[redacted]

UNIVERSITY PHYSICS

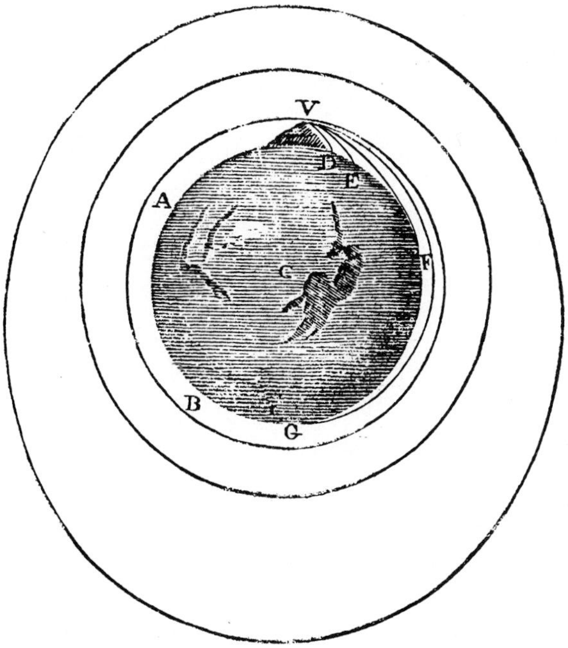

UNIVERSITY PHYSICS

George B. Arfken ▪ David F. Griffing
Donald C. Kelly ▪ Joseph Priest
Miami University, Oxford, Ohio

ACADEMIC PRESS, INC.
(Harcourt Brace Jovanovich, Publishers)
Orlando San Diego San Francisco New York
London Toronto Montreal Sydney Tokyo
São Paulo

To: Carolyn
Louise
Jane
Mary Jean

The Cover
The cover design is a modification of a sketch in Sir Isaac Newton's *Principia*. The sketch shows the transition from projectile paths on the earth to satellite orbits around the earth. Aristotle had advocated completely separate laws to describe motions on the earth and motions in the heavens. When Newton synthesized these motions under a single set of physical laws he broke a 2000-year-old tradition. With this sketch Newton made the point that his laws were universal, and from it Newtonian synthesis physics has developed as a science concerned with universal relationships.

Earth photograph courtesy of NASA.

Title page art from Isaac Newton's *Principia,* Volume 2, page 551, 1966. Reprinted by permission of the University of California Press.

Academic Press, Inc.
Orlando, Florida 32887

United Kingdom Edition Published by
Academic Press, Inc. (London) Ltd.
24/28 Oval Road, London NW1 7DX

ISBN: 0-12-059860-4

Library of Congress Catalog Card Number: 83-71920

Printed in the United States of America

Photo credits appear on page 894.

TABLE OF CONTENTS

Chapter 15

The Mechanical Properties of Matter

Chapter 16

Fluid Mechanics

Chapter 17

Wave Kinematics

Chapter 18

Mechanical Waves

Chapter 19

Special Relativity

Chapter 20

Relativistic Mechanics

Chapter 21

Temperature and Heat

Chapter 22

Thermal Properties of Matter

PREFACE

After having spent many years—more than 100, collectively—teaching introductory physics to science, engineering, and mathematics students, we have developed strong notions about the type of features a textbook should include to meet their needs. The characteristics we consider important are: (1) an authoritative treatment of physics; (2) a special sensitivity to students, particularly freshmen, who are learning physics and calculus simultaneously; (3) an emphasis on the unity of physics; (4) a sound pedagogical presentation; (5) the development of problem solving skills; and (6) a broad variety of relevant applications, problems, and questions. Some elaboration of these features should help the reader understand our intent in developing this book.

An Authoritative Treatment of Physics

A major strength of our book lies in the backgrounds and experiences of its authors. Each author has not only taught introductory physics for many years but has taught physics courses at all levels. Each has conducted physics research, and each has written successful physics textbooks. The collective wisdom of the authors, the interchange of ideas in teaching and writing, and the careful scrutiny given to each others' work have contributed to the goal of creating what we feel is an authoritative physics textbook.

For several years the manuscript was rigorously class-tested at Miami University by the authors and their colleagues. During is preparation many reviewiers, from a broad cross section of colleges and universities, have constructively criticized the manuscript. The combination of our class-testing experience and the reviewer feedback add to our confidence in the authoritative nature of the book.

Sensitivity to Students

We believe in being patient with students rather than following the sink or swim approach taken in several of the popular texts. Many capable students enter their first physics course with definite insecurities about their ability to succeed. Our goal is to create a learning environment that inspires student confidence. Advanced students do not get bored with thoughtful, accurate, and thorough explanations, and such explanations are required by average students. The liberal use of examples and illustrations, our problem solving guide, and the gradual introduction of calculus help to produce the sensitive atmosphere we are striving for.

We assume no prior knowledge of calculus, but expect that students will study it concurrently with their physics course. Particularly for freshmen, calculus should be integrated into the text carefully and gradually. Accordingly, calculus-based analyses and problems are introduced slowly.

The Unity of Physics

Newtonian mechanics, one of the foundations of physics, is normally taught first in an introductory course. To help students recognize the unity of physics, we have illustrated Newtonian mechanics with examples from such areas as astrophysics, particle physics, atomic physics, and sports, in addition to using the customary array of inclined plane, rocket, projectile, and billiard ball examples. We have constructed a bridge between classical and modern physics so students will recognize universal concepts. To reinforce this bridge, we have used extensive cross-referencing in the text.

A Sound Pedagogical Presentation

In addition to presenting the most important physical concepts in the least intimidating way, we have provided specific pedagogical aids to assist the student in using this textbook.

(1) Each chapter begins with a **preview** of what is covered in the chapter. When a chapter is the first in a sequence of related chapters, the preview introduces the entire group of chapters.

(2) A **Problem Solving Guide** is introduced early to convey the importance of problem solving skills. We know that if students cannot solve the problems, they cannot understand the physics. Therefore, this step-by-step guide

not only provides the simple mechanics of problem solving but also emphasizes the necessity for understanding the physics involved.

(3) Important principles are generously illustrated with **step-by-step examples** and clear, attractive, two-color **illustrations.** Examples provide the basis for developing problem solving skills, and illustrations portray concepts in a way words cannot express.

(4) Appropriate **end-of-chapter problems** are organized by section and keyed to specific text discussions to test the student's ability to apply physics principles.

(5) Brief **summaries** at the end of each chapter highlight significant principles, equations, and other topics covered in the chapter.

Our experience indicates that each of these pedagogical features will support the understanding of this sometimes difficult subject.

The Development of Problem Solving Skills

A step-by-step problem solving guide is introduced in Chapter 3 and extended in Chapter 6. These two strategically placed chapters provide a basis for developing the skills necessary to solve word problems in physics. Each step is carefully and systematically described to show students how to approach and solve problems. We then illustrate the use of this guide thorugh step-by-step examples. Finally, we provide problems that allow students to test their application of these skills.

Problems and Questions

To increase the value of the problems and questions, they are: (1) made appropriate to the text content; (2) graduated to challenge students' increasing abilities; and (3) varied to incorporate examples that are engaging to students. The more than 1700 class-tested problems provide the drill necessary for students to grasp the principles and concepts of physics.

In addition to the features mentioned above, the book contains sufficient material to serve either a two-, three-, or four-semester course. We use the Système International d'Unités (SI units) throughout the text, but we occasionally include pounds, miles, and other British System units for comparison.

We are indebted to many reviewers for their helpful criticisms and analyses during the development of this text. The penetrating criticism of Dr. James Smith of the University of Illinois was especially valuable. We particularly appreciate and acknowledge his contribution. A special thank you is due Kathy Civetta for her fine editorial work, and to Jane Kelly for her peerless typing.

George Arfken
David F. Griffing
Donald C. Kelly
Joseph Priest

INSTRUCTIONAL AIDS TO ACCOMPANY *UNIVERSITY PHYSICS*

The following ancillaries are available from the publisher to augment the text material.

Student Study Guide by T. William Houk, James E. Poth, and John W. Snider, Miami University

The study guide provides students with additional drill work and further review of important physics concepts. It includes brief previews of each chapter, section summaries highlighting key terms and definitions, detailed solutions of example problems, and additional practice problems for students to work.

Student Solutions Manual

This manual contains solutions or hints for over 450 of the problems in *University Physics*. Solutions are given for problems that illustrate points not covered in text examples, and problems that extend ideas developed in the text. Hints are supplied for some of the more challenging problems and in instances where the hint can shorten the solution. Following selected solutions is a collection of computer programs written in BASIC that allows students to explore certain problems in greater depth. Listings are given for Apple, IBM, and Radio Shack microcomputers.

Computer Software

An imaginative computer software package keyed specifically to sections in the text is available to students to strengthen their understanding of physics concepts while providing interest and motivation. Three computer disks, with approximately 18 programs per disk, allow students to plug in their own data and watch physics concepts in action. Seeing physics in motion is an excellent way for students to gain hands-on experience with relevant physics problems.

Instructor's Answer Book

Answers are provided to instructors for all the problems in the text.

Overhead Transparencies

Acetate transparencies of important text illustrations are available to adopters as a support to text discussions. Instructors can point out the pertinent features of text illustrations and discuss these concepts in the classroom.

chapter 1

GENERAL INTRODUCTION

1.1 The Development of Science

From earliest history people have been curious about the world around them, and have sought correlations and patterns of behavior in nature. The early Egyptians, for instance, noticed that the Nile River flooded each year when the bright star Sirius rose at dawn, and wondered whether these events were connected.

Curiosity about the motion of the planets eventually led to the Copernican revolution.* In fact, people's curiosity and their attempts to find an order to the universe led directly to the development of physics as a science. Today the patterns and the order of our physical world are expressed by physical laws and conservation principles. These laws and principles are concise expressions that enable scientists and engineers to describe many aspects of physical behavior and to predict future physical behavior. The goal of physics is often described as a search for the physical laws that govern the universe.

Physics is a diverse and evolving collection of special branches of knowledge unified by common physical laws. Table 1.1 lists a number of these branches. The division of these branches into the areas of classical physics and modern physics is arbitrary but convenient. Roughly speaking, classical physics consists of the branches that were well developed by 1900. Modern physics has emerged since that time. The distinction between classical and modern

*The Polish astronomer Nicolaus Copernicus (1473–1543) developed the theory that the sun was the center of our solar system. This theory implied a radically new view of the universe and of the place of humans in it.

Table 1.1

Areas of physics

Classical	Modern	
Mechanics Gravitation	Special relativity	

Thermodynamics (heat)	General relativity
	  Gravitation
	  Cosmology	

Electromagnetism
  (electricity, magnetism)
	Quantum mechanics
	  Atomic physics
	  Molecular physics
	  Nuclear physics
	  Solid-state physics
	  Particle physics
	  Superconductivity
	  Superfluidity
		Interdisciplinary Areas
		  Astrophysics
		  Biophysics
		  Chemical physics
		  Engineering physics
		  Geophysics
		  Medical physics
		  Physical oceanography
		  Physics of music

Optics (light)		
Acoustics (sound)		
Hydrodynamics (fluid flow)	Plasma physics	
	Magnetohydrodynamics	
	Space and planetary physics	

physics does become blurred, however, when a new device, such as a laser, rejuvenates an old field, such as optics. This text emphasizes the aspects of classical physics that lead into modern physics, and includes an introduction to two areas of the latter: special relativity and quantum theory.

Table 1.1 does not list some subjects (such as electrical engineering and computer technology) that were developed as part of physics but are now considered to be part of engineering. This movement of these topics from physics into engineering illustrates the close relationship between physics and engineering, as well as the changing nature of physics. Some old areas of physics have been revolutionized, new areas have opened up, and the existing areas continue to expand.

Physics can be defined as the study of matter and energy. As you work through this text, however, you will find that physics is also a way of thinking. You will want to master both the content of physics and this method of thinking. The problems at the end of the chapters are designed to help you do this. They are not abstract mathematical riddles, but instead explore the behavior of real physical systems.

1.2 Science and Measurement

The laws that unify physics refer to particular physical quantities—for example, forces and positions. Before we can understand the laws of physics, we must define and measure these physical quantities. In physics, the definition and measurement of a quantity are interrelated. Physics is built on **operational definitions**—definitions that are expressed in terms of measurements. For example, average speed is defined as the (measured) distance traveled divided by the (measured) time that has elasped.

There are two basic types of physical quantities—fundamental quantities and derived quantities. **Fundamental quantities** are those quantities that cannot be defined in terms of other quantities. To define a particular fundamental quantity, two steps are necessary. First, its measurement must be fully specified. Then, a standard of comparison must be established. **Derived quantities,** on the other hand, are defined in terms of fundamental quantities by means of a defining relationship that is normally an equation.

In the 19th century European scientists saw the need for a coherent system of fundamental physical quantities and their corresponding standards of measurement. This concern led to the signing of the Treaty of the Meter in 1875 and the establishment of the International Bureau of Weights and Measures. This bureau was created to "establish new metric standards, conserve the international prototypes, and carry out the comparisons necessary to insure the uniformity of measures throughout the world."

The present metric system, called the ***Système International d' Unités,*** or **SI** for short, recognizes seven fundamental physical quantities. We will consider three of these fundamental quantities—length, time, and mass—immediately. Three of the remaining four quantities will be introduced as they are needed: Temperature, measured in kelvins, first appears in Chapter 21; number of particles (such as atoms or molecules), measured in moles, in Chapter 25; and electric current, measured in amperes, in Chapter 30. The seventh fundamental SI quantity, luminous intensity, measured in candelas, is not used in this text.

Question

1. One philosophic school held that if a quantity could not be defined by an operational, or measurement–specifying, definition, then that quantity had no real meaning and no physical reality. Defend or criticize this view. (Physicists today want operational definitions for physical quantities, but they generally reject this extreme position.)

1.3 Length

Length is the first fundamental physical quantity that we will consider. An ordinary length such as the length, width, or height of a table can be measured by comparing it to a secondary standard such as a 1-foot (ft) ruler, a yardstick, or a meter stick. Indeed, we may consider such a comparison to be an operational definition of length. But what do we use as the primary standard of length? In other words, what is the meter stick measured against? In earlier times the king's foot was often the primary standard, but its length changed each time a new king appeared on the scene. Scientists need standards that do not change.

Eighteenth-century French scientists sought a natural, universal standard, free of political differences. They proposed the meter (m), a unit of length that they defined as equal to one ten-millionth of the distance from one of the earth's poles to the equator. Because this standard was tied to nature, it seemed indestructible. But although it was available to all scientists, the use of this standard was difficult. For example, what multiple of one ten-millionth of the pole–equator distance is your height? This difficulty was resolved by constructing a bar of an alloy consisting of 90% platinum and 10% iridium, and defining the standard meter as the distance between two lines engraved on the bar when the bar was at the temperature of melting ice. Accurate comparisons between the meter bar and objects of unknown length could then be made by using elaborate and very precise optical techniques. Under the auspices of the International Bureau of Weights and Measures, secondary standards of length were made and distributed to other nations of the world (see Figure 1.1).

Figure 1.1

This postage stamp was issued by France in 1954 to remind the world that France developed the metric system.

The precision of visual comparisons is limited to about one part in ten million (10^7). As science and technology developed, this precision became inadequate. In addition, there was the danger that the platinum–iridium bar might be damaged or destroyed. Consequently, the meter was redefined in 1960 by international agreement as the length equal to 1,650,763.73 wavelengths (in vacuum) of a particular color of light (a shade of orange) emitted by krypton-86 atoms. (This peculiar number was chosen in order to match the old standard meter as closely as possible.) Once again, the standard meter is tied to a property of the natural world and is thus presumably indestructible. The krypton light source is reproducible and permits convenient and precise measurements. In effect, every laboratory can now have its own primary standard of length. Remember, though, that all standards are, in principle, temporary, because they are chosen by consensus. The krypton standard quite possibly will be replaced by an even more precise laser standard within a few years.

In Table 1.2 we list the range of lengths that we are concerned with in physics, from the size of nuclear particles (about 10^{-15} m) to the distances from the Earth to far away quasars (perhaps 10^{25}–10^{26} m). When one length in this list is approximately ten times longer than the next length, we say that the first length is an **order of magnitude** larger than the second. The second length is an order of magnitude smaller than the first. Orders of magnitude, in other words, are associated with powers of 10.

For convenience in handling such a wide range of values, the set of prefixes listed in Table 1.3 is used in conjunction with SI units. The prefixes for positive powers of 10 are from Greek; those for negative powers are from Latin (except for the recently adopted prefixes **femto-** and **atto-**, which come from old Norse). All of the lengths—kilometers (km), millimeters (mm), and so on—are related to the standard meter by powers of 10. SI is thus a decimal system, which makes it easy to use. Note carefully that all

Table 1.2

Range of lengths (orders of magnitude)

Length	Meters (powers of 10)
Distance to distant quasars	10^{26}
Distance to nearest star (Proxima Centauri)	10^{16}
Radius of solar system	10^{13}
Earth–sun distance	10^{11}
Mean radius of earth	10^{7}
Height of person	10^{0}
Air-polluting aerosol particle	10^{-6}
Diameter of influenza virus	10^{-7}
Radius of a hydrogen atom	10^{-10}
Radius of a hydrogen nucleus (proton)	10^{-15}

Table 1.3

SI prefixes

Power of 10	Prefix	Symbol
10^{18}	exa	E
10^{15}	peta	P
10^{12}	tera	T
10^{9}	giga	G
10^{6}	mega	M
10^{3}	kilo	k
10^{-2}	centi	c
10^{-3}	milli	m
10^{-6}	micro	μ
10^{-9}	nano	n
10^{-12}	pico	p
10^{-15}	femto	f
10^{-18}	atto	a

Examples: 10^{6} watts = 1 megawatt (1 MW)
10^{3} meters = 1 kilometer (1 km)
10^{-3} gram = 1 milligram (1 mg)
10^{-9} second = 1 nanosecond (1 ns)

but one of the exponents are multiples of 3. The only exception, the centimeter (cm), is occasionally used when it is inconvenient to use millimeters or meters.

By law, our U.S. units of length (English system) are defined in terms of the standard meter. Thus

1 yard = 0.9144 m (exactly)

and

1 inch = 2.54 cm (exactly)

From time to time in this text we will use English units, since you are probably more familiar with these units, in order to give you a better feeling for the physical situation. In most such examples, however, the English units will be converted to SI units and the calculation carried out in SI units.

Questions

2. We are already using SI units in sports. How does the length of the 1500-m run compare with the mile?
3. Why is it desirable to have a standard tied to some natural phenomenon?
4. What is your height in centimeters? in meters?
5. Why wasn't the meter defined as 1,000,000.00, instead of the cumbersome number 1,650,763.73 wavelengths of krypton-86 light?
6. Explain how the length of an arc of a circle can be measured. (The length of the arc becomes part of the definition of an angle in Chapter 12.)
7. When the platinum–iridium meter bar was the standard meter, why was it essential to specify the temperature at which measurements were made?
8. Many length measurements are indirect. How would you measure the radius of the earth? the distance to the moon?
9. List and explain the advantages of an atomic length standard over the platinum–iridium meter bar standard.

1.4 Time

Time is another fundamental quantity, and therefore cannot be defined in terms of other quantities. However, we can measure time, and thus obtain an operational definition of time, by using a clock.

Any repetitive, periodic phenomenon can serve as a clock. The Italian astronomer Galileo Galilei used his pulse as a clock to measure the time that it took for a swinging chandelier to return to its original position. The earth, by rotating on its axis, constitutes a clock. Originally the second was defined as 1/86,400 the length of one complete revolution of the earth* (averaged over a year). The earth, however, is not a sufficiently accurate clock for today's measurements for two reasons. First, because of the effects of lunar and solar tides, the rotation of the earth is gradually slowing down. Second, the earth's rate of rotation is slightly irregular, varying with the seaons and from year to year.

By international agreement, the second was redefined in 1969 as the duration of 9,192,631,770 periods of the microwave radiation from a cesium-133 atom. A time interval in seconds is measured as the number of cesium-133 periods in that interval divided by 9,192,631,770. Thus the unit of time, like the unit of length, is now based on the behavior of atoms. This new standard of time has made it possible to measure time intervals to one part in 10^{12}. The cesium clock is reproducible and readily available to scientists and engineers the world over. The second is the unit of time in both the English and SI systems. Table 1.4 shows the range of times in orders of magnitude as expressed in seconds (powers of 10). Of course the time standard is arbitrary, just as is the standard of length. Ultraprecise hydrogen clocks may eventually replace the cesium clock as the international standard of time.

Figure 1.2 shows the variation in the rate of rotation of the earth as measured by a cesium clock. But how do we know that the irregularity is in the earth's rotation and not in the cesium clock? The way that physicists answer this

*The division of a minute (min) into 60 seconds (s), an hour (hr) into 60 min, and a day into 24 hr ($60 \times 60 \times 24 = 86,400$) began in ancient Babylon.

Table 1.4

Range of times (orders of magnitude)

Time	Seconds (powers of ten)
Time for half of a sample of neodymium-144 atoms to decay	10^{23}
Age of the earth	10^{17}
Age of the Great Pyramid	10^{11}
One year	10^{7}
One day	10^{5}
Period (reciprocal of frequency) of high-frequency sound	10^{-4}
Period of vibration of a cesium clock	10^{-10}
Time for a fast-moving nuclear particle to cross a proton diameter	10^{-23}

question illustrates how all scientists think. First, we must ask ourselves what phenomena might cause irregularities. In the case of the cesium-133 atoms, none is known. On the other hand, such phenomena as tidal friction, the seasonal motions of the winds, the melting and refreezing of the polar ice caps, and even major earthquakes could change the earth's rate of rotation. We will hypothesize that the earth's rotation is irregular. To confirm this hypothesis we look for other clocks by which to measure the rate of rotation of the earth. Vibrations of other atoms, the movement of the planets around the sun, and the movement of the moon around the earth can all serve as clocks. Although these clocks are not as accurate as the cesium clock, they do confirm that the irregularities shown in Figure 1.2 should be assigned to the earth.

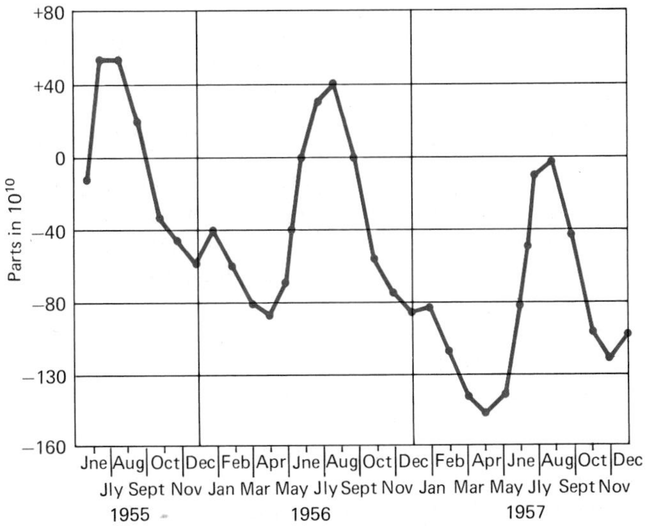

Figure 1.2

The variation in rate of rotation of the earth expressed in parts in 10^{10}. (From L. Essen, *Physics Today* **13**, 29 (July 1960).

So far we have introduced two fundamental quantities: length and time. From these fundamental quantities we can derive additional quantities, called derived quantities. Average speed is defined as the distance traveled divided by the change in time. This derived quantity is measured in a variety of units, for example, meters/second (m/s), miles/hour (mi/hr or mph), and centimeters/year (cm/yr), all with dimensions of length divided by time. As we move through our study of physics we will repeatedly derive new quantities in a similar way.

Questions

10. Why might the human pulse be an inaccurate clock?
11. The setting of 9,192,631,770 periods of cesium-133 radiation for the definition of the second seems strange. Why did scientists not pick 10,000,000,000?
12. The swinging of a pendulum could be used as a time standard. What would be the disadvantages of such a standard?
13. The standard cesium clock is said to run at a constant rate, permitting time intervals to be measured to one part in 10^{12}. What is meant by "constant rate"? How do we know that the rate of the standard cesium clock is constant?
14. Certain atomic clocks are believed to be the most accurate ever built. How could this accuracy be tested?
15. T. C. Van Flandern, in his article "Is Gravity Getting Weaker?" [*Scientific American* **236**, 44 (February 1977)], suggests that clocks governed by gravity such as the moon may differ systematically from atomic clocks. How can this difference be measured?
16. An atomic particle, assumed to be moving at approximately the speed of light, travels a distance d between its creation and decay. Explain how its lifetime (from creation to decay) could be calculated if the distance traveled and the velocity are known.
17. How could a constant speed, such as the speed of light, together with a standard of length be used to define a standard of time?
18. How could a constant speed, such as the speed of light, together with a standard of time be used to define a standard of length?

1.5 Mass

Mass, the third fundamental SI quantity, can be roughly defined as the **amount of matter.** In Chapter 6 we will discuss mass in terms of its two fundamental properties: (1) gravitational interaction and (2) resistance to acceleration; but for now a standard of mass and a method of comparison of masses is sufficient.

The SI standard of mass is a polished cylinder of platinum–iridium alloy, which was chosen for its durability and resistance to corrosion. The cylinder is kept at the International Bureau of Weights and Measures in a suburb of Paris, France. The unit of mass, the kilogram, is defined as the mass of this particular metal cylinder. Other masses can be compared to the standard kilogram by using a sensitive balance. If an unknown mass balances the standard kilogram, then the unknown mass also has a mass of 1 kilogram (kg). Objects with masses that are a fraction or a multiple of the standard kilogram can be measured as well, to a precision of one part in 10^8.

Because there is always the remote possibility that the standard platinum–iridium kilogram will be lost or destroyed, it would be convenient if we had a reproducible atomic standard of mass comparable to the atomic length and time standards. At one time, the kilogram was defined as the mass of 1000 cubic centimeters (cm^3) of water, with the water at its maximum density. But this definition turned out to be impractical. It simply was not possible to prepare and measure the water standard to the required precision. Scientists have speculated about taking the mass of a particular atom (such as the mass-12 variety of carbon, ^{12}C) as a standard, but as of now the arbitrary platinum–iridium cylinder standard kilogram is adequate and has no real competitors. Table 1.5 shows the range of masses in orders of magnitude as expressed in kilograms (powers of 10).

Questions

19. List the characteristics that would be considered desirable for a physical standard. Use the standard meter, second, and kilogram to illustrate your points.

20. What are the disadvantages or dangers of defining the kilogram in terms of a cylinder of platinum–iridium?

21. Measurements of mass, like those of length and time, are often indirect. The mass of the earth is given as 5.98×10^{24} kg. Explain how this number might be determined.

Table 1.5

Range of masses (orders of magnitude)

Mass	Kilograms (powers of ten)
Our (Milky Way) galaxy	10^{41}
The sun	10^{30}
The earth	10^{25}
A large oil tanker	10^{8}
A droplet of water 1 mm in diameter	10^{-6}
An atom of uranium	10^{-25}
A proton	10^{-27}
An electron	10^{-30}

22. The mass of a carbon-12 atom is taken as a working standard in atomic physics. How could the mass of this atom be determined in terms of the standard kilogram?

1.6 Dimensions and Units

Dimensions

In this chapter and through your own experience you have been exposed to a variety of units of length: the SI meter, kilometer, and millimeter; the English units of inches, feet, yards, and miles; and perhaps rods, fathoms, light-years, and parsecs. All of these diverse units are said to have **dimensions of length**, symbolized by L.

Likewise, all time units, such as seconds, minutes, hours, days, years, and centuries, are said to have dimensions of time, T. The kilogram and all other mass units, such as the astronomers' solar mass, have dimensions of mass, M. Generalizing, we may take a dimension (length, time, or mass) as the concept of the quantity itself.

From the three fundamental physical quantities, length, time, and mass, we mentioned that we can develop a variety of useful secondary or derived quantities. Derived quantities have different dimensions than do fundamental quantities. For instance, area, obtained by multiplying a length times a length, has the dimensions L^2. Volume, such as a cubic meter, has dimensions of L^3. Mass density is defined as mass per unit volume and has dimensions of M/L^3. The SI unit of speed is meters per second (m/s) with the dimensions L/T.

The concept of dimensions is important in understanding physics and in doing physics problems. For instance, the addition of different dimensions makes no sense; 1 m plus 2 s is meaningless. Physical equations must be dimensionally consistent. In other words, every term in an equation must be dimensionally the same. For example, the equation giving the position of a freely falling body (Chapter 4) is

$$x = v_0 t + \frac{1}{2} g t^2 \qquad (1.1)$$

where x is position (length); v_0 the initial speed (length/time), g the acceleration of gravity (length/time^2); and t time. If we analyze the equation dimensionally, we have

$$L = \frac{L}{\cancel{T}}\cancel{T} + \frac{L}{\cancel{T^2}}\cancel{T^2} = L + L$$

Note that numerical factors such as $\frac{1}{2}$ are ignored in dimensional analysis because they have no dimensions. Every term in this equation has dimensions of length, L. If we had written (by mistake)

$$x = v_0 t^2 + \frac{1}{2} g t \qquad (1.2)$$

the dimensions of the terms would be

$$L = \frac{L}{T}T^2 + \frac{L}{T^2}T$$

or

$$L = LT + \frac{L}{T}$$

Dimensionally this equation is meaningless, and thus Eq. 1.2 cannot be correct. Here, then, is a way of catching careless errors by inspection or dimensional analysis. Equation 1.1 *may* be correct, although you could still be off by a dimensionless numerical factor.

Units

In physics, and in all the sciences, we frequently have to convert a measurement from one set of units to another. The easiest way to do this is to look up the **conversion factor** if it is available. A table of conversion factors is given in Appendix 3. We can also calculate a conversion factor in terms of conversion factors that we already know. The following examples illustrate a method of converting from one set of units to another. In this method units are converted by multiplying by ratios having magnitude one.

Example 1 **Finding the Number of Seconds in a Year**

How many seconds are there in a year? We know that there are 365 days in a year, 24 hr in a day, 60 min in an hour, and 60 s in a minute. These four relationships are conversion factors. Starting with 1 yr and multiplying by these conversion factors, we obtain

$$\begin{aligned} 1\text{ yr} &= 1\,\cancel{\text{yr}} \cdot \frac{365\,\cancel{\text{days}}}{1\,\cancel{\text{yr}}} \cdot \frac{24\,\cancel{\text{hr}}}{1\,\cancel{\text{day}}} \cdot \frac{60\,\cancel{\text{min}}}{1\,\cancel{\text{hr}}} \cdot \frac{60\text{ s}}{1\,\cancel{\text{min}}} \\ &= 365 \cdot 24 \cdot 60 \cdot 60 \cdot \text{s} \\ &= 3.15 \times 10^7\text{ s} \end{aligned}$$

Appendix 3 states that

$$1\text{ yr} = 3.15569\ldots \times 10^7\text{ s}$$

Our answer is slightly off because we used 365 days as the length of a year. The year actually has 365.242 . . . days.

Example 2 **Conversion of Velocities**

Suppose that you are driving at 55 mph. How many feet per second are you traveling? How many meters per second? We start out with 55 mph and multiply by several known conversion factors:

$$\begin{aligned} 55\text{ mph} &= \frac{55\,\cancel{\text{mi}}}{1\,\cancel{\text{hr}}} \cdot \frac{5280\text{ ft}}{1\,\cancel{\text{mi}}} \cdot \frac{1\,\cancel{\text{hr}}}{3600\text{ s}} \\ &= 55 \cdot \frac{44\text{ ft}}{30\text{ s}} \\ &= 80.7\text{ ft/s} \end{aligned}$$

To convert feet per second to the SI meters per second, we look up the meter in Appendix 3.

$$1\text{ m} = 3.281\text{ ft}$$

Then

$$80.7\text{ ft/s} \cdot \frac{1\text{ m}}{3.281\text{ ft}} = 24.6\text{ m/s}$$

We must make one final point about numbers. The length standard is defined with nine significant figures. The definition of the time standard has ten significant figures. Most of the physical measurements you make, however, will not even approach this precision. As a compromise, most of the calculations in this text will carry three significant figures. Using a hand-held electronic calculator, you can obtain three-figure accuracy with no problem. Throughout this text, then, you can assume that given values for calculations are accurate to three digits unless a higher precision is specified. For example, a velocity of 1 m/s should be taken as 1.00 m/s.

1.7 Physics, Mathematics, and You

Our standards of physical quantities, both fundamental and derived, are defined with numbers. The physical laws that relate these quantities are mathematical relationships. We use mathematics to make quantitative predictions. For example, Eq. 1.1 predicts the position of a particle at time t given an initial velocity, v_0, and a constant acceleration, g. Here, in its ability to lead to a quantitative prediction, is the utility and the power of physics.

At this point we assume that you have a working knowledge of high school mathematics, especially algebra, and have just started a college course in differential and integral calculus. Calculus is a very powerful tool and we are eager to use it. But because you have just begun to study calculus, we will introduce it very slowly. Calculus will not really be needed until Chapter 6. On the other hand, an extension of algebra, called vector algebra, is needed right away and will be introduced in Chapter 2.

You should work through the examples in each chapter if you want to master physics. The questions that appear at the end of most sections are designed to stimulate you and let you check your own understanding. Finally, the end-of-chapter problems provide the crucial test.* The problems demand varying degrees of an understanding of physics and skill in mathematics.

*"Just do the exercises diligently. Then you will find out what you have understood and what you have not," Arnold Sommerfeld told his student Werner Heisenberg. Sommerfeld was one of the most famous physics teachers of all time; Heisenberg was one of the founders of modern quantum theory.

Physics can be as imaginative and wildly speculative as poetry and philosophy. But at an introductory level, mathematics and problem solving are very important. In the words of Lord Kelvin,

> *When you can measure what you are speaking about, and express it in numbers, you know something about it; but when you cannot express it in numbers, your knowledge is of a meager and unsatisfactory kind: it may be the beginning of knowledge but you have scarcely, in your thoughts, advanced to the stage of* science.
>
> William Thomson, Lord Kelvin

In addition to this text you may have lectures, demonstrations, movies, and laboratory experiments, all intended to help you learn. Don't stop there. Whether it be notes on exciting new developments in physics and related areas or comprehensive review articles, you should do some reading beyond this text. We particularly recommend *Science News* for short notes on new discoveries and new developments and *Scientific American* for broad review articles ranging from prehistoric science up to the most recent developments. References to relevant *Scientific American* articles are included in many of the chapters of this book.

Summary

Modern science is a search for order. The order that underlies our physical world is expressed by physical laws and conservation principles. These laws and principles unify widely diverse areas of physics.

Physics is built up in terms of operational definitions. These are definitions that involve measurement. At the most basic level, measurement means comparison with a fundamental standard. Standards of length, time, and mass have been chosen on the basis of criteria such as accuracy, reproducibility, and convenience. As the demands of science and technology have increased, the fundamental standards have been changed and refined. This process of refinement continues even today.

The standard of length is the meter, equal to 1,650,763.73 wavelengths (in vacuum) of a particular light emitted by krypton-86 atoms.

The standard of time is the second, equal to 9,192,631,770 periods of the microwave radiation from cesium-133 atoms.

The standard of mass is a particular cylinder of a platinum–iridium alloy. The mass of this metal cylinder is taken as the standard kilogram.

The length, time, and mass standards form part of the *Système International d' Unités* (SI). Because this system has been adopted by most of the world we shall use it almost exclusively. The system includes SI prefixes ranging from 10^{-18} to 10^{18}.

The dimension of length may be taken as the concept of length itself, a distance between two points (or two parallel lines). All the length units we use (meters, miles, etc.) have the dimension of length. From the dimensions of length, time, and mass, other quantities are derived. All physical equations must be dimensionally consistent. The conversion of units may be handled by the use of such conversion, or magnitude-one, ratios as 60 s/1 min.

Suggested Reading

A. V. Astin, Standards of Measurement, *Sci. Am.* **218**, 50–62 (June 1968).

L. Essen, Accurate Measurement of Time, *Physics Today* **13**, 26 (July 1960).

T. C. Van Flandern, Is Gravity Getting Weaker? *Sci. Am.* **236**, 44–52 (February 1977).

E. A. Mechtly, *The International System of Units: Physical Constants and Conversion Factors*, 2nd revision. NASA SP-7012 (1973).

National Bureau of Standards, *International System of Units*, Special Publication 330, 1977 ed.

Lord Ritchie-Calder, Conversion to the Metric System, *Sci. Am.* **223**, 17–25 (July 1970).

A. L. Robinson, Using Time to Measure Length, *Science* **220**, 1367 (June 1983). This article discusses the activities of scientists around the world working toward further refinement of the definition of the meter. The meter is to be defined as the distance light travels in 1/299,792,458 second.

Problems

Section 1.3 Length

1. If the distance from the north pole of the earth to the equator (measured along the surface) is taken to be 10^7 m (original definition of meter), (a) what is the circumference of the earth in meters?, (b) what is the radius of the earth in meters?

2. Astronomers, like most scientists, use units that are appropriate for the quantity being measured. For planetary distances they use the astronomical unit (AU), which is equal to the mean earth–sun distance, or 1.50×10^{11} m. For stellar distances they have the light-year, which is the distance light travels in 1 yr (3.15×10^7 s) with a speed of 3.00×10^8 m/s, and the parsec, which is equal to 3.26 light-years. Intergalactic distances might be described with megaparsecs. Convert (a) the astronomical unit, (b) the light-year, (c) the parsec, and (d) the megaparsec into meters and express each with an appropriate metric prefix.

3. The earth–moon distance (3.80×10^5 km) is regularly measured by obtaining the time of flight of a laser pulse sent to the moon and back with a large telescope. (a) Express this distance in meters. (b) What is the round trip time of flight? (Laser pulse velocity equals 3.00×10^8 m/s.)

4. If it is assumed that an atom and its nucleus are each spherical, the ratio of their radii is about 10^5. If the ratio of the radius of the orbit of the moon and the earth's radius were 10^5, how far would the moon be from the earth? How does this distance compare with the actual earth–moon distance?

5. A dime has a thickness of about 1 mm. If a dime were exactly 1 mm thick, how many wavelengths of krypton-86 radiation could you fit in the thickness of the dime?

Section 1.4 Time

6. Compare the duration of (a) a microyear and a 1-min TV commercial, (b) a microcentury and a 60-min program.

7. (a) Express your age (approximately) in seconds. (b) Is a cesium clock sufficiently precise to determine your age to within 10^{-6} s? to within 10^{-3} s?

Section 1.5 Mass

8. A carbon-12 atom (^{12}C) is found to have a mass of 1.99264×10^{-26} kg. (a) How many atoms of ^{12}C are there in 1 kg? (b) How many atoms are there in 12 grams? (This number is Avogadro's number.)

9. Density is a derived quantity defined as mass per unit volume. (a) Newton guessed that the average density of the earth was 5.5 grams/cm^3 = 5.5×10^3 kg/m^3. Taking the earth to be spherical with a radius of 6.37×10^6 m, calculate the mass of the earth. (b) The mass of the earth is actually 5.98×10^{24} kg. Calculate its average density.

Section 1.6 Dimensions and Units

10. Work out the dimensions of the following derived quantities: (a) acceleration (speed/time), (b) force (mass × acceleration), (c) work (force × distance), (d) linear momentum (mass × velocity), (e) angular momentum (distance × linear momentum).

11. Newton's law of universal gravitation (Chapter 7) is given by

$$F = G\frac{m_1 m_2}{r^2}$$

Here F is the force of attraction of one mass upon another at a distance r. F has units of kg · m · s^{-2}. What are the SI units of the proportionality constant G?

12. Consider a system in which the three fundamental quantities are the speed of light c (LT^{-1}), Planck's constant h (ML^2T^{-1}), and the mass of the proton m_p (M). (a) Develop ratios and/or products of c, h, and m_p to form a quantity with the dimension of L. (b) Develop ratios and/or products of c, h, and m_p to form a quantity with the dimension of T.

13. Imagine a poison dart experiencing a constant acceleration a over the length x of a blowgun. Assume that the speed v of the dart leaving the blowgun depends only on a and x (and possible numerical factors). From the dimensions of each quantity, prove this assumption. In other words, set up an equation with v on the left side and some combination of a and x (determined from the dimensions) on the right.

14. A planet moving in a circular orbit experiences an acceleration a. If this acceleration depends only on the linear speed v of the planet and the radius r of the orbit, show, from the dimensions, how a depends on v and r.

15. Imagine that NASA puts a series of 1000-kg satellites in various circular orbits about the earth. The energy of motion of a satellite ($K = {}^1/_2 mv^2$) is found to vary inversely as its distance from the center of the earth ($K \sim r^{-1}$). (a) How does the speed v depend on r? (b) How does the period of rotation T depend on r?

16. Using the relation 1 in. = 2.54 cm to go from the English system to the SI, calculate the number of kilometers in 1 mile.

17. Imagine that you are driving along a highway in Mexico (or in Canada). You see an 80-km/hr speed limit sign. What is this in terms of your miles-per-hour speedometer?

18. Calculate the number of cubic meters per cubic yard.

19. Determine the number of hectares in 1 acre. (1 hectare = 10^4 m^2 and 1 acre = 4840 yd^2).

20. Calculate the conversion factor for converting meters per second to miles per hour.

21. Under the influence of gravity a freely falling mass accelerates at the rate of 32.2 ft/s^2. Convert this to SI units of meters per second squared.

22. The speed of light is 3.00×10^8 m/s. Convert this to (a) miles per hour and (b) feet per nanosecond (ns).

23. A certain runner can run the mile in exactly 4 min. How long would it take this runner (running at the same speed) to cover the "metric mile" of 1500 m?

24. Chemists generally express densities in grams per cubic centimeter. The resulting numerical values are of a convenient size. The density of water, for instance, is 1 gm/cm^3. What is the density of water in SI units?

25. The head of a striking rattlesnake is reported to accelerate at the rate of 60 mph per $\frac{1}{2}$ s. Convert this acceleration to meters per second squared.

chapter 2

VECTOR ALGEBRA

The slovenliness of our language makes it easier for us to have fuzzy thoughts.

George Orwell

Preview

In this chapter we introduce vector algebra, an important tool of physics. As we explain in Section 2.1, vectors are mathematical quantities that are used to describe physical quantities having both magnitude and direction.

The operations of vector addition, subtraction, and multiplication by a number are defined in Section 2.2. Two types of vector multiplication, selected because of their usefulness in physics, are presented in Sections 2.4 and 2.5. All of these vector equations are shown from both a geometric and an algebraic point of view. At first you will probably find the geometric approach easier to understand. Later, as you gain proficiency in using vectors, you will find the algebraic approach better suited for precise computation.

2.1 Scalars and Vectors

Many quantities in our physical world are characterized by a magnitude only. These physical quantities may be described mathematically by a single number and an appropriate unit. For example, the mass of the electron is given by $m_e = 9.11 \times 10^{-31}$ kg, and the energy stored in a particular battery can be represented by $E = 2 \times 10^5$ joules (J). Physical quantities such as these, represented by a magnitude only, are called **scalars.**

Other physical quantities are characterized by a **direction** as well as a magnitude. For instance, the velocity of a car, $\mathbf{v} = 21$ m/s, northeast; or the earth's magnetic field at a particular location, $\mathbf{B} = 4 \times 10^{-5}$ tesla (T), 4° west of geographic north, are quantities having both magnitude and direction. Such quantities are called **vectors.** Indeed, we may define vectors as quantities having magnitude and direction and combining according to certain rules, which we present in the following sections.

Because a vector differs from a scalar by having a direction associated with it, we represent a vector with a distinctive symbol. In this text, vector quantities are always given in boldface: **v**, **B**. Scalar quantities, on the other hand, are given in italics: *m*, *E*. Another way to indicate that a quantity is a vector is to use a wavy underline ($\underset{\sim}{v}$, $\underset{\sim}{B}$), which is the printer's symbol for boldface type. To indicate only the magnitude of vector **v**, the mathematicians' absolute-value sign may be used, or the quantity may be printed in italics: $|\mathbf{v}| = v$.

Vectors may be represented geometrically by an arrow. The direction of the arrow is the direction of the vector, and the length of the arrow is proportional to the magnitude of the vector. Vectors may have dimensions, such as length for a position vector **r** and length/time for a velocity vector **v**. Position vectors and velocity vectors are described in the following examples.

Example 1 **The Position Vector r**

In a Cartesian (xyz) coordinate system, the distance and direction of the point (x, y, z) from the origin may be described by an arrow from the origin (0, 0, 0) to the point (x, y, z), as shown in Figure 2.1. The usual label for this position vector is **r**.

Example 2 **Velocity Vectors**

Velocities over the surface of the earth may be described by specifying the magnitude (speed)—in meters per second (m/s), miles per hour (mph), or other units—and the direction. These velocity vectors may also be presented graphically if we agree that a particular length on the graph represents a particular speed. Let us plot a velocity $\mathbf{v}_1$ of 50 m/s (105 mph), 60° east of north.

We will let 1 cm on the graph in Figure 2.2 correspond to 10 m/s. The magnitude of $\mathbf{v}_1$ is represented by an arrow from the origin to the 5-cm-radius arc. The direction of $\mathbf{v}_1$ is 60° from the y-axis (north) toward the x-axis (east).

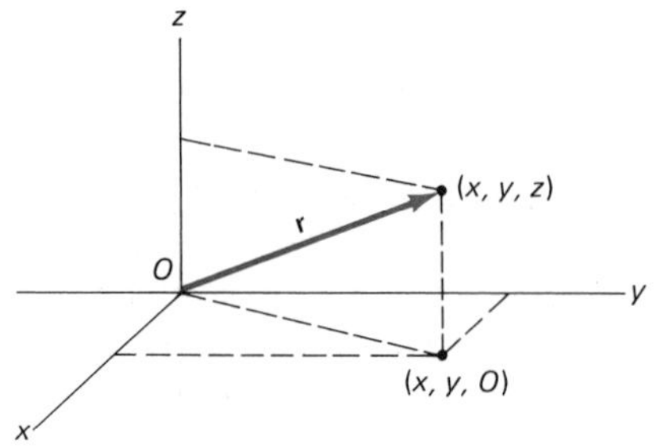

Figure 2.1

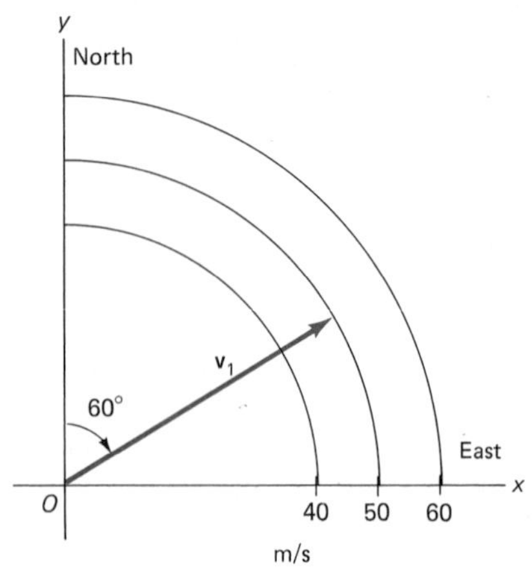

Figure 2.2

2.2 Addition and Subtraction of Vectors

Vectors can be added, subtracted, and multiplied. In this section we introduce vector addition, vector subtraction, and the multiplication of a vector by a scalar. The multiplication of two vectors is presented in Sections 2.4 and 2.5.

Addition of Vectors

In order to add vectors, we first represent them with arrows, as described in the previous section. Consider the two vectors, **A** and **B**, represented by arrows in Figure 2.3. We add these two vectors by first placing the tail end of vector **B** at the tip of vector **A**. This shift of vector **B** parallel to itself does not affect the vector, because neither the magnitude nor the direction is changed. We then draw an arrow from the tail end of **A** to the tip of **B**. This arrow represents vector **C**, the vector sum of **A** and **B**:

$$\mathbf{C} = \mathbf{A} + \mathbf{B} \tag{2.1}$$

Example 3 **Vector Addition**

A salesman leaves Dayton, Ohio, and drives 10 mi north on I-75 and then 5 mi west on I-70. Where is he relative to his starting point?

The 10-mi trip north and the 5-mi trip west are vectors, since they each have both magnitude and direction. The problem is therefore one of vector addition. We will call the 10-mi trip north vector $\mathbf{r}_1$ and the 5-mi trip west vector $\mathbf{r}_2$, and then represent these vectors with arrows (Figure 2.4). In Figure 2.4, 1 cm corresponds to 2 mi. Traveling 10 mi north (from point A to point B) and then 5 mi west (from point B to point C) is equivalent to adding the two vectors $\mathbf{r}_1$ and $\mathbf{r}_2$. The resultant change of position (from A to C) is the vector sum

$$\mathbf{r} = \mathbf{r}_1 + \mathbf{r}_2$$

Using a protractor, we find that the direction of **r** is about 26° west of north. When we measure **r** with a ruler we find that it is 5.7 cm long, which corresponds to 11.4 mi. The salesman, then, is 11.4 mi and 26° northwest of Dayton. We can also find the length of **r** algebraically by using the Pythagorean theorem. In this case, we find that the length of **r** is 11.18 mi. The graphical solution, done carefully,

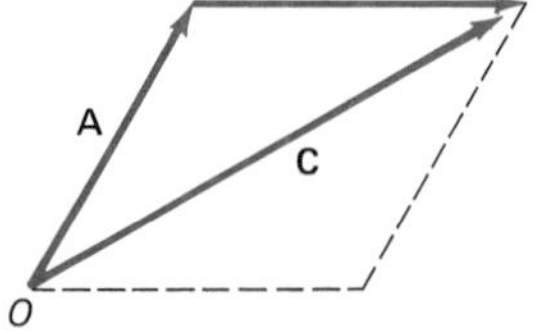

Figure 2.3

Vector addition.

Figure 2.4

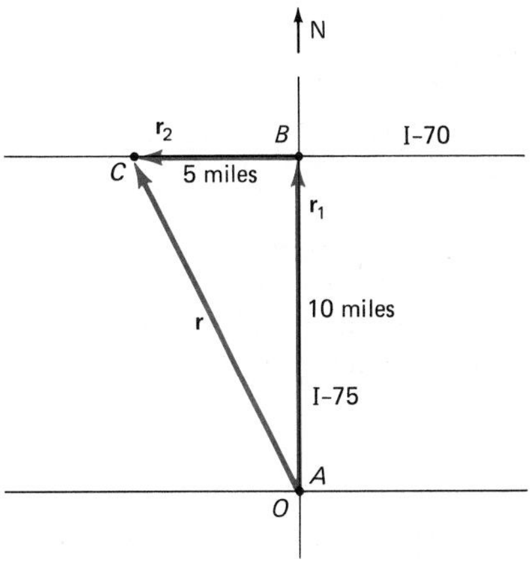

perhaps on a larger scale, may have adequate precision. It does have the advantage of giving you a picture of vectors.

The vector addition of Figure 2.3 is carried out with a triangle (*ABC*). If we complete the parallelogram (Figure 2.5) we see that

$$\mathbf{A} + \mathbf{B} = \mathbf{B} + \mathbf{A} \tag{2.2}$$

The order in which we add the two quantities does not affect the result. Vector addition is therefore commutative, or independent of order.

By adding a third vector, **D**, to **A** + **B**, as in Figure 2.6, we see that the way we group the vectors does not affect the final result.

$$(\mathbf{A} + \mathbf{B}) + \mathbf{D} = \mathbf{A} + (\mathbf{B} + \mathbf{D}) \tag{2.3}$$

With the tail of **B** at the tip of **A** and the tail of **D** at the tip of **B**, the vector sum **A** + **B** + **D** runs from the tail of **A** to the tip of **D** regardless of the order of addition of the vectors. Vector addition is therefore associative, or independent of the grouping of vectors.

The vectors $\mathbf{r}_1$ and $\mathbf{r}_2$ of Example 3 occur in sequence, one after the other. Vector quantities can also act simultaneously, as we see in Example 4.

Example 4 **Vector Addition of Forces**

Two forces act on a common point on the bumper of a car. One force of 20.0 newtons (4.50 lb) acts horizontally. A second force, also 20.0 newtons, acts vertically.

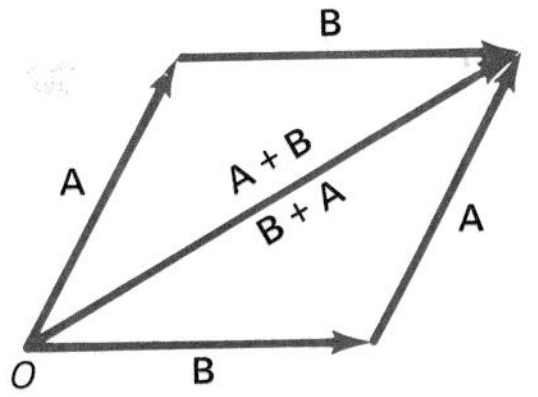

Figure 2.5

Vector addition is commutative.

$$\mathbf{A} + \mathbf{B} = \mathbf{B} + \mathbf{A}$$

Figure 2.6

Vector addition is associative.

$$(\mathbf{A} + \mathbf{B}) + \mathbf{D} = \mathbf{A} + (\mathbf{B} + \mathbf{D})$$

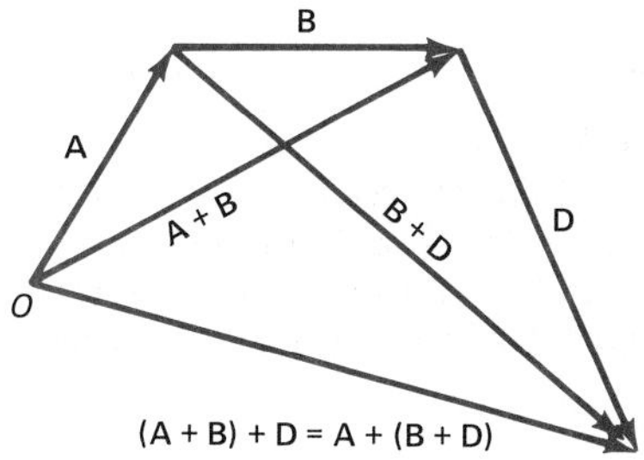

What is the single force equivalent to these two?

First, we must draw a vector diagram (Figure 2.7) to show the two forces. We shall let 1 cm of length in the diagram correspond to 4 newtons of force. Because $\mathbf{F}_1$ and $\mathbf{F}_2$ in this example are equal, the parallelogram defined by the vectors $\mathbf{F}_1$ and $\mathbf{F}_2$ is a square. The equivalent vector **F**, the vector sum of $\mathbf{F}_1$ and $\mathbf{F}_2$, is given by the diagonal. Because vectors $\mathbf{F}_1$ and $\mathbf{F}_2$ form a right angle in this case, the magnitude of **F** is given by the Pythagorean theorem:

$$|\mathbf{F}| = (F_1^2 + F_2^2)^{1/2}$$
$$= \sqrt{2(20 \text{ newtons})^2}$$
$$= 28.3 \text{ newtons}$$

Thus

$$\mathbf{F} = 28.3 \text{ newtons, } 45° \text{ above the horizontal } \mathbf{F}_1$$

In Example 4 we have *superposed* two forces. We have assumed that the result of the two forces acting together is equivalent to that of a single force vector given by the vector sum. This principle of superposition, or justification for treating forces as vectors, ultimately depends on experiment.

Remember, we specified that the two forces $\mathbf{F}_1$ and $\mathbf{F}_2$ act on the same point. By adding the forces we can form the triangle shown in Figure 2.8. To form this triangle, $\mathbf{F}_2$ is displaced sideways, parallel to the original given $\mathbf{F}_2$. This displacement is a mathematical device for adding the two vectors; *it does not imply that* $\mathbf{F}_2$ *acts on point P*. The resultant $\mathbf{F} = \mathbf{F}_1 + \mathbf{F}_2$, whether found as the hypotenuse of a triangle or the diagonal of a parallelogram, acts on the original point, *O*.

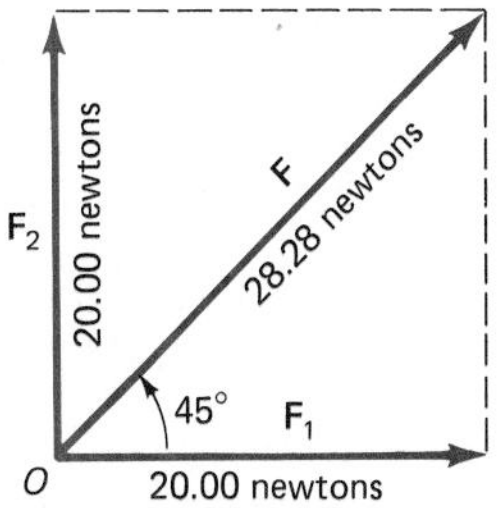

Figure 2.7

Vectors $\mathbf{F}_1$ and $\mathbf{F}_2$ acting at point *O*.

Figure 2.8

The vector F_2 is displaced to P in order to add it to F_1.

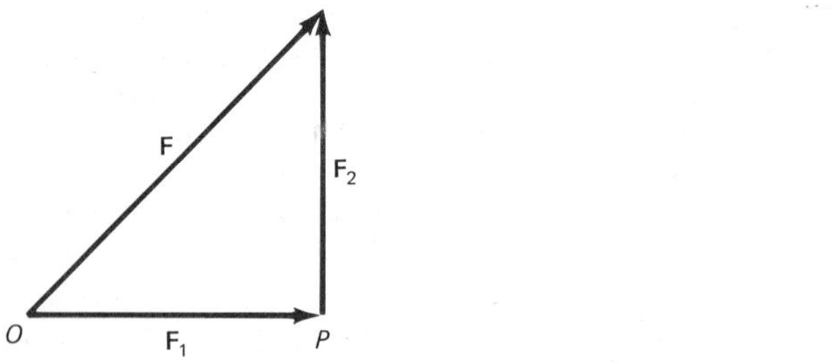

Figure 2.9

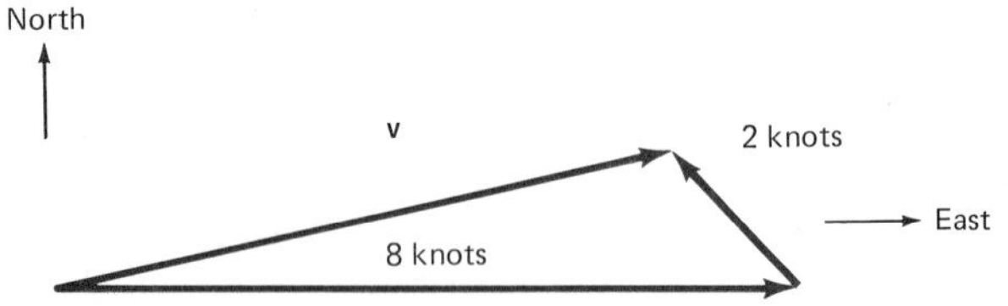

Example 5 Vector Addition of Velocities

A sailboat is moving due east with a velocity of 8 knots (4.12 m/s) relative to the water. Because of the tide, the water is moving northwest (45° west of north) with a velocity of 2 knots relative to the shore. What is the velocity of the sailboat relative to the shore?

In this problem we are asked for the vector sum of two velocities: 8 knots east and 2 knots northwest. The graphical addition of the two velocities is shown in Figure 2.9. The magnitude and direction of the resultant velocity **v** may be obtained from the diagram. Measurement with a ruler and a protractor yields 6.72 knots 12.3° north of due east.

Diagrams or graphs help us to picture what is going on and are also useful for checking calculations. However, diagrams and graphs are awkward to draw when we are dealing with three dimensions. Therefore, in Section 2.3, we will develop an algebraic method of vector addition, which is usually more accurate.

Subtraction of Vectors

If **B** is a vector shown as an arrow in Figure 2.10, then $-\mathbf{B}$ is defined to be the same vector, but in the opposite direction. The sum of a vector and its negative is zero. Therefore the subtraction of a positive vector is defined as the addition of a negative vector. For example, we write $\mathbf{A} - \mathbf{B}$ as

$$\mathbf{A} - \mathbf{B} = \mathbf{A} + (-\mathbf{B}) \tag{2.4}$$

and carry out the subtraction by adding $-\mathbf{B}$. Figure 2.11 illustrates this method for the two vectors **A** and $-\mathbf{B}$. Thus everything we have learned about vector addition may be applied to vector subtraction.

Figure 2.10

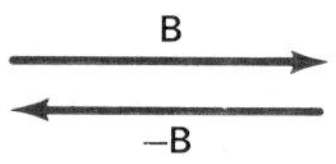

Figure 2.11

Subtraction of vectors.

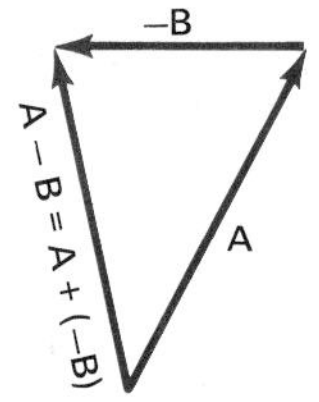

Multiplication by a Scalar

A vector **A** may be multiplied by a number or a scalar, a, to give $a\mathbf{A}$. If a is greater than 1, then this multiplication corresponds to an **expansion** of **A**. If a is less than 1, then the multiplication corresponds to a **contraction** of **A**. When a is positive, the direction of **A** is not affected. If a is negative, the direction of **A** is reversed. For instance $-2\mathbf{A}$ has double the magnitude and is in the opposite direction from **A**.

If the scalar has dimensions, multiplication will yield a completely new vector. For example, in mechanics the vector velocity **v** is often multiplied by a scalar mass m. The product $m\mathbf{v} = \mathbf{p}$, called **linear momentum**, has the dimensions mass × length/time. The new vector **p** is not at all the same as the original vector **v** with dimensions of length/time. Therefore, it would be nonsense dimensionally to try to add $\mathbf{v} + \mathbf{p}$.

At this point in our vector algebra, these important properties of ordinary algebra apply:

$$a\mathbf{v} = \mathbf{v}a \qquad \text{commutation} \tag{2.5}$$

$$a(\mathbf{v}_1 + \mathbf{v}_2) = a\mathbf{v}_1 + a\mathbf{v}_2 \qquad \text{distribution} \tag{2.6}$$

$$a(b\mathbf{v}) = (ab)\mathbf{v} \qquad \text{association} \tag{2.7}$$

We can combine the last two of these properties (distribution and association) and write

$$a(b\mathbf{v}_1 + c\mathbf{v}_2) = (ab)\mathbf{v}_1 + (ac)\mathbf{v}_2 \tag{2.8}$$

To a mathematician, Eq. 2.8 identifies vector algebra as a linear mathematical system. Since almost all of introductory physics is linear, vector algebra is used frequently.

Questions

1. The magnitudes of vectors **A** and **B** are different. Is it possible to have $\mathbf{A} + \mathbf{B} = 0$?
2. Is it possible to add three vectors of equal magnitude but different directions to get zero? If you think so, illustrate with a diagram.
3. The magnitudes of three vectors are 2 m, 4 m, and 8 m, respectively. The directions are at your disposal. Can these three vectors be added to yield zero? With a diagram, show how it is possible or why it is impossible.
4. The vector sum of **A**, **B**, and **C** equals zero. If $|\mathbf{A}| = 10$ newtons and $|\mathbf{B}| = 8$ newtons, what is the maximum possible value of $|\mathbf{C}|$? What is the minimum possible value of $|\mathbf{C}|$?
5. If **C** is the vector sum of **A** and **B**, does **C** have to lie in the plane determined by **A** and **B**?

2.3 Components

When we represent a vector by an arrow in space, a coordinate system is not needed. The graphical addition of two vectors, for instance, is independent of any coordinate system. We will now develop a new vector representation or description that requires the use of a coordinate system. The reason for developing this new representation is that it will permit us to add, subtract, and later multiply vectors analytically.

Figure 2.12 depicts an arrow in space, vector **A**, relative to a two-dimensional $x-y$ reference frame. Consider a line, drawn from the tip of **A** to a point on the x-axis, that intersects this axis perpendicularly. The distance from the origin to this intersection point represents a quantity A_x, which we call the x-component of **A**. Similarly, if we draw a line from the tip of **A** perpendicular to, and intersecting, the y-axis, the distance A_y from the intersection point to the origin is called the y-component of **A**.

You may find it helpful to think of the vector components as **projections**, or shadows, of the vector upon the axes. Imagine parallel rays of light traveling perpendicular to the x-axis (Figure 2.13). The vector **A**, blocking part of these rays, casts a shadow of length A_x on the x-axis. Similarly, if parallel rays of light travel perpendicular to the y-axis, vector **A** casts a shadow of length A_y on the y-axis (Figure 2.13).

Component A_x forms a right triangle with **A** and the x-axis, and component A_y forms a right triangle with **A** and the y-axis. The relations between the magnitude of **A** and its components A_x and A_y therefore involve trigonometric functions. As you probably know, in any triangle having a right angle, the side opposite the right angle is called the **hypotenuse.** If we call one of the two angles that are not equal to 90° α, then the side adjacent to this angle is called the **adjacent,** and the side opposite α is called the **opposite.** The ratio of the adjacent to the hypotenuse is equal to the cosine of angle α. The cosine is a function of the angle α and, as shown in Figure 2.14a, is independent of the size of the triangle.

The ratio of the opposite to the hypotenuse is the sine of α. The sine is a function of angle α and, as also shown in Figure 2.14a, is independent of the size of the triangle. The sine and cosine are considered further in Appendix 4. You might want to review this material before continuing. In particular, note that sin α is negative when α extends into the third and fourth quadrants; cos α is negative when α extends into the second and third quadrants. Looking back at Figure 2.12, we see that

$$\frac{A_x}{A} = \cos \alpha$$

and

$$\frac{A_y}{A} = \cos \beta$$

Solving for A_x and A_y, we find

$$A_x = A \cos \alpha \qquad (2.9a)$$

and

$$A_y = A \cos \beta \qquad (2.9b)$$

A component is equal to the product of the magnitude of the vector **A** and the cosine of the angle included between the positive axis and the vector. Thus, because A is positive, the components are positive or negative, depending on whether the cosines are positive or negative.

So far we have considered vector components along the coordinate axes only, but a vector can have a component in any direction. In Figure 2.13b, for example, the

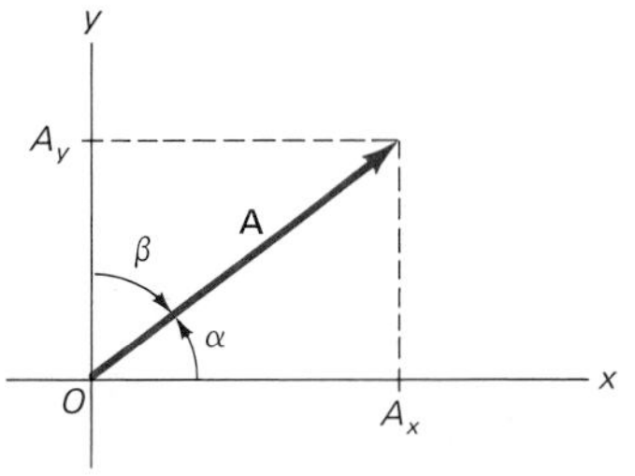

Figure 2.12

A vector and its components.

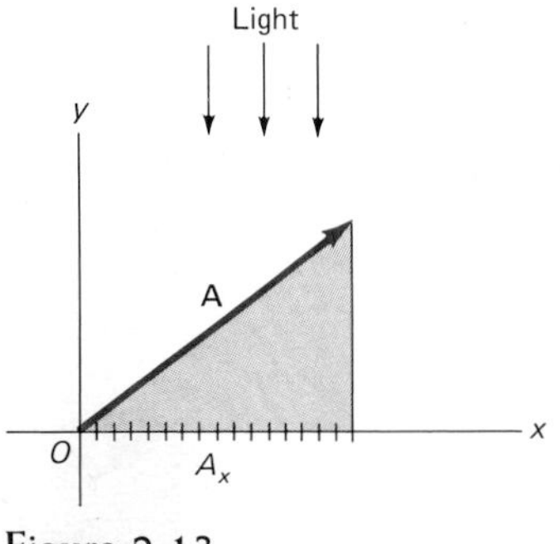

Figure 2.13

A component is a projection.

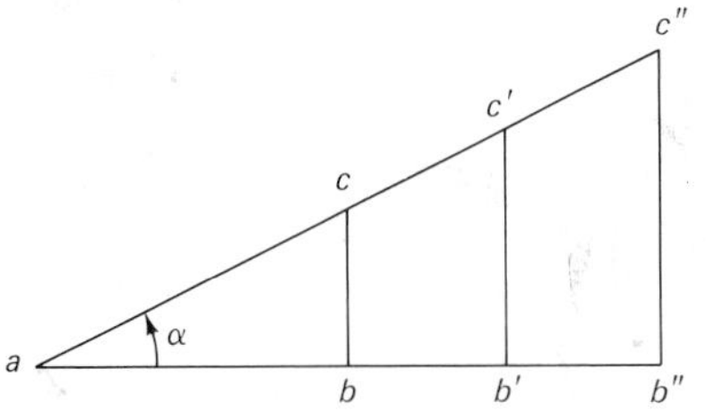

Figure 2.14a

Similar triangles. The ratio of the sides is independent of the size of the triangle.

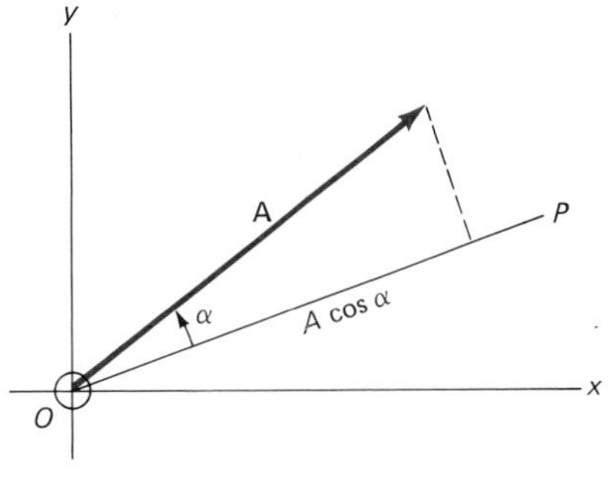

Figure 2.14b

component of **A** along the line *OP* is found by drawing a line from the tip of **A** perpendicular to *OP*; and, like components along the coordinate axes, this component is given by $A \cos \alpha$. If **A** is a force, the component along *OP* may be thought of as the effective force in the direction of *OP*.

Example 6 **Components of a Two-Dimensional Vector**

Suppose we have a velocity vector of 10 m/s, 35° north of east. What are the easterly (x) and northerly (y) components of the velocity? From Figure 2.15 we see that $\alpha = 35°$ and $\beta = 55°$. Then with a calculator we find

$$v_{\text{east}} = 10 \text{ m/s} (\cos 35°) = 10 \text{ m/s} \times 0.819$$
$$= 8.19 \text{ m/s}$$

and

$$v_{\text{north}} = 10 \text{ m/s} (\cos 55°) = 10 \text{ m/s} \times 0.574$$
$$= 5.74 \text{ m/s}$$

The component of a vector may be negative. For example, in Figure 2.16 **B** is in the second quadrant. Because the angle α running from the positive x-axis to **B** is about 140°, $\cos \alpha$, and therefore B_x, is negative. (Remember, a component is equal to the product of the vector magnitude and the cosine of the angle included between the *positive* axis and the vector.) Often, however, the measurement of the angles to the nearest axes is more convenient. For instance, if **A** is in the third quadrant, as in Figure 2.17, the angles ϕ and θ, measured from the negative axes, might be used. The x- and y-components would then be given by

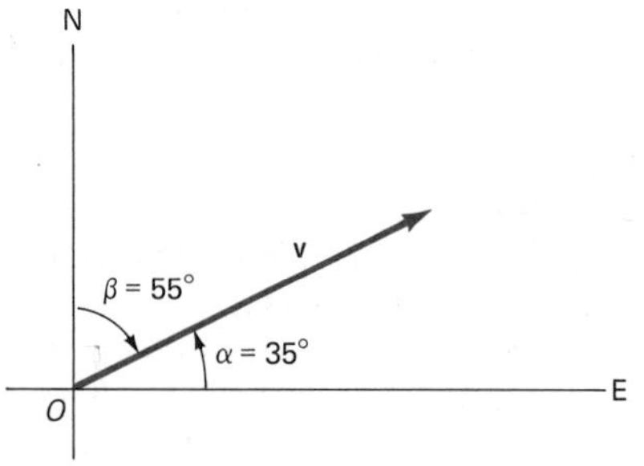

Figure 2.15

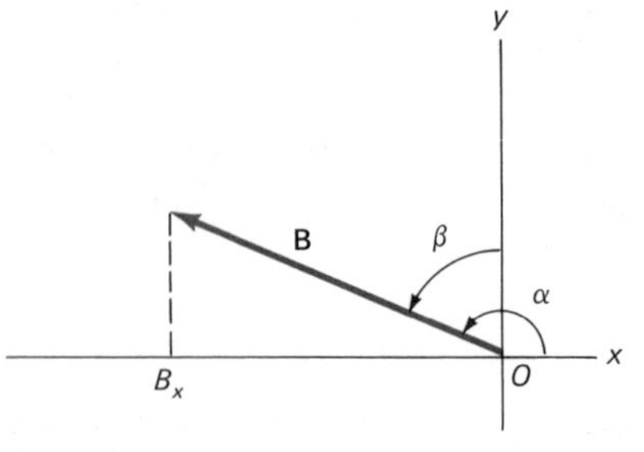

Figure 2.16

A negative x-component.

Figure 2.17

Angles may be measured from the negative axes, but note the negative signs in Eq. 2.10.

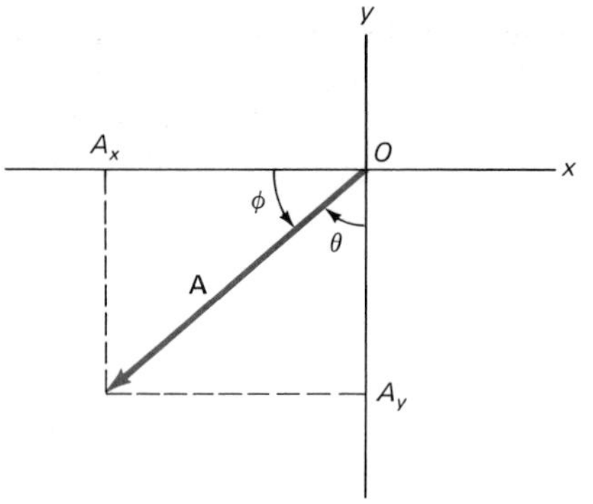

$$A_x = -A \cos \phi \tag{2.10a}$$

and

$$A_y = -B \cos \theta \tag{2.10b}$$

The negative signs are inserted because we can see that the components are negative. This "look and see" technique of component calculation can be valuable as a check on the formal calculation (Eq. 2.9), and will be used frequently in Chapter 3 and subsequent chapters.

One other alternative to Eq. 2.9 is commonly used. Returning to Figure 2.15, we see that $\alpha + \beta = 90°$ Therefore $\cos \beta = \sin \alpha$ and Eq. 2.9 may be rewritten as

$$A_x = A \cos \alpha \tag{2.11a}$$

and

$$A_y = A \sin \alpha \tag{2.11b}$$

These equations will also be used in Chapter 3 and later chapters. The relation $\alpha + \beta = 90°$ is valid, however, only if **A** is in the x–y plane. Therefore, the use of $\sin \alpha$ in place of $\cos \beta$ is limited to two-dimensional problems and should not be used in three-dimensional problems.

Example 7 **Components of Sailboat Velocity**

As in Example 5, a sailboat has a velocity $\mathbf{v}_1$ of 8 knots east relative to the water. Because of the tide the water is moving with a velocity $\mathbf{v}_2$ of 2 knots northwest (45° west of north). Find the easterly (positive x-axis) and the northerly (positive y-axis) components of each velocity. (See Figure 2.18.)

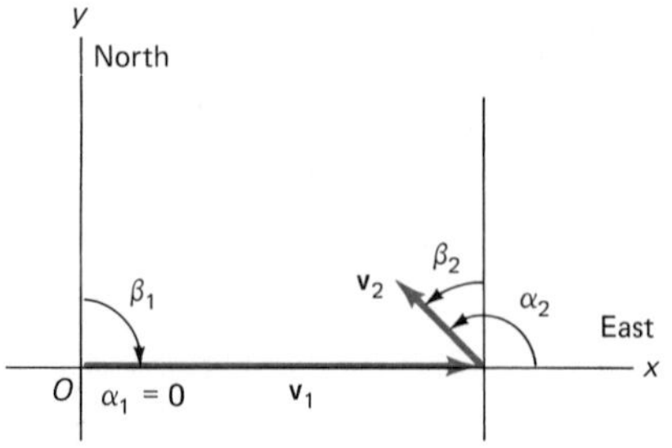

Figure 2.18

For the sailboat's motion relative to the water $v_1 = 8$ knots, $\alpha_1 = 0$, and $\beta_1 = 90°$. From Eq. 2.9, $v_{1\,(\text{east})} = 8 \cos 0°$ knots $= 8$ knots, and $v_{1\,(\text{north})} = 8 \cos 90°$ knots $= 0$ knots.

For the tidal flow, $v_2 = 2$ knots, $\alpha_2 = 135°$ and $\beta_2 = 45°$. Then $v_{2\,(\text{east})} = 2 \cos 135°$ knots $= -1.41$ knots, and $v_{2\,(\text{north})} = 2 \cos 45°$ knots $= 1.41$ knots. The minus sign for $v_{2\,(\text{east})}$ tells us that the water has a component in the westerly direction.

Three-Dimensional Vectors

So far we have used an $x-y$ coordinate system to examine vectors in two-dimensional space. In order to consider three-dimensional space, we add a z-axis perpendicular to the plane of x- and y-coordinates and then choose the positive direction of this z-axis by using what we call the **right-hand rule.** Let the fingers of your right hand curl from the positive x-axis to the positive y-axis. Your (extended) thumb indicates the positive z-direction for a right-handed system (see Figure 2.19). Figure 2.20 depicts the position of the vector **A** relative to the $x-y-z$ reference frame. As a physical example of vector **A**, think of a meter stick pointing diagonally into a room from a corner.

In Figure 2.20, vector **A** forms angles with each of the positive coordinate axes. We designate α as the angle from the x-axis to **A**, β as the angle from the y-axis to **A**, and γ as the angle from the z-axis to **A**. The projections of **A** upon the three coordinate axes are

$$A_x = A \cos \alpha \qquad (2.12a)$$

$$A_y = A \cos \beta \qquad (2.12b)$$

$$A_z = A \cos \gamma \qquad (2.12c)$$

These three projections are vector components: A_x is the x-component of **A**, A_y is the y-component of **A**, and A_z is the z-component of **A**. The three cosines of the angles that specify the direction of **A** relative to the coordinate axes are called **direction cosines.**

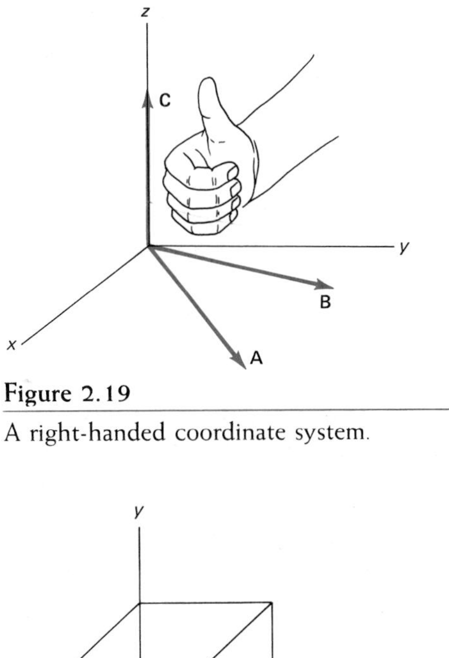

Figure 2.19

A right-handed coordinate system.

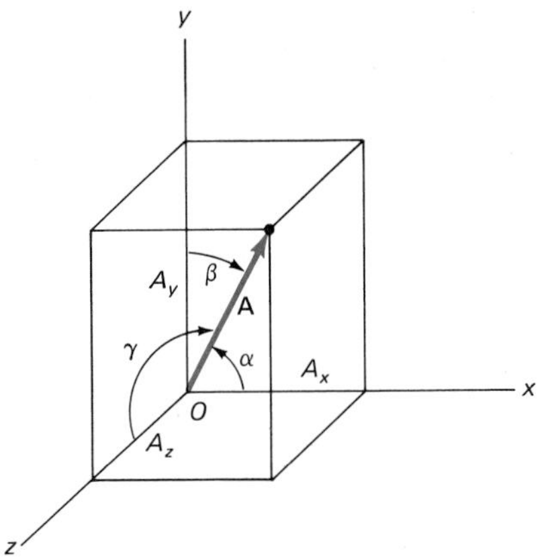

Figure 2.20

Components in three dimensions.

Example 8 **The Components of the Position Vector**

In Example 1 the position vector **r** was described by an arrow directed from the origin to a point (x, y, z) What are the components of **r**?

The magnitude of **r** is r and $\cos \alpha = x/r$. By substituting these values into Eq. 2.9a, we find

$$r_x = r\frac{x}{r}$$

or $r_x = x$. Similarly, the y-component of **r** is y and the z-component is z.

The position vector may be described either by **r** (the arrow-in-space method) or by its three components: x, y, and z.*

Equation 2.12 can be used to find the three Cartesian components A_x, A_y, and A_z from the magnitude of **A** and the direction angles α, β, and γ. Conversely, given the components, we can find the magnitude of **A**, and the direction angles α, β, and γ.

$$A = (A_x^2 + A_y^2 + A_z^2)^{1/2} \qquad (2.13)$$

Knowing the components and the magnitude A, we can then solve Eq. 2.12 for the direction angles α, β, and γ.

Unit Vectors

An advantage of using vector components is that graphical or geometric manipulation of complicated vector problems can be replaced by algebraic and numerical manipulation. To facilitate this algebraic and numerical manipulation we introduce three **unit vectors i**, **j**, and **k** (Figure 2.21). For a specific, fixed Cartesian system, **i** is in the positive x-direction, **j** is in the positive y-direction, and **k** is in the positive z-direction. These three vectors are called unit vectors because each of them has unit magnitude. They are also dimensionless. The three unit vectors **i**, **j**, and **k** do not change the magnitude or the dimensions of anything; they only indicate directions. The unit vector **i** has components 1, 0, and 0; **j** has components 0, 1, and 0; and **k** has components 0, 0, and 1.

*Our two representations, or descriptions, of vectors as arrows in space and as sets of components (in a known coordinate system) are fully equivalent. Each can be derived from the other. Each contains the same information.

Figure 2.21

Unit-magnitude vectors **i**, **j**, and **k**.

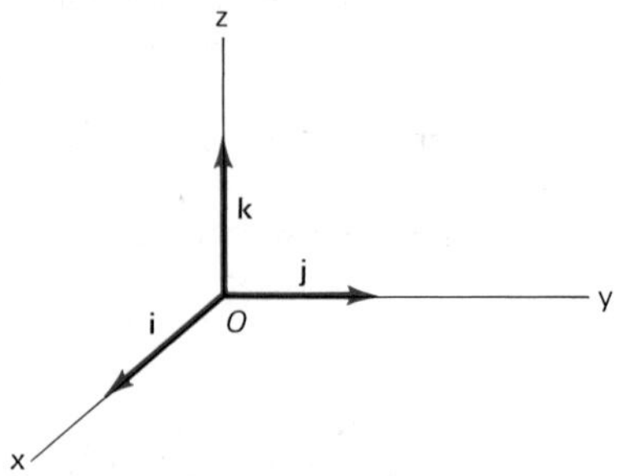

With these unit vectors we can form vectors $\mathbf{i}A_x$, $\mathbf{j}A_y$, and $\mathbf{k}A_z$ in our specific, fixed coordinate system. Then, by the definition of addition in Section 2.2,

$$\mathbf{A} = \mathbf{i}A_x + \mathbf{j}A_y + \mathbf{k}A_z \tag{2.14}$$

Because a vector is specified by its components, the collection of components (A_x, A_y, A_z) is commonly referred to as a vector.

Example 9 **From Components to Vector**

The coordinate point $(3, 2, -4)$ defines a position vector **r** from the origin with components 3, 2, and -4. Using unit vectors we can represent **r** as

$$\mathbf{r} = 3\mathbf{i} + 2\mathbf{j} - 4\mathbf{k}$$

Vector **r** can be pictured as an arrow from the origin to the coordinate point $(3, 2, -4)$. The right-hand side of this equation exhibits the components of **r** explicitly.

Equation 2.14 suggests an alternate way of looking at components. Restricting ourselves to two-dimensional space for convenience, let's ask the question, "What two vectors, one along the x-axis and the other along the y-axis, will add to yield **A**?" In two dimensions, Eq. 2.14 reduces to

$$\mathbf{A}_1 + \mathbf{A}_2 = \mathbf{A} \tag{2.15}$$

These vectors are pictured in Figure 2.22. We can say that **A** has been *resolved into its component vectors* $\mathbf{A}_1$ and $\mathbf{A}_2$. The

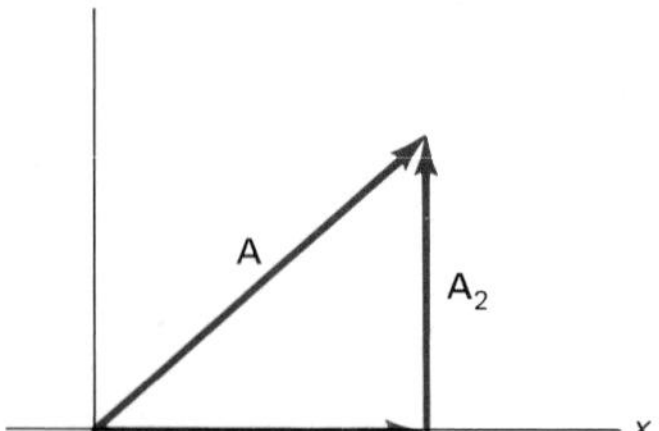

Figure 2.22

Vector addition.

length of $\mathbf{A}_1$ can be found by drawing a straight line from the tip of **A** perpendicular to the x-axis. The x-component of **A**, A_x, is thus the same as the magnitude of $\mathbf{A}_1$. Similarly, the y-component of **A**, A_y, is the same as the magnitude of $\mathbf{A}_2$.

Definitions in Component Form

Let's now redefine some vector operations in terms of unit vectors and components.

Equality. **A** = **B** when **A** and **B** have the same length, the same direction, and the same physical dimensions. We can express this equality in terms of components as follows:

$$\mathbf{i}A_x + \mathbf{j}A_y + \mathbf{k}A_z = \mathbf{i}B_x + \mathbf{j}B_y + \mathbf{k}B_z$$

This equation implies that

$$A_x = B_x \tag{2.16a}$$

$$A_y = B_y \tag{2.16b}$$

and

$$A_z = B_z \tag{2.16c}$$

If two vectors are equal, their x-components are equal to each other, their y-components are equal to each other, and their z-components are equal to each other. Here you can begin to see the power of vector algebra. The single vector equation **A** = **B** is equivalent to three numeric equations, Eqs. 2.16a, b, and c. The vector equation is thus a compact and powerful way of expressing mathematical and physical relations. You should note that it does not make sense to write an equation of the form

$$\text{vector} = \text{scalar}$$

because Eqs. 2.16a, b, and c cannot then be satisfied.

Addition. In component form, the addition of two vectors

$$\mathbf{C} = \mathbf{A} + \mathbf{B}$$

becomes

$$\begin{aligned}\mathbf{i}C_x + \mathbf{j}C_y + \mathbf{k}C_z &= (\mathbf{i}A_x + \mathbf{j}A_y + \mathbf{k}A_z) \\ &\quad + (\mathbf{i}B_x + \mathbf{j}B_y + \mathbf{k}B_z) \\ &= \mathbf{i}(A_x + B_x) + \mathbf{j}(A_y + B_y) \\ &\quad + \mathbf{k}(A_z + B_z)\end{aligned}$$

Equating corresponding components on both sides of the equation we may write

$$C_x = A_x + B_x \tag{2.17a}$$

$$C_y = A_y + B_y \tag{2.17b}$$

and

$$C_z = A_z + B_z \tag{2.17c}$$

This method of equating corresponding components can be extended to subtraction and to any number of terms. A graphical interpretation of Eqs. 2.17a and 2.17b is shown in Figure 2.23.

Figure 2.23

Vector addition by components. This figure shows two of the three dimensions.

$C_x = A_x + B_x$ and $C_y = A_y + B_y$

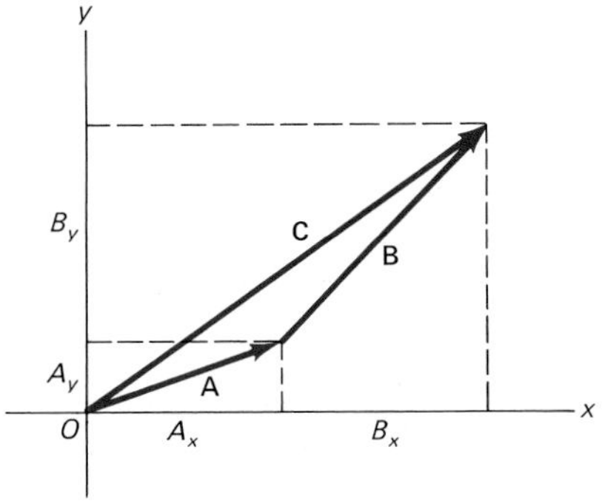

Example 10 Vector Addition by Components

A displacement vector $\mathbf{B} = 2\mathbf{i} + 4\mathbf{j} + 2\mathbf{k}$ meters is added to a displacement vector $\mathbf{A} = 1\mathbf{i} - 2\mathbf{j} + 4\mathbf{k}$ meters. What is the resultant displacement, $\mathbf{C} = \mathbf{A} + \mathbf{B}$?

Adding corresponding components, we write

$$C_x = 1 \text{ m} + 2 \text{ m} = 3 \text{ m}$$

$$C_y = -2 \text{ m} + 4 \text{ m} = 2 \text{ m}$$

$$C_z = 4 \text{ m} + 2 \text{ m} = 6 \text{ m}$$

The vector sum of **A** and **B** can now be written as

$$\mathbf{C} = 3\mathbf{i} + 2\mathbf{j} + 6\mathbf{k} \text{ m}$$

Example 10 involves the addition of numbers, and therefore might be called vector arithmetic. Vector addition by components (Eq. 2.17) also holds when the components are represented by algebraic symbols.

Example 11 Momentum Conservation in a Proton–Proton Collision

A proton with linear momentum $\mathbf{p}_i$ in the positive x-direction collides with a second stationary proton in such a way that the moving proton is deflected at an angle of 45° with the x-axis, as shown in Figure 2.24. Its linear momentum after the collision is $\frac{1}{2}p(\mathbf{i} + \mathbf{j})$. The other proton recoils with linear momentum $\mathbf{q}_f$. Determine the linear momentum $\mathbf{q}_f$ of the second particle after the collision.

Conservation of linear momentum (Chapter 9) requires that the vector sum of the linear momenta before the collision equal the vector sum of the linear momenta after the collision:

$$\mathbf{p}_i = \mathbf{p}_f + \mathbf{q}_f$$

Here the subscript i stands for initial while the subscript f stands for final. (Subscript i is not related to unit vector **i**.) Our problem is to find the final momentum of the second proton.

We have $\mathbf{p}_i = p\mathbf{i}$ and $\mathbf{p}_f = \frac{1}{2}p(\mathbf{i} + \mathbf{j})$. From the conservation of linear momentum equation we have

$$\mathbf{q}_f = \mathbf{p}_i - \mathbf{p}_f$$

Substituting the given values on the right side leads to

$$\mathbf{q}_f = p\mathbf{i} - \tfrac{1}{2}p(\mathbf{i} + \mathbf{j})$$
$$= \mathbf{i}(\tfrac{1}{2}p) + \mathbf{j}(-\tfrac{1}{2}p)$$

Our value for $\mathbf{q}_f$ shows that the second proton recoils at 45° below the x-axis. Prove this to yourself by using two nickels on a smooth table.

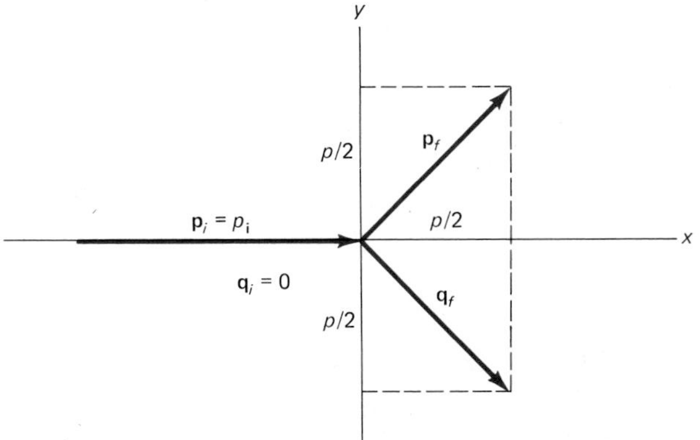

Figure 2.24

Conservation of linear momentum.

Radial and Normal Unit Vectors

In addition to **i**, **j**, and **k**, two other unit vectors are used in this text. A vector of unit magnitude in the radially outward direction is designated by $\hat{\mathbf{r}}$. The unit vector $\hat{\mathbf{r}}$ is formed by taking the position vector **r** and dividing it by its magnitude:

$$\hat{\mathbf{r}} = \mathbf{r}/r \qquad (2.18)$$

This procedure gives $\hat{\mathbf{r}}$ unit magnitude but preserves the radial direction of **r**. Unlike the three Cartesian unit vectors, $\hat{\mathbf{r}}$ may vary in direction because the position vector **r** varies in direction.

The unit vector $\hat{\mathbf{n}}$ is normal, or perpendicular, to a surface at a given point (Figure 2.25). For the special case of a spherical surface (origin at the center) the normal vector $\hat{\mathbf{n}}$ is radial and $\hat{\mathbf{n}} = \hat{\mathbf{r}}$.

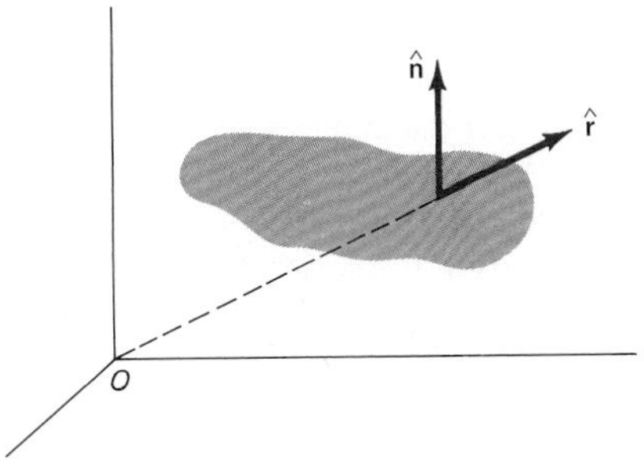

Figure 2.25

The unit vector $\hat{\mathbf{r}}$ points radially outward. The unit vector $\hat{\mathbf{n}}$ is normal (perpendicular) to the surface.

Questions

6. How can the direction cosines be negative? Describe the orientation of a vector having all three direction cosines negative.

7. The components of a vector are given by the corresponding coordinates: $A_x = x$, $A_y = y$, and $A_z = z$. What is the vector?

8. Explain why Eq. 2.11 (the replacement of $\cos \beta$ by $\sin \alpha$) is limited to two-dimensional space, that is, to a vector lying in the plane of the two axes.

9. What units are associated with the unit vectors **i**, **j**, and **k**?

10. Is the vector sum $\mathbf{i} + \mathbf{j} + \mathbf{k}$ of the three unit vectors a unit vector in the sense of having unit magnitude? Explain.

11. Explain why a vector equation may contain more information than a scalar equation.

12. At every point on a particular finite, closed surface the unit vector $\hat{\mathbf{r}}$ is equal to the unit vector $\hat{\mathbf{n}}$. What kind of a surface do you have?

13. Sketch a football. Draw the unit vectors $\hat{\mathbf{r}}$ and $\hat{\mathbf{n}}$ at several points along the surface of the football. Where is $\hat{\mathbf{r}} = \hat{\mathbf{n}}$? Where is $\hat{\mathbf{r}} \neq \hat{\mathbf{n}}$?

2.4 The Scalar, or Dot, Product

So far vector algebra has pretty much followed the rules of ordinary algebra. We can add and subtract vectors in much the same way that we add and subtract numbers. We cannot, however, multiply two vectors as we would multiply two numbers. Instead, we need a new definition of multiplication that is applicable to vectors. The scalar product and the vector product (Section 2.5) are two forms of vector multiplication. As you go through this text you will see that these two products have many applications in describing the physical world.

The **scalar product** of two vectors **A** and **B**, Figure 2.26, is indicated by a dot between **A** and **B** and is defined as

$$\mathbf{A} \cdot \mathbf{B} = AB \cos \theta \qquad (2.19)$$

the product of the two vector magnitudes and the cosine of the included angle θ. $\mathbf{A} \cdot \mathbf{B}$ is read "A dot B." $\mathbf{A} \cdot \mathbf{B}$ is called a scalar product even though it is a form of vector multiplication because the **result** is a scalar. It is also called a **dot product** because it is represented by a dot between the two vectors. In terms of components, two equivalent interpretations of $\mathbf{A} \cdot \mathbf{B}$ are possible:

$$\mathbf{A} \cdot \mathbf{B} = A(B \cos \theta)$$
$$= A \text{ times the projection of } \mathbf{B} \text{ on } \mathbf{A}$$

and

$$\mathbf{A} \cdot \mathbf{B} = B(A \cos \theta)$$
$$= B \text{ times the projection of } \mathbf{A} \text{ on } \mathbf{B}$$

Note that if **A** and **B** both have dimensions, the product $\mathbf{A} \cdot \mathbf{B}$ has dimensions different from those of either **A** or **B**.

If a vector is multiplied by itself, then $\theta = 0$, and $\cos \theta = 1$. In this case the product is the square of the magnitude of that vector: $\mathbf{A} \cdot \mathbf{A} = A^2$.

Example 12 Unit Vector $\hat{\mathbf{n}}$

In Figure 2.27 we see a position vector $\mathbf{r}$ ($\mathbf{r} \neq 0$) lying in a plane. A unit vector $\hat{\mathbf{n}}$ is combined with **r** as a scalar product. If $\hat{\mathbf{n}} \cdot \mathbf{r} = 0$ for every choice of **r**, what is the direction of $\hat{\mathbf{n}}$ relative to the plane?

$$\hat{\mathbf{n}} \cdot \mathbf{r} = (1)r \cos \theta = 0$$

Because $r \neq 0$ we must have $\cos \theta = 0$; thus $\theta = 90°$ or $270°$, which implies that $\hat{\mathbf{n}}$ is perpendicular to **r**. Because this direction holds for every position vector **r**, $\hat{\mathbf{n}}$ is perpendicular or normal to the plane.

Applying the definition of scalar product to the unit vectors **i**, **j**, and **k**, we see that when one of these unit vectors is dotted into itself the result is 1 (for instance, $\mathbf{i} \cdot \mathbf{i} = 1$), because the cosine of the included angle (0°) is unity. When one of these three Cartesian unit vectors is dotted into a different one of the three unit vectors we get 0 (for instance, $\mathbf{i} \cdot \mathbf{j} = 0$), because the cosine of the included angle (90°) is zero. We exhibit these results in Table 2.1.

The scalar, or dot, product of two vectors may be written in component form as

$$\mathbf{A} \cdot \mathbf{B} = (\mathbf{i}A_x + \mathbf{j}A_y + \mathbf{k}A_z) \cdot (\mathbf{i}B_x + \mathbf{j}B_y + \mathbf{k}B_z)$$

Applying the usual distributive and associative laws, we can expand this equation as

$$\begin{aligned}\mathbf{A} \cdot \mathbf{B} = {} & (\mathbf{i} \cdot \mathbf{i})A_xB_x + (\mathbf{i} \cdot \mathbf{j})A_xB_y + (\mathbf{i} \cdot \mathbf{k})A_xB_z \\ & + (\mathbf{j} \cdot \mathbf{i})A_yB_x + (\mathbf{j} \cdot \mathbf{j})A_yB_y + (\mathbf{j} \cdot \mathbf{k})A_yB_z \\ & + (\mathbf{k} \cdot \mathbf{i})A_zB_x + (\mathbf{k} \cdot \mathbf{j})A_zB_y + (\mathbf{k} \cdot \mathbf{k})A_zB_z\end{aligned}$$

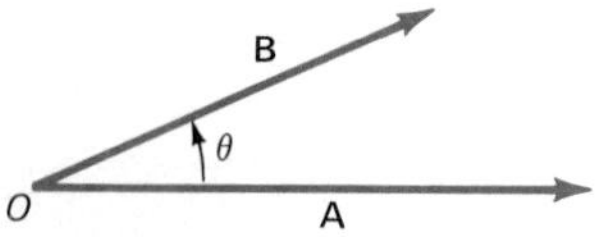

Figure 2.26

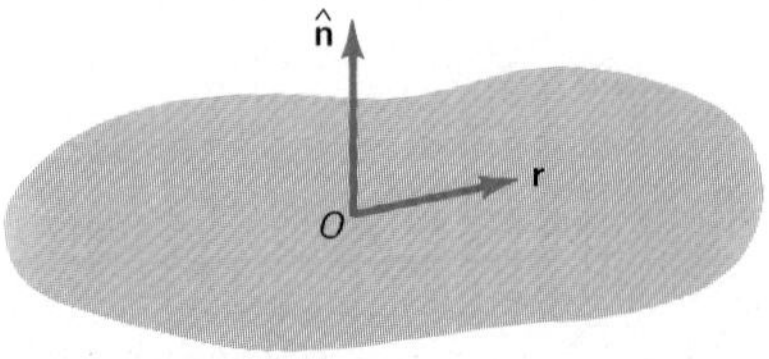

Figure 2.27

Table 2.1

Scalar products of the unit vectors

First factor	Second factor i	j	k
i	1	0	0
j	0	1	0
k	0	0	1

From the definition of scalar product the six scalar products $\mathbf{i}\cdot\mathbf{j}$, $\mathbf{i}\cdot\mathbf{k}$, $\mathbf{j}\cdot\mathbf{i}$, $\mathbf{j}\cdot\mathbf{k}$, $\mathbf{k}\cdot\mathbf{i}$, and $\mathbf{k}\cdot\mathbf{j}$ are equal to zero; thus

$$\mathbf{A}\cdot\mathbf{B} = A_xB_x + A_yB_y + A_zB_z \tag{2.20}$$

The component form of the scalar product is particularly convenient when the vectors are specified in terms of their components. From Eq. 2.20 we see that the scalar product is commutative:

$$\mathbf{A}\cdot\mathbf{B} = \mathbf{B}\cdot\mathbf{A} \tag{2.21}$$

We can also demonstrate commutivity directly from the definition, Eq. 2.19.

$$\mathbf{B}\cdot\mathbf{A} = BA\cos(-\theta) = AB\cos\theta = \mathbf{A}\cdot\mathbf{B}$$

Example 13 **Perpendicular Vectors**

A motorcyclist going across a field has a velocity $\mathbf{v} = 30\mathbf{i} - 10\mathbf{j}$ (miles per hour). The wind has a velocity of $\mathbf{V} = 6\mathbf{i} + 18\mathbf{j}$ (miles per hour). We want to find the angle between these two velocity vectors.

We use both forms of the scalar product:

$$\mathbf{v}\cdot\mathbf{V} = vV\cos\theta$$

$$\mathbf{v}\cdot\mathbf{V} = v_xV_x + v_yV_y$$

Then,

$$vV\cos\theta = v_xV_x + v_yV_y$$

Substituting numerical values for the components, we write

$$\begin{aligned} vV\cos\theta &= v_xV_x + v_yV_y \\ &= (30\times 6) + (-10\times 18) \\ &= 0\ (\text{mi/hr})^2 \end{aligned}$$

We know that neither $\mathbf{v}$ nor $\mathbf{V}$ has zero magnitude. Hence $\cos\theta$ must be zero, which means that $\theta = 90°$ or $270°$. If the dot product of two nonzero vectors is zero, the vectors are perpendicular.

One of the most important applications of the scalar product in physics is in the calculation of the work done by a force acting through a distance.

Example 14 **Work, the Scalar Product of Force and Displacement**

A parachute glider carrying an intrepid tourist is pulled across the Acapulco harbor by a power boat (Figure 2.28). The rope to the glider makes an angle of 30° with the horizontal and exerts a force of 1335 newtons (300 lb). Calculate the work done in moving the glider and tourist 100 m horizontally.

In Chapter 7 the work done by a constant force $\mathbf{F}$ acting over a displacement $\mathbf{s}$ is defined as $W = \mathbf{F}\cdot\mathbf{s}$. Using this relation we have

$$\begin{aligned} W &= 1335\cdot 100\cos 30°\ \text{newton}\cdot\text{meters} \\ &= 1335\cdot 100\cdot 0.8660\ \text{N}\cdot\text{m} \\ &= 1.156\times 10^5\ \text{N}\cdot\text{m} \end{aligned}$$

Because the definition of a scalar product, Eq. 2.19, may be interpreted as the product of one vector and the component of the other along the direction of the first, we may also write

$$\begin{aligned} W &= (F\cos\theta)\,s \\ &= (1335\cdot 0.8660)100 \\ &= 1156\cdot 100 = 1.156\times 10^5\ \text{N}\cdot\text{m} \end{aligned}$$

The component of the force that is effective in doing work is $F\cos\theta = 1156$ N.

The equivalence of the two forms of $\mathbf{A}\cdot\mathbf{B}$, that is, $AB\cos\theta$ and $A_xB_x + A_yB_y + A_zB_z$, is illustrated in the following example.

Example 15 **Work**

A force of 20 N acts on a body and moves it 2 m. The line of motion of the body and the direction of the force are shown in Figure 2.29. Calculate the work done, $W = \mathbf{F}\cdot\mathbf{s}$, by using the two forms of the scalar product.

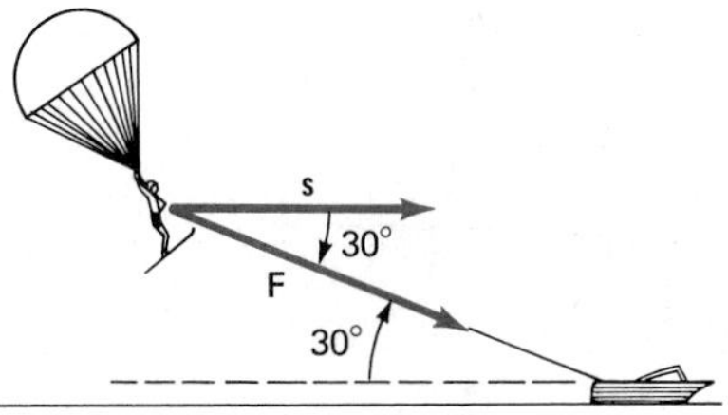

Figure 2.28

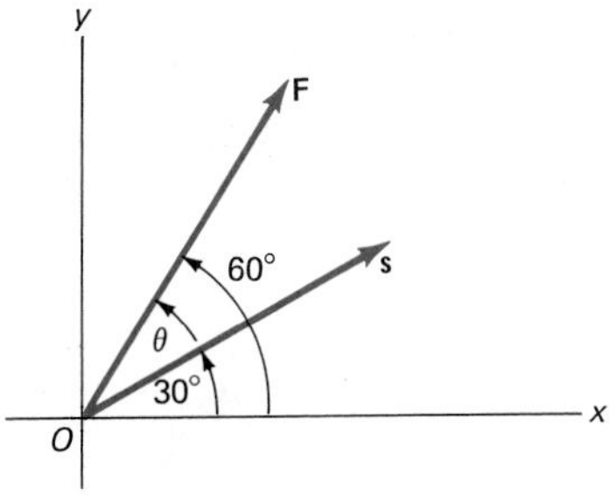

Figure 2.29

From Eq. 2.19

$$(1)\ W = Fs \cos \theta$$
$$= 20 \times 2 \cos 30° = 34.6\ \text{N} \cdot \text{m}$$

With $\mathbf{s} = \mathbf{i}x + \mathbf{j}y + \mathbf{k}z$, Eq. 2.20 yields

$$(2)\ W = F_x x + F_y y + F_z z$$
$$= (20 \cos 60° \times 2 \cos 30°)$$
$$+ (20 \cos 30° \times 2 \cos 60°) + (0 \times 0)$$
$$= (10 \times 1.73) + (17.3 \times 1)$$
$$= 34.6\ \text{N} \cdot \text{m}$$

The two formulas produce identical results, as they must, because they are mathematically equivalent.

Questions

14. Can a scalar product of two vectors be negative? If your answer is yes, give an example; if no, provide a proof.

15. You know that $\mathbf{A} \cdot \mathbf{B} = 0$. Can you conclude that **A** and **B** are perpendicular? What other information, if any, do you need to prove that they are?

16. Explain how commutivity follows from the definition of the dot product as $AB \cos \theta$.

17. The earth's gravitational force on an artificial satellite is radially inward, toward the center of the earth. For a satellite in a circular orbit, explain why the gravitational force does no work. Sketch a portion of the orbit. Draw in **F** and the distance **s** covered in some small time interval (such as a picosecond).

18. An artificial satellite is in an elliptic orbit around the earth. Pick a point on the ellipse where $\hat{\mathbf{r}}$ and $\hat{\mathbf{n}}$ are *not* parallel. ($\hat{\mathbf{n}}$ is in the plane of the ellipse but perpendicular to the curve.) At that point, is the earth's gravitational force doing work on the satellite? How does the algebraic sign of the work depend on the direction of motion of the satellite?

19. Your car has just rolled down a boat-launching ramp into a river. Water is moving past the windshield at velocity **v**. The normal to the surface of your windshield is described by the unit vector $\hat{\mathbf{n}}$. What is the relation between the statement "The water does not flow through the windshield" and the equation $\mathbf{v} \cdot \hat{\mathbf{n}} = 0$?

2.5 The Vector, or Cross, Product

The **vector product,** or **cross product**, of two vectors is a vector. If **C** is the vector product of **A** and **B**, we write

Figure 2.30

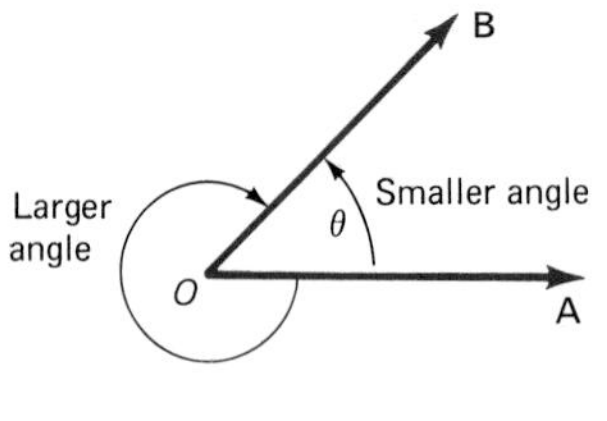

$$\mathbf{C} = \mathbf{A} \times \mathbf{B}$$

(which we read as "**C** equals **A** cross **B**"). The magnitude of **C** is defined to be equal to the product of the magnitudes of **A** and **B** and the sine of the smaller angle (see Figure 2.30) measured from the first vector **A** to the second vector **B**:

$$C = AB \sin \theta \tag{2.22}$$

The direction of vector **C** is perpendicular to both **A** and **B** and is therefore perpendicular to the plane defined by **A** and **B**. There are, however, two possible directions perpendicular to this plane. We remove this ambiguity by using the right-hand rule. As shown in Figure 2.31, let the fingers of your right hand curl from the first vector **A** to the second vector **B** through the smaller angle joining them. Your thumb will then give the direction of **C**, perpendicular to both **A** and **B**. If **A** and **B** both lie in the x–y plane, as shown in Figure 2.32, the vector product $\mathbf{C} = \mathbf{A} \times \mathbf{B}$ will be in the z-direction.

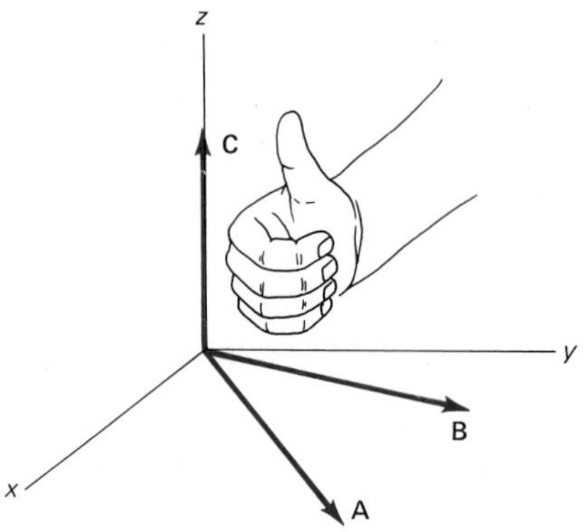

Figure 2.31

Right-hand rule, $\mathbf{A} \times \mathbf{B} = \mathbf{C}$.

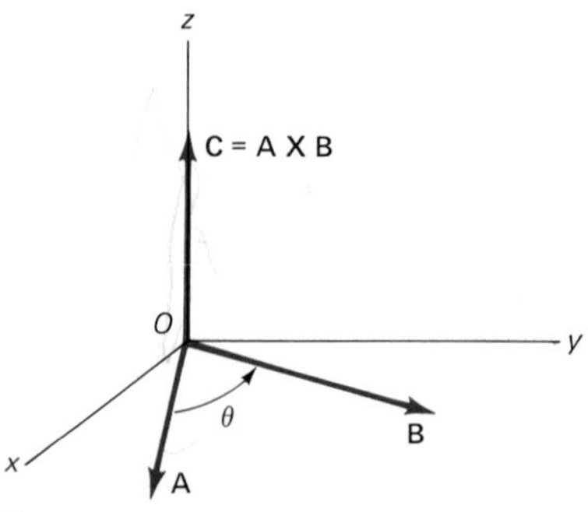

Figure 2.32

The cross-product vector $\mathbf{C} = \mathbf{A} \times \mathbf{B}$ is perpendicular to the plane defined by vectors **A** and **B**.

Figure 2.33

The vector product is anticommutative: $\mathbf{A} \times \mathbf{B} = -\mathbf{B} \times \mathbf{A}$.

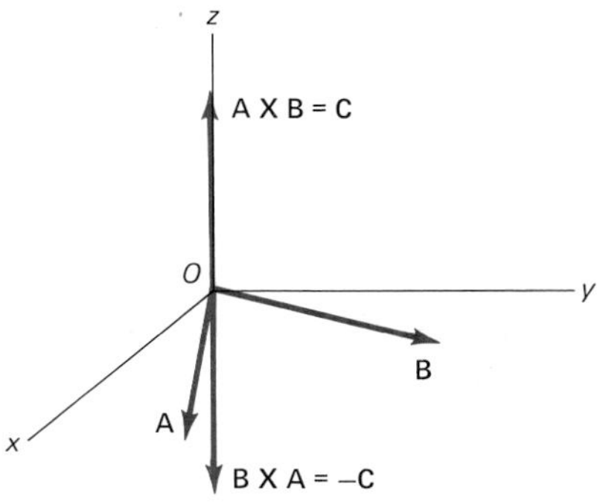

Interchanging **A** and **B** reverses the sign of the cross product. In this case, let the fingers of your right hand curl from the first vector **B** to the second vector **A** through the smaller angle. Your extended thumb, which now points along the negative z-axis, gives the direction of $-\mathbf{C}$ (see Figure 2.33). The vector product is therefore anticommutative.

$$\mathbf{A} \times \mathbf{B} = -\mathbf{B} \times \mathbf{A} \quad \text{anticommutation} \quad (2.23)$$

The vector product, like the scalar product, has survived and been included in this book because it is useful. It will aid us in describing area, torque, angular momentum, magnetic force, the flow of energy in an electromagnetic wave, and many other physical phenomena.

The vector product of two position vectors has a useful geometric interpretation: area. As shown in Figure 2.34, **A** and **B** are two vectors in the x–y plane. The magnitude of the vector product $\mathbf{C} = \mathbf{A} \times \mathbf{B}$ may be written

$$C = A(B \sin \theta)$$

But $B \sin \theta$ is the perpendicular distance from the tip of **B** to **A** (see Figure 2.34). If A is the base of the shaded parallelogram, $B \sin \theta$ is the height and C is the area. The vector representation of the area of the parallelogram defined by **A** and **B** is given by **C**. Area, in this sense, is a vector quantity.

The vector representation of an area of magnitude σ (either a plane or an area small enough so that any curvature is negligible) is $\hat{\mathbf{n}}\sigma$ with the unit normal vector $\hat{\mathbf{n}}$ selected by using the right-hand rule—the fingers curled in

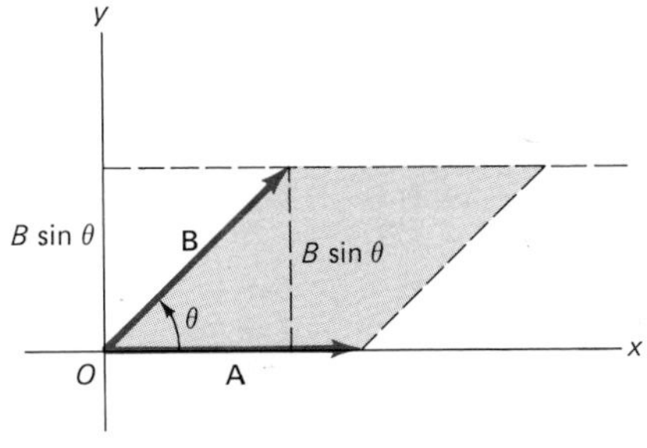

Figure 2.34

The area of the parallelogram defined by **A** and **B** is a geometric representation of the cross product $\mathbf{A} \times \mathbf{B}$.

Table 2.2

Vector product of unit vectors		
$\mathbf{i} \times \mathbf{i} = 0$	$\mathbf{i} \times \mathbf{j} = +\mathbf{k}$	$\mathbf{j} \times \mathbf{i} = -\mathbf{k}$
$\mathbf{j} \times \mathbf{j} = 0$	$\mathbf{j} \times \mathbf{k} = +\mathbf{i}$	$\mathbf{k} \times \mathbf{j} = -\mathbf{i}$
$\mathbf{k} \times \mathbf{k} = 0$	$\mathbf{k} \times \mathbf{i} = +\mathbf{j}$	$\mathbf{i} \times \mathbf{k} = -\mathbf{j}$

the direction of the circulation, or the direction in which the perimeter of σ is described (see Figure 2.35). In an actual physical situation the direction of circulation can be determined. For example, if the perimeter of σ were a wire carrying an electric current, the direction of the current would be taken as the direction of circulation. In Chapter 33 we calculate the magnetic field produced by such a current loop.

The vector product of any vector with itself is zero because the included angle is zero and $\sin 0° = 0$. Thus the vector product of any unit vector, **i**, **j**, or **k**, with itself is zero. The vector product of any one of these three unit vectors with any other one, however, is not zero because the included angle is not zero. For example, $\mathbf{i} \times \mathbf{j} = \mathbf{k}$. The included angle ($x$-axis around to y-axis) is 90° and $\sin 90° = 1$. Using the right-hand rule (the same rule we used in setting up right-handed Cartesian coordinates), we see that $\mathbf{i} \times \mathbf{j}$ points in the positive z-direction, given by unit vector **k**. The other possible unit vector products are shown in Table 2.2. Note the anticommutation: $\mathbf{i} \times \mathbf{j} = +\mathbf{k}$, but $\mathbf{j} \times \mathbf{i} = -\mathbf{k}$.

If you find Table 2.2 difficult to remember, the "cyclic order" device shown in Figure 2.36 may be helpful. Moving around the circle in the positive direction, or counterclockwise, we find that the vector product of any two successive unit vectors is the third unit vector: $\mathbf{i} \times \mathbf{j} = \mathbf{k}$. If the direction of motion is negative, or clockwise, the vector product of two successive vectors is the negative of the third vector: $\mathbf{j} \times \mathbf{i} = -\mathbf{k}$. All of the vector products listed in Table 2.2 can be found in this way.

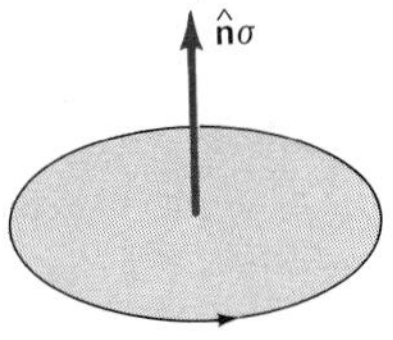

Figure 2.35

The positive direction of $\hat{\mathbf{n}}$ depends on the way the perimeter is described.

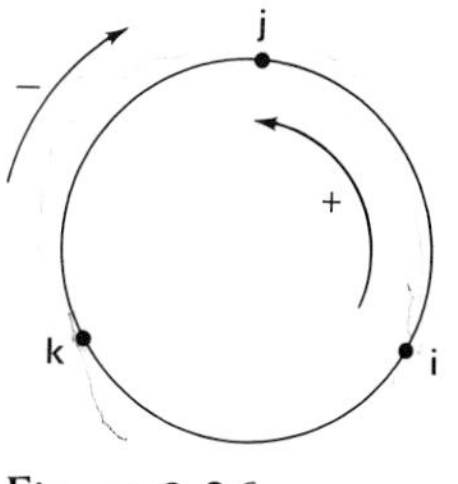

Figure 2.36

Cyclic order, $\mathbf{i} \times \mathbf{j} = \mathbf{k}$. Anticyclic order, $\mathbf{j} \times \mathbf{i} = -\mathbf{k}$.

Now let us develop the component form of the vector product

$$\mathbf{A} \times \mathbf{B} = (\mathbf{i}A_x + \mathbf{j}A_y + \mathbf{k}A_z) \times (\mathbf{i}B_x + \mathbf{j}B_y + \mathbf{k}B_z)$$

By applying the associative and distributive laws we can expand this relation as

$$\begin{aligned}\mathbf{A} \times \mathbf{B} = {} & \mathbf{i} \times \mathbf{i}A_xB_x + \mathbf{i} \times \mathbf{j}A_xB_y + \mathbf{i} \times \mathbf{k}A_xB_z \\ & + \mathbf{j} \times \mathbf{i}A_yB_x + \mathbf{j} \times \mathbf{j}A_yB_y + \mathbf{j} \times \mathbf{k}A_yB_z \\ & + \mathbf{k} \times \mathbf{i}A_zB_x + \mathbf{k} \times \mathbf{j}A_zB_y + \mathbf{k} \times \mathbf{k}A_zB_z\end{aligned}$$

Use of Table 2.2 to evaluate the vector products of the unit vectors yields

$$\begin{aligned}\mathbf{A} \times \mathbf{B} = {} & \mathbf{i}(A_yB_z - A_zB_y) + \mathbf{j}(A_zB_x - A_xB_z) \\ & + \mathbf{k}(A_xB_y - A_yB_x) \qquad (2.24)\end{aligned}$$

We equate the right-hand side of the equation to **C**. **C**, in turn, can be written in component form:

$$\mathbf{A} \times \mathbf{B} = \mathbf{C} = \mathbf{i}C_x + \mathbf{j}C_y + \mathbf{k}C_z$$

Equating x-component to x-component, y-component to y-component, and z-component to z-component, we get

$$C_x = A_yB_z - A_zB_y \qquad (2.25a)$$

$$C_y = A_zB_x - A_xB_z \qquad (2.25b)$$

and

$$C_z = A_xB_y - A_yB_x \qquad (2.25c)$$

Equations 2.25a, b, and c may seem formidable, but in many cases we can simplify the problem greatly. If **A** and **B** lie in the x–y plane (or if we orient the coordinates so that the x–y plane coincides with the plane determined by **A** and **B**), then $A_z = 0$ and $B_z = 0$. Substituting these values into Eqs. 2.25a and 2.25b, we get $C_x = 0$ and $C_y = 0$, respectively. Only the z-component, C_z, survives. Written out in component form, $\mathbf{A} \times \mathbf{B} = \mathbf{C}$ becomes

$$\begin{aligned}(\mathbf{i}A_x + \mathbf{j}A_y) \times (\mathbf{i}B_x + \mathbf{j}B_y) &= \mathbf{k}(A_xB_y - A_yB_x) \\ &= \mathbf{k}C_z \qquad (2.26)\end{aligned}$$

The cross-product vector is perpendicular to the plane of the two vectors (the x–y plane). This technique of simplifying the cross product will be used in Chapter 3 in the discussion of torques.

Example 16 **Torque**

While changing a tire, a mechanic exerts a force of 500 N (112 lb) on the end of a 0.6-m-long tire wrench, as shown in Figure 2.37. Let us calculate the magnitude of the torque at the lug nut (point 0 in Figure 2.38).

In Chapter 3 torque is defined as $\boldsymbol{\tau} = \mathbf{r} \times \mathbf{F}$. Measuring the included angle from the positive direction of **r**, the wrench, to **F**, the force exerted by the person, we have

$$\begin{aligned}\tau &= rF \sin 120^\circ \\ &= 0.6 \times 500 \times \sqrt{3}/2 = 260 \text{ N} \cdot \text{m}\end{aligned}$$

Figure 2.37

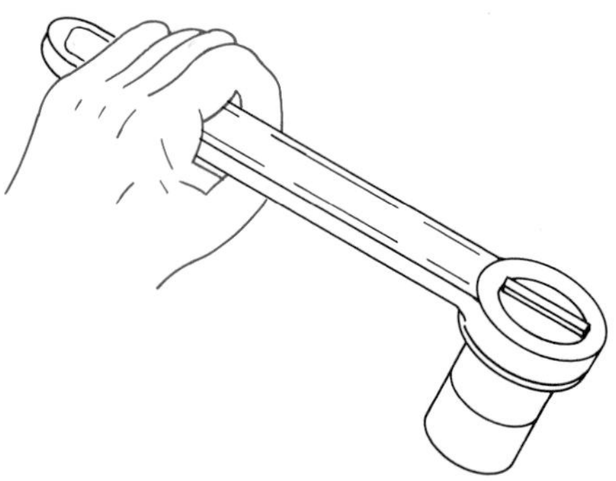

Example 17 **Torque, Component Form**

We can repeat the calculation of the torque in Example 16 by using the component form of the vector product.

From Figure 2.38 we have

$$F_x = 0 \text{ N} \qquad r_x = (0.6 \text{ m}) \cos 150^\circ = -0.520 \text{ m}$$

$$F_y = -500 \text{ N} \qquad r_y = (0.6 \text{ m}) \cos 60^\circ = 0.3 \text{ m}$$

$$F_z = 0 \text{ N} \qquad r_z = 0 \text{ m}$$

Then, from Eq. 2.25

$$\begin{aligned}\tau_x &= [0.3 \text{ m} \times 0 \text{ N}] - [0 \text{ m} \times (-0.500 \text{ N})] \\ &= 0 \text{ N} \cdot \text{m} \\ \tau_y &= [0 \text{ m} \times 0 \text{ N}] - [(-0.520 \text{ m}) \times 0 \text{ N}] \\ &= 0 \text{ N} \cdot \text{m} \\ \tau_z &= [-0.520 \text{ m} \times (-500 \text{ N})] + [0.3 \text{ m} \times 0 \text{ N}] \\ &= 260 \text{ N} \cdot \text{m}\end{aligned}$$

So the torque is 260 N · m, in agreement with Example 16. As a vector, $\boldsymbol{\tau} = +\mathbf{k}\ 260 \text{ N} \cdot \text{m}$.

As another example of the vector product, let us consider the angular momentum of an earth satellite.

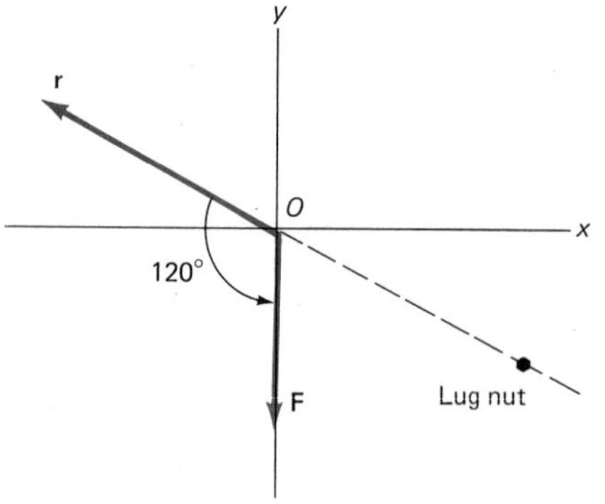

Figure 2.38

A vector diagram of the force **F** and the lever arm **r** of Figure 2.37. Note that **F** and **r** have been displaced along the lines they define to facilitate the calculation of **r** × **F**.

Figure 2.39

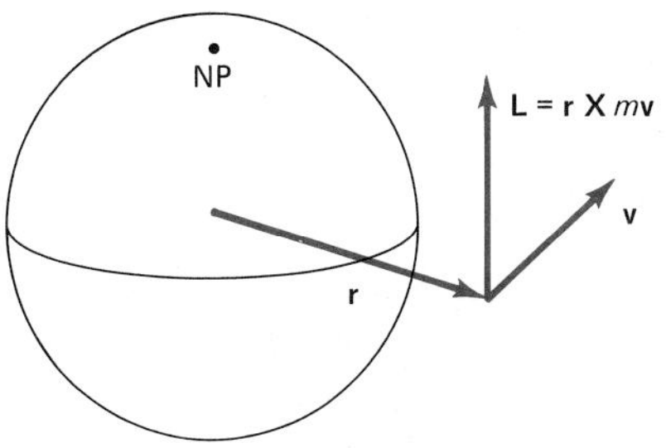

Example 18 **Angular Momentum of an Artificial Satellite**

A 200-kg artificial satellite moving eastward over the earth's equator at a constant altitude of 300 mi (6.86×10^6 m from the center of the earth) has a speed of 17,100 mph (7.63×10^3 m/s). Let's calculate the angular momentum of this satellite relative to the center of the earth.

In Chapter 11 angular momentum is defined as $\mathbf{L} = \mathbf{r} \times \mathbf{p}$ where $\mathbf{p}$ is the linear momentum $m\mathbf{v}$. The magnitude of the angular momentum is given by

$$\begin{aligned} L &= rmv \sin 90° \\ &= (6.86 \times 10^6 \text{ m})(200 \text{ kg})(7.63 \times 10^3 \text{ m/s})(1) \\ &= 1.05 \times 10^{13} \text{ kg} \cdot \text{m}^2/\text{s} \end{aligned}$$

The angle 90° is the angle between the vertical radial distance **r** and the horizontal velocity **v**. Using the right-hand rule with the fingers curled from the positive direction of **r** to the positive direction of **v**, we find that the direction of **L** is north (see Figure 2.39).

This section concludes our treatment of vectors. With vector algebra as a powerful and *precise* mathematical tool, our language can be a bit less "slovenly" than the quotation at the beginning of this chapter suggests. With vector algebra we should be able to avoid "fuzzy thoughts" a little more easily.

Questions

20. Given that $\mathbf{A} \times \mathbf{B} = 0$, can you conclude that **A** and **B** are parallel (or antiparallel)? What other information, if any, do you need to prove that **A** and **B** are parallel (or antiparallel)?

21. Suppose you are given a known nonzero vector **A**. The scalar product of **A** with an unknown vector **B** is zero. Likewise, the vector product of **A** with **B** is zero. What can you conclude about **B**?

22. How do the dimensions of torque (Example 16) compare with the dimensions of work (Example 14)? Does this mean that torque *is* work?

23. A 1-kg object has constant linear momentum **p**. Its angular momentum relative to a particular point is zero. If the object has not already passed through that point, show that if it continues on a straight line, it will pass through that point.

Summary

A scalar is a quantity that has magnitude only. The magnitude is independent of, or invariant to, the orientation of the coordinate system. Mass, time, and energy are typical scalar quantities.

A vector is a quantity that has direction as well as magnitude. Displacement, force, and velocity are examples of vector quantities.

Vectors may be represented geometrically by arrows. Addition of two vectors can be carried out by using a triangle or parallelogram diagram. Addition of more than two vectors can be represented geometrically as well. Geometric representation of vector addition gives us a picture of what is happening. The addition of two vectors is commutative. Addition of more than two vectors is associative.

A vector may be multiplied by a scalar (a number). For example, linear momentum, $\mathbf{p} = m\mathbf{v}$, is the product of mass (a scalar) and velocity (a vector).

Vectors may be resolved into components. Using the three mutually perpendicular unit vectors **i** (unit magnitude, directed along the positive x-axis), **j** (unit magnitude, directed along the positive y-axis), and **k** (unit magnitude, directed along the positive z-axis), we write

$$\mathbf{A} = \mathbf{i}A_x + \mathbf{j}A_y + \mathbf{k}A_z \tag{2.14}$$

A_x, A_y, and A_z are the three components of **A**. The advantage of using vector components lies in greater numerical precision. When combining several vectors or when using three-dimensional vectors, you will find that the component form is particularly convenient for computer calculations.

A three-dimensional vector equation is equivalent to three numeric equations. A vector may not be set equal to a scalar, nor may it be added to a scalar.

The unit vector $\hat{\mathbf{r}}$ is directed outward from the origin. The unit vector $\hat{\mathbf{n}}$ is normal to a surface.

The scalar, or dot, product is defined by

$$\mathbf{A} \cdot \mathbf{B} = AB \cos \theta \quad \text{(geometric representation)} \tag{2.19}$$

$$= A_xB_x + A_yB_y + A_zB_z \quad \text{(algebraic or component representation)} \tag{2.20}$$

where θ is the smaller angle included between the two vectors. The magnitude of the vector, or cross, product $\mathbf{C} = \mathbf{A} \times \mathbf{B}$ is defined by

$$C = AB \sin \theta \quad \text{(geometric representation)} \tag{2.22}$$

with θ measured from the first vector, **A**, to the second vector, **B**. The direction is given by the right-hand rule (see Section 2.5). In algebraic, or component, form

$$C_x = A_yB_z - A_zB_y$$

$$C_y = A_zB_x - A_xB_z$$

and

$$C_z = A_xB_y - A_yB_x \tag{2.25}$$

The scalar product is commutative: $\mathbf{A} \cdot \mathbf{B} = \mathbf{B} \cdot \mathbf{A}$. The vector product exhibits anticommutation:

$$\mathbf{A} \times \mathbf{B} = -\mathbf{B} \times \mathbf{A} \tag{2.23}$$

Problems

Section 2.1 Scalars and Vectors

1. A force of 250 N at 36.9° north of east is shown in Figure 1. Redraw the vector using axes calibrated in pounds rather than newtons (1 lb = 4.45 N).

Section 2.2 Addition and Subtraction of Vectors

2. Using a graphical technique, find the vector sum $\mathbf{A} + \mathbf{B}$ and the vector difference $\mathbf{A} - \mathbf{B}$ of the two force vectors shown in Figure 2. (Note: A graphical solution means that you draw the vectors with accurately measured magnitude and direction and then from your drawing measure the magnitude and direction of the resultant vector.)

3. A force vector $\mathbf{F}_1$ of magnitude 6 N acts on a point at the origin in a direction 30° above the x-axis in the first quadrant. A second force vector $\mathbf{F}_2$ of magnitude 5 newtons acts on the origin in the direction of the positive y-axis. Find graphically the magnitude and direction of the resultant force vector $\mathbf{F}_1 + \mathbf{F}_2$.

4. The velocity of sailboat A relative to sailboat B, $\mathbf{v}_{rel}$, is defined by the equation $\mathbf{v}_{rel} = \mathbf{v}_A - \mathbf{v}_B$ where $\mathbf{v}_A$ is the velocity of A and $\mathbf{v}_B$ is the velocity of B. Determine the velocity of A relative to B if

   $\mathbf{v}_A = 30$ km/hr east

   $\mathbf{v}_B = 40$ km/hr north

5. Two students each exert a force of 100 N on the arm of their physics professor. (a) Using the ideas of vector addition and subtraction, draw a vector diagram for the situations for the minimum and maximum net forces the students are able to exert on their professor. (b) Determine the minimum and maximum net forces.

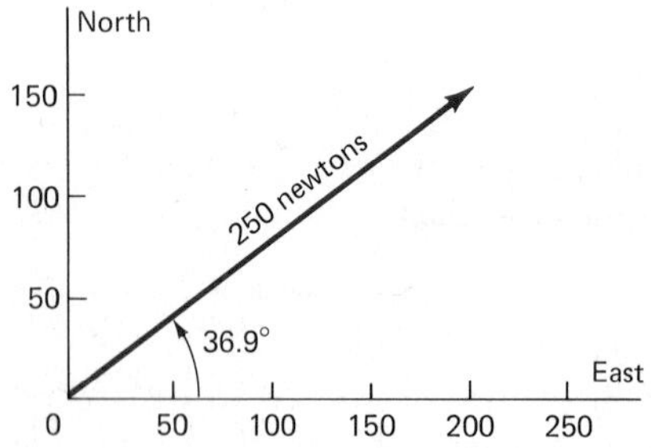

Figure 1

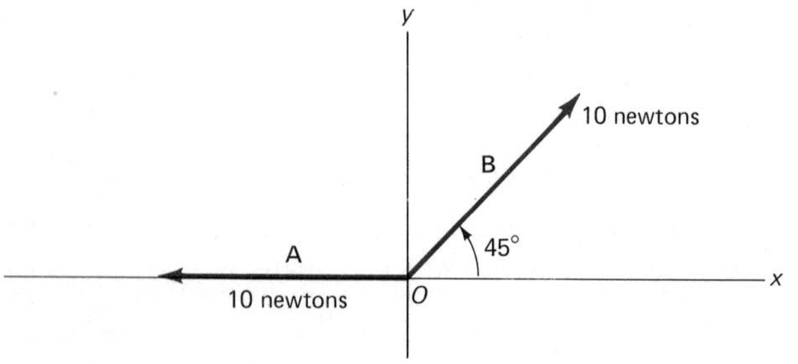

Figure 2

Section 2.3 Components

6. The driver of a car, obviously lost, drives 3 km north, 2 km northeast (45° east of north), 4 km west, and then 3 km southeast (45° east of south). Where does he end up relative to his starting point? Work out your answer graphically. Check by using components. (Assume that east–west lines and north–south lines form a Cartesian coordinate system. The car is not near either the north or the south pole.)

7. An airplane is heading 20° north of east, flying at 200 km/hr relative to the air. Simultaneously, the wind carries the plane in a direction 10° east of south at a speed of 60 km/hr. (a) Find graphically the velocity of the plane relative to the ground. (b) Find the components (east and north) of the velocity of the plane relative to the air and of the air relative to the ground. Find the plane's velocity relative to the ground by adding components.

8. Vector $\mathbf{A}$ ($|\mathbf{A}| = 2.1$) makes angles of 28.1° with the x-axis, 82.6° with the y-axis, and 63.1° with the z-axis. Calculate the components A_x, A_y, and A_z.

9. Vector $\mathbf{B}$ has x-, y- and z-components 4, 6, and 3, respectively. Calculate $|\mathbf{B}|$ and the angle $\mathbf{B}$ makes with the coordinate axes.

10. In a certain experiment two electric fields (vectors) are present. The first field $\mathbf{E}_1$ has components $E_{1x} = 21$, $E_{1y} = 68$, $E_{1z} = -18$, and the second field $\mathbf{E}_2$ has components $E_{2x} = 10$, $E_{2y} = -16$, $E_{2z} = 0$, in units of volts per meter. Find the components and magnitude of the combined electric field $\mathbf{E} = \mathbf{E}_1 + \mathbf{E}_2$.

11. A unit radial vector $\hat{\mathbf{r}}$ makes angles of $\alpha = 30°$ relative to the x-axis; $\beta = 60°$ relative to the y-axis; and $\gamma = 90°$ relative to the z-axis. Find the components of $\hat{\mathbf{r}}$.

12. A barrel of beer is rolling down a plane inclined 35° to the horizontal. The force of gravity on the barrel is 450 N vertically downward. Find the component of this gravitational force parallel to the inclined plane.

13. A sailboat sails for 1 hr at 4 km/hr on a steady compass heading of 40° east of north. The sailboat is simultaneously carried along by a current. At the end of the hour the boat is 6.12 km from its starting point. The line from its starting point to its location lies 60° east of north. (a) Find graphically the velocity of the water. (b) Find the x(easterly)- and y(northerly)-components of the water's velocity.

14. The vertices of a triangle A, B, and C are given by the points (1, 2, 1), (3, 4, 0) and (−1, 5, −1), respectively. All lengths are in meters. (a) Find the components of the vector that starts at point A and ends at point B. (b) Find the components of the vector that starts at point B and ends at point C.

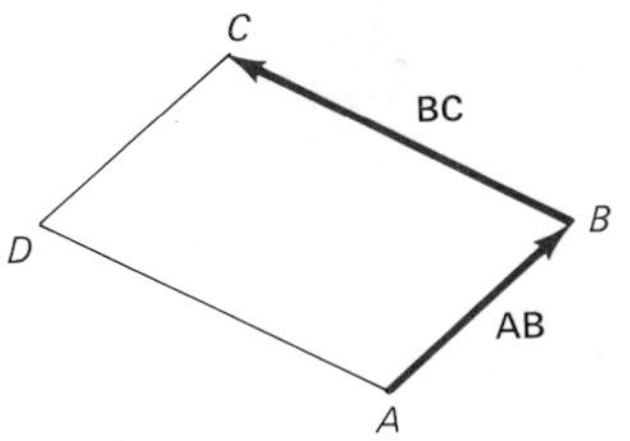

Figure 3

15. Continuing Problem 14, find the point D such that the figure $ABCD$ forms a plane parallelogram (see Figure 3).

16. A sailor aims his boat toward an island located 2 km east and 3 km north of his starting position. After an hour of rowing he sees the island due west. He then aims the boat in the opposite direction from which he was rowing, rows for another hour, and ends up 4 km east of his starting position. He correctly deduces that the ocean current is from west to east. (a) What is the ocean current speed in km/hr? (b) Give the ocean current velocity $\mathbf{v}_1$ in km/hr using unit vectors. (Let $\mathbf{i}$ be east and $\mathbf{j}$ be north.) (c) Let the sailor's rowing velocity relative to the water for the first hour be $\mathbf{v}_2 = a\mathbf{i} + b\mathbf{j}$. Show that $b/a = 1.5$. (d) Prove that $b = 3$ km/hr.

17. A vector $\mathbf{A}$ makes angles of 40° relative to the positive x-axis, 115° relative to the positive y-axis, and 61° relative to the positive z-axis. If the magnitude of $\mathbf{A}$ is 10, find the three components A_x, A_y, and A_z.

18. A given position vector $\mathbf{r} = \mathbf{i}x + \mathbf{j}y + \mathbf{k}z$ has direction cosines $\cos\alpha$, $\cos\beta$, and $\cos\gamma$. Show that

$$\cos^2\alpha + \cos^2\beta + \cos^2\gamma = 1$$

19. A football leaves the foot of a punter at an angle of 36° with respect to the vertical (positive y-direction) at a speed of 21 m/s. Determine the horizontal and vertical components of the velocity.

20. A sailboat goes 2 km east and then turns and goes along a straight line in a different direction. The sailboat arrives at a point 4 km northeast of its starting point. (a) Draw a clear diagram indicating north, east, the initial 2-km run, and the final 4-km northeast position. (b) Find the east and north components of the unknown section of travel, $\mathbf{s}$. (c) How far did the sailboat go on that unknown section of travel? (d) In what direction? relative to what?

21. A car going 55 mph along the highway is checked by radar, which measures the component of the velocity in the direction of the radar beam (Figure 4). If the horizontal radar beam makes an angle of 18° with the highway, what speed will the radar record?

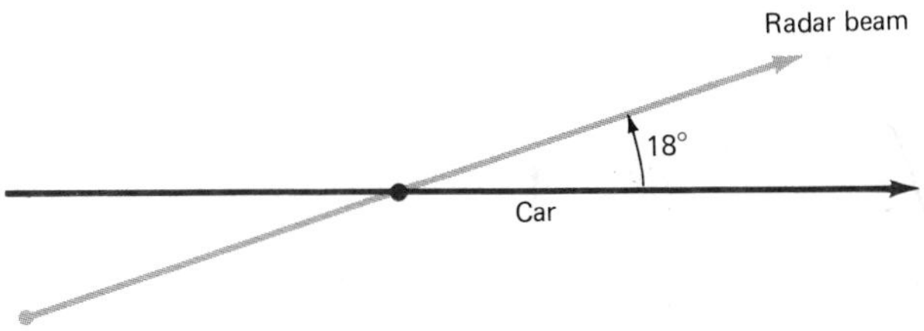

Figure 4

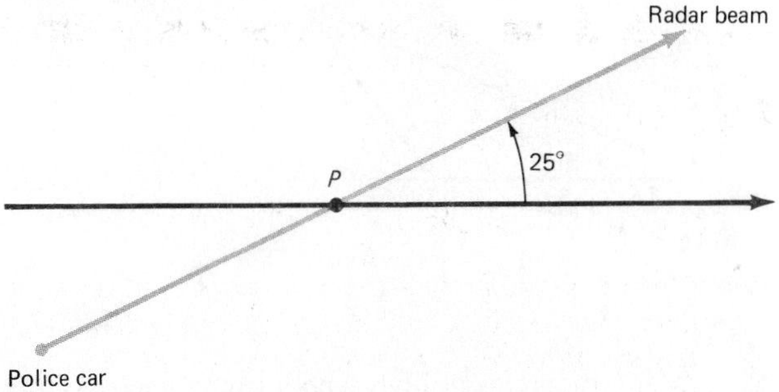

Figure 5

22. A police radar pointing 25° away from the direction of the highway records the speed of a car at point P as 50 mph (Figure 5). (This is actually the component of the car's velocity in the direction of the radar beam.) What is the actual speed of the car along the highway?

Section 2.4 The Scalar, or Dot, Product

23. Vector $\mathbf{A} = 2\mathbf{i} + 4\mathbf{j} - \mathbf{k}$ and vector $\mathbf{B} = 3\mathbf{i} - 2\mathbf{j} - 3\mathbf{k}$. Calculate $\mathbf{A} \cdot \mathbf{B}$.

24. Vector $\mathbf{A}$, having magnitude 2, lies along the positive x-axis. Vector $\mathbf{B}$, having magnitude 4, makes an angle of 45° with the positive x-axis (see Figure 6). (a) Find the scalar product $\mathbf{A} \cdot \mathbf{B}$ without using components. (b) Find $\mathbf{A} \cdot \mathbf{B}$ by using components.

25. Position vectors $\mathbf{r}_1$ and $\mathbf{r}_2$ join the origin and the points (4, −1, 2) and (2, 3, 1), respectively. Calculate the scalar product $\mathbf{r}_1 \cdot \mathbf{r}_2$.

26. Find the angle between the two vectors $\mathbf{F} = 2\mathbf{i} + 3\mathbf{j} - \mathbf{k}$ and $\mathbf{G} = 3\mathbf{j} - \mathbf{k}$. Check your result graphically.

27. Gravity does work on the beer barrel of Problem 12 as the barrel rolls down the inclined plane. Calculate the work done by gravity ($\mathbf{F} \cdot \mathbf{s}$) as the barrel moves 2 m along the inclined plane.

28. A mule pulling a barge along the old Erie Canal exerted a force of 650 N at an angle of 20° to one side of the direction of motion of the barge. Calculate the work done by the mule ($\mathbf{F} \cdot \mathbf{s}$) as it moved 100 m.

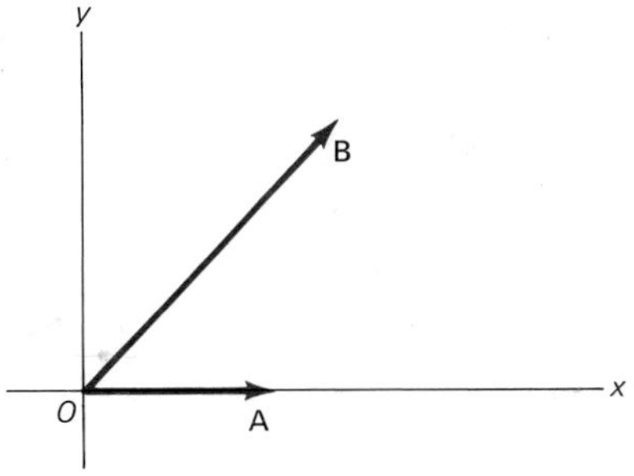

Figure 6

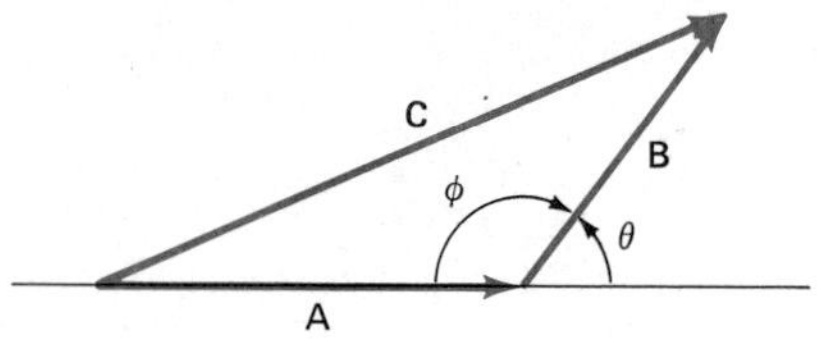

Figure 7

29. Many theorems of high school trigonometry can be rederived easily by using vector algebra. As an example, calculate $\mathbf{C} \cdot \mathbf{C} = (\mathbf{A} + \mathbf{B}) \cdot (\mathbf{A} + \mathbf{B})$ and derive the law of cosines $C^2 = A^2 + B^2 + 2AB \cos \theta$. [Note: The angle between the positive direction of **A** and **B**, θ, is an exterior angle (see Figure 7). In terms of the interior angle ϕ, the law cosines becomes $C^2 = A^2 + B^2 - 2AB \cos \phi$.]

30. Show that dotting a vector into itself leads to the Pythagorean theorem.

Section 2.5 The Vector, or Cross Product

31. Given $\mathbf{M} = 6\mathbf{i} + 2\mathbf{j} - \mathbf{k}$ and $\mathbf{N} = 2\mathbf{i} - \mathbf{j} - 3\mathbf{k}$, calculate $\mathbf{M} \cdot \mathbf{N}$ and $\mathbf{M} \times \mathbf{N}$.

32. A wrench 10 cm long has one end affixed to the head of a bolt and is oriented in a horizontal direction. If we imagine that there is an arrowhead on the "hand" end of the wrench, the wrench forms a horizontal vector 0.10 m long. If a horizontal force of 100 N is applied to the hand end, perpendicular to the wrench, determine the magnitude of the vector product of the force and the length vector.

33. Vector **T** lies along the positive y-axis and $|\mathbf{T}| = 3$. **S** is in the first quadrant of the x–y plane, making an angle of 30° with the x-axis as shown in Figure 8. $|\mathbf{S}| = 2$. Calculate $\mathbf{S} \times \mathbf{T}$.

34. Compute the area of the triangle ABC of Problem 14 as a vector product. Give the magnitude of the vector area.

35. The equation $\mathbf{F} = q(\mathbf{v} \times \mathbf{B})$ gives the force **F** on an electric charge q moving with velocity **v** through a magnetic field **B**. Calculate the (vector) force on a proton ($q_{\text{proton}} = 1.60 \times 10^{-19}$ coulombs) moving with velocity $(2.1\mathbf{i} + 0.2\mathbf{j} + 1.3\mathbf{k}) \times 10^5$ m/s in a magnetic field of $0.62\mathbf{k}$ tesla. (These SI units yield the force in newtons.)

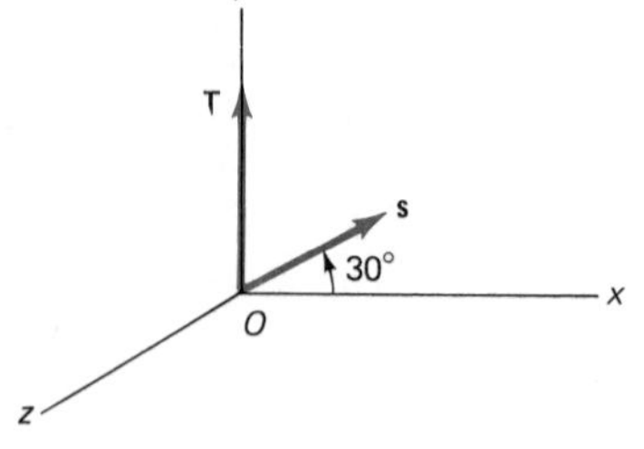

Figure 8

36. At a certain instant the position of a stone in a sling is given by $\mathbf{r} = 1.7\mathbf{i}$ m. The (linear) momentum **p** of the stone is $12\mathbf{j}$ kg · m/s. Calculate the angular momentum $\mathbf{L} = \mathbf{r} \times \mathbf{p}$ of the stone about the origin.

37. The electromagnetic Poynting vector **S** is defined by $\mathbf{S} = \mathbf{E} \times \mathbf{H}$ where **E** and **H** are electric and magnetic fields. For $\mathbf{E} = 10.1\mathbf{i} + 0.2\mathbf{j} + 0.6\mathbf{k}$ and $\mathbf{H} = -0.4\mathbf{i} + 9.8\mathbf{j} + 0.1\mathbf{k}$, calculate **S**. Disregard units (for the present).

38. If $\mathbf{P} = \mathbf{i}P_x + \mathbf{j}P_y$ and $\mathbf{Q} = \mathbf{i}Q_x + \mathbf{j}Q_y$ are any two nonparallel (also nonantiparallel) vectors in the x–y plane, show that $\mathbf{P} \times \mathbf{Q}$ is in the z-direction. Why do we have to specify "nonparallel"?

chapter 3

EQUILIBRIUM OF RIGID BODIES

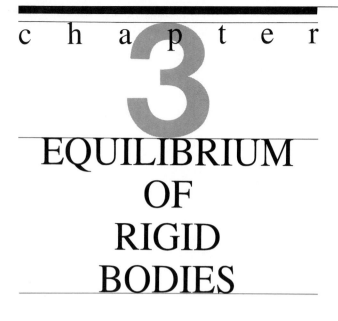

Preview

In a rigid body the distances between every pair of points in the body are constant. A rigid body can undergo translational motion, rotational motion, or a combination of translational and rotational motions. A rigid body is said to translate if every point in the body moves the same distance parallel to a line. A rigid body is said to rotate if one point in the body is fixed. When the forces acting on a rigid body combine to maintain the body in a state of rest or of motion with constant velocity the body is said to be in a state of equilibrium.

There are two conditions of equilibrium which, if satisfied, guarantee the rotational and translational equilibrium of a rigid body. In this chapter we present these two conditions and elaborate them through many examples. We develop the center of gravity concept, and classify three kinds of equilibrium according to their stability.

We also begin to answer the important questions "How do physicists think?" and "Why do they think that way?" The examples in this chapter are designed to illustrate the physicist's method of reasoning. We include specific guidelines for solving problems so that, with practice, you will learn to "think like a physicist."

Force and torque are introduced in this chapter to make it possible for us to discuss equilibrium. Their logical development involves dynamics and is completed in Chapters 6 and 11.

3.1 Force

The term *force* has many different meanings to different people. Intuitively, we regard a force as a push or a pull. A television viewer may be more likely to think of a tornado, an earthquake, or a rocket liftoff. A child might associate force with kicking a ball, pushing a swing, or pulling a wagon. Debaters use forceful arguments. Physicists must be precise in their definition and use of the term *force*. In physics **force** represents the interaction of the environment with a body In Chapter 6 we consider how force changes the state of motion of a body. In this chapter we use the deformation of a spring produced by a force to measure the force.

Four **fundamental** forces of nature studied by physicists are listed here, from the weakest to the strongest.

1 Gravitation
2 Weak nuclear force
3 Electromagnetism
4 Strong nuclear force

Much current research is aimed at increasing our understanding of these four types of force.* Other "natural" forces, such as those associated with friction, water, and wind, can be understood in terms of these four fundamental forces.

*High-energy theorists have succeeded in demonstrating that electromagnetism and the weak nuclear force are different aspects of a single more fundamental interaction. At very high energies their effects are indistinguishable. For our purposes, however, they are separate forces.

Figure 3.1

A coiled spring stretches more when a large weight is attached than when a small weight is attached.

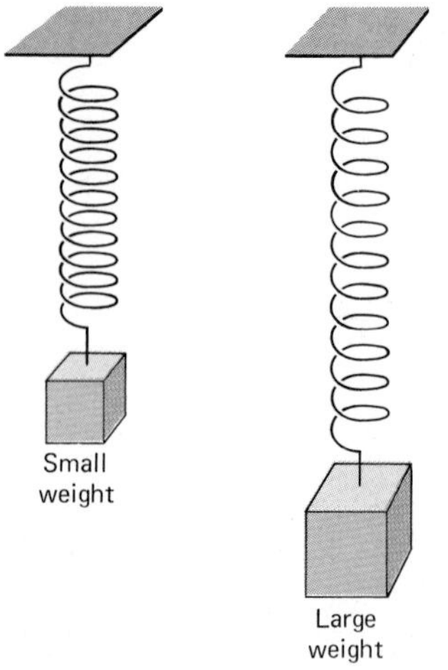

Force as a physical quantity is defined operationally in terms of its effects; we are aware of the existence of forces through these effects. When a force acts on a body, the effect is a deformation of the body and/or a change in the state of motion of the body. In Chapter 6, for instance, we develop an operational definition of force in terms of the acceleration of the "standard kilogram."** In this chapter, however, we will be content with a method of *measuring* force. Force can be measured by means of the deformation of a spring. When a coiled spring hangs vertically, its length depends on the weight* attached to the bottom of the spring, as suggested in Figure 3.1. The force that stretches the spring is a result of gravity acting on the suspended weight. Each time a particular weight is attached, the spring stretches to a definite and reproducible length that can be associated with that weight, and therefore with the force exerted by that weight on the spring.

Suppose the object we attach to the spring is the standard kilogram. Gravity acts on the standard kilogram, resulting in a force that stretches the spring. Because the spring stretches the same amount each time, we agree that any object that produces the same stretch has the same weight. Accordingly, we can create replicas of the standard kilogram. By an analogous procedure, multiples and submultiples of the standard kilogram can be produced, and we can then calibrate a scale of force. The scale relates the spring length to the applied force, as does the scale in Figure 3.2. The unit of force, F_0, is the force experienced by the spring when the standard kilogram is attached.

The size of the unit of force is determined when a value is assigned to F_0. The *SI unit of force* is the *newton* (1 pound = 4.45 newtons), abbreviated N. Setting F_0 = 9.80 newtons* calibrates the spring in newtons. In Chapter 6 we discuss the basis for this choice.

Force has both magnitude and direction. Is it then a vector quantity? Experiments show that forces can be represented by vectors. The vector character of force can be demonstrated in the following manner: A single downward force, $\mathbf{F}_2$, is applied to a vertical spring. Subsequently this single force is removed and in its place two forces, each of magnitude F_1, are applied to the same spring as shown in Figure 3.3. If force is a vector quantity, we should find $F_2 = \sqrt{2}\,F_1$ when the spring length is the same in the two cases. This relationship is depicted geometrically in Figure 3.3. Experiment confirms that $F_2 = \sqrt{2}\,F_1$, thereby showing that forces add like vector quantities.

The weight of a body is an important noncontact force. That is, gravity acts on a body whether or not that body is touching the earth. As a force, weight is a vector, and is directed toward the center of the earth. The SI unit of weight is the newton, whereas the pound is used in the British system of units. A person who weighs 132 pounds, for example, weighs 588 newtons. To express the weight

**For a discussion of the standard kilogram see Section 1.5.

*The force of gravity exerted on a body by the earth is called the **weight** of the body. The concept of weight is developed in Chapter 6.

*The spring length with the standard kilogram attached depends slightly on geographical location. We make the choice F_0 = 9.80 newtons for convenience. This choice corresponds to the average behavior of the spring over the continental United States.

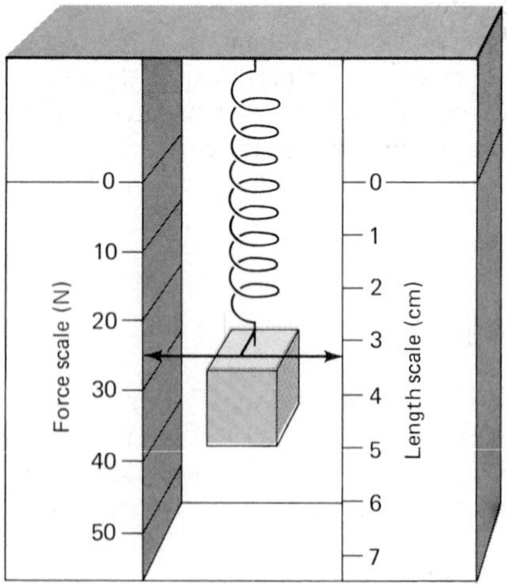

Figure 3.2

With no weight attached the spring is relaxed and both pointers give zero readings. When the spring stretches 3.3 cm the scale pointer indicates a 25 N force.

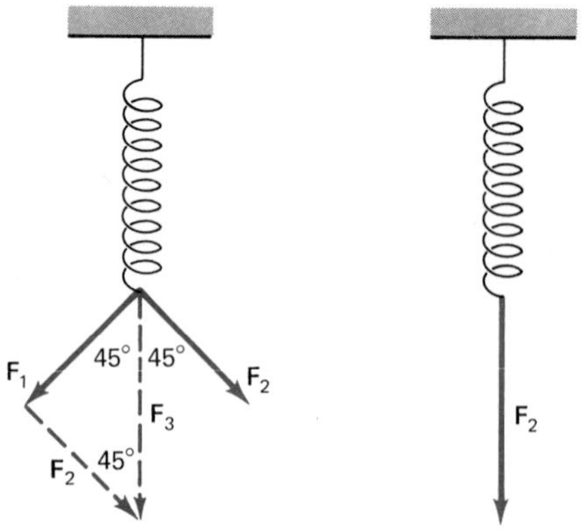

Figure 3.3

Two single forces, each of magnitude **F** applied at 45° as shown, are equivalent to the single downward force of magnitude $\mathbf{F}_3$ provided that $\mathbf{F}_3 = \sqrt{2}\,\mathbf{F}$.

in kilograms is common practice but is, strictly speaking, incorrect.* Any instrument that measures force, including a calibrated spring, may be used to measure weight.

Contact Forces

Contact forces are forces that arise when two bodies touch. A child pushes, but is unable to move, the family refrigerator. Motion of the refrigerator is opposed by contact forces between the floor and the refrigerator. Several contact forces exist in this situation—between the floor and the refrigerator, the floor and the child, the child and the refrigerator. Contact forces are electromagnetic in nature. Two electrically neutral molecules that are far apart exert little or no electromagnetic force on each other. When they come close together, however, large mutually repulsive forces are created. These electromagnetic forces prevent the molecules from coming much closer together. The same repulsive forces prevent two materials from passing through each other when contact is made. It is impractical to deal with contact forces on a molecule-by-molecule basis. Instead, we classify the net molecular contact forces exerted by one body on another into types of contact force.

The four types of contact forces that you will encounter in this chapter are

1 Tension (a pull)
2 Thrust (a push)
3 Tangential force and thrust
4 Tangential force and tension or thrust

These four kinds of contact forces are distinguished by the restrictions imposed on their directions. These forces are elaborated in the following discussion and in later examples.

*In Section 6.7 we will consider the distinction between *weight* and *mass* in detail. No such consideration is possible prior to a discussion of the law of gravity and Newton's laws of motion.

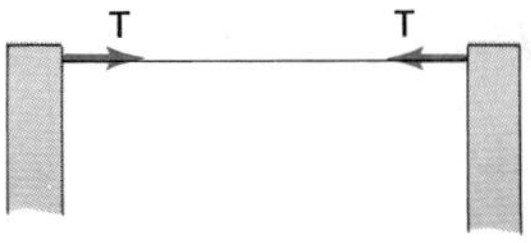

Figure 3.4a

A string stretched between posts is under tension. T is the force exerted on the posts by the tension.

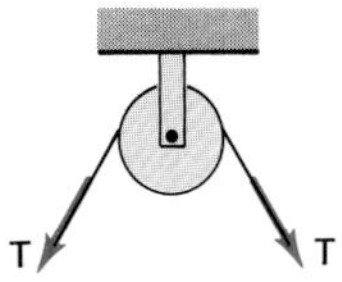

Figure 3.4b

A frictionless pulley changes the direction, but not the magnitude, of the string tension.

Figure 3.5

An ice skater coasting without friction experiences only an upward thrust by the ice.

1. Tension A light flexible string will only support a force of tension parallel to the string. Thus a dog's leash can exert a force directed away from the dog and toward the leash. We call such a force a *pull*. The string under tension in Figure 3.4a pulls on its supports at both ends. The frictionless pulley in Figure 3.4b may change the direction, but not the magnitude, of the tension.

2. Thrust An air-hockey table or sheet of ice is a good approximation to a smooth, or frictionless, surface. A smooth surface can exert a force perpendicular to and directed outward from the surface. We call such a force a *push*, or *thrust*. Thus, a sheet of ice will push upward on a skater, as shown in Figure 3.5. However, it cannot pull downward, and unless the skater cuts into the ice surface, the ice can't push sideways on the skater.

3. Tangential Force and Thrust A rough surface is one that can exert not only a thrust perpendicular to and directed outward from the surface (a push or thrust), but also a frictional force parallel to the surface. This is depicted in Figure 3.6. Like a smooth surface, however, a rough surface cannot pull downward or exert a force directed into the surface. Floors and sidewalks are good approximations to a rough surface, whereas a muddy field or quicksand is not.

4. Tangential Force and Tension or Thrust A hinge represents a fixed line or axis of contact between two solids. A force can be exerted by the hinge in any direction perpendicular to the axis of the hinge, or parallel to the hinge axis. This leaves the object free to rotate about the hinge axis. For example, consider the hinge connecting a

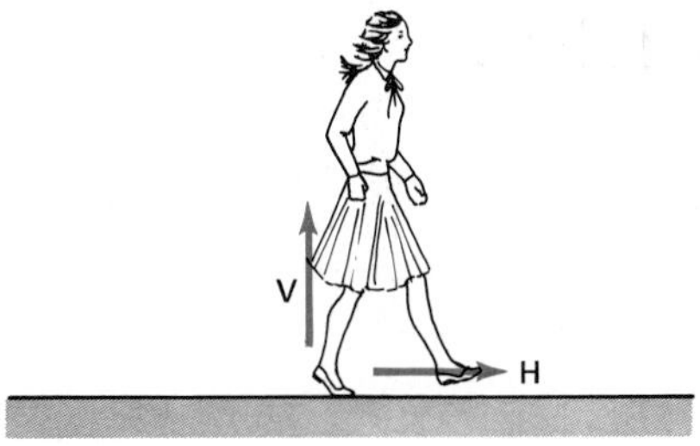

Figure 3.6

A walker may experience both a tangential frictional force, H, and a vertical thrust, V, by the surface.

Figure 3.7

The door hinge, seen from the top, may exert a force on the door tangent to the wall, such as F_1 or F_2; or perpendicular to the wall, such as F_3 or F_4.

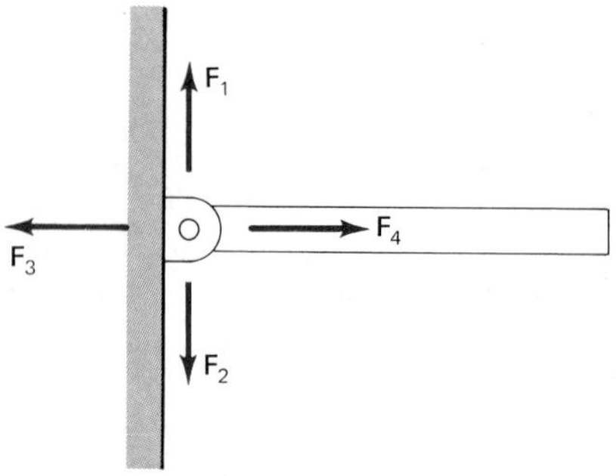

Figure 3.8

The line of action is the x-axis for A and D, the y-axis for C, and the dotted line P–P' for B. The point of application is O for A, C, and D, and O′ for B. The vectors A and D have the same magnitudes and lines of action, but differ in direction.

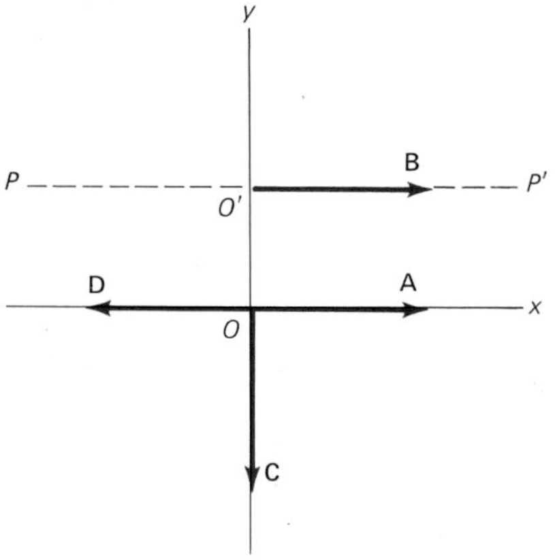

door to its frame in Figure 3.7. The force exerted by the hinge may have a component tangent to the wall, such as F_1 or F_2, and it may also have a component perpendicular to the wall, such as F_3 (tension) or F_4 (thrust).

To complete this introduction to force, we now define four terms that are useful in referring to force vectors. In Figure 3.8 vector **A** represents a force applied at the origin in the positive x-direction. The length of the vector is its **magnitude**; the entire x-axis is called the **line of action**; the origin, O, is called the **point of application**; and the **direction** is indicated by the head of the arrow. The four vectors shown in Figure 3.8 have the same magnitude. The line of action is the x-axis for **A** and **D**, the y-axis for **C**, and the dotted line P–P' for **B**. The point of application is O for **A**, **C**, and **D**, and O′ for **B**. The vectors **A** and **D** have the same magnitudes and lines of action, but differ in direction.

Questions

1. To push off, or begin skating, an ice skater turns the skate sideways, but to glide, she turns the skate forward. Contrast the contact force in the two cases.
2. List as many forces as you can that you have experienced. Divide the list into two categories—contact and noncontact forces.
3. Which kind of force, contact or noncontact, would have to be involved in psychokinesis (the use of psychic powers to produce motion of a remote object)?
4. Why doesn't quicksand qualify as a rough surface?

3.2 The First Condition of Equilibrium

When forces act on an object, the result may be a change in the object's state of motion. If certain conditions are satisfied, however, the forces may combine to maintain a state of equilibrium, that is, a state of rest or a state of motion with constant velocity. Newton's **first law of motion** describes equilibrium and the effect of force on a body that is in equilibrium It states:

> *Every body continues in its state of rest, or of constant velocity, unless it is caused to change that state by forces exerted on it by the environment.*

The **first condition of equilibrium,** a consequence of Newton's first law, may be written in vector form:

> *A body will be in translational equilibrium if and only if the vector sum of forces exerted on a body by the environment equals zero.*

For example, if three forces act on a body it is necessary that

$$\mathbf{F}_1 + \mathbf{F}_2 + \mathbf{F}_3 = 0$$

for the body to be in translational* equilibrium. This equation may be written

$$\sum_{i=1}^{3} \mathbf{F}_i = 0$$

or, for an arbitrary number of forces,

$$\sum \mathbf{F} = 0 \qquad (3.1)$$

The symbol Σ means "the sum of." This sum includes all forces exerted *on* the body *by* its environment. The vanishing of this vector sum is a necessary condition, called the first condition of equilibrium, that must be satisfied in order to ensure translational equilibrium. In three dimensions the component equations of the first condition of equilibrium are (see Section 2.3)

$$\sum F_x = 0 \quad \sum F_y = 0 \quad \sum F_z = 0 \qquad (3.2)$$

Although this condition applies to objects in motion with

*When a rigid body moves without spinning it is said to **translate.** In a translation every point in the body undergoes the same displacement.

constant velocity, which we will cover in Chapter 6, we are concerned in this chapter with statics, the equilibrium of objects at rest.

In statics we ask the question "Why are bodies at rest?" or, more specifically, "What relationship of forces maintains equilibrium?" Bodies are at rest because forces act in concert to hold them in place. That is, the conditions of equilibrium require that all forces acting on a body be mathematically related. The goal of statics is to obtain and to interpret these relationships.

In order to apply the first condition of equilibrium, we can imagine that the universe is divided into two parts. The part we choose to study is called the **body,** and everything else is called the **environment,** the part that interacts with the body. Newton's first law deals only with **external forces**—forces that are exerted on the body by the environment. Forces that are exerted on the environment by the body are not included.

A body may exhibit rotation, like a spinning top, or internal motions, like the movement of blood inside a human body. When rotation and internal motion of the body can be ignored, the body is called a **particle.** We treat a particle mathematically as a point. The size of the body is irrelevant when deciding whether or not to consider it a particle and treat it as a point. For some purposes atomic nuclei and galaxies are considered particles, whereas for other purposes both must be viewed as a collection of particles. For example, the solar system is a particle if we consider its motion as a whole through the galaxy. However, when we study planetary motion we must view the solar system as a system of particles.

When the deformation of an object is negligible the object is called a **rigid body.** More precisely, in a rigid body the distance between any two points is constant. The notion of a rigid body will prove to be useful in studies of equilibrium and rotation.

Problem Solving Guide

For many students the most difficult part of introductory physics is the formulation of the equation needed to solve a problem. You may have no trouble solving equations formulated by others, but you may experience difficulty converting "word" problems into equations. To help you attack word problems we suggest the following guide:

Guide to Problem Solving I

1 Try to guess the answer.

2 Make a simple line drawing.

3 Pick a body.

4 Draw a force diagram.

5 Choose coordinate axes.

6 Apply the first condition of equilibrium (Eq. 3.1 or 3.2).

7 Decide whether or not the solution makes sense.

If you learn this procedure and follow it faithfully, you will soon find that it helps you to develop "physical intuition." The following examples elaborate and illustrate these seven instructions.

Example 1 Weight and Cable Tension

Vertical steel cables under tension* are used to support roadbeds in suspension bridges and to anchor mobile homes to the ground. One such cable, 10 m long, weighs 500 N. How will the cable tension at the top compare with that at the bottom?

Before you attempt a formal solution, (1) try to guess the answer!** You might predict that the two tensions will be the same. Whether or not this guess is correct, a comparison with the final result will help you to develop physical insight. You should cultivate the habit of predicting the answer to a problem before completing its detailed mathematical solution. Next (2) make a simple line drawing, and (3) pick a body. The line drawing and (4) force diagram that we will use for this problem are shown in Figure 3.9. Forces acting on the cable—the body we picked—include the tension at the top ($\mathbf{F}_2$), the tension at the bottom ($\mathbf{F}_1$), and the weight ($\mathbf{W}$). (5) "Up" is selected as the positive y-direction. The first condition of equilibrium, Eq. 3.2, requires that

$$(6) \quad \sum F_y = +F_2 - F_1 - W = 0$$

In this equation F_2 appears with the plus sign and W and F_1 appear with the minus sign because the corresponding force components have directions in the plus and minus y-direction, respectively.

Our mathematical solution is a single equation, $F_2 - F_1 = W$, with two unknowns, F_1 and F_2. A complete numerical solution is not possible with the information provided. However, we can compare the tensions: F_2 exceeds F_1 by 500 N, the cable weight. If $F_1 = 1000$ N, then $F_2 = 1500$ N. Our guess that the tensions would be equal was wrong. Only if $W = 0$ would $F_2 = F_1$.

*We use compression and tension to mean push and pull, respectively. That is, they are forces.

**Wheeler's First Moral Principle, "Never make a calculation until you know the answer," is elaborated on page 60 of *Spacetime Physics* by E. F. Taylor and J. A. Wheeler (W. H. Freeman, San Francisco, 1966).

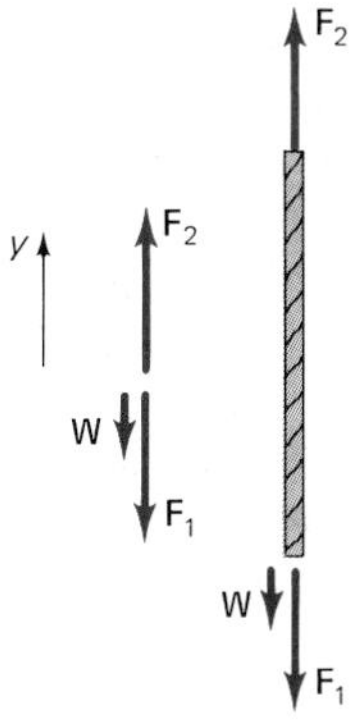

Figure 3.9

Forces acting on a section of vertical cable include its weight, $\mathbf{W}$, and the contact forces $\mathbf{F}_1$ and $\mathbf{F}_2$.

(7) Deciding whether the solution to a problem makes sense usually requires more physical intuition than is available to the beginner. Often, however, it is possible to test for "reasonableness" by considering special or limiting cases. If we set $F_1 = 0$ in our solution, then this corresponds physically to freeing the lower end of the cable, that is, disconnecting it from the roadbed. What would you expect the value of F_2 to be in that case? Our equation, $F_2 - F_1 = W$, requires that $F_2 = 500$ N when $F_1 = 0$, because we were given $W = 500$ N.

In the preceding cable example, the first condition of equilibrium led to a single component equation. Because the forces all acted vertically, the problem was one-dimensional. In a two-dimensional problem, however, the forces all lie in a plane. The forces may have horizontal and/or vertical components. In this case two component equations follow from the first condition of equilibrium, as we see in the following example.

Example 2 **Holding a Sled on a Hill**

A young girl wants to demonstrate her strength by pulling her older brother, sitting on a sled (combined weight = 785 N), up a 30° hill. Although she fails to move up the hill, she pulls with sufficient force parallel to the slope to prevent her brother and the sled from going downhill. Assuming that the sled is on an icy (frictionless) surface, let's determine the tension in the cord. Assume that the cord is weightless.

The line drawing in Figure 3.10 illustrates the problem. To determine the cord tension we must pick a body on which the force acts. Two choices exist—the girl, or the sled plus brother. The girl's weight is not given, but we *do* know the combined weight of the sled plus brother. Therefore, we will choose the sled plus brother as the body.

In this case the sled plus brother may be treated as a particle, because the two objects move (or remain at rest) as a unit. No rotation or internal motion is considered. When we draw the force vector diagram, the body, all vectors that represent forces acting on the body, dimensions, and angles should be included. In Figure 3.11 the dot represents the body, and the three vectors represent the weight (**W**), cord tension (**C**), and thrust of the hill (**B**).

Figure 3.10

A sled in equilibrium on a 30° slope.

Figure 3.11

Forces acting on the combined system of sled plus brother.

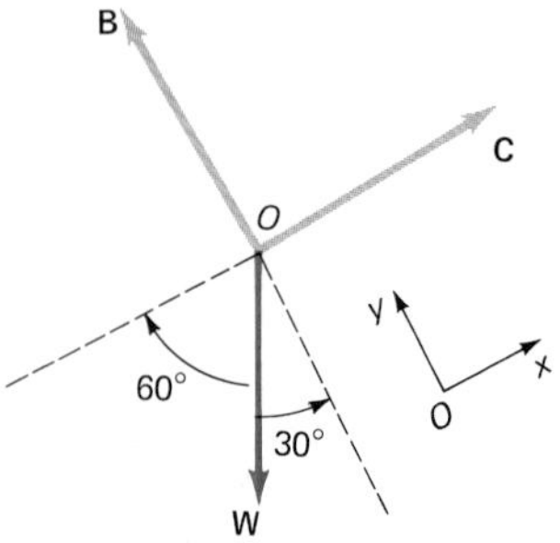

Coordinate axes may be selected in any direction, but it is best to choose axes that simplify rather than complicate the solution. For this problem, it is advantageous to choose axes parallel to and perpendicular to the slope, since in this coordinate system, **B** and **C** have only one nonzero component each. (You may find it helpful to draw arrows to define the positive directions of the x- and y-axes.) To apply the first condition of equilibrium no origin location need be specified. The first condition of equilibrium involves only the magnitude and direction of a force, and neither of these is affected by the choice of origin.

In order to obtain the equations of the first condition of equilibrium we must find the components of each vector along the chosen axes. Because **C** is parallel to the x-axis and **B** is parallel to the y-axis, we obtain the following:

$$C_x = +C \qquad B_x = 0$$

and

$$C_y = 0 \qquad B_y = +B$$

Referring to Figure 3.12 we find the components of **W** (see Section 2.3):

$$W_x = W \cos 240° = -W \cos 60°$$
$$= -W \sin 30°$$
$$W_y = W \cos 150° = -W \sin 60°$$
$$= -W \cos 30°$$

Substituting these components into Eq. 3.2 we have

$$\sum F_x = +C - W \sin 30° = 0$$

$$\sum F_y = +B - W \cos 30° = 0$$

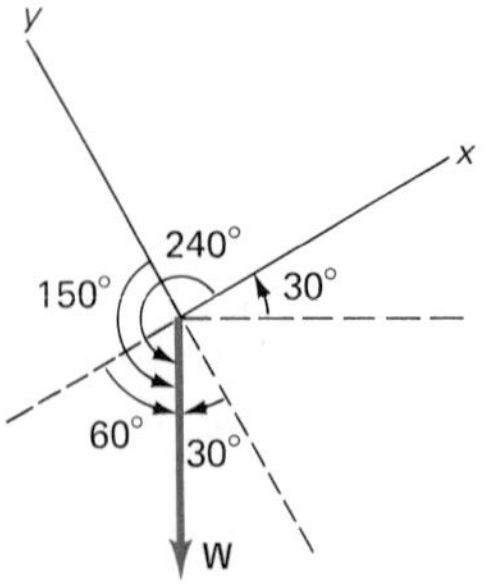

Figure 3.12

The weight vector **W** in relation to axes aligned parallel to and perpendicular to the slope.

Figure 3.13

The vectors **W**, **B**, and **C** in relation to the horizontal and vertical axes.

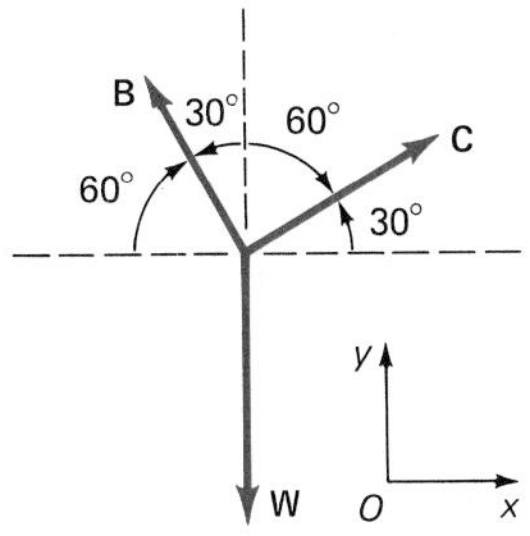

Whether the component associated with a particular vector is positive, zero, or negative in an equation depends on the relation between the vector and the chosen axes. Thus the x-component of **C** is $+C$ and the y-component of **C** is 0; the x-component of **B** is 0, and the y-component of **B** is $+B$; and the x-component of **W** is $-W \sin 30°$ and the y-component of **W** is $-W \cos 30°$. Each sign must be correct relative to the force vector drawing if the correct equations are to be obtained.

Using $W = 785$ N, we find

$$C = 392 \text{ N}$$

for the desired tension. That C is positive confirms the original direction attributed to **C** in the force vector diagram.

Other choices of axes must lead to the same numerical result, but with more difficulty. For instance, if we select horizontal and vertical axes, we obtain the force vector diagram shown in Figure 3.13. Note that these angles differ from those in Figure 3.11 as a result of the change in axes. You should verify that the equations

$$\Sigma F_x = +C \cos 30° - B \cos 60° = 0$$

$$\Sigma F_y = +C \cos 60° + B \cos 30° - W = 0$$

follow the application of the first condition of equilibrium. Then confirm the previous result by solving for C. As you will see, the solution is independent of the coordinate axes chosen. Therefore you should select axes that result in the simplest algebra. As you gain experience in problem solving, you will find it increasingly easy to select the best axes.

Newton's Third Law

When Newton's first law is satisfied for a particle, the particle is in equilibrium. However, in order for extended objects—objects whose rotation and internal motion cannot be ignored—to be in equilibrium, Newton's third law as well as Newton's first law must be satisfied. (Newton's second law, which relates to the accelerations of bodies, will not be needed until Chapter 6.)

Newton's first law has provided us with an initial insight into the nature of force. Those forces exerted on a particle by its environment determine the equilibrium of the particle in accordance with the first condition of equilibrium. But suppose that the body consists of more than one part. How are the forces between parts of a body to be handled? Is there a discernible pattern in the occurrence of forces in nature? May forces appear singly? In pairs? In groups of any size? The answers to these questions, found by observation and experiment, are summarized by Newton's third law:

To every action there is always opposed an equal reaction.

The words *action* and *reaction* mean forces. In other words, for every force there is an equal in magnitude but oppositely directed force. Newton's third law rules out the possibility of a single isolated force, because forces in nature always occur in pairs. An action–reaction pair consists of forces equal in magnitude and opposite in direction. In the mutual interaction of two bodies, the force the first body exerts on the second is equal and opposite to the force the second body exerts on the first. For example, the gravitational force exerted on the earth by the sun, $\mathbf{F}_e$, is equal in strength and opposite in direction to the gravitational force exerted on the sun by the earth, $\mathbf{F}_s$. The force exerted by the cord on the body in Example 3 is equal in magnitude and oppositely directed to the force exerted by the body on the cord. Either force of a pair may be regarded as the *action*, making the other force the *reaction*. Newton's first law involves all of the forces, regardless of number, that act on *one* body. Newton's third law, however, involves those two forces that constitute the mutual interaction of *two* bodies. The two forces constituting an action–reaction pair never act on the same body.

Example 3 **The Monkey and the Rope**

To illustrate Newton's third law and the treatment of internal forces, let's consider the 98-N monkey shown hanging on the end of a 9.80-N rope in Figure 3.14. What are the tensions, $\mathbf{F}_2$ and $\mathbf{F}_1$, at the ends of the rope?

If the rope is picked as a body, then the physical situation in Example 1 is duplicated. Only the names are changed. The forces $\mathbf{F}_1$ and $\mathbf{F}_2$ act on the rope along with $\mathbf{W}_2$ (the weight of the rope). This is shown in Figure 3.15. The result is analogous to that obtained in Example 1, $F_2 - F_1 = W_2 = 9.80$ N.

If we consider the monkey as a body, he experiences only two forces. His weight, $W_1 = 98.0$ N, acts in the negative y-direction, and the force exerted on him by the rope, $\mathbf{F}_1'$, acts in the positive y-direction. The force vector

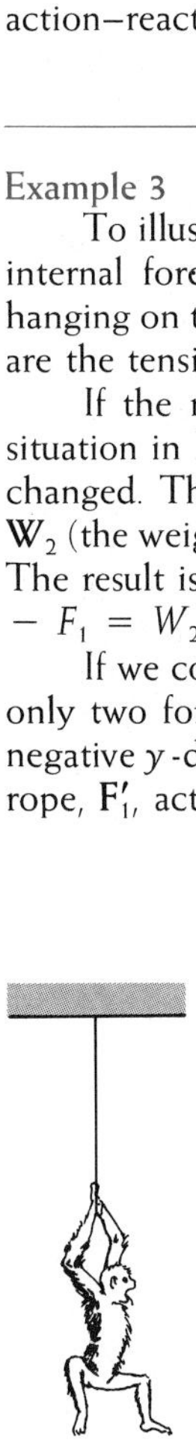

Figure 3.14

A monkey hanging on a rope.

Figure 3.15

Forces acting on the rope.

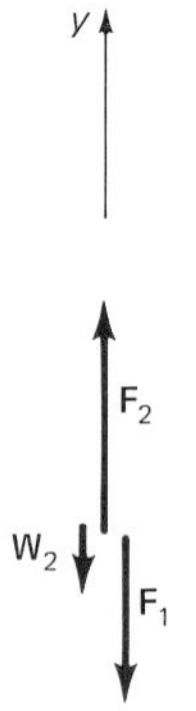

Figure 3.16

Forces acting on the monkey.

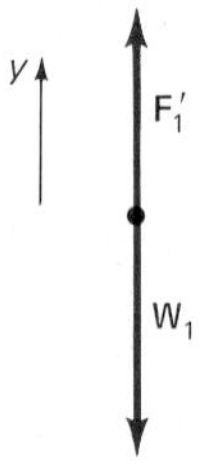

diagram is shown in Figure 3.16, with the dot representing the monkey as a particle. The first condition of equilibrium leads to

$$\sum F_y = +F_1' - W_1 = 0$$

or

$$F_1' = 98.0 \text{ N}$$

Five forces ($\mathbf{F}_1$, $\mathbf{F}_2$, $\mathbf{W}_2$, $\mathbf{F}_1'$, $\mathbf{W}_1$) and two bodies (the rope and the monkey) have been considered in this example. The force $\mathbf{F}_1$ is exerted on the rope by the monkey, while the force $\mathbf{F}_1'$ is exerted on the monkey by the rope. These two forces represent the mutual interaction of the rope and the monkey, and thus constitute an action–reaction pair of forces as described in Newton's third law. Therefore we conclude that

$$\mathbf{F}_1 = -\mathbf{F}_1'$$

so that

$$F_1 = 98.0 \text{ N} \qquad \text{and} \qquad F_2 = 108 \text{ N}$$

Note that incomplete solutions were obtained by applying our problem solving guide to the rope alone ($F_2 - F_1 = 9.80$ N) or to the monkey alone ($F_1' = 98.0$ N). The same situation occurred in Example 2. In order to obtain the complete solution we had to combine the results of the two separate problems. So if when solving a problem your first try produces an incomplete solution, then *pick another body*.

Each of the five forces ($\mathbf{F}_1$, $\mathbf{F}_2$, $\mathbf{W}_2$, $\mathbf{F}_1'$, $\mathbf{W}_1$) was considered as an external force acting either on the rope or on the monkey as a body. If the combined system consisting of the rope plus the monkey is selected as a body, however, only the two weights and $\mathbf{F}_2$ will be external. When we consider the equilibrium of the combined system, only the external forces are included in the equilibrium condition

$$\sum \mathbf{F} = 0$$

$$F_2 - W_1 - W_2 = 0$$

$$F_2 = 9.80 + 98.0 = 108 \text{ N}$$

Because $\mathbf{F}_1 + \mathbf{F}_1' = 0$, this pair of internal forces would contribute nothing to the sum $\sum \mathbf{F}$, even if included. Since internal forces occur only in action–reaction pairs, and since each pair would separately contribute zero to the sum, Newton's third law explains why internal forces should not be included when applying the first condition of equilibrium.

How can you determine whether the solution to a problem is correct? Here are some techniques employed by physicists: (1) Examine the solution for reasonableness, (2) note special symmetries, and (3) consider special cases of the solution. To illustrate these three solution checks, let's consider a two-dimensional generalization of the monkey problem.

Example 4 The Tension in a High Wire

Consider a tightrope walker ($W = 785$ N), shown in Figure 3.17, who falls off a high wire while attempting to walk across the chasm downstream from Niagara Falls. Prevented from falling to his doom only by a safety rope attached to his waist and to the high wire, he is dangling motionless in space. The wire makes the given angles, θ_1 and θ_2, with the horizontal. Let us find the tensions in the wire on each side of the point where the safety rope is attached. We will consider only the weight of the tightrope walker and not the weight of the rope.

First, we pick the tiepoint, the place where the rope and the wire come together, as the body. Because $\mathbf{W}$ is in the vertical direction, we select x- and y-axes to be positive in the horizontal and vertical directions, as drawn in the force vector diagram in Figure 3.18. The first condition of equilibrium yields the equations

$$\sum F_x = +F_1 \cos\theta_1 - F_2 \cos\theta_2 = 0$$

$$\sum F_y = +F_1 \sin\theta_1 + F_2 \sin\theta_2 - W = 0$$

One way of solving for the tensions, F_1 and F_2, is to multiply the first equation by $\sin\theta_1$ to get $F_1 \sin\theta_1 \cos\theta_1 - F_2 \sin\theta_1 \cos\theta_2 = 0$ and the second equation by $\cos\theta_1$ to get $F_1 \sin\theta_1 \cos\theta_1 + F_2 \sin\theta_2 \cos\theta_1 - W \cos\theta_1 = 0$. Then we subtract to get

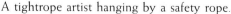

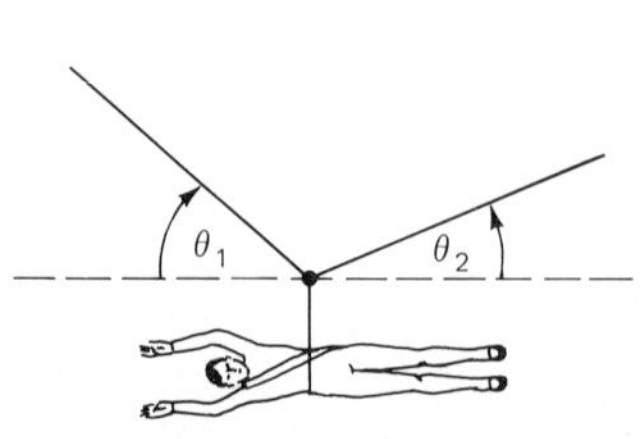

Figure 3.17

A tightrope artist hanging by a safety rope.

Figure 3.18

The forces $\mathbf{F}_1$, $\mathbf{F}_2$, and $\mathbf{W}$ act on the tiepoint where the tightrope and safety rope are attached.

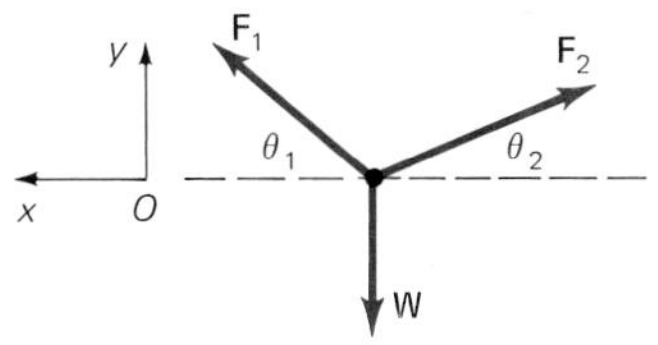

$$F_2(\sin \theta_1 \cos \theta_2 + \cos \theta_1 \sin \theta_2) = W \cos \theta_1$$

Similarly, we multiply the first equation by $\sin \theta_2$ and the second equation by $\cos \theta_2$ and add to get

$$F_1(\sin \theta_2 \cos \theta_1 + \cos \theta_2 \sin \theta_1) = W \cos \theta_2$$

These two equations can then be solved* for F_1 and F_2 to obtain

$$F_2 = \frac{W \cos \theta_1}{\sin(\theta_1 + \theta_2)}$$

$$F_1 = \frac{W \cos \theta_2}{\sin(\theta_1 + \theta_2)}$$

Let us now determine whether the solution obtained for Example 4 is correct. First, we examine the solution for reasonableness. In other words, we ask the question "Does this solution make sense?" (First verify that it is indeed a solution by substituting these values for F_1 and F_2 directly into the equations resulting from the application of the first condition of equilibrium.)

Complicated solutions, like the expressions for F_1 and F_2, generally cannot be determined to be reasonable simply by inspection. Instead we must use other methods. One powerful technique is the use of a **symmetry argument.** For example, when $\theta_1 = \theta_2$, the force vector diagram in Figure 3.18 evolves into a symmetric "Y." The solution reduces to $F_1 = F_2$ in this case. Consider whether, when $\theta_1 = \theta_2$, a symmetric solution ($F_1 = F_2$) or a nonsymmetric one ($F_1 \neq F_2$) makes more sense. If you favor the nonsymmetric solution, be prepared to say whether F_1 or F_2 should be larger.

A third method of checking solutions is to assign to the parameters (in this case the angles) special values that transform the physical situation into a much simpler problem—one that (we hope!) can be solved by inspection. For example, even though $\theta_1 = \theta_2 = 90°$ is unlikely for a tightrope, it is *possible*, and the resulting problem is easily solvable. Each rope then is vertical, as shown in Figure 3.19, and supports half the weight. Careful inspection of the solution reveals that $F_1 = F_2 = 392$ N. (Warning: Don't be too quick to substitute $\theta_1 = 90°$ and $\theta_2 = 90°$ into $\sin(\theta_1 + \theta_2)$. First let $\theta_1 = \theta_2$, then complete the algebra, and finally substitute 90° for the angle. In this way you may avoid the indeterminate result $^0/_0$!)

Special Solution Test Occasionally you may want to use other solution checks. For example, the geometric interpretation of the condition $\Sigma \mathbf{F} = 0$ requires that vectors added head-to-tail form a closed figure. Three force vectors of equal magnitude will close to form an equilateral triangle when $\Sigma \mathbf{F} = 0$. By setting $\theta_1 = \theta_2 = 30°$ in Figure 3.12, we cause the equiangular condition to be realized, and an equilateral triangle results. This requires that $F_1 = F_2 = 785$ N—a result also obtained from our original solution to Example 4.

The following example illustrates the forces exerted by a rough surface, and gives us an answer that we will then check.

*$\sin(\theta_1 + \theta_2) = \sin \theta_1 \cos \theta_2 + \cos \theta_1 \sin \theta_2$

Figure 3.19

One limiting case is that in which the tightrope is vertical and parallel to the safety rope. Here each half of the tightrope supports half of the daredevil's weight.

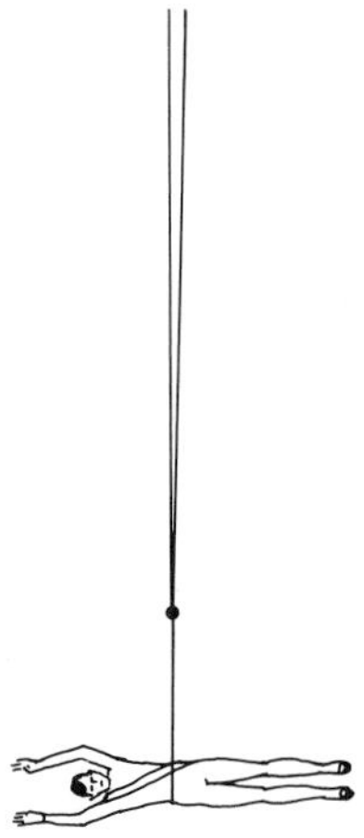

Example 5 **What Keeps a Parked Car on a Hill?**

Consider a car ($W = 9800$ N) parked on a hill inclined at the angle θ, as shown in Figure 3.20. Let us determine the forces exerted by the hill on the car. As a rough surface, the hill exerts a normal force **B** and a tangential force **C**, as drawn in Figure 3.21. Choosing x- and

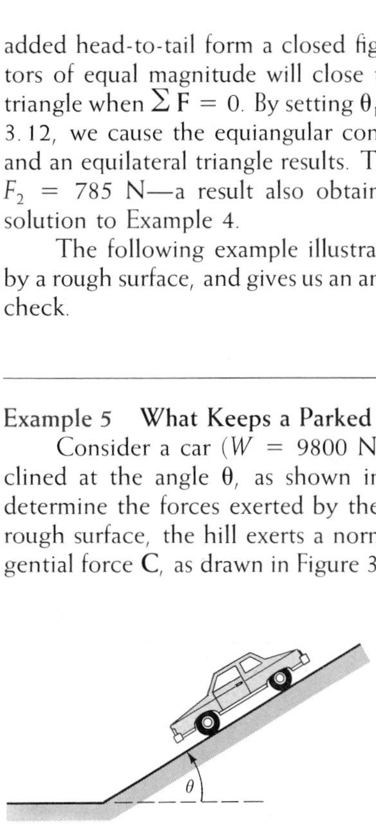

Figure 3.20

A car parked on a hill.

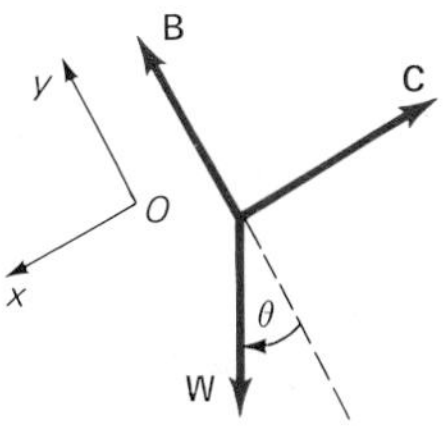

Figure 3.21

Forces acting on the car.

y-axes parallel and perpendicular to the road, we obtain the force vector diagram shown. Applying the first condition of equilibrium, we find

$$\sum F_x = -C + W \sin \theta = 0$$

$$\sum F_y = +B - W \cos \theta = 0$$

Solving for C and B we find

$$C = W \sin \theta \qquad \text{and} \qquad B = W \cos \theta$$

For example, if $\theta = 20°$, then $C = 3350$ N and $B = 9210$ N.

If you notice a resemblance between Examples 2 and 5, you are making progress!

Questions

5. A pendulum bob of weight **W** is shown in Figure 3.22 hanging on the end of a string at an angle. Give an argument based on the first condition of equilibrium showing that the pendulum ball can't be in equilibrium.
6. Two forces act on a person standing on level ground—his weight **W**, acting downward; and the normal force **P**, acting upward. In order for the person to be in translational equilibrium, the first condition of equilibrium requires that $P = W$. Explain why **P** and **W** do not constitute an action–reaction pair of forces.
7. Is the answer given in Example 5 reasonable? Use some of the solution checks discussed in this section to decide.
8. Explain why a particle experiencing only one force cannot be in equilibrium.
9. A particle is in equilibrium under the influence of two forces. Explain why the forces must be equal in magnitude and oppositely directed.

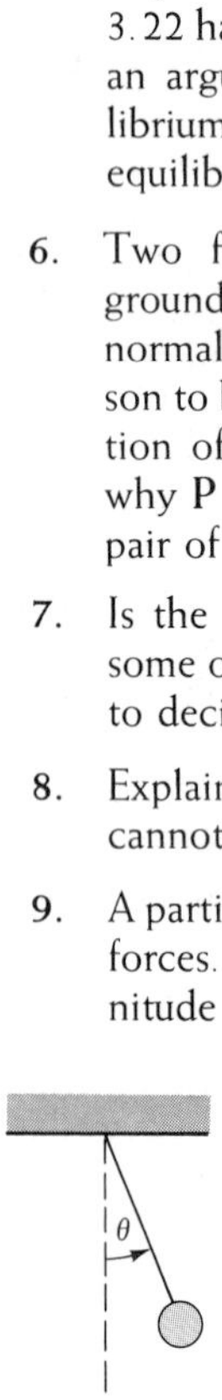

Figure 3.22

A simple pendulum.

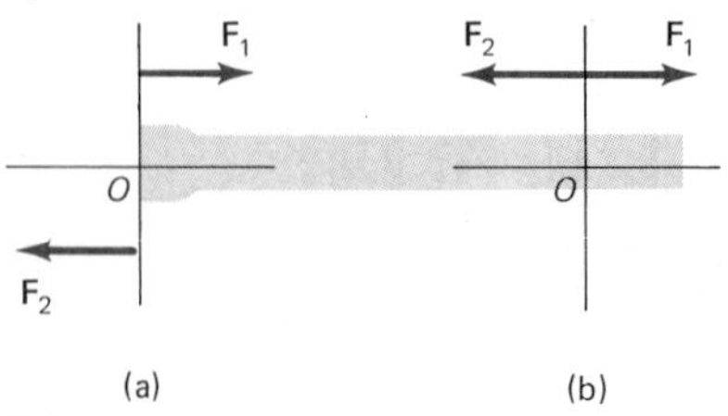

Figure 3.23

Two forces of equal magnitude and oppositely directed act on the revolving door. If the lines of action differ, as in (a), the door is not in equilibrium. If the lines of action are the same, as in (b), the door is in equilibrium.

3.3 Torque

The first condition of equilibrium must be satisfied if a body is to be in equilibrium. However, the first condition of equilibrium alone does not guarantee equilibrium. Consider, for example, the top view of a revolving door shown in Figure 3.23. In the first case, (a), one person exerts a force, $\mathbf{F}_1$, directed to the right while another person exerts a force of equal magnitude, $\mathbf{F}_2$, directed to the left along a different line of action. The first condition of equilibrium is satisfied, yet the door is not in equilibrium—it begins to revolve. In the second case, (b), $\mathbf{F}_1$ and $\mathbf{F}_2$ are the same as in (a), except that they are exerted at the same place, along the same lines of action. The door is in equilibrium, and does not begin to revolve.

Either of the forces shown in Figure 3.23 acting alone causes the door to revolve. However, if either force acts through the hinge line, it has no rotational effect on the door. Evidently the magnitude and direction of a force do not completely describe a force's effect on the equilibrium of the revolving door. The location of the line of action is also important. We can illustrate this point further with the familiar playground toy shown in Figure 3.24, the teeter-totter or seesaw.

In Figure 3.25 three forces are shown acting on a teeter-totter (seesaw) in equilibrium: $\mathbf{F}_3$ is directed upward at the fulcrum, $\mathbf{F}_1$ is directed downward at a distance l_1 to the left of the fulcrum, and $\mathbf{F}_2$ is directed downward at a distance l_2 to the right of the fulcrum. The distances l_1 and l_2 are called *lever arms* for the forces $\mathbf{F}_1$ and $\mathbf{F}_2$, respectively. If $\mathbf{F}_3 = \mathbf{F}_1 + \mathbf{F}_2$, the first condition of equilibrium is satisfied. Additionally, if $F_1 = F_2$, then $l_1 = l_2$ provides rotational equilibrium. Two children of equal weight are in equilibrium if the fulcrum is halfway between them. But if $F_1 \neq F_2$, then what must be said about the lever arms to ensure rotational equilibirum? We get no clues from the first condition of equilibrium. However, anyone familiar with teeter-totters knows that a light person can balance a

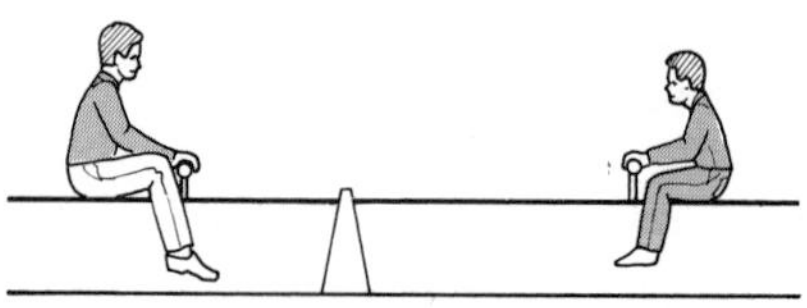

Figure 3.24

Two persons of unequal weight in balance on a teeter-totter.

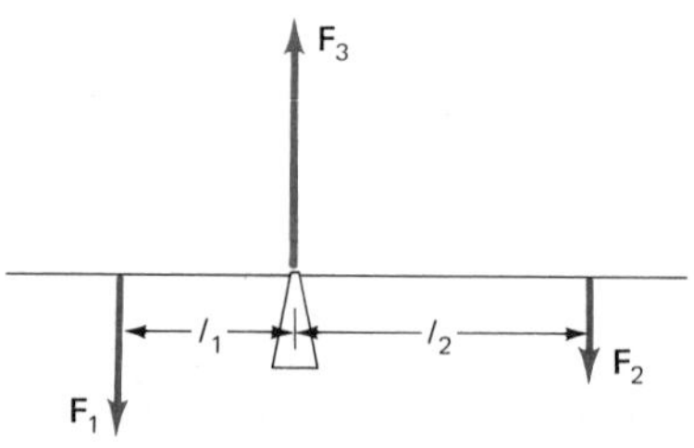

Figure 3.25

Forces acting on a balanced teeter-totter.

heavy person by moving farther away from the fulcrum. Thus if $F_1 > F_2$, then $l_2 > l_1$ for rotational equilibrium. Experimentally we find that $l_1F_1 = l_2F_2$ produces rotational equilibrium. In other words, the weight ratio determines the lever-arm ratio needed to maintain rotational equilibrium. The rotational effect of a force depends on both the force and its lever arm.

Now let us sharpen our definition of the lever arm. In Figure 3.26 a force **F** is applied at the point P at a displacement **r** from the point A. The lever arm, l, is defined as

$$l = r \sin \theta \tag{3.3}$$

Thus l is equal to the length of the perpendicular line drawn from A to the line of action of **F**.

The quantity that controls the rotational equilibrium of a body is called **torque.** The magnitude of the torque,* τ, for the force **F** is defined as the product of the magnitude of the force, F, and its lever arm, l.

$$\tau = Fl \tag{3.4}$$

Putting Eq. 3.3 into Eq. 3.4, we can write

$$\tau = Fr \sin \theta \tag{3.5}$$

for the magnitude of the torque. When the line of action of **F** in Figure 3.26 passes through the point A, the force gives rise to no rotational tendency of the body. In this case $l = 0$ and, according to Eq. 3.4, $\tau = 0$.

Equation 3.5, which gives the magnitude of the torque, can be written in the form of a vector product (see Section 2.5 for a discussion of the vector product).

$$\boldsymbol{\tau} = \mathbf{r} \times \mathbf{F} \tag{3.6}$$

We must stress that the choice of location of the axis through point A in Figure 3.26 is arbitrary. This point, once chosen, serves as a common axis for any other torques.

The vector, $\boldsymbol{\tau}$, is perpendicular to the plane of **r** and **F** and its direction is given by the right-hand rule. With the tails of the vectors **r** and **F** together, let the fingers of your right hand curl from the first vector **r** to the second vector **F** through the smaller angle joining them. Your thumb will then give the direction of $\boldsymbol{\tau}$. Thus in Figure 3.26 we see that $\boldsymbol{\tau}$ is directed into the plane of the paper.

We use the right-hand rule to interpret a torque vector. Aim the extended right-hand thumb in the direction indicated by $\boldsymbol{\tau}$. The right-hand fingers will curl in the direction in which the system would rotate from rest if $\boldsymbol{\tau}$ were the only torque acting.

In Figure 3.23a the torques assocated with both forces are directed into the figure. For each torque the right-hand rule yields a clockwise direction, which is consistent with the effect either torque acting alone would produce on the revolving door. The two forces cancel, whereas the two torques add. In Figure 3.23b the torques associated with the two forces are oppositely directed. The two forces cancel and so do the torques.

In the following example, we will determine several torques about different axes.

Example 6 Torques Acting on a Ladder

Consider a 100-N ladder 10 m long leaning against the smooth or frictionless wall illustrated in Figure 3.27. The wall exerts a force **P** perpendicular to its surface on the ladder; the floor exerts a normal or perpendicular force $\mathbf{F}_1$, and a frictional or tangential force $\mathbf{F}_2$, on the ladder. We assume that the weight, **W**, of the ladder acts downward at the ladder's midpoint. (This assumption is based on the center-of-gravity concept, which we will develop in Section 3.5.) We want to examine the torques associated with these forces about two different axes—through the points A and B.

Because the ladder is in equilibrium, we know from the first condition of equilibrium that $F_1 = W = 100$ N, and $F_2 = P$. In the next section we will see that the second

*We shall associate torque with the angular acceleration of a rigid body in Chapter 12, and with the rate of change of angular momentum of a system of particles in Chapter 11.

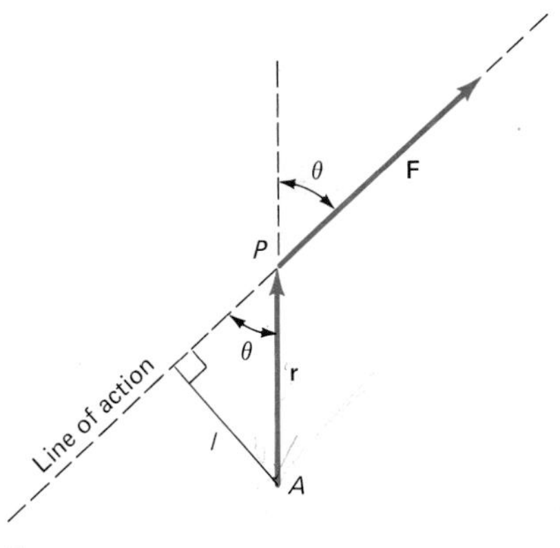

Figure 3.26

A force **F** is applied at the point P. For rotation about an axis through the point A the lever arm equals l, the perpendicular distance between the point A and the line of action of **F**.

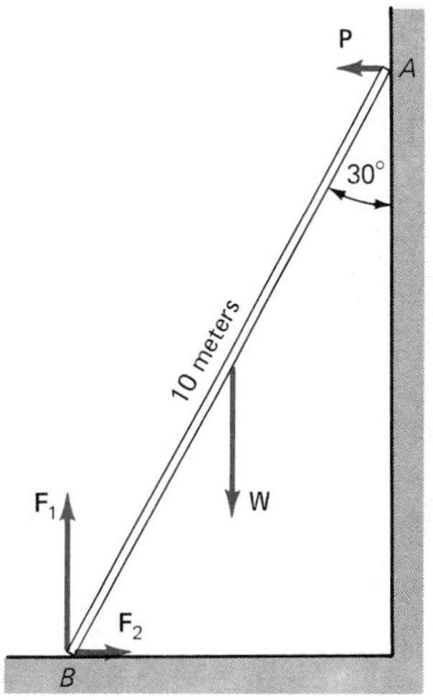

Figure 3.27

A ladder is leaning against the wall. The wall is assumed to be frictionless, and exerts a force **P** on the ladder. The frictional force, $\mathbf{F}_2$, and the normal force, $\mathbf{F}_1$, are components of the contact force exerted on the ladder by the ground. The weight of the ladder, **W**, acts at its geometric center.

condition of equilibirum requires $P = F_2 = 28.9$ N. Here, we will simply accept this value as given, since we are interested only in calculating torques. In the following table are listed the force, lever arm, and torque for each of the axes. We have designated torque vectors directed out of the figure as positive (right-hand fingers curling in a counterclockwise fashion), and torque vectors directed into the figure as negative (right-hand fingers curling in a clockwise fashion).

Axis	Force (N)	Lever Arm (m)	Torque (N · m)
A	$P = 28.9$	0	0
	$W = 100$	2.50	+250
	$F_1 = 100$	5.00	−500
	$F_2 = 28.9$	8.66	+250
B	$P = 28.9$	8.66	+250
	$W = 100$	2.50	−250
	$F_1 = 100$	0	0
	$F_2 = 28.9$	0	0

According to Eq. 3.6, $\tau = Fl$. In each case the lever arm and sign of τ must be determined in order to obtain the entry in the torque column. You should check these results by using Eq. 3.5 and the right-hand rule.

Questions

10. A right-handed screw is one that advances into the wood, metal, or other surface when a clockwise torque is applied, and a left-handed screw is one that advances into the surface when a counterclockwise torque is applied. The connectors for liquid-propane containers on travel trailers have left-handed threads. Can you think of other places where right- or left-handed threads are used?

11. To open a door that has the handle on the right and the hinges on the left a torque must be applied. Is the torque clockwise or counterclockwise when viewed from above? Does your answer depend on whether the door opens toward or away from you?

12. Explain the warning "Never use a large wrench to tighten a small bolt."

13. A central force is one that is always directed toward the same point. Can a central force give rise to a torque about that point?

14. In Figure 3.22 the weight of the pendulum bob gives rise to a torque about an axis through the point O. Is this torque vector directed into or out of the figure? Would it tend to cause rotation in a clockwise or a counterclockwise direction?

3.4 The Second Condition of Equilibrium

We saw in the previous section that the first condition of equilibrium alone cannot guarantee the equilibrium of a rigid body. Our discussion of Figure 3.23 illustrated that a revolving door may or may not be in equilibrium when the first condition of equilibrium is satisfied. We saw that the door was in equilibrium when the sum of the torques associated with $\mathbf{F}_1$ and $\mathbf{F}_2$ was zero. Similarly, you may have noticed that the ladder in equilibrium in Example 6 experienced zero *net* torque about two axes of rotation.

A body is in rotational equilibrium if its state of rotational motion does not change. In the cases of the revolving door and the ladder, rotational equilibrium corresponded to zero net torque. This correspondence holds generally, and is contained in the **second condition of equilibrium**, which states:

> *A rigid body will be in rotational equilibrium if and only if the vector sum of external torques acting on it equals zero when taken about any axis.*

That is we require that

$$\sum \tau = 0 \tag{3.7}$$

for rotational equilibrium. The choice of torque axis is arbitrary. In practice, a particular choice often simplifies the algebra involved.

Let us now consider the unequal arm balance in Figure 3.28. Unknown weights are located at a fixed distance, L_1, from the axis of rotation at O, and known weights are located at a variable distance, L_2, from the axis. We will neglect the weight of the arm itself. The weights give rise to downward forces, $\mathbf{F}_1$ and $\mathbf{F}_2$, acting on the balance arm. Observation confirms that equilibrium with respect to rotation about O is achieved when the sum of torques about any axis equals zero. Thus, for a torque axis through O

$$\sum \tau_O = F_1L_1 - F_2L_2 = 0$$

Hence the unknown weight F_1 may be written as

$$F_1 = F_2\left(\frac{L_2}{L_1}\right)$$

Six component equations constitute the necessary and sufficient conditions for equilibrium of a rigid body in three dimensions:

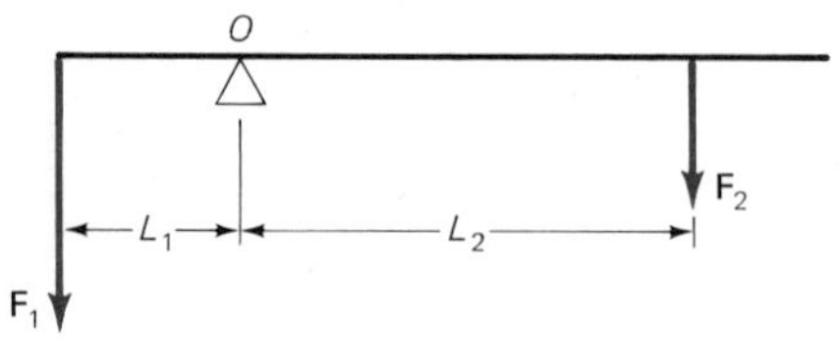

Figure 3.28

Forces acting on an unequal arm balance.

$$\sum F_x = 0 \quad \sum F_y = 0 \quad \sum F_z = 0$$
$$\sum \tau_x = 0 \quad \sum \tau_y = 0 \quad \sum \tau_z = 0 \qquad (3.8)$$

When a problem is limited to coplanar forces (forces in one plane) in the $x-y$ plane, then the x and y force and the z torque equations suffice. None of the forces have a z-component, and none of the torques have x- or y-components. Remember that counterclockwise torques are considered positive and clockwise torques negative.

At this point we will add to the problem solving guide of Section 3.2 three items that relate to the second condition of equilibrium:

Guide to Problem Solving II

8 Choose a torque axis.

9 Determine the torque for each force.

10 Apply the second condition of equilibrium.

These instructions will become clear in the following examples.

Example 7 Which Hand Exerts the Greater Force on the Pole?

A vaulter holds a 29.4-N pole in equilibrium by exerting an upward force, $\mathbf{F}_2$, with his leading hand, and a downward force, $\mathbf{F}_1$, with his trailing hand, as shown in Figure 3.29. If we assume that the weight of the pole acts at its midpoint, what are the values of F_1 and F_2?

Before reading further you should raise questions about the nature of the solution. How will F_1 compare with W? How will F_2 compare with W? Should F_2 be larger than F_1? Smaller? Could the directions of $\mathbf{F}_1$ or $\mathbf{F}_2$ be reversed and equilibrium be maintained? Think of simple ways to defend your conjectures by using the conditions of equilibrium.

Now choose a torque axis that will simplify the solution. If the axis is located through either A or B, then only one of the two forces, either $\mathbf{F}_1$ or $\mathbf{F}_2$, will appear in the torque equation. If the axis is located elsewhere along the pole, then both $\mathbf{F}_1$ and $\mathbf{F}_2$ will appear in the torque equation. We will first choose an axis through A. The second condition of equilibrium yields

$$\sum \tau_A = 0 \times F_1 + \frac{3}{4} \times F_2 - \frac{9}{4} \times 29.4 = 0$$

so that

$$F_2 = 88.2 \text{ N}$$

It is useful in this case to apply the second condition of equilibrium about more than one axis.

When we choose an axis through B we find

$$\sum \tau_B = 0 \times F_2 + \frac{3}{4} \times F_1 - \frac{3}{2} \times 29.4 = 0$$

so that

$$F_1 = 58.8 \text{ N}$$

Figure 3.29

A pole vaulter carrying a pole in equilibrium.

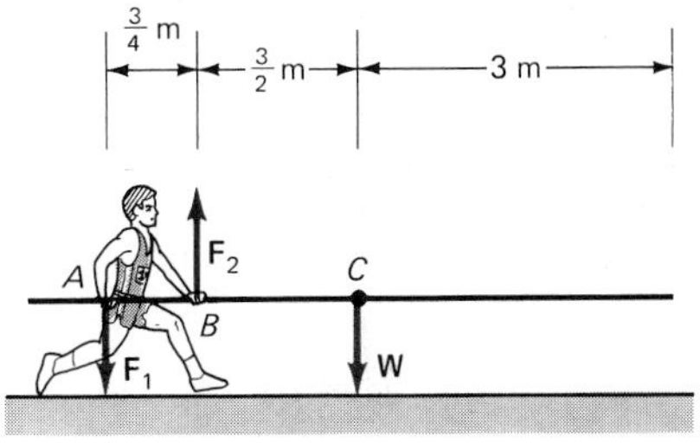

How do these results compare with your conjectures? Are the values of F_1 and F_2 compatible with the requirements of the first condition of equilibrium? Could you have determined the values of F_1 and F_2 by applying only the first condition of equilibrium?

In Example 6 we saw that the first condition of equilibrium was enough to prove the equality, but not to obtain the common value, of the forces P and F_2. The second condition of equilibrium was also needed. In Example 7 we obtained values for F_1 and F_2 without invoking the first condition of equilibrium. Often it is unclear in advance whether both conditions of equilibrium are needed, or if not, which of the two should be used. This insight comes only after solving many problems. To show how values may be found for the forces in the ladder problem of Example 6, and to illustrate the use of both equilibrium conditions to find a solution, we will now reconsider the equilibrium of the ladder.

Example 8 Equilibrium of the Ladder

Let's now reconsider the ladder in Example 6 and determine the value for F_1, F_2, and P given in the example by applying the conditions of equilibrium. The force vector diagram, the x- and y-axes, and the location of the axis of rotation (through point B) are given in Figure 3.30.

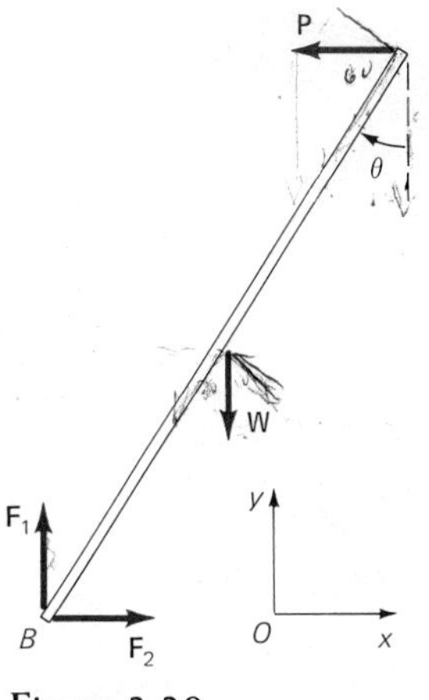

Figure 3.30

Forces acting on a ladder in equilibrium.

Recalling that the ladder is 10 m long, $W = 100$ N, and $\theta = 30°$, we write the following three equations:

$$\sum F_x = +F_2 - P = 0$$

$$\sum F_y = +F_1 - W = 0$$

$$\sum \tau_B = 0 \times F_1 + 0 \times F_2 - 5 \sin \theta \times W + 10 \cos \theta \times P = 0$$

When we solve these equations simultaneously we find

$$F_1 = 100 \text{ N} \qquad P = 28.9 \text{ N} \qquad F_2 = 28.9 \text{ N}$$

These values for P and F_2 were used in Example 6. Note how the choice of axis (through point B) eliminated two unknown forces from the torque equation.

Suppose that we want to determine a particular force **P**. Unless **P** acts on the body chosen, **P** will not appear in the equilibrium equations. Hence, to determine **P**, a body must be selected on which **P** acts. Choosing a body was simple in Examples 6–8, but as the structure becomes complex, making a good choice becomes more difficult. It requires an insight that comes only after solving many problems. To illustrate how choosing a body determines which forces appear in the equilibrium equations, we will consider an A-frame structure. In this case several body choices are possible.

Example 9 **The A-Frame**

An A-frame Swiss chalet is constructed with 10-m beams joined at the apex of a triangle in a frictionless hinge connection, as indicated in Figure 3.31. Halfway to the top a cable of negligible weight is attached to the 5000-N beams. A 1000-N chandelier is hung from the apex. If we neglect the friction of the beams with the ground, what is the force exerted on each beam by the ground and by the tension in the cable? Assume that the weight of each beam acts at its center.

First we may pick the entire assembly as a body, which gives us the force vector diagram shown in Figure 3.32. Since the A-frame is symmetric, we assume that the two forces exerted by the ground are equal. Applying the first condition of equilibrium we write

$$\sum F_y = +2P - 2W_2 - W_1 = 0$$

Substituting the given values of W_1 and W_2, we obtain

$$\sum F_y = +2P - 2(5000 \text{ N}) - 1000 \text{ N} = 0$$

$$P = 5500 \text{ N}$$

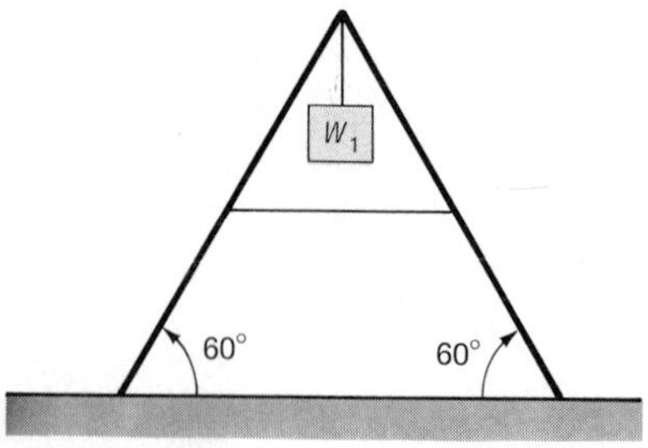

Figure 3.31

The A-frame consists of two beams hinged together at the top and connected together halfway to the ground by a steel cable.

Figure 3.32

Forces acting on the A-frame.

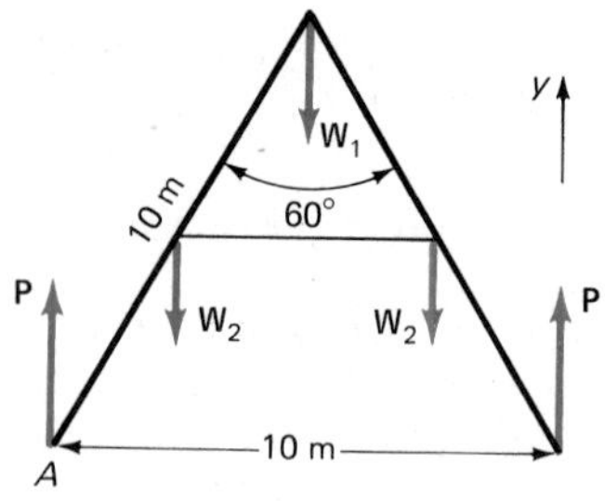

Alternately, we may apply the second condition of equilibrium by using an axis through A:

$$\sum \tau_A = P \times 0 - W_2 \times 2.5 - W_1 \times 5 - W_2 \times 7.5 + P \times 10 = 0$$

which also gives $P = 5500$ N.

No information about the cable tension may be obtained by using the full assembly as a body. Why? Because the cable tension is an internal force. To determine the cable tension we must pick a body on which this force acts. We pick the left beam as a body to obtain the force vector diagram shown in Figure 3.33. $\mathbf{F}_1$ represents the force exerted on the left beam by the cable, and $\mathbf{F}_2$ represents the force exerted on the left beam by the right beam. Why must $\mathbf{F}_2$ be drawn in the direction shown? Newton's third law requires that the forces exerted by each beam on the other be equal in magnitude and opposite in direction. Symmetry implies, therefore, that neither of these forces has a vertical component. Why is $\mathbf{W}_1/2$ (rather than $\mathbf{W}_1$) drawn acting on the left beam? By symmetry each beam supports half the weight. By choosing axes first through B and then through C, we can obtain $\mathbf{F}_1$ and $\mathbf{F}_2$ without solving simultaneous equations.

$$\sum \tau_B = 0 \times F_1 + 0 \times W_2 - 2.5P - 2.5\frac{W_1}{2} + 4.33F_2 = 0$$

$$\sum \tau_C = 0 \times F_2 + 0 \times \frac{W_1}{2} + 4.33F_1 + 2.5W_2 - 5P = 0$$

The number 4.33 comes from $5 \times \sin 60°$. Solving these equations, we find

$$F_1 = 3460 \text{ N} \qquad F_2 = 3460 \text{ N}$$

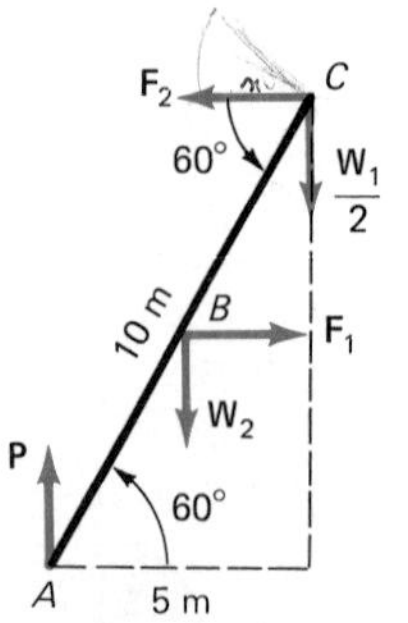

Figure 3.33

Forces acting on the left beam of the A-frame.

You should solve this problem again by using the first condition of equilibrium. Is it possible to satisfy the conditions of equilibrium with $P = 5500$ N, $W_1 = 1000$ N, $W_2 = 5000$ N, and W_1 rather than $W_1/2$ acting on the left beam? Which of the equilibrium equations, if any, would not be satisfied in this case?

In the following example we will see that each member of a truss can be taken as a body in order to determine the internal forces of the truss. A truss is an assembly of beams joined together with pins that function as hinges. The truss is often used in the construction of bridges and other structures. For simplicity we assume the beam weights to be negligible, and therefore each beam is either under compression or tension.

Example 10 **Internal Forces in a Truss**

The three-beam truss shown in Figure 3.34 rests on frictionless supports at A and C. A downward force of 1000 N acts at point B, and the assembly is supported by upward forces $\mathbf{N}_A$ and $\mathbf{N}_C$ at A and C, respectively. Suppose that we are told that $N_A = 366$ N and $N_C = 634$ N. In this case, $\mathbf{N}_A$ and $\mathbf{N}_C$ are given, and we seek the internal forces $\mathbf{C}_1$, $\mathbf{C}_2$, and $\mathbf{T}$ in the truss members.

We assume that AB and BC are in compression (forces $\mathbf{C}_1$ and $\mathbf{C}_2$), and AC is in tension (force $\mathbf{T}$). Wrong assumptions regarding compression or tension will result in negative, rather than positive, answers. Note that the compression in the beam AB means that the force on pin B is directed away from pin A and the force on pin A is directed away from pin B. Consider the truss member AB as a body. The force vector diagram is shown in Figure 3.35. Because the forces do not all act through the same point, useful equations may be obtained by applying the second condition of equilibrium. Here we calculate torques about axes through the points A and B to obtain the following equations:

$$\sum \tau_A = +C_2(10 \sin 75°) - 1000(10 \cos 30°) = 0$$

$$\sum \tau_B = +T(10 \sin 30°) - 366(10 \sin 60°) = 0$$

The quantities in parentheses are values for the lever arms. Because the lever arms for C_1 equal zero for axes through the points A and B, C_1 does not appear in the equations. When we solve these equations we find that $C_2 = 897$ N and $T = 634$ N. Now suppose that we are asked to determine C_2 and T by applying the first condition of equilibrium with the axes specified in Figure 3.35.

First, we consider the pin at B as a body. Forces acting on the pin include the two thrusts, $\mathbf{C}_1$ and $\mathbf{C}_2$, and the downward external force of 1000 N. The force vector diagram is shown in Figure 3.36. Applying the first condition of equilibrium, we have

$$\sum F_x = +C_1 \cos 30° - C_2 \cos 45° = 0$$

$$\sum F_y = +C_1 \cos 60° + C_2 \cos 45° - 1000 = 0$$

We solve these equations to find $C_1 = 732$ N and $C_2 = 897$ N. Because $\mathbf{T}$ does not act on the pin at B, it does not appear in the equations.

To obtain the tension in AC we may pick the pin at A (or C) as a body. Forces acting on the pin at A include $\mathbf{T}$, $\mathbf{C}_1$, and $N_A = 366$ N. The vector diagram is shown in Figure 3.37. Again we apply the first condition of equilibrium and obtain

$$\sum F_x = +T - C_1 \cos 30° = 0$$

$$\sum F_y = +366 - C_1 \cos 60° = 0$$

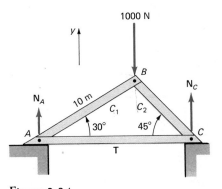

Figure 3.34

A three-member truss assembly.

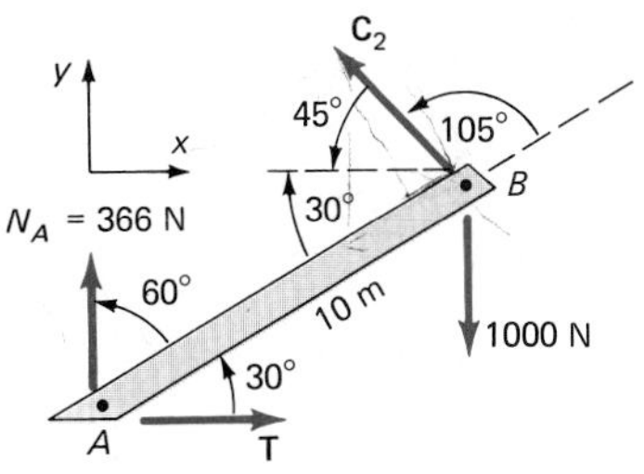

Figure 3.35

Forces acting on the upper left truss member.

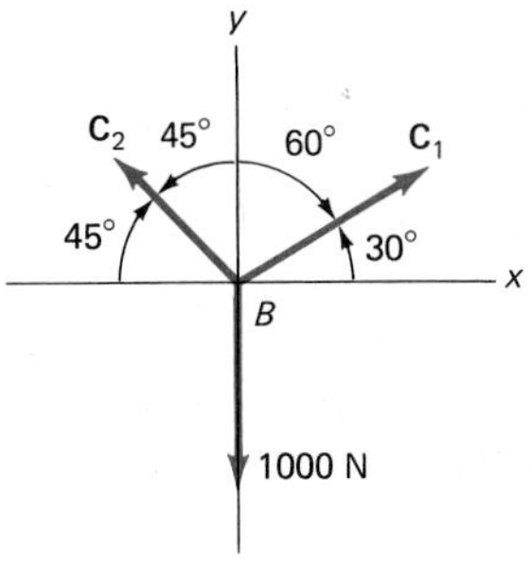

Figure 3.36

Forces acting on the truss pin at point B.

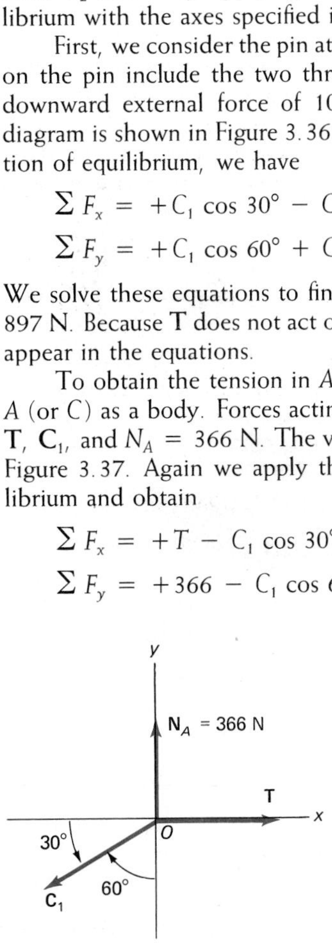

Figure 3.37

Forces acting on the truss pin at point A.

We solve these equations to find $T = 634$ N and $C_1 = 732$ N. Because C_2 does not act on the pin at A, it does not appear in the equations. Note that we had three unknowns (C_1, C_2, and T), but a total of four equations. The same value must be obtained for C_1 (732 N) by using the equations for pin A as was calculated by using the equations for pin B if the four equations are to be consistent.

Questions

15. Use an argument based on the second condition of equilibrium to show that the pendulum bob in Figure 3.22 can't be in equilibrium.

16. A uniform board attached to the ceiling by a string is shown in three positions in Figure 3.38. Assume that the weight acts at the center of the board, and determine for each orientation whether equilibrium is possible.

17. The tippie-top, when placed in the position shown in Figure 3.39, automatically rights itself. The floor exerts a normal force, **N**, directed through the center of curvature, C, and the weight, **W**, acts at the center of gravity, C'. If $N = W$, explain why the top is not in equilibrium in the position shown.

18. The upward buoyant force, **B**, acting on the swimmer is often treated as concentrated at the center of buoyancy at A, just as the weight, **W**, is viewed as concentrated at the center of gravity at C. For most people the points A and C do not coincide as shown in Figure 3.40. Determine whether an equilibrium orientation of the swimmer exists if it is assumed that $B = W$. Would your result be changed if the points A and C were reversed? Would it change if they coincided?

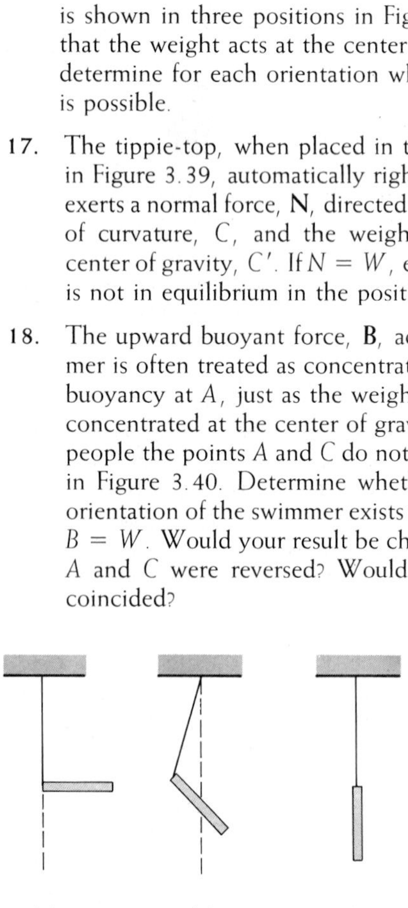

Figure 3.38

A uniform board is attached to the ceiling by a string tied to one end of the board. The board is shown (a) in a horizontal position, (b) at an angle, and (c) in a vertical position.

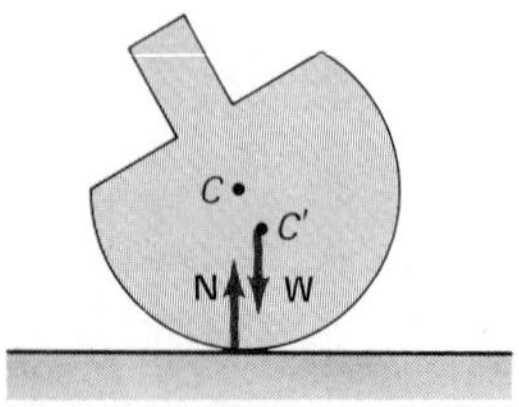

Figure 3.39

A tippie-top.

Figure 3.40

A swimmer experiences forces of buoyancy (**B**) and weight (**W**).

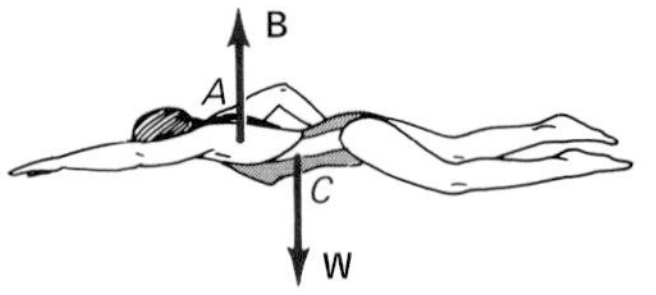

3.5 Center of Gravity

For the ladder in Examples 6 and 8, the pole in Example 7, and the A-frame in Example 9, we assumed that the total weight of an extended body acted at a particular point. This point is called the **center of gravity** of a body. Its location may be determined experimentally or can be deduced from the conditions of equilibrium. We will use the latter method before we describe an experimental procedure.

Each particle in an extended body has a weight, $\mathbf{W}_i$, which is the gravitational force exerted by the earth. The vector sum of the weights of all of the particles in a body is the total weight of the body.

$$\mathbf{W} = \sum \mathbf{W}_i \qquad (3.9)$$

It is possible to place an object in equilibrium by applying a single force, called the **equilibrant**, in just the right direction at just the right point. This point is the center of gravity. The magnitude and direction of the equilibrant, **E**, is determined by the first condition of equilibrium.

$$\mathbf{W} + \mathbf{E} = 0$$

and thus

$$\mathbf{E} = -\mathbf{W} = -\sum \mathbf{W}_i$$

The equilibrant must be equal and opposite to the weight of the object in order to satisfy the first condition of equilibrium. Also, the point of application of the equilibrant must be chosen to satisfy the second condition of equilibrium, thereby guaranteeing rotational equilibrium. To find the point of application we let the x-, y-, and z-axes be fixed in the body and oriented with the y-axis parallel to the equilibrant, as drawn in Figure 3.41. The location of the origin is arbitrary.

For rotational equilibrium

$$\sum \tau_O = \sum \mathbf{r}_i \times \mathbf{W}_i + \mathbf{r} \times \mathbf{E} = 0 \qquad (3.10)$$

where **r** is a vector from O to the point of application of **E**. We wish to determine **r**. Since $-\mathbf{E} = \mathbf{W}$, Eq. 3.10 may be written

$$\mathbf{r} \times \mathbf{W} = \sum \mathbf{r}_i \times \mathbf{W}_i$$

This equation says that the sum of all the gravitational torques is equal to the torque of the total weight acting through the center of gravity.

Figure 3.41

The element of weight, $\mathbf{W}_i$, in a body has a position vector $\mathbf{r}_i$. The equilibrant force, $\mathbf{E}$, is applied at a point with the position vector $\mathbf{r}$.

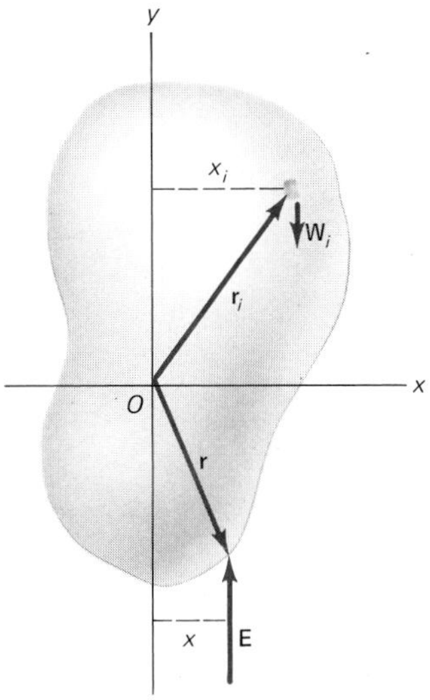

We see that $\mathbf{r}_i \times \mathbf{W}_i$ is directed into the figure (negative z-direction) for all $\mathbf{W}_i$ having $x_i > 0$, and directed out of the figure (positive z-direction) for all $\mathbf{W}_i$ having $x_i < 0$. For our choice of coordinates the component of $\mathbf{r}_i \times \mathbf{W}_i$ along the z-axis is

$$(\mathbf{r}_i \times \mathbf{W}_i)_z = -x_i \mathbf{W}_i$$

corresponding to a clockwise (negative) rotation for $x_i > 0$. Here x_i is the lever arm for the weight $\mathbf{W}_i$.*

Since $\mathbf{r} \times \mathbf{E}$ is directed out of the figure (+z-direction) as drawn, $\mathbf{r} \times \mathbf{W}$ is directed into the figure, and its z-component is

$$(\mathbf{r} \times \mathbf{W})_z = -xW$$

Thus we have

$$xW = \sum x_i W_i$$

or

$$x = \frac{1}{W} \sum x_i W_i$$

as the distance between the y–z plane and a parallel plane containing the center of gravity. We label this distance $\overline{x}$, and call it the x-coordinate of the center of gravity. Rotation of the body so that first the z-axis and then the x-axis is parallel to E establishes in a similar way values for $\overline{y}$ and $\overline{z}$. The intersection of these three planes defines the center of gravity point. Its three coordinates are

*To obtain this last result it is necessary to assume that all $\mathbf{W}_i$ are parallel to the y-axis. Only then will each x_i be the lever arm for its $\mathbf{W}_i$. The $\mathbf{W}_i$, however, converge, since they all point approximately to the center of the earth. If the dimensions are small enough, this deviation from exact parallelism may not be important. How small is small enough? It depends on how large a deviation from parallelism is tolerable. For instance, the weight vectors at opposite ends of the 2-mi linear accelerator at Stanford University fail to be parallel by an angle of only 1.7′. This deviation is unimportant.

$$\overline{x} = \frac{1}{W} \sum x_i W_i \qquad \overline{y} = \frac{1}{W} \sum y_i W_i$$
$$\overline{z} = \frac{1}{W} \sum z_i W_i \tag{3.11}$$

We may locate the center of gravity of a two-dimensional object experimentally by hanging the object by a string. If the string exerts the equilibrant force at the point A, as shown in Figure 3.42, the center of gravity must lie somewhere along the A–A' line. Repeating the procedure by using the suspension point B establishes the center of gravity along the B–B' line. The center of gravity lies at the intersection of these two lines. The three-dimensional argument is a straightforward generalization of the two-dimensional case.

The center-of-gravity concept is valuable because it simplifies many problems. Individual weights of a body may be replaced by a single weight acting at the center of gravity. You may recall that this was done in Examples 7–10, greatly simplifying our calculations.

There are several different ways to locate the center of gravity of an object. We have already seen two of them—the analytical method (using Eq. 3.11) and the experimental method illustrated by Figure 3.42. We may also find the center of gravity of an object by inspection, when the object is symmetric, or by using what we call the "negative weight" procedure. Let's examine the last two methods.

The center of gravity of a homogeneous symmetric object can be located by inspection. This is because the center of gravity of such an object coincides with the center of symmetry. To prove this, consider an origin located at the center of symmetry of an object. Let new axes be parallel to the x-, y-, and z-axes in Figure 3.41 and label them x', y', and z'. Relative to these new axes the center-of-gravity coordinates are zero, $\overline{x}' = \overline{y}' = \overline{z}' = 0$. This means, for instance, that $\sum x_i' W_i = 0$. For every weight at a positive location $(+x_i)$ there will be a corresponding weight at a negative location $(-x_i)$. Thus the sum measured relative to the symmetry center must vanish, $\overline{x} = 0$. Similarly, $\overline{y} = 0$ and $\overline{z} = 0$. Hence the center of symmetry coincides with the center of gravity.

In the following example, we use Eq. 3.11 and inspection to locate the center of gravity of a symmetric array of billiard balls.

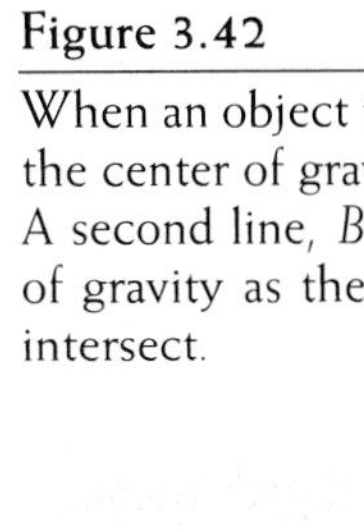

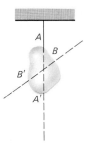

Figure 3.42

When an object is suspended by a string from the point A, the center of gravity lies below A on the vertical line AA'. A second line, BB', similarly obtained, locates the center of gravity as the point where the two lines AA' and BB' intersect.

Figure 3.43

The dots represent the centers of gravity of 15 billiard balls arranged in a triangular array.

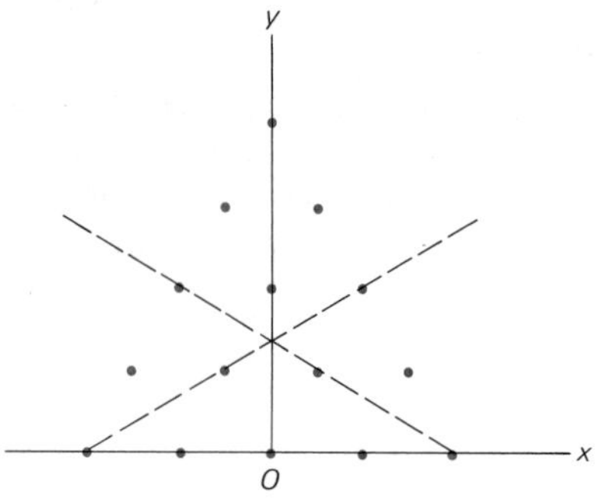

Example 11 Center of Gravity of a Billiard Ball Array

A common initial arrangement in pocket billiards has 15 object balls, each weighing 1.64 N, distributed symmetrically in a triangle, as suggested by the dot arrangement in Figure 3.43. The center-to-center distance of the billiard balls is 5.72 cm. What are the coordinates of the center of gravity of the array?

By symmetry, each ball has its center of gravity at its geometric center, so the array of centers adequately represents the balls themselves. The array itself is symmetric about the x-axis, so $\bar{x} = 0$. For $\bar{y}$, Eq. 3.11 yields

$$\begin{aligned}\bar{y} = {} & [5(1.64) \times 0 + 4(1.64) \times 4.95 \\ & + 3(1.64) \times 9.91 + 2(1.64) \times 14.9 \\ & + 1.64 \times 19.8] \\ & \div 15 \times 1.64 \\ = {} & 6.61 \text{ cm}\end{aligned}$$

Here the factor $4.95 = 5.72 \cos 30°$, the y_i for the row of four billiard balls. Symmetry indicates that $\bar{y}$ should be at the intersection of the perpendicular bisectors of the edges. We have confirmed this numerically. The location $(\bar{x}, \bar{y})$ is marked with an asterisk in Figure 3.43. You may have noted that centimeters were used in the $\bar{y}$ calculation rather than meters. Because weight appears in both the numerator and the denominator in Eq. 3.11, the units of $\bar{x}$ and $\bar{y}$ will be the units of x_i as long as W and the W_i are given the same units.

If a result can be obtained by inspection, why calculate? Whenever possible, use symmetry to locate the center of gravity of an object. Even bodies with no definite symmetry are sometimes composed of symmetric parts. Most bodies, however, lack sufficient symmetry to warrant use of the inspection method. In Example 12 we will use Eq. 3.11 to find the center of gravity of a Soma puzzle piece, an object that has too little symmetry for us to use inspection. We will also develop the negative weight procedure, which is useful in center-of-gravity calculations for objects having certain kinds of symmetry.

Example 12 The Soma Puzzle

The Soma cube puzzle consists of 27 small cubes organized into six pieces composed of 4 cubes each and one piece composed of 3 cubes. One of the pieces is L-shaped, and is illustrated (by the solid lines only) in Figure 3.44. The separate small cubes are 1 cm on an edge. What is the center of gravity of the piece?

By symmetry we know that the center of each cube is the location of its center of gravity, and $\bar{x} = \bar{y}$. Let the weight of one cube be W. From Eq. 3.11 we have

$$\bar{x} = \frac{0.5W + 0.5W + 1.5W}{3W}$$

so

$$\bar{x} = 0.83 \text{ cm} \qquad \text{and} \qquad \bar{y} = 0.83 \text{ cm}$$

Again the location of the center of gravity in Fig. 3.44 is marked with an asterisk on the figure.

Let us reconsider Example 12 in order to develop a negative weight procedure, which is useful in some center-of-gravity calculations. One more cube, placed in the position suggested by the dotted lines in Figure 3.44, would make the L-shaped piece symmetric. With this in mind, we rewrite $\bar{x}$ for Example 12:

$$\bar{x} = \frac{2 \times 0.5W + 2 \times 1.5W - 1.5W}{4W - W}$$

Terms have been added and subtracted in both numerator and denominator, leaving the value of $\bar{x}$ unchanged. A slight rearrangement gives

$$\bar{x} = \frac{1.0 \times 4W + 1.5 \times (-W)}{4W - W}$$

This formula represents a two-particle system. One particle ($W_1 = +4W$) is at $x_1 = 1.0$ cm, and the other particle ($W_2 = -W$) is at $x_2 = 1.5$ cm.

This result may be generalized as follows: If an unsymmetric object can be converted into a symmetric object by adding or subtracting one or more symmetric pieces, then the negative weight procedure will yield the correct coordinates of the center of gravity. In the case of Example 12, for instance, a symmetric one-cube object can be added to the unsymmetric object to make a symmetric four-cube object. The "positive weight" four-cube object coupled with the "negative weight" one-cube object results

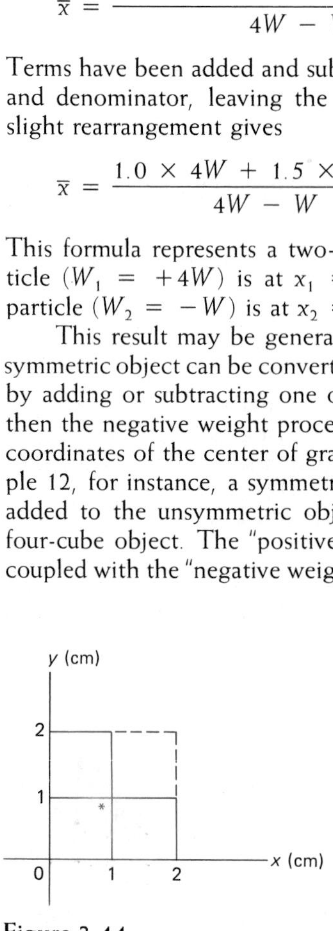

Figure 3.44

A three-cube Soma puzzle piece.

Figure 3.45

A lunate plane mass bounded by two circles.

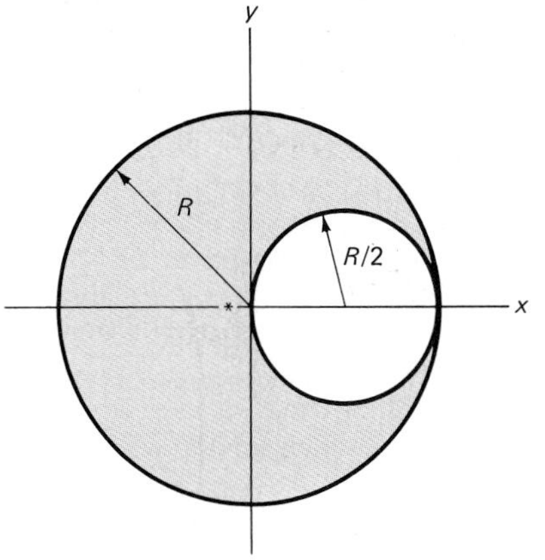

in the same center of gravity as the original three-cube object. The positive four-cube piece has a center of gravity at (1.0, 1.0), and the negative one-cube piece has a center of gravity at (1.5, 1.5). We thus calculate from these data

$$\overline{x} = \frac{1.0(4W) + 1.5(-W)}{4W - W} = 0.83 \text{ cm}$$

Again we have

$$\overline{y} = 0.83 \text{ cm} \qquad \text{by symmetry}$$

in agreement with our previous result. The following example further illustrates the negative weight procedure.

Example 13 **Center of Gravity of a Lunate Area**

Consider the lunate (crescent-shaped) area bounded by circles having radii R and $R/2$ shown in Figure 3.45. Locate the center of gravity, assuming that weight is proportional to area? By symmetry $\overline{y} = 0$, so only $\overline{x}$ need be calculated.

If the "hole" is filled in, a symmetric circle is constructed from the lunate shape. It may be viewed, therefore, as a combination of one circular area ($W_1 = +4W$) with $x_1 = 0$ and another circular area ($W_2 = -W$) with $x_2 = +R/2$. This interpretation leads to the calculation

$$\overline{x} = \frac{0 \times (+4W) + (R/2) \times (-W)}{4W - W} = -\frac{R}{6}$$

which is the x-coordinate of the center of gravity of the original lunate figure. Again an asterisk marks the spot.

Questions

19. Locate the center of gravity of some familiar objects by balancing them on your finger. Try a pencil or a book.

20. Locate the center of gravity of some familiar objects by hanging them from a string. Try scissors or a chair.

21. Normally the center of gravity of a human is about an inch below the navel in the center of the body. How is this location affected by changes in body position such as bending over? Can the center of gravity lie outside the body itself?

22. Is it possible for the center of gravity of an object to be located inside the object at a point where there is little or no matter? If your answer is yes, give an example. If you answer no, explain why not.

3.6 Stability of Equilibrium

A cone balanced on its point is in equilibrium, just as it is when firmly seated on its base. Similarly, a basketball delicately balanced atop a player's finger is in equilibrium, just as it is when motionless on the floor. In each situation two kinds of equilibrium are described, and they are not difficult to distinguish. We classify types of equilibrium according to their stability.

First, let's consider the behavior of a system in equilibrium with respect to *slight displacements* of the system away from its original equilibrium position. In its original position the sums of the external forces and torques are zero. In the displaced position the external forces and torques might differ from the original ones, and the sums may or may not be zero. When the sums of the torques and the forces are zero, the system is in equilibrium and does not move. When the sums do not equal zero the forces and/or torques will either cause the system to return to its original position or to move away from its original position.

When a body in equilibrium is given a slight displacement and the new position is an equilibrium position, then the original equilibrium position is said to be one of **neutral equilibrium.** Consider the rotor blade on a helicopter, for example. If the rotor is given a small rotation, both the new and the old position are equilibrium positions. Hence the original position is one of neutral equilibrium.

A basketball on the floor illustrates neutral equilibrium for three common types of displacements. If it slides without rotation, rolls without sliding, or rotates in place about its center of gravity, both the new and old positions in each case are equilibrium positions.

When the new position is not an equilibrium position, then one or both of the conditions of equilibrium is not satisfied by the body in the displaced position. Hence either a net force and/or a net torque acts on the body in its new position. When this net force or torque results in urging the body away from the original equilibrium position, then the original equilibrium position is called one of **unstable equilibrium.**

The gymnast shown in Figure 3.46 poised above the bar balanced on his hands exemplifies this kind of equilibrium. If he is rotated through a small angle away from equilibrium, then the net torque created as a result of the small rotation will urge him still further away.

Figure 3.46

Gymnast on a high bar.

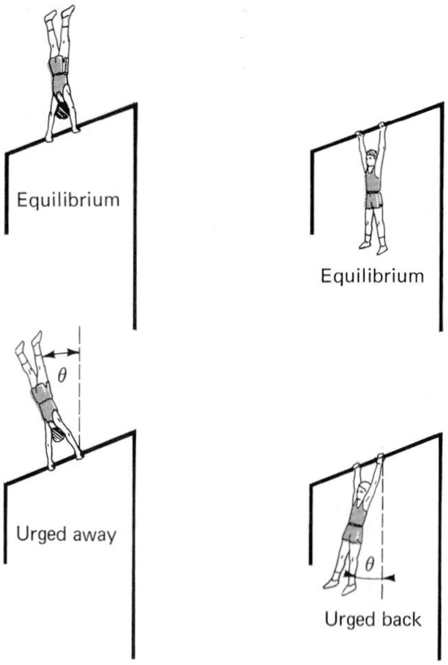

When the displaced position gives rise to a net force and/or torque that acts to urge the body back toward its original equilibrium position, then the original equilibrium position is called one of **stable equilibrium.** For example, picture a child in a swing: If the swing is rotated through a small angle away from the bottom equilibrium position, it will be urged back toward the original position by the torque generated by the weight of the child and swing.

The kind of equilibrium exhibited by a body is related to the *displacement* considered. A cone on its base is often cited as an example of stable equilibrium. If the displacements are small rotations (attempts to tip the cone over), then the cone is indeed in a position of stable equilibrium. However, if we consider small horizontal displacements of the cone then the case is one of neutral equilibrium.

Questions

23. Show that the tippie-top in Figure 3.39 is in a position of stable equilibrium if it is vertical.

24. Show that when $\theta = 0$ the pendulum bob in Figure 3.22 is in stable equilibrium.

Summary

The length of a vertical spring is used to measure force. The newton is defined by assigning 9.80 N to the force exerted by the spring when the standard kilogram is attached.

Two types of force considered in this chapter are (1) contact forces, the forces that arise when objects touch; and (2) weight, the force of gravitational attraction between the earth as a whole and the body.

The torque, τ, associated with a force $\mathbf{F}$ is defined by

$$\boldsymbol{\tau} = \mathbf{r} \times \mathbf{F} \qquad (3.6)$$

where $\mathbf{r}$ is a vector directed from the axis of rotation to the point of application of the force.

Forces and torques are called "external" when they are exerted on the body by its environment. Only the external forces determine whether a body is in equilibrium.

The necessary and sufficient conditions for a body to be in a state of equilibrium are contained in the equations

$$\sum \mathbf{F} = 0 \quad \text{first condition of equilibrium} \qquad (3.1)$$

$$\sum \boldsymbol{\tau} = 0 \quad \text{second condition of equilibrium} \qquad (3.7)$$

A body is in translational equilibrium when the first condition of equilibrium is satisfied, and it is in rotational equilibrium when the second condition of equilibrium is satisfied.

The "Guide to Problem Solving" is a list of instructions that you should follow when solving word problems concerning equilibrium.

The center of gravity is the point where the total weight of an extended body may be assumed to act. The coordinates of the center of gravity of a body are given by

$$\bar{x} = \frac{1}{W} \sum x_i W_i \qquad \bar{y} = \frac{1}{W} \sum y_i W_i \qquad \bar{z} = \frac{1}{W} \sum z_i W_i \qquad (3.11)$$

The center of gravity of a body may be located analytically, using Eq. 3.11; or experimentally; and sometimes by inspection or by using the negative weight procedure.

Equilibrium states of a body are unstable, neutral, or stable, depending on the response of the body when subjected to a small displacement from its equilibrium position.

Problems

Section 3.1 Force

1. The lengths of a spring in centimeters, and the corresponding values of the attached weight in newtons, are given in Table 3.1. (a) Construct a calibration curve for the force (F) versus the extension (x) for the spring. (b) What spring length corresponds to 0.1 N? (c) What spring length corresponds to 1.2 N?

Table 3.1

x (cm)	W (N)
10.0	0
10.5	0.5
11.0	1.0
11.5	1.5
12.0	2.0

2. The *Guinness Book of Records* (1980 edition) lists the weight of the heaviest human of all time as 1069 lb. Express this weight in newtons.

3. Experiment shows that the extension (x) of a vertically oriented spring is related to the weight (W) attached to the end of the spring by

$$W = 0.51x - 0.22$$

where W and x are in newtons and centimeters, respectively. What spring extension corresponds to a force of 0.31 N?

Section 3.2 The First Condition of Equilibrium

4. An ideal pulley changes the direction of the tension in the rope without changing its magnitude. Such a pulley weighing 100 N is used with a flexible rope of negligible weight to hoist a sailfish weighing 4000 N, as shown in Figure 1. Determine (a) the force F and (b) the tension T in the rope supporting the pulley.

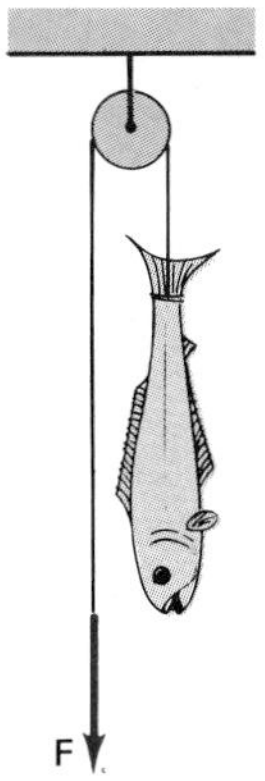

Figure 1

5. A spider descends along its self-spun thread at a constant speed. The spider weighs 0.01 N. What is the tension in the thread if the weight of the thread is negligible? (Be sure to guess at an answer first and then check it by applying the remaining steps in the problem solving guide.)

6. A 12-N weight is suspended as shown in Figure 2. The system is held motionless by the frictional force of the horizontal table top on the weight W. What is the frictional force?

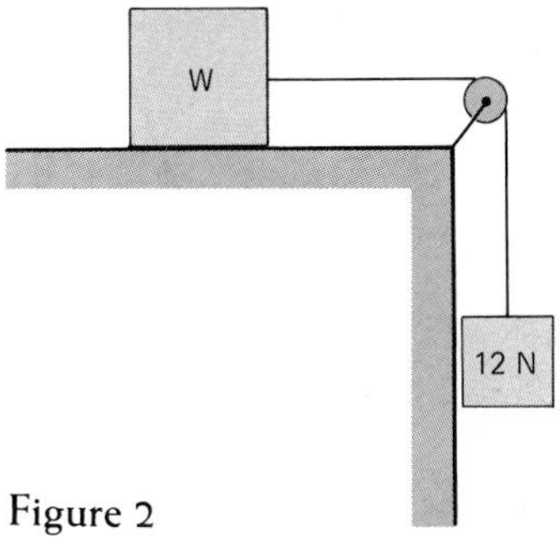

Figure 2

7. A 400-N girl sits in a playground swing of negligible weight. Normally the two supporting ropes are attached at the top (see Figure 3a) and each rope has a tension of 200 N. Suppose, however, that the two ropes pass over pulleys at the top (see Figure 3b) and the ends are held by the girl. What tension would be found in the ropes? (Assume the ropes are vertical, and ignore friction and rotational considerations.)

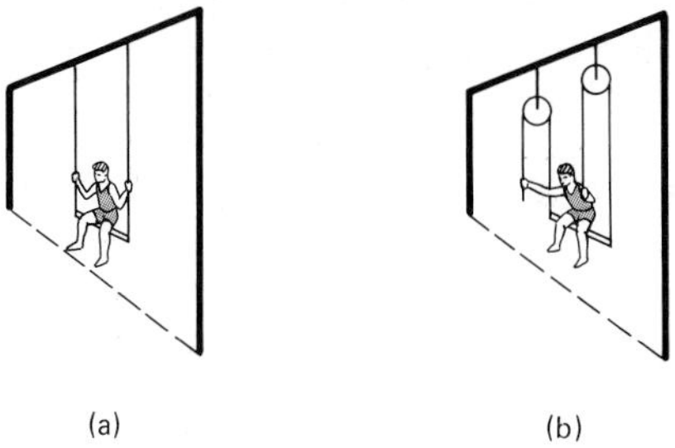

Figure 3

8. A weight W just balances the 100-N weight shown in Figure 4. The pulley and inclined plane are frictionless. (a) Obtain the components of the 100-N weight parallel and perpendicular to the plane. (b) Apply the first condition of equilibrium to the 100-N weight, and calculate the string tension, T, and the force P, exerted by the plane on the 100-N weight. (c) Calculate the weight W.

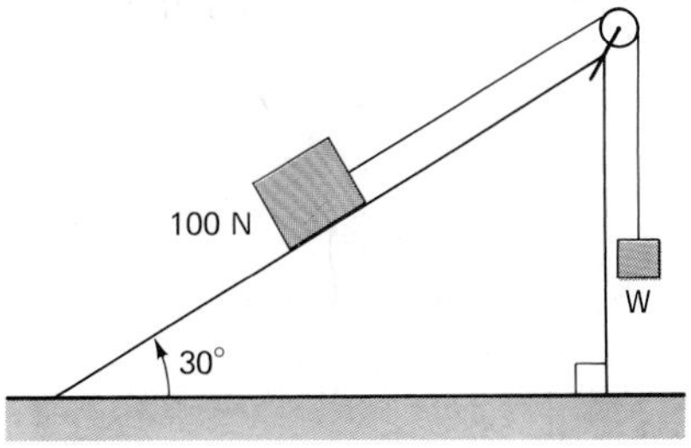

Figure 4

9. A weightless beam 4-m long is perpendicular to a frictionless wall at point C in Figure 5. The beam, in equilibrium, experiences a thrust, $\mathbf{P}$, perpendicular to the wall at C, a tension $\mathbf{T}$, at A, due to the cable AB, and a downward tension of 200 N at A due to the 200-N weight. Apply the first condition of equilibrium to the beam to determine the magnitudes of the forces $\mathbf{P}$ and $\mathbf{T}$.

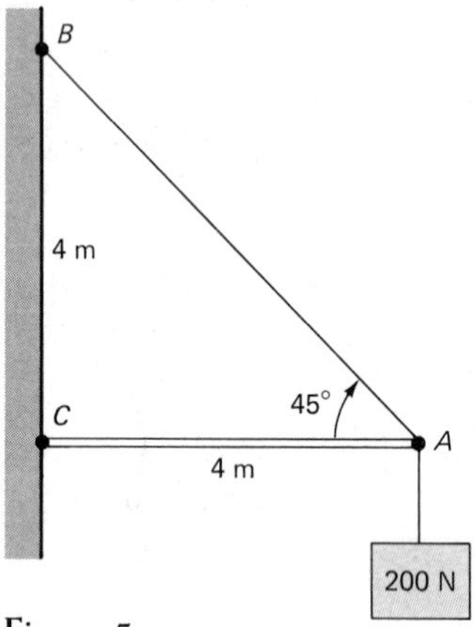

Figure 5

10. Three strings are tied at the point A as shown in Figure 6. A 39.3-N weight hangs from one string; a tension of 50 N is in the second string, which makes an angle of 37° with the horizontal; and the third string is attached to the ceiling at an angle θ with the vertical. The system is in equilibrium. (a) Obtain the horizontal and vertical components of the 50-N and 39.3-N tensions. (b) Determine the horizontal and vertical components of the tension, $\mathbf{T}$, in the third string. (c) Calculate the magnitude of $\mathbf{T}$ and the value of θ.

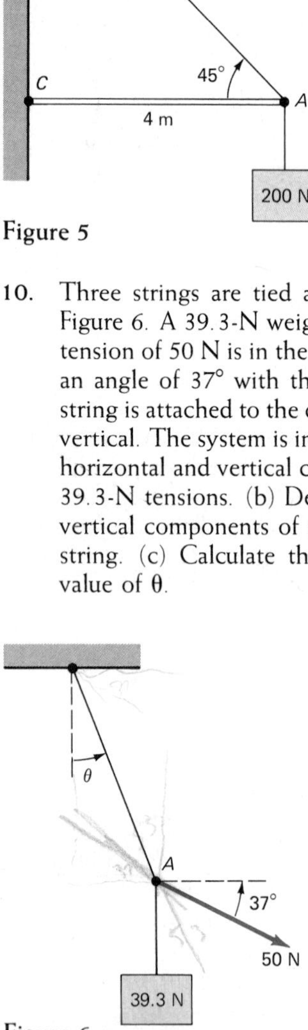

Figure 6

11. In a real pulley, shown in Figure 7, a frictional force, f, is added to the tension in the rope such that its direction opposes the impending motion of the rope. Neglecting the weight of the rope, and assuming that W_1 is on the verge of falling, (a) prove that in equilibrium $F = T_1$ and $W_1 = T_2$; (b) obtain a relation among T_1, T_2, and f. (Imagine the limiting case where the frictional force is so large that $T_1 \rightarrow 0$.) (c) How would your result in (b) change if W_1 were on the verge of rising? (d) Express T_3 in terms of the weights and F.

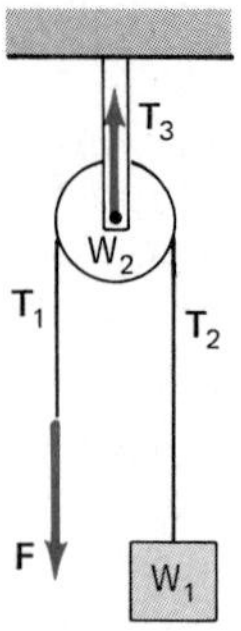

Figure 7

12. A pair of 20-N weights is connected over a pulley by a cord (whose weight may be neglected) as illustrated in Figure 8. Motion impends up the frictionless plane. (a) Prove that the frictional force introduced by the pulley is 10 N. (b) Is motion also impending down the plane? Explain.

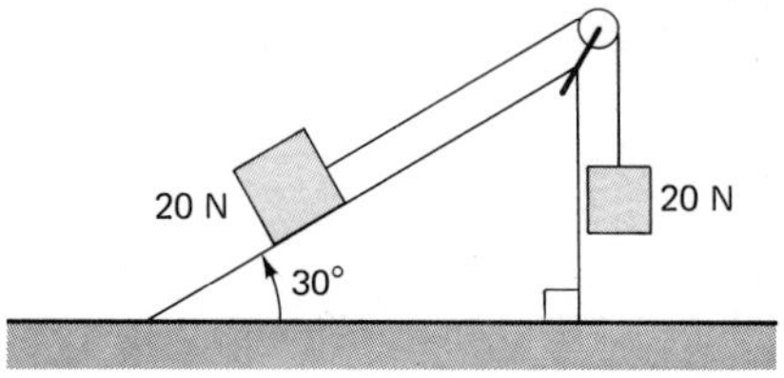

Figure 8

13. Two 200-N traffic lights are suspended from a single cable whose center section is horizontal, as shown in Figure 9. Neglect the cable weight and (a) prove that if $\theta_1 = \theta_2$, then $T_1 = T_2$. (b) Determine the three tensions if $\theta_1 = \theta_2 = 8°$.

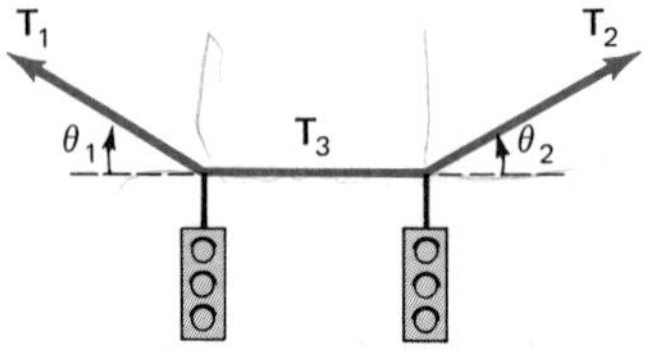

Figure 9

14. A 9000-N crate is being loaded on a ship with the symmetric cable arrangement shown in Figure 10. Assume that the weight acts at the center of the crate and neglect the cable weight. Determine the tensions.

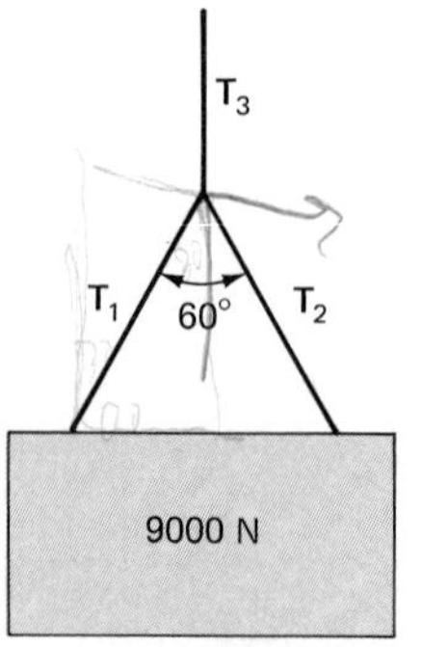

Figure 10

15. A 1.35-kg bottle of wine is stored on its side in a diamond-shaped receptacle as shown in Figure 11. Assume that the walls of the receptacle are smooth. (a) Determine the thrust exerted on the bottle by each wall when the receptacle is symmetrically oriented ($\theta_1 = \theta_2$). (b) Determine the thrust if $\theta_1 = 25°$ and $\theta_2 = 35°$.

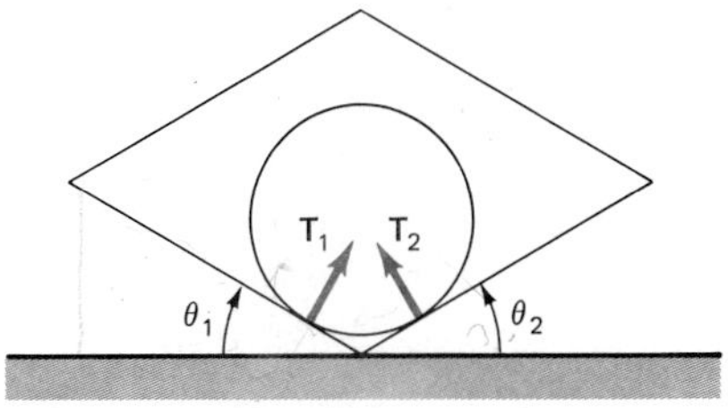

Figure 11

16. Solve the equations given at the end of Example 2 for C to confirm the numerical result obtained earlier in the example.

Section 3.3 Torque

17. Imagine you are using a crossbar wrench to change a tire. The handles are 0.4 m long and you are exerting a force of 200 N on each handle, as shown in Figure 12. Calculate the torque you are exerting.

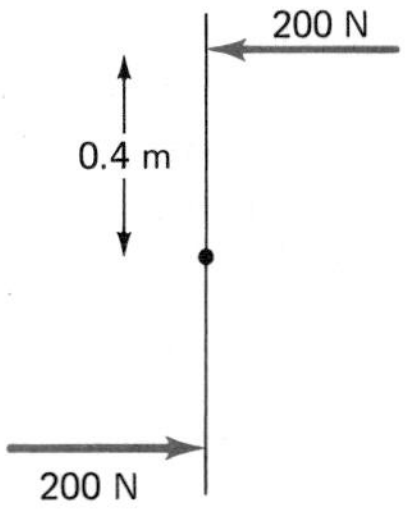

Figure 12

18. To demonstrate his strength a student holds a 50-N chair by the bottom of one leg. The weight may be considered to act along a line 20 cm to the right of the left side of the chair (Figure 13). Determine the torque of **W** about an axis through the point *O*.

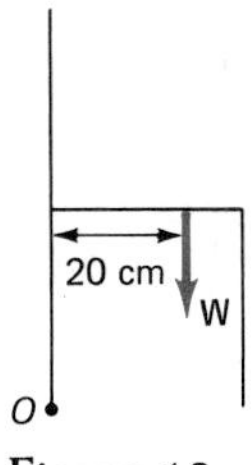

Figure 13

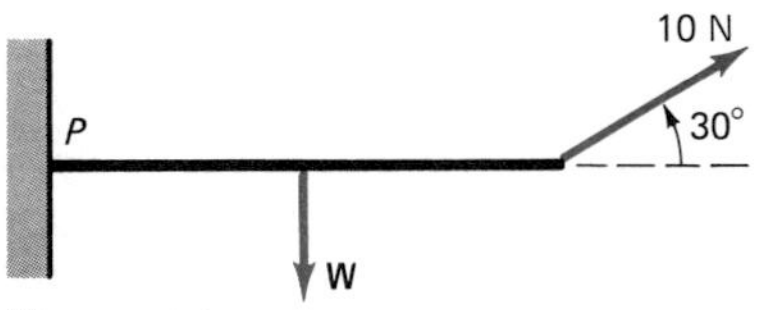

Figure 14

19. A weightless, rigid rod 4 m long is free to rotate about an axis through the point *P*. When the rod is horizontal (see Figure 14) a 10-N force is applied at 30° with the horizontal. (a) Calculate the torque of the 10-N force about *P*. (b) What weight *W* at the midpoint of the rod would give rise to a torque of equal magnitude? (c) If clockwise is negative and counterclockwise is positive, give the signs for the torques obtained in (a) and (b). (d) Give the lever arms for the two torques.

20. Determine the lever arms and torques for the three forces acting on the beam in Figure 5 (see Problem 9) for an axis through the point (a) *A*; (b) *B*.

21. The frictional force exerted on a heavy rope by the post shown in Figure 15 adds to the force, $\mathbf{F}_1$, exerted by the sailor to make the force exerted on the ship, $\mathbf{F}_2$, much larger than $\mathbf{F}_1$. Take $F_1 = 100$ N and $F_2 = 1000$ N initially. (a) Determine the net force exerted on the post by the rope. (b) If the radius of the post is 0.1 m, determine the net torque exerted on the post by the rope.

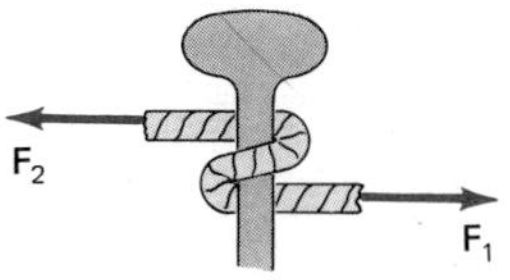

Figure 15

22. A 300-N sign is hung from the end of a 4-m uniform beam weighing 100 N. A 5-m wire connects the building to the end of the beam as shown in Figure 16. Determine the torque for each of the five forces about the axes through points *A*, *B*, and *C*. Take $H = 466\frac{2}{3}$ N, $V = 50$, and $T_1 = 583\frac{1}{3}$ N.

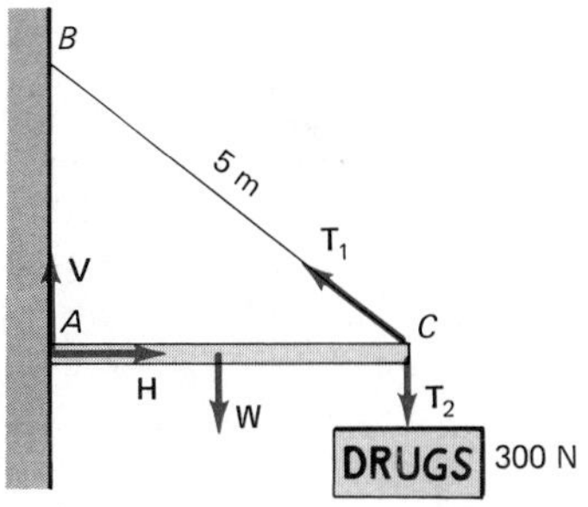

Figure 16

Section 3.4 The Second Condition of Equilibrium

23. A baseball player holds a 36-oz. bat ($W = 8.91$ N) with one hand at the point O (Figure 17). The bat weight acts along a line 60 cm to the right of *O*. Determine (a) the force and (b) the torque exerted on the bat by the player if the bat is in equilibrium.

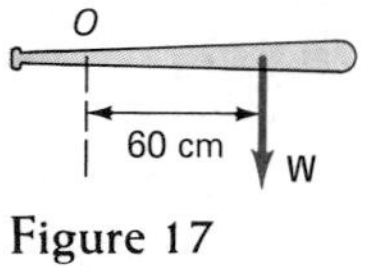

Figure 17

24. Two kids, one at each end of a teeter-totter (or seesaw) are in a state of static equilibrium. One kid, weighing 340 N, is 3 m to the right of the pivot. The other kid is 2.30 m to the left of the pivot. (a) Neglecting friction, calculate the weight of the second kid. (b) What is the upward force exerted by the pivot on the teeter-totter?

25. A weightless beam 4 m long is perpendicular to a wall. The beam, in equilibrium, is supported by the wall, a cable at 30° with the horizontal, and is pulled down by a 500-N weight hanging at the end, as shown in Figure 18. (a) Calculate the torque for each force acting on the beam about an axis through A, and apply the second condition of equilibrium. What does this imply about the vertical component of the force exerted on the beam by the wall? (b) Apply the second condition of equilibrium to the beam using an axis through the point B. What does this imply about the horizontal component of the force exerted on the beam by the wall? (c) Find the cable tension by applying the second condition of equilibrium to the beam using an axis through the point C.

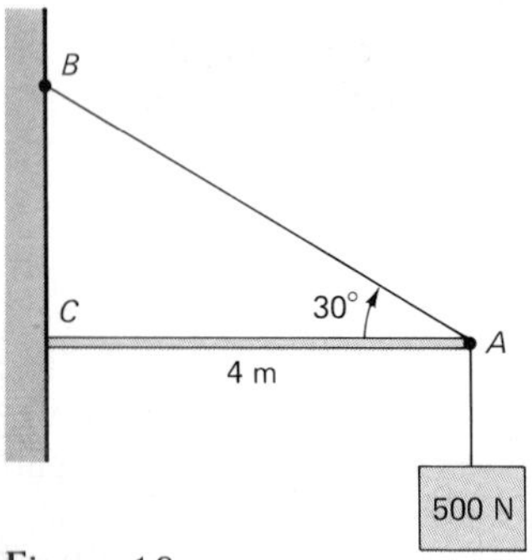

Figure 18

26. Two billiard balls are placed in a glass as shown in Figure 19. The two billiard ball centers and the point A lie on a straight line. (a) Assume that the walls are smooth, and determine P_1, P_2, and P_3. (b) Determine the force exerted on the right billiard ball by the left one. Assume that each ball weighs 1.64 N.

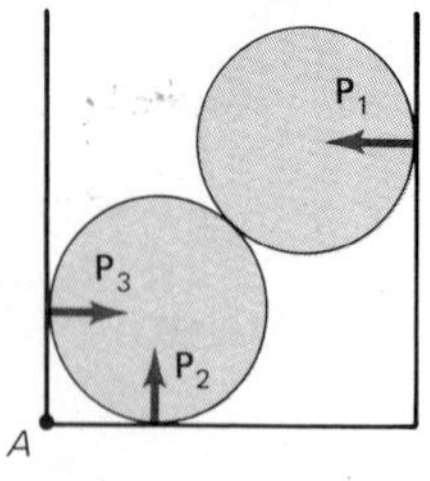

Figure 19

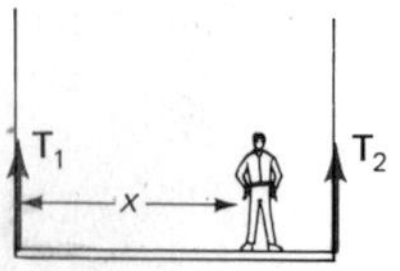

Figure 20

27. An 800-N painter is shown in Figure 20 at a distance x from the left end of a uniform 5-m board weighing 500 N. (a) Determine the maximum and minimum values of the tensions, T_1 and T_2, for values of x in the range $0 \leq x \leq 5$ m. (b) For what value of x will $T_1 = 410$ N and $T_2 = 890$ N?

28. To get some heavy barrels off the street to permit the flow of traffic, the police wrap ropes around the barrels in order to pull them up over the curbs as shown in Figure 21. If the curb height is half the barrel radius, what tension, **T**, will pull the barrel up onto the sidewalk? (Express your answer in terms of the barrel weight, W.)

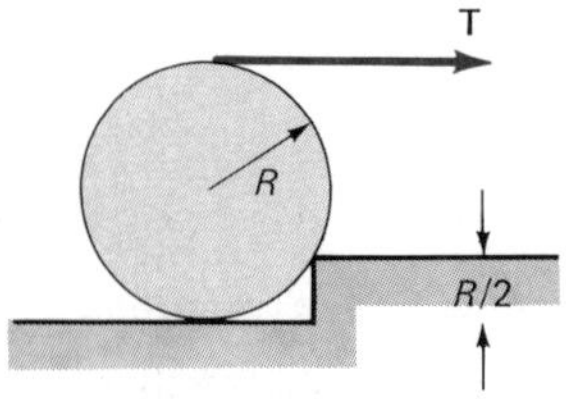

Figure 21

29. A weightless beam in a truss member experiences a single net force at each end. Prove that these two forces must be equal in magnitude and oppositely directed, and must have lines of action parallel to the beam. Thus such truss beams must be under tension or compression.

30. The weight of a pointed roof produces an outward thrust on the supporting walls. Architects used flying buttresses to counteract this thrust in Gothic cathedrals. Churches built today generally use steel rods in tension. Figure 22 shows in cross section the structural elements supporting such a roof; these elements consist of two beams symmetrically pinned at the apex and connected to each other at the bottom by a steel rod. The tension in the steel rod is adjusted until the wall experiences no net outward thrust. The share of total roof weight to be supported by each wall is 160,000 N. Determine the tension in the rod for (a) $\theta = 30°$ and (b) $\theta = 60°$.

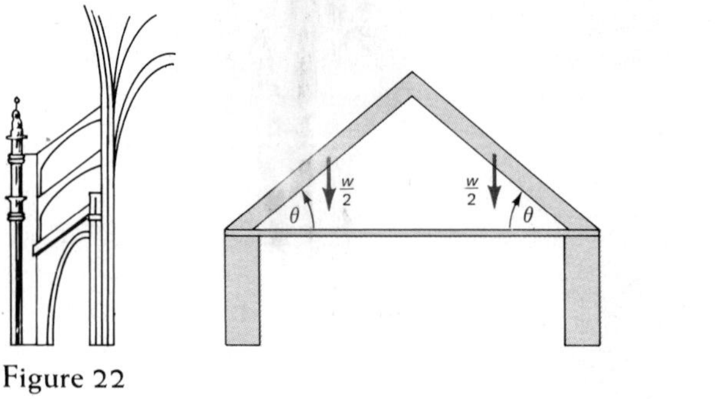

Figure 22

31. A kingpin bridge, shown in cross section in Figure 23, employs members of negligible weight consisting of two long beams under compression and two short beams and a cable under tension pinned together symmetrically as shown. The total weight supported by this section is 200,000 N. Obtain the beam compression, beam tension, and cable tension (a) if $\theta = 30°$; (b) if $\theta = 60°$.

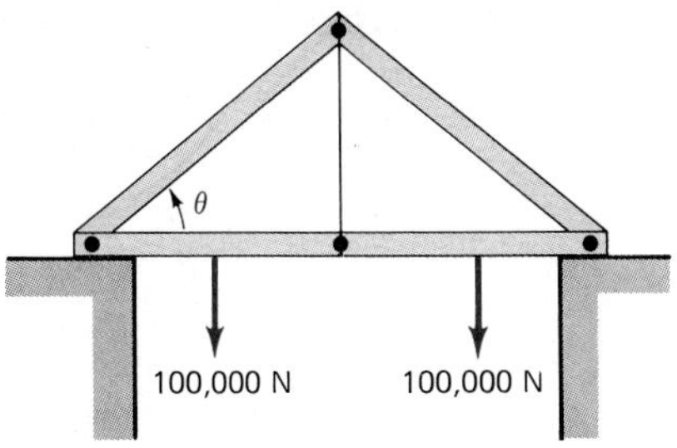

Figure 23

32. One type of automobile jack, shown in Figure 24, consists of a four-sided figure (*ABCD*) with the corners *B* and *D* connected by a screw. To raise one corner of a car, the screw is turned, shortening *DB* and raising the point *A*. Suppose the car exerts a downward force of 1000 N at *A*, and the assembly forms two equilateral triangles. Treat the assembly as a truss and determine the tension in the screw and the compression in each of the other members.

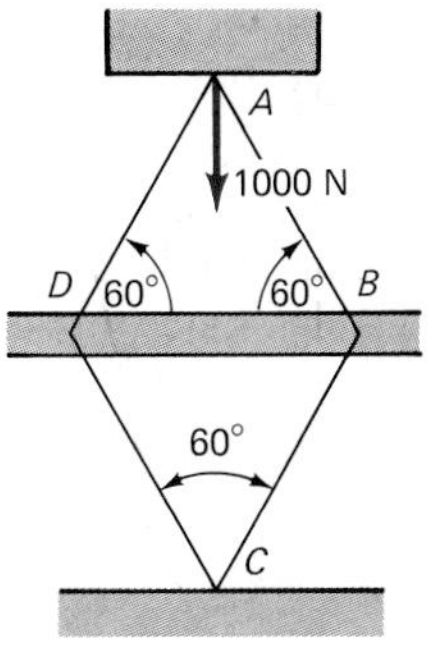

Figure 24

33. Apply the conditions of equilibrium as needed to the truss shown in Figure 3.34 to prove that $\mathbf{N}_A$ = 366 N and $\mathbf{N}_C$ = 634 N.

34. Use the first condition of equilibrium with Figure 3.35 to obtain numerical values for C_2 and T (Example 10).

35. Apply the conditions of equilibrium to Problem 22 and obtain the numerical values given for H, V, and T_1.

36. A beam with arbitrary orientation and hinged at each end is uniform and of weight **W**. Forces act on the beam at each hinge and at its center of gravity. Prove that if both conditions of equilibrium are satisfied when the beam weight is considered to act at its center of gravity, then they will also be satisfied if half the weight is considered to act at each hinge.

Section 3.5 Center of Gravity

37. A collision toy consists of five steel balls in contact (Figure 25). Each ball is of weight **W** and radius *R*. Where is the center of gravity when (a) the balls are in contact? (b) the leftmost ball is pulled a distance 4*R* to the left? (c) the first two balls on the left are pulled a distance 4*R* to the left while in contact?

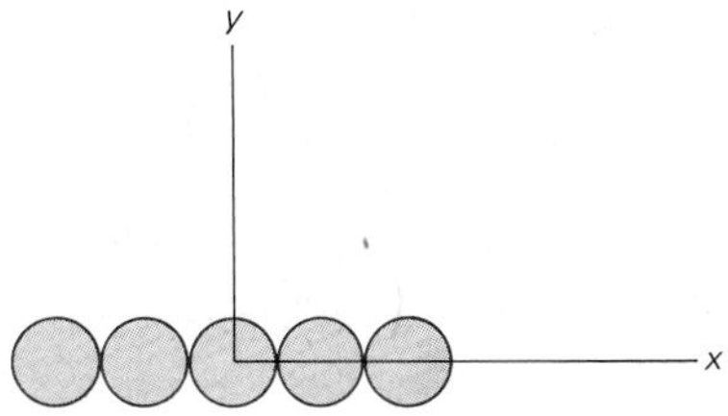

Figure 25

38. The minimum center-to-center distance between adjacent bowling pins is 12 in. When initially set up, the ten pin centers form a symmetric array that is represented in Figure 26 by dots and asterisks. A common arrangement for three pins left standing after the first ball has been bowled is that of the three dots in Figure 26. Find the *x*- and *y*-coordinates of the center of gravity of these three pins.

39. Four cubes 1 cm on a side form the Soma puzzle piece illustrated in Figure 27. The fourth cube, touching the origin, is hidden by the other three cubes in the figure. Locate the coordinates of its center of gravity.

40. Locate the coordinates of the center of gravity of the four-cube Soma puzzle piece shown in Figure 28.

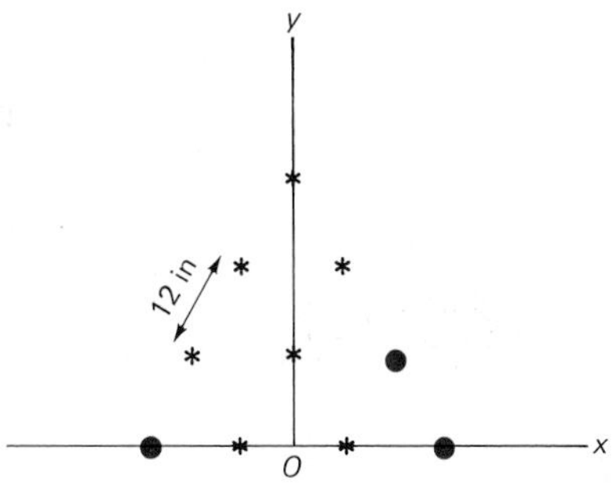

Figure 26

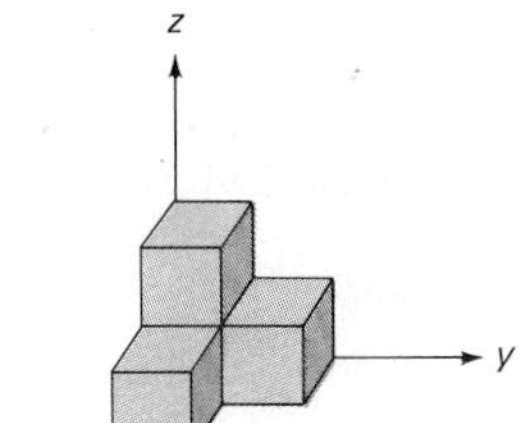

Figure 27

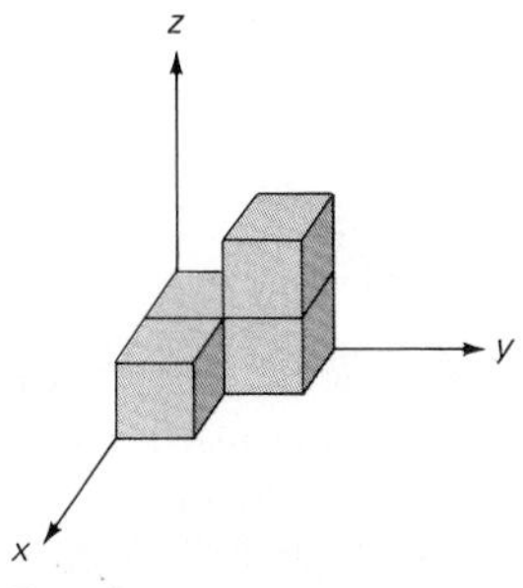

Figure 28

41. Locate the center of gravity of the plane Soma puzzle piece shown in Figure 29 by (a) inspection, (b) using Eq. 3.11, and (c) employing the negative weight procedure.

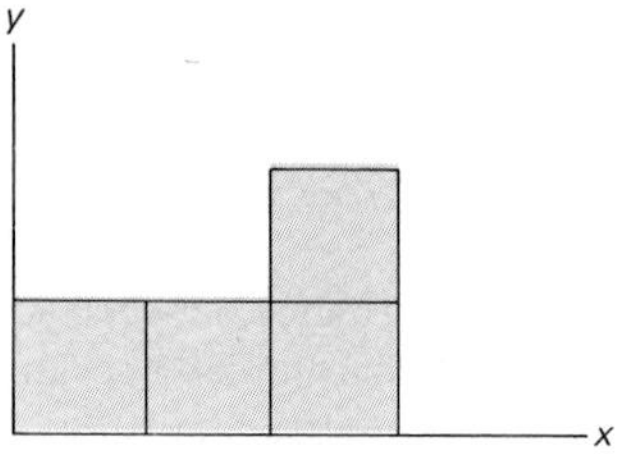

Figure 29

42. A metal wire coat hanger (without the hook) is shown in Figure 30. Assuming the wire is of uniform density, locate the center of gravity.

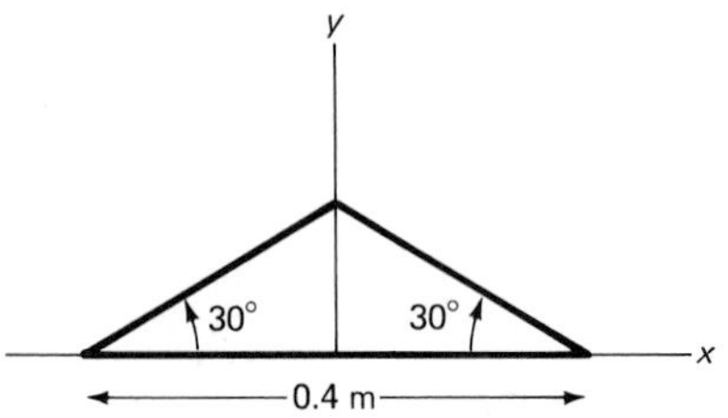

Figure 30

Section 3.6 Stability of Equilibrium

43. A rigid wooden structure consists of two uniform rods glued together at an angle of 30° (Figure 31). The longer segment has a mass of 200 gm and a length of 30 cm. The shorter segment has a mass of 100 gm and a length of 12 cm. It is carefully placed on the edge of a desk as shown. Does it remain stationary or topple off?

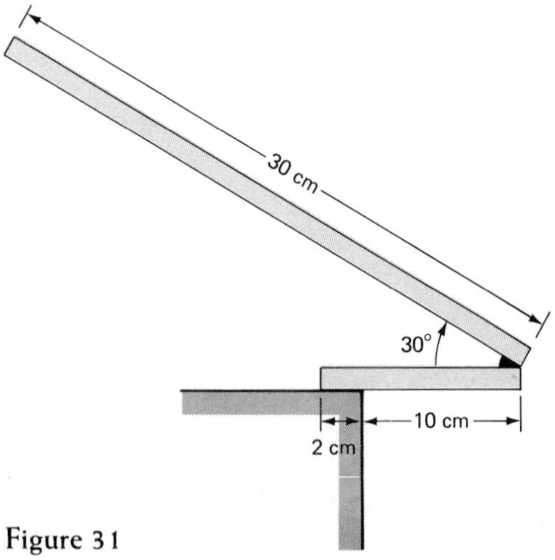

Figure 31

44. Prove that the tippie-top shown in Figure 3.39 will be in stable equilibrium with the hemisphere on the floor if C' is below C; in neutral equilibrium if C' and C coincide; and in unstable equilibrium if C' is above C.

45. The packing crate shown in Figure 32 is of weight W, height b, and width a. Its center of gravity is at its center of symmetry. Consider its stability with respect to rotation about an axis through the point O. (a) Calculate $\tan\theta$ in terms of a and b, where θ is the angle through which the crate must rotate in order to move into a position of unstable equilibrium. (b) Assume that stability increases as θ increases. Prove that if a is fixed, then stability increases as b decreases. (c) Prove that if b is fixed, stability increases as a increases. (d) Assume that large **F** (the horizontal force required to tip the crate) means large stability. Prove that if a and b are fixed, then stability increases with increasing W.

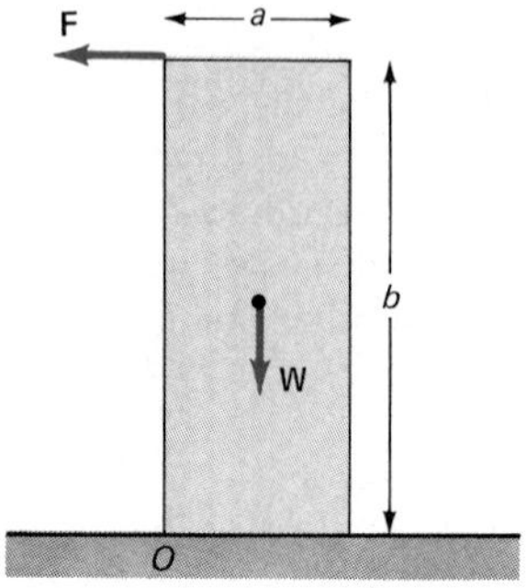

Figure 32

chapter 4

MOTION IN ONE DIMENSION

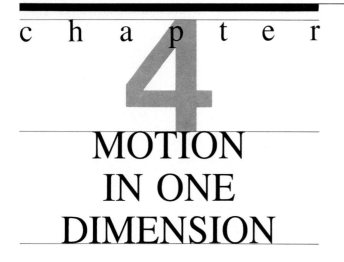

Preview

The investigations of Johannes Kepler (1571–1630) and Galileo Galilei (1564–1642) mark the beginning of modern physics. Both men made discoveries in kinematics—the study of position, time, velocity, and acceleration. Kepler discovered three laws of planetary motion and Galileo formulated the laws of falling bodies (see Figure 4.1). These two pioneering efforts provided the foundation on which Newton based his laws of motion.

The logical development of physics begins with the study of motion. In Chapters 4, 5, and 6 we follow the trail blazed by Galileo, Kepler, and Newton. In this chapter and in Chapter 5 we recreate the science of motion of Galileo and Kepler. In Chapter 6 we will use these motion ideas to formulate Newton's laws of motion.

We start with the concept of a frame of reference. Position measurements and time measurements are made relative to a frame of reference. Position and time are the bases for the two fundamental quantities of motion—displacement and the time interval. From these two quantities we derive two variables: velocity and acceleration.

To avoid calculus in giving definitions and developing equations of motion we use the concept of average velocity. Average velocity is a natural stepping-stone to instantaneous velocity, and from there to acceleration. We give general vector definitions in this chapter, but the problems and examples are limited to linear motion. In Section 4.6 we devote special attention to uniformly accelerated motion, and we treat relative motion in Section 4.7.

Once you have become familiar with the ideas of motion we will introduce some key concepts of calculus. The slope of a curve at a point, the area under a curve, and limits will all be used and discussed as they relate to the description of motion. However, we will not use integrals, and the derivative is used only in Section 4.6, where we employ it to confirm a noncalculus derivation.

If you are already familiar with motion you may be tempted to skim over the introductory material. Remember, however, that proper definitions of physical quantities require precision and attention to detail. You will profit from reading this chapter carefully.

4.1 Frames of Reference

We set the stage for the logical development of kinematics, which we defined as the study of position, time, velocity, and acceleration, by considering observers, the events they measure, and the concept of a frame of reference.

An **event** (a happening, or occurrence, at a particular place) is defined by specifying its position and time. A position vector drawn from an observer to an event gives the position of the event relative to the observer. Observers located at different points will use different vectors to

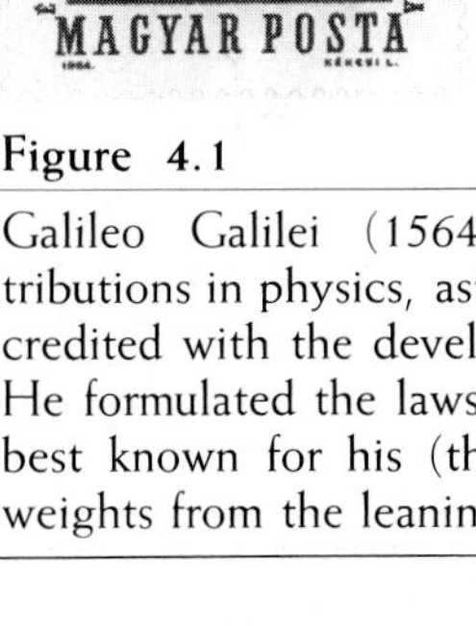

Figure 4.1

Galileo Galilei (1564–1642) made significant contributions in physics, astronomy, and mathematics, and is credited with the development of the scientific method. He formulated the laws of falling bodies, and is perhaps best known for his (thought?) experiment of dropping weights from the leaning tower of Pisa.

Figure 4.2

The three vectors in Eq. 4.1 form a closed triangle.

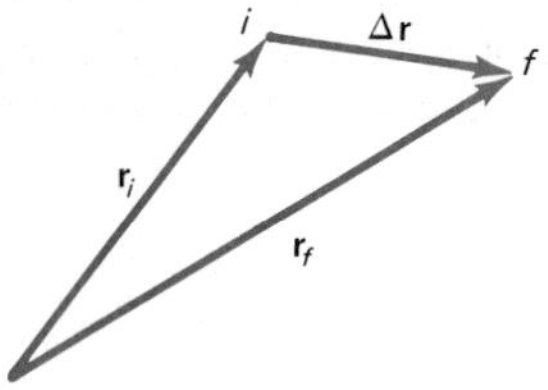

describe the position of the same event. Position measurements are therefore relative—they depend on the location of the observer. In Newtonian mechanics, however, we assume that time is absolute and universal.* Also, all observers are assumed to choose the same zero of time. With these assumptions, measurements of the time of an event will be the same for all observers.

In order to describe two events, an observer must record a distinct position and time for each one. Suppose the subscripts i and f signify initial and final events. If $\mathbf{r}_i$ and $\mathbf{r}_f$ are the position vectors, and t_i and t_f are the corresponding times of the events as recorded by observer O, then the equations

$$\Delta \mathbf{r} = \mathbf{r}_f - \mathbf{r}_i \qquad (4.1)$$

$$\Delta t = t_f - t_i \qquad (4.2)$$

define, respectively, the displacement vector and the time interval between the two events. The displacement vector $\Delta \mathbf{r}$ has a magnitude equal to the distance between events, and a direction from i to f (see Figure 4.2).

Remarkably, the displacement vector for two particular events can be the same for a large class of observers even though their respective position vectors differ. For example, consider the passing of a particular satellite first through the point i and then through the point f. Observers in Cincinnati (C), Indianapolis (I), and Dayton (D) each record pairs of position vectors for the two events, as shown in Figure 4.3. The time intervals are the same for all three observers, and as long as their relative positions are fixed, the displacements will also be the same. Indeed, every observer fixed in position relative to these three observers will obtain the same displacement and time interval separating the two events. This set of observers constitutes a **frame of reference.**

A frame of reference is a set of observers fixed in position relative to each other.

Observers in relative motion do not belong to the same frame of reference. In general, observers in relative motion record different displacements. For example, suppose an observer records a non-zero displacement of a falling apple. A second observer, falling with the apple, will record zero displacement.

A moving observer O who undergoes a displacement **s** relative to the fixed observer C in a particular frame of reference between the times t_i and t_f does not belong to the frame of reference. Compare the two displacements of an object recorded by observers O and C, $\Delta \mathbf{r}_O$ and $\Delta \mathbf{r}_C$, shown in Figure 4.4. For the stationary observer the point i represents the initial location of the object at both $t = t_i$ and $t = t_f$. For the moving observer the point i represents the initial location of the object at $t = t_i$, but, for the same observer, the point i' represents the initial location of the object at $t = t_f$. Because at $t = t_f$ the observers obtain

*This assumption is abandoned in special relativity. See Chapter 19.

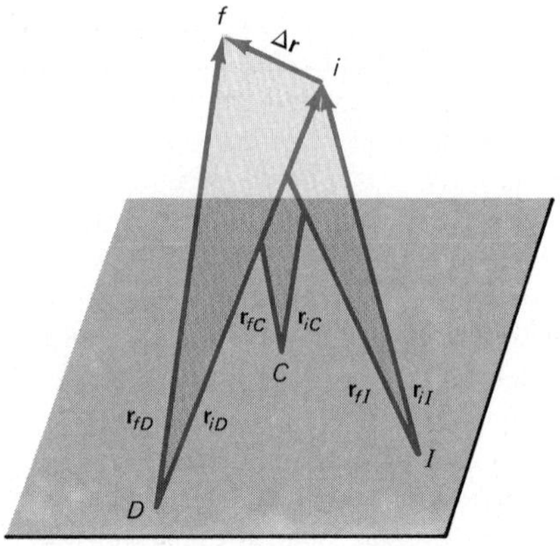

Figure 4.3

A satellite, located at point i, is seen by three observers located in Dayton (D), Cincinnati (C), and Indianapolis (I), respectively. Later the same satellite is seen at point f by the same three observers. The three position vectors at point i differ, as do the three position vectors at point f. Nevertheless the same displacement vector, $\Delta \mathbf{r}$, is obtained by each observer.

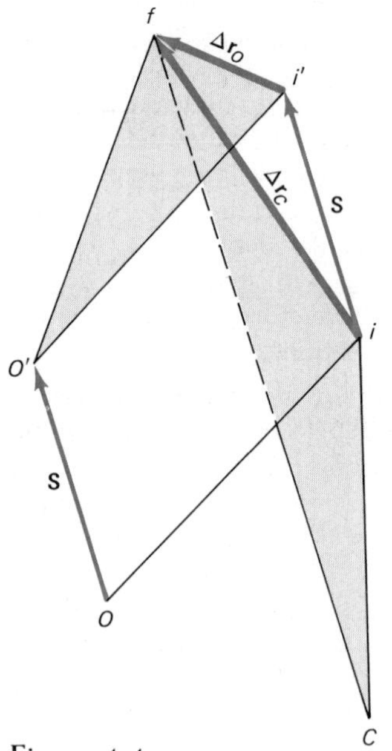

Figure 4.4

A satellite at point i is seen by a fixed observer (C) and a moving observer (O). Later the same satellite is at point f. The vector $\Delta \mathbf{r}_C$ represents the satellite displacement for C. During the time the satellite moves from i to f, the observer at O moves to O′. Hence, according to the stationary observer, the starting point for the moving observer has changed from the point i to the point i'. Thus the vector $\Delta \mathbf{r}_O$ represents the satellite displacement for O. The displacements of a moving object measured by two observers differ when the two observers are in relative motion.

different values for the initial location of the object, the two displacements differ in magnitude and direction.

A frame of reference may itself be represented by a coordinate system in which each observer is identified by a set of coordinates. If there is only one observer, this observer is usually represented by the origin. Whenever we use coordinate systems, keep in mind that they represent physical observers.

Later in this text we will introduce a particular type of reference frame called an **inertial reference frame.** Inertial frames of reference are of special interest and importance because they provide a basis for describing the laws of nature simply and elegantly.

> *A reference frame is called inertial* if every particle that is at rest in it remains at rest, and every particle that is in motion in it maintains its motion with no change in speed or direction.*

The power of this concept will be revealed gradually as you delve deeper into motion studies. Inertial reference frames surface in our discussions of acceleration in Section 4.4, of relative motion in Chapter 10, and of the postulates of special relativity in Chapter 19. With Newton's first law of motion in Section 6.2 you will begin to see the importance of inertial frames of reference in the formulation of laws of nature.

Time intervals and **displacements** are the **fundamental quantities of kinematics.** All other quantities in kinematics are defined directly or indirectly in terms of these two. The measurement of time intervals and displacements implies a frame of reference. Thus, a frame of reference, although not always explicitly stated, is always associated with a description of motion.

Relations connecting kinematic quantities add to our description of the motion of a system. For example, Kepler searched for a mathematical equation, a relation, that would describe planetary motion. The ellipse he discovered for the trajectory** or path of Mars related the two position variables in the orbital plane. Galileo considered trajectories of freely falling bodies in the uniform gravitational field of the earth, and discovered that position depended quadratically on the elapsed time. In both cases, our understanding of motion was improved because new relations were discovered among the kinematic quantities.

Trajectories are important elements in the description of motion. Nearly 2000 years ago Ptolemy obtained accurate but complex orbital trajectories for planets by using a frame of reference centered on the earth. Kepler's simpler elliptic orbits were for a frame of reference fixed in the sun. In part, the search for order in physics and astronomy has been a search for frames of reference in which the description of motion is simple.

Questions

1. If every observer in a frame of reference is displaced by the constant vector **s** (see Figure 4.4), how will their displacement measurements of an object compare?

2. A long line of cars moves at 40 km/hr down a highway. The passengers view a cow walking through a pasture of clover at the side of the road. Are all of the passengers in one frame of reference? Do the cow and the pasture belong to the same frame of reference?

*Alternatively we could define an inertial reference frame as one in which Newton's first law is valid.

**See Section 5.1 for a more detailed discussion of trajectories.

4.2 Average Velocity

Picture a bird-watcher at the origin in a particular frame of reference. The position of any bird he observes is represented by a position vector drawn from the bird-watcher to the bird. Suppose he spots a bald eagle at the position $\mathbf{r}_i$ when the time is t_i. The eagle disappears into the clouds, but later, at the time t_f, it is seen again at a new position, $\mathbf{r}_f$. During $\Delta t = t_f - t_i$, the eagle has displaced itself a $\Delta\mathbf{r} = \mathbf{r}_f - \mathbf{r}_i$. This vector relation is illustrated by the triangle in Figure 4.5. The eagle's average velocity during Δt is the ratio of the displacement, $\Delta\mathbf{r}$, to the elapsed time, Δt.

The **average velocity** of a particle, $\bar{\mathbf{v}}$, is a derived quantity. We read this term as "vee-bar." The vector $\bar{\mathbf{v}}$ is *defined* through the equation

$$\bar{\mathbf{v}} = \frac{\Delta\mathbf{r}}{\Delta t} = \frac{\text{displacement}}{\text{elapsed time}} \tag{4.3}$$

where the bar over the **v** means average. The direction of $\bar{\mathbf{v}}$ is the same as the direction of $\Delta\mathbf{r}$, regardless of the actual path from i to f. The actual path of the eagle while hidden in the clouds is unknown, and is irrelevant to determining $\bar{\mathbf{v}}$. Also, Δt need not be small. It can have any value. For example, the two eagle sightings could be an hour apart or a second apart.

The magnitude of the average velocity vector is the magnitude of the displacement divided by the elapsed time. Units for velocity are length units divided by time units. The SI unit is meters per second (m/s), but any ratio of length to time, such as mi/hr or cm/yr (for continental drift), is possible.

For linear motion in the x-direction Eq. 4.3 may be written

$$\bar{v} = \frac{\Delta x}{\Delta t} \tag{4.4}$$

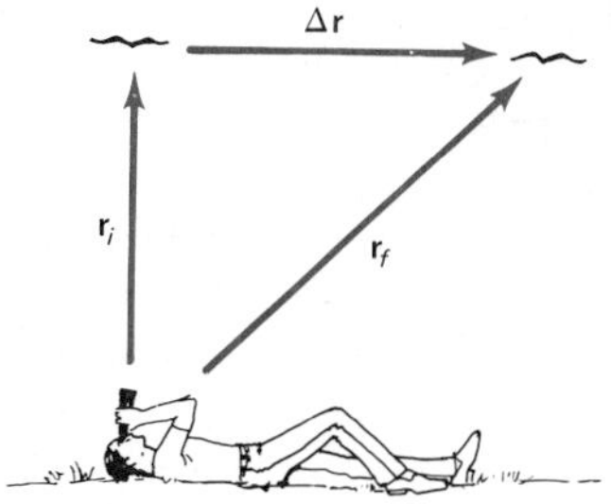

Figure 4.5

A bald eagle located initially by $\mathbf{r}_i$ and located finally by $\mathbf{r}_f$ undergoes the displacement $\Delta\mathbf{r} = \mathbf{r}_f - \mathbf{r}_i$.

where $\Delta x = x_f - x_i$ is the displacement. In the case of linear motion the only direction attributable to the average velocity is to the right (in the direction of increasing x) when $\Delta x > 0$, or to the left (in the direction of decreasing x) when $\Delta x < 0$.

The average velocity of a particle can be calculated for any time interval, large or small, for which values of Δx and Δt are known. For example, in Table 4.1 we have listed the positions, in meters, and corresponding times, in seconds, of a runner. The runner's average velocity for the complete 1.53 s can be calculated as

$$\bar{v} = \frac{9.14}{1.53} = 5.97 \text{ m/s}$$

In the same manner, the runner's average velocity between $t_i = 0.54$ s and $t_f = 0.93$ s is found to be

$$\bar{v} = \frac{4.88 - 2.44}{0.93 - 0.54} = 6.3 \text{ m/s}$$

We cannot readily calculate the average velocity for a time interval such as from $t_i = 0.50$ s to $t_f = 1.00$ s, since these times are not in the table. However, we may plot the data in Table 4.1 to produce the set of points shown in Figure 4.6. When we draw the smooth curve between data points we are extending the x versus t relationship into the region between the data points. This procedure, called *interpolation,* is useful and often necessary (although it is preferable to collect more data). We can then calculate the average velocity for the time interval $t_i = 0.50$ s, $t_f = 1.00$ s, by using this graph.

Figure 4.6

The points represent data given in Table 4.1 for a runner. The solid lines connect the points to provide a graphical interpolation of the data. The dashed line is an extrapolation of the data.

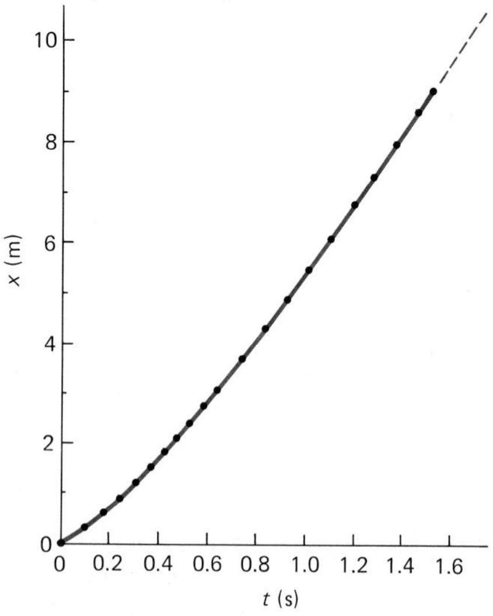

Table 4.1

Positions and times of a runner for the initial portion of a race

x (m)	t (s)
0.00	0.00
0.31	0.11
0.61	0.18
0.91	0.25
1.22	0.31
1.52	0.37
1.83	0.43
2.13	0.48
2.44	0.54
2.74	0.59
3.05	0.64
3.66	0.74
4.27	0.84
4.88	0.93
5.49	1.03
6.10	1.12
6.71	1.20
7.32	1.29
7.93	1.37
8.53	1.45
9.14	1.53

Similarly, we can calculate the average velocity for a time interval extending beyond $t = 1.53$ s. To do this we must extend the x versus t relationship into the region beyond the limits of the data. This procedure, called *extrapolation,* is also necessary at times, but is even riskier than interpolation.

Extrapolation and interpolation do not always indicate true data trends, which is why it is preferable to collect more data. Data for complex experiments are often recorded "on line" (as they occur) with a computer. A computer output is likely to be in graph form and to look much like the x versus t relation in Figure 4.7. In this graph the future ($t > 0$) is to the right. Values of x_i, x_f, t_i, t_f, Δx, and Δt are identified on the graph for a pair of events. To calculate $\bar{v}$ for the time interval represented, the values of Δx and Δt are obtained from the graph and substituted into Eq. 4.4.

You have seen that when position and time data are related, the information can be represented either in tabular or in graphical form. We can also represent the x versus t relation analytically. A theory of motion may provide an explicit form for the function $x = x(t)$. In Section 4.6, for instance, you will see that for bodies dropped from rest at $x = 0$ near the surface of the earth, the relation can be written

$$x = -\frac{1}{2}gt^2 \qquad \text{positive } x \text{ is "up"} \qquad (4.5)$$

The gravitational acceleration $g = 9.80 \text{ m/s}^2$, and x and t are expressed in meters and seconds, respectively.

Figure 4.7

The solid curve represents the relation $x = x(t)$. In particular, this means that the position x_i occurs at the time t_i and the position x_f occurs at the time t_f. Thus the displacement $\Delta x = x_f - x_i$ corresponds to the time interval $\Delta t = t_f - t_i$.

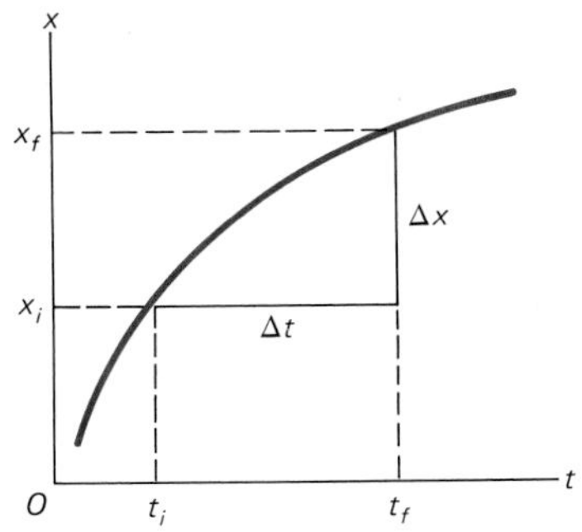

Example 1 **Average Velocity of a Falling Object**

An accurate method of measuring the position and time of an event in the trajectory of a moving object involves photocells. A photocell produces an electrical signal when an object interrupts the light beam to the cell.

Using photocells, an amateur scientist measures precisely the time it takes for a ball to drop from the roof of her building to the top and then to the bottom of her window. The ball passes the top of the window at $t_i = 2.4700$ s and it passes the bottom of the window at $t_f = 2.5300$ s. What is the average velocity during $\Delta t = t_f - t_i$? Using Eq. 4.5 we have

$$\Delta x = -\frac{9.80}{2}(2.5300)^2 + \frac{9.80}{2}(2.4700)^2$$

$$= -1.47 \text{ m}$$

and

$$\Delta t = 2.5300 - 2.4700 = 0.0600 \text{ s}$$

Using Eq. 4.4 we obtain

$$\bar{v} = \frac{-1.47}{0.0600} = -24.5 \text{ m/s}$$

Because "up" is positive, the negative answer means that the average velocity vector is downward. Note also that the original times must be measured to five significant figures if the final result is to have three-significant-figure accuracy.

To obtain average velocity, then, we must use a relationship between x and t, which may be available in tabular, graphical, or analytical form. Experimental results are often in tabular form and are useful as input data in large computers. Graphs are obtained from both theoretical calculations and laboratory experiments, and often provide an overall perspective not easily obtained in other ways. Analytical relations arise from theoretical calculations and from curve fitting of experimental data. You should become adept at handling information in all three forms.

Average Velocity and Average Speed

We have introduced average velocity for two reasons. First, the concept of average velocity provides a natural stepping-stone to the much more useful concept of instantaneous velocity, which is developed in Section 4.3. Second, as you will see in Section 4.4, average velocity is needed for our noncalculus derivation of the equations of motion for constant acceleration.

However, there are some motion problems in which the average velocity may not be useful. For example, suppose a swimmer in a 50-m pool covers 100 m in 52 s by swimming from one end to the other and back again. His average velocity, $\Delta x/\Delta t$, is not 100 m/52 s = 1.92 m/s, but instead equals 0! By swimming two lengths, the swimmer has the same starting and finishing positions; therefore $\Delta x = 0$ and $\bar{v} = 0$. The vector character of average velocity is responsible for this result. Average *speed*, a scalar, is more appropriate for such a problem.

Average speed is a term commonly used by travelers, and should not be confused with average velocity. It is determined by dividing the entire distance traveled by the elapsed time.

$$\text{Average speed} = \frac{\text{distance traveled}}{\text{elapsed time}}$$

The average speed of the swimmer who covered 100 m in 52 s is 1.92 m/s—not zero. This information is useful to the swimmer, whereas the average velocity, zero, is obviously not useful information.

Questions

3. A vulture lazily circles over an animal carcass for a long time. Explain why the magnitude of the average velocity of the vulture has a maximum value that steadily decreases as time passes.
4. Discuss the advantages and disadvantages of using average speed rather than average velocity to describe physical situations.
5. In certain circumstances the magnitude of the average velocity equals the average speed. Compare this meaning of average speed with that given in the text.
6. Would average velocity be a good variable to describe the oscillation of a child in a swing? Is your answer affected by the length of the time interval used?

4.3 Instantaneous Velocity

Driving a car properly requires a clear concept of instantaneous velocity. For example, suppose that a car with a velocity ranging from 70 km/hr north to 100 km/hr north passes through an 85-km/hr zone. Whether or not the driver gets a ticket depends on the speed of the car at the instant the police radar measures it.

Instantaneous velocity is defined at a point in time—not during a time interval. Its calculation requires evaluation of a limit. To evaluate this limit we calculate the average velocity, $\bar{v} = \Delta x/\Delta t$, using the time interval Δt and the corresponding displacement, Δx. Then we repeat the procedure using the same t with a smaller Δt and Δx. This calculation is repeated, using shorter and shorter time intervals. In the process both Δt and Δx approach zero. If the ratio $\Delta x/\Delta t$ approaches a constant value as Δt approaches zero, this constant value is called the *limit* of the ratio. This limiting value of the average velocity is defined as the instantaneous velocity at the time t. To illustrate this limiting procedure we will use two methods. We will develop a geometric interpretation with a graph, and then a numerical calculation will be given for which a hand calculator is used.

We define the **instantaneous velocity vector** as

$$\mathbf{v} = \lim_{\Delta t \to 0} \frac{\Delta \mathbf{r}}{\Delta t} \tag{4.6}$$

The notation "$\lim_{\Delta t \to 0}$" is read as "in the limit as Δt goes to zero." For linear motion this relation may be written

$$v = \lim_{\Delta t \to 0} \frac{\Delta x}{\Delta t} \tag{4.7}$$

If we compare this equation with Eq. 4.4, we see that the instantaneous velocity is the limiting value of the average velocity as the time interval approaches zero.

To develop a geometric interpretation for instantaneous velocity we first consider a geometric interpretation for average velocity. The relation $x = -\frac{1}{2}gt^2$, discussed in Section 4.2, is plotted in Figure 4.8 for the time interval $0 \le t \le 7$ s. On the graph, coordinates (t, x) are given for several points. Straight lines are drawn from several initial

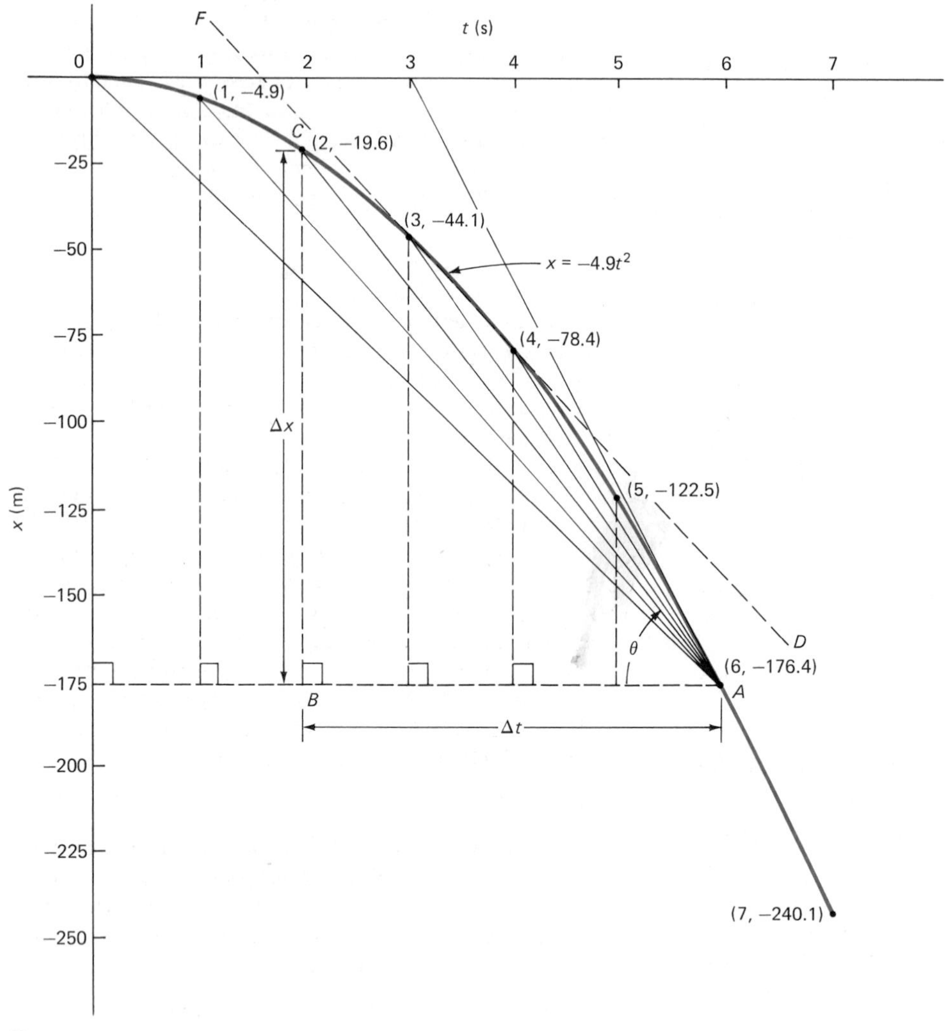

Figure 4.8

The equation $x = -4.9t^2$ for a freely falling object is shown as the solid parabolic curve. Geometrically the instantaneous velocity of the object at the point A equals the slope of the line drawn tangent to the curve at A, because the instantaneous velocity is the limit of the average velocity as $\Delta t \to 0$. To calculate the average velocity from the graph we measure a time interval $(\overline{BA})$ and displacement $(\overline{CB})$ in appropriate time and length units. The average velocity equals the ratio of $\overline{CB}$ to $\overline{BA}$. Geometrically this ratio equals the slope of the secant line $\overline{CA}$. In the limit as $\Delta t \to 0$ the point C moves along the curve until it coincides with the point A. In the process the secant line becomes the tangent line. Thus geometric evaluation of the slope of the tangent line at the point A gives the instantaneous velocity at A.

Figure 4.9

The slope of the x versus t curve equals the instantaneous velocity. Inspection of the curve determines whether the velocity is positive ($D \to B$ and $G \to E$), zero (points B and G), or negative ($B \to G$).

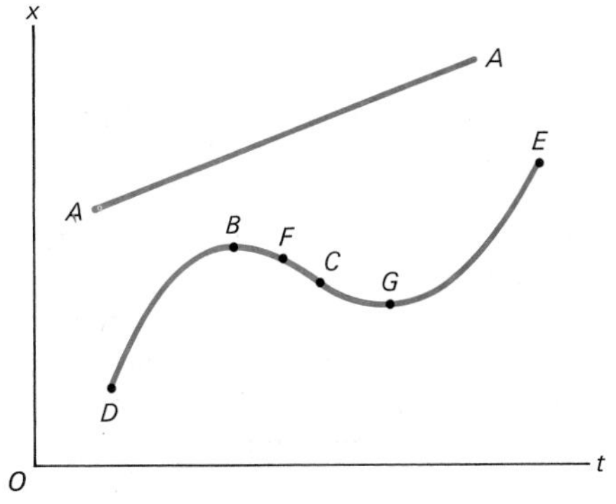

points on the curve to the single fixed point (6, −176.4). Each of these secant lines is the hypotenuse of a right triangle—such as ABC—defined by the dotted lines drawn parallel to the axes and the secant line to the curve. Geometrically, the average velocity is the ratio of the side opposite the angle θ to the side adjacent to the angle θ in the triangle ABC. That is, the average velocity is the slope of the secant line to the curve, in appropriate units.

To provide a geometric interpretation of instantaneous velocity, the limit as $\Delta t \to 0$ must be given a graphical meaning. Consider the determination of the instantaneous velocity at $t = 6$ s in Figure 4.8. Let the average velocity be evaluated for a sequence of time intervals that have the same final time ($t = 6$ s). As $\Delta t \to 0$ the point on the curve representing the beginning of the time interval "slides" along the curve toward the final point. In the process the slope of the secant line more and more nearly approximates the line tangent to the curve at $t = 6$ s, and in the limit as $\Delta t \to 0$ the slopes coincide. Thus the instantaneous velocity at any point is equal to the slope of the tangent line to the curve at the point, in appropriate units.

This geometric interpretation allows us to make some general observations:

1 A straight line, such as AA in Figure 4.9, will yield the same instantaneous velocity at all times.

2 A curve having individual points, such as those labeled B and G in Figure 4.9, with zero slope will yield zero instantaneous velocities at those points.

3 If the slope is positive (such as from D to B or from G to E), the instantaneous velocity is positive, indicating that the particle is moving in the direction of increasing x. If the slope is negative (such as from B to G), the instantaneous velocity is negative, indicating that the particle is moving in the direction of decreasing x.

Example 2 **Instantaneous Velocity: Graphical Solution**

Let us determine the instantaneous velocity at $t = 3.5$ s from the graph of the falling body in Figure 4.8.

The line DF has been constructed tangent to the curve at $t = 3.5$ s, and its slope is equal to the instantaneous velocity. We may use any time interval to evaluate the slope, since a straight line has constant slope. If we choose the time interval $t_i = 1.7$ s to $t_f = 3.5$ s, the corresponding positions are $x_i = 0$ and $x_f = -65$ m. We can write

$$\Delta t = 3.5 - 1.7 = 1.8 \text{ s}$$

and

$$\Delta x = -65 - 0 = -65 \text{ m}$$

Substituting these values into Eq. 4.7 we get

$$v = \frac{-65}{1.8} = -36 \text{ m/s}$$

For comparison, the exact value is −34.3 m/s.

The accuracy of a graphical determination is limited by the precision with which a tangent line to a curve at a point can be drawn, and by the precision with which values for Δx and Δt can be read off the graph. We saw that with tabular data, like that given in Table 4.1, the instantaneous velocity cannot be calculated at all, because the evaluation of the limit $\Delta t \to 0$ is impossible. However, when an equation relating x and t is known, this difficulty is removed and we can determine the instantaneous velocity numerically, rather than graphically.

Example 3 **Instantaneous Velocity**

Let us determine the instantaneous velocity at $t = 6$ s for the system represented exactly by the equation $x = -4.90t^2$ with x expressed in meters. The data in Table 4.2 were obtained from this equation by using a hand calculator.

Because x_i and x_f more and more nearly cancel as $\Delta t \to 0$, increased precision is required in x_i, as shown in column 3, if an accurate answer for v is to be obtained. Column 5 shows how Δt becomes progressively smaller and approaches zero. Column 6 shows that the magnitude of Δx becomes progressively smaller, and column 7 shows that the ratio $\Delta x/\Delta t$ approaches a definite limit. The average velocities in column 7 are still changing between $t = 5.99$ s and $t = 5.999$ s, although the last two entries are both very nearly −58.8 m/s. We conclude that the instantaneous velocity at $t = 6$ s is −58.8 m/s.

We saw in Section 4.2 that average speed can differ from the magnitude of the average velocity. For continuous motion, however, in the limit as $\Delta t \to 0$ the ratio $\Delta x/\Delta t$ equals the ratio of the distance covered to the elapsed time. Thus we conclude that

instantaneous speed = magnitude of the instantaneous velocity

Questions

7. If the velocity–time curve is a straight line, the average velocity is halfway between the initial and final velocities in the interval. Sketch a curve of velocity versus time where the average velocity is closer on the curve to the maximum than to the minimum velocity. Repeat for an average velocity that is closer to the minimum.

Table 4.2

Numerical evaluation of average velocity as Δt approaches zero

t_i (s)	t_f (s)	x_i (m)	x_f (m)	Δt (s)	Δx (m)	$\bar{v}$ (m/s)
0	6	0	−176.400	6	−176.4	−29.4
3	6	−44.1	−176.400	3	−132.3	−44.1
5	6	−122.5	−176.400	1	−53.9	−53.9
5.9	6.0	−170.569	−176.400	0.1	−5.831	−58.31
5.99	6.00	−175.81249	−176.40000	0.01	−0.58751	−58.751
5.999	6.000	−176.341205	−176.400000	0.001	−0.058795	−58.795
5.9999	6.0000	−176.3941200	−176.4000000	0.0001	−0.0058799	−58.799

8. What kind of velocity–time graph has average and instantaneous velocities always equal?
9. Can the velocity of a particle be equal to zero at a given instant, and yet be changing with time?
10. Describe a situation involving a particle in which both **r** and **v** are directed eastward. Repeat for the case in which **r** is eastward and **v** is westward.

4.4 Acceleration

The rate of change of velocity is called *acceleration*. A car changes its velocity when it stops, starts, or alters direction. The passengers learn to distinguish these conditions from the constant-velocity situation. Just as we introduced average and instantaneous velocity in the preceding sections, we now introduce two kinds of acceleration—average and instantaneous.

Like velocity, acceleration is a derived quantity defined in terms of other quantities. The average acceleration of a particle, $\bar{\mathbf{a}}$, is a vector defined by the equation.

$$\bar{\mathbf{a}} = \frac{\Delta \mathbf{v}}{\Delta t} \tag{4.8}$$

where $\Delta\mathbf{v} = \mathbf{v}_f - \mathbf{v}_i$ is the change in instantaneous velocity, and $\Delta t = t_f - t_i$ is the corresponding time interval. The direction of $\bar{\mathbf{a}}$ is the same as the direction of $\Delta\mathbf{v}$.

Suppose that a bluebird is observed at the locations indicated by position vectors $\mathbf{r}_i$ and $\mathbf{r}_f$ by a bird-watcher at the times t_i and t_f, as shown in Figure 4.10. The velocities of the bluebird at these times are $\mathbf{v}_i$ and $\mathbf{v}_f$, respectively. The triangle construction in Figure 4.11 illustrates the geometric relation among the three velocity vectors $\mathbf{v}_i$, $\mathbf{v}_f$, and $\Delta\mathbf{v}$. Note that the direction of $\Delta\mathbf{v}$, and hence of $\bar{\mathbf{a}}$, need not be the same as the directions of any of the vectors $\mathbf{r}_i$, $\mathbf{r}_f$, $\Delta\mathbf{r}$, $\mathbf{v}_i$, or $\mathbf{v}_f$.

For linear motion Eq. 4.8 becomes

$$\bar{a} = \frac{\Delta v}{\Delta t} \tag{4.9}$$

where Δv is the change in the instantaneous velocity. The direction of the average acceleration vector in linear motion is either to the right (in the direction of increasing x) or to the left (in the direction of decreasing x). The SI units of acceleration are m/s^2, although other units are in common use. An automobile that reaches a speed of 60 mi/hr in 10 s, for example, would be credited with an average acceleration of 6 miles per hour per second.

Figure 4.10

Positions and velocities of a bluebird at two times.

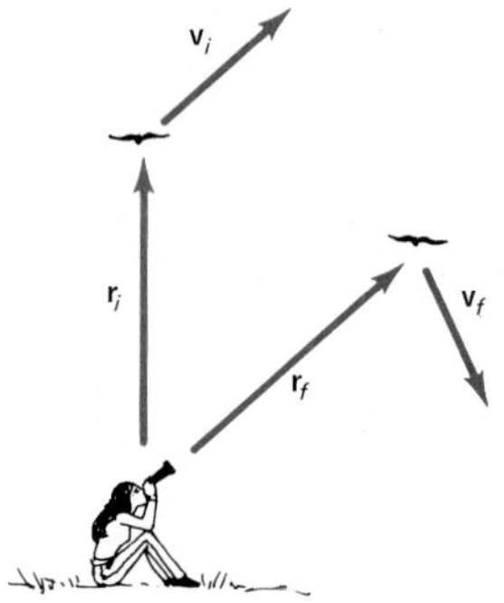

Figure 4.11

Triangle relation for velocity vectors.

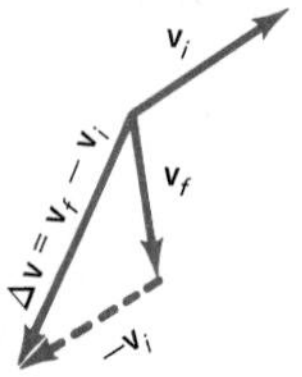

Example 4 A Drag Race

Drag racing is an acceleration competition for automobiles over a track 1320 ft long. Cars start from rest and are expected to reach the end of the quarter-mile track in the shortest possible time. In a typical race a Camaro was timed at 11.91 s and achieved a final velocity of 117 mi/hr. What was the average acceleration of the Camaro?

Using British units, we have

$$\Delta v = 117 - 0 = 117 \text{ mi/hr}$$

$$\Delta t = 11.91 \text{ s}$$

Figure 4.12

Average acceleration is the ratio of the change in velocity, Δv, to the time interval, Δt. Geometrically this equals the slope of the secant line, AB.

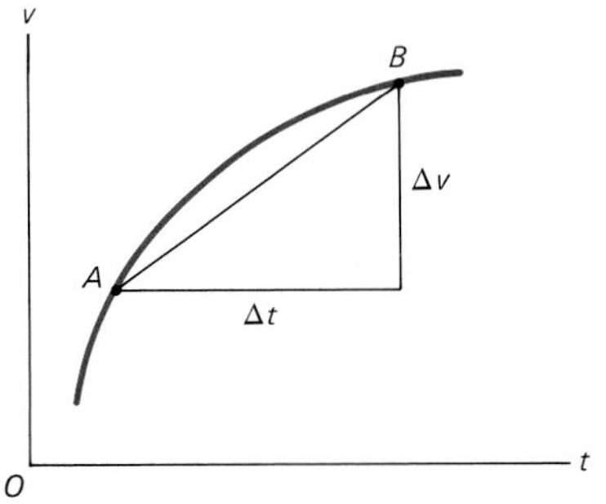

Therefore,

$$\bar{a} = \frac{117}{11.9} = 9.83 \text{ mi/hr} \cdot \text{s}$$

To convert this result to SI units we convert each unit separately and then cancel units to obtain m/s^2 in our final answer.

$$\bar{a} = 9.83 \frac{\text{mi}}{\text{hr} \cdot \text{s}} \times \frac{1.61 \times 10^3 \text{ m}}{1 \text{ mi}} \times \frac{1 \text{ hr}}{3600 \text{ s}}$$
$$= 4.40 \text{ m/s}^2$$

Just as with instantaneous velocity, the definition of instantaneous acceleration requires the use of a limit. The defining equation for **instantaneous acceleration** is

$$\mathbf{a} = \lim_{\Delta t \to 0} \frac{\Delta \mathbf{v}}{\Delta t} \qquad (4.10)$$

which becomes, for linear motion

$$a = \lim_{\Delta t \to 0} \frac{\Delta v}{\Delta t} \qquad (4.11)$$

The instantaneous acceleration is the limit of the average acceleration as the time interval shrinks to zero. We stress applications for which the acceleration is constant. For those cases, $a = \bar{a}$.

The geometric interpretation of acceleration closely parallels that of velocity. A graph giving v versus t is drawn in Figure 4.12. The average acceleration during Δt is the slope of the secant line AB in appropriate units. The instantaneous acceleration at the point B is the slope of the tangent line to the curve at the point B.

Example 5 **Acceleration of a Runner**

Figure 4.13 is the velocity versus time curve for a runner during the first 5 s of a race. What is the instantaneous acceleration of the runner at $t = 2$ s?

First we draw a line (AC) tangent to the curve at $t = 2$ s. This line becomes the hypotenuse of a triangle (ABC) that has Δv and Δt as its other two sides. From the graph we obtain

Figure 4.13

The solid curve line represents the velocity–time relation for a sprinter. The instantaneous acceleration equals the slope of the tangent line to the curve.

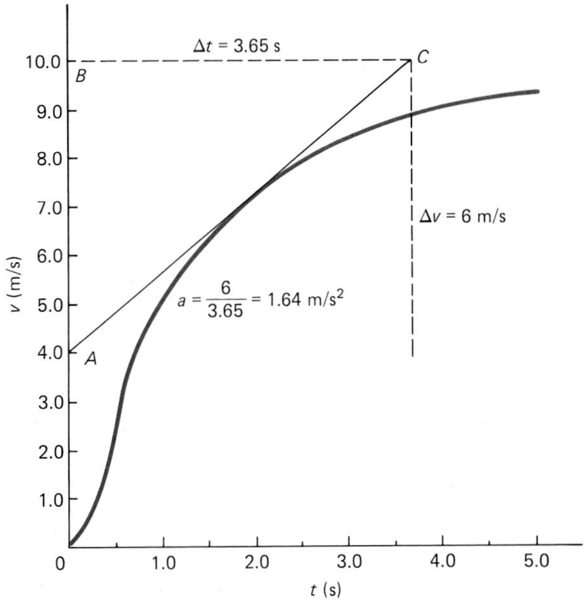

$$\Delta v = 6.00 \text{ m/s}$$
$$\Delta t = 3.65 \text{ s}$$

which gives

$$a = \frac{6.00}{3.65} = 1.64 \text{ m/s}^2$$

for the instantaneous acceleration at $t = 2$ s.

Questions

11. Give an example of an accelerating particle that has both **r** and **v** equal to zero.

12. Give an example of a moving particle that has both **r** and **a** equal to zero.

13. Give an example of a particle that is accelerating while its velocity is constant in magnitude.

14. At slow speeds air resistance is negligible, which means that at first a sky diver accelerates like any dropped object. After falling for 10 s, the diver reaches a downward velocity of constant value. Discuss her acceleration during the first 10 s. Sketch a graph of velocity versus time and one of acceleration versus time for the first 10 s.

15. A jumper leaps straight up and then returns to earth, retracing his path exactly. Sketch his velocity versus time and acceleration versus time curves for the time he is airborne. Compare this situation with that of the sky diver in Question 14.

4.5
The Program of Particle Kinematics

Kinematics is the branch of mechanics that deals with the description of motion. It does not deal with the causes of motion. In the first four sections of this chapter we introduced the variables required to describe the motion, and the changes in the motion, of a particle. When we set up definitions of velocity and acceleration we started with position and time, and then deduced the velocity and the acceleration. It is useful to pause at this point to ask the question "What is the purpose of kinematics?" We will see that the answer "To obtain future values of position and velocity of a system in terms of its present values of position and velocity and its accelerations" exactly reverses the order of the logic just outlined.

Consider the defining equation of average velocity.

$$\bar{\mathbf{v}} = \frac{\Delta \mathbf{r}}{\Delta t} \quad \text{or} \quad \Delta \mathbf{r} = \bar{\mathbf{v}}\, \Delta t \qquad (4.3)$$

Given the average velocity during the time interval Δt, this equation reveals the position $\mathbf{r}_f$ at the time t_f, providing the initial position, $\mathbf{r}_i$, and initial time, t_i, are known. Thus the average velocity, if known, may be interpreted as the quantity that provides future values of position.

Similarly, the defining equation for average acceleration

$$\bar{\mathbf{a}} = \frac{\Delta \mathbf{v}}{\Delta t} \quad \text{or} \quad \Delta \mathbf{v} = \bar{\mathbf{a}}\, \Delta t \qquad (4.8)$$

reveals the final velocity, $\mathbf{v}_f$, at the time t_f, providing the initial velocity, $\mathbf{v}_i$, initial time, t_i, and average acceleration during the time interval Δt are known. Thus the average acceleration, if known, may be thought of as a quantity that provides future values of velocity.

Kinematics starts with the acceleration of a particle and uses this acceleration to predict the future positions and velocities of the particle. But how do we determine the acceleration of a particle? The dynamic laws of nature contain the answer to this question. In Chapter 6 we will see how the acceleration of a particle is related to the external forces acting on the particle through Newton's second law. At that stage all that will be needed to complete the picture will be a knowledge of the fundamental forces of nature. In Chapter 3 these fundamental forces were listed, and much of the rest of the text will be devoted to an elaboration of them, particularly the gravitational and electromagnetic forces. For kinematic purposes, however, we will have to be content with a "given" acceleration.

In the remainder of this chapter, we will predict future positions, velocities, and displacements of a particle under various one-dimensional acceleration conditions. In the next chapter we will extend this analysis to motion in a plane.

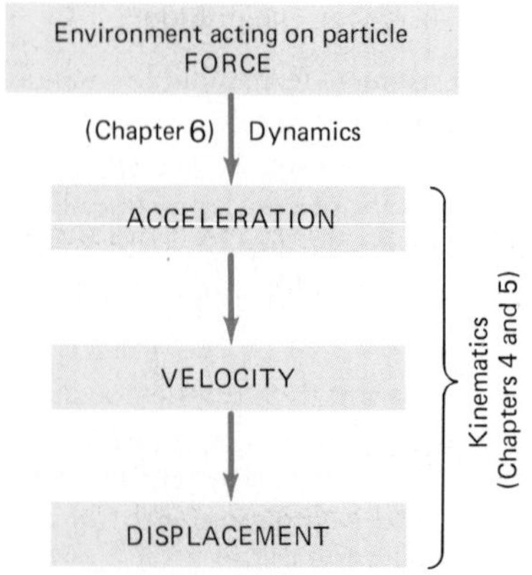

4.6
Linear Motion with Constant Acceleration

Average Velocity for Constant Acceleration

Before we develop and apply the equations for motion with constant acceleration, or uniformly accelerated motion, we will discuss a preliminary equation (Eq. 4.14) for average velocity. Previously, we have obtained values for the average velocity and instantaneous velocity from knowledge about positions and times. But as we explained in Section 4.5, the logic is reversible. In other words, if we know the instantaneous velocity at every instant in time, then we should be able to draw inferences about positions, average velocities, and displacements.

Consider a particle whose instantaneous velocity is a constant 8 m/s, and whose displacement and average velocity during a 12-s period are desired. Evidently the particle's average velocity is 8 m/s, and by using Eq. 4.4 we calculate the displacement to be 96 m. These results can be interpreted graphically. This constant instantaneous velocity plots as the straight line shown on the velocity versus time graph in Figure 4.14. The shaded area in the figure corresponds to 96 m.

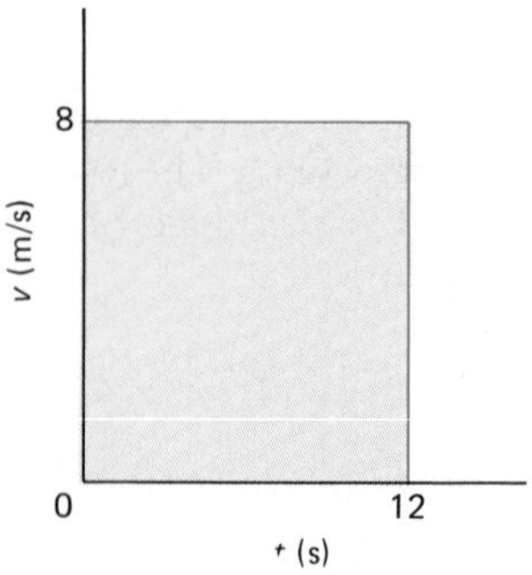

Figure 4.14

The horizontal line is the velocity–time relation for an object moving at a constant velocity of 8 m/s. In 12 s the object moves 96 m, which is also represented by the shaded area in appropriate units.

For motion with constant velocity the average velocity equals the instantaneous velocity. It is important to note that in this case the *displacement* is equal to the *area* under the velocity–time curve. But is this true of any velocity–time graph? It is; but to prove this we must use integral calculus. However, rather than attempt such a proof let's consider the important special case in which velocity varies linearly with time. We assert that the average velocity is one half the sum of the initial and final velocities, $\bar{v} = \frac{1}{2}(v_i + v_f)$, and show that this assertion (an educated guess) also leads to the interpretation of area under a velocity–time graph as displacement.

Consider a particle whose instantaneous velocity is given by the equation

$$v_f = v_i + at$$

where v_i and a are constant. The graph is given in Figure 4.15. We assert that during the time t the average velocity is given by

$$\bar{v} = \frac{1}{2}(v_i + v_f)$$

in terms of v_i and v_f. This means the average velocity is taken halfway between v_i and v_f, as indicated on the graph. According to Eq. 4.4 we have for the displacement

$$\Delta x = \bar{v}\,\Delta t$$

$$= \frac{1}{2}(v_i + v_f)\,\Delta t$$

Geometrically this is the area of a rectangle whose height is $\frac{1}{2}(v_i + v_f)$ and whose width is Δt. The corresponding area is shaded in Figure 4.15. If you study this figure you will see that this area is equal to the trapezoidal area under the line defined by the equation $v_f = v_i + at$.

We will use the expression $\frac{1}{2}(v_i + v_f)$ for **average velocity** in our consideration of uniformly accelerated motion in this section. The argument given to support the use of this expression, however, is not a formal proof.

The beauty of uniformly accelerated motion is that the equations of motion are simple and not difficult to interpret. Many examples of this type of motion are observed in nature. Freely falling bodies near the earth's surface and charged particles in a uniform electric field are examples of this kind of motion.

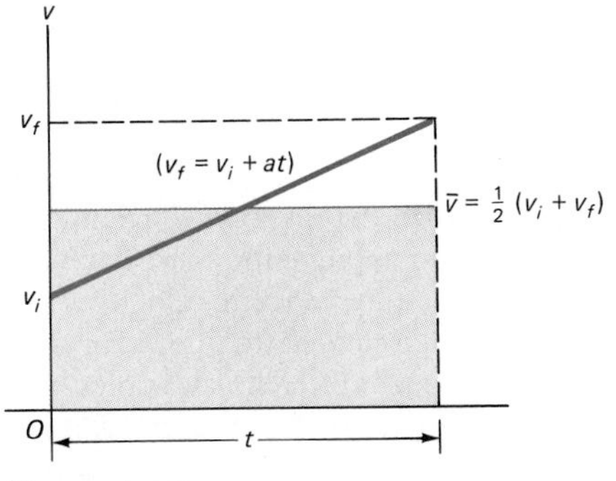

Figure 4.15

The area under the sloping line equals the shaded rectangular area. The distance covered equals the area under the velocity–time graph.

Table 4.3

Symbols for linear motion

Initial values	Final values	Intervals
$t_i = 0$	$t_f = t$	$\Delta t = t$
$x_i = 0$	$x_f = x$	$\Delta x = x$
$v_i = v_o$	$v_f = v$	$\Delta v = v - v_o$
$a_i = a$	$a_f = a$	$\Delta a = 0$

To describe uniformly accelerated motion we need a frame of reference and a set of symbols. As a frame of reference we select the positive x-axis to be in the direction of motion. The symbols we will use are listed and described in Table 4.3. The first column of the table lists the symbols for the initial time, position, instantaneous velocity, and acceleration. For simplicity, we choose $t_i = 0$ to be the initial time and $x_i = 0$ to be the initial position. The second column gives symbols for the same quantities as column 1, but at time t; and the third column lists the changes in each variable during the interval from 0 to t. The average velocity is denoted by $\bar{v}$, and the subscript o is used to distinguish between initial and final velocities. Since the acceleration does not change, the initial, final, and average accelerations are all the same.

Using these symbols, we obtain, as the defining equations (Eqs. 4.4 and 4.9) for average velocity and acceleration,

$$x = \bar{v}t \tag{4.12}$$

$$v = v_o + at \tag{4.13}$$

We argued that $\frac{1}{2}(v_i + v_f)$ was equal to the average velocity if v_f varied linearly with time. According to Eq. 4.13, v varies linearly with time. Using the notation of Table 4.3, we see that the average velocity, $\bar{v}$, may be written

$$\bar{v} = \frac{1}{2}(v_o + v) \tag{4.14}$$

The three equations 4.12, 4.13, and 4.14 include the six variables x, v, v_o, $\bar{v}$, a, and t. Other equations can be derived from these three by eliminating any two of the six variables. We will derive two particularly useful formulas in this way. First we eliminate $\bar{v}$ *and* t to obtain Eq. 4.17.

Combining Eqs. 4.12 and 4.14 to eliminate $\bar{v}$ gives us

$$\frac{1}{2}(v + v_o) = \frac{x}{t} \tag{4.15}$$

Slightly rearranging Eq. 4.13 gives

$$v - v_o = at \tag{4.16}$$

Multiplying Eqs. 4.15 and 4.16 together, and canceling a factor of t, we get

$$\frac{1}{2}(v + v_o)(v - v_o) = ax$$

which reduces to the desired result relating v and x,

$$v^2 = v_o^2 + 2ax \qquad \textit{time-independent velocity solution} \tag{4.17}$$

The utility of Eq. 4.17 is illustrated in the next three examples.

Example 6 An Automobile Collision

A driver slams on the brakes when the velocity of his car is 70 mi/hr. The car skids along the road, undergoing a constant negative acceleration of 10 mi/hr · s for 100 ft before colliding with a parked car. What is the collision velocity? We have

$$v_o = 70 \text{ mi/hr} = 103 \text{ ft/s}$$

$$a = -10 \text{ mi/hr} \cdot \text{s} = -14.7 \text{ ft/s}^2$$

$$x = 100 \text{ ft}$$

Equation 4.17 gives

$$v^2 = (103 \text{ ft/s})^2 + 2(-14.7 \text{ ft/s}^2)(100 \text{ ft})$$

$$= 10600 - 2940$$

$$= 7660 \text{ ft}^2/\text{s}^2$$

or

$$v = 87.5 \text{ ft/s} = 59.7 \text{ mi/hr}$$

This example illustrates how careful we must be with units. In the equation v^2 may have units of $(\text{ft/s})^2$ or $(\text{mi/hr})^2$, as desired, and the same units must then be used for v_o^2. The acceleration, a, and the distance, x, must combine to give compatible units. As originally defined, the units are mixed and differ from the v^2 units. Thus, we must convert units before we can obtain meaningful numerical values. A useful conversion factor is

$$15 \text{ mi/hr} = 22 \text{ ft/s}$$

In the following example we use Eq. 4.17 to deduce the acceleration of a car from its displacement, final velocity, and initial velocity.

Example 7 Acceleration in a Drag Race

In a quarter-mile drag race a contestant achieves a maximum velocity of 200 mi/hr. Assuming constant acceleration, let's find the acceleration necessary for the car to reach this maximum velocity at the moment it finishes the race.

Here $v_o = 0$, so we can write Eq. 4.17 as

$$a = \frac{v^2}{2x}$$

$$= \frac{(200 \text{ mi/hr})^2}{2(1/4) \text{ mi}}$$

$$= 80{,}000 \frac{\text{mi}}{\text{hr}^2} \times \frac{1 \cancel{\text{hr}}}{3600 \text{ s}}$$

$$= 22.2 \frac{\text{mi}}{\text{hr} \cdot \text{s}}$$

This acceleration is two to three times as large as one normally attains in an ordinary passenger car.

Police routinely measure the lengths of skid marks on the road after automobile collisions in order to estimate the speeds of the colliding cars when the brakes were applied. The acceleration limits of a skidding car are well known to the police. The acceleration and stopping distance together establish the initial velocity of a car. In Example 8 we reverse this procedure, and use a known initial velocity and acceleration to determine the stopping distance of an automobile.

Example 8 Stopping Distance of an Automobile

By applying the brakes without causing a skid, the driver of a car is able to achieve a constant deceleration of 10 m/s^2. What distance does the car travel after the brakes are applied at 50 mi/hr (22.4 m/s)? At 75 mi/hr (33.5 m/s)?

At 50 mi/hr we have

$$v_o = 22.4 \text{ m/s}$$

$$v = 0$$

$$a = -10 \text{ m/s}^2$$

Thus Eq. 4.17 gives

$$x = -\frac{v_o^2}{2a}$$

$$= -\frac{(22.4)^2}{2(-10)}$$

$$= +25.1 \text{ m}$$

or a little over 82 ft.

At 75 mi/hr we have

$$v_o = 33.5 \text{ m/s}$$

$$v = 0$$

$$a = -10 \text{ m/s}^2$$

Then Eq. 4.17 gives

$$x = -\frac{(33.5)^2}{2(-10)}$$

$$= +56.1 \text{ m}$$

or a little over 184 ft. Because x depends quadratically on the starting speed, a 50% increase in v_o results in a 125% increase (to three-figure accuracy) in stopping distance!

Dimensional Analysis

You may recall from Section 1.6 that in order to be dimensionally correct, every term in an equation must have the same dimensions. In the time-independent velocity solution, Eq. 4.17, the term v_o^2 is added to the term $2ax$. Superficially these terms are as different as apples and oranges, yet the sum is legitimate. It would be incorrect to add v to a in an equation because these terms have different dimensions. But we can add v_o^2 to $2ax$ because the dimensions of these terms are the same, as we will now show.

According to Eq. 4.4 a velocity is a length (L) divided by a time (T). The dimensions of v^2, designated $[v^2]$, in

Eq. 4.17 are thus

$$[v^2] = L^2T^{-2}$$

Evidently we also have $[v_0^2] = L^2T^{-2}$. According to Eqs. 4.4 and 4.8 acceleration is a length divided by a time squared. Clearly, x is a length. We thus have

$$[2ax] = [a][x] = (LT^{-2})(L) = L^2T^{-2}$$

Note that the factor of 2 is dimensionless, and has no effect on the dimensions of the term. All three terms in Eq. 4.17 have dimensions L^2T^{-2}. This makes the equation dimensionally correct. However, we see that we can't add v (dimensions LT^{-1}) to a (dimensions LT^{-2}).

Keep in mind the following five important points about dimensional analysis:

1 An equation that is wrong dimensionally cannot be correct.

2 Any argument (θ) of a transcendental function (like sin θ) must be dimensionless—a pure number.

3 An equation can be correct dimensionally but still incorrect. (For example, the formula for the area of a circle is $A = \pi r^2$. The formula $A = r^2$ is dimensionally correct, but is incorrect for relating the area of a circle to its radius).

4 Dimensional checks of equations ignore all pure numbers that enter as factors (but not those that enter as exponents).

5 Units are not necessary in dimensional analysis—but are necessary in numerical calculations.

Example 9 Dimensional Analysis

In this example, a represents acceleration, v stands for velocity, and x is displacement. Which of the following combinations could represent a time?

$$\text{(a)}\ \frac{x}{a} \quad \text{(b)}\ \frac{xv^2}{a} \quad \text{(c)}\ \frac{x^{1/2}}{a^{1/2}} \quad \text{(d)}\ v\left(\frac{x}{a}\right)^{1/2}$$

Option (c) is the only combination that is dimensionally a time. In Chapter 14 we will see that the equation for the period (a time) of a simple pendulum incorporates the form of option (c).

Confirm the following dimensional assignments:

$$\text{(a)}\ \frac{x}{a} = T^2 \quad \text{(b)}\ \frac{xv^2}{a} = L^2 \quad \text{(c)}\ \frac{x^{1/2}}{a^{1/2}} = T \quad \text{(d)}\ v\frac{x^{1/2}}{a^{1/2}} = L$$

The Displacement Solution

Suppose we wish to derive an equation for uniformly accelerated motion giving the displacement, x, in terms of t. To do this we write Eqs. 4.15 and 4.16 in the form

$$v + v_0 = \frac{2x}{t} \qquad (4.15)$$

$$v - v_0 = at \qquad (4.16)$$

and then subtract to eliminate v and obtain

$$2v_0 = \frac{2x}{t} - at$$

Figure 4.16

The parabola of Eq. 4.18 is represented by the solid curve. Geometrically the two terms contribute to the total displacement as indicated by the vertical dotted lines.

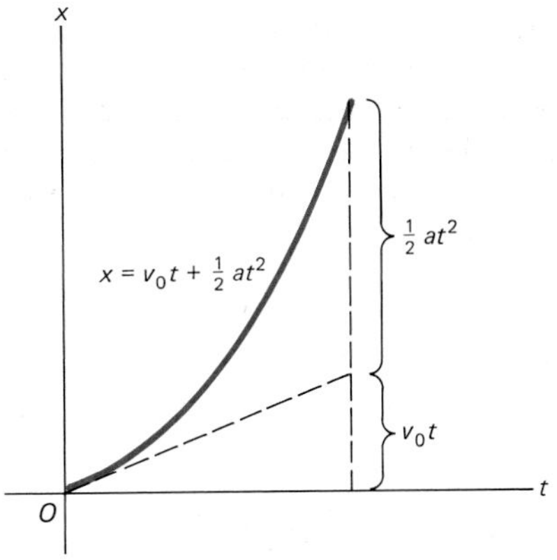

Solving for x, we find

$$x = v_0t + \frac{1}{2}at^2 \qquad (4.18)$$

The displacement at time t is expressed in terms of the time, the initial velocity, and the (constant) acceleration. This function is illustrated in Figure 4.16.

Equations 4.16 and 4.18 give future values of v and x, respectively, in terms of the initial velocity, the acceleration, and the time. Hence these two equations are a one-dimensional solution of the basic kinematics problem with constant acceleration. Before we consider applications of Eq. 4.18, let's illustrate the use of the derivative. We use differential calculus to reverse the logic and obtain Eqs. 4.16 and 4.10 from Eq. 4.18. Equation 4.7 may be written

$$v = \frac{dx}{dt}$$

Applying this to Eq. 4.18 gives us

$$v = \frac{d}{dt}\left(v_0t + \frac{1}{2}at^2\right)$$

or

$$v = v_0 + at \qquad (4.16)$$

As you can see, we have recovered Eq. 4.16 from Eq. 4.18. A second differentiation yields

$$\frac{dv}{dt} = a \qquad (4.10)$$

in agreement with the definition of instantaneous acceleration given in Section 4.4.

The acceleration due to gravity is the same for all objects at a particular location. The equations we have developed for uniformly accelerated motion apply whether the constant acceleration is at the surface of the earth or of the moon. On the moon, however, this acceleration is roughly one sixth of its value on the earth. Does this mean that an object requires six times as long to fall through 1 m on the moon as to fall through 1 m on the earth? Proper use of Eq. 4.18 in the following example will answer this question.

Example 10 **Free Fall on the Moon**

An astronaut standing on the moon drops a golf ball, letting it fall 1 m to the ground ($a = 1.62\ m/s^2$). After returning to earth, she again drops the golf ball from a height of 1 m ($a = 9.80\ m/s^2$). How long does it take the golf ball to fall to the ground in each of the situations?

In both cases $v_o = 0$, so Eq. 4.18 gives

$$t = \left(\frac{2x}{a}\right)^{1/2}$$

On the moon,

$$t = \left(\frac{2(1)}{1.62}\right)^{1/2} = 1.11\ s$$

On the earth,

$$t = \left(\frac{2(1)}{9.80}\right)^{1/2} = 0.452\ s$$

The time of free fall on the moon is longer, but not by the factor of 6, as conjectured earlier.

Often there are several ways to solve a motion problem. If the initial and final velocities and acceleration are given, then Eq. 4.13 yields the time. Alternatively, we may first obtain the distance by using Eq. 4.17, and then the time by using Eq. 4.18. We use the latter method in the next example.

Example 11 **Lunar Flight Time of a Baseball**

The best major league pitchers can throw a baseball at about 45 m/s. Suppose a baseball were thrown straight up with this speed. On the moon, how high would it rise, and what would be its total time of flight?

Using Eq. 4.17 we have $v = 0$, $v_o = +45$ m/s, and $a = -1.62\ m/s^2$. Solving for x, we obtain

$$x = -\frac{v_o^2}{2a}$$

$$= -\frac{(45)^2 2}{2(-1.62)}$$

$$= 625\ m$$

On the moon, a baseball thrown upward with an initial speed of 45 m/s would rise 625 m. To obtain t we use Eq. 4.18 to get

$$x = v_o t + \frac{1}{2}at^2$$

$$625 = 45t + \frac{1}{2}(-1.62)t^2$$

or

$$-1.62t^2 + 90t - 1250 = 0$$

This quadratic equation has the solution $t = 27.8$ s. Because of the time symmetry of the trajectory, the time of flight is just twice this time, or 55.6 s. This result (and the symmetry argument) can be checked by solving Eq. 4.18 for t with $x = 0$. This gives the equation $0 = 45t + {}^1/_2(1.62)t^2$. The two solutions are $t = 0$ (launch time) and $t = 55.6$ s (landing time). The corresponding problem on the earth is unrealistic because, for a fast baseball, air resistance is too important to neglect.

In Galileo's original work with objects undergoing uniformly accelerated motion on inclined planes he noticed a regularity that led him to the quadratic time dependence exhibited by Eq. 4.18. We follow his creative thought in the next example.

Example 12 **Galileo's Discovery**

Consider the distances traveled by a freely falling object during the 1st second of its fall, the 2nd second of its fall, the 3rd second of its fall, and so on. These distances form a regular sequence. It was Galileo's recognition of such a sequence that led him to the discovery that the distance fallen by an object is proportional to the square of the time.

Imagine dropping a ball off the Washington Monument (height 555 ft, 169 m). If down is considered positive, then $v_o = 0$, and $a = 9.80\ m/s^2$. We use Eq. 4.18 to generate the data in Table 4.4.

Table 4.4

Position and time data for a falling object in a vacuum

t (s)	x (m)	Δx (m)	$\Delta x/4.9$	$x/4.9$
0	0	—	—	—
1	4.9	4.9	1	1
2	19.6	14.7	3	4
3	44.1	24.5	5	9
4	78.4	34.3	7	16
5	122.5	44.1	9	25
6	176.4	53.9	11	36

The column headed Δx represents the displacement during the 1st second (4.9 m), 2nd second (14.7 m), 3rd second (24.5 m), and so on. Reasoning directly from similar data obtained by observing balls rolling down inclined planes, Galileo realized that the numbers formed a simple sequence, which we obtain in the fourth column of Table 4.4 by dividing each value of Δx by 4.9. Adding the integers in the fourth column produces the sequence of square numbers given in the last column. Thus $1 + 3 = 4$, $1 + 3 + 5 = 9$, and so on. This reasoning is illustrated geometrically in Figure 4.17. Each square number, $(n + 1)^2$, is seen to be the sum of the next lower square number, n^2, and the odd number $(2n + 1)$. Hence, Galileo reasoned, the position (x) varies directly with the square of the elapsed time (t). In this way Galileo arrived at Eq. 4.18.

Figure 4.17

Square numbers, n^2 with n a positive integer, are represented geometrically for $n = 1$ through $n = 5$. Each square number for n larger than 1 can be seen to be the sum of the preceding square number and a particular odd number. Thus $2^2 = 4$ is the sum of 1^2 and 3, $3^2 = 9$ is the sum of 2^2 and 5, etc. The square number, $(n + 1)^2$, is the sum of the square number, n^2, and the odd number, $(2n + 1)$.

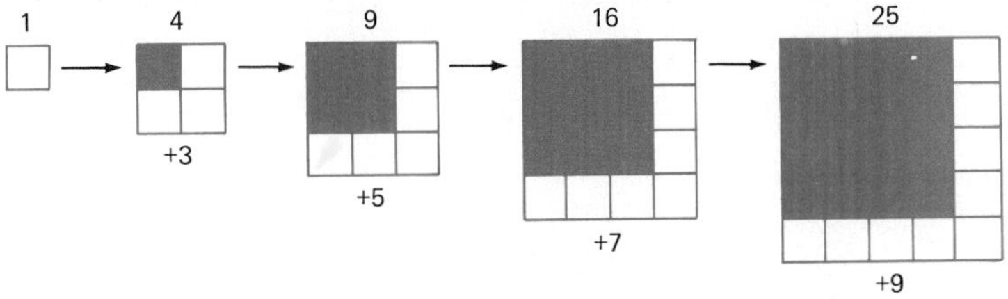

Questions

16. If we change t to $-t$ (time reversal), what changes, if any, occur in position? Displacement? Velocity? Acceleration?

17. Under time reversal what happens to Eqs. 4.17 and 4.18?

18. If up is positive for one observer, and down is positive for another observer, what effect will this have on Eqs. 4.17 and 4.18? Will their observations be the same or different?

19. Which equations in this chapter remain unchanged if the acceleration is not constant?

4.7 Relative Motion: Two Frames of Reference

So far, all of our descriptions of motion have been made relative to observers in a single frame of reference. Now let's consider the relationship between descriptions of motion made by observers in different reference frames.

Consider an automobile traveling in the $+x$-direction as shown in Figure 4.18. An observer standing on the road is in one reference frame while the passengers in T are in a second reference frame. These two frames of reference are in relative motion. The motion of a motorcycle H can be described legitimately by observers in either frame. The observer at rest on the road records the velocity of T as +15 m/s and the velocity of H as +25 m/s. However, relative to one of the passengers in T the velocity of T is 0 while the velocity of H is +10 m/s. Velocity is a relative quantity. It depends on the state of motion of the observer.

Not all quantities are relative. The number of passengers in a car, for instance, is an absolute quantity. Here absolute means "the same for all observers, independent of relative motion." Any observer in motion relative to the car will count the same number of passengers as will the car's driver.

Now let's establish the relationships between displacements, velocities, and accelerations measured relative to two different frames of reference. We use primed symbols to designate quantities measured relative to the "moving" frame of reference, T, and unprimed symbols for quantities measured relative to the "rest" frame, the ground. Thus x, v, and a denote position, velocity, and acceleration relative to the ground, while x', v', and a' refer to the "moving" frame of reference, T. It may help to imagine the axes rigidly anchored to the ground and to T, as drawn in Figure 4.19.

Let **v** denote the velocity of the primed reference frame (T) relative to the unprimed reference frame (the ground). In Figure 4.19, **v** is in the positive x-direction.

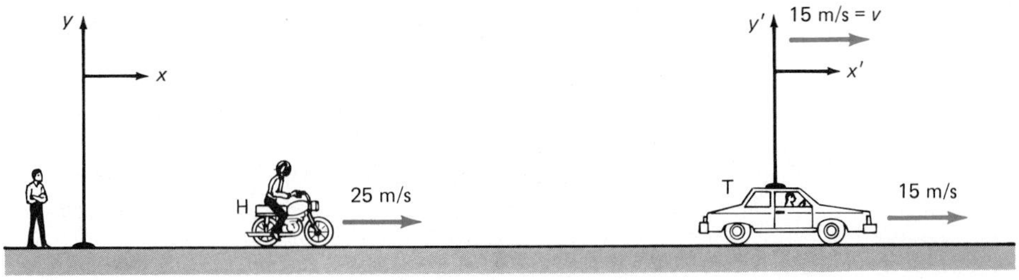

Figure 4.18

Stationary ground observers use unprimed axes affixed to the ground. Primed observers ride in the T-car and use primed axes attached to the roof of the T-car. All observers wish to describe the motion of objects such as the motorcycle H.

Figure 4.19

Unprimed (ground) observers and primed (T-car) observers can describe the motion of a jogger. The observers and the jogger are shown at the time t after the primed and unprimed axes coincided.

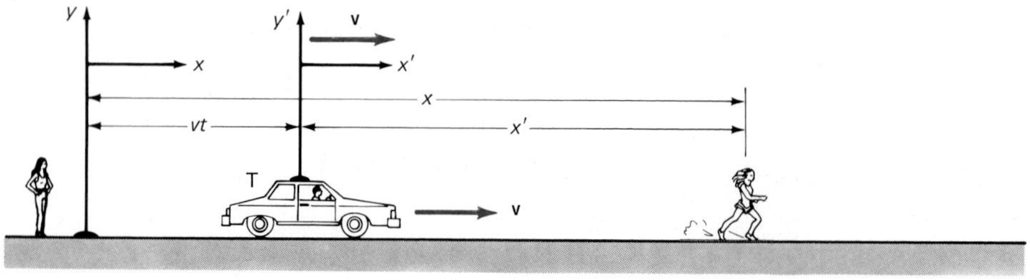

For simplicity we choose the x and x' axes to be parallel to **v**, and assume the two origins to have coincided at $t = 0$. It follows that at the time $t > 0$ the y' axis will be located at the distance vt to the right of the y-axis, as indicated in Figure 4.19.

The position of the jogger in Figure 4.19 is x, according to the unprimed observer (ground), and x', according to the primed observer (T). From Figure 4.19 we see that

$$x = x' + vt \tag{4.19}$$

is the basic **position transformation,** a relation connecting position measurements of the two observers. Note that Eq. 4.19 is consistent with our assumption that the two origins coincide at $t = 0$, because at that time the equation reduces to $x = x'$. At all other times, however, $x \neq x'$.

You may recall from Section 4.1 that the displacement of a particle is an absolute quantity, not a relative quantity, for two observers in the same frame of reference. Is this also true for two observers in different reference frames? Suppose that the jogger in Figure 4.19 were to run in the positive x-direction with speed v for a distance Δx during Δt. Because the jogger and T move in the same direction with the same speed relative to the ground, the distance between them does not change. We would have $\Delta x = v\,\Delta t$ and $\Delta x' = 0$. The displacement is not zero in the unprimed reference frame, but *is* zero in the primed reference frame. The jogger's displacement, then, is relative. It depends on the state of motion of the observer.

Now suppose the jogger runs in the positive x-direction at a speed different from v (see Figure 4.20). The velocity of the jogger relative to the ground, u, should differ from the jogger's velocity relative to T, u'. Suppose the jogger moves faster than T. During Δt the jogger will move a distance Δx, and T will move a distance $v\,\Delta t$. Meanwhile, the distance between T and the jogger will increase by the amount $\Delta x'$. Positions of T and the jogger at the times t and $t + \Delta t$ are indicated in Figure 4.20. From the figure we see that

$$\Delta x = \Delta x' + v\,\Delta t \tag{4.20}$$

Note that the velocities are defined by

$$u = \lim_{\Delta t \to 0} \frac{\Delta x}{\Delta t} \tag{4.21}$$

$$u' = \lim_{\Delta t \to 0} \frac{\Delta x'}{\Delta t} \tag{4.22}$$

From Eq. 4.20 we have

$$\frac{\Delta x}{\Delta t} = \frac{\Delta x'}{\Delta t} + v$$

In the limit as $\Delta t \to 0$, the left-hand side becomes u and the right-hand side becomes $u' + v$. Thus the velocities in the two frames are related by

$$u = u' + v \tag{4.23}$$

We can check Eq. 4.23 for special values of the variables. If the jogger is at rest relative to the ground, his velocity relative to T should be $-v$. Substitution of $u = 0$ into Eq. 4.23 yields $u' = -v$. If we substitute $u' = 0$ into Eq. 4.23, we see that $u = v$ and that the jogger is running at the same velocity as T. Equation 4.23 is the basic re-

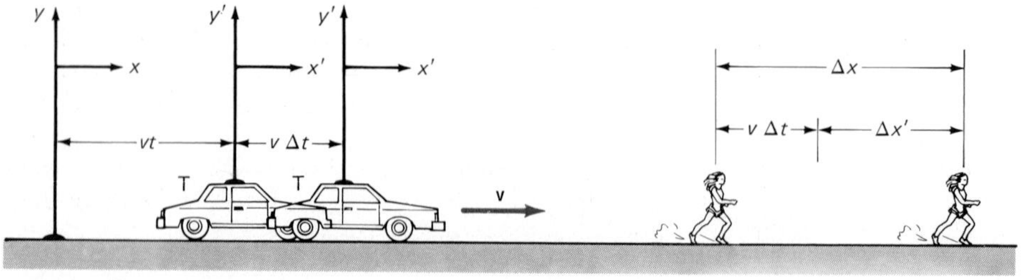

Figure 4.20

Figure 4.19, drawn at the time t, is reproduced, and the jogger and T-car are also shown at the later time $t + \Delta t$. Examination of the figure shows that $\Delta x = \Delta x' + v\,\Delta t$.

lation connecting velocity measurements of two observers in relative motion.

To illustrate the connection between Eqs. 4.19 and 4.23, we can develop a more formal derivation of Eq. 4.20. Let x_1 and x_1' denote the positions of the jogger at time t_1 and x_2 and x_2' the positions at time $t + \Delta t$. Keeping v constant, we write Eq. 4.19 at these two times:

$$x_1 = x_1' + vt$$

$$x_2 = x_2' + v(t + \Delta t)$$

Using the definition $\Delta x = x_2 - x_1$ and $\Delta x' = x_2' - x_1'$ we subtract these two equations and recover Eq. 4.20.

$$\Delta x = \Delta x' + v\,\Delta t \qquad (4.20)$$

Recall that Eq. 4.20 leads directly to Eq. 4.23.

Example 13 Relative Velocity of Two Vehicles

A sedan moving north (chosen to be the positive direction) at +50 km/hr approaches an intersection. At the same time a truck moving south at −50 km/hr approaches the intersection. Let us determine the velocity of (1) the sedan relative to the truck, and (2) the truck relative to the sedan.

1 Let $u = +50$ km/hr denote the velocity of the sedan with respect to the ground, and let $v = -50$ km/hr be the velocity of the truck with respect to the ground. Then Eq. 4.23 yields

$$u' = u - v$$

$$= +50 - (-50)$$

$$= +100 \text{ km/hr.}$$

for the velocity of the sedan relative to the truck.

2 Again associate the unprimed system with the ground, but the primed system with the sedan. Given $v = +50$ km/hr and $u = -50$ km/hr, Eq. 4.23 yields

$$u' = u - v$$

$$= -50 - (+50)$$

$$= -100 \text{ km/hr}$$

for the velocity of the truck relative to the sedan.

Throughout this section we have assumed that v is fixed, or in other words, that the relative velocity of the two frames of reference is constant. However, no such restriction applies to the velocity of objects moving relative to the two frames of reference. Both u and u' may vary with time. Let Δu and $\Delta u'$ denote the change in u and u', respectively, during Δt. Using Eq. 4.23 we have, for the relation connecting changes in velocity,

$$\Delta u = \Delta u' + \Delta v$$

However, because $\Delta v = 0$, this reduces to

$$\Delta u = \Delta u' \qquad (4.24)$$

The changes in velocity measured by observers in two different frames of reference are equal even though the velocities themselves are unequal. This remarkable result implies that the accelerations are also the same, as we will now show.

To relate the accelerations of an object measured by two observers we recall the definitions

$$a = \lim_{\Delta t \to 0} \frac{\Delta u}{\Delta t} \quad \text{and} \quad a' = \lim_{\Delta t \to 0} \frac{\Delta u'}{\Delta t}$$

Dividing Eq. 4.24 by Δt yields

$$\frac{\Delta u}{\Delta t} = \frac{\Delta u'}{\Delta t}$$

and therefore, in the limit as $\Delta t \to 0$,

$$a = a' \qquad (4.25)$$

The two observers, in relative motion, will record equal accelerations. Thus, acceleration is an absolute quantity. As such, it is said to be **invariant** with respect to the choice of reference frame (for reference frames moving with constant velocity relative to each other.)

The relations just developed for one dimension are readily extended to motion in two or three dimensions. Position, velocity, and acceleration are vectors, and Eqs. 4.19, 4.23, and 4.25 are the x-components of the vector equations

$$\mathbf{r} = \mathbf{r}' + \mathbf{v}t \qquad (4.26)$$

$$\mathbf{u} = \mathbf{u}' + \mathbf{v} \qquad (4.27)$$

$$\mathbf{a} = \mathbf{a}' \qquad (4.28)$$

Because $\mathbf{a} = \mathbf{a}'$, an object with zero acceleration in the unprimed system will have zero acceleration in the primed system. This means that if the unprimed frame of reference is an inertial frame (see Section 4.4), then the primed frame will be an inertial frame. If Newton's first law of motion holds in the unprimed system, it will also hold in the primed system. Clearly, any frame of reference moving with constant velocity relative to an inertial frame must also be an inertial reference frame.

Equations 4.26–4.28 constitute a theory of relative motion often called *Galilean relativity*. We explore this theory further in Chapters 19 and 20.

Example 14 Relative Velocity of an SST

A supersonic aircraft has a velocity directed north relative to a jet stream traveling east at 100 m/s. The SST is accelerating in its direction of motion at 10 m/s² relative to the jet stream, and at $t = 0$ is moving at 200 m/s. Let us find the velocity and acceleration of the aircraft relative to the ground.

Letting the jet-stream observer (an observer moving with the same velocity as the jet stream) be designated primed and the ground observer be designated unprimed, we can write

$$\mathbf{a}' = 10 \text{ m/s}^2 \quad \text{north}$$

$$\mathbf{u}' = (200 + 10t) \text{ m/s} \quad \text{north}$$

$$\mathbf{v} = 100 \text{ m/s} \quad \text{east}$$

Using Eq. 4.27 we have

$$\mathbf{u} = \mathbf{u}' + \mathbf{v}$$

$$\mathbf{u} = (200 + 10t) \text{ north} + 100 \text{ east m/s}$$

for the velocity of the SST relative to the ground. Using Eq. 4.28 we have

$$\mathbf{a} = \mathbf{a}' = 10 \text{ m/s}^2 \text{ north}$$

for the acceleration of the aircraft.

Questions

20. Is it possible for two observers in the same frame of reference to be in relative motion? Explain.
21. If observers in the unprimed frame of reference are inertial observers, and if **v** is constant, are observers in the primed frame of reference inertial observers? If **v** is not constant, are the primed observers inertial observers?
22. Under the transformation $\mathbf{v} \rightarrow -\mathbf{v}$, $\mathbf{r} \leftrightarrow \mathbf{r}'$, $\mathbf{u} \leftrightarrow \mathbf{u}'$, and $\mathbf{a} \leftrightarrow \mathbf{a}'$ (primed and unprimed variables exchanged) Eqs. 4.24–4.26 do not change. Discuss this result in terms of the physical situation.

Summary

A frame of reference is a set of observers fixed in position relative to each other. Each observer in a reference frame measures the same displacement vector and time interval for a pair of events. Frames of reference are often represented by coordinate systems, and individual observers by their coordinates.

Average velocity is the ratio of displacement, $\Delta\mathbf{r}$, to the time interval, Δt.

$$\bar{\mathbf{v}} = \frac{\Delta \mathbf{r}}{\Delta t} \qquad (4.3)$$

Instantaneous velocity is the limit of the average velocity as $\Delta t \rightarrow 0$.

$$\mathbf{v} = \lim_{\Delta t \rightarrow 0} \frac{\Delta \mathbf{r}}{\Delta t} \qquad (4.6)$$

For linear motion these defining equations become

$$\bar{v} = \frac{\Delta x}{\Delta t} \quad \text{and} \quad v = \lim_{\Delta t \rightarrow 0} \frac{\Delta x}{\Delta t} \qquad (4.4), (4.7)$$

The slope of the tangent line drawn to a curve of x versus t provides a geometric interpretation of the instantaneous velocity.

Average acceleration is the ratio of the change in instantaneous velocity, $\Delta\mathbf{v}$, to the time interval, Δt.

$$\bar{\mathbf{a}} = \frac{\Delta \mathbf{v}}{\Delta t} \qquad (4.8)$$

Instantaneous acceleration is the limit of the average acceleration as $\Delta t \rightarrow 0$.

$$\mathbf{a} = \lim_{\Delta t \rightarrow 0} \frac{\Delta \mathbf{v}}{\Delta t} \qquad (4.10)$$

For linear motion these defining equations become

$$\bar{a} = \frac{\Delta v}{\Delta t} \quad \text{and} \quad a = \lim_{\Delta t \rightarrow 0} \frac{\Delta v}{\Delta t} \qquad (4.9), (4.11)$$

The slope of the tangent line drawn to a curve of v versus t provides a geometric interpretation of instantaneous acceleration.

For uniformly accelerated motion five useful equations are

$$x = \bar{v}t \qquad (4.12)$$

$$v = v_0 + at \qquad (4.13)$$

$$\bar{v} = \frac{1}{2}(v_0 + v) \qquad (4.14)$$

$$v^2 = v_0^2 + 2ax \qquad (4.17)$$

$$x = v_0 t + \frac{1}{2}at^2 \qquad (4.18)$$

When one observer (primed) moves with constant velocity **v** relative to another observer (unprimed) their measurements of the position of a third object are related by the equation

$$\mathbf{r} = \mathbf{r}' + \mathbf{v}t \qquad (4.26)$$

Their velocity measurements of the third object are related by

$$\mathbf{u} = \mathbf{u}' + \mathbf{v} \qquad (4.27)$$

and their acceleration measurements by

$$\mathbf{a} = \mathbf{a}' \qquad (4.28)$$

In one dimension, these three equations are

$$x = x' + vt \qquad (4.19)$$

$$u = u' + v \qquad (4.23)$$

$$a = a' \qquad (4.25)$$

Since accelerations are observed to be the same in two different reference frames with constant relative velocity, acceleration is said to be *invariant* to the choice of reference frame.

Problems

Section 4.1 Frames of Reference

1. The coordinates (x, y, z), in meters, of two observers are $(0,0,0)$ and $(2,1,3)$. At t_i a bluebird is seen at $(10, -2, 7)$, and at t_f the bird appears at $(8, -3, 8)$. (a) Write position vectors $\mathbf{r}_i$ and $\mathbf{r}_f$ for each observer, and (b) obtain $\Delta\mathbf{r}$ for each observer.

2. Using the information given in Problem 1, assume that the first observer, $(0,0,0)$, remains fixed in location from t_i to t_f, while the second observer, at $(2,1,3)$ at $t = t_i$, moves to $(1,2,2)$ at $t = t_f$. Recalculate $\mathbf{r}_f$ and $\Delta\mathbf{r}$ for the second observer. Compare the two cases and explain any differences in the results.

3. The bases and home plate of a baseball field are situated on the corners of a square whose sides are 90 ft long. A coordinate system is established with the origin at home plate and x- and y-axes along the first and third baselines, respectively. Locate a player standing on each base with an appropriate position vector.

4. On a flat surface, the position vector of a bug for observer A is

$$\mathbf{r}_A = (1,2)\text{ m} = \mathbf{i} + 2\mathbf{j}$$

For observer B the position vector of the bug is (Figure 1)

$$\mathbf{r}_B = (3,3)\text{ m} = 3\mathbf{i} + 3\mathbf{j}$$

Determine the distance between observer A and B. (Assume their coordinate axes are parallel.)

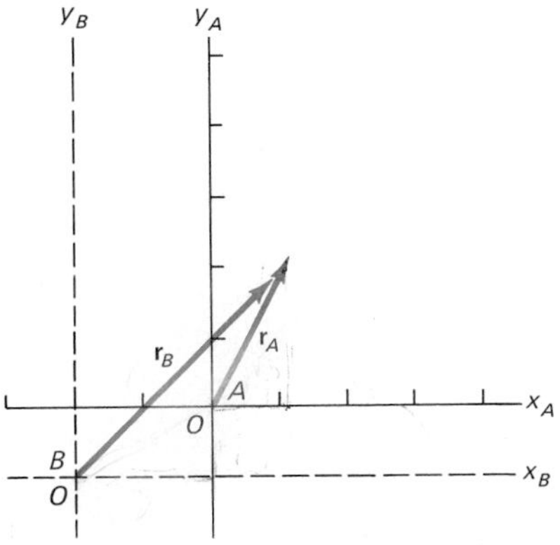

Figure 1

5. Your position riding in a bus is given by $x = 3t$, $y = 4t$ (for t in seconds and x and y in meters; see Figure 2). The position vector of a wrecked car in a field is given by $(25, 20)$. What is the position vector of the car *relative to you* at (a) $t = 4$; (b) $t = 8$?

$x = 3t$
$y = 4t$
(t in seconds; x, y in meters)

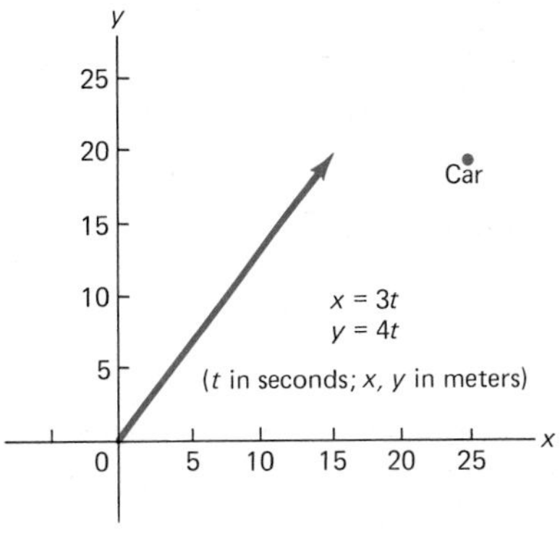

Figure 2

Section 4.2 Average Velocity

6. (a) Give the average velocity of the bluebird for either observer in Problem 1 if the two observations were 0.5 s apart. (b) What is the magnitude of this vector?

7. A car (Figure 3) travels 65 km north in 1 hr, and 150 km east in another $1\frac{1}{2}$ hr, on a trip from Cincinnati, Ohio, to Columbus, Ohio. (a) Determine the average velocity for the trip. (b) What is the magnitude of the average velocity vector? (c) Obtain the average speed for the trip. Give answers in km/hr.

8. Using data given in Table 4.1, obtain the average velocity for (a) $x_i = 0.31$ m, $x_f = 1.22$ m, (b) $x_i = 1.83$ m, $x_f = 2.74$ m, (c) $x_i = 3.66$ m, $x_f = 5.49$ m, (d) $x_i = 7.32$ m, $x_f = 9.14$ m.

9. A graph of position (in meters) versus time (in seconds) is given for a swimmer in Figure 4. Obtain the average velocity for (a) $t_i = 0$ s; $t_f = 4$ s; (b) $t_i = 3$ s, $t_f = 4$ s.

10. A freely falling ball has a position described by $x = -4.9t^2$ with x in meters and t in seconds. Determine the average velocity between $t_i = 1$ s and $t_f = 3$ s.

11. The position of a lunar rock held up and then permitted to fall freely is given by

$$y = -0.81t^2$$

with t in seconds and y in meters. Find the average velocity of the rock for $t_f = 1.00$ and (a) $t_i = 0.00$, (b) $t_i = 0.50$, (c) $t_i = 0.90$.

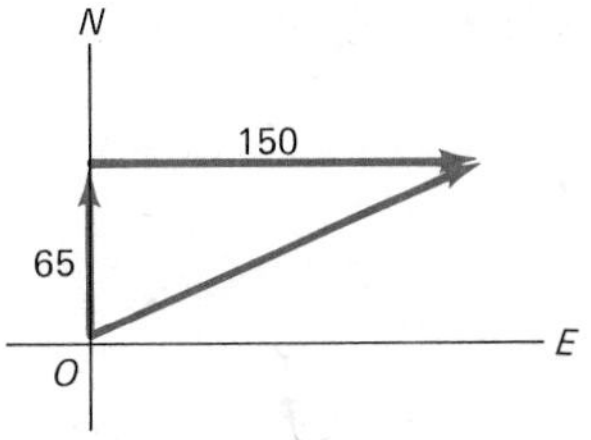

Figure 3

12. The average velocity of B is twice the average velocity of A. If it takes B 2 s to travel a given distance, how long a time is required for A to travel the same distance?

Section 4.3 Instantaneous Velocity

13. From the graph in Figure 4 (a) obtain the instantaneous velocity at $t = 3$ s. (b) By varying the slope of the line, estimate the error in your result.

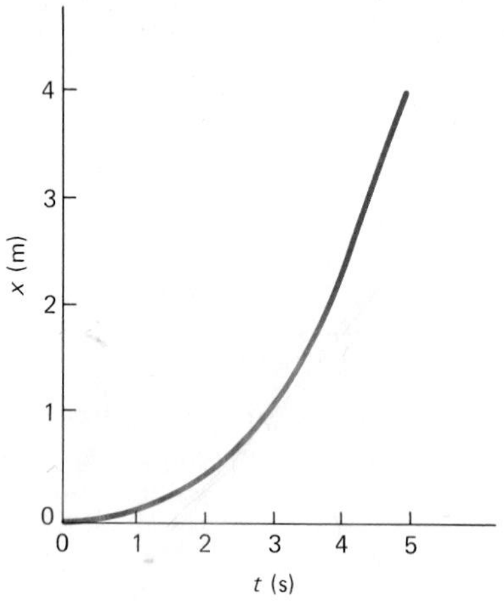

Figure 4

14. The position of a plane accelerating for takeoff down a runway is given by

$$x = 1.200t^2$$

with x in meters and t in seconds. Approximate the plane's instantaneous velocity at $t = 4$ s by taking $t_i = 4.000$ s and (a) $t_f = 4.100$ s, (b) $t_f = 4.010$ s, (c) $t_f = 4.001$ s, (d) $t_f = 4.0001$ s, (e) $t_f = 4.00001$ s.

15. From the graph in Figure 4.7 determine the instantaneous velocity at $t = 5.5$ s.

16. For a jet plane, $v = 2.4t + 30$ where t is in seconds and v is in meters per second. Plot this function on graph paper from $t = 0$ to $t = 10$, and determine the distance covered during the first 10 s geometrically by determining the area under the curve.

17. Extrapolate the average velocity graph (Figure 5) to $\Delta t = 0$ to determine the instantaneous velocity.

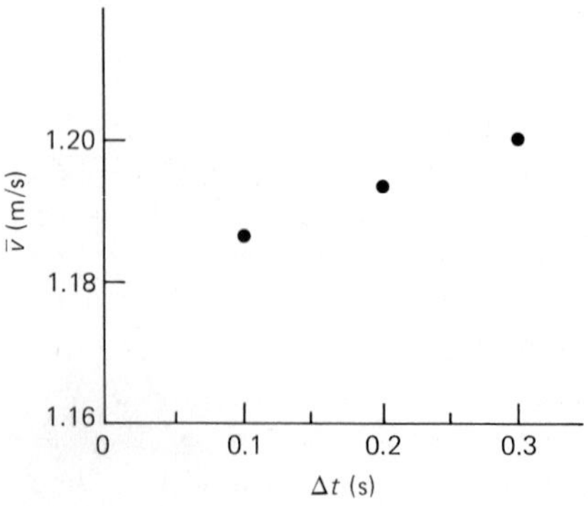

Figure 5

18. In Example 3 the position of an object is given exactly by $x = -4.9t^2$. The instantaneous velocity is approximated by letting t_i approach t_f with t_f fixed. Now we keep t_i fixed and vary t_f. Calculate the average velocity for $t_i = 6$ s and (a) $t_f = 6.1000$ s, (b) $t_f = 6.0100$ s, (c) $t_f = 6.0010$ s, (d) $t_f = 6.0001$ s.

19. A subway train moves in a straight line from one station to another. The position of the front of the train as a function of time is shown in Figure 6. Determine the instantaneous velocity at 0, 2, 4, and 6 minutes.

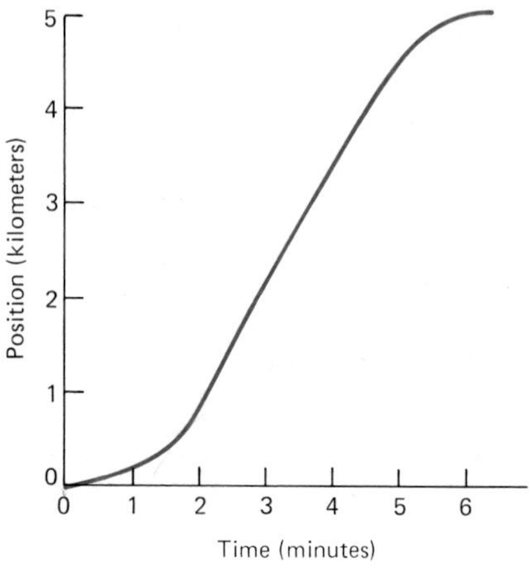

Figure 6

Section 4.4 Acceleration

20. A car moving at 30 km/hr north turns a corner and travels 30 km/hr east (Figure 7). The turn requires 4 s. Determine the average acceleration of the car during the turn. Give its magnitude and direction.

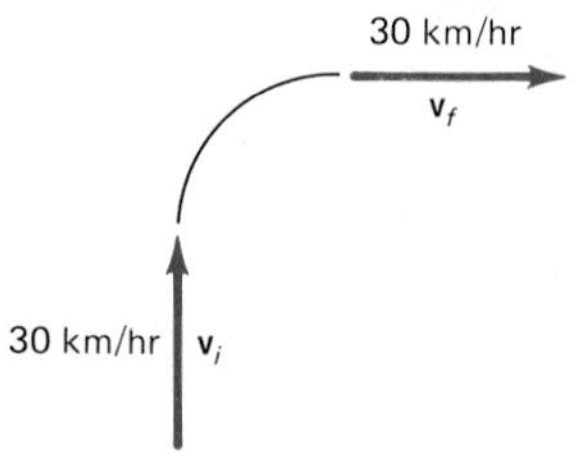

Figure 7

21. Using the data in Table 1, determine the average acceleration from (a) $t_i = 0$ s to $t_f = 2.00$ s; (b) $t_i = 2.00$ s to $t_f = 4.00$ s.

Table 1

t (s)	v (m/s)
0	0
1	2
2	8
3	18
4	32

22. From data in Table 4.1 estimate the instantaneous velocity by equating it to the average velocity for the time interval, and using as the time the middle of the time interval. From such values of instantaneous velocity estimate the average acceleration from $t_i = 0.51$ s to $t_f = 0.69$ s.

23. The velocity of a rocket during the initial launch stages (Figure 8) is given by $v = 30t - 0.5t^2$, where v is in meters per second and t is in seconds. Determine the average acceleration of the rocket from (a) $t_i = 0$ s to $t_f = 1$ s, and (b) $t_i = 5$ s to $t_f = 6$ s.

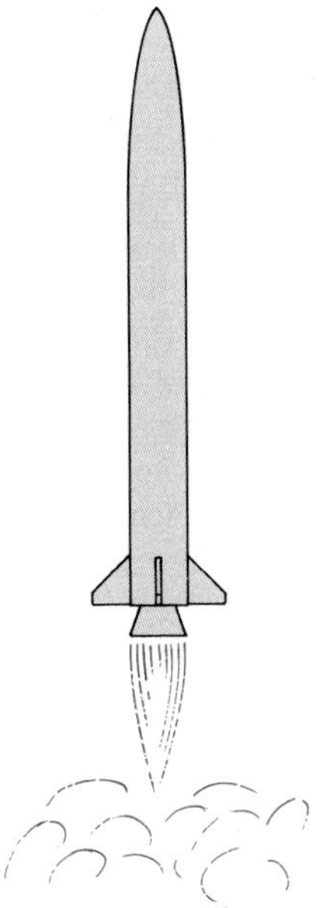

Figure 8

24. Using the formula for velocity given in Problem 23, estimate the instantaneous acceleration at $t = 5$ s by taking $t_i = 5$ s, and successively letting $t_f = 5.1$ s, $t_f = 5.01$ s, and $t_f = 5.001$ s.

25. The advertised characteristics of the 1980 Porsche Turbo sportscar are shown in Table 2. For which gear is the acceleration largest and what is the largest acceleration?

Table 2

Gear	Time starting from rest (s)	Final speed (mi/hr)
1	2.3	30
2	9.3	52
3	21.3	78
4	44.5	100
5	93.4	120

26. (a) A car moving at 10 mi/hr accelerates for 3 s. The average acceleration over the 3-s interval is 20 mi/hr · s. What is its speed at the end of the 3-s acceleration interval? (b) The car then decelerates, coming to a stop in 7 s. What is its average acceleration over the 7-s interval?

27. A car is brought to a full stop at a stop light. Forgetting that the car has automatic transmission and is in gear, the driver takes his foot off of the brake. Two seconds later the speed of the car is 1.6 m/s. What is the average acceleration of the car?

Section 4.6 Linear Motion with Constant Acceleration

28. A baseball is hit so that it travels straight up in the air (Figure 9). A fan in the stands observes that it requires 3 s to reach maximum height. (a) What was its initial vertical velocity? (b) How high did it rise? (Neglect air resistance.)

Figure 9

29. A ball is thrown vertically upward with initial speed v_0. (a) Calculate the rise time (until $v = 0$). (b) Calculate the maximum height.

With the ball at its maximum height and just starting to fall, take this position as the new origin of your coordinate system and set $t = 0$. (c) Calculate the time to fall the distance of part (b). Compare your answer with that for part (a). (d) Calculate the velocity v for the time of part (c). Compare your answer with the initial velocity v_0.

30. A student jumps straight up in the air (Figure 10) and is airborne (up plus down) for 0.6 s. (a) What was his initial vertical velocity? (b) How high did he rise? (Assume a symmetric trajectory.)

Figure 10

31. A research rocket starting from rest at $t = 0$ climbs to an altitude of 9,600 m in 80 s. (a) Calculate the rocket's average (vertical) velocity for this 80-s interval. Assuming the acceleration to be constant, calculate (b) the velocity of the rocket at $t = 80$ s and (c) the rocket's acceleration.

32. A drag racer (Figure 11) covers the quarter-mile distance in a time of 10 s. (a) What was his acceleration, assuming it was constant? (b) What was his final velocity? (Give answers in SI units.)

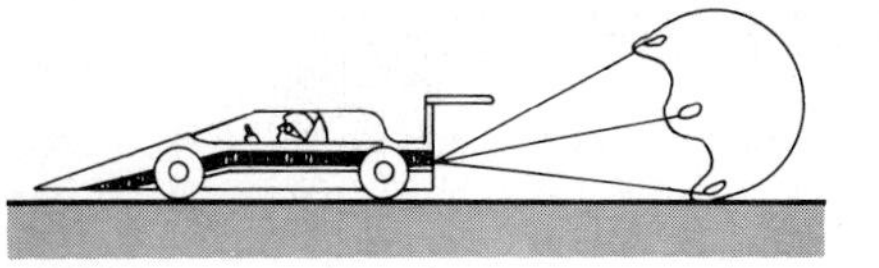

Figure 11

33. In a drag race one car reaches a maximum velocity of 90 m/s at the end of the race. The car's acceleration (constant) was 10 m/s^2. How far did this car travel?

34. In a relay race the second runner does not start from rest (Figure 12). He covers 100 m in 10 s, finishing with a maximum velocity of 12 m/s. Assuming constant acceleration, determine his initial velocity.

Figure 12

35. In the following, a, v, and x denote an acceleration, a velocity, and a displacement, respectively. A problem asks for a time of flight determination. The following multiple choice answers are presented. (a) a/v (b) ax/v (c) $x/(v - x)$ (d) x/v (e) x/a Use dimensional considerations to identify the correct answer.

36. A child drops stones from a bridge at regular intervals of 1 s (Figure 13). At the moment the fourth stone is released, the first strikes the water below. (Neglect air resistance.) (a) How high is the bridge? (b) How far above the water are each of the other falling stones at that moment?

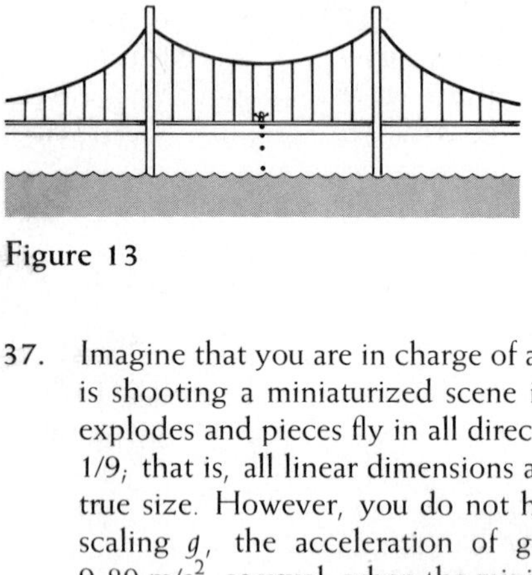

Figure 13

37. Imagine that you are in charge of a camera crew that is shooting a miniaturized scene in which a bridge explodes and pieces fly in all directions. The scale is 1/9; that is, all linear dimensions are one ninth their true size. However, you do not have the liberty of scaling g, the acceleration of gravity. It will be 9.80 m/s^2, as usual, when the miniature pieces begin flying about. If a full-scale scene were photographed at 60 frame/s, and lasted 6 s, how many frames per second would be required to make the miniature scene look natural—that is, to make it last 6 s when projected at the normal speed of 60 frames/s?

38. Two sequences of positions (in meters) and times (in seconds) are given in Table 3 for the motion of an object. Following Galileo's procedure (Example 12), determine whether these data correspond to uniformly accelerated motion. If either x_1 or x_2 does so correspond, then give the (constant) acceleration(s).

Table 3

t	x_1	x_2
0	0	0
1	3	1
2	12	2
3	27	4
4	48	6

39. A rubber ball is released from a height of 4.90 m above the floor. It falls with an acceleration of 9.80 m/s^2 and strikes the floor 1 s later. It bounces repeatedly, always rising to 81/100 of the height through which it falls. (a) Ignoring the practical fact that the ball has a finite size (in other words, treating the ball as a point mass that bounces an infinite number of times), show that its total distance of travel is 46.7 m. (Suggestion: Work out the first few distances, then try to set up and sum the infinite series. If you are unable to do this algebraically, try to sum the first several terms of the series arithmetically, perhaps using an electronic calculator.) (b) Determine the time required for the infinite number of bounces. (c) Determine the average speed. (d) Compare the result of (c) with the average speed over the first 1 s.

40. Use Eq. 4.13 to obtain the time of flight in Example 11.

41. Show that the vertical trajectory of a particle is symmetric about the maximum. That is, its height at time Δt before reaching its maximum equals its height at time Δt after reaching its maximum.

42. The horizontal velocity (assumed constant) of a punt is 12 yd/s. A player begins running in the same direction as the football with a velocity of 8 yd/s. If both start at the same time, and the player is 5 yd behind the football when it lands, how far does the football travel horizontally?

43. One car moving at a constant 50 km/hr passes a second car at rest. At the instant of passing, the second car begins accelerating at a constant 5 km/hr · s, and maintains this acceleration until it reaches the first car. Determine (a) the time required, (b) the distance covered, and (c) the speed of the second car when it reaches the first car.

44. A group of physics students devised a modified game of darts. They dropped the dart board to the ground from the window. One second after dropping the dart board they threw the darts at the board. By trial and error they found that the darts arrived at the board at the instant the board touched the ground if the dart speed initially was 20 m/s. Determine (a) the time of flight for the dart, and (b) the height of the window.

45. An object travels in a vacuum upward (Figure 14) past elevations B and A (labeled B' and A' on the trip downward). The object rises to an unknown height and then returns to its starting point, passing A' and B' on the way. The distance between the two elevations (AB or $A'B'$) is measured to be 1 m (assume that this number is exact). A laser accurately triggers a timer, which records the times when the object is located at x_1, x_2, x_3, and x_4. These times are $t_1 = 0.1982$ s, $t_2 = 0.2207$ s, $t_3 = 9.2651$ s, and $t_4 = 9.2876$ s. Treat the initial velocity and acceleration of gravity as unknowns. (a) Obtain a value for g (to three significant figures) using the given times and separation distance of 1m. (b) Obtain the initial velocity of the object $v(t = 0)$. (c) Determine the total vertical height to which the object rises relative to $x(t = 0)$.

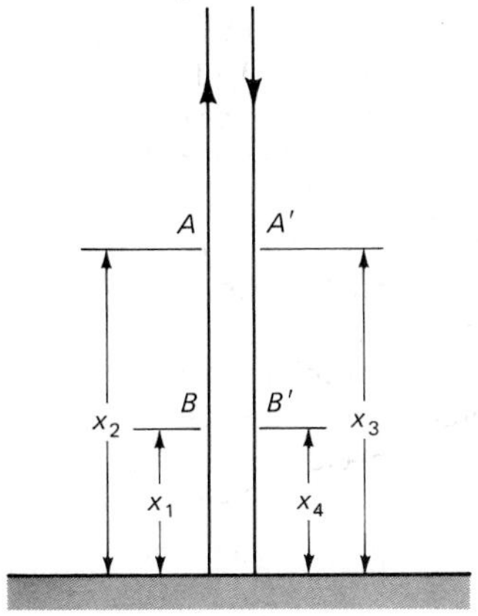

Figure 14

46. In the book *From the Earth to the Moon and a Trip Around It*, Jules Verne imagines a 900-ft cannon. The first 180 ft are filled with the propellant (gun cotton), leaving 720 ft (about 220 m) for accelerating a projectile. The projectile is to be accelerated to a speed of 1.12×10^4 m/s. Calculate the acceleration. Assume the acceleration to be constant. (*Note:* Verne's astronauts, inside the projectile, would have been squashed beyond recognition.)

Section 4.7 Relative Motion: Two Frames of Reference

47. A man standing on the shore observes that a raft floats with the river at 2 m/s downstream (Figure 15). Relative to the observer on the raft, a motorboat moves directly across the stream with the velocity $v = a_0 t$ where $a_0 = 2$ m/s^2. Determine the magnitudes and directions of the motorboat velocity and acceleration relative to the shore observer.

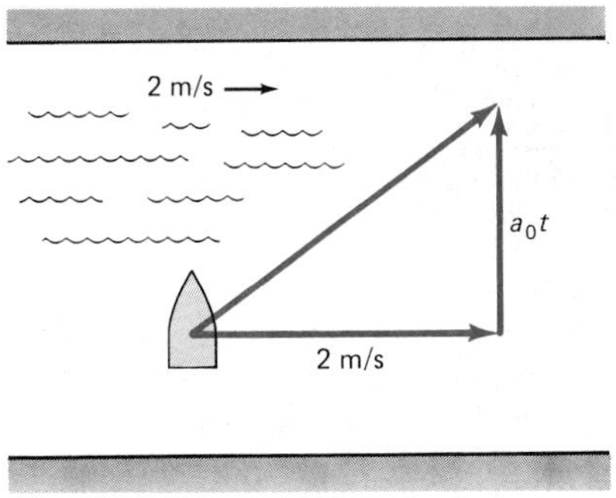

Figure 15

48. A golf club head moving at 100 mi/hr approaches a golf ball at rest (Figure 16). After the collision the club-head speed is 69 mi/hr and that of the golf ball is 139 mi/hr. All motion is in the $+x$-direction. Determine the velocity of the golf ball relative to the club head (a) before the collision; (b) after the collision. Determine the velocity of the club head relative to the golf ball (c) before the collision; (d) after the collision.

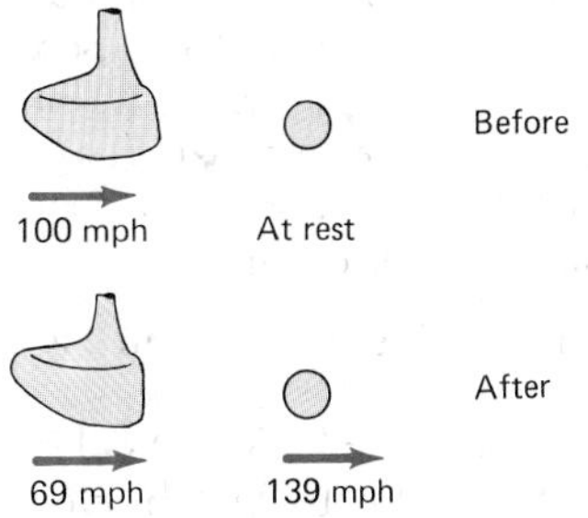

Figure 16

49. One automobile (A) travels east at 30 mi/hr while a second automobile (B) travels south at 40 mi/hr (Figure 17). Determine the velocity of (a) A relative to B, and (b) B relative to A.

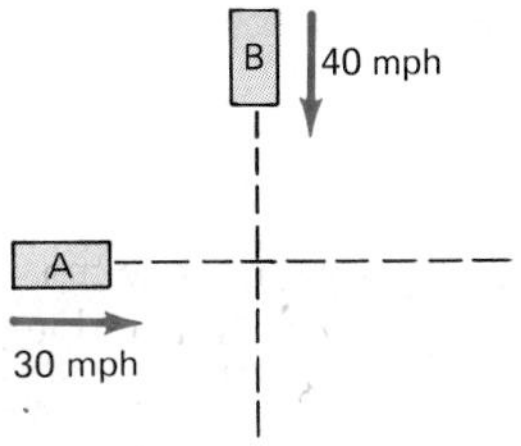

Figure 17

50. Give the velocity (see Figure 4.18) of (a) H relative to T and (b) T relative to H, and (c) explain why the answer to Problem 49 does not depend on whether A or B is approaching or leaving the intersection.

51. A flatcar travels down the railroad track at 60 mi/hr. You run from the front of the car toward the rear at a speed of 15 mph, relative to the car. What is your speed relative to the tracks?

chapter 5

MOTION IN A PLANE

Preview

In Chapter 4 we introduced two new vector quantities—velocity and acceleration. To illustrate these two concepts, and to show how equations are used to describe motion, we developed the theory of linear motion with constant acceleration, or uniformly accelerated motion.

We are now ready to consider motion in a plane. We begin this chapter with a consideration of the principle of superposition (Section 2.2) as it applies to the vector quantities position, velocity, and acceleration. Then we extend the theory of uniformly accelerated motion to motion in a plane.

Displacement, velocity, and acceleration all satisfy the principle of superposition. We use this principle as Galileo did: to consider separately the horizontal and vertical parts of projectile motion. Subsequently, the theory of linear motion developed in Chapter 4 is applied, with only minor adjustments in notation, to problems involving motion in a plane.

In uniform circular motion the speed of a particle is constant, but the direction of the velocity vector changes with time. These changes are associated with an acceleration—called centripetal—directed radially toward the center of rotation. If the speed of a particle in circular motion changes, another component of the acceleration—called tangential—is introduced, directed tangent to the path of the particle.

5.1 Motion in a Plane and the Principle of Superposition

Figure 5.1a illustrates a particle moving in the x–y plane. The heavy solid line represents the trajectory, or path, of the particle. The kinematic variables indicated on the diagram are the particle's position, **r**, velocity, **v**, and acceleration, **a**. These vector quantities can be expressed in terms of their x- and y-components as follows.

$$\mathbf{r} = \mathbf{i}x + \mathbf{j}y \tag{5.1}$$

$$\mathbf{v} = \mathbf{i}v_x + \mathbf{j}v_y \tag{5.2}$$

$$\mathbf{a} = \mathbf{i}a_x + \mathbf{j}a_y \tag{5.3}$$

In Figure 5.1b, c, and d we see the geometric representation of each of these three expressions. At the instant shown, x, y, and v_x are positive, and v_y, a_x, and a_y are negative.

The Principle of Superposition

The relation defining velocity (Section 4.3) and the relation defining acceleration (Section 4.4) are linear vector equations. These defining equations are

$$\mathbf{v} = \lim_{\Delta t \to 0} \frac{\Delta \mathbf{r}}{\Delta t} = \frac{d\mathbf{r}}{dt} \tag{5.4}$$

$$\mathbf{a} = \lim_{\Delta t \to 0} \frac{\Delta \mathbf{v}}{\Delta t} = \frac{d\mathbf{v}}{dt} \tag{5.5}$$

From the definition of vector equality in Section 2.3, each component on the left side of a vector equation must equal the corresponding component on the right side. In a plane each vector equation is equivalent to two component relations. Thus in component form Eqs. 5.4 and 5.5 can be written

$$v_x = \lim_{\Delta t \to 0} \frac{\Delta x}{\Delta t} = \frac{dx}{dt}$$
$$v_y = \lim_{\Delta t \to 0} \frac{\Delta y}{\Delta t} = \frac{dy}{dt} \tag{5.4a}$$

$$a_x = \lim_{\Delta t \to 0} \frac{\Delta v_x}{\Delta t} = \frac{dv_x}{dt}$$
$$a_y = \lim_{\Delta t \to 0} \frac{\Delta v_y}{\Delta t} = \frac{dv_y}{dt} \tag{5.5a}$$

We are interested in the special case where the acceleration is constant, a_x = constant and a_y = constant. Then these equations show that the x- and y-components of the motion are independent of each other.

As an example of this independence consider the motion of a particle in a vertical plane. If we neglect air resistance, and choose the y-axis to be in the vertical direction, we get $a_x = 0$ and a_y = constant. In this case the independence of the components of motion is confirmed by the following experiment.

Two spheres begin to fall at the same time. One is dropped ($v_{xo} = 0$), whereas the other is pushed horizontally ($v_{xo} \neq 0$). Because $a_x = 0$, the initial horizontal velocity of each sphere is retained through the fall. Both spheres have vertical motion, but only one of them moves horizontally. Figure 5.2 shows the position of each sphere at regular time intervals. Although the horizontal positions of the spheres differ, their vertical positions are the same. The horizontal motion does not affect the vertical motion, and vice versa.

Because these horizontal and vertical components of motion of a particle in a plane are uncoupled, we may apply a **principle of superposition.**

> *The separate solutions for the horizontal and vertical motions of a particle may be combined linearly to form the solution for the two-dimensional motion of the particle in a vertical plane.*

Given a_x, the solution (x, v_x) of the horizontal motion of a particle can be obtained without regard for the vertical motion; and given a_y, the solution (y, v_y) of the vertical motion of the particle can be obtained without regard for its horizontal motion. The principle of superposition applies to the linear equations $\mathbf{r} = x\mathbf{i} + y\mathbf{j}$ and $\mathbf{v} = v_x\mathbf{i} + v_y\mathbf{j}$, and ensures that $(\mathbf{r}, \mathbf{v})$ constitutes the solution to the two-dimensional motion of the particle. Keep in mind that the validity of the principle of superposition, and indeed the validity of our entire linear vector description, is based on experiment.

In summary, the principle of superposition permits us to analyze motion in a plane one component at a time, thus simplifying our task. When each component has been separately analyzed, the principle of superposition allows the results for the separate components to be combined into one path (see Figure 5.3).

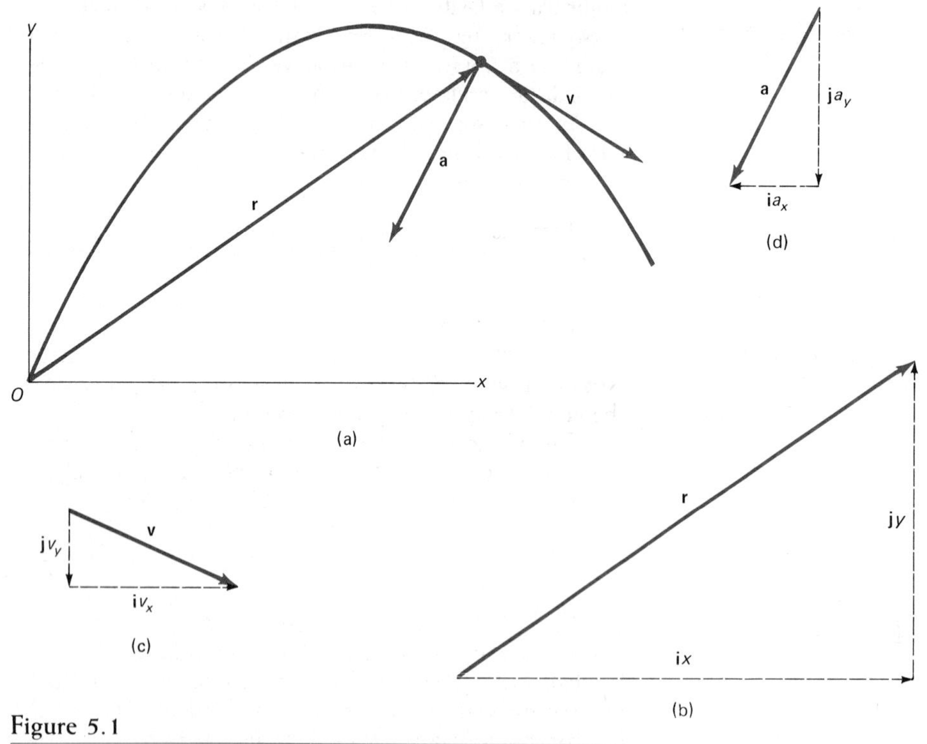

Figure 5.1

The heavy curve in (a) represents the path or trajectory traced by a particle moving in the x–y plane. At the instant shown, the position, **r**, velocity, **v**, and acceleration, **a**, vectors of the particle are as drawn. In (b), (c), and (d) the components of **r**, **v**, and **a** are shown. As drawn, x, y, and v_x are positive, whereas v_y, a_x, and a_y are negative.

Figure 5.2

The positions of two falling spheres are shown at successive equal time intervals. One sphere started to fall straight down at the same time that the other sphere started to move horizontally. Even though the horizontal motions of the spheres differ, the vertical motions are the same.

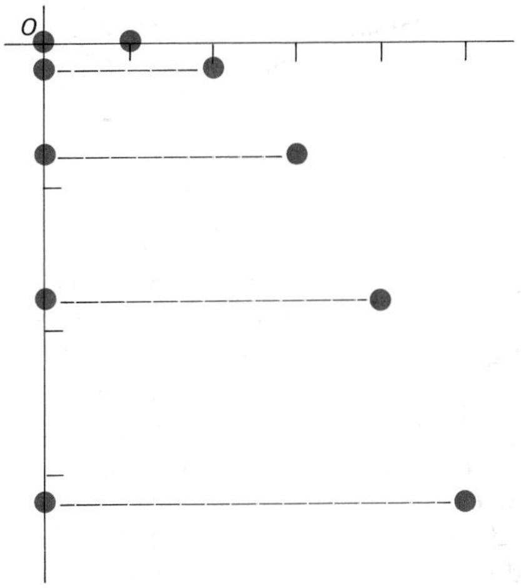

Figure 5.3

Using the principle of superposition, two one-dimensional solutions may be combined into a single two-dimensional solution. This procedure fits into the general method for attacking and solving two-dimensional problems outlined in the flowchart.

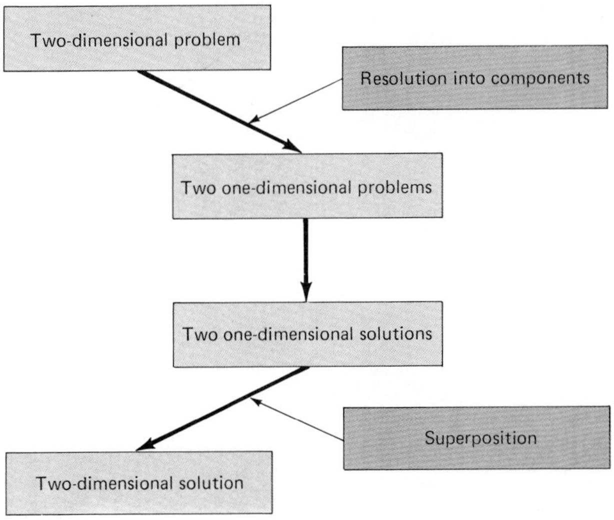

5.2 Motion in a Plane with Constant Acceleration

The equations of one-dimensional motion developed in Chapter 4 were for constant acceleration. Now we generalize this theory to motion in the x–y plane with constant acceleration. **Constant acceleration** means that neither the magnitude nor the direction of the acceleration vector varies with space or time. The motion (near the surface of the earth) of an object experiencing the force of gravity is an important example of constant acceleration. When considering the motion of an object in air, we will neglect the effects of air resistance. For simplicity we choose one axis, the y-axis, to be in the acceleration direction. This choice makes the x-component of the acceleration equal to zero.

$$a_x = 0 \tag{5.6}$$

In a plane, the **constant acceleration equations of motion,** given in Section 4.6, can be written (for $a_x = 0$) as follows:

$$x = v_{xo}t \tag{5.7}$$

$$v_x = v_{xo}, \quad \text{constant} \tag{5.8}$$

$$v_y = v_{yo} + a_y t \tag{5.9}$$

$$v_y^2 = v_{yo}^2 + 2a_y y \tag{5.10}$$

$$y = v_{yo}t + \frac{1}{2}a_y t^2 \tag{5.11}$$

You will notice that the equations for v_x and v_x^2 corresponding to Eqs. 5.9 and 5.10 contain the same information because $a_x = 0$. The following example illustrates the independence of the horizontal and vertical components of motion.

Example 1 **Vertical Drop of a Pitched Baseball**

A baseball is thrown horizontally by a pitcher toward a batter. Assuming that the ball has the velocity of a good major league fastball (90 mi/hr or 40 m/s), and travels 17 m (somewhat less than the distance between the pitcher's mound and home plate), what vertical distance does the baseball fall while en route? Ignore air resistance effects.

In the absence of horizontal acceleration the horizontal velocity will not change. We can use Eq. 5.7 to obtain the time of flight

$$x = v_{xo}t$$

so that

$$t = \frac{x}{v_{xo}} = \frac{17}{40} = 0.425 \text{ s}$$

Note that we used the horizontal motion independent of the vertical motion to get the time of flight.

The vertical velocity is zero initially and the vertical acceleration is $a_y = -g = -9.80 \text{ m/s}^2$. Thus, the equation for the vertical distance, Eq. 5.11, reduces to

$$y = -\frac{1}{2}gt^2 = -\frac{1}{2} \times 9.80(0.425)^2 = -0.885 \text{ m}$$

The baseball drops approximately 1 yard as a result of the effect of gravity alone. The minus sign indicates that the baseball arrives 0.885 m *below* its release point. Notice that we used the vertical motion of the ball independent of the horizontal motion to determine the vertical drop. The independence of the components of motion implies that if a baseball were dropped with zero vertical velocity, it would fall the same vertical distance in the same time. Comparison of the vertical distances traveled by each sphere in Figure 5.2 confirms this implication.

Trajectory

The trajectory of a particle is described by an equation, $y = y(x)$, relating values of x and y along the path traveled by the particle. To obtain this equation for a particle with constant acceleration, we solve for t in Eq. 5.7,

$$t = \frac{x}{v_{xo}}$$

and then substitute this value into Eq. 5.11:

$$y = \left(\frac{v_{yo}}{v_{xo}}\right)x + \left(\frac{1}{2}\right)\left(\frac{a_y}{v_{xo}^2}\right)x^2 \qquad (5.12)$$

You should recall from your study of algebra that every quadratic function of the form

$$y = ax^2 + bx + c$$

with $a \neq 0$ and b and c constant describes a parabola. For a constant acceleration all of the quantities in parentheses in Eq. 5.12 are constant and the trajectory is a parabola. Thus, a **parabolic trajectory** is characteristic of motion with constant acceleration.

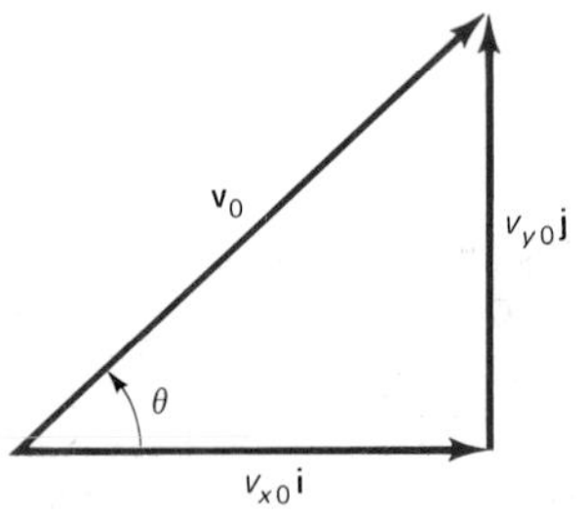

Figure 5.4

Components of the initial velocity, $\mathbf{v}_o$.

Often the launch angle, θ, and the magnitude of the initial velocity of a particle, v_o, are more likely to be known than v_{xo} and v_{yo}. Therefore, it is useful to obtain an alternate form of Eq. 5.12 in which v_0 and θ are used as variables rather than v_{xo} and v_{yo}. From Figure 5.4 we have

$$\begin{aligned} v_{xo} &= v_o \cos\theta \\ v_{yo} &= v_o \sin\theta \end{aligned} \qquad (5.13)$$

Substituting these values into Eq. 5.12 yields a new equation for the trajectory of the particle.

$$y = x \tan\theta + \left(\frac{a_y}{2v_o^2 \cos^2\theta}\right)x^2 \qquad (5.14)$$

Trajectories that correspond to several launch angles are given in Figure 5.5. These paths share the same values of v_o and a_y. The maximum height, distance from launch to landing point, and launch angle vary from one curve to the next. These paths exhibit several symmetries, which we now consider.

1 The trajectories are symmetric relative to their peaks. We can prove this as follows Let $x = x_m$ be the x-coordinate of the point of maximum height. This height is reached at the time when $v_y = 0$. From Eq. 5.9 we have for this time, t_m,

$$t_m = -\frac{v_{yo}}{a_y} \qquad (5.15)$$

Putting this result into Eq. 5.7 we have

$$x_m = -\frac{v_{xo}v_{yo}}{a_y} \qquad (5.16)$$

We know that both v_{xo} and v_{yo} are positive, while a_y is negative. Therefore, according to Eq. 5.16, x_m must be positive.

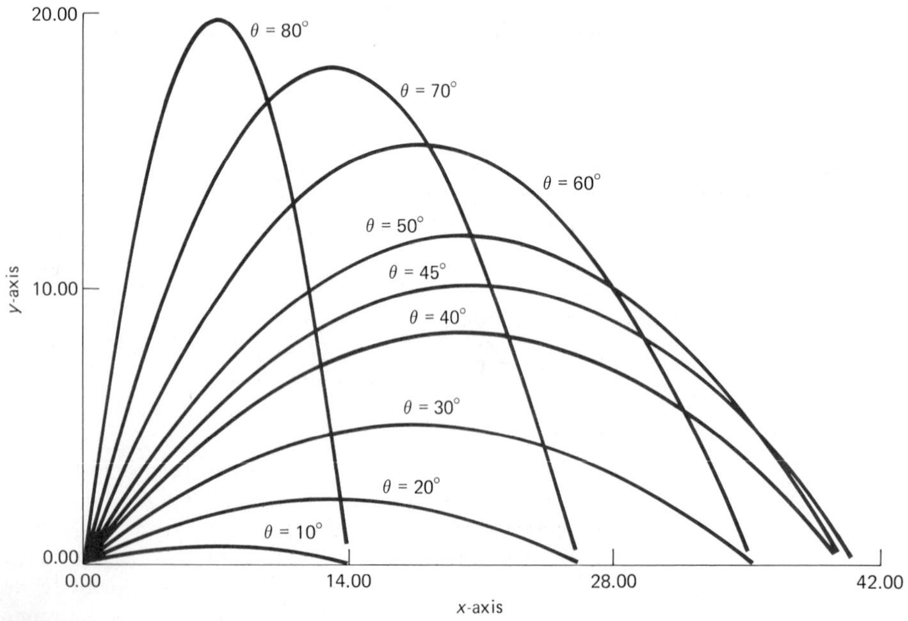

Figure 5.5

These parabolic trajectories are for the same initial speed (20 m/s) and launch angles at 10° intervals.

In order to demonstrate that the trajectory is symmetric about x_m, we must show that $y(x_m + b) = y(x_m - b)$. To do this we substitute $(x_m + b)$ for x in Eq. 5.12, using the expression in Eq. 5.16 for x_m, and then repeat the procedure with $(x_m - b)$. If the trajectory is symmetric about x_m, then y will be the same in the two cases.

Starting with Eq. 5.12, we have

$$y = \left(\frac{v_{yo}}{v_{xo}}\right)x + \frac{1}{2}\left(\frac{a_y}{v_{xo}^2}\right)x^2 \qquad (5.12)$$

Substituting $x = x_m \pm b$ into Eq. 5.12, we obtain

$$y(x_m + b) = (x_m + b)\frac{v_{yo}}{v_{xo}} + (x_m^2 + 2x_m b + b^2)\frac{1}{2}\frac{a_y}{v_{xo}^2}$$

$$y(x_m - b) = (x_m - b)\frac{v_{yo}}{v_{xo}} + (x_m^2 - 2x_m b + b^2)\frac{1}{2}\frac{a_y}{v_{xo}^2}$$

We subtract and take out a common factor of $2b$ to get

$$y(x_m + b) - y(x_m - b) = 2b\left[\frac{v_{yo}}{v_{xo}} + x_m\frac{a_y}{v_{xo}^2}\right]$$

If $x_m = -v_{xo}v_{yo}/a_y$ is substituted into this equation, the brackets vanish leaving $y(x_m + b) - y(x_m - b) = 0$, or $y(x_m + b) = y(x_m - b)$. Therefore the trajectory is symmetric relative to its peak, position x_m (Figure 5.6).

The horizontal distance covered between the two $y = 0$ points is $2x_m$, because the trajectory is symmetric. Substituting Eqs. 5.13 into Eq. 5.16 we can write

$$R = -\frac{2v_o^2 \sin\theta\cos\theta}{a_y} \qquad (5.17)$$

where $R = 2x_m$ is called the "range" of the projectile.

Now we can prove that if v_o and a_y are constant, R is a maximum when $\theta = 45°$. Equation 5.17 can be written

$$R = -\frac{v_o^2}{a_y}\sin 2\theta$$

using the relationship $\sin 2\theta = 2\sin\theta\cos\theta$. For v_o^2 and a_y constant, R will be a maximum when $\sin 2\theta$ is a maximum. This occurs when $2\theta = 90°$ or $\theta = 45°$. Note that $R = 0$ when $\theta = 0°$ and when $\theta = 90°$.

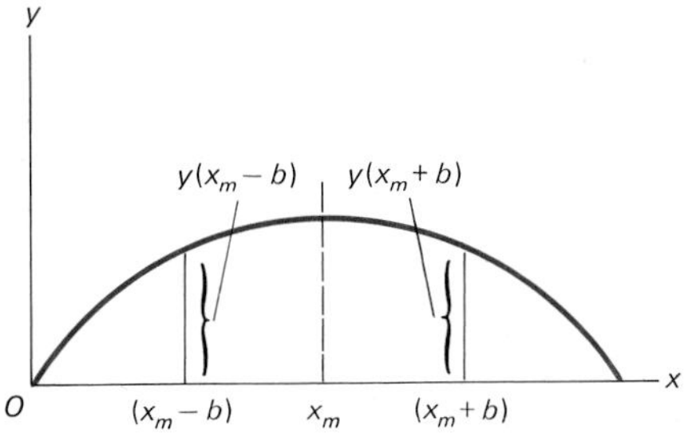

Figure 5.6

A parabolic trajectory is symmetric about its maximum at $x = x_m$. Thus y at a point a distance b to the right of x_m equals y at a point the same distance to the left of x_m.

2 Each trajectory exhibits a related symmetry. If we substitute $(90° - \theta)$ for θ in Eq. 5.17, the range remains the same. This is because

$$\sin(90° - \theta) = \cos\theta$$

$$\cos(90° - \theta) = \sin\theta$$

and so

$$\sin(90° - \theta)\cos(90° - \theta) = \sin\theta\cos\theta$$

Thus, the product $\sin\theta\cos\theta$ in Eq. 5.17 remains the same when θ is replaced by $(90° - \theta.)$ We can also say that the range is the same for positive and negative values of ϕ where $\theta = 45° \pm \phi$. The angle ϕ is measured relative to the 45° line. Thus the range for $\phi = +10°$ $(\theta = 55°)$ equals the range for $\phi = -10°$ $(\theta = 35°)$. The computer-generated trajectories exhibited in Figure 5.5 illustrate both of these symmetries.

3 Each trajectory exhibits time symmetries. It takes the same time for the projectile to rise to its maximum height after launch as it does to fall from its maximum height back to the launch elevation $(y = 0)$. Equation 5.15 gives the rise time. To prove that the total time of flight is twice the rise time, we set $y = 0$ in Eq. 5.11 and solve for t. We obtain

$$t\left(v_{yo} + \frac{1}{2}a_y t\right) = 0$$

The two solutions are $t = 0$ (the time of the launch), and $t = -(2v_{yo}/a_y)$ (the second time it reaches $y = 0$). We see that the time it takes the projectile to land is twice the time t_m needed to reach the maximum height, t_m.

4 Another kind of time symmetry exists in this parabolic motion. Suppose that we replace t by $-t$ everywhere that time appears in the original equations. This implies that $\Delta t \to -\Delta t$ and the velocities change sign but the accelerations do not. We also reverse the signs of the original velocity components. The result of reversing the sign of time and of the initial velocity is that our equations describing the motion do *not* change. Schematically, if we let

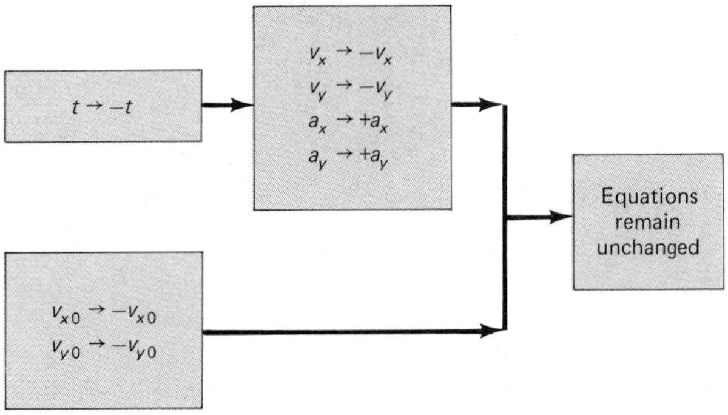

Equations that remain unchanged if the signs are changed in this way are said to exhibit **time reversal symmetry.** In the absence of friction this symmetry appears in all mechanics. It means that zero-friction mechanical experiments cannot tell us whether time is running forward or backward. Without friction Newtonian mechanics is *invariant* under time reversal.

Example 2 Initial Velocity of a Punt

A football is punted 51 m and is airborne for 4.5 s. Let us determine (a) the magnitude of the initial velocity if $\theta = 63°$, and (b) how far the football would travel horizontally if launched with the same initial velocity but with $\theta = 45°$.

(a) Solving Eq. 5.15 for v_o^2, with $a_y = -9.80$ m/s^2, we have

$$v_o^2 = \frac{51 \times 9.80}{2 \sin 63° \cos 63°} = 618 \text{ m}^2/\text{s}^2$$

or

$$v_o = 24.9 \text{ m/s}$$

for the magnitude of the initial velocity.

(b) Using Eq. 5.17, this time with $\theta = 45°$, and solving for the unknown, R, we obtain

$$R = \frac{-2 \times 618 \sin 45° \cos 45°}{-9.80}$$

The horizontal range is now

$$R = 63.1 \text{ m}$$

The horizontal distance traveled by a projectile equals the product of the projectile's horizontal velocity and time of flight. The horizontal velocity depends on the launch speed and angle, but not on the acceleration of gravity. Any change in range arising because the projectile is launched on a planet whose value for g differs from that on earth must be due to a change in the time of flight, which in turn is dependent on the acceleration of gravity.

Example 3 A Golf Drive on the Moon

In a vacuum on the earth a golf ball launched at 8° with a speed of 60 m/s (134 mi/hr) would have a range of 101 m and be airborne for 1.70 s. What range and time of flight would such a golf ball have on the moon? Take $a_y = -1.62$ m/s^2.

From Eq. 5.14 we have

$$R = \frac{-2 \times (60 \text{ m/s})^2 \sin 8° \cos 8°}{-1.62 \text{ m/s}^2}$$

Since $\sin 8° = 0.1392$ and $\cos 8° = 0.9903$,

$$R = 612 \text{ m}$$

The time of flight is equal to twice the rise time because the trajectory is symmetric with respect to time. To obtain the rise time Eq. 5.9 may be used with $v_{yo} = v_o \sin \theta$.

$$v_y = v_o \sin \theta + a_y t$$

Since the vertical velocity, v_y, is equal to zero at the highest point of the trajectory, we solve for the rise time, $t = -v_o \sin \theta / a_y$. Thus the time of flight is given by

$$\text{time of flight} = -\frac{2v_o \sin \theta}{a_y} = -\frac{2 \times 60 \text{ m/s} (\sin 8°)}{-1.62 \text{ m/s}^2} = 10.3 \text{ s}$$

Comparing moon and earth data shows that the ratios of the (1) accelerations of gravity, (2) times of flight, and (3) ranges are the same. The increase in time of flight on the moon due to the decrease in gravitational acceleration exactly accounts for the increase in range.

The maximum range of a projectile is realized with a launch angle of 45° only if the launch and landing elevations are the same. An angle greater than 45° is optimal if the landing point is above the launch point, while an angle less than 45° is optimal if the landing point is below the launch point. The following example illustrates this point.

Example 4 Optimal Launch Angle in the Shot Put

In the shot-put event, an athlete releases the shot (a 7.2 kilogram metal ball) 2 m above ground level. The athlete twice puts the shot a horizontal distance of 22 m. In one case the launch angle is 45° and in the other it is 42°. Which put requires the larger initial velocity, v_o?

To answer this question we use the trajectory relation, Eq. 5.14, to obtain the initial shot velocity in each case. Solving for v_o^2, we have

$$v_o^2 = \frac{a_y x^2}{2 \cos^2 \theta (y - x \tan \theta)}$$

Choosing the point of release as the origin, we set $y = -2$ m, $x = +22$ m, and $a_y = -9.80$ m/s^2.

For $\theta = 42°$

$$v_o^2 = \frac{(-9.80) \text{ m/s}^2 (22)^2 \text{ m}^2}{2 \cos^2 42° (-2 - 22 \tan 42°) \text{ m}}$$

$$v_o^2 = 197 \text{ m}^2/\text{s}^2$$

$$v_o = 14.0 \text{ m/s}$$

For $\theta = 45°$

$$v_o^2 = \frac{(-9.80) \text{ m/s}^2 (22)^2 \text{ m}^2}{2 \cos^2 45° (-2 - 22 \tan 45°) \text{ m}}$$

$$v_o^2 = 198 \text{ m}^2/\text{s}^2$$

$$v_o = 14.1 \text{ m/s}$$

We see that in order to travel the same distance, the shot must have a larger initial velocity at $\theta = 45°$ than at $\theta = 42°$. Hence 45° is not the optimal launch angle when the initial and final heights are different.

Questions

1. If a particle moves in a vertical plane with both its horizontal and vertical acceleration components constant (not zero), its path is parabolic. How could you orient the coordinate axes to prove this assertion?
2. Sketch the family of trajectories having the same time of flight but various ranges. What can be said about (1) the vertical component of initial velocity? (2) The height to which the projectile rises? (3) The horizontal component of initial velocity?
3. Using the trajectories for $\theta = 30°$ and $60°$, argue that with fixed initial speed $\theta = 60°$ corresponds to the greater range if the projectile lands above its launch point, where $\theta = 30°$ corresponds to the greater range if the projectile lands below its launch point.
4. For negative launch angles ($-90° < \theta < 0°$) Eq. 5.17 indicates a negative range. Explain.

5.3 Acceleration in Circular Motion

Acceleration arises in particle motion if the velocity vector changes in either magnitude or direction. Both kinds of change are important.

In our studies of linear motion in Section 4.6 we saw that acceleration occurred because the magnitude of the velocity changed with time. Another special type of acceleration occurs in uniform circular motion.

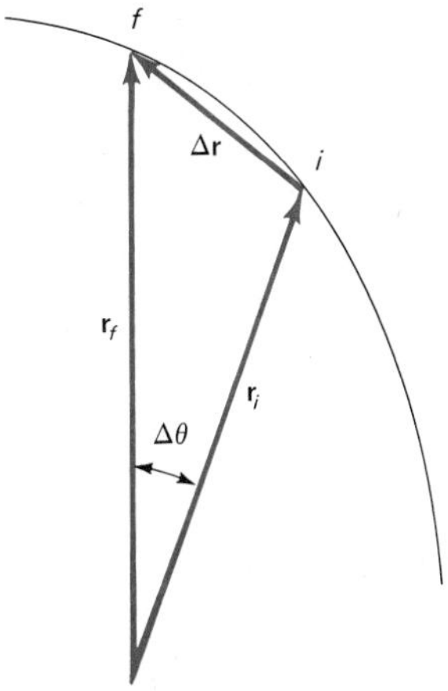

Figure 5.7

The vector $\Delta\mathbf{r} = \mathbf{r}_f - \mathbf{r}_i$ is from an initial point on the particle trajectory to a final point on the particle trajectory. For vanishingly small time intervals ($\Delta t \rightarrow 0$) the vector $\Delta\mathbf{r}$ becomes tangent to the trajectory, and this implies that $\mathbf{v}$ is tangent to the trajectory.

With circular motion the velocity vector is always tangent to the circular path. **Uniform circular motion** is circular motion at constant speed. Thus the velocity vector changes direction continuously while its magnitude remains constant. Planets in circular orbits about a star, satellites in circular orbits around a planet, and charged particles in circular orbits in accelerator storage rings are examples of uniform circular motion.

When a particle moves in a straight line, its instantaneous velocity vector, $\mathbf{v}$, is directed along the line of motion—tangent to the path. To prove that $\mathbf{v}$ is tangent to the path for motion along a curved trajectory, consider the average velocity vector, $\overline{\mathbf{v}}$, defined by (see Section 4.2)

$$\overline{\mathbf{v}} = \frac{\Delta\mathbf{r}}{\Delta t}$$

Thus the vector $\overline{\mathbf{v}}$ is parallel to the vector $\Delta\mathbf{r}$. The direction of $\Delta\mathbf{r}$, and hence of $\overline{\mathbf{v}}$, is shown in Figure 5.7. In the limit as $\Delta t \rightarrow 0$, $\Delta\theta \rightarrow 0$ and $\overline{\mathbf{v}} \rightarrow \mathbf{v}$. Accordingly, $\mathbf{v}$ is tangent to the path. In circular motion $\mathbf{v}$ is also perpendicular to $\mathbf{r}$.

Acceleration in Uniform Circular Motion

Consider a particle moving with constant speed v in a circle of radius r. In Figure 5.8a we show such a particle (1) at time t_i with position $\mathbf{r}_i$ and velocity $\mathbf{v}_i$, and (2) at time t_f with position $\mathbf{r}_f$ and velocity $\mathbf{v}_f$. Using the velocity vectors $-\mathbf{v}_i$, $\mathbf{v}_f$, and $\Delta\mathbf{v} = \mathbf{v}_f - \mathbf{v}_i$, we form the velocity triangle shown in Figure 5.8b. Because $\mathbf{r}_i$ and $\mathbf{v}_i$ are perpendicular and $\mathbf{r}_f$ and $\mathbf{v}_f$ are perpendicular, the apex angle of the velocity triangle and the angle between $\mathbf{r}_i$ and $\mathbf{r}_f$ are both $\Delta\theta$. During $\Delta t = t_f - t_i$ the equation $\Delta\mathbf{v} = \overline{\mathbf{a}}\,\Delta t$ (see Section 4.4) implies that the average acceleration and the velocity

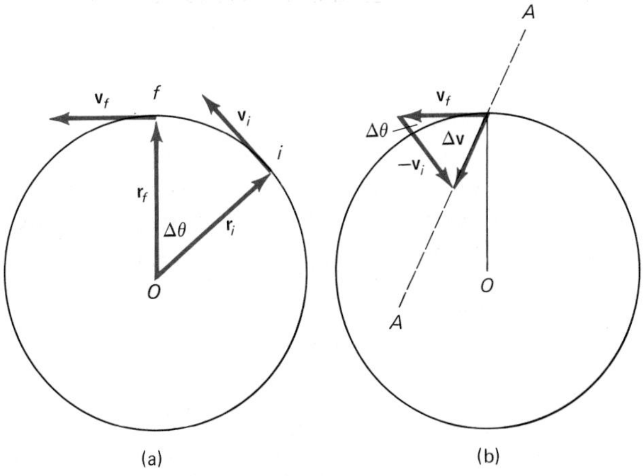

Figure 5.8

A particle moves with constant speed in a circular path. The positions, $\mathbf{r}_i$ and $\mathbf{r}_f$, and velocities, $\mathbf{v}_i$ and $\mathbf{v}_f$, are drawn in (a) at the times t_i and t_f. The equation $\Delta\mathbf{v} = \mathbf{v}_f - \mathbf{v}_i$, which gives the change in velocity, corresponds to the triangle relationship illustrated in (b). The direction of $\Delta\mathbf{v}$, indicated by the line AA, does not pass through the center, O, of the circle for finite $\Delta\mathbf{v}$. However, as $\Delta t \rightarrow 0$ both $\Delta\mathbf{v} \rightarrow 0$ and $\Delta\theta \rightarrow 0$, and the line AA passes through the point O. Thus the acceleration vector, which is parallel to $\Delta\mathbf{v}$, also passes through the center.

Figure 5.9

In uniform circular motion the magnitudes, r and v, of the position and velocity vectors are constant. Thus each of the equations $\Delta \mathbf{r} = \mathbf{r}_f - \mathbf{r}_i$ and $\Delta \mathbf{v} = \mathbf{v}_f - \mathbf{v}_i$ may be interpreted geometrically as an isosceles triangle. The space triangle is shown in (a) and the velocity triangle in (b). Because the apex angle, $\Delta\theta$, is the same, the two triangles are similar.

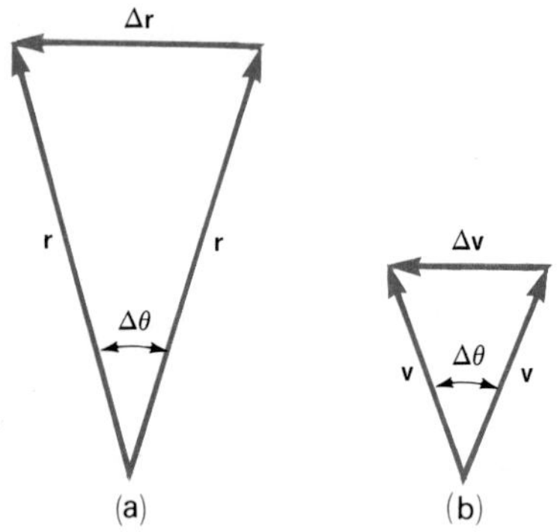

change $\Delta \mathbf{v}$ are parallel. Thus the average acceleration is directed along the broken line AA. In the limit as $\Delta t \to 0$, $\Delta\theta \to 0$, and two other things occur. The average acceleration becomes the instantaneous acceleration (see Section 4.4) and the line AA passes through the point O. Hence the instantaneous acceleration vector, $\mathbf{a}$, is directed along a radius toward the center of rotation. The vector $\mathbf{a}$ is called either a **radial** or a *centripetal* acceleration. [Centripetal, from the Latin *petere* ("to seek"), means "center seeking."]

To obtain an expression for the magnitude of $\mathbf{a}$, we let r and v denote the magnitudes of the position and velocity vectors, respectively. Notice that the space and velocity triangles in Figure 5.9 are isosceles and similar. Therefore,

$$\frac{\Delta v}{\Delta r} = \frac{v}{r}$$

Rearranging and dividing by Δt, we get

$$\frac{\Delta v}{\Delta t} = \frac{v}{r}\frac{\Delta r}{\Delta t}$$

In the limit as $\Delta t \to 0$ the ratio $\Delta v/\Delta t$ becomes a, and the ratio $\Delta r/\Delta t$ becomes v, giving

$$a = \frac{v^2}{r} \qquad (5.18)$$

for the magnitude of the **centripetal acceleration.**

We can also express a in terms of r and T where T is the time for one orbit (called the period). Using speed = distance/time we may write

$$\text{orbital speed} = \frac{\text{circumference}}{\text{time}}$$

or

$$v = \frac{2\pi r}{T} \qquad (5.19)$$

Substituting Eq. 5.19 into Eq. 5.18 we obtain

$$a = \frac{4\pi^2 r}{T^2} \qquad (5.20)$$

Example 5 **Centripetal Acceleration at the Equator**

Let us determine the centripetal acceleration of a point on the equator due to the rotation of the earth on its axis. One complete rotation requires one sidereal day, which equals 86,164 s. The radius of the earth is $R_E = 6.37 \times 10^6$ m.

Using Eq. 5.18 we can calculate directly

$$a = \frac{4\pi^2 \times 6.37 \times 10^6}{(86,200)^2}$$

$$= 3.38 \times 10^{-2} \text{ m/s}^2$$

This is roughly 0.3% of the standard value for g (9.80 m/s^2). We will see in Section 7.7 that the observed acceleration of freely falling bodies at the equator is less than we would expect without centripetal acceleration.

Many objects (celestial, terrestrial, and microscopic) move in circular paths at constant speed. Because the period and radius of the orbit are relatively easy to measure, Eq. 5.19 provides a direct method of obtaining the orbital speed of the particle.

Example 6 **Radial Acceleration of an Astronaut**

An astronaut in a circular orbit (Figure 5.10) 161 km above the earth's surface requires 5.26×10^3 s to complete one revolution. What is his orbital speed and centripetal acceleration?

An altitude of 161 km corresponds to $r = 6.53 \times 10^6$ m. Using Eq. 5.19 we obtain for the orbital speed

$$v = \frac{2\pi r}{T}$$

$$= \frac{2\pi \times 6.53 \times 10^6}{5.26 \times 10^3}$$

$$= 7.80 \times 10^3 \text{ m/s}$$

or about 17,500 mi/hr. Using Eq. 5.18 we obtain for the astronaut's centripetal acceleration

$$a = \frac{v^2}{r}$$

$$= \frac{(7.80 \times 10^3)^2}{6.53 \times 10^6}$$

$$= 9.32 \text{ m/s}^2$$

Figure 5.10

This Hungarian stamp honors astronauts McDivitt and White. The scene depicted is White's pioneering space walk.

or about 95% of the acceleration of gravity at the earth's surface.

In solving a problem, the choice of reference frame often determines the complexity of the solution. In Chapter 3 the orientation and origin of a reference frame were chosen to simplify the solution of statics problems. In Section 4.7 we examined the importance of changing reference frames.

In the following example we consider an object that is observed to be executing a combination of linear and circular motion in one reference frame. The same object, however, is observed to be in circular motion in a frame of reference at rest relative to the center of rotation. The description of motion is simpler in the second reference frame.

Example 7 **Radial Acceleration of a Tire**

What would be the radial acceleration of a point on the circumference of a bicycle tire 26 inches (0.660 m) in diameter when the bicycle moves at 13.4 m/s (30 mi/hr)? Relative to the bicycle rider, the road—and hence the point on the tire in contact with the road—moves backward at 13.4 m/s. This is the rotational speed of the point under consideration. Thus, using Eq. 5.18, we obtain

$$a = \frac{v^2}{r} = \frac{(13.4)^2}{0.330} = 544 \text{ m/s}^2$$

which is more than 55 times as large as g!

Tangential Acceleration in Circular Motion

If the speed varies in circular motion, there are changes in both the *direction* and the *magnitude* of the velocity. The particle travels on a circular path, but speeds up and slows down as it moves. The simple pendulum is an example of this type of motion.

The acceleration has components along radial and tangential directions. As before, the radial acceleration is given by Eq. 5.18:

$$a_R = \frac{v^2}{r} \tag{5.18}$$

In general, the **tangential acceleration** is determined by the rate at which the tangential component of velocity changes.

$$a_T = \lim_{\Delta t \to 0} \frac{\Delta v_T}{\Delta t} = \frac{dv_T}{dt}$$

With circular motion the velocity has no radial component, so a_T is determined by the rate at which the speed v changes.

$$a_T = \lim_{\Delta t \to 0} \frac{\Delta v}{\Delta t} = \frac{dv}{dt} \tag{5.21}$$

The sign of a_T indicates whether v is increasing or decreasing. If $a_T > 0$, then v is increasing, whereas if $a_T < 0$, then v is decreasing. In uniform circular motion, v is constant and Eq. 5.21 yields $a_T = 0$.

The magnitude and direction of the total acceleration are obtained from $\mathbf{a}_R$ and $\mathbf{a}_T$ as suggested in Figure 5.11. The vector **a** has the magnitude

$$a = \sqrt{a_T^2 + a_R^2} \tag{5.22}$$

The direction of **a** is measured relative to the center of rotation, as illustrated in Figure 5.11, and is determined by θ. The angle θ is given by

$$\tan \theta = \frac{a_T}{a_R} \tag{5.23}$$

Figure 5.11

The total acceleration, **a**, of a particle in circular motion has a tangential component, $\mathbf{a}_T$, and a radial or centripetal component, $\mathbf{a}_R$.

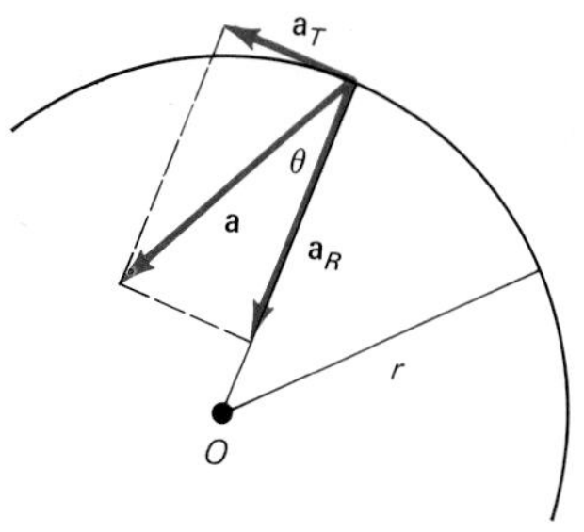

Example 8 **Acceleration of an Astronaut**

An astronaut in a circular orbit 161 km above the surface of the earth has $a_R = 9.33$ m/s². (See Example 6.) In order to leave earth's orbit and head for the moon he starts the engines, resulting in $a_T = 24.5$ m/s². What is the astronaut's total acceleration?

Using Eq. 5.22 we calculate

$$a = \sqrt{a_T^2 + a_R^2} = \sqrt{(24.5)^2 + (9.33)^2} = 26.2 \text{ m/s}^2$$

for the magnitude of the total acceleration. Using Eq. 5.23 we obtain

$$\tan \theta = \frac{a_T}{a_R} = \frac{24.5}{9.33} = 2.62$$

or $\theta = 69.1°$ for the direction of acceleration.

Objects moving in a circular path in a vertical plane undergo both tangential and radial accelerations. Three familiar examples of objects that undergo both tangential and radial accelerations are stunt planes, carnival loop-the-loop rides, and yo-yos.

Example 9 Acceleration of a Yo-yo

A yo-yo moves in a vertical circle the radius of which is 1 m. When the yo-yo's velocity is 12.6 m/s directed straight down, what is its total acceleration?

The tangential acceleration equals 9.80 m/s^2, the acceleration of gravity.

To obtain the radial acceleration we use Eq. 5.18.

$$a_R = \frac{v^2}{R} = \frac{(12.6)^2}{1} = 159 \text{ m/s}^2$$

From Eq. 5.22 we obtain

$$a = \sqrt{a_T^2 + a_R^2} = \sqrt{(9.80)^2 + (159)^2} = 159 \text{ m/s}^2$$

for the magnitude of the total acceleration. To three significant figures $a = a_R$.

Using Eq. 5.23 we obtain

$$\tan\theta = \frac{a_T}{a_R} = \frac{9.80}{159} = 0.0616$$

or $\theta = 3.53°$ for the direction of **a**. The total acceleration is at an angle of 3.53° down from the radius, as shown in Figure 5.12.

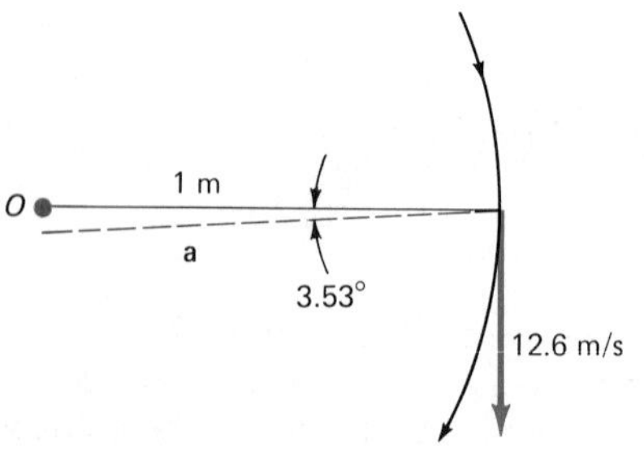

Figure 5.12

The total acceleration of a yo-yo instantaneously moving downward at 12.6 m/s has the components $a_T = 9.80$ m/s^2 and $a_R = 159$ m/s^2. The acceleration vector is angled 3.53° below the horizontal line to the center of rotation.

Questions

5. Could a particle in uniform circular motion have its acceleration vector directed radially out from the center? Tangent to the path? Explain.
6. If the acceleration of a particle is centripetal for uniform circular motion, why doesn't the particle get closer to the center?
7. Consider the motion of a child in a playground swing. Discuss the child's acceleration as motion begins. Indicate points in the trajectory where the radial and/or tangential accelerations are zero or at a maximum.
8. A particle moves in such a way that its centripetal acceleration is always zero. Describe its possible trajectories.
9. A particle moves in such a way that its tangential acceleration is always zero. Describe its possible trajectories.

Summary

The component equations of motion in a plane for constant acceleration are

$$x = v_{xo}t \qquad (5.7)$$

$$y = v_{yo}t + \frac{1}{2}a_y t^2 \qquad (5.11)$$

where $a_x = 0$ and $a_y =$ constant. Elimination of the parameter t yields the parabolic trajectory equation

$$y = x\tan\theta + \left(\frac{a_y}{2v_o^2\cos^2\theta}x^2\right) \qquad (5.14)$$

in terms of the launch velocity magnitude, v_o, and launch angle, θ.

In circular motion the acceleration, **a**, can be resolved into a radial part, a_R, and a tangential part, a_T. The radial or centripetal acceleration is directed toward the center of the circle, and has the magnitude

$$a_R = \frac{v^2}{r} \qquad (5.18)$$

For uniform circular motion a_R can also be written

$$a_R = \frac{4\pi^2 r}{T^2} \qquad (5.20)$$

where T, the period or time for one revolution, is given by

$$T = \frac{2\pi r}{v} \qquad (5.19)$$

The tangential component of the acceleration is the rate at which the particle speed changes.

$$a_T = \lim_{\Delta t \to 0}\frac{\Delta v}{\Delta t} = \frac{dv}{dt} \qquad (5.21)$$

Problems

Section 5.1 Motion in a Plane and the Principle of Superposition

1. A particle moves in a plane with the velocity components

$$v_x = 1 \text{ m/s}$$

$$v_y = -1 \text{ m/s}$$

It passes the origin ($x = 0$, $y = 0$) at $t = 0$. Sketch the particle's position versus time over the interval $t = 0$ to $t = 3$ s.

2. A marble rolls across a horizontal table at the constant speed 2.3 m/s. When it reaches the edge of the table the marble falls through a vertical distance of 1 m to the floor. (See Figure 1.) (a) For what length of time does the marble fall? (b) What is the horizontal distance between the table's edge and the impact point of the marble on the floor?

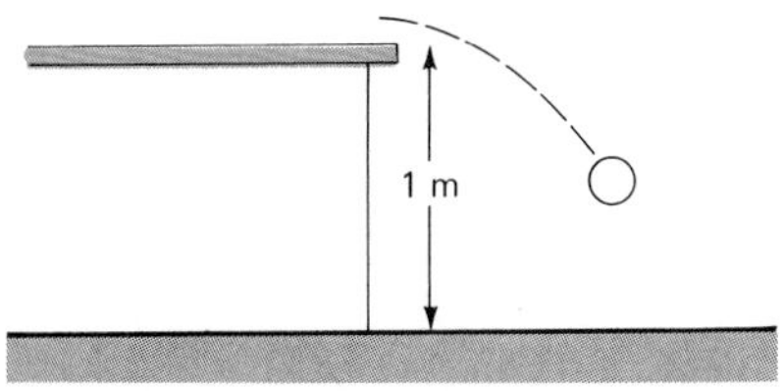

Figure 1

Section 5.2 Motion in a Plane with Constant Acceleration

3. A particle moves with the velocity components $v_x = +4$ and $v_y = -6t + 4$ with t in seconds and v_x and v_y in meters per second. Calculate the (a) magnitude $v = (v_x^2 + v_y^2)^{1/2}$ and (b) direction ($\tan\theta = v_y/v_x$) of the velocity vector at $t = 2$ s.

4. The position of a particle is given by $x = 2 + 3t$ and $y = 2 + 3t - 5t^2$ with x and y in meters and t in seconds. How far from the origin is the particle at (a) $t = 0$; (b) $t = 2$ s?

5. On a sheet of graph paper sketch the path of a particle whose position is given by (a) $x = \cos t$, $y = \sin t$; (b) $x = 4t$, $y = -5t^2$. In each case describe the trajectory.

6. Complete the algebra required to derive Eq. 5.14.

7. Show that the maximum height of a projectile is given by

$$y_{max} = \frac{v_o^2}{2g}\sin^2\theta$$

8. Show, using Eq. 5.15, that R is a maximum for $\theta = 45°$ by plotting R vs. θ for $0° \leq \theta \leq 90°$. Remember that v_o and a_y are constant.

9. What initial speed would a football need in order to travel 60 m horizontally if it is kicked at an angle of 45°?

10. A coin slides across a desk and falls to the floor 1.1 m below. (a) For what length of time does it fall? (b) If the coin strikes the floor 1.1 m beyond the edge of the desk, what was its horizontal velocity when it left the desk? (c) What is the coin's horizontal velocity just before it hits the floor?

11. Determine the range and time of flight of the golf drive in Example 3 if the ball is launched with the same speed (60 m/s) at 23° on the moon.

12. Following the procedure of Example 4, calculate v_o for every 0.25° of angle between $\theta = 40°$ and $\theta = 44°$. Plot v_o^2 versus θ, and use the graph to determine the optimal launch angle.

13. In Example 1 a baseball thrown horizontally at 40 m/s was shown to fall 0.886 m vertically while traveling 17 m horizontally (Figure 2). For the same launch speed and horizontal distance, what launch angle upward is required for the ball to arrive at its launch elevation?

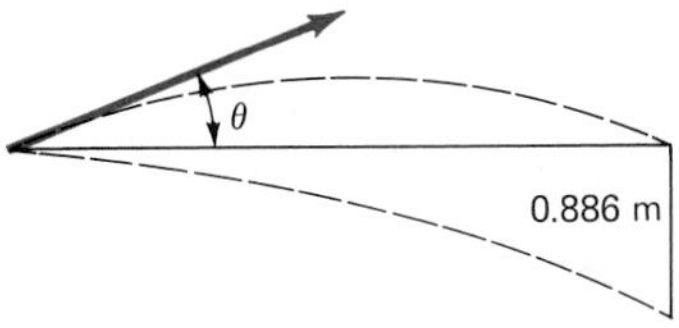

Figure 2

14. A sling is used to hurl a rock a horizontal distance of 80 m (Figure 3). The initial speed is 38 m/s. Find the two possible launch angles.

15. A football punter wants to have the football travel 60 m horizontally and stay airborne for 5 s. What combination of initial speed and launch angle will yield this result?

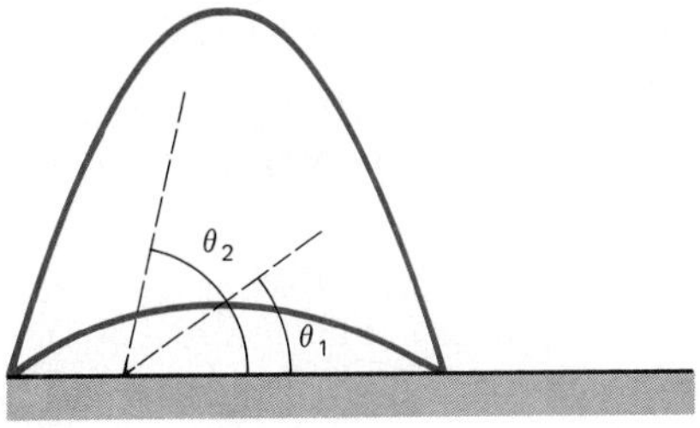

Figure 3

16. A rock is thrown through a window that is 15 m higher than the launch point and 15 m away horizontally (Figure 4). The rock is moving horizontally when it hits the window. Determine the initial speed and launch angle of the rock.

17. A batted ball is launched at 45° and just clears a fence 15 m higher than the launch and 20 m away (Figure 5). Determine (a) the launch speed, and (b) time required for the ball to reach the fence.

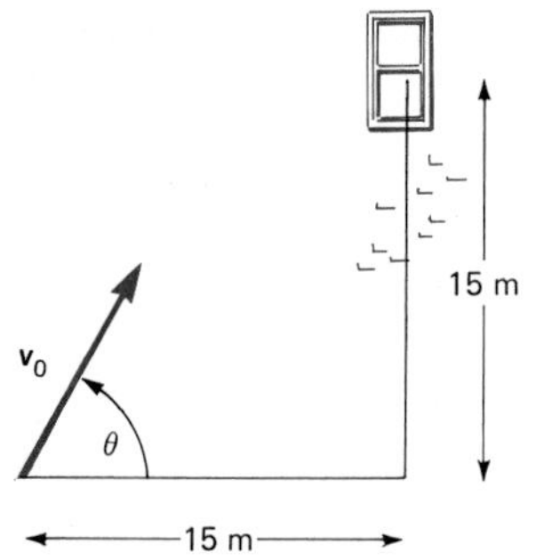

Figure 4

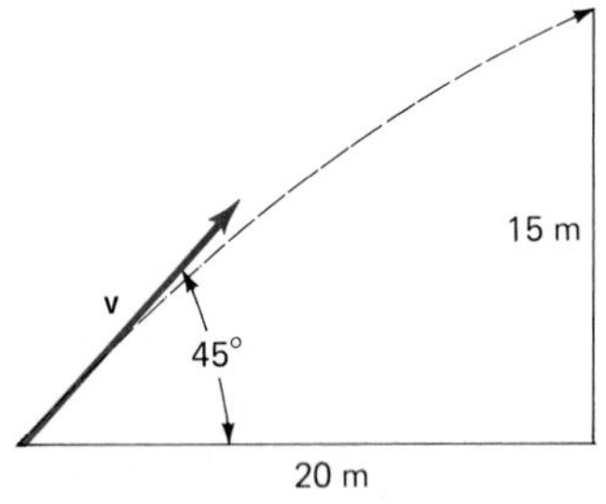

Figure 5

18. (a) Derive Eq. 5.17 for the horizontal range by requiring that $x = R$ when $y = 0$. (b) For $y = 0$, $x = R$ is one of two solutions. Interpret the second solution for $y = 0$.

19. A particle moves in a parabolic path as a result of a constant acceleration g downward. As a demonstration of time symmetry, prove that $y(t_m + t_o) = y(t_m - t_o)$ where t_m is the time of maximum altitude (Eq. 5.15) and t_o is a constant.

20. Show that the range of a projectile is the same for $\theta = 45° + \phi$ as for $\theta = 45° - \phi$ provided that $\phi < 45°$.

Section 5.3 Acceleration in Circular Motion

21. For what length of day (sidereal) would the radial acceleration at the equator have the magnitude $9.80\ \text{m/s}^2$?

22. A drop of water rides on the edge of a moving $33\frac{1}{3}$-rpm record. The diameter of the record is 30 cm. Determine the radial acceleration of the drop.

23. A student attaches a small object to a string, and then rotates the object in a circular orbit 54 cm in diameter. Assuming constant speed, calculate the object's radial acceleration for an orbital time of 1.2 s.

24. A passenger in an automobile travels around a 90° arc of a circle of radius 20 m when the car turns a corner. At a speed of 60 mi/hr what is her radial acceleration?

25. The Syncom satellite, at a distance of 4.21×10^7 m from the center of the earth, completes one orbit in one sidereal day (Figure 6). Determine its radial acceleration.

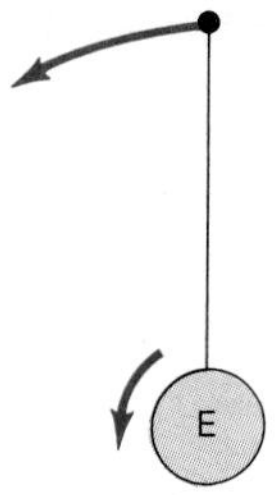

Figure 6

26. The velocity of the bob on a pendulum of length 10 m (Figure 7) is given by $v = v_o \cos(2\pi t/T)$ where $v_o = 1.00$ m/s and $T = 2\pi$ s. Determine the radial acceleration at (a) $t = 0$ s, (b) $t = \pi/4$ s, and (c) $t = \pi/2$ s.

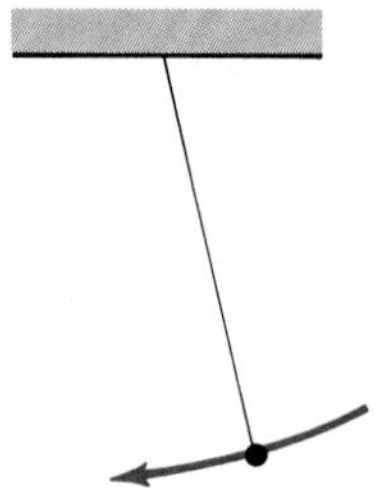
Figure 7

27. Using the limit form in Eq. 5.21, numerically obtain the tangential acceleration of the bob in Problem 26 for the three times specified. Do this by evaluating the ratio $\Delta v/\Delta t$ for successively smaller time intervals $\Delta t = 0.1,\ 0.01,\ 0.001$, and so on, with your calculator.

28. Using Eqs. 5.22 and 5.23 determine the magnitude and direction of the acceleration vector for the pendulum bob of Problems 26 and 27 for the three times specified.

29. Prove that Eq. 5.18 is dimensionally correct.

30. An astronaut encircles the earth at a low-altitude orbit (take $r = 6.37 \times 10^6$ m). (a) If his centripetal acceleration is g ($9.80\ \text{m/s}^2$), calculate his linear speed v. (b) Calculate the period T.

chapter 6

NEWTON'S LAWS

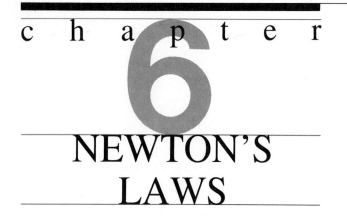

Preview

In Chapters 4 and 5 we developed kinematics—the description of motion. In this chapter we examine the causes of motion. The motion of a particle is affected by its environment and by its own properties. To account for the influence of the environment on a particle we introduce the concept of force; and to account for the response of the particle to applied forces, we refine and develop the concept of inertial mass.

Newton's first law defines the zero-force environment. In this environment a particle experiences no influence on its motion. A particle's state of being unaccelerated persists in a zero-force environment.

Newton's second law relates the inertial mass and the acceleration of a particle to the net force it experiences. Inertial mass measures the resistance of a body to an acceleration. Inertial mass is an intrinsic particle property, and it is independent of the environment. The definition of mass was discussed in Chapter 1. In this chapter we continue to develop the concept of mass by explaining how accelerations and the standard kilogram are used to assign numbers to the masses of unknown particles. We also consider quantitatively the dependence of acceleration on particle mass.

The force concept is completed by Newton's third law, which describes in part the mutual interactions of two particles. The third law stipulates that forces in nature must occur in action–reaction pairs. The treatment of internal forces is simplified as a result.

The role of inertial frames of reference is reexamined in this chapter. You will see that Newton's laws hold in inertial frames but not in noninertial frames.

The calculation of the force exerted on a particle is provided for with laws of force. Several force laws are discussed. The law of gravity is given special treatment because of its special importance. In addition, we devote considerable attention to the relationship between weight and mass. Several mechanical laws dealing with the forces exerted by solids and fluids are presented.

6.1 The Program of Particle Mechanics

The program of particle kinematics was outlined in Section 4.5. There we noted that, given the acceleration of a particle, the future positions and velocities of that particle can be determined by using the equations of motion. However, the determination of the acceleration of a particle in terms of its environment lies not in kinematics but in the larger realm of mechanics. To provide perspective we first describe the program of particle mechanics qualitatively.

In part, the acceleration of a particle is determined by its inertial mass. For a given applied force the degree to which a body resists being given an acceleration is determined by its inertial mass. Drag racing, for instance, is an acceleration competition for automobiles. Thus, for a given applied force, the smaller the mass of the car the greater its acceleration is.

The acceleration of a particle also depends on the particle's environment. Force is the term used to describe the influence of the environment on a particle. For example, we know that a baseball changes its velocity abruptly when it is hit by a bat. We account for this behavior of the particle (the baseball) by saying that the environment (the bat) exerted a force on it. Force is measured or defined by the acceleration it imparts to a known mass.

The concept of force enters mechanics in two ways. First, through Newton's second law (Section 6.4), the acceleration of a standard kilogram (Section 1.5) measures the **net force** acting on it. The net force is the vector sum of all forces acting on the object being considered. When an apple hangs on a tree its zero acceleration is the result of zero net force. As it drops to the ground the apple's acceleration is caused by a net force equal to the attractive gravitational force of the earth on the apple. The acceleration of the apple equals the net force per unit mass that it experiences. But the acceleration cannot distinguish between forces that are the same in magnitude and direction but different in nature. At this point, we encounter the second way in which force enters mechanics.

Forces are associated with particular environments. For example, frictional forces occur if bodies touch; elastic forces are exerted by springs; and viscous forces arise in fluids. Is it possible, then, to relate the force exerted on a particle to the properties of the particle and those of its environment? Laws of force make it possible. Physicists have put much effort into discovering force laws, refining our understanding of them, and investigating possible relations among them. Newton himself discovered one of these force laws—the law of gravity. (The four basic forces were listed in Section 3.1.)

Given a particle in a specified environment, how can we obtain its acceleration? The flowchart in Figure 6.1 outlines the process used in Newtonian mechanics to obtain accelerations. First, force laws are used to calculate the net force acting on the particle. Newton's second law then relates this net force to the particle's mass and acceleration. Thus, the particle acceleration is obtained in terms of its mass and the properties of its environment. As we saw in Chapter 4, knowledge of a particle's acceleration can be used to deduce future values of its position and velocity.

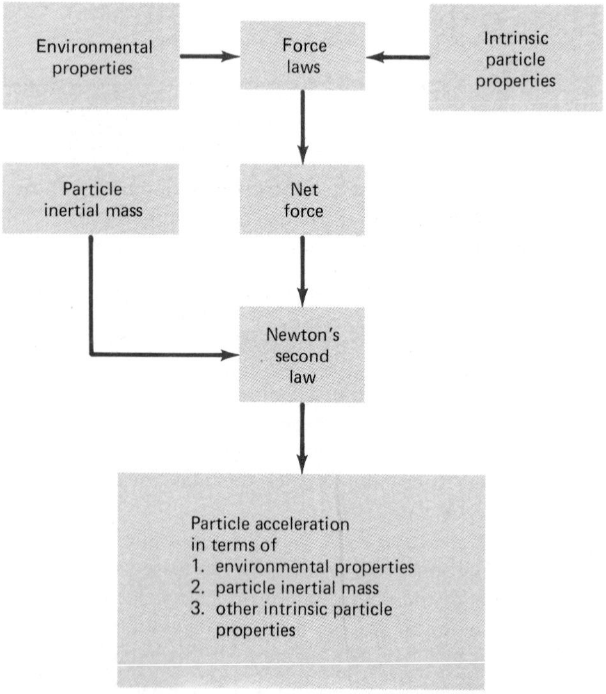

Figure 6.1

The logical structure of Newtonian mechanics is outlined in this flowchart. Force laws, such as the law of gravity, express a particular force in terms of properties of the particle and of its environment. A particle's acceleration is then explained in terms of the net force acting and the inertial mass of the particle through Newton's second law.

In order to be successful, a scientific theory must meet four criteria. It must agree with experiment, be general in scope, be fruitful, and be relatively simple. The theory of Newtonian mechanics meets all four criteria.

1 **Agreement with experiment.** Newtonian mechanics has been used to calculate the orbits of planets and astronauts in the heavens and to calculate the motions of objects on the earth.

2 **Generality in scope.** Although new theories have been developed for very large systems (general relativity), for very small objects (quantum theory), and for very fast particles (special relativity), these newer theories all reduce to Newtonian mechanics in the limited realm where Newtonian mechanics has been confirmed by experiment.

3 **Fruitfulness.** Several ideas native to Newtonian mechanics have been shown to have a wider range of validity than Newtonian mechanics itself. These ideas include the conservation principles for linear momentum, angular momentum, and energy.

4 **Simplicity.** Newtonian mechanics is routinely taught in introductory physics courses.

We can safely say, then, that Newtonian mechanics is a successful theory.

Question

1. Describe a situation where two particles differing in mass experience different accelerations when placed in the same environment.

6.2 Newton's First Law

How does a particle behave when it is unaffected by its environment? The Greeks might have phrased the question as "What is the 'natural state' of a particle?" Is it at rest? Does it move? Does it accelerate? Clearly these questions must be answered before we can begin to consider the influence of the environment on a particle.

According to the ancient Greek philosopher Aristotle, the natural state of a particle is that of rest. Although incorrect, this idea agrees very well with superficial observations. Because friction is always present in our everyday world, it appears that a force is needed in order to maintain the constant velocity of an object. For example, if a car engine breaks down on a level road, the passengers (or some other external force) must push the car to keep it moving with constant velocity. On the other hand, it seems that a force is not needed to keep an object at rest. The same car at rest on a level road remains at rest when not pushed. This kind of reasoning led Aristotle to believe that the constant velocity of a body had to be *caused*, whereas the state of rest of a body was natural. This false idea is so plausible that it was widely accepted during the two millennia between Aristotle and Galileo.

Galileo discovered the flaw in Aristotle's reasoning by extending the real situation with friction to an idealized situation without friction. He noted that when a smooth hard ball rolled down a nearly frictionless inclined plane the velocity of the ball increased. On the other hand when the same ball rolled *up* the same inclined plane its velocity decreased. By considering the limiting case with *no* friction and *no* inclination of the plane, he argued that once in motion the velocity of the ball would not change. Therefore it should be just as natural for the ball's velocity to remain constant as for the ball to remain at rest (Figure 6.2).

Galileo could imagine a frictionless environment only as a limiting case. Today, however, air tracks, magnetic suspensions, and orbiting satellites are concrete examples of nearly zero-friction situations.

Galileo's demonstration cleared the way for a deeper investigation into the natural state of a body. The state of a particle is determined by its position, velocity, and acceleration—quantities that are defined in relation to an observer. Consider the case of a drag racer. Relative to the driver, the car is at rest and has zero acceleration. Relative to a spectator, the car moves and accelerates. The motion of the car is relative, not absolute. Different choices of a frame of reference result in different descriptions of the motion.

Is there an absolute way to distinguish experimentally between an observer like the driver and an observer like the spectator? To answer this question we must reconsider the special class of observers called inertial observers.

An observer for whom every unaccelerated particle remains unaccelerated is called an **inertial observer.** According to this definition, the drag-race spectator is an inertial observer and the driver is a noninertial observer.

A two-dimensional system that is suitable for illustrating the difference between inertial and noninertial observers is an accurately level frictionless surface. This is closely approximated by a level air-hockey table rigidly attached to the floor. Air pucks at rest remain at rest, and air pucks in motion move with a constant velocity. But if we place

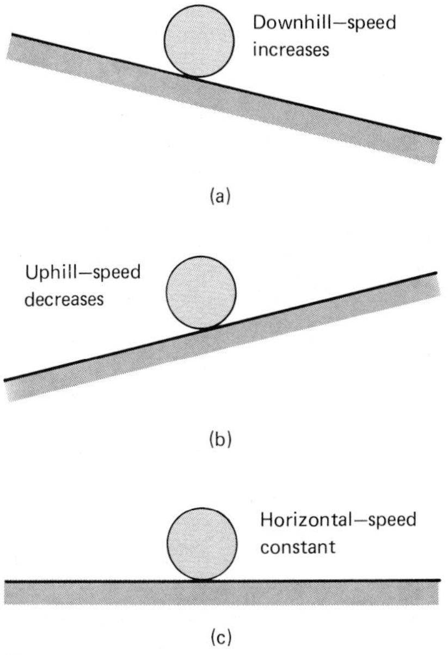

Figure 6.2

The ball moves on a perfectly flat surface without friction. The speed of the ball will (a) increase as the ball moves downhill, (b) decrease as it moves uphill, and (c) remain constant as it moves on a horizontal plane.

the table on a horizontal flatbed truck, the air puck behavior will be different. The puck will accelerate relative to the table every time the truck starts, stops, or turns. Observers can decide experimentally whether they themselves qualify as inertial observers by watching the puck behavior.

Inertial observers motionless relative to each other constitute an inertial frame of reference. We use inertial reference frames because only for inertial observers do Newton's laws have a simple mathematical form. The reference frames of special relativity (Chapter 19) are inertial frames. Indeed, special relativity is called "special" because the theory is limited to inertial reference frames.

An inertial observer can also be defined as one on whom zero force acts. Such observers are labeled inertial because Newton's first law (the law of inertia) holds for them. An accelerating drag-race driver is not an inertial observer because there is a net force that acts on him.

In an inertial reference frame the question "How does a particle behave when it is unaffected by its environment?" is answered by **Newton's first law:**

An unaccelerated body remains unaccelerated unless it is caused to change that state by forces exerted on it by the environment.

Newton's first law allows us to give an operational definition of **zero force.** (We give an operational definition of force later in this chapter.) An unaccelerated particle that remains unaccelerated experiences zero force. Newton's first law makes no distinction between a situation where force is completely absent and a situation where the vector sum of the forces (net force) acting equals zero. Both are zero-force situations.

Questions

2. A horizontal air-hockey table is used as an example of a zero-force environment. In this context, distinguish between zero force and zero net force.
3. Construct examples of objects experiencing zero net force while under the influence of (a) two forces; (b) three forces; (c) four forces.
4. Why is it easy to find examples of objects at rest experiencing zero force, but relatively difficult to find examples of objects moving with constant velocity while experiencing zero force?

6.3 Force

Intuitively we consider force to be a push or a pull. In response to a force, a body will either become deformed or will accelerate. For example, a force will cause a spring to compress or extend, while the force of gravity will cause an apple to accelerate when it falls off a tree.

Our experience is normally limited to the force of gravity, which can act remotely, and to contact forces. The force of gravity cannot be explained in terms of other forces and thus is what we call a fundamental force. We listed the four fundamental forces in Section 3.1. A contact force *can* be explained in terms of other forces and is *not* one of the fundamental forces.

In Section 3.1 we developed a calibration procedure based on a vertical spring and the standard kilogram; that procedure can be used to establish a scale of force. Then in the first two sections of this chapter we discussed qualitatively the roles that force plays in classical mechanics. We made these examinations of force without giving the term a precise meaning. Let's now develop an *operational definition* of force in terms of the acceleration of a standard kilogram.

We begin with a zero-force environment—the horizontal air-hockey table. Here the motion of the standard kilogram—a particle—is unaffected by its surroundings, and the acceleration of the kilogram is zero. An inertial observer can use Newton's first law to ensure that the environment is force free. Next we arrange for a single force to be exerted on the particle. We attach a horizontal spring to the kilogram. Any accelerations the kilogram experiences must be caused by the force exerted on it by the spring. We define the magnitude of this force to be numerically equal to the observed acceleration. The unit of force defined is the newton (N). Thus if $a = 1$ m/s^2, then $F = 1$ newton; if $a = 2$ m/s^2, $F = 2$ newtons; if $a = 3$ m/s^2, $F = 3$ newtons; and so on. We define the direction of the applied force to be in the direction of the observed acceleration of the kilogram.

The spring force is the only external force that influences the acceleration of this standard kilogram. Gravity is present, but is irrelevant because the downward force of gravity and the upward thrust of the air table add up to zero. There is no net vertical force. Therefore, the same experiment could be performed in an environment that is completely free of gravity.

The definition of force is not specially linked to a spring force. Any force causing an acceleration of 1 m/s^2 of a 1-kg mass is 1 newton.

Acceleration measurements of force, such as the one just described, are cumbersome and inconvenient at best. For this reason, secondary standards of force are nearly always used. For instance, in Section 3.1 we described a static method for defining force that involves a vertical spring attached to a standard kilogram. Because the vertical spring procedure depends on the effect of gravity on the kilogram, the experiment is not possible in an environment that is completely free of gravity. Additionally, to make the static procedure for defining force consistent with the definition in terms of acceleration, a force of 9.80 N is assigned to the force exerted on the standard kilogram by gravity at the surface of the earth. In Section 6.7 we will see how this number is chosen, and point out why it would be different in another environment, such as that on the moon.

Questions

5. Compare the utility of force-measuring schemes that use the force of gravity with those that do not.
6. A standard kilogram is dropped near the surface of the earth, and undergoes an acceleration of 9.80 m/s^2. If the same object were moved in a horizontal circle at constant speed, would a centripetal acceleration of 9.80 m/s^2 result from a centripetal force equal to its weight? Explain.

6.4 Newton's Second Law

We have seen that Newton's first law, the law of inertia, governs the behavior of a particle in a zero-force environment. Now let's consider the response of an arbitrary particle when forces are exerted on it. When a net force acts on a particle, the particle is accelerated. The acceleration is specified by **Newton's second law:**

The acceleration of a particle is equal to the ratio of the net force acting on the particle to the inertial mass of the particle.

Newton's second law can be written as an equation

$$\mathbf{a} = \frac{\mathbf{F}}{m} \quad \text{or} \quad \mathbf{F} = m\mathbf{a} \qquad (6.1)$$

where m is the inertial mass of the particle, $\mathbf{a}$ is the acceleration of the particle, and $\mathbf{F}$ is the net force—the vector sum of forces—acting on the particle. In component form, Eq. 6.1 can be written as follows.

$$F_x = ma_x \qquad F_y = ma_y \qquad F_z = ma_z$$

Equation 6.1 is a vector equation. The vector $\mathbf{a}$ is related to the vector $\mathbf{F}$ by m, the scalar proportionality constant.

The SI unit of force—the newton—is defined in terms of fundamental quantities by substituting $m = 1$ kg and $a = 1$ m/s^2 into Eq. 6.1. Thus

$$1 \text{ newton} = 1 \text{kg} \cdot \text{m/s}^2$$

A 1-kg mass accelerating at 1 m/s^2 experiences a net force of 1 newton (1 N = 0.224809 lb; 4.44822 N = 1 lb). The dimensions of force are given by

$$[F] = \frac{M \cdot L}{T^2}$$

Mass is commonly viewed as a measure of the "quantity of matter." The idea that mass measures an object's resistance to acceleration is made quantitative by Newton's second law. According to Eq. 6.1, for a fixed force, the larger the mass of a particle the smaller its acceleration.

The consistency of Eq. 6.1 with our definition of force can be shown by setting $m = 1$ kg. Force is defined in terms of the acceleration of a kilogram. With $m = 1$ kg the force and the particle acceleration have the same numerical value.

Equation 6.1 is also consistent with the law of inertia. In a zero-force environment $\mathbf{F} = \mathbf{0}$. Substituting $\mathbf{F} = \mathbf{0}$ in Eq. 6.1 gives us $\mathbf{a} = \mathbf{0}$ and the zero acceleration state of the particle will not change. Observers in such a zero-force environment are inertial observers.

Inertial Mass

The inertial mass m in Eq. 6.1 is a property of a particle associated with Newton's first law, the law of inertia. The law of inertia requires the state of motion or rest of a particle to persist or remain unchanged in a zero-force environment. This persistence, a universal behavior of matter, is explained by attributing to the particle an intrinsic property called *inertial mass*. This property, like length and time, is not derived in terms of other quantities. Hence it cannot be understood, explained, or defined in terms of other quantities.

Newton's second law provides a method for measuring the inertial mass of an arbitrary body. The method employs the horizontal frictionless surface described in Section 6.3, together with a spring scale calibrated in newtons. The body is subjected to a force large enough to induce an acceleration of 1 m/s^2. This force has a magnitude, which can be ascertained from the calibrated scale. According to Eq. 6.1, the mass of the body in kilograms is numerically equal to this scale reading:

$$m = \frac{F}{a} = \frac{F}{1} \qquad a = 1 \text{ m/s}^2$$

Newton's second law provides a second interpretation of inertial mass. Recall that inertial mass, according to the law of inertia, is associated with persistence of the state of motion (or rest) of a particle in a zero-force environment. According to Eq. 6.1, the acceleration of a particle when the net force is not zero is inversely proportional to the particle's inertial mass. If the same force is applied successively to masses m_1 and m_2, then the magnitudes of their accelerations will satisfy the equation

$$\frac{a_1}{a_2} = \frac{m_2}{m_1} \qquad (6.2)$$

The larger mass experiences the smaller acceleration, and vice versa. Hence, inertial mass is the intrinsic measure of a particle's resistance to acceleration.

Example 1 **A Sled Acceleration Race**

Identical twins plan to race their sleds across a frozen lake. Each twin is able to exert a force of 400 N on the sled. Twin A's sled, with two people on it, has a combined mass of 120 kg; twin B's sled has one person on it with a combined mass of 40 kg. Assuming that there is no friction between the ice and the sleds, find the accelerations of the two sleds, and confirm Eq. 6. 2.

From Eq. 6.1, $\mathbf{F} = m\mathbf{a}$, so for sled A

$$400 = 120a_1 \qquad a_1 = 3\tfrac{1}{3} \text{ m/s}^2$$

and for sled B

$$400 = 40a_2 \qquad a_2 = 10 \text{ m/s}^2$$

The 120-kg mass has a smaller acceleration than the 40-kg mass. Furthermore, the acceleration ratio, a_2/a_1 is the same as the inverse mass ratio, m_1/m_2, as expected.

$$\frac{a_2}{a_1} = \frac{1}{3} \qquad \frac{m_1}{m_2} = \frac{1}{3}$$

If two or more masses are considered separately, then the combined mass, considered as a particle, is the sum of the individual masses. This is what we expect of a scalar quantity. (Representative masses are given in Table 1.5.)

We often treat observers at rest on the earth's surface as inertial observers. Such observers actually undergo the centripetal accelerations associated with the earth's spin and orbital motions, and strictly speaking, do not qualify as initial observers. On a laboratory scale the effects associated with these accelerations are usually negligible, and it is legitimate to regard the earth's surface as an inertial frame of reference. But for large-scale phenomena such as ocean currents and hurricanes the acceleration caused by the earth's spin is significant and in such cases earthbound observers *cannot* be treated as inertial observers.

Questions

7. Aristotle associated "force" with velocity. Excluding gravity, give an example of an object experiencing (a) a zero force according to Aristotle's definition and a nonzero force according to Newton's definition, and (b) a nonzero force according to Aristotle and a zero force according to Newton.

8. Using Eq. 6.2 as a basis for your argument, support the contention that in collisions between football players of unequal mass the less massive player is more susceptible to injury.

6.5 Newton's Third Law

In mechanics, force represents the interaction between a particle and its environment. Newton's first two laws concentrate on the effect produced on the particle by the environment. Newton's third law completes the concept of force by taking into account the effect produced on the environment by the particle. The particle reacts to its environment as described by **Newton's third law:**

To every action there is always opposed an equal reaction.

As we discussed in Section 3.2, the words *action* and *reaction* mean forces. This law rules out the possibility of a single isolated force. Forces in nature always occur in pairs. An action–reaction pair is composed of forces equal in magnitude and opposite in direction. For example, the downward force exerted on the floor by your feet is equal in magnitude and opposite in direction to the upward force exerted on your feet by the floor. The two forces in an action–reaction pair act on different bodies. In our illustration, one force acts on your feet and the other acts on the floor.

A body cannot experience a force from its environment without exerting an equal and opposite force on its environment. We can generalize this statement and say that the net force exerted on a particle by its environment is equal in magnitude and opposite in direction to the net force exerted on the environment by the particle.

We can formalize Newton's third law as follows. Consider the interaction of the particles A and B in Figure 6.3. The force exerted on A by B, $\mathbf{F}_{AB}$, and the force exerted on B by A, $\mathbf{F}_{BA}$, satisfy the relation

$$\mathbf{F}_{AB} = -\mathbf{F}_{BA}$$

This relation holds whether A (or B) experiences other forces or not.

Figure 6.3

The force exerted on the particle at A, $\mathbf{F}_{AB}$, by the particle at B is equal in magnitude and oppositely directed to the force exerted on the particle at B, $\mathbf{F}_{BA}$, by the particle at A.

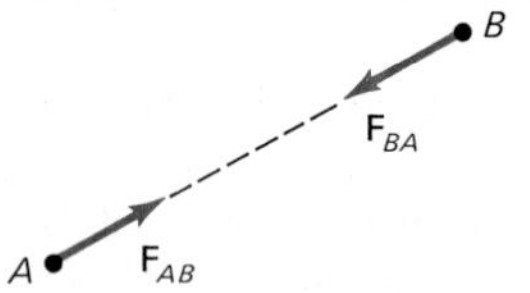

You must be careful not to confuse Newton's second and third laws. For example, how would you explain the following dilemma? The force you exert on the ground is equal in magnitude to but oppositely directed from the force exerted on you by the ground. The vector sum of these two forces is zero. If this net force is always zero, then how could you ever jump off the ground? The question accurately labels two forces as belonging to an action–reaction pair. But they do not act on the same object! Whether you jump off the ground is determined by the net force acting on *you*. The force *you exert* on the ground is not a part of the net force exerted on you.

It is also easy to confuse Newton's first and third laws. Consider a book at rest on a desk. Two forces act on the book. Its weight acts downward, and the normal force exerted on it by the desk acts upward. These forces are equal in magnitude and oppositely directed. Do they constitute an action–reaction pair? No, they do not. Both forces act on the book, whereas the forces in an action–reaction pair must act on different bodies. Yet we know that every force belongs to an action–reaction pair! Where, then, are the other two forces? The first is the downward push on the desk by the book. This contact force, together with **N** in Figure 6.4, provides one action–reaction pair of forces. The second is the gravitational attraction exerted on the earth as a whole by the book. This force, together with **W** in Figure 6.4, provides the other action–reaction pair of forces.

Newton's third law is indispensable for dealing with internal forces (Section 3.2). When two particles are part of one body, their mutual interaction (an action–reaction pair of forces) is ignored in any calculation of the net force acting on the body. This action–reaction pair of forces between the two particles adds to zero, and contributes nothing to the net force acting on the body. Because the net force is a vector sum, including all forces acting on the body, and because Newton's first two laws involve the net force, this cancellation of internal forces produces a great simplification. Imagine the confusion of applying Newton's second law to the flight of a baseball if we had to include the internal forces that bind the ball's many atoms together!

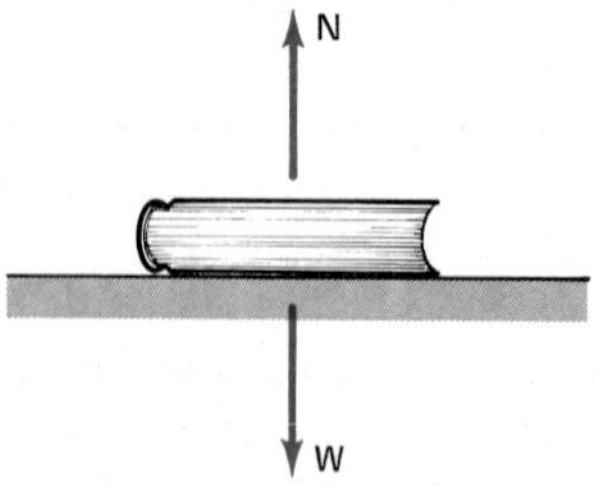

Figure 6.4

A book at rest experiences two forces. Its weight, **W**, acts downward and the contact force, **N**, acts upward. Because the book is in equilibrium, these oppositely directed forces are equal in magntiude. They act on the same body, however, and do not constitute an action–reaction pair of forces.

Example 2 **Horse and Rider**

Consider the motionless horse and rider shown in Figure 6.5 as a composite particle. Is the force exerted on the horse by its environment affected by the forces that the horse and rider exert on each other? Let the force exerted on the rider by the horse be $\mathbf{N}_{RH}$, the force exerted on the

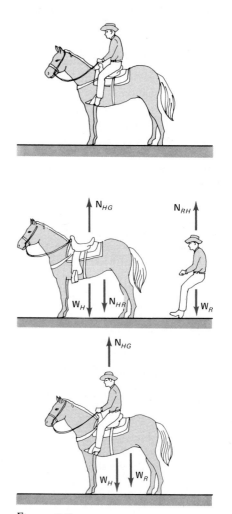

Figure 6.5

The horse and rider, considered as a single body, experience three forces. These are the two weights, $\mathbf{W}_H$ and $\mathbf{W}_R$, and the upward contact force, $\mathbf{N}_{HG}$, exerted by the ground. The rider, considered separately, experiences his weight, $\mathbf{W}_R$, and an upward contact force, $\mathbf{N}_{RH}$, exerted by the horse. The horse, considered separately, experiences his weight, $\mathbf{W}_H$, the ground contact force, $\mathbf{N}_{HG}$, and a downward contact force, $\mathbf{N}_{HR}$, exerted by the rider. The two contact forces, $\mathbf{N}_{HR}$ and $\mathbf{N}_{RH}$, constitute an action–reaction pair.

horse by the rider be $\mathbf{N}_{HR}$, the force exerted on the horse by the ground be $\mathbf{N}_{HG}$, and their weights be $\mathbf{W}_R$ and $\mathbf{W}_H$, respectively. The net force, **F**, acting on the composite body is

$$\mathbf{F} = \sum \mathbf{F}_i = \mathbf{N}_{RH} + \mathbf{N}_{HR} + \mathbf{N}_{HG} + \mathbf{W}_R + \mathbf{W}_H$$

$$= 0$$

The net force must equal zero if the horse and rider are to be in equilibrium. The first two forces in the sum constitute an action–reaction pair, and their sum is zero.

$$\mathbf{N}_{RH} + \mathbf{N}_{HR} = 0$$

Therefore, we conclude that

$$\mathbf{N}_{HG} + \mathbf{W}_R + \mathbf{W}_H = 0$$

According to this equation, the combined weight of the horse and the rider is equal to the force exerted on the horse by the ground. In other words, the ground supports both horse and the rider, and the forces that the rider and horse exert on each other do not affect this conclusion.

Questions

9. A man whose feet touch the floor experiences two forces. Gravity acts downward, and the floor pushes upward on him. A student states, "According to Newton's third law, these forces must be equal and opposite regardless of the acceleration of the man." Criticize this statement on the basis of Newton's second law.
10. Criticize the statement in Question 9 on the basis of Newton's third law.
11. Two stars, A and B, that are gravitationally bound together form a binary star system. Qualitatively discuss the motion of a binary star system if star A exerts an attractive force on star B that is larger than the attractive force exerted on star A by star B.
12. The book in Figure 6.4 experiences two forces. These are its weight, **W**, and the normal force, **N**, exerted by the desk. Where are the other members of the action–reaction pairs of forces to which **W** and **N** belong?

6.6 The Universal Force of Gravitation

In Newton's day the word *gravity* referred to the force exerted on objects by the earth. Gravity was thought to be a local force, and only those objects near the earth were believed to be affected by it. Galileo had already formulated the laws of kinematics, which described the motion of objects influenced by gravity. However, the explanation of these gravitational effects remained for Newton to discover.

An apparently unrelated problem of 17th-century science involved celestial motion. The unknown force guiding planets in their orbits acted over great distances. Kepler had studied the planets and discovered three laws describing their motion. However, Kepler failed in his attempt to find this long-range guiding force. This, too, remained for Newton to discover.

Prior to Newton, scientists believed that physics on the earth differed from physics in the heavens. The motion of falling apples, for example, was never connected to the motion of an orbiting planet. In Newton's hands, however, *universal* gravitation and a *universal set* of dynamic laws evolved together. Newton saw the orbiting moon as the limiting case of a falling object possessing a large horizontal velocity. For example, as the horizontal velocity of a falling apple increases, it reaches a point where, because of the curvature of the earth, the earth recedes from the apple at the same rate that the apple falls toward the earth. The apple is then in orbit! Newton recognized that the same force that acts on the apple—gravity—also acts on the moon.

Newton's realization that gravity is universal had important implications. Newton recognized that the laws of motion must also be universal in character. In other words, the same laws of motion—Newton's laws of motion—govern both celestial and terrestrial motion.

According to Newton's second law, the net force acting on a particle "causes" it to accelerate. Thus a falling apple accelerates as a result of the force exerted on it by the earth. The acceleration of the moon is "explained" in a similar way. However, Newton's second law does not describe any specific force in terms of the properties of the particle and of its environment. In this sense Newton's second law is incomplete. Newton filled this gap by formulating the universal law of gravity. He built this law on the foundation provided by Kepler's laws of planetary motion and Galileo's studies of terrestrial motion. To provide insight we now introduce the universal law of gravity by exhibiting its relationships to the results of Kepler and Galileo.

In 1619, Johannes Kepler culminated 30 years of research (Kepler's first and second laws were published in 1609) with the publication of the last of his **three laws of planetary motion:**

1 Planets move in elliptic orbits with the sun located at one focus.

2 During equal time intervals, equal areas are swept out by the line connecting the planet and the sun.

3 The square of the planetary period (time for one complete orbit) is proportional to the cube of the mean planetary radius (semimajor axis of the ellipse).

The elliptic orbit specified in Kepler's first law was much simpler than the complex system previously used by astronomers. It was also a break with 2,000 years of Greek astronomy. Kepler's predecessors had believed that all celestial motions were composed of circles.

Kepler's second law relates the orbital position and planetary speed. In Figure 6.6a for example, assume that the times required for a planet to move from a to b, c to d, and e to f are the same. Kepler's second law requires the three shaded areas, swept out by the line between the

Figure 6.6a

Suppose the times required for a planet to travel along the ellipse from a to b, c to d, and e to f are equal. Then, according to Kepler's second law, the shaded areas are equal.

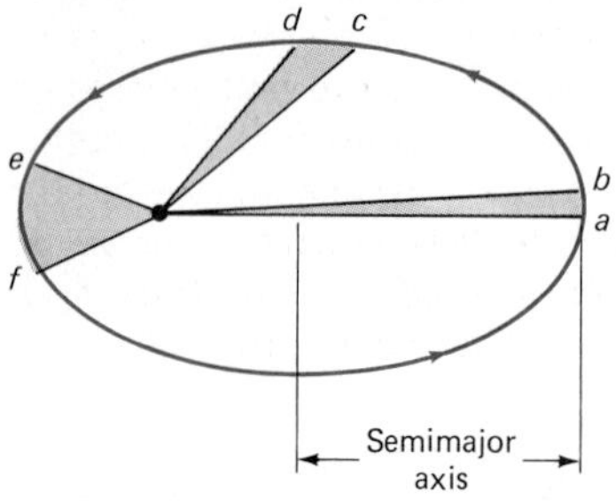

Figure 6.6b

The two ellipses are possible planetary orbits about the sun drawn with the semimajor axis, r_1, of the smaller ellipse one fourth the length of that of the larger ellipse. According to Kepler's third law, the planetary "year" for a planet moving along the large ellipse is eight times larger than the planetary "year" for a planet moving along the small ellipse.

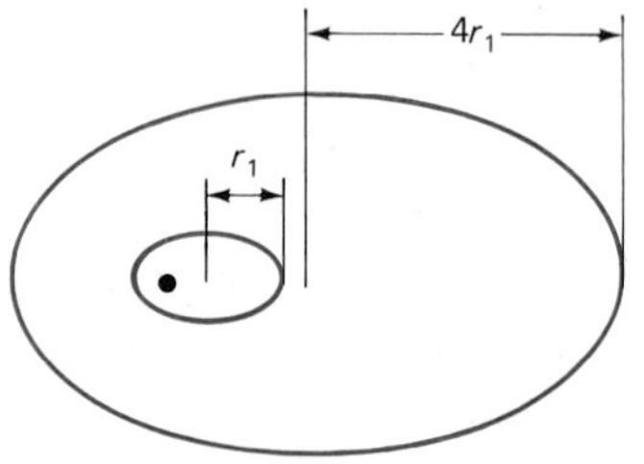

planet and the sun, to be the same. Hence the speed of the planet increases as it moves form a to c to e. (In Chapter 11 we will consider Kepler's second law as an example of conservation of angular momentum.)

Unlike his first two laws, Kepler's third law involves more than one planetary orbit. For example, the mean radii of the two ellipses shown in Figure 6.6b are in the ratio 4:1. Kepler's third law requires that the ratio of the corresponding periods be 8:1. Thus different orbits are "related."

Newton recognized that any theory of planetary motion should account for Kepler's three laws. Newton was the first to prove rigorously that an inverse square law force ($F \sim 1/r^2$) follows for a planetary orbit in the shape of a conic section (hyperbola, parabola, ellipse). Thus the ellipse specified in Kepler's first law requires an inverse square law force to be exerted on the planet by the sun. This could be written

$$F = \frac{c_1}{r^2}$$

where F is the force, r is the sun–planet distance, and c_1 is the proportionality constant. To determine c_1, we must measure both F and r. The force exerted by the sun on a planet cannot be measured, and consequently c_1 could not be determined. Hence this approach was limited.

What was the nature of this force acting on a body? Newton assumed that it was proportional to the mass of the body. We can write this assumption as

$$F_g = c_2 m$$

where c_2 is the proportionality constant. If we apply this force law to a freely falling apple and use Newton's second law, we obtain

$$F_g = c_2 m = ma$$

Hence

$$a = c_2$$

The acceleration of the apple is independent of its mass. This was Galileo's observation: In the absence of air resistance, all freely falling bodies near the earth's surface undergo the same acceleration. For a universal force this result would also apply to celestial objects. The acceleration of a planet at a given point in space does not depend on its mass. These results strongly support Newton's conjecture that the force of gravity is proportional to the mass of the particle.

In order to arrive at Newton's form of the law of gravity, let us consider the interaction between the earth and an apple (Figure 6.7). The earth exerts a force on the apple proportional to the apple's mass ($F_1 \sim m_1$). The apple exerts a force on the earth proportional to the mass of the earth ($F_2 \sim m_2$). Newton's third law (Section 6. 5) requires these forces to be equal in magnitude ($F_1 = F_2$). This requirement is satisfied by Newton's assumption that force is proportional to the product of the two masses ($F_1 \sim m_1 m_2$ and $F_2 \sim m_1 m_2$). If we include the inverse square law condition, we obtain*

$$F_g = G\frac{m_1 m_2}{r^2} \qquad (6.3)$$

*In vector form the force of gravity may be written

$$\mathbf{F}_g = -\frac{Gm_1 m_2}{r^2}\hat{\mathbf{r}}$$

where $\hat{\mathbf{r}}$ is a unit vector directed outward, away from the origin. If the origin is chosen at m_1, $\mathbf{F}_g$ represents the force on m_2. If the origin is chosen at m_2, $\mathbf{F}_g$ represents the force on m_1. Because the force is attractive, the minus sign is needed to indicate that $\mathbf{F}_g$ and $\hat{\mathbf{r}}$ are oppositely directed.

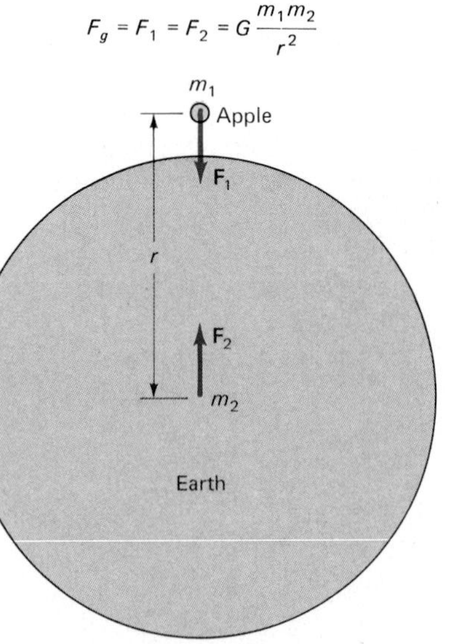

Figure 6.7

An apple (mass = m_1) is at a distance r from the center of the earth (mass = m_2). Each object attracts the other one with a force of gravity. The attractive force, $\mathbf{F}_1$, exerted on the apple is equal in magnitude and opposite in direction to the force of attraction, $\mathbf{F}_2$, exerted on the earth.

Here G is a universal constant, and r is the distance between the two "point masses." The force is attractive—directed along the line joining the two particles, as shown in Figure 6.7. For two objects having spherical symmetry, r is the center-to-center distance. (This can be proved with integration, but a more elegant method, which we will see in Chapter 27, involves Gauss' law.) The gravitational force of attraction between point particles varies directly with the product of the masses and inversely with the square of the distance between them.

Newton's law of gravity, Eq. 6.3, is not derived from some more fundamental law. It *relates* to the work of Kepler and Galileo, but is a *new* law (see Figures 6.8 and 6.9).

Example 3 Kepler's Third Law and the Force of Gravity

We can perform an independent confirmation of Newton's law of gravity by using Kepler's third law. Consider a planet of mass m moving in a circular orbit of radius r about the sun. What is the mathematical form of the gravitational force exerted on the planet?

The acceleration of the planet is centripetal, and thus may be written (See Section 5.2)

$$a = \frac{4\pi^2 r}{T^2}$$

where T is the period of the orbit. We write Kepler's third law in the form

$$T^2 = Kr^3$$

where K is the constant for satellites of the sun. Inserting this expression for T^2 into $a = 4\pi^2 r/T^2$ gives

$$a = \frac{4\pi^2}{Kr^2}$$

for the acceleration of the planet. From Newton's second law, the force acting on the planet equals ma, so we obtain for the force

$$F = ma = \frac{4\pi^2 m}{Kr^2}$$

This equation displays the inverse square dependence of Newton's law of gravity (Eq. 6.3), and provides additional support for its form.

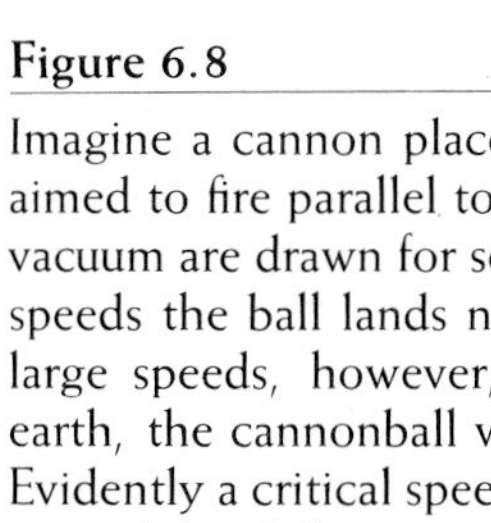

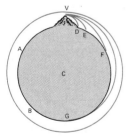

Figure 6.8

Imagine a cannon placed on top of a high mountain and aimed to fire parallel to the earth's surface. Trajectories in vacuum are drawn for several cannonball speeds. For small speeds the ball lands near the mountain, as expected. At large speeds, however, because of the curvature of the earth, the cannonball will land partway around the earth. Evidently a critical speed exists at which the ball will travel around the globe in a circle. (That is, it would be in orbit.) With this clever drawing Newton bridged the gap between Kepler's celestial orbits and Galileo's terrestrial trajectories.

Figure 6.9

Sir Isaac Newton (1642–1727) discovered the law of universal gravitation. He incorporated this law, and his three laws of motion, into a comprehensive dynamic theory that explained celestial and terrestrial motion. He also invented the calculus and made important contributions to other areas of physics and mathematics. In 1687 Newton published the *Principia*. This treatise includes (1) Newton's three laws of motion, (2) his law of universal gravitation, and (3) his mathematical discoveries of calculus. The *Principia* is recognized as one of man's supreme achievements and is commemorated on this Nicaraguan stamp.

Gravitational Mass and Inertial Mass

Mass, as it appears in the law of gravity, measures the gravitational interaction of a particle with other objects. We refer to this property as **gravitational mass.** Mass, as it appears in Newton's second law, measures the particle's resistance to acceleration. We refer to this property as **inertial mass.** The acceleration, a, of an apple that is subject only to the gravitational force of the earth is described by using the law of gravity together with Newton's second law. Thus

$$F_1 = G\frac{m_1 m_2}{r^2} = m_1' a$$

or

$$a = \left(\frac{m_1}{m_1'}\right)\frac{Gm_2}{r^2}$$

where m_1 and m_1' denote, respectively, the gravitational and inertial masses of the apple. There is no reason to suspect that the ratio m_1/m_1' should be the same for all objects. Yet Galileo argued, and experiment has confirmed, that this ratio is the same. All freely falling objects at a given point undergo the same acceleration.*

*Newton thought that this was a coincidence. Einstein recognized it as a profound principle. He made it his *principle of equivalence,* the starting point for his theory of general relativity (1915).

For convenience, we take gravitational mass and inertial mass to be identical:

$$m = m'$$

The constant G was offered as universal, but Newton didn't know its numerical value. Measurements of G require separating two known masses a known distance and measuring the force with which one mass attracts the other. Everything in Eq. 6.3 would then be known except G. The needed technology (a torsion balance) wasn't invented until about 110 years after Newton advanced his law of gravitation. However, Newton was able to estimate the value of G, as the following example shows.

Example 4 Estimation of G

Consider an apple at the surface of the earth. We can apply Newton's second law to relate the acceleration of gravity to the gravitational constant.

$$F_A = G\frac{m_A M_E}{R_E^2} = m_A g$$

Here g is the acceleration of gravity at the surface of the earth, and the subscript A denotes the apple, whereas the subscript E denotes the earth. We assume that the force of attraction of the earth is the same as it would be if the mass of the earth were concentrated at its center. Solving for G, we have

$$G = \frac{gR_E^2}{M_E} \tag{6.4}$$

Newton did not know the mass of the earth, but could estimate its average density, ρ_E, the ratio of its total mass to its total volume (Section 15.1).

$$M_E = \frac{4}{3}\pi R_E^3 \rho_E \tag{6.5}$$

Substituting the value of m_E in Eq. 6.5 into Eq. 6.4 gives

$$G = \frac{3g}{4\pi R_E \rho_E} \tag{6.6}$$

Both g and R_E were known to Newton. Let us assume, along with Newton, that ρ_E must lie between the value for water (1 gm/cm^3 or 1000 kg/m^3) and ten times this value. Using $R_E = 6.37 \times 10^6$ m and $g = 9.80\ \text{m/s}^2$, we have the limits for G:

$$3.67 \times 10^{-11}\frac{\text{N}\cdot\text{m}^2}{\text{kg}^2} < G < 3.67 \times 10^{-10}\frac{\text{N}\cdot\text{m}^2}{\text{kg}^2}$$

If we use Newton's value for the density of rock, $\rho \sim 5500\ \text{kg/m}^3$, we get

$$G \sim 6 \times 10^{-11}\frac{\text{N}\cdot\text{m}^2}{\text{kg}^2}$$

which is very close to the presently accepted value.

The constant G was measured accurately* for the first time in 1798 by Henry Cavendish. Using his newly developed torsion balance (see Figure 6.10), Cavendish found $G = 6.75 \times 10^{-11}\ \text{N}\cdot\text{m}^2/\text{kg}^2$. The presently accepted value is

Figure 6.10

To determine G, a torsion balance is used to measure the gravitational force between lead spheres. Two identical lead spheres (B) are connected to a light rod, forming an assembly that is free to rotate about an axis through the point O. Two torques act on the assembly. A clockwise torque is produced by the gravitational forces between identical fixed large lead spheres (A) and the small spheres (B). A counterclockwise torque is produced by the supporting fiber as it attempts to untwist. (The figure shows the view looking along the twisted fiber.) In equilibrium, these counteracting torques are equal in magnitude. Calibration of the fiber permits the torque, and hence the gravitational force, to be measured. The use of Eq. 6.3 then yields G.

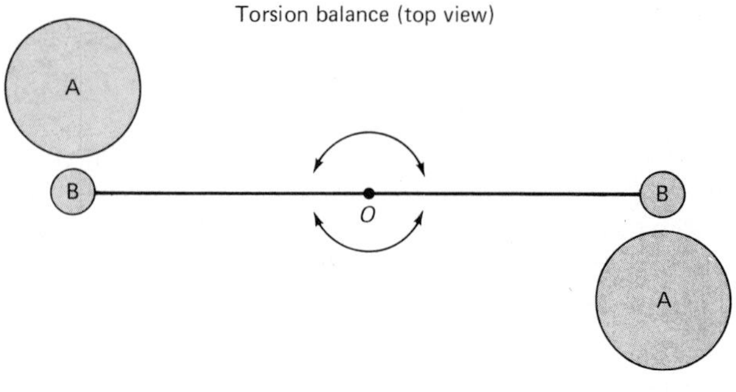

$$G = 6.6704 \times 10^{-11}\ \text{N}\cdot\text{m}^2/\text{kg}^2$$

We have emphasized that Newton's law of gravity is *universal*. This one force law can explain both falling apples and orbiting planets. Newton's laws of motion also are not restricted to terrestrial phenomena, but are universally valid. As the first scientist to advance universal laws of nature, Newton established a goal for scientists involved in all areas of fundamental research—namely, to seek out *universally valid* relationships.

Questions

13. The force of gravity is proportional to the mass of a body, yet all freely falling bodies near the surface of the earth experience the same acceleration. Explain.

14. Two satellites whose masses are m_1 and m_2 circle the earth in an orbit of radius R. Compare (a) their centripetal accelerations, and (b) the forces of gravity exerted on each by the earth.

15. Increases in altitude are accompanied by decreases in the force of gravity. How would the motion of bodies at the surface of the earth differ if the force of gravity increased with altitude?

16. Discuss the motion of freely falling bodies on the surface of a fictitious planet whose force of gravity is independent of the mass of the falling body. Compare your conclusion with Galileo's conclusion that all freely falling objects on the surface of the earth undergo the same acceleration.

17. In what way would the force of gravity have to depend on mass for the acceleration of a body to be proportional to its weight, as Aristotle contended?

*Accuracy is relative here. Today G is the least accurately known of all fundamental constants.

6.7 Weight and Mass

A particle at a distance r from the center of the earth experiences a gravitational force, according to Eq. 6.3, that is equal to

$$F_g = G\frac{m_1 M_E}{r^2} \qquad \text{for} \qquad r \geq R_E$$

where m_1 and m_E are the masses of the particle and the earth, respectively, and R_E is the radius of the earth. The local **weight** of the object (see Section 3.1) is defined to be this force. Hence we may write

$$W = G\frac{m_1 M_E}{r^2} \qquad \text{for} \qquad r \geq R_E \tag{6.7}$$

for the weight, W, of the object, showing that weight varies with the distance of the object from the center of the earth.

Example 5 **Variation of Weight with Altitude**

How much weight does an object "lose" when it is carried 555 ft to the top of the Washington Monument? This altitude is about 169 m. Let $W(R_E)$ be the weight of the object at the bottom. If R_E is the earth's radius and h denotes the altitude, then

$$r = R_E + h$$

From Eq. 6.7 we have

$$\frac{W(r)}{W(R_E)} = \frac{R_E^2}{r^2}$$

Thus

$$W(r) = W(R_E)\frac{R_E^2}{(R_E + h)^2}$$

$$= W(R_E)\frac{1}{(1 + h/R_E)^2}$$

Although you should have a hand calculator, we stress the binomial theorem and other analytical techniques to exhibit general characteristics. To show the dependence of weight on altitude, we use the binomial theorem (Appendix 4):

$$\left(1 + \frac{h}{R_E}\right)^{-2} \cong 1 - \frac{2h}{R_E} \qquad \text{for} \qquad \frac{h}{R_E} \ll 1.$$

Then

$$W(R_E + h) = W(R_E)\left(1 - \frac{2h}{R_E}\right)$$

The weight loss of the object is

$$W(r) - W(R_E) \cong W(R_E)\left(-\frac{2h}{R_E}\right)$$

$$= -5.31 \times 10^{-5}\, W(R_E)$$

where $2h/R_E = 5.31 \times 10^{-5}$ is the fractional weight loss. If $W(R_E) = 200$ lb, the weight loss is

$$5.31 \times 10^{-5} \cdot 200 \text{ lb} = 1.06 \times 10^{-2} \text{ lb}$$

which is about 0.17 ounce.

The weight of a particle near the surface of the earth is approximately constant in magnitude over the surface of the earth, and is directed nearly toward the center of the earth. If the distances covered in particle motion are small compared with the radius of the earth, we may assume the weight to be constant in magnitude and direction.

$$\mathbf{F}_g = \mathbf{W} \tag{6.8}$$

Consider a particle under the influence of its own weight alone—a freely falling body. Substituting $\mathbf{F} = \mathbf{W}$ in Eq. 6.1 gives

$$\mathbf{a} = \frac{\mathbf{W}}{m} \tag{6.9}$$

for the particle's acceleration. We might expect this acceleration to vary with mass, just as it did in Eq. 6.1, where $\mathbf{F} \neq \mathbf{W}$. However, experimentally $\mathbf{a}$ is found to be the same for all m in Eq. 6.9. This remarkable result, first proclaimed by Galileo, indicates that $\mathbf{W}$ is different from other forces (compare the discussion of this point in Section 6.6). Specifically, the weight of a body is proportional to its mass. We assign a special symbol to $\mathbf{a}$ for the case of a freely falling body near the surface of the earth.

$$\mathbf{a} = \mathbf{g} \tag{6.10}$$

The symbol $\mathbf{g}$ represents the acceleration due to gravity. We can therefore rewrite Eq. 6.9 as

$$\mathbf{W} = m\mathbf{g} \qquad \textit{Weight–mass relation} \tag{6.11}$$

The acceleration of gravity is not completely uniform. Local irregularities in the acceleration of gravity result from high- and low-density regions in the earth, which affect the weight slightly. These effects are used by geologists in their search for minerals and petroleum. The value of g on the surface of the earth depends on latitude and local geology. This acceleration has been measured precisely at specific locations on the earth. Its *standard value* in SI units is $g = 9.80665$ m/s^2. The value $g = 9.80$ m/s^2 is used in our examples and problems.

Equation 6.11 is the source of the confusion between weight and inertial mass. Any measurement of mass as a ratio will automatically give the corresponding weight ratio. That is,

$$\frac{m_1}{m_2} = \frac{m_1 g}{m_2 g} = \frac{W_1}{W_2}$$

But mass and weight are different. Mass, the amount of matter, is measured in kilograms. Weight, the gravitational force on mass, is measured in newtons. The confusion between these two different quantities often results in the incorrect quotations of weight in kilograms and mass in pounds (the British unit of force). For example, a can of peanuts might have "Net Weight 5 lb = 2.27 kg" printed on its label.

Because mass is an intrinsic property of a body, it does not vary with the environment. Weight is a specific force and *is* dependent on the environment. The weight of an astronaut on the moon is about one sixth of his weight on the earth, but his mass is the same on the moon as it is on the earth. Both **g** and **W** in Eq. 6.11 vary with the environment, whereas m is the same for a particle everywhere.

Example 6 **Weight of the Standard Kilogram**

Let us use Eq. 6.11 to find the weight of the standard kilogram on the earth. We have on the earth

$$W = mg$$

$$W = 1\text{kg} \times 9.80665\ \text{m/s}^2$$

$$W = 9.80665\ \text{N}$$

To determine this weight in pounds we need the conversion factor from newtons to pounds. In SI units 1 lb is defined to be a force equal to the weight of a body whose mass is 0.453592 kg at a point where $g = 9.80665\ \text{m/s}^2$. Hence

$$1\ \text{lb} = 0.453592\ \text{kg} \times 9.80665\ \text{m/s}^2$$

$$= 4.44822\ \text{N}$$

We then have, for the weight of the standard kilogram,

$$W = 9.80665\ \text{N}\frac{1\ \text{lb}}{4.44822\ \text{N}}$$

$$= 2.20462\ \text{lb}$$

Questions

18. A calibrated vertical spring or an equal-arm balance may be used to measure weight. Explain why the spring measurement varies with altitude, latitude, and local geographic conditions, whereas the equal-arm balance always gives the same result.

19. Discuss the two methods for measuring weight given in Question 18. Consider their use for measuring mass, and state which method you would like to see used if your were buying a specified quantity of gold bullion, and why.

6.8 Mechanical Force Laws

Through Newton's second law the acceleration of a particle is expressed in terms of its inertial mass and the net force it experiences. The acceleration of a 1-kg mass equals the net force acting on it but does not distinguish among different kinds of forces. For example, a 10-N force produces the same acceleration of the 1-kg mass whether the force is electrical or gravitational in nature. Laws of force, on the other hand, permit the calculation of a force in terms of the specific properties of a particle and of its environment. We are able to distinguish electric forces from gravitational forces by the differences in their force laws.

Fundamental forces of nature—for example, gravity—have no known explanation in terms of other forces. Mechanical force laws—the linear spring force law, for instance—are empirical, and may be understood in terms of other forces, in this case the electric forces between molecules. In this section we consider several examples of mechanical force laws.

Linear Spring

The calibration of a vertical spring in newtons was described in Section 3.1, and for a horizontal spring, calibration in newtons was described in Section 6.3. Calibration implies a relation between the change in the spring length and the restoring force exerted by the spring. For many materials this relation is linear provided that the extension of the spring does not exceed the latter's "elastic limit." In this case we can write

$$F = -kx \qquad \textit{Hooke's law} \quad (6.12)$$

where x is the change in length of the spring from its equilibrium length, F is the restoring force, and k is the spring constant, a positive quantity. Equation 6.12 is often referred to as Hooke's law (see Section 15.3). Figure 6.11 shows why F is referred to as a *restoring* force. The direction of the spring force is always opposite to the displacement, and thus serves to restore the mass to its original position. Note from Eq. 6.12 that because the spring constant, k, is positive, the sign of x is always opposite to that of F.

In order to calibrate a linear spring, we must evaluate the constant k. We can do this by using the static procedure outlined in Section 3.1 or the dynamic method described in Section 6.3.

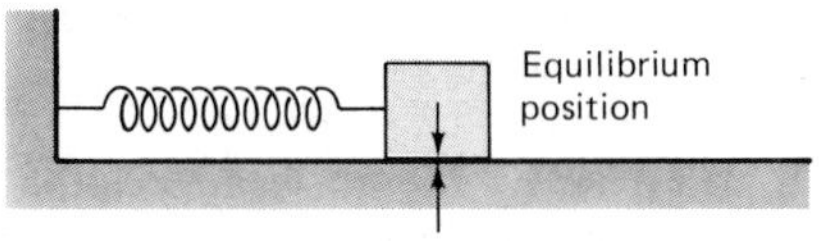

Figure 6.11a

A horizontal spring in its equilibrium position.

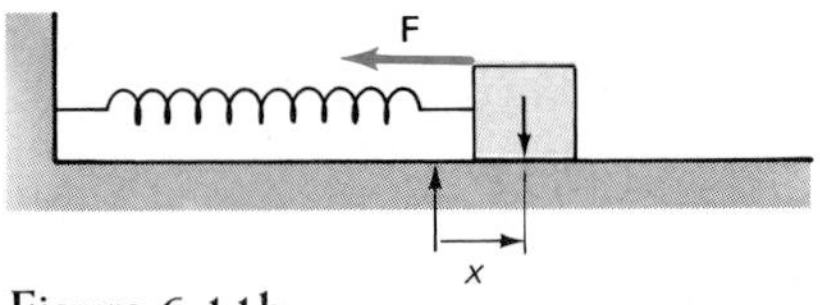

Figure 6.11b

The horizontal spring exerts a force directed to the left if the spring is stretched.

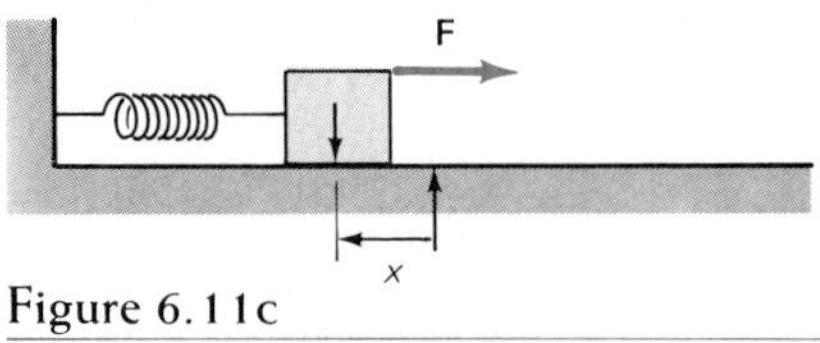

Figure 6.11c

The horizontal spring exerts a force directed to the right if the spring is compressed.

Example 7 **Accelerating a Standard Kilogram**

A horizontal spring satisfying Hooke's law ($k = 1000$ N/m) is used to accelerate a standard kilogram across a horizontal frictionless table. During acceleration the spring extension is $x = +0.01$ m. What is (a) the force exerted by the spring, and (b) the acceleration of the kilogram?

(a) To determine the force exerted by the spring, we use Hooke's law (Eq. 6.3).

$$F = -kx = -\left(1000 \frac{\text{N}}{\text{m}}\right)(0.01 \text{ m})$$
$$= -10 \text{ N}$$

(b) To determine the acceleration we use Newton's second law (Eq. 6.1).

$$a = \frac{F}{m} = \frac{-10 \text{ N}}{1 \text{ kg}}$$
$$= -10 \text{ m/s}^2$$

Because F and a have the same algebraic sign, they are in the same direction.

Sliding Friction

In Section 3.1 we discussed the contact forces that arise when solid objects touch. Such forces have one component, called the normal force (N), perpendicular to the surface, and one component, called the frictional force (f), parallel to the surface. The normal force inhibits the penetration of one object into another, and the frictional force opposes relative motion of the objects parallel to the contacting surfaces. The rubbing of surface irregularities is responsible for kinetic friction. Thus when object A in Figure 6.12 moves to the right with the velocity $\mathbf{v}_A$ relative to object B while A and B are in contact, the forces exerted on A by B (f_A and N_A) will be directed as indicated. The frictional force depends on the condition of the two surfaces, the material, and the normal force. For a great variety of materials and velocities f and N are related approximately by

$$f = \mu_k N \qquad \textit{Kinetic force of friction} \qquad (6.13)$$

where μ_k is the coefficient of kinetic friction. Within broad limits this relation is independent of contact area and velocity. Typical values are given in Table 6.1. (Note that the coefficient of kinetic friction may exceed 1.)

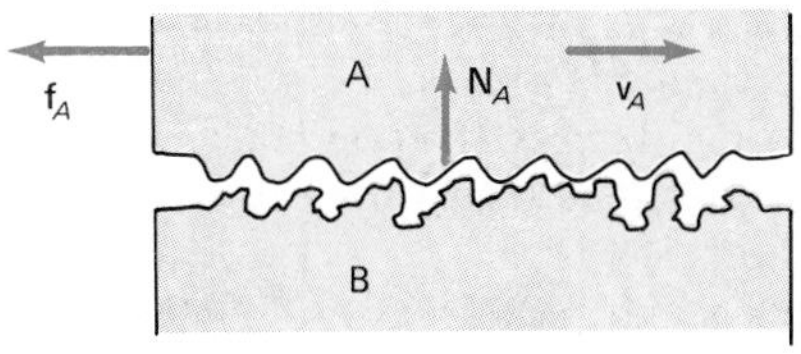

Figure 6.12

If block A moves to the right relative to block B, a kinetic frictional force is exerted on block A. This force is directed opposite to this motion, or to the left.

Table 6.1

Coefficient of kinetic friction (Handbook of Chemistry and Physics)

Material	T (°C)	μ_k
waxed ski on dry snow	0	0.04
waxed ski on dry snow	−10	0.18
waxed ski on dry snow	−40	0.40
ice on ice	−10	0.035
ebonite on ice	−10	0.05
brass on ice	−10	0.075
rubber on dry cement	—	1.02
rubber on wet cement	—	0.97

Static Friction

If no relative motion occurs between contacting surfaces, then we speak of static friction. The static frictional force can take on all values from 0 up to a certain maximum, f_{max}, which is determined by the nature of the contacting surfaces and the normal force. The maximum static frictional force is related to the normal force by $f_{max} = \mu_s N$, where μ_s is the coefficient of static friction.

As noted, the static force of friction may be less than this maximum force, $\mu_s N$. For example, suppose that a book is at rest on a level table (see Figure 6.13). If no horizontal force is exerted on the book, the static friction force also must be zero in order to ensure that equilibrium be preserved. Suppose a student pushes the book to the right with a force F that is less than the maximum frictional force, $\mu_s N$. The book remains at rest. The frictional force assumes a value just large enough to prevent acceleration of the book. The frictional force may equal zero, may equal $\mu_s N$, or may lie between these extremes. Therefore we may write

$$f_s \leq \mu_s N$$

In most situations $\mu_s > \mu_k$. This relation has implications in braking an automobile, as we will now show.

Consider a car of weight W moving on a horizontal road (see Figure 6.14). Vertical equilibrium requires $N = W$. What happens when the brakes are applied? If the car skids, then the retarding frictional force $f_k = \mu_k W$. If the

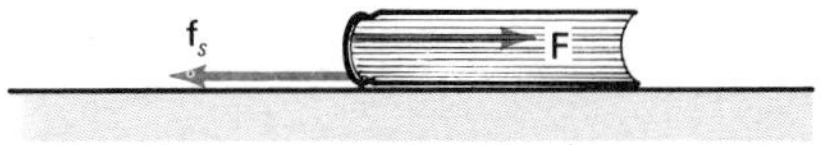

Figure 6.13

An external horizontal force **F** exerted on the book is too weak to cause the book to move. The opposing static frictional force, $\mathbf{f}_s$, must be equal in magnitude to **F** because the book remains in horizontal equilibrium.

Figure 6.14

Two identical cars brake to a stop. The maximum frictional force exerted on the nonskidding car is larger than the frictional force exerted on the skidding car.

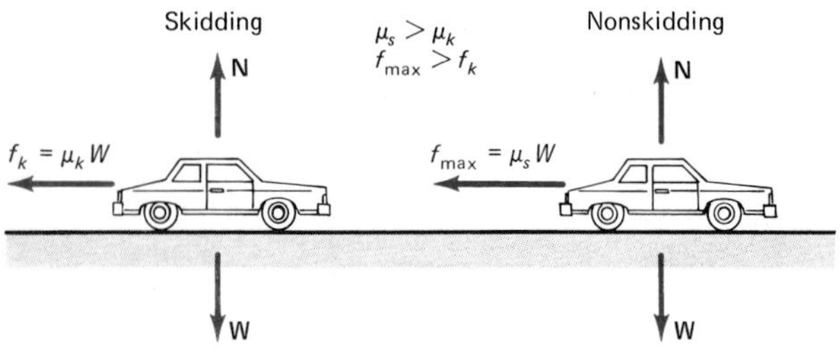

car does not skid, then the frictional force has a maximum given by $f_{max} = \mu_s W$. Because $\mu_s > \mu_k$, $f_{max} > f_k$. The maximum effect of braking without skidding exceeds the effect of braking with skidding.

Example 8 **Stopping Distance**

At $-10°C$ a skier moving horizontally at 20 m/s coasts to a stop as a result of the frictional force exerted on her skis by the snow (see Figure 6.15). What is the stopping distance? (Take $m = 80$ kg and $N = 784$ N, and neglect air resistance.)

We first determine the acceleration of the skier. The only horizontal force acting is f, so using Table 6.1 we have

$$f = -\mu_k N = -0.18 \times 784 \text{ N} = -141 \text{ N}$$

The minus sign is needed because f is in the negative x-direction. Using Newton's second law, we obtain for the acceleration ($F = f$)

$$a = F/m = \frac{-141 \text{ N}}{80 \text{ kg}} = -1.76 \text{ m/s}^2$$

Because the force is constant, the acceleration is constant, and we can use the results developed in Chapter 4 (Eq. 4.17).

$$v^2 = v_o^2 + 2ax \qquad (4.17)$$

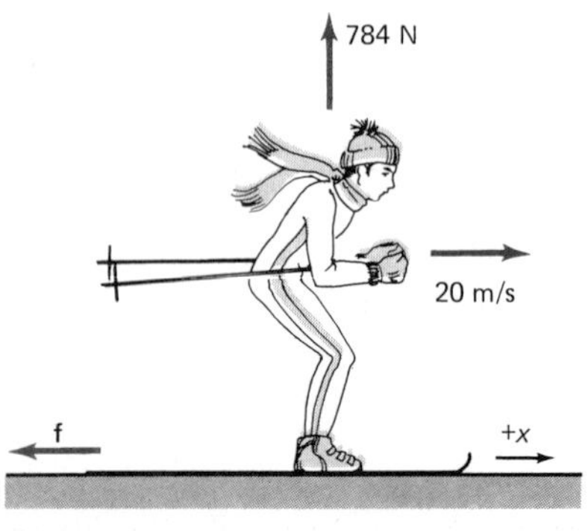

Figure 6.15

A skier coasting horizontally to the right experiences a kinetic frictional force directed to the left.

With $v = 0$ and $v_o = +20$ m/s we have

$$x = -\frac{(20 \text{ m/s})^2}{2(-1.76 \text{ m/s}^2)} = +114 \text{ m}$$

Because of the relatively low coefficient of kinetic friction, the skier coasts 114 m before coming to a stop.

Example 8 illustrates the program of particle mechanics discussed in Section 6.1. The law of force ($f = \mu_k N$) is used to obtain the net force. This is substituted into Newton's second law to obtain the acceleration in terms of the properties of the environment (friction) and of the particle (mass). From the acceleration, we obtain the displacement corresponding to zero velocity (stopping distance) and the future location of the particle.

Fluid Friction

Next we consider a mechanical force law for fluids. (By fluid, we mean a gas or a liquid.) This law describes a force exerted on an object that is moving through a fluid. Like the laws for solids, fluid laws are empirical. The physics of fluids will be developed in Chapter 16. Here we are interested only in the motion of objects subject to forces exerted by fluids. We will take this law and explore its consequences using Newton's laws.

The size, shape, and orientation of an object determine the fluid force on that object. Swimmers and sky divers change their shape and orientation by bending, twisting, and moving their arms and legs. This allows them to manipulate the fluid forces and consequently to control their speed and direction of motion. The expressions for fluid laws are simplest for spheres. Therefore we will limit our considerations to the study of fluid forces acting on a sphere.

Fog and mist are collections of tiny water droplets. Close examination reveals that these droplets fall very slowly. The effect of drag on the droplets' speed is large in comparison with drag's effect on Newton's falling apple. At the low speeds of water droplets the drag force is well represented by *Stokes' law*. This law expresses the drag, F_D, on a sphere of radius r moving with speed v as

$$F_D = 6\pi\eta r v \qquad (6.14)$$

where η (Greek eta) is an empirical quantity called the *viscosity*. Viscosity, which we will discuss in Section 15.7, is the fluid analogue of the coefficient of kinetic friction. The SI unit of viscosity is kilograms per meter per second (kg/m · s).

Consider the vertical motion of a fog droplet subject to the forces of gravity and drag. The droplets begin to move from rest. After a time they acquire a speed v downward. Two forces act on the droplet. Its weight, mg, is downward, and the drag, $6\pi\eta r v$, is upward. The droplet's acceleration downward is determined from Newton's second law

$$F = +mg - 6\pi\eta r v = ma$$

Dividing by m we obtain

$$a = g - \frac{6\pi\eta r v}{m} \qquad (6.15)$$

Figure 6.16

The acceleration of a fog droplet, a, is plotted versus its velocity, v. Initially $v = 0$ and the droplet acceleration equals g. When the velocity reaches the terminal velocity, v_T, the acceleration has been reduced to zero.

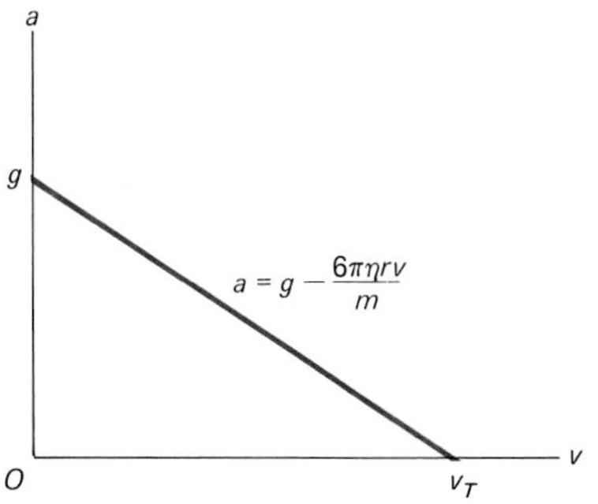

Perhaps the most striking feature of this result is the velocity dependence. Freely falling objects share a common acceleration g. The droplet acceleration (see Figure 6.16) is g initially ($v = 0$), but then falls to zero. The velocity increases until the two forces become equal in magnitude. At this point the velocity is at a maximum called the terminal velocity, v_T. Setting $a = 0$ in Eq. 6.15 we obtain for the terminal velocity

$$v_T = \frac{mg}{6\pi\eta r} \qquad (6.16)$$

We see from this relation that the terminal velocity of an object is proportional to the object's mass! The more massive an object, the faster it falls through a fluid.

To determine the size dependence of the terminal velocity we introduce the mass density, ρ,

$$m = \left(\frac{4}{3}\pi r^3\right)\rho$$

where the term in parentheses is the volume of the droplet. Substituting this expression for m into Eq. 6.16 yields

$$v_T = \left(\frac{4}{3}\pi r^3\rho\right)\frac{g}{6\pi\eta r}$$

or

$$v_T = \frac{2\rho g r^2}{9\eta} \qquad (6.17)$$

The terminal velocity of a sphere of given material (fixed ρ) varies directly with the square of the radius. For example, doubling the radius produces a fourfold increase in terminal velocity.

Example 9 **Velocity of a Falling Fog Droplet**

Using a microscope we find that the radius of a small fog droplet is 5.1×10^{-6} m, or about five thousandths of a millimeter (0.005 mm). (This radius, typical for droplets found in fog and clouds, is roughly one tenth of the radius of the smallest droplet visible to the human eye.) We can use this measurement to obtain the settling speed of the droplet, assuming that Stokes' law holds (for air, $\eta = 1.90 \times 10^{-5}$ kg/m · s).

We start with Eq. 6.17 for v_T.

$$v_T = \frac{2\rho g r^2}{9\eta}$$

Using $\rho = 10^3$ kg/m^3 for water, we obtain

$$v_T = 2.7 \times 10^{-2} \text{ m/s}$$

A droplet falling with this speed requires 37 s to fall 1 m.

Questions

20. Would you expect rubber bands to obey Hooke's law? Explain.
21. A newspaper article reports that a material has been discovered that is useful for the manufacture of springs that will exert a restoring force proportional to the square of their displacement from equilibrium. Could a spring have this property? Explain.
22. Explain why the braking action of a car is less if the car skids than it is if the car does not skid but is on the verge of skidding.
23. What methods do highway departments use to change the coefficient of friction between the road and automobile tires? What effect does the weather have on the coefficient of friction? What methods are available to car owners to modify the coefficient of friction?
24. What generally prevents raindrops from becoming as large as hailstones?

6.9 Applications of Newton's Laws

To help you solve word problems involving Newton's laws we now extend the *Problem Solving Guide* given in Section 3.2.

1 Try to guess the answer.
2 Make a simple line drawing.
3 Pick a body.
4 Draw a force diagram.
5 Choose coordinate axes.
6 Resolve vectors along chosen axes.
7 Apply Newton's laws of motion.
8 Decide, if possible, whether or not the solution makes sense.

In selecting axes, try to choose one axis in the direction of the acceleration. This may simplify the algebra. Let's now use this guide to solve the word problems in this section.

Figure 6.17

Two ice skaters pushing each other apart.

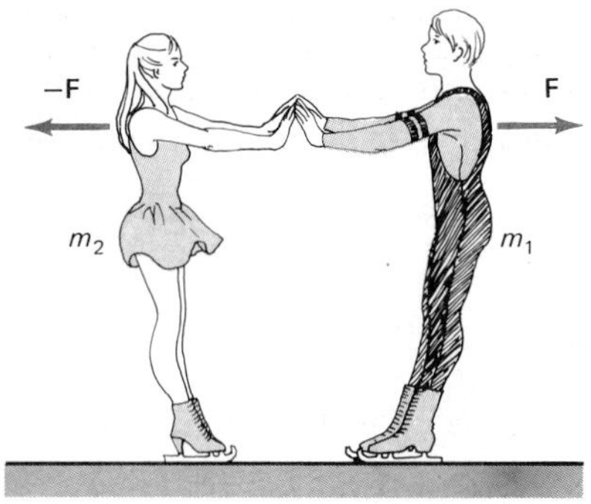

Figure 6.18

The two blocks (masses m_1 and m_2) move without friction on a horizontal surface. The connecting string is under tension T_1 and an external force T_2 is exerted on the right block.

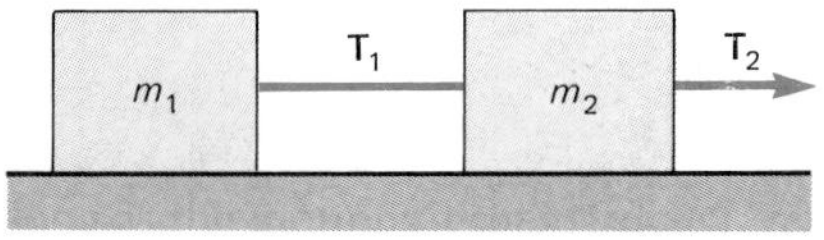

Consider two ice skaters (see Figure 6.17) whose masses are m_1 and m_2. Initially they are standing at rest on frictionless horizontal ice. Each skater pushes horizontally on the other. Assume that the forces are constant and act for a time t. What conclusions may be drawn if no other horizontal forces act?

According to Newton's third law, the two forces must be equal and opposite. Using this law together with Newton's second law for each skater, we have

$$-F = m_1 a_1$$

$$+F = m_2 a_2$$

so

$$\frac{a_1}{a_2} = -\frac{m_2}{m_1}$$

The accelerations of the skaters are inversely proportional to their masses. This relation is readily tested in the laboratory by using air tracks. This inverse relationship between mass and acceleration may be interpreted analogously to Eq. 6.2. The difference is in the minus sign here, which indicates that the skaters accelerate in opposite directions.

If a force is constant, a_1 and a_2 are also constant. In this case we may write for the final velocities

$$v_1 = a_1 t$$

$$v_2 = a_2 t$$

and the velocity ratios are the same as the acceleration ratios. We have

$$\frac{v_1}{v_2} = -\frac{m_2}{m_1} \tag{6.18}$$

This equation can be rearranged to $m_1v_1 + m_2v_2 = 0$. We will see in Section 9.3 that the two ice skater problem is an example of linear momentum conservation.

Acceleration of Coupled Objects

Two objects having masses m_1 and m_2 move horizontally on a straight frictionless air track. One string (under tension T_1) connects the objects, and a second string (under tension T_2) is used to accelerate the system to the right, as shown in Figure 6.18. The acceleration, a, and masses are given. The problem is to determine the two tensions in terms of m_1, m_2, and a. You might first want to guess the answers for T_1 and T_2. In particular, does your intuition tell you that $T_1 = T_2$?

Suppose we now choose m_2 as a body. Two horizontal forces act on this body. The force vector diagram is given in Figure 6.19. Choosing the x-axis as drawn and applying Newton's second law, we obtain

$$\sum F_x = +T_2 - T_1 = m_2 a$$

This tells us that T_2 exceeds T_1 by the product m_2a. The two tensions are not equal. Because neither T_1 nor T_2 can be separately determined, the solution is incomplete. We must pick another body.

We select m_1 as a body. One horizontal force acts on m_1. The corresponding force vector diagram is shown in Figure 6.20. This time Newton's second law yields

$$\sum F_x = T_1 = m_1 a$$

This equation gives T_1 in terms of m_1 and a, as desired. However, we still do not have any information about T_2. This is because T_2 does not act on m_1, and can't appear in the equation for Newton's second law. This solution is also incomplete. We obtain the complete solution by combining the results of choosing m_1 as a body with the results of choosing m_2.

$$T_1 = m_1 a$$

$$T_2 = (m_1 + m_2)a$$

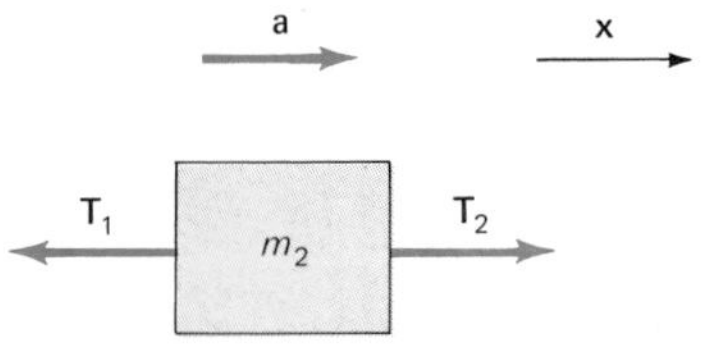

Figure 6.19

Horizontal forces acting on the right block in Figure 6.18 are the tensions $\mathbf{T}_1$ and $\mathbf{T}_2$.

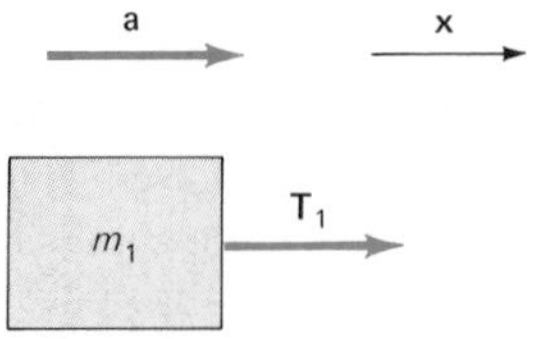

Figure 6.20

The only horizontal force acting on the left block in Figure 6.18 is the tension $\mathbf{T}_1$.

Figure 6.21

The only horizontal force acting on the pair of blocks in Figure 6.18 is the tension T_2.

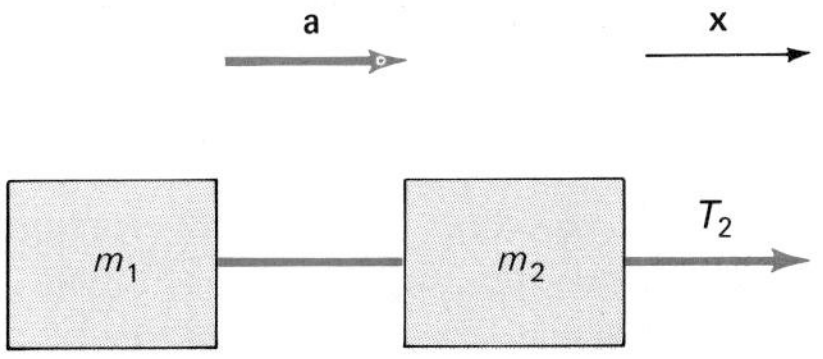

It might not occur to you to pick the combined system of m_1 plus m_2 plus the connecting string as a body. Yet, provided all parts of an object move as a unit (have the same acceleration) such a choice is possible. One horizontal force acts on this composite body. The force vector diagram is shown in Figure 6.21. Applying Newton's second law, we have

$$\sum F_x = T_2 = (m_1 + m_2)a$$

This is a solution for T_2. No information about T_1 is obtained because T_1 is an internal force.

Insight is needed in order to decide how to choose a body. In the problem just described, three choices were available. Any two of these choices *combined* leads to a complete solution. As you work through examples and end-of-chapter problems, this sort of insight will come more easily.

The Accelerometer

A device that measures acceleration is called an accelerometer. A plumb bob, for instance, can be used to measure horizontal accelerations in a van (Figure 6.22) and is therefore an accelerometer. The angle between the unaccelerated plumb line and its direction when attached to the roof of the accelerating van can be related to the acceleration. In both cases, accelerated and unaccelerated, the plumb bob must be motionless relative to the van. To achieve this in practice it may be necessary to hold the bob in position momentarily in order to eliminate oscillations.

To find the van's acceleration, we choose the x-axis as horizontal. With T as the tension in the plumb line we write

$$F_x = +T \sin \theta$$

$$F_y = +T \cos \theta - W$$

for the components of the net force acting on the plumb bob (Figure 6.23). According to Newton's second law we have

$$T \sin \theta = ma_x$$

$$T \cos \theta - W = ma_y$$

Substituting $a_x = a$, $a_y = 0$, and $W = mg$, we obtain for a the expression

$$a = g \tan \theta$$

This is the desired relation. From a measurement of the angle θ we calculate the acceleration of the bob, and therefore of the van. If the plumb line hangs at $\theta = 45°$, the van's horizontal acceleration equals g.

Figure 6.22

A van moving horizontally to the right at constant speed (a), or at rest (b), is unaccelerated. In either case, a plumb bob hangs straight down from the van ceiling. If the van undergoes a constant acceleration to the right (c), the plumb bob will hang down from the ceiling at an angle θ with the vertical.

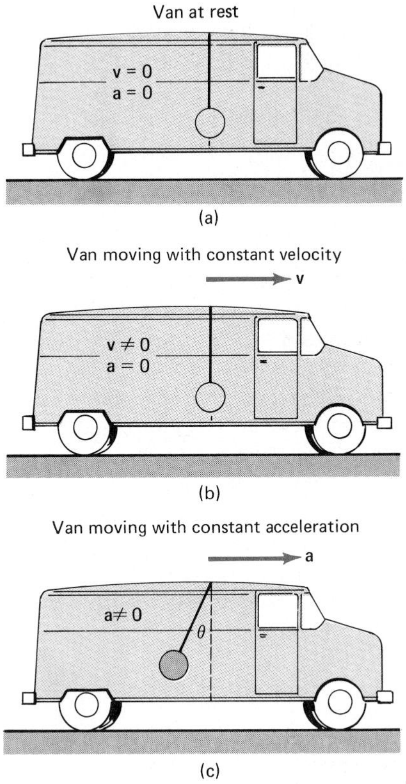

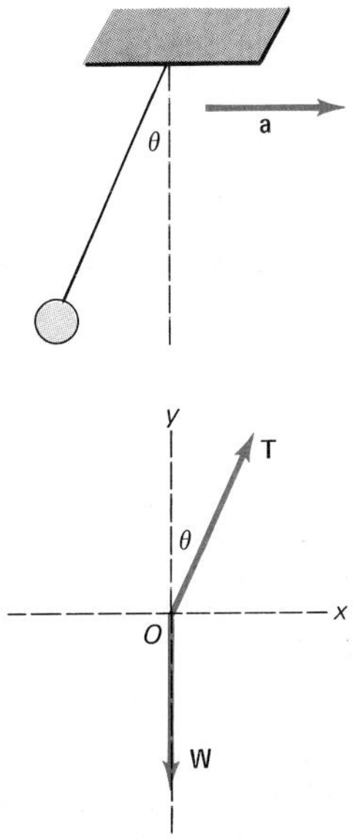

Figure 6.23

Picture the plumb bob hanging at the angle θ from the accelerating van in Figure 6.22c. Forces acting on the plumb bob are its weight, **W**, and the string tension, **T**.

Accelerated Motion on a Hill

A van of mass m is moving downhill and encounters an icy stretch in the road where the friction is negligibly small. We wish to determine the acceleration of the van and the force exerted on it by the road in terms of the inclination angle θ, g, and m.

We pick the van as a body. The two forces acting on the van are the normal force, exerted perpendicular to the road, and the weight, directed downward. The van is constrained to move parallel to the road, and therefore axes should be selected parallel and perpendicular to the road, as shown in Figure 6.24. This choice of axes simplifies the solution by making $a_y = 0$. Applying Newton's second law, we get

$$\sum F_x = +mg \sin\theta = ma_x$$

$$\sum F_y = +N - mg\cos\theta = 0$$

Solving for a_x in the first equation and N in the second equation, we get

$$a_x = g\sin\theta \qquad \text{and} \qquad N = mg\cos\theta$$

We should check the plausibility of this solution. For $\theta = 0$ the road is horizontal and supports the weight, and no horizontal acceleration is expected. Thus if $\theta = 0$ we expect $a_x = 0$ and $N = mg$. This conclusion is confirmed by setting $\theta = 0$ in the solution.

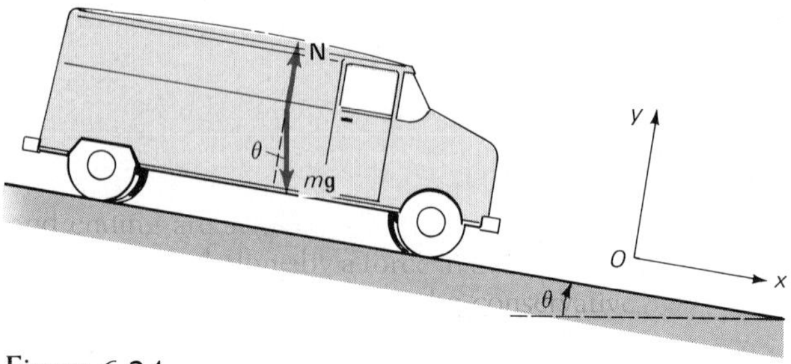

Figure 6.24

A van skidding on an icy (frictionless) hill experiences two forces. These are its weight, $m\mathbf{g}$, and the normal force, **N**, exerted perpendicular to the slope.

When $\theta = 90°$ the van becomes a freely falling object. We expect it to fall with the acceleration of gravity. Therefore, if $\theta = 90°$ we expect $a_x = g$ and $N = 0$. This too is confirmed by setting $\theta = 90°$ in the solution. Although the $\theta = 90°$ plausibility test may seem a bit unrealistic to you, both situations ($\theta = 0$ and $\theta = 90°$) are possible and are relatively quick tests of the solution.

We can generalize the preceding problem by considering the van moving down the same hill, but in the presence of friction. At a moment when the van velocity is v, the driver sees a deer ahead and slams on the brakes. The van begins to skid. The available stopping distance is x_o, and the coefficient of kinetic friction is μ_k. Will the deer—which is paralyzed by fright—be hit or not?

We pick the van as a body and choose axes as drawn in Figure 6.25 because the acceleration is known to be parallel to the plane. We write the component equations of Newton's second law, taking advantage of the fact that $a_y = 0$ (the van stays on the road):

$$\sum F_x = -f + mg\sin\theta = ma_x$$

$$\sum F_y = +N - mg\cos\theta = 0$$

In addition, we have

$$f = \mu_k N$$

We multiply the y-component equation by μ_k and add it to the x-component equation to eliminate both f and N to obtain

$$a_x = g(\sin\theta - \mu_k\cos\theta)$$

For a given angle θ and coefficient of kinetic friction μ_k this will be the actual acceleration of the van. The van has an initial velocity v_{xo}, and must stop in a distance $x \leq +x_o$ to avoid a collision. To calculate the stopping distance we use the equation (Section 4.6)

$$v_x^2 = v_{xo}^2 + 2a_x x$$

and set $v_x = 0$ to obtain

$$x = -\frac{v_{xo}^2}{2a_x}$$

The minus sign results in a positive value for x because the acceleration is directed up the plane of the hill (negative).

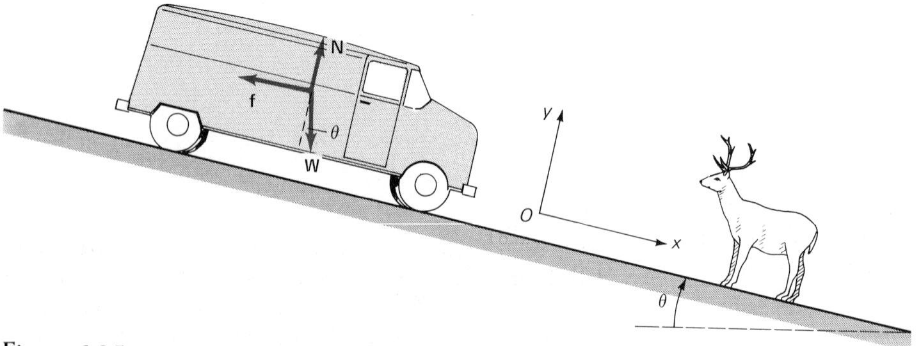

Figure 6.25

The van skids down the hill toward the deer. Forces acting on the van include friction, **f**, the normal force, **N**, and the van weight, **W**.

We require $x \le x_o$ for no collision. This gives the condition

$$-\frac{v_{xo}^2}{2a_x} \le x_o$$

Substituting for a_x yields

$$\frac{-v_{xo}^2}{2g(\sin\theta - \mu_k\cos\theta)} \le x_o$$

Rearranging, we obtain the condition

$$v_{xo}^2 \le 2gx_o(\mu_k\cos\theta - \sin\theta)$$

which must be met if the deer is not to be struck.

Example 10 **Was the Deer Hit?**

Let the deer be located 30 m from the car. From Table 6.1, $\mu_k = 1.02$ for rubber on dry concrete. Therefore, for $\theta = 5°$ and $v_{xo} = 80$ km/hr (22.2 m/s), we have for the stopping distance of the car

$$x = +\frac{(22.2\text{ m/s})^2}{2 \times 9.80(1.02\cos 5° - \sin 5°)}$$

$$= +27.1\text{ m}$$

Because the deer was located 30 m from the van when the driver applied the brakes, the van stops 2.9 m short of the deer.

Weight at the Equator and in an Accelerating Rocket

The reading on a spring scale does not disclose the weight (mg) of a person who is accelerating. Consider an astronaut standing on a scale in a rocket that is leaving the launching pad. We pick the astronaut as a body and write Newton's second law.

$$F = ma$$

Two forces contribute to the net force, F. One is the weight, mg, of the astronaut, and the other is the normal force, N, exerted on the astronaut by the scale. Thus we have

$$F = N - mg$$

Combining the two expressions for F gives the result

$$N = m(g + a) \tag{6.19}$$

for the force exerted on the astronaut by the scale. The scale reading equals N. When the rocket is at rest on the launching pad, mg is the scale reading and is equal to the weight of the astronaut. When the rocket accelerates upward, $a > 0$ and $N > mg$; the scale reading exceeds the weight of the astronaut. We see that

if $a > 0$ then $N > mg$

if $a = 0$ then $N = mg$

if $a < 0$ then $N < mg$

If the astronaut experiences $a = -g$, then $N = 0$. This condition is often referred to as weightlessness. If the astronaut stands on the scale, it will read zero.

In Figure 6.26 the acceleration of the astronaut is linear. However, the effect just described is also observed with centripetal acceleration. For example, in Figure 6.27 a person is shown standing at the equator. The diagram is drawn from the viewpoint of someone who is looking at the earth from the North Star. The earth spin is counterclockwise, and the person on the equator travels in a circular orbit whose radius is the earth's radius. Two forces act on this person. The weight is directed toward the center of the earth, and the normal force is directed outward. Applying Newton's second law gives

$$\sum F = mg - N = ma_c$$

where a_c is the person's centripetal acceleration. We solve for N to get

$$N = m(g - a_c)$$

Clearly N, equal to the scale reading, is less than the weight. In Section 5.3 we calculated $a_c = 3.39 \times 10^{-2}$ m/s^2. This value is much smaller than $g = 9.80$ m/s^2. The scale reading is less than the weight because part of the weight must provide for the centripetal acceleration. Only the remainder of the weight is available to produce the scale reading. For a 200-lb man the scale reads about $199\frac{1}{3}$ lb because of the man's centripetal acceleration.

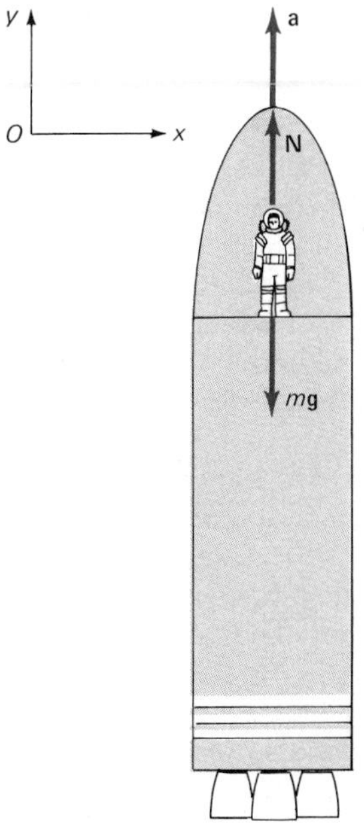

Figure 6.26

The rocket accelerates upward. Forces acting on the astronaut inside the rocket include his weight, $m\mathbf{g}$, and the normal force, **N**. When the astronaut stands on a scale, the scale must exert the force **N**. The force exerted on the scale by the astronaut is equal in magnitude to **N**, and the scale reading is also equal to **N**. With any acceleration, therefore, the scale reading will differ from his weight.

Figure 6.27

As viewed from the North Star, a person standing on the equator moves counterclockwise on a circular path due to the spin of the earth. This person would experience a centripetal acceleration directed toward the center of the earth. The net force acting is the vector sum of the person's weight and the normal force.

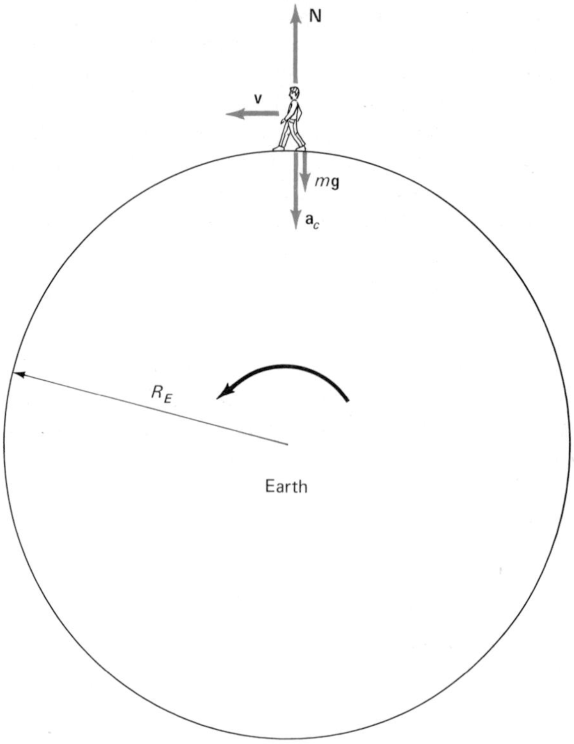

Kepler's Third Law

We now reconsider Kepler's third law in order to (1) exhibit its derivation from the law of gravity and Newton's second law, (2) illustrate its use in obtaining mean planetary or satellite orbit radii in terms of orbital periods, and (3) show how the mass of a central body can be obtained if it has a satellite whose mass is relatively small.

1 Consider an object of mass m_1 moving at constant speed v in a circular orbit of radius r about an object of mass m (Figure 6.28). We assume that $m \gg m_1$. Newton's second law requires that the force exerted on m_1 equal the mass of m_1 times its centripetal acceleration. Thus

$$F = m_1 a_c$$

The only force acting is gravity, and so we have

$$F = G\frac{mm_1}{r^2}$$

Substituting $a_c = v^2/r$ in the first equation gives

$$F = m_1\frac{v^2}{r}$$

By eliminating F we obtain

$$m_1\frac{v^2}{r} = G\frac{mm_1}{r^2}$$

Canceling m_1 gives us

$$\frac{v^2}{r} = \frac{Gm}{r^2}$$

To obtain Kepler's third law, we substitute $v = 2\pi r/T$ where T is the period. This gives

$$\frac{1}{r}\left(\frac{2\pi r}{T}\right)^2 = \frac{Gm}{r^2}$$

After canceling r and solving for T^2 we get

$$T^2 = \left(\frac{4\pi^2}{Gm}\right)r^3 \qquad (6.20)$$

The square of the period is proportional to the cube of the radius. Equation 6.20 is Kepler's third law. This law may be generalized to elliptic orbits in which r is the semimajor axis (often called the mean radius). Note that this derivation reverses the logic used in Example 3.

2 If the same central body (m fixed) has two satellites, then Eq. 6.20 may be written once for each satellite. Upon division of one equation by the other, the constant $(4\pi^2/Gm)$ cancels, leaving

$$\left(\frac{T_1}{T_2}\right)^2 = \left(\frac{r_1}{r_2}\right)^3 \qquad (6.21)$$

Generally the periods are relatively easy to measure. Thus using Eq. 6.22 enables us to calculate r_2 in terms of r_1 and the period ratio. In this way the mean radii of all the planet orbits can be expressed in terms of one standard. For the solar system the standard used is 1 astronomical unit (AU)—the mean radius of the earth's orbit.

3 We may solve Eq. 6.20 for m to obtain

$$m = \frac{4\pi^2 r^3}{GT^2} \qquad (6.22)$$

This equation expresses the mass of the attracting body in terms of the period and mean radius of one of its satellites. The mass of any celestial object with a small satellite can be determined in this way. Astronomers have used this technique to determine the masses of moons, planets, stars, galaxies, and galactic clusters.

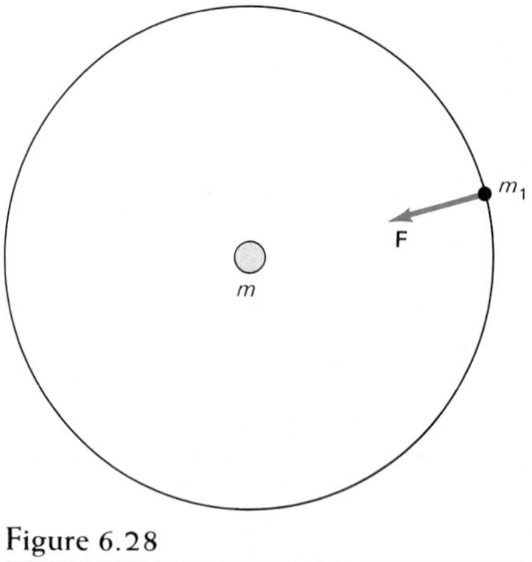

Figure 6.28

An object (mass $= m_1$) in a circular orbit about an object of mass m.

Example 11 **Mass of the Earth**

An artificial satellite placed in orbit about the earth has a period of 90 minutes and a mean radius of 6.65×10^6 m. Use these data to determine the mass of the earth ($G = 6.67 \times 10^{-11}$ N · m²/kg²).

We use Eq. 6.22.

$$M_E = \frac{4\pi^2 r^3}{GT^2}$$

$$= \frac{4\pi^2(6.65 \times 10^6 \text{ m})^3}{(6.67 \times 10^{-11} \text{ N} \cdot \text{m}^2/\text{kg}^2)(5.4 \times 10^3 \text{ s})^2}$$

$$= 5.98 \times 10^{24} \text{ kg}$$

From Newton's second law and his law of gravity, we derived Kepler's third law. Now, using the experimental value of the gravitational constant, G, we have calculated the mass of the earth.

Questions

25. A single force **F** acts on a particle moving with velocity **v**. What can be said about the relative directions of (a) **F** and **v**; (b) **a** and **v**; (c) **F** and **a**?

26. If according to Newton's third law forces must always occur in pairs, how could a particle experience a single force?

27. In space an astronaut is said to be weightless. What actions of the astronaut would reveal that his mass is not zero? How could the astronaut tell which of two objects had the larger mass?

28. In an elevator, a bathroom scale may read your weight correctly, or the reading may be too low or too high. Describe the state of motion of the elevator for each case.

29. At an amusement park, passengers ride a vertical loop-the-loop moving with constant speed. If a passenger is seated on a bathroom scale, how would the reading compare at the top, the bottom, and halfway down the ride?

Summary

An observer for whom every unaccelerated particle remains unaccelerated is called an inertial observer.

Newton's first law states that every body continues in its state of rest, or of constant velocity, unless it is caused to change that state by forces exerted on it.

A body is unaffected by its environment if it moves with constant velocity, or is at rest, relative to an inertial observer. Such a particle experiences zero net force.

The net force exerted on the standard kilogram is in the direction of its acceleration, and numerically equal to it (in SI units). This fact can be used to calibrate a horizontal spring force scale in newtons. Experiment confirms that forces may be added as vector quantities.

Newton's second law states that the acceleration of a particle is equal to the ratio of the net force acting on the particle to the inertial mass of the particle. Symbolically,

$$\mathbf{a} = \frac{\mathbf{F}}{m} \quad \text{or} \quad \mathbf{F} = m\mathbf{a} \qquad (6.1)$$

where m is the inertial mass, **a** is the acceleration of the particle, and **F** is the net force acting.

Newton's second law may be taken as an operational definition of (a) *force,* based on the acceleration of a standard kilogram, and of (b) *mass*, based on the acceleration produced by a known force.

Newton's third law states that to every action there is always opposed an equal reaction. This means that forces in nature occur in pairs.

Laws of force provide for the calculation of the force on a particle in terms of the properties of the particle and of its environment. One fundamental force law is the law of gravity. Between point masses the gravitational attraction is

$$F = G\frac{m_1 m_2}{r^2} \qquad (6.3)$$

Weight is the force of gravity exerted on a body. The weight of a body depends on its location. Mass is an intrinsic property of a body, and does not depend on its environment.

Empirical force laws for solids and fluids include the following.

Linear spring	$F = -kx$	(6.12)
Sliding friction	$F = \mu_k N$	(6.13)
Low-velocity drag (Stokes' law)	$F_D = 6\pi\eta r v$	(6.14)

Suggested Reading

R. H. Dicke, The Eötvös Experiment, *Sci. Amer.* 205, 84 (December 1961).

T. S. Kuhn, *The Copernican Revolution,* Random House, New York, 1959.

Problems

Section 6.2 Newton's First Law

1. A particle moves in a straight line. Its positions (x) at various moments of time are displayed in Table 1. Determine whether or not the particle remains free of a net force over the entire interval.

Table 1

x (m)	t (s)
1.1	0
2.2	1
3.4	2
4.8	3
6.0	4
7.0	5

2. The position (x) is shown plotted versus time (t) for three types of "motion" of the same particle (Figure 1). Which type(s) of motion indicate that zero net force acts on the particle?

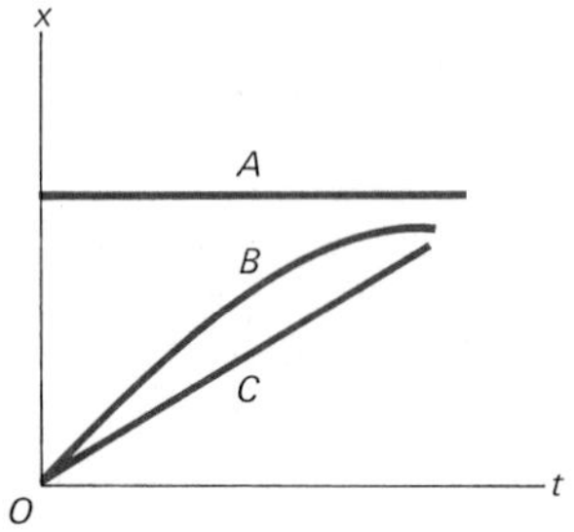

Figure 1

Section 6.3 Force

3. The discus used by women athletes has a weight of 2.2 lb. What is its weight in newtons?

4. By pulling on a string attached to a standard kilogram, a student gives it the acceleration 2.3**j** m/s^2. Determine the force exerted on the string by the standard kilogram.

5. A standard kilogram is observed to accelerate horizontally at 10 m/s^2 at 30° north of east (Figure 2). One of the two forces acting on the kilogram is of magnitude 5 N and is directed north. Obtain the magnitude and direction of the second force acting on the kilogram.

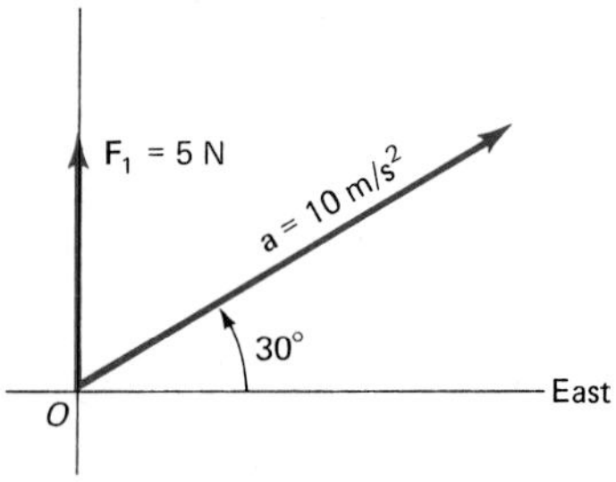

Figure 2

6. Forces $\mathbf{F}_1$ (east) and $\mathbf{F}_2$ (north) are applied to a standard kilogram (Figure 3). As a result, its acceleration is 10 m/s^2 at 30° north of east. (a) Give the magnitude and direction of the net force exerted on the standard kilogram. (b) Determine $\mathbf{F}_1$ and $\mathbf{F}_2$.

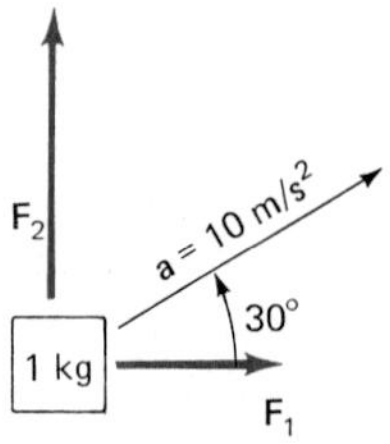

Figure 3

7. Horizontal forces of magnitude 10 N and 15 N are applied to the standard kilogram. (a) Determine the maximum and minimum accelerations for arbitrary horizontal directions of the 15-N force if the 10-N force is always east. (b) Determine the direction(s) of the 15-N force if the acceleration is 10 m/s^2. (c) Determine the direction(s) of the acceleration in (b).

Section 6.4 Newton's Second Law

8. A 1000-kg auto accelerates at 4 m/s^2. What is the net force on it?

9. A 0.2-kg mass attached to the end of a string is whirled in a vertical circle by a student. At some position the mass experiences a downward force of -62**j** N and a horizontal force of 38**i** N. Determine the acceleration at this position.

10. Forces of 10 N north, 20 N east, and 15 N south are simultaneously applied to a 4-kg mass (Figure 4). Obtain its acceleration.

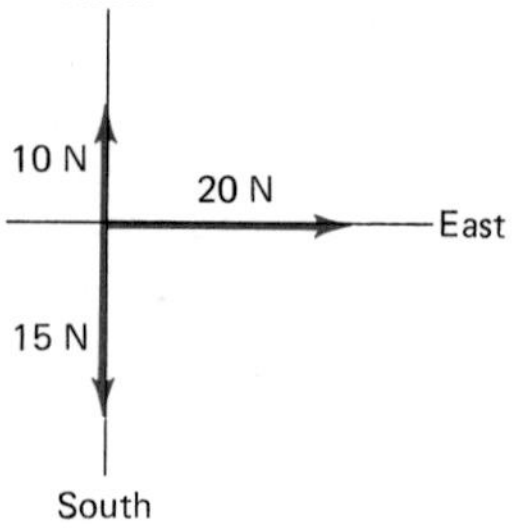

Figure 4

11. A force is applied to a 2-kg mass giving it an acceleration of 3 m/s^2 east. The same force results in a 1-m/s^2 east acceleration of the mass m. (a) Determine the applied force. (b) Determine m.

12. What force would be needed (in Problem 11) to give m an acceleration of 3 m/s^2 east?

13. Forces $\mathbf{F}_1$ (east) and $\mathbf{F}_2$ are simultaneously applied to a 3-kg mass. When $\mathbf{F}_2$ is east, $\mathbf{a} = 5$ m/s^2 east, and when $\mathbf{F}_2$ is west, $\mathbf{a} = 1$ m/s^2 east. Determine the magnitudes of F_1 and F_2.

14. A constant force $\mathbf{F}$ changes the velocity of an 80-kg sprinter from 3 m/s to 4 m/s in 0.5 s. (a) Calculate the acceleration of the sprinter. (b) Obtain the magnitude of the constant force. (c) Determine the acceleration of a 50-kg sprinter experiencing the same force. (Assume linear motion.)

15. A 1000-kg automobile moving at 20 m/s is brought to rest in a distance of 80 m by the constant net force $\mathbf{F}$. Determine (a) the acceleration of the automobile, and (b) the magnitude of $\mathbf{F}$.

16. An air-hockey puck ($m = 0.1$ kg) moves without friction in a circle of radius 0.5 m on a horizontal table (Figure 5). The puck is attached to the center of rotation by a string of negligible mass. The speed of the puck is 6 m/s. Determine (a) the centripetal acceleration of the puck, and (b) the string tension. (c) The 0.1-kg puck is replaced by a 0.4-kg puck. The speed of the new puck is adjusted until the string tension equals the tension in parts (a) and (b). What is the speed of the new puck?

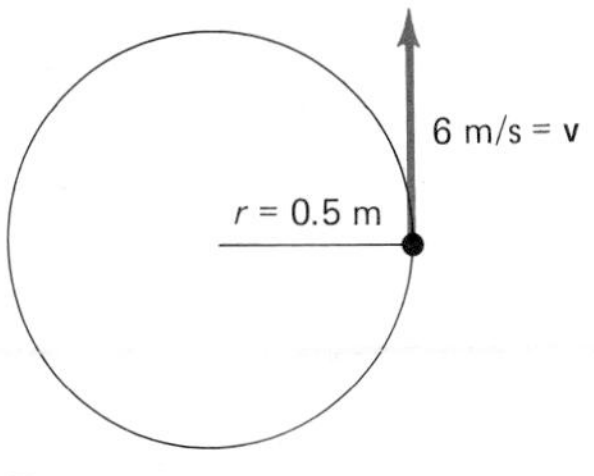

Figure 5

Section 6.5 Newton's Third Law

17. The mass of the earth is 5.98×10^{24} kg. If the earth's gravitational force causes a 60-kg student to accelerate "downward" at 9.8 m/s^2, calculate the "upward" acceleration of the earth resulting from the student's gravitational reaction force acting on the earth.

18. The weight of an 80-kg person has a magnitude of 176 lb directed radially toward the center of the earth. Give the magnitude and direction of the (gravitational) force exerted on the earth by this person which, together with the weight, forms an action–reaction pair.

Section 6.6 The Universal Force of Gravitation

19. Determine the acceleration of gravity at an altitude of 100 mi.

20. Determine the gravitational force exerted by the earth on an 80-kg man at a distance of 250,000 miles from the center of the earth (roughly the distance to the moon).

21. At the moment of a total solar eclipse the moon lies along a line from the earth to the sun (Figure 6). Your weight, as indicated by a scale, is affected by the gravitational pulls of the sun and moon. (a) Compare the gravitational forces exerted on you by the moon and sun with the gravitational force exerted on you by the earth. Express your answers as ratios of the forces rather than as the forces themselves. (b) If a total eclipse occurs at high noon, do the combined gravitational pulls of the moon and sun increase or decrease the scale reading? If your weight is 800 N, what is the combined pull of the moon and sun?

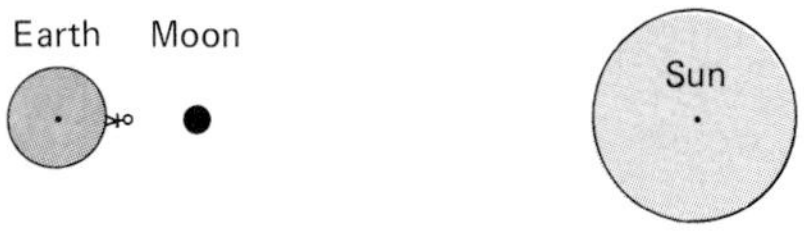

Figure 6

22. Calculate the gravitational force between two bowling balls, each with a mass of 7.27 kg, that are 1 m apart. Compare this force with the weight of one bowling ball.

23. At a point along the line joining the centers of the earth and moon their gravitational forces exerted on a third body cancel. (a) Determine the distance of this point from the center of the earth. (b) Repeat the calculation for the earth–sun system.

24. The ratio of the mean radii is 4:1 for the two planetary ellipses shown in Figure 6.6b. Determine the period ratio for the planets in these orbits.

25. A hypothetical planet has five moons. The moons move in circular orbits. Table 2, listing orbit radii and periods, contains one erroneous *radius* entry. Use Kepler's third law to discover which entry is in error and determine the correct value.

26. A lead sphere has a radius of 0.05 m. (a) Determine the force of gravity exerted on the sphere by the earth at the earth's surface. (b) Calculate the gravitational force exerted by the sphere on an identical sphere if the two spheres just touch. (Take the density of lead to be 11,300 kg/m^3.)

27. On the earth a weight lifter can lift a mass of 200 kg. If we assume no change in his strength, what mass could he lift on the moon?

28. According to Eq. 6.4, $g = GM_E/R_E^2$ where $g = 9.80$ m/s^2. Confirm this equation explicitly by evaluating GM_E/R_E^2 numerically.

Table 2

Moon	r (10^6 m)	T (days)
I	$\frac{1}{4}$	$\frac{1}{8}$
II	1	1
III	2	2.83
IV	3	6
V	4	8

29. Some stars evolve into small dense objects called neutron stars. Consider a neutron star whose mass is $m = 2.8 \times 10^{30}$ kg and whose radius is $r = 1.2 \times 10^4$ m. (a) Calculate the average density of the star (for comparison, atomic nuclei have $\rho \sim 2 \times 10^{17}$ kg/m^3). (b) Calculate the acceleration of gravity at the surface of the star. (c) Assume the star spins with period T. Determine the value of T for which the acceleration of gravity equals the centripetal acceleration for a particle on the equator. (d) Obtain the speed of this particle (for comparison, the speed of light $c = 3 \times 10^8$ m/s).

Section 6.7 Weight and Mass

30. Determine the weight of an 80-kg man in newtons and in pounds.

31. If you weigh 120 lb at the surface of the earth, how much do you weigh at an altitude of 6370 km? (The radius of the earth is 6370 km.)

32. Matter in a neutron star may have a mass density $\rho \cong 10^{17}$ kg/m^3. Consider the science fiction situation of a 75-kg person lying on the surface of a 5-m-radius sphere of this material (neutronium). (a) Calculate the local gravitational acceleration at the surface of the sphere. (b) How much does the person weigh (in newtons)? (c) How much does the person weigh in tons?

33. A group of people decide to construct a brick tower to the sky. As the tower rises, it shares in the rotation of the earth. At what altitude will the ratio of the weight of a brick to its mass be equal to its centripetal acceleration?

34. (a) From your weight in pounds on the earth calculate your mass in kilograms. (b) Obtain your weight in pounds on the moon from your mass in kilograms.

35. (a) At what altitude above the surface of the moon would the acceleration of gravity be 1 m/s^2? (b) A very heavy man weighs 40 lb on the moon's surface. What will he weigh at the altitude determined in part (a)?

36. At what altitude above the earth is a person's weight reduced by (a) 0.1%? (b) 1%? (c) 10%?

Section 6.8 Mechanical Force Laws

37. A horizontal spring with an unstretched length of 0.20 m stretches to a length of 0.25 m (Figure 7) when a force of 5 N is applied. (a) Determine the force constant, k, of the spring. (b) Obtain the acceleration of a 5-kg mass when the spring length is 0.15 m. (c) Repeat part (b) for 0.50 m.

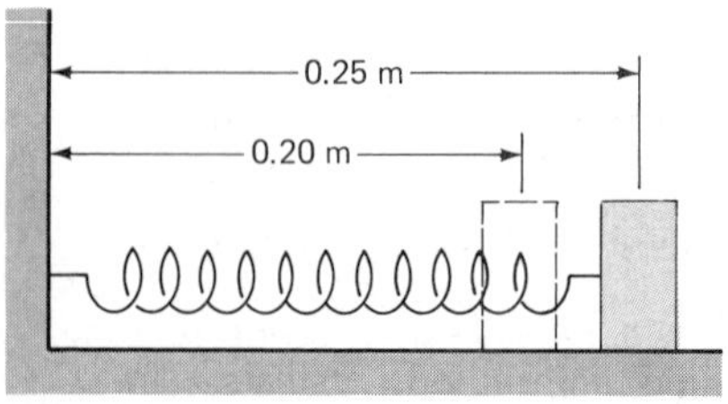

Figure 7

38. A Slinky is a common toy consisting of a coiled spring about 8 cm in diameter. When hung vertically, a 100-g mass attached to its free end stretches the Slinky 6.2 ft. Determine its spring constant in newtons per meter.

39. A spring of negligible mass has a spring constant $k = 500$ N/m. One end of the spring is attached to the ceiling. A slab of beef is attached to the other end. The weight stretches the spring 42 cm. What is the weight of the beef? (Answer in newtons and pounds.)

40. A 10-kg mass is attached to a horizontal spring obeying Hooke's law (Eq. 6.12) with $k = 200$ N/m. The mass is pulled to $x = +0.25$ m and released, and it moves back and forth in the interval $-0.25 \text{ m} \leq x \leq +0.25$ m. Neglect friction and restrict your answers to the given interval. Give the value of x for which (a) the force is a maximum in the $+x$-direction; (b) the force is zero; (c) the acceleration is a maximum in the $-x$-direction; (d) the acceleration is zero. Sketch (e) the force (F) versus the displacement (x) and (f) the acceleration (a) versus the displacement (x).

41. A spring obeying Hooke's law has $k = 50$ N/m. When suspended vertically (Figure 8) its equilibrium length is 0.5 m. Determine the new equilibrium length when a 0.5-kg object is attached.

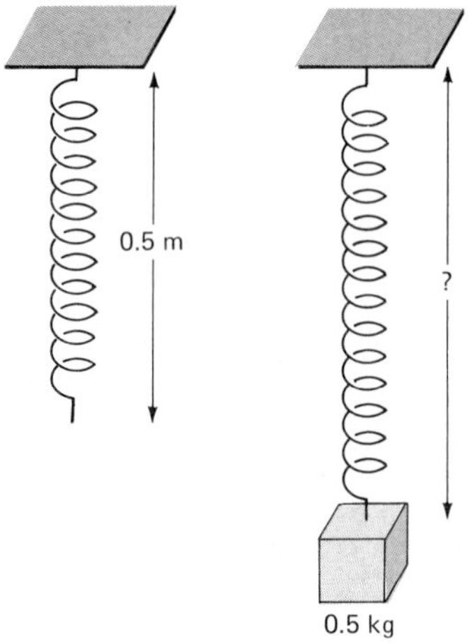

Figure 8

42. A spring satisfying Hooke's law has spring constant k. Determine the spring constant of a new spring constructed by putting two of the original springs (a) end to end; (b) side by side.

43. Determine the stopping distance of the skier in Example 8 if her initial speed is (a) 10 m/s or (b) 30 m/s.

44. A fireman whose mass is 75 kg slides down a 6-m pole in a time of 2 s. The force of friction exerted on the fireman by the pole is constant. Determine (a) the acceleration of the fireman, and (b) the force of friction.

45. A crate of mass m slides across the warehouse floor (Figure 9). The acceleration, a, of the crate is opposite in direction to its motion. Assume the force of friction is constant and prove that $\mu_k = a/g$.

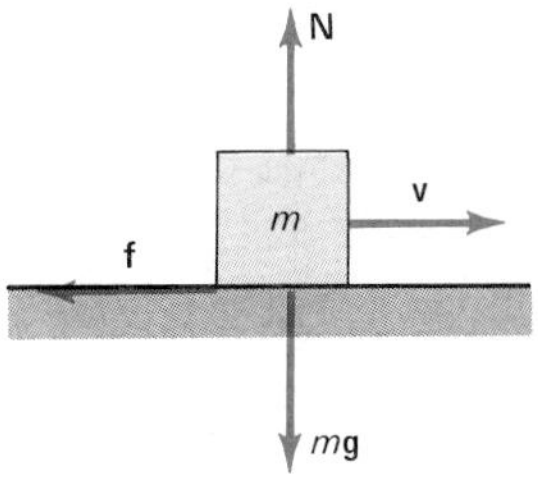

Figure 9

46. About 200 years ago C. A. Coulomb invented the tribometer, a device employed to investigate static friction. The instrument is represented schematically in Figure 10. To determine the coefficient of static friction, the hanging mass m_2 is increased or decreased as necessary until m_1 is on the verge of sliding. Prove that $\mu_s = m_2/m_1$.

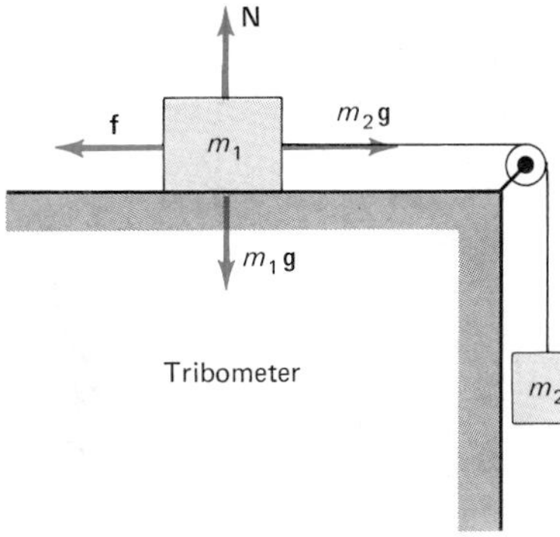

Figure 10

47. On a dry road a 1000-kg car moving at 30 m/s requires a minimum distance of 50 m to come to rest without skidding after the brakes are applied (Figure 11). Determine (a) the car's acceleration, (b) the frictional force exerted by the road, and (c) the coefficient of static friction. If the road is icy, then $\mu_s = 0.11$. If the car moves initially at 30 m/s on an icy road, determine (d) the maximum frictional force, and (e) the minimum stopping distance.

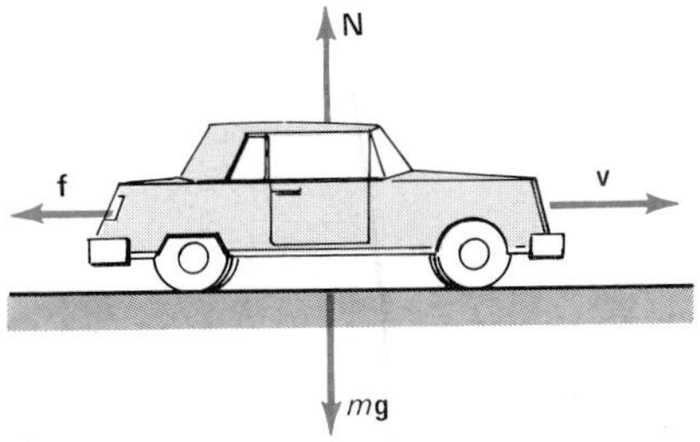

Figure 11

48. The minimum stopping distance is 35 m for a particular 1200-kg car after the brakes have been applied if the initial speed is 24.6 m/s (55 mph). (a) Determine the time required for the car to come to rest. Tests show that the engine is capable of accelerating the car from rest to 24.6 m/s in 12 s. (b) Calculate this accelerating distance. If road friction rather than the engine limited the acceleration, then (c) what would be the minimum time to accelerate from rest to 24.6 m/s? (Assume no skidding and constant acceleration.)

49. An automobile moving at 20 m/s brakes to a stop in the minimum distance without skidding. The driver of a following car moving at 30 m/s sees the brake lights, applies the brakes after a 0.1 s delay to react, and brakes to a stop in the minimum distance, also without skidding. Assume that $\mu_s = 0.75$ for both cars. Calculate the minimum distance between cars at the instant the driver of the lead car applies the brakes if a collision from the rear is to be avoided. (Assume constant acceleration.)

50. An automobile driver applies the brakes when moving at 20 m/s. The coefficients of static and kinetic friction are $\mu_s = 0.90$ and $\mu_k = 0.75$. (a) How much farther will the car travel skidding to a stop than if the car comes to rest on the verge of skidding? (b) Repeat for an initial speed of 30 m/s.

51. A spherical oil droplet ($\rho = 900$ kg/m^3) whose radius is 5×10^{-8} m is emitted from the top of a smokestack 350 m above the ground. Estimate the time required for the oil droplet to settle to the ground if its descent is governed by Stokes' law. (Take $\eta = 1.9 \times 10^{-5}$ kg/m · s for air and $g = 9.80$ m/s^2.)

52. Particulates emerging from a smokestack settle vertically in accord with Stokes' law as they are carried horizontally by a 10-mph wind (Figure 12). Consider two spherical particulates ($\rho = 2000$ kg/m^3) after they emerge from a 350-m smokestack. Assume a perfectly flat earth and estimate the horizontal distance from the smokestack where the spheres (radii are 1.5×10^{-6} m and 2×10^{-4} m) strike the ground. Use $\eta = 1.9 \times 10^{-5}$ kg/m · s for air.

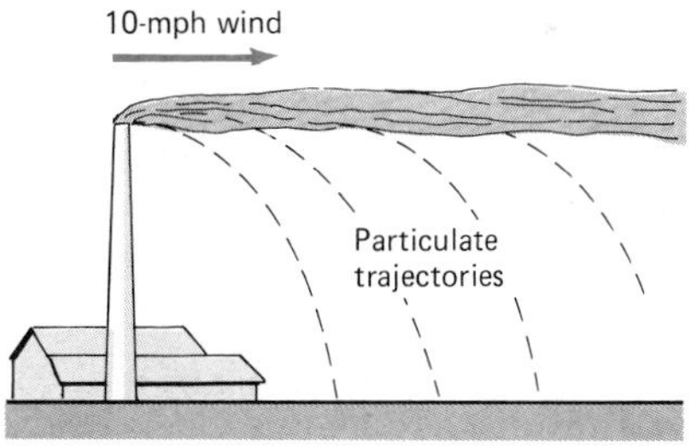

Figure 12

53. To determine the viscosity of a liquid, a steel sphere 1 mm in radius is timed as it falls through a 0.1-m distance in the liquid at terminal velocity. To account for buoyancy, the weight of the sphere in the liquid is used. Assume that Stokes' law is obeyed and determine the viscosities of the three liquids listed in the table below.

Liquid	Weight in liquid	Time to fall 0.1 m
Water	2.9×10^{-4} N	6.5×10^{-3} s
Glycerin	2.8×10^{-4} N	5.0 s
Sucrose	2.7×10^{-4} N	1.9×10^{4} s

Section 6.9 Applications of Newton's Laws

54. Two ice skaters are at rest on a frozen canal 50 m wide, and stand a distance x from one side (Figure 13). They push each other apart, and then coast without friction, arriving at the same time on opposite sides of the canal. Their masses are 80 kg and 50 kg, and during the acceleration period of 0.5 s they push on each other with a constant force of 400 N. Determine the (a) acceleration of each skater during the push; (b) speeds v_1 and v_2; and (c) accelerating distance for each skater. (d) Locate the starting point. (e) Calculate the total time (accelerating plus coasting time) required for either skater to reach the canal side.

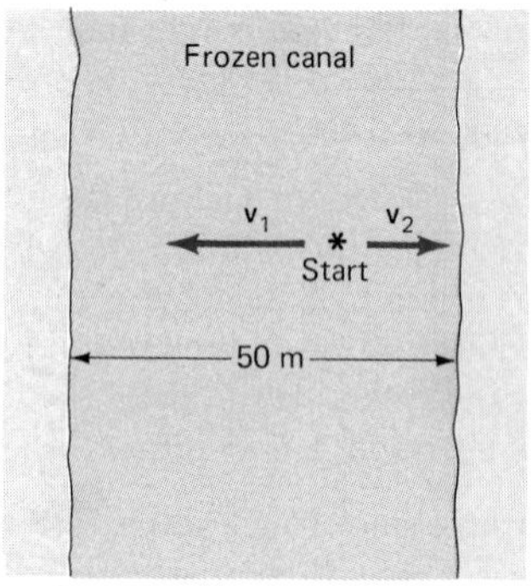

Figure 13

55. Two blocks connected by a string move as a unit horizontally on a frictionless surface (Figure 14). The acceleration $a = 3$ m/s^2 with $m_1 = 2$ kg and $m_2 = 4$ kg. (a) Determine each of the tensions T_1 and T_2. (b) For the same acceleration, repeat the calculation if the two blocks are interchanged.

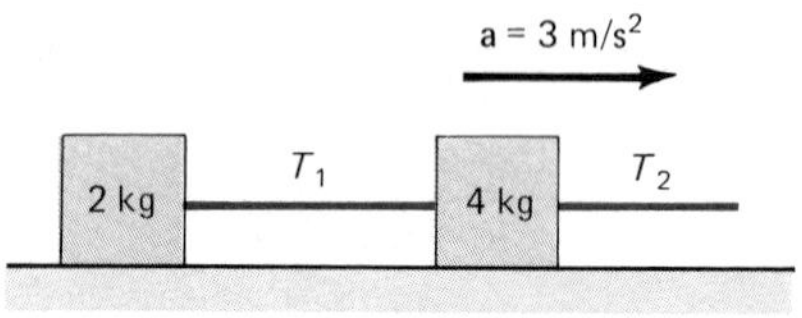

Figure 14

56. The driver of a car traveling 60 km/hr applies the brakes and skids 30 m to a stop. Determine the coefficient of kinetic friction between the tires and the road.

57. In an acceleration test a van (Figure 15) is brought from rest to 25 m/s in 10 s. At constant acceleration (a) what angle with the vertical will a cord make that is hanging from the rear view mirror? (b) What acceleration time for the same final speed would give rise to an angle of 30°?

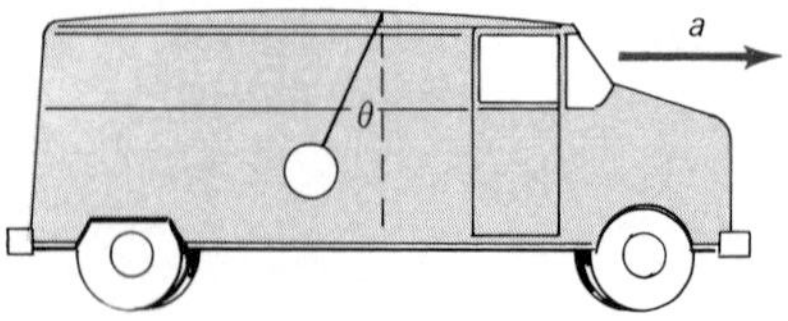

Figure 15

58. A 1000-kg automobile is moving up a 10° hill at 25 m/s (Figure 16). It strikes an icy stretch and coasts to a stop without friction in a distance x. Determine the (a) force N exerted by the road, (b) the acceleration down the hill, (c) the time required to come to rest, and (d) the stopping distance. (Compare with Problem 60.)

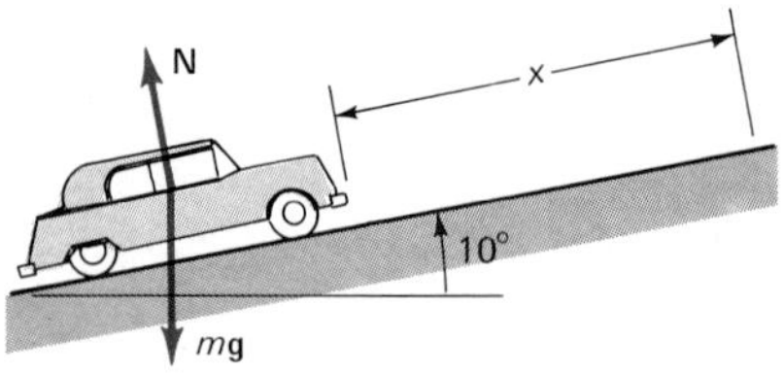

Figure 16

59. A worker piles one box on top of another and pushes on the bottom box (Figure 17). The two boxes move ahead together with an acceleration of 1 m/s^2. What horizontal force does the bottom box exert on the upper box?

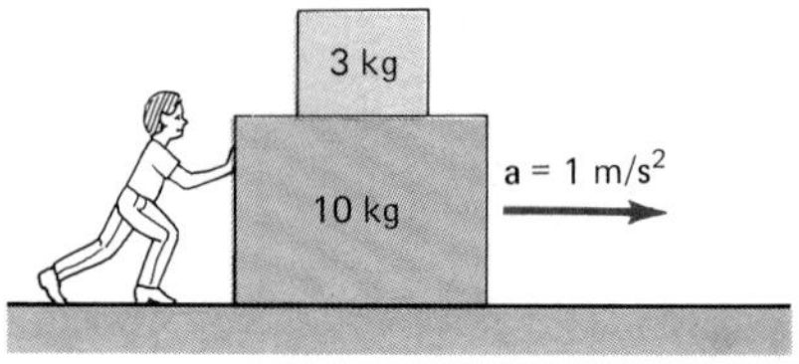

Figure 17

60. An 80-kg skier (Figure 18) moving initially at 10 m/s glides up a 15° slope ($\mu_k = 0.18$). He coasts up the slope to a stop, and then glides back to the starting point. Determine each of the following quantities for the trip up and for the trip back: (a) the forces f and N; (b) the acceleration of the skier; (c) the distance traveled; and (d) the time required. (Compare with Problem 58)

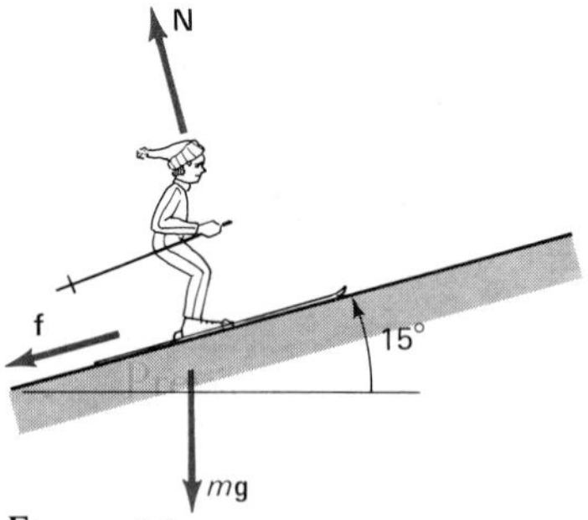

Figure 18

61. A block slides down a plane at constant speed (Figure 19). Prove that the angle of the plane, θ, and the coefficient of kinetic friction, μ_k, are related according to $\mu_k = \tan\theta$.

62. Imagine an 80-kg man standing on a spring scale (calibrated in newtons) in an elevator. (a) The elevator accelerates downward at 3 m/s^2. What is the scale reading? (b) When the scale reading is 1000 N, what is the acceleration of the elevator?

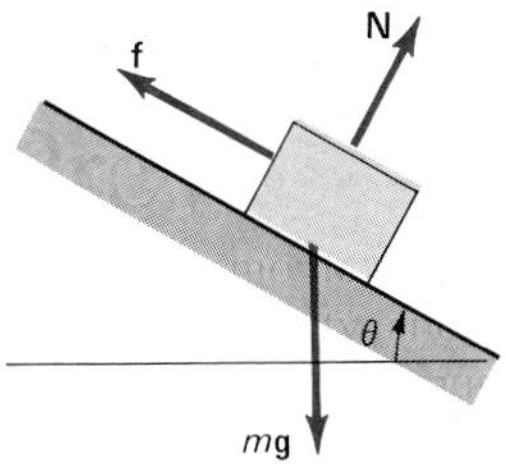

Figure 19

63. A person stands on a scale in an elevator (Figure 20). The maximum and minimum scale readings are 591 N and 391 N, respectively. Assume the magnitude of elevator acceleration is the same during starting and stopping, and determine (a) the weight of the person, (b) the person's mass, and (c) the elevator acceleration.

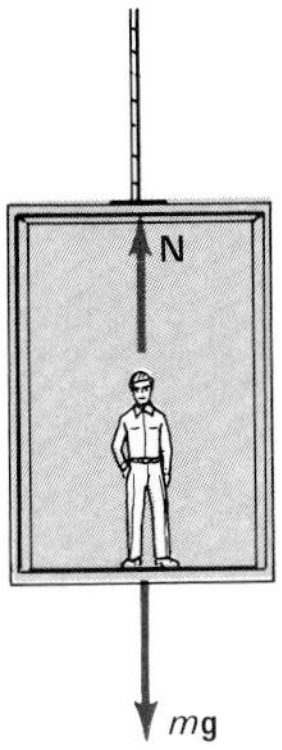

Figure 20

64. Look up the period of the earth going around the sun (in seconds) and the mean radius of the earth's orbit (in meters). Using Kepler's third law, calculate the mass of the sun.

65. Look up the period of the moon going around the earth and the mean radius of the moon's orbit. Using Kepler's third law, calculate the mass of the earth.

66. The mean orbital distance of Pluto from the sun is 39.44 AU. Calculate its period in years.

67. Use Kepler's third law to calculate the altitude of a synchronous satellite (a satellite having a period of 1 sidereal day).

68. With the aid of a binomial expansion, show that Newton's law of gravity leads to a constant acceleration

$$a = \frac{GM_E}{R_E^2}$$

as long as the height h above the earth's surface is very small compared to R_E ($2h/R_E \ll 1$).

69. In Larry Niven's science fiction story *Ringworld,* a solid ring of material rotates about a star (Figure 21). The rotational speed of the ring is 1.25×10^6 m/s. Its radius is 1.53×10^{11} m. The inhabitants of this ring world experience a normal contact force, N, that produces an inward acceleration of 9.90 m/s^2. Additionally, the star at the center of the ring exerts a gravitational force on the ring and its inhabitants. (a) Show that the total centripetal acceleration of the inhabitants is 10.2 m/s^2. (b) The difference between the total acceleration and the acceleration provided by the contact force is due to the gravitational attraction of the central star. Show that the mass of the star is approximately 10^{32} kg.

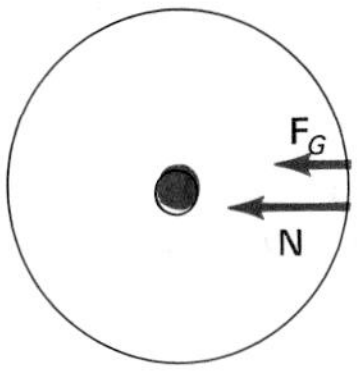

Figure 21

70. The diagram in Figure 22 is of Atwood's machine. Assume that the pulley is frictionless and has no effect other than to change the direction of the string. (a) Treat the string together with the two masses as a single body, and prove that the system acceleration (assume $m_2 > m_1$) is

$$a = g\frac{m_2 - m_1}{m_2 + m_1}$$

(b) Determine the tension in the connecting cord.

71. A conical pendulum is a ball moving in a horizontal circle at the end of a long wire (Figure 23). The angle between the wire and the vertical does not change. Consider a conical pendulum (m = 80 kg) with a 10-m wire making a 5° angle with the vertical. (a) Determine the tension in the wire and its horizontal and vertical components. (b) Determine the radial acceleration of the ball. (c) Determine the total force exerted on the wire by the ball. (Use Newton's third law, and neglect the mass of the wire.)

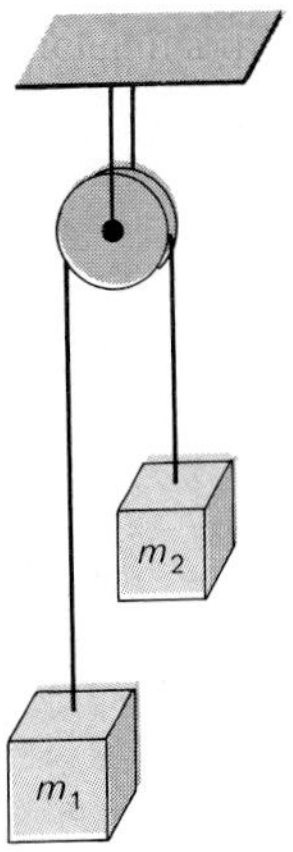

Figure 22

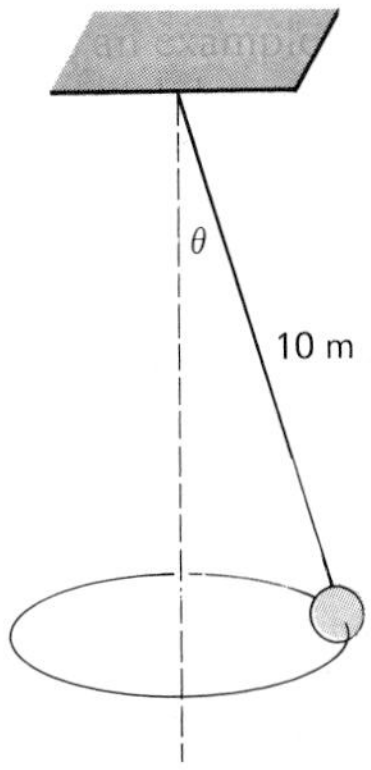

Figure 23

chapter 7

WORK, ENERGY, AND POWER

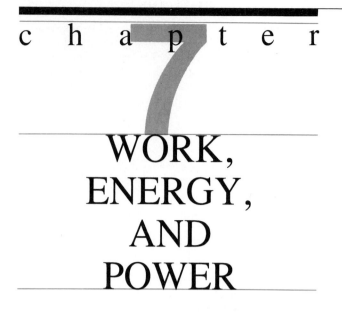

Preview

In Chapter 6 we learned how Newton's laws of motion enable us to predict future values of the position and velocity of a particle. In this chapter we introduce a second way of relating force to particle motion. The scalar product of force and displacement defines work, and half the product of the mass of a particle and the square of its velocity defines its kinetic energy. Using the concepts of work and kinetic energy, we derive the work–energy principle in Section 7.4. The work–energy principle is the central idea of this chapter, and plays a role analogous to that of Newton's second law.

The work–energy principle is of special importance because it leads to the law of conservation of energy (Chapter 8). A conservation law states that under specified conditions a particular quantity (energy in this case) remains constant.

In later chapters we will consider two additional ways of relating force to motion that also lead to conservation laws. The product of force and time defines linear impulse, and linear impulse is a prelude to the law of conservation of linear momentum (Chapter 9). The vector product of position and force defines torque, and torque is a prelude to the law of conservation of angular momentum (Chapter 11).

7.1 Work Done by a Constant Force

Force represents the influence of the environment on a particle (Section 6.3). In response to the **net force,** a particle accelerates. We now extend this idea by introducing a quantity—the work done by a force—that measures the effect of the force during the displacement of the particle.

Consider the special case of a body that experiences a constant force **F** while undergoing a displacement $\Delta\mathbf{s}$ (see Figure 7.1) as it moves from O to P. The force that the body experiences has the same magnitude and direction at every point between O and P.

We define ΔW, the work done by the force **F** in displacing the body $\Delta\mathbf{s}$, as the scalar product of **F** and $\Delta\mathbf{s}$. (See Section 2.4 for a discussion of the scalar product of two vectors.)

$$\Delta W = \mathbf{F} \cdot \Delta\mathbf{s} = F \cos\theta \, \Delta s \qquad (7.1)$$

The sign of ΔW is plus or minus, depending on the sign of $\cos\theta$ where θ is the angle between **F** and $\Delta\mathbf{s}$. We see, then, that the work done can be either positive or negative. Note also that $F\cos\theta$ is the projection of **F** on $\Delta\mathbf{s}$; hence an alternative form of ΔW is

$$\Delta W = F_s \, \Delta s \qquad (7.2)$$

where $F_s = F\cos\theta$.

Because work is a derived quantity, its units are determined by the defining equation (Eq. 7.1). According to Eq. 7.1 with **F** and $\Delta\mathbf{s}$ parallel, one unit of work is the

Figure 7.1

The constant force **F**, drawn acting on an object at points O and P, has the same magnitude and direction at all points. This force **F** will do an amount of work equal to $F \, \Delta s \cos \theta$ on an object displaced $\Delta \mathbf{s}$ from point O to point P.

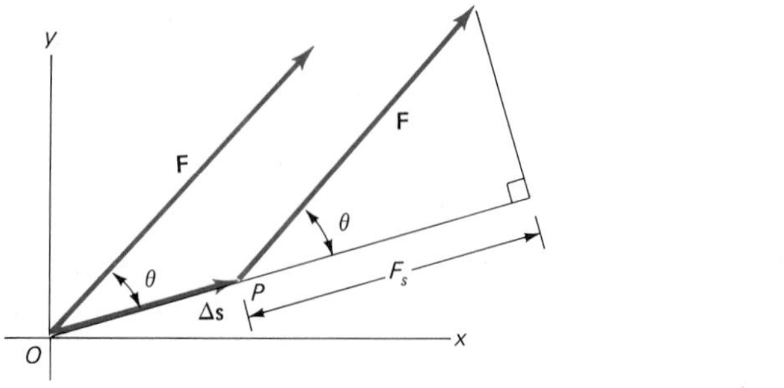

Figure 7.2

For a particle moving at constant speed in a circle the displacement, $\Delta \mathbf{s}$, is tangent to the circumference, whereas the net force **F** is directed toward the center. The work done by **F** in this case equals zero.

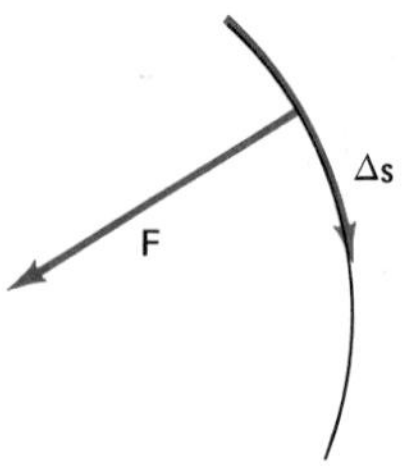

product of a force of 1 newton and a displacement of 1 meter. Therefore, the SI unit of work is 1 newton · meter, which we call the joule;

$$1 \text{ joule} = 1 \text{ newton} \cdot \text{meter} \quad \text{or} \quad 1 \text{ J} = 1 \text{ N} \cdot \text{m}.$$

Because $1 \text{ N} = 1 \text{ kg} \cdot \text{m/s}^2$ we may also write

$$1 \text{ J} = 1 \text{ kg} \cdot \text{m}^2/\text{s}^2$$

Example 1 **Work Done by Gravity**

Consider a sky diver falling through a distance L. What is the work done on the sky diver by the force of gravity (which is presumed constant) during the fall?

The magnitude of the force vector is

$$F = mg$$

and the magnitude of the displacement vector is

$$\Delta s = L$$

These two vectors are parallel; therefore $\theta = 0$ and $\cos \theta = +1$. Using Eq. 7.1 the work done by gravity becomes

$$\Delta W = F \, \Delta s$$

or

$$\Delta W = +mgL$$

For a sky diver having $m = 66$ kg and falling $L = 1000$ m,

$$\Delta W = +66 \times 9.80 \times 1000 \text{ kg} \cdot \frac{m}{s^2} \cdot m$$

$$\Delta W = 6.47 \times 10^5 \text{ J}$$

Whether or not other forces act on the sky diver is irrelevant to this calculation. The work done on the sky diver by the force of gravity does not depend on her velocity or on the presence or absence of other forces.

According to Eq. 7.1, $\Delta W = 0$ if any of the factors (F, Δs, $\cos \theta$) equals zero. That $\Delta W = 0$ if $F = 0$ is not surprising when we recall that work was defined to represent one way that force influences the motion of a particle. In the absence of force there is no influence, and hence, no work.

If $\Delta s = 0$, then $\Delta W = 0$ regardless of the size of the force that is acting. In other words, if a force does not cause a displacement of a particle, then it does not do work on that particle. This idea may contradict your everyday understanding of work. Consider, for example, a weightlifter exerting an upward force on a barbell. This force does work while he lifts the barbell from the floor ($\Delta s \neq 0$), but it does no work while he holds the barbell motionless above his head after the lift ($\Delta s = 0$). This does not mean that holding a barbell overhead is an easy task! It does mean that work, as we have defined it, is not involved. Now imagine a stalemated tug-of-war game. Neither team does work through the force exerted on the rope because the displacement of the rope is zero. It is possible to become exhausted by exerting a force that does no work!

When $\cos \theta = 0$, $\Delta W = 0$. This means that $\theta = 90°$, and **F** is perpendicular to $\Delta \mathbf{s}$. A charged particle moving in a circular orbit under the influence of a magnetic force, and a planet moving in a circular orbit under the influence of the gravitational force of the sun, are examples of motion with **F** perpendicular to $\Delta \mathbf{s}$. The displacement is along the circumference of the orbit while the force is radial (see Figure 7.2). Clearly the force influences the motion in these cases, and yet the work done is zero. This result is a direct consequence of our definition of work, and nicely illustrates the fact that work does not *completely* represent the effect of the environment on particle motion. The work done can be zero even when the environment influences the particle motion.

Example 2 **Work Done on a Truck**

Consider a truck of mass m moving up a hill as shown in Figure 7.3. What is the work done by gravity? The angle between the force, $m\mathbf{g}$, and $\Delta \mathbf{s}$ is $90° + \theta$, and we have from Eq. 7.1

$$\Delta W = mg \, \Delta s \cos(90° + \theta)$$

or

$$\Delta W = -mg \, \Delta s \sin \theta$$

From the geometry of Figure 7.3 the angle θ is both the angle between $m\mathbf{g}$ and the normal to the road, and the angle of the hill. Hence we may write $\Delta s \sin \theta = \Delta y$, which gives

$$\Delta W = -mg \, \Delta y$$

Figure 7.3

The displacement, $\Delta\mathbf{s}$, of the truck makes the angle $(90° + \theta)$ with the weight $m\mathbf{g}$.

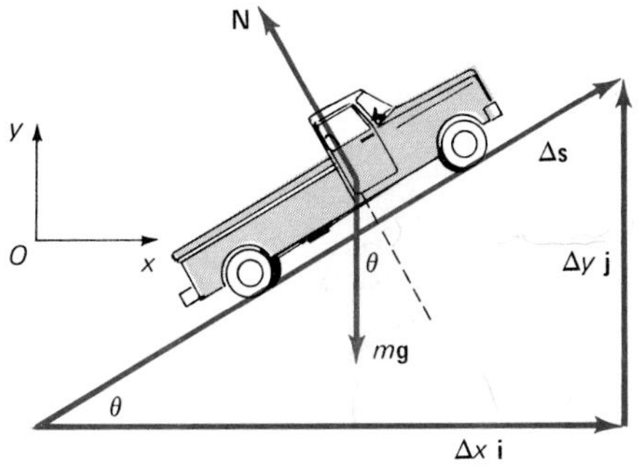

This is consistent with the result of Example 1 except for the minus sign. Here the vertical part of the displacement is opposite in direction to the force, whereas in Example 1 they were parallel. To bring out this aspect of the calculation we resolve $\Delta\mathbf{s}$ into components parallel and perpendicular to $m\mathbf{g}$, and recall (Section 2.4) that the scalar product can be expressed in component form. This gives

$$\begin{aligned}\Delta W &= \mathbf{F} \cdot \Delta\mathbf{s} \\ &= (\mathbf{i}F_x + \mathbf{j}F_y + \mathbf{k}F_z) \cdot (\mathbf{i}\,\Delta x + \mathbf{j}\Delta y + \mathbf{k}\,\Delta z) \\ &= F_x\,\Delta x + F_y\,\Delta y + F_z\,\Delta z\end{aligned}$$

In this case we have

$$\mathbf{F} = -mg\,\mathbf{j}$$

and

$$\Delta\mathbf{s} = \mathbf{i}\,\Delta x + \mathbf{j}\,\Delta y$$

so that

$$\begin{aligned}\Delta W &= (-mg\mathbf{j}) \cdot (\mathbf{i}\,\Delta x + \mathbf{j}\,\Delta y) \\ &= -mg\,\Delta x\,(\mathbf{j} \cdot \mathbf{i}) - mg\,\Delta y\,(\mathbf{j} \cdot \mathbf{j})\end{aligned}$$

The first term equals zero and the second term gives the previous result.

Note that the work done by the normal force $\mathbf{N}$ must equal zero since $\mathbf{N}$ and $\Delta\mathbf{s}$ are perpendicular. If we write

$$\mathbf{F}_{\text{net}} = \mathbf{N} + m\mathbf{g}$$

then the work done by $\mathbf{F}_{\text{net}}$ equals the work done by the weight, since $\mathbf{N}$ does no work. The work done by the net force is important in the work–energy principle, discussed in Section 7.4.

Example 3 Walking Down a Flight of Stairs

Suppose you walk down a flight of 12 stairs to a landing located at a distance L below the starting elevation, as shown in Figure 7.4. Normally the trip would consist of 12 falls through a vertical distance $(L/12)$ alternating with 12 horizontal moves through the same distance. What is the work done on you by gravity in terms of L, g and your mass m?

According to Eq. 7.1, $\Delta W = \mathbf{F} \cdot \Delta\mathbf{s}$. In each of the 12 horizontal moves $\mathbf{F}$ and $\Delta\mathbf{s}$ are perpendicular. Hence, the work done by gravity during the horizontal part of the trip is zero, or $\Delta W = 0$.

In each of the 12 vertical falls, $\mathbf{F} = -mg\mathbf{j}$ and $\Delta\mathbf{s} = -(L/12)\mathbf{j}$. Hence the work done by gravity for one of the falls is

$$\begin{aligned}W &= (-mg\mathbf{j}) \cdot \left(-\frac{L}{12}\mathbf{j}\right) \\ &= +\frac{1}{12}\,mgL\end{aligned}$$

You make 12 vertical falls going from A to B. Therefore, the work done by gravity for the complete trip is

$$W = +mgL$$

Note that the total work done by gravity is independent of the number of steps in the flight of stairs. For instance, the same result $(W = +mgL)$ would have been obtained with one very large step $(A \rightarrow B)$ or with four large steps $(A \rightarrow D \rightarrow E \rightarrow F \rightarrow B)$ Figure 7.5. Indeed, the work done by gravity is independent of the path taken from A to B. We return to this very important point in Section 8.1.

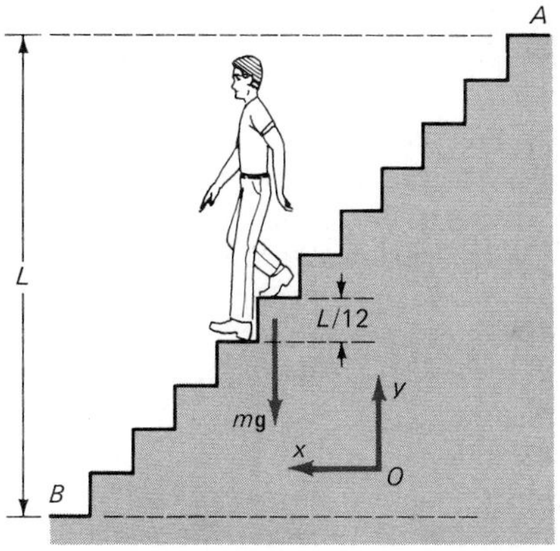

Figure 7.4

The motion is from A to B along the stairlike path.

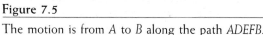

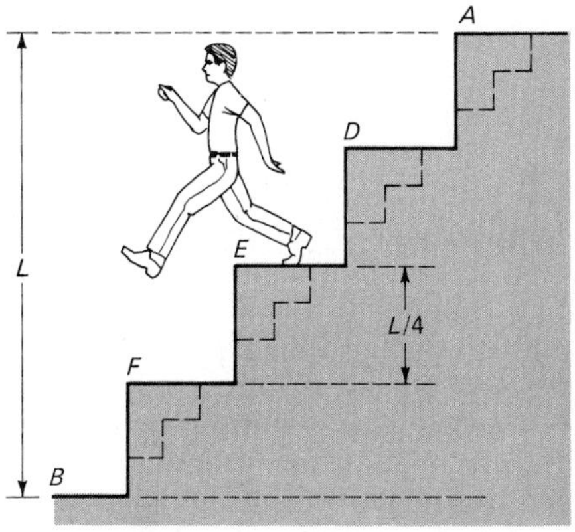

Figure 7.5

The motion is from A to B along the path $ADEFB$.

Questions

1. In order to strengthen their wrists and hands, athletes repeatedly squeeze soft rubber balls. Is the work done on a ball during compression positive, zero, or negative? Explain.
2. The word *work* means one thing in everyday language and another thing in physics. Construct two situations where the two meanings would agree. Construct two situations where the two meanings would disagree.
3. The work done by gravity on a horizontally moving object is zero. The work done on a swinging child by supporting chains also equals zero. Cite two other situations where the work equals zero because force and displacement are perpendicular.
4. Describe a situation where the work done by a constant force is increased by allowing the force to act over a larger distance.

7.2 Work Done by a Variable Force

As long as the force acting on a particle is constant, Eq. 7.2 can be used to calculate the work done by that force whether the displacement of the particle is large or small. This equation can also be used to estimate the work done by a variable force, providing the displacement is sufficiently small. For large displacements involving variable forces, we must generalize the definition of Eq. 7.2 by using the definite integral.

To visualize the difficulty posed by variable forces and large displacements consider the work done by the gravitational force of the earth on an 80-kg astronaut in orbit when he is displaced 1 m farther from the earth. The force of gravity, discussed in Section 6.6, is an inverse square law. When the separation of two particles is tripled, for example, the force is reduced by a factor of 3^2, or 9. At the surface of the earth the astronaut weighs 785 N, and we estimate the work due to the 1-m upward displacement of the astronaut to be -785 J. The minus sign indicates that the force and displacement are oppositely directed. At a distance of 1.27×10^7 m (an altitude equal to twice the earth's radius) the astronaut weighs only 87 N. Therefore, the 1-m upward displacement of the astronaut results in only -87 J of work. If the astronaut moves continuously from the surface of the earth to an altitude of 1.27×10^7 m, then how much work would be done by the gravitational force of the earth in this large displacement? In the first meter of the astronaut's displacement there is a contribution of -785 J to the total work, whereas in the final meter of displacement there is a contribution of -87 J of work. In each intervening meter there is a contribution to the total work of an amount that falls somewhere between -785 J and -85 J. The total work done will be closely

Figure 7.6

Work done by a constant force equals the shaded rectangular area.

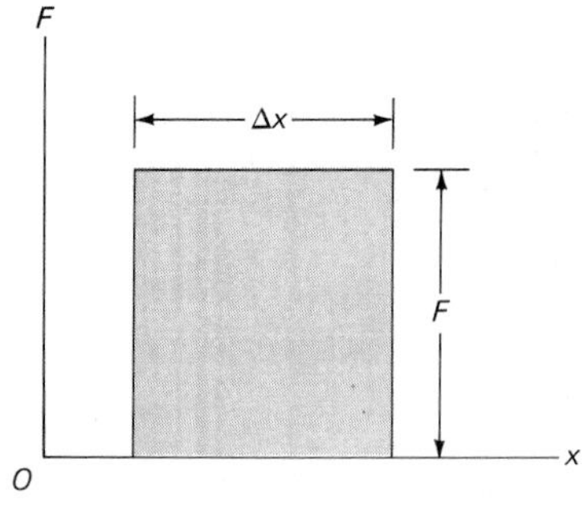

Figure 7.7

Work done by a variable force may be estimated geometrically by constructing a set of rectangles. Error in the estimate is caused by the shaded-in corner areas. The exact area under the curve equals the work done.

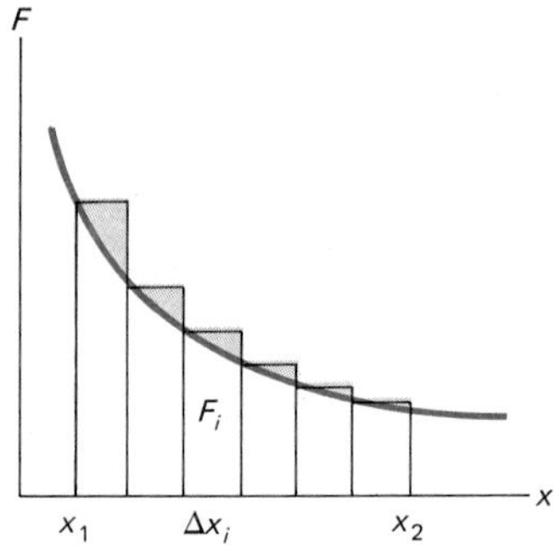

approximated by the sum of all of these amounts. Each amount will be in the form

$$\Delta W_i = -F_i \times (1 \text{ m})$$

where F_i ranges from 785 N to 87 N. We wish to find the value of the sum—the total work done.

$$W = \Sigma\, \Delta W_i = -\Sigma\, F_i \times (1 \text{ m})$$

This sum includes 1.27×10^7 terms! Yet you must learn how to evaluate sums like this one if you are to treat variable forces and large displacements. First let's see how such a sum can be evaluated geometrically.

Following the procedure suggested above in the continuous movement of an astronaut away from the earth, we can develop a geometric interpretation of the work done by the force F. This development parallels that used in Section 4.6. Figure 7.6 shows the graph for the one-dimensional case where F is constant. Equation 7.2 requires

$$\Delta W = F\, \Delta x$$

Geometrically this is the area of the shaded rectangle in Figure 7.6, and equals the area "under the curve."

If F varies with x, the area under the curve is still equal to the work done. We show a displacement curve for varying F in Figure 7.7. In order to find the area under the curve, we divide the region between x_1 and x_2 into a series of vertical strips. The area of the ith strip is $F_i\, \Delta x_i$, and should equal the work done by the force F_i for the displacement Δx_i. This is not exactly equal to the area under the curve, however, or to the work done for the displacement Δx_i, because of the small triangular area that does not lie under the curve. In Figure 7.7 you can see that for each vertical strip there is a triangular area—shaded in the figure—that does not lie under the curve. Therefore, the sum of the area of these vertical strips is an estimate, and not an exact measurement, of the area under the curve. However, in the limit as all $\Delta x_i \to 0$ the number of strips increases, and the combined areas of the small triangles approach zero, until the area under the curve exactly equals the work done by F for the finite displacement $\Delta x = x_2 - x_1$.

For example, in Figure 7.8 the gravitational force acting on the 80-kg astronaut is plotted versus the astronaut's distance from the center of the earth. In a displacement of the astronaut from the surface of the earth ($r = R_E$ to $r = 3R_E$) the work done by gravity equals the area labeled *ABCDEA*. We can determine this area by successive approximation using narrower and narrower strips, or rectangles.

First we divide the interval $R_E \leq r \leq 3R_E$ into six rectangles. The combined area of the six rectangles constitutes our first estimate of the area *ABCDEA*. Each rectangle has an area larger than the corresponding area bounded by the curve. Hence the estimated area is too large. The excess area is shaded in Figure 7.8. To reduce the error in our estimate we double the number of rectangles by halving the width of each one. This second approximation reduces the excess area and improves the estimate. Repeating this procedure further reduces the excess area until, in the limit as the width of the rectangles approaches zero, the estimate becomes the exact area *ABCDEA*.

The approach of the estimated area to the exact area is illustrated geometrically in Figure 7.8. The section of graph labeled *ABF* is drawn in four successive approximations. This same point is demonstrated numerically in Table 7.1. The combined areas of the rectangles in the first

Table 7.1

Number of Rectangles	$W \times 10^{-9}$ (J)
6	−4.16
12	−3.72
24	−3.52
48	−3.43
⋮	⋮
∞	−3.34

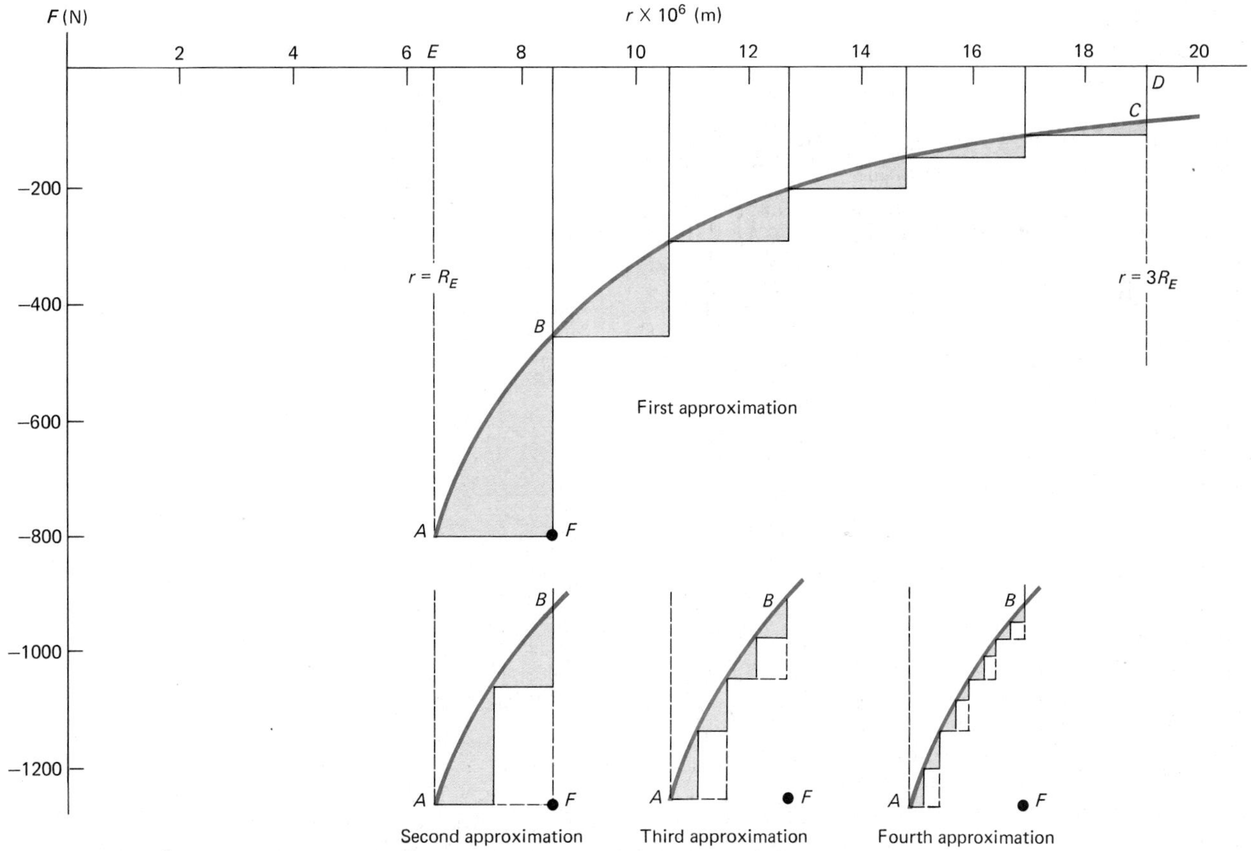

Figure 7.8

The gravitational force acting on a person who weighs 784 N at the earth's surface is shown versus distance from the center of the earth. To estimate the area, and thus the work done, for a displacement of the person from $r = R_E$ to $r = 3R_E$, six rectangles are drawn. The shaded area *ABF* is the error associated with the left rectangle. Doubling the number of rectangles reduces the error of this estimate as is shown by the second, third, and fourth approximations.

four approximations are calculated and displayed in the table, together with the exact result, -3.34×10^9 J, for comparison. Both the graph and the table suggest the approach to the limit. Evaluation of the area in appropriate units yields the work in joules. Later in this chapter you will learn how to obtain the exact result.

Equation 7.2 yields the correct result for the work done by a constant force. However, Eq. 7.2 does not, in general, give the correct result for the work done by a variable force. With a variable force, Eq. 7.2 gives only an estimate of the work done. Therefore we must derive a new equation that holds for both constant and variable forces. (This new equation is a generalization of Eq. 7.2.)

Let F_i represent the force exerted in the x-direction at coordinate x_i, and let Δx_i be a small displacement at x_i. We can use Eq. 7.2 to estimate the work done.

$$\Delta W \cong F_i \, \Delta x_i \tag{7.3}$$

in this displacement. The smaller we choose Δx_i, the more accurate will be our estimate. To determine the work done, W_{12}, in a finite displacement, $\Delta x = x_2 - x_1$, from the point x_1 to the point x_2, we must add all ΔW_i between x_1 and x_2. This gives

$$W_{12} \cong \sum \Delta W_i \cong \sum F_i \, \Delta x_i \tag{7.4}$$

as our estimate for the work done in the displacement from x_1 to x_2. To obtain an exact result instead of this estimate, two things are necessary. The approximation $W_{12} \cong \sum \Delta W_i$ must be made exact, and Eq. 7.3 must be made exact. Both objectives are attained by letting all $\Delta x_i \rightarrow 0$ while their number is increased without limit in such a way as to maintain exactly $\Delta x = \sum \Delta x_i$ at every stage of the limiting process. The definite integral from x_1 to x_2 is defined to be the limit of the sum

$$\int_{x_1}^{x_2} F \, dx = \lim_{\Delta x_i \rightarrow 0} \sum F_i \, \Delta x_i \tag{7.5}$$

The general definition of work for one-dimensional motion is

$$W_{12} = \int_{x_1}^{x_2} F \, dx \tag{7.6}$$

The integral symbol $\int$, can be thought of as a stretched s, for *sum*. Equation 7.6 is read as "The work done by the force F in going from x_1 to x_2 equals the integral of F from x_1 to x_2."

Geometrically the definite integral equals the area under the F versus x curve, and thus equals the work done. The work done is positive, zero, or negative, according to the sign of the area. The area above the x-axis is positive (Figure 7.9a), giving $W > 0$; that below the x-axis is negative (Figure 7.9b), giving $W < 0$; and if equal positive and negative areas occur (Figure 7.9c), then $W = 0$.

When the force is constant, F can be factored out of the integral in Eq. 7.6. Thus

$$\begin{aligned} W_{12} &= \int_{x_1}^{x_2} F \, dx \\ &= F \int_{x_1}^{x_2} dx \\ &= F(x_2 - x_1) \\ &= F \, \Delta x \end{aligned}$$

This shows that the definition of work given in Eq. 7.6 includes Eq. 7.2 as a special case. Equation 7.6 is a general equation, whereas Eq. 7.2 is the definition of work for the special case where the force is constant. Some useful integrals are listed in Appendix 4.

Example 4 Work Done on an Astronaut

What is the work done by the force of gravity on a 80-kg astronaut in a displacement from the surface of the earth (point A in Figure 7.10) to a point whose altitude is 2 earth radii (point B in Figure 7.10)? We estimated this value earlier, using Figure 7.8. Using 7.6 we have

$$W = -\int_{R_E}^{3R_E} \frac{GM_E m}{r^2} \, dr$$

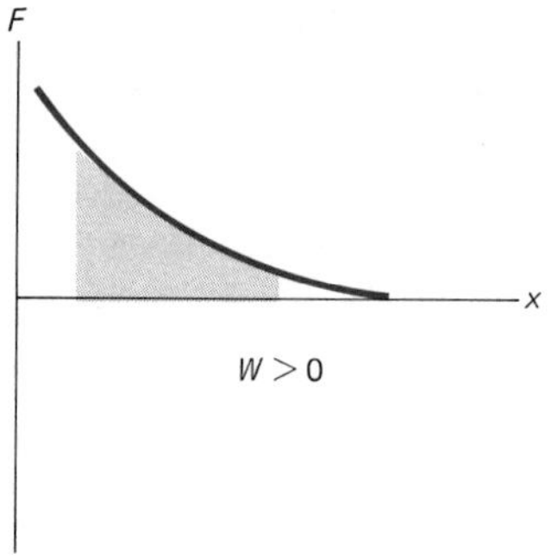

Figure 7.9a

When $F > 0$ the work and area are positive.

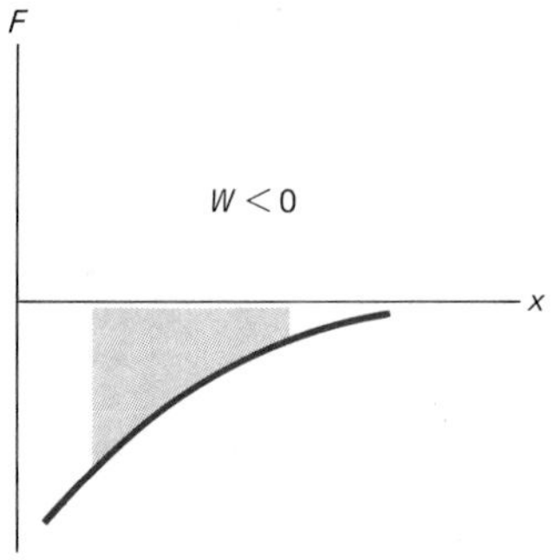

Figure 7.9b

When $F < 0$ the work and area are negative.

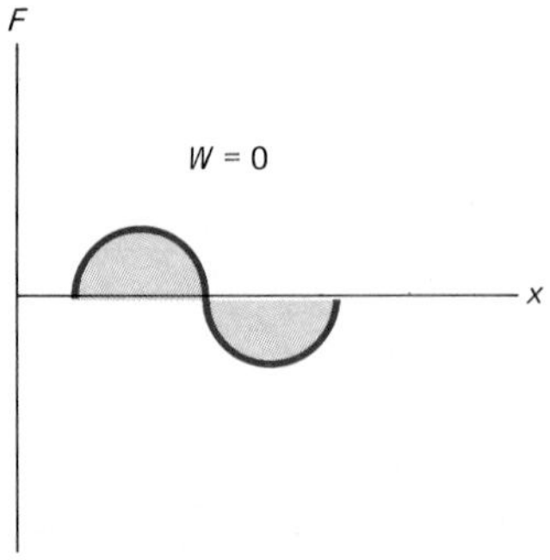

Figure 7.9c

Plus and minus areas can lead to zero net work.

where $x_1 = R_E$ and $x_2 = 3R_E$. Here

$$F = -\frac{GM_E m}{r^2}$$

from Section 7.6 (positive r-direction is *up*; F is *down* and thus negative), and $dx = dr$. From Appendix 4 we have

$$\int \frac{dr}{r^2} = -\frac{1}{r}$$

so

$$W = +\frac{GM_E m}{r}\Bigg|_{R_E}^{3R_E} = GM_E m\left(\frac{1}{3R_E} - \frac{1}{R_E}\right)$$

or

$$W = -\frac{2}{3}\frac{GM_E m}{R_E}$$

The work done is negative because the force is directed toward the earth (down) while the displacement is directed away from the earth (up).

Solving this equation for the 80-kg astronaut, we have

$$W = -\frac{2}{3}\,\frac{6.67 \times 10^{-11}\ \mathrm{N \cdot m^2/kg^2} \times 5.98 \times 10^{24}\ \mathrm{kg} \times 80\ \mathrm{kg}}{6.37 \times 10^6\ \mathrm{m}}$$

$$= -3.34 \times 10^9\ \mathrm{J}$$

This is the limiting, or exact, value given earlier in Table 7.1. Alternatively, using $g = GM_E/R_E^2$, we can write

$$W = -\frac{2}{3}mgR_E$$

which gives the same numerical result.

The force is plotted versus r in Figure 7.11. Geometrically the work equals the shaded area. The negative result in Example 4 corresponds to the area below the r-axis. If the astronaut in Example 4 had moved from B to A instead of from A to B, the work done on him would have been numerically the same, only positive: $+3.34 \times 10^9$ J. Thus, an object falling toward the earth has positive work done on it by gravity.

Example 5 Work Done by a Spring

Consider the work done by a spring force while the spring is stretching. The form of the spring force is

$$F = -kx \qquad (6.12)$$

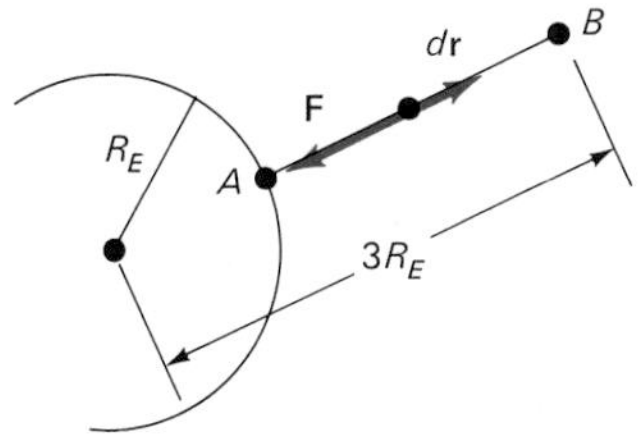

Figure 7.10

In a displacement $d\mathbf{r}$ away from the earth the force and displacement are opposite in direction. The work is negative.

The work done as the spring stretches from $x = 0$ to $x = x_1$ is the shaded area in Figure 7.12. This area equals

$$\text{area} = \frac{1}{2}(\text{base})(\text{altitude})$$

where the base is x_1 and the altitude is $-kx_1$. Therefore

$$W = \frac{1}{2}(x_1)(-kx_1)$$

$$W = -\frac{1}{2}kx_1^2$$

Again the minus sign indicates that the area is below the x-axis. The work done by the spring is negative because the force and the displacement are oppositely directed.

Because we knew the formula for the area of a triangle we were able to determine the work done without using integration. More generally, however, the area under a

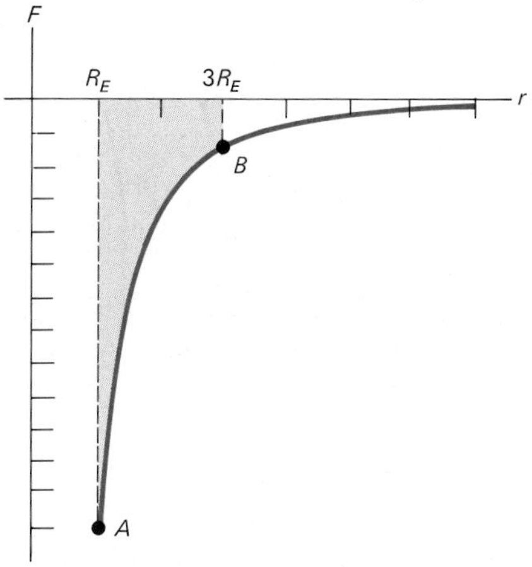

Figure 7.11

The force acting on a object that is moving away from the earth yields the negative shaded area, which corresponds to the negative work done in the displacement shown in Figure 7.10.

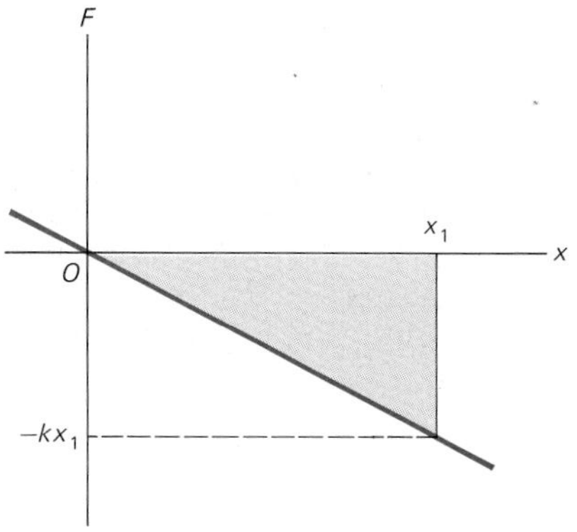

Figure 7.12

The force, F exerted by a spring has this linear form. In stretching, the spring does work equal to the shaded negative area.

curve $y(x)$ from $x = a$ to $x = b$ can be taken as a geometric representation of the definite integral. Therefore, if $y(x) = df/dx$, then

$$\int_a^b y(x)\, dx = f(b) - f(a)$$

In the case of the spring, $y(x) = -kx = \frac{d}{dx}\left(-\frac{1}{2}kx^2\right)$, so we have

$$\int_0^{x_1} -kx\, dx = -\frac{1}{2}kx_1^2 - \left(-\frac{1}{2}k0^2\right)$$

which gives

$$W = -\frac{1}{2}kx_1^2$$

Questions

5. The area under a curve of F versus x can be positive, zero, or negative. Sketch a curve corresponding to each case for $F = -kx$.
6. A force F acts on a body displaced an amount x. What would it mean physically if the work done were negative? zero? positive?
7. A child bounces up and down repeatedly to the same height on a pogo stick. Consider the work done on the child by gravity from the peak of one jump to the next peak. Is it positive, negative, or zero? How would your answer change if the height steadily decreased?

7.3 Work in Three Dimensions

When external forces act on a particle, the particle does not always move in a straight line. Therefore, we must extend our definition of work to include motion on a curved path. Consider a particle moving along the curved path shown in Figure 7.13. We can take a path segment, $\Delta \mathbf{s}_i$, small enough so that Eq. 7.1 approximates the work done on the particle for that segment.

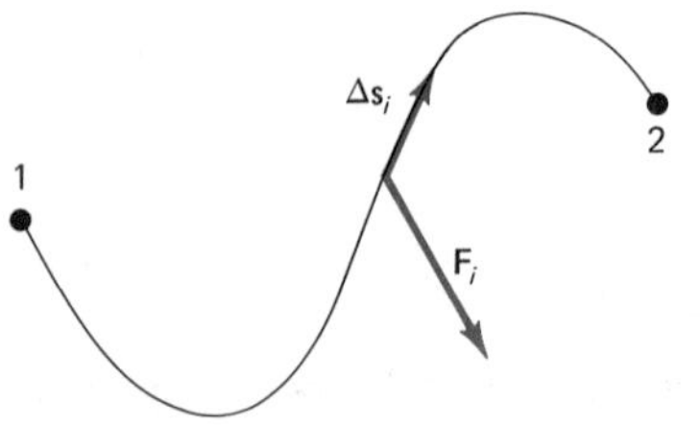

Figure 7.13

For an arbitrary path connecting points 1 and 2 we approximate the total work done by considering a small displacement $\Delta \mathbf{s}_i$. The work done in this displacement is approximately $\mathbf{F}_i \cdot \Delta \mathbf{s}_i$ where $\mathbf{F}_i$ is the force. The total work done equals the sum of similar contributions over the complete path.

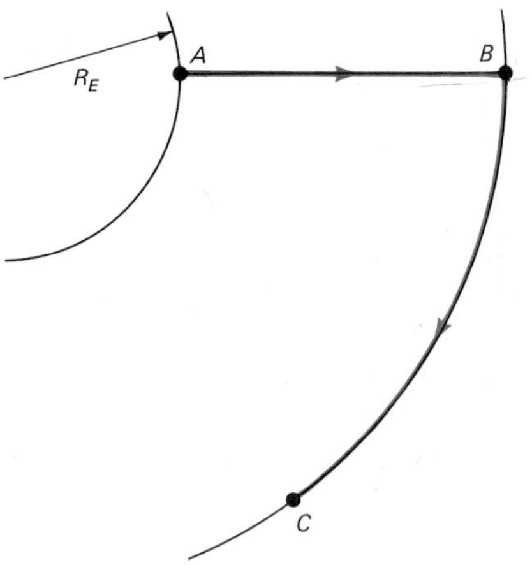

Figure 7.14

An astronaut moves radially out from the earth from A to B during launch. Then he enters a circular orbit, moving from B to C.

$$\Delta W_i = \mathbf{F}_i \cdot \Delta \mathbf{s}_i$$

The smaller $\Delta \mathbf{s}_i$, the more accurate will be the approximation of ΔW_i. We divide the path from the initial position of the particle, labeled 1, to its final position, labeled 2, into a large number of segments. Each segment contributes a term like ΔW_i to the work. Using a limiting procedure analogous to the one used in Section 7.2, we obtain the general definition of work in three dimensions.

$$W = \int_1^2 \mathbf{F} \cdot d\mathbf{s} \qquad (7.7)$$

Equation 7.7 is the natural extension of Eq. 7.6, the general definition of work for one-dimensional motion, to three dimensions.

Both the integral in Eq. 7.7 and the integral in Eq. 7.6 are called *path integrals* or *line integrals*, because they are evaluated along a path or line. Equation 7.7 is similar in principle to, but more general than, Eq. 7.6 because it includes both straight and curved paths. The following example illustrates the use of path integrals.

Example 6 The Astronaut Revisited

Let's return to the astronaut described in Example 4 as he enters a circular orbit (Figure 7.14). The astronaut is launched at A, rises to an altitutde of $2R_E$ at B, and then enters the orbit. We can find the work done by gravity along the path $A \to B \to C$ by using Eq. 7.7.

First we break the integral of Eq. 7.7 into two parts corresponding to the segments $A \to B$ and $B \to C$.

$$W = \int_A^B \mathbf{F} \cdot d\mathbf{s} + \int_B^C \mathbf{F} \cdot d\mathbf{s}$$

In Example 4 we determined that

$$W_{A\to B} = -\frac{2}{3}\frac{GM_E m}{R_E}$$

By inspection, $W_{B\to C} = 0$, since **F** is perpendicular to $d\mathbf{s}$ at every point of the arc. Hence we have

$$W_{A\to C} = -\frac{2}{3}\frac{GM_E m}{R_E}$$

which, as before, equals -3.34×10^9 J for the 80-kg astronaut.

Questions

8. Give an argument showing that the net force acting on a particle in uniform circular motion does zero work.
9. A moving charged particle in the vicinity of a magnet experiences a force that is perpendicular to its velocity. Is the work done on the charge necessarily zero?

7.4 Kinetic Energy and the Work–Energy Principle

When a particle experiences a net force it accelerates. If the net force does positive work, the speed of the particle increases, whereas with negative work its speed decreases. Although we might expect the work done on a particle by the net force to be measured by the change in speed, it is not. Instead, the **kinetic energy** of a particle is defined in such a way that the work done by the net force equals the change in the kinetic energy of the particle. Kinetic energy is a quantity that involves the speed of a particle and its mass.

A particle of mass m moving at speed v has a kinetic energy K defined by

$$K = \frac{1}{2}mv^2 \qquad (7.8)$$

The SI unit of kinetic energy is the joule, the same as the SI unit of work.

Example 7 **Kinetic Energy of a Skier**

A 52-kg skier moves down a slope at a speed of 14.0 m/s (31.3 mph). Determine the kinetic energy of the skier.

Using Eq. 7.8 we have

$$K = \frac{1}{2}mv^2$$

$$= \frac{1}{2} \times 52 \text{ kg} \times (14.0)^2 \frac{\text{m}^2}{\text{s}^2}$$

$$= 5096 \text{ J}$$

Similarly, a 52-kg platform diver falling 10 m down to the water is accelerated by gravity from rest to a speed of 14 m/s. Thus the kinetic energy of the 52-kg platform diver at impact is also 5096 joules.

We have seen that the work done by the net force can be calculated by using laws of force. The change in kinetic energy of a particle can be expressed in terms of the particle's mass and its speed just before and just after the performance of the work. The **work–energy principle** states:

The work done on a particle by the net force equals the change in kinetic energy of the particle.

Formally

$$W_{\text{net}} = \Delta K \qquad (7.9)$$

where

$$W_{\text{net}} = \int \mathbf{F}_{\text{net}} \cdot d\mathbf{s} \qquad (7.10)$$

is the work done by the net force.

The work–energy principle is an *alternative* to Newton's second law; both describe the influence of the environment on the particle. Let us derive the work–energy principle and then illustrate it with examples. We will begin with a simplified development that is valid for one-dimensional motion if the force is constant, and then follow with a general derivation.

For a constant force acting over the distance x, Eq. 7.2 gives

$$W = Fx$$

In this equation, F can represent any force. Only if F represents the *net* force may we use Newton's second law, $F = ma$, to obtain

$$W_{\text{net}} = max$$

Here a is constant because F is constant. Therefore, we can find ax by using Eq. 4.17

$$v^2 = v_0^2 + 2ax$$

where v_0 is the initial speed and v is the speed after a displacement x. Solving for ax, we write

$$ax = \frac{1}{2}v^2 - \frac{1}{2}v_0^2 \qquad (7.11)$$

Substituting this value of ax into the equation for work, $W_{\text{net}} = max$, we get

$$W_{\text{net}} = \frac{1}{2}mv^2 - \frac{1}{2}mv_0^2 \qquad (7.12)$$

Then, from the definition of kinetic energy (Eq. 7.8), we see that this becomes

$$W_{\text{net}} = \Delta K \qquad (7.9)$$

where

$$\Delta K = \frac{1}{2}mv^2 - \frac{1}{2}mv_0^2 \qquad (7.13)$$

The quantity $K = \frac{1}{2}mv^2$ is the final kinetic energy and $K_0 = \frac{1}{2}mv_0^2$ is the initial kinetic energy. Thus W_{net} determines the change in a particle's kinetic energy.

This derivation demonstrates that the integral in Eq. 7.10 equals the change in kinetic energy of the particle. To complete the derivation of the work–energy principle, we should evaluate this integral explicitly in terms of properties of the particle and its environment. However, such an evaluation requires knowledge of all the force laws con-

Figure 7.15

A rocket launched from point A at the surface of the earth coasts radially away from the earth along the line AB. Point B is at a distance $3R_E$ from the center of the earth.

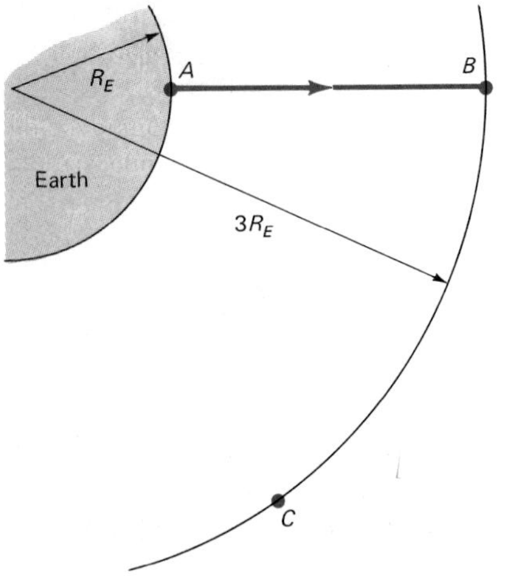

tributing to $\mathbf{F}_{net}$. Because such knowledge may or may not be available, even if the environment of the particle is completely defined, an explicit evaluation of the integral is generally not possible. For this reason the work–energy principle is stated in the form of Eq. 7.9:

$$W_{net} = \Delta K$$

When a constant force is the only force acting on a particle, the net work is readily calculated. For example, in Galileo's celebrated thought experiment, cannonballs are dropped through a vertical distance h from the leaning tower of Pisa. Only the constant force of gravity, mg, acts on a cannonball during the fall. The work done by the net force thus equals the work done by gravity, $W_{net} = +mgh$. With zero initial speed the change in kinetic energy of the cannonball is $\Delta K = {}^1\!/_2\, mv^2$, where v is the impact speed. Substituting for W_{net} and ΔK in Eq. 7.9, we obtain

$$mgh = \frac{1}{2} mv^2$$

or

$$v = \sqrt{2gh}$$

This expression for v is exactly what we would obtain by using Eq. 4.17 with $a = g$, $v_o = 0$, and $x = h$. This shows that Eq. 4.17 is a special form of the work–energy theorem.

Example 8 **A Rocket Ride**

A rocket is launched at point A on the surface of the earth and coasts radially outward to point B (Figure 7.15). If air resistance, the rotation of the earth about its own axis, and the acceleration distance during the launch are neglected, what is the speed of the rocket at point B? Take the launch speed to be 1.12×10^4 m/s (about 25,000 mph).

As we saw in Example 4, the work done by gravity is

$$W_{net} = -\frac{2}{3}\frac{GM_E m}{R_E}$$

Substituting this value of W_{net} into Eq. 7.12, we write

$$-\frac{2}{3}\frac{GM_E m}{R_E} = \frac{1}{2} mv_B^2 - \frac{1}{2} mv_A^2$$

Multiplying by 2 and canceling m yields

$$-\frac{4}{3}\frac{GM_E}{R_E} = v_B^2 - v_A^2$$

Finally, solving for v_B^2, we obtain

$$v_B^2 = v_A^2 - \frac{4}{3}\frac{GM_E}{R_E}$$

Using $G = 6.67 \times 10^{-11}\ \mathrm{N \cdot m^2/kg^2}$, $M_E = 5.98 \times 10^{24}$ kg, and $R_E = 6.37 \times 10^6$ m from Appendix 2, we obtain

$$v_B^2 = \left(1.12 \times 10^4\ \frac{\mathrm{m}}{\mathrm{s}}\right)^2 - \frac{4}{3}\,\frac{6.67 \times 10^{-11}\ \mathrm{N \cdot m^2/kg^2} \times 5.98 \times 10^{24}\ \mathrm{kg}}{6.37 \times 10^6\ \mathrm{m}}$$

$$= 4.19 \times 10^7\ \frac{\mathrm{m}^2}{\mathrm{s}^2}$$

$$v_B = 6480\ \mathrm{m/s}$$

Comparing v_B with v_A, we see that the speed of the rocket has been cut almost in half. This is due to the negative work done on the rising rocket by the force of the earth's gravity.

Although Newton's second law and the work–energy principle both describe the influence of the environment on a particle, these relations are not identical. Newton's second law is in vector form, whereas the work–energy principle is in scalar form. Information about the speed, but not about the direction of the velocity, emerges from the work–energy principle.

General Work–Energy Principle

We now rederive the work–energy principle by starting with the general definition of work given in Eq. 7.7. This removes the limitations of constant force and one-dimensional motion. Using Newton's second law,

$$\mathbf{F}_{net} = m\mathbf{a} \qquad (6.1)$$

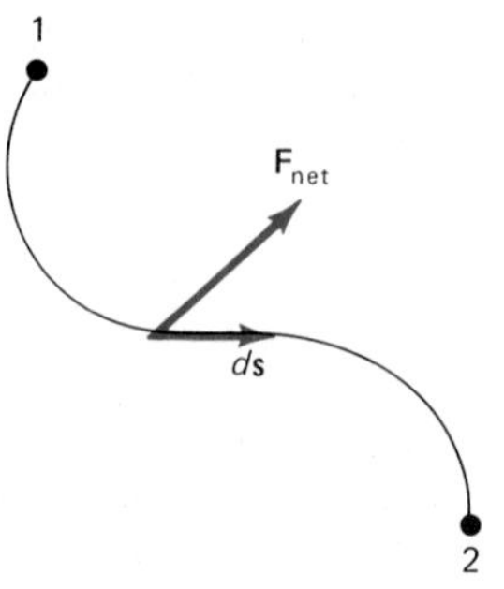

Figure 7.16

A particle moves from point 1 to point 2 along an arbitrary path. The displacement $d\mathbf{s}$ and net force $\mathbf{F}_{net}$ will, in general, vary from point to point along the path.

we obtain for Eq. 7.10

$$W_{net} = \int_1^2 m\mathbf{a} \cdot d\mathbf{s}$$

Here the acceleration, **a**, may be variable because $\mathbf{F}_{net}$ need not be constant. Also, the path may curve from the initial position 1 to the final position 2 in Figure 7.16. We must convert the integrand ($m\mathbf{a} \cdot d\mathbf{s}$) into a form that can be evaluated. We start with the defining relation for velocity (see Section 4.3):

$$d\mathbf{s} = \mathbf{v}\,dt$$

so that

$$\mathbf{a} \cdot d\mathbf{s} = \mathbf{a} \cdot (\mathbf{v}\,dt)$$

But as we saw in Section 4.4 from the defining relation for acceleration

$$d\mathbf{v} = \mathbf{a}\,dt$$

Therefore, substituting $d\mathbf{v}$ for $\mathbf{a}\,dt$ in our equation, we get

$$\mathbf{a} \cdot d\mathbf{s} = (d\mathbf{v}) \cdot \mathbf{v}$$
$$= \mathbf{v} \cdot d\mathbf{v}$$

Thus the integral now has the form

$$W_{net} = \int_1^2 m\mathbf{v} \cdot d\mathbf{v}$$

The limits now represent velocities rather than positions because we have changed the variable of integration from position ($d\mathbf{s}$) to velocity ($d\mathbf{v}$). Using the identity

$$\mathbf{v} \cdot d\mathbf{v} = \frac{1}{2}\, d(\mathbf{v} \cdot \mathbf{v}) = \frac{1}{2}\, d(v^2)$$

we have for the integral

$$W_{net} = \frac{m}{2} \int_{v_o}^{v} d(v^2)$$

Here v_o and v in the limits represent initial and final speeds, respectively. This integral may be evaluated at once to give

$$W_{net} = \frac{1}{2} mv^2 - \frac{1}{2} mv_o^2$$

or, by using Eq. 7.13,

$$W_{net} = K - K_o \qquad (7.14)$$

Equation 7.14 is the general form of the work–energy principle, where W_{net} is calculated by using Eq. 7.10.

The significance of the more general derivation is that it removes the limitations of constant force and one-dimensional motion. As a result, the work–energy principle is generally applicable.

Example 9 **Speed of a Basketball**

A basketball player shoots a free throw from a point 1 m below and 15 ft horizontally away from the basket. The initial speed of the basketball is 10 m/s. What is its speed as it passes through the basket? Using Eq. 7.2, we get

$$W_{net} = -mgh = -m \times 9.80 \times (1.00)$$
$$= -9.80\, m$$

where m is the mass of the basketball. Using Eq. 7.13 we obtain

$$\Delta K = \frac{1}{2} mv^2 - \frac{1}{2} m(10)^2$$
$$= \frac{1}{2} mv^2 - 50.0\, m$$

Hence

$$-9.80\, m = \frac{1}{2} mv^2 - 50.0\, m$$

or

$$v^2 = 80.4$$
$$v = 8.97 \text{ m/s}$$

Note that the result is independent of the mass of the ball and its horizontal displacement! Also, because W and K are scalars, only the magnitude of **v**, not its direction, is obtained.

The work done on a particle will be negative if the force is opposite in direction to the displacement. When we solve Eq. 7.14 for the final kinetic energy

$$K = K_o + W_{net}$$

we see that if W_{net} is negative, then the kinetic energy of the particle must decrease ($K < K_o$ when $W_{net} < 0$).

Example 10 **Driving into a Brick Wall**

An automobile ($m = 10^3$ kg) is driven into a brick wall in a safety test. The bumper behaves like a spring ($k = 5 \times 10^5$ N/m), and is observed to compress a distance of 0.1 m while the car is brought to rest. What was the initial speed of the automobile?

The work by the bumper is negative (see Example 5 in this chapter), and is given by

$$W_{net} = -\frac{1}{2} kx^2$$
$$= -\frac{1}{2} \times 5 \times 10^5 \frac{\text{N}}{\text{m}} \times (0.1 \text{ m})^2$$
$$= -2.5 \times 10^3 \text{ J}$$

The final kinetic energy equals zero, and the initial kinetic energy is given by

$$K_o = \frac{1}{2} mv_o^2$$
$$= \frac{1}{2} \times (10^3 \text{ kg}) v^2$$

Substituting these expressions for K_o and W_{net} into Eq. 7.14, we get

$$W_{net} = K - K_o$$

$$-2.5 \times 10^3 \text{ J} = 0 - \frac{1}{2} \times (10^3 \text{ kg}) v^2$$

or

$$v_o^2 = \frac{2 \times 2.5 \times 10^3}{10^3}$$

$$v_o^2 = 5$$

$$v_o = 2.24 \text{ m/s}$$

or about 5 mph.

The work–energy principle relates property of a particle, its increase in kinetic energy, to the influence of the environment on the particle motion, the work done on the particle by the net force acting. In this sense the work–energy principle plays a role in mechanics analogous to Newton's second law, from which it was derived. However, the work–energy principle is a scalar principle, whereas Newton's second law is a vector law. Also, the integral work–energy principle relates initial and final values only, whereas the differential Newton's second law provides information at intermediate times also. When deciding whether or not to use the work–energy principle, we weigh its simplicity and generality against the information loss in comparison with the Newton's second law approach.

Questions

10. Explain why positive work results in an increase in velocity and negative work results in a decrease in velocity. Consider all possible relative directions of **v** and **F**.

11. The magnetic force on a charged particle is perpendicular to its velocity. Can the speed of the particle be changed by the force? How about the velocity?

12. Give an example of motion where no work is done by the net force, yet the particle accelerates.

13. A body falls through a height h under the influence of gravity (Figure 7.17). During its fall the body is pushed upward by a spring. At the beginning and the end of the fall the velocity of the body is zero. Discuss the work done by gravity, by the spring, and by the net force at various stages of the fall.

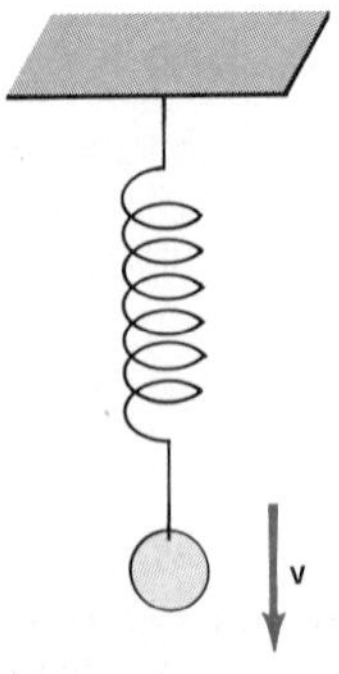

Figure 7.17

As the object falls it is subject to the forces exerted by gravity and the spring.

7.5 Power

Often we are interested not only in the amount of work being done, but in the rate at which it is done. **Power,** P, is defined as the rate at which work is performed.

$$P = \frac{dW}{dt} \tag{7.15}$$

If a total amount of work ΔW is performed in a time interval Δt, then the average power $\overline{P}$ during Δt is given by

$$\overline{P} = \frac{\Delta W}{\Delta t} \tag{7.16}$$

The SI unit of power is the watt. From Eq. 7.15

1 watt = 1 joule/second or 1 W = 1 J/s

Any power unit multiplied by a time unit is a valid unit of energy or work. For example, electric energy is often measured in kilowatt hours (kW · hr, or kWh).

Example 11 Energy Unit Conversion

Let's determine the number of joules in 1 kWh. Using the unit conversion factors (Section 1.6) we can write

$$1 \text{ kWh} = 1 \cancel{\text{k}}\cancel{\text{W}} \cdot \cancel{\text{hr}} \times \frac{10^3 \text{ W}}{1 \cancel{\text{kW}}} \times \frac{3600 \text{ s}}{1 \cancel{\text{hr}}}$$

$$= 3.6 \times 10^6 \text{ W} \cdot \text{s}$$

Since 1 W · s = 1 J, then

$$1 \text{ kWh} = 3.6 \times 10^6 \text{ J}$$

Because power is defined by Eq. 7.15 as work (or energy) divided by time, the performance of work or the transfer of energy in a small amount of time may result in a large instantaneous power. This burst of power occurs in lasers and electron accelerators, for example. An electron accelerator accelerates electrons until they have large kinetic energies. In one electron accelerator, high-velocity electrons give up 60,000 J of kinetic energy to the target in an impact time of 24 nanoseconds (ns). The average power is

$$\overline{P} = \frac{6 \times 10^4 \text{ J}}{24 \times 10^{-9} \text{ s}} = 2.5 \times 10^{12} \text{ W}$$

or 2.5 terawatts (TW).

Another useful expression for power follows from Eq. 7.1. We have

$$\Delta W = \mathbf{F} \cdot \Delta \mathbf{s}$$

Dividing by Δt and taking the limit as $\Delta t \to 0$ gives

$$\frac{dW}{dt} = \mathbf{F} \cdot \frac{d\mathbf{s}}{dt}$$

Recognizing dW/dt and ds/dt as the definitions of power

and velocity, respectively, we can write

$$P = \mathbf{F} \cdot \mathbf{v} \tag{7.17}$$

Because power depends on velocity, P is of little value for rating rocket engines. These engines are instead rated by their thrust—the force they deliver via Newton's third law.

Example 12 **Horsepower**

A common but archaic unit of power is the horsepower (hp). It is approximately equal to 746 W. When an automobile moves with constant velocity, the power developed by the engine is used to overcome friction and air resistance. If the power developed is 50 hp, what total frictional force acts on a car at 55 mph (24.6 m/s)? From Eq. 7.17

$$F = \frac{P}{v} = \frac{50 \text{ hp} \times 746 \text{ W/hp}}{24.6 \text{ m/s}} = 1520 \text{ N}$$

or about 340 lb. This is the frictional drag due to the tires and air resistance.

Questions

14. The words *powerful*, *energetic*, and *strong* mean one thing to a physicist and another to a layman. Compare the two meanings of each word.
15. In a time T (1 yr, for instance) an electric utility company delivers a total energy E to its customers. During this time the maximum power required by the customers is P. Discuss the power company situation if the generators are designed to produce a maximum power equal to (a) E/T, (b) P. (c) Discuss the economics of the two cases.
16. The world is said to be experiencing an "energy crisis," although presumably the power-generating-capacity problem is solvable. Contrast this situation with one in which ample energy is available, but the world experiences a "power crisis."
17. Using Eq. 7.17 as a guide describe situations for which the power is positive, negative, or zero.

7.6 Applications of the Work–Energy Principle: Simple Machines

Machines normally serve either of two functions. They either increase the applied force, or they increase the applied velocity. For example, by using a claw hammer a carpenter easily pulls nails that he couldn't move with his bare hands. The claw hammer increases the applied force. The velocity of a tennis racket in the hand of a tennis player is much greater than the velocity of the player's hand. The tennis racket increases the applied velocity. In a simple machine, the increase or multiplication of force or velocity is governed by a work principle, which we will now derive from the work–energy principle.

If the kinetic energy of a machine does not change, $K = K_o$, then Eq. 7.12 reduces to $W_{net} = 0$. It is useful to distinguish between the positive and negative parts of W_{net}, the work done on the machine. Let W_i (i for input) be the positive work, and $-W_o$ (o for output) be the negative work ($W_{net} = W_i - W_o$). The relation $W_{net} = 0$ can then be written as work input = work output, or

$$W_i = W_o \tag{7.18}$$

In most simple machines we deal with a single input force and a single output force. Schematically we can think of the machine as a **work transmitter** (see Figure 7.18). Any work done on the machine by the environment, $+W_i$, is transmitted, or returned, to the environment as work done by the machine, $+W_o$. Following Eq. 7.1 this can be written

$$F_i \,\Delta s_i = F_o \,\Delta s_o \tag{7.19}$$

where F_i is the input force, Δs_i is the input displacement, F_o is the output force, and Δs_o is the output displacement. During the time that the input force acts through Δs_i, the output force acts through Δs_o. Because Δs_i and Δs_o occur in the same time, we can divide Eq. 7.19 by Δt to obtain

$$F_i v_i = F_o v_o \tag{7.20}$$

where $v_i = \Delta s_i/\Delta t$ and $v_o = \Delta s_o/\Delta t$ are the input and output velocities, respectively. Equations 7.18–19 can be read as "Work in equals work out." Equation 7.20 can be read as "Power in equals power out."

The **ideal mechanical advantage**, or I.M.A., of a machine is the force multiplication achieved in the absence of friction. Using Eqs. 7.19 and 7.20, we can write this as

$$\text{I.M.A.} = \frac{F_o}{F_i} = \frac{\Delta s_i}{\Delta s_o} = \frac{v_i}{v_o} \tag{7.21}$$

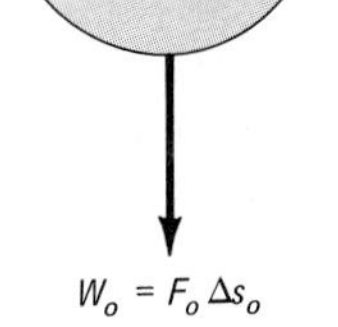

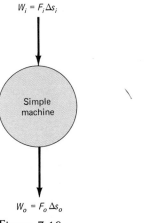

Figure 7.18

Simple machines transmit work. The input force F_i acts through the input displacement Δs_i to develop an input work W_i. The output force F_o acts through the output displacement Δs_o to develop an output work W_o. For ideal simple machines no friction is present and $W_i = W_o$.

Figure 7.19

In a wheel and axle the input force $\mathbf{F}_i$ is applied to a handle moving in a circle of radius r_2. The output force $\mathbf{F}_o$ is the tension in a rope wound about a drum in a circle of radius r_1.

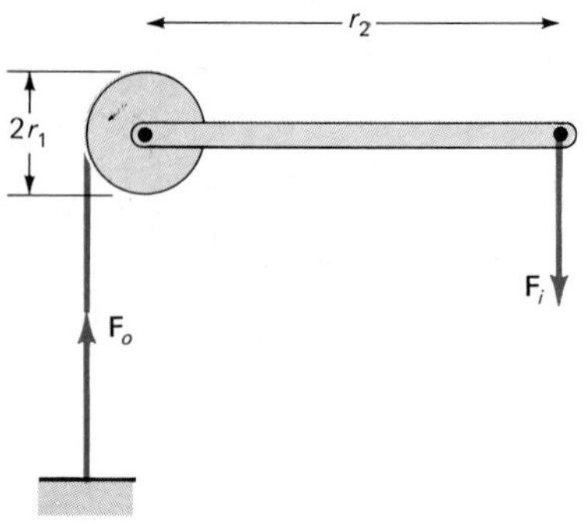

Equation 7.21 shows that force multiplication is attained at the expense of velocity (or distance), and vice versa. The easiest way to determine the I.M.A. is generally to use the ratio $\Delta s_i/\Delta s_o$. The actual ratio of output to input force will differ from F_o/F_i because of friction. To obtain the actual mechanical advantage of the machine, we must measure the actual ratio of output force to input force. Because this ratio is less than F_o/F_i as a result of the effects of friction, the actual mechanical advantage of a machine will be less than its I.M.A.

Example 13 **Wheel and Axle**

For centuries the wheel and axle has been used to raise water from a well. This system (see Figure 7.19) consists of a drum of radius r_1 and a handle that moves in a circle of radius r_2. As the handle is turned, the rope supporting the bucket wraps around the drum. What is the ideal mechanical advantage of this system?

The input force is the force applied by the handle and the output force is the force applied by the drum. For one complete revolution of the drum and handle

$$\Delta s_i = 2\pi r_2$$

$$\Delta s_o = 2\pi r_1$$

Hence

$$\text{I.M.A.} = \frac{r_2}{r_1}$$

After making I.M.A. = 10 for the wheel and axle, a feeble physics professor is able to raise a 40-lb bucket of water (with no friction) by applying a 4-lb force. Unfortunately, because the well is 40 ft deep, the professor must turn his hand through a total distance of 400 ft! Also, he must turn the handle at ten times the speed of the bucket.

A **lever** is a simple machine that consists of a rigid body pivoted on a fixed point called the *fulcrum*. The human body is a complex system of levers that determine the way forces can be applied. Some are designed to multiply the force and others to multiply the velocity. The Greek mathematician Archimedes was so impressed by the virtue of the lever that he once said "Give me a place to stand and I will move the earth."

Example 14 **The Elbow as a Lever**

A schematic representation of the human arm is shown in Figure 7.20. The shoulder, elbow, and hand are points A, O, and C, respectively. The forearm pivots about O as a result of the force, $\mathbf{F}_i$, exerted by the muscle at point B. Let us calculate the I.M.A. for this system.

When the forearm, OC, rotates about O, the distances moved by the points B and C are proportional to their radii. Thus we have

$$\frac{\Delta s_i}{\Delta s_o} = \frac{r_i}{r_o}$$

so that

$$\text{I.M.A.} = \frac{r_i}{r_o}$$

You can estimate this ratio by looking at your own elbow. Suppose the I.M.A. $= \frac{1}{8}$. The muscle in your forearm must exert a 400-lb force if your hands are to support 50 lb. Clearly the arrangement of your elbow is designed as a velocity multiplier, not as a force multiplier. (Note that forces applied at the pivot do not require consideration, since no work is done by them.)

Questions

18. A baseball bat or golf club can be viewed as an extension of an athlete's hands. What kind of multiplication is provided by the club or bat?

19. Explain the term *leverage*.

20. A simple machine can be called a *work transmitter*. Explain why the term *velocity transmitter* or *force transmitter* would be wrong.

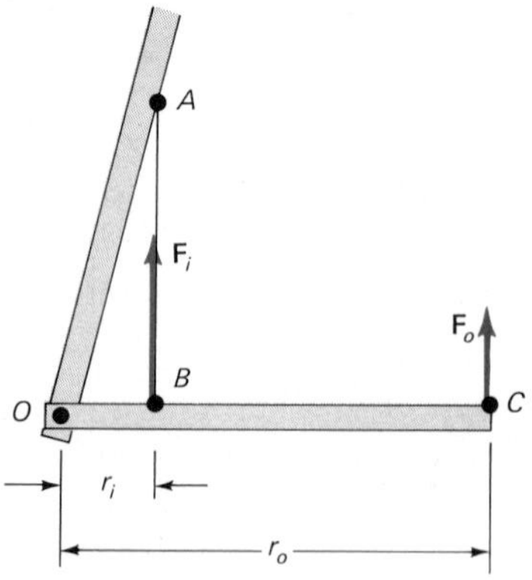

Figure 7.20

The elbow joint is a lever that can be modeled as two sticks hinged at O with an input force supplied by the biceps muscle at B. The output force is exerted by the hand at C, which moves in the circle of radius r_o.

Summary

The work done by a constant force, **F**, is given by

$$\Delta W = \mathbf{F} \cdot \Delta \mathbf{s} \qquad (7.1)$$

where $\Delta \mathbf{s}$ is the displacement. If the force varies, we must use integration to calculate the work. In one dimension the work can be expressed as

$$W = \int_{x_1}^{x_2} F \, dx \qquad (7.6)$$

where F is the component of force in the x-direction, and where x_1 and x_2 are the initial and final points along the path of integration. In three dimensions the work is defined by the path integral.

$$W = \int_1^2 \mathbf{F} \cdot d\mathbf{s} \qquad (7.7)$$

Here 1 and 2 represent the initial and final locations on the path of integration.

A particle has energy by virtue of its motion. This property is called kinetic energy, K, and is defined by

$$K = \frac{1}{2} m v^2 \qquad (7.8)$$

The fundamental scalar relation between the change in kinetic energy of a particle and the effect of its environment on its motion is the work–energy principle. It states that

$$W_{net} = \Delta K \qquad (7.9)$$

where W_{net} is the work done on the particle by the net force acting on it, and $\Delta K = K - K_o$ is the change in kinetic energy of the particle.

Power is the rate of performance of work,

$$P = \frac{dW}{dt} \qquad (7.15)$$

An alternative expression for power is

$$P = \mathbf{F} \cdot \mathbf{v} \qquad (7.17)$$

where **F** is the applied force and **v** is the velocity associated with the application of the force.

When the kinetic energy of a body does not change, the work–energy principle requires that

$$W_{net} = 0$$

For simple machines with one input and one output force, labeled F_i and F_o, respectively, this equation can be written

$$F_i \, \Delta s_i = F_o \, \Delta s_o \qquad (7.19)$$

The work input equals the work output.

Because input and output work are done in the same time, this equation can also be written as a power relation:

$$F_i v_i = F_o v_o \qquad (7.20)$$

Simple machines serve to increase the applied force, velocity or distance. The ideal mechanical advantage, I.M.A., of a simple machine is

$$\text{I.M.A.} = \frac{F_o}{F_i} = \frac{\Delta s_i}{\Delta s_o} = \frac{v_i}{v_o} \qquad (7.21)$$

Problems

Section 7.1 Work Done by a Constant Force

1. A raindrop ($m = 3.35 \times 10^{-5}$ kg) falls vertically at constant speed under the influence of the forces of gravity and drag (Figure 1). In falling through 100 m, what work is done by (a) gravity and (b) drag?

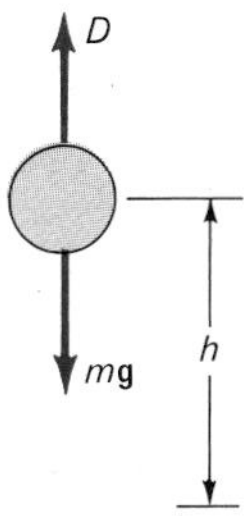

Figure 1

2. A 4-kg bowling ball skids horizontally for the first 4 m of its path. Forces acting on the ball include **N**, **f**, and $m\mathbf{g}$, as shown in Figure 2. If $\mu_k = 0.1$, what work is done by (a) gravity, (b) **N**, and (c) friction?

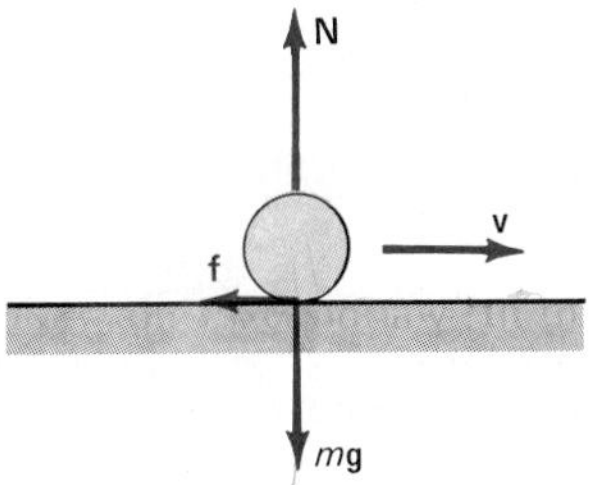

Figure 2

3. A group of students pushes a stalled car (m = 1000 kg) up a 3° incline at constant speed for a distance (measured parallel to the incline) of 100 m. How much work is done by (a) the normal force during the trip; (b) gravity during the trip; (c) gravity, if the car retraces its path at constant speed?

4. A child pulls a sled carrying his friend at constant velocity up a snow-covered slope inclined at an angle of 23° relative to the horizontal. If the force provided by the child is parallel to the slope and the sled–passenger system has a combined mass of 83 kg, how much work is done by the child in moving the sled 16 m up the slope? Assume that the frictional force is negligible.

5. Two 4-kg objects are connected by a string of negligible mass (Figure 3). The vertically moving object falls at constant speed through a distance of 2 m. Determine the work done by (a) gravity on m_1, (b) gravity on m_2, (c) tension on m_1, (d) tension on m_2, (e) friction on m_1.

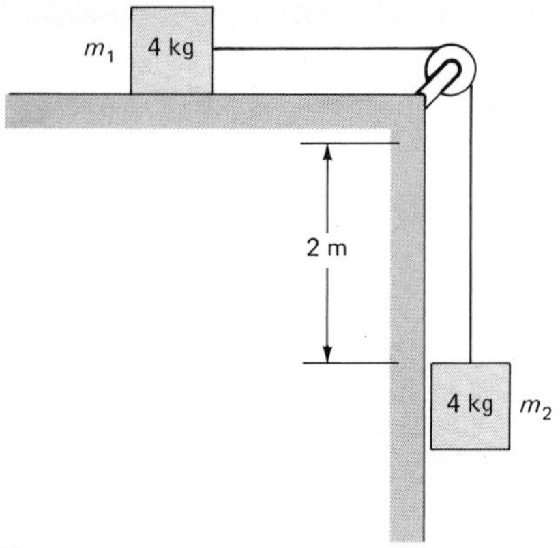

Figure 3

6. A weightlifter elevates at constant speed a mass of 180 kg through 2 m. Calculate the work done on the mass by (a) gravity; (b) the weightlifter.

Section 7.2 Work Done by a Variable Force

7. A spring pushes with a force of $+500$ N when it is compressed a distance of 0.1 m, and pulls with a force of -500 N when it is stretched a distance of 0.1 m (see Figure 4). How much work is done by the spring when its extension changes from (a) $x = -0.1$ m to $x = 0$ m; (b) $x = 0$ m to $x = +0.1$ m; (c) $x = -0.1$ m to $x = +0.1$ m?

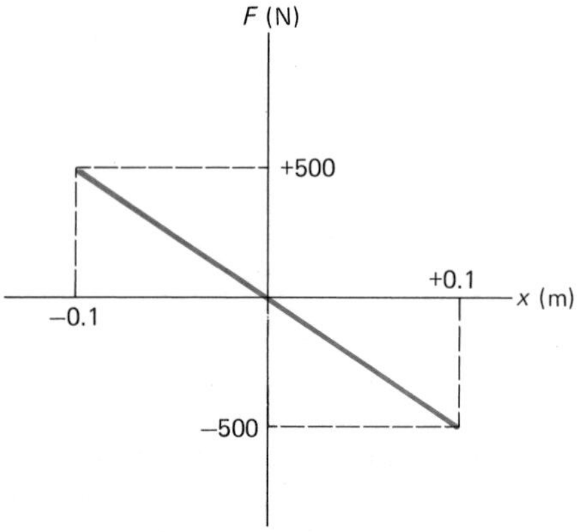

8. When a pendulum bob is displaced horizontally a small distance x, the horizontal force on the pendulum may be approximated by the expression $F = -m(g/l)x$ where m is the mass of the bob and l is the length of the pendulum. Calculate the work done by this force as the pendulum bob changes from position $x = x_0$ to position $x = 0$.

9. In Figure 5 the force exerted by gravity on a 945-kg object is plotted versus distance. For an upward displacement of 8 m, determine the work done by gravity (a) graphically (by measuring the area), and (b) analytically.

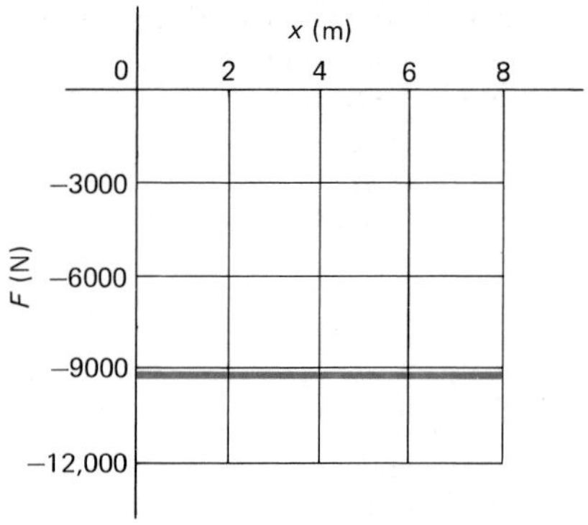

Figure 5

10. The free end of a linear spring ($k = 10$ N/m) is pulled 10 cm from its equilibrium position. Calculate the work done by the spring in the displacement from (a) $x = 0$ to $x = 5$ cm, and (b) $x = 5$ cm to $x = 10$ cm.

11. The force in newtons exerted on a 10-gm mass by a spring is plotted versus displacement in meters in Figure 6. Use the area under the curve to determine the work done from (a) $x = -0.3$ m to $x = -0.1$ m; (b) $x = -0.2$ m to $x = +0.1$ m; (c) $x = -0.3$ m to $x = 0$; (d) $x = -0.2$ m to $x = +0.2$ m.

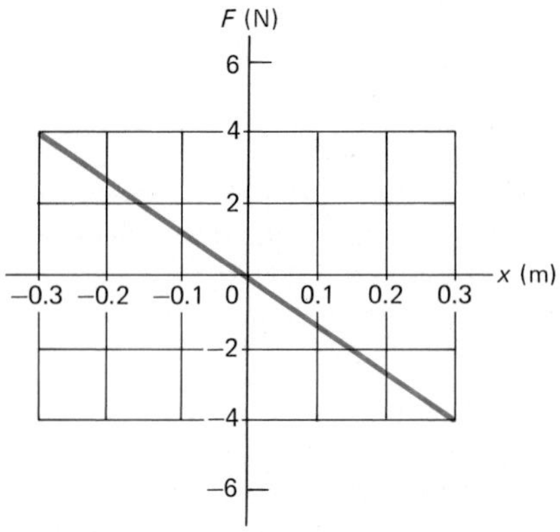

Figure 6

12. An athlete exercises his legs on a machine by repeatedly extending them against a large resistive force. The force he exerts is represented in Figure 7. From the graph determine the work done by his legs during one extension.

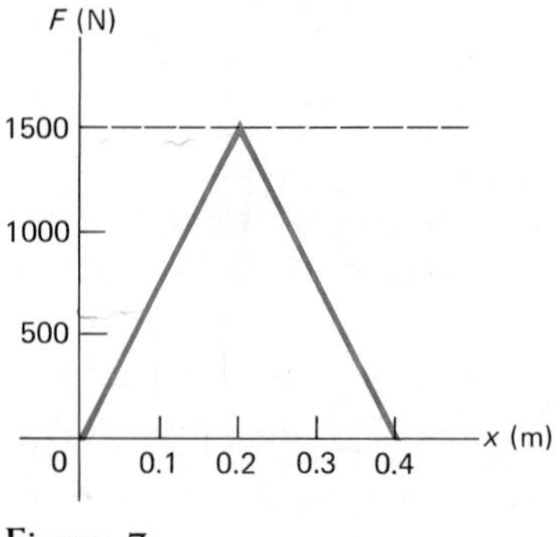

Figure 7

13. The force acting on a particle is shown in Figure 8 for the interval $-4 \leq x \leq +4$. Determine the work done on the particle for displacements defined by the motion from (a) $x = -4$ m to $x = -2$ m; (b) $x = -4$ m to $x = 0$ m; (c) $x = 0$ m to $x = +4$ m; (d) $x = -4$ m to $x = +4$ m; and (e) $x = -2$ m to $x = +2$ m.

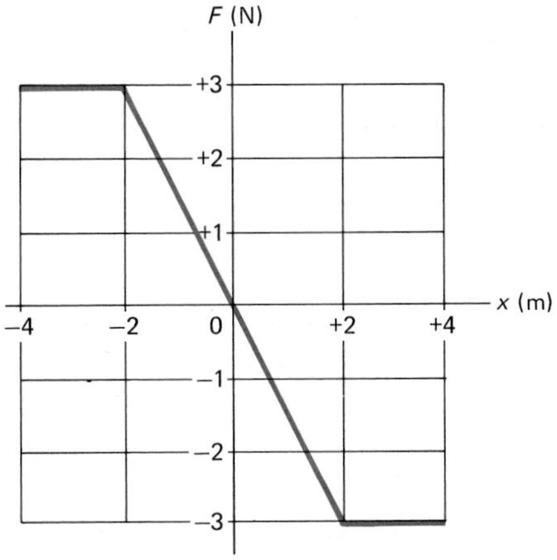

Figure 8

14. The repulsive force exerted on an alpha particle by a gold nucleus is given by

$$F = +\frac{A}{r^2}$$

where F is in newtons, r is in meters, and $A = +3.65 \times 10^{-26}$ N · m^2. This function is plotted in Figure 9. From the graph (a) determine the work (area under the curve) in joules from $r_1 = 4 \times 10^{-14}$ m to $r_2 = 12 \times 10^{-14}$ m. (b) Determine the work done over the same interval by integrating the function.

15. Diatomic molecules are formed when the constituent atoms exert attractive forces at large distances and repulsive forces at short distances. The Lennard–Jones force is a good approximation for many molecules, and is given by

$$F = F_o\left[2\left(\frac{\sigma}{r}\right)^{13} - \left(\frac{\sigma}{r}\right)^{7}\right]$$

where r is the center-to-center distance of the atoms in the molecule. For an oxygen molecule $F_o = 9.6 \times 10^{-11}$ N and $\sigma = 3.5 \times 10^{-10}$ m. This function is plotted in Figure 10. (a) From the graph determine the area bounded by the curve from $r = 4 \times 10^{-10}$ m to $r = 9 \times 10^{-10}$ m. (b) Determine the work done over the same interval by integrating the function.

16. A certain spring exerts a restoring force if it is displaced from equilibrium. Suppose this force is represented by

$$F = -kx + k'x^3$$

with $k = 10$ N/m and $k' = 100$ N/m^3. Calculate the work done by this force when the spring is displaced from $x = 0$ to $x = 0.1$ m.

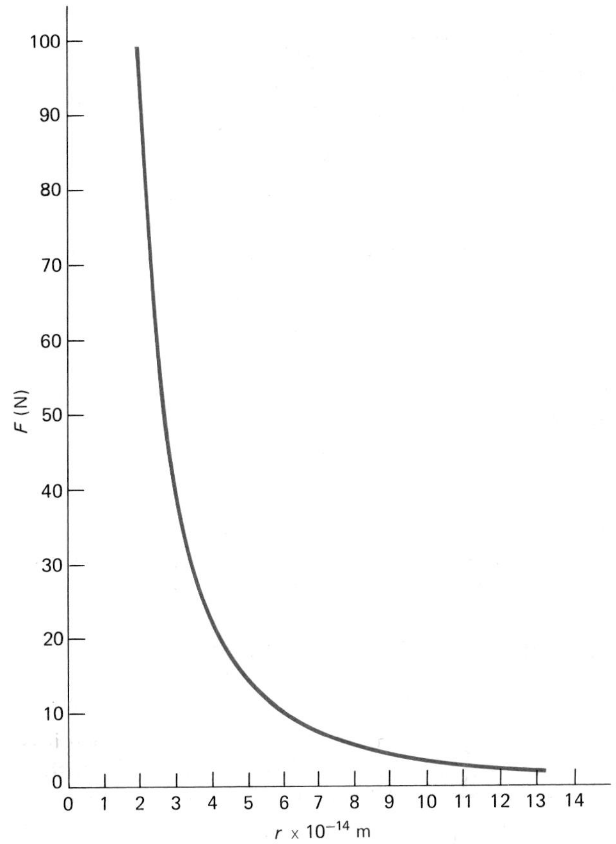

Figure 9

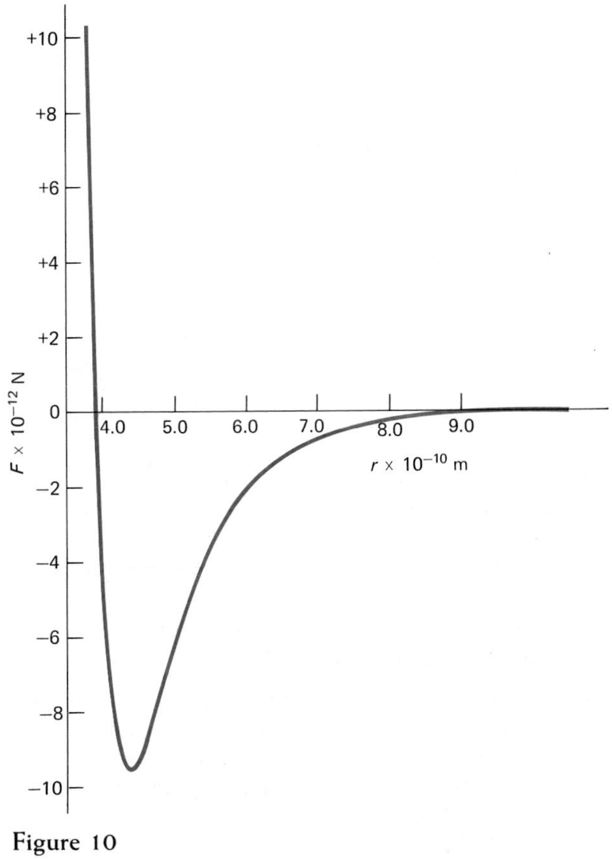

Figure 10

Section 7.3 Work in Three Dimensions

17. Imagine a very large frictionless plane tangent to the surface of the earth at point C, as shown in Figure 11. The points A and B are equidistant from C. Consider the block (mass $= m$) as it slides along the plane. Will the work done by gravity be positive, zero, or negative if the block moves from (a) A to C; (b) C to B; (c) A to B?

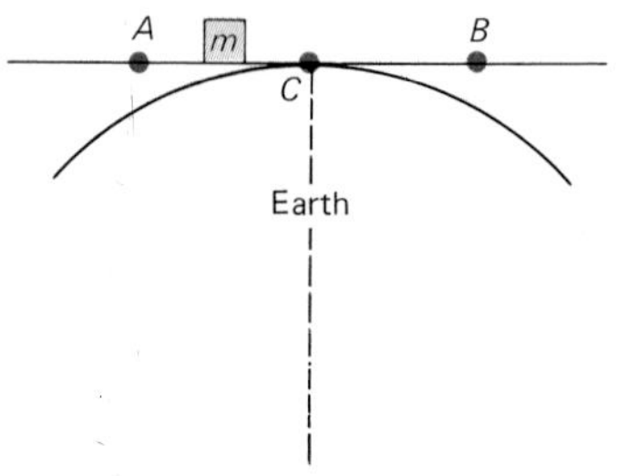

Figure 11

18. A skateboard rider moves on the inside of a bowl-shaped surface (Figure 12). Determine the work done by gravity if the rider moves (a) down a vertical distance y; (b) up a vertical distance y.

Figure 12

19. Assume that the point mass m_1 is fixed in position at the center of concentric circles of radii r_1 and r_2 (Figure 13). A second point mass, m_2, is attracted to m_1 by the universal force of gravity. In terms of m_1, m_2, r_1, r_2, and G as necessary, calculate the work done by gravity on m_2 for a displacement (a) radially outward from A to B; (b) radially inward from B to A; (c) tangentially along the circumference from B to D. (d) Prove that the work done along a tangential path following the circumference from A to C equals the work done along the alternate path $A \rightarrow B \rightarrow D \rightarrow C$.

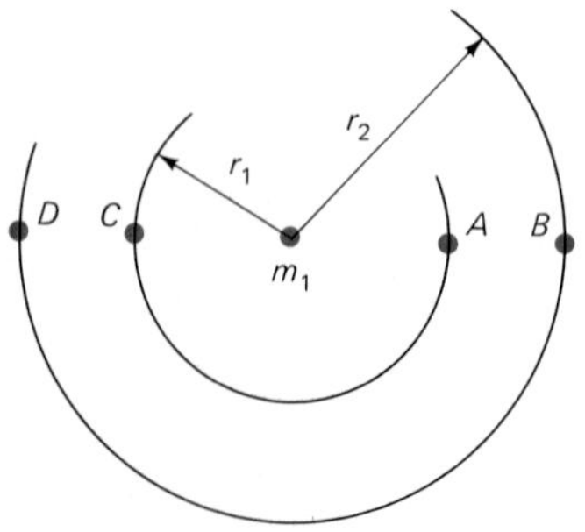

Figure 13

20. A 50-N weight is taken on a closed rectangular path $1 \rightarrow 2 \rightarrow 3 \rightarrow 4 \rightarrow 1$, as shown in Figure 14. The height of the rectangle is 2 m and its width is 3 m. Obtain numerical values for the work done by gravity for a displacement of the weight from (a) 1 to 2; (b) 2 to 3; (c) 3 to 4; (d) 4 to 1.

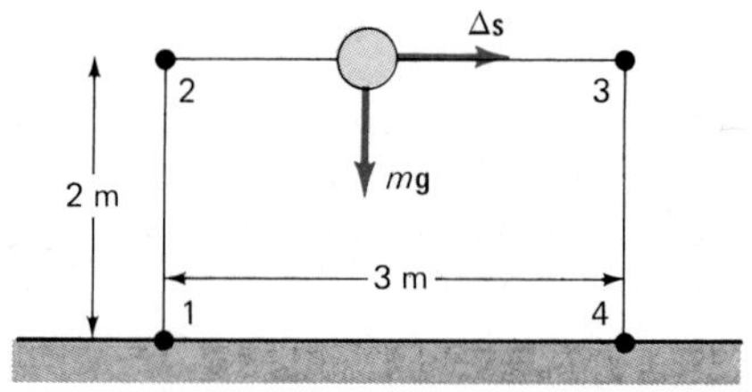

Figure 14

21. In two dimensions a spring force may be represented by the equation

$$\mathbf{F} = -k(\mathbf{i}x + \mathbf{j}y)$$

with x and y in meters and $k = 5$ N/m. Evaluate the work done by this force on a particle by using Eq. 7.7 with $d\mathbf{s} = \mathbf{i}\,dx + \mathbf{j}\,dy$. Consider the three paths shown in Figure 15: (a) $(0,0) \rightarrow (0,1) \rightarrow (1,1)$; (b) $(0,0) \rightarrow (1,0) \rightarrow (1,1)$; and (c) $(0,0) \rightarrow (1,1)$ along the line $x = y$.

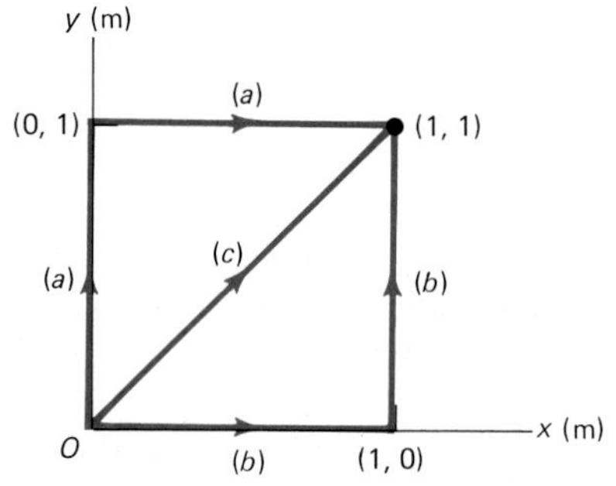

Figure 15

22. A hydrodynamic force is described by the equation

$$\mathbf{F} = -k(\mathbf{i}y + \mathbf{j}a)$$

with y in meters $a = 1$ m, and $k = 3$ N/m. Use Eq. 7.7 to evaluate the work done by this force on a particle with $d\mathbf{s} = \mathbf{i}\,dx + \mathbf{j}\,dy$. Consider the two paths shown in Figure 16, (a) $(0,0) \rightarrow (0,2) \rightarrow (3,2)$ and (b) $(0,0) \rightarrow (3,0) \rightarrow (3,2)$.

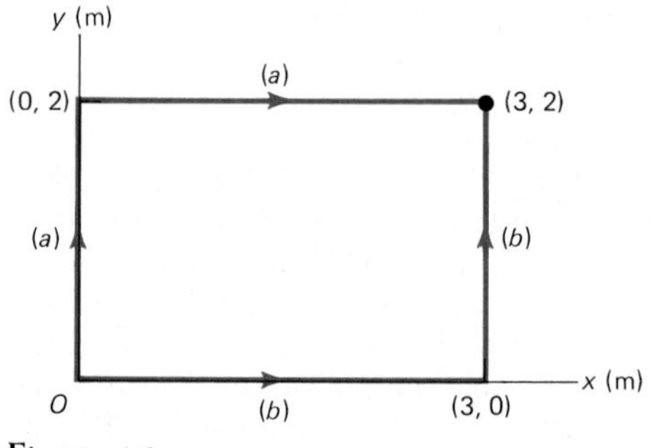

Figure 16

Section 7.4 Kinetic Energy and the Work–Energy Principle

23. Two cars have masses of 1500 kg and 3000 kg. If both cars have the same kinetic energy and the 1500-kg car is moving at 55 km/hr, what is the speed of the 3000-kg car?

24. A well-conditioned sprinter achieves speeds of about 10 m/s. For comparison, this is roughly equal to the speed acquired by a dropped object in 1 s. Determine the kinetic energy of an 80-kg sprinter running at 10 m/s.

25. Electrons accelerated through 10,000 V in a TV picture tube have a kinetic energy of 1.60×10^{-15} J. Calculate the speed of such electrons. (Do not consider relativistic corrections; the velocity of light is 3×10^8 m/s).

26. According to kinetic theory (Section 25.3), a typical gas molecule in thermal equilibrium at room temperature has a kinetic energy $K = 6 \times 10^{-21}$ J, regardless of mass. Estimate the speed at room temperature of (a) a hydrogen molecule ($m = 3.3 \times 10^{-27}$ kg) and (b) a xenon atom ($m = 2.0 \times 10^{-25}$ kg).

27. Treat the earth as a particle and calculate its orbital kinetic energy about the sun. Use $r = 1.50 \times 10^{11}$ m, $T = 3.16 \times 10^7$ s, and $m = 5.98 \times 10^{24}$ kg.

28. In Olympic competition, an athlete throws a 16-lb hammer as far horizontally as possible. Assume a 45° launch angle and an 80-m throw (near the world record). Determine the initial (a) speed, and (b) kinetic energy of the hammer. For comparison, the kinetic energy of a runner ranges approximately from 1000 J in the marathon to 4000 J in a sprint.

29. An astronaut fires a distress signal packet ($m = 0.5$ kg) upward from the lunar surface. The position of the packet is described by

$$x = v_0 t + \frac{1}{2} a t^2$$

with $v_0 = +32.4$ m/s and $a = -1.62$ m/s^2. Calculate the (a) velocity of the packet at $t = 0$, 10, 20, 30, and 40 s; and (b) the kinetic energy of the packet at $t = 0$, 10, 20, 30, and 40 s.

30. A spring ($k = 19.60$ N/m) hanging vertically has an equilibrium length of 0.25 m (Figure 17). A 100-gm particle is carefully attached to the spring and then allowed to drop. During the particle's fall the spring increases in length from 0.25 m to 0.35 m. The particle is brought to rest momentarily when the spring attains the 0.35-m length, and subsequently moves upward as the spring contracts. Calculate the work done on the particle during the fall by (a) gravity, (b) the spring, and (c) the net force.

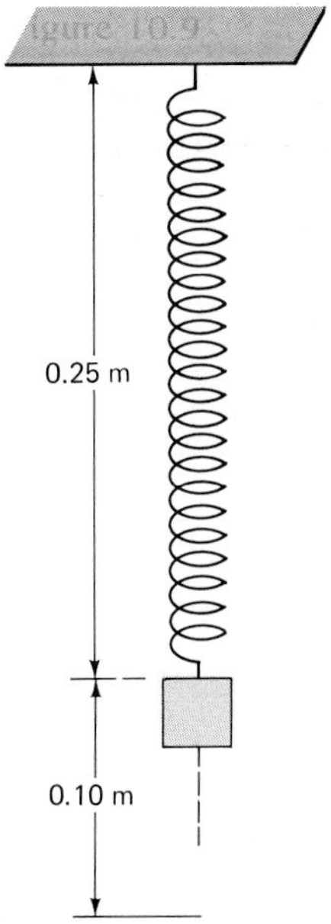

Figure 17

31. With the 100-gm particle attached, the equilibrium length of the vertical spring in Problem 30 differs from its original equilibrium length of 0.25 m (Figure 18). (a) Calculate this new equilibrium length. Note that the force exerted on the spring is up if the spring is longer than 0.25 m. Now let the particle be pushed 0.05 m above the new equilibrium position determined in (a) and then released. It returns to and moves through the new equilibrium position. Calculate the work done on the particle for the first 0.05 m of this downward move by (b) gravity, (c) the spring, and (d) the net force. (e) Calculate the kinetic energy of the particle as it moves through its new equilibrium position.

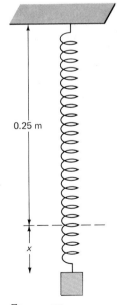

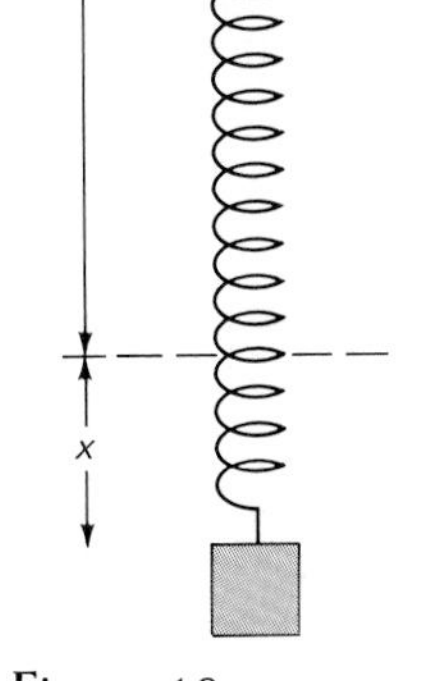

Figure 18

32. A 50-gm mass is at rest on a horizontal frictionless airtrack. This mass is held against a spring compressed a distance of 10 cm. The mass is released, allowing the spring to expand toward its relaxed position. Determine the velocity of the mass at the moment the spring is relaxed (k = 100 N/m).

33. The force $F = A/r^2$ is exerted on an alpha particle with $A = 3.65 \times 10^{-26}$ N · m^2 and r in meters. Suppose the speed of the alpha particle at a great distance from the force center (nucleus) is 1.70×10^7 m/s. As the alpha particle approaches the nucleus, the repulsive force F does work on the alpha particle, reducing its speed. Either graphically (using Figure 10) or by integration, calculate the work done by F for motion of the alpha from $r \rightarrow \infty$ to $r = r_o$. Use the work–energy principle to obtain the alpha-particle speed when r_o equals (a) 10×10^{-14} m; (b) 8×10^{-14} m; (c) 6×10^{-14} m. (d) Determine the value of r_o for which the alpha-particle speed equals zero.

34. A skier glides a distance x down a slope at a constant speed (Figure 19). Forces acting on the skier include the normal force, gravity, and friction. Use the work–energy principle to determine the angle θ of the slope (μ_k = 0.18).

35. Determine the stopping distance x for a skier with a speed of 20 m/s. Assume μ_k = 0.18 and θ = 5° (see Figure 19).

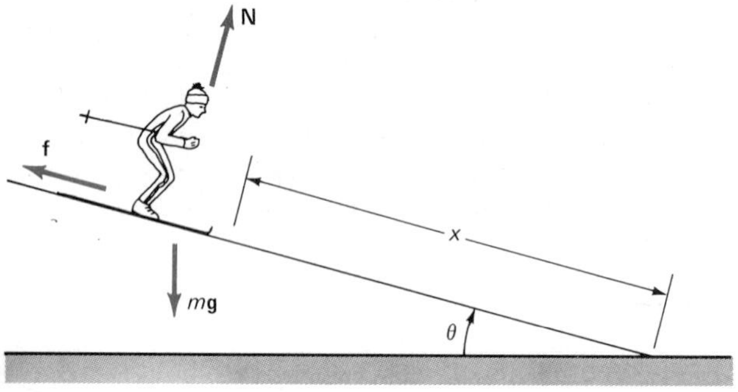

Figure 19

36. The force F exerted by one oxygen atom on another as a function of their separation is represented by the equation in Problem 15 and graphically in Figure 10. (a) Either graphically or by integration, determine the area bounded by the curve in Figure 10 between the minimum and $r \rightarrow \infty$. (b) Determine the change in kinetic energy of one oxygen atom as two oxygen atoms move from an infinite distance to a separation of 4.35×10^{-10} m. (Assume the two atoms share the kinetic energy equally.) (c) Using $m = 2.70 \times 10^{-26}$ kg for an oxygen atom, calculate the speed of one of the atoms at a separation of 4.35×10^{-10} m, assuming the kinetic energy at infinity to be negligible.

37. A 50-gm yo-yo moves in a vertical circle 1 m in radius (Figure 20). If, when the yo-yo is at point A, the string tension equals the weight of the yo-yo, (a) calculate the yo-yo's speed at point A. (b) Use the work–energy principle to determine its speed at point B. (c) If the string breaks when the yo-yo is at point A, determine the yo-yo's speed when it arrives at the elevation of point B. (d) Determine the string tension at point B for the circular motion case.

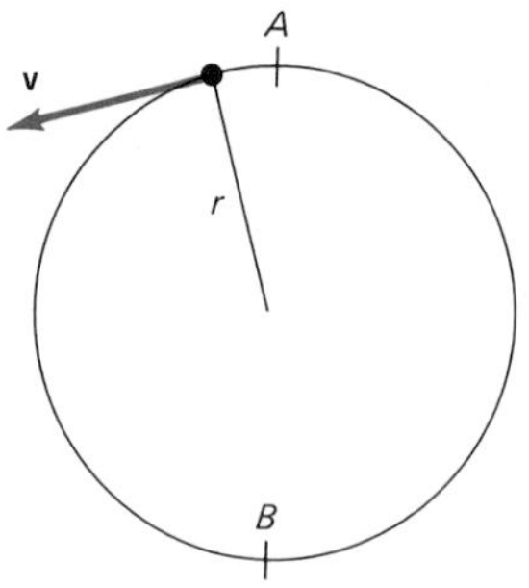

Figure 20

38. A mass m falls freely through a distance y, and then strikes a spring having spring constant k (Figure 21). The spring is compressed a distance x as the mass is brought to rest. Use the work–energy principle to show that (a) the speed of m at the moment it strikes the spring is $v = \sqrt{2gy}$; (b) the speed is reduced to zero at $x = (mg/k)[1 + \sqrt{1 + (2ky/mg)}]$, and (c) the maximum speed attained is given by

$$v = \sqrt{2gy}\left(1 + \frac{mg}{2ky}\right)^{1/2}$$

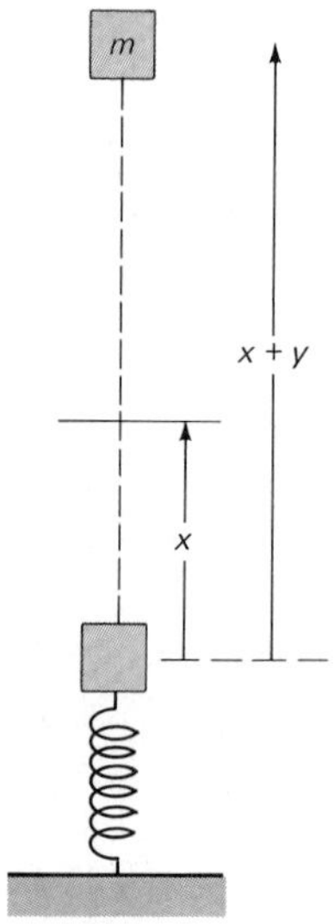

Figure 21

39. A simple pendulum of mass m moves in a circular arc of radius R in a vertical plane (Figure 22). (a) Prove that in a downward swing from the angle θ to 0° the work done by gravity is $mgR\,(1 - \cos\theta)$. (b) For the same swing, prove that the work done by the tension

T equals zero. Use the work–energy principle to determine the speed of the pendulum bob at the bottom for (c) $\theta = 5°$ and (d) $\theta = 90°$. (e) Compare the answer to part (d) with the speed acquired by the same bob in falling through a vertical distance R. (Use $R = 10$ m.)

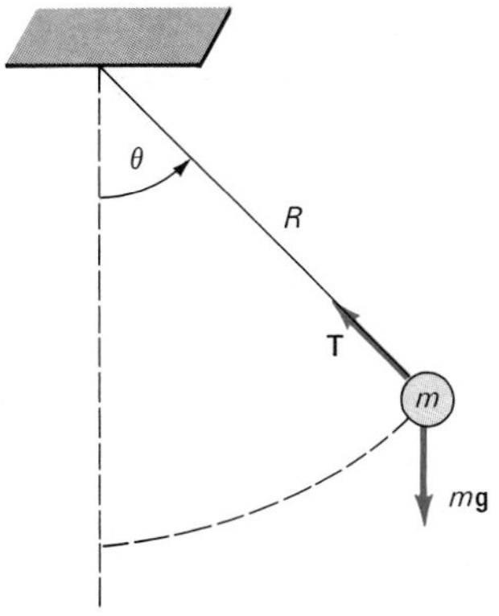

Figure 22

Section 7.5 Power

40. A person derives energy from food at the rate of about 2500 kcal/day. Convert the 2500 kcal/day to a power in watts.

41. How many joules of energy are used by a homeowner whose electric bill is \$65.00, and who pays \$0.10 per kilowatt hour?

42. What average power is developed by a gymnast ($m = 80$ kg) who pulls himself up a 6-m rope at constant velocity in 6 s?

43. Determine the average power (in units of hp) required to accelerate a 1000-kg automobile from rest to 60 mph in 10 s.

44. At Lawrence Livermore Laboratory the SHIVA system is designed to produce a laser fusion pulse with 10^4 J of energy in a time of 10^{-9} s. (a) What is the average power delivered in 10^{-9} s? (b) A thermal energy of 2.26×10^6 J will vaporize 1 kg of water. How many times would SHIVA need to be fired in order to boil away 1 kg of water? (c) What average power would be delivered by SHIVA if (b) occurred in 1 s?

45. A 1000-kg automobile undergoes constant acceleration from rest to 25 m/s in 10 s. (a) Obtain the speed as a function of time. (b) Use Eq. 7.17 to calculate the accelerating power at $t = 4$ s. (c) Calculate the kinetic energy as a function of time. (d) Deduce the work done by the accelerating force using the work–energy principle, and employ Eq. 7.15 to calculate the power developed by this force at $t = 8$ s. (e) Calculate the power at $t = 4$ s by using the method in (d), and (f) calculate the power at $t = 8$ s by using the method in (b).

46. When falling through the air at a constant speed of 45 m/s the drag force (Section 7.8) equals the weight of an 80-kg sky diver. (a) How much power is developed by the drag force? (b) How many 100-W light bulbs could be lit simultaneously with this much power?

47. A horse walks at about 1 m/s while pulling a plow. To develop 1 hp, what force must the horse exert on the plow? Express your answer both in newtons and in pounds.

48. To operate one of the generators in the Grand Coulee Dam 7.24×10^8 W of mechanical power are required. This power is provided by gravity as it performs work on the water while it falls 87 m down to the generator. The kinetic energy acquired during the fall is given up turning the generator. (a) Prove that the available water power is mgh/t where m is the mass of water falling through the height h during the time t. (b) Calculate the flow rate in kilograms per second. (c) Determine the volume of water needed to power this generator for one day. (d) If the water for one day were stored in a circular lake 10 m deep, what lake radius would be required?

49. Power windmills turn in response to the force of high-velocity drag (Section 6.8). For a sphere, $F_D \sim r^2v^2$ where r is the radius of the sphere and v is the fluid speed. Thus the power developed $P \sim r^2v^3$, according to Eq. 7.17. For a windmill the power delivered by the wind can be represented by $P = ar^2v^3$ where r is the windmill radius, v is the wind speed, and $a = 2\ \text{W} \cdot \text{s}^3/\text{m}^5$. For a home windmill with $r = 1.5$ m, calculate the power delivered to the generator (this representation ignores windmill efficiency—about 25%) if (a) $v = 8$ m/s; (b) $v = 24$ m/s. For comparison, a typical home needs about 3 kW of electric power.

Section 7.6 Applications of the Work–Energy Principle: Simple Machines

50. A man ($m = 80$ kg) raises himself at constant velocity by pulling downward on a rope that passes around a fixed pulley overhead and is tied around his waist. (a) How much work is done by the man if a 10-m length of the rope passes through his hands? (b) What power is expended if 4 s are required?

51. A simple pulley system for lifting a weight is shown in Figure 23. Prove that the ideal mechanical advantage of this system is 2.

52. A jackscrew is part of the automobile jack, a once-common device used to lift a car while changing a tire. Each revolution of the screw raises the screw a distance equal to its pitch (0.5 cm in Figure 24). If the handle of the automobile jack moves in a circle of 25-cm radius, then (a) determine I.M.A. for this machine. (b) What force is required (neglecting friction) to raise 1000-lb weight?

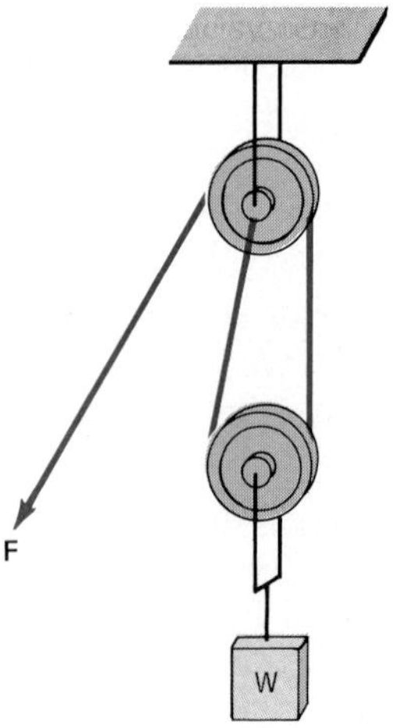

Figure 23

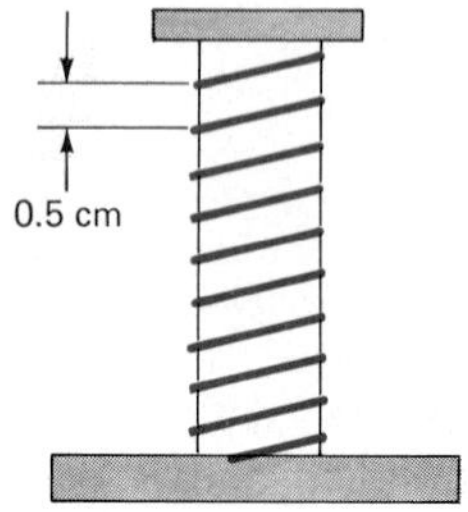

Figure 24

53. What force F is required to lift the weight using the block and tackle shown in Figure 25? (Assume no friction.)

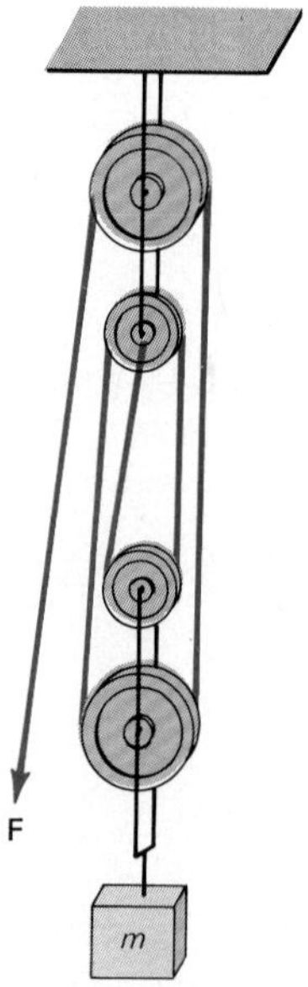

Figure 25

54. The inclined plane can be treated as a simple machine which enables a workman to raise an object whose weight is much greater than the maximum force he can exert (Figure 26). If the maximum force a workman can exert is 500 N, (a) what is the largest angle θ that permits him to push a 25,000-N crate up the incline? (b) For what angle θ is the I.M.A. of the plane 100?

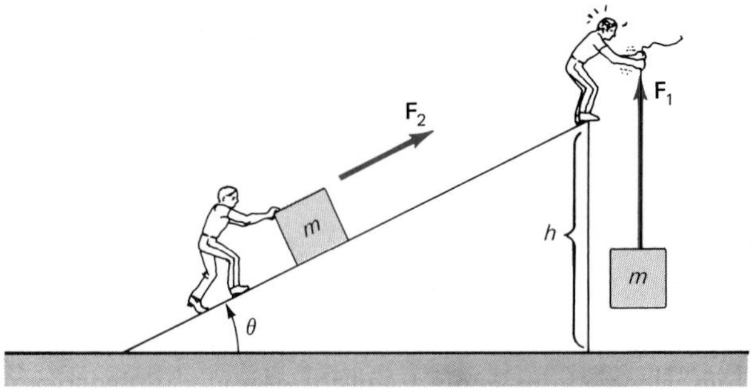

Figure 26

55. An external force $F = 50$ N raises a mass $m = 12$ kg (Figure 27) at constant velocity. Neglecting friction, calculate r_2 if $r_1 = 0.1$ m.

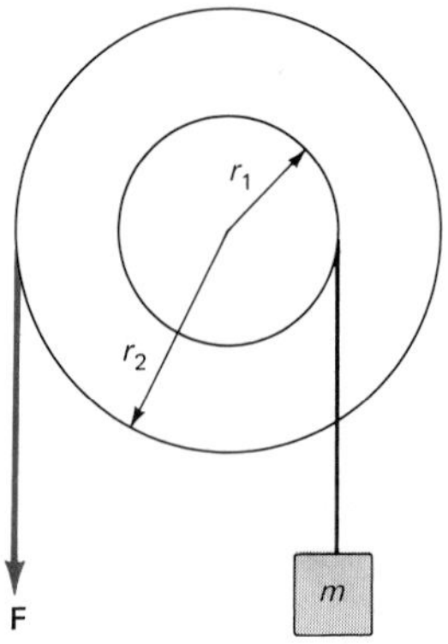

Figure 27

56. A lever 2 m long is used to lift one edge of a 9000-N commercial refrigerator (Figure 28). The maximum available external force F is 500 N. (a) Assuming the lever supports half the weight, how far from the point A should the fulcrum be placed? (b) For what fulcrum distance from A will the I.M.A. $= 49$?

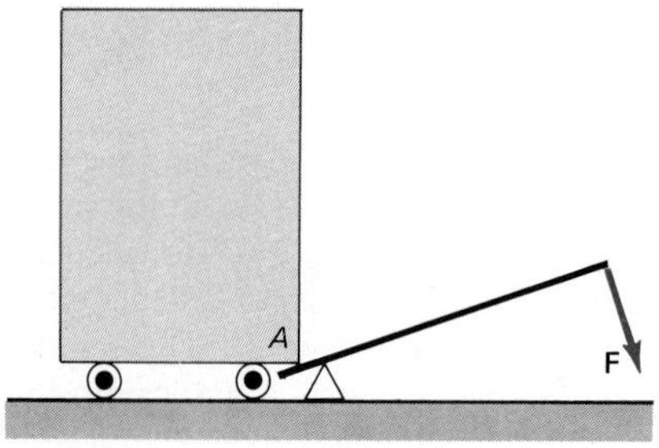

Figure 28

c h a p t e r

8

CONSERVATION OF ENERGY

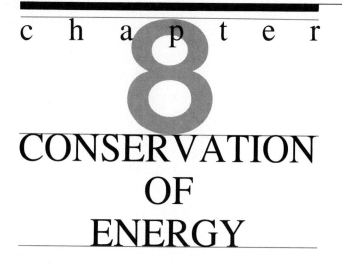

Preview

To the layman the word *conservation* often means the wise use of resources. To the scientist, however, the conservation of a physical quantity means that it *does not change with time.*

The law of conservation of energy is one of the most fruitful results to emerge from Newtonian mechanics. Since its origin the law has been extended to include thermal and electromagnetic energy and forms of nuclear energy.

Physically, energy is identified with the capacity to perform work. We saw in Chapter 7 that a particle possessing kinetic energy is capable of performing work—converting kinetic energy into work. In this chapter we enlarge the law of conservation of energy to include a new form of energy, *potential energy*. A system possessing potential energy is also capable of performing work.

To lay the foundation for a definition of potential energy we introduce the notion of a *closed path.* A closed path may be of arbitrary shape, but its beginning and ending must be at the same point. We also distinguish two types of forces, conservative and nonconservative. If the work done by a particular force in every closed path equals zero, then that force is called *conservative.* Forces not having this property are called *nonconservative.*

While under the influence of a conservative force a particle can do work. We say the particle possesses potential energy because it has this capacity to do work.

If the work done on a particle by the *net* force equals zero, then the kinetic energy of the particle is unchanged —it is conserved. In this chapter we generalize this kinetic energy conservation principle. The total energy of a particle is the sum of its kinetic and potential energies. If only conservative forces act, then the total energy of the particle is conserved.

8.1 Conservative and Nonconservative Forces

We generally divide forces into two classes, conservative and nonconservative. The distinction between a conservative and a nonconservative force is based on the work the force does over a closed path. For example, the work done by gravity on a skier being pulled to the top of a slope by a tow rope is negative. On her downhill run the work done by gravity is positive. If the skier returns to her exact starting point, then zero work was done by the force of gravity over the round trip. This holds true regardless of the paths up and down the mountain—they need not be the same. As long as the skier returns to her exact starting point, the path is called a **closed path.** We find that *the net work done by the force of gravity around any closed path is zero.* Not all forces in nature do zero work around a closed path, but forces that do are called **conservative.**

The force of friction is an example of a **nonconservative** force. The frictional force exerted on a pair of skis is always directed opposite to the ski displacement. Hence the work done by friction is negative, both as the skier is pulled to the top of the slope and as she skis to the bottom. Thus, for the closed path, the work done on the skier by friction is negative rather than zero.

We define a conservative force as one for which the work done around every closed path equals zero. Formally,

$$\oint \mathbf{F} \cdot d\mathbf{s} = 0 \tag{8.1}$$

where the small circle around the integral sign means that the path is closed.

Example 1 Work Done on an Astronaut

An astronaut leaves the earth at point A and rises to an altitude equal to twice the earth's radius at point B. He travels in orbit along an arc to point C, and then returns to earth at point D. Finally he returns to his starting point by surface transportation. We show his trip schematically in Figure 8.1. How much work is done on the astronaut by gravity in the round trip?

In Example 4 of Chapter 7 the work done in going from A to B was determined to be

$$W_{AB} = -\frac{2}{3}\frac{GM_E m}{R_E}$$

Because the force and displacement are perpendicular from B to C, $\mathbf{F} \cdot d\mathbf{s} = 0$ and

$$W_{BC} = 0$$

To calculate the work from C to D we note that

$$W_{CD} = -W_{DC}$$

because the force and displacement are parallel along $C \rightarrow D$, whereas they are oppositely directed along $D \rightarrow C$. For the path $D \rightarrow C$ the calculation is identical with that in Example 4 of Chapter 7, so we have

$$W_{DC} = -\frac{2}{3}\frac{GM_E m}{R_E}$$

and

$$W_{CD} = +\frac{2}{3}\frac{GM_E m}{R_E}$$

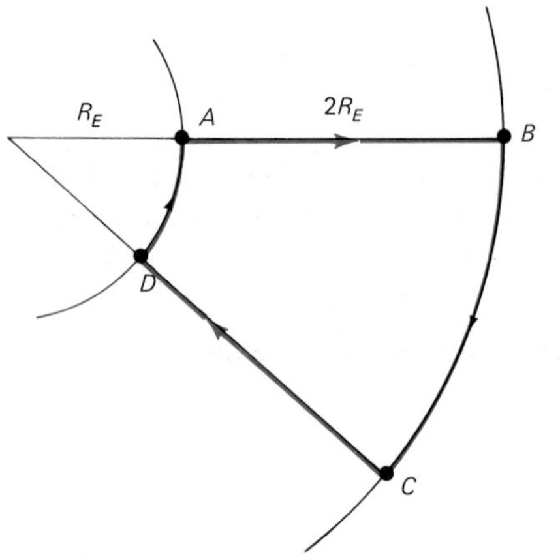

Figure 8.1

An astronaut travels radially outward from the surface of the earth ($A \rightarrow B$), along a circular path ($B \rightarrow C$), radially inward toward the center of the earth ($C \rightarrow D$), and back to the start along the surface of the earth ($D \rightarrow A$). For this closed path the work done on the astronaut by gravity equals zero.

Because the force and displacement are perpendicular from D to A, again $\mathbf{F} \cdot d\mathbf{s} = 0$ and

$$W_{DA} = 0$$

For the complete path, $A \rightarrow B \rightarrow C \rightarrow D \rightarrow A$, we thus have

$$W = W_{AB} + W_{BC} + W_{CD} + W_{DA}$$

$$= -\frac{2}{3}\frac{GM_E m}{R_E} + 0 + \frac{2}{3}\frac{GM_E m}{R_E} + 0 = 0$$

The net work done on the astronaut by the force of gravity is zero. In Section 28.4 a more general proof of the conservative nature of this type of inverse square law force is presented by using the electrostatic force as an example.

Why should a single force satisfying Eq. 8.1, $W = \oint \mathbf{F} \cdot d\mathbf{s} = 0$, be characterized as conservative? In physics a property that is conserved is one that does not change. According to the work–energy principle, the change in kinetic energy of a particle is zero if the work done by the net force is zero. When only one force acts, that force is the net force, and if it satisfies Eq. 8.1, kinetic energy is conserved and the force is conservative.

We can define a conservative force acting in a closed path as a force that does zero work, or produces no change in particle kinetic energy. But there is a second, and completely equivalent, definition of conservative force, which we will use to introduce the concept of potential energy.

Let us consider the skier on the mountain. She can travel from the mountaintop to the ski lodge along many routes. She can take the steepest trail and arrive in a short time; or she can take a circuitous and scenic route that requires more time. She might even break her leg and be brought back to the lodge by the ski patrol! Yet for each of these possible trips, the skier starts at the same place on the mountain top and finishes at the same place at the bottom. The difference in elevation, h, is the same between the starting and finishing points. Hence the work done by the force of gravity is $+mgh$ for each trip. This bring us to the second way of defining conservative force. A conservative force is one for which the work done by the force on a particle over every path connecting two points is the same. The work done between two fixed points, A and B, is independent of the path. Formally, we write the second definition of conservative force as

$$\int_A^B \mathbf{F} \cdot d\mathbf{s} = \text{constant} \tag{8.2}$$

independent of the path from A to B.

Equivalence of the Two Definitions of Conservative Force

Equation 8.1, the closed-path definition of conservative force, and Eq. 8.2, the path-independent definition of conservative force, are equivalent. To demonstrate this equivalence, we can use Eq. 8.2 to prove Eq. 8.1. Keeping in mind that other forces may also act, we have, as the work done by a single force, $\mathbf{F}$, from A to C to B, (Figure 8.2)

Figure 8.2

For a particle to move from point A to point B many paths are possible (such as $A \to C \to B$ or $A \to D \to B$). If the work done by a particular force is the same for all paths connecting A and B, the force is conservative. Equivalently, a force is conservative if the work equals zero for all closed paths (such as $A \to C \to B \to D \to A$).

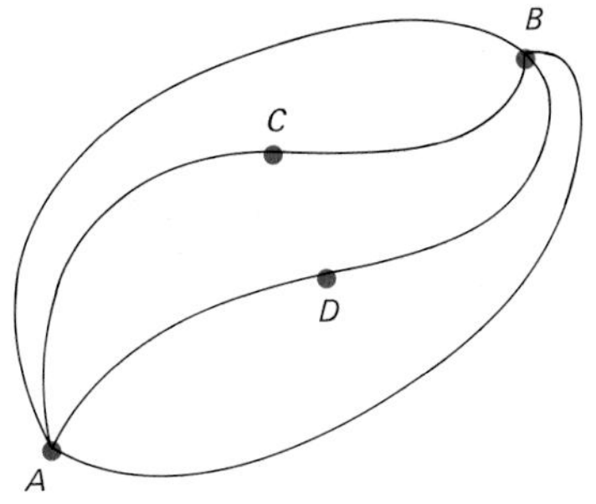

$$W(A \to C \to B) = \int_A^C \mathbf{F} \cdot d\mathbf{s} + \int_C^B \mathbf{F} \cdot d\mathbf{s}$$

Equation 8.2 requires

$$W(A \to C \to B) = W(A \to D \to B) \qquad (8.3)$$

If we reverse direction along one of the paths we have

$$W(A \to C \to B) = -W(B \to C \to A) \qquad (8.4)$$

because **F** is unaffected, whereas $d\mathbf{s}$ changes sign. Thus Eq. 8.3 can be written

$$-W(B \to C \to A) = W(A \to D \to B)$$

which in turn becomes

$$W(A \to D \to B) + W(B \to C \to A) = 0$$

But this is nothing more than Eq. 8.1—the work done in a closed path equals zero. The proof can be carried out in reverse order, and therefore the two definitions are completely equivalent. Note that Eq. 8.4 does not hold for the force of friction. If $d\mathbf{s}$ reverses, then **F** reverses too.

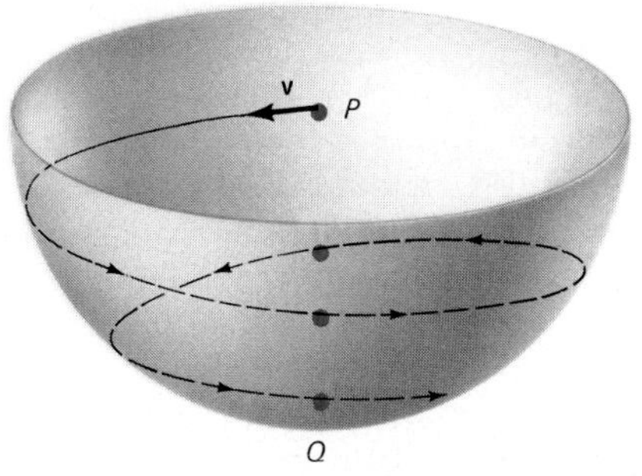

Figure 8.3

A steel sphere is given an initial kinetic energy at P such that the sphere subsequently moves along the inside surface of the water-filled hemisphere. Eventually the sphere comes to rest at Q on the bottom. For all starting speeds and directions the work done (1) by the normal force equals zero, (2) by gravity is the same positive quantity, and (3) by the drag force is a negative quantity whose value depends on the path.

One situation merits special mention. A force that is always perpendicular to the particle displacement does zero work on the particle for any path, including a closed path. Such forces do have an effect on the particle trajectory, but not on the particle energy. For example, the tension of the supporting wire of a pendulum restricts the motion of the pendulum bob, but does not affect the total energy of the bob. Similarly, the gravitational force of the earth may hold a satellite in a circular orbit, but this radial force does no work as long as the orbit is circular. Thus, forces perpendicular to the path of a particle can be ignored in considerations of work and energy.

Example 2 **A Steel Sphere in a Glass Bowl**

To illustrate further the nature of a conservative force, let's consider the motion of a steel sphere in a hemispherical bowl filled with water (Figure 8.3). Three forces act on the steel sphere. The bowl exerts a force, **N**, directed toward the center of the hemisphere and perpendicular to the velocity of the sphere. The water exerts a drag force, $\mathbf{F}_D$, opposite in direction to the velocity of the sphere. The force of gravity, $\mathbf{F}_g$, is downward.

Let the steel sphere be given an initial velocity **v** at the point P. The magnitude and direction of **v** are arbitrary except that (a) **v** must be tangent to the surface of the bowl, and (b) the subsequent trajectory of the sphere must lie on the bowl's surface. Many trajectories can be imagined. For every trajectory, however, the following statements are true:

1 The sphere starts at P and comes to rest at the bottom of the bowl at point Q.

2 No work is done by **N** because it is perpendicular to $d\mathbf{s}$ along the path—**N** is a conservative force.

3 The work done by gravity is positive and the same for all paths—gravity is a conservative force.

4 The work done by the drag force is negative and path dependent—drag is a nonconservative force.

Questions

1. A boomerang is thrown in such a way that it returns to the thrower at the same height from which it was thrown. Would you expect the boomerang to have the same speed returning as when launched? Greater speed? Less speed? Explain.
2. Cite two nonconservative forces acting on a moving automobile.

8.2 Potential Energy

The kinetic energy of a particle is one measure of the particle's ability to do work on its environment. A particle may also be able to do work on its environment when its kinetic energy is zero. For example, a particle falling through a distance h in a vacuum (so we can ignore air resistance) at the earth's surface acquires a speed $v = \sqrt{2gh}$ and a kinetic energy $K = \frac{1}{2}mv^2 = mgh$. As the particle falls, it gains kinetic energy and the ability to do work. Did the earth–particle system have this ability to do work at the start of the fall? It would certainly appear so.

To focus more sharply on this question we consider the analogous process of stretching a spring—that is, doing work on it (see Figure 8.4). Let the relaxed length of the spring be l_o, its spring constant k, and its mass negligible. Consider first a situation where the kinetic energy of a particle (mass $= m$) attached to the spring disappears as work is done on the spring (Figure 8.4a).

Let the spring be horizontal with the attached particle free to move in one dimension on a horizontal frictionless surface. When the spring length is l_o the particle has a velocity v_o directed from left to right. As the spring stretches from l_o to $l_o + x_1$ the velocity of the particle is gradually reduced to zero. The initial kinetic energy, $K_i = \frac{1}{2}mv_o^2$, is reduced to zero as the particle does work on the spring. Clearly the kinetic energy of the particle enabled it to stretch the spring.

If the spring now contracts, pulling the particle back to its starting position, the particle recovers its original kinetic energy. Evidently the original kinetic energy was

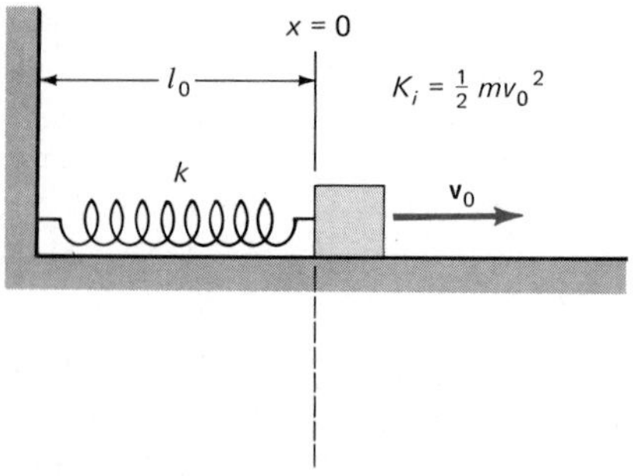

Figure 8.4a

At the instant drawn, $x = 0$, the spring is relaxed, and the block moves from left to right with velocity $\mathbf{v}_o$. The kinetic energy of the block, $\frac{1}{2}mv_o^2$, decreases as the block moves from $x = 0$ to the right.

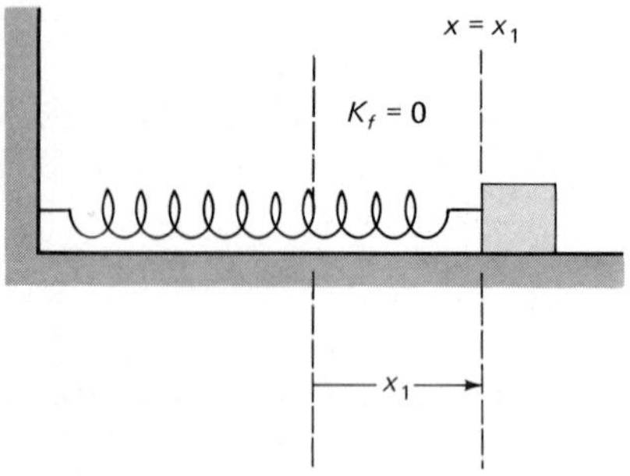

Figure 8.4b

The original kinetic energy of the block represented a capacity to perform work. The block stretched the spring as its kinetic energy decreased. At $x = x_1$ the kinetic energy is zero. The "lost" kinetic energy is completely recovered, in the absence of friction, if the spring contracts back to $x = 0$. Therefore at $x = x_1$ the original kinetic energy of the block may be viewed as stored in the spring.

Figure 8.5

The block is held motionless at $x = 0$ and dropped (a). Initially its speed and kinetic energy are zero, and the spring is relaxed. Because of gravity the block falls from $x = 0$ to $x = x_2$, coming momentarily to rest at $x = x_2$ (b). During this drop the block's kinetic energy increases from zero to a maximum and then returns to zero. Because originally the block could fall from $x = 0$ to $x = x_2$, it originally possessed the capacity to perform work. The block stretched the spring during the drop.

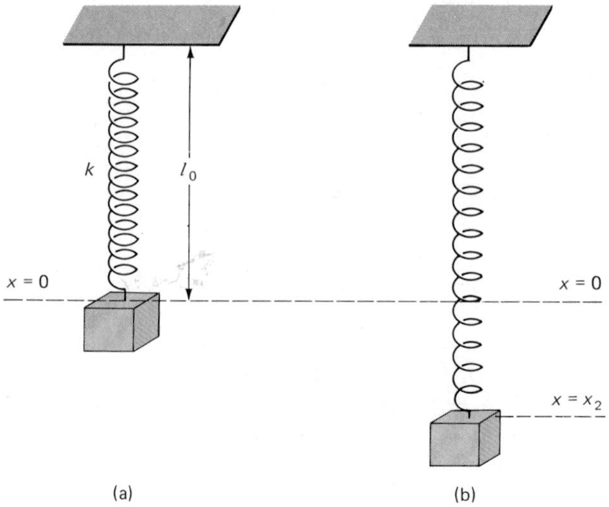

not lost in stretching the spring but was somehow *stored* in the spring. This stored energy is called *potential energy* because its release can produce kinetic energy.

Next, consider a situation where there is no net change in the kinetic energy of the particle. If the particle does work on the spring under these conditions, then this ability to do work under these conditions is not associated with the particle kinetic energy.

We mount the spring vertically, as shown in Figure 8.5. When the spring length is l_o we attach the particle, hold it motionless, and then drop it. At first it speeds up, but finally it slows down and momentarily comes to rest when the spring length is $l_o + x_2$. No net change in kinetic energy has occurred. Because gravity did work on the particle as it descended, the particle was able to stretch the spring. In its original position the particle possessed a capacity to do work on the spring—it could fall.

In the first case the spring was stretched a distance x_1 because the particle's energy allowed it to do work on the spring. In the second case the spring was stretched because the particle was able to fall a distance x_2 under the influence of gravity—the particle was located in an advantageous position. A change in the particle's position, rather than in its kinetic energy, enabled it to do work on the spring. The energy of position is called **potential energy.** Now let us define potential energy quantitatively by using our second definition of conservative force (Section 8.1).

When a conservative force does positive work on a particle, the kinetic energy of the particle increases. At the same time, the capacity of the particle to do work on its environment by virtue of particle *motion* increases. However, the ability of the particle to do work on its environment by virtue of particle *position* decreases. Thus gravity does positive work on a falling ball whose kinetic energy increases during the fall, but the ability of the ball to do work by falling decreases during the fall. Observation shows that the total ability of the ball to do work is not affected by falling. This observation suggests the following definition of the change in potential energy, $\Delta U = U_f - U_i$, of a particle moving under the influence of the single force **F**.

Let a particle be displaced from the initial point i to the final point f. We define the change in potential energy to be

$$\Delta U = U_f - U_i = -\int_i^f \mathbf{F} \cdot d\mathbf{s} \qquad (8.5)$$

The integral, $\int_i^f \mathbf{F} \cdot d\mathbf{s}$, is the work done by the single conservative force **F**. Hence ΔU is the negative work done by the same force. The minus sign is needed so that ΔU will be negative if the integral is positive. This choice of sign helps us generalize the conservation of energy laws. Because the force **F** depends on properties of the environment through force laws (see Section 6.8), the change in potential energy ΔU likewise depends on properties of the environment. Quantitative expressions for U and ΔU are derived below for several laws of force.

Note that Eq. 8.5 defines a difference in potential energy, but says nothing about the absolute value of the potential energy, because the concept of absolute value here does not have physical significance. Each time Eq. 8.5 is used, the location of the zero of potential energy can be chosen for convenience. For example, for gravitational potential energy we might choose the floor or desk top for $U = 0$. Common choices are the origin of coordinates and the point at infinity.

Quantitative Relations for Potential Energy

The force **F** in Eq. 8.5 refers to a single conservative force. Each different law of force will produce a particular functional form of ΔU when used in Eq. 8.5. Several force laws are given in Section 6.8, and of these only the frictional force is nonconservative. For each conservative force law we have discussed so far, we now develop expressions for ΔU by using Eq. 8.5. In each case we give the most common choice for $U = 0$, and the corresponding equation for U. We also give a graph for each U for specific values of the constants in the force law.

Linear Spring

In Figure 8.6 the force exerted by a spring stretched a distance x is

$$F = -kx \qquad (8.6)$$

The change in potential energy of a particle moved from 0 to x is the negative of the work done by the spring on the particle. Thus we have

$$\Delta U = -\int_0^x (-kx)\, dx = +\frac{1}{2}kx^2$$

We take $U_i = 0$ at the origin, $x = 0$, and $U_f = U$ to obtain

$$U = \frac{1}{2}kx^2 \qquad (8.7)$$

Figure 8.6

As the block moves to the right from the origin it exerts the force $-kx$ directed to the left. Negative work is done on the block by the spring. In the process, the capacity of the spring to perform work by contracting increases, and the energy stored in the spring increases.

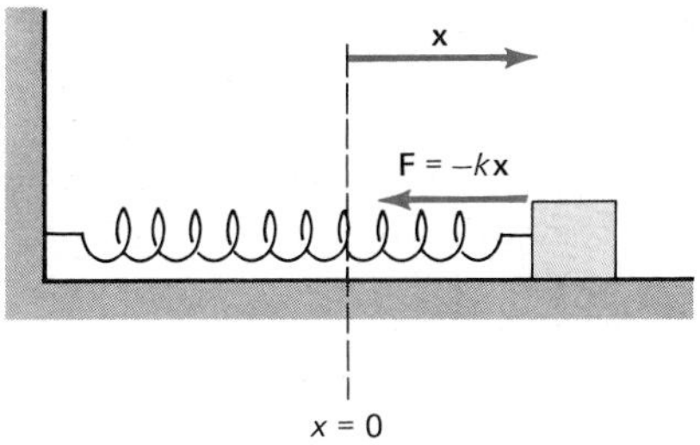

Figure 8.7

The potential energy $U = \frac{1}{2}kx^2$ of a spring is shown versus spring extension x for a spring constant $k = 10$ N/m.

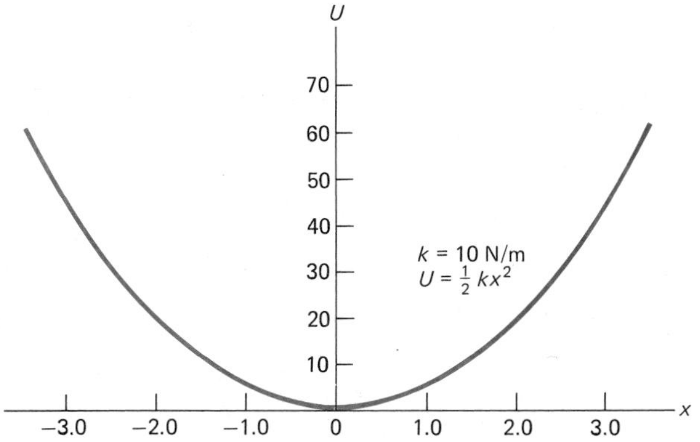

for the potential energy of the spring. This also equals the external work done on the spring in stretching it as given in Example 4 of Section 7.2. This function is plotted in Figure 8.7. Note that the spring stores energy in both the stretched ($x > 0$) and compressed ($x < 0$) condition.

Constant Gravitational Force

The constant force of gravity near the surface of the earth is given by

$$\mathbf{F} = -mg\mathbf{i} \qquad (8.8)$$

where the positive x-axis is directed upward. The change in potential energy of the mass when it is raised a distance x is given by Eq. 8.5 as follows (using $d\mathbf{s} = \mathbf{i}\, dx$):

$$\Delta U = -\int_0^x (-mg\mathbf{i}) \cdot (\mathbf{i}\, dx)$$

Because $\mathbf{i} \cdot \mathbf{i} = +1$, the right side of this equation becomes

$$+mg \int_0^x dx = +mgx$$

To evaluate the left side of the same equation we recall that

$$\Delta U = U(x) - U(0)$$

If $U(0)$ is chosen equal to zero, then we obtain

$$U(x) = +mgx \qquad (8.9)$$

as the expression for the potential energy.

Figure 8.8

The potential energy $U = mgx$ of an 80-kg mass is shown versus height x near the surface of the earth.

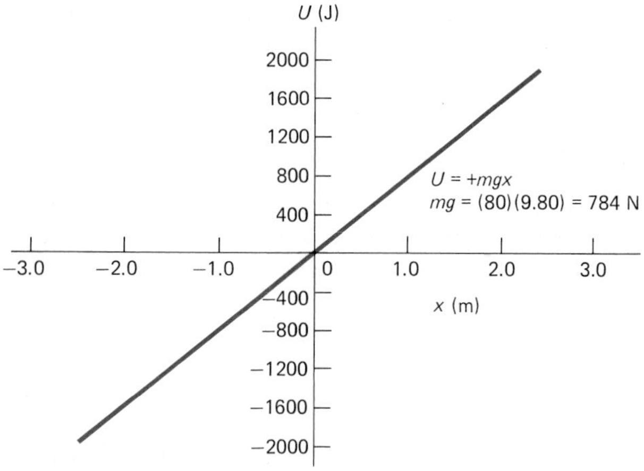

Note that if $x > 0$, then $U > 0$. A raised weight has the capacity to do work by falling to a lower level. The potential energy function described by Eq. 8.8 is shown in Figure 8.8 for $m = 80$ kg ($mg = 784$ N).

Universal Force of Gravitation

The magnitude of the attractive force of gravity between point masses m_1 and m_2 (Section 6.6) is

$$F = G\frac{m_1 m_2}{r^2} \tag{8.10}$$

where r is the distance between m_1 and m_2. If the distance between m_1 and m_2 is increased from r to r_o, the change in potential energy is

$$\Delta U = -\int_r^{r_o} \mathbf{F} \cdot d\mathbf{s}$$

$$\Delta U = U(r_o) - U(r) = -\int_r^{r_o} (-1)\left(G\frac{m_1 m_2}{r^2}\right) dr$$

$$= (Gm_1 m_2)\int_r^{r_o} \frac{1}{r^2}\, dr$$

$$= -Gm_1 m_2 \left.\frac{1}{r}\right|_r^{r_o}$$

The minus sign in the integrand is required because force and displacement are oppositely directed ($\mathbf{F} \cdot d\mathbf{s} = F\,ds \cos 180°$; $ds = dr$).

$$U(r_o) - U(r) = -Gm_1 m_2\left(\frac{1}{r_o} - \frac{1}{r}\right)$$

For simplicity, we take $U(r_o) = 0$ when $r_o \to \infty$. The choice $U(r_o \to \infty) = 0$ makes $U = 0$ correspond to $F = 0$. This choice gives the gravitational potential energy of a mass m_1 in the gravitational field of a mass m_2 as

$$U(r) = -\frac{Gm_1 m_2}{r} \tag{8.11}$$

This expression is symmetric in m_1 and m_2—a consequence of Newton's third law. Hence Eq. 8.11 is also the potential energy of m_2 in the field of m_1. Gravitational potential energy is a mutual property of m_1 and m_2. Because of the $1/r$ and the minus sign, the maximum value of U is zero. For all finite distances U is negative. However, U increases if r increases. Positive work must be done against gravity in order to separate two masses. This positive work appears as an increase in gravitational potential energy. The function given in Eq. 8.11 is shown in Figure 8.9 for $Gm_1 m_2 = 3.19 \times 10^{16}$ J · m. This is the constant for an 80-kg object interacting with the earth.

Figure 8.9

The mutual gravitational potential energy $U = -Gm_1 m_2/r$ between an 80-kg object and the earth is shown versus their separation r.

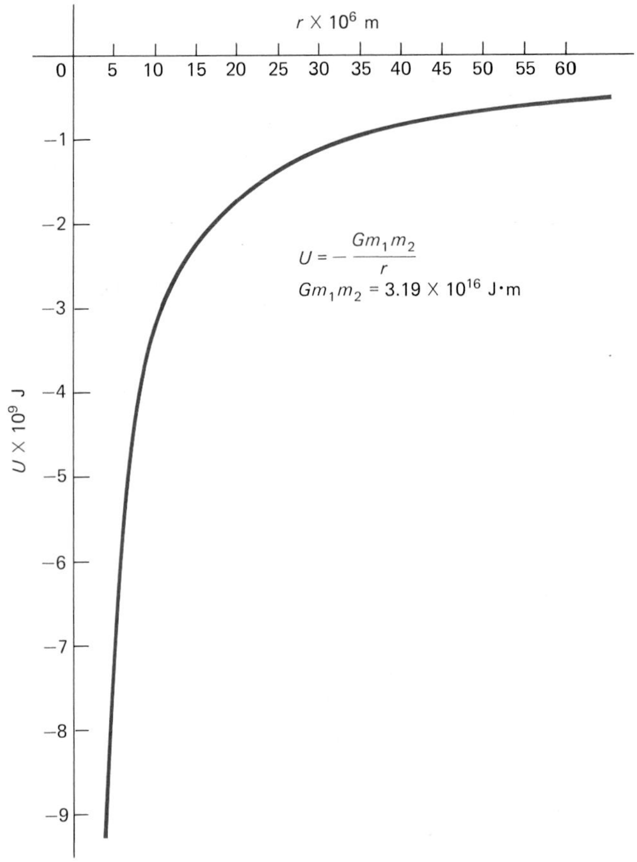

Repulsive Electrostatic Force

The fundamental law for the magnitude of the force between charged particles, Coulomb's law, takes a form similar to that of the gravitational law. Here the form is

$$F = k\frac{q_1 q_2}{r^2} \tag{8.12}$$

where k is a universal constant, q_1 and q_2 are quantities called electric charges (discussed more fully in Chapter 26), and r is the distance between the two point charges. For the particular case of an alpha particle and gold nucleus the expression $kq_1 q_2$ has the value

$$A = kq_1 q_2 = 3.65 \times 10^{-26} \text{ N} \cdot \text{m}^2$$

Coulomb's law thus takes the form

$$F = \frac{A}{r^2} \tag{8.13}$$

Like Newton's force of gravity, F, the force between charged particles varies inversely with the square of the separation between interacting particles, and acts through empty space. However, for the alpha-nuclear interaction F is repulsive.

If the separation between the two particles is increased from r to r_o, the change in potential energy is

$$\Delta U = U(r_o) - U(r) = -\int_r^{r_o} (+1)\frac{A}{r^2}\,dr$$

The plus sign in the integrand is a reminder that the force and displacement vectors are parallel. ($\mathbf{F} \cdot d\mathbf{s} = F\,ds \cos 0°$; $ds = dr$). Carrying out the integration gives

$$U(r_o) - U(r) = A\left(\frac{1}{r_o} - \frac{1}{r}\right)$$

Again choosing $U(r_o) = 0$ for $r_o \to \infty$ we have

$$U(r) = +\frac{A}{r} \tag{8.14}$$

for the potential energy. Here U decreases as the separation increases, and U is positive over the available range of r. The function is shown in Figure 8.10.

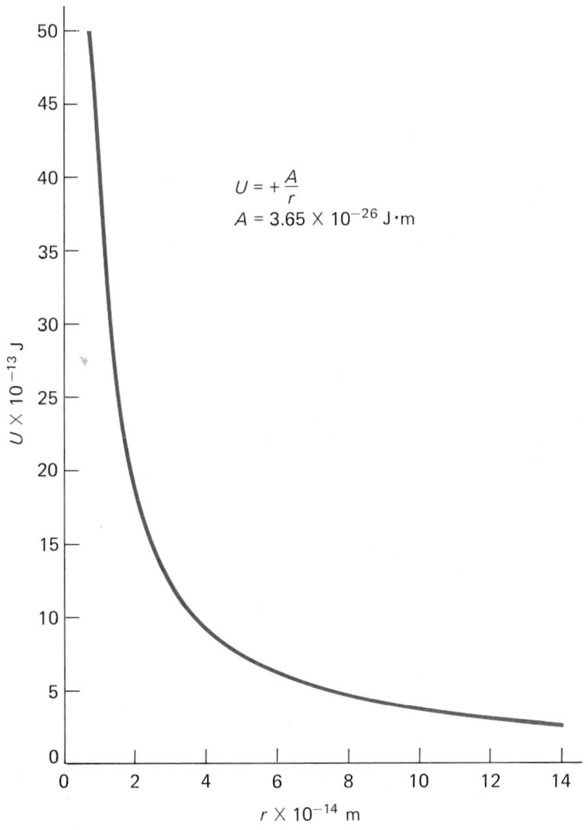

Figure 8.10

The mutual electrostatic potential energy $U = +A/r$ between an alpha particle and a gold nucleus is shown versus separation r.

Interatomic Force: The Lennard–Jones Force

In Problem 15 of Chapter 7 the force between atoms in an oxygen molecule was given as

$$F = F_o\left[2\left(\frac{\sigma}{r}\right)^{13} - \left(\frac{\sigma}{r}\right)^{7}\right] \tag{8.15}$$

where $F_o = 9.6 \times 10^{-11}$ N and $\sigma = 3.5 \times 10^{-10}$ m. If we increase the separation of two atoms from a distance r to a distance r_o, the change in potential energy is

$$\Delta U = -\int_r^{r_o} F_o\left[2\left(\frac{\sigma}{r}\right)^{13} - \left(\frac{\sigma}{r}\right)^{7}\right] dr$$

$$= +F_o\left[\frac{2\sigma}{12}\left(\frac{\sigma}{r}\right)^{12} - \frac{\sigma}{6}\left(\frac{\sigma}{r}\right)^{6}\right]_r^{r_o}$$

$$= -\frac{F_o\sigma}{6}\left[\left(\frac{\sigma}{r}\right)^{12} - \left(\frac{\sigma}{r_o}\right)^{12} - \left(\frac{\sigma}{r}\right)^{6} + \left(\frac{\sigma}{r_o}\right)^{6}\right]$$

Again picking $U(r_o) = 0$ when $r_o \to \infty$ we have

$$U(r) = \frac{F_o\sigma}{6}\left[\left(\frac{\sigma}{r}\right)^{12} - \left(\frac{\sigma}{r}\right)^{6}\right]$$

or

$$U(r) = U_o\left[\left(\frac{\sigma}{r}\right)^{12} - \left(\frac{\sigma}{r}\right)^{6}\right] \tag{8.16}$$

where

$$U_o = \frac{F_o\sigma}{6} = 5.60 \times 10^{-21}\ \text{J}$$

The function $U(r)$ is shown in Figure 8.11 using the above value of U_o.

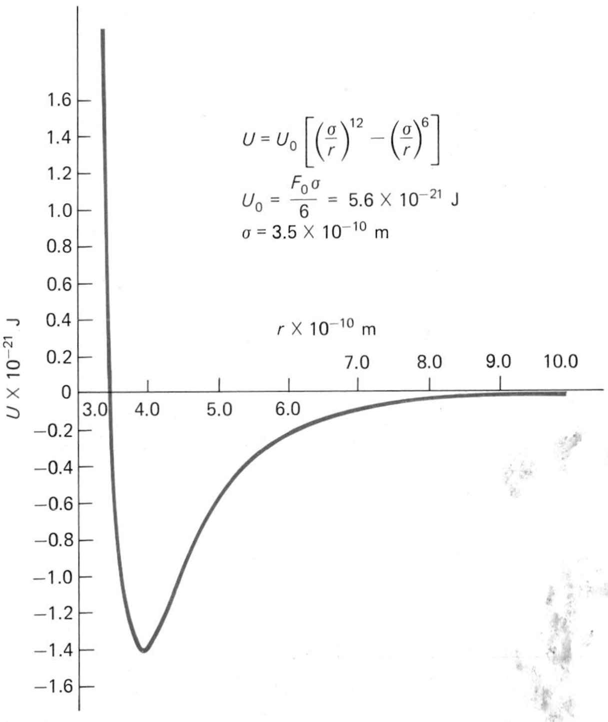

Figure 8.11

The mutual interatomic potential energy U between two oxygen atoms is shown versus separation r as given by the Lennard–Jones force.

Table 8.1

Summary of force and potential energy relations

Type of Force	Force (N)	Potential Energy (J)	Zero of Potential Energy	Value of Constant
Spring	$-kx$	$+\frac{1}{2}kx^2$	$x = 0$	k depends on the spring
Gravity (constant)	$-mg$	$+mgx$	$x = 0$	mg depends on the mass.
Gravity (universal)	$G\frac{m_1m_2}{r^2}$	$-G\frac{m_1m_2}{r}$	$r \to \infty$	$G = 6.67 \times 10^{-11} \frac{\text{N}\cdot\text{m}^2}{\text{kg}^2}$
Electrostatic for gold nucleus–alpha particle	$\frac{A}{r^2}$	$+\frac{A}{r}$	$r \to \infty$	$A = 3.65 \times 10^{-26}\ \text{N}\cdot\text{m}^2$
Lennard–Jones	$F_o\left[2\left(\frac{\sigma}{r}\right)^{13} - \left(\frac{\sigma}{r}\right)^{7}\right]$	$U_o\left[\left(\frac{\sigma}{r}\right)^{12} - \left(\frac{\sigma}{r}\right)^{6}\right]$	$r \to \infty$	$F_o = 9.6 \times 10^{-11}$ N $\sigma = 3.5 \times 10^{-10}$ m $U_o = 5.6 \times 10^{-21}$ J

If the value of the integral (Eq. 8.5) were different for different paths, then ΔU would not be well defined. This would be the case for a nonconservative force like friction, which is why we can't define ΔU for such a force. For a conservative force the integral in Eq. 8.5 is the same for all paths connecting i and f. The change in potential energy, ΔU, is well defined for conservative forces and depends on the initial and final positions of the particle, but not on the path joining them.

Questions

3. A swimmer experiences a drag force that increases as the swimmer's speed increases. Explain why no potential energy is associated with this drag force.
4. Potential energy is energy of position. Does this mean that for a given force every position has a potential energy different from that of other positions? Explain.
5. Give an example of a force having the property that no change in potential energy occurs for (a) a circular path; (b) a straight-line path.

8.3 The Law of Conservation of Energy

Our understanding of energy is widened when we consider both potential energy and kinetic energy. Our starting point is the work–energy principle:

$$W_{net} = \Delta K \tag{8.17}$$

The work done on a particle by the sum of all forces—conservative and nonconservative—acting on that particle is included in W_{net}. To introduce potential energy we limit our consideration to systems experiencing conservative forces only. With this restriction we can state that W_{net} equals the work done by the sum of all conservative forces acting.

The total work done by all of the conservative forces can be written

$$W_{net} = W_1 + W_2 + \cdots$$

Using Eq. 8.5, we write

$$W_1 = -\Delta U_1,\ W_2 = -\Delta U_2, \ldots \tag{8.18}$$

where W_1, W_2, . . . individually represents the work done by the conservative forces, and ΔU_1, ΔU_2, . . . are the associated potential energy changes. Introducing

$$U = U_1 + U_2 + \cdots \tag{8.19}$$

and putting Eq. 8.18 and 8.19 into Eq. 8.17, $W_{net} = \Delta K$, yields $-\Delta U = \Delta K$, which can be written

$$\Delta U + \Delta K = 0 \tag{8.20}$$

This equation represents the law of conservation of mechanical energy, which can be stated as follows:

If only conservative forces act on the system, the sum of changes in the kinetic and potential energies of a particle equals zero. The sum of the kinetic and potential energies of the particle is conserved.

Alternatively we can write this law as

$$K_i + U_i = K_f + U_f \tag{8.21}$$

The sum of the initial kinetic and potential energies equals the sum of the final kinetic and potential energies.

Do not confuse the criterion with the law. The law of conservation of energy is, The energy of the particle does not change. The criterion is, If only conservative forces act on the system. . . .

Example 3 **Escape Velocity of a Projectile**

Many decades before our astronauts landed on the moon, Jules Verne wrote a tale called *De la Terre à la Lune* (*From the Earth to the Moon*) about a voyage to the moon and back. Verne's fictitious spaceship was launched vertically upward out of a huge cannon buried in the ground. Neglecting air resistance, let us find the minimum launch velocity that a projectile requires in order to escape the earth's gravity. By "escape" we mean that the projectile velocity and kinetic energy approach zero as $r \to \infty$.

Take the initial point of the spaceship to be at the earth's surface and the final point at infinity. Let v_i and v_f be the corresponding projectile speeds. We have

$$K_i = \frac{1}{2} m v_i^2 \quad \text{and} \quad K_f = \frac{1}{2} m v_f^2$$

From Eq. 8.11 we have

$$U_i = -\frac{GM_E m}{R_E}$$

$$U_f = -\frac{GM_E m}{\infty} = 0$$

for the potential energies of the projectile.

The law of conservation of energy requires

$$K_i + U_i = K_f + U_f$$

or

$$\frac{1}{2} m v_i^2 - \frac{GM_E m}{R_E} = \frac{1}{2} m v_f^2$$

Solving this equation for the launch speed, we get

$$v_i = \sqrt{\frac{2GM_E}{R_E} + v_f^2}$$

The minimum launch speed is called the *escape velocity*, and corresponds to $v_f = 0$. Using

$$G = 6.67 \times 10^{-11} \frac{\text{N} \cdot \text{m}^2}{\text{kg}^2}$$

$$M_E = 5.98 \times 10^{24} \text{ kg}$$

and

$$R_E = 6.37 \times 10^6 \text{ m}$$

we have

$$v_i^2 = \frac{2\left(6.67 \times 10^{-11} \frac{\text{N} \cdot \text{m}^2}{\text{kg}^2}\right)(5.98 \times 10^{24} \text{ kg})}{6.37 \times 10^6 \text{ m}}$$

$$= 1.25 \times 10^8 \text{ m}^2/\text{s}^2$$

so that

$$v_i = 1.12 \times 10^4 \text{ m/s}$$

or just over 25,000 mi/hr.

In terms of the work–energy principle (Section 7.4) the earth's gravitational force does negative work on a projectile, reducing its kinetic energy to zero. In terms of energy conservation the kinetic energy of the projectile is converted into gravitational potential energy. The kinetic energy corresponding to the minimum launch velocity is just sufficient to raise the initially negative potential energy to zero (as $r \to \infty$).

Total Energy

The total energy, E, of a particle is defined as the sum of the particle's kinetic energy, K, and potential energy, U.

$$E = K + U \tag{8.22}$$

Strictly speaking, neither U nor E is a property of the particle alone, but is a property shared jointly by the particle and its environment. Using E we can write Eq. 8.20 as

$$\Delta E = 0 \qquad E_f = E_i \tag{8.23}$$

or

$$\Delta K + \Delta U = 0 \tag{8.24}$$

or

$$K_i + U_i = K_f + U_f \tag{8.25}$$

The total energy of the particle is conserved. Kinetic energy and potential energy change in such a way that the total energy is constant. Kinetic energy is converted to potential energy and vice versa. Each kind of energy conversion provides us with an opportunity to define a new kind of energy and thus to generalize further the law of conservation of energy. This is precisely what is done in the case of thermal, nuclear, electric, magnetic, gravitational, and rest mass energies.

Questions

6. Construct an example of a system for which (1) the net work done is not zero; (2) the work done by the conservative forces is zero; and (c) the kinetic energy of the system is reduced to zero.

7. The words *conservative* and *conservation* are both important in science. Discuss and compare their meanings in the phrases *conservative force* and *conservation of energy*.

8.4 Force from Potential Energy

The equation defining the change in potential energy

$$\Delta U = -\int_i^f \mathbf{F} \cdot d\mathbf{s} \tag{8.5}$$

can be used to calculate ΔU when the force is given. Equation 8.5 can be evaluated for a single force described by a law of force. This was done in Section 8.3 to find expressions of U and ΔU for several force laws.

There may be occasions where U is known and we want to determine the force from it. To do this we need an equation that is solved for $\mathbf{F}$ in terms of U rather than solved for ΔU in terms of $\mathbf{F}$. For example, the potential energy associated with a spring is given by

$$U = \frac{1}{2} kx^2 \tag{8.7}$$

and the force exerted by the spring is

$$F(x) = -kx \tag{8.6}$$

In this case the formula

$$F(x) = -\frac{dU}{dx} \tag{8.26}$$

gives the correct expression for the force, Eq. 8.6, if the known potential energy function, Eq. 8.7, is used.

To prove Eq. 8.26 more generally, let us consider a one-dimensional system in which U depends only on x, $U = U(x)$. For $\Delta x = x_f - x_i$, Eq. 8.5 can be written approximately as

$$\Delta U(x) \cong -F(x)\, \Delta x$$

where Δx is a small change in position associated with the change in potential energy, $\Delta U(x)$, and $F(x)$ is the force. Dividing by Δx and taking the limit as $\Delta x \to 0$, we have

$$F(x) = -\frac{dU}{dx} \tag{8.27}$$

giving Eq. 8.26.

In the case where the displacement is radial and the potential energy depends only on the radial coordinate, $U = U(r)$, we can write

$$F(r) = -\frac{dU}{dr} \tag{8.28}$$

where $F(r)$ is the component of force in the radial direction.

Example 4 Alpha Particle Potential Energy

Returning once again to the alpha-particle–gold-nucleus problem (Section 8.2), let us determine F_r, the component of force in the r-direction. Using Eq. 8.14 for U, we have from Eq. 8.28

$$F_r = -\frac{dU}{dr} \quad \text{and} \quad U = +\frac{A}{r}$$

or

$$F_r = -\left(-\frac{A}{r^2}\right) = +\frac{A}{r^2}$$

in agreement with Eq. 8.16.

Geometric Analysis of Potential Energy

When U depends on only one variable (x, for instance) a graph can be drawn of U versus x. Equation 8.27 can then be interpreted as follows: The component of F in the x-direction, $F(x)$, is the negative of the slope of the U versus x curve. In Figure 8.12 we have plotted a possible potential-energy curve for an alpha particle that is very close to a nucleus. We can obtain useful information about force from such a curve by considering the behavior of the slope:

1 At $x = x_B$ and $x = x_D$ the slope vanishes; hence $F(x) = 0$ at these two points.

2 If $dU/dx \to 0$ as $x \to \infty$, then $F(x) \to 0$ as $x \to \infty$

3 $F(x) > 0$ for $0 < x < x_B$ because the slope is negative in this interval. The same remark holds for $x > x_D$. In these regions the force is repulsive.

4 $F(x) < 0$ for $x_B < x < x_D$ since the slope is positive. The force is attractive in this region.

Recall that in one dimension the direction of the force vector is determined by its sign. Thus for $F(x) > 0$ the force is in the positive x-direction, and for $F(x) < 0$ the force is in the negative x-direction. Moreover, in this analysis the origin is taken to be at the force center. Thus attractive forces are directed toward the origin and repulsive forces are directed away from the origin.

The behavior of a ball rolling on a hilly surface provides an easy way to remember the force–potential energy relationship. Suppose that the ball in Figure 8.12 is free to roll under the influence of gravity on a contour having the U versus x shape. The U-axis is "up." At $x = x_C$ the ball is urged to the left, and $F(x) < 0$; at $x = x_B$ the ball is urged neither right nor left, and $F(x) = 0$; and at $x = x_A$ the ball is urged to the right, and $F(x) > 0$. This intuitive image of a ball on the curve gives correct results for the sign of $F(x)$, and is completely in agreement with points 1–4.

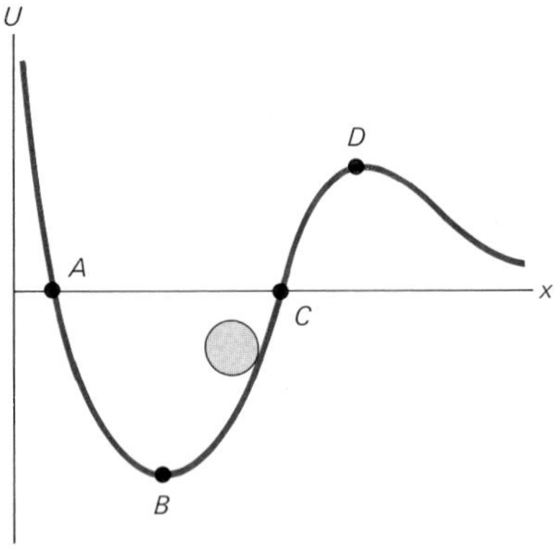

Figure 8.12

A potential energy U is shown plotted versus x. Geometrically the force F is the negative of the slope of the curve. From B to D the slope is positive, so the force is negative or directed to the left. To remember the force–potential-energy relation, imagine a ball free to move on a surface having the shape of the graph. Such a ball would be pushed in the direction of F. When located between B and D the ball is urged to the left.

Figure 8.13

The potential energy $U = \frac{1}{2}kx^2$ is shown for a spring. When the spring is compressed, $x < 0$, the slope is negative, and the force is positive. When the spring is stretched, $x > 0$, the slope is positive, and the force is negative.

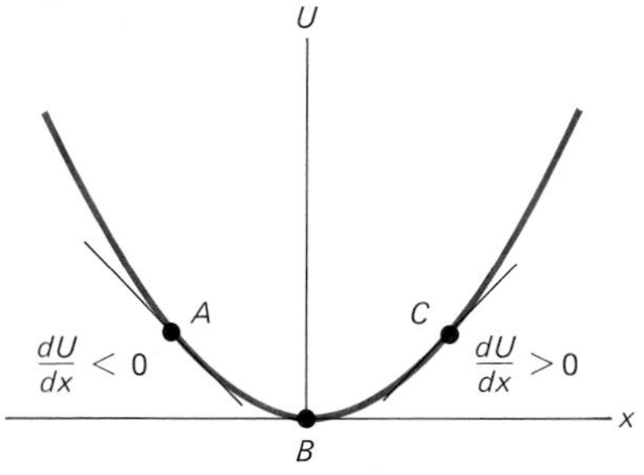

Example 5 Potential Energy of a Spring

The potential energy of a spring, defined by

$$U = \frac{1}{2}kx^2 \tag{8.7}$$

is shown in Figure 8.13. At the origin, point B,

$$\frac{dU}{dx} = 0 \quad \text{hence } F(x) = 0$$

When x is positive, as at point C,

$$\frac{dU}{dx} > 0 \quad \text{hence } F(x) < 0$$

and the force is attractive. When x is negative, as at point A,

$$\frac{dU}{dx} < 0 \quad \text{hence } F(x) > 0$$

and the force is attractive.

Using the image of a rolling ball, we see that for any displacement from point B the ball is urged back to point B. This fits in well with our knowledge that a stretched or compressed spring will seek to return to its equilibrium length.

Question

8. Equation 8.27 expresses force as the negative derivative of the potential energy. How does the picture of a ball rolling on a slope "explain" this relation, as suggested in the discussion of Figure 8.12?

8.5 Stability of Equilibrium

In Chapter 3 we considered the equilibrium and stability of a particle, using force as the central theme. In one dimension, the first condition of equilibrium can be written

$$F_{net} = 0 \tag{8.29}$$

Relabeling F_{net} as $F(x)$, we have

$$F(x) = 0 \tag{8.30}$$

Our studies of potential energy and its connection with force make it possible to reconsider stability and equilibrium from an energy viewpoint. Using Eq. 8.27 we see that Eq. 8.30 is equivalent to

$$\frac{dU}{dx} = 0 \tag{8.31}$$

as the condition for the equilibrium of a particle. In more than one dimension Eq. 8.31 must hold for all choices of displacement. With regard to a graph of U versus x equilibrium points are those having zero slope.

Equilibrium points occur in the four situations depicted in Figure 8.14. In the case of a relative minimum in the U versus x curve, like point A, the equilibrium is stable. An infinitesimal displacement to the right places the particle on a point with positive slope. As we saw in Example 5, this corresponds to a negative force tending to urge the particle back to point A. An infinitesimal displacement to the left places the particle on a point with negative slope. This slope corresponds to a positive force which also urges the particle back toward point A. If the particle is displaced in either way from point A, a restoring force will arise that urges the particle back toward point A. This is our definition of **stable behavior.**

Points of stable equilibrium are particularly interesting and are considered further in the next section. An example of a stable equilibrium situation is a particle subject to the force of a spring. Whether stretched or compressed, the spring urges the particle toward its equilibrium position. The potential energy for this case is shown in Figure 8.7, with $x = 0$ as a point of stable equilibrium.

In the case of a relative maximum, like point B in Figure 8.14, the equilibrium is unstable. Whether the particle is displaced to the left of point B or to the right, a force arises that urges the particle still farther away from B. This is **unstable behavior.**

In the case of a horizontal point of inflection like point C, the equilibrium can't be characterized as either stable or unstable. When the particle is displaced to the right its behavior is unstable, whereas when it is displaced to the left its behavior is stable. A three-dimensional exam-

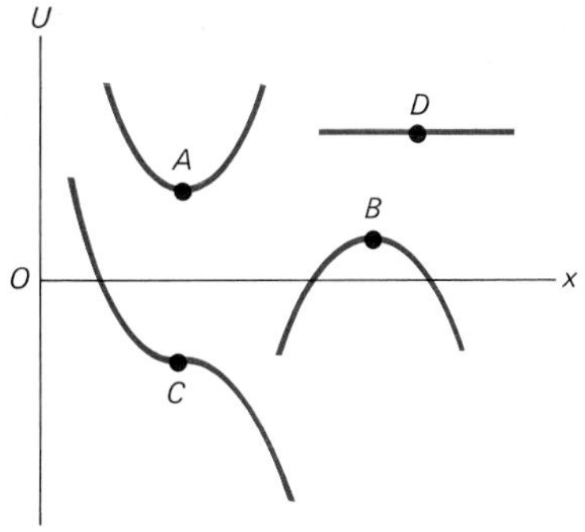

Figure 8.14

Points (A, B, C, D) on the potential-energy versus position graph with zero slope are equilibrium positions. The equilibrium is characterized as stable (A), unstable (B), or neutral (D), according to the graph in the neighborhood of the equilibrium point. For a curve like that near point C the equilibrium cannot be characterized simply as stable, unstable, or neutral.

Figure 8.15

The surface near a pass between mountain peaks is saddle shaped. A ball resting at point B is in equilibrium. For displacements along the line $A \rightarrow B \rightarrow C$ connecting the pass with peaks the equilibrium is stable. For displacements perpendicular to this line the equilibrium is unstable.

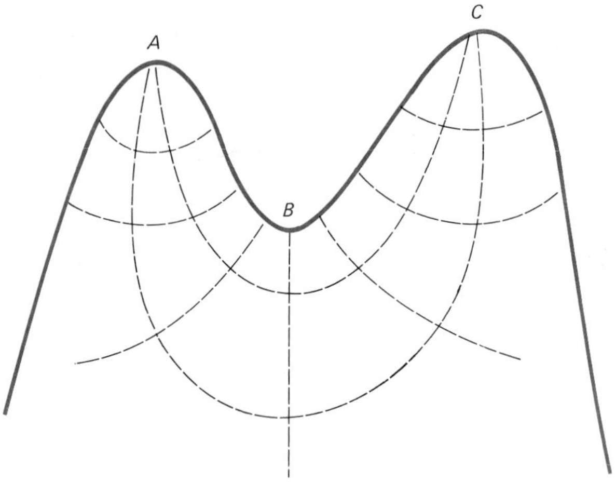

ple is a ball rolling on a saddle-shaped surface (Figure 8.15) like that of a pass between mountain peaks. This type of equilibrium is discussed in Chapters 27 and 28.

In the case of a point on a horizontal line of zero slope, like point D, the equilibrium is neutral. Infinitesimal displacements in either direction place the particle in a new equilibrium position. This is called **neutral behavior.**

Questions

9. A gymnast rotating in a vertical circle on a high bar is in equilibrium at the top and bottom of his swing. Examine the stability of each of these equilibrium positions.

10. What effect will the addition of a constant to the energy of a particle have on the particle's equilibrium? What affect will this addition have on its stability?

8.6 Energy Graphs

In addition to the location of the equilibrium point, much can be learned about the motion of a particle by considering a graph of potential energy versus position. Let's consider the case of a potential energy that depends on a single variable. We start with an alternate form of Eq. 8.21

$$E = K + U \tag{8.21}$$

for the total energy, E, of the particle—which we assume to be constant (conservative forces only).

The potential energy, U, may be positive, negative, or zero, but, the kinetic energy, $K = \frac{1}{2}mv^2$, must be positive or zero. From Eq. 8.21 we see that we must have $E \geq U$ in order to have $K \geq 0$. To identify regions where $K \geq 0$ and regions where $K < 0$ we superimpose the graphs of E versus x (a horizontal straight line for constant E) and U versus x. Regions where $E < U$ correspond to negative kinetic energies, and therefore represent inaccessible regions. Regions where $E \geq U$ correspond to non-negative kinetic energies, and therefore represent accessible regions.

To illustrate these regions consider the potential-energy graph for a spring shown in Figure 8.16. The horizontal dashed line labeled E is the graph of E versus x. As long as the motion of the particle is restricted to the interval $x_A \leq x \leq x_B$, the total energy E is greater than U, and the kinetic energy is non-negative. Such motion is possible. However, motion of the particle corresponding to either $x > x_B$ or $x < x_A$ requires $K < 0$. Such motion is not possible. The shaded regions in Figure 8.16 represent the values of x "forbidden" to the particle if its energy is E. Changing E will change the boundary between the allowed and forbidden regions.

Clearly the points A and B, determined by the intersection of the E and U curves, are special. At these points $E = U$ so that $K = 0$. The particle is motionless. Because F is a restoring force at both points, the particle acceleration is directed toward $x = 0$, the equilibrium point. If x is slightly less than x_B, for example, the particle is slowing down if moving to the right or speeding up if moving to the left. Its direction of motion changes at $x = x_B$. A similar argument holds for x_A, so these points are called *turning points* of the trajectory.

When the intersection of the E and U curves is in the neighborhood of a relative maximum rather than of a relative minimum, the situation shown in Figure 8.17 results. The points A and B are still turning points, but it is the forbidden region that lies between them rather than the allowed region. The two allowed regions are separated by a forbidden region through which no particle with this energy may pass. Increasing the energy of the particle reduces (or can eliminate) the forbidden region, while decreasing the particle energy increases the size of the forbidden region. The following example further illustrates the use of energy diagrams.

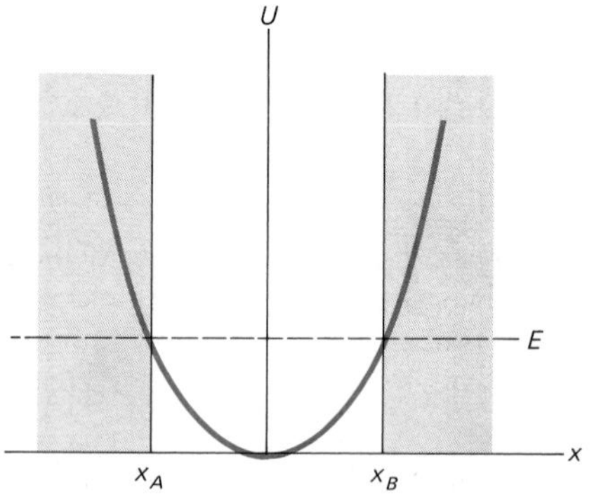

Figure 8.16

The heavy solid curve is the potential energy, U, versus position, x, of a particle whose total energy $E = K + U$ is constant. Because the kinetic energy is nonnegative, motion is not possible for regions where the E-line lies below the U-curve. These regions are shaded. Positions x_A and x_B separate the "allowed" and "forbidden" regions. Such positions are called turning points because the direction of motion reverses at these points. Here an allowed region separates two forbidden regions.

Figure 8.17

Here a forbidden region separates two allowed regions. With the total energy E a particle moving in one allowed region can never move to the other allowed region.

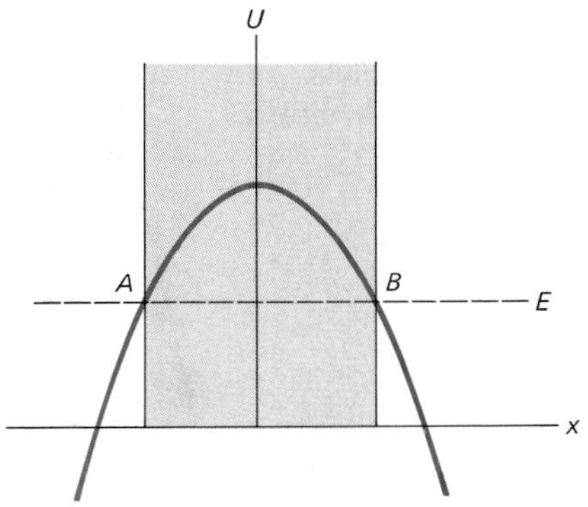

Example 6 **A Sounding Rocket**

Consider a 2000-kg rocket that is 60 km above the earth's surface and coasting radially outward. Its speed is 8.66×10^3 m/s. Let us graphically determine the maximum altitude of the rocket.

The potential energy is given by

$$U = -\frac{GM_E m}{R_E + h}$$

$$= -\frac{\left(6.67 \times 10^{-11}\,\frac{\text{N}\cdot\text{m}^2}{\text{kg}^2}\right)(5.98 \times 10^{24}\text{ kg})(2000\text{ kg})}{6.37 \times 10^6\text{ m} + 6 \times 10^4\text{ m}}$$

$$= -1.24 \times 10^{11}\text{ J}$$

Its kinetic energy is

$$K = \frac{1}{2}mv^2$$

$$= \frac{1}{2}(2000\text{ kg})\left(8.66 \times 10^3\,\frac{\text{m}}{\text{s}}\right)^2$$

$$= 0.75 \times 10^{11}\text{ J}$$

Hence its total energy is $E = K + U = -0.49 \times 10^{11}$ J.

In Figure 8.18 E and U are plotted versus r. The turning point of the potential energy curve is at $r = 16 \times 10^6$ m. Hence the maximum altitude of this rocket is $r = 9.6 \times 10^6$ m or about 1.5 earth radii.

Figure 8.18

The gravitational potential energy U of a sounding rocket ($m = 2000$ kg) is shown for a total energy $E = -0.49 \times 10^{11}$ J. The turning point is at an altitude of about 1.5 earth radii.

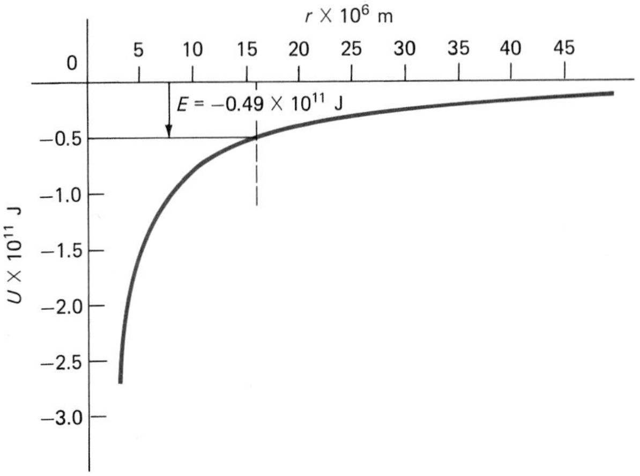

The gravitational potential energy, $U(r)$, was chosen to be zero (Section 8.2) in the limit as $r \to \infty$. This choice makes $U(r)$ negative for finite values of r. We can make some generalizations about the behavior of such a particle in terms of its total energy, E. A particle with $E > 0$ is said to be an **unbound state.** In an unbound state, $K > 0$ as $r \to \infty$, and thus $E = K + U$ is positive. Unbound states are associated with positive total energy ($E > 0$), and **bound states** are associated with negative total energy ($E < 0$). The same remarks apply, for example, to the electron–proton system, because the electrostatic potential energy (see Section 28.4) is analogous to the gravitational potential energy. Thus the hydrogen atom is a bound state with $E < 0$, whereas an electron scattering off a proton constitutes an unbound state with $E > 0$.

Finally, let us consider the interaction between an alpha particle and a nucleus. This interaction can be represented by the potential-energy curve shown in Figure 8.19. The radial position of the alpha relative to the center of the nucleus is specified by r. We consider four total energies of the system, labeled E_1, E_2, E_3, and E_4.

For all values of r we have $E_1 < U$, so we know that the alpha cannot exist with an energy of E_1. With the negative energy E_2, one turning point is located at r_A. The alpha may have values of r only in the interval $0 \leq r \leq r_A$. Because the alpha is trapped in this case, the alpha is said to be in a *bound state.*

If the alpha has the energy E_4, there are no turning points. The alpha may have any value of r. This corresponds to the alphas approaching the nucleus, passing

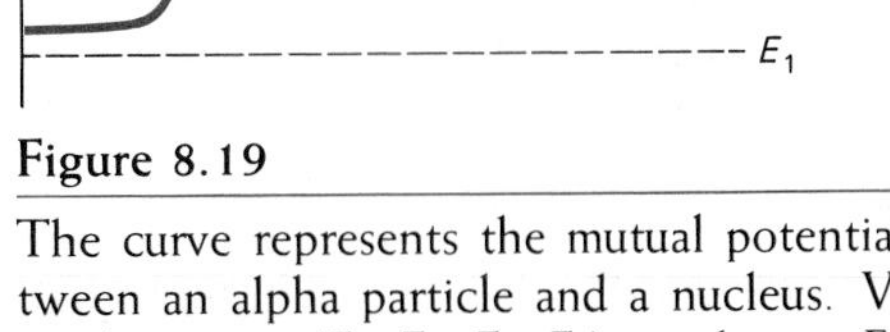

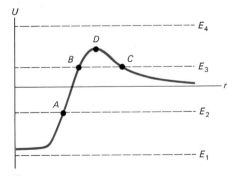

Figure 8.19

The curve represents the mutual potential energy U between an alpha particle and a nucleus. Various constant total energies (E_1, E_2, E_3, E_4) are shown. For $E = E_1$ there is no allowed region. For $E = E_2$ the single allowed region is separated from the forbidden region by the turning point at A. For $E = E_3$ two allowed regions and one forbidden region (turning points B and C) are defined. For $E = E_4$ one allowed region and no forbidden regions occur.

through it, and then going off to infinity again. The alpha is said to be in an *unbound state*.

When the alpha energy is E_3, the situation is complex. According to Newtonian mechanics, turning points at r_B and r_C define one forbidden region ($r_B < r < r_C$) and two allowed regions ($r \leq r_B$ and $r \geq r_C$). If the alpha is at a point r such that $r \geq r_C$, the trajectory is unbound. If the alpha is at a point r such that $r \leq r_B$, the trajectory is bound, and the case is similar to that of an alpha having energy $E = E_2$.

According to Newtonian mechanics, an alpha having energy E_3 may be in either a bound state or an unbound state, but it cannot move from one state to the other. Here we encounter one of the limitations of Newtonian mechanics discussed in Section 6.1. Quantum mechanics permits an alpha particle bound in the ^{238}U nucleus to make a transition to an unbound state. The probability of such a transition is small. Roughly once in 10^{40} encounters with the nuclear surface a bound alpha particle in a ^{238}U nucleus "tunnels" through the forbidden region and becomes unbound.

It is remarkable how much information can be obtained from a study of the potential energy of a system. Its scalar nature makes potential energy easier to use than force, which is a vector concept. In advanced classical and quantum mechanics potential energy is used almost exclusively, and force is greatly de-emphasized.

Question

11. Describe the motion of a particle moving under the influence of the potential energy given in Figure 8.11. Give as much detail as possible for (a) positive and (b) negative total particle energies.

Summary

A closed path is a path of arbitrary shape whose beginning and ending are at the same point.

If the work done by a force in every closed path equals zero, then the force is said to be conservative.

$$\oint \mathbf{F} \cdot d\mathbf{s} = 0 \qquad (8.1)$$

Forces that are not conservative are called nonconservative. Friction is an example of a nonconservative force. A conservative force can be used to define a potential-energy change, ΔU, according to

$$U_f - U_i = -\int_i^f \mathbf{F} \cdot d\mathbf{s} \qquad (8.5)$$

The zero of potential energy is selected for convenience. Common choices are the origin and the point at infinity.

If no nonconservative forces act on a particle, then its total energy is conserved. Formally conservation of total energy can be expressed as

$$K_i + U_i = K_f + U_f \qquad (8.20)$$

The force in the x-direction, $F(x)$, can be calculated from the potential energy by using

$$F(x) = -\frac{dU}{dx} \qquad (8.26)$$

Points of equilibrium are those for which $dU/dx = 0$. The negative slope of the U versus x graph is $F(x)$. Relative minimums are stable and relative maximums are unstable points of equilibrium.

Using $E = K + U$ together with $K \geq 0$ and $E =$ constant, we can make trajectory conclusions from the graph of U versus x. Points with $E = U$ and $K = 0$ are turning points. Turning points divide space into allowed and forbidden regions for particle motion.

Problems

Section 8.1 Conservative and Nonconservative Forces

1. Reverse the logic of the proof given in the text and prove that if the work done by a force in every closed path equals zero, then the work done by that force is path independent.

2. A conservative force performs -2.2 J of work over the path $A \rightarrow B \rightarrow C$ in Figure 1. How much work is done by the force (a) along the path $A \rightarrow D \rightarrow E \rightarrow C$; (b) along the path $C \rightarrow E \rightarrow B \rightarrow A$?

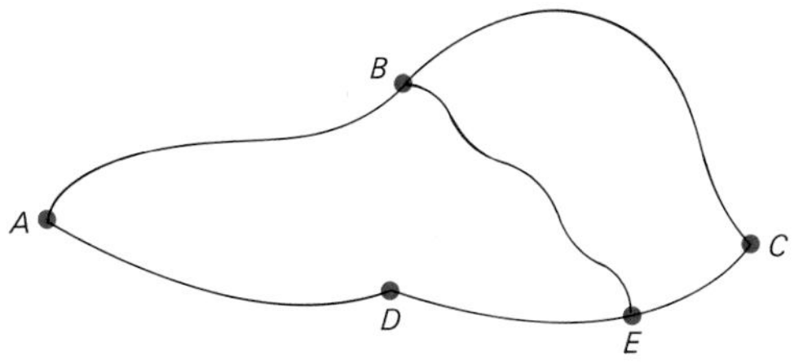

Figure 1

3. A particle can move along any of the paths drawn in Figure 2. The force is nonconservative along the path $C \rightarrow E$ and conservative along all other paths. The work done equals zero for paths $A \rightarrow D$, $D \rightarrow E$, and $B \rightarrow C$; $+10$ J for the path $A \rightarrow B$; and -20 J for the path $C \rightarrow E$. How much work is done along the path (a) $A \rightarrow B \rightarrow C \rightarrow D \rightarrow A$? (b) $C \rightarrow D$; (c) $A \rightarrow D \rightarrow C$; (d) $C \rightarrow E \rightarrow D \rightarrow C$?

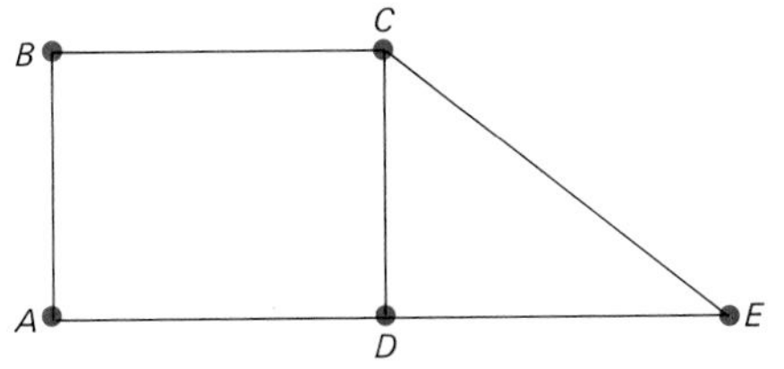

Figure 2

4. A crane lifts a 450-kg load straight upward through 22 m, moves the load sideways for a distance of 10 m, lowers the load 22 m, and then moves the load sideways 10 m to its original position. Determine the work done by the force of gravity during the entire trip.

5. A constant force of $+275$ N acts in the $+x$-direction on an object moving along the x-axis. If the position of the object varies with time according to the equation

$$x = -3t^2 + 6t$$

where x is in meters and t is in seconds, how much work is done by the constant force during the time interval (a) $t = 0$ s to $t = +1$ s; (b) $t = +1$ s to $t = +2$ s; (c) $t = 0$ s to $t = +2$ s?

6. The force on an object constrained to move along the x-axis is represented by the equation

$$F = 100x - 100$$

where F is in newtons and x is in meters. (a) Starting at the origin, through what distance would the object have to travel in order that the total work done on the object be equal to zero? (b) Is this "zero work" proof that the given force is conservative?

Section 8.2 Potential Energy

7. A spring is stretched 0.1 m in excess of its equilibrium length. The potential energy stored is 5 J. What is the spring constant?

8. Determine the increase in potential energy of a mass attached to the spring in Problem 7 when its length is increased from 0.05 to 0.20 m in excess of its equilibrium length.

9. Three identical springs are attached to three identical masses as shown in Figure 3. The system of springs and masses is free to move on a circle of radius R without friction. The springs are relaxed when the arrangement is symmetric. Determine the potential energy of the system as drawn. (Take $k = 1000$ N/m and $R = 0.1$ m.)

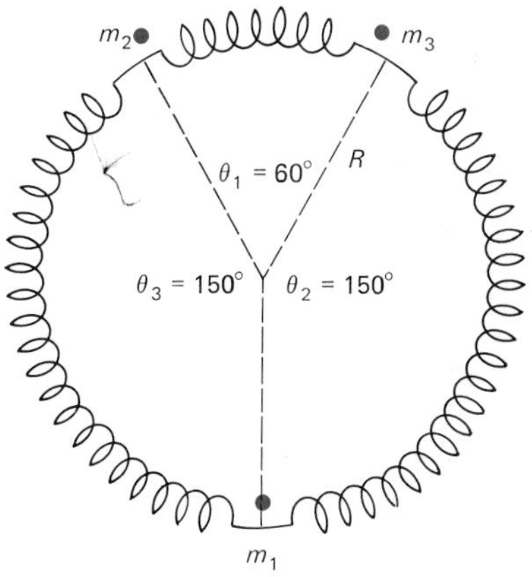

Figure 3

10. An object is connected to a pair of identical springs ($k = 500$ N/m), each 0.2 m in length. As drawn in Figure 4, the springs are relaxed. Calculate the potential energy of the system if the object is displaced 0.1 m from its equilibrium position (a) in the $+x$-direction; (b) in the $+y$-direction.

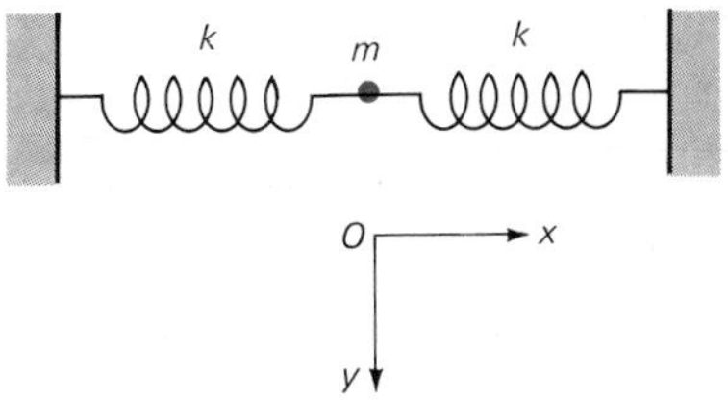

Figure 4

11. An 80-kg sky diver jumps at an altitude of 2300 m and opens the parachute at an altitude of 450 m. Determine the change in the gravitational potential energy of the sky diver during this 1850-m fall. (Use $g = 9.80\ m/s^2$.)

12. A 5-kg mass and a 10-kg mass are connected by an inextensible string of negligible mass (Figure 5). The connecting string passes over a pulley so that a rise by one mass is accompanied by an equal fall of the other. Calculate the change that occurs in the total potential energy of the system when the 10-kg mass falls through 2 m.

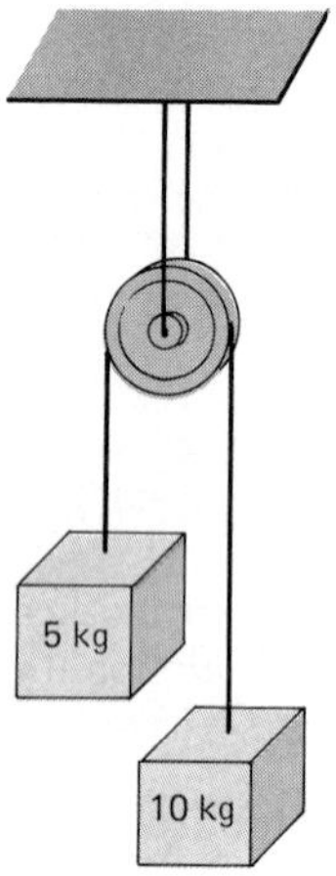

Figure 5

13. A pendulum bob ($m = 70$ kg) is pulled from its equilibrium position ($\theta = 0°$) until the supporting wire ($l = 13$ m) makes the angle θ with the vertical (Figure 6). (a) Prove the the change in potential energy of the bob equals

$$\Delta U = mgl(1 - \cos\theta)$$

Determine the resulting change in potential energy for (b) $\theta = 5°$; (c) $\theta = 45°$; (d) $\theta = 90°$.

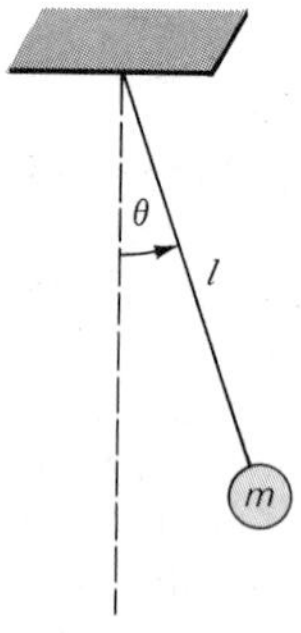

Figure 6

14. Trams ($m = 2000$ kg) that carry passengers between two mountain peaks are suspended from steel cables strung between the peaks. Pairs of trams operate together, so that while one travels up from A to B (Figure 7) the other travels down from B to A. If B is 1 km higher than A, calculate the change in potential energy (a) of the rising empty tram in a trip from A to B; (b) of the falling empty tram in a trip from B to A; (c) of the pair of empty trams in the trips defined in (a) and (b). Twenty passengers ride from A to B while five ride from B to A. If $m = 70$ kg for all passengers, calculate the change in the potential energy (d) of the system for the trip.

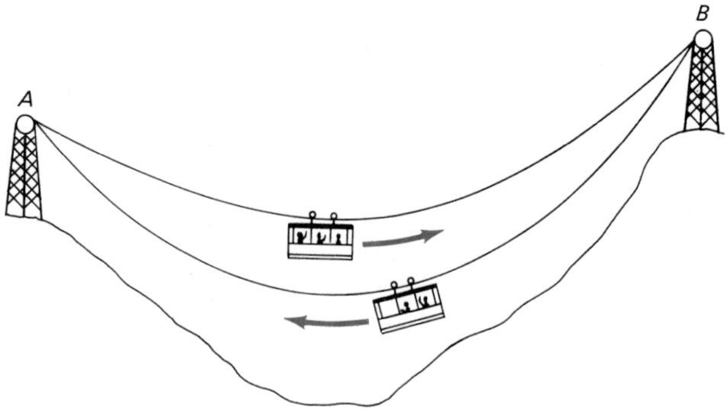

Figure 7

15. Use Eq. 8.11 to calculate the gravitational potential energy of the electron–proton system in the hydrogen atom. Assume that $r = 0.529 \times 10^{-10}$ m. (For comparison, the electrostatic potential energy of the electron–proton system at the same separation is -4.36×10^{-18} J.)

16. An astronaut of mass m moves from a point P (at a distance r from the center of the earth) to a point Q (at a distance $r + x$ from the center of the earth; see Figure 8). (a) Prove that the change in the gravitational potential energy of the astronaut equals

$$\Delta U = \frac{mgx}{1 + \frac{x}{r}} \left(\frac{R_E}{r}\right)^2$$

(b) Show that $\Delta U \cong mgx$ if $r = R_E$ and $x \ll r$, as we expect from Eq. 8.9. For $x = 1$ m, $m = 80$ kg, and the radius of the earth $R_E = 6.37 \times 10^6$ m, calculate ΔU; for (c) $r = R_E$; (d) satellite altitudes, $r = (41/40)R_E$; and (e) $r = 2R_E$.

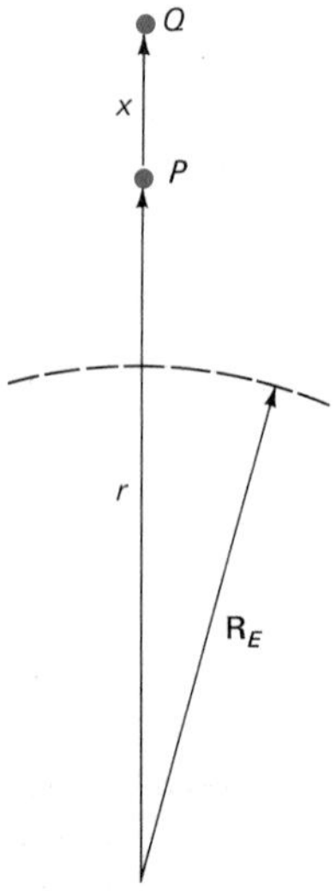

Figure 8

17. A rocket (m = 1000 kg) is launched from the earth's surface and moves completely out of the solar system. Consider only the earth–rocket system, and (a) prove that by the time the rocket's distance from the center of the earth is r its gravitational potential-energy change equals ($R_E = 6.37 \times 10^6$ m, radius of the earth)

$$\Delta U = mgR_E\left(1 - \frac{R_E}{r}\right)$$

(b) Calculate ΔU for the complete flight (to $r \to \infty$); (c) determine the value of r for which the potential-energy change is half its final value.

18. At the distance of closest approach, the potential energy of an alpha particle in the neighborhood of a gold nucleus is 12 MeV (1 MeV = 1.60×10^{-13} J). Determine this distance by using (a) Figure 8.10 and (b) Eq. 8.14.

19. Relative to a gold nucleus an alpha particle is located initially at r_A and finally at r_B (Figure 9). In the move through the distance $r_A - r_B$ the potential energy increases 3×10^{-12} J. Determine the final particle separation, r_B, and the distance moved by the alpha, $r_A - r_B$, if (a) $r_A \to \infty$; (b) $r_A = 10^{-13}$ m; (c) $r_A = 2 \times 10^{-14}$ m. Obtain the answers both graphically (Figure 8.10) and analytically (Eq. 8.14).

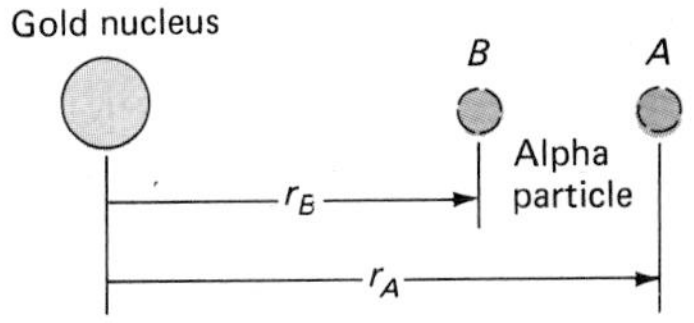

Figure 9

20. Determine the change in potential energy of two oxygen atoms whose separation is increased from (a) 3.6×10^{-10} m to 3.9×10^{-10} m; (b) 3.5×10^{-10} m to 4.95×10^{-10} m; and (c) 3.95×10^{-10} m to ∞. Obtain the answers both graphically (Figure 8.11) and analytically (Eq. 8.16).

21. Determine the two values of r for which the potential energy of two oxygen atoms is -1.0×10^{-21} J. Obtain the result both graphically (Figure 8.11) and analytically (Eq. 8.16).

22. A student whose mass is 59 kg climbs three flights of stairs, thereby raising her weight through a height of 10.6 m. Determine the change in her gravitational potential energy, taking care to note whether it increases or decreases.

Section 8.3 The Law of Conservation of Energy

23. A block (m = 2 kg) collides with a horizontal spring (k = 10 N/m). The spring (Figure 10) is compressed a distance x = 0.2 m before bringing the block to rest. Determine the original speed of the block.

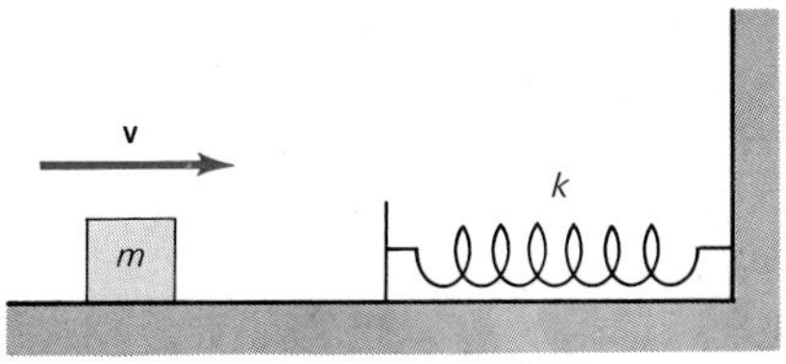

Figure 10

24. A 15-kg mass (Figure 11) is attached to a vertical spring (k = 1000 N/m). It is held motionless with the spring in its relaxed position, and then dropped. How far will it drop before coming to rest momentarily?

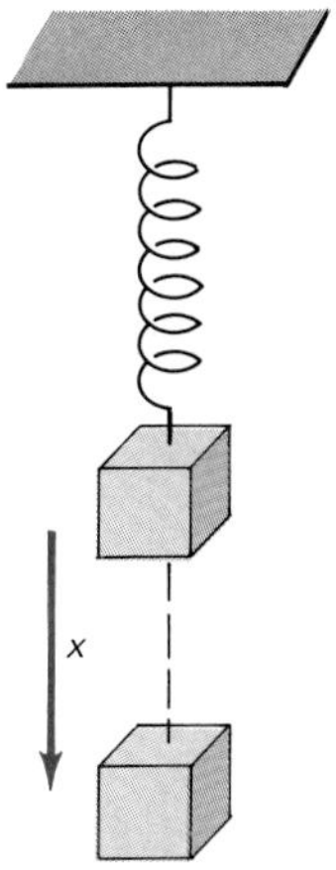

Figure 11

25. The system of identical springs (k = 1000 N/m) and masses ($m_1 = m_2 = m_3 = 0.1$ kg) in Figure 3 is held at rest in the configuration drawn, and abruptly released. (a) Assume that m_1 remains at rest after the release, and determine the speed of m_2 at the instant when all three angles are 120°. (See Problem 9.) (b) Determine a second set of values for θ_1, θ_2, and θ_3 for which the three masses are at rest momentarily. Do this by inspection, and analytically.

26. A child's pogo stick (Figure 12) stores energy in a spring ($k = 2.5 \times 10^4$ N/m). At A ($x_1 = -0.1$ m) the spring compression is a maximum, and the child is momentarily at rest. At B (x = 0) the spring is relaxed, and the child is moving upward. At C the child is again momentarily at rest at the top of the jump. Assume that the combined mass of the child and the pogo stick is 25 kg, and (a) calculate the total energy of the system if both potential energies are zero at x = 0. (b) Determine x_2. (c) Calculate the speed of the child at x = 0. (d) Determine the value of x for which the kinetic energy of the system is a maximum. (e) Obtain the child's maximum speed.

27. If the two masses in Problem 12 start from rest, what is their speed at the moment the 10-kg mass has dropped 2 m?

28. A pendulum bob (Figure 13) is dropped when the supporting wire (l = 5 m) is horizontal. Determine the speed of the bob (a) when $\theta = 0°$; (b) when $\theta = 45°$

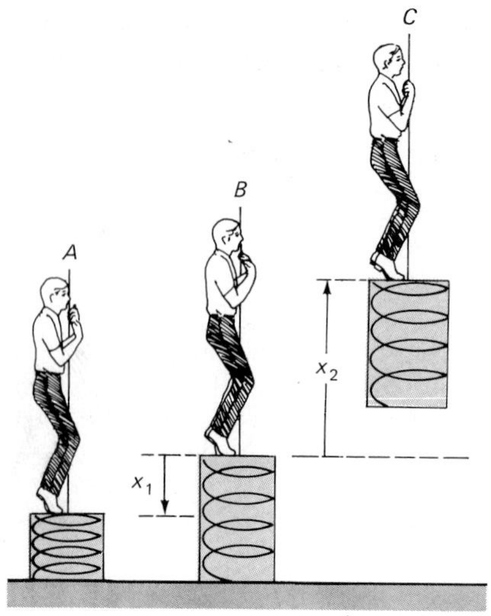

Figure 12

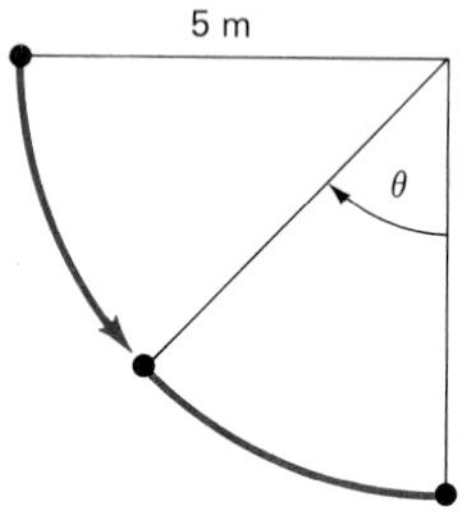

Figure 13

29. The engine of a car that is moving up a hill at 20 m/s suddenly fails. How much farther up the hill will the car (Figure 14) coast (neglect friction) before coming to rest? The slope of the hill is $\theta = 8°$. Express your answer (a) as an elevation and (b) as the distance measured parallel to the road.

Figure 14

30. A 2000-kg steel beam falls 100 m to the ground during the construction of a tall building. At the start of its fall its potential energy is zero. Calculate (a) the total energy of the beam during the fall; (b) the beam potential energy just before impact; (c) the beam kinetic energy just before impact; (d) the beam speed just before impact.

31. An object is whirled in a vertical circle 1 m in radius. The tension in the string is 6 times the weight of the object at the top and 12 times the weight of the object at the bottom of the swing. Determine the speed of the object at (a) the top and (b) the bottom of the swing.

32. In springboard diving competition the centers of gravity of divers are elevated about 2 m by the spring. Assume that a diver falls into the water from a maximum height of 3 m in the 1-m springboard competition and from a maximum height of 5 m in the 3-m springboard competition. Calculate the speed of entry into the water for the (a) 1-m competition and (b) 3-m competition. In platform diving the diver falls rather than springs into the water. Calculate the speed of entry into the water of a platform diver from (c) 10 m and (d) 30 m.

33. The roller coaster in Figure 15 moves at 5 m/s. Determine the speed of the roller coaster at the (a) top of the loop-the-loop (point C) and (b) bottom of the loop-the-loop (point B). The loop-the-loop has a radius $R = 8$ m. (Neglect friction.)

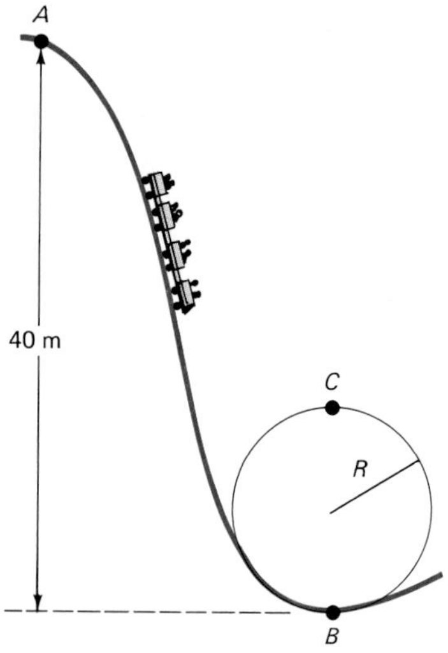

Figure 15

34. A meteorite with negligible initial velocity starts to accelerate toward the earth from a great distance. What is the impact speed of the meteorite at the earth's surface? (Neglect atmospheric effects and the earth's spin.)

35. A satellite of mass m is in a circular orbit of radius r about the earth. Use Newton's second law together with Eqs. 7.8, 8.11, and 8.22 to prove that the kinetic energy and total energy of the satellite can be expressed in the form

$$K = \frac{1}{2}\frac{GM_E m}{r} \qquad E = -\frac{1}{2}\frac{GM_E m}{r}$$

36. An 80-kg astronaut is in a circular orbit about the earth at an altitude of 200 km. (a) Neglecting the spin of the earth, calculate the energy the astronaut must lose in order to arrive at the earth's surface with no kinetic energy. (b) How much energy must the astronaut gain in order to move from the original orbit to another circular orbit at an altitude of 400 km? (c) Assume that the energy increase in (b) is realized by a constant force equal to the astronaut's weight pushing through a distance x. Determine x (see Problems 16 and 35).

37. A scientific satellite is sent radially toward the sun from a point on the earth's orbit. Assume that the satellite's initial speed is 30 km/s, and calculate its impact speed at the sun's surface. ($M_s = 1.99 \times 10^{30}$ kg; $R_s = 6.96 \times 10^8$ m.)

38. Determine the kinetic energy of the alpha particle in Problem 18 (a) as $r \to \infty$ and (b) at $r = 8 \times 10^{-14}$ m.

39. Two oxygen atoms approach each other and collide head on (see Figure 8.11). The maximum potential energy of the system, 1.0×10^{-21} J, is attained when each atom comes to rest. What is the kinetic energy of the system when (a) $r = 3.6 \times 10^{-10}$ m; (b) $r = 3.95 \times 10^{-10}$ m; (c) $r \to \infty$?

40. A block of wood ($m = 0.22$ kg) falls vertically through a distance of 12 m. It starts from rest and ends up moving at 14.1 m/s. Is mechanical energy conserved during the process?

Section 8.4 Force from Potential Energy

41. Determine $F(x)$ for the spring potential energy given in Figure 8.7 by (a) measuring the slope of the tangent line to the curve, and (b) using the formula (Eq. 8.26) for the points $x = -3.0$ m, $x = +1.0$ m, and $x = +2.0$ m.

42. Determine $F(r)$ (a) from the slope of the tangent line to the curve and (b) using Eq. 8.28 for the potential energy in Figure 8.9 at the points $r = 6.4 \times 10^6$ m, $r = 10^7$ m, and $r = 35 \times 10^6$ m.

43. Determine $F(r)$ (a) from the slope of the tangent line to the curve and (b) using Eq. 8.28 for the potential energy in Figure 8.11, for $r = 3.5 \times 10^{-10}$ m, $r = 4.4 \times 10^{-10}$ m, and $r = 6 \times 10^{-10}$ m.

44. The potential energy of a particle is shown in Figure 16 as a function of r. (a) Give the value of r for which the attractive force acting on the particle is a maximum. (b) Determine the maximum attractive force graphically. (c) Obtain the force at $r = 1$ m. (d) Obtain the force at $r = 5$ m.

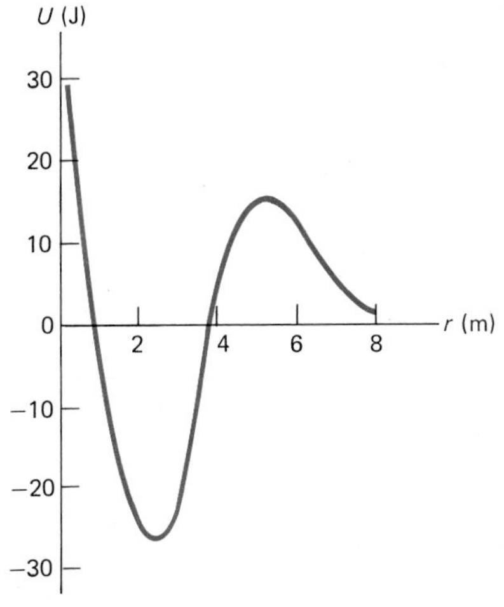

Figure 16

45. A linear spring ($k = 0.72$ N/m) is stretched 0.12 m. (a) What is its elastic potential energy? (b) What force is required to stretch it?

46. The potential energy of a sky diver is described by the equation

$$U = Ax + B$$

with U in joules, x in meters, $A = +588$ N, and $B = 3200$ J. What force acts on the sky diver at (a) $x = 5000$ m; (b) $x = 0$?

47. A 10-kg block is free to slide on a frictionless plane elevated 40° (Figure 17) above the horizontal. From the potential energy of the block, calculate the gravitational force on the block directed along the plane.

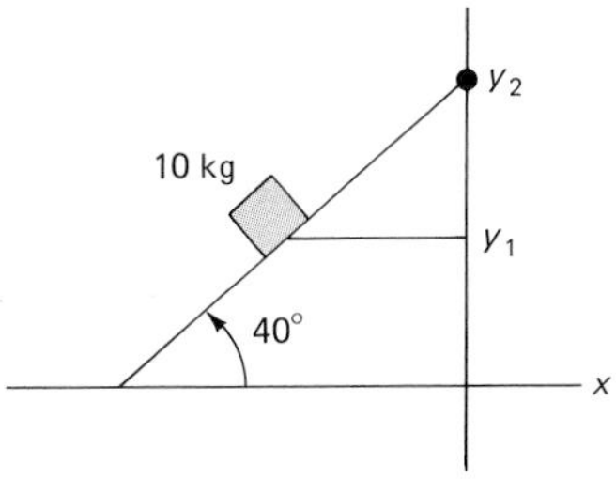

Figure 17

48. A particular nonlinear spring has an elastic potential energy $U = Cx^3$ where C is a constant. Find the restoring force exerted by this spring.

49. A certain electric force gives rise to a potential energy term $U = a/r^2$. Calculate the force.

Section 8.5 Stability of Equilibrium

50. The graph in Figure 18 represents the potential energy of a particle versus x. (a) Give the intervals for which the force is positive. (b) List the points where the force is zero. (c) Give the intervals for which the force is negative. (d) List the points of stable equilibrium and (e) list the points of unstable equilibrium.

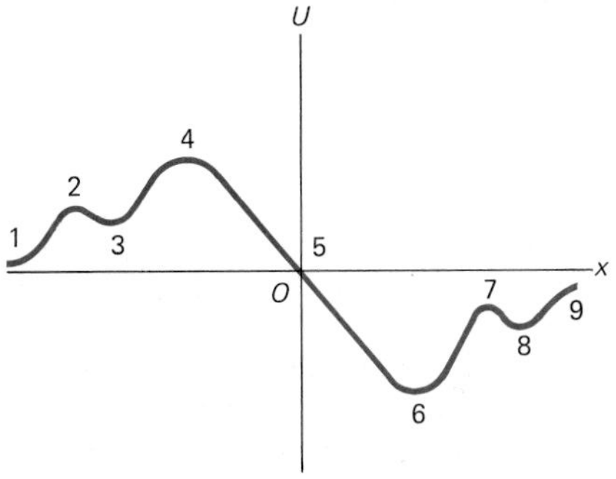

Figure 18

51. Determine the value of r for stable equilibrium for two oxygen atoms (a) graphically and (b) from Eq. 8.15.

52. Draw a graph of U versus θ for a simple pendulum (Figure 19) covering the interval $-250° < \theta < +250°$. Locate all points of equilibrium and determine the nature of their stability. (Take $m = 10$ kg and $R = 2$ m.)

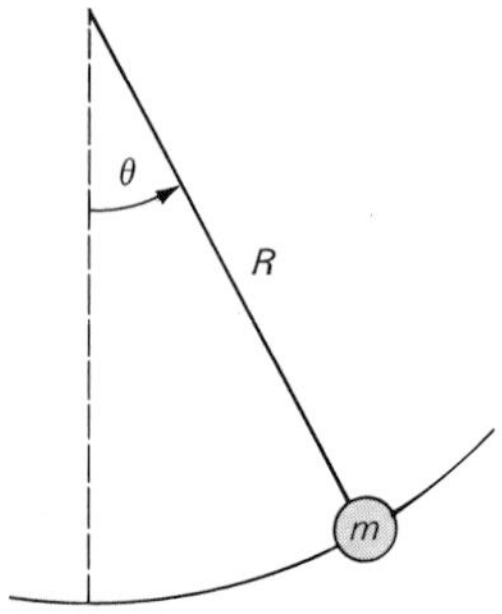

Figure 19

53. Consider the particle subject to the potential energy drawn in Figure 16. Locate all points of equilibrium and determine the nature of their stability.

54. The potential energy of a particle satisfies the equation

$$U = \frac{1}{2}(\sqrt{1 + x^2} - 1)^2$$

with U in joules and x in meters. (a) Prove that $x = 0$ is an equilibrium position. (b) Characterize the stability of this equilibrium position by plotting U versus x and examining the graph.

55. A potential energy curve is described by $U(x) = x^3$. Find a point of equilibrium and determine whether the term *stable* or *unstable* applies.

56. The potential energy of a system is described by $U(x) = x^4 - 4x^2$. (a) Find the points of equilibrium. (b) Determine which of the points of equilibrium are stable and which are unstable.

Section 8.6 Energy Graphs

57. A particle subject to the potential energy in Figure 16 has a total energy $E = -10$ J. Determine the turning points and the maximum kinetic energy.

58. The potential energy of a 10-kg mass attached to a spring is shown in Figure 8.7. For a total energy $E = 20$ J, determine (a) the turning points; (b) the maximum kinetic energy; (c) the maximum speed; (d) the acceleration at the turning points.

59. A particle subject to the potential energy in Figure 16 has a total energy $E = +10$ J. (a) Determine the turning points; (b) give the intervals defining allowed and forbidden regions of motion.

60. A particle whose potential energy U is described by the equation given in Problem 54 has a total energy $E = 1$ J. (a) What is the minimum value of U? (b) What is the maximum value of the kinetic energy K? (c) Calculate the turning points of the motion. (d) Determine K for $x = \sqrt{3}$ m.

61. A particle with a total energy E (see Figure 20) is released from rest at $x = x_D$. (a) Why does it move? (b) Describe its motion. (c) Where does it reach its maximum speed? (d) What are the turning points?

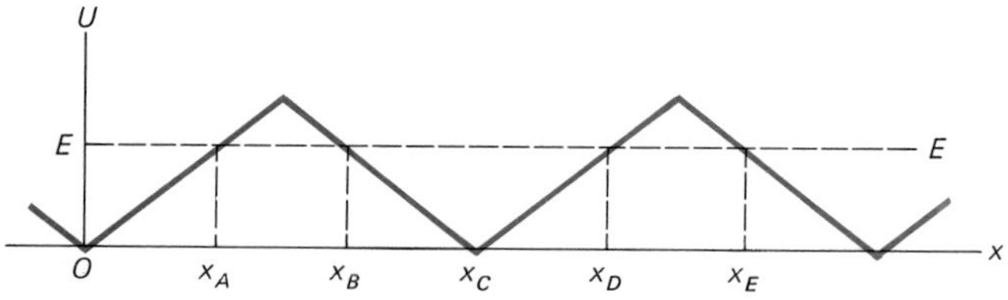

Figure 20

62. For the potential energy $U = x^4 - 5x^2$ determine the turning points and the allowed regions if the total energy E is (a) 36; (b) -4. (*Hint:* Let $y = x^2$ and use the quadratic formula. The turning points correspond to $x = \pm\sqrt{y}$ for y positive.)

chapter 9

CONSERVATION OF LINEAR MOMENTUM

Preview

The linear momentum of a particle is the product of that particle's velocity and mass. Linear momentum deserves special attention because, like energy, it is conserved under certain conditions. To provide a basis for the law of conservation of linear momentum we will introduce the concept of linear impulse and use it to reformulate Newton's second law. In this new formulation the net linear impulse represents the influence of the environment on the particle. Initially, we consider how the linear momentum of a particle changes in response to an external linear impulse. In order for linear momentum to be conserved, the net external linear impulse must vanish. We then extend this idea to interactions involving two particles, and illustrate this vector law with several examples.

9.1 Linear Impulse

When a golf club strikes a golf ball a large force of short duration is exerted on the ball by the club. Similarly, when an alpha particle enters a thin gold foil and strikes a gold nucleus, a large force of short duration is exerted on the alpha particle by the nucleus. Observable changes in the motions of the golf ball and the alpha particle occur as a result of these forces.

In order to deal with sudden or abrupt changes in the motion of particles we introduce **linear impulse,** the physical quantity that relates the initial and final states of motion of a particle. The scattering of alpha particles by gold nuclei in a thin foil (Figure 9.1) is an example of an impulse situation. The force that causes an alpha particle to accelerate is exerted only while the particle is inside the foil. Trajectories of the alpha particle inside the foil, which are deduced by using the concepts of force and acceleration, cannot be directly tested, but we can measure the positions and velocities of the alpha particle as it enters and leaves the gold foil.

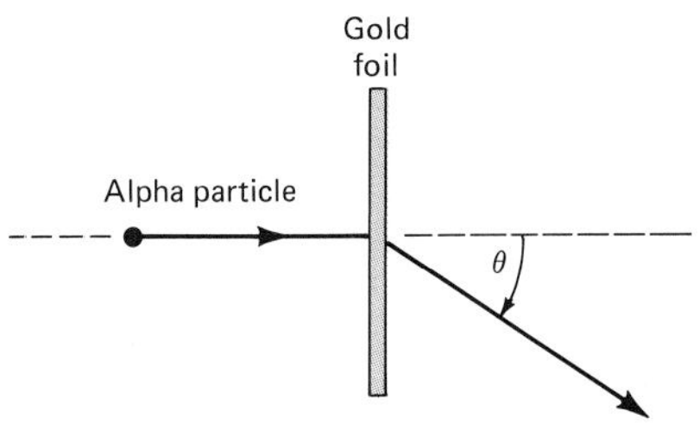

Figure 9.1

An alpha particle approaches a very thin gold foil from the left, passes through the foil, and emerges at an angle θ. This scattering of the alpha particle is caused when it strikes a gold nucleus inside the foil.

Linear impulse is the second of three physical quantities that represent the influence of the environment on particle motion. Each of these three quantities, work (Chapter 7), linear impulse (this chapter), and torque (Chapter 11), is defined in terms of force, and each leads to a conservation law.

Linear Impulse in One Dimension

For a constant force, F, in one dimension the linear impulse, ΔJ, experienced by a particle is defined to be the product of the force and the time interval, Δt, during which the force acts.

$$\Delta J = F\,\Delta t \tag{9.1}$$

This defining equation is similar in form to the corresponding equation for work. In one dimension the work done by a constant force F exerted through a displacement Δx is $\Delta W = F\,\Delta x$. Because of the close analogy between these two definitions, many of the equations given in Chapter 7 that relate to work also apply (with some modifications) to linear impulse.

Equation 9.1 defines linear impulse dimensionally as the product of a force and a time. Thus the units of linear impulse are newton-seconds (N · s).

When a constant force F is plotted versus time, a horizontal straight line results. The area bounded by the curve equals $F\,\Delta t$, as shown in Figure 9.2. The linear impulse, ΔJ, is equal to the shaded area under the F versus t graph.

If the force varies during the time interval, then the linear impulse is obtained by using a limiting procedure analogous to that used to obtain work in Section 7.2. In Figure 9.3 a plot of a variable force, F, versus t is shown for the time interval $\Delta t = t_f - t_i$. We divide this time interval into many small intervals. During the time interval Δt_j Eq. 9.1 can be used to approximate the corresponding linear impulse, ΔJ_j:

$$\Delta J_j \cong F_j\,\Delta t_j$$

This is the shaded rectangle in Figure 9.3. Summing the areas of all of the small intervals in the interval from t_i to t_f gives the expression

$$\sum \Delta J_j \cong \sum F_j\,\Delta t_j$$

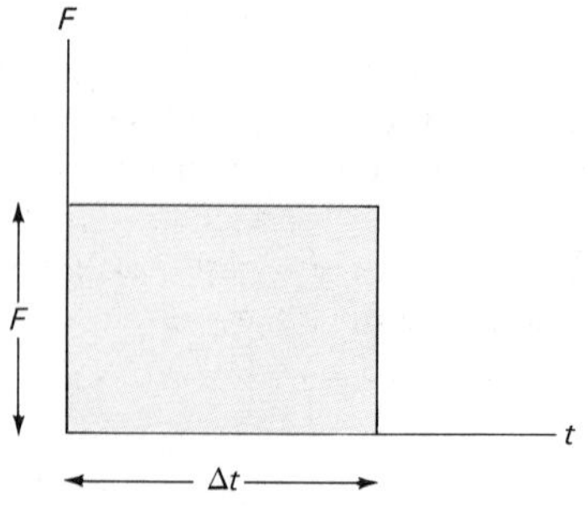

Figure 9.2

A constant force F acting for the time interval Δt develops the linear impulse $\Delta J = F\,\Delta t$. Geometrically this impulse equals the shaded area.

Figure 9.3

The linear impulse associated with a variable force $F(t)$ equals the area under the F versus t curve. This area is approximated by adding together areas of many narrow rectangles. In the limit as the width of the rectangles approaches zero the area equals the definite integral $\int_{t_i}^{t_f} F\,dt$.

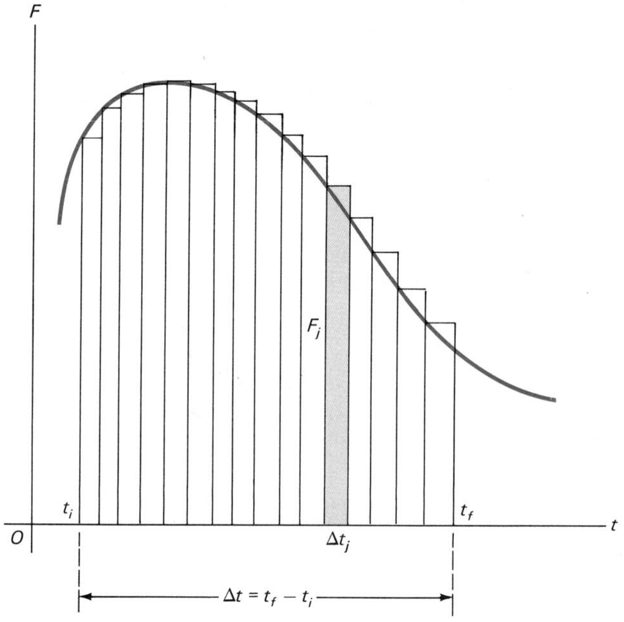

In the limit as all $\Delta t_j \to 0$ in such a way that the time interval $t_f - t_i = \sum \Delta t_j$ remains fixed, the left side of this relation becomes the linear impulse, J, and the right side becomes the area bounded by the curve. In other words, as the small rectangles become smaller and smaller, the sum of their areas more closely approximates the area under the curve. Thus we arrive at the defining equation for linear impulse for a variable force acting in one dimension:

$$J = \int_{t_i}^{t_f} F\,dt \tag{9.2}$$

Note that the work done in a displacement from x_i to x_f by a variable force F acting in one dimension is $W = \int_{x_i}^{x_f} F\,dx$. Thus the analogy between work and linear impulse in one dimension holds for variable forces.

Example 1 Deceleration of a Pole Vaulter

The oscillation of a mass that is subject to the spring force $F = -kx$ will be considered in Chapter 14. For time intervals that are short compared with the period of oscillation the position, x, is proportional to the time, t, and the force can be approximated as $F \cong -ct$ with c constant. The foam rubber pit into which a pole vaulter falls after his vault can be viewed as such a spring. We want to find the linear impulse delivered to an 82-kg pole vaulter during the first 0.010 s of his contact with a pit that has $c = 6.32 \times 10^5$ N/s.

In Figure 9.4 we show F versus t with $c = 6.32 \times 10^5$ N/s. The force exerted upward on the vaulter by the pit builds from 0 to -6320 N between $t = 0$ s and $t = 0.010$ s, respectively. The linear impulse is given by the shaded

Figure 9.4

Between $t = 0$ and $t = 0.01$ s the force exerted by a foam pole-vaulting pit is linear with time. The linear impulse $J = -31.6$ N · s equals the shaded area.

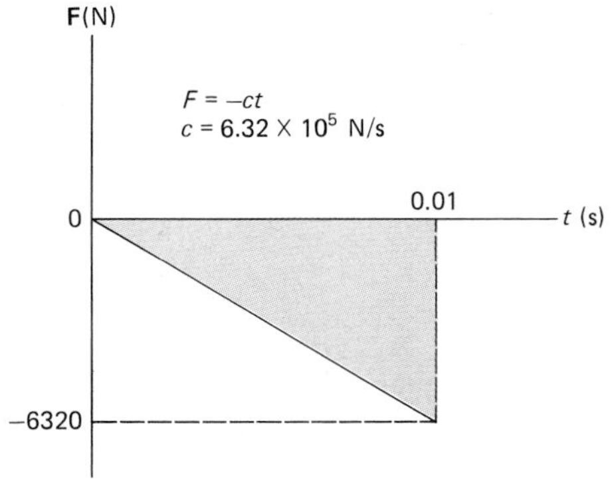

Figure 9.5

To reduce the linear momentum of the falling glass to zero a particular linear impulse $\mathbf{J} = \overline{\mathbf{F}}\,\Delta t$ is needed. With the carpet a small average force $\overline{\mathbf{F}}$ acts for a long time interval and the glass survives. With the cement a large average force must act during a short time, shattering the glass.

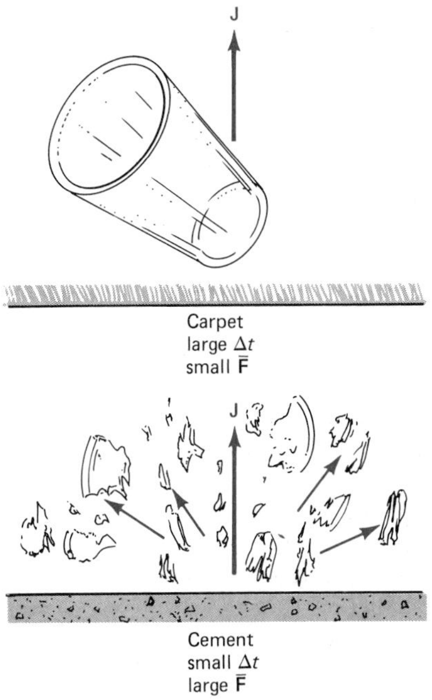

triangular area. This "area" equals half the product of the base (0.010 s) times the altitude (6320 N) of the triangle, giving

$$J = -\frac{1}{2}(0.010 \text{ s})(6320 \text{ N}) = -31.6 \text{ N} \cdot \text{s}$$

Alternatively we can use Eq. 9.2 to obtain

$$\begin{aligned} J &= -6.32 \times 10^5 \int_0^{0.010} t\,dt \\ &= -6.32 \times 10^5 \left(\frac{t^2}{2}\right)\Bigg|_0^{0.010} \\ &= -31.6 \text{ N} \cdot \text{s} \end{aligned}$$

Because the force varies, the acceleration is not constant, and the constant-acceleration equations of Section 4.6 are not applicable. However, knowing the linear impulse does permit us to draw conclusions about the motion of the vaulter, as we shall see in Section 9.2.

Average Force

When dealing with a variable force, it is sometimes helpful to introduce an **average force,** $\overline{F}$, defined by

$$\overline{F} = \frac{J}{\Delta t} \qquad (9.3)$$

Here $\overline{F}$ is the constant force that delivers the same linear impulse as the actual variable force during $\Delta t = t_f - t_i$.

Because J is the area under the F versus t curve, the average force is defined as the area divided by the total time interval.

$$\overline{F} = \frac{1}{t_f - t_i}\int_{t_i}^{t_f} F\,dt \qquad (9.4)$$

We needed a similar definition of "average," as applied to velocity, to obtain the relation $\overline{v} = \frac{1}{2}(v_i + v_f)$, which was used in Section 4.6 for the special case of constant acceleration. Note that there is an inverse relationship between $\overline{F}$ and Δt. We can illustrate this relationship by dropping a glass tumbler onto a carpet and then onto a cement floor (Figure 9.5). In both cases, the glass is brought to rest as the result of the upward linear impulse, J, delivered by the floor. The glass survives a fall onto the carpet (large $\Delta t \rightarrow$ small $\overline{F}$), whereas it is broken when it hits the cement floor (small $\Delta t \rightarrow$ large $\overline{F}$).

Example 2 **Pole Vaulter**

What average force acts on the pole vaulter in Example 1 during the time interval $\Delta t = 0.010$ s? Putting the value of J obtained in Example 1 into Eq. 9.3, we get

$$\overline{F} = \frac{J}{\Delta t} = \frac{-31.6 \text{ N} \cdot \text{s}}{0.010 \text{ s}} = -3160 \text{ N}$$

The shaded rectangle in Figure 9.6 has a length equal to the time interval ($\Delta t = 0.010$ s) and a width equal to the average force ($\overline{F} = -3160$ N). Geometrically, the "area" of this rectangle is the product of these two numbers. This area, -31.6 N · s, is the same as the area of the shaded triangle in Figure 9.4. This must be the case, since the linear impulse, J, is the same in both examples.

Questions

1. A Superball collides briefly with the floor. Consider vertical motion and choose up as the positive direction. (a) Sketch F_s versus t where F_s is the contact force exerted on the superball by the floor. (b)

Figure 9.6

The force (Figure 9.4) exerted by the vaulting pit develops the linear impulse $J = -31.6$ N · s during 0.01 s. This impulse equals the triangular area OPQ. The average force $\bar{F} = -3160$ N acting for 0.01 s develops the same impulse, shown as a shaded rectangular area. Thus $\bar{F} = -3160$ N is the average force exerted by the foam pit during this time interval.

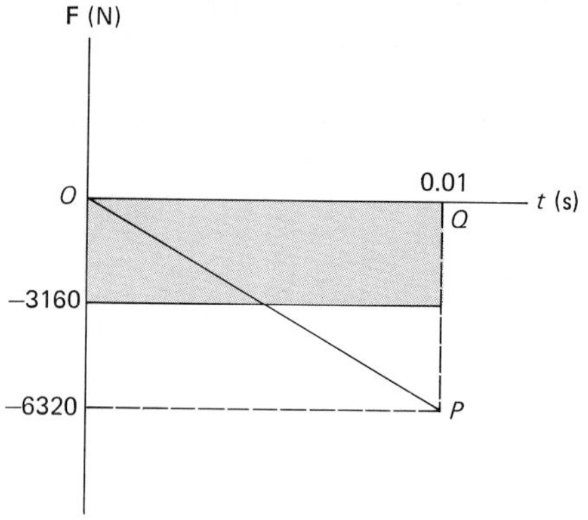

Sketch F_f versus t where F_f is the contact force exerted on the floor by the Superball. (c) Must the two linear impulses associated with F_s and F_f be exact opposites? Explain.

2. A particle of mass m experiences the constant downward force mg due to gravity. Choose up to be positive. (a) Can the linear impulse due to gravity for a given time interval $\Delta t = t_f - t_i$ be positive? (b) Zero? (c) Negative? Explain.

3. A particle of mass m is constrained to move from right to left at constant velocity **v** on a line parallel to the x-axis. An identical particle, fixed at the origin, gives rise to an attractive force **F** directed along the line joining the two particles and dependent on the separation. If $t = 0$ at the moment the particle crosses the y-axis, then for the time interval

$$\frac{-b}{v} \le t \le +\frac{b}{v}$$

(a) sketch the x-component of **F** versus t, and (b) sketch the y-component of **F** versus t. (c) In which direction would the particle at the origin be urged as a result of the impulse during this time interval?

9.2 Linear Momentum and the Linear Impulse–Momentum Principle

We have introduced linear impulse as a second way of describing the effect of the environment on a particle's motion. Through Newton's second law the net force exerted on a particle determines the product of the particle's mass and acceleration. Through the work–energy principle the work done by the net force determines the change in the kinetic energy of the particle. Now we will see that through the linear impulse–momentum principle, the linear impulse delivered by the net force determines the change in linear momentum of the particle.

Before we define linear momentum and derive the linear impulse–momentum principle, let's reconsider Newton's second law, $\mathbf{F} = m\mathbf{a}$. Using $\mathbf{a} = d\mathbf{v}/dt$ for acceleration (Section 4.4), we can write

$$\mathbf{F} = m\frac{d\mathbf{v}}{dt} = \frac{d}{dt}(m\mathbf{v}) \tag{9.5}$$

where $\mathbf{v} = d\mathbf{r}/dt$ is the instantaneous velocity (Section 4.3) and m is constant.

According to Eq. 9.5, the net force acting on a particle equals the change in the product $m\mathbf{v}$ per unit time. This product was called "motion" by Newton. The modern term is *linear momentum*, and it is assigned the symbol **p**. Its defining equation is

$$\mathbf{p} = m\mathbf{v} \tag{9.6}$$

Note that because **v** is a vector and m is a positive scalar, **p** is a vector parallel to **v**. We can therefore rewrite Eq. 9.5 as

$$\mathbf{F} = \frac{d\mathbf{p}}{dt} \tag{9.7}$$

The two forms of Newton's second law—$\mathbf{F} = m\mathbf{a}$ and $\mathbf{F} = d\mathbf{p}/dt$—are equivalent if the mass is constant. In our notation, Newton's second law was given as

$$\mathbf{F} = \frac{d}{dt}(m\mathbf{v})$$

by Newton. If the mass varies, as with a rocket, then $\mathbf{F} = m\mathbf{a}$ does not apply, and we turn to $\mathbf{F} = d\mathbf{p}/dt$.

Linear Impulse–Momentum Principle

In order to derive the linear impulse–momentum principle we must begin with the defining equation for linear impulse,

$$\mathbf{J} = \int_{t_i}^{t_f} \mathbf{F}\, dt \tag{9.8}$$

Provided that **F** represents the net force, we can write, for Newton's second law,

$$\mathbf{F} = \frac{d\mathbf{p}}{dt}$$

This equation can be written in the differential form

$$\mathbf{F}\, dt = d\mathbf{p}$$

which converts the time integral in Eq. 9.8 into the expression

$$\mathbf{J} = \int_{\mathbf{p}_i}^{\mathbf{p}_f} d\mathbf{p}$$

Here $\mathbf{p}_f$ is the linear momentum at t_f and $\mathbf{p}_i$ is the linear momentum at t_i. This integrates to

$$\mathbf{J} = \mathbf{p}_f - \mathbf{p}_i = \Delta\mathbf{p} \tag{9.9}$$

which we call the **linear impulse–momentum principle.** Stated in words, we have

The net linear impulse delivered to a particle equals the change in linear momentum of the particle.

Just as the change in the kinetic energy of a particle measures the net work done on the particle, so the change in linear momentum of a particle measures the linear impulse delivered to the particle by the net force. We see that the vector linear impulse–momentum principle, Eq. 9.9, is analogous to the scalar work–energy principle (Eq. 7.9). Both principles involve integration and connect the initial and final values of quantities without involving the values of these quantities at intermediate times.

Example 3 **Bouncing a Superball**

A Superball ($m = 60$ gm) dropped from a height of 2 m rebounds from the sidewalk to a height of 1.8 m. What linear impulse is delivered to the Superball by the sidewalk?

We neglect air resistance and calculate the linear momentum of the Superball immediately before and after impact. Applying the conservation of energy law to the fall we obtain

$$\frac{1}{2}mv_i^2 = mgh$$

Solving for v_i, the velocity immediately before impact, we obtain

$$v_i = (2gh)^{1/2} = (2 \times 9.80 \text{ m/s}^2 \times 2 \text{ m})^{1/2}$$
$$= +6.26 \text{ m/s}$$

where we have selected down to be positive. Similarly, we obtain for the rebound velocity (the velocity after impact)

$$v_f = -(2 \times 9.80 \times 1.8)^{1/2} = -5.94 \text{ m/s}$$

Thus the change in linear momentum is

$$\Delta p = p_f - p_i$$
$$= mv_f - mv_i$$
$$= (0.060)(-5.94) - (0.060)(+6.26)$$
$$= -0.732 \frac{\text{kg} \cdot \text{m}}{\text{s}}$$

Accordingly we obtain for J, the linear impulse,

$$J = -0.732 \frac{\text{kg} \cdot \text{m}}{\text{s}}$$

Because we selected down to be positive, the minus sign means that the linear impulse (and the average force) is directed upward. Because J was obtained by using Eq. 9.9, it represents the linear impulse produced by the net force, not by one of the specific forces acting on the Superball.

Impacts like the Superball–sidewalk collision occur during short time intervals. Consequently, a large average acceleration may result even for modest velocity changes. The golf drive discussed in Example 4 makes this point.

Example 4 **The Golf Drive**

A golf ball ($m = 0.0460$ kg) initially at rest is struck by a golf club and leaves the face of the club at a horizontal velocity of 60 m/s.

1 What is the net linear impulse delivered to the golf ball during the impact?

2 Estimate the average net force and the golf ball acceleration if the impact time is 5×10^{-4} s.

1 Again we use the change in linear momentum to obtain the net linear impulse. We have

$$p_i = mv_i = 0$$
$$p_f = mv_f = 0.0460 \text{ kg} \times 60 \text{ m/s}$$
$$= 2.76 \frac{\text{kg} \cdot \text{m}}{\text{s}}$$

so

$$\Delta p = p_f - p_i = 2.76 \frac{\text{kg} \cdot \text{m}}{\text{s}}$$

Thus from Eq. 9.9 we obtain

$$J = 2.76 \frac{\text{kg} \cdot \text{m}}{\text{s}} = 2.76 \text{ N} \cdot \text{s}$$

2 To determine the average force, $\overline{F}$ we need to know the impact time. High-speed photography indicates that

$$\Delta t \sim 5 \times 10^{-4} \text{ s}$$

Hence, using Eq. 9.3, we obtain

$$\overline{F} = \frac{2.76 \text{ N} \cdot \text{s}}{5 \times 10^{-4} \text{ s}} \sim 6 \times 10^3 \text{ N}$$

This force is slightly more than 0.5 ton! Another way to look at this result is to compare the average acceleration of the golf ball with the acceleration of gravity. We have

$$\overline{a} = \frac{\overline{F}}{m} = \frac{6 \times 10^3}{0.046} \sim 10^5 \text{ m/s}^2$$

which is about 10,000 times the acceleration of gravity.

The Superball bounce and the golf drive provide one-dimensional illustrations of the linear impulse–momentum principle. To bring out the multi-dimensional, or vector, nature of this principle we consider the collision of a baseball with a bat.

Example 5 **Batting a Baseball**

In this example we will consider only the horizontal components of the motion of a baseball ($m = 0.145$ kg). The first baseline is the x-axis and the third baseline is the y-axis (Figure 9.7). A baseball is moving with a velocity $\mathbf{v}_i = -31.8(\mathbf{i} + \mathbf{j})$ m/s when it reaches the batter. (The unit vectors $\mathbf{i}$ and $\mathbf{j}$ have their usual meanings.) After the baseball is hit by the batter it travels with a horizontal velocity

Figure 9.7

A baseball travels from pitcher to catcher along the line AA. The batter strikes the ball, causing it to travel along the third-base foul line.

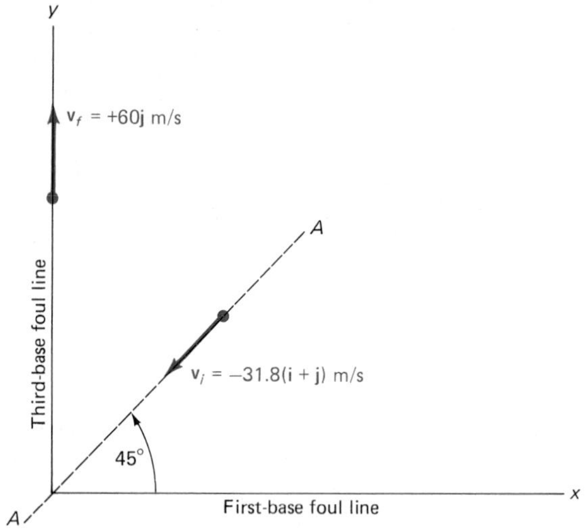

Figure 9.8

Initially the baseball approaches the catcher with linear momentum $\mathbf{p}_i$. Then the batter strikes the baseball with the bat, delivering the linear impulse $\mathbf{J}$ to the ball. Finally, the baseball travels away from the batter with linear momentum $\mathbf{p}_f$.

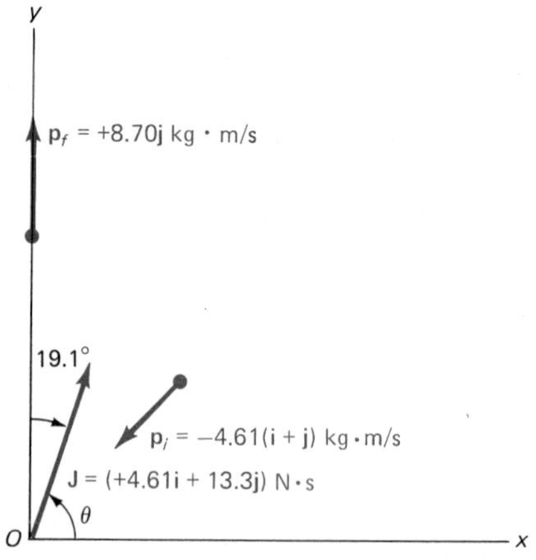

$\mathbf{v}_f = +60\mathbf{j}$ m/s along the third baseline. What linear impulse was delivered to the baseball by the batter?

We have

$$\mathbf{p}_i = 0.145 \text{ kg} \times (-31.8)(\mathbf{i} + \mathbf{j}) \text{ m/s}$$
$$= -4.61(\mathbf{i} + \mathbf{j}) \text{ kg} \cdot \text{m/s}$$

and

$$\mathbf{p}_f = 0.145 \text{ kg} \times 60\mathbf{j} \text{ m/s}$$
$$= +8.70\mathbf{j} \text{ kg} \cdot \text{m/s}$$

Hence

$$\Delta\mathbf{p} = \mathbf{p}_f - \mathbf{p}_i$$
$$= +8.70\mathbf{j} - [-4.61(\mathbf{i} + \mathbf{j})]$$
$$= (+4.61\mathbf{i} + 13.3\mathbf{j}) \text{ kg} \cdot \text{m/s}$$

Again using $\mathbf{J} = \Delta\mathbf{p}$ we have

$$\mathbf{J} = (+4.61\mathbf{i} + 13.3\mathbf{j}) \text{ N} \cdot \text{s}$$

for the net linear impulse delivered to the baseball.

The magnitude of $\mathbf{J}$ is obtained from its components. Thus we have

$$J = \sqrt{J_x^2 + J_y^2}$$
$$= \sqrt{(4.61)^2 + (13.3)^2}$$
$$= 14.1 \text{ N} \cdot \text{s}$$

The direction of $\mathbf{J}$ is given by

$$\tan\theta = \frac{J_y}{J_x}$$
$$= \frac{13.3}{4.61}$$
$$= 2.88$$

Thus $\theta = 70.9°$ and $\mathbf{J}$ makes an angle of about 19.1° with the third baseline (y-axis) as shown in Figure 9.8.

Questions

4. A particle of mass m and speed v approaches a wall. Consider two possible collisions: (a) The particle sticks to the wall; and (b) the particle rebounds and leaves the wall with the initial speed v. The interaction time in case (b) is twice what it is in case (a). Compare the two cases from the standpoint of the linear impulses and average forces.

5. Two players exert the same linear impulse, $\mathbf{J}$, in order to catch a baseball. Player A allows his hands to recoil with the baseball while bringing it to rest. Player B holds his hand rigid during the catch. Which player exerts the larger average force? Explain by using Eq. 9.3.

9.3 Conservation of Linear Momentum

According to the linear impulse–momentum principle,

if the net linear impulse exerted on a particle equals zero, then the change in linear momentum of the particle equals zero.

Formally, if

$$\mathbf{J} = 0$$

then

$$\Delta\mathbf{p} = 0 \quad \text{or} \quad \mathbf{p}_f = \mathbf{p}_i \tag{9.10}$$

The linear momentum of the particle is conserved. This is a special case of the law of conservation of linear momentum. As a special case, if $\mathbf{F}_{\text{net}} = 0$, then $\Delta\mathbf{p} = 0$ or $\mathbf{p}_f =$

$\mathbf{p}_i$. This "special case" is actually the case we usually encounter.

Much of our discussion so far has dealt with the motion of a single particle and with the laws governing the relationship between the particle and its surroundings. We have just seen how linear momentum is conserved for an isolated particle. Now let us generalize this principle to a composite system of two particles isolated from their surroundings, but able to interact with each other.

Consider two air-hockey pucks that are free to move on a horizontal frictionless surface. Their paths intersect as shown in Figure 9.9. Their masses are m_1 and m_2, and they each exert a force on the other when they touch. The only horizontal linear impulse experienced by either mass is that produced on it by the other mass. If $\mathbf{F}_2$ is the force exerted on m_2 by m_1, and $\mathbf{F}_1$ is the force exerted on m_1 by m_2, then Newton's third law requires that $\mathbf{F}_1 = -\mathbf{F}_2$. For the time interval $\Delta t = t_f - t_i$ we have

$$\mathbf{J}_1 + \mathbf{J}_2 = 0 \tag{9.11}$$

where $\mathbf{J}_1$ is the linear impulse experienced by m_1 and $\mathbf{J}_2$ is the linear impulse experienced by m_2. According to Eq. 9.9 we must have

$$\mathbf{J}_1 = \Delta\mathbf{p}_1 \tag{9.12}$$

and

$$\mathbf{J}_2 = \Delta\mathbf{p}_2 \tag{9.13}$$

where $\Delta\mathbf{p}_1$ is the change in linear momentum of m_1 and $\Delta\mathbf{p}_2$ is the change in linear momentum of m_2. Substituting Eqs. 9.12 and 9.13 into Eq. 9.11 gives

$$\Delta\mathbf{p}_1 + \Delta\mathbf{p}_2 = 0 \tag{9.14}$$

If we define the combined linear momentum, $\mathbf{p}$, of the pair of particles to be the vector sum of the momenta of the individual particles,

$$\mathbf{p} = \mathbf{p}_1 + \mathbf{p}_2 \tag{9.15}$$

then Eq. 9.14 becomes

$$\Delta\mathbf{p} = 0 \quad \text{or} \quad \mathbf{p}_f = \mathbf{p}_i \tag{9.16}$$

In the absence of a net external linear impulse, the total linear momentum $\mathbf{p}$ of the system of two particles is conserved independent of the internal forces. Equation 9.16 is identical in form to Eq. 9.10. Note carefully, however, that Eq. 9.16 refers to a composite system of two particles, whereas Eq. 9.10 refers to one particle.

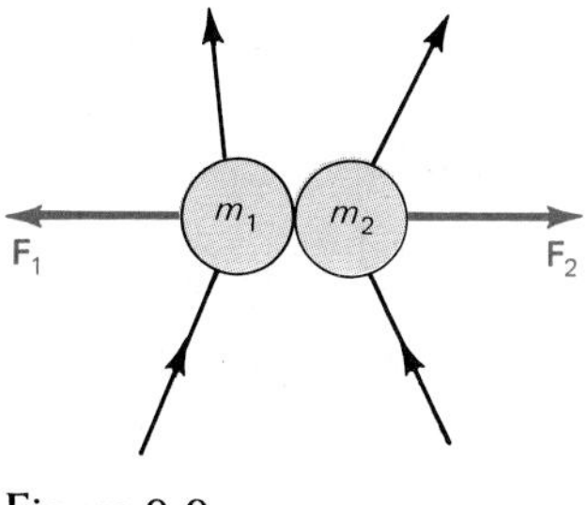

Figure 9.9

Two air-hockey pucks, moving horizontally on a frictionless surface, approach each other, collide, and recede from each other. During impact the horizontal forces that each puck exerts on the other are $\mathbf{F}_1$ and $\mathbf{F}_2$.

By assuming the validity of Newton's third law for a pair of mutually interacting particles that are otherwise isolated, we have demonstrated that the linear momentum of the combination is conserved. We could have reversed the logic by assuming conservation of linear momentum and proving that Newton's third law holds. In fact, Newton arrived at the third law while studying such pairs of interacting particles.

No exceptions to the law of conservation of linear momentum have been discovered. It has been observed to hold even in circumstances where the force is unknown. For this reason, the law of conservation of linear momentum is considered to be more fundamental than Newton's third law.

Example 6 **Two Ice Skaters**

Two ice skaters ($m_1 = 60$ kg, $m_2 = 80$ kg) face each other on the ice. They stand motionless and exchange angry words. Without warning the skater whose mass is m_1 pushes the skater whose mass is m_2, delivering a linear impulse of $J_2 = 50.0$ N · s to m_2. What happens?

First we say that m_2 acquires a backward velocity, v_2, given by

$$v_2 = \frac{J_2}{m_2} = \frac{50.0 \text{ N}\cdot\text{s}}{80.0 \text{ kg}} = 0.625 \text{ m/s}$$

The backward direction is taken to be positive. To calculate v_1, the velocity of m_1, we note that the linear momentum of the pair of skaters is conserved. From Eq. 9.16

$$m_1v_1 + m_2v_2 = 0$$

or

$$v_1 = -\frac{m_2v_2}{m_1} = -\frac{80 \text{ kg} \times 0.625 \text{ m/s}}{60 \text{ kg}}$$

$$= -0.833 \text{ m/s}$$

The negative sign means that m_1 moves in the forward direction. The same result can be obtained by using Newton's third law. According to Eq. 9.11, m_2 delivers a linear impulse $J_1 = -50.0$ N · s to m_1. Hence we obtain

$$v_1 = \frac{J_1}{m_1} = \frac{-50.0 \text{ N}\cdot\text{s}}{60 \text{ kg}} = -0.833 \text{ m/s}$$

Although the "little guy" pushed the "big guy," we see that, without any overt action, the big guy pushed the little guy as well.

A common device in science fiction stories is an "advanced" spacecraft engine for interstellar travel that operates without the mass ejection necessary with a rocket engine. The Dean Space Drive, discussed in the following example, was seriously proposed as such an engine. However, it would contradict linear momentum conservation, as we will now see.

Example 7 **The Dean Space Drive**

Norman Dean, an inventor, obtained a U.S. patent in 1959 for a "system for converting rotary motion into unidirectional motion." The basic element in this system was a symmetrically mounted pair of counter-rotating cylinders. Two centers of rotation were provided for each cylinder. One of these centers was at the center of symmetry of the cylinder and the other center was located to one side. During rotation it was possible to switch each cylinder back and forth between these two centers of rotation. When the two cylinders counter-rotate as cams, the entire device moves forward and backward along a straight line. By appropriately shifting the centers of rotation twice per revolution, Dean claimed to have eliminated the backward, but not the frontward, motion. Thus the device was supposed to surge forward (or upward) in jerks. A power supply was required, but there was no need for an external force.

This machine was advertised to be the answer to space propulsion, because it would make the mass ejection required in rocket engines unnecessary. Is this device, or any device based on the same premise, consistent with linear momentum conservation?

According to Dean, even if the device were enclosed in a giant plastic bag through which no force acts, and through which no matter passes, the device would accelerate. Its linear momentum would change with time. But zero force implies that the external linear impulse acting equals zero.

$$\mathbf{J} = 0$$

Equation 9.9 gives us

$$\mathbf{p}_f = \mathbf{p}_i$$

If the net external linear impulse is zero, no change in linear momentum is permitted by the law of conservation of linear momentum. This is a clear contradiction of Dean's claims. If a confirmed working model of such a device were ever made, linear momentum conservation would be violated. However, the Dean Drive did not work and the conservation law has survived.

The conservation of linear momentum is an important physical principle that finds wide application in science. One example of its application is its use in the interpretation of irregularities in the motion of Barnard's star. An isolated star cannot exhibit such irregularities and also conserve linear momentum. Thus the irregularities imply the existence of an invisible companion—perhaps a planet. In the continuing study of nuclear physics, the neutrino (Section 43.2) was postulated to exist partly because experimental studies of radioactive decay involving electron or positron emission seemed to imply violation of linear momentum conservation. The missing linear momentum was attributed to the neutrino. This elusive particle was first observed in 1956, two decades after its prediction by Wolfgang Pauli.

These two examples illustrate the scope of the law of conservation of linear momentum. It is more than just a detail of a particle's behavior; it is one of the fundamental principles of dynamics.

Questions

6. An astronaut is stranded motionless 10 m from his spacecraft. To generate motion toward the spacecraft he decides to throw a piece of equipment. Use conservation of linear momentum to establish the correct direction in which to throw the equipment. How would the speed of the astronaut toward the spacecraft compare with the speed of the thrown equipment? Explain.

7. An ice skater moving north also faces north. Suddenly the ice skater jumps and lands facing south. If the horizontal linear impulse exerted by the ice is negligible, then must the skater continue to move north? Explain.

9.4 Two-Particle Collisions

When a bowling ball collides with a bowling pin each object experiences an abrupt change in velocity. The reason for this characteristic behavior of colliding particles seems clear. Each body exerts a large force on the other for a short time. The sudden changes in the velocity of colliding objects are associated with their mutual interaction, and not with the influence of the rest of their environment. A collision between two particles is an interaction where the linear impulse between particles during impact is large compared with all other linear impulses acting on either particle. Under these conditions the colliding particles can be considered to be isolated from their environment.

One of the cornerstones of nuclear physics was laid when Geiger and Marsden performed their famous alpha-particle scattering experiment (see Section 42.2). Under the supervision of the English physicist Rutherford, they directed a beam of alpha particles toward a thin gold foil. They were able to observe the deflection of the alpha particles resulting from their collisions with gold atoms. The gold was deliberately formed into extremely thin foils so that any deflection of a particular alpha particle resulting from its traversal of the foil could be interpreted as a collision with a single gold atom.

In a collision between two particles the external linear impulse can be assumed to be approximately zero, and the combined linear momentum is conserved. Equation 9.16 then applies, and can be written

$$\Delta\mathbf{p} = m_1\,\Delta\mathbf{v}_1 + m_2\,\Delta\mathbf{v}_2 = 0 \qquad (9.17)$$

or

$$\Delta\mathbf{v}_1 = -\frac{m_2}{m_1}\,\Delta\mathbf{v}_2 \qquad (9.18)$$

where m_1 and m_2 are the masses, and $\Delta\mathbf{v}_1$ and $\Delta\mathbf{v}_2$ the velocity changes, of the colliding bodies. You should consider why Eq. 9.18 so closely resembles Eq. 6.2. In Eq. 9.18, the larger velocity change is associated with the smaller mass, and in Eq. 6.2 the larger acceleration is associated with the smaller mass. This corresponds well with our common-sense expectations. For example, in a collision between a compact car and a cement truck, we would expect the compact car to undergo the larger velocity change.

Example 8 A High Jumper

Let us consider the earth and a high jumper ($m = 81$ kg) as a two-body system. Both are at rest initially, and when the jumper leaps he elevates his center of gravity a distance of 1 m. What is the recoil velocity of the earth associated with the jump?

We use Eq. 4.17 to estimate the launch velocity of the high jumper.

$$v^2 = 2gx = 2 \times 9.80 \text{ m/s}^2 \times 1.00 \text{ m}$$

$$v = 4.43 \text{ m/s}$$

For the high jumper $\Delta v_1 = v = 4.43$ m/s. Thus, we can obtain the recoil velocity of the earth from Eq. 9.18.

$$\begin{aligned} \Delta v_2 &= -\frac{m_1}{m_2}\Delta v_1 \\ &= -\frac{81.0 \text{ kg}}{5.98 \times 10^{24} \text{ kg}}(4.43 \text{ m/s}) \\ &= -6.00 \times 10^{-23} \text{ m/s} \end{aligned}$$

To help you to see how small this number is, imagine that the earth traveled at this speed for a time equal to 1000 times the age of the earth (4.6×10^9 y). In this time the earth would move a distance of only 9 mm. Because the recoil velocity of the earth is very small it can be ignored in most situations.

Perfectly Elastic Collisions

If the combined kinetic energy of two colliding particles is conserved, their collision is termed **perfectly elastic.** The conservation equations for linear momentum and for kinetic energy in a one-dimensional perfectly elastic collision (Figure 9.10) can be written

$$m_1v_1 + m_2v_2 = m_1v_1' + m_2v_2' \tag{9.19}$$

$$\frac{1}{2}m_1v_1^2 + \frac{1}{2}m_2v_2^2 = \frac{1}{2}m_1v_1'^2 + \frac{1}{2}m_2v_2'^2 \tag{9.20}$$

Here the subscripts label the particles, unprimed quantities denote values before the collision, and primed quantities denote values after the collision.*

These equations can be solved to express v_1' and v_2' (normally the unknown quantities) in terms of the masses and initial velocities.

$$v_1' = \frac{m_1 - m_2}{m_1 + m_2}v_1 + \frac{2m_2}{m_1 + m_2}v_2 \tag{9.21}$$

$$v_2' = \frac{2m_1}{m_1 + m_2}v_1 - \frac{m_1 - m_2}{m_1 + m_2}v_2 \tag{9.22}$$

In Section 10.4 we will see that the derivation is simpler when the problem is recast in terms of a center-of-momentum reference frame.

*You may notice that $v_2 = v_2'$ and $v_1 = v_1'$ constitutes a mathematical solution to these equations. Physically this solution means that no collision occurred because each particle has the same linear momentum and kinetic energy after the "collision" that it had before. If we want to describe a collision, we reject this "noncollision" solution, and note that it is sometimes necessary to reject a correct mathematical solution on physical grounds.

Figure 9.10

Two particles, with masses m_1 and m_2, move along the same straight line and collide. Before colliding the particles have velocities $\mathbf{v}_1$ and $\mathbf{v}_2$, respectively; after the collision their velocities are $\mathbf{v}_1'$ and $\mathbf{v}_2'$.

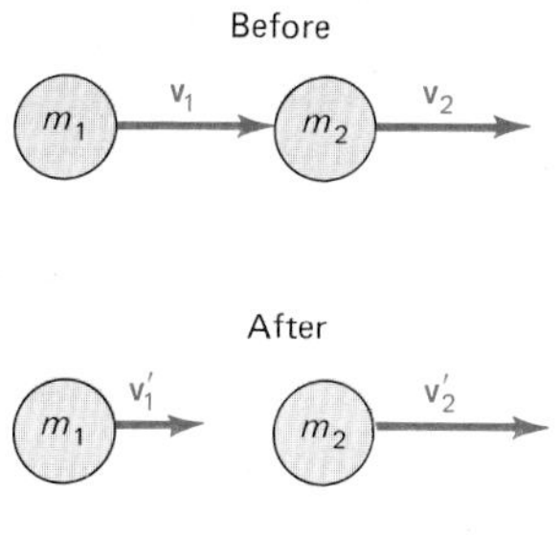

In nuclear scattering events, billiard ball collisions, and other types of elastic collisions it is common to have one of the two particles—called the *target*—at rest. Setting $v_2 = 0$ in Eqs. 9.21 and 9.22, we obtain

$$v_1' = \frac{m_1 - m_2}{m_1 + m_2}v_1 \tag{9.23}$$

$$v_2' = \frac{2m_1}{m_1 + m_2}v_1 \tag{9.24}$$

Let the initial direction of motion of m_1 be positive ($v_1 > 0$). Then, according to Eqs. 9.23 and 9.24, the final velocities are conditioned by the mass relationship as summarized in Table 9.1.

To help visualize the limiting cases in Table 9.1, imagine the head-on collision of a bowling ball with a stationary marble (case 3), a marble with a stationary bowling ball (case 4), and a marble with a stationary marble (case 5). In the bowling ball–marble collisions the velocity of the bowling ball changes very little, whereas the velocity of the marble changes markedly. In Example 9 we consider the important case of two equal masses ($m_1 = m_2$).

Table 9.1

1. If $m_1 > m_2$,	then $v_1' > 0$	and	$v_2' > 0$.
2. If $m_1 < m_2$,	then $v_1' < 0$	and	$v_2' > 0$.
3. If $m_1 \gg m_2$,	then $v_1' \approx +v_1$	and	$v_2' \approx +2v_1$.
4. If $m_1 \ll m_2$,	then $v_1' \approx -v_1$	and	$v_2' \approx 0$.
5. If $m_1 = m_2$,	then $v_1' = 0$	and	$v_2' = +v_1$.

Example 9 Proton–Proton Collision

Physicists who study the interactions of elementary particles often make use of proton projectiles. The protons are accelerated electrically and caused to collide with the nuclei of target atoms. In one type of experiment the target chamber is filled with liquid hydrogen and the projectile protons collide with target protons—the hydrogen nuclei. In such collisions the two particles have equal masses ($m_1 = m_2$). According to Eq. 9.23, when $m_1 = m_2$ we have

$$v_1' = 0$$

while according to Eq. 9.24

$$v_2' = v_1$$

The projectile proton comes to a halt, and all its velocity is given to the target proton. While this conclusion was derived from Newtonian mechanics, it also holds true in special relativity for elastic collisions of identical particles.

A common illustration of this type of collision is the "stop shot" in billiards. A cueball skidding without rotation at impact will stop after colliding head on with an object ball of equal mass.

In an elastic collision maximum kinetic energy transfer from incident (incoming) to struck particle occurs for particles of equal mass. In the proton–proton collision of Example 9, all of the kinetic energy of the incident proton was transferred to the struck proton. If one particle is much more massive than the other particle, however, we will find that the kinetic energy transfer is inefficient.

Equation 9.18 shows that in a two-particle collision conservation of linear momentum requires that the smaller velocity change be experienced by the larger mass. An extreme example of such a collision is between a molecule, m_1, and a wall, m_2. We can draw more specific conclusions about this collision by writing Eqs. 9.23 and 9.24 in the following form.

$$v_1' = \frac{\dfrac{m_1}{m_2} - 1}{\dfrac{m_1}{m_2} + 1} v_1 \tag{9.23}$$

$$v_2' = \frac{2\left(\dfrac{m_1}{m_2}\right)}{\dfrac{m_1}{m_2} + 1} v_1 \tag{9.24}$$

Because the mass of the wall is enormous compared with that of the molecule, we consider the limit as $m_1/m_2 \to 0$ to obtain

$$v_1' \to -v_1$$

$$v_2' \to 0$$

The wall does not move, whereas the molecule rebounds with the same speed it had originally, but in the opposite direction. Also, in this case the molecule retains all of its original kinetic energy, but with its linear momentum reversed in sign. What is the final linear momentum of the wall? We have

$$\begin{aligned} p_2' &= m_2 v_2' \\ &= \frac{2m_1 v_1}{\dfrac{m_1}{m_2} + 1} \\ &\cong 2m_1 v_1 \\ &\cong 2p_1 \end{aligned}$$

The wall ends up with twice the linear momentum originally possessed by the molecule, although the wall's velocity approaches zero. Note that even though the wall acquires twice the linear momentum of the molecule, the net change in the total linear momentum of the molecule–wall system is zero (because the particle's momentum is reversed in direction) and the total linear momentum is conserved.

Consider the kinetic energy ratio for particle 1 hitting a stationary target

$$\frac{K_1'}{K_1} = \frac{(m_1 - m_2)^2}{(m_1 + m_2)^2} \tag{9.25}$$

where K_1' is the final kinetic energy of m_1 and K_1 is its initial kinetic energy. In the limit where either mass is large compared with the other ($m_1 \ll m_2$ or $m_1 \gg m_2$), $K_1' \cong K_1$. The incident particle retains its kinetic energy and transfers very little kinetic energy to the target particle. If the masses are equal ($m_1 = m_2$), then $K_1' = 0$, and there is a maximum transfer of kinetic energy. To transfer kinetic energy efficiently, the masses of colliding particles should be as nearly equal as possible.

Example 10 **Kinetic Energy Transfer in a Nuclear Reactor**

In a nuclear reactor certain heavy nuclei capture neutrons and then undergo fission, emitting high-speed neutrons (Section 43.5). These emitted neutrons go on to initiate additional fission reactions, and the process continues in a chain reaction.

In *thermal* nuclear reactors low-speed neutrons are used to initiate fission. Therefore, the high-speed neutrons that are emitted in fission must slow down or transfer most of their kinetic energy in collisions before they can initiate another fission reaction. Thus, elements of small mass, like hydrogen and helium, are concentrated in the thermal reactor core to maximize the chance of a neutron collision with a particle of nearly equal mass. This makes the energy transfer as efficient as possible. In a *fast-breeder* nuclear reactor high-speed neutrons are used to initiate fission (Section 43.5). Designers strive to minimize the energy transfer by neutrons emitted in fission reactions. Thus elements of large mass are concentrated in the reactor core. For example, liquid sodium, rather than water, is used as a coolant in order to avoid having hydrogen (from water) in the reactor core.

Perfectly Inelastic Collisions

When the relative velocity of two colliding particles is zero after the collision—in other words, if they stick together—the collision is termed *perfectly inelastic*. While the linear momentum of a pair of isolated colliding particles is still conserved in such a collision, kinetic energy is not conserved—some is converted to other forms of energy in the process. We may use the same symbols that were used for perfectly elastic collisions to rewrite the linear momentum conservation equation, Eq. 9.19, as

$$m_1 v_1 + m_2 v_2 = (m_1 + m_2)v \tag{9.26}$$

where we have introduced $v = v_1' = v_2'$ as the common final velocity.

Example 11 Inelastic Collision of Identical Cars

Driving a new car off the assembly line, a careless worker hits the stopped car ahead and the two cars stick together as a result of the impact. Setting $v_2 = 0$ in Eq. 9.26, we see that if $m_1 = m_2$, we obtain $v = v_1/2$ for the combined velocity of identical colliding particles. We seek the loss in kinetic energy of the two cars as a fraction of the original value, or $\Delta K/K_1$. We have

$$K_1 = \frac{1}{2}m_1v_1^2$$

and

$$K' = \frac{1}{2}(m_1 + m_2)v^2$$

Since $v = v_1/2$ and $m_1 = m_2$, we can write

$$K' = \frac{1}{2}(2m_1)\left(\frac{v_1}{2}\right)^2 = \frac{1}{4}m_1v_1^2$$

Hence

$$\Delta K = K_1 - K' = \frac{1}{2}m_1v_1^2 - \frac{1}{4}m_1v_1^2$$

$$= \frac{1}{4}m_1v_1^2$$

Thus

$$\frac{\Delta K}{K_1} = 0.5$$

or 50% of the original kinetic energy is lost. It is converted into other forms of energy.

In the inelastic collision of the previous example each car does work on the other, producing costly damage and some heat. Similarly, in a collision between two football players, the work each does on the other in slowing down converts their kinetic energy into other forms of energy. If atoms or nuclei collide, the kinetic energy may put the colliding particles into excited states, with the energy stored internally. For velocities approaching the velocity of light the relativistic kinetic energy may be converted to the rest mass energy of new particles that are created by the collision.

Two-Dimensional Collisions

The law of conservation of linear momentum, Eq. 9.10, is a vector relation. If the linear momentum vector lies in the x–y plane, then the two component equations are

$$J_x = \Delta p_x = 0 \qquad (9.27)$$

and

$$J_y = \Delta p_y = 0 \qquad (9.28)$$

It is important to note that the component equations are independent. Thus if $J_x = 0$, then $\Delta p_x = 0$ whether $J_y = 0$ or not. For example, a molecule striking a "frictionless" wall will have its component of linear momentum parallel to the wall conserved even if the component perpendicular to the wall is suddenly reversed.

Figure 9.11

A kaon at rest decays spontaneously into a pair of oppositely charged pions. The pions move apart with momenta of equal magnitude but the momenta are oppositely directed.

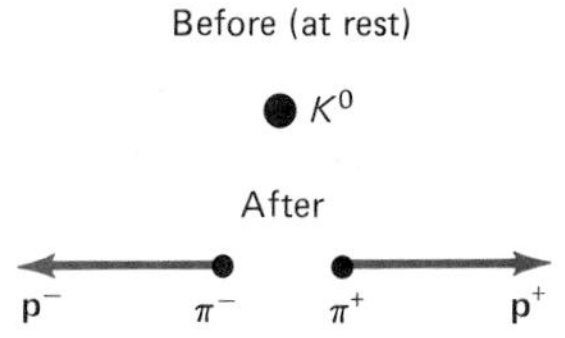

Example 12 Decay of the Kaon at Rest

A meson is a nuclear particle that is more massive than an electron but less massive than a proton or neutron. A type of meson called the neutral kaon decays into two charged (+ and −) pions that are equal in mass. A pion is a particle associated with the strong nuclear force that binds protons and neutrons together in the nucleus. If the kaon is initially at rest, we want to prove that the pion linear momenta after the decay of the kaon must be equal in magntiude and opposite in direction.

The decay of the kaon, represented in Figure 9.11 can be written

$$K^0 \rightarrow \pi^+ + \pi^-$$

We write $\mathbf{p}_i$ for the kaon, K^0, linear momentum; $\mathbf{p}^+$ for the positive pion, π^+, linear momentum; and $\mathbf{p}^-$ for the negative pion, π^-, linear momentum. The final linear momentum, $\mathbf{p}_f$, of the system is given by

$$\mathbf{p}_f = \mathbf{p}^+ + \mathbf{p}^-$$

Because the kaon is initially at rest, $\mathbf{p}_i = 0$. Since $\mathbf{J} = 0$, $\mathbf{p}_f = 0$. We write

$$0 = \mathbf{p}^+ + \mathbf{p}^-$$

which gives

$$\mathbf{p}^+ = -\mathbf{p}^-$$

The two linear momentum vectors are equal in magnitude and oppositely directed.

Questions

8. The purpose of a moderator in a nuclear reactor is to reduce the speed of neutrons. A neutron has roughly the same mass as a proton. Explain why graphite (pure carbon) would be a better moderator than lead.

9. A particle of mass m moving with speed v collides perfectly elastically with a particle of mass M at rest. Discuss the mass ratio necessary for maximum (a) kinetic energy transfer, (b) linear momentum transfer, and (c) speed of the struck particle. (Assume a one-dimensional collision.)

10. A spaceship initially at rest is suddenly attacked. Will the spaceship move if it fires a missile? Explain. (Assume that the spaceship's engines are turned off.)

Summary

Linear impulse, a measure of the effect of a force over a period of time, is defined in one dimension by

$$J = \int_{t_1}^{t_2} F\,dt \qquad (9.2)$$

The average force $\overline{F}$ acting over the time interval $\Delta t = t_2 - t$ is defined in terms of J.

$$\overline{F} = \frac{J}{\Delta t} \qquad (9.3)$$

In one dimension J is the area under the F versus t curve. In three dimensions the vector **J** is defined by

$$\mathbf{J} = \int_{t_1}^{t_2} \mathbf{F}\,dt \qquad (9.8)$$

The linear momentum of a particle is the vector

$$\mathbf{p} = m\mathbf{v} \qquad (9.6)$$

Newton's second law for variable force is written

$$\mathbf{F} = \frac{d\mathbf{p}}{dt} \qquad (9.7)$$

or, in its impulsive form, as the linear impulse–momentum principle,

$$\mathbf{J} = \Delta\mathbf{p} \qquad (9.9)$$

where **F** and **J** are the net external force and net external linear impulse, respectively.

If the net external linear impulse equals zero, $\mathbf{J} = 0$, then the linear momentum of a particle is conserved:

$$\Delta\mathbf{p} = 0 \quad \text{or} \quad \mathbf{p}_f = \mathbf{p}_i \qquad (9.10)$$

The same relations hold for a pair of particles providing **p** is the combined linear momentum of the system.

Perfectly elastic collisions are those in which kinetic energy is conserved, and perfectly inelastic collisions are those in which the colliding particles stick together after the collision. Linear momentum is conserved in both types of collision. Kinetic energy is not conserved in an inelastic collision.

Problems

Section 9.1 Linear Impulse

1. A punter's foot is in contact with a football for 3×10^{-3} s. During this time the average force exerted on the football by the foot is 4000 N (about 900 lb). What linear impulse does the player deliver to the football?

2. A weightlifter holds a 200-kg barbell overhead for 0.80 s. Determine the linear impulse he delivers to the barbell during that time interval.

3. During an impact time of approximately 5×10^{-4} s a golf club exerts an average force of about 5000 N on the golf ball. What linear impulse is delivered to the ball as a result?

4. A child bounces a Superball on the sidewalk. The linear impulse delivered by the sidewalk to the Superball is 2 N · s during the 1/800 s of contact. What is the average force exerted on the Superball by the sidewalk? What is the direction of this average force?

5. Figure 1 is a graph showing the force exerted by a boxer as he punches a bag. Determine the linear impulse delivered over the time inteval from $t = 0$ to $t = 0.10$ s.

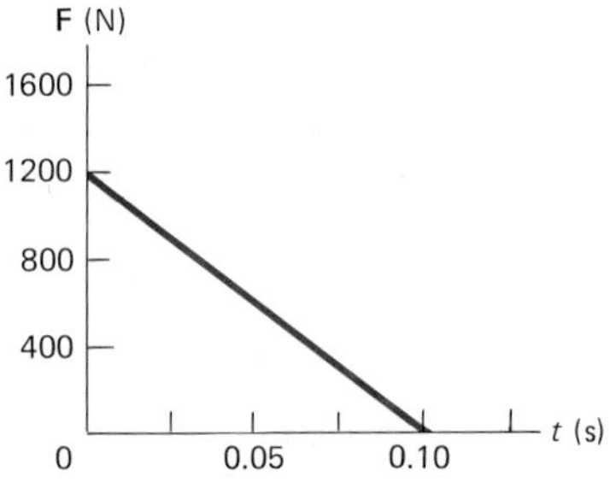

Figure 1

6. In a classroom demonstration a linear impulse of 0.25 N · s is delivered electromagnetically to a metallic ring in a time of 0.01 s, causing the ring to jump. What average force acts on the ring?

7. A graph of force versus time is given in Figure 2 for the force exerted on a sprinter's feet by the starting blocks. Graphically determine the total linear impulse associated with this force.

8. A sharply peaked force is described by

$$F(t) = \begin{cases} 1000\,t & 0 \le t \le 0.01 \text{ s} \\ 10 & 0.01 \le t \le 0.02 \text{ s} \\ 1000(0.03 - t) & 0.02 \le t \le 0.03 \text{ s} \\ 0 & t \le 0 \text{ and } t \ge 0.03 \text{ s} \end{cases}$$

where F is expressed in newtons if t is in seconds. (a) Sketch $F(t)$ versus t. (b) Calculate the linear impulse J. (c) Calculate the average force over the interval $0 \le t \le 0.03$ s.

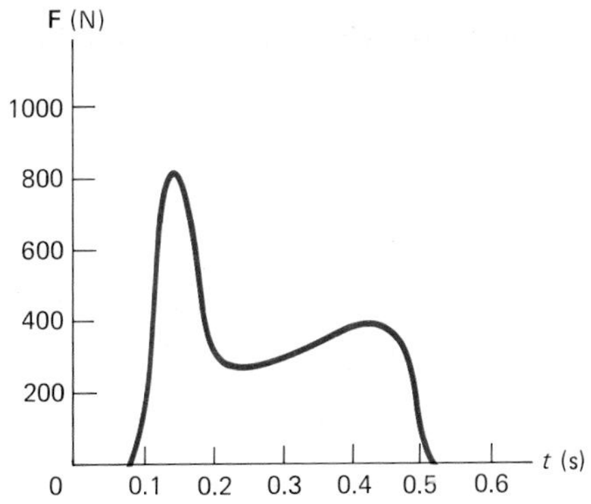

Figure 2

9. A spring exerts a force described by $F = -kx$, where $x = A \sin \omega t$, and A and ω are constants. Determine the linear impulse delivered by the spring during the interval $0 \leq \omega t \leq 90°$.

10. During liftoff of the spaceship Columbia the linear impulse associated with the upward thrust for the first 30 s was 8.5×10^8 N · s. Assuming a constant thrust in this 30-s period, calculate the ratio of this thrust to the 102-ton weight of the Columbia.

11. A positively charged cosmic ray of charge q_1 passes within a distance b of a negative charge q_2 as it passes through matter. The electrical force between the two charges is attractive and acts along the line joining the charges (Figure 3). Assume the cosmic ray speed is c, that it travels in a straight line, that it is not deflected, and that it loses a very tiny fraction of its speed in the interaction. Taking $F = kq_1q_1/r^2$, prove that (a) the component of the linear impulse delivered to the cosmic ray along its line of travel equals zero; (b) the component of linear impulse perpendicular to its line of travel is $J = 2kq_1q_2/bc$; and (c) the answer given in (b) is dimensionally correct.

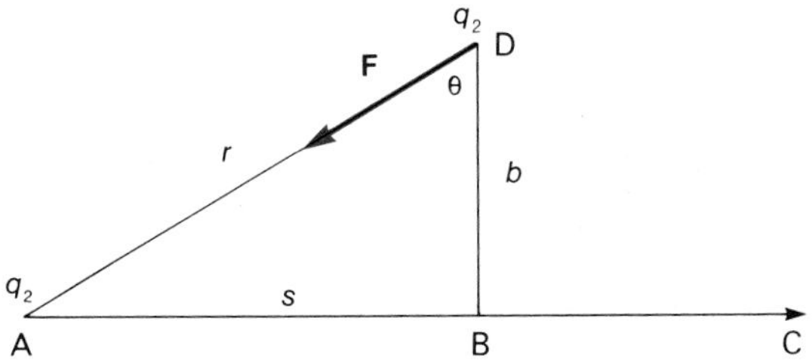

Figure 3

12. A mass moving on the x-axis is subjected to a force $\mathbf{F} = 530\mathbf{i}$ N. The velocity of the mass as a function of its position x (in meters) is represented by

$$\frac{dx}{dt} = 5x^2$$

Separate the variables and integrate this equation to show that the mass moves from $x = 2$ m to $x = 4$ m in a time interval of 0.05 s. Calculate the linear impulse associated with this force as the position changes from $x = 2$ m to $x = 4$ m.

13. When used to define average velocity rather than average force Eq. 9.4 can be written

$$\bar{v} = \frac{1}{t_f - t_i}\int_{t_i}^{t_f} v\, dt$$

Prove that this reduces to $\bar{v} = \frac{1}{2}(v_i + v_f)$ if the acceleration is constant.

Section 9.2 Linear Momentum and the Linear Impulse–Momentum Principle

14. A golf ball (mass = 0.046 kg) receives a linear impulse of 2.72 kg · m/s when it is struck with a driving iron. At what speed does it leave the tee?

15. A mass m experiencing a spring force has a linear momentum given by $p = mA \cos \omega t$ where A and ω are constants. Give the force acting at (a) $\omega t = 90°$; (b) $\omega t = 0°$.

16. A hockey puck whose mass is 0.1 kg is struck by a hockey stick, which changes the puck's velocity from zero to 45 m/s in a contact time of 1 ms. (a) What linear impulse was delivered to the puck? (b) What was the average force?

17. An air-hockey puck with velocity $\mathbf{v} = \mathbf{i}v_x + \mathbf{j}v_y$ ($v_x > 0$, $v_y < 0$) hits the side of the table ($y = 0$) and rebounds (Figure 4) with no change in the *magnitude* of its velocity. The linear impulse $\mathbf{J}$ is perpendicular to the table edge, $\mathbf{J} = \mathbf{j}J$. Show that the effect of $\mathbf{J}$ is to (a) leave the x-component of $\mathbf{v}$ unchanged and (b) reverse the y-component of $\mathbf{v}$. This means an exit velocity $\mathbf{v} = \mathbf{i}v_x - \mathbf{j}v_y$.

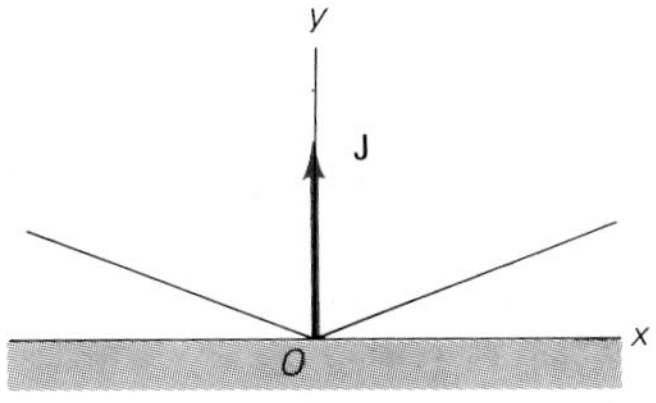

Figure 4

18. A golf ball ($m = 0.046$ kg) struck by a driver acquires a velocity of 60 m/s in an impact time of 5×10^{-4} s. Determine (a) the linear impulse delivered to the golf ball and (b) the average force exerted on the golf ball.

19. While accelerating from a stopped position, a 1000-kg car is subjected to a constant force of 6100 N. Using the linear impulse–momentum principle, determine the speed of the car 8 s after start.

20. An 75-kg ice skater moving at 10 m/s crashes into a stationary skater of the same mass. The two skaters move as a unit after the collision at 5 m/s. The

average force the skater can tolerate without breaking a bone is 1000 lb. If the impact time is 0.1 s, does a bone break?

21. Two balls collide in such a way that J is the linear impulse associated with the contact force. Show that the ratio of masses determines the ratio of velocity changes provided that other impulses are negligible.

22. A baseball ($m = 0.15$ kg) moves downward at 40 m/s toward a spectator. To catch the ball the spectator uses his bare hands. (a) Calculate the linear impulse required to bring the baseball to rest. What average force does the spectator exert if the baseball is brought to rest in (b) 0.015 s; (c) 0.0015 s? (d) Calculate the linear impulse delivered to the baseball by gravity during 0.015 s and compare the result with the answer to part (a).

23. For 6 s a 5-kg particle experiences the force shown graphically in Figure 5. (a) Calculate the linear impulse delivered during the first 6 s. (b) Determine the particle velocity at $t = 6$ s if it started from rest.

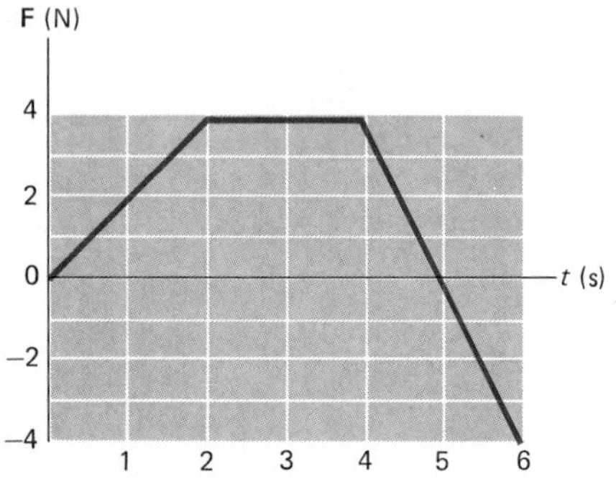

Figure 5

24. A baseball of mass m has an initial velocity $+v\mathbf{i}$ (toward the batter). After the bat strikes the baseball its velocity is $+v\mathbf{j}$ (straight up). In terms of m, v, and the unit vectors $\mathbf{i}$ and $\mathbf{j}$, determine the linear impulse $\mathbf{J}$ delivered to the baseball by the bat.

25. An alpha particle, initially moving in the negative x-direction, is deflected by a gold nucleus at the origin (Figure 6). The magnitude of the linear impulse delivered to the alpha particle is J. Express this linear impulse in vector form if the alpha particle emerges from the collision with its speed unchanged and at an angle of 120° with the x-axis.

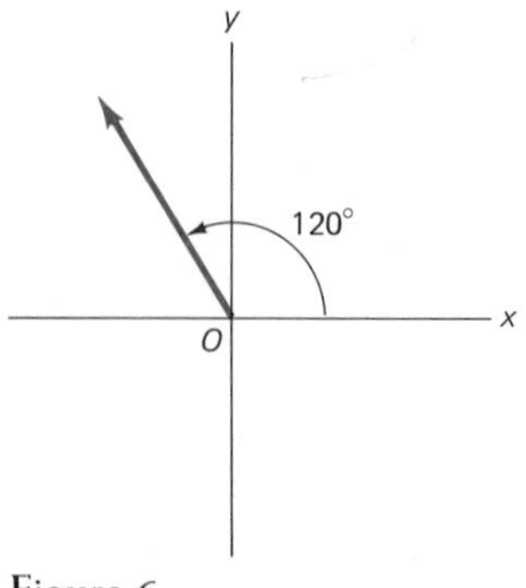

Figure 6

Section 9.3 Conservation of Linear Momentum

26. An ice skating couple push each other apart. He leaves with a speed of 2 m/s. Neglecting external impulses, (a) what impulse did she (60 kg) deliver to him (80 kg) and (b) what was her speed?

27. A 70-kg astronaut is floating free in space several meters form her spacecraft. She has a 1.60-kg wrench, which she can hurl with a speed of 22.0 m/s. (a) In what direction should she hurl the wrench so that she will move toward the spacecraft? (b) What will be her speed toward the spacecraft?

28. When ^{238}U (atomic mass $\cong 238$) decays emitting an alpha particle, the residual ^{234}Th ($m \cong 234$) nucleus recoils. If the original uranium nucleus is at rest, and the alpha particle ($m \cong 4$) leaves with a speed of 1.5×10^7 m/s, determine the recoil speed of the thorium nucleus.

29. A highly excited carbon-12 nucleus (^{12}C) is at rest. Having a large amount of internal energy it spontaneously disintegrates into three alpha particles. If in the disintegration of this ^{12}C nucleus we observe an alpha particle moving in the $+x$-direction with speed v and an alpha particle moving in the $+y$-direction with speed $2v$, what is the velocity of the third alpha particle?

30. Three ice skaters of equal mass move north as a unit at 6 m/s. The second skater pushes the third skater southward at a certain moment. Subsequently the third skater moves north at 2 m/s slower than the first two skaters. Assuming that no horizontal external impulses act, calculate the velocity of (a) the third skater and (b) the first two skaters.

31. Two hockey players ($m_1 = 70$ kg, $m_2 = 90$ kg) move together with velocity $\mathbf{v}$ as they fight. At a certain moment they push each other apart. Then the 70-kg player velocity is $+15\mathbf{i}$ m/s and the 90-kg player velocity is $+10\mathbf{j}$ m/s. Determine the magnitude and direction of $\mathbf{v}$.

32. A high jumper and the earth exert linear impulses on each other whose magnitudes are both equal to 400 N · s when the high jumper springs over the bar. Calculate the velocity of the high jumper ($m = 80$ kg) and of the earth as they each move away from the contact point.

Section 9.4 Two-Particle Collisions

33. A train with 90 freight cars backs up at a speed of 5 km/hr to connect to a string of 10 freight cars at rest. What is the velocity of the resulting 100-car train just after impact? Take the mass of each of the cars to be the same.

34. The danger to an automobile driver involved in a collision with a stationary car is related to the loss of speed of the moving vehicle. To assess the role of the masses of the cars, consider a perfectly inelastic col-

lision of a moving car with a car initially at rest. Calculate the loss of speed Δv for these conditions:

	Moving Car	Stationary Car
(a)	mass M, speed v	mass $2M$
(b)	mass $2M$, speed v	mass M

35. Starting with Eqs. 9.23 and 9.24, derive Eq. 9.25.

36. An outfielder stands at the base of the centerfield fence watching a long drive head for the seats. He throws his glove (mass = 0.60 kg) straight upward. Just as the glove reaches its peak, it "catches" the ball. The ball (mass = 0.15 kg) is traveling 32 m/s at the moment it strikes the glove. At what speed does the ball–glove system move just after the "catch?"

37. For a perfectly elastic collision, prove that the sign of the velocity of particle 1 relative to particle 2 reverses, and that the magnitude of this velocity does not change.

38. Imagine that you have two masses, m_1 and m_2. They are together and motionless but free to move apart. The two masses suddenly acquire kinetic energy as a result of a mutual interaction. Show that the total kinetic energy is divided inversely as the ratio of the masses; that is,

$$\frac{K_1}{K_2} = \frac{m_2}{m_1}$$

39. In one nuclear fission reaction a nucleus of mass 5 splits into a neutron (mass = 1) and a helium nucleus (mass = 4). The total kinetic energy available is 14 units. How many of these 14 units does the neutron get?

40. Prove that the statements in Table 9.1 about v_1' and v_2' are correct.

41. A neutron (m = 1) collides perfectly elastically with a nucleus initially at rest. Calculate the fractional energy loss of the neutron, $(K_1' - K_1)/K_1$, in a head-on collision if the struck nucleus is (a) a proton (m = 1); (b) a deuteron (m = 2); (c) an alpha particle (m = 4); (d) carbon (m = 12); (e) sodium (m = 23).

42. A 3000-kg rhinoceros running north at 10 m/s collides head on with an 800-kg Land Rover (a type of car) moving south at 25 m/s. A perfectly inelastic collision occurs. Determine the (a) final velocity of the combined system and (b) linear impulse delivered to the rhinoceros during impact.

43. A 10-gm bullet is stopped in a block of wood (m = 5 kg). The combined speed of the bullet plus wood immediately after the perfectly inelastic collision is 0.6 m/s. What was the original speed of the bullet?

44. A child sits motionless in a playground swing, and her sister (same mass) runs at 5 m/s and leaps onto the swing. How fast does the pair move initially? (Include only the girls' mass.)

45. A proton (m = 1) and a triton (m = 3) approach each other with equal speeds and undergo a perfectly elastic collision (Figure 7). In terms of the initial speed v, determine the final speed of (a) the proton and (b) the triton.

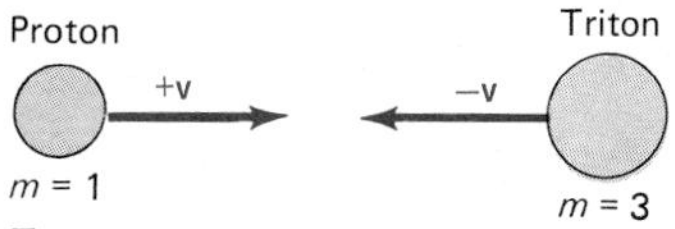

Figure 7

chapter 10

MANY-PARTICLE SYSTEMS

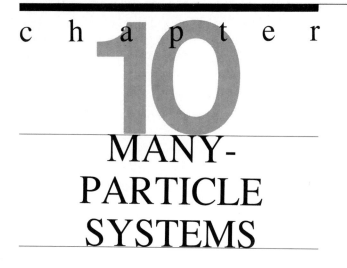

Preview

We have developed Newton's laws of motion and the conservation laws for energy and linear momentum, and applied them to one- and two-particle systems. To extend this treatment on a particle-by-particle basis to a system of many particles would require the use of a large electronic computer, but when the center-of-mass concept is used to describe the dynamic behavior of a many-particle system, the description is no more complex than that of a single particle.

In this chapter, we develop the center-of-mass concept and apply it to several systems. We then introduce the center-of-momentum reference frame, which is a generalization of the center-of-mass concept. We illustrate the utility of this reference frame with examples, including several drawn from atomic and nuclear physics.

10.1 Center of Mass

In Section 3.2 we saw that in order to apply Newton's first law, we must conceptually divide the universe into two parts. We used the same division in Chapter 6 to apply Newton's second law. In this division one part of the universe is called the *body* and the other part is called its *environment*. When rotation can be ignored, the body is called a *particle*.

For example, every atom of the diver shown in Figure 10.1 experiences the same displacement in any given time interval. Every atom of the diver's body traverses a parabolic path of the same shape. Rotation is absent, and the body can be viewed as a particle.

In Figure 10.2 another diver is shown executing a forward one-and-a-half somersault dive. Note that one point associated with the diver follows the same parabolic path that we discussed in relation to Figure 10.1. This point is called the *center of mass* of the diver. Although none of the atoms in the diver's body necessarily follow this exact parabolic path, and although the diver's motion is complex and difficult to describe quantitatively, one simplifying feature stands out. The center of mass of the diver moves in the same manner as would a particle subject to the same force and the same initial conditions. Therefore, we can effectively replace a very complicated set of calculations for every atom of the diver with a much simpler equation for one particle.

This qualitative idea—that the center of mass is a simplifying concept—can be made quantitative. In Section 10.2 special forms of the conservation laws for linear momentum and energy and for Newton's second law will be derived that apply to the center of mass of a body.

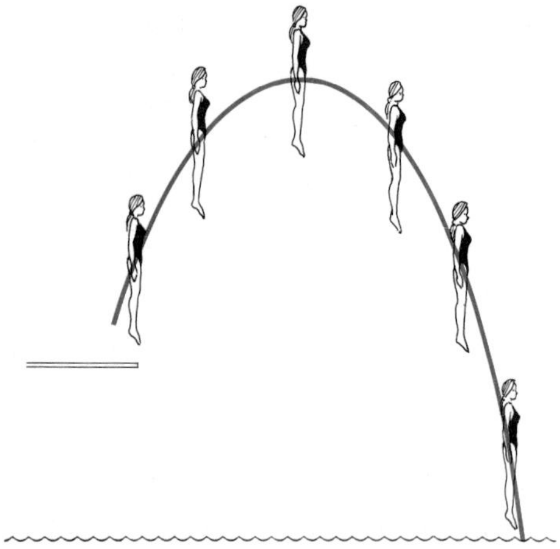

Figure 10.1

When a rigid diver moves through the air without rotation, the same parabola describes the trajectory of each part of the diver's body.

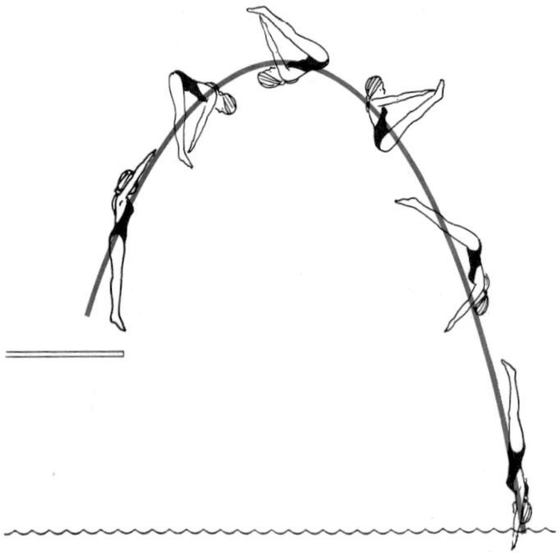

Figure 10.2

One point, the center of mass of the diver, follows a parabolic path even though the diver rotates while moving through the air.

We begin quantifying the center of mass by stating the definition of the center of mass for a collection of particles. Later we will extend this definition to cover the center of mass for a continuous mass distribution.

For a discrete set of particles, the *center-of-mass position vector*, $\mathbf{r}_{cm}$, is defined by

$$\mathbf{r}_{cm} = \frac{1}{M}(m_1\mathbf{r}_1 + m_2\mathbf{r}_2 + \cdots)$$

$$= \frac{1}{M}\sum m_i\mathbf{r}_i \qquad (10.1)$$

Here, the total mass of the set is

$$M = \sum m_i \qquad (10.2)$$

and $\mathbf{r}_i = x_i\mathbf{i} + y_i\mathbf{j} + z_i\mathbf{k}$ is the position vector of the ith particle, whose mass is m_i (Figure 10.3). Interpreted roughly, the center-of-mass vector starts at the origin and ends in the center of the set of particles. This sum includes every particle in the set.

Figure 10.3

In a collection of particles the position vector, $\mathbf{r}_i$, of the ith particle (mass $= m_i$) has components x_i, y_i, and z_i.

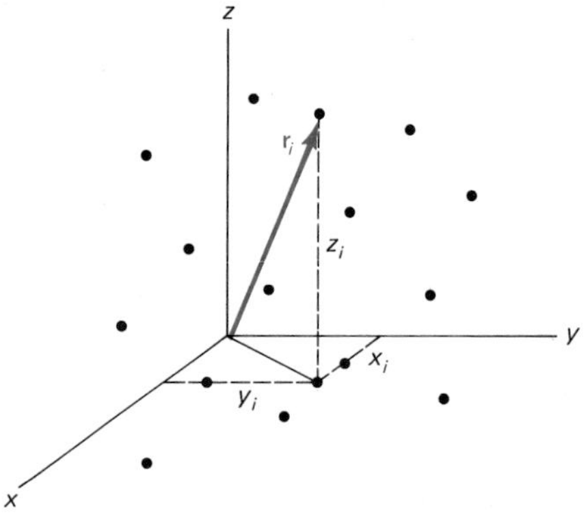

In terms of unit vectors the right side of Eq. 10.1 can be written

$$\sum m_i\mathbf{r}_i = \left(\sum m_ix_i\right)\mathbf{i} + \left(\sum m_iy_i\right)\mathbf{j} + \left(\sum m_iz_i\right)\mathbf{k}$$

The components of $\mathbf{r}_{cm}$, x_{cm}, y_{cm}, z_{cm}, are defined by

$$\mathbf{r}_{cm} = x_{cm}\mathbf{i} + y_{cm}\mathbf{j} + z_{cm}\mathbf{k}$$

and so we can write Eq. 10.1 in component form as the set of equations

$$x_{cm} = \frac{1}{M}\sum m_ix_i \qquad y_{cm} = \frac{1}{M}\sum m_iy_i$$

$$z_{cm} = \frac{1}{M}\sum m_iz_i \qquad (10.3)$$

For symmetric systems the center of mass is at the center of symmetry. For example, in a binary star system composed of equal-mass stars, the center of mass of the system will be halfway between the two star centers.

Numerical values obtained for the center-of-mass coordinates have meaning only relative to the origin chosen.

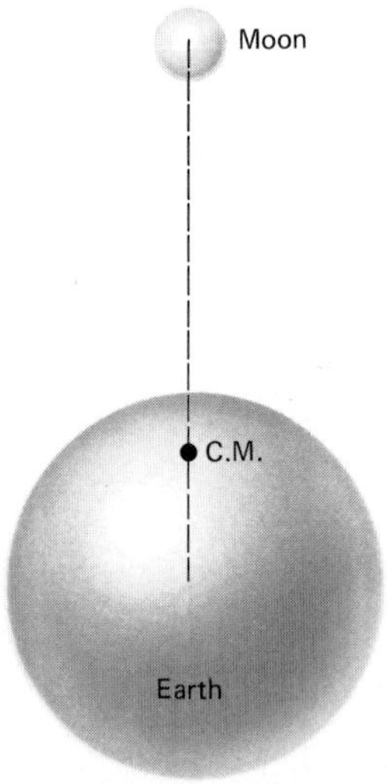

Figure 10.4

The center of mass (C.M.) of the earth–moon system is located about 1100 miles below the earth's surface.

Figure 10.5

Two hydrogen atoms (H) and an oxygen atom (O) form a water molecule.

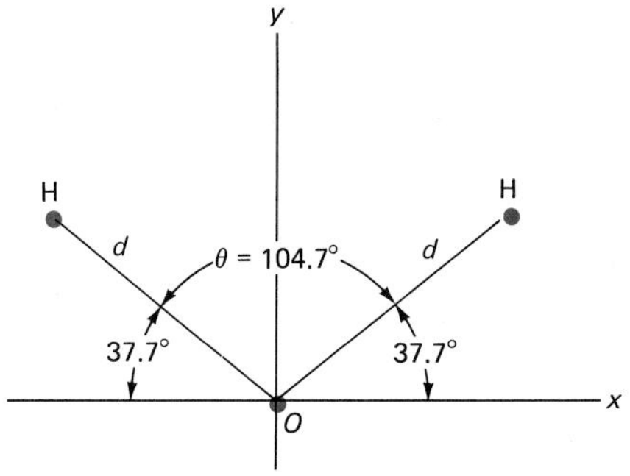

The center of mass of the earth–moon system is located at a distance 4.6×10^6 m from the center of the earth—one choice for origin—or about 1.8×10^6 m beneath the surface of the earth—a second choice for origin (Figure 10.4). The physical location of the center of the mass is the same in the two cases, but the numbers used to describe its position differ. Often it is useful to place the origin at the center of mass itself. In this case all components of $\mathbf{r}_{cm}$ equal zero.

Example 1 **Center of Mass of a Water Molecule**

The water molecule, H_2O, is an arrangement of two hydrogen atoms and an oxygen atom distributed as shown in Figure 10.5. Each of the hydrogen atoms is a distance d from the oxygen atom, and their position vectors make the angle θ. Where is the center of mass of the system? In Figure 10.5 the axes are selected to be symmetric. By symmetry $x_{cm} = 0$, which leaves y_{cm} to be evaluated. Using Eq. 10.3 we write

$$y_{cm} = \frac{\sum m_i y_i}{\sum m_i}$$

$$= \frac{m_H d \sin 37.7° + m_O \cdot 0 + m_H d \sin 37.7°}{2m_H + m_O}$$

For the water molecule $d = 1.0 \times 10^{-10}$ m, and using $m_H = 1.0$ u* and $m_O = 16$ u gives

$$y_{cm} = 6.8 \times 10^{-12} \text{ m}$$

The center of mass is at the point $(0, 6.18 \times 10^{-12}$ m).

Relation of Center of Mass and Center of Gravity

In Section 3.5 we saw that the components of the center-of-gravity vector of a body are given by

*Carbon 12 (^{12}C) is assigned a mass of exactly 12 u (12 atomic mass units). The masses of hydrogen and oxygen atoms are determined experimentally in atomic mass units by comparison with the ^{12}C standard.

$$\bar{x} = \frac{1}{W}\sum W_i x_i \qquad \bar{y} = \frac{1}{W}\sum W_i y_i$$

$$\bar{z} = \frac{1}{W}\sum W_i z_i \tag{10.4}$$

where $W = \sum W_i$ is the total weight of the body. This set of equations bears a striking similarity to the corresponding relations for the center of mass in Eqs. 10.3. Are Eqs. 10.3 and 10.4 indeed the same?

The weight and mass of the ith particle are related according to

$$\mathbf{W}_i = m_i \mathbf{g}_i \tag{10.5}$$

where $\mathbf{g}_i$ is the value of the acceleration of gravity at the location of the ith particle. If $\mathbf{g}_i$ is constant in magnitude and direction throughout the body, then

$$\mathbf{W} = \sum \mathbf{W}_i = \left(\sum m_i\right)\mathbf{g} = m\mathbf{g}$$

relates the total mass and total weight of the body. Let us apply this result to $\bar{x}$, the x coordinate of the center of gravity, in Eq. 10.4. We have

$$\bar{x} = \frac{\sum W_i x_i}{\sum W_i} = \frac{(\sum m_i x_i)g}{(\sum m_i)g} = \frac{\sum m_i x_i}{\sum m_i} = x_{cm}$$

proving that $\bar{x} = x_{cm}$. Proofs for $\bar{y}$ and $\bar{z}$ are analogous. We conclude that if $\mathbf{g}$ is constant in magnitude and direction, the center of mass and center of gravity of a body are located at the same point.

Now let's see what happens if $\mathbf{g}$ is not constant for all particles in the system.

Example 2 **Center of Mass and Center of Gravity in a Nonuniform Gravitational Field**

Two astronauts of equal mass are located at the points A and B at opposite ends of a spacecraft near the earth (Figure 10.6). The *center of mass* of the two astronauts is at the midpoint of the line joining them. The center of mass is a geometric property of a body or system whose location depends on the mass distribution but not on the environment. Because the astronaut at A is closer to the earth than the astronaut at B, the astronaut at A weighs more. Hence the *center of gravity* of the two astronauts is closer to point

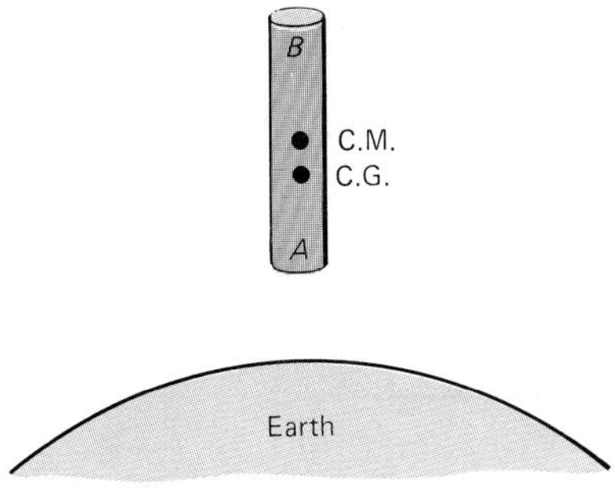

Figure 10.6

Two identical astronauts are located at opposite ends (points A and B) of a space capsule. The center of mass (C.M.) of the astronauts is halfway between them, whereas their center of gravity (C.G.) is closer to the astronaut at A.

A than to point B. We see, then, that the location of the center of gravity of a body or system depends on both the mass distribution and the environment.

Continuous Mass Distribution

Equations 10.1–10.3, which are valid for sets of particles, can be extended to a continuous distribution by subdividing the continuous distribution into elements of mass small enough to be considered point masses. In practice this means replacing the summation over discrete masses, $\sum m_i$, with an integration over the continuous mass, $\int dm$, whenever $\sum m_i$ occurs. Thus we obtain the definitions for continuous systems.

$$\mathbf{r}_{cm} = \frac{\int \mathbf{r}\, dm}{\int dm} \tag{10.6}$$

with

$$M = \int dm \tag{10.7}$$

and

$$x_{cm} = \frac{1}{M}\int x\, dm \qquad y_{cm} = \frac{1}{M}\int y\, dm$$

$$z_{cm} = \frac{1}{M}\int z\, dm \tag{10.8}$$

where all integrals extend over the body in question.

We used Eqs. 10.1–10.3 to find the center of mass of a set of three particles (H, H, and O) in Example 1, and to find the center of mass of a set of two particles (the astronauts) in Example 2. Now we will use these equations in their extended forms (Eqs. 10.6–10.8) to find the center of mass of a continuous object—a straight rod.

Example 3 Center of Mass of a Straight Rod

Consider a uniform homogeneous rod of length l and cross-sectional area A. Coordinate axes are selected as drawn in Figure 10.7. We want to locate the center of mass of the rod (a) by inspection and (b) with integration.

(a) By symmetry, the center of mass is at the geometric center of the rod, $x_{cm} = +l/2$. This conjecture is confirmed if we observe that

1 the center of mass and center of gravity are at the same point; and

2 the center of gravity is at the balance point, $\bar{x} = +l/2$.

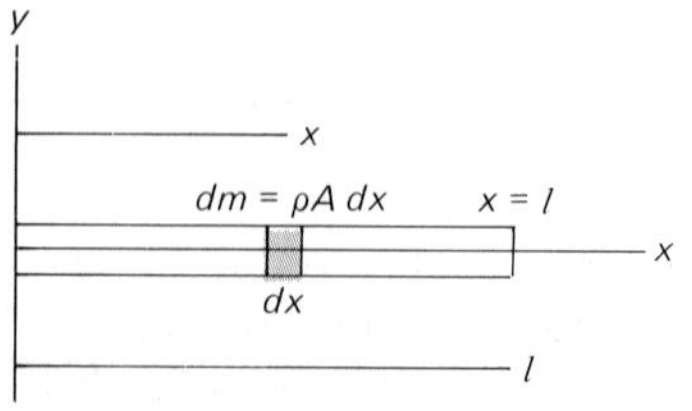

Figure 10.7

The shaded region represents an element of mass dm at a distance x from the left end of the thin homogeneous rod.

A ruler or meter stick balances at its midpoint.

(b) To find the center of mass by integration we use Eq. 10.8:

$$x_{cm} = \frac{\int x\, dm}{\int dm}$$

We choose $dm = \rho A\, dx$ for the element of mass where ρ is the mass density, A is the cross-sectional area, and dx is the width of the mass element (Figure 10.7). With limits from 0 to l we obtain

$$x_{cm} = \frac{\int_0^l \rho A x\, dx}{\int_0^l \rho A\, dx} = \frac{\rho A l^2/2}{\rho A l} = \frac{l}{2}$$

Example 4 Center of Mass of a Triangular Lamina

Consider the thin plate, or lamina, shown in Figure 10.8. This plate is in the shape of an isosceles triangle, and is of thickness t. Let's find its center of mass by using a procedure similar to that used in Example 3. Only x_{cm} needs to be calculated because by symmetry, $y_{cm} = 0$ for the axes chosen.

We choose the shaded section in Figure 10.8 as an element of mass. The volume of this section is

$$dV = t\left(\frac{xb}{a}\right) dx$$

where t is the thickness, xb/a the length, and dx the width. This gives

$$dm = \rho\, dV = \rho\left(\frac{tb}{a}\right) x\, dx$$

Putting this value of dm into Eq. 10.8 and integrating between the limits of 0 and a, we have

$$x_{cm} = \frac{\int x\, dm}{\int dm} = \frac{\int_0^a x\rho\left(\frac{tb}{a}\right) x\, dx}{\int_0^a \rho\left(\frac{tb}{a}\right) x\, dx}$$

$$= \frac{a^3/3}{a^2/2}$$

$$= \frac{2}{3}a$$

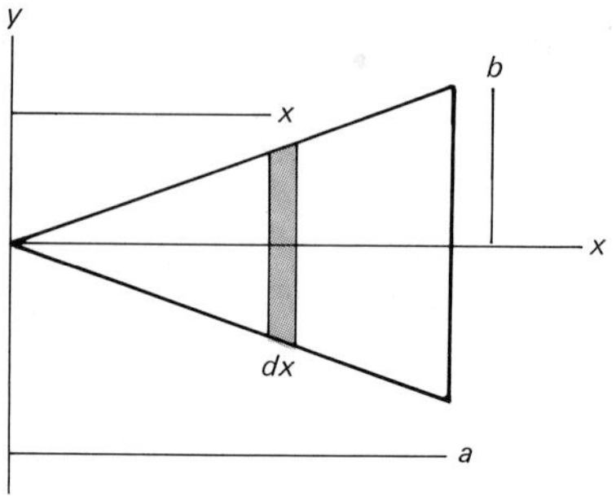

Figure 10.8

The shaded region represents an element of mass dm at a distance x from the apex of the thin triangular lamina.

The center of mass is located at two thirds of the altitude of the triangle measured from the triangle's apex. You can check this result experimentally. Cut a triangle out of paper and locate the center of mass by balancing the paper on the tip of your finger. The balance point should be at a spot two thirds of the distance from the apex to the base of the triangle.

Questions

1. To complete a successful high jump, the jumper must pass over the crossbar without knocking it off. How might this occur without the jumper's center of mass attaining the crossbar height?
2. A runner runs in such a fashion that she raises and lowers her head with each stride. Does this mean that the runner raises and lowers her center of mass with each stride? Explain.
3. Can a body's center of mass lie outside the body? Explain.

10.2 Dynamics of the Center of Mass

In many situations it is legitimate and desirable to ignore the rotation and internal motions of a system. In these situations the center-of-mass concept greatly simplifies our analysis of the motion, because the many-particle system can be treated as a single particle located at the system's center of mass. For example, if we want to examine the motions of galaxies, it is often helpful to view a galaxy of 100 billion stars as a single particle located at its center of mass. We will begin with the defining equation for the center of mass, Eq. 10.1, and derive a series of simplifying equations that show how a collection of particles can be treated as a single particle.

Velocity of the Center of Mass

The center-of-mass position, $\mathbf{r}_{cm}$, is given by

$$\mathbf{r}_{cm} = \frac{1}{M}\sum m_i \mathbf{r}_i \qquad (10.1)$$

where

$$M = \sum m_i \qquad (10.2)$$

The center-of-mass velocity is defined by the relation

$$\mathbf{v}_{cm} = \frac{d\mathbf{r}_{cm}}{dt} \qquad (10.9)$$

Figure 10.9

A hockey center skates at 8 m/s toward a goalie at rest.

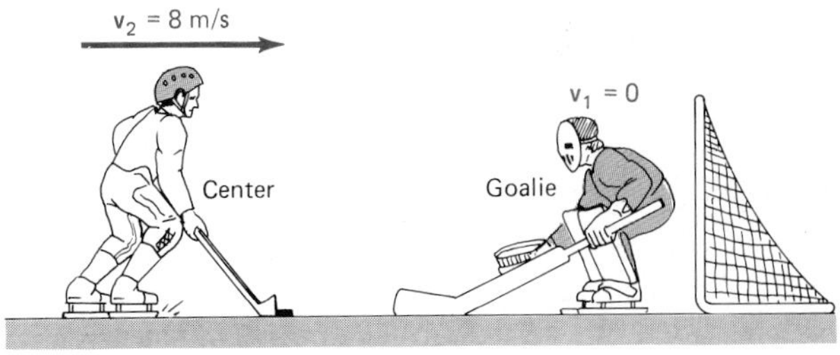

This equation is in the same form as the definition of the velocity of a single particle, Eq. 4.6 in Section 4.3. Differentiating Eq. 10.1, we obtain

$$\mathbf{v}_{cm} = \frac{1}{M}\sum m_i \mathbf{v}_i \qquad (10.10)$$

which expresses $\mathbf{v}_{cm}$ in terms of the individual particle velocities, $\mathbf{v}_i = d\mathbf{r}_i/dt$.

Example 5 Center-of-Mass Velocity of Two Hockey Players

A 90-kg goalie stands at rest while a 110-kg center skates toward him at 8 m/s (Figure 10.9). What is the velocity of the center of mass of the pair?

Let the subscript 1 label the goalie and the subscript 2 label the center. Then, according to Eq. 10.10, we can write

$$v_{cm} = \frac{1}{M}(m_1 v_1 + m_2 v_2)$$

$$= \frac{1}{200}(90 \times 0 + 110 \times 8)$$

$$= 4.4 \text{ m/s}$$

The center moves toward the goalie at 8 m/s. The center of mass moves toward the goalie at 4.4 m/s.

Linear Momentum

The linear momentum of a system of particles, $\mathbf{p}$, is the vector sum of the individual linear momenta.

$$\mathbf{p} = \sum \mathbf{p}_i = \sum m_i \mathbf{v}_i \qquad (10.11)$$

Substituting $\mathbf{p}$ for $\sum m_i \mathbf{v}_i$ in Eq. 10.10, we can write

$$\mathbf{p} = M\mathbf{v}_{cm} \qquad (10.12)$$

This equation can be interpreted as follows. If the mass of the system were concentrated at the center of mass and moved with the velocity of the center of mass, then the linear momentum of the concentrated mass would equal the system linear momentum. Here is one justification for treating a system—a galaxy, for instance—as a particle located at the center of mass.

Acceleration of the Center of Mass

The acceleration of the center of mass, $\mathbf{a}_{cm}$, is defined by the equation

$$\mathbf{a}_{cm} = \frac{d\mathbf{v}_{cm}}{dt} \qquad (10.13)$$

Again, this equation is identical in form to the definition of the acceleration of a single particle, Eq. 4.10 in Section 4.4. Differentiating Eq. 10.10 and combining Eqs. 10.10 and 10.13, we obtain

$$\mathbf{a}_{cm} = \frac{1}{M} \sum m_i \mathbf{a}_i \qquad (10.14)$$

This result helps pave the way for the application of Newton's second law to the center of mass.

Newton's Second Law

Newton's second law can be written for a particle as

$$\mathbf{F}_i = m_i \mathbf{a}_i = \frac{d\mathbf{p}_i}{dt}$$

where $\mathbf{F}_i$ is the net force acting on the ith particle. If we sum $\mathbf{F}_i$ over all particles in the system, we obtain

$$\mathbf{F}_{ex} = \sum \mathbf{F}_i \qquad (10.15)$$

as the net external force acting on the system. In this sum all internal forces between the particles of the system cancel in pairs because of Newton's third law, as discussed in Section 3.2. Combining Eqs. 10.14 and 10.15, we obtain

$$\mathbf{F}_{ex} = M\mathbf{a}_{cm} \qquad (10.16)$$

According to this equation, the net external force exerted on a system equals the total mass of the system multiplied by the acceleration of the center of mass. With respect to its environment, the system as a whole moves like a particle of mass M that is always located at the center of mass. When $\mathbf{F}_{ex} = 0$ the motion of the center of mass is unaccelerated.

One point should be emphasized. In the summation over all $\mathbf{F}_i$ in Eq. 10.15 *all internal forces between the particles of the system cancel*. Hence $\mathbf{F}_{ex}$ includes only *external* forces acting on the system.

Example 6 The Siamese Twin Balloons

Two unstretchable Mylar balloons of equal fixed volumes are joined by a thin membrane (Figure 10.10). Initially one balloon is filled with gas while the other is evacuated and collapsed. The mass of the balloons is negligible in comparison with the mass of the gas. At a certain moment the membrane ruptures, allowing the gas to fill the two balloons equally. Neglecting friction, where is the final location of the balloons if only horizontal motion is permitted?

Because in this case $\mathbf{F}_{ex} = 0$, we conclude from Eq. 10.16,

$$\mathbf{F}_{ex} = M\mathbf{a}_{cm}$$

Figure 10.10

Identical unstretchable balloons are connected by a thin membrane. Initially the left balloon is inflated and the right balloon is collapsed. The membrane breaks, producing a final situation where the original gas fills both balloons. If the balloon mass is neglected, the center of mass of the gas will not move.

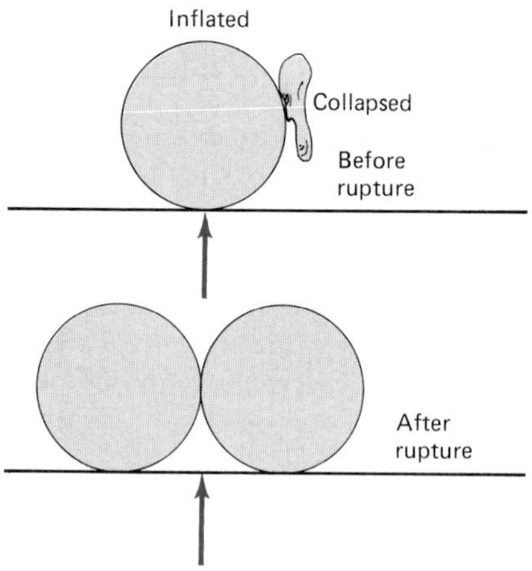

that $\mathbf{a}_{cm} = 0$. But we know that

$$\mathbf{a}_{cm} = \frac{d\mathbf{v}_{cm}}{dt} \qquad (10.13)$$

so this means that $\mathbf{v}_{cm}$ is constant. Since initially $\mathbf{v}_{cm} = 0$ (the balloons were motionless) and because we know that $\mathbf{v}_{cm}$ is constant, the final value of $\mathbf{v}_{cm}$ must be zero as well. We also know that

$$\mathbf{v}_{cm} = \frac{d\mathbf{r}_{cm}}{dt} \qquad (10.9)$$

Therefore, because $\mathbf{v}_{cm} = 0$, $\mathbf{r}_{cm}$ is constant. This means that the position of the center of mass remains the same before, during, and after rupture of the membrane between the balloons. The arrows in Figure 10.10 indicate the center-of-mass positions before and after rupture. Because (as we said earlier) the only mass in the system is that of the gas, the balloons must move.

Conservation of Linear Momentum

If we differentiate Eq. 10.12

$$\mathbf{p} = M\mathbf{v}_{cm}$$

we obtain

$$\frac{d\mathbf{p}}{dt} = M\frac{d\mathbf{v}_{cm}}{dt}$$

Table 10.1

Single-particle–Many-particle analogy

Single Particle	Many-Particle System
$\mathbf{p} = m\mathbf{v}$	$\mathbf{p} = M\mathbf{v}_{cm}$
$\mathbf{F}_{net} = m\mathbf{a}$	$\mathbf{F}_{ex} = M\mathbf{a}_{cm}$

or, from Eq. 10.13,

$$\frac{d\mathbf{p}}{dt} = M\mathbf{a}_{cm}$$

Because $\mathbf{F}_{ex} = M\mathbf{a}_{cm}$ (Eq. 10.16), we can write

$$\mathbf{F}_{ex} = \frac{d\mathbf{p}}{dt} \tag{10.17}$$

This means that the net external force acting on the system as a whole equals the time rate of change of the total linear momentum of the system. When $\mathbf{F}_{ex} = 0$, the total linear momentum of the system does not change—it is conserved. That is, if

$$\mathbf{F}_{ex} = 0$$

then

$$\mathbf{p}_i = \mathbf{p}_f \quad \text{or} \quad \mathbf{p} = \text{constant} \tag{10.18}$$

We see, then, that conservation of linear momentum applies to an isolated system of particles just as it does to an isolated particle.

Because $\mathbf{F}_{ex}$ includes no internal forces (remember, they cancel), the internal forces in a complex system have no influence on the time rate of change of the momentum of the system. Put simply, it is *not* possible to lift yourself by your own bootstraps! An external force is required.

Conservation of Kinetic Energy

In Section 7.4 the increase in kinetic energy of a particle was defined as the work done on the particle by the net force, $\mathbf{F}_i$, acting on it. Thus

$$W_i = \Delta K_i \tag{10.19}$$

where W_i is the work done by the net force acting on the ith particle. Here we are looking at just one particle. The net force $\mathbf{F}_i$ could include forces internal to the system. K_i is the change in kinetic energy of the ith particle. If we sum Eq. 10.19 over all particles in the system, we obtain

$$W_{total} = \Delta K \tag{10.20}$$

where $W_{total} = \sum W_i$ and $\Delta K = \sum \Delta K_i$. When the total work done on the system equals zero, there is no change in the kinetic energy of the system—it is conserved; that is, if

$$W_{total} = 0$$

then

$$K_i = K_f \quad \text{or} \quad \Delta K = 0 \tag{10.21}$$

It is important to note that W_{total} is not necessarily equal to the work done by the net external force. It is quite possible for the internal forces to do work, and therefore to contribute to W_{total}, although they would not contribute to $\mathbf{F}_{ex}$. Hence the kinetic energy of a system may not be conserved even when its linear momentum is conserved.

Consider two automobiles of equal mass moving head on toward each other at equal speeds from opposite directions (Figure 10.11). There are no external forces acting on the automobiles in the horizontal direction. A collision occurs, leaving the wreckage at rest in a heap. Because $\mathbf{F}_{ex} = 0$, linear momentum is conserved. Clearly, $\mathbf{p} = 0$ both before and after the collision. However, $W_{total} \neq 0$, because the contact forces—which are internal to the system—do work in bringing the automobiles to rest. Thus kinetic energy is not conserved. The initial kinetic energy is converted to other forms of energy in the collision. The final kinetic energy is zero.

Figure 10.11

Identical cars approaching each other with the same speed strike in a perfectly inelastic collision.

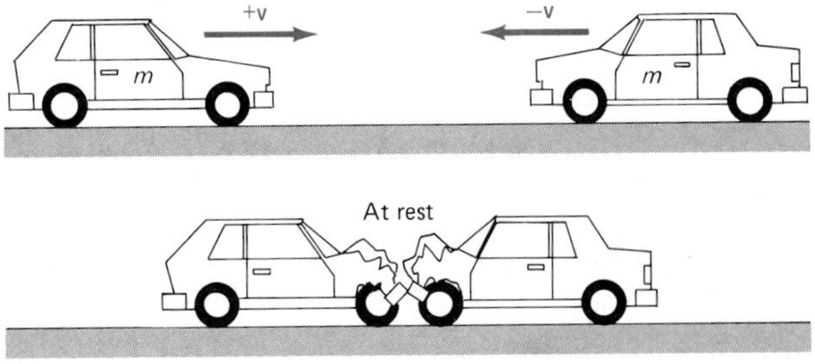

We call a collision in which kinetic energy is not conserved *inelastic*. One example of an inelastic collision occurs when two or more particles combine to form a single system at rest. In the process the kinetic energy of the particles is converted into some form of internal energy (gravitational, thermal, rest mass, etc.) of the system.

It is also possible for a system at rest to fragment inelastically into two or more parts. In this process, internal energy of the system is converted into the kinetic energy of the fragments. Whether the inelastic process involves combining parts into a whole or fragmenting a whole into parts, the internal forces of the whole perform work. In the inelastic collision of two automobiles (Figure 10.11) it is not clear how the internal work is performed. In Example 7, however, we consider an inelastic situation where the internal work is performed explicitly.

Example 7 **Internal Work Done by a Spring**

Consider two identical air carts free to move without friction on a straight horizontal air track (Figure 10.12). The carts are held together by a thread and a spring is compressed between them. If the system is initially at rest, what happens when the thread breaks? Neglect the masses of the thread and spring.

Again $\mathbf{F}_{ex} = 0$, and so the linear momentum is conserved. The carts move apart with equal speeds in opposite directions. The spring does work as it expands, accelerating each cart to speed v. In the process the internal

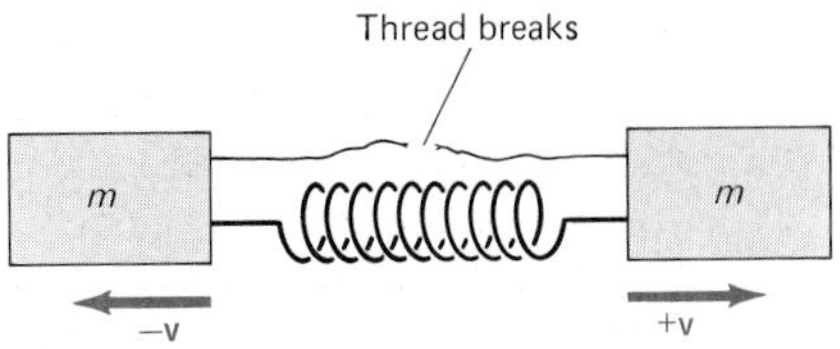

Figure 10.12

A thread holds identical objects together, compressing a spring between them. When the thread breaks they move apart with equal speeds.

energy of the system (potential energy stored in the compressed spring) is converted into the kinetic energy of the two carts. The initial kinetic energy of the system is zero but the final kinetic energy is not zero. Kinetic energy is not conserved because work is done by the internal force—the spring force.

A rigid body was defined in Section 3.2 as one for which the distance between any two points is constant. If the distance between two points is constant, however, the corresponding displacement, $\Delta \mathbf{s}$, is zero. The work done by a force (we assume central forces) is $\Delta W = \mathbf{F} \cdot \Delta \mathbf{s}$, and vanishes if $\Delta \mathbf{s}$ is zero. Hence the work done by internal forces in a rigid body is zero.

Questions

4. An astronaut is isolated (no external forces act) from his spacecraft, and at rest relative to it. Can he change the location of his center of mass relative to the spacecraft? Can he start his body moving toward the spacecraft? Explain.
5. An artillery shell explodes into many fragments while in flight. Neglecting air resistance, is the (a) total linear momentum changed; (b) total kinetic energy changed? Explain.
6. Two objects are isolated from their environment but exert forces on each other. Discuss the relation of Newton's third law to conservation of linear momentum deduced from the motions of the two objects.

10.3 Galilean Relativity

Relativity concerns the laws of physics as they are formulated by observers in relative motion. For example, consider two observers: an elk standing motionless on a snow-covered field and a crewman in a helicopter that is flying at a moderate altitude with some constant velocity (Figure 10.13). The crewman drops a bale of hay to feed the snowbound elk. The elk sees the bale fall in a graceful parabola. The helicopter crewman sees it drop straight down as he continues along at constant velocity. The elk and the crewman observe different positions and different velocities, but both trajectories *hold relative to each individual reference frame,* as predicted by Newton's law of motion. Even though the observers *disagree* on the actual trajectory, they do agree that Newton's laws are applicable.

Experiments show that if Newtonian mechanics is valid in one reference frame, it is valid in all other reference frames moving at constant velocity relative to the first. This is a principle of relativity. The key requirement here is the absence of any significant acceleration of the reference system. A convenient interpretation is no significant acceleration (including rotation) with respect to the "fixed" stars or with respect to distant galaxies. These constant-velocity reference frames are called **inertial systems** because (as we saw in Chapter 6) Newton's first law, which describes the property of inertia, holds in these systems. In modern terminology we say that Newtonian mechanics is **invariant** to the choice of inertial system.

Figure 10.13

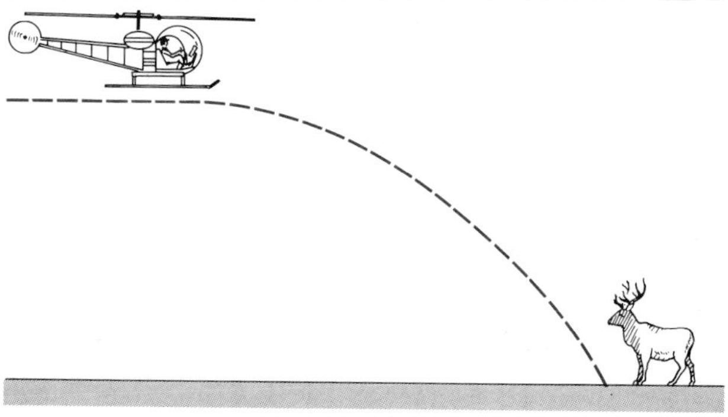

The invariance of Newtonian mechanics to the choice of inertial reference frame is called *Galilean relativity.*

The term *invariance* means that observers in different inertial systems *agree*—they agree that Newtonian mechanics adequately describes the observed motions. Newtonian mechanics, then, is invariant to the choice of inertial reference frame. However, observers in different inertial systems *disagree* on the actual positions and velocities in the observed trajectories: The helicopter crewman in the example cited saw the bale of hay drop vertically, whereas the elk saw it fall in a graceful parabola.

Let's study the helicopter and the elk more carefully. The helicopter constitutes one reference frame. The elk standing on the earth constitutes a second reference frame, as described in Section 4.1. Let's attach a coordinate system (Figure 10.14) to each: coordinate system S to the stationary elk and S' to the helicopter moving to the elk's right with velocity $\mathbf{v}$. To simplify matters we take the x'-axis of system S' and the x-axis of system S to be parallel to each other. Further, let's start measuring time when the S and S' origins coincide.

As developed in Section 4.7, points x' in system S' are related to points x in system S by

$$x = x' + vt \qquad (10.22)$$

Also

$$x' = x - vt \qquad (10.23)$$

Here we're following Newton's assumption of one univeral time. In other words, the elk and the crewman see their own clocks and each other's clocks all running at the same rate (Figure 10.15). We will refer to equations 10.22 and 10.23 as the **Galilean transformations.**

Now suppose that the crewman moves forward through the helicopter with a speed u' (in the $+x$-direction). This is his velocity relative to the helicopter. Referring back to Section 4.7 again, we have, for his velocity relative to the ground, u, the expression

$$u = u' + v \qquad (10.24)$$

This is the "commonsense" rule for velocity addition. Because we will refer to this velocity addition formula again, we'll give it a name: **Galilean velocity addition.**

Figure 10.14

Two reference frames. The elk is stationary in system S. The helicopter is stationary in system S'.

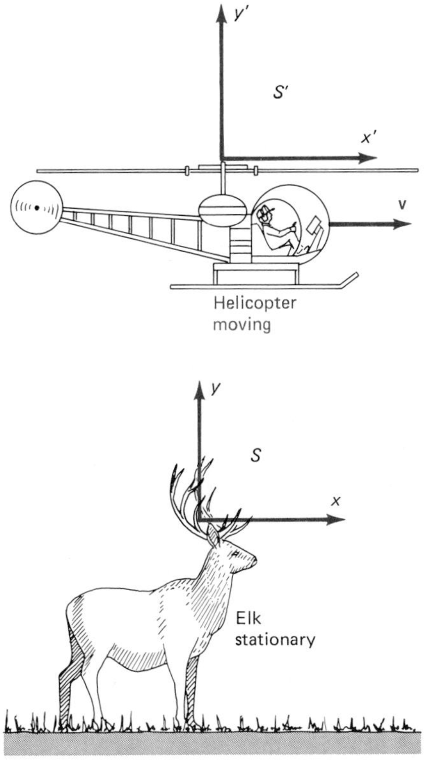

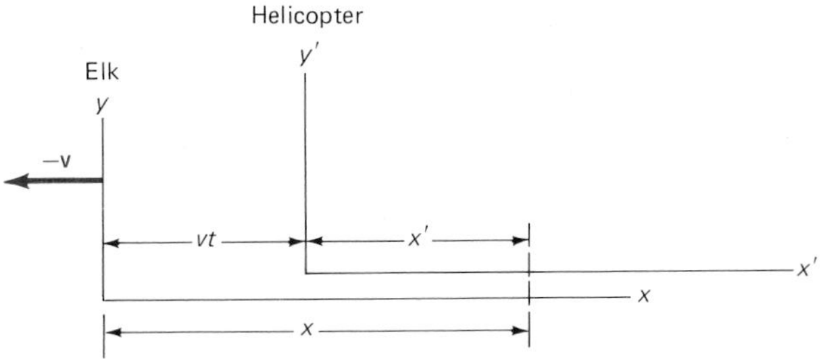

Figure 10.15a

Two reference frames from the elk's viewpoint.

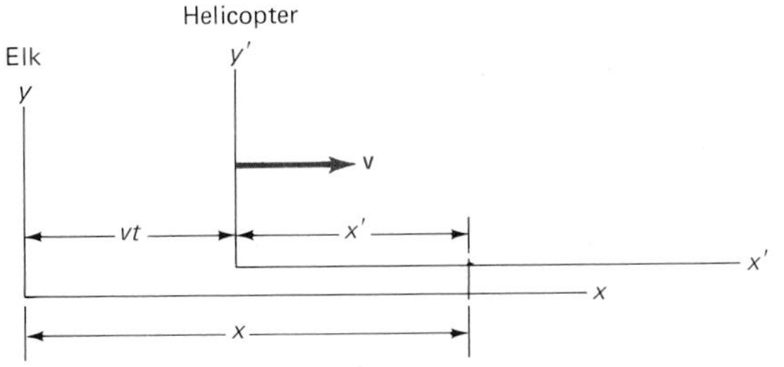

Figure 10.15b

Two reference frames from the helicopter pilot's viewpoint.

If observers at rest in S and S' disagree on position and velocity, will they also disagree on acceleration? No, they will not! They each will see the same acceleration because they move at a *constant* relative velocity.

$$a = a' \tag{10.25}$$

The acceleration is invariant to the choice of inertial system. The acceleration of the falling bale of hay is the same for the elk as it is for the helicopter pilot, even though the positions, velocities, and trajectories differ. The different inertial observers agree on Eq. 10.25.

Generalizing, we assert that there exists a system of mechanics, based on Newton's laws of motion and called *Newtonian mechanics*, that describes particle motions in our everyday work to high accuracy. A principle of relativity goes with it: Galilean relativity. For Newtonian mechanics, there is no preferred inertial reference system.

10.4 Center-of-Momentum Reference Frame

We can analyze the dynamics of processes like collisions or the decay of unstable nuclear particles more easily when we do so relative to the center-of-momentum reference frame. The center-of-momentum reference frame is defined as the reference frame in which the total linear momentum of the particles of the system equals zero. Thus observers for whom

$$\mathbf{p} = \sum_i \mathbf{p}_i = 0 \tag{10.26}$$

constitute a center-of-momentum reference frame.

According to Eq. 10.9, the velocity of the center of mass,

$$\mathbf{v}_{cm} = \frac{1}{M}\sum_i m_i \mathbf{v}_i$$

equals zero in this reference frame. Consequently, the center-of-momentum reference frame can also be called the center-of-mass reference frame.*

Newtonian Mechanics

For a many-particle system the center-of-mass position is given by Eq. 10.1:

$$\mathbf{r}_{cm} = \frac{1}{M}\sum_i m_i \mathbf{r}_i$$

*Since Eq. 10.9 is Newtonian or nonrelativistic, it holds in Newtonian mechanics. However, the center-of-momentum reference frame and the center-of-mass reference frame do not necessarily coincide if any of the particles have relativistic velocities. Indeed, if photons or other zero-rest-mass particles are involved, a center-of-mass system would be meaningless.

Here, $\mathbf{r}_{cm}$ is the position vector from the origin to the center of mass. Now suppose that we choose the origin to be located at, and moving with, the center of mass. (This choice is often a great convenience.) Then, with primes denoting variables relative to an origin taken at the center of mass, $\mathbf{r}'_{cm}$ is equal to zero. Hence

$$\mathbf{r}'_{cm} = \frac{1}{M} \sum_i m_i \mathbf{r}'_i = 0 \qquad (10.27)$$

where $\mathbf{r}'_i$ is the position of the ith particle. Similarly, $\mathbf{v}'_{cm}$, the velocity of the center of mass relative to an origin taken at the center of mass, also equals zero, giving

$$\mathbf{v}'_{cm} = \frac{1}{M} \sum_i m_i \mathbf{v}'_i = 0 \qquad (10.28)$$

For a two-particle system, Eqs. 10.27 and 10.28 require

$$\mathbf{r}'_1 = -\frac{m_2}{m_1} \mathbf{r}'_2 \qquad (10.29)$$

$$\mathbf{v}'_1 = -\frac{m_2}{m_1} \mathbf{v}'_2 \qquad (10.30)$$

The two particles are located in opposite directions from the center of mass at distances inversely proportional to the corresponding masses. The particles move in opposite directions with speeds inversely proportional to their masses.

Example 8 An Exercising Astronaut

While in orbit, astronauts film one another moving back and forth between the ends of the Skylab satellite. Can such motion form the basis for space propulsion engines?

Consider what happens when an astronaut pushes off from the bottom end of the space capsule, as shown in Figure 10.16. If m_a and $\mathbf{v}'_a$ are the mass and velocity of the astronaut relative to the center of mass of astronaut and satellite combined, and m_c and $\mathbf{v}'_c$ are the capsule quantities, then Eq. 10.30 requires that

$$\mathbf{v}'_c = -\frac{m_a}{m_c} \mathbf{v}'_a$$

When the astronaut moves upward, the capsule must move downward. When the astronaut lands at the top end of the capsule, both astronaut and capsule stop. During the process, the capsule has moved downward a certain distance.

To return to the starting point the astronaut must push upward on the capsule, and the capsule will also return to its starting point. In one round trip no progress has been made, and consequently the device fails as a space propulsion engine.

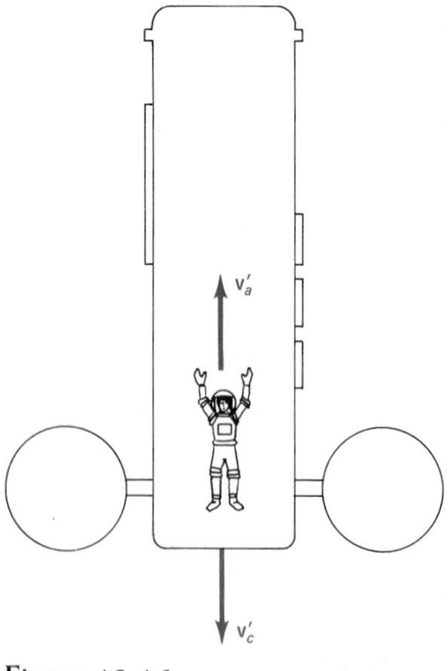

Figure 10.16

An astronaut and space capsule are initially at rest. The astronaut jumps upward at a velocity v'_a, causing the capsule to move downward with a velocity v'_c.

One-Dimensional Collisions

The center-of-mass reference frame is usually the simplest reference frame to use for the theoretical analysis of a collision. Experiments are performed in laboratories, however, and the laboratory frame of reference may differ from the center-of-mass reference frame. Therefore, both reference frames are often used to compare theory with experiment. The theoretical solution is obtained in the center-of-mass frame, and this solution is transformed into a solution in the laboratory frame for comparison with experiment.

Converting velocities back and forth between reference frames requires a law of transformation of velocities. The Galilean velocity transformation law, discussed in Section 5.3, holds in Newtonian mechanics. For motion along the x-axis (+ is to the right and − is to the left) the Galilean velocity addition law is

$$\mathbf{v} = \mathbf{v}' + \mathbf{v}_{cm} \qquad (10.31)$$

Here $\mathbf{v}$ is the laboratory velocity of an object (Figure 10.17). $\mathbf{v}'$ is the center-of-mass velocity of the same object, and $\mathbf{v}_{cm}$ is the laboratory velocity of the center of mass.

To illustrate the making of a laboratory problem into a center-of-mass problem, let's take a problem that can be solved by inspection—one involving a *perfectly inelastic collision*. Consider two objects (Figure 10.18) of mass m_1 and m_2, respectively, initially moving in the laboratory with respective velocities $\mathbf{v}_{1i}$ and $\mathbf{v}_{2i}$. The two objects collide

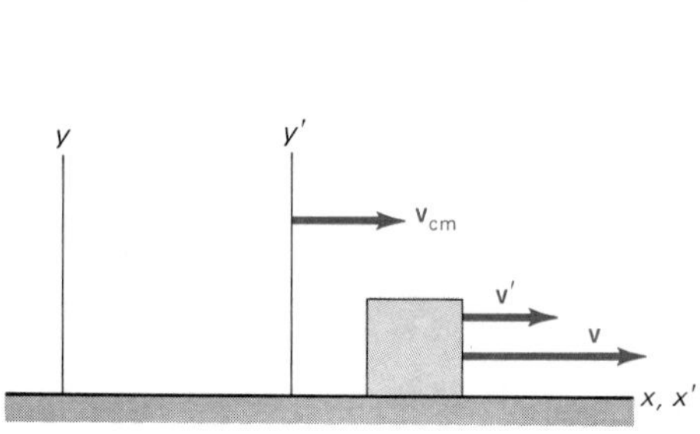

Figure 10.17

The primed axis (at rest with the center of mass) moves to the right with speed $\mathbf{v}_{cm}$ in the unprimed system. The velocity of an object is $\mathbf{v}$ (unprimed system) or $\mathbf{v}'$ (primed system).

Figure 10.18

Two blocks, seen in the unprimed system, move prior to colliding. The center of mass and the primed system move with the velocity $\mathbf{v}_{cm}$.

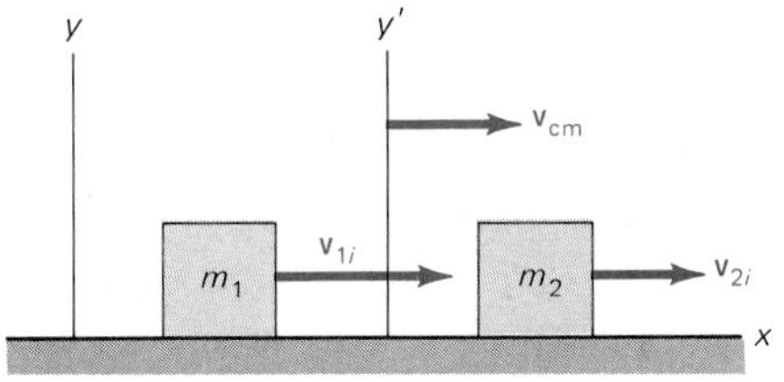

Figure 10.20

In the center-of-mass system two colliding blocks are at rest after colliding (perfectly) inelastically.

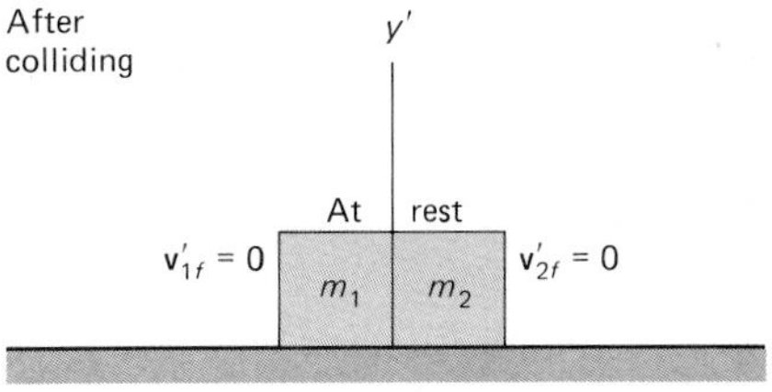

inelastically, and we seek their velocities after the collision.

A perfectly inelastic collision is characterized by the fact that the colliding objects move as a unit after the collision. Their common velocity is the velocity of the center of mass. According to Eq. 10.10 we have

$$\mathbf{v}_{cm} = \frac{m_1\mathbf{v}_{1i} + m_2\mathbf{v}_{2i}}{m_1 + m_2}$$

for the center-of-mass velocity, and hence for the final velocities, $\mathbf{v}_{1f}$ and $\mathbf{v}_{2f}$, of the two objects. This is the desired solution, which we have obtained by inspection.

Now, let us solve the problem again by using the center-of-mass frame of reference (Figure 10.19). First we obtain the initial velocities in this reference frame by using Eq. 10.31.

$$\mathbf{v}'_{1i} = \frac{m_2(\mathbf{v}_{1i} - \mathbf{v}_{2i})}{m_1 + m_2} \qquad (10.32)$$

$$\mathbf{v}'_{2i} = \frac{m_1(\mathbf{v}_{2i} - \mathbf{v}_{1i})}{m_1 + m_2} \qquad (10.33)$$

A little algebra reveals that

$$\mathbf{p}'_i = m_1\mathbf{v}'_{1i} + m_2\mathbf{v}'_{2i} = 0$$

or the linear momentum of the system equals zero initially. Linear momentum is conserved in the collision, so $\mathbf{p}'_f = 0$ also. In an inelastic collision the particles move as a unit after colliding. Hence $\mathbf{v}'_{1f} = 0$ and $\mathbf{v}'_{2f} = 0$. This is the solution in the center-of-mass reference frame—both particles are at rest (Figure 10.20).

To obtain the solution in the laboratory we use Eq. 10.31 to obtain $\mathbf{v}_{1f}$ and $\mathbf{v}_{2f}$. This gives

$$\mathbf{v}_{1f} = \mathbf{v}_{2f} = \frac{m_1\mathbf{v}_{1i} + m_2\mathbf{v}_{2i}}{m_1 + m_2}$$

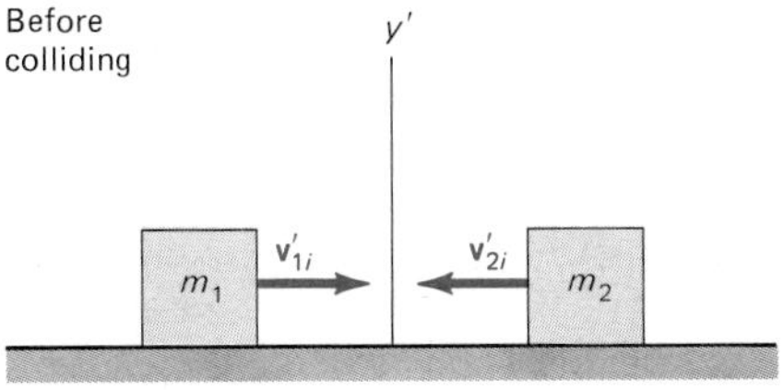

Figure 10.19

Two blocks, seen in the primed or center-of-mass system, move prior to colliding.

as before. Note that kinetic energy is not conserved in perfectly inelastic collisions, although linear momentum is conserved.

Now let us consider a *perfectly elastic collision* (one in which both linear momentum and kinetic energy are conserved) in one dimension (Figure 10.21). We start with the same physical situation and the same initial conditions as in the previous example. Thus, $\mathbf{v}_{1i}$ and $\mathbf{v}_{2i}$ are the initial velocities in the laboratory, and in the center-of-mass reference frame the initial velocities are given by Eqs. 10.32 and 10.33.

We must obtain the solution—the final velocities—in the center-of-mass reference frame. Recall that in Section 9.4 we obtained expressions for the final velocities in a perfectly elastic collision. In the center-of-mass reference frame these expressions reduce to

$$\mathbf{v}'_{1f} = -\mathbf{v}'_{1i} \qquad (10.34)$$

$$\mathbf{v}'_{2f} = -\mathbf{v}'_{2i} \qquad (10.35)$$

As illustrated in Figure 10.21, the final velocities are the negatives of the initial velocities!

As a final step, we again apply Eq. 10.31 to obtain the laboratory solution.

$$\mathbf{v}_{1f} = \mathbf{v}'_{1f} + \mathbf{v}_{cm} \qquad (10.36)$$

$$\mathbf{v}_{2f} = \mathbf{v}'_{2f} + \mathbf{v}_{cm} \qquad (10.37)$$

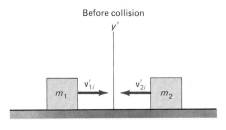

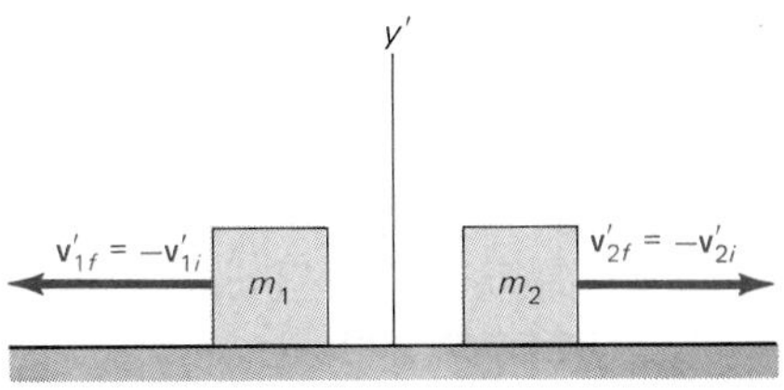

Figure 10.21

A perfectly elastic collision in the center-of-mass reference frame.

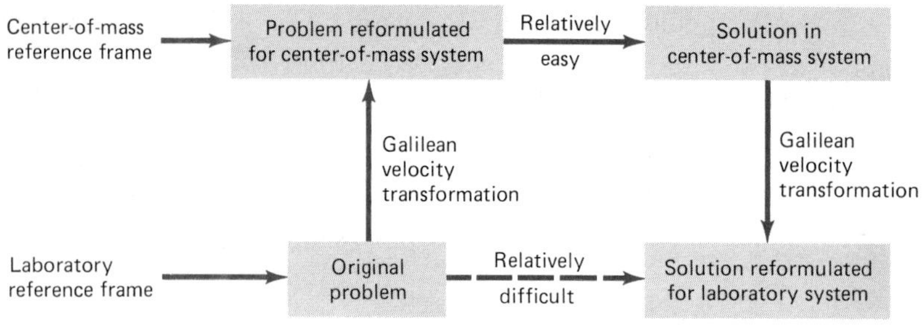

Use of the center of mass to solve problems

Diagram 10.1

It is not difficult to prove that if $\mathbf{v}'_{1f}$, $\mathbf{v}_{cm}$, and $\mathbf{v}'_{2f}$ are expressed in terms of the initial laboratory velocities, then Eqs. 10.36 and 10.37 can be written as

$$\mathbf{v}_{1f} = \frac{m_1 - m_2}{m_1 + m_2}\mathbf{v}_{1i} + \frac{2m_2}{m_1 + m_2}\mathbf{v}_{2i} \qquad (10.38)$$

$$\mathbf{v}_{2f} = \frac{2m_1}{m_1 + m_2}\mathbf{v}_{1i} + \frac{m_1 - m_2}{m_1 + m_2}\mathbf{v}_{2i} \qquad (10.39)$$

This method of solution is outlined in Diagram 10.1, and leads to the same solution as the method used in Section 9.4. You can confirm the equivalence of these two methods by comparing Eqs. 10.38 and 10.39 with Eqs. 9.21 and 9.22 in Section 9.4.

Example 9 **Center-of-Mass Collision Analysis**

Two objects ($m_1 = 8$ kg and $m_2 = 2$ kg) collide in one dimension in a perfectly elastic manner (Figure 10.22). Initially their velocities are $v_{1i} = +6$ m/s and $v_{2i} = -4$ m/s. We wish to calculate (a) the velocity of the center of mass, (b) the initial velocities in the center-of-mass reference frame, (c) the final velocities in the center-of-mass frame, and (d) the final velocities in the laboratory reference frame.

(a) Using Eq. 10.10 we obtain

$$v_{cm} = \frac{1}{M}(m_1 v_{1i} + m_2 v_{2i})$$

$$= \frac{8\text{ kg} \times 6\text{ m/s} + 2\text{ kg} \times (-4\text{ m/s})}{8\text{ kg} + 2\text{ kg}}$$

$$= +4\text{ m/s}$$

for the center-of-mass velocity relative to the laboratory.

(b) Using Eq. 10.32 gives

$$v'_{1i} = \frac{m_2(v_{1i} - v_{2i})}{m_1 + m_2}$$

$$= \frac{2\text{ kg}[6\text{ m/s} - (-4\text{ m/s})]}{8\text{ kg} + 2\text{ kg}}$$

$$= +2\text{ m/s}$$

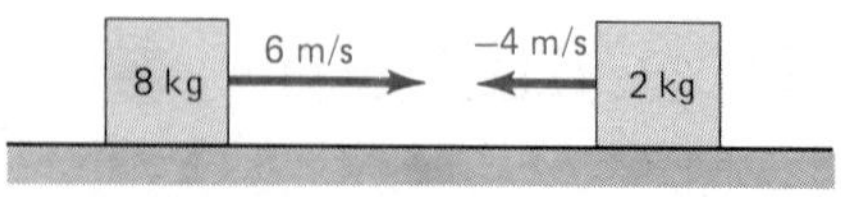

Figure 10.22

The collision of blocks unequal in mass seen in the laboratory reference frame.

for the initial velocity of the 8-kg mass relative to the center of mass, and using Eq. 10.33 gives

$$v'_{2i} = \frac{m_1(v_{2i} - v_{1i})}{m_1 + m_2}$$

$$= \frac{8\text{ kg}(-4\text{ m/s} - 6\text{ m/s})}{8\text{ kg} + 2\text{ kg}}$$

$$= -8\text{ m/s}$$

for the initial velocity of the 2-kg mass relative to the center of mass.

(c) Using Eqs. 10.34 and 10.35 we obtain

$$v'_{1f} = -v'_{1i} = -2\text{ m/s}$$

and

$$v'_{2f} = -v'_{2i} = +8\text{ m/s}$$

for the final velocities relative to the center-of-mass reference frames.

(d) Using Eqs. 10.36 and 10.37 yields

$$v_{1f} = v'_{1f} + v_{cm}$$

$$= -2 + 4$$

$$= +2\text{ m/s}$$

and

$$v_{2f} = v'_{2f} + v_{cm}$$

$$= +8 + 4$$

$$= +12\text{ m/s}$$

for the final velocities relative to the laboratory reference frame. If we know the final velocities in the center-of-mass reference frame we can find the final velocities in the laboratory by a series of simple steps.

Collisions in Two Dimensions

When $m_1 = m_2$, Eqs. 10.29 and 10.30 require $\mathbf{r}'_1 = -\mathbf{r}'_2$ and $\mathbf{v}'_1 = -\mathbf{v}'_2$. For a collision between two electrons or between two protons, for example, the two particles will always be equidistant from the center of mass and will be located in opposite directions from it. Also, the particles will be moving in opposite directions. For convenience the x- and x'-axes are chosen parallel to the line joining the two particles. A scattering event like this can be represented by the scattering angle θ', as shown in Figure 10.23. These conclusions regarding the position and ve-

Figure 10.23

In two-dimensional collisions in the center-of-mass system, the emerging particles move along a line making the angle θ' with the original direction of motion.

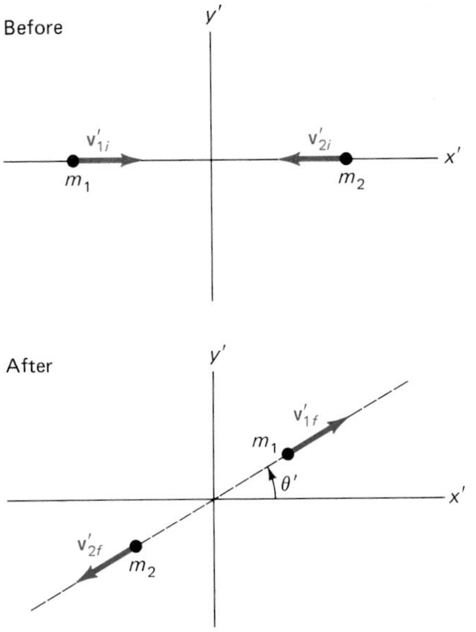

locity vectors relative to the center of mass hold for any angle ($0 \leq \theta' \leq 180°$).

When a collision is studied in the laboratory, often one of the two interacting particles, called the *target particle*, is at rest. The masses may be equal, but more frequently $m_{\text{target}} > m_{\text{projectile}}$. In a collision between a moving particle and a stationary target the center of mass moves relative to the laboratory. Particle positions, $\mathbf{r}_i$, and velocities, $\mathbf{v}_i$, relative to the laboratory are related to the corresponding quantities in the center-of-mass reference frame by the equations

$$\mathbf{r}_i = \mathbf{r}_{\text{cm}} + \mathbf{r}'_i \tag{10.40}$$

$$\mathbf{v}_i = \mathbf{v}_{\text{cm}} + \mathbf{v}'_i \tag{10.41}$$

From the geometry of Figure 10.24 we can write

$$\tan\theta = \frac{v'_{1f}\sin\theta'}{v'_{1f}\cos\theta' + v_{\text{cm}}} \tag{10.42}$$

as the relation between the scattering angle (θ) in the laboratory and the scattering angle (θ') in the center-of-mass reference frame. The x-axis is taken to be in the direction of the center-of-mass motion relative to the laboratory and to the center-of-mass reference frames, respectively. If linear momentum and kinetic energy are conserved in the collision, then we can prove that (see Problem 20)

$$v_{\text{cm}} = \frac{m_1 v'_{1f}}{m_2} \tag{10.43}$$

Putting Eq. 10.43 into Eq. 10.42 gives

$$\tan\theta = \frac{\sin\theta'}{\cos\theta' + (m_1/m_2)} \tag{10.44}$$

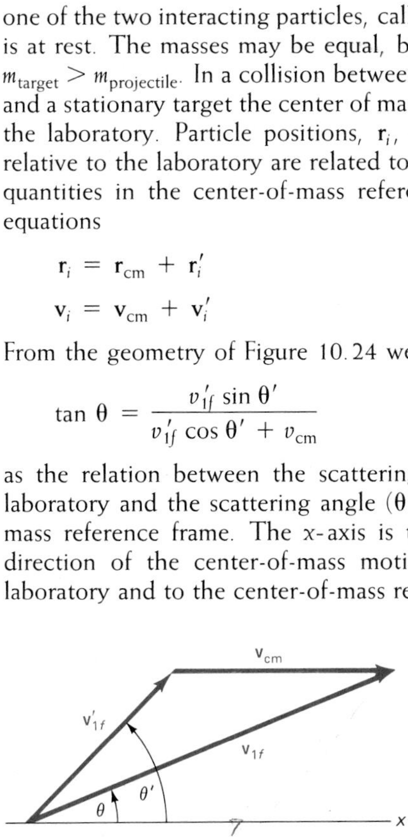

Figure 10.24

Two particles collide and scatter through the angle θ in the laboratory or θ' in the center-of-mass system. Geometrically the equation $\mathbf{v}_{1f} = \mathbf{v}_{\text{cm}} + \mathbf{v}'_{1f}$ relating the laboratory and center-of-mass velocities of the scattered particle is a triangle relation.

Example 10 Alpha-Particle Scattering

An alpha particle ($m = 4$ u) strikes a uranium nucleus ($m = 238$ u) at rest. Determine the scattering angle in the center-of-mass reference frame if the laboratory scattering angle is observed to be 90°. Using Eq. 10.44 we have

$$\tan\theta = \frac{\sin\theta'}{\cos\theta' + (4/238)}$$

Since $\tan\theta \to \infty$, we require $\cos\theta' = -4/238$. This gives $\theta' = 90.96°$ as the scattering angle in the center-of-mass reference frame. This is very close to 90°, the laboratory scattering angle. We might have expected this, because $m_u \gg m_a$.

When the projectile mass m_1 is equal to the target mass m_2 we call this special situation the *equal-mass case*. In this event $\theta = \theta'/2$ (see Problem 21). When $m_1 \ll m_2$ we have an important limiting case called the *heavy-target case*. In this case, $\theta \to \theta'$. Scattering of alpha particles by heavy nuclei is an example of the heavy-target case, and proton–proton scattering is an example of the equal-mass case.

Physicists make measurements in the laboratory system, but they often perform calculations in the center-of-momentum system, where calculations are simpler. Equations such as Eq. 10.42 are used to transform back and forth from one system to another.

Questions

7. Two particles having masses m_1 and m_2 (with $m_2 > m_1$) approach each other with speeds v_1 and v_2. (a) If their combined linear momentum is zero, are their kinetic energies equal? If not, which is larger? (b) If their kinetic energies are equal, are the magnitudes of their linear momenta equal? If not, which is larger?

8. A heavy nucleus at rest emits an alpha particle at high velocity. Use the center-of-momentum reference frame to prove that the residual nucleus can't be at rest.

10.5 Rocket Propulsion

Two conditions must be satisfied in order for Newton's laws of motion to hold when applied to a system. First, the observations must be made by an inertial observer. Second, dividing the universe into a system, or "body," and its

Figure 10.25

Robert H. Goddard (1882–1945) pioneered in rocket research and development. He is credited with building and successfully launching the first liquid-fueled rocket.

environment must be given careful attention. For example, a rocket engine transfers mass from the rocket to its surroundings. Suppose we choose the system to be the rocket (including its unburned fuel) at some instant. At that instant all parts of the system have the same velocity. A moment later some of the fuel will have been burned and ejected. The two parts of the system, the rocket and the ejected gases, have different velocities. Correct application of Newtonian mechanics requires that we keep track of the momentum of the ejected gases as well as of the rocket momentum (Figure 10.25).

Example 11 **How to Accelerate a Flatcar**

A railroad brakeman (Figure 10.26) of mass m stands on a flatcar of mass M at rest. He runs on the flatcar with a speed u_o relative to the flatcar, and then jumps off. What is the resulting flatcar recoil velocity, assuming that friction is negligible?

Because $\mathbf{F}_{ex} = 0$, linear momentum is conserved. We must be careful to measure velocities relative to an inertial observer, and so we consider the situation relative to the ground. We have

initially $\quad p_i = 0$

finally $\quad p_f = Mv - m(u_o - v)$

Conservation of linear momentum requires $p_f = 0$, giving

$$v = \frac{mu_o}{M + m}$$

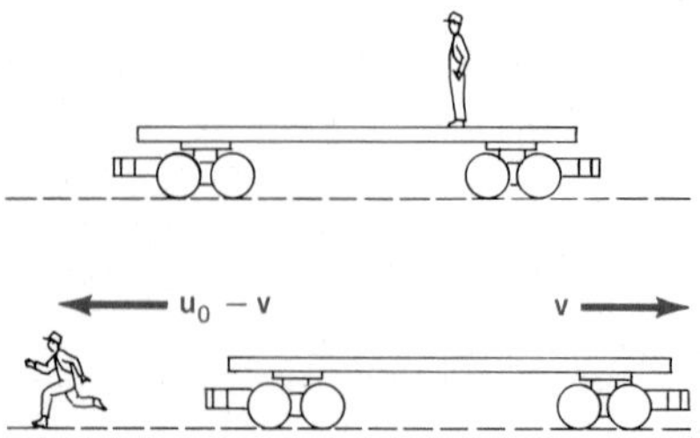

Figure 10.26

A brakeman and flatcar are initially at rest. When the brakeman accelerates to the left by running on the flatcar, the flatcar accelerates to the right.

The flatcar velocity depends on u_o, the velocity of the brakeman relative to the flatcar. The denominator is the initial (and final) mass of the system, and in the numerator m is the mass ejected by the flatcar.

Like the flatcar-and-brakeman system just considered, a rocket ejects mass. In the ejection process the rocket exerts a force on the mass being ejected. From Newton's third law this mass must exert a reaction force back on the rocket. Hence the rocket accelerates, and in a direction opposite to that in which the mass is ejected.

A Force-Free Rocket

Consider a rocket (Figure 10.27) whose mass at time t is m and that ejects mass from the engine at the constant rate of α (in kg/s). The velocity of escaping gases is constant relative to the rocket, and is equal to u_o. We want to determine the rocket acceleration in the absence of any net external force.

At the time t we observe the rocket of mass m moving with velocity v. Thus initially the linear momentum of the "system" is given by

$$p_i = mv$$

Later, at the time $t + \Delta t$, we observe the rocket mass to be $m + \Delta m$ and its velocity to be $v + \Delta v$. Additionally a mass $(-\Delta m)$ of gas has been ejected by the engine and moves with velocity $v + \Delta v - u_o$. Note that Δm is negative.

In these circumstances (1) the combined mass at $t + \Delta t$ is equal to m, just as it is at time t. We are considering one system at two different times. (2) We expect Δm to be negative (so m decreases with time) and Δv to be positive (the rocket accelerates). At the time $t + \Delta t$ the linear momentum of the system

$$\begin{aligned} p_f &= (m + \Delta m)(v + \Delta v) + (-\Delta m)(v + \Delta v - u_o) \\ &= mv + m\,\Delta v + u_o\,\Delta m \end{aligned}$$

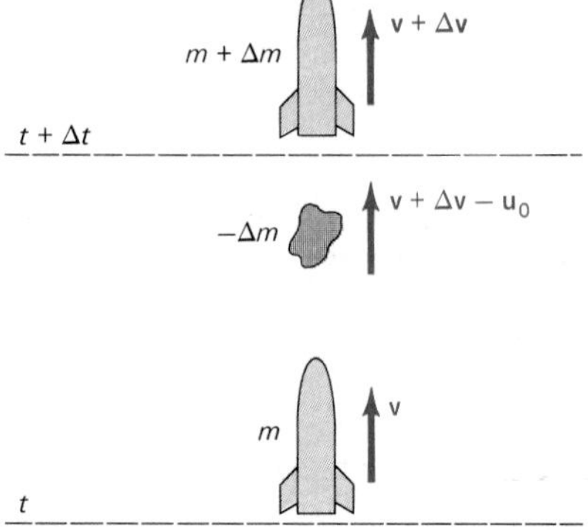

Figure 10.27

A system consists of a rocket of mass m moving with velocity v at time t. At time $t + \Delta t$ the same system includes a rocket of mass $m + \Delta m$ moving with velocity $v + \Delta v$, and the gas of mass $-\Delta m$ moving with velocity $v + \Delta v - u_o$.

where u_o is the velocity of escaping gases relative to the rocket. Using $\Delta p = p_f - p_i$, we obtain the expression

$$\Delta p = m\,\Delta v + u_o\,\Delta m \tag{10.45}$$

for the change in the linear momentum of the system during the time interval Δt.

We can use Eq. 10.45 to draw inferences about the functioning of a rocket engine. In the force-free case we know that the linear momentum of the rocket–gas system is conserved, or $\Delta p = 0$. Thus we can write, from Eq. 10.45,

$$\Delta v = -\frac{u_o}{m}\,\Delta m \tag{10.46}$$

We see from this equation that the change in velocity of the rocket is proportional to the mass ejected and to the velocity of escaping gases, and is inversely proportional to the rocket mass. This is exactly the same situation as in the flatcar-and-brakeman example. Note that Δv is positive if Δm is negative. Consequently, if an astronaut found himself separated from his spacecraft, for example, he could acquire a velocity toward the spacecraft by throwing something in the opposite direction. Similarly, a rocket propels itself by literally throwing part of itself (the escaping gases) in a direction opposite to that in which the rocket gains velocity. In Eq. 10.46, $\Delta m/m$ is the fractional mass ejection. To produce a large Δv this mass ratio must be large.

Note that rocket engines are self-contained—both fuel and oxygen are on board ready for combustion. On the other hand, a jet engine does not carry oxygen. The oxygen must come from the air. Consequently, a jet engine will not work in a vacuum. A rocket engine, however, functions most efficiently in a vacuum because air exerts a frictional drag that slows the rocket. The mass of the rocket during the burn decreases with time, as suggested in Figure 10.28.

If we divide Eq. 10.46 by Δt and take the limit as $\Delta t \to 0$ we obtain

$$\frac{dv}{dt} = -\frac{u_o}{m}\left(\frac{dm}{dt}\right) \tag{10.47}$$

Introducing $a = \dfrac{dv}{dt}$ and $\alpha = -\dfrac{dm}{dt}$ (so that α is positive), we have

$$a = \frac{u_o \alpha}{m} \tag{10.48}$$

for the acceleration of the rocket. Note that a depends on both α, the rate of mass ejection, and u_o, the exhaust gas velocity. These two quantities are determined by the design of the rocket engine. For maximum acceleration, both must be large. The mass ejection per second must be large, and when the mass is ejected it should leave with the largest possible velocity.

In Eq. 10.48 the quantity $T = u_o\alpha$ is called the thrust of the rocket engine. Using T we can write Eq. 10.48 in the form

$$T = ma \tag{10.49}$$

which resembles Newton's second law. To the astronaut in the rocket his acceleration is "explained" by Eq. 10.49; but because the astronaut is not an inertial observer, Eq. 10.49 cannot be Newton's second law.

Figure 10.28

A rocket whose initial total mass is m_o loses mass as the engine burns fuel. The detailed shape of the curve depends on the burning rate versus time.

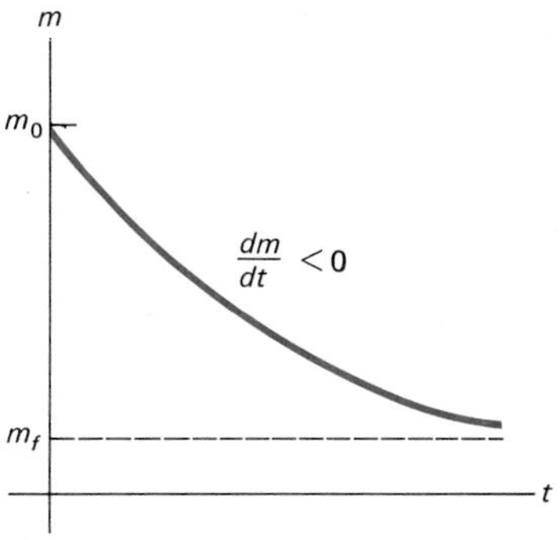

As discussed in Section 7.5, power is an inappropriate concept for rating a rocket engine. Rockets are usually rated in terms of the thrust that they deliver. The Saturn V rocket that carried American astronauts to the moon had a thrust of about 8 million pounds.

We obtain the velocity of the rocket as a function of its mass by integrating Eq. 10.46. We have

$$\int_{v_o}^{v} dv = -u_o \int_{m_o}^{m} \frac{dm}{m}$$

which leads to the result

$$v = v_o + u_o \ln \frac{m_o}{m} \tag{10.50}$$

Here v_o and m_o are the initial values, and v and m are the values at time t, of the velocity and mass, respectively, of the rocket. Because $m_o > m$, $v > v_o$.

Example 12 **The Force-Free Rocket**

Let us consider the velocity change of a rocket in the absence of external forces (a) when the fractional mass change is -0.01 and (b) when the fractional mass change is -0.99. We take $u_o = 2500$ m/s.

(a) When $\Delta m/m$ is small ($\Delta m/m \ll 1$) Eq. 10.46 can be used as an approximation. We obtain

$$\Delta v = v - v_o = -u_o\left(\frac{\Delta m}{m}\right)$$
$$= -2500(-0.01)$$
$$= +25 \text{ m/s}$$

The fraction, $\Delta v/u_o$, is the same as $-(\Delta m/m)$. A small mass ejection leads to a small change in velocity.

(b) With $\Delta m/m_o = -0.99$ we must use the integrated form—Eq. 10.50. For this value of $\Delta m/m_o$ we have an initial-to-final mass ratio $m_o/m = 100$, and so we obtain

$$v = +2500 \ln 100$$
$$= 11{,}500 \text{ m/s}$$

Here $\Delta v/u_o = 4.6$, which is much larger than the value of 0.99 we would obtain with the approximate equation, Eq. 10.46.

Equation 10.50 can be used to answer other questions. In the next example, we want our rocket to take us through interstellar space at velocities approaching the velocity of light.

Example 13 A Rocket to the Stars

What mass ratio, m_o/m, would be required to accelerate a rocket from rest to $v = (1/10)c$? Suppose that $u_o = 5000$ m/s (a very large figure for a chemical rocket engine). According to Eq. 10.50 we have

$$\ln \frac{m_o}{m} = \frac{v}{u_o} = \frac{3 \times 10^7}{5 \times 10^3} = 6000$$

and

$$\frac{m_o}{m} = e^{6000} = 10^{2600}$$

This ratio is so large that much less than one molecule of rocket would be left, regardless of the rocket's initial mass! Chemical rockets do not develop a sufficiently large exhaust velocity to be viable options for interstellar travel.

Questions

9. Consider the case of two men on the flatcar in Example 11. If the men jump off one at a time, each with speed u_o relative to the flatcar, the flatcar speed is greater than if they jump off simultaneously. Explain.

10. In a two-stage rocket (Figure 10.29) two rocket engines and two fuel supplies are launched as a unit. The first-stage engine alone lifts the entire assembly at launch. When all of the fuel in the first stage has been burned, the considerable mass of the first stage (engine, fuel container, controls, etc.) is jettisoned. Then the second-stage (usually a small fraction of the mass originally launched) engine is ignited. Explain qualitatively the advantages and disadvantages of the two-stage rocket in relation to the one-stage rocket.

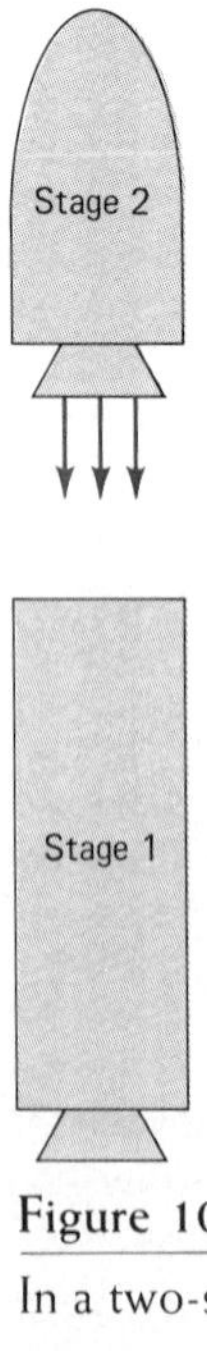

Figure 10.29

In a two-stage rocket both stages have an engine and fuel.

Summary

The center of mass of a collection of particles is defined by

$$\mathbf{r}_{cm} = \frac{1}{M} \sum m_i \mathbf{r}_i \qquad (10.1)$$

where $M = \sum m_i$ is the total mass of the collection. The components of the center-of-mass vector are given by

$$x_{cm} = \frac{1}{M} \sum m_i x_i \qquad y_{cm} = \frac{1}{M} \sum m_i y_i$$

$$z_{cm} = \frac{1}{M} \sum m_i z_i \qquad (10.3)$$

For a continuous distribution of mass the sums over particles become integrals over the distribution. Thus the center-of-mass position vector is given by

$$\mathbf{r}_{cm} = \frac{\int \mathbf{r}\, dm}{M} \qquad (10.6)$$

where $M = \int dm$. In component form $\mathbf{r}_{cm}$ is given by

$$x_{cm} = \frac{1}{M} \int x\, dm \qquad y_{cm} = \frac{1}{M} \int y\, dm$$

$$z_{cm} = \frac{1}{M} \int z\, dm \qquad (10.8)$$

If the acceleration of gravity is constant throughout the mass distribution, then the center of mass and center of gravity are located at the same point.

To describe the dynamics of the center of mass, we define the center-of-mass velocity

$$\mathbf{v}_{cm} = \frac{1}{M} \sum m_i \mathbf{v}_i \qquad (10.10)$$

and the center-of-mass acceleration

$$\mathbf{a}_{cm} = \frac{1}{M} \sum m_i \mathbf{a}_i \qquad (10.14)$$

Galilean relativity is a principle of relativity for Newtonian mechanics. Newtonian mechanics has no preferred reference frame; it is invariant under the Galilean transformations.

The center-of-momentum reference frame for a collection of particles is defined by the condition

$$\sum m_i \mathbf{v}_i = 0 \qquad (10.26)$$

If all particles in the system are non-relativistic, this reference system is equivalent to the center-of-mass system for which $\mathbf{v}_{cm} = 0$.

A rocket in a force-free environment has a velocity given by

$$v = v_o + u_o \ln \frac{m_o}{m} \qquad (10.50)$$

Problems

Section 10.1 Center of Mass

1. The three dots in Figure 1 represent equal masses. The points labeled 1, 2, and 3 are "prospects" for the center of mass. Identify the center of mass and give a *qualitative* reason why each of the other two points *cannot* be the center of mass.

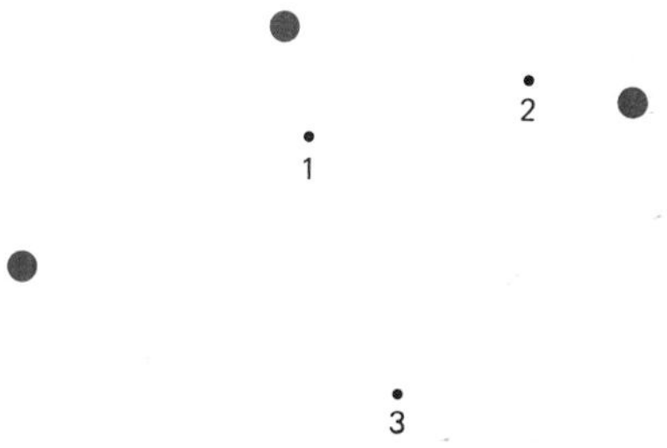

Figure 1

2. Using the center of the earth as an origin, calculate the center of mass of the earth–moon system (Figure 2), treating both objects as points. Use the average center-to-center distance $\bar{x} = 3.84 \times 10^8$ m and the ratio of the mass of the moon to that of the earth $M_{moon}/M_{earth} = 0.0123$.

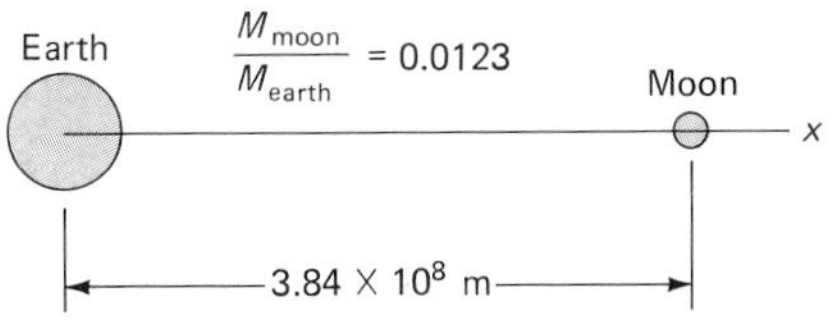

Figure 2

3. Two 5-kg masses and a 10-kg mass are on the x-axis as drawn in Figure 3. Locate the center of mass of the system (a) as drawn; (b) if the two 5-kg masses exchange positions; and (c) if the 10-kg mass exchanges position with the 5-kg mass at $x = +2$ m.

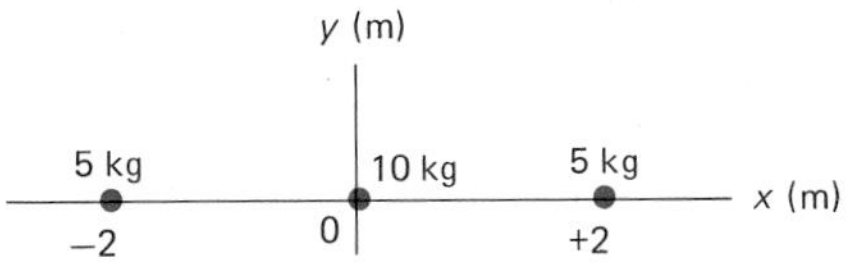

Figure 3

4. Three objects are located in the x–y plane as shown in Figure 4. Determine the location of a fourth object having mass $4M$ such that the center of mass of the four objects is at the origin.

5. Using the proton as origin, locate the center of mass of the proton–electron system in a hydrogen atom. Use the center-to-center distance $r = 0.529 \times 10^{-10}$ m, $m_e = 9.11 \times 10^{-31}$ kg, and $m_p = 1.67 \times 10^{-27}$ kg.

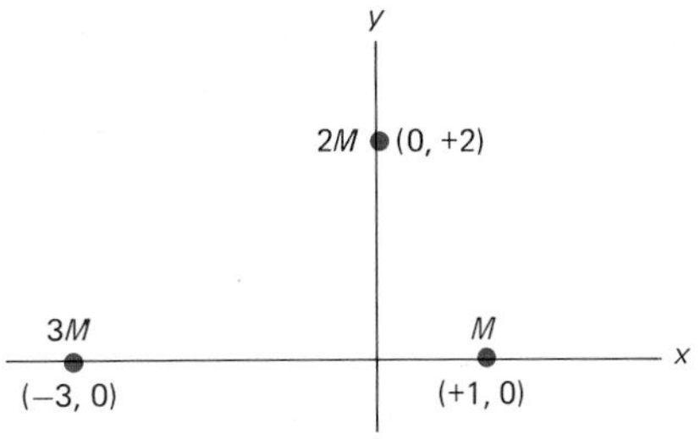

Figure 4

6. Two point masses lie on the x-axis. Prove that the center of mass (a) lies on the x-axis and (b) is located between the two given point masses.

7. Using the axes given in Figure 5, determine the coordinates of the center of mass of the water molecule.

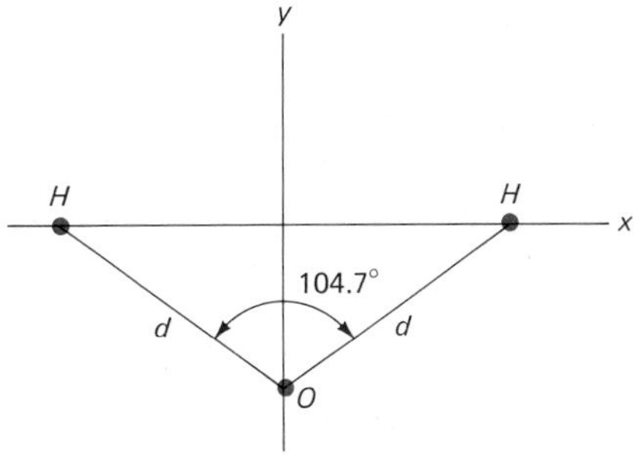

Figure 5

8. Using a graphical technique, locate the center of mass of the figure drawn in Figure 6. Assume that its mass density is the same throughout. (*Hint:* Draw his outline on graph paper, and take its mass to be proportional to its area. A scale marked in feet is given for your reference.)

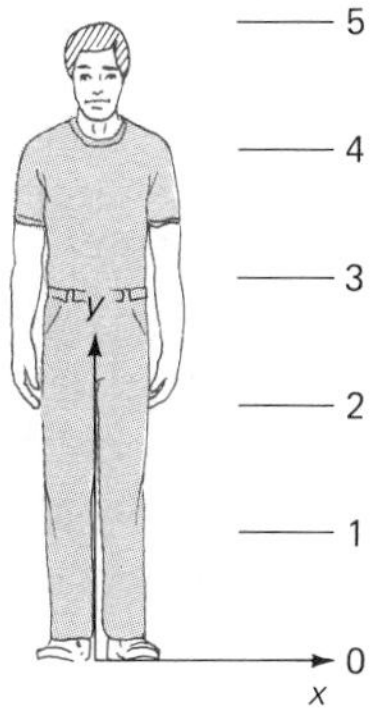

Figure 6

9. A wrench consists of a uniform steel rod 0.40 m long with a second steel rod, similar to the first but 0.60 m long, welded to the first rod to form a "T" (Figure 7). Find the center of mass of this T wrench relative to the place where the rods are welded together. Neglect the thickness of the rods.

10. A boomerang (Figure 8) is constructed in the shape of a 90° angle with each side 0.4 m in length. Locate the center of mass if the mass density is constant by using (a) a symmetry argument and (b) integration.

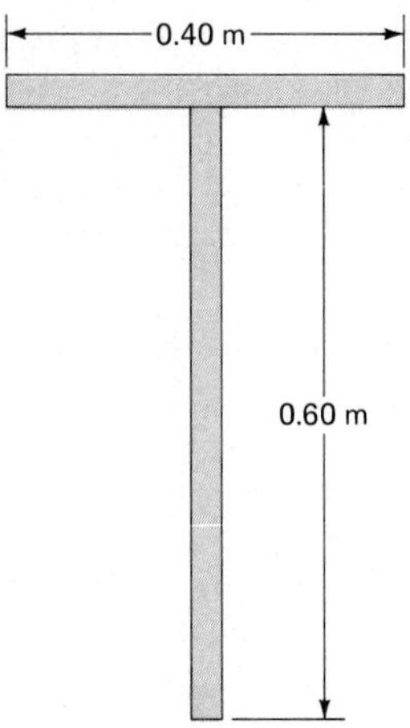

Figure 7

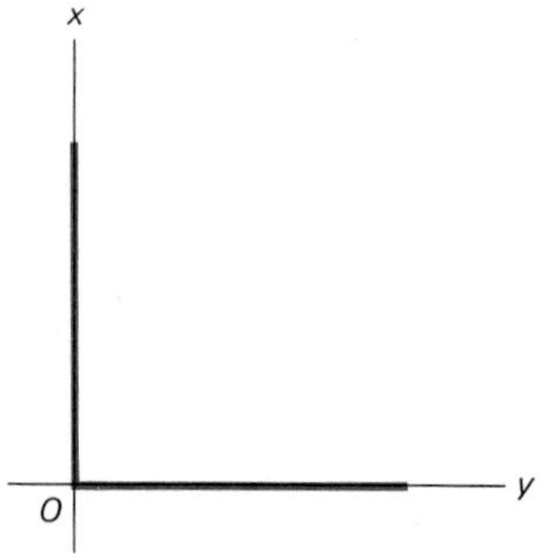

Figure 8

11. The ammonia molecule consists of three hydrogen atoms and one nitrogen atom situated on the corners of a pyramid. When the hydrogen atoms are located at $x = +0.814 \times 10^{-10}$ m, $y = 0$, $z = 0$; $x = -0.814 \times 10^{-10}$ m, $y = 0$; $z = 0$, and $x = 0$, $y = +1.41 \times 10^{-10}$ m, $z = 0$, the nitrogen atom is located at $x = 0$, $y = 0.470 \times 10^{-10}$ m, $z = 0.383 \times 10^{-10}$ m. Taking the ratio of the nitrogen mass to the hydrogen mass to be 14, locate the center of mass of the ammonia molecule.

12. A steel bar (Figure 9) is bent in the shape of one fourth of a circle of radius R. Determine the center-of-mass coordinates by using (a) the given axes and (b) symmetry axes.

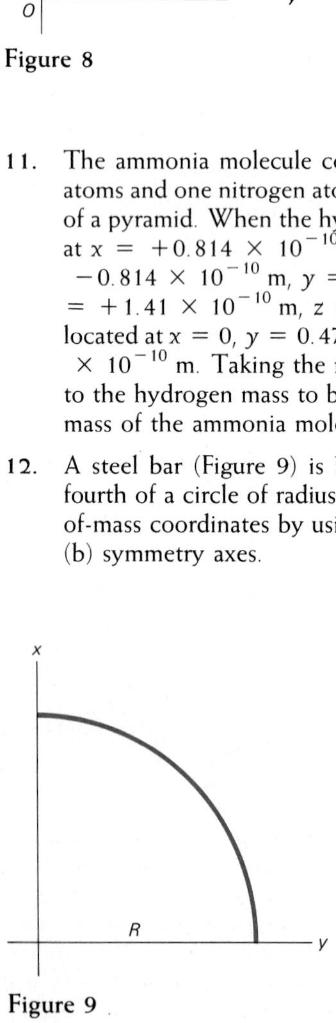

Figure 9

13. Locate the center of mass of a homogeneous right circular cone (Figure 10) having altitude a and a base of radius b. Note that Figure 10.8 can be interpreted helpfully as a side view of such a cone. However, the expression for dm employed in Example 4 does not apply to the cone.

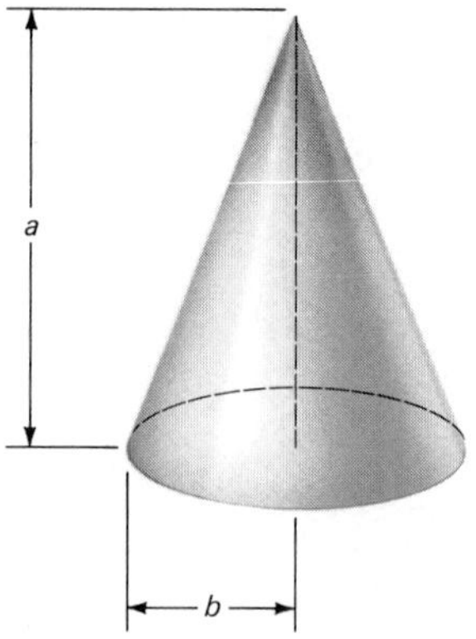

Figure 10

14. The water in a glass of water forms a 12-cm-high truncated cone (Figure 11) that has a 2-cm radius at the base and a 3-cm radius at the top. Find the height of the center of mass of the water above the base.

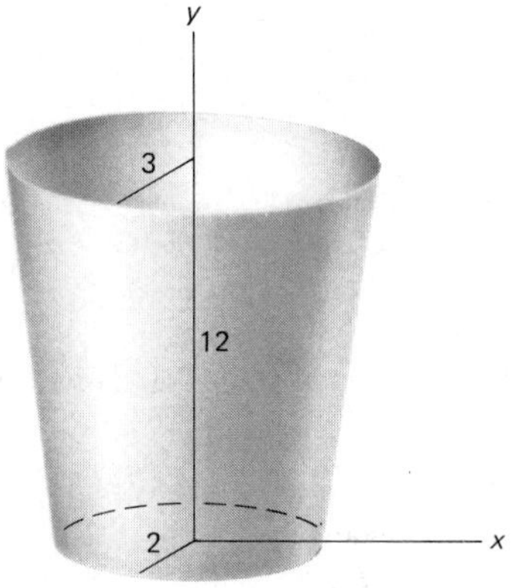

Figure 11

Section 10.2 Dynamics of the Center of Mass

15. A 100-kg man stands motionless at point A on a 900-kg raft at rest in the water (Figure 12). He goes across the raft, coming to rest at B. Neglecting water resistance, determine how far (a) the man and (b) the raft moved from their initial locations. Take $\overline{AB} = 10$ m.

Figure 12

16. Astronaut A ($m = 60$ kg) is motionless a distance x from one end of a space capsule 20 m long (Figure 13). She is holding an 8-kg mass in one hand. If she throws the mass to astronaut B, it arrives at B simultaneously with the arrival of A at the other end of the capsule. Determine the value of x.

Figure 13

17. Astronaut B (see Problem 16) jumps from one end of the capsule to the other. His mass is 80 kg. While he moves, the space capsule is observed to move 2 cm in the opposite direction. What is the mass of the space capsule?

18. A man whose mass is 100 kg stands at rest on a motionless 50-kg boat. Then he begins to walk at a speed of +2 m/s relative to the boat. Determine the velocity of the man and of the boat relative to the shore. Neglect friction between the water and the boat.

19. Two identical air cars equipped with identical linear springs (Figure 14) move toward each other on an air track. They collide, compressing the springs, and come to rest momentarily before rebounding. Find the maximum compression of a spring ($m = 0.20$ kg, $k = 3000$ N/m, $v = 3.0$ m/s).

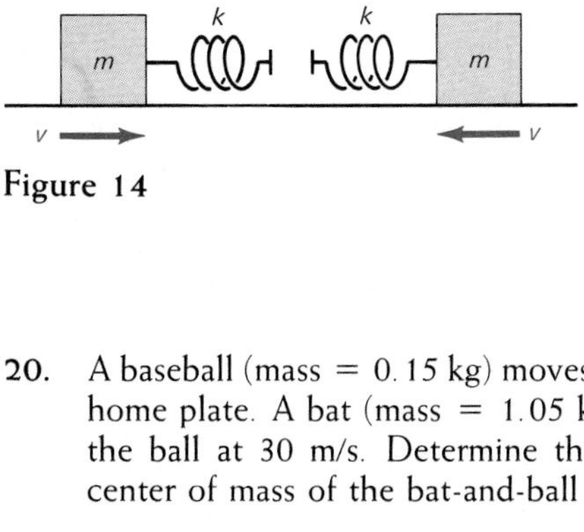

Figure 14

20. A baseball (mass = 0.15 kg) moves at 40 m/s toward home plate. A bat (mass = 1.05 kg) moves toward the ball at 30 m/s. Determine the velocity of the center of mass of the bat-and-ball system.

21. A uranium nucleus ($m = 232$ u) at rest emits an alpha particle ($m = 4$ u) with a speed $v = +1.32 \times 10^7$ m/s relative to the laboratory. Determine the velocity of the residual nucleus ($m = 228$ u) relative to the laboratory.

22. If the residual nucleus in Problem 21 were at rest relative to the laboratory after the alpha decay, what would have been (a) the alpha velocity relative to the laboratory and (b) the uranium nucleus velocity? (c) Determine the velocity of the alpha relative to the residual nucleus.

23. A plutonium nucleus ($m = 239$ u), initially at rest at the origin of a coordinate system, disintegrates into a uranium nucleus ($m = 235$ u) and an alpha particle ($m = 4$u). Some short time later the alpha particle passes the point $x = 0.1$ m moving with a speed of 1.82×10^7 m/s relative to the laboratory. What is the (a) speed and (b) location of the uranium nucleus when the alpha particle passes $x = 0.1$ m?

24. Three 1-kg objects, which are connected by rigid rods of negligible mass, are located at $x = 40$ cm, $x = -40$ cm, and $y = 70$ cm (see Figure 15). A net force of $\mathbf{F} = 12\mathbf{i} + 12\mathbf{j}$ N is applied to the object initially located at $x = 40$ cm. Locate the center of mass of the system 1 s after this force is applied.

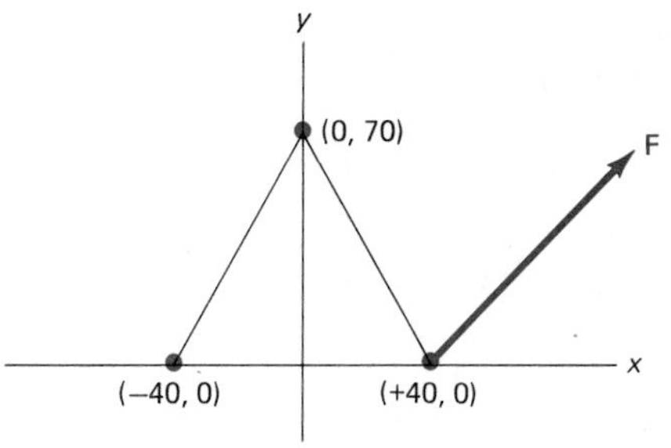

Figure 15

Section 10.4 Center-of-Momentum Reference Frame

25. Determine the initial velocities of the two hockey players (see Example 5) in the center-of-mass reference frame.

26. A 1200-kg car traveling 70 km/hr due east approaches a 1500-kg car traveling 80 km/hr due west. Determine the velocity of the center of mass of this two-car system.

27. Calculate the combined kinetic energy of the two objects in Example 9 before and after the collision (a) in the laboratory reference frame and (b) in the center-of-mass reference frame. [The numerical values obtained in (a) differ from those obtained in (b). That $\Delta K = 0$ in both cases illustrates the principle that conservation of energy is *invariant* under a Galilean transformation.]

28. A 1000-kg car (A) traveling due east at 60 km/hr approaches an identical car (B) waiting for a red light to turn green. (a) Determine the velocities of the cars in the center-of-momentum reference frame. (b) Calculate the total kinetic energy in the laboratory and center-of-momentum reference frames, and comment on any differences.

29. Determine the center-of-mass velocity in (a) Problem 21 and (b) Problem 22.

30. A railroad boxcar (mass M, speed 5.0 m/s) overtakes and couples with a second boxcar (mass $0.8M$) that had been traveling in the same direction at 3.5 m/s. (a) Calculate the velocity of the center-of-momentum reference frame before the collision. (b) What is the velocity of two coupled cars after collision? (c) Calculate the net linear momentum relative to the center-of-momentum system before collision. (d) What is the linear momentum of the two coupled cars relative to the center-of-momentum system after collision?

31. Calculate the fraction of the initial kinetic energy that is lost in the inelastic collision of the two railroad boxcars of Problem 30. (*Note:* Calculate the kinetic energy relative to the earth.)

32. Demonstrate explicitly that the linear momentum of two particles equals zero in the center-of-mass reference frame by proving that $p_i' = m_1 v_{1i}' + m_2 v_{2i}' = 0$. For v_{1i}' and v_{2i}' use Eqs. 10.32 and 10.33.

33. Derive Eqs. 10.38 and 10.39.

34. Prove the assertion that in a perfectly elastic collision in the center-of-mass frame of reference the final velocities are negatives of the initial velocities. In other words, derive Eqs. 10.34 and 10.35.

35. Derive Eq. 10.43.

36. Starting with Eq. 10.44 prove that $\theta = \theta'/2$ when $m_1 = m_2$.

37. An alpha particle is scattered through an angle of 60° in the laboratory by a collision with a cobalt nucleus (m = 60 u). Determine the scattering angle in the center-of-mass system.

Section 10.5 Rocket Propulsion

38. A boy and a girl each of mass m are standing on a small stationary rail car, mass $2m$. They jump off to the rear with a speed of u_o relative to the rail car. Neglecting friction, consider two distinct cases: (a) The boy and the girl jump off simultaneously. Calculate the resultant forward speed of the rail car. (b) The boy jumps off first. Calculate the resultant forward speed of the rail car. Then the girl jumps. Calculate the change in speed of the rail car and the final speed of the rail car. (c) Compare the final speed of the rail car in cases (a) and (b).

39. Continuing Problem 38, calculate the linear momentum of the boy and of the girl as they jump, and of the rail car (all relative to the ground), for the second case, (b). Show that linear momentum has been conserved.

40. What mass ratio m_o/m would be required to accelerate a force-free rocket from rest to $v = 3 \times 10^4$ m/s $= 10^{-4}\,c$ with the exhaust velocity u_o = 5000 m/s?

41. A rocket in outer space (force free) starts from rest and ejects mass at a constant rate with speed u_o relative to the rocket. When the rocket achieves a speed $v = u_o$, what mass fraction has been ejected?

42. A large rocket with an exhaust velocity u_o = 3000 m/s develops a thrust of 24×10^6 N (about 5.4×10^6 lb). How much mass is being blasted out the rocket exhaust per second?

43. What is the maximum velocity a rocket starting from rest in a force-free environment can acquire if $u_o = 3 \times 10^3$ m/s and 90% of the initial mass of the rocket is fuel?

44. A force-free two-stage rocket (see Question 10) is constructed so that 90% of the total original mass is fuel for the first stage. Of the remaining 10% of the original mass, half is the total mass of the second stage and half is mass to be jettisoned when the first-stage burn is complete. The second-stage rocket separately is also 90% (by mass) fuel. Each rocket engine has an exhaust velocity of 2500 m/s. Determine the maximum speed of the second stage, assuming that its initial speed equals the maximum speed of the first stage and that the first stage starts from rest.

chapter 11

CONSERVATION OF ANGULAR MOMENTUM

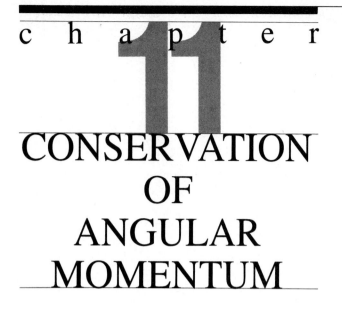

Preview

The concept of linear momentum, described in the preceding chapter, is special because, like energy, it is associated with a conservation principle. Angular momentum is the rotational analog of linear momentum and is also associated with a conservation principle. Kepler's second law of planetary motion is an example of angular momentum conservation. An ice skater's rotational motion also demonstrates the conservation of angular momentum. When unstable nuclei decay with the emission of particles, angular momentum is conserved. The conservation of angular momentum is a universal principle.

In this chapter we introduce the law of conservation of angular momentum, consider the criteria for its validity, and illustrate its scope with varied examples. We begin with the definition of angular momentum for a particle, and then extend this definition to a system of particles. Using the center-of-mass concept, we then express the angular momentum of a system of particles as a sum of two types of angular momenta, spin and orbital. This separation of angular momentum into two types simplifies drawing conclusions about the rotational motion of the system as a whole.

11.1 Angular Momentum of a Particle

For a particle having linear mometum **p** located at the point Q (Figure 11.1a) with position vector **r** we define the angular momentum vector, **L**, by the equation

$$\mathbf{L} = \mathbf{r} \times \mathbf{p} \tag{11.1}$$

The angular momentum vector **L** is perpendicular to the x–y plane determined by **r** and **p**, and in a direction determined by the right-hand rule for the cross product (see Section 2.5). Thus **L** is directed along the positive z-axis, as shown in Figure 11.1b. If the particle has mass m and is nonrelativistic, $\mathbf{p} = m\mathbf{v}$. Therefore, we can write

$$\mathbf{L} = m\mathbf{r} \times \mathbf{v} \tag{11.2}$$

From this equation we see that the units of **L** are kilogram · meters^2 per second, or kg · m^2/s. We will use these units throughout this text. Because 1 joule · second equals 1 kilogram · meter^2 per second (1 J · s = 1 kg · m^2/s), an alternate unit of angular momentum is the joule · second. This unit is used for angular momentum in modern physics.

We can see from Eqs. 11.1 and 11.2 that the angular momentum vector of a particle depends on the location of the origin through the dependence of **L** on **r**. Thus the angular momentum vector, **L**′, of the particle in Figure 11.2 relative to the origin O′ differs, in general, from the

Figure 11.1

A particle whose linear momentum is $\mathbf{p}$ is located at the point Q. The x-axis is chosen to be parallel to $\mathbf{r}$, and the x–y plane is chosen to be the plane determined by the vectors $\mathbf{r}$ and $\mathbf{p}$. According to the right-hand rule, the angular momentum vector $\mathbf{L} = \mathbf{r} \times \mathbf{p}$ for the particle is directed along the positive z-axis.

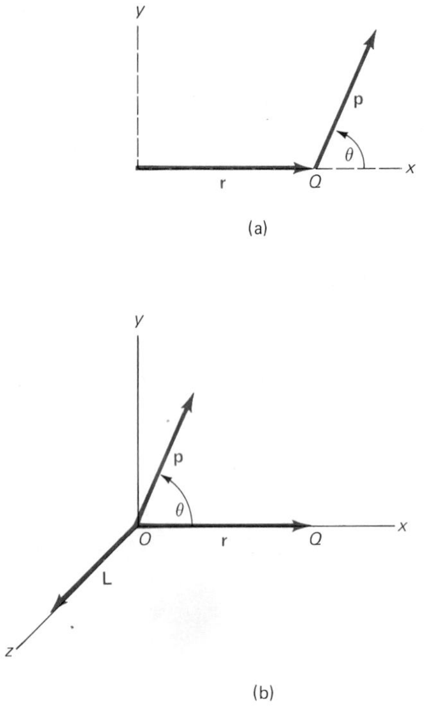

Figure 11.2

A particle whose linear momentum is $\mathbf{p}$ is located at the point Q. The two position vectors $\mathbf{r}$ and $\mathbf{r}'$ are for the two origins. In general the two angular momentum vectors $\mathbf{L} = \mathbf{r} \times \mathbf{p}$ and $\mathbf{L}' = \mathbf{r}' \times \mathbf{p}$ will not be equal.

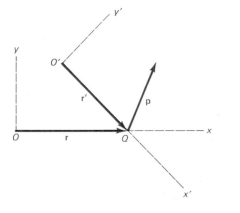

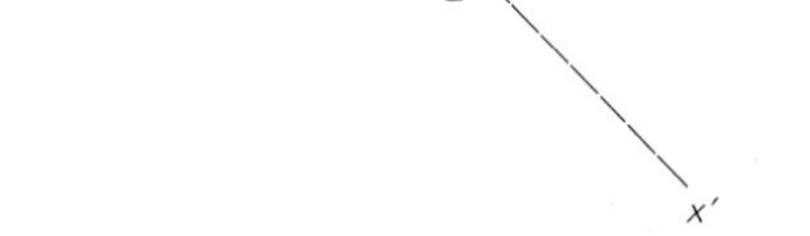

angular momentum vector, $\mathbf{L}$, of the same particle relative to the origin at O. Note that if $\mathbf{p}$ is directed toward or away from the origin, then the angular momentum equals zero relative to this origin.

The magnitude of the angular momentum vector is given by

$$L = rp \sin \theta \qquad (11.3)$$

Figure 11.3

A particle with linear momentum $\mathbf{p}$ is located by the position vector $\mathbf{r}$. The component of $\mathbf{p}$ perpendicular to $\mathbf{r}$ is labeled $p_\perp$ and the component of $\mathbf{r}$ perpendicular to $\mathbf{p}$ is labeled $r_\perp$. The magnitude of the angular momentum vector $\mathbf{L}$ may be written $L = r_\perp p = rp_\perp$.

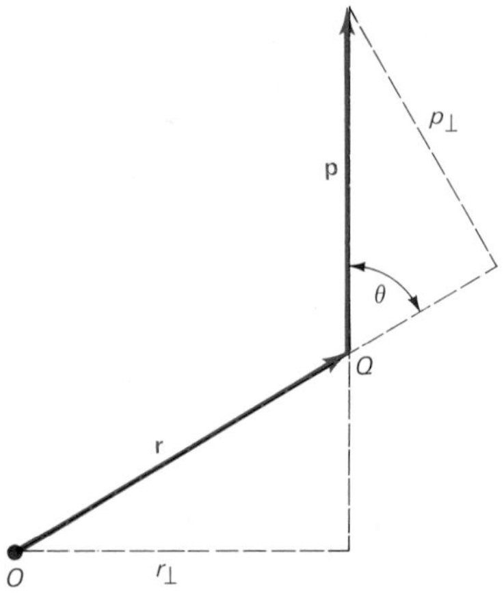

in accordance with the cross-product definition. This can also be written as

$$L = r_\perp p = rp_\perp \qquad (11.4)$$

where $r_\perp = r \sin \theta$ is the projection of $\mathbf{r}$ on a line perpendicular to $\mathbf{p}$, and $p_\perp = p \sin \theta$ is the projection of $\mathbf{p}$ on a line perpendicular to $\mathbf{r}$. The geometry of these relations is illustrated in Figure 11.3.

Motion with Constant Velocity

A particle of mass m moves in a straight line with constant speed v. We want to prove that such a particle has constant angular momentum relative to an origin fixed at O. First we will show the direction of $\mathbf{L}$ is constant.

Consider the particle at the two positions described by $\mathbf{r}_1$ and $\mathbf{r}_2$. We take $\mathbf{r}_1$ to denote the position along the particle's path nearest to O, and $\mathbf{r}_2$ to denote an arbitrary position along the path. We assign the symbol b to the magnitude of the vector $\mathbf{r}_1$. Because the particle has constant velocity, $\mathbf{v}_1 = \mathbf{v}_2$ and we have $\mathbf{p}_1 = \mathbf{p}_2$. From the definition of angular momentum (Eq. 11.1), $\mathbf{L}_1 = \mathbf{r}_1 \times \mathbf{p}_1$ and $\mathbf{L}_2 = \mathbf{r}_2 \times \mathbf{p}_2$. Using the right-hand rule, we find that both $\mathbf{L}_1$ and $\mathbf{L}_2$ are directed into the page. Indeed, $\mathbf{L}$ is directed into the page for all particle positions along this path, and thus $\mathbf{L}$ is constant in direction.

Now we consider the magnitude of $\mathbf{L}$. To obtain the magnitude of $\mathbf{L}_1$ and of $\mathbf{L}_2$ we use Eq. 11.3 to get

$$L_1 = r_1 p_1 \sin 90° = mv_1 r_1$$

$$L_2 = r_2 p_2 \sin \theta = mv_1 r_2 \sin \theta$$

From Figure 11.4, $r_1 = r_2 \sin \theta = b$. Therefore we can substitute b into both equations to get

$$L_1 = mvb$$

$$L_2 = mvb$$

and

$$L_1 = L_2 = mvb$$

Figure 11.4

A particle of mass m moves with constant speed v in a straight line. The perpendicular distance beween the origin O and this straight line is called the impact parameter b. As drawn, the angular momentum vector **L** is directed into the figure and has the magnitude $L = mvb$.

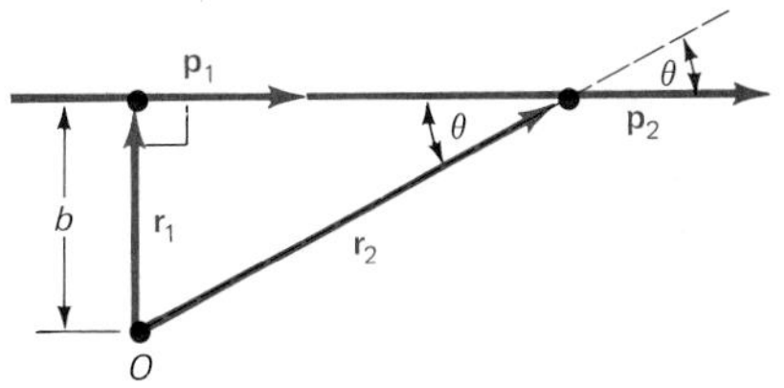

where v is the constant speed of the particle. Hence the magnitude of **L** is constant.

Because both the magnitude and direction of the angular momentum are constant, we have shown that the vector **L** is constant. Note that if the linear momentum of this particle varies in magnitude while maintaining a fixed direction, then the angular momentum of the particle is no longer constant.

Uniform Circular Motion

We have just proved that a particle moving in a straight line with constant speed v has a constant angular momentum relative to a fixed origin. Now we consider a particle moving in a circular path at constant speed (Figure 11.5). Such a particle will have constant angular momentum relative to an origin at the center of the circle. Both **r** and **p** have constant magnitudes, and vary in direction with a constant angle of 90° between them. Thus the vector $\mathbf{L} = \mathbf{r} \times \mathbf{p}$ has the constant direction perpendicular to the plane of the path. The direction of **L** will be into the page for clockwise motion and out of the page for counterclockwise motion. The constant magnitude of **L** is given by

$$L = rp = mvr \tag{11.5}$$

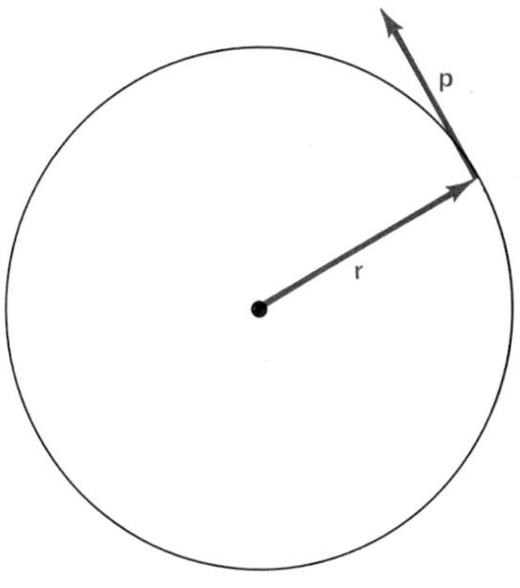

Figure 11.5

A particle moves in a circular path with constant speed v. As drawn, the angular momentum vector **L** is directed out of the figure and has the magnitude $L = mvr$.

Example 1 **Orbital Angular Momentum of the Moon**

We can use the angular momentum calculation for circular motion to evaluate the angular momentum of the moon relative to the earth. We take $r = 3.84 \times 10^8$ m, $T = 2.36 \times 10^6$ s, and $m = 7.35 \times 10^{22}$ kg.

The speed of the moon in its orbit is given by

$$v = \frac{2\pi r}{T}$$

$$= \frac{2\pi \times 3.84 \times 10^8}{2.36 \times 10^6}$$

$$v = 1.02 \times 10^3 \text{ m/s}$$

Thus the linear momentum of the moon has a magnitude

$$p = mv$$

$$= (7.35 \times 10^{22})(1.02 \times 10^3)$$

$$= 7.50 \times 10^{25} \text{ kg} \cdot \text{m/s}$$

The angular momentum therefore has a magnitude

$$L = rp$$

$$= (3.84 \times 10^8)(7.50 \times 10^{25})$$

$$= 2.88 \times 10^{34} \text{ kg} \cdot \text{m}^2\text{/s}$$

and, according to the right-hand rule, is directed out of the page (Figure 11.6). This corresponds to a direction toward the north polar star relative to the solar system.

We have just considered two examples of motion in which the angular momentum is constant, or conserved. Angular momentum is conserved in planetary motion, in the motion of spinning figure skaters, and in the scattering of alpha particles by atomic nuclei. Yet angular momentum is not always conserved. What criterion determines whether the angular momentum will be constant or will change with time? To answer this question we must determine what governs the change in angular momentum with time. We start with the defining equation for **L** for a particle;

$$\mathbf{L} = \mathbf{r} \times \mathbf{p} \tag{11.1}$$

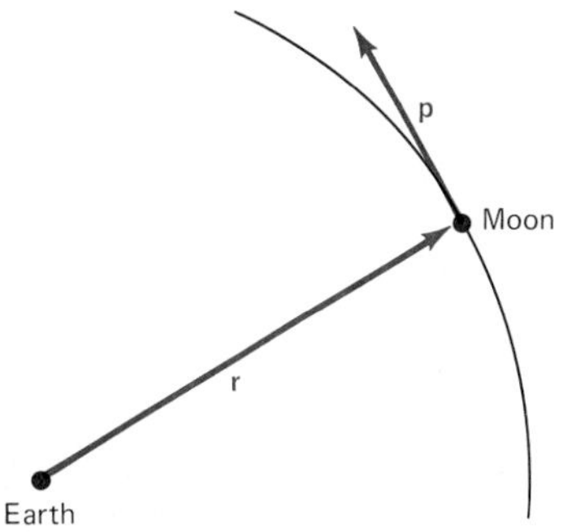

Figure 11.6

For circular motion with a linear momentum of constant magnitude p the moon has an angular momentum vector **L** directed out of the figure and of magnitude $L = rp$.

and differentiate it to get $d\mathbf{L}/dt$, the time rate of change of angular momentum.

$$\frac{d\mathbf{L}}{dt} = \mathbf{r} \times \frac{d\mathbf{p}}{dt} + \frac{d\mathbf{r}}{dt} \times \mathbf{p} \qquad (11.6)$$

Because $\mathbf{v} = d\mathbf{r}/dt$ and $\mathbf{p} = m\mathbf{v}$, we can write the second term in this equation as

$$\frac{d\mathbf{r}}{dt} \times \mathbf{p} = \mathbf{v} \times (m\mathbf{v})$$

But $\mathbf{v} \times (m\mathbf{v}) = m(\mathbf{v} \times \mathbf{v}) = 0$, so the second term equals zero. Equation (11.6) then becomes

$$\frac{d\mathbf{L}}{dt} = \mathbf{r} \times \frac{d\mathbf{p}}{dt}$$

Now we want to convert Eq. 11.6 from a form that simply relates particle properties to a form that also includes the influence of the environment on the particle behavior. To do this we use Newton's second law,

$$\mathbf{F} = \frac{d\mathbf{p}}{dt} \qquad (11.7)$$

where $\mathbf{F}$ represents the net force acting on the particle. Substituting Eq. 11.7 into Eq. 11.6 gives

$$\frac{d\mathbf{L}}{dt} = \mathbf{r} \times \frac{d\mathbf{p}}{dt} = \mathbf{r} \times \mathbf{F}$$

From the definition of torque in Section 3.3 this equation becomes

$$\frac{d\mathbf{L}}{dt} = \boldsymbol{\tau}_{ex} \qquad (11.8)$$

where $\boldsymbol{\tau}_{ex}$ is the net torque acting on the particle.

Equation 11.8 is the rotational analog of Newton's second law of motion. Just as the net force in Eq. 11.7 determines the rate of change of the linear momentum of a particle, so the net torque in Eq. 11.8 determines the rate of change of the angular momentum of a particle. The net force represents the influence of the environment on the linear momentum of the particle, and the net torque represents the influence of the environment on the particle's angular momentum. This is why torque is a useful variable, and is introduced in mechanics. The analogy between the linear and angular forms of Newton's second law is summarized in Table 11.1.

Conservation of Angular Momentum

The law of conservation of angular momentum follows from Eq. 11.8.

Table 11.1

Linear			Angular
$\mathbf{p}$	momentum		$\mathbf{L}$
$\mathbf{F}$	force	torque	$\boldsymbol{\tau}$
$\mathbf{F} = \frac{d\mathbf{p}}{dt}$	law of motion		$\boldsymbol{\tau} = \frac{d\mathbf{L}}{dt}$

When the net torque equals zero, $d\mathbf{L}/dt$ is zero; the angular momentum of the particle does not change—it remains constant, and is thus conserved.

If

$$\boldsymbol{\tau} = 0$$

then

$$\mathbf{L}_i = \mathbf{L}_f \quad \text{or} \quad \mathbf{L} = \text{constant} \qquad (11.9)$$

This is the law of conservation of angular momentum for a particle. The criterion that must be satisfied if angular momentum is to be conserved is that the net torque equals zero.

The angular momentum of a particle is a particle property in the same sense as is its position, velocity, or linear momentum. If the position and linear momentum vectors of a particle are known at all times, then the equation $\mathbf{L} = \mathbf{r} \times \mathbf{p}$ can be used to calculate the particle's angular momentum at all times. The calculation will reveal whether $\mathbf{L}$ is constant, and therefore whether the angular momentum is conserved. Although nothing was explicitly stated about the net torque in Example 1, the angular momentum was found to be conserved. Equation 11.8 implies that the net torque must be zero.

The case where the net force is a central force merits special consideration. A central force is one whose direction is always toward or away from a fixed point. If $\mathbf{r}$ is the vector from the force center to the particle, then $\mathbf{F}$ is along $\mathbf{r}$. Electrical and gravitational forces provide examples of central forces. When $\mathbf{F}$ and $\mathbf{r}$ are parallel (or antiparallel), their cross product is zero. Therefore, for a central force, $\mathbf{r} \times \mathbf{F} = 0$ and $\boldsymbol{\tau} = 0$. The angular momentum of a particle moving under the influence of a central force is conserved (Figure 11.7).

Example 2 Alpha-Particle Scattering

In Section 9.4 we discussed scattering experiments in which alpha particles of mass m and velocity $\mathbf{v}$ were directed at thin gold foils. As a result of their collisions with the gold nuclei, the alpha particles were scattered through various angles. A typical scattering event is depicted in Figure 11.8. The alpha-particle trajectory is the heavy curved line AB, and the nucleus is located at the point P. The position vector $\mathbf{r}$ and the repulsive electrical force $\mathbf{F}$ are drawn in Figure 11.8 for the alpha particle when it is

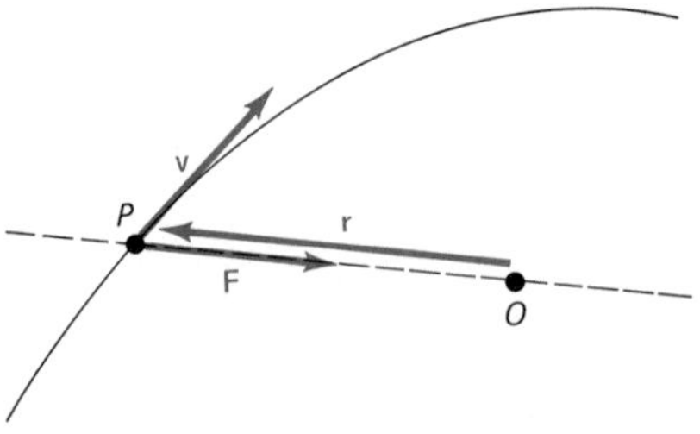

Figure 11.7

The force $\mathbf{F}$, a central force, is always directed toward (or away from) the fixed point O. The torque $\boldsymbol{\tau} = \mathbf{r} \times \mathbf{F}$ equals zero because $\mathbf{r}$ and $\mathbf{F}$ are antiparallel (parallel), so the angular momentum vector $\mathbf{L}$ does not change.

Figure 11.8

An alpha particle approaches a nucleus located at P along the line AA', and is deflected by the force **F** so that the particle leaves the impact region along the line BB'. The heavy curved path AQB represents the trajectory of the alpha particle while it is near the nucleus. The impact parameter b is the distance between the line AA' and a parallel line CC' drawn through the point P. The angle θ between the incoming and outgoing directions is the scattering angle.

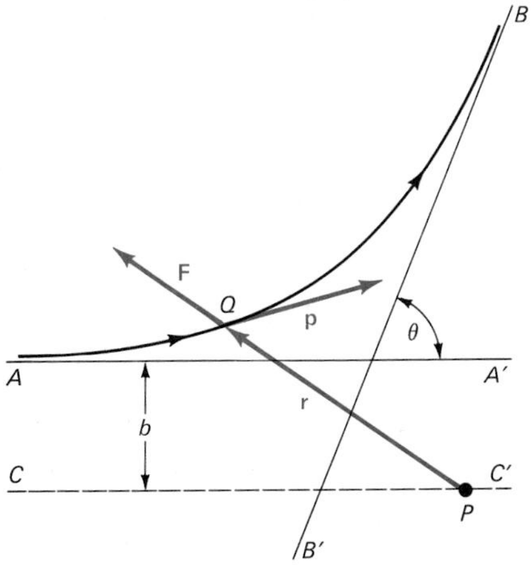

located at point Q. The angular momentum of the alpha particle can't be calculated directly from **r** and **p** because their values are not given. Since the electrical force is a central force, we know that **F** is parallel to **r**, and so $\boldsymbol{\tau} = \mathbf{r} \times \mathbf{F} = 0$. Thus we infer that **L** is constant.

The astronomer Tycho Brahe recorded precise naked-eye observations of planetary positions. Working with Brahe's data for Mars, Johannes Kepler (Section 6.6) fitted the trajectory of this planet to an ellipse with the sun at one focus. He generalized this result to all planets, and we know it today as Kepler's first law of planetary motion: Planets move in elliptic orbits with the sun located at one focus. Kepler's second law of planetary motion states that in equal time intervals the radius vector from the sun to a planet sweeps out equal areas. We can now show how Kepler's first two laws lead to a proof that the angular momentum of a planet is conserved.

Consider the part of the planetary orbit exhibited in Figure 11.9. When the planet moves from P to Q with velocity **v** the area swept out by **r** is ΔA. For a very small time

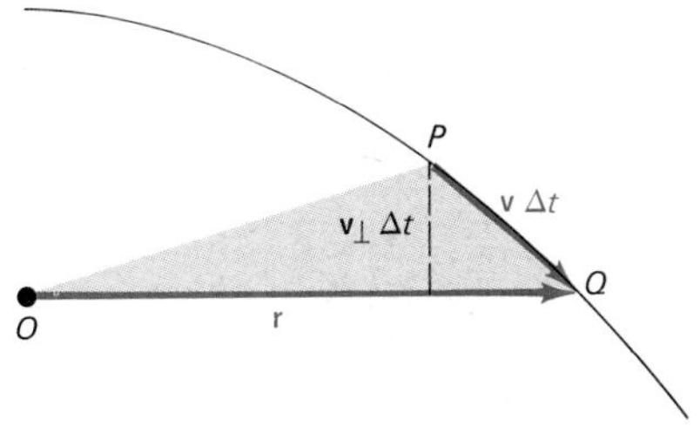

Figure 11.9

While a planet moves from P to Q the radius vector **r** from the sun sweeps out the shaded area.

interval, Δt, this area (shaded in Figure 11.9) approximates a triangle. The area of a triangle is $^1/_2$ (base)(altitude), so we have approximately

$$\Delta A \cong \frac{1}{2}(\text{base})(\text{altitude})$$

The base is approximately equal to r. To obtain the length of the altitude we note that the arc length PQ is approximately $v\ \Delta t$. If we project **v** on a line perpendicular to **r**, then the projected length of $(v\ \Delta t)$ will be the altitude of the triangle. Using this projected length, $v_\perp\ \Delta t$, we write for ΔA the expression

$$\Delta A \cong \left(\frac{1}{2}\right) r v_\perp\ \Delta t$$

Now we divide by Δt and take the limit as $\Delta t \rightarrow 0$ to obtain

$$\frac{dA}{dt} = \left(\frac{1}{2}\right) r v_\perp \tag{11.10}$$

In this limit the approximations made earlier become exact; dA/dt is the time rate at which the area is swept out by the radius vector.

Now we calculate the magnitude of **L**. Using Eq. 11.4 we can write

$$L = r p_\perp = r m v_\perp$$

so

$$r v_\perp = \frac{L}{m} \tag{11.11}$$

Substituting Eq. 11.11 into Eq. 11.10 gives

$$\frac{dA}{dt} = \frac{L}{2m} \tag{11.12}$$

Kepler's second law is a statement that dA/dt is constant, and is, by Eq. 11.12, equivalent to requiring that the magnitude of the angular momentum of the planet be constant.

Kepler's first law requires the planet to move in an elliptic orbit with the sun at one focus. Thus **r** and **p** are always in this orbital plane. The vector $\mathbf{L} = \mathbf{r} \times \mathbf{p}$ is perpendicular to this plane, as is consistent with the right-hand rule for the cross product. Hence **L** is always in the same direction. With both the magnitude and direction of **L** fixed, the angular momentum is constant. Taken together, Kepler's first two laws are equivalent to the law of conservation of angular momentum. We expect the angular momentum of a planet to be constant because the gravitational force is a central force.

Example 3 **Halley's Comet**

Halley's comet moves about the sun in an elliptic orbit (Figure 11.10) with perihelion distance $R_p = 0.59$ AU and an aphelion distance $R_a = 35$ AU (1 AU $= 1.50 \times 10^{11}$ m). Can we find the aphelion speed if we know the perihelion speed is $v_p = 5.4 \times 10^4$ m/s?

Because the sun's force of gravity acts along the line joining the sun and the comet, the angular momentum of the comet is conserved. The angular momentum at perihelion must equal the angular momentum at aphelion.

Figure 11.10

The angular momentum of Halley's comet may be written $L = mR_a v_a$ in terms of aphelion quantities or $L = mR_p v_p$ in terms of perihelion quantities. The two expressions for L are equal because angular momentum is conserved.

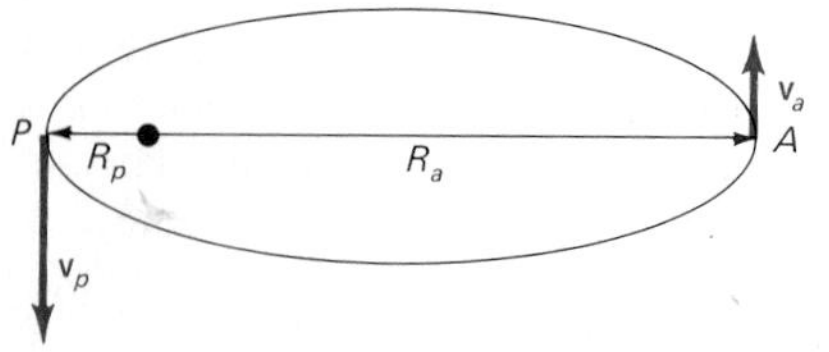

Applying 11.11 to perihelion and aphelion we write

$$mR_p v_p = mR_a v_a$$

or

$$v_a = v_p\left(\frac{R_p}{R_a}\right)$$

$$= \frac{5.4 \times 10^4 \times 0.59}{35}$$

$$= 9.1 \times 10^2 \text{ m/s}$$

The data are accurate to two significant figures, so the result is accurate to two significant figures.

To illustrate a physical situation where the angular momentum of a particle may or may not be conserved we next consider the motion of an ancient sling, such as that used by David to slay Goliath. For centuries this weapon competed favorably with the bow and arrow.

Example 4 **The Sling**

The sling consists of a stone and string. The geometry of this device is shown in Figure 11.11. Here the stone moves in a horizontal circle. Initially we consider motion such that the string passes through a fixed vertical axis of rotation. The thrower holds his hand motionless overhead, as suggested in Figure 11.11. In this case the force is central and the angular momentum is constant. To prove this, consider the forces acting on the stone. The horizonal component of the string tension, $\mathbf{T}_H$, is the only horizontal force acting on the stone. Only horizontal forces contribute to the vertical component of the torque. Thus $\mathbf{T}_H$ is the only force potentially contributing to the vertical component of the torque. But because $\mathbf{T}_H$ is a central force, the vertical component of torque equals zero. Because of the vector nature of Eq. 11.8 this implies that the vertical component of angular momentum is conserved.

Two vertical forces, $\mathbf{T}_v$ and $\mathbf{W}$, act on the stone. Because the motion of the stone is horizontal we have equilibrium in the vertical direction, and so these vertical forces are equal in magnitude and opposite in direction. The resultant torque of $\mathbf{T}_v$ and $\mathbf{W}$ is zero, and therefore the total angular momentum of the stone is also conserved.

Now let the stone move horizontally in a circle of the same radius, but with the string no longer directed toward the center of the circle at O. The thrower moves his hand in a horizontal circle of radius b overhead, as indicated in Figure 11.12. Thus the end of the string, labeled P, also moves in a small horizontal circle of radius b. This is the procedure that would be used to increase the speed of the stone in its rotation. Note that T_H is not a central force in this case. The external torque due to T_H no longer vanishes, but is given by

$$\tau = T_H b$$

Figure 11.11

The sling thrower's hand is held motionless, so the stone in the sling continues to move with constant speed in a horizontal circle.

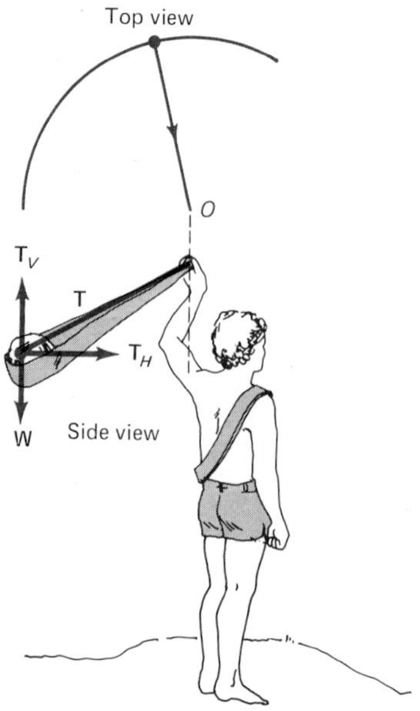

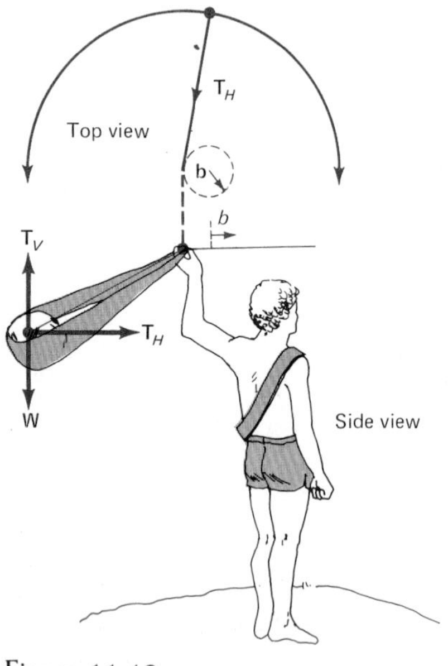

Figure 11.12

If the sling thrower moves his hand in a small circle (radius b), a torque is developed that changes the angular momentum of the stone in the sling. The stone then moves in a circular path with increasing speed.

Thus the vertical component of L changes with time—it is no longer conserved. This increase in L appears as an increase in the speed of the stone.

Questions

1. In nuclear interactions the distance b (Figure 11.4) is called the *impact parameter*. An accelerator produces particles of fixed (nonrelativistic) energy E. How will the angular momentum of the beam of particles depend on mass for a given b and E?
2. Two moons of identical mass are in stable circular orbits of different radii about the same planet. Which of the two moons has the larger angular momentum, the one closer to the planet or the one farther away? Explain.
3. The centripetal acceleration of an artificial satellite as it orbits the moon is caused by the force of gravity. Can this same force change the angular momentum of the moon? Explain.
4. Why is dA/dt equal to the ratio of the area of the ellipse to the orbital period for a Kepler ellipse?

11.2 Angular Momentum of a System of Particles

In the preceding section we defined angular momentum for the special case of a single particle. Now let us extend the angular momentum principles to a system of particles. We define the angular momentum of a system of particles to be the vector sum of the individual angular momentum vectors. Thus

$$\mathbf{L} = \sum \mathbf{L}_i = \sum \mathbf{r}_i \times \mathbf{p}_i \qquad (11.13)$$

defines the total angular momentum of a system of particles, where $\mathbf{r}_i$ and $\mathbf{p}_i$ are the position and linear momentum vectors, respectively, of the ith particle in the system.

For a single particle we derived the equation

$$\tau_{ex} = \frac{d\mathbf{L}}{dt} \qquad (11.8)$$

relating the external torque to the rate of change of angular momentum. To extend this relation to a system of particles requires a somewhat involved and abstract derivation. To simplify the algebra without compromising the argument we will consider a system having only two particles for part of the derivation, and then generalize the result to a system having any number of particles.

Let τ_i be the net torque acting on the ith particle. Then, according to Eq. 11.7, $\boldsymbol{\tau}_i = d\mathbf{L}/dt$, where $\mathbf{L}_i = \mathbf{r}_i \times \mathbf{p}_i$ is the angular momentum of the ith particle. The resultant torque, τ_{sys}, acting on the system is given by

$$\tau_{sys} = \sum \tau_i \qquad (11.14)$$

We will show that τ_{sys} has no net contribution from any internal torques. An internal torque is a torque exerted by one particle of the system on another particle of the system. If, for an isolated system, the sum of internal torques is not zero, then the system's angular momentum will change. An isolated system could set itself into rotation in these circumstances. This behavior is not observed in nature, however, and so we infer that the sum of internal torques is always zero. If the sum of internal torques equals zero, then $\tau_{sys} = \tau_{ex}$ where τ_{ex} includes the sum of all external torques. As a check on our inference, let us prove that the sum of internal torques is zero for a system of two particles interacting through a central force.

Suppose $\mathbf{F}_1$ is the central force exerted on particle 1 by particle 2, and $\mathbf{F}_2$ is the central force exerted on particle 2 by particle 1. For the sum of internal torques we write,

$$\tau_1 + \tau_2 = \mathbf{r}_1 \times \mathbf{F}_1 + \mathbf{r}_2 \times \mathbf{F}_2$$

Newton's third law requires that $\mathbf{F}_1 = -\mathbf{F}_2$. Using the fact that $\mathbf{F}_1 = -\mathbf{F}_2$, we obtain for the internal torque sum

$$\tau_1 + \tau_2 = (\mathbf{r}_2 - \mathbf{r}_1) \times \mathbf{F}_2$$

As shown in Figure 11.13, the vector $(\mathbf{r}_2 - \mathbf{r}_1)$ is along the line joining the particles, and because the forces are central, is therefore parallel to $\mathbf{F}_2$. As a consequence, the vector product $(\mathbf{r}_2 - \mathbf{r}_1) \times \mathbf{F}_2$ equals zero, so the sum of internal torques equals zero.

Observation shows that a body does not change its angular momentum spontaneously when it experiences zero net external torque. This observation confirms zero as the sum of internal torques acting on the body.

We have shown that if Newton's third law is obeyed, and if the forces are central, then the sum of internal torques equals zero for the case of two particles. We assert that this is true for any number of particles, and therefore $\tau_{sys} = \tau_{ex}$. (Though not explicitly stated, this was assumed to be true in Section 3.4.) Thus we can rewrite Eq. 11.14 as

$$\tau_{ex} = \sum \tau_i \qquad (11.15)$$

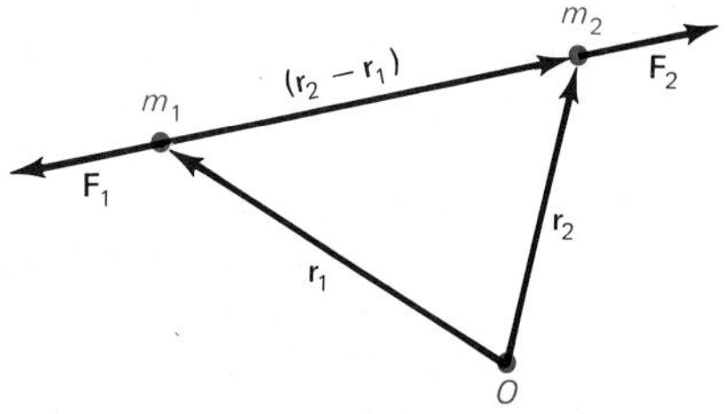

Figure 11.13

Particle 1 experiences a force $\mathbf{F}_1$ due to particle 2, and particle 2 experiences a force $\mathbf{F}_2$ due to particle 1.

To complete the derivation of Eq. 11.8 we substitute for τ_i in Eq. 11.15, using $\tau_i = d\mathbf{L}_i/dt$ for each particle. Thus

$$\tau_{ex} = \sum \tau_i = \sum \frac{d\mathbf{L}_i}{dt}$$

$$= \frac{d}{dt} \sum \mathbf{L}_i$$

Hence

$$\tau_{ex} = \frac{d\mathbf{L}}{dt} \qquad (11.8)$$

because $\mathbf{L} = \sum \mathbf{L}_i$. The net external torque acting on a system of particles determines the time rate of change of the total angular momentum of the system.

From Eq. 11.8 we can now extend the principle of angular momentum conservation to a system of particles.

If the net external torque acting on a system of particles equals zero, then the total angular momentum of the system is conserved.

In equation form, if

$$\tau_{ex} = 0$$

then

$$\mathbf{L} = \text{constant} \qquad (11.16)$$

This is the law of conservation of angular momentum for a system of particles.

A system experiencing zero external torque can produce striking changes in its own rotational motion through the action of its internal forces, yet the angular momentum of such a system is conserved. Insight into this behavior is provided by the expressions for the angular momentum of a particle:

$$L = rp_\perp = mrv_\perp \qquad (11.4)$$

where $v_\perp$ is the particle's tangential velocity and $p_\perp$ is the tangential component of linear momentum. For example, a figure skater begins her final spin about a vertical axis with her arms and legs as far from the axis of rotation (large r) as possible (Figure 11.14). If vertical torques are negligible, her angular momentum about a vertical axis is constant. As she draws her arms and legs closer to the spin axis, reducing r, her average tangential velocity must increase. Her speed of rotation increases just enough to keep her angular momentum constant.

The conservation of the linear momentum of a system and the conservation of the angular momentum are mutually independent. It is possible for a system to have its linear momentum conserved while its angular momentum changes, and vice versa. Example 5 illustrates these points.

Figure 11.14

To initiate her final spin a figure skater extends both arms and one leg as far from the axis as possible. To complete the spin she brings all parts of her body as close to the axis as possible.

Figure 11.15

A playground merry-go-round on which six children are riding is shown from above.

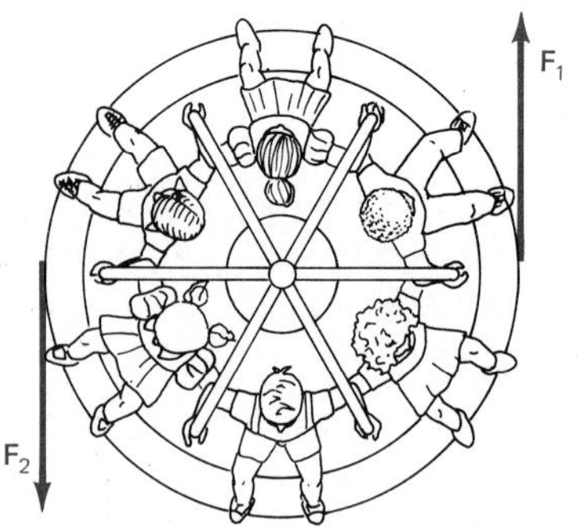

Example 5 **A Playground Merry-Go-Round**

The top view of a small playground merry-go-round with six passengers is depicted in Figure 11.15. To set it into rotation, two boys on opposite sides of the merry-go-round exert the forces $\mathbf{F}_1$ and $\mathbf{F}_2$ on it. Because $\mathbf{F}_1$ and $\mathbf{F}_2$ are equal in magnitude but oppositely directed, the net external force equals zero. Hence the linear momentum of the merry-go-round and its riders is conserved. However, the corresponding torque vectors, τ_1 and τ_2, are also equal in magnitude, and they are in the same direction. Thus the sum of external torques is not equal to zero. The angular momentum of the merry-go-round and its occupants changes with time.

Questions

5. Can an external force give rise to a torque that will change the relative angular momentum of two particles?

6. Suppose the action–reaction forces exerted on each other by a pair of particles resulted in a net torque. Describe a possible result.

11.3 Spin and Orbital Angular Momentum

Many systems exhibit two distinctive types of angular momentum. The electron in a hydrogen atom, for example, may possess two kinds of angular momentum. One is associated with the electron's orbital motion about the nucleus and the other is an intrinsic property of the electron that is unrelated to its motion about the nucleus. Analogously, the earth has two kinds of angular momentum. The angular momentum associated with the earth's orbital motion about the sun is called its *orbital angular momentum*. The angular momentum associated with the spinning of the earth about its own axis is called *spin angular momentum*. It is useful to be able to refer separately to these two angular momenta.

Our goal is to show formally that for any mechanical system

$$\mathbf{L} = \mathbf{L}_o + \mathbf{L}_s \qquad (11.17)$$

where $\mathbf{L}$ is the total angular momentum of the system, $\mathbf{L}_o$ is the orbital angular momentum of the system, and $\mathbf{L}_s$ is the spin angular momentum of the system. To do this, we need to express $\mathbf{L}_o$ and $\mathbf{L}_s$ in terms of the position and linear momentum vectors of the individual particles of the system.

We begin with the defining equation for $\mathbf{L}$.

$$\mathbf{L} = \sum \mathbf{r}_i \times \mathbf{p}_i \qquad (11.13)$$

Next we introduce the center-of-mass variables (see Section 10.1). Let $\mathbf{r}_i$ and $\mathbf{p}_i$ be the position and linear momentum vectors of the ith particle relative to O', the center of mass. The position and velocity of the center of mass relative to O, an arbitrary origin, are $\mathbf{r}_{cm}$ and $\mathbf{v}_{cm}$, respectively. The position vectors are related according to

$$\mathbf{r}_i = \mathbf{r}_{cm} + \mathbf{r}_i' \qquad (11.18)$$

as shown in Figure 11.16. Differentiating Eq. 11.18 with respect to time gives

$$\mathbf{v}_i = \mathbf{v}_{cm} + \mathbf{v}_i' \qquad (11.19)$$

as the relation among velocity vectors. Multiplying Eq. 11.19 by the mass of the ith particle, m_i, leads to

$$\mathbf{p}_i = m_i\mathbf{v}_{cm} + \mathbf{p}_i' \qquad (11.20)$$

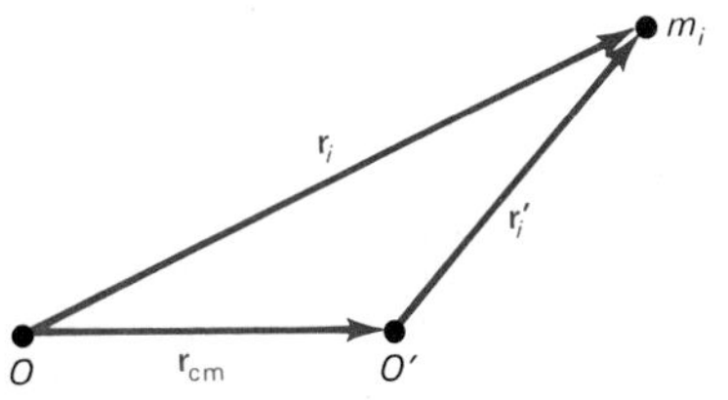

Figure 11.16

The triangle relation represents the vector equation $\mathbf{r}_i = \mathbf{r}_{cm} + \mathbf{r}_i'$.

where we have used $\mathbf{p}_i = m_i\mathbf{v}_i$ and $\mathbf{p}_i' = m_i\mathbf{v}_i'$. Now substitute for $\mathbf{r}_i$ and $\mathbf{p}_i$ in Eq. 11.13, using the expressions given for them in Eqs. 11.18 and 11.20, respectively. This results in the expression

$$\mathbf{L} = \sum (\mathbf{r}_{cm} + \mathbf{r}_i') \times (m_i\mathbf{v}_{cm} + \mathbf{p}_i') \qquad (11.21)$$

If we carry out the indicated cross multiplication (remember that $\mathbf{r}_{cm}$ and $\mathbf{v}_{cm}$ are not summed) we obtain four terms:

$$\mathbf{L} = \mathbf{r}_{cm} \times \left(\sum m_i\right)\mathbf{v}_{cm} + \mathbf{r}_{cm} \times \left(\sum \mathbf{p}_i'\right) + \left(\sum m_i\mathbf{r}_i'\right) \times \mathbf{v}_{cm} + \sum \mathbf{r}_i' \times \mathbf{p}_i' \qquad (11.22)$$

The second and third terms equal zero, as we now prove.

In the third term of Eq. 11.22 we have the sum $\sum m_i\mathbf{r}_i'$. By definition (see Section 10.1) this can be written

$$\sum m_i\mathbf{r}_i' = M\mathbf{r}_{cm}' \qquad (11.23)$$

where $\mathbf{r}_{cm}'$ is the position vector of the center of mass relative to an origin at the center of mass. Clearly $\mathbf{r}_{cm}'$ equals zero.

To prove that the second term in Eq. 11.22 equals zero, differentiate Eq. 11.23 to obtain

$$\sum m_i\mathbf{v}_i' = \sum \mathbf{p}_i' = M\mathbf{v}_{cm}'$$

Here $\mathbf{v}_{cm}'$ is the velocity of the center of mass relative to the center of mass at the origin. Clearly $\mathbf{v}_{cm}'$ also vanishes.

In the first term of Eq. 11.22 we note that $\sum m_i = M$, the total mass of the system. Thus Eq. 11.13 can be written

$$\mathbf{L} = M\mathbf{r}_{cm} \times \mathbf{v}_{cm} + \sum \mathbf{r}_i' \times \mathbf{p}_i' \qquad (11.24)$$

We define the orbital and spin angular momentum vectors as

$$\mathbf{L}_o = M\mathbf{r}_{cm} \times \mathbf{v}_{cm} \qquad (11.25)$$

and

$$\mathbf{L}_s = \sum \mathbf{r}_i' \times \mathbf{p}_i' \qquad (11.26)$$

respectively. Substituting these two equations into Eq. 11.24 shows that we can write $\mathbf{L} = \mathbf{L}_o + \mathbf{L}_s$. Notice that using the center-of-mass variables allowed us to resolve $\mathbf{L}$ into $\mathbf{L}_o$ and $\mathbf{L}_s$. This is another major reason why the center-of-mass concept is useful.

In Figure 11.17 a spinning planet is shown orbiting a star that is located at point O. The trajectory or orbit of the planet's center of mass is indicated by the dotted line. The orbital angular momentum vector, $\mathbf{L}_o = \mathbf{r} \times \mathbf{p}$, is perpendicular to the orbital plane as drawn. The spin angular momentum vector, $\mathbf{L}_s$, could be in any direction relative to $\mathbf{L}_o$, but in Figure 11.17 the planet is spinning in such a way that $\mathbf{L}_s$ and $\mathbf{L}_o$ are parallel.

To determine the direction of $\mathbf{L}_s$ for a spinning body in general, consider the earth as seen by someone looking down on the North Pole from the North Star (see Figure

Figure 11.17

From above the planet spins counterclockwise as it orbits its sun counterclockwise.

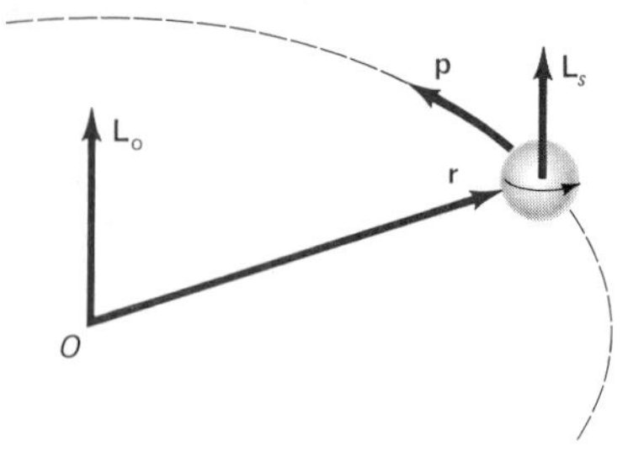

Figure 11.18

From the North Star the earth is seen to spin counterclockwise.

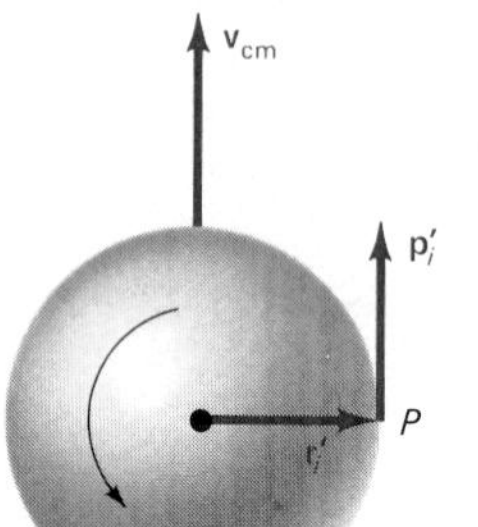

11.18). The spin direction of the earth is counterclockwise. Now choose a particle at the equator, and label its position $\mathbf{r}_i'$ and its linear momentum $\mathbf{p}_i'$ (relative to the center of mass). The direction of $\mathbf{L}_s$ for the earth is the direction of $\mathbf{r}_i' \times \mathbf{p}_i'$ of the particle. Thus $\mathbf{L}_s$ points out of the plane of this page or in the direction of the North Star from the earth.

The angular momentum of a system due to the motion of its center of mass is $\mathbf{L}_o$, the orbital angular momentum. Compare the defining equation, $\mathbf{L}_o = M\mathbf{r}_{cm} \times \mathbf{v}_{cm}$, with Eq. 11.2, $\mathbf{L} = m\mathbf{r} \times \mathbf{v}$, for the angular momentum of a particle. To calculate the orbital angular momentum of a system of particles, we imagine all of the mass of the system to be concentrated at the center of mass and moving with the center-of-mass velocity. We then calculate the angular momentum of the resulting "particle."

The angular momentum of a system due to its motion relative to an origin at the center of mass is $\mathbf{L}_s$, the spin angular momentum.

When the external torque acting on a system of particles equals zero, the angular momentum of the system is conserved. This means that $\Delta\mathbf{L} = 0$ and, according to Eq. 11.17,

$$\Delta\mathbf{L}_s = -\Delta\mathbf{L}_o \tag{11.27}$$

Therefore, any change in the spin angular momentum of a system whose total angular momentum does not change must be accompanied by an equal and opposite change in the orbital angular momentum of the system.

Angular Momentum in the Earth–Moon System

Consider the earth and its moon as an isolated system. The earth spins on its axis and the moon orbits the earth. The earth and moon each orbit about the system center of mass, C, as suggested by the dotted trajectories of their centers of mass (E and M) drawn in Figure 11.19. Additionally, the earth and moon each spin about their own axes. The total angular momentum of the system thus has two spin components and two orbital components. The spin angular momentum of the moon and the orbital angular momentum of the earth are small, and can be neglected. Therefore the system angular momentum, $\mathbf{L}$, can be represented by

$$\mathbf{L} = \mathbf{L}_o + \mathbf{L}_s$$

where $\mathbf{L}_o$ is the oribital angular momentum of the moon and $\mathbf{L}_s$ is the spin angular momentum of the earth.*

According to Eq. 11.27 any change in the spin angular momentum of the earth is accompanied by an equal and opposite change in the orbital angular momentum of the moon, $\Delta\mathbf{L}_s = -\Delta\mathbf{L}_o$. The tides on the earth are a mechanism for changing $\mathbf{L}_s$, and hence for changing $\mathbf{L}_o$.

The gravitational force, $\mathbf{F}_g$, exerted by the moon on the earth gives rise to earth tides. A tide is a large-scale bulge in the atmosphere, crust, or ocean resulting from the variation of $\mathbf{F}_g$ with distance. The mechanism that produces tides can be understood by considering Figure 11.20. Let point 1 be near the sublunar point, point 2 the center of mass of the earth, and point 3 the far sublunar point on the earth; and suppose that F_1, F_2, and F_3 each represent the gravitational force exerted by the moon on a 1-kg mass located at the respective points. Because $F_g \propto$

*Our treatment of the earth–moon system in this text holds providing that we use the component of the earth's spin angular momentum that is perpendicular to the moon's orbital plane. In reality, however, the spin axis of the earth is not perpendicular to the moon's orbital plane.

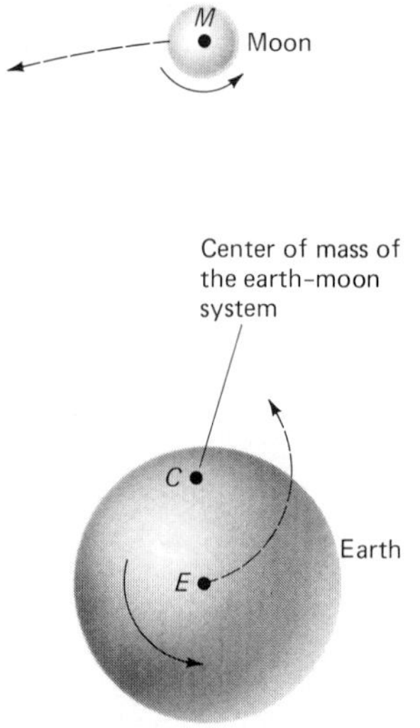

Figure 11.19

The earth and moon each orbit about their common center of mass at C. Additionally, each spins about its own axis.

$1/r^2$, we have $F_1 > F_2 > F_3$. *Relative to point 2*, the kilogram at point 1 experiences a slight attraction toward the moon and the kilogram at point 3 experiences a slight repulsion away from the moon. As a result, tidal bulges are produced on opposite sides of the earth, as suggested in Figure 11.20. At mid-ocean the bulges are of the order of 1 ft in height.

Without earth spin the tidal bulges would be located along the line joining the earth and moon, as suggested by Figure 11.20. When the effects of earth spin are included, however, two important changes occur: (1) The tides are swept eastward by the daily rotation of the earth, as shown in Figure 11.21. Thus the tidal bulges no longer align with the earth–moon direction as in Figure 11.20. (2) The earth rotates, whereas the tidal bulges do not. The relative motion between the tides and the earth acts as a brake on the earth's rotation. This tidal friction gives rise to a torque acting on the earth that reduces the spin angular momentum of the earth. In order to keep the total angular momentum of the system constant, this decrease in the spin angular momentum of the earth must be accompanied by an increase in the orbital angular momentum of the moon.

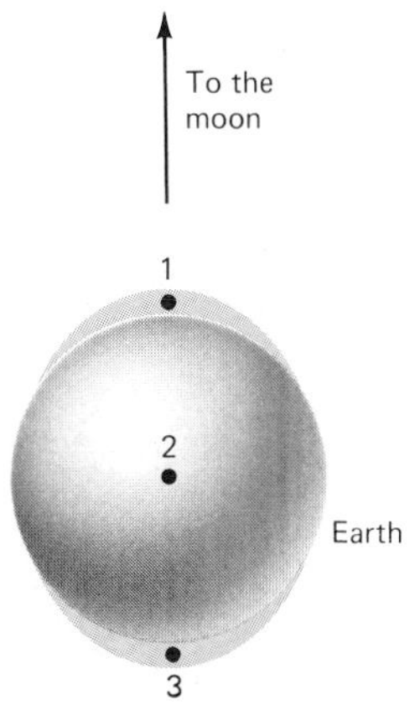

Figure 11.20

The gravitational attraction of the moon gives rise to tidal bulges on the near side (1) and far side (3) of the earth.

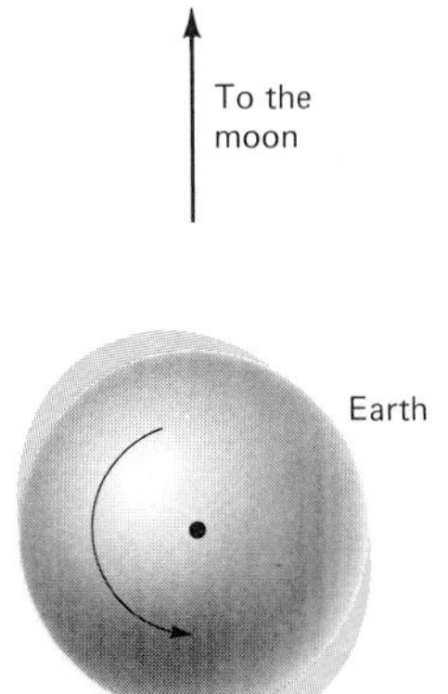

Figure 11.21

Friction between the earth and its tidal bulges causes the bulges to be swept around with the earth, to some extent, as it rotates.

Figure 11.22

The orbital angular momentum vector of the moon $\mathbf{L}_o$ and the spin angular momentum vector of the earth come out of the figure.

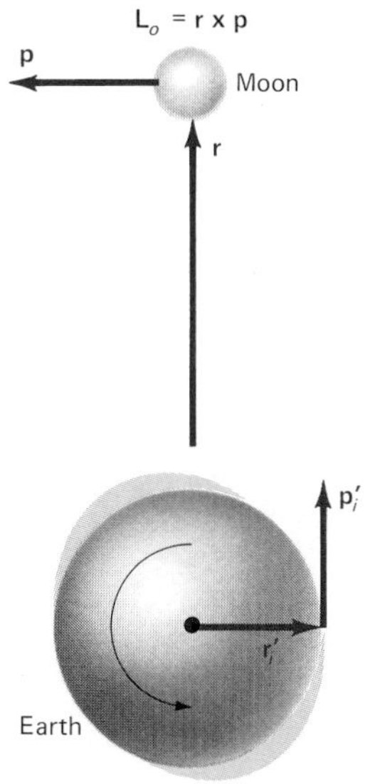

In Figure 11.22 the motions of the earth–moon system are shown as they would appear from the North Star. Evidently the orbital angular momentum of the moon, $\mathbf{L}_o = \mathbf{r} \times \mathbf{p}$, is directed out of the figure. The angular momentum of the ith particle on the equator, $\mathbf{r}_i' \times \mathbf{p}_i'$, is directed out of the figure; hence the spin angular momentum of the earth is also directed out of the figure. The vectors $\mathbf{L}_o$ and $\mathbf{L}_s$ are parallel. The total angular momentum of the system, $\mathbf{L} = \mathbf{L}_o + \mathbf{L}_s$, is constant. The decrease in $\mathbf{L}_s$ due to tidal action must result in an increase in $\mathbf{L}_o$ (note that here we refer to the magnitudes of $\mathbf{L}_o$ and $\mathbf{L}_s$). In Example 6 we explore the consequences of this increase.

Example 6 The Earth–Moon System

We take the angular momentum of the earth–moon system to include the spin angular momentum of the earth and the orbital angular momentum of the moon about the earth. We assume that the system is isolated so that $\tau_{ex} = 0$, and the total angular momentum is conserved. There are known tidal friction torques associated with the atmosphere, the ocean, and the crust of the earth. We postulate that the spin angular momentum L_s decreases as a result of these internal torques. Then the orbital angular momentum of the moon L_o must increase according to Eq. 11.27. What are the implications of this increase?

Applying Newton's second law and the law of gravity to the motion of the moon, we can write

$$\frac{GM_mM_E}{r^2} = \frac{M_mv^2}{r}$$

This is a statement that the force of gravity provides the centripetal force holding the moon in its orbit. Multiplying by M_mr^3 we obtain

$$M_m^2r^2v^2 = GM_m^2M_Er$$

Figure 11.23

Coral-band studies indicate that the day was shorter in length in the past than it is today. The day was only 20 hours long 700 million years ago.

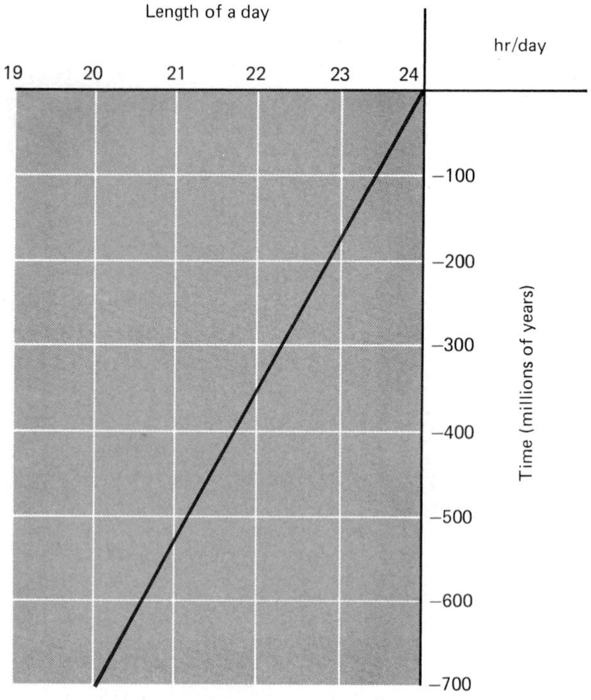

The left side of this equation is simply L_o^2, the square of the orbital angular momentum of the moon. Therefore we have

$$L_o^2 = (GM_m^2M_E)r$$

If L_o^2 increases, then r must increase, because the quantity in parentheses is constant. Consequently, if L_s is reduced, L_o increases and r increases. Or simply, as the spin of the earth slows down, the moon orbits the earth at a greater distance. The moon–earth distance in the past was less than it is today. Studies of coral bands (see Figure 11.23) indicate that in the past the earth's spin angular momentum was greater than it is today. The earth's spin is decreasing and the length of the day is said to be increasing at the rate of 0.0015 s/century.

Let us take another look at Eqs. 11.24–11.26, and consider their implications.

$$\mathbf{L} = M\mathbf{r}_{cm} \times \mathbf{v}_{cm} + \sum \mathbf{r}_i' \times \mathbf{p}_i' \qquad (11.24)$$

$$\mathbf{L}_o = M\mathbf{r}_{cm} \times \mathbf{v}_{cm} \qquad (11.25)$$

$$\mathbf{L}_s = \sum \mathbf{r}_i' \times \mathbf{p}_i' \qquad (11.26)$$

First we note that if $\mathbf{r}_{cm} = 0$ or $\mathbf{v}_{cm} = 0$, then $\mathbf{L}_o = 0$ and $\mathbf{L} = \mathbf{L}_s$, according to Eq. 11.25. In either case the total angular momentum of the system equals the spin angular momentum. When $\mathbf{r}_{cm} = 0$ the origin is located at the center of mass itself. If $\mathbf{r}_{cm} = 0$, then $\mathbf{L} = \mathbf{L}_s$. When $\mathbf{v}_{cm} = 0$ the origin may be located anywhere, but the center of mass is at rest. If $\mathbf{v}_{cm} = 0$, then $\mathbf{L} = \mathbf{L}_s$. The angular momentum of a system whose center of mass is at rest is the same for all possible locations of the origin.

When isolated unstable particles decay, $\mathbf{F}_{ex} = 0$ and $\tau_{ex} = 0$. Thus both linear and angular momenta are conserved in the decay. The conservation principles of linear and angular momentum are particularly useful in deciding among possible modes of the decay of particles and in assigning values of L_s to the particles. These points are illustrated in Example 7.

Example 7 Decay of Positive Kaon

The positive kaon is an unstable particle with $\mathbf{L}_s = 0$ that decays into a neutral pion and a positive pion, both of which also have $\mathbf{L}_s = 0$. When the pion masses are taken to be equal, linear momentum conservation requires that the two pions emerge from the decay with equal speeds at a 180° angle, as illustrated in Figure 11.24. Because the pions travel away from the decay along the same straight line, this mode of decay has zero orbital angular momentum ($\mathbf{L}_o = 0$) and angular momentum is conserved.

A hypothetical alternate mode of decay is illustrated in Figure 11.25. In this hypothetical decay the pions travel

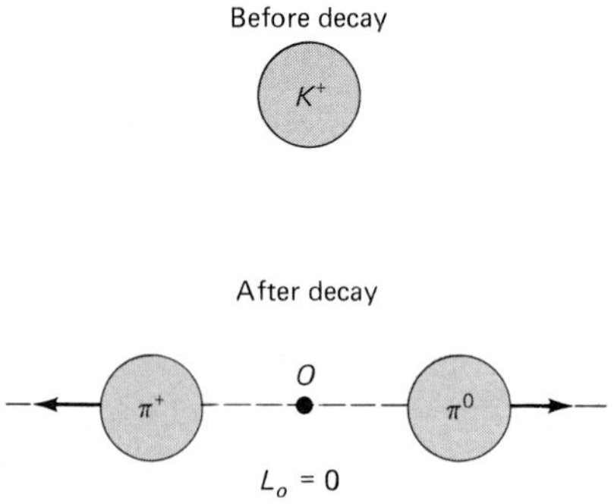

Figure 11.24

The kaon (K^+) is an unstable particle that spontaneously decays into a pion pair (π^+ and π^0). Here the pions emerge from the decay in opposite directions along the same line. This decay is observed.

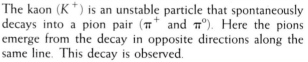

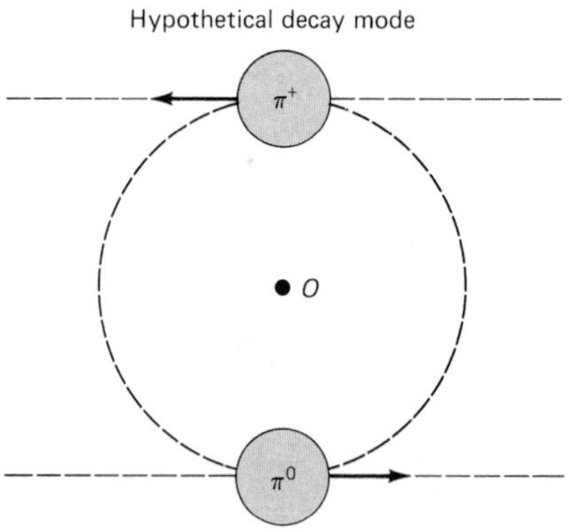

Figure 11.25

In this hypothetical decay of the kaon (see Figure 11.24) the pions emerge from the decay in opposite directions along different parallel lines. This decay is not observed.

away from the decay location along different parallel lines. The orbital angular momentum of the pion pair is not zero ($\mathbf{L}_o \neq 0$), whereas the spin angular momentum of the pion pair equals zero ($\mathbf{L}_s = 0$). Thus the total angular momentum of the pion pair is not zero. Because the original kaon angular momentum equals zero, the total angular momentum of the system changes in this hypothetical decay. Angular momentum is not conserved in such a decay.

Questions

7. Two ice skaters circle about a point while holding hands (Figure 11.26). At a certain moment both let go and move along straight lines. Are the two straight lines the same? Explain.
8. Two astronauts are connected by a rope, and each orbits about the center of mass. Assume that no forces act other than that of the rope. By pulling the rope in, the astronauts reduce their separation. Is the total angular momentum conserved? Explain.
9. Is the kinetic energy of the astronauts in Question 8 conserved? Explain.

Summary

The angular momentum of a particle is defined by

$$\mathbf{L} = \mathbf{r} \times \mathbf{p} \tag{11.1}$$

where $\mathbf{r}$ is the position vector and $\mathbf{p}$ is the linear momentum vector of the particle. The magnitude of $\mathbf{L}$ can be written

$$L = rp \sin\theta = rp_\perp = r_\perp p \tag{11.4}$$

where $r_\perp$ is the projection of $\mathbf{r}$ on a line perpendicular to $\mathbf{p}$, and $p_\perp$ is the projection of $\mathbf{p}$ on a line perpendicular to $\mathbf{r}$.

The rate of change of $\mathbf{L}$ with time is governed by

$$\tau_{ex} = \frac{d\mathbf{L}}{dt} \tag{11.8}$$

where τ_{ex} is the net torque acting on the particle. When $\tau_{ex} = 0$, then the angular momentum of the particle is conserved. If

$$\tau_{ex} = 0$$

then

$$\Delta\mathbf{L} = 0 \qquad \mathbf{L} = \text{constant}$$

The total angular momentum of a system of particles is defined by

$$\mathbf{L} = \sum \mathbf{L}_i = \sum \mathbf{r}_i \times \mathbf{p}_i. \tag{11.13}$$

The net internal torque due to the interparticle forces sums to zero, and the net external torque determines the time rate of change of angular momentum for the system of particles. Thus Eq. 11.8 is also satisfied for a system of particles where τ_{ex} is the net external torque, and $\mathbf{L}$ is total angular momentum of the system. Furthermore, if $\tau_{ex} = 0$, then the angular momentum of the system is conserved.

The total angular momentum of a system can be expressed as a sum of spin and orbital angular momentum. Thus

$$\mathbf{L} = \mathbf{L}_s + \mathbf{L}_o \tag{11.17}$$

where the spin angular momentum

$$\mathbf{L}_s = \sum \mathbf{r}_i' \times \mathbf{p}_i' \tag{11.26}$$

is the angular momentum relative to the center of mass, and the orbital angular momentum

$$\mathbf{L}_o = M\mathbf{r}_{cm} \times \mathbf{v}_{cm} \tag{11.25}$$

is the angular momentum of the total mass of the system located at the center of mass and moving with the center-of-mass velocity. Here $\mathbf{r}_{cm}$ and $\mathbf{v}_{cm}$ are the center-of-mass position and velocity vectors, respectively, and primes denote quantities relative to the center of mass.

If the angular momentum is conserved, $\Delta\mathbf{L} = 0$, and

$$\Delta\mathbf{L}_s = -\Delta\mathbf{L}_o \tag{11.27}$$

Any change in the spin angular momentum of a system whose total angular momentum is conserved is matched by an equal and opposite change in the orbital angular momentum of the system.

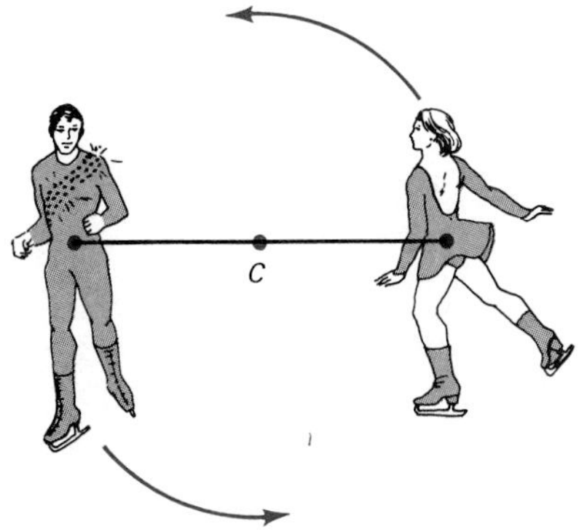

Figure 11.26

Two ice skaters.

Problems

Section 11.1 Angular Momentum of a Particle

1. A 1000-kg automobile travels on a circular racetrack 4 km in circumference. The driver completes one lap at 300 km/hr. What is the magnitude of the angular momentum of the car relative to the center of the racetrack?

2. A 1500-kg car travels at constant velocity (v = 80 km/hr) down an interstate highway. A farmer located at a point 100 m off the highway observes the car. Determine the angular momentum of the car (a) with respect to a car ahead, which is moving along the same line at 90 km/hr; and (b) with respect to the farmer.

3. For the car described in Problem 2 and shown in Figure 1, determine r and L when (a) $\theta = 90°$; (b) $\theta = 45°$; (c) $\theta = 0°$.

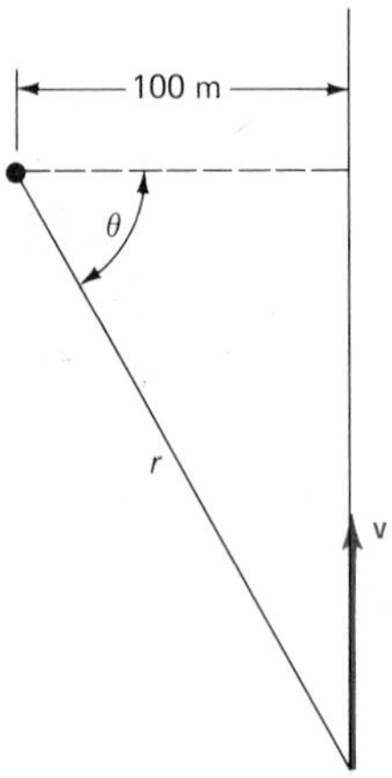

Figure 1

4. The alpha particle in Example 1 has an angular momentum equal to 3.17×10^{-34} J · s. Determine the value of the impact parameter (b in Example 1) if $v = 1.5 \times 10^7$ m/s and $m = 6.68 \times 10^{-27}$ kg.

5. An alpha particle has a kinetic energy of 6.60 MeV. It approaches a nucleus with an impact parameter of 3.12×10^{-15} m. Determine the magnitude of its angular momentum. Express your result as a multiple of $\hbar$ ($\hbar = 1.0546 \times 10^{-34}$ J · s)

6. Assume the orbit of the earth to be circular. Use values for the earth's orbital speed and radius to determine the area per second swept out by the earth's orbital radius.

7. Use the result of Problem 6 to calculate the orbital angular momentum of the earth. Check your answer by using Eq. 11.2.

8. The ball shown in Figure 2 is allowed to swing in a vertical plane like a simple pendulum. At noon the plane of vibration is north–south. Ignore the rotation of the earth. (a) Is the angular momentum of the ball conserved? (b) Calculate the direction of **L** at some time. Does it change? (c) What force acting on the pendulum always gives zero torque about an axis perpendicular to the motion plane and through the point of support? (d) Calculate the torque due to the weight of the ball about this axis at the angle θ. (Let l = pendulum length.) (e) Calculate the magnitude of the rate of change of angular momentum of the pendulum bob at θ.

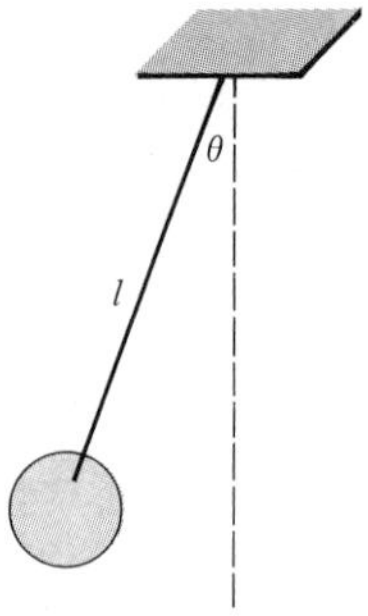

Figure 2

9. Prove that the orbit of a particle lies in a plane if its angular momentum is constant. (*Hint:* Evaluate $\mathbf{r} \cdot \mathbf{L}$.)

10. Using the definition $\mathbf{L} = \mathbf{r} \times \mathbf{p}$, prove that the direction of **L** is constant for the alpha particle in Example 2.

11. In Figure 11.9 call the vector from O to Q $(\mathbf{r} + \Delta\mathbf{r})$, the vector from O to P $\mathbf{r}$, and the vector from P to Q $\Delta\mathbf{r} = \mathbf{v}\,\Delta t$. Use the cross-product interpretation of area to prove that

$$\frac{dA}{dt} = \frac{1}{2}|\mathbf{r} \times \mathbf{v}|$$

from which Eq. 11.10 readily follows.

12. A comet orbits the sun in an elliptic orbit. The ratio of aphelion radius to perihelion radius is $r_a/r_p = 10$. If the perihelion speed is $v_p = 3 \times 10^5$ m/s, what is its aphelion speed?

13. A pendulum bob (Figure 3) of mass m is set into motion in a circular path in a horizontal plane as shown. During the motion, the supporting wire of

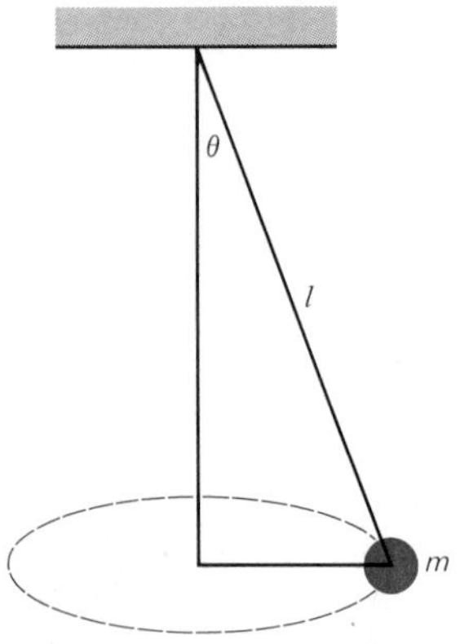

Figure 3

length l maintains the constant angle θ with the vertical. Prove that

$$L^2 = m^2 g l^3 \frac{\sin^4 \theta}{\cos \theta},$$

where L is the magnitude of the angular momentum vector of the bob about a vertical axis through the point of support.

14. A 50-gm particle has the position $x = +2$ m and $y = -3$ m and velocity $v_x = 6$ m/s and $v_y = +2$ m/s. What is the particle's angular momentum?

15. A 1.5-kg particle moves with velocity $\mathbf{v} = 4.2\mathbf{i} - 3.6\mathbf{j}$ m/s. Determine its angular momentum relative to the origin when its position is $\mathbf{r} = 1.5\mathbf{i} + 2.2\mathbf{j}$ m.

16. A particle ($m = 1.2$ kg) located at $x = 2.5$ m, $y = 4.3$ m, $z = 0$ moves in the $+x$-direction at 12 m/s. Determine its angular momentum relative to the point $x = 5.0$ m, $y = -5.0$ m, $z = 0$.

17. The pendulum of Problem 13 consists of a 0.402-kg bob on a wire of length 1.23 m. For $\theta = 10°$ calculate (a) the angular momentum, L, and (b) the linear velocity, v.

Section 11.2 Angular Momentum of a System of Particles

18. Calculate the orbital angular momentum of the moon and of the earth as they orbit their common center of mass. Take each body to be a point mass located at its own center of mass. (The common center of mass is located at 4.66×10^6 m from the center of the earth.)

19. Two 80-kg ice skaters (Figure 4) move together along parallel lines separated by a distance of 2 m. Their speeds are 15 m/s. Determine the angular momentum of the system relative to an origin through points A, B, and C.

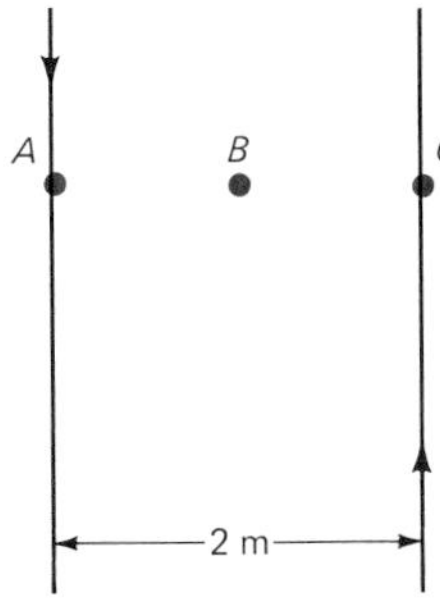

Figure 4

20. Two astronauts (Figure 5), each having a mass of 75 kg, are connected by a (massless) rope 10 m in length. They are isolated and orbiting their common center of mass with an orbital speed of 5 m/s. (a) Calculate the angular momentum of the system by treating the astronauts as particles. (b) Calculate the kinetic energy of the system. By pulling together on the rope, the astronauts shorten it to 5 m. (c) What is the new angular momentum of the system? (d) What is the new orbital speed? (e) What is the new kinetic energy of the system? (f) How much work (internal) was done by the astronauts in shortening the rope?

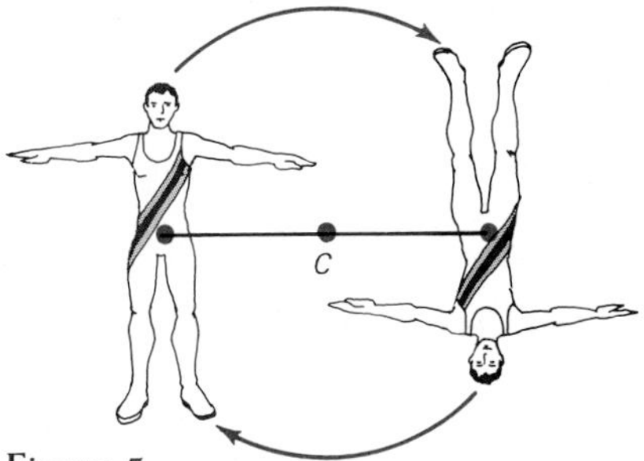

Figure 5

21. Two identical masses move in a circular orbit about a common center at opposite ends of a diameter of the circle. Show that

$$L^2 = \frac{G}{4} m^3 r$$

where r is the radius of the circular orbit. (*Hint:* Equate gravitational force and centripetal force.)

22. Consider a system of three identical particles. Each particle has an angular momentum of magnitude $\hbar = 1.0546 \times 10^{-34}$ J · s. Draw figures showing how the three angular momenta can be oriented to give a total angular momentum (vector sum) with magnitudes of 0, $\hbar$, $2\hbar$, and $3\hbar$.

23. Two 2-kg particles have the positions at $+2$ and -2 (on $x(m)$ axis) and velocities of $2\mathbf{j}$ m/s and $-2\mathbf{j}$ m/s, respectively. Calculate the angular momentum of the system about an axis parallel to the z-axis and passing through (a) the origin; (b) the point $x = +1$ m, $y = 0$; (c) the point $x = +2$ m, $y = 0$.

24. Two particles confined to the same plane move in circular paths about a common center. One particle ($m = 5.2 \times 10^{-3}$ kg) has an orbital speed of 31.0 m/s and an orbital radius of 2.1 m. The other particle ($m = 8.1 \times 10^{-3}$ kg) has an orbital speed of 26 1 m/s. (a) Determine the orbital radius of the "other particle" if the angular momentum of the system equals zero. (b) Determine the magnitude of the system angular momentum if the particles orbit in the same direction with radii and speeds the same as in part (a).

25. The velocity of a particle is $\mathbf{v} = 3.12\mathbf{i} + 4.82\mathbf{j}$ m/s. When the particle crosses the x-axis at $x = 2.13$ m, another identical particle traveling parallel to the x-axis crosses the y-axis at $y = 1.32$ m. What is the velocity of the second particle if the angular momentum of the system is zero?

Section 11.3 Spin and Orbital Angular Momentum

26. Treat the earth–moon system as isolated with all spin and orbital angular momentum vectors parallel. Neglect the spin angular momentum of the moon

and the orbital angular momentum of the earth. (a) Calculate the orbital angular momentum of the moon. (b) Take $L_s = 5.9 \times 10^{33}$ kg · m²/s for the earth today, and calculate the radius of the moon's orbit back when the length of the day was 10 hr instead of 24 hr. Assume that $L_s \propto 1/T$, where T is the length of the day. (*Hint:* Use Example 6.)

27. Continuing Problem 26 (with the same assumptions stated there), what will be the length of the day when the moon is 5% farther away than it is now?

28. The electron has a spin angular momentum $L_s = h/4\pi = 0.527 \times 10^{-34}$ kg · m²/s. (a) Determine its orbital angular momentum about the proton in hydrogen if $r = 0.529 \times 10^{-10}$ m and its tangential velocity is 2.20×10^6 m/s. (b) If spin and orbital angular momenta for the electron must be either parallel or antiparallel, what are possible values for the total angular momentum?

29. The orbital angular momentum of the earth, relative to the center of mass of the earth–moon system, is given by

$$L_E = M_E v r = \frac{2\pi M_E r_E^2}{T}$$

where M_E is the mass of the earth, r_E is the distance of the center of the earth from the center of mass of the earth–moon system, and T is the orbital period of the earth–moon system about its center of mass. A similar expression can be written for L_E, the orbital angular momentum of the moon. Show that

$$\frac{L_E}{L_M} = \frac{M_M}{M_E}$$

and use the tabulated masses of the moon and earth to evaluate the ratio.

30. A student attaches a string 1 m in length to each of two 0.1-kg particles. Holding the free ends of the string in one hand, the student walks at 0.5 m/s in a straight line while rotating the particles so that the strings form a straight line (Figure 6). If the center of mass of the strings is always 1.5 m above the floor, and if the orbital speed of each mass about the center of mass is 1.2 m/s, what is the total angular momentum of the system about an axis in the floor?

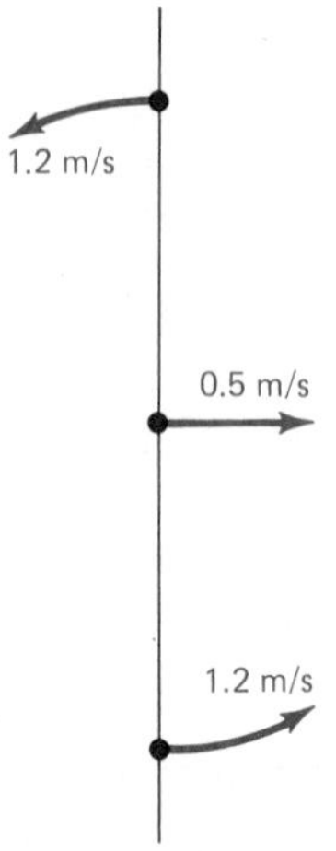

Figure 6

31. Two crates of mass m are connected by a (massless) rope of length h, and orbit the earth in a circular path aligned as shown in Figure 7. Let r_1 be the orbital radius for the crate nearest the earth, and (a) use the binomial theorem with the law of gravity to obtain the approximate expression.

$$T \cong \frac{2GM_E mh}{r_1^3}$$

for the tension, T, in the rope where M_E is the mass of the earth. (b) Calculate T if $m = 1$ kg, $h = 20$ m, and the orbit is 200 km in altitude. (c) For comparison, calculate the gravitational attractive force between the two crates. (d) Assume that the crates and earth form an isolated system, and determine the orbital radius for which the tidal repulsion [part (b)] equals the gravitational attraction [part (c)].

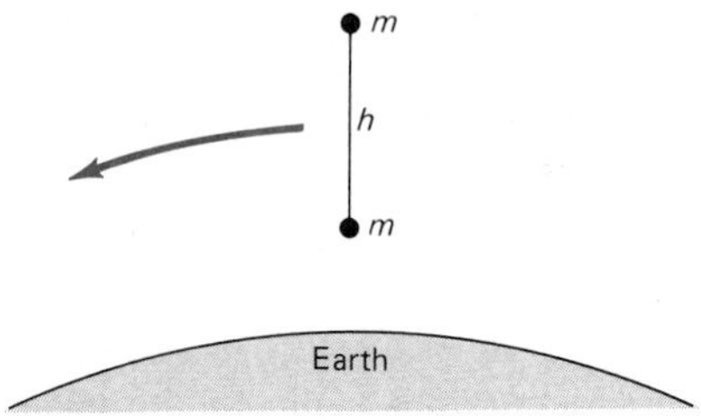

Figure 7

32. A pair of 2-kg particles are attached by a light string so that they rotate about their center of mass as they move along. At a certain moment their positions and velocities are as drawn in Figure 8. Calculate the (a) spin, (b) orbital, and (c) total angular momenta about the z-axis.

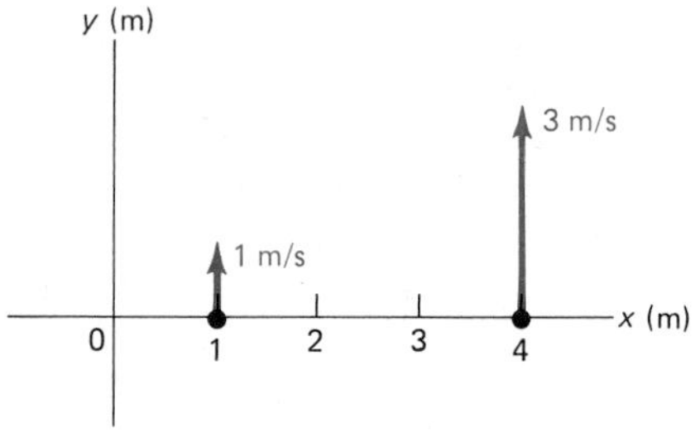

Figure 8

chapter 12

ROTATION OF A RIGID BODY

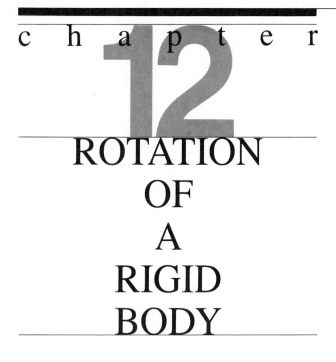

Preview

In Chapters 10 and 11 we developed general principles for treating the motion of mechanical systems that consist of many particles. A rigid body is an example of such a system. Many objects—such as nuclei, gyroscopes, bicycle wheels, and heavenly bodies—can be approximated as rigid. The motion of a rigid body reduces to a translation of its center of mass plus a rotation relative to its center of mass, and is therefore comparatively easy to describe. For these reasons rigid bodies merit special attention.

We can discover many aspects of rigid body motion by studying the motion of rigid bodies that have an axis of rotation that is fixed in direction. Spinning electrical machinery that is anchored in position and a majorette's airborne whirling baton are examples of this type of motion. In this chapter we introduce special variables and relations that are useful for dealing with rigid body motion where the axis of rotation is fixed in direction. We also derive a rotational form of Newton's second law applicable to this kind of motion.

12.1 Angular Kinematics

We have seen that the center-of-mass concept is useful in describing the linear momentum (Section 10.2) and angular momentum (Section 11.3) of a system of particles. The descriptions are simplified because motion relative to the center of mass is separated from the motion of the system "as a whole." For an arbitrary collection of particles no additional simplification is possible. However, when the particles form a rigid body, no internal motion occurs, and additional simplification *is* feasible. In this case, the only possible motion of the system is a motion of the center of mass plus a rotation relative to the center of mass. We have already developed the variables needed to describe the center-of-mass motion in Chapter 10. Now we introduce the variables required to describe the rotation of a rigid body relative to an axis through its center of mass.

Radian Measure

When dealing with the rotation of rigid bodies we must measure angles. Two common methods are used to establish a unit of angle. One method divides a circle into an integral number of equal parts. Thus one *revolution* and 360 *degrees* establish the revolution and degree, respectively, as units of angle. The second method defines one unit of angle to be the angle subtended by an arc equal in length to the radius, as shown in Figure 12.1. Because the arc length for a complete circle equals $2\pi r$, it follows that the radius divides the circumference into 2π parts. The value of this unit of angle, called the *radian*, is given in degrees by

$$1 \text{ radian} = \frac{360}{2\pi} \text{ degrees} \cong 57.3 \text{ degrees}$$

Figure 12.1

The angle subtended by an arc equal in length to the radius equals one radian.

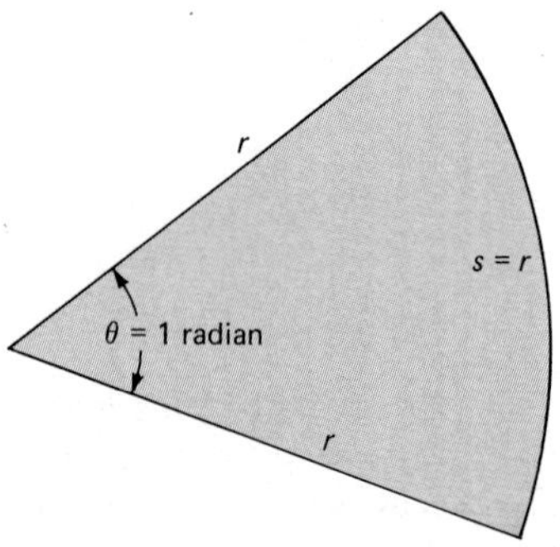

Many pocket calculators incorporate this conversion factor as a feature.

For a circle the arc length, s, is proportional to the angle it subtends (is opposite to), θ, and to the radius of the circle, r. If the angle is expressed in radians, this relation connecting s and r can be written

$$s = r\theta \qquad \text{or} \qquad \theta = \frac{s}{r} \tag{12.1}$$

Thus $s = r$ means $\theta = 1$ radian. If θ is expressed in degrees, this same relation must be written

$$s = \frac{\pi}{180} r\theta \tag{12.2}$$

The simpler form of Eq. 12.1 singles out the radian as the "natural" unit of angle. Test these equations yourself to determine whether the formula for the circumference of a circle, $C = 2\pi r$, is recovered.

By convention, we measure an angle counterclockwise relative to its axis. In Figure 12.2, θ_2 is a positive angle and θ_1 is a negative angle.

Angular displacement, $\Delta\theta$, is the difference between two angles measured relative to the same axis and origin.

$$\Delta\theta = \theta_2 - \theta_1 \tag{12.3}$$

A positive angular displacement (see Figure 12.3) corresponds to $\theta_2 > \theta_1$, and a negative angular displacement to $\theta_2 < \theta_1$. The angular displacement $\Delta\theta$ and arc length Δs subtended by $\Delta\theta$ are related according to Eq. 12.1. That is, $\Delta s = r\,\Delta\theta$.

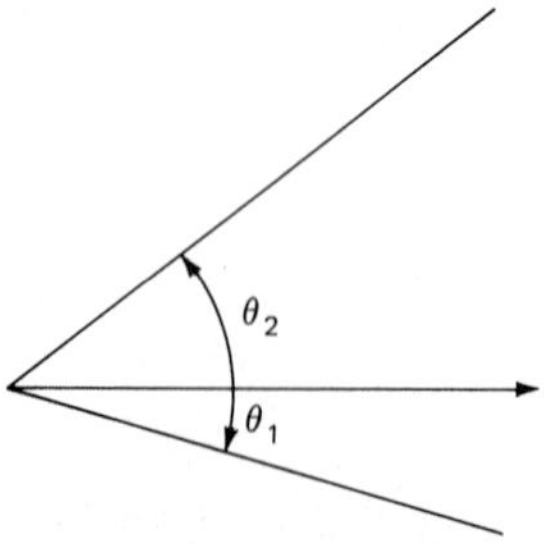

Figure 12.2

An angle is positive, like θ_2, when measured counterclockwise, and negative, like θ_1, when measured clockwise from a fixed line.

Figure 12.3

Angular displacement, $\Delta\theta = \theta_2 - \theta_1$, equals the difference between two angles measured from the same fixed line.

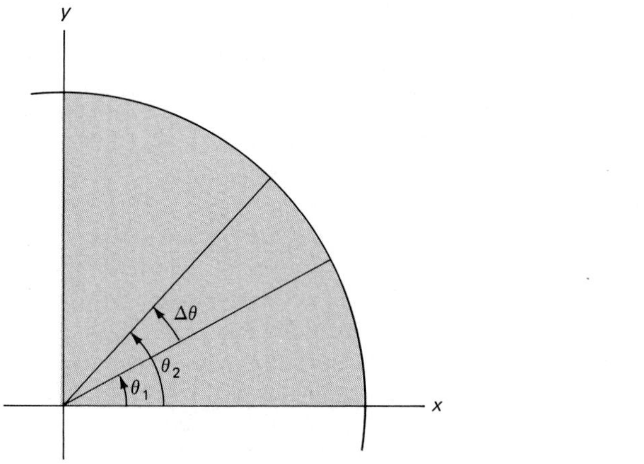

In Section 4.3 we defined the instantaneous linear velocity in terms of the linear displacement by using the equation

$$v = \frac{dx}{dt} \tag{4.7}$$

Analogously, we define the instantaneous angular velocity, ω,* in terms of the angular displacement by the equation

$$\omega = \frac{d\theta}{dt} \tag{12.4}$$

If time is expressed in seconds and angular displacement in radians, then the units for ω are radians per second (rad/s). Radians have no dimensions, as you can see from the definition, Eq. 12.1. "Rad" carries no dimensions; it is simply a reminder of how ω is measured.

Equation 12.4 can be written in the integral form

$$\int_{\theta_1}^{\theta_2} d\theta = \int_{t_1}^{t_2} \omega\, dt$$

For constant ω this equation can be integrated to give

$$\Delta\theta = \omega\,\Delta t \tag{12.5}$$

where $\Delta\theta = \theta_2 - \theta_1$ is the angular displacement during the time interval $\Delta t = t_2 - t_1$.

For a particle moving in a circular path a fundamental relation connects its tangential velocity with its angular velocity (Figure 12.4). To derive this relation we differentiate the equation $s = r\theta$ (keeping r constant) to get

$$\frac{ds}{dt} = r\frac{d\theta}{dt}$$

With the definitions $v_T = \frac{ds}{dt}$ and $\omega = \frac{d\theta}{dt}$ this relation can be written

$$v_T = r\omega \tag{12.6}$$

Because v_T is tangent to the circle it is called the *tangential velocity*.

*A direction will later be attributed to the quantity ω.

Figure 12.4

The tangential velocity vector is tangent to the circle centered at point O.

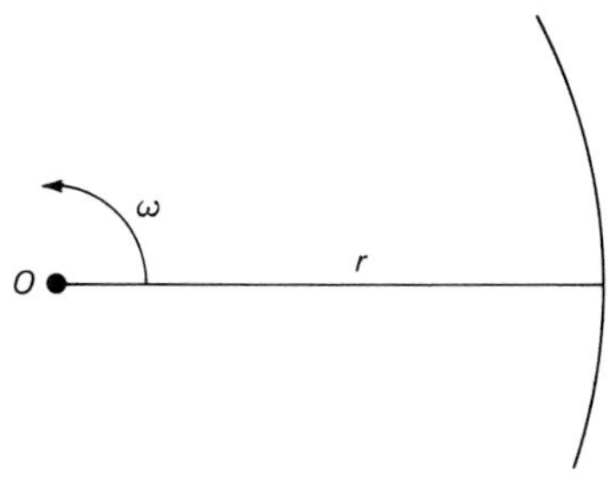

Example 1 An Orbiting Racing Car

During an automobile race on a circular track 2 km in circumference the lead car requires 30 s to complete one lap. Assuming that the speed of the car is constant, we want (a) to calculate the angular velocity of the car, (b) to calculate its tangential velocity, and (c) to determine whether the results are consistent with Eq. 12.6.

(a) To obtain ω we use Eq. 12.5. For one revolution we have $\Delta\theta = 2\pi$ rad and $\Delta t = 30$ s. Because the speed of the car is constant, we have $\omega = \Delta\theta/\Delta t = 0.209$ rad/s.

(b) Because the speed is constant, we obtain V_T by using $v_T = \Delta s/\Delta t$. For one revolution we have $\Delta s = 2000$ m and $\Delta t = 30$ s. Therefore $v_T = 66.7$ m/s.

(c) To check Eq. 12.6 for consistency we first calculate the radius of the circular racetrack.

$$r = \frac{2000}{2\pi} = 318 \text{ m}$$

According to Eq. 12.6 we should have

$$v_T = r\omega$$

From part (b) $v_T = 66.7$ and $r\omega = (318)(0.209) = 66.5$. The discrepancy of 0.2 is caused by rounding off the values for v_T, r, and ω. If we carry four significant figures, the comparison is $66.67 = (318.3)(0.2094) = 66.65$, and the discrepancy is only 0.02.

An object may possess an angular velocity because it is in orbit, like the car in Example 1 that orbits the center of the racetrack. An object can also possess an angular velocity because it spins on its own axis, as in the following example.

Example 2 The Spinning Earth

The earth spins about its own axis, completing one revolution relative to the "fixed" stars in one sidereal day (Section 5.3). We want to determine the angular velocity of the earth associated with this spinning motion. We have, for one revolution,

$$\Delta\theta = 2\pi \text{ rad}$$

$$\Delta t = 1 \text{ sidereal day} = 86,164 \text{ s}$$

Hence, to three significant figures,

$$\omega = \frac{\Delta\theta}{\Delta t} = \frac{2\pi}{8.62 \times 10^4} = 7.29 \times 10^{-5} \frac{\text{rad}}{\text{s}}$$

How do we determine ω for a particle if the motion is not circular? Consider a particle moving along the radius of a circle. Such a particle sweeps out zero angular displacement in any time interval. Hence $\omega = 0$ because $\Delta\theta = 0$. Thus the component of velocity in the radial direction, v_r, contributes nothing to ω. The component of velocity perpendicular to the radius, v_T, sweeps out an angular displacement, $\Delta\theta$, and an arc length, Δs, in time Δt that are related approximately according to

$$\Delta s \sim r\,\Delta\theta \sim v_T\,\Delta t$$

as shown in Figure 12.5. After dividing by Δt and taking the limit as $\Delta t \rightarrow 0$ this approximation becomes exact

$$v_T = r\omega \qquad (12.6)$$

Thus the angular velocity of an object that does not move in a circle is determined by the radius and the component of velocity perpendicular to the radius.

Example 3 Angular Velocity with Linear Motion

An automobile travels in a straight line down an interstate highway at 50 mph, as a farmer, located 100 meters from the highway, watches. Let's determine the angular velocity of the car relative to the farmer.

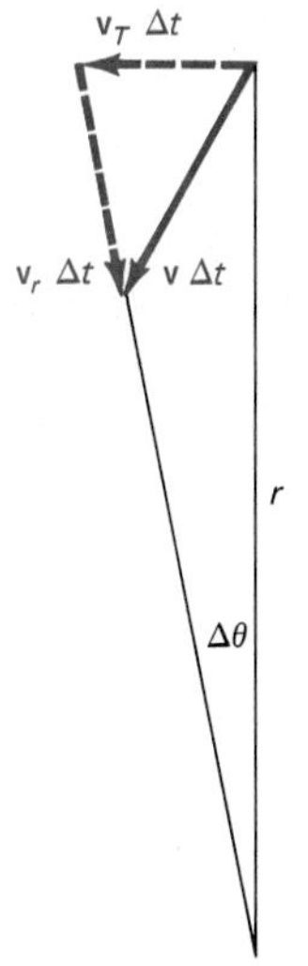

Figure 12.5

During Δt a particle moves through the displacement $\mathbf{\Delta s} = \mathbf{v}\ \mathbf{\Delta}t$ where $\mathbf{v}$ is the particle's velocity vector. The component of displacement perpendicular to the radius equals $v_T\ \Delta t$ where v_T is the tangential velocity component, and the component of displacement parallel to the radius equals $v_r\ \Delta t$ where v_r is the radial velocity component.

Figure 12.6

For a particle that does not move in a circle the angular velocity is determined by the radius r and the tangential velocity $\mathbf{v}_T$.

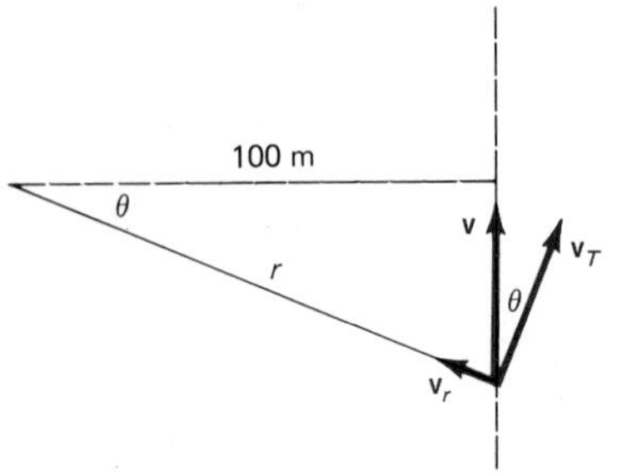

When the car is at an angle θ, as shown in Figure 12.6, its velocity is resolved into components parallel and perpendicular to the line joining the car with the farmer. The component of v perpendicular to this line is the instantaneous tangential velocity, and with r it determines ω. The component parallel to this line is the instantaneous radial velocity and contributes nothing to ω. Thus

$$v_T = v \cos \theta$$

so

$$\omega = \frac{v \cos \theta}{r}$$

and since $r = 100$ m/cos θ, and $v = 26.8$ m/s, we obtain

$$\omega = \frac{26.8}{100} \cos^2 \theta = 0.268 \cos^2 \theta \frac{\text{rad}}{\text{s}}$$

You can check this solution at $\theta = 0$ and $\theta = \pi/2$ rad by solving the problem directly at these two angles and comparing the results with the solution above.

Analogously to the linear acceleration a, defined by $a = \frac{dv}{dt}$, the *angular acceleration*, α, is defined by the equation

$$\alpha = \frac{d\omega}{dt} \tag{12.7}$$

The SI units for α are radians per second squared (rad/s^2).

Equation 12.7 can be cast into the integral form

$$\int_{\omega_0}^{\omega} d\omega = \int_0^t \alpha\, dt$$

For constant α this equation can be integrated to give

$$\Delta\omega = \omega - \omega_o = \alpha t \tag{12.8}$$

Equation 12.8 is analogous to the relation developed for one-dimensional linear motion with constant acceleration $\Delta v = v - v_o = at$ (Section 4.6).

For a particle moving in a circular path, Eq. 12.6 relates tangential and angular velocities. To obtain the corresponding equation relating the tangential and angular accelerations of such a particle we differentiate this equation

$$v_T = r\omega$$

to get

$$\frac{dv_T}{dt} = r\frac{d\omega}{dt}$$

Using the definitions $a_T = \frac{dv_T}{dt}$ and $\alpha = \frac{d\omega}{dt}$, we obtain the desired result:

$$a_T = r\alpha \tag{12.9}$$

The tangential acceleration, a_T, of a particle moving in a circular path is associated with a change in the magnitude of the tangential velocity of the particle. If this magnitude is constant, then $a_T = 0$.

You may recall (Section 5.3) that a particle moving in a circle experiences a radial acceleration $a_R = v^2/r$ associated with the changing direction of the velocity vector. This radial acceleration should not be confused with the tangential acceleration of a_T of this particle arising because the magnitude of the velocity vector changes.

Example 4 A Bicycle Wheel

A point on the circumference of a bicycle tire that is 30 inches in diameter and has its axle fixed in one position undergoes a constant angular acceleration from rest to 35 rad/s in 15 s. Find the tangential and angular accelerations of this point during the 15-s interval.

To determine the tangential acceleration we first determine the final tangential speed. According to Eq. 12.6

$$v_T = r\omega$$

We know that $r = 30/2$ in. $= 0.381$ m and $\omega = 35$ rad/s. Therefore,

$$v_T = 0.381 \times 35.0 = 13.3 \text{ m/s}$$

The tangential acceleration is calculated by using the relation

$$a_T = \frac{\Delta v_T}{\Delta t} = \frac{13.3 - 0}{15} = 0.887 \text{ m/s}^2$$

To determine the angular acceleration we use a form of Eq. 12.8.

$$\alpha = \frac{\omega - \omega_o}{t} = \frac{35 - 0}{15} = 2.33 \text{ rad/s}^2$$

We have calculated a_T and α for the point on the circumference of the tire, and the radius, r, was given. We

Table 12.1

Quantity	Tangential Defining Equation	Relation for Circular Motion	Angular Defining Equation
Displacement	$\Delta s = s_2 - s_1$	$\Delta s = r\,\Delta\theta$	$\Delta\theta = \theta_2 - \theta_1$
Velocity	$v_T = \dfrac{ds}{dt}$	$v_T = r\omega$	$\omega = \dfrac{d\theta}{dt}$
Acceleration	$a_T = \dfrac{dv_T}{dt}$	$a_T = r\alpha$	$\alpha = \dfrac{d\omega}{dt}$

can now determine whether these numerical values of a_T, r, and α are consistent with Eq. 12.9. We have

$$a_T = r\alpha$$

$$= 0.381 \times 2.33 = 0.888$$

The discrepancy of 0.001 is due to round-off error, and is not significant. Our values are consistent with Eq. 12.9.

Note that Eqs. 12.1, 12.6, and 12.9 are valid only if the angles are expressed in radians. The radian, from Eq. 12.1, is a dimensionless unit of angle. Often we include it to remind us of the units, but it may be dropped at any time. In the calculation in Example 4 the radian unit disappears (a_T is in m/s^2, r is in m, α is in rad/s^2), although it does not cancel.

In linear motion the linear velocity and linear acceleration of a particle may be of the same or opposite sign. For example, a car traveling north (with positive linear velocity) may be speeding up (positive linear acceleration) or slowing down (negative linear acceleration). Similarly, in circular motion the angular velocity and angular acceleration of a particle may be of the same or opposite sign.

Example 5 A Spinning Basketball

A familiar trick is to spin a basketball and balance the spinning ball on the tip of one finger. The balancing finger and axis of rotation of the basketball are both vertical. While the basketball is being set in rotary motion, its angular velocity and angular acceleration have the same sign. Once the ball is started and balanced on the player's finger, the angular velocity of the basketball will decrease as a result of friction. Thus $\Delta\omega = \omega_2 - \omega_1$ will be negative, and the angular acceleration will therefore be negative. While the basketball is slowing down, the angular velocity and angular acceleration are opposite in sign.

We have introduced angular displacement, angular velocity, and angular acceleration. In each case the defining equation was analogous to the corresponding defining equation for the linear or tangential quantity. Additionally, corresponding tangential and angular variables are related for a particle in circular motion. In order to emphasize this analogy further, we have collected these results in Table 12.1.

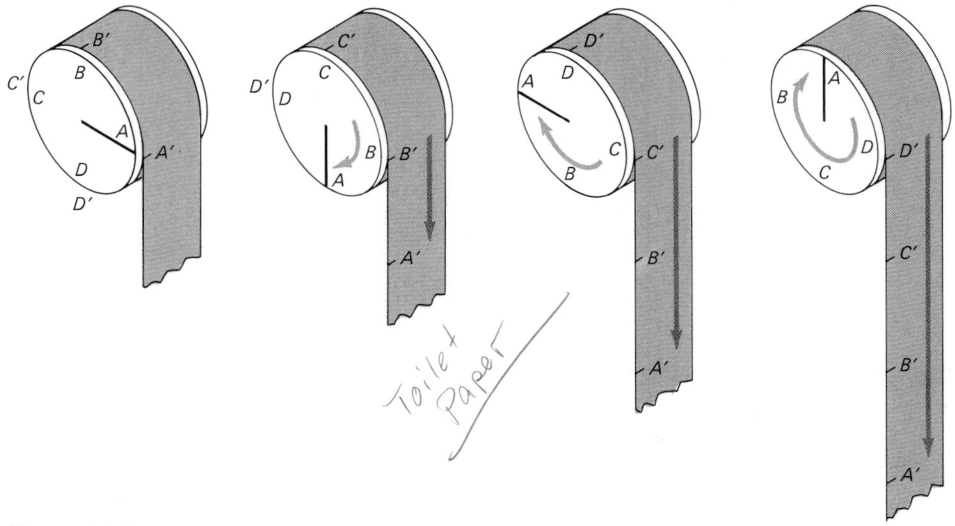

Figure 12.7

As the cylinder rotates a paper tape unwinds from it, forming a straight section that hangs down. Points on the tape labeled A', B', C', D' had been located on the cylinder opposite corresponding points labeled A, B, C, D.

We have seen that angular variables are analogous both to tangential variables (for a particle in circular motion) and to linear variables (for a particle in straight-line motion). To illustrate these analogies consider the motion of points on a paper tape as it unwinds from a rotating cylinder. The effect of one revolution is depicted in Figure 12.7. Points on the tape (A', B', C', D') were initially in contact with the corresponding points on the cylinder (A, B, C, D). Clearly the tangential velocities and accelerations for points on the circumference are the same as the linear velocities and accelerations for points on the tape. Given r, ω, and α for the cylinder, we can calculate the linear and tangential quantities through the relations

$$v_T = r\omega \quad \text{and} \quad v = v_T$$

and

$$a_T = r\alpha \quad \text{and} \quad a = a_T$$

In Section 4.6 we derived a set of equations relating linear quantities for the case where the linear acceleration is constant. Rather than derive the corresponding equations relating angular quantities when the angular acceleration is constant, we use the analogy, and display two sets of variables and equations in Table 12.2.

Example 6 The Bicycle Wheel Revisited

Through what angle did the bicycle wheel in Example 4 turn? According to the solution given in the example, $\omega = 35$ rad/s, $\omega_o = 0$, and $\alpha = 2.33$ rad/s^2. Using Eq. 12.12 we have

$$\theta = \frac{\omega^2 - \omega_o^2}{2\alpha} = 263 \text{ rad}$$

or about 42 revolutions.

Let us make one final comment regarding angular variables of motion. It is possible and useful to define ω and α to be vectors. Their magnitudes have already been defined according to $\omega = \dfrac{d\theta}{dt}$ and $\alpha = \dfrac{d\omega}{dt}$.

To complete the vector definitions we need to assign directions to ω and α. For a particle circling the z-axis in the x–y plane (see Figure 12.8) ω is defined to be in the $+z$-direction. Let the fingers of your right hand curl in the direction of rotation and your extended thumb will point in the direction of ω.

Figure 12.8

A particle moves in a circular path in the x–y plane. The angular velocity vector ω for this particle is in the positive z-direction.

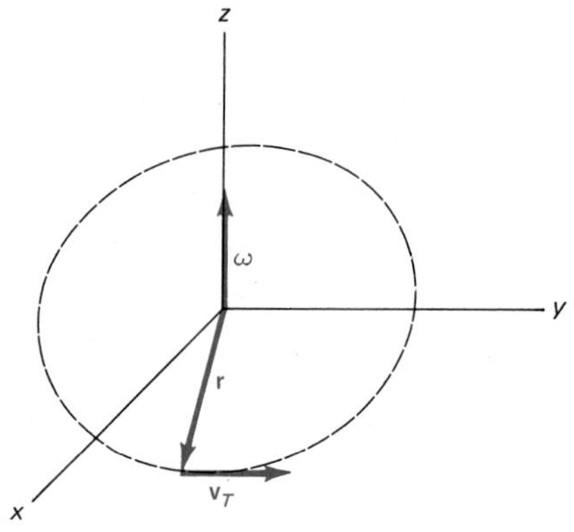

If the radius and velocity of the particle are labeled $\mathbf{r}$ and $\mathbf{v}_T$, respectively, then the three vectors are related according to

$$\mathbf{v}_T = \boldsymbol{\omega} \times \mathbf{r} \qquad (12.13)$$

This is a vector form of Eq. 12.6. You should check this equation for vector consistency.

Questions

1. Vectors satisfy the commutative law of addition. The displacement vector $\mathbf{s}_1$ followed by the displacement vector $\mathbf{s}_2$ leads to the same total displacement as when the displacement $\mathbf{s}_2$ occurs first and is followed by the displacement $\mathbf{s}_1$. We describe this equality with the equation $\mathbf{s}_1 + \mathbf{s}_2 = \mathbf{s}_2 + \mathbf{s}_1$.

   Three views of a die are shown in Figure 12.9. The number of dots on the three hidden sides in each view of the die can be inferred from the fact that the sum of the dots on opposite faces is seven. In Figure 12.9a the die is shown in its initial orientation relative to the axes; in Figure 12.9b the same die is shown after a rotation $\theta_1 = 90°$ about the x-axis; in Figure 12.9c the same die is shown following a second rotation of $\theta_2 = 90°$ about the y-axis. Repeat

Table 12.2

Linear		Angular	
s	displacement	θ	
v	velocity	ω	
a	acceleration	α	
$v = at + v_o$		$\omega = \alpha t + \omega_o$	(12.10)
$s = s_o + v_o t + \frac{1}{2}at^2$		$\theta = \theta_o + \omega_o t + \frac{1}{2}\alpha t^2$	(12.11)
$v^2 = v_o^2 + 2as$		$\omega^2 = \omega_o^2 + 2\alpha\theta$	(12.12)

Figure 12.9

Dots are arranged on the faces of the die so that the sum of dots for each opposing pair of faces equals seven. We show the die (a) aligned with the axes, (b) after a 90° rotation about the x-axis, and then (c) after a 90° rotation about the y-axis.

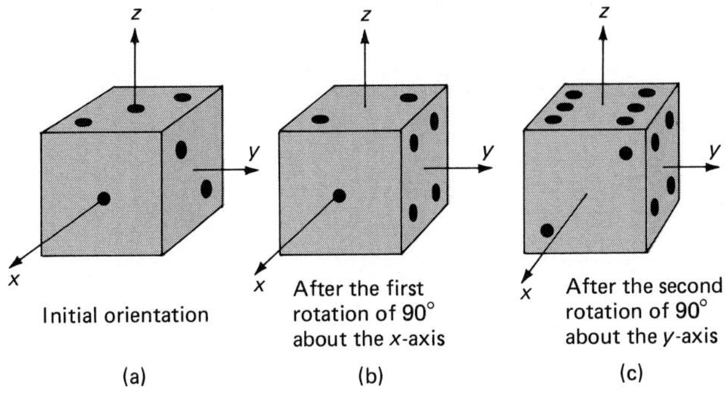

the procedure in reverse order. Is the result of θ_2 followed by θ_1 the same as when θ_1 precedes θ_2? Does this experiment suggest that rotations through large angles behave like vectors?

2. Determine the form of Eq. 12.6 if ω is expressed in degrees per second rather than radians per second. Which form is more convenient?

3. How many angles are required in order to specify the orientation of a rigid body with one point fixed? Why are two angles (longitude and latitude) sufficient for ocean and aircraft navigation?

4. A particle moving in a circular path has constant linear speed. What can be said about its angular velocity? angular acceleration? centripetal acceleration? tangential acceleration?

5. A particle moving in a circular path undergoes a constant tangential acceleration. What can be said about its linear speed? angular velocity? angular acceleration? centripetal acceleration?

12.2 Angular Momentum and Rotational Kinetic Energy

For linear motion the linear velocity, v, of a particle is related to its linear momentum, p, and linear kinetic energy, K, by $p = mv$ and $K = \frac{1}{2}mv^2$. Equations analogous to these can be derived that relate the angular velocity, ω, of a body to its angular momentum, L, and rotational kinetic energy, K, provided that (1) the direction of the axis of rotation is fixed, and (2) the body is rigid (Section 3.2). The analogous relations are $L = I\omega$ and $K = \frac{1}{2}I\omega^2$. These useful equations apply to a wide variety of rotational phenomena. Complications arise, however, if the direction of the axis of rotation varies. Except for the treatment of the gyroscope and symmetric top (Section 13.3), we will assume a fixed direction for the axis of rotation.

Angular Momentum of a Rigid Body

Let a rigid body be free to rotate about an axis that is fixed in direction. We choose the z-axis to coincide with the axis of rotation. If L denotes the z-component of the angular momentum, then we want to prove that L and ω are related by the equation

$$L = I\omega \tag{12.14}$$

where I is called the *moment of inertia* of the body about the specified rotation axis. The proof leads to a definition of I (Eq. 12.18) in terms of the mass distribution of the body relative to the fixed axis of rotation.

We begin by considering a particle rotating about the z-axis with angular velocity ω. Let its path lie in the x–y plane, as shown in Figure 12.10. In terms of its position—$\mathbf{R}_i$*—and linear momentum—$\mathbf{p}_i$—its angular momentum (see Section 11.1) is

$$\mathbf{L}_i = \mathbf{R}_i \times \mathbf{p}_i$$

The vector $\mathbf{R}_i$ lies in a plane perpendicular to the z-axis. This vector specifies the position of the particle relative to the axis of rotation. The linear momentum vector, $\mathbf{p}_i$, lies in the same plane and is perpendicular to $\mathbf{R}_i$. Hence we can write

$$L_i = R_i p_i = R_i(m_i v_i)$$

for the magnitude of $\mathbf{L}_i$. Here $\mathbf{v}_i$ is the particle velocity and m_i is its mass. Using Eq. 12.6, $v_i = R_i\omega$, we can express L_i as

$$L_i = m_i R_i^2 \omega$$

or

$$L_i = I_i \omega \tag{12.15}$$

*The vector $\mathbf{R}_i$ used here is perpendicular to the z-axis by definition. This vector should not be confused with the radius vector $\mathbf{r}_i$, which locates the particle relative to the origin.

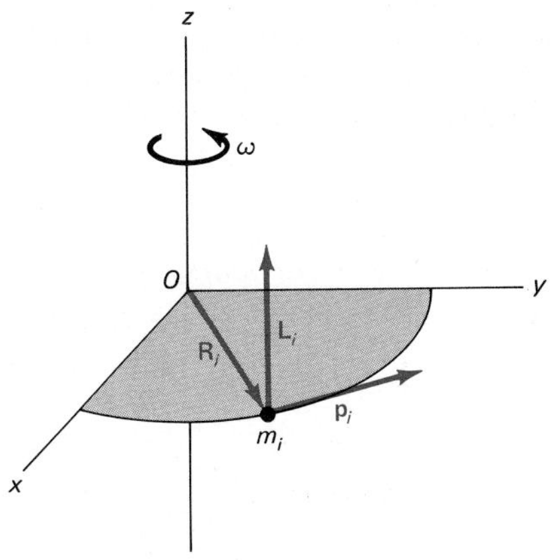

Figure 12.10

A particle moving in the x–y plane has the linear momentum $\mathbf{p}_i$ and moves in a circle of radius $\mathbf{R}_i$. With the directions shown for $\mathbf{p}_i$ and $\mathbf{R}_i$, the angular momentum vector $\mathbf{L}_i$ is in the positive z-direction.

where we have introduced

$$I_i = m_i R_i^2 \qquad (12.16)$$

as the moment of inertia of the particle about the axis of rotation.

Note that Eq 12.15 has the same form as Eq. 12.14, and Eq. 12.16 expresses I in terms of the mass distribution relative to the axis of rotation. Both of these equations, however, were derived for a particle. Now we will generalize these equations for the case of a rigid body.

The z-component of angular momentum of a rigid body, L, is the sum of z-components of angular momenta, L_i, taken over all particles (m_i) in the body.

$$L = \sum L_i \qquad (12.17)$$

Before expressing L_i in terms of the angular velocity we note that some care must be exercised in defining ω for a rigid body. Such a body is said to have an angular velocity vector ω at some time t if

1 the ith particle in the body moves tangent to a circle of radius $\mathbf{R}_i$, and

2 all such circles, regardless of radius, have centers on the same straight line, called the *axis of rotation*. In this case $v_i = \omega R_i$ for the ith particle with ω the same for all particles, and ω is directed along the axis of rotation.

Now we use Eq. 12.15 to write $L_i = m_i R_i^2 \omega$ so that Eq. 12.17 becomes

$$L = \sum m_i R_i^2 \omega = \left(\sum m_i R_i^2\right)\omega$$

Here we have factored ω out of the sum because it is the same for all particles in the body. This equation can be written

$$L = I\omega \qquad (12.14)$$

with

$$I = \sum m_i R_i^2 \qquad (12.18)$$

defined to be the moment of inertia of the rigid body relative to the axis of rotation.

In Eq. 12.18 m_i is the mass of the ith particle and $R_i = (x_i^2 + y_i^2)^{1/2}$ is the radius of the circle in which it moves. The moment of inertia is a property of the body that depends on both the mass and the way the mass is distributed relative to the center of mass. The moment of inertia also depends on the location and orientation of the axis of rotation. These points will be elaborated in Section 12.3, where the calculation of the moment of inertia is discussed and illustrated.

Example 7 Moment of Inertia of a Hoop

We want to find the moment of inertia of a hoop of mass M and radius R about an axis that is perpendicular to the plane of the hoop and goes through its center (Figure 12.11). We will derive an expression for I in terms of M and the hoop geometry (radius).

Figure 12.11

For the hoop every mass element m_i is located at the same distance $R_i = R$ from an axis perpendicular to the plane of the hoop through the point O.

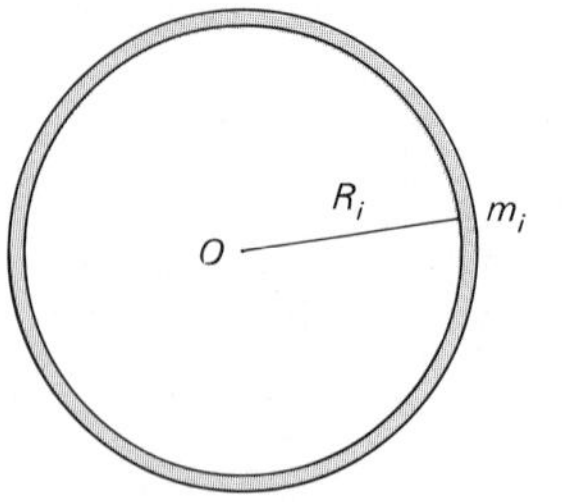

Our starting point is Eq. 12.18.

$$I = \sum m_i R_i^2$$

Evidently $R_i = R$, the hoop radius, for all particles in the hoop. This gives us

$$I = \sum m_i R^2$$

$$= \left(\sum m_i\right) R^2$$

$$= MR^2$$

where $M = \sum m_i$ is the total mass of the hoop. This expression for the moment of inertia, $I = MR^2$, is identical to Eq. 12.16, the expression for the moment of inertia of a particle of mass M at a distance R from the axis of rotation.

Equation 12.14, $L = I\omega$, relates L to ω analogously to the way $p = mv$ relates p and v. Linear and angular momentum and linear and angular velocity are pairs of analogous variables. Perhaps, then, the moment of inertia I is the rotational analog of mass. The following energy derivation reinforces this suggestion.

Rotational Kinetic Energy

Using the moment-of-inertia concept, we derived the equation $L = I\omega$ for the angular momentum of a rigid body. A similar simplification is possible for the rotational kinetic energy of a rigid body. We will now derive such an expression for a rigid body rotating about an axis that is fixed in position and direction.

For a collection of particles, the kinetic energy is defined to be

$$K = \sum \frac{1}{2} m_i v_i^2 \qquad (12.19)$$

where $\frac{1}{2} m_i v_i^2$ is the kinetic energy of the ith particle. Each particle moves with the same angular velocity ω, so

$$v_i = R_i \omega$$

Table 12.3

Linear			Angular	Relationship
Δs	displacement		$\Delta\theta$	$\Delta\theta = \frac{\Delta s}{r}$
v	velocity		ω	$\mathbf{v} = \boldsymbol{\omega} \times \mathbf{r}$
a	acceleration		α	$\mathbf{a}_T = \boldsymbol{\alpha} \times \mathbf{r}$
p	momentum		L	$\mathbf{L} = \mathbf{r} \times \mathbf{p}$
F	force	torque	τ	$\boldsymbol{\tau} = \mathbf{r} \times \mathbf{F}$
M	mass	moment of inertia	I	$I = \sum m_i R_i^2$

according to Eq. 12.6, where R_i is the radius of the circle on which the particle moves. Substituting Eq. 12.6 into Eq. 12.19 gives, after factoring,

$$K = \frac{1}{2}\left(\sum m_i R_i^2\right)\omega^2$$

The quantity in parentheses is simply the moment of inertia, I, and so

$$K = \frac{1}{2} I \omega^2 \qquad (12.20)$$

for the rotational kinetic energy of the body.

Comparison of Eq. 12.20 with the corresponding equation for the kinetic energy of a particle, $K = \frac{1}{2}mv^2$, again brings out the analogy between v and ω and between m and I. This analogy between linear and angular quantities is often useful. The two sets of variables and their relationships are displayed in Table 12.3

The primary applications of the moment-of-inertia concept are to rigid bodies having continuous mass distributions. We consider these in detail in later sections. In some physical situations, however, several discrete masses may move in such a way that their interparticle distances do not change. Such a system of particles can also be treated as a rigid body. We consider one of these systems in the following example.

Example 8 The Diatomic Molecule

A diatomic molecule consists of two atoms (masses m_1 and m_2 in Figure 12.12) bound together at a specific distance from each other, called the *equilibrium separation* and

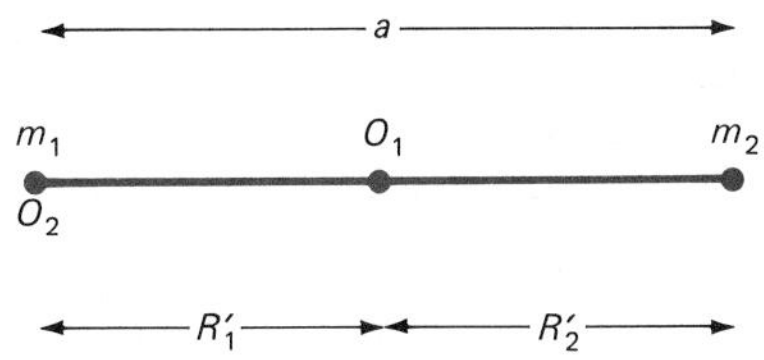

Figure 12.12

Two atoms with masses m_1 and m_2 are bound together to form a diatomic molecule. The center of mass at O_1 is located at a distance R_2' from m_2 and R_1' from m_1.

denoted by a. Diatomic molecules may translate, and may also undergo both rotational and vibrational motion. These kinds of motion are important in studies of the properties of gases. Here we consider only the rotational motion of a diatomic molecule. We calculate the angular momentum and kinetic energy of rotations of the molecule for an angular velocity ω.

Consider rotation about an axis through the center of mass at O_1. The distances of m_1 and m_2, respectively, from O_1 are R_1' and R_2' (primes denote center-of-mass coordinates).

$$R_1' = \frac{m_2 a}{m_1 + m_2}$$

$$R_2' = \frac{m_1 a}{m_1 + m_2}$$

These distances are the radii, R_1' and R_2', required in Eq. 12.18 to determine the moment of inertia of the molecule about an axis perpendicular to the figure through O_1. Thus we obtain for I the expression

$$I = m_1 R_1'^2 + m_2 R_2'^2$$

$$= m_1\left(\frac{m_2 a}{m_1 + m_2}\right)^2 + m_2\left(\frac{m_1 a}{m_1 + m_2}\right)^2$$

$$= \frac{m_1 m_2 a^2}{m_1 + m_2}$$

Using Eq. 12.14 we have

$$L = \frac{m_1 m_2 a^2}{m_1 + m_2}\omega$$

for the angular momentum of the rotation. Using Eq. 12.18 we have

$$K = \frac{1}{2}\frac{m_1 m_2}{m_1 + m_2} a^2 \omega^2$$

for the kinetic energy of rotation.

Combined Translation and Rotation

In Section 11.3 the total angular momentum of a system of particles was expressed as a sum of two terms. One term is the angular momentum of the center of mass and the other is the angular momentum relative to the center of mass. Similarly, it is useful to separate the kinetic energy of a rigid body into two terms. We shall prove that the kinetic energy can be written

$$K = \frac{1}{2}Mv_{cm}^2 + \frac{1}{2}I_c\omega_c^2 \qquad (12.21)$$

The first term, $\frac{1}{2}Mv_{cm}^2$, is the kinetic energy the body would have if its mass were all to move with the center-of-mass velocity. The second term, $\frac{1}{2}I_c\omega_c^2$, represents the rotational kinetic energy of the body relative to an axis through the center of mass.

To prove Eq. 12.21 we begin with the definition of the kinetic energy of a collection of particles:

$$K = \sum \frac{1}{2}m_i v_i^2$$

where $\mathbf{v}_i$ is the particle velocity relative to an arbitrary origin. Following the same procedure used in Section 11.3, we introduce the center-of-mass velocity, $\mathbf{v}_{cm}$, using Eq. 11.19

$$\mathbf{v}_i = \mathbf{v}_{cm} + \mathbf{v}_i'$$

to relate $\mathbf{v}_i$ to the particle velocity, $\mathbf{v}_i'$, relative to the center of mass as origin. Noting that $v_i^2 = \mathbf{v}_i \cdot \mathbf{v}_i$, we can write the kinetic energy

$$K = \sum \frac{1}{2}m_i(\mathbf{v}_{cm} + \mathbf{v}_i') \cdot (\mathbf{v}_{cm} + \mathbf{v}_i')$$

$$= \frac{1}{2}\sum m_i(v_{cm}^2 + v_i'^2 + 2\mathbf{v}_{cm} \cdot \mathbf{v}_i')$$

$$= \frac{1}{2}\left(\sum m_i\right)v_{cm}^2 + \frac{1}{2}\sum m_i v_i'^2 + \mathbf{v}_{cm} \cdot \left(\sum m_i \mathbf{v}_i'\right) \qquad (12.22)$$

Here we have used the fact that $\mathbf{v}_{cm}$ is a factor common to each term in the third sum.

The third term in Eq. 12.22 equals zero. We can prove this just as we did in Section 11.3. That is,

$$\mathbf{v}_{cm}' = \frac{1}{M}\sum m_i \mathbf{v}_i'$$

represents the velocity of the center of mass relative to the center of mass. Hence, $\sum m_i \mathbf{v}_i' = 0$.

To simplify the form of the second term we introduce ω_c, the angular velocity of the rigid body relative to an axis through the center of mass. Thus

$$v_i' = R_i'\omega_c \qquad (12.6)$$

where R_i' is the radius of the circle on which the ith particle moves. The second term in Eq. 12.22 thus becomes

$$\frac{1}{2}\sum m_i v_i'^2 = \frac{1}{2}\sum m_i R_i'^2 \omega_c^2$$

$$= \frac{1}{2}\left(\sum m_i R_i'^2\right)\omega_c^2$$

$$= \frac{1}{2}I_c\omega_c^2$$

where I_c is the moment of inertia of the body about an axis through the center of mass.

The first term in Eq. 12.22 is simplified if we recall that $M = \sum m_i$ is the total mass of the body. Hence

$$\frac{1}{2}\left(\sum m_i\right)v_{cm}^2 = \frac{1}{2}Mv_{cm}^2$$

completing the derivation, and confirming the expression

$$K = \frac{1}{2}Mv_{cm}^2 + \frac{1}{2}I_c\omega_c^2 \qquad (12.21)$$

for the kinetic energy of a rigid body.

We have derived significant results (Eqs. 12.14, 12.20, and 12.21) for the angular momentum and kinetic energy of a rigid body. The key assumption that made these derivations possible was that of a fixed direction of the axis of rotation. A general analysis of rigid body motion without a limitation on the direction of the axis of rotation is beyond the scope of this text. (In Section 13.4, however, we will present a *restricted* treatment of gyroscopic motion—an important example of rigid body motion where the axis of rotation is not fixed in direction.)

A second element that makes possible the results obtained in this section is the introduction of the moment of inertia of the rigid body. In Section 12.3 we illustrate procedures for calculating moment of inertia; in general, however, the moment of inertia of an object must be determined experimentally. The basis of experimental methods is developed in Section 12.4. In Chapter 14 we describe a procedure for obtaining the moment of inertia of a body by observing its behavior while it undergoes rotational oscillations.

Example 9 **Kinetic Energy of the Earth–Moon System**

We assume that the earth and moon maintain constant separation as they orbit about their center of mass, and we treat them as point masses (Figure 12.13). This allows us to neglect the rotational energy due to spin. The center of mass, we assume, moves with constant speed in a circular orbit about the sun. To illustrate the use of Eq. 12.21 we calculate the contribution of these two motions to their total kinetic energy. We need to evaluate

$$K = \frac{1}{2}Mv_{cm}^2 + \frac{1}{2}I_c\omega_c^2 \qquad (12.21)$$

To evaluate the first term we have

$$M = M_E + M_m$$
$$= 5.98 \times 10^{24} + 7.35 \times 10^{22}$$
$$= 6.05 \times 10^{24} \text{ kg}$$

and

$$v_{cm} = 2.98 \times 10^4 \text{ m/s}$$

This gives

$$\frac{1}{2}Mv_{cm}^2 = \frac{1}{2} \times 6.05 \times 10^{24} \times (2.98)^2 \times 10^8$$
$$= 2.69 \times 10^{33} \text{ J}$$

Figure 12.13

The center of mass of the earth–moon system is located at a distance R_E from the earth and at a distance R_M from the moon. The system rotates about the center of mass with angular velocity ω_{cm}. The center of mass moves wih linear velocity v_{cm}.

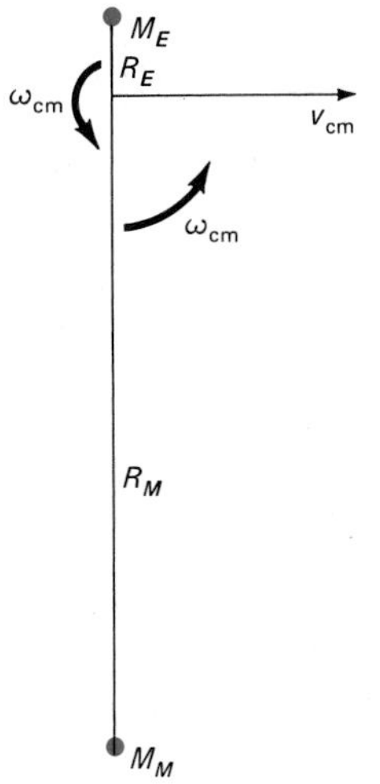

To evaluate the second term we have to obtain I_c and ω_c For I_c we have

$$\begin{aligned} I_c &= M_E R_E^2 + M_m R_m^2 \\ &= (5.98 \times 10^{24})(4.67 \times 10^6)^2 \\ &\quad + (7.35 \times 10^{22})(3.79 \times 10^8)^2 \\ &= 1.30 \times 10^{38} + 1.06 \times 10^{40} \\ &= 1.07 \times 10^{40} \text{ kg} \cdot \text{m}^2 \end{aligned}$$

For ω_c we have

$$\begin{aligned} \omega_c &= \frac{2\pi}{T} \\ &= \frac{2\pi}{27.3} \text{ days} \\ &= 2.66 \times 10^{-6} \text{ rad/s} \end{aligned}$$

Hence

$$\begin{aligned} \frac{1}{2} I_c \omega_c^2 &= \frac{1}{2}(1.07 \times 10^{40})(2.66 \times 10^{-6})^2 \\ &= 3.79 \times 10^{28} \text{ J} \end{aligned}$$

Here is a case where the kinetic energy associated with the center-of-mass motion $\frac{1}{2}Mv_{cm}^2$ is large compared with the kinetic energy associated with the orbital motion relative to the center of mass $\frac{1}{2}I_c\omega_c^2$.

Questions

6. In Example 9 the moment of inertia of the earth–moon system is calculated relative to the system center of mass on the assumption that both objects are point masses. This assumption means the spin kinetic energy of the system is neglected. Explain why we expect the spin kinetic energy of the earth to exceed that of the moon. (*Hint:* The same hemisphere of the moon always faces the earth.)

7. For an origin at the center of rotation (Figure 12.10) the angular velocity and angular momentum vectors of a particle are parallel. Will this be true for an origin elsewhere along the z-axis in Figure 12.10? What happens to the direction of ω in this case as rotation proceeds? What happens to the direction of $\mathbf{L}_i$?

8. Convince yourself that any displacement of this textbook can be reduced to a single translation of the center of mass followed by a rotation through an angle about a single axis. Is this reduction related to Eq. 12.21?

12.3 Calculation of the Moment of Inertia

The moment of inertia for a collection of point masses is defined by

$$I = \sum m_i R_i^2 \qquad (12.18)$$

To convert this equation into an expression for the moment of inertia of a body whose mass is distributed continuously we follow the rule of replacing the sum over particle masses, $\sum m_i$, by an integral over the entire body, $\int_M dm$. Thus we get

$$I = \int_M R^2 \, dm \qquad (12.23)$$

for the definition of I for a continuous body. The M on the integral sign represents the total mass of the body, and is there to emphasize that the integration is over the total mass of the body.

Before proceeding formally with the calculation of I for specific mass distributions, let us consider qualitatively some implications of the defining equations (Eqs. 12.18 and 12.23). First, the moment of inertia is proportional to the mass of the body. All other things being equal, the moment of inertia increases or decreases in direct proportion to the body's mass. We can prove that I and M are proportional if we can show that increasing one by a factor of 2 increases the other by the same factor. Imagine every m_i to increase by a factor of 2. This causes M to increase by 2, and increases I by 2, as Eq. 12.18 shows. Hence M and I are directly proportional. This will become more clear later in this section when every calculation is seen to lead to an answer proportional to the mass.

Sometimes it is convenient to formalize the fact that M and I are directly proportional by writing I as the product of the total body mass and the square of a length, s, called the *radius of gyration*.

$$I = Ms^2 \qquad (12.24)$$

Figure 12.14

The thin parallelepiped lies in the $x-y$ plane with the origin located at its center of symmetry.

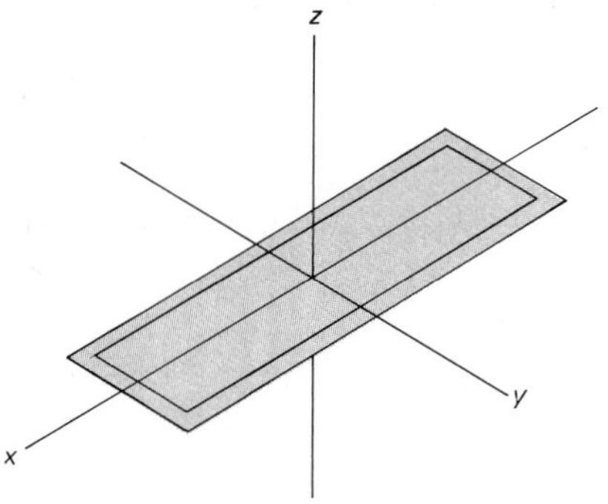

Equations 12.18 and 12.24 together define s^2 according to

$$s^2 = \frac{1}{M} \sum m_i R_i^2 \qquad (12.25)$$

If all of the body mass is located a distance R from the axis of rotation, then $R_i = R$ for all particles, and $I = MR^2$, giving $s = R$. For this special case the radius of gyration equals the radius of the circle on which all of the particles move.

The moment of inertia also depends on the distribution of mass relative to the chosen axis. Consider the thin parallelepiped shown symmetrically located in the $x-y$ plane of Figure 12.14. We can consider the x- or the z-axis to be an axis of rotation. Each of these axes passes through the center of mass of the rectangle, and in each case the total mass of the rectangle is the same. Yet very different results are obtained for the moments of inertia, I_x and I_z, about these axes. Intuitively, we expect I_x to be relatively small and I_z to be relatively large, because the mass is relatively close to the x-axis and relatively far from the z-axis. This turns out to be correct.

When an object is sufficiently symmetric, it is possible to calculate its moment of inertia analytically. In a calculation the idea is to express I in terms of the total mass and geometry of the object. We start with the defining equation for a continuous body, Eq. 12.23:

$$I = \int_M R^2 \, dm$$

where R (Figure 12.15) is the perpendicular distance of the mass element, dm, from the axis of rotation, $A-A$. This integral is a definite integral taken over the body. To carry out the integration it is necessary to express dm in terms of the corresponding volume element, dV, of the object. This entails introducing the mass density, ρ, defined by

$$dm = \rho \, dV \qquad (12.26)$$

and expressed in units of kilograms per cubic meter (kg/m^3). Substituting for dm in Eq. 12.23 and using Eq. 12.26, we write for I the integral

$$I = \int_V R^2 \rho \, dV \qquad (12.27)$$

This integral also is definite and taken over the body. In most of our applications ρ is constant, and in this case it is given by

$$\rho = \frac{M}{V} \qquad (12.28)$$

where M is the total mass and V is the total volume of the body.

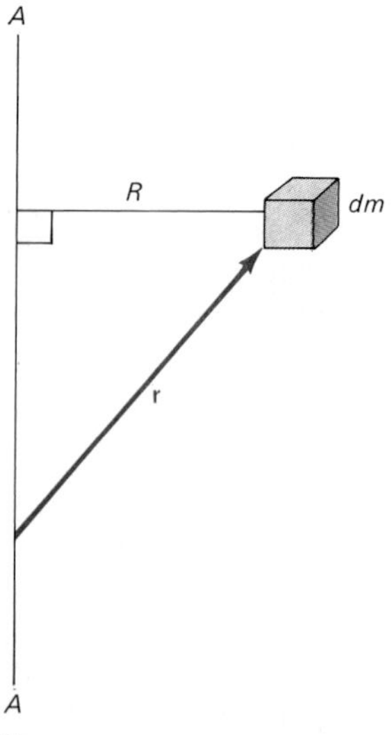

Figure 12.15

The mass element dm is located at a distance R from the axis of rotation.

Example 10 The Long Thin Rod

We want to calculate I for a rod of length L about the y-axis through one end (Figure 12.16). Assume the rod to have cross-sectional area A and constant mass density ρ.

We can write for a volume element dV

$$dV = A \, dx$$

Putting this into Eq. 12.27 we write

$$I = \int_0^L x^2(\rho A \, dx)$$

where we have set $R = x$, and used limits from 0 to L to integrate over the entire rod. Since ρ and A are both constant, we obtain

$$I = \frac{1}{3}\rho A x^3 \Big|_0^L = \frac{1}{3}\rho A L^3$$

Since $V = AL$, we have

$$\rho = \frac{M}{AL}$$

giving for I the expression

$$I = \frac{1}{3} M L^2$$

Comparison with Eq. 12.24 shows that we obtain

$$s = \frac{L}{\sqrt{3}}$$

Figure 12.16

A thin rod of length L lies along the x-axis with one end of the rod at the origin. A small section of the rod of width dx is located at a distance x from the origin.

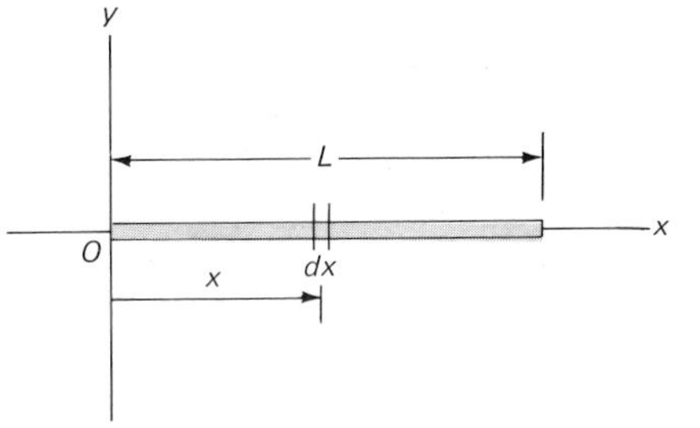

for the radius of gyration of the rod about an axis through one end of the rod.

Parallel Axis Theorem

The moment of inertia for an axis through the center of mass of a body is related to the moment for any parallel axis by a theorem. First we prove the theorem for an axis perpendicular to a lamina. Let I_c be the moment of inertia about an axis through the center of mass of the lamina, I be the moment of inertia of the same lamina about a parallel axis, and h be the perpendicular distance between the two axes. The theorem—called the parallel axis theorem—states that

$$I = I_c + Mh^2 \tag{12.29}$$

To prove this theorem, consider the arbitrarily shaped lamina in Figure 12.17. The points O and O′ lie on parallel axes through the origin and center of mass, respectively, indicated by the lines in the figure. The vectors $\mathbf{R}_i$ and $\mathbf{R}_i'$ are from O and O′ to an element of mass m_i, and the vector $\mathbf{h}$ is drawn perpendicular to the two axes. By definition

$$I = \sum m_i R_i^2 \tag{12.18}$$

We see from the figure that $\mathbf{R}_i = \mathbf{h} + \mathbf{R}_i'$, and since $R_i^2 = \mathbf{R}_i \cdot \mathbf{R}_i$ we have

$$R_i^2 = (\mathbf{R}_i' + \mathbf{h}) \cdot (\mathbf{R}_i' + \mathbf{h})$$

or

$$R_i^2 = R_i'^2 + h^2 + 2\mathbf{R}_i' \cdot \mathbf{h} \tag{12.30}$$

Putting Eq. 12.30 into Eq. 12.18 we obtain

$$I = \sum m_i R_i'^2 + \sum m_i h^2 + \sum 2m_i \mathbf{R}_i' \cdot \mathbf{h}$$

The last term is zero, because

$$\sum 2m_i \mathbf{R}_i' \cdot \mathbf{h} = 2\left(\sum m_i \mathbf{R}_i'\right) \cdot \mathbf{h}$$

and

$$\mathbf{R}_{cm}' = \frac{1}{M} \sum m_i \mathbf{R}_i' = 0$$

The vector from the center of mass to the center of mass equals zero.

The second term equals Mh^2, and the first term defines I_c, thus proving the parallel axis theorem for a lamina.

To generalize this theorem, let us consider the arbitrarily shaped three-dimensional object shown in Figure 12.18. If this object consists of a stack of parallel laminae, then the theorem separately holds for each lamina in the stack. Because each term in the theorem (I, I_c, and Mh^2)

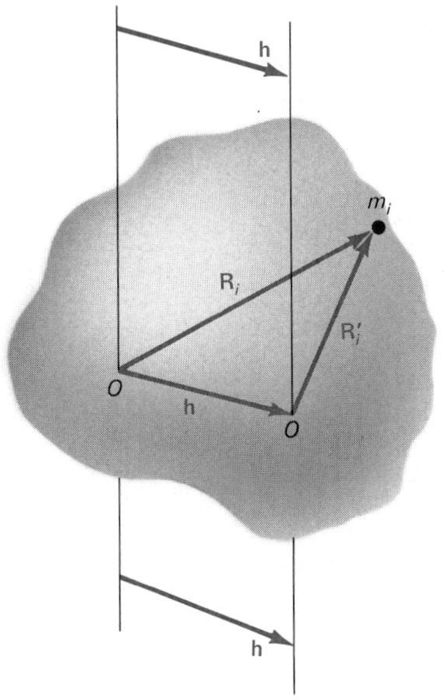

Figure 12.17

Two axes are shown perpendicular to the lamina. One axis intersects the lamina at O and the other axis intersects the lamina at the center of mass at O′. The two axes are separated by the vector displacement $\mathbf{h}$. A mass m_i has the radius vectors $\mathbf{R}_i$ and $\mathbf{R}_i'$, respectively, relative to the two axes.

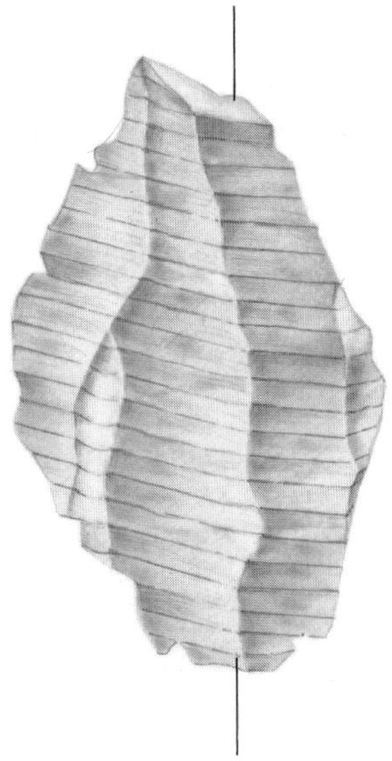

Figure 12.18

A three-dimensional object of arbitrary shape is considered to be a stack of parallel laminae.

Figure 12.19

A thin rod is perpendicular to two axes, labeled y and y'.

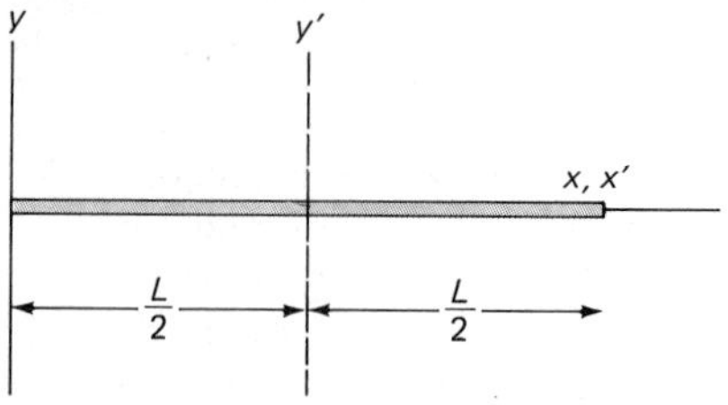

is proportional to the mass, it follows that the theorem also holds for the stack of laminae.

Because the term Mh^2 is not negative, it is clear that the moment of inertia is a minimum for the axis passing through the center of mass.

Example 11 **Long Thin Rod—Again**

In Example 10 the moment of inertia of a thin rod of mass M and length L is calculated to be $\frac{1}{3}ML^2$ about an axis through one end. What is the moment of inertia of the same rod about a parallel axis through the center of mass?

We can use the parallel axis theorem to find the moment of inertia. Referring to Figure 12.19, we have

$$I = \frac{1}{3}ML^2 \qquad \text{and} \qquad h = \frac{L}{2}$$

Using Eq. 12.29 we have

$$I_c = I - Mh^2$$
$$= \frac{1}{3}ML^2 - M\left(\frac{L}{2}\right)^2$$
$$= \frac{1}{12}ML^2$$

Alternately, we can note that a single rod of length L has the same I_c as the combined moments of inertia of two rods of length $L/2$ and mass $M/2$ about axes through their ends. The moments of inertia add algebraically. For a rod of length $L/2$, Example 10 shows that $I_c = (1/3)(M/2)(L/2)^2 = (1/24)ML^2$. Doubling this value for two rods confirms our previous result, $I_c = (1/12)ML^2$.

In general, a moment-of-inertia calculation that uses Eq. 12.27 involves a three-dimensional integration. (The calculation done in Example 10 reduced to one dimension as a consequence of the "thin rod" assumption. Integration in the plane perpendicular to the rod was avoided.) As you will see in the following example, it is sometimes better to solve such a problem as three one-dimensional problems rather than as a single three-dimensional problem.

Example 12 **Moment of Inertia of a Solid Sphere**

We want to determine I_c for a solid sphere of radius R and mass M. Rather than attack the problem directly, using a three-dimensional form of Eq. 12.27, let's divide the problem into three stages. We start with the moment of inertia, I_c, of a circular hoop obtained in Example 7; then we use this result to obtain the moment of inertia, I_c, of a solid disk; and finally the disk result is used to determine I_c for the solid sphere.

(a) *Hoop*. From Example 7 a hoop of mass M and radius r has a moment of inertia about an axis through its center of mass equal to

$$I_c = Mr^2$$

(b) *Solid Disk*. To determine I about an axis perpendicular to the figure for the disk (see Figure 12.20) we view the disk as an assembly of concentric hoops. From part (a) we know that a hoop of radius r and mass dm will have a moment of inertia

$$dI_c = r^2\, dm$$

Assuming constant density, we write

$$dm = \rho\, dV = \rho(2\pi r\, dr\, \Delta Z)$$

where ΔZ is the disk width and the quantity in parentheses is the disk volume element dV. The limits on ring radius go from 0 to R, giving

$$I_c = \int dI_c = \int_0^R r^2 \rho(2\pi r\, dr\, \Delta Z)$$

or

$$I_c = \frac{\pi}{2}\rho\, \Delta Z\, R^4$$

Using $M = \pi R^2 \Delta Z\, \rho$ for the disk to eliminate ρ gives the result

$$I_c = \frac{1}{2}MR^2$$

(c) *Solid Sphere*. To determine the moment of inertia of the solid sphere about an axis through its center we imagine the sphere to be constructed of many solid disks, as suggested in Figure 12.21. A typical disk of radius y and thickness dz, and located a distance z from the center, is singled out. According to part (b) this disk has a moment of inertia

$$dI_c = \frac{1}{2}y^2\, dm$$

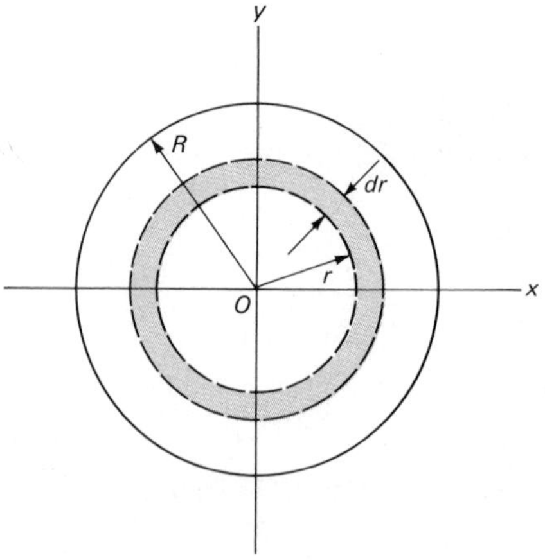

Figure 12.20

The disk is considered to be a set of nested rings. A representative ring of radius r and thickness dr is drawn.

Figure 12.21

The sphere is considered to be a stack of disks of varying radii. A representative disk of thickness dz and radius y is shown.

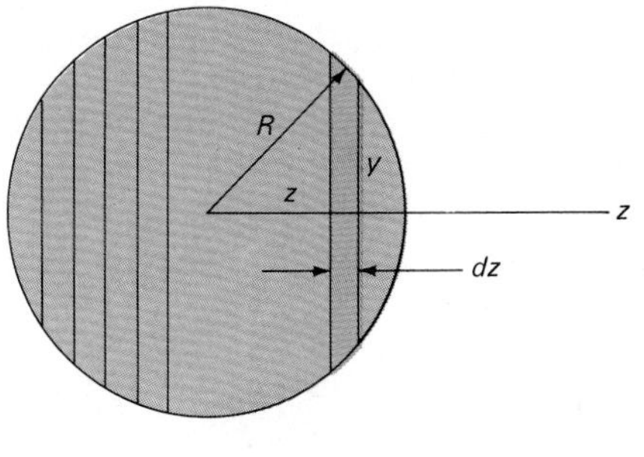

where dm is its mass. For dm we substitute

$$dm = \rho\, dV = \rho(\pi y^2\, dz)$$

which gives

$$dI_c = \frac{\pi}{2}\rho y^4\, dz$$

To include the entire sphere we add contributions from $z = -R$ to $z = +R$, obtaining the integral

$$I_c = \int_{-R}^{R} \frac{\pi}{2}\rho y^4\, dz$$

To eliminate y from this integral we note from Figure 12.21 that $y^4 = (R^2 - z^2)^2$. Putting this term into the expression for I_c gives

$$I_c = \int_{-R}^{R} \frac{\pi}{2}\rho(R^2 - z^2)^2\, dz$$

which leads to the expression

$$I_c = \frac{8\pi}{15}\rho R^5$$

Eliminating ρ by using $M = (4/3)\pi R^3\rho$ produces the final result

$$I_c = \frac{2}{5}MR^2$$

This method of solution has yielded the results

$I_c = MR^2$ for the hoop,

$I_c = \frac{1}{2}MR^2$ for the disk,

and

$I_c = \frac{2}{5}MR^2$ for the sphere

for an axis through the center of mass. Use of the parallel axis theorem provides the moment of inertia I for any of these bodies about a parallel axis.

For an axis perpendicular to the plane of the disk and through its center of mass we find in Example 12 that $I_c = {}^1/_2MR^2$. This answer is independent of disk thickness, and applies to any right circular cylinder. A similar generalization applies to any solid whose cross-sectional area does not vary with depth.

Questions

9. Two aluminum rods circular in cross section are of equal mass and length. One is hollow and the other is solid. Compare their moments of inertia about an axis (a) through one end perpendicular to the rod, and (b) that coincides with the axis of the rod. Explain the difference in comparisons.

10. Two flywheels having the same mass and made of the same material are designed to store energy. Explain why, for the same ω, the flywheel storing the greater energy is more likely to fracture.

11. A cylinder, sphere, and cube are solid and of equal mass. The dimensions are as shown in Figure 12.22. For the axes AA through the center of mass, which has the largest moment of inertia? Which has the smallest?

Table 12.4

Moments of inertia about axes through the center of mass		
Hoop or cylindrical shell (Axis through the center along the axis of the hoop)	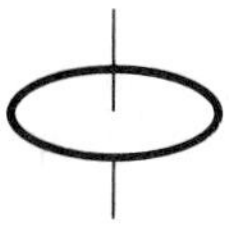	MR^2
Hoop (Axis along a diameter)		$\frac{1}{2}MR^2$
Disk or solid cylinder (Axis through the center, perpendicular to the plane of disk)	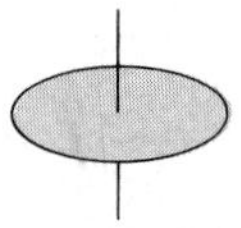	$\frac{1}{2}MR^2$
Thin disk (Axis along a diameter)	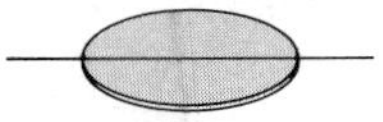	$\frac{1}{4}MR^2$
Spherical shell (Axis along a diameter)	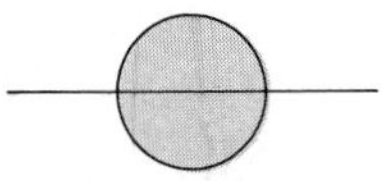	$\frac{2}{3}MR^2$
Solid sphere (Axis along a diameter)	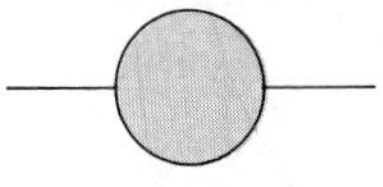	$\frac{2}{5}MR^2$
Thin rod (Axis through the midpoint, perpendicular to the rod)	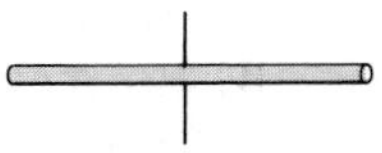	$\frac{1}{12}ML^2$
Rectangular lamina (Axis perpendicular to figure, and through the center of mass)		$\frac{M}{12}(a^2 + b^2)$

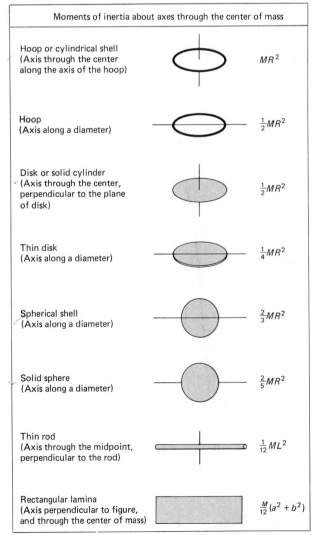

Figure 12.22

The cylinder, sphere, and cube have an axis drawn through the center of mass of each. In each case the axis intersects the surface at a 90° angle.

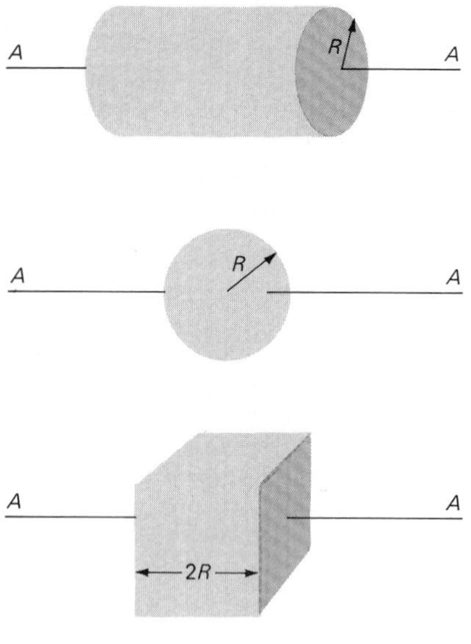

12. The moment of inertia of the earth about its axis of rotation is

$$0.3306MR^2 = 0.8265 \times {}^2/_5MR^2$$

Why should the moment of inertia of the earth be some 17% less than that of a sphere of uniform mass density? What does the factor of 0.8265 tell you (qualitatively) about the mass distribution within the earth?

12.4 Dynamics of Rigid Body Motion

Dynamics is the study of motion and the causes of changes in motion. In the first three sections of this chapter we introduced the variables required to describe the motion of a rigid body. In this section we consider how the interaction of a body with its environment causes its motion to change. The linear acceleration of a particle is determined by its mass and by the net external force. Newton's second law, $\mathbf{F} = m\mathbf{a}$, expresses this relationship. In an analogous way, the equation $\tau = I\alpha$ describes the angular acceleration of a rigid body in terms of its moment of inertia and the net external torque. We begin by deriving this equation.

In Section 11.2 the time rate of change of the angular momentum of a system of particles was shown to equal the net external torque acting on the system.

$$\boldsymbol{\tau}_{ex} = \frac{d\mathbf{L}}{dt} \tag{11.12}$$

This relation holds for an inertial observer* with the forces between particles satisfying Newton's third law. In particular, Eq. 11.12 holds for a rigid body. This equation is the starting point for developing the dynamics of rigid body motion.

We want to obtain a form of Eq. 11.12 that applies to rigid body motion when the axis of rotation is fixed in direction. Let us choose the z-axis as parallel to the fixed direction. In terms of its angular velocity and moment of inertia, the angular momentum of the rigid body can be written (see Section 12.2)

$$L = I\omega \tag{12.14}$$

where L, I, and ω all refer to the given axis of rotation. Differentiating Eq. 12.14 with respect to time gives

$$\frac{dL}{dt} = I\frac{d\omega}{dt} = I\alpha \tag{12.31}$$

where α is the angular acceleration relative to the axis of rotation. Note that I is constant in time because the axis of rotation is fixed in direction and because the body is rigid. Substituting Eq. 12.31 into Eq. 11.12 gives

$$\tau = I\alpha \tag{12.32}$$

where τ represents the external torque. Note the analogy between $F = ma$ and Eq. 12.32. Just as force causes the linear acceleration of a particle, so does torque cause the angular acceleration of a rigid body.

The inertial attribute of the moment of inertia is revealed by Eq. 12.32. Consider two rigid bodies subjected to the same external torque. According to Eq. 12.32, the ratio of their angular accelerations is inversely related to the ratio of their moments of inertia.

$$\frac{\alpha_1}{\alpha_2} = \frac{I_2}{I_1} \tag{12.33}$$

Comparison of this equation with Eq. 6.2 strengthens the analogy between the linear and angular forms of Newton's second law. Just as m is the property of a body that determines how well it "resists being given a linear acceleration" ($a = F/m$), so too I is the property of a rigid body that determines how well it "resists being given an angular acceleration."

To calculate the moment of inertia of a body, the mass, the way the mass is distributed, and the location and orientation of the axis of rotation must all be considered. These same factors affect the angular acceleration of the body when it experiences an external torque. The mass alone does not determine the response of a rigid body to an external torque.

Equation 12.32 is the basis of any dynamic determination of the moment of inertia of a rigid body. Such methods require measurement of the angular acceleration of the body and of the net external torque applied to it.

*Equation 11.12 holds for an observer located at the center of mass whether or not the observer is inertial.

Example 13 Measurement of the Moment of Inertia of a Disk

The moment of inertia of a disk can be determined in a simple experiment. A sensitized tape is wrapped around the disk (see Figure 12.23), and a small mass m is attached to it. As the tape unwinds, a constant torque is applied to the disk, causing it to undergo an angular acceleration. Position measurements on the tape at regular time intervals allow measurement of its acceleration. Measurement of the external torque is accomplished indirectly, permitting I to be expressed in terms of measured quantities.

Let T be the tension in the tape. Using Newton's second law, we write

$$\sum F_i = -T + mg = ma$$

where a is the linear acceleration of m, the disk circumference, and the tape. A second equation results when we apply Eq. 12.32 to the disk.

$$\sum \tau_i = TR = I\alpha$$

where α is the angular acceleration of the disk. Using Eq. 12.9 to relate a and α gives

$$a = R\alpha$$

Combining these three equations, we obtain the equation for I,

$$I = mR^2\left(\frac{g - a}{a}\right) \qquad (12.34)$$

in terms of the measured quantities m, R, g, and a.

In Example 12, taking advantage of the symmetry of the disk, we calculated $I = {}^1\!/_2 MR^2$ where M is the mass of the disk. With this equation, a measurement of M and R gives one value for I that can be compared with the result obtained by using Eq. 12.34. The two results should be equal.

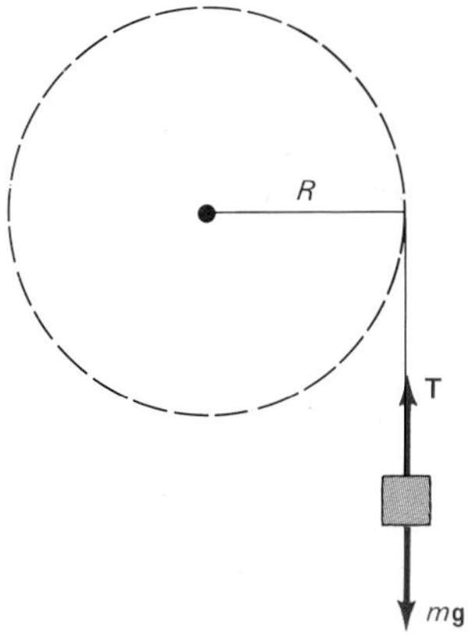

Figure 12.23

A particle of mass m is attached to the paper tape that unwinds from the cylinder.

Figure 12.24

A yo-yo unwinds under the influence of two forces. The string exerts an upward tension **T**, and gravity exerts a downward force $m\mathbf{g}$.

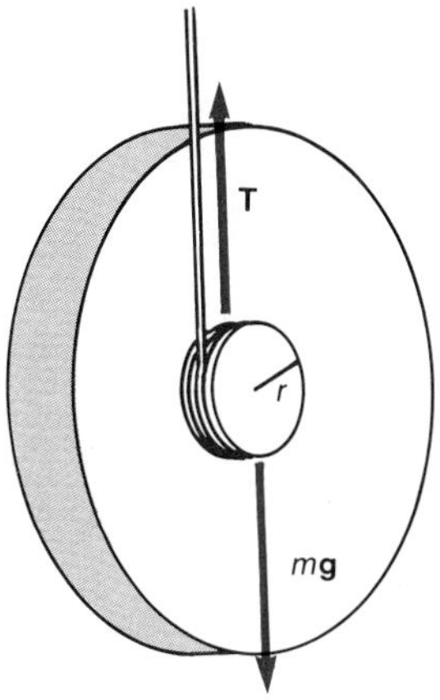

The cumulative nature of physics is illustrated in Example 14. Both the linear and rotational forms of Newton's second law are used, together with several of the results in rotational kinematics developed in Section 12.1.

Example 14 The Yo-Yo

A yo-yo is set into rotation by dropping it (Figure 12.24). At the same time that the yo-yo is dropped the string is pulled upward with a constant tension **T** so that the center of mass of the yo-yo remains fixed in position as the string unwinds. What is the angular velocity when the 1-m string has unwound, assuming that the initial angular velocity is zero? The yo-yo has $r = 9.00 \times 10^{-3}$ m, $I = 1.60 \times 10^{-5}$ kg · m^2, and $m = 4.00 \times 10^{-2}$ kg.

For zero vertical acceleration the net force must equal zero. Hence the string tension equals the yo-yo weight.

$$\begin{aligned} T &= mg \\ &= 0.0400 \times 9.81 \\ &= 0.392 \text{ N} \end{aligned}$$

The net torque relative to the center of mass is due to the tension alone. Hence

$$\begin{aligned} \tau &= Tr \\ &= 3.92 \times 10^{-1} \times 9.00 \times 10^{-3} \\ &= 3.53 \times 10^{-3} \text{ N} \cdot \text{m} \end{aligned}$$

Using Eq. 12.32 we obtain, for the angular acceleration relative to the center of mass,

$$\begin{aligned} \alpha &= \frac{\tau}{I} \\ &= \frac{3.53 \times 10^{-3}}{1.60 \times 10^{-5}} \\ &= 2.21 \times 10^2 \text{ rad/s}^2 \end{aligned}$$

The angle in revolutions through which the yo-yo has turned during acceleration is the string length divided by the circumference.

$$\theta = \frac{1 \text{ m}}{2\pi \times 9 \times 10^{-3}}$$

$$= 17.7 \text{ rev}$$

$$= 111 \text{ rad}$$

Using Eq. 12.12 in Table 12.2 we have, with $\omega_o = 0$,

$$\omega^2 = 2\alpha\theta$$

$$= 2 \times 2.21 \times 10^2 \times 111$$

$$= 4.91 \times 10^4$$

or

$$\omega = 221 \text{ rad/s}$$

or about 35 rev/s.

The angular velocity may be increased by giving the yo-yo an initial downward velocity. The tension will have to exceed mg in order to decelerate the yo-yo, causing an increase in τ, α, and ω.

Changing the Moment of Inertia with Internal Forces

When a diver goes from the layout position to the tuck position she changes her moment of inertia (Figure 12.25). This maneuver can reduce her moment of inertia sevenfold. There are many similar situations where a body can be treated as rigid before and after internal forces act to change its moment of inertia relative to the center of mass. During the change, of course, the body is not rigid, the moment of inertia is not constant, and so Eq. 12.31 does not apply. Because Eq. 11.12 is not subject to this restriction we use it to analyze this kind of situation.

Consider a system isolated in the sense that the net external torque acting equals zero. We start with the equation

$$\boldsymbol{\tau}_{ex} = \frac{d\mathbf{L}}{dt} \qquad (11.12)$$

and because $\tau_{ex} = 0$, we conclude that

$$\mathbf{L}_i = \mathbf{L}_f$$

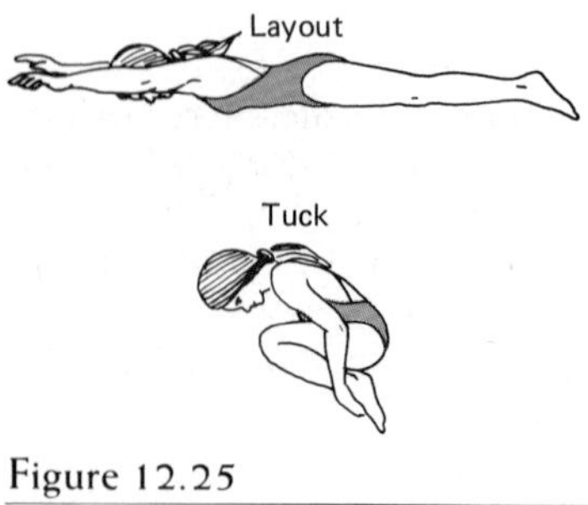

Figure 12.25

A diver is shown in the layout and tuck positions.

where $\mathbf{L}_i$ is the initial and $\mathbf{L}_f$ the final angular momentum of the system. The angular momentum is conserved. Before and after the change the body is presumed rigid. For a rigid body we can write

$$L_i = I_i\omega_i$$

$$L_f = I_f\omega_f$$

Putting these expressions into the conservation of angular momentum equation gives

$$I_i\omega_i = I_f\omega_f \qquad (12.35)$$

Because $\omega = 2\pi/T$ where T is the time for one complete revolution (the period), we can write Eq. 12.35 in the alternate form

$$\frac{I_i}{T_i} = \frac{I_f}{T_f} \qquad (12.36)$$

A flexible object can use internal forces to vary I. The value of ω (or T) will automatically adjust to maintain the product, $I\omega$, constant. This result holds for an inertial observer or for an observer at the center of mass.

Example 15 Creation of a Neutron Star

In the last stages of the life of a large star, the outer portion explodes while the interior part implodes. Let's consider a star whose core has a mass of 4.2×10^{30} kg and a radius of 8.0×10^8 m rotating with a period of 30 days. The angular momentum of the imploding core equals 3.6×10^{41} kg · m²/s. Under the influence of the implosion and gravity the core is compressed to a neutron star configuration with a radius of 13 km. As a neutron star its density is approximately constant throughout. We seek (a) the moment of inertia of the stellar core before collapse, (b) the moment of inertia in the neutron star configuration, and (c) the period of rotation, T, of the neutron star.

(a) Using Eq. 12.14 to describe the angular momentum of the core before the implosion we write

$$L_i = I_i\omega_i = I_i\left(\frac{2\pi}{T_i}\right)$$

Thus the moment of inertia before the implosion is

$$I_i = \frac{T_i L_i}{2\pi}$$

$$= \frac{30 \times 24 \times 3600}{2\pi} \times 3.6 \times 10^{41}$$

$$= 1.5 \times 10^{47} \text{ kg} \cdot \text{m}^2$$

(b) From Table 12.4, $I_c = (2/5)MR^2$ is the moment of inertia of a uniform sphere about an axis through its center. Thus we have, for the moment of inertia of the neutron star after the implosion,

$$I_f = \frac{2}{5}(4.2 \times 10^{30})(1.3 \times 10^4)^2$$

$$= 2.8 \times 10^{38} \text{ kg} \cdot \text{m}^2$$

(c) Angular momentum is conserved, so Eq. 12.35 applies with the given value for L_i used for $I_i\omega_i$.

$$L_i = L_f = I_f\omega_f = I_f\left(\frac{2\pi}{T_f}\right)$$

Solving for the rotational period of the neutron star we have

$$T_f = \frac{2\pi I_f}{L_i}$$
$$= \frac{2\pi(2.8 \times 10^{38})}{3.6 \times 10^{41}}$$
$$= 4.9 \times 10^{-3} \text{ s}$$

The newly formed neutron star rotates at about 200 rev/s!

The large speed of rotation obtained for the neutron star is consistent with observations. A value of about 30 rev/s has been measured for a neutron star in the Crab nebula. This star is believed to have been formed in a supernova observed in 1054 A.D. Since then its angular velocity has been reduced significantly as its kinetic energy of rotation has been converted into other forms of energy.

As discussed in Section 11.3, tidal friction causes the spin angular momentum of the earth to decrease. This decrease is accompanied by an increase in the length of the day and by an increase in the earth–moon distance (see Example 6 in Chapter 11). In the earth–moon system this conversion of spin angular momentum of the earth into lunar orbital angular momentum stops when the earth reaches the condition called *tidal lock*. In Example 16 we consider the future tidal lock of the earth.

Example 16 Tidal Lock of the Earth

Let's assume that the earth–moon system is free of external torque so that the total angular momentum of the system does not change. The system's angular momentum includes that due to the spin of each body together with that associated with their orbital motion about the center of mass (C in Figure 12.26). The period of the spin and orbital motion of the moon are the same—the same hemisphere of the moon faces the earth at all times. An orbiting satellite that always presents the same hemisphere to the attracting body is in tidal lock. Thus the moon has reached a state of tidal lock.

Tidal friction will slow down the spin of the earth until its rotational period (now 1 day) equals the orbital period of the moon (now 27.3 days). To conserve angular momentum, the loss of spin angular momentum must be compensated by an increase in orbital angular momentum. As a result, the length of the day then will equal the length of the lunar month. The moon recedes as this slowdown proceeds, eventually reaching a distance 1.5 times its present distance (now 354,000 km). Let us determine the length of the day when tidal lock of the earth is achieved.

The spin angular momentum of the earth is

$$L_s = \frac{2}{5}MR^2\left(\frac{2\pi}{T_s}\right)$$

where T_s is the spin period. The orbital angular momentum of the moon about the center of mass of the earth–moon system is

$$L_o = mvr = mr^2\left(\frac{2\pi}{T_o}\right)$$

Figure 12.26

The earth and moon orbit their common center of mass at C. Additionally, each body spins about its own center of mass.

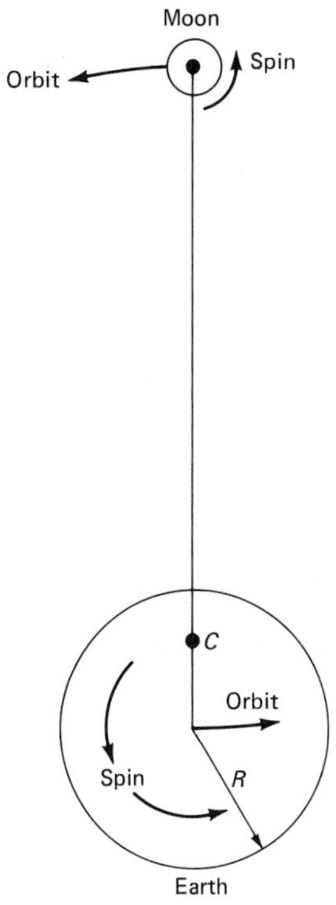

where T_o is the orbital period. We ignore the spin angular momentum of the moon and orbital angular momentum of the earth as negligible (see Problems 18 and 26 in Chapter 11 and Problem 18 at the end of this chapter). The total angular momentum is therefore

$$L_{tot} = L_s + L_o$$
$$= \frac{2}{5}MR^2\left(\frac{2\pi}{T_s}\right) + mr^2\left(\frac{2\pi}{T_o}\right)$$

Conservation of angular momentum requires L_{tot} to be the same today as when tidal lock of the earth is achieved. Thus

$$\frac{2}{5}MR^2\left(\frac{2\pi}{T}\right) + m(1.5r)^2\left(\frac{2\pi}{T}\right) = \frac{2}{5}MR^2\left(\frac{2\pi}{T_s}\right) + mr^2\left(\frac{2\pi}{T_o}\right)$$

where T represents both the spin period of the earth and the lunar (or earth) orbital period. For the orbital term on the left side of this equation the moon's distance from the system center of mass has been set equal to $1.5r$.

Note that a period T occurs in the denominator of every term in the equation. Provided the same time unit is used with each period, the time units cancel. Taking T, T_s, and T_o to be in days, we have

$$\frac{0.4}{T} + \frac{2.25}{T}\left(\frac{m}{M}\right)\left(\frac{r}{R}\right)^2 = \frac{0.4}{1} + \frac{1}{27.3}\left(\frac{m}{M}\right)\left(\frac{r}{R}\right)^2$$

Solving for T yields

$$T = \frac{0.4 + 2.25\left(\frac{m}{M}\right)\left(\frac{r}{R}\right)^2}{0.4 + 0.0366\left(\frac{m}{M}\right)\left(\frac{r}{R}\right)^2}$$

Now the moon–earth mass ratio is $m/M = 0.0123$, and the radius ratio $r/R = 60.3$, so we obtain

$$T = 49.6 \text{ days}$$

If the external torque (primarily due to the sun) could be neglected, the earth would reach a tidal lock with the day and lunar month each equal to 49.6 present earth days.

Questions

13. While airborne a long jumper (Figure 12.27) rotates his trunk forward, bending at the waist. Simultaneously his legs must rotate forward. This effect is sometimes referred to as "rotational action and reaction, a rotational version of Newton's third law." Justify this interpretation on the basis of angular momentum conservation.

14. Explain how a cat can change the orientation of its body (assumed initially at rest) in midair without changing its angular momentum.

Summary

One radian is the angle subtended by an arc equal in length to the radius, r. The circular arc length, s, and angle, θ, are related by

$$s = r\theta \qquad (12.1)$$

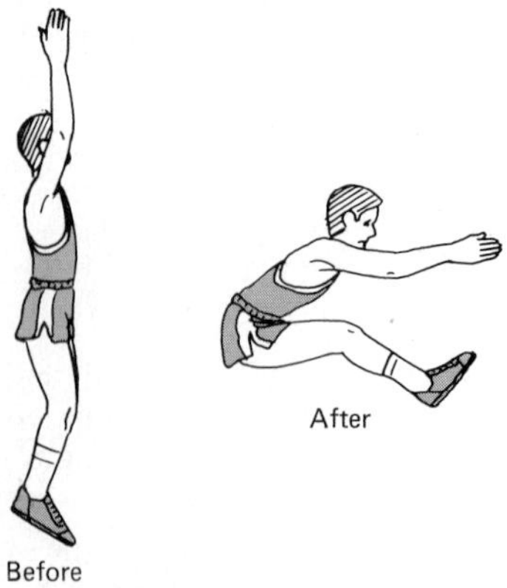

Figure 12.27

While airborne, a long jumper's body is straight (before) early in the jump and bent at the waist (after) late in the jump.

provided the angle is expressed in radians. The angular velocity, ω, and angular acceleration, α, are defined by

$$\omega = \frac{d\theta}{dt} \qquad (12.4)$$

and

$$\alpha = \frac{d\omega}{dt} \qquad (12.7)$$

The quantities ω and α are related to the tangential linear velocity, v_T, and tangential linear acceleration, a_T, according to

$$v_T = r\omega \qquad (12.6)$$

and

$$a_T = r\alpha \qquad (12.9)$$

For constant angular acceleration three equations, analogous to their linear counterparts, are useful.

$$\omega = \omega_0 + \alpha t \qquad (12.10)$$

$$\theta = \theta_0 + \omega_0 t + \frac{1}{2}\alpha t^2 \qquad (12.11)$$

$$\omega^2 = \omega_0^2 + 2\alpha\theta \qquad (12.12)$$

Angular velocity may be assigned a direction. This direction is along the axis of rotation, in accord with a right-hand rule. The position vector, $\mathbf{r}$, and velocity vector, $\mathbf{v}_T$, are related by the equation

$$\mathbf{v}_T = \boldsymbol{\omega} \times \mathbf{r} \qquad (12.13)$$

When the axis of rotation is fixed in direction, the components of angular momentum, L, and angular velocity, ω, parallel to this axis are related according to

$$L = I\omega \qquad (12.14)$$

where I is the moment of inertia of the body about the same axis.

The rotational kinetic energy of the system can be written

$$K = \frac{1}{2}I\omega^2 \qquad (12.20)$$

The total kinetic energy of the body can be expressed as a sum

$$K_{tot} = \frac{1}{2}Mv_{cm}^2 + \frac{1}{2}I_c\omega_c^2 \qquad (12.21)$$

where the first term is the kinetic energy of the center of mass and the second term is the kinetic energy relative to the center of mass.

For a discrete collection of particles the moment of inertia relative to a particular axis of rotation is defined by

$$I = \sum m_i R_i^2 \qquad (12.18)$$

Here, m_i is the mass of the ith particle, and R_i is the perpendicular distance to the axis of rotation. When the mass of the body is continuously distributed, the moment of inertia is calculated from the equation

$$I = \int_M R^2\, dm \qquad (12.23)$$

For any rigid body the parallel axis theorem relates the moment of inertia about an axis through the center of mass, I_c, to the moment of inertia, I, about a parallel axis. The relation takes the form

$$I = I_c + Mh^2 \qquad (12.29)$$

where M is the total mass of the body and h is the perpendicular distance between the axes.

For a rigid body the rotational form of Newton's second law (Eq. 11.12)

$$\tau_{ex} = \frac{d\mathbf{L}}{dt}$$

takes the form

$$\tau = I\alpha \qquad (12.32)$$

This equation reveals the inertial attribute of the body's moment of inertia. It is that property of a body governing its resistance to being given an angular acceleration.

Because angular momentum is conserved in a deformable body when the external torque equals zero, the initial and final states of a system deformed through the action of its internal forces are related according to

$$I_i\omega_i = I_f\omega_f \qquad (12.35)$$

providing the body in its initial and final states may be regarded as rigid.

Problems

Section 12.1 Angular Kinematics

1. Determine the difference in angle between 180° and 3 rad in (a) degrees and (b) radians.

2. What length of arc subtends an angle of 1° if the radius is 1 m?

3. Express 60 rev/s in radians per second.

4. You walk once around a circle (radius 10 m) in a time of 200 s. What is your angular velocity about the center of the circle?

5. A watch with an hour hand, minute hand, and second hand is at rest on a table (but with the mechanism running). Calculate the angular velocity of each hand in radians per second.

6. The moon always presents the same hemisphere to an observer on the earth. Use this fact to determine the spin angular velocity of the moon on its axis relative to the fixed stars. (The sidereal period of the moon is 27.3 days.)

7. A model airplane is flying in a horizontal circle (R = 40 m) at 40 mi/hr when the engine stops. The plane travels through half the circumference of the circle while coming to rest. Assume that the angular acceleration during this coasting period is constant, and calculate (a) the angular velocity of the plane when the engine stops, (b) the angular acceleration of the plane while it coasts, and (c) the time required to coast to a stop.

8. A major league pitcher releases a fast ball at a speed of 40 m/s. At the moment of release the ball is traveling along a circular path of radius is 0.80 m. Determine the angular velocity of the ball at that moment.

9. A ray of light strikes a plane mirror (Figure 1). The incident and reflected rays make equal angles (θ) with the normal. What is the *angle of deviation* (the angle through which the ray is "bent")?

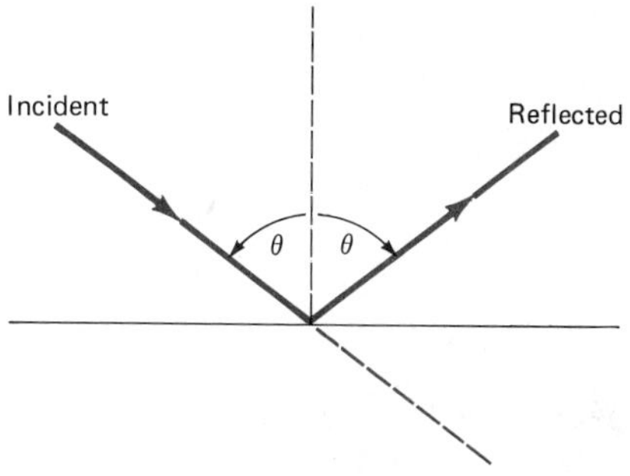

Figure 1

10. The angular velocity of the earth about the sun can be expressed as 1 rev/yr. (a) Express ω for the earth in rad/s. (b) Using $r = 1.50 \times 10^{11}$ m for the mean orbital radius, determine the orbital velocity of the earth about the sun.

11. An ant is perched on the rim of a 10-inch-diameter phonograph record rotating at $33\frac{1}{3}$ revolutions per minute (rpm). Determine the tangential velocity of the ant.

12. A stock car averages 175 mi/hr in a race on a circular track 2 km in circumference. If the car velocity is reduced to zero in a time of 20 s, determine the angular acceleration.

13. A student uniformly accelerates her sports car from rest. After 5 s of motion the tachometer indicates an engine rotation of 1500 rpm. Determine the angular acceleration of the engine.

14. A discus thrower accelerates the discus from rest to a speed of 25 m/s by whirling it through 1.25 rev (Figure 2). Assume the discus moves on the arc of a circle 1 m in radius, and calculate (a) the final angular velocity of the discus, (b) the average angular acceleration of the discus, and (c) the acceleration time.

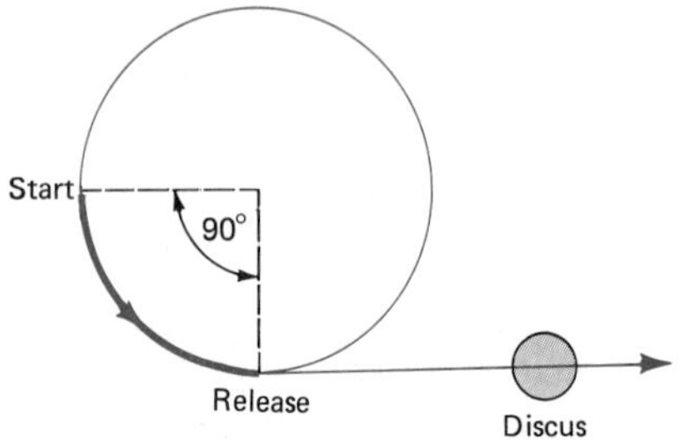

Figure 2

15. The circular wheel of a tool grinder moving at 26 rev/s comes to rest in 8.1 s. How many revolutions occurred while the wheel was coming to rest?

16. An airplane is observed directly overhead by a physics student who wonders how high it is. She assumes it is a jet traveling at 700 mi/hr, and measures its angular velocity to 2 deg/s. What is its altitude? (Give the answer in both meters and feet.)

17. In Example 3 the angular velocity of a car moving with a constant speed of 26.8 m/s in a straight line is shown to be

$$\omega = 0.268 \cos^2 \theta \text{ rad/s}$$

relative to a farmer located 100 m off the road. Prove that the angular acceleration is $\alpha = -0.144 \sin \theta \cos^3 \theta \cdot \text{rad/s}^2$.

Section 12.2 Angular Momentum and Rotational Kinetic Energy

18. The moment of inertia of the earth about its spin axis is 8.04×10^{37} kg · m^2. Use the fact that the earth rotates through 2π rad in 1 sidereal day to determine its (a) angular velocity, (b) angular momentum (magnitude), and (c) rotational kinetic energy.

19. A pitched baseball (r = 3.8 cm) moves with v_{cm} = 38 m/s and ω_c = 125 rad/s. Calculate the ratio of the spin kinetic energy to the translational kinetic energy of the baseball.

20. A popular (to some) amusement park ride consists of a rotating cylinder (R = 3 m). The passengers stand inside with their backs to the wall and rotate with the cylinder. If the rotational speed is 25 rpm, what is (a) the angular momentum and (b) the rotational kinetic energy of an 80-kg person? (Treat the person as a point mass.)

21. An electron ($m = 9.11 \times 10^{-31}$ kg) completes one revolution about the proton in hydrogen in a time of 1.52×10^{-16} s. Assume a circular orbit of radius $r = 5.29 \times 10^{-11}$ m, and determine the orbital angular momentum of the electron.

22. Two point masses m_1 and m_2 rotate about a common center of mass. Show that the ratio of the individual kinetic energies of rotation vary inversely as the ratio of the masses:

$$\frac{K_1}{K_2} = \frac{m_2}{m_1}$$

From the data of Example 9 calculate K_{moon}/K_{earth}.

23. An ammonia molecule consists of three hydrogen atoms and one nitrogen atom assembled into a pyramid-like structure with the hydrogen atom at the corners of the base and the nitrogen atom at the top. Each hydrogen atom is located 1.628×10^{-10} m from the two neighboring hydrogen atoms. Calculate the moment of inertia of the ammonia molecule about an axis passing through the nitrogen atom and the midpoint of the base of the pyramid.

24. A pair of ice skaters holding hands spin about their common center of mass completing 2 rev each second. One has a mass of 80 kg and the other a mass of 50 kg. Treat each skater as a point mass with their separation equal to 2 m, and determine their combined angular momentum.

25. Determine the rotational kinetic energy of the ice skaters in Problem 24 by using (a) Eq. 12.19 and then (b) Eq. 12.20.

26. An oxygen molecule has a mass of 5.31×10^{-26} kg and a moment of inertia of 1.95×10^{-46} kg · m^2 about an axis perpendicular to the line joining the oxygen atoms and through the center of mass. If the total angular momentum of the molecule is 1.05×10^{-34} J · s, then (a) determine the angular velocity relative to the center of mass, (b) determine the kinetic energy relative to the center of mass, and (c) find the value for the center-of-mass velocity that would make the kinetic energy of the center of mass equal to the kinetic energy relative to the center of mass.

27. Take the earth and moon to be homogeneous spheres with $I = \frac{2}{5}MR^2$ (see Example 12), and calculate their spin kinetic energies. For comparison (see Example 9), the orbital kinetic energies are $K = 4.60 \times 10^{26}$ J and $K = 3.75 \times 10^{28}$ J for the earth and moon, respectively.

28. Treat the earth–moon system as isolated, and the earth and moon as homogeneous spheres (see Problem 27), and calculate the (a) spin and (b) orbital

angular momentum of the earth. Also calculate the (c) spin and (d) orbital angular momentum of the moon. (See Example 6 and Problem 18 in Chapter 11.)

29. Three identical particles of mass m are situated on the corners of an equilateral triangle in the x–y plane (Figure 3). The triangle has a side of length a. In terms of m and a determine the moment of inertia of the three particles about an axis through the center of mass and parallel to the (a) x-axis; (b) y-axis; (c) z-axis.

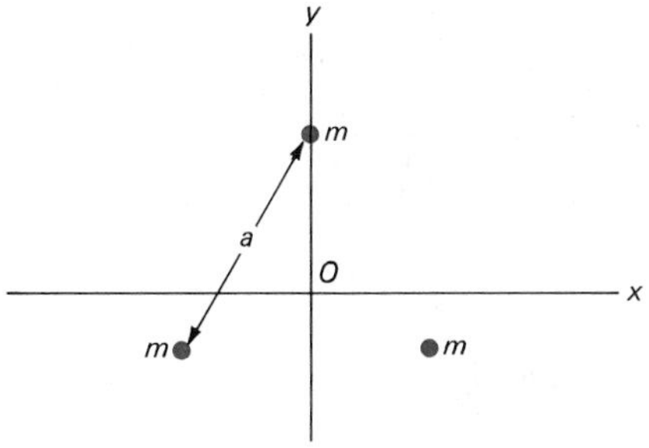

Figure 3

30. Three identical particles of mass m are located a distance a apart on the x-axis, as shown in Figure 4. In terms of m and a determine the moment of inertia of the three particles about an axis through the center of mass and parallel to the (a) x-axis; (b) y-axis; (c) z-axis.

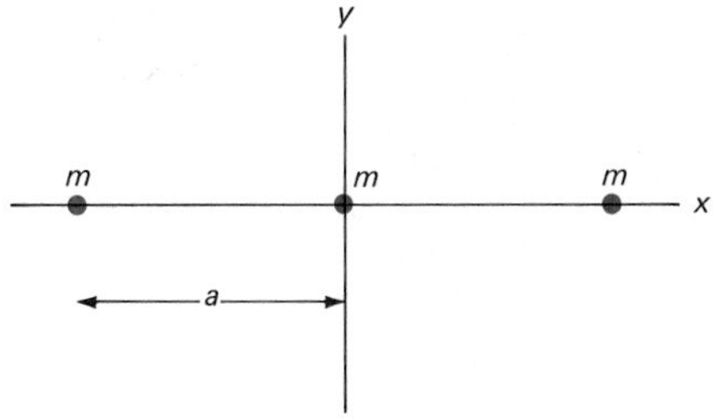

Figure 4

31. A 1500-kg car that had been traveling at 16 m/s comes to a stop at a stop sign. Imagine that all of its kinetic energy has been transformed into the rotational kinetic energy of a flywheel. The flywheel is a disk with a radius of 0.40 m and whose mass is 50 kg; it is rotating about the axis of rotational symmetry. Calculate its angular velocity ω.

Section 12.3 Calculation of the Moment of Inertia

32. The moment of inertia I of a long thin rod (mass $= M$, length $= L$) is $\frac{1}{3}ML^2$ for an axis perpendicular to the rod and passing through one end. Determine I for a parallel axis a distance x from the center of the rod by (a) treating the rod as two rods of length $(L/2) - x$ and $(L/2) + x$ rotating about a common axis, and (b) using the parallel axis theorem.

33. Determine the moment of inertia of the carpenter's square (Figure 5) and the cross (Figure 6) about an axis through O (perpendicular to the plane of the square). Treat each of the arms as a long thin rod of length L and mass M.

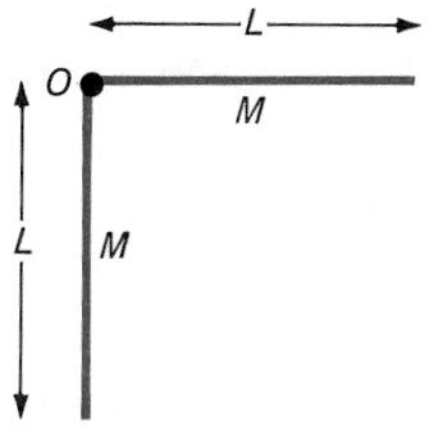

Figure 5

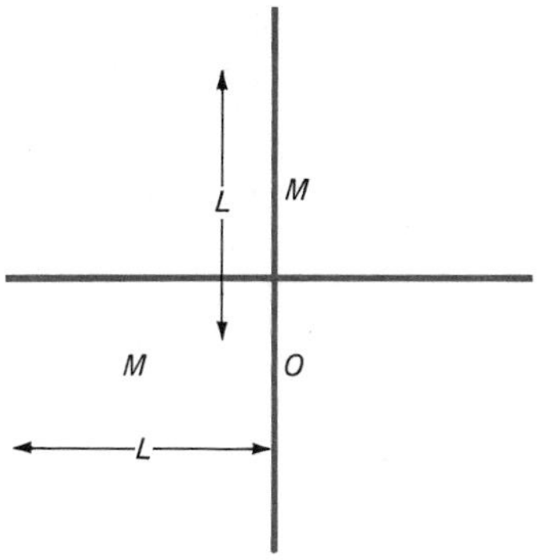

Figure 6

34. A solid double cone is a toy usually set into rotation with a string. In terms of the total mass M and cone geometry calculate the moment of inertia about the axis AA (Figure 7).

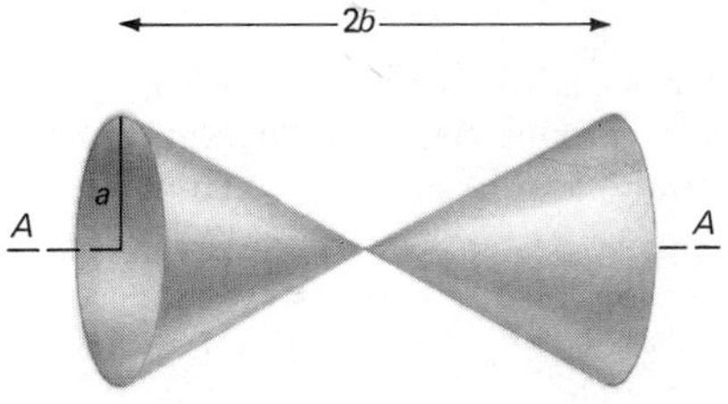

Figure 7

35. Obtain the moment of inertia of a hoop of mass M and radius R about an axis perpendicular to the plane of the hoop and passing through the circumference.

36. A wheel is formed from a hoop and 12 spokes (Figure 8). The mass of the hoop is six times the mass of a spoke. Calling m and r the mass and length, respectively, of a spoke, determine the moment of inertia of the wheel rotating about an axis through its center and perpendicular to a plane containing the wheel.

37. Determine the moment of inertia of the earth–moon system relative to their center of mass. Solve their problem (a) by treating the earth and moon as point masses (this approximation means that $I_c = 0$) located at their geometric center, and (b) by treating each as a homogeneous solid sphere and using the parallel axis theorem.

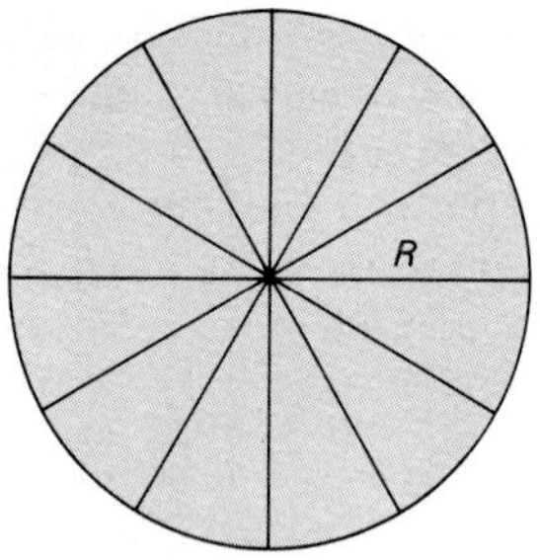

Figure 8

38. Obtain the moment of inertia of a hoop of mass M and radius R about an axis in the plane of the hoop tangent to the circumference.

39. Derive the expression given in Table 12.4 for the moment of inertia of rectangular lamina.

40. Use the parallel axis theorem to deduce the moment of inertia of a solid sphere of mass M and radius R about an axis tangent to its surface.

41. A thin rod of mass M is bent into the shape of a semicircle of radius R (Figure 9). Determine the three moments of inertia about axes through the center of mass that are parallel to the x-, y-, and z-axes.

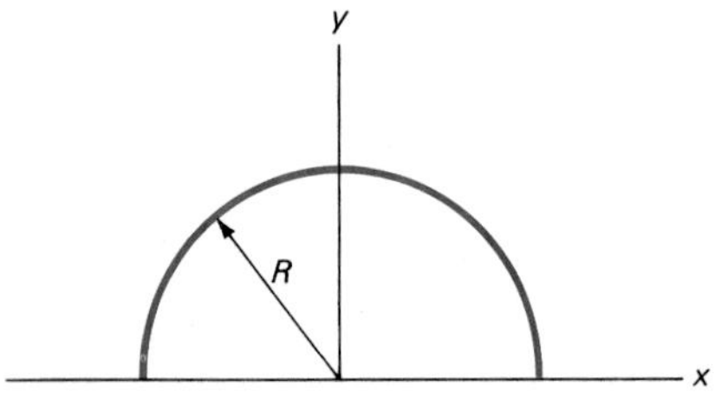

Figure 9

42. A block of mass M has edges a, b, and c. For an axis parallel to edge c through the center of mass use the procedures employed in Example 11 to prove that

$$I_c = \frac{M}{12}(a^2 + b^2)$$

43. A thin spherical shell of mass M and radius R has a moment of inertia $I_c = \frac{2}{3}MR^2$ about an axis through its center of mass. Derive this result by starting with the result for a solid sphere. Imagine the spherical shell to be created by subtracting from the solid sphere of radius R a solid sphere with a slightly smaller radius.

44. For a thin spherical shell of thickness dr and mass dM the result of Problem 43 can be rewritten

$$dI = \frac{2}{3}r^2\,dm = \frac{8\pi}{3}\rho r^4\,dr$$

For a spherical body like a planet or a star the density is a function of radial position: $\rho = \rho(r)$. With the moment of inertia written as aMR^2, show that

$$aR^2 = \frac{\int_0^R \frac{8\pi}{3}\rho r^4\,dr}{\int_0^R 4\pi\rho r^2\,dr}$$

45. The density of the earth is given very approximately by

$$\rho(r) = \left(14.2 - 11.6\frac{r}{R}\right) \times 10^3\,\frac{\text{kg}}{\text{m}^3}$$

Show that this density leads to a moment of inertia

$$I_E = 0.330MR^2$$

The accepted value (from astronomical observation) is $I_E = 0.331MR^2$.

Section 12.4 Dynamics of Rigid Body Rotation

46. A bicycle wheel ($M = 18$ kg, $R = 0.4$ m) is set into rotation by applying a force of 150 N tangent to the circumference. Treat the wheel as a hoop and calculate its angular acceleration.

47. A long thin rod of mass M and length L is hinged at one end (Figure 10). It is released and allowed to rotate about the hinged end. Show that its angular acceleration is

$$\frac{3}{2}\frac{g}{L}\cos\theta$$

where θ is the angle between the rod and the horizontal. Show also that the tangential acceleration of the free end exceeds g when θ falls below $\cos^{-1}(2/3)$.

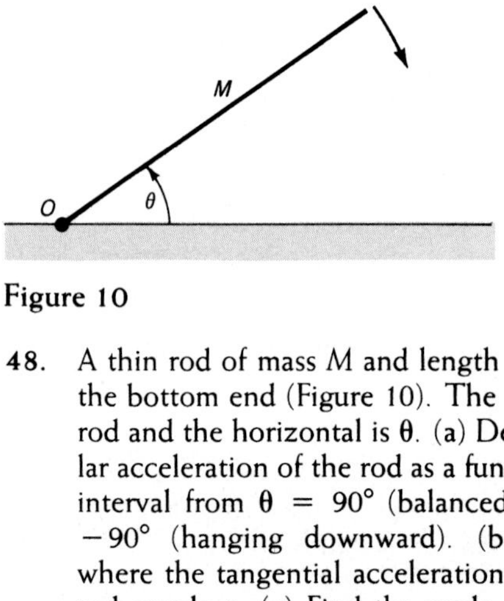

Figure 10

48. A thin rod of mass M and length L is pivoted about the bottom end (Figure 10). The angle between the rod and the horizontal is θ. (a) Determine the angular acceleration of the rod as a function of θ over the interval from $\theta = 90°$ (balanced upward) to $\theta = -90°$ (hanging downward). (b) Find the angle where the tangential acceleration of the end of the rod equals g. (c) Find the angle where the vertical component of the tangential acceleration of the end of the rod equals g.

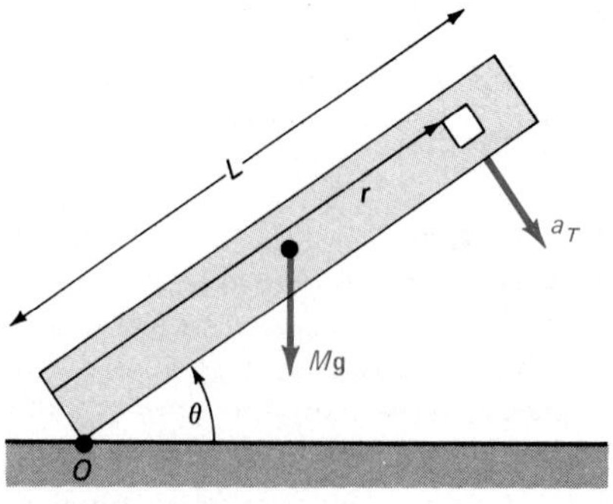

Figure 11

49. A disk is set in rotation as described in Example 13. The disk has a radius of 0.10 m and a mass of 2.10 kg. The mass $m = 0.16$ kg experiences an acceleration of 1.30 m/s^2 downward. Calculate I_{disk} (a) from the geometry of the disk and (b) by using the acceleration of the small mass.

50. Referring again to Example 13, show that the acceleration of m is independent of the radius of the disk and depends on the ratio of the masses. Specifically, show that

$$\begin{aligned} a &= \frac{g}{\frac{1}{2}M/m + 1} \\ &= g\,\frac{2m}{M + 2m} \end{aligned}$$

Check the limiting values of a for $m \to 0$ and $m \to \infty$ and argue that these limits are to be expected.

51. When a brick smokestack (Figure 11) is toppled to the ground it invariably fractures before the angle θ reaches zero. Why? Consider a brick whose tangential acceleration is a_T. If the vertical component of a_T exceeds g, the brick mortar will be under tension and will fracture. Show that this occurs for the top brick when

$$\cos^2\theta \geq \frac{2}{3}$$

52. A baseball batter accelerates the bat ($m = 1$ kg and $s = 0.5$ m) through 120° from rest to 30 rad/s. The accelerating torque is supplied by equal and opposite forces exerted on the bat by his hands (this is the extreme case of a "wrist hitter"), which are separated a distance of 0.1 m. Determine (a) the angular acceleration of the bat, (b) the tangential acceleration of the point 0.5 m from the axis of rotation between his hands, (c) the net torque, and (d) the force each hand exerts.

53. A disk (mass = M, radius = R) rotates about an axis perpendicular to the disk and through its center. When its angular velocity equals ω a particle of mass M, initially at rest, becomes attached to a point on the rim of the disk. While becoming attached the particle is accelerated by the disk and the disk is slowed by the particle. If no external torques act, calculate (a) the new angular velocity of the disk and (b) the new kinetic energy of the system.

54. A child rolls a solid disk at a constant linear velocity v. What fraction of the total kinetic energy of the disk is due to rotational motion?

55. An 80-kg diver has a moment of inertia of 30 kg · m^2 in the layout position and 4 kg · m^2 in the tuck position. During the dive his angular momentum is such that he has 0.6 s in which to complete three somersaults in the tuck position. Determine (a) the maximum angular velocity he attains, (b) his angular momentum, and (c) the time needed to complete one revolution in the layout position.

56. A student working on an upended bicycle sets a wheel in motion and then slows it down by applying a force on the tire. The force is applied in such a way that it increases with time (in seconds) according to $F = 1.21t$ N. If the moment of inertia of the wheel is 0.2 kg · m^2 and the radius is 0.3 m, what is the angular acceleration 2 s after application of the force?

57. Prove the assertion that the spin angular momentum of the moon and the orbital angular momentum of the earth are negligible compared with other contributions to the total angular momentum of the earth–moon system.

chapter 13

MOTION OF A RIGID BODY

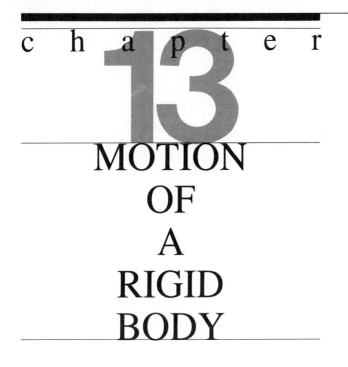

Preview

In Chapter 12 we introduced the variables needed to describe rigid body rotation about an axis that is fixed in direction. We also drew an analogy between linear and rotational motion which culminated in the rotational analog of Newton's second law. In this chapter we extend these ideas in several directions.

First, we introduce rotational forms for work and impulse. Then we use these forms as a basis for developing a rotational treatment of energy and power, and for dealing with impact situations that involve rotation. When a rigid body possesses linear and rotational motion, we can simplify the description of its motion by choosing the origin to be in one of two special locations. The first is a point in the body that is fixed, even if only for an instant, and the second is the center of mass.

We also introduce one of the most striking types of rigid body rotational motion, called *precession*. Precession involves a spin axis that varies in direction, and is found in everything that spins—atoms, neutron stars, planets, and toy tops, for example. Many of the ideas that have been developed so far in this text help to explain precession.

13.1 Work, Energy, and Power in Rotation

The work–energy principle, $W_{net} = \Delta K$, relates the change in kinetic energy of a system of particles to the work done by the net force that is acting (see Section 10.2 and note Eq. 10.20). In particular, the work–energy principle applies to the special case of a rigid body. To restate this principle using variables suitable for a rigid body (torque rather than force, for instance), we must first derive an expression for the work done on a rigid body by a single force **F** in terms of the torque associated with **F**. Remember that the axis of rotation is assumed to be fixed in location and direction. This means that only forces that are perpendicular to the axis, and torques that are parallel or antiparallel to the axis, need to be considered.

Referring to Figure 13.1, we see that the work done by the force **F** in producing the displacement $d\mathbf{s}$ of a point as it rotates about an axis through O is

$$\begin{aligned} dW &= \mathbf{F} \cdot d\mathbf{s} \\ &= F\, ds \cos\theta \end{aligned}$$

Substituting $ds = r\,d\theta$ into this equation, we get

$$dW = (Fr\cos\theta)\, d\theta = (Fr\sin\phi)\, d\theta$$

The quantity in parentheses is the magnitude of the torque, τ, about an axis through O. Therefore, we have the result

$$dW = \tau\, d\theta \tag{13.1}$$

This is the rotational analog of the one-dimensional relation $dW = F\,ds$.

Figure 13.1

The tangential displacement ds of a point subtends the angle $d\theta$. The angles θ and ϕ are complementary.

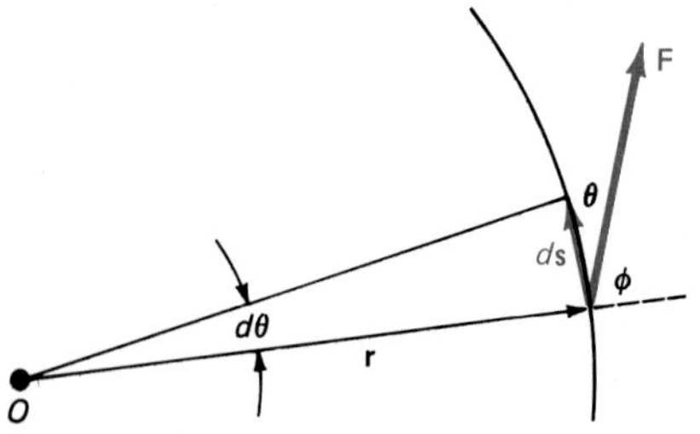

For a finite rotation about a fixed axis, Eq. 13.1 becomes

$$W = \int_{\theta_1}^{\theta_2} \tau\, d\theta \qquad (13.2)$$

which is the rotational analog of Eq. 7.6 for one-dimensional motion of a particle.

If more than one external torque acts, then the resultant torque is the sum of individual torques.

$$\tau_{ex} = \sum \tau_i$$

Because the axis of rotation is fixed in direction (along the z-axis), only the z-component of torque is included in the sum. Each torque, τ_i, will be directed one way or the other along the fixed axis of rotation. Therefore, we must be careful to account correctly for the directions of the torques when performing this sum.

In an infinitesimal displacement $d\theta$, the work done by the net external torque is given by

$$dW_{ex} = \tau_{ex}\, d\theta = \sum \tau_i\, d\theta$$

and so the work done in the finite displacement is

$$W_{ex} = \int_{\theta_1}^{\theta_2} \tau_{ex}\, d\theta$$

In Section 10.2 we saw that the work done on a system of particles by the net force equals the increase in kinetic energy of the system providing the internal forces do no work. In a rigid body the interparticle distances are fixed, guaranteeing that internal forces do no work. Thus we can write

$$W_{ex} = \text{change in rotational kinetic energy}$$
$$= K_f - K_i = \Delta K$$

In Section 12.2 we derived the expression for the rotational kinetic energy of a rigid body, $K = {}^1\!/_2 I\omega^2$, by starting with the expression for the kinetic energy of a system of particles. Thus

$$W_{ex} = \frac{1}{2}I\omega_f^2 - \frac{1}{2}I\omega_i^2$$

where $K_f = {}^1\!/_2 I\omega_f^2$ and $K_i = {}^1\!/_2 I\omega_i^2$ represent the final and initial rotational kinetic energies, respectively.

To stress the analogy between linear and rotational motion, we now give an alternate derivation of the equation $K = {}^1\!/_2 I\omega^2$ that parallels the derivation (Section 7.4) of the linear kinetic energy of a particle, $K = {}^1\!/_2 mv^2$. We start with Eq. 13.2

$$W_{ex} = \int \tau_{ex}\, d\theta$$

where τ_{ex} is the net torque acting. Using the rotational form of Newton's second law,

$$\tau_{ex} = \frac{dL}{dt}$$

we can write

$$W_{ex} = \int_i^f \frac{dL}{dt}\, d\theta$$

The angular momentum is $L = I\omega$ with I constant. Thus

$$\frac{dL}{dt} = I\frac{d\omega}{dt}$$

This gives

$$W_{ex} = \int_i^f I\frac{d\omega}{dt}\, d\theta$$

Rewriting the integrand, we obtain

$$W_{ex} = I\int_i^f \frac{d\theta}{dt}\, d\omega$$
$$= I\int_i^f \omega\, d\omega$$

Integrating and substituting in the limits yields the result

$$W_{ex} = \frac{1}{2}I\omega_f^2 - \frac{1}{2}I\omega_i^2 \qquad (13.3)$$

which completes the derivation. Equation 13.3 explains why ${}^1\!/_2 I\omega^2$ is the rotational analog of ${}^1\!/_2 mv^2$. We designate Eq. 13.3 as the rotational analog of the work–energy principle. If the work done is positive, then the kinetic energy of the body increases. If the work done is negative, then the kinetic energy of the body decreases. If $W_{ex} = 0$, then the kinetic energy does not change—it is conserved.

A common example of negative work done on rotating systems is provided by the frictional torques. If these are the only torques acting, the angular velocity of the body decreases. The tidal torques acting on the earth as it rotates provide another illustration of negative work. These torques cause a decrease in the earth's angular velocity (see Example 6 in Chapter 11), and thus in its rotational kinetic energy about its spin axis.

Example 1 **An Automobile Flywheel**

Among several energy storage devices being investigated for use in automobiles are high-speed flywheels. According to Post and Post (see "Suggested Reading"), about 30 kWh of energy can be stored in a quartz flywheel weighing 130 lb (a mass of 59.1 kg). If we assume that the wheel is a uniform cylinder with a radius of 0.5 m, what angular velocity would correspond to this energy storage?

The rotational kinetic energy stored is given by $K = {}^1\!/_2 I\omega^2$. We can find the angular velocity, ω, by using this equation. However, first we must express K and I in SI units. For K we have

$$K = 30\ \text{kW} \cdot \text{h} = 3 \times 10^4\ \text{W} \cdot \text{h} = 1.08 \times 10^8\ \text{J}$$

For a cylinder, Table 12.3 gives us

$$I = \frac{1}{2}MR^2$$

$$= \frac{1}{2} \times 59.1\ \text{kg} \times (0.5\ \text{m})^2$$

$$= 7.39\ \text{kg} \cdot \text{m}^2$$

Thus we have

$$K = \frac{1}{2}I\omega^2$$

or

$$\omega^2 = \frac{2K}{I}$$

$$= \frac{2 \times 1.08 \times 10^8}{7.39}$$

$$= 2.92 \times 10^7\ \text{rad}^2/\text{s}^2$$

Hence

$$\omega = 5.41 \times 10^3\ \text{rad/s}$$

Because 1 rev/s corresponds to 2π rad/s, the flywheel must rotate at about 861 rev/s!

Rotational Power

The rate at which work is performed, dW/dt, is called *power* (Section 7.5). A rotational form of the equation for power delivered to a body by a torque τ follows directly from Eq. 13.1,

$$dW = \tau\, d\theta$$

which gives the work done by the torque τ acting through the angle $d\theta$. Thus

$$\frac{dW}{dt} = \tau \frac{d\theta}{dt}$$

so that the power can be written

$$P = \tau\omega \qquad (13.4)$$

Equation 13.4 is the rotational analog of the linear relation, $P = Fv$, given for a particle in Section 7.5.

With a fixed axis of rotation, the directions of $\boldsymbol{\tau}$ and $\boldsymbol{\omega}$ must be either "in" or "out" of Figure 13.1. Let "out" be designated the positive direction. Clearly τ and ω can separately be positive or negative. Hence the sign of P in Eq. 13.4 is positive or negative according to whether τ and ω are of the same sign or of opposite signs, respectively.

Example 2 **An Electrical Power Generator**

Electrical power is generated by applying to a large generator an external torque that causes the generator to turn at a predetermined angular velocity. Such generators are so efficient ($\gtrsim$99%) that virtually all of the mechanical power input is realized as electrical power output. A typical generator running at 1800 rpm yields an electrical power of 0.840×10^9 watts (W). What is the torque required?

Corresponding to 1800 rpm we have $\omega = 188$ rad/s. Using Eq. 13.4 we obtain

$$\tau = \frac{P}{\omega}$$

$$= \frac{0.840 \times 10^9}{188}$$

$$= 4.46 \times 10^6\ \text{N} \cdot \text{m}$$

In power stations this torque is often supplied by a large steam turbine, which is equivalent, in its effect, to a wheel and axle. What driving force would be needed to develop this torque by using a wheel and axle comparable in size to the turbine? For a radius of 2 m (see Figure 13.2) the torque of 4.46×10^6 N · m corresponds to a force of 2.23×10^6 N, or roughly 250 tons.

Figure 13.2

The wheel section represents the rotor of a large steam turbine.

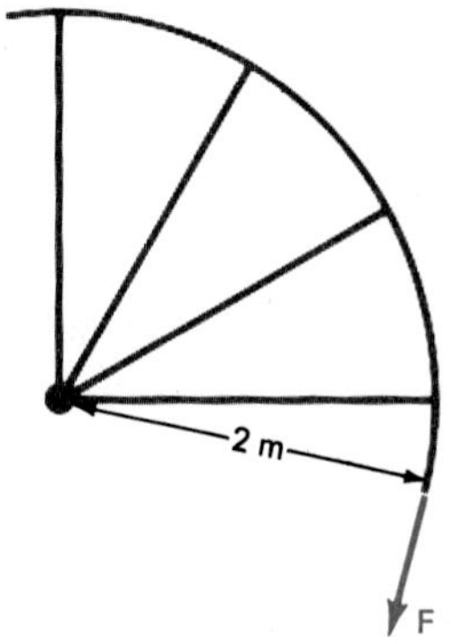

Questions

1. Why does a helicopter need a small propeller having an axis perpendicular to that of the large rotor?
2. A cyclist maintains a constant angular velocity of the bicycle pedals while going both uphill and down. How does the power delivered to the pedals vary with the inclination of the road?
3. If the cyclist in Question 2 exerts a constant downward force with one leg, how does the power delivered to the cycle depend on pedal angle? Consider only the downward motion.
4. A certain rigid body moves with its axis of rotation fixed in direction. In analyzing the motion of this body, only those forces perpendicular to the axis of rotation and only those torques parallel to the axis of rotation need to be considered. What is the justification for this simplifying assumption?

13.2 Rotation with Translation

In a pure rotation, which we considered in Section 12.4, points of the rigid body that are on the axis of rotation do not move. A different type of rotational motion is typified by a billiard ball that *rolls without slipping*. In this case, not only is there motion relative to the center of mass, but the center of mass itself moves. In Figure 13.3, a circle is shown rolling from left to right on a flat surface without slipping. The center of the object moves in a straight line parallel to the surface. A point on the object's circumference traces out a cycloid curve as shown. This point is at rest for the instant that it touches the flat surface. In other words, there is no *relative* motion between the surface and the point of contact on the wheel. The lack of relative motion is what makes this motion that we call *rolling without slipping*.

It seems natural to consider rolling-without-slipping motion as a combination of translation and rotation. We adopted this point of view in Section 12.2, where the kinetic energy of a rigid body was shown to be

$$K = \frac{1}{2}Mv_c^2 + \frac{1}{2}I_c\omega_c^2 \qquad (12.21)$$

Here the first term represents the translational kinetic energy of the body as a whole, and the second term represents the rotational kinetic energy of the body relative to an axis through the center of mass.

Curiously, this rolling-without-slipping motion can also be treated instantaneously as a pure rotation of the rigid body about an axis through the point of contact. This greatly simplifies calculations, and for this reason we will treat the specific case of rolling without slipping before we consider rotation with translation in general.

To compare pure rotational motion with rolling-without-slipping motion consider Figure 13.4. In Figure 13.4a rolling without slipping is depicted. As we have seen, points on the circumference of the circular object trace out cycloid curves as the motion proceeds. The cycloid curves shown in the figure give portions of the trajectories for circumferential points to the left of the diameter PP'. The center of mass at C moves on a straight line to the right with speed v_c. The point of contact P is instantaneously at rest. Hence the point C moves with an

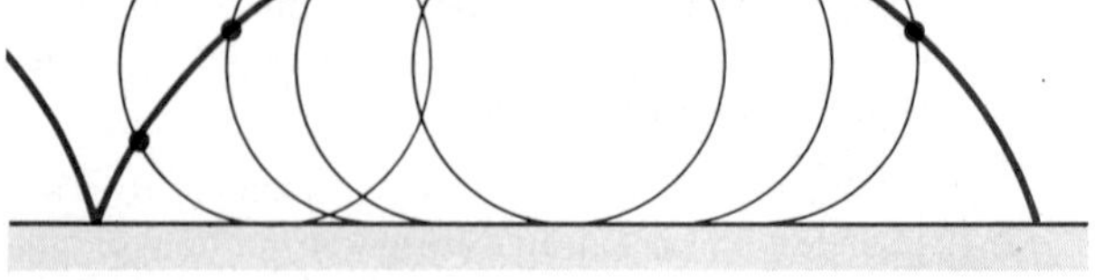

Figure 13.3

When a circle rolls without slipping on a flat surface, each point on the circumference traces out a cycloid curve.

Figure 13.4a

Portions of the various cycloids for points on the circumference produce the family of curves drawn when the circle rolls without slipping. The point P is instantaneously at rest.

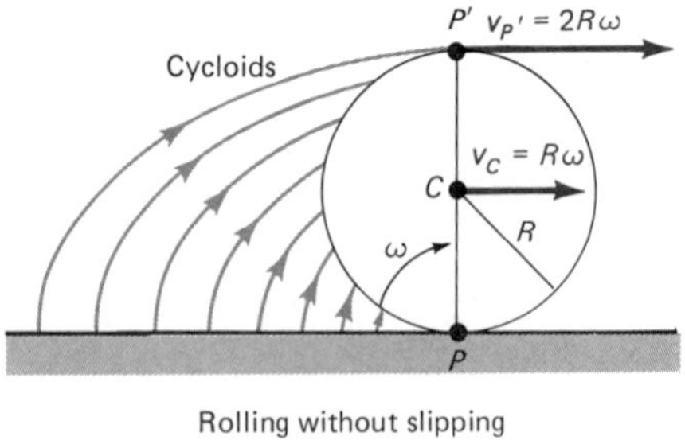

Figure 13.4b

The circle in Fig. 13.4b is drawn as it would look if rotating about an axis through P.

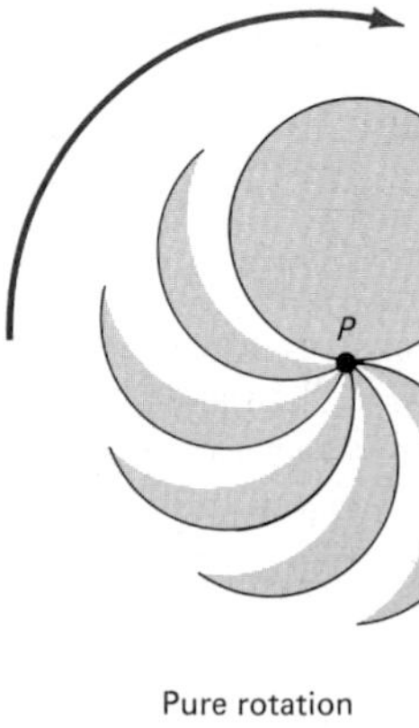

Figure 13.4c

At the instant drawn, the circle moves so that all points have the same velocities as those depicted in Fig. 13.4a. Thus the circle in Fig. 13.4a can be viewed as rotating about an axis through P at the instant drawn.

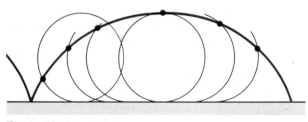

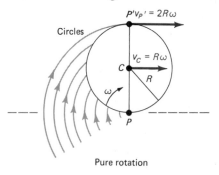

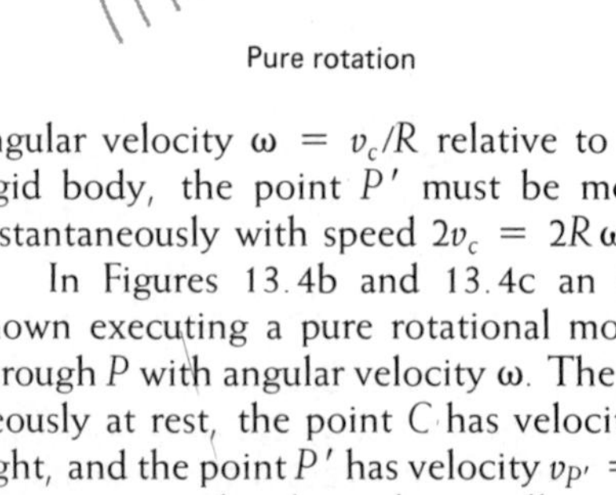

angular velocity $\omega = v_c/R$ relative to the point P. For a rigid body, the point P' must be moving to the right instantaneously with speed $2v_c = 2R\omega$.

In Figures 13.4b and 13.4c an identical object is shown executing a pure rotational motion about an axis through P with angular velocity ω. The point P is instantaneously at rest, the point C has velocity $v_c = R\omega$ to the right, and the point P' has velocity $v_{P'} = 2R\omega$ to the right. Every point in the object that is rolling without slipping (a) has the same instantaneous velocity, at the instant drawn, as the corresponding point in the object that is rotating (b). Thus a rolling-without-slipping motion on a flat surface is equivalent instantaneously to a pure rotation about an axis through the point of contact, P.

To illustrate the dynamic relationship between these two viewpoints we start with the expression for the kinetic energy of a rigid body that was obtained by considering the body's motion to be combined rotation and translation. Then we transform the kinetic energy into the form obtained by considering the body's motion to be pure rotation.

In Section 12.2, and again in this section, we showed that the kinetic energy of a rigid body can be expressed as the sum

$$K = \frac{1}{2}Mv_c^2 + \frac{1}{2}I_c\omega_c^2 \qquad (12.21)$$

where v_c is the center-of-mass speed, I_c is the moment of inertia about the center of mass, and ω_c is the angular velocity of the body about its center of mass. The first term, $\frac{1}{2}Mv_c^2$, is kinetic energy associated with translation of the center of mass, and the second term, $\frac{1}{2}I_c\omega_c^2$, is kinetic energy associated with rotation about the center of mass. This relation applies, in particular, to the object shown rolling without slipping in Figure 13.5a. An observer at rest relative to point P will attribute to the center of mass C a speed v_c directed to the right. An observer at rest relative to the center of mass will attribute to the object a pure rotational motion with angular velocity ω_c about the center of mass, as shown in Figure 13.5b. Using v_c and ω_c along with M and I_c, we can obtain the total kinetic energy of the object with Eq. 12.21.

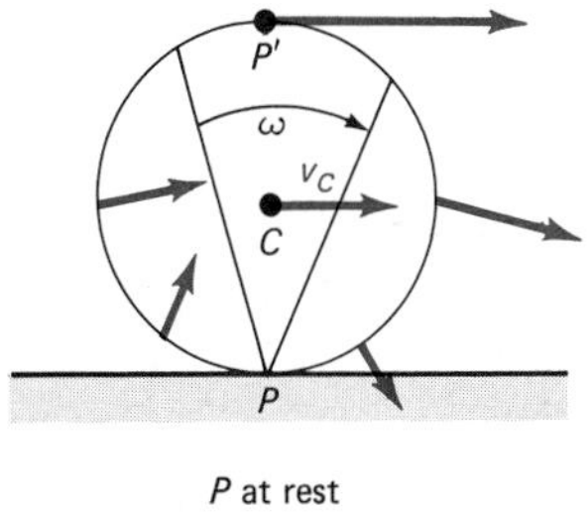

Figure 13.5a

At the instant drawn, the circle (which is rolling to the right without slipping) is viewed as rotating about P with angular velocity ω. The point P is at rest.

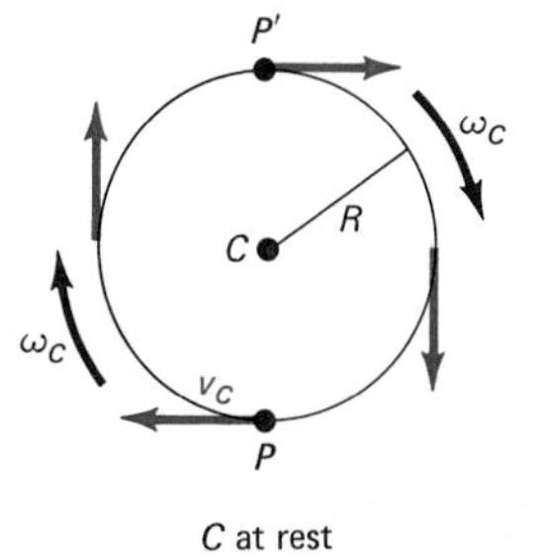

Figure 13.5b

Viewed from the point C, the point P moves to the left with speed v_c and the circle motion is a pure rotation.

The kinetic energy of the object shown rolling without slipping in Figure 13.5a can also be interpreted as a pure rotation about an axis through the point P. This alternate interpretation is revealed by an equation that we can derive from Eq. 12.21. For the observer at rest relative to the point P, the center of mass (and the entire object) has an angular velocity $\omega = v_c/R$. For the observer at rest relative to the center of mass the object has an angular velocity $\omega_c = v_c/R$. Clearly, $\omega_c = \omega$, and so we can write

$$K = \frac{1}{2}MR^2\omega^2 + \frac{1}{2}I_c\omega^2$$

$$= \frac{1}{2}(MR^2 + I_c)\omega^2$$

Using the parallel axis theorem (see Eq. 12.29)

$$I_p = I_c + MR^2$$

is the moment of inertia of the object about the point P. This gives the result

$$K = \frac{1}{2}I_p\omega^2 \qquad (13.5)$$

for the kinetic energy, and this is exactly the expression for K that we would have written had the body originally been viewed as rotating about an axis through the point P. Each of the two viewpoints developed is useful. Sometimes one and sometimes the other is more convenient.

Example 3 The Great Race

A solid disk, a hoop, and a solid sphere (Figure 13.6) roll, without slipping, downhill. They start from rest and have equal radii and masses. Which one reaches the bottom of the hill first?

We take the finish line to be at a vertical distance y below the starting line. Evidently the object with the largest linear velocity down the hill wins. We use the law of conservation of energy to write

$$Mgy = \frac{1}{2}I_p\omega^2 = \frac{1}{2}I_p\left(\frac{v_c}{R}\right)^2$$

where the axis of rotation instantaneously is through the contact point. Then

$$v_c^2 = \frac{2MR^2gy}{I_p} = \frac{2gy}{I_p/MR^2}$$

$$= \frac{2gy}{1 + (I_c/MR^2)}$$

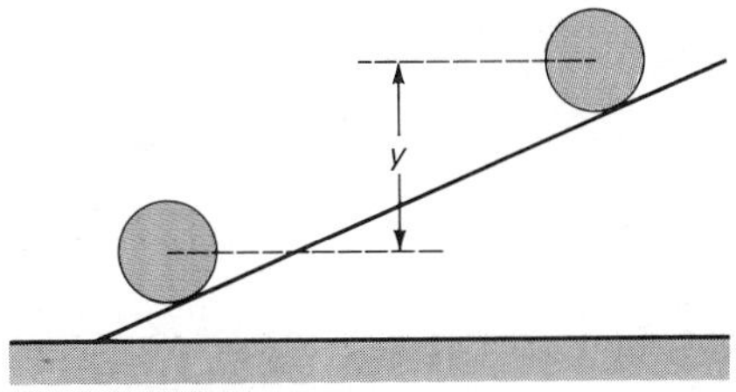

Figure 13.6

An object of circular cross section rolls down the plane through a vertical distance y.

Figure 13.7

When given the proper spin a Superball will bounce back and forth several times.

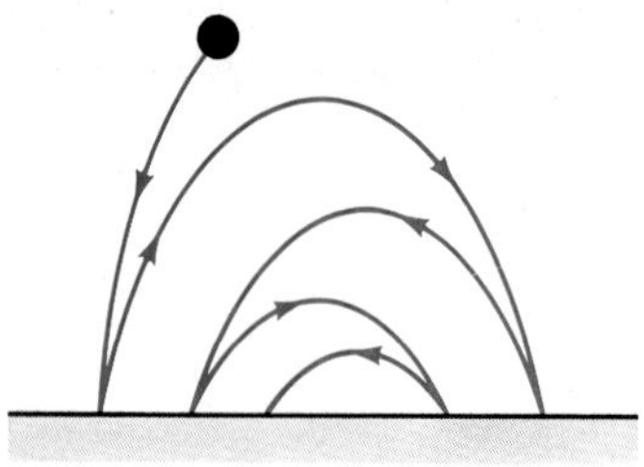

where in the last step we use the parallel axis theorem. Reference to Table 12.3 shows that I_c/MR^2 is smallest for the solid sphere. Hence v_c^2 is largest for the solid sphere, and the sphere wins the race.

Physically the smallest I_c presents the smallest resistance to angular acceleration. Thus the angular acceleration will be largest for an object with minimum I_c. This in turn implies that the linear acceleration, and therefore V_c, will be largest. From an energy viewpoint the smallest I_c accompanies the smallest rotational energy. This, coupled with a constant total energy, implies a larger translational energy, and hence a larger v_c.

Questions

5. A child in a swing that is already moving can increase the amplitude of motion in the swing without being pushed. How is this done?
6. A Superball dropped with the "right" spin and direction of velocity will bounce back and forth as suggested in Figure 13.7. Explain.
7. Can the path of a bowling ball be a curve if the ball rolls on a flat surface without slipping? Explain.

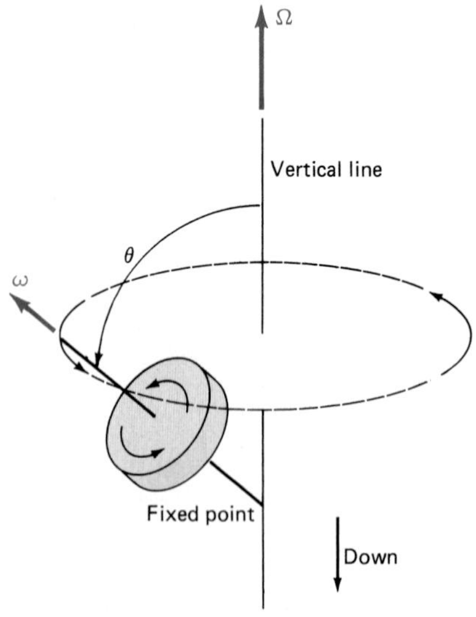

Figure 13.8

As the toy gyroscope spins with angular velocity **ω** its spin axis rotates about the vertical line, tracing out the surface of a cone. This regular rotation of the spin axis is called precession, and the precessional angular velocity is labeled **Ω**.

13.3 Precession

We now consider motion where the axis of rotation changes direction. Spinning symmetric bodies such as gyroscopes, toy tops, and the earth exhibit **precession**—a regular rotation of the spin axis. In Figure 13.8 we show a toy gyroscope executing precessional motion. One end of the spin axis of the gyroscope is fixed in position, but otherwise it is free to move. The spin axis and spin angular velocity ω make a fixed angle θ with the vertical. In this pure precessional motion the spin axis traces a cone and the precessional angular velocity **Ω** is directed upward. To begin with, we consider precession for which $\theta = 90°$ and the spin axis is horizontal.

In this section we will examine an important special case of precession—the motion of a "fast top." The fast top approximation entails neglecting all angular momenta other than spin angular momentum, and assuming a steady precession of the spin axis.

Consider a spinning bicycle wheel such as is often used in classroom demonstrations of precession. One end of the axle is supported, with a string for example, while the other end is left free to move. In Figure 13.9 such a wheel is shown with its spin angular velocity vector, **ω**, along the $+x$-axis. The z-axis is vertical and passes through the supported end of the axle. If the wheel is properly started, and **ω** is large, then the spin axis will rotate in the x–y (horizontal) plane. This rotation of the spin axis is called precession, and the angular velocity, **Ω**, associated with this rotation is called the *precessional angular velocity*.

The vector relationship between **ω** and **Ω** is shown in Figure 13.10. Here **ω** lies in the x–y plane and rotates from the $+x$-axis toward the $+y$-axis. The **Ω** associated with this precession is along the $+z$-axis according to the right-hand rule.

The spin angular momentum vector, L, is parallel to **ω**. As **ω** rotates, L rotates with it in the x–y plane. In a

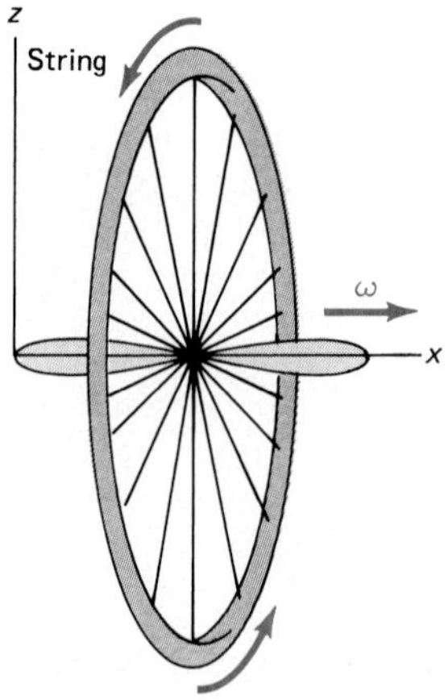

Figure 13.9

The bicycle-wheel axle is attached at one end to a string. The spin angular velocity **ω** precesses in the horizontal plane.

Figure 13.10

The precession angular velocity vector $\boldsymbol{\Omega}$ is directed upward along the positive z-axis, and the spin angular velocity vector $\boldsymbol{\omega}$ rotates in the x–y plane. As precession proceeds, the tip of the spin angular velocity vector moves as suggested from 1 to 2 to 3 to 4.

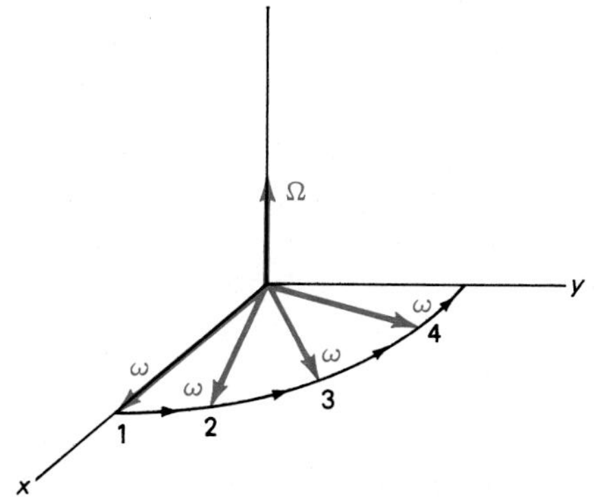

time interval Δt, $\mathbf{L}$ will rotate through an angle $\Delta\phi$. The change in angular momentum is related to the initial ($\mathbf{L}_i$) and final ($\mathbf{L}_f$) angular momenta according to

$$\Delta \mathbf{L} = \mathbf{L}_f - \mathbf{L}_i \qquad (13.6)$$

This equation is exhibited geometrically in Figure 13.11a.

To produce such a change in angular momentum an angular impulse is required. The dynamics are governed by the equation

$$\Delta \mathbf{L} = \boldsymbol{\tau}\,\Delta t \qquad (13.7)$$

where $\boldsymbol{\tau}$ is the net applied torque. Note that $\Delta\mathbf{L}$ and $\boldsymbol{\tau}$ are in the same direction. If $\boldsymbol{\tau}$ is parallel to $\mathbf{L}_i$, the magnitude, but not the direction, of $\mathbf{L}_i$ changes. The wheel spins faster or slower, but its axis of rotation is not altered. If $\boldsymbol{\tau}$ is perpendicular to $\mathbf{L}_i$, the direction, but not the magnitude, of $\mathbf{L}_i$ changes. In precession $\boldsymbol{\tau}$ and $\mathbf{L}_i$ are perpendicular.

The direction relations for the position vector $\mathbf{r}$, the weight vector $\mathbf{W}$, and the torque vector $\boldsymbol{\tau}$ are exhibited in Figure 13.11b. For $\mathbf{r}$ in the positive x-direction and $\mathbf{W}$ in

Figure 13.12

The gyroscope axis is parallel to the x-axis with $\boldsymbol{\omega}$ in the positive x-direction. The torque $\boldsymbol{\tau}$ due to gravity is in the positive y-direction. For these directions of $\boldsymbol{\omega}$ and $\boldsymbol{\tau}$ the precession angular velocity $\boldsymbol{\Omega}$ is in the positive z-direction.

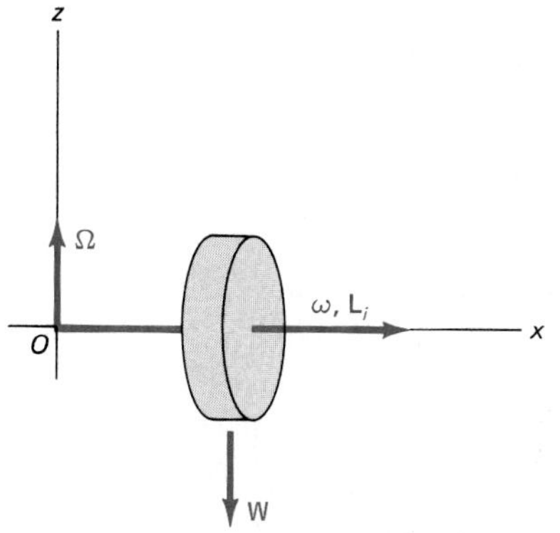

the negative z-direction, $\boldsymbol{\tau}$ will be in the positive y-direction in accordance with $\boldsymbol{\tau} = \mathbf{r} \times \mathbf{W}$.

The precessional behavior just outlined is a steady-state motion. This behavior will occur only if the system is properly started in its motion. We will assume this steady-state motion, and derive a relation among the external torque, $\boldsymbol{\tau}$, the angular momentum, $\mathbf{L}$, and the precessional angular velocity, $\boldsymbol{\Omega}$.

Consider the body in Figure 13.12, shown spinning instantaneously about the $+x$-axis. The end of the axle is attached to the point O. Both $\boldsymbol{\omega}$ and $\mathbf{L}_i$ lie along the $+x$-axis. The weight, $\mathbf{W}$, gives rise to a torque, $\boldsymbol{\tau} = \mathbf{r} \times \mathbf{W}$, about the $+y$-axis. The angular impulse delivered by $\boldsymbol{\tau}$ during Δt is given by

$$\mathbf{A} = \boldsymbol{\tau}\,\Delta t \qquad (13.8)$$

The vectors $\boldsymbol{\tau}$ and $\mathbf{A}$ stay 90° ahead of the vectors $\boldsymbol{\omega}$ and $\mathbf{L}$. Combining Eqs. 13.6, 13.7, and 13.8, we have

$$\mathbf{L}_f = \mathbf{L}_i + \boldsymbol{\tau}\,\Delta t \qquad (13.9)$$

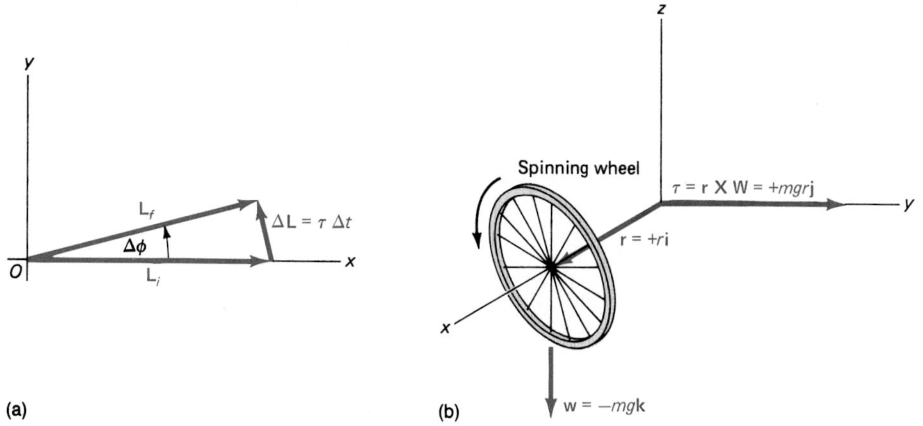

Figure 13.11

When the spin angular velocity vector is in the positive x-direction, the vector $\mathbf{r}$ from the origin to the center of gravity is $\mathbf{r} = r\mathbf{i}$. The weight vector $\mathbf{W}$ is directed downward along the negative z-axis, and is given by $\mathbf{W} = -mg\mathbf{k}$. Hence the torque vector $\boldsymbol{\tau} = \mathbf{r} \times \mathbf{W} = +mgr\mathbf{j}$ is directed along the positive y-axis.

Figure 13.13

The vector equation $\mathbf{L}_f = \mathbf{L}_i + \boldsymbol{\tau}\,\Delta t$ corresponds to the triangle relation drawn.

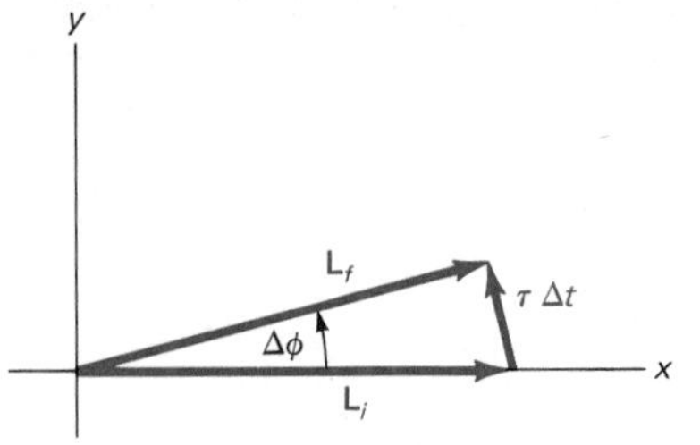

This equation is interpreted geometrically in Figure 13.13. Because $\boldsymbol{\tau}\,\Delta t$ is in the $+y$-direction and $\mathbf{L}_i$ is in the $+x$-direction, $\mathbf{L}_f$ lies in the x–y plane rotated through $\Delta\phi$ counterclockwise relative to $\mathbf{L}_i$. According to the right-hand rule, this establishes the direction of the precession angular velocity, $\boldsymbol{\Omega}$, to be "out" of Figure 13.13, or along the $+z$-direction, as shown in Figure 13.12.

The magnitude of $\boldsymbol{\Omega}$ is defined by the equation

$$\Omega = \lim_{\Delta t \to 0} \frac{\Delta\phi}{\Delta t}$$

Let L represent the common magnitude of $\mathbf{L}_i$ and $\mathbf{L}_f$. From the triangle in Figure 13.11a we have for small $\Delta\phi$

$$\tau\,\Delta t \approx L\,\Delta\phi$$

Dividing by Δt and taking the limit as $\Delta t \to 0$ makes the expression become exact:

$$\tau = \Omega L \qquad \text{or} \qquad \Omega = \frac{\tau}{L} \tag{13.10}$$

for the magnitude of Ω. When we combine Eq. 13.10 with our considerations of direction, we are able to write the result as the vector product

$$\boldsymbol{\tau} = \boldsymbol{\Omega} \times \mathbf{L} \tag{13.11}$$

This relation holds for a top or gyroscope that is "fast," that is, one for which the spin angular velocity is large compared with the precessional angular velocity ($\omega \gg \Omega$). We can see that Eq. 13.11 could not be valid for a slow top if we consider a top having $\omega = 0$. Referring to Figure 13.12, we see that the torque due to gravity in the $+y$-direction causes the top to fall. The top doesn't precess at all.

A top or gyroscope whose moment of inertia about its spin axis is I_c has a spin angular momentum

$$L = I_c\omega$$

If the center of gravity precesses in a horizontal plane (Figure 13.14) in a circle of radius R, the torque due to gravity is

$$\tau = MgR$$

Substituting these expressions for L and τ into Eq. 13.10 gives for the precessional angular velocity

$$\Omega = \frac{MgR}{I_c\omega} \tag{13.12}$$

The inverse relation between spin and precessional angular velocities means that a large spin angular velocity (ω) accompanies a small precessional angular velocity (Ω). We say that a top is fast if

$$\omega \geq 10\Omega$$

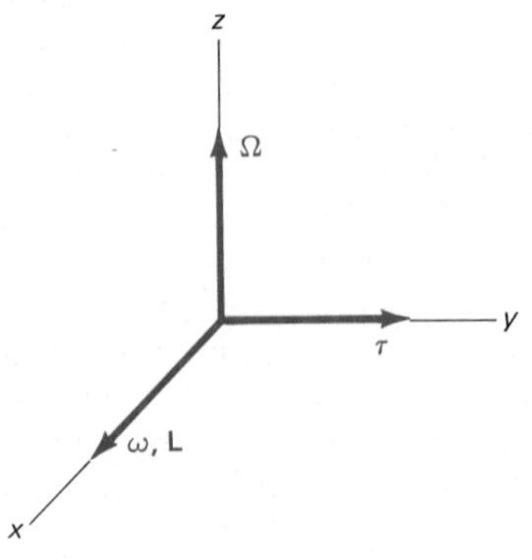

Figure 13.14

When a fast top precesses with its spin axis moving in a horizontal plane the vectors $\boldsymbol{\omega}$ (or L), $\boldsymbol{\Omega}$, and $\boldsymbol{\tau}$ are mutually perpendicular.

Example 4 A Fast Toy Gyroscope

A toy gyroscope is made in the shape of a uniform disk with an axle of negligible mass (see Figure 13.15). For simplicity, let the distance from the point of support to the center of gravity and the radius of the disk both equal R. What is the minimum spin angular velocity for which the gyroscope is fast?

With the axle horizontal, and supported at the end point O, the only torque acting about an axis through O is that due to gravity. With $I_c = {}^1\!/_2 MR^2$ Equation 13.12 gives

$$\Omega = \frac{MgR}{\frac{1}{2}MR^2\omega} = \frac{2g}{R\omega}$$

for the precessional angular velocity. A typical toy gyroscope might have $R = 0.04$ m, giving

$$\Omega = \frac{490}{\omega}$$

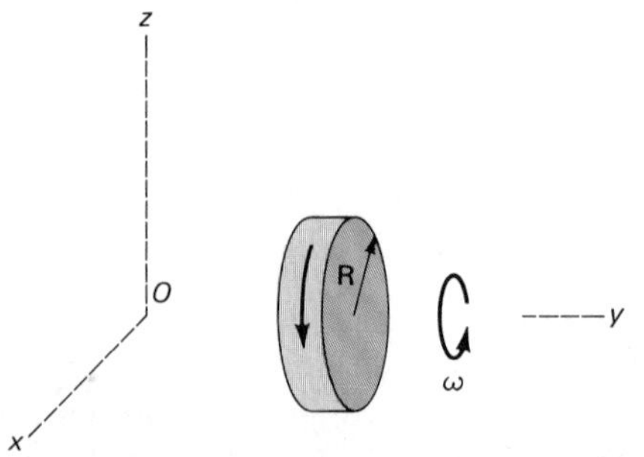

Figure 13.15

A toy gyroscope is shown with its spin angular velocity $\boldsymbol{\omega}$ horizontal and in the positive y-direction.

For the gyroscope to be fast, the minimum spin angular velocity is given by

$$\omega \geqq 10\Omega$$

Eliminating Ω, we obtain

$$\omega \geqq \frac{10 \cdot 490}{\omega}$$

or

$$\omega^2 \geqq 4900$$

This gives

$$\omega \geqq 70 \text{ rad/s} = 11.1 \text{ rev/s}$$

as the minimum spin angular velocity. For this value of ω we have $\Omega = 7$ rad/s $= 1.11$ rev/s. Larger values of ω would correspond to smaller values of Ω, and would be consistent with the pure precession of a fast top.

The reciprocal relation between ω and Ω indicates that the rate of precession increases as ω decreases. As energy is lost as a result of friction, ω will decrease, which explains why a "dying" top or gyroscope precesses faster and faster as it dies.

In the derivation of the equation $\boldsymbol{\tau} = \boldsymbol{\Omega} \times \mathbf{L}$ we assumed that the spin angular velocity vector $\boldsymbol{\omega}$ maintained the fixed angle of 90° with the z-axis during precession. However, a fast top or gyroscope can precess steadily for fixed angles other than 90° between $\boldsymbol{\omega}$ and the z-axis. In Figure 13.16 a steady precession is depicted for an arbitrary angle θ. We assert that the equation $\boldsymbol{\tau} = \boldsymbol{\Omega} \times \mathbf{L}$ can be generalized to include this type of steady precession provided that the spin is fast.

When a toy top is initially set into rotation on a rough horizontal surface it will often begin to precess at an angle θ. However, the top will gradually right itself until the symmetry axis is vertical. It is not difficult to understand qualitatively how the action of friction to right the top fits into our discussion of precession.

Imagine the point of the top, shown in Figure 13.17 in contact with the floor at the point P, to be moving perpendicular to and into the figure as a result of the spin about the symmetry axis of the top. The frictional force exerted on the top at P by the floor will oppose this motion, and thus be directed out of the figure. Relative to the center of mass at C this frictional force gives rise to the torque τ_1 in the plane of the figure. The angular impulse associated with this torque will change the angular momentum of the top and will produce a change in the direction of the spin axis toward the vertical. Since this motion of the axis vanishes only if the top is in the "sleeping," or vertical, position, the effect of friction will be to cause the spin axis of the top to rotate until it is vertical.

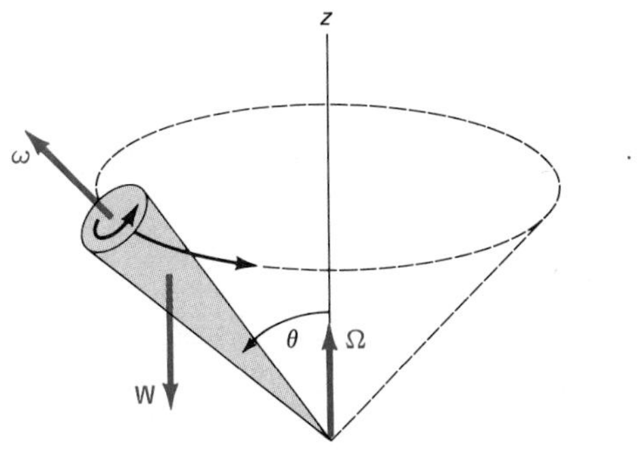

Figure 13.16

The spin axis of a top or gyroscope traces the surface of a cone (angle θ) during precession.

Figure 13.17

The tip of a spinning top is shown when the spin axis of the top is not vertical. When the spin is fast precession causes the spin axis of such a top to rotate into a vertical or "sleeping-top" orientation.

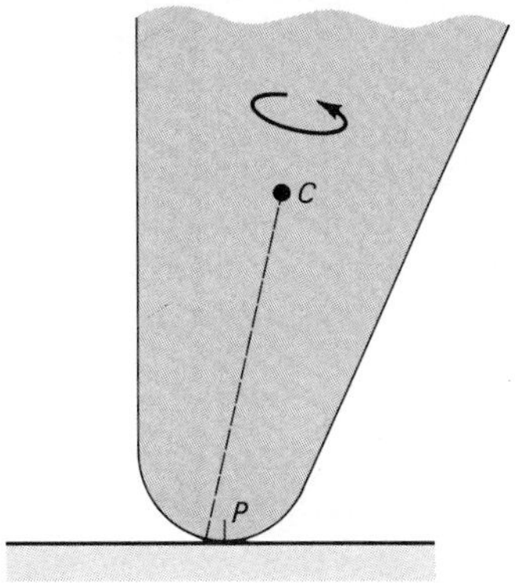

Precession of the Equinoxes

Precession is not limited to small tops and gyroscopes. Planets and stars can spin, and thus will precess if subjected to an angular impulse. More than 2000 years ago Hipparchus found that the equinoctal points disagreed with measurements made by earlier astronomers, and in this way discovered the precession of the earth.

Equinoctal points are positions in the earth's orbit about the sun at which day and night are of equal length. That the spin axis of the earth is inclined at an angle of about 23.5° with the perpendicular to the earth's orbital plane accounts for the difference in length of day and night, and for the existence of seasons. During one revolution about the sun the orientation of the earth's spin axis is nearly constant. When the North Pole is tilted toward the sun, day is longer than night and it is summer in the northern hemisphere. When the North Pole is tilted away from the sun, day is shorter than night and it is winter in the northern hemisphere. Halfway between these orbital extremes day and night are of equal length. These orbital positions are the two equinoctal points.

As the spin axis precesses the equinoctal points move synchronously, giving rise to the "precession of the equinoxes." Today the earth's spin axis is known to precess once roughly every 26,000 years, tracing a cone with an angular radius of 23.5°. Thus experimentally we have for the precession of the equinoxes of the earth

$$\omega = 1 \text{ rev/day} = 7.27 \times 10^{-5} \text{ rad/s}$$

$$\Omega = 1 \text{ rev/26,000 yr} = 7.66 \times 10^{-12} \text{ rad/s}$$

Quite clearly the "fast" criterion $\omega \geqq 10\Omega$ is satisfied.

The physical origin of the torque that gives rise to this precession is the gravitational interaction between the moon (and the sun) and the equatorial "bulges" associated with the oblate (spheroidal) shape of the earth. If the force exerted by the moon on every unit mass in the earth were the same, then no torque would be exerted on the earth. Because of the inverse-square nature of gravity, the force on unit mass increases as the mass gets closer to the moon. Imagine that the force exerted on unit mass at the earth's center by the moon is subtracted from the moon's force on every unit mass in the earth. A net attractive force, $+\mathbf{F}$, will survive on the near bulge, and a net repulsive force, $-\mathbf{F}$, will survive on the far bulge. This situation is depicted in Figure 13.18.

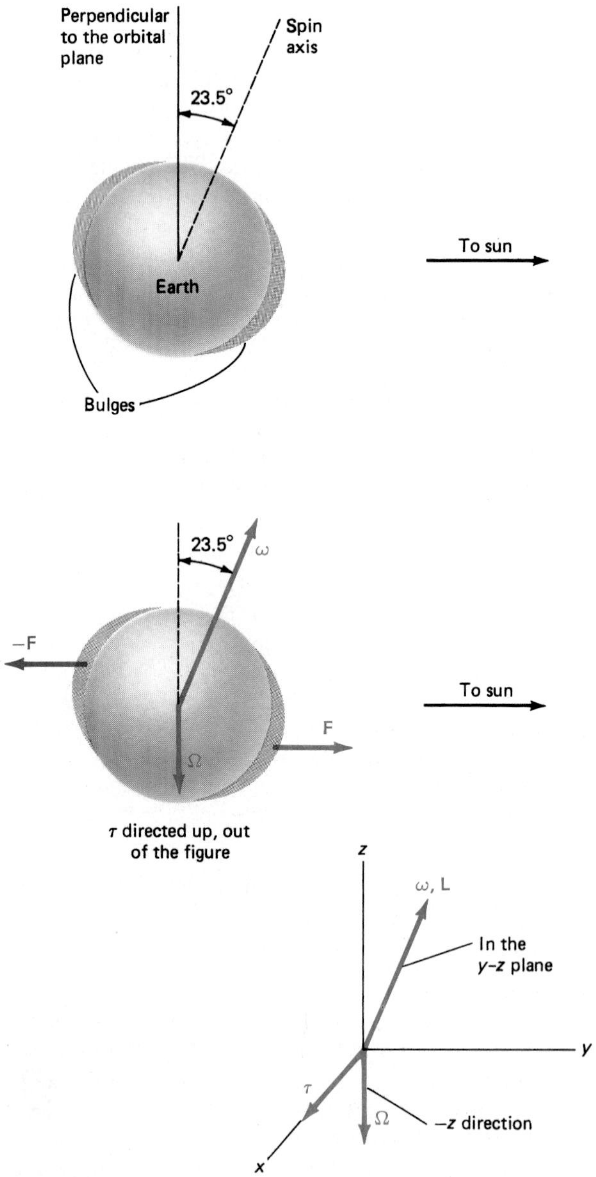

Figure 13.18

The spin axis of the earth is tilted at 23.5° to the normal of the earth's orbital plane. This tilt gives rise to gravitational torques due to the sun and moon causing the spin axis of the earth to precess.

Relative to the center of mass of the earth, the forces $+\mathbf{F}$ and $-\mathbf{F}$ give rise to a torque directed out of Figure 13.18. The angular momentum vector is parallel to $\boldsymbol{\omega}$. Thus $\boldsymbol{\Omega}$ is directed as drawn. This analysis is confirmed by observation. Note that gravity acts to tip the spin axis of the gyroscope away from the vertical. Gravity acts on the earth to align the spin axis with the perpendicular to the earth's orbital plane. As a result, the precessional angular velocity, $\boldsymbol{\Omega}$, is reversed for the earth relative to the gyroscope.

Question

8. Could the moon exert a torque on the earth if the earth were a perfect homogeneous sphere? Does the shape of the moon matter? Explain.

Summary

The work done on a rigid body by a torque $\boldsymbol{\tau}$ in a finite rotation is

$$W = \int_{\theta_1}^{\theta_2} \tau \, d\theta \qquad (13.2)$$

If the torque is the net external torque, then the work–energy principle requires this to equal the increase in kinetic energy of the body. Hence we have

$$W_{\text{ex}} = \frac{1}{2}I\omega_f^2 - \frac{1}{2}I\omega_i^2 \qquad (13.3)$$

The power developed by a torque τ acting at an angular velocity ω is given by

$$P = \tau\omega \qquad (13.4)$$

An object rolling on a flat surface without slipping may be viewed instantaneously as rotating about P, the contact point. In this case the object's kinetic energy is

$$K = \frac{1}{2}I_p\omega^2 \qquad (13.5)$$

The object can also be viewed as both translating and rotating. In this case its kinetic energy is

$$K = \frac{1}{2}Mv_c^2 + \frac{1}{2}I_c\omega_c^2 \qquad (12.21)$$

Precession is a regular rotation of an object's spin axis. When a top or gyroscope is precessing, the torque $\boldsymbol{\tau}$, angular momentum $\mathbf{L}$, and precessional angular velocity $\boldsymbol{\Omega}$ are related by

$$\boldsymbol{\tau} = \boldsymbol{\Omega} \times \mathbf{L} \qquad (13.11)$$

Suggested Reading

R. F. Post and S. F. Post, Flywheels, *Sci. Am.*, 229(6), 17–23 (December 1973).

Problems

Section 13.1 Work, Energy, and Power in Rotation

1. A constant force of 150 N is applied tangentially to the circumference of a bicycle wheel (M = 8 kg, R = 0.4 m) whose angular velocity increases from zero while the wheel turns through an angle of 90°. Treat the wheel as a hoop and calculate (a) the applied torque, (b) the angle in radians, (c) the work done by the applied torque, and (d) the angular velocity reached after the 90° turn.

2. In some ancient power mills, a horse (m = 800 kg) was made to walk continuously on an inclined wheel (θ = 10°). The wheel turned as the horse walked (Figure 1) so that the position of the horse relative to the ground was fixed. Take the radius of the horse's path to be 2 m, and determine (a) the walking speed of the horse and (b) the torque exerted on the wheel by the horse if the animal is able to sustain a power output of 1 hp (1 hp = 746 W).

Figure 1

3. The energy that can be stored in a rotating disk is limited by the tensile strength of the construction material. For an energy-storing disk made of fiberglass the maximum energy per kilogram of material is 238 Wh/kg. How long would it take a 5-hp motor to give a 200-kg disk its maximum energy?

4. A flywheel energy storage unit being proposed for a small bus consists of 6 disks, each of radius 0.48 m and mass 15 kg. When moving at maximum speed (fully charged), the flywheel rotates at 16,000 revolutions per minute (rev/min). If the bus requires 4.1×10^6 J for each kilometer of travel, how many miles can the bus travel on a single "charge" of the flywheel?

5. Gasoline has an energy content of 1.33×10^8 J per gallon. What is the angular speed of a 200-kg disk with a radius of 1 m if its kinetic energy is equivalent to 1 gallon of gasoline?

6. A father hears a cry for help and discovers that his daughter has fallen into a well. He tells her to hold on to the bucket so that he can pull her up with the hand winch. If the father is capable of a maximum power of 60 W and the mass of the daughter is 40 kg, (a) what is the maximum upward velocity of the daughter? (b) If the father rotates the handle at 1 rev/s, what torque does he exert?

7. A sling accelerates a 0.10-kg projectile from rest to a speed of 70 m/s. (a) Determine the work performed on the projectile. (b) The projectile experiences a constant torque and the acceleration takes place over an angular displacement of $\pi/3$ radians. Determine the magnitude of the torque that acts. (c) The angular velocity of the projectile just before the moment of release is 100 rad/s. Determine the power delivered at that moment.

8. Flywheels weighing several hundred tons have been proposed as mechanical energy storage devices. Consider a 100-ton flywheel that stores 10^4 kWh of energy at an angular speed of 3500 rev/min. Determine (a) the moment of inertia of the flywheel in kg · m², (b) the time required to bring it to rest while it delivers 3000 kW of power, and (c) the initial torque delivered externally by the flywheel while it is losing energy at the rate of 3000 kW.

9. The friction of the tidal bulges exerts a retarding torque tending to slow the earth's rate of rotation. Suppose that one year from now a day is 10^{-5} s longer than it is today. Calculate (a) the present angular velocity of the earth and (b) the change in the angular velocity a year from now. (*Note:* $d\omega/\omega = -dT/T$.)

10. Calculate (a) the present rotational kinetic energy of the earth and (b) the change in the rotational energy a year from now (see Problem 9). (*Note:* $dK/K = 2\,d\omega/\omega$; $I_{\oplus} = 0.3306 \times 2.432 \times 10^{38}$ kg · m².)

11. (a) From this decrease (Problem 10) in the rotational energy of the earth of 4.95×10^{19} J in 1 yr, calculate the average torque of the tidal bulges. (b) Calculate the rate dW/dt at which rotational energy is being lost (converted into thermal energy by friction).

Section 13.2 Rotation with Translation

12. A basketball rolls without slipping across the floor. (a) Calculate the ratio of its rotational kinetic energy (relative to the center of mass) to its translational kinetic energy. (b) Would this ratio be the same for a solid sphere?

13. A bowling ball rolls without slipping down the bowling lane. Determine the ratio of rotational kinetic energy to translational kinetic energy.

14. A pool ball (radius 0.0288 m, mass 0.165 kg) rolls without slipping. Its center of mass moves at a speed of 2.1 m/s. Determine (a) the translational kinetic energy ($\frac{1}{2}Mv_c^2$), (b) the rotational kinetic energy as reckoned from the center of mass ($\frac{1}{2}I_c\omega_c^2$), and (c) the total kinetic energy.

15. A bowling ball is started down a bowling lane without spin at a speed v_0. What fraction of the ball's original kinetic energy is lost by the time it starts to roll without slipping? (*Hint:* Use both linear and rotational forms of Newton's second law in impulsive form.)

16. A solid sphere with a radius of 15 cm is given an angular speed of 25 rad/s. It rolls without slipping up an inclined plane. (a) Through what vertical distance will the center of mass of the sphere rise before it comes to rest momentarily? (b) What is the speed of the center of mass when the sphere returns to its starting point?

17. A solid sphere of radius r and mass m rolls without slipping down the track shown in Figure 2. At the end of its run at point Q its center-of-mass velocity is directed upward. To what height above the base of the track will the sphere rise?

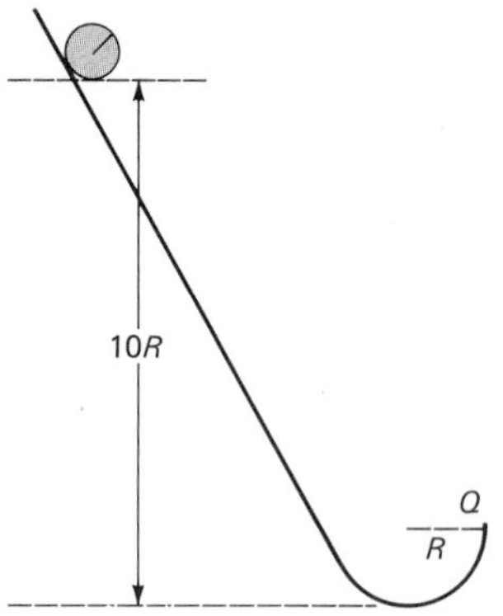

Figure 2

18. A marble starting from rest rolls without slipping on the track of a "loop-the-loop" as shown in Figure 3. (a) Explain why the net force on the marble at position A is $N + mg$, where N is the normal force exerted by the track on the marble. (b) Show that $N = 0$ when the marble starts from a height given by $H = 2.7R$.

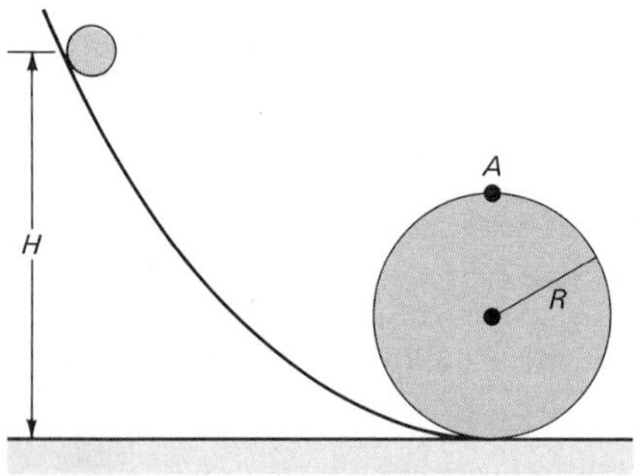

Figure 3

19. A billiard ball is given a linear velocity of 10 m/s without spin. It begins to roll without slipping after traveling a distance of 2 m. Determine the coefficient of friction between the table and the billiard ball.

20. Consider an object whose moment of inertia is described by $I_{cm} = aMR^2$ where a is a constant. Given the relation $v_c = R\boldsymbol{\omega}_c$, show that when the object is rolling without slipping, the ratio of rotational kinetic energy to translational kinetic energy is equal to the constant a.

21. Prove that all solid spheres, independent of mass or radius, arrive at the bottom of the incline simultaneously under the conditions of the race described in Example 3.

22. A uniform ladder of mass M and length L rests against a wall making an angle θ with the horizontal. The wall and floor are frictionless, and the ladder is held in place with a string. At a certain moment the string breaks and the ladder slides down to the floor. Determine, at the moment of impact with the floor, (a) the total kinetic energy of the ladder, (b) the magnitude of its center-of-mass velocity, and (c) its angular velocity relative to its center of mass.

Section 13.3 Precession

23. Determine the spin angular velocity of the gyroscope in Example 4 if $\Omega = 0.2$ rev/s.

24. Suppose that the toy gyroscope in Example 4 (mass = 0.5 kg) spins at 20 rev/s. Calculate (a) the spin angular momentum; (b) the torque due to gravity; (c) the precessional angular momentum (take $I = (5/4)MR^2$ about the z-axis); (d) the angular impulse delivered by gravity while the top precesses through an angle of 0.1 rad.

25. A bicycle wheel ($I = 1.50 \text{ kg} \cdot \text{m}^2$ and $M = 8$ kg) spins at 25 rad/s with its axle horizontal. If the axle is pivoted about a point O (Figure 4) located 0.1 m from the plane of the wheel, calculate the precession rate of the axle.

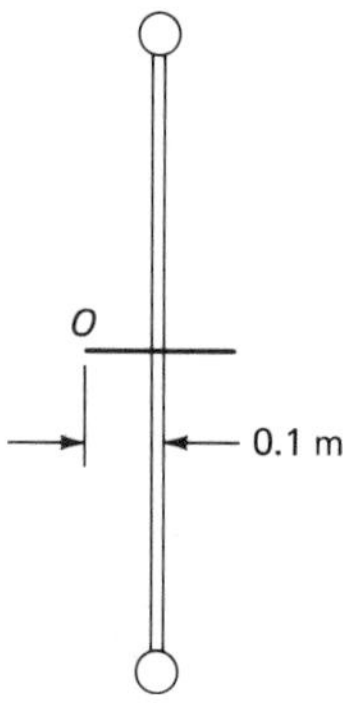

Figure 4

26. A bicycle-wheel "gyroscope" used to demonstrate fast precession has a *precessional* period of 28 s. The center of gravity moves in a horizontal circle whose radius is 35 cm. The mass of the wheel is 3.9 kg and its moment of inertia about its spin axis is $1.3 \text{ kg} \cdot \text{m}^2$. Determine its spin period (time to complete one revolution about its spin axis). Does the precession satisfy the fast condition, $\omega \geqq 10\Omega$?

27. A fast top (Figure 5) has its spin axis tilted at an angle θ with the vertical. The distance from the point of support, O, to the center of gravity, C, is l. The top executes a precessional motion with its center of gravity moving in a horizontal circle of radius R.

Prove that the precessional angular velocity is given by

$$\Omega = \frac{mgl}{I_c \omega}$$

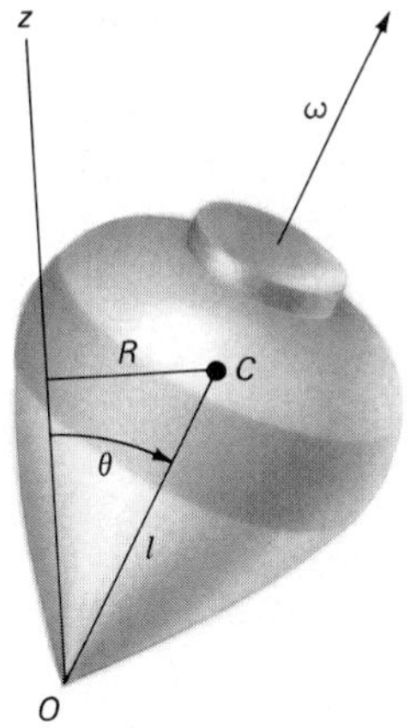

Figure 5

28. Prove that for a fast top the spin kinetic energy is large compared to the maximum decrease in potential energy were the top to fall over.

chapter 14

OSCILLATORY MOTION

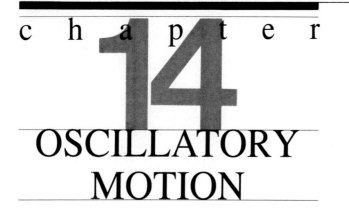

Preview

Vibrational, or oscillatory, motion is important not only because it is common, but because it is a basic constituent of wave motion. This chapter forms a bridge between the mechanics of particles and rigid bodies, discussed earlier, and the physics of wave motion which is discussed in Chapters 17 and 18.

We limit our considerations largely to "one-dimensional" systems—systems whose motion can be described by a single linear or angular variable. In each system that we analyze, the displacement, whether linear or angular, varies sinusoidally, or approximately so, with time. The sine and cosine functions are called harmonic functions, and this type of motion is therefore referred to as simple harmonic motion (SHM).

First, we demonstrate that a system executes SHM if its acceleration and displacement are directly proportional but in opposite directions. We also show that SHM describes the motion, projected on a diameter, of an object engaged in uniform circular motion. Then we describe the transient behavior of a harmonic oscillator that is subjected to a dampening mechanism. (A damping mechanism converts the mechanical energy of an oscillator into thermal or other nonmechanical forms of energy.) Finally, we consider the steady-state response of a damped oscillator to a sinusoidal driving force.

14.1 Simple Harmonic Motion

Oscillatory Motion

Periodic motion is regular, and repeats with a characteristic time. The motions of a jump rope, a clock pendulum, and an orbiting planet are all periodic. An oscillation, or vibration, is a special type of periodic motion in which the system moves back and forth through an equilibrium position. The motion of a point on the hand of a clock is periodic but not oscillatory, whereas the motion of a clock pendulum is periodic and oscillatory.

The most striking feature of oscillatory motion is its regularity. This regularity is characterized by the time between vibrations. The time required for one complete oscillation of a system or body is called the *period*, T, of the motion. The frequency of the motion is the number of complete oscillations of the system per unit time, and is designated by the Greek letter ν (nu). By definition, then,

$$\nu = \frac{1}{T} \tag{14.1}$$

If time is expressed in seconds, frequency is expressed in reciprocal seconds (s^{-1}). This frequency unit is called the *hertz** and is abbreviated as Hz.

$$1 \text{ Hz} = 1 \text{ oscillation/s}$$

*This unit is named after Heinrich Hertz, who in 1887 generated and detected the electromagnetic waves that were predicted in 1865 by James Clerk Maxwell.

When a system oscillates it does so about an equilibrium position. Two physical conditions must be met in order for a system to oscillate.

1 The net torque or force exerted on the body when it is not in its equilibrium position must be *restoring* in character. A restoring torque or force is one that urges a body toward its equilibrium position, and such equilibrium positions are stable (see Section 8.5). For example, a simple pendulum is urged back toward its equilibrium position regardless of the direction in which it was displaced.

2 In order that its motion through the equilibrium position persists, the oscillating object must possess *inertia*. Because a moving pendulum has rotational inertia, and therefore angular momentum, a net torque is necessary to bring it to rest. At the moment the pendulum arrives at its equilibrium position, however, the net torque equals zero. Because of its angular momentum the pendulum does not stop there, but continues moving through the equilibrium position.

Important aspects of motion can be deduced from an energy diagram, as discussed in Section 8.6. The force exerted on a particle is determined from the particle's potential energy by the relation $F = -dU/dx$. Geometrically, the force equals the negative slope of the U versus x graph. Consider the behavior of a particle that is subject to a potential energy function with a stable equilibrium position. Such a function is depicted in Figure 14.1. Using slopes determined from this graph, we see that when $x > 0$, $F < 0$; and when $x < 0$, $F > 0$. Thus F is directed toward the equilibrium position (the origin) for all such values of x and qualifies as a restoring force. The mass of the particle provides the inertia. This particle executes oscillatory motion between the turning points at x_a and x_b.

In Section 8.6 we discussed the potential energy $U = {}^1\!/_2 kx^2$ of a particle of mass m attached to a spring with spring constant k. This potential energy function is shown in Figure 14.2. Imagine this curve to be a frictionless track with gravity acting vertically. A mass released at rest at $x = x_b$ will slide down to the origin and up the other side, stopping at $x = x_a$. It will then retrace its path back to $x = x_b$, then back to $x = x_a$, and so on, moving back and forth. The sliding mass executes oscillatory motion. Several other systems that exhibit oscillatory motion are depicted in Figure 14.3.

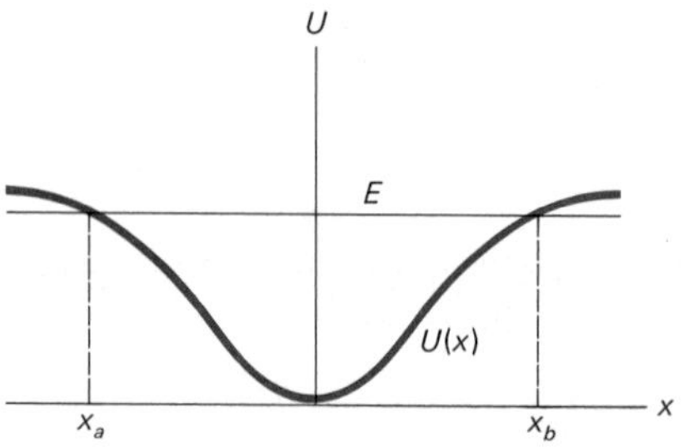

Figure 14.1

Near a minimum, $x = 0$, in the potential energy, $U(x)$, the force $F = -dU/dx$ is directed toward the minimum point.

Figure 14.2a

A particle executes simple harmonic motion (SHM) if the potential energy $U(x) = \frac{1}{2}kx^2$.

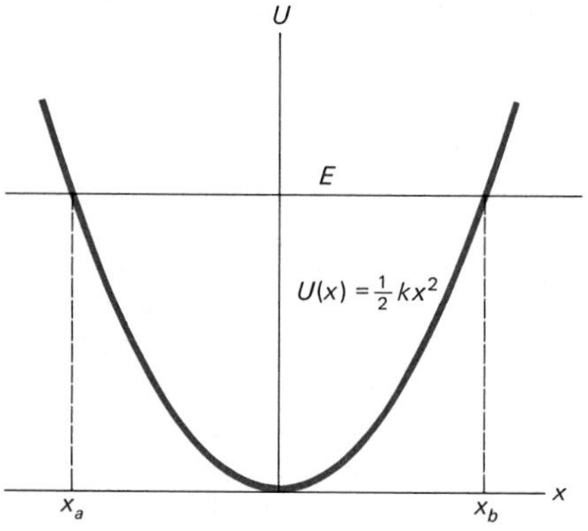

Figure 14.2b

SHM is exhibited by the horizontal component of the motion of a particle sliding without friction on a parabolic hill.

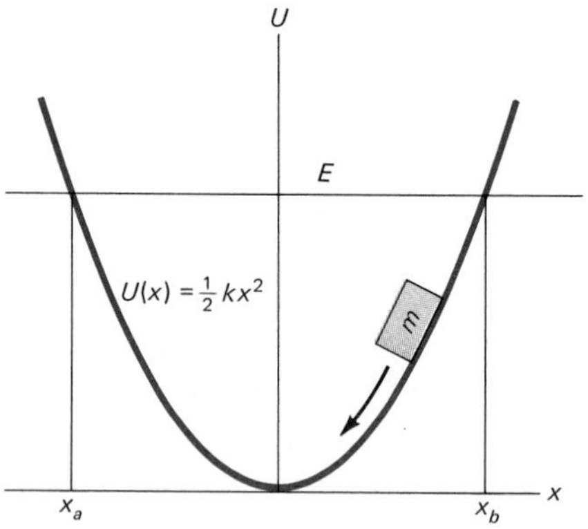

Simple Harmonic Motion

Simple harmonic motion (SHM) is a particularly important kind of oscillatory motion. *Physically* we define a body to be executing SHM if its acceleration is proportional to its displacement and of opposite sign. A body executing SHM is called a *harmonic oscillator.* In this chapter we limit our analysis of oscillating systems to harmonic oscillators. As a simple example or prototype of SHM we will use a mass–spring system on a horizontal frictionless surface.

In Figure 14.4 a body of mass m is attached to a spring that obeys Hooke's law. Such a spring exerts on the mass m a restoring force that is proportional to x—the displacement of the mass from equilibrium. For a displacement $+x$ the force is $-kx$, where k is called the *spring constant.* Using Newton's second law, we can write

$$F = -kx = ma$$

so that we have

$$a = -\left(\frac{k}{m}\right)x \tag{14.2}$$

Figure 14.3

The two mass–spring systems, the fluid in the U-tube, and the pendulum all illustrate SHM. The pendulum exhibits SHM provided that the approximation $\sin\theta \cong \theta$ holds.

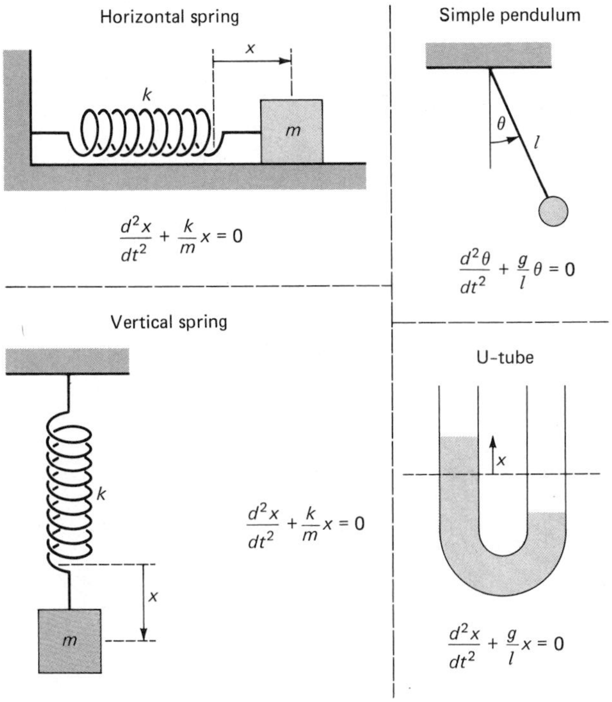

Figure 14.4

The particle's horizontal motion is SHM if the spring is linear and friction is absent.

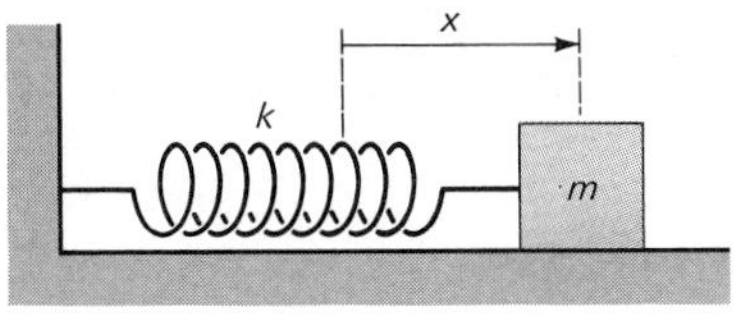

The ratio k/m is constant. Hence the acceleration is proportional to the displacement and of opposite sign, and the body executes SHM. We can use this same procedure to determine whether any system is in simple harmonic oscillation.

Example 1 The Simple Pendulum

A simple pendulum, like the one illustrated in Figure 14.5, moves back and forth in a plane. Does this motion constitute SHM?

Consider the pendulum bob as the body. Forces acting on the bob are its own weight, **W**, and the tension, **T**, in the supporting string. We view the motion of the bob as rotational, and apply the rotational form of Newton's second law (Section 12.4), $\tau = I\alpha$, about an axis through O and perpendicular to the plane of the figure. Treating the bob as a particle, we have for the moment of inertia I

Figure 14.5

Two forces act on the pendulum bob. These are the string tension **T** and the weight **W** of the bob. For small angles the bob executes SHM.

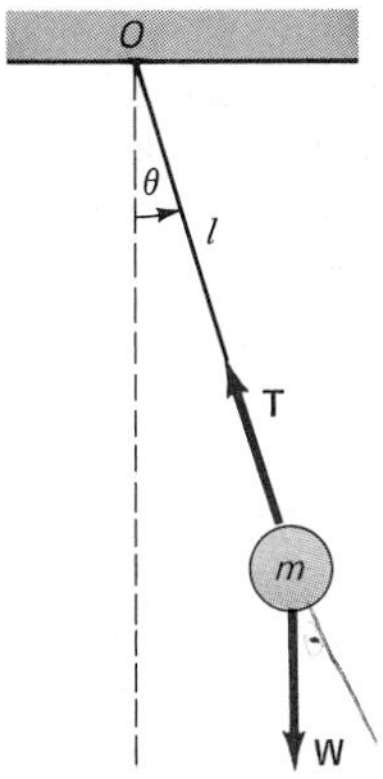

$= ml^2$. The torque due to **T** vanishes, leaving only the torque due to the weight, $W = mg$. Therefore, the net torque is $\tau = -mgl \sin\theta$. The minus sign is necessary because $\tau < 0$ for $\theta > 0$ and $\tau > 0$ for $\theta < 0$. Putting this all together, we get

$$-mgl \sin\theta = ml^2\alpha$$

or, solving for the angular acceleration, α, we have

$$\alpha = -\left(\frac{g}{l}\right)\sin\theta$$

As it stands this equation does not represent SHM. However, if θ is small, we can use the approximation $\sin\theta \approx \theta$ to write

$$\alpha = -\left(\frac{g}{l}\right)\theta \qquad (14.3)$$

The angular acceleration, α, is proportional to the angular displacement, θ, and of opposite sign. For small angles the simple pendulum executes SHM.

To arrive at a general form for the equation of motion of a harmonic oscillator we consider Eqs. 14.2 and 14.3. Recalling that $a = d^2x/dt^2$ and $\alpha = d^2\theta/dt^2$, we can rewrite Eqs. 14.2 and 14.3 as

$$\frac{d^2x}{dt^2} = -\left(\frac{k}{m}\right)x \quad \text{and} \quad \frac{d^2\theta}{dt^2} = -\left(\frac{g}{l}\right)\theta$$

These are special forms of the general equation

$$\frac{d^2x}{dt^2} = -\omega^2x \qquad (14.4)$$

Here ω is constant, and is related to the physical constants in Eqs. 14.2 and 14.3 according to

$$\omega^2 = \frac{k}{m} \qquad (14.5)$$

and

$$\omega^2 = \frac{g}{l} \qquad (14.6)$$

respectively. The unit for ω is reciprocal seconds (s^{-1}), not hertz (Hz), which is reserved for frequency, ν. In Section 14.2 we will explain why the symbol ω, which is used for angular velocity in rotational motion, is also used here.

Mathematically we define SHM to be motion for which the displacement satisfies Eq. 14.4, the harmonic oscillator equation of motion. We examine its solution in Section 14.2.

Energy Analysis

We can deduce some aspects of SHM without obtaining the explicit solution (x as a function of t) to Eq. 14.4. For example, we will see that the energy equation (see Section 8.6) for the harmonic oscillator can be used to derive Eq. 14.4. We can then freely draw inferences from the energy equation by using methods developed in Sections 8.4–8.6.

The constant total energy, E, of the system is

$$E = K + U = \frac{1}{2}mv^2 + \frac{1}{2}kx^2 \qquad (14.7)$$

where $K = {}^1\!/_2 mv^2$ is the kinetic energy, and $U = {}^1\!/_2 kx^2$ is the potential energy. Dividing by m and letting $\omega^2 = k/m$, we can write Eq. 14.7 as

$$\frac{E}{m} = \frac{1}{2}v^2 + \frac{1}{2}\omega^2x^2$$

Because E/m is constant, its derivative with respect to time is zero. Thus, if we differentiate this equation with respect to time, we obtain

$$0 = v\frac{dv}{dt} + \omega^2x\frac{dx}{dt}$$

Using the definitions $\frac{dx}{dt} = v$ and $\frac{dv}{dt} = \frac{d^2x}{dt^2}$, we obtain

$$v\left(\frac{d^2x}{dt^2} + \omega^2x\right) = 0$$

After dividing by v we have

$$\frac{d^2x}{dt^2} = -\omega^2x \qquad (14.4)$$

This is the harmonic oscillator equation of motion. If we wished, we could reverse this argument, and derive Eq. 14.5 by integrating Eq. 14.4.

During each cycle of oscillator motion, the speed and displacement reach definite maxima. These maxima are related to the total energy, E. We will let A, called the *amplitude*, represent the maximum value of the displacement, x. From Eq. 14.7 we see that x is a maximum when $v = 0$. Setting $x = A$ and $v = 0$ in Eq. 14.7 gives

$$E = \frac{1}{2}kA^2 \qquad \text{or} \qquad A = \left(\frac{2E}{k}\right)^{1/2} \qquad (14.8)$$

The total energy of a harmonic oscillator is determined by the maximum displacement and the spring constant (see Eq. 14.8). Therefore, one way of starting a harmonic oscillator (giving it energy) is to pull it a distance A from equilibrium and release it there from rest. At this point the oscillator begins the conversion of potential energy into kinetic energy as it approaches the equilibrium position. This conversion is completed when the oscillator reaches its equilibrium position and achieves maximum speed.

Equation 14.7 shows that maximum speed occurs when $x = 0$. If v_{max} represents maximum speed ($x = 0$), then

$$E = \frac{1}{2}mv_{max}^2 \qquad \text{or} \qquad v_{max} = \left(\frac{2E}{m}\right)^{1/2} \qquad (14.9)$$

According to Eq. 14.9, the total energy of a harmonic oscillator is determined by its maximum speed and mass. Consequently, another way to start a harmonic oscillator is to give it an initial velocity while it is in its equilibrium

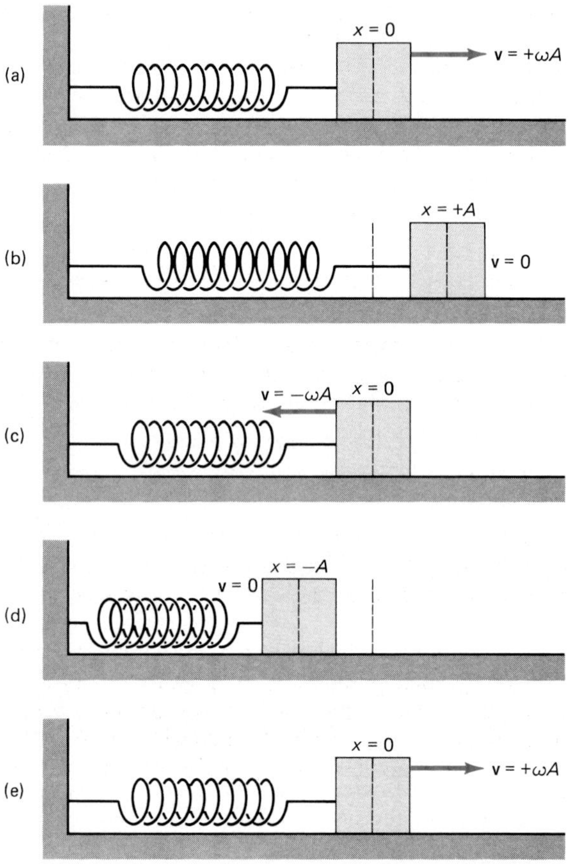

Figure 14.6

One complete cycle of SHM is illustrated by the mass–spring system. In (a) the particle is at the origin, $x = 0$, moving to the right with velocity $v = +\omega A$. In (b) the particle velocity has been reduced to zero at its maximum displacement, $x = +A$, to the right. In (c) the particle's velocity has changed to $v = -\omega A$ when it passes through the origin moving to the left. In (d) the particle velocity has again been reduced to zero at its maximum displacement, $x = -A$, to the left. In (e) the particle is back to its starting position with velocity $v = \omega A$ to the right, completing the cycle.

position—by hitting it sharply, for instance. In this case the oscillator converts kinetic energy into potential energy as it moves away from its equilibrium position. This conversion is completed when the oscillator reaches its maximum displacement.

The total energy may be expressed in terms of the maximum speed (Eq. 14.9) or maximum displacement (Eq. 14.8). This suggests that a relationship exists between maximum speed and maximum displacement. To obtain this relationship we divide Eq. 14.9 by Eq. 14.8 to get

$$\frac{v_{max}}{A} = \left(\frac{k}{m}\right)^{1/2}$$

From Eq. 14.5, $(k/m)^{1/2} = \omega$, giving

$$v_{max} = A\omega \qquad (14.10)$$

The relationship between the maximum speed and the maximum displacement of a harmonic oscillator is summarized by Eq. 14.10.

To see how these ideas apply to a physical system consider the horizontal spring–mass system shown in Figure 14.6. The equilibrium position ($x = 0$) is labeled with a dotted line. Each of the drawings in Figure 14.6 represents a stage in one complete oscillation of the system.

(a) The object is in its equilibrium position ($x = 0$) and is moving to the right with velocity $v = +\omega A$. The energy is all kinetic ($K = \frac{1}{2}m\omega^2A^2$; $U = 0$).

(b) The object is at its maximum displacement ($x = +A$) to the right and is at rest ($v = 0$). The energy is all potential ($U = \frac{1}{2}kA^2$; $K = 0$).

(c) The object is passing through its equilibrium position ($x = 0$) and is moving to the left with velocity $v = -\omega A$. Again the energy is all kinetic ($K = \frac{1}{2}m\omega^2A^2$; $U = 0$).

(d) The object is at its maximum displacement ($x = -A$) to the left and instantaneously at rest ($v = 0$). The energy is all potential ($K = 0$).

(e) The object has returned to conditions that are the same as those in Figure 14.6a. The position ($x = 0$) and velocity ($v = +\omega A$) are identical and the cycle is completed.

From an energy viewpoint the oscillation is a continuing transformation of kinetic and potential energy, one into the other, while the total energy doesn't change. Because the points $x = \pm A$ are located where the velocity changes in direction, they are called *turning points* of the motion.

Questions

1. According to Eq. 14.4, $\omega^2 = k/m$ for a mass attached to a spring. Explain qualitatively why, for this oscillator, you might expect ω to be smaller for a larger mass and to be larger for a stiffer spring. Try to use *physical* reasoning.
2. A mass on a spring oscillates with $\omega = \sqrt{k/m}$. The spring breaks and the same mass is attached to the remaining portion of the original spring. Compare the old and new values of ω. Explain any difference.

14.2 Solution of the Harmonic Oscillator Equation

Now let's find a solution for the harmonic oscillator equation

$$\frac{d^2x}{dt^2} = -\omega^2 x \qquad (14.4)$$

which we discussed in the preceding section. This equation involves the second derivative, d^2x/dt^2. Its solution, $x(t)$, will contain two constants of integration. Calling A and ϕ the constants of integration, we have the following solution of the harmonic oscillator equation:

$$x = A\cos(\omega t + \phi) \qquad (14.11)$$

To verify that Eq. 14.11 does in fact satisfy Eq. 14.4, we differentiate $x = A\cos(\omega t + \phi)$ twice. The first derivative is

$$\frac{dx}{dt} = -\omega A\sin(\omega t + \phi)$$

The second derivative is

$$\frac{d^2x}{dt^2} = -\omega^2 A\cos(\omega t + \phi) \qquad (14.12)$$

Substituting Eqs. 14.11 and 14.12 into Eq. 14.4 gives

$$-\omega^2 A\cos(\omega t + \phi) = -\omega^2 A\cos(\omega t + \phi)$$

This equation is of course correct, confirming that $x = A\cos(\omega t + \phi)$ is a solution of Eq. 14.4, the harmonic oscillator equation.

We have seen that oscillatory motion is periodic with a period T. To obtain the relation between T and ω we consider the periodicity of the cosine function. The cosine function is periodic, with a period of 2π rad. That is,

$$\cos(\theta + 2\pi) = \cos\theta$$

If we increase the time by T in our solution, $x = A\cos(\omega t + \phi)$, we must get the same displacement. Hence,

$$A\cos[\omega(t + T) + \phi] = A\cos(\omega t + \phi)$$

Comparison of these two periodic conditions yields

$$\omega(t + T) + \phi = \omega t + \phi + 2\pi$$

or

$$\omega T = 2\pi \qquad (14.13)$$

This equation relates the constant ω (called the *radian* or *angular frequency*), appearing in Eq. 14.4, to the period, T, of the motion. We can also relate frequency, ν, to T by using Eq. 14.1,

$$\nu = \frac{1}{T}$$

Therefore

$$\omega = \frac{2\pi}{T} = 2\pi\nu^* \qquad (14.14)$$

We see then that the three quantities ω, T, and ν contain the same information. We may use whichever quantity is convenient.

Example 2 A Seconds Pendulum

Many antique clocks use a pendulum to keep time. The bob of a "seconds" pendulum requires 1 s for each swing, so that its period is 2 s. What is the length of a simple pendulum that has a period of 2 s, and the corresponding values of ω and ν?

We know that $T = 2$ s. Therefore, according to Eq. 14.14, we have

$$\omega = \frac{2\pi}{T} = 3.14 \text{ rad/s}$$

and

$$\nu = \frac{1}{T} = 0.500 \text{ Hz}$$

To obtain the pendulum's length we use Eq. 14.6 to get

$$l = \frac{g}{\omega^2} = \frac{9.80}{(3.14)^2} = 0.994 \text{ m}$$

You may want to construct such a pendulum for yourself, and see how close you can come to a period of 2 s.

---

*The unit for ν, the frequency, is the hertz (reciprocal second), whereas ω, the radian frequency, is measured in radians per second.

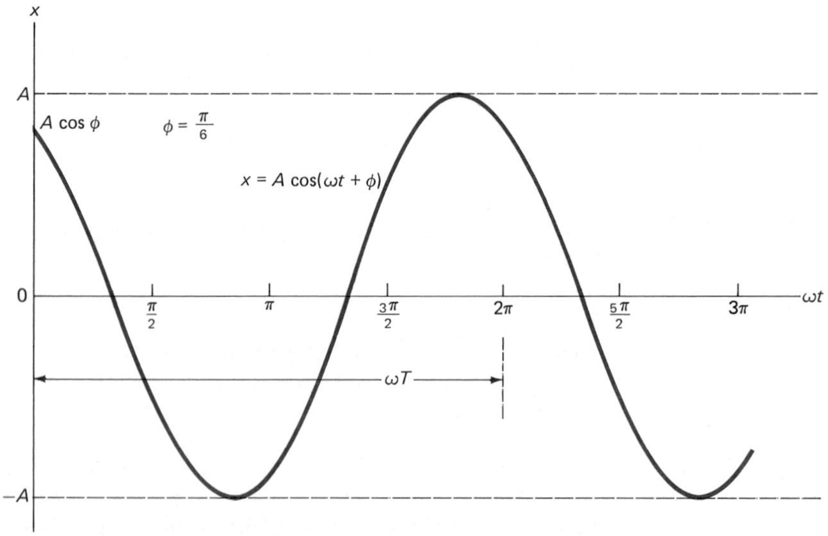

Figure 14.7

The function $x = A\cos(\omega t + \phi)$ describes the motion of a harmonic oscillator. The position $x = A\cos\phi$ at $t = 0$ is determined by the initial phase ϕ and amplitude A. One complete cycle occurs in an angle of 2π radians or in a time $T = 2\pi/\omega$ seconds.

In Figure 14.7 the function $x = A\cos(\omega t + \phi)$ is plotted versus ωt. The connection between ω and T is emphasized by designating $\omega T = 2\pi$ rad for one complete oscillation.

The argument of the cosine function, $\omega t + \phi$, is called the *phase angle* or *phase*. Because the phase equals ϕ when $t = 0$, ϕ is called the *initial phase*. The value of x at $t = 0$ is $x = A\cos\phi$. Thus, for a specified amplitude, the initial value of x determines ϕ. Alternatively we could say that ϕ determines where in the motion we start the clock (set $t = 0$). With $\phi = 0$, for instance, $x = +A$ at $t = 0$. Because we do not lose any generality by choosing $\phi = 0$, we will do so, and write

$$x = A\cos\omega t \qquad (14.15)$$

This form of the solution is shown in Figure 14.8.

In the previous section we saw that $x = +A$ and $x = -A$ represent the turning points of the motion. In both Figure 14.7 and Figure 14.8 the curve lies between the dotted lines ($x = \pm A$). These dotted lines represent the maximum and minimum values of x, and physically correspond to the maximum excursion from the equilibrium position ($x = 0$).

With $\phi = 0$ we have for the velocity and acceleration the expressions

$$v = \frac{dx}{dt} = -\omega A\sin\omega t \qquad (14.16)$$

and

$$a = \frac{d^2x}{dt^2} = -A\omega^2\cos\omega t \qquad (14.17)$$

Figure 14.8

The function $x = A \cos \omega t$ describes the motion of a harmonic oscillator having the initial phase $\phi = 0$. Such an oscillator begins its motion with zero velocity and a displacement $x = +A$.

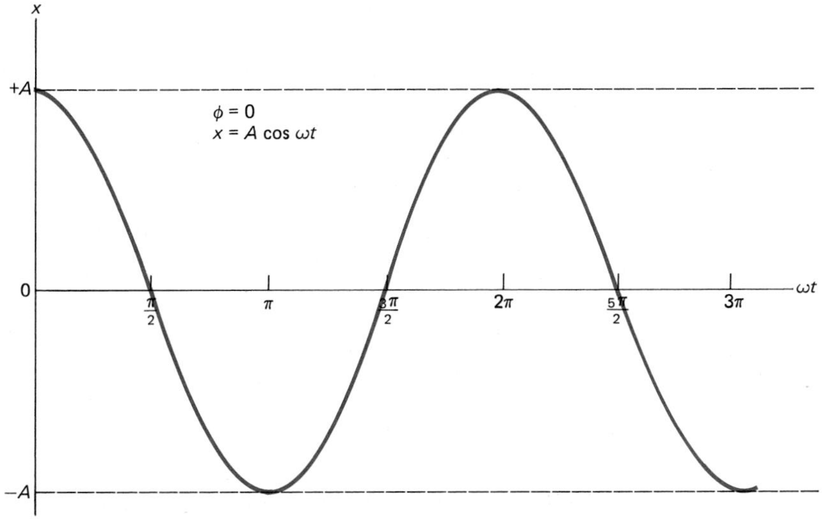

obtained by differentiating Eq. 14.15. These expressions are the same as those obtained at the beginning of the section, but now with $\phi = 0$. From Eq. 14.16 we see that ωA is the maximum value of v, as discussed in Section 14.1.

Equations 14.15–14.17 show that the velocity is a maximum when the displacement and acceleration are zero. The converse is also true. The maximum displacement and acceleration occur when the speed is zero. We can write, for these maximum values,

$$x_{max} = A \tag{14.18}$$

$$v_{max} = \omega A \tag{14.19}$$

$$a_{max} = \omega^2 A \tag{14.20}$$

The Energy Equation

In Section 14.1 we used the energy equation for the harmonic oscillator (Eq. 14.7) to draw inferences about the oscillator's motion and to obtain the equation of motion (Eq. 14.4) by using differentiation. Here we reverse the argument by starting with Eq. 14.4 and deriving the energy equation by using integration.

To integrate the equation

$$\frac{d^2x}{dt^2} + \omega^2 x = 0 \tag{14.4}$$

we use the chain rule of calculus to write

$$\frac{d^2x}{dt^2} = \frac{dv}{dt} = \frac{dv}{dx}\frac{dx}{dt} = \frac{dv}{dx}v$$

so that we can rewrite Eq. 14.4 in the form

$$\frac{dv}{dx}v + \omega^2 x = 0$$

That this relation is an expansion of the equation

$$\frac{d}{dx}\left(\frac{1}{2}v^2 + \frac{1}{2}\omega^2 x^2\right) = 0$$

is readily verified by carrying out the indicated differentiation. To integrate, we first observe that because the derivative of the expression $(\frac{1}{2}v^2 + \frac{1}{2}\omega^2 x^2)$ equals zero, the expression itself must be constant. We choose to write this constant in the form E/m with both E and m constant and E/m having units of velocity squared. This gives

$$\frac{1}{2}v^2 + \frac{1}{2}\omega^2 x^2 = \frac{E}{m}$$

This choice for the form of the constant leads to an expression that is identical to the energy equation, Eq. 14.7. To see this multiply by m to obtain

$$\frac{1}{2}mv^2 + \frac{1}{2}m\omega^2 x^2 = E$$

Now introduce $k = m\omega^2$ to get the energy equation

$$\frac{1}{2}mv^2 + \frac{1}{2}kx^2 = E \tag{14.7}$$

Because $\frac{1}{2}mv^2$ is the kinetic energy and $\frac{1}{2}kx^2$ is the potential energy, E represents the total mechanical energy of the oscillator. The energy equation is sometimes called the first integral of the equation of motion.

We have verified that $x = A \cos \omega t$ is a solution of the simple harmonic equation of motion, Eq. 14.4. Now we wish to verify that it is also a solution of the energy equation, Eq. 14.7. Starting with

$$x = A \cos \omega t$$

we differentiate once to obtain

$$v = \frac{dx}{dt} = -\omega A \sin \omega t$$

Substituting these expressions into Eq. 14.7 yields

$$\frac{1}{2}m(-\omega A \sin \omega t)^2 + \frac{1}{2}k(A \cos \omega t)^2 = E$$

or

$$\frac{1}{2}m\omega^2 A^2 \sin^2 \omega t + \frac{1}{2}kA^2 \cos^2 \omega t = E$$

Using $k = m\omega^2$, Eq. 14.5, this equation can be reduced to

$$\frac{1}{2}m\omega^2 A^2(\sin^2 \omega t + \cos^2 \omega t) = E$$

or

$$\frac{1}{2}m\omega^2 A^2 = E \qquad (14.9)$$

where we have used the identity $\sin^2 \omega t + \cos^2 \omega t = 1$. In a similar manner $m\omega^2 = k$ can be used to eliminate $m\omega^2$, and we obtain

$$\frac{1}{2}kA^2 = E \qquad (14.8)$$

These two relations, which connect E to m and ω and to k and A, are identical to those obtained in Section 14.1. Hence if $\omega = \sqrt{k/m}$, $x = A \cos \omega t$ is also a solution of the energy equation.

Questions

3. According to Eqs. 14.15 and 14.16, at $t = 0$, $x = A$ and $v = 0$. If the velocity is zero, how does the oscillator get back to its equilibrium position at $x = 0$?
4. If Eq. 14.16 is substituted into $K = \frac{1}{2}mv^2$, the kinetic energy of the oscillator is seen to vary as $\sin^2 \omega t$. How can the total energy be constant when K varies with time?
5. Because $U = \frac{1}{2}kx^2$ and $F_x = -dU/dx$, $x = 0$ is an equilibrium position. The force and acceleration are both zero at $x = 0$. Can the oscillator ever leave $x = 0$? How?

14.3 Applications of Simple Harmonic Motion

The period of simple harmonic motion depends on the physical properties of the oscillating system and its environment. A precise measurement of the period is possible by counting many vibrations, and often provides a dynamic method of determining characteristics of the system or of its environment. For instance, using Eq. 14.5

$$\omega^2 = \frac{k}{m}$$

and Eq. 14.14

$$\omega = \frac{2\pi}{T}$$

we can write the period for a horizontal spring–mass system as

$$T = 2\pi\left(\frac{m}{k}\right)^{1/2} \qquad (14.21)$$

If we use a calibrated spring (Section 3.1), then the value of k is known. Thus a measurement of T can be used to measure m. This experiment is of particular interest because it is only the inertial mass that is involved. This technique was used in Skylab for measuring mass. Common mass measurements using balances or vertical springs involve the force of gravity.

In Example 2 the measured period of a seconds pendulum (2 s) together with the known value of g enabled us to determine the pendulum length l. We can reverse this experiment and use a pendulum of measured length with a measured period to infer a value for g. Using Eq. 14.6

$$\omega^2 = \frac{g}{l}$$

and Eq. 14.14, we can express g in terms of l and T:

$$g = \frac{4\pi^2 l}{T^2} \qquad (14.22)$$

In Newton's day the pendulum was used to obtain values for g at different latitudes over the surface of the earth. Newton's conjecture that the earth was an oblate spheroid was confirmed in this way.

For a simple pendulum, T can be determined with accuracy, but it is not possible to obtain an equally accurate value of l. This in turn limits the accuracy of the values of g obtained by using a simple pendulum. The state of the art in the measurement of g is represented by the method used at the National Bureau of Standards. An object is projected upward in a vacuum and allowed to retrace its path as suggested in Figure 14.9. It rises, passing levels I

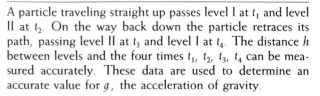

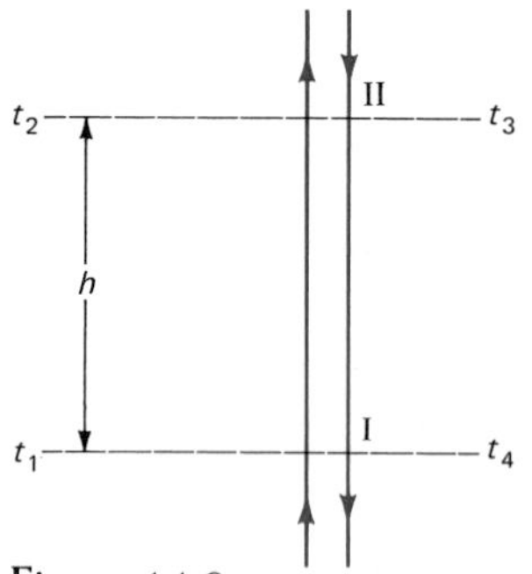

Figure 14.9

A particle traveling straight up passes level I at t_1 and level II at t_2. On the way back down the particle retraces its path, passing level II at t_3 and level I at t_4. The distance h between levels and the four times t_1, t_2, t_3, t_4 can be measured accurately. These data are used to determine an accurate value for g, the acceleration of gravity.

Figure 14.10

The period T of a simple pendulum of length l may be used to measure g, the acceleration of gravity.

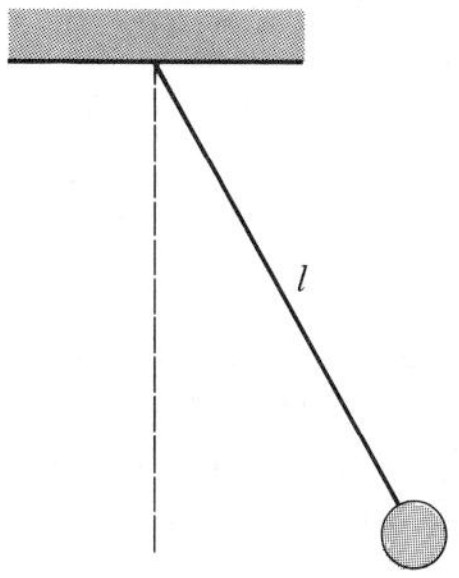

and II at the times t_1 and t_2, and then falls, passing levels II and I at times t_3 and t_4. Precise measurements of h, the vertical distance between levels I and II, and of the four times are made, permitting the computation of an accurate value for g (see Chapter 4, Problem 44).

Example 3 **Measurement of g with a Simple Pendulum**

A simple pendulum (Figure 14.10) can be used to determine the acceleration of gravity, g, at a particular location. The expression for the period is

$$T = 2\pi\sqrt{\frac{l}{g}}$$

Thus we have

$$g = \frac{4\pi^2 l}{T^2}$$

for the acceleration of gravity in terms of the measured period, T, and length, l, of the pendulum.

A pendulum that is 1.00 m in length is observed to require 100.60 s to execute 50 complete oscillations. These data give a period

$$T = \frac{100.60}{50}$$

$$= 2.01 \text{ s}$$

Thus we calculate

$$g = \frac{4\pi^2 \times 1}{(2.01)^2}$$

$$= 9.77 \text{ m/s}^2$$

for the acceleration of gravity.

Measurement of the Moment of Inertia

A torsion pendulum (see Sections 6.6 and 26.3) consists of a body suspended on the end of a wire or rod that can be set into rotational oscillations. The period of such oscillations depends on the moment of inertia of the system. A torsion pendulum can be used to measure the unknown moment of inertia of an object. A vertical torsion pendulum, arranged to measure the moment of inertia of a standing person about an axis through the center of mass, is shown in Figure 14.11. We will show how a measurement of T for torsional vibrations can be used to obtain I for the person.

First, we must calibrate the torsion pendulum. A known torque, τ, is applied to twist the rod until it comes to equilibrium at the angle θ. We assume that Hooke's law holds for the rod, so that the ratio of applied torque to twist angle is constant. This assumption permits us to write

$$\tau = -\beta\theta \qquad (14.23)$$

The minus sign is needed because τ is a restoring torque. A static measurement of τ and θ establishes the value of the twist constant of the rod, β.

Next the pendulum—without the person in place—is twisted through the angle θ and released. If I_o is the moment of inertia of the empty pendulum, the rotational form of Newton's second law enables us to write

$$\tau = -\beta\theta = I_o\alpha$$

or

$$\alpha = -\left(\frac{\beta}{I_o}\right)\theta$$

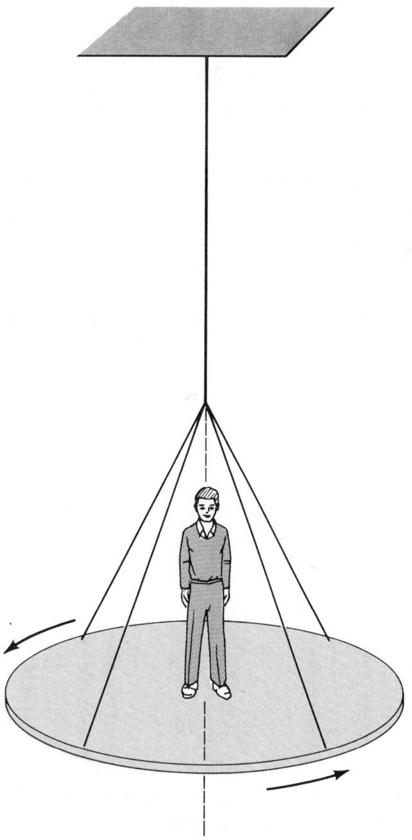

Figure 14.11

A torsion pendulum is an object free to rotate when suspended by a vertical wire or rod. If the wire is twisted and released, it unwinds and the object undergoes rotational oscillations. Here the torsion pendulum period is used to measure the moment of inertia of a person.

showing that the pendulum executes SHM. Thus we have for the period

$$T_o = 2\pi\left(\frac{I_o}{\beta}\right)^{1/2} \quad \text{or} \quad I_o = \frac{\beta T_o^2}{4\pi^2} \qquad (14.24)$$

Finally, the person climbs onto the pendulum platform, and the pendulum is again twisted through the angle θ and released. If I_p is the moment of inertia of the person alone, we can follow the same procedure to write

$$\tau = -\beta\theta = (I_o + I_p)\alpha$$

or

$$\alpha = -\left(\frac{\beta}{I_o + I_p}\right)\theta$$

showing that we again have SHM. Thus we can write for the period

$$T = 2\pi\left(\frac{I_o + I_p}{\beta}\right)^{1/2} \quad \text{or} \quad I_o + I_p = \frac{\beta T^2}{4\pi^2} \qquad (14.25)$$

Substituting the value of I_o obtained in Eq. 14.24 into Eq. 14.25, we can solve for the moment of inertia of the person, I_p,

$$I_p = \frac{\beta}{4\pi^2}(T^2 - T_o^2) \qquad (14.26)$$

in terms of the three measured quantities β, T, and T_o. By having the person assume various positions we can measure any desired moment of inertia of the person.

Example 4 **Moment of Inertia of a Person**

Let us use the torsion pendulum method to determine the moment of inertia of a person. The person stands vertically with the symmetry line of the apparatus passing through his or her center of mass. We measure $\beta = 125$ N·m/rad, $T_o = 2.000$ s, and $T = 2.152$ s. What is the person's moment of inertia?

Using Eq. 14.26 we have

$$I_p = \frac{125}{4\pi^2}[(2.152)^2 - 2.000^2]$$

$$= 2.00 \text{ kg} \cdot \text{m}^2$$

for the moment of inertia of the person. Note that, because of the cancellation in Eq. 14.26, T_o and T must be measured to four significant figures in order to obtain I_p to three significant figures.

Questions

6. If oscillations through large angles are permitted, the period of a simple pendulum depends on amplitude. Will the period corresponding to large amplitude be greater than or less than the period corresponding to small amplitude? Give a physical argument in support of your answer.
7. Explain why it is important for the person to stand at the exact center of the platform in Example 4.

14.4 Uniform Circular Motion and Simple Harmonic Motion

On a clear night, through a good pair of binoculars, you can see the four Galilean moons of Jupiter. Galileo watched these moons through his telescope, and was fascinated as he saw them change their positions night after night. He believed that Jupiter and the four orbiting moons constituted an analog of the solar system.

We now know that the orbits of the Galilean moons are nearly circular. However, from the earth their orbital plane is seen nearly edge on (Figure 14.12), so that their motion appears to be linear. Observations of this projected motion were used to prove that the orbits are actually circular. Here, we are interested in the inverse relationship. If the actual motions are circular, then what kind of motion is exhibited by the projections?

To answer this question we consider a particle moving with constant angular velocity ω in a circle of radius r. At $t = 0$ the particle is at $x = +r$ with $y = 0$ and $\theta = 0$. At time t it is located at $\theta = \omega t$, as shown in Figure 14.13. The point P marks the location of the particle. The point P' is the projection of P on the x-axis. The projected position is thus x, which is given as a function of time by the equation

$$x = r \cos \omega t \qquad (14.27)$$

Equation 14.27 is a special case of Eq. 14.11, and is obtained by setting $A = r$ and $\phi = 0$. Except for a difference

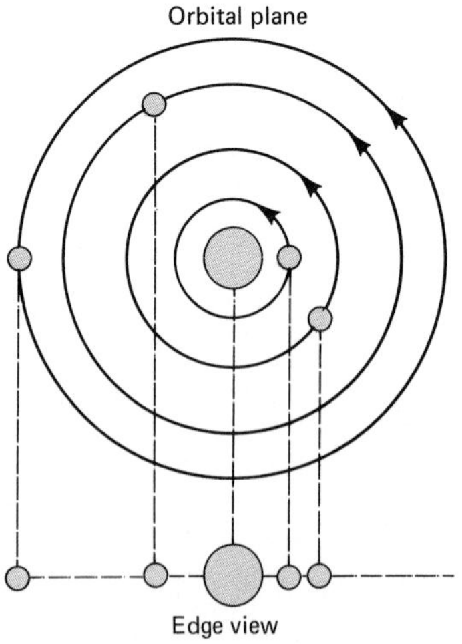

Figure 14.12

The four Galilean moons of Jupiter move about the parent planet in nearly circular orbits. From the earth this circular motion is viewed from the edge. This projected motion is simple harmonic.

in the interpretation of the constant ω,* the two expressions for x both describe SHM.

One point remains to be clarified. According to Eq. 14.14,

$$\omega = \frac{2\pi}{T} = 2\pi\nu$$

and ω, ν, and T are different ways of describing the time for one complete oscillation of a harmonic oscillator in SHM. In this section so far, however, ω has represented the constant angular velocity of a point P moving in circular motion. To relate these two ideas we ask, "What time is required for one revolution of the point P?" Call this time T_p. The angle swept out in time T_p is 2π rad, so that

$$\omega = \frac{2\pi}{T_p}$$

or

$$T_p = \frac{2\pi}{\omega} \tag{14.28}$$

In spite of the difference in concept, the period of SHM equals the period of the uniform circular motion. That is, $T = T_p$. A glance at Figure 14.13 shows why. In the same time that is required for P to complete one revolution, P' completes one "back and forth" motion.

The motion of the Galilean moons of Jupiter as seen in projection is simple harmonic. This gives us good reason to believe that we are observing the projected uniform circular motion of these moons as they orbit the planet.

What are the maximum values of x, v_x, and a_x? Consider Eq. 14.27,

$$x = r \cos \omega t$$

*The radian frequency ω of this chapter is the same as the angular velocity ω used in previous chapters. The connection between uniform circular motion and SHM developed in this section explains why the symbol ω is used for both concepts.

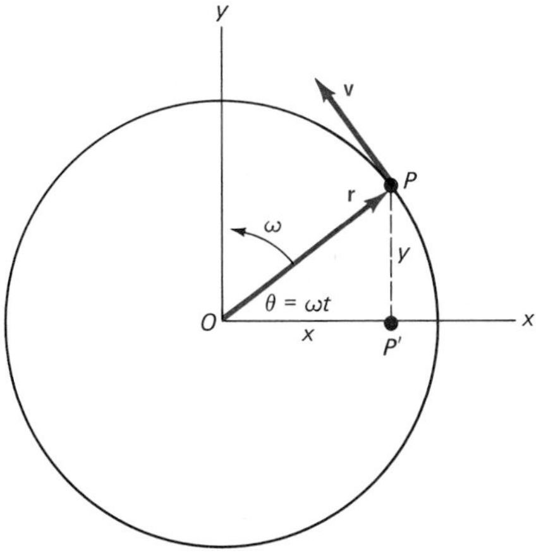

Figure 14.13

The particle at point P moves around the circle with constant speed $v = r\omega$. The point P' is the projection of P on the x-axis. As P moves counterclockwise in uniform circular motion the point P' executes SHM.

Differentiating twice, we obtain the following expressions for v_x and a_x:

$$v_x = -\omega r \sin \omega t$$

$$a_x = -\omega^2 r \cos \omega t$$

Because the maximum value of $\sin \omega t$ and $\cos \omega t$ is unity, we can assert that

$$x_{max} = r \tag{14.29}$$

$$v_{max} = r\omega \tag{14.30}$$

$$a_{max} = r\omega^2 \tag{14.31}$$

Alternatively these equations can be deduced by using knowledge of uniform circular motion. Referring to Figure 14.13 we see that if $t = 0$, then $\theta = 0$ and $x = r$. Evidently this x is x_{max}, and $x_{max} = r$. This confirms Eq. 14.29. Again referring to Figure 14.13 we see that if $t = \pi/2\omega$, then $\theta = \pi/2$ and $|v_x| = v_T$. Apparently this is the largest magnitude of v_x, which means that $v_{max} = v_T$. In Section 12.1 we proved that the tangential velocity of a particle in uniform circular motion is given by $v_T = r\omega$. Therefore $v_{max} = r\omega$, confirming Eq. 14.30. In Section 5.2 we proved that the radial acceleration of a particle in uniform circular motion may be expressed as $a_R = v_T^2/r$, where v_T is the tangential velocity. Because $v_T = r\omega$ we can write $a_R = (r\omega)^2/r$, which gives us $a_R = r\omega^2$. Referring to Figure 14.13 for the third time, we see that if $t = 0$, then $\theta = 0$ and $|a_x| = a_R$. This is the largest magnitude for a_x, which means that $a_{max} = a_R$. Therefore, $a_{max} = r\omega^2$ and Eq. 14.31 is confirmed.

Note that the components of the vector **r** are projections on the coordinate axes. When we write $\mathbf{r} = x\mathbf{i} + y\mathbf{j}$ we are expressing the position vector in terms of its projections. Thus

$$\mathbf{r} = r \cos \omega t\,\mathbf{i} + r \sin \omega t\,\mathbf{j}$$

Similarly, the components of the velocity and acceleration vectors, **v** and **a**, can be written in terms of their projections

$$\mathbf{v} = v_x\mathbf{i} + v_y\mathbf{j}$$

$$\mathbf{a} = a_x\mathbf{i} + a_y\mathbf{j}$$

Example 5 Derivation of Centripetal Acceleration from SHM

We have used the expression for radial or centripetal acceleration, $a_R = r\omega^2$, to draw inferences about SHM. Now let us do the reverse in order to illustrate the power of a vector argument. Beginning with the position vector, $\mathbf{r} = x\mathbf{i} + y\mathbf{j}$, we will assume that $x = r \cos \omega t$ and $y = r \sin \omega t$ with r and ω constant. Thus we can assume that x and y execute SHM, so that **r** rotates with constant angular velocity ω, and executes uniform circular motion. We will then show that the vector acceleration, **a**, is centripetal.

Differentiating x and y twice with respect to t, we write

$$\frac{d^2x}{dt^2} = -\omega^2 r \cos \omega t \quad \text{and} \quad \frac{d^2y}{dt^2} = -\omega^2 r \sin \omega t \tag{14.32}$$

Figure 14.14

As the wheel rotates, the right end of the shaft, point A, moves in a circular path. The left end of the shaft, point B, can move only in the horizontal direction.

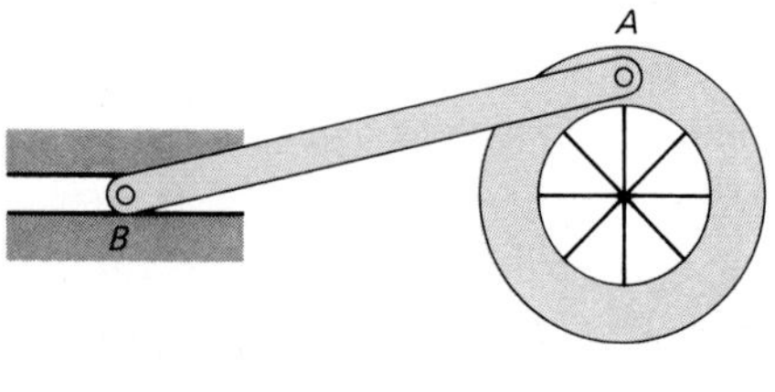

Substituting x for $r \cos \omega t$ and y for $r \sin \omega t$, we have

$$\frac{d^2x}{dt^2} = -\omega^2 x \qquad \text{and} \qquad \frac{d^2y}{dt^2} = -\omega^2 y$$

Hence

$$a_x = -\omega^2 x \tag{14.33}$$

$$a_y = -\omega^2 y \tag{14.34}$$

To prove that **a** is centripetal we express **a** in terms of its projections:

$$\mathbf{a} = a_x\mathbf{i} + a_y\mathbf{j}$$

Substituting the expressions for a_x and a_y from Eqs. 14.33 and 14.34 yields

$$\mathbf{a} = -\omega^2(r \cos \omega t\mathbf{i} + r \sin \omega t\mathbf{j})$$

Using $x = r \cos \omega t$ and $y = r \sin \omega t$, we can write

$$\mathbf{a} = -\omega^2(x\mathbf{i} + y\mathbf{j})$$

But the quantity in parentheses is **r**, which gives

$$\mathbf{a} = -\omega^2\mathbf{r} \tag{14.35}$$

This relation shows that because **r** is directed away from the center, **a** is directed toward the center, and is therefore centripetal (see Section 5.2).

Questions

8. A cyclist pedals a bicycle at constant velocity. Compare the pedal motion as it is viewed from the side with the motion as it appears from front or rear.

9. One end of a drive shaft (point A in Figure 14.14) moves in a circular path as the wheel rotates. The other end of the drive shaft (point B in Figure 14.14) moves back and forth in a straight line. Is the end at B executing simple harmonic motion? Does your answer depend on the kind of motion at A?

14.5 Damped Harmonic Motion

When a simple pendulum is set into oscillation, friction from a resistance gradually causes this oscillation to stop. So far we have neglected this effect in oscillatory motion. We now consider this kind of "decaying vibration," which we called *damped harmonic motion*.

To account for friction analytically we need an expression for the frictional force. We will limit our consideration of frictional force to linear motion in one dimension along the x-axis. Experiment shows that for small velocities, the viscous frictional force exerted on an object—such as a falling fog droplet—is proportional to the object's velocity (see Section 16.6 for a discussion of viscosity). Such a force can be represented by the equation

$$\mathbf{f} = -\gamma\mathbf{v} \tag{14.36}$$

where γ is an empirically determined constant. The minus sign indicates that f and v are of opposite sign. The frictional force tends to reduce the velocity as suggested by Fig. 14.15.

Let's add the frictional force given in Eq. 14.36 to our treatment of the harmonic oscillator. Our prototype oscillator will now consist of a mass m subjected to the forces $-kx$ and $-\gamma v$. Applying Newton's second law, we write

$$F = -kx - \gamma v = ma$$

With $v = \dfrac{dx}{dt}$ and $a = \dfrac{d^2x}{dt^2}$ this equation can be written in the form

$$m\frac{d^2x}{dt^2} + \gamma\frac{dx}{dt} + kx = 0 \tag{14.37}$$

Solutions of this equation describe damped harmonic motion. But before we obtain the solutions, let's consider three different types of damped harmonic motion.

1 For small values of γ the frictional force on a harmonic oscillator will be small, and we expect the system to approximate SHM. The system should oscillate with a steadily decreasing amplitude. This motion—illustrated in Figure 14.16—is called *underdamped harmonic motion*.

2 For large values of γ the frictional force will nearly balance the restoring force of the spring as the mass returns to its equilibrium position. As Eq. 14.37 indicates, this means that the acceleration will be small throughout the motion. If displaced, we expect the mass to creep back to equilibrium. In other words, no oscillation will occur. This motion—illustrated in Figure 14.17—is called *overdamped harmonic motion*.

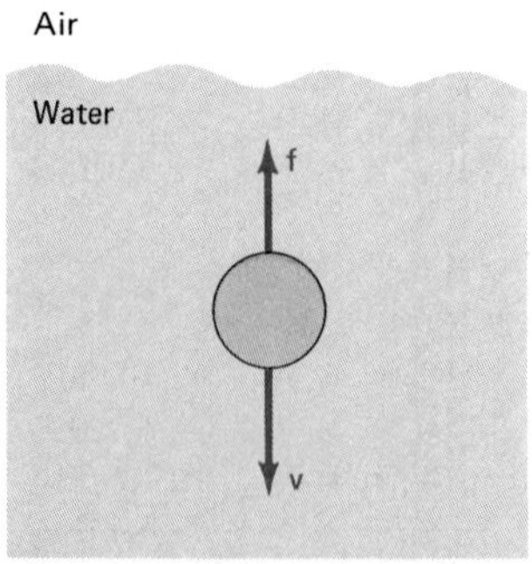

Figure 14.15

A sphere falling in water experiences an upward viscous drag.

Figure 14.16

In underdamped harmonic motion the particle oscillates with a steadily decreasing maximum displacement.

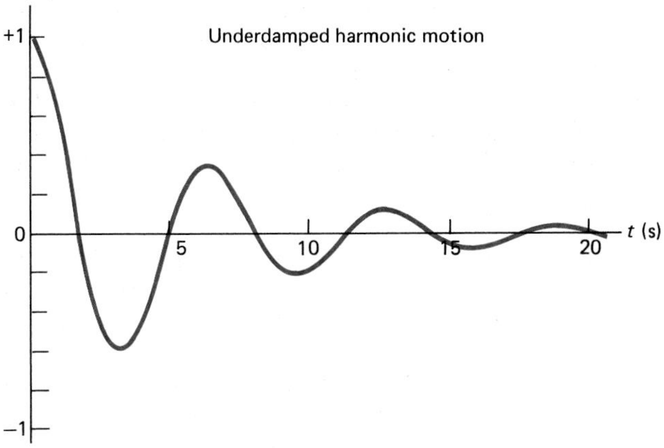

Figure 14.17

With overdamped harmonic motion no oscillation occurs. The displacement steadily decreases to zero.

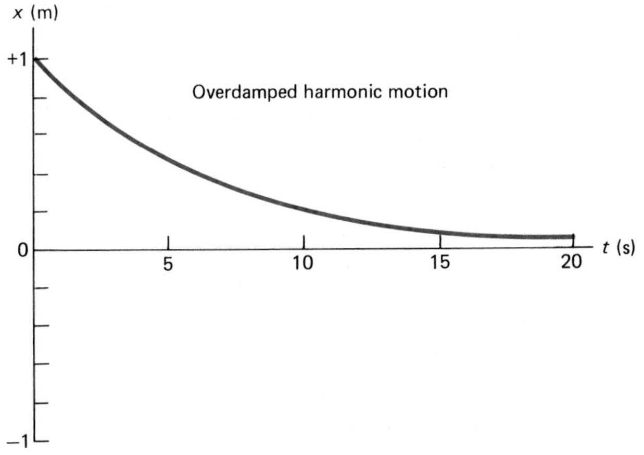

3 Evidently, if γ is too large, underdamped harmonic motion becomes impossible. Similarly, if γ is too small, overdamped harmonic motion becomes impossible. There is a critical value of γ that separates these two types of motion. This critical value is

$$\gamma = 2(km)^{1/2} \tag{14.38}$$

With this value of γ, the system executes *critically damped motion*. Any smaller value of γ leads to oscillatory motion.

Relations like Eq. 14.37 arise in many physical situations. For example, the consideration of transient currents in a circuit containing an inductor, resistor, and capacitor (see Section 35.4) leads to a similar equation. Therefore, it is convenient to develop a standardized form of Eq. 14.37 that will apply to many situations. If we divide the equation by m we get

$$\frac{d^2x}{dt^2} + \frac{\gamma}{m}\frac{dx}{dt} + \frac{kx}{m} = 0$$

Then, using Eq. 14.5

$$\omega^2 = \frac{k}{m} \tag{14.5}$$

and introducing

$$\tau = \frac{2m}{\gamma} \tag{14.39}$$

we have for Eq. 14.37

$$\frac{d^2x}{dt^2} + \frac{2}{\tau}\frac{dx}{dt} + \omega^2 x = 0 \tag{14.40}$$

Here ω is the radian frequency of an undamped oscillator and τ is a parameter that characterizes the time needed for the oscillations to "decay."

Underdamped Oscillations

We can consider the important features of the solution to Eq. 14.40 without deriving the general solution. We begin by discussing the underdamped case. In previous sections you learned that ω is the natural frequency of a harmonic oscillator. We now wish to develop an interpretation of τ. First, looking at Eq. 14.40, we observe that τ is dimensionally a time. We can gain additional insight into τ by comparing the motion of the underdamped oscillator with SHM.

In Section 14.2 the equation describing SHM was shown to be

$$x = A \cos \omega t \tag{14.15}$$

This function is plotted in Figure 14.18 with $A = 1$ m and $\omega = 1$ rad/s. The curve begins at $x = +1$ m and oscillates between constant limits set by $x = \pm 1$ m. The time between alternate zeros, called the *period*, is related to ω by $T = 2\pi/\omega$. In this case $T = 2\pi$ s $= 6.28$ s.

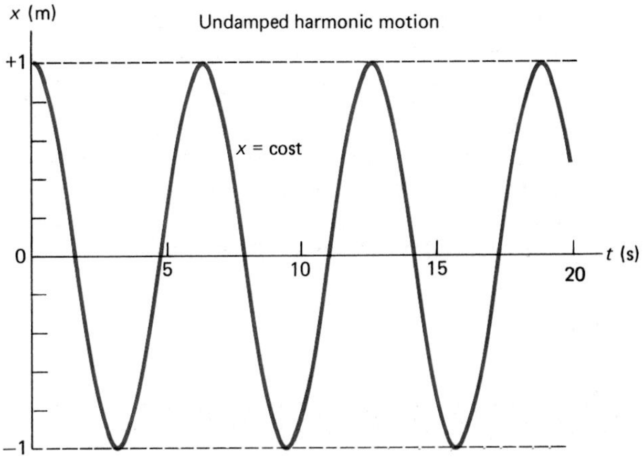

Figure 14.18

When there is no damping the maximum displacement of a harmonic oscillator started at $x = +1$ m with zero speed is bounded by the two straight lines $x = +1$ m and $x = -1$ m.

The equation that describes underdamped harmonic motion is

$$x = Ae^{-t/\tau} \cos \omega_1 t \qquad (14.41)$$

where

$$\omega_1 = \omega\left(1 - \frac{1}{\omega^2\tau^2}\right)^{1/2}$$

is the radian frequency of the cosine function, with ω_1 real and less than ω because $\omega\tau > 1$. This function is plotted in Figure 14.19 (with $A = 1$ m and $\tau = 6$ s). The curve begins at $x = +1$ m and oscillates between decreasing limits set by the envelope curves $x = \pm e^{-t/\tau}$. In this case the time between alternate zeros is

$$T = \frac{2\pi}{(\omega^2 - 1/\tau^2)^{1/2}} \qquad (14.42)$$

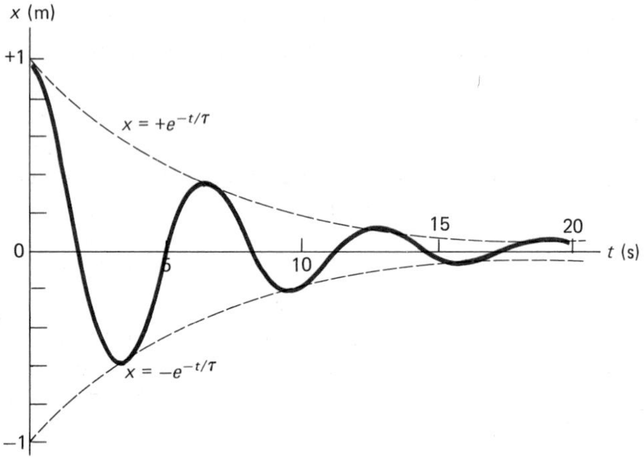

Figure 14.19

With underdamped harmonic motion the maximum displacement of an oscillator started at $x = +1$ m with zero speed is bounded by the two exponentials $x = +e^{-t/\tau}$ and $x = -e^{-t/\tau}$ (here $\tau = 8$ s).

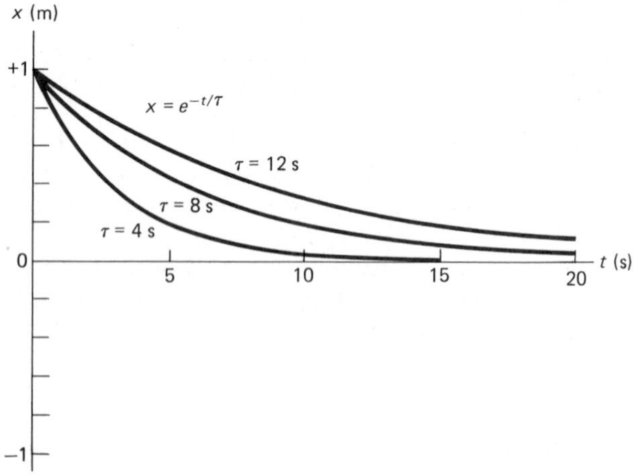

Figure 14.20

The mean time, τ, determines the rate at which the exponential $e^{-t/\tau}$ decreases to zero. The greater the value of τ, the more slowly the exponential is reduced to zero.

To see how τ controls the "amplitude," $Ae^{-t/\tau}$, in Eq. 14.41, we have plotted the exponential, $e^{-t/\tau}$, versus t for three values of τ in Figure 14.20. All of the curves approach zero as $t \rightarrow \infty$. The time, τ, determines how quickly this occurs. For large τ the approach to zero is slow. The reverse is also true. For small τ the approach to zero is rapid. The time τ determines how rapidly the exponential approaches zero, and therefore how rapidly underdamped oscillations die out.

With functions that approach zero exponentially with time, like those in Figure 14.20, it is customary to assign to each curve a characteristic time called the *mean*, which describes the rate of approach to zero of that curve. We define the mean time to be the time required for the function to fall to $1/e$ of its initial value.

For instance, consider the amplitude, $Ae^{-t/\tau}$, of the underdamped oscillator at the times $t = 0$ and $t = \tau$. At $t = 0$,

$$Ae^{-0/\tau} = A$$

and at $t = \tau$,

$$Ae^{-\tau/\tau} = \frac{A}{e}$$

From the definition of mean time and from these expressions, it is clear that the mean time for the underdamped oscillator is τ. Thus the amplitude of the underdamped oscillator approaches zero with a characteristic time τ.

For large τ, γ is small and the damping force is small. Underdamped motion occurs only for small values of γ. But how small must these values be? Consider Eq. 14.42, which gives the period of underdamped oscillations. If $\omega\tau > 1$, this equation gives a real value for the period, and underdamped motion will be the solution to Eq. 14.39. The condition $\omega\tau > 1$ is equivalent to the condition $\gamma < 2\sqrt{mk}$.

Example 6 A Damped Simple Pendulum

A long pendulum with a heavy bob will oscillate for several hours before coming to rest. This sytem is a good illustration of underdamped harmonic motion. Consider a pendulum 20 m long that requires 1 hr for its amplitude to be reduced to half its original value. What are the values of ω and τ?

We assume that $\omega\tau \gg 1$ so that we can use the approximation

$$T = \frac{2\pi}{(\omega^2 - 1/\tau^2)^{1/2}} \approx \frac{2\pi}{\omega}$$

From the simple pendulum expression

$$T = 2\pi\sqrt{\frac{l}{g}}$$

we calculate

$$T = 2\pi\sqrt{\frac{20}{9.80}} = 8.98 \text{ s}$$

Therefore

$$\omega = \frac{2\pi}{T} = 0.700 \text{ rad/s}$$

To evaluate τ we compare the maximum values of displacement at $t = 0$ s and at $t = 3600$ s (1 hr). From Eq. 14.41 we have

$$x_{max}(0) = Ae^0 = A$$

$$x_{max}(3600) = Ae^{-3600/\tau}$$

Using

$$\frac{x_{max}(0)}{x_{max}(3600)} = 2$$

we obtain the equation

$$2 = e^{+3600/\tau}$$

Solving for τ we obtain

$$\tau = \frac{3600}{\ln 2} = 5190 \text{ s}$$

$$= 1.64 \text{ hr}$$

Overdamped Harmonic Motion

A solution to Eq. 14.40 for overdamped motion ($\omega\tau < 1$) is

$$x = \frac{1}{\alpha_2 - \alpha_1}(\alpha_2 e^{-\alpha_1 t} - \alpha_1 e^{-\alpha_2 t}) \qquad (14.43)$$

where α_1 and α_2 are roots of the quadratic equation $\alpha^2 - (2/\tau)\alpha + \omega^2 = 0$. This particular solution satisfies the initial conditions $x = 1$ m and $dx/dt = 0$ at $t = 0$.

In Figure 14.21 we show several plots of x versus t for Eq. 14.43 with $\omega = 1$ rad/s. The greater the damping (large γ and small τ), the slower the approach of x to zero. In the limiting case, $\omega\tau = 0$, the system does not move, and x maintains a value of 1. The largest value $\omega\tau$ may have without oscillation is unity, corresponding to the limit of critically damped harmonic motion.

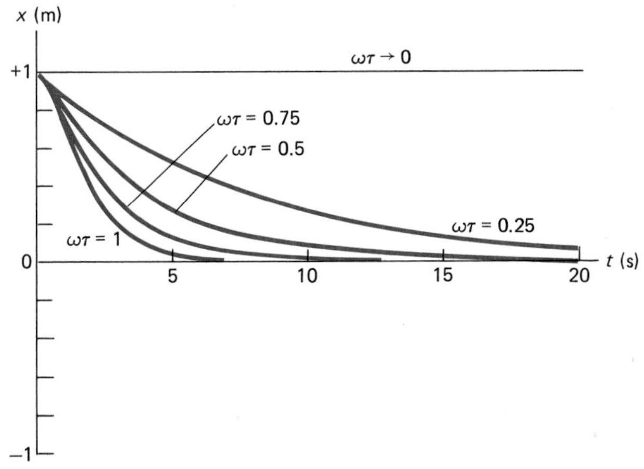

Figure 14.21

These curves for overdamped harmonic motion are for $\omega = 1$ rad/s with $\tau = 0$, $\tau = 0.25$ s, $\tau = 0.5$ s, and $\tau = 0.75$ s. In each case the particle is released from $x = +1$ m with zero velocity.

Critically Damped Motion

With underdamped motion the most rapid approach to zero of the maximum displacement corresponds to the smallest τ or the largest γ—the maximum damping. With overdamped motion the most rapid approach to zero corresponds to the largest τ or smallest γ—the minimum damping. Critically damped motion, $\omega\tau = 1$ or $\gamma = 2(mk)^{1/2}$, evidently produces the most rapid return to zero (Figure 14.21). This fact is used to advantage in some electrical instruments.

Energy Considerations in Damped Motion

The kind of motion represented by any of the three solutions to Eq. 14.40 is transient. The motion dies out exponentially. Mechanical energy is not conserved, but is converted to thermal energy. To show this, we write the total mechanical energy, $E = K + U$, in the form

$$E = \frac{1}{2}mv^2 + \frac{1}{2}kx^2 \qquad (14.44)$$

Differentiating, and using $v = dx/dt$ and $a = dv/dt$, we can write the rate of change of E as

$$\frac{dE}{dt} = (ma + kx)v \qquad (14.45)$$

With the harmonic oscillator, $(ma + kx) = 0$, so $dE/dt = 0$. Mechanical energy is conserved. With the damped harmonic oscillator, however, we have $(ma + kx) = \gamma v$, so Eq. 14.45 can be written

$$\frac{dE}{dt} = -\gamma v^2 \qquad (14.46)$$

The rate at which the total mechanical energy changes is negative, $dE/dt < 0$. Therefore, the mechanical energy decreases with time. We can assume that the mechanical energy is converted into thermal energy by viscous friction.

Although we used the linear motion of a horizontal spring as the model for a damped harmonic oscillator, our discussion is valid for any damped harmonic oscillator subject to a damping force in the form given in Eq. 14.38. The resistance in electrical circuits introduces a damping term in the circuit equation just as the frictional force in the horizontal spring introduces a damping term in Eq. 14.39. The flow of charge through a resistor converts electrical energy into thermal energy. The theory of damped harmonic motion applies equally well to transient currents in electrical circuits. These damped solutions are called *transient* (meaning temporary) because all of them go to zero with time.

Question

10. Shock absorbers are placed in cars to provide damping. Should the shock absorbers be designed to result in underdamped, critically damped, or overdamped motion? Explain the disadvantages of the two wrong options.

14.6 Forced Harmonic Motion: Resonance

In transient motion, no external forces act to replace the mechanical energy, which is continuously being converted to thermal energy through friction. Consequently, the motion eventually stops in all three cases of damping. In Section 14.5 we assumed that the moving body experienced an elastic force ($-kx$) and a frictional force ($-\gamma v$). We now study the behavior of a damped harmonic oscillator that is also subjected to a third force, which varies sinusoidally with time. We call such a system a *forced oscillator*.

Our prototype of a forced oscillator consists of a mass m that is subjected to a spring force, $-kx$, a frictional force, $-\gamma v$, and a driving force, $F_o \cos \omega' t$. The motion is governed by the equation

$$ma + \gamma v + kx = F_o \cos \omega' t \qquad (14.47)$$

Making the same substitutions as in the previous section, and letting $a_o = F_o/m$, we obtain a standardized form of Eq. 14.47:

$$\frac{d^2x}{dt^2} + \frac{2}{\tau}\frac{dx}{dt} + \omega^2 x = a_o \cos \omega' t \qquad (14.48)$$

In this equation $\omega^2 = k/m$ and $\tau = 2m/\gamma$, as before, and ω' is the radian frequency of the impressed force, $F_o \cos \omega' t$.

This differential equation has a general solution consisting of two parts. The transient part of the solution involves underdamped, critically damped, or overdamped motion, depending on the value of $\omega\tau$, as discussed in Section 14.5. The other part of the general solution to Eq. 14.48, called the *steady-state solution*, is our concern here.

At the onset of forced vibration an oscillator will usually exhibit a mixture of transient and steady-state motions. The particular mixture obtained depends on how and where the system started to move. Any transient motion initially present will die out exponentially with time, but the steady-state motion will persist. After a sufficiently long time ($t \gg \tau$) the motion will be steady state. We assume here that the system executes steady-state motion without the presence of transients.

In the steady state the system is forced to execute SHM with a frequency, ω', determined by the driving force. Therefore we write the solution in the form

$$x = A \cos(\omega' t + \phi) \qquad (14.49)$$

to resemble the harmonic oscillator solution given in Eq. 14.11. There are, however, significant differences between these two solutions. In Eq. 14.11, ω is determined by k and m, whereas in Eq. 14.49, ω is determined by the driving force. Additionally, A and ϕ are constants of integration in Eq. 14.11, but in Eq. 14.49 A and ϕ are determined by ω, ω', and τ. Much of our interest in the forced oscillator relates to the "response" of the system. By response we mean the dependence of A and ϕ on ω'.

The variation of A with ω' is illustrated by the following example. A car with nylon cord tires is not driven for a prolonged period of time, and consequently the tires develop flat spots. Then, when the car is first driven, the passengers feel a rhythmic bumping each time that these flat spots hit the road. The amplitude of vibration of the car is generally small. However, the driver discovers that this bumping is tolerable if the speed of the car is either slow or fast, but that at 38 mph the amplitude becomes large and the car threatens to shake itself to pieces. This large amplitude at a particular frequency is an example of resonance. We now demonstrate that Eq. 14.49 exhibits resonance by solving Eq. 14.48 when $\omega' = \omega$.

Analytic Solution for $\omega' = \omega$

We must evaluate the first and second derivatives of Eq. 14.49, and then substitute these values into Eq. 14.48. We have

$$x = A \cos(\omega' t + \phi) \qquad (14.49)$$

$$\frac{dx}{dt} = -\omega' A \sin(\omega' t + \phi)$$

$$\frac{d^2x}{dt^2} = -\omega'^2 A \cos(\omega' t + \phi)$$

First we note that

$$\frac{d^2x}{dt^2} + \omega^2 x = (-\omega'^2 + \omega^2) A \cos(\omega' t + \phi)$$

equals zero if $\omega' = \omega$. Hence Eq. 14.48 becomes

$$\frac{-2\omega' A}{\tau}\sin(\omega' t + \phi) = a_o \cos \omega' t$$

Expanding $\sin(\omega' t + \phi)$ we have

$$\frac{-2\omega' A}{\tau}(\sin \omega' t \cos \phi + \cos \omega' t \sin \phi)$$

$$= a_o \cos \omega' t$$

This expression must be identically valid for all values of t. Hence the coefficient of the $\sin \omega' t$ term must equal zero and the coefficients of $\cos \omega' t$ must be equal. For the coefficient of $\sin \omega' t$ to equal zero we have $\cos \phi = 0$. Choosing $\phi = -\pi/2$ rad we obtain

$$x = A \sin \omega' t \qquad (14.50)$$

For the coefficients of $\cos \omega' t$ to be equal we have

$$A = \frac{\tau a_o}{2\omega'} \qquad (14.51)$$

as our steady-state solution.

Recall from Section 14.5 that large τ accompanies small γ and weak damping. This, in turn, produces a large amplitude. In the limit of zero damping, Eq. 14.51 shows $A \rightarrow \infty$. Energy is being fed into the system by the external force. Our model for forced harmonic motion provides for energy removal from the system by the damping mechanism. Without damping the energy of the system increases without limit, and the amplitude goes to infinity.*

*This unlimited increase in energy is based on the assumption that Hooke's law ($F = -kx$) holds. In practice Hooke's law will fail before $A \rightarrow \infty$.

We observe that $\phi = -\pi/2$ rad if $\omega' = \omega$. The driving force reaches its maximum at $t = 0$ because $\cos \omega' t = +1$ when $\omega' t = 0$. The displacement reaches its maximum at $t = \pi/2\omega'$ because $\sin \omega' t = +1$ when $\omega' t = \pi/2$. The driving force peaks earlier than the displacement by a phase angle of $\pi/2$ rad. Both of these results are characteristic of resonant behavior at $\omega' = \omega$.

Approximate Solution for Small ω'

When Eq. 14.49 is substituted into Eq. 14.48 we obtain

$$-A\left[\omega'^2 \cos(\omega' t + \phi) + \frac{2\omega'}{\tau} \sin(\omega' t + \phi) - \omega^2 \cos(\omega' t + \phi)\right] = a_0 \cos \omega' t$$

For sufficiently small ω' both $\omega'^2 \ll \omega^2$ and $2\omega'/\tau \ll \omega^2$. Hence the first two terms on the left side of the equation are negligible, provided that $\omega'/\omega \ll 1$ and $\omega'/\omega \ll \omega\tau/2$. This means we are left with

$$\omega^2 A \cos(\omega' t + \phi) = a_0 \cos \omega' t$$

Following the same procedure as we did in the $\omega' = \omega$ case, we find the solution

$$x = A \cos \omega' t \tag{14.52}$$

with

$$A = \frac{a_0}{\omega^2} \tag{14.53}$$

In this limit the amplitude is constant and independent of τ and ω'. Here the displacement and driving force are in phase. They both peak when $t = 0$, and the solution requires $\phi = 0$.

Approximate Solution for Large ω'

When Eq. 14.49 is substituted into Eq. 14.48, the second and third terms on the left side of Eq. 14.48 may be neglected if ω' is sufficiently large. This means that $\omega'^2 \gg \omega^2$ and $\omega' \gg 2/\tau$. Thus we have

$$-\omega'^2 A \cos(\omega' t + \phi) = a_0 \cos \omega' t$$

Following the same procedure as before, we obtain the solution

$$x = -A \cos \omega' t \tag{14.54}$$

with

$$A = \frac{a_0}{(\omega')^2} \tag{14.55}$$

Here the amplitude approaches zero as $\omega' \to \infty$. Additionally, $\phi = \pi$ rad, and the displacement peaks in one direction at the same time that the driving force peaks in the opposite direction.

We have been concerned with resonant behavior that occurs when the damping is small. Small damping can lead to very large amplitudes, as we have seen. In Figures 14.22 and 14.23 we display curves of ϕ and A as a function of ω' for both large and small damping. Note that these curves are consistent with our analytical results in the two limiting cases and at $\omega' = \omega$.

Figure 14.22

The relative phase angle, ϕ, between the displacement and driving force is shown versus driving frequency ω'. At resonance, $\omega' = \omega$ and $\phi = \pi/2$ radians.

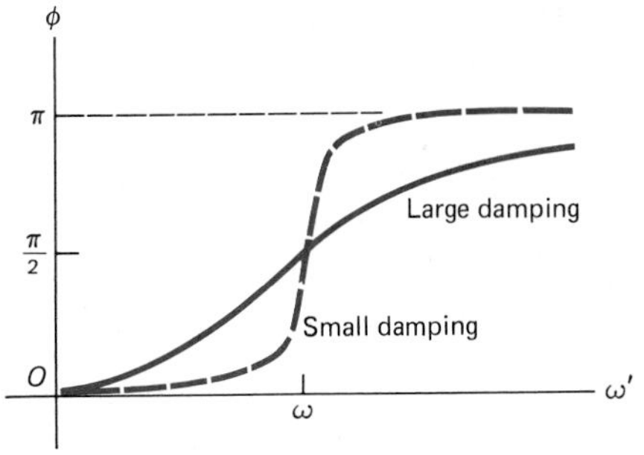

Figure 14.23

The amplitude A of the motion is shown versus the driving frequency ω'. Near resonance the amplitude becomes large. The effect becomes more pronounced as the damping is reduced.

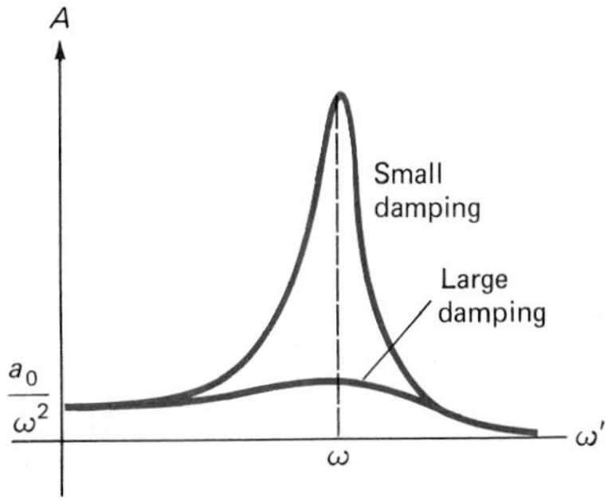

Questions

11. Why would it be sensible to design machinery so that all resonant frequencies are higher than any frequencies encountered in normal use?
12. Some opera stars are said to be able to shatter wineglasses simply by singing. Explain. [See Haym Kruglak and René Pittet in *Physics Teacher*, 17(1) 49 (January 1979).]
13. Are there any examples of forced oscillations in a children's playground? Are these examples of resonance?

Summary

The period, T, of harmonic motion is the time required for one complete oscillation. It is the reciprocal of the frequency, ν, which denotes the number of oscillations per unit time.

$$T = \frac{1}{\nu} \tag{14.1}$$

Simple harmonic motion (SHM) is an oscillatory motion for which the acceleration and displacement are proportional, but of opposite sign. The equation of motion of a harmonic oscillator is

$$a = -\omega^2 x \quad \text{or} \quad \frac{d^2x}{dt^2} + \omega^2 x = 0 \qquad (14.4)$$

where

$$\omega = \frac{2\pi}{T} = 2\pi\nu \qquad (14.14)$$

is constant. The solution to the harmonic oscillator equation is

$$x = A\cos(\omega t + \phi) \qquad (14.11)$$

where A is the amplitude and ϕ is the initial phase.

A simple pendulum approximates SHM with a period given by

$$T = 2\pi\left(\frac{l}{g}\right)^{1/2} \qquad (14.6)$$

For a horizontal spring the period is

$$T = 2\pi\left(\frac{m}{k}\right)^{1/2} \qquad (14.5)$$

The mechanical energy of the oscillator is given by

$$E = \frac{1}{2}mv^2 + \frac{1}{2}kx^2 \qquad (14.7)$$

The projection on a diameter of a point engaged in uniform circular motion moves with SHM. The maximum displacement, velocity, and acceleration are given by

$$x_{max} = r \qquad (14.29)$$

$$v_{max} = r\omega \qquad (14.30)$$

$$a_{max} = r\omega^2 \qquad (14.31)$$

respectively, where r is the radius of the circle and ω is the angular velocity.

In the presence of a frictional force that is proportional to the velocity, the harmonic oscillator equation becomes

$$\frac{d^2x}{dt^2} + \frac{2}{\tau}\frac{dx}{dt} + \omega^2 x = 0 \qquad (14.37)$$

where τ governs the extent of friction. Motion may be overdamped ($\omega\tau < 1$) critically damped ($\omega\tau = 1$), or underdamped ($\omega\tau > 1$). In all cases motion approaches zero for large t. In such motion mechanical energy is converted to thermal energy.

When an external sinusoidal force acts on a damped harmonic oscillator the equation of motion becomes

$$\frac{d^2x}{dt^2} + \frac{2}{\tau}\frac{dx}{dt} + \omega^2 x = a_o \cos \omega' t \qquad (14.48)$$

where a_o is a measure of the external force. The displacement and driving force differ in phase by an angle ϕ between 0 and π rad. The amplitude, A, and phase, ϕ, depend on the driving frequency.

A damped harmonic oscillator is said to be in resonance when its amplitude becomes large. With small damping the amplitude is large when $\omega' = \omega$. At this frequency the phase equals $\pi/2$ rad.

Problems

Section 14.1 Simple Harmonic Motion

1. When a mass of 5 gm is attached to a vertical spring the spring's length increases by 2 cm. The same spring is now arranged to oscillate horizontally. A mass of 2 gm is attached and the spring is pulled a distance x_0 from its equilibrium position and then released from rest. The initial acceleration is 50 m/s^2. Determine (a) the spring constant of the spring; (b) the radian frequency, ω, for the oscillation; and (c) the initial displacement x_0.

2. A particle executing simple harmonic motion along the x-axis returns to its starting position every 0.25 s. Determine the period, frequency, and radian frequency.

3. Determine the mass that must be attached to the horizontal spring in Problem 1 in order for ω to equal 2π rad/s.

4. Give the frequency in hertz of (a) the second hand, (b) the minute hand, and (c) the hour hand of a watch.

5. A simple pendulum that is 1 m in length has an angular acceleration given by $\alpha = -(g/l)\sin\theta$ that becomes $\alpha = -(g/l)\,\theta$ for small angles. Determine the percentage error resulting from the use of this approximation when θ equals (a) 1°; (b) 10°; (c) 90°.

6. A simple harmonic oscillator in the form of a linear spring and a 0.150-kg mass has a period of 2.10 s. Determine the spring constant.

7. A mass m is oscillating freely on a linear spring (Figure 1) that obeys Hooke's law. When $m = 0.81$ kg the period is 0.91 s. An unknown mass on the same spring is observed to have a period of oscillation of 1.16 s. Determine (a) the spring constant k of the spring and (b) the unknown mass.

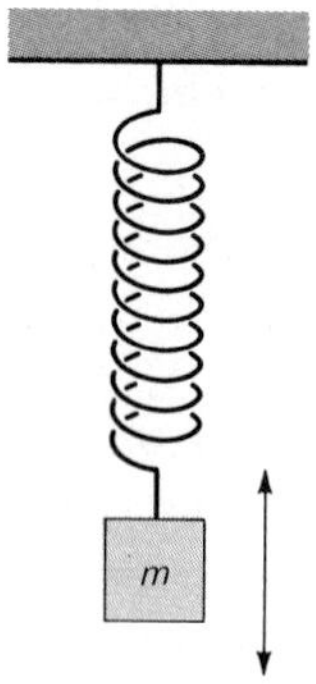

Figure 1

8. A mass of 0.79 kg oscillates freely at the end of a linear spring that obeys Hooke's law. The period of oscillation is 2.1 s. Calculate (a) the radian frequency, ω, and (b) the spring constant, k. (c) If the

0.79-kg mass is replaced by a 0.52-kg mass, what will be the new period of oscillation?

9. Determine the displacement of a horizontal spring at which the potential and kinetic energies are the same. Express your answer in terms of the amplitude A.

10. Reasoning directly from Eq. 14.4, prove that the constant ω has dimensions of inverse time.

11. Starting with the equation of motion for the harmonic oscillator (Eq. 14.4), integrate once to obtain the energy equation (Eq. 14.7).

Section 14.2 Solution of the Harmonic Oscillator Equation

12. The angular position of the string of a pendulum is represented by the equation $\theta = 0.32 \cos(4.43\ t)$. Determine the period, frequency, radian frequency, and length of the pendulum.

13. The velocity of a 0.3-kg mass attached to the end of a spring is represented by $v = 1.60 \sin 2.83t$ m/s. Calculate the total energy of the system.

14. Prove by direct substitution that $x = A \sin(\omega t + \phi)$ satisfies Eq. 14.4 for arbitrary A and ϕ.

15. Prove by direct substitution that $x = c_1 \sin \omega t + c_2 \cos \omega t$ satisfies Eq. 14.4 for arbitrary values of the constants c_1 and c_2.

16. A spring–mass simple harmonic oscillator is described by the equation

$x = 0.073 \cos 2t$ m

(a) Where are the turning points? (b) What is the period?

17. For the equation $x = 4.2 \cos(t/2 + \pi)$ give (a) the amplitude, A, (b) the constant, ω, (c) the period, T, and (d) the initial phase angle, ϕ.

18. Sketch the equation $x = A \sin(\omega t + \phi)$ for two complete cycles by using $A = 2$ m, $\omega = 2$ rad/s, and (a) $\phi = -\pi/4$ rad; (b) $\phi = 0$; (c) $\phi = +\pi/4$ rad.

Section 14.3 Applications of Simple Harmonic Motion

19. A meter stick is set into oscillation about an axis that passes through a point 20 cm from one end. Determine the (a) period and (b) length of the equivalent simple pendulum.

20. (a) A meter stick oscillates about a horizontal axis perpendicular to the stick. Where should the axis be located to achieve the minimum period? (b) What is this period?

21. A hoop (Figure 2) of mass M and radius R is set into oscillation about an axis that passes through O and is perpendicular to the plane of the figure. Derive an expression for the period. Give the period for $R = 0.1$ m and $R = 10$ m.

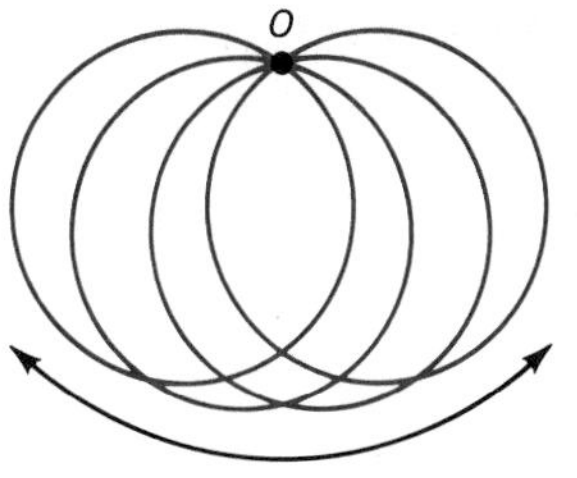

Figure 2

22. Determine the length of the simple pendulum that has the same period as the hoop in Problem 21. Do this for both values of R.

23. The equation of motion of a simple pendulum executing small oscillations is

$$\frac{d^2\theta}{dt^2} = -\frac{g}{l}\theta$$

where θ is the angle between the string and the vertical. Show that for small oscillations this equation can also be written as

$$\frac{d^2x}{dt^2} = -\frac{g}{l}x$$

where x is the horizontal distance between the bob and the vertical.

24. The equation of motion of a simple pendulum is found to be $d^2\theta/dt^2 = -9.0\ \theta$ rad/s^2. Show that a solution of this equation is $\theta = 0.10 \cos 3t$ rad.

25. A 0.12-kg particle attached to the end of a spring oscillates horizontally on a frictionless surface. Its position as a function of time is represented by $x = 0.35 \cos \omega t$ m. Determine the spring constant if the particle's maximum acceleration equals g.

26. A spring having an unstretched length L is cut in half to make two springs of length $L/2$. When a mass m_1 is connected to the original spring (length L), the vibrational frequency is the same as for a mass m_2 connected to a spring of length $L/2$. How are m_1 and m_2 related?

27. (a) Explain why you would expect the period of a simple pendulum to increase when it is moved to an elevated position above the earth's surface. (b) If a simple pendulum has a period of 1 s when on the earth's surface, at what distance above the earth's surface is its period 2 s?

28. A simple pendulum of length l has a period T_0. If the length is increased by a small amount Δl, show that the fractional change in the period is

$$\frac{\Delta T}{T} = \frac{1}{2}\left(\frac{\Delta l}{l}\right)$$

29. A particular simple pendulum has a period of 1.0 s. Find the frequency, ν, and the length of the pendulum.

30. A simple pendulum of length 1.45 m is observed to take 242 s to complete 100 full oscillations. Calculate g. (*Note:* The experiment is carried out at high altitude.)

31. Prove that a physical pendulum (Figure 3) has a minimum period when the distance between the axis of rotation and the center of mass equals the radius of gyration, $h = s$ (see Section 12.3).

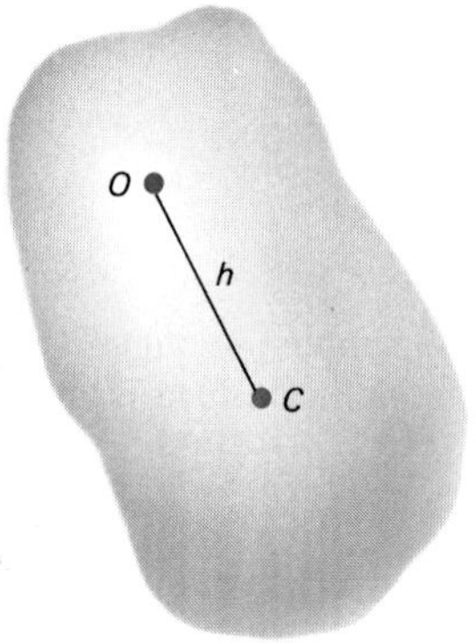

Figure 3

32. An athlete of mass 80 kg assumes a prone position with his center of mass on the symmetry line of the torsion pendulum shown in Figure 14.11. The measured period of oscillation of the pendulum with the athlete is $T = 3$ s and without the athlete is $T_o = 2$ s. Using the value of β given in Example 4, determine (a) the moment of inertia of the athlete about a vertical axis through his center of mass and (b) the moment of inertia for rotation about a parallel axis through a point 1.5 m above his center of mass.

Section 14.4 Uniform Circular Motion and Simple Harmonic Motion

33. An alien traveling in a spacecraft views the earth's orbital motion edge on, and concludes that the earth is in a circular orbit. From measurements of the amplitude of the motion, $A = 9.3 \times 10^7$ miles, and the period of the oscillation, $T = 1$ year, the alien obtains values for the orbital velocity of the earth, v, and its centripetal acceleration. What are these values?

34. Saturn's largest moon, Titan, is observed to execute uniform circular motion about Saturn. Its position relative to Saturn may be described by

$$r = r_o \cos \frac{2\pi t}{T}$$

with $r_o = 1.22 \times 10^6$ km and $T = 15.95$ days. (a) Find the frequency, ν, and (b) the angular frequency, ω. (c) Assuming that Titan is bound to Saturn by gravity, calculate the mass of Saturn.

35. Equation 14.35 gives the magnitude of the centripetal acceleration as $\omega^2 r$. The earlier calculation in Section 5.2 led to v^2/r. Reconcile these two results.

36. An object undergoing uniform circular motion in a plane has a displacement given by

$$\mathbf{r} = \mathbf{i}\, 0.31 \cos 4t + \mathbf{j}\, 0.31 \sin 4t \text{ m}$$

(a) Determine the radius of the circle. (b) Determine the period of the motion.

37. For Io, one of the Galilean moons of Jupiter, the time for one revolution is 1.77 days and the orbital radius is 4.22×10^8 m. Using these data, express the edge on motion as the equation $x = A \cos \omega t$, giving numerical values for A and ω.

38. A cyclist rides directly toward an observer who views only the projected up and down motion of points on the circumference of the wheels. The wheels are 0.33 m in radius and spin with an angular velocity of 50 rad/s about their centers of mass. Calculate the observed projected maximum (a) velocity and (b) acceleration of a point on the circumference of a wheel.

Section 14.5 Damped Harmonic Motion

39. A simple pendulum with a 70-kg bob oscillates as an underdamped harmonic oscillator. It has a period of 2π s and a damping constant of 0.01 kg/s. Determine (a) the pendulum length; (b) the maximum bob velocity if its original maximum amplitude is 3°; (c) the original maximum kinetic energy; (d) the maximum kinetic energy 2 hr after it starts oscillation; (e) the original maximum power dissipated by friction and (f) the maximum power dissipated 2 hr after it starts to oscillate.

40. The oscillation of a simple pendulum with a 10-kg bob is described by the equation

$$10\frac{d^2x}{dt^2} + 6.25 \times 10^{-3}\frac{dx}{dt} + 23.5x = 0$$

Calculate (a) the angular frequency, ω; (b) the period, T; (c) the time for the amplitude to fall to half of its original value.

41. A damped simple harmonic oscillator has

$\gamma = 0.11$ kg/s

$k = 1.20$ N/m

$m = 0.130$ kg

(a) Is its motion overdamped or underdamped? (b) Determine the value of γ for critically damped motion.

42. Prove that as $t \to \infty$, $e^{-(\omega^2\tau t/2)} \to 0$ more slowly than $e^{-t/\tau}$ if $\omega\tau \ll 1$.

43. Prove that $\gamma < 2(mk)^{1/2}$ is equivalent to $\omega\tau > 1$ as a criterion for underdamped motion to occur.

Section 14.6 Forced Harmonic Motion: Resonance

44. Prove that with sufficiently small damping, A at resonance is large compared with A in the low-frequency limit.

45. Two particles of equal mass are connected to separate springs having spring constants of k and $4k$. With one end of each spring fixed, the particles are given identical displacements and then released into

oscillatory motion. If the smaller period of the two springs is 1 s, (a) what is the minimum time required for the two particles to return simultaneously to their starting positions? (b) How many oscillations has each particle made during this time?

46. A poorly designed piece of electrical equipment is being driven at its natural (resonant) frequency of 60 Hz. The parameters are $m = 1.21$ kg, $k = 1.72 \times 10^5$ N/m, $\gamma = 3.10 \times 10^{-2}$ kg/s, and $F_o = 3.68 \times 10^{-3}$ N. Calculate (a) a_{oi} (b) τ_i (c) the resonant amplitude, A.

47. A spring–mass oscillator has $k = 4.72$ N/m, $m = 0.93$ kg. What must be the period of a simple harmonic driving force if it is to excite resonance?

chapter 15

THE MECHANICAL PROPERTIES OF MATTER

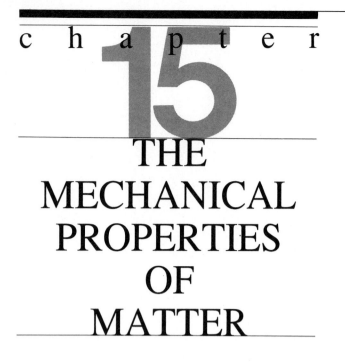

Preview

Man's successes in exploiting the properties of matter have been chronicled in history. Historians speak of the Stone Age, the Bronze Age, and the Atomic Age. During the first half of the 20th century man's ability to make the most of Nature's raw materials hinged on a grasp of the basic laws governing matter in bulk form, that is, in quantities of macroscopic size. In this chapter and in Chapter 16 we study the mechanical properties of matter in bulk form. This chapter is devoted primarily to solids, whereas Chapter 16 is devoted to the mechanics of fluids. The microscopic properties of matter—its thermal, electrical, and magnetic properties—will be treated in later chapters.

In this chapter the familiar words *stress* and *strain* are given precise definitions. An empirical relation between stress and strain—Hooke's law—is formulated. The three primary elastic moduli are introduced. A number of examples are presented that show how the elastic moduli are used in various practical and theoretical calculations.

15.1 States of Matter

In this chapter, our analysis of the properties of matter is at the macroscopic level—the level at which our measuring instruments and senses can respond. A *macroscopic* quantity registers the combined actions of many atoms or molecules. In contrast, a *microscopic* quantity describes a single atom or molecule. The mass of your body, for example, is a macroscopic quantity; the mass of a single carbon atom in your body is a microscopic quantity.

The physical states of matter can be divided into two broad classes—solids and fluids. A solid maintains its shape: It resists the actions of external forces that tend to change its shape or volume, and it is not easily "sliced" or compressed. A fluid, on the other hand, can easily change shape and flow when subjected to a force. A fluid is easily sliced.

Gases and liquids are fluids. The distinction between a gas and a liquid is made on the basis of compressibility. Under ordinary conditions, gases are relatively easy to compress, whereas liquids are virtually incompressible. A qualitative distinction between solids, liquids, and gases can be made as follows: A gas can be sliced and compressed, a liquid can be sliced but is not easily compressed, and a solid is not easily sliced or compressed.

Solids, liquids, and gases can also be distinguished at the microscopic level. A crystalline solid is a highly ordered array of atoms that are bound by mutual electrical forces. Individual atoms are not free to roam about. Their motions are restricted to vibrations about positions of equilibrium. A liquid, on the other hand, is a crowded assembly of mobile atoms. Each atom is in contact with several

Table 15.1

Physical properties of solids*

Material	Density (units of 10^3 kg/m^3)	Young's Modulus (Y)	Shear Modulus (μ)	Bulk Modulus (B)
		(units of 10^{10} Pa)		
C (diamond)	3.25	83	34	55
Al	2.70	6.9	2.6	7.2
Fe	7.87	19	719	14
Cu	8.96	12	4.6	13
Pb	11.36	1.8	0.55	6
U	19.07	17	7.6	11
Steel	7.8	21	8	16
Glass	2.5	6.9	2.4	4.3
Brass	8.6	9.0	3.5	6.0

*Values are approximate. The quantities Y, μ, and B are defined in Sections 15.3, 15.6, and 15.5, respectively. The pascal (abbreviated as Pa) is the SI unit of force per unit area, 1 Pa = 1 N/m^2.

neighboring atoms, but is not bound securely to any of them. As an atom moves about in a liquid, it collides frequently with its neighbors. In a gas, the average distance between atoms is so great that collisions are infrequent in comparison with those in a liquid. The atoms in a gas spend most of their time in free flight.

The mass density, ρ, is useful in describing both solids and fluids. The mass density is defined as the mass per unit volume. If a mass m is distributed uniformly over volume V, the mass density is

$$\rho = \frac{m}{V} \qquad (15.1)$$

If the mass is not uniformly distributed, Eq. 15.1 defines the average density. The SI unit of density is kilograms per cubic meter (kg/m^3). The densities of several materials are listed in Table 15.1. Also included in Table 15.1 are other mechanical properties that will be introduced later.

Example 1 Average Mass Density of the Earth and Sun

The matter in the earth and sun is not uniformly distributed. We can use Eq. 15.1 to calculate the average mass density for these two bodies.

The mass and radius of the earth are

$$m_E = 5.98 \times 10^{24}\ \text{kg} \qquad R_E = 6.37 \times 10^6\ \text{m}$$

The volume of the earth is

$$V_E = \frac{4}{3}\pi R^3 = 1.08 \times 10^{21}\ \text{m}^3$$

The average mass density is

$$\rho_E = \frac{5.98 \times 10^{24}\ \text{kg}}{1.08 \times 10^{21}\ \text{m}^3} = 5.52 \times 10^3\ \text{kg/m}^3$$

By comparison, the density of water is 1.00×10^3 kg/m^3. The mass and radius of the sun are

$$m_S = 1.99 \times 10^{30}\ \text{kg} \qquad R_S = 6.95 \times 10^8\ \text{m}$$

The average mass density is

$$\rho_S = \frac{1.99 \times 10^{30}\ \text{kg}}{1.41 \times 10^{27}\ \text{m}^3} = 1.41 \times 10^3\ \text{kg/m}^3$$

Although the mass of the earth is much smaller than that of the sun, the volume of the earth is smaller as well—so much so that the average density of the earth is nearly four times as great as the average density of the sun.

Questions

1. Classify the following quantities as microscopic or macroscopic:
   (a) the mass of an oxygen molecule;
   (b) the weight of the air in your classroom;
   (c) your body temperature;
   (d) the kinetic energy of a water molecule in your body;
   (e) your height;
   (f) the diameter of a calcium atom in your femur (thighbone).
2. The characteristic diameter of an atom is 10^{-10} m. Using this value, approximately how many atoms tall are you?
3. An optical microscope can distinguish objects with sizes as small as 10^{-6} m. Approximately how many atoms having a diameter of 10^{-10} m would fit inside a cube with a side of length 10^{-6} m?
4. As salt dissolves in water, does it cause the density of the solution to increase, decrease, or remain unchanged? Explain. (You might want to try adding a spoonful of salt to a *full* glass of water without causing it to run over.)

15.2 Stress and Strain

In the absence of external forces a solid maintains its shape. A solid can be deformed by applying external forces. A material is said to be *elastic* if it reverts to its original shape when the deforming forces are removed. *Plastic* materials, on the other hand, remain permanently distorted when the deforming forces are removed. There are a number of *elastic moduli* that measure the response of an elastic solid when it is subjected to a deforming force. The deforming force is described in terms of the *stress* that it exerts, and the deformation of the solid is described in terms of the *strain* that results. In this section we will define and illustrate the primary types of stresses and strains. Then, in the next section, we will see how stress and strain are related by an elastic modulus.

A log easily splits when an ax strikes it parallel to the grain. The same blow applied across the grain fails to split

Figure 15.1

The equal but opposite forces F_t applied to the ends of the rod cause its length to increase from l to $l + \Delta l$. The rod is said to be in *tension*.

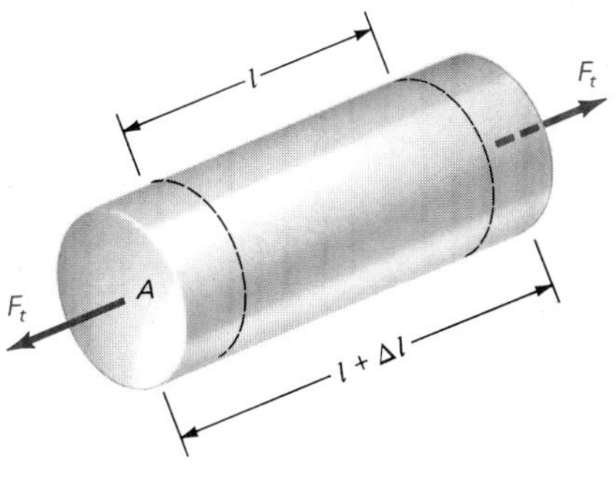

the wood. We describe this type of directional dependence by saying that wood is *anisotropic*. In other words, the physical properties of wood have a *directional dependence*. Many crystalline solids display anisotropies in their mechanical, thermal, and electromagnetic properties. The directional dependence of these properties of bulk matter can be traced to the anisotropic crystalline structure at the atomic level. Materials that do not exhibit any such directional preferences are said to be *isotropic*. Gases and most liquids are isotropic because the motions of their atoms are not strongly correlated. Noncrystalline solids, called *amorphous* solids, are isotropic. We limit our discussion in this text to isotropic materials.

Consider a rod of length l and cross-sectional area A. Suppose that a force F_t is applied at both ends, parallel to the length, so as to stretch the rod by Δl (Figure 15.1). The rod is said to be *under tension*. The **tensile stress** is defined as the ratio of the force of tension to the cross-sectional area:

$$\text{Tensile stress} = \frac{F_t}{A} \qquad (15.2)$$

The resulting **tensile strain** is defined as the ratio of the change in length to the original length:

$$\text{Tensile strain} = \frac{\text{change in length}}{\text{original length}} = \frac{\Delta l}{l} \qquad (15.3)$$

Note that tensile stress has the dimensions of force per unit area, whereas tensile strain is dimensionless. The **pascal** (abbreviated Pa) is the SI unit of stress.

$$1 \text{ Pa} = 1 \text{ N/m}^2$$

Example 2 **Stress and Strain**

A television commercial shows a 47,000-lb (209,400-N) railroad car being lifted by a cord made of a material used to strengthen automobile tires. The diameter of the cord is 9.1 cm. The tensile stress is

$$\frac{F_t}{A} = \frac{209{,}400 \text{ N}}{\frac{\pi}{4}(0.091)^2 \text{ m}^2} = 3.22 \times 10^7 \text{ N/m}^2$$

$$= 3.22 \times 10^7 \text{ Pa}$$

The unstretched length of the cord is 8 m. The length increases by 2.60 cm when the railroad car is raised. The tensile strain is

$$\frac{\Delta l}{l} = \frac{2.60 \text{ cm}}{800 \text{ cm}} = 0.00325$$

The large applied stress produces a relatively small change in length—slightly over 0.3%.

An object that is submerged in a fluid experiences compressive forces over its entire surface. If these compressive forces change the object's volume from V to $V + \Delta V$, the **volume strain** is defined as $\Delta V/V$. Because the compressive forces cause the volume to decrease, ΔV is negative. Consequently, the volume strain is negative. The **compressive stress** associated with volume strain is called *pressure*.* If F_p denotes the compressive force normal to a surface element of area A (Figure 15.2) the pressure is defined as

$$\text{pressure} = P = \frac{F_p}{A} \qquad (15.4)$$

Several units of pressure are commonly encountered in the sciences. In addition to the pascal (Pa) there are pounds per square inch (psi) and atmospheres (atm). The **atmosphere** is defined as

$$1 \text{ atm} = 1.01325 \times 10^5 \text{ Pa}$$

The weight of Earth's atmosphere exerts a pressure of approximately 1 atm on Earth's surface. The relation between pounds per square inch and pascals is

$$1 \text{ psi} = 6.89476 \times 10^3 \text{ Pa}$$

You can verify that

$$1 \text{ atm} = 14.696 \text{ psi}$$

If the ends of a rod are subjected to equal pushes, the rod is said to be *under compression*. The compressive stress and strain are defined in precise analogy to the tensile stress and strain. For many solids, experiment shows that the ratio of compressive stress to strain equals the ratio of tensile stress to strain.

$$\frac{\text{Compressive stress}}{\text{Compressive strain}} = \frac{\text{tensile stress}}{\text{tensile strain}}$$

Generally, this equality persists only for small elastic strains, extending roughly up to 0.01.

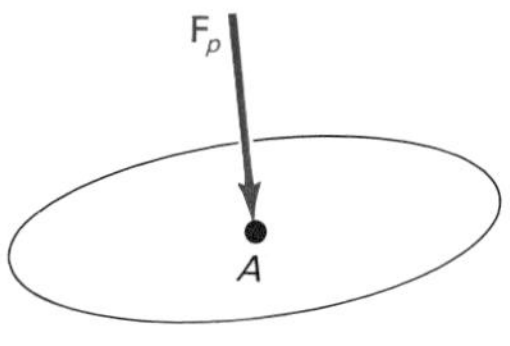

Figure 15.2

The pressure P is defined as the ratio of the magnitude of the normal force $\mathbf{F}_p$ to the area A on which the force acts: $P = F_p/A$.

*For liquids the terminology *hydrostatic pressure* is often used, even in situations where the pressure is not exerted by water.

The pressure of a confined gas results from impacts of the gas molecules on the container walls. In dealing with gas pressures, we generally measure the so-called *gauge pressure*, which is the difference between the total, or *absolute*, pressure and the surrounding atmospheric pressure. Thus, for example, if the gauge pressure for an auto tire reads 32 psi at a time when the atmospheric pressure is 15 psi, the absolute pressure in the tire is 47 psi.

Example 3 **How to Weigh Your Car with a Ruler**

The four tires of an automobile are inflated to a gauge pressure of 2.0×10^5 Pa (29 psi). Each of the four tires has an area of 0.024 m^2 that is in contact with the ground. We wish to determine the mass and weight of the automobile.

The weight, Mg, is balanced by the contact forces on the tires. The total contact force is $4PA$, where P is the gauge pressure and A is the contact area for one tire. We have for the weight

$$Mg = 4PA = 4 \cdot (2.0 \times 10^5 \text{ N/m}^2) \cdot (0.024 \text{ m}^2)$$
$$= 19,200 \text{ N}$$

which is 4310 lb. The mass is

$$M = \frac{19,200 \text{ N}}{9.80 \text{ m/s}^2} = 1960 \text{ kg}$$

Why is the gauge pressure used instead of the absolute pressure?

Shear Stress

The compressive stress that we call pressure changes the volume but not the shape of an object. A cube remains cubical when it is compressed by a uniform pressure. A *shear stress*, on the other hand, tends to change the shape of a body. Referring to Figure 15.3, we see that a force F_s parallel to the surface of area A deforms the rectangular area xy into a parallelogram. The shape of the body is altered. The shear stress is defined as

$$\text{shear stress} = S = \frac{F_s}{A} \tag{15.5}$$

and the shear strain is defined as

$$\text{shear strain} = \frac{\Delta x}{y} = \tan\theta \simeq \theta \tag{15.6}$$

The approximation $\tan\theta \simeq \theta$ is valid for small strains.*

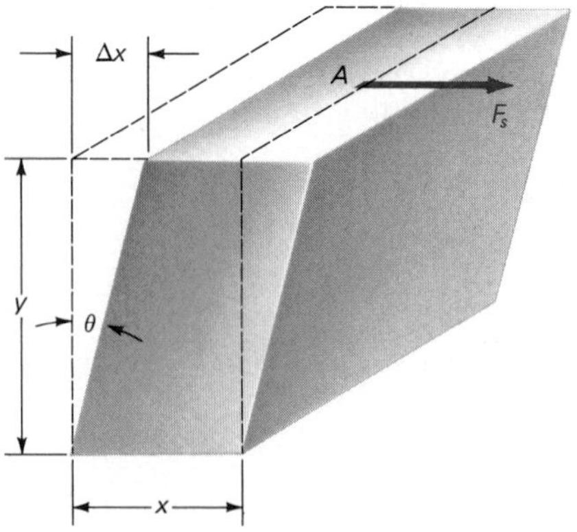

Figure 15.3

The shear stress S is defined as the ratio of the magnitude of the tangential force $\mathbf{F}_s$ to the area A on which the force acts: $S = F_s/A$.

Example 4 **Thumbs-Up Shear Stress**

A physics professor tells his class that human skin cannot withstand a shear stress in excess of 10^6 Pa. A group of students decide to check this assertion. They invite their physics professor to help. Moments after accepting the invitation, the professor finds himself suspended by his thumbs. His weight is distributed evenly over two strips of adhesive tape. The total contact area between his thumbs and the tape is 0.001 m^2. His mass is 70 kg. The shear stress on the skin is

$$S = \frac{F_s}{A} = \frac{70 \text{ kg} \cdot 9.80 \text{ m/s}^2}{0.001 \text{ m}^2} = 6.86 \times 10^5 \text{ Pa}$$

This is less than 10^6 Pa, and the professor stays suspended. The adhesive on the tape shears at a stress of 8.0×10^5 Pa. Can the students verify the professor's claim by cutting one strip of tape, thereby transferring all of his weight to the other strip?

Questions

5. Cite an example of a structure whose components are subject to tensile, compressive, and shear stresses. Identify the type of stress experienced by each component.
6. *Estimate* the compressive stress (pressure) on the soles of your shoes when you stand erect, with your weight evenly distributed on both feet. Express the pressure in pascals, pounds per square inch and atmospheres.

15.3 Hooke's Law and Young's Modulus

Many everyday situations show that stress and strain are related. For example, as a balloon is inflated, two quantities increase: the balloon's volume and the pressure of the air inside. The change in volume is a measure of the strain,

*See Appendix 4.

and the air pressure is a measure of the stress. An increase in the stress (as you put air in the balloon) results in an increase in the strain (the balloon expands). Similarly, the extension of a spring balance is a measure of its strain. The suspended weight establishes the stress. The greater the weight, the longer the extension. Again we see that increased stress leads to increased strain.

For many elastic solids, stress and strain are linked by a simple linear relation known as *Hooke's law.** Hooke's law states that stress is directly proportional to strain. The proportionality factor between stress and strain is called an *elastic modulus*. Hooke's law may be expressed as an equation:

$$\frac{\text{Stress}}{\text{Strain}} = \text{elastic modulus} \qquad (15.7)$$

In the particular case of tensile stress and strain, the elastic modulus is known as *Young's modulus,*† Y. Thus,

$$\frac{\text{tensile stress}}{\text{tensile strain}} = \frac{F_t/A}{\Delta l/l} = Y \qquad (15.8)$$

To emphasize the fact that the larger Y is, the more rigid or unyielding is the material, Hooke's law may be expressed as

$$\frac{\Delta l}{l} = \frac{F_t/A}{Y} \qquad (15.9)$$

This form of Hooke's law shows that for a fixed stress, the larger Y, the smaller the strain. Table 15.1 presents values of Y for several solids. Other elastic constants that will be introduced later are also included in Table 15.1.

Example 5 Stretching Steel

A Foucault pendulum has an 80-kg mass suspended from a steel wire. The wire is 18 m long and 3 mm in diameter. We want to determine the increase in the length of the wire when the pendulum is not in motion. In this situation, the tensile force equals the weight of the suspended mass

$$F_t = mg = 80\ \text{kg} \cdot 9.80\ \text{m/s}^2 = 784\ \text{N}$$

The cross-sectional area is

$$A = \frac{\pi}{4}(0.003)^2\ \text{m}^2 = 7.07 \times 10^{-6}\ \text{m}^2$$

The tensile stress is

$$\frac{F_t}{A} = \frac{784\ \text{N}}{707 \times 10^{-6}\ \text{m}^2} = 1.11 \times 10^8\ \text{Pa}$$

From Table 15.1, Young's modulus for steel is 21×10^{10} Pa. Using Eq. 15.9, we get the tensile strain:

$$\frac{\Delta l}{l} = \frac{F_t/A}{Y} = \frac{1.11 \times 10^8}{21 \times 10^{10}} = 0.00053$$

With $l = 18$ m we find $\Delta l = 9.5$ mm. The wire is stretched nearly 1 cm.

Hooke's law is valid only for very small strains. The linear relation between stress and strain generally fails in the range of strain between 0.001 and 0.01. For example, if a metal rod is subjected to increasing tension, it eventually undergoes a plastic deformation—it "gives"—at a characteristic **yield stress.** Further increases in the strain eventually lead to *fracture*—the metal "snaps." The **fracture stress** lies in the range of 10^8–10^9 Pa for many solids. Figure 15.4 shows the nature of the stress–strain relation for a ductile metal.

Man-made structures cover a wide range of sizes, from tiny solid-state electronic circuits to the Great Wall of China. The shapes of most solids are maintained by electrical forces between atoms. At the macroscopic level these electrical forces are referred to as *tensile* forces. The tensile forces act to oppose any change in shape—compression, extension, shearing, and the like. The gravitational forces acting between different segments of typical man-made objects are negligible in comparison to the tensile forces. Tensile forces, not gravity, hold together desks and chairs and other relatively small objects. But *self-gravity*, the mutual gravitational attraction of the atoms, holds together the sun and the planets and the moons of some planets. If an object is large enough, gravity not only holds it together, but also determines its shape. The crush of gravity overwhelms the tensile forces and compacts the matter into a nearly spherical form.*

We want to show that the limiting size for a body whose shape is determined by tensile forces is a *few hundred kilometers*. The limiting size may be inferred by using a dimensional argument and noting that 10^9 Pa is a typical fracture stress for solids. Gravity will fracture a solid whenever the gravitational stress exceeds the fracture stress.

*Rotation causes deviations from a spherical shape. Earth, for example, is an oblate spheroid because of its rotation.

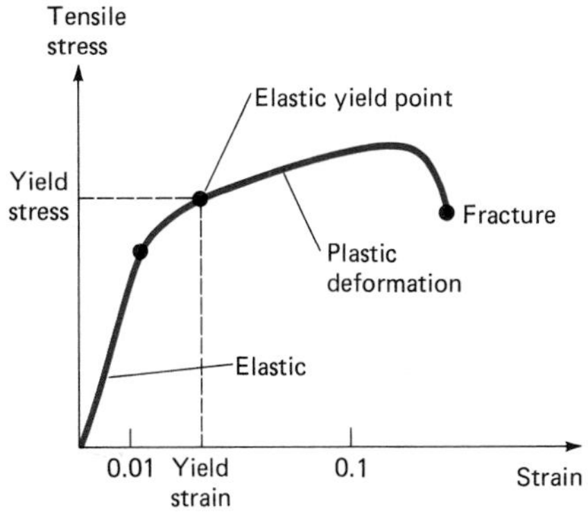

Figure 15.4

Stress–strain relation for a metal rod.

*Robert Hooke (1635–1703), an English scientist, first announced the law that now bears his name in an anagram at the end of his book on helioscopes. Hooke feared that others would claim to have made the discovery first, so he presented his "law of springs" as "ceiiinosssttuv." Two years later, he translated this into "ut tension, sic vis," Latin for "as the stretch, so the force." Hooke's law of springs is presented in Section 6.8 in the form $F = -kx$.

†Thomas Young (1773–1829), an English physician, is most famous for his discovery of the interference of light.

Figure 15.5

The two hemispheres exert equal and opposite gravitational forces on each other, and thereby set up a gravitational pressure within the solid.

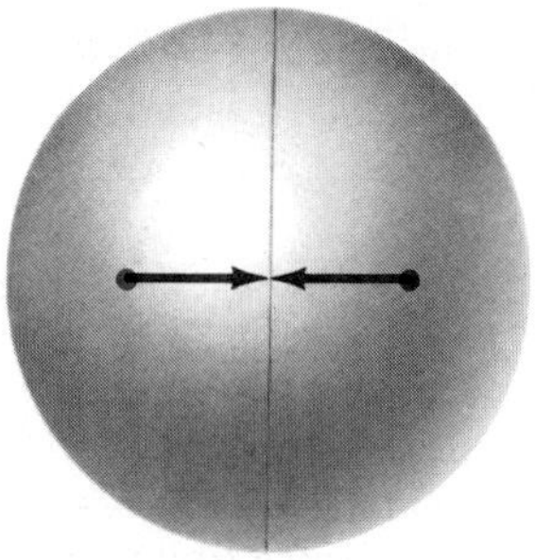

Consider a sphere of mass M and radius R (Figure 15.5). In Section 6.6 you learned that the attractive gravitational force between two particles having masses m_1 and m_2 and separated by a distance r is

$$F = \frac{Gm_1m_2}{r^2} \tag{15.10}$$

G is the gravitational constant,

$$G = 6.67 \times 10^{-11}\ \text{N} \cdot \text{m}^2/\text{kg}^2$$

On purely dimensional grounds the characteristic gravitational force that the two hemispheres exert on each other is roughly GM^2/R^2. This force acts over an area of approximately πR^2. Our order-of-magnitude approach does not justify inclusion of factors of π, 2, and the like, so we take R^2 as the area over which the gravitational force acts. The gravitational stress, or pressure, is therefore

$$P_{\text{gravity}} = \frac{F_{\text{gravity}}}{\text{area}} \approx \frac{GM^2/R^2}{R^2} = \frac{GM^2}{R^4}$$

We can eliminate M in favor of the mass density ρ. The volume is roughly R^3, so that

$$M \simeq \rho R^3$$

In terms of ρ and R,

$$P_{\text{gravity}} \approx GR^2\rho^2 \tag{15.11}$$

When $P_{\text{gravity}} > P_{\text{fracture}}$, gravity will determine the shape of the object. Taking $\rho = 10^4\ \text{kg/m}^3$ and $P_{\text{fracture}} = 10^9$ Pa as typical of solids, we find that gravity "wins" when

$$GR^2\rho^2 > 10^9$$

or

$$R \gtrsim \frac{1}{\rho}\sqrt{\frac{P_{\text{fracture}}}{G}} \simeq \frac{1}{10^4}\sqrt{\frac{10^9}{6.67 \times 10^{-11}}}\ \text{m} \simeq 400\ \text{km}$$

The radius of the smallest planet, Mercury, is 2432 km. Thus, gravity rather than tensile forces determines the shape of all the planets in our solar system. Most asteroids have sizes of a few kilometers. We therefore expect their shapes to be determined by tensile forces, and if this is so, it is unlikely that they are spherical. This expectation is borne out by observations of the sunlight that is reflected by asteroids. The light intensity varies as the asteroids rotate, and the shapes of the asteroids can be determined by studying light intensity patterns. Most asteroids are found to be oblong bodies. Phobos, one of the two moons of Mars, has an irregular shape—one determined by tensile forces, not gravity (Figure 15.6). This is not surprising because its largest dimension is less than 30 km.

Figure 15.6

A view of Phobos from Viking Orbiter 1 (range 612 km). The largest crater at the top lies near the North Pole. (NASA photo)

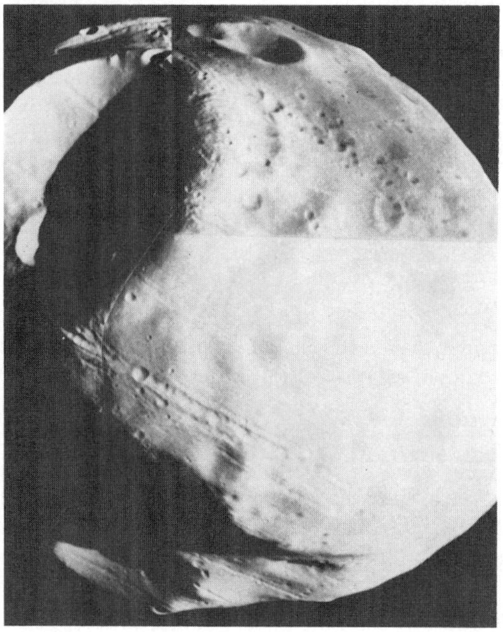

In arriving at our estimate of 400 km, we assumed that the object was spherical. But why is it that gravity tends to fashion an object into a spherical shape? One argument is that when an object has a uniform mass density and no rotation, there is no preferred direction in the object. It follows that the distance from the center of mass to the surface must be the same for all directions. Consequently the object is a sphere.

Question

7. The strain at which a solid yields and undergoes plastic deformation is typically 0.01. This may seem like a small strain. In fact, it is quite large. To sense just how large, imagine that you experience a tensile strain of 0.01. How much taller would you be? Do you think your body could withstand such a stress without undergoing a plastic deformation?

15.4 Elastic Energy and the Spring Constant

When an elastic solid is deformed, work is done by the applied force. This work is stored as elastic energy and is released when the solid reverts to its unstrained shape. We encountered an example of elastic potential energy in Chapter 8, when we considered the linear spring (see Section 8.2). The external force required to change the length of the spring from an initial length l to a length $l + x$ is

$$F_{spring} = kx \tag{15.12}$$

The quantity k is the spring constant. The potential energy of the deformed spring is (Section 8.2).

$$U_{spring} = \frac{1}{2}kx^2 \tag{15.13}$$

The tensile force required to change the length of an elastic solid from l to $l + \Delta l$ is obtained from Eq. 15.9,

$$F_{tensile} = \left(\frac{YA}{l}\right)\Delta l \tag{15.9}$$

If we compare Eq. 15.12 to Eq. 15.9 we see that x in Eq. 15.12 corresponds to Δl in Eq. 15.9 and k in Eq. 15.12 corresponds to (YA/l). Thus, the spring constant of an elastic solid is

$$k = \frac{YA}{l} \tag{15.14}$$

It follows from Eqs. 15.13 and 15.14 that the elastic potential energy stored in a solid under tension or compression is given by

$$U_{elastic} = \frac{1}{2}\left(\frac{YA}{l}\right)(\Delta l)^2 \tag{15.15}$$

In Section 14.3 we saw that a mass m attached to the end of a linear spring can execute simple harmonic motion (SHM). The period (T) of the oscillations is given by

$$T_{spring} = 2\pi\sqrt{\frac{m}{k}} \tag{15.16}$$

If a mass m is attached to the end of a wire, the mass can be made to execute SHM. Inserting the effective spring constant $k = YA/l$ given by Eq. 15.14 into Eq. 15.16 gives the period of the oscillations:

$$T = 2\pi\sqrt{\frac{ml}{YA}} \tag{15.17}$$

Equation 15.17 provides a basis for measuring Young's modulus.

Example 6 A Dynamic Method for Measuring Young's Modulus

A 20-kg mass is suspended from a wire that is 15 m long and 3.1 mm in diameter. The vertical oscillations of the mass have a period of 0.11 s. Using Eq. 15.17 we can determine Young's modulus:

$$Y = \frac{4\pi^2 ml}{T^2 A} = \frac{4(3.14)^2 \cdot 20 \cdot 15}{(0.11)^2 \cdot \frac{3.14}{4}(3.1 \times 10^{-3})^2}$$

$$= 1.3 \times 10^{11} \text{ Pa}$$

An alternate form of Eq. 15.15 incorporating the mass of the elastic solid is useful. To derive this form we note that the volume of a solid of cross-sectional area A and length l is $V = Al$. The mass m of the solid is given by

$$m = \rho V = \rho A l \tag{15.18}$$

We can rearrange this equation to get $A = m/\rho l$. Then, substituting this value for A into Eq. 15.15, we obtain the following for the elastic potential energy.

$$U_{elastic} = \frac{1}{2}m\frac{Y}{\rho}\left(\frac{\Delta l}{l}\right)^2 \tag{15.19}$$

Equation 15.19 can be used to establish an important limit on shock absorbers. A conventional shock absorber converts the kinetic energy of the object being protected into elastic potential energy of the shock absorber. Suppose an object of mass M moving at speed v is to be brought to rest, safely, by using a shock absorber of mass m whose elastic potential energy is given by Eq. 15.19. If we assume that the mass of the object is much greater than the mass of the shock absorber ($M \gg m$), we can ignore the kinetic energy of the shock absorber itself. The conversion of the kinetic energy of the moving object into the elastic energy of the shock absorber is expressed by

$$\frac{1}{2}Mv^2 = \frac{1}{2}m\frac{Y}{\rho}\left(\frac{\Delta l}{l}\right)^2$$

This gives

$$v = \left(\frac{m}{M}\frac{Y}{\rho}\right)^{1/2}\left(\frac{\Delta l}{l}\right) \tag{15.20}$$

In order to protect the object, the shock absorber itself must survive.* We observe from Eq. 15.20 that v is directly proportional to $\Delta l/l$, the strain of the shock absorber. The maximum "safe" speed at which the object can move can be estimated by setting $\Delta l/l$ equal to the yield strain. Greater speeds would result in plastic deformation or fracture of the shock absorber. Of course, survival of the shock absorber does not guarantee that the object will survive undamaged. The point is that if the shock absorber cannot withstand the strain, the objects stands little chance of doing so. We will take $\Delta l/l = -0.02$. As a reasonable figure we specify $m/M = 1/50$ (5 lb of shock-absorbing material to protect a 250-lb object).

Finally we note that

$$\frac{Y}{\rho} \simeq 10^7 \text{ m}^2/\text{s}^2$$

for many different materials (see Table 15.1). Inserting these values into Eq. 15.20 gives the maximum "safe" speed at which the object can travel,

$$v_{max} = \frac{1}{50} \cdot 10^7 \cdot 4 \times 10^{-4} \text{ m/s} = 30 \text{ m/s} \simeq 68 \text{ mph}$$

This value of v_{max} means that an auto passenger, for example, could conceivably survive a crash at speeds up to 68 mph. At greater speeds, the materials currently available for use in elastic shock absorbers cannot withstand the strain.

*Our estimates apply to *elastic* shock absorbers. Devices like auto air bags are not elastic and can withstand greater shocks.

8. If a linear spring is cut into two pieces of equal length, how is the spring constant (k) changed?

9. The box springs of a bed consist of a network of uniformly spaced identical springs. A rectangular package weighing 800 N is placed on the springs. Each end of the package has an area of 0.84 m^2, and each side has an area of 0.96 m^2. The top and bottom of the package each have an area of 1.12 m^2. Will the compression of the springs be greatest when the package rests on its end, on its side, or on its bottom?

15.5 Bulk Modulus and Compressibility

For tensile stress and strain Hooke's law takes the form given by Eq. 15.8, tensile stress/tensile strain = Y, where Y is Young's modulus. We now examine Hooke's law for volume stress and strain, which is written

$$-\frac{\text{volume stress}}{\text{volume strain}} = \frac{-P}{\Delta V/V} = B \qquad (15.21)$$

The quantity B is called the *bulk modulus*. The change in volume, ΔV, is negative. The minus sign has been inserted in Eq. 15.21 to make B positive. The units of bulk modulus, like those of Young's modulus, are pascals. Values of B for several materials are given in Table 15.1. Typically, values of B for solids are in the neighborhood of 10^{11} Pa, or 10^6 atm. Knowing this, we can use Eq. 15.21 to find the pressure required to decrease the volume of a solid by some specified percentage and vice versa. For example, the pressure needed to decrease the volume of a solid by 1% follows from Eq. 15.21 if we set $B = 10^6$ atm and $\Delta V/V = -0.01$. The result is 10^4 atm.

The reciprocal of B is sometimes a more convenient parameter. It is called the *compressibility* and is denoted by the symbol K:

$$K = \frac{1}{B} = -\frac{\Delta V/V}{P} \qquad (15.22)$$

The smaller the value of K, the larger must be the pressure needed to produce a specified volume strain. The compressibility is a meaningful quantity for fluids as well as solids. In fact, liquids can transmit hydraulic pressure effectively because of their very small compressibilities.

Example 7 Compression of Water and Steel

From experiment, the compressibilities of water and steel are found to be

$$K_{water} = 5 \times 10^{-5}/\text{atm} = 5 \times 10^{-10}/\text{Pa}$$

and

$$K_{steel} = 6 \times 10^{-7}/\text{atm} = 6 \times 10^{-12}/\text{Pa}$$

We see that the compressibility of water is about 80 times that of steel. In other words, it is 80 times easier to compress water than it is to compress steel. However, both water and steel are difficult to compress. To show this, we rewrite Eq. 15.22 as

$$-\frac{\Delta V}{V} = KP \qquad (15.23)$$

A 1% decrease in volume ($\Delta V/V = -0.01$) requires a pressure of

$$P = \frac{0.01}{K}$$

Solving this equation for water we obtain

$$P_{water} = \frac{0.01}{5 \times 10^{-5}} \text{ atm} = 200 \text{ atm}$$

Thus, to compress water by 1% we must apply a pressure of 200 atm, which is about 100 times the pressure in an automobile tire. To compress steel by the same amount requires 80 times as great a pressure.

$$P_{steel} = \frac{0.01}{6 \times 10^{-7}} \text{ atm} \simeq 16{,}000 \text{ atm}$$

One of the consequences of the small compressibility of water is a slight increase in the density of water with depth. At great depths in the oceans, water is compressed by the weight of the water above it, and its density is thereby increased.

The bulk modulus, B, is defined by Eq. 15.21 in terms of the volume strain, $\Delta V/V$. The volume strain $\Delta V/V$ is related to the fractional change in mass density through the conservation of mass. Therefore, an alternate definition relates B to the fractional change in mass density. Thus, if the density and volume change by $\Delta\rho$ and ΔV, respectively,

$$\rho V = \text{mass} = \text{constant} = (\rho + \Delta\rho)(V + \Delta V)$$

By ignoring the relatively small product $\Delta\rho\,\Delta V$, we can rearrange this equation to obtain

$$-\frac{\Delta V}{V} = \frac{\Delta\rho}{\rho} \qquad (15.24)$$

Inserting this relation into Eq. 15.21, $-P(\Delta V/V) = B$, gives us an alternate definition of B in terms of the pressure and fractional density change,

$$B = \frac{P}{\Delta\rho/\rho} \qquad (15.25)$$

15.6 Shear Modulus

So far we have introduced Hooke's law for tensile stress and strain, which involves Young's modulus, and Hooke's law for volume stress and strain, which involves the bulk modulus. When we deal with shear stress and strain, Hooke's law takes the form

$$\frac{\text{shear stress}}{\text{shear strain}} = \frac{F_s/A}{\theta} = \mu \qquad (15.26)$$

Figure 15.7

Shear stress and strain for a cylindrical rod. The base of the rod is anchored. The vertical strip (shaded area) is distorted when a tangential force $\mathbf{F}_s$ is applied. The shear strain is $\Delta x/y \simeq \theta$.

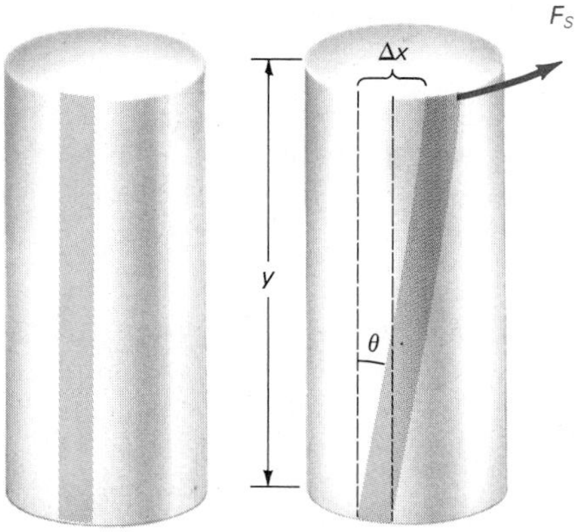

where μ is called the *shear modulus*. As Table 15.1 reveals, values of μ for many materials are comparable to those of Y and B.

Shear stresses and strains arise when two surfaces slide over one another. They also occur in rotating or twisted structures. For example, the twisted rod in Figure 15.7 exhibits a shear strain, as does the shot-putter's shoe sole described in Example 8.

Example 8 The Shear Strain of a Shot-Putter's Shoe Sole

A shot-putter exerts a shear stress on the sole of her shoe as she drives her body forward (Figure 15.8). What is the resulting shear strain on the sole? The athlete exerts a shearing force of 700 N, which is distributed over an area of 0.015 m^2. The shear modulus of the rubber sole is μ = 1.5 × 10^6 Pa. We substitute these values into Eq. 15.26 and solve for θ, the shear strain.

Figure 15.8

As the shot-putter drives herself forward the sole of her shoe experiences a shear stress. The resulting shear strain can be estimated by using Eq. 15.26.

$$\theta = \frac{F_s/A}{\mu} = \frac{700\ \text{N}/0.015\ \text{m}^2}{1.5 \times 10^6\ \text{N/m}^2} = 0.03$$

This strain is rather large because the shear modulus μ for rubber is much smaller than it is for typical solids.

Summary

Stress and strain are related by Hooke's law,

$$\frac{\text{stress}}{\text{strain}} = \text{constant} = \text{elastic modulus} \qquad (15.7)$$

Hooke's law is an empirical relation, and is valid only for very small strains ($\lesssim 0.01$). Three types of stresses and strains are encountered frequently:

$$\frac{\text{Tensile stress}}{\text{Tensile strain}} = \frac{\begin{array}{c}\text{tensile force}\\ \text{per unit area}\end{array}}{\begin{array}{c}\text{fractional change}\\ \text{in length}\end{array}} = \frac{F_t/A}{\Delta l/l}$$

$$= Y\ \text{(Young's modulus)} \qquad (15.8)$$

$$-\frac{\text{Pressure or compressive stress}}{\text{Volume strain}}$$

$$= -\frac{\begin{array}{c}\text{normal force}\\ \text{per unit area}\end{array}}{\begin{array}{c}\text{fractional change}\\ \text{in volume}\end{array}}$$

$$= \frac{-P}{\Delta V/V} = B\ \text{(bulk modulus)} \qquad (15.21)$$

$$\frac{\text{Shear stress}}{\text{Shear strain}} = \frac{\begin{array}{c}\text{tangential}\\ \text{force/area}\end{array}}{\begin{array}{c}\text{angular displacement}\\ \text{of stressed face}\end{array}}$$

$$= \frac{F_s/A}{\theta} = \mu\ \text{(shear modulus)} \qquad (15.26)$$

Suggested Reading

Materials, W. H. Freeman, San Francisco, 1967.

Problems

Section 15.1 States of Matter

1. The SI unit of mass density is kilograms per cubic meter. The corresponding cgs unit is grams per cubic centimeter. Derive the conversion factor in the equation

$$1\ \mathrm{kg/m^3} = a\ \mathrm{gm/cm^3}$$

2. Lead bricks like the one in Figure 1 are used to shield persons from radioactive materials. Calculate the (a) mass and (b) weight of such a brick.

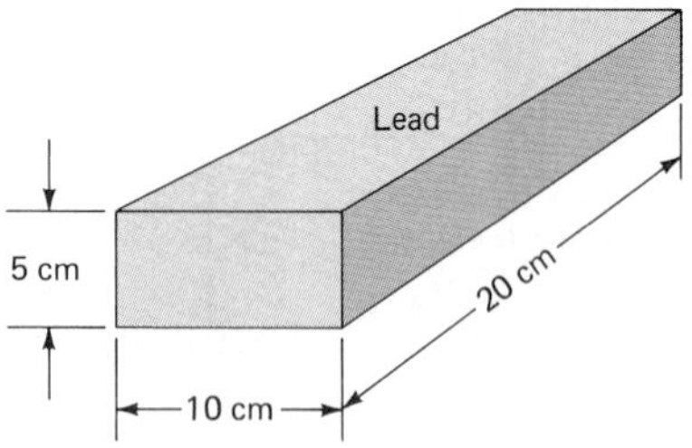

Figure 1

3. For the college men's shot-put event the mass of the shot is 7.27 kg. Assume that the shot is a uniform iron sphere and calculate its (a) volume and (b) radius.

4. Listed in Table 1 are mass and radius values for the planets in the solar system. (Pluto is not included because its radius is uncertain.) Determine the mass density of each planet. The four inner planets (Mercury, Venus, Earth, and Mars) are called the terrestrial, or Earthlike, planets. The outer four are referred to as the Jovian, or Jupiter-like, planets. What property suggests these groupings?

Table 1

Planet	Mass (kg)	Radius (m)
Mercury	3.3×10^{23}	2.43×10^{6}
Venus	4.87×10^{24}	6.06×10^{6}
Earth	5.98×10^{24}	6.37×10^{6}
Mars	6.4×10^{23}	3.38×10^{6}
Jupiter	1.90×10^{27}	6.97×10^{7}
Saturn	5.69×10^{26}	5.79×10^{7}
Uranus	8.69×10^{25}	2.37×10^{7}
Neptune	1.03×10^{26}	2.25×10^{7}

5. Use your mass and approximate density to estimate your volume.

Section 15.2 Stress and Strain

6. (a) Calculate the total force exerted by the atmospheric pressure (1 atm) on 1 $\mathrm{cm^2}$ of your skin. (b) If an object that weighed 10.1 N were placed on top of your head, would you notice it? (c) Why don't you ordinarily notice the forces exerted on your body by the atmosphere?

7. The gauge pressure of the air in a car tire is found to be 29 psi. The atmospheric pressure is 14.8 psi. Calculate the absolute pressure of the air in the tire in (a) pounds per square inch; (b) atmospheres; (c) pascals.

8. A single ice skate contacts the ice over an area 10 cm long and 3 mm wide. Assume that the weight of the skater (mass = 60 kg) is uniformly distributed over this area, and calculate the resulting pressure exerted on the ice (a) in pascals and (b) in atmospheres.

9. To pluck a certain steel guitar string (length = 1 m), a musician pulls it sideways a distance of 2 mm, forming the triangle shown in Figure 2. If the point B is originally 10 cm from one end of the string, calculate the tensile strain produced by plucking. (*Warning:* Watch significant figures!)

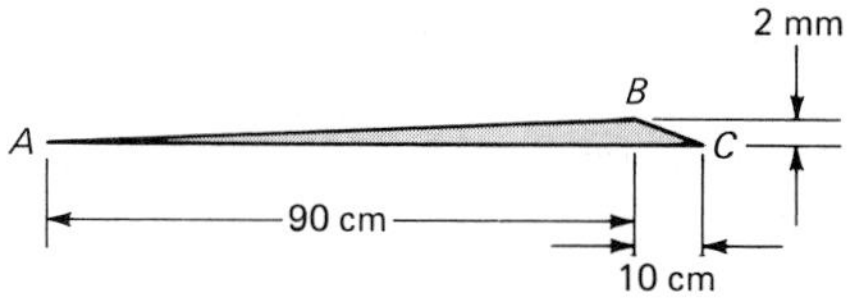

Figure 2

10. A mass of 4 kg is suspended from the end of a copper wire that has a diameter of 1 mm. What tensile stress does the wire experience?

11. A cube is subjected to a hydrostatic pressure that shortens each edge of the cube by 1%. What is the approximate volume strain?

12. A physics textbook weighing 15 N slides across a desk. The coefficient of sliding friction is 0.40 and the area of the book that is in contact with the desk is 0.05 $\mathrm{m^2}$. Determine the shear stress exerted on the book.

13. The radius of the earth is approximately 6300 km. The pressure of the atmosphere at the surface of the earth is approximately 10^5 Pa. The thickness of the atmosphere is small compared to the radius of the earth. Consequently the acceleration of gravity may be treated as a constant ($g = 9.80\ \mathrm{m/s^2}$) throughout the atmosphere. (a) What is the surface area of the earth? (b) What is the weight of the earth's atmosphere? (c) What is the mass of the earth's atmosphere? (d) If the average mass of one atmospheric molecule is 4.8×10^{-26} kg, how many molecules are there in the earth's atmosphere?

Section 15.3 Hooke's Law and Young's Modulus

14. A student records the following data from an experiment designed to determine Young's modulus for a steel wire.

    Initial (unstressed) length = 2.06 m
    Stressed length = 2.10 m
    Diameter = 0.22 mm
    Tensile force = 133 N

    Using these data, determine the value of Young's modulus.

15. A metal wire 3 m long with a diameter of 5 mm is stretched 2 mm under a tension of 10,000 N. Determine Young's modulus for the wire.

16. A copper wire 2 m long and 1 mm in diameter is subjected to a tensile force of 125 N. How much will it stretch?

17. A steel wire 100 m long and 3 mm in diameter hangs down alongside a skyscraper under construction. How much would the wire stretch if one of the workers (mass = 100 kg) grabbed the bottom of the wire and hung freely?

18. Estimate the gravitational stress exerted between the right and left sides of your body. Compare the result with the characteristic fracture stress for solids (10^9 Pa).

Section 15.4 Elastic Energy and the Spring Constant

19. A linear spring has a spring constant of 21.2 N/m. (a) What force is required to increase its length by 1.20 cm? (b) How much elastic potential energy is stored in the spring when its length is increased 1.20 cm?

20. An aluminum cylinder used in an attempt to detect gravitational radiation is 4 m long and 0.80 m in diameter. How much elastic energy is stored in this rod when it is compressed so that $\Delta l = 10^{-16}$ m?

21. A 1-m-long copper rod (diameter = 4 mm) is stretched 1.5 mm by pulling on each end. (a) What tensile force acts on each end? (b) How much elastic energy is stored?

22. Determine the elastic potential energy stored in the stretched steel wire of Problem 14.

23. A pendulum bob of mass m is suspended from a cord of length l (unstretched). The bob is released with the cord in a horizontal position. Show that the increase in length of the cord at the bottom of the swing (position V in Figure 3) is given by

    $$\Delta l = \frac{3mg}{YA} l$$

    where Y is Young's modulus for the cord, A is its cross-sectional area, and g is the acceleration of gravity. (*Hint:* The law of conservation of mechanical energy will help you to relate the speed of the bob to the distance it has "fallen." Ignore the elastic potential energy change.) Newton's second law will let you relate the tension in the cord to the centripetal acceleration of the bob.

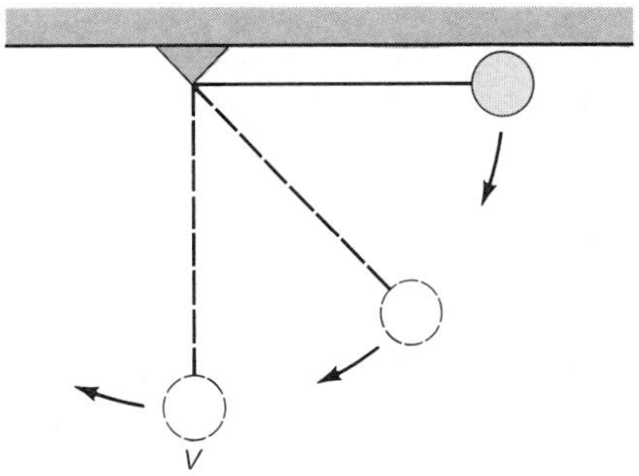

Figure 3

24. A cylindrical aluminum rod has a length of 120 cm and a diameter of 3 cm. The rod is held vertically with its lower end 2 m above a rigid steel plate and then released. Assuming that the steel plate does not deform, estimate the maximum strain experienced by the rod as it strikes the plate. (*Hint:* Gravitational potential energy is converted into elastic energy. Show that $(\Delta l/l)^2 = 2\rho gh/Y$, where h is the distance the rod falls.)

25. For the rod in Problem 24, show that the strain-producing force is more than 1000 times its weight.

26. Two mountain climbers are connected by a nylon safety rope (Figure 4). One climber loses his footing and plummets over the edge of an overhang. His quick-thinking companion winds their connecting line around a tree stump, and thus avoids being

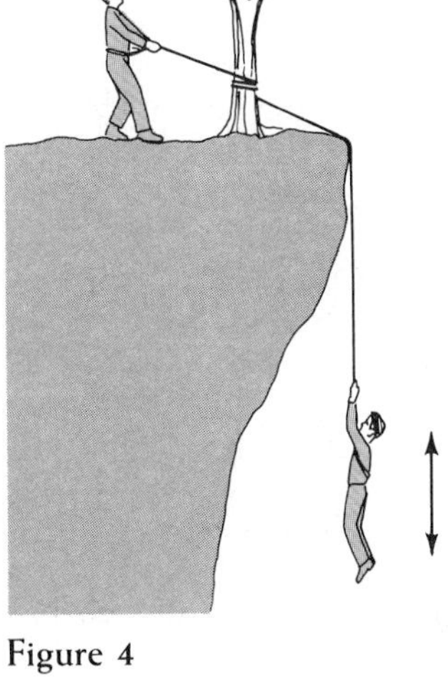

Figure 4

pulled over the cliff. The first climber ends up oscillating at the end of the line, which is elastic. The mass of the oscillating climber is 75 kg. The *unstretched* length of the line from tree to climber is 30 m. The diameter of the line is 1 cm and its Young's modulus is 10^8 Pa. (a) Calculate the "spring constant" (k) for the line. (b) Determine the maximum length of the line. (*Hint:* The maximum length of the line is the distance the climber falls. The overall result of this fall is to convert gravitational potential energy into elastic potential energy. A mathematical statement of this fact leads to a quadratic equation for the "stretch" of the line.) (c) Even though the restoring force (tension) in the line follows Hooke's law ($F = -kx$), the climber will not undergo *simple* harmonic motion. Explain why not.

Section 15.5 Bulk Modulus and Compressibility

27. A pressure of 100 atm is applied to a metal canister. The canister volume decreases by 3%. Determine the bulk modulus and compressibility of the canister.

28. The pressure on an object submerged in seawater increases in direct proportion to its depth. For every 10-m increase in depth, the pressure on the object increases by about 1 atm. At what depth would a typical solid be compressed to 99.9% of its volume at the surface? (Volume strain = -0.001.) Consult Table 15.1 and choose a reasonable value for the bulk modulus of a "typical solid."

29. The pressure of seawater increases by about 1 atm for each 10-m increase in depth. By what percentage is the density of water increased in the deepest oceanic trenches (which have depths of around 12 km)? The compressibility of water is 50×10^{-6}/atm.

30. A woman who weighs 500 N wears spike heels. The contact area of one heel is $2 \times 10^{-6}\ m^2$. Her weight results in the distribution over this area of a force of 250 N. (a) What pressure does one heel exert on the floor? (b) A linoleum floor has a bulk modulus of 10^{10} Pa and is damaged by volume strains larger than -0.003. Will the woman's spike heel damage the floor?

31. In an experiment 750 cm^3 of water expands to 765 cm^3 when heated. What hydrostatic presssure is required in order to squeeze the water back to its original volume? (The compressibility of water is 50×10^{-6}/atm.)

Section 15.6 Shear Modulus

32. From the definition of the shear modulus μ, show that it has units of pascals.

33. A pedestrian leaps sideways to avoid being run down by an automobile. Her rubber shoe sole experiences a shear force of 800 N on an area of 0.021 m^2. The shear modulus of the sole is 1.8×10^7 Pa. Determine the shear strain of the sole.

34. To polish a rectangular block of brass the polishing tool moves back and forth repeatedly, pushing to the right (or left) on the top surface (AB) (Figure 5). During the polishing process the bottom surface (CD) is held rigidly fixed in position. If polishing gives rise to a shear strain of 0.001, calculate (a) the accompanying shear stress and (b) the shear force exerted on the polished surface by the polisher. (Assume a top surface area equal to 10 cm^2.)

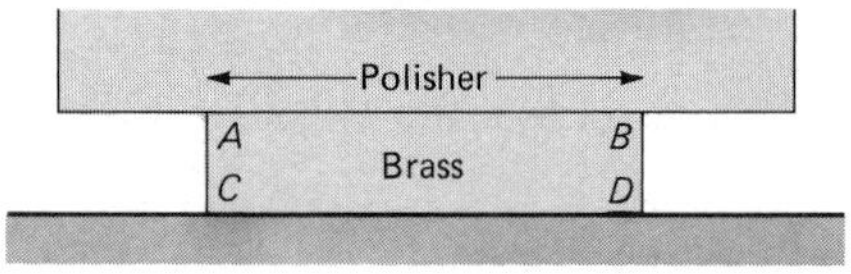

Figure 5

chapter 16

FLUID MECHANICS

Preview

Fluid mechanics is the analysis of the behavior of fluids. The description of fluid motion relies on Newton's laws of motion, formulated in terms of measurable properties of the fluid, such as density, pressure, and flow velocity. Many ideas carry over from our earlier study of Newtonian mechanics. A fluid has weight and therefore exerts a pressure on submerged objects. A fluid in motion transports momentum and energy.

We deal first with fluid statics and learn about the buoyant force exerted by fluids (Archimedes' principle). Next, we derive and apply the pressure–depth relation, which relates fluid pressure to depth below the fluid surface.

We deal with two basic equations: (1) the equation of continuity, which is a consequence of mass conservation; and (2) Bernoulli's equation, a fluid version of the work–energy principle.

The final two sections of this chapter provide an introduction to viscosity (fluid friction) and turbulence (unstable fluid motion).

16.1 Fluid Statics: Archimedes' Principle

A *fluid* is a substance that cannot support a shearing stress (Section 15.2). When a fluid is subjected to a shearing stress the fluid layers slide relative to one another. This characteristic gives fluids the ability to flow or change shapes. Both gases and liquids are fluids. Fluid mechanics is concerned with the behavior of fluids at the *macroscopic* level—the level at which measurements are made with pressure gauges, thermometers, and flow meters. Accordingly, when we speak of a **fluid particle** or **fluid element** we are not referring to a single molecule. Rather, we mean some volume of fluid that is small by macroscopic standards, but nonetheless contains many molecules. The properties of such a fluid particle are insensitive to the behavior of any one molecule.

Fluid statics is concerned with fluids in which the center of mass of each fluid particle has zero velocity and zero acceleration. In ordinary language we would say that a static fluid is "at rest," even though the individual molecules move about incessantly. A fluid that is at rest is said to be in *hydrostatic equilibrium*.*

An important property of fluids in hydrostatic equilibrium is described by *Archimedes' principle:*

An object that displaces a fluid experiences a buoyant force equal to the weight of the fluid that it displaces.

Archimedes' principle can be expressed in equation form as

$$B = W_{\text{displaced}} \tag{16.1}$$

*Hydrostatic equilibrium is not restricted to water. The terminology is used for all liquids.

Figure 16.1

The net effect of the surface forces is a buoyant force, B, equal to the weight of the displaced fluid.

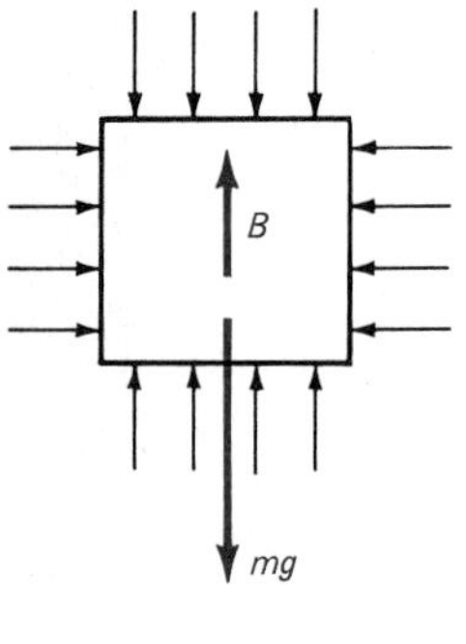

Figure 16.2

An object floats provided that the buoyant force, $\rho_{fl}V_s g$, equals the weight of the object, $\rho_o Vg$. This equality will be satisfied if the average density, ρ_o, of the object is less than the density of the surrounding fluid, ρ_{fl}.

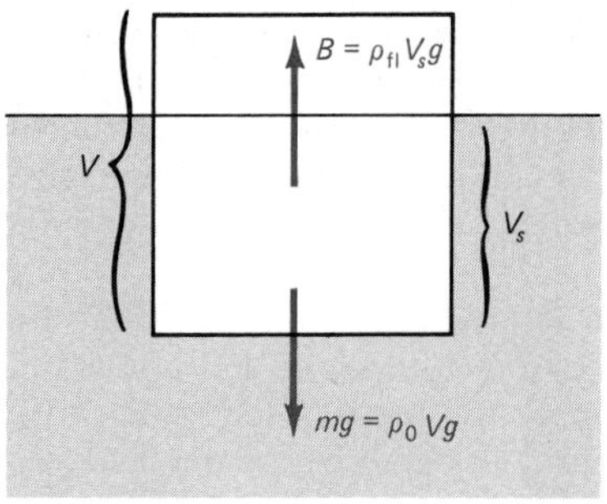

where B is the buoyant force and $W_{\text{displaced}}$ is the weight of the displaced fluid (Figure 16.1). It is this buoyant force that supports a floating object. To understand this principle, consider some arbitrary particle of a fluid in hydrostatic equilibrium. The net force on the fluid particle vanishes when it is in hydrostatic equilibrium. Two types of forces act on the particle: a gravitational force equal to the weight of the particle, and the forces exerted on the particle's surface by the surrounding fluid. In equilibrium, the net surface force balances the weight of the fluid particle. This surface force is a buoyant force—it prevents the fluid particle from sinking. If a solid object displaces the fluid particle, the same forces act on the object's surface as once acted on the now displaced fluid. Hence, the object experiences a buoyant force equal to the weight of the fluid it displaced.

In a swimming pool, you may have noticed that it is relatively easy to carry someone who is partially submerged. This is because you must support only part of the person's weight. The buoyant force supports the remainder. Water-skiers notice a change in buoyancy as they go from a nearly submerged starting position to an upright stance. While submerged a skier experiences a large buoyant force, and the vertical force on the skis is small. In an upright skiing position the skier is no longer submerged. The buoyant force is nil and the skis must support nearly all the skier's weight.

We can use Archimedes' principle to prove that an object will float in any liquid whose density is greater than the density of the object. Ice floats in water, for example, because the density of ice is less than that of water ($\rho_{\text{water}} = 10^3$ kg/m^3). Let ρ_0 denote the average density of an object of volume V that floats in a fluid of density ρ_{fl}. In order for the object to float, the fraction of the volume V that is submerged must be enough to make the buoyant force equal to the weight of the object. The weight of the object is related to its volume and density (ρ_0) by

$$\text{weight of object} = mg = \rho_0 Vg$$

If V_s is the volume submerged, the mass of the displaced fluid is $\rho_{fl}V_s$ and the weight of the displaced fluid is $\rho_{fl}V_s g$. The buoyant force, B, is therefore

$$B = W_{\text{displaced}} = \rho_{fl}V_s g \qquad (16.2)$$

In order for the object to float, the buoyant force must equal the weight of the floating object (Figure 16.2). Therefore, we write

$$\rho_{fl}V_s g = \rho_0 Vg$$

Solving this equation for the fraction of the object that is submerged, V_s/V, we get

$$\text{fraction submerged} = \frac{V_s}{V} = \frac{\rho_0}{\rho_{fl}} \qquad (16.3)$$

Equation 16.3 shows that the object floats ($V_s < V$) provided that its density is less than that of the fluid ($\rho_0 < \rho_{fl}$).

Example 1 Just the Tip of the Iceberg

The density of ice is 920 kg/m^3 and the density of seawater is 1020 kg/m^3. From Eq. 16.3, the submerged fraction of an iceberg is

$$\frac{V_s}{V} = \frac{920 \text{ kg/m}^3}{1020 \text{ kg/m}^3} = 0.90$$

Thus, only about 10% of an iceberg—its tip—is above the surface of the water (Figure 16.3). The expression, "That's just the tip of the iceberg," has come to mean that the enormity of a situation is suspected, even though only a tiny part is evident.

Two toys rely on Archimedes' principle. One, known as the "Cartesian diver," consists of a hollow glass or plastic "diver" placed in a water-filled bottle (see Figure 16.4) that is fitted with a stopper. Small holes in the side of the diver allow it to be partially filled with water, to the point where it has a density slightly less than the density of the water. This enables the diver to float with a small amount of air trapped inside. The bottle is filled, leaving little or no air space. If the bottle is squeezed, the diver plummets to the bottom. This is because squeezing the bottle increases the pressure in the water. This forces water into the diver, increasing the diver's average density. The diver sinks when its average density rises above that of the water. If the squeeze pressure is relaxed, water is forced out of the diver and it rises.

The "lava lamp" also relies on Archimedes' principle. In this device, a large globule of paraffin is submerged in a cylinder that is filled with machine oil. At room tem-

Figure 16.3

Only about 10% of the volume of an iceberg is above water.

perature the density of the paraffin is just slightly greater than that of the oil. An electric bulb illuminates and heats the base of the cylinder. The heat causes the paraffin to expand more than the oil until, having become less dense than the oil, it rises toward the surface. As the paraffin rises, its temperature is lowered by the cooler oil toward the top of the cylinder. It contracts until it becomes more dense than the surrounding oil, whereupon it sinks. The cohesive nature of the paraffin prevents the globule from breaking up, and thus it repeatedly rises and falls.

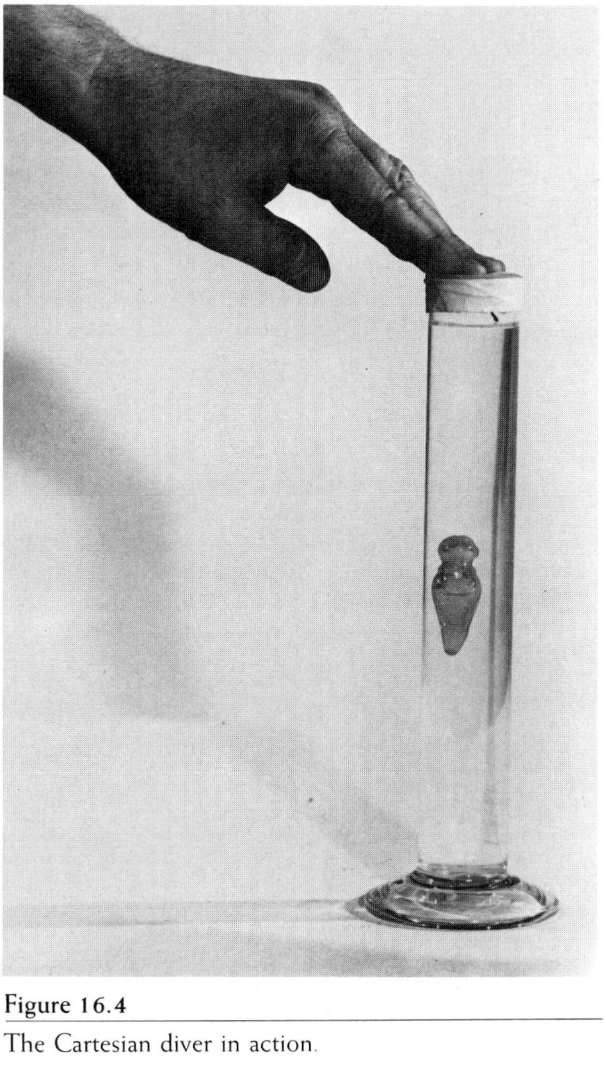

Figure 16.4

The Cartesian diver in action.

Questions

1. How could you estimate the volume of your body without using any length measurements?
2. Why is it easier for a person to float in the Great Salt Lake than in Lake Superior?
3. A block of wood is held beneath the surface of a bucket of water by a string. The bucket rests on a scale. The string breaks and the wood floats to the surface. After things settle down, has the scale reading increased, decreased, or remained unchanged?
4. Could you float in a pool of mercury? What might make it difficult to walk across a pool of mercury without falling?

16.2 Pressure–Depth Relation

Each square meter of the earth's surface experiences a force of approximately 10^5 N (over 11 tons!). This force is the weight of the atmosphere above each square meter of the surface. The atmospheric weight per square meter of surface area is called the *atmospheric pressure.* The atmospheric pressure at the surface of the earth is approximately 10^5 Pa. If you were to ascend through the atmosphere, let's say by climbing a mountain, you would find that the atmospheric pressure *decreases* as you move upward. The atmospheric pressure at any level reflects the weight of the air above that level. At an altitude of 5 km the pressure is approximately one-half of the pressure at the surface. In other words, approximately one-half of the earth's atmosphere lies within 5 km of the surface. At an altitude of 31 km the pressure is less than 1% of the pressure at the surface. More than 99% of the earth's atmosphere lies within 31 km of the surface. We literally live at the bottom of an ocean of air, the weight of which is responsible for the atmospheric pressure.

When divers descend in the ocean they experience a pressure increase. The pressure increases by about 10^5 Pa for every 10-m increase in depth. The pressure at any depth reflects the weight of the fluids—water and air—above each square meter at that level.

We can derive the pressure–depth relation by considering the forces acting on a thin slab of fluid at a depth z (see Figure 16.5). Let A denote the cross-sectional area of the slab and dz its thickness. The volume of the slab is $A\,dz$ and its mass is $dm = \rho A\,dz$, where ρ is the mass density of the fluid. The weight of the fluid slab is

$$dW = g\,dm = \rho g A\,dz \qquad (16.4)$$

For a fluid in hydrostatic equilibrium, in order for the downward force dW to be compensated by the net fluid pressure force (the buoyant force), the net force on the slab must be zero.

The upward pressure on the bottom face exceeds the downward pressure by an amount dP. In hydrostatic equilibrium the buoyant force, $A\,dP$, just balances dW. Thus,

$$A\,dP = dW = \rho g A\,dz$$

Canceling the common factor A gives

$$dP = \rho g\,dz \qquad (16.5)$$

This equation relates the changes in pressure and depth. If P_0 denotes the pressure at the surface ($z = 0$), then the pressure P at a depth z is determined by integrating Eq. 16.5:

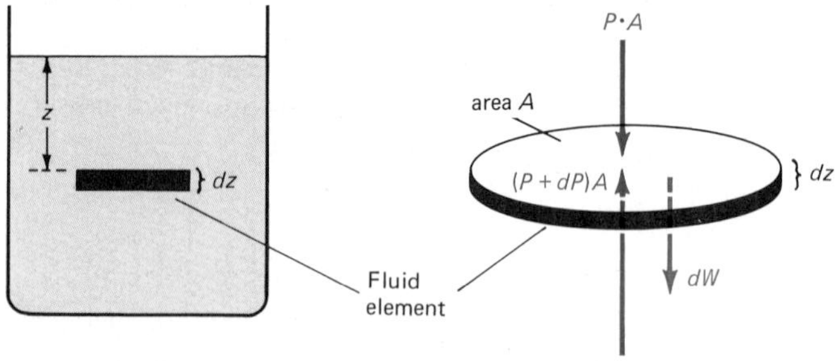

Figure 16.5

Three vertical forces act on the fluid element. The weight of the element is denoted as dW. The fluid force on the top is $P \cdot A$. The fluid force on the bottom is $(P + dP)A$. The difference between these two fluid forces, $dP \cdot A$, constitutes the buoyant force. For hydrostatic equilibrium the buoyant force balances the weight.

Figure 16.6

The pressure–depth relation. The pressure at the level $z = 0$ is P_0. At a depth z the pressure is $P_0 + \rho g z$.

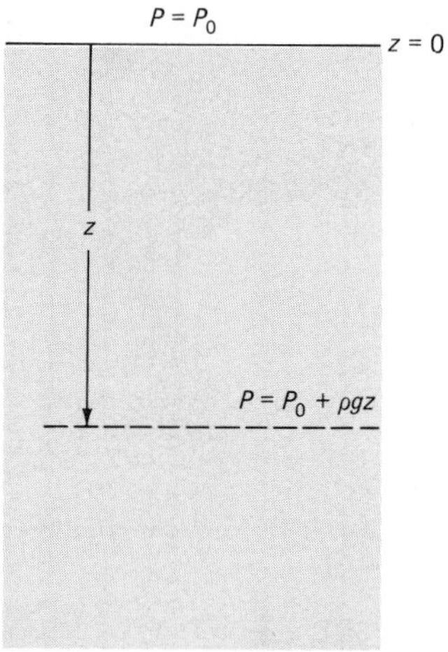

$$\int_{P_0}^{P} dP = \int_0^z \rho g\,dz$$

The left side gives $P - P_0$. For an incompressible fluid ρ is constant. If g is also taken to be constant, the right side gives $\rho g z$. We now have $P - P_0 = \rho g z$. Rearranging gives

$$P = P_0 + \rho g z \qquad (16.6)$$

This is the *pressure–depth relation* (Figure 16.6). Note that if the pressure at the surface (P_0) is changed, an equal change in pressure is felt at all depths, that is, throughout the fluid. This fact was stated by the French scientist Blaise Pascal (1623–1662) and is called *Pascal's principle:*

Pressure applied to an enclosed fluid is transmitted undiminished to every portion of the fluid and the walls of the containing vessel.

Pressure Measurement

A barometer is a device used to measure the pressure of the atmosphere. The conventional mercury-filled barometer consists of a narrow tube that is closed at one end. The tube is first filled with mercury, then inverted in a mercury reservoir (Figure 16.7). The pressure at the surface of the

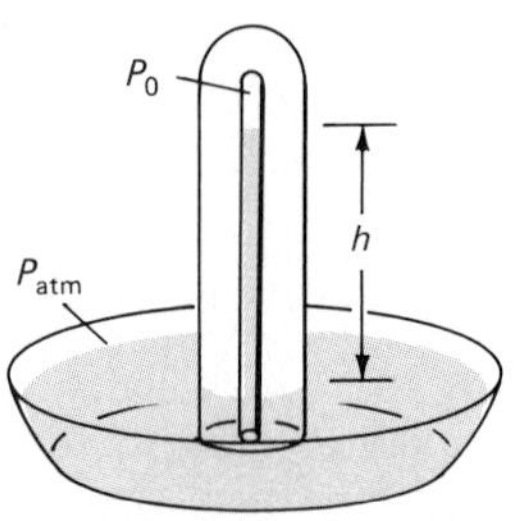

Figure 16.7

In an ordinary mercury barometer, the atmospheric pressure exceeds $\rho g h$ slightly, because of the vapor pressure P_0 in the region above the fluid.

reservoir is atmospheric (P_{atm}). Inside the tube the pressure at the level of the reservoir is $P_o + \rho g h$, where h is the height of the mercury column. The pressure P_o at the top of the mercury column is not quite zero because a small amount of mercury vapor fills the otherwise evacuated space. This vapor produces the pressure P_o, which is negligible at ordinary temperatures, except in precise scientific work. Hydrostatic equilibrium is achieved when the pressure at the reservoir surface ($P_o + \rho g h$) equals the atmospheric pressure (P_{atm}):

$$P_{atm} = P_o + \rho g h \qquad (16.7)$$

For mercury, $\rho = 13.6 \times 10^3$ kg/m^3, $g = 9.80$ m/s^2, and $P_{atm} = 1.01 \times 10^5$ Pa. Substituting these values into Eq. 16.7, we get $h = 0.758$ m $= 75.8$ cm. Thus, the atmosphere exerts a pressure that can support a column of mercury over 75 cm in height.

For a fixed pressure difference ($P_{atm} - P_o$), Eq. 16.7 shows that h is inversely proportional to the density of the barometric fluid. Thus, at a fixed pressure, the higher the density of the liquid, the lower the height of the column. The high density of mercury, therefore, suits it to barometers of modest size. Other fluids give unwieldy heights for pressure differences on the order of 1 atm.

Example 2 **The Wine Barometer**

The barometer was invented by Torricelli, one of Galileo's students (Figure 16.8). Torricelli used many different liquids to fill his barometers, including wine. He found that at a fixed pressure, the lower the density of the fluid, the greater the height of the column. We already have seen that mercury, with a density of 13.6×10^3 kg/m^3, rises to a height of 0.758 m at atmospheric pressure. Therefore, if we take the density of wine to be 1.02×10^3 kg/m^3, the height of the column in a wine barometer at atmospheric pressure is

$$h_{wine} = 0.758 \text{ m} \cdot \frac{13.6}{1.02} = 10.1 \text{ m}$$

which converts to 33 ft 1 in. Because this is such an unwieldy height, we do not generally use wine barometers to measure the atmospheric pressure.

Figure 16.8

This Russian postage stamp honors Torricelli, inventor of the barometer.

Equation 16.5 relates the changes in pressure and depth whether or not ρ and g are constant. Equation 16.6, on the other hand, is restricted to systems wherein ρ and g are constant. In the case of planetary atmospheres, for example, both ρ and g decrease as the altitude increases, and Eq. 16.6 cannot be used. The variation of ρ with altitude can seldom be ignored. However, in liquids and solids, the density variations are usually minor enough so that Eq. 16.6 is applicable. In particular, the pressure at modest depths in the earth's crust can be estimated by using Eq. 16.6. This estimation enables us to answer a fascinating question: How deep can you dig a hole?

Example 3 **How Deep Can You Dig a Hole?**

There is a limit to how far we can dig—or drill—a hole in the ground, because the pressure of the earth increases with depth. At a sufficient depth the pressure becomes so great that it will crush any material, whether man-made steel or nature's most rigid rock.

We can estimate the maximum depth possible for a hole in the ground by combining the pressure–depth relation, Eq. 16.6,

$$P = P_o + \rho g z$$

with Hooke's law for volume stress and strain (Section 15.5),

$$P = -B\left(\frac{\Delta V}{V}\right) \qquad (15.21)$$

Ignoring P_o (1 atm), we have

$$P = \rho g z = B\left(-\frac{\Delta V}{V}\right)$$

Solving for the depth, z, we get

$$z = \frac{1}{\rho g} B\left(-\frac{\Delta V}{V}\right)$$

Because we want an estimate of the maximum depth, we set $\Delta V/V$ equal to the volume strain at which solids fracture. Values appropriate to the crust of the earth are as follows:

$\rho = 3 \times 10^3$ kg/m^3	density of crustal rocks
$g = 9.8$ m/s^2	acceleration of gravity
$B = 6 \times 10^{10}$ Pa	bulk modulus of crustal rocks
$\dfrac{\Delta V}{V} = -0.02$	volume strain at which crust crumbles (estimated)

Substituting these values into our equation, we get

$$z = \frac{(6 \times 10^{10})(0.02)}{(3 \times 10^3)(9.8)} \text{ m} = 40 \text{ km} \simeq 25 \text{ mi}$$

If we compare this value of z with the mean radius of the earth ($R_E = 6371$ km), we see that the deepest hole possible would extend less than 1% of the way to the center of the earth.

Our rough calculation shows that rigid material cannot exist at depths greater than about 40 km. It follows that this same depth, 40 km, should represent the thickness of the earth's rigid crustal layer. Seismic studies bear out this expectation. The earth's crustal thickness ranges from about 5 km in the ocean floors to 35 km in the continental blocks. At such depths the earth's makeup changes from a rigid solid to a plastic-like material.

Questions

5. Deep-sea divers have air pumped to them through hoses. Why wouldn't a long, flexible tube, open to the atmosphere, serve just as well?
6. A student decides to make a barometer by simply placing a tube in a mercury reservoir, without first filling the tube with mercury. He has heard his instructor explain that the barometer operated because the external (atmospheric) pressure supports the weight of the mercury column. The student argues that if the atmosphere is willing to support the column, it should not mind forcing the mercury upward into the column as well. Explain why this will not happen.

16.3 Fluid Dynamics

In Chapter 10 we showed that the motion of a system of particles is determined by the net external force acting on the system. (See Section 10.2.) The internal forces cancel by virtue of Newton's third law. The results derived in Section 10.2 are applicable to a fluid particle, because such a particle is composed of many molecules and is itself a system of particles. We can describe the macroscopic motions of a fluid in terms of the motions of a fluid particle. In particular, the velocity **u** of the fluid is described as the center-of-mass velocity of a *fluid particle* (Section 16.1). The fluid velocity may vary both with the position of the fluid particle and with time. Thus, in general

$$\mathbf{u} = \mathbf{u}(x, y, z, t)$$

We limit our study of fluid dynamics to a special, but very important, class of fluid flows, described in the next four paragraphs. This lets us deal with the more important concepts of fluid dynamics without using sophisticated mathematical techniques.

First, fluid flow may be *rotational* or *irrotational*. If we place a small paddle wheel in a fluid where the flow is irrotational, the paddle wheel does not rotate (Figure 16.9a). In a rotational flow it rotates (Figure 16.9b). We restrict ourselves to *irrotational* flows.

Second, ordinary liquids exhibit *viscosity*, a fluid form of friction. In many situations viscosity can be ignored, just as rolling or sliding friction often can be neglected in treating the motions of rigid bodies. We consider only *nonviscous* flow at this time. (Viscosity is introduced in Section 16.6.)

Figure 16.9

(a) A small paddle wheel placed in an irrotational flow undergoes translation but not rotation. (b) In a rotational flow the paddle wheel rotates.

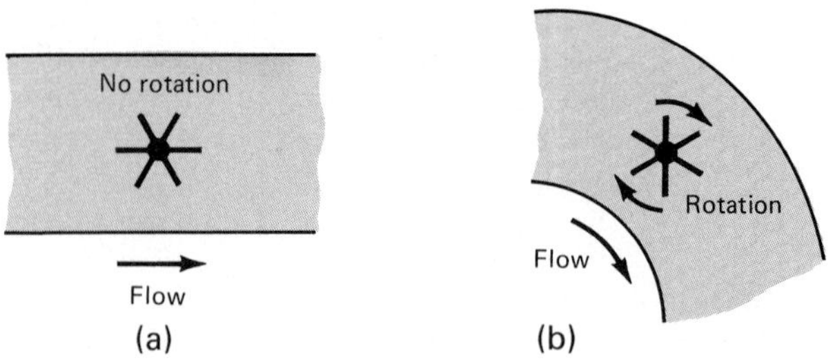

Third, we distinguish gases and liquids on the basis of their abilities to withstand compressive stresses. Ordinarily, a gas is readily compressed, while a liquid is compressed much less readily. However, there are many situations in which the flow of a gas can be regarded as incompressible. These are flows where the gas is not subjected to large *changes* in pressure. We will deal only with *incompressible* fluid flow. In an incompressible fluid, the *mass density* (ρ) is a *constant*:

$$\rho = \text{constant} \rightarrow \text{incompressible flow}$$

Finally, fluid flow is termed *steady* if the fluid velocity **u** does not change with time. Note that this does not require that the fluid velocity be the same at all positions. For example, consider the steady flow of water in a stream. The flow velocity is greatest at the surface. The velocity decreases with depth, falling to zero at the bottom, where the water adheres to the stream bed. The velocity is different at different positions; but at a given position, **u** does not change. We restrict our analysis to *steady flows*.

To begin our study of steady, irrotational, nonviscous, incompressible flow we must introduce the concept of a *streamline*. If we draw the velocity vector **u** of a fluid particle at successive points (A, B, C, etc.) and connect these points, we construct a line—which we call a *streamline* (see Figure 16.10). The fluid particles move along streamlines. If we spilled colored ink into a liquid, for example, the ink would trace out the streamlines. Streamlines have interesting and useful properties. By definition, they are parallel to the fluid velocity at all points. As a consequence, no two streamlines can cross. We can prove this by assuming that two streamlines *do* cross (see Figure 16.11). *If* two streamlines crossed, the fluid velocities associated with them would be in different directions at the point of intersection. This in turn would imply that the fluid velocity has two different values at one point. This is clearly impossible. Hence, streamlines never cross.*

The fact that fluid particles move along streamlines which never cross enables us to keep track of any given volume of fluid. We need only sketch a few representative streamlines on the periphery of the fluid volume of interest. These streamlines outline a surface. The region bounded by this surface is called a *tube of flow* (see Figure 16.12). Fluid particles inside a tube of flow remain inside. Fluid particles outside never enter. Neither sort of crossing

*An exception occurs at *stagnation points*, which are positions where the fluid velocity is zero. Streamlines can "meet" at a stagnation point.

Figure 16.10

The fluid velocity at each point is tangent to the streamline at that point.

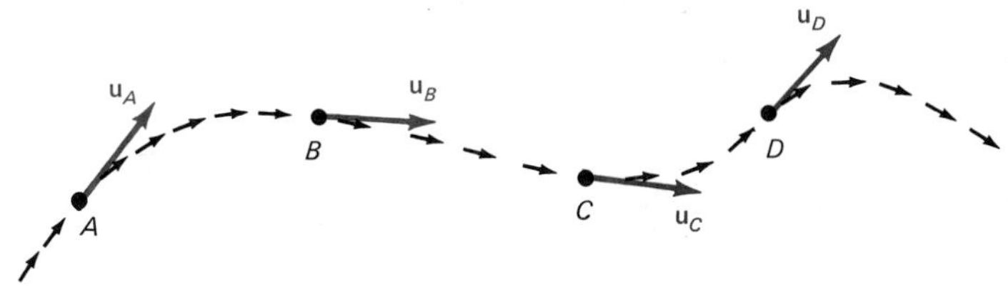

can occur, for both would entail the crossing of streamlines.

In the next two sections we use the concepts of the fluid particle and the tube of flow to derive two important fluid dynamic equations: the equation of continuity and Bernoulli's equation.

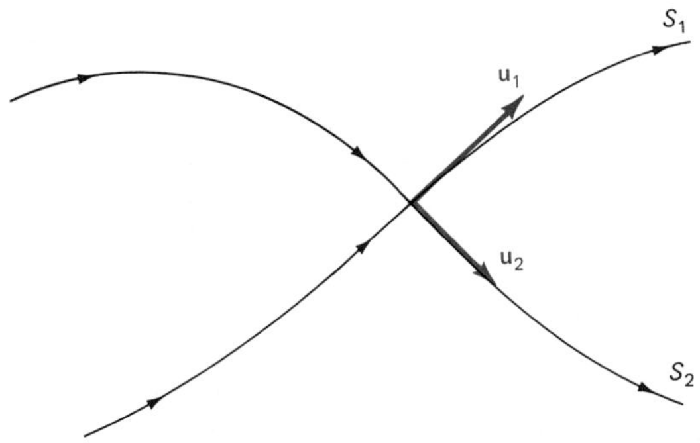

Figure 16.11

Streamlines never cross. *If* the streamlines S_1 and S_2 crossed, the velocities associated with the two ($\mathbf{u}_1$ and $\mathbf{u}_2$) would be in different directions at the point of intersection. This in turn would imply that the average fluid velocity has two different values at one point. This is clearly impossible. Hence, streamlines never cross.

16.4 The Equation of Continuity

We can use the tube-of-flow concept to derive the equation of continuity, which amounts to a statement of the conservation of mass. Let us focus our attention on one segment of a tube of flow (see Figure 16.13). For steady flow the streamlines are fixed and the volume of the tube segment remains constant. For incompressible flow the density remains constant. The total mass in the segment, the product of density and volume, is therefore constant. It follows that equal masses must flow into and out of the ends of the segment.

Consider points ① and ② at the ends of the segment, where the flow speeds are u_1 and u_2 and the cross-sectional areas are A_1 and A_2, respectively (see Figure 16.13). In a time t, a fluid volume A_1u_1t flows past point ① into the segment. A volume A_2u_2t flows past point ② and out of the segment. The mass of each volume is equal to ρ times the volume. Thus,

$$m_1 = \rho A_1u_1t = \text{mass into segment}$$

$$m_2 = \rho A_2u_2t = \text{mass out of segment}$$

In order for the mass inside the segment to remain constant, the mass flowing into the segment must equal the mass flowing out. Setting $m_1 = m_2$ and canceling common factors, we get

$$A_1u_1 = A_2u_2 \tag{16.8}$$

This is one form of the *equation of continuity* for steady incompressible flow. According to this equation, if the cross-sectional area of the flow is decreased, the fluid ve-

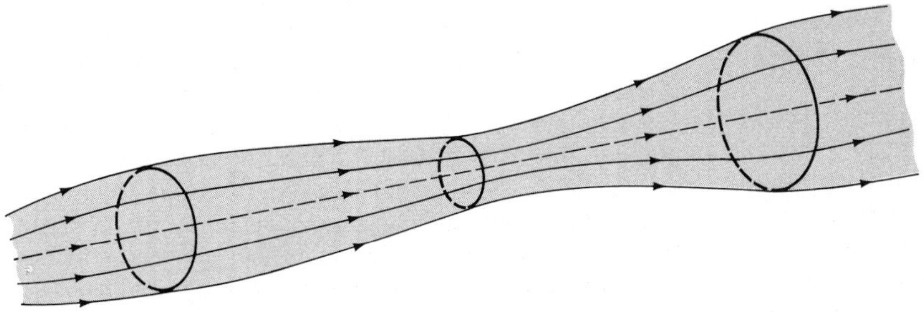

Figure 16.12

Segment of a tube of flow generated by streamlines.

Figure 16.13

For steady incompressible flow, mass in equals mass out.

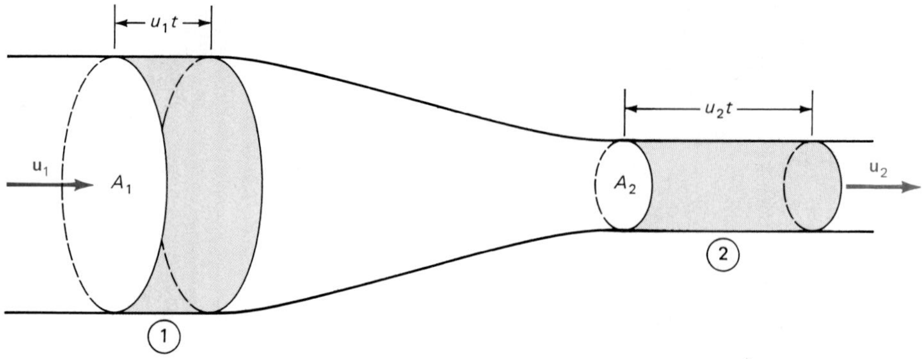

locity is increased. Many common devices illustrate the equation of continuity. For example, when you slowly squeeze the trigger of a water pistol, water is forced through a very small opening, and it emerges at a high speed. Likewise, in a syringe, the plunger has a relatively large cross-sectional area and moves slowly. At the tiny opening in the tip of the needle the fluid emerges at a much higher speed.

The quantity Au is called the *discharge rate*, Q, and gives the volume of fluid that crosses the area A per unit time:

$$\text{Discharge rate} \equiv Q = Au \qquad (16.9)$$

Thus, for example, if gas flows at a speed of 10 m/s through a pipe with a cross-sectional area of 2 m^2, the discharge rate is 20 m^3/s.

Multiplying the discharge rate by the mass density gives the rate of mass flow. Thus,

$$\text{rate of mass flow} \equiv \alpha = \left(\frac{\text{mass}}{\text{volume}}\right) \cdot \left(\frac{\text{volume}}{\text{second}}\right)$$

$$= (\text{mass density}) \cdot (\text{discharge rate})$$

or

$$\alpha = \rho A u = \rho Q \qquad (16.10)$$

Let's use Eq. 16.10 to consider the flow of natural gas through a pipeline.

Example 4 **Flow Speed in a Pipeline**

A natural gas pipeline has a diameter of $D = 0.25$ m. The rate of mass flow is $\alpha = 1.4$ kg/s. The density of the gas is $\rho = 0.90$ kg/m^3. We wish to find the flow speed, u. Using Eq. 16.10 for the rate of mass flow, we write

$$\alpha = \frac{\rho \pi D^2}{4} u$$

where $\pi D^2/4 = A$. Solving for the flow speed and substituting the values given, we get

$$u = \frac{4\alpha}{\rho \pi D^2} = \frac{4 \cdot 1.4 \text{ kg/s}}{0.90 \text{ kg/m}^3 \cdot 3.14(0.25)^2 \text{ m}^2}$$

$$= 32 \text{ m/s}$$

Questions

7. The distance a stream of water travels after it emerges from a hose can be increased by closing off part of the exit. Explain.
8. A reservoir of gravy breaks through a small rift in the surrounding dam of mashed potatoes. The flow of gravy down the side of the potato dam is quite rapid, but the flow of gravy within the reservoir is barely noticeable. Explain.

16.5 Bernoulli's Equation

Because fluid particles obey Newton's second law, the work–energy principle can be applied to their motions (see Section 7.4). For nonviscous flow, the work–energy principle leads to Bernoulli's equation. This equation is named after Daniel Bernoulli, who in 1738 wrote the first "modern" monograph on hydrodynamics. In it he derived the equation and applied it to several problems of fluid flow. The derivation of Bernoulli's equation that we present here generalizes the analysis used to obtain the pressure–depth relation in Section 16.2.

Consider a section of a tube of flow (see Figure 16.14). Let's apply the work–energy principle to the fluid that, at some instant, occupies the region between the cross sections labeled ① and ②. We consider a flow that is steady, incompressible, and nonviscous. We follow the motion of the fluid over a short time interval Δt during which the configuration of the system changes. The fluid advances as indicated in Figure 16.14. The change in configuration takes place through the combined actions of the fluid forces P_1A_1 and P_2A_2 and of gravity. Each fluid particle in the system advances along the tube of flow, but net work and energy transfers involve only the mass elements at the ends. The energy of the intervening mass is unchanged because the density is constant and the flow is steady. For a steady flow of constant density, equal masses (m) flow past ① and ② during the interval Δt. The net result of the flow during the interval Δt is to transfer a mass

Figure 16.14

The system consists of the fluid between points ① and ② at time t. In a time Δt the system moves forward under the influence of the forces P_1A_1, P_2A_2, and gravity. Applying the work–energy principle to the system relates the work done by these forces to the change in kinetic energy of the fluid.

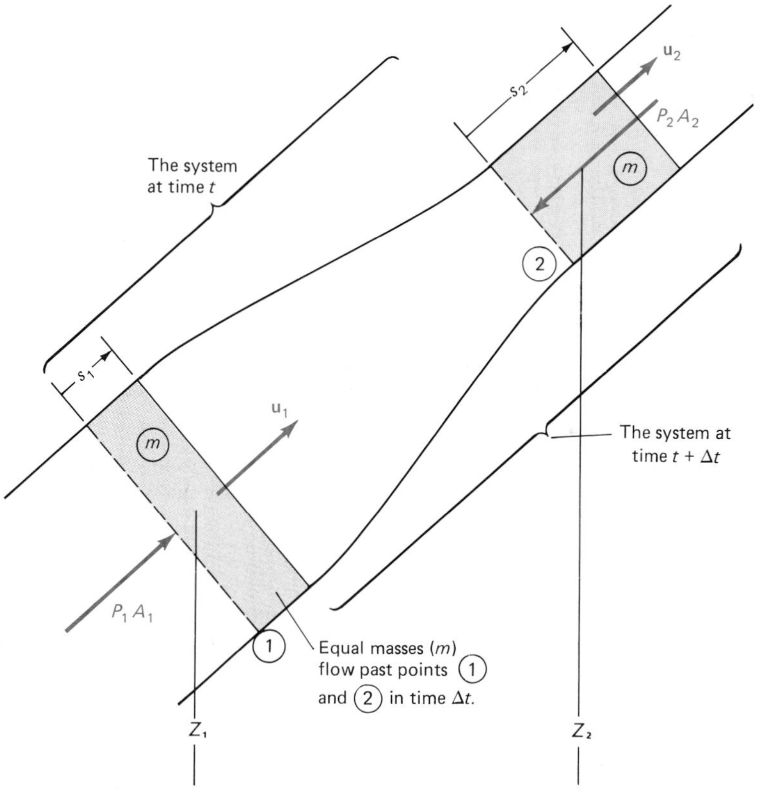

m from the left end of the tube to the right end. The change in the kinetic energy of the system is the change in the kinetic energy of this mass m. Thus,

$$\text{change in kinetic energy} = \frac{1}{2}mu_2^2 - \frac{1}{2}mu_1^2 \qquad (16.11)$$

Gravitational work is performed because the mass m moves through a vertical distance $z_2 - z_1$. The net work done on the system *by* gravity is thus

$$W_g = -mg(z_2 - z_1) \qquad (16.12)$$

At the left end, the fluid force P_1A_1 acts through the displacement s_1 and does work $P_1A_1s_1$. At the right end, the fluid force P_2A_2 and the displacement s_2 are in opposite directions. The work done by P_2A_2 is thus $-P_2A_2s_2$. The net work done on the system by the fluid forces P_1A_1 and P_2A_2 is

$$W_{fl} = P_1A_1s_1 - P_2A_2s_2 \qquad (16.13)$$

Note that A_1s_1 and A_2s_2 are the volumes of the equal masses m, and that the density is constant. Because volume is equal to mass divided by density, we can substitute m/ρ for both A_1s_1 and A_2s_2:

$$A_1s_1 = A_2s_2 = \frac{m}{\rho} \qquad (16.14)$$

Substituting m/ρ into Eq. 16.13, we get

$$W_{fl} = (P_1 - P_2)\frac{m}{\rho} \qquad (16.15)$$

for the net work done on the system by the fluid. A verbal statement of the work–energy principle for the system is

$$\begin{array}{c}\text{net work done}\\ \text{on system by}\\ \text{external forces}\end{array} = \begin{array}{c}\text{change in}\\ \text{kinetic energy}\\ \text{of system}\end{array} \qquad (16.16)$$

Collecting the results of Eqs. 16.11, 16.12, and 16.15, we can translate this verbal statement into

$$\underbrace{(P_1 - P_2)\frac{m}{\rho}}_{\text{work by fluid}} + \underbrace{[-mg(z_2 - z_1)]}_{\text{work by gravity}} = \underbrace{\frac{1}{2}mu_2^2 - \frac{1}{2}mu_1^2}_{\text{change in kinetic energy}}$$

which can be rearranged to read

$$P_2 + \rho g z_2 + \frac{1}{2}\rho u_2^2 = P_1 + \rho g z_1 + \frac{1}{2}\rho u_1^2 \qquad (16.17)$$

This is a standard form of Bernoulli's equation. An equivalent statement is

$$P + \rho g z + \frac{1}{2}\rho u^2 = \text{constant} \qquad (16.18)$$

that is, the quantity $P + \rho g z + \frac{1}{2}\rho u^2$ has the same value at all points along a streamline.

Each term in Bernoulli's equation has the dimensions of energy per unit volume. The term $\frac{1}{2}\rho u^2$ is the kinetic energy per unit volume associated with the *macroscopic* fluid motions.* The term $\rho g z$ is the gravitational potential energy per unit volume. The pressure, P, represents the *flow energy* per unit volume. Flow energy is work done by the fluid against its surroundings as a result of its motions. The sum of the three terms, $P + \rho g z + \frac{1}{2}\rho u^2$, is called the *available energy* per unit volume. Bernoulli's theorem is therefore a statement that the available energy per unit volume remains constant along any given tube of flow. The various forms of available energy (kinetic energy, gravitational potential energy, and flow energy) can be transformed into one another, but the total available energy remains constant. We now consider three examples of Bernoulli's equation.

Example 5 Hydrostatic Limit

When a fluid is at rest, $u_1 = u_2 = 0$. In this special case, which we call the *hydrostatic limit*, Bernoulli's equation reduces to the pressure–depth relation,

$$P_1 = P_2 + \rho g (z_2 - z_1) \qquad (16.6)$$

This equation was derived earlier (in Section 16.1). Here, however, we see it in a new light. We see that change in flow energy is balanced by an opposite change in the gravitational energy, and thus this equation is an example of energy conservation. Energy can be changed from one form to another, but there can be no change in the total amount of energy.

In certain circumstances, the available energy per unit volume will have the same value along *all* streamlines, and thus is a constant throughout the fluid. The following example illustrates this point.

Example 6 Flow Speed from a Reservoir (Torricelli's Theorem)

Let us determine the speed at which water escapes through a small orifice at the base of a reservoir. Both the top of the reservoir and the orifice are open to the atmosphere (Figure 16.15). Bernoulli's equation is applied at the upper surface (point 1) and in the stream emerging from the orifice (point 2). The available energy is the same at all points on the upper surface. Because all tubes of flow originate at this surface, the available energy is the same along all streamlines. We can therefore regard the entire volume of water between 1 and 2 as a single tube of flow.

The Bernoulli equation, Eq. 16.17, is simplified by noting that the height of the water column is

$$h = z_1 - z_2$$

Figure 16.15

Torricelli's theorem follows from Bernoulli's equation when the speed of the surface, u_1, is ignorable compared to the exit speed, u_2.

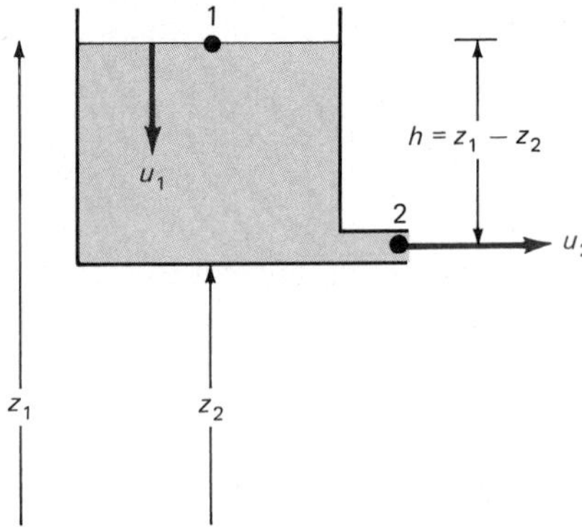

and that $P_1 = P_2 = P_{atm}$, since both points 1 and 2 are open to the atmosphere. Bernoulli's equation reduces to

$$\frac{1}{2}u_2^2 - \frac{1}{2}u_1^2 = gh \qquad (16.19)$$

This form of Bernoulli's equation illustrates the transformation of gravitational potential energy into kinetic energy. Thus, if we envision a mass m as falling along a streamline through a distance h, its gravitational potential energy decreases by mgh. If its speed increases from u_1 to u_2, its kinetic energy rises from $\frac{1}{2}mu_1^2$ to $\frac{1}{2}mu_2^2$. Energy conservation dictates that the potential energy decrease be matched by an equal increase in kinetic energy. In equation form, energy conservation requires

$$\frac{1}{2}mu_2^2 - \frac{1}{2}mu_1^2 = mgh$$

This equation is identical to Eq. 16.19 except for the factor m. We could eliminate u_1 in our equation by using the equation of continuity,

$$u_1A_1 = u_2A_2 \qquad (16.8)$$

Instead, however, we make use of the fact that the surface area A_1 is much greater than the cross-sectional area of the emerging stream, A_2. This makes u_1 very small in comparison with u_2. Therefore, we can drop u_1^2 from the right side of Eq. 16.19 to get

$$u_2 = \sqrt{2gh} \qquad (16.20)$$

Notice that the speed of flow is the same as would be acquired in free fall from the surface of the reservoir to the level of the orifice. This result, Eq. 16.20, is known as *Torricelli's theorem*.

There are many other situations in which the gravitational potential energy term is ignorable. In these situations Bernoulli's equation relates pressure changes and kinetic energy changes, as we see in Example 7.

*For a hydrostatic system, $u = 0$, and the macroscopic kinetic energy is zero. However, there is a considerable amount of kinetic energy associated with the molecular motions at the *microscopic* level.

Example 7 Fluid Dynamics of Vaccination

Vaccinations are now routinely administered by using high-pressure "guns" rather than hypodermic needles. The guns force the vaccine through a tiny opening in a sapphire tip. Typically guns exert a pressure of 500 psi to force vaccine through an orifice 0.005 in. in diameter. We can use Bernoulli's equation to determine the speed at which the vaccine emerges. We ignore the gravitational energy term in Bernoulli's equation, and approximate the flow speed *inside* the gun as zero. In this case, Bernoulli's equation describes the transformation of flow energy into kinetic energy. Equation 16.17 assumes the form

$$P_i - P_o = \frac{1}{2}\rho u_o^2$$

(where the subscripts i and o mean inside and outside, respectively) with

$$\rho = 1.1 \times 10^3 \text{ kg/m}^3$$

$$P_i = 500 \text{ psi} = 3.45 \times 10^6 \text{ N/m}^2$$

$$P_o = 15 \text{ psi} = 0.10 \times 10^6 \text{ N/m}^2$$

We get for the speed of the emerging vaccine

$$u_o = \sqrt{\frac{2(P_i - P_o)}{\rho}} = \sqrt{\frac{2(3.35 \times 10^6)}{1.1 \times 10^3}} \text{ m/s} = 78 \text{ m/s}$$

which is approximately 175 mph, or about 0.25 of the speed of sound in air.

Questions

9. An insulated water jug has an airtight cover and a spigot. If the spigot is opened, water flows for a short time and then stops. In order to maintain the flow it is necessary to loosen the airtight cover, thereby maintaining atmospheric pressure inside the jug. Explain why the flow stops when the top is securely tightened. (*Hint:* The pressure of the confined air is inversely proportional to its volume.) How might you modify the construction of the jug to make it unnecessary to loosen the top?

Figure 16.16

A siphon.

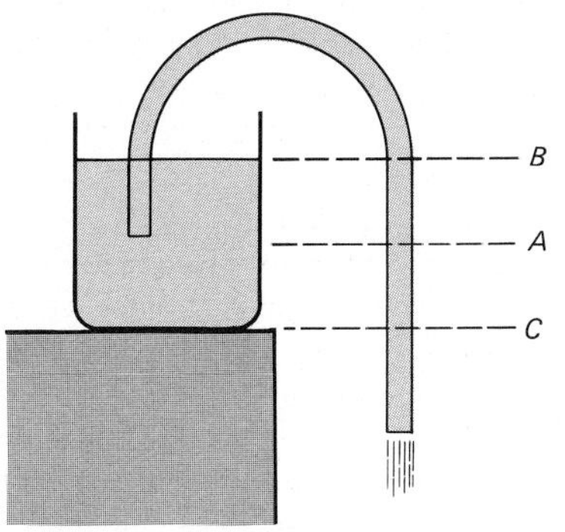

10. Figure 16.16 shows a siphon. If the submerged end is at A and the tube is filled with water, at what level must the other end be in order for the siphon to begin operation: just below level B, just below level A, or just below level C? Explain briefly. If the siphon is not initially filled with water, why isn't it self-starting?

16.6 Viscosity

By definition, a fluid cannot support a shearing stress (Section 16.1). Nevertheless, fluids do *resist* shearing motions. The property of a fluid that measures its resistance to shearing motions is called *viscosity*.

Viscosity is the fluid analog of the shear modulus for an elastic solid (Section 15.6). Recall our discussion of the shear stress–strain relationship for a solid. We can think of a solid as a set of adjacent layers. From our discussion of the shear stress–strain relationship for a solid (Section 15.2) you may recall that a shearing stress produces a relative displacement of these layers (see Figure 16.17).

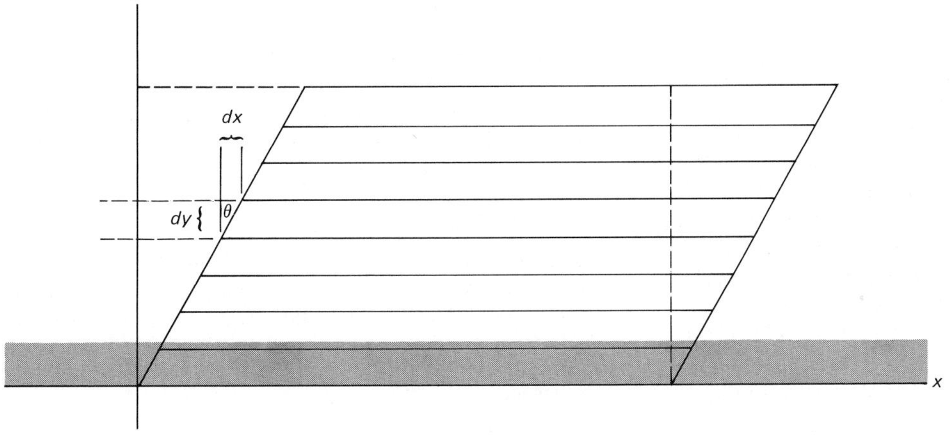

Figure 16.17

In a solid, a shearing stress results in the relative displacement of adjacent layers of the solid. The strain, $dx/dy \approx \theta$, is a measure of the distortion produced by the stress.

Figure 16.18

In a liquid, a shearing stress causes adjacent layers to be in relative motion. The rate of strain, du/dy, is a measure of the distortion produced by the stress.

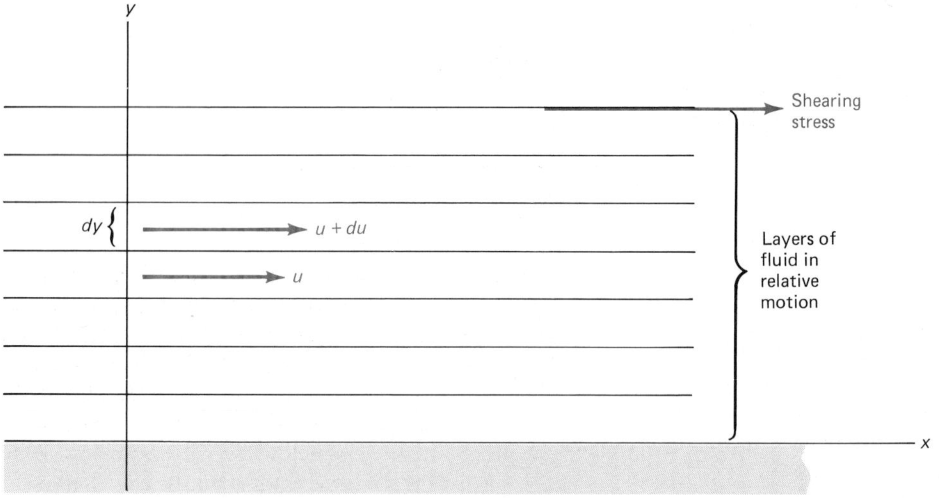

Figure 16.18 shows the analogous situation for a fluid. A shearing stress (F/A) sets the fluid in motion and produces *relative motion* of the fluid layers. The relative displacement of adjacent layers increases indefinitely and is not an appropriate quantity for describing shear effects in a fluid. Instead, we use the *relative velocity* of adjacent fluid layers. The relative velocity of fluid layers can be inferred from the *velocity profile*, a graph or equation that relates the fluid velocity to position. Figures 16.19a and 16.19b show the velocity profiles typical of flow in a pipe and of flow in an open channel (a river, for example). If two layers separated by a distance dy have velocities u and $u + du$, their relative velocity is du. The *rate of strain* is defined as du/dy. The quantity du/dy is often referred to as the *velocity gradient*; it measures how the flow velocity changes with position. Geometrically du/dy is the slope of the velocity profile. In an elastic solid the ratio (shear stress/shear strain) is defined as the shear modulus (Section 15.2):

$$\mu = \frac{\text{shear stress}}{\text{shear strain}} = \frac{F/A}{\theta} \qquad (15.26)$$

For small strains $\theta \simeq dx/dy$ and

$$\mu = \frac{F/A}{dx/dy}$$

In fluids, the analog of the shear modulus for solids is the ratio of the shear stress to the rate of strain. This ratio (η) is termed the *dynamic viscosity* or simply the viscosity:

$$\eta = \frac{\text{shear stress}}{\text{rate of shear strain}} = \frac{F/A}{du/dy} \qquad (16.21)$$

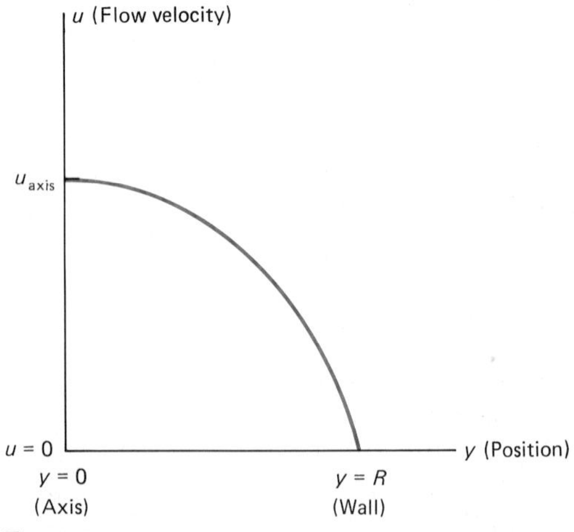

Figure 16.19a

Velocity profile for a pipe of radius R. The fluid velocity is zero at the wall and rises to a maximum along the central axis.

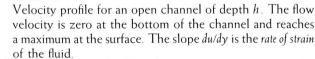

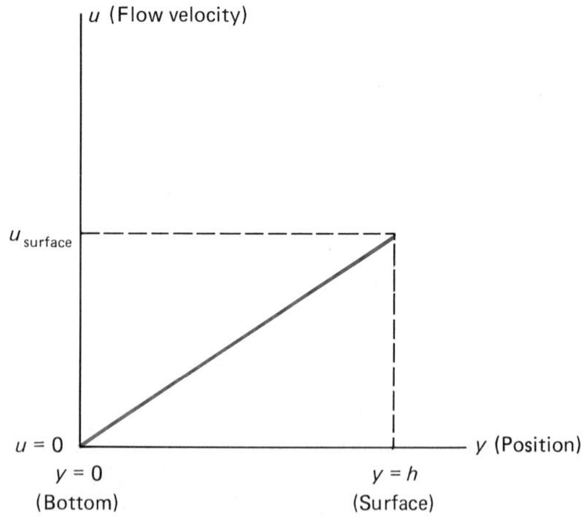

Figure 16.19b

Velocity profile for an open channel of depth h. The flow velocity is zero at the bottom of the channel and reaches a maximum at the surface. The slope du/dy is the *rate of strain* of the fluid.

In Eq. 16.21, F/A is the shearing stress between fluid layers, which produces a rate of strain du/dy. We can rewrite Eq. 16.21 to display the dynamic viscosity as the proportionality factor relating shear stress and rate of strain

$$\frac{F}{A} = \eta \frac{du}{dy} \qquad (16.22)$$

The SI unit of viscosity is the (N · s/m^2). This unit suffers the indignity of having no name. The cgs unit of viscosity is the dyne · s/cm^2, and is called the *poise:*

$$1 \text{ poise} = 1 \text{ dyne} \cdot \text{s/cm}^2$$

The centipoise is one one-hundredth of a poise. We can verify that 1 dyne/cm^2 = 0.1 N/m^2, so that

$$1 \text{ poise} = 0.1 \text{ N} \cdot \text{s/m}^2$$

The viscosities of liquids and gases typically differ by a factor of roughly 100. For example, the viscosities of water and air at 20°C are 1.0 centipoise (cP) and 1.8×10^{-2} centipoise, respectively. Table 16.1 contains a list of viscosities for several fluids at 20°C. Keep in mind that in general the viscosity decreases as the temperature of the fluid increases.

The viscous force F in Eq. 16.21 is the fluid analog of the sliding friction force between two solid surfaces. For this reason, viscosity is often referred to as *fluid friction*. Like other frictional forces, viscous forces oppose the relative motion of adjacent fluid layers. Whereas solid frictional forces are approximately independent of velocity, viscous forces increase with velocity, as shown in Example 8.

Figure 16.20

The metal plate slides down the incline, reaching a limiting speed.

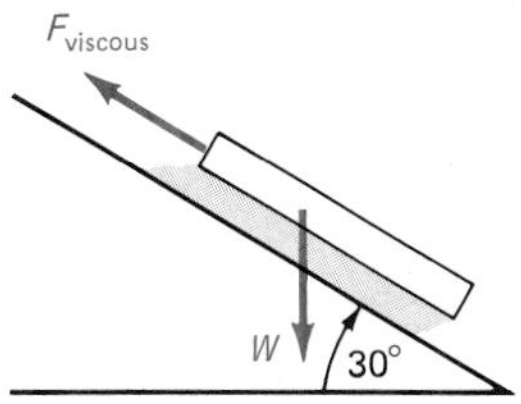

Example 8 **Fluid Friction**

A flat steel plate starts from rest and slides down an inclined plane under the combined action of gravity and the viscous force exerted by a thin film of oil (Figure 16.20). What is the limiting speed of the plate, assuming that the rate of strain, du/dy, is constant across the thickness of the film? The plane is inclined at 30° to the horizontal. The plate weighs 0.5 N, and its contact area with the oil is 0.02 m^2. The viscosity of the oil is 0.05 N · s/m^2, and the film thickness is 0.2 mm.

The oil in contact with the incline is at rest. The fluid speed increases across the film, reaching a maximum, v, at the surface in contact with the plate. The speed at which the plate moves is the maximum fluid speed (v). With a constant rate of strain

$$\frac{du}{dy} = \frac{v}{h}$$

Table 16.1

Viscosities of selected fluids*

Fluid	Dynamic Viscosity (η) (N · s/m^2)	Kinematic Viscosity (ν) (m^2/s)
Liquids		
Benzene	6.6×10^{-4}	7.5×10^{-7}
Gasoline	2.9×10^{-4}	4.3×10^{-7}
Jet Fuel (JP-4)	8.7×10^{-4}	1.1×10^{-6}
Crude Oil	7.2×10^{-3}	8.4×10^{-6}
Water	1.0×10^{-3}	1.0×10^{-6}
Glycerin	1.5	1.2×10^{-3}
Gases		
Hydrogen	0.90×10^{-5}	1.1×10^{-4}
Helium	1.97×10^{-5}	1.2×10^{-4}
Methane	1.34×10^{-5}	2.0×10^{-5}
Air	1.81×10^{-5}	1.5×10^{-5}
Oxygen	2.01×10^{-5}	1.5×10^{-5}
Carbon Dioxide	1.47×10^{-5}	0.80×10^{-5}

*Values refer to a temperature of 20°C and a pressure of 1 atm. Adapted from *Elementary Fluid Mechanics* by J. K. Vannard and R. L. Street, 5th ed., with permission of the publisher, John Wiley & Sons, New York.

where h, the film thickness, is 2 mm. This viscous force is

$$F_{vis} = \eta \frac{v}{h} A$$

Notice that the viscous force increases as the speed of the plate increases. The speed of the plate increases until F_{vis} equals $W \sin 30°$, the component of the weight acting along the surface. When $F_{vis} = W \sin 30°$, the net external force acting parallel to the surface is zero and the plate then moves with a constant velocity. Taking

$$F_{vis} = \eta \frac{v}{h} A = W \sin 30°$$

gives

$$v = \frac{hW \sin 30°}{\eta A}$$

Inserting numerical values gives

$$v = \frac{0.5\ \text{N}\,(0.5)(2 \times 10^{-4}\ \text{m})}{(0.05\ \text{N}\cdot\text{s/m}^2)\,(0.02\ \text{m}^2)} = 0.05\ \text{m/s}$$

The viscosity is often referred to as the *absolute*, or dynamic, viscosity to distinguish it from a related quantity, the *kinematic viscosity*. The kinematic viscosity, ν, is defined as the ratio of the dynamic viscosity and the mass density,

$$\nu \equiv \frac{\eta}{\rho} \tag{16.23}$$

Both ν and η measure the resistance to shearing motions. The kinematic viscosity is introduced because in many applications it is the ratio η/ρ rather than η alone that determines the importance of viscous forces. The cgs unit of kinematic viscosity is the *stoke:*

$$1\ \text{stoke} = 1\ \text{cm}^2/\text{s}$$

The kinematic viscosity η/ρ has many applications. We will encounter one such application in the next section, where we study fluid turbulence.

Molecular Origin of Viscosity

The origin of the shearing force in a fluid is not immediately evident. Why should there be a shearing force between two layers of fluid in relative motion? There are, in fact, two sources of viscous forces. This fact is suggested by the observation that the viscosity of a liquid *decreases* as temperature increases, whereas the viscosity of a gas *increases* as its temperature rises. In liquids, viscosity originates with the cohesive forces between molecules in adjacent layers. The viscous force results because the attractive forces between molecules tend to prevent the relative motion of molecules. In low-density gases cohesive forces make only a minor contribution to the viscosity. The molecules are widely separated and exert very weak attractive forces, and a different mechanism is responsible for viscosity. In a gas, molecules constantly are being exchanged between adjacent layers. If molecules of mass m are traded between layers where the flow speeds are u and $u + du$, there will be a net transfer of momentum from the faster layer to the slower layer. The slower layer is accelerated—it receives .an impulse of $+m\,du$ when it catches a molecule from the faster layer. The faster layer receives a negative impulse, $-m\,du$, in the trade. Overall, the momentum transfers tend to reduce the relative velocity of the layers. At the macroscopic level this momentum transfer is the viscous force that opposes relative motion.

16.7 Turbulence

The stable streamline flow in a viscous fluid is called *laminar* flow. Adjacent layers, or laminae, of fluid slide smoothly over each other. The stability of laminar flow is maintained by viscous forces. However, experiment shows that laminar flow is disrupted if the speed of flow is sufficiently great. A random and irregular motion of the fluid replaces the smooth laminar flow. Such an irregular motion is referred to as *turbulent flow*, or turbulence. The flow speed at which turbulence develops depends on the viscosity of the liquid, and on the geometry of the flow.

There are countless examples of turbulence around us. There may be turbulence in the smoke rising from a cigarette or a smokestack, or in the buildup of a cumulus cloud on a summer afternoon. The twinkle of starlight is the result of turbulence in our atmosphere. The Gulf Stream is a turbulent flow. The wakes left by autos, ships, and airplanes are turbulent.

We can think of fluid turbulence as groups of swirling eddies or currents moving along with the average fluid velocity. Viscous forces gradually dissipate the energy of individual eddies, causing them to die out. However, new eddies will form as long as the basic fluid instability persists.

Turbulence can be "good" or "bad," depending on the circumstances. Turbulent flows are often undesirable because they invariably "waste" energy. As eddies form and decay, energy is dissipated uselessly as heat. Thus, for example, when the flow through a pipe changes from laminar to turbulent, the pressure must be increased in order to maintain a constant flow speed. On the other hand, turbulence can be very desirable because it causes rapid mixing and increases rates of transfer of mass, momentum, and energy. For example, the blades of an eggbeater promote turbulent flow and rapid mixing.

Reynolds Number

The Reynolds number is a useful parameter that allows us to decide—or at least estimate—whether a flow will be steady or turbulent. The **Reynolds number** is defined as

$$R = \frac{\rho u L}{\eta} \tag{16.24}$$

where ρ is the fluid density, u is the flow speed, η the dynamic viscosity, and L some *characteristic length* associated with the flow. For flow past a sphere, L would be the diameter of the sphere. For flow through a pipe, L would be the diameter of the pipe.

The Reynolds number is dimensionless. In a sense, it is the "simplest" dimensionless quantity that can be formed from ρ, u, L, and η. The Reynolds number is often written as

$$R = \frac{uL}{\nu} \qquad (16.25)$$

where

$$\nu = \frac{\eta}{\rho} \qquad (16.23)$$

is the kinematic viscosity.

Experiment shows that geometrically similar flows become turbulent at the same value of R. For example, oil and water have different kinematic viscosities; but if we pump oil and water through geometrically similar pipes (pipes with the same type of cross section), the flows become unstable and turbulent at approximately the same Reynolds number, called the *critical Reynolds number*. Although it varies with geometry, we know that for many fluids the critical Reynolds number lies between 1000 and 10,000. (Engineers frequently take 2300 as the critical Reynolds number.)

Example 9 **A Classroom Determination of the Critical Reynolds Number**

The critical Reynolds number for water can be measured in the following simple experiment. The flow velocity of tap water is increased until the stream changes its appearance. Laminar flow is characterized by a smooth, transparent stream. When turbulence sets in at the critical velocity, the stream becomes twisted and translucent. The volume of water that flows out of the pipe in a time t is $V = ut \cdot \pi L^2/4$, where L is the diameter of the pipe. If the water is collected in a cylindrical vessel of diameter D, the volume V can be expressed as

$$V = \frac{h\pi D^2}{4}$$

where h is the height of the water. Equating these two expressions for V gives

$$uL = \frac{D^2 h}{Lt}$$

The Reynolds number is $R = uL/\nu$. Thus,

$$R = \frac{D^2 h}{\nu L t}$$

The following data were obtained in a classroom experiment:

$$D = 0.167 \text{ m} \qquad h = 0.089 \text{ m} \qquad t = 40 \text{ s}$$

$$L = 0.014 \text{ m}$$

Taking $\nu = 1.0 \times 10^{-6}$ m²/s gives

$$R = \frac{(0.167)^2(0.089)}{(1.0 \times 10^{-6})(0.014)(40)} = 4400$$

To check the precision of this result, try the experiment yourself and then compare your result with this one.

We have stressed the idea that turbulence results from a fluid instability. The stability of fluid particles moving along a tube of flow depends on the viscous friction exerted by the surrounding fluid. If the viscous friction is disturbed, instability results. The following analogy animates the idea of unstable flow and illustrates the significance of the Reynolds number. Consider a group of students running to their physics class on a cold winter morning. They run at a speed u, taking strides of length L. The coefficient of friction between their shoes and the sidewalk is μ. We define the analog of the Reynolds number for this "flow" of students as

$$R_s = \frac{uL}{\mu}$$

Consider what happens when the students move from a bare sidewalk to one covered with ice. The value of μ decreases abruptly, causing the student Reynolds number to increase. Whereas friction prevents any slipping on the bare sidewalk, the students must be careful on the ice—any slight imbalance may result in *unstable* motion. To recover stability, the students slow down and take shorter strides. By decreasing u and L, the students lower their Reynolds number and thereby tend to stabilize their motions.

Summary

Archimedes' principle states that an object experiences a buoyant force (B) equal to the weight of the fluid it displaces:

$$B = \underbrace{\rho_{fl} V_s}_{\text{mass of fluid displaced}} g \qquad (16.2)$$

where ρ_{fl} is the mass density of the displaced fluid and V_s is the volume of fluid displaced.

The hydrostatic pressure in a fluid increases with depth. For a fluid of constant density ρ the pressure at a depth z is given by the pressure–depth relation

$$P = P_o + \rho g z \qquad (16.6)$$

where P_o is the pressure at the surface ($z = 0$).

For *steady, irrotational, nonviscous* flow in an *incompressible* fluid the conservation of mass principle leads to the equation of continuity,

$$A_1 u_1 = A_2 u_2 \qquad (16.8)$$

which states that the *discharge rate, Au,* is constant along any *streamline*.

The work–energy principle of mechanics leads to Bernoulli's equation,

$$P_2 + \rho g z_2 + \frac{1}{2}\rho u_2^2 = P_1 + \rho g z_1 + \frac{1}{2}\rho u_1^2 \qquad (16.17)$$

which states that the available energy per unit volume ($P + \rho g z + \frac{1}{2}\rho u^2$) is constant along any tube of flow.

Viscosity is the property of a fluid that measures its resistance to shearing forces. The *dynamic viscosity* (η) is defined by

$$\eta = \frac{\text{shear stress}}{\text{rate of shear strain}} = \frac{F/A}{du/dy} \qquad (16.21)$$

The *kinematic viscosity* (ν) is defined by

$$\nu = \frac{\eta}{\rho} \qquad (16.23)$$

where ρ is the mass density of the fluid.

Unstable fluid flow produces turbulent eddies. The transition from stable streamline flow to turbulent flow is related to the Reynolds number,

$$R = \frac{uL}{\nu} \qquad (16.25)$$

where u is the flow speed, L is a characteristic length, and ν is the kinematic viscosity. Fluid flow becomes turbulent when R exceeds a critical value. For many fluids the critical Reynolds number lies between 1000 and 10,000.

Suggested Reading

J. D. Bernal, The Structure of Liquids, *Sci. Am.* Vol. 203, 124 (August 1960).

V. P. Starr and N. E. Gant, Negative Viscosity, *Sci. Am.* Vol. 223, 72 (July 1970).

Problems

Section 16.1 Fluid Statics: Archimedes' Principle

1. One mole of water contains 6.02×10^{23} molecules and occupies a volume of 18×10^{-6} m^3. (a) Determine the average volume per molecule. (The smallest length that can be resolved by an optical microscope is about 10^{-6} m. A cube with a volume of 10^{-18} m^3 is therefore near the limit of visibility. A cube with a volume of 10^{-21} m^3—a thousand times smaller than the measurable limit—can safely be regarded as "small" by macroscopic standards.) (b) How many water molecules would occupy a volume of 10^{-21} m^3, on the average? (c) Would you expect the measurable properties of the fluid to deviate significantly from their average values if the number of molecules in the "sample" volume (10^{-21} m^3) were to increase by 10 molecules? What if the number of molecules decreased by 10 molecules?

2. A 425-N pig marooned on a wooden board floats down a flooded river. If the board is 15 cm thick and 3 m long, and has a density of 600 kg/m^3, what is the width of the board if the top surface is level with the water?

3. A body of mass m and density ρ_o is completely submerged in a fluid of density ρ_{fl}, where $\rho_{fl} < \rho_o$. Show that the apparent weight of the submerged body (mg − buoyant force) is

$$W_{app} = mg\left(1 - \frac{\rho_{fl}}{\rho_o}\right)$$

4. A 10-kg rock ($W = 98$ N) is submerged in water. If the density of the rock is 3.1 gm/cm^3, what is its apparent weight under water? (See Problem 3.)

5. A pousse-café is a drink made of six different liqueurs, which lie in layers one on top of the other. The colors and densities of these liqueurs are indicated in Figure 1. Each layer is 2 cm thick. A small plastic cube with a volume of 1 cm^3 and a density of 1.056 g/ml^3 is carefully lowered into the drink. (a) Between which layers does it float? (b) Show that five-sevenths of its volume is in one layer and two-sevenths is in another.

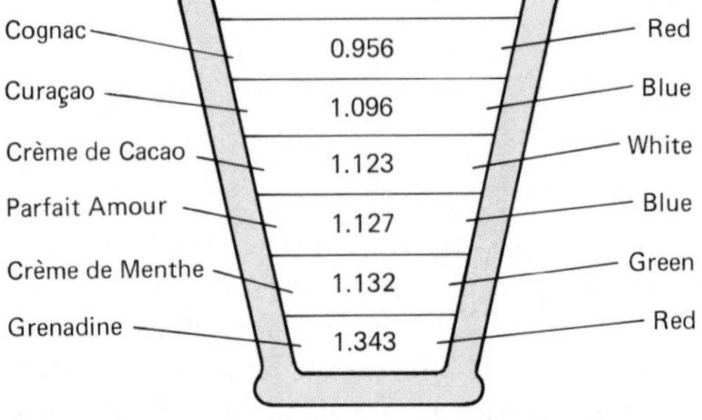

Figure 1

6. A plastic sphere floats in water with 0.5 of its volume submerged. This same sphere floats in oil with 0.4 of its volume submerged. Determine the densities of the oil and the sphere.

7. The molecular weight of helium is 4.00 gm/mole. The molecular weight of air is 28.9 gm/mole. A helium-filled balloon rises because it displaces a weight of air in excess of its own weight. The molar volume is approximately the same for all gases, 0.0224 m^3 at a temperature of 0°C and a pressure of 1 atm. (a) Estimate the mass and volume of helium required to lift a mass of 100 kg. (b) In view of your answer for part (a), would it be feasible to market miniblimps? In particular, would a one-man blimp fit into a one-car garage?

8. Estimate the volume of your body without using any length measurements.

Section 16.2 Pressure–Depth Relation

9. An oil well 1600 m deep (about 1 mi) is filled with a fluid of density $\rho = 950$ kg/m^3. What is the pressure at the bottom of this well?

10. Experimentalists measuring pressures in the range of a few millimeters of Hg sometimes use a barometer (Figure 16.7) filled with oil rather than mercury. If the density of the oil is 920 kg/m^3, what is the change in the level of oil when the pressure changes by 1 mm Hg?

11. The gravitational acceleration at the surface of the earth's moon is 1.62 m/s^2. The average mass density of the moon is 3.33×10^3 kg/m^3 and the bulk modulus of its crustal rocks is comparable to the bulk modulus of the earth's crustal rocks. Estimate the depth to which you could dig a hole in the moon.

12. The pressure at the surface of a swimming pool is 1.02×10^5 Pa. Calculate the pressure at a depth of 1.63 m. Take the density of water to be 1.00×10^3 kg/m^3.

13. A diver descends to a depth of 20 m in search of abalone. What pressure acts on her at that depth? Express your result in both pascals and atmospheres. Is your answer an absolute pressure or a gauge pressure?

14. The pressure at the surface of a swimming pool is 1 atm. At what depth does the pressure equal 1.2 atm? Take the density of water to be 1.00×10^3 kg/m^3.

15. Torricelli was the first to realize that we live at the bottom of an ocean of air. He correctly surmised that the pressure of our atmosphere is due to the weight of the air. The density of air at the earth's surface is 1.3 kg/m^3. The density decreases with increasing altitude as the atmosphere "thins out." Suppose that the density were a constant (1.3 kg/m^3) up to some altitude h, and zero above. Then h would represent the depth of the ocean of air, or the "thickness" of our atmosphere. Determine the value of h that gives a pressure of 1.01×10^5 Pa (= 1 atm) at the surface of the earth. Would the peak of Mt. Everest rise above the surface of such an ocean?

Section 16.4 The Equation of Continuity

16. A particular garden hose has an inside diameter of 5/8 in. The nozzle opening has a diameter of 1/4 inch. When the flow speed of the water in the hose is 0.43 m/s, what is the flow speed through the nozzle?

17. An electromagnet used in a research laboratory is cooled with water flowing through copper tubes (inside diameter = 0.5 cm). A pump that maintains the circulation is connected to the copper tubes by a flexible tube having an inside diameter of 1 cm. The flow speed of water in the copper tube is 5 m/s. Determine (a) the flow speed in the flexible tubing, (b) the discharge rate of the pump, and (c) the rate of mass flow from the pump.

18. A water pipe narrows from a diameter of 9.7 cm to 2.1 cm. The flow speed in the wider section is 0.10 m/s. Determine the flow speed in the narrower section.

19. A gas jet has a diameter of 2 mm. It is fed by a line with a diameter of 1 cm. The flow speed through the jet is 10 m/s. Determine the flow speed along the line.

20. Oil flows out of a pipe at a speed of 1.15 m/s. The diameter of the pipe is 5.12 cm. Determine the discharge rate.

21. Determine the discharge rate for gas flow in Example 4.

Section 16.5 Bernoulli's Equation

22. Water flows through a horizontal pipe that gradually narrows so that the final inside diameter is one-half the original diameter. In the wide section the flow speed is 4.1 m/s and the pressure is 10.4×10^2 kPa. Determine (a) the flow speed in the narrow section and (b) the pressure in the narrow section.

23. A J-shaped glass tube is connected to a pipe as shown in Figure 2. The tube is filled with mercury and the open end is exposed to the atmosphere where the pressure is $P_0 = 1.05 \times 10^5$ Pa. If the difference in levels of the mercury is 28 cm, what is the pressure of the air in the pipe?

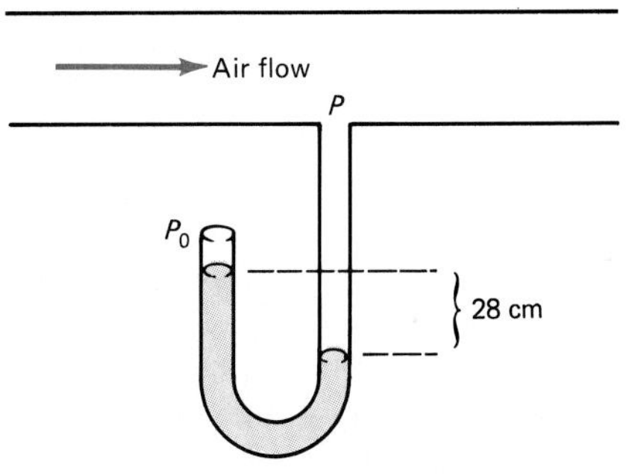

Figure 2

24. When a water tower is filled, the surface is 14 m above the exit pipe. (a) Assume that Torricelli's theorem (Eq. 16.20) applies, determine the exit flow speed. (b) How far above the exit pipe must the surface be in order to give an exit flow speed of 8.3 m/s?

25. A fountain in a shopping mall is designed to give a stream of water that rises to a height of 10 m. What is the gauge pressure in the pipe that feeds the fountain?

26. Imagine an open water tank filled to a depth H (Figure 3). If a hole is punched in the tank just slightly below the surface of the water, the stream of water through this hole will have a low speed and will strike the ground relatively close to the base of the tank. If a hole is punched in the tank near its base, the water will emerge from this hole with a higher speed, but will still strike the ground relatively close to the base of the tank. Presumably there is a point at which a hole can be punched so as to maximize the horizontal distance that the water travels. (a) Show that the horizontal distance the water travels is a maximum when the hole is punched at a point $H/2$ above the base of the tank. (b) Show that the maximum horizontal distance is equal to H. [*Hint:* Observe that $R = vt$ expresses the horizontal distance of travel in terms of the speed of the emerging water, v, and the time of flight, t, of the water. It is algebraically advisable to start with the equation $R^2 = v^2t^2$, and then use Bernoulli's equation and the free-fall equation (relating the time of flight, t, and the distance of vertical fall, y) to obtain the equation $R^2 = 4(H - y)y$. The condition

$$\frac{d(R^2)}{dy} = 0 \qquad \text{for maximum } R^2$$

leads to the value of y that maximizes R.

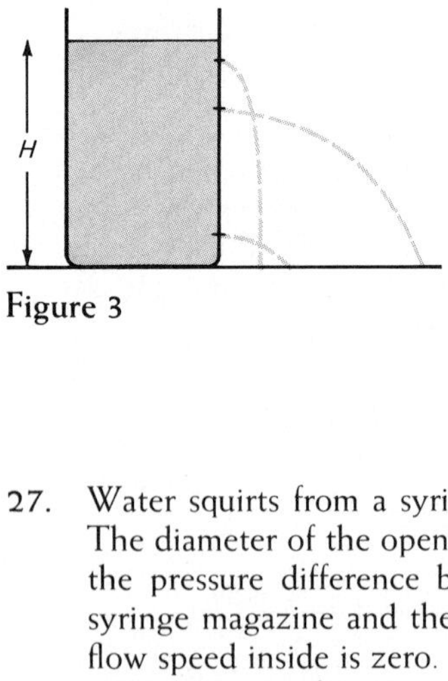

Figure 3

27. Water squirts from a syringe at a speed of 30 m/s. The diameter of the opening is 0.01 cm. Determine the pressure difference between the inside of the syringe magazine and the outside. Assume that the flow speed inside is zero. Express your result in pascals and atmospheres.

28. A toy squirt gun ejects water with a nozzle speed of 10 m/s. The diameter of the nozzle is 1 mm. The diameter of the water magazine is 1 cm. Determine the internal pressure exerted on the water. Perform *two* determinations. In the first, ignore the flow speed inside the magazine. Express the pressure in pascals and in atmospheres. In the second, include the internal flow speed in your calculation. Do the results differ if you work to three-significant-figure accuracy?

Section 16.6 Viscosity

29. The velocity profile for laminar flow in a pipe of radius R is given by

$$u(y) = u_o\left(1 - \frac{y^2}{R^2}\right)$$

where y is the distance from the axis of the pipe and u_o is the flow speed along the axis. (a) Where is the fluid velocity greatest? (b) At what position(s) does the velocity gradient have the largest magnitude? (c) Where is the shearing stress greatest?

30. A child tows a thin piece of wood through a water puddle at a speed of 15 cm/s. The depth of the water is 2 mm. The surface area of contact between water and wood is 200 cm^2. The viscosity of water is 1.0 centipoise. Assume that the velocity gradient (du/dy) is a constant from bottom to surface. Determine the horizontal component of the force exerted by the tow cord.

31. A sphere of radius r moving through a fluid of dynamic viscosity η at a speed v experiences a viscous force given by Stokes' law,

$$F_{\text{viscous}} = 6\pi\eta r v$$

Show that this equation is dimensionally consistent.

32. A particle falling through air accelerates until it reaches what is called the *terminal speed*. The terminal speed is reached when the downward gravitational force on the particle (its weight) equals the upward viscous force. Take the particle to be a sphere of radius r and assume that the viscous force is given by Stokes' law (see Problem 31). Show that the terminal speed is given by

$$v = \frac{2}{9}\frac{\rho g r^2}{\eta}$$

where ρ is the mass density of the particle and η is the viscosity of air.

33. When coal is burned in an electric power plant, particulate matter is released into the atmosphere. The particulates rise to heights of a kilometer or more and then gradually settle back to earth. Consider two particulates with diameters of 1 μm and 100 μm. The terminal (settling) speed of the 100-μm particulate is 25 cm/s. The terminal speed is proportional to the square of the particulate diameter (see Problem 32). Determine (a) the terminal speed of the 1-μm particulate and (b) the settling time of the 1-μm particulate, assuming it reaches its terminal speed at a height of 1 km.

Section 16.7 Turbulence

34. Prove that the Reynolds number is dimensionless.

35. The flow of helium through a particular pipe becomes turbulent at a flow speed of 30 m/s. At what

speed will the flow become turbulent for oxygen (through the same pipe at the same temperature and pressure)?

36. Oil flows through a pipe at a speed of 6 m/s. The kinematic viscosity of the oil is 8.0×10^{-6} m^2/s. The diameter of the pipe is 12 cm. (a) Determine the Reynolds number for the flow. (b) Would you expect the flow to be streamline or turbulent?

37. A sphere with a diameter of 36 mm moves at a speed of 3 m/s through oil with a density of 900 kg/m^3 and a viscosity of 0.10 kg/m · s. Determine the Reynolds number. Is the flow laminar or turbulent?

38. Air flows through a tube at a speed of 10 m/s. If the flow must be laminar, what is the maximum diameter that the pipe can have? (Take the critical Reynolds number to be 2300.)

39. The Reynolds number for flow through a pipe can be related to the discharge rate Q (see Eq. 16.9). Show that, for a pipe having a circular cross section of diameter d,

$$R = \frac{4Q}{\pi \nu d}$$

40. A golf ball with a radius of 2 cm moves through air at 55 m/s. Calculate the Reynolds number. Would you expect the flow to be laminar or turbulent?

41. Calculate the Reynolds number for the flow of water in the copper pipe of Problem 17. Is the flow laminar or turbulent?

chapter 17

WAVE KINEMATICS

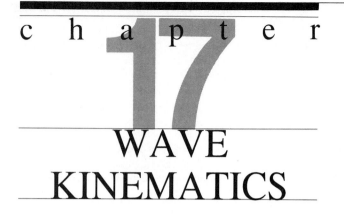

Preview

We learn about waves at an early age. As children we watch water waves sweep across ponds and see flags waving in the breeze. We are taught that the sun sends out light waves that illuminate our planet and provide the energy to nourish life. Shock waves from supersonic aircraft may rattle our windows. Seismic waves produced by earthquakes travel thousands of miles along the surface of the earth and through its interior. Physicists actively search for gravitational waves—faint "wrinkles" in the strength of gravity. Evidently, waves abound in nature.

This chapter is about the *kinematics* of waves. You will learn how to describe wave motions without reference to the particular physical nature of the wave. Sound waves, light waves, and water waves are quite different physically. Yet, each displays certain kinematic features that are identical or very similar.

In later chapters you will learn about specific types of waves. In Chapter 18 we study waves in solids and fluids, such as water waves and sound waves. In Chapter 38 we develop various aspects of electromagnetic waves. In Chapter 42 we encounter perhaps the most exotic waves of all—the "matter waves" or "probability waves" that are used to describe the atomic and nuclear structure of matter.

17.1 Wave Characteristics

A wave is a disturbance that travels at a definite speed. For example, a stone thrown into a pond disturbs the smooth pond surface. This disturbance moves across the surface of the pond as a circular wave.

Waves are of great interest to scientists and engineers because they transfer energy and momentum. In the example just mentioned, the waves transfer energy and momentum from the point where the stone strikes the surface to the edge of the pond. The surface of the pond returns to its undisturbed condition after the waves pass. There is a transfer of energy and momentum but no transfer of mass.

A water wave is one type of *mechanical* wave. In a mechanical wave matter experiences an oscillatory displacement from equilibrium. Mechanical forces acting on and between adjacent elements of the matter cause the disturbance to advance at a definite speed. The matter is the *medium* through which the wave travels.

There are waves that do not require a material medium to support their travel. For example, light waves (electromagnetic waves) travel through the reaches of empty space between the earth and the stars. In electromagnetic waves the "disturbances" are oscillatory *electric* and *magnetic fields* (Chapters 27 and 32). No mechanical forces need act in order to propagate electromagnetic waves. Gravitational waves can also travel through empty space. Gravitational waves are oscillatory variations in a *gravitational field,* and they travel at the speed of light. As yet there is no clear-cut experimental evidence that gravitational waves exist. A number of scientific research teams are currently trying to detect gravitational waves, which presumably are generated in various astrophysical events, such as the supernova explosion of an aging star.

Figure 17.1a

In a transverse mechanical wave the oscillatory displacement of the medium is perpendicular to the direction of wave travel.

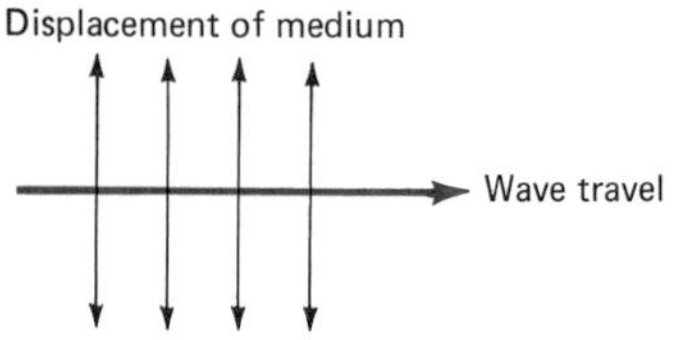

Figure 17.1b

In a longitudinal mechanical wave the oscillatory displacement of the medium is parallel to the direction of wave travel.

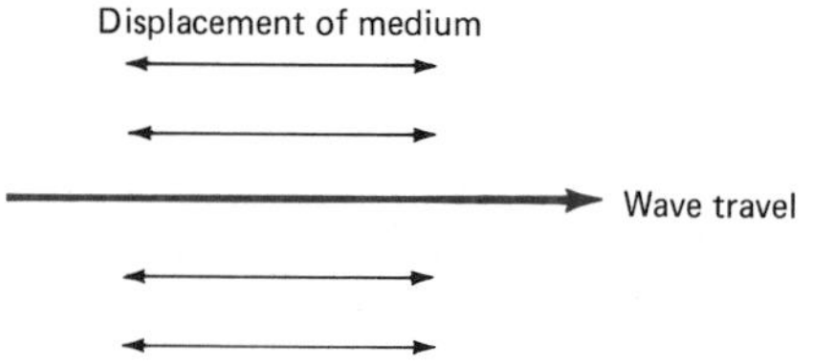

Many waves may be classified as either *transverse* or *longitudinal*. In a transverse mechanical wave the displacement of the medium is perpendicular to the direction of wave travel (Figure 17.1a). In our example, the wave moving across the pond appears to be transverse. We see the surface of the water rise and fall as the waveform moves horizontally. However, water waves are not purely transverse. The surface of the water also moves to and fro in the horizontal direction, although this component of the motion is not usually apparent to the casual observer. In a longitudinal wave the displacement is parallel to the direction of wave travel (Figure 17.1b). Sound waves in air are longitudinal. The "displacements" are variations in the air pressure. These pressure variations are parallel to the direction in which the sound travels. A helical spring can be used to demonstrate longitudinal wave motion. The wave is seen as moving variations in the spacing of the turns of the helix (Figure 17.2).

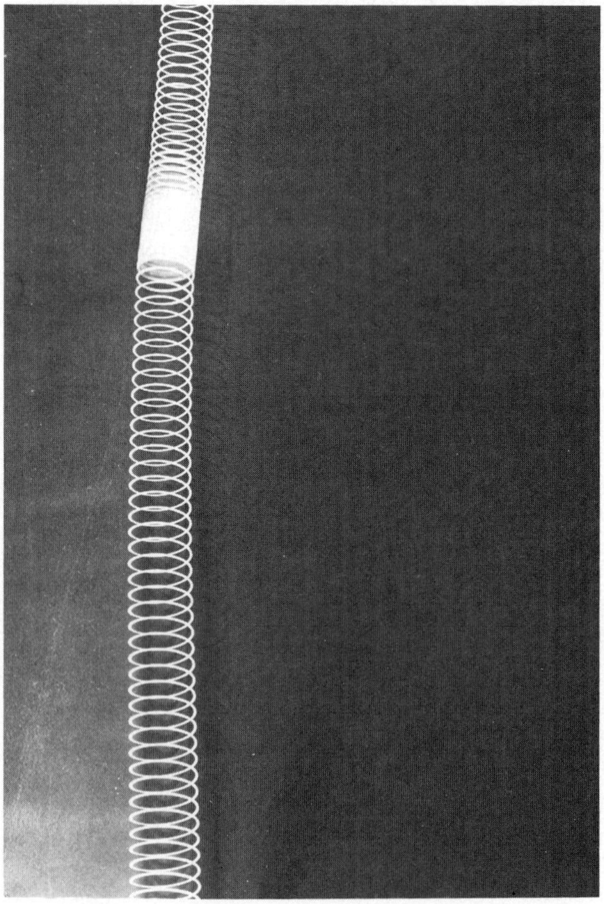

Figure 17.2

A longitudinal wave travels along the helical spring of a Slinky.

Another aspect of a wave is its *polarization*. A wave is said to be polarized along the direction of the disturbance. Therefore, the polarization of a longitudinal wave is parallel to the direction of wave motion. The polarization of longitudinal waves is of little practical importance because it cannot be changed or controlled. In a transverse wave the disturbance (and thus the polarization) lies in a plane perpendicular to the direction of wave travel. The direction of polarization can be controlled in some transverse waves. For example, a transverse wave can be generated by shaking one end of a flexible cord or string (Figure 17.3). For a wave on a string advancing along the x-axis the polarization lies in the y–z plane. Figure 17.3 shows two possible polarizations, one along the z-axis and one along the y-axis. That the polarization of light waves can be controlled has many practical applications. For example, sunglasses that reduce the glare of reflected light function because such reflected light is selectively polarized. The polarization of light is dealt with in Chapter 38 (Section 38.6).

To avoid making our study of wave motion unnecessarily abstract, we frequently refer to the type of wave called a *wave on a string*. This is the kind of wave that is generated by shaking one end of a flexible cord or string (Figure 17.3). Waves on a string propagate in one dimension, that is, along the length of the string. The description of one-dimensional waves requires just one space variable. In contrast, the water waves generated by throwing a stone into a pond are two-dimensional waves—they spread out over a surface. The flash of light generated by a firefly travels outward in all directions as a three-dimensional pulse. In general, the mathematical description of two- and three-dimensional waves requires two and three space variables, respectively.

Consider a wave on a string traveling in the positive x-direction at a constant speed (see Figure 17.4). Let $\psi(x,t)$* denote the height of the wave at a point x at time t. The shape of the waveform describes how ψ depends on x and t. In this discussion we ignore the effects of absorption and dispersion. Absorption reduces the wave height, and dispersion changes the wave shape. Thus, we treat a wave that maintains its shape and it height.

We can imagine all sorts of wave shapes, each expressing ψ in terms of x and t. However, there is one very special relationship between x and t for all waves: For a

*The Greek character ψ (psi, pronounced sī) is the 23rd letter of the Greek alphabet.

Figure 17.3

A wave on a string is transverse. The motion of points along the string is perpendicular to the direction of wave travel. We describe this by saying that the *polarization* of the wave is in the plane perpendicular to the direction of wave travel. In particular, the polarization can be along either of two mutually perpendicular axes which are themselves perpendicular to the direction of wave travel. In (a) and (b) the wave travels along the x-axis. In (a) the wave is polarized along the z-axis. In (b) the wave is polarized along the y-axis.

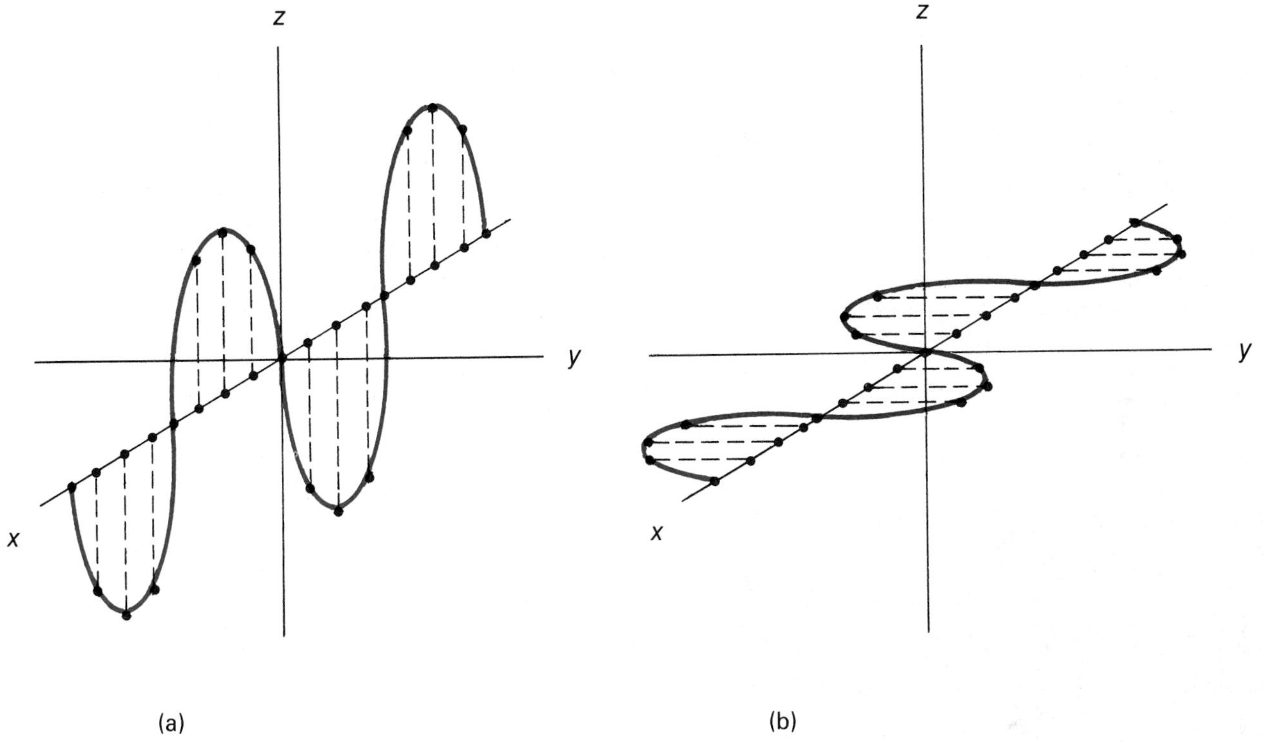

Figure 17.4

The waveform advances at speed v. If the crest is at x_0 at time $t = 0$, it will advance to $x_0 + vt$ in time t.

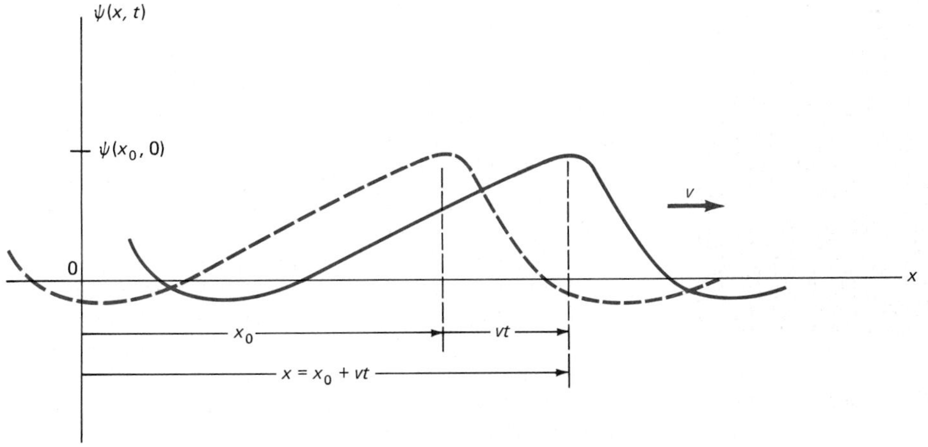

Figure 17.5

The waveform $\psi(x,t) = \dfrac{a^3}{a^2 + (x - vt)^2}$ with $a = 1$ cm and $v = 3$ cm/s is shown at three times, $t = 0$, $t = 1$ s, and $t = 2$ s. The wave moves to the right without change of shape.

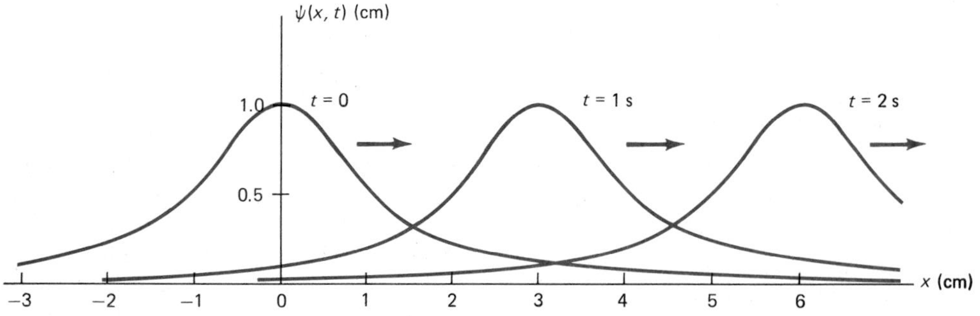

wave traveling in the positive x-direction at speed v the variables x and t *must* occur in $\psi(x,t)$ *only* as the combination $x - vt$. Symbolically,

$$\psi(x,t) = \psi(x - vt) \qquad (17.1)$$

Figure 17.5 illustrates how the $x - vt$ combination results in wavelike behavior for

$$\psi(x,t) = \frac{a^3}{a^2 + (x - vt)^2} \qquad (17.2)$$

where $a = 1$ cm and $v = 3$ cm/s. Note that x and t occur as the combination $x - vt$. Figure 17.5 shows that the crest of the wave is at $x = 0$ at $t = 0$. The height of the crest is 1 cm.

$$\psi(x = 0,\ t = 0) = \frac{a^3}{a^2 + (0 - 0)^2} = a = 1 \text{ cm}$$

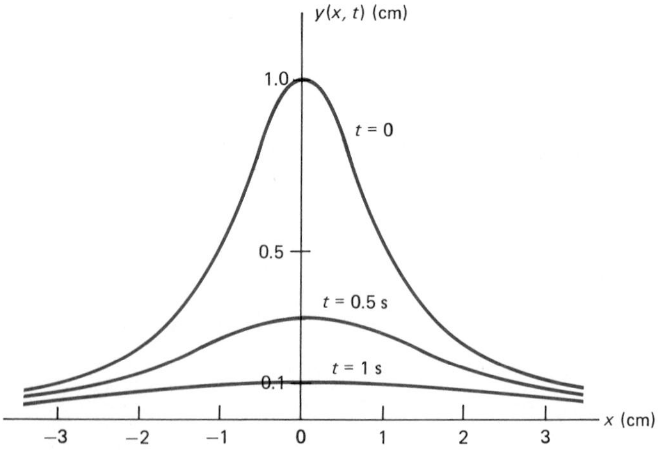

Figure 17.6

The function $y(x,t) = \dfrac{a^3}{a^2 + x^2 + v^2t^2}$ for $a = 1$ cm, $v = 3$ cm/s is shown at three times, $t = 0$, $t = 0.5$ s, and $t = 1$ s. The peak remains at $x = 0$ and "flattens out" as t increases. The function $y(x,t)$ does not describe a wave.

At a speed of $v = 3$ cm/s the entire waveform—crest included—has moved 3 cm to the right at $t = 1$ s. Thus, setting $x = 3$ cm, $v = 3$ cm/s, and $t = 1$ s in Eq. 17.2 gives

$$\psi(x = 3 \text{ cm},\ t = 1 \text{ s}) = \frac{a^3}{a^2 + (3 - 3)^2} = a = 1 \text{ cm}$$

At $t = 2$ s the crest is at $x = 6$ cm. In general, the crest is at $x = vt$ at time t.

For contrast, consider the behavior of

$$y(x,t) = \frac{a^3}{a^2 + x^2 + v^2t^2} \qquad (17.3)$$

where again $a = 1$ cm and $v = 3$ cm/s. The form of $y(x,t)$ looks similar to that of $\psi(x,t)$, as given by Eq. 17.2. However, $y(x,t)$ does not describe a wave: The variables x and t do not occur as the combination $x - vt$. Figure 17.6 shows $y(x,t)$ for $t = 0$, 1 s, 2 s. The peak of $y(x,t)$ remains at $x = 0$. Clearly, $y(x,t)$ does not describe a wave.

We can test any proposed wavelike disturbance by checking to see if it involves x and t *only* as the combination $x - vt$. Possible waveforms *include*

$$\psi(x,t) = A \sin[k(x - vt)]$$

and

$$\psi(x,t) = \frac{a^5}{(x - vt)^4 + a^4}$$

but *exclude*

$$\psi(x,t) = A \sin\left(\frac{kx^2}{vt}\right)$$

and

$$\psi(x,t) = \frac{a^3}{xvt + a^2}$$

The first two equations describe waves that advance at speed v; the latter two do not. When we say that ψ is a function of $x - vt$, we mean that x and t can appear only in the combination $x - vt$. No other combination can describe a wave traveling in the positive x-direction at speed v. A similar argument can be used to show that the height of a wave traveling in the negative x-direction at speed v must be described by a function of the combination $x + vt$.

Questions

1. Cite examples illustrating that waves transfer energy without any transfer of matter.
2. Water waves are very nearly transverse. What is their direction of polarization? Is it possible to control their polarization?
3. Waves may lose energy to the medium through various absorption processes. How would absorption affect the amplitude of a wave?
4. A particle carries energy and momentum. How does a particle differ from a wave?
5. A raindrop strikes the surface of a puddle and generates a circular wave. Trace the kinetic and potential energy transfers, starting with the falling drop and ending with the beaching of the wave.

17.2 Sinusoidal Waves

A *waveform* is described by the function $\psi(x,t)$. Sinusoidal waveforms are particularly important. They have the forms

$$\psi(x,t) = A \sin\left[\frac{2\pi}{\lambda}(x - vt)\right] \quad (17.4)$$

$$\psi(x,t) = B \cos\left[\frac{2\pi}{\lambda}(x - vt)\right] \quad (17.5)$$

The quantity λ is the wavelength, and is defined below. The importance of sinusoidal waveforms stems from two facts:

1 Sinusoidal waveforms adequately describe many types of waves observed in nature.

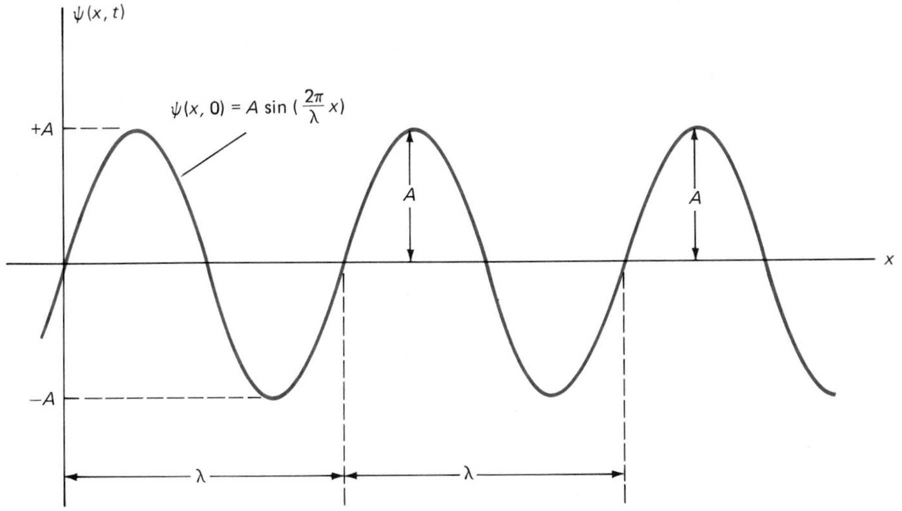

Figure 17.7

The waveform $\psi(x,t) = A \sin\left[\frac{2\pi}{\lambda}(x - vt)\right]$ at time $t = 0$ has the form $\psi(x,0) = A \sin(2\pi x/\lambda)$.

2 More complicated waveforms can be synthesized through the superposition (addition) of sinusoidal waves.

Our immediate concern is with the kinematic aspects of sinusoidal waves. Figure 17.7 shows the waveform

$$\psi = A \sin\left[\frac{2\pi}{\lambda}(x - vt)\right]$$

at $t = 0$. From the figure, or the equation

$$\psi(x,t = 0) = A \sin\frac{2\pi x}{\lambda} \quad (17.4a)$$

we see that A is the maximum value of ψ. We define A to be the **amplitude** of the wave. Geometrically, A is the height of the wave crests above the x-axis. The **wavelength,** λ, is the distance over which the waveform repeats itself. The **period,** T, of the wave can be defined as the time between the passage of successive crests, or equivalently, as the time required for the waveform to advance through a distance λ. Since the waveform moves at speed v, it advances a distance vT in time T. Thus, the wavelength λ is related to the speed and period by

$$\lambda = vT \quad (17.6)$$

Equation 17.6 is the basic kinematic relation for sinusoidal waves.

Example 1 Determining the Specific Form of $\psi(x,t)$

A transverse sinusoidal wave travels along a rope at a speed of 5 m/s. The wavelength is 4 m and the amplitude is 0.05 m. The height of the wave is zero at $x = 0$, $t = 0$. We can insert this information ($v = 5$ m/s, $\lambda = 4$ m, $A = 0.05$ m) into Eq. 17.4 to obtain

$$\psi(x,t) = A \sin\left[\frac{2\pi}{\lambda}(x - vt)\right]$$

$$= 0.05 \sin\left[\frac{\pi}{2}(x - 5t)\right] \text{ m}$$

Table 17.1

Relationships among wave properties

Name	Wavelength	Wave Number	Period	Frequency	Radian Frequency	Velocity
Symbol	λ	k	T	ν	ω	v
Definition	$\frac{2\pi}{\lambda} = k$		$\frac{1}{T} = \nu$		$2\pi\nu = \omega$	
Basic Kinematic Relation (Text Eq.)		$\lambda = vT$ (17.6)		$\nu\lambda = v$ (17.8)	$\frac{\omega}{k} = v$ (17.13)	
Equivalent Expressions for the Sinusoidal Waveform $\psi(x,t)$	$\psi(x,t) = A \sin\left[\frac{2\pi}{\lambda}(x - vt)\right]$					(17.4)
	$\psi(x,t) = A \sin\left[2\pi\left(\frac{x}{\lambda} - \frac{t}{T}\right)\right]$					(17.9)
	$\psi(x,t) = A \sin(kx - \omega t)$					(17.12)

The height of the wave at any point and time is found by inserting the appropriate values of x and t. Thus, at $x = 6$ m, $t = 1$ s,

$$\psi(6\text{ m}, 1\text{ s}) = 0.05 \sin\left[\frac{\pi}{2}(6 - 5)\right] = 0.05\text{ m}$$

which shows that there is a crest at $x = 6$ m at $t = 1$ s.

The **frequency** of the wave, ν, is the number of crests that pass an observer per second. If the time between crests (T) is 0.1 s, ten crests pass per second. Thus, the frequency is the reciprocal of the period:

$$\nu = \frac{1}{T} \qquad (17.7)$$

The SI unit of frequency, 1 cycle per second (cps), is called the *hertz* (abbreviated Hz), in honor of Heinrich Hertz, a German physicist.* We can substitute the reciprocal of the frequency, $1/\nu$, for T in the basic kinematic relation $\lambda = vT$. This gives us a second form of the kinematic relation,

$$\nu\lambda = v \qquad (17.8)$$

The two forms of the kinematic relation can be used to represent sinusoidal waves in alternate forms. For example, using Eq. 17.6 converts Eq. 17.4 into

$$\psi(x,t) = A \sin\left[2\pi\left(\frac{x}{\lambda} - \frac{t}{T}\right)\right] \qquad (17.9)$$

Other useful representations of the sinusoidal waveform are obtained by introducing the **wave number** (k),

$$k = \frac{2\pi}{\lambda} \qquad (17.10)$$

measured in radians per meter, and the **radian frequency** (ω),

$$\omega = \frac{2\pi}{T} = 2\pi\nu \qquad (17.11)$$

measured in radians per second.* Expressed in terms of k and ω, Eq. 17.9 becomes

$$\psi(x,t) = A \sin(kx - \omega t) \qquad (17.12)$$

Example 2 **Determining T, ν, k, and ω for the Wave of Example 1**

For the wave of Example 1 we can use Eq. 17.6 to establish the period

$$T = \frac{\lambda}{v} = \frac{4\text{ m}}{5\text{ m/s}} = 0.8\text{ s}$$

The frequency of the wave is

$$\nu = \frac{1}{T} = 1.25\text{ cps} = 1.25\text{ Hz}$$

The wave number, k, is

$$k = \frac{2\pi}{\lambda} = 1.57\text{ rad/m}$$

The radian frequency, ω, is

$$\omega = 2\pi\nu = 7.85\text{ rad/s}$$

According to Eq. 17.6 the wave speed is given by $v = \lambda/T$. By combining Eqs. 17.10 and 17.11 we obtain yet another form of the kinematic relation, namely,

$$v = \frac{\omega}{k} \qquad (17.13)$$

*In 1887, Hertz generated and detected the electromagnetic waves predicted in 1865 by James Clerk Maxwell.

*Radian frequency was introduced in Chapter 14, in connection with harmonic motions.

Table 17.1 summarizes the relationships among the many wave properties.

So far we have focused our attention on the wave—the disturbance that moves through the medium. What about the medium? What sort of motion does it undergo? In particular, if a wave having the form $\psi = A\sin(kx - \omega t)$ moves along a rope, what sort of motion does the rope execute? The answer is that each point along the rope undergoes simple harmonic motion (see Chapter 14). This is most readily established by observing the point $x = 0$. The motion of the point $x = 0$ is given by

$$\psi(0,t) = A\sin(-\omega t) = -A\sin\omega t \qquad (17.14)$$

This equation describes simple harmonic motion at radian frequency ω and amplitude A.

It is also important to distinguish between the speed at which the wave moves and the velocity of the rope. The wave in Example 1 moves in the x-direction at a constant speed of 5 m/s. Points along the rope move perpendicular to the x-direction with a velocity that varies with time. The vertical velocity of the rope at a position x is given by $\partial\psi/\partial t$, the time derivative of $\psi(x,t)$ with x held constant (This time derivative is computed according to the usual rules. The independent variable x is treated as a constant.) Thus,

$$v_{\text{rope}} = \frac{\partial\psi}{\partial t} = -\frac{2\pi v}{\lambda}A\cos\left[\frac{2\pi}{\lambda}(x - vt)\right]$$

Example 3 **Wave Speed and Rope Velocity**

The velocity of the rope in Example 1 at the position $x = 0$ at the time $t = 0$ is

$$v_{\text{rope}} = -\frac{2\pi\cdot 5}{4}(0.05)\cos[0] = -0.393 \text{ m/s}$$

The wave speed is $v = 5$ m/s. Thus at $x = 0$, $t = 0$ the rope is moving downward at 0.393 m/s while the waveform is moving forward at 5 m/s.

Questions

6. How could you determine the frequency of water waves in a harbor by observing a floating bottle?
7. A sinusoidal wave has a frequency of 60 Hz. What is its period?

17.3 Phase and Phase Difference

For the sinusoidal waveform $\psi(x,t) = A\sin(kx - \omega t)$, the argument of the sine function

$$kx - \omega t \equiv \phi \qquad (17.15)$$

is referred to as the *phase angle*, or more briefly as the *phase*. At any given time the phase, ϕ, is different for each point along the waveform. For example, at a fixed time t, the points x_1 and x_2 differ in phase by

$$\begin{aligned}\text{phase difference} &= \phi_2 - \phi_1\\ &= (kx_2 - \omega t) - (kx_1 - \omega t)\\ &= k(x_2 - x_1)\end{aligned}$$

or

$$\begin{aligned}\text{phase difference} &= k(x_2 - x_1)\\ &= \frac{2\pi(x_2 - x_1)}{\lambda}\end{aligned} \qquad (17.16)$$

If we let $\Delta\phi$ stand for the phase difference ($\phi_2 - \phi_1$, in radians) and let Δx stand for the spatial separation along the waveform ($x_2 - x_1$), we can write

$$\Delta\phi = 2\pi\left(\frac{\Delta x}{\lambda}\right) \qquad (17.17)$$

This relationship between phase difference and spatial separation along the waveform is used extensively in the study of optics (Chapters 39–41).

Example 4 **Phase Difference and Path Difference**

A tuning fork generates sound waves with a frequency of 256 Hz. The waves travel in opposite directions along a hallway, are reflected by walls, and return. What is the phase difference between the reflected waves when they meet? The corridor is 47 m long and the tuning fork is located 14 m from one end. The speed of sound in air is 330 m/s.

The kinematic relation $\nu\lambda = v$ lets us determine the wavelength of the sound waves.

$$\lambda = \frac{v}{\nu} = \frac{330 \text{ m/s}}{256 \text{ c/s}} = 1.29 \text{ m}$$

The phase difference between the reflected waves is given by

$$\Delta\phi = 2\pi\left(\frac{\Delta x}{\lambda}\right)$$

where Δx is the *difference* in *path length* for the two waves. One wave travels 14 m, is reflected, and returns 14 m. The other travels 33 m, is reflected, and returns 33 m. The difference in their path lengths is

$$\Delta x = 66 \text{ m} - 28 \text{ m} = 38 \text{ m}$$

The phase difference between the two is

$$\Delta\phi = 2\pi\left(\frac{38 \text{ m}}{1.29 \text{ m}}\right) = 2\pi(29.5) = 185 \text{ rad}$$

It follows from the relation $\Delta\phi = 2\pi(\Delta x/\lambda)$ that two points separated by one wavelength differ in phase by 2π (Figure 17.8). However, because the waveform repeats over a distance of one wavelength, the corresponding

Figure 17.8

A physical separation of one wavelength is equivalent to a phase difference of 2π radians.

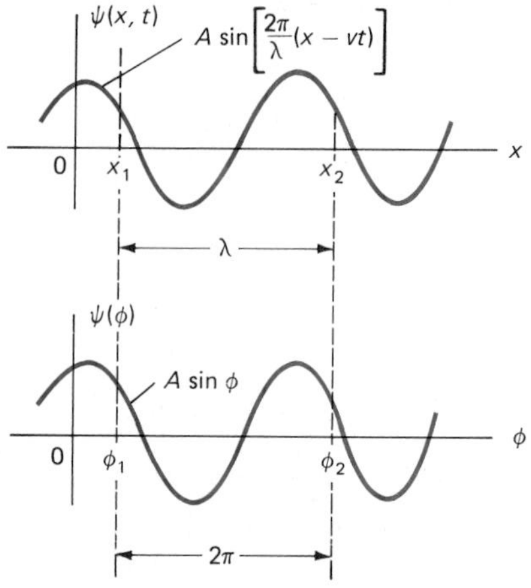

phase difference of 2π is equivalent to zero phase difference. In other words, a displacement of one full wavelength or a phase difference of 2π leaves the waveform unchanged. This reasoning can be extended to show that

1 phase differences that are *integral multiples* of 2π are equivalent to zero phase difference, and

2 any phase angle or phase difference can be increased or reduced by an integral multiple of 2π without affecting subsequent calculations and results.

For example, the phase difference in Example 4 was $2\pi(29.5)$. We can ignore the $2\pi(29)$ and concern ourselves only with $2\pi(0.5) = \pi$, without affecting subsequent calculations involving $\Delta\phi$.

Example 5 **How to Find a Wounded Elephant**

Imagine that you are an African game warden searching for a wounded elephant. You wander aimlessly until you hear the elephant's distress call. Then, confident that you will find the animal, you set off in the direction from which the sound came. If you have normal hearing, you will be able to determine the direction of a source of sound rather accurately. When you face the source, sound waves entering your ears have zero phase difference. If you turn your head as little as 3° to either side of the direction of the source there is a noticeable phase difference (Figure 17.9). If your ears are separated by a distance L, then rotating your head through a small angle θ will result in a path difference of $L\theta$. The corresponding phase difference for waves of length λ is

$$\text{phase difference} = 2\pi\left(\frac{L\theta}{\lambda}\right)$$

If we take $\theta = 3°$ we get the minimum detectable phase difference. If the elephant trumpets at a wavelength of 1 m (frequency = 330 Hz) and we take $L = 20$ cm, $\theta = 3° = 0.052$ rad, we find that the minimum detectable phase

Figure 17.9

Turning the head through a small angle θ results in a path difference of approximately $L\theta$ between waves reaching the two ears simultaneously. For sound with a frequency of 330 Hz the minimum detectable phase difference corresponds to an angle $\theta \simeq 3°$.

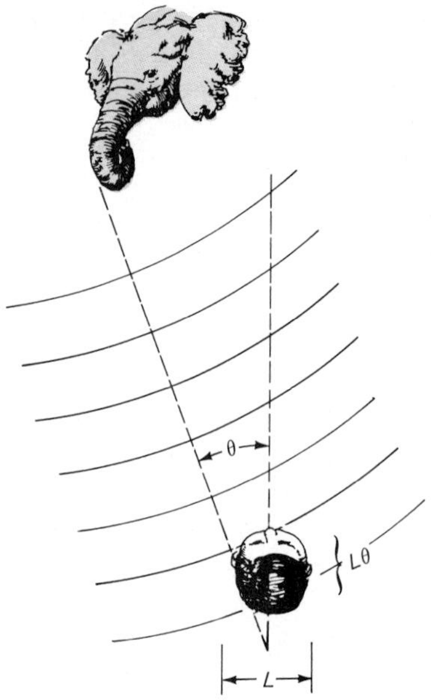

difference is 0.068 rad. An alternate way of emphasizing this high degree of directional sensitivity is to note that $L\theta/\lambda$ is the path difference expressed in wavelengths. For these data, $L\theta/\lambda = 0.0104$. In other words, at a wavelength of 1 m, your ears can detect a path difference of about 1 cm.

Questions

8. What is the smallest positive phase angle that is equivalent to $\phi = (9/2)\pi$?
9. Waves travel along a string at a speed v. If you were to run alongside the string at speed v, in the same direction as the waves travel, how would the waveform appear to you?

17.4 The Principle of Superposition for Waves

There are many instances in which two or more waves meet. For example, raindrops strike the surface of a pond and generate circular waves that meet each other. The meetings do not alter the individual waves. The waves meet, and then travel on unaffected. In other words, the individual waves move independently. We express this independence by saying that the waves obey a **principle of superposition.** As you may recall from Chapter 5, we used the superposition of x and y motions to build up the parabolic path of a uniformly accelerated particle. We also encountered a principle of superposition for forces in Chapters 2 and 3. The principle of superposition for forces states that the simultaneous application of two or more forces is equivalent to a single force equal to the vector sum of the individual forces. The principle of superposition for waves can be phrased in similar terms:

A meeting of two or more waves produces a waveform that is the algebraic sum of the waveforms produced by each wave acting separately.

The principle of superposition can be stated mathematically as follows:

If ψ_1 and ψ_2 are the waveforms produced by two sources of waves acting separately, then $\psi_1 + \psi_2$ is the waveform produced when the two sources act together.

Standing Waves

Figure 17.10 illustrates the principle of superposition for waves on a string. For the superposition of sinusoidal waveforms, we can use the trigonometric addition relations to determine the resultant waveform. Consider, for example,

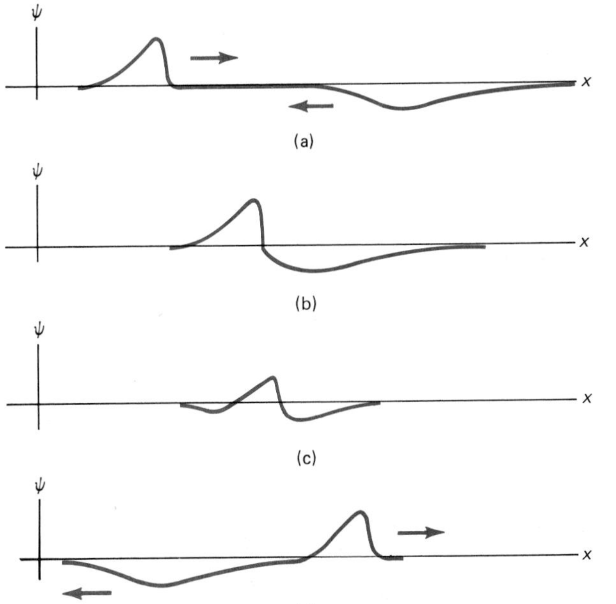

Figure 17.10

The superposition of waves on a string. (a) Waveforms traveling in OPPOSITE directions meet. (b) (c) Their superposition (c) gives a resultant waveform that is the algebraic sum of the two individual waveforms. (d) The waveforms separate; their amplitudes and speeds are unchanged by the superposition.

the superposition of two sinusoidal waveforms having equal amplitudes, wavelengths, and frequencies, but traveling in opposite directions.

$$\psi_1 = A\sin(kx - \omega t) \qquad \psi_2 = A\sin(kx + \omega t)$$

The trigonometric relations

$$\sin(kx - \omega t) = \sin kx \cos \omega t - \sin \omega t \cos kx$$

$$\sin(kx + \omega t) = \sin kx \cos \omega t + \sin \omega t \cos kx$$

give

$$A\sin(kx - \omega t) + A\sin(kx + \omega t)$$

$$= 2A \sin kx \cos \omega t \qquad (17.18)$$

The superposition of ψ_1 and ψ_2 therefore results in

$$\psi = \psi_1 + \psi_2 = 2A \sin kx \cos \omega t \qquad (17.19)$$

Figure 17.11 shows the result of this superposition. It is an oscillation ($\cos \omega t$) with an amplitude ($2A \sin kx$) that varies with position. This wave is called a *standing wave* because there is no motion of the disturbance along the x-direction. A standing wave is distinguished by stationary positions of zero displacement called *nodes*. In Figure 17.11 there are nodes at both ends and at the center. A more detailed study of standing waves is presented in Chapter 18.

Not all waves obey the principle of superposition. For example, large-amplitude waves in fluids and solids interact in a complicated nonlinear fashion. Thus, shock waves generated by jet aircraft and ocean surf do not obey the principle of superposition. In general, only experiment can determine whether or not the principle of superposition is applicable to a given situation. In general, we can say that the smaller the amplitudes of the waves, the more precisely valid is the principle of superposition. This is true because a small-amplitude wave alters the properties of the medium so slightly that the change does not affect a second small-amplitude wave, and the superposition principle applies.

Interference of Waves

The superposition of waves can result in what we call *interference*. At points where superposition causes the waves to oppose one another we say there is *destructive interference*. At points where superposition causes the waves to reinforce one another we say there is *constructive interference*.

Various types of mirrors and lenses, both optical and acoustic, focus waves at a point (ideally) where they exhibit constructive interference. For example, the parabolic "dishes" in evidence along the sidelines of televised football games reflect sound waves to a microphone. The microphone is mounted at a point where the waves interfere constructively. Similar reflecting dishes are used to collect television signals beamed from communication satellites (Figure 17.12). Waves reflected from the dish interfere constructively at the point where the receiving horn is positioned.

The interference of light waves has important applications. We will make frequent use of the ideas of constructive and destructive interference when we study optical interference in Chapters 40 and 41.

Figure 17.11

Superposition of sinusoidal waves traveling in opposite directions results in a standing wave. The wave $\psi_1 = A \sin(kx - \omega t)$, Figure 17.11a, travels to the right. The wave $\psi_2 = A \sin(kx + \omega t)$, Figure 17.11b, travels to the left. The "multiple exposure," Figure 17.11c, shows $\psi_1 + \psi_2$ at different times. Their superposition, $\psi_1 + \psi_2 = 2 \sin kx \cos \omega t$, is called a standing wave because there is no motion of the disturbance along the x-direction.

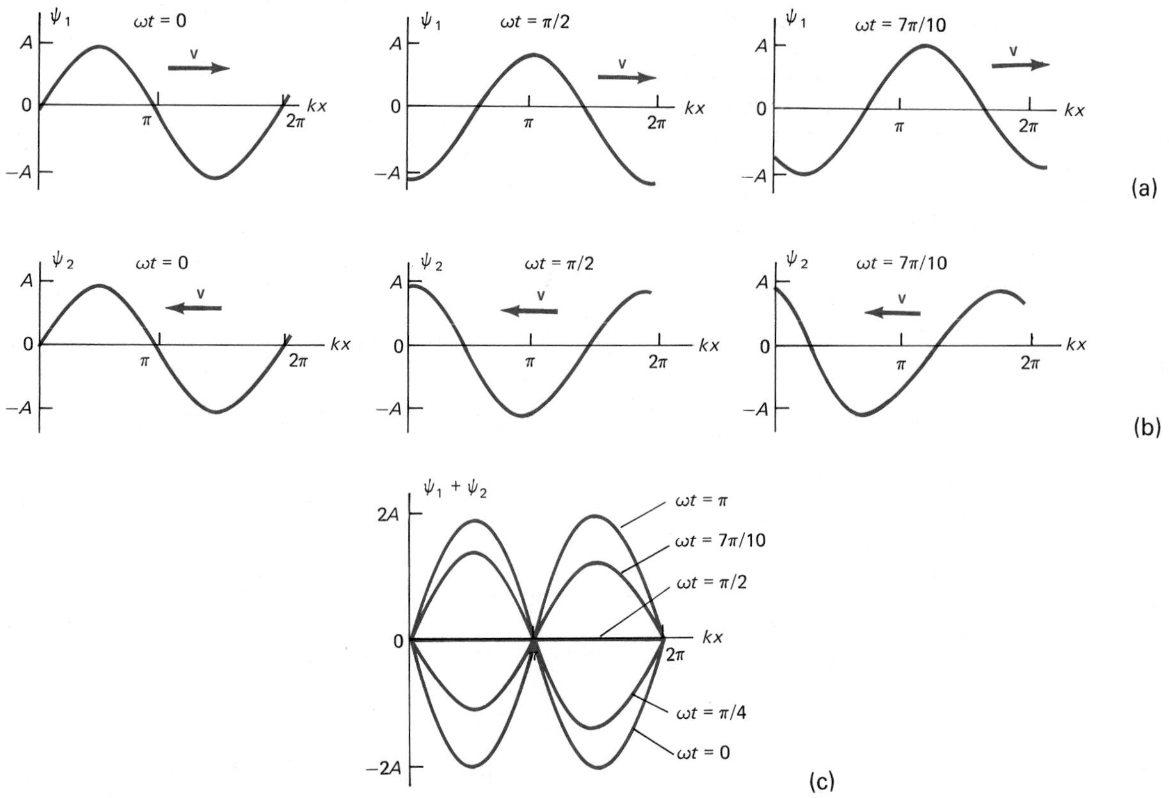

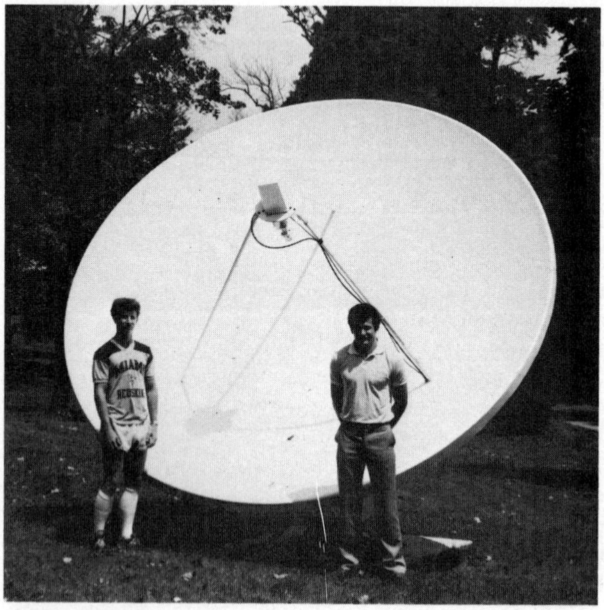

Figure 17.12

A receiver for television signals transmitted from a communications satellite. Waves reflected by the "dish" interfere constructively at the horn.

Questions

10. Explain the periodic fading of a television signal that sometimes results when an airplane passes near the receiving antenna.

11. Does destructive interference result in a loss of energy? If so, where does the energy go?

12. A principle of superposition does not hold for all waves (e.g., shock waves and surf). Can such waves still exhibit interference effects?

17.5 Beats

The superposed sound waves produced by two identical tuning forks are heard by the human ear as a single pure tone. However, if the frequency of one fork is altered slightly by "loading" it with a piece of wax, the resulting sound exhibits periodic variations in loudness called *beats*. (Ask your instructor to demonstrate this, or do it yourself.) The beats are a result of the periodic swings between constructive interference (louder sound) and destructive interference (fainter sound). Beat phenomena are not restricted to sound waves; they are produced whenever two waves with slightly different frequencies are superposed.

We can analyze the beat phenomenon by considering the superposition of the waves

$$\psi_1 = A\cos(k_1x - \omega_1t) \text{ and } \psi_2 = A\cos(k_2x - \omega_2t)$$

At the fixed point $x = 0$ (chosen for convenience), the waveform is

$$\psi = \psi_1 + \psi_2 = A(\cos\omega_1t + \cos\omega_2t) \qquad (17.20)$$

From this equation we can see that at $t = 0$ the waves interfere constructively and give a wave of amplitude $2A$. If we suppose that $\omega_2 > \omega_1$, then the phase angle ω_2t increases more rapidly than the phase angle ω_1t. The *phase difference* of the waves is $(\omega_2 - \omega_1)t$. When the phase difference reaches π (or 180°) the waves interfere destructively and $\psi = 0$. For example, if $\omega_2t = 2\pi$ and $\omega_1t = \pi$, then $(\omega_2 - \omega_1)t = \pi$ and

$$\psi = A(\cos\pi + \cos 2\pi)$$
$$= A(-1 + 1) = 0$$

The phase difference increases linearly with time. At the instants when the value of the phase difference is 0, 2π, 4π, 6π, and so on, there is maximum constructive interference. At those instants when the phase difference has the values π, 3π, 5π, 7π, and so on, there is complete destructive interference ($\psi = 0$). We can demonstrate this mathematically by writing

$$\omega_2t = \omega_1t + (\omega_2 - \omega_1)t \qquad (17.21)$$

Then

$$\psi = A\{\cos\omega_1t + \cos[\omega_1t + (\omega_2 - \omega_1)t]\} \qquad (17.22)$$

The quantity

$$\Delta\phi = (\omega_2 - \omega_1)t \qquad (17.23)$$

is the phase difference between ψ_1 and ψ_2. This phase difference increases linearly with time. Maximum amplitude occurs at times when

$$(\omega_2 - \omega_1)t = 0, 2\pi, 4\pi, \ldots \qquad \psi = 2A\cos\omega_1t$$

Destructive interference occurs at those instants for which

$$(\omega_2 - \omega_1)t = \pi, 3\pi, 5\pi, \ldots \qquad \psi = 0$$

The time between successive moments of constructive interference is called the **beat period,** T_{beat}. In a time T_{beat} the phase difference increases by 2π. Thus

$$(\omega_2 - \omega_1)T_{beat} = 2\pi \qquad (17.24)$$

In general, if the two radian frequencies differ by $\Delta\omega = \omega_2 - \omega_1$, we can write

$$T_{beat} = \frac{2\pi}{\omega_2 - \omega_1} = \frac{2\pi}{\Delta\omega} \qquad (17.25)$$

The number of beats per second, called the **beat frequency** (ν_{beat}), is the reciprocal of the beat period

$$\text{beat frequency} = \nu_{beat} = \frac{\omega_2 - \omega_1}{2\pi}$$

Recall that $\omega = 2\pi\nu$ relates the ordinary and radian frequencies. This relation allows us to replace $(\omega_2 - \omega_1)/2\pi$ by $\nu_2 - \nu_1$. Thus,

$$\nu_{beat} = \nu_2 - \nu_1 \qquad (17.26)$$

The beat frequency equals the difference between the two superposed frequencies.

An alternate way of displaying the beat oscillations makes use of the trigonometric equations

$$\cos(x + y) = \cos x\cos y - \sin x\sin y$$
$$\cos(x - y) = \cos x\cos y + \sin x\sin y$$

Using these equations we can express $\psi = A(\cos\omega_1t + \cos\omega_2t)$ as

$$\psi = 2A\cdot\cos\left(\frac{\omega_2 - \omega_1}{2}\right)t\cdot\cos\left(\frac{\omega_2 + \omega_1}{2}\right)t \qquad (17.27)$$

When the difference in frequencies is small compared to the sum of the frequencies,

$$\omega_2 - \omega_1 \ll \omega_2 + \omega_1$$

the time behavior of ψ is that of a nearly sinusoidal oscillation at the average frequency $\frac{1}{2}(\omega_2 + \omega_1)$, as incorporated in the $\cos[(\omega_2 + \omega_1)/2]t$ factor. The oscillation is not precisely sinusoidal because the factor $2A\cos[(\omega_2 - \omega_1)/2]t$ also varies with time. The factors $\pm 2A\cos[(\omega_2 - \omega_1)/2]t$ describe a slowly varying *amplitude envelope* that modulates the more rapid oscillations described by the $\cos[(\omega_1 + \omega_1)/2)]t$ factor. Figure 17.13 shows the superposition of waves with $\omega_2 = 20\pi$ rad/s and $\omega_1 = 18\pi$ rad/s. The oscillations are contained in the amplitude envelope and give rise to beats with a frequency of 1 Hz.

$$\nu_{beat} = \frac{20\pi - 18\pi}{2\pi}\text{ Hz} = 1\text{ Hz}$$

As Figure 17.13 shows, the amplitude envelope repeats every 1 s.

Equation 17.26 shows that the beat frequency equals the difference between the frequencies of the superposed waves. This fact makes it possible to measure an unknown frequency by superposing a wave of known frequency and then measuring the beat frequency. The advantage of this method is that the beat frequency can be made much lower than either of the two superposed frequencies. This relatively low beat frequency can then be measured by audio or electronic methods that are "too slow" to measure the individual frequencies. For example, guitar players who lack a sure sense of pitch rely on beats to tune their instrument. Police radar units also operate on a beat signal.

Figure 17.13

The amplitude envelope $\pm 2A\cos\left(\frac{\omega_2 - \omega_1}{2}\right)t$ limits the amplitude of the sinusoidal oscillations, which proceed at the average frequency $\frac{\omega_2 + \omega_1}{2}$

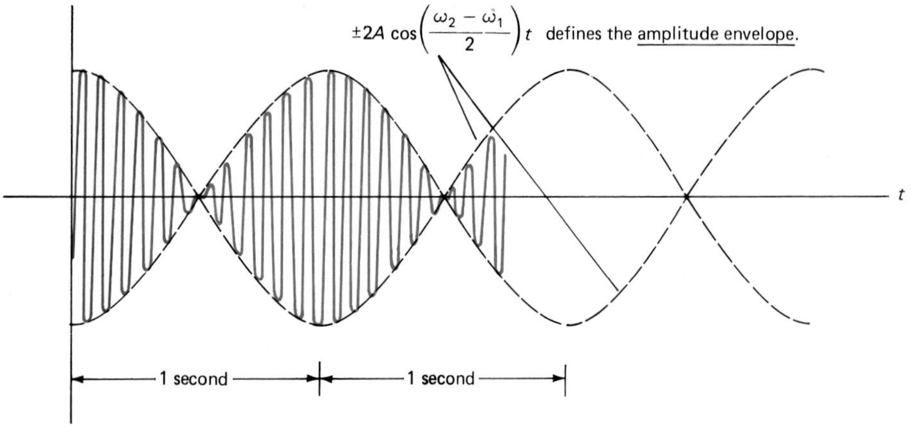

17.6 The Doppler Effect

Relative motion between a source of waves and an observer of waves results in the **Doppler effect**, a change in the observed frequency or wavelength. We measure this change by comparing the frequency or wavelength when there is relative motion to the frequency or wavelength when there is no relative motion.

For example, a motorist waiting at a railroad crossing can observe the Doppler effect for sound waves. As the train passes, the motorist notices a sudden decrease in the frequency of the train whistle. If the motorist then speeds away from the crossing in an effort to make up for the delay he may discover an application of the Doppler effect—police radar. A police radar unit sends out a beam of microwaves. These are electromagnetic waves with a wavelength of a few centimeters. An approaching vehicle reflects portions of the microwave beam back to the radar unit, which also acts as a receiver. Portions of the transmitted and received waves are superposed, giving a signal at the beat frequency—the difference between the two frequencies. As we will show, the difference between the two frequencies is directly proportional to the speed of the vehicle. Hence, a frequency meter capable of responding at the beat frequency is readily converted into a "speed meter."

Consider a tuning fork that emits waves of frequency ν_o. If the waves travel at speed v, an observer at rest with respect to the fork would find that their wavelength is

$$\lambda_o = \frac{v}{\nu_o} = vT_o \qquad (17.28)$$

It is helpful to think of T_o, the period of the wave, as the time between the emission of successive wave crests. If the fork *recedes* from the observer at speed u, the distance between successive crests is increased by uT_o—the distance the fork moves between the emission of successive crests (see Figure 17.14). The wavelength is thereby increased from λ_o to $\lambda_o + \Delta\lambda$, where

$$\Delta\lambda = uT_o \qquad (17.29)$$

Combining Eqs. 17.28 and 17.29 gives us one version of the Doppler formula:

$$\frac{\Delta\lambda}{\lambda_o} = \frac{u}{v} = \frac{\text{speed of recession}}{\text{speed of wave}} \qquad (17.30)$$

Note that the wavelength *increases* when the relative motion *increases* the distance between the source of waves and the observer.

Waves pass the observer at a constant speed v whether or not the fork is in motion, because v is determined by the medium, not by the motion of the source. Consequently, the increased wavelength also registers as a decreased frequency. To show how the frequency is reduced we note that the speed of the wave can be expressed by

$$v = \nu(\lambda_o + \Delta\lambda)$$

where ν is the frequency corresponding to the wavelength $\lambda_o + \Delta\lambda$. The wave speed is also given by

$$v = \nu_o\lambda_o$$

Equating these two expressions for v gives

$$\nu = \frac{\nu_o\lambda_o}{\lambda_o + \Delta\lambda} = \frac{\nu_o}{1 + (\Delta\lambda/\lambda_o)}$$

Using Eq. 17.30 to replace $\Delta\lambda/\lambda_o$ by u/v, we obtain an expression for the Doppler-shifted frequency ν:

$$\nu = \frac{\nu_o}{1 + (u/v)} \qquad (17.31)$$

Figure 17.14

When the source of waves recedes from the observer the wavelength is increased, resulting in a decrease in frequency for the observer.

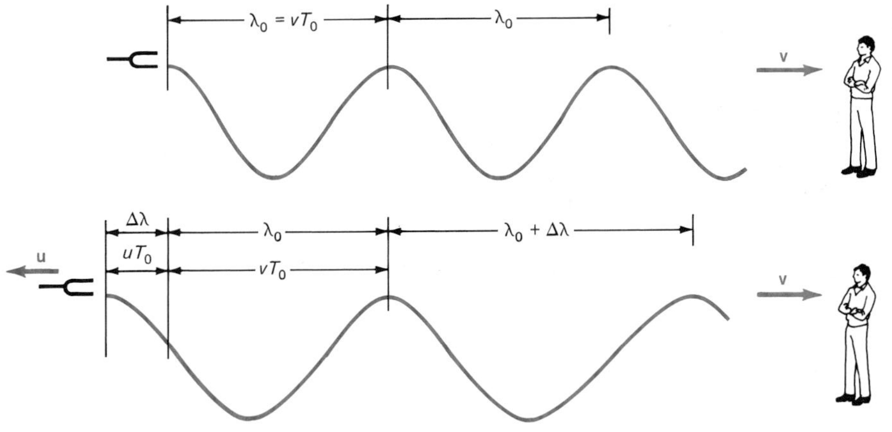

When the source moves toward the observer the wavelength is decreased and the frequency is increased. The equation relating ν and ν_o for when the source moves toward the observer is obtained by replacing u by $-u$ in Eq. 17.31.

$$\nu = \frac{\nu_o}{1 - (u/v)} \tag{17.32}$$

The following example illustrates the Doppler effect for a source that first moves toward an observer and then moves away from the observer.

Example 6 **Doppler Shift of the Frequency of an Air Horn**

A student standing at the side of a roadway hears the air horn of an approaching truck. She notices that the frequency of the horn suddenly decreases as the truck passes her. What frequencies does she hear as the truck approaches and recedes? The truck moves at a speed of 90 km/hr (25 m/s) and the speed of sound is 330 m/s. The frequency of the horn, as observed by the truck driver, is 1200 Hz (ν_o). Equation 17.32 gives the frequency heard by the student as the truck approaches:

$$\nu = \frac{\nu_o}{1 - (u/v)} = \frac{1200 \text{ Hz}}{1 - (25/330)} = 1298 \text{ Hz}$$

Equation 17.31 gives the frequency she hears as the truck moves away:

$$\nu = \frac{\nu_o}{1 + (u/v)} = \frac{1200 \text{ Hz}}{1 + (25/330)} = 1115 \text{ Hz}$$

As the truck moves toward her and then away from her the student hears the frequency drop from 1298 Hz to 1115 Hz. You may have observed a similar decrease in the frequency of the whine of automobile tires as cars move past you.

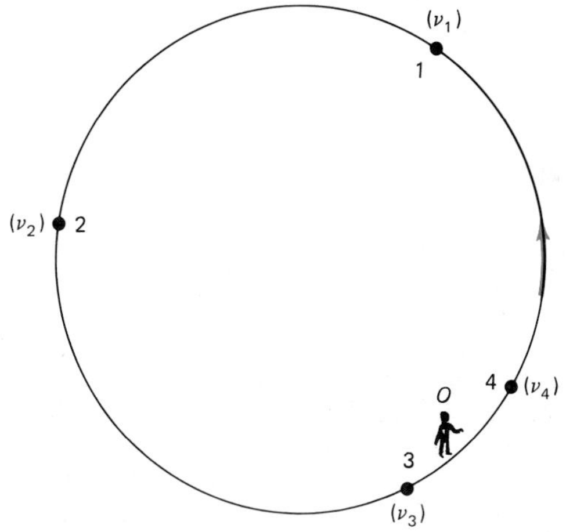

Figure 17.15

Question

13. A toy train moves counterclockwise around a circular track at a constant speed. The train generates sound waves that reach a child sitting at O (Figure 17.15). The four frequencies received by the child when the train is at positions 1, 2, 3, and 4 are labeled ν_1, ν_2, ν_3, and ν_4, respectively. Arrange the frequencies in order of increasing magnitude, with the lowest frequency first.

Summary

This chapter deals with the kinematic aspects of waves, particularly *sinusoidal* waves having the form

$$\psi(x,t) = A \sin\left[\frac{2\pi}{\lambda}(x - vt)\right] \tag{17.4}$$

The basic kinematic relation between wavelength (λ), frequency (ν), and phase velocity (v) is

$$\nu\lambda = v \tag{17.8}$$

Other useful wave variables include the wave number

$$k = \frac{2\pi}{\lambda} \tag{17.10}$$

and radian frequency

$$\omega = 2\pi\nu \tag{17.11}$$

In terms of ω and k, the phase velocity is given by

$$v = \frac{\omega}{k} \tag{17.18}$$

For sinusoidal waves we define the *phase* (ϕ) by the relations

$$\phi = \frac{2\pi}{\lambda}(x - vt) = kx - \omega t \tag{17.15}$$

Any phase can be increased or reduced by an integral multiple of 2π without affecting subsequent conclusions.

Empirically, it is found that a *principle of superposition* is valid for small-amplitude waves:

A meeting of two or more waves produces a waveform that is the algebraic sum of the waveforms produced by each wave acting separately.

In short, small-amplitude waves move independently.

The principle of superposition leads to the notion of wave *interference*. Constructive interference results when superposition causes the waves to reinforce one another. Destructive interference results when superposition causes the waves to oppose one another. *Beats* are the periodic variations between constructive and destructive interference that are evident when two waves of slightly different frequencies are superposed. The beat frequency equals the difference between the two superposed frequencies,

$$\nu_{beat} = \nu_2 - \nu_1 \tag{17.26}$$

Relative motion between a source of waves and an observer results in the *Doppler effect*, a change in the measured frequency or wavelength. The Doppler effect is kinematic in origin and shows itself as an *increase* in wavelength when the relative motion *separates* the source of waves and the observer.

Problems

Section 17.1 Wave Characteristics

1. Which of the following are functions of $x - vt$? (a) $e^{a(x-vt)}$, (b) $(x - vt)^3$, (c) $x^2 - 2xvt + v^2t^2$ (d) $x^2 - v^2t^2$, (e) $x^4 - v^4t^4$, (f) $\sin(kx) \cdot \cos(kvt)$, (g) $\cos[k(x + vt)]$

2. Ocean waves with a crest-to-crest distance of 10 m can be described by

   $$\psi(x,t) = 0.8 \sin 0.63(x - 1.2\ t)\ \text{m}$$

   (a) Sketch $\psi(x,t)$ at $t = 0$. This corresponds to a snapshot of the waves at $t = 0$. (b) Sketch $\psi(x,t)$ at $t = 2$ s. Note how the entire waveform has shifted 2.4 m in the positive x-direction.

Section 17.2 Sinusoidal Waves

3. A sinusoidal waveform has a maximum displacement (ψ_{max}) of 1.80 cm at $x = 2.11$ m and a (nearest) minimum displacement of -1.80 cm at $x = 2.62$ m. Determine its amplitude and wavelength.

4. A waveform is described by

   $$\psi = 2.0 \sin(2.11x - 3.62t)\ \text{cm}$$

   where x is the position along the waveform (in meters) and t is the time (in seconds). Determine the amplitude, wave number, wavelength, radian frequency, and speed of the wave.

5. A sinusoidal wave on a string is described by $y(x,t) = 0.51 \sin(3.1x - 9.3t)$ cm, where x is measured in centimeters and t is measured in seconds. How far does a wave crest move in 10 s? Does it move in the $+x$-direction or in the $-x$-direction?

6. Light travels through a vacuum at a speed of 3.00×10^8 m/s. Light with a wavelength of 650 nm (1 nm $= 10^{-9}$ m) is red in color. Determine the frequency of red light waves.

7. Light waves with a frequency of 7.00×10^{14} Hz have a frequency that is too high for most human eyes to detect. Calculate the corresponding wavelength in nanometers.

8. A sinusoidal wave has a wavelength of 6 m and a speed of 10 m/s. Determine its period, frequency, wave number, and radian frequency.

9. A wave of the form $\psi(x,t) = B\cos(kx - \omega t)$ has a wave number equal to 1.25 rad/m and a radian frequency of 3 rad/s. Determine the wavelength and the velocity of the wave.

10. A transverse sinusoidal wave travels to the right along a rope at a speed of 10 m/s. The wavelength is 2 m and the amplitude is 0.1 m. The height of the wave is zero at $x = 0$, $t = 0$. (a) Write an expression for the displacement $\psi(x,t)$. (b) Determine the

height at $x = 0$, $t = 0.1$ s. (c) Determine the velocity of the *rope* at $x = 0$, $t = 0.1$ s. (d) Determine the period of the wave motion.

11. Use Eqs. 17.6, 17.10, and 17.11 to show that the wave speed can be written as $v = \omega/k$.

12. If the waveform describing the displacement of the medium is given by $\psi(x,t) = B\cos(kx - \omega t)$, show that at any particular point (x_0) the medium executes simple harmonic motion at radian frequency ω with amplitude B.

13. A sinusoidal waveform has the form

$$\psi = A\sin(2.11x - 3.62t)$$

where x is in meters and t is in seconds. Determine the phase of the wave at $x = 1.16$, $t = 0.73$ and at $x = 2.27$, $t = 0.83$

14. A sinusoidal wave has a wavelength of 2.28 m. At some fixed moment in time, how far apart are points that differ in phase by π rad?

Section 17.3 Phase and Phase Difference

15. A calculation shows a phase difference of 63.7 rad. What is the smallest (positive) *equivalent* phase difference?

16. Use the addition theorem,

$$\sin(x + y) = \sin x \cos y + \sin y \cos x,$$

to prove that

$$\sin(\phi + 2\pi N) = \sin\phi$$

where N is an integer. This shows that any phase angle can be altered by an integral multiple of 2π without changing the value of a sinusoidal wave property.

17. Red light ($\lambda = 650$ nm) moves in the $+x$-direction. Calculate Δx corresponding to a phase difference of π.

18. A light wave undergoes a phase change of $\pi/4$ rad in passing through a thin film of oil. If the wavelength of the light as it moves through the film is 520 nm, how thick is the film?

19. A wave has the form

$$y(x,t) = 0.26\sin(2.3x - 8.1t + 1.6)\ \text{cm}$$

where x is measured in centimeters and t in seconds. Determine how the origin on the x-axis can be shifted so that the wave has the form

$$y(x',t) = 0.26\sin(2.3x' - 8.1t)\ \text{cm}$$

20. Two cellos in an orchestra are lined up, one behind the other, with a listener. When they sound a note ($\nu = 220$ Hz) simultaneously, the sounds arrive at the listener with a phase difference of 180°. Determine the smallest separation of the cellos that will give this phase difference (sound speed = 344 m/s).

Section 17.4 The Principle of Superposition for Waves

21. Two sinusoidal waves of equal amplitude are superimposed. The resultant wave is

$$\psi(x,t) = 3\sin\left(\frac{\pi x}{L}\right)\cos\left(\frac{\pi t}{K}\right)$$

(a) Find the forms of the two sinusoidal waves that are superimposed. (b) If $L = 3$ m and $T = 2$ s, determine the wavelengths (λ) and frequencies (ν) of the two waves. (c) Find their velocities.

22. (a) Use the principle of superposition to show that the two waves

$$\psi_1 = 2\cos(2\pi x - 6\pi t)$$

$$\psi_2 = 2\cos(2\pi x + 6\pi t)$$

give rise to a standing wave. (b) What are the period and wavelength of the resulting standing wave?

23. Two traveling waves are represented by

$$y_1 = A\sin(kx + \omega t)$$

$$y_2 = A\sin\left(kx + \omega t + \frac{\pi}{2}\right)$$

Show that the amplitude of the superimposed waves $(y_1 + y_2)$ is $\sqrt{2}A$.

Section 17.5 Beats

24. A pair of tuning forks are slightly out of tune: One vibrates at 513 Hz, the other at 511 Hz. What is the beat frequency of their superimposed waves?

25. Two tuning forks exhibit beats at a beat frequency of 3 Hz. The frequency of one fork is known to be 256 Hz. The frequency of this fork is then lowered slightly by adding a bit of tacky wax to one tine. The two forks then exhibit a beat frequency of 1 Hz. Determine the final frequencies of the two forks.

26. Determine the beat frequency resulting from the superposition of two waves represented by

$$y_1 = \sin(3.1x - 4320t)$$

$$y_2 = \sin(3.1x - 4330t)$$

where t is measured in seconds.

27. Beat frequencies in the range of 2–8 Hz are considered musically pleasant. To take advantage of this, an organ designer will sometimes include two pipes constructed to play the same note, and then deliberately detune one of them to produce beats in the pleasant range. If one pipe is to produce an A ($\nu = 440$ Hz), what frequencies are available for the second pipe if pleasant beats are to result?

Section 17.6 The Doppler Effect

28. A police radar unit emits waves with a frequency of 4.80×10^9 Hz. A motorist moves toward the radar unit at a speed of 100 km/hr. What is the difference

between the broadcast frequency of 4.80×10^9 Hz and the frequency his "fuzz-buster" radar warning unit detects?

29. A train whistle generates sound waves with a frequency of 486 Hz. (a) What frequency will you hear if the train moves toward you at a speed of 30.1 m/s? Take the speed of sound to be 331 m/s. (b) What frequency will you hear if the train moves away from you at 30.1 m/s?

30. A physics professor, willing to go to any limit to provide memorable illustrations, goes one step too far in his demonstration of the Doppler effect. With vibrating tuning fork in hand, he invites his students to note its frequency. A music major who possesses perfect pitch correctly identifies the frequency as $\nu = 300$ Hz. The professor then runs away from his class at top speed, urging them to "listen to the Doppler sh . . . ." His last word is only partly audible because his last step carries him out a window. Unfortunately, the classroom is located on the 25th floor. The students rush to the window, eager to observe the Doppler shift. (a) Do the students hear a frequency greater than or less than 300 Hz? (The professor remains silent so as not to spoil the demonstration.) (b) Does the frequency they hear remain constant throughout the demonstration? Explain. (c) The acceleration of gravity is 9.80 m/s^2, and the professor demonstrates for 5.50 s. Determine the approximate frequency that will be heard during the final moments of the demonstration. The speed of sound in air is 330 m/s. (d) One student, named Spock, has very "sharp" ears. He is able to hear the waves arriving directly from the tuning fork and those that travel to the ground and are reflected upward. What beat frequency does Spock hear as the demonstration concludes?

31. (a) Consider the situation in which an observer moves at speed u toward a stationary source of sound waves of frequency ν_0. Show that the Doppler effect increases the frequency by $\nu - \nu_0 = \Delta\nu$, where

$$\frac{\Delta\nu}{\nu_0} = \frac{u}{v}$$

(b) Compare $\Delta\nu$, as given in (a), with $\Delta\nu$ as it would be determined by using Eq. 17.32 for $\nu_0 = 256$ Hz, $u = 60$ mph, and $v = 760$ mph.

32. Starting with Eq. 17.31, show that the Doppler shift $\Delta\nu \equiv \nu - \nu_0$ is given by

$$\Delta\nu = \frac{-(u/v)\nu_0}{1 + \dfrac{u}{v}}$$

When the speed of the source (u) is small compared to the speed of the waves (v), this reduces to $\Delta\nu = -(u/v)\nu_0$, showing that the Doppler shift is directly proportional to the speed of the source.

chapter 18

MECHANICAL WAVES

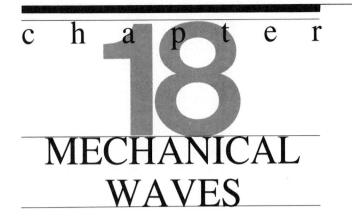

Preview

In this chapter we are concerned primarily with *mechanical waves*. These are waves—such as sound waves and water waves—that involve the displacement of ordinary matter. We treat transverse waves on a string first because they are conceptually simple. We also introduce the wave equation, which governs the propagation of mechanical waves. We trace the origin of the wave equation to Newton's second law and the empirical equations describing the elastic properties of the medium. We describe several different kinds of mechanical waves and show how their speeds are related to the elastic and inertial properties of the medium.

The concept of intensity is introduced as a measure of the rate at which a wave transports energy. We study standing waves to illustrate how material boundaries give rise to boundary conditions that must be satisfied by solutions of the wave equation. Boundary conditions enable us to identify the physically acceptable solutions of the wave equation contained within the larger group of mathematically acceptable solutions.

18.1 Waves on a String

We can send a mechanical wave along a flexible string by anchoring one end of the string and shaking the other end. This wave on a string is a transverse wave. The variable that we use to describe this wave is the displacement of the string from its equilibrium shape. We call this displacement the *height* of the wave and denote it symbolically as $\psi(x,t)$ (Figure 18.1).

Let us now derive the wave equation for waves on a string—the equation governing $\psi(x,t)$. The wave equation for $\psi(x,t)$ emerges from Newton's second law, applied to a small segment of the string. We must make several assumptions in the course of this derivation. At the outset we assume that the tension, T, in the string remains constant even when waves are present. Figure 18.2 shows the three external forces acting on the segment located between x and $x + \Delta x$. These three forces are the weight of the segment and the tensile forces exerted on the ends of the segment. We ignore the weight of the segment. Thus, the dynamics of the string are controlled by the tension T, not by gravity. As Figure 18.3 indicates, the tensile forces "pulling" on the ends of the segment have equal magnitudes. The directions of these two tensile forces are different (in general), and so a net vertical force acts on the segment. This net vertical force $T_v(x + \Delta x) - T_v(x)$ results in a vertical acceleration of the segment, as prescribed by Newton's second law.

Figure 18.1

$\psi(x,t)$ denotes the displacement of the string from its equilibrium shape (horizontal). The figure represents a "snapshot" view of a wave; that is it depicts $\psi(x,t)$ for different values of x at a fixed moment of time t.

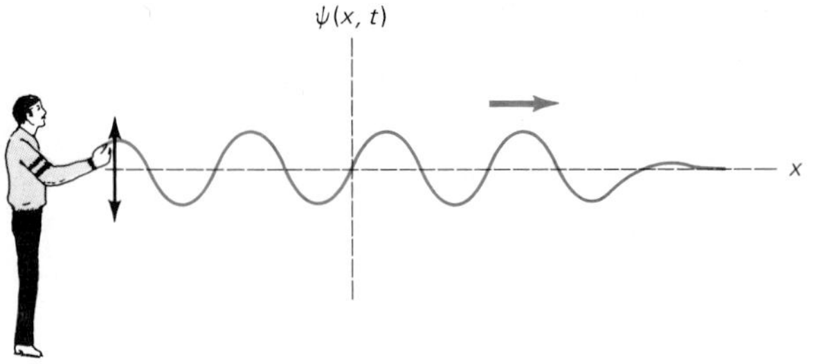

Figure 18.2

Three external forces act on the segment located between x and $x + \Delta x$. These three forces are the weight, W, of the segment, and the tensile forces exerted on the ends of the segment.

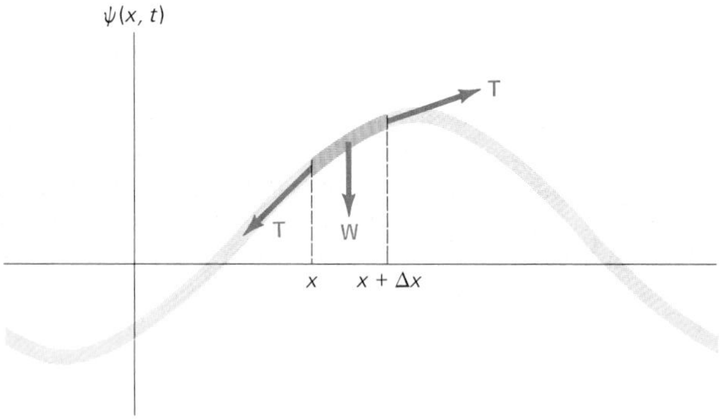

Figure 18.3

The tensile forces acting on the ends of the segment have equal magnitudes but different directions. As a result there is a net vertical force, $T_v(x + \Delta x) - T_v(x)$, on the segment. The slope $\partial\psi/\partial x$ is related to the angle θ by the relation $\tan\theta = \partial\psi/\partial x$.

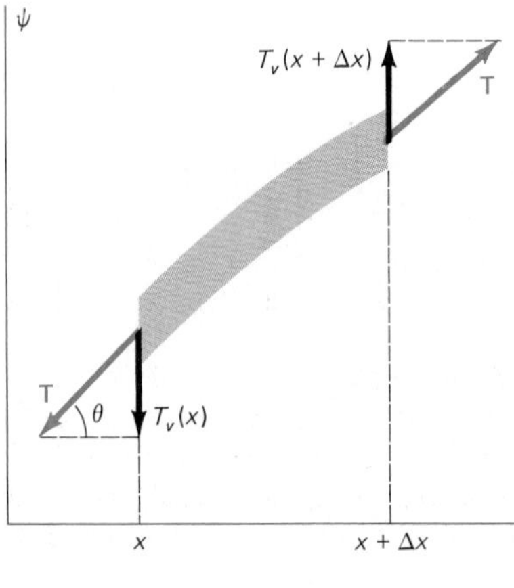

We consider a uniform string of mass M and length L. The mass per unit length is M/L,

$$\frac{M}{L} = \frac{\text{mass of string}}{\text{length of string}} \tag{18.1}$$

The mass of the segment of length Δx is $(M/L)\,\Delta x$. The vertical acceleration of the segment is

$$a_{\text{vertical}} = \frac{\partial^2\psi}{\partial t^2} \tag{18.2}$$

Newton's second law for the segment is

$$\frac{M}{L}\Delta x \cdot \frac{\partial^2\psi}{\partial t^2} = T_v(x + \Delta x) - T_v(x) \tag{18.3}$$

Next, we divide through by Δx and consider the limit as $\Delta x \to 0$. The right side becomes the derivative of T_v:

$$\lim_{\Delta x \to 0} \frac{T_v(x + \Delta x) - T_v(x)}{\Delta x} = \frac{\partial T_v}{\partial x} \tag{18.4}$$

Newton's second law is thereby transformed into

$$\frac{M}{L}\frac{\partial^2\psi}{\partial t^2} = \frac{\partial T_v}{\partial x} \tag{18.5}$$

The vertical component T_v is related to the tension T by

$$T_v = T\sin\theta \tag{18.6}$$

At this point we make a second important assumption—namely, that the angle θ is small ($\theta \ll 1$). As long as the height of the wave remains small compared to the wavelength, the assumption that θ is small is legitimate, and it permits us to replace $\sin\theta$ by $\tan\theta$. Physically, this assumption restricts our analysis to small-amplitude waves. From Figure 18.3 we note that $\tan\theta$ is the slope, $\partial\psi/\partial x$. Thus,

$$\sin\theta \simeq \tan\theta = \frac{\partial\psi}{\partial x} \tag{18.7}$$

and

$$T_v = T\sin\theta \simeq T\frac{\partial\psi}{\partial x} \tag{18.8}$$

Replacing T_v in Eq. 18.5 by $T\,\partial\psi/\partial x$ gives

$$\frac{M}{L}\frac{\partial^2\psi}{\partial t^2} = T\frac{\partial^2\psi}{\partial x^2} \tag{18.9}$$

Dividing both sides by M/L produces

$$\frac{\partial^2\psi}{\partial t^2} = \frac{T}{M/L}\frac{\partial^2\psi}{\partial x^2} \tag{18.10}$$

The quantity $\sqrt{\dfrac{T}{M/L}}$ has the dimensions of *speed*. We therefore write

$$v = \sqrt{\frac{T}{M/L}} \tag{18.11}$$

whereupon Eq. 18.10 becomes

$$\frac{\partial^2\psi}{\partial t^2} = v^2\frac{\partial^2\psi}{\partial x^2} \tag{18.12}$$

Equation 18.12 is the standard form of the wave equation for small-amplitude waves on a string. These waves travel at the speed v given by Eq. 18.11.

Example 1 Sinusoidal Waves on a String

In Chapter 17 we showed that sinusoidal waves of the form

$$\psi(x,t) = A \sin\left[\frac{2\pi}{\lambda}(x - vt)\right] \tag{18.13}$$

travel at a speed v. We now show that this sinusoidal waveform satisfies the equation describing waves on a string, Eq. 18.10, provided that the speed v in Eq. 18.13 equals $\sqrt{\frac{T}{M/L}}$, the speed defined by Eq. 18.11. From Eq. 18.13,

$$\frac{\partial\psi}{\partial t} = A \cos\left[\frac{2\pi}{\lambda}(x - vt)\right] \cdot \frac{2\pi}{\lambda}(-v)$$

and

$$\frac{\partial^2\psi}{\partial t^2} = -A \sin\left[\frac{2\pi}{\lambda}(x - vt)\right] \cdot \left(\frac{2\pi}{\lambda}\right)^2 \cdot v^2$$

which can be expressed as

$$\frac{\partial^2\psi}{\partial t^2} = -v^2\left(\frac{2\pi}{\lambda}\right)^2 \psi(x,t) \tag{18.14}$$

Similarly,

$$\frac{\partial\psi}{\partial x} = A \cos\left[\frac{2\pi}{\lambda}(x - vt)\right] \cdot \left(\frac{2\pi}{\lambda}\right)$$

and

$$\frac{\partial^2\psi}{\partial x^2} = -A \sin\left[\frac{2\pi}{\lambda}(x - vt)\right] \cdot \left(\frac{2\pi}{\lambda}\right)^2$$

$$= -\left(\frac{2\pi}{\lambda}\right)^2 \psi(x,t)$$

Thus,

$$v^2\frac{\partial^2\psi}{\partial x^2} = -v^2\left(\frac{2\pi}{\lambda}\right)^2 \psi(x,t) \tag{18.15}$$

By comparing Eqs. 18.14 and 18.15 we see that the relation $\psi(x,t) = A \sin[(2\pi/\lambda)(x - vt)]$ satisfies the following equation:

$$\frac{\partial^2\psi}{\partial t^2} = v^2\frac{\partial^2\psi}{\partial x^2} \tag{18.12}$$

By comparing Eq. 18.12 with Eq. 18.10, the equation describing waves on a string, we see that they are identical provided that the wave speed v is given by Eq. 18.11:

$$v = \sqrt{\frac{T}{M/L}} \tag{18.11}$$

Thus, sinusoidal waves can travel along a string at a speed given by Eq. 18.11.

Example 2 Clothesline Waves

A clothesline has a mass of 0.80 kg. The line stretches between two supports that are 12 m apart. The tension in the line is 100 N. What is the time required for a wave to travel from one end of the line to the other and then return? The mass per unit length is

$$\frac{M}{L} = \frac{0.80 \text{ kg}}{12 \text{ m}}$$

The speed of the waves is

$$v = \sqrt{\frac{100}{0.80/12}} \text{ m/s} = 38.7 \text{ m/s}$$

The round trip travel time is

$$t = \frac{24 \text{ m}}{38.7 \text{ m/s}} = 0.620 \text{ s}$$

We observe that the larger T is, the larger is the force required to alter the shape of the string. In other words, the tension in a string is a measure of the string's stiffness or elasticity. The mass per unit length of the string is an inertial factor. Therefore, according to Eq. 18.11, the speed of a wave on a string depends on the ratio of an elastic factor (T) and an inertial factor (M/L). In fact, Eq. 18.11 is one example of the general relation

$$v = \sqrt{\frac{\text{elastic factor}}{\text{inertial factor}}} \tag{18.16}$$

We will see that Eq. 18.16 also holds true for other types of mechanical waves.

Our derivation shows that Eq. 18.10 describes small-amplitude waves on a string. However, the wave equation, Eq. 18.12, also describes many other types of mechanical waves, such as water waves and sound waves. In Section 18.2 we trace the origin of this general form of the wave equation to Newton's second law.

18.2 Origin of the Wave Equation for Mechanical Waves

We stated without proof that many other types of mechanical waves are described by the same form of wave equation,

$$\frac{\partial^2\psi}{\partial t^2} = v^2\frac{\partial^2\psi}{\partial x^2} \tag{18.12}$$

We can uncover the origin of Eq. 18.12 without performing detailed derivations for each type of wave. The derivations begin by our applying Newton's second law to a small-mass element that is displaced from its equilibrium position.

The mass of the element is proportional to some intrinsic inertial property of the medium—some sort of "inertial factor." For sound waves this inertial factor is the mass density. For waves on a string the inertial factor is the mass per unit length. If ψ denotes the displacement of the element from equilibrium, its acceleration is $\partial^2\psi/\partial t^2$. The displacement gives rise to a restoring force because the elastic properties of the medium attempt to restore equilibrium. For small displacements, the restoring force is proportional to $\partial^2\psi/\partial x^2$. Establishing this proportionality is usually the most complicated step in the derivation. The proportionality constant incorporates an "elastic factor"—a quantity that measures the "stiffness" of the medium. For example, in our derivation of the equation describing sound waves, the bulk modulus B (Section 15.5) and the shear modulus μ (Section 15.6) enter as elastic factors. For waves on a string the elastic factor is the tension in the string. In general the restoring force takes the form

$$\text{restoring force on element} = (\text{elastic factor})\,\frac{\partial^2\psi}{\partial x^2} \qquad (18.17)$$

Newton's second law, applied to the displaced element, takes the form

$$\begin{pmatrix}\text{mass of}\\ \text{element}\end{pmatrix}\begin{pmatrix}\text{acceleration}\\ \text{of element}\end{pmatrix} = \begin{pmatrix}\text{restoring force}\\ \text{on element}\end{pmatrix} \qquad (18.18)$$

The left-hand side is converted to

$$\begin{pmatrix}\text{mass of}\\ \text{element}\end{pmatrix}\begin{pmatrix}\text{acceleration}\\ \text{of element}\end{pmatrix} = (\text{inertial factor})\,\frac{\partial^2\psi}{\partial t^2} \qquad (18.19)$$

The right-hand side is transformed by using Eq. 18.17. These manipulations leave Newton's second law in the form

$$(\text{inertial factor})\cdot\frac{\partial^2\psi}{\partial t^2} = (\text{elastic factor})\cdot\frac{\partial^2\psi}{\partial x^2}$$

Rearranging this as

$$\frac{\partial^2\psi}{\partial t^2} = \left(\frac{\text{elastic factor}}{\text{inertial factor}}\right)\frac{\partial^2\psi}{\partial x^2}$$

we find that it is the wave equation, $\partial^2\psi/\partial t^2 = v^2\,\partial^2\psi/\partial x^2$, the wave speed being given by Eq. 18.16:

18.3 Mechanical Waves: A Sampling

In this section we present descriptions and discussions of several types of mechanical waves, both to expose you to unfamiliar types of waves and to reveal some interesting features of familiar waves. The presentations are brief and most results are simply stated without proof.

All of the waves studied in this section can be described by the wave equation, Eq. 18.12:

$$\frac{\partial^2\psi}{\partial t^2} = v^2\,\frac{\partial^2\psi}{\partial x^2} \qquad (18.12)$$

However, the physical significance of $\psi(x,t)$ and the dependence of the wave speed on the properties of the medium differ from one type of wave to the next.

Figure 18.4

In water waves, the fluid particles execute circular motions. The radii of the circular motions decrease with depth, becoming negligible at a depth of a few times the wavelength. Thus, the wave motions are confined to layers within a few wavelengths of the surface. We therefore classify water waves as *surface* waves.

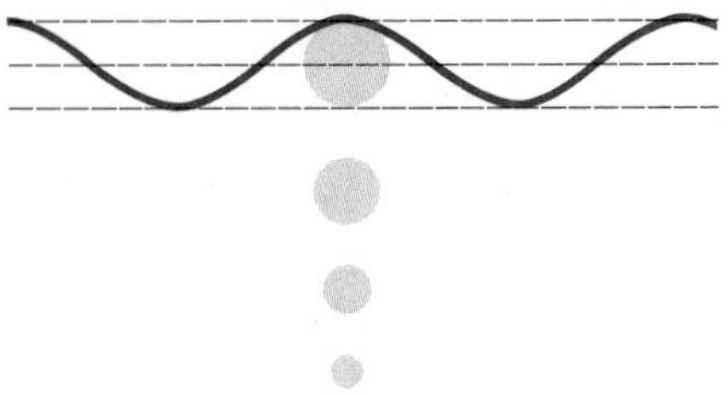

1. Waves on a Liquid Surface (Water Waves)

Water waves are among the more familiar waveforms encountered in our everyday experience. These waves appear to be transverse; the liquid rises and falls as the waveform travels along the surface. But in fact water waves have both transverse and longitudinal components. You will see this if you watch a small floating object such as a cork or bottle. You will notice that the cork not only bobs up and down but also moves back and forth along the surface. A detailed theoretical analysis reveals that the fluid particles (Section 16.1) execute *circular* motions. The radius of the circular paths (the wave amplitude) is greatest at the surface. At a depth of a few wavelengths, the wave amplitude has diminished to a negligible value (see Figure 18.4). Water waves are therefore classified as *surface* waves.

Two types of forces, gravity and surface tension, act to restore displacements of the water. This means that two "elastic factors" will enter the equation for wave speed. There are also two characteristic lengths to be considered. These are the wavelength and the depth of the water. When the depth (h) is much smaller than the wavelength (λ), we are dealing with *shallow-water waves*. The speed of shallow-water waves is

$$v = \sqrt{gh} \qquad h \ll \lambda \qquad (18.20)$$

Equation 18.20 can be viewed as an expression of the equation

$$v = \sqrt{\frac{\text{elastic factor}}{\text{inertial factor}}} \qquad (18.16)$$

in which gravity supplies the elastic factor ($\rho g h$) and the mass density (ρ) serves as the inertial factor.

When the depth of the water is much greater than the wavelength, the waves are called *deep-water waves*. The speed of deep-water waves is given by

$$v = \sqrt{\frac{g\lambda}{2\pi} + \frac{2\pi S}{\rho\lambda}} \qquad h \gg \lambda \qquad (18.21)$$

where S is the surface tension,* g is the acceleration of gravity, and ρ is the density of the liquid. This expression

*The surface tension S is related to the force required to "stretch" a surface, thereby increasing its surface area. The units of S are newtons per meter (N/m).

shows that waves of very short wavelength (ripples) are controlled by the surface tension, whereas those of long wavelength are gravity controlled. There is a minimum speed for deep-water waves. We can determine this minimum by setting $dv/d\lambda = 0$.

$$\frac{dv}{d\lambda} = \frac{\frac{1}{2}\left(\frac{g}{2\pi} - \frac{2\pi S}{\rho\lambda^2}\right)}{\sqrt{\frac{g\lambda}{2\pi} + \frac{2\pi S}{\rho\lambda}}} = 0$$

Solving the equation for λ, we find that the minimum speed occurs for the wavelength

$$\lambda = 2\pi\sqrt{\frac{S}{\rho g}} \qquad (18.22)$$

We then substitute this expression for λ into Eq. 18.21 to obtain the equation for minimum speed.

$$v_{\text{min}} = \left(\frac{4gS}{\rho}\right)^{1/4} \qquad (18.23)$$

The surface tension of water varies slightly with temperature, ranging from 0.0756 N/m at 0°C to 0.0589 N/m at 100°C. At 15°C (59°F),

$$S = 0.07359 \text{ N/m} \quad \text{and} \quad \rho = 1002 \text{ kg/m}^3$$

Taking $g = 9.80$ m/s^2, we get 0.232 m/s for the minimum speed. We find the wavelength at which this minimum speed occurs by inserting these same values of S, ρ, and g into Eq. 18.22. This gives a wavelength of 1.72 cm.

The existence of this minimum wave speed for deep-water waves has an interesting consequence. When air moves over the surface of water at a speed less than the minimum wave speed, it is unable to transfer energy to the water. Any aspiring wave finds itself fighting a headwind and dies out. Ripples are generated only for wind speeds greater than $(4gS/\rho)^{1/4}$ (Equation 18.22). (You can demonstrate this by blowing on the surface of a glass of water.) The fact that ripples do not arise for low air speeds can be used to measure the surface tension of a liquid.

2. Acoustic Gravity Waves

The density of our atmosphere decreases with altitude. This variation in density allows the propagation of transverse density variations called *acoustic gravity waves*. Such waves could not exist in a homogeneous fluid. In our heterogeneous atmosphere, the acoustic gravity waves travel parallel to the earth's surface. The density variation is vertical, and the waves are transverse.

Two forces combine to produce atmospheric gravity waves. These are the force of gravity and a buoyancy force (Chapter 16). Consider a parcel of air—some arbitrary chunk of the atmosphere. In equilibrium, a parcel of air experiences a buoyant force equal to its weight. If the parcel is displaced upward, the buoyant force acting on it is reduced because the parcel displaces air that is less dense. Because the parcel's weight is unchanged, the net force is downward, and the air parcel is urged downward toward its equilibrium altitude. If the parcel is displaced downward, the buoyant force increases because the parcel displaces air that is more dense. The buoyant force then exceeds the weight and the net force is an upward restoring force.

Acoustic gravity waves are of considerable meteorological interest because they appear to be related to clear-air turbulence. Strong winds can cause the waves to become unstable. Their amplitudes grow until they break like surf on a beach. The breaking of an acoustic gravity wave leaves a wake of turbulent eddies. These swirling eddies produce the annoying, sometimes dangerous, "bumps" that are felt by airplane passengers. Thus, a vertical displacement either upward or downward gives rise to a restoring force. Parcels of air can undergo oscillations about equilibrium—just as segments of a string can oscillate vertically about equilibrium. These oscillatory displacements of air parcels propagate horizontally as acoustic gravity waves.

3. Longitudinal and Transverse Elastic Waves in Solids and Fluids

A homogeneous isotropic solid can support both longitudinal and transverse elastic waves. The longitudinal waves in a solid correspond to the acoustic waves in a fluid. Both are traveling pressure and density variations. Transverse shear waves can exist because a solid can successfully resist shearing stresses (Section 15.6).

The speeds of the longitudinal and transverse waves are given by the following equations:

$$v_{\text{long}} = \sqrt{\frac{B + \frac{4}{3}\mu}{\rho}} \qquad (18.24)$$

$$v_{\text{trans}} = \sqrt{\frac{\mu}{\rho}} \qquad (18.25)$$

In these equations ρ is the mass density, B is the bulk modulus, and μ is the shear modulus (Sections 15.5 and 15.6). The longitudinal waves travel at a greater speed than the transverse waves. This can be seen by comparing Eq. 18.24 with Eq. 18.25, keeping in mind that B and μ are always positive. Typical values of v_{long} and v_{trans} fall in the range from 1 km/s to 10 km/s. Values of B, μ, and ρ for several solids and liquids are displayed in Table 18.1, along with values of v_{trans} and v_{long} computed from Eqs. 18.24 and 18.25.

In an ideal fluid there is no resistance to shearing motions and the shear modulus (μ) is zero. The speed of longitudinal waves in an ideal fluid is given by Eq. 18.24, with $\mu = 0$. The speed of transverse waves drops to zero when $\mu = 0$, showing that transverse waves do not propagate in an ideal fluid.

$$v_{\text{long}} = \sqrt{\frac{B}{\rho}} \qquad (18.26)$$

(ideal fluid, $\mu = 0$)

$$v_{\text{trans}} = 0 \qquad (18.27)$$

Example 3 The Speed of Underwater Sound

We can use Eq. 18.26 to calculate the speed of sound in water. The bulk modulus of water at 20°C is 2.18×10^9

Table 18.1*

Material	ρ (10^3 kg/m^3)	B (10^{10} Pa)	μ (10^{10} Pa)	v_{long} (km/s)	v_{trans} (km/s)
Aluminum (Al)	2.70	7.2	2.6	6.3	3.1
Copper (Cu)	8.96	13.0	4.6	4.6	2.3
Lead (Pb)	11.36	6.0	0.55	2.4	0.7
Steel	7.8	16.0	8.0	5.8	3.2
Glass	2.5	4.3	2.4	5.5	3.1
Benzene	0.88	0.221	0	1.6	0
Ethyl Alcohol	0.79	0.090	0	1.1	0
Water	0.998	0.218	0	1.5	0
Mercury	13.45	0.240	0	0.42	0

Solids: $v_{long} = \sqrt{\dfrac{B + \frac{4}{3}\mu}{\rho}} \qquad v_{trans} = \sqrt{\dfrac{\mu}{\rho}}$

Fluids: $v_{long} = \sqrt{\dfrac{B}{\rho}} \qquad v_{trans} = 0$

*The units are shown in parentheses following the symbol. Thus, for example, the density of aluminum is 2.7×10^3 kg/m^3.

Pa. The density of water at 20°C is 998 kg/m^3. The speed of sound in water at 20°C is therefore

$$v = \sqrt{\frac{2.18 \times 10^9 \text{ Pa}}{998 \text{ kg/m}^3}} = 1.48 \text{ km/s}$$

This is nearly five times the speed of sound in air at 20°C.

In more sophisticated treatments of acoustic waves, the ratio B/ρ appearing in Eq. 18.26 is replaced by the partial derivative $\partial P/\partial\rho$, where P denotes the fluid pressure. The wave speed as given by Eq. 18.26 then becomes

$$v_{long} = \sqrt{\frac{\partial P}{\partial \rho}} \tag{18.28}$$

Example 4 The Speed of Sound in Air

For ordinary sound waves in a gas, the variations in pressure and density are related by

$$P = C\rho^{\gamma} \qquad \gamma, C = \text{constants}$$

We can use this relation and Eq. 18.28 to determine the speed of sound in air. The constant γ (Greek letter gamma) has the value 1.40 for air. From $P = C\rho^{\gamma}$ we find

$$\frac{\partial P}{\partial \rho} = \gamma C \rho^{\gamma - 1} = \frac{\gamma P}{\rho}$$

Thus, from Eq. 18.28 the speed of sound in air is

$$v_{sound} = \sqrt{\frac{\gamma P}{\rho}}$$

Under standard conditions (0°C, 1 atm)

$$P = 1 \text{ atm} = 1.01 \times 10^5 \text{ Pa} \qquad \rho = 1.29 \text{ kg/m}^3$$

$$\gamma = 1.40$$

we find

$$v_{sound} = 331 \text{ m/s}$$

This is approximately 740 mi/hr.

In real fluids, particularly in liquids, viscosity (fluid "friction") gives the medium some resistance to shear. This is because transverse waves produce relative motion of fluid layers, which brings viscous shear stresses into play. Whereas a *static* shear stress would lead to continued relative motion, the *oscillatory* shear stresses characteristic of wave motions produce displacements that average to zero. Thus, transverse (shear) waves can propagate in viscous fluids. In particular, transverse ultrasonic waves can travel considerable distances through liquids.

Viscosity has another important effect on waves in fluids. Viscous stresses convert the energy of a wave into heat—just as frictional forces convert the kinetic energy of a skidding auto into heat. As a result, the amplitude decreases steadily as the wave progresses. We speak of this dwindling of the wave amplitude as wave *attenuation*.

Ultrasonic waves are used widely in the study of solids and liquids. Measurements of the wave speeds (v_l, v_t) yield values of the elastic constants (B, μ) of solids and liquids via Eqs. 18.24, 18.25, and 18.26. Ultrasonic waves have many practical applications. Reflections of ultrasound by cracks and impurities make such defects "visible," just as reflected light reveals a crack in a window pane. Medical applications of ultrasonic waves include the inspection of a fetus within its mother's body and the detection of tumors. The mechanical properties of metal castings can be improved by passing ultrasonic waves through the molten material. The waves pulverize tiny impurities, remove gas bubbles, and homogenize the molten metal.

The solid portions of the earth support both longitudinal and transverse elastic waves. Much of our knowledge about the interior of the earth has been inferred from measurements on *seismic* waves generated by earthquakes or man-made explosions. An earthquake generates both longitudinal and transverse waves. Seismologists refer to the faster-moving longitudinal disturbances as primary, or "P" waves. The slower transverse waves are called secondary, or "S" waves. Because of the difference in their speeds, the P waves reach a seismic station before the S waves. The time delay between their arrival can be used to determine the distance from the station to the epicenter.*

Studies of the reflection and transmission of S and P waves have given us the following picture of the internal structure of the earth: The outer layer, called the *crust*, is relatively thin. The thickness of the crust is approximately 50 km, less than 1% of the radius of the earth (6.37×10^3 km). Beneath the crust, extending to a depth of 2900 km, is the *mantle*. Both S and P waves travel freely through the mantle, indicating that it is a solid.

The central portion of the earth is called the *core*. It consists of a solid portion, the inner core, surrounded by a fluid, the outer core. The fluid nature of the outer core is inferred from observations of the transverse S waves. The S waves are strongly reflected at the mantle–core boundary. The transmitted S wave travels at a greatly reduced speed in the outer core and is severely attenuated. These characteristics lead us to conclude that the outer core is a dense liquid.

4. Elastic Surface Waves (Rayleigh Waves)

In addition to the longitudinal and transverse waves that can propagate through the body of an elastic solid, various types of *surface* waves can also be transmitted across the surface of a solid. Among these are what we call *Rayleigh waves*. The amplitude of a Rayleigh wave diminishes with depth below the surface, and is ignorably small at depths of more than a few wavelengths. The disturbance is therefore confined to the surface layers of a solid. The surface layers experience both transverse and longitudinal components of motion. In this respect, Rayleigh waves are similar to water waves.

Rayleigh waves are well suited for use in a variety of transducers† which process radar, television, and radio signals. Of primary importance is the fact that, for a given frequency, the wavelength of an electromagnetic wave is about 10^5 times greater than the wavelength of a Rayleigh wave. This follows from the basic kinematic relation (Section 17.2)

$$\nu\lambda = v \tag{17.8}$$

Electromagnetic waves travel at $v = 3 \times 10^8$ m/s. Rayleigh waves, on the other hand, travel at "acoustic" speeds of a few kilometers per second. Thus, for equal frequencies, and a Rayleigh wave speed of 3 km/s,

$$\frac{\lambda_{\text{electromagnetic}}}{\lambda_{\text{Rayleigh}}} = \frac{3 \times 10^8 \text{ m/s}}{3 \times 10^3 \text{ m/s}} = 10^5$$

If we want to transmit a wave without causing severe attenuation, the length of the transmitting device must be comparable to or greater than the wavelength. Because Rayleigh waves of a given frequency are 10^5 times shorter than electromagnetic waves of the same frequency, devices that transmit Rayleigh waves can be made much smaller than their electromagnetic counterparts. This reduction in size reduces the weight, which is an important consideration for applications in aircraft and space vehicles. Another desirable property of Rayleigh waves is that they are *nondispersive*—that is, their speed is independent of frequency. Therefore, a Rayleigh-wave device can process a multifrequency signal without any change in the waveform (zero distortion).

*The epicenter is the point on the surface of the earth vertically above the origin of the earthquake.

†A transducer is a device that converts electrical energy into mechanical energy and/or vice versa.

Questions

1. A stampeding herd of cattle can be "felt" before it is heard. Explain.
2. A rule of thumb for estimating the distance between yourself and a lightning stroke is this: Count the seconds that elapse between the flash and the arrival of the thunder, and divide this number by three. The result is the distance in kilometers. For example, a 6-s delay indicates a distance of 2 km. Explain the basis for this rule.
3. Suppose that you are studying in your room with the door closed and the window open. When someone opens the door and enters your room, you notice the curtains trying to "leap out" the open window. Explain. What will the curtains do when your guest leaves?
4. The sound speed in a solid is inversely proportional to the square root of the density. The densities of solids are typically 1000 times greater than the density of a gas like air. Why are sound speeds in most solids greater than the speed of sound in air?

18.4 Energy Flow and Wave Intensity

An important feature of wave motion is the transfer of energy. In this section we introduce the concept of *intensity*, a measure of the energy flow in a wave.

Suppose that a steady progression of waves emanates from a source. The wave energy E falling on a detector of collecting area A perpendicular to the direction of energy flow in a time t is proportional to both A and t. If we divide E by t and A, we get a quantity that is independent of A and t. Thus, the quantity E/tA is an intrinsic measure of energy flow in the wave. The rate of energy flow per unit area is E/tA. We call E/tA the **intensity** of the wave, and represent it by the symbol I:

$$\text{Intensity} = I \equiv \frac{E}{tA}$$

$$= \text{energy per second per unit area} \tag{18.29}$$

The SI unit of intensity is the watt per square meter. The intensity of sunlight reaching the upper levels of the earth's atmosphere is $1.38 \times 10^3\ \text{W/m}^2$. The minimum intensity of sound waves (at a frequency of 1000 Hz) to which the human ear is sensitive is approximately $10^{-12}\ \text{W/m}^2$.

We obtain a very useful expression for I by multiplying the numerator and denominator of the right side of Eq. 18.29 by the wave velocity, v. The result is

$$I = \left(\frac{E}{Avt}\right)v \tag{18.30}$$

As Figure 18.5 shows, the quantity Avt is the *volume* that contains the wave energy E. The quantity E/Avt, then, is the wave energy per unit volume. We call this quantity the *energy density* of the wave, u.

$$u = \text{wave energy per unit volume} = \frac{E}{Avt} \tag{18.31}$$

The intensity can now be written as

$$I = uv \tag{18.32}$$

We can use Eq. 18.32 to relate the intensity of a sinusoidal wave to the wave amplitude and frequency. Consider a sinusoidal wave traveling through an elastic medium—for example, a longitudinal sound wave traveling along a metal rod. The wave reveals itself as a traveling series of compressions and rarefactions. The atoms that make up the metal rod execute simple harmonic motion about their equilibrium positions. Their total energies are a sum of kinetic and potential energies. The total energy of an individual particle undergoing simple harmonic motion is a constant, equal to the maximum kinetic energy. For sinusoidal waves of the form

$$\psi(x,t) = D\cos(kx - \omega t)$$

the particle velocity is

$$\frac{\partial\psi}{\partial t} = \omega D\sin(kx - \omega t)$$

This particle velocity has a maximum value of ωD. If m is the mass of some segment of the metal rod undergoing this motion, the maximum kinetic energy is $\frac{1}{2}m(\omega D)^2$. Dividing by the volume of the segment gives the energy density, u. The result is

$$u = \frac{1}{2}\rho\omega^2 D^2 \tag{18.33}$$

where ρ is the mass density of the metal rod. Thus, the intensity for a sinusoidal wave in an elastic medium is

$$I = \frac{1}{2}\rho\omega^2 D^2 v \tag{18.34}$$

Note that the intensity is proportional to D^2, the *square* of the wave amplitude.

Wave Fronts

When you watch water waves travel across a pond your eye tends to follow a wave crest. The visual impression of a wave crest leads to the idea of a **wave front.** A wave front is defined as a surface over which the phase of the wave is constant. In a particular wave front, at a given moment of time, all particles of the medium are undergoing the same motion. Two types of wave fronts are particularly important. They are *plane* wave fronts and *spherical* wave fronts. If a wave travels in one fixed direction, we call it a *plane wave.* For example, a sound wave in which the pressure differences $P(x,t)$ vary sinusoidally,

$$P(x,t) = P_0\sin(kx - \omega t)$$

is a plane wave traveling in the x-direction. The wave fronts of plane waves are planes that are perpendicular to the direction of propagation (Figure 18.6a).

Spherical wave fronts can be produced by a point source—a source whose dimensions are small compared with the distance to an observer. If waves travel outward in all directions from a point source, the phase of one of these waves will be the same at points equidistant from the source, that is, on the surface of a sphere centered on the source (see Figure 18.6b). Therefore, the wave fronts are spherical surfaces and the waves are called *spherical waves.*

If a wave front diverges, the energy carried by the wave spreads over a larger and larger area, and consequently the intensity decreases. The intensity of a plane wave is constant because plane wave fronts neither converge nor diverge.* The intensity of a spherical wave, on the other hand, is inversely proportional to the square of the distance from the source. This is so because the out-

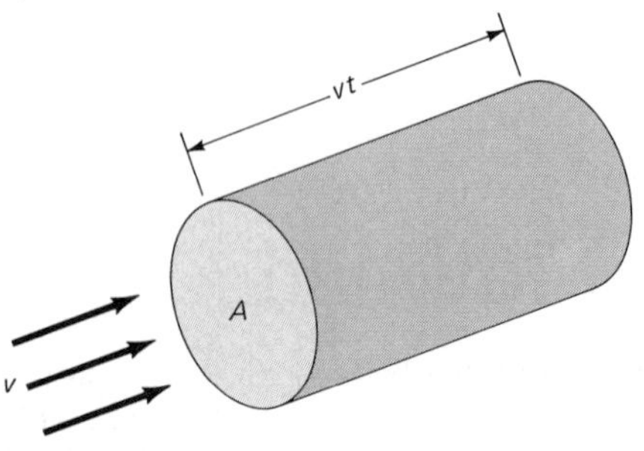

Figure 18.5

The waves that carry energy E across area A travel a distance vt in time t and thus "fill" a volume Avt.

*This assertion assumes that the medium through which the wave travels does not absorb energy from the wave and does not feed energy into the wave.

Figure 18.6a

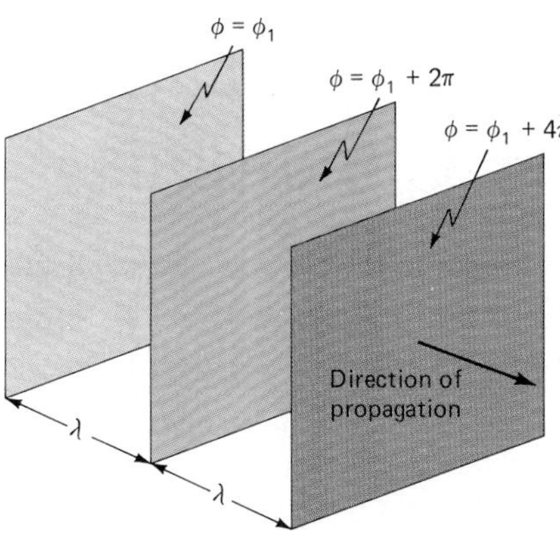

Figure 18.6b

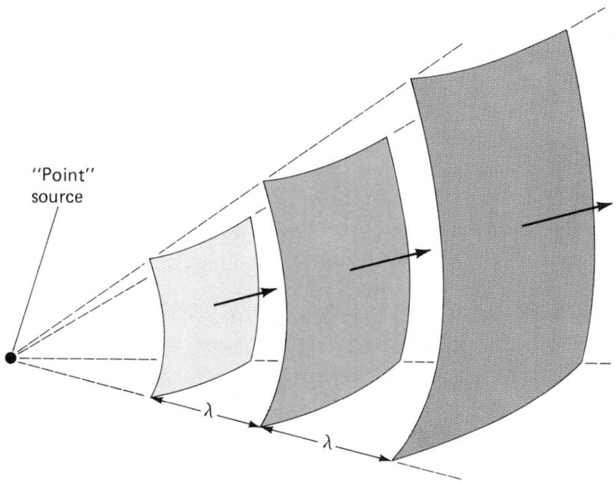

Figure 18.7

The same energy passes through all spherical surfaces. The area of the spherical surfaces is directly proportional to r^2. Thus, the energy per unit area is inversely proportional to r^2.

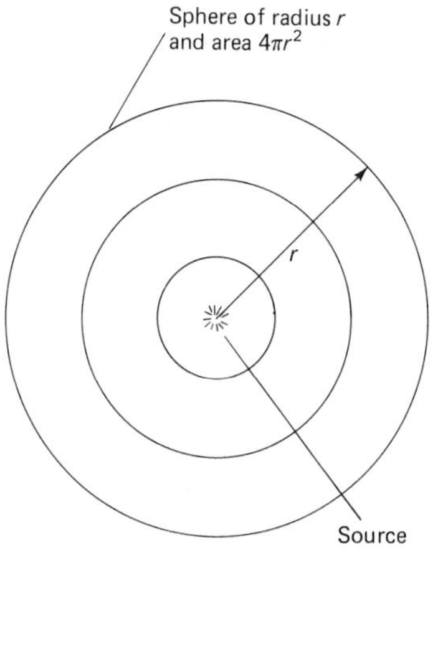

going waves must spread over larger and larger surface areas, while obeying the energy conservation law. To understand this inverse square law variation of intensity, consider spherical waves leaving the point source of Figure 18.7. Next, imagine a series of spherical surfaces concentric with the source (*geometric* surfaces, not real physical surfaces). The rate of energy flow across each such spherical surface must be the same because energy is conserved. This presumes that no energy is absorbed or released between surfaces. Thus, the rate of energy flow, E/t, is the same for each spherical surface. The surface area of a sphere of radius r is $4\pi r^2$. The rate of energy flow is proportional to the product of the intensity (I) and the surface area,

$$I \cdot 4\pi r^2 = \frac{E}{t} = \text{constant}$$

Because E/t is constant, it follows that the intensity of a spherical wave varies inversely as the square of the distance from the source

$$I_{\text{spherical wave}} \propto \frac{1}{r^2} \qquad (18.35)$$

Finally, recall that for sinusoidal waves the intensity is proportional to the square of the wave amplitude (Eq. 18.34). As a consequence the amplitude of a spherical wave is inversely proportional to the distance from the source,

$$\text{amplitude of spherical wave} \propto \frac{1}{r} \qquad (18.36)$$

Sound Wave Intensity and Intensity Level

The intensity of ordinary sound waves in air and the auditory sensation of loudness are not linearly related. In other words, doubling the intensity does not double the loudness of the sound. Instead, there is a roughly *logarithmic* relation between intensity and the sensation of loudness. For frequencies near 1000 Hz the loudness seems twice as great if the intensity is increased by a factor of about 10. This qualitative notion of loudness is measured by a quantity called the *intensity level*, which *by definition*, exhibits a logarithmic relation to the intensity. Specifically, the intensity levels of two waves of intensity I and I_o are defined to differ by β decibels (dB), where

$$\beta = 10 \log_{10}\left(\frac{I}{I_o}\right) \text{ dB} \qquad (18.37)$$

Thus, for example, if I is twice I_o, then

$$\log_{10}\left(\frac{I}{I_o}\right) = \log_{10} 2 = 0.301$$

and their intensity levels differ by 3.01 dB. In the case of ordinary sound waves, I_o is taken to be 10^{-12} W/m^2. This serves as a standard of intensity that is near the threshold of hearing for most people (at 1000 Hz). The intensity of ordinary speech is on the order of 10^{-6} W/m^2, and gives intensity levels in the range of 60 dB.

The intensity of sinusoidal waves is proportional to the square of the wave amplitude. For sound waves in air, the wave amplitude is the maximum pressure, measured relative to the undisturbed atmospheric pressure. Thus,

$$\frac{I}{I_o} = \frac{P^2}{P_o^2}$$

where P and P_o are the pressure amplitudes. An alternate definition of the intensity level for sound waves is therefore

$$\beta = 20 \log_{10}\left(\frac{P}{P_o}\right) \text{ dB} \qquad (18.38)$$

The standard for sound wave pressure is $P_o = 2 \times 10^{-5}$ Pa, the threshold for hearing at 1000 Hz. Compared to the typical atmospheric pressure (1 atm $= 1.013 \times 10^5$ Pa), P_o is less than one billionth of an atmosphere. Ordinary sound waves are indeed small-amplitude waves.

Example 5 **Sound Intensities at Rock Concerts**

The intensity level of the sound generated by some rock musicians frequently reaches 120 dB. The intensity level of ordinary speech is typically 60 dB. The ratio of the actual intensities of rock music and ordinary speech follows from Eq. 18.37:

$$\beta_{rock} - \beta_{speech} = 10 \log_{10}\left(\frac{I_R}{I_o}\right) - 10 \log_{10}\left(\frac{I_s}{I_o}\right)$$

Recalling that $\log a - \log b = \log(a/b)$ gives

$$120 - 60 = 10 \cdot \log_{10}\left(\frac{I_R}{I_s}\right)$$

and thus

$$\log_{10}\left(\frac{I_R}{I_s}\right) = 6$$

Therefore

$$\frac{I_R}{I_s} = 10^6$$

The intensity of rock music can be a million times greater than that of ordinary speech. Since medical studies show that prolonged exposure to high-intensity sound is dangerous, the conclusion should be obvious. Don't sit in the front row at a rock concert!

Questions

5. Suppose that you must convince a 10-year-old child that waves transfer energy. What are some examples of wave energy transfer that might make sense to the child? Allow for the likelihood that the child has only a fuzzy notion of the concept of energy.

6. Doubling the intensity of a sound wave raises the intensity level by 3.01 dB. What change in intensity level results if the intensity is quadrupled?

7. Approximately how many radios (each delivering sound of the same intensity) are needed to sound twice as loud as a single radio?

18.5 Standing Waves and Boundary Conditions

In Chapter 17 we introduced the basic kinematic relation for sinusoidal waves,

$$\nu\lambda = v \qquad (17.8)$$

As we have seen, the speed v is determined by the medium. Depending on the physical constraints imposed on the medium, either ν or λ may be regarded as the independent variable. Up to now, we have envisioned situations where a source drives a steady succession of waves through some medium. The frequency of the source can be set at any value, within limits. We have therefore made frequency the independent variable. The wavelength becomes the dependent variable, and its value is dictated by Eq. 17.8. Let us now turn to a different situation, where physical constraints on the medium fix the value of λ. The frequency, as given by Eq. 17.8, thereby becomes the dependent variable.

The physical constraints on a medium are generally referred to as *boundary conditions*. To envision the idea of boundary conditions, consider a uniform string stretched between two rigid supports a distance L apart (Figure 18.8). No matter how the string moves, its displacement must be zero at both ends, where the string is anchored. We can describe these boundary conditions mathematically as follows: Label the positions of the ends of the string $x = 0$ and $x = L$, and let $\psi(x,t)$ denote the transverse displacement of the string. The boundary conditions are then

$$\psi(0,t) = 0 \qquad (18.39)$$
$$\psi(L,t) = 0$$

Now let's see if sinusoidal waveforms can satisfy these boundary conditions. The boundary conditions can be satisfied by a superposition of two sinusoidal waves with equal amplitudes, traveling in opposite directions. Such a superposition results in *standing waves*, like those shown in Figure 18.9. At any given position x the string undergoes simple harmonic motion in the vertical direction, just as it does in a traveling wave. However, in a standing wave the amplitude varies with position along the string. The wave-

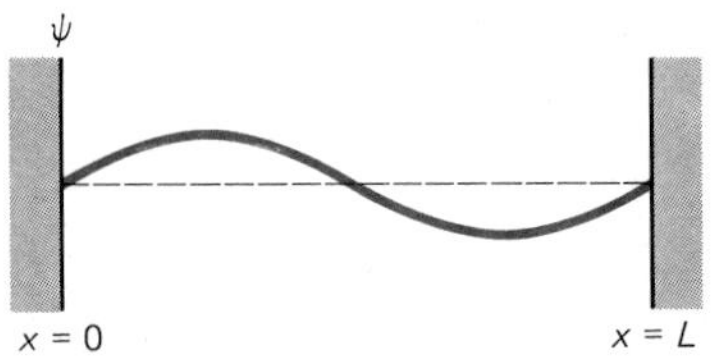

Figure 18.8

A standing wave on a string. The only physically acceptable solutions of the wave equation for the string are those for which the displacement is zero at both ends.

Figure 18.9

Standing waveforms of a string.

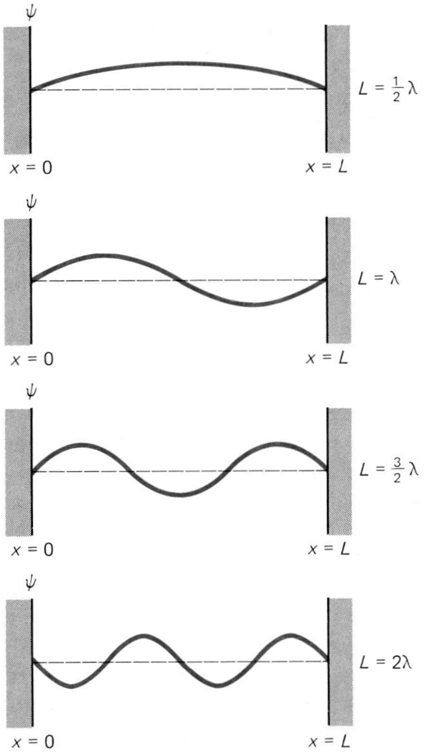

form gives no impression of advancing crests—hence the designation *standing* wave.*

Positions of zero displacement in a standing wave are called *nodes*. Only sinusoidal waveforms that have nodes at $x = 0$ and $x = L$, *for all values of* t, can satisfy the boundary conditions. If we were to examine a sinusoidal waveform, we would discover that the distance between adjacent nodes is one-half wavelength. Therefore, acceptable sinusoidal waveforms are those for which an *integral* number of half wavelengths "fit" into the distance L,

$$L = \frac{1}{2}\lambda,\ \lambda,\ \frac{3}{2}\lambda,\ 2\lambda,\ \ldots \qquad (18.40)$$

or, in a more succinct notation,

$$L = n \cdot \frac{1}{2}\lambda \qquad n = \text{positive integer} \qquad (18.41)$$

Figure 18.9 shows the waveforms at one instant of time for $n = 1, 2, 3, 4$.

The condition imposed by Eq. 18.41 can be derived more formally. It is worthwhile to learn how this formal derivation is carried out for standing waves on a string because the techniques carry over to many types of boundary value problems. First, we note that no single traveling wave can satisfy the boundary conditions. For example, the traveling wave $\psi = A\sin(kx - \omega t)$ would satisfy the condition $\psi(0,t) = 0$ at time $t = 0$ but not at an instant before or at an instant later. However, a standing waveform that does satisfy the boundary conditions can be built up by a superposition of two waves traveling in opposite directions. As you saw in Section 17.4, the superposition of the two waves

$$\psi_R(x,t) = A\sin(kx - \omega t)$$

$$\psi_L(x,t) = A\sin(kx + \omega t)$$

gives a standing wave of the form

$$\psi(x,t) = \psi_R + \psi_L = 2A\sin kx \cdot \cos\omega t \qquad (18.42)$$

Equation 18.42 is the mathematical description of a standing wave. It describes a simple harmonic motion ($\cos\omega t$) with an amplitude ($2A\sin kx$) that varies with position along the string. This waveform clearly satisfies the boundary condition at $x = 0$ because $\sin 0 = 0$. The boundary condition at $x = L$,

$$\psi(L,t) = 0 = 2A\sin kL \cdot \cos\omega t$$

is also satisfied for all times t, provided that

$$\sin kL = 0 \qquad (18.43)$$

The condition $\sin kL = 0$ is satisfied for special values of kL, namely

$$kL = n\pi \qquad n = 1, 2, 3, \ldots \qquad (18.44)$$

Recalling that $k = 2\pi/\lambda$, we can convert Eq. 18.44 into Eq. 18.41, which is the condition deduced earlier:

$$L = n \cdot \frac{1}{2}\lambda \qquad n = \text{positive integer} \qquad (18.41)$$

Equation 18.41 illustrates our earlier statement that "physical constraints on the medium fix the value of λ." We can see from the equation that only definite values of λ satisfy the boundary conditions and produce a standing waveform.

We can find the frequencies of the standing waves by using Eq. 18.41 and the kinematic relation, $\nu\lambda = v$. Thus,

$$\nu = \frac{v}{\lambda} = n\left(\frac{v}{2L}\right) \qquad n = 1, 2, 3, \ldots \qquad (18.45)$$

For waves on a uniform string the speed v is given by the square root of the ratio of the tension and mass per unit length, as shown by Eq. 18.11:

$$v = \sqrt{\frac{T}{M/L}}$$

Substituting this value into Eq. 18.45 we get

$$\nu = \frac{n\sqrt{T/(M/L)}}{2L} \qquad n = 1, 2, 3, \ldots \qquad (18.46)$$

for the standing-wave frequencies. Equation 18.46 shows that the *spectrum* of standing-wave frequencies consists of integral multiples of the *fundamental frequency* ($n = 1$),

$$\nu_{\text{fund}} = \frac{\sqrt{T/(M/L)}}{2L} \qquad (18.47)$$

The guitar provides a familiar example of this relation. Increasing the tension, T, in a guitar string raises the string's fundamental frequency. We hear this increase in frequency as a "higher" note when we pluck the string. Decreasing the length, L, through fingering accomplishes the same result.

*If equipment is available, standing waves on a string can be demonstrated. Stroboscopic illumination makes it possible to "stop" the string and reveal forms like those shown in Figure 18.9.

Example 6 Standing Waves on a String

A uniform string of mass 0.1 kg is stretched between two supports separated by a distance of 3 m. The tension in the string is 20 N. What are the speed, the fundamental frequency, and the corresponding wavelength of the waves along the string? The fundamental frequency corresponds to the longest wavelength, which is twice the length of the string; $\lambda = 6$ m. The wave speed is given by Eq. 18.11,

$$v = \sqrt{\frac{T}{M/L}} = \sqrt{\frac{20}{(0.1)/3}} \text{ m/s} = 24.5 \text{ m/s}$$

The fundamental frequency follows from Eq. 18.47,

$$\nu_{\text{fund}} = \frac{v}{2L} = \frac{24.5 \text{ m/s}}{6 \text{ m}} = 4.08 \text{ Hz}$$

Our treatment of standing waves on a string illustrates a general consequence of boundary conditions:

Boundary conditions identify the physically acceptable solutions contained within the larger group of mathematically acceptable solutions of the wave equation.

To see this more clearly, consider the wave equation

$$\frac{\partial^2 \psi}{\partial t^2} = v^2 \frac{\partial^2 \psi}{\partial x^2}$$

One set of mathematically acceptable solutions to this equation has the form

$$\psi(x,t) = A \sin(kx - \omega t) + B \sin(kx + \omega t)$$

The only constraint imposed by the wave equation is that $\omega/k = v$. In particular, A and B can have any values. However, for standing waves on a string only solutions with $A = B$ are admissible, and the values of k are restricted by the condition $kL = n\pi$.

These restrictions do not limit the physically acceptable solutions to sinusoidal waveforms. For example, a triangular waveform can be set up by plucking the string at its midpoint. Such a triangular waveform can be synthesized by a *superposition* of sinusoidal waveforms. In Figure 18.10 the dashed line shows the single sinusoidal waveform

$$\psi_1 = \frac{8}{\pi^2} \sin \frac{\pi x}{L}$$

The dotted line shows the sum (superposition) of three sinusoidal waveforms

$$\psi_3 = \frac{8}{\pi^2} \left[\sin \frac{\pi x}{L} - \frac{1}{9} \sin\left(\frac{3\pi x}{L}\right) + \frac{1}{25} \sin\left(\frac{5\pi x}{L}\right) \right]$$

Even better "fits" to the triangular waveform can be achieved by adding properly chosen sinusoidal waveforms to ψ_3. Figure 18.10 shows how the superposition of one (dashed line) and three (dotted line) sinusoidal waveforms approximates the triangular waveform. The technique of building up waveforms by superposition of sinusoidal waveforms is called *Fourier synthesis*. The sum of sinusoidal waveforms that describes the synthesis is called a *Fourier series*. Fourier synthesis can be used to synthesize various types of waveforms in electrical circuits.

The effect of boundary conditions—which is to sort out physically acceptable solutions—is not limited to waves on a string, nor to waves in material media. Electromagnetic waves and the probability waves of quantum mechanics are frequently subjected to boundary conditions. Thus, the ideas and techniques developed in connection with standing waves on a string can be extended to many other types of wave phenomena.

You don't need to be a mathematical wizard to gain insight into standing waves. In fact, it is possible to *estimate* the fundamental frequency for any oscillatory system with-

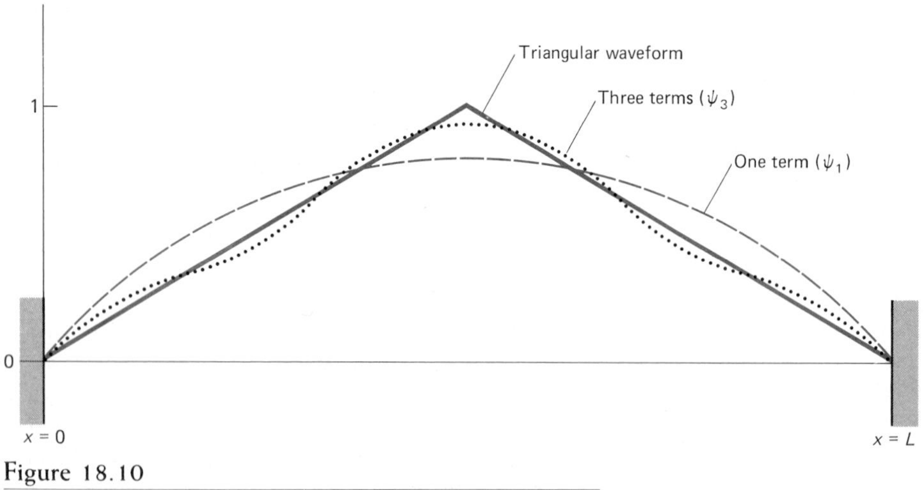

Figure 18.10

The triangular waveform can be synthesized by superimposing standing waveforms. The dashed line indicates the approximation achieved with a standing wave of the fundamental frequency. The dotted line shows the approximation obtained by superimposing three standing waves having appropriate amplitudes.

out solving the wave equation.* The fundamental frequency is the *lowest* frequency for a standing waveform. From $\nu = v/\lambda$ it follows that the fundamental frequency corresponds to the *maximum* wavelength. Thus,

$$\nu_{\text{fund}} = \frac{v}{\lambda_{\text{max}}} \qquad (18.48)$$

where λ_{max} is the maximum standing wavelength and v is the speed of the wave. The precise value of λ_{max} depends on geometric factors and the boundary conditions. However, we can estimate λ_{max} quite simply, and thereby obtain an approximate value for ν_{fund}. The estimate for λ_{max} is

$$\lambda_{\text{max}} \simeq \text{largest dimension of system} \equiv \pounds \qquad (18.49)$$

This estimate is readily justified because it is the only possible choice. (What other length could enter the problem?)

With $\lambda_{\text{max}} \approx \pounds$ we have

$$\nu_{\text{fund}} \approx \frac{v}{\pounds} \qquad (18.50)$$

$$\pounds = \text{largest dimension of system}$$

This estimate for the fundamental frequency is generally within a factor of 2 or 3 of the precise value obtained by solving the wave equation. Thus, Eq. 18.50 gives an *order-of-magnitude estimate*† for the fundamental frequency of standing waves. You can use this equation to double-check your formal solution of a given wave equation. If your solution indicates a fundamental frequency that differs by a factor of 1000 from your order-of-magnitude calculation, go back and check your mathematics.

Example 7 **Seismic-Wave Frequencies**

Earthquakes can excite standing waves that involve the entire earth. The approximate fundamental frequency of such waves is given by Eq. 18.50,

$$\nu_{\text{fund}} = \frac{v_{\text{seismic}}}{D}$$

where v_{seismic} is the speed of the seismic wave and D is the diameter of the earth. The seismic wave speed varies with depth and direction. However, an estimate of the seismic wave speed, which is typical of the speed of sound in solids, is

$$v_{\text{seismic}} \approx 11 \text{ km/s}$$

Taking $D \simeq 12{,}000$ km gives

$$\nu_{\text{fund}} \approx 10^{-3} \text{ Hz}$$

The reciprocal of ν_{fund}, the *period* of the fundamental mode of vibration, is about 1000 s, or 16 min. The motion to which this period refers is called the *breathing mode* because the radius of the earth expands and contracts slightly, as if the planet were breathing. How good is our estimate? A rigorous calculation sets the period at 1228.8 s. Experimental data fix the period at 1227.6 s. Because our value, 1000 s, is within a factor of 2 of these figures, it is a satisfactory order-of-magnitude estimate.

---

*P. A. M. Dirac, a Nobel Prize winner in physics, said, "I understand what an equation means if I can figure out the characteristics of its solution without actually solving it."

†"The only thing that matters in physics is the order of magnitude—except in cases of salary, where factors of two become important," stated Henry Fairbank.

Questions

8. Is there a transfer of energy along a standing wave on a string? Look to *symmetry* for support of your answer.

9. When standing waves are set up in a pan of water, the ends of the pan are called *antinodes*—positions of maximum displacement. For a rectangular pan, what is the relationship between the length of the pan and the wavelength of the fundamental mode (longest wavelength)? It may be helpful to sketch the profile of the water surface versus position along the pan.

Summary

For a uniform string under tension, small displacements from equilibrium propagate as waves with a speed

$$v = \sqrt{\frac{T}{M/L}} \qquad (18.11)$$

where T is the tension in the string and M/L the mass per unit length. The wave equation governing the propagation of such waves has the form

$$\frac{\partial^2\psi}{\partial t^2} = v^2\frac{\partial^2\psi}{\partial x^2} \qquad (18.12)$$

Many other types of waves can be described by the same equation, but with different physical significances for the variable $\psi(x,t)$ and different values for the wave speed, v. In general, the speed of a mechanical wave is determined by the ratio of an elastic factor, which measures the "stiffness" of the medium, and an inertial factor

$$v = \sqrt{\frac{\text{elastic factor}}{\text{inertial factor}}} \qquad (18.16)$$

Wave intensity is a measure of the energy flow in a wave. The intensity, I, defined as the time rate of energy flow per unit area or as the rate of energy flow per unit area, can be expressed as the product of the energy density (u) and the wave velocity (v):

$$I = uv \qquad (18.32)$$

A wave front is a surface over which the phase of a wave is constant. For plane waves the wave fronts are planes perpendicular to the direction of propagation. For spherical waves the wavefronts are spheres centered on the source of the waves.

Standing waves can arise in situations where boundary conditions dictate the possible wavelengths. For standing waves on a string, the boundary conditions require that the length of the string equal an integral number of half wavelengths. The corresponding wave frequencies are integral multiples of the lowest (or fundamental) frequency. In general, an order-of-magnitude estimate of the fundamental frequency is given by

$$\nu_{\text{fund}} = \frac{v}{\pounds} \tag{18.50}$$

where £ is the largest dimension of the system and v is the wave speed.

Suggested Reading

D. M. Boore, The Motion of the Ground in Earthquakes, *Sci. Am.* 237, 68 (December 1977).

A. A. Few, Thunder, *Sci. Am.* 233, 80 (July 1975).

G. S. Kino and J. Shaw, Acoustic Surface Waves, *Sci. Am.* 227, 50 (September 1972).

Problems

Section 18.1 Waves on a String

1. Verify that the following waveforms satisfy the wave equation, Eq. 18.12.
(a) $\psi(x,t) = (x - vt)^2$
(b) $\psi(x,t) = \dfrac{1}{a^2 + (x - vt)^2}$ $\quad a = \text{constant}$
(c) $\psi(x,t) = e^{j(kx-\omega t)}$ $j, k, \omega = \text{constants}, v = \omega/k$
2. Verify that $\psi(x,t) = B\cos(kx - \omega t)$ satisfies Eq. 18.12, provided that $v = \omega/k$. (B is a constant.)
3. (a) The tension in a cord is 60 N. The mass per unit length is 0.01 kg/m. What is the wave speed along the rope? (b) If the rope is 8 m long, what time is required for a wave to travel from end to end?
4. A clothesline has a mass of 0.75 kg. It is stretched between two posts that are 10 m apart. What tension in the line will enable a wave to travel its length in 0.11 s?
5. A violin string has a length of 0.76 m and a mass of 6 gm. The tension in the string is 104 N. Determine the speed of waves along the string. The speed of sound in air at 0°C is 331 m/s. Is the wave speed significantly greater or smaller than the speed of sound in air?
6. A bucket of water weighing 80 N hangs at the end of a rope. The rope is 6 m long and has a mass of 0.60 kg. (a) Determine the tension in the rope at both ends. (b) Determine the wave speed at both ends.
7. A uniform rope of weight Mg and length L hangs vertically. The tension at the free end is zero. The tension increases linearly with the distance above the free end, reaching the value Mg at the point of support. (a) Show that the wave speed a distance x above the free end is given by

$$v = \sqrt{gx}$$

(b) Show that the time required for a wave to travel the length of the rope is

$$t = 2\sqrt{\frac{L}{g}}$$

(*Hint:* The speed varies along the length of the rope because the tension varies. The tension equals the weight at one end and falls to zero at the free end. Start with

$$\frac{dx}{dt} = \sqrt{gx}$$

and separate variables before integrating to relate t and L.)

Section 18.3 Mechanical Waves: A Sampling

8. A jet plane takes off in Hawaii and flies to California at an average speed of 800 km/hr. At the moment of

takeoff a volcanic eruption occurs, generating a tidal wave, or tsunami, in the ocean and sound waves in the air. Determine the order of arrival in California for the plane, tsunami, and sound waves. Take h = 5 km for the depth of the Pacific Ocean. Take the pressure and density of air to be 1.06×10^5 Pa and 1.31 kg/m^3, respectively.

9. The speed of sound in air equals $\sqrt{1.4P/\rho}$, an expression that is independent of frequency. As a consequence, sound waves should not exhibit dispersion. Waves of all frequencies should travel at the same speed. What elementary observations on sound are consistent with the absence of dispersion?

10. Do you think that you can hear sound whose wavelength is larger than the size of your ear? Calculate the wavelengths corresponding to the auditory limits of 20 Hz and 20 kHz.

11. The bulk modulus for mercury at 40°C is 2.40×10^9 Pa. The density of mercury at 40°C is 13,450 kg/m^3. Calculate the speed of sound in mercury at 40°C.

12. Determine the speed of deep-water waves having a wavelength of 20 m. Take $g = 9.80$ m/s^2. Does your answer differ significantly if you ignore the surface tension term in Eq. 18.21? ($S = 0.0736$ N/m; ρ = 1002 kg/m^2.)

13. Imagine that you are standing near the shore in water with a depth of 1.5 m. Ocean waves near the shore are coming in. A surfer, riding a wave, collides with you. What is the speed of impact in meters per second and in miles per hour?

14. Plot v versus h for shallow-water waves, Eq. 18.20, and for deep-water waves, Eq. 18.21, taking λ = 10 m. (Ignore surface tension.) Let h range from 0 to 20 m. Indicate on your plot where you expect each curve to give a valid description of the wave speed.

15. Using Eqs. 18.24 and 18.25, prove that the speed of a longitudinal elastic wave in a solid is at least 15% greater than the speed of a transverse elastic wave in the same material. Make use of Table 18.1 to evaluate $(v_{\text{long}} - v_{\text{trans}})/v_{\text{trans}}$ for any two solids listed therein. Do the tabulated values conform to the theoretical limit imposed by Eqs. 18.24 and 18.25?

16. The speed of P waves and S waves in the outer core of the earth is 8.15 km/s and 1.40 km/s, respectively. The density is 9400 kg/m^3. Determine the bulk modulus (B) and the shear modulus (μ) for the outer core.

17. Ultrasonic waves can be generated with frequencies ranging from 20,000 Hz to 10^{10} Hz. Assuming a speed of 1.47 km/s (water), determine the corresponding range of wavelengths.

18. One end of a long steel beam is struck with a hammer. The blow generates longitudinal and transverse waves that travel the length of the beam, are reflected, and return. The round trip times for the longitudinal and transverse waves are 3.2 ms and 5.8 ms, respectively. Determine the *ratio* B/μ for steel. Compare your answer with the value of B/μ computed from the data of Table 18.1.

19. A Rayleigh wave with a frequency of 100 MHz, travels at a speed of 5.2 km/s. Determine its wavelength.

Section 18.4 Energy Flow and Wave Intensity

20. A laser beam transfers an energy of 6 J across an area of 2×10^{-6} m^2 in a time of 10^{-9} s. What is the intensity of the beam?

21. Calculate the energy density for the laser beam in Problem 20, assuming that the wave travels at the speed of light, 3×10^8 m/s.

22. The intensity of a sound wave is raised, first from 10^{-12} W/m^2 to 6×10^{-12} W/m^2 and then to 10^{-11} W/m^2. Determine the sequential changes in intensity level (in decibels) for the two changes.

23. Your noisy neighbor creates a sound wave that has an intensity of 100 dB. (a) What is the intensity in watts per square meter? (b) What is the wave energy per cubic meter?

24. The pressure amplitude of a wave is raised from 2×10^{-5} Pa to 4×10^{-5} Pa. (a) By how many decibels is the intensity level raised? (b) By what *factor* is the intensity increased? (c) If the pressure amplitude is raised by an additional 2×10^{-5} Pa (to 6×10^{-5} Pa), will the intensity level rise by the same number of decibels as in part (a)? Back up your answer with an explanation and a calculation.

25. The threshold of auditory pain sets in at an intensity level of about β = 140 dB. (a) What is the corresponding sound wave intensity? (b) Determine the corresponding sound pressure in pascals and in atmospheres.

26. An SST produces sonic booms in which the pressure changes by about 100 Pa. *Estimate* the intensity level of the boom in decibels. (The sonic boom is a shock wave, a far cry from the sinusoidal waveforms for which Eq. 18.38 [$\beta = 20 \log_{10}(P/P_o)$] strictly applies. Thus Eq. 18.38 can be relied on only for an estimate of the intensity level.)

27. The intensity of sunlight incident at the top of earth's atmosphere is 1.38 kW/m^2. (This is called the *solar constant*.) (a) The *absolute luminosity* of a star is defined as the total energy it radiates per second. Show that the absolute luminosity of the sun is 3.9×10^{26} W. (The earth–sun distance is 1.496×10^{11} m.) (b) Of the incident solar energy, about 39% is reflected back into space. The remainder is absorbed by the atmosphere and surface. Calculate the total solar energy absorbed by the earth in 1 yr. What fraction of the solar energy would have to be utilized in order to match the annual U.S. electrical energy "consumption" of about 1.5×10^{12} kW · h? The radius of the earth is 6.37×10^6 m.

28. A surface wave that spreads out from a line source generates what are called cylindrical wave fronts. Present an argument that shows that the intensity of a *cylindrical* wave varies inversely as the first power of

the distance from the source ($I_{cyl} \propto 1/r$). Earthquakes generate surface waves (in addition to the P and S waves that travel through the body of the earth.) The surface waves are particularly destructive, even at great distances from the origin of the disturbance, because their intensity diminishes only as $1/r$ rather than $1/r^2$.

29. A radio antenna radiates energy with a power of 50,000 W. What is the intensity of the radio wave at a distance of 100 km, assuming that the energy spreads uniformly in all directions, with (a) the half that strikes the ground being completely absorbed; (b) the half that strikes the ground being completely reflected, thereby enhancing the energy radiated directly outward?

30. The intensity level of an orchestra is 85 dB. A single violin achieves an intensity level of 70 dB. How does the intensity of the full orchestra compare with that of the violin?

31. The intensity of sunlight incident on the upper atmosphere of the earth is 1.38 kW/m^2. How many people, each generating sound at an intensity 10,000 times the threshold level, I_o, would be needed to produce an intensity equal to that of the sun? Compare your result with the earth's population ($\approx 5 \times 10^9$).

32. A light meter records an intensity of 16 W/m^2 at an unknown distance from a point source. Increasing the distance by 1 m reduces the intensity to 9 W/m^2. What is the distance from the source to the point where the intensity is 9 W/m^2?

Section 18.5 Standing Waves and Boundary Conditions

33. Verify that the condition $kL = n\pi$ of Eq. 18.44 reduces to $L = n \cdot \frac{1}{2}\lambda$, Eq. 18.41.

34. A student wants to establish a standing wave on a wire 1.8 m long that is clamped at both ends. (The wave speed is 509 m/s.) (a) What is the minimum frequency disturbance that she should apply to the wire? (b) What next higher frequency should she apply to create a (different) standing wave?

35. Standing waves having a frequency of 9.9 Hz are established on a rope 3.2 m long. The wavelength is twice the length of the rope ($n = 1$). Determine the wave speed.

36. Estimate the fundamental frequency for longitudinal sound waves (standing waves) in your skull. Assume a speed of 4 km/s.

37. Make an order-of-magnitude estimate of the fundamental frequency for standing waves on a lake 6 km long and 12 m deep. (Assume shallow-water waves.)

chapter 19

SPECIAL RELATIVITY

Preview

It is difficult to get rid of deep-rooted prejudices, but there is no other way.

A. Einstein and L. Infeld

Albert Einstein's theories of special relativity (1905) and general relativity (1915) together represent one of mankind's greatest intellectual achievements (Figure 19.1). In this chapter, we develop the theory of special relativity, or measurement in constant-velocity reference frames. We present special relativity as a generalization of Galilean relativity, a generalization that is required for speeds comparable to the speed of light but that reduces to Galilean relativity for speeds small compared to the speed of light. The basic concepts of length and time are re-examined.

This chapter is optional, which means that you can skip it without any loss of continuity. So, you may ask, why do we include it here at all? There are two major reasons: First, this chapter is an introduction to 20th-century physics. In the preceding chapters on dynamics you have been studying physics that is valid for ordinary velocities—meters per second or kilometers per second. This optional chapter, however, sets the foundation for developing the physics that give agreement with experiment for phenomena involving velocities up to the velocity of light. Chapter 20 uses the appropriate relativistic expressions for energy and momentum. Special relativity enables us to study dynamics from a 20th-century viewpoint.

The second reason for including this chapter is that special relativity is one of the most fascinating topics in physics. It involves re-educating your common sense to bring it into agreement with experimental facts. We think that you will find the strangeness, the basic simplicity, and the profound consequences of special relativity exciting.

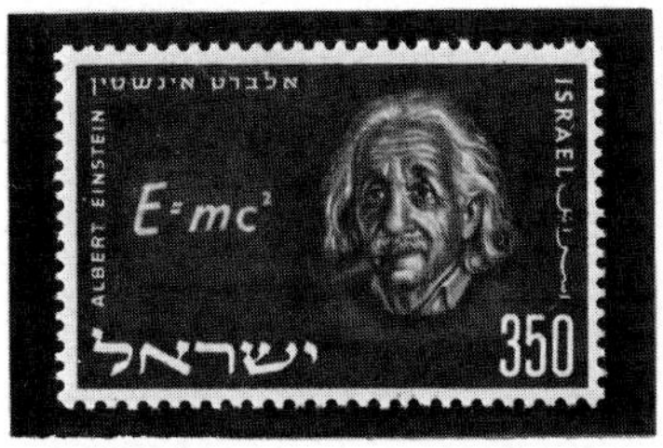

Figure 19.1

Albert Einstein (1879–1955). German–Swiss physicist. Famous for the development of the theory of special relativity in 1905 and the theory of general relativity in 1915, was awarded the Nobel Prize in 1921 for his work on the photoelectric effect.

19.1 Einstein's Postulates of Special Relativity

In the 19th century scientists had recognized that light had wave properties, and were searching for the hypothetical medium that propagated the light waves. Astronomical observations were made, repeated, and confirmed, and a variety of optical experiments were carried out. At every turn the hypothetical medium seemed to elude the investigator. The French mathematician Henri Poincaré spoke of a "conspiracy of nature" to conceal the propagating medium. It remained for the genius of Albert Einstein to realize that this *conspiracy* of nature was actually a *law* of nature. In 1905, recognizing that reasonable postulates had failed, Einstein postulated:

1 The laws of nature have the same form in all inertial reference frames.

2 The speed of light, is constant, the same in all inertial reference frames, independent of any relative motion of the source and of the observer.

In his first postulate Einstein asserted the existence of a single principle of relativity that is valid for mechanics, electromagnetism, and all of nature.

In his second postulate Einstein dispensed with the medium. Light is not a classical wave and does not need a medium to transmit it. Einstein's second postulate is incompatible with Galilean velocity addition.

Figure 19.2

Einstein as a boy. "Common sense is that thin veneer of prejudice we acquire before the age of 17."

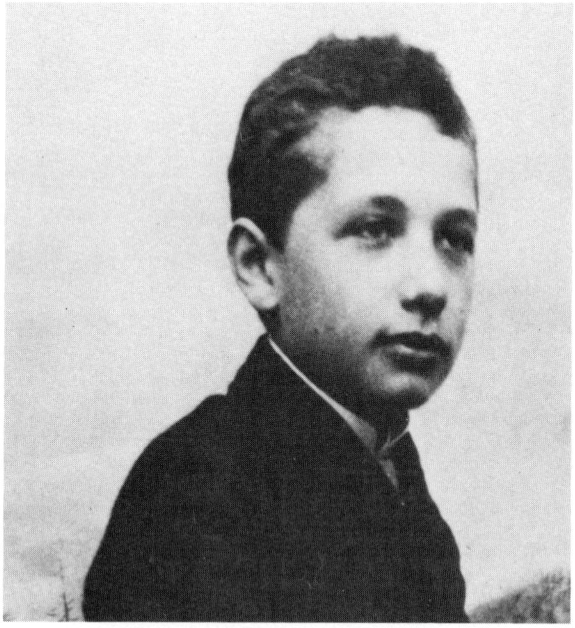

The validity of Einstein's postulates is judged on a very pragmatic basis. We ask, are their consequences in agreement with experiment? Although the postulates defy "common sense," their consequences are verified by experimental evidence and we must accept them as valid. Also, we should remember that our common sense has been formed in a world where velocities of most objects are on the order of meters per second, but light travels over a million times faster than do these everyday objects. The extension of everyday concepts to light is an extrapolation of over six orders of magnitude. Extrapolations are often necessary, but in this case extrapolated common sense fails. Einstein's postulates are verified by the experimental evidence.

This pragmatic judgment of Einstein's postulates on the basis of experiment is not the whole story. Einstein's postulates have great scope and generality. The "laws of nature" referred to in the first postulate include all the laws of our natural world. This universality appealed to many scientists who expected perfection in the natural laws much as the Greeks expected perfection in natural phenomena. Quite literally, the great esthetic appeal of Einstein's theory gained it a sympathetic hearing. Its ultimate acceptance as valid, however, is based on its agreement with experimental evidence.

Question

1. A lighthouse sends out a well-focused beam of light resulting in a spot of light seen on clouds at the horizon. Suppose the light rotates once each second, and let the horizon be arbitrarily far away. Can the spot of light on the clouds move at a speed greater than c? Does this motion of the spot of light on the clouds violate Einstein's second postulate? Explain.

19.2 The Relativity of Time

Before we begin the derivation of Einstein's relativity transformations, it is helpful to see how Einstein's postulates force scientists to reconsider the nature of space and time. Consider the following example.

A rocket ship flashes past an asteroid at half the speed of light (Figure 19.3). An alien stationed on the asteroid sees the ship go by with speed $v = 0.5c$. The alien also sees a pulse of light, going in the same direction, overtake and pass the ship. The alien observes the speed of the light pulse to be c.

The pilot of the rocket ship knows that she has a speed $v = 0.5c$ relative to the asteroid. In accordance with Einstein's second postulate, she sees the light pulse pass her with speed c relative to the rocket. Keep in mind that all of these speeds (rocket relative to asteroid, light relative to rocket, and light relative to asteroid) are measured speeds.

Suppose that each of the two observers, the rocket pilot and the alien, uses the ordinary Galilean addition of speeds, $u = u' + v$ (Eq. 4.23 of Section 4.7), to predict the speed of light measured by the other observer. The alien calculates the speed of the light pulse relative to the rocket as c minus the rocket's speed of $0.5c$ or $c - 0.5c = 0.5c$. The rocket pilot has observed the light pulse to have speed c. This is a contradiction. The pilot calculates the speed of light relative to the asteroid as her speed plus that of the light, or $0.5c + c = 1.5c$. The alien has observed a speed of c. Again we have a contradiction.

Clearly the commonsense Galilean relativity is inconsistent with Einstein's second postulate. Evidently the addition of velocities must be re-examined. This in turn means that we must reconsider length and time measurements.

As a second example, consider the sequence of events shown in Figure 19.4. A moving train is struck by lightning in the front and in the rear. The lightning marks the train and jumps to the track, marking the track also. The flash from each lightning stroke is seen from inside the train and also from the track. A repairman sitting by the track exactly halfway between the two lightning strokes sees the two flashes simultaneously. In the diner, exactly halfway between the two lightning marks, the railroad president sees a flash at the front of the train. A very short time later (10^{-3} sec for a high-speed train of reasonable length) he sees a flash at the rear. The track repairman saw the flashes as simultaneous; the railroad president did not. There is no paradox here. Both observers give correct descriptions of

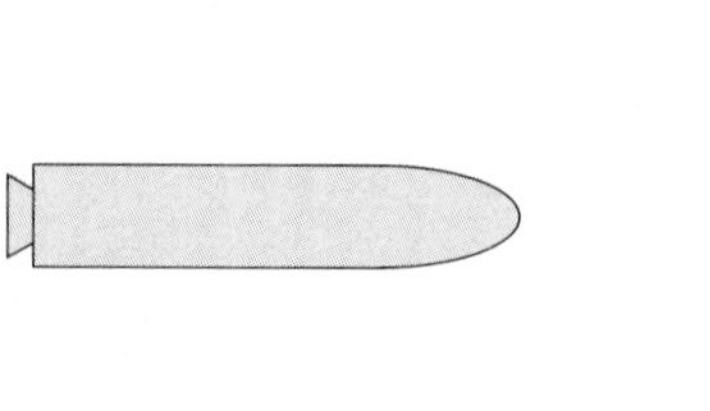

Figure 19.3

Figure 19.4

1. Lightning strikes the train at A' and B' and the track at A and B.
2. The light signal from A' reaches the President.
3. The signals from A and B reach the track repairman.
4. The signal from B' reaches the President.

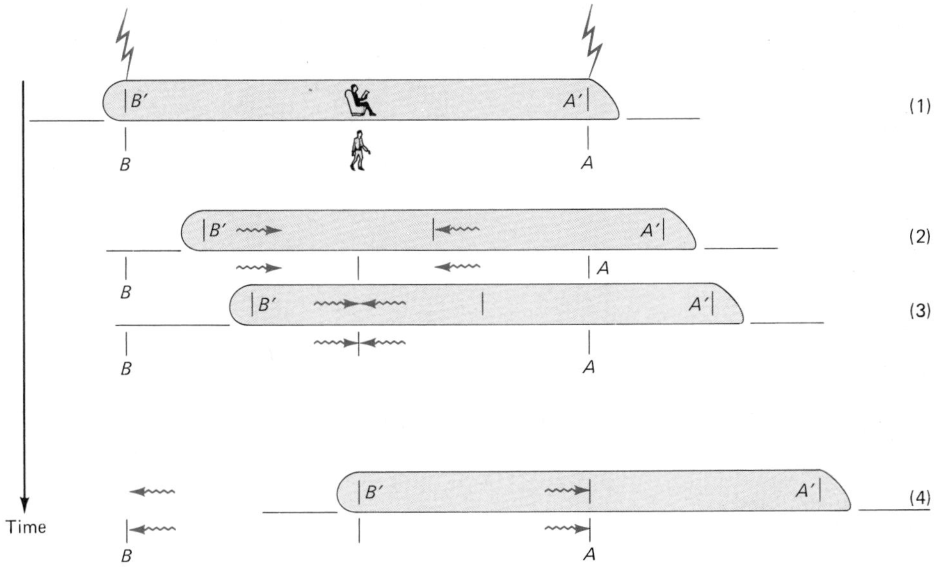

the sequence of light flashes. We see, then, that *simultaneity* (meaning an instance of events' occurring at the same time) is relative, not absolute. Because the velocity of light is finite, simultaneity depends on the motion of the observer. This in turn implies that time may not be an absolute that all observers agree upon, but may depend on the motion of the observer.

As you will see from the diagrams and the description, because the track repairman sees both lightning flashes simultaneously, points A and A' seem to coincide at the same time that points B and B' coincide, which means that the track length AB appears equal to the train length $A'B'$. The railroad president, however, views things a bit differently. Since he sees the lightning flash from the front of the train before he sees the one from the rear, it seems to him that A and A' coincide before B and B' do. He thus asserts that the track length AB is shorter than the train length $A'B'$. The two observers, each supported by his measurements, disagree about length as well as time.

Both the spaceship–asteroid example and the lightning–struck train example are farfetched. We do not have spaceships that can travel at $v = c/2$ nor are passenger trains likely to be struck by lightning in this fashion. But we can use these examples as thought experiments. We can think through the effects consistently, analyze the systems, and derive the logical consequences. As stated at the beginning of this chapter, "It is difficult to get rid of deep-rooted prejudices, but there is no other way."

Questions

2. Suppose the lightning flashes in Figure 19.4 occur simultaneously for the railroad president. Would they be simultaneous for the track repairman? Which flash would he see first, the front or the rear?

3. The railroad president saw the front of the train hit before the rear was hit. The track repairman saw the two events as simultaneous. Consider a third observer in some third inertial reference frame. Observer 3 passes the track repairman at the time that the lightning flashes strike (time in the track repairman's reference frame). Is it possible for observer 3 to see the flash that hit the rear of the train before he sees the flash that hit the front? Explain.

4. Notice that Figure 19.4 is drawn from the point of view of the track repairman. Now sketch the lightning-struck train, the president, the track repairman, and the light flashes at various times from the point of view of a helicopter pilot moving along with the train. Explain how the railroad president can argue that because event AA' happened before event BB', length AB is shorter than length $A'B'$ as seen by the railroad president and the helicopter pilot.

19.3 The Lorentz Transformations

We now present an algebraic derivation of the transformation equations implied by Einstein's postulates. These equations relate the position and time measurements of one observer to the corresponding measurements of another observer in motion relative to the first.

We associate the reference frames S and S' with two observers (Figure 19.5). The Galilean transformation equations are

$$x' = x - vt \tag{19.1}$$

and

$$x = x' + vt' \tag{19.2}$$

The final t' takes into account the possibility of different time scales. Next, because the length scales may change, we introduce a scale factor, γ (independent of the position and time but possibly dependent on the speed):

$$x' = \gamma(x - vt) \tag{19.3}$$

$$x = \gamma(x' + vt') \tag{19.4}$$

In accordance with Einstein's first postulate, the second scale factor must be the same as the first in order to avoid singling out one reference frame as a preferred system. Remember that γ is being introduced as a mathematical possibility. The only restriction on γ is that it must approach unity as v approaches zero. This is because, experimentally, Galilean relativity holds for velocities that are small compared to the velocity of light.

We now have two inertial reference frames (S and S'), with the corresponding observers equipped with meter sticks and clocks so that positions and times can be measured in each frame. Our problem is to determine the scale factor γ by making use of Einstein's postulates. To do this, we consider the following physical situation. When the origins coincide we start our clocks, one clock in S keeping t time and an identical clock in S' keeping t' time. At ($x = 0$, $t = 0$) and ($x' = 0$, $t' = 0$) there is a flash of light from a source at the coincident origins (Figure 19.6). The light moves out along the x-axis with velocity c measured by an observer in S. The light also moves out along the x'-axis with velocity c as measured by an observer in S'. (The same velocity c in both systems is required by Einstein's second postulate.) Then

$$x = ct \tag{19.5}$$

$$x' = ct' \tag{19.6}$$

Substituting into Eqs. 19.3 and 19.4 we obtain

$$ct' = \gamma(c - v)t \tag{19.7}$$

$$ct = \gamma(c + v)t' \tag{19.8}$$

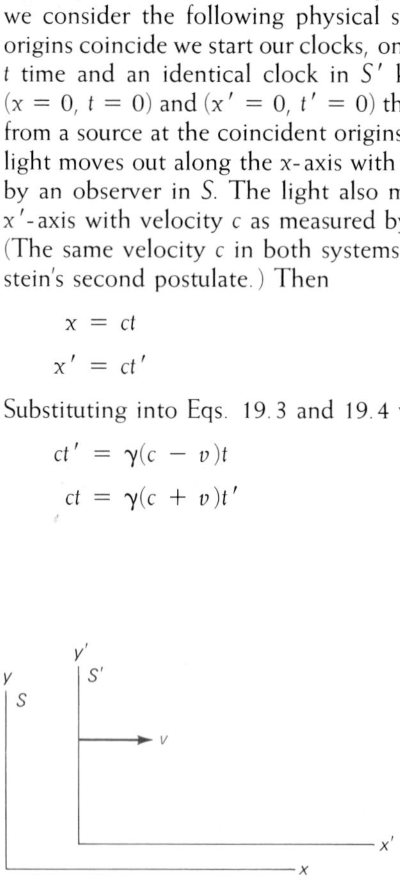

Figure 19.5

Two inertial reference frames. Consider the x'-axis to be sliding along the x-axis at a constant speed v.

Figure 19.6

A flash of light travels with the same speed in both S and S'.

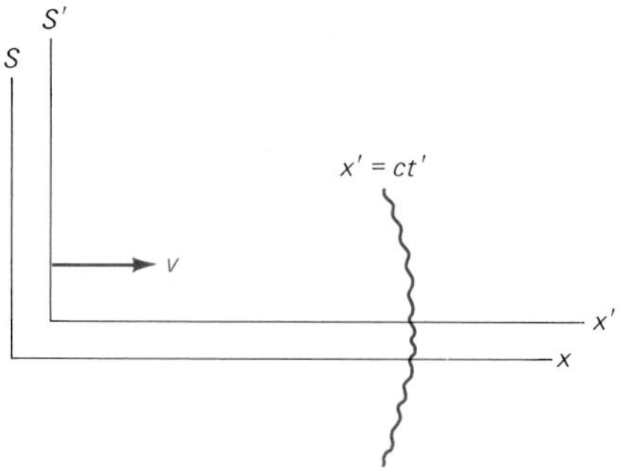

We can eliminate t and t' by multiplying these two equations and dividing each side by tt'. This yields

$$c^2 = \gamma^2(c^2 - v^2) \tag{19.9}$$

or

$$\gamma = \pm\frac{1}{\sqrt{1 - v^2/c^2}} \tag{19.10}$$

We choose the plus sign on physical grounds. Consider the limiting case of zero relative velocity, $v = 0$. Then $\gamma = \pm 1$. Choosing the plus sign leads to $x' = x$ in Eqs. 19.3 and 19.4, as we would expect. With zero relative velocity the two length coordinates coincide. On the other hand, the choice of the minus sign leads to $x = -x'$. We reject this solution. With this choice of $+\gamma$ in the the modified Galilean transformations, Eq. 19.3 and 19.4, we have

$$x' = \frac{x - vt}{\sqrt{1 - v^2/c^2}} \tag{19.11}$$

$$x = \frac{x' + vt'}{\sqrt{1 - v^2/c^2}} \tag{19.12}$$

By eliminating either x or x', we can solve these equations for t' and t:

$$t' = \frac{t - vx/c^2}{\sqrt{1 - v^2/c^2}} \tag{19.13}$$

$$t = \frac{t' + vx'/c^2}{\sqrt{1 - v^2/c^2}} \tag{19.14}$$

These transformation equations, consequences of Einstein's postulates of special relativity, have been named after H. A. Lorentz, a Dutch theoretical physicist who had used these expressions earlier, although in a different connection and with a different interpretation.

Equation 19.13 has a profound implication. Suppose that all of the clocks in S have been synchronized. Time t is the same for all x; but t' is different for each different x. The clocks in S' as seen by an observer in S are *not* synchronized. Einstein's time is relative. We have lost Newton's absolute, universal time.

Symmetry Relations

Interchanging the S and S' systems is equivalent to interchanging x and x', t and t', and replacing v by $-v$. This substitution converts Eq. 19.11 into 19.12 and vice versa. Likewise, Eq. 19.13 becomes 19.14 and vice versa. Physically this means that experiment cannot decide which observer moves and which observer is at rest. The observer at rest in system S claims that system S' is moving to his right. The observer at rest in S' claims system S is moving to his left. Both are correct.

Small-Velocity Limit

For velocities much less than the velocity of light we can expect the Lorentz transformation equations to reduce to the Galilean expressions. Why? Because the Galilean equations have been verified for $v \ll c$ by experiment.

For small velocities, $v \ll c$, v^2/c^2 is negligible compared to 1, and $1 - v^2/c^2$ reduces to 1. Equation 19.13 can be rewritten as

$$ct' = ct - \frac{v}{c}x \tag{19.15}$$

For small velocities and for distances that are small compared to the distance light can travel in the time t ($v/c \ll 1$ and $vx/c \ll ct$), this equation becomes

$$t' = t \tag{19.16}$$

and we're back to one universal time scale, which is characteristic of Newtonian mechanics and Galilean relativity. Equation 19.12 reduces to

$$x = x' + vt \tag{19.17}$$

identical to Eq. 4.19 of Galilean relativity (see Section 4.7). We see that classical Galilean relativity, extensively verified by experiment, provides a low-velocity limit for Einstein's special relativity. Note that the same mathematical reduction would take place if c increased without limit: $c \to \infty$. If light had an infinite velocity, time would indeed be absolute, as Newton postulated, and there would be no Lorentz transformations and no Einstein relativity. But the velocity of light, although large, is finite and, in vacuum, is always equal to c.

Einstein's two postulates of special relativity, which we presented in Section 19.1, can now be restated:

All laws of physics must be invariant under Lorentz transformations.

Here we see special relativity elevated to the position of a supertheory. All other theories in physics must be consistent with it. This is an extremely stringent restriction on physical theories, but in advanced theoretical physics, including electromagnetism, atomic and particle physics, and gravitation, this restriction has been very fruitful.

Question

5. Explain in your own words why we pick the positive solution for γ in Eq. 19.10 and why we reject the negative solution.

19.4 The Einstein Velocity Addition Formula

All the speeds considered in this chapter represent motion along the x- and x'-axes, which are parallel to each other. If this is understood, we can speak of the "addition of velocities." The special relativistic addition of nonparallel velocities is more complicated.

Let us return to the "paradox" of Section 19.2, in which the two observers, the rocket pilot and the alien, disagreed on their prediction of the velocity of light. We define S and S' as the alien's and the pilot's reference frame, respectively, and x and x' as axes parallel to the direction of motion. The pilot (S') sees a light flash moving along the x'-axis with a velocity c, and, using Galilean relativity, predicts that the velocity of this light signal as seen in S will be

$$u = c + v \tag{19.18}$$

This is inconsistent with Einstein's second postulate and with the speed that the alien actually observes.

We now take the Lorentz transformation equations and derive a new velocity addition law. First we rewrite Eqs. 19.12 and 19.13 for the difference of positions and difference of times:

$$\Delta x = \frac{\Delta x' + v\,\Delta t'}{\sqrt{1 - v^2/c^2}} \tag{19.19}$$

and

$$\Delta t = \frac{\Delta t' + (v/c^2)\,\Delta x'}{\sqrt{1 - v^2/c^2}} \tag{19.20}$$

Then dividing Eq. 19.19 by Eq. 19.20, we get

$$\frac{\Delta x}{\Delta t} = \frac{\Delta x' + v\,\Delta t'}{\Delta t' + (v/c^2)\,\Delta x'} = \frac{(\Delta x'/\Delta t') + v}{1 + (v/c^2)(\Delta x'/\Delta t')} \tag{19.21}$$

Since $\Delta x/\Delta t = u$ and $\Delta x'/\Delta t' = u'$, Eq. 19.21 becomes

$$u = \frac{u' + v}{1 + u'v/c^2} \tag{19.22}$$

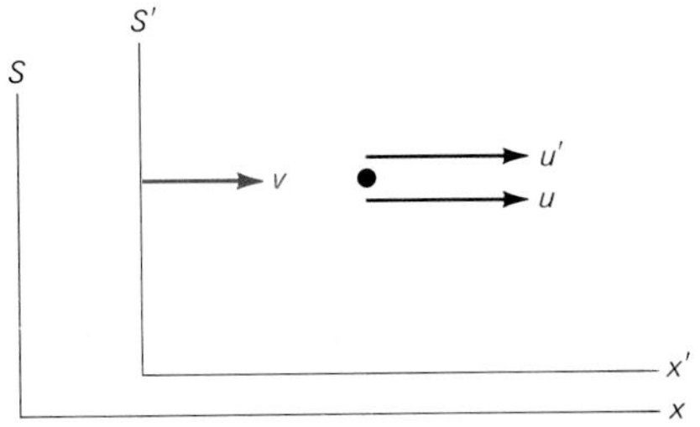

Figure 19.7

A particle has speed u relative to S and speed u' relative to S'. Velocities $\mathbf{u}$, $\mathbf{u}'$, and $\mathbf{v}$ all are directed parallel to the x- and x'-axes.

which is the Einstein velocity addition formula. It differs from Galilean velocity addition by that all-important denominator.

The Einstein velocity addition formula can be interpreted with the aid of Figure 19.7. Here u' is the velocity of some particle relative to S' and u is the velocity of the same particle relative to S. We have insisted that v, the velocity of S' relative to S, be constant, although the particle velocities u and u' may change with time.

Example 1 **Relativistic Addition of Velocities**

A rocket moving at $0.8c$ away from the earth shoots a projectile forward (away from the earth) at $0.9c$ relative to the rocket. What is the velocity of the projectile relative to the earth?

We have

$$u' = 0.9c$$

$$v = 0.8c$$

Galilean velocity addition leads to

$$u = 0.9c + 0.8c = 1.7c$$

which is 70% faster than the speed of light. By contrast, Einstein's velocity addition formula, Eq. 19.22, gives us

$$u = \frac{0.9c + 0.8c}{1 + (0.9c)(0.8c)/c^2} = \frac{1.70c}{1.72} = 0.9884c$$

Not only is the resultant velocity, u, not equal to $1.70c$, it's not even as great as c. In fact, for any velocities less than the velocity of light, the Einstein sum will never reach the velocity of light (Figure 19.8).

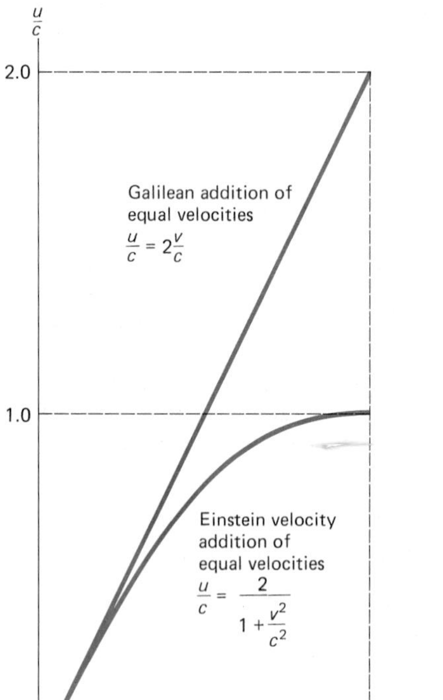

Figure 19.8

A comparison of Galilean and Einstein addition of equal velocities, $u' = v$. The Einstein addition leads to an upper limit of $u = c$.

Now let's use the Einstein velocity addition formula to resolve the contradictory conclusions made by the rocket ship pilot and the alien on the asteroid.

Example 2 **The Rocket Ship and the Asteroid: Relativistic Resolution**

The rocket ship pilot of Section 19.2 was traveling at a velocity $v = 0.5c$ (Figure 19.9). Labeling the pilot's reference frame S' and the alien's reference frame (the asteroid) S, we have, for the velocity of S' relative to S, $v = 0.5c$. The pilot saw light passing her at a velocity of $u = c$. Upon addition of velocities according to Eq. 19.22, the velocity of light seen by the alien becomes

$$u = \frac{c + 0.5c}{1 + 0.5c^2/c^2} = \frac{1.5c}{1.5} = c$$

The alien's observation of a velocity of c was correct. The rocket pilot's Galilean calculation of what she thought the alien saw was wrong, as was the alien's calculation of what he (it) thought the pilot saw.

The Einstein sum of c and any other velocity always equals c. This result should not be too surprising. It was built in from the beginning by Einstein's second postulate, which stated that the velocity of light was a constant for all inertial observers, independent of inertial reference frames.

The Einstein velocity addition formula is subject to observational test. For example, astronomers can observe regularly pulsing x-ray sources in binary star systems. These x rays are a form of electromagnetic radiation (as is light) and should travel with the velocity of light, c. If it is assumed that the x rays are emitted by a source moving at velocity v toward us, Galilean velocity addition predicts that the x rays should arrive here with velocity

$$c' = c + v$$

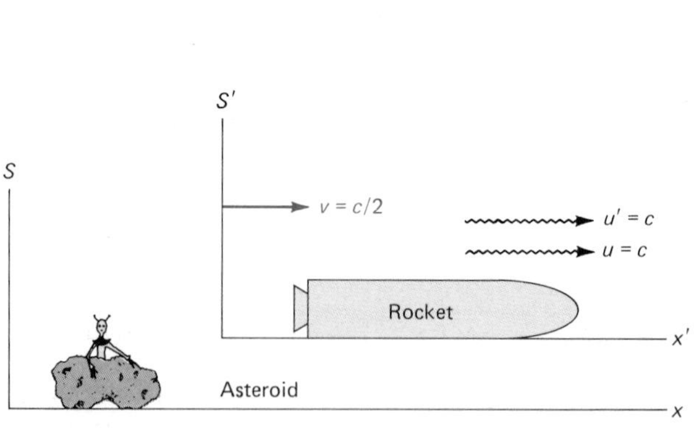

Figure 19.9

However, observation shows that the arrival velocity is actually

$$c' = c + kv \qquad (19.23)$$

where k is less than 2×10^{-9} and possibly 0.* This result contradicts Galilean velocity addition but is consistent with Einstein velocity addition.

Reduction to Galilean Velocity Addition

In Section 19.3 we saw that the Lorentz transformations reduce to the Galilean expressions for $v \ll c$. An analogous result holds for the Einstein velocity addition theorem. If the relative velocity v and the particle velocity u' are *both* small compared to the velocity of light,

$$v \ll c \quad \text{and} \quad u' \ll c$$

then the denominator reduces to 1 and the Einstein velocity addition formula, Eq. 19.22, reduces to Galilean velocity addition, $u = u' + v$.

19.5 Lorentz–FitzGerald Contraction

Length Parallel to Velocity

Einstein forced us to reconsider and refine our concepts of length and time. In this section we will see that length depends on motion.

Suppose that we take a meter stick that has been carefully calibrated by the National Bureau of Standards and set it at rest in the reference frame S', which is moving relative to us (see Figure 19.10). The stick's end positions are x_2' and x_1'. Next we measure x_2 and x_1, the corresponding positions in S. The measurements must be made simultaneously in S if $x_2 - x_1$ is to be interpreted as a length.† Using the Lorentz transformation equations, we subtract x_1' from x_2'. Because the measurements in S are made simultaneously, $t_1 = t_2$, and the terms involving the time cancel out, we are left with

$$x_2' - x_1' = \frac{x_2 - x_1}{\sqrt{1 - v^2/c^2}} \qquad (19.24)$$

or

$$L = L'\sqrt{1 - \frac{v^2}{c^2}} \qquad (19.25)$$

*K. Brecher, Is the Speed of Light Independent of the Velocity of the Source? *Phys. Rev. Lett.* 39, 1051 (1977).

†This is a definition of length. If the two position measurements are not simultaneous, the "length" measurement may well be complete nonsense. Question 8 is a specific example of this nonsense.

Figure 19.10

System S' is attached to the meter stick.

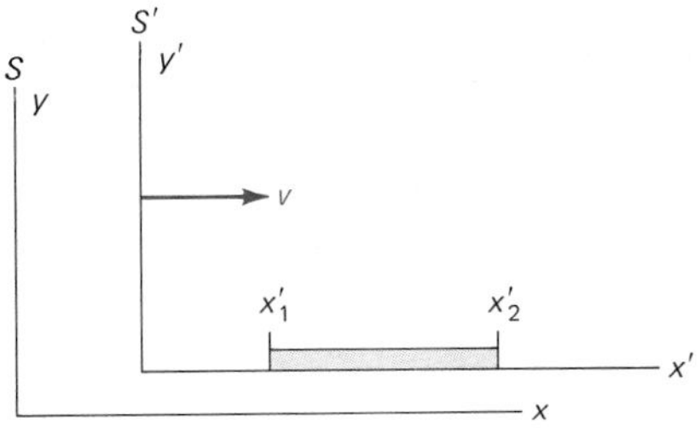

where L' is the length of a meter stick measured by an observer in S', at rest relative to the stick; and L is the length of a meter stick measured by an observer in S. (This stick is moving relative to this observer with speed v.)

The length of our meter stick in $S' = 1$ m. Since $\sqrt{1 - v^2/c^2} < 1$ (for $0 < v \le c$), our *measured* length in S is less than 1 m. A moving length contracts in the direction of motion. The length of the stick is greatest for the observer who sees it at rest.

Again this is a very high-velocity phenomenon. For small v/c, ($v/c \ll 1$), the $\sqrt{1 - v^2/c^2}$ reduces to 1 and the lengths are the same—to within an error that is proportional to v^2/c^2. FitzGerald's name is often attached to this contraction because of his 1892 suggestion that lengths parallel to the motion of the medium contracted by a factor of $\sqrt{1 - v^2/c^2}$. (But remember that Einstein discarded the medium.) In this section we are concerned only with relative motion, the motion of the meter stick relative to the observer.

Example 3 **A Moving Klingon Ship Contracts**

A Klingon D7 class battle cruiser measures 216.408 earth meters in length when it is at rest. What length would a Federation observer measure it to be as it flashes past him at $v = 2.0 \times 10^8$ m/s?

Attach the inertial system S' to the Klingon ship. Then $L' = 216.408$ m. From Eq. 19.25, with $c = 2.9979 \times 10^8$ m/s, the length measured by the Federation observer (reference frame S) is

$$L = 216.408 \text{ m}\sqrt{1 - \left(\frac{2. \times 10^8 \text{ m/s}}{2.9979 \times 10^8 \text{ m/s}}\right)^2}$$

$$= 216.408 \text{ m}\sqrt{0.5549} = 161.2 \text{ m}$$

(If you use $c = 3.0 \times 10^8$ m/s, you'll get 161.3 m.)

Example 3 illustrates the direct calculation of the Lorentz contraction for a high velocity. If the velocity v is very small compared to c, a different mathematical technique is useful. The term $(1 - v^2/c^2)^{1/2}$ can be expanded by the binomial theorem, Appendix 4.

Example 4 Calculation of a Small Contraction via the Binomial Theorem

A rod at rest in system S' has a length L' in S'. The rod moves past you (system S) with velocity v. We want to calculate the contraction $L - L'$.

From Eq. 19.25,

$$L = L'\left(1 - \frac{v^2}{c^2}\right)^{1/2}$$

By the binomial theorem

$$\left(1 - \frac{v^2}{c^2}\right)^{1/2} = 1 - \frac{1}{2}\frac{v^2}{c^2} \qquad \text{for } \frac{v^2}{c^2} \ll 1$$

$$L = L' - \frac{L'}{2}\frac{v^2}{c^2}$$

The contraction is

$$L' - L = \frac{L'}{2}\frac{v^2}{c^2}$$

proportional to v^2/c^2. This is an approximation. The error, for $v^2/c^2 \ll 1$, depends primarily on the next term in the binomial expansion. Again, from Appendix 4, the next term (which we omitted) is $-\frac{1}{8}(v^2/c^2)^2$. If $v^2/c^2 \ll 1$, then $(v^2/c^2)^2$ will be negligible.

Now let's apply this analysis to a meter stick. Suppose the meter stick flies past you at 40 m/s (89.5 mi/hr). What contraction, $L' - L$, could you expect to see? From the result above we have

$$L' - L = \frac{1.0 \text{ m}}{2} \cdot \frac{(40)^2 \text{ (m/s)}^2}{(3 \times 10^8)^2 \text{ (m/s)}^2}$$

$$= 8.89 \times 10^{-15} \text{ m}$$

For this speed the contraction is tiny; it is on the order of the size of an atomic nucleus.

Length Perpendicular to Velocity

We must emphasize that the Lorentz–FitzGerald contraction applies only to lengths or components of length oriented in the direction of the velocity v. Lengths perpendicular to the velocity are *not* affected.

As a physical argument supporting this assertion, imagine two identical rocket ships in outer space passing each other at high relative speed v as indicated in Figure 19.11. Each ship is equipped with two special lasers that will mark the other rocket as it passes. Let's assume that lengths perpendicular to the velocity (such as the separation of the lasers) do contract and then show that this assumption leads to a contradiction. (The contradiction means that the assumption is false.) You are in rocket ship S. If you see L', the distance between lasers on rocket ship S', as contracted ($L' < L$), then your counterpart on rocket S' should see $L < L'$. If either of you sees a contraction, the other should also. Remember, we have reciprocity. There must be no preferred coordinate system. If you see L' contracted so that $L' < L$, you expect that at the instant of passing the other rocket you will get marked with the laser beams. On the other hand, you expect that your laser beams will straddle and miss rocket S'. Your counterpart on S' predicts just the opposite—that when you pass, you will mark his ship and he will miss yours. This is a contradiction; the assumption of contraction is false. At the same instant, you cannot be both marked and missed. We conclude that the lengths perpendicular to the direction of the motion are not affected by the motion. So we can add to the transformation equations of both Einstein's special relativity and Galilean relativity

$$y' = y$$

and

$$z' = z \tag{19.27}$$

Questions

6. Explain why we are unaware of special relativistic effects in our everyday lives.
7. Scientists generally agree that if they can measure a quantity, it is real. In this sense, is the Lorentz–FitzGerald contraction real? Does the moving meter stick really contract?
8. An object is moving relative to us. We define its length as the difference in position of the two end points. Explain why we demand that the two endpoint positions be determined simultaneously. As an example, consider measuring the position of the tail of an airplane as it leaves New York, and then measuring the position of the nose of the plane some hours later as it approaches Los Angeles.
9. In this section we presented an argument and a thought experiment demonstrating that lengths perpendicular to the relative velocity are invariant. Why doesn't this argument apply to lengths parallel to the relative velocity? (*Hint:* Reconsider the train described in Section 19.2.)

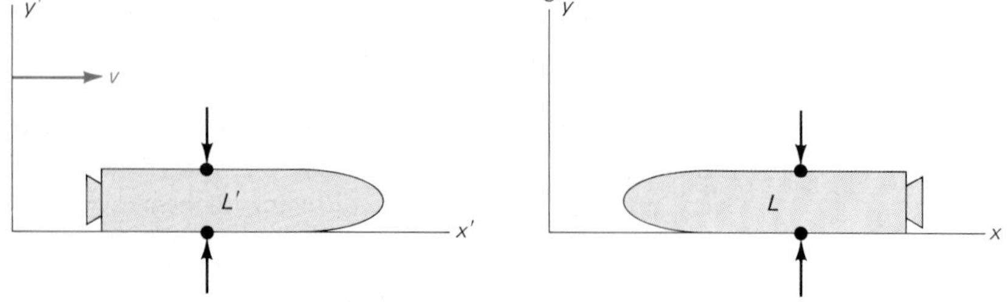

Figure 19.11

10. Our derivation of the Lorentz–FitzGerald contraction utilized the Lorentz transformation, Eq. 19.11. Why didn't we use the other Lorentz transformation, Eq. 19.12? (*Hint:* What would we do with $\Delta t'$?)

19.6 Time Dilation

In place of the meter stick of Section 19.5 let's now put a clock at a fixed point x_0' in system S' moving relative to us with velocity v, as shown in Figure 19.12. In S' the clock records a time interval $\Delta t'$. The distance interval in S', $\Delta x'$, is 0 because the clock stays at x_0'. In system S we have two clocks identical to the S' clocks. The S clocks are synchronized with each other and are running at the same rate. When the S' clock passes the S clock at x_1, the S' clock time t_1' and the S clock time t_1 are recorded. When the S' clock passes the S clock at x_2, t_2' and t_2 are recorded. The S' time interval is $\Delta t' = t_2' - t_1'$. The time interval in S is $\Delta t = t_2 - t_1$. From Eq. 19.20

$$\Delta t = \frac{\Delta t' + (v/c^2)\,\Delta x'}{\sqrt{1 - v^2/c^2}}$$

Because the clock is stationary in S', $\Delta x' = 0$. Then

$$\Delta t = \frac{\Delta t'}{\sqrt{1 - v^2/c^2}} \qquad (19.28)$$

where $\Delta t'$ is the time interval measured by clocks moving past us with velocity v and Δt is the time interval measured by clocks at rest relative to us. Because $\sqrt{1 - v^2/c^2}$ is less than 1 (for $0 < v \le c$), the time interval Δt measured in S is longer than the time interval $\Delta t'$ measured by the moving clock (moving relative to us). A moving clock runs slow. This means quite literally that the rate of passage of time depends on the relative motion.

The effect is reciprocal. It must be reciprocal if we are to avoid having a reference frame singled out for special or preferred treatment. An observer who is stationary in S' finds his clock keeping perfect time but observes the clock in S running slow. So who is right? Both are correct! Read on.

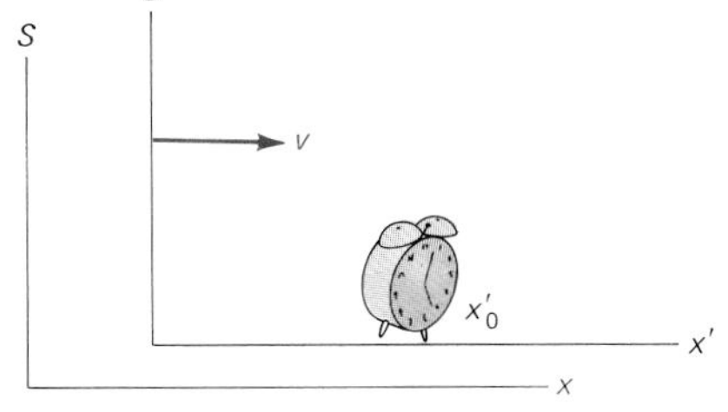

Figure 19.12

System S' is attached to the clock.

Experimental Evidence: Muon Decay

Most of the evidence that supports Einstein's special relativity comes from relativistic momentum and energy, which we will consider in Chapter 20. But there is evidence that bears directly on moving clocks and moving lengths.

Muons are unstable (radioactive) particles with a mean lifetime of about 2 microseconds (μs), measured by clocks at rest relative to the muons. Some muons are created by the bombardment of the upper atmosphere, 20 miles or more above the earth, with high-energy cosmic rays. A classical calculation suggests that even if muons traveled downward at almost the velocity of light, they could only travel a distance of about 3×10^8 m/s $\times\ 2 \times 10^{-6}$ s or 600 m before decaying. But many muons do reach the surface of the earth. Our time dilation calculation recognizes the muon to be a moving clock that is running slowly; and, running slowly, it may well have time enough to come down 20 miles to the surface before it decays.

Einstein's relativity gives us another way of looking at this phenomenon. Imagine that you are riding with the muon. Now there is no time dilation. You have only 2 μs to get down to the earth. However, you see the earth moving up toward you. The distance to the surface is moving. Hence, this distance is contracted and may well be short enough to traverse in 2 μs.

Example 5 Muon Decay

One rather fast muon is created 30 km above the earth's surface. It moves toward the earth at $v = 0.9999c$. Can it reach the surface in 2 μs (measured by a clock traveling with the muon)? Let's consider that $\Delta t'$ is 2×10^{-6} s. By Eq. 19.28

$$\Delta t = \frac{2 \times 10^{-6}\ \text{s}}{\sqrt{1 - (0.9999)^2}} = 141 \times 10^{-6}\ \text{s}$$

The modern pocket electronic calculator can handle this problem quite easily but let's try to solve it analytically. (Sometimes there is more to be learned by looking at the general behavior of a function than at a set of illuminated numbers.) Let's see how the time dilation effect enters the problem. For $v \approx c$, we'll write $v/c = 1 - \delta$. Hence $\delta = 0.0001$. Then

$$\frac{1}{\sqrt{1 - v^2/c^2}} = \frac{1}{\sqrt{1 + \dfrac{v}{c}}}\,\frac{1}{\sqrt{1 - \dfrac{v}{c}}}$$

$$= \frac{1}{\sqrt{2 - \delta}\,\sqrt{\delta}} \cong \frac{1}{\sqrt{2}\,\sqrt{\delta}}$$

For very small δ, δ is negligible compared to 2 and has been dropped from the first radical. We see that Δt is proportional to $1/\sqrt{\delta}$. Because $\delta \ll 1$, Δt increases by a large factor. At velocity $0.9999c$ (or essentially c, because δ is negligible compared to 1) the muon can go a distance of $c\,\Delta t$ or

$$3 \times 10^8\ \text{m} \times 141 \times 10^{-6}\ \text{s} = 423 \times 10^2\ \text{m}$$
$$= 42\ \text{km}$$

Yes, this muon can reach the earth's surface in 2 μs.

The same time dilation analysis applies to the beams of very short-lived particles created in modern high-energy particle accelerators such as the one at Fermi National Laboratory in Illinois. The fast-moving radioactive particles behave like moving clocks, and in agreement with Einstein's relativity, go many times further before they decay than we would otherwise expect. Measurements of the distance traveled have provided very precise confirmation of the time dilation formula.

Question

11. Our derivation of the time dilation formula utilized Eq. 19.14 as the Lorentz transformation. Why didn't we use the other Lorentz transformation, Eq. 19.13? (*Hint:* What would we do with Δx?)

19.7 The Twin Effect

Newton considered time to be absolute, flowing uniformly on, independent of reference frame. Most of us have consciously or unconsciously adopted this view. It is in matters of time that Einstein's relativity most radically defies our Newtonian common sense. This problem is brought into sharp focus in a thought experiment called the *twin effect.*

Imagine that astronaut O.a.B. Smith climbs into his rocket ship (Figure 19.13), quickly accelerates to 80% the velocity of light, and coasts to Alpha Centauri, 4 light-years away.* With the help of the gravitational pull of Alpha Centauri O.a.B. circles around and heads back to earth again at $v = 0.8c$.

Meanwhile, O.a.B.'s twin brother S.a.H. Smith has remained on earth. He knows that 10 yr have elapsed between O.a.B.'s takeoff and return. This is the time measured by S.a.H.'s earthbound clocks and is consistent with the distance-to-velocity ratio

$$\frac{8 \text{ light-years (round trip)}}{0.8c} = 10 \text{ yr}$$

*A light-year is the distance light can travel in 1 yr: $3.156 \times 10^7 \text{ s} \times 2.998 \times 10^8 \text{ m/s} = 9.46 \times 10^{15}$ m, about 5.88×10^{12} mi.

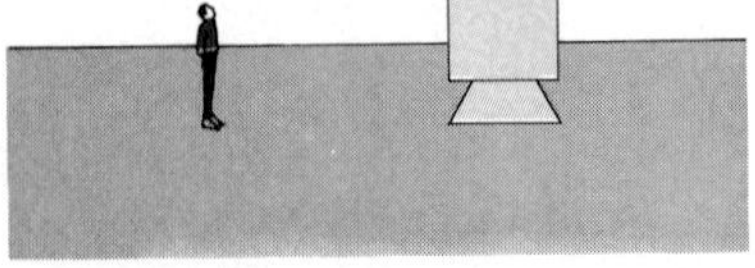

Figure 19.13

S.a.H. Smith also knows that his twin has been a moving clock and hence has aged more slowly by a factor of $\sqrt{1 - (0.8)^2} = 0.6$. O.a.B. Smith steps out of the rocket. Since the takeoff, S.a.H. has aged 10 yr. O.a.B. has aged only 6 and is now 4 yr younger than his twin brother!

This may seem paradoxical if we argue from the astronaut O.a.B.'s point of view. To O.a.B. the 4-light-year distance to Alpha Centauri was contracted, and he saw the distance as $0.6 \times 4 = 2.4$ light-years. Over the round trip, O.a.B. ages 6 yr. But the motion was relative; O.a.B. saw his twin brother S.a.H. as a moving clock. So why didn't S.a.H. Smith age less? Why don't we have reciprocity?

The answer is that we don't have symmetry between the twins. Stay-at-Home Smith, on earth, is always in the same inertial system. Out-and-Back Smith is in one inertial reference frame at first (velocity $+v$ relative to the earth) and then in a second inertial reference frame (velocity $-v$ relative to the earth). This change makes all the difference. S.a.H. Smith's time dilation calculation is valid, but O.a.B. Smith's claim for reciprocity is not. We do not have symmetry and we do not have reciprocity.

This situation is shown schematically in Figure 19.14. Plotting position x against time t, S.a.H. Smith remains on the same horizontal line, $x = 0$. O.a.B. Smith's outward journey appears as a straight line (velocity v) through the origin. His return trip is a different straight line (velocity $-v$) representing a different inertial system. The space-time tracks of O.a.B. Smith and S.a.H. Smith are not symmetric with each other.

Could the twin effect really happen? Wouldn't the twins really age equally? We've not yet sent space travelers to distant stars, and considering the energy requirement, perhaps we never will. The experiment has been carried out, however, with radioactive atoms and with atomic clocks.* A number of commercial atomic clocks were synchronized with each other. Half of the clocks were placed in a commercial airplane and sent around the world eastward, the airplane's speed adding to the speed of the earth's rotation. The other atomic clocks were placed on a commercial airplane and sent around the world westward, the airplane's speed subtracting from the speed of the earth's

*J. C. Hafele and R. E. Keating, Around-the-World Atomic Clocks: Predicted Relativistic Time Gains; and Around-the-World Atomic Clocks: Observed Relativistic Time Gains, *Science* 177, 166; 168 (1972).

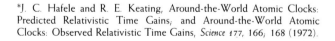

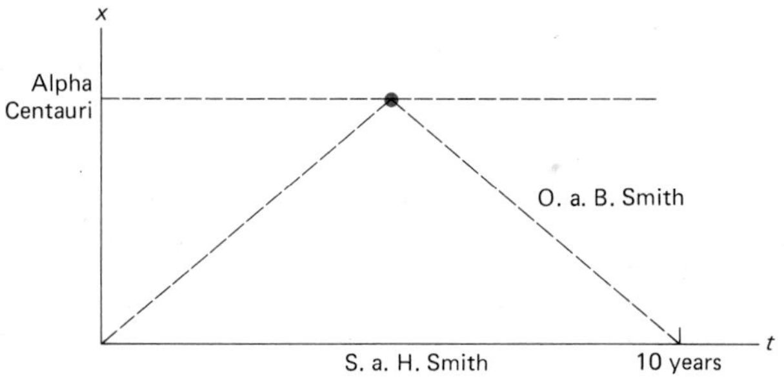

Figure 19.14

A plot of position versus time for the astronaut and for the earthbound observer. The earthbound observer stays in one inertial system. The astronaut does not.

rotation. When the planes returned to their starting point, the clocks were compared. The experimental results were in full agreement with Einstein's relativity: A moving clock runs slow. The astronaut twin going into space and coming back (O. a. B) at high velocity will age less than his stay-at-home earthbound sibling (S. a. H).

A number of modern science fiction writers, acquainted with special relativity, have used the twin effect as a sort of suspended animation. In Poul Anderson's *Tau Zero*, for instance, astronauts live centuries longer than other people because of the many years they spend at a speed close to that of light. Sometimes the space travelers return to earth to find their great-grandchildren grown, married, and physically more aged than they are!

Einstein's Importance in Physics and Beyond

Einstein's special relativity has had a major impact on the field of physics, in the calculation and understanding of high-velocity phenomena, and an even more important effect on our ways of thinking. Our understanding of space and time is much greater now than it was at the turn of the century. Special relativity has also had consequences that go beyond physics. The concept that measurements, such as measurements of length and time, are relative to an observer rather than independent of the observer, has influenced philosophy, religion, and our society in general. One of special relativity's more interesting (and more trivial) consequences has been its effect on pseudoscientific or "crackpot" literature.* Before Einstein, Newton was the figure of authority in physics, and the automatic target for attack. Since the theory of special relativity appeared, crackpots have declared Newton to be correct and have concentrated their criticism and vituperation on Einstein.

Summary

In 1905, Einstein proposed special relativity, based on two postulates:

1 The laws of nature are the same (invariant) in all inertial reference frames.

2 The speed of light is a constant, the same in all inertial reference frames, independent of the motion of the source and of the observer.

This second postulate does away with the need for a medium.

Einstein's postulates forced a re-examination of the concepts of space and time. Simultaneity is not an absolute but depends on the motion of the observer.

Einstein's postulates led to the Lorentz transformations:

$$x' = \frac{x - vt}{\sqrt{1 - v^2/c^2}} \qquad (19.11)$$

$$t' = \frac{t - vx/c^2}{\sqrt{1 - v^2/c^2}} \qquad (19.13)$$

*The classic reference on crackpots is Martin Gardner's *Fads and Fallacies in the Name of Science*, Dover Publications (1957). We recommend it.

These transformations relate inertial systems and establish a new relativity, which we call *special relativity*. Special relativity agrees with Galilean relativity for $v \ll c$ but is significantly different for velocities comparable to c.

The Lorentz transformations led to a new form of velocity addition:

$$u = \frac{u' + v}{1 + u'v/c^2} \qquad (19.22)$$

This is the Einstein velocity addition formula. Here, v is the speed of the inertial system S' relative to inertial system S. u' is the speed of some object relative to S', while u is its speed relative to S. The special relativistic sum of any two velocities less than c is always less than c. The sum of c and any velocity is c.

A moving length contracts in the direction of the motion by a factor of $\sqrt{1 - v^2/c^2}$. However, a moving length perpendicular to the direction of motion is invariant.

A moving clock runs slow by a factor of $\sqrt{1 - v^2/c^2}$. Muons and beams of short-lived particles from accelerators confirm time dilation.

An astronaut who went into space and then returned to earth would age less than his stay-at-home twin. Experimental confirmation is provided by radioactive atoms and by atomic clocks.

Suggested Reading

J. Bronowski, The Clock Paradox, *Sci. Am.* 206, 134 (February 1962).

R. S. Shankland, The Michelson–Morley Experiment, *Sci. Am.* 211, 107 (November 1964).

R. S. Shankland, Michelson and His Interferometer, *Physics Today* 230, 36 (April 1974).

A. B. Stewart, The Discovery of Stellar Aberration, *Sci. Am.* 210, 100 (March 1964).

See also *Special Relativity Theory: Selected Reprints*, American Institute of Physics, New York, 1963.

Problems

Section 19.2 Relativity of Time

1. In the example of the train struck by lightning front and rear, the railroad president sees first the flash from the front. The flash from the rear reaches him a short time later. For a train 180 m long traveling at 40 m/s (about 90 mi/hr), calculate this "short time." (Here is another reason for considering this example to be a thought experiment rather than an actual experiment.)

Section 19.3 Derivation of the Lorentz Transformations

2. Because of the form of the Lorentz transformation equations, all of our special relativistic expressions will involve a factor $\sqrt{1 - v^2/c^2}$ in the numerator or denominator. (a) Expand $\sqrt{1 - v^2/c^2}$ by the binomial expansion through the term proportional to v^4/c^4. (b) If $v^2/c^2 = 10^{-4}$, what is the fractional error in approximating $\sqrt{1 - v^2/c^2}$ by $1 - \frac{1}{2}(v^2/c^2)$?

3. (a) Show that $(1 - v^2/c^2)^{-1/2} \approx 1 + \frac{1}{2}(v^2/c^2) + \frac{3}{8}(v^4/c^4)$. (b) How small must v/c be for the approximation $(1 - v^2/c^2)^{-1/2} \approx 1 + \frac{1}{2}(v^2/c^2)$ to be correct to within 0.0001? (c) From part (b), what is v in miles per hour?

4. (a) With the help of the binomial expansion show how the Lorentz transformation, Eq. 19.12, reduces to the Galilean transformation, $x = x' + vt$, for very small speeds, $v \ll c$. (b) What do you have to neglect to get the Galilean equation?

5. Show that if the velocity of light were infinite, all of the Lorentz transformation equations (19.11–19.14) would reduce to the corresponding Galilean transformation equations, and we would have no special relativity.

Section 19.4 The Einstein Velocity Addition Formula

6. Calculate the Einstein sum of (a) $u' = 0.4c$ and $v = 0.6c$; (b) $u' = 0.9c$ and $v = 0.9c$; (c) $u' = 0.99c$ and $v = 0.99c$.

7. Two jets of matter move in opposite directions away from the center of a radio galaxy at speeds of 0.75c. Determine the speed at which the two jets recede from each other. That is, determine their relative velocity. Do this first according to Galilean relativity, then again using the Einstein velocity addition relation.

8. Show that the special relativistic sum of c and *any* speed equals c.

9. A Klingon space ship moves away from the earth at a speed of 0.8c. The Starship Enterprise pursues at a speed of 0.9c relative to the earth. Earthbound observers see the Enterprise overtaking the Klingon ship at a relative speed of 0.1c. With what speed is the Enterprise overtaking the Klingon ship as seen by the crew of the Enterprise? Note that $\Delta v = 0.1c$ is not a *directly* measured velocity.

10. Einstein's velocity addition formula yields

$$u = \frac{v + v}{1 + v^2/c^2}$$

for the addition of two equal velocities v and v to form a third velocity, u. Solving for v, we have

$$\frac{v}{c} = \frac{1 - \sqrt{1 - u^2/c^2}}{u/c}$$

Using a binomial expansion, show that this formula for v is consistent with Galilean velocity addition for $u/c \ll 1$.

Section 19.5 Lorentz–FitzGerald Contraction

11. Rederive the Lorentz–FitzGerald contraction starting from the difference form of the Lorentz transformation:

$$\Delta x' = \frac{\Delta x - v\Delta t}{\sqrt{1 - v^2/c^2}}$$

Define your terms clearly.

12. For $v/c \ll 1$, show that the Lorentz–FitzGerald contraction may be approximated by $L = L'[1 - \frac{1}{2}(v^2/c^2)]$.

13. A plane is measured in the reference system in which the plane is at rest. The length is found to be 30 m. Observers on the ground make a precise measurement of the plane's length as the plane flies past at 1000 mi/hr. What length do they measure? By what fraction of a millimeter has the plane contracted?

14. In the Stanford Linear Accelerator (SLAC) electrons are accelerated to an energy of 20 GeV (a speed of 0.999,999,999,67c). The evacuated tube through which the electrons are accelerated is about 2 mi (3.2 km) long. How long will the tube appear to an observer "sitting" on the 20-GeV electron?

15. The moving meter stick ($L' = 1$ m) described in this section is measured to be 50 cm long ($L = 50$ cm). What is its speed in terms of the speed of light?

Section 19.6 Time Dilation

16. Rederive the time dilation formula, Eq. 19.28, by working from the Lorentz transformations, Eqs. 19.11–19.14. Define your terms clearly.

17. On a rocket ship a pulse of light bouncing up and down between two mirrors constitutes a clock. The light path seen by observers in the rocket is perpendicular to the direction of motion of the rocket. An

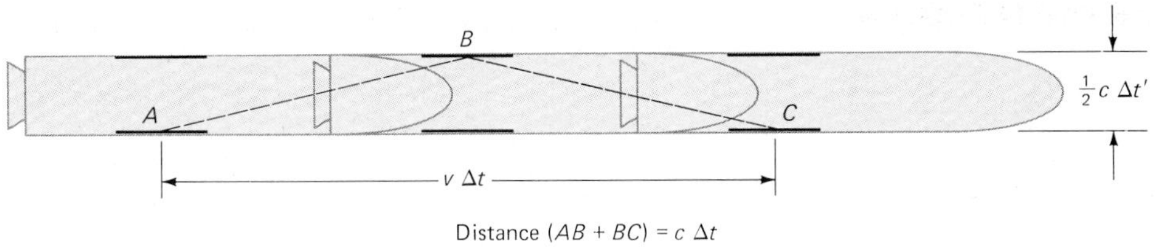

Figure 1

observer on some planet sees the rocket zoom past with constant velocity, v. He sees the light pulse following a *diagonal* path. He sees the light-pulse clock running slow because the light pulse traverses a longer path. With the help of Figure 1, rederive the time dilation formula from this light-pulse clock.

18. Suppose that you are sitting on a muon 30 km (measured by an observer on earth) above the earth's surface. The earth appears to be moving up toward you at $v = 0.9999c$. (a) What distance do you see between yourself and the earth? (b) Will the earth reach you in 2 μs (measured by your clock)?

19. A very accurate atomic clock (S') moves at 1000 km/hr for 1 hr as measured by an identical earthbound atomic clock (S). How many nanoseconds slow will the moving atomic clock be after the 1 hr earth time (S)? [*Hint:* Use a binomial expansion of $(1 - v^2/c^2)^{-1/2}$.]

20. A jet plane is traveling at about 2200 mi/hr or 1 km/s. As viewed by observers on the ground, at what rate, in seconds per year, are clocks on the plane running slow (because of the time dilation effect)?

21. An astronaut circling the earth at low altitude has a speed of about 18,000 mi/hr. For 1 s of earth time consider the astronaut to be an inertial system S' moving along the x'-axis. As seen by ground observers, by how much will the astronaut's watch run slow in this 1 s of t time? How many seconds slow would this be per year?

22. A muon rushing down toward the earth is at rest in inertial system S'. In S' this muon exists for 2.2×10^{-6} s (time measured by clocks at rest in S'). How long does the muon exist as measured by earthbound clocks (internal system S) if its approach velocity is (a) $0.9c$; (b) $0.99c$; (c) $0.999c$; (d) $0.9999c$?

23. The muon in Problem 22 ($v = 0.99c$) is created 20 km above the earth's surface. Will it reach the surface of the earth?

24. The Klingons in Problem 19 were being overtaken by the Starship Enterprise ($v_{K-\text{earth}} = 0.8c$, $v_{E-\text{earth}} = 0.9c$). The Klingons departed 1 yr (earth reckoning) before the Enterprise. (a) How much time had elapsed on the Klingon clocks when the Enterprise departed? (b) What was the distance between earth and the Klingons (Klingon reckoning) when the Enterprise departed? (c) The Klingons saw the Enterprise overtaking their ship at a speed $0.357c$. How long (Klingon time) is required for the Enterprise to overtake them? (d) Check your results by using the Lorentz transformation equations.

chapter 20

RELATIVISTIC MECHANICS

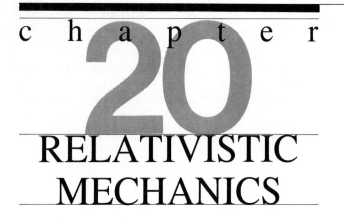

Preview

In Chapter 19 we introduced Einstein's theory of special relativity for motion with speeds comparable to the speed of light. We presented Einstein's two postulates and derived the kinematic consequences. These include the Lorentz transformations, length contraction, time dilation, and the Einstein velocity addition law. Earlier, in Chapters 6–10, we considered speeds much less than the speed of light and developed Newtonian mechanics. Specifically, in Chapters 6–10 force and mass were introduced and two very important conserved quantities, linear momentum and kinetic energy, were described.

The Newtonian mechanics of Chapter 6–10 was developed on the basis of the kinematics of Chapters 4 and 5, a Galilean relativity. The logical relationships between special relativity, Galilean relativity, and Newtonian mechanics are indicated in Figure 20.2. Galilean relativity is a special case of special relativity for speeds very much less than the speed of light, $v \ll c$. Newtonian mechanics is consistent with Galilean relativity but is not consistent with special relativity. Light propagation and electromagnetic theory in general are consistent with special relativity.

In 1905 scientists were faced, then, with two broad physical theories—Newtonian mechanics and electromagnetism—each with its own theory of relativity. Scientists recognized that there should be only one relativity, and they began to search for it. Their solution was to develop a relativistic mechanics that is consistent with special relativity. This relativistic mechanics incorporates Newtonian mechanics as a special case for $v \ll c$. Relativistic mechanics has been amply confirmed by experiment.

Chapter 20, like Chapter 19, is optional. In this chapter, two key dynamic concepts, linear momentum and kinetic energy, are generalized so that the conservation laws of linear momentum and energy hold for particle speeds up to the speed of light. The relativistic expression for kinetic energy leads directly to the famous mass–energy relation, $E = mc^2$.

These relativistic conservation laws are tested in Section 20.4 when the conservation laws are applied to systems in which particles are created or annihilated.

Figure 20.1
Albert Einstein on an Israeli banknote.

Figure 20.2

Relativity, mechanics, and electromagnetism.

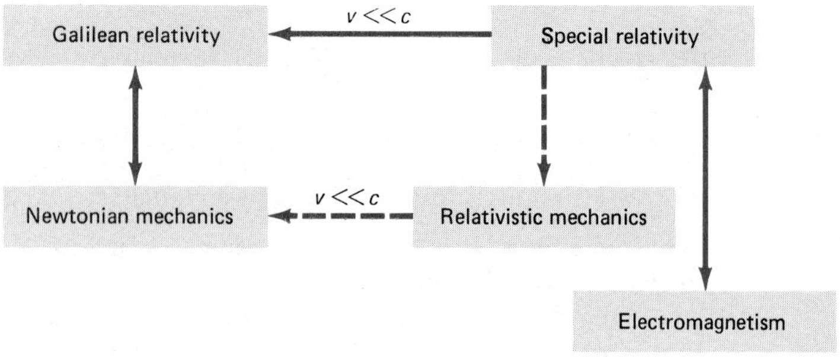

20.1 Introduction

Beginning in Chapter 6, our study of Newtonian mechanics led to the very useful laws of conservation of energy (Section 8.3) and conservation of linear momentum (Section 9.3). For speeds that are very much less than the speed of light ($v \ll c$) these conservation laws are absolute, and hold without exception. They have been repeatedly confirmed by experiment.

As particle speeds approach the speed of light, however, experimental verification disappears, and the Newtonian conservation laws do not hold. The enormity of the failure of Newtonian mechanics as v approaches c is indicated by particle reactions involving the annihilation or creation of particles. For example, suppose that an electron and a positron (the antimatter equivalent of an electron) come together with negligible linear momenta and kinetic energies. The electron and positron annihilate each other and are replaced by two gamma photons moving in opposite directions at $v = c$. These gamma photons are bundles of radiant (electromagnetic) energy. There is a 100% conversion of particle mass into energy. Radiant energy is not conserved; it is created. Another case where the conservation laws fail is in the creation of particle mass by proton–antiproton pair production, which we consider in detail in Section 20.4. The kinetic energy of high-velocity protons is not conserved.

The development of relativistic mechanics starts with the generalization of the classical expressions for linear momentum and kinetic energy. We must impose two theoretical restrictions:

1 The relativistic linear momentum and relativistic energy are defined in such a way that linear momentum and energy are conserved. We recognize that the conservation laws are extremely useful and we want to maintain them within the framework of special relativity.

2 For $v \ll c$, the relativistic linear momentum expression must reduce to the Newtonian expression $\mathbf{p} = m\mathbf{v}$ and the relativistic kinetic energy expression must likewise reduce to $K = \frac{1}{2}mv^2$. The reason for this restriction is simply that for $v \ll c$, Newtonian mechanics works.

We introduce the relativistic form of linear momentum in Section 20.2. Then, in Section 20.3, we use the work–energy theorem to derive relativistic energy from the relativistic momentum. Finally, in Section 20.4 we apply the relativistic momentum and energy to describe proton–antiproton pair production.

20.2 Relativistic Linear Momentum

The form of the relativistic linear momentum was originally developed by Einstein on a purely theoretical basis as part of a comprehensive relativity theory for mechanics. Modern experiments measuring the deflection of charged particles in magnetic fields (see Section 32.3) yield values that are in excellent agreement with Einstein's expression for relativistic momentum. On the basis of theory and experiment we define the relativistic linear momentum as

$$\mathbf{p} = \frac{m\mathbf{v}}{(1 - v^2/c^2)^{1/2}} \qquad (20.1)$$

Figure 20.3 shows both the relativistic form, Eq. 20.1, and the classical Newtonian form, $p = mv$, plotted against v/c. Note how the ever-present variable $(1 - v^2/c^2)^{1/2}$ in the denominator raises the value of the relativistic momentum above the Newtonian momentum and drives the relativistic momentum to infinity in the limit as v approaches c. Once again, as in Chapter 19, the speed of light is an upper limit.

For smaller speeds, such as $v/c < 0.2$, the two curves of Figure 20.3 merge, indicating that the relativistic momentum reduces to the classical momentum. The formal mathematical reduction of Eq. 20.1 to the Newtonian form is accomplished by letting v/c be so small that v^2/c^2 is negligible compared to 1. (This does *not* mean that $v = 0$.) Then the denominator goes to 1 and $\mathbf{p} = m\mathbf{v}$. For $v/c \ll 1$ the special relativistic linear momentum reduces to the classical Newtonian form.

Figure 20.3

A comparison of momentum in relativistic mechanics and in Newtonian mechanics. Note how the relativistic momentum goes to infinity as $v \rightarrow c$.

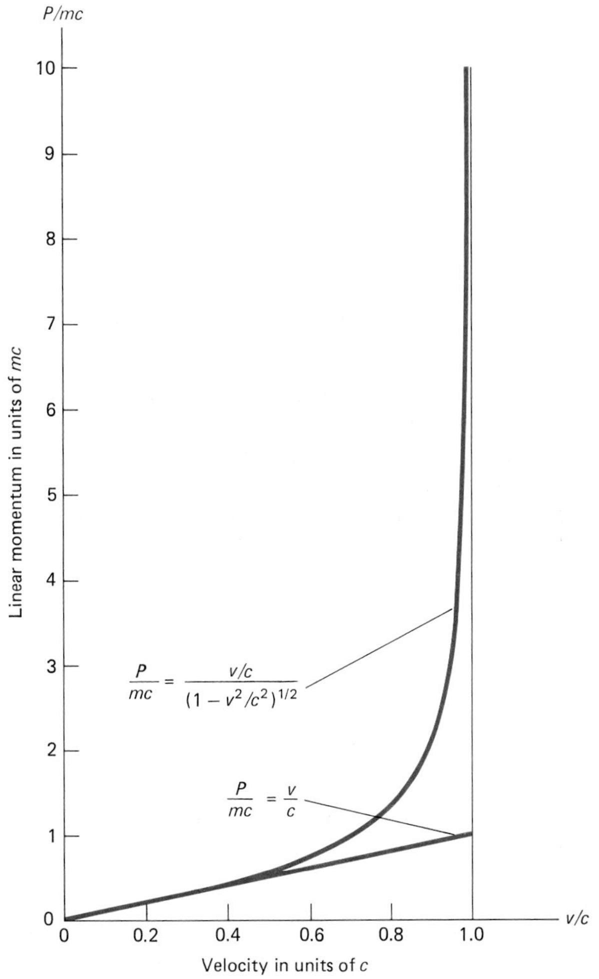

One further point should be made regarding Eq. 20.1. In Newtonian mechanics the mass of a particle is taken to be a constant, independent of the state of motion. We use this interpretation here. Throughout this text the mass, m, is the mass measured when the particle is at rest. It does *not* change with velocity. Some physicists speak of a relativistic increase in mass, but we reject this idea for the following three reasons:

1 There is no way of directly measuring such an increase in mass.

2 All the observed effects are actually covered by relativistic momentum, as given by Eq. 20.1.

3 In mathematical developments of relativistic mechanics beyond the scope of this text, it is much more convenient to hold the mass invariant (constant).

Now that we have defined relativistic momentum, we can write an equation in the form of Newton's second law:

$$\frac{d\mathbf{p}}{dt} = \mathbf{F} \tag{20.2}$$

This, in effect, defines a relativistic force, **F**.

Motion with Constant Force

Consider a particle of mass m that is accelerated from rest by a force of constant magnitude and direction. According to Newtonian mechanics its velocity increases without limit. We have $v = at = (F/m)t$. As time, t, approaches infinity, so does the velocity, v.

Using the relativistic version of Newton's second law we can show that the particle speed approaches c in the limit as $t \rightarrow \infty$. With F = constant, we can integrate Eq. 20.2,

$$\frac{dp}{dt} = F = \text{constant}$$

to obtain

$$\int dp = \int F\,dt$$

or

$$p = Ft$$

Substituting

$$p = \frac{mv}{(1 - v^2/c^2)^{1/2}}$$

and solving for v/c gives

$$\frac{v}{c} = \frac{Ft/mc}{[1 + (Ft/mc)^2]^{1/2}} \tag{20.3}$$

As usual, we look at limiting values, both to check the derivation and to get a better feeling for the meaning of the result. For small values of time we have

$$\lim_{Ft/mc \ll 1} \frac{v}{c} = \frac{Ft}{mc}$$

In the nonrelativistic limit, $F/m = a$ (acceleration), and we have

$$v = at$$

in agreement with nonrelativistic, or Galilean, kinematics (Section 4.4). For large values of time Eq. 20.3 yields

$$\lim_{Ft/mc \gg 1} \frac{v}{c} = 1$$

No matter how long the force is applied and no matter how long mass m accelerates, the speed of the mass will never exceed c. Note particularly that constant force does not imply constant acceleration.*

*If we retain the definition of acceleration as $a = dv/dt$, then $a = (F/m)(1 - v^2/c^2)^{3/2}$.

Example 1 Acceleration Toward the Stars

Imagine a rocket, filled with passengers, that experiences a constant force mg for a time of 1 yr. What speed does it achieve, starting from rest? Calculate the speed first according to Newtonian kinematics and then according to the relativistic equation 20.3.

From the classical definition of acceleration with $F/m = a = g$,

$$\begin{aligned} v &= at \\ &= 9.80\ \text{m/s}^2 \times 3.15 \times 10^7\ \text{s} \\ &= 3.09 \times 10^8\ \text{m/s} = 1.03c \end{aligned}$$

We see that Newtonian kinematics gives a speed that is 3% greater than the speed of light. This result, $v > c$, is a direct contradiction of the second postulate of special relativity (Section 19.4).

However, using Eq. 20.3, which is derived from relativistic momentum, we get

$$\frac{Ft}{mc} = \frac{gt}{c} = \frac{9.80\ \text{m/s}^2 \times 3.15 \times 10^7\ \text{s}}{3.00 \times 10^8\ \text{m/s}} = 1.029$$

and

$$\frac{v}{c} = \frac{1.029}{[1.0 + (1.029)^2]^{1/2}} = 0.717$$

Here our result, $v < c$, agrees with special relativity. Figure 20.4 exhibits the Newtonian and relativistic speeds as a function of time.

Equation 20.3 can be integrated to obtain the distance traveled by a mass, m. With $v = dx/dt$, Eq. 20.3 becomes

$$\frac{1}{c}\int_0^x dx = \int_0^t \frac{(Ft/mc)\,dt}{[1 + (Ft/mc)^2]^{1/2}}$$

Integrating, we obtain

$$\frac{1}{c}x = \frac{mc}{F}\left|\left[1 + \left(\frac{Ft}{mc}\right)^2\right]^{1/2}\right|_0^t$$

or

$$x = \frac{mc^2}{F}\left\{\left[1 + \left(\frac{Ft}{mc}\right)^2\right]^{1/2} - 1\right\} \qquad (20.4)$$

Again consider the limiting values. For small values of time, $Ft/mc \ll 1$, the binomial expansion leads to

$$\lim_{Ft/mc \ll 1} x = \frac{1}{2}\frac{FT^2}{m} = \frac{1}{2}at^2$$

in full agreement with the Newtonian (Galilean) form (Section 4.6). For $Ft/mc \gg 1$,

$$\lim_{Ft/mc \gg 1} x = ct$$

For $Ft/mc \gg 1$, the mass m spends most of the time traveling at almost $v = c$, and so the distance x is approximated by ct.

Example 2 Are We There Yet?

How far does the rocket in Example 1 travel in 1 yr? From Eq. 20.4 with $F/m = 9.80\ \text{m/s}^2$,

$$\begin{aligned} x &= \frac{3.00 \times 10^8\ \text{m/s}}{9.80\ \text{m/s}^2}\{[1 + (1.029)^2]^{1/2} - 1\} \\ &= 3.99 \times 10^{15}\ \text{m} \end{aligned}$$

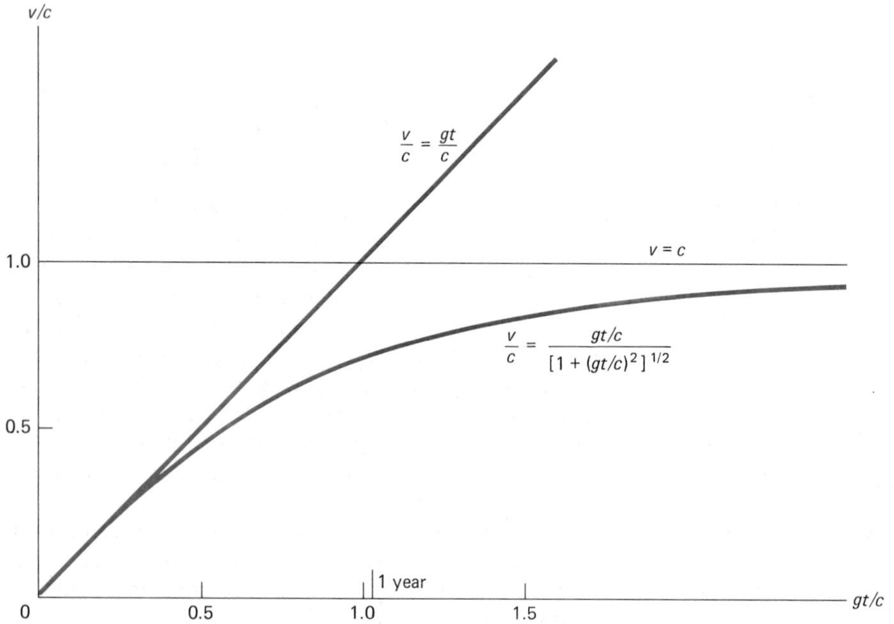

Figure 20.4

A comparison of Newtonian and relativistic speeds as a function of time. The force (dp/dt) is mg, constant.

Figure 20.5

A comparison of Newtonian and relativistic distance calculations for a rocket starting from rest and accelerating under a constant force, mg.

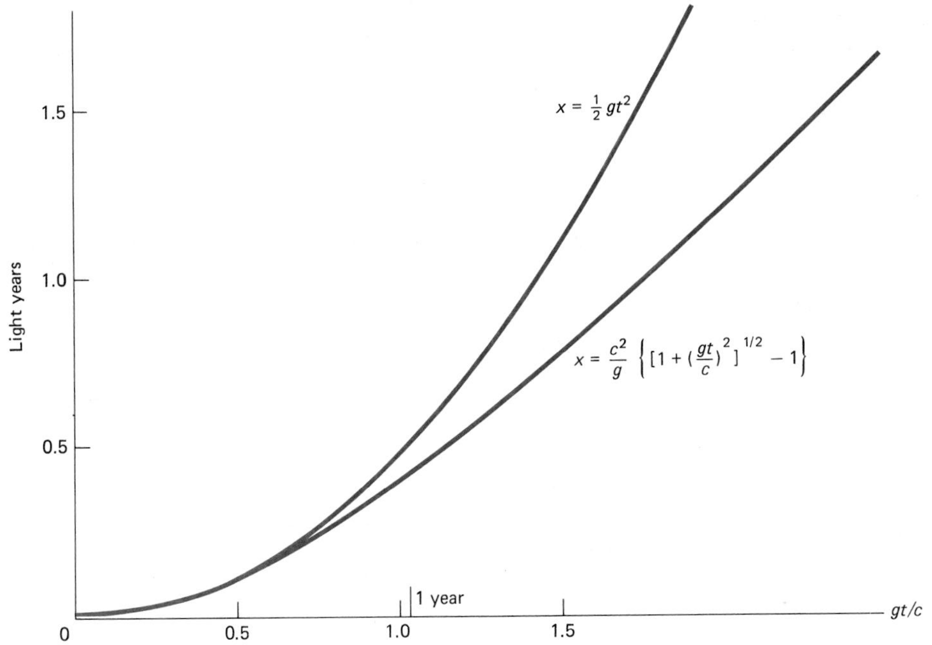

Meters don't mean very much on an interstellar scale. Let's convert meters to light-years. (One light-year is the distance light travels in 1 yr.)

$$1 \text{ light-year} = 3.00 \times 10^8 \frac{\text{m}}{\text{s}} \times 3.15 \times 10^7 \text{ s}$$
$$= 9.45 \times 10^{15} \text{ m}$$

Then

$$x = \frac{3.99 \times 10^{15}}{9.45 \times 10^{15} \text{m/light years}}$$
$$= 0.423 \text{ light years}$$

Unfortunately, the nearest star is about 4 light-years away, and so our rocket is only one tenth of the way there. Figure 20.5 gives a graphical comparison of the Newtonian and relativistic distance calculations.

Questions

1. A particle moves with a speed close to the speed of light. A constant force F is applied for a time t in the direction of motion. Would you expect the speed change predicted by relativistic mechanics to be greater or less than that predicted by Newtonian mechanics? Explain.
2. During the time t a constant force acts on a particle that is moving at a speed close to the speed of light. Qualitatively compare the change in linear momentum of the particle according to (a) Newtonian and (b) relativistic mechanics.
3. Discuss the appropriateness of the term *accelerator* used to describe large machines whose purpose it is to produce beams of high-energy particles.
4. In Newtonian mechanics doubling the velocity will cause the momentum to be doubled. Discuss the effect on p (relativistic momentum) of doubling the velocity of a particle. (Take $v/c < 0.5$.)

20.3 Relativistic Energy

In Section 7.4 the work done on a particle by a net force F was calculated as $\int F\,dx$. The result is the change in the kinetic energy, $\frac{1}{2}mv^2$. The basic relationship is given by the work–energy theorem. When we repeat this calculation in a relativistic framework, we obtain the relativistic kinetic energy. The calculation illustrates the parallel relationships between Newtonian concepts of linear momentum and kinetic energy and the corresponding relativistic concepts. However, the establishment of the resultant expression (Eq. 20.10) as the appropriate relativistic form depends on the criteria of Section 20.1.

As a special case of the work–energy theorem for a particle starting from rest, the kinetic energy of the particle is the work done in accelerating the particle from rest. As a mathematical equation,

$$K = \int_0^x F\,dx \tag{20.5}$$

Using $F = dp/dt$ and $dx = v\,dt$ gives

$$K = \int_0^t \frac{dp}{dt} v\,dt = \int_0^p v\,dp \qquad (20.6)$$

In Section 7.4 we integrated with respect to v. Here we select p as the variable of integration. This choice makes our calculations somewhat easier. Now we must express v in terms of p. If we use the nonrelativistic relation $v = p/m$, Eq. 20.6 gives $K = p^2/2m$. With $p = mv$ this is $K = \frac{1}{2}mv^2$. Using the relativistic relation, Eq. 20.1,

$$p = \frac{mv}{(1 - v^2/c^2)^{1/2}}$$

we can solve for v in terms of p to obtain

$$v = \frac{p/m}{[1 + (p/mc)^2]^{1/2}} \qquad (20.7)$$

which is the relativistic analog of $v = p/m$. Equation 20.6 becomes

$$K = \int_0^p \frac{(p/m)\,dp}{[1 + (p/mc)^2]^{1/2}} \qquad (20.8)$$

Analogous to the integration of Eq. 20.3 to obtain x, Eq. 20.8 can be integrated to give the result

$$K = mc^2 \left\{ \left[1 + \left(\frac{p}{mc} \right)^2 \right]^{1/2} - 1 \right\} \qquad (20.9)$$

Using Eq. 20.1 to express p in terms of v, we can rewrite Eq. 20.9 as

$$K = \frac{mc^2}{(1 - v^2/c^2)^{1/2}} - mc^2 \qquad (20.10)$$

This is the relativistic generalization of $K = \frac{1}{2}mv^2$. We will offer an interesting explanation of the terms in this equation after we look at how Eq. 20.10 reduces to the Newtonian kinetic energy.

Reduction to Newtonian Form

The expression for relativistic kinetic energy, Eq. 20.10, holds for all speeds $0 \leq v < c$. Thus it must reduce to the Newtonian form for small speeds ($v \ll c$), since Newtonian mechanics has been verified by experiment for $v \ll c$. The $(1 - v^2/c^2)^{-1/2}$ term can be expanded by the binomial theorem, Appendix 4. This expansion leads to

$$K = mc^2 \left[\not{1} + \frac{1}{2}\frac{v^2}{c^2} + \frac{3}{8}\left(\frac{v^2}{c^2}\right)^2 + \cdots - \not{1} \right]$$

If v^2/c^2 is small enough that the v^4/c^4 term can be neglected, then

$$K = mc^2 \left[\frac{1}{2}\frac{v^2}{c^2} + \cancel{\frac{3}{8}\left(\frac{v^2}{c^2}\right)^2} + \cdots \right]$$

or

$$K = \frac{1}{2}mv^2$$

in agreement with the Newtonian form. Criterion 2 of Section 20.1 is satisfied.

Figure 20.6

A comparison of kinetic energy in relativistic mechanics and in Newtonian mechanics. Note how the relativistic kinetic energy goes to infinity as $v \rightarrow c$.

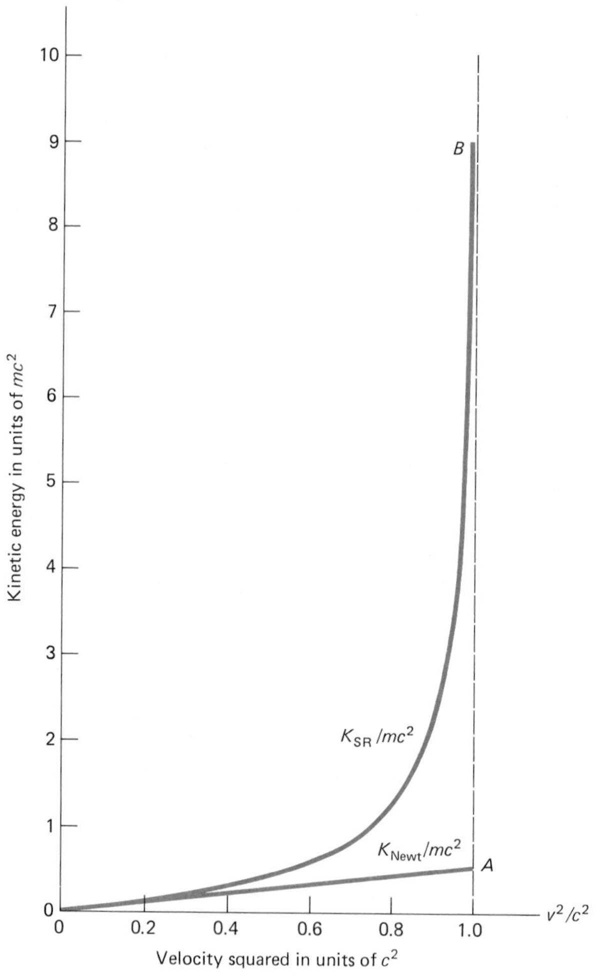

In Figure 20.6 the Newtonian and special relativistic kinetic energy curves appear to merge for small v. The reason $\frac{1}{2}mv^2$ works as well as it does for most purposes is that v/c (and v^2/c^2) is very small in everyday applications. However, in problems involving atoms and nuclei v^2/c^2 is often not small compared to 1, and the relativistic forms, Eqs. 20.10 and 20.12, must be used.

Example 3 **Newtonian and Relativistic Kinetic Energies at Small Speeds, $v \ll c$**

Consider an 80-kg astronaut whose rocket moves at 1.12×10^4 m/s (~25,000 mi/hr). What error do we make in using the Newtonian expression for the kinetic energy of the astronaut? From Eq. 7.9 we have

$$K_{Newt} = \frac{1}{2}mv^2$$

$$= \frac{1}{2} \times 80 \text{ kg} \times (1.12 \times 10^4 \text{ m/s})^2$$

$$= 5.02 \times 10^9 \text{ J}$$

For the astronaut, $v/c = 3.73 \times 10^{-5}$ and $v^2/c^2 = 1.39 \times 10^{-9}$. With such small values for v^2/c^2, the relativistic equation,

$$K_{SR} = \frac{mc^2}{(1 - v^2/c^2)^{1/2}} - mc^2$$

should not be evaluated directly, because too many significant figures would be needed. The two terms in Eq. 20.10 give nearly equal values, and many significant figures cancel in subtraction. Instead we use the binomial series expression (see Section 19.8 for similar use of the binomial theorem in length contraction) to calculate the difference between the relativistic kinetic energy, K_{SR}, and K_{Newt}. We have

$$K_{SR} - K_{Newt} = mc^2\left(\frac{1}{2}\frac{v^2}{c^2} + \frac{3}{8}\frac{v^4}{c^4} + \frac{5}{16}\frac{v^6}{c^6} + \cdots\right) - mc^2\,\frac{1}{2}\frac{v^2}{c^2}$$

Recall that the zero-order terms (the 1's) canceled in the development of the small-velocity approximation of K_{SR}. Here we have exact cancellation of the v^2/c^2 terms. For a v/c that is small compared to 1, so that $(5/16)(v^6/c^6)$ can be neglected, the energy difference is proportional to v^4/c^4.

$$K_{SR} - K_{Newt} \cong mc^2 \cdot \frac{3}{8}\frac{v^4}{c^4} = \frac{1}{2}mv^2 \cdot \frac{3}{4}\frac{v^2}{c^2}$$

$$= K_{Newt} \cdot \frac{3}{4}\frac{v^2}{c^2}$$

Substituting numerical values, we write

$$K_{SR} - K_{Newt} = 5.02 \times 10^9\ \text{J} \times \frac{3}{4} \times 1.39 \times 10^{-9}$$

$$= 5.25\ \text{J}$$

At a velocity of 25,000 mi/hr the error in using Newtonian kinetic energy instead of the "exact" relativistic kinetic energy is about 1 part in 10^9.

To obtain the result 5.25 J by a direct calculation of K_{SR} and K_{Newt}, we would need to calculate both values to 12 significant figures.

Rest Mass Energy and Relativistic Total Energy

Let's return to the expression for relativistic kinetic energy:

$$K = \frac{mc^2}{(1 - v^2/c^2)^{1/2}} - mc^2 \qquad (20.10)$$

The second term, mc^2, is independent of the velocity and is labeled the *rest mass energy*, E_o:

$$E_o = mc^2 \qquad (20.11)$$

Here we are associating an energy with a mass m. This does not mean, however, that particle mass can always be converted into energy, or vice versa. Equation 20.11 simply expresses an equivalence. Whether or not mass (rest mass) can be transformed to energy or energy to mass depends on the physics of the specific situation. Transformation of mass into energy and of energy into mass is considered at the end of this section.

The first term in Eq. 20.10, $mc^2/(1 - v^2/c^2)^{1/2}$, is *interpreted* as the *total* relativistic energy, E (kinetic energy plus rest mass energy), of a mass m moving with velocity **v**.

$$E = \frac{mc^2}{(1 - v^2/c^2)^{1/2}} \qquad (20.12)$$

$$= K + E_o$$

In Section 20.4 we will see that it is this total relativistic energy E that is conserved in particle interactions.

The Newtonian kinetic energy $K = \frac{1}{2}mv^2$ and the relativistic kinetic energy, Eq. 20.10, are plotted in Figure 20.6. As with the relativistic momentum (see Figure 20.3), the relativistic kinetic energy is greater than the Newtonian kinetic energy for a given velocity, and indeed approaches infinity as $v \rightarrow c$. Here is a physical reason for regarding c as a limiting speed: It would take an infinite amount of work to give a nonzero mass m a speed equal to the speed of light.

Now let's move from the small velocity in Example 3 (25,000 mi/hr is slow compared to the speed of light) to velocities that are comparable to c, the speed of light.

Example 4 **Kinetic Energy and Rest Energy**

At what velocity will the kinetic energy of a particle be equal to its rest energy? To answer this question we set

$$K = E_o$$

or

$$\frac{mc^2}{(1 - v^2/c^2)^{1/2}} - mc^2 = mc^2$$

This reduces to the expression

$$\left(1 - \frac{v^2}{c^2}\right)^{1/2} = \frac{1}{2}$$

or

$$\frac{v}{c} = \frac{\sqrt{3}}{2} = 0.866$$

Substituting $c = 3 \times 10^8$ m/s in the previous equation, we get $v = 2.60 \times 10^8$ m/s. At a velocity of 2.60×10^8 m/s the particle's relativistic kinetic energy, K_{SR}, is equal to its rest mass energy, E_o

We can use Figure 20.6 to check this calculation. $K_{SR} = E_o$ is equivalent to $K_{SR}/mc^2 = 1$. This corresponds to $v^2/c^2 = 0.75$ or $v/c = 0.866$, as found by direct calculation.

As a particle's velocity approaches c, the relativistic energy E approaches infinity. In this extreme relativistic range the problem of finding the energy that corresponds to a given velocity or the velocity that corresponds to a given energy calls for the technique of Section 19.9, as we see in the following example.

Example 5 Extreme Relativistic Velocity, $v \approx c$

An electron in the Stanford Linear Accelerator is accelerated to a speed that differs from c by only 3 parts in 10^{10}:

$$\frac{c - v}{c} = \frac{3}{10^{10}}$$

or

$$1 - \frac{v}{c} = 3.0 \times 10^{-10}$$

We want to calculate the ratio of the total relativistic energy to the rest mass energy, E/E_0.

Now algebraically

$$1 + \frac{v}{c} = 2 - \left(1 - \frac{v}{c}\right)$$
$$= 2 - 3 \times 10^{-10}$$
$$\approx 2$$

Then

$$1 - \frac{v^2}{c^2} = \left(1 + \frac{v}{c}\right) \times \left(1 - \frac{v}{c}\right) \cong 6 \times 10^{-10}$$

and

$$\left(1 - \frac{v^2}{c^2}\right)^{1/2} \cong 2.45 \times 10^{-5}$$

From their definitions, Eqs. 20.11 and 20.12,

$$\frac{E}{E_0} = \frac{1}{(1 - v^2/c^2)^{1/2}} \cong \frac{1}{2.45 \times 10^{-5}}$$
$$\approx 40,000$$

This electron is in the extreme relativistic range, with its total relativistic energy 40,000 times greater than its rest mass energy.

If you wonder why we used such an indirect technique in Example 5, try punching

$$\frac{v}{c} = 0.9999999997$$

into your electronic calculator and carrying out a direct calculation. If your calculator carries ten or more significant figures you can get away with a direct calculation, but if your calculator carries nine figures or less, you'll get nonsense.

An Energy–Momentum Relation

Relativistic energy and relativistic linear momentum are related by the useful identity

$$E^2 = (pc)^2 + (mc^2)^2 \qquad (20.13)$$

This is derived directly from Eq. 20.9 and can be verified by substituting the value of **p** from Eq. 20.1 and the value of E from Eq. 20.12 into Eq. 20.13. The identity is in the form of a Pythagorean equation and can be interpreted by using the Pythagorean theorem, as shown in Figures 20.7a and 20.7b. Figure 20.7a shows the energy–momentum relationship corresponding to small speeds—the Newtonian case. Figure 20.7b presents the extreme relativistic case with $v \approx c$ and $E \gg mc^2$.

Figure 20.7a

An energy–momentum relation. For small speeds, the linear momentum is small and $E \approx mc^2$.

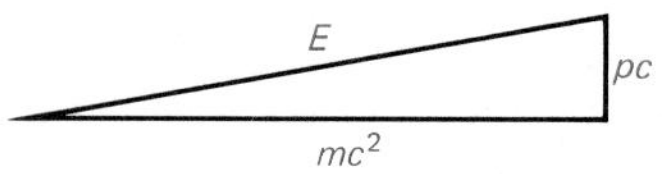

Figure 20.7b

In the extreme relativistic case with speeds approaching c, $E \approx pc$.

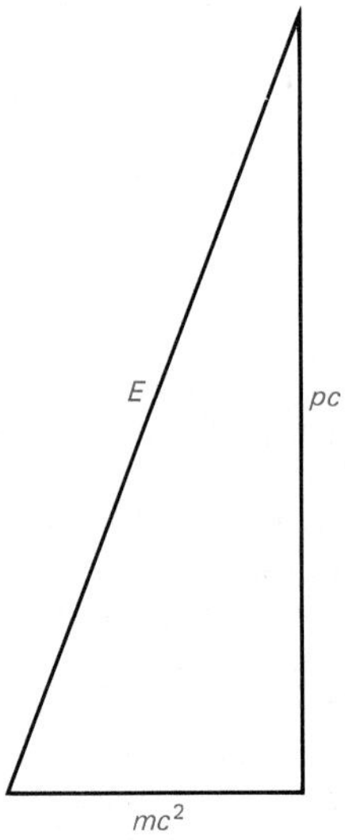

Because $(mc^2)^2$ is a constant, an invariant that is independent of the motion of the observer, then $E^2 - (pc)^2$ is a constant. In terms of the moving reference frames of Chapter 19

$$(E')^2 - (p'c)^2 = E^2 - (pc)^2$$
$$= (mc^2)^2 \quad \text{invariant} \qquad (20.14)$$

Because they are energies, E and E_0 can be properly expressed in joules. Actually, however, joules are seldom used in relativistic mechanics. Special relativity is important primarily in describing electrons, protons, and other subatomic particles. In accelerators these electrically charged particles are given energy by electrical forces that do work on them. As we will explain further in Section 28.1, it has become customary to describe these particle energies in electron volts (eV) or in metric multiples thereof, such as MeV (million, or 10^6, electron volts), GeV (billion, or 10^9, electron volts), or TeV (trillion, or 10^{12}, electron volts). For MeV the conversion relation is

$$1 \text{ MeV} = 1.60219 \times 10^{-13} \text{ J}$$

Because pc has the same dimensions as mc^2, the relativistic momentum is often given in units of mc where m is the rest mass of the particle under discussion. Alternatively, if the energies are given in units of MeV, it is convenient to give the momentum p in units of MeV/c.

$$1\,\frac{\text{MeV}}{c} = \frac{1.60219 \times 10^{-13}\,\text{J}}{2.99793 \cdot 10^{+8}\,\text{m/s}}$$
$$= 5.34432 \times 10^{-22}\,\text{kg} \cdot \text{m/s}$$

For example, we can find the momentum of an electron with a kinetic energy of 3 MeV as follows. Using Eq. 20.13 we write

$$pc = [E^2 - (mc^2)^2]^{1/2}$$

With the value of mc^2 for an electron equal to 0.511 MeV and $E = 3.511$ MeV, we have

$$pc = [(3.511)^2 - (0.511)^2]^{1/2}$$
$$= 3.47 \text{ MeV}$$

and

$$p = 3.47 \text{ MeV/c}$$

Zero-Mass Particles

One of the fundamental properties of a particle is its mass. We have seen that a particle of mass m at rest possesses a definite rest mass energy, mc^2. A particle in motion possesses kinetic energy in addition to its rest mass energy.

Although we usually think of the mass of a particle as a number greater than 0, physicists consider three types of particles that have zero rest mass. These are the photon, the neutrino,* and the as yet unobserved graviton. The photon is associated with light. We can view a beam of light as a stream of photons, each with a definite energy, and each traveling at the speed of light. The photon has zero rest mass—it does not exist as an object at rest. The neutrino is produced in certain types of radioactive decay reactions (Section 43.2). Like the photon, the neutrino is a zero-mass particle, and does not exist as an object at rest. The graviton is a hypothetical zero-mass particle that is associated with the úniversal force of gravitation. Photons and neutrinos have been detected, but the existence of gravitons has yet to be experimentally verified. (Because of the relative weakness of the gravitational force, individual gravitons should carry very little energy. The effects of the absorption of an individual graviton are far too small to permit detection with contemporary techniques.)

The photon, neutrino, and graviton all share two important characteristics. Each has zero mass and each travels at the speed of light. In fact, we can prove that a zero-mass particle carrying energy must travel at the speed of light. The energy of a zero-mass particle is kinetic in nature—the particle has no mass and thus no rest mass energy.

To prove that $v = c$ for any particle with $m = 0$, $E \neq 0$, we first rewrite Eq. 20.12 for the total energy as

$$E \cdot \left(1 - \frac{v^2}{c^2}\right)^{1/2} = mc^2 \qquad (20.15)$$

*Some experiments have suggested that at least one form of neutrino may have a very small but nonzero rest mass. The rest mass energy equivalent might be approximately 20–40 eV.

Figure 20.8

Before and after for the decay of a neutral pion into two gamma-ray photons.

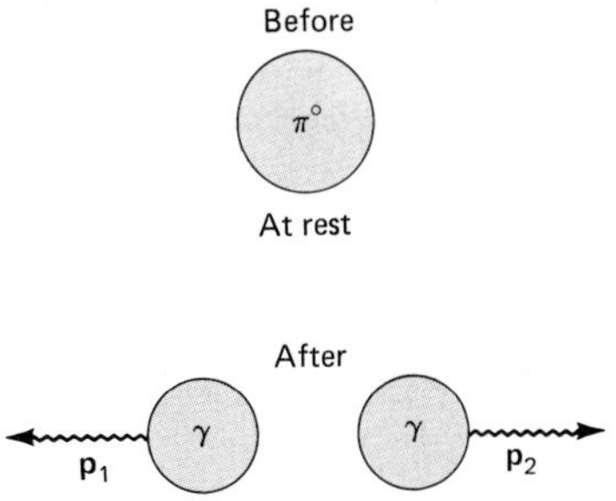

If $m = 0$, we are left with

$$E \cdot \left(1 - \frac{v^2}{c^2}\right)^{1/2} = 0 \qquad (20.16)$$

But because $E \neq 0$, $(1 - v^2/c^2)^{1/2}$ must equal zero, which means that $v = c$. Zero-mass particles carrying energy must travel at the limiting speed, the speed of light. Conversely, a particle moving at the speed of light (but with only a finite amount of energy) must have zero mass. If its mass were not zero, its energy would be infinite according to Eq. 20.12.

Example 6 **Decay of a Neutral Pion**

One unstable elementary particle is the pion. The pion has a mass intermediate between that of an electron and that of a proton. Pions may carry a positive or a negative electric charge, or be neutral (uncharged). A neutral pion decays at rest into two photons, as shown in Figure 20.8. Conservation of linear momentum requires that the photon linear momenta be equal in magnitude and opposite in direction. Thus Eq. 20.14 requires that the two photons have the same energy. Hence the rest energy of the pion must be equally divided between the two photons.

The neutral pion rest mass m is equal to 264 electron rest masses, m_e. Then, in terms of electron rest mass units, the energy of each photon is

$$E_\gamma = 1/2 \times 264 m_e c^2$$
$$= 132 m_e c^2$$

From Eq. 20.14 the magnitude of the momentum of each photon is

$$p = \frac{E}{c}$$
$$= 132 m_e c$$

Can you imagine zero-rest-mass "particles" carrying energy and momentum? The sale of sunburn remedies testifies to the transport of energy from the sun to the skin by photons. The linear momentum of light was first measured early in this century.

Questions

5. In Newtonian mechanics, doubling the velocity of a particle will cause its kinetic energy to increase fourfold. What would be the effect on K (the relativistic kinetic energy) of doubling the velocity of a particle? (Take $v/c < 0.5$.)

6. If the total relativistic energy E is conserved in a nuclear reaction, why isn't the relativistic kinetic energy also conserved?

7. Physicists generally ignore rest mass energy when analyzing collisions between particles moving at low velocities. Why?

8. Physicists who apply special relativity to high-velocity collisions concentrate their attention on the total relativistic energy E and tend to ignore the explicit form (Eq. 20.10) of relativistic kinetic energy. Why?

9. The net work done on a particle equals the particle's increase in kinetic energy, in both Newtonian and relativistic mechanics. Compare the changes in the kinetic energy, total energy, and rest mass energy of a particle that are associated with the net work done.

20.4 Conservation of Energy and Momentum in Particle Interactions

One spectacular aspect of the interaction of very high-energy subatomic particles is the "creation" of new particles, or more properly, the conversion of kinetic energy into particle mass. One example of this phenomenon is the creation of a proton–antiproton pair in the high-energy collision of two protons. We will calculate the minimum energy necessary—the threshold—for this proton–antiproton pair production.

Proton–Antiproton Pair Production in the Center-of-Momentum System

We start our calculation in the center-of-momentum (center-of-mass) reference frame* because the calculation much easier here than in the laboratory reference frame. As indicated schematically in Figure 20.9, the initial state (before collision) consists of two protons with equal speeds but opposite direction that are about to collide head on. The final state (after collision) consists of the two incident protons plus a third proton and an antiproton (proton mass but negative electric charge). At the threshold, the total kinetic energy of the two colliding protons is just sufficient to create the proton–antiproton pair. There is no energy left over for any motion. The four particles are stationary in the center-of-momentum system.

*The center-of-momentum reference system is one in which the net linear momentum equals zero. For two particles of equal mass this system coincides with the center-of-mass system.

Figure 20.9

Production of a proton–antiproton pair. Before and after as seen in the center-of-momentum system.

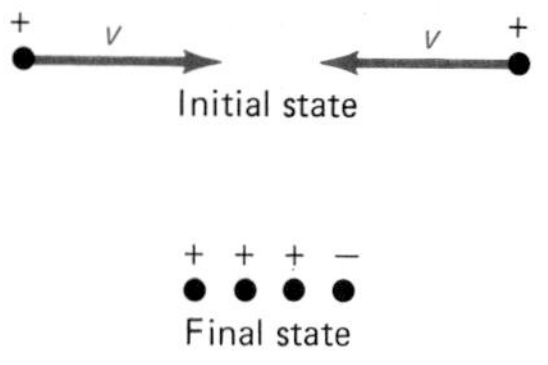

The initial state has two protons, each with speed v, and so the total initial relativistic energy is

$$E_{\text{initial}} = \frac{2m_p c^2}{(1 - v^2/c^2)^{1/2}} \qquad (20.17)$$

The final state has three protons, each with mass m_p, and an antiproton, which also has mass m_p. At the threshold, or minimum energy, condition these four particles are at rest. The final total energy is

$$E_{\text{final}} = 4m_p c^2 \qquad (20.18)$$

By applying conservation of total relativistic energy, we get

$$\frac{2m_p c^2}{(1 - v^2/c^2)^{1/2}} = 4m_p c^2 \qquad (20.19)$$

This yields

$$\left(1 - \frac{v^2}{c^2}\right)^{1/2} = \frac{1}{2}$$

or

$$v = \frac{\sqrt{3}}{2} c = 0.866c \qquad (20.20)$$

Compare this result with the result in Example 4. The numerical equality is not just coincidence. This is the velocity at which the kinetic energy is equal to the rest mass energy. With each of the two incoming protons having this velocity, there is enough kinetic energy to create a proton–antiproton pair. We have applied conservation of relativistic energy in order to calculate the threshold velocity.

Proton–Antiproton Pair Production in the Laboratory System

The use of the center-of-momentum system simplifies our calculation of the kinetic energy threshold, but the original experiments were carried out in a laboratory by having high-energy protons from an accelerator strike stationary protons in a target. Therefore, we need to transform velocities from the center-of-momentum reference frame to the laboratory frame. In other words, we convert our solutions to the reference frame in which the right-hand proton in Figure 20.9 is at rest.

Figure 20.10

Velocity transformation from the center of momentum system S' back to the laboratory system S. The primed and unprimed velocities are related by the Einstein velocity addition formula, Eq. 19.29 of Section 19.7.

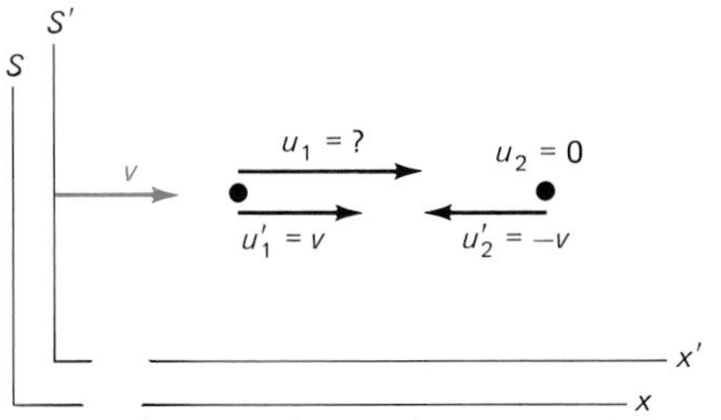

The velocity relationships are shown in Figure 20.10 (The coordinates and the velocities are labeled in the same style as in Chapter 19.) System S' is the center-of-momentum system; S is the laboratory system. The velocity of a particle is u' relative to S', while u is the corresponding velocity relative to the laboratory. We seek u_1, the velocity of the incident proton relative to the laboratory.

We may *not* use the Galilean velocity addition formula to obtain $u_1 = u_1' + v = 2v$ because the velocities approach that of light. Instead, velocity addition must be carried out by means of the Einstein velocity addition formula, Eq. 19.29 of Section 19.7. With $u_1' = v$ this formula yields

$$u_1 = \frac{v + v}{1 + v^2/c^2} = \frac{2v}{1 + v^2/c^2}$$

Substituting $v = (\sqrt{3}/2)\,c$ from Eq. 20.20, we get

$$u_1 = \frac{\sqrt{3}c}{7/4} = \frac{4\sqrt{3}}{7}c = 0.9897c \qquad (20.21)$$

The total relativistic energy of the incident particle in the laboratory system is

$$E = \frac{m_pc^2}{(1 - u_1^2/c^2)^{1/2}} = 7m_pc^2 \qquad (20.22)$$

The incident proton has a rest mass energy $E_o = m_pc^2$. The accelerator must supply a kinetic energy of $6m_pc^2$. With $m_pc^2 = 0.938 \times 10^9$ eV $= 0.938$ GeV, the required minimum kinetic energy for proton–antiproton pair production is 5.6 GeV. In 1955 the first proton–antiproton pairs were produced—at a kinetic energy threshold of 5.6 GeV, exactly as predicted by relativistic calculations.

Notice that the threshold kinetic energy is much higher in the laboratory system ($6m_pc^2$) than it is in the center-of-mass system ($2m_pc^2$). In the extreme relativistic range ($E \gg mc^2$), the energy available in the center-of-mass system rises only as the square root of the energy in the laboratory system. This places a very severe restriction on physicists' attempts to achieve higher and higher energies in the center-of-mass system by using a high-speed projectile and a stationary target. The alternative, already in use around the world, is to create collisions between beams of particles moving in opposite directions. Then for equal-mass, equal-speed projectiles the laboratory system is the center-of-momentum system. Devices that operate by this principle are called *storage rings*, because the accelerated particles are stored in oppositely directed beams before they are allowed to collide.

Conservation of Linear Momentum in the Laboratory System

At the beginning of this chapter we emphasized that the relativistic momentum and energy are defined in such a way that they are conserved in the absence of external forces. Let's check to see whether linear momentum is conserved in the laboratory reference frame for proton–antiproton pair production. With

$$u_1 = \left(\frac{4\sqrt{3}}{7}\right)c$$

we have

$$p_i = \frac{m_pu_1}{(1 - u_1^2/c^2)^{1/2}} = 4\sqrt{3}m_pc$$

for the initial momentum. After collision there are four particles with a velocity $v = (\sqrt{3}/2)c$. Then

$$p_f = \frac{4m_pv}{(1 - v^2/c^2)^{1/2}} = 4\sqrt{3}m_pc$$

The relativistic linear momentum is conserved.

As a further example of energy and momentum conservation in particle interactions, we now consider the radioactive decay of a subatomic particle called the *kaon*.

Example 7 Radioactive Decay of a Neutral Kaon

The neutral kaon is an unstable particle that has been produced by high-energy accelerators. A neutral kaon at rest can decay into a pair of oppositely charged pions, as shown in Figure 20.11. The kaon mass is 3.566 times as large as the mass of a charged pion. We seek the linear momentum and energy of each pion, and we want to know whether it is necessary to use relativistic mechanics rather than Newtonian mechanics.

The pions must be oppositely directed, as shown in Figure 20.11. This follows from linear momentum conservation. For the kaon, $\mathbf{p} = 0$, and hence $\mathbf{p}_1 + \mathbf{p}_2 = 0$ for the pion pair. The linear momenta of the pions must be equal in magnitude and opposite in direction.

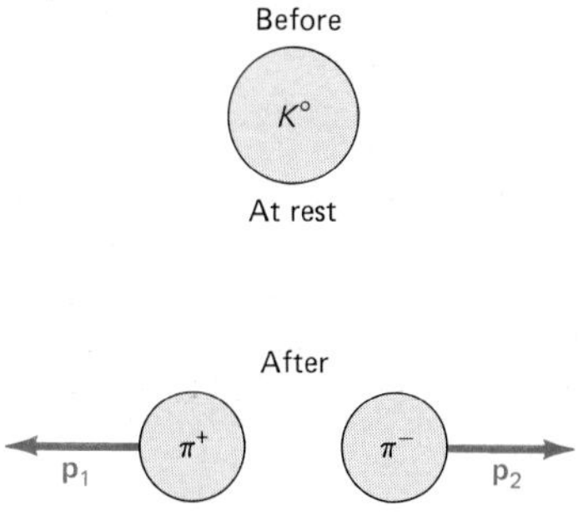

Figure 20.11

The decay of a neutral kaon into two pions.

From Eq. 20.14 the total energy of a pion is given by

$$E_1 = [(p_1c)^2 + (m_\pi c^2)^2]^{1/2}$$

Because p_1 and m_π are the same for each pion, it follows that their total energies must also be the same. Conservation of energy requires that the sum of the total energies for the pions, $E_1 + E_2$, be equal to the original rest energy of the kaon. Thus we can write

$$3.566m_\pi c^2 = E_1 + E_2 = 2E_1$$

or

$$E_1 = 1.783m_\pi c^2$$

From Eq. 20.14, the pion momentum is given by

$$p_1 = \frac{1}{c}[(1.783m_\pi c^2)^2 - (m_\pi c^2)^2]^{1/2}$$

or

$$p_1 = 1.476m_\pi c$$

You may ask, is relativistic mechanics necessary? If you try to interpret $p = 1.476m_\pi c$ in terms of a Newtonian mv, you will get $v = 1.476c$, which is nonsense. The answer, then, is yes, relativistic mechanics is necessary. As a further indication of the necessity of relativistic mechanics, you can show from the pion energy that the pion speed is 0.83c. The factor $(1 - v^2/c^2)^{1/2}$ is 0.561, significantly different from unity.

Mass–Energy Conversion

Example 7 and numerous other examples in this text illustrate the transformation of rest mass (particles at rest) into energy and energy into rest mass. Because we want energy to do useful work, we are vitally interested in the transformation of mass into energy. From the fundamental relation, Eq. 20.11,

$$E_o = mc^2$$

we see that a small number of kilograms corresponds to a large number of joules.

$$1 \text{ kg} \rightarrow c^2 \text{ J}$$
$$= 9 \times 10^{16} \text{ J}$$

The complete conversion of 1 kg of mass into electrical energy could provide electrical power at the rate of 2.86×10^3 MW for 1 yr.

$$2.86 \times 10^3 \times 10^6 \frac{\text{J}}{\text{s}} \times 3.15 \times 10^7 \text{ s}$$
$$= 9.0 \times 10^{16} \text{ J}$$

For comparison, the electricity-generating capacity of our entire country is about 400×10^3 MW.

Rewriting Eq. 20.11 as $m = E_o/c^2$, we see that the proportionality constant $1/c^2$ is so small that in chemical reactions the change in the rest mass of the particles is undetectable (Table 20.1). In nuclear reactions, however, the mass change, Δm, of the particles is often a much larger fraction of the initial rest mass. The initial and final masses can be measured (by magnetic techniques if the particles carry charge—see Section 32.3) and Δm can be calculated.

Nuclear fission is the process of splitting a heavy element such as uranium into two fragments. The fractional rest mass loss, $\Delta m/m$, is small (about 0.001), but thanks to the factor c^2, the energy release is large. This is the source of the electrical energy produced by nuclear reactors. Nuclear fusion is the combining of light elements, usually forms of hydrogen, to build up heavier elements. Here, too, the fractional rest mass loss, $\Delta m/m$, is small (about 0.006). Whereas a practical nuclear fission reactor is a working reality, a nuclear fusion reactor is only a hope. Further discussion of both of the reactions involved appears in Chapter 43.

In extreme cases, the fractional rest mass loss, $\Delta m/m$, may be unity; 100% of rest mass is converted into radiant energy. An electron–positron pair annihilate each other, giving rise to radiant energy in the form of two gamma photons. The decay of the pion, π^o, into two gamma photons was considered in detail in Example 6.

Table 20.1

Reaction	$\Delta m/m$
Chemical	$\sim 10^{-9}$
Nuclear	10^{-3} and higher
Fission	1×10^{-3}
Fusion	6×10^{-3}
π^o decay and	
e^-–e^+ annihilation	1.00

m is the total initial particle rest mass; Δm is the change in the total particle rest mass. The energy released (radiant, kinetic, thermal, etc.) is $\Delta E = \Delta mc^2$

Question

10. The energy threshold for the creation of a proton–antiproton pair is higher in the laboratory system than it is in the center-of-momentum system. Explain.

Summary

The relativistic generalization of linear momentum is given by

$$\mathbf{p} = \frac{m\mathbf{v}}{(1 - v^2/c^2)^{1/2}} \qquad (20.1)$$

The following form of Newton's second law

$$\mathbf{F} = \frac{d\mathbf{p}}{dt} \qquad (20.2)$$

is used to define a relativistic force.

From the work–energy principle the relativistic generalization of kinetic energy is

$$K = \frac{mc^2}{(1 - v^2/c^2)^{1/2}} - mc^2 \qquad (20.10)$$

The quantity

$$E = \frac{mc^2}{(1 - v^2/c^2)^{1/2}} \qquad (20.12)$$

is identified as the total relativistic energy; E includes the rest mass energy

$$E_o = mc^2 \qquad (20.11)$$

The mass m is taken to be the mass measured at rest.

The relativistic linear momentum **p** and the relativistic kinetic energy K each reduce to the corresponding Newtonian form for $v \ll c$.

The relativistic linear momentum **p** and the relativistic total energy E each satisfy a conservation law—in the absence of external forces. These conservation laws hold even though the particles in a given isolated system may be created or annihilated.

Energies can be expressed in joules, in electron volts (or metric multiples such as MeV), and in mc^2 units using the mass–energy equivalence. Corresponding units for linear momenta are kg · m/s, MeV/c, and mc.

Relativistic total energy and relativistic linear momentum are related by

$$E^2 = (pc)^2 + (mc^2)^2 \qquad (20.13)$$

Zero-mass particles (photons, neutrinos, and gravitons) carry energy and momentum and travel at the speed of light.

Suggested Reading

A. S. Goldhaber and M. M. Nieto, The Mass of the Photon, *Sci. Am.*, *234*, 86–96 (May 1976).

C. K. O'Neill, Particle Storage Rings, *Sci. Am.*, *215*, 107–116 (November 1966).

E. Segrè and C. C. Wiegand, The Antiproton, *Sci. Am.*, *194*, 37–41 (June 1956).

Problems

Section 20.2 Relativistic Linear Momentum

1. A particular very high-energy proton has a speed of 0.9999995 c. Calculate the ratio of the relativistic momentum to the Newtonian momentum.

2. If the relativistic momentum of a particle is given by $p = 10mc$, what is the ratio v/c?

3. A hypothetical rocket experiences a force-to-mass ratio of $3g$ ($g = 9.80$ m/s^2) for a time of 60 days. Starting from rest, what speed does it achieve according to (a) Newtonian kinematics; (b) relativistic mechanics?

4. (a) What constant F/m ratio would get a space probe to Alpha Centauri (4.3 light-years away) in 10 yr? (b) How fast would the space probe be going as it flashed past Alpha Centauri?

5. An electron, mass 9.11×10^{-31} kg, experiences a force of 1.602×10^{-13} N in an accelerator. (a) Calculate the initial acceleration (F/m) of this electron. (b) If the electron, starting from rest, were to accelerate for 10 ns, how fast would it be traveling?

6. How far does the electron of Problem 5 travel?

7. An electron achieves a speed of $0.9c$ in 1 ms. Assuming F/m to be constant, calculate F/m.

8. Show that the expression for distance, Eq. 20.4, is dimensionally consistent by showing that both sides have dimensions of L.

9. Relativistic mechanics reduces to Newtonian mechanics if the relativistic momentum is much less than mc: $p \ll mc$. Prove that this statement implies that $v/c \ll 1$, thus justifying itself.

Section 20.3 Relativistic Energy

10. Calculate the kinetic energy in joules of an electron and a proton moving with $v = (\sqrt{3}/2)c$.

11. Determine the speed of a proton whose kinetic energy is 400 GeV (1 GeV = 1.602×10^{-10} J).

12. A particle moves at a speed that is one third the speed of light. Determine the ratio of its kinetic energy and rest mass energy, K/mc^2, by using (a) Newtonian mechanics and (b) relativistic mechanics.

13. A particle of mass m has a kinetic energy that is eight times its rest mass energy. Calculate the ratio of its speed to the speed of light by using (a) relativistic mechanics (Eq. 20.10) and (b) Newtonian mechanics.

14. An electron is moving with a speed $v = 0.01c$. Calculate the relative error, $(K_{Newt} - K_{SR})/K_{SR}$, made in using the Newtonian form of the kinetic energy rather than the special relativistic form.

15. Calculate the maximum velocity (as a ratio v/c) a particle may have and still permit a calculation of the kinetic energy by the Newtonian $\frac{1}{2}mv^2$ with a relative error of magnitude no larger than 0.005.

16. A proton in the Fermilab accelerator has a v/c ratio of 0.999998. Calculate the ratio of its total relativistic energy E to its rest mass energy E_o (a) directly and (b) by use of the technique of Example 5.

17. Use the binomial expansion to show that Eq. 20.13 reduces to the Newtonian result $K = p^2/2m$ in the limit as $p \ll mc$.

18. Verify the relativistic energy–momentum relation

$$E^2 = (pc)^2 + (mc^2)^2$$

19. An electron has 1 MeV of kinetic energy. Calculate its linear momentum.

20. A fast-moving meson has a total relativistic energy of 400 MeV and a relativistic momentum of 375 MeV/c. Calculate its rest mass (in electron mass units); $m_e c^2 = 0.511$ MeV.

21. A proton has a relativistic momentum of 1500 MeV/c. (a) Calculate the proton's total relativistic energy E. (b) Calculate the proton's speed (as v/c).

22. The magnitude of the linear momentum of a particle of mass m is $0.60mc$. Determine the ratio of its kinetic energy and rest mass energy (K/mc^2) by using (a) Newtonian mechanics and (b) relativistic mechanics.

23. Use Eqs. 20.1 and 20.12 to show that the velocity, momentum, and total energy of a relativistic particle are related by

$$\mathbf{v} = \frac{\mathbf{p}c^2}{E}$$

Show that this equation reduces to the nonrelativistic relation $\mathbf{v} = \mathbf{p}/m$ when the kinetic energy is negligibly small compared with the rest mass energy.

24. Use Eq. 20.15 to prove the statement, "A particle of nonzero energy moving at the speed of light must have zero mass."

25. Assume that a neutrino is found to have a mass such that its energy equivalent $m_\nu c^2$ is 20 eV. Calculate the value of $1 - v/c$ for this neutrino when its total energy is 1 MeV.

26. Prove that relativistic kinetic energy reduces to Newtonian kinetic energy if the relativistic kinetic energy is much less than the rest mass energy: $K \ll mc^2$.

27. From the work–energy theorem for a particle initially at rest, kinetic energy is given by $K = \int F\,dx = \int v\,dp$. Using $v = p/m$, integrate and show that the kinetic energy is $K = \frac{1}{2}(p^2/m) = \frac{1}{2}mv^2$ in agreement with Section 7.4.

Section 20.4 Conservation of Energy and Momentum in Particle Interactions

28. Show that the total relativistic energy is conserved in the laboratory system for the proton–antiproton pair production reaction at threshold.

29. A fast-moving proton collides with a stationary proton. The result of the collision is two protons plus a positive-pion–negative-pion pair. Calculate the minimum kinetic energy of the incident proton (in the laboratory reference frame) necessary for this reaction to occur. The pion mass $m_\pi = 0.1488m_p = 273m_e$. Also, $m_\pi c^2 = 139.58$ MeV.

30. A proton with 10 GeV of kinetic energy ($10 \text{ GeV} = 10.66m_p c^2$) smashes into a stationary proton. How much kinetic energy is available in the center-of-momentum system for the creation of new particles?

31. A proton incident upon a stationary proton target has an energy E_{lab} that is much greater than the proton rest mass energy, $E_{lab} \gg m_p c^2$. In this extreme relativistic range show that the energy in the center-of-momentum system rises only as the square root of the energy in the laboratory system: $E_{cm} \sim (E_{lab})^{1/2}$.

32. Prove that if an electron and a positron at rest annihilate each other, creating two photons in the process, the two photons must (a) be equal in energy and (b) emerge in opposite directions.

33. The Λ^o is a neutral unstable particle whose rest mass is $7.993m_\pi$ (m_π = pion rest mass). The Λ^o at rest decays into a proton ($m_p = 6.722m_\pi$) and a negative pion as shown in Figure 1. (a) Set up the equations for conservation of relativistic linear momentum and for conservation of total relativistic energy. (b) From part (a) you have a pair of simultaneous nonlinear equations in the particle speeds. Assume that $[1 - (v_p/c)^2]^{1/2} \approx 1$ and solve the energy equations numerically for v_π/c. (c) With the pion speed v_π obtained in part (b), calculate v_p/c from the momentum equation and check the assumption you made in part (b). (d) Calculate the relativistic kinetic energy of the pion and of the proton. Notice that the less massive particle gets the greater share of the kinetic energy.

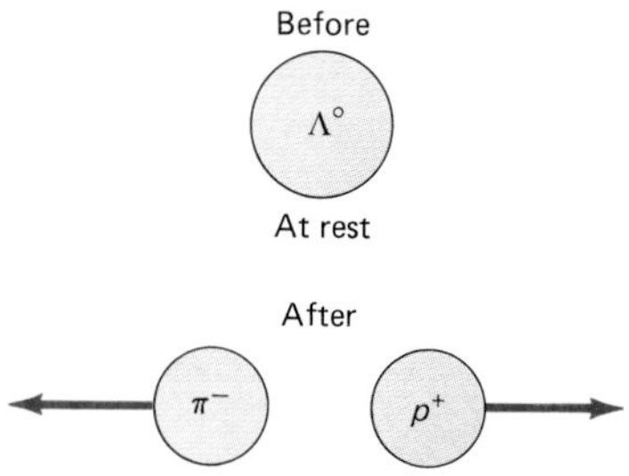

Figure 1

34. (a) As an alternative solution to Problem 33 show that conservation of total relativistic energy can be written

$$[(p_\pi c)^2 + (6.722m_\pi c^2)^2]^{1/2} + [(p_\pi c)^2 + (m_\pi c^2)^2]^{1/2} = 7.993m_\pi c^2$$

(b) Solve for p_π (numerically) in units of $m_\pi c$ and for v_π/c. (c) From the equality of linear momenta, calculate the numerical value of v_p/c. (d) Calculate the relativistic kinetic energy of the pion and of the proton.

c h a p t e r 21

TEMPERATURE AND HEAT

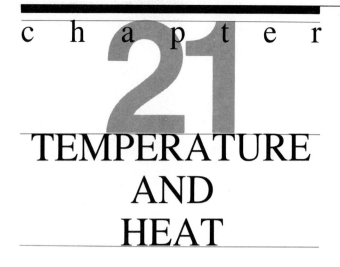

A Preview of Thermodynamics

Chapters 21–25 are devoted to thermodynamics. Thermodynamics deals with the properties of bulk matter under conditions where the effects of heat and temperature are significant. The laws of thermodynamics comprise the essence of 200 years of experimentation and theoretical interpretation. Thermodynamics often seems rather formal, partly because of the universal character of its laws. The generality of the laws of thermodynamics requires that they be independent of the detailed workings of any particular physical system. Thermodynamics did not evolve from atomic models of matter that let you see "how it works." Such insights came later with the development of kinetic theory and statistical mechanics. We will present thermodynamics in much the same way that it developed historically.

In the following five chapters we concern ourselves with three laws of thermodynamics: the zeroth law, the first law, and the second law.

The zeroth law formalizes an important experimental fact, namely, that thermal equilibrium between two systems demands the equality of one property, the property we call *temperature*.

The first law recognizes heat as a form of energy and states that energy is conserved in all processes.

The second law places limits on the extent to which heat can be converted into other forms of energy, such as electric energy. In effect, the second law says that the conversion of heat into other forms of energy is incomplete: Some fraction of the heat must be ejected to the environment.

In this chapter we study both the zeroth law and the first law of thermodynamics. In the process we develop operational definitions of temperature and heat.

21.1 Temperature and the Zeroth Law of Thermodynamics

In mechanics we rely on only three fundamental quantities: mass, length, and time. Other quantities, such as kinetic energy, moment of inertia, and mass density, are derived. In thermodynamics, we must introduce a fourth fundamental quantity—temperature. In this chapter we simply give temperature an operational definition, so that we can measure it. In Chapter 25 we will uncover the kinetic interpretation of temperature, showing that it measures the average kinetic energy of molecular motions.

Temperature is a physical quantity based on our subjective sense of hot and cold. We quantify this sense with temperature. An object that feels hot is said to be at a higher temperature than one that feels cold. It is also common experience that if a hot object is placed in contact with a cold object, their temperatures change. Eventually this change in temperature stops and the objects are said to be in *thermal equilibrium*. We find it convenient to say that their *temperatures are equal*. This definition of temperature equality is not tied to any particular scale of temperature. At this point we have no operational way of assigning temperatures. We have simply recognized a new property, called *temperature*, that indicates when two objects are in thermal equilibrium.

Suppose a warm watermelon is placed in a mountain stream. It cools off and comes to thermal equilibrium with the stream. A bottled soft drink placed in the stream also comes into thermal equilibrium with the stream. When the watermelon and the soft drink are placed in contact, neither warms up or cools down: They are already in thermal equilibrium. This result may not be surprising, but it is of profound significance. It is called the **zeroth law of thermodynamics** and may be stated formally as

Two objects (A and B), each in thermal equilibrium with a third object (C), are in thermal equilibrium with each other.

The zeroth law is significant because it recognizes that equality of a single physical property—temperature—is both necessary and sufficient to ensure thermal equilibrium. In the next section we study methods used to measure temperature.

Figure 21.1

An elementary form of optical pyrometer.

21.2 Temperature Measurement

The invention of the thermometer and the development of the concept of temperature mark the beginnings of the science of thermodynamics. Early in the 18th century Joseph Black, a Scottish physician, used the thermometer to identify the tendency for objects to achieve thermal equilibrium. Temperature became a quantitative concept and the significance of temperature equality was recognized. In principle, any system whose properties change with temperature can be used as a thermometer. In the optical pyrometer, the property is color (Figure 21.1). Other properties used to indicate temperature are the electrical resistance of a wire, the pressure of a confined gas, and the length of a column of alcohol or mercury. In recently developed medical thermometers the color of a liquid crystal is used to register body temperature.

The most fundamental scale of temperature, called the *Kelvin scale,* is based on the laws of thermodynamics. On the Kelvin scale, temperature is defined in terms of heat. A temperature defined in this way is independent of the properties of the material used to construct the thermometer. The Kelvin scale is discussed further in Section 24.4.

An empirical temperature scale is obtained by simply *defining* a temperature in terms of some measurable property. For example, in the constant-volume gas thermometer (Figure 21.2), temperature (T) is defined in terms of gas pressure (P) by

$$T = bP \tag{21.1}$$

where b is a proportionality constant. The choice of a linear relation between temperature and pressure is arbitrary but convenient. Equal changes of pressure correspond to equal changes of temperature. The scale defined by Eq. 21.1 is one example of an *absolute* scale. On an absolute scale, the lowest temperature is numerically zero.*

*The fact that temperature has a lower limit is embodied in the third law of thermodynamics. See, for example, J. Wilks, *The Third Law of Thermodynamics,* Oxford Univ. Press, London and New York, 1961.

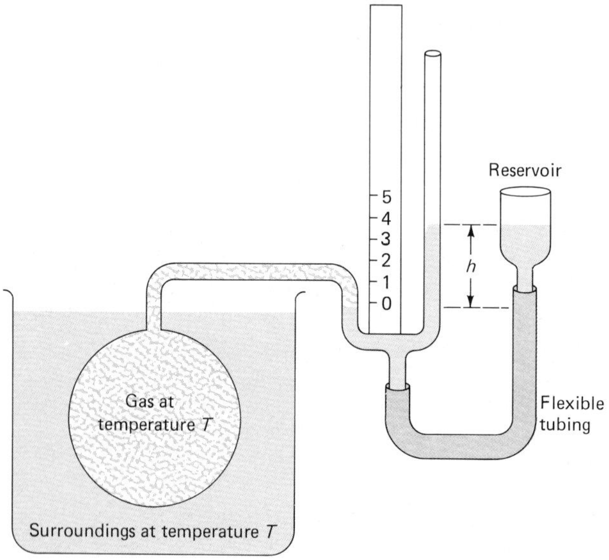

Figure 21.2

Schematic of a constant-volume gas thermometer. The gas is maintained at a constant volume by means of the mercury reservoir. If the temperature T of the bath rises, the level of the reservoir must be increased to maintain the same volume. In practice, several refinements and corrections are necessary. Among these are accounting for the change in volume of the gas container and the fact that the gas in the tube leading to the pressure gauge is not at the same temperature as the gas in the main vessel.

The value of the constant b in Eq. 21.1 is fixed by assigning a temperature to some standard *fixed point*. By international agreement, the temperature of water at its triple point* has been assigned the value 273.16 K, where K is the symbol for the temperature unit called the *kelvin*.† Accessibility and ease of reproduction are the primary reasons for using the triple point of water as the fundamental fixed point.

If P_3 denotes the gas pressure at the triple point of water, we have

$$273.16 \text{ K} = bP_3$$

or

$$b = \frac{273.16 \text{ K}}{P_3}$$

The constant-volume gas temperature scale is now defined by

$$T = 273.16 \text{ K} \left(\frac{P}{P_3}\right) \quad (21.2)$$

The boiling of water is a familiar phenomenon. The *boiling point* is the temperature at which a liquid and its vapor are in thermal equilibrium. The *normal boiling point* is the equilibrium temperature when the pressure of the vapor is exactly 1 atm. The normal boiling point of pure water is called the *steam point*.

Example 1 **Steam-Point Measurement**

Using a constant-volume gas thermometer of modest precision, a student finds the values $P_3 = 2.68 \times 10^4$ Pa and $P_{\text{steam}} = 3.67 \times 10^4$ Pa for the triple-point and steam-point pressures. The corresponding steam-point temperature is

$$T_{\text{steam}} = 273.16 \text{ K} \frac{3.67 \times 10^4}{2.68 \times 10^4} = 374 \text{ K}$$

The constant-volume gas thermometer is of special interest because it can be used to establish the Kelvin scale over an important range of temperatures. The temperature recorded by a constant-volume gas thermometer depends on the amount and the type of gas. Thermometers that use different gases generally do not agree, except at the triple point of water. Furthermore, if we remove some of the gas in a thermometer and then repeat the measurements, we record a lower triple-point pressure and a different temperature. A sequence of such measurements in which progressively lower triple-point pressures are used enables us to extrapolate the results to the point where $P_3 = 0$. In this limit, the gas density equals zero; that is, $P_3 = 0$ corresponds to the removal of all gas. The amazing feature of constant-volume gas thermometers is that this extrapolated temperature is *the same for all gases*. Figure 21.3 shows how sequences of measurements using different gases and different triple-point pressures extrapolate to a common value for the steam point.

Figure 21.3

The steam-point temperature recorded by a constant-volume gas thermometer. For any particular gas, the steam-point temperature depends on the pressure of the triple point of water. However, when the data are extrapolated to zero pressure, the unique steam point 373.15 K is indicated. Adapted from *Heat and Thermodynamics*, 5th ed., by Mark W. Zemansky; McGraw-Hill, New York, 1968. Used with permission of McGraw-Hill Book Company.

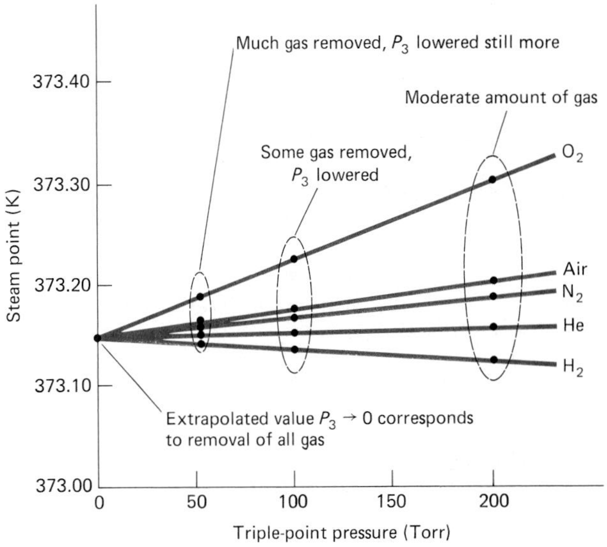

The fact that the extrapolated temperature is the same for all gases leads to the important concept of an *ideal gas*. The extrapolated temperature is called the **ideal gas temperature,**

$$T = 273.16 \text{ K} \lim_{P_3 \to 0} \left(\frac{P}{P_3}\right) \quad (21.3)$$

Other features of the ideal gas are developed in subsequent chapters. The ideal gas scale coincides with the Kelvin scale of temperature (Section 24.4). In principle, its application is restricted only by the requirement that the substance in the thermometer be a gas. At sufficiently low temperatures, all gases liquefy. With gaseous helium, at low pressure, the constant-volume gas thermometer can be used to measure temperatures as low as about 0.3 K. In practice, the constant-volume gas thermometer serves primarily as a standard. Other types of thermometers offer the advantages of portability and flexibility.

In addition to the Kelvin scale, three other scales are in widespread use today. These are the Celsius, Fahrenheit, and Rankine scales. On the Celsius scale, temperature t_C is related to the Kelvin temperature T by

$$t_C = (T - 273.15)°\text{C} \quad (21.4)$$

where °C is the abbreviation for degrees Celsius. Thus, for example, the Celsius temperature of the triple point of water is 0.01°C, because $T_3 = 273.16$ K. Notice that the

*The solid, liquid, and gaseous phases of water can coexist in thermodynamic equilibrium at a unique pressure (611 Pa) and temperature. This is referred to as the *triple point of water*.

†At one time temperatures on the Kelvin scale were designated °K, read "degrees kelvin." In 1968 the designation was changed to K, read "kelvins." Thus, 100 K is the temperature "one hundred kelvins," *not* "one hundred degrees kelvin."

Celsius degree and the kelvin have the same size. The scales differ only in the location of their zeros.*

The symbol °F stands for degrees Fahrenheit. Temperatures on the Fahrenheit scale are related to those of the Celsius scale by the equation

$$t_F = \left(32 + \frac{9}{5} t_C\right)°\text{F} \tag{21.5}$$

This places the ice and steam points at 32°F and 212°F. The Fahrenheit and Celsius scales evidently differ in the size of their units and in the location of their zeros.

Example 2 **The Celsius and Fahrenheit Temperatures of Liquid Nitrogen**

The normal boiling point of nitrogen is 77.4 K. The corresponding Celsius temperature follows from Eq. 21.4,

$$t_C = (77.4 - 273.15)°\text{C} = -195.8°\text{C}$$

We can insert this result into Eq. 21.5 to find the equivalent Fahrenheit temperature,

$$t_F = \left[32 + \frac{9}{5}(-195.8)\right]°\text{F} = -320.4°\text{F}$$

Thus, 77.4 K, −195.8°C, and −320.4°F are equivalent temperatures expressed on different scales.

The Rankine temperature scale is widely used in engineering. The Rankine and Kelvin scales have a common zero, but the size of their units differs. The Rankine degree is equal to the Fahrenheit degree, and thus is only 5/9 as large as the Celsius degree and the kelvin,

$$1\ \text{R}° = \frac{5}{9}\ \text{K} \tag{21.6}$$

Consequently, a temperature on the Rankine scale is 9/5 as great as the corresponding Kelvin temperature

$$t_R = \left(\frac{9}{5} T\right)°\text{R} \tag{21.7}$$

For example, the Rankine temperature of the ice point of water is

$$\frac{9}{5}(273.15)°\text{R} = 491.67°\text{R}$$

The Rankine scale bears the same relationship to the Fahrenheit scale as does the Kelvin scale to the Celsius scale. That is, the temperature units are of equal size but the locations of the origins differ. Figure 21.4 compares temperatures on the four scales.

*It has become standard practice to distinguish the units of *temperature* and the units of *temperature change* on the Celsius scale. We denote temperatures on the Celsius scale by the symbol °C (read "degrees Celsius"). Temperature changes and temperature differences on the Celsius scale are expressed in C° (read "Celsius degrees"). Thus, for example, 20°C is a temperature and 20 C° is a temperature difference.

Figure 21.4

A comparison of four temperature scales. Numerical values indicate the temperatures of several standard fixed points. NBP stands for *normal boiling point* (the boiling point at a pressure of 1 atm). NMP stands for *normal melting point*.

	Kelvins	°R	°C	°F
NMP of gold	1337.65	2407.77	1064.45	1948.1
NMP of zinc	692.66	1246.8	419.51	787.13
Steam point	373.15	671.67	100.00	212.00
Triple point of water	273.16	491.69	0.01	32.02
NBP oxygen	90.17	162.3	−110.9	−297.37
NBP hydrogen	20.26	36.47	−252.89	−423.20
Absolute zero	0	0	−273.15	−459.67

Phenomena of scientific interest span an enormous range of temperature (Figure 21.5). For example, research refrigerators using liquefied helium operate in the millikelvin (0.001 K) range. Attempts to achieve controlled thermonuclear fusion reactions have produced temperatures in excess of 10^6 K. The temperature at the center of our sun is over 10^7 K, while the temperature at the center of an exploding hydrogen bomb is over 10^8 K.

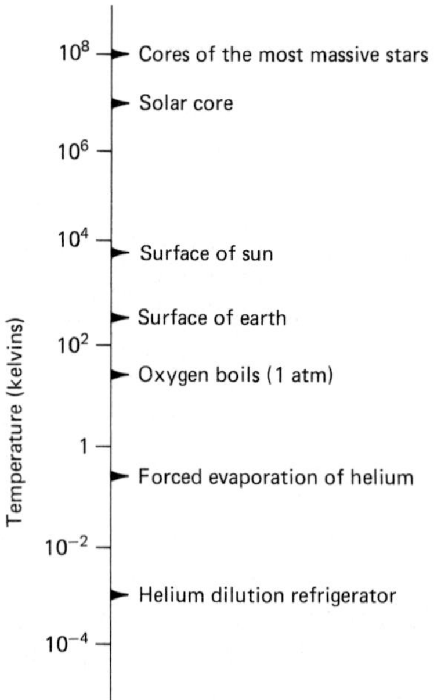

Figure 21.5

A sampling of the temperatures encountered in nature (kelvin scale). Note that the scale is logarithmic.

Questions

1. On the Celsius scale, the ice point is 0°C and the steam point is 100°C. If you had the opportunity to assign new values to these two fixed points, what values would you choose? What would be the primary advantage of your choices?
2. When asked to give the temperature at the center of an exploding hydrogen bomb, you reply that it is 10^8°C. A fellow student objects, saying that the correct answer is 10^8 K. How would you defend your answer?
3. On the Kelvin scale a piece of steel is twice as hot as an ice cube at 0°C. What is the Kelvin temperature of the steel?

21.3 Energy and the First Law of Thermodynamics

Energy is a multifaceted concept. Mechanics has acquainted us with the kinetic energy of a moving object, the gravitational potential energy of a raised weight, and the elastic potential energy of a stretched spring. A charged battery stands ready to convert chemical energy into the energy of motion of electric charges.

Our notion of energy is unlikely to change much from the elementary idea that energy measures the capacity to perform work. But our appreciation of energy grows as we find it thrust forward as a unifying concept in all areas of physics, and in fact, throughout all science. Two primary attributes of energy are important: Energy can be converted from one form to another, and energy is conserved. The first law of thermodynamics quantifies these two attributes of energy. All forms of energy (mechanical, electrical, chemical, nuclear, mass, etc.) are recognized, together with the possibility of conversions from one type to another. In particular, the recognition of heat as a form of energy is necessary in order to preserve the conservation of energy. A statement of the first law of thermodynamics adequate for our purposes is this:

Energy is conserved in all physical processes. It cannot be created or destroyed.

The experimental foundation of the first law is the fact that no one has been able to build a cyclic device that can produce more energy than it takes in. Such a device, called a *perpetual motion machine of the first kind,* would—if it existed—violate the first law of thermodynamics.

In the next three sections of this chapter we analyze the three forms of energy related by the first law of thermodynamics. These are heat, thermodynamic work, and internal energy.

21.4 Heat and Specific Heat Capacity

Today we recognize heat as a form of energy transferred by virtue of a temperature difference. We can set up an operational definition of heat that relates it to other forms of energy. This modern view of heat replaces an earlier theory of heat developed during the 17th and 18th centuries. This early theory—the *caloric theory of heat*—viewed heat as a fluid. Called *caloric,* this fluid was invisible and weightless. If the temperature of a body increased, it was because caloric had been absorbed. Cooling was accompanied by the loss of caloric. In order to explain certain experiments it was necessary to postulate that caloric was conserved, that is, that it could be neither created nor destroyed. The caloric theory was immensely successful because it could explain a great many thermal phenomena, such as the expansion of heated objects. However, the caloric theory was eventually discarded because it could not offer satisfactory explanations of certain phenomena involving friction. For instance, if two ice cubes are rubbed together in a vacuum, they gradually melt. This phenomenon cannot be explained by the caloric theory because it requires the *creation* of caloric.

Today we understand that heat is a form of energy rather than a substance. However, the language of thermodynamics contains many reminders of the caloric theory. Among these is the unit of heat, the kilocalorie. The definition of the kilocalorie* is the following:

One kilocalorie (1 kcal) is the amount of heat required to raise the temperature of 1 kg of water from 14.5°C to 15.5°C.

Because heat is a form of energy, the kilocalorie can be expressed in kilojoules (kJ). The presently accepted equivalence is

$$1.0000 \text{ kcal} = 4.1858 \text{ kJ} \qquad (21.8)$$

We will follow the trend away from caloric units and use the joule or the kilojoule as the unit of heat.†

Although the caloric theory did not recognize heat as a form of energy, two significant attributes of heat were recognized:

1 The amount of heat required to produce a specified change of temperature is directly proportional to the mass of the material.

2 For a given mass of material the amount of heat absorbed is directly proportional to the temperature increase. This direct proportionality holds true only for small changes of temperature and only as long as the heated substance does not undergo a change of phase, such as melting.

*In dietetics, the kilocalorie is generally referred to as the Calorie, abbreviated Cal, 1 kcal ≡ 1 Cal.

†Many engineering applications of thermodynamics employ the *British thermal unit* (Btu) of heat. The equivalences are 1 Btu = 0.252 kcal; 1 Btu = 1.05 kJ.

Table 21.1

Specific heat capacity at constant pressure (C_p)[a]

Substance	C_p (kJ/kg · C°)	Substance	C_p (kJ/kg · C°)
SOLIDS		LIQUIDS	
Silver (Ag)	0.234	Benzene (C_6H_6)	1.62
Aluminum (Al)	0.900	Bromine (Br)	0.448
Gold (Au)	0.130	Ethyl alcohol (C_2H_5OH)	2.43
Bismuth (Bi)	4.31	Methyl alcohol (CH_3OH)	2.52
Copper (Cu)	0.385	Water (H_2O)	4.19
Iron (Fe)	0.448	-GASES-	
Brick	0.837	Argon (Ar)	0.523
Concrete	0.879	Carbon dioxide (CO_2)	0.95
Magnesium (Mg)	1.04	Chlorine (Cl_2)	0.481
Sodium (Na)	1.23	Helium (He)	5.24
Nickel (Ni)	0.431	Hydrogen (H_2)	14.2
Lead (Pb)	0.130	Nitrogen (N_2)	1.04
Zinc (Zn)	0.352	Neon (Ne)	1.03
Wood (Pine)	2.81	Oxygen (O_2)	0.917

[a]Values refer to a pressure of 1 atm and a temperature of 25°C for solids and liquids; 15°C for gases. Adapted with permission from *International Critical Tables of Numerical Data: Physics, Chemistry, and Technology*, National Research Council, Washington, DC, 1923–1933, and E. H. Kennard, *Kinetic Theory of Gases*, McGraw-Hill, New York, 1938.

These attributes led to the concept of specific heat capacity.

Specific Heat Capacity

Let Q stand for the heat absorbed or rejected when a body of mass m undergoes a temperature change ΔT. The observation that Q is proportional to m and ΔT lets us introduce a proportionality factor (C) and write the equation

$$Q = mC\,\Delta T \qquad \textit{specific heat} \qquad (21.9)$$

The factor C is called the *average specific heat capacity* over the temperature range ΔT. The value of the average specific heat capacity depends on the physical and chemical composition of the material. Where no confusion can result we often shorten "average specific heat capacity" to "specific heat." The specific heats for a number of substances are presented in Table 21.1. The tabulated values refer to measurements made with the pressure held constant at 1 atm. In Table 21.1 the subscript p on C_p signifies the constant pressure.

Example 3 The Specific Heat of Water

Table 21.1 lists the specific heat of water as 4.19 kJ/kg · C°. To see that this value conforms to the definition of 1 kcal we use Eq. 21.9 with m = 1 kg, C = 4.19 kJ/kg · C°, and ΔT = 1 Celsius degree. This gives the amount of heat needed to raise the temperature of 1 kg of water by 1 degree:

$$Q = (1\text{ kg})(4.19\text{ kJ/kg}\cdot\text{C°})(1\text{ C°}) = 4.19\text{ kJ}$$

According to Eq. 21.8, 4.19 kJ equal 1 kcal (to three-significant-figure accuracy).

We can think of Eq. 21.9 as an operational definition of heat, provided we define the average specific heat of water to be 4.19 kJ/kg · C° over the range from 14.5°C to 15.5°C. Although Eq. 21.9 can define heat, it cannot answer the question, "What is heat?" It tells us how to measure heat, but not what heat itself measures. Today we recognize that *heat is energy transferred by virtue of a temperature difference*. We often refer to heat as *thermal energy* in order to emphasize its basic character.

Equation 21.9 can be expressed in differential form as

$$dQ = mC\,dT \qquad (21.10)$$

In this equation dT is the small temperature change resulting from the exchange of the small amount of heat dQ. The total heat absorbed or rejected in a finite change of temperature is the sum (integral) of the small amounts

$$Q = \int_{T_i}^{T_f} dQ = m\int_{T_i}^{T_f} C\,dT \qquad (21.11)$$

In general, the specific heat varies with T and other thermodynamic variables. This is why Eq. 21.9 can define only an average specific heat. If the specific heat capacity does not change significantly over the temperature range T_i to T_f, C can be treated as a constant in the integration. Performing the integration results in

$$Q = mC(T_f - T_i) = mC\,\Delta T \qquad (21.12)$$

which is identical with Eq. 21.9. Happily, many substances have nearly constant specific heat capacities over sizable temperature ranges, permitting us to use the relation $Q = mC\,\Delta T$ instead of the more involved integral relation, Eq. 21.11.

Example 4 **Constant Specific Heat**

The specific heat of zinc is 0.352 kJ/kg · C° for temperatures near 25°C. We want to determine the amount of heat required to raise the temperature of 0.50 kg of zinc from 20°C to 30°C. Taking the specific heat to be a constant, we can use Eq. 21.12 to find

$$Q = mC\,\Delta T = (0.50\text{ kg})\left(0.352\,\frac{\text{kJ}}{\text{kg}\cdot\text{C}^\circ}\right)(10\text{ C}^\circ)$$

$$= 1.76\text{ kJ}$$

Equation 21.12 is not applicable when the specific heat changes significantly over the temperature range of interest. For a variable C, we must use Eq. 21.11 to determine Q.

Example 5 **Variable Heat Capacity**

At sufficiently low temperatures the heat capacities of many crystalline solids are proportional to T^3. For zinc at temperatures below 100 K, the specific heat is given by

$$C = 1.02 \times 10^{-6}\,T^3\,\frac{\text{kJ}}{\text{kg}\cdot\text{K}^4}$$

We want to determine the amount of heat required to raise the temperature of 0.50 kg of zinc from 10 K to 20 K. This is the same temperature change as in Example 4. In this case, however, we must use Eq. 21.11 because the heat capacity varies with temperature.

$$Q = m\cdot\int_{T_i}^{T_f} C\,dT$$

$$= 0.50\text{ kg}\left(1.02 \times 10^{-6}\,\frac{\text{kJ}}{\text{kg}\cdot\text{K}^4}\right)\int_{T_i}^{T_f} T^3\,dT$$

giving

$$Q = 1.28 \times 10^{-7}(T_f^4 - T_i^4)\,\frac{\text{kJ}}{\text{K}^4}$$

$T_f = 20$ K and $T_i = 10$ K gives $Q = 0.0192$ kJ. The heat required in this example is more than a hundred times smaller than the heat required for the same change of temperature in Example 4. This is because the average heat capacity is far lower over the range from 10 K to 20 K than it is over the range from 20°C to 30°C (293–303 K).

Heat Capacity

The product of the mass of a substance and its specific heat capacity is called the **heat capacity.**

$$\text{Heat capacity} = \text{mass} \times \text{specific heat capacity} = mC$$

Thus, in Example 4, the heat capacity of the 0.50-kg sample of zinc is the product

$$\text{heat capacity} = (0.50\text{ kg})(0.352\text{ kJ/kg}\cdot\text{C}^\circ)$$

$$= 0.176\text{ kJ/C}^\circ$$

Note that the heat capacity and specific heat capacity have different units.

The Method of Mixtures

A basic technique for measuring specific heats is called the *method of mixtures*. Suppose that we place 0.60 kg of lead, initially at a temperature of 25°C, in an aluminum can that holds 0.10 kg of water. We call the can a *calorimeter*. The initial temperature of the water and calorimeter is 80°C. The mass of the calorimeter is 0.020 kg and its specific heat is 0.900 kJ/kg · C°. The final temperature of the mixture is 72°C. We can use these data to determine the specific heat of lead.

A basic assumption in the method of mixtures is that all of the heat rejected by the initially warmer material is absorbed by the initially cooler material. In other words, the combined system (the calorimeter and its contents) neither gains nor loses heat. Heat rejected by one part of the system is absorbed by another part. This ignores heat transfer to or from the system—via radiation, for example. The heat absorbed by the lead is

$$Q_A = m_{\text{Pb}}C_{\text{Pb}}\,\Delta T_{\text{Pb}} = (0.60\text{ kg})(47\text{ C}^\circ)C_{\text{Pb}}$$

The heat rejected by the water and calorimeter is

$$Q_R = (m_{\text{Al}}C_{\text{Al}} + m_W C_W)\,\Delta T \qquad \Delta T = 8\text{ C}^\circ$$

$$= \left[(0.020\text{ kg})\left(0.900\,\frac{\text{kJ}}{\text{kg}\cdot\text{C}^\circ}\right) + (0.10\text{ kg})\left(\frac{4.19\text{ kJ}}{\text{kg}\cdot\text{C}^\circ}\right)\right](8\text{ C}^\circ) = 3.50\text{ kJ}$$

Equating Q_A and Q_R gives for the specific heat of lead

$$C_{\text{Pb}} = \frac{3.50\text{ kJ}}{(0.600\text{ kg})(47\text{ C}^\circ)} = 0.124\,\frac{\text{kJ}}{\text{kg}\cdot\text{C}^\circ}$$

At ordinary temperatures the specific heats of solids and liquids depend only slightly on the pressure and volume. (Representative values are given in Table 21.1.) However, the specific heats of gases depend strongly on pressure and volume, and it is therefore important to distinguish the conditions under which heat is absorbed or rejected. For example, the specific heat of a gas maintained at a constant pressure is greater than the specific heat measured when the gas is held at a constant volume. This curious fact led Robert Mayer* to hypothesize that

*Julius Robert Mayer (1814–1878), a German physician, conceived the idea of the equivalence of heat and work while serving as a ship's physician.

heat is a form of energy and that it can be converted into other forms of energy. In 1842 he observed that the amount of heat required to change the temperature of a gas was greater when the pressure of the gas was held constant than when the volume of the gas was held constant. When heat is added at constant pressure the gas expands and performs work. When heat is added at constant volume no work is performed. Mayer realized that the "extra" heat required under constant-pressure conditions was converted into the work of expansion. For this singular stroke of genius, Mayer shares (with James Prescott Joule) the credit for formulating the first law of thermodynamics. Heat is one of the three forms of energy that are related by the first law. The other two are *work* and *internal energy*. Before stating the first law in equation form, we will review the concept of work in the next section.

Questions

4. The sands of a desert get very hot during the day and very cool at night. How does the specific heat capacity of sand compare to that of water?
5. Suppose that the specific heat capacity of water were ten times its actual value. Would more or less time be required to boil water in a teakettle?
6. Does the heat capacity of a substance depend on how much of that substance is present? Does its specific heat depend on how much is present?

21.5 Thermodynamic Work

A set of values of the thermodynamic variables defines an equilibrium state for a system. For a gas enclosed in a cylinder fitted with a piston, a set of values of pressure, volume, and temperature (P, V, T) specifies an equilibrium state. If we use a Cartesian coordinate system to plot sets of values of P, V, and T, different equilibrium states correspond to different points (Figure 21.6).

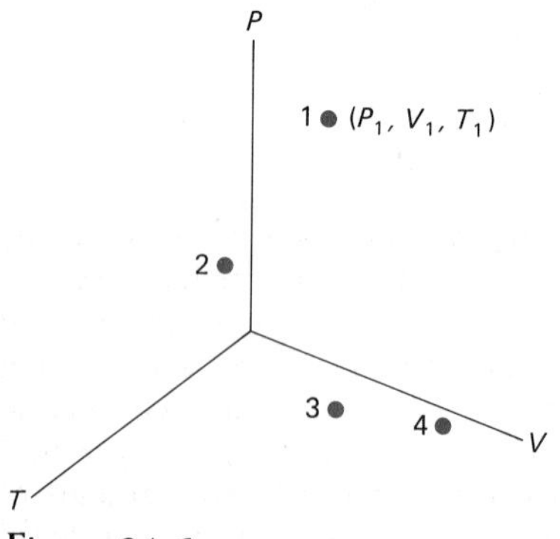

Figure 21.6

Different equilibrium states of a gas can be represented by points defined by values of pressure (P), volume (V), and temperature (T).

Figure 21.7

A quasi-static process may be represented by a path that joins the initial and final states by a succession of intermediate equilibrium states.

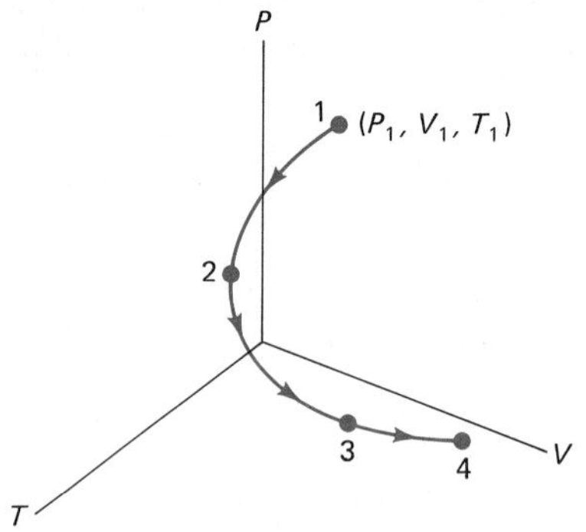

In a thermodynamic process, one or more of the thermodynamic variables change. A *quasi-static* process is one in which the surroundings and thermodynamic variables of the system change so slowly that the process can be viewed as one in which the system passes through a succession of equilibrium states. Geometrically, a quasi-static process can be represented by a path that joins the initial and final equilibrium states by a succession of intermediate equilibrium states (Figure 21.7). Figure 21.8 suggests how a gas might be compressed quasi-statically. A cylinder filled with gas has a piston at one end. Sand piled on the piston helps to compress the gas. By adding sand slowly, one grain at a time, the gas can be compressed slowly.

We can use the example of a gas being compressed quasi-statically to introduce the important concept of a *reversible process*. Consider first the behavior of the gas–piston system when there is no friction between the piston and the cylinder walls. Adding or removing a grain of sand constitutes an *infinitesimal change* in the surroundings. At any stage we can halt the compression and initiate expansion

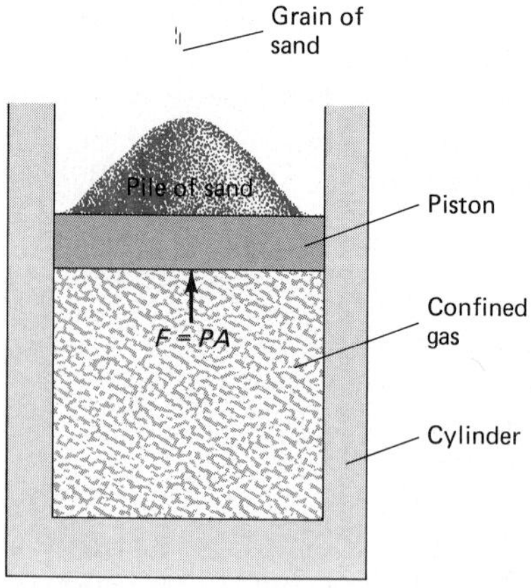

Figure 21.8

The gas confined in the piston-fitted cylinder is compressed quasi-statically. This can be achieved by slowly piling sand on the piston. If the piston moves without friction, the removal of one grain of sand (an infinitesimal change) causes the piston to reverse its motion. If friction is present, the removal of one grain need not produce a reversal of the piston.

by removing a grain of sand, provided there is no friction. If the sand is removed grain by grain, so as to reverse the order in which it was added, the gas retraces the thermodynamic path (the same sequence of equilibrium states) it followed during the compression. A significant point is that the reversal can be initiated by an infinitesimal change in the surroundings.

The quasi-static, frictionless compression just described is an example of a reversible process. A general definition is this:

A **reversible process** is a quasi-static process that can be reversed by an infinitesimal change in the surroundings.

Please note that a reversible process does not necessarily go one way and then back again. The word *reversible* implies that the process *can* be reversed, not that it is reversed.

In contrast, an irreversible process is one that cannot be reversed by an infinitesimal change in the surroundings. Examples of irreversible processes include the following:

(a) a chemical reaction in which a precipitate forms;

(b) the explosive combination of hydrogen and oxygen to form water;

(c) the diffusion of perfume vapor into air.

Irreversible processes occur spontaneously within a system that is not in thermodynamic equilibrium. Heat flow is another type of irreversible process. Heat flows spontaneously (that is, without any outside help) from regions of higher temperature to regions of lower temperature. Observation reveals that irreversible processes tend to produce a state of mutual equilibrium between the system and its surroundings. Once attained, an equilibrium state persists as long as the surroundings remain unchanged.

Friction destroys reversibility. For example, if there is friction between the piston and the cylinder shown in Figure 21.8, we can still compress the gas quasi-statically by adding sand slowly. However, the compression cannot be reversed by an infinitesimal change in the surroundings. Removing one grain of sand—an infinitesimal change—does not reverse the process. A finite change in the surroundings is needed to revert from compression to expansion.

Although friction cannot be eliminated completely, the frictional force can often be rendered negligible by comparison with the force exerted by the gas. In such circumstances compressions and expansions may be regarded as reversible processes.

An important consequence of reversibility is this: The work done by the system during a reversible process can be expressed in terms of its thermodynamic variables and the changes in these variables. We can illustrate this point by means of a gas–piston system (Figure 21.9). If P is the gas pressure and A the cross-sectional area of the piston, the force exerted on the piston by the gas is

$$F = PA \tag{21.13}$$

Suppose that the gas expands, pushing the piston outward a distance ds. The work done by the gas is

$$dW = F\,ds = PA\,ds \tag{21.14}$$

Figure 21.9

When the piston moves upward a distance ds the volume of the gas increases by $dV = A\,ds$ and the gas does work $P\,dV$.

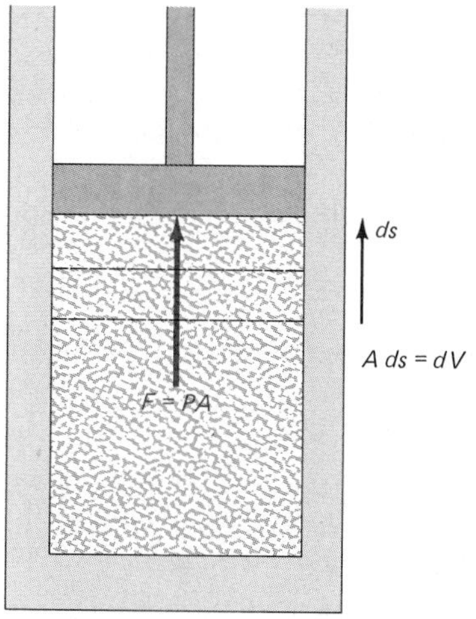

However, $A\,ds = dV$ is the *change* in the volume of the gas. Thus, the work done by the gas, dW, during a quasi-static change of volume, dV, is

$$dW = P\,dV \tag{21.15}$$

We have succeeded in expressing the work solely in terms of the thermodynamic variables of the system. The nature of the external force and other characteristics of the surroundings do not appear in the expression for dW. This remarkable result is possible only to the extent that the compression is reversible—that is, quasi-static and frictionless. Friction would introduce into Eq. 21.15 an additional term representing work done by frictional forces. The work $dW = P\,dV$ is often referred to as *thermodynamic work*, because it is expressed entirely in terms of the thermodynamic variables. By contrast, the work done by a frictional force would not be classified as thermodynamic work because such a force is not a thermodynamic variable of the system. However, we will deal only with thermodynamic work, and therefore will refer to it simply as work.

Returning to the gas–piston system shown in Figure 21.9, we find the total work done during a finite change of volume by integrating Eq. 21.15 between some initial state (i) and some final state (f):

$$W = \int_i^f dW = \int_{V_i}^{V_f} P\,dV \tag{21.16}$$

V_i and V_f denote the initial and the final volume of the gas, respectively. In order to evaluate the integral it is necessary to specify a relationship between P and V. An equilibrium state of a gas is specified by values of P, V, and T. Using a P–V diagram (Figure 21.10) to label the sequence of states through which the gas passes, we get a graphical description of the process. Geometrically we can think of the sequence of points on the P–V diagram as a "path" from the initial state to the final state.

Let's look at this path more closely. The P–V diagram in Figure 21.11 shows the path of the expansion process as well as the initial and final pressures and vol-

Figure 21.10

$P-V$ diagram showing the expansion of a gas.

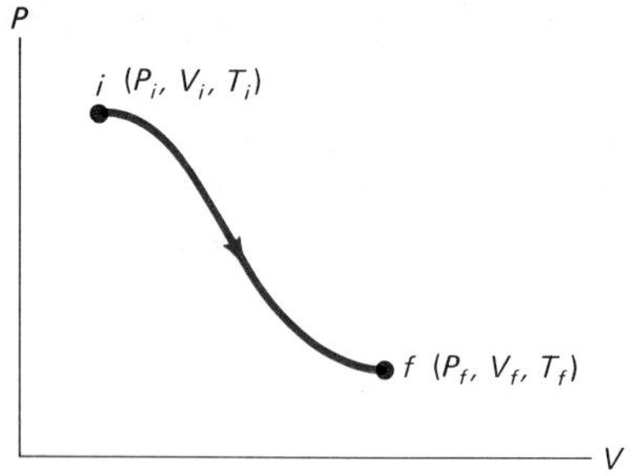

Figure 21.11

Geometrically, the work $dW = P\,dV$ corresponds to the cross-hatched area. The total work W done by the gas during the expansion from V_i to V_f corresponds to the shaded area beneath the $P-V$ path.

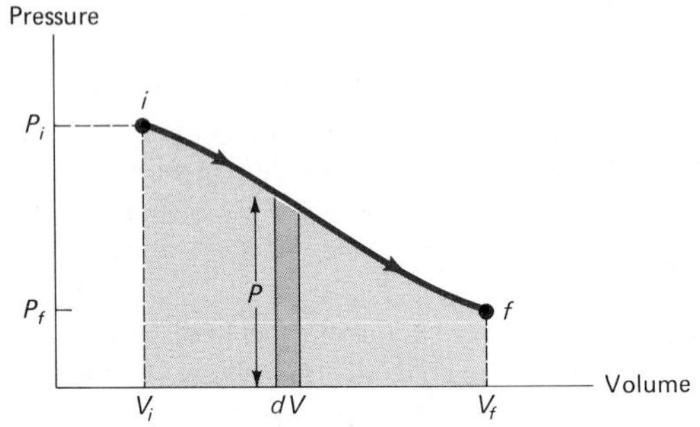

umes. As the expansion proceeds the pressure changes. The quantity $P\,dV$ is the work done during the volume change dV and is represented by the cross-hatched area under the $P-V$ path. (Keep in mind that although $P\,dV$ can be represented by this area, its units are those of work.) The total work, W, done by the gas corresponds to the shaded area beneath the $P-V$ path. Therefore, different paths between V_i and V_f could result in different amounts of work, even though they connect the same initial and final states. We say that the work, W, is a *path-dependent* quantity because it depends on the nature of the path as well as on the initial and final states. Figure 21.12 shows three different paths connecting the same initial and final states. These different paths can be achieved by controlling the temperature of the gas during the expansion. The work done by the gas is greatest for the path marked "1," which encloses the greatest area. The work done by the gas is smallest for the path marked "3."

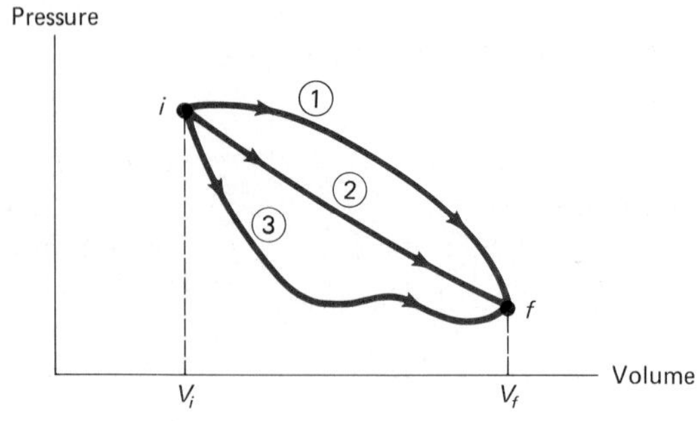

Figure 21.12

Three different paths connecting the same initial and final states. The work done by the gas is greatest for the path marked ① which encloses the greatest area.

Figure 21.13

The work performed along path ① is greater than the work performed along path ②.

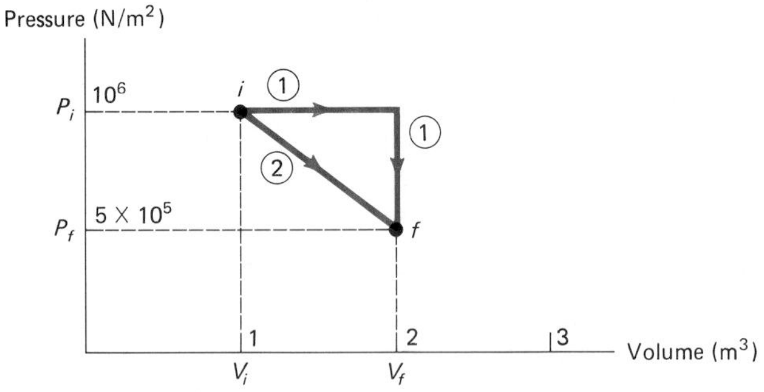

Example 6 Path-Dependent Work

We can demonstrate the path dependence of thermodynamic work numerically as follows: Consider a gas that starts at a pressure of 10^6 Pa and a volume of 1 m^3 and expands to a final pressure of 5×10^5 Pa and a final volume of 2 m^3. In other words, its volume doubles and its pressure is halved. Figure 21.13 shows two possible paths connecting the initial and final states. The work (W_1) performed along path 1 corresponds to a rectangular area with a height P_i and work (W_2) performed along path 2 corresponds to the trapezoidal area. It is clear from the figure that $W_1 > W_2$. Direct calculation verifies this. W_1 corresponds to the area of a rectangle of height P_i and width $V_f - V_i$:

$$W_1 = P_i(V_f - V_i) = 10^6 \text{ J}$$

W_2 corresponds to the area of a trapezoid of average height $\frac{1}{2}(P_i + P_f)$ and width $V_f - V_i$:

$$W_2 = \frac{1}{2}(P_i + P_f)(V_f - V_i) = 7.5 \times 10^5 \text{ J}$$

We see then that the thermodynamic work done by a system depends on the path of the process connecting the initial and final states.

The work integral can be negative. For example, if the expansion described by Figure 21.11 is reversed, the magnitude of the work done is still represented by the area under the $P-V$ path. However, the work is negative because dV is negative—the volume is decreasing. Alternately, we can see the work of compression as amounting to an interchange of the limits on the work integral $\int_{V_i}^{V_f} P\,dV$, which thereby changes its sign.

The mechanical work performed by a force is the most familiar form of work. However, the work performed

Figure 21.14

Path 1 is longer than path 2. For which path is the thermodynamic work greater?

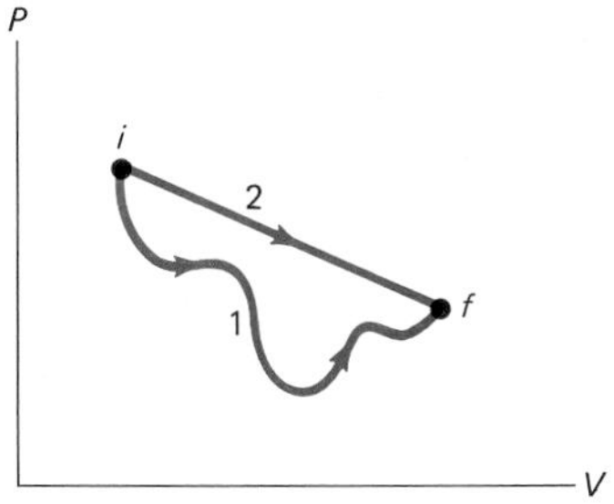

on a thermodynamic system can take other forms. In fact, in thermodynamics, work includes all forms of energy transfer except heat. For example, when a battery is being charged, electrical work is performed and stored in chemical form. Another nonmechanical form of work of importance is magnetic work. When a piece of steel is magnetized, magnetic work is performed.

Just as there are nonmechanical forms of thermodynamic work, the energy a system stores internally need not be purely mechanical. Chemical and magnetic energy can be stored internally as well. The concept of *internal energy* is introduced in thermodynamics to cover the many possible forms of energy that can be stored within a system. In the next section we show how internal energy is operationally defined.

Questions

7. Cite an example of a quasi-static process, other than those considered in the text. Is this process also reversible? Explain.
8. When a book slides across a desk top and comes to rest, some of its initial kinetic energy is converted into heat. Is this conversion a reversible or an irreversible process? Explain.
9. Paths for two different processes are shown on the $P-V$ diagram in Figure 21.14. The length of path 1 is greater than the length of path 2. Is the thermodynamic work greater for path 1 or for path 2?

21.6 Internal Energy

Consider a system that undergoes a transformation during which no heat is either absorbed or rejected. Such a process is called an **adiabatic process.** In an adiabatic process a system can perform work only at the expense of its "stored" energy. If the system performs a positive amount of work on the environment, it seems reasonable that the energy stored in the system has decreased. This observation can be used to set up an operational definition of *internal energy,* the energy "stored" within a system. Let W denote the work done by a system during an adiabatic process that carries the system from some initial equilibrium state to some final equilibrium state. Let U designate the internal energy of the system—the quantity we wish to define. We postulate that energy is conserved by requiring that

$$\left\{\begin{matrix}\text{the change in the}\\ \text{internal energy of}\\ \text{the system}\end{matrix}\right\} + \left\{\begin{matrix}\text{the work done by}\\ \text{the system on its}\\ \text{environment}\end{matrix}\right\} = 0$$

Internal energy: definition

In symbolic form this equation becomes

$$U_f - U_i + W = 0 \tag{21.17}$$

The subscripts refer to the initial and final states. We already have an operational definition of work. Equation 21.17 therefore serves to define the change in the internal energy of the system, $U_f - U_i$. This equation is often referred to as the *adiabatic form of the first law of thermodynamics.* Note that only the *change* in internal energy, $U_f - U_i$, is defined. This feature of internal energy is similar to potential energy, as defined for conservative forces in Section 8.2.

The adiabatic form of the first law agrees with our earlier observation about work and internal energy. It shows that if W is positive, the change in U is negative and therefore $U_f < U_i$. In other words, the system performs work at the expense of its internal energy. If W is negative, then the change in U is positive and the internal energy increases ($U_f > U_i$). For example, the adiabatic compression of a gas increases the internal energy of the gas.

Example 7 **Adiabatic Compression of a Gas**

A person exerts an average force of 1000 N in compressing the gas in a pump cylinder. The force acts over a distance of 0.500 m. Gases are notoriously poor heat conductors. For this reason the rapid expansion or compression of a gas can be regarded as an adiabatic process—that is, it is over before the gas has time to absorb or eject a significant amount of heat. Assuming that the compression is adiabatic, we want to calculate the work done by the gas and the change in its internal energy.

The work done *on* the gas by the person is positive. The work done *by* the gas (force × displacement) is negative,

$$W = -(1000\text{ N})(0.500\text{ m}) = -500\text{ J}$$

The increase in the internal energy of the gas follows from the adiabatic form of the first law, Eq. 21.17:

$$U_f - U_i = -W = +500\text{ J}$$

The temperature of the gas rises as a result of this 500-J increase in the internal energy. If you have ever pumped up a tire by means of a hand pump you may have noticed the temperature rise that results from an adiabatic compression.

Our operational definition of internal energy does not give us much of a feel for the quantity. This is one of the drawbacks of the fact that thermodynamics operates at the macroscopic level. If we want to "see inside" a system and understand what makes it tick, we must construct a *microscopic model* and a theory to go with it. We must believe in atoms and puzzle over the forces that bind them together to form molecules and solid objects. The challenge of understanding thermodynamics from a microscopic viewpoint inspired men like Rudolph Clausius, James Clerk Maxwell, Ludwig Boltzmann, and Josiah Willard Gibbs to develop the kinetic theory of gases and statistical mechanics. These two disciplines permit events to be followed at the microscopic level in a statistical fashion. We will get a taste of the ideas involved when we study kinetic theory in Chapter 25.

Questions

10. Water drips from a faucet into a partially filled basin. Trace the energy transfers for a drop of water from the moment it begins its fall until it loses its identity in the basin.
11. Think of a way to (a) change the temperature of an object without adding heat to it; (b) add heat to an object without changing its temperature. The object can be any system you choose.
12. The internal energy of a system is doubled as the result of a thermodynamic process. Is there any way to decide whether the increase came about as a result of the absorption of heat or through the performance of work?

21.7 The First Law of Thermodynamics

In Section 21.3 we stated the first law of thermodynamics:

Energy is conserved in all physical processes. It cannot be created or destroyed.

In Sections 21.4, 21.5, and 21.6 we studied the three general forms of energy that we deal with in thermodynamics: heat, thermodynamic work, and internal energy.

Heat is energy transferred by virtue of a temperature difference. Thermodynamic work is energy—exclusive of heat—transferred between a system and its surroundings. Internal energy is energy stored within a system.

We are now in a position to formulate the first law of thermodynamics in terms of these three forms of energy.

Credit for formulating the first law of thermodynamics is shared by Robert Mayer and James Prescott Joule. Working independently and in response to quite different motivations, both realized that heat is a form of energy and that energy is conserved.

In 1842, Mayer presented the first determination of the mechanical equivalent of heat, the amount of mechanical energy equivalent to 1 kcal. Mayer's value was within 20% of the presently accepted value of

$$1.0000 \text{ kcal} = 4.1858 \text{ kJ} \tag{21.8}$$

James Prescott Joule based his belief in the conservation of energy on religious grounds.* In a series of experiments spanning 40 years Joule measured the mechanical equivalent of heat. He demonstrated that the loss of a given amount of mechanical or electrical energy always resulted in the gain of a proportional amount of heat.

A quantitative formulation of the first law of thermodynamics recognizes that a change in the internal energy of a system can be accomplished by the performance of work, by the exchange of heat, or by a combination of the two. Let Q denote the quantity of heat added to a system during a process that carries it from some initial equilibrium state to some final equilibrium state. In general such a process changes the internal energy of the system and results in the performance of work. The *first law of thermodynamics* can be stated as

$$\begin{Bmatrix}\text{heat added}\\ \text{to a system}\end{Bmatrix} \text{ appears as } \begin{Bmatrix}\text{a change in the}\\ \text{internal energy}\end{Bmatrix} \text{ and/or } \begin{Bmatrix}\text{is used to}\\ \text{do work}\end{Bmatrix}$$

The equation expressing this statement, a quantitative formulation of the first law of thermodynamics, is

$$Q = U_f - U_i + W \tag{21.18}$$

where Q is the name for the net amount of heat *added* to the system. If Q is negative, it simply means that the system *ejected* a net amount of heat. In an adiabatic process $Q = 0$ and Eq. 21.18 reverts to the adiabatic form, Eq. 21.17.

Example 8 **Internal Energy Change**

A piston-fitted cylinder contains superheated steam. The steam absorbs 1.2×10^7 J of heat as it expands, driving the piston. The expanding steam performs 1.7×10^7 J of work against its environment. We want to determine the change in the internal energy of the steam. First, solve Eq. 21.18 for $U_f - U_i$.

$$U_f - U_i = Q - W$$

Because heat is absorbed by the steam,

$$Q = +1.2 \times 10^7 \text{ J}$$

Positive work is done by the steam against the environment:

$$W = +1.7 \times 10^7 \text{ J}$$

Inserting these values of Q and W gives

$$\begin{aligned} U_f - U_i &= 1.2 \times 10^7 \text{ J} - 1.7 \times 10^7 \text{ J} \\ &= -5.0 \times 10^6 \text{ J} \end{aligned}$$

*James Prescott Joule (1818–1889), an English brewer, saw energy as one of God's grandest creations. As such it was not something that mortals could create or destroy. Joule wrote, "The grand agents of nature are, by the Creator's fiat, indestructible;" [*Philosophical Magazine*, Series 3, 435 (1843)]. He began his investigations of the transformations of electrical and mechanical energy into heat in 1840. Results of his determination of the mechanical equivalent of heat were published in 1843.

The minus sign means that the final internal energy is 5.0×10^6 J less than the initial internal energy. Internal energy decreased because the work performed exceeded the heat absorbed.

We saw in Section 21.5 that work is a path-dependent quantity. In contrast, the internal energy depends on the particular thermodynamic state of the system and is independent of the path by which the system arrived at that state. For example, the internal energy of a given amount of gas depends on its pressure and temperature. It makes no difference how the gas achieved a particular pressure and temperature. All processes (all paths) leading to a particular pressure and temperature leave the gas with the same internal energy. It follows that the change in internal energy $U_f - U_i$ depends only on the initial and final states and not on the sequence of intermediate states. We say that $U_f - U_i$ is path independent. From the first law,

$$Q = U_f - U_i + W$$

we can see that Q is the sum of a path-independent quantity, $U_f - U_i$, and a path-dependent quantity, W. Therefore Q is a path-dependent quantity. This means that the amount of heat absorbed or ejected by a system depends on the nature of the path it follows between the initial and final states. In summary, heat and work are path-dependent quantities. Internal energy is a property of a system. Stated differently, heat and work represent two ways in which a system can exchange energy with its environment, and both heat and work depend on the details of the exchange process. Internal energy describes the energy "possessed" by a system.

When a thermodynamic process proceeds smoothly we can think of it as a continuous sequence of small changes. Mathematically we can represent the quantities appearing in the first law as infinitesimals. The differential form of the first law is

$$dQ = dU + dW \tag{21.19}$$

The physical content of the first law is unaltered by rewriting it in this form. The symbol dQ denotes an infinitesimal amount of heat absorbed by a system. The term dU represents the infinitesimal change in its internal energy and dW is the infinitesimal work done by the system. However, Eq. 21.19 presents a pitfall of which you must be aware. Representing heat and work as dQ and dW does not imply the existence of properties—Q and W—that measure the heat and work content of a system. There are no such properties. In other words, dQ and dW are mathematically infinitesimal, but they are not true differentials: They denote small amounts of heat and work. There are no functions Q and W that can be differentiated.

The first law serves two purposes: In its adiabatic form (Eq. 21.17) it gives an operational definition of internal energy change. In the general form (Eq. 21.18) it recognizes heat as a form of energy. The first law of thermodynamics is a statement of energy conservation. The internal energy of a system changes if it absorbs a net amount of heat or if it performs a net amount of work against its environment.

Questions

13. Heat and work are path-dependent quantities. Is their difference path dependent? Explain.

14. If all of the heat absorbed by a system in some process is converted into work, what must be true about the internal energy of the system?

15. A gas is confined in a cylinder by a piston. A sudden (adiabatic) compression causes the temperature to rise by 20 degrees (Celsius) and the volume to decrease by 30%. This process causes the internal energy to increase ($\Delta U > 0$). Negative work is done by the gas ($W < 0$) and there is no heat exchange ($Q = 0$). Describe a different process that results in the *same change* of state for the gas. Indicate whether ΔU, W, and Q are positive, negative, or zero.

21.8 Applications of the First Law

Let's see how the first law applies to various types of thermodynamic processes.

1. Cyclic Process

In a cyclic process (Figure 21.15), the initial and final states of the system are the same, so that $U_f = U_i$, and the first law reduces to

$$Q = W \tag{21.20}$$

In this form the first law states that the net work done by the system over the cycle equals the net heat absorbed over the cycle.

In the case of a fluid system, where $dW = P\,dV$, we can give the net work a geometric interpretation. The net work done by the pressure forces over one complete cycle can be written

$$W_{\text{cycle}} = \oint P\,dV \tag{21.21}$$

The symbol $\oint$ means that the integration is over a full cycle. Geometrically, $\oint P\,dV$ is represented by the area enclosed by the path describing the cycle on a $P-V$ dia-

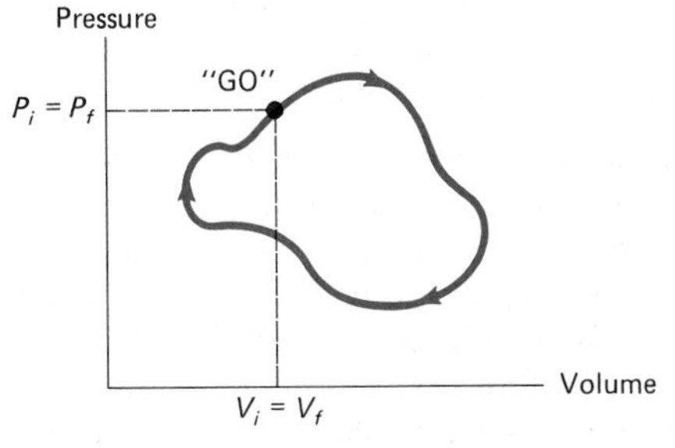

Figure 21.15

$P-V$ diagram for a cyclic process.

Figure 21.16

(a) The work of expansion is positive; it corresponds to the hatched area. (b) The work of compression is negative; its magnitude corresponds to the cross-hatched area. (c) The net work is positive. It corresponds to the area enclosed by the clockwise path. (d) A cycle for which the net work done is negative; its magnitude corresponds to the enclosed area. Note that the cycle generates a counterclockwise path.

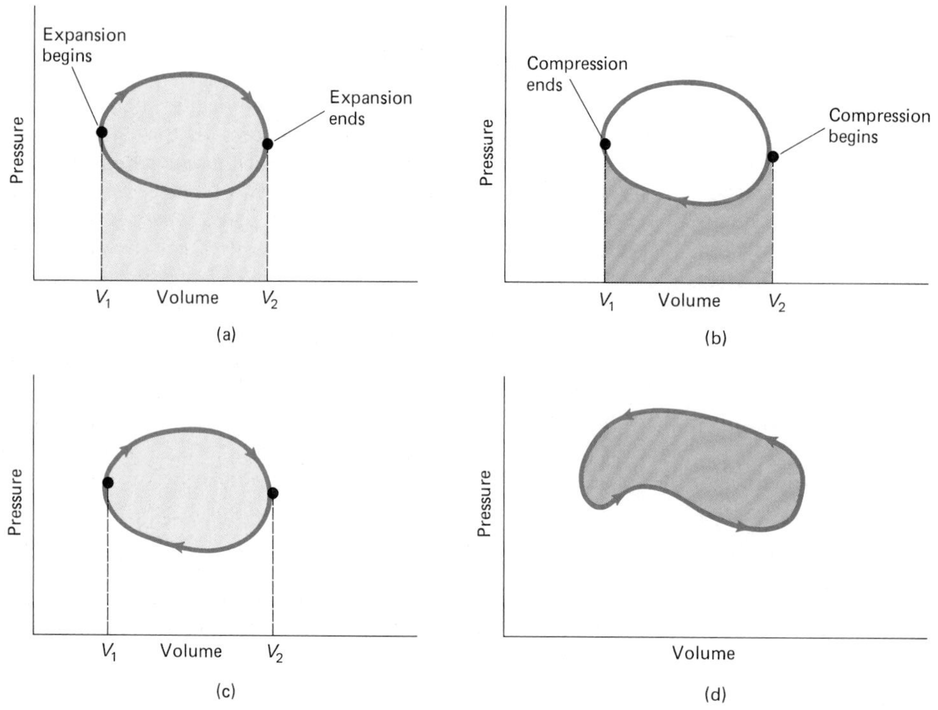

gram. Figure 21.16 illustrates this point. During expansion the volume increases so that dV and thus $dW = P\,dV$ are positive. The total work of expansion is thus positive and its magnitude corresponds to the hatched area of Figure 21.16a. The compression from V_2 to V_1 reduces the volume so that dV and dW are negative. The total work of compression is therefore negative, and its magnitude is given by the cross-hatched area of Figure 21.16b. The net work done over a full cycle is evidently the sum of the positive work of expansion and the negative work of compression. Therefore, the net work corresponds geometrically to the *difference* between the corresponding areas. Figure 21.16c shows that this difference is the *area enclosed* by the path of the cycle. In Figure 21.16c the net work done by the system is positive because the pressure during expansion is greater than the pressure during compression. Note that the cyclic process in Figure 21.16c generates a clockwise path on the $P-V$ diagram. A clockwise path on a $P-V$ diagram always corresponds to positive work.

Figure 21.16d shows a cyclic process for which the net work is negative. The work of compression is negative and its magnitude *exceeds* the positive work of expansion. Note that the cycle traces out a counterclockwise path. On a $P-V$ diagram, a counterclockwise path always corresponds to negative work. The magnitude of the net work corresponds to the enclosed area.

Example 9 Cyclic Work

Example 6 illustrates the path dependence of work. Now consider a cycle that uses the two paths of that example (Figure 21.17). Starting at i the cycle proceeds in clockwise fashion along the two-legged path 1 to f, and returns to i along the path labeled "2." The net work performed over the cycle is positive (the cycle traces a clockwise path) and corresponds geometrically to the triangular area enclosed. Using the numerical values given in Example 6, we find that the net work performed is

$$W = \frac{1}{2}(P_i - P_f)(V_f - V_i) = 2.5 \times 10^5 \text{ J}$$

In this cycle the net heat absorbed also equals 2.5×10^5 J. This is not a coincidence: The first law requires that $Q = W$ for a cyclic process.

2. Adiabatic Process

By definition, in an adiabatic process no heat is absorbed or ejected by the system. For an adiabatic process the first law of thermodynamics states that

$$U_f - U_i + W = 0 \qquad (21.17)$$

Problems

Section 21.2 Temperature Measurement

1. Using a constant-volume gas thermometer, a student finds values of 1.374, 1.370, and 1.368 for the ratio of the steam-point pressure to the triple-point pressure. (a) Determine the corresponding steam points. (b) The three values correspond to triple-point pressures of 2.667×10^4 Pa, 1.333×10^4 Pa, and 6.667×10^3 Pa, respectively. Extrapolate the data (graphically or otherwise) to determine the $P_3 = 0$ limit of the steam point.
2. Derive the inverse of Eq. 21.5, that is, the equation expressing Celsius temperature in terms of Fahrenheit temperature. Show that your result gives the correct temperature for the steam point.
3. Determine the Fahrenheit temperature of absolute zero (0°R).
4. The normal melting point of gold is 1064.50°C. Determine its value on the Kelvin scale.
5. Determine the temperature for which the Fahrenheit and Kelvin scales coincide.
6. Express a body temperature of 98.6° F on the Celsius and Kelvin scales.
7. In Eq. 21.3 the definition of the ideal gas temperature involves the limit

$$\lim_{P_3 \to 0} \left(\frac{P}{P_3}\right)$$

It *appears* as though the limit corresponds to division by zero. Why does the limit approach a definite value?

Section 21.3 Energy and the First Law of Thermodynamics

8. Electric utilities charge for energy usage on the basis of the kilowatt hour (kWh). Show that 1 kWh is approximately equal to 3600 kJ.
9. The *quad* is an energy unit.

$$1 \text{ quadrillion Btu} = 10^{15} \text{ Btu} \equiv 1 \text{ quad}$$

The *megaton* is another energy unit. One megaton is the energy released in the explosion of 1 million tons of TNT.

$$1 \text{ megaton} = 4 \times 10^{17} \text{ J}$$

The yearly energy consumption in the United States totals nearly 100 quads. Show that the yearly energy consumption is approximately 250 megatons.

Section 21.4 Heat and Specific Heat Capacity

10. The specific heat capacity of gold is 0.130 kJ/kg · C°. Determine the heat required to raise the temperature of 100 gm of gold from 270 K to 300 K.
11. The temperature of a silver bar rises by 10.0 C° when it absorbs 1.23 kJ of heat. The mass of the bar is 525 gm. Determine the specific heat of silver.
12. The specific heat of aluminum is 0.900 kJ/kg · C°. (a) What is the heat capacity of 3 kg of aluminum? (b) How much heat must be added to 3 kg of aluminum to raise its temperature from 27°C to 37°C?
13. Lake Erie contains roughly 4×10^{11} m^3 of water. How much heat is required to raise the temperature of that volume of water from 62°C to 63°C? How long would it take to supply this amount of heat by using the full output of a 1000-MW (10^9 W) electric power plant?
14. Determine the heat capacity of 1 gal of water.
15. The specific heat of a substance is given by

$$C = a + bT$$

where $a = 1.12$ kJ/kg·C° and $b = 0.016$ kJ/kg·C° K. Use Eq. 21.11 to determine the amount of heat required to raise the temperature of 1.20 kg of the material from 280 K to 312 K.

16. A student obtains the following data in a method-of-mixtures experiment designed to measure the specific heat of aluminum:

Initial temperature of water and calorimeter: 70°C

Mass of water: 0.400 kg

Mass of calorimeter: 0.040 kg

Specific heat of calorimeter: 0.63 kJ/kg · C°

Initial temperature of aluminum: 27°C

Mass of aluminum: 0.200 kg

Final temperature of mixture: 66.3°C

Use these data to determine the specific heat of aluminum. Your result should be within 15% of the value listed in Table 21.1.

17. The air temperature above coastal areas is profoundly influenced by the large specific heat of water (4.19 kJ/kg · C°). One reason is that the heat released when 1 m^3 of water cools by 1 C° will heat an enormously greater volume of air by this same amount. Estimate the volume of air that can be heated by 1 C° by the heat released by the cooling of 1 m^3 of water by 1 C°. The specific heat of air is approximately 1.0 kJ/kg · C°. Take the density of air to be 1.3 kg/m^3.
18. The Joule constant, $J = 4.1858$ kJ/kcal, is often referred to as the *mechanical equivalent of heat*. At the end of his 1842 essay Mayer expresses his results for the mechanical equivalent of heat by stating that the warming of a given weight of water from 0°C to 1°C

is equivalent to the mechanical energy acquired by the same weight of water in falling from a height of 365 m. Show that the value of J for these data is approximately

$$J = 3.58 \text{ kJ/kcal}$$

Assume that the specific heat of water is 1 kcal/kg · C° and take the acceleration of gravity to be 9.80 m/s^2.

19. A stuntman who weighs 735 N falls from a height of 25 m into a pool of water that is 15 cm deep, 4 m wide, and 4 m long. Assume that all of his initial gravitational potential energy is converted into heat and that this heat is absorbed by the water. Determine the temperature increase of the water. In fact, not all of the energy is converted into heat upon impact. In what other ways would the energy be distributed?

20. Water going over Angel Falls in Venezuela (the world's highest waterfall) drops 1000 m. Calculate the increase in temperature from the conversion of gravitational potential energy into thermal energy. (Actually, other effects—such as air resistance and evaporative cooling—significantly change this result.)

21. Soil has a specific heat of 0.80 kJ/kg · C°. Equal masses of soil and water absorb equal quantities of heat. The temperature of the water increases by 4 C°. Determine the temperature increase of the soil.

Section 21.5 Thermodynamic Work

22. A block of aluminum in the form of a cube with a volume of 1 m^3 expands. The surrounding pressure is 1 atm (1.013×10^5 Pa). The length of each edge increases by 1 mm. Determine the work by the block in expanding against its surroundings.

23. A fluid expands at a constant pressure of 6.2×10^5 Pa, increasing its volume by 3.1 m^3. Determine the work done by the fluid.

24. A gas expands from an initial volume of 1.0 m^3 to a final volume of 2.0 m^3. The pressure falls as shown in Figure 1. Determine the work done by the gas.

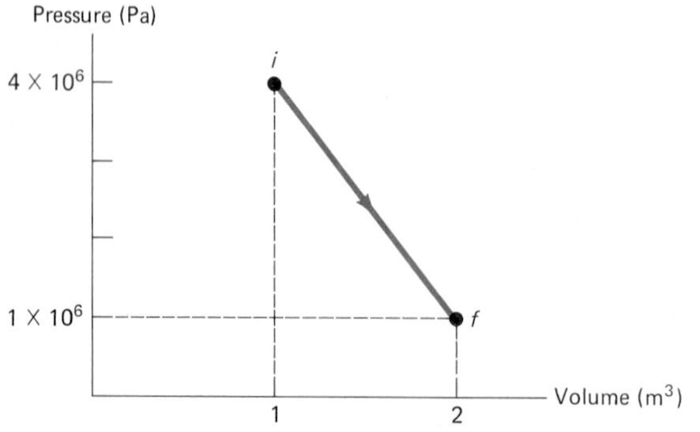

Figure 1

25. Determine the work performed by a fluid that expands from i to f as indicated in Figure 2. How much work is performed by the fluid if it is compressed from f to i along the same path?

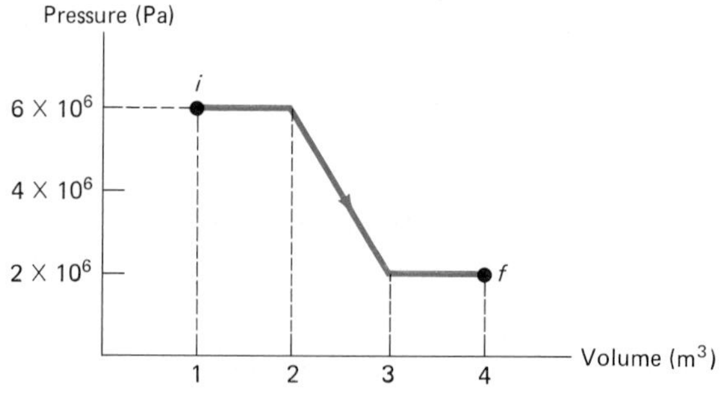

Figure 2

26. A certain gas obeys the equation of state

$$\frac{PV}{T} = \text{constant}$$

(a) Use a $P-V$ diagram to trace the path of processes in which the gas expands, doubling its volume, (i) isobarically and (ii) isothermally. (b) Use the geometric interpretation of work to determine in which case the gas performs more work.

27. The pressure of a gas (in pascals) is related to its volume (in cubic meters) by

$$P = 10^5 V^{3/2}$$

Calculate the work done by the gas as its volume increases from 1 m^3 to 2 m^3.

28. Helium with an initial volume of 1 liter (10^{-3} m^3) and an initial pressure of 10 atm expands to a final volume of 1 m^3. The relationship between pressure and volume during the expansion is PV = constant. Calculate the (a) value of the constant, PV; (b) final pressure; and (c) work done by the helium during the expansion.

Section 21.7 The First Law of Thermodynamics

29. The environment performs 1.7×10^7 J of work *on the system* of Example 8. If the system absorbs 5.0×10^6 J of heat, compute the change in its internal energy.

30. For many years the U.S. Patent Office was deluged with applications for patents on various types of perpetual motion machines. This flood subsided abruptly and completely when the rules were amended to require that a working model accompany each application. Suppose that you were employed in the Patent Office prior to this ruling and that you received an application for a patent on a heat engine that claimed the following performance figures for the engine:

Net heat consumption per cycle: 40 kJ

Mechanical work output per cycle: 100 kJ

Electrical work output per cycle: 10^5 W · s

Would it be necessary to give detailed consideration to the application? Explain.

31. When the helium of Problem 28 expanded, its temperature and internal energy remained constant. Heat was supplied to hold the temperature constant and maintain the PV = constant relationship. How much heat was absorbed by the expanding helium?

Section 21.8 Applications of the First Law

32. A fluid is carried through the cycle indicated in Figure 3. How much work (in kilojoules) is done by the fluid during (a) the expansion from 1 to 2; (b) the constant-volume decompression from 2 to 3; (c) the compression from 3 to 4? (d) What is the net amount of heat (in kJ) transformed into work during the cycle? (1 atm = 1.013×10^5 Pa.)

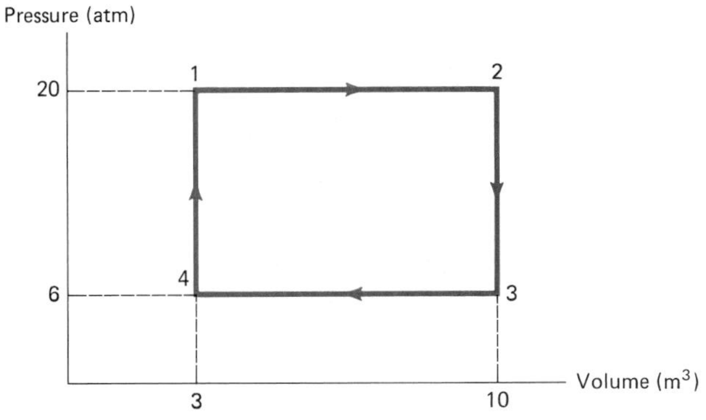

Figure 3

33. The paths for three cyclic processes are indicated in Figure 4. Indicate for each part whether the net work done is positive, negative, or zero. Explain the basis for your choice.

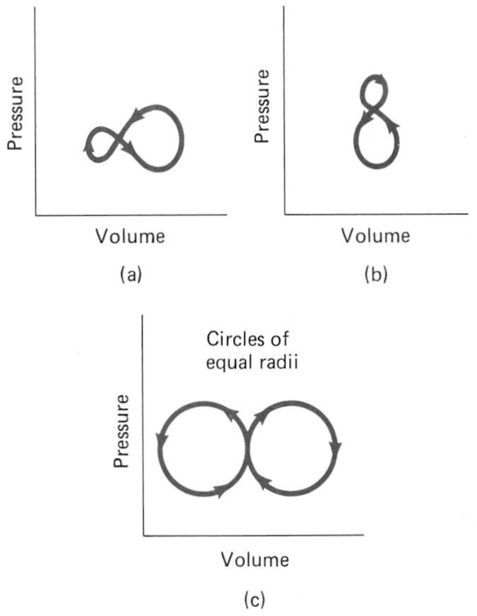

Figure 4

34. A fluid undergoes a cyclic process. The $P-V$ path is a "circle," as indicated in Figure 5. Determine the net work performed by the fluid and the net heat absorbed. (*Hint:* If it bothers you to see a "circle" with one "diameter" drawn vertically as 9×10^5 Pa and the other "diameter" drawn horizontally as 2 m^3, think of the figure as an ellipse. A small amount of research will uncover the formula for the area of an ellipse.)

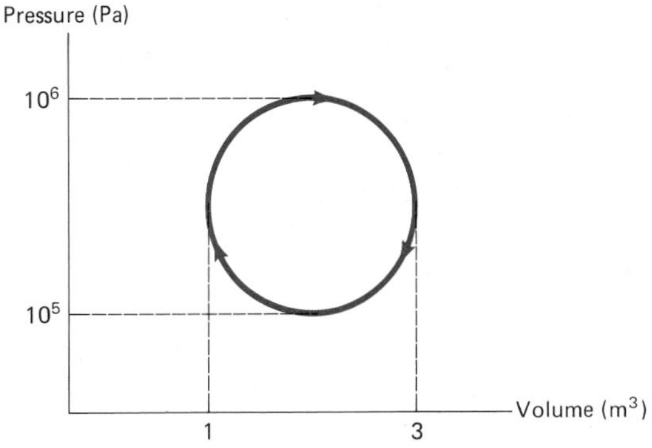

Figure 5

chapter 22

THERMAL PROPERTIES OF MATTER

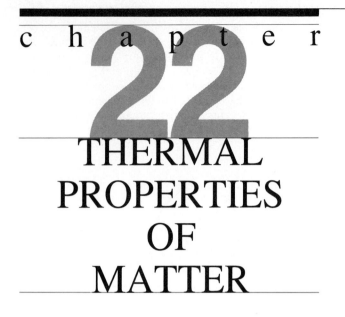

Preview

Certain basic thermal properties of matter are familiar to us through our everyday experiences. We have all seen ice melt and water boil. We also know that heated solids expand. That is the reason that telephone lines sag most noticeably in hot weather. The thermal expansion of solids and liquids is analyzed in Section 22.1. Changes of phase, which include melting and boiling, are discussed in Section 22.4.

Additionally, this chapter develops the concept of an ideal gas, first mentioned in Section 21.2. The ideal gas is a generalized model of a gas that incorporates the basic properties common to all real gases. This model offers both practical and conceptual advantages. The engineer who wants a rough idea how a temperature change will affect the pressure of a particular gas can rely on the ideal gas model for an approximate answer. A precise answer would depend on factors such as the chemical composition of the gas. By using the ideal gas model, the engineer can avoid minor details and quickly arrive at an approximate answer. The ideal gas model not only makes calculations simpler but also makes it easier to grasp certain thermodynamic concepts. The ideal gas plays the same role in thermodynamics as does the "point mass" or "particle" in mechanics. It can be used to illustrate the principles of thermodynamics at an intuitive level.

22.1 Thermal Expansion

Most of us are familiar with the fact that solids and liquids expand or contract with temperature changes. Concrete highway blocks are separated by tar-filled gaps to allow for expansion during the summer. Automobile engines are fitted with expansion plugs. Changes in the volume of the fluid in a household thermometer signal temperature changes.

There are many situations that require a quantitative description of thermal expansion. These range from the design of buildings and bridges to achieving a tight-fitting joint between two different metals. A mathematical description of thermal expansion will enable us to analyze such situations.

If a rod of length L and temperature T experiences a small change in temperature, ΔT, its length changes by an amount proportional to both ΔT and L (Figure 22.1). The change in length, ΔL, is proportional to the product $L\,\Delta T$. We introduce a proportionality factor, α, characteristic of the solid, and write

$$\Delta L = \alpha L\,\Delta T \tag{22.1}$$

The quantity α is called the *coefficient of linear expansion*. Note that Eq. 22.1 describes expansion when ΔT is positive and contraction when ΔT is negative. If we set* $\Delta T = 1\ \mathrm{C}^\circ$, we can see from Eq. 22.1 that α represents the fractional change in length for a 1-degree change in temperature

$$\left(\frac{\Delta L}{L}\right)_{\Delta T = 1\,\mathrm{C}^\circ} = \alpha \cdot 1\ \mathrm{C}^\circ \tag{22.2}$$

*Temperature *changes* are expressed in units of Celsius degrees, abbreviated C°.

Figure 22.1

A temperature change, ΔT, results in a change of length, ΔL. Experiment shows that ΔL is directly proportional to ΔT and directly proportional to L.

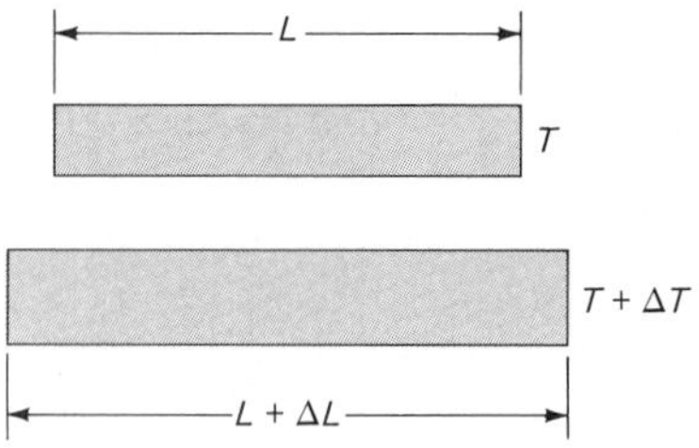

Table 22.1 presents values of α for a variety of materials. All of these values are close to 10^{-5}/C°. We can get a feeling for thermal expansion by noting that a solid 1 km long with $\alpha = 10^{-5}$/C° increases its length by just 1 cm for each 1-degree rise in temperature.

The smaller the value of α, the less thermal expansion will occur. Materials with near-zero values of α therefore maintain their dimensions and their shapes very well in spite of temperature changes. Such materials can be very useful. For example, the mirrors used in astronomical telescopes must have a paraboloidal reflecting surface in order to focus light sharply. Thermal expansion or contraction alters the shape of the surface. The glass used for the 200-in. Hale telescope on Mount Palomar has $\alpha = 0.325 \times 10^{-5}$/C°. The glass to be used in the mirror for the proposed Space Telescope has a value of α at least 100 times smaller than that of the Hale mirror.

Example 1 Why Telephone Calls Travel Farther in Warm Weather

The value of α for copper is 1.7×10^{-5}/C°. A copper telephone wire 10 km long at a temperature of 60°F (16°C) lengthens when the temperature rises. At 96°F (36°C), the increase in length is

$$\Delta L = \alpha L \, \Delta T = (1.7 \times 10^{-5}/\text{C}°)(10^4 \text{ m})(20 \text{ C}°)$$

$$= 3.4 \text{ m}$$

Differences in the thermal expansion of two solids can be demonstrated by using a bimetallic strip. This consists of two pieces of metal, each having a different value of α, bound together (Figure 22.2). An increase in temperature produces unequal changes in their lengths, and results in a curvature of the bimetallic strip. A decrease in temperature results in a curvature in the opposite direction. Bimetallic strips are used as temperature change sensors in home thermostats.

Temperature changes that alter the dimensions of an object also change its volume. If a substance of volume V experiences a small change in temperature, ΔT, its volume changes by an amount proportional to V and ΔT. We express the change in volume as

$$\Delta V = \beta V \, \Delta T \qquad (22.3)$$

where β is called the *coefficient of volume expansion*. Setting $\Delta T = 1$ C° in Eq. 22.3 shows that β is numerically equal

Table 22.1

Coefficients of expansion for selected liquids and solids*

Solids	α (units of 10^{-5}/C°)	Liquids	β (units of 10^{-3}/C°)
Silver (Ag)	1.89	Acetic acid	3.21
Aluminum (Al)	2.80	Acetone	4.47
Gold (Au)	1.42	Benzene	3.72
Calcium (Ca)	2.50	Carbon tetrachloride	3.72
Iron (Fe)	1.17	Ether	4.98
Sodium (Na)	7.1	Mercury	0.54
Lead (Pb)	2.91	Pentane	4.83
Rubidium (Rb)	9.0	Water	0.63
Zinc (Zn)	3.3		
Carbon steel	1.15		
Glass	0.90		

*The column entry for α expresses the coefficient of linear expansion in units of 10^{-5}/C°. Thus, the value of α for silver is 1.89×10^{-5}/C°. For the coefficient of volume expansion, β, the units are 10^{-3}/C°. Thus, the value of β for acetic acid is 3.21×10^{-3}/C°. Values refer to measurements at a pressure of 1 atm and for a temperature near 20°C. Values for liquids are adapted from *Chemical Engineer's Handbook* by J. H. Perry, McGraw-Hill, New York, 1950. Used with permission of McGraw-Hill Book Company. Values for solids are adapted from *International Critical Tables of Numerical Data: Physics, Chemistry, and Technology*, with permission of the National Academy of Sciences, Washington, DC.

Figure 22.2

A bimetallic strip is used to demonstrate differences in thermal expansion. Two pieces of metals with different values of α are bound together. An increase in temperature produces unequal changes in their lengths and results in a curvature. A decrease in temperature produces a curvature in the opposite direction.

Figure 22.4

An isotropic solid expands when heated. Each dimension increases in length by an amount proportional to $\alpha \, \Delta T$. Because $\alpha \, \Delta T \ll 1$ the resulting change of volume is also proportional to ΔT.

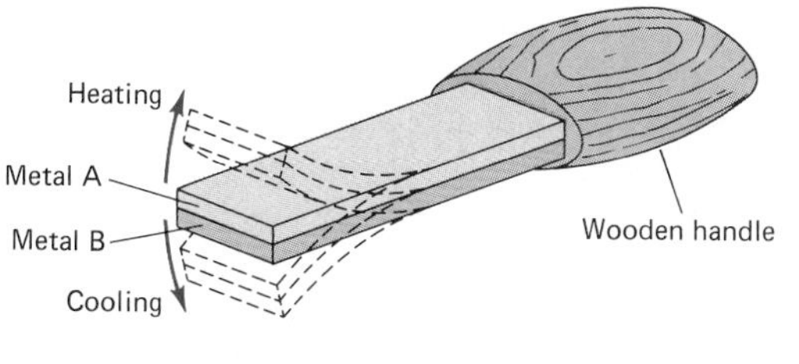

to the fractional change in volume for a 1-degree change in temperature

$$\left(\frac{\Delta V}{V}\right)_{\Delta T = 1\,C^\circ} = \beta \cdot 1 \; C^\circ \qquad (22.4)$$

Typical values of β for liquids are listed in Table 22.1. The tabulated values of β for liquids cluster near $10^{-3}/C^\circ$. This means that a liquid expands by about one part in 10^3 for each degree of change in temperature. A few substances exhibit negative values of β over certain ranges of temperature. Most notable of these is water, for which β is negative over the narrow range of temperature from 0°C to 3.98°C. Thus, if water at 1°C is warmed to 2°C, its volume decreases and its density increases. The maximum density of water occurs at a temperature of 3.98°C (at a pressure of 1 atm). Figure 22.3 shows a graph of the density of water over a narrow range of temperature.

Table 22.1 lists values of β only for liquids. For isotropic solids we can show that β and α are simply related. Figure 22.4 shows a solid in the form of a rectangular parallelepiped. If the temperature rises by ΔT, the lengths of the sides increase from L_1, L_2, and L_3 to

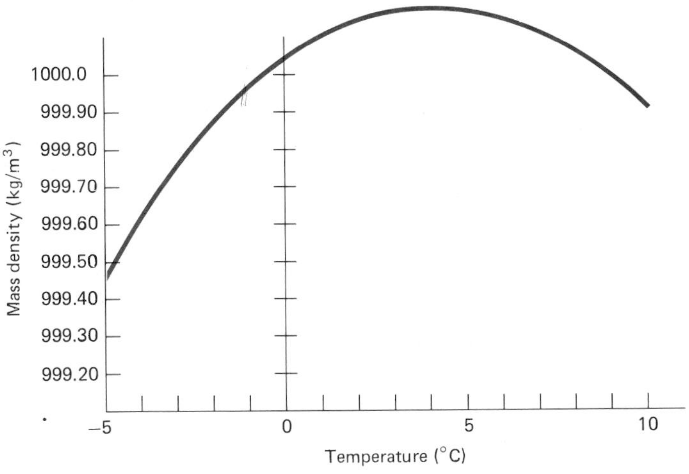

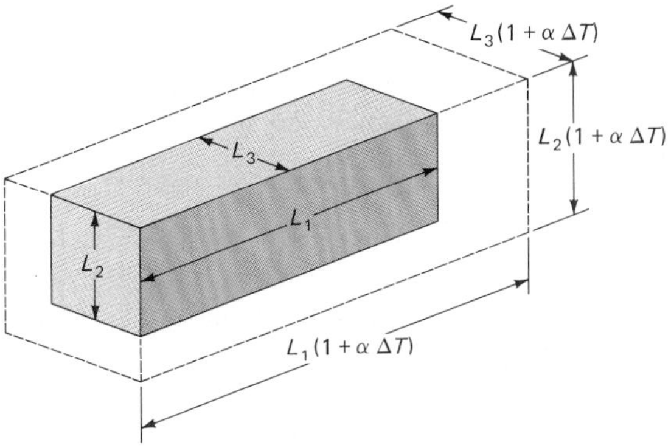

$$L_1(1 + \alpha \, \Delta T) \quad L_2(1 + \alpha \, \Delta T) \text{ and } L_3(1 + \alpha \, \Delta T).$$

The volume changes from $V = L_1 L_2 L_3$ to $V + \Delta V = L_1 L_2 L_3 (1 + \alpha \, \Delta T)^3$. Expanding $(1 + \alpha \, \Delta T)^3$ gives for the change in volume

$$\Delta V = V \cdot [1 + 3\alpha \, \Delta T + 3(\alpha \, \Delta T)^2 + (\alpha \, \Delta T)^3] - V$$

Because $\alpha \, \Delta T \ll 1$ we can ignore the terms $3(\alpha \, \Delta T)^2$ and $(\alpha \, \Delta T)^3$ in comparison to the term $3\alpha \, \Delta T$. This leaves

$$\Delta V = 3\alpha V \cdot \Delta T \qquad (22.5)$$

When we compare this expression for ΔV with Eq. 22.3 ($\Delta V = \beta V \, \Delta T$), we see that β and α are related by

$$\beta = 3\alpha$$

Figure 22.3

The density of water versus temperature at a pressure of 1 atm. The maximum density of 10^3 kg/m^3 occurs at 3.98°C. (The range from -5°C to 0°C is for supercooled liquid water. The density of ice is 920 kg/m^3, far below the lowest value on the graph.)

Questions

1. Explain why ice forms first at the surface of a lake rather than at the bottom.
2. Suppose that ice did not float. What would be the consequences to life on earth?
3. A steel washer has inner diameter of 1 cm and an outer diameter of 2 cm. A temperature increase causes the outer diameter to increase by 0.01%. Does the inner diameter increase or decrease? By what percentage?
4. The listed height of the Washington Monument is 555 ft 5 in. Use the fact that $\alpha \approx 10^{-5}/\text{C}°$ for solids to *estimate* the change in height of the structure when the temperature rises by 10 degrees.

22.2 The Ideal Gas

Thermal expansion is one of the basic thermal properties of solids and liquids. The properties of a confined gas are strongly temperature dependent. Let us examine the basic thermal properties of gases from an elementary viewpoint.

The quantitative study of gases was made possible by von Guericke's invention of the vacuum pump in 1650. In 1662 Robert Boyle used a crude vacuum pump to discover the law that bears his name. Boyle found that the volume (V) of a fixed mass of gas held at a constant temperature varied inversely as its pressure (P). This relation can be written as an equation:

$$PV = \text{constant (for fixed mass of gas at constant temperature)} \qquad (22.6)$$

Nearly a century later, Charles and Gay-Lussac independently found that equal changes in temperature produce equal changes in volume for a fixed mass of gas held at a constant pressure. On an absolute temperature scale the law of Charles and Gay-Lussac can be expressed as

$$\frac{V}{T} = \text{constant (for fixed mass and pressure)} \qquad (22.7)$$

When we combine Eqs. 22.6 and 22.7 we get

$$\frac{PV}{T} = \text{constant (for a fixed mass)} \qquad (22.8)$$

All gases at low pressures ($P < 1$ atm) and high temperatures ($T > 200$ K) obey this equation closely. That all gases behave similarly allows us to make certain general statements about gases and to develop a general model of a gas. This model is known as the *ideal gas* model. The ideal gas is characterized by three relations: (1) an equation of state, (2) an equation for the internal energy, and (3) an equation for the heat capacity.

1 Equation of State In general, an equation of state relates two or more measurable properties of a system. The *ideal gas equation of state* is

$$PV = nRT \qquad (22.9)$$

where P, V, and T are the pressure, volume, and temperature, respectively, of the gas, and n denotes the number of moles* of gas. R is the universal gas constant,

$$R = 8.31434 \text{ joules/kelvin} \qquad (22.10)$$

For a fixed amount of gas (n = constant) the ideal gas equation of state shows that only two of the three variables P, V, and T are independent. For example, if P and T are used as independent variables, then V is determined by Eq. 22.9. To illustrate this, suppose that we have 1 mole of gas at the standard temperature and pressure,

$$P = 1 \text{ atm} = 1.01325 \times 10^5 \text{ Pa}$$

$$T = 0°\text{C} = 273.16 \text{ K}$$

The volume occupied by 1 mole of a gas at this pressure and temperature is given approximately by the ideal gas equation of state,

$$V = \frac{nRT}{P} = \frac{1(8.3143)(273.16)}{1.01325 \times 10^5} \text{ m}^3$$

$$= 2.2414 \times 10^{-2} \text{ m}^3$$

or

$$V = 22{,}414 \text{ cm}^3$$

This value is approximate because real gases do not conform precisely to the ideal gas equation of state under standard conditions. Real gases are better described by more complicated equations of state. However, even in these equations only two of the three variables P, V, and T are independent.

Example 2 **Deciding that Helium Is a Precious Natural Resource**

A cylinder containing gaseous helium is delivered to a low-temperature laboratory. Two students working in the lab are told that the cost of the helium is \$35. They decide to figure out the cost of 1 kg of helium. They reason that the ideal gas law ($PV = nRT$) can be used to calculate n, the number of moles of gas, provided they can determine P, V, and T. Knowing n, they can determine the mass of gas because they know that 1 mole of helium has a mass of 4 gm. The pressure gauge on the cylinder reads 2200 lb/in.^{2} when the temperature is 70°F. They are not certain whether the gauge shows the absolute pressure or whether the absolute pressure is 1 atm (14.7 lb/in.^{2}) higher. They realize that this slight difference will not seriously affect their estimate and so take P = 2200 lb/in.^{2}. They measure the dimensions of the cylinder and estimate the volume of the gas to be 0.011 m^3. Converting the pressure to pascals gives

$$P = 2200 \text{ lb/in.}^2 \frac{6.89 \times 10^3 \text{ Pa}}{1 \text{ lb/in.}^2} = 1.52 \times 10^7 \text{ Pa}$$

*By definition, 1 mole is the amount of a substance that contains the same number of molecules (or atoms in the case of monatomic substances) as there are carbon atoms in 12 gm of carbon 12. The mass of 1 mole of a substance is called its *molecular weight* (see Section 25.1).

The Kelvin temperature corresponding to 70°F is

$$T = 273 + \frac{5}{9}(70 - 32)\text{ K} = 294\text{ K}$$

The ideal gas law gives

$$n = \frac{PV}{RT} = \frac{1.52 \times 10^7 \cdot 0.011}{8.31 \cdot 294} = 68.4\text{ moles}$$

which converts to 0.274 kg. They conclude that, at \$128 per kilogram, helium is a precious natural resource.

---

2 Internal Energy Equation The second relation that characterizes the ideal gas is the equation for the internal energy, U. Experiments by Joule and others show that the internal energy of a gas depends strongly on temperature (T) and the amount of gas (n) but is nearly independent of volume. Here, we regard pressure as the dependent variable. Through this section we consider a fixed amount of gas so that n is a constant. The *internal energy of the ideal gas* is taken to be independent of volume

$$U_{\text{ideal gas}} = U(T) \tag{22.11}$$

3 Heat Capacity Equation Finally, the third relation that characterizes the ideal gas is the equation for the specific heat. Experiment shows that the specific heat of a gas is nearly constant over a moderate range of temperature. Therefore the specific heat of the ideal gas is taken to be *independent of temperature.*

We now assume a constant specific heat and use the first law of thermodynamics to show that the internal energy of the ideal gas increases linearly with its Kelvin temperature. The differential form of the first law of thermodynamics is (Section 21.7)

$$dQ = dU + dW \tag{22.12}$$

where dQ is the heat absorbed by the gas, dU is the change in the internal energy of the gas, and dW is the work performed by the gas. We can express dQ as (see Eq. 21.10)

$$dQ = m\,C dT \tag{22.13}$$

where m is the mass of the gas, C is its specific heat, and dT is the change in its temperature. The work done by the gas is given by (see Eq. 21.15)

$$dW = P\,dV \tag{22.14}$$

Substituting these expressions for dQ and dW into Eq. 22.12 gives

$$m\,C dT = dU + P\,dV \tag{22.15}$$

If heat is added with the volume of the gas held constant, then $dV = 0$, in which case Eq. 22.15 reduces to

$$mC_v\,dT = dU \tag{22.16}$$

The subscript v is a reminder that Eq. 22.16 applies to the condition of heat exchange at constant volume. In general, the value of the specific heat depends on the heat exchange. For example, we show later in this section that the specific heat for a constant-pressure heat exchange (C_p) is greater than the specific heat for a constant-volume heat exchange (C_v). For the ideal gas the heat capacity C_v is independent of temperature. This permits us to integrate Eq. 22.16 to obtain

$$U = mC_vT + \text{constant} \tag{22.17}$$

This is the desired result: The internal energy of the ideal gas increases linearly with its temperature. The constant of integration in Eq. 22.17 reflects the fact that only changes in internal energy can be measured. In subsequent work we take the integration constant to be zero and write

$$U = mC_vT \qquad C_v = \text{constant} \tag{22.18}$$

At the outset we assumed that the internal energy U of a given mass of an ideal gas depends on its temperature but not on its volume. Equation 22.18 shows explicitly how U depends on T.

Heat added to a gas at constant volume goes entirely into the increase of internal energy—no work is done. The same amount of heat added at constant pressure produces a smaller rise in temperature because the gas expands and converts some of the heat into work. Consequently, the specific heat is larger at constant pressure than at constant volume. We can show this as follows: If C_p denotes the specific heat at constant pressure and $(dQ)_p$ is the heat absorbed by the gas at constant pressure, then

$$(dQ)_p = mC_p\,dT \tag{22.19}$$

For heat added at constant pressure the first law has the form

$$mC_p\,dT = dU + P\,dV \tag{22.20}$$

With pressure constant we can form the differential $P\,dV$ using the ideal gas equation of state, Eq. 22.9 ($PV = nRT$). This gives

$$P\,dV = nR\,dT \tag{22.21}$$

Inserting Eqs. 22.16 and 22.21 into Eq. 22.20 and canceling a factor of dT leaves

$$mC_p = mC_v + nR \tag{22.22}$$

We can conclude from this equation that $C_p > C_v$, as we asserted earlier. Equation 22.22 also shows that C_p is a constant for an ideal gas, because C_v and R are constants. For 1 mole of gas ($n = 1$, $m \equiv M$, the molecular weight), Eq. 22.22 shows that the *molar heat capacities*, MC_p and MC_v, differ by a constant amount of $R = 8.314$ J/K.

$$MC_p - MC_v = R \tag{22.23}$$

Table 22.2 shows that the molar heat capacities of various real gases are in good agreement with Eq. 22.23.

To summarize, basic properties of real gases are described by a generalized model called the ideal gas. The ideal gas is characterized by three relations:

1	Equation of state	$PV = nRT$	(22.9)
2	Internal energy	$U = mC_vT$	(22.18)
3	Constant specific heat	C_v = constant, independent of temperature	

Table 22.2

Molar heat capacities of gases				
Gas	MC_p (J/K)	MC_v (J/K)	$MC_p - MC_v$ (J/K)	$\gamma = C_p/C_v$
Helium (He)	20.93	12.61	8.32	1.66
Argon (Ar)	20.93	12.53	8,40	1.67
Hydrogen (H_2)	28.76	20.39	8.37	1.41
Nitrogen (N_2)	29.05	20.75	8.30	1.40
Oxygen (O_2)	29.43	21.02	8.41	1.40

Adiabatic Process

An adiabatic process is one in which there is no exchange of heat. Many processes occur so rapidly that there is little heat transfer. Such processes are often considered to be adiabatic. The rapid expansions and compressions of gases in internal combustion engines constitute one example. Let's consider the adiabatic transformation of an ideal gas. The fact that there is no heat exchange will let us establish another relation among P, V, and T, in addition to the equation of state. An adiabatic process is defined by the condition $dQ = 0$, and the first law becomes

$$dU + P\,dV = 0 \qquad (22.24)$$

For the ideal gas we know that

$$dU = mC_v\,dT \qquad (22.16)$$

and

$$P = \frac{nRT}{V}$$

Inserting these expressions into Eq. 22.24 and dividing through by mC_v gives

$$dT + \frac{nRT}{mC_v}\frac{dV}{V} = 0 \qquad (22.25)$$

From Eq. 22.22 (with $n = 1$ and $m = M$) we find

$$\frac{nR}{mC_v} = \frac{M(C_p - C_v)}{mC_v} = \frac{C_p}{C_v} - 1 \qquad (22.26)$$

The ratio of specific heats C_p/C_v is denoted by γ (the Greek letter gamma):

$$\frac{C_p}{C_v} = \gamma \qquad (22.27)$$

For an ideal gas C_p and C_v are constants, so γ is a constant. In terms of γ,

$$\frac{nR}{mC_v} = \gamma - 1 \qquad (22.28)$$

We can see from this relation that $\gamma - 1$ is a positive number. Replacing nR/mC_v by $\gamma - 1$ in Eq. 22.25 and multiplying through by $V^{\gamma-1}$ produces

$$V^{\gamma-1}\,dT + T(\gamma - 1)V^{\gamma-2}\,dV = 0 \qquad (22.29)$$

Recalling the product rule for differentials,

$$d(xy) = y\,dx + x\,dy$$

we see that the left-hand side of Eq. 22.29 is the differential of $TV^{\gamma-1}$.

$$d(TV^{\gamma-1}) = V^{\gamma-1}\,dT + T(\gamma - 1)V^{\gamma-2}\,dV$$

Thus, Eq. 22.29 shows that the differential of $TV^{\gamma-1}$ is zero during an adiabatic process. It follows that $TV^{\gamma-1}$ is a constant for the adiabatic transformation of an ideal gas:

$$TV^{\gamma-1} = \text{constant} \qquad (22.30)$$

This relation may also be expressed as

$$TV^{\gamma-1} = T_0V_0^{\gamma-1} \qquad (22.31)$$

where T_0 and V_0 are the temperature and volume for some particular state of the gas during the adiabatic process. Equation 22.31 can be used to show in a quantitative way how an adiabatic compression raises the temperature of a gas. Thus, let (T_0, V_0) denote the initial values of temperature and volume and let (T_f, V_f) denote the final values of temperature and volume. The final temperature T_f follows from Eq. 22.31 as

$$T_f = T_0\left(\frac{V_0}{V_f}\right)^{\gamma-1} \qquad (22.32)$$

For a compression V_0/V_f is greater than unity, and so T_f exceeds T_0.

Example 3 Just Hot Air

Adiabatic compression can lead to a substantial increase in temperature. The piston-fitted cylinder device shown in Figure 22.5 can be used to demonstrate this fact. Forcing the piston downward rapidly compresses the air adiabatically and raises its temperature in accord with Eq. 22.32. A small piece of tissue placed at the bottom of the cylinder flashes to incandescence, revealing the rapid increase in temperature.

In a typical compression the volume decreases by a factor of 20. Both nitrogen and oxygen, the primary constituents of air, have $\gamma = 1.4$. If we take $T_0 = 300$ K, a typical room temperature, Eq. 22.32 gives

$$T_f = 300\text{ K} \cdot (20)^{0.4} = 994\text{ K}$$

a temperature well above the flash point of tissue. Note that absolute temperatures must be used in Eq. 22.32.

In an ordinary automobile engine, an air–fuel mixture is ignited by a spark. In a diesel engine there are no spark plugs. The air–fuel mixture is raised to ignition tem-

Figure 22.5

A device for demonstrating the heating that occurs during an adiabatic compression. A glass tube serves as a transparent cylinder in which air is compressed by a plunger fitted with O rings. A metal cylinder with an oval viewing port at the bottom serves as a safety shield. A small piece of tissue paper placed in the cylinder will flash to incandescence when the air is suddenly compressed.

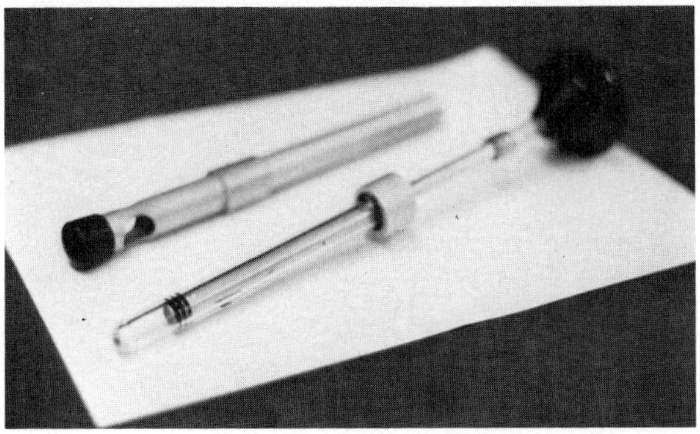

perature by a nearly adiabatic compression. The high temperatures at which ordinary automobile engines operate often cause "dieseling"—the engine continues to run for a few moments after the ignition has been turned off.

Equation 22.31 relates the temperature and volume during an adiabatic process. We can obtain a relationship between the pressure and volume during an adiabatic process by noting that the ideal gas equation of state can be expressed as (n = constant)

$$\frac{PV}{T} = \frac{P_o V_o}{T_o} \tag{22.33}$$

Multiplying Eq. 22.31 through by PV/T gives

$$PV^{\gamma} = P_o V_o^{\gamma} \tag{22.34}$$

which is often expressed as

$$PV^{\gamma} = \text{constant} \tag{22.35}$$

A third form of the adiabatic equation, which relates T and P, can be obtained by using Eq. 22.33 to eliminate the volume from Eq. 22.34. The resulting relation is

$$\frac{T}{T_o} = \left(\frac{P}{P_o}\right)^{(\gamma-1)/\gamma} \tag{22.36}$$

Example 4 Adiabatic Expansion and Gas Liquefaction

One method used to liquefy a gas lowers the temperature of the gas by causing it to expand and thereby to perform "$P\,dV$" work against its surroundings. If the expansion is adiabatic or nearly so, the work is done at the expense of the internal energy of the gas, and its temperature decreases. That is, $U = mC_vT$ so that if U decreases so must T. We can get a rough idea of the cooling produced by treating the gas as ideal. We take $\gamma = 1.40$, a value appropriate for diatomic gases (Table 22.2). For initial state values we set

$$P_o = 50 \text{ atm} \qquad T_o = 300 \text{ K}$$

For the final state pressure we take

$$P = 1 \text{ atm}$$

The final temperature follows from Eq. 22.36 as

$$T = T_o\left(\frac{P}{P_o}\right)^{(\gamma-1)/\gamma} = 300 \text{ K}\left(\frac{1}{50}\right)^{0.283} \approx 99 \text{ K}$$

Real gases depart significantly from ideal gas behavior at 50 atm. Nevertheless the result is not unrealistic. Adiabatic expansion "engines" are prominent features of commercial gas liquefiers.

Questions

5. Why does a playground basketball bounce noticeably higher on a hot day than it does on a cold day?
6. Why does the air escaping from an inflated tire feel cold?
7. Suppose that you want to double the temperature of an ideal gas. How would you accomplish this (a) without changing the volume of the gas; (b) without changing the pressure of the gas? Which requires the addition of more heat, (a) or (b)?

22.3 *P–V–T* Surfaces

A *pure* substance is one that is composed of a single chemical species rather than a mixture of chemically distinct components. Water, for example, consists only of the compound H_2O, and is a pure substance. The equilibrium states of a pure substance can be described by three variables, pressure, molar volume,* and temperature. The equation of state specifies one relation among P, V, and T so that only two are independent variables. For example, the equation of state for an ideal gas is $PV = RT$. From a geometric viewpoint the equation of state defines a *surface* in a space in which P, V, and T are the coordinates. Figure 22.6 shows the $P–V–T$ surface for an ideal gas. Figures 22.7 and 22.8 show the $P–V–T$ surfaces for carbon dioxide and water. Such surfaces are helpful in following the progress of various processes. For example, a closed path $A \rightarrow B \rightarrow C \rightarrow A$ is shown on the ideal gas surface of Figure 22.6. The path $A \rightarrow B$ describes an isothermal compression (T = constant). The path $B \rightarrow C$ describes an isobaric expansion (P = constant). The segment $C \rightarrow A$ describes a process in which the volume remains constant while pressure and temperature decrease.

*The volume divided by the number of moles is called the *molar volume*. Throughout this section we denote the molar volume by V.

Figure 22.6

The P–V–T surface for an ideal gas. The path from A to B describes an isothermal (constant-temperature) compression of the gas. The path from B to C describes an isobaric (constant-pressure) expansion of the gas. The path from C to A describes an isovolumic (constant-volume) decompression.

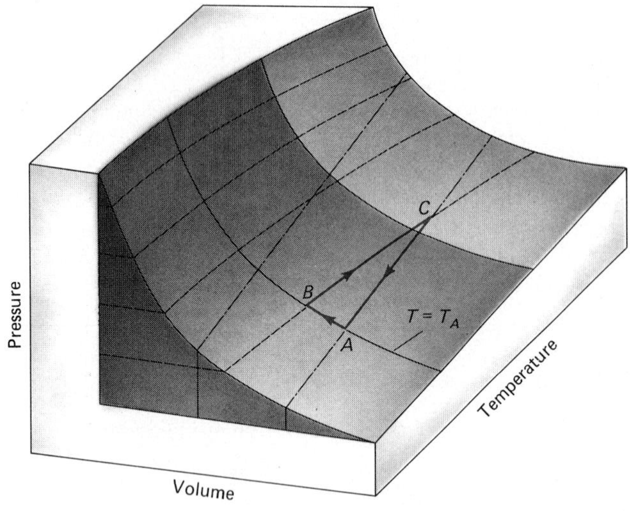

Figure 22.7

The P–V–T surface for carbon dioxide. The dashed line marks the critical isotherm. The appearance of the surface, when viewed along the volume axis, is shown in Figure 22.9.

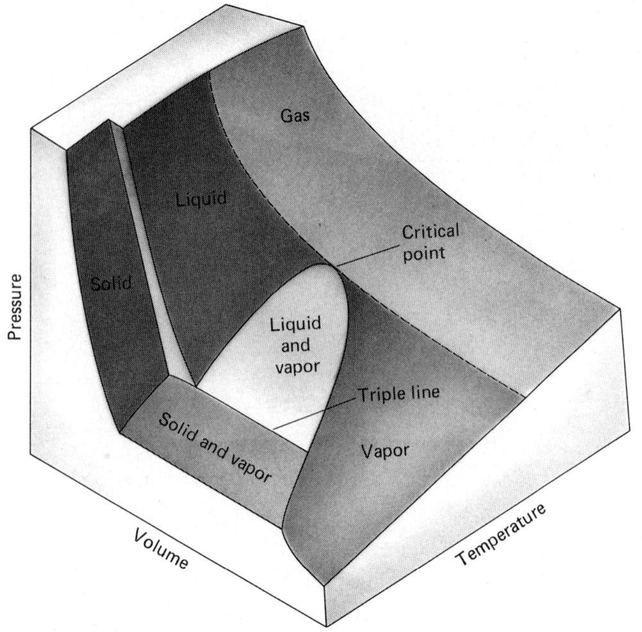

Example 5 **Ideal Gas Cycle**

Consider 1 mole of air (treated as an ideal gas) that is carried through the cycle of Figure 22.6. Let point A correspond to the conditions

$$P_A = 1 \text{ atm} \qquad T_A = 273 \text{ K} \qquad V_A = 10 \text{ m}^3$$

Suppose that the isothermal compression $A \rightarrow B$ reduces the volume to 5 m^3. It follows from the equation of state $PV = RT$ that the pressure must double if V is cut in half while T remains constant. Thus, the conditions at point B are

$$P_B = 2 \text{ atm} \qquad V_B = 5 \text{ m}^3 \qquad T_B = T_A = 273 \text{ K}$$

The path from B to C proceeds at constant pressure. From Figure 22.6 we see that the volume doubles from B to C, returning to its original value of 10 m^3. The equation of state shows that the temperature must also double if pressure remains constant. Thus,

$$P_C = 2 \text{ atm} \qquad V_C = 10 \text{ m}^3 \qquad T_C = 546 \text{ K}$$

Finally, $C \rightarrow A$ proceeds at a constant volume of 10 m^3. Heat is rejected and the pressure and temperature return to their original values of 1 atm and 273 K.

Figures 22.7 and 22.8 reveal that certain portions of the P–V–T surface correspond to states where two phases coexist in thermal equilibrium. An ice cube in equilibrium with water is a familiar example of such a condition. The *triple line* marks states where liquid, solid, and vapor* simultaneously exist in equilibrium. The *critical point* marks the upper terminus of the liquid–vapor coexistence region. It is located at a unique pressure, temperature, and molar volume characteristic of the substance (P_C, T_C, V_C). For example, for water $P_C = 217.7$ atm, $T_C = 647.3$ K, and $V_C = 5.7 \times 10^{-5}$ m^3. If you look at a liquid confined in a test tube, the meniscus marks the liquid–vapor boundary. This meniscus disappears at the critical point. If a gas is compressed at a temperature in excess of the critical temperature, it never undergoes condensation; it remains a

*The word *vapor* is often used to describe a gas that is in equilibrium with the liquid or solid phase. We use the words *vapor* and *gas* interchangeably.

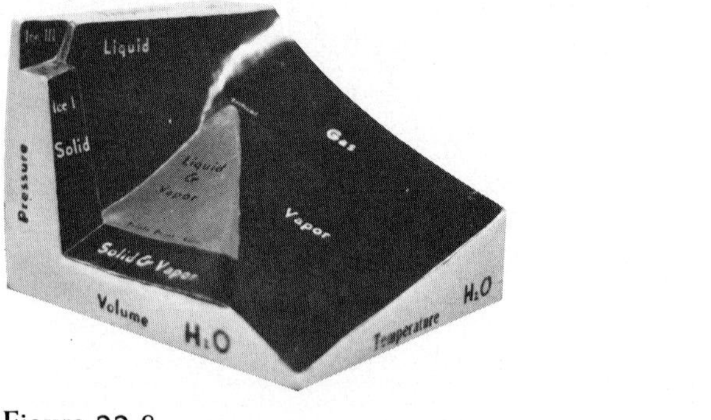

Figure 22.8

A model of the P–V–T surface for water.

gas. If a gas is compressed at a temperature less than the critical temperature, it separates into liquid and vapor phases.

In many situations it is convenient to view a projection of the P–V–T surface. Figure 22.9 shows projections of a P–V–T surface onto the P–T plane and onto the P–V plane. Figure 22.10 shows the P–T diagram for carbon dioxide. The triple line projects onto the P–T diagram as the triple point. The portion of the P–V–T surface labeled "liquid and vapor" in Figure 22.9 appears as a line on the P–T diagram—the vaporization curve. The other coexistence regions likewise appear as lines on the P–T plane. The solid–liquid region projects onto the P–T plane as the line designated "melting curve," or the fusion line. The solid–gas region projects onto the P–T plane as the sublimation curve. The physical significance of these various lines on the P–T diagram is discussed in the next section.

Questions

8. Two different gases (which behave as ideal gases) occupy equal volumes and have equal temperatures. The pressure of one gas is twice that of the other. It follows from the ideal gas equation of state

$$PV = nRT$$

that there must be twice as many moles of one gas as there are moles of the other. Suppose that you have

Figure 22.10

P–T diagram for carbon dioxide (not to scale). The lines define conditions for stable equilibrium between two phases. Their intersection is the triple point (5.1 atm, 216.6 K). Carbon dioxide is typical of most pure substances in that it expands upon melting. One consequence is that the slope dP/dT of its melting curve is positive ($dP/dT > 0$). The horizontal arrow at the normal sublimation point indicates that CO_2 sublimes at 1 atm (194.6K). If the pressure is raised above 5.1 atm, heating the solid may result in melting. The values at the critical point are 72.9 atm and 309.1 K.

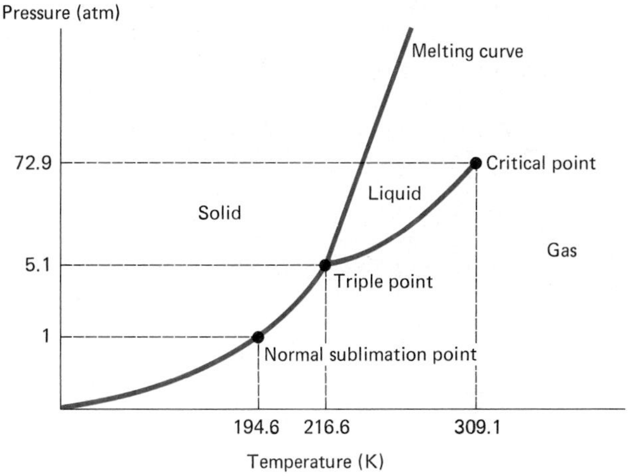

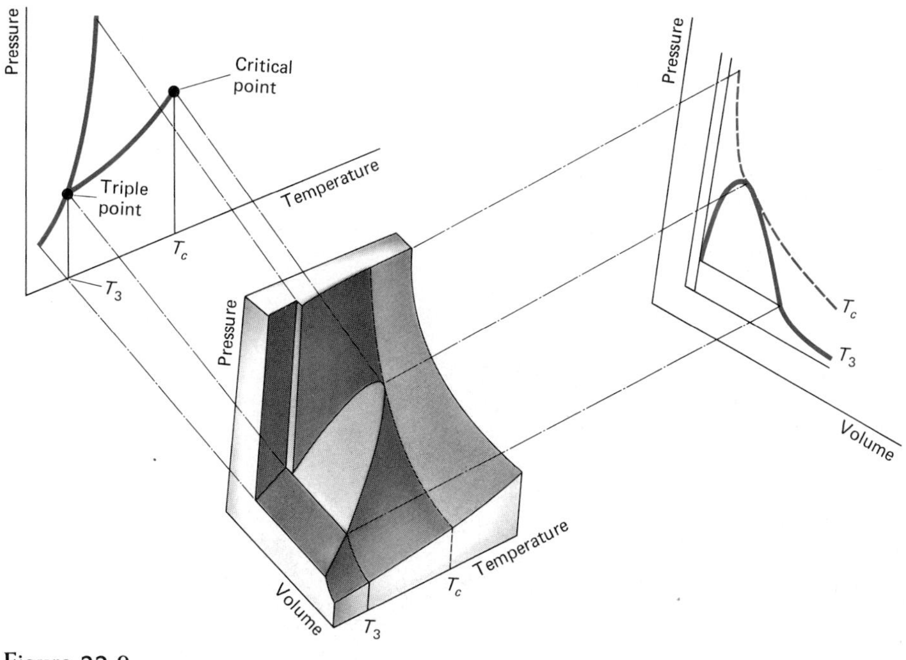

Figure 22.9

Projections of the P–V–T surface onto a P–V diagram and a P–T diagram.

2 moles of one solid at a given volume and temperature and 1 mole of another solid at the same volume and temperature. Would you expect the pressures of the solids to differ by a factor of 2? Would you expect the volume of either solid to change significantly if you doubled its pressure?

9. Make a sketch of the P–T diagram for carbon dioxide (Figure 22.10). Use it to show how you could start with gaseous CO_2, at a temperature slightly above the triple-point temperature, and transform it first into a liquid and then into a solid by (a) compressing it isothermally or (b) cooling it isobarically.

22.4 Change of Phase

A solid can be heated until it melts and becomes a liquid. The liquid in turn can be heated until it reaches its boiling point and vaporizes. For pure substances, such phase changes occur at a fixed temperature. There is an exchange of heat between the substance and its environment, but the temperature remains constant during the change of phase. The *heat of fusion*, L_f, is the amount of heat per kilogram absorbed in converting a solid to a liquid. The *heat of vaporization*, L_v, is the amount of heat per kilogram absorbed in converting a liquid to a vapor. The process called *sublimation* is one in which a solid is converted to a vapor without passing through the liquid phase. The sublimation of "dry ice" (solid CO_2) is a familiar example. The sublimation process involves a *heat of sublimation*, L_s, the amount of heat per kilogram absorbed in the solid-to-vapor conversion. Table 22.3 lists latent heats for various substances.

Figure 22.11

The heating curve for a sample of m kg of ice. The initial temperature (point A) is -5°C. Heat is added at a constant rate, making the absorbed heat (Q) proportional to t, the heating time. At B the ice has reached the melting point. During the change of phase (melting) the temperature remains constant (0°C). At C the ice has melted completely. The specific heat of ice is about half that of liquid water. This is reflected in the slope of the heating curve. The slope from A to B is approximately twice that from C to D.

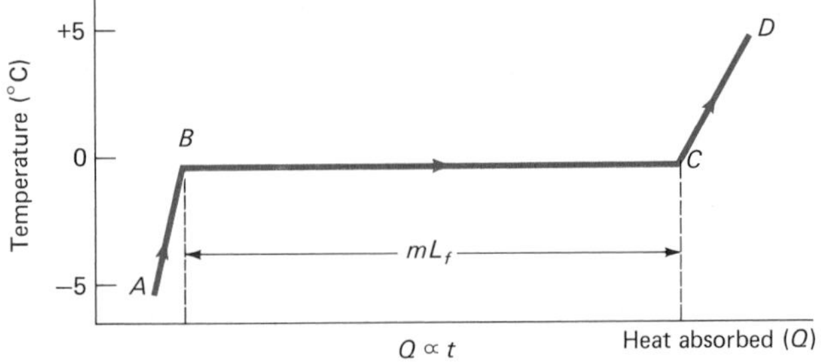

A heating curve illustrates the relationship between temperature and heat absorbed. Figure 22.11 shows the heating curve for ice and water near the melting point. It is a plot of temperature versus Q, the amount of heat absorbed. If heat is added at a constant rate, the amount of heat absorbed is directly proportional to the heating time, t. The heating curve shows that the sample starts as ice at -5°C. The addition of heat at a constant rate results in a uniform temperature rise until the ice reaches 0°C. No further increase of temperature results until all of the ice has melted. For a sample of m kg, the total heat absorbed during the phase change is mL_f, as indicated in Figure

Table 22.3
Latent heats

Substance	Normal Melting Point (K)	Latent Heat of Fusion (kJ/kg)	Normal Boiling Point (K)	Latent Heat of Vaporization (kJ/kg)
Water	273.16	334	373.16	2,260
Silver	1234	111	2485	2,360
Aluminum	931.7	399	2600	10,500
Gold	1336.16	64.4	2933	62,300
Cesium	301.9	15.7	963	514
Copper	1356.2	205	2868	4,800
Germanium	1232	479	2980	3,920
Lithium	459	416	1640	19,600
Sodium	371	115	1187	4,260
Lead	600.6	23.1	2023	859
Silicon	1683	1650	2750	10,600
Zinc	692.7	102	1180	1,760

22.11. Once all of the ice has melted, the continued addition of heat causes the temperature rise to resume. The slope of the heating curve is inversely proportional to the specific heat. The temperature rises most slowly when the specific heat is greatest.

Example 6 **The Heat of Fusion for Ice**

A student performs an experiment in which a heating curve is used to determine the latent heat of ice. Heat is added at a constant rate of $dQ/dt = 100$ W to $m = 20$ gm of ice. The time during which the temperature remains constant (melting time) is 66 s. The total heat absorbed during melting may be expressed as

$$mL_f = \left(\frac{dQ}{dt}\right) t_{\text{melt}}$$

which gives

$$L_f = \frac{(100 \text{ J/s})(66 \text{ s})}{0.02 \text{ kg}} = 330 \text{ kJ/kg}$$

This compares favorably with the accepted value of 333.6 kJ/kg.

The temperature at which a solid melts is called the *melting point.* The melting point varies with pressure. For example, Figure 22.10 shows that the melting point of carbon dioxide increases with pressure. The melting point at a pressure of 1 atm is called the *normal* melting point. Similar terms are used for boiling and sublimation. The temperature at which a liquid boils is called the *boiling point* and the temperature at which a solid sublimes is called the *sublimation point.* The boiling point at a pressure of 1 atm is called the *normal* boiling point. The heats of fusion, vaporization, and sublimation also vary with pressure. For example, as a substance approaches the critical point, two phases become one and the heat of vaporization drops to zero (see Figure 22.7).

Example 7 **A Heat Calculation Involving Melting**

A mixture composed of 0.160 kg of ice and 0.320 kg of water is initially at 0°C. What is the amount of heat required to melt the ice and then raise the water to a temperature of 16°C? The latent heat of fusion of ice is 334 kJ/kg (Table 22.3). The heat required to melt the ice is

$$Q_{\text{melt}} = m_{\text{ice}} L_f = 0.160 \text{ kg} \cdot 334 \frac{\text{kJ}}{\text{kg}} = 53.4 \text{ kJ}$$

The heat required to raise the 0.480 kg of water from 0°C to 16°C is

$$Q_{\text{water}} = m_{\text{water}} C_{\text{water}} \Delta T$$

$$= 0.480 \text{ kg } 4.19 \frac{\text{kJ}}{\text{kg} \cdot \text{C}^\circ} \cdot 16\text{C}^\circ = 32.2 \text{ kJ}$$

The total heat required is 85.6 kJ.

Most of the heat absorbed during a phase change is used to alter the microscopic structure of the substance. For example, when a solid melts most of the heat absorbed (the heat of fusion) is used to break the bonds between atoms. A tiny fraction of the heat of fusion is converted into the $P\,dV$ work associated with a change of volume. In sublimation and vaporization, however, the volume of the substance increases substantially and a significant portion of the heat absorbed is used to perform the work of expansion.

Example 8 **$P\,dV$ Work during Vaporization**

The heat of vaporization for water at a pressure of 1 atm is 2256 kJ/kg. The vaporization of water at the boiling point (373.15 K) increases its volume by a factor of 1761. Thus, 1 kg of liquid, which occupies 0.001 m^3, expands as a vapor to a volume of approximately 1.76 m^3. A portion of the heat of vaporization goes into the $P\,dV$ work of expansion. The remainder is used to break molecular bonds and shows up as a change of internal energy. We apply the first law of thermodynamics,

$$Q = U_f - U_i + W \tag{22.37}$$

to the process in which 1 kg of water at its boiling point absorbs 2256 kJ and is vaporized:

$$Q = mL_v = 2256 \text{ kJ} = \text{heat of vaporization of 1 kg}$$

$$U_f - U_i = U_{\text{vapor}} - U_{\text{liquid}} = \text{internal energy change between liquid and vapor phases}$$

$$W = \int_{V_l}^{V_v} P\,dV = \text{work of expansion against atmosphere}$$

Since the vapor expands at constant pressure the work expansion is given by

$$W = \int_{V_l}^{V_v} P\,dV = P(V_v - V_l)$$

Taking $P = 1 \text{ atm} \simeq 10^5 \text{ N/m}^2$ and $V_v - V_l \simeq 1.76 \text{ m}^3$ gives

$$W \simeq 176 \text{ kJ}$$

Comparing $W = 176$ kJ with $mL_v = 2256$ kJ, we conclude that about 8% of the heat of vaporization goes into $P\,dV$ work. The remaining 92% is accounted for by the increase in internal energy.

Vaporization

The vaporization of a liquid can be readily understood from the microscopic viewpoint. By virtue of their kinetic energy some molecules are able to overcome the attractive intermolecular forces, escape the liquid, and become part of the vapor atmosphere above the liquid surface. When the vapor and liquid are in equilibrium we say the vapor is *saturated.*

In general, the pressure just above the surface of a liquid that is open to the atmosphere is partially due to the surrounding air and partially due to the vapor. For example, suppose you set a small pan of water on a heating element in a room where the atmospheric pressure is 1 atm, and slowly heat the water. The pressure throughout the room, including the region just above the surface of the water, will remain 1 atm. At 60°C the pressure of the water vapor is about $\frac{1}{4}$ atm. At this temperature, water vapor exerts a pressure of $\frac{1}{4}$ atm and air molecules exert a pressure of $\frac{3}{4}$ atm. At 82°C the vapor pressure of water is about $\frac{1}{2}$ atm. Thus, when the temperature of the water reaches 82°C the 1-atm pressure just above the liquid surface is due in equal parts to the water vapor and the air molecules. As the temperature approaches 100°C, the water vapor contributes an increasingly larger fraction of the total pressure of 1 atm. When the water reaches 100°C, the vapor pressure equals 1 atm. The atmosphere immediately above the liquid contains only water vapor, and we say that the water has reached its normal boiling point. If heating continues, the water will "boil away." In other words, the liquid will continue to escape into the surrounding atmosphere, while maintaining a pressure of 1 atm at the surface of the liquid.

The saturated vapor pressure depends on two things: (1) the physical nature of the molecules, and (2) the temperature. The lower the temperature, the lower is the saturated vapor pressure. In general, the boiling point of a liquid is the temperature at which the vapor pressure equals the surrounding atmospheric pressure. Consequently, if the atmospheric pressure is reduced, the boiling point is reduced. For example, if a pan of water is placed in a sealed container and the air is pumped out, the water begins to boil even though no heat is added. Pumping away the air reduces the surrounding pressure until it equals the saturated vapor pressure, at which point boiling occurs. One consequence of this pressure dependence of the boiling point is that mountain climbers may find it impossible to cook certain foods properly. The atmospheric pressure decreases as they climb upward, and at high elevations the boiling point may be reduced to a temperature too low to cook foods properly.

Cooling via Evaporation

The relationship between vapor pressure and temperature is important in cryogenics, the science of low-temperature phenomena. If the vapor above a cryogenic liquid is pumped away, liquid evaporates in an attempt to maintain equilibrium. But vaporization requires heat and the only source of the heat of vaporization is the liquid itself. The vapor carries away thermal energy and the liquid is cooled. Thus, by using a pump to maintain a low vapor pressure, we can lower the temperature of a liquid—that is, we can cool it by forced evaporation. For example, the normal boiling point of liquid helium is 4.2 K. The temperature of liquid helium may be lowered to less than 1 K by forced evaporation.

The motion of the "drinking duck" (Figure 22.12) relies on cooling by evaporation and on the fact that the saturated vapor pressure depends strongly on temperature.

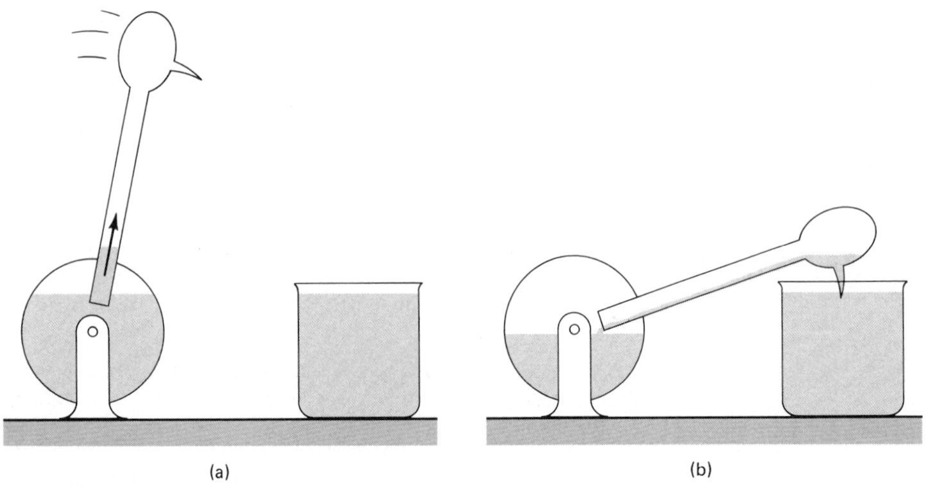

Figure 22.12

(a) Evaporation cools the head of the duck, causing the vapor pressure in the head and stem to decrease. The vapor pressure above the fluid in the body of the duck thereby becomes greater than the vapor pressure in the stem and head. This greater pressure in the body forces liquid upward in the stem and causes the center of gravity of the duck to rise. (b) The duck eventually dips forward. This allows it to wet its beak again. It also exposes the lower end of the stem, allowing fluid in the head to be replaced by vapor from the body. Thus, the center of gravity is lowered and the duck tips upward and the vapor pressures in the head and body are once again in near equilibrium. The "drinking" action continues as long as the beak remains moist and evaporation continues.

The beak of the duck is covered with a layer of an absorptive fabric, much like an ink blotter. If the beak is moistened and atmospheric conditions are such as to permit the liquid to evaporate, the duck begins to sway on its perch, and finally tips forward to "drink" from a cup of water. It then rights itself. So long as the beak is moist the drinking action continues. The fluid filling the duck is highly volatile (usually ethyl ether). The duck's behavior is a consequence of the cooling action of evaporation. Initially, the vapor trapped in the body of the duck is at the same temperature and pressure as that in the head. Evaporation cools the head and the ensuing condensation reduces the vapor pressure there. The vapor pressure in the body is then greater than the vapor pressure in the head. The greater pressure in the body forces liquid upward in the stem and causes the center of gravity of the duck to rise. As the duck loses its balance and tips forward, the lower end of the stem breaks contact with the fluid in the body. The vapor itself then rises through the tube, allowing the liquid to drain and permitting the duck to right itself. The action of the duck continues as long as the beak remains moist.

Plastic Flow, Melting, and Mountains

In Section 16.3 we showed that there is a limit to the depth of a hole in the ground. This limit exists because the pressure increases with depth, eventually reaching a value sufficient to crush any material. A similar effect limits the height of mountains. The limiting mountain height can be estimated by the following energy argument: Imagine that you set out to build a mountain by stacking blocks of granite one on top of the other. As the height of your mountain increases, the pressure and temperature at its base increase. Eventually, the pressure and temperature reach a value at which plastic flow occurs—the bottom block turns into something resembling "silly putty." Once plastic flow occurs at the base, your mountain has achieved its maximum height.

The amount of energy required to induce plastic flow is roughly the same as the heat of fusion, and we shall assume that they are equal. The energy required to melt a block of mass m is mL_f. The work done to lift a block of mass m to a height h is mg. So long as

$$mgh < mL_f \qquad (22.38)$$

it is energetically favorable to stack up blocks. Nature favors the configuration of minimum energy, and so the height of the mountain can be increased so long as the inequality of Eq. 22.38 is satisfied. The limiting height is therefore set by the condition that the work done to lift a block to the top equals the energy required to melt a block at the base. Thus, we set

$$mgh_{\text{max}} = mL_f \qquad (22.39)$$

The heat of fusion for rocks is roughly 250 kJ/kg. The resulting limit is

$$h_{\text{max}} = \frac{L_f}{g} = \frac{250\ \text{kJ/kg}}{9.80\ \text{m/s}^2} \simeq 25\ \text{km}$$

The actual limit is about one half this figure. Our calculated limit is too high, primarily because the energy required to produce plastic flow is less than L_f.

Questions

10. Explain why you may experience a chill upon emerging from a shower.

11. The gravitational acceleration on the moon is 1.6 m/s^2. How would this affect the maximum height of lunar mountains? Try to find out how heights of the taller lunar mountains compare with their terrestrial counterparts.

Summary

The coefficient of linear expansion (α) relates the change in length (ΔL) to the length (L) and the temperature change (ΔT):

$$\Delta L = \alpha L\, \Delta T \qquad (22.1)$$

The coefficient of volume expansion (β) relates the change in volume (ΔV) to the volume (V) and the temperature change:

$$\Delta V = \beta V\, \Delta T \qquad (22.3)$$

For an isotropic solid $\beta = 3\alpha$. Typically,

$$\alpha_{\text{solid}} \simeq 10^{-5}/\text{C}^\circ$$

$$\beta_{\text{liquid}} \simeq 10^{-3}/\text{C}^\circ$$

Many basic properties of real gases are described by the ideal gas model. The ideal gas is characterized by three relations: (1) the equation of state relates the pressure (P), volume (V), Kelvin temperature (T), and the number of moles of gas (n):

$$PV = nRT \qquad (22.9)$$

(2) the internal energy (U) is directly proportional to the Kelvin temperature

$$U = mC_vT \qquad (22.18)$$

(3) the specific heat at constant volume (C_v) is independent of temperature,

$$C_v = \text{constant}$$

The adiabatic (no heat exchange) transformations of an ideal gas are governed by the following (equivalent) relations:

$$TV^{\gamma-1} = T_oV_o^{\gamma-1} \qquad (22.31)$$

$$PV^{\gamma} = P_oV_o^{\gamma} \qquad (22.34)$$

$$\frac{T}{T_o} = \left(\frac{P}{P_o}\right)^{(\gamma-1)/\gamma} \qquad (22.36)$$

The constant γ is the ratio of the specific heats at constant pressure and constant volume

$$\gamma = \frac{C_p}{C_v} \qquad (22.27)$$

The quantities T_o, P_o, and V_o denote the temperature, pressure, and volume of any particular state of the gas during the adiabatic process.

The change of phase of a pure substance involves the absorption (or ejection) of heat. Most of the heat is accounted for as a change of internal energy, being used to rearrange the internal (molecular) structure of the material. A small fraction is transformed into the $P\,dV$ work of expansion.

Suggested Reading

H. C. Jensen, Freezing Nitrogen: A Modification, *Am. J. Phys.* 36, 919 (1968).

G. H. Heilmeier, Liquid-Crystal Display Devices, *Sci. Am.* 222, 100 (April 1970).

R. L. Wild and D. C. McCollum, Dramatic Demonstration of Change of Phase, *Am. J. Phys.* 35, 540 (1967).

M. Zemansky, Why Does the Slush Form? *The Physics Teacher* 14, 517 (November 1976).

Problems

Section 22.1 Thermal Expansion

1. A gold ring has an inner diameter of 2.168 cm at a temperature of 37°C. Determine its diameter at 122°C.

2. A steel tape used by a surveyor is made of Invar ($\alpha = 0.7 \times 10^{-6}$/C°). The tape was calibrated at a temperature of 25°C. The markings on the tape have a width of 0.04 mm. On a day when the temperature is 85°F the tape indicates a distance of 61.16 m. Is the actual distance greater or smaller? By what amount? Compare the thermal expansion distance with the width of the scale ruling.

3. A brass sphere ($\alpha = 1.9 \times 10^{-5}$/C°) has a diameter of 5.000 cm at 25°C. A steel ring ($\alpha = 1.1 \times 10^{-5}$/C°) has a diameter of 4.995 cm at 25°C. At what temperature will the sphere just be able to pass through the ring (a) if the ring but not the sphere, is heated; (b) if both ring and sphere have the same temperature?

4. Figure 1 shows a circular iron casting with a gap. If the casting is heated, the iron expands ($\alpha = 1.17 \times 10^{-5}$/C°). Does the width of the gap increase or decrease? The gap width is 1.600 cm when the temperature is 30°C. Determine the gap width when the temperature rises to 190°C.

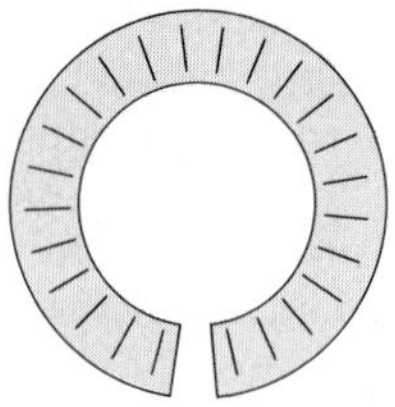

Figure 1

5. A bimetallic strip consisting of strips of aluminum and zinc is riveted together as shown in Figure 2. When the strip is heated does it curl upward or downward? Explain.

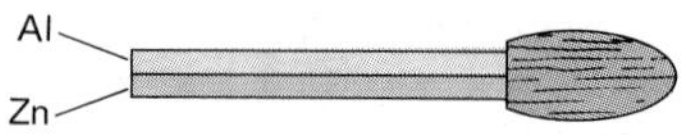

Figure 2

6. (a) The 200-in. diameter Hale telescope mirror is made of a borosilicate glass with an expansion coefficient 3.25×10^{-6}/C°. Calculate the change in diameter (in millimeters) for a 10-C° change in temperature. (b) A titanium silicate glass being used for a space telescope mirror has an expansion coefficient of less than 0.03×10^{-6}/C°. Calculate the maximum change in diameter (in millimeters) of a 200-in. mirror of this glass for a 10-C° change in temperature.

7. The period of a pendulum is proportional to the square root of its length. A "seconds pendulum" has a period of precisely 1 s and executes 86,400 complete oscillations in 1 day. The pendulum coefficient of linear expansion is 1.2×10^{-5}/C°. The pendulum is calibrated at a temperature of 20°C. How many oscillations does it execute in 1 day when the temperature is held fixed at 40°C?

Section 22.2 The Ideal Gas

8. A gas at a pressure of 2.11×10^5 Pa occupies a volume of 0.116 m^3. The gas temperature is 284 K. Assuming the gas is ideal, how many moles are present?

9. One mole of an ideal gas expands isothermally at a temperature of 294 K. The initial pressure is 2.20×10^5 Pa. The final pressure is 1.24×10^5 Pa. Determine the initial and final volumes.

10. Two moles of an ideal gas occupy a volume of 0.017 m^3 at a temperature of 288 K. (a) Determine the gas pressure. (b) The gas is compressed isobarically to half its initial volume. Determine its final temperature. (c) Does the internal energy increase or decrease?

11. A mole of helium expands from 2 atm pressure to 1 atm pressure. The initial temperature is 20°C. Treat the helium as an ideal gas. (a) Calculate the initial volume. Assuming the expansion to be isothermal, (b) calculate the final volume. Assuming the expansion to be adiabatic, (c) calculate the final volume and (d) the final temperature.

12. Assume that helium behaves like an ideal gas. Consult Table 22.2 for values of MC_v and MC_p. Determine the amount of heat that 1 mole of helium must absorb in order to raise its temperature from 280 K to 290 K when (a) the pressure is held constant; (b) the volume is held constant. (c) Determine the increases in internal energy for (a) and (b).

13. Air expands adiabatically from $T_0 = 300$ K, $P_0 = 100$ atm to a final pressure of 1 atm. Treat the gas as ideal and determine the final temperature ($\gamma = 1.40$). Compare your result with the normal boiling points of oxygen and nitrogen (90.1 K and 77.3 K, respectively). Does your result necessarily mean that some of the air would liquefy?

14. Equal amounts of heat are added to 1 mole of helium and to 1 mole of hydrogen, both at constant pressure. Which gas undergoes the larger temperature change? Would your answer be different if equal amounts of heat were added to the helium at constant pressure and to the hydrogen at constant volume? (*Hint:* Consult Table 22.2)

15. Air initially at 27°C expands adiabatically, tripling its volume. Its initial pressure is 4 atm. Treat the gas as ideal and determine its final pressure and temperature ($\gamma = 1.40$).

16. An ideal gas expands, doubling its volume. Among the many ways this can be accomplished are (i) isobarically, (ii) isothermally, and (iii) adiabatically. Determine which of these types of expansion (a) results in the largest and smallest changes in the gas temperature (regard a temperature decrease as smaller than no change); (b) results in the largest and smallest amounts of work being performed against the surroundings; (c) requires the largest and smallest heat absorption by the gas. Where possible, use qualitative arguments based on the P–V diagram, knowledge of U, and so on, rather than quantitative calculations.

17. Use Eqs. 22.33 and 22.34 to derive the result

$$\frac{T}{T_0} = \left(\frac{P}{P_0}\right)^{(\gamma-1)/\gamma} \qquad (22.36)$$

describing the path of an adiabatic process for the ideal gas.

18. Use the pressure and temperature data given in Example 4 to calculate the final temperature for an ideal gas with $\gamma = 1.67$, a value appropriate for monatomic gases. (a) Is the cooling larger or smaller for a monatomic gas than it is for a diatomic gas (for diatomic gases $\gamma = 1.4$)? (b) What would the final temperature be in the limit as $\gamma \to 1$?

Section 22.3 P–V–T Surfaces

19. Figure 3 shows the P–T diagram for water. Use the diagram to explain how ice at −20°C can melt under pressure even though its temperature remains constant. (This is the mechanism by which glaciers "flow" around large obstacles.)

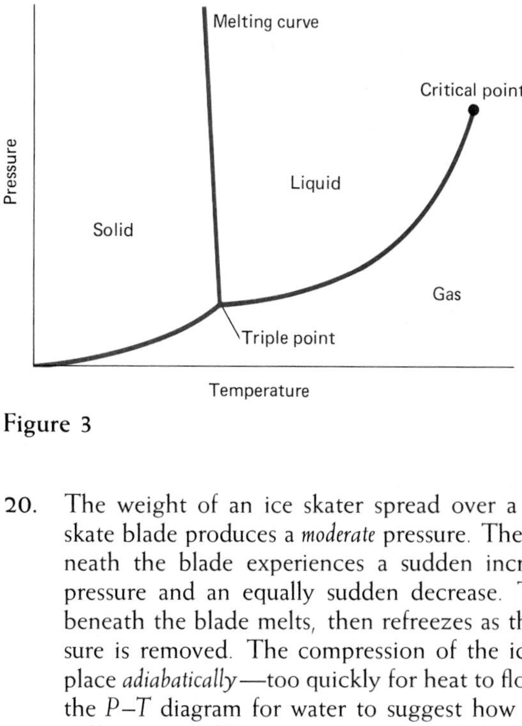

Figure 3

20. The weight of an ice skater spread over a narrow skate blade produces a *moderate* pressure. The ice beneath the blade experiences a sudden increase in pressure and an equally sudden decrease. The ice beneath the blade melts, then refreezes as the pressure is removed. The compression of the ice takes place *adiabatically*—too quickly for heat to flow. Use the P–T diagram for water to suggest how ice initially at −20°C can melt under adiabatic compression. (*Hint:* The process is *not* isothermal.) Would you expect the temperature to rise or fall during the adiabatic compression?

Section 22.4 Change of Phase

21. An immersion heater is placed in a mixture of 100 gm of ice and 150 gm of water (at 0°C). The heater converts electrical energy into heat at a rate of 1000 W. Assume that all of the heat is transferred to the water–ice mixture. (This ignores the heat absorbed by the heater coil and container.) Determine the time intervals required (a) to melt the ice (L_f = 333.6 kJ/kg); (b) to bring the water to its normal boiling point (C_p = 4.19 kJ/kg · C°); (c) to completely evaporate the water (L_v = 2256 kJ/kg).

22. A block of ice in a picnic cooler maintains a temperature of 0°C. The ice melts at the rate of 6 gm/min. At what rate (watts) is thermal energy leaking into the cooler?

23. An insulated glass holds 300 gm of water and 100 gm of ice. When the glass is left undisturbed, it is found that the ice melts completely in 31 min. Calculate the thermal power input to the ice.

24. A mixture composed of 0.10 kg of ice and 0.25 kg of water is initially at a temperature of 0°C. Enough heat is added to completely melt the ice and raise the 0.35 kg of water to a temperature of 15°C. Determine the amount of heat added.

25. A block of aluminum with a mass of 2.20 kg is heated until it reaches its melting point. How much additional heat must it absorb in order to melt?

26. A drop of liquid sodium with a mass of 0.130 gm is heated until it reaches its boiling point. How much additional heat must it absorb in order to vaporize?

27. A sample of dry ice at the normal sublimation point of 194.6 K absorbs heat at a rate of 100 W. Its mass decreases via sublimation from 200 gm to 100 gm in 9 min 21 s. Determine the heat of sublimation (tabulated value is 573 kJ/kg).

28. A company ships meats in insulated chests packed with dry ice. Heat "leaks" into the chests at a rate of 0.7 W. Estimate the amount (in kg) of dry ice needed to protect the meat for 1 week. The heat of sublimation of dry ice is 573 kJ/kg.

29. Comets are believed to be composed largely of ice and frozen organic molecules. Assume that a comet composed entirely of ice strikes the earth when moving at a speed of 30 km/s (the orbital speed of the earth). Assume further that one half of its kinetic energy is converted into heat that is absorbed by the comet. (a) Determine the amount of heat absorbed by 1 kg of the comet. The temperature of the comet before impact is −20°C. Is the heat absorbed sufficient (b) to melt the comet; (c) to "boil" it? (The average specific heat of ice is 2.0 kJ/kg · C°; the latent heat of fusion of ice is 334 kJ/kg.) Describe what you would expect to happen if a comet 10 km in radius struck a populated area of the earth.

chapter

23 HEAT TRANSFER

Preview

A transfer of heat occurs whenever an object is not in thermal equilibrium with its surroundings. Applications of heat transfer are found in virtually every phase of engineering and in many areas of the life sciences. The three forms of heat transfer are conduction, convection, and thermal radiation.

Conduction is a transfer of heat without any flow of matter. For example, the handle of a spoon is heated by conduction when the bowl of the spoon is submerged in hot coffee. Heat conduction is well understood and can be given a concise mathematical formulation.

Convection is a form of heat flow that can occur only in fluids. Heat is transported by currents of fluid that circulate between hotter and cooler regions. The convection is called free or natural when the convection currents are driven by internal forces arising from a lack of thermal equilibrium. For example, free convection can arise when a pan of water is heated on a stove. As the water at the bottom of the pan becomes warm it expands and becomes less dense than the cooler water above. As this less dense water rises it is replaced by the cooler water above. The rising warmer water and the descending cooler water constitute a convection current. Heat is transferred between the currents and their surroundings by conduction. The convection is called forced when the convective currents are driven by external forces. The circulation of air by a furnace blower is an example of forced convection.

Because convection involves both fluid motions and heat conduction its mathematical description is very complicated. We will avoid most of the mathematical complications by restricting ourselves to the description of convection that is known as Newton's law of cooling.

The third type of heat transfer, thermal radiation, involves the transfer of energy in the form of electromagnetic waves. The rate at which thermal radiation is emitted depends strongly on the temperature of the emitting object. In this chapter, we introduce the Stefan–Boltzmann law, which governs thermal radiation, and use it to show how the temperature of the sun can be determined by using only an "eyeball" measurement and an ordinary thermometer.

23.1 Conduction

The flow of heat via conduction through a brick wall or through a pane of glass is an example of heat flow in one dimension. We can let the x-axis denote the direction of heat flow and consider the flow of heat between the faces of a flat plate of thickness Δx and face area A (Figure 23.1). Let ΔT denote the temperature difference that is maintained across the thickness Δx. This temperature difference is what gives rise to the heat flow. Let q denote the rate of heat flow across the slab (q is measured in watts). We might expect that the thicker the slab, the smaller q. For small Δx, experiment bears out this expectation and shows that

$$q \propto \frac{1}{\Delta x}$$

Likewise, we would expect q to increase if the temperature difference increases. Again, experiment shows that our expectation is correct. For small ΔT

$$q \propto \Delta T$$

Finally, experiment confirms the expectation that the rate of heat flow should be directly proportional to the face area:

$$q \propto A$$

Thus, for small ΔT and small Δx, the rate of heat flow is proportional to $A\ \Delta T/\Delta x$:

$$q \propto A\frac{\Delta T}{\Delta x}$$

Figure 23.1

Heat flow by conduction. For small ΔT and Δx the rate of heat flow is proportional to $A\ \Delta T/\Delta x$. The rates of heat flow across the two faces are equal provided that the temperatures T and $T + \Delta T$ are maintained constant. For heat flow in the direction shown, ΔT is negative.

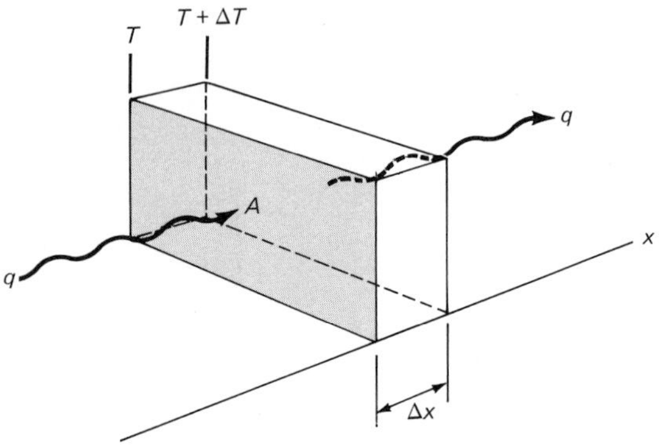

This relation can be converted into an equation by introducing a proportionality factor, k, called the *thermal conductivity*.

$$q = -kA\frac{\Delta T}{\Delta x} \tag{23.1}$$

Equation 23.1 is known as **Fourier's law.*** The minus sign accounts for the fact that heat flows from a higher temperature to a lower temperature. Thus, for example, in Figure 23.1, heat flows in the positive x-direction. For this to happen the temperature must decrease as x increases. Thus, ΔT is negative in Figure 23.1. The units of k are W/m · C°. The value of the thermal conductivity depends primarily on the physical composition of the material. A material with a large value of k would be described as a good thermal conductor. Metals are among the best conductors. A material with a small value of k would be classified as a poor thermal conductor, or a good thermal insulator. Gases and various porous plastics such as styrofoam are among the best insulators. Table 23.1 presents values of the thermal conductivities of many different materials.

Throughout this section we assume that the rate of heat flow and the temperature do not change with time. That is, we assume a *steady-state* condition. Under steady-state conditions the slab does not absorb a net amount of heat. (If it did, its temperature would change.) It follows then that the value of q must be the same for both faces. In other words, in a steady-state condition the rate of heat flow into one face of the slab is equal to the rate of heat flow out of the opposite face.

Table 23.1

Thermal conductivity*

Material	$k\left(\frac{W}{m \cdot C°}\right)$	Material	$k\left(\frac{W}{m \cdot C°}\right)$
Gases		Solids	
Air	0.0237	Glass	0.760
Carbon dioxide	0.0145	Ice	2.21
Oxygen	0.0246	Steel	16.3
Hydrogen	0.167	Iron	72.7
Helium	0.142	Aluminum	228
Methane	0.0305	Copper	386
		Silver	417
Liquid			
Water	0.566		
Solids			
Corrugated cardboard	0.064		
Asbestos	0.150		
Wood	0.190		

*Values refer to a temperature of 0°C and a pressure of 1 atm.)

Example 1 Heat Flow through a Window

We want to calculate the rate of conductive heat flow through a large glass window pane. The glass measures 1.00 m × 1.30 m × 0.50 cm. The temperature of the inner surface of the glass is 68°F (20°C). The outer surface of the glass is at 5°F (−15°C). The thermal conductivity of glass is 0.76 W/m · C°. The temperature difference is $\Delta T = -35$ C°. Using Eq. 23.1 we get

$$q = -0.76\frac{\text{W}}{\text{m}\cdot\text{C}°}\frac{(1.00 \times 1.30\ \text{m}^2)(-35\ \text{C}°)}{5 \times 10^{-3}\ \text{m}}$$

$$= 6.9\ \text{kW}$$

This enormous rate of heat flow shows that glass is not a very good thermal insulator.

The rate of heat flow through a glass window can be reduced substantially by using a thin layer of air sandwiched between two layers of glass. The trapped layer of air provides better thermal insulation than an equal thickness of glass.

Many types of insulation make use of the low thermal conductivity of air. The fluffy insulation used in the walls of homes traps air. This air provides a thick layer of material with a low thermal conductivity. The fluffy composition also restricts convective heat flow by limiting circulation of the trapped air. Even thermal underwear makes use of the low thermal conductivity of air. The numerous small holes in this underwear trap air between the body and outer garments. The hobo who sleeps beneath the pages of a newspaper also makes use of the insulating quality of air. The paper traps an insulating layer of air and reduces convection currents.

The Differential Form of Fourier's Law

We obtain the differential form of Fourier's law by replacing $\Delta T/\Delta x$ in Eq. 23.1 by the derivative dT/dx. This replacement produces

$$q = -kA\frac{dT}{dx} \tag{23.2}$$

*Jean-Baptiste-Joseph Fourier (1768–1830) is best known for the mathematical theory that introduced the Fourier series. His research on heat conduction was published in 1822 in *Analytic Theory of Heat*.

Figure 23.2

The temperature gradient dT/dx is the slope of a graph of temperature versus position. For one-dimensional heat flow $dT/dx = \Delta T/\Delta x$ is constant and the T versus x graph is a straight line.

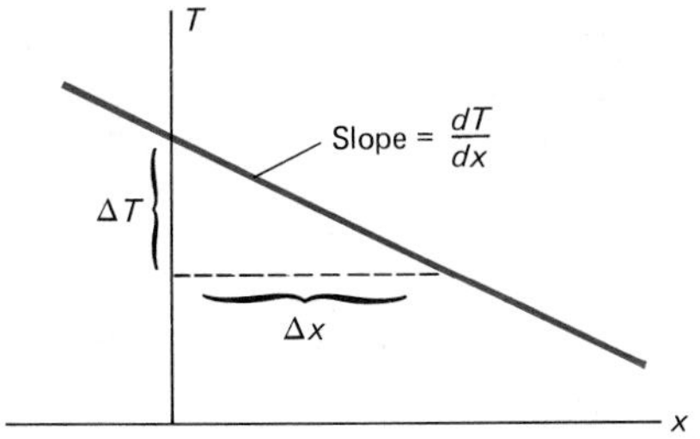

The derivative dT/dx is called the *temperature gradient* and measures how temperature changes with position. Geometrically, dT/dx is the slope of a graph of T versus x. The heat flow shown in Figure 23.1 is one-dimensional—along the x-direction. For one-dimensional heat flow the temperature gradient is a constant, provided that the material is uniform (k = constant).

$$\frac{dT}{dx} = -\frac{q}{kA} = \text{constant}$$

Consequently, the temperature decreases *linearly* with distance along the direction of heat flow, giving a straight-line graph of T versus x (Figure 23.2).

Example 2 **Temperature Variation for One-Dimensional Heat Flow**

The temperature gradient for the window in Example 1 is

$$\frac{dT}{dx} = \frac{\Delta T}{\Delta x} = -\frac{35\ \text{C}^\circ}{0.005\ \text{m}} = -7000\ \text{C}^\circ/\text{m}$$

This equation can be integrated to give

$$T = -7000x + \text{constant}$$

where x is in meters and T is in degrees Celsius. If the inner face of the window is taken to be at $x = 0$, the constant of integration must equal the temperature at that position. Thus $T_{\text{inner}} = 68°\text{F} = 20°\text{C}$ gives

$$T = (20 - 7000x)°\text{C}$$

We can check to see that this gives the correct temperature for the outer face. Taking $x = 0.5\ \text{cm} = 5 \times 10^{-3}$ m gives

$$T_{\text{outer}} = (20 - 7000 \cdot 5 \times 10^{-3})°\text{C}$$
$$= -15°\text{C} = +5°\text{F}$$

This is the outer temperature that was given in Example 1.

Heat Flow in Two Dimensions

We can apply Fourier's Law to situations where the heat flow is two-dimensional or three-dimensional by allowing for variation in the area A. For example, consider a steam

Figure 23.3

Geometry for cylindrical heat flow. The inner radius of the pipe is r_i. The cylindrical sleeve has a radius r, thickness dr, and length L. Its inner surface area is $2\pi rL$.

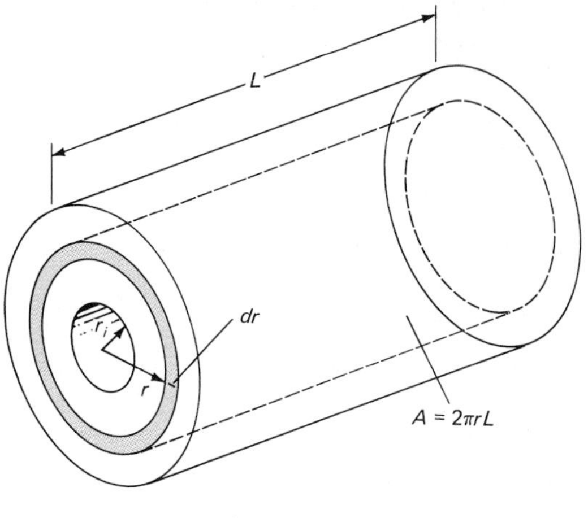

pipe in which heat flows radially outward. This type of heat flow is called *cylindrical* heat flow. The geometry of cylindrical heat flow is shown in Figure 23.3. We can conceptually divide the cylinder into a series of concentric cylindrical sleeves. The rate of heat flow through a cylindrical sleeve of radius r and thickness dr is given by Fourier's law:

$$q = -kA\frac{dT}{dr} \tag{23.3}$$

where

$$A = 2\pi rL \tag{23.4}$$

is the surface area of the cylindrical sleeve. Thus, for cylindrical heat flow, Fourier's law has the form

$$q = -2\pi kLr\frac{dT}{dr} \tag{23.5}$$

For steady-state conditions (q constant) we can solve Eq. 23.5 to find how T varies with r. Writing Eq. 23.5 as

$$dT = -\frac{q}{2\pi kL}\frac{dr}{r} \tag{23.6}$$

we integrate Eq. 23.6 from a point r_i where the temperature is T_i to some arbitrary point r where the temperature is $T(r)$,

$$\int_{T_i}^{T} dT = -\frac{q}{2\pi kL}\int_{r_i}^{r}\frac{dr}{r} \tag{23.7}$$

The result is

$$T(r) - T_i = -\frac{q}{2\pi kL}\ln\left(\frac{r}{r_i}\right) \tag{23.8}$$

This equation shows that the temperature decreases *logarithmically* with r for cylindrical heat flow.

Example 3 **Reducing Heat Loss with Insulation**

A stainless steel pipe has an inner radius of 2.0 cm and an outer radius of 2.5 cm (Figure 23.4a). The pipe carries

Figure 23.4a

Cross section of steel pipe (r_i = 2.0 cm; r = 2.5 cm).

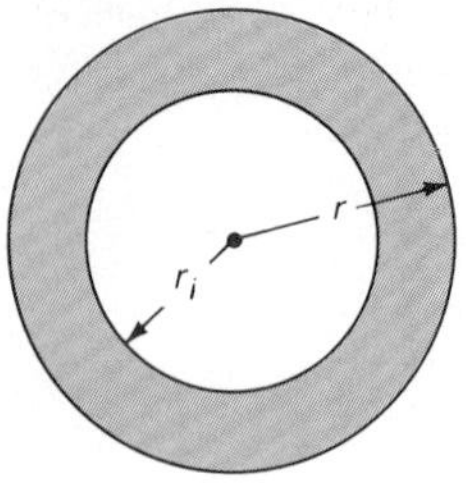

Figure 23.4b

Cross section of pipe wrapped with insulation. The values of r and r_i differ from those of Figure 23.4a. Here r_i denotes the inner radius of the insulating sleeve (r_i = 2.5 cm) and r denotes the outer radius of the sleeve.

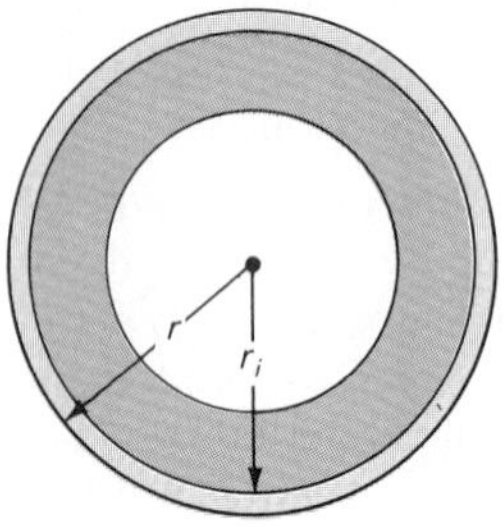

hot water at a temperature of 130°F (54.4°C). The temperature at the pipe's outer surface is 126°F (52.2°C). What is the rate of heat flow per unit length of the pipe? Solving Eq. 23.8 for q/L gives

$$\frac{q}{L} = \frac{-2\pi k[T(r) - T_i]}{\ln(r/r_i)}$$

Setting

$$r = 2.5 \text{ cm} \qquad r_i = 2.0 \text{ cm}$$

$$T(r) = 52.2°\text{C} \qquad T_i = 54.4°\text{C}$$

$$k = 19 \text{ W/m} \cdot \text{C}° \qquad \text{(for stainless steel)}$$

gives

$$\frac{q}{L} = \frac{-2\pi\ 19(52.2 - 54.4) \text{ W/m}}{\ln(2.5/2.0)} = 1200 \text{ W/m}$$

Now let's determine the thickness of insulation required to reduce this heat loss by a factor of 10 and achieve a temperature of 100°F (37.8°C) at the outer surface of the insulation. As far as the pipe is concerned, a reduction in q/L by a factor of 10 requires that the temperature difference between the inner and outer surfaces be reduced by the same factor. Thus, the original 4 F° difference is reduced to 0.4 F° so that the outer surface of the pipe is at 129.6°F (54.2°C). We are now in a position to use Eq. 23.8 again, this time for a cylindrical shell of insulation, with an inner radius of 2.5 cm and an outer radius (r) that remains to be determined (Figure 23.4b). Solving Eq. 23.8 for $\ln(r/r_i)$ gives

$$\ln\left(\frac{r}{r_i}\right) = \frac{-2\pi k[T(r) - T_i]}{q/L}.$$

Setting

$$T(r) = 37.8°\text{C} \qquad T_i = 54.2°\text{C}$$

$$q/L = 120 \text{ W/m}$$

$$k = 0.15 \text{ W/m} \cdot \text{C}° \qquad \text{(typically)}$$

gives

$$\ln\left(\frac{r}{r_i}\right) = \frac{-2\pi(0.15)(37.8 - 54.2)}{120} = 0.129$$

and

$$\frac{r}{r_i} = 1.14 \qquad r_i = 2.5 \text{ cm}$$

The required thickness of the insulation is $r - r_i$:

$$r - r_i = 1.14r_i - r_i = 3.5 \text{ mm}$$

Questions

1. For slabs of equal thickness (Δx) and equal heat flow per unit area (q/A), which of the following materials has the largest temperature difference between opposite faces: air, glass, or steel?
2. For slabs of equal thicknesses (Δx) and equal temperature differences (ΔT), which material has the largest heat flow per unit area (q/A) across it: air, glass, or steel?
3. Explain why inserting an aluminum nail into a large potato decreases the potato's baking time.
4. The ability of building materials to restrict the conductive flow of heat between two regions of different temperature is measured by a quantity called the *R-value*. As the R-value increases the heat flow decreases. How would you expect the R-value to depend on (a) thermal conductivity and (b) thickness of material?

23.2 Convection

As we have just seen, conduction can take place in both solids and fluids. Convection, on the other hand, occurs only in fluids. Unlike conduction, convection involves the flow of matter. In convective heat flow, circulating currents transfer parcels of fluid between regions of different temperature. The parcels from a higher temperature region are gradually cooled by conduction as they move through fluid with lower temperatures. If the forces that drive the convection arise from a lack of thermal equilibrium, the convection is called *free* or *natural*. A *forced* convection, on the other hand, is a convective flow that is produced by a pump or some other external agent.

Radiation, conduction, and convection all compete as heat transfer mechanisms. If radiation and conduction transport heat swiftly enough, thermal instabilities do not occur and there is no natural convection. If radiation and conduction cannot remove heat fast enough, a thermal instability arises and convection results.

Convection is commonplace in our atmosphere because gases are poor heat conductors. We can compare atmospheric convection to convection in a pan of water. Water in a pan is heated from below by a flame or by electric heating coils. Our atmosphere is also heated from below by the earth. The heating of both the water and the atmosphere is uneven. This causes some parcels of fluid to become hotter than the surrounding fluid. Conduction cannot transfer this heat through the fluid quickly enough, and so these parcels expand, lowering their density. This lowered density makes the parcels buoyant and they begin to rise, passing through the cooler, more dense fluid above them. As the warmer parcels rise, cooler parcels descend and take their places, and we have convection. Because gases are poor conductors, thermal conduction is not very effective in suppressing convection in the atmosphere. However, another factor can suppress convection. The density and temperature in our atmosphere decrease with altitude. Consequently, a hot parcel of material may become slightly less dense than its surroundings and yet still be more dense than the fluid above it. In this case, the parcel begins to rise, but when it reaches the less dense fluid above, it stops. Convection develops only when the density inside the parcel *remains* less than that of the surrounding fluid.

Ordinarily, convection helps to rid the lower atmosphere of pollutants. Convection currents carry hot polluted air upward, where it is dispersed. However, when atmospheric conditions are such that convection is inhibited, the pollutants may linger near the earth's surface as smog. The condition under which convection is inhibited is called a *temperature inversion*. A temperature inversion takes place when the normal decrease of temperature with altitude is inverted—the temperature increases with altitude. The cooler air near the earth's surface is more dense than the air above it and consequently does not rise.

Newton's Law of Cooling

Many situations of practical interest involve convective heat transfer via a fluid that is in contact with a solid surface. For example, a highway may be cooled via convection through the air currents in contact with the highway surface. Figure 23.5 suggests how the temperature changes near the surface. Most of the temperature change takes place via conduction across a thin *boundary layer* of the fluid. This boundary layer is like a fluid "skin" that clings to the solid surface. The thickness of the boundary layer is not a fixed quantity. It depends on such things as viscosity, flow speed, and whether the fluid flow is laminar or turbulent. For flowing air or water a typical boundary layer thickness is about 0.1 mm. Beyond the boundary layer the temperature changes more gradually. Convective mixing maintains a nearly uniform temperature in the bulk of the fluid.

The fact that most of the temperature change occurs by conduction through the boundary layer allows us to make a very useful, though approximate, analysis of convective heat transfer. Let's assume that the full surface-to-fluid temperature change $(T_s - T_f)$ occurs across the

Figure 23.5

The solid line suggests the actual temperature–position relation for convection near a solid surface. The dashed line shows the approximate temperature–position relation embodied in Newton's law of cooling.

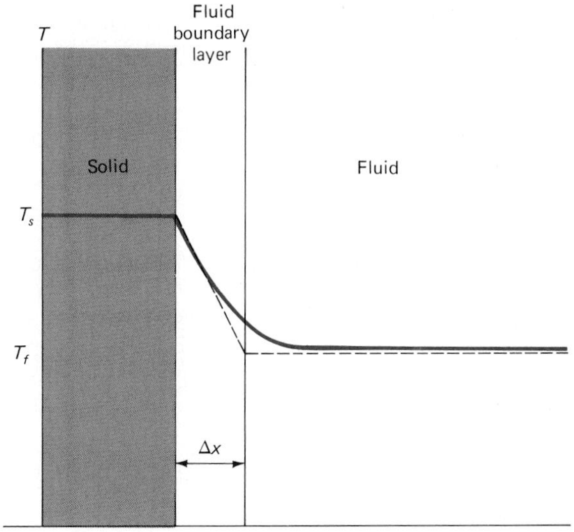

boundary layer. Such heat transfer could be described by Fourier's law:

$$q = A\frac{k}{\Delta x}(T_s - T_f) \tag{23.9}$$

where Δx is the boundary layer thickness. However, as we already noted, the thickness of the boundary layer is not a fixed quantity. For this reason the combination $k/\Delta x$ in Eq. 23.9 is replaced by a single empirical factor, h, called the *heat transfer coefficient* or *film coefficient*. The units for h are $W/m^2 \cdot C°$. The resulting equation for the rate of convective heat flow is

$$q = Ah(T_s - T_f) \tag{23.10}$$

Equation 23.10 is known as *Newton's law of cooling*. When it was originally proposed by Isaac Newton, this law was meant to describe all cooling, because at that time there was no distinction between convection and conduction. Even though Eq. 23.10 is only an approximate description

Table 23.2

Representative values of convective heat transfer coefficients (approximate)*

Type of Convection, Fluid	h ($W/m^2 \cdot C°$)
Free, air	5–25
Forced, air	10–500
Forced, water	100–15,000
Forced, boiling water	2500–25,000
Forced, condensing steam	5000–100,000

*From *Heat Transfer*, 4th ed., by J. P. Holman; McGraw-Hill, New York, 1976. Used with permission of McGraw-Hill Book Company.

of convective heat flow, it is useful in many applications where great precision is not required.

Values of h can be determined experimentally. The precise value of h depends on the type of fluid, the nature of the surface, the speed of fluid flow, and whether the flow is laminar or turbulent. Some representative values of h are given in Table 23.2. The wide variations in these values reflect the wide range of flow conditions encountered in convective heat transfer.

Example 4 **Highway Heat Flux**

A highway surface is heated to a temperature of 140°F (60°C) by the summer sun. The air temperature several feet above the surface is 88°F (31°C). The heat transfer coefficient for air flowing over the pavement is 25 W/m² · C°. What is the rate of heat flow per square meter of highway surface? From Eq. 23.10,

$$\frac{q}{A} = h(T_s - T_f) = \frac{25\ \text{W}}{\text{m}^2 \cdot \text{C}^\circ}(60 - 31)\text{C}^\circ$$

$$= 700\ \text{W/m}^2$$

Cooling Time

We can use Newton's law of cooling to study how an object tends toward thermal equilibrium with its environment. Consider an object that no longer receives heat and that loses heat to its environment through convection. Let dQ denote the amount of heat convected from the object in a time dt. The rate of convective heat flow (q) is related to dQ and dt by

$$dQ = q\, dt \qquad (23.11)$$

Using Newton's law of cooling, Eq. 23.10, we can replace q with $Ah(T_s - T_f)$ to get

$$dQ = Ah(T_s - T_f)\, dt \qquad (23.12)$$

The heat loss dQ is related to the decrease in surface temperature dT_s by

$$dQ = -mC\, dT_s \qquad (23.13)$$

where m is the mass of the object and C is its specific heat. Combining Eqs. 23.12 and 23.13 gives

$$\frac{dT_s}{T_s - T_f} = -\frac{Ah}{mC}\, dt \qquad (23.14)$$

which can be integrated* to give the following equation:

$$\ln\left(\frac{T_s - T_f}{T_o - T_f}\right) = -\frac{t}{mC/Ah} \qquad (23.15)$$

$$^*\int_{T_o}^{T_s} \frac{dT_s}{T_s - T_f} = \ln\left(\frac{T_s - T_f}{T_o - T_f}\right) = -\frac{Ah}{mC}\int_0^t dt = -\frac{t}{mC/Ah}$$

Note: It is assumed that the temperature of the surrounding fluid (T_f) remains constant.

Figure 23.6

Surface temperature (T_s) versus time (in units of mC/Ah) according to Newton's law of cooling. Cooling begins at $t = 0$ with the surface at a temperature T_o. As cooling proceeds the surface temperature approaches T_f, the temperature of the surrounding fluid.

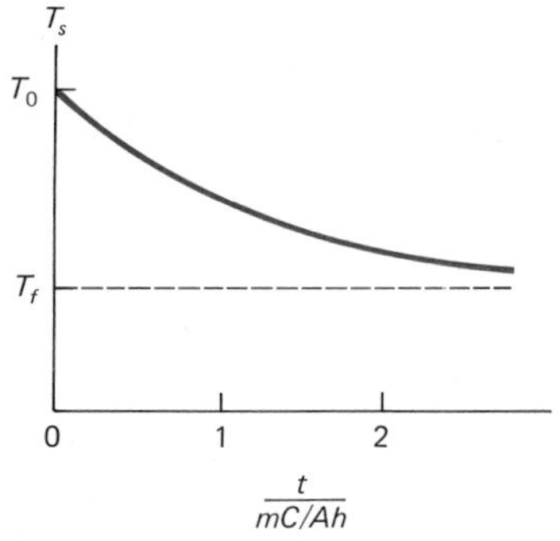

This equation relates the temperature of the surface (T_s) to the time (t) that it has been cooling via convection. An equivalent expression is

$$\frac{T_s - T_f}{T_o - T_f} = e^{-\left(\frac{t}{mC/Ah}\right)} \qquad (23.16)$$

where T_o is the initial (time = 0) temperature of the surface and T_f is the temperature of the fluid that carries away the heat. As Figure 23.6 shows, the temperature of the surface decreases steadily toward T_f.

Both sides of Eq. 23.15 are dimensionless, and so the quantity mC/Ah must have the dimensions of time. We refer to mC/Ah as the *cooling time:*

$$t_{\text{cooling}} = \frac{mC}{Ah} \qquad (23.17)$$

As Eq. 23.16 and Figure 23.6 show, t_{cooling} is the time required for the difference $T_s - T_f$ to be reduced by a factor of $e^{-1} = 0.367$.

The cooling time dependence on m, C, A, and h is reasonable: The larger the mass, m, and specific heat, C, of an object, the more slowly the object cools. Equation 23.17 verifies this by showing that $t_{\text{cooling}} \propto mC$. The larger the surface area, A, the more rapidly an object cools. Equation 23.17 also verifies this because it shows $t_{\text{cooling}} \propto 1/A$. Finally, Eq. 23.17 shows that $t_{\text{cooling}} \propto 1/h$. We expect this because the heat transfer coefficient is a measure of how quickly convection transfers heat. Thus, the larger h is, the more quickly cooling proceeds and the smaller t_{cooling} is.

Example 5 **Cooling Fins**

A power transistor is mounted on a base equipped with cooling fins. Heat is convected to the surrounding air. How much time is required for the fins to cool to a temperature of 32°C? The temperature of the fin when the transistor is switched off is 80°C. The temperature of the surrounding air is 27°C. The other data are as follows:

surface area	$A = 6.0\ \text{cm}^2 = 6.0 \times 10^{-4}\ \text{m}^2$
heat transfer coefficient	$h = 38\ \text{W/m}^2 \cdot \text{C}°$
mass	$m = 80\ \text{gm} = 0.080\ \text{kg}$
specific heat	$C = 1.2\ \text{kJ/kg} \cdot \text{C}°$

The characteristic cooling time, given by Eq. 23.17, is

$$t_{\text{cooling}} = \frac{mC}{Ah} = \frac{(0.08)(1.2)}{(6.0 \times 10^{-4})(38)} = 4.2\ \text{s}$$

Before we perform the calculation, we can anticipate that the time for the fins to cool from 80°C to 32°C will be only a few times 4.2 s. We base this expectation on the *exponential* decrease of $T_s - T_f$ with time, as shown by Eq. 23.16 and by Figure 23.6. Setting $t_{\text{cool}} = 4.2$ s, $T_s =$ 32°C, $T_f = 27$°C, and $T_o = 80$°C in Eq. 23.15 gives

$$t = -4.2 \ln\left(\frac{32 - 27}{80 - 27}\right)\ \text{s} = 9.9\ \text{s}$$

This time is only slightly more than two times 4.2 s, as we expected.

Questions

5. Certain types of solar energy collectors have a transparent glass plate above a black absorbing surface. A portion of the incident sunlight is reflected by the plate and never reaches the absorbing surface. Nevertheless, the presence of the glass plate raises the temperature of the absorber and increases the overall efficiency of the collector. What is the purpose of the plate?
6. Wood is a better insulator than glass, and yet fiberglass, a material consisting of glass fibers in resin, is used extensively in home insulation. Why aren't wood shavings used in place of fiberglass?
7. An electric fan compresses air very slightly, thereby *warming* it. Why then are you cooled when this heated air passes you?

23.3 Thermal Radiation

Every object loses energy by the emission of electromagnetic waves, at a rate that depends strongly on the object's temperature. This type of energy transfer is called *thermal radiation*. An object also absorbs a fraction of the thermal radiation incident from its environment. The temperature of an object tends to rise when the rate of absorption exceeds the rate of emission. The temperature tends to drop when the rate of emission exceeds the rate of absorption. Radiative equilibrium exists when the rates of absorption and emission are equal.

Thermal radiation has special significance in the history of science because Max Planck's theory of thermal radiation, published in 1900, gave birth to quantum theory (Chapter 42). In this section we concentrate simply on presenting and illustrating the basic law governing thermal radiation.

An ordinary object that absorbs most of the visible light reaching it appears black. A *blackbody* is an idealized system that absorbs all incident radiant energy. The rate at which a blackbody emits thermal radiation is specified by the Stefan–Boltzmann law. This law states that the rate at which energy is radiated per unit surface area is proportional to the fourth power of the Kelvin temperature of the blackbody (Figure 23.7). In equation form the *Stefan–Boltzmann law* reads

$$F = \sigma T^4 \tag{23.18}$$

where F is the energy radiated per second per square meter and is called the *surface flux*, T is the Kelvin temperature, and σ is a universal constant called *Stefan's constant*,

$$\sigma = 5.6703 \times 10^{-8}\ \frac{\text{W}}{\text{m}^2 \cdot \text{K}^4} \tag{23.19}$$

This value of Stefan's constant is calculated on the basis of Planck's theory and agrees with the experimentally determined value of σ.

The surface flux of thermal radiation from an object that is not a blackbody is usually expressed as

$$F = \epsilon\sigma T^4 \tag{23.20}$$

where ϵ is called the *emissivity* of the surface. The emissivity depends on the physical composition of the surface and on its temperature. Because a blackbody is a perfect absorber it is also a perfect emitter. In other words, its emissivity is greater than that of an object that is not a blackbody. Thus, $\epsilon = 1$ for a blackbody, but for all other objects $0 < \epsilon < 1$.

As we mentioned earlier, an object also absorbs thermal radiation from its environment (Figure 23.8). The combined effects of absorption and emission act to establish thermal equilibrium between the object and its environment. One portion of the incident flux is reflected and makes no net contribution to the total flux near the object. However, the remaining portion of the incident flux is absorbed by the object and reduces the total flux. If the environment is at a temperature T_e, then the surface flux incident on the object is $\epsilon_e \sigma T_e^4$, where ϵ_e is the emissivity

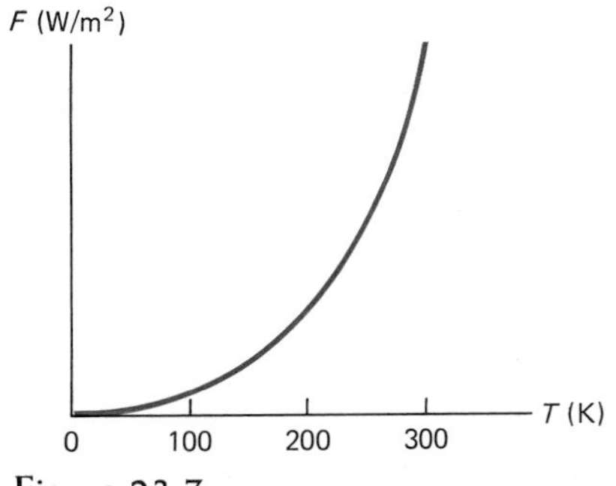

Figure 23.7

The surface flux (F) of a blackbody is related to the surface temperature (T) by the Stefan–Boltzmann law, $F = \sigma T^4$.

Figure 23.8

An object at a temperature T exchanges thermal energy with its environment, which has a temperature T_e. The rate at which thermal energy is emitted by the object per unit area is $\epsilon\sigma T^4$, where ϵ is the emissivity of the object. Thermal energy reaches the object from the environment, and some of it is absorbed. The net rate at which thermal energy leaves the object per unit area is $\epsilon\sigma(T^4 - T_e^4)$.

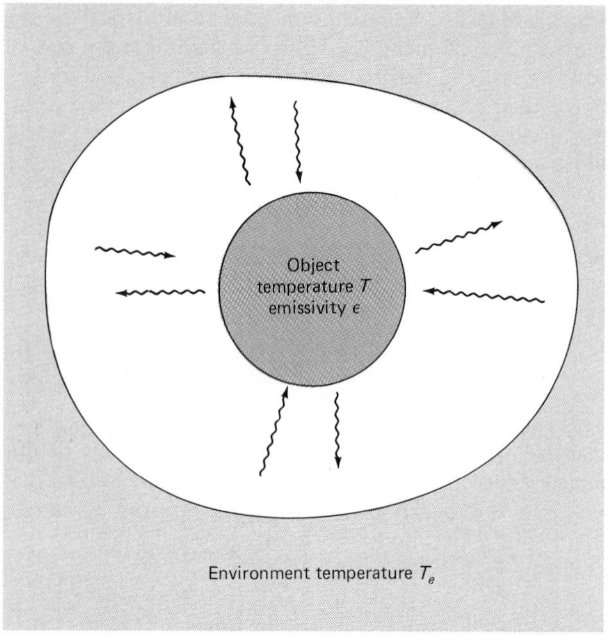

of the environment. If a fraction α of this incident flux is absorbed, the net thermal radiation flux is

$$F_{net} = \epsilon\sigma T^4 - \alpha\epsilon_e\sigma T_e^4$$

However, when $T = T_e$, the object is in thermal equilibrium with its environment. In this case F_{net} must be zero. This can be true only if $\alpha\epsilon_e = \epsilon$ for all temperatures. Thus, the net thermal radiation flux for an object at temperature T radiating energy into an environment at temperature T_e is

$$F_{net} = \epsilon\sigma(T^4 - T_e^4) \qquad (23.21)$$

Note that a positive value for F_{net} indicates a net *loss* of energy by the object ($T > T_e$).

The total rate at which an object radiates energy is called its *luminosity*, L. The luminosity of a body that radiates energy according to Eq. 23.20 is

$$L = FA = \epsilon\sigma T^4 A \qquad (23.22)$$

where A is the radiating area.

Example 6 Stellar Luminosity

The star Antares is one of a class of stars called *red giants*, so named because of their red color and enormous size. The radius of Antares is 500 times that of our sun. The surface temperature of Antares is 3800 K. Antares and other stars are among nature's closest approximations to blackbodies. Therefore we can take $\epsilon = 1$ for Antares. What is the luminosity of Antares? The luminosity is related to its surface flux and radius (R) by Eq. 23.22:

$$L = FA = F \cdot 4\pi R^2$$

The radius of our sun is 6.96×10^8 m. Because the radius of Antares is 500 times that of our sun, we set $R = 500 \times (6.96 \times 10^8 \text{ m})$. Substituting this value of R into Eq. 23.22 we get

$$L = (5.67 \times 10^{-8})(3800)^4 \cdot 4\pi(500 \times 6.96 \times 10^8)^2 \text{ W} = 1.8 \times 10^{31} \text{ W}$$

This figure is truly astronomical. The luminosity of our sun is small in comparison—it is 3.0×10^{26} W. Antares emits thermal radiation—which we see as starlight—at a rate 44,000 times that of our sun.

If an object absorbs and radiates energy at equal rates it is in *radiative equilibrium* with its environment. However, radiative equilibrium generally does not require that the object have the same temperature as the source from which it receives radiant energy.* For example, the earth is in radiative equilibrium, emitting and absorbing thermal radiation at equal rates. However, the temperature of the earth differs greatly from that of the sun, the source of the radiation that the earth absorbs. We can use the concept of radiative equilibrium to estimate the sun's surface temperature.

Example 7 Estimate of the Surface Temperature of the Sun

The rate at which the earth receives energy from the sun depends on the temperature of the sun. The rate at which the earth emits thermal radiation depends on the temperature of the earth. The condition of radiative equilibrium relates these two temperatures. Treating both the earth and the sun as blackbodies, estimate the sun's surface temperature. The luminosity of the sun is given by Eq. 23.22, with $A = 4\pi R_S^2$ for the surface area of the sun.

$$L_S = \sigma T_S^4 \cdot 4\pi R_S^2 \qquad (23.23)$$

T_S is the solar surface temperature and R_S is the solar radius. The energy radiated by the sun spreads out uniformly in all directions. A tiny fraction is intercepted by our planet (Figure 23.9). As viewed from the sun, the earth appears as a circular disk. Its absorbing area is πR_E^2, where R_E is the earth's radius. At a distance r from the sun the radiation is spread uniformly over a sphere of area $4\pi r^2$. If we take r equal to the earth-to-sun distance, the fraction of the solar luminosity absorbed by the earth is $\pi R_E^2/4\pi r^2$. The energy absorbed per second is therefore

$$\sigma T_S^4 \cdot 4\pi R_S^2 \cdot \frac{\pi R_E^2}{4\pi r^2} = \text{energy absorbed per second} \qquad (23.24)$$

*If the source is maintained at a fixed temperature, and completely surrounds the object, then radiative equilibrium does require that the object have the same temperature as the source.

Figure 23.9

Earth intercepts a small fraction of the energy radiated by the sun.

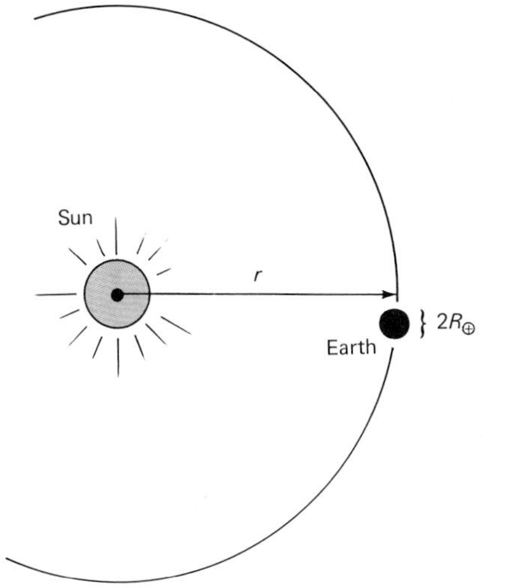

The temperature variations over the earth are only a small fraction of the average earth temperature. We therefore regard the earth as a spherical blackbody of temperature T_E. It emits thermal radiation at a rate

$$\sigma T_E^4 \cdot 4\pi R_E^2 = \text{energy emitted per second} \qquad (23.25)$$

For radiative equilibrium the rates of absorption and emission are equal. Therefore we can equate the expressions in Eqs. 23.24 and 23.25.

$$\sigma T_E^4 \cdot 4\pi R_E^2 = \sigma T_S^4\, 4\pi R_S^2 \frac{\pi R_E^2}{4\pi r^2} \qquad (23.26)$$

Solving for T_S, the temperature of the sun, we get

$$T_S = \frac{2T_E}{\sqrt{2R_S/r}} \qquad (23.27)$$

We can determine the numerator, $2T_E$, simply by using a thermometer. Let's take $T_E = 285$ K (65°F) as a representative value of earth's temperature. The denominator can be determined using an eyeball measurement because the quantity $2R_S/r$ is the *angular diameter* of the sun as viewed from the earth (Figure 23.10). The angular diameter is

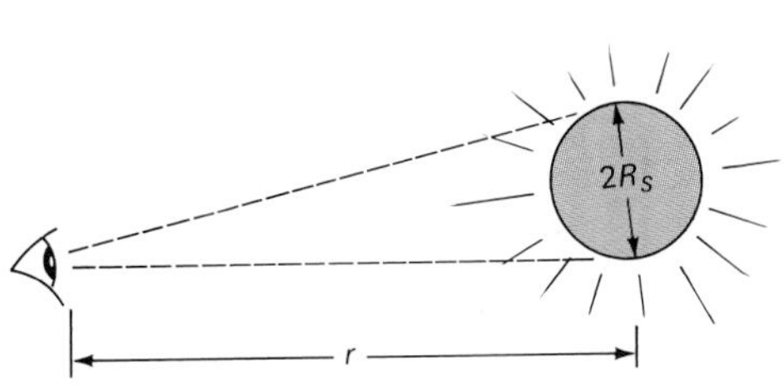

Figure 23.10

The angular diameter of the sun is $2R_S/r$.

$2R_S/r \simeq 0.00934$ rad. Using Eq. 23.27, our estimate of the surface temperature of the sun is

$$T_S = \frac{2(285\text{ K})}{\sqrt{0.00934}} \simeq 5900\text{ K}$$

quite close to the accepted figure of 5800 K.

In some instances, thermal radiation represents an undesirable energy loss. Engineers have devised ingenious ways to reduce such losses. The Thermos® bottle is a prime example (Figure 23.11). The space between the double glass walls of the Thermos bottle is evacuated, and the two surfaces facing the evacuated space are silvered. These silvered surfaces reflect thermal radiation, thereby preventing its inward flow. In addition, because the space between the walls is a vacuum, there can be no heat flow across it by convection and conduction. The primary heat flow takes place via conduction through the stopper and the glass.

The Dewar flasks used in cryogenics (Figure 23.12) are essentially open Thermos® bottles. Those flasks are designed to insulate cryogenic fluids (liquid He, H_2, N_2, etc.) against the inward flow of heat. The Dewar flask has no stopper because if the flask were closed, the slight influx

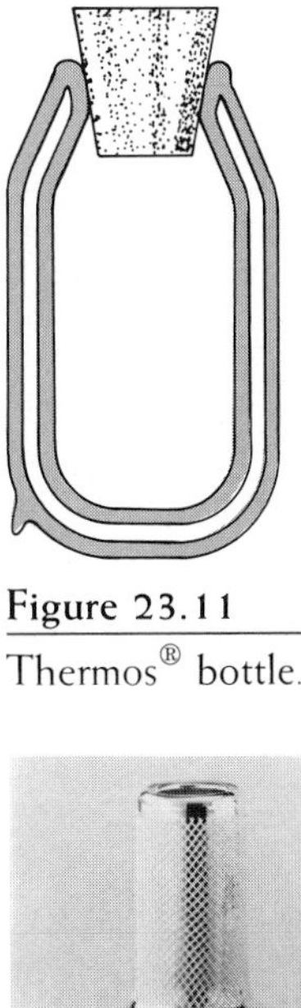

Figure 23.11

Thermos® bottle.

Figure 23.12

A Dewar flask used in handling cryogenic fluids.

of heat would build up a dangerous pressure above the liquid surface. Instead, Dewar flasks are either left open to the atmosphere or connected to a pump in order that a low pressure be maintained above the liquid surface.

Questions

8. A familiar highway sign warns Bridge Surface Freezes Before Roadway. Why does the bridge lose heat more rapidly than the rest of the highway?

9. Two sheets of paper, one black, the other white, are placed on a snow-covered surface. On a sunny day, will the snow melt more rapidly beneath the white paper or the black paper?

10. The insulation used in the walls of homes often has a reflecting surface. When installed, this surface faces outward. What is the purpose of this reflecting surface?

11. Explain how stirring increases the rate at which a cup of hot coffee cools. How would the rate of cooling be increased by combining the stirring with removing spoonfuls of coffee and allowing them to fall back into the cup? (Consider all three modes of heat transfer.)

Summary

Conduction is a transfer of heat without any flow of matter, and is governed by Fourier's law,

$$q = -kA\frac{\Delta T}{\Delta x} \qquad (23.1)$$

where q is the rate of heat flow through an area A and thickness Δx, across which the temperature change is ΔT. The quantity k is the thermal conductivity of the material. The differential form of Fourier's law replaces $\Delta T/\Delta x$ by dT/dx.

$$q = -kA\frac{dT}{dx} \qquad (23.2)$$

In convection, heat is transported by parcels of fluid that circulate between hotter and cooler regions. The forces driving convective flow can arise from a thermal instability (free convection) or can be external (forced convection). A useful, though approximate, description of convection is contained in Newton's law of cooling:

$$q = Ah(T_s - T_f) \qquad (23.10)$$

The value of the heat transfer coefficient h depends on properties of the fluid and on the nature of the fluid flow.

The temperature of an object that cools in accord with Newton's law of cooling approaches the temperature of its environment in a time

$$t_{\text{cooling}} = \frac{mC}{Ah} \qquad (23.17)$$

The variables m, C, and A denote the mass, specific heat, and surface area of the object, respectively; h is the heat transfer coefficient.

Every object emits energy in the form of electromagnetic waves at a rate that depends strongly on the object's temperature. This type of energy transfer is called thermal radiation. The surface flux of thermal radiation (F is the energy emitted per second per unit area) is given by

$$F = \epsilon\sigma T^4 \qquad (23.20)$$

where ϵ is the emissivity of the object. The limiting case $\epsilon = 1$ defines a blackbody. A blackbody is a perfect absorber and emitter. For all other objects, $0 < \epsilon < 1$.

The absorption of thermal radiation tends to drive an object toward thermal equilibrium with its environment. The net surface flux is

$$F_{\text{net}} = \epsilon\sigma(T^4 - T_e^4) \qquad (23.21)$$

where T is the temperature of the object and T_e is the temperature of its environment.

The luminosity of an object is the total rate at which it emits energy. An object emitting thermal radiation in accord with Eq. 23.20 through an area A has a luminosity

$$L = \epsilon\sigma T^4 A \qquad (23.22)$$

Suggested Reading

H. N. Pollack and D. S. Chapman, The Flow of Heat from the Earth's Interior, *Sci. Am.* 237, 60 (August 1977).

Problems

Section 23.1 Conduction

1. A temperature difference of 110 C° is maintained across a copper plate 2 cm thick. The plate is 12 cm long and 8 cm wide. Calculate the rate at which heat is conducted across the plate.

2. A large plate glass window measures 3.0 m × 2.5 m × 0.80 cm. Determine the rate of heat flow through the window when the temperature drop across it is 10 C°.

3. A large picture window consists of a sandwich of air between two panes of glass. The thermal conductivity of air is much lower than that of glass, and so a thin layer of air provides better insulation than an equal thickness of glass. What thickness of glass would give the same heat flux (q/A) and temperature drop (ΔT) as a 0.50-cm thickness of air?

4. A window 1.2 cm thick has an inside temperature of 27°C and an outside temperature of 15°C. Determine the temperature gradient across the window.

5. A pair of metal plates having equal areas and equal thicknesses are in thermal contact (Figure 1). One is made of iron, and the other is aluminum. Assume that the thermal conductivity of aluminum is exactly three times that of iron. The outer face of the iron plate is maintained at 0°C. The outer face of the aluminum plate is maintained at 60°C. Determine the temperature of the surface where the two plates are in contact.

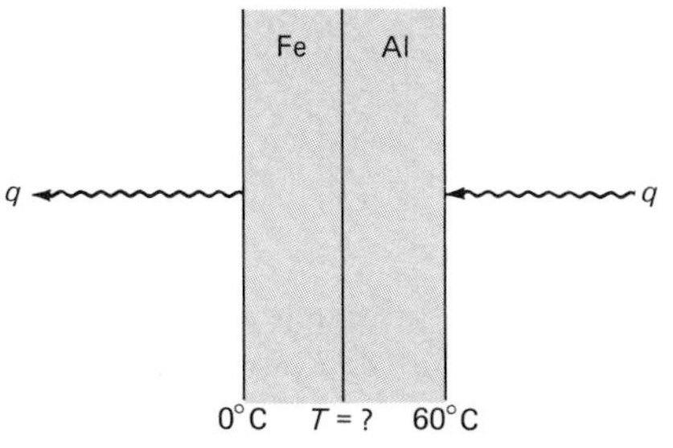

Figure 1

6. Homeowners are urged to lower their thermostats in winter in order to conserve energy. If the outside temperature is 5°C, how does the conductive heat flow through a window compare for thermostat settings of 21°C and 20°C?

7. The windows in a house consist of a single thickness of glass. One window has an area of 3 m^2 and a thickness of 0.6 cm. An adjacent window has an area of 2 m^2 and a thickness of 1.3 cm. Compare the conductive heat flow through the two windows, assuming that they have equal thermal conductivities.

8. A pipe with an outer radius of 1 cm carries a hot fluid. With a 1-cm thickness of insulation the temperature difference across the insulation is 100 C°. The rate of heat loss per meter length of the pipe (q/L) is found to be 100 W/m. Calculate the value of k, the thermal conductivity of the insulation.

9. The temperature difference across the insulation is held constant for the pipe considered in Problem 8. What happens to the rate of heat loss per meter if the insulation thickness is increased from 1 cm to (a) 2 cm; (b) 3 cm; (c) 4 cm?

10. The heat flow through the steel pipe in Example 3 is 120 W/m. The pipe is wrapped with a cylindrical sleeve of insulation (k = 0.15 W/m · C°) 1 cm thick. If the temperature of the inner surface of the insulation is 129.6°F (54.2°C), determine the temperature of the outer surface.

11. A steel pipe with an outer radius of 1 cm is wrapped with insulation (k = 0.15 W/m · C°). The temperature difference across the insulation is 40 C°. Compare the rate of heat flow per meter (q/L) via conduction for insulation thicknesses of 1, 3, and 10 cm.

12. (a) Show that the flow of heat in a system with spherical symmetry (radially outward from a spherical source) is governed by the equation

$$q = -4\pi k r^2 \frac{dT}{dr}$$

where r is the distance from the center of the source to a point where the temperature is T. (b) Rearrange this equation and integrate to show that the temperature difference between the points r_1 and r_2 is given by (Figure 2)

$$T_1 - T_2 = \frac{q}{4\pi k}\left(\frac{1}{r_1} - \frac{1}{r_2}\right)$$

(c) A spherical pressure vessel made of steel surrounds a small nuclear reactor. The inner and outer diameters of the vessel are 0.88 m and 1.22 m. The outward rate of heat flow is 3700 W. The temperature of the inner surface is 650°F (343°C). Determine the temperature of the outer surface.

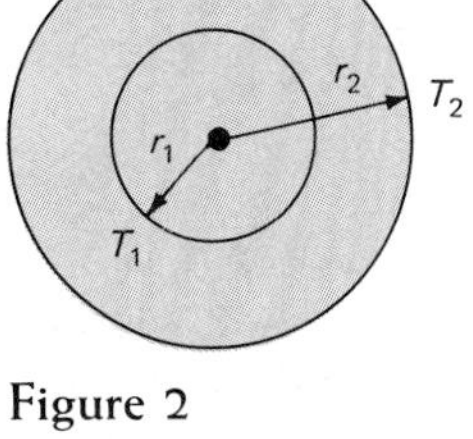

Figure 2

Section 23.2 Convection

13. The heat transfer coefficient for boiling water in a teakettle is h = 1500 W/m^2 · C°. The temperature difference across the boundary layer is 0.50 C°. Find the rate of heat flow through an area of 0.12 m^2.

14. How long will it take the power transistor of Example 5 to cool to (a) 30°C; (b) 28°C?

15. Long-distance swimmers stop at regular intervals to consume high-caloric foods (typically, several thousand calories each hour). A significant fraction of the energy obtained from these foods is used to compensate the body heat loss to the water. Assume that the heat transfer coefficient for the swimmer is 100 $W/m^2 \cdot C°$. Take the surface area to be 1.0 m^2. If the temperature of the water is 21°C and the body temperature is 37°C, determine the heat loss in Calories in 1 hr (1 Calorie = 10^3 calorie = 4.19×10^3 J). Round your answer to the nearest 100 Cal.

16. A swimmer emerges from the water and experiences a chill because a breeze increases the convection heat flow. If $h = 250\ W/m^2 \cdot C°$ and the temperature difference between the air and wet skin is 4.4 C°, determine the rate of heat flow from a swimmer with an exposed area of 0.62 m^2.

Section 23.3 Thermal Radiation

17. A wire carrying an electric current dissipates energy in the form of heat. The heat is radiated from the surface of a ceramic material encasing the wire. Will the surface temperature of the ceramic be greater if it is coated with silver-colored paint ($\epsilon = 0.30$) or with black paint ($\epsilon = 0.90$)? If the surface temperature is 360 K with black paint, determine its value for silver paint. Take the temperature of the surroundings to be 310 K.

18. A lamp filament reaches 2500 K. The surrounding glass bulb is at 400 K. The emissivity of the filament is 0.32 and its surface area is 0.04 cm^2. At what rate is electrical energy supplied to the filament?

19. Assume that a nudist's body radiates like a blackbody at a temperature of 32°C and calculate the *net* radiation loss per day to an environment at 27°C. Compare this with the typical caloric intake of 3000 kcal/day. Take the radiating area to be 0.84 m^2.

20. The heating element of a large electric heater radiates energy at the rate of 1500 W. Its temperature is 2500 K. Assuming an emissivity of 0.35, what is its surface area?

21. A room in a house is kept at a temperature of 20°C. The ceiling ($\epsilon = 0.40$) is maintained at a temperature of 30°C by electric heating. Calculate the net thermal power radiated into the room per square meter of ceiling.

22. A solar collector used for heating water has an area of 3.0 m^2 and an emissivity of 0.86. Determine the rate at which it radiates energy when its temperature is 120°C.

23. A 100-W light bulb emits approximately 10 W of radiation. The wire filament of the bulb is 2.5 cm long and 0.1 mm in diameter. Estimate the temperature of the filament, assuming it behaves as a blackbody.

24. The star Betelgeuse is a red giant with a surface temperature of 3700 K. Its luminosity is $10^5 L_S$, where $L_S = 3.9 \times 10^{26}$ W is the luminosity of our sun. Assume that $\epsilon = 1$ (blackbody). (a) Calculate the surface flux, F. (b) Calculate the radius of Betelgeuse. Express your result as a multiple of the solar radius, $R_S = 7.0 \times 10^8$ m.

25. The star Sirius B is a white dwarf with a surface temperature of 6600 K. Its luminosity is $1.1 \times 10^{-3} L_S$, where $L_S = 3.9 \times 10^{26}$ W is the luminosity of our sun. Assuming that $\epsilon = 1$ (blackbody), calculate the radius of Sirius B. Express your result as a multiple of the earth's radius ($R_E = 6.37 \times 10^6$ m).

26. Pluto is roughly 40 times as far from the sun as is the earth. Estimate the surface temperature of Pluto.

27. A sphere of radius R is maintained at a surface temperature T by an internal heat source (Figure 3). The sphere is surrounded by a thin concentric shell of radius $2R$. Both objects absorb and emit as blackbodies. Show that the temperature of the shell is $T/(8^{1/4}) = 0.595$ T. (*Hint:* Both the inner and outer surfaces of the shell emit as blackbodies.)

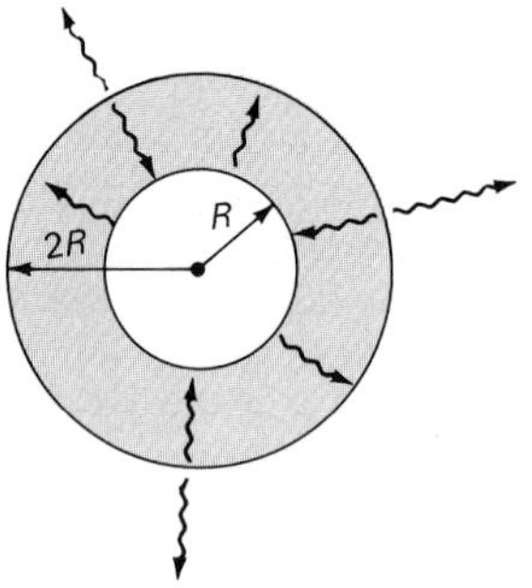

Figure 3

28. In Section 23.2 we carried through an analysis that showed how the temperature decreases steadily for a body cooled by convection (Eqs. 23.15 and 23.16). Perform a similar analysis for a body that cools by thermal radiation alone. (a) Specifically, show that the temperature T at time t after the cooling begins is given by

$$\frac{1}{T^3} - \frac{1}{T_o^3} = \left(\frac{3A\epsilon\sigma}{mC}\right)t$$

where T_o is the initial ($t = 0$) temperature and m is the mass, C the specific heat, ϵ the emissivity, and A the surface area. Assume that the body radiates but does not absorb. (b) An iron ingot cools initially primarily by radiation. Its initial temperature is 2000 K. How much time is required for the ingot to cool to a temperature of 1400 K? Take $m = 2 \times 10^4$ kg; $C = 0.45\ kJ/kg \cdot C°$; $\epsilon = 0.09$; $A = 12\ m^2$.

29. An object is completely surrounded by a source of radiant heat (Figure 4). A heating element maintains the source at a fixed temperature, T_e. Use Eq. 23.21 to prove that the temperature of the object equals T_e when radiative equilibrium is achieved.

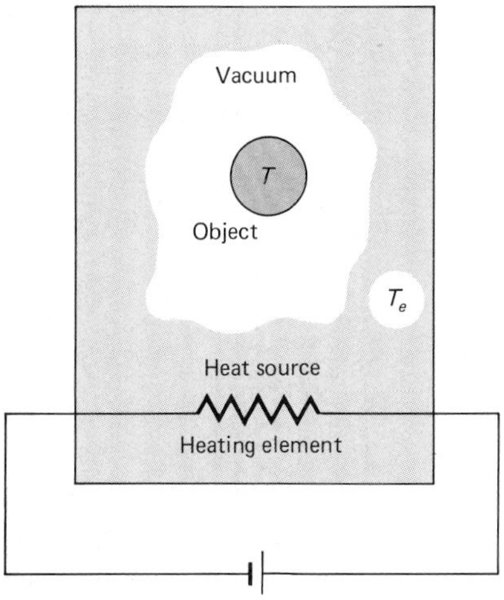

Figure 4

chapter 24

THE SECOND LAW OF THERMODYNAMICS

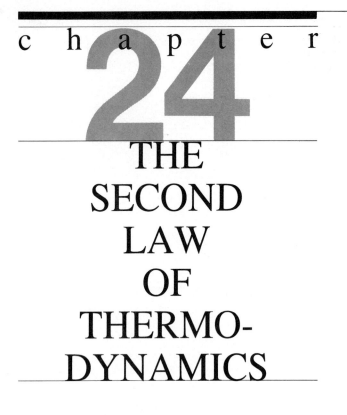

Preview

Most of the laws of physics take the form of equations that describe precisely what must happen in a given situation. For example, Newton's second law states that an object subjected to a net force must experience an acceleration, and the law determines precisely the magnitude and direction of the acceleration. The first law of thermodynamics states that energy is conserved—energy remains precisely constant. But the second law of thermodynamics, introduced in this chapter, is a different kind of physical law. It is an inequality that describes what cannot happen. With certain qualifications, the second law recognizes that heat cannot be converted completely into work. The second law places restrictions on heat engines—devices that transform heat into electrical, mechanical, and chemical energy. Not all of the heat intake of a heat engine can be converted into work; a significant fraction must be ejected to the environment.

We begin this chapter by describing heat engines. We then introduce the Carnot cycle, the most efficient type of cyclic process for converting heat into work. A heat engine that uses the Carnot cycle provides a standard of comparison for all heat engines. For many reasons, few practical heat engines operate close to the ultimate Carnot efficiency. We use the Carnot cycle to define the Kelvin temperature scale, an absolute scale that is independent of the materials used to actually construct a thermometer.

We present the Kelvin and Clausius statements of the second law of thermodynamics and show their equivalence. The last two sections of this chapter are devoted to the concept of entropy. We show by example that the entropy of a system is a measure of its disorder, or randomness, at the microscopic level. Processes that increase disorder at the molecular level increase entropy. Finally, we show how the second law of thermodynamics can be stated in terms of entropy changes.

24.1 Heat Engines and Thermodynamic Efficiency

There is a sharp distinction between the first and second laws of thermodynamics. The first law states that energy is conserved, but places no other restriction on processes that convert heat into work. The second law of thermodynamics recognizes further limits on processes that convert heat into work. The upshot of the second law is that the complete conversion of heat into work is not possible. The second law is of great practical importance because the conversion of heat into work is a matter of vital importance to technology.

We must introduce the concept of a heat reservoir* before we formulate the second law. A heat reservoir is a body of uniform temperature throughout, the mass of which is sufficiently large that its temperature is unchanged by the absorption or ejection of heat. For example, if an ice cube is tossed into the Atlantic Ocean, it produces no observable change in the temperature of the ocean. Therefore the Atlantic Ocean qualifies as a heat reservoir. It is possible for a particular object to be a heat reservoir in some instances but not in others. For example, if an ice cube is dropped into a cup of hot tea, there is a noticeable change in the temperature of the tea. But if a snowflake falls into the tea, no significant temperature change occurs for the tea. The tea therefore qualifies as a heat reservoir in the latter case, but not in the former.

*The terminology *heat reservoir* is somewhat imprecise. A heat reservoir can accept and eject heat, but it contains energy. We often refer to heat as *thermal energy*—energy transferred by virtue of a temperature difference.

Figure 24.1 Thermodynamic aspect of steam turbine operation.

(a) Water is converted to steam by absorbing heat from a coal-fired or nuclear heat reservoir. (b) Steam expands and does work on turbine blades. Heat is converted into rotational kinetic energy. (c) Expanded steam ejects heat to river or ocean water and condenses. (d) Cycle is completed by pumping water back to boiler.

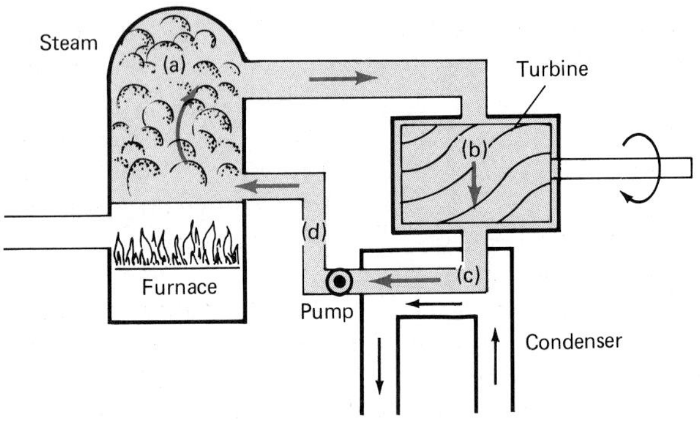

Figure 24.2

Energy flow for a heat engine. Heat Q_H is absorbed from a high-temperature reservoir. A portion of Q_H is converted into work, W, and the remainder, Q_L, is ejected to a low-temperature reservoir. The first law of thermodynamics requires that $W = Q_H - Q_L$.

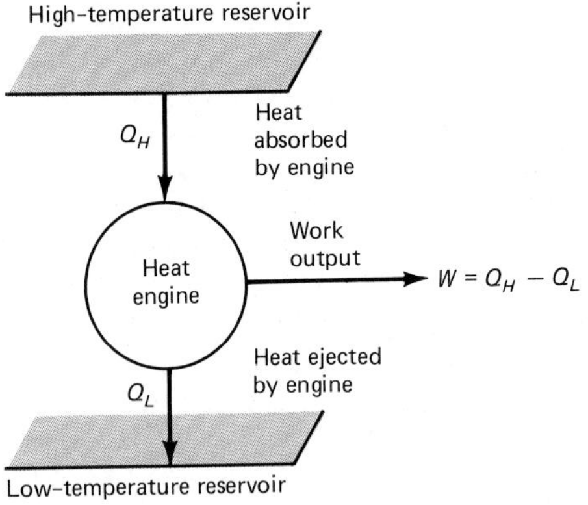

Our industrialized society is powered by many types of *heat engines*. A heat engine is any device, operating in a cycle, that absorbs heat from a high-temperature reservoir, converts part of it into work, and ejects the remainder into a low-temperature reservoir. Some sort of "working substance" is used to exchange heat and perform work. For example, in the steam turbine, water is the working substance. Water is heated and converted into high-pressure steam in a boiler. The steam expands and does work on the turbine blades, is condensed by ejecting heat into the atmosphere or a river, and is then pumped back to the boiler. These steps are shown in Figure 24.1.

The fact that the earth's supply of fossil fuels is dwindling, and the environmental dangers of thermal pollution have made many people energy conscious. The ejection of heat to the environment seems wasteful, and perhaps harmful. In recent years, engineers have searched for and found a variety of beneficial uses for the thermal discharges from heat engines. For example, the warm water ejected by electric power plants can be used to heat homes, and the hot air ejected by refrigeration units can heat the interiors of supermarkets.

However, no clever or innovative engineering design can overcome the fundamental limitation imposed on heat engines by the second law of thermodynamics. The practical thrust of the second law is that no heat engine can absorb a given amount of heat and convert it completely into work. (A more precise statement of the second law is given in Section 24.3.)

Thermodynamic Efficiency

Figure 24.2 shows the energy flows common to all types of heat engines. Heat Q_H is absorbed from a high-temperature reservoir. A portion of Q_H is converted into work, and the remainder, Q_L, is ejected to a low-temperature reservoir.* The working substance is carried through a cycle and is left unchanged. For a cyclic process there is no change in the internal energy, and the first law of thermodynamics requires that the net work done, W, equal the net heat absorbed:

$$W = Q_H - Q_L \tag{24.1}$$

A heat engine is efficient if it converts a large fraction of its heat intake into work. The thermodynamic efficiency of a heat engine, denoted by the Greek letter eta (η), is defined as the fraction of Q_H converted into work:

$$\eta \equiv \frac{\text{work out}}{\text{heat absorbed}} = \frac{W}{Q_H} \tag{24.2}$$

An alternate expression for η follows if we use Eq. 24.1 to replace W by $Q_H - Q_L$,

$$\eta = \frac{Q_H - Q_L}{Q_H} \tag{24.3}$$

Insofar as the first law of thermodynamics is concerned, η can take on any value between 0 ($Q_L = Q_H$; $W = 0$) and 1 ($Q_L = 0$; $W = Q_H$). The actual values of η are well below unity in all areas of energy conversion technology. For example, at electric power plants, steam turbines convert thermal energy into work, which in turn is converted into rotational kinetic energy of the turbine. The thermo-

Table 24.1

Thermodynamic efficiencies

Device	η
Liquid-fuel rocket	0.48
Steam turbine	0.46
Diesel engine	0.37
Automobile engine	0.25
Steam locomotive	0.08
Thermocouple	0.07

*Both Q_H, heat absorbed, and Q_L, heat ejected, are *positive* quantities.

dynamic efficiency for these steam turbines is approximately 0.46. This means that only 46% of the heat absorbed by the water is converted into work. Over half of the heat absorbed is ejected. Other heat engines in widespread use today, like the automobile engine, have even smaller efficiencies. Table 24.1 lists thermodynamic efficiencies for several contemporary heat engines.

Example 1 Thermodynamic Efficiency of a Steam Turbine

In one day the boilers of a typical electric power plant supply 3.0×10^{14} J of heat to the steam that drives the turbines. The daily mechanical work output from the turbines is 1.26×10^{14} J. The thermodynamic efficiency for conversion of heat into work is given by Eq. 24.2,

$$\eta = \frac{W}{Q_H} = \frac{1.26 \times 10^{14}\ \text{J}}{3.0 \times 10^{14}\ \text{J}} = 0.42$$

The thermodynamic efficiency, η, refers only to processes in which heat is transformed into work. Many other energy conversion processes are important. In an electric motor, for example, electric energy is converted into rotational kinetic energy. The conversion efficiency (mechanical energy output/electric energy input) of such processes is generally quite high (over 0.90) in comparison with typical thermodynamic efficiencies.

Refrigerators, Heat Pumps, and Coefficient of Performance

A heat engine extracts heat from a high-temperature reservoir, converts part of it into work, and ejects the remaining heat to a low-temperature reservoir, normally our environment. A refrigerator is a device that extracts heat from a low-temperature reservoir, has positive work performed on it by the environment, and ejects heat at a higher temperature. It therefore pumps heat uphill—that is, from a lower to a higher temperature.

Because a refrigerator ejects heat, one method of heating a building is to refrigerate the surrounding ground or atmosphere. A variety of devices called *heat pumps* are now commercially available. They are reversible heat engines that cool a home in the summer (by ejecting heat to the atmosphere) and warm it in the winter (by refrigerating the atmosphere).

A useful measure of the capabilities of a refrigerator or heat pump is the coefficient of performance, or COP. It is defined as follows:

$$\text{COP} \equiv \frac{\text{heat extracted from low-temperature reservoir}}{\text{net work performed on refrigerant to extract heat}}$$

$$= \frac{Q_L}{W} = \frac{Q_L}{Q_H - Q_L} \qquad (24.4)$$

where Q_H is the heat ejected to the high-temperature reservoir. Optimally, a refrigerator extracts a sizable amount of heat from a low-temperature reservoir with a minimal expenditure of work; that is, it has a large coefficient of performance. Typical values of the COP for refrigerators and heat pumps range from 2 to 5.

The economic appeal of heat pumps is apparent when we consider the following: Suppose that COP = 2, a modest value. We can replace Q_L by $Q_H - W$ in Eq. 24.4 to obtain

$$\text{COP} = \frac{Q_H - W}{W} = 2$$

from which it follows that $Q_H/W = 3$. This means that the amount of heat ejected to the high-temperature reservoir is three times the work done. Suppose you were given the choice of heating your home with (a) a heat pump that supplies 3 kJ of heat to your home for every 1 kJ of electric energy used, or (b) electric heating elements that supply 1 kJ of heat for every 1 kJ of electric energy used. Which unit would you choose?

Several generations of intelligent, dedicated people have worked to improve the efficiency of heat engines. Why, then, are the efficiences listed in Table 24.1 still under 50%? The answer is that nature has established a fundamental limit on all processes that convert heat into work. The second law of thermodynamics is a formal statement that such a limit exists, and the relation $\eta = (Q_H - Q_L)/Q_H$ is a quantitative measure of this limit. This limit is *not related to friction* within a heat engine. It is true that friction reduces the actual work output of heat engines and thereby reduces their thermodynamic efficiencies.* However, even if friction were eliminated completely, the thermodynamic efficiency would still be well below unity. For example, in a steam turbine, friction reduces the thermodynamic efficiency by less than 1%. The fundamental limit to efficiency is thermodynamic in origin.

What then is the optimal design for a heat engine? This question was answered by the French engineer Sadi Carnot† (Figure 24.3). Because of his brilliant analysis of the problem, Carnot is given credit for discovering the second law of thermodynamics.

*Some of the work is converted into heat via friction and ejected. Thus in Eq. 24.3, $\eta = (Q_H - Q_L)/Q_H$, friction increases Q_L and lowers η.

†Nicolas Leonard Sadi Carnot (1796–1832) published his monumental paper ("Reflections on the Motive Power of Heat") in 1824. Carnot's work is especially noteworthy because it was completed in 1824, nearly 20 years before the Mayer–Joule formulation of the first law of thermodynamics. However, because Carnot used the caloric theory of heat, his applications of the second law were not always sound. Carnot's notebooks, published long after his death, revealed his suspicion of the caloric theory and his resolve to perform experiments similar to those carried out by Joule.

Figure 24.3

French stamp honoring Sadi Carnot.

There are two remarkable aspects of Carnot's research. First, he found that the efficiency with which heat can be converted into other forms of energy depends on the nature of the cyclic process used but not on the working substance. Second, Carnot discovered the most efficient cyclic process for converting heat into other forms of energy. We will discuss this cycle, now known as the *Carnot cycle,* in Section 24.2. Carnot's achievements are truly monumental. Without benefit of a complete theory of heat, he set forth the ultimate standard of heat engine efficiency.

24.2 The Carnot Cycle

Consider a system composed of an ideal gas confined in a cylinder fitted with a piston (Figure 24.4). The ideal gas is the working substance; the piston and cylinder walls are its environment. The gas can exchange heat with its surroundings via conduction through the walls of the cylinder. Heat exchange can be eliminated if desired by suitably insulating the cylinder.

Figure 24.5 shows a $P-V$ diagram for a Carnot cycle in which an ideal gas serves as the working substance. The Carnot cycle consists of two adiabatic paths (no heat exchange) connected by a pair of isothermal paths (constant temperature). We can analyze the Carnot cycle shown in Figure 24.5 by using the following four relations:

1 the equation of state for an ideal gas (Section 22.2),

$$PV = nRT \qquad (24.5)$$

2 the equation for the internal energy (U) of an ideal gas,

$$U = mC_vT \qquad (24.6)$$

3 the relation describing the adiabatic transformation of an ideal gas,

$$TV^{\gamma-1} = \text{constant} \qquad \gamma = \frac{C_p}{C_v} > 1 \qquad (24.7)$$

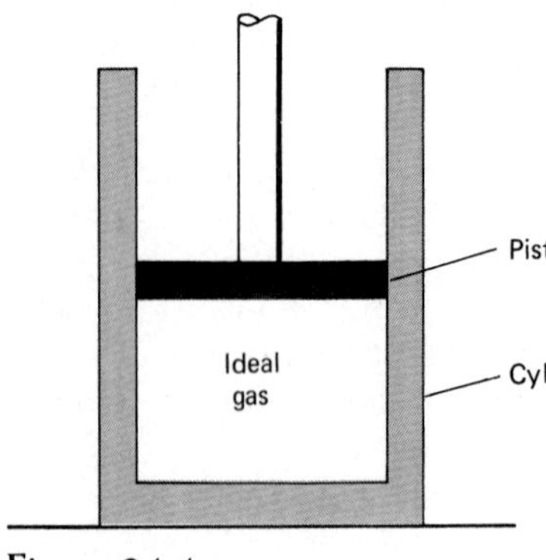

Figure 24.4

An ideal gas is confined in a cylinder fitted with a piston. The ideal gas is the working substance. The piston and cylinder walls are its environment.

Figure 24.5

The Carnot cycle for a gas as it appears on a $P-V$ diagram, where T_H and T_L denote, the high- and low-temperature reservoirs, respectively. Heat Q_H is absorbed from a reservoir as the gas expands isothermally from 1 to 2. The expansion from 2 to 3 is performed adiabatically. Heat Q_L is ejected during the isothermal compression from 3 to 4. The cycle is completed by an adiabatic compression from 4 to 1.

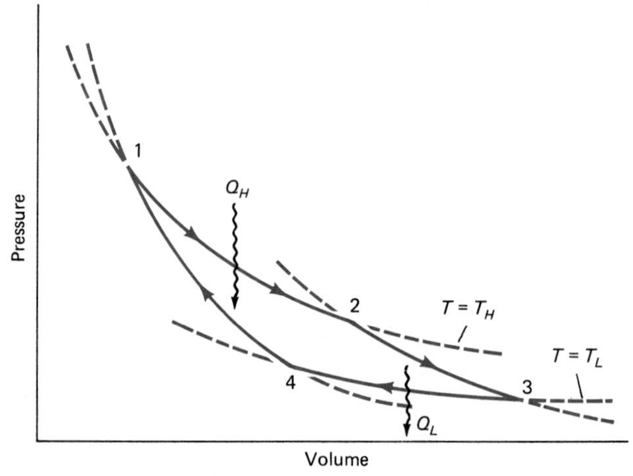

4 the first law of thermodynamics,

$$Q = \Delta U + W \qquad (24.8)$$

The four legs of the Carnot cycle described in Figure 24.5 consist of an isothermal expansion ($1 \rightarrow 2$), an adiabatic expansion ($2 \rightarrow 3$), an isothermal compression ($3 \rightarrow 4$), and an adiabatic compression ($4 \rightarrow 1$). All four of these processes must proceed *reversibly* (Section 21.5). During the isothermal expansion from 1 to 2, heat Q_H is absorbed from a high-temperature reservoir (Figure 24.6). The expanding gas performs a positive amount of work W_{12} against its environment. Because $U = mC_vT$ for the ideal gas, $\Delta U = 0$ along an isotherm. Thus, there is no change in the internal energy over the path from 1 to 2, and the first law reduces to

$$Q_H = W_{12}$$

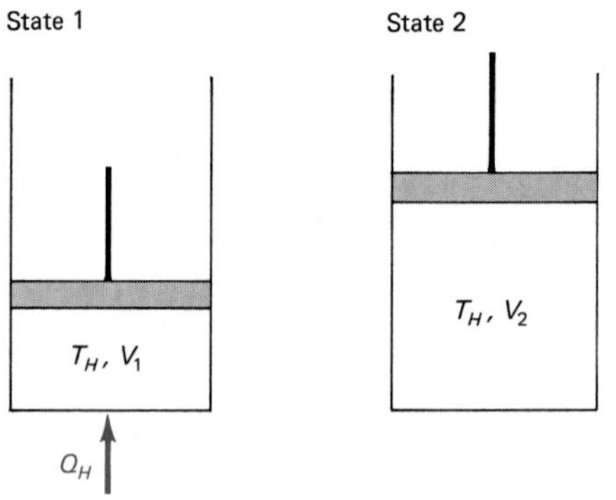

Figure 24.6

The gas absorbs heat Q_H and expands from a volume V_1 to a volume V_2. The process proceeds at a rate that maintains the gas at a constant temperature, T_H.

Figure 24.7

An adiabatic expansion from V_2 to V_3 results in a drop in temperature for the gas.

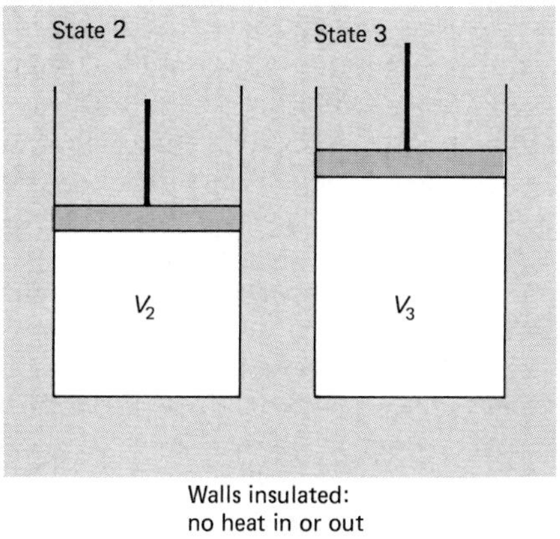

All of the heat absorbed from 1 to 2 is used to perform work against the environment.

The expansion of the gas continues from 2 to 3 under adiabatic conditions; that is, the gas is insulated from its surroundings during the expansion (Figure 24.7). During the adiabatic expansion the gas performs a positive amount of work, W_{23}. This work is done at the expense of the internal energy of the gas because no heat is received. The internal energy decreases during the adiabatic expansion (2 to 3). The logic is

$$Q = 0 \qquad \text{(adiabatic process)}$$

$$W_{23} > 0 \qquad \text{(expansion)}$$

$$\Delta U = U_3 - U_2 = -W_{23} < 0 \qquad \text{(first law; work performed at expense of internal energy)}$$

$$U_3 < U_2 \qquad \text{(internal energy decreases from 2 to 3)}$$

Because $U = mC_vT$, a decrease in U means a decrease in temperature, $T_L < T_H$. Therefore, the gas cools during the adiabatic expansion. We can also use Eq. 24.7, $TV^{\gamma-1}$ = constant, to reach the same conclusion. The volume increases as the gas expands, so the temperature must decrease to satisfy Eq. 24.7.

The foregoing arguments are reversed for the isothermal compression from 3 to 4 (Figure 24.8) and the adiabatic compression from 4 to 1 (Figure 24.9), completing the cycle. During the isothermal compression from 3 to 4 the environment (piston) performs a positive amount of work on the gas. The gas must eject heat (Q_L) to the low-temperature reservoir in order to avoid a temperature increase. The adiabatic compression from 4 to 1 completes the cycle. There is no heat transfer during this compression, and the positive work performed by the environment therefore increases the internal energy of the gas. This increase shows up as a temperature rise from T_L to T_H.

We chose an ideal gas as the working substance in the Carnot cycle simply for convenience. We could have used any number of other working substances. The *essential* features of the Carnot cycle are that (1) it consists of two isothermal paths connected by two adiabatic paths, and (2) the adiabatic and isothermal processes proceed *reversibly*.

Figure 24.8

Heat Q_L is ejected during the isothermal compression from V_3 to V_4. The gas temperature remains fixed at T_L during the compression.

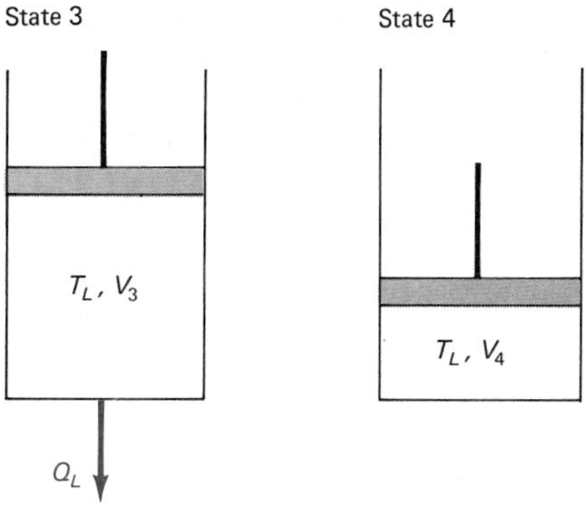

Figure 24.9

An adiabatic compression from V_4 to V_1 causes the temperature of the gas to rise.

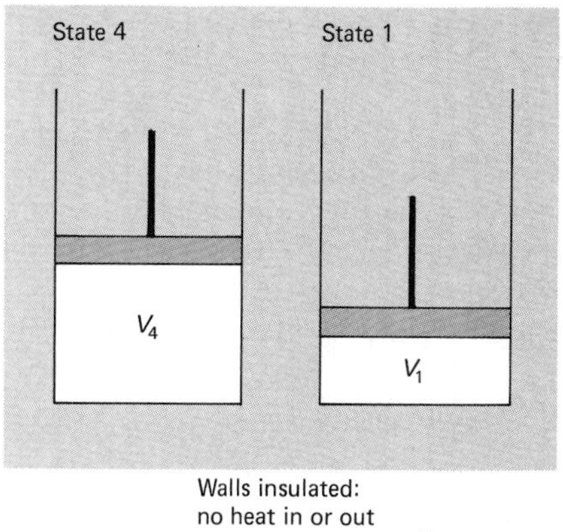

The overall result of the Carnot cycle is the conversion into work of some of the heat extracted from the high-temperature reservoir. The particular Carnot cycle just described acts as a heat engine, converting the net heat $Q_H - Q_L$ into work. Therefore, we often refer to such a cycle as a *Carnot engine.*

Perhaps the most important aspect of the Carnot engine is that it is the most efficient heat engine operating between the two heat reservoirs. This fact, known as *Carnot's theorem,* is proved in Section 24.3 and means that the Carnot engine is the ultimate standard of comparison for all heat engines.

Another special aspect of the Carnot engine is that heat is absorbed and ejected only during isothermal processes. Thus, a single temperature is associated with Q_H and likewise with Q_L. This makes it possible to define temperature in terms of heat. In Section 24.4 we show how the efficiency of the Carnot engine is used to define the Kelvin temperature scale.

Questions

1. Can you air condition your kitchen by opening the refrigerator door? How would you explain your answer to (a) your physics classmates; (b) a classmate who has never studied physics?

2. A ship driven by steam turbines consumes fuel at a lower rate when it is in the cold North Atlantic Ocean than when it is in the warm South Pacific Ocean, even though its speed is the same in both. Suggest an explanation based on thermodynamic efficiency.

3. Why is it necessary to condense the steam on the low-temperature side of a steam turbine? It seems somewhat wasteful of energy to cool the water, return it to the boiler, and then convert it back into superheated steam. Why not just let the steam move through the turbine without condensing?

24.3 The Second Law of Thermodynamics

Precise statements of the second law of thermodynamics were formulated by William Thomson, who later became Lord Kelvin,[*] and by Rudolph Clausius[†] in the early 1850s. Although worded differently, their statements are equivalent.

The Kelvin statement of the second law is this:

It is impossible to devise a process whose only result is to convert heat, extracted from a single reservoir, entirely into work.

The word *only* in Kelvin's statement is an important qualifier. Many processes can be performed whereby a system converts heat completely into work. However, all such processes result in a final state that differs from the initial state. Thus, there is some other result besides the conversion of heat into work. For example, in the isothermal expansion of the ideal gas considered in the Carnot cycle, the heat absorbed was converted entirely into work. At the end of the expansion, however, the thermodynamic state of the gas had changed. Its pressure had decreased and its volume had increased. The word *single* is also a key word in Kelvin's statement. Heat engines convert heat into work. A heat engine is a cyclic device that extracts heat from a high-temperature reservoir, converts part of it into work, and ejects the remainder to a low-temperature reservoir. Converting the net heat absorbed into work does not violate the second law, and is in fact demanded by the first law. A violation of Kelvin's statement of the second law would result if a heat engine could extract heat from a single reservoir and convert it entirely into work without the ejection of a net amount of heat to other reservoirs.

*Lord Kelvin (William Thomson, 1824–1907) was a major scientific figure. He published over 600 scientific papers. In addition to his formulation of the second law, Kelvin was the first to propose an absolute temperature scale (Section 24.4).

†In addition to stating his version of the second law of thermodynamics, Clausius (1822–1888) was one of the founders of the kinetic theory of gases (Chapter 25).

The Clausius statement of the second law is this:

It is impossible to doubt a process whose only result is to extract heat from a reservoir and eject it to a reservoir at a higher temperature.

According to the Clausius statement, the spontaneous flow of heat from a low-temperature reservoir to a high-temperature reservoir is impossible. In other words, heat flow is an irreversible process. Heat flows from higher to lower temperatures. It *never* spontaneously flows in the other direction. Of course, a refrigerator transfers heat from a low-temperature reservoir to a high-temperature reservoir, but work must be performed to accomplish this transfer—it is not spontaneous.

Heat flow is one very obvious example of the intimate relation between irreversible processes and the second law. The second law is a consequence of irreversibility. It is possible, although not always enlightening, to show that *if* a particular irreversible process were to become reversible, *then* the reversed process would violate the second law.

The evidence supporting the second law of thermodynamics is the failure of all attempts to construct a perpetual motion machine of the second kind,[*] a machine that would contradict the second law. If such a machine could be constructed, however, the benefits would be enormous. For example, the crust of our earth, its atmosphere, and its oceans are giant heat reservoirs. If an engine could be constructed that could extract heat from the atmosphere and convert it into mechanical energy, the mechanical energy could be used to operate machinery and to generate electric currents. Air taken in and cooled by the extraction of heat could be used as a refrigerant. Air conditioning systems would be a by-product of such an engine. The supply of energy in the atmosphere is virtually limitless, because it is maintained by the sun. It would not be necessary to mine coal or drill for gas and oil—*if* there were such engines. The oceans of our earth also contain enormous quantities of thermal energy. If a ship's engine could extract heat from seawater, the heat could be converted into work and used to propel the ship. However, such wondrous heat engines violate the second law of thermodynamics and seem destined to remain figments of the imagination.[†]

Equivalence of Kelvin and Clausius Formulations

At the beginning of this section, we said that the Kelvin and Clausius formulations of the second law are equivalent. This equivalence is demonstrated by showing that if Kelvin's statement is false, then so is Clausius', and vice versa.

*Recall that a perpetual motion machine of the first kind is one that would manufacture energy—it would violate the first law of thermodynamics (Section 21.4).

†Our Earth's interior is hotter than its surface. The interior is a high-temperature reservoir and is now being tapped at several geothermal power plants. There have been serious attempts to extract heat from the ocean by making use of the temperature difference between the warmer surface layers and the cooler, deeper layers. See "Suggested Reading" at end of this chapter.

Let us suppose that Kelvin's statement is false. This would mean that a process is possible whose only result is to convert heat extracted from a reservoir completely into work. The work obtained could then be completely converted back into heat at any desired temperature, via friction, for example. In particular, it could be deposited in a heat reservoir whose temperature exceeds that of the reservoir from which the heat was extracted. The only result of this process would be to transfer a quantity of heat from a lower temperature to a higher temperature. But this is a violation of Clausius' statement. Thus, if Kelvin's statement is false, so is Clausius' statement.

Carnot's Theorem

Carnot's theorem provides the basis for a thermodynamic temperature scale—a scale independent of the thermometric substance. Carnot's Theorem is this:

No heat engine operating between two heat reservoirs can be more efficient than a Carnot engine operating between the same two reservoirs.

The proof of the theorem follows these lines: One assumes that the second law is valid. Next, two reversible heat engines are introduced. One is a Carnot engine and the other is an engine assumed to be more efficient than the Carnot engine. If Carnot's theorem were false it would be possible to use the two engines in a cycle that violates the second law. By assumption it is not possible to circumvent the second law. Thus, Carnot's theorem is not false.

Carnot's Theorem is important. It imposes a fundamental limit on the conversion of heat into work. No amount of ingenuity can devise a heat engine more efficient than the Carnot engine.

Question

4. Suppose that you dipped your finger into a glass of cold water and subsequently noticed that your finger became warmer and warmer and that the water began to freeze. (a) Would this necessarily violate the first law of thermodynamics? (b) Would this necessarily violate the second law of thermodynamics?

24.4 The Kelvin Temperature Scale

We saw in Section 24.3 that the efficiency of a Carnot engine is independent of the working substance used in the engine. In Chapter 21 we noted that the Kelvin scale of temperature is based on the laws of thermodynamics—it is independent of the thermometric substance used to construct a thermometer. The fact that the efficiency of a Carnot engine is independent of the working substance makes it possible to use the Carnot efficiency to define the Kelvin temperature scale.

Figure 24.10

A Carnot cycle with one isotherm fixed at the triple point of water can be used to define the Kelvin temperature scale.

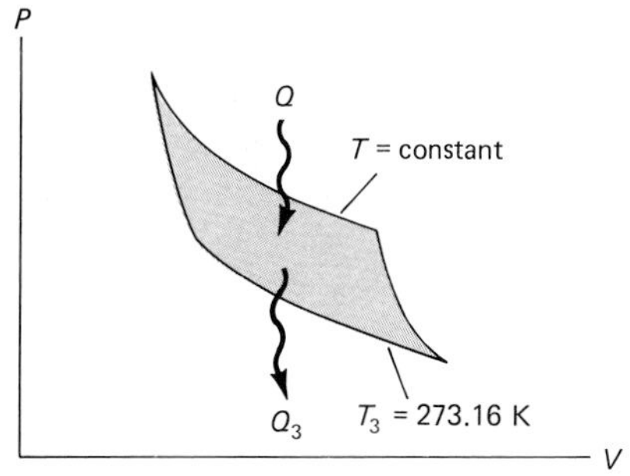

Consider a Carnot engine that absorbs heat Q_H along an isothermal path and ejects Q_L along an isothermal path. A thermodynamic temperature T is *defined* by taking $Q_L/Q_H = T_L/T_H$ or, equivalently,

$$\frac{Q_H}{T_H} = \frac{Q_L}{T_L} = \text{constant} \qquad (24.9)$$

We emphasize that this is an arbitrary, but operational, definition of temperature. This definition tells us that the thermodynamic temperature T is proportional to Q, but leaves open the size of the temperature unit. As noted in Chapter 21, the temperature of the triple point of water has been assigned the value 273.16 K. Suppose that we operate a Carnot engine with the low-temperature reservoir at 273.16 K. Let Q_3 denote the heat ejected to the low-temperature reservoir (Figure 24.10). If the heat absorbed from the high-temperature reservoir is Q, its Kelvin temperature T is given by Eq. 24.9,

$$\frac{Q}{T} = \frac{Q_3}{273.16}$$

or

$$T = 273.16\left(\frac{Q}{(Q_3)}\right)\text{K} \qquad (24.10)$$

Equation 24.10 defines the Kelvin temperature scale.

The efficiency of a Carnot engine operating between reservoirs at temperatures T_L and T_H can be expressed in terms of T_L and T_H. We start by rearranging

$$\eta = \frac{Q_H - Q_L}{Q_H} = 1 - \frac{Q_L}{Q_H} \qquad (24.11)$$

and use Eq. 24.9 to write

$$\frac{Q_L}{Q_H} = \frac{T_L}{T_H} \qquad (24.12)$$

Replacing Q_L/Q_H by T_L/T_H in Eq. 24.11 gives an equation that expresses the efficiency of a Carnot engine in terms of T_L and T_H,

$$\eta = 1 - \frac{T_L}{T_H} = \eta_C \qquad (24.13)$$

We call $1 - T_L/T_H$ the Carnot efficiency and denote it by η_C.

Example 2 Carnot Efficiency of a Steam Turbine

A steam turbine operates between a boiler temperature of 550°F (561 K) and a condenser temperature of 100°F (311 K). We want to determine the Carnot efficiency,

$$\eta_C = 1 - \frac{T_L}{T_H}$$

The actual thermodynamic efficiency is

$$\eta = \frac{W}{Q_H} = \frac{\text{work output}}{\text{heat absorbed}}$$

The Carnot efficiency is an *upper limit* for η—the second law guarantees that $\eta \leq \eta_C$. Taking $T_H = 551$ K and $T_L = 311$ K in Eq. 24.13 gives

$$\eta_C = 1 - \frac{311}{511} = 0.446$$

This efficiency of less than 50% is a consequence of the second law of thermodynamics. Some of the absorbed heat must be ejected and consequently is not available for conversion into work.

A glance at the relation $\eta_C = 1 - T_L/T_H$ suggests that to raise the efficiencies of heat engines we must lower T_L and/or raise T_H. But the only convenient low-temperature reservoirs that nature provides (large bodies of water and the atmosphere) have $T_L \simeq 290$ K. Despite the best efforts of materials scientists and engineers there is still a disappointingly low limit on T_H. The thermal properties of working substances and materials used to fabricate high-temperature reservoirs have so far prevented significant increases in T_H. In particular, prolonged operation at high temperatures reduces the strength of metals and increases their rate of corrosion.

As reserves of abundant and cheap fuels dwindle, scientists and engineers look for new sources of fuel, strive to increase the efficiency of existing heat engines, and seek to develop new, more efficient heat engines. In the quest for new and better heat engines, the second law of thermodynamics provides both a goal and a limit: For given high and low temperatures, the maximum thermodynamic efficiency is given by Eq. 24.13.

24.5 Entropy

In this section we introduce a new physical property—entropy. The concept of entropy may not seem intuitively clear to you at first. This is partly because entropy is a new concept and partly because thermodynamics does not suggest any microscopic interpretation of quantities like temperature and entropy. We introduce entropy because it is a very useful thermodynamic property. For example, we can state the second law of thermodynamics in terms of entropy (Section 24.6).

In Chapter 21 we introduced the differential form of the first law of thermodynamics:

$$dQ = dU + dW \tag{24.14}$$

The internal energy is a property of a physical system, and this makes it possible to express dU in terms of appropriate thermodynamic variables and their changes. The work dW can also be expressed in terms of thermodynamic variables and their changes. For example, for a fluid system undergoing a reversible change of volume, dW can be expressed in terms of the pressure (P) and the change of volume (dV),

$$dW = P\,dV \tag{24.15}$$

Is it possible to express dQ in a similar fashion? Yes, provided that the heat exchange is reversible. For a reversible process dQ can be expressed in terms of the temperature (T) and the entropy change dS. The relation between dQ and dS is

$$dQ = T\,dS \tag{24.16}$$

We can rewrite Eq. 24.16 as

$$dS = \frac{dQ}{T} \tag{24.17}$$

to stress that it is a *definition* of the entropy change of a system that absorbs heat dQ. Remember that Eqs. 24.16 and 24.17 are valid only for reversible processes. The total entropy change in a reversible process is obtained by integrating Eq. 24.17 to obtain the total entropy change, ΔS.

$$\Delta S \equiv S_f - S_i = \int_i^f \frac{dQ}{T} \tag{24.18}$$

where S_f and S_i are the entropies of the final and initial states, respectively. A special case of Eq. 24.18 is noteworthy. If the process is isothermal, the right-hand side of Eq. 24.18 becomes

$$\int_i^f \frac{dQ}{T} = \frac{1}{T}\int_i^f dQ = \frac{Q}{T} \tag{24.19}$$

where Q is the total heat absorbed by the system. Thus, for an isothermal process the entropy change is

$$\Delta S = \frac{Q}{T} \tag{24.20}$$

The units of entropy and entropy change are joules per kelvin (J/K). As the following example demonstrates, entropy is a property of a system.

Example 3 Entropy Change for Carnot Cycle

If entropy is a property of a system it must be unchanged in any *cyclic* process. Let's check to see if entropy is unchanged in the Carnot cycle. Figure 24.5 shows the Carnot cycle for a gas–piston system. There is no entropy change along either of the adiabats (S = constant) because $dQ = 0$ and thus

$$dS = \frac{dQ}{T} = 0$$

Along the high-temperature isotherm ($1 \rightarrow 2$), heat Q_H is absorbed. The entropy change is given by Eq. 24.20, with $Q = Q_H$ and $T = T_H$.

$$\text{Entropy change along } T_H \text{ isotherm} = \frac{Q_H}{T_H}$$

Along the low-temperature isotherm ($3 \rightarrow 4$), heat Q_L is ejected at $T = T_L$. Equation 24.20 applies, with $T = T_L$ and $Q = -Q_L$. The minus sign enters because an amount of heat Q_L is ejected along the low-temperature isotherm.

$$\text{Entropy change along } T_L \text{ isotherm} = -\frac{Q_L}{T_L}$$

The minus sign enters because the entropy decreases when heat is ejected. The total entropy change over the Carnot cycle is

$$\Delta S_{\text{cycle}} = \oint \frac{dQ}{T} = \frac{Q_H}{T_H} - \frac{Q_L}{T_L}$$

However, Eq. 24.9 relates Q_H/T_H and Q_L/T_L for the Carnot cycle,

$$\frac{Q_H}{T_H} = \frac{Q_L}{T_L} = \text{constant} \qquad (24.9)$$

Therefore,

$$\frac{Q_H}{T_H} - \frac{Q_H}{T_L} = 0$$

and

$$\Delta S_{\text{cycle}} = 0 \qquad (24.21)$$

The entropy change for the Carnot cycle is zero. This is because entropy is a property of a physical system. The entropy depends on the thermodynamic state and hence is unchanged in any cyclic process.

Entropy and Disorder

Eq. 24.18 provides a way to measure entropy changes, but does not explain what entropy itself measures. Entropy is important because it is useful, but it will help our understanding if we can relate entropy to some aspect of our day-to-day experience. Entropy is a measure of disorder. If the entropy of a system increases, then its microscopic structure becomes more disordered. The examples that follow should help you to understand the idea that entropy measures disorder.

Example 4 **Entropy Change for Melting**

Consider a 1-kg chunk of ice at its normal melting point. If we add the heat of fusion, 334 kJ, the ice melts. What is the change in entropy? Because this change of phase takes place at constant temperature and is reversible, we can use Eq. 24.20 to obtain

$$\Delta S = S_{\text{liquid}} - S_{\text{solid}} = \frac{L_f}{T_m}$$

Taking $L_f = 334$ kJ, $T_m = 273.15$ K gives

$$\Delta S = 1.22 \text{ kJ/K}$$

The value of ΔS is positive showing that melting results in an *entropy increase*. This entropy increase is also in accord with the idea that entropy measures disorder. The molecular structure of the final state (liquid) is more disordered than the molecular structure of the initial state (solid). In a crystalline solid the molecules form an ordered, periodic structure. In a liquid only a short-range order is evident. By short-range order we mean that any given molecule is able to influence only the behavior of nearby molecules. Melting results in an entropy increase and molecular disordering. Entropy measures disorder.

The computed value, $\Delta S = 1.22$ kJ/K obtained in Example 4 will mean more if we compare it with the entropy changes for other processes. To provide a basis for comparison we compute entropy changes for two other processes involving 1 kg of water: (a) raising its temperature isobarically to the boiling point and (b) vaporizing it.

Example 5 **Isobaric Entropy Increase**

If heat is added to a liquid at constant pressure, the liquid's temperature rises until it reaches the boiling point. The heat added in changing the temperature by an amount dT is

$$dQ = mC_p\, dT$$

where C_p is the specific heat at constant pressure. The corresponding entropy change is

$$dS = \frac{dQ}{T} = \frac{mC_p\, dT}{T}$$

For water C_p is nearly constant between the melting and boiling points. If C_p is treated as constant, the total entropy change between the melting and boiling points is

$$\Delta S = mC_p \int_{T_m}^{T_b} \frac{dT}{T} = mC_p \ln\left(\frac{T_b}{T_m}\right) \qquad (24.22)$$

For 1 kg of water, $mC_p = 4.19$ kJ/K and $\ln(373/273) \simeq 0.312$. These data give

$$\Delta S \simeq 1.31 \text{ kJ/K}$$

This change in entropy is only slightly larger than the entropy change of melting. Raising the temperature of liquid water increases the molecular disorder by increasing the average molecular speed. The chance that a given molecule will influence the behavior of a nearby molecule decreases as their relative speeds increase—the molecules pass too swiftly to communicate effectively. Thus, as the temperature increases, molecular interactions diminish and disorder rises.

Once water is brought to its boiling point, it can be vaporized by adding still more heat. The entropy change in going from the liquid to the vapor phase is readily found by the same arguments that gave us $\Delta S = L_f/T$ in Example 4. In place of L_f we now have L_v, the heat of vaporization, and the temperature T is the boiling point, T_b. The entropy change is

$$\Delta S = S_{vapor} - S_{liquid} = \frac{L_v}{T} \tag{24.23}$$

For 1 kg of water at 1 atm, $L_v = 2260$ kJ, $T_b = 373$ K, and $\Delta S \simeq 6.05$ kJ/K. The entropy change for the liquid–vapor transition is roughly five times the entropy change for melting. What disruption at the molecular level has caused this entropy increase? We can answer this question in part by noting that at the normal boiling point, 1 kg of liquid water occupies a volume of about 10^{-3} m^3. At the same pressure and temperature the vapor occupies a volume of approximately 1.7 m^3—1700 times the volume of the liquid! The collection of water molecules, which started as a comparatively dense liquid, has spread out into a less dense vapor. The molecules in the vapor phase are greatly disorganized because they now have a larger volume in which to roam. This is somewhat like the disorder that would result if a swarm of bees left the confines of its hive and filled a room with a volume 1700 times that of the hive.

Ideal Gas Entropy

An expression for the entropy of an ideal gas (Section 22.2) is readily derived and is very useful. We first obtain an expression for $dS = dQ/T$ for an ideal gas. Using the first law we can write $dQ = dU + P\,dV$, obtaining for dS

$$dS = \frac{dQ}{T} = \frac{dU + P\,dV}{T} \tag{24.24}$$

For an ideal gas

$$dU = mC_v\,dT \tag{24.25}$$

and

$$PV = nRT \tag{24.26}$$

The specific heat C_v is a constant. The expression for dS converts to

$$dS = mC_v\frac{dT}{T} + nR\frac{dV}{V} \tag{24.27}$$

This relation can be integrated to give the entropy change in a process that carries the ideal gas from (T_1, V_1) to (T_2, V_2). Thus,

$$\int_{S_1}^{S_2} dS = mC_v\int_{T_1}^{T_2}\frac{dT}{T} + nR\int_{V_1}^{V_2}\frac{dV}{V}$$

By carrying out the indicated integrations, we get

$$S_2 - S_1 = mC_v \ln\left(\frac{T_2}{T_1}\right) + nR\ln\left(\frac{V_2}{V_1}\right) \tag{24.28}$$

Notice that the entropy change has been expressed in terms of the temperatures and volumes of the two states, showing that entropy is a property of a physical system. This equation is valid for any series of reversible processes in which the system starts at (T_1, V_1) and ends at (T_2, V_2). An example of such a process would be an isothermal volume change ($V_1 \rightarrow V_2$ with T_1, constant) followed by an isovolumic temperature change ($T_1 \rightarrow T_2$ with V_2 constant).

We also have stressed that the relation

$$dS = \frac{dQ}{T} \tag{24.17}$$

is valid only for reversible processes. The following discussion of the free expansion of a gas illustrates this fact, and the idea that entropy measures disorder.

Free Expansion of a Gas

Consider the experiment represented in Figure 24.11. A vessel with rigid insulated walls is divided by a partition. Initially, one side contains an ideal gas and the other side is a vacuum. Consider what happens when the partition ruptures, allowing the gas to expand and fill the entire container. The insulation prevents heat exchange. Thus, $dQ = 0$ for each stage of the expansion. The rigid walls guarantee that no work is done on the system, that is, there can be no change in the volume of the system and thus no $P\,dV$ work by the surroundings. The process is called a *free expansion* because of the absence of any external work as gas moves into the evacuated region. We therefore have $dQ = 0$ and $dW = 0$ at each stage of the expansion. It follows from the first law that $dU = 0$; the internal energy of the gas does not change. For an ideal gas $U = mC_vT$; the internal energy depends on T alone. If U remains constant, so must the temperature; thus $T_2 = T_1$. The final volume of the gas V_2 is clearly greater than the initial volume V_1, so that $\ln(V_2/V_1)$ is a positive number. From Eq. 24.28 we then have

$$S_2 - S_1 = nR\ln\left(\frac{V_2}{V_1}\right) > 0$$

That is, the final entropy exceeds the initial entropy. The gas has expanded but its temperature has not changed. The gas spreads out over a larger volume, resulting in further disorder. The predicted entropy increase is in agreement with the idea that entropy measures disorder. It all seems very straightforward, but is it? Let us retreat momentarily to the point where we prescribed that the container was

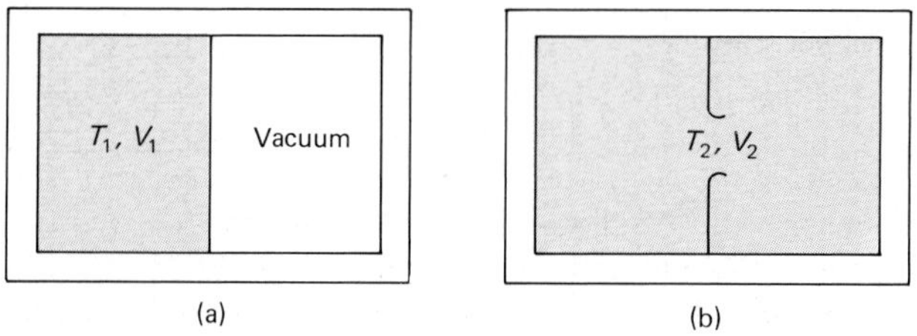

Figure 24.11

Initially, (a) one portion of the rigid insulated container is a vacuum and an ideal gas occupies the other section. After the partition ruptures, (b) the gas expands to fill the container. The insulation and rigid walls guarantee that the system is isolated.

insulated. We noted that the insulation guaranteed that $dQ = 0$ *at each stage* of the expansion. From

$$dS = \frac{dQ}{T}$$

it might seem that we should conclude that $dS = 0$ at each stage, and therefore that the entropy of the gas remains constant. But this is not so! The entropy *does* change. The expression $dS = dQ/T$ *does not apply* to the free expansion, because the free expansion is an irreversible process. It cannot be reversed by any infinitesimal change in the surroundings; and as we said earlier, the relation $dS = dQ/T$ holds only for reversible transformations.

The result

$$S_2 - S_1 = nR \ln\left(\frac{V_2}{V_1}\right) > 0$$

obtained via Eq. 24.28 *is* valid here even though the process leading from state 1 to state 2 is irreversible. This is because entropy is a property of the thermodynamic state. The entropy change $(S_2 - S_1)$ is independent of the process that carries the system from one state to another.

Questions

5. (a) Give an example of an irreversible process that occurs spontaneously in nature (other than the ones mentioned in the text). (b) Indicate how the process you chose increases disorder.

6. A piece of steel absorbs 10 kJ of heat and undergoes a 10-C° increase in temperature. It then returns to its initial state by ejecting 10 kJ of heat. (a) What is the change in entropy for the overall process? (b) Is the process necessarily reversible?

24.6 Entropy Formulation of the Second Law

We noted in Section 24.3 that *if* an irreversible process, like the flow of heat, were to become reversible, *then* the reversed process could be used to violate the second law. The second law is a consequence of irreversibility. Although the energy of an isolated system remains constant, its entropy need not remain constant. The free expansion of an ideal gas considered in the preceding section shows that the entropy of an isolated system increases as the result of an irreversible process. Entropy, unlike energy, is not a conserved quantity. In general, we can say that irreversible processes generate entropy. For example, the diffusion of perfume vapor is an irreversible process that leads to increased disorder—an entropy increase. Evidently there is an "irreversibility connection" between entropy and the second law. This connection led Rudolph Clausius to an entropy formulation of the second law:

The entropy of an isolated system never decreases.

This statement is generally referred to as the *entropy principle*. Stated mathematically, this version of the second law is

$$\Delta S_{\text{isolated system}} \geq 0 \qquad (24.29)$$

where ΔS denotes the change in entropy brought about by any thermodynamic processes that connect equilibrium states of an isolated system. If an isolated system is not in equilibrium, irreversible processes operate to drive it toward equilibrium. Irreversible processes generate entropy. They mix, disorder, or randomize. The inequality in Eq. 24.29 refers to nonequilibrium situations in which irreversible processes generate entropy. From these remarks it follows that equilibrium corresponds to a state of maximum entropy—the entropy increases until the system attains equilibrium, after which the entropy remains constant.

Keep in mind that the entropy principle applies *only* to isolated systems. For example, a healthy 10-yr-old boy converts mashed potatoes into muscle—disorder into order. But the boy is not an isolated system, and the entropy principle does not apply.

Questions

7. Which is more disordered, ice or liquid water? Does the entropy of the water increase or decrease when the water freezes? Does the spontaneous freezing of a pond in winter violate the second law of thermodynamics?

8. Many students find that their rooms tend toward a highly disordered state as final exams approach. Is this a consequence of the second law of thermodynamics?

Summary

Heat engines are cyclic devices that extract heat from a high-temperature reservoir and convert part of it into work. The fraction of heat converted is defined as the thermodynamic efficiency:

$$\eta = \frac{\text{work output}}{\text{heat absorbed}} = \frac{W}{Q_{\text{abs}}} \qquad (24.2)$$

The second law of thermodynamics places limits on the conversion of heat into work. In effect, the second law states that $\eta < 1$. In other words, the complete conversion of heat into work is not possible. The second law is a consequence of the irreversibility of processes that occur spontaneously, such as the flow of heat.

The Carnot engine is the most efficient form of heat engine. A Carnot engine that extracts heat from a high-temperature reservoir and ejects heat to a low-temperature reservoir has an efficiency that is independent of the working substance but dependent on the temperatures of the reservoirs. If a Carnot engine operates with one reservoir at the triple point of water, the Kelvin temperature of the other reservoir is given by

$$T = 273.16 \left(\frac{Q}{Q_3}\right) K \qquad (24.10)$$

where Q_3 is the heat exchanged with the reservoir at the triple point of water and Q is the heat exchanged with the reservoir at temperature T.

The efficiency of a Carnot engine that operates between reservoirs at temperatures T_H and T_L is

$$\eta_C = 1 - \frac{T_L}{T_H} \tag{24.13}$$

No heat engine operating between T_H and T_L can have an efficiency greater than η_C.

Entropy measures the randomness, or disorder, of a system at the molecular level. The entropy change dS during a reversible process in which a system at temperature T exchanges heat dQ is defined by

$$dS = \frac{dQ}{T} \tag{24.17}$$

The entropy change during a finite reversible process is

$$\Delta S \equiv S_f - S_i = \int_{T_i}^{T_f} \frac{dQ}{T} \tag{24.18}$$

Entropy, unlike energy, is not a conserved quantity. Irreversible processes lead to increased disorder and hence generate entropy. The second law of thermodynamics can be written in terms of entropy:

The entropy of an isolated system never decreases.

Irreversible processes drive an isolated system toward equilibrium while increasing its entropy. The equilibrium state corresponds to a state of maximum entropy.

Suggested Reading

S. W. Angrist and L. G. Helper, *Order and Chaos*, Basic Books, New York, 1967.

L. Bryant, Rudolf Diesel and His Rational Engine, *Sci. Am.* 221, 108 (August 1969).

O. V. Lounasmaa, New Methods for Approaching Absolute Zero, *Sci. Am.* 221, 26 (December 1969).

C. M. Summers, The Conversion of Energy, *Sci. Am.* 225, 148 (September 1971).

G. Walker, The Stirling Engine, *Sci. Am.* 229, 80 (August 1973).

Problems

Section 24.1 Heat Engines and Thermodynamic Efficiency

1. An inventor claims that his heat engine absorbs 38 kJ of heat, performs 42 kJ of work, and ejects 4 kJ of heat per cycle. Would you be inclined to invest money in this engine? Explain.

2. A heat engine absorbs 18 kJ of heat each cycle. Its thermodynamic efficiency is 0.36. (a) How much heat does this engine eject per cycle? (b) How much work does it do each cycle?

3. A heat engine performs 24 kJ of work and ejects 36 kJ of heat during one cycle. (a) Determine the heat absorbed by the heat engine over one cycle. (b) Determine the thermodynamic efficiency.

4. Determine the thermodynamic efficiency for the cycle shown in Figure 1.

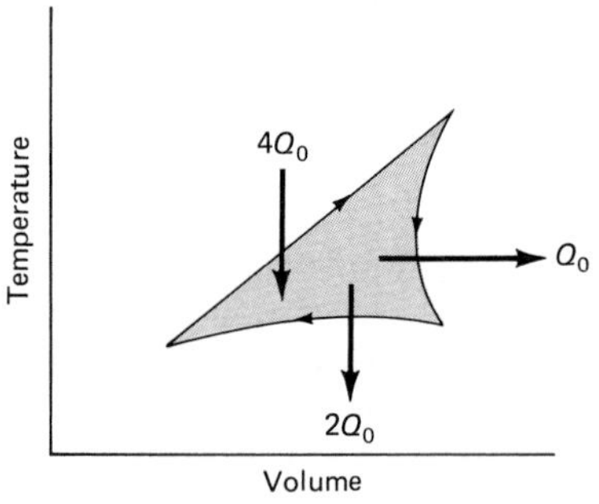

Figure 1

5. Show that the efficiency of the cycle shown in Figure 2 is $\frac{1}{3}$.

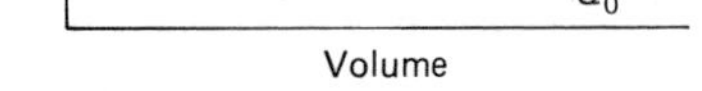

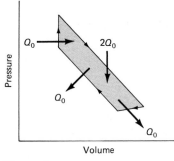

Figure 2

6. (a) Calculate the net work done over the cycle shown in Figure 3. The expansion from 2 to 3 and the compression from 4 to 1 are adiabatic. The arrows indicate the absorption of heat from 1 to 2 and the ejection of heat from 3 to 4. (b) Calculate the thermodynamic efficiency.

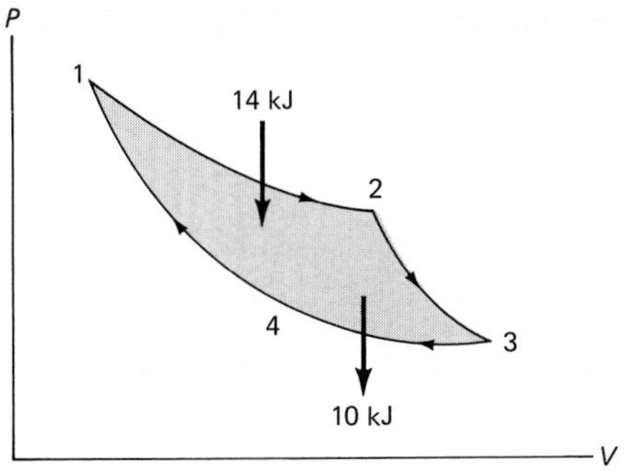

Figure 3

7. A heat engine operates on the cycle shown in Figure 4. It absorbs heat along two legs and ejects heat along two legs, as shown. It performs 10 kJ of work in each cycle. (a) Determine the heat ejected along the lower leg of the cycle. (b) Determine the efficiency of the engine.

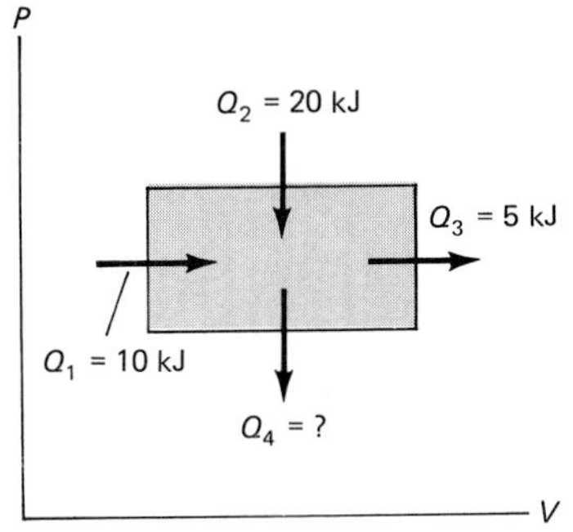

Figure 4

8. Which has the greater thermodynamic efficiency, a heat engine for which (a) the heat absorbed from the high-temperature reservoir is three times the work output, or one for which (b) the heat absorbed from the high-temperature reservoir is twice the heat ejected to the low-temperature reservoir?

9. A heat pump has a COP = 2.7. The net work performed on the refrigerant in 2 hr is 7 kWh. Determine the heat extracted from the low-temperature reservoir in 2 hr.

10. Prove that the coefficient of performance (COP) and thermodynamic efficiency (η) are related by

$$\eta = \frac{1}{1 + \text{COP}}$$

11. The coefficient of performance for a refrigerator is 3.7. Determine its thermodynamic efficiency.

12. The thermodynamic efficiency of an air conditioner is 0.23. (a) Determine its coefficient of performance. (b) If the air conditioner performs 3×10^7 J of work in 1 hr, how much heat does it extract from the low-temperature reservoir in that time?

13. During each cycle of its operation, a refrigerator extracts 480 kJ of energy from the low-temperature reservoir and ejects 360 kJ of energy to the high-temperature reservoir. Determine (a) the work required and (b) the coefficient of performance.

Section 24.2 The Carnot Cycle

14. Suppose that an ideal gas is used as the working substance in the Carnot cycle depicted in Figure 24.5. Remember that the internal energy of an ideal gas depends on the temperature alone. (a) If the gas absorbs Q_H = 3 kJ during the isothermal leg from 1 to 2, how much work is performed by the gas? (b) If the gas performs 1 kJ of work during the adiabatic expansion from 2 to 3, what is the change in its internal energy? (c) How much work must be performed on the gas from 3 to 4 for it to eject Q_L = 2 kJ of heat? (d) How much work must be performed on the gas to take it from 4 to 1? (e) What is the net amount of work performed by the gas during the cycle? (f) What is the thermodynamic efficiency of the cycle?

15. An ideal gas undergoes the Carnot cycle shown in Figure 24.5. The P, V, and T values at points 1, 2, 3, and 4 are as follows.

1:	$P = 10 \times 10^5$ N/m^2	$V = 1.$ m^3	$T = 400$ K
2:	$P = 5 \times 10^5$ N/m^2	$V = 2.$ m^3	$T = 400$ K
3:	$P = 2.44 \times 10^5$ N/m^2	$V = 3.08$ m^3	$T = 300$ K
4:	$P = 4.87 \times 10^5$ N/m^2	$V = 1.54$ m^3	$T = 300$ K

Section 24.3 The Second Law of Thermodynamics

16. One mole of an ideal gas expands isothermally. (a) If the gas doubles its volume, show that the work of expansion is

$$W = RT \ln 2$$

(b) Since the internal energy of an ideal gas depends solely on its temperature, there is no change in U during the expansion. It follows from the first law that the heat absorbed by the gas during the expansion is converted completely into work. Why does this *not* violate the second law?

Section 24.4 The Kelvin Temperature Scale

17. Which change would produce a greater increase in the Carnot efficiency, (a) a 25-K increase in the temperature of the high-temperature reservoir, or (b) a 25-K decrease in the temperature of the low-temperature reservoir? Explain.

18. A Carnot cycle operates between reservoirs at temperatures of 180°C and 30°C. Determine its Carnot efficiency.

19. A heat engine absorbs heat from a reservoir at 60°C and ejects heat to the surroundings at 20°C. Its thermodynamic efficiency is 0.08. Determine its Carnot efficiency.

20. There are serious proposals to tap the thermal energy of the ocean by operating a heat engine utilizing $T_H = 32°C$ of warm surface water in the tropics and $T_L = 14°C$ of cooler deep water. Calculate the Carnot efficiency of a heat engine operating between these temperatures.

21. An inventor comes to you with the claim that her heat engine (which employs water as a working substance) has a thermodynamic efficiency of 0.61. She explains that it operates between heat reservoirs at 4°C and 0°C. It is a very complicated device, with many pistons, gears, and pulleys, and the cycle involves freezing and melting. Does her claim that $\eta = 0.61$ warrant serious consideration? Explain.

22. The Carnot efficiency is given by Eq. 24.13. For the same value of T_L, which produces the larger increase in the Carnot efficiency, a rise in T_H from 500 K to 510 K or a rise in T_H from 1000 K to 1010 K?

23. If T_L for a Carnot cycle is limited to 290 K, what value of T_H is needed for a Carnot efficiency of 0.70?

24. A nuclear reactor heats pressurized water to a temperature of 313°C. Heat is ejected to the condenser at a temperature of 38°C. Determine the Carnot efficiency for the cycle.

Section 24.5 Entropy

25. At a pressure of 1 atm, liquid helium boils at 4.2 K. The latent heat of vaporization is 20.5 kJ/kg. Determine the entropy change (per kilogram) resulting from vaporization.

26. The specific heat of copper (at constant pressure) is 0.39 kJ/kg · C°. (a) If the temperature of 20 gm of copper is raised from 0°C to 20°C, determine its entropy change. (b) How much heat is absorbed by the copper in the process? (c) If all of the heat computed in part (b) were absorbed by the copper at the average temperature (10°C), what would be the entropy change?

27. The molar heat of fusion of lead (Pb) is 4.69 kJ/mole. The normal melting point is 327.5°C. Determine the entropy change of 100 gm of lead when it melts at a pressure of 1 atm. The atomic weight of lead is 207.2.

28. The heat of vaporization of ethyl alcohol is 1120 kJ/kg at 27°C. Determine the entropy change of 1 gm of ethyl alcohol when it evaporates at room temperature (27°C).

29. How much water must be vaporized at the normal boiling point to produce the same entropy change as occurs in the melting of 1 kg of ice?

30. A substance undergoes a Carnot cycle during which its temperature changes by a factor of 2 and its entropy changes by a factor of 4. (a) Sketch a graph of the cycle using temperature–entropy (T–S) axes. (b) What is the physical significance of the "area" enclosed by the graph?

31. The isothermal expansion of an ideal gas carries it from $P = 10 \times 10^5$ N/m², $V = 1.$ m³ to $P = 5 \times 10^5$ N/m², $V = 2.$ m³ at a temperature of 400 K. Calculate the entropy change.

32. The rigid insulated container used in the free expansion experiment (Section 24.6) is prepared with 1 mole of chlorine gas on one side of the partition and 1 mole of bromine gas on the other. The partition ruptures, allowing the two gases to mix. The volumes V_1 and V_2 are equal. Treat the two gases as ideal and show that the entropy increase of the system is

$$\Delta S = 2R \ln 2$$

This quantity is known as the *entropy of mixing*. How has disorder increased?

33. One kg of water at 10°C is mixed with 1 kg of water at 30°C. The process proceeds at constant pressure. When the mixture has reached equilibrium, (a) what is the final temperature? (b) Take $C_p = 4.19$ kJ/kg · C° for water and show that the entropy of the system increases by

$$\Delta S = 4.19 \ln\left(\frac{293}{283} \cdot \frac{293}{303}\right) \text{ kJ/K}$$

(c) Verify numerically that $\Delta S > 0$. (d) Is the mixing process irreversible?

34. It is an empirical fact, known as *Trouton's rule*, that the entropy change per mole associated with boiling is roughly the same for many liquids. Test this rule for the liquids listed below by calculating the entropy change per *kilomole*.

Liquid	Latent Heat of Vaporization (kJ/kg)	Molecular Weight (kg/kmol)	Normal Boiling Point (K)
H_2O	2256	18	373
O_2	213	32	90
^4He	20.5	4	4.2
C_2H_5 (ethyl alcohol)	854	46	351
C_6H_6 (benzene)	402	78	353

chapter 25

KINETIC THEORY

Preview

Kinetic theory is based on an atomic model of matter. The basic assumption of kinetic theory is that the measurable properties of gases, liquids, and solids reflect the combined actions of countless numbers of atoms and molecules. For example, the pressure exerted on the walls of a bicycle tire is produced by the impacts of an enormous number of air molecules. Kinetic theory relates the microscopic properties of atoms and molecules, which are not directly measurable, to measurable properties of matter, like temperature and pressure.

We begin this chapter by showing that the characteristic size of an atom is about 10^{-10} m. We then use this knowledge to show that the atoms in a gas spend most of their time in free flight. The important concept of a collision cross section is introduced as a measure of the likelihood of collisions between atoms.

Next we use the atomic model of a gas to derive the ideal gas equation of state. The analysis of this equation leads to a kinetic interpretation of temperature: Temperature measures the average kinetic energy of atomic and molecular motions.

In the final section we introduce the distribution function concept. The Maxwellian distribution of molecular speeds is presented and used to determine the most probable molecular speed.

25.1 The Atomic Model of Matter

A model is a conceptual picture coupled with a set of assumptions. The atomic model pictures matter as composed of small particles called *atoms*. In many instances the basic unit of matter is the *molecule*, formed by two or more atoms held together by interatomic forces. For example, the air we breathe is made up largely of oxygen molecules (O_2) and nitrogen molecules (N_2). The water we drink is composed of molecules, each containing two hydrogen atoms and one oxygen atom (H_2O). The primary assumption we make is that atoms and molecules obey Newton's laws of motion.

One of the seven fundamental quantities in the SI system (Section 1.1) is related to the atomic model of matter. This quantity is called the *mole*. One mole is the amount of a substance that contains Avogadro's number (N_A) of atoms or molecules:

$$N_A = 6.02205 \times 10^{23} \tag{25.1}$$

Avogadro's number is defined as the number of atoms in 12 gm of carbon 12 (^{12}C).*

*Note that the mass is specified in grams rather than in the SI unit, the kilogram. One kilomole is defined as the number of atoms in 1 kg of carbon 12.

The mass of 1 mole of a substance is called its *gram molecular weight*. By definition, then, the gram molecular weight of ^{12}C is 12 gm. Some approximate gram molecular weights are the following: hydrogen (H_2), 2.016 gm; helium (He), 4.003 gm, water (H_2O), 18.02 gm, oxygen (O_2), 32 gm. Thus, 2.016 gm of hydrogen contains approximately 6.022×10^{23} hydrogen molecules (H_2), and 4.003 gm of helium contains approximately 6.022×10^{23} helium atoms (He).

A monatomic gas is a gas composed of single atoms. Helium (He) is a monatomic gas. A diatomic gas is a gas composed of molecules containing two atoms. Hydrogen (H_2) and carbon monoxide (CO) are diatomic gases. A polyatomic gas is a gas composed of molecules containing more than two atoms. Methane (CH_4) and steam (water vapor) are polyatomic gases.

We cannot see the molecules of a gas because atoms and molecules are very small. But just how small are they? For example, what is the size of a typical atom? We do not need sophisticated equipment or an understanding of quantum mechanics to answer this question. A reliable estimate of atomic size can be made, based on the observation that solids and liquids are virtually incompressible. Extremely high pressures are required to alter the volume of a solid or liquid by a few percent. By contrast, a gas is readily compressed. These facts suggest that the atoms in a gas are relatively far apart whereas those in liquid or solid form are packed rather tightly.

We can estimate atomic sizes by assuming that the volume of a solid or liquid equals the total volume of the constituent atoms. That is, we ignore the space between atoms. This is a reasonable assumption, because if there were much space between atoms in a solid or liquid it would be possible to squeeze them closer together by applying pressure. However, as we already noted, squeezing a solid or a liquid has virtually no effect on its volume.

Let us first consider 1 mole of a solid. Let V denote its volume and let M signify its gram molecular weight. The number of molecules in this molar volume is $N_A \simeq 6 \times 10^{23}$. Now imagine that the volume V is composed of N_A cubical subvolumes of equal size—one for each molecule. If the length of the edge of such a cube is d, the cube's volume is d^3. Thus d is a measure of molecular size. Taken collectively, the N_A cubes comprise the total volume V and

$$V = N_A d^3$$

Next we observe that the mass density ρ may be expressed by

$$\rho = \frac{\text{mass of 1 mole}}{\text{volume of 1 mole}} = \frac{M}{V} = \frac{M}{N_A d^3}$$

Solving for d gives

$$d = \left(\frac{M}{N_A \rho}\right)^{1/3} \qquad (25.2)$$

We can use Eq. 25.2 to find the molecular size of solids and liquids. For example, let's use this equation to find the size of a water molecule. For water, M = 18 gm and ρ = 1 gm/cm^3, and Eq. 25.2 gives

$$d_{\text{water}} = \left(\frac{18 \text{ gm}}{6 \times 10^{23} \times 1 \text{ gm/cm}^3}\right)^{1/3}$$
$$= 3.1 \times 10^{-8} \text{ cm}$$

Table 25.1

Characteristic atomic dimensions*

Element	M (gm)	ρ (gm/cm^3)	$(M/N_A\rho)^{1/3}$ (Å)
Solids			
Li	6.9	0.53	2.8
Ne	20.2	1.00	3.2
Na	23.0	0.97	3.4
Al	27.0	2.70	2.6
K	29.1	0.86	4.2
Ar	39.9	1.65	3.4
Zn	65.4	7.1	2.5
Rb	85.5	1.53	4.5
Ag	107.9	10.5	2.6
Cs	132.9	1.87	4.9
Pt	195.2	21.4	2.5
Au	197.0	19.3	2.6
Pb	207.2	11.3	3.1
Liquids			
He	4.0	0.12	3.8
Ne	20.2	1.21	2.9
Ar	39.9	1.39	3.6
Kr	83.8	2.61	3.7
Xe	131.3	3.06	4.1
Hg	200.6	13.6	2.9

*The second column gives the gram molecular weight; the third column gives the mass density in grams per cubic centimeter; the fourth column gives the characteristic atomic dimension $d = (M/N_A\rho)^{1/3}$ in angstroms (1 Å = 10^{-10} m).

The angstrom (abbreviated Å) is a convenient unit of length for atomic and molecular sizes.

$$1 \text{ angstrom} \equiv 10^{-8} \text{ cm} = 10^{-10} \text{ m} = 1 \text{ Å}$$

Thus, the size characteristic of a water molecule is 3.1 Å.

When Eq. 25.2 is used to evaluate d for many solids and liquids an interesting pattern emerges. As Table 25.1 shows, $d = (M/N_A\rho)^{1/3}$ is very near 3 Å for many elemental solids and liquids, even though the values for M and ρ differ widely. Other materials furnish similar results. Measurements of atomic spacing made by scattering x rays from solids give values within 10% of those in Table 25.1. We conclude that the angstrom characterizes molecular and atomic sizes.

Example 1 **Sizing Up Ethyl**

Ethyl alcohol has the chemical formula C_2H_5OH. Its gram molecular weight is 46.1 gm and its density is 0.789 gm/cm^3. Inserting these figures into Eq. 25.2 gives

$$d = \left(\frac{46 \text{ gm}}{6 \times 10^{23} \times 0.789 \text{ gm/cm}^3}\right)^{1/3} = 4.6 \text{ Å}$$

Although the ethyl alcohol molecule is not cubic, d = 4.6 Å is a rough measure of its size.

Questions

1. This textbook contains approximately 1000 pages. Use this fact to estimate the thickness of one page.
2. In making an estimate of atomic size, we treated the atoms as cubes. Had we treated them as rigid spheres, would our computed estimate of their size have been larger, smaller, or the same?

25.2
Mean Free Path and Cross Section

The atoms of a gas are constantly in motion. Atoms collide with each other, and with the walls of their container, and are deflected. As a result, they travel in irregular zigzag paths. The distance that an atom travels between collisions is called its free *path length*. Since collisions occur randomly, the length of individual free paths cannot be predicted. However, a useful measure of free path length is the average, or mean, distance between collisions, the so-called mean free path (λ). We can estimate λ as follows: Let the effective radius of an atom be denoted by R_0. By effective we mean that a collision occurs if parallel lines through the centers of two approaching atoms are separated by less than $2R_0$ (Figure 25.1). The effective radius R_0 is a length that measures the range of interatomic forces, that is, the distance over which the atoms exert forces on one another. As we might expect, R_0 is typically a few angstroms.

Corresponding to R_0 there is an area, πR_0^2. Geometrically, πR_0^2 represents the *cross-sectional area* of a sphere of radius R_0. Physically, πR_0^2 is a measure of the target area that each atom presents to other atoms. This area is called the *atomic collision cross section*, or simply the cross section, and is denoted by the Greek letter sigma (σ). If we take $R_0 \simeq 3$ Å $= 3 \times 10^{-10}$ m, the cross section is

$$\sigma = \pi R_0^2 \simeq 3 \times 10^{-19}\ \text{m}^2 \qquad (25.3)$$

As atoms move about between collisions, their circular cross sections sweep out cylindrical volumes. On the average the cross-sectional area σ sweeps out a volume $\sigma\lambda$, where λ is the mean free path (Figure 25.2). For N atoms in a container of volume V, the volume per atom is V/N. We can estimate λ by equating $\sigma\lambda$ and V/N:

$$\sigma\lambda = \frac{V}{N} \qquad (25.4)$$

The number of atoms per unit volume is called the *number density, n*:

$$n = \frac{N}{V} \qquad (25.5)$$

Setting $V/N = 1/n$, we can rewrite Eq. 25.4 as

$$n\sigma\lambda = 1 \qquad (25.6)$$

This is one of the most useful of all kinetic theory relations: The product of (number density) × (cross section) × (mean free path) equals unity. Equation 25.6 can be applied to many processes other than the simple "billiard-ball" collision of atoms.

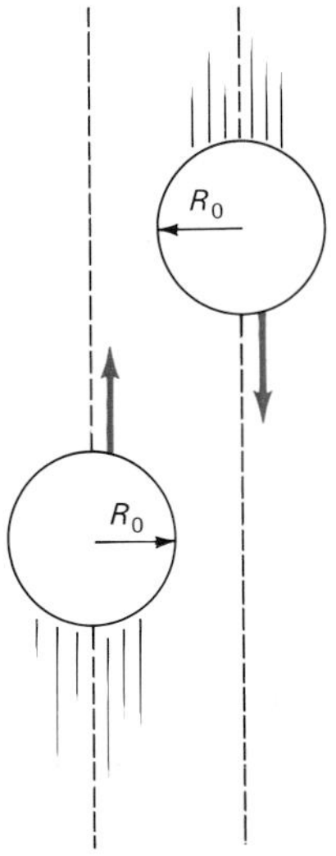

Figure 25.1

The effective radius of an atom (R_0) is a measure of the range of interatomic forces. A collision occurs if parallel lines through the centers of two approaching atoms are separated by less than $2R_0$.

Figure 25.2

The cross-sectional area, σ, sweeps out a cylindrical volume. Over one mean free path length, λ, the volume swept out is $\sigma\lambda$.

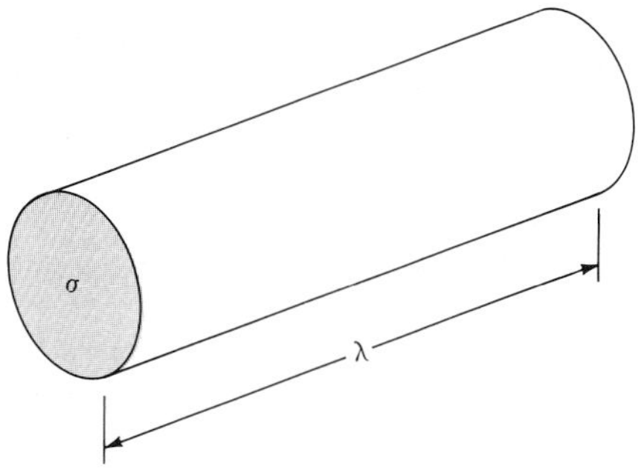

Example 2 Mean Free Path for Atmospheric Molecules
Estimate the mean free path for a molecule under typical atmospheric conditions. The number density can be estimated by using the ideal gas law (Section 22.2). For 1 mole of ideal gas

$$V = \frac{RT}{P}$$

The number density is

$$n = \frac{N_A}{V} = \frac{N_A P}{RT}$$

We take

$$P = 1 \text{ atm} \simeq 10^5 \text{ N/m}^2$$

$$T = 300 \text{ K } (27°\text{C})$$

With $N_A = 6.02 \times 10^{23}$ and $R = 8.31$ J/K we find

$$n = 2.4 \times 10^{25}/\text{m}^3$$

For the cross section we take

$$\sigma = 3 \times 10^{-19} \text{ m}^2$$

Inserting these values of n and σ into Eq. 25.6 we get for the mean free path

$$\lambda = \frac{1}{n\sigma} = 1.4 \times 10^{-7} \text{ m} = 1400 \text{ Å}$$

Our estimate for σ used $R_o = 3$ Å. The mean free path of 1400 Å is more than 400 times as large, showing that an atmospheric molecule spends most of its time in free flight.

Questions

3. The pressure of a gas is doubled by compressing it at constant temperature. Assuming that the gas obeys the ideal gas equation of state ($PV = nRT$), what change, if any, is there in the mean free path (λ) of the gas molecules?
4. Complete this statement and justify your choice. Your mean free path through a crowd is ____ the number of people per square meter in the crowd.
   (a) directly proportional to
   (b) directly proportional to the square of
   (c) inversely proportional to the square of
   (d) inversely proportional to
   (e) independent of
5. The number of densities of two gases are the same. The atoms of the gases have diameters of 2 Å and 4 Å. If the mean free path for the smaller atoms is 2000 Å, what is it for the larger atoms?
6. How does the total volume of the atoms compare with the volume of the container for (a) a gas; (b) a liquid?

25.3 The Ideal Gas: Kinetic Interpretation of Temperature

We have made frequent use of the ideal gas equation of state (Section 22.2),

$$PV = nRT \tag{25.7}$$

We will now rewrite this equation to obtain a form that emphasizes the microscopic viewpoint. We express nR as

$$nR = nN_A \cdot \frac{R}{N_A} \equiv Nk \tag{25.8}$$

where N_A is Avogadro's number and

$$nN_A = N \tag{25.9}$$

is the total number of molecules in n moles. The ratio

$$\frac{R}{N_A} \equiv k = 1.38062 \times 10^{-23} \text{ J/K} \tag{25.10}$$

is called the *Boltzmann constant.** Using Eq. 25.8 we replace nR by Nk in Eq. 25.7 to obtain the alternate form of the ideal gas equation of state,

$$PV = NkT \tag{25.11}$$

This equation of state can be derived by introducing a microscopic model of an ideal gas. This model makes four basic assumptions:

1 The gas consists of a large number of particles (atoms†) that obey Newton's laws of motion.

2 The motion of the atoms is random.

3 The volume of the atoms themselves is negligible by comparison with the volume of the container. The atoms of a monatomic gas are treated as point masses.

4 The atoms collide elastically with the container walls, but do not collide with one another.

For a monatomic gas, this means that the atoms possess only translational kinetic energy, except during the instant of impact in a wall collision.

Let's consider these four assumptions one by one and see what evidence supports each:

1 Experimental evidence for assuming that the gas contains many atoms is related to determinations of Avogadro's number, $N_A \simeq 6 \times 10^{23}$. At 0°C and 1 atm, 1 cm^3 of gas contains approximately 3×10^{19} atoms, an enormous number by any standard.

The assumption that the atoms obey Newton's laws of motion is adequate for the development of most aspects of kinetic theory. The internal structure of atoms can be determined by applying modern quantum theory. However, the most basic features of kinetic theory are independent of the internal structure of the atoms and molecules. Historically, kinetic theory evolved within a classical framework, and we follow this historical path.

2 Experimental evidence for the random motion of atoms is based on observations of what we call *Brownian motion.* In 1827, the Scottish botanist Robert Brown used a microscope to observe tiny spores of pollen suspended in water. The pollen appeared to dance about in erratic fashion. At first, Brown thought that the pollen was alive and that its motion was some sort of fertility dance! Subsequent studies of liquid suspensions of various inanimate particles convinced observers that the liquid itself was responsible for the erratic motions. Eventually, the irregular motion of the suspended particles was explained in detail by assuming that the liquid was composed of molecules in random motion. The random motion of atoms in a gas was demon-

*The units for k are joules per Kelvin (J/K). The product kT has the units of energy.

†For the present we consider a monatomic gas.

strated by Fletcher and Millikan in 1911. They observed the Brownian motion of electrically charged oil droplets, using many of the same techniques developed by Millikan to determine the charge of the electron (Section 27.6).

3 An atom behaves like a point mass in only a few gross respects. However, it is these same few characteristics that are crucial to determining the equation of state. The translational kinetic energy of the atom can be written as $\frac{1}{2}mv^2$, where m is the mass of the atom and v is the speed of the center of mass. The linear momentum is given by $m\mathbf{v}$. With respect to translational kinetic energy and linear momentum, the structure of the atom is irrelevant—it behaves like a point mass.

Liquid-to-vapor transition data justify the neglect of the atomic volume in a rarified gas. For example, 1 gm of water at the normal boiling point occupies a volume of approximately 1 cm^3. After vaporization it occupies a volume of over 1700 cm^3, suggesting that each molecule in the vapor has about 1700 times as much room to move about in as it had in the liquid state.

4 The assumption of elastic collisions is a simplification. A more realistic description of collisions is much more complicated. For example, some atoms collide inelastically with the walls of the container. An inelastic collision may result in the transfer of energy from the atom to the wall, or vice versa. However, the assumption of elastic collisions does not lead to any error so long as the gas is in thermal equilibrium with the container walls. Thermal equilibrium requires that no *net* transfer of energy occur between the gas and the walls. When thermal equilibrium prevails, the inelastic collisions collectively do not lead to any net transfer of energy to or from the gas. Hence, ignoring inelastic collisions does not affect our conclusions about the equilibrium states of the gas.

Perhaps the most significant part of the fourth assumption is that the atoms do not collide with one another. This assumption that the atoms do not collide with one another is supported by our estimate of the mean free path, λ, in Example 2. There we found that $\lambda = 1400$ Å. Compared with an atomic size of 3 Å it is evident that atoms spend most of their time in free flight. Thus, we ignore collisions between atoms; however, even in a rarified gas, atoms experience mutual interatomic forces, and it is these interatomic forces that cause the characteristics of real gases to differ from those of an ideal gas.

The equation of state,

$$P = \frac{NkT}{V}$$

expresses the gas pressure in terms of other thermodynamic variables. Therefore, to derive this equation of state we must use the atomic model to determine the gas pressure. Collisions of the atoms with the container walls give rise to a force and thus to a pressure. To determine the pressure we calculate the average momentum transferred to a wall per second. According to Newton's second law, the momentum change per second equals the force. This force divided by the area of the wall is the gas pressure.

1 Momentum Transfer

We first calculate the momentum transfer. Let v_x denote the x-component of velocity of an atom that subsequently collides with the wall (see Figure 25.3). The atom rebounds with an x-component of velocity equal to $-v_x$. The change in its x-component of linear momentum is

$$(p_x)_{\text{after}} - (p_x)_{\text{before}} = -2mv_x$$

A momentum change of equal magnitude and opposite direction is transferred to the container wall. The x momentum transferred to the wall is

$$\begin{matrix}\text{momentum transfer to} \\ \text{wall in one collision}\end{matrix} \equiv \Delta p_x = +2mv_x \qquad (25.12)$$

2 Time Between Collisions

Next we must determine the time between collisions. The gas is enclosed in a rectangular container having sides of length L_x, L_y, and L_z (Figure 25.4). An atom with velocity

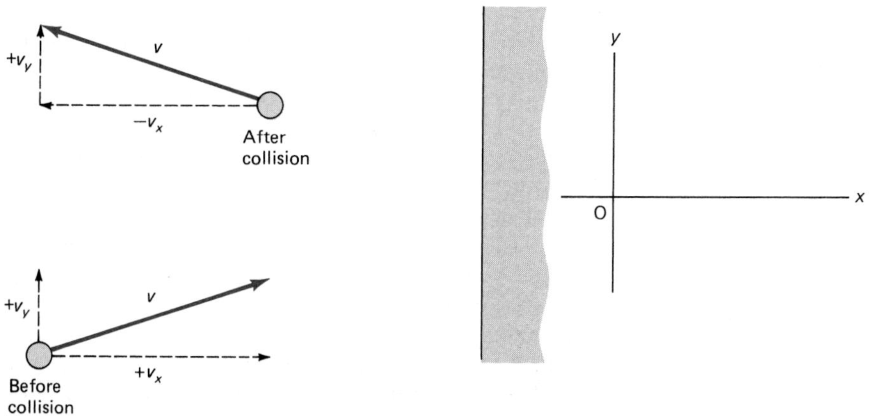

Figure 25.3

An atom undergoes an elastic collision with the wall. The normal component of its velocity (v_x) changes sign. Its speed (v) is unchanged.

Figure 25.4

The rectangular container has a volume $L_xL_yL_z$. The area A of the face perpendicular to the x-direction is L_yL_z.

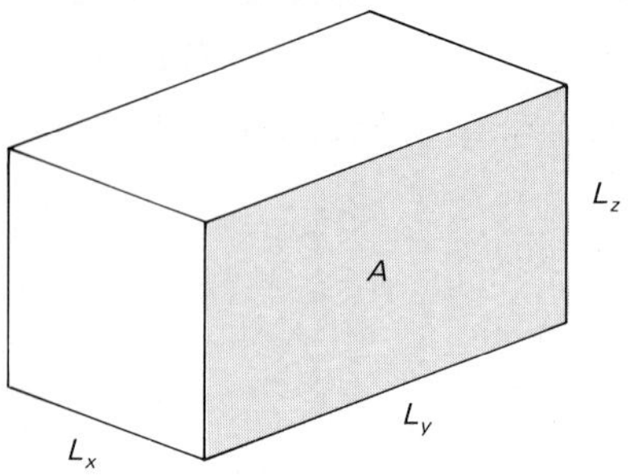

component v_x travels the length L_x of the container in a time L_x/v_x. The time Δt between collisions with a particular wall is twice as great:

$$\Delta t = \frac{2L_x}{v_x} \tag{25.13}$$

3 Average Force

Next, we use Newton's second law to determine the average force exerted on the wall. The average rate at which an atom transfers momentum to the wall is

$$\frac{\Delta p_x}{\Delta t} = \frac{2mv_x}{2L_x/v_x} = \frac{mv_x^2}{L_x} \tag{25.14}$$

By Newton's second law, the rate of change of momentum equals force:

$$\frac{\Delta p_x}{\Delta t} = f_x \tag{25.15}$$

Thus, the average force exerted by the impacts of a single atom having an x velocity component v_x is

$$f_x = \frac{mv_x^2}{L_x} \tag{25.16}$$

4 Pressure

The force exerted by a single atom is not observable: A pressure gauge responds to the combined effects of many such collisional forces. The total force on the wall, F_x, is obtained by summing over all atoms.

$$F_x = \sum_{\text{aa}} f_x = \sum_{\text{aa}} \frac{mv_x^2}{L_x} \tag{25.17}$$

where "aa" stands for "all atoms." The area of the container wall perpendicular to the x-axis is $A = L_yL_z$ (Figure 25.4). The pressure P on this area is the force F_x divided by the wall area $A = L_yL_z$.

$$P = \frac{F_x}{A} = \sum_{\text{aa}} \frac{mv_x^2}{L_xL_yL_z} \tag{25.18}$$

Although Eq. 25.18 is an expression for the pressure, we must eliminate the reference to direction to put it in a useful form. Because $L_xL_yL_z$ is the volume (V) of the container, we can write

$$PV = \sum_{\text{aa}} mv_x^2 \tag{25.19}$$

Now, the pressure of the gas is the same on every wall; there is *no preferred direction of motion*. Thus we can also write

$$PV = \sum_{\text{aa}} mv_y^2 = \sum_{\text{aa}} mv_z^2 \tag{25.20}$$

Further, since

$$v_x^2 + v_y^2 + v_z^2 = v^2 \tag{25.21}$$

we can write

$$PV = \frac{1}{3}\sum_{\text{aa}} m(v_x^2 + v_y^2 + v_z^2)$$

$$= \frac{1}{3}\sum_{\text{aa}} mv^2 \tag{25.22}$$

We know that the total kinetic energy K_{tot} of the gas is

$$K_{\text{tot}} = \sum_{\text{aa}} \frac{1}{2} mv^2 \tag{25.23}$$

Thus, comparing Eq. 25.22 and 25.23 shows that

$$PV = \frac{2}{3} K_{\text{tot}} \tag{25.24}$$

5 Equation of State

By introducing the average kinetic energy per atom we can maintain the particle point of view and express Eq. 25.24 in a form directly comparable with the equation of state, $PV = NkT$. We write K_{tot} as N times the average kinetic energy per atom, $\langle K\rangle$:

$$K_{\text{tot}} = N\langle K\rangle$$

whereupon Eq. 25.24 becomes

$$PV = \frac{2}{3} N\langle K\rangle \tag{25.25}$$

Equation 25.25 is the equation of state that the ideal gas model has produced. Experimentally, the equation of state that is approximately obeyed by all gases is

$$PV = NkT \tag{25.11}$$

Comparing Eqs. 25.11 and 25.25, we can conclude that the temperature of an ideal gas is proportional to the average kinetic energy of the atoms. Specifically, we must have

$$\langle K\rangle = \left\langle \frac{1}{2} mv^2 \right\rangle = \frac{3}{2} kT \tag{25.26}$$

if the model is to reproduce the empirical form of the equation of state. We emphasize that Eq. 25.26 is *not* an operational definition of temperature, because we cannot measure the average kinetic energy of an atom. The Kelvin temperature T is defined operationally in terms of heat and the Carnot cycle (Section 24.4). However, our model of the ideal gas does give us a microscopic *interpretation* of temperature. It takes us beyond the semiquantitative statement, "Equality of temperature is a condition for thermal equilibrium," and gives us a mechanical feel for temperature. It shows us that the temperature of a gas measures the translational kinetic energy of the thermal motions of its atoms.

Our derivation of the equation of state applies equally well to diatomic and polyatomic gases, for the following reason. Only considerations involving the translational kinetic energy and linear momentum of the center of mass enter the pressure calculation. Because we follow only the center-of-mass motion, any structure of the atoms or molecules is irrelevant insofar as the equation of state is concerned. Thus, ideal gases obey the same equation of state regardless of any microscopic structure attributed to their atoms and molecules.

Example 3 **Average Thermal Energy of Atoms**

What is the average kinetic energy for atoms in a gas at room temperature ($T = 300$ K)? Using Eq. 25.26 we have

$$\langle K\rangle = \frac{3}{2}kT = \frac{3}{2}\left(1.38 \times 10^{-23}\ \frac{\text{J}}{\text{K}}\right)(300\ \text{K})$$

$$= 6.21 \times 10^{-21}\ \text{J}$$

This value of $\langle K\rangle$ is more conveniently expressed in the widely used energy unit, the *electron volt* (eV).

$$1\ \text{eV} \equiv 1.6 \times 10^{-19}\ \text{J}$$

Thus

$$\langle K\rangle = 6.21 \times 10^{-21}\ \text{J} = 0.0388\ \text{eV}$$

or approximately $\frac{1}{25}$ eV. Note that the mean kinetic energy is independent of the mass of the atom.

RMS Speed

The speed characteristic of molecular motions can be inferred from Eq. 25.26, $\frac{1}{2}m\langle v^2\rangle = \frac{3}{2}kT$. The *mean square* speed is

$$\langle v^2\rangle = \frac{3kT}{m} \tag{25.27}$$

The square root of $\langle v^2\rangle$ is called the *root-mean-square* (rms) speed. Thus,

$$v_{\text{rms}} \equiv \sqrt{\langle v^2\rangle} = \sqrt{\frac{3kT}{m}} \tag{25.28}$$

For helium atoms at $T = 300$ K ($m = 6.66 \times 10^{-27}$ kg)

$$v_{\text{rms}} = \sqrt{\frac{12.40 \times 10^{-21}}{6.66 \times 10^{-27}}}\ \text{m/s} = 1360\ \text{m/s}$$

The foregoing calculation shows that atoms and molecules move swiftly. In fact, the rms speed is comparable to the *speed of sound*. We can demonstrate this fact by recalling from Section 18.3 that acoustic waves in a fluid travel at a speed given by

$$v_{\text{sound}} = \sqrt{\left(\frac{\partial P}{\partial \rho}\right)} \tag{25.29}$$

Sound wave compressions and rarefactions take place *adiabatically* at typical audio frequencies. Therefore, the derivative $\partial P/\partial\rho$ must be evaluated for adiabatic conditions. For an ideal gas (Section 22.2) the equation

$$PV^{\gamma} = \text{constant} \tag{25.30}$$

governs adiabatic transformations. Because $\rho = M/V$, Eq. 25.30 may be rewritten

$$P = C\rho^{\gamma} \qquad C = \text{constant} \tag{25.31}$$

It follows that

$$\frac{\partial P}{\partial \rho} = \gamma C\rho^{\gamma-1} = \gamma\frac{P}{\rho} \tag{25.32}$$

Finally, we rearrange the equation of state

$$P = \frac{NkT}{V} = \frac{mN}{V}\frac{kT}{m} = \rho\frac{kT}{m}$$

to find

$$\frac{P}{\rho} = \frac{kT}{m} \tag{25.33}$$

Using Eq. 25.33 in Eq. 25.32 gives

$$\frac{\partial P}{\partial \rho} = \gamma\frac{kT}{m}$$

which converts Eq. 25.29 to

$$v_{\text{sound}} = \sqrt{\frac{\gamma kT}{m}} \tag{25.34}$$

The near equality of $v_{\text{rms}} = \sqrt{3kT/m}$ and $v_{\text{sound}} = \sqrt{\gamma kT/m}$ is not accidental because sound waves travel by virtue of collisions between atoms.

It is often desirable to estimate the energy or speed characteristics of thermal motions. In such instances, we can disregard factors of 3/2 or γ. We take

$$\text{thermal energy} \approx kT \tag{25.35}$$

$$\text{thermal speed} \approx \sqrt{\frac{kT}{m}} \tag{25.36}$$

Figure 25.5 indicates the wide range of temperatures and thermal energies encountered in nature.

Example 4 **Thermal Energies and Speeds for the Sun**

At the solar surface $T \simeq 5800$ K, and the thermal energy and speed of a proton (the nucleus of a hydrogen atom) are

$$kT = 1.38 \times 10^{-23} \times 5800\ \text{J} \times \frac{1\ \text{eV}}{1.6 \times 10^{-19}\ \text{J}}$$

$$\simeq 0.50\ \text{eV}$$

$$\sqrt{\frac{kT}{m}} = \sqrt{\frac{1.38 \times 10^{-23} \times 5800}{1.67 \times 10^{-27}}}\ \text{m/s} \simeq 6900\ \text{m/s}$$

In the core of the sun, where the temperature is approximately 16×10^6 K,

$$kT \approx 1000\ \text{eV}$$

$$\sqrt{\frac{kT}{m}} \approx 350{,}000\ \text{m/s}$$

Figure 25.5

Characteristic temperatures and thermal energies. Note that the thermal energy (kT) is expressed in *electron volts*, a convenient unit of energy. One electron volt equals 1.60219×10^{-19} joule.

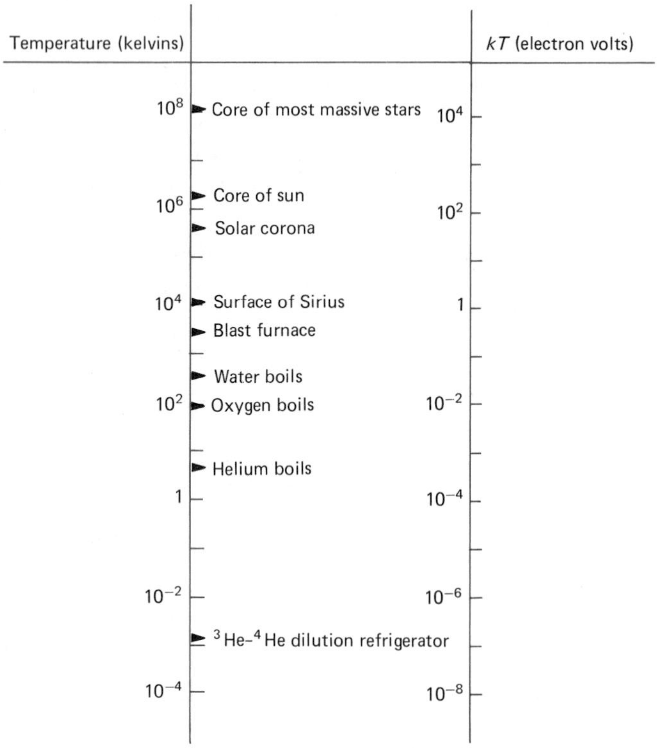

The thermal energy of the solar protons is maintained by thermonuclear fusion reactions in the core of the sun.

Effects of Gravity

Gravity alters the speeds of gas molecules only very slightly under ordinary laboratory conditions. For example, suppose that a nitrogen molecule falls vertically through a distance of 5.0 m. We take its initial speed (downward) to be 516 m/s, a value characteristic of room temperature conditions. The speed (v) after falling a distance h is given by

$$v^2 = v_o^2 + 2gh$$

where

$$v_o = \text{initial speed} = 516 \text{ m/s}$$

$$g = \text{gravitational acceleration} = 9.80 \text{ m/s}^2$$

With $h = 5.0$ m we find the change in speed is

$$v - v_o = 0.095 \text{ m/s}$$

a change of well under 0.1%.

Gravity becomes significant in kinetic theory only when it changes the speeds of molecules by an amount comparable to the rms speed. For example, if the nitrogen molecule falls a distance of 5000 m, then we find that $v = 604$ m/s, a significant change of over 17%.

The atmospheres of planets and stars are among the places where atoms and molecules can move through long distances while subject to gravitational accelerations. As Example 5 shows, we can combine our knowledge of kinetic theory and gravity to estimate the thickness of the earth's atmosphere.

Example 5 How High Can a Molecule Jump? (How Thick Is Our Atmosphere?)

A small fraction of the molecules in the earth's atmosphere have kinetic energies that are 10 times the average value. These exceptional molecules can be thought of as forming the upper limit of the earth's atmosphere because virtually no other molecule can jump higher. We can estimate the thickness of the earth's atmosphere by determining how high these exceptional molecules jump against gravity. If a molecule starts at the earth's surface with a speed v_i, it can rise freely to an altitude h, given by

$$2gh = v_i^2$$

For v_i^2 we take 10 times the mean square value ($v_{\text{rms}}^2 = 3kT/m$). This gives

$$v_i^2 = \frac{30kT}{m} = \frac{30RT}{mN_A} = \frac{30RT}{M}$$

where $M = mN_A$ is the molecular weight. With $T = 300$ K and $M = 28.9$ gm (molecular weight of air) and $R = 8.31$ J/mol · K we get

$$h = \frac{30RT}{2gM} = \frac{30(8.31)(300)}{2(9.80)(28.9 \times 10^{-3})} = 132 \text{ km}$$

which is approximately 80 miles.

The earth's atmosphere does not have a sharp upper boundary. However, virtually all of the atmospheric molecules lie below an altitude of 132 km. The actual mass density of the atmosphere at the earth's surface is approximately 1.29 kg/m^3. At an altitude of 132 km it is less than 10^{-8} kg/m^3, a decrease by a factor of over 100 million. Thus, our value of 132 km is a reasonable estimate for the "thickness" of the atmosphere because virtually all of the atmospheric mass lies below that altitude.

Quantum Zero-Point Energy

There is a potential pitfall in the relation

$$\langle K \rangle = \frac{3}{2} kT \qquad (25.26)$$

Taken *alone* it implies that the kinetic energy drops to zero as the temperature approaches absolute zero. This in turn suggests that absolute zero is a state in which there is no atomic or molecular motion. However, *such an inference is not valid*. The kinetic energy does *not* drop to zero as the temperature approaches absolute zero because there is *another* contribution to the kinetic energy of the atoms and molecules besides thermal kinetic energy. This contribution to the total kinetic energy, ignorably small at ordinary temperatures and densities, is quantum mechanical in nature. The quantum contribution to the kinetic energy is called the *zero-point energy* because it is the energy that remains at $T = 0$.

There are several physical systems wherein the zero-point energy outweights the ordinary thermal kinetic energy. Liquid helium is perhaps the best-known example of such a system. At atmospheric pressure most liquids solidify as the temperature is reduced. The only exception is helium, which remains liquid even at the lowest temperature. The attractive force between He atoms is too weak to overcome the quantum zero-point motion.

Questions

7. A molecule bounces off a wall. There is a transfer of momentum of $2mv_x$ to the wall. Explain why we can neglect any transfer of energy.

8. Rocket fuels are selected so that the exhaust gases (a) are at a very high temperature and (b) have low molecular weight. Explain why these two features are desirable.

9. Is it meaningful to assign a temperature to a single molecule? Discuss.

10. Suggest a kinetic argument to explain why the speed of sound in solids and liquids is considerably higher than the speed of sound in air.

11. A furnace heats the air in a home by 5 C°, without any change in pressure. How is this accomplished without violating the ideal gas law? (Assume that air does obey the ideal gas law!)

12. Does the presence of water vapor tend to increase or decrease the atmospheric pressure?

13. Give a kinetic theory argument in support of the statement, "There is no such thing as an object *at rest*."

14. The mean velocity of a group of atoms is zero. Does this require that their mean speed be zero? Explain.

25.4 The Distribution of Molecular Speeds

While deriving the ideal gas equation of state in Section 25.3, we dealt with quantities such as the average kinetic energy and the average force per unit area. These average values enter the kinetic theory because there is a wide range of molecular speeds. Let's now investigate this distribution of molecular speeds for a gas in thermal equilibrium.

The quantity used to describe the distribution of molecular speeds is called a *distribution function*. We define the distribution function, $f(v)$, by the statement

$$\begin{array}{c}\text{number of molecules}\\ \text{with speeds in the}\\ \text{range } v \text{ to } v + dv\end{array} = dN = f(v)\,dv \qquad (25.37)$$

Figure 25.6 is a graph of $f(v)$ versus v that describes the distribution of molecular speeds for 1 mole of oxygen molecules. The molecules are in thermal equilibrium at a temperature of 293 K (68°F). The distribution function for the oxygen molecules in your classroom is likely to be very similar to the one shown in Figure 25.6. The quantity $dN = f(v)\,dv$ has a geometric interpretation—it corresponds to the "area"* under $f(v)$ between v and $v + dv$ (Figure 25.7). The number of molecules with speeds in the range from v_1 to v_2 is given by the integral

$$\int_{v_1}^{v_2} f(v)\,dv = \begin{array}{c}\text{number of molecules}\\ \text{with speeds in the}\\ \text{range from } v_1 \text{ to } v_2\end{array} \qquad (25.38)$$

In Figure 25.7 we see that this integral corresponds to the area under $f(v)$ between v_1 and v_2. The total "area" under the $f(v)$ graph corresponds to the total number of molecules in the gas. Thus the integral of $f(v)\,dv$ over the full

*Dimensionally, the quantity $f(v)\,dv$ is not an area, it is a pure number.

Figure 25.6

The distribution function $f(v)$ for 1 mole of oxygen molecules in thermal equilibrium at a temperature of 293 K (68°F).

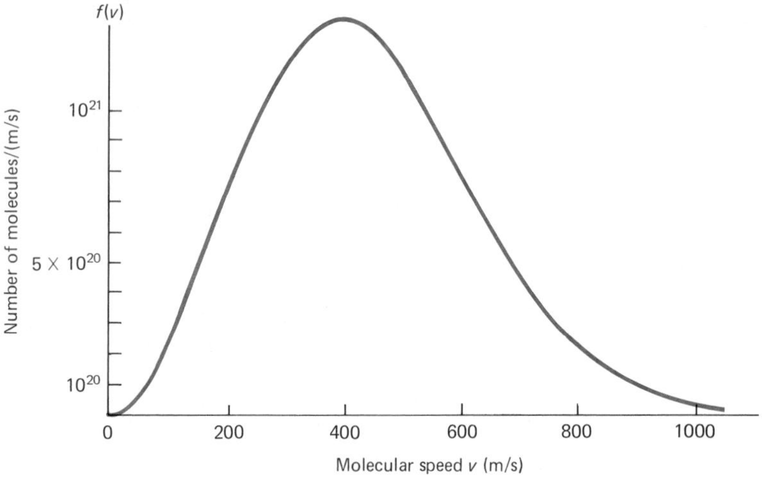

Figure 25.7

The number of molecules with speeds between v and $v + dv$ corresponds to the "area" under the $f(v)$ graph between the speeds v and $v + dv$. The number of molecules with speeds between v_1 and v_2 corresponds to the area under the $f(v)$ graph between v_1 and v_2. The total number of molecules in the gas corresponds to the total area under the $f(v)$ graph.

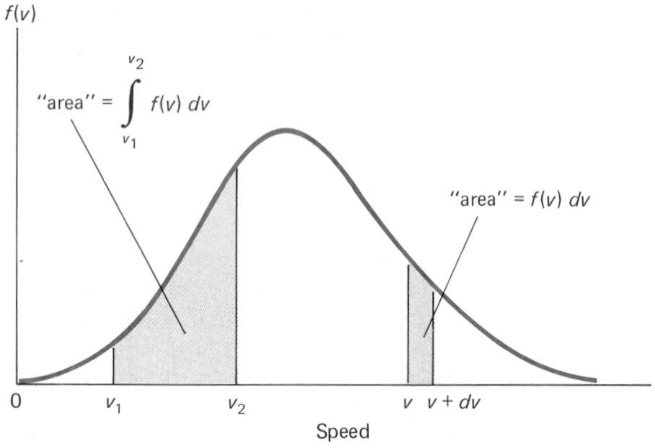

range of molecular speeds* equals N, the total number of molecules:

$$N = \int_0^\infty f(v)\, dv \tag{25.39}$$

Over 100 years ago James Clerk Maxwell determined the form of $f(v)$ that describes a gas in thermal equilibrium. The Maxwellian distribution function for N molecules, each of mass m, in thermal equilibrium at a temperature T is*

$$f(v) = 4\pi N\left(\frac{m}{2\pi kT}\right)^{3/2} v^2 e^{-(mv^2/2kT)} \tag{25.40}$$

The only intrinsic property of the molecules that appears in the Maxwellian $f(v)$ is their mass m. In particular, the Maxwellian distribution is independent of the forces that the molecules exert on one another in the process of establishing equilibrium. Nevertheless, such intermolecular forces are important. In a very real sense, intermolecular forces "drive" a gas toward thermal equilibrium and subsequently maintain the equilibrium state.

Example 6 Maxwellian Distribution of Oxygen Molecule Speeds

Let's consider a gas composed of 10^6 oxygen molecules with a Maxwellian distribution of speeds at a temperature of 300 K. Determine the number of molecules with speeds between (a) 100 m/s and 101 m/s; (b) 300 m/s and 301 m/s; (c) 1000 m/s and 1001 m/s; (d) 3000 m/s and 3001 m/s.

For the small range of speeds considered (1 m/s) we can use Eq. 25.37, which relates the number of molecules (dN) for a specified range of speeds (dv) to the distribution function $f(v)$:

$$dN = f(v)\, dv = \text{number of molecules with speeds in range } v \text{ to } v + dv$$

Here we have $dv = 1$ m/s as the range of speeds. The Maxwellian distribution is given by Eq. 25.40:

$$f(v) = 4\pi N\left(\frac{m}{2\pi kT}\right)^{3/2} v^2 e^{-(mv^2/2kT)}$$

*The actual upper limit of molecular speeds is the speed of light, c. A relativistic kinetic theory takes this fact into account. Taking the upper limit of molecular speeds to be infinite rather than equal to c has a negligible effect under ordinary conditions.

*$e^{-x} = \frac{1}{e^x}$ denotes the exponential function.

For 10^6 oxygen molecules at a temperature of 300 K

$$N = 10^6 \qquad T = 300\ K$$

$$m = 5.31 \times 10^{-26}\ \text{kg}$$

Inserting these values of N, T, and m into the Maxwellian form of $f(v)$ and setting $dv = 1$ m/s gives

$$dN = f(v)\,dv = (3.67 \times 10^{-2} v^2) e^{-(6.42 \times 10^{-6} v^2)}$$

By consistently using SI units throughout, we make certain that the exponent will be dimensionless provided the speed v is expressed in the SI unit of m/s. To find dN for the range $v = 100$ m/s to 101 m/s we set $v = 100$ m/s to find

$$dN = 3.67 \times 10^{-2}(100)^2 e^{-[6.42 \times 10^{-6}(100)^2]} = 344$$

We have rounded the numerical result to two significant figures. In the same way we can evaluate dN for $v = 300$ m/s, 1000 m/s, and 3000 m/s to find (rounded to two significant figures)

$$dN = 1900 \qquad (v = 300 \text{ m/s to } 301 \text{ m/s})$$

$$dN = 60 \qquad (v = 1000 \text{ m/s to } 1001 \text{ m/s})$$

$$dN = 0 \qquad (v = 3000 \text{ m/s to } 3001 \text{ m/s})$$

Even though we considered equal ranges of speeds ($dv = 1$ m/s) for each of the four intervals, the numbers of molecules are quite different. We can trace the variations in dN to the $v^2 e^{-(mv^2/2kT)}$ dependence of the Maxwellian distribution function. At low speeds, $mv^2/2kT \ll 1$ and the exponential factor is nearly 1:

$$v^2 e^{-(mv^2/2kT)} \simeq v^2 e^0 = v^2$$

At high speeds $mv^2/2kT \gg 1$, and the exponential factor is very important:

$$e^{-(mv^2/2kT)} \ll 1$$

and so

$$v^2 e^{-(mv^2/2kT)} \ll v^2$$

Thus, with $v = 1000$

$$e^{-(6.42 \times 10^{-6} v^2)} = e^{-6.42} = 1.63 \times 10^{-3}$$

and with $v = 3000$

$$e^{-(6.42 \times 10^{-6} v^2)} = e^{-57.78} = 8.06 \times 10^{-26}$$

It is the exponential factor $e^{-(mv^2/2kT)}$ that causes the Maxwellian distribution function to drop sharply toward zero at high speeds.

Figure 25.6 shows that $f(v)$ has a maximum value. The speed for which $f(v)$ is a maximum is called the *most probable speed*. Mathematically the peak of $f(v)$ is determined by the condition $df(v)/dv = 0$. For the Maxwellian distribution the condition $df/dv = 0$ leads to

$$\frac{d}{dv}\left[v^2 e^{-(mv^2/2kT)}\right] = 0$$

This gives

$$2v\left(1 - \frac{mv^2}{2kT}\right) e^{-(mv^2/2kT)} = 0 \tag{25.41}$$

There are three solutions to this equation. Two of them ($v = 0$ and $v = \infty$) correspond to minima of $f(v)$. The third solution comes from

$$1 - \frac{mv^2}{2kT} = 0 \tag{25.42}$$

and it determines the *most probable speed*, v_{mp}

$$v_{mp} = \sqrt{\frac{2kT}{m}} \tag{25.43}$$

Example 7 Most Probable Speed for Oxygen Molecules

For oxygen molecules ($m = 5.31 \times 10^{-26}$ kg) at a temperature of 293 K the most probable speed is

$$v_{mp} = \sqrt{\frac{2(1.38 \times 10^{-23})(293)}{5.31 \times 10^{-26}}}\ \text{m/s} = 390 \text{ m/s}$$

This is comparable to the speed of sound in air.

The distribution function is a central concept in more advanced treatments of kinetic theory. Such advanced theories relate deviations of $f(v)$ from the Maxwellian form to phenomena such as conductive heat flow and molecular viscosity.

Questions

15. On the average, which move faster, the oxygen molecules or the nitrogen molecules in the air of your classroom?
16. From Figure 25.6, would you expect that the fraction of oxygen molecules with speeds greater than the most probable speed is greater than $\frac{1}{2}$, less than $\frac{1}{2}$, or exactly equal to $\frac{1}{2}$?
17. We mentioned that the Maxwellian distribution is independent of the intermolecular forces that come into play when molecules in a gas collide. Consider a mixture of two or more gases in thermal equilibrium. Give an argument to support the statement that each gas will have the same Maxwellian distribution of speeds that it would have if the other gases were not present.

Summary

Kinetic theory is based on an atomic model of matter. It relates the microscopic properties of atoms and molecules, which are not directly measurable, to measurable properties of matter, like temperature and pressure.

The characteristic atomic dimension is the angstrom (1 Å $= 10^{-10}$ m). The key observation that reveals this fact is that solids and liquids are nearly incompressible.

The collision cross section (σ) and mean free path (λ) of an atom are related to the number density (n = number of particles per unit volume) by the relation

$$n\sigma\lambda = 1 \tag{25.6}$$

A calculation of gas pressure leads to the ideal gas equation of state,

$$PV = NkT \tag{25.11}$$

and reveals the kinetic interpretation of temperature: Temperature measures the average kinetic energy of atomic and molecular motions.

$$\left\langle \frac{1}{2} mv^2 \right\rangle = \frac{3}{2} kT \tag{25.26}$$

A gas in thermal equilibrium has a distribution of molecular speeds described by the Maxwellian distribution function,

$$f(v) = 4\pi N \left(\frac{m}{2\pi kT} \right)^{3/2} v^2 e^{-(mv^2/2kT)} \tag{25.40}$$

Suggested Reading

M. Born, *The Restless Universe*, Chapter 1, Dover, New York, 1951.

D. K. C. MacDonald, *Near Zero*, Chapter 4, Anchor Books, New York, 1961.

Problems

Section 25.1 The Atomic Model of Matter

1. Calculate the mass in kilograms of (a) one atom of ^{12}C; (b) one atom of hydrogen.

2. The size characteristic of the neutron is 10^{-15} m. The mass of the neutron is 1.67×10^{-27} kg. A star composed entirely of closely packed neutrons has a mass of 2×10^{30} kg (the mass of our sun). Estimate the radius of such a *neutron star*.

3. (a) Estimate your mass and mass density and then use the values to compute your volume. (b) Use your computed volume and knowledge of the characteristic atomic size to estimate the number of atoms in your body. (c) Devise a method for estimating the number of atoms in your body that does not require knowledge of the characteristic atomic size. Compare the result with that of part (b).

4. A laboratory meter rod has a width of 2.5 cm and a thickness of 0.70 cm. Assuming that the characteristic dimension of a molecule is 3 Å, estimate the number of molecules in the rod.

Section 25.2 Mean Free Path and Cross Section

5. A small vacuum chamber has a width of 20 cm. If the mean free path in the chamber is 1000 Å when the pressure is 10^5 Pa, at what pressure will the mean free path equal 20 cm? (Assume that the temperature remains constant.)

6. In some regions of intergalactic space the density of molecules (atoms, ions) may be $n = 1.0/\text{m}^3$. Calculate the mean free path in kilometers. Take $R_o = 3$ Å in estimating the cross section, σ.

7. "Bumper" cars are a favorite attraction at many amusement parks. One park has 22 cars, each with an effective width of 2 m, free to roam about an area of 300 m^2. Estimate the mean free path of a car, assuming the cars move randomly.

8. The standard billiard table measures 4 ft × 8 ft. The diameter of a billiard ball is 2.25 in. Estimate the mean free path lengths for collisions between billiard-balls when there are four balls moving about on the table.

9. *Photons* are particle-like quanta of light. Photons produced by nuclear reactions in the core of the sun are scattered by electrons. The scattering cross section is 2.4×10^{-29} m^2. If there are 10^{31} electrons per cubic meter in the sun, what is the photon mean free path?

10. In using Eq. 25.4 to estimate λ we ignored the atomic volume ($\frac{4}{3}\pi R_o^3$), claiming that it is small compared to V/N. Use the data of Example 2 to justify this step. (Show that $V/N \gg \frac{4}{3}\pi R_o^3$.)

11. Three atoms move at speeds of 1, 2, and 3 km/s. Determine the mean speed and the rms speed of the trio.

12. You are given two particles, one with speed v_1 and the second with speed v_2, $v_1 \neq v_2$. Prove that $\sqrt{\langle v^2 \rangle} > \langle v \rangle$; that is, prove that

$$\sqrt{\frac{1}{2}(v_1^2 + v_2^2)} > \frac{1}{2}(v_1 + v_2)$$

for $v_1 \neq v_2$.

Section 25.3 · The Ideal Gas: Kinetic Interpretation of Temperature

13. The "escape speed" from earth's gravity is approximately 11 km/s. Compare this with the rms speed of a hydrogen atom at a temperature of 200 K. Over many millions of years, would you expect any particular hydrogen atom in the upper atmosphere to have a good chance to escape earth's gravity?

14. Approximately how long does it take a nitrogen molecule moving at the rms speed to travel across your classroom?

15. A neutron emitted in a nuclear fission reaction has a kinetic energy of 2 MeV (1 MeV $= 1.6 \times 10^{-13}$ J). (a) If the neutron mass is 1.67×10^{-27} kg, what is the neutron speed? Fission neutrons transfer their kinetic energy to moderating materials via collisions. Their "ordered" energy is converted into "random" thermal energy. A *thermal* neutron is one whose kinetic energy equals the mean thermal energy of the atoms in the material through which it moves. (b) Determine the speed of a thermal neutron moving through a gas at $T = 320$ K.

16. The average kinetic energy for an electron in a good electrical conductor is 3×10^{-19} J. What would be the temperature of such an "electron gas" if this energy stemmed from purely thermal motions?

17. Thermonuclear fusion reactors will operate at a temperature of 10^8 K. (a) What is the rms speed of protons at this temperature? (b) What is the rms speed of a *deuteron* at this temperature? (A deuteron is the nucleus of a "heavy" hydrogen atom, and is very nearly twice as massive as a proton.) Compare the speeds with the speed of light. (c) Would you expect to need special relativistic formulas to adequately describe the particles?

18. Estimate the number of air molecule impacts each second on 1 cm^2 of your forehead.

19. In our treatment of the ideal gas, we ignored the effects of gravity. Compare the mean thermal energy of an oxygen molecule with the change in its gravitational potential energy in falling from the ceiling to the floor in a classroom. Use your own values for room temperature and room height. Explain how your calculation justifies ignoring gravity in our treatment of the ideal gas.

20. Particles in a plasma (an ionized gas) have an average kinetic energy of E electron volts. Assume that the average kinetic energy of a plasma particle is given by $\frac{3}{2}kT$ and derive the conversion factor that will let you express this energy in terms of temperature.

$$E(\text{eV}) = (\text{conversion factor}) \cdot T(\text{K})$$

Check to see if your conversion factor is correct by using the fact that a temperature of 7752 K gives an average energy of 1 eV.

Section 25.4 The Distribution of Molecular Speeds

21. For a gas in thermal equilibrium we considered three characteristic thermal speeds, the rms speed, the speed of sound in the gas, and the most probable speed. For a diatomic gas, arrange these three speeds in order, with the lowest speed first. (*Hint:* Consult the discussion of rms speed in Section 25.3. The value of γ for a diatomic gas is approximately 1.40.)

22. Gaseous helium is in thermal equilibrium with liquid helium at a temperature of 4.20 K. Determine the most probable speed of a helium atom (mass $= 6.70 \times 10^{-27}$ kg).

23. A gas composed of 1000 oxygen molecules has a Maxwellian distribution of speeds. Use graph paper to plot the Maxwellian distribution function for two temperatures:

$$T_1 = 100 \text{ K} \qquad T_2 = 300 \text{ K}$$

Make your plots on the same speed axis, and answer the following questions: (a) Which temperature has the larger most probable speed? (b) Which temperature has the larger maximum value of $f(v)$?

24. Smoke particles with a mass of 1.0×10^{-17} kg are suspended in an oxygen atmosphere. The oxygen molecules and the smoke particles are in thermal equilibrium. Calculate the ratio of the most probable speeds of an oxygen molecule and a smoke particle.

25. Figure 25.6 shows the Maxwellian distribution function for 1 mole of molecular oxygen at a temperature of 293 K. How would it compare to the Maxwellian distribution function for (a) 1 mole of atomic sulfur vapor at a temperature of 293 K; (b) 1 mole of acetylene (C_2H_2) at a temperature of 238 K; (c) 1 mole of methane (CH_4) at a temperature of 146.5 K?

26. Figure 25.6 shows the distribution of molecular speeds for oxygen molecules. Using the fact that area under the $f(v)$ graph corresponds to the number of molecules, estimate the fraction of oxygen molecules with speeds that are two or more times the speed of a molecule traveling at the most probable speed.

27. A gas composed of 10^6 carbon 12 atoms (mass of one carbon 12 atom $= 1.99 \times 10^{-26}$ kg) has a Maxwellian distribution of speeds at a temperature of 300 K. Determine the number of atoms with speeds between (a) 100 m/s and 101 m/s; (b) 300 m/s and 301 m/s; (c) 1000 m/s and 1001 m/s; (d) 3000 m/s and 3001 m/s. Round your calculated number to two significant figures.

chapter 26

ELECTRIC CHARGE

Preview

In this chapter, we begin a study of the second of the four basic physical forces listed in Chapter 3, the electrical force. Electrical forces play a dominant role in the structure of atoms and molecules and an important role in the structure of nuclei.

Unlike mechanical phenomena, electrical phenomena are often beyond the direct perception of our senses. For example, we can see a baseball arching through the air, but we cannot see electric charge flowing through a wire, nor can we feel its magnetic effects. Consequently, electrical phenomena are more difficult to comprehend than mechanical phenomena. In fact, the development of our understanding of electricity lagged more than a century behind the development of Newtonian mechanics, largely because of this difficulty. We cannot rely on common sense or intuition when studying electricity. Instead we rely on experimental evidence and insist on operational definitions of quantities.

To place electrical phenomena in a proper perspective, we begin this chapter with the historical background of electricity. We then present electric charge in terms of what it does as we discuss electrostatic attraction and electrostatic repulsion. Experiments show that this electrical force is described quantitatively by the expression we call Coulomb's law. Experiment also shows that two or more electrical forces may be combined vectorially. Thus there is a principle of superposition for electrical forces just as there is a principle of superposition for mechanical forces.

This chapter begins a study of electromagnetism (electrical and magnetic phenomena) that continues through Chapter 38. These 13 chapters are tied together by the progressive development of two central themes: fields and energy, including the principle of conservation of energy.

26.1 Discovery of Electricity

Mankind was introduced to electricity by a variety of natural phenomena. Lightning was one of these. Another was the strange blue glow that early sailors sometimes saw atop their ship's masts, which in Christian times was called *St. Elmo's fire* (Figure 26.1). (We now know this blue glow to be an electric discharge.) Fishermen in the Mediterranean Sea sometimes caught electric fish; and, finally, there was the amber effect. It was found that if amber, a fossilized resin, was rubbed with fur, the amber could then attract bits of papyrus and chaff. This peculiar property of amber was known at least as far back as the time of the Greek philosopher Thales of Miletus (640–456 B.C.). Today we recognize this amber effect as an example of a particular aspect of electricity—static electricity.

Although you may not have experimented with amber, you have probably experienced static electricity. On a cold winter day, if you shuffle across a rug and then bring your finger close to a friend's ear, there will be a small spark and perhaps a large reaction. A balloon rubbed against your hair will cause your hair to stand on end, and when released, the balloon itself may adhere to the ceiling temporarily.

The first scientific investigation of this electrostatic attraction is attributed to William Gilbert.* Gilbert was primarily interested in magnetism (to help the English sea

*William Gilbert (1544–1603) was an English physicist and court physician to Queen Elizabeth I. Gilbert's notable work on magnetism is characterized by originality and reliance upon experimental evidence.

Figure 26.1

St. Elmo's fire.

captains navigate across unknown oceans), but he also devoted some attention to electrification by friction. Gilbert gave us our word *electricity,* from the Greek *elektron,* meaning amber. He found that a delicately balanced wooden pointer would point to amber if the amber was first rubbed with fur. Many other substances, such as glass, porcelain, and gems, also made the pointer rotate toward them after they were rubbed. Gilbert called these substances "electrics." Today we know that these substances, unlike metals, do not readily permit electricity to flow through them. We call these substances *insulators.*

Over a century later, in 1731, the English scientist Stephen Gray reported to the Royal Society his findings that the ability to produce the amber effect could be transmitted by some substances. These substances included metals and were called *conductors.* Gray's observations were explained and generalized by the French scientist Charles-François Dufay (1698–1739), who clearly recognized and distinguished conductors and insulators.*

Dufay also observed that electrified, or charged, objects sometimes attracted each other and sometimes repelled each other. He proposed two kinds of electricity: vitreous, found on glass that had been rubbed with silk; and resinous, found on amber that had been rubbed with fur. Later Dufay's theories were reformulated in terms of two kinds of fluids. Each form of electricity, vitreous and resinous, was assumed to consist of a continuous, invisible fluid. When amber was rubbed with fur it was said that the resinous fluid flowed into the amber. The amber had acquired an electric "charge." Gradually the amber lost its ability to attract chaff, and this loss was explained by saying that the invisible fluid had leaked out.

The American scientist Benjamin Franklin* recognized that it was not necessary to have two distinct electric fluids in order to explain the behavior of charged objects. Franklin proposed that there was only one electric fluid, and that electrification, or charging, is a redistribution of this fluid. When rubbed with silk, glass acquired an excess of the electric fluid, while the silk was left with a deficiency. Franklin called the glass *positive* and the silk *negative.* These terms correspond to Dufay's vitreous and resinous electricity.

The physicists of the late 18th century were unable to decide between the one-fluid theory of Franklin and the two-fluid theory of Dufay. Indeed, our modern view of electricity incorporates some elements of each.

By the mid-1700s effective frictional techniques for generating electricity had been developed (Figure 26.2). There was tremendous popular interest in electricity, and people flocked to formal lecture-demonstrations of electrostatics and played electrostatic parlor games in their homes. This interest created strong encouragement and support for scientists.

26.2 Electric Charge

Quantization

The modern view of electric charge is that it is a basic property of atoms, the fundamental particles of which all matter is made. Atoms themselves are composed of three different types of particles—protons, neutrons, and electrons. A proton carries one unit of positive† charge (Dufay's vitreous electricity) and an electron carries one unit of negative charge (Dufay's resinous electricity). A neutron carries no charge; it is neutral.

*Today we recognize a third class of materials, the semiconductors, intermediate between the conductors and the insulators. Conductors, semiconductors, and insulators do not actually form three sharply defined classes but rather offer an almost continuous distribution in their ability to conduct. Conduction is discussed in Chapter 30.

*Benjamin Franklin (1706–1790) was an American statesman, philosopher, and scientist whose accomplishments in the study of electricity include the discovery of the identity of lightning and the invention of the lightning rod.

†The labels *positive* and *negative* are just that—labels. They are convenient in that they suggest and facilitate algebraic addition of charge. See Examples 2 and 3.

Figure 26.2

Playing with static electricity.

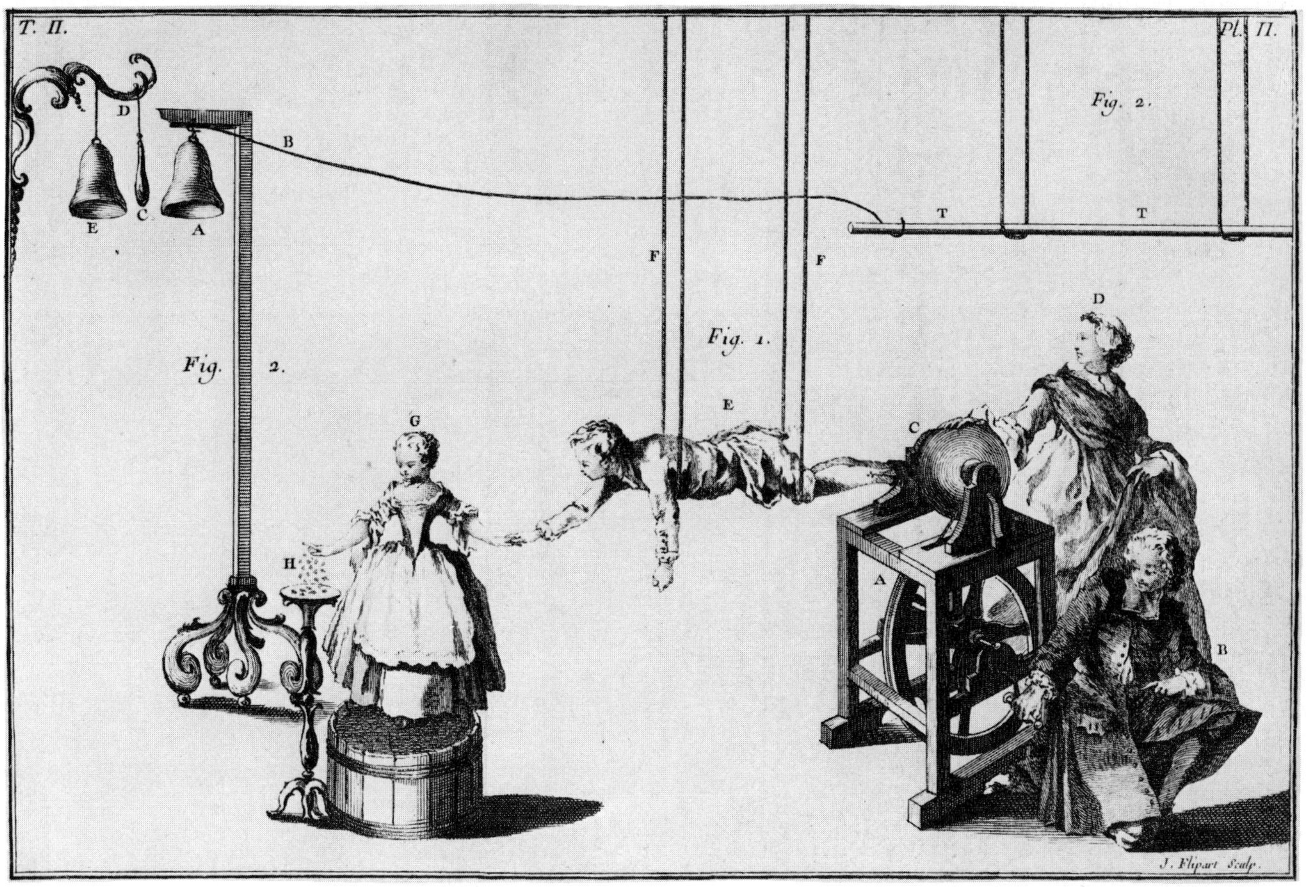

The mass of a proton is almost exactly equal to the mass of a neutron. An electron is much less massive than a proton or neutron. However, the magnitude of the electron charge is exactly equal to the magnitude of the proton charge to within the experimental error of one part in 10^{22}. Protons, neutrons, and electrons are arranged within atoms as shown in Figure 26.3.

Eighteenth-century scientists thought of electricity as a continuous fluid. By talking about unit charges on electrons and protons, we have made a profound shift from this idea. From a wide variety of evidence, including electrochemistry and high-energy particle physics, we now know that electric charge is not a continuous fluid, but occurs only in certain discrete amounts. We describe this situation by saying that electric charge is *quantized*. In particular, all electric charge is an integral multiple of the electron (or proton) charge. Mass is also quantized, or present only in discrete amounts. It is not continuous, but instead consists of small particles (atoms, neutrons, protons, electrons).

Table 26.1

Particle	Symbol	Charge	Mass
Proton	p	$+e$	$1.6726485 \times 10^{-27}$
Neutron	n	0	$1.6749543 \times 10^{-27}$
Electron	e	$-e$	9.109534×10^{-31}

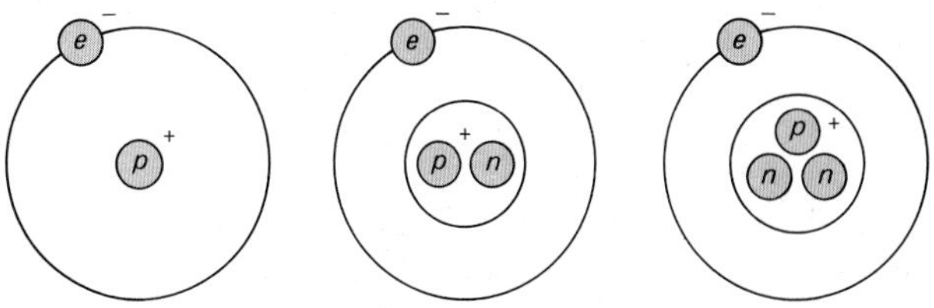

Figure 26.3

This is a diagram of hydrogen atoms. It is a schematic diagram, not a literal picture. The proton in hydrogen, proton plus neutron in deuterium, and proton plus two neutrons in tritium form massive central nuclei for these hydrogen atoms. Outside this nucleus there is one electron.

Figure 26.4

Benjamin Franklin.

Figure 26.5

Separation of electric charge when a glass rod is rubbed with a silk cloth.

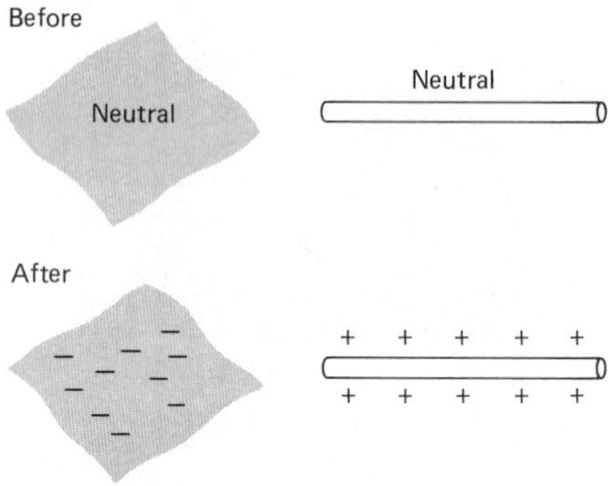

Conservation of Charge

The concept that electric charge can never be created or destroyed, but merely rearranged, goes back at least to Benjamin Franklin (Figure 26.4) and his one-fluid theory of electricity. Franklin's most important contribution to our understanding of electricity is the idea that electric charge is conserved. In the case of the glass rod rubbed with silk, electrons are transferred from the glass to the silk, giving the silk a net negative charge and leaving the glass rod with an equal positive charge. This transfer of electrons is illustrated in Figure 26.5. No change in the total charge of glass plus silk occurs. Conservation of charge has been tested repeatedly in the realm of high-energy physics and has been found to hold in all circumstances. The principle of conservation of electric charge can be stated as follows:

In any interaction the net algebraic amount of electric charge remains constant.

The following three examples illustrate conservation of electric charge on an elementary-particle basis. Example 1 is a redistribution of positive charge. Examples 2 and 3 emphasize the net (algebraic) feature of conservation of charge.

Example 1 Uramium 238 Alpha Decay

The element uranium 238, ^{238}U, decays by emitting an alpha particle (helium nucleus). This nuclear reaction may be written

$$^{238}_{92}U \rightarrow {}^{4}_{2}He + {}^{234}_{90}Th$$

The subscripts give the number of protons in each nucleus and are therefore a measure of positive nuclear charge. Notice that the number of protons (and the amount of positive charge on the left-hand side of the equation) is the same as the total number of protons on the right-hand side of the equation. The balancing of the subscripts represents *exact* conservation of charge in this nuclear (radioactive) disintegration.

In Example 1 nuclear charge in the form of protons is simply rearranged. Sometimes electric charge is created: Positive and negative charge are created in equal amounts.

Example 2 Carbon 14 Beta Decay

Carbon 14, $^{14}_{6}C$, has six protons in its nucleus and is formed in our atmosphere by cosmic ray bombardment of nitrogen 14, $^{14}_{7}N$. Carbon 14 is unstable and undergoes radioactive decay by emitting a beta particle (an electron) and an antineutrino (zero mass, zero charge) and becoming nitrogen (seven protons):

$$^{14}_{6}C \rightarrow {}_{-1}e + {}^{14}_{7}N + \bar{\nu}$$

In this process a neutral particle in the ^{14}C nucleus, called a *neutron*, is transformed into three particles—a positively charged proton, a negatively charged electron, and a neutral antineutrino. The proton, the electron, and the antineutrino are *created* from the neutron in the reaction. But although both positive and negative charges are created, the net charge ($+6$ before $= -1 + 7$ after $= +6$ after) remains the same. Again the subscripts represent exact conservation of electric charge.

In Example 2 we saw the creation of positive and negative electric charge. Now let's consider the reverse of charge creation: charge annihilation.

Example 3 Electron–Positron Annihilation

An ordinary electron, unit negative charge, is brought into contact with an antielectron,* or positron, unit positive charge. Both are at rest. The electron (particle) and the positron (antiparticle) literally annihilate each other,

$${}_{-1}e + {}_{+1}e \rightarrow 2\gamma$$

producing two gamma rays. The energy of each gamma ray, E_γ, equals the rest energy of the electron or positron, m_ec^2. We have a 100% conversion of matter into energy (electromagnetic radiation) and exact conservation of electric charge. There is zero net charge before the annihilation and zero net charge after annihilation.

---

*For each type of elementary particle (matter) there is a corresponding antiparticle (antimatter). For the electron there is the antielectron, with a positive charge. For the proton there is the antiproton, with a negative charge. For the neutron there is the antineutron. Modern particle theory is concerned with the problem of understanding why we live in a matter-dominated universe rather than in an antimatter-dominated universe.

Questions

1. Electric charge is quantized. All electric charges are integral multiples of the electron charge. Is mass quantized in the same way? For instance, can all masses be expressed as a sum of various numbers of elementary-particle masses?

2. Discuss the electron–positron annihilation reaction in terms of these conservation laws: (a) charge, (b) linear momentum, (c) energy, and (d) angular momentum (spin).

26.3 Coulomb's Law

Quantitative work in electrostatics began in 1785 when the French scientist Charles Augustin de Coulomb* (Figure 26.6) introduced the torsion balance.† This was a technological breakthrough comparable to Galileo's invention of the telescope. Earlier work by Priestley and Cavendish in England had suggested that the electrostatic force obeyed a law similar to Newton's law of gravitation. With the torsion balance Coulomb was able to establish this law, called the *electrostatic force law* or *Coulomb's law,* with much higher precision than had earlier workers.

Coulomb found that the force between two small‡ electrically charged spheres at rest was *inversely proportional to the square of the distance* between them (center-to-center distance if the small spheres were uniformly charged).

$$F \propto \frac{1}{r^2} \qquad (26.1)$$

By varying the electric charge on the two spheres, Coulomb further found that the force was *proportional to the product of the charges.* He expected this by analogy with Newtonian gravitation (Section 6.6). This result is known as *Coulomb's law of electrostatic interaction* and may be written

$$F = k_e\left(\frac{q_1 q_2}{r^2}\right) \qquad (26.2)$$

where F is the magnitude of the force exerted on charge q_1 by charge q_2 separated by a distance r. By Newton's third law, this quantity equals the magnitude of the force exerted by charge q_1 on charge q_2 (Figure 26.7).

$ke = 8.98755 \times 10^9 \frac{Nm^2}{C^2}$ (26.4)

*Charles Augustin de Coulomb (1736–1806), a French physicist, independently invented the torsion balance and used it to make a series of precise measurements that laid the foundation for a mathematical theory of electricity.

†The torsion balance had actually been developed earlier by John Mitchell in England and used by Henry Cavendish to determine the constant G in Newton's law of universal gravitation. These results had not been published in 1785. For a diagram and a description of the operation of the torsion balance see Section 6.6.

‡"Small" means small compared to the distance between the spheres. As a mathematical ideal we shall assume that the spheres have zero radius. In this limit the charges become point charges.

Figure 26.6

Charles Augustin Coulomb.

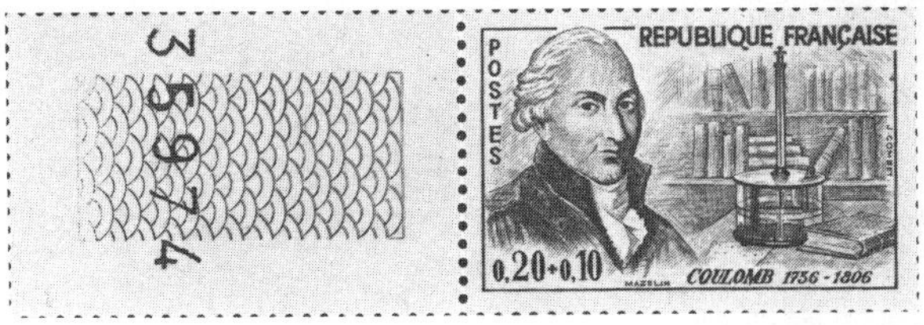

Figure 26.7

As indicated by the arrows, $\mathbf{F}_1$ is the force exerted on q_1 by q_2 and $\mathbf{F}_2$ is the force exerted on q_2 by q_1 for the case where q_1 and q_2 have the same sign, repulsion. If q_1 and q_2 have opposite signs, the arrows are reversed. The force is then attractive.

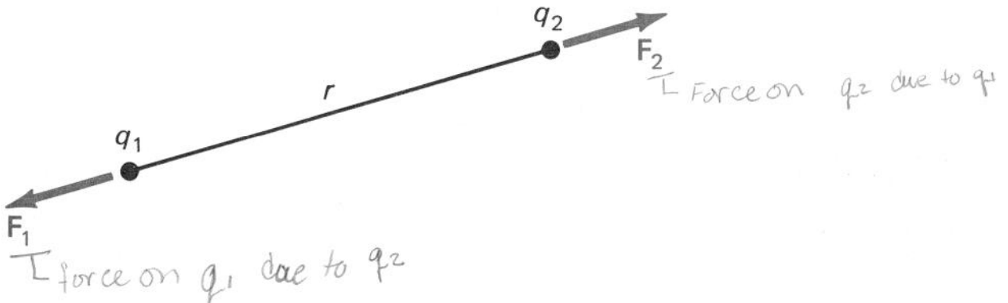

The quantity k_e is a proportionality constant. The choice of the numerical value and dimensions of k_e is arbitrary. However, choosing k_e in Coulomb's law determines the unit of electric charge. For instance, choosing $k_e = 1$ (with no dimensions) yields the electrostatic definition of charge. We set $q_1 = q_2$ with $r = 1$ cm. If the (repulsive) force is 1 dyne (1 dyne $= 10^{-5}$ newtons), then q_1 is 1 electrostatic unit (esu) of charge, also known as 1 statcoulomb. This electrostatic system of units is still in common use in some areas of advanced physics, but we will not use it here.

Alternatively, if we choose the unit of charge, q, then the proportionality constant, k_e, can be determined by experiment by using Coulomb's law. The *Système International d' Unités* (SI) follows this second option. The unit of charge is taken to be the coulomb and k_e is determined by experiment. The SI unit of charge, the coulomb, is actually defined indirectly by making use of the magnetic effects of charges in motion—electric current. The reason for this indirect approach is that the magnetic measurement is easier and more precise. We will explore this definition in detail in Section 33.5.

Coulomb's law becomes

$$F = k_e\left(\frac{q_1 q_2}{r^2}\right) \qquad (26.3)$$

with q_1 and q_2 in coulombs (C), r in meters, and F in newtons. Then k_e, no longer dimensionless, is given by

$ke = 8.98755 \times 10^9 \frac{N\,m^2}{C^2}$ (26.4)

The force between two point charges* at rest is directly proportional to the product of the charges and inversely proportional to the square of the distance between them.

*"Point charge" means that the electric charge resides on a geometric point. Coulomb's law holds for charged objects whose sizes or spatial dimensions are much smaller than the distance between the objects.

This electrostatic Coulomb force is repulsive for like charges and attractive for unlike charges, as had been noted by Dufay. For comparison of units of charge, the charge on the proton is

$$+e = 4.8033 \times 10^{-10} \text{ esu} = 1.60219 \times 10^{-19} \text{ C}$$

The charge on the electron is the negative of the proton charge.

$$-e = -4.8033 \times 10^{-10} \text{ esu}$$
$$= -1.60219 \times 10^{-19} \text{ C}$$

Example 4 **Coulomb Repulsion**

Calculate the force of repulsion between two 1-C charges 1 m apart. From Coulomb's law, Eq. 26.3, we have

$$F = 8.99 \times 10^9 \text{ N} \cdot \text{m}^2/\text{C}^2 \times \frac{1 \text{ C} \times 1 \text{ C}}{1^2 \text{ m}^2}$$
$$= 8.99 \times 10^9 \text{ N}$$

In British units this is equal to about 1 million tons. This force is far larger than the forces involved in electrostatic demonstrations or laboratory experiments. Clearly, by electrostatic–Coulomb law standards, a coulomb of charge is a tremendously large amount of charge. In fact, this example is completely imaginary and could not actually be done. There is no way that we would get a net charge of 1 C to stay on any small surface. For applications involving electric currents (Section 30.1), however, the coulomb is a reasonably sized unit.

Coulomb's law of electrostatic interaction, Eq. 26.3, and Newton's law of universal gravitation, Eq. 6.3, have the same mathematical form. But what about the relative strength of these two fundamental forces? Let's calculate the ratio of electrostatic attraction to the gravitational attraction of an electron and a proton on each other. Combining Coulomb's law, Eq. 26.3, and Newton's law of universal gravitation from Section 6.6, we have

$$F_{\text{grav}} = G\left(\frac{m_e m_p}{r^2}\right) \qquad (6.3)$$

$$\frac{F_{\text{elec}}}{F_{\text{grav}}} = \frac{k_e\left(\frac{q_e q_p}{r^2}\right)}{G\left(\frac{m_e m_p}{r^2}\right)}$$

$$= \frac{8.99 \times 10^9 \text{ N} \cdot \text{m}^2/\text{C}^2}{6.67 \times 10^{-11} \text{ N} \cdot \text{m}^2/\text{kg}^2} \left(\frac{q}{m}\right)_e \frac{\text{C}}{\text{kg}} \left(\frac{q}{m}\right)_p \frac{\text{C}}{\text{kg}}$$

The distance factor cancels out.* Inserting values for the charge-to-mass ratio for the electron (1.75880×10^{11} C/kg) and for the proton (9.57895×10^7 C/kg), we obtain for the ratio of electrical force to gravitational force

$$\frac{F_{\text{elec}}}{F_{\text{grav}}} = 2.27 \times 10^{39} \qquad (26.5)$$

*That Newton's law of gravitation and Coulomb's law have the same $1/r^2$ distance dependence has impressed many scientists, including Einstein, as more than mere coincidence. So far, no profound relationship or common origin has been discovered.

The number 10^{39} is a tremendously large number. Imagine grains of sand so fine that you can pack 10^6 grains in 1 cm^3: 10^{39} of these grains would occupy the volume of a million earths! Clearly the Coulomb electrical force is far stronger than gravity, at least for electrons and protons. (Gravity dominates when the masses are large—like planetary or stellar masses—and the net electric charges are small.) A consequence of the strength of the electrical force is that the universe must be almost exactly neutral.

There is a second and more immediate consequence of the strength of the electrical force. Suppose that we have 5 gm of copper (equal in mass to a nickel), and that we place 1 millicoulomb (mC) of negative charge (electrons) on this copper. Because of the great strength of the electrical force, this 1-mC charge can exert a very significant force. But what of the effect of the charge on the chemical nature* of the copper? Our 5 gm of copper contains about 1.37×10^{24} electrons (and an equal number of protons). The 1 mC we add amounts to 6.24×10^{15} electrons. There are 220 million electrons in our 5 gm of copper for every one electron we add. The relative change in number of electrons in tiny, and so the chemical nature of the copper remains unchanged.

Electrical Force and Newtonian Mechanics

In this Section we have introduced a new force, the Coulomb electrical force. This force describes a way in which the environment (q_2) may influence a particle (q_1). The behavior of the particle (q_1 in this case) is given by Newtonian mechanics, introduced in Chapter 6. Newton's second law still applies in the case of electrical forces: $\mathbf{F}_{\text{net}} = d\mathbf{p}/dt$. The Coulomb force can do work, $W = \int \mathbf{F} \cdot d\mathbf{s}$, just like any other force.

Keep in mind that throughout this chapter we are considering charges at rest relative to each other or moving at a very low speed ($v \ll c$).

Questions

3. A philosophy major asks you to explain what electric charge "really" is. How would you explain the concept of electric charge to a student who has not studied physics?
4. What difference would it make if we called the electron positive and the proton negative, in effect reversing Franklin's choice?
5. Is Coulomb's law of electrostatic interaction consistent with Newton's third law? Explain.
6. How could you show that the electrostatic force on a charged object is independent of the mass of the object?
7. A scientist suggests that the proton charge may be slightly larger than the electron charge, perhaps by one part in 10^{18}. What consequences would this difference have for (a) the number of electrons in an atom; (b) the net charge on a planet, a star, a galaxy;

*The chemical nature of a substance depends essentially on the number of electrons in a neutral atom of the substance.

(c) the interaction of galaxies? (This suggestion has been rejected based on laboratory experiments.)

8. A hydrogen atom consists of one proton and one electron bound together by the attractive Coulomb force. Assume that the proton charge exceeds the electron charge by one part in 10^{18}. How does the resulting electrostatic repulsion between two hydrogen atoms compare qualitatively with the gravitational attraction?

9. Equation 26.5 shows that the electrostatic force is stronger than the gravitational force by a factor on the order of 10^{39}. From this, argue that our galaxy must be electrically neutral or almost electrically neutral.

26.4 Superposition

Vector Form of Coulomb's Law

The Coulomb electrical force, like all forces, is a vector quantity—it has direction as well as magnitude. From the full statement of Coulomb's law in the preceding section we can write Coulomb's law in vector form.

$$\mathbf{F}_{12} = k_e\left(\frac{q_1 q_2}{r_{12}^2}\right)\hat{\mathbf{r}}_{12} \qquad (26.6)$$

Except for sign, Eq. 26.6 has the same mathematical form as the vector form of Newton's law of gravity (Section 6.6). The subscripts 1 and 2 on $\mathbf{F}_{12}$ are included so that $\mathbf{F}_{12}$ will be understood as the force on charge 1 due to charge 2. Similarly, $\mathbf{F}_{21}$ is the force on charge 2 due to charge 1. Here $\hat{\mathbf{r}}_{12}$ is a unit vector (unit magnitude) pointing toward q_1 from q_2, as shown in Figure 26.8. The unit vector in the $\mathbf{r}_{12}$ direction is given by

$$\hat{\mathbf{r}}_{12} = \frac{\mathbf{r}_{12}}{|\mathbf{r}_{12}|} \qquad (26.7)$$

If the charges q_1 and q_2 are both positive or both negative, the force on q_1 exerted by q_2, symbolized by $\mathbf{F}_{12}$, is repulsive, parallel to $\mathbf{r}_{12}$, and directed away from q_2. The fact that electrostatic force can be repulsive differentiates it from gravitation. Gravitational forces are always attractive.

If one of the charges is negative, then $\mathbf{F}_{12}$ will be negative, which indicates that q_1 is pulled in toward q_2, opposite to $\hat{\mathbf{r}}_{12}$. In other words, like charges repel and unlike charges attract.

Superposition

Suppose that we have more than two charges. Will the presence of other charges affect the interaction of a given pair of charges? Consider the following example.

Example 5 Superposition

Charges of 3.0×10^{-6} C, 4.0×10^{-6} C, and 6.0×10^{-6} C are placed along a line (Figure 26.9). Let us calculate the two separate forces on the 6.0-μC charge. In each case we will calculate the force as if the two interacting charges were isolated (as if they were the only two charges in the universe). First consider the force exerted by the 3.0-μC charge. From Coulomb's law,* Eq. 26.3, the force exerted on the 6.0-μC charge by the 3.0-μC charge is

$$F_{63} = 8.99 \times 10^9 \times \frac{3 \times 10^{-6} \times 6 \times 10^{-6}}{3^2}$$

$$= 8.99 \times 10^9 \times 2 \times 10^{-12}$$

$$= 1.80 \times 10^{-2} \text{ N (directed to the right)}$$

Next, we consider the force exerted by the 4-μC charge.

$$F_{64} = 8.99 \times 10^9 \times \frac{4 \times 10^{-6} \times 6 \times 10^{-6}}{2^2}$$

$$= 8.99 \times 10^9 \times 6 \times 10^{-12}$$

$$= 5.39 \times 10^{-2} \text{ N (directed to the right)}$$

Because F_{63} and F_{64} are parallel, we obtain for the total force on the 6.0-μC charge

$$F_6 = F_{63} + F_{64} = 7.19 \times 10^{-2} \text{ N}$$

To within the limits of experimental accuracy the total force on the 6-μC charge has been confirmed to be the sum of F_{63} and F_{64}, or 7.19×10^{-2} N. In other words,

*If we take the three charges to lie on the x-axis, then the unit vectors $\hat{\mathbf{r}}_{63} = \mathbf{i}$ and $\hat{\mathbf{r}}_{64} = \mathbf{i}$. The force vectors are parallel and we may simply add magnitudes.

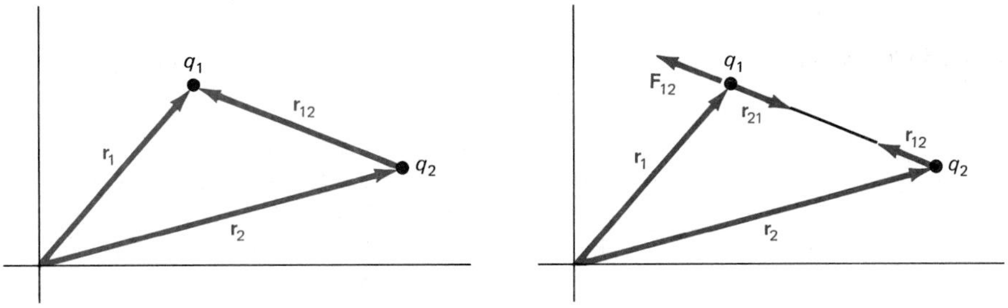

Figure 26.8

Vector diagrams.

Figure 26.9

Vector diagram.

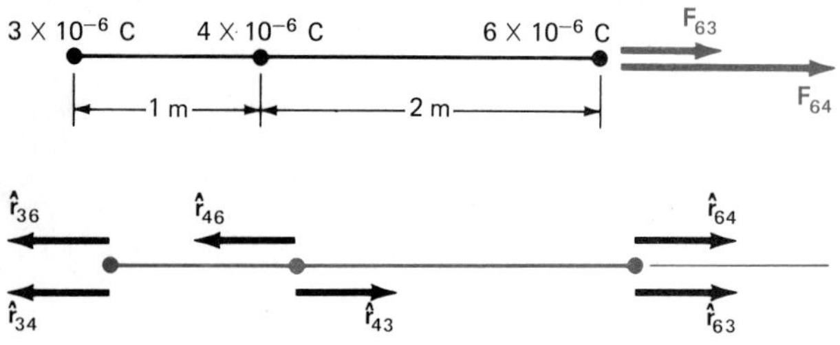

experiment shows that the presence of a third charge does not influence the Coulomb force between the other two charges.

We can generalize the experimental result stated in Example 5 by saying that electrical forces obey the *principle of superposition*.

The net force exerted by two or more charges on a single charge Q is the vector sum of the individual Coulomb forces acting on Q.

Keep in mind that this principle is the result of experiment.* It is not obvious or automatic.

Mathematically, the principle of superposition means that our electrostatic system is linear. if one charge had influenced the force that another charge exerted on the 6-μC charge, our mathematical description would have been nonlinear.

*Example 5 is a relatively simple and artificial system. Much better experimental proof comes later when our electrostatic theory is extended to include time dependence, combined with magnetism and applied to physical situations such as electromagnetic waves (Chapter 38) and atomic systems (Chapter 42).

Example 5 involves only parallel and antiparallel forces. Now let's apply the principle of superposition to a system where the forces are nonparallel. Let's also build in enough symmetry so that we can check our results.

Example 6 A Triangle of Charges—Vector Addition

Consider three 1-μC charges at the vertices of an equilateral triangle, 1 m on a side. What is the net force that two of the charges exert on the third?

First, let's inspect the system. This array of three equal charges has left–right symmetry relative to a vertical line through q_1. We know from this symmetry that the net force of q_2 and q_3 on q_1 will be vertical, and in the upward direction. (All charges have the same sign; all forces are repulsive.) Then the net force on q_1 is the vertical component of the force exerted by q_2 plus the vertical component of the force exerted by q_3. Since the two vertical components are equal (by symmetry), then

$$F = 2k_e\left(\frac{q_1q_2}{r^2}\right)\cos 30^\circ$$

$$= \frac{2 \times 8.99 \times 10^9 \times 10^{-6} \times 10^{-6} \times 0.866}{1^2}$$

$$= 1.56 \times 10^{-2}\text{ N}$$

Now, let's go through the calculation in detail using the vector form of Coulomb's law, Eq. 26.6. The force vectors and unit vectors are illustrated in Figure 26.10. By the principle of superposition we can apply Coulomb's law to charges q_1 and q_2 alone. For $\mathbf{F}_{12}$, the force that q_2 exerts on q_1, we have

$$\mathbf{F}_{12} = k_e\left(\frac{q_1q_2}{r_{12}^2}\right)\hat{\mathbf{r}}_{12}$$

$$= (8.99 \times 10^9 \times 10^{-12})\hat{\mathbf{r}}_{12}$$

$$= (8.99 \times 10^{-3})\hat{\mathbf{r}}_{12}\text{ N}$$

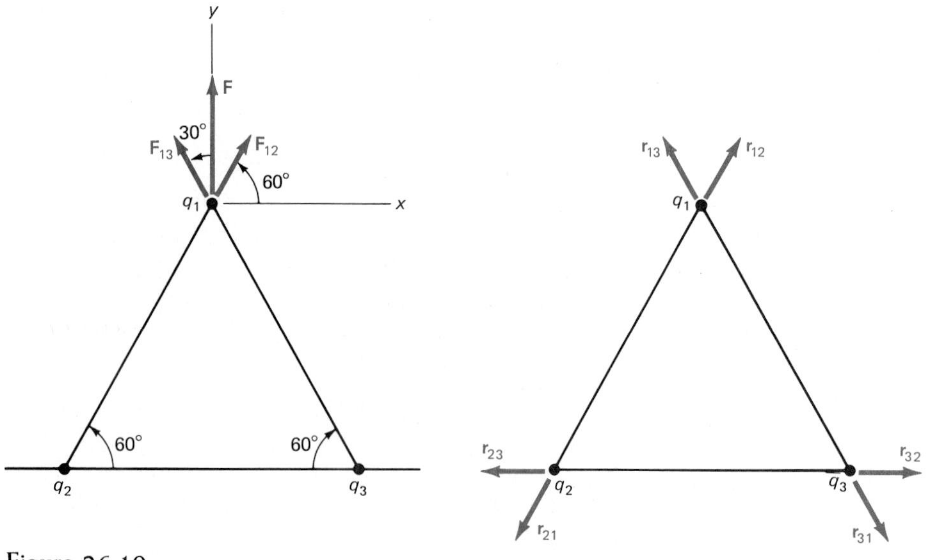

Figure 26.10

Vector diagrams.

Let's resolve $\hat{\mathbf{r}}_{12}$ into its Cartesian components (Section 2.3). For this choice of coordinates and this configuration of charges we have $\hat{\mathbf{r}}_{12} = \cos 60° \, \mathbf{i} + \cos 30° \, \mathbf{j}$. Then

$$\mathbf{F}_{12} = (8.99 \times 10^{-3})\cos 60° \, \mathbf{i}$$
$$+ (8.99 \times 10^{-3})\cos 30° \, \mathbf{j}$$
$$= (4.49 \times 10^{-3})\mathbf{i} + (7.78 \times 10^{-3})\mathbf{j} \text{ N}$$

In terms of Cartesian components

$$\hat{\mathbf{r}}_{13} = \cos 120° \, \mathbf{i} + \cos 30° \, \mathbf{j}$$

Then

$$\mathbf{F}_{13} = (-4.49 \times 10^{-3})\mathbf{i} + (7.78 \times 10^{-3})\mathbf{j} \text{ N}$$

The net force on q_1 is the vector sum of $\mathbf{F}_{12}$ and $\mathbf{F}_{13}$.

$$\mathbf{F} = \mathbf{F}_{12} + \mathbf{F}_{13}$$
$$= (15.6 \times 10^{-3})\mathbf{j} \text{ N}$$

This solution is in agreement with the first calculation. The horizontal components cancel each other, as a consequence of the symmetry of the system.

We see from Example 6 that when symmetry is present in a system the solution of the net force is much easier. If symmetry is not present, however, we can find the net force by using the vector form of Coulomb's law and the principle of superposition.

Continuously Distributed Charge

All electric charge is actually a collection of discrete charges such as electrons and protons. However, when we consider a large number of charges from a large distance away, the distribution of charges appears to be continuous and we can treat the distributed discrete charges as continuous. The principle of superposition may be applied as before. The summation over the discrete charges becomes an integration over the distributed charge.

The following is an example of continuously distributed charge that requires consideration of only two dimensions.

Example 7 Force Exerted on a Point Charge by a Line of Charge

A thin rod (a line) extends from $-x_1$ to $+x_1$ as shown in Figure 26.11. This rod has an electric charge of λ coulombs per unit length, where λ is a constant. We wish to calculate the force on a charge q_o at position $(0, y_o)$. Here we have the same left–right symmetry as we did in Example 6, and just as in Example 6, the net force exerted by the rod on q_o will be vertical.

Consider an element of charge $dq = \lambda \, dx$ where dx is small enough so that $\lambda \, dx$ acts as a point charge. The vertical component of the force exerted on q_o by this element of charge $dq = \lambda \, dx$ is given by

$$dF = k_e \left[\frac{q_o \lambda \, dx}{(x^2 + y_o^2)} \right] \cos \theta \qquad (26.8)$$

Because $\cos \theta = y_o/(x^2 + y_o^2)^{1/2}$,

$$dF = k_e q_o \lambda y_o \frac{dx}{(x^2 + y_o^2)^{3/2}}$$

Figure 26.11

Line charge.

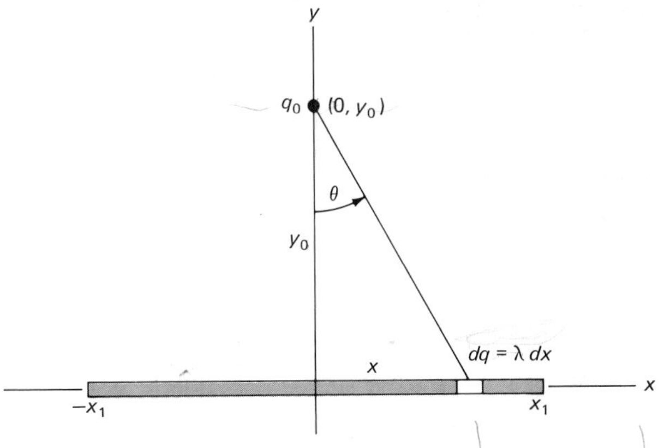

We integrate from $-x_1$ to $+x_1$ to cover all of the charge on the rod.

$$F = k_e q_o \lambda y_o \int_{-x_1}^{x_1} \frac{dx}{(x^2 + y_o^2)^{3/2}}$$

From Appendix 4 the integral is

$$\int \frac{dx}{(x^2 + y_o^2)^{3/2}} = \frac{x}{y_o^2 (x^2 + y_o^2)^{1/2}}$$

(We can check this by differentiating the right-hand side.) The net force F exerted by the charged rod on q_o becomes

$$F = k_e q_o \lambda \frac{2x_1}{y_o(x_1^2 + y_o^2)^{1/2}}$$

If we write $Q = 2x_1\lambda$ for the total charge on the rod, then

$$F = k_e \frac{q_o Q}{y_o(x_1^2 + y_o^2)^{1/2}} \qquad (26.9)$$

Example 7 is essentially a two-dimensional problem (x and y) with an integral that must be looked up in a table. Now we turn to a three-dimensional example of continuously distributed charge, with an integration that can be done by inspection.

Example 8 A Ring of Charge

Consider a circular ring with a uniform distribution of charge, and a total charge Q (Figure 26.12). Calculate the force on a charge q_o on the symmetry axis perpendicular to the plane of the ring passing through the center of the ring.

Every element of charge dq_1 can be paired with an equal element of charge dq_2 on the opposite side of the ring. The y-component of force exerted by dq_1 just balances (cancels) the y-component of force by dq_2. Only the x-component, dF_x, survives. This means that we need only calculate the x-component of the Coulomb force exerted by each dq.

$$dF_x = k_e q_o \left(\frac{dq}{r^2} \right) \cos \theta \qquad (26.10)$$

Figure 26.12

Ring charge.

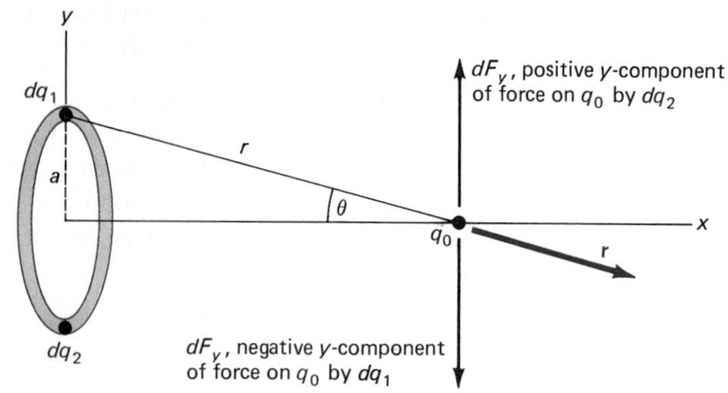

We can obtain the total force F_x by integrating over all elements of charge dq over the entire ring. Now the proportionally constant k_e and the test charge q_o are constants, independent of the choice and location of dq_i. For the very special case of q_o on the axis of rotational symmetry, r and $\cos\theta$ are also constant. Therefore

$$F_x = \int dF_x = \int k_e q_o \frac{\cos\theta}{r^2}\,dq$$

$$= k_e q_o \frac{\cos\theta}{r^2}\int dq$$

Since $\int dq = Q$, where Q is the total charge, we have

$$F_x = k_e q_o Q\frac{\cos\theta}{r^2}$$

$$= k_e q_o Q\frac{x}{(x^2+a^2)^{3/2}} \qquad (26.11)$$

This is the Coulomb force exerted by the ring charge Q on the charge q_o (on the symmetry axis).

Now let's consider a couple of limiting cases, both to check Eq. 26.11 and to better understand the behavior of this force. First, what happens when $x = 0$? From Eq. 26.11, if $x = 0$, then $F_x = 0$. We get this result from Figure 26.12 by inspection. The point $x = 0$ on the x-axis is the geometric center of the ring. The force $d\mathbf{F}_1$ exerted by dq_1 on a test charge is exactly canceled by the force $d\mathbf{F}_2$ exerted by dq_2. Continuing around the ring, we find that the force exerted by any dq is exactly canceled by the force exerted by a dq on the opposite side of the ring. The net force is zero. Second, what happens as q_o goes far out on the x-axis, $x \gg a$? When this happens, Eq. 26.11 reduces to

$$F_x = \frac{k_e q_o Q}{x^2}$$

Is this reasonable? Yes; if you are a large distance x away from the ring ($x \gg a$), the ring will look like a point charge.

Finally, what happens if q_o is not on the x-axis? If q_o is not on the x-axis, then r and $\cos\theta$ are not constants. They do not come out of the integral and we are left with a messy problem. Such a problem can be solved only with an electronic computer or with advanced mathematics.

Coulomb's law (Section 26.3) and the principle of superposition (introduced in this section) cover all of electrostatics. No additional new forces or principles are needed. However, we introduce electric fields in Chapter 27, and the electric potential in Chapter 28, for convenience (very great convenience) in calculation and understanding.

Questions

10. Two equal positive charges, q_1 and q_2, are fixed in position 1 m apart. Where could you place a third charge of the same sign, q_3, so that the net force on it vanishes? Is q_3 then in stable equilibrium? What happens if q_3 has opposite sign?

11. Electric charge is distributed uniformly over a spherical surface and is fixed in position. One electron is placed at the center of the sphere. Is this one electron in equilibrium? Explain.

Summary

Electrical insulators were recognized 400 years ago in terms of the amber effect of electrostatic attraction. Electrical conductors were recognized a century later. Conductors readily permit electric charge to flow through them. Insulators do not readily permit the flow of electric charge.

Electric charge occurs only in discrete amounts—it is quantized. All electric charge is an integral multiple of the electron or proton charge (Section 27.6).

Electric charge is a scalar quantity that can be positive or negative. It is always exactly conserved. In any interaction the algebraic sum of the charges before the interaction equals the algebraic sum of the charges after the interaction.

Experimentally two electric charges are found to attract or repel each other with a force that is (a) inversely proportional to the square of the distance between them, (b) proportional to the product of the charges, and (c) along the line joining the charges (a central force). These findings are summarized by Coulomb's law:

$$\mathbf{F} = k_e\left(\frac{q_1 q_2}{r_{12}^2}\right)\hat{\mathbf{r}}_{12} \qquad (26.6)$$

For practical reasons the SI unit of charge, the coulomb, is defined in terms of the magnetic effects of electric currents. (Electric currents are charges in motion.) Then k_e is determined by experiment.

$$k_e = 8.98755 \times 10^9\ \mathrm{N\cdot m^2/C^2} \qquad (26.4)$$

The electrostatic force is much stronger than the gravitational force. The ratio of electrical force to gravitational force for both an electron and a proton is 2.27×10^{39}.

Coulomb forces may be added vectorially. There is a principle of superposition and a linear mathematical system for Coulomb forces.

Problems

Section 26.2 Electric Charge

1. A uranium nucleus (92 protons) undergoes nuclear fission. Emerging from the fission event are a krypton fragment (36 protons), a barium fragment (56 protons), and some "other particles." Use conservation of charge to prove that if only two "other particles" emerge, they must be either uncharged or oppositely charged.

Section 26.3 Coulomb's Law

2. (a) From Coulomb's law in SI units show that the proportionality constant k_e has units of $N \cdot m^2/C^2$. (b) With the coulomb defined as an ampere second ($C = A \cdot s$), show that the units of k_e may also be written $kg \cdot m^3/A^2 \cdot s^4$.

3. An electroscope can detect the presence of as little as 10^{-10} C of electric charge. Approximately how many electrons could be present without being detected by the electroscope?

4. A mole of molecular hydrogen, H_2, contains two times Avogadro's number (Section 25.1) of protons and an equal number of electrons. How many coulombs of charge do these protons provide?

5. (a) Calculate the number of electrons in a small silver pin, electrically neutral, with a mass of 10 gm. Silver has 47 electrons per atom. The atomic mass of silver is 107.87. (b) Electrons are added to the silver pin until the net charge is 1 mC. How many electrons are added for every billion (10^9) electrons already present?

6. Two small neutral spheres are 0.12 m apart. Electrons are removed from one sphere and deposited on the other sphere. The result is an attractive force exerted by each sphere on the other of 1.03×10^{-3} N. How many electrons were transferred?

7. Two identical small metal spheres *attract* each other with a force of 8.53×10^{-2} N. The distance between the spheres is 1.19 m. The spheres are brought into electrical contact with each other so that the net charge is shared equally. When returned to a distance of 1.19 m, the spheres are found to *repel* each other with a force of 1.96×10^{-2} N. Find the charge on each sphere.

8. Example 4 shows the tremendous force between two 1.0-C charges 1 m apart. What is the repulsive force for a separation distance of (a) 1 km; (b) the earth–moon distance; (c) the earth–sun distance?

9. A pair of equal charges ($q_1 = q_2 = q$) are separated by 1 m. The Coulomb force exerted by each charge on the other is 1 N. Determine the magnitude of q.

10. A pair of equal charges ($q_1 = q_2 = q$) are separated by 1 m. The Coulomb force exerted by each charge on the other is 9800 N. Determine the magnitude of q.

11. A pair of equal charges ($q = 10^{-4}$ C) exert forces of 90 N on each other. How far apart are they?

12. Atomic nuclei contain protons and neutrons. A typical proton–proton separation is 10^{-15} m. Calculate the repulsive force between two protons at this distance. (In stable nuclei this disruptive Coulomb force is counterbalanced by the strong attractive nuclear force.)

13. A hydrogen atom may be pictured (Section 42.5) as an electron revolving around a fixed proton at a distance of 0.529×10^{-10} m. (a) Calculate the attractive electrical force the proton exerts on the orbiting electron. (b) This electrical force is the centripetal force. Knowing this, calculate the speed of the electron.

14. In the ground state of hydrogen the electron–proton separation is 0.529×10^{-10} m. In the "first excited state" of hydrogen the electron–proton separation is four times as large. Calculate the electrostatic force on the hydrogen electron in the (a) ground state and (b) first excited state.

15. In the novel *Ringworld*, science fiction writer Larry Niven postulates a device that separates the positive charges from the negative charges of atoms. Suppose that the electrons and the protons of 1 microgram (μg) of hydrogen were separated into two small spheres 0.1 m apart. What would be the attractive force exerted by each sphere on the other? Express your result in newtons and in tons. (*Hint:* Avogadro's number is $N_A = 6.022 \times 10^{23}$ atoms/mole.)

16. A crystal of ordinary table salt consists of sodium ions with 1 electron unit of positive charge and chlorine ions with 1 electron unit of negative charge. Calculate the electrostatic force that an adjacent sodium ion and chlorine ion exert on each other for a separation distance of 2.82×10^{-10} m.

17. In a typical example of nuclear fission, a uranium nucleus splits into a krypton nucleus ($Z = 36$, or 36 protons in the nucleus) and a barium nucleus ($Z = 56$, or 56 protons in the nucleus). For a separation distance of 2×10^{-14} m, what electrostatic force does each bare nucleus fission fragment (krypton and barium) experience?

18. Two very small spheres 4 cm apart carry equal negative charges. Each sphere experiences a repulsive force of 2×10^{-6} N. (a) What is the charge on each sphere? (b) How many extra electrons does each sphere have (to give rise to this charge?)

19. Imagine that the sun carries a net excess positive charge while the earth carries a net negative charge. If the excess charge is proportional to the mass, $q_{sun} = km_{sun}$, $q_{earth} = -km_{earth}$, (a) What charges would be needed on the earth and on the sun in order to provide electrostatic attraction equal to the gravitational attraction? (b) what is the value of the proportionality constant k (in coulombs per kilogram)?

20. The man-made substance plutonium 239, $^{239}_{94}Pu$, decays into uranium 235, $^{235}_{92}U$, emitting an alpha particle ($^{4}_{2}He$) in the process. (*Note:* The alpha particle is a bare helium nucleus with a net charge of $+2e$.) (a) Calculate the force on the alpha particle exerted by the uranium 235 nucleus when the center-to-center distance is 1.0×10^{-14} m. (b) What is the alpha particle's acceleration at this separation distance? (*Note:* $m_\alpha = 4.00260u = 6.646 \times 10^{-27}$ kg; $m_U = 235.044u = 3.903 \times 10^{-25}$ kg.)

21. Two completely ionized carbon 12 atoms ($q = 6e$) are brought to a distance of 1×10^{-14} m of each other. (a) What is the force of repulsion exerted by each carbon ion on the other? (b) If the carbon 12 ions are free to move, what is the initial acceleration of each?

22. Two small insulating spheres suspended on 50-cm-long insulating cords each carry a net charge $+q$ uniformly distributed. Each sphere has a mass of 20 gm. When the spheres are hung from the same point, it is found that each cord makes an angle of 5° with the vertical. (a) Find q, the charge on each sphere. (b) What angle will each cord make with the vertical if one charge is halved and the other charge doubled?

23. The points of suspension of the two spheres in the preceding problem are moved apart until the distance between the two spheres has doubled. Show that the cords now make an angle of 1.25° with the vertical.

24. Two small charged spheres contain charges of $+Q$ and $+q$, respectively. A charge $+\Delta q$ is removed from one sphere and transferred to the other. Show that the electrostatic force between the two spheres is a maximum when each sphere possesses a charge $(Q + q)/2$.

25. Suppose that there is a difference of 10^{-42} C in the magnitudes of the charges of the electron and proton. Each atom of the earth and its moon would then have a net electric charge. Compare the *ratio* of the electrical force between two hydrogen atoms, one on the earth and the other on the moon, and the gravitational force between the same two atoms.

Section 26.4 Superposition

26. Three positive point charges are placed on a line as shown in Figure 1. Redraw the diagram and show the six $\hat{\mathbf{r}}_{ij}$ unit vectors. Find the total force (magnitude and direction) on each charge. Check your answers by showing that the sum of all the forces is zero. Why zero?

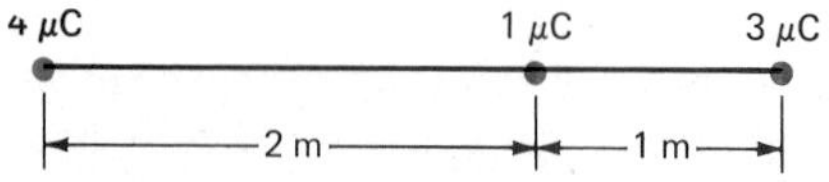

Figure 1

27. Four equal charges ($Q = 10^{-6}$ C) are located as shown in Figure 2. Determine the *net* electrical force on the charge at position 3 (midway between the charges at positions 2 and 4).

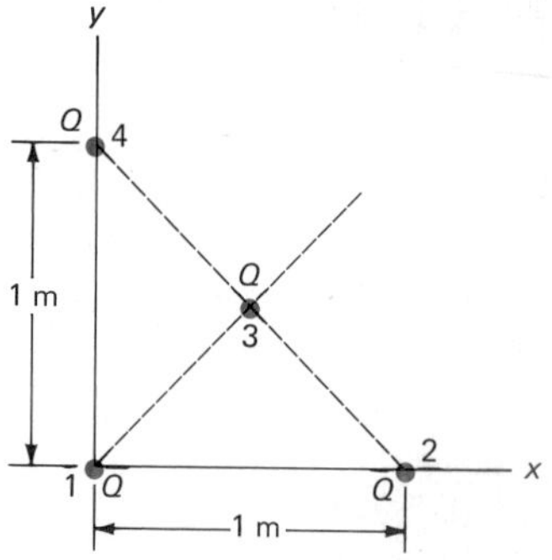

Figure 2

28. A charge of $+1$ μC is located 10 m from a charge of $+81$ μC (Figure 3). (a) Where, along the line joining the two charges, could you place a negative charge $-q_0$ so that the net electrostatic force on $-q_0$ would vanish? (b) Is this a position of stable equilibrium? Explain. (*Note:* This is an electrostatic analog of the gravitational forces on a spacecraft between the earth and the moon.)

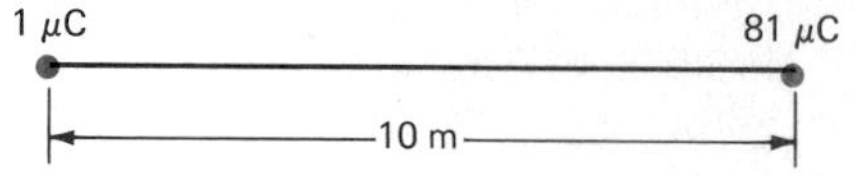

Figure 3

29. A positive charge of 9 μC is fixed at the origin, $x = 0$, in Figure 4. A second charge, q, is fixed on the x-axis at $x = a$. A positive charge q_0 on the x-axis at $x = 3a$ does not feel any net electrostatic force. (a) Find the value of the charge q. (b) By displacing q_0 a small distance Δx along the x-axis, show whether the equilibrium is stable or unstable (with respect to displacement along the x-axis).

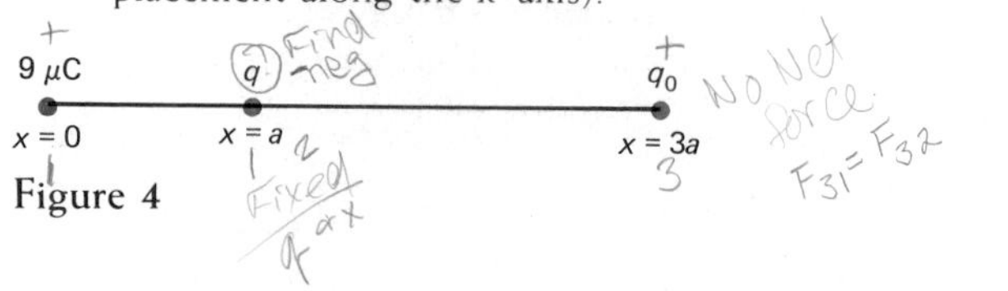

Figure 4

30. A charge of $+1.0 \times 10^{-6}$ C is at the origin. A second charge, $+3.0 \times 10^{-6}$ C, is on the x-axis at $x = 2$. A third charge, Q, is introduced. All three charges are in (unstable) equilibrium. (a) Find the location of Q. (b) Determine Q (magnitude and sign).

31. The pair of charges shown in Figure 5 forms an electric dipole. (a) Show that the force these charges exert on a test charge q_0 on the x-axis is given by

$$\mathbf{F} = \mathbf{i}\, 4k_e\left(\frac{qq_0a}{x^3}\right) \text{ N} \quad \text{for} \quad x \gg a$$

(*Note:* This problem calls for binomial expansions. Remember that $x \gg a$ does *not* mean that x becomes infinite; it means only that $(a/x)^2$ is negligible compared to a/x.) (b) Show that the force exerted by the charges in Figure 5 on a test charge q_0 (with q_0 on the y-axis) is given by

$$\mathbf{F} = -\mathbf{i}\ 2k_e\left(\frac{qq_0a}{y^3}\right)\ \text{N} \quad \text{for} \quad y \gg a$$

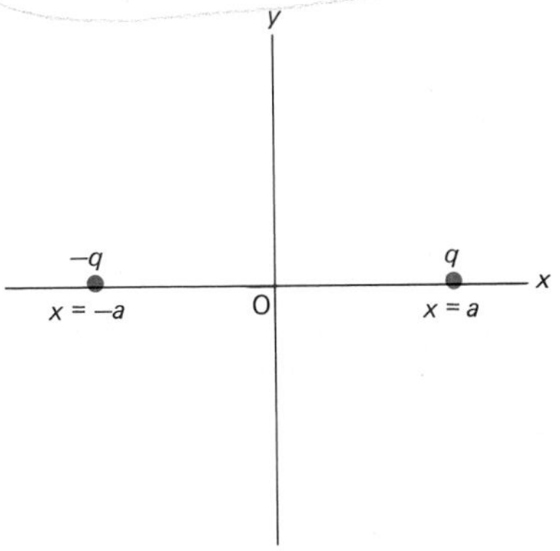

Figure 5

32. A charge of 2 μC is at point (0, 1 m) in Figure 6. An identical charge is at point (0, − 1 m). Calculate the force on a 1-μC charge for (a) points on the y-axis (excluding $y = \pm 1$ m); (b) points on the x-axis; (c) Locate the points on the x-axis for which F_x is a maximum.

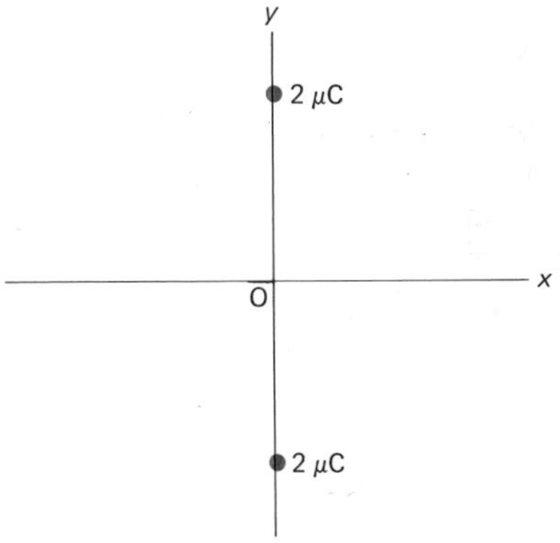

Figure 6

33. A charge of 2 μC is at point (0, 1 m) in Figure 7. A negative charge, −2 μC, is at point (0, − 1 m). Calculate the force on a 1-μC charge for (a) points on the y-axis (excluding $y = \pm 1$ m); (b) points on the x-axis.

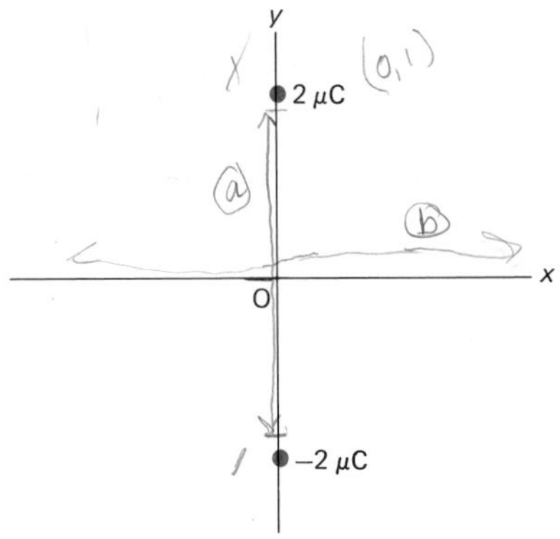

Figure 7

34. Three identical charges q lie on the corners of an equilateral triangle, as shown in Figure 8. A fourth charge, q_0, is free to move on the x-axis under the influence of the electrical repulsive forces of the other three charges. Prove that an equilibrium position for q_0 exists at $x = +a/\sqrt{3}$, $y = 0$.

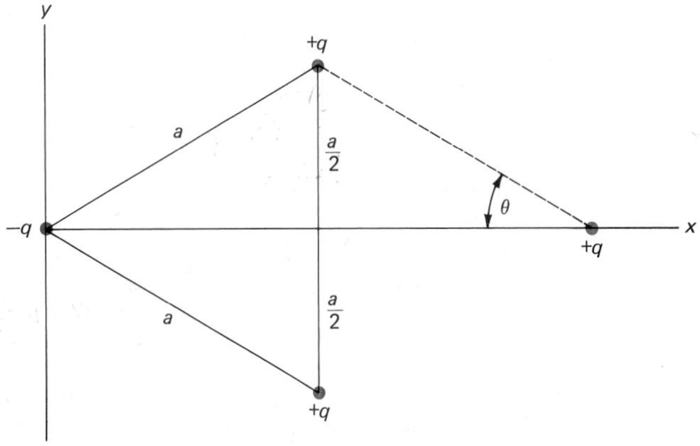

Figure 8

35. Three charges of equal magnitude q lie on the corners of an equilateral triangle. Two are positive and the third, at the origin in Figure 9, is negative. A fourth charge, $+q_0$, is free to move on the x-axis under the influence of the two repulsive forces and one attractive force exerted by the other three charges. Find an equilibrium position on the x-axis for q_0 for $x > 0$. (*Hint:* Find an equation for θ and determine θ numerically by using your pocket calculator.)

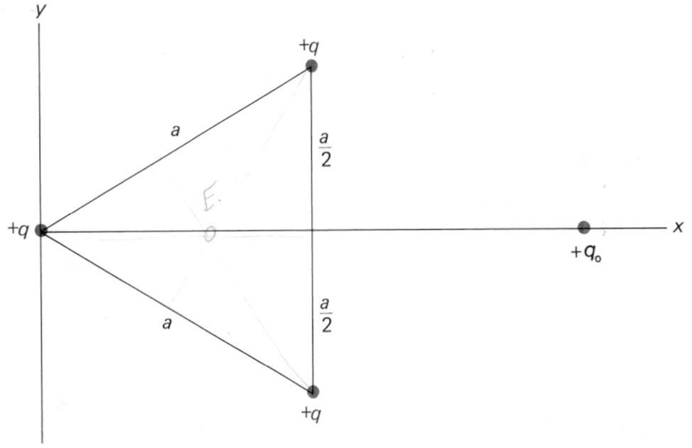

Figure 9

36. Each of the four corners of a square 2 m on a side is occupied by a 1-μC charge (positive), as in Figure 10. (a) Calculate the force on a 1-μC positive charge at an arbitrary point on the x-axis. (b) Verify that the points (±0.773, 0) are equilibrium points.

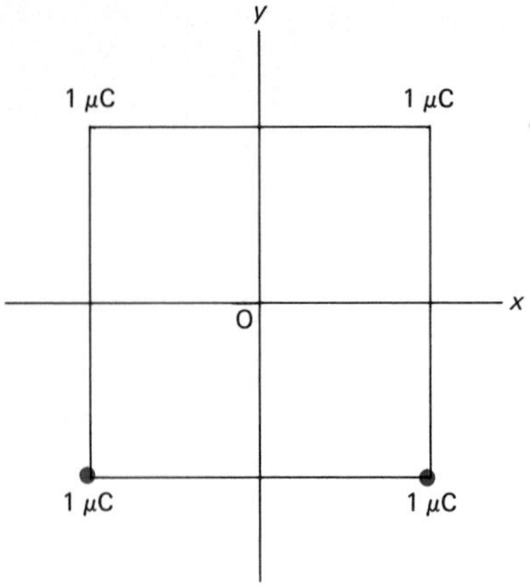

Figure 10

37. As an extension of Problem 36 a positive charge q_o (mass m) is constrained to stay on the x-axis but is free to move along the x-axis. (a) For $|x| \ll 1$ show that

$$F_x = -[k_e 2^{-1/2} q_o (1 \times 10^{-6})]x$$

and that q_o will therefore undergo simple harmonic motion. (b) Calculate the period of oscillation of q_o. (*Note:* This simple harmonic behavior is not some rare accident. Almost every physically significant potential can be approximated by $a_o + a_2 x^2$ in the vicinity of a minimum. It is this parabolic form near a minimum that leads to simple harmonic oscillation.)

38. A line of positive charge, λ C/m, is formed in the shape of a semicircle of radius a and placed in the x–y plane with the ends at $x = a$ and $x = -a$ and the open end facing the positive y-direction. Calculate the force on a charge q_o at the origin.

39. A thin charged rod of length L is placed on the x-axis with one end at the origin and the other end at $x = L$. The charge per unit length on the rod, λ, is constant. Determine the force on a positive charge q_o located at a position $x = a$ $(a > L)$.

40. Referring back to Example 7, Section 26.4, find the expression for the force on q_o as q_o is moved far out on the y-axis, $y_o \gg x_1$. Compare your result with the force that would be exerted on q_o by a point charge Q that is y_o m distant.

41. In Example 7 the force on a charge q_o at $(0, y_o)$ exerted by a line charge ($-x_1$ to $+x_1$), λ C/m, was found to be

$$F = k_e q_o \lambda \left[\frac{2x_1}{y_o (x_1^2 + y_o^2)^{1/2}} \right]$$

Find the expression for the force on q_o as the rod becomes infinitely long (λ constant).

42. Let's change Example 7 by putting the charge q_o at a point x on the x-axis, $x > x_1$ (see Figure 11). (a) Show that the electrical force exerted by the charged rod on q_o is

$$F = k_e \left(\frac{q_o 2\lambda x_1}{x^2 - x_1^2} \right) = k_e \left(\frac{q_o Q}{x^2 - x_1^2} \right)$$

(b) Find the limiting form of the force on q_o for $x \gg x_1$ and compare with the force between q_o and a point charge Q at the origin. (*Note:* $x \gg x_1$ does *not* mean setting x equal to infinity.)

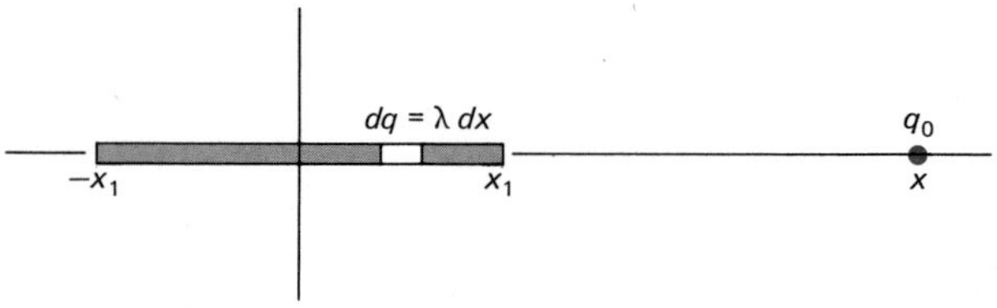

Figure 11

43. In Section 26.4 we presented a calculation of the force exerted by a ring of charge on a point charge located along the axis of the ring. (a) Show that this force is proportional to x for a displacement x from the origin very small compared to the radius of the ring. (b) The charge q_o is constrained to stay on the x-axis. If the ring is positive and the charge q_o is negative, what is the period of oscillation of q_o $(|x| \ll a)$?

44. Suppose that you have a charge of 2 μC at the origin and a charge of −1 μC on the x-axis at $x = -a$, as shown in Figure 12. Select a charge q and place it on the x-axis at $x = b$. Then the force on a test charge q_o far out on the x-axis $(x \gg a,\ x \gg b)$ can be written

$$F = \frac{a_2}{x^2} + \frac{a_3}{x^3} + \frac{a_4}{x^4} + \cdots$$

(a) Choose q so that $a_2 = 0$. (b) Choose b so that $a_3 = 0$. You have now constructed a linear electric quadrupole.

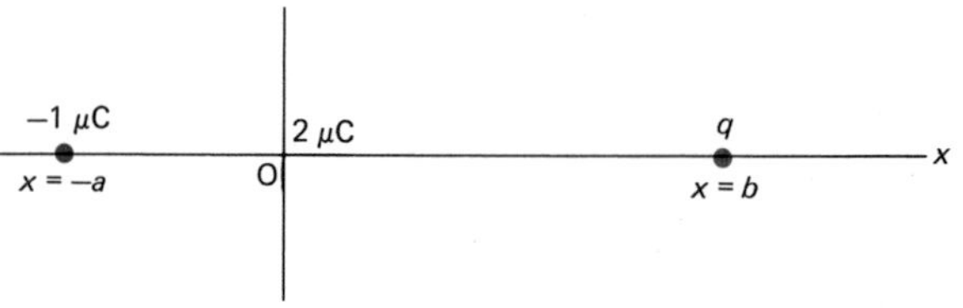

Figure 12

c h a p t e r 27

ELECTRIC FIELD AND GAUSS' LAW

Preview

Coulomb's law of electric force, like Newton's law of universal gravitation, left many scientists uneasy. How, they asked, was the force exerted through empty space? The tremendous success of the law of gravity in explaining terrestrial and celestial motion dispelled all objections to gravitation. In the case of Coulomb forces, however, the objections were overcome when the British experimental physicist Michael Faraday proposed the electric field as an alternative to the concept of force acting through empty space. The electric field became the mechanism for explaining the Coulomb force.

In this chapter we introduce the concept of an electric field in terms of Faraday's electric field lines. Although not necessary, these electric field lines are extremely helpful in visualizing, correlating, and understanding a wide range of electrical phenomena.

Gauss' law, which relates electric fields to charges, is developed in terms of these electric field lines. We then apply Gauss' law to a variety of physical situations.

Finally we consider the influence of an external electric field on point charges—both individual charges and charges in positive–negative pairs. The action of an electric field on individual charges supplies evidence that electric charge is quantized, and leads to a measurement of the fundamental discrete amount of charge.

27.1 Electric Field

Coulomb's law (Section 26.3) precisely describes the forces between electric charges. But Coulomb's law offers no explanation or *mechanism* for these forces. To many scientists this situation was unacceptable. The forces with which they were familiar required *contact* between interacting objects, such as the contact of two colliding billiard balls. If a charge experienced a force, they reasoned, then there should be something acting directly at the location of the charge. In Chapter 6 we discussed the action of the environment on a particle. In the case of electric charge, scientists wanted to describe or define an environment that would result in the Coulomb force on the charge.

It was already known that iron filings would trace out a pattern in the vicinity of a magnet. Gilbert had shown this two centuries earlier. Fine, particle-like sawdust exhibited a similar pattern in the vicinity of electric charges. These patterns and the search for a mechanism for Coulomb's law led Michael Faraday* to propose the concept of an *electric field*.

The electric field $\mathbf{E}$ is defined at each point in space as the electric force $\mathbf{F}$ on a small, stationary† positive test charge q_0 at that point divided by the magnitude of the test charge q_0.

$$\mathbf{E} = \frac{\mathbf{F}}{q_0} \text{ newtons/coulomb} \qquad (27.1)$$

*Michael Faraday (1791–1867) was an English physicist and chemist whose experiments covered a wide range of electrical, magnetic, optical, and chemical phenomena. He is best known for his discovery of electromagnetic induction.

†If the test charge q_0 is in motion, it will experience the same electric force but in addition there may be a magnetic force, proportional to the velocity, if magnetic fields are present. See Chapter 32.

Mathematically, because the force **F** is directly proportional to q_o (as in Coulomb's law), the division by q_o results in a quantity that is independent of the magnitude and sign of the charge q_o. We can think of the electric field **E** as the "environment" of the "particle" q_o. The force $\mathbf{F} = q_o\mathbf{E}$ is described as a combination of properties of the particle (q_o) and its environment (**E**). Since force is a vector quantity, **E** must be a vector quantity (taking q_o as a scalar). Equation 27.1 is an operational definition of electric field. The force **F** on the test charge q_o indicates the existence of the electric field. A measurement of the force, together with a knowledge of q_o, yields the magnitude and direction of the field. The test charge q_o must be sufficiently weak so that it does not disturb the distribution of electric charges that are producing the field that q_o measures. From the definition, Eq. 27.1, the SI units of the electric field are newtons per coulomb. In Section 28.2 we will introduce alternate SI units of volts per meter.

One change in notation is in order. In Chapter 26 we wrote Coulomb's law with a proportionality constant, k_e. In this chapter we will follow the standard (SI) usage and introduce another constant, ε_o, defined by

$$k_e = \frac{1}{4\pi\varepsilon_o}$$

$$\varepsilon_o = 8.85419 \times 10^{-12}\ \mathrm{C^2/N \cdot m^2} \qquad (27.2)$$

This new constant, ε_o, is called the *electric permittivity* of empty space. (In the presence of a medium, the proportionality constant will change. We consider this in Section 29.4.)

Electric Field of a Point Charge

From Coulomb's law, the force exerted on a small test charge q_o by a point charge q is

$$\mathbf{F} = \left(\frac{1}{4\pi\varepsilon_o} \cdot \frac{qq_o}{r^2}\right)\hat{\mathbf{r}} \qquad (27.3)$$

By convention the unit vector $\hat{\mathbf{r}}$ is radially outward from q (Figure 27.1). If q and q_o are taken to be positive, the force on q_o is along the line joining q_o and q and directed away from q. Applying the definition of electric field,

$$\mathbf{E} = \frac{\mathbf{F}}{q_o}$$

we get the field of a point charge:

$$\mathbf{E} = \left(\frac{1}{4\pi\varepsilon_o} \cdot \frac{q}{r^2}\right)\hat{\mathbf{r}} \qquad (27.4)$$

The field **E** of a positive charge is directed outward, away from the positive charge. The field created by a negative charge is directed inward, toward the negative charge.

There are two ideas that you should keep in mind: First, Eq. 27.4 is *not* the definition of an electric field. It merely describes the electric field for the specific case of a point charge. Second, because Coulomb's law and Newton's law of gravitation have the same mathematical form, Faraday's field concept can also be applied to gravitation. The earth, for example, can be said to move in the gravitational field of the sun, which can be written

$$\mathbf{E}_{\text{grav}} = \frac{\mathbf{F}_{\text{grav}}}{m} = -G\left(\frac{M_{\text{sun}}}{r^2}\right)\hat{\mathbf{r}}$$

where m is a "test mass" analogous to the test charge q_o.

The gravitational force of the earth on a mass m at the earth's surface is given by $F = mg$. The gravitational force per unit mass F/m is simply g. We can refer to the gravitational acceleration g as the earth's gravitational field. The direction of the field is downward.

Faraday's concept of electric field allows us to shift our attention from the electric charges to the medium between the charges, and to look at the medium (which may be a vacuum) as a mechanism for transmitting the electrical force. Although physicists today are cautious about giving mechanical properties to electromagnetism, the concept of electric fields has survived because, as we will soon see, it is very useful.

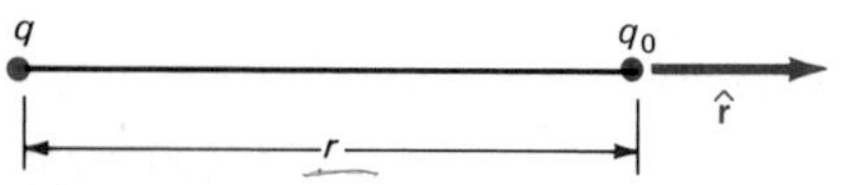

Figure 27.1

27.2 Electric Field Lines

From the definition of electric field we can *imagine* the empty space between electric charges to be filled with little arrows, or vectors. Each of these vectors gives the magnitude and the direction of the electric field at that point. In fact, we can *assign* a vector to *every* point in space. However, rather than draw an arrow at every point in space, which would be an impossible task, Faraday suggested *picturing* an electric field by drawing a few representative lines that follow the curvature mapped out by the arrows, as shown in Figure 27.2. Faraday's electric field lines have the following properties:

1 Electric field lines originate on positive charges and terminate on negative charges (or go on to infinity). They never originate or terminate on a charge-free point in finite space.

2 At each point in space, the field line through that point is tangential to the electric field vector **E** at that point. (When making a field line drawing, make the number of field lines originating on a positive charge or terminating on a negative charge proportional to the magnitude of the charge.)

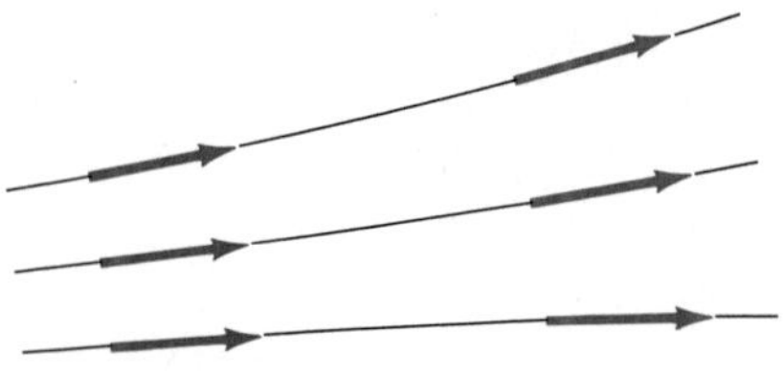

Figure 27.2

Electric field lines tangent to the electric field vectors.

Example 1 **A Positive Point Charge**

The Coulomb force exerted by a positive charge q on a positive test charge q_o is directed radially outward from q. Three such force vectors are shown in Figure 27.3. From Eq. 27.4 we obtain the electric field for the point charge q.

$$\mathbf{E} = \left(\frac{q}{4\pi\varepsilon_o} \cdot \frac{1}{r^2}\right)\hat{\mathbf{r}}$$

The electric field vectors also point radially out from the positive charge, q. Hence the field lines must be radially outward, as shown in Figure 27.4. If q were the only charge in the universe (other than our small test charge q_o), the lines would continue radially outward to infinity.

If q is negative, the pattern is the same except that all the electric field lines are reversed in direction. They continue radially inward, terminating on the negative charge, as shown in Figure 27.5.

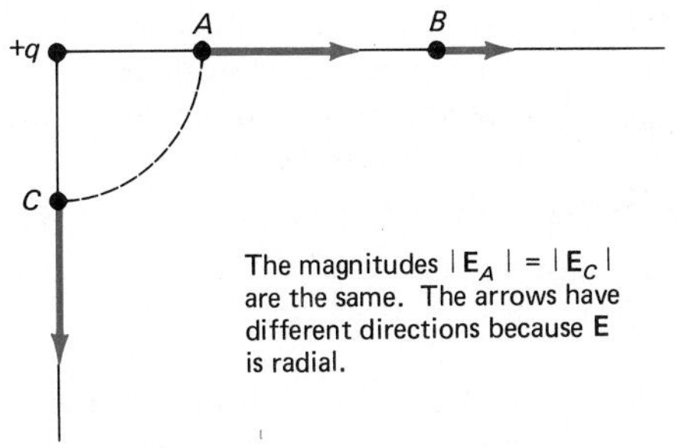

Figure 27.3

The magnitudes $|\mathbf{E}_A| = |\mathbf{E}_C|$ are the same. The arrows have different directions because **E** is radial

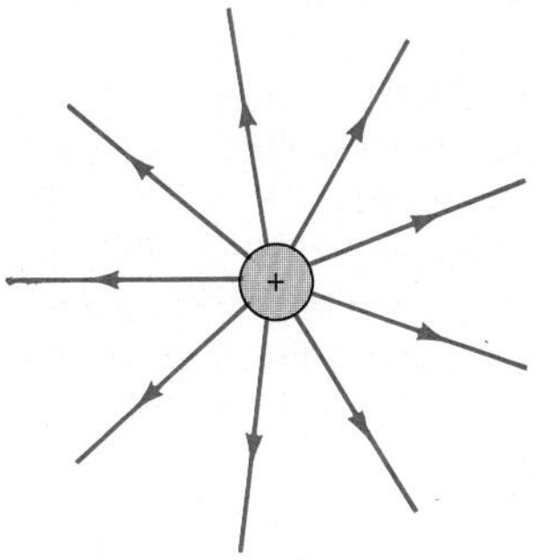

Figure 27.4

Electric field lines representing the electric field of an isolated positive charge. Note the closer spacing where the field is relatively strong and the wider spacing where the field is relatively weak.

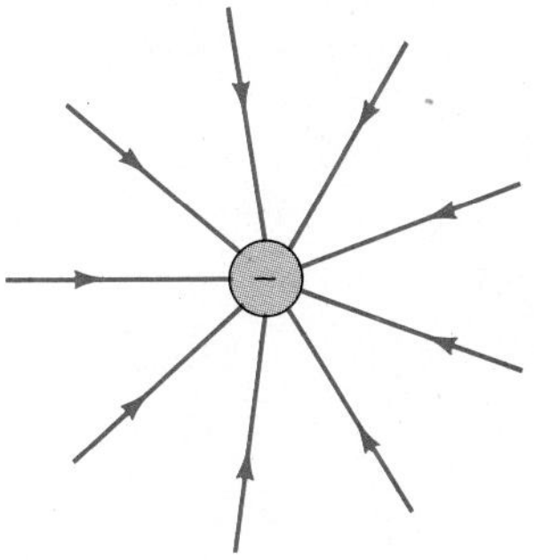

Figure 27.5

Electric field lines in the vicinity of a negative charge.

Principle of Superposition

When we consider two or more charges, we can use the same principle of superposition that was introduced in Section 26.4 to determine **E**. To prove this let's consider again Example 6 of Chapter 26, where **F** represents the total force exerted on a charge q_1 by each of two charges, q_2 and q_3. Superposing forces (which is justified by experiment), we can write

$$\mathbf{F} = \mathbf{F}_{12} + \mathbf{F}_{13}$$

The net force on q_1 is the vector sum of the forces of the separate charges acting individually on q_1. Now let's regard q_1 as a positive test charge and divide the equation by q_1:

$$\frac{\mathbf{F}}{q_1} = \frac{\mathbf{F}_{12}}{q_1} + \frac{\mathbf{F}_{13}}{q_1}$$

This gives

$$\mathbf{E} = \mathbf{E}_2 + \mathbf{E}_3$$

This last equation proves that the electric fields satisfy the principle of superposition. The net electric field is the vector sum of the fields produced by q_2 and q_3 individually. Superposition of Coulomb forces has become superposition of electric fields.

Suppose that we have two $+1$-μC charges 2 m apart on the x-axis. Let's determine the net electric field at a point P between them, x m to the right of the left charge (see Figure 27.6). The field created by the charge at the origin, $x = 0$ is

$$\mathbf{E}_1 = \left(\frac{1}{4\pi\varepsilon_o}\right)\left(\frac{1 \times 10^{-6}}{x^2}\right)\mathbf{i} \text{ N/C}$$

The field created by the charge at 2 m is

$$\mathbf{E}_2 = \frac{1}{4\pi\varepsilon_o}\left[\frac{1 \times 10^{-6}}{(2-x)^2}\right](-\mathbf{i}) \text{ N/C}$$

Note that for $0 < x < 2$, the two fields $\mathbf{E}_1$ and $\mathbf{E}_2$ are oppositely directed. Adding $\mathbf{E}_1$ and $\mathbf{E}_2$ by the principle of

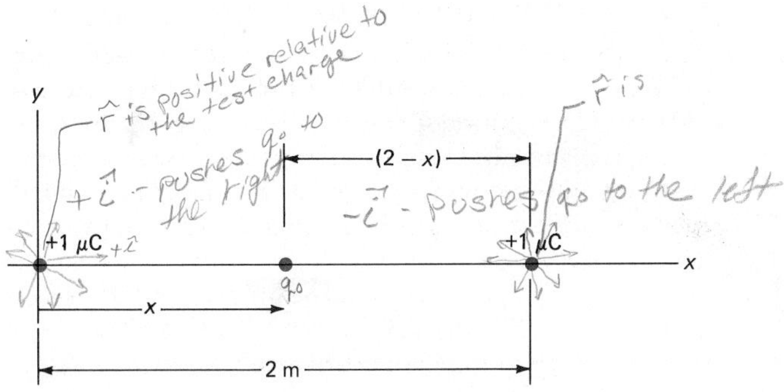

Figure 27.6

Superposition of the fields of two point charges.

superposition, we obtain the resultant electric field E at point P:

$$\mathbf{E} = \mathbf{E}_1 + \mathbf{E}_2$$

$$= \left(\frac{1}{4\pi\varepsilon_0}\right)10^{-6}\left[\frac{1}{x^2} - \frac{1}{(2-x)^2}\right]\mathbf{i}\ \text{N/C} \qquad (27.5)$$

For the special case where P is at the midpoint, $x = 1$, the superposed individual Coulomb fields cancel and $\mathbf{E} = 0$. The net force on a test charge placed at $x = 1$ is zero. Such a test charge would be in equilibrium. If this test charge is positive, its equilibrium will be stable for small displacements along the x-axis but *unstable* for small displacements perpendicular to the x-axis.

Example 2 **The Electric Field between Two Positive Charges of Equal Magnitude**

Consider again the two positive charges shown in Figure 27.6. Let's calculate the field E at the point where $x = 0.5$ and at the point where $x = 1.5$. From Eq. 27.5

$$\mathbf{E}(0.5) = \mathbf{i}(1 \times 10^{-6})(8.99 \times 10^9) \times \left[\frac{1}{(0.5)^2} - \frac{1}{(1.5)^2}\right]\text{N/C}$$

The numbers in brackets yield 3.56, and solving the equation gives us

$$\mathbf{E}(0.5) = \mathbf{i}(3.20 \times 10^4)\ \text{N/C}$$

To find $\mathbf{E}(1.5)$ we again use Eq. 27.5.

$$\mathbf{E}(1.5) = \mathbf{i}(1 \times 10^{-6})(8.99 \times 10^9)\left[\frac{1}{(1.5)^2} - \frac{1}{(0.5)^2}\right] = -\mathbf{i}(3.20 \times 10^4)\ \text{N/C}$$

The field E at $x = 1.5$ has the same magnitude as E at $x = 0.5$. We would expect this from the symmetry of the system. (The plane through $x = 1$, perpendicular to the x-axis, is a plane of symmetry.) However, the sign is reversed. $\mathbf{E}(0.5) = -\mathbf{E}(1.5)$. At $x = 0.5$ the charge at $x = 0$ dominates and pushes a test charge q_0 to the right ($+\mathbf{i}$). At $x = 1.5$ the charge at $x = 2.0$ dominates and pushes a test charge q_0 to the left ($-\mathbf{i}$).

The Electric Field of an Electric Dipole

We have just seen how the electric field between two positive charges of equal magnitude can be obtained by using the principle of superposition. Now let's use this same principle to examine two point charges of equal magnitude but opposite sign. This combination of charges is called an *electric dipole*. Figure 27.7 shows an electric dipole consisting of charges q and $-q$. The product of the positive charge (q) and the distance ($2a$ in Figure 27.7) between charges q and $-q$ is called the *electric dipole moment*, $2qa$.

Let's calculate the electric field produced by the dipole in Figure 27.7. We will call the field produced by the positive charge $\mathbf{E}_+$ and the field produced by the negative charge $\mathbf{E}_-$. Then, by the principle of superposition, the total electric field at any point in space is

$$\mathbf{E} = \mathbf{E}_+ + \mathbf{E}_-$$

The individual fields $\mathbf{E}_+$ and $\mathbf{E}_-$ can each be calculated from Eq. 27.4. Using Cartesian coordinates we obtain

$$\mathbf{E} = \frac{q}{4\pi\varepsilon_0}\left[\frac{\hat{\mathbf{r}}_+}{x^2 + y^2 + (z-a)^2} - \frac{\hat{\mathbf{r}}_-}{x^2 + y^2 + (z+a)^2}\right] \qquad (27.6)$$

where $\hat{\mathbf{r}}_+$ is radially outward and away from the positive charge ($z = a$), and $\hat{\mathbf{r}}_-$ is radially outward and away from the negative charge ($z = -a$).

Let's examine Eq. 27.6 extremely close to each of the charges. In the vicinity of the positive charge, x, y, and $z - a$ are small, while $z + a \approx 2a$ and is not small. Therefore in magnitude, $E_+ \gg E_-$. This means that very close to the positive charge, the field is dominated by the positive charge. The electric field lines are radially outward, as for an isolated positive charge. Close to the negative charge, the reverse is true; and field lines are radially inward.

Figure 27.7

An electric dipole: charge $+q$ at $z = a$ and charge $-q$ at $z = -a$. Zero net charge.

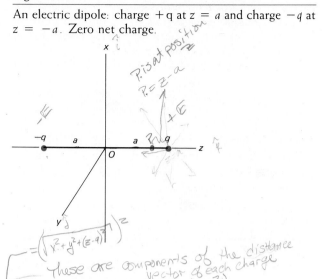

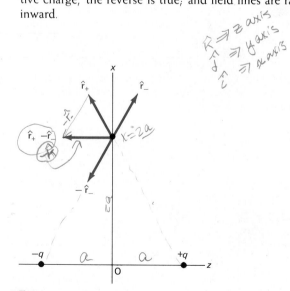

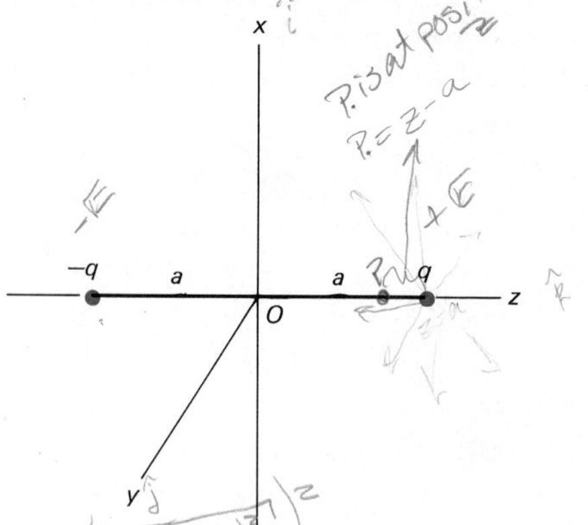

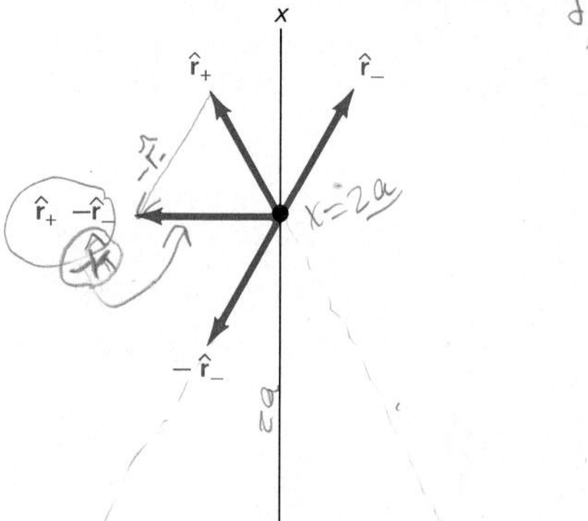

Figure 27.8

Radial unit vectors $\hat{\mathbf{r}}_+$ and $\hat{\mathbf{r}}_-$ giving the direction $\hat{\mathbf{r}}_+ - \hat{\mathbf{r}}_-$ of the resultant field for a point on the x-axis.

At other points the fields of *both* charges contribute; both terms in Eq. 27.6 are important. For instance, at a point $x = 2a$ on the x-axis,

$$E = \frac{q}{4\pi\varepsilon_o}\left(\frac{\hat{r}_+}{4a^2 + 0 + a^2} - \frac{\hat{r}_-}{4a^2 + 0 + a^2}\right)$$

$$= \frac{q}{4\pi\varepsilon_o} \cdot \frac{1}{5a^2}(\hat{r}_+ - \hat{r}_-)$$

The vector subtraction of the two unit vectors is shown in Figure 27.8. Along the x-axis (or anywhere in the $x - y$ plane) the electric field is in the $-\mathbf{k}$-direction.

Example 3　The Dipole Field along the Dipole Axis, I

To get a better feeling for the form of the dipole field let's calculate the field for points far out on the dipole axis (z-axis), as in Figure 27.9. This means that $x = y = 0$ and $z \gg a$. On the z-axis the unit vectors of Eq. 27.5 may be rewritten $\hat{r}_+ = \hat{r}_- = \mathbf{k}$, the Cartesian unit vector. Then Eq. 27.5 becomes

Field line points in neg. dir.

$$E = \mathbf{k}\left(\frac{q_s}{4\pi\varepsilon_o}\right)\left[\frac{1}{(z - a)^2} - \frac{1}{(z + a)^2}\right]$$

Because $z > a$ we can take out a factor of z^2 from each denominator in the brackets and then expand each of these terms by the binomial expansion. We have　*like taylors*

$$E = \mathbf{k}\left(\frac{q}{4\pi\varepsilon_o z^2}\right)\left[\left(1 - \frac{a}{z}\right)^{-2} - \left(1 + \frac{a}{z}\right)^{-2}\right]$$

$$= \mathbf{k}\left(\frac{q}{4\pi\varepsilon_o z^2}\right)\left[\left(1 + \frac{2a}{z} + \cdots\right)\right.$$
$$\left. - \left(1 - \frac{2a}{z} + \cdots\right)\right]$$

The 1's cancel but the $2a/z$ terms add, resulting in

$$E = \mathbf{k}\left[\frac{2(2aq)}{4\pi\varepsilon_o z^3}\right] \qquad (z \gg a)$$
characteristic of dipole

This field along the z-axis is proportional to the electric dipole moment $2aq$ and falls off as the cube of the distance. This inverse cube dependence is characteristic of a dipole.

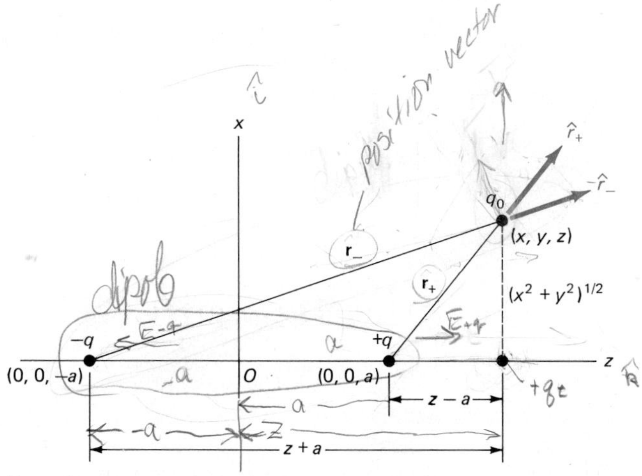

Figure 27.9

An electric dipole acting on a test charge q_0 at (x, y, z).

Figure 27.10

$a = 1. \times 10^{-10} \, m$

$20 \times 10^{-10} \, m$

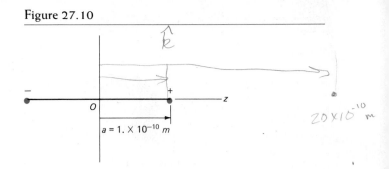

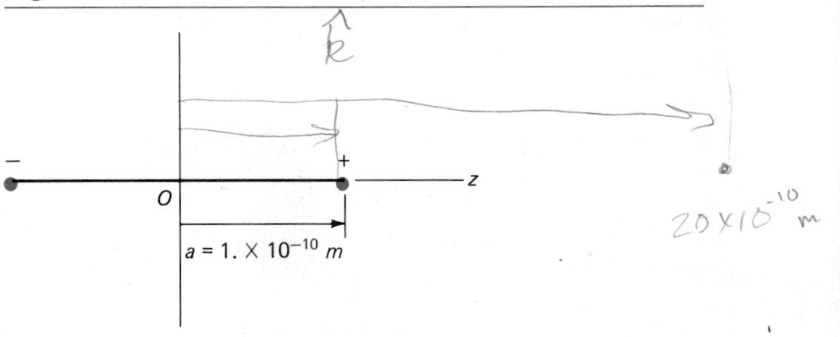

Example 3 and the discussion preceding it are algebraic in nature. Now let's calculate a dipole field using actual numbers.

Example 4　The Dipole Field along the Dipole Axis, II

A proton and an electron separated by 2.0×10^{-10} m form an electric dipole, as shown in Figure 27.10. Calculate the electric field $\mathbf{E}$ at $z = 20 \times 10^{-10}$ m. From Example 3 we have

$$E = \mathbf{k}(8.99 \times 10^9)$$
$$\times \frac{2(2 \times 1 \times 10^{-10}) \times (1.602 \times 10^{-19})}{(20.0 \times 10^{-10})^3}$$
$$= \mathbf{k}(7.20 \times 10^7) \text{ N/C} \qquad (z \gg a)$$

This is an approximation based on the assumption that $z \gg a$. Let's check this approximation by recalculating $\mathbf{E}$ using the exact equation 27.6. With $\hat{r}_+ = \hat{r}_- = \mathbf{k}$ as in Example 3, Eq. 27.6 becomes　*vector addition of $\hat{r}_+ + \hat{r}_- = 1 + (-1)$*

$$E = \mathbf{k}\left(\frac{q}{4\pi\varepsilon_o}\right)\left\{\frac{1}{[(20-1)\times 10^{-10}]^2} - \frac{1}{[(20+1)\times 10^{-10}]^2}\right\}$$
$$= \mathbf{k}(8.99 \times 10^9) \times (1.602 \times 10^{-19})$$
$$\times [(2.770 \times 10^{17}) - (2.268 \times 10^{17})]$$
$$= \mathbf{k}(7.23 \times 10^7) \text{ N/C}$$

Clearly the calculation of Example 3 is a rather good approximation for $z/a = 20$.

We might ask what effect this dipole field would have on a proton at $z = 20 \times 10^{-10}$ m. Imagine that the dipole charges are "nailed down," or fixed in place, but that the proton at $z = 20 \times 10^{-10}$ m is free to move. The force on the proton is

$$F = qE = \mathbf{k}(1.602 \times 10^{-19}) \times (7.23 \times 10^7)$$
$$= \mathbf{k}(1.16 \times 10^{-11}) \text{ N}$$

Under the influence of this force the proton at $z = 20 \times 10^{-10}$ m will have an initial acceleration of

$$a = \frac{F}{m}$$
$$= \mathbf{k}\left(\frac{1.16 \times 10^{-11} \text{ N}}{1.67 \times 10^{-27} \text{ kg}}\right)$$
$$= 6.95 \times 10^{15} \text{ m/s}^2$$

We see then that this dipole electric field has quite an effect on a proton at $z = 20 \times 10^{-10}$ m.

Faraday's Electric Field Lines

Faraday's picture of the electric field lines for an electric dipole is shown in Figure 27.11. The lines originate on the positive charge, curve in accordance with the direction of the electric field, and terminate on the negative charge. The electric field lines for two equal positive charges are shown in Figure 27.12. The field lines originate on the positive charges and run out to infinity. Note in both Figure 27.11 and Figure 27.12 that the field lines are close together near the charges, where the field is strong.

The field lines are relatively far apart away from the charges, where the field is weak. In the next section we will take the number of field lines per unit area of cross section as a measure of the strength of the electric field.

At the beginning of this section, we described two properties of Faraday's electric field lines. A great deal of electrostatics can be covered in general terms (qualitatively) if we attribute two *additional* properties to electric field lines:

3 The field lines tend to assume a shape as though they were under tension, each line resembling a stretched rubber band pulling on the charge on which it originates and on the charge on which it terminates.

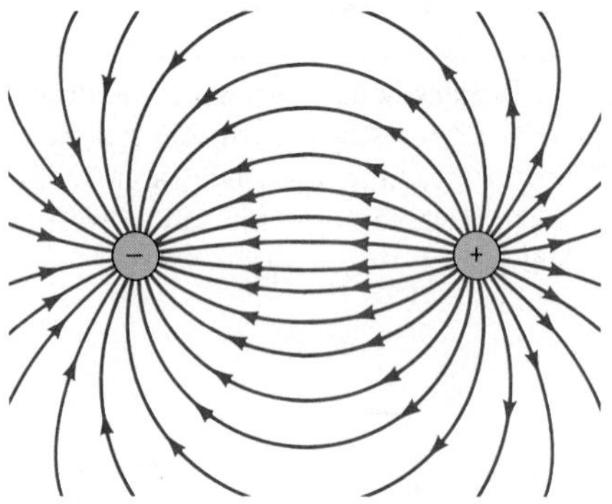

Figure 27.11
The electric field of an electric dipole.

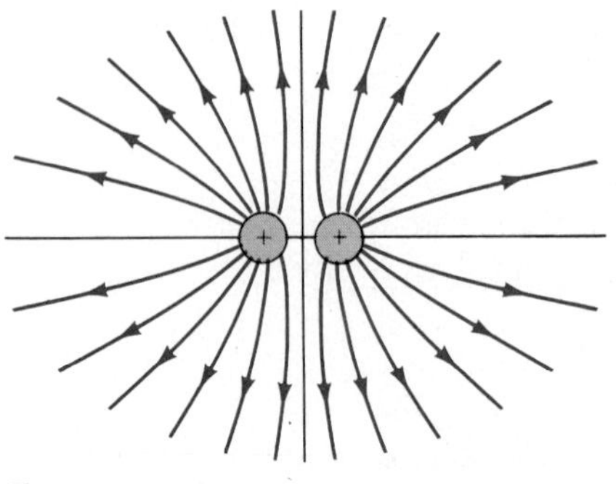

Figure 27.12
The electric field of two closely spaced positive charges.

(When applied to the electric dipole, this property gives us a way of thinking of the attractive force between the two charges.)

Tension alone is not sufficient to account for Figure 27.11. To account for the fact that the field lines do not all lie along the line joining the two charges we postulate a fourth property:

4 The field lines tend to spread out as though they repelled each other.

Remember, Faraday's electric field lines are not real. They are a representation, or picture, of the Coulomb force on a test charge in space. You can ignore the field lines and work exclusively with the mathematical equations if you wish. Or you may want to use the field lines as an aid to understanding electric fields and as a guide in applying the mathematical equations.

Questions

1. In the definition of **E** (Eq. 27.1), how do we know that "the division by q_0 results in a force that is independent of the magnitude (size) of the charge q_0"?
2. The electric field is a vector field. Name two other examples of vector fields.
3. Is it possible for the electrostatic field lines to intersect? Explain how this can happen or why it may not happen.
4. The electric field lines, by definition, give the direction of the electrostatic force on a static positive charge. Do they also give the direction of acceleration of a charge free to move? Do they give the direction of the velocity?
5. Using the electric field lines and the four properties ascribed to field lines, explain (qualitatively) the mutual repulsion of two positive charges.
6. The electric field above an infinite plane that is uniformly charged is found to be independent of the distance from the plane. Sketch a field line picture and explain this independence of distance in terms of the geometry of the field lines. (*Hint:* Read the caption of Figure 27.4.)
7. In Chapter 26 we introduced a principle of superposition of Coulomb forces, and in this chapter we introduced a principle of superposition of electric fields. Are these two principles of superposition independent of each other? Explain.

27.3 Electric Flux

The following analogy is very helpful in understanding electric flux. Imagine a field of wheat with tall, straight, uniformly spaced parallel stalks (Figure 27.13). Suppose that we place a hoop, like a barrel hoop or a hula hoop, down over the wheat stalks. How many wheat stalks are enclosed by the hoop?

Figure 27.13

Wheat stalks and a circular hoop seen edge on.

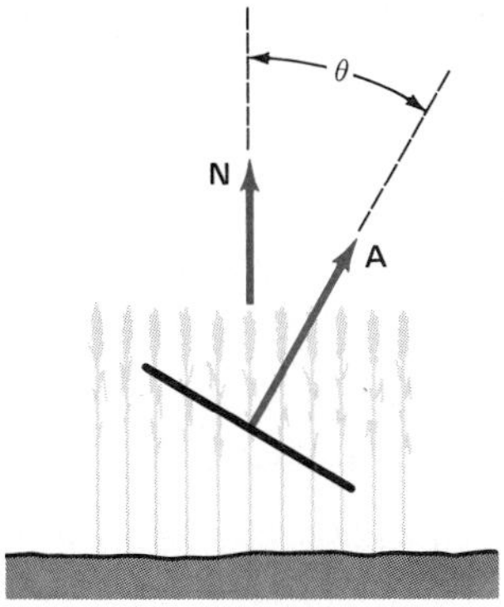

The number of wheat stalks enclosed depends on the number of wheat stalks per square meter, the area of the hoop, and one other factor: the orientation of the hoop relative to the wheat stalks. Let's introduce the vector **N** to describe the wheat stalks. The magnitude of **N** is defined as the *area density* of stalks. The direction of **N** is up, and the units of **N** are stalks per square meter. Now let's introduce the vector **A** to describe the hoop: **A** is defined as the area of the hoop, and its direction is *perpendicular* to the circular surface defined by the hoop. If the hoop is placed on end (hoop plane vertical, **A** horizontal), then no wheat stalks pass through the hoop. If the hoop is horizontal and **A** is vertical, the number of stalks enclosed is a maximum equal to NA. For any orientation of the hoop we define the number of wheat stalks enclosed as the wheat *flux*.

Wheat flux = number of wheat stalks enclosed

$$= NA \cos \theta = \mathbf{N} \cdot \mathbf{A}$$

Now imagine that the stalks of wheat are actually electric field lines. If we replace the wheat density vector **N** by the electric field vector **E**, and replace the area of the hoop by any area **A**, then we have the definition of *electric flux*.

the No. of field lines enclosed by Area

For a uniform field (independent of position) and a plane surface, the electric flux, Φ_E, is defined to be the scalar product of the electric field vector **E** and the plane surface area **A**:

$$\Phi_E = \mathbf{E} \cdot \mathbf{A} \qquad (27.7)$$

Area vector is Normal to the surface.

For nonuniform fields and curved surfaces we rewrite this definition of electric flux in differential form:

$$d\Phi_E = \mathbf{E} \cdot d\mathbf{A} \qquad (27.8)$$

In both definitions of electric flux we use a vector quantity **A**, for area. We are justified in doing this because, as you may recall from Section 2.5, area is the geometric representation of a vector product. The *magnitude* poses no problem; the length of the vector **A** is defined as equal to the number of square meters. The *direction* of **A** is normal (perpendicular) to the surface. Two conventions specify which normal direction is positive:

1 If the surface is *open*, that is, if it has a perimeter like a sheet of paper or a flat dish, let the fingers of your right hand point in the direction in which the perimeter is de-

Figure 27.14

The perpendicular vector $d\mathbf{A}$ represents the area.

scribed (Figure 27.14*). Then your thumb gives the direction of the positive normal. (This is very similar to the right-hand rule for the vector or cross product, and with good reason. Area is a geometric representation of the vector product—Section 2.5.)

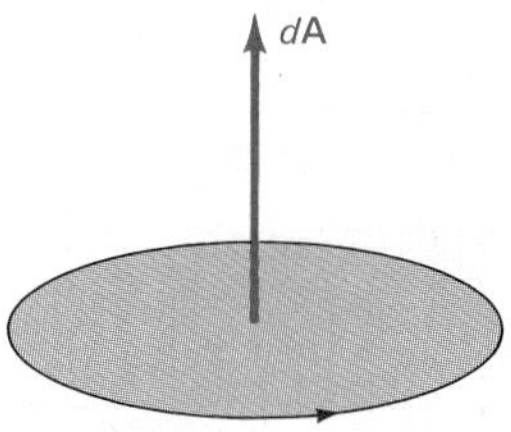

2 If the differential area $d\mathbf{A}$ is a part of a closed surface (like a basketball, for instance), the positive normal is taken to be *outward* from the volume being enclosed. Frequently the vector area differential $d\mathbf{A}$ will be written

$$d\mathbf{A} = \hat{n}\, dA \qquad (27.9)$$

where $\hat{n}$ is a unit vector normal to the surface in the positive direction and dA is the scalar magnitude. The advantage of the $\hat{n}\, dA$ form of the vector is that it separates the direction $\hat{n}$ of the vector $d\mathbf{A}$ from the magnitude dA. In this chapter, the two forms $d\mathbf{A}$ and $\hat{n}\, dA$ are used interchangeably.

$$d\vec{A} = \hat{n}\, dA$$

Electric Field Line Interpretation

Faraday's electric field lines provide a convenient picture of electric flux. We can interpret Φ_E as proportional to the number of field lines intersecting the given area (Figure 27.15[†]). Then, for a surface area perpendicular to the field lines ($\hat{n}$ and **E** parallel),

$$d\Phi_E = E\, dA$$

or

$$E = \frac{d\Phi_E}{dA} \qquad (27.10)$$

*In some physical problems the positive direction of the perimeter is suggested by the positive direction of an electric current around the perimeter. In the absence of physical clues you may choose the positive direction as you wish, as long as you are consistent in your choice.

[†]Keep in mind that electric field lines do not actually exist.

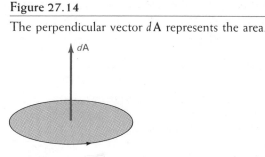

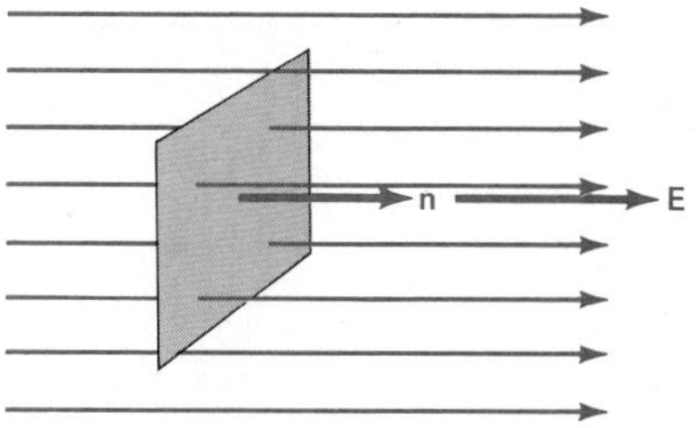

Figure 27.15

E perpendicular to the surface; **E** and $\hat{n}$ parallel.

Figure 27.16

Electric field lines entering the surface are counted as negative: $\mathbf{E} \cdot \hat{\mathbf{n}} < 0$. Electric field lines leaving the surface are counted as positive: $\mathbf{E} \cdot \hat{\mathbf{n}} > 0$.

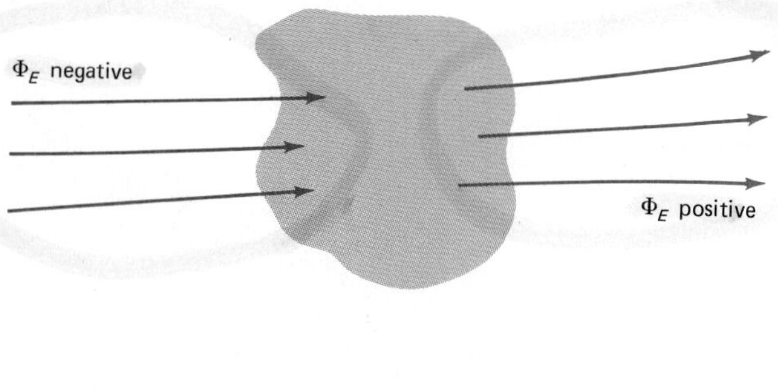

We can picture the magnitude of the electric field as proportional to the number of field lines per unit area.

According to the convention for the positive normal of a closed surface, when a field line intersects a closed surface, the electric flux is negative as the field line enters the surface and positive as the field line leaves the surface (Figure 27.16).

Example 5 **A Point Charge at the Center of Concentric Spherical Shells**

As a further illustration of the relation between a field **E** and the flux Φ_E consider the field produced by a positive point charge at the origin. The electric field lines run radially outward and are assumed to be continuous (Figure 27.17). In the absence of other charges, they continue into infinity. Then the number of field lines passing through a given spherical shell is independent of the radius of the shell. From our picture of electric flux as proportional to the number of field lines passing through a given surface, the flux is independent of the radius of the shell; see Figure 27.17.

But the area of a shell of radius r is $4\pi r^2$. So the number of field lines per unit area, and therefore E, falls off as the inverse square of the distance—in agreement with Eq. 27.4 in Example 1.

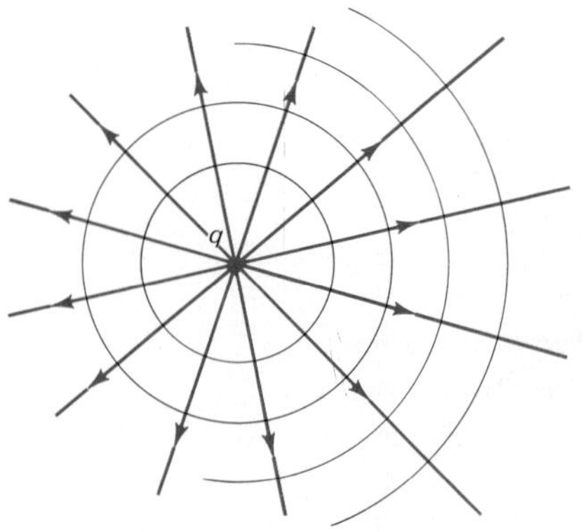

Figure 27.17

Field lines directed radially outward from a positive point charge intersecting concentric spherical shells.

Figure 27.18

The same number of field lines intersect the irregular surface S_2 as intersect the spherical surface S_1.

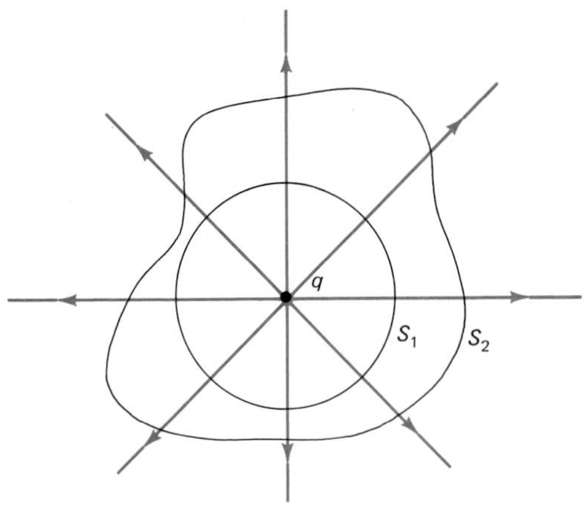

In this example we assume that the field lines are continuous (in the absence of charge). This leads to E proportional to $1/r^2$. Conversely, knowing that E is proportional to $1/r^2$, we find that the field lines must be continuous.

Note that the spherical shells of Example 5 are mathematical shells. In other words, they are not necessarily physical in nature. Also because the field lines are continuous (on out to infinity) they will intersect any closed surface of any shape as long as the surface encloses the point charge. This fact is illustrated in Figure 27.18.

If we were to draw $1/\varepsilon_0 = 1.12941 \times 10^{11}$ electric field lines per coulomb, the electric field E would be numerically *equal* to the number of field lines per square meter. To see this, take the source q to be 1 C and the surface to be a shell of radius r with q at the center. Then

$$E = \frac{1}{4\pi\varepsilon_0} \cdot \frac{1}{r^2} = \frac{1}{4\pi r^2} \cdot \frac{1}{\varepsilon_0} \text{ N/C}$$

Since the field E is constant for fixed r and is everywhere normal to the spherical surface, the flux is given by

$$\Phi_E = E \cdot 4\pi r^2 = \frac{1}{\varepsilon_0}$$
$$= 1.12941 \times 10^{11} \text{ N} \cdot \text{m}^2/\text{C}$$

If we were to draw $1/\varepsilon_0 = 1.12941 \times 10^{11}$ electric field lines per coulomb, the electric flux Φ_E would be numerically equal to the number of field lines. The electric field $E = \Phi_E/4\pi r^2$ would be numerically equal to the number of field lines per square meter.

Example 6 **Uniform Field, Plane Surface**

For an area $\mathbf{A}_1$ with positive normal parallel to **E** the electric flux is

$$\Phi_E = \mathbf{E} \cdot \mathbf{A}_1 = EA_1$$

This is the magnitude of the electric field times the magnitude of the area. (The vector representation of the surface is parallel to **E**, but the surface itself is perpendicular

Figure 27.19

A few of the possible orientations of the electric field **E** and the area **A**.

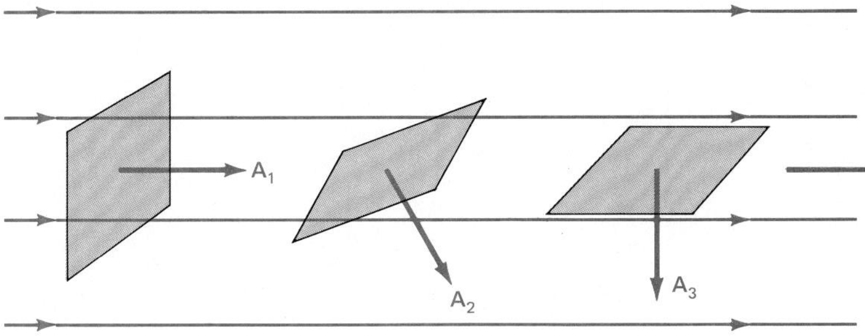

to **E**. See Figure 27.19.) For the tilted surface $\mathbf{A}_2$ shown in Figure 27.19, the flux is given by the scalar product.

$$\Phi_E = \mathbf{E} \cdot \mathbf{A}_2 = EA_2 \cos \theta$$

The flux is reduced by the cos θ factor. For $\theta = \pi/2$, cos θ vanishes and the flux is zero. The field lines are parallel to the surface (perpendicular to $\hat{\mathbf{n}}$) and do not intersect the surface.

In terms of field lines, the flux is the number of field lines passing through a given surface. The tilted surface intercepts fewer field lines for a given surface area, and the flux is therefore reduced.

Questions

8. A certain closed surface contains no net charge. In terms of the electric field lines from external sources, what is the net electric flux through the closed surface?
9. What would the answer to Question 8 be if the closed surface enclosed a net positive charge?

27.4 Gauss' Law

Most of the physical laws that we have seen so far are expressed in differential form. In this section we develop an *integral* relation, Gauss' law (Figure 27.20). The differential laws, such as Newton's second law, are sometimes called *local* laws because they indicate what happens to a particle at a point. Certain integral laws, on the other hand, can be termed *global* because they are concerned with what is happening, or with what exists, over some wide range. Gauss' law is a very powerful, useful global law.

We begin our development of Gauss' law with the very simple case of a positive point charge q at the center of a spherical shell. From Coulomb's law and the definition of the electric field **E**, the charge q produces an electric field **E**. We want to integrate **E** over the surface of the sphere, which means that we must calculate the flux. Calculating the flux is equivalent to counting all the electric field lines that emerge from the shell, or in other words, equivalent to evaluating the integral $\int \mathbf{E} \cdot d\mathbf{A}$. For the case of a spherical shell, the electric field **E** and the area $d\mathbf{A}$ are both radial and therefore parallel, as shown in Figure 27.21. The integral becomes $\int E\, dA$.

The surface integral $\int E\, dA$ is developed similarly to the Riemann integral of ordinary calculus:

$$f(x)\, dx = \lim_{\Delta x \to 0} \sum_i f(x_i)\, \Delta x$$

Figure 27.20

Karl Friedrich Gauss (1777–1835), German mathematician and physicist. Gauss made major contributions to the theory of electromagnetism.

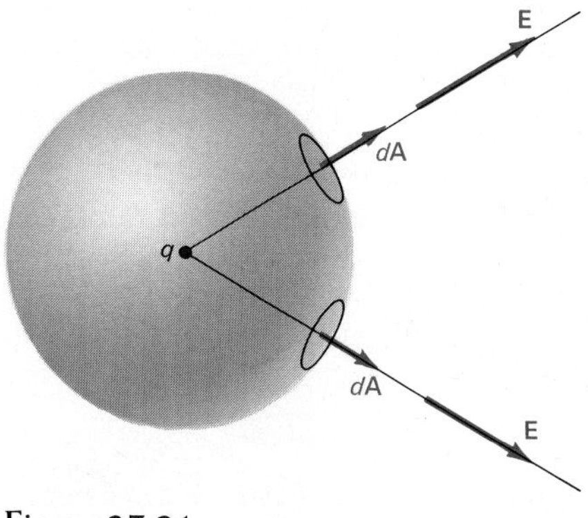

Figure 27.21

For q at the center of the sphere $d\mathbf{A}$ and **E** are always parallel.

Geometrically, the integral of $f(x)$ is the limit of a large number of rectangles, of width Δx, as the width shrinks to zero. A surface integral can be approached in the same way. It is the limit of the sum of small elements of area ΔA, each multiplied by the value of the integrand, E in this case, at that area element.

$$\int E\, dA = \lim_{\Delta A \to 0} \sum_i E_i\, \Delta A \qquad (27.11)$$

For our spherical shell the magnitude of **E** is constant over the surface. Then

$$\int E\, dA = E \int dA \quad \text{if } E \text{ is constant (uniform)}$$

$$= \underbrace{\frac{1}{4\pi\varepsilon_o} \cdot \frac{q}{r^2}}_{E} \int dA \qquad (27.12)$$

The integral of dA is simply the area of the spherical surface:

$$\int dA = 4\pi r^2 \qquad (27.13)$$

Finally,

$$\int_{\text{sphere}} \mathbf{E} \cdot d\mathbf{A} = \frac{1}{4\pi\varepsilon_o} \cdot \frac{q}{r^2} \cdot 4\pi r^2 = \frac{q}{\varepsilon_o} \qquad (27.14)$$

— Always flux

Equation 27.14 is Gauss' law for a spherical shell.

The electric field integral in Eq. 27.14 is related to electric flux by the defining equation, Eq. 27.8:

$$\Phi_E = \int d\Phi_E = \int \mathbf{E} \cdot d\mathbf{A}$$

This is the total flux through the spherical surface. In terms of the field lines, the integral $\int \mathbf{E} \cdot d\mathbf{A}$ is proportional to the number of field lines originating on q (positive charge).

We can generalize Gauss' law for a spherical shell to any charge q anywhere inside a shell of any shape. If we hold the magnitude of the charge constant, the number of field lines going outward is fixed. Because the lines are continuous, we can say the following: *lines per shell Area*

1 All of the electric field lines that intersect the spherical shell when the charge q is at the center continue to intersect the shell when the charge is moved off center (Figure 27.22). The total flux over the enclosing surface is invariant to the location of the enclosed charge.

2 All of the field lines that intersect the spherical shell must intersect any other closed surface that encloses the charge q. The total flux is invariant to the shape and size of the enclosing surface (Figure 27.18).

This analysis can be justified by formal mathematics, but it's really just a matter of counting Faraday's field lines (flux). The result is that Eq. 27.14 holds for any closed surface about any distribution of charge:

$$\int_{\substack{\text{closed} \\ \text{surface}}} \mathbf{E} \cdot d\mathbf{A} = \frac{q_{\text{interior}}}{\varepsilon_o} \qquad (27.15)$$

total enclosed charge for any closed surface

Equation 27.15 is a generalized mathematical statement of Gauss' law.

We have taken the enclosed charge q_{interior} in Eq. 27.15 to be a point charge, but this restriction is not

Figure 27.22

The electric charge q is centered relative to a spherical shell S_1 but off center relative to a second, identical shell S_2. All of the field lines from q that intersect shell S_1 also intersect S_2.

necessary. We can sum over many enclosed point charges and integrate over an enclosed continuous charge distribution. The result is that q_{interior} is simply the enclosed charge.

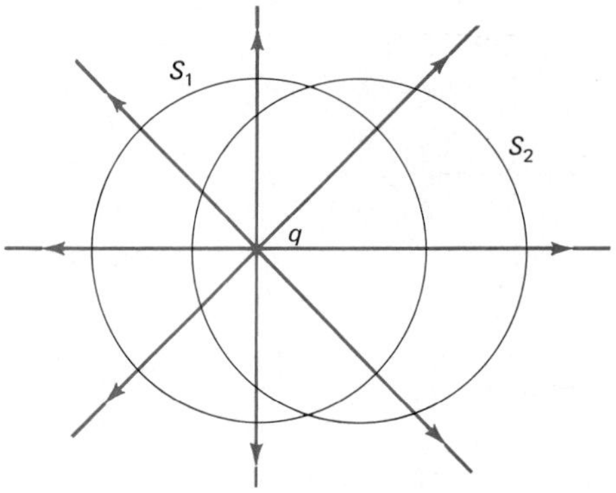

Example 7   From Gauss to Coulomb

Just as we derived Gauss' law from Coulomb's law, we can derive Coulomb's law from Gauss' law, as we now show. For a point charge q at the origin, and a spherical surface, we have spherical symmetry—no preferred direction. Therefore, we assert that the electric field **E** must be radial in direction and that its magnitude is the same everywhere on the surface. Then at the position of an element of area $d\mathbf{A}$ on the surface, **E** is parallel to $d\mathbf{A}$ and $\mathbf{E} \cdot d\mathbf{A} = E\, dA$.

$$\int_{\text{sphere}} \mathbf{E} \cdot d\mathbf{A} = \int E(r)\, dA$$

Because the magnitude of the field is constant over the surface of the sphere

$$\int \mathbf{E} \cdot d\mathbf{A} = E(r) \int dA = E(r) \cdot 4\pi r^2$$

From Gauss' law

$$\int \mathbf{E} \cdot d\mathbf{A} = \frac{q}{\varepsilon_o}$$

$E(r) \cdot 4\pi r^2 = \dfrac{q}{\varepsilon_o}$

$E(r) = \dfrac{q}{\varepsilon_o} \dfrac{1}{4\pi r^2}$

Equating these two results and solving for $E(r)$, we obtain

$$E(r) = \frac{1}{4\pi\varepsilon_o} \cdot \frac{q}{r^2}$$

Then

$$\mathbf{E} = \hat{r} E(r) = \frac{1}{4\pi\varepsilon_o} \left(\frac{q}{r^2}\right) \hat{r}$$

This is the electric field produced by q. From the definition of **E**, the force on the test charge q' is

$$\mathbf{F} = \frac{1}{4\pi\varepsilon_o} \left(\frac{qq'}{r^2}\right) \hat{r}$$

which is Coulomb's law.

Questions

10. Gauss' law has been derived by assuming that the enclosed charge is positive. If q is negative, what happens to the field lines? What happens to the flux?

11. What relation is there between a Coulomb force of $1/r^2$ and the assumption that electric field lines are continuous (originating and terminating only on charge)?

12. We assumed that electric field lines were continuous in our development of Gauss' law. Now reverse this assumption. Start with Gauss' law and show that it implies continuity of the electric field lines.

13. Why does spherical symmetry demand that E be constant over the surface of the sphere? (See Example 5.)

27.5 Applications of Gauss' Law

Gauss' law has many applications in both elementary and advanced physics. We describe several of these applications in this section. We will use Gauss' law first to show that the interior of a conductor in electrostatic equilibrium contains no net charge. Then we will use Gauss' law to determine the electric field produced by various charge distributions. Each of these charge distribution–electric field calculations can be carried out by using Coulomb's law directly with integration, but provided there is sufficient symmetry, it is simpler to use Gauss' law. Gauss' law is used in two important ways:

1 Certain general results may be derived, as illustrated in Example 8.

2 **E** can be calculated for a given charge distribution. This is illustrated in Examples 9, 10, and 11.

In all of these cases we can ask, "How do we choose the surface for calculating the electric flux?" The immediate answer is to choose the Gaussian surface in such a way that we can calculate the flux. In Examples 9, 10, and 11 the choice is dictated by the symmetry of the physical system. The choice in Example 8 seems almost trivial, but it's a valid choice and it leads to an important physical result.

Example 8 **Absence of Net Charge Inside a Conductor**

A conductor consists of an array of positively charged atomic nuclei bound in place relative to one another. Electrons provide an equal amount of negative electric charge, and so the material is electrically neutral. Most of the electrons are bound to the nuclei, but in a metallic conductor some electrons (perhaps one per atom) are free to move through the metal. The electrons move in response to an electric field, and in motion they constitute an electric current.

Figure 27.23

A Gaussian surface just inside the actual surface of the conductor.

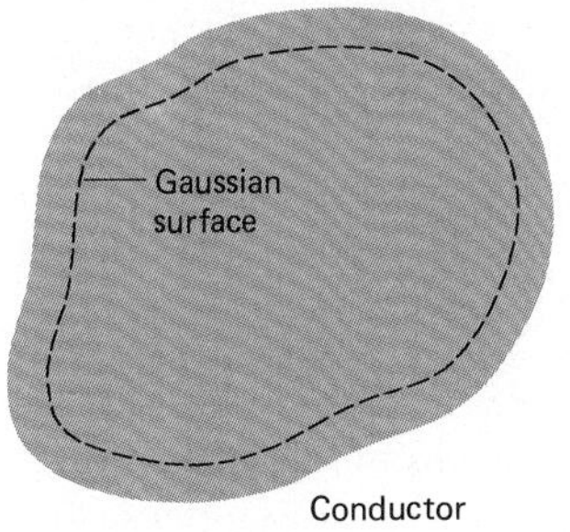

We want to show that there is no net charge in the interior of a conductor that is in electrostatic equilibrium. Electrostatic equilibrium means that there are no electric currents. Let's consider a solid conductor, such as a chunk of metal, that has a net charge. In electrostatic equilibrium the electric field inside the conductor must be zero. To understand why **E** must be zero remember that the conductor has a large amount of both positive and negative charge in its interior and that some of this charge is free to move. If **E** were not zero, the electrical forces would give rise to a motion of some of the charge. We would then have electric currents, and consequently we would not have electrostatic equilibrium.

We choose our Gaussian surface just beneath the actual physical surface of the conductor (see Figure 27.23). The reason for this choice of Gaussian surface is that we know $\mathbf{E} = 0$ over this entire surface. Integrating over this surface we get

$$\int \mathbf{E} \cdot d\mathbf{A} = \frac{q}{\varepsilon_0}$$

But the integral is zero because **E** is zero everywhere on this interior surface. Therefore the enclosed charge q equals zero. In a state of static equilibrium there is no net excess charge in the interior of a conductor. (The positive and negative charges present occur in equal amounts.) Any excess charge must lie on the surface of the conductor.

This result was found experimentally before Gauss' law was formulated.

As long as the enclosed volume is very much larger than the individual atoms or molecules, the Gaussian surface may be chosen to be any closed surface in the interior of the conductor. Example 8 shows that there can be no net charge in any such volume in the interior of a conductor. That is, in the interior of a conductor in electrostatic equilibrium the net charge density is zero. Note that charge on the surface of an isolated conductor is actually trapped on the surface. The charge is bound to the surface and is *not* free to wander off into space. In fact, if we want to get electrons off the surface of an isolated conductor in a vacuum, we must exert an extremely large electrical force (see Section 28.6).

The following are guidelines for applying Gauss' law to calculate the electric field of a given charge distribution. There are two major steps:

I Evaluate the electric flux, $\Phi_E = \int \mathbf{E} \cdot d\mathbf{A}$.
1 From the symmetry of the charge distribution determine the pattern of the field lines.
2 Construct a Gaussian surface to take advantage of this symmetry. This means that, ideally, you should arrange the surface so that **E** is either parallel to $d\mathbf{A}$ or perpendicular to $d\mathbf{A}$ everywhere over the surface. Then for any given surface element $d\mathbf{A}$, either $\mathbf{E} \cdot d\mathbf{A} = 0$ or $\mathbf{E} \cdot d\mathbf{A} = E\,dA$.
3 Still exploiting the symmetry of the charge distribution, arrange the Gaussian surface so that **E** is constant over that portion where **E** and $d\mathbf{A}$ are parallel. Then $\int \mathbf{E} \cdot d\mathbf{A} = \int E\,dA = E \int dA$.
4 Finally, the electric flux is given by $\Phi_E = EA$ where A is the area over which **E** and $d\mathbf{A}$ are parallel. (Remember that vector $d\mathbf{A}$ is perpendicular to the surface!)

II Evaluate q, the charge enclosed by your choice of Gaussian surface. By Gauss' law, Eq. 27.15, q/ε_0 is equal to the flux calculated in step I.

Strong emphasis is placed on symmetry in step I. Indeed, the flux calculation is reasonable only when we have a high degree of symmetry. The charge distributions in Examples 9, 10, and 11 are all highly symmetric.

Example 9 **Electric Field at a Charged Conducting Plane**

Consider a conducting plane—a metal plate that is in static equilibrium with positive electric charge distributed uniformly over its surface. We will assume that the conducting plane is infinite in extent, and thus avoid the question of what happens at the edge of a plane. Since the plane is conducting, the tangential component of **E** at the surface must vanish. Otherwise, we would have currents on the surface, contrary to our assumption of static equilibrium. Then if **E** ($E \neq 0$) has no tangential component, it must be normal. (This argument will be developed in more detail in Section 28.5.)

Because the charge distribution is uniform on a uniform conducting plane we can argue that the electric field must be normal to the surface by symmetry. Apart from the normal there is no reason to single out any one angle for **E**, or any one direction.

For the Gaussian surface of integration we take a short squat cylinder of height h and radius r (Figure 27.24a). The bottom surface of the cylinder is inside the conductor (where $\mathbf{E} = 0$). Since **E** is tangential to the curved surfaces of the cylinder, the surface integral receives no contribution from the curved sides of the cylinder. This contribution to the flux is zero, $\mathbf{E} \cdot d\mathbf{A} = 0$ (Figure 27.24b). The only remaining contribution is from the top surface of the short squat cylinder. Here **E** is parallel to $d\mathbf{A}$ (perpendicular to the top surface) and constant in magnitude over the surface. Following step I of our guidelines for using Gauss' law, we write

$$\Phi_E = \int \mathbf{E} \cdot d\mathbf{A} = E \int dA = E\pi r^2$$

Figure 27.24a

A cylindrical Gaussian surface.

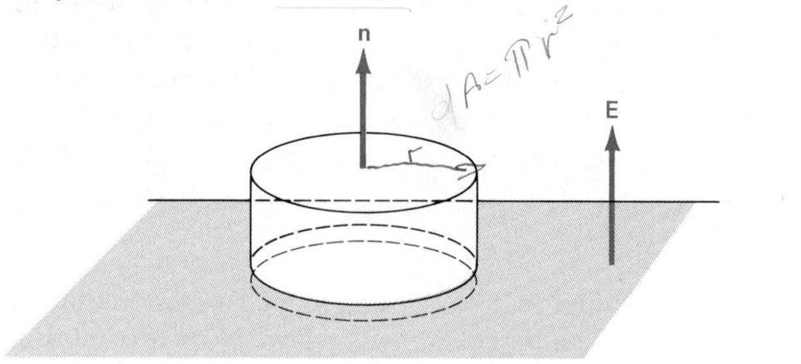

Figure 27.24b

A cross-sectional view.

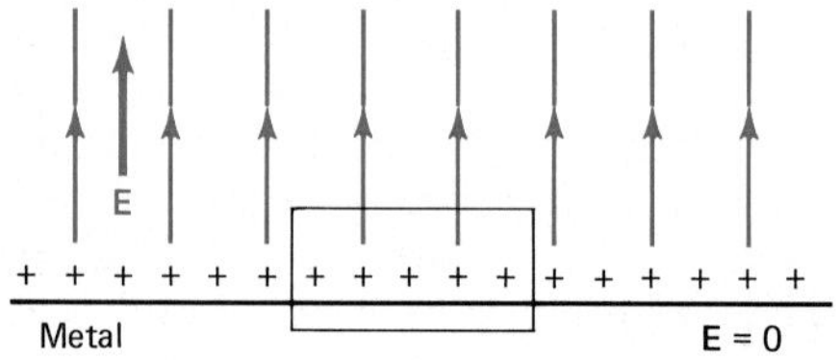

By Gauss' law (step II) the electric flux is equal to q/ε_0. As usual q is the enclosed charge. Equating these two forms of Φ_E we get

$$E\pi r^2 = \frac{q}{\varepsilon_0}$$

It is convenient to describe the charge q in terms of a surface charge density σ, having units of coulombs per square meter. Then $q = \sigma\pi r^2$ and

$$E = \frac{\sigma}{\varepsilon_0} \qquad (27.16)$$

The field directly above a charged conducting plane is normal to the plane. The field strength is proportional to the surface charge density. Note carefully that we have used a metal plate as a conducting plane. All of the field lines go in the same direction—out from the metal surface. If we had instead considered a charged mathematical surface, half of the field lines would have gone in one direction (perpendicular to the surface) and half in the opposite direction. For a mathematical charged plane Eq. 27.16 would require a factor of 2 in the denominator.

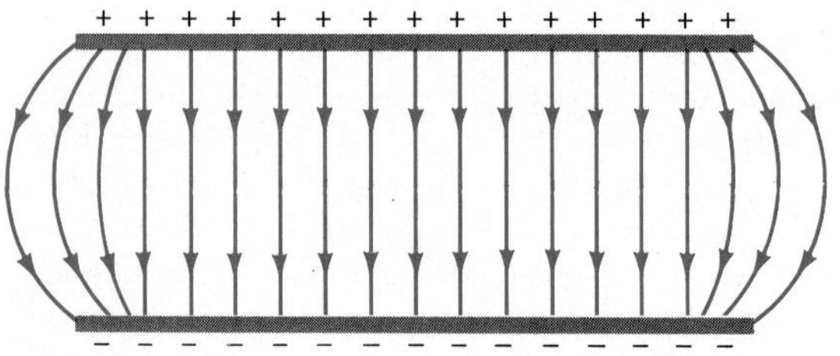

Figure 27.25

The electric field of two parallel charged plates.

A system of two parallel conducting planes, one with a uniform positive charge and the other with an equal negative charge, results in an electric field between the parallel planes. The resulting field is very uniform—as long as we don't get too close to the edges of the planes. At the edges of the planes, the field lines become curved due to tension and mutual repulsion. (See Figure 27.25.) In Section 29.1 we will discuss the use of this system of two parallel planes as a device for storing electric charge.

Example 10 A Uniformly Charged Spherical Shell— Exterior Field

Let us find the electric field outside a uniformly charged spherical shell, total charge Q (Figure 27.26). First, note the symmetry that we have imposed by using the words *uniform* and *spherical*. The electric charge is spread out uniformly—there is a constant number of coulombs per square meter over the spherical surface. There is complete spherical symmetry—there are no angles defined. There is no preferred reference line or axis. Thus, by symmetry, the field lines must go radially outward. This means that $\mathbf{E}$ must be radial in direction. The magnitude of $\mathbf{E}$ depends only on the distance r from the center of the sphere. We write

$$\mathbf{E} = \hat{\mathbf{r}} E(r)$$

To apply Gauss' law, we choose for the Gaussian surface an imaginary second spherical shell that is outside of and concentric with the charged shell. This is exactly what we did in Example 6 (Section 27.4). With $\mathbf{E}$ and $d\mathbf{A}$ parallel over the surface, the integral becomes

$$\int \mathbf{E} \cdot d\mathbf{A} = \int E \, dA \;=\; E_{(r)}\, dA \;=\; E(r)\cdot A \;=\; E(r)\cdot 4\pi r^2$$

As in Example 6, the magnitude E is a function only of r and thus is constant over the spherical surface. Then the flux integral may be written

$$\int \mathbf{E} \cdot d\mathbf{A} = E(r) \cdot 4\pi r^2$$

gaussian surface

Electric field lines

Q

R

Charged sphere

r

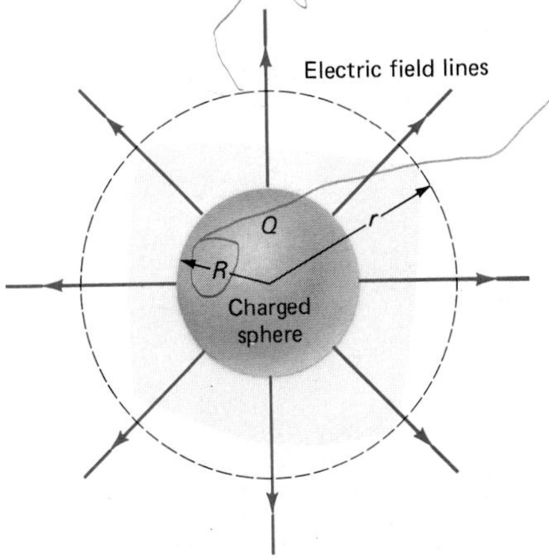

Figure 27.26

The electric field of a charged conducting sphere.

Figure 27.27

The (radial) electric field produced by a uniformly charged spherical shell of radius R.

at surface of sphere

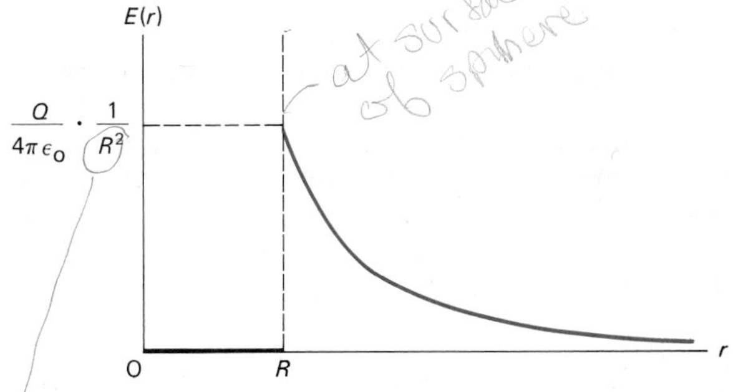

This completes step I of our guidelines for applying Gauss' law. Step II is the evaluation of the enclosed charge. The charge is given as Q, so

$$E(r) \cdot 4\pi r^2 = \frac{Q}{\varepsilon_o}$$

With unit vector $\hat{\mathbf{r}}$ to indicate direction

$$\mathbf{E} = \frac{1}{4\pi\varepsilon_o}\left(\frac{Q}{r^2}\right)\hat{\mathbf{r}} \qquad (27.17)$$

Equation 27.17 is the electric field *outside* the uniformly charged spherical shell ($r > R$). The electric field outside the uniformly charged spherical shell would be the same if all the charge were concentrated at the center of the shell. This result is shown in Figure 27.27 ($r > R$).

Just as a uniformly charged spherical shell acts as if all the charge were concentrated at the center, the earth attracts Newton's apple as if all the mass of the earth were concentrated at the center of the earth.

Example 11 A Uniformly Charged Spherical Shell— Interior Field

Now let's find the electric field *inside* the charged spherical shell. Again, by symmetry, if any field exists at all, it must be radial. For an imaginary sphere that is inside and concentric with our charged sphere we have

$$\int \mathbf{E} \cdot d\mathbf{A} = 0$$

by Gauss' law. No charge is enclosed. Again, writing $\mathbf{E} = \hat{\mathbf{r}} E(r)$ (because symmetry demands this) and integrating over the imaginary sphere (r is constant), we get

$$E(r) \cdot 4\pi r^2 = 0$$

or

$$E(r) = 0$$

hollow Not solid sphere

Hence $\mathbf{E} = 0$. We conclude that *there is no electric field inside a uniformly charged spherical shell*. This result is represented in Figure 27.27 by the heavy line on the r-axis for $0 \le r < R$.

Table 27.1

Source	Electric Field Dependence on Distance	Reference
Point	$\propto 1/r^2$	Eq. 27.4
Line	$\propto 1/r$	Problem 44
Plane	constant	Example 9

A uniformly charged spherical shell is approximated rather well by an automobile with a fallen electric power line stretched across its roof. A passenger inside the car will be unharmed because there is no electric field inside a uniformly charged spherical shell.

We have illustrated plane symmetry and spherical symmetry in Examples 9, 10, and 11. Cylindrical symmetry is illustrated in Problems 44 and 45 (see Table 27.1).

Gravitational Implications

The results of Examples 10 and 11, shown in Figure 27.27, may be applied to the gravitational force exerted by spherically symmetric mass distributions. For instance, imagine that you are in a deep hole at a distance a from the center of the earth. If gravitational force behaves just like electrical force, then the results of Example 11 mean that the mass at a distance $r > a$ exerts no net gravitational force on you. (This means that if the earth were hollow and inhabited, as sometimes described in science fiction, then the people inside would be weightless, apart from small centripetal effects.) The results of Example 10 indicate that the gravitational force on you would be that of the mass for $r < a$, as though that mass were concentrated at the center of the earth, $r = 0$. Astrophysicists must take the gravitational analogs of Examples 10 and 11 into account in developing models of the structure of planets, stars, and galaxies.

Questions

14. The electric field **E** is zero at every point over the surface of a sphere. What is the *net* electric charge inside the sphere? Is it possible to have any charges at all inside the sphere?

15. The net electric flux through a closed surface is zero. Does this mean that **E** vanishes everywhere on the surface? What is the net charge within the surface?

16. A charge q is at the geometric center of a cube. What is the electric flux through *one* of the cube faces? Is the flux through one particular face independent of the location of the charge?

17. If you know the value of the net charge Q within some closed surface, is Gauss' law sufficient to enable you to calculate the electric field? What additional information might you need?

18. Does the converse of Example 8 hold? That is, if there is no net charge within a closed Gaussian surface, does this mean that **E** vanishes over the entire surface? Explain.

19. Suppose that you have a spherical distribution of positive electric charge. The charge density (charge per unit volume) is spherically symmetric, meaning that it may vary with radial distance but is independent of angular position. Will the resulting electric field take on a maximum value at the surface of your distribution? Explain.

27.6 Motion of Point Charges in a Static Electric Field

A charged particle such as a proton (charge e) in an electric field **E** experiences a force $\mathbf{F} = e\mathbf{E}$. In a uniform static electric field ($\mathbf{E}$ = constant) the motion of the proton is like the motion of a shot (shot put) in a uniform gravitational field, with one important difference. In the case of a shot put in a gravitational field, gravitational force and resistance to acceleration are both proportional to m (gravitational mass and inertial mass, Section 6.6) and the mass m cancels out. However, the electric force on a charged particle is proportional to the charge of the particle and independent of the mass. Newton's second law for a charged particle becomes

$$\mathbf{F} = q\mathbf{E} = m\mathbf{a} \tag{27.18}$$

or

$$\mathbf{a} = \left(\frac{q}{m}\right)\mathbf{E} \tag{27.19}$$

The charge/mass ratio q/m therefore remains in our equation. The acceleration depends on particle mass.

Example 12 Motion (Anti)Parallel to E

A proton, charge e, with a velocity of 10^6 m/s enters a field **E** of 10,000 N/C. The proton's velocity **v** and the field **E** are antiparallel (Figure 27.28). How far into this field will the proton penetrate?

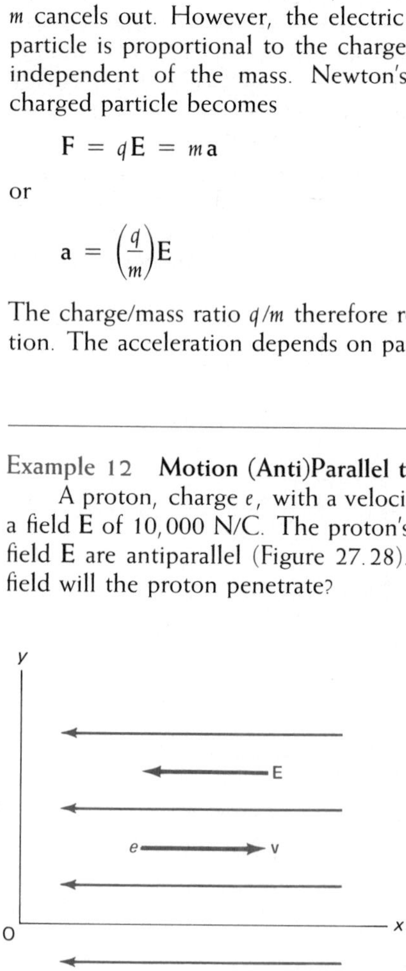

Figure 27.28
Velocity **v** and electric field **E** antiparallel.

Let us choose the positive x-axis in the direction of motion. The force equation is

$$ma = -eE \quad \text{or} \quad a = \frac{-eE}{m}$$

The negative sign appears because the electric field is in the negative x-direction. Physically the field is retarding the motion of the proton. Now that we have identified acceleration, our problem is reduced to one of elementary mechanics. From Eq. 4.17 of Section 4.6, we can write

$$v^2 - v_i^2 = 2ax$$

Here we want to find x for $v = 0$. This leads to

$$x = -\frac{v_i^2}{2a}$$

$$= \frac{(10^6)^2}{(2 \times 10^4)(e/m)} = \frac{10^8}{2e/m} \text{ meters}$$

From Appendix 2, $e/m_{\text{proton}} = 9.57895 \times 10^7$ C/kg and the penetration distance is $x = 0.522$ m. The logic could be reversed. We could measure x and thus determine the charge-to-mass ratio. We will present a better method for determining this ratio in Section 32.3.

Example 13 **Determination of e: The Millikan Oil Drop Experiment**

The American physicist R. A. Millikan (1868–1953) observed the motion of electrically charged droplets of oil as they fell, under the influence of gravity, and rose, under the influence of a uniform electric field (produced by horizontal parallel plates). In his analysis, Millikan considered the buoyancy of the oil droplets in the air and the very significant effect of low-velocity air drag (Stokes's law for the motion of a sphere in a viscous medium—air, Section 6.8—and a correction to Stokes's law).

Because of the air drag, Millikan's droplets quickly reached a terminal velocity. Falling at terminal velocity v_1, the net gravitational force, $(m_{\text{oil}} - m_{\text{air}})g$, was balanced by the viscous drag, Kv_1 (see Figure 27.29).

$$(m_{\text{oil}} - m_{\text{air}})g = Kv_1$$

Applying an electric field **E**, Millikan introduced an upward force $q\mathbf{E}$. Then the force balance for the upward-moving drop (terminal velocity v_2 *upward*) was

$$qE - (m_{\text{oil}} - m_{\text{air}})g = Kv_2$$

with K the Stokes viscous drag coefficient. Upon addition of these two equations, the charge q on the oil drop is found to be

$$q = \frac{K}{E}(v_1 + v_2)$$

Repeating the experiment many times, Millikan found that the values of q clustered closely about 1.6×10^{-19} C, 3.2×10^{-19} C, 4.8×10^{-19} C, ... These data led Millikan to conclude that (1) the magnitude of the electronic charge is $(1.603 \pm 0.002) \times 10^{-19}$ C. (The modern value is 1.60219×10^{-19} C.) (2) The charges on all the oil drops are always integral multiples of this value. The electric charge occurs only in discrete amounts. In other words, charge is quantized.

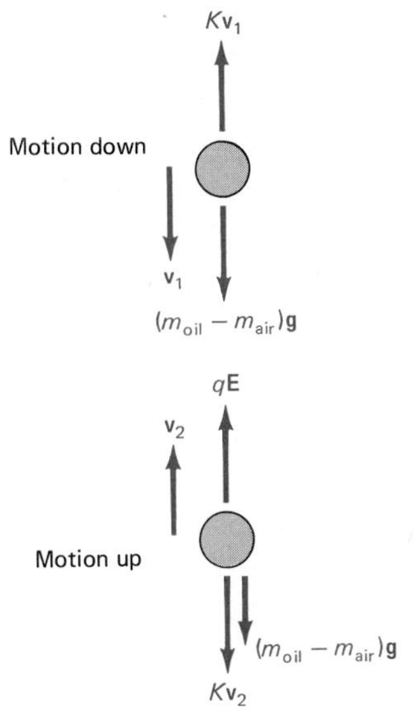

Figure 27.29

Forces on the moving droplet.

Motion Perpendicular to a Field

The motion of a charged particle initially moving perpendicular to an electric field is like that of a ball thrown horizontally in a uniform gravitational field. The velocity perpendicular to the electric field is unchanged; no electric force is acting in this direction. There is an acceleration parallel to the field, however, as shown in the preceding examples. The parallel and perpendicular motions may be superposed (as in Section 5.1). For a uniform electric field, the trajectory of the charged particle is a parabola—just as in Section 5.1.

The motion of electrons in a perpendicular field, which we will now describe, is illustrated by the modern oscilloscope. We choose the $+x$-axis in the direction of **E**. Electrons are first accelerated to a velocity v_z and then passed between a pair of conducting plates. An electric field E_x has been created in the region between the plates. The electron experiences an acceleration a_x and acquires an x-component of velocity, $v_x = a_x \, \Delta t$ where Δt is the time spent between the plates. If the length of the plates is l, then $\Delta t = l/v_z$ and

$$v_x = \frac{-e}{m} \cdot \frac{l}{v_z} \cdot E_x$$

The minus sign is a consequence of the electron's negative charge. The acquired transverse velocity v_x is opposite to the direction of **E**. The angle of deflection θ is given by

$$\tan\theta = \frac{v_x}{v_z} = -\frac{e}{m} \cdot \frac{l}{v_z^2} \cdot E_x$$

The deflected electrons signal their presence (and deflection) by producing a flash of light when they are absorbed by a special material (phosphor) covering the screen. The deflection is proportional to the electric field across the deflecting plates.

Question

20. A point charge starting from rest is permitted to move in response to a uniform electric field. Will the trajectory of the charge coincide with a single field line? Would the trajectory of the charge coincide with a single field line if the field were nonuniform? (*Hint:* Consider Newton's first law.)

27.7 Electric Dipole in an Electric Field

Uniform Electric Field

As you may recall from Section 27.2, an electric dipole consists of two charges equal in magnitude but of opposite sign that are separated by some distance. In Section 27.2 we investigated the electric field *produced by* an electric dipole. We will now consider the influence of an external electric field *on* a dipole. (Figure 27.30 shows an electric dipole in a uniform external electric field **E**.) The dipole consists of two charges, $-q$ and q, separated by a certain distance. We'll denote the vector displacement of q relative to $-q$ by **l**. In a uniform external electric field **E** there is a force on q, $\mathbf{F}_+ = q\mathbf{E}$. Similarly, there is a force on $-q$, $\mathbf{F}_- = -q\mathbf{E}$. The net force on the dipole is therefore

$$\mathbf{F} = \mathbf{F}_+ + \mathbf{F}_- = q\mathbf{E} - q\mathbf{E} = 0 \qquad (27.20)$$

A *uniform* electric field exerts no net force on an electric dipole. But the forces $\mathbf{F}_+$ and $\mathbf{F}_-$ do not have the same line of action. Therefore, there is a net torque τ. Let's try to predict what will happen to the dipole. Clearly the torque due to the force $\mathbf{F}_+$ on q will attempt to align **l** in the direction of the external field **E**. The torque due to $\mathbf{F}_-$ on $-q$ will also attempt to line **l** up with **E**. The two torques are acting in the same manner. They do not *cancel*.

Now, let's do a formal vector analysis of the torque exerted on **l**. Using the equations for torque from Section 3.3 we get

$$\tau_+ = qE\left(\frac{l}{2}\right)\sin\theta \qquad \text{clockwise}$$

$$\tau_- = qE\left(\frac{l}{2}\right)\sin\theta \qquad \text{clockwise} \qquad (27.21)$$

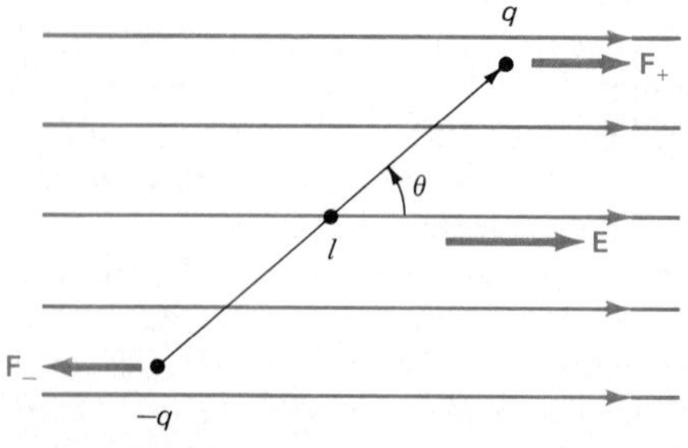

Figure 27.30

Forces on an electric dipole in a uniform electric field.

The total torque is

$$\tau = qlE\sin\theta \qquad \text{clockwise} \qquad (27.22)$$

This torque is proportional to the field **E** and to the product ql, which is called the *dipole moment*. The forces $\mathbf{F}_+$ and $\mathbf{F}_-$ form a couple and, following Section 3.3, the torque is invariant to the choice of axis (perpendicular to the plane of $\mathbf{F}_+$ and $\mathbf{F}_-$).

This torque may cause the dipole to rotate and line up parallel to the electric field, just as a compass needle lines up in the magnetic field of the earth. Or if the dipole has angular momentum and is behaving like an isolated gyroscope, the torque will cause the dipole to precess (Section 13.4) about the direction of the field.

Note that a measurement of the torque and of the field gives information only about the product ql and not about the factors q and l individually. This is a general characteristic of dipoles and explains why the dipole moment ql is important in its own right.

The torque on a dipole is frequently given in vector form. The vector dipole moment is defined by the vector equation

$$\mathbf{p} = q\mathbf{l} \qquad (27.23)$$

With the vector displacement **l** directed from the negative charge toward the positive charge, the vector torque is

$$\boldsymbol{\tau} = \mathbf{p} \times \mathbf{E} \qquad (27.24)$$

This yields Eq. 27.22 for the magnitude of $\boldsymbol{\tau}$ and, by the right-hand rule, gives the direction of the torque as into the paper. Note that **p** is the dipole moment and *not* linear momentum.

Example 14 Torque on a Water Molecule

The electric dipole moment of an isolated water molecule, H_2O, is $ql = 6.17 \times 10^{-30}$ coulomb-meters (C · m). Let's calculate the maximum torque on this molecule when an external field of 10^5 N/C is applied.

For the electric field perpendicular to the dipole moment ($\sin\theta = 1$), Eq. 27.22 leads to

$$\tau = 6.17 \times 10^{-30}\ \text{C}\cdot\text{m} \times 10^5\ \text{N/C}$$
$$= 6.17 \times 10^{-25}\ \text{newton-meters}$$

Nonuniform Electric Field

Suppose that a dipole is aligned parallel to an electric field but that the field is *nonuniform*, that is, not constant in both magnitude and direction. Taking the x-axis to be in the direction of **E** (see Figure 27.31), we have for the force on the negative charge at x

$$F_- = -qE(x)$$

The force on the positive charge at $x + l$ in the field $E(x + l)$ is

$$F_+ = qE(x + l)$$
$$= q\left[E(x) + l\frac{dE(x)}{dx} + \cdots\right] \qquad (27.25)$$

Figure 27.31

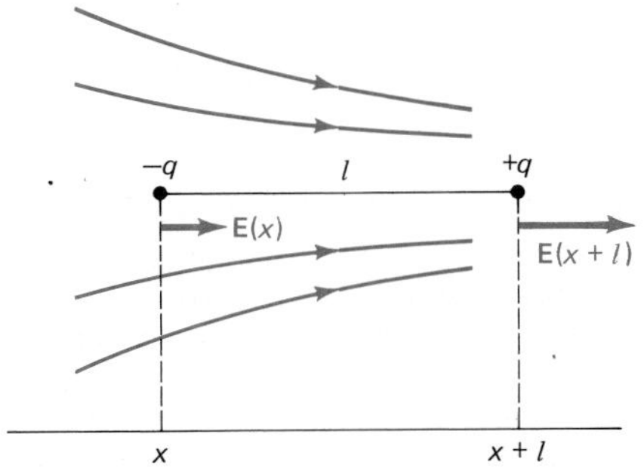

Figure 27.32

The nonuniform field of a charged amber rod creates and attracts an electric dipole.

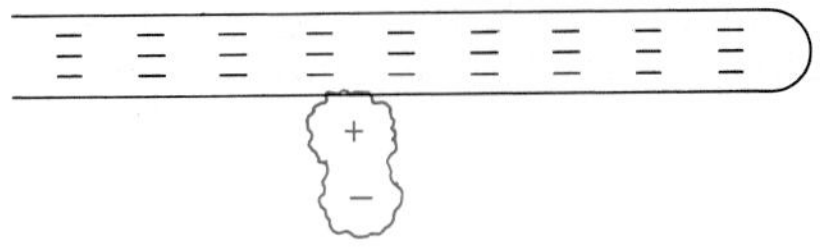

Here we used the first two terms of a Taylor expansion of $E(x + l)$ (see Appendix 4). Adding the two forces F_+ and F_-, we obtain

$$F = F_+ + F_- = ql\frac{dE(x)}{dx} \tag{27.26}$$

There is a net force on the dipole that is proportional to the dipole moment ql and to the space *derivative* of the external (nonuniform) electric field. The force is directed toward the stronger electric field.

This is the basis of the amber effect, which we described in Section 26.1. The negatively charged amber repels the electrons in a small piece of paper. There is a deformation of the electron distribution, and a small dipole moment is induced in the bit of paper. The electric field of the amber is nonuniform and the paper is attracted to the amber (Figure 27.32).

Questions

21. An uncharged conducting sphere is placed in the field of a point charge q_1. Sketch the field lines in the vicinity of the conducting sphere. Will there be a net force on the neutral sphere? Explain.

22. Suppose that the conducting sphere described in Question 21 carries a small net charge q_2. How will the total force exerted on the sphere by q_1 compare with the Coulomb force between q_1 and q_2 (with q_2 at the center of the sphere) if q_1 and q_2 have the same sign? If q_1 and q_2 have opposite signs?

23. What are the implications of the preceding two questions for the use of q_1 as a test charge to measure the electric field of the charge (q_2) on the conducting sphere?

Summary

The electric field is defined as the quotient of the force (apart from gravity and any forces of constraint) on a small, stationary test charge and the magnitude of the test charge.

$$\mathbf{E} = \frac{\mathbf{F}}{q} \tag{27.1}$$

The electric field is a vector quantity with magnitude and direction. Its units are newtons per coulomb (or volts per meter). The concept of an electric field, initially introduced to describe electrostatic interactions across empty space, has developed into one of the most fruitful concepts of modern physics.

The electric field can be illustrated with electric field lines. Electric field lines originate on positive charges, terminate on negative charges, and are tangential to the field vector **E** at every point.

Electric flux is defined as the scalar product of the electric field **E** and area $d\mathbf{A}$:

$$d\Phi_E = \mathbf{E}\cdot d\mathbf{A} \tag{27.8}$$

and can be pictured as the number of electric field lines passing through a given area.

Gauss' law

$$\int_{\text{closed surface}} \mathbf{E}\cdot d\mathbf{A} = \frac{q_{\text{interior}}}{\varepsilon_0} \tag{27.15}$$

relates the charge inside a closed surface to the integral of **E** over that surface. Applications of Gauss' law include

1 the demonstration of the absence of excess charge inside a conductor;

2 the relation of the normal electric field immediately above a plane surface to the surface density of electric charge on that surface, $E = \sigma/\varepsilon_0$;

3 proof that the field outside a uniformly charged spherical shell is the same as the field of equivalent charge concentrated at the center of the sphere;

4 proof that the field inside a uniformly charged spherical shell vanishes.

Electric charges move in response to an unbalanced electric force in accordance with Newton's second law:

$$\mathbf{a} = \left(\frac{q}{m}\right)\mathbf{E}$$

From observation of the motion of charged oil droplets under the influence of gravity and electric fields, R. A. Millikan determined that (a) the value of the electron charge is -1.60219×10^{-19} C (modern value), and (b) all droplets have an integral multiple of this electron charge. (Electric charge is quantized.)

Electric dipoles in a uniform electric field experience zero net force. They *do* experience a net torque, which tends to align the dipole with the field. An electric dipole in a nonuniform electric field experiences a net electric force proportional to the space derivative of the field. Two examples of this force are the amber effect of Chapter 26 and the van der Waals forces between molecules in gases.

Suggested Reading

W. P. Dyke, Advances in Field Emission, *Sci. Am.*, 210, 108–118 (January 1964).

H. W. Lissmann, Electric Location by Fishes, *Sci. Am.*, 208, 50–59 (March 1963).

A. D. Moore, Electrostatics, *Sci. Am.*, 226, 46–58 (March 1972).

H. A. Pohl, Nonuniform Electric Fields, *Sci. Am.*, 203, 106–116 (December 1960).

Problems

Section 27.1 Electric Field

1. A vertical electric field is adjusted to just counteract the force of gravity. What field (magnitude and direction) will be required for (a) an electron; (b) a proton; (c) a neutron?

2. An electric field of 100 N/C acts on an alpha particle (a doubly ionized helium atom, net charge = $2e$, $\text{mass}_\alpha = 4.0015$ u). (a) What force is exerted on the alpha particle? (b) What is the acceleration of the alpha particle?

3. A small crystal of diamond (carbon) has an excess of electrons. In a horizontal electric field of 1.5×10^5 N/C the diamond experiences an acceleration of 0.3 m/s^2 horizontally. (a) What is the charge/mass ratio of the charged diamond crystal? (b) On the average, how many excess electrons are there per carbon atom?

4. Determine the magnitude of a point charge that produces an electric field of 1.0 N/C at points 1 m away.

5. An electron inside a TV picture tube experiences a force of 8.0×10^{-14} N directed toward the front of the tube ($+x$-direction). What is the (a) magnitude and (b) direction of the electric field that produces this force?

6. Calculate the gravitational field (N/kg) of the earth (a) at the surface of the earth and (b) at the location of the moon.

7. Calculate the gravitational field of the sun at the location of the earth. (This is the field that holds the earth in its orbit about the sun.)

8. An electric dipole is formed with 1 μC at $x = 1$ m and -1 μC at $x = -1$ m. Calculate the electric field at $x = 0$, 0.1, 0.2, 0.3, 0.4, 0.5, 0.6, 0.7, 0.8, and 0.9 m. Plot $E(x)$ against x for your 10 points. What field do you expect for the corresponding negative values of x? Why was the point $x = 1$ omitted from the list?

9. (a) From Example 8 of Section 26.4 show that the electric field produced by the charged ring is

$$\mathbf{E} = \mathbf{i}\left(\frac{Q}{4\pi\varepsilon_0}\right)\frac{x}{(x^2 + a_0^2)^{3/2}}$$

for points on the x-axis. (b) Where does this field have a maximum?

10. A charge of 1.5 μC and a charge of 2.8 μC are 1.62 m apart. (a) Find the electric field each charge produces at the position of the other charge. (b) What force does each charge exert on the other charge? Is Newton's third law satisfied?

Section 27.2 Electric Field Lines

11. Sketch out the electric field lines of an array of equal charges on the vertices of a square (Figure 1). Pay careful attention to the space within the square.

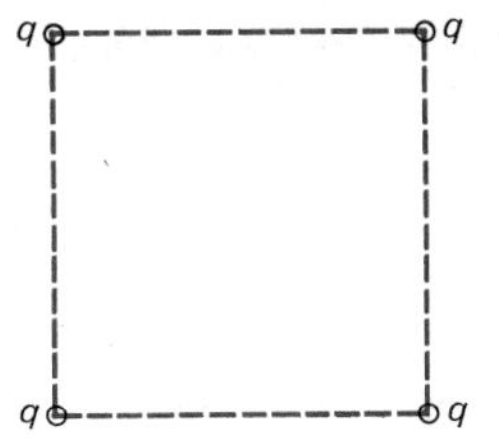

Figure 1

12. When an isolated charge $+q$ is placed at $x = -a$, one of its electric field lines is directed along the $+x$-axis. Similarly, an isolated charge $+q$ located at $y = -a$ has an electric field line directed along the $+y$-axis. Sketch these two lines when both charges are present, one at $x = -a$ and the other at $y = -a$.

13. (a) Using Figure 27.3 as a guide, draw the vector that represents the electric field at a point midway between A and C on the dotted line joining A and C. (b) How does the magnitude of the electric field determined in part (a) compare with the field at point B in Figure 27.3?

14. Problem 28 of Chapter 26 is an electrostatic analog of the earth–moon–spaceship gravitational problem. Sketch the resultant field lines for the 81-μC and 1-μC charges of that problem.

15. In Example 3 the field of an electric dipole along the dipole axis was found to be

$$E = k\left(\frac{q}{4\pi\epsilon_0}\right)\left[\frac{1}{(z-a)^2} - \frac{1}{(z+a)^2}\right]$$

Example 3 exhibits the leading term for $z \gg a$:

$$E \approx k\left[\frac{2(2aq)}{4\pi\epsilon_0 z^3}\right]$$

Show that the next nonvanishing term is

$$k\left(\frac{8a^3 q}{4\pi\epsilon_0 z^5}\right)$$

For $(z/a)^2 \gg 1$ this correction term is negligible.

16. Suppose that you have an electric dipole consisting of a charge $+q$ at $z = a$ on the z-axis and a charge $-q$ at $z = -a$ on the z-axis. See Figures 27.7, 27.8, and 27.9. Investigate the resulting electric field on the x-axis as follows:

(a) Show that

$$E(x, 0, 0) = \frac{q}{4\pi\epsilon_0} \cdot \frac{\hat{r}_+ - \hat{r}_-}{x^2 + a^2}$$

(b) Show that

$$\hat{r}_+ - \hat{r}_- = -k\left[\frac{2a}{(x^2 + a^2)^{1/2}}\right]$$

The second equation is a mathematical statement that the x-components of the two individual fields cancel and that the resultant field is in the negative z-direction. This is indicated in Figure 27.9.

(c) Show that

$$E(x, 0, 0) = -k\left[\frac{(2aq)}{4\pi\epsilon_0}\right]\frac{1}{(x^2 + a^2)^{3/2}}$$

(d) Finally, for $x \gg a$, show that

$$E(x, 0, 0) \approx -k\left[\frac{(2aq)}{4\pi\epsilon_0}\right]\frac{1}{x^3}$$

Again we have the $1/r^3$ dependence characteristic of a dipole.

17. Two oppositely directed dipoles are placed side by side (Figure 2). This is one form of an electric quadrupole. Sketch the electric field lines for this configuration of charges. Include the interior of the square and a portion of the exterior space.

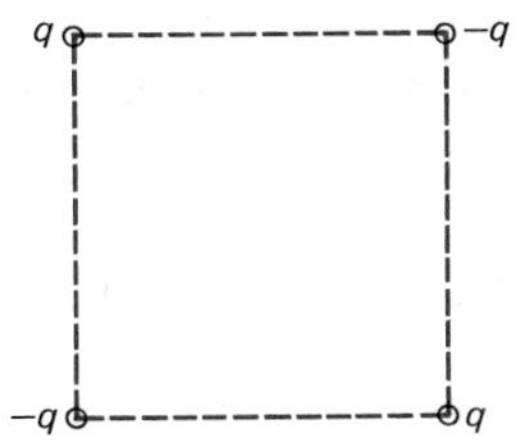

Figure 2

18. Two dipoles that are back to back form a linear electric quadrupole (Figure 3). Sketch the electric lines of force of this quadrupole configuration of charges.

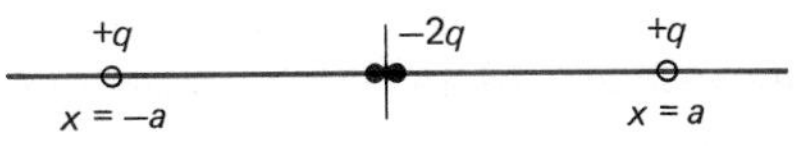

Figure 3

19. For the electric quadrupole of Problem 18 (a) calculate E_x for points on the x-axis, $x \gg a$; (b) calculate E_y for points on the y-axis, $y \gg a$. Do your calculated results agree with your field line picture?

20. Three identical positive point charges lie on the corners of an equilateral triangle whose sides are of length a. Axes are symmetrically oriented with the origin at the geometric center of the three charges (Figure 4). Prove that the electric field on the axis is

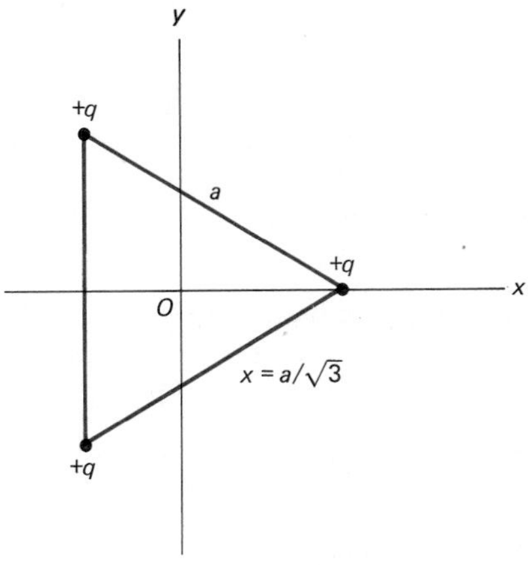

Figure 4

(a) zero at the origin, (b) directed to the right if $x > a/\sqrt{3}$, and (c) directed to the left if $x < -a/2\sqrt{3}$.

Section 27.3 Electric Flux

21. A hoop of area 4.0 m^2 encircles 832 vertical wheat stalks when the *normal* to the plane of the hoop is oriented at an angle of 60° to the vertical. Assuming that the wheat stalks are distributed uniformly, how many would be enclosed when the normal to the plane of the hoop is vertical?

22. A uniform electric field is directed from left to right along the $+x$-axis. If the magnitude of the field is 5.0×10^4 N/C, what flux passes through a circular loop with an area of 0.2 m^2 if the "normal" to the loop is (a) in the $+x$-direction, (b) in the $-x$-direction, (c) in the $+y$-direction, and (d) in the $-y$-direction? (e) For what direction of the normal will the flux be half of the answer to part (a)?

23. An area of 0.05 m^2 is tilted so that its normal makes an angle of 80° with a uniform electric field of 5×10^5 N/C. Calculate the electric flux, Φ_E, through this area.

24. An electric field is produced by a charge of 0.4 μC. Calculate the electric flux, Φ_E, through (a) a sphere 0.6 m in radius concentric with the charge; (b) an arbitrary closed cylindrical surface enclosing the charge.

25. A nonuniform electric field described by $\mathbf{E} = \mathbf{i}(E_o + kx)$ passes through a cube. The cube, 0.53 m on a side, lies along the coordinate axes as shown in Figure 5. Calculate the net electric flux through this cube. (The constant k has units of N/C · m.)

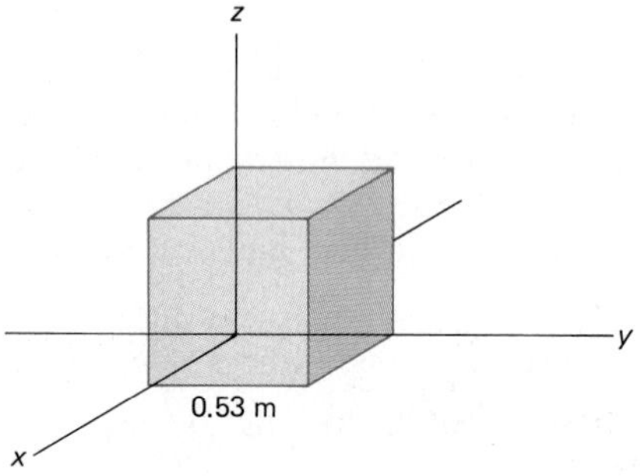

Figure 5

26. A conducting sphere 0.16 m in radius has a surface charge density of 4.2×10^{-8} C/m^2. (a) Calculate the electric field **E** just outside the spherical surface. (b) What is the electric flux, Φ_E, through a closed surface enclosing this sphere?

27. Consider the electric dipole shown in Figure 6. Calculate the electric flux, Φ_E, through the y–z plane as follows: (a) Show that

$$E_x = \frac{qa}{2\pi\epsilon_o} \cdot \frac{1}{r^3}$$

on the y–z plane a distance y from the origin. (b) Taking dA to be a ring of radius y and width dy in the y–z plane, $dA = 2\pi y\, dy$, integrate $d\Phi_E = E_x\, dA$ and show that

$$\Phi_E = \frac{q}{\epsilon_o}$$

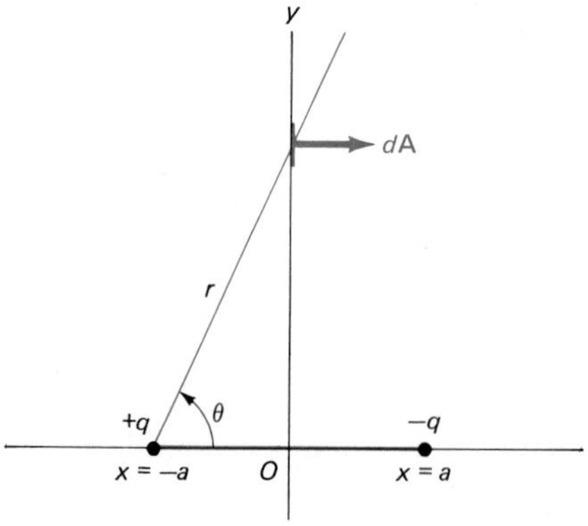

Figure 6

28. Determine the electric flux through the surfaces A and B in Figure 7.

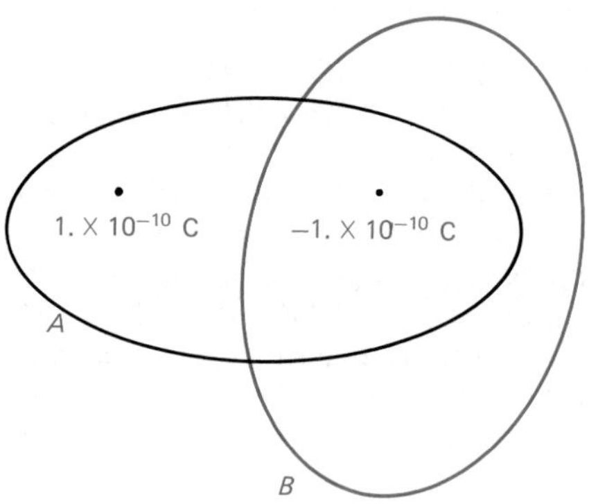

Figure 7

29. A +10-μC charge located at the origin of a Cartesian coordinate system is surrounded by a nonconducting hollow sphere of radius 10 cm. A drill with a radius of 1 mm is aligned along the z-axis, and a hole is drilled in the sphere. Calculate the electric flux in the hole.

Section 27.4 Gauss' Law

30. A point charge of 4.62×10^{-8} C is at the geometric center of a cube. Find the total electric flux through the faces of the cube.

31. The electric flux threading a spherical Gaussian surface is 1.53×10^4 N·m^2/C. What is the net charge residing within this surface?

32. Consider the irregularly shaped closed surface pictured in Figure 8. The only flux emerging from the surface is due to the positive point charge at P. Prove that this flux is the same whether the surface boundary is the solid curve or the dotted curve by (a) reasoning directly from the electric field lines, and (b) considering the flux at the surface areas F, G, and H.

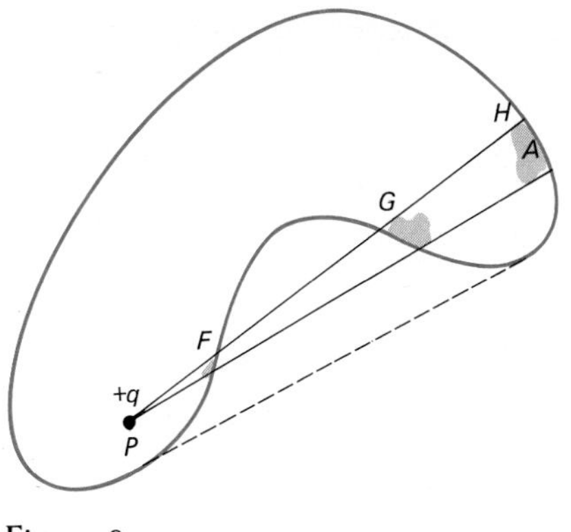

Figure 8

Section 27.5 Applications of Gauss' Law

33. A dry cell is connected across two parallel metal plates 1 cm apart to produce an electric field across the plates of 150 N/C. (a) Find the surface charge density σ (in C/m^2) on the negative plate. (b) How many excess electrons per square centimeter are represented by your value of σ?

34. Two horizontal parallel metal plates are separated by a small distance (Figure 9). Positive electric charge in the amount of 10.0 $\mu C/m^2$ is distributed uniformly over the upper surface of the lower plate. On the lower surface of the upper plate there is a charge equal in magnitude but negative. (a) Calculate the magnitude of the electric field produced by this charge (far from the plate edge). (b) Is the field directed upward or downward?

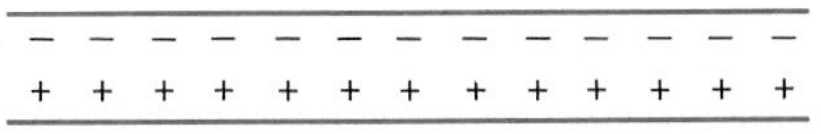

Figure 9

35. Apply Gauss' law to a uniformly charged mathematical plane (not conducting) and show that

$$E = \frac{\sigma}{2\epsilon_0}$$

36. Consider two horizontal, parallel, uniformly charged mathematical planes, one plane with a surface charge density of $+\sigma$ and the other with a surface charge density of $-\sigma$. Draw a diagram showing the electric fields above the two planes, between the two planes, and below the two planes. Show that the fields above and below the two planes are zero while the field between them is given by

$$E = \frac{\sigma}{\epsilon_0}$$

37. The nucleus of lead 208, $^{208}_{82}Pb$, has 82 protons within a sphere of radius 6.34×10^{-15} m or 6.34 fermis (fm). Assume that the protons are uniformly distributed throughout the spherical volume. Calculate the electric field produced by this nuclear charge (a) at the surface of the lead nucleus and (b) at a distance of 10^{-12} m.

38. Suppose that you have two concentric conducting spherical shells. A charge of $+q_1$ is placed on the inner shell (radius a), $-q_2$ on the outer shell (radius b). Take $q_1 > q_2$. (a) Develop an electric field line picture to enable you to predict, at least qualitatively, the electric field for $r < a$, for $a < r < b$, and for $r > b$. (b) Using Gauss' law, calculate the electric field for $r < a$, for $a < r < b$, and for $r > b$.

39. Imagine a solid sphere, radius R, uniformly charged throughout its volume. The total charge is Q. Using Gauss' law, show that the field *inside the sphere* a distance r from the center is

$$E = \hat{r}\left(\frac{Q}{4\pi\epsilon_0}\right) \cdot \frac{1}{R^2} \cdot \frac{r}{R} \qquad 0 \le r \le R$$

40. An early model of the atom consisted of a spherical, uniformly distributed positive charge with one or more (point) electrons within the positive charge. Consider such a model of hydrogen having total positive charge $= e$, radius, $r_0 = 0.5 \times 10^{-10}$ m, and one electron. (a) From Gauss' law (Examples 9 and 10) show that the Coulomb force on the electron, distance r from the center, is

$$F = -\frac{1}{4\pi\epsilon_0} \cdot \frac{e^2 r}{r_0^3}$$

(b) Assuming that the result of part (a) is the net force, show that the electron can oscillate radially with a radian frequency

$$\omega = 4.50 \times 10^{16} \text{ rad/s}$$

41. The charge density inside a sphere of radius R varies as $\rho = kr$ C/m^3. Use Gauss' law to deduce the electric field inside and outside the sphere.

42. A charge of 6.0×10^{-10} C is spread uniformly over the surface of a hollow metal sphere. The sphere has a radius of 0.10 m. Determine the magnitude of the electric field (a) 0.05 m from the center of the sphere; (b) 0.20 m from the center of the sphere.

43. A uniformly charged spherical surface whose radius is 0.73 m has a surface charge density of 6.3×10^{-8} C/m^2. What is the total electric flux leaving this spherical surface?

44. A long straight conducting wire carries a charge of λ C/m (Figures 10 and 11). (a) Develop a symmetry argument to demonstrate that the electric field must be radial, perpendicular to the conductor. (b) Using Gauss' law, show that the electric field at a distance r out from the conductor is

$$E = \frac{\lambda}{2\pi\epsilon_0} \cdot \frac{1}{r}$$

(c) Show that this same expression for E can also be obtained from Problem 41 of Chapter 26. *Note:* If a problem has the symmetry necessary to apply Gauss' law, the solution will be easier. The Gauss' law solution is also more general in the sense that it applies to an extended cylindrical distribution. Problem 41 of Chapter 26 is limited to a line charge.

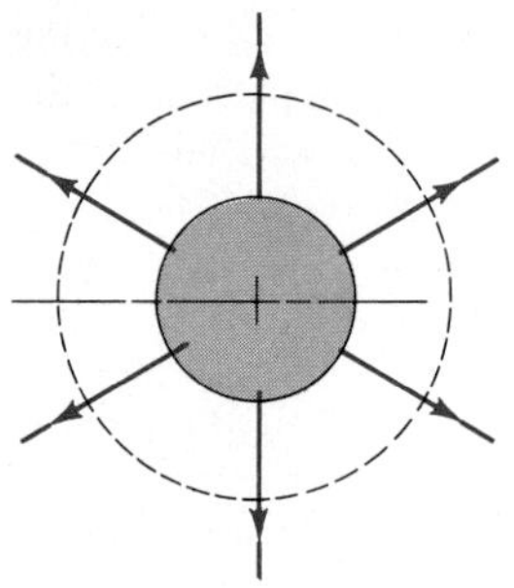

Figure 10 End view.

45. Imagine an indefinitely long solid cylinder of radius R that is uniformly charged throughout its volume. The charge is λ C/m. Using Gauss' law, show that the electric field *inside the cylinder* a distance r from the axis is

$$E = \hat{r}\left(\frac{\lambda}{2\pi\epsilon_0}\right) \cdot \frac{1}{R} \cdot \frac{r}{R} \qquad 0 \leq r \leq R$$

(This is the two-dimensional analog of Problem 39.)

46. A long straight wire has a charge of 6.21×10^{-9} C/m. Using Gauss' law (and an appropriate Gaussian surface) calculate the electric field **E** at a distance 0.182 m away from the wire. Explain carefully what you are doing.

47. A coaxial cable consists of a cylindrical inner conductor surrounded by a conducting cylindrical shell, as suggested in Figure 12. At a given moment the net charge on the two conductors equals zero. Assume that the charge per unit length on the wire is λ (positive). (a) Use Gauss' law to prove that the charge per unit length on the shell is $-\lambda$, and that this charge resides on the inner surface of the shell. (b) Calculate the electric field in the region between the conductors in terms of r.

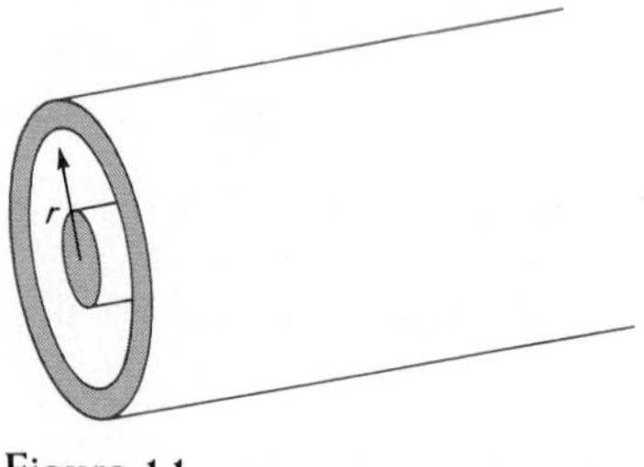

Figure 11

48. A cubical Gaussian surface, side length 1 m, is centered on a $+5.0$ μC charge. Determine the average value of the perpendicular component of the electric field on one of the faces of this Gaussian surface. (*Hint*: Calculate flux through one face of the cube.)

Section 27.6 Motion of Point Charges in a Static Electric Field

49. An electron placed in an electric field experiences an upward acceleration of 3.6×10^{13} m/s^2. Determine the acceleration of a proton placed in the same electric field.

50. In a particular Van de Graaff accelerator, a proton starts from rest and is accelerated by an electric field of 1.2×10^6 N/C over a distance of 4 m. (a) What is the acceleration of the proton? (b) What is the velocity of the proton after it has been accelerated by this field over a length of 4 m? (c) How long does it take the proton to travel this 4-m length?

51. An electron, initially at rest, is accelerated over a length of 1 m by an electric field of E N/C. (a) Find the final velocity of the electron as a function of the accelerating field E. (b) Treating the electron nonrelativistically, what value of the electric field will accelerate it to a velocity of $0.1c$ in this 1-m length? (When the electron velocity becomes a significant fraction of the velocity of light, calculations of velocity, momentum, and energy must be made relativistically.)

52. An electron with a velocity of 3×10^6 m/s moves into a uniform electric field of 10^3 N/C. The field is parallel to the electron's velocity and acts to decelerate the electron. How far will the electron go before it is brought to a halt?

53. Generalizing Problem 52, let the electron have a velocity v (nonrelativistic), and the retarding electric field have a magnitude E. Calculate the distance, s, that the electron travels before being brought to rest as a function of v and E.

54. A photon of blue light transfers its energy, 5.78×10^{-19} J, to an electron on the surface of a piece of cesium. The electron loses 3.04×10^{-19} J by escaping from the surface. Emerging from the cesium surface, the electron moves directly against a retarding field of 100 N/C. How far will the electron travel before its speed is reduced to zero?

55. An electron is ejected with a kinetic energy of 1.6×10^{-19} J from a metal surface. The metal is left with a net positive charge e, which we may consider to be a point charge located at a distance r beneath the surface where r equals the distance of the electron above the surface. Assuming that the force attracting the electron back to the surface begins to act when $r = 5 \times 10^{-11}$ m, determine the maximum distance of the electron from the surface.

Section 27.7 Electric Dipole in an Electric Field

56. A water molecule in the vapor state has an electric dipole moment of 6.17×10^{-30} C · m. What is the maximum torque experienced by a water molecule in an electric field of 5.0×10^5 N/C?

57. The dipole moment of a carbon monoxide molecule is 4.0×10^{-31} C · m. Determine the maximum torque that can be exerted on such a molecule in a uniform electric field with $E = 3.0 \times 10^{10}$ N/C.

58. An electric dipole, with dipole moment p and moment of inertia I, undergoes small-amplitude oscil-

lations in a uniform electric field E. Show that the oscillation frequency ν is

$$\nu = \frac{1}{2\pi}\sqrt{\frac{pE}{I}}$$

59. A particular electric dipole consists of a 1-μC charge and a $-$1-μC charge 1 m apart. The dipole is lined up with an electric field of 10^4 N/C (Figure 12). (a) Calculate the restoring torque on this dipole when it is rotated to an angle θ. (b) How much work would you have to do to rotate this dipole through 180°?

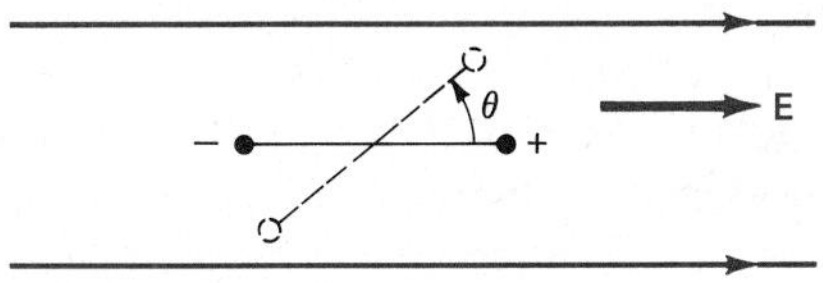

Figure 12

60. Generalize Problem 59 by calculating the work required to rotate an electric dipole (dipole moment **p**) 180° out of its stable equilibrium position in a uniform electric field E.

61. A water molecule, mass 18u = 2.99×10^{-26} kg, electric dipole moment 6.17×10^{-30} C · m, is on the x-axis at the point $x = 2$ m. The water molecule is exposed to an electric field with an x-component $E_x = 10^4(1 + 0.2x^2)$ N/C. (a) Calculate the force on the water molecule. (b) What will be its acceleration?

62. A geiger counter consists of a cylindrical wire (radius $= 10^{-3}$ m) surrounded by a cylindrical conducting shell. If the charge per unit length on the wire is 10^{-8} C/m, determine (a) the electric field at the surface of the wire, (b) the electric field gradient, dE/dr, at the surface of the wire, and (c) the force exerted on a water molecule ($p = 6.17 \times 10^{-30}$ C · m) at the surface of the wire. (*Hint:* See Problem 44.)

chapter

28 ELECTRIC POTENTIAL

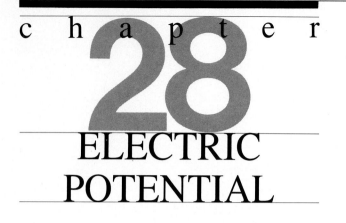

Preview

In Chapter 27 we saw that the electric field vector **E** provides us with a mechanism for the exertion of electric force. In this chapter we will build on the mechanical concepts of work and energy, and define a scalar potential V as the work done per unit charge by the electric field **E**.

The potential V is the direct electrostatic analog of the gravitational potential energy per unit mass. That the potential V is a scalar quantity makes V easier to work with than the electric field **E**, which is a vector.

28.1 Potential Difference

Electric fields are vector fields with a direction as well as a magnitude at every point of the field. Here, in close analogy to mechanics, we are about to introduce a scalar potential to replace the vector **E**, with a considerable gain in mathematical simplicity. This potential (or voltage) will be used extensively in later chapters when we analyze the behavior of various electrical circuits. Also, in troubleshooting electrical circuits, a measurement of the potential difference between two points is by far the easiest and most common electrical measurement made.

Uniform Electric Field

Consider a uniform electric field **E**, constant in direction and constant in magnitude. Suppose that we place a positive charge q into this field. The charge experiences a force

$$\mathbf{F} = q\mathbf{E} \qquad (28.1)$$

We can think of this situation as the action of the environment on a particle. The influence of the environment is represented by the electric field **E**, which acts on the particle charge q, resulting in a force **F** on the particle. When the particle moves through some displacement $\Delta\mathbf{s}$, from point A to point B (Figure 28.1), we say that the environment (the electric field) does *work* on the particle,

$$W_{AB} = \mathbf{F} \cdot \Delta\mathbf{s} = q\mathbf{E} \cdot \Delta\mathbf{s} \qquad (28.2)$$

When we considered gravity in Section 8.2, we defined the change in gravitational potential energy as the *negative* of the work done on a mass by the gravitational

Figure 28.1

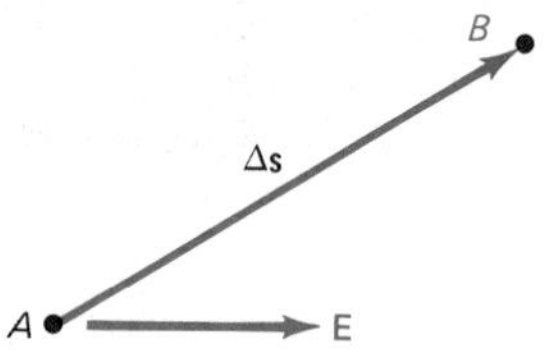

Figure 28.2

Electrostatic and gravitational force and potential energy: an exact analogy

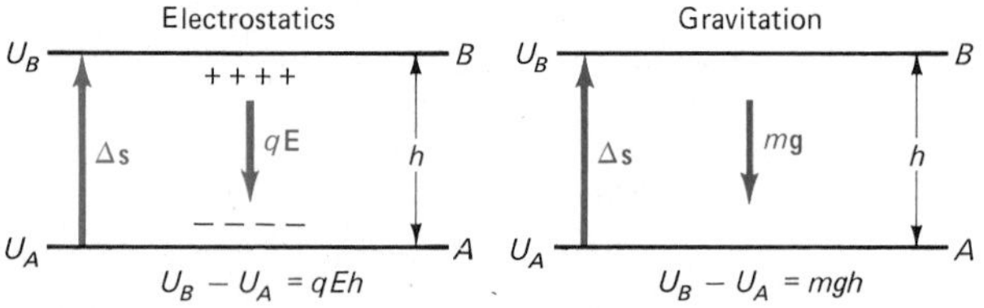

Table 28.1

Typical electric potential differences (volts)	
biochemical	10^{-3}
common dry cell	1.5
automobile battery	12
standard household	110–120
transmission lines	
local	4,400
cross country	120,000
high voltage	500,000–1,000,000
lightning	10^8–10^9

force. We will follow the same procuedure here and define the change in electrostatic potential energy to be the *negative* of the work done by the electric force:

$$\Delta U = U_B - U_A = -W_{AB} \tag{28.3}$$

so that

$$U_B - U_A = -q\mathbf{E} \cdot \Delta\mathbf{s} \tag{28.4}$$

The analogy between electrostatics and gravitation is exhibited in Figure 28.2. Note that in both cases the potential energy U_B at the higher level is greater than the potential energy U_A at the lower level. Also note that in both cases the force vector is directed from the level of higher potential energy toward the level of lower potential energy.

If the concept of potential energy is to be meaningful (uniquely defined), it is necessary that the work done by the field be independent of the path joining the points A and B. This path independence was demonstrated for gravitational forces (force radial and proportional to $1/r^2$) in Section 8.1. Because the Coulomb electrostatic force has the same mathematical form as the gravitational force, we expect that the work done by the field of a point charge will be independent of the path. In Section 28.3 we will show this explicitly. The Coulomb electrostatic force, and therefore the Coulomb electric field, is conservative. By superposition of point charge fields this result can be extended to any electrostatic field.

From Eq. 28.2 the work done by **E** is proportional to the magnitude of q. To eliminate this dependence on a particle property and obtain a quantity characteristic of the field, we divide by q and consider the *work done per unit positive test charge*;

$$\frac{W_{AB}}{q} = \mathbf{E} \cdot \Delta\mathbf{s} = -\Delta V \tag{28.5}$$

$$= V_A - V_B \tag{28.6}$$

Equation 28.5 defines the *potential difference* between A and B. The usefulness of this quantity stems from the fact that it is independent of the test charge q and is therefore a property of the electric field alone. By definition the units of the potential V are joules per coulomb (J/C); 1 J/C has been named the *volt*, (V), after Alessandro Volta (Figure 28.3). Typical values of potential differences for a variety of systems are presented in Table 28.1

You may notice that our discussion here closely parallels the discussion of gravitational potential energy for a uniform gravitational force in Section 8.2. There the potential energy difference was work, or mgh. In Section 8.2 we emphasized that only a difference in gravitational potential energy is physically meaningful. The same arguments apply here to both potential energy and potential. Only a *difference* in potential is physically meaningful. The zero of potential is arbitrary. Any point in the electric field (or electrical circuit) may be assigned zero potential. Using the common terminology for an electric circuit, we say that this point is *grounded*.*

Figure 28.3

Alessandro Volta (1745–1827), Italian physicist. Volta discovered how to convert chemical energy into electric energy. This made it possible for scientists to create and maintain steady electric currents as contrasted with intermittent spark discharges.

*Modern household wiring includes a third wire, a ground wire. Appliances with three-pronged plugs inserted in a properly wired three-hole electric outlet are automatically grounded. (The ground wire is often connected to metal water pipes, which are in contact with the earth.) Alternately, the appliance may have a polarized connector that can be inserted into the socket only one way, so that the ground terminal of the power cord always connects to the ground connection on the appliance.

Figure 28.4

Electrically charged, parallel conducting plates.

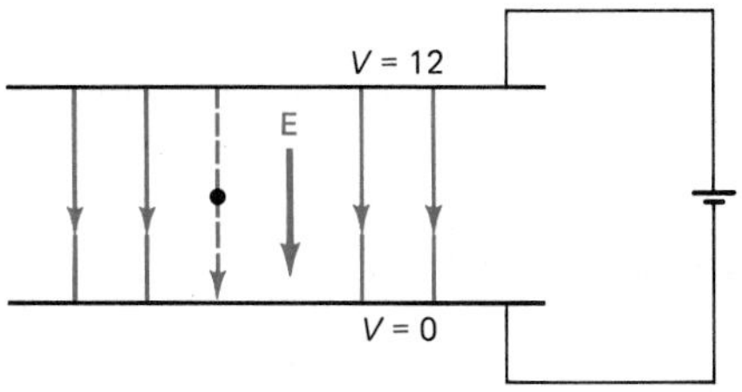

Example 1 Work Done in a Uniform Electric Field

Two parallel metal plates are connected to a 12-V automobile battery as shown in Figure 28.4. The lower plate is assigned a potential equal to zero and, in this case, is connected to the earth. (In cars made in the United States it is conventional to ground the negative terminal of the battery to the chassis.)

After the battery is disconnected, one proton moves from the upper plate (V = 12 volts) to the lower plate (V = 0 volts). How much work does the field do on this charge? Using Eq. 28.5 and replacing the unit volts by its equivalent, joules per coulomb, we get

$$\Delta W = -\Delta V q$$
$$= \frac{12\ \mathrm{J}}{\mathrm{C}} \times 1.60 \times 10^{-19}\ \mathrm{C}$$
$$= 1.92 \times 10^{-18}\ \mathrm{J}$$

This is the work done by the field **E**. If the proton can move freely, this work will show up as kinetic energy of the proton.

Nonuniform Electric Field

If the field **E** varies in either magnitude or direction, the work done by the field is described in infinitesimal steps $d\mathbf{s}$ (instead of a possibly large $\Delta\mathbf{s}$). We have, for the work done in moving a charge q a distance $d\mathbf{s}$ by a field **E**,

$$dW = \mathbf{F} \cdot d\mathbf{s} = q\mathbf{E} \cdot d\mathbf{s} \qquad (28.7)$$

Then the work done by the field moving q from A to B (Figure 28.5) is the integral of Eq. 28.7:

$$W_{AB} = q \int_A^B \mathbf{E} \cdot d\mathbf{s} \qquad (28.8)$$

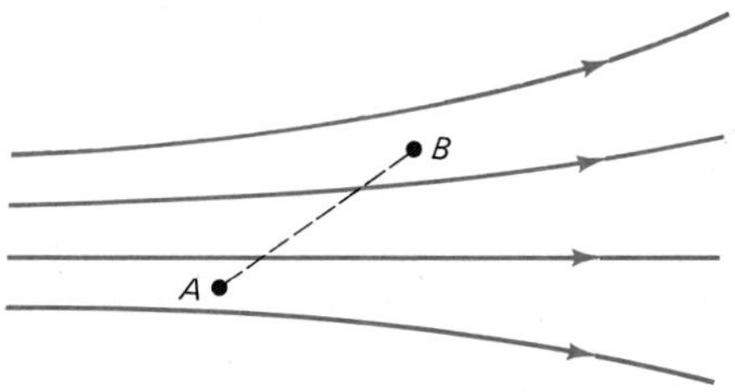

Figure 28.5

A displacement in a nonuniform electric field.

If the path of integration is along the x-axis, $d\mathbf{s}$ becomes $\mathbf{i}\,dx$ and the scalar product $\mathbf{E} \cdot d\mathbf{s}$ becomes $E_x\,dx$. We illustrate Eq. 28.8 in the following example, when we consider the work done by the nonuniform field of a ring of charge.

Example 2 Work Done by a Nonuniform Field

In Section 26.4 we found the force exerted by a ring of charge on a test charge q_o on the x-axis to be

$$\mathbf{F} = \mathbf{i}F_x = \left(\frac{1}{4\pi\epsilon_o}\right)Qq_o\left[\frac{x}{(x^2 + a^2)^{3/2}}\right]\mathbf{i}$$

The electric field on the x-axis becomes

$$\mathbf{E} = \mathbf{i}E_x = \frac{Q}{4\pi\epsilon_o} \cdot \frac{x}{(x^2 + a^2)^{3/2}} \cdot \mathbf{i}$$

Let's calculate the work done by this nonuniform field in moving a test charge q_o from $x = b$ (Figure 28.6) out to infinity, $x \rightarrow \infty$. From Eq. 28.8,

$$W_{b\infty} = q_o \int_{x=b}^{\infty} \mathbf{E} \cdot d\mathbf{s}$$
$$= \frac{Qq_o}{4\pi\epsilon_o} \int_b^{\infty} \frac{x}{(x^2 + a^2)^{3/2}}\,dx$$

This integrates to

$$W_{b\infty} = \frac{Qq_o}{4\pi\epsilon_o}\left| -\frac{1}{(x^2 + a^2)^{1/2}} \right|_b^{\infty}$$

Hence, the work done by the field of the ring charge Q on q_o is

$$W_{b\infty} = \frac{Qq_o}{4\pi\epsilon_o}\left[\frac{1}{(a^2 + b^2)^{1/2}}\right]$$

Note that this result is finite. If the field **E** had been uniform, the work done would have been infinite. (Try integrating a constant over the range from b to infinity.)

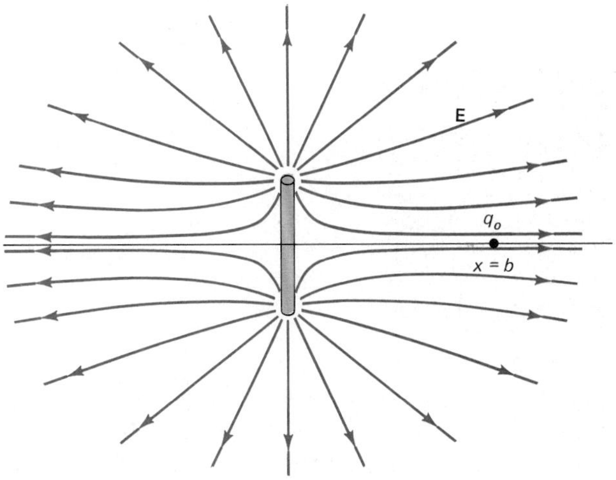

Figure 28.6

The electric field of a uniformly charged ring.

Recalling that the potential difference is defined as the negative of the work per unit charge W_{AB}/q, we have

$$V_B - V_A = -\int_A^B \mathbf{E} \cdot d\mathbf{s} \qquad (28.9)$$

If $\mathbf{E}$ is known, Eq. 28.9 provides a means of calculating $V_B - V_A$. Compare Eq. 28.9 with Eq. 8.5 of Section 8.2. As with the uniform field, the choice of the zero of potential difference is arbitrary. For instance, if the potential of the ring charge of Example 2 is taken to be zero at infinity, the potential on the x-axis ($x = b$) becomes

$$V_b = \frac{Q}{4\pi\epsilon_0}\left[\frac{1}{(a^2 + b^2)^{1/2}}\right]$$

The integral in Eq. 28.9 is a line integral. In general, the value of a line integral depends on the path—here the path that the charge q follows in moving from initial point A to final point B. We show in Section 28.3 that for the electrostatic field, the value of this line integral is independent of the choice of path joining initial point A and final point B.

Effects of Potential Difference

A potential difference of several thousand volts between you and the earth can result from combing your hair, taking off a T-shirt, or shuffling your feet across a rug. In the case of the latter, touching a doorknob can result in an electrostatic discharge. The amount of charge involved is very low (fractions of microcoulombs), and you feel only a minor shock. The lethal effects of electricity come not from potential difference but from *current*, or charge in motion, passing through the body (see Chapter 30). The effect of even a small current (millicoulombs per second) passing through the heart can be lethal.

To illustrate the importance of potential difference, consider a squirrel on a moderately high-voltage power line. As long as the squirrel stays on one wire, everything is fine. The potential of the one wire is irrelevant. There is no potential difference across any part of the squirrel. But if the squirrel moves from one wire to another across a large potential difference, there may be a flow of charge through its body, electrocuting the squirrel.

28.2 Conservation of Energy

At the beginning of Section 28.1 we considered an electric charge q in an electric field. The electric field acted on this charge, exerted a force on it, and did work on it. Let's assume that no other forces act on q. Then

$$\mathbf{F} = q\mathbf{E}$$

The charge will be accelerated as a result of this net force and will acquire kinetic energy. By repeating the work–energy analysis of Section 7.3 we obtain

$$W_{AB} \equiv q\int_A^B \mathbf{E} \cdot d\mathbf{s} = \int_A^b \mathbf{F} \cdot d\mathbf{s}$$

From Newton's second law

$$\mathbf{F} = m\frac{d\mathbf{v}}{dt}$$

where m is the mass of the charge. Substituting for $\mathbf{F}$, we have

$$W_{AB} = m\int_A^B \frac{d\mathbf{v}}{dt} \cdot d\mathbf{s} = m\int_A^B d\mathbf{v} \cdot \frac{d\mathbf{s}}{dt}$$

$$= m\int_A^B \mathbf{v} \cdot d\mathbf{v}$$

Integrating, we obtain

$$W_{AB} = \frac{1}{2}mv_B^2 - \frac{1}{2}mv_A^2$$

$$= \Delta K$$

where ΔK is the change in the kinetic energy. Because $W_{AB} = -\Delta U$ (Eq. 28.3), we arrive at

$$\Delta K = -\Delta U$$

where ΔU is the change in potential energy. Since $\Delta K = K_B - K_A$ and $\Delta U = U_B - U_A$, we may also write

$$K_A + U_A = K_B + U_B \qquad (28.10)$$

Just as in mechanics, we have conservation of energy. The kinetic energy plus electrostatic potential energy at point A equals the kinetic plus electrostatic potential energy at point B. (The electrostatic potential energy is defined so that it works out this way.)

A New Energy Unit, the Electron Volt

Suppose that an electric field accelerates an intially stationary proton (charge e). The work done by the field shows up as the kinetic energy of the proton. When the proton moves through a potential difference of (minus) 1 volt the proton kinetic energy is 1 electron volt (eV). This is the operational definition of an electron volt. The work done by the field is given by Eq. 28.5,

$$\Delta W = e(-\Delta V)$$

$$= 1.60219 \times 10^{-19}\ \text{C} \times 1\ \text{V}$$

$$= 1.60219 \times 10^{-19}\ \text{J}$$

From Eq. 28.10 we see that this is the change in kinetic energy,

$$\Delta K = +\Delta W$$

$$= 1.60219 \times 10^{-19}\ \text{J}$$

Therefore

$$1\ \text{eV} = 1.60219 \times 10^{-19}\ \text{J}$$

The electron volt is not a metric or an SI unit. However, it is widely used because it facilitates calculations: the atomic and nuclear particles have charges that are integral multiples of the electron charge. The standard SI prefixes (Table 1.2) are in common use: 10^3 eV = 1 keV, 10^6 eV = 1 MeV, and 10^9 eV = 1 GeV. Several modern particle accelerators operate in the tens- or hundreds-of-GeV range.

Example 3 Speed of a 1-eV Electron

An electron on the lower plate of two parallel plates has a potential energy of 1 eV relative to the upper plate. When the electron is released from the lower plate the electric field causes it to accelerate (Figure 28.7). The electron hits the upper plate with 1 eV of kinetic energy. What is its speed of impact?

There are two possible methods for finding the speed of impact. We will show both of them here.

Method 1 Using conservation of energy, we can write

$$1\ \text{eV} = 1.602 \times 10^{-19}\ \text{J} = \frac{1}{2} m_e v^2$$

With the electron mass $m_e = 9.1096 \times 10^{-31}$ kg the speed at impact becomes

$$v = \left(\frac{2 \times 1.602 \times 10^{-19}}{9.11 \times 10^{-31}}\right)^{1/2}$$

$$= 5.93 \times 10^5\ \text{m/s}$$

(For comparison, this is 20 times the orbital speed of the earth going around the sun and 1/500th the speed of light.)

Although Method 1 is perfectly correct, it may be inconvenient to look up the electron mass m and the joule equivalent of 1 eV. Method 2 is an alternative.

Method 2 Using the mass–energy relation for an electron, $mc^2 = 0.511$ MeV (Section 29.3), we write

$$1\ \text{eV} = \frac{1}{2} m v^2 = \frac{1}{2}(mc^2\ \text{eV})\frac{v^2}{c^2}$$

$$= \frac{1}{2} \times 0.511 \times 10^6 \left(\frac{v^2}{c^2}\right)$$

With eV units on both sides of the equation, the units cancel and

$$\frac{v^2}{c^2} = \frac{2}{511{,}000}$$

Then

$$\frac{v}{c} = 1.98 \times 10^{-3}$$

and

$$v = 5.93 \times 10^5\ \text{m/s}$$

Figure 28.8

The relative error in using Newtonian kinetic energy instead of relativistic kinetic energy for v/c up to 0.15.

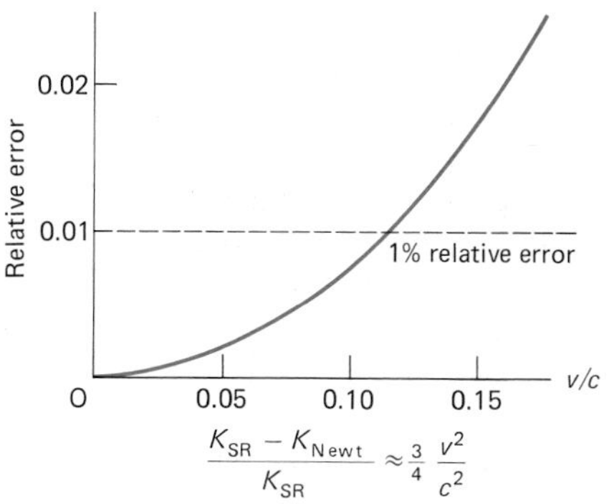

Since $v/c \approx 2 \times 10^{-3}$, our 1-eV electron is nonrelativistic.* But what about a 10,000-eV electron (= 10 keV) or a 10^6-eV electron (1 MeV)? Are they relativistic? A convenient standard for comparison is the rest mass energy of the electron (already used in Example 3),

$$E = mc^2 = 8.188 \times 10^{-14}\ \text{J}$$
$$= 0.511 \times 10^6\ \text{eV}$$
$$= 0.511\ \text{MeV} \qquad (28.11)$$

If the kinetic energy of the electron is comparable to its rest mass energy (0.511 MeV), relativistic mechanics (Section 20.3) will be needed (see Table 28.2).

Figure 28.8, based on Section 20.3, shows the relative error made in using the Newtonian form of kinetic energy rather than the special relativistic form for a given v/c. For the 3-keV electron, $v/c \approx 0.11$, and the relative error is 0.9%.

Questions

1. Why are electron volts more convenient than the SI joule?
2. An electron is accelerated through a potential difference of 100,000 V. Which will yield a higher value for its speed, Newtonian mechanics or relativistic mechanics?

*Note that although this calculation uses $E_0 = mc^2$ as a convenient way of handling units, the calculation itself is nonrelativistic, based on Newtonian kinetic energy.

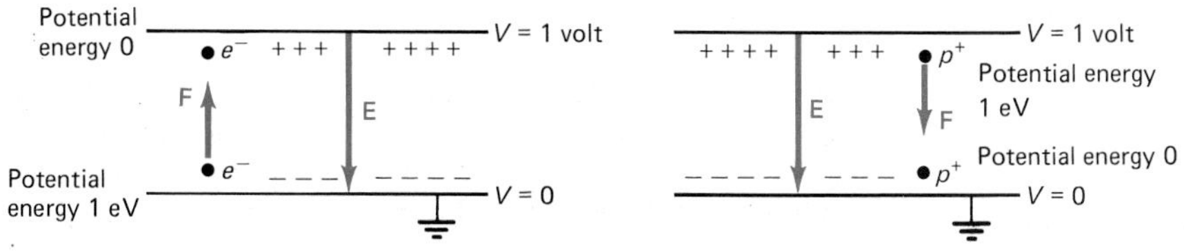

Figure 28.7

Electric force and potential energy for an electron and a proton in a uniform electric field.

3. A single electric force acts on a point charge. Explain why its kinetic energy and electric potential energy cannot simultaneously increase.

28.3 Electric Potential

We have defined the potential difference between points A and B, $V_B - V_A$, as the negative of the work done by the field **E** per unit charge in moving a charge from A to B. This definition leads to

$$V_B - V_A = -\int_A^B \mathbf{E} \cdot d\mathbf{s} \qquad (28.12)$$

When the potential at point A is chosen equal to zero, $V_A = 0$, V_B becomes the *absolute* potential or simply the potential.* One point in the system may be assigned zero potential. Often this point is physically connected to the earth—grounded. For fields decreasing as $1/r^2$ or faster, it is convenient to take ∞ as zero potential (Section 28.4). These are the same choices that we make for the gravitational potential, and we make our choices here for the same reasons. Both force laws have the same $1/r^2$ dependence.

Given a field **E**, Eq. 28.12 permits us to calculate the potential difference by integration. Electric potential is an integral of the electric field. Even if the integral cannot be evaluated by being looked up in integral tables, it can be evaluated numerically with an electronic computer. Since we can calculate V from **E** by using integration,

$$\mathbf{E} \xrightarrow{\text{integration}} V$$

we should be able to calculate **E** from V by using differentiation.

$$V \xrightarrow{\text{differentiation}} \mathbf{E}$$

For a uniform electric field the relation of field **E** and potential V given by Eq. 28.5 reduces to

$$\Delta V = -E\,\Delta s \qquad (\text{for } \mathbf{E} \text{ and } \Delta\mathbf{s} \text{ parallel})$$

or

$$E = -\frac{\Delta V}{\Delta s}\,\frac{\text{volts}}{\text{meter}} \qquad (28.13)$$

In this form the electric field E can be calculated from a known potential V.

Example 4 A Uniform Electric Field

In Example 1 the potential difference between two parallel metal plates was fixed at 12 V by a battery. Now, given that the plate separation is 1 cm, let us calculate the electric field (which we are assuming to be constant). From Eq. 28.12 we have

$$\Delta V = 12 = -\int_{\text{bottom plate}}^{\text{top plate}} \mathbf{E} \cdot d\mathbf{s}$$

Taking the x-axis to be vertical and positive upward, as shown in Figure 28.9, we have

$$\int \mathbf{E} \cdot d\mathbf{s} = \int E_x\,dx \qquad (\textit{with } E_y \textit{ and } E_z = 0)$$

Then

$$12 = -\int E_x\,dx = -E_x\,\Delta x$$

Solving for E_x with $\Delta x = 1$ cm yields

$$E_x = \frac{12\text{ V}}{1\text{ cm}} = -1200\text{ V/m}$$

We obtain units of volts per meter for **E**, equivalent to the newtons per coulomb of Section 28.1. E_x is negative because the vector **E** points downward, in the negative x-direction.

Equivalently, from Equation 28.13

$$E = -\frac{\Delta V}{\Delta s} = -\frac{12\text{ V}}{1\text{ cm}} = -1200\text{ V/m}$$

The minus sign in this equation reminds us that increasing V, ΔV positive, means a displacement in the negative **E**-direction.

---

*The word *absolute* is seldom used. Generally, people simply say "potential." It is assumed that a zero of potential has been specified and agreed upon.

Table 28.2

Particle	Rest Mass Energy (MeV)
electron	0.511
proton	938.26
neutron	939.55

Electric fields and electric potentials are relatively abstract concepts. Therefore you may find it helpful to think in terms of gravitational analogies. The earth's gravitational force is directed downward. If we define a gravitational field as force divided by mass, this also is downward. But gravitational potential energy increases upward. Likewise, the gravitational potential (potential energy divided by mass) increases upward. A comparison of gravitational and electric potential *energy* is presented in Figure 28.2.

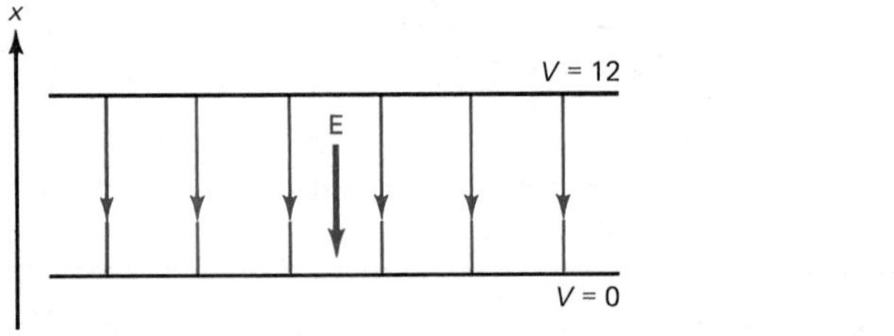

Figure 28.9

Example 5 A Uniform Electric Field Revisited

Returning to the parallel plates of Examples 1 and 4, let us recalculate the work done by the field in moving one proton from the top plate to the bottom plate. We have

$$W = F\,\Delta x = eE\,\Delta x$$
$$= 1.60 \times 10^{-19}\text{ C} \times 1200\text{ N/C} \times 10^{-2}\text{ m}$$
$$= 1.92 \times 10^{-18}\text{ N}\cdot\text{m (joules)}$$

in agreement with Example 1.

Nonuniform Electric Field

Now let's consider the potential difference in a nonuniform electric field (Figure 28.10). Once again we can use Eq. 28.12, only now points A and B are on the x-axis,

$$\Delta V = -\int_A^B \mathbf{E}\cdot d\mathbf{s} = -\int_A^B E_x\,dx \qquad (28.14)$$

For a displacement $\Delta\mathbf{s} = \mathbf{i}\,\Delta x$ so small that E_x does not vary significantly, the integral can be replaced by the product

$$\Delta V \cong -E_x\,\Delta x \qquad (28.15)$$

(You will find it useful here to review the mean value theorem of integral calculus.) Equation 28.15 becomes exact for the case of a uniform field. We have

$$E_x \cong -\frac{\Delta V}{\Delta x} \qquad (28.16)$$

In the limit as $\Delta x \to 0$, this approximation becomes exact:*

$$E_x = -\frac{dV}{dx} \qquad (28.17)$$

Equation 28.17 may also be applied in other coordinate systems. For example, in the circular cylindrical coordinate system $r = (x^2 + y^2)^{1/2}$, the perpendicular distance from the z-axis. When the potential V is a function only of r, we apply Eq. 28.17 to get

$$E_r = -\frac{dV(r)}{dr} \qquad (28.18a)$$

*If V is a function of other variables (y and z) as well as x, the ordinary derivative of Eq. 28.17 is replaced by a partial derivative: $E_x = -\partial V/\partial x$. This notation says, "Differentiate with respect to x but hold y and z constant."

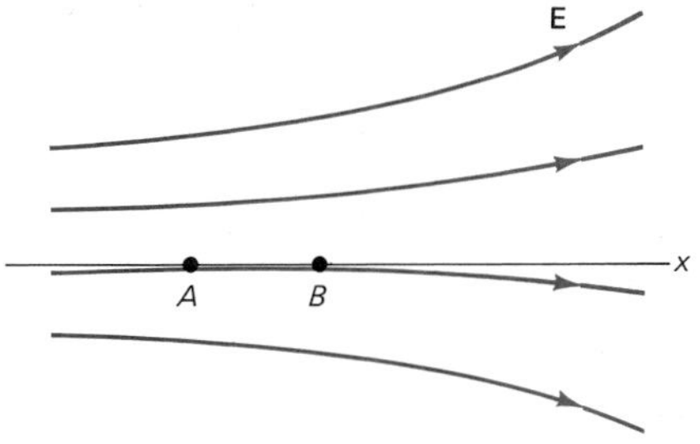

Figure 28.10

A displacement in a nonuniform electric field.

In the spherical polar coordinate system r is measured radially outward from the origin, $r = (x^2 + y^2 + z^2)^{1/2}$. When the potential V depends only on this radial distance, we can write

$$E_r = -\frac{dV(r)}{dr} \qquad (28.18b)$$

Keep in mind that all of the foregoing derivatives are derivatives with respect to a *length*. If we differentiate with respect to an angle (the polar angle θ of spherical polar coordinates), then

$$E_\theta \neq -\frac{\partial V}{\partial \theta}$$

Instead

$$E_\theta = -\frac{1}{r}\frac{\partial V}{\partial \theta}$$

Questions

4. What is the difference between electric potential energy and electric potential?
5. Do protons tend to accelerate from a region of high potential to a region of low potential or vice versa? What is the tendency for electrons?
6. An electric charge q is moved from point A to point B in an electric field. At point B, the electric potential is lower than it is at point A, but the electric potential energy of q is higher. Is this possible? Explain.
7. When we assign the earth a potential of zero, does this imply that the earth has no net charge?
8. If the earth had a net charge, could we still use the earth as a zero of potential?
9. If the electric field **E** is zero at a point or in a region, does this imply that the potential V is zero at that point or in that region?
10. If the potential V is zero at a point, does this imply that **E** is zero at that point? What if V is zero in an extended region?
11. Someone claims that a measurement of the potential at a point automatically gives the electric field **E** at that point. Show, from the scalar nature of V and the vector nature of **E**, that this claim is ridiculous.
12. The success of a certain experiment depends on holding the potential constant over a volume of about 1 m^3. How could you hold the potential constant?
13. Imagine that you are sitting on a metal chair on an insulating platform. An electrostatic generator is connected to the chair and both you and the chair are raised to a potential of 15,000 V above ground. What effect will this have on you? (See Figure 28.11.) Explain, in terms of forces, why this might happen.

Figure 28.11

Electrostatic repulsion. (See Question 13.)

Figure 28.12

Integration along a radial line in the electric field of a point charge at the origin.

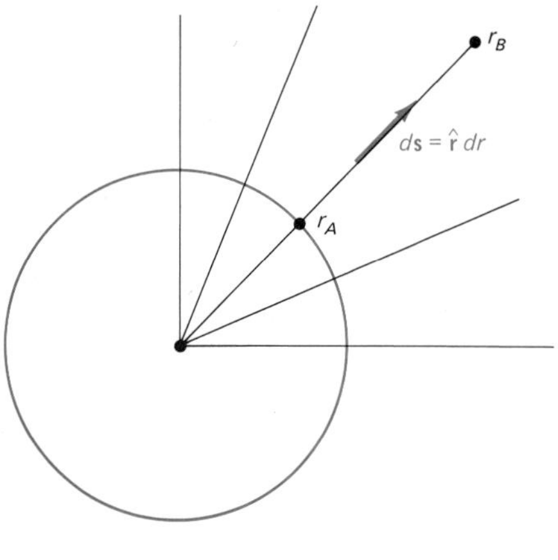

28.4 Point Charge Potential

The electric field of a point charge q is given by (Eq. 27.4)

$$\mathbf{E} = \frac{q}{4\pi\epsilon_0} \cdot \frac{1}{r^2} \cdot \hat{\mathbf{r}} \qquad (28.19)$$

We can obtain the electrostatic potential difference between positions labeled r_A and r_B (Figure 28.12) by integrating this field.

$$V(r_B) - V(r_A) = -\int_{r_A}^{r_B} \mathbf{E} \cdot d\mathbf{s} \qquad (28.20)$$

Here $V(r_A)$ and $V(r_B)$ denote the potentials at distances r_A and r_B from the point charge. Integrating outward from r_A to r_B along a radial line with $d\mathbf{s} = \hat{\mathbf{r}}\,dr$ (Figure 28.12), we have

$$V(r_B) - V(r_A) = -\frac{q}{4\pi\epsilon_0}\int_{r_A}^{r_B} \frac{dr}{r^2}$$

or

$$V(r_B) - V(r_A) = \frac{q}{4\pi\epsilon_0}\left(\frac{1}{r_B} - \frac{1}{r_A}\right) \qquad (28.21)$$

We've previously emphasized that the zero of potential is completely arbitrary. However, when we set $V(r_B \rightarrow \infty) = 0$ and $r_A = r$ we get the simplest formula:

$$V(r) = \frac{q}{4\pi\epsilon_0} \cdot \frac{1}{r} \qquad (28.22)$$

for the potential at a distance r from a point charge q (relative to zero potential at an infinite distance). If we agree to call $V(r_B \rightarrow \infty) = 0$, then Eq. 28.22 can be taken as the absolute potential, or simply the potential.

Coulomb Potential Energy

If a point charge q' is located a distance r from a point charge q, the electrostatic potential energy of the charges is

$$U(r) = V(r) \cdot q' = \frac{1}{4\pi\epsilon_0} \cdot \frac{qq'}{r} \qquad (28.23)$$

Now we must decide how to interpret the right-hand side of Eq. 28.23. In Section 28.2 we intepreted potential energy in terms of a charged particle and its environment (the electric field). It is correct to say that q sets up a field and provides the environment for q'. But we must also recognize that q' creates a field and provides an environment for q. The solution is to recognize that the potential energy is a *mutual* property depending on the two charges q and q' and the distance between them. The potential energy is a property of the two-charge system and not of either charge alone.

An important example of a two-charge system is the hydrogen atom, which consists of an electron and a proton separated by a small distance. In Example 6 we calculate the potential energy of the hydrogen atom.

Example 6 **Potential Energy of the Hydrogen Atom**

According to the planetary model of the hydrogen atom, an electron circles a proton at a distance $r = 5.29 \times 10^{-11}$ m. Let us calculate the electrostatic potential set up by the proton at this distance and the corresponding potential *energy* of the atom.

From Eq. 28.23

$$V(r = 5.29 \times 10^{-11}\text{ m})$$
$$= (8.99 \times 10^9)\,\frac{1.60 \times 10^{-19}}{5.29 \times 10^{-11}}$$
$$= 27.2 \text{ V}$$

The potential energy of the atom is -1.60219×10^{-19} C times V, or

$$U = -1.60 \times 10^{-19} \times 2.72 \times 10^{+1}$$
$$= -4.76 \times 10^{-18}\text{ J}$$

In this instance the electron volt is a more convenient unit than joules (or attojoules). Because 1 eV = 1.60×10^{-19} J

$$U = -27.2 \text{ eV}$$

The negative sign is characteristic of the bound state and the choice of $V = 0$. The electron is bound to the proton, and we would have to do positive work in order to free the electron.

A pair of protons in a nucleus has a positive electrostatic potential energy. The approximate size of this energy and the ramifications of the energy's being positive are discussed in Example 7.

Example 7 **Proton–Proton Electrostatic Interaction**
Two protons in a nucleus experience both an electrostatic and a nuclear force. The nuclear force is significant only when the center-to-center distance between the two protons has been reduced to about 2.8×10^{-15} m. Let us calculate the electrostatic potential energy of two protons that are 2.8×10^{-15} m apart. From Eq. 28.23

$$U = (8.99 \times 10^{9})\frac{(1.60 \times 10^{-19})^2}{2.8 \times 10^{-15}}$$

$$= 8.24 \times 10^{-14} \text{ J}$$

$$= 0.514 \text{ MeV}$$

If the two protons are released, they will repel each other. The mutual potential energy will be converted into kinetic energy. This is essentially what happens in nuclear fission. The mutual electrostatic potential energy of the fission fragments is converted into kinetic energy. In a fission reactor this kinetic energy is quickly randomized, and we speak of it as thermal energy.

Path Independence

In Section 28.1 we emphasized that the definition of potential difference, Eq. 28.12, was meaningful only if the integral was independent of the path of integration. Here, with the field of point charge, we are in a position to prove that the integral is indeed independent of the path of integration. We repeat the development of path independence given in Chapter 8 in an electrostatic context. While Chapter 8 deals with force and potential energy, here we use a field (force per unit charge) and potential (potential energy per unit charge).

Consider two points, A and B, in the Coulomb field of a point charge, with A and B not necessarily on the same radial line. The potential difference is

$$V_B - V_A = -\int_A^B \mathbf{E} \cdot d\mathbf{s} \qquad (28.24)$$

as before. Now let us compare the integral over paths 1 and 2 of Figure 28.13 with the integral over paths 3 and 4.

Figure 28.13

Integration along radial and tangential paths in the electric field of a point charge at the origin.

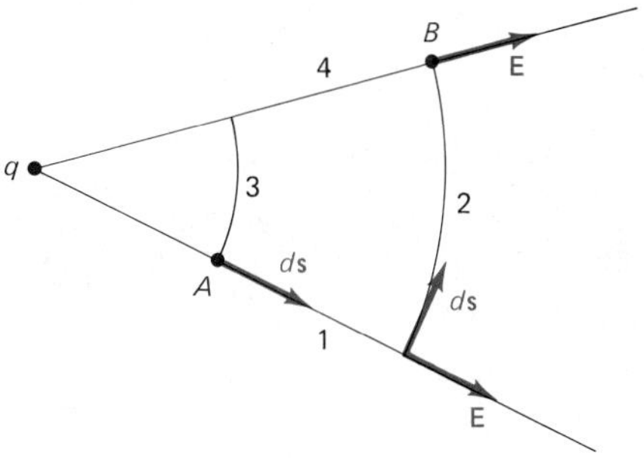

Over path 1, $d\mathbf{s} = \mathbf{r}\,dr$, and $\mathbf{E}$ and $d\mathbf{s}$ are parallel. Over path 2, $\mathbf{E}$ and $d\mathbf{s}$ are perpendicular, and $\mathbf{E} \cdot d\mathbf{s} = 0$. Then

$$-\int_A^B \mathbf{E} \cdot d\mathbf{s} = -\frac{q_0}{4\pi\epsilon_0}\int_{r_A}^{r_B}\frac{dr}{r^2} + 0$$

$$= \frac{q_0}{4\pi\epsilon_0}\left(\frac{1}{r_B} - \frac{1}{r_A}\right) \qquad (28.25)$$

For paths 3 and 4 we obtain the same results: Over path 3, $\mathbf{E} \cdot d\mathbf{s} = 0$, while over path 4, $\mathbf{E} \cdot d\mathbf{s} = (q/4\pi\epsilon_0)(dr/r^2)$. The integral leads to the same expression as that given in Eq. 28.25.

Any path joining A and B may be broken into radial parts and tangential parts. Always, $\mathbf{E} \cdot d\mathbf{s} = 0$ over the tangential segments (Figure 28.14). No work is done when moving perpendicular to the field. Only the radial segments contribute. The potential difference is independent of the path chosen. The electrostatic field is a conservative field.

Since every electrostatic field is the result of a superposition of the fields of point charges, this result—the potential difference of the two points A and B—is independent of the path joining the points. In other words, it holds for every electrostatic field.

This path independence may be restated as follows.

$$\underset{\text{paths 1 and 2}}{\int_A^B \mathbf{E} \cdot d\mathbf{s}} - \underset{\text{paths 3 and 4}}{\int_A^B \mathbf{E} \cdot d\mathbf{s}} = 0$$

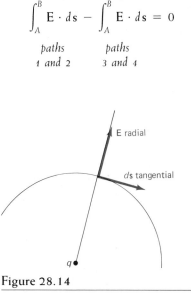

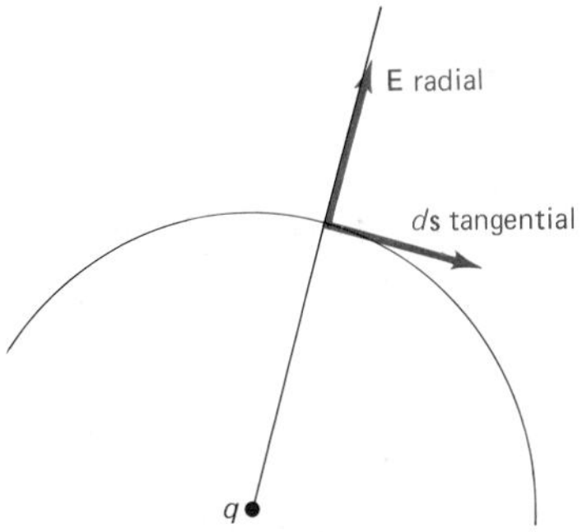

Figure 28.14

When $\mathbf{E}$ is radial and $d\mathbf{s}$ is tangential, $\mathbf{E} \cdot d\mathbf{s} = 0$.

or

$$\underbrace{\int_A^B \mathbf{E}\cdot d\mathbf{s}}_{\substack{\text{paths}\\ \text{1 and 2}}} + \underbrace{\int_B^A \mathbf{E}\cdot d\mathbf{s}}_{\substack{\text{paths}\\ \text{3 and 4}}} = 0$$

This is the same as integrating around the closed loop 1–2–4–3, or

$$\oint \mathbf{E}\cdot d\mathbf{s} = 0 \tag{28.26}$$

The line integral of the electrostatic field around a closed loop is zero. (As in Section 8.1, the circle on the integral sign indicates that the integral is over a closed path.)

According to Eq. 28.26, the work done in moving a charge around a closed loop in an electrostatic field is zero. When the charge returns to the starting point there is no net loss of energy and no net gain. Energy has been conserved. The electrostatic force, like the gravitational force, is a *conservative* force. Because we're describing the electrostatic force as an electrostatic field, we say that the electrostatic field is *conservative*. It should be emphasized that this result is for static fields, where there is no time dependence. A time-varying field is not necessarily conservative.

Questions

14. In terms of work done on a positive test charge by a field, what is meant by a negative potential in the vicinity of an isolated negative point charge?

15. A bound system is characterized by a *negative* potential energy. Explain what this means in terms of the work that would have to be done in order to send a rocket to the moon or to pull an electron away from an atom.

28.5 Multiple Charge Potentials

There are two general ways to calculate electrostatic potentials.

1 If the field E is known, the potential difference may be calculated as the negative of the work done per unit charge by the field E in moving a test charge from point A to point B. In Section 28.3 we found the potential of a point charge in this way.

2 Alternatively, the potential due to a given charge distribution may be obtained as the sum of the potentials due to the individual charges (or the integral over an assumed continuous charge). In this case, we invoke the same *superposition principle* that was found experimentally to hold for electrostatic forces and electric fields.

Let's now consider the second method. The development of the superposition principle for potentials begins with Eq. 28.20.

$$V = -\int \mathbf{E}\cdot d\mathbf{s}$$

To simplify the presentation, we have omitted the Δ's and the end points. If **E** is the resultant of two or more discrete or distributed charges, then from the superposition principle for fields, Section 27.2, we can write

$$V = -\int (\mathbf{E}_1 + \mathbf{E}_2 \cdots)\cdot d\mathbf{s}$$

Integration is a linear mathematical operation, which means that the integral of the sum equals the sum of the integrals. Therefore,

$$V = -\int \mathbf{E}_1\cdot d\mathbf{s} - \int \mathbf{E}_2\cdot d\mathbf{s} - \cdots$$

or

$$V = V_1 + V_2 + \cdots$$

The resultant potential V is the algebraic sum of the individual potentials due to the discrete or distributed charges.

Discrete Charges

Using this superposition principle to justify taking the potential due to a number of discrete charges as the algebraic sum of the potentials due to the individual charges, we write

$$V = \frac{1}{4\pi\epsilon_0}\left(\frac{q_1}{r_1} + \frac{q_2}{r_2} + \cdots\right)$$

$$= \frac{1}{4\pi\epsilon_0}\sum_i \frac{q_i}{r_i} \tag{28.27}$$

Here q_i is the ith charge and r_i is its distance away from our point of measurement. (For scalars, direction is irrelevant.)

Electric Dipole Potential

Consider a charge q at $x = a$ and a charge $-q$ at $x = -a$, which form an electric dipole with dipole moment $p = 2aq$. From Eq. 28.22 the potential is

$$V = \frac{q}{4\pi\epsilon_0}\left(\frac{1}{r_1} - \frac{1}{r_2}\right) \tag{28.28}$$

where r_1 is the distance from the charge q to the point where V is determined and r_2 is the distance from the charge $-q$ to the point where V is determined (Figure 28.15). For $r_1 = r_2$ we get $V = 0$ by inspection. We can interpret this result as a consequence of the symmetry of the dipole system. Any given point on the y–z plane ($x = 0$) is equidistant from the two charges. The $+q$ charge creates a positive potential. The $-q$ charge creates an equal but negative potential. The two potentials cancel and yield zero.

For any position with $r_1 \neq r_2$, we must express r_1 and r_2 in terms of a common coordinate system. If we locate the point where V is determined with Cartesian coordi-

Figure 28.15

An electric dipole.

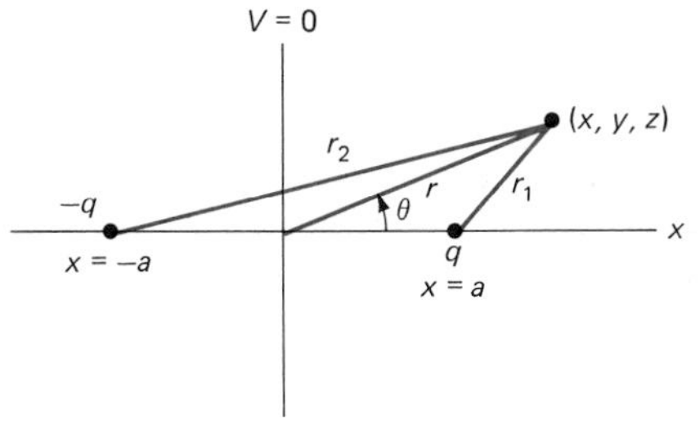

nates (x, y, z), it follows that $r_1 = [(x - a)^2 + y^2 + z^2]^{1/2}$ and $r_2 = [(x + a)^2 + y^2 + z^2]^{1/2}$. Hence

$$V(x,y,z) = \frac{q}{4\pi\epsilon_0}\left\{\frac{1}{[(x - a)^2 + y^2 + z^2]^{1/2}} - \frac{1}{[(x + a)^2 + y^2 + z^2]^{1/2}}\right\}$$

Let's look at a special case, the potential far out on the x-axis, $x \gg a$. With $y = z = 0$, we have

$$V(x,0,0) = \frac{q}{4\pi\epsilon_0}[(x - a)^{-1} - (x + a)^{-1}]$$

$$= \frac{q}{4\pi\epsilon_0}\left[\left(1 - \frac{a}{x}\right)^{-1} - \left(1 + \frac{a}{x}\right)^{-1}\right] \qquad (28.29)$$

A binomial expansion of each of the terms in the outer brackets yields

$$V(x,0,0) = \frac{q}{4\pi\epsilon_0 x}\left[\left(1 + \frac{a}{x} + \cdots\right) - \left(1 - \frac{a}{x} + \cdots\right)\right]$$

$$= \frac{2aq}{4\pi\epsilon_0 x^2} + \cdots \qquad (28.30)$$

Thus the electric dipole potential (on the x-axis) is proportional to the dipole moment $2aq$ and inversely proprotional to the square of the distance. The terms omitted are proportional to x^{-4} and higher negative powers of x, and are negligible for $x \gg a$.

The potential for $r_1 = r_2$ $(V = 0)$ and the equation for the potential far out on the x-axis (Eq. 28.30) are special cases of a general expression for the potential that can be written in spherical polar coordinates as

$$V(r,\theta) = \frac{2aq}{4\pi\epsilon_0}\cdot\frac{\cos\theta}{r^2} \qquad (r \gg a) \qquad (28.31)$$

The angle θ and the distance r are shown in Figure 28.15.

This general form is convenient because it is much simpler to use. Note that the dipole potential falls off as r^{-2}. Just as an r^{-1} potential is characteristic of a point charge (an electric monopole), so an r^{-2} potential is characteristic of an electric dipole. Note also the lack of spherical symmetry. The orientation of a dipole is important, whether the dipole is a radio transmitting antenna or the rabbit ears on a home TV set.

Radio antennas are often electric dipoles with a radio-frequency potential applied, so that the sign of the charge at a given end of the dipole reverses rapidly or oscillates. Much atomic radiation (radiation from individual atoms) is conveniently treated as coming from an oscillating electric dipole of atomic dimensions. As we noted in Section 27.7, the water molecule H_2O has an electric dipole moment. It is this dipole moment that is responsible for many of the unique properties of water. Dipole forces also play a significant role in the adherence of neutral atoms and molecules to each other. Since our bodies are composed of neutral atoms and molecules, dipole forces help hold us together.

Distributed Charges

In Section 26.4 we calculated the electrostatic force exerted by a ring of charge on a test charge located on the symmetry axis of the ring. We will now calculate the potential for a ring of charge. This is an easier task, because the electrostatic potential is a scalar quantity, whereas the force **F** is a vector quantity. For a ring of radius a in the y–z plane (Figure 28.16), the potential due to a charge element dq is

$$dV = \frac{1}{4\pi\epsilon_0}\left(\frac{dq}{r}\right) = \frac{1}{4\pi\epsilon_0}\left[\frac{dq}{(x^2 + a^2)^{1/2}}\right] \qquad (28.32)$$

Summing up the contributions from all of the charge elements is the same as integrating around the ring. Thus

$$V(x) = \frac{1}{4\pi\epsilon_0}\int\frac{dq}{(x^2 + a^2)^{1/2}}$$

$$= \frac{1}{4\pi\epsilon_0(x^2 + a^2)^{1/2}}\int dq \qquad (28.33)$$

This last equation holds because in this case x and a are constant for all charge elements dq. Then

$$V(x) = \frac{q}{4\pi\epsilon_0}\cdot\frac{1}{(x^2 + a^2)^{1/2}} \qquad (28.34)$$

for the potential on the x-axis created by the ring of charge. At the center of the ring, the electric field **E** is zero. But from Eq. 28.34 we see that the potential at the center of the ring is not zero.

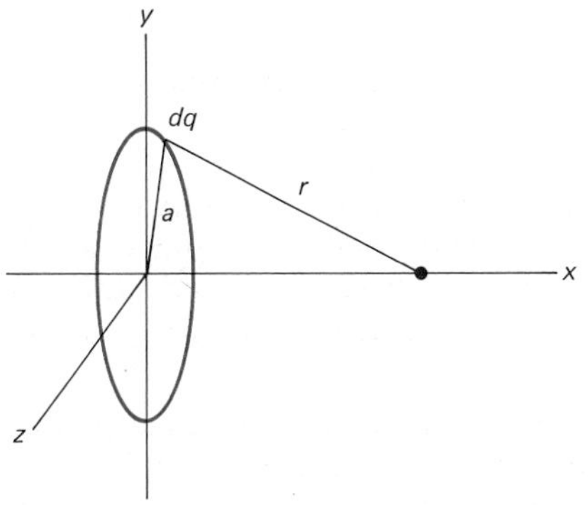

Figure 28.16

Calculation of the electrostatic potential of a ring of charge along the axis of rotational symmetry.

Example 8 The Potential of a Ring of Charge at Large Distances

The electrostatic potential of a ring of charge is given by Eq. 28.34 for all points on the symmetry axis. Let's explore the behavior of this potential for points far out on the x-axis, specifically for $x \gg a$.

First, we can use an analytic approach. For $x \gg a$, x^2 is much larger than a^2 and we can simply drop the a^2 in the sum $x^2 + a^2$. Then Eq. 28.34 becomes

$$V(x) = \frac{q}{4\pi\epsilon_0} \cdot \frac{1}{(x^2)^{1/2}} = \frac{q}{4\pi\epsilon_0} \cdot \frac{1}{x}$$

This is exactly what we would get for a point charge q at the origin. At large distances the ring charge yields the same potential as a point charge.

Now let us use a numerical approach. How do a point charge potential and a ring charge potential compare at $x = 20a$? For the point charge,

$$V_{pt} = \frac{q}{4\pi\epsilon_0} \cdot \frac{1}{20a} = \frac{q}{4\pi\epsilon_0 a} \times 0.05000$$

For the ring charge,

$$V_{ring} = \frac{q}{4\pi\epsilon_0} \cdot \frac{1}{[(20a)^2 + a^2]^{1/2}}$$

$$= \frac{q}{4\pi\epsilon_0 a} \times 0.04994$$

The ratio V_{pt}/V_{ring} is 1.0012, very close to unity. In terms of potential, at $x = 20a$ the ring looks like a point.

In Section 27.5 we used Gauss' law to show that the electric field **E** outside a uniformly charged spherical shell is the same as it would be if all of the charge were concentrated at the origin. Since the electrostatic potential $V(r)$ is an integral of **E**, the potential outside the uniformly charged spherical shell (radius a) is the same as that set up by a point charge at the origin.

$$V(r) = \frac{q}{4\pi\epsilon_0} \cdot \frac{q}{r} \qquad (r \geq a) \qquad (28.35)$$

Inside the uniformly charged spherical shell, $\mathbf{E} = 0$, and therefore $V =$ constant. Because the potential is continuous, the constant must be the value of $V(r)$ at $r = a$,

$$V(r) = \frac{q}{4\pi\epsilon_0} \cdot \frac{q}{a} \qquad (0 \leq r \leq a) \qquad (28.36)$$

(If the potential were discontinuous at $r = a$, then the discontinuity would imply a finite amount of work done over a zero distance and therefore an infinite force.)

Now let's consider a sphere that is uniformly charged throughout its volume. (This is a fairly reasonable model of the proton distribution in an atomic nucleus.) Outside the sphere, the electrostatic potential is the same as it would be if all the charge were concentrated at the origin.

$$V(r) = \frac{1}{4\pi\epsilon_0} \cdot \frac{q}{r} \qquad (r \geq a) \qquad (28.37)$$

Inside the sphere at radius r_0 (Figure 28.17a) only the charge at a smaller radial distance $r < r_0$ exerts a net force,

Figure 28.17a

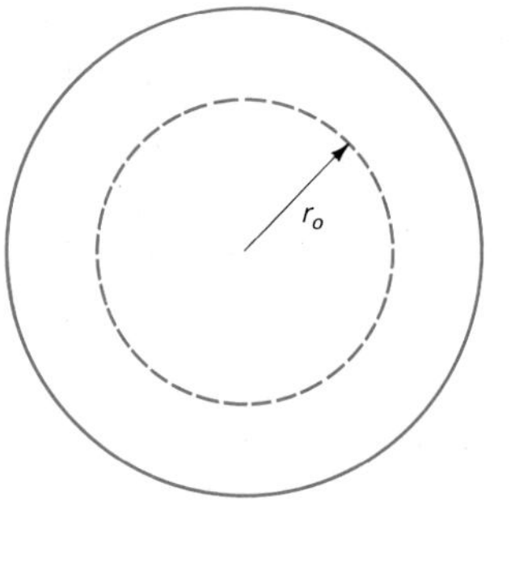

Figure 28.17b

The electrostatic potential of a sphere of radius a uniformly charged throughout its volume.

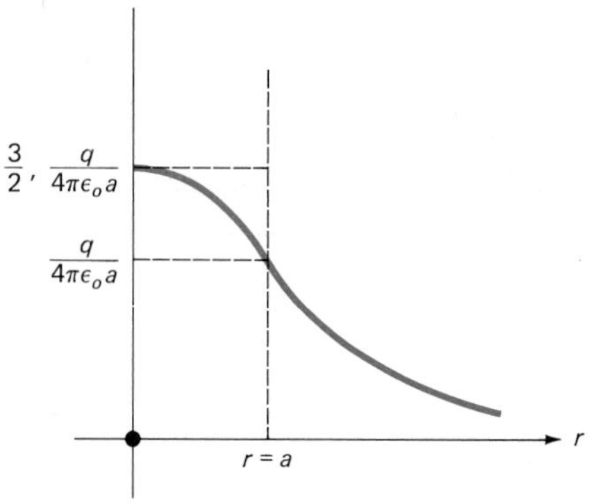

but all of the charge in the space $r_0 \leq r \leq a$ contributes to the potential. The result is a parabolic dependence of the interior potential.

$$V(r) = \frac{q}{4\pi\epsilon_0 a}\left(\frac{3}{2} - \frac{1}{2}\frac{r^2}{a^2}\right) \qquad (0 \leq r \leq a) \qquad (28.38)$$

The potential $V(r)$ of a uniformly charged sphere is plotted in Figure 28.17b.

Example 9 Electrostatic Potential of a Lead Nucleus

A nucleus of the lead isotope, lead 208, has a radius of 6.34×10^{-15} m and contains 82 protons, each with a charge of 1.60×10^{-19} C. Let us calculate the electrostatic potential at the surface of this nucleus, assuming that the charge is distributed uniformly throughout a sphere of radius 6.34×10^{-15} m.

We locate the origin of a coordinate system at the center of the nucleus. The total charge in the nucleus is $82(1.60 \times 10^{-19})$ C. Hence from Eq. 28.38 the electrostatic potential at the surface ($r = a$) is

$$V_{surface} = \frac{4}{4\pi\epsilon_0 a}\left(\frac{3}{2} - \frac{1}{2}\frac{r^2}{a^2}\right)$$

$$= 8.99 \times 10^9 \cdot \frac{82 \times 1.60 \times 10^{-19}}{6.34 \times 10^{-15}}$$

$$= 1.86 \times 10^7 \text{ V}$$

Table 28.3

Charge Distribution	Electric Field (magnitude)	Potential	Reference for Potential
monopole	$\sim 1/r^2$	$\sim 1/r$	Eq. 28.22
dipole	$\sim 1/r^3$	$\sim 1/r^2$	Eq. 28.31
quadrupole	$\sim 1/r^4$	$\sim 1/r^3$	Problem 40

At the center of the nucleus the potential is higher than it is at the surface by a factor of $\frac{3}{2}$. Hence

$$V_{center} = \frac{3}{2} V_{surface} = 2.79 \times 10^7 \text{ V}$$

In Chapter 27 we investigated the electric fields produced by single charges (monopoles) and by dipole and quadrupole arrays of charges. In this chapter we have determined or will determine the corresponding potentials. These results are summarized in Table 28.3.

Questions

16. The calculation of the potential V of a ring of charge is asserted to be easier than the calculation of the electric field **E**. Explain.
17. The calculation of the potential of a ring of charge would not be so easy if (a) the charge were distributed nonuniformly or (b) we went off the axis. Show, from the equations, what problems would arise in these circumstances.
18. Imagine two spheres of equal radius and equal total charge. One sphere is uniformly charged throughout its volume. The second sphere has all of its charge distributed uniformly over its surface. How do the potentials at the surface compare?

28.6 Equipotential Surfaces

Any surface, planar, or curved, over which the potential is constant is called an *equipotential surface.* In Section 28.3 we considered the uniform electric field between two parallel metal plates that are 1 cm apart. Let us consider this system again. The potential relative to the lower plate is given by

$$V_B - V_{bottom\ plate} = -\int_{x=0}^{x_B} \mathbf{E} \cdot d\mathbf{s} \qquad (28.39)$$

$$= E x_B \qquad (0 \leq x \leq x_{top\ plate})$$

For a fixed value of x_B (the height above the bottom plate) the potential V_B is constant—independent of coordinates parallel to the plates; x_B = constant defines a plane of constant potential (Figure 28.18).

Figure 28.18

Equipotential planes in a uniform electric field.

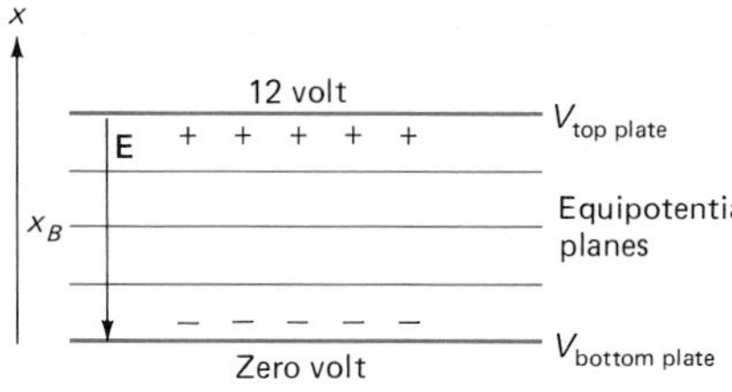

Because the potential is constant on an equipotential surface, the change in potential, ΔV, between two points on the surface is zero. The general relation between potential and field is

$$\Delta V = -\mathbf{E} \cdot \Delta \mathbf{s}$$

For $\Delta\mathbf{s}$, an increment of length on the equipotential surface, $\Delta V = 0$. Therefore

$$\mathbf{E} \cdot \Delta \mathbf{s} = 0$$

Because this equation holds for all choices of $\Delta\mathbf{s}$ on the equipotential surface, **E** has no component parallel to the surface and hence is perpendicular to the surface. The Faraday electric field lines are normal to an equipotential surface and directed from higher potential to lower potential. In terms of work done

$$dW = q\, dV = 0$$

for any diplacement $\Delta\mathbf{s}$ on the equipotential surface. No work is done in moving a charge on an equipotential surface.

For a point charge q the potential is

$$V = V(r) = \frac{q}{4\pi\epsilon_0} \cdot \frac{1}{r} \qquad (28.22)$$

The spherical surface defined by $r = (x^2 + y^2 + z^2)^{1/2}$ = constant becomes an equipotential surface. No work is done in moving a charge over any such spherical surface, as we saw in Section 28.3.

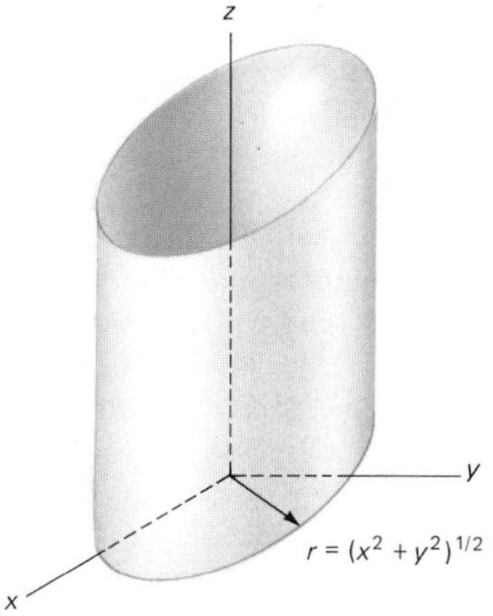

Figure 28.19

Cylindrical surface defined by $r = (x^2 + y^2)^{1/2}$ = constant.

Figure 28.20

Electric dipole field lines (solid) and equipotential surfaces (dotted).

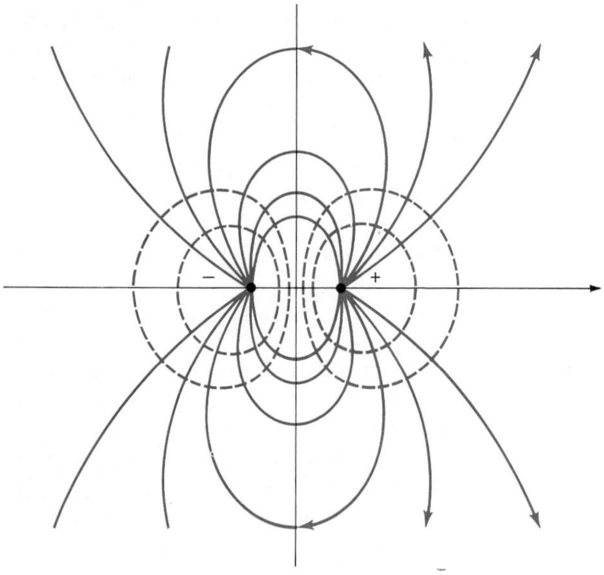

A line charge gives $V(r) \sim \ln(r/r_0)$. Then $V(r)$ is constant over the cylindrical surface defined by r = constant as shown in Figure 28.19. (Here we are using circular cylindrical coordinates. If z is the axis of the cylinder, then $r = (x^2 + y^2)^{1/2}$ is the perpendicular distance from the z-axis.)

More complicated charge systems will have more complicated systems of field lines and equipotential surfaces. The field lines and equipotential surfaces for a dipole are shown in Figure 28.20.

Conducting Surfaces

Any mathematical equipotential surface may be replaced by a physical conducting surface without altering the electrostatic field (except in the interior of the actual physical conductor).

Conversely, in our electrostatic case, any conducting surface is automatically an equipotential surface. When a conducting surface is placed in an electrostatic field (and does *not* coincide with a previously existing equipotential surface), charge will flow until a steady state is reached—that is, until the field produced by the new charge distribution exactly cancels the external applied field on and within the conducting surface. The introduction of a conducting surface (that does not coincide with some mathematical equipotential surface) distorts the original field. Figure 28.21 shows the field lines and equipotential surfaces that result from placing a conducting sphere in a previously uniform electric field.

Notice that in Figure 28.21 the sphere constitutes a new equipotential surface (after the charges in the conducting sphere have reached an equilibrium distribution). The field lines touching the sphere are perpendicular to the spherical surface. The field lines do not penetrate the interior of the conductor (Section 27.5). The external field induces a net negative charge on the upper surface of the sphere and a net positive charge on the lower surface. The sphere becomes an (induced) electric dipole.

Figure 28.21

Electric field lines and equipotential surfaces near a conducting sphere. The electric field is uniform at large distances.

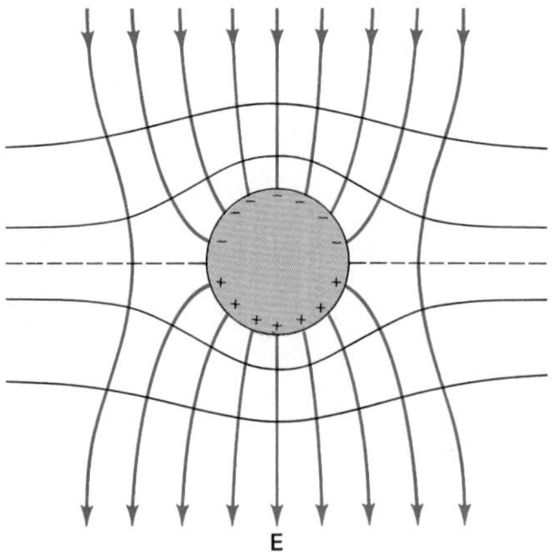

Question

19. A diagram of equipotential surfaces shows the intersection of two surfaces with different potentials. Is this reasonable? What would happen if the surfaces were conductors? Interpreting potential in terms of work, what is the implication for the force at the point of intersection?

28.7 Corona Discharge

Corona Discharge

When a conductor in air has such a large surface charge that the electric field at the surface exceeds 3×10^6 V/m, the oxygen and nitrogen molecules in the air will be ionized and an electric discharge will occur. This is a *corona discharge*. The St. Elmo's fire mentioned at the beginning of Chapter 26 is an example of a corona discharge. St. Elmo's fire has a certain visual beauty, but corona discharges in electrical equipment may do physical damage and are to be avoided.

In terms of a charged conducting sphere (radius a), we have the following relations.

$$E_{\text{surface}} = \frac{q}{4\pi\epsilon_0} \cdot \frac{1}{a^2} \tag{28.40}$$

$$V_{\text{surface}} = \frac{q}{4\pi\epsilon_0} \cdot \frac{1}{a} \tag{28.41}$$

and

$$q = 4\pi a^2 \sigma \tag{28.42}$$

with σ the surface charge density (its units are coulombs per square meter). Not all objects are spherical, and so we

Figure 28.22

Two conducting spheres, each at potential V, joined by a long fine wire. The distance between the spheres is large compared to the sphere radii.

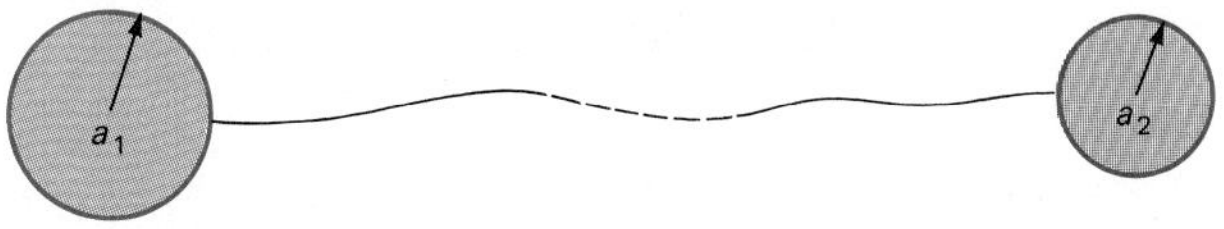

must extend these relations. Consider two spheres, each at potential V. For the first sphere, radius a_1,

$$V = \frac{q_1}{4\pi\epsilon_o} \cdot \frac{1}{a_1} = \frac{\sigma_1 a_1}{\epsilon_o}$$

while for the second sphere, radius a_2,

$$V = \frac{q_2}{4\pi\epsilon_o} \cdot \frac{1}{a_2} = \frac{\sigma_2 a_2}{\epsilon_o}$$

Then, since the potentials are equal,

$$\sigma_1 a_1 = \sigma_2 a_2 \qquad (28.43)$$

which means that the surface charge density is inversely proportional to the sphere radius for a fixed potential. Suppose that these two spheres are separated by a distance that is very large compared to the sphere radii. Then the potential V and the charge distribution of each sphere are that of an isolated sphere, unaltered by the other sphere. Now, imagine that the spheres are joined by a long, very fine wire (Figure 28.22). The flow of charge onto the very fine wire is negligible. The two spheres, each at the same potential V, retain the same potential and same charge distribution; but with the spheres joined, we have *one* body with a blunt end, radius a_1, and a sharp end, radius a_2. This suggests that on an irregularly shaped charged conductor a portion of the surface that can be described with radius r will have a surface charge density $\sigma \sim 1/r$.

We see the importance of this result when we consider Eqs. 28.40 and 28.42. They yield

$$E_{\text{surface}} = \frac{\sigma}{\epsilon_o} \qquad (28.44)$$

and therefore $E_{\text{surface}} \sim 1/r$. This $1/r$ dependence for E_{surface} also followed directly from Eqs. 28.40 and 28.41 with the potential V held fixed. *A small radius may lead to a large electric field.*

Example 10 **Electric Field of a Small Conducting Sphere**

A small conducting sphere is charged to a potential of 1000 V. How small must the sphere be in order to have an electric field of 3×10^6 V/m at its surface. We take the ratio of Eqs. 28.40 and 28.41:

$$\frac{E}{V} = \frac{1}{r}$$

$$r = \frac{1000 \text{ V}}{3 \times 10^6 \text{ V/m}} = 3.3 \times 10^{-4} \text{ m} = 0.33 \text{ mm}$$

The sharp tip of a lightning rod will set up a strong electric field at relatively low voltage. By ionizing, or breaking down, the molecules of the air, the lightning rod transfers charge to the atmosphere and thereby lowers the potential difference between the building it is protecting and a charged cloud overhead. The function of the lightning rod is to reduce a dangerously high potential difference, not to "catch" lightning. Airplanes often have similar devices to help eliminate any excess charge.

If the electric field at the surface of a conductor is high enough, it can literally pull electrons out from the surface of the conductor. This phenomena is called *field emission*. The electric field required for field emission to occur depends on the nature of the substance, and on the temperature. For a substance like tungsten at room temperature, the required field is a few million volts per centimeter, which is roughly a hundred times the electric field required to ionize the molecules of oxygen and nitrogen in the air. Such high fields can be achieved at pointed surfaces (in a vacuum). Field emission has been used in microscopes to study the nature of surfaces. By using magnification of over 1 million, we can see the outlines of heavy atoms on a pointed emitting surface.

Questions

20. An isolated metal cube is given an electrostatic charge Q. Qualitatively, how will this charge distribute itself over the cube? Consider the plane faces, the edges, and the corners.

21. If the charge of the cube in Question 20 is made steadily larger, what will happen? Where on the cube will this most likely happen?

Summary

The electrostatic *potential energy* at point B relative to point A is defined as the negative of the work done on the charge by the electric field when the charge moves from point A to point B:

$$\Delta U = U_B - U_A = -W_{AB} \qquad (28.3)$$

The electrostatic *potential difference* between points B and A is defined as the negative of the work done on the charge by the electric field when the charge moves from A to B, divided by the charge:

$$\Delta V = V_B - V_A = -\frac{W_{AB}}{q} \qquad \begin{matrix}(28.5)\\(28.6)\end{matrix}$$

The units are joules per coulomb, or equivalently, volts. The change in potential ΔV and the change in potential energy are related by

$$\Delta V = \frac{\Delta U}{q}$$

In the general case, the potential difference is the negative of the line integral of the electric field

$$V_B - V_A = -\int_A^B \mathbf{E} \cdot d\mathbf{s} \qquad (28.12)$$

The choice of zero of potential is arbitrary. For point charges it is often convenient to take the zero of potential at $r \to \infty$. Integrating the Coulomb field yields

$$V(r) = \frac{q}{4\pi\epsilon_0} \cdot \frac{1}{r} \tag{28.22}$$

for the potential due to a point charge q at a distance r from the charge.

From a known potential V, the electric field **E** may be obtained by differentiation. In the one-dimensional case

$$E(x) = -\frac{dV}{dx} \tag{28.17}$$

For convenience in atomic and nuclear physics problems, the electron charge and the volt are combined to construct an energy unit—the electron volt—with 1 eV $= 1.60219 \times 10^{-19}$ J. Particle energies are measured in kilo-electron-volts, mega-electron-volts, giga-electron-volts and tera-electron-volts.

The work done by the electrostatic field **E** on a charge moving from point A to point B is independent of the path. The electrostatic field is conservative. (An electric field that varies with time may not be conservative.)

There is a principle of superposition for electrostatic potentials. This is an extension of the principle of superposition for electrostatic fields. Since the potentials are scalars, superposition means direct algebraic addition as opposed to vector addition, which is required for the electric field.

An electric monopole (single charge) produces a characteristic $(1/r)$-dependent potential in the space around it. Electric dipoles produce a characteristic $1/r^2$ potential.

Any surface over which the potential is constant is called an equipotential surface. The electric field lines are perpendicular (normal) to an equipotential surface.

For a spherical surface at a fixed potential, the surface charge density (and therefore the electric field) is inversely proportional to the radius of the spherical surface.

Suggested Reading

G. Giannini, Electrical Propulsion in Space, *Sci. Am.*, 204, 57–65 (March 1961).

P. H. Rose and A. B. Wittkower, Tandem Van de Graaff Accelerators, *Sci. Am.*, 223, 24–33 (August 1970).

Problems

Section 28.1 Potential Difference

1. A charge of 2×10^{-7} C moves from the coordinate origin to the point (4, 4), with coordinates measured in meters (Figure 1). Calculate the work done by the uniform electric field $\mathbf{E} = (1 \times 10^4)\mathbf{i}$ N/C for each of three paths: (a) $(0,0) \to (4,0) \to (4,4)$; (b) $(0,0) \to (4,4)$, along diagonal; (c) $(0,0) \to (0,4) \to (4,4)$. (d) What is the electrostatic potential at point (4, 4) relative to the origin?

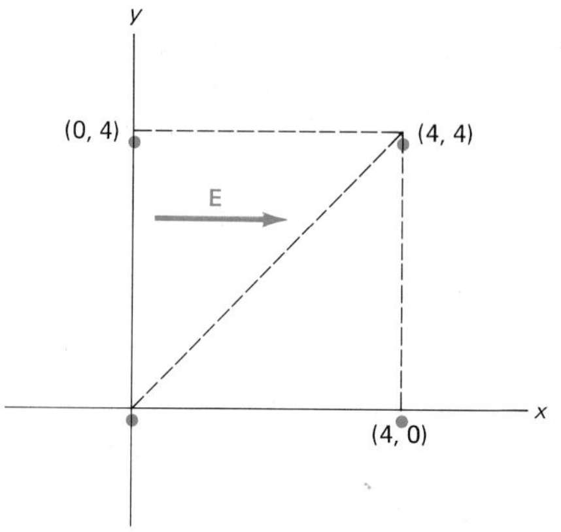

Figure 1

2. Repeat problem 1 using $\mathbf{E} = (1/\sqrt{2})(-\mathbf{i} + \mathbf{j}) \times 10^4$ N/C.

Section 28.2 Conservation of Energy

3. The relative error made in using the Newtonian form of kinetic energy rather than the special relativistic form is

$$\frac{K_{SR} - K_{Newton}}{K_{SR}}$$

Show, by binomial expansion, that this ratio is approximately $\frac{3}{4}(v^2/c^2)$.

4. (a) Calculate v/c for a 10-keV electron, assuming that Newtonian mechanics applies. (b) Calculate v/c for a 10-keV electron, assuming that special relativity applies.

5. A rigid electric dipole, dipole moment $2aq$, is in a uniform electric field **E**. Calculate the electrostatic potential energy of the dipole in this external field (a) when the dipole is parallel to the field (Figure 2a); (b) when the dipole is antiparallel to the field (Figure 2b).

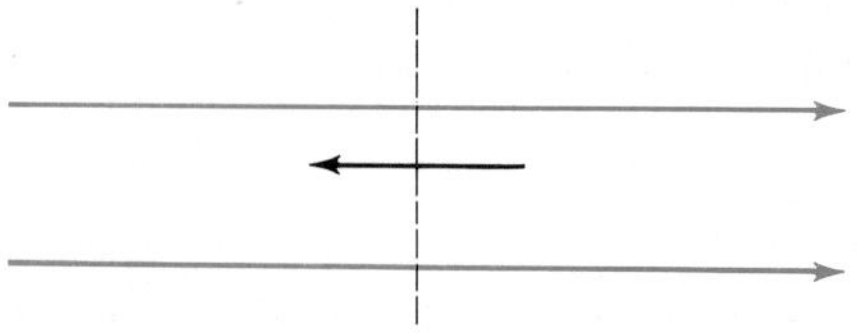

Figure 2a

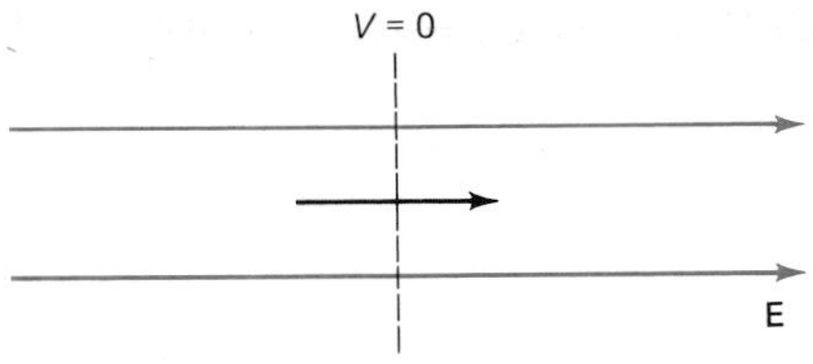

Figure 2b

6. To move 6.0 C of charge through an electrical resistance, 72.0 J of work are required. What is the potential difference between the ends of the resistor?

7. An electron is released from the negative plate of two parallel plates that are 2.0 cm apart. Accelerated by the uniform field **E** between the parallel plates, the electron strikes the positive plate with a speed of 3.0×10^6 m/s. (a) What is the magnitude of the electric field **E**? (b) What is the potential difference between the plates?

Section 28.3 Electric Potential

8. A potential difference of 10,000 V is established across two parallel metal plates, creating a uniform electric field between them. (a) Calculate the strength of the electric field **E** if the plates are d meters apart. (b) When the electric field strength rises to about 30,000 V/cm, the oxygen and nitrogen molecules of the air between the plates are ionized and electrical breakdown (arcing) occurs. What plate separation will give rise to this critical field?

9. Two parallel metal plates (Figure 3) large enough so that edge effects are negligible, are charged. The lower plate has a charge density $\sigma = +1.2 \times 10^{-6}$ C/m^2, and the upper plate has a charge density $\sigma = -1.2 \times 10^{-6}$ C/m^2. (a) Calculate the electric field between the two plates. (b) If the potential difference between these two plates is 440 V, what is the distance between the plates?

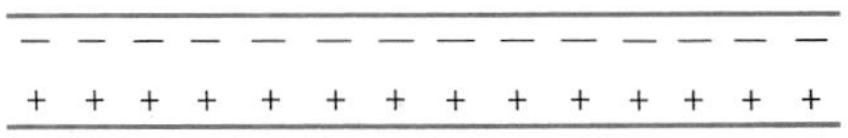

Figure 3

10. Point charges $+q$ and $-q$ are located on the x-axis as drawn in Figure 4. Three points (A, B, C) are indicated on the line joining the two charges. If $V_B = 0$, and V_A is 12 V higher than V_B and 24 V higher than V_C, determine V_A and V_C.

Figure 4

11. The electric field along a section of the x-axis is known to be in the x-direction: $\mathbf{E} = \mathbf{i}E(x)$. Calculate $E(x)$ for an electric potential (a) $V(x) = -10^4 x$; (b) $V(x) = +10^3 x^2$.

12. Using results from Problem 11, calculate the work done by the electric field when a test charge q_0 moves from $x = 2$ m to $x = 6$ m. Then determine $\Delta V = V(6) - V(2)$ and compare with $\Delta V = \Delta V(6) - V(2)$ calculated from the expressions for $V(x)$ given in Problem 11.

13. In the Fermilab accelerator protons are accelerated at 500 GeV. If 10^6 protons per second hit, and stop in, a particular target, how many joules per second are delivered to that target? (*Note:* At this energy most of the protons will actually continue on through the target.)

14. In a tandem Van de Graaff accelerator a proton is accelerated through a potential difference of 14×10^6 V. Assuming that the proton starts from rest, calculate its (a) final kinetic energy in joules; (b) final kinetic energy in MeV; (c) final velocity.

15. Starting from rest, an electron is accelerated by an electric field and moves through a potential difference V. Calculate the electron's final kinetic energy (in joules) and velocity (in meters per second) for (a) $V = 10$ V; (b) $V = 10^6$ V.

16. Starting from rest, a proton is accelerated by an electric field and moves through a potential difference V. Calculate the proton's final kinetic energy (in joules) and velocity (in meters per second) for (a) $V = 10$ V; (b) $V = 10^6$ V.

17. An electron in the 2-mile-long Stanford Linear Accelerator acquires an energy of 20 GeV (20×10^9 eV). (a) Calculate the ratio of the final speed of the electron to the speed of light. (b) If the actual accelerator, with its carefully timed impulses, were replaced by a uniform electric field, what strength field (in volts per meter) would be required to accelerate an electron to this energy in this distance? (*Note:* This problem requires the relativistic expressions presented in Chapter 20.)

Section 28.4 Point Charge Potential

18. At a distance r from a particular isolated point charge Q the electric field is 1250 N/C. The electrostatic potential is 5000 V. (a) Determine the distance r. (b) What is the magnitude of the point charge Q?

19. A cube, 1 m on a side, is centered on a $+50$-μC charge. How much work is required to move a $+5.0$-μC charge between any two corners of the cube?

20. A point charge of 2.3×10^{-7} C is at the origin (Figure 5). Calculate the electrostatic potential difference between the points (a) $x = 4$ m ($y = 0$, $z = 0$) and $x = 2$ m ($y = 0$, $z = 0$); (b) $y = 4$ m ($x = 0$, $z = 0$) and $z = 2$ m ($x = 0$, $y = 0$).

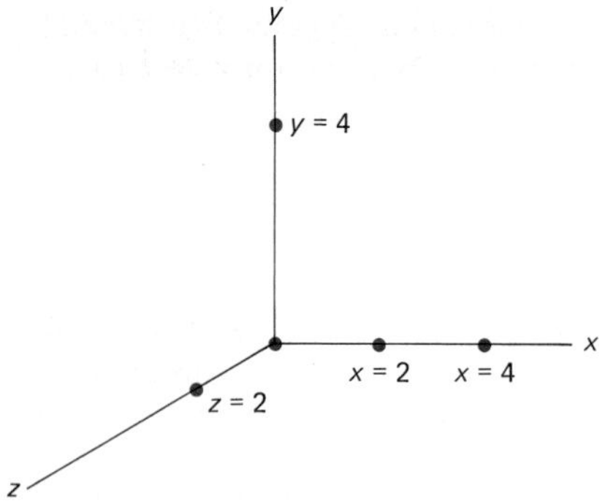

Figure 5

21. A point charge $q = 5.0\ \mu C$ is located at the origin. Determine the potential difference between the point (2.5 m, 0) and the points (a) (0, 2.5 m); (b) (−2.5 m, 0); and (c) (2.5 m, 2.5 m).

22. A charge $+q$ is at the origin. A charge $-2q$ is at $x = 2.0$ m on the x-axis. (a) For what finite value(s) of x is $\mathbf{E} = 0$? (b) For what finite value(s) of x is $V = 0$? (*Note:* The values of x for $\mathbf{E} = 0$ do not coincide with the values for $V = 0$.)

23. The two protons in a helium nucleus ^4He are 2×10^{-15} m apart on the average (time average). Calculate the mutual electrostatic potential energy of this pair of protons corresponding to this separation distance. (*Note:* As the number of protons in the nucleus increases, the electrostatic potential energy increases and is ultimately responsible for instability and nuclear fission; see Section 43.5.)

24. A proton is fixed in place. A second proton is brought to within a distance r of the first proton and then released. When the potential energy of the second proton in the field of the first proton is converted entirely into kinetic energy (at infinite distance), what speed will the second proton have for (a) $r = 1.0$ m; (b) $r = 3. \times 10^{-15}$ m?

25. The (mutual) potential energy of charges Q and q, a distance r apart, is

$$\frac{1}{4\pi\epsilon_0} \cdot \frac{Qq}{r}$$

Show that this energy is equal to the work that would have to be done on q against the Coulomb force of Q to bring q in from infinity to a distance r from Q, with Q fixed in space.

Section 28.5 Multiple Charge Potentials

26. A -1-μC charge is located at $x = 1$ m and a $+2$-μC charge is located at $x = -2$ m. Make a sketch of the electric potential for positions on the y-axis and show that the electric potential is a maximum at $y = \pm 2.03$ m.

27. A spherical conductor of radius a carries a uniformly distributed charge $-Q$. A proton moving radially out from the conductor is retarded by the electric field of the conductor, but reaches an arbitrarily large distance before its speed is reduced to zero $[v(r \to \infty) = 0]$. What is the proton's initial velocity, $v(r = a)$, in terms of the charge $-Q$ and the radius a? (*Note:* This is the electrostatic analog of the escape velocity of a particle in a gravitational field.)

28. Given two 2-μC charges, as shown in Figure 6, and a positive test charge $q_o = 1.28 \times 10^{-18}$ C at the origin. (a) What is the net electrostatic force exerted on q_o by the two 2-μC charges? (b) What electrostatic field $\mathbf{E}$ do the two 2-μC charges produce at the origin? (c) What is the electrostatic potential V produced by the two 2-μC charges at the origin? (d) What is the electrostatic potential energy of q_o?

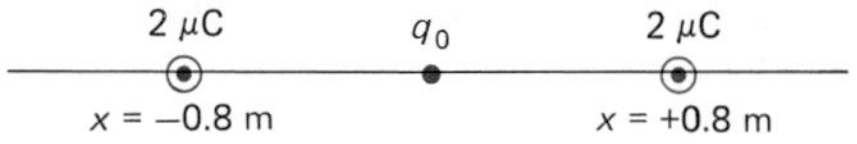

Figure 6

29. Each of the vertices of an equililateral triangle 1 m on a side carries a 1-μC (positive) charge. Calculate the potential energy of this configuration by calculating the work done in bringing the electric charges in form infinity (one at a time). (*Hint:* The first charge comes in free: No work is required. Bring the second charge in along the x-axis, and the third charge in along the y-axis; see Figure 7.)

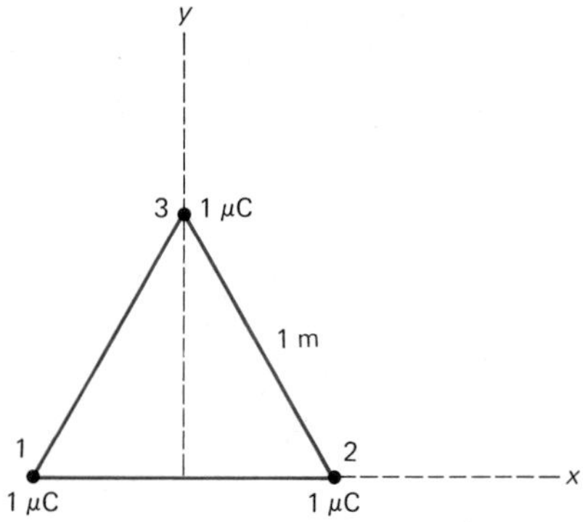

Figure 7

30. Four equal positive charges, q, are located on the vertices of a square of side a. How much work had to be done to assemble these four charges, bringing them in from infinity?

31. The potential on the x-axis of a ring of charge is found to be

$$V(x) = \frac{Q}{4\pi\epsilon_0}\left[\frac{1}{(x^2 + a^2)^{1/2}}\right]$$

Show that this is consistent with the electric field produced by the ring of charge. (The Coulomb force exerted by the ring of charge was calculated in Section 26.4.)

32. Calculate the electrostatic potential at the center of a uniform ring of charge with radius a and charge Q. Take the zero of potential to be at infinity.

33. The axis of rotational symmetry of a ring charge Q is taken to be the x-axis. On this axis the potential is

$$V(x) = \frac{Q}{4\pi\epsilon_0}\left[\frac{1}{(x^2 + a^2)^{1/2}}\right]$$

Determine the potential of a *uniformly* charged disk, radius a_0, by integrating this result. (*Note:* Q will depend on the radius of the ring.)

34. (a) Derive the expression for the electric field of a uniformly charged disk along the rotational symmetry axis of the disk. (*Hint:* Differentiate.) (b) Show that the field **E** has the expected behavior at $x = 0$, and for $x \gg a_0$.

35. The calcium nucleus $^{40}_{20}Ca$ is nearly spherical, with a radius of 3.66×10^{-15} m. Calculate the electrostatic potential at the surface of the nucleus.

36. What is the electrostatic potential at the midpoint of an electric dipole ($+q$ at $x = +a$ and $-q$ at $x = -a$)? Check your answer by calculating the work required to bring a test charge in from infinity.

37. In spherical polar coordinates the radial component of the electric field is given by

$$E_r = -\frac{\partial V}{\partial r}$$

The $\partial/\partial r$ symbol means "Differentiate with respect to r, holding the angles θ and ϕ constant. (a) Calculate E_r for an electric dipole (at large distances) given

$$V(r,\theta) = \frac{2aq\cos\theta}{4\pi\epsilon_0 r^2}$$

Compare your result with the dipole electric field calculated in Section 27.2. (b) This potential has no azimuthal (ϕ) dependence. Why not?

38. (a) From Section 28.5 the potential of an electric dipole far out on the dipole (x) axis is

$$V(x) = \frac{2aq}{4\pi\epsilon_0}\cdot\frac{1}{x^2}$$

From this equation derive directly the corresponding electric field (E_x) far out on the dipole axis. (b) The electric field far out on the dipole axis (see Example 3, Section 27.2) is

$$\mathbf{E} = \mathbf{i}\left(\frac{2aq}{2\pi\epsilon_0}\right)\cdot\frac{1}{x^3}$$

Use this result to derive the dipole potential far out on the dipole (x) axis.

39. An electric quadrupole creates a potential $V(x) = K/x^3$ along the x-axis. (K is a constant, x is large compared to the quadrupole dimensions.) (a) Show that the electric field along the x-axis is $E_x = 3K/x^4$. (b) Integrating the field from x to ∞, verify the given form of the quadrupole potential.

40. Suppose that you have a charge of $-2q$ at the origin, a charge of $+q$ at $x = -a$, and a charge of q' at $x = b$, as shown in Figure 8. The charge q' and the value of b are at your disposal. Assume that the electrostatic potential along the x-axis for $x \gg b$ can be written

$$V(x) = \frac{a_1}{x} + \frac{a_2}{x^2} + \frac{a_3}{x^3} + \frac{a_4}{x^4} + \cdots$$

(a) Choose q' so that $a_1 = 0$. (b) Choose b so that $a_2 = 0$. With these choices you have constructed a linear electric quadrupole.

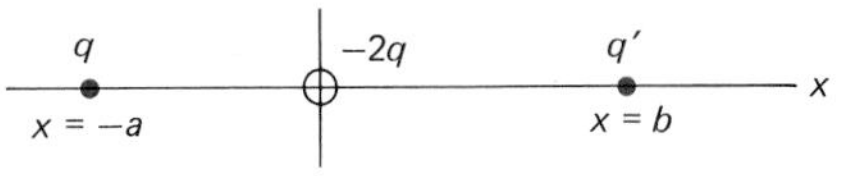

Figure 8

41. From Gauss' law, the electric field of the uniform line charge, λ C/m is $\mathbf{E} = \hat{\mathbf{r}}(\lambda/2\pi\epsilon_0 r)$ where $\hat{\mathbf{r}}$ points radially outward, perpendicular to the line source. (a) Calculate the potential difference between $r = r_1$ and $r = r_2$. (b) For a point source we take the zero of potential to be at $r \to \infty$. Is this choice appropriate for a line source? Explain.

42. A molecule has an electric dipole moment $p = 1.0 \times 10^{-30}$ C · m. Calculate the potential at a distance $r = 5.0 \times 10^{-9}$ m from the molecule at the angles (a) $\theta = 0$; (b) $\theta = 45°$; and (c) $\theta = 90°$. (θ is the angle measured relative to the dipole axis.)

Section 28.6 Equipotential Surfaces

43. The equipotential surfaces of an isolated point charge are concentric spheres. For an isolated point charge of 3.0×10^{-8} C calculate the radius of the 100.0 V equipotential surface.

44. A long straight wire carries a charge of 1.0×10^{-8} C/m (Figure 9). A coaxial cylindrical equipotential surface, radius 1.0 m, has a potential of 100.0 V. (a) What is the radius of the 1000.0 V equipotential cylinder? (b) What is the potential of the coaxial equipotential cylinder of radius 2.0 m?

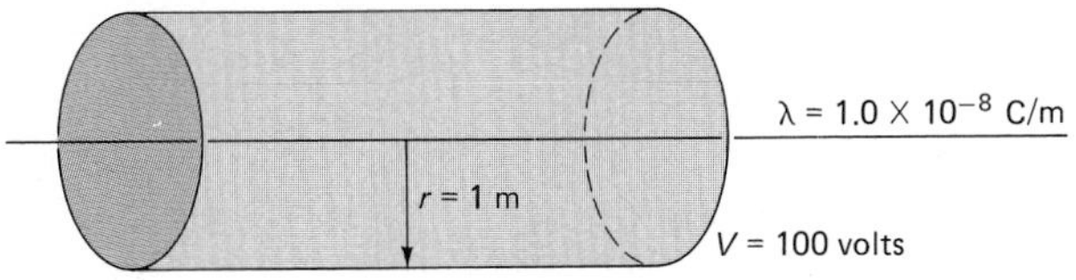

Figure 9

45. A point charge $+q$ is located at $x = -l$ and a point charge $-2q$ is located at the origin. Prove that the equipotential surface that has zero potential is a sphere centered at $(-\frac{4}{3}l, 0, 0)$ whose radius $r = \frac{2}{3}l$.

46. Using Eq. 28.31, show that the equipotential surfaces of an electric dipole are described by the equation $r^2 = k\cos\theta$ where k is a constant.

Section 28.7 Corona Discharge

47. A very large conducting sphere carries a charge of 1 C. The electrostatic potential at the surface of the

sphere is 1 V. (a) Calculate the radius of the sphere. How does this compare with the radius of the earth? (b) Calculate the surface charge density in C/m^2 and in proton charges/m^2.

48. A 1-μC charge is placed on an isolated 1-cm diameter steel ball bearing. Calculate the electric field at the surface of the sphere. (*Note:* This is an enormously strong electric field that would lead to corona discharge. For this physical system 1-μC is a very large charge.)

49. (a) The surface of a sphere 0.5 m in radius is at a potential of 100 V. What is the surface charge density, σ? (b) What is the electric field just outside the surface of the sphere?

chapter 29

CAPACITANCE AND CAPACITORS

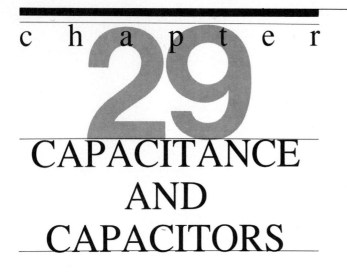

Preview

No single electronic component plays a more important role in our electronic age than a charge-storing mechanism called a capacitor. Capacitors help us to understand the energy aspects of electric fields and the properties of insulators. They are used to establish electric fields, to minimize voltage variations in electronic power supplies, to increase the efficiency of electric power transmission, and to provide energy for certain types of nuclear fusion energy devices and nuclear particle accelerators. Capacitors are used in electronic circuits that detect and generate electromagnetic waves, and as components of electronic circuits used to measure time.

We begin our discussion with a type of capacitor consisting of two conductors separated by an insulator. We show how this capacitor can accumulate and store a net amount of electric charge. Capacitance is introduced as a quantitative measure of this charge-storing ability. After calculating the capacitance of some symmetric capacitors of interest, we determine the equivalent capacitance of series and of parallel combinations of capacitors. We examine the energy aspects of capacitors and conclude with a discussion of the role of the insulator in a capacitor.

29.1 Capacitance and Capacitors

Both conductors and insulators are electrically neutral in their natural state. Subjecting either to electric forces can redistribute charges and produce localized regions having a net charge. However, the electric fields required to remove charges from conductors are much smaller than those needed to remove the charges from insulators. If two conductors separated by an insulator are connected to an energy source such as a battery, electric forces cause electrons to flow from the conductor connected to the positive terminal to the conductor connected to the negative terminal (Figure 29.1). Electron flow stops when the potential difference between the conductors equals the potential difference across the battery. Conservation of charge demands that the amount of negative charge removed from one conductor (leaving this conductor with a net positive charge) equal the amount of negative charge accumulated on the other conductor. If the battery is removed, the charge remains stored on the two conductors. Because of this charge-storing ability, this configuration of two conductors is termed a **capacitor.** A capacitor in a circuit is represented by the symbol —||—. The parallel lines represent the conductors and the lines connected to the parallel lines represent electrical contacts to the conductors.

The ability of a capacitor to store charge is measured by a quantity we call **capacitance.** We call any configuration of two conductors separated by an insulator a capacitor, and we define the capacitance of a capacitor as

$$C = \frac{Q}{V} \tag{29.1}$$

Figure 29.1

When a battery is connected to two conductors separated by an insulator, electric forces remove electrons from one conductor and store them on the other.

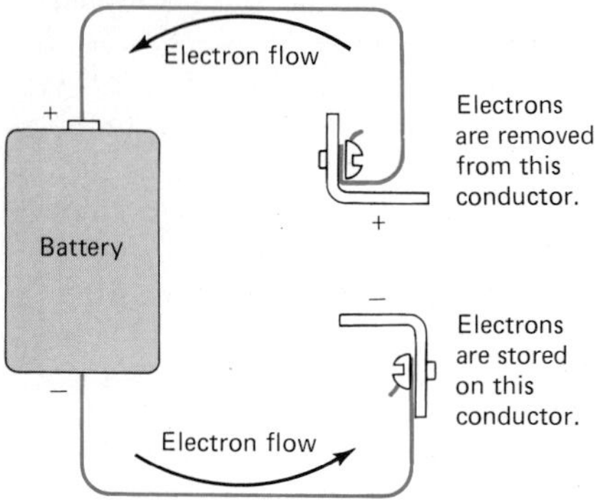

Figure 29.2

Two oppositely charged concentric spheres separated by a vacuum. The radii of the spheres are r_1 and r_2 ($r_1 < r_2$). A charge $+Q$ is distributed uniformly on the surface of the inner sphere. A charge $-Q$ is distributed uniformly on the surface of the outer sphere.

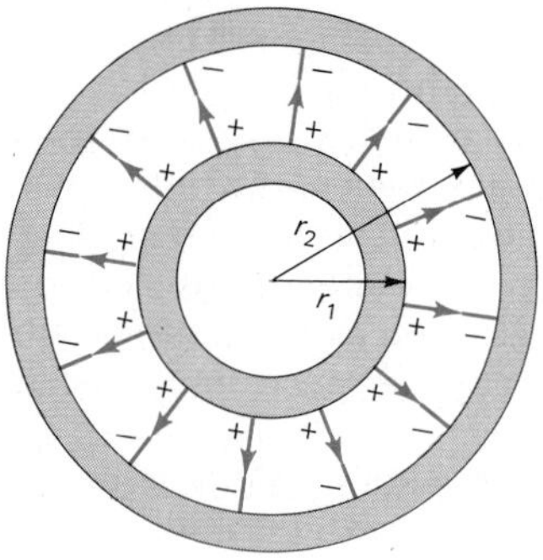

where V is the potential difference* between the two conductors and Q is the magnitude of the charge on either conductor. From Eq. 29.1 capacitance has units of coulombs per volt, or farads (abbreviated F). Because a conductor is an equipotential surface (Section 28.6), V can be measured between any two points on the oppositely charged conductors. Accurate measurements of Q and V allow us to determine the capacitance of capacitors having any geometric shape. For a few capacitors we can determine the relation between Q and V theoretically and then identify the ratio Q/V as the capacitance.

In practice, determining capacitance is not always simple. Consider, for example, the intertwining, insulated conducting strips in the integrated electronic circuit of a pocket calculator. In such a system it is impossible to completely isolate the two conductors of a particular capacitor from the other conductors in the environment. Some capacitance will arise from these other conductors as well as from the conductors of the capacitor. These capacitances are called **stray capacitances.** Stray capacitances are usually negligible, however, and are therefore omitted from our analysis of capacitance.

Spherical Capacitor

To illustrate one method of calculating capacitance, we consider two oppositely charged, concentric, hollow spherical conductors separated by a vacuum (Figure 29.2). Let us first imagine that the outer sphere is absent and focus our attention on the inner sphere with radius r_1. Electric field lines extend radially outward from positive charges distributed uniformly over the surface of the inner sphere. The electric potential is constant on the surface and throughout the inside of the sphere. Outside the sphere, the equipotential surfaces are concentric spheres (Section 28.5). The potential difference between any equipotential surface of radius r and the spherical conductor of radius r_1 is (Section 28.4)

$$V = \frac{1}{4\pi\epsilon_0}\left(\frac{Q}{r_1}\right) - \frac{1}{4\pi\epsilon_0}\left(\frac{Q}{r}\right) \tag{29.2}$$

*In Chapter 28 we used the notation ΔV to denote a potential difference. Because only potential differences are involved in determining capacitance, we simplify the notation by omitting the Δ.

If we now introduce the outer sphere of radius r_2, we know from arguments presented in Section 28.6 that the electric field line pattern is unaltered between the spheres. Field lines emanating from positive charges on the inner sphere will terminate on negative charges distributed uniformly on the outer sphere. The potential difference between the spheres is

$$\begin{aligned} V &= \frac{1}{4\pi\epsilon_0}\left(\frac{Q}{r_1}\right) - \frac{1}{4\pi\epsilon_0}\left(\frac{Q}{r_2}\right) \\ &= Q\frac{1}{4\pi\epsilon_0}\left(\frac{1}{r_1} - \frac{1}{r_2}\right) \end{aligned} \tag{29.3}$$

Solving for the ratio Q/V, we find the capacitance,

$$C = \frac{Q}{V} = 4\pi\epsilon_0\left(\frac{r_1 r_2}{r_2 - r_1}\right) \tag{29.4}$$

Note that C depends on geometric factors and on the insulator (a vacuum in this case), but is independent of the values of Q and V.

Although this method of determining the potential difference between the spheres illustrates nicely the utility of symmetry, it is not the only method available. Using Gauss' law to determine the electric field between the spheres, we could have calculated the potential difference from its defining relation (Eq. 28.9). We will use this second method in some forthcoming examples.

The effect of the outer sphere on the capacitance diminishes as its radius increases. At some point (formally for $r_2 \to \infty$) the capacitance is determined solely by r_1 and we have the capacitance for an isolated spherical capacitor, $C = 4\pi\epsilon_0 r_1$. A spherical satellite in space is a good approximation to an isolated spherical capacitor.

Example 1 Capacitance of the Echo Satellite

The Echo I satellite was a metal-coated plastic sphere, 30 m in diameter, that orbited the earth at a mean altitude of 1600 km. Because its altitude was large compared to its radius, it was essentially isolated. It had a capacitance

$$\begin{aligned} C &= 4\pi\epsilon_0 \cdot 15 \\ &= 1.7 \times 10^{-9} \text{ farad} \end{aligned}$$

To produce a capacitance of 1 F the satellite would require a radius of 8.85×10^9 m. This is about 1400 times the radius of the earth! Practical capacitances are therefore usually many orders of magnitude smaller than a farad, and it is customary to use units of microfarads (10^{-6} F) and picofarads (10^{-12} F), abbreviated μF and pF, respectively.

A variation of this concentric sphere calculation involves shifting the spheres from the concentric position (Figure 29.3). The capacitance is still given by $C = Q/V$ but the calculation of the ratio Q/V is much more difficult. The initial symmetry is lost and the electric field lines are no longer radial. The outer sphere no longer corresponds to an equipotential surface for the inner sphere. Charge has redistributed itself on both spheres in an undetermined way, making it impossible to determine the potentials of the two spheres by a symmetry argument. Because the electric field lines are no longer radial, we cannot use Gauss' law to obtain the electric field, which, in turn, is needed to calculate the potential difference from its defining relation. Therefore we will consider only situations in which symmetry simplifies the calculations. The capacitance of the nonsymmetrical cases may be determined by experimental measurements.

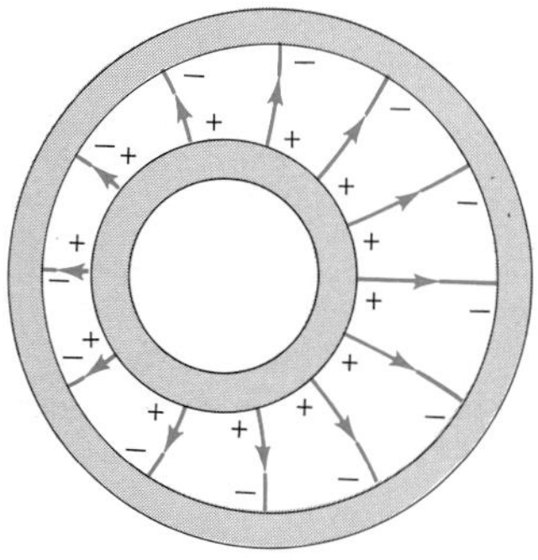

Figure 29.3

Two nonconcentric spheres separated by an insulator constitute a capacitor. Calculation of the capacitance is difficult because the electric field lines do not extend radially out from the inner sphere.

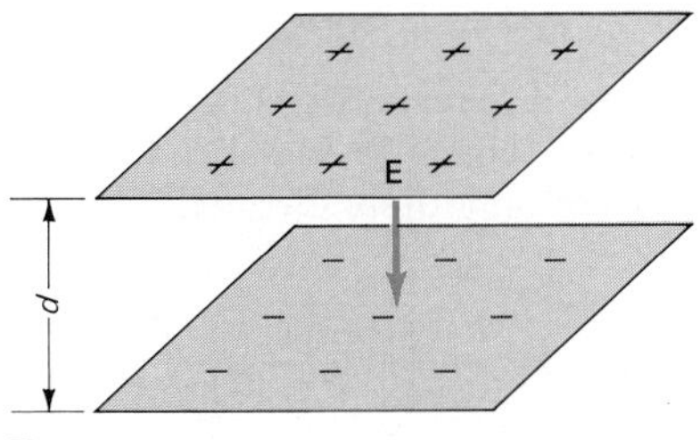

Figure 29.4

Two parallel, conducting plates separated by an insulator (a vacuum here) constitute a capacitor. When the plate dimensions are large compared to the distance between the plates, the electric field is nearly constant between the oppositely charged plates.

Parallel-Plate Capacitor

A parallel-plate capacitor consists of two plane-parallel conductors of equal area separated by an insulator (Figure 29.4). For simplicity, let's assume that the insulator is a vacuum. If the capacitor plates were of infinite size, the electric field between the plates would be $E = \sigma/\epsilon_0$ (Section 27.5) where σ is the charge per unit area. Practical parallel-plate capacitors are not infinitely large, but the dimenisons of the area A of one of the plates are usually very large compared to the plate separation d. When this is the case, the electric field can be approximated by the infinite-plate expression. Using this expression, we obtain for the potential difference between the plates of the capacitor in Figure 29.4

$$V = Ed = \left(\frac{\sigma}{\epsilon_0}\right)d \tag{29.5}$$

The magnitude of the total charge on either plate is $Q = \sigma A$ and the capacitance is

$$C = \frac{Q}{V} = \frac{\sigma A}{\sigma d/\epsilon_0} = \frac{A\epsilon_0}{d} \tag{29.6}$$

Just as with spherical capacitors, C depends only on ϵ_0 and geometric factors. If lengths are in meters C is in farads. Also note that as d decreases, C increases. From Eq. 29.6 we see that alternate units for ϵ_0 are farads per meter.

Although Eq. 29.6 pertains only to parallel-plate capacitors, its implications are much broader. The capacitance of *any* capacitor increases as the distance between the conductors decreases. Decreasing d is an effective way of increasing capacitance.

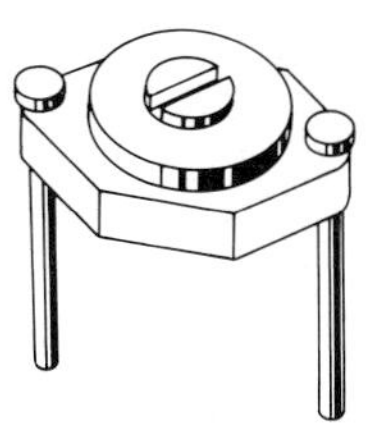

Figure 29.5

The capacitance of this capacitor is varied by adjusting the plate separation.

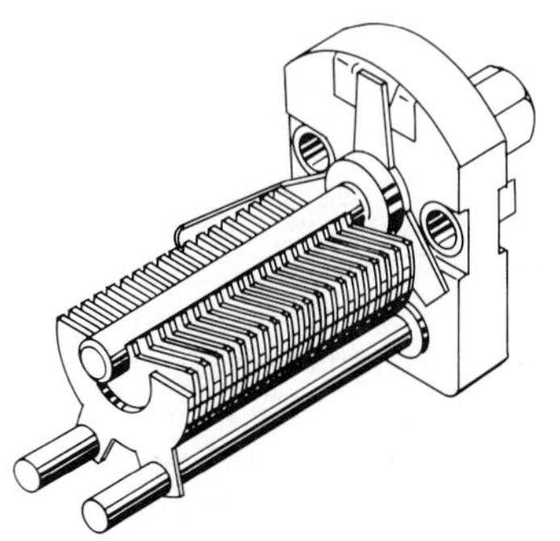

Figure 29.6

The capacitance of this capacitor is varied by adjusting the effective area of the plates.

Many electronic devices require a capacitance that is variable over a prescribed range. For example, the selection of a radio station on a transistor radio often is done by varying a capacitance in a tuning circuit. A variable capacitor, symbolized by ⇥, exploits the dependency of capacitance on the separation between, and the area of, the plates. One type of variable capacitor (Figure 29.5) maintains constant plate area but allows the distance between plates to be varied with a screw adjustment. A second type of variable capacitor (Figure 29.6) consists of a set of plates that is free to rotate between a second set of fixed plates. The effective plate area of the capacitor, and consequently its capacitance, is varied by rotating the plates.

Example 2 **Comparison of a Capacitor with a Battery**

A high-quality 12-V automobile storage battery is rated at 100 ampere hours (A·hr). This means that it can transfer a total of 3.6×10^5 C of charge. Because capacitors are also charge-storing devices, you may wonder why they are not used in automobiles. The answer lies in their size. Let's determine the physical size of a parallel-plate capacitor that stores 3.6×10^5 C when the potential difference across it is 12 V. The desired capacitance would be $C = Q/V = 3 \times 10^4$ F. The capacitance depends on both the areas of and the separation between the plates. For a plate separation of 0.1 mm the area is

$$A = \frac{Cd}{\epsilon_0} = \frac{3 \times 10^4 \times 10^{-4}}{8.85 \times 10^{-12}}$$
$$= 3.39 \times 10^{11} \text{ m}^2$$

This area corresponds to a square of about 5.82×10^5 m (about 360 miles) to a side! Because of its size this parallel-plate capacitor cannot be used in place of an automobile storage battery. However, we will see later that capacitors perform many important functions that batteries cannot.

Coaxial Transmission Line

The transmission of electric currents between electronic equipment is often accomplished with coaxial cables. These cables consist of concentric cylindrical conductors separated by an insulator (Figure 29.7). These cables thus have the structure of a cylindrical capacitor (Figure 29.8). The cable capacitance is important because it influences the current. To determine the capacitance of a cylindrical capacitor we use the parallel-plate capacitor example as a guide. If the inner and outer conductors contain uniformly distributed charges $-Q$ and $+Q$ and the length of the capacitor is large compared to the separation of the conductors, then symmetry arguments suggest that the electric field is radial. We can then use Gauss' law (Section 27.5) to determine the electric field. We surround the inner conductor with a concentric cylindrical Gaussian surface of radius r. Because of the radial nature of the field, no electric flux passes through the ends of the Gaussian surface. Assuming that the conductors are separated by a vacuum, we can write for Gauss' law

$$\int \mathbf{E} \cdot \mathbf{n}\, dA = \frac{-Q}{\epsilon_0}$$

where integration is over the cylindrical surface and Q is the net charge residing within the surface. From symmetry arguments stated in Chapter 27, the integral reduces to

$$-\int E\, dA = -E\int dA = -E\,\underbrace{2\pi rL}$$

because **E** is antiparallel to **r**; because E is constant on the Gaussian surface; area of the cylindrical Gaussian surface

Hence

$$E2\pi rL = \frac{Q}{\epsilon_0} \qquad (29.7)$$

To determine the potential difference we choose a radial path between the conductors (Figure 29.9). Then $\mathbf{E} \cdot d\mathbf{r} = -E\,dr$ because **E** and $d\mathbf{r}$ are oppositely directed, and the potential difference is

$$V = -\int_{r_1}^{r_2} -\frac{Q}{2\pi\epsilon_0 Lr}dr = \frac{Q}{2\pi\epsilon_0 L}\ln\left(\frac{r_2}{r_1}\right)$$

The capacitance is

$$C = \frac{Q}{V} = \frac{2\pi\epsilon_0 L}{\ln(r_2/r_1)} \qquad (29.8)$$

As for the parallel-plate capacitor, C depends only on geometric quantities and ϵ_0.

Figure 29.7

Section of a type of coaxial cable. Material has been stripped from the central conductor to show the construction.

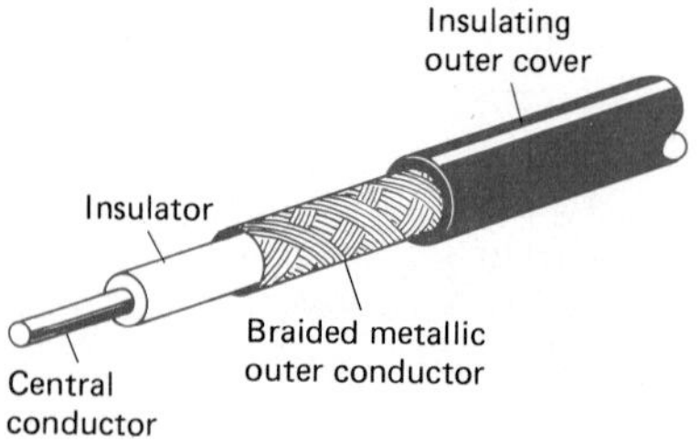

Figure 29.8

Two concentric cylindrical conductors separated by an insulator (a vacuum here) constitute a cylindrical capacitor. The electric field lines are radial in the region between the conductors.

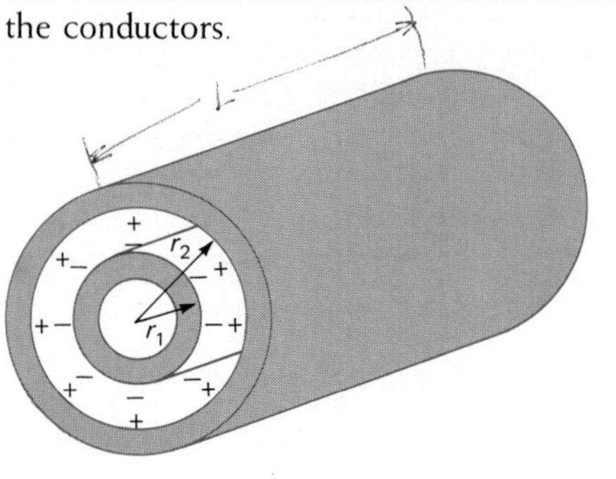

Figure 29.9

End view of the cylindrical capacitor shown in Figure 29.8. The path of integration used to determine the potential difference between the cylinders is a line extending radially out from the axis of the capacitor.

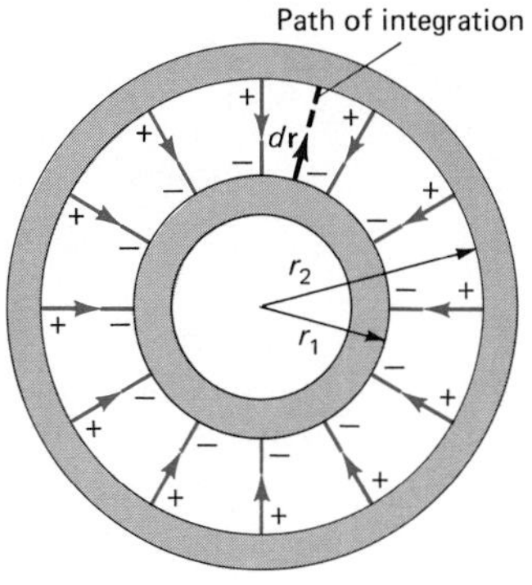

Questions

1. Capacitance is a measure of the ability of a capacitor to store charge. Similarly, volume is a measure of the ability of a container to store a liquid. Name some other systems that have an ability to store or hold something. Just from the meaning of capacitance, is it surprising that it depends only on geometric factors? Explain.
2. Rubbing a glass rod with a piece of silk produces a net positive charge on the rod. How could you use the positively charged rod to produce a positive or negative charge on a metallic conductor? (You may temporarily connect the metallic conductor to another conductor.)
3. Examine the two wires connecting a TV antenna to a TV set. Would you expect these two wires to have capacitance?
4. How might a moving conductor like an airplane acquire a charge as it flies through the air?
5. A commercial airplane flying at about 40,000 ft approximates an isolated conductor. Discuss the difficulties involved in a theoretical calculation of the capacitance of this airplane.
6. Two long, separated parallel cylinders of equal radii possess uniform charge distributions of $+\sigma$ and $-\sigma$ C/m^2. Sketch the electric field lines between the two cylinders and comment on the problems involved in calculating the capacitance of the configuration.
7. What are some ways that a measure of capacitance could be used as an indirect method for measuring a length or a thickness?
8. Describe the charging of a capacitor by a battery, assuming that positive charges are transferred.
9. Commercial capacitors are usually cylindrical in shape. However, the capacitance of a commercial capacitor is normally determined by the parallel-plate capacitor formula. How might cylindrical capacitors be constructed from long flexible strips of metal and insulator?

29.2 Capacitors in Series and in Parallel

Capacitors are often connected or combined in electronic systems. If we understand how various combinations of capacitors behave, we can then understand how they affect the performance of these systems. If we wish, we can even replace combinations of capacitors with single equivalent capacitors without changing the system.

Capacitors can be combined in many ways, but the two combinations called *series* and *parallel* are of special interest. Two or more capacitors connected to the same battery (or other source of potential difference) in the way shown in Figure 29.10 are said to be connected in series. Two or more connected as shown in Figure 29.11 are connected in parallel.

Whether we are charging a single capacitor, as in Section 29.1, or a combination of capacitors, the principle of charge conservation is always satisfied. Charges are redistributed, but no charge is ever added to or subtracted from the total system. For example, consider the charging of two capacitors in series (Figure 29.12). In Figure 29.12a, all conductors of the capacitors are electrically neutral. They are not yet connected to the battery because the switch, S, is open. When the switch is closed (Figure 29.12b), the battery is connected and negative charge flows from the upper conductor of C_1 to the lower conductor of C_2. No charge is removed from the inner conductors of C_1 and C_2 because these conductors are not connected to the battery. However, a separation of charge does occur on the inner conductors (Figure 29.12c). Overall electrical neutrality is maintained. Capacitor C_2 charges *as if* electrons were removed from its upper conductor, making the lower conductor of C_1 charged negatively and the upper conductor of C_2 charged positively. Because the connected

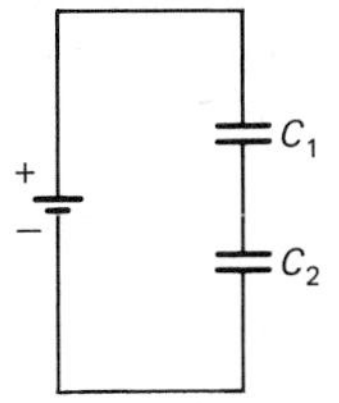

Figure 29.10

A series combination of two capacitors.

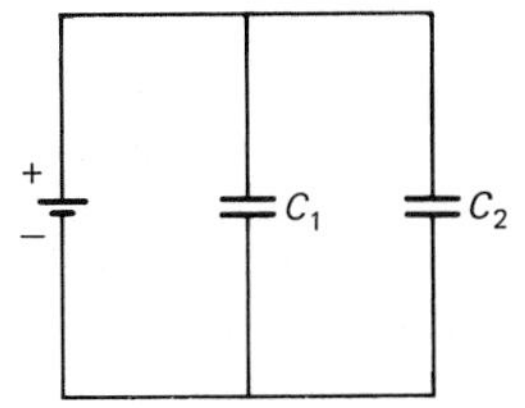

Figure 29.11

A parallel combination of two capacitors.

Figure 29.12

The charging of two capacitors in series by a battery. (a) The two capacitor plates outlined in dashed lines are connected electrically with a wire. However, there is no direct connection between these plates and the rest of the circuit. The *net* charge is always zero within the boundary shown. (b) Closing the switch (S) completes the electrical circuit, and electrons are extracted from the top plate of C_1 and transferred to the bottom plate of C_2. (c) Electric fields produced by the accumulation of charge on capacitors C_1 and C_2 cause a charge redistribution on the plates within the boundary shown. However, the *net* charge on the combined parts of the circuit within the boundary is always zero. (d) The potential differences, V_1 and V_2, sum to V.

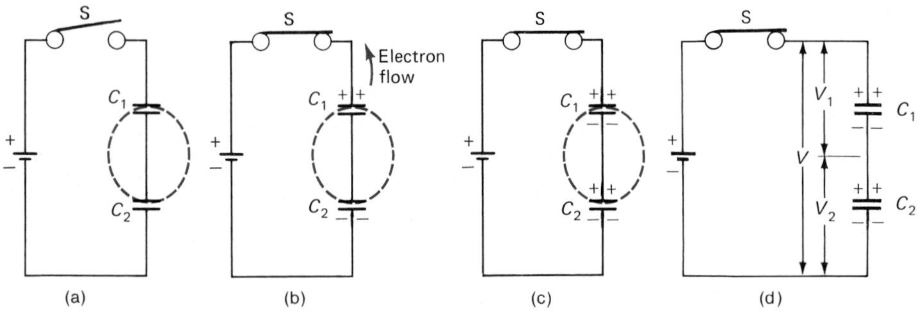

inner conductors of the capacitors must be electrically neutral overall, the magnitude of the charge on both inner conductors must be equal. Thus each plate of the two capacitors contains the same magnitude of charge. Relating potential difference to charge and capacitance, we have

$$V_1 = \frac{Q}{C_1} \quad \text{and} \quad V_2 = \frac{Q}{C_2}$$

where V_1 is the potential difference across capacitor 1, which has capacitance C_1, and V_2 is the potential difference across capacitor 2, which has capacitance C_2. The charge Q is the same for both capacitors because they are connected in series. Potential differences V_1 and V_2 sum to V, the potential difference across the battery (Figure 29.12d).

$$V = V_1 + V_2$$

Now suppose that we want to replace C_1 and C_2 with a single equivalent capacitor. We must ask, "What single capacitor connected to the same potential difference V will store the same magnitude of charge Q on each plate?" That is, $V = Q/C$. The capacitance C is thus equivalent to the series combination of C_1 and C_2. Because $V = V_1 + V_2$

$$\frac{Q}{C_1} + \frac{Q}{C_2} = \frac{Q}{C}$$

or

$$\frac{1}{C_1} + \frac{1}{C_2} = \frac{1}{C} \qquad (29.9)$$

from which

$$C = \frac{C_1C_2}{C_1 + C_2} \qquad (29.10)$$

Note that the equivalent capacitance is less than either C_1 or C_2. For an arbitrary number of series-connected capacitors the charge is the same on each capacitor and the equivalent capacitance is determined from

$$\frac{1}{C} = \frac{1}{C_1} + \frac{1}{C_2} + \frac{1}{C_3} + \cdots + \frac{1}{C_N}$$

or

$$C = \frac{1}{\frac{1}{C_1} + \frac{1}{C_2} + \frac{1}{C_3} + \cdots + \frac{1}{C_N}} \qquad (29.11)$$

The reciprocal of the equivalent capacitance equals the sum of the reciprocals of the series capacitances.

Example 3 **Capacitors in Series**

Two 10-μF series-connected capacitors are connected to a 10-V potential difference. Using symmetry arguments, let us determine the charge on each capacitor. Because both capacitors have the same capacitance, symmetry suggests that equal potential differences exist across each. Because the total potential difference across the combination is 10 V, the potential difference across each capacitor must be 5 V. Hence the charge on each capacitor is

$$Q = CV = 10 \times 10^{-6}\ \text{F} \cdot 5\ \text{V}$$
$$= 50\ \mu\text{C}$$

Alternatively, we can determine the equivalent capacitance from

$$\frac{1}{C} = \frac{1}{C_1} + \frac{1}{C_2}$$

$$\frac{1}{C} = \frac{1}{10} + \frac{1}{10} = \frac{2}{10}$$

$$C = 5\ \mu\text{F}$$

Thus the charge on the equivalent capacitance is

$$Q = CV = 5 \times 10^{-6}\ \text{F} \cdot 10\ \text{V}$$

$$= 50\ \mu\text{C}$$

The charge is the same on each capacitor in a series combination. Hence the charge on each capacitor is 50 μC.

When capacitors are connected in parallel, the potential difference for each capacitor is the same (Figure 29.11). The equivalent capacitance for N capacitors in parallel equals the sum of the capacitances in the combination.

$$C = C_1 + C_2 + C_3 + \cdots + C_N \qquad (29.12)$$

The determination of this equivalent capacitance is straightforward and is outlined in the problems for this section.

Example 4 Capacitors in Parallel

A combination of 10 identical capacitors connected to a 6-V potential difference causes a total of 120 μC of charge to be transferred. Let's determine the capacitance of each capacitor if the capacitors are all in parallel. The equivalent capacitance is

$$C = \frac{Q}{V} = \frac{120 \times 10^{-6}\ \text{C}}{6\ \text{V}} = 20\ \mu\text{F}$$

For a parallel combination, the equivalent capacitance is the sum of the individual capacitances. If C_i is the capacitance of a single capacitor, then

$$10C_i = 20\ \mu\text{F}$$

and each capacitor has a capacitance

$$C_i = 2\ \mu\text{F}$$

When a combination of capacitors is connected to a potential difference we often need to know the charge on each capacitor plate and the potential difference across each capacitor. The following example illustrates how our knowledge of series and parallel combinations of capacitors allows us to determine these quantities.

Example 5 Combinations of Series and Parallel Capacitor Connections

Three capacitors are connected to a 10-V potential difference as shown in Figure 29.13. We want to calculate the charge on each plate and the potential difference across each capacitor when fully charged. Capacitors C_2 and C_3 are in parallel and can be replaced by a single capacitor having capacitance $C_4 = C_2 + C_3 = 10\ \mu\text{F}$. The actual circuit is equivalent to

Figure 29.13

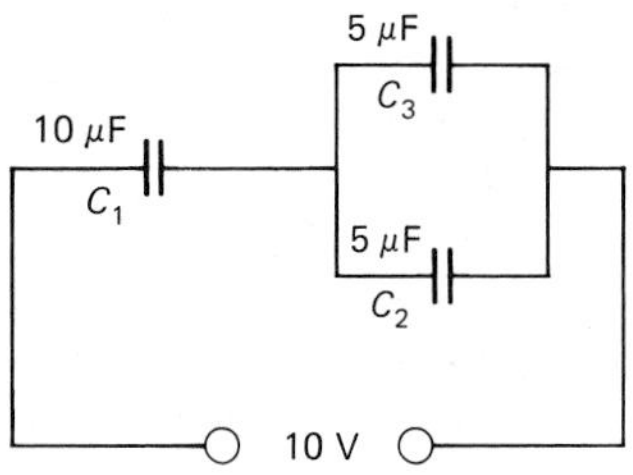

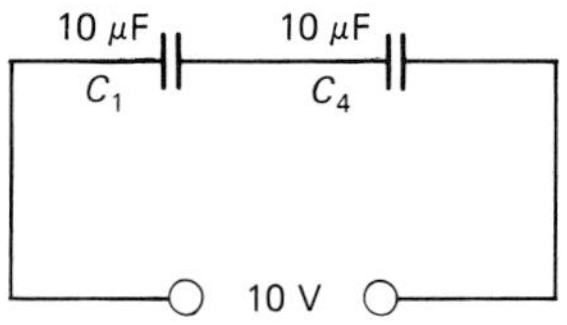

Capacitors C_1 and C_4 are in series and can be replaced with a single capacitor having capacitance

$$C_5 = \frac{C_1 C_4}{C_1 + C_4} = \frac{10 \times 10}{20} = 5\ \mu\text{F}$$

Thus the circuit is equivalent to

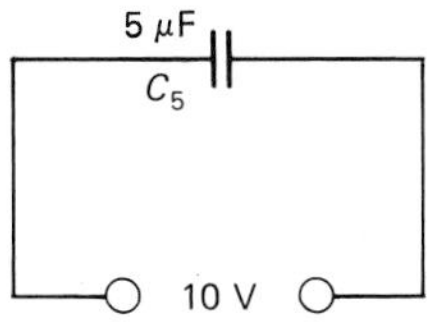

The charge on C_5 is $Q_5 = C_5 V_5 = 5\ \mu\text{F} \cdot 10\ \text{V} = 50\ \mu\text{C}$. Because the charge is the same on each capacitor in a series connection, the charges on C_1 and C_4 (which are equivalent to C_5) are also 50 μC. The potential differences across C_1 and C_4 are

$$V_1 = \frac{Q_1}{C_1} = \frac{50\ \mu\text{C}}{10\ \mu\text{F}} = 5\ \text{V}$$

$$V_4 = \frac{Q_4}{C_4} = \frac{50\ \mu\text{F}}{10\ \mu\text{F}} = 5\ \text{V}$$

Note that $V_1 + V_4 = 10$ V, as required.

Because the potential difference is the same across each capacitor in a parallel connection, the potential difference across C_2 and C_3 (which together are equivalent to C_4) is 5 V. Hence the charges on C_2 and C_3 are

$$Q_2 = C_2 V_2 = 5\ \mu\text{F} \cdot 5\ \text{V} = 25\ \mu\text{C}$$

$$Q_3 = C_3 V_3 = 5\ \mu\text{F} \cdot 5\ \text{V} = 25\ \mu\text{C}$$

Note that $Q_2 + Q_3 = 50\ \mu\text{C}$, as required.

Questions

10. Why is the equivalent capacitance of a parallel connection of capacitors always *greater* than the capacitance of any single capacitor in the connection?

11. Why is the equivalent capacitance of a series connection of capacitors always *less* than the capacitance of any single capacitor in the connection?

29.3 Electrostatic Energy of a Charged Capacitor

Figure 29.14 symbolizes a circuit consisting of a battery, a switch S, and a capacitor C. The battery maintains a constant potential difference V across its terminals and provides an energy source for charging the capacitor when the switch S is closed. As we described in Section 29.1, the top plate of the capacitor acquires a positive charge as electrons are removed from it. In principle, the capacitor charges *as if* positive charges were pulled off the bottom plate and transferred directly to the upper plate (Figure 29.15). This concept of positive charge movement is very useful when discussing the behavior of capacitors in electrical circuits.

Before the switch in Figure 29.14 is closed, there is no potential difference across the capacitor because there is no charge on the capacitor plates. When the switch is closed, charge begins accumulating on the plates. For each increment of positive charge dq transferred from the lower plate to the upper plate the work required is

$$dW = V_C\, dQ$$

$$= \frac{q}{C}\, dq \qquad (29.13)$$

where q is the magnitude of the charge on each capacitor plate at the instant the charge dq is transferred, and $V_C = q/C$ is the potential difference between the capacitor plates.

Charge transfer continues until the potential difference across the capacitor equals the potential difference across the battery. When fully charged, each plate contains charge $Q = CV$. The total work required to increase the stored charge from zero to some amount Q is found by integrating.

$$W = \int_0^Q \frac{q}{C}\, dq = \frac{1}{2}\frac{Q^2}{C} \qquad (29.14)$$

This work is stored as electrostatic potential energy U.

$$U = \frac{1}{2}\frac{Q^2}{C} \qquad (29.15)$$

Because $Q = CV$ the electrostatic potential energy can also be written as

$$U = \frac{1}{2}CV^2 \qquad (29.16)$$

or

$$U = \frac{1}{2}QV \qquad (29.17)$$

We use whichever of these three equations is most convenient for solving the problem at hand. In the following example we consider some of the energy aspects of charging a capacitor.

Example 6 Energy Aspects of Charging a Capacitor

We have shown that when a charge Q is removed from one conductor of a capacitor and stored on the opposite conductor, work equal to $\frac{1}{2}QV$ is required. This work is performed by the energy source (a battery) connected to the capacitor. During the charging, is any energy expended other than this $\frac{1}{2}QV$?

To calculate the total work done by the battery we use the definition of potential difference. A total charge Q is moved through a *constant* potential difference V provided by the battery (Figure 29.15). Hence the total work done is

$$W = QV$$

This is twice the energy actually expended to charge the capacitor. But where did the other $\frac{1}{2}QV$ of energy go? In Chapter 30 we will see that work is also required to force the charges through the wires (conductors) connecting the capacitor plates to the potential difference. This work produces heat, and accounts for the remaining $\frac{1}{2}QV$ of the work done by the battery.

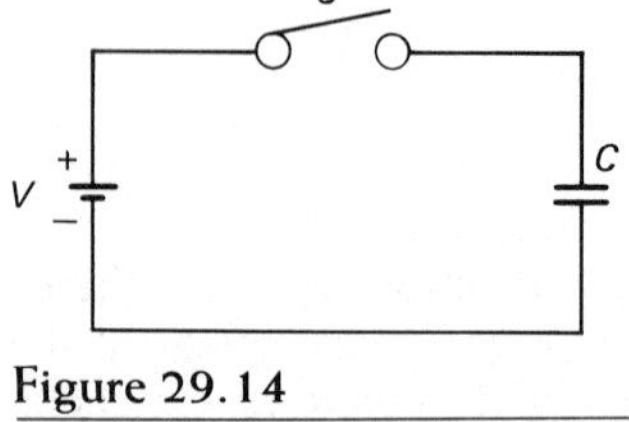
S
V
+
−
C

Figure 29.14

A scheme for charging a capacitor with a battery. No charge accumulates until the switch is closed. Then electric forces cause a redistribution of charge on the capacitor plates.

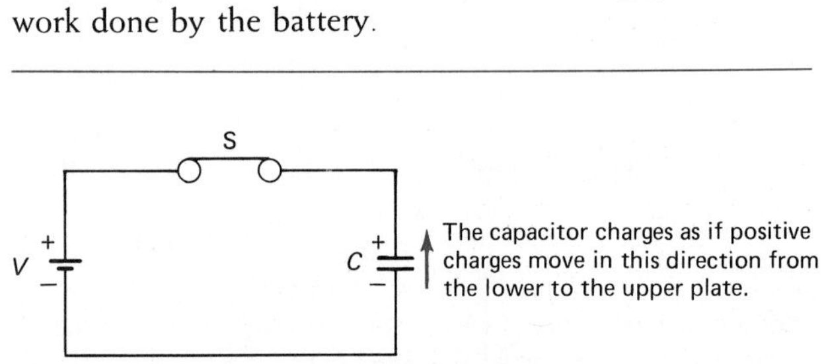
S
V
+
−
C
+
−
The capacitor charges as if positive charges move in this direction from the lower to the upper plate.

Figure 29.15

Closing the switch in the circuit shown in Figure 29.14 allows charges to accumulate on the plates of the capacitor. Positive charge accumulates on the plate connected to the positive terminal of the battery; negative charge accumulates on the plate connected to the negative terminal of the battery. The capacitor charges *as if* positive charges moved directly from the lower plate to the upper plate.

In Example 2 we compared the charge-storing abilities of a capacitor and a battery. Now let us compare the energy-storing capability of these two devices.

Example 7 **Comparison of a Capacitor with a Battery as an Energy-Storing Device**

Twelve-volt lead–acid batteries convert chemical energy into electric energy in the electrical systems of nearly all contemporary automobiles. Since capacitors also store electric energy, let us determine the feasibility of using capacitors rather than lead–acid batteries as electric energy sources in automobiles. The largest capacitance available in a 12-V commercial capacitor is about 10^{-1} F. Such a capacitor will store 7.2 J of energy. But a good 12-V lead–acid battery can deliver 4.3×10^6 J (see Example 2). This means that it would take 6×10^5 capacitors to provide the same amount of energy as one 12-V battery. Even if each capacitor cost only 10¢, the cost of 6×10^5 capacitors would be about \$60,000. Each of these capacitors would occupy about 20 cm^3 of space. The total volume of all 6×10^5 capacitors corresponds to a cube about 2 m on a side, which is roughly the size of a contemporary car. Thus, we conclude that capacitors cannot compete with lead–acid storage batteries as energy sources in contemporary automobiles.

Energy Density

A charge in an electric field experiences an electric force $\mathbf{F} = q\mathbf{E}$. If the charge is free to move, then work is done on it by the electric force. If we think of the electric field as the environment that provides the force, then we may also identify the electric field as the energy source for the work done by the field. We can see that this interpretation of electric field is meaningful when we consider the energy stored by a parallel-plate capacitor.

For any capacitor the total energy stored is

$$U = \frac{1}{2}CV^2$$

For a parallel-plate capacitor $C = A\epsilon_0/d$ and $V = Ed$. Hence

$$U = \frac{1}{2}\left(\frac{A\epsilon_0}{d}\right)(Ed)^2$$

$$= \frac{1}{2}(Ad)\epsilon_0 E^2 \qquad (29.18)$$

Note that Ad is the volume between the plates of the capacitor and that it coincides with the volume where the electric field exists. By dividing both sides of Eq. 29.18 by this volume, we arrive at quantity u. This quantity has units of joules per cubic meter, independent of the geometric aspects of the capacitor.

$$u = \frac{U}{Ad} = \frac{1}{2}(\epsilon_0 E^2) \qquad (29.19)$$

We interpret this relation as an energy density associated with the electric field between the capacitor plates. This interpretation suggests that there is an energy density proportional to the square of the electric field where an electric field exists. Although we have actually shown this to be true only for a parallel-plate capacitor, it is a general result and is true of all electric fields. Because the strength of an electric field can vary with position, the energy density can also vary with position. This happens, for example, between the conductors of spherical and cylindrical capacitors. To obtain the total energy in situations like these, we can integrate the energy density throughout the volume of interest, as we do in the following example.

Example 8 **Energy of a Spherical Capacitor**

We know that the energy stored in any capacitor is $U = \frac{1}{2}(Q^2/C)$ (Eq. 29.15). We also know that the capacitance of a spherical capacitor (Eq. 29.4) is

$$C = 4\pi\epsilon_0\left(\frac{r_1 r_2}{r_2 - r_1}\right)$$

and that the electric field between the conductors (Eq. 27.17) is

$$E = \frac{1}{4\pi\epsilon_0}\left(\frac{Q}{r^2}\right)$$

Let us take the energy density of a spherical capacitor to be $u = \frac{1}{2}\epsilon_0 E^2$, and show by direct integration that this leads to the known relation for the energy of a spherical capacitor.

We note that u depends only on the radial distance from the center of the capacitor. Therefore if we pick a spherical shell of thickness dr, volume $4\pi r^2\, dr$, the energy density is the same at any position within this shell (Figure 29.16). The energy within this shell is

$$dU = u\, 4\pi r^2\, dr = \frac{1}{2}\epsilon_0\left[\frac{1}{4\pi\epsilon_0}\left(\frac{Q}{r^2}\right)\right]^2 4\pi r^2\, dr$$

The total energy is then

$$U = \int dU = \frac{1}{2}\left(\frac{Q^2}{4\pi\epsilon_0}\right)\int_{r_1}^{r_2}\frac{dr}{r^2}$$

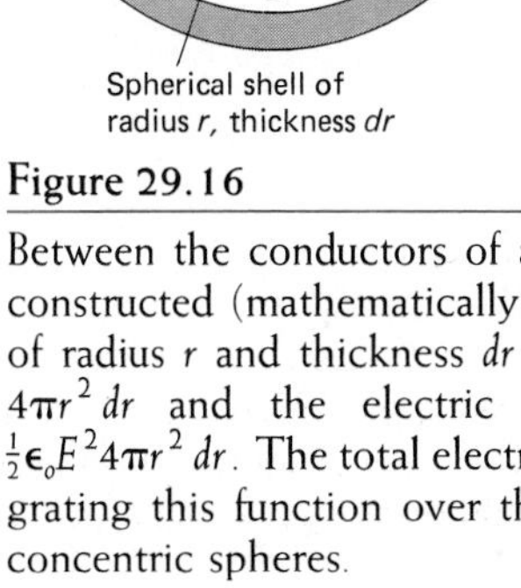

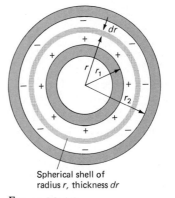

Figure 29.16

Between the conductors of a spherical capacitor there is constructed (mathematically) a concentric spherical shell of radius r and thickness dr. The volume of the shell is $4\pi r^2\, dr$ and the electric energy in this volume is $\frac{1}{2}\epsilon_0 E^2 4\pi r^2\, dr$. The total electric energy is obtained by integrating this function over the total volume between the concentric spheres.

Integrating, we obtain

$$= \frac{1}{2}\left(\frac{Q^2}{4\pi\epsilon_0}\right)\left(-\frac{1}{r}\right)\Bigg|_{r_1}^{r_2}$$

$$= \frac{1}{2}\left(\frac{Q^2}{4\pi\epsilon_0}\right)\left(\frac{1}{r_1} - \frac{1}{r_2}\right) = \frac{1}{2}\left(\frac{Q^2}{4\pi\epsilon_0}\right)\left(\frac{r_2 - r_1}{r_1 r_2}\right)$$

If we identify the capacitance as

$$C = 4\pi\epsilon_0\left(\frac{r_1 r_2}{r_2 - r_1}\right)$$

it follows that

$$U = \frac{1}{2}\frac{Q^2}{C}$$

Thus we have verified the energy density formula for the spherical capacitor. We can also verify the energy density formula for a cylindrical capacitor by using a similar calculation.

Questions

12. The electric energy for a certain nuclear fusion energy system is supplied by a group of capacitors. The system requires a large amount of electric charge. Why would these capacitors be connected in parallel rather than in series?
13. The addition of each bowling ball lifted from the floor to a tabletop increases the gravitational potential energy of the system of bowling balls. How is this process similar to and different from transferring charges from one capacitor plate to another?
14. How might the electrostatic potential energy of a capacitor be converted to mechanical energy? thermal energy? chemical energy? light energy?

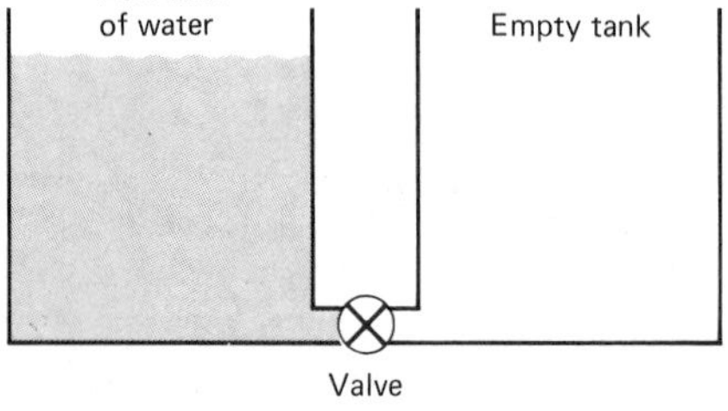

Figure 29.17a

Water flows from the full tank to the empty tank when the valve is opened.

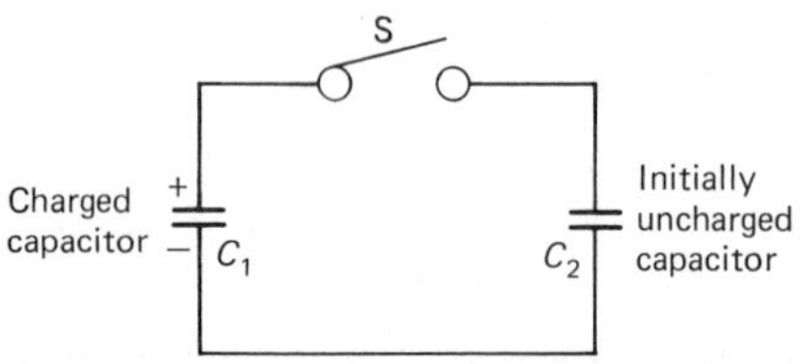

Figure 29.17b

Charge flows from C_1 to C_2 when switch S is closed.

15. Water in a tank is allowed to drain into an adjacent tank (Figure 29.17a). What determines when the flow stops? What effect does the size of the connecting pipe have on the final levels of the water in the two tanks? How is this situation like the "draining" of charge from a capacitor to an initially uncharged capacitor connected to it (see Figure 29.17b)?
16. How would the variation of electrostatic energy density along a radius in a cylindrical or spherical capacitor differ from that in a parallel-plate capacitor?
17. Speculate on the disposition of the energy of a charged capacitor when a wire is connected across the plates of the capacitor.

29.4 Effect of an Insulator on Capacitance

Earlier, when we discussed capacitors of varying geometric shapes, we noted that capacitance depends on the geometry, area, conductor separation, and the insulator between the conductors. We will now discuss in more detail the role of the insulator.

Figure 29.18 shows an apparatus for investigating capacitance changes when an insulator is inserted between the plates of a parallel-plate capacitor. The effect of the insulator is represented by the dielectric constant, κ (Greek kappa), which is defined by the following experiment.

An electric energy source such as a battery is connected to a parallel-plate capacitor in a vacuum. A potential difference, V_0, equal to the potential difference across

Figure 29.18

A type of parallel-plate capacitor used to investigate the effect of an insulator on capacitance.

Figure 29.19

(a) A parallel-plate capacitor with vacuum for an insulator is charged by being connected to a battery. When fully charged, the potential difference is the same (V_o) across the battery terminals and capacitor plates. (b) Because there is no way for charge to be removed from the capacitor, the potential difference across the capacitor remains at V_o when the capacitor is isolated from the battery. (c) When an insulator such as glass is inserted between the plates of the isolated capacitor, the potential difference across the capacitor decreases to a value $V < V_o$.

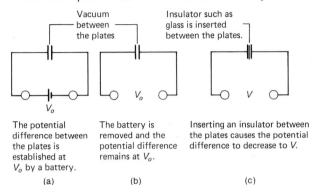

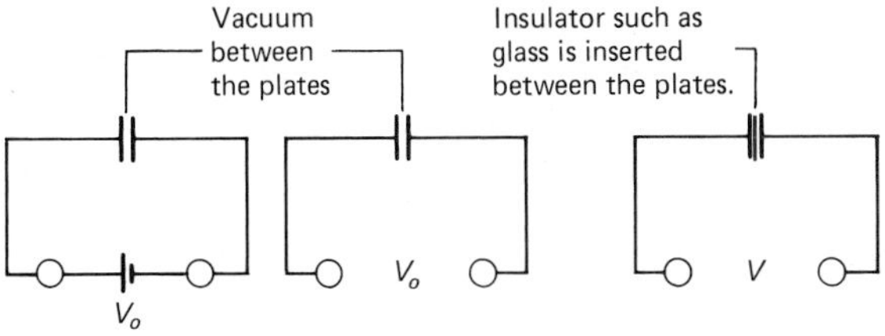

The potential difference between the plates is established at V_o by a battery.
(a)

The battery is removed and the potential difference remains at V_o.
(b)

Inserting an insulator between the plates causes the potential difference to decrease to V.
(c)

Table 29.1

Representative dielectric constants of some common materials*

Material	κ
Water	78.3
Air	1.000590
Lucite	2.84
Plexiglas	3.12
Polystyrene	2.55
Neoprene	6.60
Polyethylene	2.26
Pyrex	4–6†
Teflon	2.1
Titanates (Ba, Sr, Ca, Mg, and Pb)	15–12,000

*For a more complete listing see *Handbook of Chemistry and Physics*, CRC Press, 2000 N.W. 24th Street, Boca Raton, FL 33431.

†Depending on type.

the energy source develops across the capacitor when it is fully charged (Figure 29.19a). The energy source is then removed and the capacitor is isolated so that charge flow between the plates can no longer occur (Figure 29.19b). If an insulator such as glass is inserted between the plates of the isolated capacitor, the potential difference across the plates decreases (Figure 29.19c). We call the new value of the potential difference V, and define the dielectric constant κ as*

$$\kappa = \frac{V_o}{V} \tag{29.20}$$

Although this ratio is affected to some extent by external influences such as temperature, systematic experiments reveal that it does *not* depend on plate size, plate separation, or initial potential difference. The ratio depends almost entirely on the type of insulator used. In fact, the ratio is a property of the insulator. The dielectric constant is unity for a vacuum and is a number always greater than 1 for any material medium. Some representative dielectric constants are listed in Table 29.1.

Because potential difference and charge are related by $V = Q/C$, we might suspect that the reduction in V is due to a loss of charge. However, since the plates are not connected electrically (remember, the battery has been removed), there is no way for charge to leave the plates except by jumping from one plate to the other. Indeed, when the insulator is removed from the isolated capacitor the potential difference returns to the initial value, indicating that the presence of the insulator does *not* affect the charge stored on either plate. We conclude that the reduction in potential difference is due to an *increase in capacitance*. With Q_o, V_o, and C_o the charge, potential difference, and capacitance when there is a vacuum between the plates, and Q, V, and C the charge, potential difference, and capacitance when there is an arbitrary insulator between the plates, it follows from $\kappa = V_o/V$ that C is related to C_o by

$$C = \frac{Q_o}{V} = \frac{Q_o}{V_o/\kappa} = \kappa\left(\frac{Q_o}{V_o}\right) = \kappa C_o \tag{29.21}$$

The capacitance has increased by a factor of κ.

Now suppose that we insert an insulator between the capacitor plates while the battery is still connected. If the battery maintains a constant potential difference V_o across the capacitor, then according to Eq. 29.1 the charge stored on the capacitor is

$$Q = CV_o$$

where C is the capacitance when the insulator is present. Because $C = \kappa C_o$ and $Q_o = C_oV_o$, Eq. 29.1 becomes

$$Q = \kappa C_o V_o$$
$$= \kappa Q_o \tag{29.22}$$

Measurements confirm that the charge stored with the insulator present has increased by a factor of κ.

Equation 29.21 is a general equation. It shows that an insulator alters the capacitance of any capacitor. We can rewrite Eq. 29.21 specifically for a parallel-plate capacitor as follows.

$$C = \kappa C_o$$
$$= \kappa\epsilon_o\left(\frac{A}{d}\right) \tag{29.23}$$

Because κ is dimensionless, $\kappa\epsilon_o$ has the same dimensions as ϵ_o. This quantity, denoted by ϵ, is called the **permittivity**.

$$\epsilon = \kappa\epsilon_o \tag{29.24}$$

*In Section 29.5 we will see why the term **dielectric** is used.

Table 29.2

Dielectric strengths of some common insulators*

	Dielectric strength (kV/mm)
Air	3
Lucite	20
Plexiglas	20
Polystyrene	20
Neoprene	12
Polyethylene	18
Pyrex	14
Teflon	19
Titanates (Ba, Sr, Ca, Mg, and Pb)	2–12

*The dielectric strengths of plastics will vary with thickness. For this reason, there is some variation in the values quoted in the literature. Those listed here are representative. For a more complete listing see *Handbook of Chemistry and Physics*, CRC Press, 2000 N.W. 24th Street, Boca Raton, FL 33431.

Permittivity is a property of the insulator. Because $\epsilon = \epsilon_0$ for a vacuum, ϵ_0 is appropriately called the **permittivity of free space.** In terms of permittivity, the capacitance of a parallel-plate capacitor becomes

$$C = \epsilon\left(\frac{A}{d}\right) \tag{29.25}$$

Similarly, the energy density in an electric field in a material medium becomes

$$u = \frac{1}{2}(\epsilon E^2) \tag{29.26}$$

Dielectric Strength

Capacitors for electronic applications vary in capacitance from about 10^{-3} μF to about 10^5 μF. Manufacturers achieve this wide range of values by taking advantage of the dependence of capacitance on geometry and dielectric constant. When a chosen capacitor is used in a circuit, it is charged by an energy source that provides a potential difference across its terminals. This external energy source establishes an electric field between the capacitor plates. For any insulator there is a maximum electric field that can be maintained without ionizing atoms in the insulator and causing it to be conductive. This maximum electric field, called the **dielectric strength,** depends on the physical structure of the insulator. Representative values of the dielectric strength of some common insulators are shown in Table 29.2. Once the plate separation of a capacitor has been determined, there is a maximum potential difference that can be applied across its terminals to avoid breakdown of the insulator. (This breakdown potential difference is quoted by commercial manufacturers and must be observed in circuit applications of capacitors.) For a parallel-plate capacitor the breakdown potential difference ($V_{\text{breakdown}}$) and the dielectric strength ($E_{\text{breakdown}}$) are related by

$$E_{\text{breakdown}} = \frac{V_{\text{breakdown}}}{d} \tag{29.27}$$

where d is the plate separation.

Example 9 **Breakdown Potential Difference of a Representative Capacitor**

A certain parallel-plate capacitor has a plate separation of 0.01 mm and uses Teflon as an insulator. What is the maximum potential difference that can be applied to the terminals of the capacitor? From Table 29.2 the dielectric strength of Teflon is 19 kV/mm. Thus the maximum applied potential difference is

$$\begin{aligned} V_{\text{breakdown}} &= E_{\text{breakdown}} \cdot d \\ &= 19\,\frac{\text{kV}}{\text{mm}} \cdot 0.01\text{ mm} \\ &= 0.19\text{ kV} \\ &= 190\text{ V} \end{aligned}$$

If a potential difference larger than 190 V is applied, this capacitor will probably be destroyed.

Electrolytic Capacitors

Decreasing the distance between the conductors of a capacitor is an effective way of increasing the capacitance. Therefore a thin insulator is desirable in a capacitor, although the insulator must never be so thin that the conductors are allowed to touch. An **electrolytic capacitor** is specially constructed with an extremely thin insulator between two conducting media.

One of the two conductors in an electrolytic capacitor is a metal foil, usually made from tantalum or aluminum. On the surface of this conductor is a very thin nonconducting oxide of the metal. This metal oxide serves as the insulator. The other conductor is a conducting paste or a liquid that makes intimate contact with the metal oxide. Because the metal oxide layer is very thin, capacitances as large as 500,000 μF can be attained.

The physical size of a capacitor depends on both the capacitance and the breakdown potential difference. An electrolytic capacitor having a capacitance of 50,000 μF and a breakdown potential difference of 15 V is about 5 cm in diameter and 15 cm long. If this capacitor were to have a breakdown potential difference of 40 V, the diameter would have to increase to about 8 cm. A 5-μF electrolytic capacitor having a breakdown potential difference of 15 V would have a diameter of about 0.5 cm and length of about 1.3 cm. While electrolytic capacitors have many modern electronic applications, their utility is severely limited by the fact that the polarity of the metal conductor must always be positive. Otherwise chemical (electrolytic) reactions occur that break down the oxide layers.

18. A certain capacitor has a breakdown potential of 10 V. How would you construct from 20 such capacitors a capacitor that had a breakdown potential of 200 V? What is the equivalent capacitance of your grouping?

19. List three possible ways of changing the capacitance of a parallel-plate capacitor.

20. Apart from frictional effects, why would you have to do work to pull an insulator from between the plates of an isolated charged parallel-plate capacitor?

21. Why does the energy of an isolated charged capacitor decrease when a material of larger dielectric constant is inserted between the conductors?

29.5
Atomic Viewpoint of the Effect of an Insulator on Capacitance

So far, we have established that the insertion of an insulator between the conductors of an isolated charged capacitor diminishes the potential difference between the conductors (Section 29.4). Let's now try to interpret this effect in terms of the atoms and molecules within the insulator. In doing so, we will show why the value of the dielectric constant is always greater than 1.

A sufficiently strong electric field applied to an insulator may free electrons from atoms and create an electric current, but ordinarily the electric fields used are only capable of displacing an electron a fraction of an atomic distance from its host atom. Such a small separation of positive and negative charges produces an electric dipole at the atomic level. In some molecules—for example, H_2O and N_2O—internal electric fields cause a charge separation that gives rise to a permanent molecular dipole moment. Such molecules are termed **polar**. Other molecules, like H_2, N_2, and O_2, do not possess a permanent dipole moment, but when they are placed in an external electric field an ensuing charge separation gives rise to an induced electric dipole moment. Such molecules are termed **nonpolar**. Insulators containing either polar or nonpolar molecules are called **dielectrics**. When an electric dipole is placed in a uniform electric field it experiences a torque that aligns the axis of the dipole along the electric field lines (Section 27.7). The positive component of the dipole tends to move in the direction of $\mathbf{E}$, and the negative component tends to move opposite to $\mathbf{E}$ (Figure 29.20). The electric field lines of the dipole tend to point opposite to the applied electric field. Thus the net electric field, which is the superposition of $\mathbf{E}$ and $\mathbf{E}_{dipole}$, is diminished in the

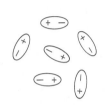

Figure 29.20

Alignment of nonpolar and polar molecules by an electric field.

In the absence of an applied electric field, the charge distribution is spherically symmetric around nonpolar molecules. The molecule exhibits no net electric dipole moment.

In the presence of an applied electric field, the positive and negative charges tend to separate, producing molecular dipoles. The dipoles then align along the electric field lines.

Polar molecules have a net dipole moment in the absence of an applied electric field. Because they are randomly oriented, there is no net electric dipole field in a macroscopic size sample.

In the presence of an applied electric field, the permanent molecular dipoles experience a torque and orient along the electric field lines.

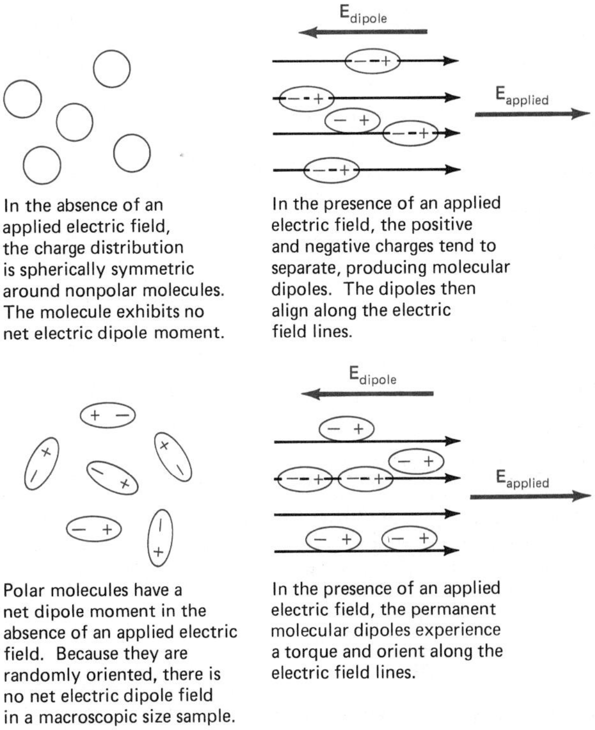

Figure 29.21

Alignment of molecular dipoles in an insulator between the plates of a capacitor.

Capacitor plates with equal but opposite charge

Within the insulator molecular dipoles are aligned by the electric field produced by charges on the plates of the capacitor.

vicinity of the dipole. When an insulator is inserted in the uniform electric field between the plates of a parallel-plate capacitor, its molecular constituents, whether polar or nonpolar, tend to align along the electric field lines (Figure 29.21). As a result of the dipole alignment, the net electric field in the insulator is diminished (Figure 29.22). It is this dipole alignment that gives rise to the observed reduction in potential difference between the plates of an isolated capacitor when an insulator is inserted between the plates. Since the potential difference of the charged isolated capacitor (V_0) is always greater than the potential difference (V) produced when an insulator is inserted between the plates of the same isolated capacitor, the dielectric constant ($\kappa = V_0/V$) will always be a number greater than 1. This conforms with the experimental observation.

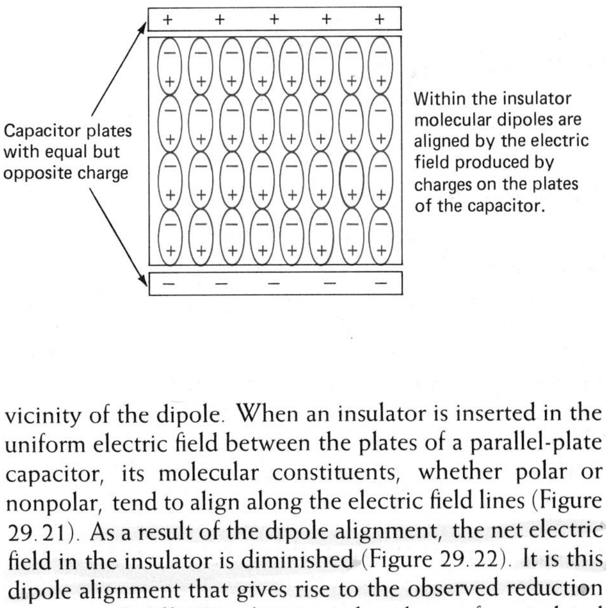

Figure 29.22

Depiction of the electric fields in an insulator between the plates of a capacitor.

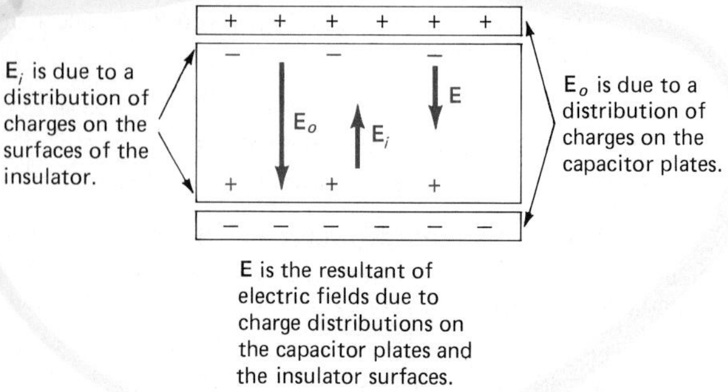

E_i is due to a distribution of charges on the surfaces of the insulator.
E_o
E_i
E
E_o is due to a distribution of charges on the capacitor plates.
E is the resultant of electric fields due to charge distributions on the capacitor plates and the insulator surfaces.

Questions

22. A tiny bit of paper having no net charge can be attracted to a charged comb. Explain the physical mechanism for this.
23. A tiny bit of paper attracted to a charged comb will often be repelled by the comb after touching it. Explain.
24. A stream of water from a tap will be deflected when a charged rod is brought close to the stream. Remembering that water molecules are polar, explain why the stream experiences an electric force, and indicate the direction in which the water moves relative to the rod.
25. Knowing the charge of an electron and the approximate size of an atom, why would you expect the permanent dipole moment of a polar molecule to be of the order of 10^{-29} C·m?

Summary

Any configuration of conductors separated by an insulating material is capable of storing charge. This ability to store charge is measured as capacitance and has units of coulombs per volt, called farads and symbolized by F.

A capacitor consists of two conductors separated by an insulating material. Normally, a capacitor contains a charge $+Q$ on one conductor and a charge $-Q$ on the other conductor. The capacitance of a capacitor is defined as

$$C = \frac{Q}{V} \qquad (29.1)$$

where Q is the magnitude of the charge on either conductor and V is the magnitude of the potential difference between the two conductors.

The capacitance of a parallel-plate capacitor having a vacuum between the plates is

$$C = \frac{A\epsilon_0}{d} \qquad (29.6)$$

where A is the area of either plate, d is the separation of the plates, and ϵ_0 is the permittivity of free space.

The equivalent capacitance of N capacitors in series is

$$\frac{1}{C} = \frac{1}{C_1} + \frac{1}{C_2} + \frac{1}{C_3} + \cdots + \frac{1}{C_N} \qquad (29.11)$$

The equivalent capacitance of N capacitors in parallel is

$$C = C_1 + C_2 + C_3 + \cdots + C_N \qquad (29.12)$$

The potential energy of charge stored on a capacitor is equal to the work done in charging the capacitor and is given by

$$U = \frac{1}{2}\frac{Q^2}{C} \qquad (29.15)$$

or

$$U = \frac{1}{2}CV^2 \qquad (29.16)$$

or

$$U = \frac{1}{2}QV \qquad (29.17)$$

Here Q is the charge on either plate, C is the capacitance, and V is the potential difference across the capacitor.

Alternatively, we can view the energy as being stored in the electric field established between the plates. The energy density at a given position is related to the electric field at that position by

$$u = \frac{1}{2}\epsilon E^2 \qquad (29.19)$$

The insertion of an insulator between the plates of an isolated capacitor reduces the electric field and potential difference between the plates and thereby increases the capacitance. The ratio of the potential difference without the insulator to the potential difference with the insulator is a property of the insulator. This property is called the dielectric constant and is defined as

$$\kappa = \frac{V_0}{V} \qquad (29.20)$$

The reduction of the electric field is due to an electrical alignment of electric dipoles by the electric field produced by charges on the capacitor plates.

Problems

Section 29.1 Capacitance and Capacitors

1. Two separated metallic coffee cans (Figure 1) are connected to the terminals of a 10-V battery. A sensitive electrometer indicates that each has accumulated a charge of 10^{-12} C. What is the capacitance of the configuration?

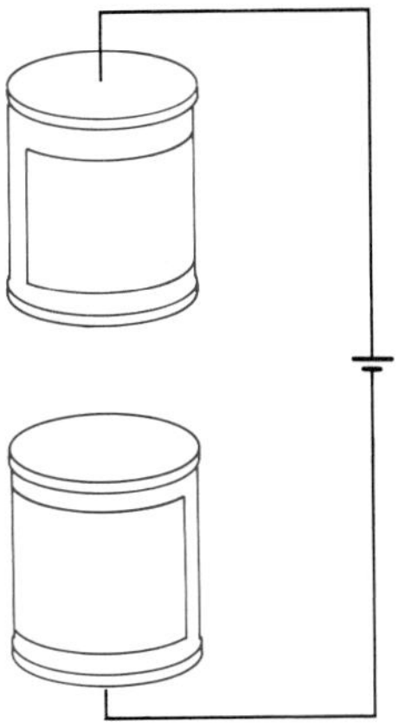

Figure 1

2. Two coins carry equal but opposite charges of 6.0×10^{-8} C. The capacitance of the combination is 300 pF. Determine the potential difference between the coins.

3. A capacitor with a charge of 10^{-4} C has a potential difference of 100 V. What charge is required to produce a potential difference of (a) 1 V; (b) 5 V; (c) 25 V?

4. Aluminum has density 2.7 gm/cm^3, atomic number 13, and atomic weight 27. Compare the total amount of charge of the electrons in a cubic centimeter of aluminum with the charge accumulated on either plate of a 1-F capacitor charged to 1000 V.

5. Calculate the capacitance of an isolated sphere having a radius equal to that of the earth.

6. (a) Calculate the capacitances of two parallel-plate capacitors, each having a plate area of 1 m^2 but plate separations of 0.1 mm and 0.01 mm. (b) Determine the percentage change in each of these capacitances if each separation is decreased 0.005 mm.

7. Light flash units for cameras often discharge a capacitor across the lamp to produce the light flash. (a) How much charge can be stored on a 300-μF photoflash capacitor when the potential difference across it is 300 V? (b) What is the maximum potential difference attainable across the capacitor when connected to a power supply providing 300 V? (c) Why would one choose to power the photoflash lamp with a capacitor rather than just connecting the lamp across the power supply that charges the capacitor?

8. A certain parallel-plate capacitor has a plate area of 1 m^2, plate separation of 0.1 mm, and a vacuum separating the plates. Determine the surface charge density on each plate and the electric field between the plates when the potential difference across the capacitor is 10 V.

9. Calculate the area of each plate of a parallel-plate capacitor having a capacitance of 1 F, a plate separation of 1 mm, and a vacuum separating the plates.

10. The electric field outside an isolated spherical conductor of radius r is given by $E = (1/4\pi\epsilon_0)(Q/r^2)$ where Q is the total charge on the sphere. Following the Gauss law treatment in Section 29.1 for determining the capacitance of two concentric cylinders, show that the potential difference between the conductors of two concentric spheres of a spherical capactor is

$$V = \frac{Q}{4\pi\epsilon_0}\left(\frac{r_2 - r_1}{r_1 r_2}\right)$$

and that the capacitance is

$$C = 4\pi\epsilon_0\left(\frac{r_1 r_2}{r_2 - r_1}\right)$$

11. The earth and the ionosphere may be considered to be a spherical capacitor with a separation $r_2 - r_1$ of about 100 km. Calculate the capacitance of this earth–ionosphere system.

12. The capacitance of two concentric spheres was shown to be

$$C = 4\pi\epsilon_0\left(\frac{r_1 r_2}{d}\right)$$

where $d = r_2 - r_1$ is the separation of the spheres. If both radii become very large, the surface becomes very "flat" over a local region and the surfaces become parallel plates. For this situation the expressions for the capacitance of two concentric spheres and a parallel-plate capacitor should be the same. Show that this is indeed the case.

13. The capacitance of two spheres having radii a and b whose surfaces are a distance c apart (Figure 2) is

$$C = \frac{4\pi\epsilon_0}{\frac{1}{a} + \frac{1}{b} - \frac{2}{c}}$$

if

$c \gg b$ or $c \gg a$

(a) Is this expression correct dimensionally? (b) Why does letting $c \to \infty$ correspond to a situation in-

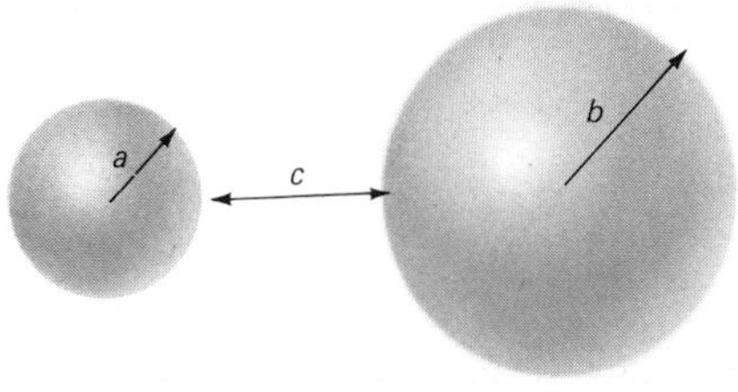

Figure 2

volving two isolated spheres? (c) For c very large, show that the capacitance of the system is equivalent to that of two isolated spherical capacitors in series.

14. The capacitance of two parallel wires of length l and radius a whose axes are separated by a distance d such that $d \gg a$ is

$$C = \frac{\pi \epsilon_o l}{\ln(d/a)}$$

If a charge per meter of $+\lambda$ C/m is placed on one wire and a charge per meter of $-\lambda$ C/m is placed on the other wire, show that the potential difference between the two wires is

$$V = \frac{\lambda \ln(d/a)}{\pi \epsilon_o}$$

15. Some variable capacitors have nonparallel plane plates, as shown in Figure 3. The capacitance is varied by changing the tilt angle of the upper plane with a screw adjustment. To calculate the capacitance, we can divide the plates into strips of width dx and length l. This divides the capacitor into a large number of capacitors in parallel, each having a capacitance $dC = \epsilon_o l(dx/y)$. Since the capacitance of a parallel combination is the sum of the individual capacitances, the total capacitance is

$$C = \epsilon_o l \int \frac{dx}{y}$$

Show that the capacitance is

$$C = \frac{A\epsilon_o}{\Delta} \ln\left(1 + \frac{\Delta}{d}\right)$$

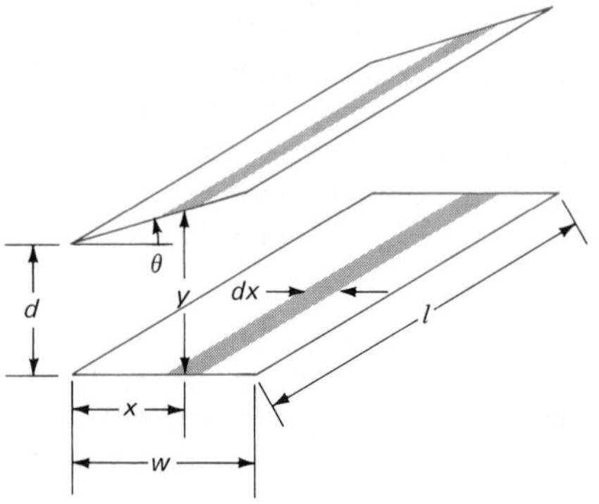

Figure 3

where $\Delta = w \tan \theta$ and A is the area of the bottom plate. Note that if $\Delta = 0$, this expression should reduce to the parallel-plate expression $C = A\epsilon_o/d$. Show that this is indeed the case.

Section 29.2 Capacitors in Series and in Parallel

16. Figure 4 shows a parallel connection of an arbitrary number (N) of capacitors. (a) Why is the potential difference the same across any capacitor in the circuit? (b) Calling V the potential difference across any capacitor, the charge on any capacitor is $Q = CV$. What is the sum total of all the charges stored in the parallel combination? (c) Show that a single capacitor having capacitance $C = C_1 + C_2 + C_3 + \cdots + C_N$ would store the same amount of charge if connected to the same potential difference V.

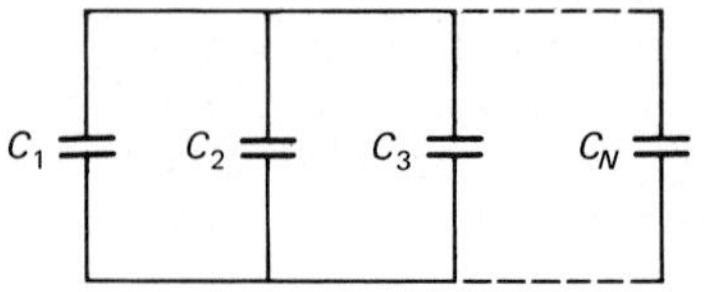

Figure 4

17. Two capacitors in series have an effective capacitance of

$$C = \frac{C_1 C_2}{C_1 + C_2}$$

Three capacitances in series have an effective capacitance of

$$C = \frac{C_1 C_2 C_3}{C_1 C_2 + C_1 C_3 + C_2 C_3}$$

Using mathematical induction, write the equivalent capacitance of four capacitors in series. Check your result by using the general relationship (Eq. 29.11) for the equivalent capacitance for resistors in series.

18. Determine the equivalent capacitance for each of the capacitor configurations in Figure 5.

19. Points A and B are joined by an array of five capacitors, as shown in Figure 6. The numbers give the capacitances in microfarads. Find the equivalent capacitance between A and B.

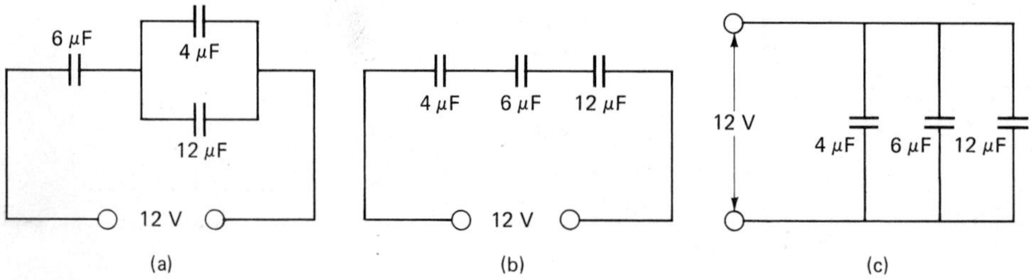

Figure 5

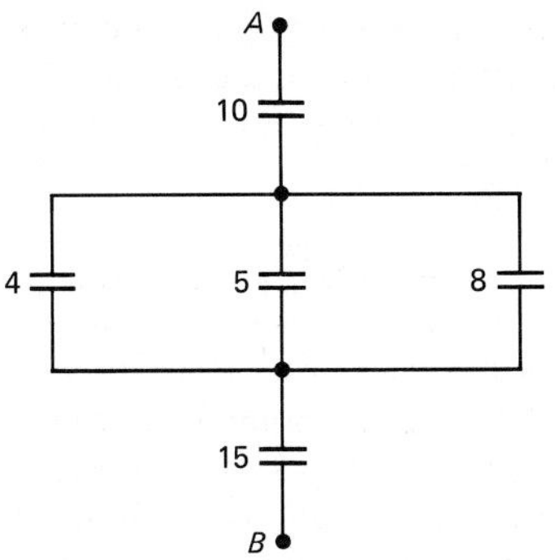

Figure 6

20. Determine the charge stored on the 6-μF capacitor in each of the configurations in Figure 5.

21. A 5-μF capacitor is charged by connecting it to a 10-V battery. The capacitor is removed and connected in parallel with an unknown, uncharged capacitor. At equilibrium, a potential difference of 1 V is established across each capacitor. Determine the unknown capacitance.

22. Determine the capacitances for the combinations shown in Figure 7.

23. How many 0.25-pF capacitors must be connected in parallel in order to store 1.2 μC of charge when connected to a battery providing a potential difference of 10 V?

24. A 4.0 μF capacitor and a 6.0 μF capacitor are charged in series by a 200-V source. The charged capacitors are disconnected (without discharging them) and reconnected in parallel. (a) What is the charge on each capacitor while they are in series? (b) What is the final charge on each capacitor after they are connected in parallel? (c) What is the potential difference across the capacitors when they are in parallel?

25. A group of identical capacitors is connected first in series and then in parallel. The combined capacitance in parallel is measured to be 100 times larger than that for a series connection. How many capacitors are in the group?

26. If you have a 4.0-μF capacitor and a 6.5-μF capacitor, each of which has stored charge of 10.0 μC, what will the charge stored on each capacitor be when they are connected electrically (Figure 8)?

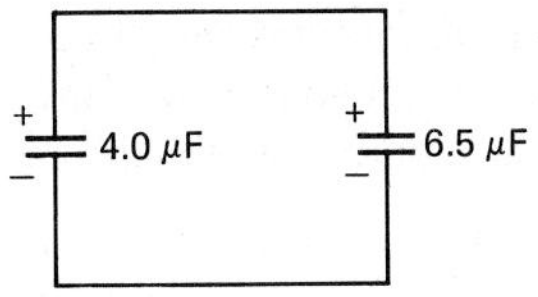

Figure 8

27. Suppose that you have six 12-pF capacitors. Determine the minimum and the maximum equivalent capacitance of connections involving all six capacitors.

28. Two variable area parallel-plate capacitors having the same plate separation have a total plate area A. Show that when connected in series the equivalent capacitance is a maximum when the area of each plate is $A/2$.

29. (a) Using a symmetry argument, determine the potential difference across each capacitor in Figure 9. (b) Determine the total amount of charge redistributed by the battery. (c) Using the result of part (b), calculate an equivalent capacitance for the circuit. Check your result by calculating the equivalent capacitance with the formulas for the equivalent capacitance of series and parallel combinations.

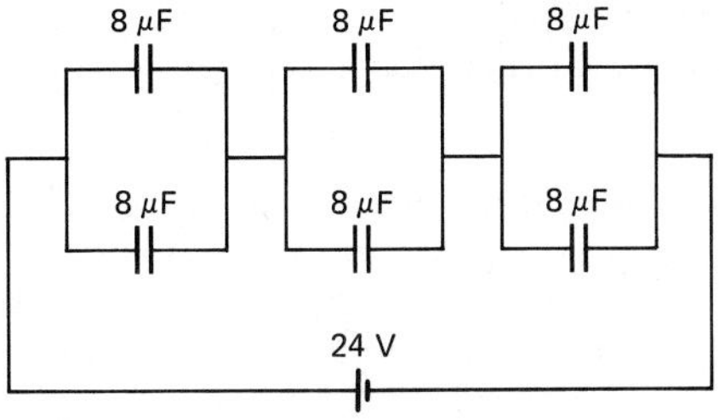

Figure 9

30. During the charging of the capacitors in Figure 10 the battery redistributes a total of 3.1 μC of charge. Determine the equivalent capacitance of the configuration. If you knew the capacitance of each

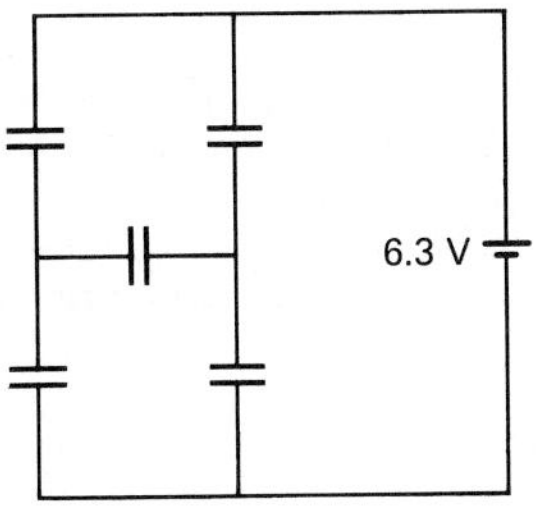

Figure 10

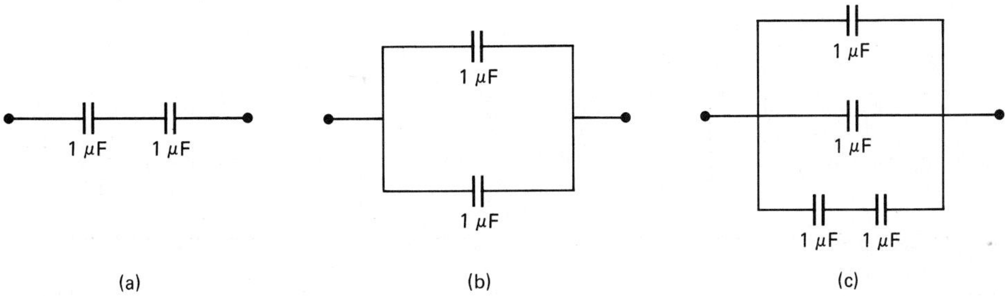

Figure 7

capacitor, why couldn't you determine the equivalent capacitance by using the formulas for the equivalent capacitance of series and parallel combinations of capacitors?

Section 29.3 Electrostatic Energy of a Charged Capacitor

31. Determine the electrostatic energy of the 4-μF capacitor in Figure 5.

32. The two capacitors shown in Figure 11 are uncharged when the switch is closed. Determine the ratio of the energies (U_2/U_1) stored in the two capacitors after the switch S is closed and the capacitors have become fully charged.

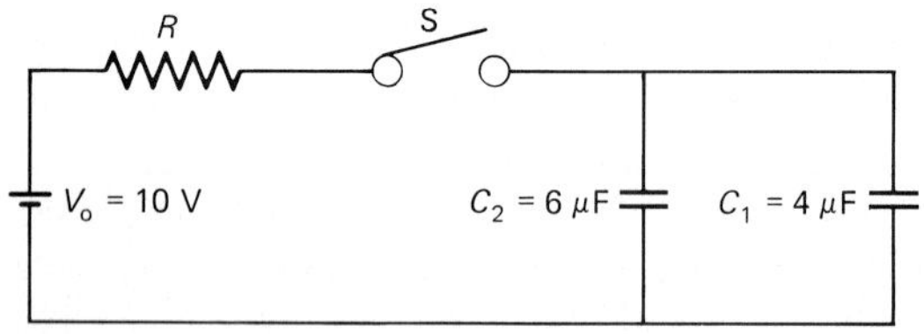

Figure 11

33. The energy stored in a particular capacitor is increased fourfold. What is the accompanying change in the (a) charge of and (b) potential difference across the capacitor?

34. The potential difference across a charged 5-μF capacitor is 10 V. If the plates of this capacitor are connected electrically to the plates of an uncharged 5-μF capacitor, determine the total electrostatic energy of the combination and comment on any energy difference between the initial and final systems.

35. The electrostatic energy of a charged 12-μF capacitor is 1.3×10^{-4} J. Determine the charge on the capacitor plates and the potential difference across the capacitor.

36. A parallel-plate capacitor with air as its dielectric is charged and then separated from the charging source. If the plate separation is doubled, determine the change in the electrostatic potential energy of the capacitor and account for its increase.

37. (a) Show that the potential energy of a charged parallel-plate capacitor can be written

$$U = \frac{1}{2}\left(\frac{Q^2}{\epsilon_0 A}\right)x$$

where x is the plate separation. (b) Show that the pressure (force per unit area) on a plate is

$$P = -\frac{1}{2}\sigma E = -\frac{1}{2}\epsilon_0 E^2$$

38. A battery having a 12-V potential difference across its terminals causes 8 μC of charge to be transferred between the plates of a capacitor connected to its terminals. (a) How much work is done by the battery? (b) How much electrostatic energy is stored by the capacitor? (c) Why is there a difference between the work done by the battery and the energy stored by the capacitor?

39. A 16-μF capacitor is connected to a battery. When fully charged, the potential difference across the capacitor is 24 V. How much additional work is required to charge the capacitor to 72 V?

40. When considering the energy supply for an automobile, the energy per unit mass of the supply is an important point. Using the data at the bottom of the page, compare the energy per unit mass (J/kg) for gasoline, lead–acid batteries, and capacitors.

41. Assume that the energy density for a charged capacitor can be written $u = \frac{1}{2}(\epsilon_0 E^2)$. For a cylindrical capacitor the electric field between the cylinders is $E = Q/2\pi\epsilon_0 Lr$ where L is the length of the cylinder, Q is the total charge on either cylinder, and r is the radial distance from the axis of the cylinders. Show by direct integration that the total energy is $U = \frac{1}{2}(Q^2/C)$, thus supporting the concept that the energy may be considered to be stored in the electric field.

42. If the earth–ionosphere capacitor of Problem 11 is charged to 250,000 V (by thunderstorms), what is the electric energy density just above the surface of the earth?

43. Calculate the electric field and the electric energy density in the vicinity of a proton. Take $r = 2.8 \times 10^{-15}$ m.

44. A capacitor (C_1) is charged to a potential difference V_1 and then connected to an initially uncharged capacitor C_2 (Figure 12). (a) What quantity is conserved in this process? (b) Show that the total energy of the system after equilibrium is established is

$$U_{\text{final}} = U_{\text{initial}} \cdot \frac{C_1}{C_1 + C_2}$$

Gasoline	Lead–acid battery	Capacitors
126,000 Btu/gal	voltage = 12 V	breakdown potential difference = 12 V
1 Btu = 1055 J	ampere hours = 100	capacitance = 10^{-1} F
1 gal = 231 in.^{3}	mass = 16 kg	mass = 0.1 kg
density = 0.67 gm/cm^3		

(c) In a perfectly inelastic collision of a mass m_1, whose velocity is v_1, with a mass m_2 that is initially at rest, what quantity is conserved? (d) Show that the total kinetic energy after the perfectly inelastic collision is

$$E_{\text{final}} = E_{\text{initial}} \cdot \frac{m_1}{m_1 + m_2}$$

(e) Comment on the analogies between the collision and capacitor problems.

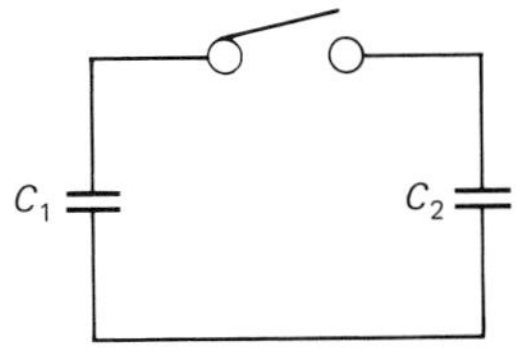

Figure 12

45. How would the analysis of Problem 44 change if capacitor C_2 possessed an initial charge before the switch was closed? What is the collision analog?

Section 29.4 Effect of an Insulator on Capacitance

46. A 0.001-μF parallel-plate capacitor is to be made with a Teflon dielectric ($\kappa = 2.1$, dielectric strength 19 kV/mm) 0.005 mm thick. Determine (a) the required plate area and (b) the breakdown potential difference.

47. A parallel-plate capacitor is to be made with a polystyrene dielectric ($\kappa = 2.55$, dielectric strength 20 kV/mm), plate area 0.001 m^2, and breakdown potential difference of 12 V. Determine the maximum possible capacitance consistent with the required breakdown potential difference.

48. Each plate of a parallel-plate capacitor has an area of 1.00 m^2. The plate separation is 0.80 cm. For a potential difference of 10.0 V, (a) calculate the electric field between the plates. The vacuum between the plates is filled with an insulating plastic, κ (kappa) = 3. Calculate (b) the potential difference with the dielectric between the plates; (c) the electric field with the dielectric between the plates; and (d) the ratio of capacitances

$$\frac{C}{C_0} = \frac{\text{capacitance (plates + dielectric)}}{\text{capacitance (plates + vacuum)}}$$

49. A potential difference of 9.6 V exists across a parallel-plate capacitor having a plate separation of 0.001 mm. If the dielectric constant of the medium between the plates is 2.6, determine the electric energy density of the capacitor.

50. Two parallel-plate capacitors have the same plate area. One has a Teflon dielectric ($\kappa = 2.1$) and a plate separation of 0.12 mm. The other has a neoprene dielectric ($\kappa = 6.6$) and a plate separation of 0.32 mm. Which has the larger capacitance?

51. A certain electronic circuit calls for a capacitor having a capacitance of 1.2 pF and a breakdown potential difference of 1000 V. If you have a supply of 6-pF capacitors having a breakdown potential difference of 200 V, how could you meet this circuit requirement?

52. Each capacitor in the configuration shown in Figure 13 has a breakdown potential difference of 15 V. What is the breakdown potential difference of the equivalent capacitance?

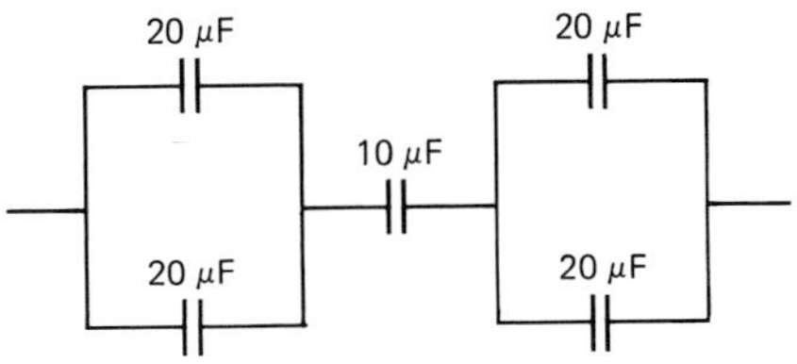

Figure 13

53. A parallel-plate capacitor having air between the plates is charged to 31.5 V. The capacitor is isolated from the charging source and the volume between the plates is filled with Plexiglas. Determine the potential difference across the capacitor.

54. Two capacitors with the same plate area and plate separation, one having air as an insulator and the other having a Teflon insulator, are connected to a 15-V battery. Compare the charges stored on the capacitors and the potential energies of the capacitors.

55. A disk-shaped parallel-plate capacitor having a diameter of 25 cm, plate separation 0.1 mm, and air as an insulator is connected to a 9.3-V battery. While the capacitor is still connected to the battery, a 0.1-mm-thick sheet of glass ($\kappa = 4.3$) is inserted between the capacitor plates. Determine the change in the charge stored by the capacitor and the change in the electric potential energy of the capacitor.

56. The capacitor shown in Figure 29.18 has a potential difference of 10,376 V across it when it is in air at a pressure of 1 atm. It is then placed in a vacuum chamber and the air is removed. In the vacuum the potential difference rises to 10,382 V. Determine the dielectric constant of air.

57. A charged capacitor stores 50 J of energy when the dielectric constant equals 1. A new dielectric is added whose dielectric constant equals 10. What energy is stored in the presence of the new dielectric if it is added (a) while the battery is connected to the capacitor and (b) while the battery is disconnected from the capacitor?

58. A potential difference of 12 V exists across an isolated parallel-plate capacitor having a capacitance of 15 pF and a material between the plates having a dielectric constant of 2.5. Determine the electrostatic energy of the capacitor after the insulator is pulled away from the capacitor.

59. A charged isolated parallel-plate capacitor contains an insulator having dielectric constant κ_1. If this dielectric is replaced with one having dielectric con-

stant κ_2, show that the initial electrostatic potential energy U_1 is related to the final electrostatic energy U_2 by

$$\frac{U_2}{U_1} = \frac{\kappa_1}{\kappa_2}$$

60. A cylindrical capacitor can be constructed by forming metal foils on the inside and outside walls of a cylindrically shaped glass tube. If the walls of a Pyrex tube ($\kappa = 6.0$) 25 cm long that has an inside diameter of 2.5 cm and a wall thickness of 2 mm are covered with aluminum foil, what is its capacitance?

61. The manufacturer of a type of coaxial cable used for stereo systems quotes the capacitance per unit length as 18 pF/ft. Verify this value, knowing that the radius of the inner and outer conductors are 0.0254 and 0.242 in. respectively, and are separated by polyethylene having dielectric constant $\kappa = 2.62$.

62. A slab of material having dielectric constant κ is placed between the plates of a parallel-plate capacitor as shown in Figure 14. (a) Explain why this configuration is equivalent to two capacitors in series and then show that the capacitance is

$$C = \frac{A\epsilon_0}{d}\left[\frac{\kappa}{(x/d) + \kappa(1 - x/d)}\right]$$

(b) Show that the capacitance expression reduces to the expected results for $x = 0$, $x = d$, and $\kappa = 1$.

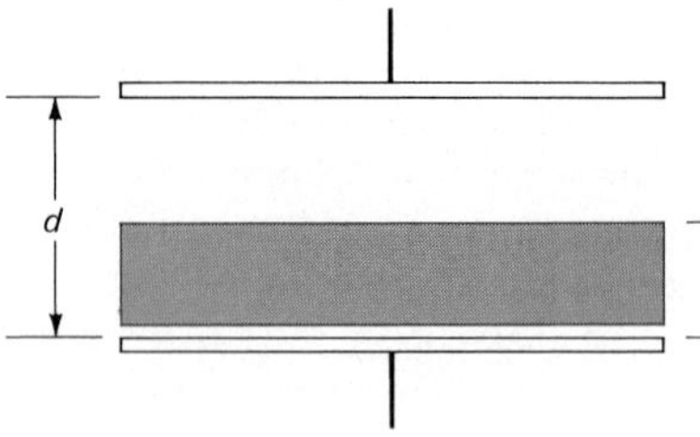

Figure 14

63. A slab of material having dielectric constant κ is placed between the plates of a parallel-plate capacitor as shown in Figure 15. (a) Explain why this configuration is equivalent to two capacitors in parallel and then show that the capacitance is

$$C = \frac{A\epsilon_0}{d}\left[\frac{\kappa x}{W} + \left(1 - \frac{x}{W}\right)\right]$$

(b) Show that the capacitance expression reduces to the expected results for $x = 0$, $x = W$, and $\kappa = 1$.

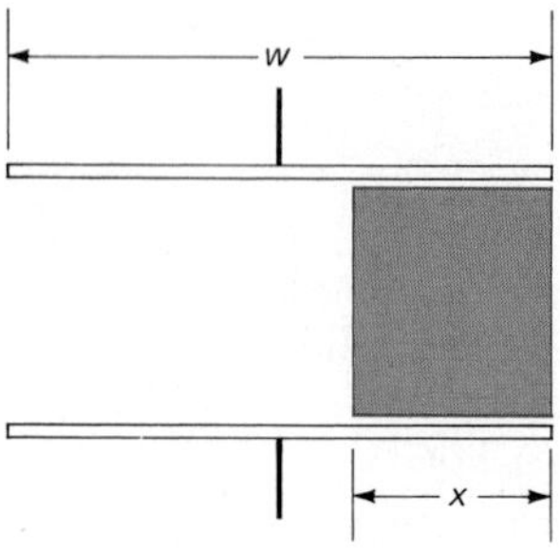

Figure 15

chapter 30

ELECTRIC CURRENT

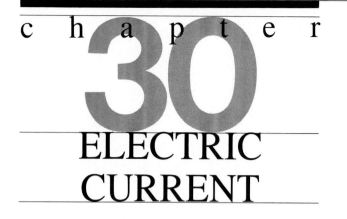

Preview

A net flow of charge constitutes an electric current. It is as electric currents that charges are of greatest practical use. Around the home, food mixers, refrigerators, light bulbs, and stoves, for example, are powered by electric currents. Computers and calculators could not perform their high-speed calculations without the flow of charge through them. Nuclear fusion energy involves charges moving at very high speed in ionized gases. The intense magnetic fields used to contain these charges are produced by electric currents in conductors.

We begin the study of electric currents by introducing the ampere as the unit of electric current. The concepts of resistance, resistivity, and conductivity describe how the physical characteristics of a conductor limit the rate of charge flow. Electrical measurements on a metallic conductor lead to Ohm's law, which relates the electric current in the conductor and the potential difference between two points along the conductor.

30.1 Electric Current

Free-Electron Model of Conduction

An electric current is a net flow of charge. Currents most often occur in conductors, although they can also exist in a vacuum. For example, circulating protons in the evacuated channels of the accelerator at the Fermilab in Batavia, Illinois, are an electric current. The currents that we deal with every day however, like the current in a light bulb, occur in solid metallic conductors.

We can think of a metallic conductor as an arrangement of positive ions and free electrons. The ions are in relatively fixed positions, whereas the electrons are much more mobile. Some electrons are free to roam about within the array, or *lattice,* of ions. In this sense the electrons in a conductor are like the molecules of a gas confined to a container. Each atom in a copper conductor, for example, contributes about one electron to the electron "gas." The electrons in the conductor are distributed uniformly, like the molecules in the gas of an inflated balloon. The motion is random and there is no net charge flow across any imaginary area within the conductor. As many electrons per second cross an area from right to left as from left to right. If an electric field $\mathbf{E}$ is established in the conductor, each free (conduction) electron experiences a force $\mathbf{F} = -e\mathbf{E}$, causing a net flow of electrons in the direction opposite to $\mathbf{E}$. This description of electric current in a metallic conductor is known as the **free-electron model of conduction.**

Figure 30.1

In this diagram, electric charges move to the right in the conductor. The current is the rate of flow of charge through any cross section of the conductor.

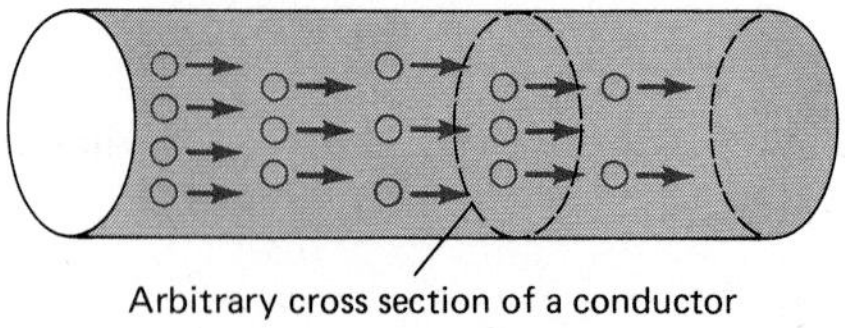

The value of the electric current in a conductor is measured by the net rate of flow of charge through a cross section of the conductor (Figure 30.1). If a net charge dq flows through a cross section in time dt, the current I is

$$I = \frac{dq}{dt} \tag{30.1}$$

The units for q are coulombs and the units for t are seconds. Thus, the units of current are coulombs per second, called *amperes** and abbreviated A. If the current is *constant*, it is given by

$$I = \frac{Q}{t} \tag{30.2}$$

where Q is the net charge flowing through a cross section of the conductor in time t. A constant current of 1 A means that during each second, 1 C of charge flows through a cross section of the conductor. In this chapter we consider only constant currents.

Note carefully that the definition of electric current does not specify the algebraic sign of the moving charges. This is because both positive and negative charges can produce electric currents. For example, the storage battery in an automobile involves the motion of both positive and negative charges in a liquid electrolyte. A radioactive particle produces both positive ions and free negative electrons in the gas of the detecting head of a Geiger counter, and both the positive charges and the negative charges give rise to electric currents. In the semiconducting materials from which transistors are made, electric currents are due to positive or negative charges, or both. Charges of opposite sign move in opposite directions in a given electric field. How then can we assign a direction to a current that is due to both positive and negative charges? It happens that, with very few exceptions (such as the Hall effect, discussed in Section 32.3), *the effects produced by negative charges moving in one direction are identical to those produced by positive charges moving in the opposite direction*. Therefore, for simplicity, we can assume that electric current is due to positive charges and that the direction of a current is the direction in which positive charges move. We make this assumption by convention. If the charges are actually negative, as electrons are, then they move opposite to the direction of the current (Figure 30.2).

*Although this is a meaningful interpretation of current and its units, the ampere is defined operationally in terms of the force exerted on a current-carrying wire in a magnetic field (Section 33.5). Then Eq. 30.1 is used to define the coulomb. The ampere is named for André Marie Ampère, a French mathematician and physicist (1775–1836).

Figure 30.2

When a battery establishes a potential difference between the ends of a conductor with the left end positive, then an electric field is directed to the right. *If present*, positive charges would be urged to the right. Thus the *current* direction would be to the right. Free electrons in the conductor would move to the left. Nevertheless, by convention, the current direction is to the right, from high potential to low potential.

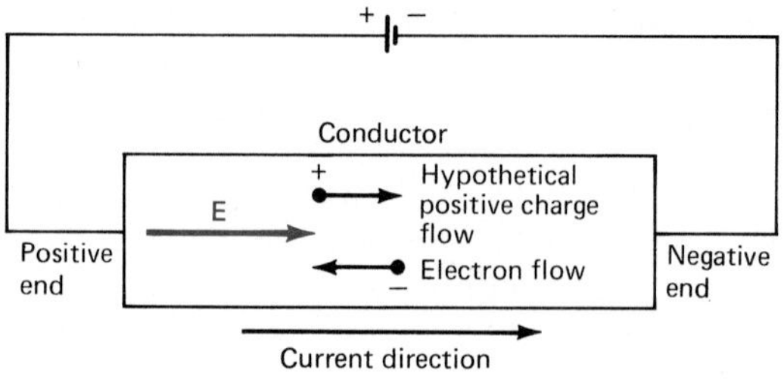

Figure 30.3

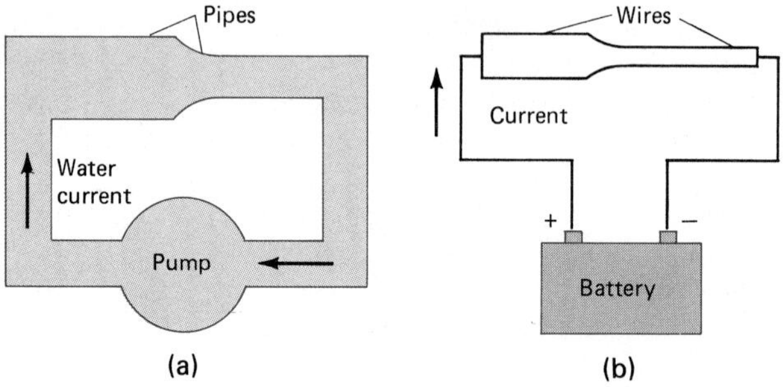

Conservation of Charge

In many ways the flow of charge in a conductor is analogous to the flow of water through pipes. For example, in the cooling system of an automobile engine, the flow of water in kilograms per second is the same through any portion of the system (Figure 30.3). The plumbing may be circuitous, it may contain internal obstructions, and the cross section of the pipes may vary, but as long as there are no holes or other mechanisms for removing or adding water, the rate of mass movement is constant (Figure 30.3a). There is conservation of mass. Similarly, a conductor may have distortions of its structure, it may be bent in a peculiar shape, and it may have varying cross-sectional area, but the current is the same in any cross section of the conductor because there are no mechanisms for removing or adding charge (Figure 30.3b). Charge can be neither created nor destroyed; it is conserved.

Drift Velocity

When an electric field is established in a conductor, each conduction electron (or free electron) experiences a force $\mathbf{F} = -e\mathbf{E}$, and an instantaneous acceleration $\mathbf{a} = \mathbf{F}/m$. An electron flow is established with a direction opposite to the electric field (Figure 30.4). During acceleration, the kinetic energy of an electron increases. The electron does not achieve unlimited speed, however, because it collides

Figure 30.4

Electrons moving through a conductor under the influence of an electric force are analogous to marbles dropping vertically in a chute with barriers made of nails. The marbles change direction as a result of collisions with the nails, but they acquire an average velocity in the direction of the gravitational force.

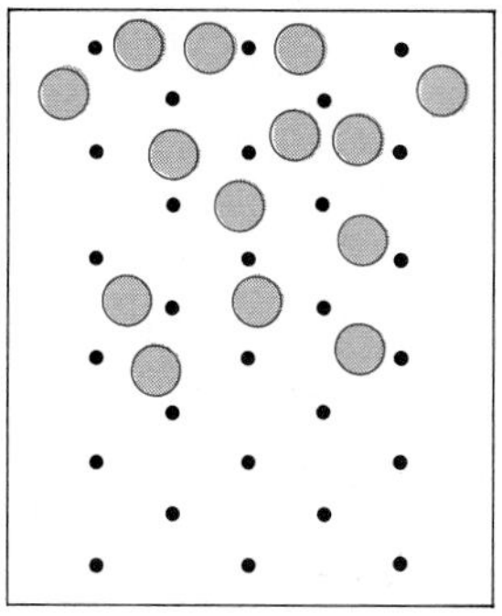

with the ions or atoms that make up the conductor. On the average, kinetic energy is lost in such collisions. The resulting motion of the electron is zigzag, but the average velocity is in the direction opposite to the electric field. This resultant average electron velocity is called the **drift velocity** and is denoted by v_D. The motion of an electron through a conductor is analogous to the motion of a person through a crowd. The person may collide with others or be forced to change direction, but manages to drift through the crowd with some net average velocity in the desired direction.

What is the relation between the drift velocity and the electric current? Because current is the rate of flow of charge, we expect the current to increase as the drift velocity increases. Consider a conductor of uniform circular cross section A (m^2) and free-electron density n (electrons/m^3; Figure 30.5). If we assume that each electron has the drift velocity v_D, the current is determined by counting the number of charges that flow through any cross-sectional area in a time t. During the time interval t all electrons move a distance $v_D t$. This means that all those electrons within a distance $v_D t$ to the left of the counting position pass through the cross-sectional area. All electrons in the cylinder of cross-sectional area A, length $v_D t$,

Electron flow

$v_D t$

Count the number of electrons passing this area in time t. Then all electrons within a distance $v_D t$ flow past this area.

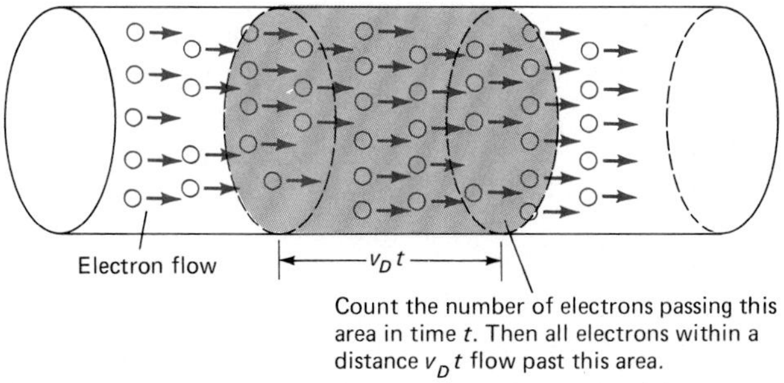

Figure 30.5

Electron flow in a conductor is similar to traffic flow on a road. By determining the amount of charge flowing through a cross section in some time interval we can calculate the current.

and volume $A v_D t$ pass through the right end of this cylinder. The number passing through is then given by

$$N(\text{electrons}) = \left(\frac{\text{electrons}}{\text{m}^3}\right) \cdot \text{m}^3$$

$$= n A v_D t$$

Because each electron carries a charge of $-e$, the amount of charge passing through the cross section is

$$Q = Ne = e n A v_D t$$

and the current is

$$I = \frac{Q}{t} = e n A v_D \qquad (30.3)$$

Free electron density
area
drift velocity
charge in coulombs

The current is *directly proportional* to the drift velocity. Remember, however, that if the electron flow is to the right, as it is in this example, the direction of the current, *by convention*, is to the left.

The drift velocity of electrons in a conductor can be estimated by using Eq. 30.3 if we make some reasonable assumptions. We find that it is about 10^{-4} m/s (Example 1). This means that it takes about 10,000 s or about 3 hr for an electron to travel 1 m! The drift velocity is very small compared to the average speed of an electron in the electron "gas." The situation is analogous to that of air moving through a pipe. At a temperature of 300 K, the average speed of a molecule due to its thermal motion is large compared to the speed of the gas moving in the pipe.

Example 1   Drift Velocity of Conduction Electrons in a Metal

Let us estimate the drift velocity of the conduction electrons in a copper wire that is 1 mm in diameter and carries a current of 1 A.

We begin with Eq. 30.3, which relates current and drift velocity. Solving this equation for v_D, we have

$$v_D = \frac{I}{enA}$$

We need to find n, the free-electron density.

We will assume that each copper atom contributes one conduction electron, with the result that the number of conduction electrons per cubic meter equals the number of copper atoms per cubic meter. We know that a mole of copper contains Avogadro's number of copper atoms. To get the free-electron density n, we need to know the volume occupied by the mole of copper. We calculate the volume from a knowledge of the density of copper.

$$\text{Volume of a mole} = \frac{\text{mass of a mole}}{\text{density}}$$

$$V = \frac{M}{\rho}$$

$$\text{Number of copper atoms per cm}^3 = \frac{\text{Avogadro's number}}{\text{volume of a mole}}$$

$$n = \frac{N_A}{V} = \frac{N_A \rho}{M}$$

Knowing that $N_A = 6.022 \times 10^{23}$ atoms/mole, $\rho = 8.89$ gm/cm^3, and $M = 63.54$ gm/mole, we have

$$n = \frac{8.89 \text{ gm/cm}^3 \cdot 6.022 \times 10^{23} \text{ atoms/mole}}{63.54 \text{ gm/mole}}$$
$$= 8.43 \times 10^{22} \text{ electrons/cm}^3$$
$$= 8.43 \times 10^{28} \text{ electrons/m}^3$$

The cross-sectional area of the conductor is

$$A = \frac{\pi d^2}{4} = \frac{\pi(10^{-3})^2}{4} \text{ m}^2$$

Hence the drift speed is

$$v_D = \frac{I}{neA}$$
$$= 1 \text{ A} \div \left(8.43 \times 10^{28} \frac{\text{electrons}}{\text{m}^3}\right) \cdot \left(1.602 \times 10^{-19} \frac{\text{C}}{\text{electron}}\right) \cdot \frac{\pi(10^3)^2}{4} \text{ m}^2$$
$$= 9.4 \times 10^{-5} \text{ m/s}$$

At this speed, about 10 s are required for an electron to drift a distance of 1 mm.

Current Density

The relation between electric current and drift velocity involves quantities (I, area) that are characteristic of the conductor as a whole, and quantities (q, n, v_D) that are characteristic of a point inside a conductor rather than of the conductor as a whole. Separating these macroscopic and microscopic variables in Eq. 30.3 and replacing e by q, we can write

$$\underset{\text{macroscopic}}{I \div \text{area}} = \underset{\text{microscopic}}{qnv_D} \tag{30.4}$$

The quantity $I \div$ area measures the concentration of the current. We call this the **current density** and denote it by J. For example, a wire in a household electrical circuit usually has a diameter of about 10^{-3} m, and may carry a current of 1 A. The current density in this case is

$$J = I \div \text{area} = \frac{1 \text{ A}}{\dfrac{\pi(0.001)^2}{4} \text{ m}^2} = \frac{4}{\pi} \times 10^6 \text{ A/m}^2$$

The same current density in a wire 1 m^2 in cross section would produce a current of $(4/\pi) \times 10^6$ A.

In general, the current density is a vector quantity. For our model of conduction, where each charge has the same drift velocity, we write

$$\mathbf{J} = qn\mathbf{v}_D \tag{30.5}$$

Like q, n, and $\mathbf{v}_D$, $\mathbf{J}$ is microscopic.

The drift velocity vector $\mathbf{v}_D$ has the direction of the moving charge. If q is negative, as it is for electrons, then $\mathbf{J}$ is directed opposite to $\mathbf{v}_D$. Situations arise where currents involve charges of different types, different concentrations, and different drift velocities. The net current density in this case is the vector sum of the current density for each individual component contributing to the current.

In some situations, the density of charges varies with position. This means that current density also varies with position. Therefore, to obtain the total current through some area of interest, we must integrate the current density over this area.

$$I_{\text{total}} = \int \mathbf{J} \cdot d\mathbf{A} \tag{30.6}$$

Note that since the current I is obtained as a scalar product of current density and area, I is a scalar quantity.

Example 2 Current in a Plasma

An ionized gas of equal numbers of positive and negative charges is called a *plasma*. In a plasma having a circular cross section with radius 10 cm, the current density changes with the radial distance from the axis of the current direction according to the relation

$$\mathbf{J} = 100r\mathbf{k} \text{ A/m}^2$$

The charges move in the positive z-direction. Let us determine the total current in the plasma.

We pick an area dA having the geometry of a ring of radius r and width dr (Figure 30.6). The area element is $d\mathbf{A} = 2\pi r\, dr\, \mathbf{k}$ and

$$I = \int \mathbf{J} \cdot d\mathbf{A} = \int_0^{0.1} (100r\mathbf{k}) \cdot (2\pi r\, dr\, \mathbf{k})$$
$$= 2\pi(100) \int_0^{0.1} r^2\, dr$$
$$= 2\pi(100)\left(\frac{r^3}{3}\right)\Bigg|_0^{0.1}$$

The total current in the plasma is

$$I = 0.209 \text{ A}$$

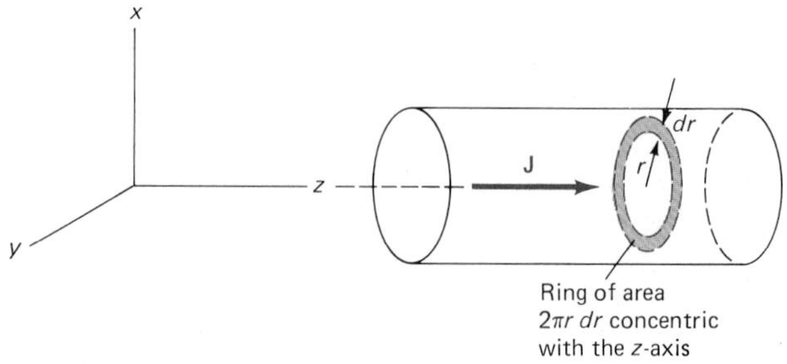

Figure 30.6

Current in the plasma is directed along the positive z-axis. We construct a ring of radius r, area $2\pi r\, dr$, concentric with the cylindrical axis of the plasma. Then the current in this ring is $J 2\pi r\, dr$. We obtain the total current by integrating $\int J 2\pi r\, dr$ over the cross section of the plasma.

Although current density is probably new to you, it is related to the notion of energy intensity introduced in Section 18.4. Dimensionally an energy intensity is energy/(time · area). It is a measure of the rate of flow of energy through a unit area. The intensity of a traveling wave is

$$I = \left(\frac{E}{\text{volume}}\right)v$$

This intensity is an energy per unit volume multiplied by the velocity of the wave. Current density has the same characteristics as energy intensity with charge analogous to energy.

$$J = \frac{\text{charge}}{\text{time} \cdot \text{area}}$$

Using Eq. 30.5 for the current density, we have $J = (nq)v_D$ and we see that this is a charge per unit volume (nq) multiplied by the velocity of the charge.

$$J = \left(\frac{\text{charge}}{\text{volume}}\right) \cdot v_D$$

You should try to seek similar relationships; doing so will make your study of physics more meaningful.

Questions

1. A beam of protons emerging from a particle accelerator such as a cyclotron constitutes an electric current. If this current is used to charge a capacitor, how could a measure of the charge on the capacitor be used to determine the average rate at which protons emerge from the accelerator?
2. If the positive and negative connections to the terminals of an automobile battery were reversed, would you expect the automobile lights to function?
3. Explore the analogy between the equation relating electric field, electric flux, and area and the equation relating current density, current, and area.

30.2 Electrical Resistance and Ohm's Law

In the previous section we compared electric current in a conductor to the flow of water in a pipe in order to explain conservation of charge. Let's continue with this analogy.

An open pipe permits water to flow through it but there is some resistance to the flow due to friction. An electrical conductor, by definition, permits charge to flow through it, but again there is some resistance to the flow. This resistance is a property of the conductor and is called the conductor's **electrical resistance,** or simply the resistance. When we compare the parameters that determine the flow of water in a pipe with those determining the flow of charge in a conductor, we arrive at a quantitative definition of resistance, as we will now show.

In order for water to flow through the cooling system of an automobile, a pressure difference must be maintained between the inlet to the system and the outlet to the system. This pressure difference is maintained by a pump driven by the automobile engine. The rate at which water flows depends on the pressure difference and the nature of the pipes through which the water flows. Obstructions in the pipes or other limitations in the openings through which water flows tend to diminish the flow rate. Similarly, in order for electrons to flow through the wires of an electrical circuit, a potential difference must be maintained between the ends of the wires. This potential difference can be maintained by a battery. The rate at which charge flows (i.e., the current) depends on the potential difference and the nature of the wires through which charge flows. Obstructions or constrictions in these wires tend to diminish the current.

The electrical resistance (R) of a conductor is defined as the potential difference between the ends (denoted by a and b) of the conductor (V_{ab}) divided by the current (I) in the conductor.

$$R = \frac{V_{ab}}{I} \qquad (30.7)$$

Electrical resistance has units of $\frac{\text{volts}}{\text{amperes}}$, called *ohms,** which are symbolized by Ω. If a potential difference of 12.1 V across a conductor causes a current of 0.25 A in the conductor, its resistance is

$$R = \frac{12.1 \text{ volts}}{0.25 \text{ amperes}} = 48.4\ \Omega$$

For many conductors, particularly those made of metal, the electrical resistance is approximately the same, regardless of the applied potential difference used to measure it. Such a conductor is said to obey Ohm's law. If we plot the current of a conductor that obeys Ohm's law against the potential difference, we get a straight line (Figure 30.7). The slope of the line is the reciprocal of the resistance. Resistors are conductors obeying Ohm's law that we include in a circuit deliberately to impede or limit the current or to produce heat. In electrical circuits, resistors are represented by ─╱╲╱╲─. If we know the resistance

*Named for Georg Simon Ohm, a German physicist (1787–1854).

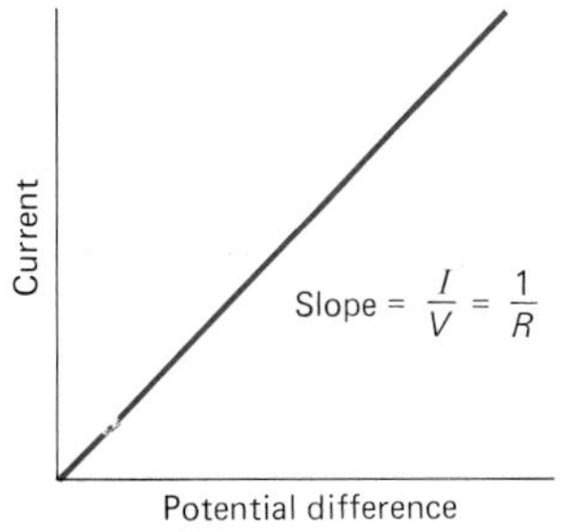

Figure 30.7

For an element obeying Ohm's law, a plot of current versus potential difference yields a straight line. The slope of the straight line yields the reciprocal of the resistance.

Figure 30.8

Shown here is a plot of the current and potential difference for a solid-state diode. The solid-state diode does not obey Ohm's law because a plot of the current versus potential difference does not yield a straight line.

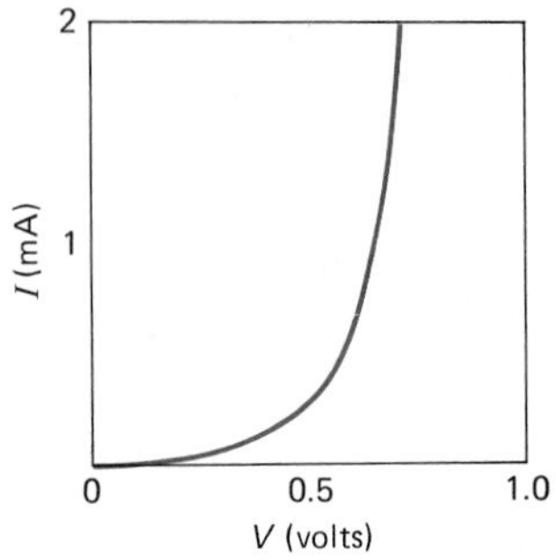

of a resistor and the current in it, then Eq. 30.7 allows us to calculate the potential difference across it ($V = IR$). Or if we know the resistance and the potential difference across a resistor, then Eq. 30.7 allows us to calculate the current in it ($I = V/R$). In this sense, $V = IR$ plays the same role for resistors that the capacitance equation $Q = CV$ plays for capacitors. In the next chapter we will see how these two equations help us to analyze electrical circuits that have resistors and capacitors as components.

For many conductors, the resistance is sensitive to the potential difference across it and the current in it (Figure 30.8). These conductors do *not* obey Ohm's law and are called *nonohmic conductors*. The concept of resistance as defined by Eq. 30.7 has little practical usefulness for nonohmic conductors. To find the current in such a conductor, we must either know the mathematical relationship between potential difference and current or resort to a graph displaying the current–potential difference relationship. In Section 30.4 we will discuss the concept of *dynamic* resistance, which *does* have utility for nonohmic conductors.

Questions

4. How would you define a fluid resistance for a water-carrying pipe?
5. Why would you expect the resistance of a wire (a) to decrease as its cross-sectional area increases, and (b) to decrease as its length decreases?
6. An electrical resistance exists between any two areas on a human body. This resistance figures into the effects of electric shock on a person. How might you measure the electrical resistance between your hands? What difficulties do you foresee in trying to calculate this resistance?
7. Why is a thin region of a light bulb filament more vulnerable to "burning out" than a region with greater cross-sectional area?

30.3 Electrical Conductivity and Electrical Resistivity

The fluid resistance of a pipe depends on its length and cross-sectional area. Long, small-diameter pipes have large resistances. Short, large-diameter pipes with no internal obstructions have small resistances. Analogously, the electrical resistance of a conductor of uniform cross section will depend on its length and cross-sectional area. As the length (L) increases, the resistance increases. As the cross-sectional area (A) increases, the resistance decreases. Additionally, the resistance of a conductor depends on the type of material from which the conductor is made. Incorporating these ideas, we write the resistance of a conductor as

$$R = \rho\left(\frac{L}{A}\right) \tag{30.8}$$

where ρ is called the **electrical resistivity.** Resistivity has units of ohm meters ($\Omega \cdot m$) and is an intrinsic property of the material used in making the conductor. It is independent of cross-sectional area and length. The reciprocal of electrical resistivity has units of reciprocal ohm meters $[(\Omega \cdot m)^{-1}]$, is called **electrical conductivity,** and is denoted by σ.

$$\sigma = \frac{1}{\rho} \tag{30.9}$$

Using the relation $R = (1/\sigma)(L/A)$, we can now put $V_{ab} = IR$ into a more fundamental form involving electrical conductivity, current density, and electric field.

For a conductor of uniform cross-sectional area A we can write

$$V_{ab} = IR$$
$$= I \cdot \frac{L}{\sigma A}$$

Rearranging terms, we have

$$\frac{I}{A} = \sigma\left(\frac{V_{ab}}{L}\right) \tag{30.10}$$

Here I/A is the current density (J) and V_{ab}/L is the electric field (E) established by the potential difference V_{ab}. Hence Eq. 30.10 can be written

$$J = \sigma E \tag{30.11}$$

or

$$\rho J = E \tag{30.12}$$

The quantities J, σ, and E are all characteristic of a point in the conductor; they are microscopic quantities. It is appropriate to think of $J = \sigma E$ as a microscopic form of $V_{ab} = IR$.

Table 30.1 shows the conductivities of some common materials. Note that the largest and smallest conductivities listed differ by about 20 orders of magnitude. Conductors are customarily categorized as materials with conductivities greater than $10^5\ (\Omega \cdot m)^{-1}$. Insulators are materials with conductivities less than $10^{-5}\ (\Omega \cdot m)^{-1}$ (Figure 30.9). Materials with conductivities between 10^{-5} and $10^5\ (\Omega \cdot m)^{-1}$ are called *semiconductors.* The elements germanium and silicon and the compound gallium arsenide are important semiconductors used in the fabrication of diodes, transistors, and integrated circuits.

Electrical conductivity is an important property of a conductor that experimental physicists measure and theoretical physicists try to calculate from first principles of physics. Electrical conductivity also has practical value when we want to determine the resistance of a conductor. For example, an electromagnet consisting of a length of wire wound as a series of coils onto a cylindrical core requires a certain current for proper operation. Knowing the conductivity (or resistivity) and dimensions of the wire, we can compute the resistance from Eq. 30.8. Using $V_{ab} = IR$, we can then determine the potential difference between the ends of the wire that is needed to produce the correct current (Example 3).

Figure 30.9

Categorization of materials according to electrical conductivity. The boundaries between the categories are not sharply defined.

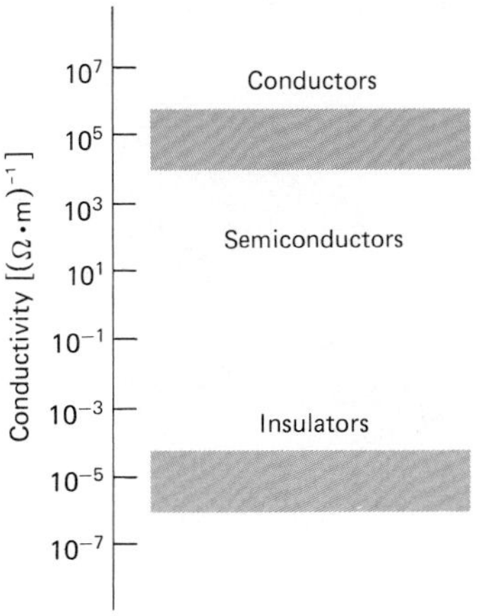

Example 3 **Electrical Resistance of a Solenoid**

A solenoid (Figure 30.10) is constructed from 500 m of copper wire having a diameter of 1.5 mm. We will determine its electrical resistance by using Eq. 30.8. The

Table 30.1

Electrical resistivity, electrical conductivity, and temperature coefficient of resistivity of some common materials*

	Resistivity $(\Omega \cdot m)$	Conductivity $(\Omega \cdot m)^{-1}$	Temperature Coefficient (/°C)
Aluminum	2.824×10^{-8}	3.541×10^7	0.0039
Brass†	7×10^{-8}	1.4×10^7	0.002
Constantan‡	49×10^{-8}	2.0×10^6	0.000002
Copper	1.724×10^{-8}	5.800×10^7	0.00393
Gold	2.44×10^{-8}	4.10×10^7	0.0034
Iron	10.0×10^{-8}	1.00×10^7	0.0050
Nichrome§	100×10^{-8}	1×10^6	0.0004
Silver	1.59×10^{-8}	6.29×10^7	0.0038
Carbon	3500×10^{-8}	2.86×10^4	-0.0005
Polyethylene	10^8–10^9	10^{-9}–10^{-8}	
Polystyrene	10^7–10^{11}	10^{-11}–10^{-7}	
Neoprene	10^9	10^{-9}	
Teflon	10^{14}	10^{-14}	
Glass	10^{10}–10^{14}	10^{-14}–10^{-10}	
Porcelain	10^{10}–10^{12}	10^{-12}–10^{-10}	

*Measurements were made at 20° C. For a more complete listing see *Handbook of Chemistry and Physics,* CRC Press, 2000 N.W. 24th Street, Boca Raton, FL 33431.

†Average values; depends on type.

‡An alloy of equal parts of nickel and copper having a very low temperature coefficient of resistivity.

§An alloy of nickel and chromium often used to make heating elements.

Figure 30.10

A solenoid consisting of closely-spaced turns of wire on a cylindrical insulating form.

Figure 30.11

Electrical resistivity of copper as a function of temperature. The dashed line represents Eq. 30.13 with $T_o = 20°C$, $\rho_o = 1.724 \times 10^{-8}\ \Omega \cdot m$, and $\alpha = 3.93 \times 10^{-3}/°C$.

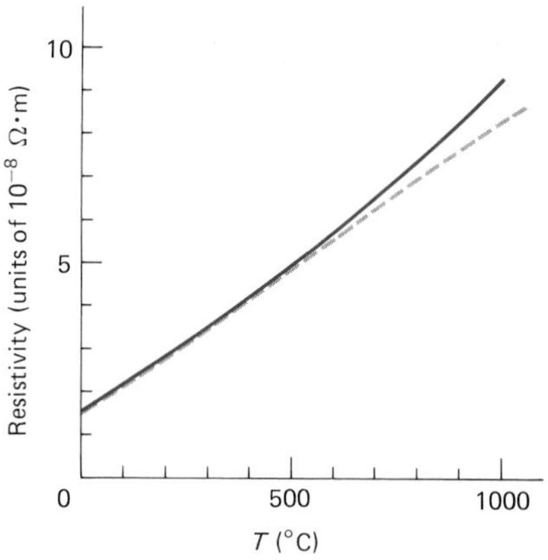

cross-sectional area of the wire is

$$A = \frac{\pi d^2}{4} = \frac{\pi(1.5 \times 10^{-3})^2}{4} = 1.767 \times 10^{-6}\ m^2$$

From Table 30.1 we obtain the electrical resistivity of copper.

$$\rho = 1.724 \times 10^{-8}\ \Omega \cdot m$$

Hence the resistance of the solenoid is

$$R = \rho\left(\frac{L}{A}\right) = \frac{1.724 \times 10^{-8} \cdot 500}{1.767 \times 10^{-6}} = 4.88\ \Omega$$

If the solenoid requires a current of 1.25 A, then the potential difference between the ends of the wire must be

$$V_{ab} = IR = 1.25 \times 4.88 = 6.10\ V$$

Experiments show that electrical resistivity depends on the temperature of the material. This is illustrated in Figure 30.11, which shows the resistivity of copper for temperatures between 0°C and 1000°C. The resistivity of copper and most metals increases as the temperature increases. With fairly good accuracy, the data in Figure 30.11 can be represented by a straight line over the entire temperature range shown. For temperature intervals of a few hundred degrees, a straight line is an excellent representation. This means that the data can be represented by a straight-line formula relating electrical resistivity to temperature T. We can write this expression as

$$\rho = \rho_o[1 + \alpha(T - T_o)] \qquad (30.13)$$

where ρ_o is the resistivity at temperature T_o. Because we usually are interested in the electrical resistivity of materials near room temperature, T_o is generally taken to be 20°C. The symbol α, with units of 1/°C, is called the **temperature coefficient of resistivity.** Given the resistivity and temperature coefficient of resistivity at temperature T_o, we can compute the resistivity for some different temperature T. The temperature coefficient of resistivity for several materials measured at 20°C is shown in Table 30.1.

The change of resistivity with temperature makes possible electrical thermometers. Platinum is commonly used in these thermometers. Commercial temperature-dependent resistors called *thermistors* are made of semiconducting materials, because the resistivity of a semiconductor is particularly sensitive to temperature.

Example 4 **Change of Resistivity with Temperature**

The temperature coefficient of resistivity of copper at 20°C is $3.9 \times 10^{-3}/°C$. Let us determine the percentage increase in resistivity when the temperature increases to 220°C. The fractional change is

$$\frac{\rho - \rho_o}{\rho} = \alpha(T - T_o) = 3.9 \times 10^{-3}/°C \cdot (220 - 20)°C = 0.78$$

Therefore, the percentage increase is 78%. This is a significant change.

Superconductivity

In 1911, the Dutch physicist J. Kammerlingh Onnes discovered that the resistivity of mercury decreased abruptly to zero at a temperature of 4.153 K (modern value). The

Figure 30.12

The resistance of mercury as a function of temperature in kelvins as reported by Kammerlingh Onnes.* The abrupt decrease at about 4.20 K denotes the onset of superconductivity. Kammerlingh Onnes reported "the experiment left no doubt about the disappearance of the resistance of mercury."

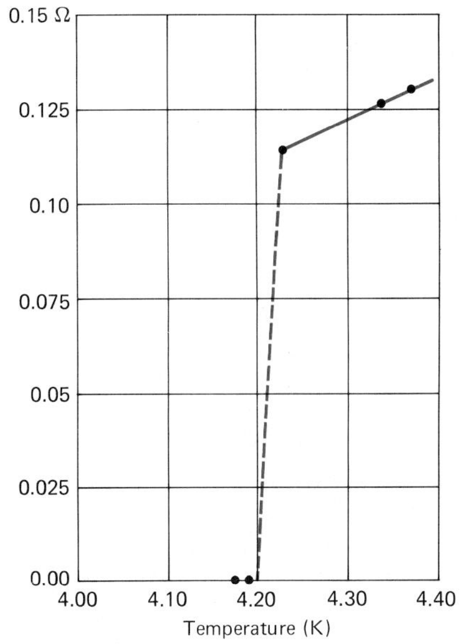

resistivity remained zero as long as the temperature remained below 4.153 K (Figure 30.12). The total absence of resistivity is called **superconductivity** and materials showing this behavior are termed **superconductors.** Once a current is initiated in a ring of superconducting material, the current will persist undiminished as long as the material is maintained in the superconducting state. Since Onnes' discovery of mercury's superconductivity, numerous other metals and alloys have been found to be superconductors (see Section 42.7). But a satisfactory theory of superconductivity was not developed until 1957. This theory is based on quantum mechanical ideas, which we will discuss in Chapter 42. Many experimental devices requiring large magnetic fields utilize superconducting materials in the magnets, as we will see in that chapter.

Questions

8. Two copper wires having the same length but different cross-sectional area are connected to the same battery. How does the current density in the two wires compare?

9. The current in an electric light bulb rises to a maximum almost at the instant the bulb is turned on. The current then decreases significantly. Why does the current decrease?

*Commun. Kammerlingh Onnes Lab. Univ. Leiden (Suppl. 34b, 1913).

10. Table 30.1 shows that copper and carbon have positive and negative temperature coefficients of resistivity, respectively. How could you use copper and carbon resistors to produce a resistor having zero temperature coefficient of resistivity?

30.4 Dynamic Resistance

Because many materials obey Ohm's law, the relation $R = V/I$ is enormously useful. One determination of the resistance of a conductor establishes the relation between V and I. Then the current can be calculated for a given applied potential difference, or if the potential difference is known, the current can be computed. But for conductors that do *not* obey Ohm's law, the resistance as defined by $R = V/I$ has little practical use. In this case the relation between V and I is more complex, and must be determined experimentally. For example, a solid-state tunnel diode does not obey Ohm's law. The relation between V and I for a typical tunnel diode is illustrated in Figure 30.13. The value of dV/dI, called the **dynamic resistance,*** is an important characteristic of this curve.

For nonohmic conductors, the dynamic resistance (dV/dI) is likely to have more utility than the resistance (V/I) because the dynamic resistance reflects how the potential difference changes for small changes in current. Of particular interest are those conditions for which dV/dI is *negative,* as it is for a solid-state tunnel diode (Figure 30.13). A negative dynamic resistance corresponds to instability—a decrease in potential difference produces an increase in current.

To illustrate the negative resistance characteristic, let us consider the neon glow lamp shown in Figure 30.14.

*For an ohmic material $dV/dI = R =$ constant.

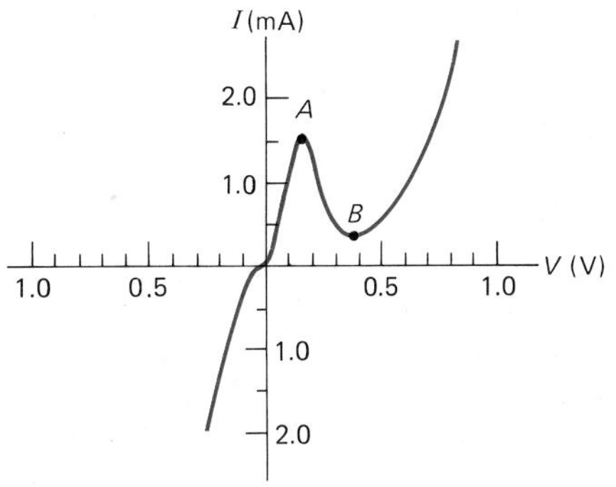

Figure 30.13

A tunnel diode is an electronic device having two electrical connections. The current in the device depends on the potential difference between the two electrical connections. The current–potential-difference relationship for a typical tunnel diode is shown in this figure. dV/dI is negative between points A and B.

Figure 30.14

A neon glow lamp connected in series with a resistor and a source of variable emf. A neon glow lamp consists of two metal electrodes in a glass envelope containing neon gas at a low pressure. An electric field of sufficient magnitude between the electrodes will produce an electric discharge in the gas. The discharge gives rise to a characteristic reddish glow.

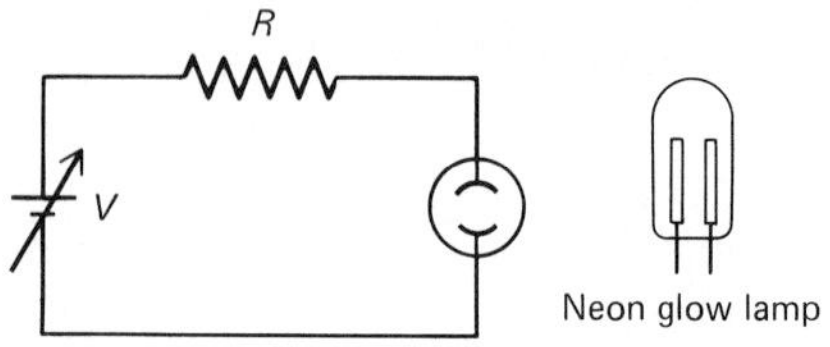

Figure 30.15a

As the potential difference across the lamp (Figure 30.14) increases, there is no current until a value V_f is achieved.

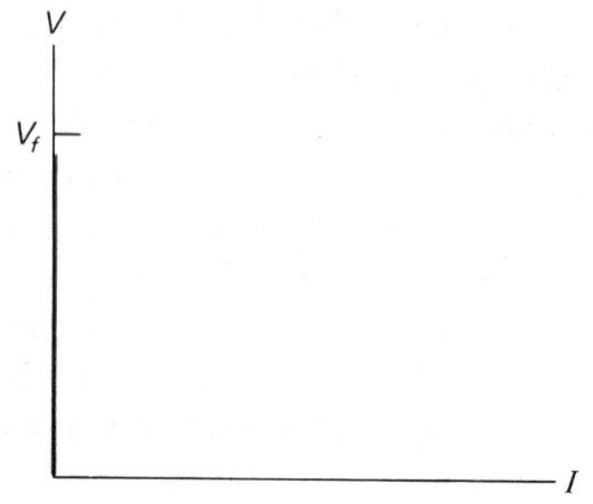

Figure 30.15b

Once the potential difference V_f is reached, the lamp becomes conductive and the potential difference across the lamp decreases rapidly to a value V_e.

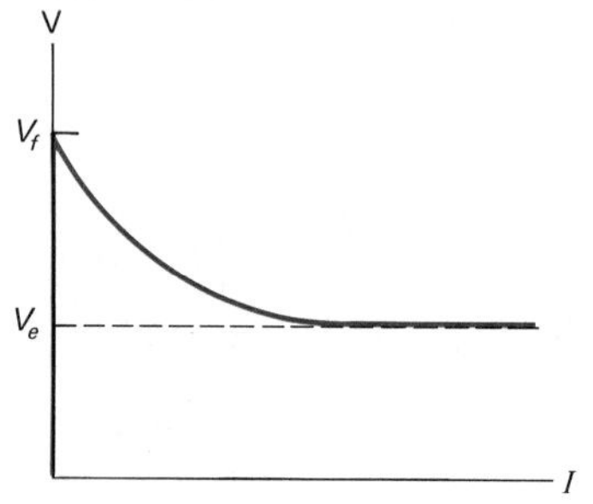

(These reddish-glowing lamps are often used as bedroom night-lights.) We will connect the lamp in series with a resistor and a variable potential difference, and then determine the potential difference across the glow lamp and the current in the glow lamp. When we start with zero potential difference across the lamp, an increase in potential difference produces zero current until the critical value labeled V_f in Figure 30.15a is achieved. In the potential difference range from 0 to V_f, $dV/dI \to \infty$; the dynamic resistance is infinite. Once V_f is achieved, the lamp suddenly becomes conductive and glows. The potential difference across the lamp decreases as the current in it increases. In the potential difference range $V < V_f$, $dV/dI < 0$; the dynamic resistance is small, negative, and rapidly approaches a near-zero value. The potential difference across the lamp decreases and approaches the value labeled V_e in Figure 30.15b. The circuit current is given by

$$I = \frac{V_{\text{source}} - V_e}{R} \qquad (30.14)$$

where V_{source} is the potential difference of the battery.

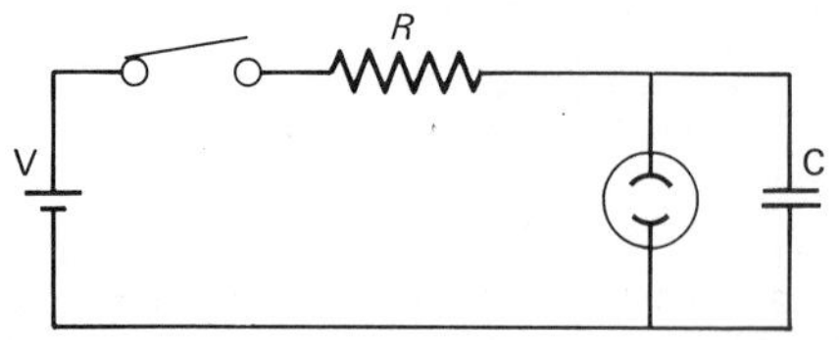

Figure 30.16

The circuit in Figure 30.14 is modified to include a switch in series with the resistor and a capacitor in parallel with the neon glow tube. Because the capacitor and neon glow tube are connected in parallel, the potential difference is the same across both of them.

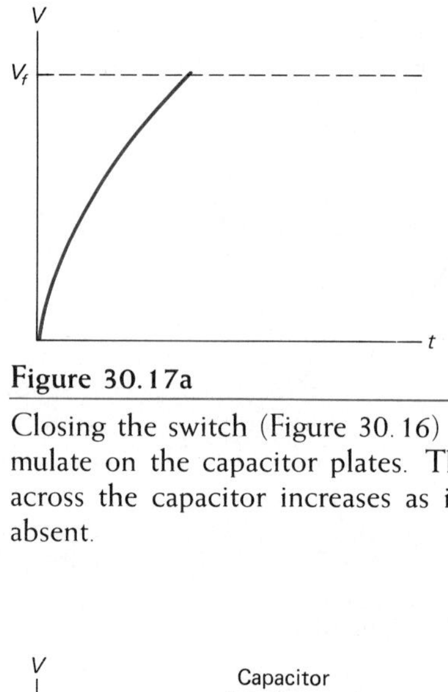

Figure 30.17a

Closing the switch (Figure 30.16) allows charge to accumulate on the capacitor plates. The potential difference across the capacitor increases as if the glow lamp were absent.

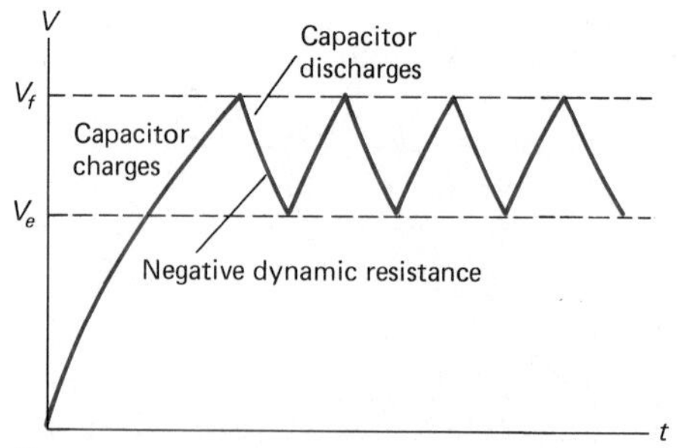

Figure 30.17b

Once the potential difference V_f is reached, the glow lamp becomes conductive, short-circuiting the capacitor. Charge leaves the capacitor plates and the potential difference decreases rapidly to V_e, whereupon the glow lamp becomes nonconductive.

Figure 30.18

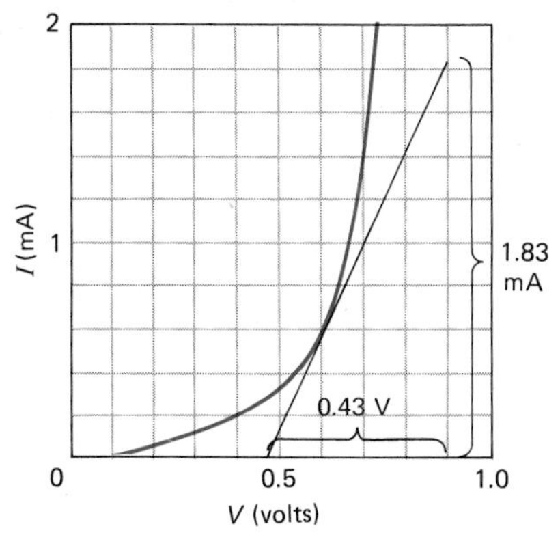

Figure 30.19

A light bulb connected to a battery. The potential difference (V_{ab}) between the battery terminals is the same as that across the leads to the light bulb.

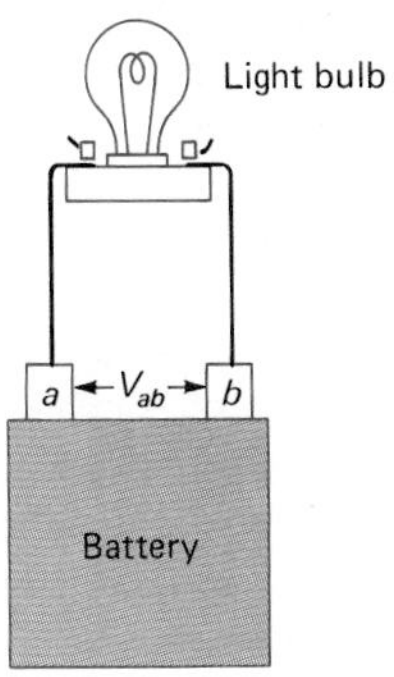

The glow lamp can be made to blink on and off by connecting a capacitor in parallel with the lamp (Figure 30.16). Initially there is no charge on the capacitor. Closing the switch completes the circuit and charge accumulates on the capacitor plates. No charge flows through the lamp until the critical potential V_f is reached. Then it becomes conductive and its dynamic resistance becomes very low. The low resistance of the bulb allows the capacitor to discharge rapidly. When the potential difference across the capacitor falls below V_e, the bulb becomes nonconductive; its dynamic resistance becomes very large and the capacitor recharges, beginning a new cycle. The potential difference across the bulb and capacitor varies with time as shown in Figure 30.17.

Example 5 **Determination of Dynamic Resistance**

Figure 30.18 shows the $I-V$ relation for a solid-state diode. Let us estimate the dynamic resistance when the potential difference across the diode is 0.6 V.

By definition, the reciprocal of the dynamic resistance is dI/dV, the slope of the I versus V curve. For $V = 0.6$ V we draw a tangent to the curve. Measuring the slope, we find

$$\frac{dI}{dV} = \frac{1.83 \times 10^{-3}}{0.43} \frac{\text{A}}{\text{V}} = 4.26 \times 10^{-3} \frac{\text{A}}{\text{V}}$$

Hence the dynamic resistance is

$$\frac{1}{4.26 \times 10^{-3}} = 230\ \Omega$$

30.5 Energy Conversion and Electric Power

The storage battery in an automobile maintains a constant potential difference (usually 12 V) across two metallic terminals. When an electrical device such as a light bulb is connected to these terminals (Figure 30.19), electrons flow from the negative terminal to the positive terminal through the device, and lose electric potential energy as they do so. The transformation of energy takes many forms. If the electrical device is a radio, some of the electric energy is changed to sound energy; if the device is a cigarette lighter, electric energy is transformed to heat; if the device is the starter of an engine, electric energy is converted to mechanical energy; and if the device is a headlamp, electric energy becomes light. The rate at which an electrical device uses energy is practical information. With it, we can determine how much energy is used in a given time, and then, knowing the price of energy, we can compute the cost of operating the device.

Regardless of the nature of the connected device or the manner of producing the potential difference V_{ab}, the electric energy converted by a charge dq moving to a lower potential is

$$dU = V_{ab}\, dq \qquad (30.15)$$

Equation 30.15 follows directly from the definition of potential difference (Section 28.1). In terms of the current in the element

$$dq = I\, dt$$

Hence

$$dU = V_{ab} I\, dt \qquad (30.16)$$

Finally, the rate at which energy is converted is

$$\frac{dU}{dt} = V_{ab} I \qquad (30.17)$$

By definition (Section 7.5), the rate of converting energy is power. The rate of converting electric energy, as in this case, is called *electric power.*

$$P = V_{ab} I \qquad (30.18)$$

The unit of electric power is a volt · ampere = watt = joule per second. You may recall from Chapter 7 that the units of mechanical power are also watts.

Equation 30.18 is a general relation for any type of device that is connected to a constant potential difference. We need not know how V_{ab} is produced. All we need to know to determine the power are V_{ab} and I. If some special relation between V_{ab} and I exists for a particular device, then this relation can be incorporated in Eq. 30.18. For example, if the connected device is a resistor, then we can replace V_{ab} with IR and obtain

$$P = I^2R \tag{30.19}$$

or we can replace I with V_{ab}/R and obtain

$$P = \frac{V_{ab}^2}{R} \tag{30.20}$$

Electrons migrating through a resistor lose energy through collisions with atoms in the structure, increasing the internal energy of the resistor. As a result, the temperature of the resistor rises just as if the resistor had been placed in contact with an object at a higher temperature. This effect is called the *Joule effect* and the thermal energy produced is called **Joule heat.** Electric stoves, toasters, and clothes dryers, for example, use Joule heat.

Example 6 Electric Water Heater

Heaters used to warm small volumes of water consist of a coil of resistance wire that is immersed in the water to be heated. One such heater produces 600 W of power when connected to a 120-V source. Let us determine how long it would take for this heater to bring 0.5 liter of water (enough for two cups of coffee) to a boil if the initial temperature of the water was 27°C, assuming that all the Joule heat of the current-carrying wire is absorbed by the water.

The Joule energy produced by the heater in time t is

$$Q = Pt = 600t$$

The energy required to warm m gm of water by ΔT degrees is (Section 22.5)

$$Q_{H_2O} = mC_{H_2O}\,\Delta T$$

where C_{H_2O} is the specific heat of water. Inserting numerical values, we have

$$Q_{H_2O} = \frac{10^{-3}\ \text{kg}}{\text{cm}^3} \cdot 500\ \text{cm}^3 \cdot \frac{4.19\ \text{kJ}}{\text{kg} \cdot \text{C}^\circ} \cdot (100 - 27)\text{C}^\circ$$
$$= 153\ \text{kJ}$$

Hence

$$t = \frac{153{,}000}{600} = 255\ \text{s}$$
$$= 4.3\ \text{min}$$

What would the cost of heating the water be if the company that supplies the electric power charges 8¢ per kilowatt hour? The energy, expressed in kilowatt hours, is

$$Q = 0.6\ \text{kW} \cdot 4.3\ \text{min} \cdot \frac{1\ \text{hr}}{60\ \text{min}}$$
$$= 0.043\ \text{kWh}$$

At a cost of 8¢ per kilowatt hour, the total cost is

$$\text{cost} = 0.043\ \text{kWh}\ \frac{8¢}{\text{kWh}}$$
$$= 0.34¢$$

Figure 30.20

The size of the potential difference (V) shown here changes with time. A negative value for V indicates a change in polarity for the potential difference.

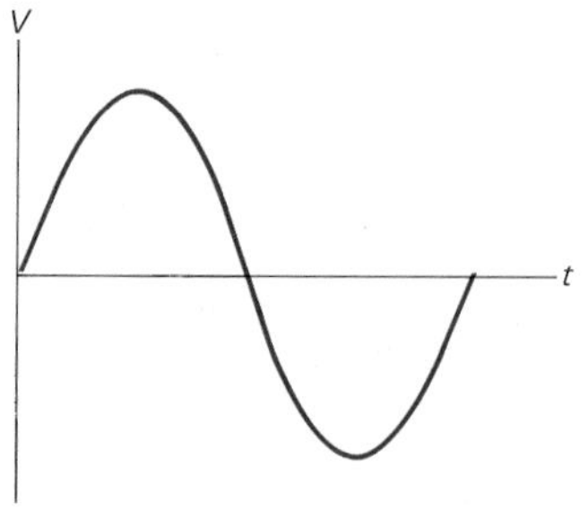

Questions

11. If the rate of production of Joule heat in a resistor is $P = I^2R$, why do heating elements in electric stoves, for example, have resistances of only about 10 Ω?
12. Why is current density fundamental to determining the current-carrying capacity of a wire?
13. In the normal operation of a resistor, what limits the temperature rise due to Joule heating?
14. If for some reason the potential difference across a resistor were to vary as shown in Figure 30.20, would you expect the resistor to produce heat?
15. The heating coils of most household electric stoves require 240 V. If the potential difference were reduced to 120 V while the resistance was kept constant, how much change would there be in the thermal power produced?
16. Describe qualitatively what would happen to the system in Question 15 if the resistance increased with temperature.

Summary

A net flow of charge constitutes an electric current (I). The numerical value of the electric current is equal to the rate of flow of charge.

$$I = \frac{dq}{dt} \tag{30.1}$$

Electric current has units of amperes (A). One ampere equals a rate of charge flow of 1 C/s.

The effect on energy conversion of negative charges moving in one direction is indistinguishable from the effect of positive charges moving in the opposite direction. By convention we assume that an electric current is always due to the motion of positive charges. Consequently, the movement of negative electrons through a conductor is opposite in direction to the current.

Electrons moving through a conductor under the influence of an applied electric field collide with atoms and ions in the conductor. As a result, electrons achieve an average velocity in the direction of the net force. This average velocity is called the drift velocity $\mathbf{v}_D$.

Current density J, expressed in amperes per square meter, measures the concentration of current in a conductor. It is a microscopic quantity that may vary with position. Current density is related to charge density and drift velocity by

$$\mathbf{J} = nq\mathbf{v}_D \qquad (30.5)$$

where nq is the conduction charge density measured in coulombs per cubic meter.

Electrical resistance measures the resistance offered by a conductor to charge flow through it. It is defined as

$$R = \frac{V_{ab}}{I} \qquad (30.7)$$

where I is the current produced when a potential difference V_{ab} exists across the conductor.

For many conducting elements, particularly those made of metal, the electrical resistance is essentially independent of potential difference and current. An element having this characteristic is said to obey Ohm's law.

The resistance of a conductor of length L and cross-sectional area A can be written

$$R = \rho\left(\frac{L}{A}\right) \qquad (30.8)$$

where ρ is the electrical resistivity. In terms of electrical resistivity the equation $V_{ab} = IR$ can be written

$$E = \rho J \qquad (30.12)$$

where E is electric field and J is current density.

Electrical conductivity, σ, is defined as

$$\sigma = \frac{1}{\rho} \qquad (30.9)$$

In terms of electrical conductivity, the equation $V_{ab} = IR$ can be written

$$J = \sigma E \qquad (30.11)$$

For metals, the variation of electrical resistivity with temperature can be described by

$$\rho = \rho_o[1 + \alpha(T - T_o)] \qquad (30.13)$$

where ρ_o is the electrical resistivity at temperature T_o and α is the temperature coefficient of resistivity.

For a class of metals called superconductors the electrical resistivity vanishes at a sufficiently low temperature.

The dynamic resistance, defined as dV/dI, is a useful concept for devices that do not obey Ohm's law. Conditions for negative dV/dI are of special interest because of their utility in oscillators.

The electric power produced when a potential difference V_{ab} is connected across a conductor is

$$P = V_{ab}I \qquad (30.18)$$

where I is the current produced.

Thermal energy produced in a resistor is called Joule heat. The thermal power is given by

$$P = I^2R \qquad (30.19)$$

or

$$P = \frac{V_{ab}^2}{R} \qquad (30.20)$$

Problems

Section 30.1 Electric Current

1. How many electrons per second flow through any cross section of a wire carrying a current of 1 A?

2. To start a car the starter delivers a 200-A current to the car engine for 2 s. How much charge passes through the starter?

3. How much charge flows each minute through any cross section of a wire carrying a current of 2.53 A?

4. In a high-energy particle accelerator protons are moving with almost the speed of light. (a) If the electric current delivered to the target by the protons is 1 mA, how many protons hit the target per second? (b) How many protons are there in a 1.0-m length of the proton beam?

5. On a worldwide basis, lightning transfers a net charge of 1.5×10^8 C from the surface of the earth to the "sky" each day. (The earth acts like one plate of a spherical capacitor. The other "plate" is about 50 km above the surface of the earth. The atmosphere between these two plates is not a perfect insulator, and charge can flow between the two plates.) Determine the average current.

6. Alpha particles from a Van de Graaff accelerator deposit 0.153 mC of charge in a target in 5 min. Determine (a) the number of alpha particles impinging on the target and (b) the current (assumed constant) produced by the alpha particles.

7. Protons emerging from an accelerator produce a constant current of 1.35×10^{-7} A. If the beam impinges on a target, how many protons strike the target each second?

8. (a) What is the velocity of electrons accelerated through a potential difference of 19 kV in a cathode-ray oscilloscope? (b) If the diameter of the electron beam is 1 mm and the current produced is 130 μA, what is the number density of electrons in the beam?

9. A fuse in an electrical circuit is designed to open the circuit like a switch when the current exceeds some preset value. If a fuse is made of a material that melts when the current density reaches 400 A/cm^2, what diameter wire is needed to limit a current to 0.25 A?

10. The current-carrying capacity of a copper wire 0.13 cm in diameter is 6 A. Excluding any differences in heat transfer characteristics, what diameter copper wire is required to have a current-carrying capacity of 20 A?

11. An electric discharge between the ends of two cylindrically shaped conductors produces a cylindrically shaped charge flow. The current density at a radial distance r m from the axis of the cylindrical conducting path can be represented by

$$J = -15r + 0.15 \text{ A/m}^2 \quad \text{for} \quad r \leq 0.01 \text{ m}$$

and

$J = 0$ for $r > 0.01$ m

Determine the total current in the discharge.

Section 30.2 Electrical Resistance and Ohm's Law

12. An orbital electron in a hydrogen atom has a kinetic energy of 13.6 eV. If the radius of the orbit is 5.3×10^{-11} m, what is the electric current produced by the orbiting electron?

13. Suppose that the current in a wire varies with time according to the relation $I = 2.5e^{-1.2t}$ mA when t is expressed in seconds. Determine the total amount of charge that has flowed through the wire by the time the current has diminished to zero.

14. In Example 1 the electron drift speed in a copper wire 1 mm in diameter carrying a current of 1 A was shown to be about 10^{-4} m/s. (a) If the diameter was increased to 10 mm and the current remained the same, what change would occur in the drift velocity? (b) If the diameter remained the same and the current increased to 10 A, what change would occur in the drift velocity?

15. Suppose that the southbound traffic on a straight north–south interstate highway moves with an average speed v. If the number of vehicles per mile of highway is n and the average mass of a vehicle is M, derive an expression for the rate at which mass moves by any point in the southbound lane. What is the relationship between this exercise and that leading to Eq. 30.3?

16. Electrons are "sprayed" onto a moving thread as shown in Figure 1. Calling v the speed of the thread and λ the charge per unit length on the thread, deduce a relation for the current produced by the moving charges.

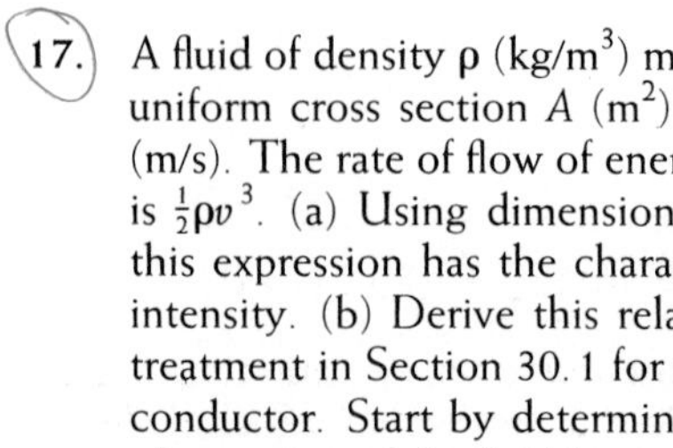

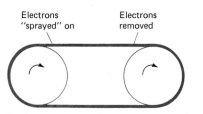

Figure 1

17. A fluid of density ρ (kg/m^3) moves through a pipe of uniform cross section A (m^2) at a constant speed v (m/s). The rate of flow of energy through a unit area is $\frac{1}{2}\rho v^3$. (a) Using dimensional analysis, show that this expression has the characteristics of an energy intensity. (b) Derive this relation by following the treatment in Section 30.1 for the flow of charge in a conductor. Start by determining the kinetic energy of a portion of the fluid having a cross-sectional area A and width d.

18. A geologist measures a potential difference of 3 mV between two probes planted 6 m apart in the earth. The current flowing between the probes is 10^{-12} A. Determine the resistance between the probes.

19. An automobile starter delivers 200 A when driven by a 12-V battery. Assuming that all electrical resistance in the circuit is in the starter, what is the starter resistance?

Section 30.3 Electrical Conductivity and Electrical Resistivity

20. A current of 1 A exists in a copper wire 1 m long and 0.1 mm in diameter. Determine (a) the electric field within the wire and (b) the potential difference between its ends.

21. A Nichrome wire 2 m long and 0.2 mm in diameter is connected to a 100-V battery. Determine the current in the wire.

22. Determine the resistance of a 10-m-long copper conductor having an inside diameter of 0.4 mm and outside diameter of 0.5 mm.

23. The resistivity ρ of copper is 1.724×10^{-8} $\Omega \cdot$ m; that of aluminum is 2.824×10^{-8} $\Omega \cdot$ m. What diameter aluminum wire will have the same resistance per unit length as a copper wire 1 mm in diameter?

24. A wire is stretched to twice its original length by drawing it through a tiny hole. If the wire assumes a uniform cross-sectional area in the stretched configuration and its volume is unchanged by the stretching, show that its electrical resistance increases by a factor of 4.

25. Two cylindrical resistors made from the same material have the same resistance. Prove that their diameters differ by a factor of 2 if their lengths differ by a factor of 4.

26. The conductors that carry current from the generators of large electric power plants are often in the form of a hollow metallic pipe several inches in diameter. Calculate the resistance of a 2-cm-thick aluminum pipe 3 m long with an outside diameter of 20 cm.

27. The copper wire leads for a $\frac{1}{8}$-W commercial resistor are 0.635 mm in diameter and 8 cm long. Determine the resistance of the leads. Does your calculation show that the lead resistance can normally be neglected in most applications?

28. A current of 0.32 A exists in a wire-wound resistor of unknown composition when it is connected to a 10-V potential difference. If the length of the wire is 2 m and its diameter is 0.2 mm, determine the electrical conductivity and determine a possible composition by using Table 30.1 as a guide.

29. A solenoid 10 cm long is formed by winding 25 layers of coil from a continuous length of copper wire

1.25 mm in diameter onto an insulating core 2 cm in diameter. Determine the resistance of the wire.

30. For a research project a coil is to be wound of #44 copper wire (the wire's diameter is 8×10^{-5} m and its resistivity is $1.77 \times 10^{-8}\ \Omega \cdot$ m). The coil is to have 5000 turns, with an average turn diameter of 2×10^{-2} m. What is the expected coil resistance?

31. A current density of 10^{-12} A/m^2 carries charges from the surface of the earth to a region approximately 50 km above the earth's surface. The electric field in this region is 100 V/m. Determine the electrical conductivity of the earth's atmosphere.

32. A 10-cm-long copper bar with a rectangular cross section of dimensions 1 cm $\times$ 2 cm is to be used to conduct 100 A of current between two opposite faces. (a) Through which two opposite faces should the charge flow in order to minimize the potential difference between the faces? (b) What is the current density in the bar? (c) What is the potential difference between the faces?

33. A copper wire [its diameter is 2×10^{-3} m and its conductivity is $5.65 \times 10^{7}\ (\Omega \cdot \text{m})^{-1}$] carries a 200-A current. Determine the electric field inside the wire.

34. A conductor of uniform cross section is constructed from a material whose resistivity varies along its length. If the resistivity can be represented by $\rho = \rho_o + \rho_o m x$, where x denotes the position in meters along the wire ($x = 0$ denotes one end) and m is a constant, derive an expression for the resistance of a wire of length L and cross-sectional area A.

35. Determine from Figure 30.11 the temperature coefficient of resistivity for copper at 20°C. Check your result with that presented in Table 30.1.

36. Figure 30.11 shows how the equation $\rho = \rho_o[1 + \alpha(T - T_o)]$ with $\alpha = 0.00393/°\text{C}$ and $\rho_o = 1.724 \times 10^{-8}\ \Omega \cdot$ m represents the electrical resistivity of copper as a function of temperature. Estimate the percentage difference between the actual and calculated values of resistivity for temperatures of 20°C, 500°C, and 1000°C.

37. Figure 30.11 shows that the resistivity of copper plotted as a function of temperature is nearly linear. This implies that the data can be represented by a linear function of the form

$$\rho = mT + b$$

where m is the slope and b is the intercept. (a) If we choose to write the slope as $m = \alpha\rho_o$, where ρ_o is the resistivity at some temperature T_o, what would the units for the parameter α be? (b) Using this definition for α, show that the intercept is

$$b = \rho_o(1 - \alpha T_o)$$

and that

$$\rho = \rho_o[1 + \alpha(T - T_o)]$$

38. Assuming that the resistivity of copper is linearly related to temperature between 500°C and 1000°C and using data from Figure 30.11, show that the resistivity as a function of temperature can be written as follows.

$$\rho = (5.00 \times 10^{-8})[1 + 0.00172(T - 500)]\ \Omega \cdot \text{m}$$

39. In Section 30.3 we saw that the temperature variation of the resistivity of copper can be approximated by a linear function of the form $\rho = \rho_o[1 + \alpha(T - T_o)]$. (a) Why would a function of the form $\rho = \rho_o[1 + \alpha(T - T_o) + \beta(T - T_o)^2]$ provide a somewhat better representation of the temperature variation of the resistivity? (b) What units would be assigned to the β term? (c) Taking $T_o = 20°\text{C}$ and $\rho_o = 1.724 \times 10^{-8}\ \Omega \cdot$ m for copper, we find that $\alpha = 0.00393/°\text{C}$ for the linear relationship. If $\rho = 5.000 \times 10^{-8}\ \Omega \cdot$ m at 500°C, determine the value of β assuming that $\alpha = 0.00393/°\text{C}$.

40. A wire has a current-carrying capacity that must be considered in the design of an electrical circuit. (a) Why does the current-carrying capacity of a conductor increase as the temperature of the conductor decreases? (b) Assuming that the current-carrying capacity is directly related to resistivity, determine the temperature change required to produce a 10% increase in the current-carrying capacity of a copper wire.

41. A resistor at 20°C has resistance R_o. (a) Show that if this temperature is changed by some small amount ΔT, its resistance changes by $\Delta R = \alpha R_o\ \Delta T$. (b) If two resistors in series have positive and negative temperature coefficients of resistivity, then the change in resistance of one resistor tends to be compensated by the opposite resistance change of the other. Show that the compensation for two series resistors R_1 and R_2 is exact if $R_1/R_2 = -\alpha_2/\alpha_1$. (c) Determine the values of a carbon resistor and an iron resistor if it is desired to maintain $R_{\text{carbon}} + R_{\text{iron}} = 100\ \Omega$.

42. A more general definition of the temperature coefficient of resistivity is

$$\alpha = \frac{1}{\rho}\frac{d\rho}{dT}$$

where ρ is the resistivity at temperature T. (a) Assuming that α is constant, show that

$$\rho = \rho_o e^{\alpha(T - T_o)}$$

where ρ_o is the resistivity at temperature T_o. (b) Using the series expansion for e^x, show that $\rho = \rho_o[1 + \alpha(T - T_o)]$ for $\alpha(T - T_o) \ll 1$.

Section 30.4 Dynamic Resistance

43. The current–potential difference relationship for a solid-state germanium point-contact signal diode is shown in Figure 2. Determine the resistance and dynamic resistance for currents of 0.5 and 1.5 mA.

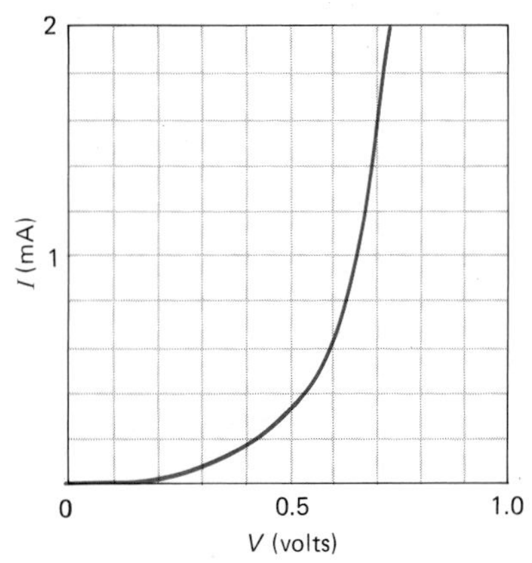

Figure 2

44. A common two-terminal electronic element called a *Zener diode* has a potential difference–current relationship closely approximated by Figure 3. Make a sketch of the resistance of the Zener diode versus the potential difference.

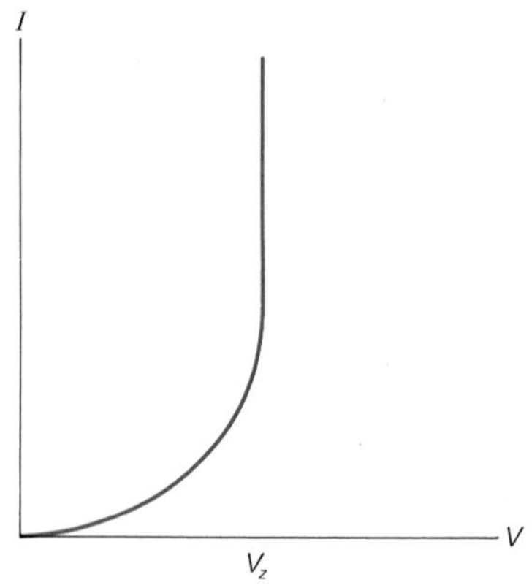

Figure 3

45. A device has a potential difference–current characteristic described by the equation

$$V = V_1 + \frac{P}{I}$$

where $V_1 = 9 \times 10^4$ V and $P = 10^5$ W. When the current $I = 10$ A, calculate (a) the ratio V/I, and (b) the dynamic resistance. Express your answers in ohms.

46. Figure 30.13 shows the characteristics of a tunnel diode. Estimate the average dynamic resistance for the diode over the region between *A* and *B*.

Section 30.5 Energy Conversion and Electric Power

47. An electron beam machine delivers 200 A at 250 kV for 1.10 μs. (a) What is the power in watts during this time interval? (b) How much energy in joules is delivered in this time interval?

48. A 10-W light bulb is connected to a 12-V battery having negligible internal resistance. Determine (a) the current in the bulb and (b) the resistance of the bulb.

49. A heating coil is constructed from Nichrome wire 0.62 mm in diameter. If the coil is to produce 225 W of thermal power when connected to a 24-V potential difference, what length of wire is required?

50. A 10^4-W heater operates with 120 V. What is the heater resistance?

51. Suppose that you want to install a heating coil that will convert electric energy to heat at a rate of 300 W for a current of 1.5 A. (a) Determine the resistance of the coil. (b) The resistivity of the coil wire is 10^{-6} Ω · m. The diameter of the wire is 0.3 mm. Determine its length.

52. The bulb in a two-cell flashlight is rated at 2 W. If the flashlight is left on continuously, the batteries go dead in about 1 hr. (a) Determine the number of kilowatt hours of energy produced during the lifetime of the batteries. (b) If the batteries cost 25¢ each, what is the cost per kilowatt hour of energy? Compare this with a cost of 5¢ per kilowatt hour from an electric utility company. What do you really buy when you use batteries to produce light?

53. An electric heater with a resistance of 20 Ω requires 100 V across its terminals. A built-in switching circuit repetitively turns the heater on for 1 s and off for 4 s. (a) How much energy is produced by the heater in 1 hr? (b) What is the average power delivered by the heater?

54. For the heater in Problem 53, what would the average power be if the potential difference were uniformly increased from 0 to 100 V during the "on" part of the cycle?

55. An aluminum wire 2 m long and 2 mm in diameter is surrounded by a cylindrical copper covering 1 mm thick. A 1-V potential difference is placed between the ends of the wire. Determine the current and thermal power in each component of the wire.

56. Electrons in a color TV are accelerated through a potential difference of 25 kV. If the current produced is 0.25 mA, how many watts of power are dissipated in the TV screen by the electrons?

57. The proton beam of a 6-MeV Van de Graaff generator constitutes an electric current of 1.0 μA delivered to the target. (a) How many protons hit the target per second? (b) With each proton having 6 MeV of kinetic energy, how much energy (in million electron volts and in joules) is delivered to the target in 1 s? (c) What is the power input to the target?

58. A circuit breaker inserted in the wiring of an electrical circuit opens the circuit when the current achieves some maximum value. An electrical circuit in an automobile powered by a 12-V battery utilizes a circuit breaker designed to open when the current reaches 3 A. (a) What is the maximum electric power that can be delivered by the battery? (b) How many 5-W parallel-connected light bulbs could be connected in this circuit protected by a 3-A circuit breaker?

59. If the power dissipated in a 1000-Ω resistor cannot

exceed 0.25 W, what is the maximum tolerable current in the resistor?

60. An electric water heater has a 60-gal capacity and a 5000-W heating element. Assuming that all the electric energy is transferred to the water, determine (a) the cost of heating a 60-gal charge from 70°F to 135°F if the electric energy costs 4¢ per kilowatt hour, and (b) the time required.

61. A 1000-Ω carbon resistor used in an electronic circuit is rated at 0.5 W. Determine the limit of the current in the resistor and the limit for the potential difference across the resistor.

62. A 1000-W electric heater is designed to operate with 24 V. What is the resistance of the heater and the operating heater current?

63. If the thermal power in a 12-m-long copper wire cannot exceed 10 W when a potential difference of 10 V exists across its end, what is the maximum diameter of the wire?

64. A pan containing 1.2 liters of water rests on a 1000-W heating element of an electric stove. If the water is initially at a temperature of 25°C and is brought to a boil in 10.2 min, what is the energy efficiency of this arrangement for boiling water?

65. A current density J exists in a conductor having resistivity ρ and uniform cross-sectional area. Show that the thermal power produced per unit volume of the conductor is ρJ^2.

66. Calculate the thermal power per unit volume in an aluminum wire 2 mm in diameter carrying a current of 0.83 A.

67. A potential difference established between the ends of a conductor of uniform cross section establishes a uniform electric field E in the conductor. Show that the thermal power produced per unit volume of the conductor is σE^2, where σ is the conductivity of the conductor.

68. A potential difference of 0.132 V is established between the ends of a 47.5-cm length of copper wire. Determine the thermal power produced per unit volume of the conductor.

chapter 31

DIRECT-CURRENT CIRCUITS

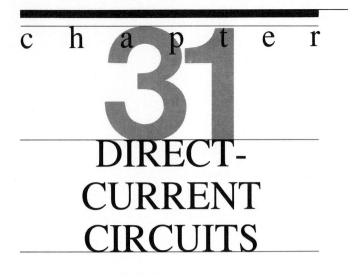

Preview

The commonplace hand-held calculator is a marvel of modern electronics. It performs arithmetical operations with speed and accuracy. Although the design of electronic devices like these calculators requires a wide variety of expertises, an essential one is an understanding of the laws that govern the flow of charge through resistors and capacitors in electronic circuits. This chapter is an introduction to these laws. We start with a discussion of emf (pronounced like the series of letters *e*, *m*, *f*)—the energy source for driving charges through circuits. The principles of energy conservation and charge conservation then lead to Kirchhoff's rules for circuit analysis. By applying these rules we learn how to analyze circuits involving combinations of resistors, capacitors, and sources of emf. Finally, we apply circuit analysis to the electrical instruments used to measure currents and potential differences.

31.1 Source of Electromotive Force (emf)

Contrary to its name, a *source of electromotive force* is a source of energy, not of force. To avoid confusion, we use the term *source of emf* or, in most cases, just emf. Specifically, an emf is a source of electric energy that provides a potential difference between two terminals. An emf provides the electric energy needed to move charges through a conductor that is connected to the terminals. A flashlight battery is an example of an emf. When connected to a light bulb, the battery provides the electric energy to move charges through the filament of the bulb. This electric energy is converted to thermal and radiant energy in the bulb when the charges (assumed positive) move from the higher (positive) potential to the lower (negative) potential. In energy terms, this process is analogous to the situation shown in Figure 31.1, where a ski lift (a source of energy) raises skiers to an elevated position, thereby increasing their gravitational potential energy. The acquired gravitational energy is converted to thermal energy when the skiers slide down and come to rest at the bottom of the slope. In a flashlight battery chemical energy is converted to electric energy.

Figure 31.1

Skiers gain potential energy as the lift raises them to the top of the slope. This potential energy is converted to thermal energy when the skiers slide down the slope and come to rest at the bottom of the slope.

It is possible for emf's to produce currents that vary in both direction and magnitude. For example, the electric generator in a commercial electric power plant produces currents that change direction 120 times a second. [We will study these alternating currents (ac) in Chapter 37.] The emf's considered in this chapter produce currents having only one direction. These currents are called *direct currents*, abbreviated dc.

The variety of useful emf's is large. Some important types are (1) batteries used to convert chemical energy to electric energy (found in automobiles and electronic calculators); (2) biological emf's that convert chemical energy to electric energy in nerve and muscle cells; (3) thermocouples, important temperature-measuring devices consisting of a junction of two dissimilar metals in which thermal energy is converted to electric energy; (4) photovoltaic cells that convert radiant energy to electric energy (used in the light-metering systems of cameras and the electric power supplies of satellites); and (5) electric generators that convert mechanical energy to electric energy (used in electric power plants). For the present we are not concerned with the way in which emf's like these operate. Rather we will study the characteristics of an emf that give it its ability to provide energy to charges.

Work must be done to move positive charges to a region of higher potential. In an emf, this work is accomplished by forces characteristic of that particular emf. In a battery, for example, the forces are electrical. In a generator in an electric power plant the force is provided by a steam turbine. The strength ($\mathscr{E}$) of an emf is defined as the work done per unit of positive charge moved from the lower potential to the higher potential. If work dW is done in moving charge dq, then

$$\mathscr{E} = \frac{dW}{dq} \tag{31.1}$$

Here (and throughout this chapter) we are speaking of conventional currents involving the flow of positive charges. If W and q are measured in joules and coulombs, respectively, then $\mathscr{E}$ is measured in volts: 1 volt = 1 joule per coulomb.

Example 1 Work Done by an emf

A flashlight battery is an emf that develops a constant potential difference of 1.5 V between its positive and negative terminals. Let us calculate the work done in moving

0.1 C of charge from the low-potential (negative) region to the high-potential (positive) region.

Because the strength of the emf is constant (1.5 V), the work done is

$$W = \mathscr{E}q$$
$$= 1.5(0.1)$$
$$= 0.15 \text{ J}$$

Chemical energy is converted to electric energy to account for the work.

Not all of the energy, chemical or otherwise, converted to electric energy by a source of emf is available to external devices. Some of this electric energy is converted internally into thermal energy. This thermal energy is accounted for in circuit analysis by attributing an internal resistance to the emf. The symbol used to denote both the strength ($\mathscr{E}$) and the internal resistance (r) of an emf is

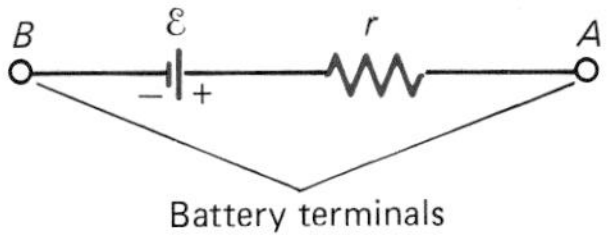

If the emf is connected to a resistor, the equivalent circuit is as shown in Figure 31.2. Electric energy will be converted to heat in both the emf and the external resistor (R). Current is directed out of the positive terminal (A), through the resistor, and into the negative terminal (B). The charges will collide with atoms in the resistor, but there is no mechanism for removing or storing charge anywhere in the circuit. Charge is conserved. Hence the current, dq/dt, is the same at all points in the circuit. From Eq. 31.1, the rate at which electric energy is delivered by the emf is

$$\frac{dW}{dt} = \frac{dW}{dq} \cdot \frac{dq}{dt} = \mathscr{E}I \qquad (31.2)$$

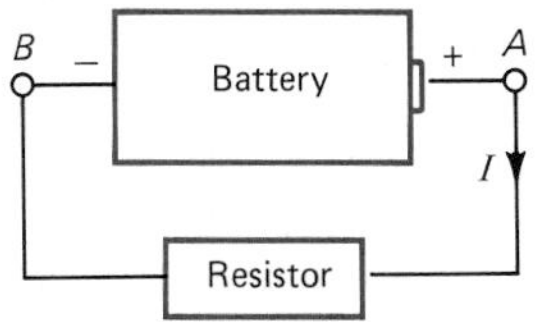

Figure 31.2a

A battery connected to a resistor such as an electric heater or a light bulb.

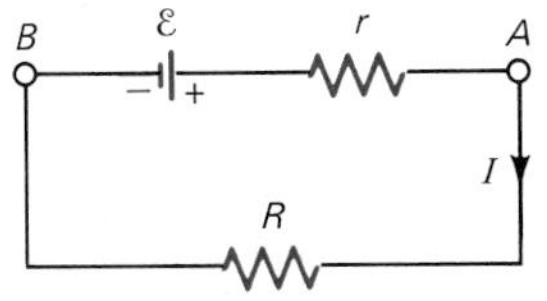

Figure 31.2b

Schematic representation of an emf connected to a resistor.

Experiment shows that there is a conservation of energy and that the thermal power produced internal to and external to the emf equals the electric power supplied by the emf. Hence

$$I^2r + I^2R = \mathscr{E}I$$

Here I^2r and I^2R are the rates at which electric energy is converted to thermal energy in the emf and external resistor, and $\mathscr{E}I$ is the rate at which electric energy is delivered by the emf.* Solving for the current (I), we have

$$I = \frac{\mathscr{E}}{r + R} \qquad (31.3)$$

Equation 31.3 illustrates the important role of the internal resistance, r. If $r \ll R$, as it is for an automobile battery in good condition, then $I \simeq \mathscr{E}/R$, and the internal resistance has little effect on the current produced. But if $r \gg R$, then $I \simeq \mathscr{E}/r$, and the current in the external circuit is determined largely by the internal resistance. When a battery "goes dead" and is unable to deliver sufficient current for some external device to function, it is not necessarily because the strength of the emf decreases. Rather, it is probably because the internal resistance *increases* and prevents adequate current in the external circuit. In subsequent discussions we will assume that the strength ($\mathscr{E}$) and internal resistance (r) are constant for a given emf and that the emf is essentially a source of unlimited energy.

With a current I in the circuit shown in Figure 31.2, the potential difference across the terminals of the emf is

$$V = \mathscr{E} - Ir \qquad (31.4)$$

Thus only if there is no current delivered by the emf will the terminal potential difference be equal to the strength of the emf, that is,

$$V = \mathscr{E} \qquad (31.5)$$

If current *is* delivered by the emf, then $V < \mathscr{E}$. The difference between V and $\mathscr{E}$ increases as the current delivered increases.

Example 2 **Experimental Determination of the Strength ($\mathscr{E}$) and Internal Resistance (r) of a Source of emf**

When a 5-Ω resistor is connected to a flashlight battery, a current of 0.25 A is produced. Removing the 5-Ω resistor and connecting a 9-Ω resistor produces a current of 0.15 A. Let us determine the internal resistance and the strength of the emf of the battery from these measurements.

Using $\mathscr{E} = Ir + IR$ (Eq. 31.3), we can write

$$\mathscr{E} = 0.25r + 0.25(5) \qquad \text{5-}\Omega \text{ resistor connected}$$

$$\mathscr{E} = 0.15r + 0.15(9) \qquad \text{9-}\Omega \text{ resistor connected}$$

*We assume here and in all subsequent analyses that the resistance of any connecting lead is negligible, so that no electric energy is dissipated in the leads. This assumption is valid as long as the resistance of the connecting leads is small compared to the resistance of the elements connected.

Solving these two equations, we find

$r = 1\ \Omega$

$\mathcal{E} = 1.5\ \text{V}$

This method is sometimes used experimentally to determine the strength and the internal resistance of an unknown emf.

Questions

1. A copper and an aluminum electrode stuck into a grapefruit will often produce an emf of 1 V. Why will a light bulb that requires 1 V not light when it is connected to the aluminum and copper electrodes?
2. Describe the energy transformations for charges when an emf is connected to a light bulb.
3. Denote the energy transformations for charges when a household food blender is connected to an electric outlet.
4. How would you revise the model in Figure 31.1 to include an internal resistance for the source of energy for the skiers?
5. Why do commercial testers of transistor radio batteries require that an electrical load, such as a resistor, be connected across the terminals of the battery during the test?
6. A plane carrying a parachutist flies to a certain altitude and remains there for a certain length of time. The parachutist jumps from the plane, floats gracefully to the ground, and walks back to the initial takeoff position at the airport. What energy transformations are involved in the circuitous trip of the parachutist? What energy analogies are there between this example and the path taken by an electric charge in a circuit containing an emf and a resistor?

31.2 Kirchhoff's Loop Rule for Potential Differences

Consider a positive charge q as it moves clockwise around the circuit shown in Figure 31.3. The charge gains electric potential energy in going from lower $(-)$ potential to higher $(+)$ potential in the emf, and loses energy in moving through the resistors. When all the energy changes in the complete circuit or loop are accounted for, experiment reveals that

the net energy gained by the charge =
the net energy lost by the charge

Figure 31.3

Energy aspects of a positive charge making a complete traversal of an electrical circuit.

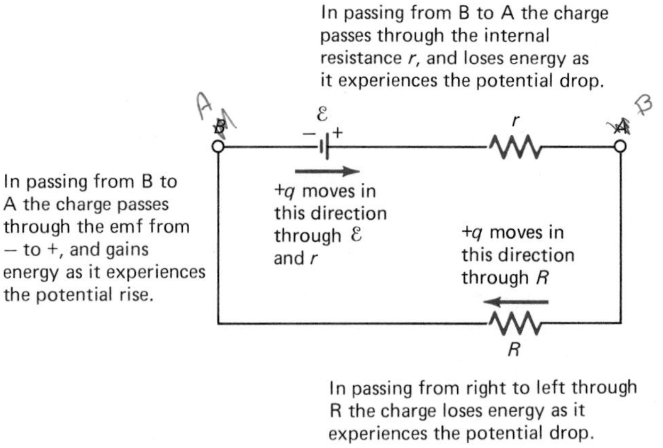
In passing from B to A the charge passes through the internal resistance r, and loses energy as it experiences the potential drop.
A
B
$\mathcal{E}$
r
In passing from B to A the charge passes through the emf from $-$ to $+$, and gains energy as it experiences the potential rise.
$+q$ moves in this direction through $\mathcal{E}$ and r
$+q$ moves in this direction through R
R
In passing from right to left through R the charge loses energy as it experiences the potential drop.

In other words, there is a conservation of energy. This equation is equivalent to

the net energy gained by the charge $-$
the net energy lost by the charge $= 0$

If we take an energy gain as positive and an energy loss as negative, this equation can be rephrased as follows:

The algebraic sum of the energy changes is zero for a positive charge moving around a complete circuit.

An energy gain means that the positive charge has experienced an increase, or positive change, in potential; and an energy loss means that the positive charge has experienced a decrease, or negative change, in potential. Therefore an equivalent statement in terms of potential is the following:

The algebraic sum of the potential changes is zero for a positive charge moving around a complete circuit.

This statement is called **Kirchhoff's* loop rule** and is a result of the conservation of energy in electrical circuits. This result is anticipated from our discussion in Section 28.3 of the conservative nature of electrostatic fields. We are dealing here with conservative forces, namely, electrostatic forces. The work done around a closed path by conservative forces is zero. Energy is conserved.

Kirchhoff's loop rule is applicable only to a complete electrical circuit. This means that to use the rule we must pick some starting point in an electrical circuit, follow a possible path for a positive charge, and then return ultimately to the starting point. A charge returning to its starting point in an electrical circuit experiences no net change in its electric potential energy. This is analogous to starting at the base of a mountain and walking up the mountain, around the mountain, and through crevices, but eventually returning to the starting point. On your return, there is no change in your gravitational potential energy. When applying Kirchhoff's loop rule to a circuit we need only remember that

1 a positive charge experiences a potential rise (positive quantity) when moving from $-$ to $+$ through an emf, and

*Named for German physicist Gustave Robert Kirchhoff (1824–1887), who contributed greatly to our understanding of electricity, optics, and heat.

2 a positive charge experiences a potential drop (negative quantity) when moving through a resistor, for example, in the direction of the current.

Example 3 **An Application of Kirchhoff's Loop Rule**
A 1.5-V battery having a 1-Ω internal resistance is connected to a 10-Ω resistor as shown in Figure 31.4. Let us use Kirchhoff's loop rule to determine the current in the circuit.

Following convention, we label the current direction to correspond to positive charge flowing out of the positive terminal of the emf. A positive charge passing point A moves through the emf, flows through resistors r and R, and returns to point A. This constitutes a complete circuit, or loop, for the charge. Let us start at A and proceed clockwise. As the charge crosses the emf from negative to positive, it experiences a potential rise. As it crosses r and R the charge experiences potential drops. Applying the loop rule, we have

$$\mathscr{E} - Ir - IR = 0$$

Solving for I yields

$$I = \frac{\mathscr{E}}{r + R} = \frac{1.5}{1.0 + 10} = \frac{1.5}{11.0} = 0.136 \text{ A}$$

Note that the equation

$$I = \frac{\mathscr{E}}{r + R}$$

is identical to the equation we obtained by using conservation of energy (Eq. 31.3).

We obtain the same result if we start at A and proceed counterclockwise around the circuit:

$$IR + Ir - \mathscr{E} = 0$$

$$I = \frac{\mathscr{E}}{r + R}$$

Figure 31.5 shows schematically a series connection of two emf's and a resistor. This circuit could be used to charge the battery in a golf cart, for example; $\mathscr{E}_1$ could represent the "charger" and $\mathscr{E}_2$ could represent the battery. Convention dictates that the current direction be out of the positive terminal of an emf. Because there are two emf's in this circuit, we might be inclined to label a current direction out of each emf. However, electric charge is

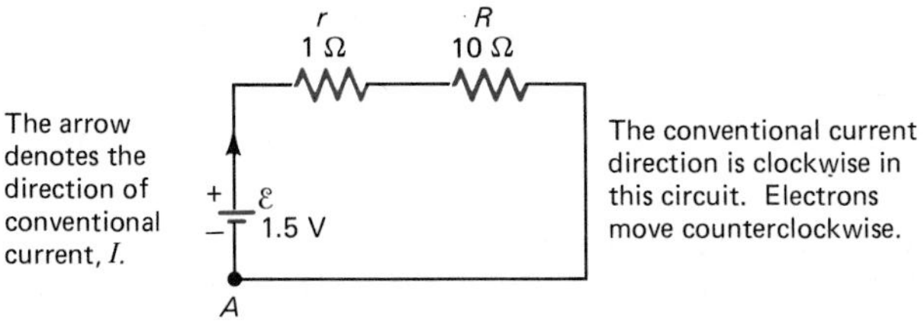
r
1 Ω
R
10 Ω
The arrow denotes the direction of conventional current, I.
+
−
ℰ
1.5 V
A
The conventional current direction is clockwise in this circuit. Electrons move counterclockwise.

Figure 31.4

Schematic diagram of a battery (strength 1.5 V, internal resistance 1 Ω) connected to a 10-Ω resistor.

Figure 31.5

A circuit diagram of two emf's connected by a resistor. The current direction is *assumed* to be counterclockwise in the circuit. We choose arbitrarily to proceed clockwise from G around the circuit when applying Kirchhoff's loop rule.

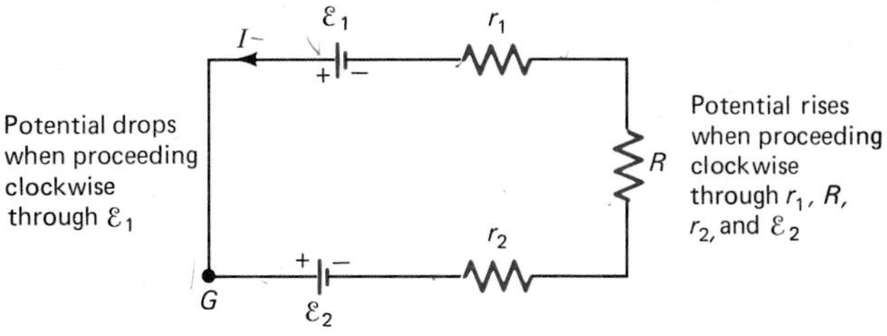
ℰ1
r1
I
+
−
Potential drops when proceeding clockwise through ℰ1
Potential rises when proceeding clockwise through r1, R, r2, and ℰ2
R
r2
G
ℰ2

conserved. Hence, the current in each circuit element (emf or resistor) must be the same. Suppose the circuit conditions are such that if charge flows out of $\mathscr{E}_1$, then the same charge must flow into $\mathscr{E}_2$. Physically, this means that if the emf's are reversible devices like the batteries used in automobiles, then $\mathscr{E}_1$ is being discharged (converting chemical energy into electric energy) and $\mathscr{E}_2$ is being charged (converting electric energy into chemical energy).

If we assume a given direction for the current and this direction proves to be wrong, then the current calculated by using the loop rule will be negative. Thus, we need not know the direction of the current before solving the problem because our answer will tell us if we guessed correctly. To start, then, let us assume that current direction is counterclockwise, as shown. Starting at G and proceeding clockwise around the circuit, we have

$$\underbrace{-\mathscr{E}_1}_{\text{potential drop}} + \underbrace{Ir_1 + IR + Ir_2 + \mathscr{E}_2}_{\text{potential rises}} = 0$$

Note that Ir_1, IR, and Ir_2 are potential rises because we elected to go in the opposite direction to the assumed direction of I. If we had gone counterclockwise, these three terms would have been potential drops. Rearranging terms, we have

$$\mathscr{E}_1 - \mathscr{E}_2 = Ir_2 + IR + Ir_1$$

Solving for the current I, we have

$$I = \frac{\mathscr{E}_1 - \mathscr{E}_2}{r_1 + r_2 + R} \qquad (31.6)$$

Thus if $\mathscr{E}_1 > \mathscr{E}_2$, I is positive and the charge flows in the assumed direction. If $\mathscr{E}_1 < \mathscr{E}_2$, then I is negative and the charge flows opposite to the assumed direction. If $\mathscr{E}_1 = \mathscr{E}_2$, then $I = 0$.

Experimental measurements provide a way of confirming the theoretical results deduced from Kirchhoff's rules. In principle, we have a choice of measuring currents or potential differences. In order to measure a current in a circuit we must break the circuit and insert a meter. However, we can measure a potential without breaking the circuit. We need only attach the meter connections to the points of interest in the circuit. Generally, potentials are measured relative to some point in the circuit that is taken to have zero potential. It is customary to call this point *ground* (Section 28.1) and denote it by the symbol ⏚. The

Figure 31.6

The symbol ⏚ in this circuit designates ground. It denotes a point where the potential is taken arbitrarily to be zero. All other potentials in the circuit are then measured relative to ground. The potential values shown are measured relative to ground.

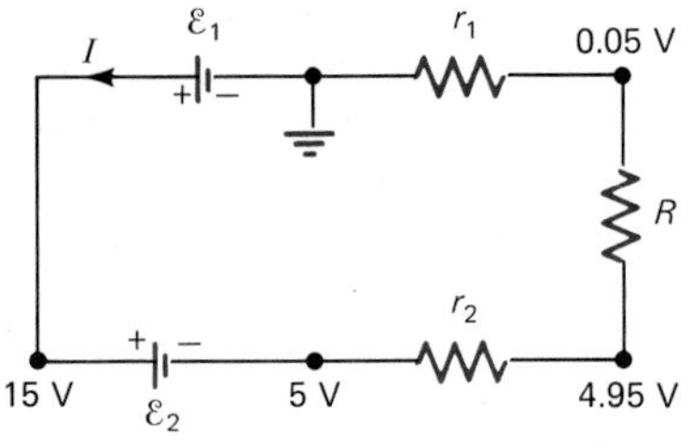

potentials shown in Figure 31.6 are all measured relative to the point representing ground. We can show that these potentials are consistent with the analysis done for the same circuit (Figure 31.5). Starting at ground and going counterclockwise around the circuit, the potential changes from 0 to +15 V in crossing $\mathcal{E}_1$; this is a potential rise of +15 V. Crossing $\mathcal{E}_2$, the potential changes from 15 V to 5 V; this is a potential drop of −10 V. Crossing r_2, the potential changes from +5 V to 4.95 V; this is a potential drop of −0.05 V. Likewise, crossing R and r_1 gives potential drops of −4.9 and −0.05 V. Summing the potential changes, we have

$$\begin{array}{cccccc} +15 & -10 & -0.05 & -4.9 & -0.05 & = 0 \\ \mathcal{E}_1 & \mathcal{E}_2 & r_2 & R & r_1 & \end{array}$$

The algebraic sum of the potential changes equals zero, in accordance with Kirchhoff's loop rule. Given the value of an external resistor and having determined the potential difference across it, we can calculate the current by using Eq. 30.7.

Questions

7. What is the net change in potential energy for the parachutist in Question 6 when he returns to the starting position? Devise a scheme for summing the potential energy changes for the parachutist, and deduce a rule similar to Kirchhoff's loop rule for potential differences.

8. Draw a circuit for the battery, horn, and horn switch in an automobile. Apply Kirchhoff's loop rule to this circuit when (a) the horn switch is not depressed (that is, when it is open) and (b) the horn switch is depressed (closed).

9. Draw a circuit for the battery, headlights (high beam and low beam), light switch, and high-beam switch in an automobile. Apply Kirchhoff's loop rule to a couple of circuits in this system.

10. How would the loop rule for potential differences be changed if we interpreted the current as a flow of negative charges?

31.3 Application of Kirchhoff's Loop Rule: Resistors in Series

A series connection of resistors is shown in Figure 31.7. We can use Kirchhoff's loop rule to replace such a combination of resistors with a single resistor that produces no change in the circuit current. The key to doing this is to recognize that charge is conserved, and that therefore the current is the same at all points in the series connection. Applying Kirchhoff's loop rule for potential differences in the circuit in Figure 31.7 and solving for the current, we have

$$I = \frac{\mathcal{E}}{r + R_1 + R_2 + R_3} \qquad (31.7)$$

If a single resistor R having a value $R = R_1 + R_2 + R_3$ were connected to the terminals of the emf, the current would be

$$I = \frac{\mathcal{E}}{r + R} \qquad (31.8)$$

and the current would be unchanged. In general, any series combination of resistors can be replaced by a single resistor having resistance equal to the sum of the resistances without changing current in the circuit. For N resistors in series the equivalent resistance (R) is

$$R = R_1 + R_2 + R_3 + \cdots + R_N \qquad (31.9)$$

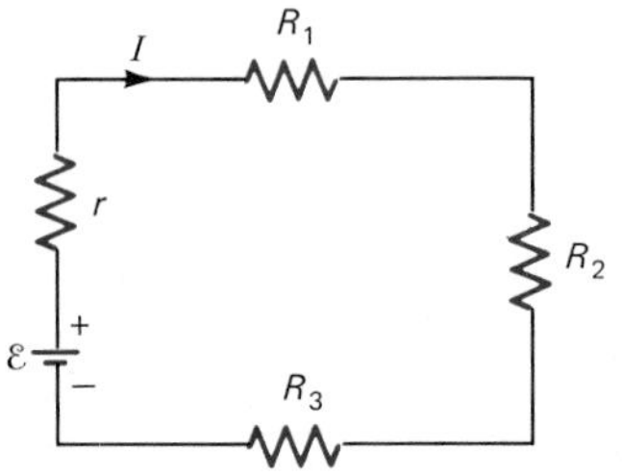

Figure 31.7

Resistors R_1, R_2, and R_3 in this circuit are said to be connected in series.

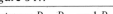

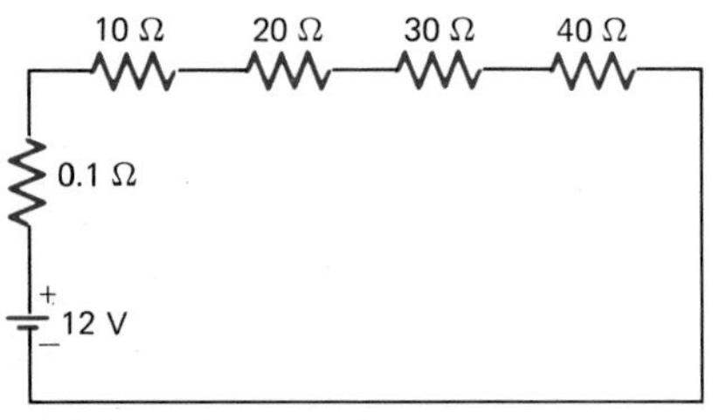

Figure 31.8

A series connection of four resistors and a battery.

Note that determining the equivalent resistance of resistors in *series* is like computing the equivalent capacitance of capacitors in *parallel*. This is because for a resistor $RI = V$ and for a capacitor $(1/C)Q = V$.

Example 4 **Resistors in Series**

A 10-Ω, a 20-Ω, a 30-Ω, and a 40-Ω resistor are connected in series to a battery having an emf of 12 V and an internal resistance of 0.1 Ω (Figure 31.8). What is the current in the circuit? The equivalent resistance of the resistors is

$$R = 10\ \Omega + 20\ \Omega + 30\ \Omega + 40\ \Omega$$
$$= 100\ \Omega$$

Effectively, these resistors are connected in series with the internal resistance of the battery, and so the resistance "felt" by the emf is

$$R + r = 100\ \Omega + 0.1\ \Omega$$
$$= 100.1\ \Omega$$

Hence the current is

$$I = \frac{\mathcal{E}}{r + R} = \frac{12}{100.1} = 0.12\ \text{A}$$

In this example $r \ll R$. Therefore, even if we had neglected the internal resistance, the current calculation would be unchanged, since the calculation is accurate to only two digits.

$$I = \frac{12}{100} = 0.12\ \text{A}$$

Questions

11. An electric power plant is linked to a community by transmission lines held off the ground by utility poles or towers. Why is it appropriate to say that these pole-to-pole connections are connected in series?

12. Suppose that the traffic on a highway consisted of two parallel lanes of cars moving with the same speed in the same direction. If the highway narrowed to one lane and the rate of flow of traffic, and the vehicle speed, remained the same, what must have occurred to the arrangement of the vehicles? If both lanes initially were filled bumper to bumper and the rate of flow of traffic remained constant when the two lanes narrowed to one, what must have happened to keep the rates constant? How are these examples similar to the flow of electric charges in a circuit?

13. Suppose that you and a friend are stationed at opposite ends of a tunnel and are measuring the rate of flow of vehicles through the tunnel. Comparing notes, you find that the rates are identical. What conclusions can you draw about events inside the tunnel? How is this example analogous to the flow of charges through a resistor?

31.4 Kirchhoff's Junction Rule for Currents

The intersection of electrical conductors such as metal wires is called a *junction* (Figure 31.9). When charges arrive at a junction they divide and leave the junction through the available conductors. Because experiment shows that charge is conserved, the charge flowing into a junction per second must equal the charge flowing out of the junction per second. For example, if the current is 1.0 A directed toward the junction in the horizontal branch of Figure 31.10 and the current is 0.3 A directed away from the junction in the upward branch, then the current must be 0.7 A directed away from the junction in the downward branch. This observation is the basis for **Kirchhoff's junction rule:**

The total current directed toward a junction equals the total current directed away from a junction.

If we adopt the convention that a current directed toward a junction is positive and a current directed away from a junction is negative, then Kirchhoff's junction rule can be stated as follows:

The algebraic sum of the current at any junction is zero.

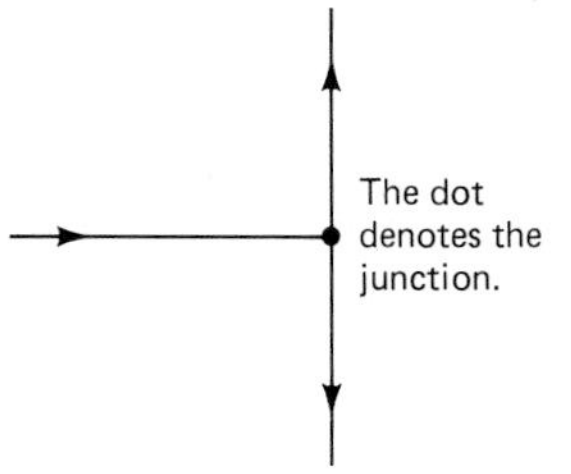

Figure 31.9

A physical connection of two or more conductors forms a current junction. Current entering a junction will divide.

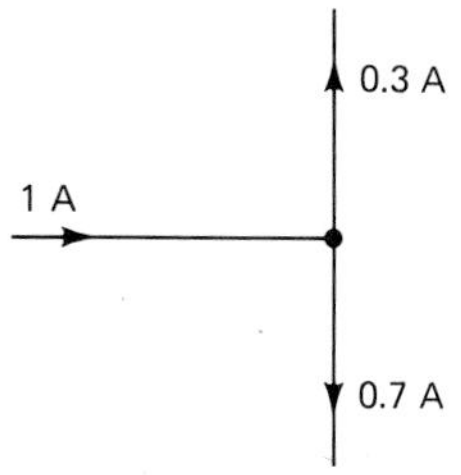

Figure 31.10

The total current directed toward a junction must equal the total current directed away from the junction. Here, 1.0 A is directed toward the junction and 0.3 A + 0.7 A = 1.0 A is directed away from the junction.

Figure 31.11

A parallel combination of an emf and two 100-Ω resistors.

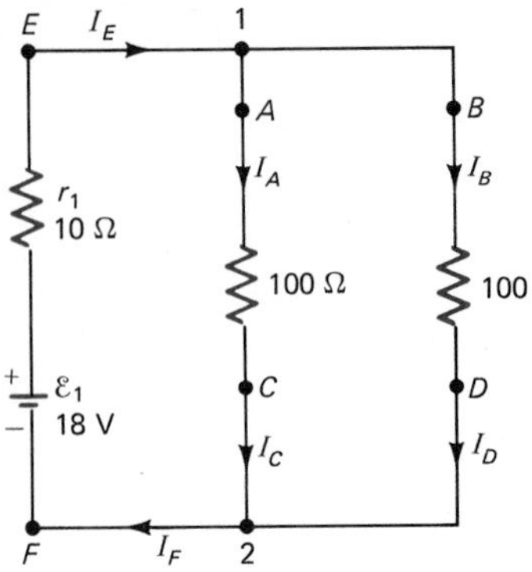

Example 5 **Charge Conservation and Kirchhoff's Junction Rule**

An emf having a 10-Ω internal resistance and a strength of 18 V is connected to two 100-Ω resistors as shown in Figure 31.11. Experimental measurements indicate that the currents at points A and B in the 100-Ω resistors are each 0.15 A. What are the current values at the points labeled C, D, E, and F?

Current is constant in any continuous segment of a circuit. Hence the current at C is the same as the current at A, and the current at D is the same as the current at B:

$$I_A = I_C = 0.15 \text{ A}$$

$$I_B = I_D = 0.15 \text{ A}$$

Convention has it that current is directed out of the positive terminal of the emf. Hence charge passing through point E flows into the junction labeled 1. Charges flowing through points A and B flow out of the junction labeled 1. Kirchhoff's junction rule requires that at junction 1,

$$I_E = I_A + I_B$$

We know that

$$I_A = 0.15 \text{ A}$$

and

$$I_B = 0.15 \text{ A}$$

Hence

$$I_E = 0.15 + 0.15$$

$$= 0.30 \text{ A}$$

Charges passing through points C and D flow into the junction labeled 2. Applying Kirchhoff's junction rule at junction 2, we have

$$I_C + I_D = I_F$$

Again, we know that

$$I_C = +0.15 \text{ A}$$

and

$$I_D = +0.15 \text{ A}$$

Hence

$$I_F = 0.15 + 0.15$$

$$I_F = 0.30 \text{ A}$$

A current-measuring meter at points E and F will register the same values for the current. This is to be expected from the principle of charge conservation.

We can illustrate both the Kirchhoff junction rule and the Kirchhoff loop rule by considering the circuit in Figure 31.12a. Let us assume that the values of the resistors and the characteristics of the emf's are known, and that we want to determine the unknown currents. The first step is to label the junctions and unknown currents in the circuit (Figure 31.12b). There are junctions at A and B, and three currents, I_1, I_2, and I_3. The arrows indicate the assumed directions of these three currents. These directions need not be correct, since our calculations will yield a negative value for any current whose assumed direction is incorrect. The next step is to identify the circuit loops (Figure 31.12c). One loop can be formed by starting at junction A and proceeding first to junction B, then to point C, then to point D, and returning to A (Figure 31.12c). Two other loops are present, one involving $A \rightarrow B \rightarrow F \rightarrow E \rightarrow A$ and the other $A \rightarrow D \rightarrow C \rightarrow B \rightarrow F \rightarrow E \rightarrow A$. If we consider the two junctions and three loops, Kirchhoff's rules will yield five equations. We need three independent equations in order to solve for the three currents, but we cannot arbitrarily pick any three. Let us see why. Applying the junction rule at A and B, we have

$$I_1 + I_2 = I_3 \quad \text{at } A \qquad (31.10)$$

$$I_3 = I_1 + I_2 \quad \text{at } B \qquad (31.11)$$

These two equations are algebraically identical and therefore are not independent. Both equations equal $I_1 + I_2 - I_3 = 0$. Applying the loop rule, we obtain three more equations:

$$-I_3R + \mathscr{E}_1 - I_1r_1 = 0 \qquad (31.12)$$
for loop 1 $A \rightarrow B \rightarrow C \rightarrow D \rightarrow A$

$$-I_3R + \mathscr{E}_2 - I_2r_2 = 0 \qquad (31.13)$$
for loop 2 $A \rightarrow B \rightarrow F \rightarrow E \rightarrow A$

$$I_1r_1 - \mathscr{E}_1 + \mathscr{E}_2 - I_2r_2 = 0 \qquad (31.14)$$
for loop 3 $A \rightarrow D \rightarrow C \rightarrow B \rightarrow F \rightarrow E \rightarrow A$

However, note that the loop 3 equation can be reproduced by subtracting the loop 1 equation from the loop 2 equation. Thus only two of the three loop equations are independent. We are left with only the required three independent equations involving the unknown currents I_1, I_2, and I_3. Note that the equations are linear; doubling the sizes of $\mathscr{E}_1$ and $\mathscr{E}_2$ causes all currents to double. Because the equations are linear, their solution is straightforward. Picking either one of the two junction equations (Eq. 31.10 or Eq. 31.11) and any two of the loop equations, (Eq. 31.12, 31.13, or 31.14), we can solve for I_1, I_2, and I_3.

Figure 31.12

The steps involved in the application of Kirchhoff's rules are illustrated in this series of diagrams.

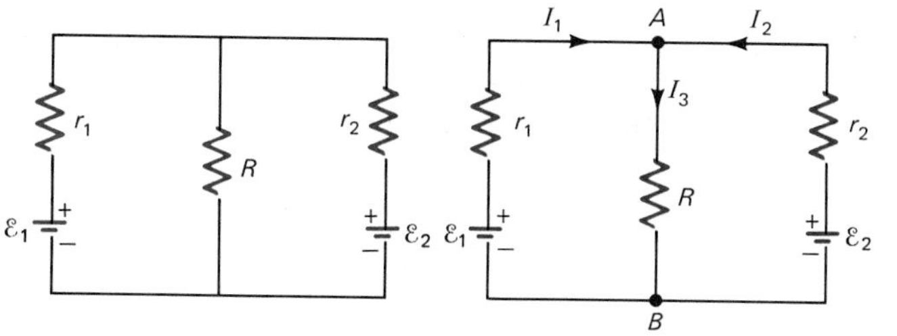

(a)

We start with the circuit diagram.

(b)

Then we label the junctions and identify the currents to be determined. Convince yourself, using Example 5 if you like, that the currents at junction *A* are the same as those at junction *B*.

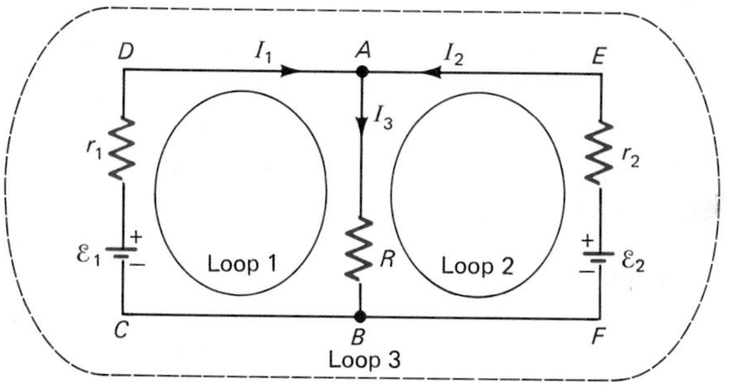

(c)

The next step is to identify the loops in the circuit.

$$I_1 = \frac{\mathscr{E}_1(r_2 + R) - \mathscr{E}_2 R}{r_1 r_2 + r_1 R + r_2 R} \qquad (31.15)$$

$$I_2 = \frac{\mathscr{E}_2(r_1 + R) - \mathscr{E}_1 R}{r_1 r_2 + r_1 R + r_2 R} \qquad (31.16)$$

$$I_3 = \frac{\mathscr{E}_1 r_2 + \mathscr{E}_2 r_1}{r_1 r_2 + r_1 R + r_2 R} \qquad (31.17)$$

Before checking the algebra, let's see if the answers are reasonable. First, dimensions and units must be consistent. This holds true for our three equations. The left-hand side of each equation has units of amperes. The right-hand side of each equation has units of volts per ohm, which are also equal to amperes.

Second, it may be possible to reduce the circuit to some situation where the answer is obvious or known from a previous calculation. For example, if loop 2 is open-circuited by the severing of a wire, then no charge flows through r_2 and the circuit reduces to that in Figure 31.2. The answer from a previous calculation (Eq. 31.4) is $I_1 = I_3 = \mathscr{E}_1/(r_1 + R)$. Severing the wire is equivalent to letting r_2 approach infinity. Thus Eqs. 31.15 and 31.17 should yield $I_1 = I_3 = \mathscr{E}_1/(r_1 + R)$ for the limiting case in which $r_2 \to \infty$. You can verify this by letting r_2 approach infinity in Eqs. 31.15 and 31.17. Instead of loop 2, suppose that the wire connecting R to junction A is severed. Then the current in R must vanish. Severing the wire is equivalent to letting R approach infinity. Using Eq. 31.17, convince yourself that $I_3 \to 0$ if $R \to \infty$. With R removed, the circuit reduces to that shown in Figure 31.5. Drawing on the solution for the current in that circuit (Eq. 31.6), we expect

$$I_1 = -I_2 = \frac{\mathscr{E}_1 - \mathscr{E}_2}{r_1 + r_2}$$

You can convince yourself that this is so by letting R approaach infinity in Eqs. 31.15 and 31.16.

Kirchhoff's loop and junction rules always yield exactly the number of independent equations needed to solve for the circuit currents. But it is a tedious job to sort out and solve the resulting simultaneous equations if there are more than three unknown currents. There are several systematic approaches to solving this problem that are beyond the scope of our interests here. A particularly attractive method takes advantage of the ease of solving simultaneous linear algebraic equations with modern digital computers. Specially designed computer programs for circuit analysis are readily available.* You might want to become familiar with these possibilities if you have access to a computer.

Questions

14. What analogies, if any, exist between the flow of charges up to and away from a junction of two wires and the flow of cars up to and away from a junction of two highways?

*See, for example, *Fortran for Engineering Physics—Electricity, Magnetism, and Light*, by Alan B. Grossberg; McGraw-Hill, New York, 1973.

15. Equations 31.15, 31.16, and 31.17 are the solutions for the unknown currents I_1, I_2, and I_3 for the circuit depicted in Figure 31.12. Under what conditions will the current I_1 be opposite to that designated in Figure 31.12b? How is it possible to have I_2 or I_3 equal to zero? How is it possible to have I_3 equal to zero?

31.5 Application of Kirchhoff's Junction Rule: Resistors in Parallel

The terminals of a number of light bulbs in a house trailer are all connected to the same battery. We say that these bulbs are connected in parallel. The lights in an automobile and the lights in a house are also connected in parallel. A parallel connection of resistors is shown schematically in Figure 31.13a. All three resistors, R_1, R_2, and R_3, are connected across the emf. If we know the resistance values and the characteristics of the emf, we can determine all currents and potential differences by using Kirchhoff's rules. This can prove very useful. Also, we may want to determine what single resistance could replace the parallel combination and leave the total current (I) in the circuit unchanged (Figure 31.13b). Or we may want to know how I is changed when more resistors are placed in parallel. If we neglect the resistance of the conductors that connect the resistors, then the potential difference across each resistor in the parallel combination is the same. Let us call this potential difference V. From the junction rule for currents and the equation $I = V/R$, it follows that

$$I = I_1 + I_2 + I_3$$

$$= \frac{V}{R_1} + \frac{V}{R_2} + \frac{V}{R_3} \qquad (31.18)$$

If the resistor R connected between the same potential V leaves the current I unchanged, then

$$I = \frac{V}{R} \qquad (31.19)$$

Combining Eqs. 31.18 and 31.19 yields

$$\frac{V}{R} = \frac{V}{R_1} + \frac{V}{R_2} + \frac{V}{R_3}$$

Canceling V, we have

$$\frac{1}{R} = \frac{1}{R_1} + \frac{1}{R_2} + \frac{1}{R_3} \qquad (31.20)$$

$$R = \frac{1}{\dfrac{1}{R_1} + \dfrac{1}{R_2} + \dfrac{1}{R_3}} \qquad (31.21)$$

Figure 31.13a

Three resistors connected in parallel to the terminals of a battery. The terminals are labeled A and B.

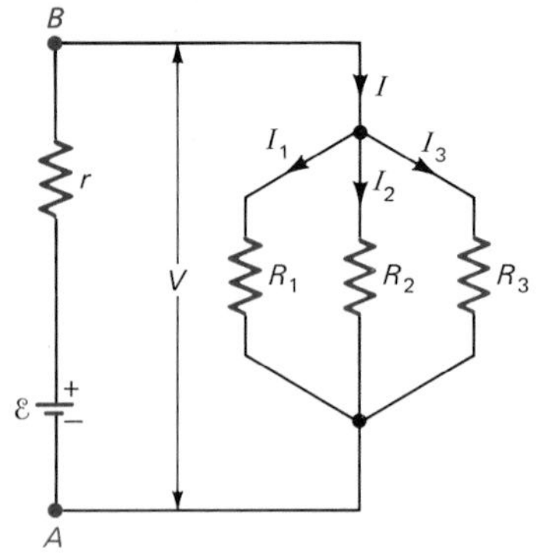

Figure 31.13b

The single resistor R causes the same current I as the three resistors R_1, R_2, and R_3 in Figure 31.13a.

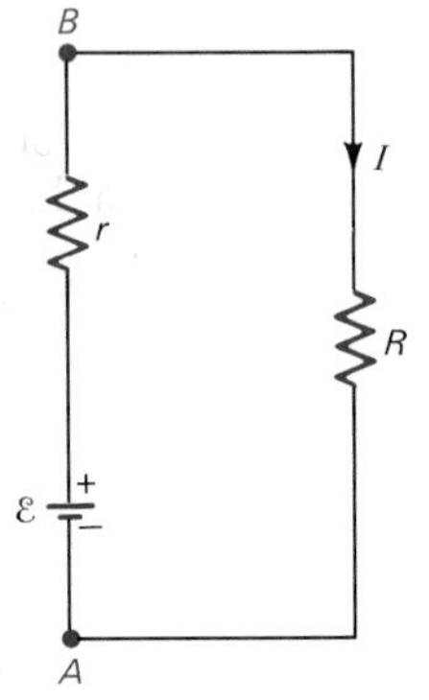

R is the equivalent resistance of R_1, R_2, and R_3 connected in parallel. The same total current results in the circuit when either the single resistor R, or the three resistors R_1, R_2, R_3 connected in parallel, are present. Generalizing Eqs. 31.20 and 31.21 for N resistors in parallel, we have

$$\frac{1}{R} = \frac{1}{R_1} + \frac{1}{R_2} + \frac{1}{R_3} + \cdots + \frac{1}{R_N} \qquad (31.22)$$

$$R = \frac{1}{\dfrac{1}{R_1} + \dfrac{1}{R_2} + \dfrac{1}{R_3} + \cdots + \dfrac{1}{R_N}} \qquad (31.23)$$

Note that determining the equivalent resistance of resistors in *parallel* is like computing the equivalent capacitance of capacitors in *series*. The effective resistance of any number of resistors in parallel is obtained by first adding the reciprocals of all the resistances and then taking the reciprocal of the sum. For instance, the effective resistance of 20 5-Ω resistances in parallel is

$$\underbrace{\frac{1}{R} = \frac{1}{5} + \frac{1}{5} + \frac{1}{5} + \cdots + \frac{1}{5}}_{\text{20 terms}}$$

$$= 20\left(\frac{1}{5}\right) = 4$$

$$R = \frac{1}{4}\ \Omega$$

Some current exists in each resistor in a parallel combination. Therefore the total current in a parallel combination has to exceed the current in any single resistor in the set. Hence the equivalent resistance of the parallel combination is always smaller than the resistance of any single resistor in the set.

Example 6 **Resistors in Parallel**

What is the equivalent resistance of a parallel combination of 5-Ω, 10-Ω, and 20-Ω resistors? Using Eq. 31.20 we write

$$\frac{1}{R} = \frac{1}{5} + \frac{1}{10} + \frac{1}{20}$$

$$= \frac{7}{20}$$

$$R = \frac{20}{7} = 2.86\ \Omega$$

Note that the equivalent resistance is less than the smallest resistance in the combination.

If this parallel combination is connected to an emf of 12 V and internal resistance $\frac{1}{7}\ \Omega$, what is the current in the battery? (Figure 31.14)? The equivalent circuit for this situation is shown in Figure 31.15. The total current I is

$$I = \frac{\mathscr{E}}{r + R}$$

$$= \frac{12}{\frac{1}{7} + \frac{20}{7}}$$

$$= 4\ \text{A}$$

A good battery provides a terminal potential difference that is relatively insensitive to the resistance connected between its terminals. A parallel combination of identical resistors has a significant effect on the current delivered by such a battery, but not on the battery's terminal potential difference.

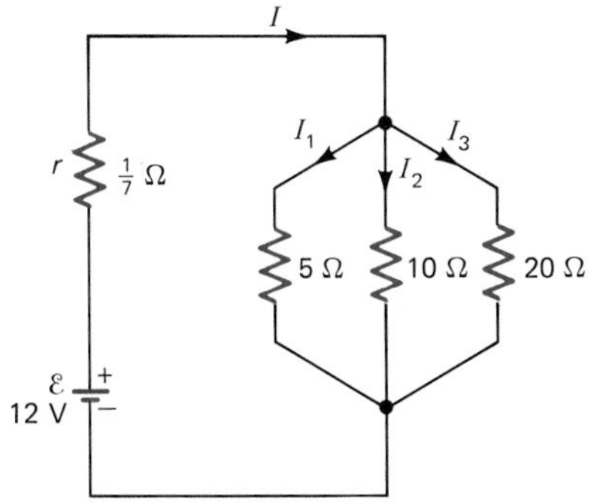

Figure 31.14

A parallel combination of a 5-Ω resistor, a 10-Ω resistor, and a 20-Ω resistor connected to a battery. The total current delivered by the battery is labeled I.

Example 7 **Turning on More Lights**

Consider a house trailer supplied with a 12-V automobile battery. Each light in the trailer is identical, and can be connected to the battery terminals (Figure 31.16). What is the effect on the battery current (I) of the number of lights turned on?

If n lights are connected, the equivalent resistance is R/n. Thus the current I is given by

$$I = \frac{\mathscr{E}}{r + R/n} = \frac{n\mathscr{E}}{nr + R}$$

If nr is negligible in comparison with R, then $I \cong n\mathscr{E}/R$.

The battery current is proportional to the number of lights turned on. Because the potential difference is constant, this means that the power is also proportional to the number of lights turned on (as we saw in Section 30.5).

The circuit in Figure 31.17 involves both series and parallel combinations of resistors. By applying either Kirchhoff's rules or the methods for reducing series and parallel combinations of resistors that we derived by using Kirchhoff's rules, we can determine the currents and potential differences. We will use the latter method. The R_2, R_3 and R_4, R_5 combinations are in series. Therefore, R_2 and R_3 can be replaced with a resistor equal in value to the sum of the resistances of R_2 and R_3, while R_4 and R_5 can be replaced with a resistor equal in value to the sum of the

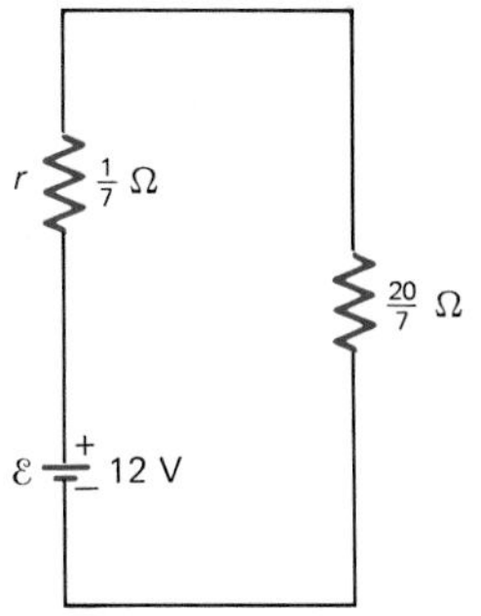

Figure 31.15

This circuit is equivalent to the one shown in Figure 31.14 in the sense that the battery delivers the same total current I.

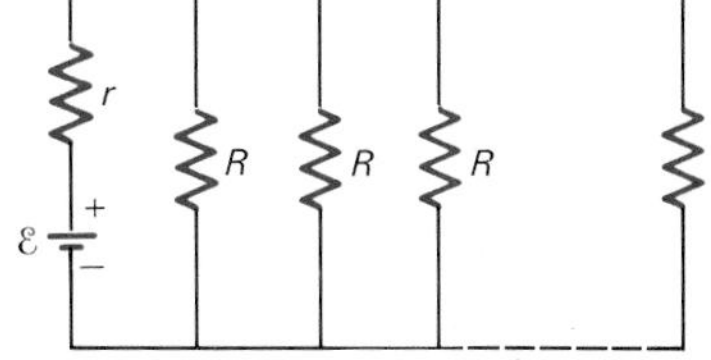

Figure 31.16

In this circuit, R represents the resistance of a light connected to the battery. An arbitrary number n of lights is connected.

Figure 31.17

A circuit having a combination of series and parallel connections of resistors.

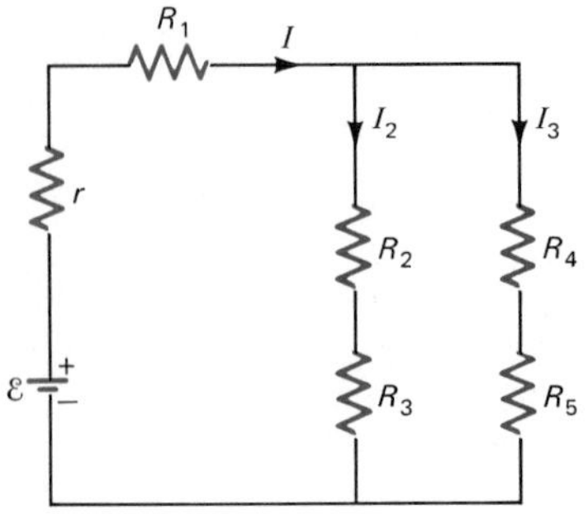

Figure 31.18

In this circuit, R_6 and R_7 replace resistors R_2, R_3, R_4, and R_5 in Figure 31.17. The new circuit leaves the total current I unchanged.

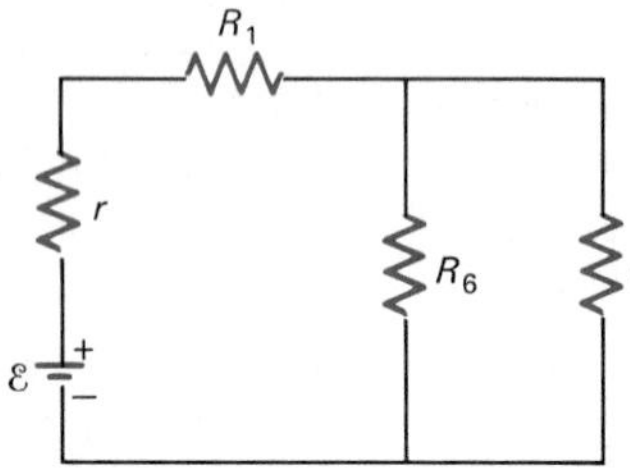

Figure 31.19

In this circuit, R_8 replaces the parallel combination of R_6 and R_7 in Figure 31.18.

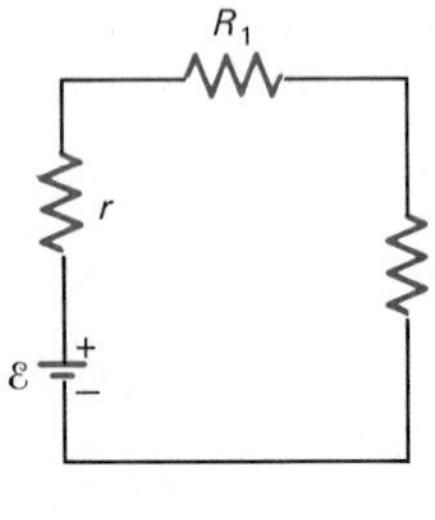

Figure 31.20

In this circuit, R replaces the series combination of r, R_1, and R_8 in Figure 31.19. The total current I is the same in all the circuits shown in Figures 31.17–31.20.

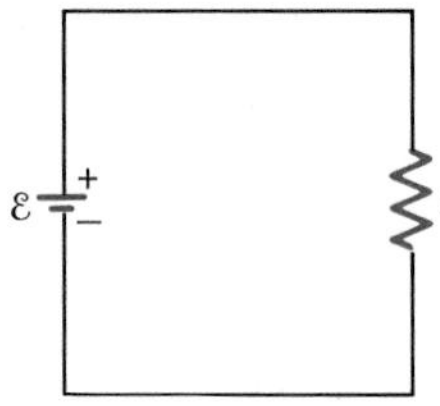

resistances of R_4 and R_5. Designating these equivalent resistances as R_6 and R_7, we have

$$R_6 = R_2 + R_3$$

and

$$R_7 = R_4 + R_5$$

As far as the emf is concerned, the circuit reduces to that shown in Figure 31.18. The equivalent resistances R_6 and R_7 are in parallel and can be replaced with a single resistance (R_8) that we can compute from the relation

$$\frac{1}{R_8} = \frac{1}{R_6} + \frac{1}{R_7}$$

The circuit now reduces to the one shown in Figure 31.19. We now see that the equivalent resistance R_8 is in series with r and R_1. These three resistances can be replaced with a single resistance (R) equal to the sum of the resistances of r, R_1, and R_8.

$$R = r + R_1 + R_8$$

As far as the emf is concerned, the current I is determined by a single resistor R connected to the emf, as shown in Figure 31.20.

If we know the numerical values of the resistances and the emf, we can compute the currents and potential differences of the circuit. Suppose that

$$\mathcal{E} = 2\text{ V} \qquad r = 1\ \Omega \qquad R_1 = 4\ \Omega$$

$$R_2 = R_5 = 7\ \Omega \qquad R_3 = R_4 = 3\ \Omega$$

Then the equivalent resistances R_6, R_7, and R_8 are

$$R_6 = R_2 + R_3$$
$$= 7 + 3 = 10\ \Omega$$
$$R_7 = R_4 + R_5$$
$$= 3 + 7 = 10\ \Omega$$
$$\frac{1}{R_8} = \frac{1}{R_6} + \frac{1}{R_7}$$
$$= \frac{1}{10} + \frac{1}{10}$$
$$= \frac{2}{10}$$
$$R_8 = 5\ \Omega$$

Because R_6, R_7, and R_8 are in series, their equivalent resistance is

$$R = r + R_1 + R_8$$
$$= 1 + 4 + 5 = 10\ \Omega$$

Thus the current in the series combination of r, R_1, and R_8 is

$$I = \frac{\mathcal{E}}{R}$$
$$= \frac{2}{10} = 0.2\text{ A}$$

The potential difference across R_8 is

$$V_8 = IR_8$$
$$= 0.2(5) = 1\text{ V}$$

Because R_8 represents the parallel combination of R_6 and R_7, the potential difference across R_6 and R_7 is also 1 V. Thus the current in R_6 is

$$I_6 = \frac{V_8}{R_6}$$

$$= \frac{1}{10} = 0.1 \text{ A}$$

and the current in R_7 is

$$I_7 = \frac{V_8}{R_7}$$

$$= \frac{1}{10} = 0.1 \text{ A}$$

Since R_6 is the equivalent resistance of R_2 and R_3 in series, the current (I_2) in R_2 and R_3 is 0.1 A. Similarly, the current (I_3) in R_4 and R_5 is 0.1 A. Note that $I = I_2 + I_3$, as required by Kirchhoff's junction rule. The potential difference across r and R_1 is

$$V_r = Ir$$

$$= 0.2(1) = 0.2 \text{ V}$$

$$V_{R_1} = IR_1$$

$$= 0.2(4) = 0.8 \text{ V}$$

Starting at the negative terminal of the emf (Figure 31.17) and proceeding clockwise, we find

$$\underset{\mathcal{E}}{2} - \underset{Ir}{0.2} - \underset{IR_1}{0.8} - \underset{IR_2}{0.7} - \underset{IR_3}{0.3} = 0$$

Thus Kirchhoff's loop rule is satisfied.

Suppose that the circuit shown in Figure 31.17 is altered to include an additional resistor placed between the connection of R_2 and R_3 and the connection of R_4 and R_5 (Figure 31.21). Can we compute series and parallel equivalent resistances to help solve for the currents and potential differences? Generally, the answer is no. Only direct application of Kirchhoff's loop and junction rules will work in such a case. This is because the presence of the resistor R_0 gives an additional current pathway for the currents out of R_2 and R_4. Thus neither R_2 and R_3 nor R_4 and R_5 are connected in series. The current in a series combination must be the same throughout, but only in special circumstances will the current through R_2 be the same as the current through R_3, or the current through R_4 the same as the current through R_5.

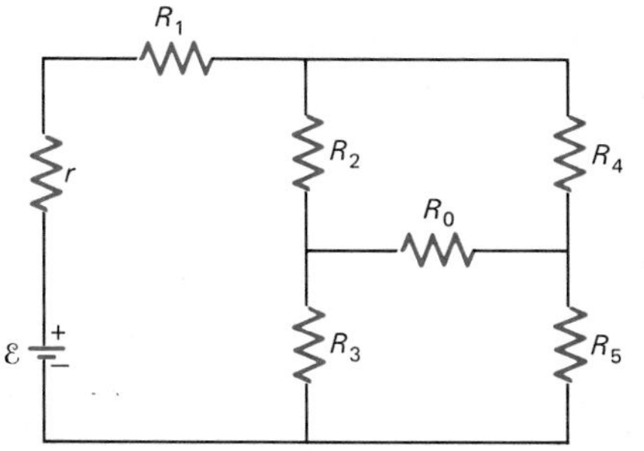

Figure 31.21

The current in each resistor of this circuit can be determined by using Kirchhoff's rules. However, we cannot solve for the currents by computing series and parallel equivalent resistances because there are no such combinations in the portion involving R_0, R_2, R_3, R_4, and R_5.

Figure 31.22

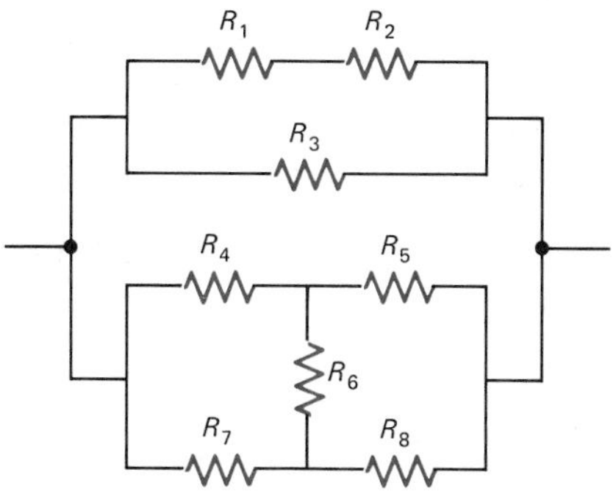

Questions

16. If a second transmission line from an electric power plant is required to supply electricity to a city, should the two lines be connected to each other in series or in parallel? Explain.

17. Are the electrical appliances in a household kitchen connected in series or in parallel with the household electric outlets?

18. What series and parallel connections are involved in the United States interstate highway system?

19. Connecting a resistor in parallel with another resistor provides an additional path through which charge can flow. Just from the meaning of electrical resistance, why would you expect the overall resistance of the parallel combination to be less than the resistance of either resistor in the parallel combination?

20. Identify the series and parallel connections in the circuit shown in Figure 31.22.

31.6 Charging a Capacitor: The *RC* Circuit

In this section we examine circuits that contain capacitors as well as resistors. Figure 31.23 shows a capacitor connected in series with a resistor and an emf. Understanding the way that charge accumulates on such a capacitor is vital to understanding the general behavior of capacitors in electrical circuits.

In Chapter 29 we examined the charging of a capacitor connected to an emf such as a battery. Our primary interest then was in the total charge stored by the capacitor. Now as we analyze the charging of a capacitor connected in series with an emf *and a resistor,* we will be interested mainly in the influence of the resistor. This

Figure 31.23

A series connection of a battery, a switch, a resistor, and a capacitor. We have neglected the internal resistance of the emf. Initially, the switch is open and there is no charge stored by the capacitor. Closing the switch allows for charge to be stored by the capacitor.

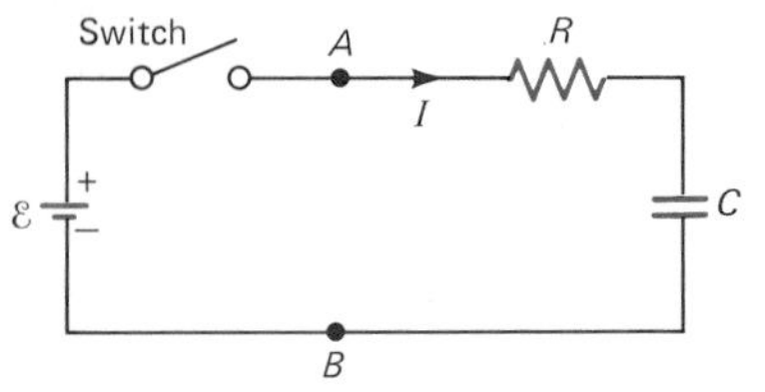

combination of emf, resistor, and capacitor is called an **RC circuit.**

Whether or not the resistor is present, charge does not begin accumulating on the capacitor at the instant that the switch is closed. Instantaneous movement of a charge requires an infinite force, which means that we would need an infinite electric field—a physical impossibility. Instead, the accumulation takes time, although only a very short time. Once the capacitor does begin storing charge, however, the rate of accumulation is affected by the size of the series resistance. We can better understand this situation if we think of a pump used to force water through a pipe and into an elevated tank. Because of its mass, the water does not move at the instant the pump is energized. Such a movement would require an infinite force, which is a physical impossibility. Once the water does begin moving, however, the pipe impedes the flow of water just as a resistor impedes the flow of charge. It takes time to store water in the tank just as it takes time to store charge in a capacitor. The rate at which the pump forces water into the tank is determined in large part by the resistance offered by the connecting pipe. Similarly, the rate at which a capacitor accumulates charge is affected by the resistance of the series resistor. Ultimately the potential difference across the capacitor equals the strength of the emf, and charge stops accumulating. If the maximum potential difference across the capacitor is $\mathcal{E}$, then we assert that the maximum charge stored by the capacitor is $q_{max} = C\mathcal{E}$.

Let us analyze the energy transformations involved in removing a positive charge dq from one plate of the capacitor in Figure 31.23 and storing it on the other plate. To move the charge dq through the potential difference $\mathcal{E}$, work $\mathcal{E}\,dq$ must be done by the emf. The flow of the charge in the resistor produces a current I and thermal power I^2R. In time dt the thermal energy produced in the resistor is $I^2R\,dt$. The electrostatic energy added to the capacitor by storing the charge dq is

$$V\,dq = \frac{q}{C}\,dq$$

We can use the principle of conservation of energy to equate the work done by the emf to the sum of the thermal energy produced and the additional electrostatic energy stored. We have

$$\mathcal{E}\,dq = I^2R\,dt + \frac{q}{C}\,dq \qquad (31.24)$$

Keep in mind that both current (I) and charge (q) may depend on time. Charge conservation requires that $dq = I\,dt$. Thus we can write Eq. 31.24 as

$$\mathcal{E}I\,dt = I^2R\,dt + \left(\frac{q}{C}\right)I\,dt$$

and

$$\mathcal{E}I = I^2R + \left(\frac{q}{C}\right)I \qquad (31.25)$$

Equation 31.25 states that the power ($\mathcal{E}I$) delivered by the emf is equal to the sum of the thermal power (I^2R) produced and the rate of energy storage [$(q/C)I$] on the capacitor.

Canceling the current I in Eq. 31.25 yields

$$\mathcal{E} = IR + \frac{q}{C}$$

$$\mathcal{E} = R\frac{dq}{dt} + \frac{q}{C} \qquad (31.26)$$

This equation is based on physical arguments (conservation of charge and conservation of energy) and describes the circuit behavior only *after* the switch is closed.

Equation 31.26, written as

$$\mathcal{E} - R\frac{dq}{dt} - \frac{q}{C} = 0$$

can also be viewed as a result of Kirchhoff's loop rule. As we proceed clockwise around the circuit (Figure 31.23), we find that $\mathcal{E}$ represents a potential rise (positive quantity) and IR and q/C represent potential drops (negative quantities). These potential changes sum to zero, in accordance with Kirchhoff's loop rule. Even though charge on the capacitor and current in the circuit may vary with time, the conservation of charge and energy holds at any instant. Thus Kirchhoff's loop and junction rules are valid at any instant. Because they are so general, we will apply Kirchhoff's rule to other circuits of interest.

The general solution of Eq. 31.26 requires mathematical, not physical, techniques. Rewriting the equation so that each composite term has units of charge, we have

$$RC\frac{dq}{dt} + q = C\mathcal{E} \qquad (31.27)$$

Because each of these terms in Eq. 31.27 has units of charge, resistance multiplied by capacitance (RC) must have dimensions of time. When R and C are expressed in ohms and farads, the units of RC are seconds. This RC combination is called the *time constant,* and as we shall see, it plays a very important role in the charging and discharging of a capacitor.

Equation 31.27 is a differential equation having variables of charge (q) and time (t). It is much like the differential equations involving velocity and time that we encountered in Chapter 4. There we wanted to determine velocity as a function of time. Here we want to determine the charge on the capacitor as a function of time. Like all problems of this type, we must establish the initial conditions. In effect this means deciding when to start measuring time. It is convenient to start when there is no charge stored by the capacitor and to choose the initial time ($t = 0$) to be the instant the switch is closed. Because charge cannot accumulate instantaneously on the capacitor, $q = 0$ when $t = 0$. To solve Eq. 31.27, we first rewrite it as

$$\frac{dq}{q - C\mathcal{E}} = -\frac{1}{RC}\,dt \qquad (31.28)$$

We have placed terms involving q on the left-hand side of the equation and terms involving t on the right-hand side of the equation. Both sides of the equation can now be integrated.

$$\int_0^q \frac{dq}{q - C\mathcal{E}} = -\frac{1}{RC}\int_0^t dt$$

$$\ln\left(\frac{q - C\mathcal{E}}{-C\mathcal{E}}\right) = -\frac{t}{RC} \qquad (31.29)$$

By using the definition of the natural logarithm, we can rewrite Eq. 31.29 as

$$q = C\mathcal{E}(1 - e^{-t/RC}) \qquad (31.30)$$

Here, $e = 2.71828\ldots$ denotes the base number for natural logarithms, and *not* electronic charge.

As required, $q = 0$ when $t = 0$; there is no charge stored by the capacitor at the instant the switch is closed. But as time progresses, charge accumulates on the capacitor, and the maximum charge stored on either plate is $C\mathcal{E}$. After a time $t = RC$ has elapsed, the charge on the capacitor is $q = C\mathcal{E}(1 - e^{-1}) = 0.632\ C\mathcal{E}$. Thus about 63% of the maximum charge that can accumulate on either capacitor plate has done so in a time equal to the time constant. This sequence of events is shown graphically in Figure 31.24.

By differentiating Eq. 31.25 we obtain the current

$$I = \frac{dq}{dt} = \left(\frac{\mathcal{E}}{R}\right)e^{-t/RC} \qquad (31.31)$$

The current in the circuit immediately after the switch is closed is $\mathcal{E}/R$. Charge accumulating on the capacitor tends to reduce the further flow of charge. As time progresses, the current decreases and approaches zero as the capacitor becomes fully charged. After a time $t = RC$ has elapsed, the current has decreased to $I = (\mathcal{E}/R)e^{-1} = 0.37\mathcal{E}/R$, or about 37% of its initial value. This sequence of events is presented graphically in Figure 31.25.

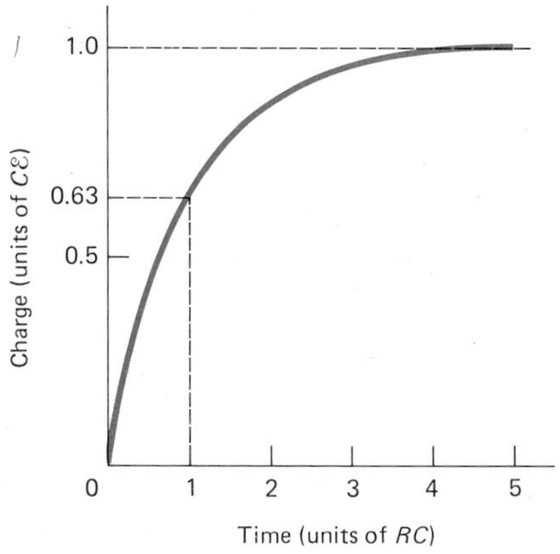

Figure 31.24

A generalized plot of the charge stored by a capacitor connected in series with a resistor and a battery. Time is expressed in units of RC, the circuit time constant. Charge is expressed in units of $C\mathcal{E}$, the maximum charge stored by the capacitor. In principle, it takes an infinite amount of time to store the maximum amount of charge; 63% is stored in a time equal to one time constant.

Figure 31.25

A generalized plot of the current in a resistor–capacitor–battery series circuit. Current is expressed in units of $\mathcal{E}/R$, the maximum current in the circuit. The current drops exponentially after the switch is closed, reaching 37% of its initial value in a time equal to one time constant.

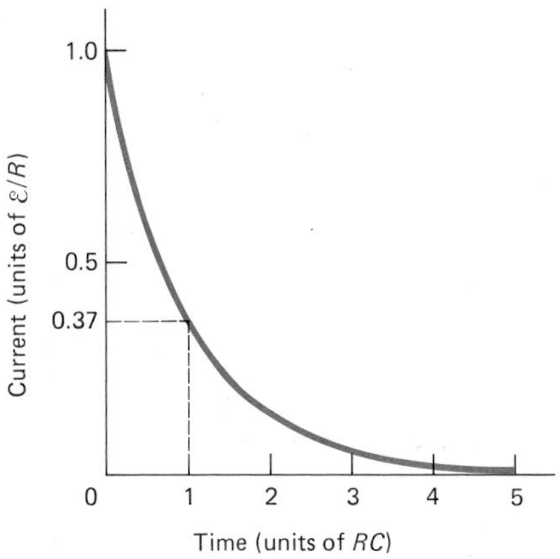

Example 8 Making Use of the Time Required to Charge a Capacitor

Figure 31.26 represents an unspecified electronic circuit having a very large resistance between its input terminals connected in parallel with a capacitor. This circuit is designed to respond when the potential difference across the capacitor rises to 1 V after the switch is closed. What possible values of R and C would result in a time delay of 1 s between responses? Using Eq. 31.30 we have

$$V_C = \frac{q}{C} = \mathcal{E}(1 - e^{-t/RC})$$

Substituting values yields

$$1 = 10(1 - e^{-1/RC})$$

$$e^{-1/RC} = \frac{9}{10}$$

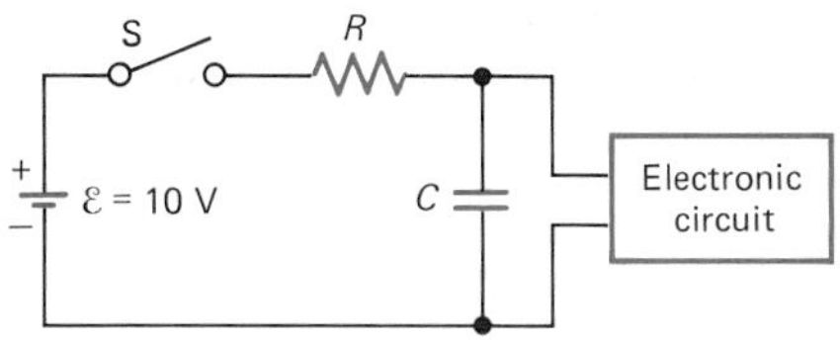

Figure 31.26

The circuit in Figure 31.23 is modified to include an electronic circuit connected in parallel with the capacitor. There is negligible current in the wires leading to the electronic circuit. We want the electronic circuit to respond when the potential difference across the capacitor rises to 1 V.

Taking the natural logarithm of each side of this equation, we have

$$-\frac{1}{RC} = -0.1054$$

or

$$RC = 9.49 \text{ s}$$

Any combination of R and C that has the product $RC = 9.49$ s would result in a 1-s time delay. For example, if $C = 10\ \mu$F, then

$$R = 0.949 \times 10^6\ \Omega$$
$$= 0.949 \text{ megohms}$$
$$= 0.949\ \text{M}\Omega$$

Energy Stored by the Capacitor

Referring to either Eq. 31.30 or Figure 31.24, we see that the maximum charge stored on the capacitor is $q_{max} = C\mathscr{E}$, independent of the value of the series resistance. This means that the maximum energy stored by the capacitor

$$U_{max} = \frac{1}{2}C\mathscr{E}^2$$

is also independent of the resistance. Thus the resistor affects the rate at which the capacitor charges but has no effect on the ultimate value of the charge or the energy stored. Using Eq. 31.31 for the charging current, we can calculate the thermal power developed in the resistor as

$$P = I^2R = \left(\frac{\mathscr{E}^2}{R}\right)e^{-2t/RC} \qquad (31.32)$$

and the thermal energy produced during the charging as

$$\text{thermal energy} = \frac{\mathscr{E}^2}{R}\int_0^t e^{-2t/RC}\,dt$$

$$= \frac{C\mathscr{E}^2}{2}(1 - e^{-2t/RC}) \qquad (31.33)$$

Hence, during a *complete* charging ($t \to \infty$), the thermal energy produced, $C\mathscr{E}^2/2$, is exactly equal to the energy stored on the capacitor. Regardless of the rate at which the capacitor charges, the total amount of energy converted by the emf is $C\mathscr{E}^2$. One half of this energy is converted to thermal energy, and the other half is stored by the capacitor.

Note that although the charge on the capacitor is zero at the instant the switch is closed (that is, $t = 0$), the circuit current at $t = 0$ is $I = \mathscr{E}/R$. This reflects the fact that before the switch is closed the potential difference across the capacitor is zero. To remove charge instantaneously from one plate and store it on the other would require an infinite force, which is an impossibility. If the capacitor were short-circuited, then the circuit current would be $I = \mathscr{E}/R$ for all time. Because $I = \mathscr{E}/R$ at $t = 0$, this means that *instantaneously* the capacitor behaves as if it is short-circuited (zero resistance). This idea is useful in determining the instantaneous current in more complex resistive–capacitive circuits.

Figure 31.27

In this circuit, it is assumed that the switch connecting terminals 1 and 3 has been closed long enough so that the capacitor is fully charged. Then the switch is actuated, connecting terminals 3 and 2, and the capacitor starts discharging through the resistor. We have neglected the internal resistance of the emf.

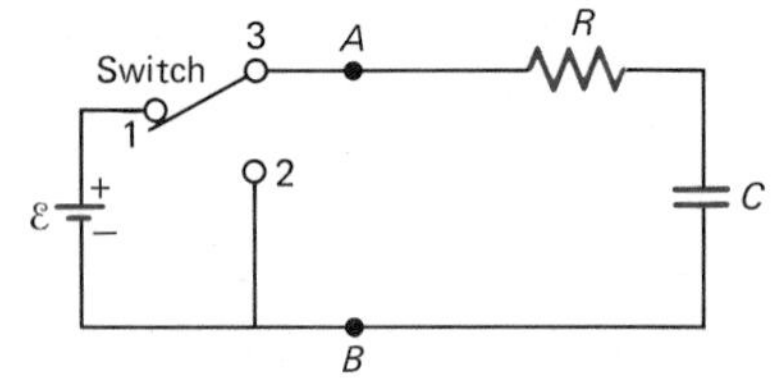

Discharging the Capacitor

If the switch in the circuit in Figure 31.27 is moved to position 2, the potential difference between A and B drops from $\mathscr{E}$ to zero, and Eq. 31.27 reduces to

$$q + RC\frac{dq}{dt} = 0$$

To solve this differential equation, let us restart the clock and call zero the time that the switch is moved to position 2. If we call the charge on the capacitor q_o when $t = 0$, and separate the charge and time variables, we have

$$\int_{q_o}^{q}\frac{dq}{q} = \int_o^t -\frac{dt}{RC}$$

which integrates to

$$\ln\left(\frac{q}{q_o}\right) = -\frac{t}{RC}$$

Using the definition of the natural logarithm, we obtain

$$q = q_o e^{-t/RC} \qquad (31.34)$$

Differentiating this expression to get the current, we have

$$I = \frac{dq}{dt} = \left(-\frac{q_o}{RC}\right)e^{-t/RC} \qquad (31.35)$$

The negative sign indicates that the current direction is opposite to the assumed direction. From the point of view of the charge on the capacitor, the negative sign is a consequence of the decrease of charge q with time. Note that the magnitude of both q and I falls to $1/e$ of their initial value in a time $t = RC$.

The deduction of Eq. 31.26 by using conservation principles allowed us to analyze the charging and discharging process in a physically meaningful way. Alternatively, we could have deduced this equation by using Kirchhoff's loop rule, because this rule is a reflection of energy conservation. Indeed, the practical analysis of circuits containing resistors and capacitors normally proceeds from the application of Kirchhoff's rules. However, unlike the coupled linear algebraic equations that evolve in purely resistive circuits, circuits containing resistors and capacitors yield coupled differential equations. The algebraic signs of the derivative terms then become very important and often are not obvious. We will not dwell on this

Figure 31.28

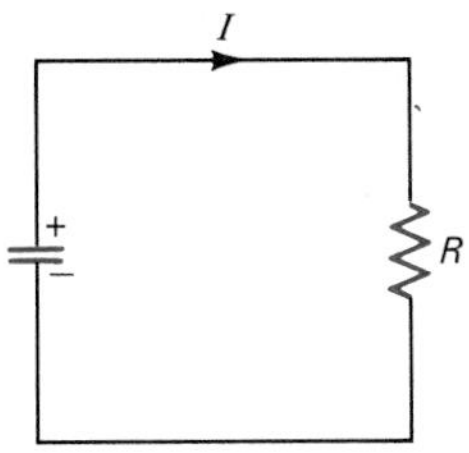

I
+
R

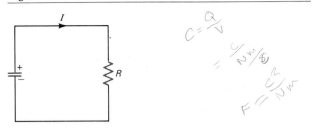

problem here, but you might consider the consequences of choosing the current direction opposite to that shown in Figure 31.28 when the capacitor is discharging through the resistor. If you use $I = -dq/dt$, everything works out fine.

Questions

21. When a capacitor, a resistor, and a battery are connected in series, the ultimate charge stored on the capacitor is independent of the resistance of the resistor. What role *does* the resistor play in the charging process?

22. How can a measure of the rate of discharge of a capacitor through a resistor be used to determine the amount of resistance?

23. A capacitor discharges through a resistance, as shown in Figure 31.28. The positive direction of the current is taken to indicate positive charge flowing from the positive plate of the capacitor. Why should we take $I = -dq/dt$?

31.7 Current and Potential Difference Measurements

So far, we have devised ways of calculating currents and potential differences in electrical circuits but have said very little about measuring these quantities. Not only do measurements give us confidence in the analysis procedure, but they sometimes provide the only way of determining currents and potential differences. An instrument for measuring current is called an *ammeter*. An instrument for measuring potential difference is called a *voltmeter*.

In order for an ammeter to measure the size of a current, charge must flow *through* it. Therefore, an ammeter has two terminals that allow the current to enter and to leave. The circuit must be broken at the point of interest, and the terminals of the ammeter connected to recomplete the circuit (Figure 31.29).

A voltmeter measures the potential difference *between* two points. A voltmeter has two terminals that can be connected at any two points in a circuit. The circuit does not have to be broken in order to connect a voltmeter (Figure 31.30), and for this reason, it is more convenient to measure potential difference than it is to measure current.

All ammeters and voltmeters rely on some effect that is produced by a current or a potential difference. For example, one of the most versatile modern voltmeters, the oscilloscope (Figure 31.31), relates the amount of deflection of an electron beam to a potential difference established between two parallel plates that the electron beam passes between. A digital ammeter or voltmeter (Figure 31.32) that visually displays the numerical value of a current or potential difference measurement is based on the principle that a transistor or electronic circuit will produce a response that is related to the current or potential difference that it is measuring. A versatile and widely used type of meter (Figure 31.33) that requires the experimenter to read the position of a pointer on a calibrated scale relies on the principle that a current-carrying coil of wire between the poles of a permanent magnet will experience a torque that rotates the coil (this effect will be studied in Chapter 32).

When an ammeter or voltmeter is inserted into an electrical circuit for the purpose of measuring a current or potential difference, the meter will generally alter the circuit to some extent. This is because meters require energy for their operation and they extract this energy from the circuit. The energy requirement is associated with the electrical resistance (R_m) appearing between the two terminals of the meter (Figure 31.35). Resistance R_m is the equivalent of the resistance of the active element and of any resistance that is connected in parallel or in series with the active element. For example, a conventional moving-coil ammeter has an intrinsic resistance associated with the wire constituting the moving coil (the active element) and a parallel resistance (commonly called a *shunt*) that allows

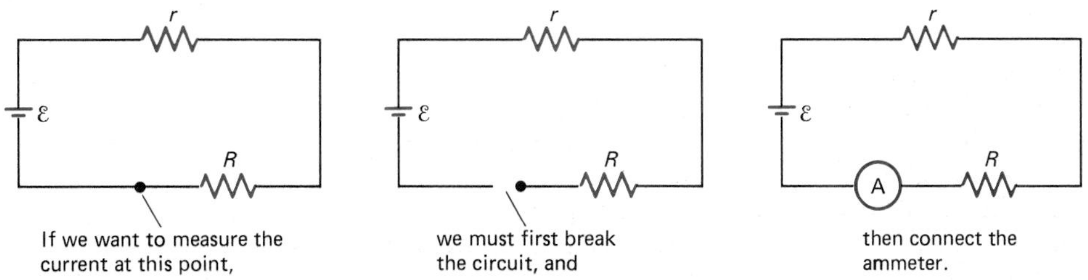

r
ℰ
R
If we want to measure the current at this point,
r
ℰ
R
we must first break the circuit, and
r
ℰ
A
R
then connect the ammeter.

Figure 31.29

The steps involved in connecting an ammeter in an electrical circuit.

Figure 31.30

The steps involved in connecting a voltmeter in an electrical circuit.

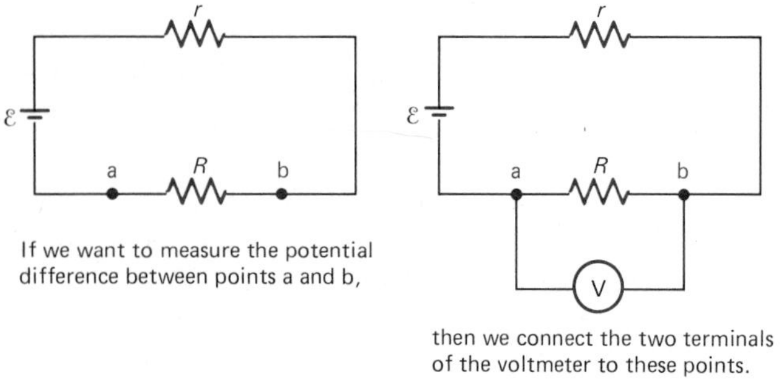

Figure 31.31

An oscilloscope. The vertical deflection of the beam is related to the voltage applied to the input.

Figure 31.32

A digital voltmeter.

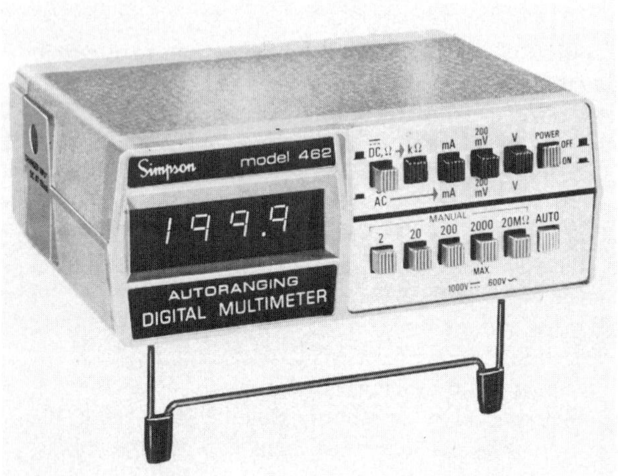

Figure 31.33

A conventional moving-coil ammeter.

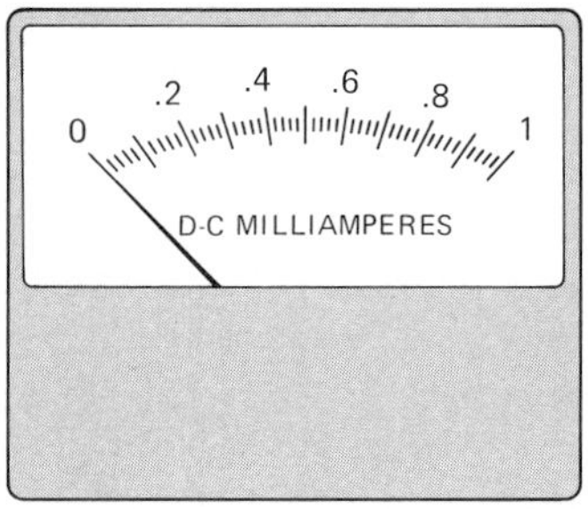

Figure 31.34

The components of a moving-coil ammeter. The active element is the moving coil. The deflection of the meter pointer is proportional to the current in the moving coil. The resistance of the meter is due to the resistances of the shunt and the moving coil.

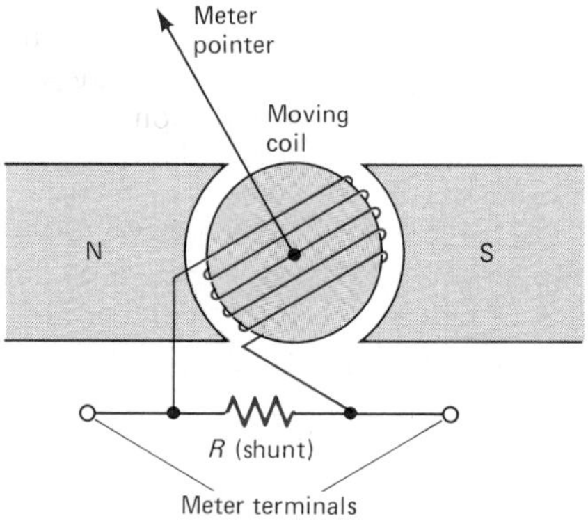

Figure 31.35

The ammeter in this circuit introduces a resistance R_m in series with the circuit resistors r and R.

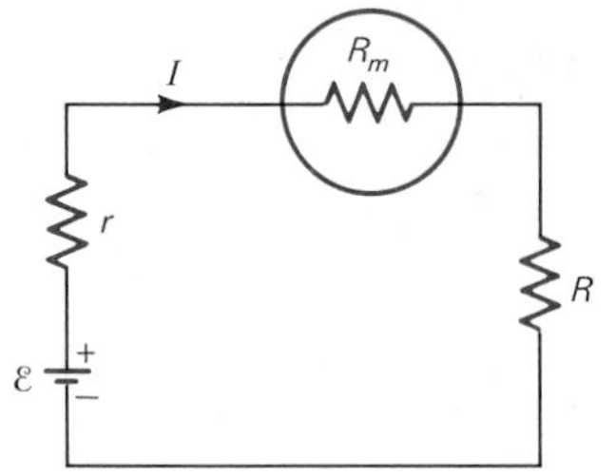

current to bypass the moving coil (Figure 31.34). To illustrate the effect of the measuring instrument, we can consider the circuit in Figure 31.35, which shows an ammeter with resistance R_m connected to measure a current through the resistance designated R. The ammeter is connected in series with R because the charge that flows through R must also flow through the ammeter. In the absence of the ammeter the current I is

$$I = \frac{\mathscr{E}}{r + R} \tag{31.36}$$

With the meter present the current is

$$I_m = \frac{\mathscr{E}}{r + R + R_m} \tag{31.37}$$

Clearly, $I \neq I_m$. However, if $R_m \ll R$, then $I \cong I_m$. When this is the case, R_m can be neglected. Generally, an ammeter will not significantly affect a current measurement, but we should always be aware of the possibility that it will, and be prepared to make corrections for it. When a voltmeter is used to measure the potential difference across a resistor (Figure 31.30), the meter will not significantly affect the measurement if $R_m \gg R$.

Summary

An emf is a two-terminal energy source capable of driving charges through some externally connected circuit. A battery is an example of an emf.

By convention, a positive charge experiences a potential rise when moving from − to + across an emf and a potential drop when moving through a resistor in the direction of the current.

Energy conservation in electrical circuits leads to Kirchhoff's loop rule, which states that in any complete electrical circuit (called a loop) the algebraic sum of the potential changes is zero.

The equivalent resistance (R) of N series-connected resistors is

$$R = R_1 + R_2 + R_3 + \cdots + R_N \qquad (31.9)$$

Charge conservation in electrical circuits leads to Kirchhoff's junction rule, which states that the sum of the currents directed toward any junction is equal to the sum of the currents directed away from the junction.

The equivalent resistance (R) of N parallel-connected resistors is determined from

$$\frac{1}{R} = \frac{1}{R_1} + \frac{1}{R_2} + \frac{1}{R_3} + \cdots + \frac{1}{R_N} \qquad (31.22)$$

The charge stored by a capacitor being charged by an emf $\mathcal{E}$ that is in series with a resistor and the capacitor is represented by

$$q = C\mathcal{E}(1 - e^{-t/RC}) \qquad (31.30)$$

The current in the resistor–capacitor–emf series circuit is

$$I = \left(\frac{\mathcal{E}}{R}\right)e^{-t/RC} \qquad (31.31)$$

Resistance multiplied by capacitance (RC) has dimensions of time. Of the maximum amount of charge that will accumulate on the capacitor, 63% will do so in a time equal to RC.

For a current to be measured with an ammeter, the circuit must be broken and the ammeter connected to close the circuit. The ammeter will produce negligible effect on the current if the internal resistance of the meter is negligible compared with the resistance of the circuit.

For measurement of a potential difference across a resistor with a voltmeter, the circuit need not be broken in order to connect the voltmeter. The voltmeter will give an accurate reading if the internal resistance is very large compared with the resistance of the resistor.

Suggested Reading

B. Chalmers, The Photovoltaic Generation of Electricity, *Sci. Am.*, Vol. 235, 34–43 (October 1976).

E. W. McWhorter, The Small Electronic Calculator, *Sci. Am.*, Vol. 234, 88–98 (March 1976).

R. N. Noyce, Microelectronics, *Sci. Am.* (September 1977). The entire issue is devoted to microelectronics.

Problems

Section 31.1 Source of Electromotive Force (emf)

1. A battery having an emf of 7 V produces a current of 0.5 A when connected to an external resistance. If 150 J of energy are produced in 1 min in the external resistor, determine the values of the internal and external resistances.

2. An old 12-V automobile battery with an internal resistance of 1 Ω delivers 1 A of current to a resistor for 10 min. Determine (a) the chemical energy converted by the battery, (b) the energy produced in the resistor, and (c) the resistance of the resistor.

3. A battery (emf = 12 V) of unknown internal resistance delivers 22 W to an external resistance when the current is I. If the battery current doubles, the external power delivered becomes 40 W. Determine (a) the original current I and (b) the internal resistance of the battery.

4. A 12-V storage battery delivers a steady current of 2.3 A. At what rate does the battery deliver energy?

5. A high-quality 12-V automobile battery with negligible internal resistance can deliver a current of 5 A for 20 hr. (a) How much energy can the battery provide? (b) If a copper bar whose dimensions are 2 cm × 2 cm × 10 cm accidentally fell across the terminals of the battery, and all the energy of the battery were dissipated into the bar, what would happen to the bar? (The specific heat of copper is 0.39 kJ/kg · C°.) The density of copper is 8.9×10^3 kg/m^3. (This problem should give you some indication of the concentration of energy in an automobile battery.)

Section 31.2 Kirchhoff's Loop Rule for Potential Differences

6. A 12-V storage battery has an internal resistance of 0.30 Ω. It is placed in series with a resistor whose resistance is 3.20 Ω. Determine the current in the circuit.

7. A nonohmic resistance of initially unknown characteristics is connected in series with a resistor and a battery that has negligible internal resistance (Figure 1). (a) If V_x is the potential difference across the device, show that the current in the circuit is $I = (\mathcal{E} - V_x)/R$. (b) If it is found experimentally that the potential difference V_x is related to the current by $V_x = kI^2$, what are the units of k? (c) With the knowledge that $V_x = kI^2$, derive a relation for the current.

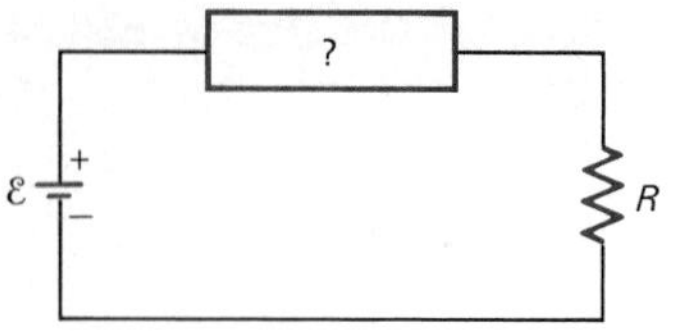

Figure 1

8. A battery with a strength of 12 V and internal resistance of 0.1 Ω is connected to a resistance that can be varied continuously from 0 to 1 Ω. (a) Determine the thermal power developed in the external resistor for several values of the external resistance and make a plot of the thermal power in the resistor versus the value of the external resistance. (b) Compare the thermal power in the resistor with the power developed by the emf and comment on why there is a difference.

9. (a) For the circuit in Figure 31.2 show that

$$\frac{1}{I} = \frac{R}{\mathscr{E}} + \frac{r}{\mathscr{E}}$$

(b) The data below were recorded in the circuit shown in Figure 31.2 with an ammeter of negligible internal resistance. Using these data, make a plot of $1/I$ versus R and determine the emf and internal resistance of the battery.

R (Ω)	I (A)
1	4.8
4	2.0
8	1.1
12	0.8
15	0.6

Sections 31.3 and 31.5 Resistors in Series and in Parallel

10. Show that the equivalent resistance of any series combination of resistors is greater than the resistance of any single resistor in the combination.

11. Show that the equivalent resistance of any number of resistors in parallel is always less than the smallest resistance in the combination.

12. Show how four 6-Ω resistors can be connected to produce equivalent resistances of 1.5, 6, 8, 15, and 24 Ω.

13. When two resistors are connected in series with a battery having an emf of 10 V and negligible internal resistance, a current of 1 mA is produced. When only one resistor is connected to the battery the current is 3 mA. What are the values of the resistors?

14. When two resistors are connected in series with a battery having an emf of 10 V and negligible internal resistance, a current of 1 mA is produced. When the resistors are connected in parallel to the same battery, a total current of 4 mA is produced. What are the resistances of the resistors?

15. Five 1000-Ω resistors are connected in series. Each of five identical resistors of unknown resistance is connected in parallel with the series combination of five 1000-Ω resistors. The equivalent resistance of the combination is 2500 Ω. What is the unknown resistance?

16. Five 1000-Ω resistors in series are connected in series with five identical parallel-connected resistors of unknown resistance. If the equivalent resistance is 10,000 Ω, what is the resistance of the unknown resistors?

17. Two light bulbs, each having a resistance of 10 Ω, are connected in series with a 12-V storage battery having negligible internal resistance. When a third identical light bulb is connected in parallel with one of the original bulbs, which bulbs will brighten and which will dim?

Sections 31.2 and 31.4 Kirchhoff's Rules

18. Two batteries and two resistors are connected in the single loop shown in Figure 2. The potential at point D equals zero. Determine the potentials at points (a) A, (b) B, and (c) C.

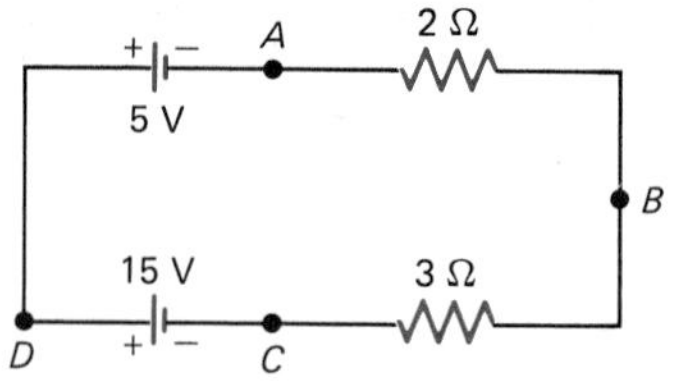

Figure 2

19. N identical resistors in parallel have an equivalent resistance of R/N where R is the resistance of a single resistor. (a) If an additional identical resistor is connected in parallel, show that the percentage change in the equivalent resistance is

$$\frac{100}{N + 1}$$

(b) Evaluate the percentage change for $N = 1$, $N = 10$, and $N = 100$.

20. Batteries in a two-cell flashlight are connected in series (Figure 3). (a) What advantage has a series connection over a parallel connection? (b) Determine the current in the circuit.

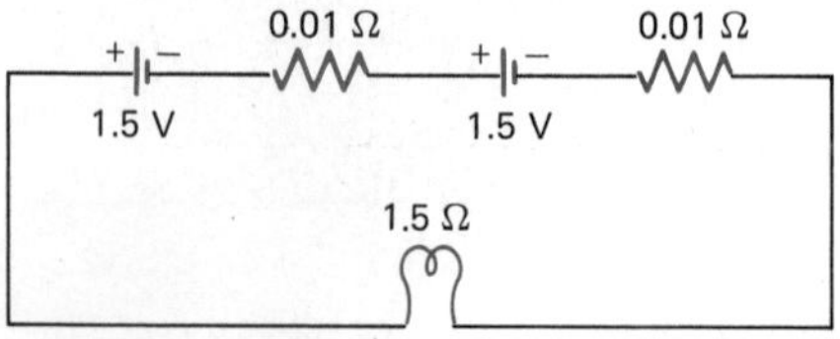

Figure 3

21. A dead battery is boosted by connecting it to the live battery of another car (Figure 4). Determine the current in the starter and in the dead battery.

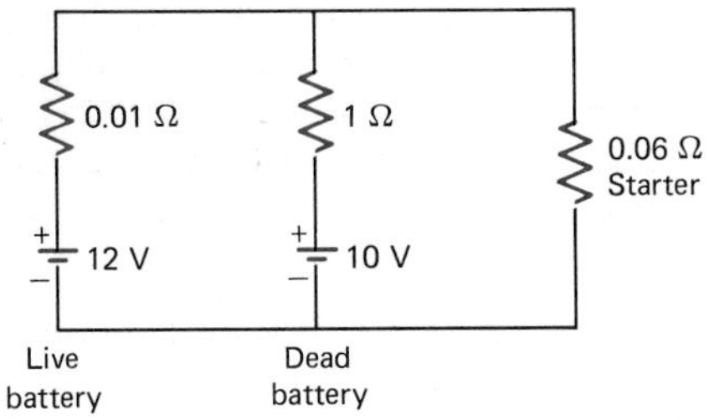

Figure 4

22. Determine the current I_2 in the circuit shown in Figure 5.

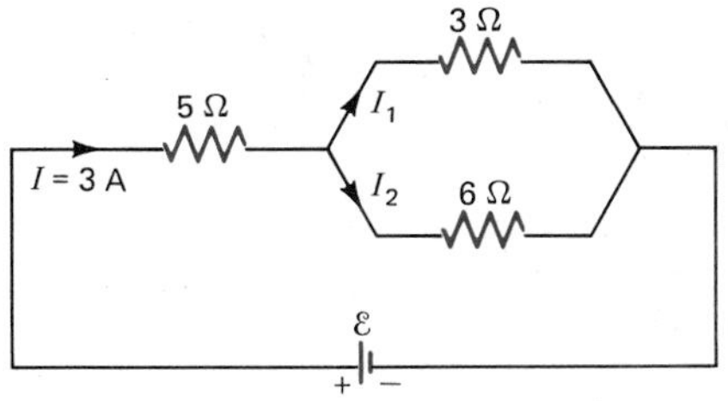

Figure 5

23. Determine the current in the 10-Ω resistor in Figure 6 by using (a) Kirchhoff's rules and (b) equivalent resistances of series and parallel combinations of resistors.

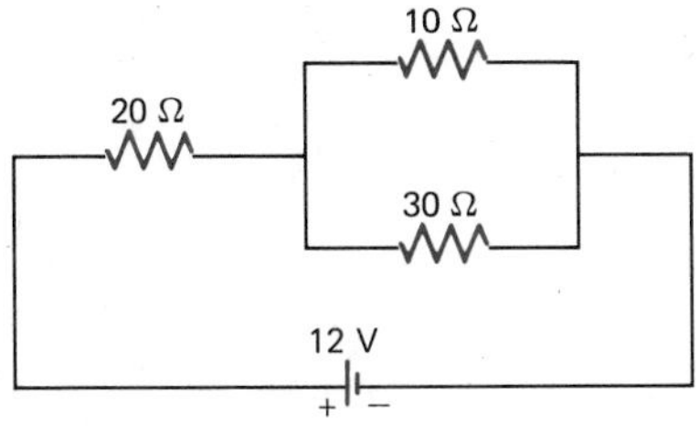

Figure 6

24. Referring to Figure 7, (a) calculate (1) the current in R_1, (2) the potential drop across R_1, and (3) the thermal power developed in R_1 when S_1 is closed and S_2 is *open*. (b) Repeat for R_1 with *both* switches closed.

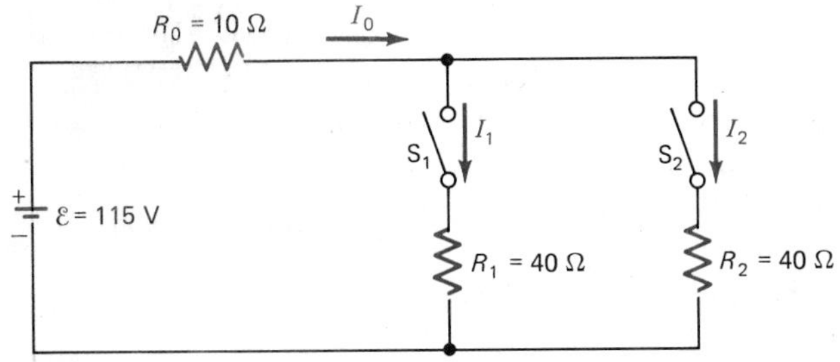

Figure 7

25. Determine I_2, I_4, I_5, and I_7 as shown in Figure 8.

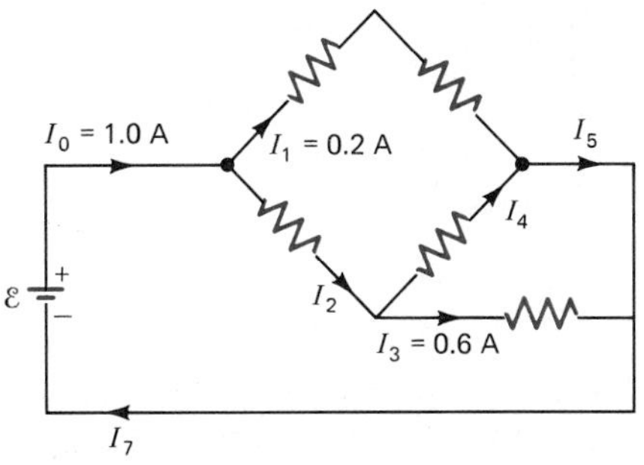

Figure 8

26. An ideal on–off switch has infinite resistance when off and zero resistance when on. Assuming an ideal switch in the automobile circuit shown in Figure 9, determine the potential difference across the switch when it is on and when it is off.

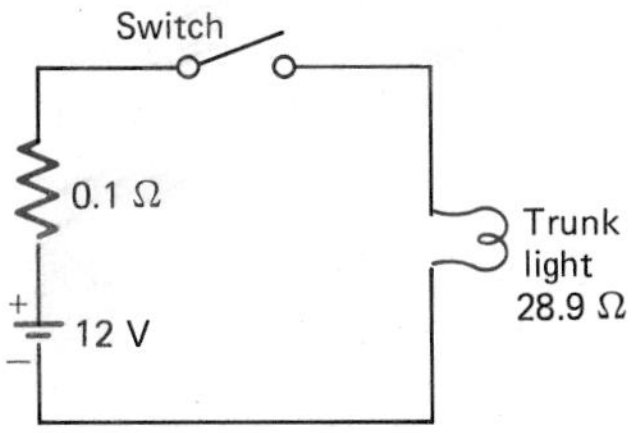

Figure 9

27. A 20-W bulb designed to operate with a 12-V potential difference across its terminals is connected to a 12-V battery having an internal resistance of 0.6 Ω. (a) Determine the potential difference across the bulb. (b) Determine the power converted in the bulb. (c) Determine the change in the potential difference across the bulb and the change in the power converted by the bulb when a second 20-W bulb is connected in parallel with the battery.

28. A current I dissipates 25 W in the resistor R_1. When connected in series with the resistor R_2 the same current will dissipate 75 W in the two resistors. Calculate the ratio R_2/R_1 of the two resistors.

29. A battery with a constant emf $\mathscr{E}$ has an internal resistance r. It is placed in series with a variable resistor, as suggested in Figure 10. (a) Show that the power delivered to the resistor can be expressed by

$$P = \frac{R\mathscr{E}^2}{(R + r)^2}$$

(b) Make a sketch of P versus R for $R = 0$–$10\ \Omega$. Take $\mathscr{E} = 1$ V, $r = 3\ \Omega$. (c) Using the expression (a) for P, prove analytically that the maximum value of P occurs for $R = r$. Does your sketch in (b) show that P is a maximum at $R = 3\ \Omega$?

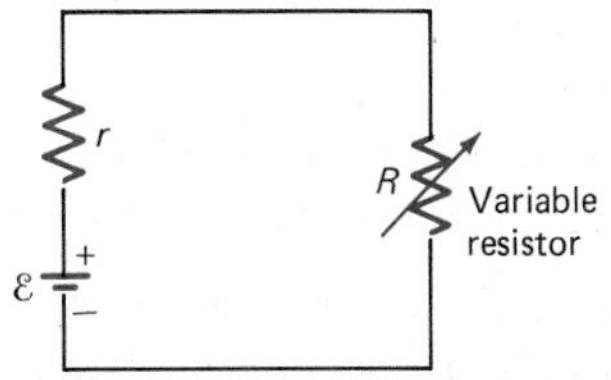

Figure 10

30. When two unknown resistors are connected in series, 225 W are dissipated with a total current of 5 A. For the same total current 50 W are dissipated when the resistors are connected in parallel. Determine the values of the two resistors.

31. An electrical circuit is somewhat like a jigsaw puzzle in which each independent loop outlines a piece of the puzzle. If you cut the puzzle into sections each of which contains a single loop, then the number of sections equals the number of independent loop equations. For example, the circuit in Figure 11 has two sections. (a) Label the unknown currents in the circuit. (b) How many current equations are required in addition to the two equations obtainable from the two loops? (c) Apply Kirchhoff's junction and loop rules and solve for the circuit currents.

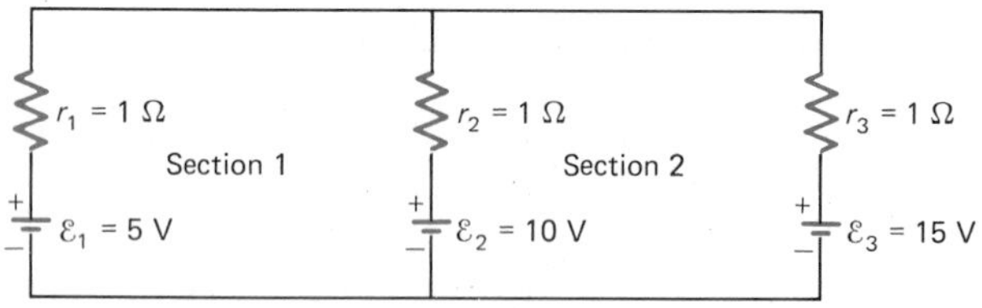

Figure 11

32. Determine the number of independent potential difference and current equations in the circuit in Figure 12 and show that they are sufficient for calculating the circuit currents.

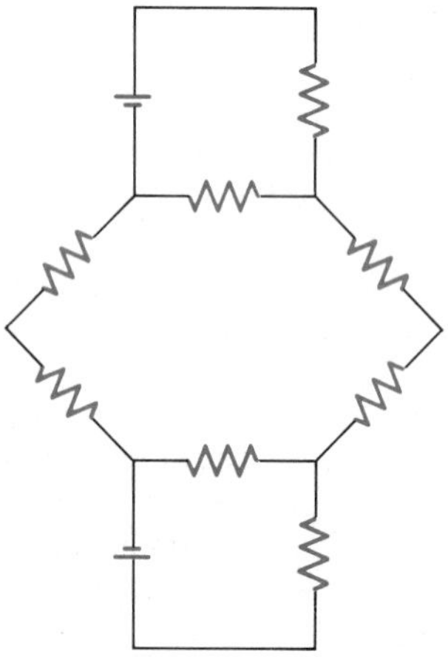
Figure 12

33. For the circuit shown in Figure 31.12, take $\mathcal{E}_1 = 12.0$ V, $\mathcal{E}_2 = 6.0$ V, $r_1 = 4.0\ \Omega$, $r_2 = 2.0\ \Omega$, and $R = 5.0\ \Omega$. (The resistances r_1 and r_2 are no longer identified as internal resistances.) (a) Calculate the three currents I_1, I_2, and I_3. (b) Calculate the potential drop across each resistance. (c) Calculate the power dissipated in each resistance. (d) Show that the total thermal power developed in the resistors equals the input electric power.

34. (a) Using Kirchhoff's loop law, show that the current in resistor R in the circuit shown in Figure 13 is

$$I = \frac{\mathcal{E}_1 - \mathcal{E}_2}{r_1 + r_2 + R}$$

(b) You can think of the current in R as the superposition of currents provided independently by $\mathcal{E}_1$ and $\mathcal{E}_2$. Calculate the current provided by $\mathcal{E}_1$ when $\mathcal{E}_2$ is short-circuited. Do the same for $\mathcal{E}_2$ assuming that $\mathcal{E}_1$ is short-circuited. Superpose these two currents and show that the result is identical to that obtained in part (a).

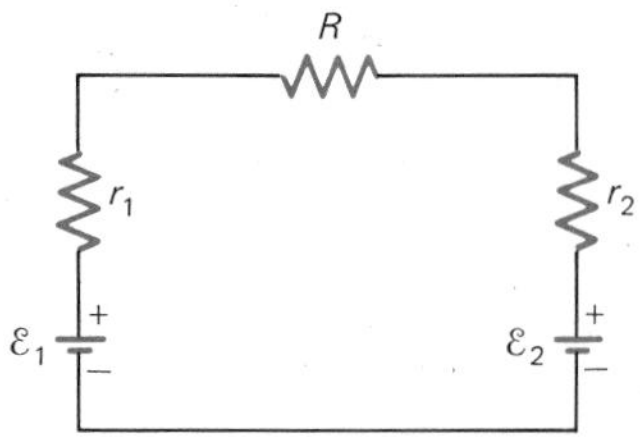

Figure 13

35. A circuit for an unbalanced Wheatstone bridge is shown in Figure 14. Currents are labeled for each branch in the circuit. If $I_1 = 10$ A, $I_2 = 6$ A, and $I_3 = 2$ A, use Kirchhoff's junction rule to obtain (a) I_4, (b) I_5, and (c) I_6.

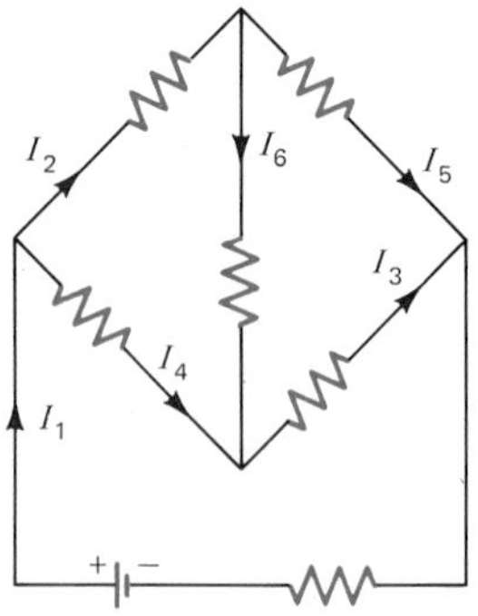

Figure 14

36. A sealed box contains two wires leading to an unknown combination of resistors inside the box. When a potential difference of 1.51 V is connected across the leads, a current of 0.206 A is produced. (a) What is the effective resistance of the combination of resistors in the box? (b) Opening the box reveals the combination of resistors shown in Figure 15. Determine the equivalent resistance and verify the calculation in part (a).

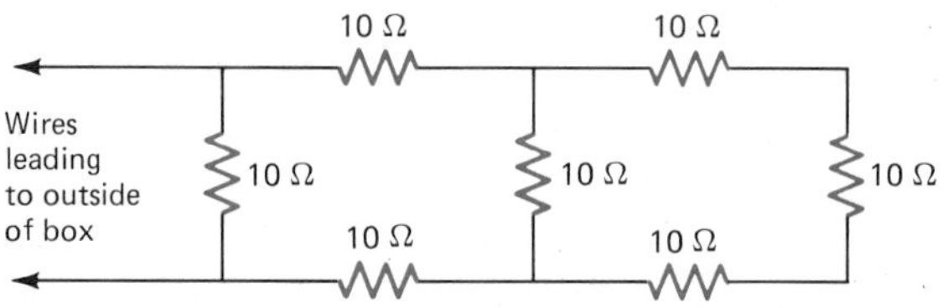

Figure 15

37. Determine the equivalent resistances of the configurations of the equal-valued resistors shown in Figure 16.

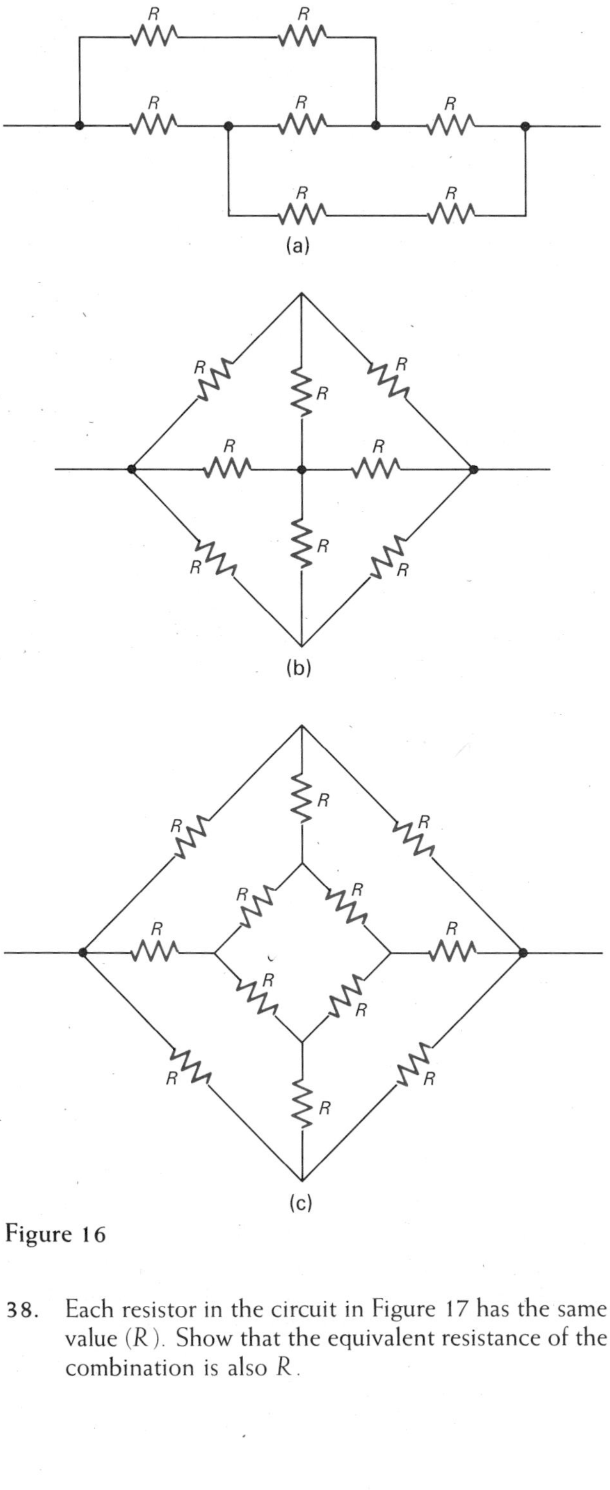

Figure 16

38. Each resistor in the circuit in Figure 17 has the same value (R). Show that the equivalent resistance of the combination is also R.

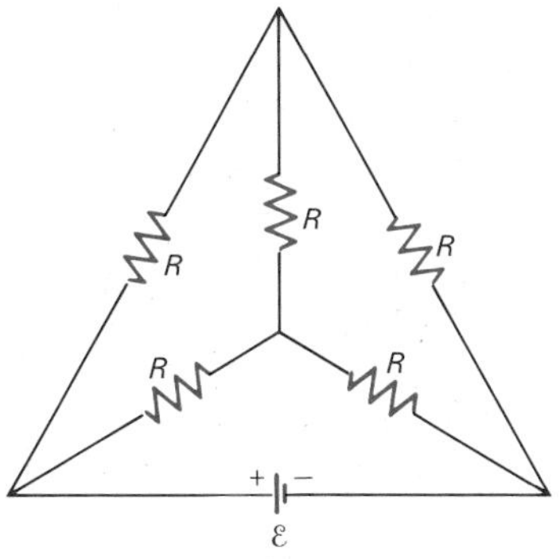

Figure 17

39. (a) Using symmetry arguments, show that the current through any resistor in the configuration of Figure 18 is either $I/3$ or $I/6$. (b) Show that the equivalent resistance between points A and B is $(5/6)r$.

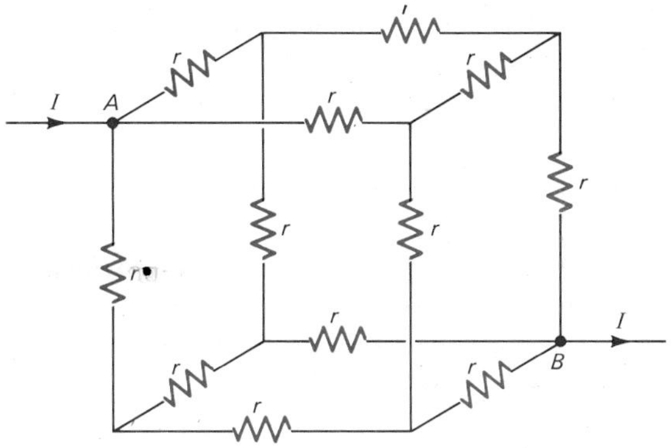

Figure 18

40. A symmetric infinite array of equal-valued resistors is shown in Figure 19. (a) Why is the equivalent resistance of this array equal to or greater than $2r$ but less than $3r$? (b) If the current in r_1 is I, why must the current in r_3 also be equal in magnitude to I? (c) If the current in r_4 is kI (k is a positive constant), why must the current in r_2 be $(1 - k)I$ and the current in r_5 be $k(1 - k)I$? (d) By applying Kirchhoff's loop rule to the loop containing r_2, r_4, r_5, and r_6, show that $k = 2 - \sqrt{3}$. (e) Given that $k = 2 - \sqrt{3}$, apply Kirchhoff's loop rule to the loop containing V, r_1, r_2, and r_3 and show that the equivalent resistance is $(1 + \sqrt{3})r$.

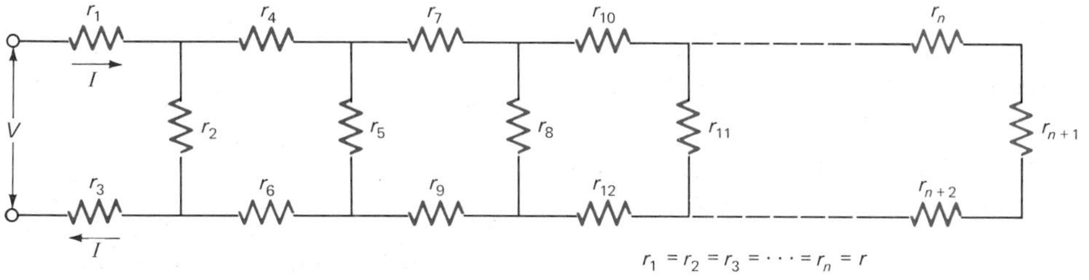

Figure 19

Section 31.6 Charging a Capacitor: The RC Circuit

41. A 1-μF capacitor in a circuit has an initial charge of 10 μC. Suppose that we want another circuit to respond when the potential difference across the capacitor falls to 1 V as it discharges through a 100-kΩ resistor. How much time is required for this circuit to respond?

42. Using Ohm's law and the definition of capacitance, show that RC has dimensions of time.

43. In an elementary circuit consisting of a battery, a resistor, a capacitor, and a switch, about 63% of the maximum charge that can be accumulated by the capacitor will be done in a time $t = RC$ after the switch is closed. At this time, what percentage of the maximum electrostatic energy is stored by the capacitor?

44. A 12.0-V battery is charging a 10,000-μF capacitor through a 1.0-MΩ resistor. How long will it take to raise V_C, the potential difference across the capacitor, to 6.0 V?

45. An engineer wants to design a circuit that will produce a blinking light, and decides to use an RC series combination as part of the circuit. The time constant is to be 0.30 s. If the resistor has $R = 47$ kΩ, what capacitance is required?

46. For a resistor, capacitor, and battery in series, we showed that

$$IR + \frac{q}{C} = \mathscr{E}$$

at any instant. (a) Differentiate this equation with respect to time to obtain a differential equation involving the current (I). (b) Solve the differential equation for the current as a function of time.

47. At $t = 0$, the switch in Figure 31.23 is closed and the capacitor starts to charge. Show that the potential difference across the capacitor, V_C, is

$$V_C = \mathscr{E}(1 - e^{-t/RC})$$

For $t \ll RC$ show that V_C is approximated by

$$V_C = \left(\frac{\mathscr{E}}{RC}\right)t$$

(*Hint:* For $|x| \ll 1$, $e^x \cong 1 + x$.)

48. In Section 31.6 we showed that the current in a series circuit consisting of an emf, a resistor, and a capacitor is $I = (\mathscr{E}/R)e^{-t/RC}$. If the power developed by the emf is $\mathscr{E}I$, show that the energy converted by the emf during the complete charging of the capacitor is $C\mathscr{E}^2$.

49. A fully charged capacitor has 12 J of energy stored. How much energy remains stored when its charge has decreased to half its original value during a discharge?

50. Following the analysis of the charging of a capacitor, derive a relation for the charge on the capacitor as a function of time, given that the capacitor has an initial charge q_0.

51. Figure 20 shows an oscilloscope trace of the discharge of a 1-pF capacitor through a parallel-connected resistor. The time base on the horizontal axis is 1 μs/division. Estimate the resistance of the resistor.

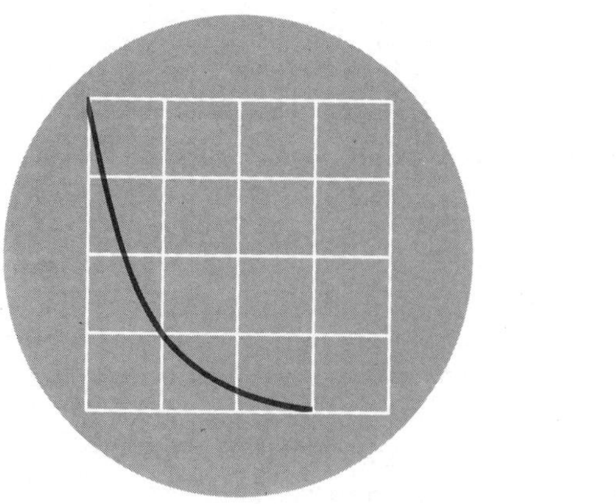

Figure 20

52. Show by substitution that

$$R\frac{dq}{dt} + \frac{q}{C} = 0$$

is solved by $q = q_0e^{-t/RC}$

53. Figure 21 shows a series connection of a charged capacitor, a resistor, and an open switch. When the switch is closed, the capacitor will discharge through the resistor. (a) Without doing any calculations, make a qualitative sketch of the charge on the capacitor and current in the circuit as a function of time. Assume that $t = 0$ when the switch is closed and that the initial charge on the capacitor is q_0. (b) Convince yourself that Kirchhoff's loop rule for this circuit yields $-IR + q/C = 0$ if you assume a clockwise current indicating discharge of the capacitor. Then from your sketches in part (a) and the definition of current, convince yourself that $I = -dq/dt$. (c) Substitute $I = -dq/dt$ in the loop equation and derive relations for the charge on the capacitor and current in the circuit as a function of time. Show that your relations are consistent with your sketches for part (a).

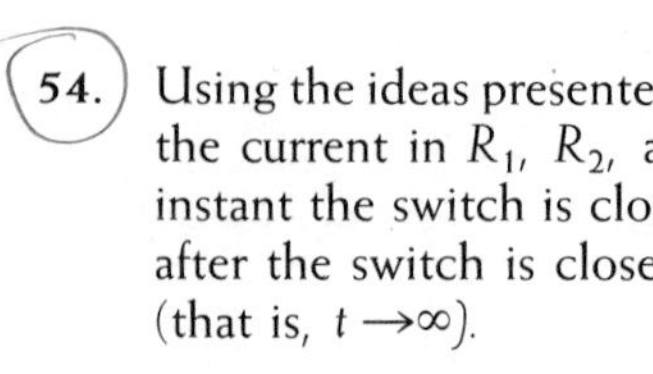

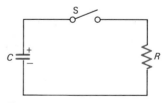

Figure 21

54. Using the ideas presented in Section 31.5, determine the current in R_1, R_2, and C (Figure 22) (a) at the instant the switch is closed (that is, $t = 0$), and (b) after the switch is closed for a long period of time (that is, $t \to \infty$).

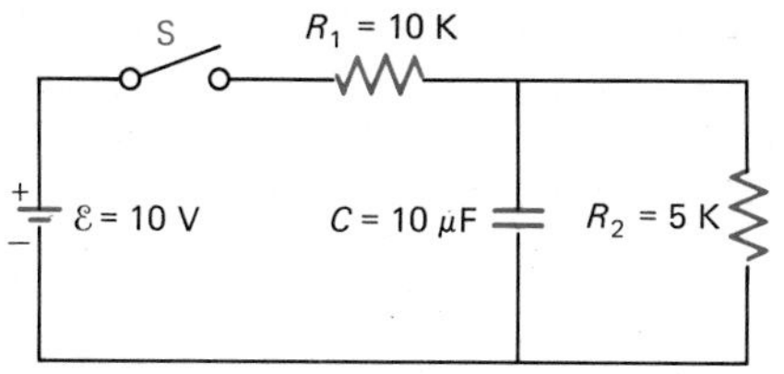

Figure 22

55. (a) Show that the ratio of the potential difference across the capacitor and the current in the circuit in Figure 31.23 can be written as $V_C/I = R(e^{t/RC} - 1)$. (b) If we interpret V_C/I as the "resistance" of the capacitor, what is the resistance for $t = 0$ and $t \to \infty$? Do these results make sense physically in light of the way that the capacitor is charged by the emf?

Section 31.7 Current and Potential Difference Measurements

56. A battery having an emf of 12 V and internal resistance r of 0.2 Ω is connected in series with two 18-Ω resistors. A voltmeter is used to measure the potential difference across one of the two resistors. What potential difference will the voltmeter measure if its internal resistance equals (a) 18 Ω; (b) 180 Ω; (c) 1800 Ω; (d) infinity?

chapter 32

MAGNETISM AND THE MAGNETIC FIELD

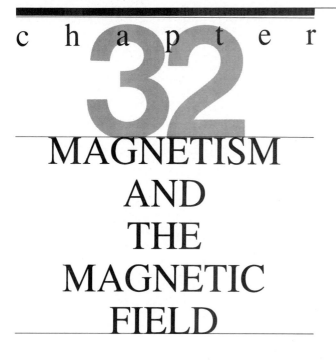

Preview

Magnetic phenomena have engaged the human imagination for about 2000 years, and numerous "explanations" of magnetism have evolved. We now believe that magnetism is associated with electric charges. James Clerk Maxwell incorporated this premise in an exquisite theory of electricity and magnetism. This chapter begins our study of the magnetic aspects of Maxwell's electromagnetic theory.

We start by describing some 19th-century ideas that led to Maxwell's description of magnetism. This approach reveals the analogy between electric and magnetic phenomena. The concept of magnetic field is introduced as a mechanism for exerting a magnetic force on a moving charge. An operational definition of the magnetic field is expressed in terms of three measured quantities: (1) the velocity of a charged particle, (2) the electric charge, and (3) the magnetic force experienced by the particle. We discuss several applications that exploit our detailed understanding of the magnetic force on moving charges, and then examine the force exerted by a magnetic field on a current-carrying conductor. This force is utilized in an electric motor. We conclude the chapter by deriving formal relations for the torque on, and the potential energy of, a magnetic dipole in a magnetic field. Both of these relations are important in magnetic phenomena.

32.1 Magnetism

Most of us have at some time been curious about magnets. If you've ever played with a magnetic toy, used a compass, or attached notes to your refrigerator with a magnet, you've probably wondered how magnets work. Mankind's interest in magnetism can be traced to the Greek's discovery, some 2000 years ago, that lodestones from Magnesia (called *magnetite*) exert forces on each other. Countless technological applications involving magnetic phenomena have evolved since then. One of the most important of these is the magnetic navigation compass first used in China in about 1000 A.D. and still in use today.

Magnets hold a certain fascination for many of us, perhaps because it is possible to actually *feel* magnetic force. For example, if you hold a magnet in your hand while standing near an iron or steel object, you feel your hand being pulled toward the object. If you hold a magnet in each hand, you sense forces exerted by one magnet on the other, even when the magnets are not in contact; and if you place an insulating material such as glass between the two magnets, the forces persist. In fact, the forces exist even if the magnets are in a vacuum.

If you continue to experiment with magnets in this way, you soon learn that the sources of the magnetic force in a magnet are concentrated in regions called *poles*. You also find that forces between magnets can be attractive or repulsive. We can account for this attraction and repulsion by defining two types of poles, N and S. Two N or two S poles repel each other, but an N pole and an S pole attract

Figure 32.1

Two N poles or two S poles are urged apart by magnetic forces, but an N pole and an S pole are attracted by magnetic forces.

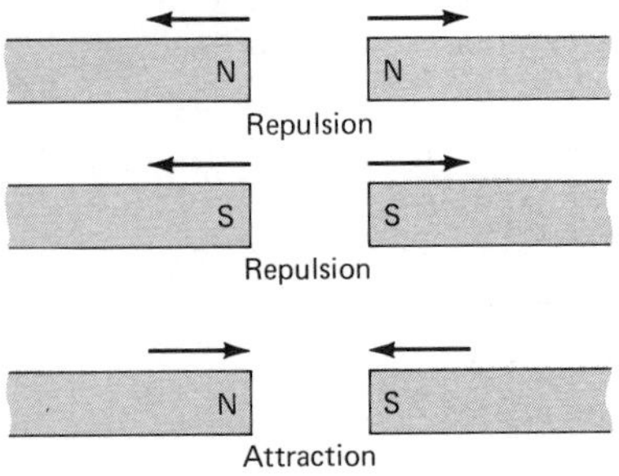

each other (Figure 32.1). Because the earth behaves as a giant magnet with poles located near its geographic poles, we define an N pole as that pole of a freely suspended magnet that is attracted to, and therefore points toward, the earth's magnetic pole, which is located near the North geographic pole. Since only *unlike* poles attract each other, the magnetic pole near the North geographic pole must be an S pole.

If we try to isolate a magnetic pole by cutting a magnet into two pieces, we do not obtain a separate N pole and a separate S pole, but instead get two smaller magnets, each having an N pole and an S pole (Figure 32.2). This happens no matter how many times we cut the magnets, and *an isolated magnetic pole is never obtained.* Despite rare claims that a magnetic monopole has been discovered,* exhaustive experiments with magnets lead us to conclude that the elementary magnetic entity is a magnetic dipole having one N pole and one S pole.

*Magnetic monopoles have been incorporated in some theories of magnetism. Encouraged by these speculations, many researchers have undertaken experimental searches. Generally, their results have been negative. For a claim (so far unconfirmed and unaccepted) of discovery, see P. B. Price, E. K. Shirk, W. Z. Osborne, and L. S. Pinsky, Evidence for Detection of a Moving Magnetic Monopole, *Phys. Rev. Lett.* 35, 487 (August 1975).

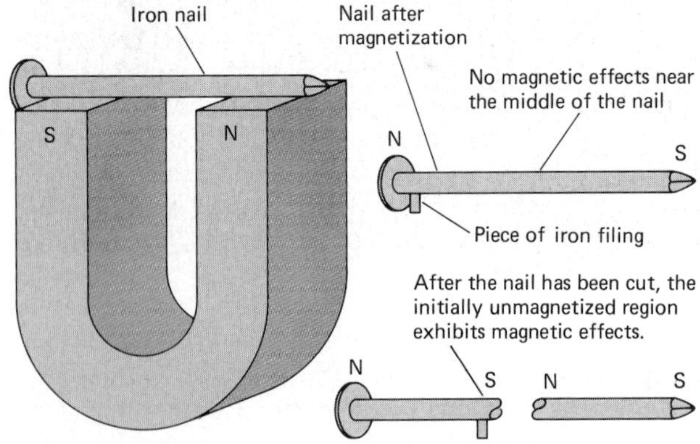

Figure 32.2

An iron nail becomes magnetized when laid across the pole faces of a magnet. A piece of iron filing is attracted to the ends of the magnetized nail but there is no attraction at the midpoint. Cutting the nail in half produces two magnets. The ends that were near the middle of the unsevered nail are now magnetic poles.

If we sprinkle iron filings in the vicinity of a bar magnet, the filings become magnetized and produce a pattern remarkably similar to that produced by tiny bits of thread scattered in the vicinity of an electric dipole (Figure 32.3a, b). The needle of a magnetic compass placed in the vicinity of the magnet will align itself with the iron filings (Figure 32.3c). The poles of the bar magnet produce this alignment by exerting forces on the poles of the magnetized iron filings. We can envision these forces as being transmitted to the iron filings and compass via a magnetic field created by the magnet. But because isolated poles are nonexistent, we cannot explore magnetic fields as we explored electric fields with static charges. It is not possible to investigate either the force between the isolated poles or the magnetic field produced by an isolated pole. Instead we consider the forces and torques associated with magnetic dipoles and moving charges.

Although many similarities exist between electric and magnetic phenomena, no physical connection was inferred until July 21, 1820. During a lecture demonstration on this date, the Danish scientist Hans Christian Oersted accidentally discovered that a compass beneath a current-carrying wire orients itself perpendicular to the wire (Figure 32.4). Oersted reversed the current in the wire, and observed that the N and S positions of the compass needle interchanged. He established for the first time an interaction between moving charges (electric current) and a magnetic dipole. Had Oersted sprinkled bits of iron on a sheet of paper and inserted the current-carrying wire through the center and perpendicular to the plane of the paper, he would have seen the bits line up in distinct circular patterns centered on the axis of the wire (Figure 32.5), indicating the existence of a magnetic field. A compass needle on any circle would have aligned perpendicular to the radius. The torque experienced by the compass needle results from an interaction of the magnetic needle

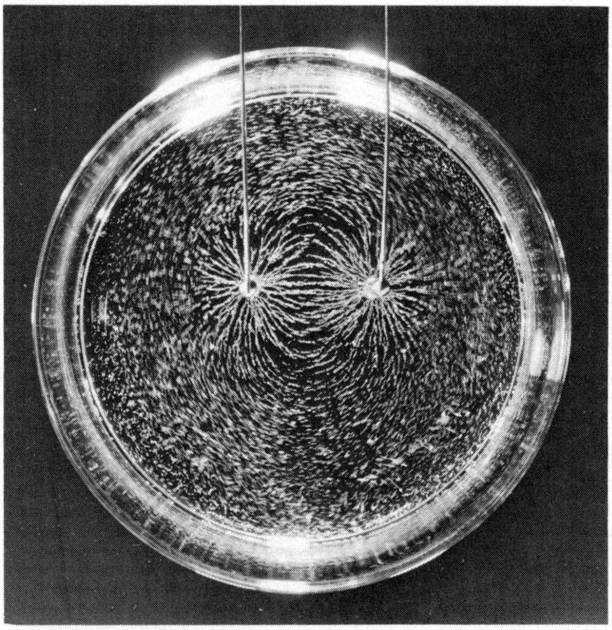

Figure 32.3a

The electric field line pattern of an electric dipole as revealed by the alignment of tiny bits of thread.

Figure 32.3b

The magnetic field line pattern of a magnetic dipole as revealed by the alignment of tiny bits of iron.

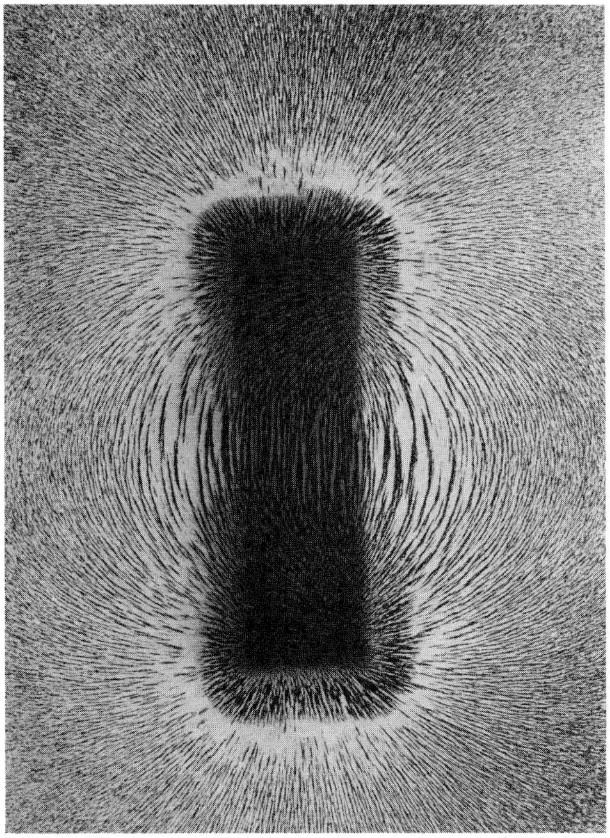

Figure 32.3c

A compass aligns tangentially to a magnetic field line.

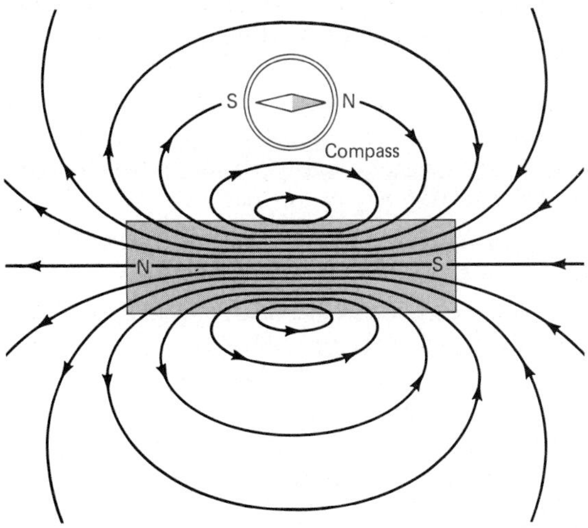

and the magnetic field produced by the current in the wire. A moving charge producing a magnetic field is called the **Oersted effect.** We will consider the production of magnetic fields by moving charges in the next chapter. Now we turn to the force an existing magnetic field exerts on a moving charge.

Figure 32.4

A compass beneath a current-carrying wire experiences a torque and orients perpendicular to the wire.

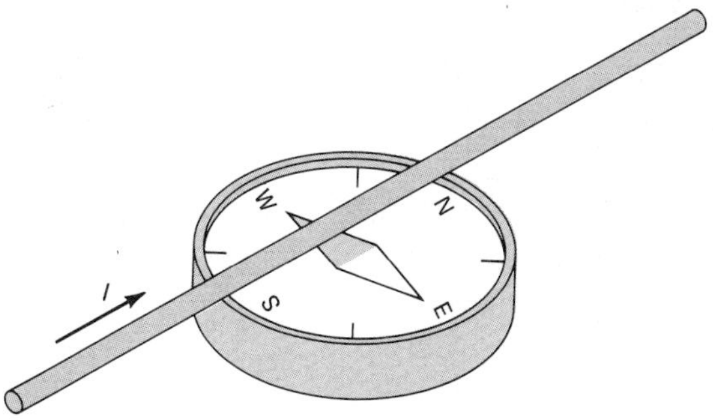

Figure 32.5

A current-carrying wire extends through a flat plastic sheet oriented perpendicular to the wire. Iron filings sprinkled on the sheet orient along the circular magnetic field lines produced by the current in the wire.

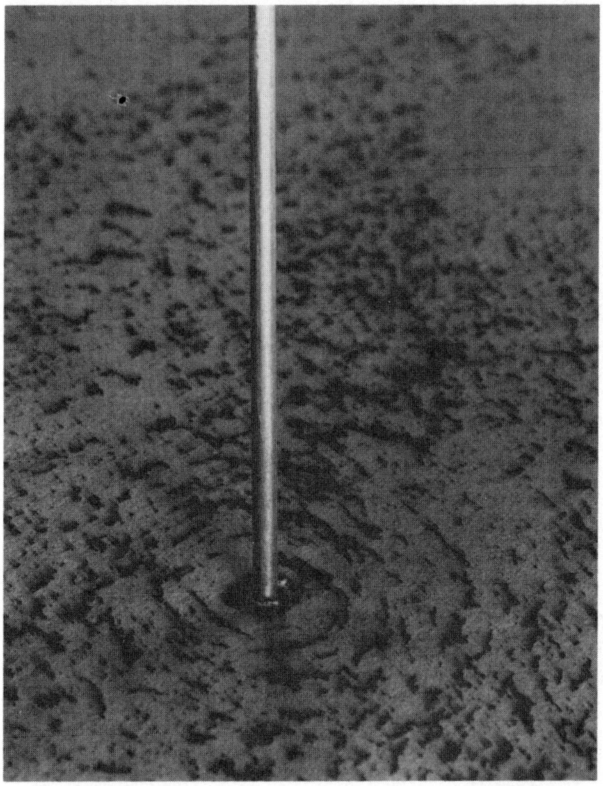

Questions

1. Two pieces of iron are identical in appearance. However, one is a magnetic dipole and the other is not. By observing the forces that the two pieces exert on each other, how can you tell which one is not magnetized?
2. What advantage might there be in calling the poles of a magnet positive and negative rather than N and S?

3. Suppose that you had a large number of tiny permanent magnets. How could you assemble these so as to produce a long, thin bar magnet? How could you use a permanent bar magnet to produce the alignment for you?

4. How can iron nails having no initial magnetism be made to hang end-to-end from a permanent magnet?

5. Simple magnetic compasses are free to rotate about one axis only. What changes would occur in the orientation of a compass needle if the needle were free to rotate about any axis through its center?

32.2 Magnetic Field

Magnetic Force on a Moving Charge

Because a properly oriented magnet experiences a force when placed in a magnetic field of a current-carrying conductor, we might expect from Newton's third law that a force is experienced by the moving charges. Experiment confirms this expectation. Moving charges, unlike isolated magnetic poles, do exist and they can be used to explore magnetic fields.

Experimentally, moving charges can be produced in a variety of ways. Certain radioactive materials emit both positive and negative charges. Cathode-ray tubes (like those used in oscilloscopes—Figure 32.6) can be used to generate a beam of electrons. This technique is attractive because it permits the experimenter to control the speed of the charges by accelerating them through a known potential difference. Additionally, the termination of the electron trajectories can be monitored by observing the spot of light produced by the charges as they strike a phosphorescent screen.

Now that we have a qualitative feeling for what magnets and magnetic fields are, let's consider a beam of electrons produced by a cathode-ray tube, and use it to explore a uniform magnetic field. (Such a uniform field filling a suitable volume of space exists between the pole faces of many types of large research magnets, as shown, for example, in Figure 32.17.) This will lead us to an operational definition of the magnetic field. We can check the uniformity of a magnetic field by observing the alignment of a compass as it is moved around between the pole faces. This exploration shows that the compass always aligns perpendicular to the faces of the poles of the magnets. No electric field exists in the region of interest. Therefore, electrons in the cathode-ray tube experience no electric force.

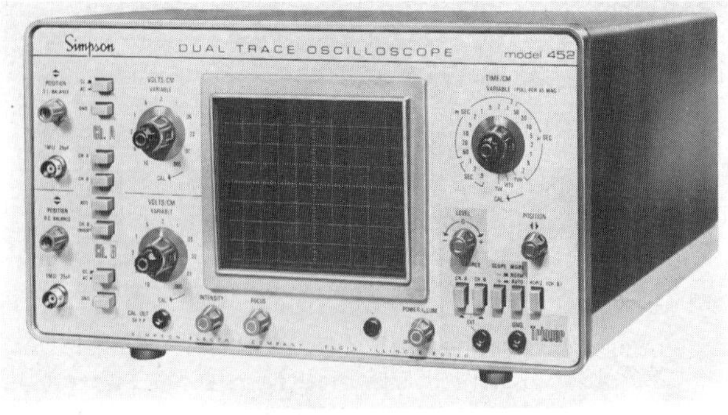

Figure 32.6

A modern cathode-ray oscilloscope used in physics research.

Figure 32.7a depicts a beam of electrons striking the center of the screen of a cathode-ray tube removed from a magnetic field of interest. Suppose that the tube is inserted into the magnetic field as shown in Figure 32.7b. For reference, we locate a coordinate system with origin at the midpoint between the pole faces, and orient the axes of the poles along the z-axis. Then, if the electrons are directed along the y-axis, we find that the beam is deflected in the x-direction; the electrons experience a force perpendicular to their direction of motion. Interchanging the N and S poles reverses the deflection; the beam is deflected in the negative x-direction. The force increases as the speed of the electrons increases, but the deflection decreases because the electrons move faster through the field. The force also depends on the direction of the beam of electrons. For example, if the cathode-ray tube is rotated as shown in Figure 32.7c, the beam is still deflected along the x-axis but the amount of deflection is diminished. If the electrons move parallel to the z-axis, the deflection vanishes. Similar experiments with positive charges reveal that for a given setup the force on positive charges is directed opposite to the force on negative charges. For example, if the beam of electrons in Figure 32.7b were instead a beam of protons, it would be deflected along the negative x-axis.

Exhaustive experiments disclose four facts about the magnetic force on a moving charge in a magnetic field:

1 There is a unique direction in space along which the moving charges experience no magnetic force. This direction, which we call the *zero-force direction*, lies along a line perpendicular to the pole faces (the z-axis in Figure 32.7).

2 The magnitude of the magnetic force is directly proportional to the product of the charge, the speed, and the sine of the angle between the velocity and the zero-force direction ($F \propto qv \sin \theta$).

3 The direction of the magnetic force is perpendicular to both the velocity and the zero-force direction.

4 Negative and positive charges moving in the same direction are deflected in opposite directions.

We can summarize these four facts by defining a magnetic field vector **B** to be along the zero-force direction, and by writing the force as

$$\mathbf{F} = q\mathbf{v} \times \mathbf{B} \tag{32.1}$$

This definition of the magnetic field **B** is similar to the definition of the electric field **E** in that the field is defined in terms of the *force* it produces. It is an *operational* definition. To find the field, we must measure the force; but the mathematical structure of the definition of **B** is quite different from that of the definition of **E**. The magnetic field **B** is contained in a vector product, and its definition is therefore indirect.

Figure 32.7

(a) The stream of electrons in the cathode-ray tube experiences no magnetic force when the tube is outside the magnetic field produced by the large magnet. The undeflected electron beam produces a spot of light in the center of the cathode-ray tube screen. (b) When the cathode-ray tube is moved horizontally into the magnetic field, the electron stream is deflected by a magnetic force. If a positive z-axis is directed from the S pole to the N pole, and the electrons move in the positive y-direction, then the magnetic force on the electrons is in the positive x-direction. (c) When the cathode-ray tube is tilted up toward the positive z-axis, the electron stream is still deflected in the positive x-direction but the force on the electrons is diminished. When the electron stream moves in the z-direction, it is undeflected; the force on the electrons is zero.

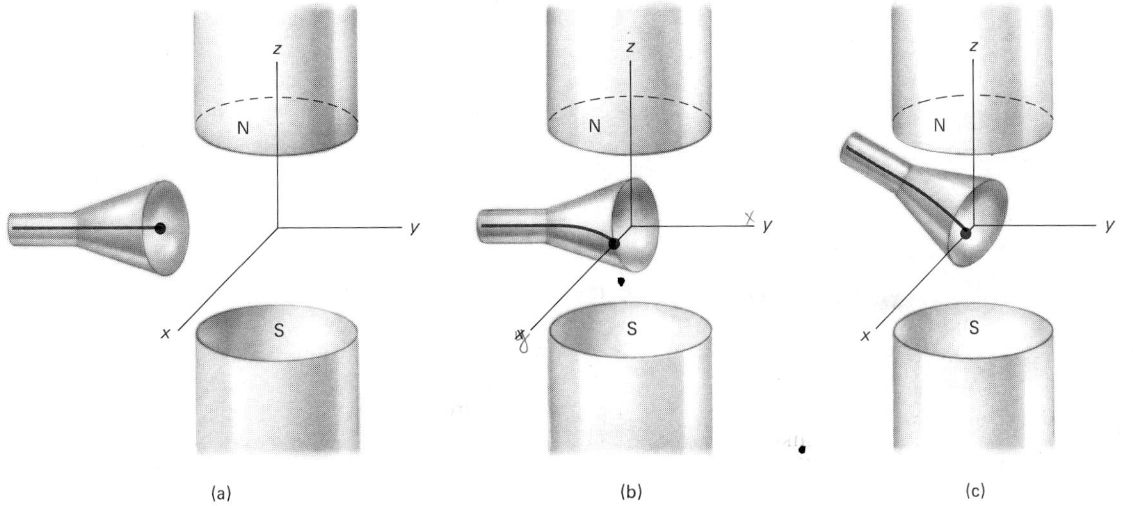

Let's now prove to ourselves that this definition of **B** agrees with the four experimental facts about the force on a moving charge in a magnetic field. We will let the magnetic field **B** be unknown but constant, and let the velocity **v** vary in direction.

Fact 1 When **v** is parallel or antiparallel to **B** (which is perpendicular to the pole faces)* the vector product $q\mathbf{v} \times \mathbf{B}$ vanishes, and consequently the force also vanishes. Thus our definition of **B** accounts for the experimentally observed zero-force direction. According to $\mathbf{F} = q\mathbf{v} \times \mathbf{B}$, **B** is in the zero-force direction.

Fact 2 From the meaning of a vector product, the magnitude of $\mathbf{F} = q\mathbf{v} \times \mathbf{B}$ is $F = qvB \sin\theta$ where θ is the angle ($\leq 180°$) between **v** and **B**, the zero-force direction. This agrees with the experimental observation that $F \propto qv \sin\theta$.

Fact 3 The direction of the force as determined from the vector product $q\mathbf{v} \times \mathbf{B}$ is perpendicular to **v** and **B**. Thus the equation $\mathbf{F} = q\mathbf{v} \times \mathbf{B}$ agrees with experimental fact.

Fact 4 Because $-q\mathbf{v} \times \mathbf{B}$ is directed opposite to $q\mathbf{v} \times \mathbf{B}$, the magnetic force on negative charges is opposite to the magnetic force on positive charges. Because opposite forces must deflect charges in opposite directions, our definition of **B** agrees with the fact that negative and positive charges moving in the same direction are deflected in opposite directions.

According to Eq. 32.1

$$\mathbf{F} = q\mathbf{v} \times \mathbf{B}$$

Hence the magnetic field **B** has units of

$$\frac{\text{force}}{\text{charge} \cdot \text{velocity}} = \frac{\text{newtons}}{\text{coulomb}\ \dfrac{\text{meter}}{\text{second}}}$$

Because the ratio coulombs/second is equal to the unit of current, the ampere, the units of magnetic field can also be expressed as

$$\frac{\text{newtons}}{\text{ampere meter}}$$

This unit is called a *tesla* and its symbol is T. The range of magnetic field strength encountered in nature is shown in Table 32.1 in teslas.

*It would be legitimate to define **B** via the relation $\mathbf{F} = q\mathbf{B} \times \mathbf{v}$. Such a choice would merely reverse the direction of **B**. The choice actually adopted, $\mathbf{F} = q\mathbf{v} \times \mathbf{B}$, thus arbitrarily defines the sense of **B**.

Table 32.1

Neutron star, pulsar	10^8 T
Largest man-made fields from short bursts of electric current	10^3 T
Very strong laboratory superconducting magnet	10^1 T
Strong bar magnet	10^{-1} T
Earth's magnetic field	10^{-5} T
Interplanetary magnetic fields	10^{-9} T
Magnetic field associated with the human body	10^{-12} T

Because many magnetic field strengths of interest tend to be only a fraction of a tesla, you will often see magnetic field strengths expressed in the unit of gauss (G). For conversion purposes, $1\ \text{T} = 10^4$ G.

An electric field and a magnetic field are both defined operationally in terms of the force on a test charge. In principle, a measurement of the force on a known static test charge is sufficient to determine **E**. It is somewhat more difficult to determine **B** because **B** is buried in a vector product ($\mathbf{F} = q\mathbf{v} \times \mathbf{B}$), requiring measurements of both the size and velocity of the test charge.

Example 1 Operational Determination of a Magnetic Field

We can determine a magnetic field as follows. We first vary the direction of a moving test charge in the magnetic field until the zero-force direction is found. Because **B** is in the zero-force direction we now know the line along which the magnetic field vector lies. Let us orient the z-axis of a coordinate system along this line (Figure 32.8). The test charge is directed perpendicular to this direction (the z-axis) and the force is recorded. If we know the sign of the test charge, we can infer the direction of **B** from the direction of the force. For example, if a positive test charge moves in the positive x-direction and experiences a force in the positive y-direction, then the magnetic field is directed in the negative z-direction. Knowing the speed and charge, and measuring the force, we obtain for the magnitude of **B**

$$B = \frac{F}{qv \sin\theta} = \frac{F}{qv}$$

because $\theta = \pi/2$. If the moving test charge is a proton ($+q = 1.602 \times 10^{-19}$ C) moving with a speed of 2.0×10^6 m/s, and the force is measured to be 3.84×10^{-13} N, then the magnitude of the field is

$$B = \frac{F}{qv}$$

$$= \frac{3.84 \times 10^{-13}}{1.60 \times 10^{-19}(2.0 \times 10^6)}$$

$$= 1.20\ \text{T}$$

Figure 32.8

A positive charge moving with velocity **v** along the positive x-axis experiences a force in the positive y-direction. Because $\mathbf{F} = q\mathbf{v} \times \mathbf{B}$, the magnetic field **B** is in the negative z-direction.

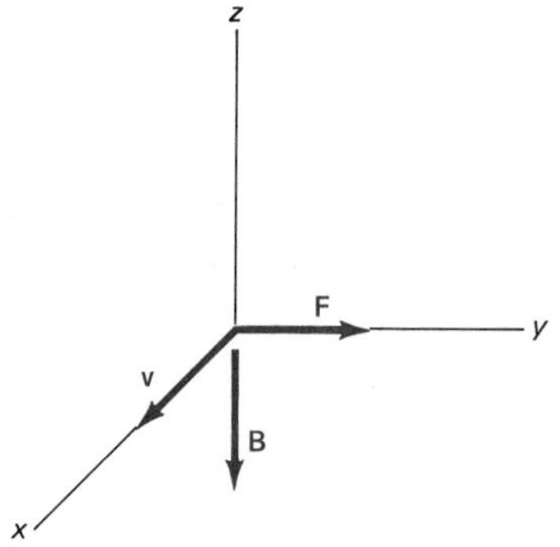

Incorporating the direction, we write

$$\mathbf{B} = -1.20\mathbf{k}\ \text{T}$$

Although this scheme can be used in principle to determine an unknown magnetic field, it is not a practical method because of the difficulty of measuring **v** and **F**. We will note how magnetic fields are actually measured as we proceed. The Hall effect, described in Section 32.3, is one practical method.

Magnetic Field Lines

Recall from Section 27.2 that electric field lines, although they do not actually exist, provide a visual picture of an electric field. They allow us, for example, to determine

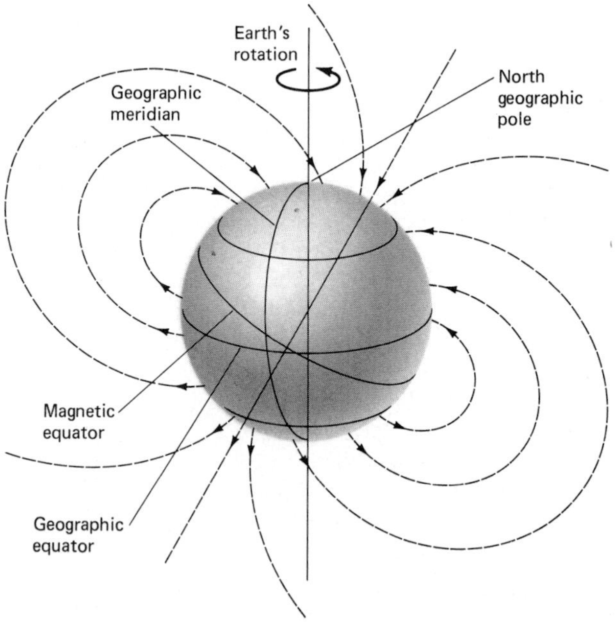

Figure 32.9

The magnetic field line pattern of Earth's magnetic field. By convention the pole near the North geographic pole is an S pole.

qualitatively the motion of a charge under the influence of the field. The pattern of iron filings that forms around a bar magnet suggests that this same concept can be extended to magnetic fields. The magnetic field vector **B** is defined at every point in a magnetic field. Thus, in principle, we could draw an infinite number of **B** vectors denoting the magnitude and direction of **B** at every point in the magnetic field. For simplicity we will instead represent the field with a few selected lines. At each point in the field, the magnetic field vector **B** is tangential to the magnetic field line through that point. Figure 32.9 shows the magnetic field line pattern for earth's magnetic field. To a first approximation this field is characteristic of a magnetic dipole with an S pole near the North geographic pole and an N pole near the South geographic pole. (We will discuss Earth's magnetic field in more detail in Section 36.6.)

Magnetic Flux

We construct a magnetic flux Φ_B in much the same way that we construct an electric flux Φ_E. For a constant magnetic field **B** and a plane surface A, the magnetic flux is defined as the scalar product of the magnetic field and the area.

$$\Phi_B = \mathbf{B} \cdot \mathbf{A} \tag{32.2}$$

The flux unit is a tesla meter^2, which in the SI system is called a weber (Wb). The flux can be interpreted as *the number of magnetic field lines passing through the area A*. For a nonuniform magnetic field and a curved surface we use a differential form of Eq. 32.2:

$$d\Phi_B = \mathbf{B} \cdot d\mathbf{A} \tag{32.3}$$

By convention, we find the direction of the area vector by using the right-hand rule (Figure 32.10), just as we did in Section 27.3. For the special case of **B** and $d\mathbf{A}$ parallel we can write

$$B = \frac{d\Phi_B}{dA} \tag{32.4}$$

Thus we can infer the magnitude of the magnetic field at a given position from the number of magnetic field lines per square meter at that position. The greater the concentration of magnetic field lines, the larger the magnetic field (Figure 32.11).

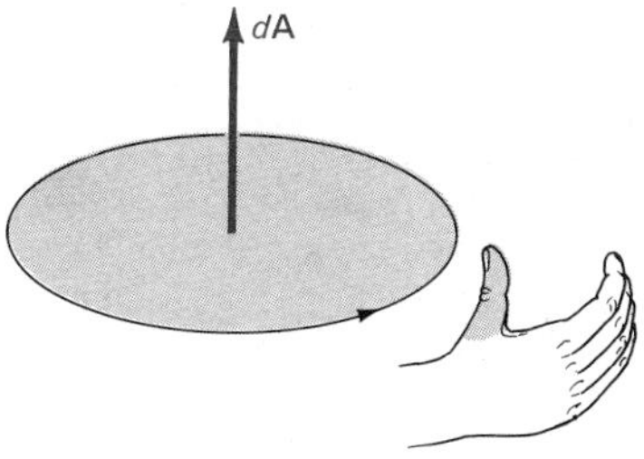

Figure 32.10

The number of magnetic field lines threading the area $d\mathbf{A}$ is $\mathbf{B} \cdot d\mathbf{A}$. The direction of $d\mathbf{A}$ is determined by the direction given for the perimeter of the area. Curl the fingers of your right hand in the direction of the perimeter. Then your thumb points in the direction of $d\mathbf{A}$. For a closed surface, $d\mathbf{A}$ always points outward from the surface.

Figure 32.11

The number of lines threading dA in the $\mathbf{B}_1$ field is greater than the number of lines threading dA in the $\mathbf{B}_2$ field. Hence $B_1 > B_2$.

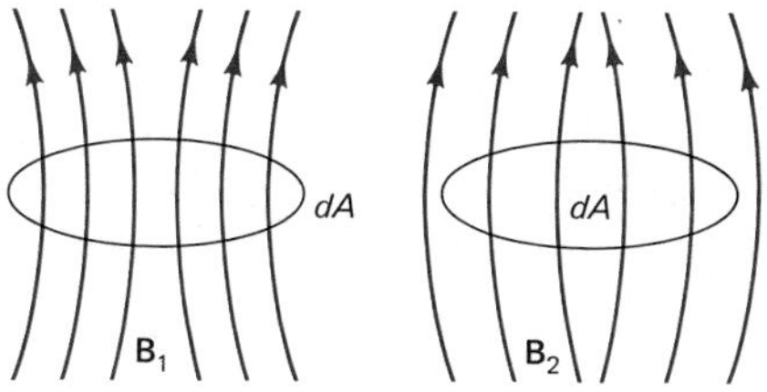

The *net* number of magnetic field lines, or net magnetic flux penetrating a closed surface, is obtained by integrating Eq. 32.3 over the surface.

$$\Phi_B = \int_{\substack{\text{closed}\\\text{surface}}} \mathbf{B} \cdot d\mathbf{A} \tag{32.5}$$

Similarly, we found the amount of electric flux threading a closed surface to be

$$\Phi_E = \int_{\substack{\text{closed}\\\text{surface}}} \mathbf{E} \cdot d\mathbf{A} = \frac{q}{\epsilon_0}$$

where q is the *net* charge residing inside the surface. This net charge can be positive, negative, or zero. In magnetism there is no magnetic counterpart of *net* charge. In other words, there is no proof of the existence of magnetic monopoles. In the absence of magnetic monopoles the magnetic flux threading a closed surface is zero.

$$\int_{\substack{\text{closed}\\\text{surface}}} \mathbf{B} \cdot d\mathbf{A} = 0 \tag{32.6}$$

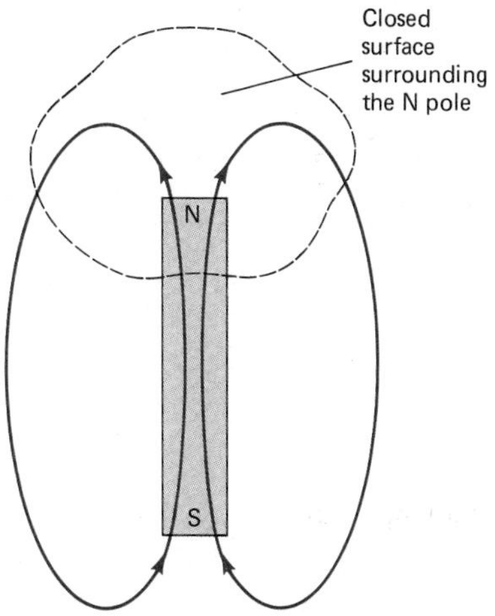

Figure 32.12

The Gaussian surface surrounds only the N pole of the magnet. Because the field lines are continuous, any line passing from the inside to the outside must pass back into the surface. Hence the net number of lines penetrating a closed surface is zero.

Unlike electric field lines, which *originate* on a positive charge, curve in accordance with the direction of the electric field, and *terminate* on a negative charge, magnetic field lines have no origin and no termination. They always form closed paths. The magnetic field lines of a magnetic dipole emerge from the N pole, curve in accordance with the direction of the magnetic field, and pass into and through the S pole (Figure 32.12). If any part of the magnetic field line pattern is surrounded by a closed surface, then any line entering the surface must pass back out of the surface (Figure 32.12), so that the net number of lines threading the closed surface is zero.

Questions

6. How could you determine whether or not there is a magnetic field in the area you are now in?
7. Why is it that a static charge is adequate for exploring an electric field, but a moving charge is needed to investigate a magnetic field?
8. A charged particle moves with velocity **v** in a region where both an electric field **E** and a magnetic field **B** exist. What orientation of **v**, **E**, and **B** would make the net force on the charge equal zero?
9. A compass needle is deflected when a charged plastic rod is held near it. What is the origin of the force that produces the deflection?
10. The northern lights, or aurora borealis, are produced by charged particles of cosmic origin interacting with Earth's atmosphere near the North Pole. These particles approach Earth in a nearly due south direction. Why aren't these particles deflected away from Earth by a magnetic interaction with Earth's magnetic field?
11. A charged particle traveling due east enters a chamber in which the path of the particle can be observed. The chamber is located in a uniform magnetic field that is directed due north. Looking into the chamber, we see that the particle executes a clockwise circular motion. What is the sign of the charge on the particle?
12. How do the units of electric field **E** and $\mathbf{v} \times \mathbf{B}$ compare? Justify your answer.

32.3 Applications of Moving Charges in a Magnetic Field

In addition to its fundamental importance as a definition of the magnetic field vector **B**, the relation $\mathbf{F} = q\mathbf{v} \times \mathbf{B}$ has many practical applications and is involved in a variety of natural phenomena. Let us consider a few of the more important applications.

Charged-Particle Linear Momentum Analyzer

When a net force **F** causes a particle to displace $d\mathbf{s}$, the net work dW done is (Section 7.1)

$$dW = \mathbf{F} \cdot d\mathbf{s} = F \cos\theta \, ds$$

where θ is the angle between **F** and $d\mathbf{s}$. If the net work is zero, then according to the work–energy principle there is no change in the kinetic energy. There will be zero net work if the force is perpendicular to the displacement. We see this in the circular motion of a satellite about the earth. No work is done on the satellite by the gravitational force because the force and instantaneous displacement are perpendicular. Although the force instantaneously changes the direction of the satellite and keeps it moving in a circle, there is no change in kinetic energy or speed. The magnetic force on a charged particle is always perpendicular to its instantaneous velocity. Relating the displacement to velocity by

$$d\mathbf{s} = \mathbf{v}\, dt$$

we see that there is no work done by the magnetic force because the magnetic force and displacement are perpendicular. If the magnetic force is the net force on the particle, then there will be no change in the particle's kinetic energy. The magnetic force can change the *direction* of the velocity but *not* the speed.

Consider a positively charged particle moving in a uniform magnetic field directed perpendicular to the velocity of the charge (Figure 32.13).* The magnetic force does no work on the particle. Therefore, if the magnetic force is the net force, the particle executes uniform circular motion with radial acceleration v^2/r. Using Newton's second law to relate the magnetic force, Bqv, to mass times acceleration, we have

$$Bqv = \frac{mv^2}{r} \tag{32.7}$$

Solving for the linear momentum p, we obtain

$$p = mv = Bqr \tag{32.8}$$

*It is sometimes difficult to view the three-dimensional aspects of the magnetic force. When we are interested in the force and velocity, it is common to draw **F** and **v** in the plane of the paper and to orient the magnetic field perpendicular to the plane. We use a dot to denote a magnetic field line coming out of the paper. This is like looking into the pointed head of the arrow representing the direction of vector **B**. We use a cross to denote a magnetic field line directed into the paper. This is like looking into the tail feathers of the arrow representing the direction of vector **B**. A uniform magnetic field directed out of the paper is shown as

```
• • • • • • •
• • • • • • •
• • • • • • •
• • • • • • •
```

A uniform magnetic field directed into the paper is shown as

```
x x x x x x
x x x x x x
x x x x x x
x x x x x x
```

Figure 32.13

A charged particle in a uniform magnetic field moves in a circle when its velocity is directed perpendicular to the magnetic field lines.

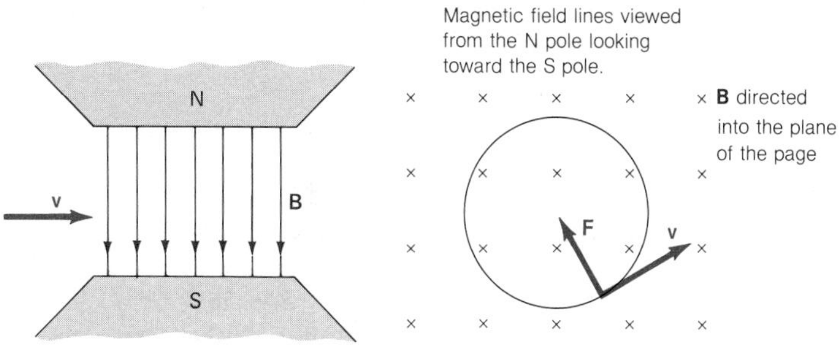

Although we have derived Eq. 32.8 by using Newtonian mechanics, it is also valid in relativistic mechanics if we use the relativistic form for linear momentum:

$$p = \frac{mv}{\sqrt{1 - \dfrac{v^2}{c^2}}}$$

If we know the strength of the magnetic field and the charge of the particle, we can measure the radius of the circular path of the particle in order to determine its momentum. This principle is used to measure the linear momentum of charged particles. The circular trajectories of charged particles are made visible in a device called a *liquid-hydrogen bubble chamber* (Figure 32.14).

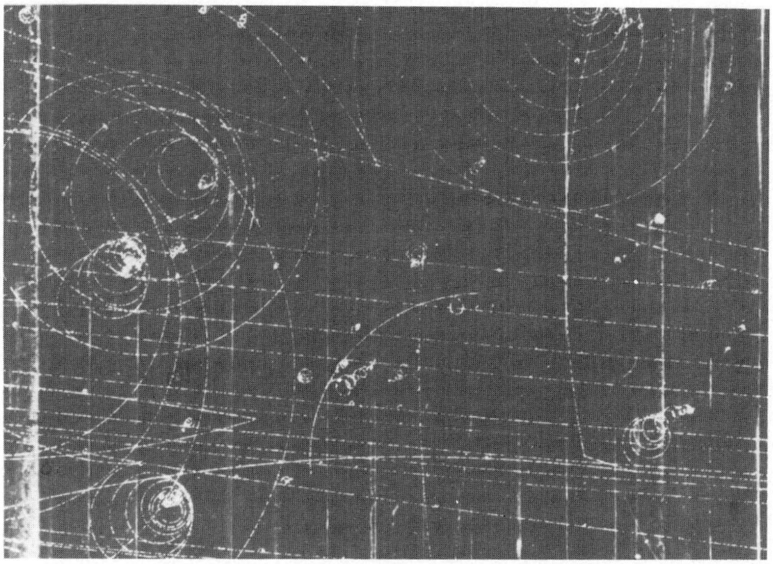

Figure 32.14

Spiral path of an electron in a liquid-hydrogen bubble chamber that has been placed in a magnetic field. Continuous collisions with hydrogen atoms cause the electron to lose speed, thereby continually decreasing the radius of the electron orbit (Eq. 32.8).

Figure 32.15

Schematic drawing of a mass spectrometer. Positively charged particles from the source are accelerated by an applied potential difference and enter a uniform magnetic field. The particles move in a semicircle of radius r and then strike the detector.

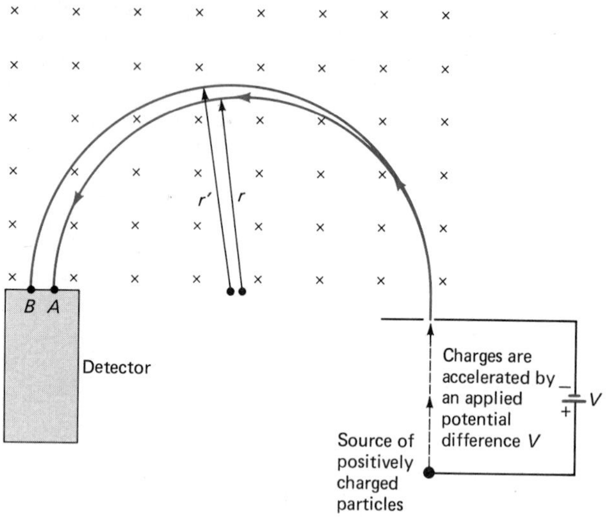

r′
r
B A
Detector
Charges are accelerated by an applied potential difference V
V
Source of positively charged particles

Mass Spectrometer

A mass spectrometer is an instrument used to measure the mass of a particle. One type of mass spectrometer enables us to measure the ratio of mass and charge by using magnetic principles. In this spectrometer, a charged particle of known velocity enters a uniform magnetic field with direction perpendicular to the magnetic field direction (Figure 32.15). The magnetic force on the particle causes it to move in a circular trajectory of radius r. According to Eq. 32.7, the mass-to-charge ratio is given by

$$\frac{m}{q} = \frac{Br}{v} \qquad (32.9)$$

By placing an appropriate charged-particle detector in the spectrometer we are able to determine the radius r. For example, the orbit of a particle that hits the detector at A must have radius r, while the orbit of a particle that hits the detector at B has radius r' (Figure 32.15). The velocity of a particle can be determined by accelerating it through a known potential difference V. (The velocity can also be determined with a velocity selector, described later in this section.) Using the work–energy principle, we can write

$$\text{net work done on the charge} = \text{change in kinetic energy}$$

$$qV = \frac{1}{2}mv^2 \qquad (32.10)$$

Squaring the terms of Eq. 32.9 and using Eq. 32.10 to eliminate the velocity, we obtain for the mass-to-charge ratio

$$\frac{m}{q} = \frac{B^2r^2}{2V} \qquad (32.11)$$

Thus measurements of B, r, and V allow us to determine m/q. If the charge of the particle is known, then the mass can be computed.

Although an atom is electrically neutral in its natural state, it is relatively easy to remove one or more of its electrons, leaving it with a net positive charge. If two atoms with the same charge but different masses are accelerated through the same potential difference and passed into the same magnetic field, then the ratio of their masses is related to their radii of curvature by

$$\frac{m_1}{m_2} = \left(\frac{r_1}{r_2}\right)^2 \qquad (32.12)$$

It is customary to measure atomic masses relative to the most abundant isotope* of carbon. On the atomic mass scale this atom of carbon is assigned a relative mass of exactly 12 atomic mass units (u). On an absolute mass scale

$$1 \text{ u} = 1.66053 \times 10^{-27} \text{ kg}$$

The relative masses of atoms are often measured to an accuracy of one part in 10^8.

*Isotopes are discussed in Chapter 43. Basically, isotopes are atomic species having the same chemical properties (same number of electrons and protons) but differing masses (different number of neutrons).

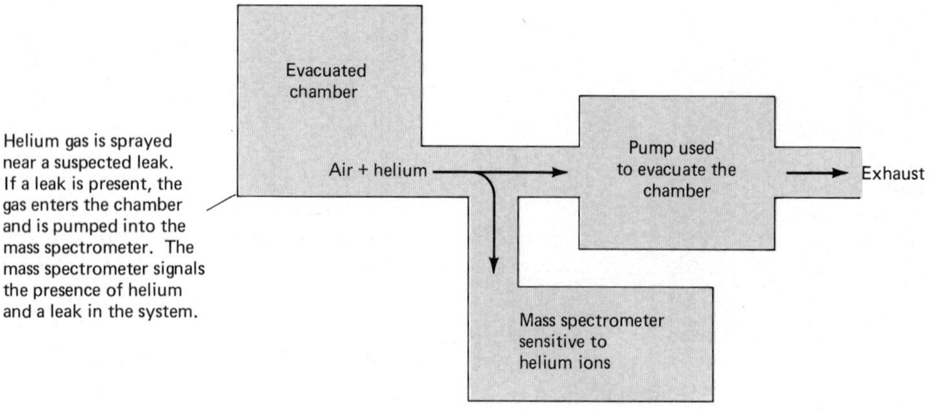

Helium gas is sprayed near a suspected leak. If a leak is present, the gas enters the chamber and is pumped into the mass spectrometer. The mass spectrometer signals the presence of helium and a leak in the system.
Evacuated chamber
Air + helium
Pump used to evacuate the chamber
Exhaust
Mass spectrometer sensitive to helium ions

Figure 32.16

Leaks in the plumbing of research vacuum systems are located with a helium leak detector.

Sometimes a mass spectrometer is used to study only a single type of charged particle. When this is the case, the potential difference of the spectrometer can be set permanently for this particle. (Such a scheme is used in a popular apparatus for detecting leaks in research systems requiring a vacuum chamber; see Figure 32.16.)

Figure 32.17

The cyclotron principle. Positively charged particles released near the center of the dees (labeled D_1 and D_2) begin their journey by being accelerated across the gap from D_1 to D_2. Once inside D_2 the particle feels no electric force and is bent into a circular path by the magnetic force. When it reenters the gap between the dees the polarity between the dees has changed and the particle gets a second acceleration across the gap. This process continues until the particles emerge from the left dee.

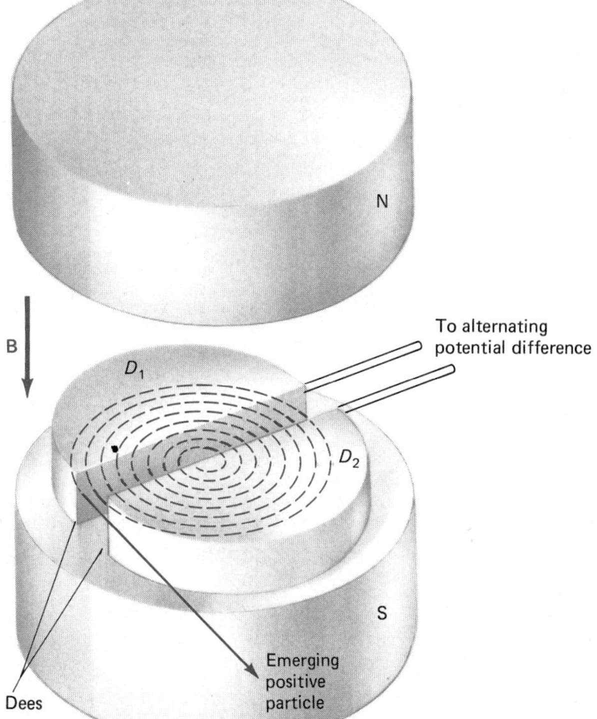

Cyclotron

You may recall that the speed of a particle executing uniform circular motion is related to orbital frequency and radius by

$$v = 2\pi\nu r$$

Equation 32.7 can therefore be written as

$$\frac{Bq}{2\pi m} = \nu \qquad (32.13)$$

Note that the *frequency of revolution (and thus the period) is independent of the radius of the orbit and the speed of the charge.* This frequency is called the *cyclotron frequency* because the operation of a device called a *cyclotron* is based on this relation.

A cyclotron consists of two hollow, D-shaped metal vessels called *dees* in an evacuated chamber (Figure 32.17). A potential difference, the polarity of which varies sinusoidally with time, is connected between the dees. The frequency of the oscillations is the same as the cyclotron frequency. The dees are placed in a uniform magnetic field directed perpendicular to the plane of the dees, and a source of charged particles is placed near the center of the dees. The charged particles are then accelerated across the gap between the dees. Once inside the dees, the particles are shielded electrically, and coast at constant speed while being bent in a circular path by the magnetic force $F = Bqv$. When they reenter the gap between the dees the polarity of the potential difference between the dees has changed and the particles are again accelerated, which increases their speed. Again they coast and bend in a circle. However, the radius of this second circle is larger than that of the first because the speed is greater. The particles are repeatedly accelerated and bent in this way until they emerge from the dees. Each acceleration of a charge q through a potential difference ΔV imparts an energy $q\,\Delta V$ to the charge. Because two accelerations are produced for each complete revolution, an energy $2q\,\Delta V$ is produced per turn. Typically, the potential difference is about 70,000 V, and a particle executes about 200 revolutions before emerging from the dees. This means that a particle emerges with an energy of $200(2)(70{,}000)q$ joules. If the particle is a proton, for example, then $q = 1.60 \times 10^{-19}$ C, and its energy upon emerging from the cyclotron is 28 MeV. This energy is equivalent to that acquired by a proton accelerated through a potential difference of 28 MV. But setting up such a potential difference is very difficult, which is why the cyclotron is such a useful experimental device for nuclear physicists.

As the velocity of the accelerated particles increases, relativistic effects are encountered, so that the cyclotron frequency can no longer be considered independent of speed. If we insert the relativistic momentum expression (Eq. 20.1) into Eq. 32.7 to account for these effects, we obtain

$$\nu = \frac{Bq}{2\pi m}\sqrt{1 - \frac{v^2}{c^2}} \qquad (32.14)$$

Thus we see that the cyclotron frequency decreases as the speed increases. A modified version of the cyclotron called the *synchrocyclotron* compensates for this effect by accelerating particles in bunches and decreasing the oscillator frequency after each half revolution to ensure that the particles are accelerated when they pass between the dees.

Contemporary high-energy particle accelerators called *synchrotrons* utilize varying magnetic fields, varying frequency, and a fixed circular path. The synchrotron at the Fermi National Laboratory in Batavia, Illinois, utilizes a particle orbit 2 km in diameter and achieves proton energies of over 500×10^9 eV (500 GeV).

Charged-Particle Velocity Selector

An electric field can coexist with a magnetic field in a region of space. If a moving charge is in motion in such a region, it experiences an electric force $\mathbf{F}_E = q\mathbf{E}$ as well as a magnetic force $\mathbf{F}_B = q\mathbf{v} \times \mathbf{B}$. The total force is the superposition of the electric and magnetic forces, and so

$$\begin{aligned}\mathbf{F} &= \mathbf{F}_E + \mathbf{F}_B \\ &= q(\mathbf{E} + \mathbf{v} \times \mathbf{B})\end{aligned} \tag{32.15}$$

The combination of electric and magnetic forces in Eq. 32.15 is commonly called the *Lorentz* force*. A charged-particle velocity selector, often used in conjunction with magnetic mass spectrometers, exploits the Lorentz force. A pencil-like stream of particles having a distribution of speeds is formed by allowing particles from some source to pass through holes in two separated plates. If a Cartesian coordinate system is oriented with the x-axis along a line connecting the two small holes, then the velocity of a particle passing through the second plate has the form $\mathbf{v} = v\mathbf{i}$. If the particles pass through an electric field $E\mathbf{j}$ established between the plates, then each particle experiences a force $qE\mathbf{j}$. If a magnetic field $+B\mathbf{k}$ is superimposed on the electric field, each particle experiences a magnetic force $qv\mathbf{i} \times (+B\mathbf{k}) = -qvB\mathbf{j}$. For those particles still passing through the holes, the net force must be zero. Hence

$$\begin{aligned}\mathbf{F}_{\text{net}} &= qE\mathbf{j} - qvB\mathbf{j} \\ &= 0\end{aligned}$$

Therefore

$$E = vB$$

or

$$v = \frac{E}{B} \tag{32.16}$$

Measurement of the electric field E and magnetic field B allows us to determine the velocity of the particles passing through the holes.

Force on a Current-Carrying Conductor

By converting electric energy into mechanical energy, electric motors are able to perform countless tasks. Regardless of their size or complexity, most electric motors operate on the principle that a current-carrying conductor experiences a force when placed in a magnetic field.

According to the free-electron model of conduction (Section 30.1), conduction electrons in an isolated conductor move randomly about, like molecules in a gas. Just as a closed container confines gas molecules, the conductor confines the electrons within its boundaries. If the isolated conductor is placed in a magnetic field, each conduction electron experiences a force, $\mathbf{F} = q\mathbf{v} \times \mathbf{B}$, because conduction electrons are moving charged particles. But because the electron motion is random, the vector sum of the forces on all the electrons is zero, and *the conductor as a whole experiences no net force*. If, however, a current is established in the conductor, then an ordered motion among the conduction electrons is created, and the vector sum of the forces on all the electrons is not zero, but is instead related to the electrons' drift velocity. Because the electrons are confined to the conductor, the conductor as a whole experiences a net force.

*Hendrik Antoon Lorentz is known also for the Lorentz transformation in the theory of special relativity (Section 19.3).

Figure 32.18

A straight wire of uniform cross section carrying a current I is placed in a uniform magnetic field. The magnetic field has no special direction.

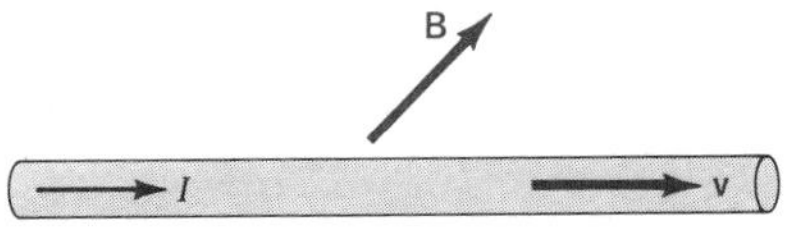

To show this quantitatively, suppose that we establish a current I in a straight wire of uniform cross section, and place the wire in a uniform magnetic field $\mathbf{B}$ (Figure 32.18). In any time t the amount of charge flowing through any cross section of the wire is $q = It$. This charge experiences a force

$$\begin{aligned}\mathbf{F} &= q\mathbf{v} \times \mathbf{B} \\ &= It\mathbf{v} \times \mathbf{B}\end{aligned} \tag{32.17}$$

where $\mathbf{v}$ represents the drift velocity of the charges. If t represents the time required for the charges to travel the length (l) of the wire, then it follows that

$$l = vt \tag{32.18}$$

By defining a vector $\mathbf{l} = \mathbf{v}t$ having the same direction as $\mathbf{v}$ (and the current) we can write

$$\mathbf{F} = I\mathbf{l} \times \mathbf{B} \tag{32.19}$$

Because the charges are confined to the wire, this force is exerted on the wire as a whole. Note that if the magnetic field is parallel (or antiparallel) to the length of the wire, there is no force on the wire. If the magnetic field is perpendicular to the length of the wire, the force is a maximum, and has magnitude

$$F = IlB \tag{32.20}$$

Principle of a Direct-Current (dc) Electric Motor

As we have said, most electric motors operate on the principle that a current-carrying conductor experiences a force when placed in a magnetic field. Let's examine this principle further by using a battery-powered motor as an example (Figure 32.19). A coil of wire wound onto a shaft

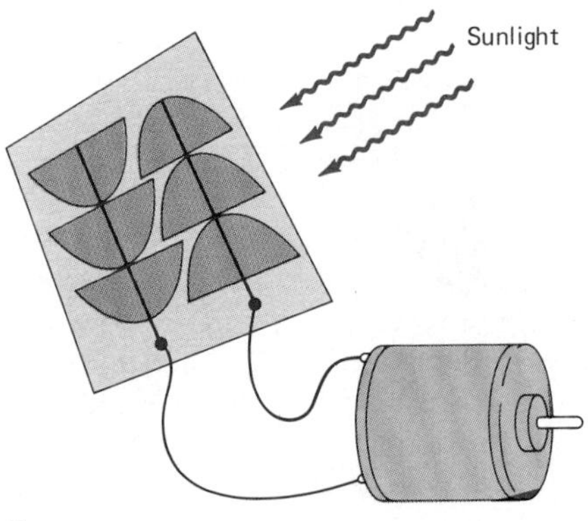

Figure 32.19

The emf that powers this small dc motor is derived from a solar (photovoltaic) cell.

rotates as a result of a torque exerted on the coil. We can find the origin of the torque by examining a single loop of a current-carrying wire in the magnetic field of a permanent magnet (Figure 32.20). Sides 1 and 2, which are the same length (b) and carry the same current (I), experience equal and opposite forces ($\mathbf{F}_1 = -\mathbf{F}_2$) with the same lines of action. Hence they can neither translate nor rotate the loop. Sides 3 and 4 also experience equal but opposite forces, and therefore produce no translational motion. But because their lines of action are not colinear there is a net torque on the loop.* Because the angle between side 3 (or side 4) and B is 90°, the magnitude of the force on either side as determined from the equation $\mathbf{F} = I\mathbf{l} \times \mathbf{B}$ is

$$F_3 = BIa = F_4$$

where a is the length of the side. Choosing an axis of rotation (OO') passing through the midpoints of sides 1 and 2, we have for the net torque on the loop

$$\tau = F_3\left(\frac{b}{2}\right)\sin\theta + F_4\left(\frac{b}{2}\right)\sin\theta$$

$$= BIab \sin\theta \qquad (32.21)$$

*Torque is discussed in Chapter 3.

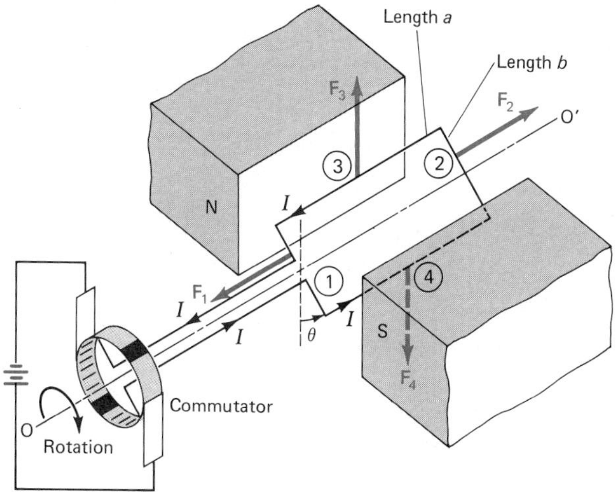

Viewed from the commutator end, the current in side 3 is directed toward the observer. In side 4, the current is directed away from the observer.

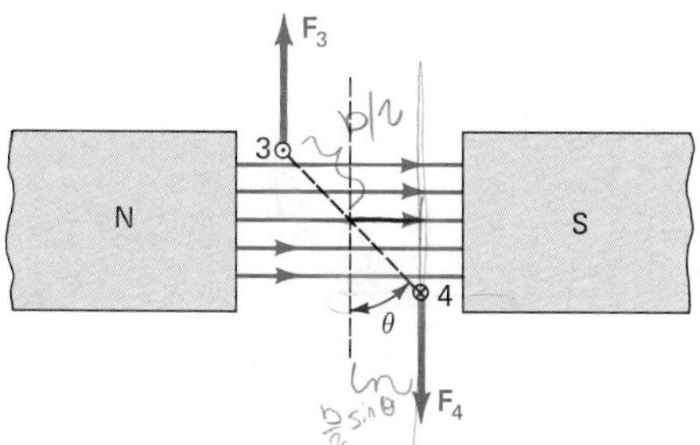

The force ($\mathbf{F} = I\mathbf{l} \times \mathbf{B}$) is up on side 3, down on side 4.

Figure 32.20a

Figure 32.20b

When the coil rotates to the upright position, the current drops instantaneously to zero as the metal contacts on the commutator touch the insulating position.

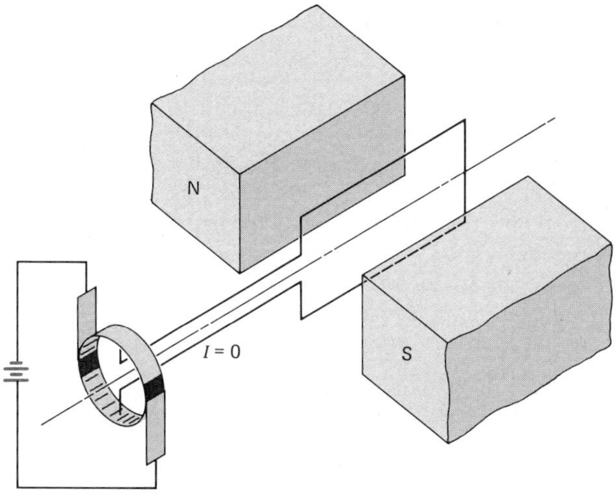

In the upright position the current drops instantaneously to zero. Hence the forces on the coil drop instantaneously to zero.

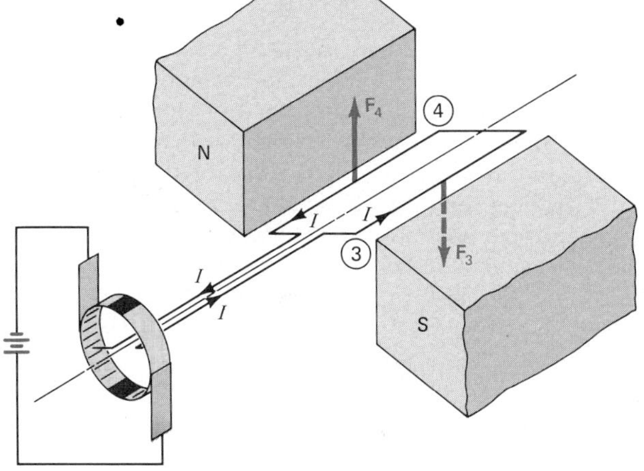

When the coil rotates past the upright position, the current direction changes in sides 3 and 4 of the coil. Viewed from the commutator end, the current in side 3 is directed away from the observer. In side 4, the current is directed toward the observer.

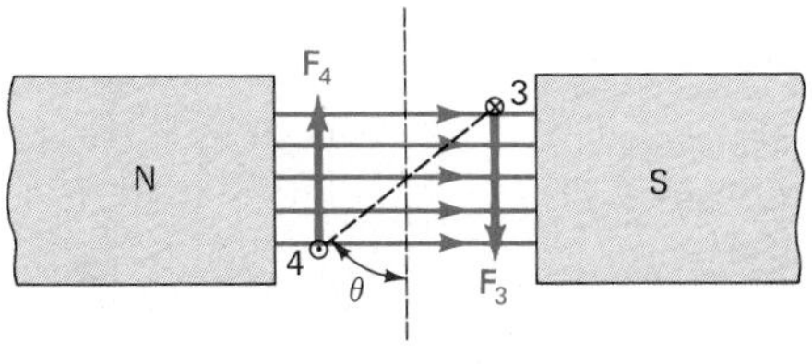

The force ($\mathbf{F} = I\mathbf{l} \times \mathbf{B}$) is down on side 3, up on side 4.

Figure 32.20c

When the coil rotates past the upright position, the commutator changes the direction of the current in sides 3 and 4. This causes the magnetic force on these two sides to change direction. The force is down on side 3, up on side 4, and the coil continues to rotate clockwise.

This torque rotates the loop in a clockwise direction. In the upright position (Figure 32.20b) the torque is zero because the lines of action of $\mathbf{F}_3$ and $\mathbf{F}_4$ coincide ($\theta = 0$ in Eq. 32.21). The motion of the loop carries it past this equilibrium position, at which time the current direction in the loop changes by the action of the commutator (Figure 32.20). The coil continues to rotate clockwise because of a clockwise torque. At the two equilibrium positions of the coil, the current direction changes in such a way as to always provide a clockwise torque. Consequently, the coil rotates in a continuous fashion and produces a simple motor. This is called a *direct-current (dc) motor* because the current directions into and out of the battery never change. A practical motor contains many loops of wire in order to increase the torque on the coil.

Equation 32.21 reveals that the torque on a current-carrying loop is proportional to the current in the loop. Thus a measure of the torque can be used to determine the current, as shown in Example 2.

Example 2 Principle of a D'Arsonval Galvanometer

For n closely wound loops having the same size, orientation, position, and current, the net torque on the coil is n times the torque on a single loop (see Eq. 32.21).

$$\tau = nBIab \sin \theta \tag{32.22}$$

This principle is used in the D'Arsonval galvanometer, which is the basic meter in most mechanical ammeters and voltmeters. A coiled spring attached to the rotational axis of the coil exerts a torque counter to that produced by current in the coil. The pole faces of the permanent magnet providing the magnetic field are designed so that the magnetic field lines are always parallel to the plane of the coil. The angle θ in Eq. 32.22 is then $\pi/2$, and the torque due to the current I is $\tau = nBIab$. The spring is designed so that it exerts a restoring torque directly proportional to the angular displacement (ϕ) of the meter needle attached to the axis of the coil.

$$\tau_{\text{restoring}} = k\phi \tag{32.23}$$

where k is the proportionality constant that depends on the strength of the spring. A current in the coil gives rise to a torque that rotates the coil. The coil rotates until the restoring torque of the spring balances the torque produced by the current. Then

$$k\phi = nBIab$$

$$I = \frac{k\phi}{nBab} \tag{32.24}$$

Thus, if k, n, B, a, and b are known, then a measurement of the angular deflection ϕ determines the current in the coil.

The Hall Effect

Figure 32.21a depicts a current in a thin, flat strip of metal connected to a battery. Electrons flow with drift speed v from the negative connection to the positive connection on the conducting strip. If a magnetic field is oriented perpendicular to the flat face of the strip, electrons feel a transverse force $-evB$ and are deflected from their previous course (Figure 32.21b). Because the electrons cannot escape from the conductor, negative charges accumulate on one side of the strip, leaving a net positive charge on the opposite side. This separation of charges produces a transverse electric field E_H, with H standing for Hall, for whom this effect is named.* As a result of this electric field, the

*E. H. Hall was an American physicist (1855–1938).

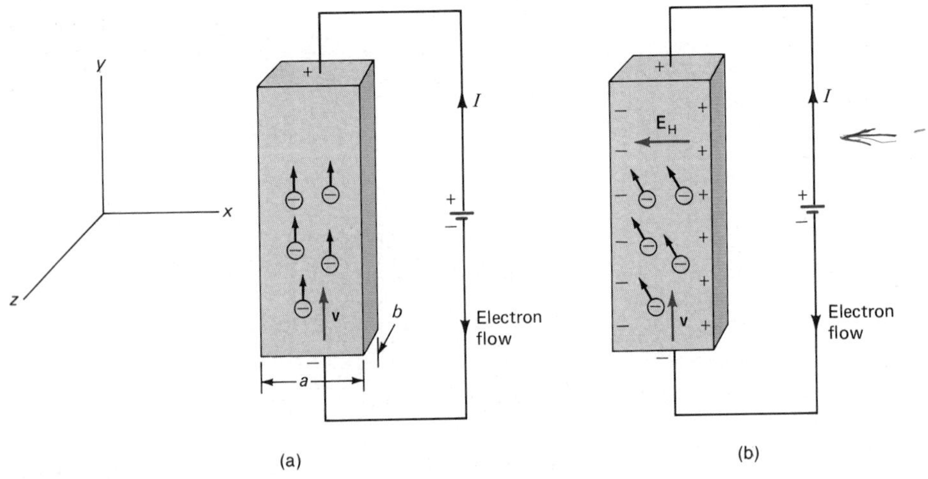

Figure 32.21

(a) A ribbonlike conductor is oriented with its long axis along the y-axis. The drift velocity of the electrons is in the positive y-direction. (b) A magnetic field is oriented perpendicular (positive z-direction) to the face of the "ribbon." Electrons experience a magnetic force in the negative x-direction.

electrons experience an electric force that opposes the magnetic force. Charges accumulate, and E_H increases, until the electric force on an electron cancels the magnetic force and equilibrium is established. Equating the electric force on an electron to the magnetic force on an electron, we have

$$-eE_H = -evB \tag{32.25}$$

In terms of the potential difference (called the *Hall voltage*, V_H) across the strip, the electric field can be written

$$E_H = \frac{V_H}{a} \tag{32.26}$$

where a is the width of the strip. The drift speed v of the charges is related to the current density J by

$$J = nev \tag{30.5}$$

In terms of the cross-sectional area of the strip, ab, and the current, I, the current density (amperes per square meter) can also be written

$$J = \frac{I}{ab}$$

Using Eq. 30.5 to relate the drift speed to charge density (n) and current densitv (J), we have

$$v_D = \frac{J}{ne} = \frac{I}{abne} \tag{32.27}$$

By combining Eqs. 32.25, 32.26, and 32.27 we obtain the Hall voltage:

$$V_H = \frac{I}{bne} \cdot B \tag{32.28}$$

Here $e = 1.60 \times 10^{-19}$ C. Measurements of I, B, b, and V_H allow us to calculate the charge density n. Typically, V_H is in the microvolt range for metals and in the millivolt range for semiconductors. Its measurement requires sensitive measuring instruments. Because V_H is inversely proportional to thickness, the sample is made as thin as possible in order to enhance the Hall voltage.

Note carefully that the Hall voltage uniquely determines the algebraic sign of the charge carriers. In some materials, the polarity of the Hall voltage indicates the movement of *positive* charges! This seems to contradict our free-electron model of conduction, in which the charge carriers are negative electrons. However, solid-state physics offers an explanation.

The Hall effect can be used to measure magnetic fields. For a given strip of material, the Hall voltage is directly proportional to the magnetic field. By placing the strip in a known magnetic field, and measuring V_H, we can determine the constant of proportionality between V_H and B. Unknown magnetic fields can then be determined by measuring the Hall voltage when the strip is placed in a magnetic field of interest. This is an electrical method requiring no mechanical parts, unlike the flip-coil method described in Section 34.4. These Hall probes are widely used to measure magnetic fields. Semiconducting materials with relatively low charge-carrier density are generally used for the probes so that the Hall voltage will be as large as possible.

Example 3 Estimation of a Typical Hall Voltage for a Copper Strip

The measured value of $1/ne$ for copper is reported to be 5.4×10^{-11} m^3/C. How large is the Hall voltage in a copper strip 2 mm wide and 0.05 mm thick, if the strip carries a current of 100 mA and is placed in a magnetic field of 1 T? From Eq. 32.28

$$V_H = \frac{IB}{bne}$$

$$= \frac{(10^{-1}\ \text{A})(1\ \text{T})(5.4 \times 10^{-11}\ \text{m}^3/\text{C})}{5 \times 10^{-5}\ \text{m}}$$

$$= 1.1 \times 10^{-7}\ \text{V}$$

The Hall voltage is 0.11 μV. Measurement of this potential difference requires a sensitive measuring instrument.

Questions

13. How could you use a magnetic field to distinguish between positive and negative particles ejected by radioactive nuclei?
14. The resonant absorption of energy by systems executing some sort of oscillatory motion is an extremely important physical phenomenon. If a metal is placed in a uniform magnetic field, then electrons will resonantly absorb electromagnetic radiation if the frequency of the radiation equals the cyclotron frequency. How does a measurement of this resonant frequency allow us to determine the mass of an electron?
15. If the magnet of a D'Arsonval galvanometer lost some of its magnetism over a period of time, would you expect the reading recorded for a given current to be less than or greater than the true value of the current?
16. It was shown that the current in the coil of a D'Arsonval galvanometer is directly proportional to the angular deflection of a pointer attached to the coil. How can this proportionality constant be determined experimentally?

32.4 Magnetic Dipole in a Magnetic Field

Magnetic Dipole Moment

A magnetic dipole in a uniform magnetic field experiences a torque, just as an electric dipole experiences a torque in a uniform electric field (Section 27.7). By calculating the torque on an electric dipole we obtained the relation

$$\boldsymbol{\tau} = \mathbf{p} \times \mathbf{E} \tag{27.24}$$

Figure 32.22

Two equal but opposite polarity charges a distance l apart constitute an electric dipole having a dipole moment $p = ql$. In a uniform electric field, as shown here, the dipole experiences a clockwise torque; $\boldsymbol{\tau} = \mathbf{p} \times \mathbf{E}$ points into the plane of the paper.

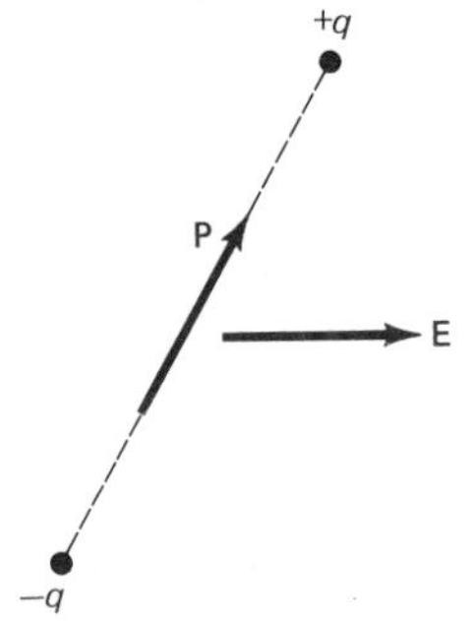

Figure 32.23

The magnetic dipole as shown here experiences a clockwise torque. The vector representing the torque points into the plane of the paper. Measurements of the magnitude and direction of the torque ($\boldsymbol{\tau}$) and magnetic field (**B**) allow $\boldsymbol{\mu}$ to be determined.

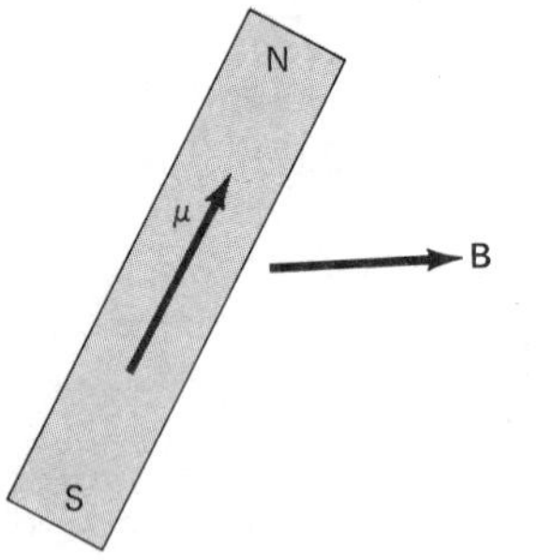

where **p** is the electric dipole moment, which has magnitude equal to the product of one of the dipole charges and the length of the dipole, and direction pointing from the negative to the positive charge (Figure 32.22). Analogously, the torque on a magnetic dipole in a magnet field and the magnetic field itself are related by a property of the dipole called the **magnetic dipole moment** ($\boldsymbol{\mu}$) that is defined by the relation

$$\boldsymbol{\tau} = \boldsymbol{\mu} \times \mathbf{B} \qquad (32.29)$$

We cannot relate the magnetic dipole moment to a dipole length because the poles cannot be located precisely; but just as magnetic field is defined operationally by Eq. 32.1, the magnetic dipole moment is operationally defined by Eq. 32.29. We measure $\boldsymbol{\tau}$ and **B**, and deduce $\boldsymbol{\mu}$. Magnetic moment has units of

$$\frac{\text{torque}}{\text{magnetic field}} = \frac{\text{newton meters}}{\text{newtons/ampere meter}}$$

$$= \frac{\text{J}}{\text{T}} \quad \text{or} \quad \text{A}\cdot\text{m}^2$$

The magnetic dipole moment points from the S pole to the N pole (Figure 32.23).

Magnetic Dipole Moment of a Current-Carrying Loop

A current in a circular loop produces a magnetic field pattern like that produced by a magnetic dipole having N and S poles (Figure 32.24). When a current-carrying loop is placed in a magnetic field, the loop experiences forces and a torque much like those experienced by a magnetic dipole such as a compass. We can make this analogy between current loops and magnetic dipoles quantitative as follows. If we compare the expression for the magnitude of the torque on a magnetic dipole with the expression for the magnitude of the torque experienced by a current-carrying loop (Eq. 32.21), we are then able to assign a magnetic dipole moment to the current-carrying loop. The torque on a magnetic dipole in a uniform magnetic field is given by Eq. 32.29:

$$\boldsymbol{\tau} = \boldsymbol{\mu} \times \mathbf{B}$$

The magnitude of the torque is

$$\tau = \mu B \sin\theta \qquad (32.30)$$

where θ is the angle between $\boldsymbol{\mu}$ and **B**. The magnitude of the torque on a rectangular current loop in a uniform magnetic field is

$$\tau = Blab \sin\theta \qquad (32.21)$$

which can be written as

$$\tau = IAB \sin\theta \qquad (32.31)$$

where $A = ab$ is the area of the rectangular loop. Here, as when we calculated magnetic flux in Section 32.2, we can treat area as a vector having magnitude A and direction perpendicular to the plane containing the area. If we take the direction of the current as the positive direction around the perimeter of the loop, then the direction of the area

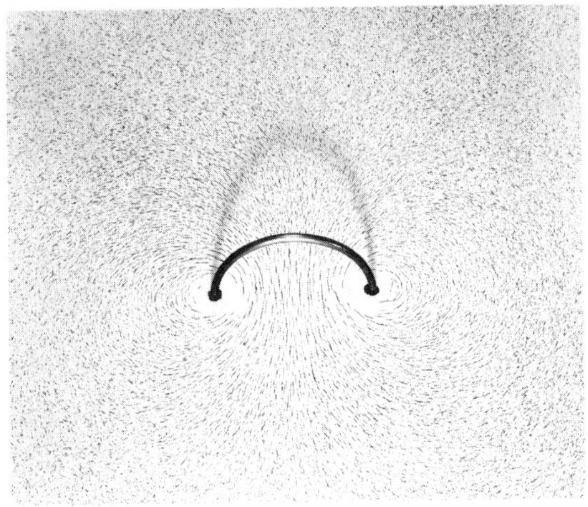

Figure 32.24

The arrangement of iron filings on a sheet of paper through which a circular loop of electric current passes shows a magnetic field pattern similar to that produced by a magnetic dipole. (The other half of the current-carrying loop extends below the sheet of paper that supports the iron filings.)

Figure 32.25

A current loop is represented by an area vector **A** having a magnitude equal to that of a flat surface with perimeter identical to the loop. To get the direction, curl the right-hand fingers in the direction of I; then the thumb denotes the direction.

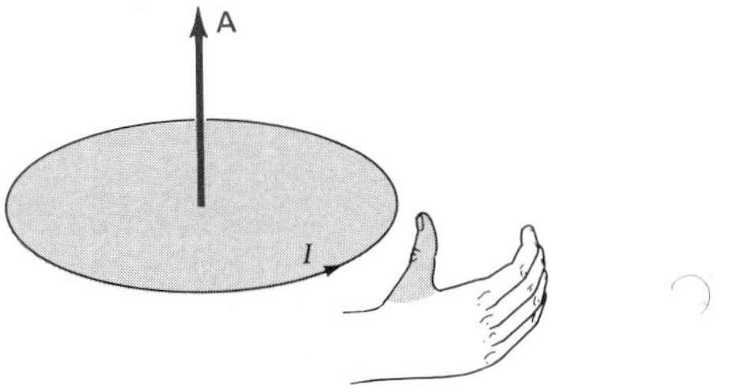

vector can be determined by the right-hand rule (Figure 32.25). Thus we define the magnetic dipole moment of a loop to be

$$\boldsymbol{\mu} = I\mathbf{A} \quad (32.32)$$

Substituting $\boldsymbol{\mu}$ for IA in Eq. 32.31, we can write the torque on the loop:

$$\tau = \mu B \sin\theta \quad (32.30)$$

and

$$\boldsymbol{\tau} = \boldsymbol{\mu} \times \mathbf{B} \quad (32.29)$$

The equation $\boldsymbol{\tau} = \boldsymbol{\mu} \times \mathbf{B}$ is appropriate for either a magnetic dipole moment associated with a magnet such as a compass or a current-carrying loop. Although we assumed a rectangular current-carrying loop, this derivation applies to a current loop of arbitrary shape.

The magnetic dipole moment $\boldsymbol{\mu}$ of a circulating current is of fundamental significance. For example, in Chapter 36 we will see that circulating electron currents in atoms contribute to the magnetic properties of materials

Potential Energy of a Magnetic Dipole

Work is done by a torque when the torque causes an object such as a wheel to rotate. In Section 13.1 we saw that the work done by a mechanical torque is

$$dW = \tau\, d\phi$$

where $d\phi$ is the angular diplacement of the rotating object (Figure 32.26). Similarly, work is done when a magnetic field exerts a torque on a magnetic dipole and causes it to rotate (Figure 32.27). Just as for a mechanical torque, the work done by a magnetic torque is

$$dW = \tau\, d\phi$$
$$= \mu B \sin\theta\, d\phi$$

To relate the work to only a single angular variable we note from Figure 32.27 that the angle θ between $\boldsymbol{\mu}$ and **B** and the angle ϕ together equal 90°, or

$$\phi = \frac{\pi}{2} - \theta$$

Figure 32.26

A torque τ producing an angular displacement $d\phi$ does work $dW = \tau\, d\phi$.

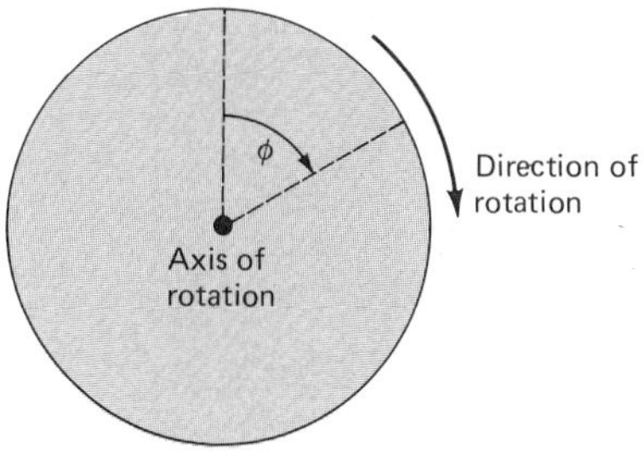

Figure 32.27

A magnetic dipole in a uniform magnetic field feels a torque and rotates. Work $dW = \tau\, d\phi$ is done by the torque.

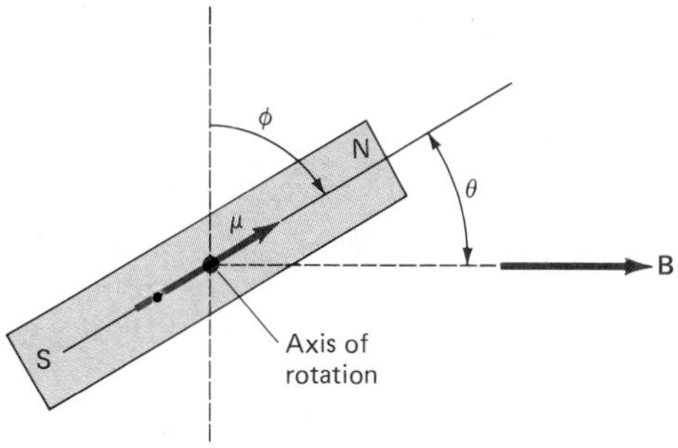

Hence $d\phi = -d\theta$, and in terms of the angle θ, the work done is

$$dW = -\mu B \sin\theta\, d\theta$$

The work done on the magnetic dipole by the magnetic field is negative and the magnitude increases as the angle θ between $\boldsymbol{\mu}$ and **B** increases.

When the magnetic dipole moment and the magnetic field vectors are not parallel, the dipole is capable of doing external work and hence possesses potential energy. As a convenience, we choose the zero of potential energy to be when the axis of the dipole is perpendicular to the magnetic field ($\theta = \pi/2$). By definition (Section 8.2) the potential energy relative to this position is

$$U = -\int_{\pi/2}^{\theta} dW = -\int_{\pi/2}^{\theta} -\tau\, d\theta = \mu B \int_{\pi/2}^{\theta} \sin\theta\, d\theta$$
$$U = \mu B(-\cos\theta)\Big|_{\pi/2}^{\theta}$$
$$U = -\mu B \cos\theta$$

The quantity $\mu B \cos\theta$ can be represented by the scalar product of the vectors $\boldsymbol{\mu}$ and **B**. Thus

$$U = -\boldsymbol{\mu} \cdot \mathbf{B} \quad (32.33)$$

Notice in Figure 32.28 that the potential energy U of the magnetic dipole is at a maximum when $\boldsymbol{\mu}$ and **B** are antiparallel,

$$U_{max} = +\mu B$$

Figure 32.28

Orientations of magnetic moment and magnetic field for minimum (a) and maximum (b) potential energy.

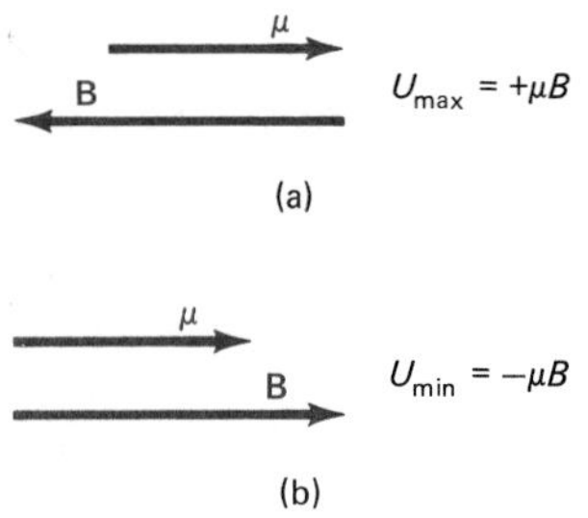
μ
B
U_max = +μB
(a)
μ
B
U_min = −μB
(b)

and at a minimum when μ and $\mathbf{B}$ are parallel,

$$U_{\text{min}} = -\mu B$$

The torque on the dipole ($\boldsymbol{\tau} = \boldsymbol{\mu} \times \mathbf{B}$) is zero whether $\boldsymbol{\mu}$ and $\mathbf{B}$ are parallel or antiparallel. Therefore both orientations are equilibrium configurations. The parallel configuration is one of stable equilibrium, and the antiparallel configuration is one of unstable equilibrium. For example, the magnetic moment of a compass needle is in stable equilibrium when its magnetic moment is parallel to the earth's magnetic field. If the needle is rotated through 180°, it remains in unstable equilibrium until some disturbance alters its alignment. The needle is then urged back toward the stable configuration.

Example 4 Calculating the Work Required to Rotate a Coil in a Uniform Magnetic Field

A current of 0.5 A exists in a circular coil of wire situated in a uniform magnetic field of 0.5 T. The coil has a mean radius of 5 cm and contains 250 turns of wire. How much work is done by magnetic forces in rotating the coil from a position where $\boldsymbol{\mu}$ is perpendicular to $\mathbf{B}$ to a position where $\boldsymbol{\mu}$ is parallel to $\mathbf{B}$? The magnitude of the magnetic moment is

$$\begin{aligned}\mu &= nIA = 250(0.5)\pi(0.05)^2\\ &= 0.982 \text{ A}\cdot\text{m}^2\end{aligned}$$

Using the definition of work, we write

$$\begin{aligned}W &= -\int_{\pi/2}^{0} \tau\, d\theta = -\mu B \int_{\pi/2}^{0} \sin\theta\, d\theta\\ &= -\mu B(-\cos\theta)\Big|_{\pi/2}^{0}\\ &= \mu B\\ &= 0.982(0.5)\\ &= 0.491 \text{ J}\end{aligned}$$

Alternatively, we can calculate the change in potential energy between the two positions and then take the negative of this change to arrive at the work done by magnetic forces.

$$\begin{aligned}U_{\text{initial}} &= -\mu B \cos\frac{\pi}{2} = 0\\ U_{\text{final}} &= -\mu B\cos\theta = -\mu B\\ \Delta U &= U_{\text{final}} - U_{\text{initial}} = -\mu B\\ W &= -\Delta U = \mu B\\ &= 0.491 \text{ J}\end{aligned}$$

Hydrogen atoms have magnetic dipoles whose configurations also change from unstable to stable orientation, as described in the following example.

Example 5 Magnetic Dipoles in Astrophysics

The hydrogen atom consists of an electron and a proton, both of which have an intrinsic magnetic moment. The magnetic moment of the electron sets up a magnetic field that affects the magnetic moment of the proton. There are two equilibrium configurations of the moments. In the stable configuration, the magnetic moment and magnetic field are parallel (Figure 32.29a). In the unstable configuration, the magnetic moment and magnetic field are antiparallel (Figure 32.29b). The energies associated with these two configurations are

$$\begin{aligned}U_- &= -\mu_p B_e &&\text{(stable configuration)}\\ U_+ &= +\mu_p B_e &&\text{(unstable configuration)}\\ U_\pm &= \pm\mu_p B_e && (32.34)\end{aligned}$$

where μ_p is the magnetic moment of the proton and B_e is the magnetic field produced by the electron.

There are vast clouds of hydrogen in our galaxy. Most of the individual hydrogen atoms are in the lower energy state. Occasionally, a collision will supply enough energy to boost an atom into the higher energy configuration. The energy required to do this is the energy difference between the stable and unstable states.

$$\text{Energy required} = U_+ - U_- = 2\mu_p B_e$$

The atom can then return to the stable configuration by *emitting* energy in the form of an electromagnetic wave.

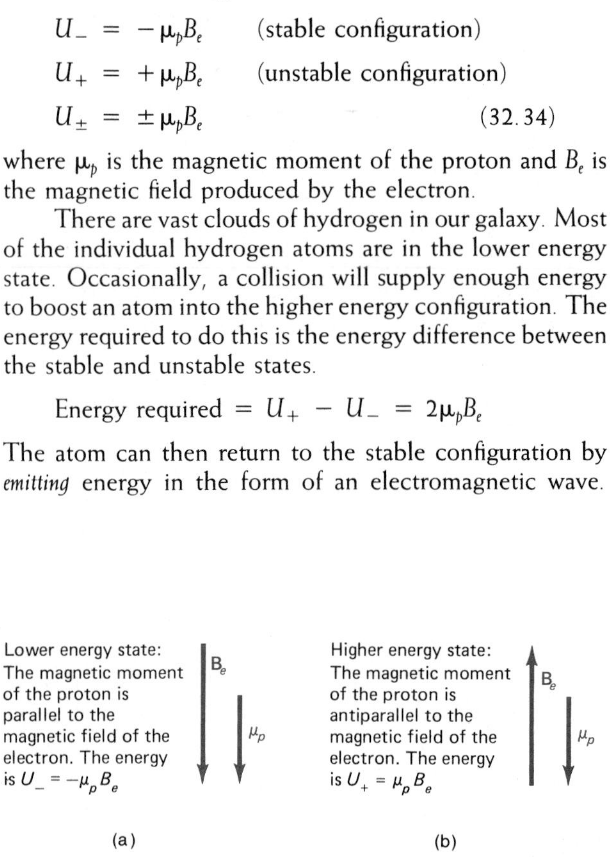
Lower energy state: The magnetic moment of the proton is parallel to the magnetic field of the electron. The energy is U_ = −μ_p B_e
B_e
μ_p
Higher energy state: The magnetic moment of the proton is antiparallel to the magnetic field of the electron. The energy is U_+ = μ_p B_e
B_e
μ_p
(a)
(b)

Figure 32.29

A schematic representation of magnetic energy states in the hydrogen atom. Electromagnetic radiation is emitted when the atom changes from the higher energy state (a) to the lower energy state (b). The energy difference is $2\mu_p B_e$.

This radiation can be detected by a radio telescope. The direction and intensity of this radiation have made it possible for radio astronomers to locate and "map" hydrogen-rich portions of our galaxy.

Magnetic Mirrors

In a constant magnetic field, a magnetic dipole experiences zero net force. This behavior is explained when we observe that the forces on the N pole and on the S pole are equal in magnitude and oppositely directed (Figure 32.30a). However, a magnetic dipole experiences a force if it is placed in a magnetic field that is not constant, but instead varies spatially (Figure 32.30b). This behavior is identical to that of an electric dipole in an electric field (Section 27.7). For an electric field varying only along the x-axis, the net force on the electric dipole is

$$F = p\frac{dE}{dx} \tag{27.26}$$

where dE/dx is the spatial variation of the electric field. For a *magnetic* field varying only along the x-axis, the net force on the magnetic dipole is

$$F = \mu\frac{dB}{dx} \tag{32.35}$$

where dB/dx is the spatial variation of the magnetic field. Positive dB/dx means that B increases as x increases. Positive μ in Eq. 32.35 means that μ points in the positive x-direction. Thus a positive force (that is, in the positive x-direction) occurs when both $\boldsymbol{\mu}$ and dB/dx are positive *or* when both μ and dB/dx are negative. A negative force (a force in the negative x-direction) occurs when one or the other of the quantities μ and dB/dx is negative. It is possible to design a magnetic field in such a way that this principle can be used to make a magnetic dipole move back and forth within a selected region of the field. To see how this works, we can consider the following situations.

We have shown that a charged particle executes circular motion when it enters a uniform magnetic field with its velocity *perpendicular* to the field (Section 32.3). The magnetic dipole moment produced by such a circulating charge is oriented antiparallel to the magnetic field (Figure 32.31). If, on the other hand, the particle enters the field with its velocity *parallel* to the magnetic field, then it experiences no magnetic force, and moves parallel to the magnetic field lines (Figure 32.32). If the particle enters

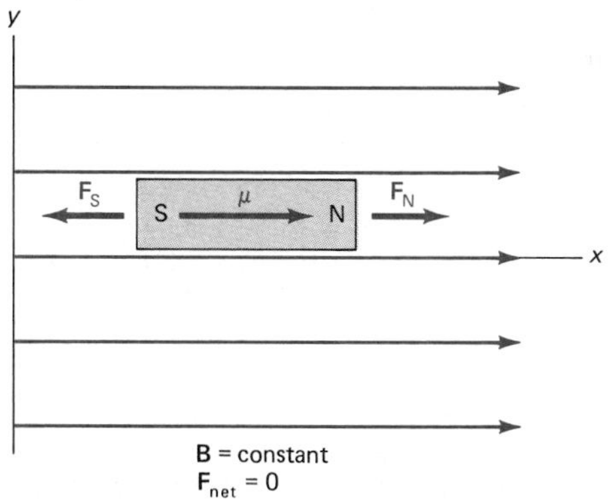

Figure 32.30a

In a uniform magnetic field the forces on the poles of a magnetic dipole are equal in magnitude but opposite in direction; the net force is zero.

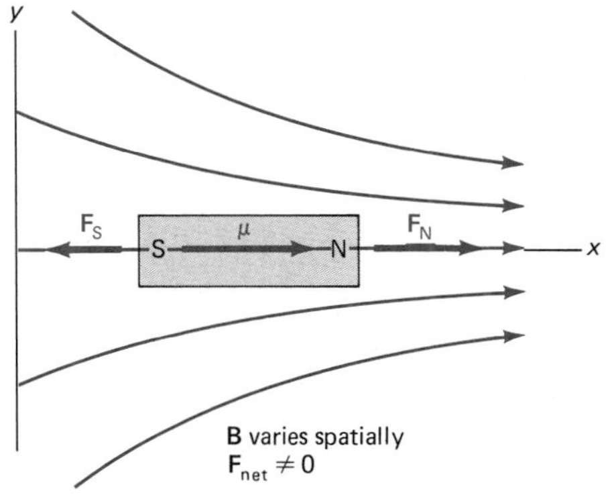

Figure 32.30b

In a nonuniform magnetic field the forces on the magnetic poles are unequal. For the situation shown here the N pole is in a region of larger magnetic field than the S pole. Hence there is a net force on the dipole to the right (positive x-direction).

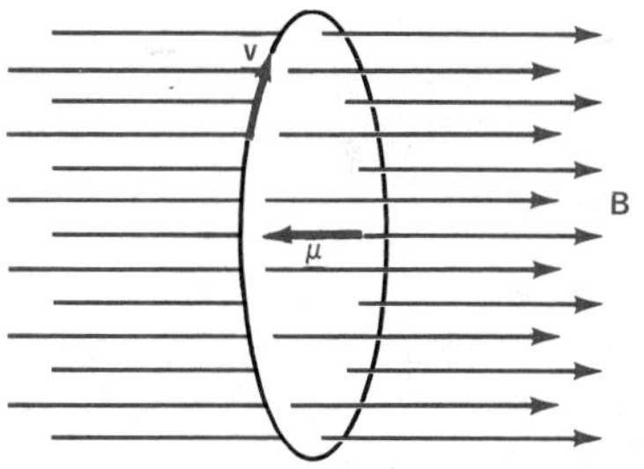

Figure 32.31

If a charge enters a magnetic field with $\mathbf{v} \perp \mathbf{B}$, then the charge executes circular motion, and the magnetic dipole moment $\boldsymbol{\mu}$ of the circulating charge is oriented antiparallel to the magnetic field.

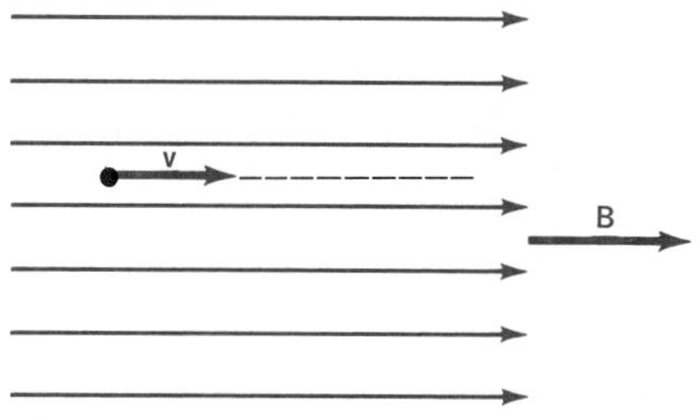

Figure 32.32

If a charge enters a magnetic field with $\mathbf{v} \parallel \mathbf{B}$, then the charge experiences no magnetic force; it moves in a straight line in the direction of $\mathbf{v}$.

Figure 32.33

If a charge enters a magnetic field with a velocity component ($\mathbf{v}_\perp$) perpendicular to **B** and a velocity component ($\mathbf{v}_\parallel$) parallel to **B**, then the charge spirals around the magnetic field lines.

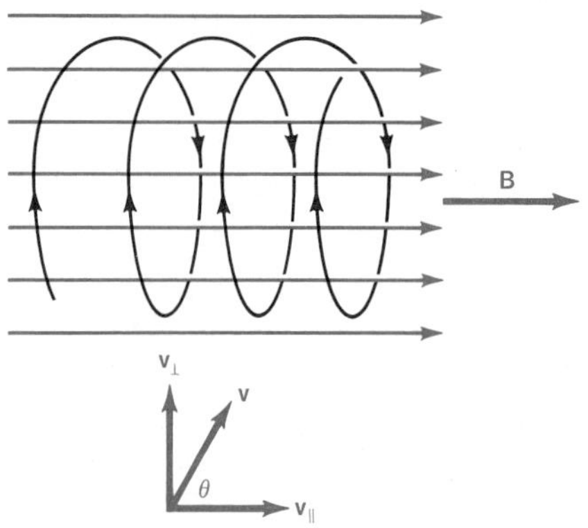

the field with *components of velocity perpendicular and parallel* to the field, its motion is a superposition of the motions due to the two components. The perpendicular component causes the particle to execute circular motion; the parallel component causes it to move in the direction of the field lines. Consequently, the particle moves in a spiral path (Figure 32.33). In effect, Figure 32.33 is a combination of Figures 32.31 and 32.32. The circular orbit and associated magnetic moment of Figure 32.31 move along the magnetic field lines with the velocity of Figure 32.32 (Figure 32.33). If the uniform field were of infinite extent, then the charge would spiral indefinitely. But let's suppose that in its motion the dipole encounters a region of increased concentration of magnetic field lines; $dB/dx > 0$ (Figure 32.34). Because the dipole moment is antiparallel to **B**, the dipole experiences a net force to the *left*. It is *decelerated*, and eventually changes direction and moves out of the nonuniform magnetic field. If in its motion to the left (negative x-direction) the dipole encounters a region of increased concentration of magnetic field lines ($dB/dx < 0$), it experiences a net force to the right (positive x-direction). Again, the dipole is decelerated, and eventually changes direction and moves out of the nonuniform magnetic field. It is trapped between the regions having larger concentrations of magnetic field lines. A magnetic field configuration like this is called a *magnetic mirror* because it reflects electric

Figure 32.35

Electrons entering the earth's magnetic field experience a magnetic force that traps them in spiraling paths around the magnetic field lines. The concentrated field lines near the poles function as magnetic mirrors that reflect the electrons.

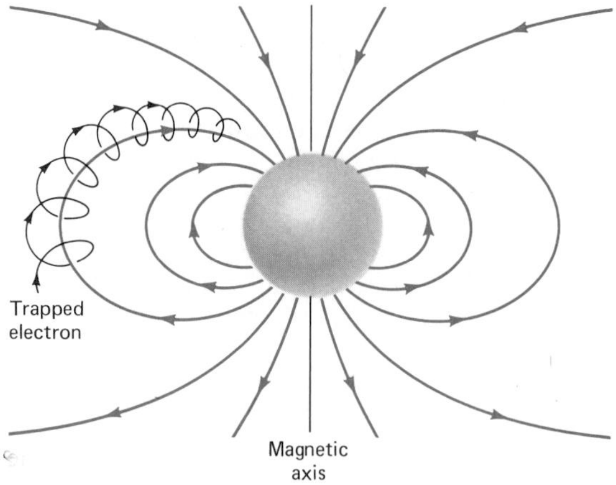

charges. In the future, nuclear fusion devices for producing electricity will use magnetic fields to contain charged particles.

The earth's magnetic field traps charged particles approaching the earth from outer space, as illustrated in Figure 32.35 (see the article by O'Brian under Suggested Reading).

Example 6 **A Neutron Storage Ring**

The neutron has a magnetic moment of $\mu = 9.66 \times 10^{-27}$ J/T. By using carefully shaped magnets, scientists have been able to hold very low-energy neutrons ($v = 20$ m/s) in a circular orbit of radius 0.6 m. Let us determine the required spatial variation of the magnetic field dB/dx.

While orbiting, the neutrons experience a radial acceleration

$$a = \frac{v^2}{r} = \frac{20^2}{0.6}\ \text{m/s}^2$$

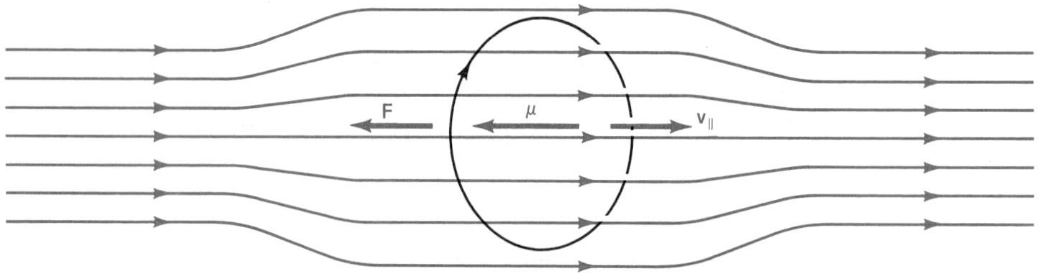

Figure 32.34

As the particle spirals to the right toward a region of large magnetic flux ($dB/dx > 0$) there is a force on the charge that changes its original direction of motion.

and a net radial force

$$F = m\left(\frac{v^2}{r}\right) = 1.675 \times 10^{-27}\left(\frac{20^2}{0.6}\right)$$
$$= 1.12 \times 10^{-24}\ \text{N}$$

Using Eq. 32.31, we obtain for the required field gradient

$$\frac{dB}{dx} = \frac{F}{\mu} = \frac{1.12 \times 10^{-24}}{9.66 \times 10^{-27}} = 116\ \text{T/m}$$

Questions

17. A child playing in a moving car twirls a mass on a string. What is the shape of the path of the mass as seen by (a) a person in the car and (b) a person standing on the ground?
18. When considering the forces between two magnetic dipoles, why do we not specifically account for the forces between the poles of a given dipole?
19. Why is the net force on a magnetic dipole zero in a uniform magnetic field?
20. Why is the torque not necessarily zero on a magnetic dipole in a uniform magnetic field?

Summary

The sources of magnetic force between magnets are concentrated in regions near the two ends of the magnets. These two regions are termed N and S poles.

An N pole is defined to be that pole of a freely suspended magnet that points toward the geographic north.

Like magnetic poles repel each other; unlike magnetic poles attract each other.

The magnetic field **B** is defined in terms of the force experienced by a moving charge. Specifically,

$$\mathbf{F} = q\mathbf{v} \times \mathbf{B} \qquad (32.1)$$

where **F** is the force experienced, q is the size of the charge (including algebraic sign), and **v** is the velocity of the charge.

The tesla (T) is the SI unit of magnetic field. Magnetic field lines are used in the same sense as are electric field lines. At any point in space, the tangent to the magnetic field line gives the direction of the magnetic field.

The magnetic flux $d\Phi$ threading an area element $d\mathbf{A}$ is defined as

$$d\Phi = \mathbf{B} \cdot d\mathbf{A} \qquad (32.3)$$

where **B** is the magnetic field vector at the location of $d\mathbf{A}$.

Because magnetic monopoles do not exist, magnetic field lines do not terminate, and for any closed surface

$$\int_{\substack{\text{closed}\\\text{surface}}} \mathbf{B} \cdot d\mathbf{A} = 0 \qquad (32.6)$$

A conductor of length l carrying a current I may experience a force when placed in a magnetic field. Defining a vector **l** having length equal to that of the wire and direction the same as that of the current, the force is represented by

$$\mathbf{F} = I\mathbf{l} \times \mathbf{B} \qquad (32.19)$$

This principle is used in electric motors and galvanometers.

The magnetic moment $\boldsymbol{\mu}$ of a magnetic dipole is defined by the equation

$$\boldsymbol{\tau} = \boldsymbol{\mu} \times \mathbf{B} \qquad (32.29)$$

where $\boldsymbol{\tau}$ is the torque experienced by the dipole in a magnetic field **B**.

The magnetic dipole moment of a current loop carrying a current I is identified as

$$\boldsymbol{\mu} = I\mathbf{A} \qquad (32.32)$$

where **A** is the area determined by the boundaries of the loop. The direction of **A** is determined by the right-hand rule.

The potential energy of a magnetic dipole in a magnetic field is given by

$$U = -\boldsymbol{\mu} \cdot \mathbf{B} \qquad (32.33)$$

A magnetic dipole in a nonuniform magnetic field experiences a net force

$$F = \mu \frac{dB}{dx} \qquad (32.35)$$

where dB/dx is the spatial variation of the magnetic field.

Suggested Reading

H. P. Furth, M. A. Levine, and R. W. Waniek, Strong Magnetic Fields, *Sci. Am.*, 198, 28–33 (February 1958).

W. C. Gough and B. J. Eastlund, The Prospects for Fusion Power, *Sci Am.*, 224, 50–64 (February 1971).

H. Kohn, J. Oberteuffer, and D. Kelland, High-Gradient Magnetic Separation, *Sci. Am.*, 233, 46–54 (November 1975).

J. E. Kunzler and M. Tanenbaum, Superconducting Magnets, *Sci. Am.*, 206, 60–67 (June 1962).

B. J. O'Brian, Radiation Belts, *Sci. Am.*, 208, 84–96 (May 1963).

G. E. Pake, Magnetic Resonance, *Sci. Am.*, 199, 58–66 (August 1958).

W. B. Sampson, P. P. Craig, and M. Strongin, Advances in Superconducting Magnets, *Sci. Am.*, 216, 114–123 (March 1967).

R. A. Carrigan, Jr. and W. Peter, Superheavy Magnetic Monopoles, *Sci. Am.*, 246, 106 (April 1982).

K. W. Ford, Magnetic Monopoles, *Sci. Am.*, 209, 122–131 (December 1963).

Problems

Section 32.2 Magnetic Field

1. An electron having velocity $10^6\mathbf{i}$ m/s experiences a maximum force of $1.6 \times 10^{-14}\mathbf{k}$ N when it enters a uniform magnetic field. What is the magnitude and direction of the magnetic field?

2. A proton moving perpendicular to a magnetic field of 10^{-4} T experiences a magnetic force that is 10 billion (10^{10}) times its weight. Determine the speed of the proton.

3. A proton having a speed of 2×10^6 m/s enters a uniform magnetic field of 0.5 T. The angle between the velocity and magnetic field vectors is 96°. Determine the magnitude of the force on the proton.

4. Electrons from outer space enter the earth's magnetic field with velocities on the order of 10^7 m/s. (a) Assuming the earth's magnetic field to be of the order of 10^{-5} T, estimate the magnetic force on a cosmic electron. (b) Compare this force with the weight of an electron at the earth's surface.

5. A stream of particles, some positive, some negative, and some neutral, are directed along the east line in Figure 1. The magnetic field is directed into the plane containing the N–S, E–W lines (as represented by the crosses). Draw the possible trajectories for the three types of particles.

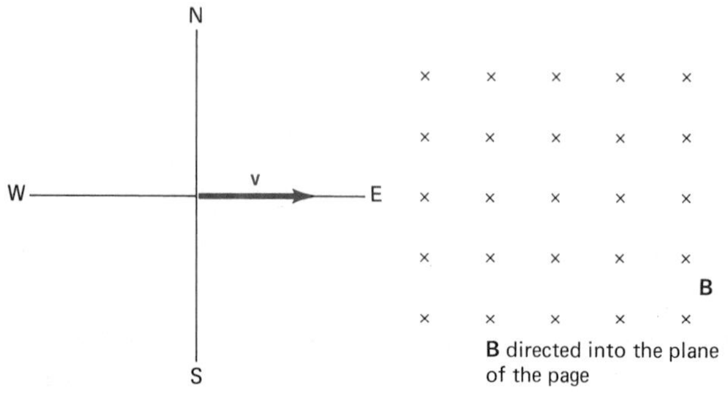

Figure 1

6. An electron with a velocity of $10^6\mathbf{i}$ m/s enters a uniform magnetic field. One microsecond later it emerges from the magnetic field with a velocity of $10^6\mathbf{j}$ m/s. (a) What is the direction of the magnetic field? (b) What is the average force experienced by the electron?

7. A positively charged particle traveling due east enters a uniform magnetic field and experiences a force in a due north direction. Give a possible direction for the magnetic field.

8. A positive charge ($Q = 1.6 \times 10^{-19}$ C) moves through a magnetic field ($B = 2$ T). If the velocity is in the $\pm z$-direction, then the magnetic force equals zero. If the velocity is in the $+y$-direction, then the force $\mathbf{F} = 6.4 \times 10^{-12}\mathbf{i}$ N. (a) What is the direction of **B**? If the charge moves parallel to the direction ($\mathbf{i} + \mathbf{j}$) with the same speed as in part (a), what is the (b) magnitude and (c) direction of the force?

9. A proton having velocity $\mathbf{v} = 10\mathbf{j}$ m/s enters a uniform magnetic field represented by $\mathbf{B} = 0.5\mathbf{i} - 0.8\mathbf{k}$ T. Determine (a) the force and (b) the acceleration experienced by the proton when it enters the field.

10. A proton ($q = 1.60 \times 10^{-19}$ C) moves along the positive x-axis at 1.0×10^8 m/s. The magnetic field B is in the x–y plane, and makes an angle of 20° with the x-axis. Calculate the magnetic force on the proton when $B = 3 \times 10^{-5}$ T.

11. Electrons in a color TV set are often accelerated through a potential difference of 10 kV. Calculate the maximum force that an electron experiences when it enters a magnetic field of 3000 G after being accelerated through a potential difference of 10 kV.

12. A free neutron exists for about 15 min and then disintegrates into a proton, an electron, and an antineutrino (which has no charge). If the magnitudes of the linear momenta of the proton and electron are equal, describe the subsequent motions in the neutron disintegration shown in Figure 2 if the proton initially is directed along the $+x$-axis and the electron initially is directed along the $-x$-axis.

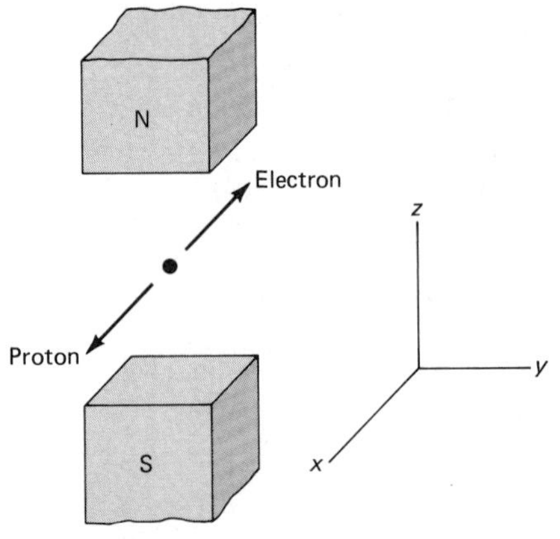

Figure 2

Section 32.3 Applications of Moving Charges in a Magnetic Field

13. A cosmic ray proton traveling at three fourths the speed of light is heading directly toward the center of the earth in the plane of the earth's equator. Will it hit the earth? As an estimate, assume that the earth's magnetic field is 5×10^{-5} T and extends out one earth diameter, or 1.3×10^7 m. Calculate the radius of curvature of the proton in this magnetic field. Will the proton hit the earth?

14. A velocity "filter" (Figure 3) for charged particles can be made by using constant electric and magnetic fields oriented so that **E**, **B**, and the velocity **v** are mutually perpendicular. Positively charged particles

moving too slowly will be deflected downward. Those moving with a particular speed will experience zero net force and will travel through the small hole in the baffle, and thereby be filtered out of the beam. (a) Determine the direction of **B** (into or out of the page) that will give a force opposing the electric field force. (b) If $E = 100$ V/m and $B = 0.01$ T, what is the speed of particles that pass undeflected through the fields?

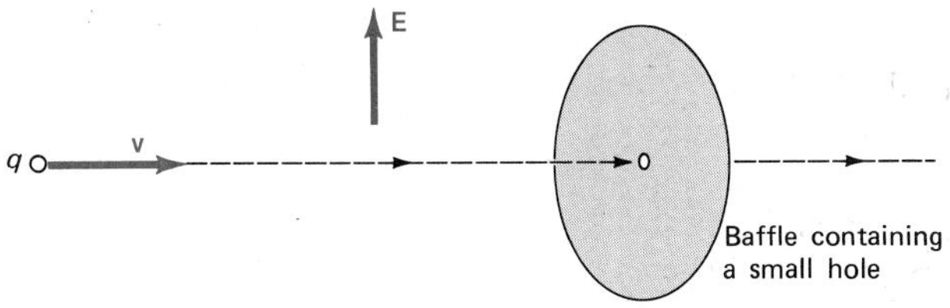

Figure 3

15. An electron passes through a velocity selector having $E = 10^4$ V/m, $B = 2 \times 10^{-2}$ T. Determine the electron speed.

16. A proton having kinetic energy of 100 keV enters a uniform magnetic field such that its velocity **v** is perpendicular to the magnetic field **B**. What is the strength of the magnetic field if its radius of curvature is 2 m?

17. Electrons are accelerated through a potential difference of 1000 V and then steered into a mass spectrometer having a magnetic field strength of 0.0025 T. Determine the radius of curvature of the electrons in the mass spectrometer.

18. In a certain mass spectrometer the radius of curvature of all charged particles investigated is 0.75 m. This constant radius of curvature is achieved by controlling the potential difference through which the particles are accelerated. If the magnetic field strength is 0.5 T, determine the potential difference required for singly charged ^{238}U and ^{235}U.

19. Consider an electron moving in a circular path in the Large Electron Positron (LEP) which is located in Western Europe. Assume that $p = 2.67 \times 10^{-17}$ kg · m/s (for 50-GeV electrons) with a path diameter of 27 km, and calculate the bending magnetic field if it is assumed to be spread uniformly over the path.

20. Charged particles all having the same linear momentum enter a uniform magnetic field ($\mathbf{v} \perp \mathbf{B}$). Compare the radii of curvature of protons and electrons, protons and deuterons, and deuterons and alpha particles.

21. Protons of different linear momentum enter a uniform magnetic field such that their velocity vector is always perpendicular to the direction of the magnetic field. It is desired to have the radii of curvature differ by at least 0.1%. Determine the corresponding percentage differences in (a) linear momentum and (b) kinetic energy. You may use the nonrelativistic expression for kinetic energy.

22. Protons from a cyclotron are steered between the pole faces of a magnet, as shown in Figure 4. We want to deflect the protons to the right of their path from the cyclotron (that is, out of the paper, toward you). Determine the required direction of the magnetic field in the bending magnet and label the N and S poles of the magnet.

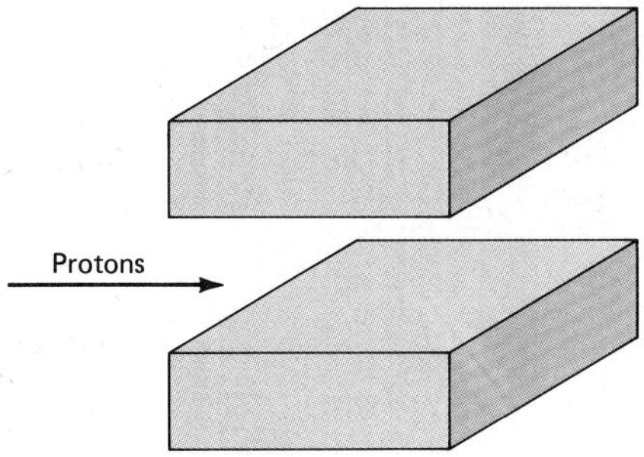

Figure 4

23. In a certain magnetic field, a proton and an alpha particle are observed to revolve in circles having the same radii. Compare their (a) linear momentum, (b) velocity, and (c) kinetic energy.

24. A charged particle executes circular motion between the pole faces of a permanent magnet, as shown in Figure 5. Is the charge positive or negative?

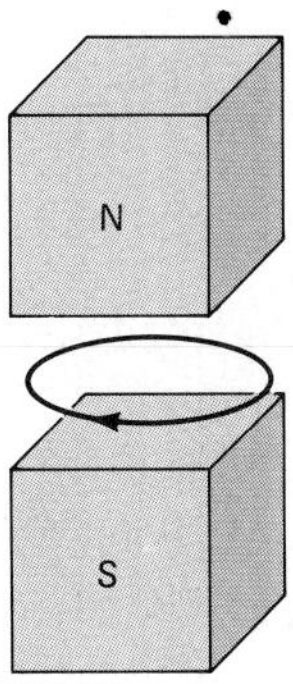

Figure 5

25. Calculate the cyclotron frequency of electrons in a magnetic field of 10 T.

26. A horizontal length of copper wire ($\rho = 8960$ kg/m^3 and $A = 8.37 \times 10^{-6}$ m^2) experiences an upward magnetic force due to an imposed horizontal magnetic field ($B = 2$ T). What electric current is needed to support the weight of the wire exactly?

27. A coil consisting of 300 turns of wire has a mean radius of 10 cm and carries a current of 1.5 A. Initially its dipole moment is aligned parallel to a uniform magnetic field of 0.75 T. Calculate the minimum amount of work that must be done by the field in order to rotate the coil into a position where its magnetic moment is antiparallel to the magnetic field.

28. A thin ribbon of copper is shown in Figure 6. On the diagram show the following clearly and unambiguously: (a) the direction of motion and deflection of the charge carriers; (b) the resulting polarity (+, −) of the opposite edges of the ribbon; and (c) the Hall electric field, $\mathbf{E}_H$.

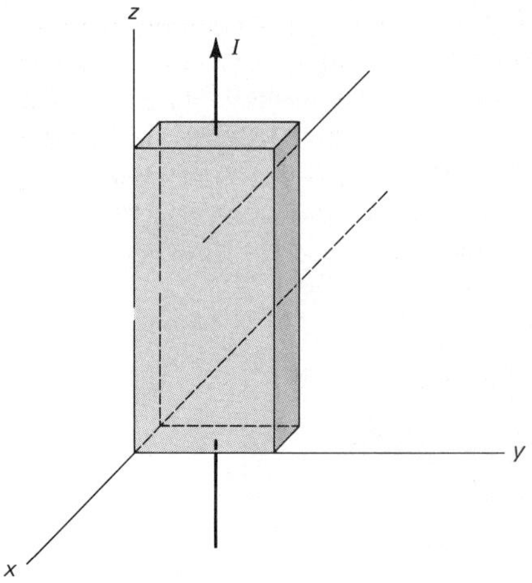

Figure 6

29. A Hall probe used to measure magnetic fields has a thickness of 0.15 mm and is made from a material having a charge carrier density of $10^{24}/m^3$. For a current of 10 mA, a Hall voltage of 42 μV results when the probe is placed in a magnetic field of unknown strength. What is the value of the magnetic field?

30. In an undergraduate physics laboratory, students are required to use the Hall effect to measure the density of charge carriers in a metal. The metal sample, measuring 0.13 mm thick, is placed in a uniform magnetic field of 0.30 T. When a current of 0.10 A is established in the sample, a Hall voltage of 1.6×10^{-6} V is measured. What value did the students obtain for the density of charge carriers?

31. A commercial Hall effect probe used to measure magnetic fields is designed to operate with a 200-mA current in it. When placed in a magnetic field of 10 kG the probe produces a Hall voltage of 510 mV. When placed in an unknown magnetic field it produces a Hall voltage of 235 mV. What is the strength in teslas of the unknown field?

32. In a Hall effect experiment a potential difference will often develop across the potential difference contacts, even if no magnetic field is present. This is an *IR*-type potential difference resulting from misalignment of the contacts. Although the polarity of the Hall voltage depends on the direction of the impressed magnetic field, the *IR* potential difference does not. Show that if two potential difference measurements are taken with oppositely directed magnetic fields, then the true Hall voltage is

$$V_H = \frac{V_1 - V_2}{2}$$

33. The copper strip described in Example 3 is to be used to measure a magnetic field. For a Hall voltage $V_H = 1.35 \times 10^{-7}$ V and current $I = 50 \times 10^{-3}$ A, what size field is employed?

34. Table 1 shows measurements of a Hall voltage and corresponding magnetic field for a probe used to measure magnetic fields. (a) Make a plot of these data and deduce a relationship between the Hall voltage and magnetic field. (b) If the measurements were taken with a current of 0.2 A and the sample is made from a material having a charge-carrier density of $10^{26}/m^3$, what is the thickness of the sample?

Table 1

V_H (μV)	B (T)
0	0
11	0.1
19	0.2
28	0.3
42	0.4
50	0.5
61	0.6
68	0.7
79	0.8
90	0.9
102	1.0

35. A thin ribbon of a silver alloy 2.00 cm wide and 1.50×10^{-3} cm thick carries a current of 6.98 A. The Hall potential difference is found to be 1.24×10^{-3} V when the magnetic field is perpendicular to the face of the ribbon (Figure 6) is 25.0 T. (a) Starting with Figure 6, show the direction of motion of the charge carriers (electrons), the polarity of the opposite sides of the strip, and the direction of the Hall electric field, E_H. (b) Calculate the drift velocity, v_D. (c) Calculate n, the number of charge carriers per cubic meter.

Section 32.4 Magnetic Dipole in a Magnetic Field

36. The needle of a Boy Scout compass experiences a maximum torque of 5.2×10^{-6} N·m when placed in a region where the earth's magnetic field has a strength of 2×10^{-5} T. Determine the magnetic dipole moment of the compass.

37. A certain nonuniform magnetic field is represented by $\mathbf{B} = 0.01x^2\mathbf{i}$ T where x is the distance in meters from the origin of a Cartesian coordinate system. Determine the force on a magnetic dipole having a dipole moment of 2.8 J/T at a position 2 m from the origin.

38. A superconducting solenoid produces a field described by

$$\mathbf{B} = (+16.4\mathbf{j})\ \text{T}$$

In this field there is a magnetic dipole described by $\boldsymbol{\mu} = 3.18 \times 10^{-2}\hat{\mathbf{n}}$ A·m² where $\hat{\mathbf{n}} = 0.611\mathbf{j} - 0.792\mathbf{k}$. (a) Calculate the torque on this dipole. (b) Calculate the potential energy of this dipole, taking the potential energy to be zero when the dipole and field are mutually perpendicular.

39. A magnetic dipole of strength 10 J/T is placed in a uniform magnetic field of strength 0.5 T. Calculate the work done by the field on the dipole when the angle between **μ** and **B** changes from 127° to 35°.

40. The maximum torque on a rectangular current loop rotating in a magnetic field $B = 10^{-2}$ T is 4.6×10^{-4} N·m. The loop area is 2×10^{-3} m^2. Determine the current in the loop.

41. A spring is connected to the end of a magnetic dipole that is pivoted about its midpoint as shown in Figure 7. The torque exerted on the dipole by the spring can be represented by $\tau_s = k\theta$, where k is a constant and θ is a small angle expressed in radians. (a) Assuming that only the spring and the magnetic field exert forces on the dipole, show that the equilibrium angular position of the dipole is determined by the relation

$$\mu B \cos \theta = k\theta$$

(b) Determine the equilibrium angular position if $\mu = 10$ J/T, $B = 0.025$ T, and $k = 1$ N·m/rad.

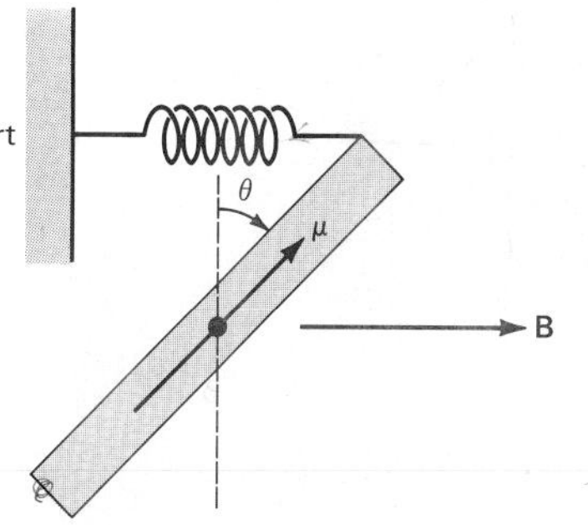

Figure 7

42. The proton has a magnetic moment of $\mu = 1.41 \times 10^{-26}$ J/T. When the proton "flips" from the unstable equilibrium position to the stable equilibrium position in the magnetic field of the electron, 9.47×10^{-25} J of energy are released as 21-cm electromagnetic radiation. Calculate the magnetic field experienced by the proton.

43. Typically, magnetic dipole moments of atoms and molecules are on the order of one *Bohr magneton* ($\mu_B = 9.2741 \times 10^{-24}$ J/T). These atomic and/or molecular magnetic moments may experience magnetic fields on the order of 1 T. Show that the corresponding magnetic potential energy is roughly one ten thousandth of an *electron volt*.

44. Consider a copper rod 1 cm in diameter. A cross-sectional disk that is one atom in thickness contains about 6.2×10^{14} copper atoms. Each copper atom has 27 electrons, which means that there are about 1.7×10^{16} electrons in the disk. These electrons are tiny magnets, each having a spin magnetic moment $\mu_s = 9.3 \times 10^{-24}$ A·m^2. Suppose these elementary magnets are all parallel to the axis of the rod. (a) What is the disk's magnetic moment? (b) What circumferential current would give rise to the same magnetic moment for the disk?

45. A pattern of magnetic lines corresponding to a nonuniform magnetic field is shown in Figure 8. Several small bar magnets are shown in the figure and numbered for reference. (a) Which magnets will experience a net force that urges them along a field line from left to right? (Ignore forces between magnets.) (b) Which magnets experience the least torque? (Ignore torques between magnets.)

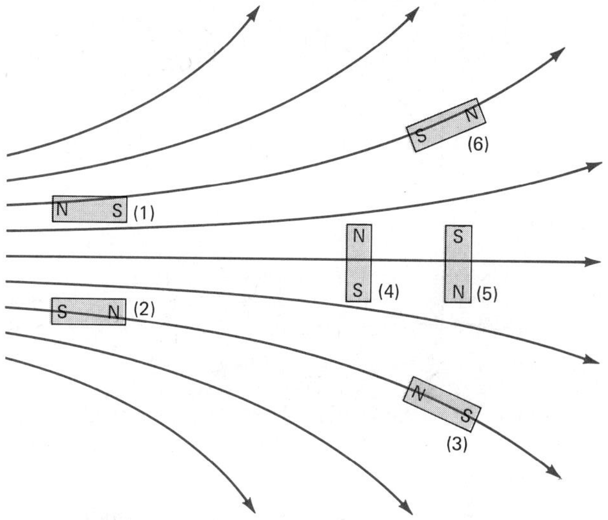

Figure 8

chapter 33

MAGNETIC FIELD OF ELECTRIC CURRENT

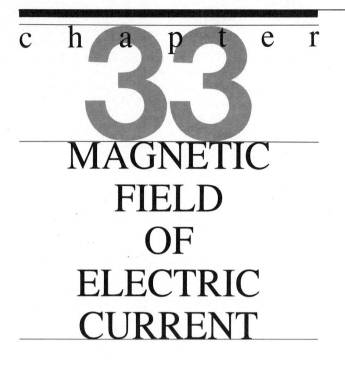

Preview

Electric currents create magnetic fields. This discovery by Oersted in 1820 opened up a new field of research. As we will see in this chapter, Oersted's discovery has been described quantitatively in two distinct ways: (1) Biot and Savart proposed a relation in differential form that is somewhat analogous to a differential form of Coulomb's law of electrostatics, and (2) Ampère devised a law—an integral equation—that differs markedly in form from the law of Biot and Savart.

The law of Biot and Savart can be applied to a straight-line segment of an electric current circuit to yield the magnetic field produced by that portion of the circuit. A second application of the law of Biot and Savart to a circular current loop gives the magnetic field produced by the electric current loop. If we compare this magnetic field with the electric field of an electric dipole, we can then assign a magnetic dipole moment to the current loop, consistent with the assignment (based on torque) of Section 32.5.

We apply Ampere's law to calculate the magnetic field of a long straight wire, and obtain the same result obtained with the law of Biot and Savart. This demonstrates the consistency of the two laws. We then use Ampere's law to calculate the magnetic field of a solenoid. To provide a picture and a qualitative understanding of the magnetic field, we develop Faraday's interpretation of magnetic field lines.

Finally, by drawing on results from Chapter 32 we calculate the force exerted on one electric current by the magnetic field of a second current. This result is used to define the ampere in terms of the force between two parallel current-carrying conductors.

33.1 Biot and Savart's Law

As we mentioned in Section 32.1, Oersted discovered that an electric current in a wire produces a magnetic field. Soon after this discovery, Jean Baptiste Biot and Felix Savart* developed what is known as the *law of Biot and Savart.* This law, which we will now describe, gives a quantitative description of Oersted's magnetic field in terms of the electric current.

Because a current-carrying wire can be bent into an arbitrary shape, we start with the differential contribution to the magnetic field $d\mathbf{B}$ produced by the current in a differential length element $d\mathbf{s}$. The product $I\,d\mathbf{s}$ is called a *differential current element.* Here I is the electric current, expressed in amperes, and $d\mathbf{s}$ is the vector length of an element of the circuit. The length $d\mathbf{s}$ is taken parallel to the direction of the current. For a steady state (no time-dependent fields or currents) Biot and Savart proposed that the current element $I\,d\mathbf{s}$ generated a magnetic field $d\mathbf{B}$ given by

$$d\mathbf{B} = k_m \frac{I\,d\mathbf{s} \times \hat{\mathbf{r}}}{r^2} \tag{33.1}$$

This result, the law of Biot and Savart, is illustrated in Figure 33.1. In this figure, $\hat{\mathbf{r}}$ is a unit vector directed from the current element $I\,d\mathbf{s}$ (called the *source point*) toward the point P at which $\mathbf{B}$ is to be calculated (the field point). The differential magnetic field $d\mathbf{B}$ given by the vector product is perpendicular to $I\,d\mathbf{s}$ and to $\hat{\mathbf{r}}$.

*Jean Baptiste Biot (1774–1862), a French physicist, was noted for his investigations of the polarization of light as well as for the analysis of the magnetic field of electric current. Felix Savart (1791–1841) was also a French physicist.

Figure 33.1

The current element $I\,d\mathbf{s}$ generates a contribution $d\mathbf{B}$ to the magnetic field at P a distance r away.

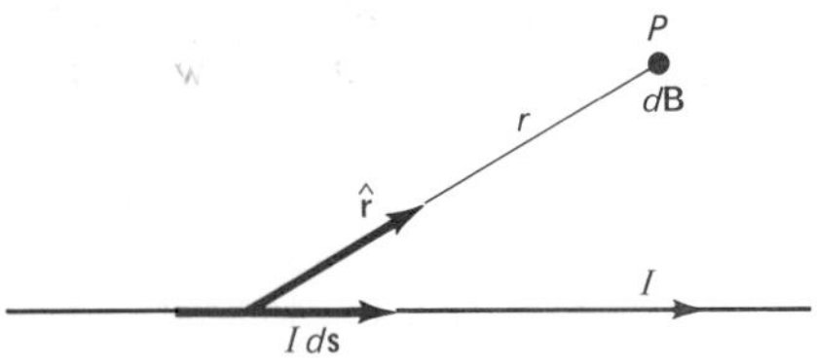

The proportionality constant k_m is assigned the value of 10^{-7} (SI units) in a vacuum:

$$k_m = 10^{-7}\ \mathrm{T \cdot m/A} \tag{33.2}$$

[The tesla (T), as you will recall from Section 32.2, is a unit of magnetic field strength.] Keep in mind that, unlike the Coulomb proportionality constant k_e (Section 26.3), which is determined by experiment, the magnetic proportionality constant k_m is *assigned* the numerical value of 10^{-7} (SI units). The assignment of a value to k_m leads to the determination of the size of the unit of current, the ampere. We return to this point in Section 33.5, where the ampere is defined operationally.

In Section 27.1, we replaced the Coulomb proportionality constant k_e by $1/4\pi\epsilon_0$. When discussing magnetism, it is customary to make a similar substitution. We introduce the magnetic permeability constant μ_0, which is defined by

$$k_m = \frac{\mu_0}{4\pi} \tag{33.3}$$

or

$$\mu_0 = 4\pi \times 10^{-7}\ \mathrm{T \cdot m/A} \tag{33.4}$$

The subscript zero is a reminder that this is the value of the magnetic permeability constant in a *vacuum*. Note carefully that μ_0 is used as a proportionality constant. We replace the magnetic proportionality constant k_m with $\mu_0/4\pi$. Despite its appearance, μ_0 is *not* a magnetic moment.

It is useful to compare the law of Biot and Savart with Coulomb's law:

Biot and Savart	Coulomb
$d\mathbf{B}$ (current element $I\,d\mathbf{s}$, a vector)	$d\mathbf{E}$ (charge dq, a scalar)
$\frac{1}{r^2}$ distance dependence	$\frac{1}{r^2}$ distance dependence
proportional to electric current	proportional to electric charge
tangential, perpendicular to the $\mathbf{r}$ direction	radial, in the $\mathbf{r}$ direction

Integration over the Circuit

We can use the law of Biot and Savart to obtain the magnetic field produced by an extended circuit. We simply add the differential field contributions $d\mathbf{B}$ from the current elements $I\,d\mathbf{s}$. In this way, we apply a principle of superposition for magnetic fields analogous to the superposition of electric fields (Section 27.2). For a complete circuit (Figure 33.2) we integrate Eq. 33.1 and obtain

Figure 33.2

Integration of $I\,d\mathbf{s}$ around the electric circuit.

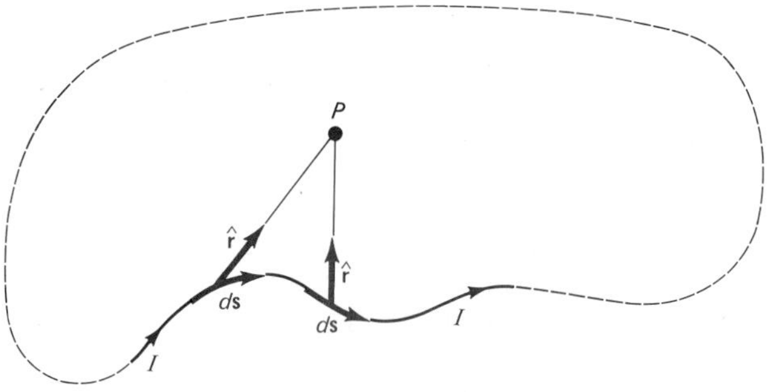

$$\mathbf{B} = \frac{\mu_0 I}{4\pi} \oint \frac{d\mathbf{s} \times \hat{\mathbf{r}}}{r^2} \tag{33.5}$$

The circle through the integral sign emphasizes that our path of integration (around the circuit) is a closed loop. The integration of current elements $I\,d\mathbf{s}$ here is directly analogous to the integration of charge elements dq over some continuous distribution (Section 26.4). Equation 33.5 is a line integral. (You may recall that line integrals also appeared in Section 8.1 in our calculations of work.)

Question

1. In Figure 33.1 how does the contribution $d\mathbf{B}$ vary with the direction of $\hat{\mathbf{r}}$ (r fixed)? (Consider P to move on a circle of radius r about $I\,d\mathbf{s}$.) What current element $I\,d\mathbf{s}$ along the straight wire will make the largest contribution to the net field $\mathbf{B}$?

33.2 Magnetic Field of Electric Current

The line integral equation 33.5 can be calculated analytically in situations where symmetry simplifies the problem. In this section, we will consider two highly symmetric cases, that of a straight segment of wire and that of a circular current loop.

A Straight Wire

We can apply the law of Biot and Savart to calculate the magnetic field produced by the electric current I in a straight wire. (The current is continuous and the complete circuit forms a closed loop, but we focus attention on a *straight segment* of this complete loop.)

Figure 33.3

Calculation of the magnetic field generated by the current in a finite section of a long straight wire. $d\mathbf{s}$ is integrated from $-s_o$ to $+s_o$.

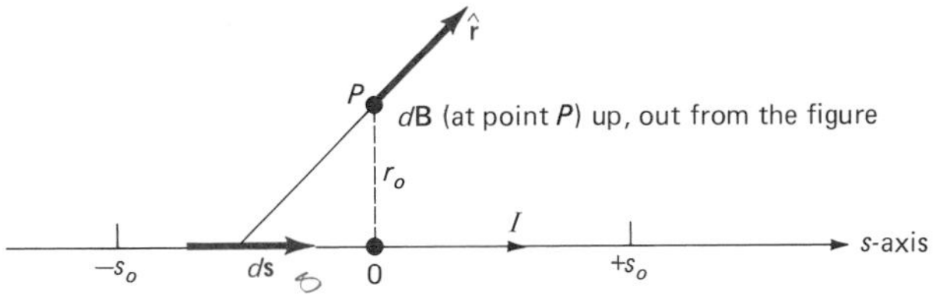

Figure 33.4

Coordinate relations for the finite line segment.

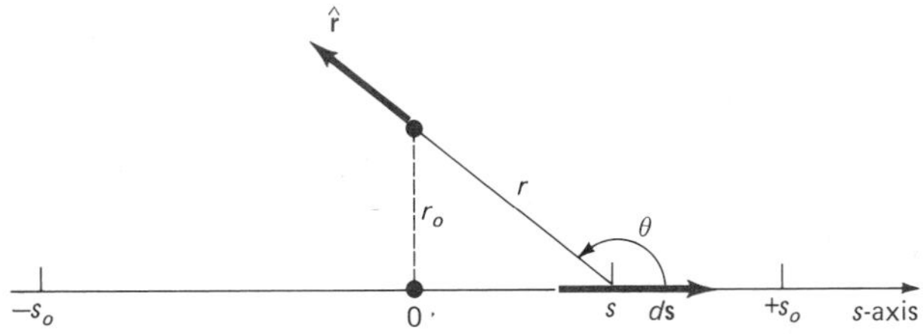

The differential field $d\mathbf{B}$ at point P produced by $I\,d\mathbf{s}$ is given by

$$d\mathbf{B} = \frac{\mu_o I}{4\pi} \cdot \frac{d\mathbf{s} \times \hat{\mathbf{r}}}{r^2} \tag{33.6}$$

Let us consider the directional relationships first, as shown in Figure 33.3. For every element $d\mathbf{s}$, the cross product $d\mathbf{s} \times \hat{\mathbf{r}}$ is perpendicular to the plane of the figure and, by the right-hand rule, directed up from the page toward you. This means that all the $d\mathbf{B}$'s are parallel. Invoking the principle of superposition as we did in electrostatics, we can add (integrate) magnitudes, as follows.

We first consider a finite length of wire running from $-s_o$ to $+s_o$. The current element $I\,d\mathbf{s}$ (Figure 33.4) creates a magnetic field

$$dB = \frac{\mu_o I}{4\pi} \cdot \frac{\sin\theta\, ds}{r^2}$$

Integrating from $-s_o$ to $+s_o$, we obtain for the magnetic field of the line segment

$$B = \frac{\mu_o I}{4\pi} \int_{-s_o}^{s_o} \frac{\sin\theta\, ds}{r^2} \tag{33.7}$$

The integral in Eq. 33.7 can be evaluated in three different ways. We can integrate with respect to s, with respect to θ, or with respect to r. Because the variable s may be a little more meaningful physically, we will express $\sin\theta$ and r^{-2} in terms of s.

From Figure 33.4

$$r^2 = s^2 + r_o^2$$

and

$$\sin\theta = \sin(\pi - \theta) = \frac{r_o}{r}$$

With these substitutions Eq. 33.7 becomes

$$B = \frac{\mu_o I r_o}{4\pi} \int_{-s_o}^{s_o} \frac{ds}{(s^2 + r_o^2)^{3/2}} \tag{33.8}$$

From Appendix 4

$$\int \frac{ds}{(s^2 + r_o^2)^{3/2}} = \frac{s}{r_o^2(s^2 + r_o^2)^{1/2}} \tag{33.9}$$

(You can check this equation by differentiating the right-hand side.) Substituting in the limits s_o and $-s_o$, we obtain

$$B = \frac{\mu_o I}{2\pi r_o} \cdot \frac{s_o}{(s_o^2 + r_o^2)^{1/2}} \tag{33.10}$$

This result, Eq. 33.10, is the magnetic field produced by current I in a finite current segment extending from $-s_o$ to $+s_o$. But suppose that we want to know the magnetic field of a *long* straight wire. We say that a wire is long if the length $2s_o$ is very large relative to the distance from the wire (r_o) at which we are evaluating B. If we let s_o become large so that $s_o \gg r_o$, then the factor $s_o/(s_o^2 + r_o^2)^{1/2}$ on the right-hand side of Eq. 33.10 approaches 1. Thus at a distance $r_o = r$ the magnetic field produced by a current I in a long straight wire is

$$B = \frac{\mu_o I}{2\pi r} \tag{33.11}$$

Thus far, we have emphasized the $1/r^2$ dependence of the Biot and Savart law (Eq. 33.1). But now we find in Eq 33.11 a $1/r$ dependence (r perpendicular to the wire). This is a consequence of integrating over a *long* straight wire, where $s_o \gg r_o$. (For $s_o \ll r_o$, Eq. 33.10 brings us back to a $1/r^2$ dependence. See Problem 2.) Equation 33.11 was discovered experimentally by Biot and Savart before the differential form, Eq. 33.1, was developed. Some physicists therefore refer to Eq. 33.11 as the law of Biot and

Figure 33.5

The iron filing pattern outlining the magnetic field about a long straight current-carrying wire.

Figure 33.6

The right-hand rule relating the direction of the current and the direction of the magnetic field.

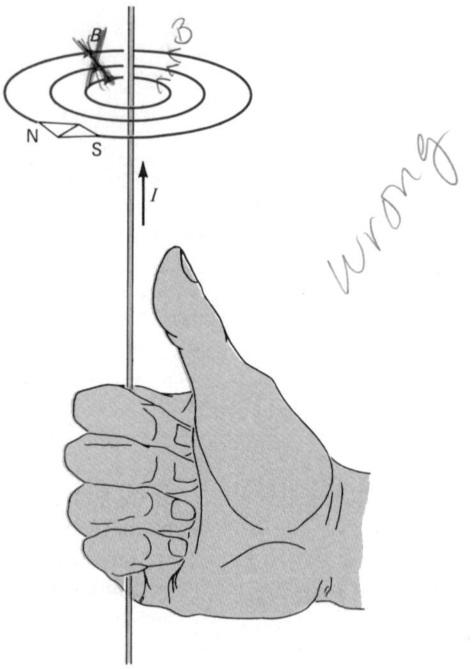

Savart. But in the sense that the differential form applies to *any* circuit, Eq. 33.1 is clearly more fundamental.

Equation 33.11 is the mathematical representation of the magnetic field that Oersted discovered. The field lines form circles about the current I (s-axis), and the intensity of the field falls off as the reciprocal of the distance from the wire (Figure 33.5).

A convenient way to remember the direction of the magnetic field of a current-carrying wire is to use the *right-hand rule*. If you let the extended thumb of your right hand point in the direction of the current, then the fingers of your right hand, when curled inward, give the direction in which the magnetic field lines encircle the wire (Figure 33.6).

Example 1 Calculating the Magnetic Field at the Center of a Square Current Loop

Let's use Eq. 33.10 to calculate the magnetic field at the center of a square loop that is 1 m on a side, and carries 1 A of current (Figure 33.7). First, we must consider the directions of the **B** vectors. Vector $\mathbf{B}_1$, produced by the current in line segment 1, is perpendicular to the plane containing line 1 and the origin. In other words, $\mathbf{B}_1$ is perpendicular to the $x-y$ plane and therefore lies in either the positive or negative z-direction. By using the right-hand rule, we find that $\mathbf{B}_1$ is in the positive z-direction:

$$\mathbf{B}_1 = \mathbf{k}B_1$$

The same argument applies to the remaining three **B** vectors. All are in the positive z-direction. The four **B** vectors

Figure 33.7

The magnetic field at the center of a square current loop.

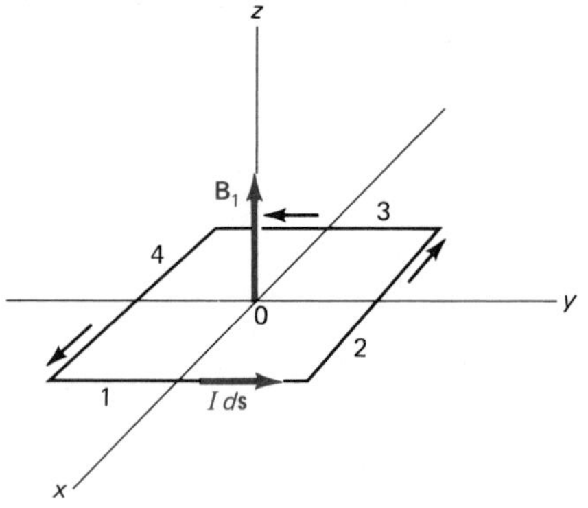

may be added algebraically.

$$\begin{aligned}\mathbf{B} &= \mathbf{B}_1 + \mathbf{B}_2 + \mathbf{B}_3 + \mathbf{B}_4\\ &= \mathbf{k}(B_1 + B_2 + B_3 + B_4)\end{aligned}$$

Further, since all four sides are the same length and are the same distance from the origin, the four **B** vectors all have the same magnitude. Therefore

$$\mathbf{B} = \mathbf{k}4B_1$$

To find the magnitude B_1 we substitute values into Eq. 33.10. We have $s_o = 0.5$ m, $r_o = 0.5$ m, and $I = 1.0$ A. Therefore

$$\begin{aligned}B_1 &= \frac{4\pi \times 10^{-7} \times 1.0}{2\pi \times 0.5} \cdot \frac{0.5}{[(0.5)^2 + (0.5)^2]^{1/2}}\\ &= 2\sqrt{2} \times 10^{-7} = 2.83 \times 10^{-7}\ \mathrm{T}\end{aligned}$$

Multiplying this value of B_1 by 4 to include all four line segments, and by **k** to include the direction, we obtain **B**, the magnetic field at the center of this 1-m-square loop:

$$\mathbf{B} = \mathbf{k}(1.13 \times 10^{-6})\ \mathrm{T}$$

Comparison with an Electric Line Charge

Let us compare the magnetic field of a long straight current-carrying wire with the electric field produced by a uniform line charge. In Chapter 27, Problem 44, we found the electric field to be

$$E = \frac{1}{2\pi\epsilon_o}\left(\frac{\lambda}{r}\right) \tag{33.12}$$

for a uniform charge distribution of λ C/m. Both the electric field strength and the magnetic field strength vary inversely with the perpendicular distance r from the source. This $1/r$ dependence of the electric and magnetic fields is characteristic of a line source. The field *directions* are different, however: **E** is radially out from the line source (for λ positive), as in Figure 33.8a, whereas the lines of **B** encircle the current (Figure 33.8b).

Figure 33.8a

The electric field lines go radially outward from a positively charged line source. The line of charge in this figure is perpendicular to the plane of the paper.

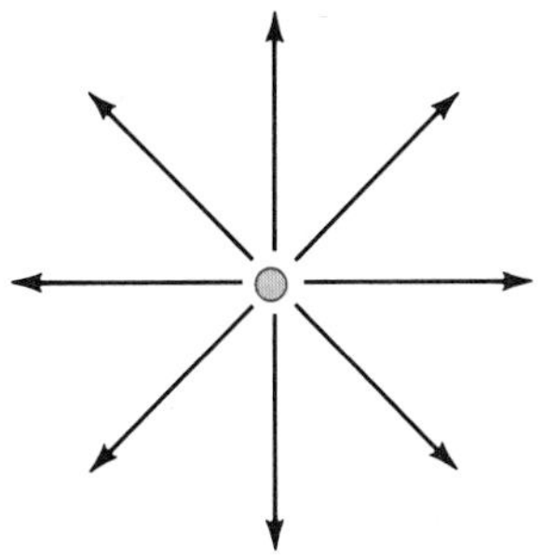

Figure 33.8b

The magnetic field lines encircle an electric current. The wire carrying the current is perpendicular to the plane of the paper, and the current is directed toward the observer.

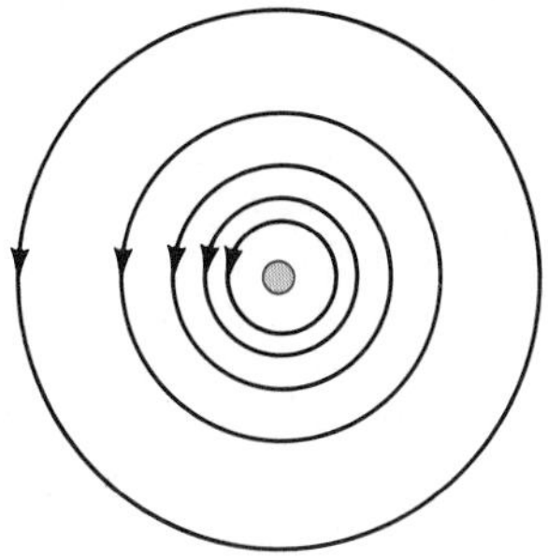

Magnetic Field of a Circular Current Loop

So far we have used the law of Biot and Savart to calculate the magnetic field produced by electric current in a straight wire. Let's now use the same law to calculate the magnetic field produced along the axis of a circular current loop. Again we do the vector part of the problem first, using symmetry to help us. One current element, $I\,d\mathbf{s}_1$, produces a magnetic field $d\mathbf{B}_1$, as shown in Figure 33.9. The direction of $d\mathbf{B}_1$, given by the right-hand rule for the cross product $d\mathbf{s}_1 \times \hat{\mathbf{r}}_1$, is perpendicular to both $d\mathbf{s}_1$ and $\hat{\mathbf{r}}_1$. In Figure 33.9 $d\mathbf{s}_1$ is parallel to the z-axis; and since $d\mathbf{B}_1$ is perpendicular to $d\mathbf{s}_1$, it also is perpendicular to the z-axis, and must therefore be in the x–y plane. Because $d\mathbf{B}_1$ is perpendicular to $\hat{\mathbf{r}}_1$, $d\mathbf{B}_1$ makes an angle α relative to the x-axis. Another current element, $I\,d\mathbf{s}_2$, on the opposite side of the circular loop will produce magnetic field $d\mathbf{B}_2$, also shown in Figure 33.9. The direction of $d\mathbf{B}_2$, given by the right-hand rule for the cross product $d\mathbf{s}_2 \times \hat{\mathbf{r}}_2$, is perpendicular to both $d\mathbf{s}_2$ and $\hat{\mathbf{r}}_2$. Because $d\mathbf{s}_2$ is parallel to the negative z-axis, as shown in the figure, $d\mathbf{B}_2$ is perpendicular to the negative z-axis and is therefore also in the x–y plane. Magnetic field $d\mathbf{B}_2$ is also perpendicular to $\hat{\mathbf{r}}_2$, and makes an angle α relative to the x-axis. For equal-length current elements ($ds_1 = ds_2$) the two differential

Figure 33.9

The magnetic field on the axis of a circular current loop.

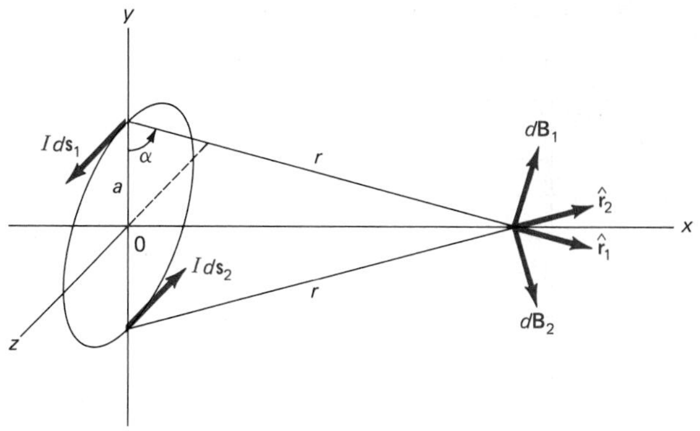

magnetic fields are equal in magnitude, and are inclined at the same angle α to the x-axis. This means that when we add contributions for all elements in the circular current loop, the $d\mathbf{B}$ form a cone about the x-axis with α constant. The components of $d\mathbf{B}$ that are perpendicular to the x-axis, dB_y and dB_z, will average out to zero, and only the component along the x-axis, dB_x, will remain. This symmetry argument converts the vector integral problem to an algebraic one.

Now we can drop the subscripts, since they are no longer necessary. Because $d\mathbf{s}$ and $\hat{\mathbf{r}}$ are at right angles, the sine factor from the vector product is unity, giving for the magnitude dB:

$$dB = \frac{\mu_o I}{4\pi} \cdot \frac{ds}{r^2}$$

The x-component is

$$dB_x = dB \cos\alpha$$

$$= \frac{\mu_o I}{4\pi} \cdot \frac{\cos\alpha}{r^2} \cdot ds$$

Then integrating ds around the loop, we get the total field, B_x.

$$B_x = \frac{\mu_o}{4\pi} \cdot \frac{I\cos\alpha}{r^2} \oint ds$$

$$= \frac{\mu_o}{4\pi} \cdot \frac{I}{r^2} \cdot \frac{a}{r} \cdot 2\pi a = \frac{\mu_o I}{2} \cdot \frac{a^2}{r^3} \qquad (33.13)$$

In terms of the distance x from the center,

$$B_x = \frac{\mu_o I}{2} \cdot \frac{a^2}{(a^2 + x^2)^{3/2}} \qquad (33.14)$$

At the center of the loop, $x = 0$ and $r = a$ (see Figure 33.10) and Eq. 33.14 becomes

$$B_x = \frac{\mu_o I}{2a} \qquad (33.15)$$

The field pattern for a current loop can be exhibited by using iron filings (Figure 33.11) or can be calculated on an electronic computer.

Figure 33.10

The magnetic field at the center of a circular current loop.

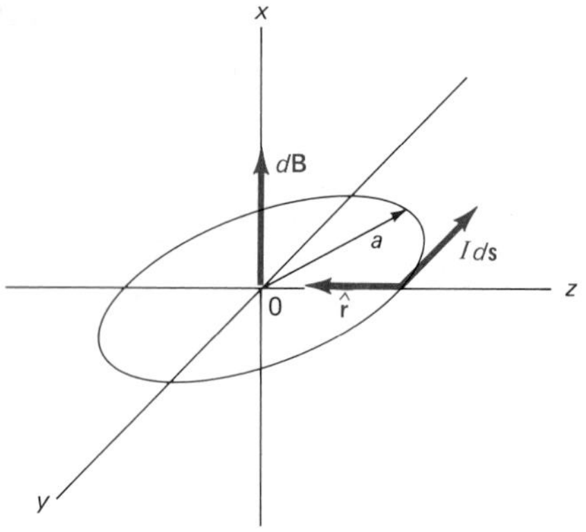

Figure 33.11

The iron filing pattern outlining the magnetic field of a circular current loop.

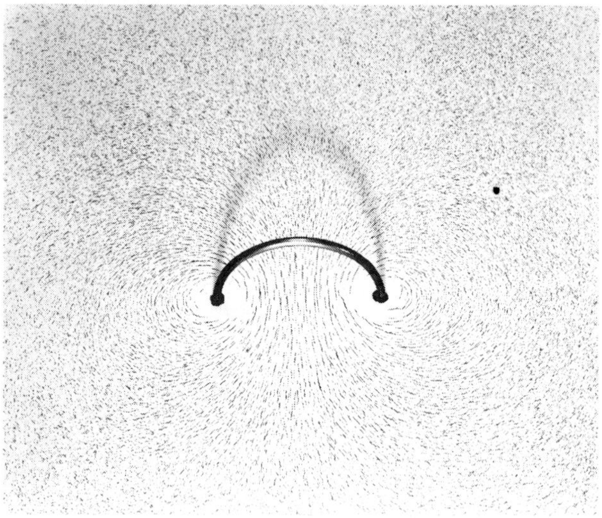

Example 2 Magnetic Field of a Hydrogen Electron

The planetary model of a hydrogen atom consists of a negatively charged electron in a circular orbit about a positive proton (Figure 33.12). The motion of the electron constitutes an electric current. Let's calculate the magnetic field produced by the orbiting electron at the location of the proton.

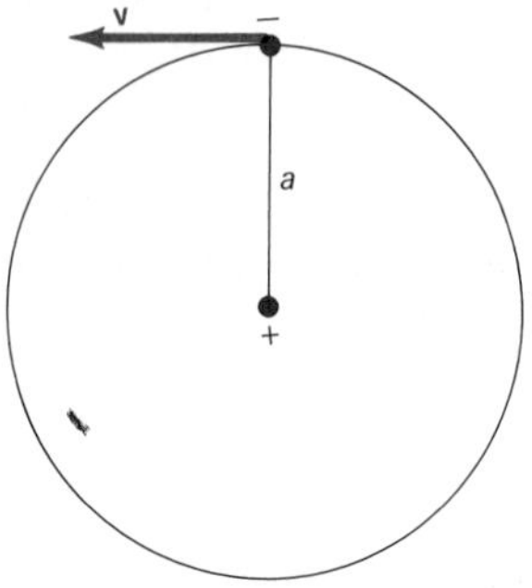

Figure 33.12

The planetary model of hydrogen: an electron in orbit about a proton.

The radius of the electron orbit (described in Section 42.3) is $a = 5.29 \times 10^{-11}$ m. The electron's orbital speed v is 2.19×10^{6} m/s. This means that the time it takes to complete one orbit (the period) is

$$T = \frac{2\pi a}{v}$$

$$= 1.52 \times 10^{-16} \text{ s}$$

The electric current, I, is equal to the electron charge, e, divided by this time.

$$I = \frac{e}{T} = 1.06 \times 10^{-3} \text{ A}$$

Then from Eq. 33.15 the magnetic field at the center of the electron orbit is

$$B = \frac{4\pi \times 10^{-7} \times 1.06 \times 10^{-3}}{2 \times 5.29 \times 10^{-11}}$$

$$= 12.6 \text{ T}$$

This magnetic field is quite strong. In fact, it is 30,000 times stronger than the earth's magnetic field at the equator.

Magnetic Dipole Moment of a Circular Current Loop

The magnetic field produced on the axis of a circular current loop at a distance x from the plane of the loop is given by Eq. 33.14. For large values of x ($x \gg a$) this equation is approximated by

$$B_x \cong \frac{\mu_0}{4\pi} \frac{2(I\pi a^2)}{x^3} \qquad (x \gg a)$$

We have said that, by symmetry, B_y and B_z both equal zero, and only B_x remains. Therefore, we can write for **B**

$$\mathbf{B} = \mathbf{i}B_x = \frac{\mu_0}{4\pi}\left[\frac{2(I\pi a^2)}{x^3}\right]\mathbf{i} \qquad (33.16)$$

In Section 27.2, Example 3, we saw that the electric field of an electric dipole (Figure 33.13) is

$$\mathbf{E} = \mathbf{i}E_x = \frac{1}{4\pi\epsilon_0}\left[\frac{2(ql)}{x^3}\right]\mathbf{i} \quad (x \gg l) \qquad (33.17)$$

This is the electric field at a distance x along the x-axis from an electric dipole at the coordinate origin, pointing in the positive x-direction. In Section 32.5 we identified the product (current × area) as the magnetic dipole mo-

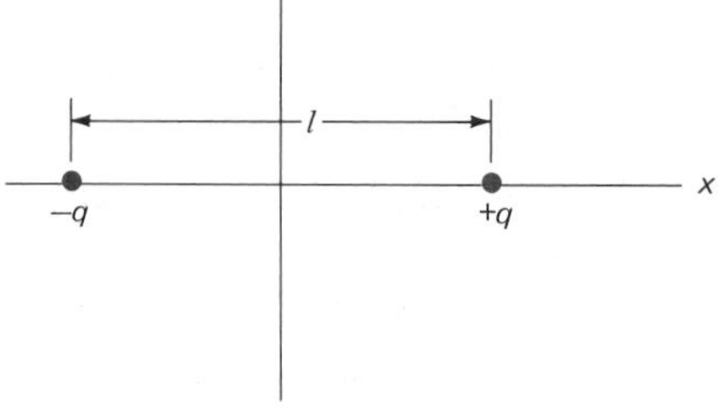

Figure 33.13

An electric dipole.

Figure 33.14

In order to describe the magnetic field and magnetic interactions, you may imagine the current-carrying loop to be replaced by a bar magnet, which is a magnetic dipole.

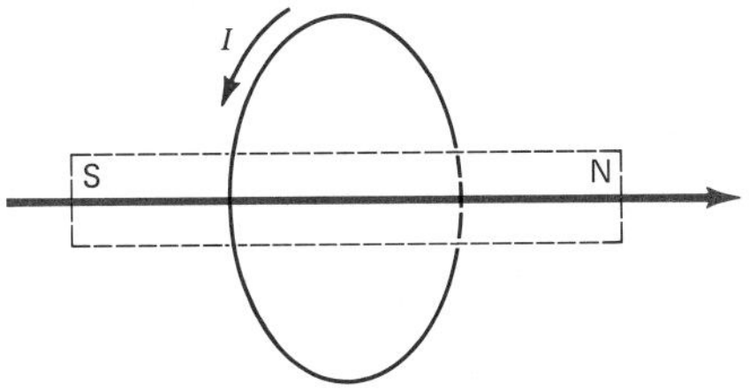

ment μ of the current loop, specifically for a square loop. If we make this same identification here for the circular loop,

$$\mu = I\pi a^2 \qquad (33.18)$$

then the magnetic field of the circular current loop and the electric field of the electric dipole have the same mathematical form at large distances. In particular, both show an x^{-3} dependence, which is characteristic of a dipole field.* The similarity between the circular current loop magnetic field (Eq. 33.16) and the electric dipole electric field (Eq. 33.17, with $\mu_0/4\pi$ replacing $1/4\pi\epsilon_0$) confirms the identification of current times area as the magnetic moment of the loop.

The direction of the dipole moment of the current loop is given by the right-hand rule. If you let the fingers of your right hand curl around the loop in the direction of the current, then your extended thumb points in the direction of the dipole moment. Figure 33.14 shows the magnetic dipole equivalent to the current loop dipole moment as a bar magnet.

The earth has a magnetic dipole moment of 7.94×10^{22} A · m^2. Geophysicists speculate that the earth's magnetic field is produced by a number of electric current loops in the interior. The energy source of these electric currents must ultimately be the rotation of the earth on its axis and the thermal energy in the earth's core. We will discuss the earth's magnetic field further in Section 36.6.

Example 3 Van Allen Belt Current Loop

Charged particles (mostly electrons and protons) trapped in the earth's magnetic field make up what we call the *Van Allen belts* (Figure 33.15; see Section 32.5). Because of nonuniformities in the magnetic field, the charged particles migrate—the protons move westward and the electrons move eastward. This migration of charges constitutes a tremendous electric current encircling the earth. Let us approximate the very complex electron and proton distributions by a current of 100,000 A at an altitude of 1600 km encircling the earth westward above the geomagnetic equator. Despite the size of this current, we can demonstrate that the resulting magnetic field is small compared to the earth's magnetic field. For example, we can calculate

*The magnetic field of any current loop will decrease at least as rapidly as 1 over the cube of the distance for large distances.

Figure 33.15

The Van Allen radiation belts.

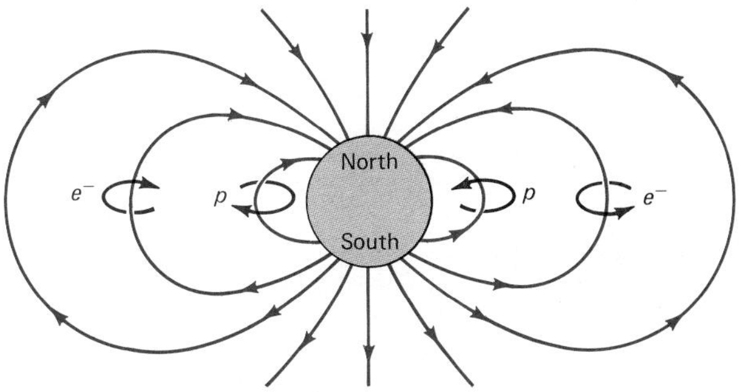

the magnetic field that this Van Allen current produces at the earth's center.

From Eq. 33.15

$$B_x = \frac{\mu_0 I}{2a} = \frac{4\pi \times 10^{-7} \times 10^5}{2 \times 8 \times 10^6}$$

taking a, the distance to the earth's center, to be 8×10^6 m. The result is

$$B_x = 0.8 \times 10^{-8}\ \text{T}$$
$$= 0.8 \times 10^{-4}\ \text{G}$$

By comparison, the earth's magnetic field is a few tenths of a gauss in magnitude. Thus, although the electric current produced by the Van Allen belts is tremendous when compared with ordinary household or laboratory currents, the resulting magnetic field is only a small disturbance in the earth's magnetic field.

Questions

2. A proton moves parallel to a long straight wire, in the direction of the current carried by the wire. What is the direction of the force on the proton?
3. The magnetic field of the earth is thought to be produced by a large current within the earth, more or less through the equatorial plane. Would this current have an eastward or a westward direction?
4. The sealed wooden box shown in Figure 33.16 has two wire leads conducting current into and out of the box. (The leads are twisted around each other in order that their magnetic effects will cancel out.) A careful investigation shows that there is a magnetic field whose strength at large distances varies inversely as the cube of the distance from the box. What sort of a current distribution inside the box does this imply?

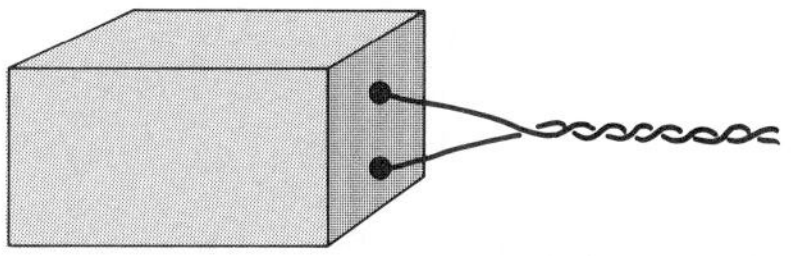

Figure 33.16

5. Suppose that you have made the magnetic field measurements at large distances from a sealed box like that in Figure 33.16 and found that $B \sim r^{-2}$. Is this possible?

33.3 Ampère's Law

The work of Oersted and of Biot and Savart was continued by André Marie Ampère. Ampère found that any line integral of the magnetic field around a closed path was proportional to the current encircled by the path. In SI units Ampère's result becomes

$$\oint \mathbf{B} \cdot d\mathbf{s} = \mu_0 I_{\text{encircled}} \qquad (33.19)$$

Again we use the symbol $\oint$ to indicate that the path is a closed path. There is a sign associated with each encircled current. For the area defined by the path of integration, there is a positive normal ($\hat{\mathbf{n}}$), given by the right-hand rule (Section 2.5). If the current through the area bounded by the path of integration is in this positive direction, we count it as positive. If the current is in the reverse direction, we count it as negative (Figures 33.17 and 33.18). Equation 33.19 is Ampère's law, and holds for steady-state conditions (no time-dependent fields or currents or changes in the path of integration). The extension of Ampère's law to time-dependent fields and currents was made by Maxwell. Indeed, Ampère's law, generalized for time dependence, became one of Maxwell's equations.

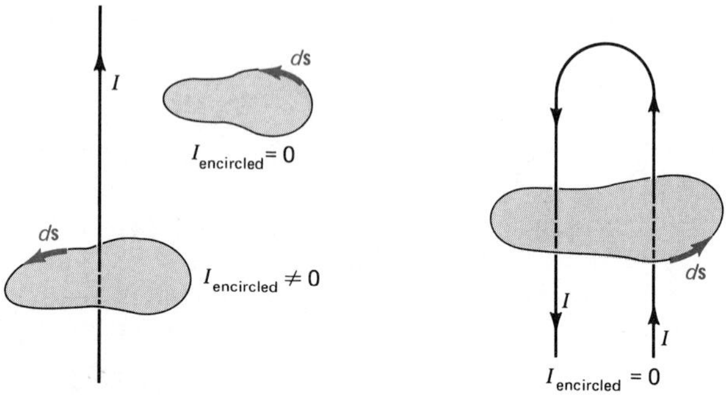

Figure 33.17

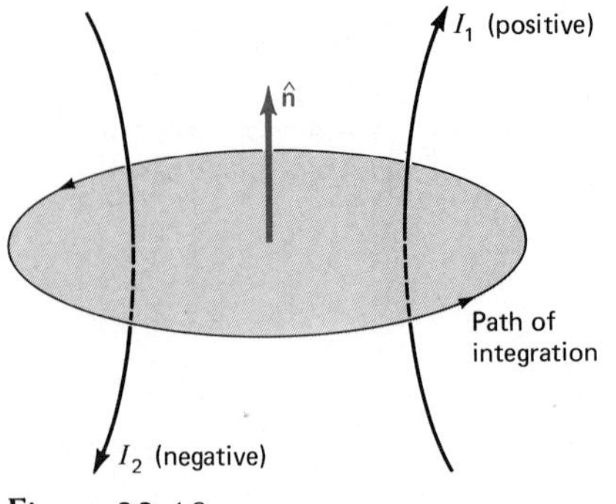

Figure 33.18

Positive and negative contributions of enclosed current.

We must emphasize that the path of integration is an *arbitrary* closed path in space. Unlike the integrated law of Biot and Savart, the path of integration does *not* follow the current. The path may or may not follow a magnetic field line. How, then, do we choose the path of integration? You may recall that we faced a similar problem in choosing a surface of integration for Gauss' law in Section 27.5. Our solution then and now is to exploit the symmetry of the physical system. We may choose a path so that

(a) the component of $\mathbf{B}$ parallel to $d\mathbf{s}$ is constant;

(b) $\mathbf{B} \cdot d\mathbf{s} = 0$ because $\mathbf{B} = 0$;

(c) $\mathbf{B} \cdot d\mathbf{s} = 0$ because $\mathbf{B}$ is perpendicular to the path; or

(d) a combination of these three options is present.

For example, in the investigation of the magnetic field of a solenoid (Section 33.4) we use all three options. For the field of a long straight current-carrying wire we use option (a), as we now show.

The Field of a Long Straight Current-Carrying Wire

We will apply Ampère's law (Eq. 33.19) to a long straight current-carrying wire for the following four reasons: By doing so, we (1) illustrate the application of symmetry principles; (2) describe the field that Oersted discovered, and (3) reach the same result that we did by using the law of Biot and Savart, and thus show the consistency of these two approaches. Also, (4) the result will be useful in Section 33.5, when we consider the current balance and finally define the ampere.

As a guide in selecting a path of integration, we consider the general symmetry of the field. The long straight wire is an axis of rotational symmetry. Any direction outward and perpendicular to the wire is the same as any other direction perpendicular to the wire. This cylindrical symmetry invites the use of circular cylindrical coordinates. In Figure 33.19 we see such a system oriented so that the z-axis lies along the wire with the positive direction in the direction of the current. The magnetic field may have a radial component B_r, an axial component B_z, and an azimuthal component B_ϕ. Because of the rotational sym-

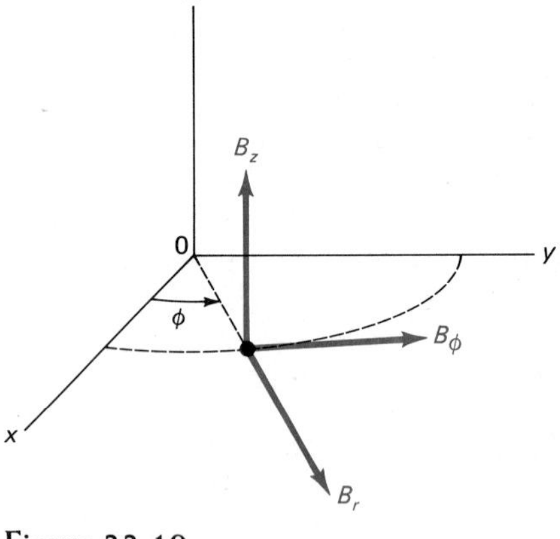

Figure 33.19

Circular cylindrical components of **B**.

Figure 33.20

A cylindrical surface length L, radius r for the calculation of possible magnetic flux: $\Phi_B = 2\pi r L B_r$. Since $\Phi_B = 0$ for a closed surface, we conclude that $B_r = 0$.

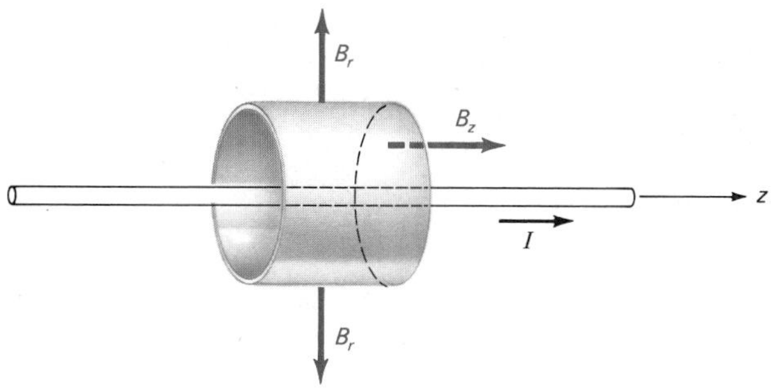

metry, none of these three field components can depend on the azimuthal angle ϕ. Nor can they depend on z, since we have a *long* straight wire, and every value of z is like every other value (translational symmetry). Thus, by symmetry, the components can depend only on r. Let us consider each of these possible components in turn.

First, could we have a radial field, B_r? (Such a field is illustrated in Figure 33.20.) This would mean that the magnetic field lines either originate or terminate at the wire. The answer is no, because the magnetic field lines form closed loops and thus never terminate in finite space (see Section 32.2). Therefore we conclude that $B_r = 0$. This result can be confirmed by integrating the magnetic field over the cylindrical surface shown in Figure 33.20.

Could we have an axial field, $B_z(r)$? If we did, it would have to decrease with increasing radial distance* and, at large enough distances, become negligible. To determine whether or not this is so, we take a $B_z(r)$ that *does* decrease with increasing radial distance, and apply Ampère's law to the circuit shown in Figure 33.21. We select the path segments so that $\mathbf{B} \cdot d\mathbf{s} = 0$ over the radial parts [option (c)] while $B_z(r)$ is constant over each of the segments parallel to the z-axis [option (a)]. From Ampère's law

$$\oint \mathbf{B} \cdot d\mathbf{s} = B_z(r) \int_E^F dz + \int_F^G B_r\, dr + B_z \int_G^H dz + \int_H^E B_r\, dr = 0 \qquad (33.20)$$

The zero appears because zero current is enclosed. The two radial integrals ($F \rightarrow G$ and $H \rightarrow E$) vanish because $B_r = 0$. If the $G \rightarrow H$ integral is taken far out so that B_z along that segment is negligible, then that integral vanishes. We are left with

$$B_z(r) \int_E^F dz = B_z(r)\overline{EF} = 0$$

The length $\overline{EF}$ is not zero. Hence

$$B_z = 0$$

and there is no axial field, $B_z(r)$.

We are left with only an azimuthal field, $B_\phi(r)$, a field whose pattern consists of circular loops about the current. We can drop the subscript ϕ, since we know that no other components exist, and apply Ampère's law once again. This time we select a path of integration so that B remains constant in magnitude [option (a) earlier in this section]—a circle concentric with the z-axis (Figure 33.22). Then

$$\oint \mathbf{B} \cdot d\mathbf{s} = B \int_0^{2\pi} r\, d\phi = 2\pi r B \qquad (33.21)$$

By Ampère's law this equals $\mu_0 I$ or

$$B = \frac{\mu_0 I}{2\pi r} \qquad (33.22)$$

in complete agreement with the result obtained in Section 33.2 from the law of Biot and Savart.

Notice here that by using symmetry arguments we were able to show without formal, detailed calculations that $B_r = 0$ and $B_z = 0$. Then Eq. 33.21 for the total field followed very quickly. With practice these and other symmetry arguments will become more familiar to you.

*As with the electric field, an energy density may be associated with the magnetic field (Section 35.3). If any nonzero axial magnetic field did not decrease with distance, this failure to decrease would imply infinite energy per unit length of the wire.

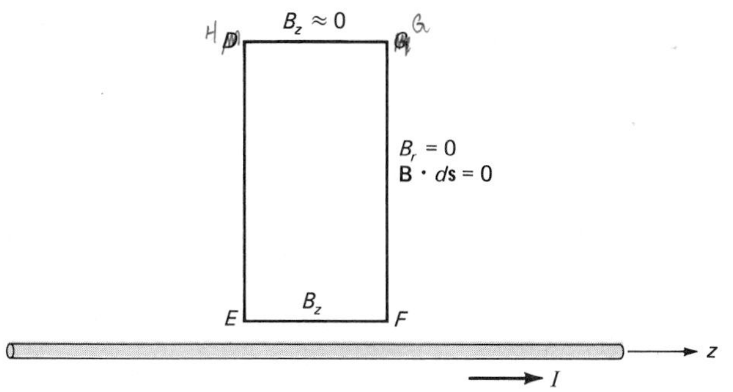

Figure 33.21

A path of integration for the calculation of a possible axial magnetic field B_z.

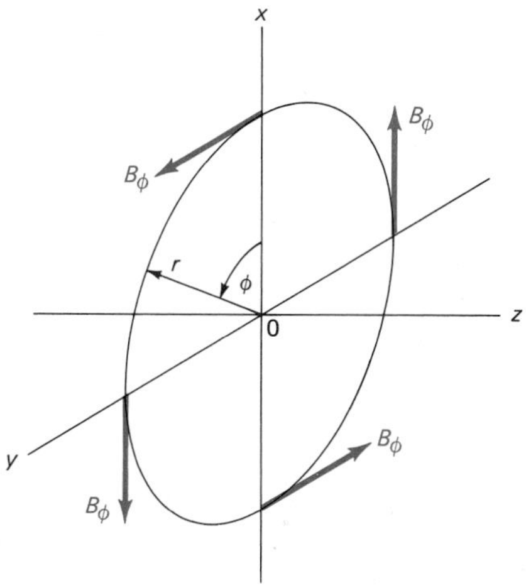

Figure 33.22

A path of integration for the calculation of the azimuthal magnetic field B_ϕ.

Figure 33.23

A path of integration for calculating the magnetic field inside a cylindrical distribution of current.

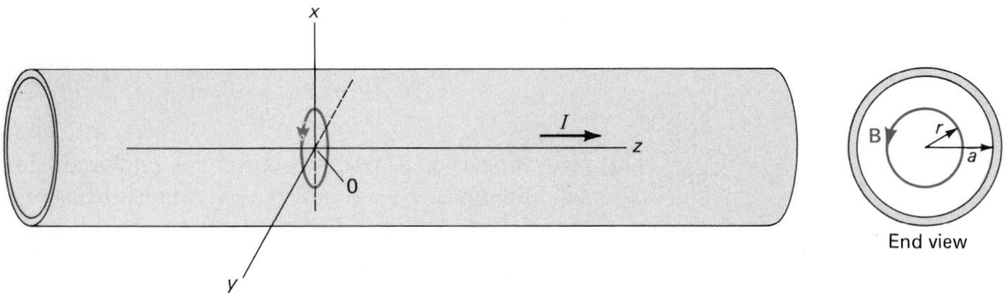

Consider Eq. 33.22 for a moment. What happens to the field as $r \rightarrow 0$? Does $\mathbf{B}$ really become infinite? The following example answers these questions and provides a further illustration of the use of Ampère's law.

Example 4 Magnetic Field in the Interior of a Current-Carrying Wire

A round wire with radius a carries a total current I_o. Calculate B_ϕ in the interior of the wire.

Let us assume that the current density (amperes per square meter) is constant over the cross section of the wire.* The current density, J, is

$$J = \frac{I_o}{\pi a^2}$$

where I_o is the current in the wire and πa^2 is the cross-sectional area of the wire. Then, upon application of Ampère's law to a circle of radius r, $r \leq a$, the line integral becomes

$$\oint \mathbf{B} \cdot d\mathbf{s} = 2\pi r B_\phi$$

exactly as it does in Eq. 33.21. The circular path inside† the round wire (Figure 33.23) gives us the same rotational symmetry that we have for the circular path outside the round wire. The source term (μ_o × encircled current) is

$$\mu_o I = \mu_o J \pi r^2 = \mu_o I_o \left(\frac{\pi r^2}{\pi a^2}\right)$$

knowing that $\oint \mathbf{B} \cdot d\mathbf{s} = 2\pi r B_\phi$ and $\mu_o I = \mu_o I_o(\pi r^2/\pi a^2)$, and then using Ampère's law (Eq. 33.19), we have

$$2\pi r B_\phi = \mu_o I_o \left(\frac{\pi r^2}{\pi a^2}\right)$$

*This is true for steady currents (direct currents). It is not true for high-frequency alternating currents. When the frequency is sufficiently high, the rapidly varying magnetic fields tend to drive the moving charges toward the surface of the conductor.

†In the interior of a material medium we should take the magnetic properties of the medium into account and replace μ_0 by μ, analogous to $\epsilon_0 \rightarrow \epsilon$ for dielectrics (Section 29.4). Actually the effect is generally very small *except* in ferromagnetic media. For copper the effect is only a few parts per million (Section 36.3).

or

$$B_\phi = \frac{\mu_o I_o}{2\pi} \cdot \frac{r}{a^2} \qquad (33.23)$$

Inside the round wire, the field is directly proportional to the distance out from the center line. The azimuthal field B_ϕ does not go to infinity as r goes to zero. Instead B_ϕ goes to zero.

The strength of the azimuthal magnetic field described in Example 4 is plotted in Figure 33.24 as a function of r. Equation 33.23 provides the linear portion, $0 \leq r \leq a$, and Eq. 33.22 yields a hyperbolic curve for $r \geq a$. Note very carefully that we may *not* use Eq. 33.22 all the way in to $r = 0$. Conditions change when $r < a$ (we're inside the conductor) and Eq. 33.22 no longer applies.

Ampère's Law and Gauss' Law

Both Ampère's law and Gauss' law are useful in developing general proofs and in setting up differential equations. But each of these two laws is useful as a device for calculating fields *only when the physical system has sufficient symmetry*. Mathematically Ampère's law and Gauss' law are quite distinct.

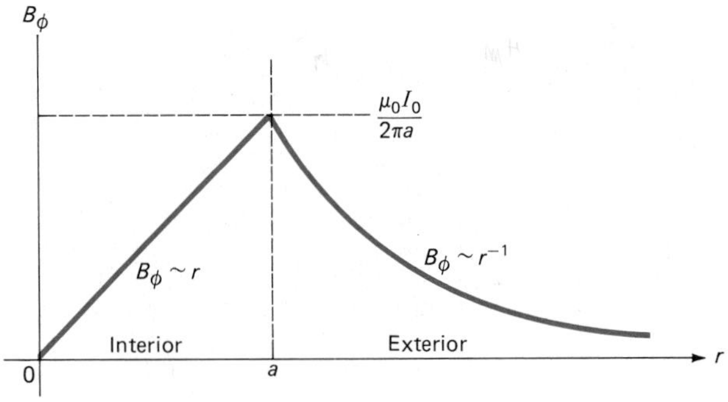

Figure 33.24

The magnetic field inside and outside a uniform cylindrical distribution of current.

Ampere's Law	Gauss' Law	
$\oint \mathbf{B} \cdot d\mathbf{s} = \mu_o I_{encircled}$	$\int_{surface} \mathbf{E} \cdot d\mathbf{A} = \frac{q_{enclosed}}{\epsilon_o}$	(33.24)
integration of **B** along a closed *path*	integration of **E** over a closed *surface*	
proportional to the current encircled	proportional to the charge enclosed	

Questions

6. When using Ampère's law, how is the path of integration chosen?
7. Suppose that you are building a piece of electronic equipment that must contain two wires carrying equal currents in opposite directions. If you want to minimize the magnetic effects of the two wires, should you place them close together or far apart? Explain.

33.4 Magnetic Field of a Solenoid

A *solenoid* is a long current-carrying wire tightly wound into a helix, or coil (Figure 33.25). Usually the length of the coil is larger than the radius. Solenoids are quite interesting because the magnetic field inside a solenoid can be very strong and very uniform over a large volume. In this section we investigate the magnetic field of a solenoid. First we develop an approximate picture to enable us to determine the symmetry that allows Ampère's law to be used. Then we calculate the magnetic field quantitatively, using Ampère's law.

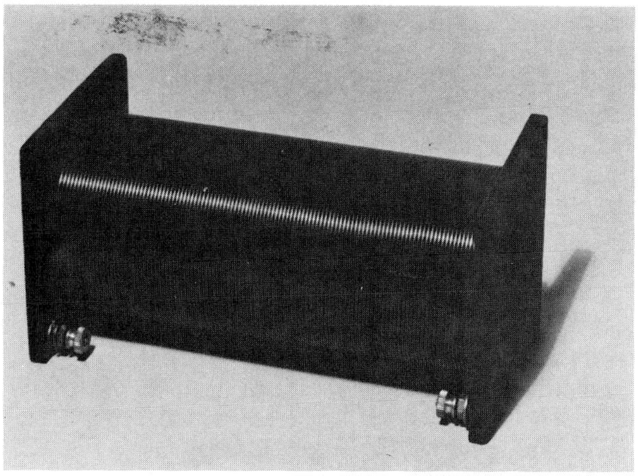

Figure 33.25

A solenoid.

An Approximate Picture

A solenoid is approximately a series of circular current loops. We calculated the field of a single circular current loop in Section 33.2. If the wire is loosely wound (Figure 33.26a), some magnetic field lines will encircle the individual turns of the coil. Other field lines will combine to form the axial magnetic field of the solenoid. Here again we use the principle of superposition, adding the (vector) fields of the individual circular current loops. If the wire is wound more tightly, the number of field lines encircling each individual turn is reduced and the desired axial field is strengthened (Figure 33.26b).

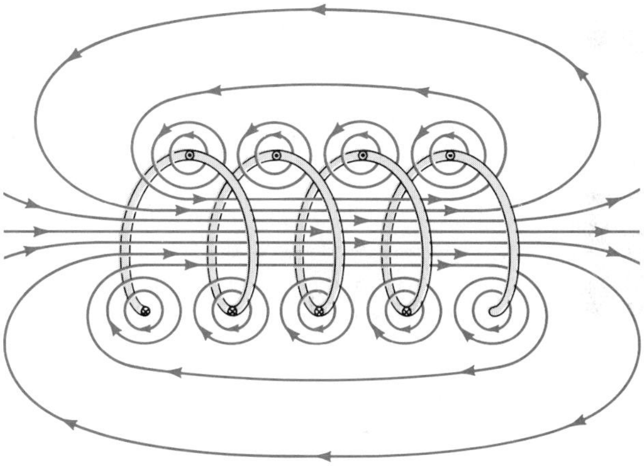

(a)

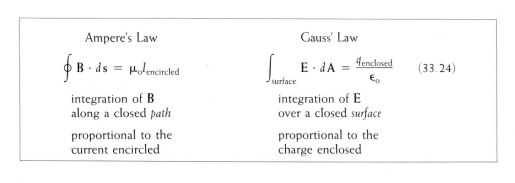

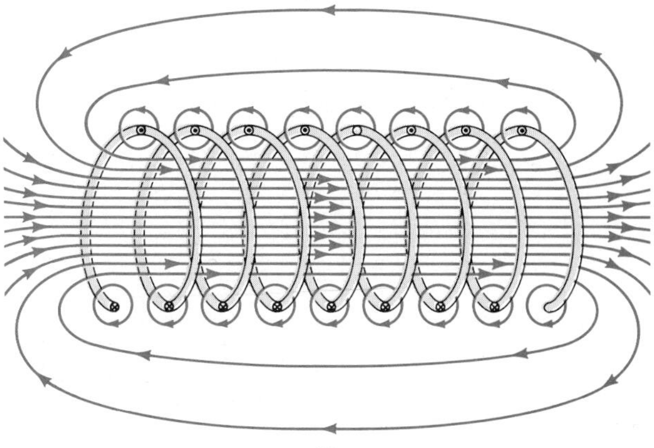

(b)

Figure 33.26

(a) Magnetic field lines for a loosely wound solenoid. (b) Magnetic field lines for a tightly wound solenoid.

Solenoid Flux by Ampère's Law

An ideal solenoid is infinitely long and so tightly wound that the current is essentially a cylindrical sheet (Figure 33.27). When applying Ampère's law, the problem again is to find an appropriate path of integration—a path of integration over which the integral can be evaluated, and that yields a useful result. First we select a rectangular path like that shown in Figure 33.28.

The line segments EF and GH are parallel to the z-axis, and FG and HA are perpendicular to the z-axis. Therefore the integral around the closed path can be expressed as a sum of four integrals:

$$\oint \mathbf{B} \cdot d\mathbf{s} = \int_E^F B_z\,dz + \int_F^G B_r\,dr + \int_G^H B_z\,dz + \int_H^E B_r\,dr \qquad (33.25)$$

The equation is just like Eq. 33.20 (except for the "= 0"). In fact, the Ampère law investigation of the field of a solenoid is very similar to the determination of the field of a long straight wire that we saw in Section 33.3.

We've taken the solenoid to be infinitely long. This means that, since the ends are infinitely far away, nothing, including the radial field B_r, can depend on z. This independence of z is really an expression of another symmetry property: translational symmetry. If the solenoid is infinitely long, then there is nothing physical to single out any one value of z from any other value of z. Then

$$B_r(z) = B_r(z + l)$$

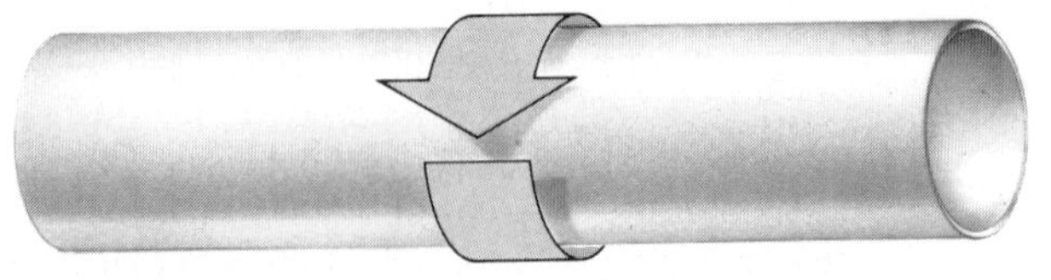

Figure 33.27

The current in very closely spaced turns may be approximated as a sheet of current. If n is the number of turns per meter length of the solenoid and each turn carries current I, then nI is the current per meter length of the solenoid.

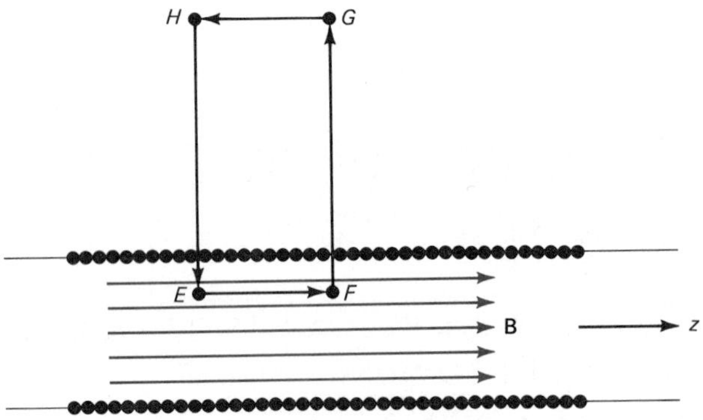

Figure 33.28

A path of integration for the calculation of the axial magnetic field inside a solenoid.

Since the two radial integrals ($F \rightarrow G$ and $H \rightarrow E$) are in opposite directions, they exactly cancel each other. (Actually $B_r = 0$, by the arguments used in Section 33.3, and each integral vanishes.)

Next we must consider the integral

$$\int_G^H B_z\,dz$$

involving the value of B_z outside the solenoid. Let us assume that if there is a nonzero field B_z outside the solenoid, this field falls off with distance. Again, if such an axial field exists, it must decrease with increasing distance, or else the magnetic energy stored in the field per unit length of solenoid would be infinite. Now, because the line segment GH is taken arbitrarily far out, the field B_z is arbitrarily small, and the integral over GH vanishes. This means that *all* $B_z = 0$ (exterior to an infinitely long solenoid).

Inside the solenoid there can be no radial field. This is the same symmetry argument that we applied to the long straight wire. There is no azimuthal field with field lines forming circular loops like the field produced by a long straight current-carrying wire. From both of these arguments, which are essentially symmetry arguments, we conclude that the magnetic field inside the solenoid is purely axial, or in other words, parallel to the axis of the solenoid. Because this is our z-axis, $\mathbf{B} = \mathbf{k}B_z$, with B_z independent of z (and independent of any azimuthal angle ϕ).

Then

$$\oint \mathbf{B} \cdot d\mathbf{s} = \int_E^F B_z\,dz = B_z l$$

For a current I, and a winding of N turns in the distance $\overline{EF} = l$ m, the current enclosed by the path of integration is IN. Ampère's law becomes

$$B_z l = \mu_0 IN \qquad (33.26)$$

If we let $N = nl$, where n is the number of turns per meter, the axial field of the solenoid is

$$B_z = \mu_0 In \qquad (r < a) \qquad (33.27)$$

Note carefully that we have not specified where the line segment EF is located within the solenoid. This result holds inside the solenoid for any distance out from the z-axis, $0 \leq r < a$. Therefore the axial field B_z is uniform, constant across the cross section of the solenoid, and independent of r as well as of z. This is why solenoids are so useful in scientific work: They can produce strong, uniform magnetic fields. Also, because of the strength of the axial field, solenoids have important everyday applications. For example, solenoids are used in the starters of cars.

Once again we can use a right-hand rule: If the fingers of your right hand curl in the direction of the electric current in a solenoid, then your extended thumb points in the positive direction of the axial field.

Because a solenoid is actually a helix, rather than a series of perfect circular loops, there is a net flow of charge in the direction of the axis of the solenoid. This is the current I. The result is a small tangential field B_ϕ outside the solenoid. By applying Ampère's law to the circular path shown in Figure 33.29, we get

$$B_\phi = \frac{\mu_0 I}{2\pi r} \qquad (r > a) \qquad (33.28)$$

The Current Balance

The force that one current-carrying conductor exerts on a second, parallel, current-carrying conductor is used to define the unit of current, the ampere. The same current is sent through two parallel conductors. The force per unit length on conductor 2 (from Eq. 33.31) is

$$\frac{dF_2}{ds_2} = \frac{\mu_o I^2}{2\pi d} \tag{33.33}$$

This force is measured with a very sensitive, accurate balance.* The conductor length and the separation distance d are also measured accurately. We *assign* μ_o the value of $4\pi \times 10^{-7}$ (tesla meters per ampere, newtons per ampere squared, or other equivalent units) and then substitute these values of dF_2, d, ds_2, and μ_o into Eq. 33.33. We can then solve this equation for current I in order to find the value of the current in amperes. *The ampere can therefore be defined as follows:*

One ampere is the electric current in each of two parallel conductors separated by 1 m that gives rise to a force per unit length of 2×10^{-7} N/m on each conductor.

With the ampere defined we can now define the coulomb as the quantity of electric charge that passes a given point in a conductor in 1 s when the current is 1 A. Or

$$1 \text{ A} = 1 \text{ C/s}$$

We have now defined all of the terms in Coulomb's law (Section 26.3).

When we use this definition of the coulomb, Coulomb's law, in principle, provides us with an experimental determination of the proportionality constant k_e and the electric permittivity ϵ_o. Scientists could have first assigned values to k_e and ϵ_o, then measured the coulomb by Coulomb's law, and finally measured μ_o. But instead, by international agreement, μ_o is assigned the value $4\pi \times 10^{-7}$ (SI units) and the ampere is defined and measured by current balances. The reasons for this choice are strictly pragmatic. Scientists can control and measure electric current far more precisely and accurately than they can control and measure electric charges.

Questions

10. Current is established in a loosely wound helix. Do the magnetic forces cause the helix to elongate or to contract?

11. A plasma inside a hollow cylinder carries a current and thereby produces a magnetic field. Assuming uniform current density across the cylinder, what is the direction of the magnetic force on the outer edge of the plasma?

12. Equations 33.31 and 33.32 (for the force per unit length between parallel wires) include a factor $I_1 I_2$. This factor $I_1 I_2$ appears in the expression for the force between any two coils, independent of shape or orientation. Explain why you would expect the force to be proportional to the product of the currents, $I_1 I_2$.

*The "parallel" conductors are not straight but are actually bent into a circular shape. However, the radius of curvature of the circles is large compared to the separation distance d, and we can continue to think of them as parallel conductors.

Summary

The law of Biot and Savart gives the magnetic field $d\mathbf{B}$ at a field point produced by a current element $I\,d\mathbf{s}$ at a source point:

$$d\mathbf{B} = \frac{\mu_o}{4\pi} \frac{I\,d\mathbf{s} \times \hat{\mathbf{r}}}{r^2} \tag{33.1),(33.6}$$

The unit vector $\hat{\mathbf{r}}$ is directed from the source point to the field point. The total field $\mathbf{B}$ is obtained by integrating over the entire circuit.

The proportionality constant $\mu_o/4\pi$ is assigned the value 10^{-7}. The units are T · m/A or N/A^2.

When applied to a long straight current-carrying wire, the law of Biot and Savart yields

$$B = \frac{\mu_o I}{2\pi r} \tag{33.11}$$

at a perpendicular distance r from the wire. The field lines form concentric circles about the wire, with the direction given by the right-hand rule.

When applied to a circular current loop of radius a, the law of Biot and Savart yields

$$B_x = \frac{\mu_o}{4\pi} \frac{2(I\pi a^2)}{r^3} \tag{33.13}$$

where the x-axis is the axis of rotation of the loop. The quantity $I\pi a^2 = I \cdot$ area is the magnetic dipole moment of the current loop. At the center of the loop the field strength is

$$B = \frac{\mu_o I}{2a} \tag{33.15}$$

Ampère's law,

$$\oint \mathbf{B} \cdot d\mathbf{s} = \mu_o I_{\text{encircled}} \tag{33.19}$$

relates the line integral of the magnetic field to the current encircled. It is convenient to use the law of Biot and Savart, or Ampère's law, only when the current element has a high degree of symmetry.

Ampère's law permits us to calculate the interior axial field of a long solenoid,

$$B_z = \mu_o I n \tag{33.27}$$

independent of position (as long as end effects can be neglected).

The magnetic field created by one current, I_1, will exert a force on a second current, I_2. For parallel currents a distance d apart the magnetic force per unit length exerted on one current by the other is

$$\frac{dF}{ds} = \frac{\mu_o}{2\pi} \frac{I_1 I_2}{d} \tag{33.31),(33.32}$$

For parallel currents this force is attractive. For antiparallel currrents this force is repulsive.

The magnetic force between two parallel currents is used to define the ampere:

One ampere is the electric current in each of two parallel conductors separated by 1 m that gives rise to a force per unit length of 2×10^{-7} N/m on each conductor.

The ampere is used to define the coulomb.

$1\ C = 1\ A \cdot s$

One coulomb is the charge that passes a given point in 1 s when the current is 1 A.

Problems

Section 33.1 Biot and Savart's Law

1. Show that the product $\mu_0 \epsilon_0$ has dimensions of (velocity)$^{-2}$, that is, T^2/L^2.

2. An element of a conductor ($ds = 10^{-3}$ m) carries a current of 100 A. This current element is at the origin and is in the positive z-direction: $I\,d\mathbf{s} = \mathbf{k}I\,ds$. Find the contribution to the magnetic field, $d\mathbf{B}$, of this current element at each of the following points (distances in meters): (a) (1, 0, 0); (b) (0, 0, 1); (c) (1, 1, 1).

Section 33.2 Magnetic Field of Electric Current

3. Equation 33.10 gives the magnetic field produced by the current in a straight wire segment of length $2s_0$. Let s_0 shrink to $ds_0/2$ and show that Eq. 33.10 reduces to the law of Biot and Savart (Eq. 33.1).

4. Consider a segment of current extending along the x-axis from 0 to $+s_0$ (Figure 1). Calculate the magnetic field that this current segment will contribute to the magnetic field at point P on the y-axis.

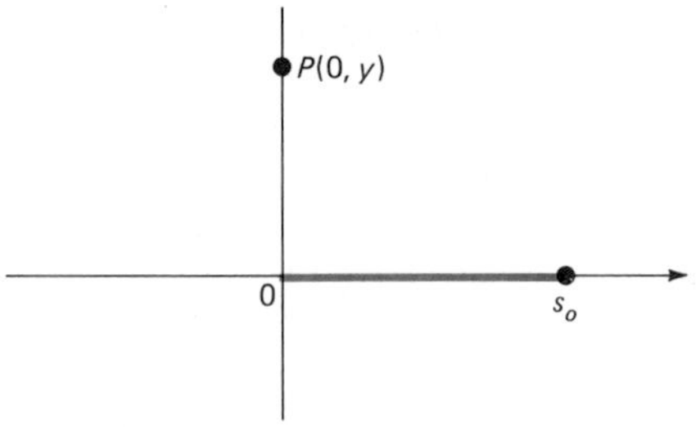

Figure 1

5. (a) A square loop $2a$ m on a side conducts a current I (Figure 2). Calculate the magnetic field at the center of the square. (*Hint:* Use the principle of superposition and the expression derived for the field of a finite current-carrying line segment.) (b) Compare this result with the field at the center of a circular loop (1) of radius a and (2) of radius $2a$.

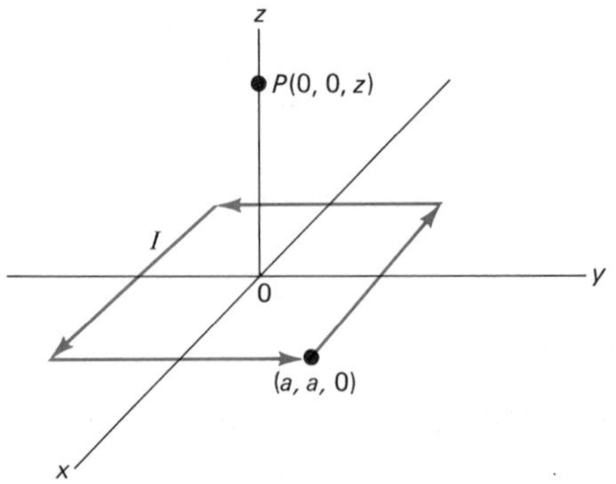

Figure 2

6. Calculate the magnetic field for a point $P\ (0, 0, z)$ on the axis of a square loop of current (Figure 2). (*Hint:* Superpose components but note that the four vectors are *not* parallel.)

7. (a) Using the result of Problem 6, calculate the magnetic field $\mathbf{B}$ for a point on the z-axis with $z \gg a$ (asymptotic form). (b) Compare your result with the field of a magnetic dipole and assign a magnetic moment to the square loop.

8. The length of wire in the square loop in Problems 5–7 is $8a$ m. (a) For the same current, how does the magnetic field at the center of the square loop compare with that at the center of a rectangle $a \times 3a$, (Figure 3)? (b) For the same current, how does the magnetic field of the square loop compare with that at the center of a circle of the same perimeter, $8a$ m?

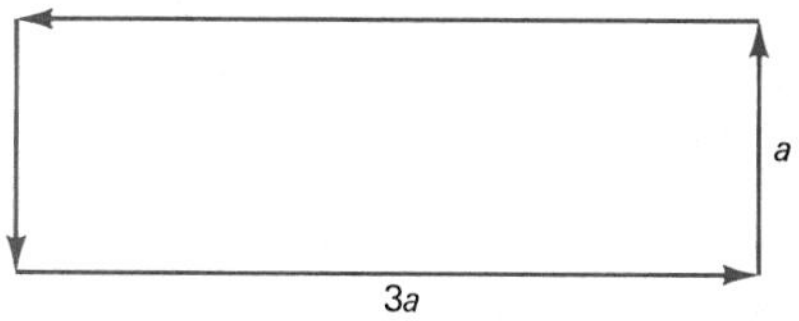

Figure 3

9. A flat current loop is in the shape of a rectangle. Prove that for a given fixed perimeter, the magnetic field at the center of the loop is smallest for the case of a rectangle having four equal sides—that is, a square. In your proof use (a) a symmetry argument, and then (b) the fact that the derivative dB/ds_0 must vanish at a relative minimum.

10. A ribbon-like conductor of width $2a$, infinite length, and carrying a current I lies in the $x-y$ plane with the x-axis oriented along the center line of the length of the ribbon (Figure 4). Using Eq. 33.11 for the magnetic field a distance R from an infinitely long current-carrying wire, calculate the magnetic field at a position $y = R$ on the y-axis.

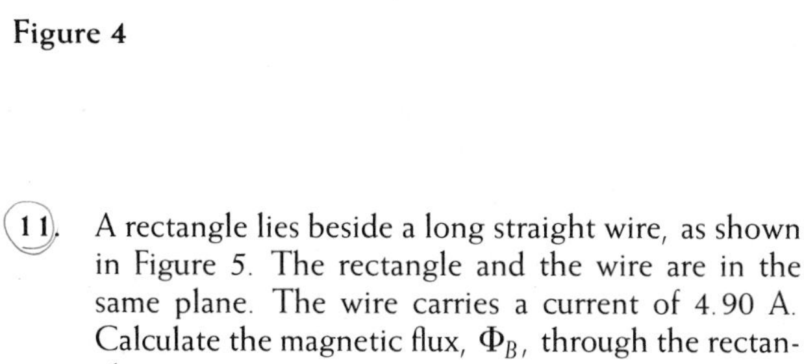

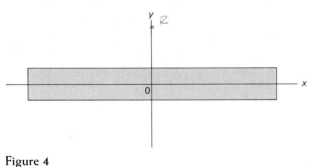

Figure 4

11. A rectangle lies beside a long straight wire, as shown in Figure 5. The rectangle and the wire are in the same plane. The wire carries a current of 4.90 A. Calculate the magnetic flux, Φ_B, through the rectangle.

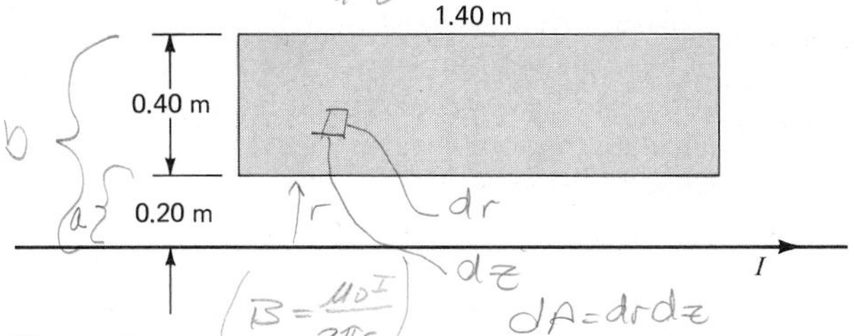

Figure 5

12. Continuing Problem 11, we add a second long straight wire in the plane of the rectangle and 0.20 m above it. The two wires carry the same number of amperes of current. Calculate the magnetic flux through the rectangle if (a) the two currents are parallel; (b) the two currents are antiparallel.

13. A circular coil of wire whose radius is 10 cm has 100 turns. The current through the coil is 0.3 A. (a) Calculate the magnetic field at the center of the coil. (b) At what points along the axis of the circular coil has B dropped to 50% of its value at the center of the coil?

14. A circular coil (radius 0.50 m) consists of 180 turns. Each turn carries a current of 4.21 A. (a) Calculate the magnitude of the magnetic field at the center of the loop. At the center of this large coil is a small coil of 20 turns (radius 1.00 cm), each turn carrying 0.680 A. (b) Calculate the magnetic moment of this small coil. (c) Assuming that the magnetic field of the large coil is uniform over the small coil, calculate the maximum torque on the small coil. (d) Calculate the difference in potential energy between the unstable and the stable orientations of the small coil in the field of the large coil.

15. Three circular loops of wire of radius R, $2R$, and $3R$, respectively, are made into concentric circles. A clockwise current I exists in the innermost and outermost loops. Determine the magnitude and direction of the current in the middle loop if the net magnetic field at the center is zero.

16. A long straight wire is bent as shown in Figure 6 to form a circle 1 m in radius and a smaller circle of radius r_0. The direction of the current I along the wire is shown in the diagram. The magnetic field B at the common center point is measured and found to be zero. Calculate the value of r_0. (*Hint:* Use the principle of superposition. Treat the straight segment as a wire of infinite length.)

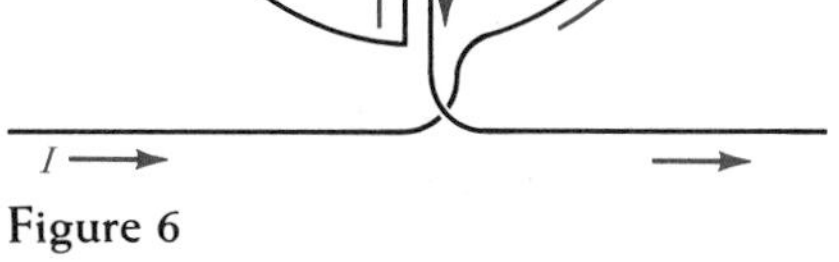

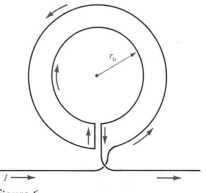

Figure 6

17. (a) Calculate the magnetic field **B** produced at the origin by a current I flowing in a semicircular arc of radius a. (b) The complete circuit is shown in Figure 7. Calculate the total magnetic field at the origin, P (by inspection).

Figure 7

18. Helmholtz coils consisting of two circular conductors, each of radius a, are placed parallel to each other with axes coinciding, a distance a apart (Figure B). Each loop carries a current I in the same sense of circulation. (a) Sketch in the field lines you expect for a cross-sectional plane containing the common axis (the x–y plane). (b) Calculate the magnetic field **B** on the common axis at the midpoint between the coils. (c) Demonstrate that this field of the Helmholtz coils is relatively uniform at the midpoint by showing that

$$\left.\frac{dB_x}{dx}\right|_{x=0} = 0 \quad \text{and} \quad \left.\frac{d^2B_x}{dx^2}\right|_{x=0} = 0$$

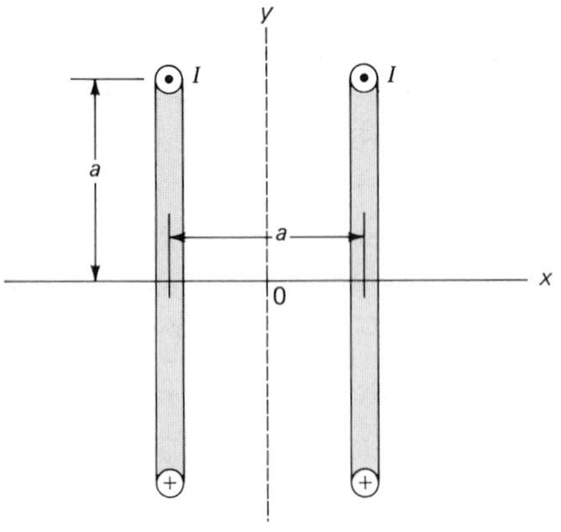

Figure 8

19. Evaluate the integral of Eq. 33.8 by using r as the variable of integration. (See Figure 9.) Calculate the magnetic field B. Extend the limits of integration as appropriate for a long wire and compare your calculated B with Eq. 33.11.

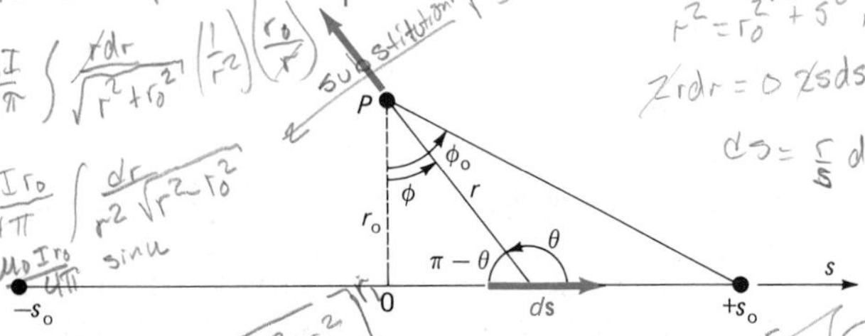

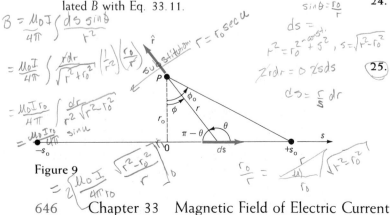

Figure 9

20. Repeat Problem 19, this time using θ (or ϕ) as the variable of integration.

21. Calculate the magnetic dipole moment of the Van Allen belt current of Example 3. Compare this magnetic dipole moment with the magnetic dipole moment of the earth (7.94×10^{22} A·m^2).

22. The magnetic field produced by a straight wire at a perpendicular distance r_o from the center of the wire can be expressed as (see Eq. 33.10)

$$B = \left(\frac{\mu_o I}{2\pi r_o}\right)\sin\phi_o$$

where $2\phi_o$ is the angle subtended by the wire from the field point P. (a) Show that the field produced at the center of a current loop in the form of an N-gon (a regular polygon with N sides) can be expressed as

$$B_N = \left(\frac{\mu_o I}{2r_o}\right)\frac{\sin(\pi/N)}{\pi/N}$$

(b) Show that for $N = 4$, B_N gives the result for a square loop derived in Example 1. (c) Obtain the field at the center of a circular loop of radius r_o by considering a limiting form of B_N.

Section 33.3 Ampère's Law

23. A section of a circuit running to and from a battery consists of two parallel wires (Figure 10). The current in each wire is 2.0 A. Evaluate the loop integral $\oint \mathbf{B} \cdot d\mathbf{s}$ for (a) a path encircling the top wire only, as shown; (b) a path encircling the bottom wire only, as shown; and (c) a path encircling both wires, as shown.

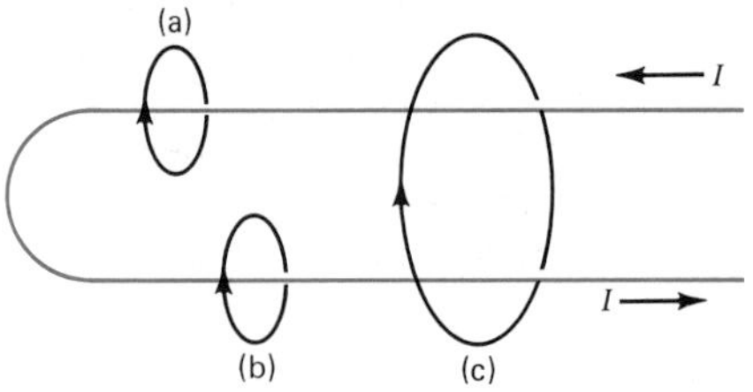

Figure 10

24. An evaluation of the loop integral in Ampère's law around a current-carrying conductor yields a value of 0.314×10^{-6} T·m. Determine the current in the wire.

25. A long wire lying on the line $x = a$ in the x–z plane carries a current I in the positive z-direction. A second long wire, parallel to the first on the line $x = -a$, carries a current I in the negative z-direction (Figure 11). (a) Sketch the magnetic field lines you expect in the x–y plane. (b) Calculate the resultant magnetic field **B** at point P on the y-axis as a function of y.

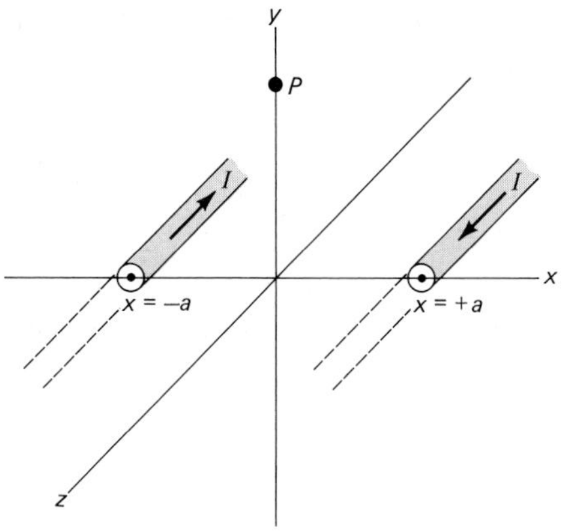

Figure 11

26. Show that the expression for the magnetic field (Problem 25) on the y-axis at large distances ($y \gg a$) reduces to

$$\mathbf{B} = -\mathbf{j}\,\frac{\mu_o I}{\pi}\cdot\frac{a}{y^2}$$

27. Repeat Problem 25 but with *both* currents in the positive z-direction.

28. Show that the expression for the magnetic field (Problem 27) on the y-axis at large distances ($y \gg a$) reduces to

$$\mathbf{B} = -\mathbf{i}\,\frac{\mu_o I}{\pi}\cdot\frac{1}{y}$$

29. In an attempt to thaw a frozen copper water pipe, a welding generator is connected to send 120 A through the pipe (a circular cylindrical shell). The copper pipe has an inner radius r_1 and an outer radius r_2. Assume that the current is uniformly distributed in the copper of the pipe. Calculate B for (a) $r < r_1$, inside the pipe; (b) $r_1 < r < r_2$, in the copper; (c) $r_2 < r$, outside the pipe.

30. A coaxial cable (Figure 12) consists of an inner solid conductor $r = a$ and an outer concentric conductor, inner radius $r = b_1$ and outer radius $r = b_2$. For the regions designated, calculate the magnetic field when the inner conductor carries a current I to the right (into the paper) and the outer conductor carries a current I to the left (out of the paper; assume that the current density is independent of position in each conductor): (a) $r < a$ inside the inner conductor; (b) $a < r < b_1$ between the two conductors; (c) $b_1 < r < b_2$ in the interior of the outer conductor; (d) $r < b_2$ beyond the outer conductor.

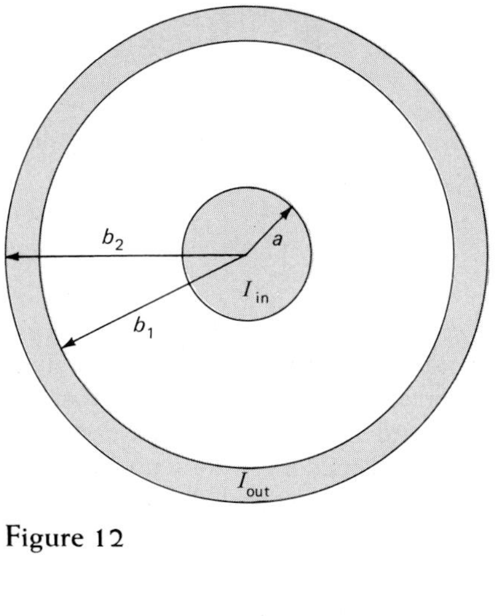

Figure 12

31. Lead, at 4.3 K, is superconducting. At this temperature the superconductivity is effectively destroyed by a magnetic field of 0.052 T or stronger at the surface. Consider a wire of superconducting lead, at 4.3 K, 1.00 mm in radius. How much current can it transmit (uniformly distributed) without destroying its own superconductivity by its own magnetic field?

32. The current density inside a particular cylindrical wire (radius $= R$) is given by $J = (3I/2\pi R^3)r$ where r is the distance from the wire axis and I is the total wire current. (a) Prove that I is the total current by evaluating the integral $\int_0^R J\,dA$ over the wire cross section. (b) Evaluate B inside the wire by using Ampère's law.

33. A current in a long straight wire produces a magnetic field. Show that the radial component, B_r, is zero. (*Hint:* Integrate over the cylindrical surface shown in Figure 33.20. Justify and then use (1) B_z independent of z and (2) B_r constant for fixed r.)

34. Electric charge flows along a long wide sheet of metal. The metal sheet is in the x–y plane. The charge flows in the positive x-direction (Figure 13). The current is perfectly uniform, independent of position in the x–y-plane. Show, by Ampère's law, that the magnetic field is given by

$$\mathbf{B} = \begin{cases} -\mathbf{j}\,\frac{1}{2}\mu_o K & z > 0 \\ +\mathbf{j}\,\frac{1}{2}\mu_o K & z < 0 \end{cases}$$

Here K is the *surface* current density in amperes per meter. Assume that $|z| \ll$ width and length of the current sheet.

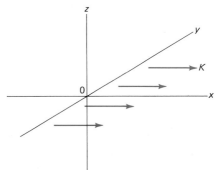

Figure 13

35. From Example 1 the magnetic field produced by a circular current loop (radius a) is

$$B_x(x,0,0) = \frac{\mu_o I}{2}\cdot\frac{a^2}{(a^2 + x^2)^{3/2}}$$

along the axis of the loop. Evaluate $\int_{-\infty}^{\infty} B_x(x,0,0)\,dx$ and interpret your result in terms of Ampère's law. [*Note:* In this situation Ampère's law can serve as a partial check on your result (Example 1). However, it is useless in finding B_x initially.]

Section 33.4 Magnetic Field of a Solenoid

36. At room temperature the superconducting solenoid of Example 5 has a resistance of $10^4\ \Omega$. A 12-V potential difference is applied to the solenoid terminals. Calculate the magnetic field produced in the interior of the solenoid under these conditions.

37. A 12-ft bubble chamber is in a solenoid of niobium–titanium superconducting coils. The interior field of the actual solenoid is roughly equivalent to that of a simple coil of 796 turns per meter, each turn carrying a current of 1800 A. Calculate the magnetic field inside this equivalent solenoid.

38. Two long solenoids are coaxially mounted, one inside the other. The solenoids are wound in the opposite direction with the same number of turns per meter, n, and the same current I. If the radii are R_1 and R_2 with $R_1 < R_2$, determine the magnetic field (a) for $R < R_1$; (b) $R_1 < R < R_2$; (c) $R > R_2$.

39. For a research project, a student needs a solenoid that produces an interior magnetic field of 0.03 T. She decides to use a current of 1.0 A and a wire 0.5 mm in diameter, and to wind the wire as layers onto an insulating form 1.0 cm in diameter and 10.0 cm long. Determine the number of layers of wire needed and the total length of the wire.

40. Picture a long solenoid curved to form a toroid (that is, a doughnut shape). Figure 14 shows a cross section of this toroidal solenoid. The radius of the individual turns, a, is much less than the radius b. Using Ampère's law, integrate around the circle of radius b by following the geometric center of the individual turns. Show that

$$B = \mu_0 I n$$

where n is the number of turns per unit length of the coil.

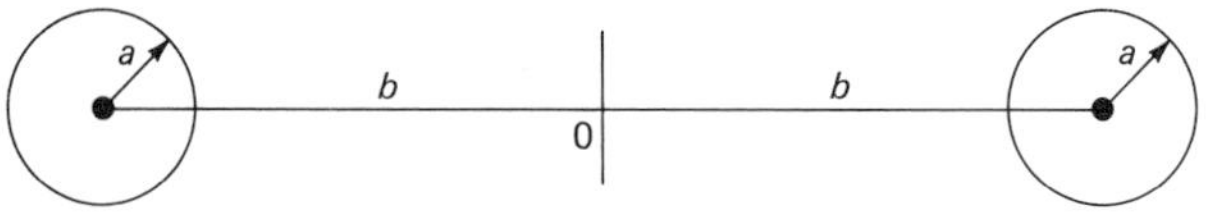

Figure 14

41. Using Ampère's law, show that there is no azimuthal (tangential) field B_ϕ in the interior of a long solenoid. (*Hint:* Try the path of integration shown in Figure 15.)

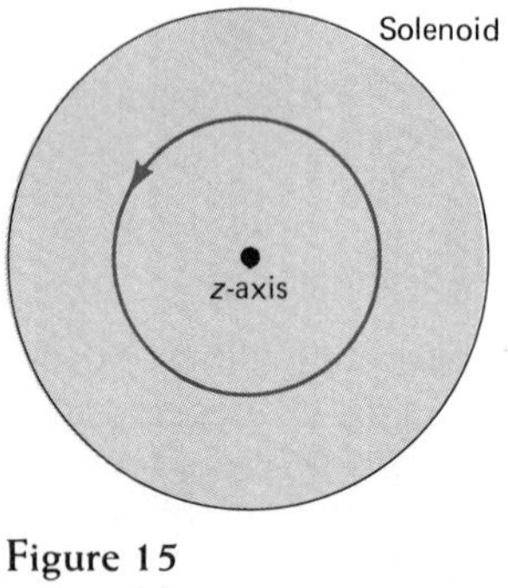

Figure 15

42. Assuming that the magnetic field is uniform in the interior of a solenoid, evaluate the integral $\oint \mathbf{B} \cdot d\mathbf{s}$ around a rectangular path in the interior. Orient one side of the rectangle parallel to the axis of the solenoid.

Section 33.5 Definition of the Ampere

43. From Eq. 33.31, show that μ_0 must have units of N/A^2.

44. Three parallel straight wires lie in a plane and are equally spaced (Figure 16). The center wire is a distance d from the outside wires. The magnitude of the current in each wire is I. Give the direction and magnitude of the force per unit length on the center wire, B, if its current is in the $+x$-direction, and (a) the currents in A and C are both in the $+x$-direction; (b) the current in A is in the $+x$-direction and the current in C is in the $-x$-direction.

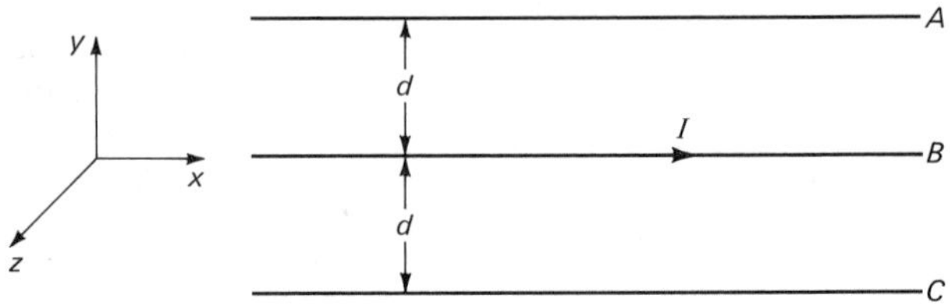

Figure 16

45. A rectangular conducting loop, 0.1 m × 0.8 m, carrying 2.0 A is placed 0.1 m away from a long straight conductor carrying 6.0 A (Figure 17). (a) Calculate the net magnetic force on the rectangular loop. (b) Calculate the net torque on the rectangular loop about an axis through the center of the loop parallel to the longer sides.

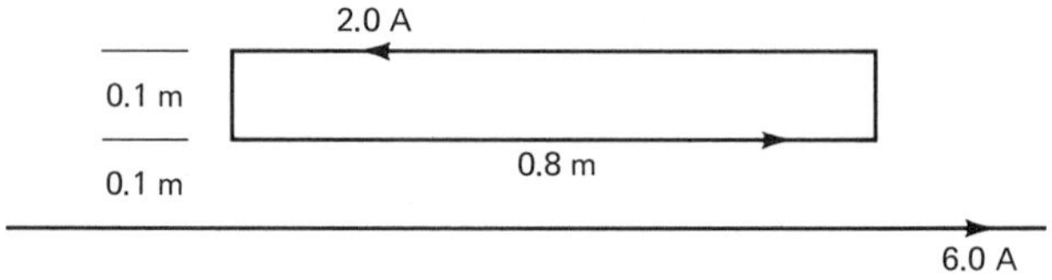

Figure 17

46. A long straight wire is carrying a current I (Figure 18). A small section of the wire is displaced sideways slightly. Will the magnetic forces tend to increase the displacement or to reduce it? Answer by calculating the force on a current element $I\,ds$ displaced parallel to the wire by an amount Δx.

Figure 18

47. The movable conductor of a freshman laboratory current balance has an effective length of 0.8 m. The fixed parallel conductor is a distance 0.01 m away. Both the movable and the fixed conductor carry the same current. If the balance is sensitive to 10^{-3} gm, what is the minimum current that the balance can detect?

48. (a) Two circular loops are parallel, coaxial, and almost in contact, 1 mm apart (Figure 19). Each loop is 10 cm in radius. The top loop carries a current of 14 A clockwise. The bottom loop carries a current of 14 A counterclockwise. Calculate the force that the bottom loop exerts on the top loop. (*Hint:* For a separation distance much smaller than the radius, each small section of the upper loop "sees" a long straight (anti-) parallel current. (b) The upper loop has a mass of 0.021 kg. Calculate its upward acceleration, assuming it is free to move.

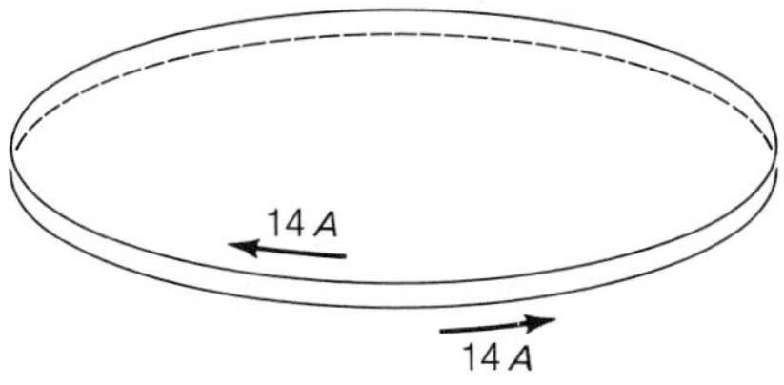

Figure 19

49. To get a feeling for the magnetic forces on the coils of a large solenoid, calculate the magnetic force on a 4-m length of conductor (diameter of the solenoid) that is carrying a current of 1800 A perpendicular to the field of 1.8 T.

chapter 34

ELECTROMAGNETIC INDUCTION

Preview

After Oersted's discovery in 1820 that a steady electric current produced a steady magnetic field, experimentalists eagerly sought to demonstrate the reverse: that a steady magnetic field could create a steady electric current. But all attempts to do so failed. This is not surprising, from the modern point of view of conservation of energy. An electric current is a source of energy—it can run motors or produce heat—and thus if we created an electric current with a steady magnetic field, we would be producing energy without putting energy into the system. We now know that this would violate the law of conservation of energy.

In 1831 Michael Faraday accidentally discovered electromagnetic induction. When Faraday closed a switch to send a current through one circuit, a galvanometer needle in a nearby circuit moved and then returned to its equilibrium position. The motion of the galvanometer needle indicated that an electric current had been induced in the second circuit. It was a mark of Faraday's genius that he recognized the significance of the motion of the galvanometer needle and followed it up. At the time, electric currents were produced by chemical reactions in batteries. Faraday's discovery offered an alternative method of producing electric current.

In this chapter we start with a somewhat simpler physical system than Faraday's, namely, a conductor moving across a uniform magnetic field. The electric charges in the moving conductor experience a magnetic force (Chapter 32). This is the basic physics of the electric generator that provides electricity for our modern society. Generalizing this special result, we find that a changing magnetic field creates an electric field. This is Faraday's law of induction.

The directions of the induced electric field and the induced electric current (if there is a complete circuit) are governed by Lenz's law. Lenz's law is a law of opposition, and essentially a law of conservation of energy.

In the last section of this chapter we apply Faraday's law to a few circuits.

34.1 Motional Electromotive Force

Faraday's* discovery of electromagnetic induction in 1831 can be demonstrated in several different ways. Let's investigate an example of Faraday's law of electromagnetic induction that we can develop and analyze in detail.

Consider a conductor of length l sliding along and making electrical contact with parallel metal rails. The conductor is moving in a direction perpendicular to its length with constant velocity **v**, which is perpendicular to a constant magnetic field **B**, as shown in Figures 34.1 and 34.2. The voltmeter shown in Figure 34.1 indicates that an emf is being generated. What is the origin of this emf? Upon what physical parameters does it depend? In this section we develop the answer to these two questions and relate the emf to a time rate of change of magnetic flux.

*Michael Faraday (1791–1867), an English physicist and chemist, was a genius in experimentation, making significant contributions in electricity, magnetism, optics, and various areas of chemistry.

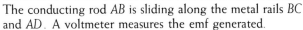

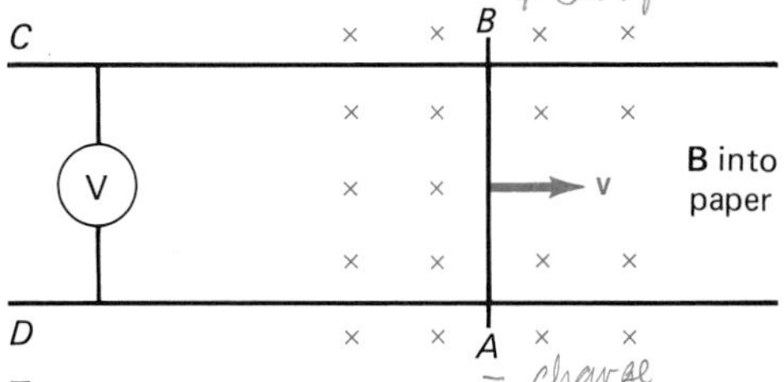

Figure 34.1

The conducting rod AB is sliding along the metal rails BC and AD. A voltmeter measures the emf generated.

Figure 34.2

A length of conductor, AB, is moving across a magnetic field.

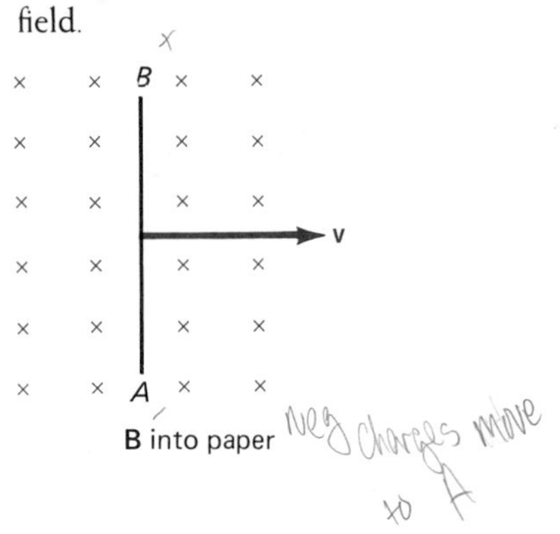
B
v
A
B into paper
Neg charges move to A

For purposes of interpretation the frame of reference must be specified carefully. We and the magnetic field **B** are stationary in the laboratory, our reference frame. The conductor is moving relative to our laboratory reference. The charges in the moving conductor move relative to the field **B** and therefore experience a force

$$\mathbf{F} = q\mathbf{v} \times \mathbf{B} \tag{34.1}$$

This is the same equation that we used to define **B** in Section 32.2. The positive charges and most of the negative charges are bound in the solid structure. However, some of the electrons are free to flow through the conductor. In response to the force given by Eq. 34.1, these electrons by virtue of their negative charge move to the lower part of the conductor, point A in Figure 34.3a. The vector relations are shown in Figure 34.3b. The result of this movement is that a net negative charge accumulates at A (and in the rail AD in Figure 34.1). A net positive charge accumulates at B (and on the rail BC in Figure 34.1). This separation of charge opposes the continued flow of electrons. The electron concentration at A increases until equilibrium is achieved. At equilibrium, the magnetic force is balanced by the electrostatic force set up by the charge separation. This magnetically induced separation of charge may be called a *motional emf*, analogous to the chemical emf of the flashlight dry cell or the automobile battery.

We introduce the vector **l** directed from A to B to describe the length of the moving conductor. In the case we are considering, **v**, **B**, and **l** are mutually perpendicular. We have

$$F = qvB \tag{34.2}$$

Recalling that emf, $\mathscr{E}$, is work per unit charge (Section 31.1), to transfer charge from point A to point B we have

$$\mathscr{E} = \frac{Fl}{q} = Bvl \tag{34.3}$$

for the induced emf of this moving conductor. Note that $\mathscr{E} = 0$ when $v = 0$. This motional emf is not present in a static system. Additionally, the emf depends on the size of the system ($\mathscr{E} \propto l$) and on the strength of the field ($\mathscr{E} \propto B$). These features are generally true of motional emf's.

The moving conductor gives rise to an emf, but no current is present in the conductor after the initial redistribution of charge is complete. The charge has no place to go. Now let's complete the circuit. Imagine that the conducting rails are joined at the left edge (CD) by a resistance R to form a complete circuit (Figure 34.4). The electrons accumulating on the rail AD are free to move through the circuit. Positive charge would flow in the direction opposite to the electron flow, as indicated in Figure 34.4. In terms of potential differences, the induced emf creates a potential difference along the moving rod, $V_B > V_A$. This potential difference is fully equivalent to the potential difference created by the chemical reactions in a battery; but whereas a battery is a *chemical* device for generating electric current, the system shown in Figure 34.4 is a *mechanical* device for generating current (as long as the rod AB is moving through the magnetic field **B**).

You will note that a resistor has been included in the circuit. For purposes of analysis in this chapter we are holding the induced current I to a small value so that we can neglect the magnetic effects of this induced current. These magnetic effects (self-induction) are considered in Chapter 35.

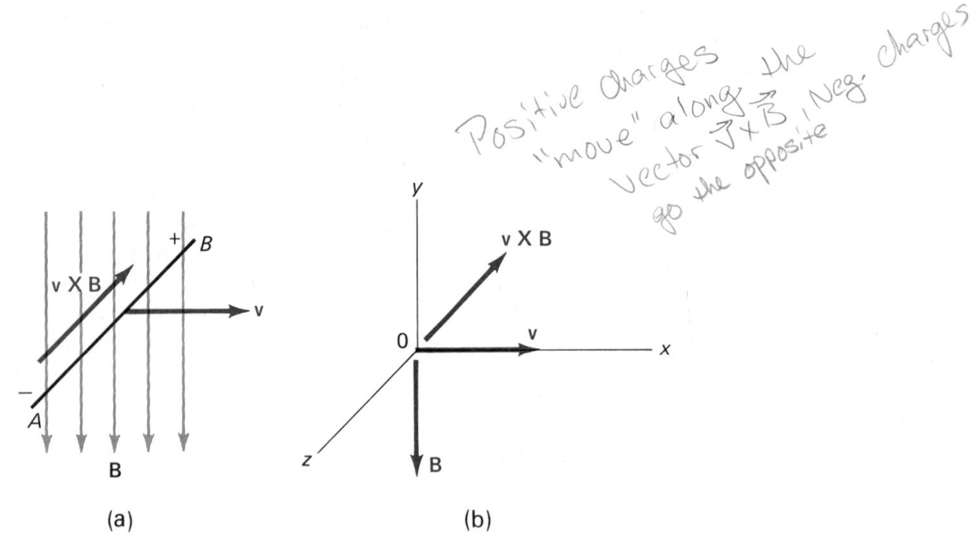
Positive charges "move" along the vector v×B, Neg. charges go the opposite
v X B
+
B
v
−
A
B
(a)
y
v X B
0
v
x
z
B
(b)

Figure 34.3

(a) The magnetic force on charges is proportional to $\mathbf{v} \times \mathbf{B}$. (b) For $\mathbf{v} = \mathbf{i}v$ and $\mathbf{B} = -\mathbf{j}B$, $\mathbf{v} \times \mathbf{B} = -\mathbf{k}bB$.

Figure 34.4

The moving conductor AB makes a sliding electrical contact at A and B. We have a complete electric circuit, $ABCDA$.

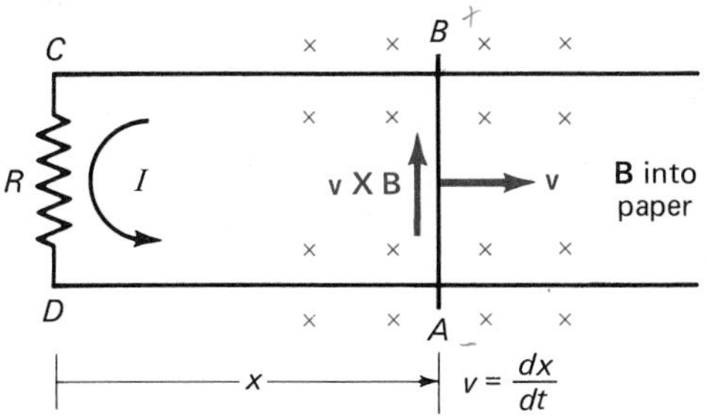

Cutting Magnetic Field Lines

Let's now give ourselves a picture of electromagnetic induction in terms of Faraday's magnetic field lines, and at the same time relate our mechanically generated emf to the time rate of change of magnetic flux. Consider the conductor AB of Figure 34.4 to be moving to the right. As it moves it passes across, or "cuts," the magnetic field lines. The number of field lines passing through a given area is the magnetic flux, Φ_B. For the area bounded by our circuit $ABCD$, the rate at which the magnetic flux changes (which is the rate at which the moving conductor cuts the field lines) is

$$\frac{d\Phi_B}{dt} = \frac{d}{dt}(BA) = \frac{d}{dt}(Bxl)$$

$$= B\frac{dx}{dt}l$$

$$= Bvl \qquad (34.4)$$

But as we see from Eq. 34.3, this result is equal to the motional emf $\mathscr{E}$. Therefore, we can picture the emf, or induced potential difference, as *the time rate of cutting the magnetic field lines.* (We return to this point in Section 34.2, where we express Faraday's law in terms of Φ_B.)

Example 1 A Moving Conductor in a Magnetic Field
The conductor in Figure 34.4 is 0.1 m long, and moves with a speed of 10.0 m/s across a magnetic field of 1.5 T. Let's calculate the emf and the force per coulomb of charge.

From Eq. 34.3

$$\mathscr{E} = Bvl$$

$$= 1.5 \times 10.0 \times 0.1 = 1.5 \text{ V}$$

This is the emf of the moving conductor. The force per coulomb of charge in this linear conductor is

$$\frac{F}{q} = Bv$$

$$= 1.5 \times 10.0 = 15 \text{ T} \cdot \text{m/s}$$

The direction of the magnetic force $\mathbf{F}$ in $\mathbf{F} = q\mathbf{v} \times \mathbf{B}$ is along the conductor (upward).

Note that F/q may also be written $\mathscr{E}/l$. In this form we get units of volts per meter.

34.2 Faraday's Law of Induction

A Different Observer, an Induced Electric Field

In the preceding section we saw that the moving conductor of Figure 34.4 cut magnetic field lines at the rate of Bvl. This is the rate of change of magnetic flux. We have

$$\mathscr{E} = Bvl = \frac{d\Phi_B}{dt} \qquad (34.5)$$

Faraday's law, which we are about to develop, is usually written with a line integral of an induced electric field, $\oint \mathbf{E} \cdot d\mathbf{s}$. So far, we do not have an induced electric field, but Example 1 provides a clue. There we mentioned that F/q can be obtained in volts per meter. All of Section 34.1 is developed relative to a laboratory frame of reference. Although a stationary observer in the lab sees a *magnetic* force $\mathbf{F} = q\mathbf{v} \times \mathbf{B}$ on the charges in the moving rod, this magnetic force is equivalent to an electric force exerted by an electric field.

$$\mathbf{E} = \mathbf{v} \times \mathbf{B} \qquad (34.6)$$

Another way to look at this induced field $\mathbf{E}$ is to consider the reference frame in which the rod AB is stationary, as shown in Figure 34.5. An observer sitting on the rod sees the same emf and the same current that we see from our laboratory perspective. He must, because special relativity insists that there be no preferred reference frame. However, the observer moving with the rod AB has a different interpretation. He measures no magnetic field; the velocity of his conductor is zero. He does see electrons accelerated from B to A and, by the definition of an electric field (Section 27.1), he ascribes the force on the charges to an electric field. He sees the electric field $\mathbf{E}$ on the left-hand side of Eq. 34.6. Observers stationary in the laboratory see the $\mathbf{v} \times \mathbf{B}$ on the right-hand side.

For the observer sitting on the rod AB

$$\mathscr{E} = \int_A^B \mathbf{E} \cdot d\mathbf{s} = El \qquad (34.7)$$

There is no emf or induced field in any other portion of our circuit. Then, remembering that we are talking about the induced field, we can write

$$\mathscr{E} = \oint \mathbf{E} \cdot d\mathbf{s} \qquad (34.8)$$

Figure 34.5

This is Figure 34.4 from the viewpoint of an observer sitting on the rod AB. The rod AB is stationary. The rest of the circuit, $BCDA$, and the magnetic field **B** are moving to the left with velocity **v**.

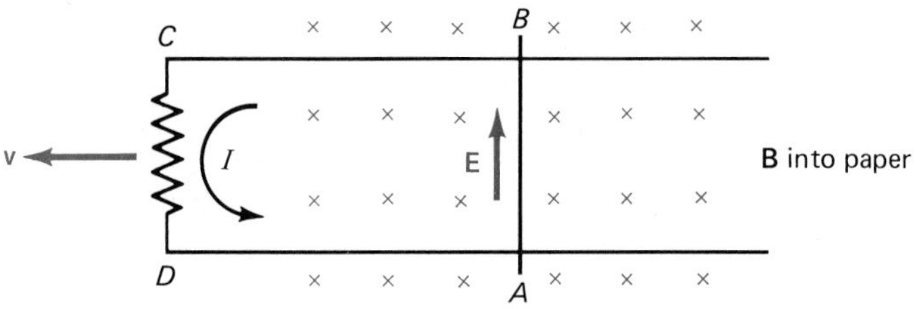

The induced emf, $\mathscr{E}$, is expressed as a line integral of the field E around the entire circuit.

Faraday's Law

Equation 34.3, $\mathscr{E} = Bvl$, allows us to relate $d\Phi_B/dt$ to $\oint \mathbf{E} \cdot d\mathbf{s}$. But before we write this equation, let us consider the sign. First let's look again at the sliding conductor in Figure 34.4. The external magnetic field **B** is directed into the paper. The charge (assumed positive) flows around the conducting loop as shown. Taking the loop $ABCDA$ to be in the direction of E and the current I, we see that E and $d\mathbf{s}$ are parallel, making the line integral $\oint \mathbf{E} \cdot d\mathbf{s}$ positive. But is the magnetic flux Φ_B positive or negative? To find the answer, consider Figure 34.6 and 34.7. The magnetic flux consists of the field lines passing through the surface defined by the current loop $ABCDA$, and our analysis so far has ignored positive and negative directions of this surface. The positive direction of the surface outlined by $ABCDA$ is up, out from the paper. This makes the magnetic flux *negative*. We have

$$\Phi_B = \int_{\text{surface}} \mathbf{B} \cdot \hat{\mathbf{n}}\, dA \qquad (34.9)$$

with **B** and $\hat{\mathbf{n}}$ in opposite directions, $\hat{\mathbf{n}}$ up and **B** down. So $\mathbf{B} \cdot \hat{\mathbf{n}} = -B$. The motion of the conductor means that Φ_B is increasing in magnitude, becoming more *negative*. The time derivative of the flux, $d\Phi_B/dt$, is negative

$$\frac{d\Phi_B}{dt} = -Bvl \qquad (34.10)$$

We can now write the equation relating $d\Phi_B/dt$ to $\oint \mathbf{E} \cdot d\mathbf{s}$.

$$\oint \mathbf{E} \cdot d\mathbf{s} = -\frac{d\Phi_B}{dt} \qquad (34.11)$$

This is Faraday's law of electromagnetic induction. In words:

The emf induced in a circuit is equal to the negative of the total time rate of change of the magnetic flux of the circuit.

The crucial *negative sign* in Eq. 34.11 is interpreted as Lenz's law in the next section. The particular mathematical form was developed long after Faraday's work, but the law is named after Faraday because he made the key discovery.

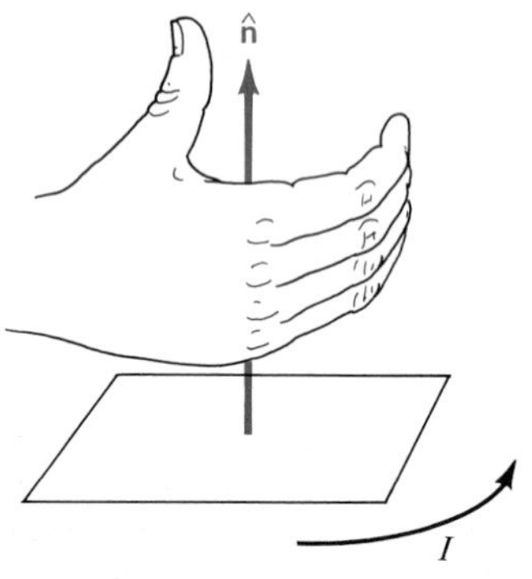

Figure 34.6

The positive normal for this rectangular area is given by the right-hand rule.

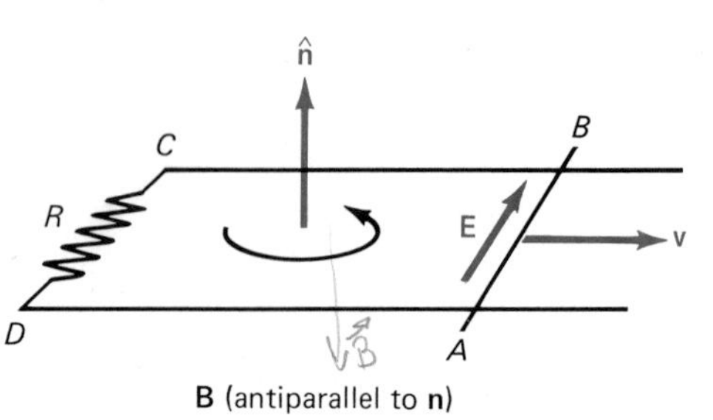

Figure 34.7

Directional relations. The magnetic field **B** and the unit normal vector $\hat{\mathbf{n}}$ are oppositely directed, antiparallel.

Generalization

Faraday's law is far more general than our derivation suggests. First, the field **E** of our derivation is not produced by a static distribution of positive or negative charge. In the case considered, **E** was created by the motion of a conductor through a magnetic field. Second, the field **E** of Eq. 34.7 can exist in entirely empty space, and in empty space the field lines of **E** form closed loops (Figure 34.8). There are no electric charges on which the electric field lines can begin or end. Finally, since the integral $\oint \mathbf{E} \cdot d\mathbf{s}$ does not vanish in a closed loop, this induced field is *not* a conservative field (Section 28.3).

So far, we have approached Faraday's law by analyzing (1) *a moving conductor and a stationary field* and (2) *a stationary conductor and a moving field*. Actually, Eq. 34.11 does not require relative motion of the conductor and field. The time rate of change of magnetic flux, $d\Phi_B/t$, may

Figure 34.8

The electric field induced in a loop by a changing magnetic field.

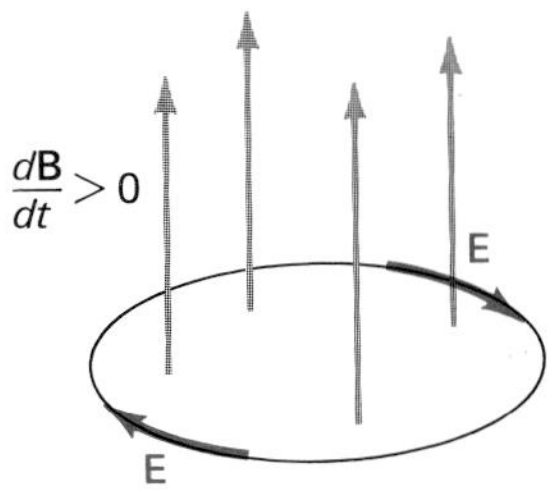

come from a field that is stationary relative to the conductor but that changes in intensity with time. There may be (3) *no relative motion, but a time-dependent field.*

This last case is what Faraday initially observed. An increasing current through one coil creates an increasing magnetic field through an adjacent coil. This induces an emf and thus a current in the second coil. (This is the basic principle of the transformer,* and we return to it in Chapter 37.)

From the third case mentioned—the time-dependent magnetic field—we have another form of Faraday's law:

A changing magnetic field produces an electric field.

This formulation of Faraday's law is entirely in terms of fields. There is no reference to electric charges or magnetic dipoles. In Chapter 38 we will see this form of Faraday's law become one of Maxwell's equations.

The word *changing* in the restatement of Faraday's law is particularly significant. Faraday's discovery shifted the attention of physicists from electrostatics to electrodynamics, and time dependence thus assumed a major role.

Example 2 **Electromagnetic Induction by a Time-Dependent Field**

Consider a long solenoid with a cross-sectional area of 8 cm^2 (Figures 34.9 and 34.10). A time-dependent current in the wire winding creates a time-dependent magnetic field $B(t) = B_0 \sin 2\pi\nu t$ T. Here B_0 is constant, and equal to 1.2 T. The quantity ν is the frequency of the magnetic field. With $\nu = 60$ Hz and the ring resistance $R = 1.0\ \Omega$, we want to calculate the emf and the current (I)

*Transformers are widely used as devices to change the size of time-varying voltage. For instance, an electric power company will use a transformer to reduce the high voltage of the power lines down to 240 and 120 V for household use.

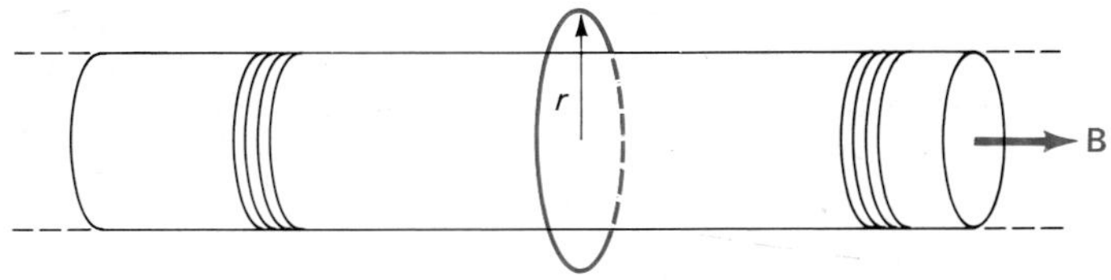

Figure 34.9

A long solenoid and a concentric ring outside the solenoid.

Figure 34.10

Solenoid and concentric ring: a cross-sectional view.

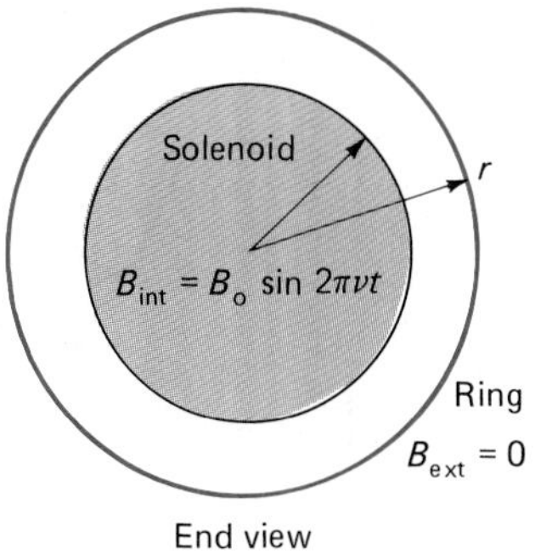

induced in a ring of radius r concentric with the axis of the solenoid.

The magnetic flux is equal to

$$\Phi_B = B_0 \sin 2\pi\nu t \cdot A$$

and so

$$\mathscr{E} = -\frac{d\Phi_B}{dt} = 2\pi\nu\, A\, B_0 \cos 2\pi\nu t$$

$$= -2\pi \cdot 60\ s^{-1} \cdot 8 \times 10^{-4}\ m^2 \cdot 1.2\ T \cdot \cos 2\pi\nu t$$

$$= -0.362 \cos 2\pi\nu t\ V$$

The current in the ring is $I = \mathscr{E}/R$. Therefore

$$I = \frac{-0.362 \cos 2\pi\nu t\ V}{1.0\ \Omega}$$

$$= -0.362 \cos 2\pi\nu t\ A$$

The induced current oscillates with frequency ν and an amplitude of 0.362 A.

Questions

1. Imagine that you are in a region of space filled with a uniform magnetic field in the positive z-direction. Do you detect an electric field, and if so, what is its direction, if you and some electric charges are moving with constant speed in (a) the positive z-direction; (b) the positive x-direction; (c) the positive y-direction? (Note carefully that you are no longer at rest in the laboratory. You are moving with the electric charges through the magnetic field.)

Figure 34.11

A bar magnet moving away from a metal ring.

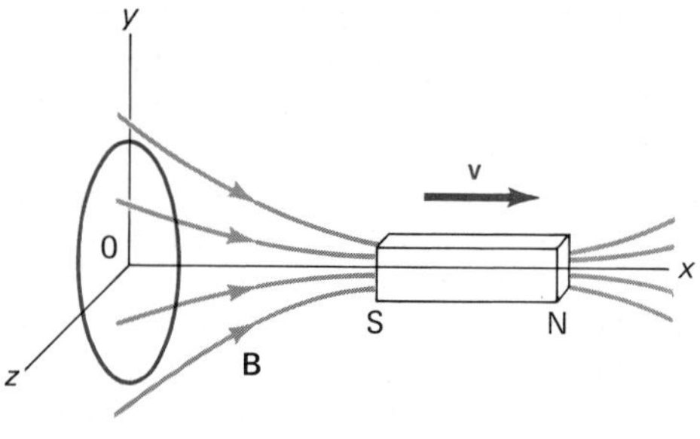

2. A wire loop is displaced (translation, no rotation) but stays entirely within a uniform magnetic field. What emf, if any, is induced in the loop?

3. A bar magnet moves away from a metal ring as shown in Figure 34.11. In what direction will current be induced in the metal ring?

4. A positively charged particle passes through a circular conducting loop along the rotational symmetry axis of the loop. Is an emf induced in the loop? What is the direction of the induced current (if any)?

5. The two lead wires attached to a loudspeaker are held close together in a plastic coating. Does being close together rather than separated make the leads more vulnerable or less vulnerable to magnetic induction effects (stray pickup)?

6. A bar magnet moves inside a long solenoid along the solenoid axis. What emf is induced in the solenoid?

34.3 Lenz's Law

Let's look again at Faraday's law, Eq. 34.11. Why is the minus sign there? What are its implications?

To answer these questions we will consider two physical situations. First, let's take another look at the moving rod of the preceding two sections. As the rod moves across the magnetic field, an emf $\mathscr{E}$ is created in the rod. As long as the rod is isolated so that there is no current, the motion of the rod across the magnetic field is unrestrained. There is no opposing force. However, when the rod slides on conducting rails and we have a current, an opposing force appears. From Section 32.4 the magnetic force on the current-carrying rod is

$$\mathbf{F} = I\mathbf{l} \times \mathbf{B} \tag{34.12}$$

Vector $\mathbf{l}$ has a magnitude equal to the length of the rod and has the direction of the current I. By the right-hand rule for the direction of a cross product, $\mathbf{F}$ is in the negative v-direction (Figures 34.12 and 34.13), opposing the motion of the rod. The interaction of the external magnetic

Figure 34.12

Directional relations. The current in the moving conductor results in a retarding force, $\mathbf{F}$.

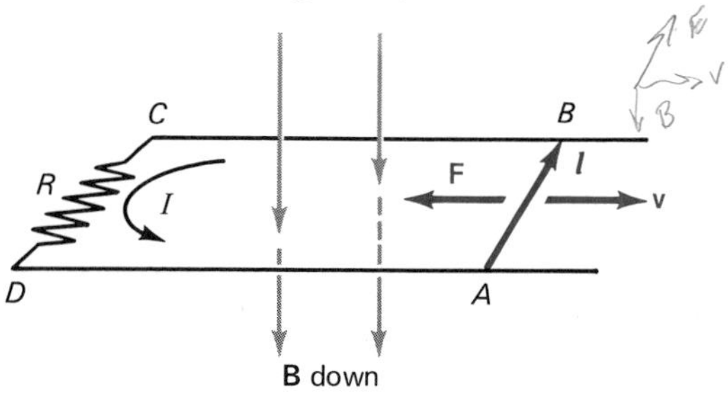

Figure 34.13

Relative to the usual xyz (Cartesian) coordinates, $\mathbf{l} = -\mathbf{i}l$, $\mathbf{B} = -\mathbf{k}B$. Then $\mathbf{l} \times \mathbf{B} = (-\mathbf{i}) \times (-\mathbf{k})lB = -\mathbf{j}lB$, in the negative y-direction.

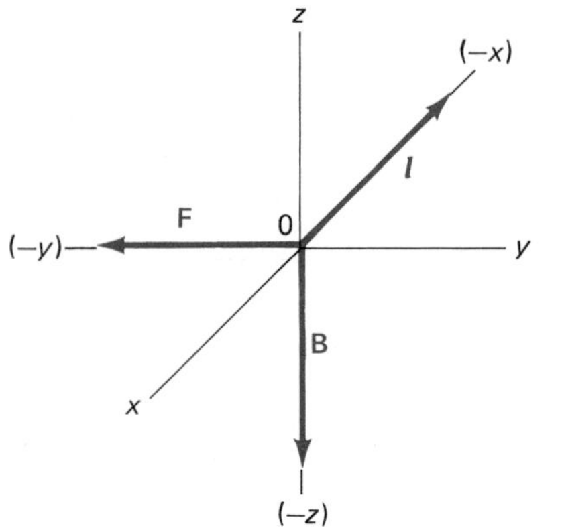

field and the current leads to a magnetic force on the rod opposing the motion that is inducing the current.

As a second physical example, consider a bar magnet approaching a conducting ring (Figure 34.14a). To apply Faraday's law to this system we first choose a positive direction around the ring. Let's take the direction from z to x as positive. (Either choice is legitimate, as long as we are consistent.) For our choice the positive normal for the area of the ring is in the y-direction and the magnetic flux is negative. As the distance between the conducting ring and the N pole of the bar magnet decreases, more and more field lines go through the ring, making the flux more and more negative. Thus $d\Phi_B/dt$ is negative. By Faraday's law $\oint \mathbf{E} \cdot d\mathbf{s}$ is positive relative to our chosen direction. The current I is directed as shown.

The current induced in the ring creates an induced secondary magnetic field that is opposite to the original field inside the ring. This induced magnetic field is similar in form to the field of a second bar magnet, shown dotted in Figure 34.14b. This induced magnet repels the incoming magnet. This opposition is a consequence of the minus sign in Faraday's law, and is formalized as *Lenz's law:**
When a current is induced in a conductor, the direction of the current will be such that the current's magnetic effects oppose the change that induced it.

*Heinrich Friedrich Emil Lenz (1804–1865) was a German physicist. In 1834 he proposed that an induced current produces effects opposing the forces that induced it.

Figure 34.14a

A bar magnet approaching a metal ring.

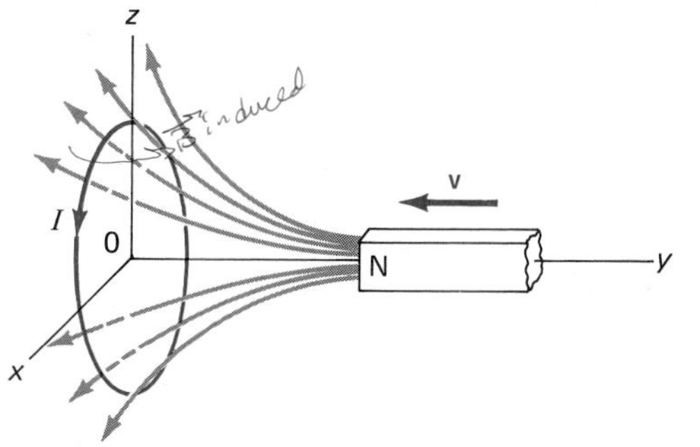

Figure 34.14b

The magnetic field of the induced current opposes the approaching bar magnet.

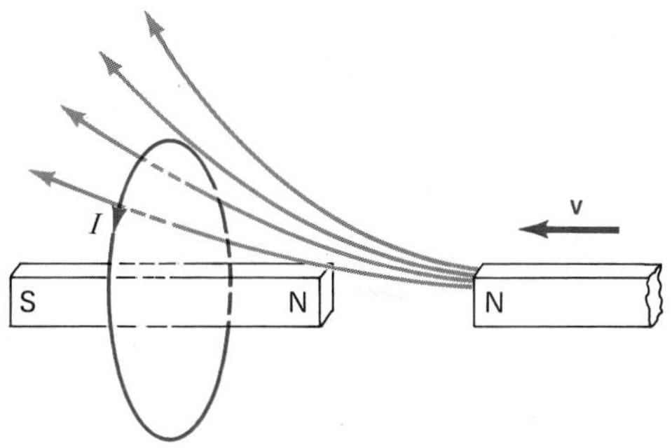

The key word in this statement is *oppose*. Lenz's law is one of opposition; it is a declaration that we are not going to get something for nothing. When the bar magnet is pushed toward the ring, the current induced in the ring creates a magnetic field that opposes the change in flux. The magnetic field produced by the induced current repels the incoming magnet. If we push the magnet toward the ring, we must do work on the magnet to move it inward. The work we do shows up as electrical energy in the ring. Lenz's law thus becomes a law of conservation of energy.

Suppose that Faraday's law did not contain that minus sign, and that the ring current was in the reverse direction. Then the secondary magnetic field produced by the ring current would *reinforce* the original field, and $d\Phi_B/dt$ would be larger, the induced emf would be larger, and the induced current would rise. The equivalent induced magnet would be reversed, and the incoming magnet would therefore be attracted. In short, the magnet would do mechanical work, and we would get electric current in the conducting ring. We'd have an energy generator that violated the law of conservation of energy. However, the world doesn't work this way. Faraday's law does have a minus sign. Lenz's law says, in effect, that if we want to get electric energy out of this sort of an inductive system, we'll have to do work. We will give the quantitative form of Lenz's law in the next section.

Magnetic Field Lines Picture

The concept of opposition, which is central to Lenz's law, may be pictured very nicely by using Faraday's field lines. For example, in Figure 34.15a we see an external field **B** directed vertically down, and a conducting rod moving to the right with velocity **v**. If the circuit is complete, the current I is directed into the paper, creating a secondary magnetic field encircling the induced current (by the right-hand rule). You may recall that this is the same situation as that in Figure 34.12. The two magnetic fields combine as vectors to give the resultant field sketched in Figure 34.15b. Here again we have invoked the principle of superposition. The magnetic field in front of the moving conductor is strengthened, and the field behind the conductor is weakened. If we think of the field lines as being under tension and exhibiting mutual repulsion, then we can picture the field as exerting a net retarding force on the moving conductor. The field opposes the change—as Lenz's law predicts.

In mathematical terms, the retarding force on the moving conductor involves two cross products. First, the force on a moving charge (in the conductor) is a cross product. From Eq. 34.1

$$\mathbf{F}_1 = q\mathbf{v} \times \mathbf{B}$$

This force $\mathbf{F}_1$, shown in Figure 34.16a, gives rise to the induced emf and a current I (if there is a complete circuit and the charge can flow). Then the external field **B** exerts a force on the induced current I. From Eq. 34.12 the force on the moving conductor is given by

$$d\mathbf{F}_2 = I\,d\mathbf{s} \times \mathbf{B}$$

In our example (Figure 34.15), with **v** perpendicular to **B** and the conductor (and $d\mathbf{s}$) perpendicular to both **v** and **B**, the force $d\mathbf{F}_2$ is in the opposite direction from **v** (Figure 34.16b). Force $d\mathbf{F}_2$ is the retarding force on the moving conductor.

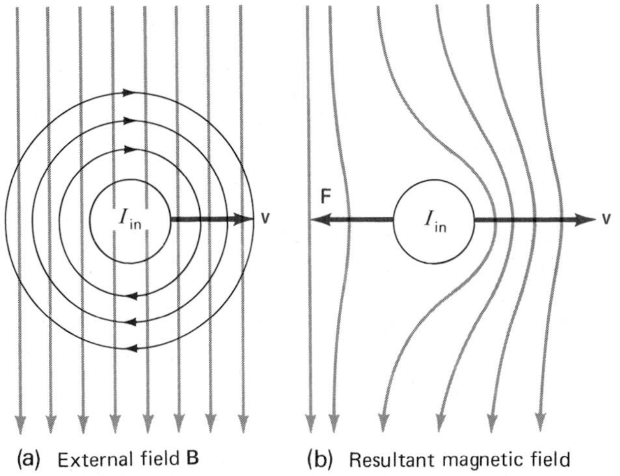

Figure 34.15

(a) Two magnetic fields: the uniform external field (vertical here); the circular field of the current, I. (b) Superposition of the external magnetic field and the field of the current.

Figure 34.16

(a) Vector relations for $\mathbf{v} \times \mathbf{B}$. (b) Vector relations for $d\mathbf{s} \times \mathbf{B}$.

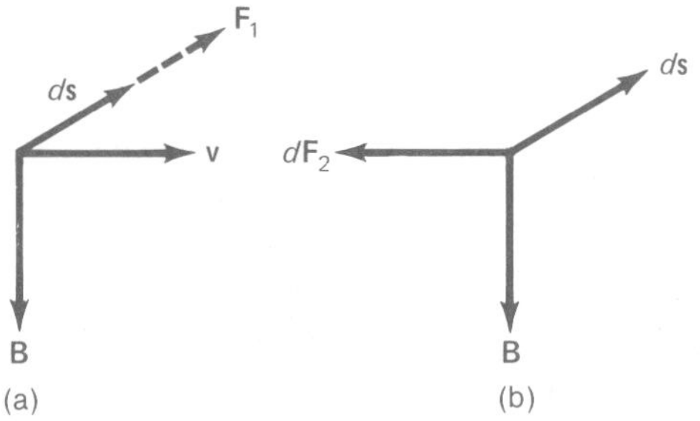

Example 3 Lenz's law for a Rotating Loop

Consider a rectangular loop of wire rotating in a magnetic field. A cross-sectional view of the wire is shown in Figure 34.17. The upper and lower conductors and the axis of rotation run perpendicular to the page. The magnetic field $\mathbf{B}$ is in the page, directed to the right. The loop is completed by two conductors parallel to the page. With the loop rotating counterclockwise as shown, the problem is to predict the direction of the current in the upper conductor. There are *four* ways to solve this problem:

1 Use Faraday's law. Choose a positive normal $\hat{\mathbf{n}}$ for the loop. Either choice is acceptable; just be consistent with your choice. For the vector $\hat{\mathbf{n}}$ shown in Figure 34.17 the magnetic flux $\Phi_B = \mathbf{B} \cdot \hat{\mathbf{n}}\, A$ is positive but decreasing; $d\Phi/dt$ is negative. From Faraday's law, Eq. 34.11, $\mathscr{E}$ is positive *relative* to the perimeter corresponding to the choice of $\hat{\mathbf{n}}$. By the right-hand rule, this means that $\mathscr{E}$, and therefore I, are both positive (out of the paper for the upper conductor).

2 Use magnetic moment. The current in the rectangular loop will create a magnetic moment $I\,A$, which is equivalent to a bar magnet. From Lenz's law this equivalent bar magnet should experience a restoring torque $\boldsymbol{\tau} = \boldsymbol{\mu} \times \mathbf{B}$. In the case shown in Figure 34.17 this means that the upper right end must be an N pole, the lower left end an S pole. The magnetic moment $\boldsymbol{\mu}$ points from S to N. Then, by the right-hand rule (with your thumb parallel to $\boldsymbol{\mu}$), the current in the upper conductor is directed out of the paper.

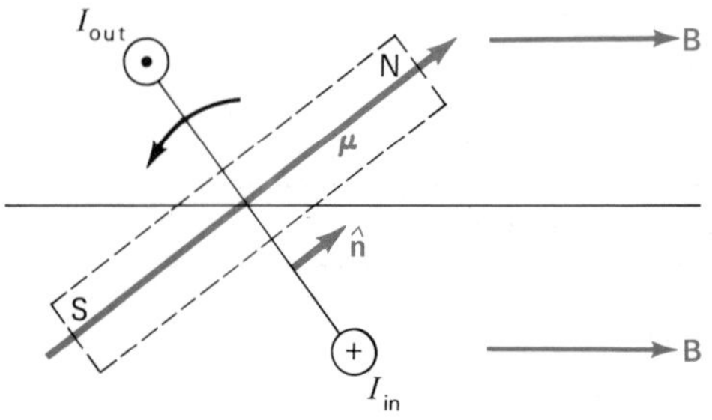

Figure 34.17

The magnetic moment and equivalent magnetic dipole resulting from the induced current.

Figure 34.18

Two magnetic fields: reinforcement in front of the moving wire, cancellation behind.

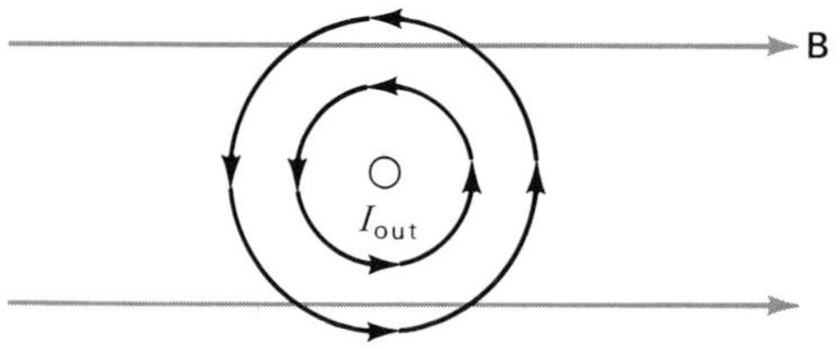

Figure 34.19

Vector relations for $\mathbf{v} \times \mathbf{B}$.

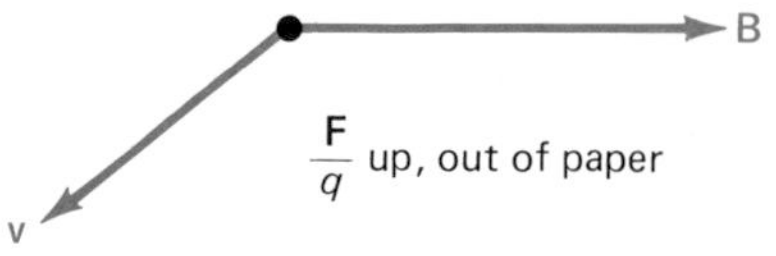

3 Use field lines. To get the opposition that Lenz's law demands, the current in the upper conductor (the conductor is moving down) must create a field that reinforces the external field below the conductor (Figure 34.18). By the right-hand rule (for a long straight wire), the current in the upper conductor is directed out of the paper.

4 Use the definition of magnetic field. Direct application of Eq. 34.1, $\mathbf{F} = q\mathbf{v} \times \mathbf{B}$, yields a force on the electrons in the conductor, and therefore yields a current I out of the paper for the upper conductor (Figure 34.19).

All four methods give the same answer. Therefore, when you are faced with a similar problem, use whichever method you find most convenient.

Questions

7. A bar magnet (Figure 34.20) moves vertically down, N pole first, approaching a circular conducting loop in the x–y plane. What is the direction of the induced current in the loop?

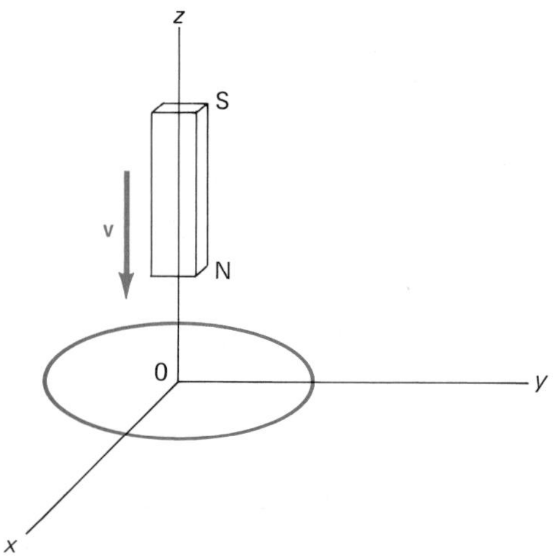

Figure 34.20

Figure 34.21

Parallel coaxial conducting loops.

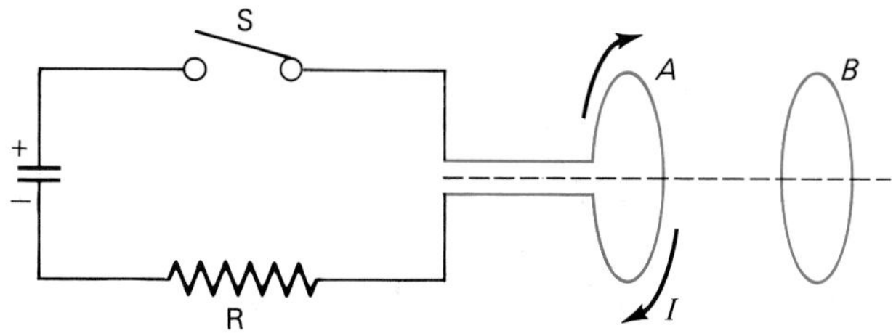

8. Loops A and B are two parallel, coaxial conducting loops perpendicular to the plane of the paper (Figure 34.21). When switch S is closed, the capacitor discharges through loop A. Will current be induced in loop B? In which direction?

34.4 Applications of Faraday's Law

Flip-Coil Measurement of Magnetic Field

An old but still useful way of measuring a magnetic field is to rapidly rotate a coil of wire (Figure 34.22) and measure the electric charge that is driven through a circuit connected to the coil. Consider a coil with area A whose plane is perpendicular to a uniform magnetic field **B**. The flux through the coil is

$$\Phi_B = NBA \tag{34.13}$$

The flux through one turn, BA, is multiplied by N, the number of turns of wire in the coil. If the coil is flipped over (rotated 180°), the change in flux is $\Delta\Phi_B = -2NBA$ in the time interval Δt. The magnitude of the induced current is

$$I = \frac{\mathscr{E}}{R} = -\frac{1}{R} \cdot \frac{d\Phi_B}{dt} \tag{34.14}$$

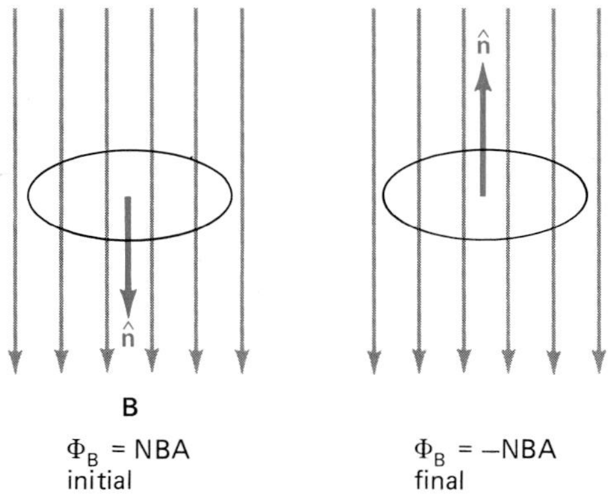

Figure 34.22

Positive and negative magnetic flux in a flip coil.

Figure 34.23

The galvanometer records the charge Q passing through it when the coil is rotated. Then the magnetic field **B** is calculated by Eq. 34.16.

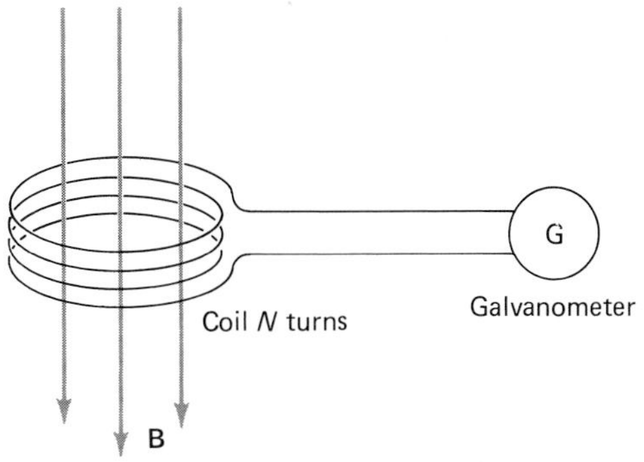

where R is the total resistance of the circuit, which consists of the coil, leads, and a galvanometer. We integrate Eq. 34.14 over the time interval Δt during which charge flows. The integral equals Q, the total charge passing through the galvanometer.

This last integral yields $\Delta\Phi_B$, so

$$Q = -\frac{1}{R}\,\Delta\Phi_B$$

or

$$Q = \frac{2NBA}{R} \tag{34.15}$$

Presumably N, A, and R are known or can be measured. Then a measurement of the charge Q yields the value of B from the equation (Figure 34.23).

$$B = \frac{RQ}{2NA} \tag{34.16}$$

A Rectangular Loop Generator

A rectangular loop of wire of length l and height h is rotated about an axis perpendicular to an external field **B**, as shown in Figure 34.24. The angular velocity is a constant ω. The magnetic flux through the loop is

$$\Phi_B = \mathbf{B} \cdot \hat{\mathbf{n}}\, A = Blh \cos \omega t \tag{34.17}$$

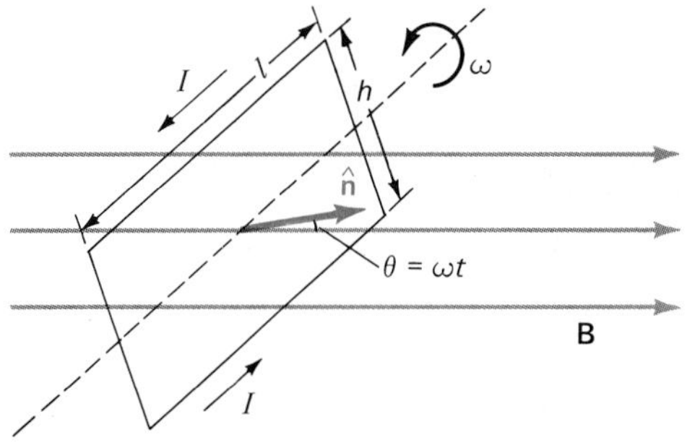

Figure 34.24

A rotating rectangular loop. The axis of rotation is perpendicular to the magnetic field.

Figure 34.25

A rotating loop. The time dependence of the magnetic flux and its time derivative.

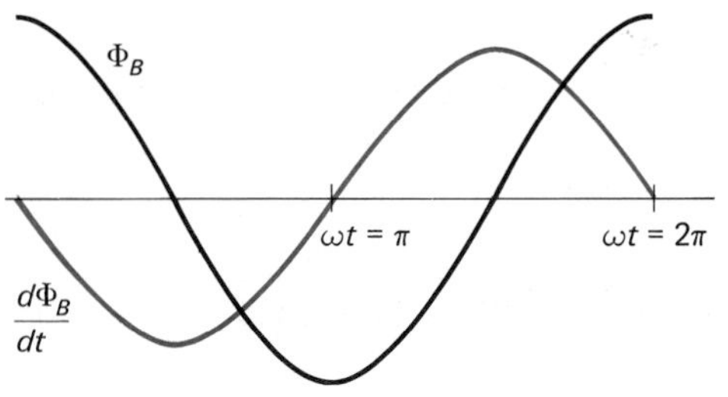

Figure 34.26

A rotating rectangular loop: cross-sectional view.

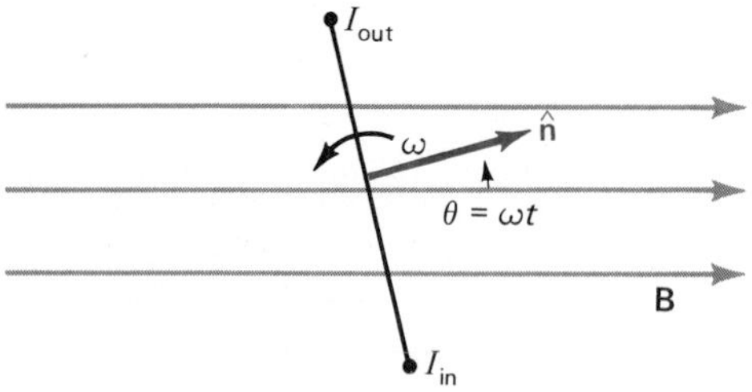

for the case with the loop vertical ($\hat{n}$ horizontal) at $t = 0$. Then

$$\frac{d\Phi_B}{dt} = -\omega Blh \sin \omega t = -\mathscr{E} \quad (34.18)$$

by Faraday's law. From Eq. 34.17 the magnetic flux Φ_B is proportional to cos ωt. The cosine comes from the scalar product and is a function of time because $\hat{n}$ is rotating. Then, by differentiation, $d\Phi_B/dt$ is proportional to $-\sin \omega t$. The time relations are shown in Figure 34.25.

The minus sign in Eq. 34.18 reminds us that the charge will flow (if the circuit is complete) in such a direction as to *oppose the change that induced it*. In this case the induced current will produce a magnetic field to oppose the decrease in the flux ($d\Phi_B/dt$ is negative). The induced magnetic moment of the loop will be in the positive $\hat{n}$-direction. By the right-hand rule, the charge flows out along the top conductor and in along the bottom conductor, as indicated by I_{out} and I_{in} in Figure 34.26. (See Example 3 in Section 34.3.) This holds for $0 < \theta \le \pi/2$, the situation illustrated in Figure 34.26. For $-\pi/2 \le \theta < 0$ the current is reversed, directed in along the top conductor and out along the bottom conductor.

Let us compare the rate at which electrical energy is dissipated in the loop with the rate at which we must do work to turn the loop. If the loop has a total resistance R, the current is

$$I = \frac{\mathscr{E}}{R} = \frac{\omega Blh}{R} \sin \omega t \text{ A} \quad (34.19)$$

The rate of dissipation of electrical energy as heat becomes

$$P_{elec} = I^2 R = \frac{\omega^2 B^2 l^2 h^2}{R} \sin^2 \omega t \text{ W} \quad (34.20)$$

From Eq. 7.17 (in Section 7.5) the power (rate of doing mechanical work) is

$$P_{mech} = \frac{dW}{dt} = \mathbf{F} \cdot \mathbf{v} \quad (34.21)$$

Note that **v** is the velocity of the conductor itself, not the velocity of charges through the conductor. Consider the magnetic force on the top conductor, whose length is l. From Eq. 32.19 of Section 32.4,

$$\mathbf{F}_{mag} = I\mathbf{l} \times \mathbf{B}$$

$$F_{mag} = \frac{\omega Blh}{R} \sin \omega t \cdot lB \quad (34.22)$$

This magnetic force is vertically up, perpendicular to **l** and **B**. Resolving $\mathbf{F}_{mag}$ into radial and tangential components, we must exert a force to overcome F_{tan} to turn the loop (Figure 34.27).

$$F_{tan} = F_{mag} \sin \omega t$$

$$= \frac{\omega B^2 l^2 h}{R} \sin^2 \omega t \quad (34.23)$$

The radial component of F_{mag} may be ignored here. It does no work because it is always perpendicular to the displacement. Likewise, the forces on the front and back sides of the loop do no work, and may be ignored.

The rate at which we must do mechanical work to turn the loop is $\mathbf{F} \cdot \mathbf{v}$, where the force we exert, **F**, has a magnitude equal to F_{tan} but is directed in the positive v-direction, or tangential direction.

$$P_{mech} = 2F_{tan}v = \frac{2\omega B^2 l^2 h}{R} \sin^2 \omega t \cdot \frac{\omega h}{2}$$

$$= \frac{\omega^2 B^2 l^2 h^2}{R} \sin^2 \omega t \quad (34.24)$$

The 2 in the numerator is included to take into account the bottom conductor as well as the top conductor. By comparing Eq. 34.24 with Eq. 34.20 we find that $P_{elec} = P_{mech}$. Mechanical energy is transformed into electrical energy and then into thermal energy. Overall, energy is conserved.

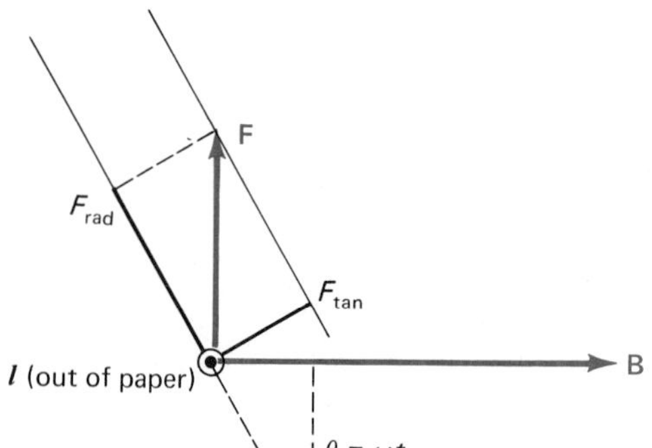

Figure 34.27

Tangential and radial components of the magnetic retarding force.

Figure 34.28

(a) Slip rings. (b) Commutator.

What we have just described are the basic principles of the electric generator. The loop in Figure 34.24 may be opened, and the circuit completed to the outside world, by using sliding contacts for alternating current or commutators for direct current (Figures 34.28a and 34.28b). When the circuit is made to the outside world, the loop becomes a generator. Today almost all the electricity we use comes from generators operating on the principles of this rotating loop and Faraday's law.

Example 4 **A Small Generator**

Imagine the single loop of Figure 34.24 replaced by a coil of 240 turns (so the induced emf is multiplied by a factor of 240). The coil is square in shape (0.08 m on a side) and is rotated at a frequency $\nu = 5$ Hz. Let's calculate the maximum induced emf for a field $B = 0.25$ T.

From Eq. 34.14

$$\mathscr{E} = \omega B l^2 \sin \omega t \text{ V}$$

Replacing ω by $2\pi \cdot 5$, multiplying by 240 for the number of loops (series circuit), and setting $\sin \omega t = 1$, we obtain

$$\mathscr{E}_{max} = 240 \cdot 2\pi \cdot 5 \cdot 0.25 \cdot (0.08)^2$$
$$= 12 \text{ V}$$

The maximum induced emf, proportional to the frequency of rotation, the area of a loop, the number of turns, and the field strength, is 12 V.

The Spin–Echo Magnetometer

The spin–echo magnetometer is a magnetic field measuring device. One of its many applications is in measuring magnetic anomalies in the ocean floor. This device consists of a source of protons (hydrogen nuclei), such as a bottle of distilled water, surrounded by a coil of heavy wire able to carry a large current (a solenoid; see Figure 34.29). The magnetometer is lowered to the ocean floor, and a strong pulse of current is then sent through the coil, or solenoid. This large current creates a strong magnetic field in the solenoid (Section 33.4). Each proton has an intrinsic magnetic moment, or in other words is a small magnetic dipole. Therefore, the protons align themselves in the strong magnetic field of the solenoid.

The current is then dropped to zero. Acting like tiny gyroscopes (Section 15.4), the aligned protons tend to precess about the weak magnetic field that remains (Figure 34.30). This magnetic field is the magnetic field of the

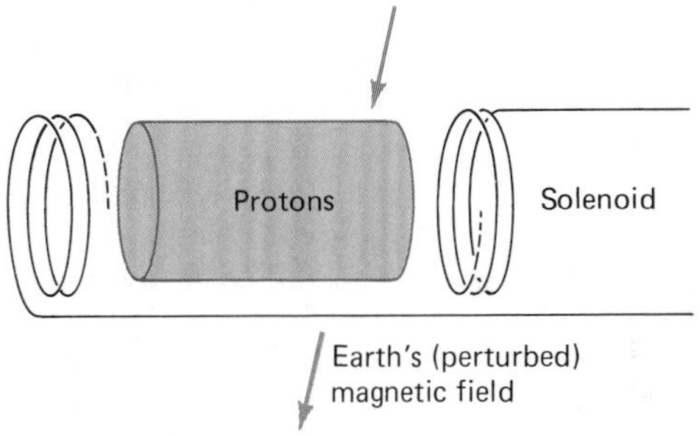

Figure 34.29

A strong magnetic field resulting from a large pulse of current in the solenoid aligns the proton magnetic dipoles. When the solenoidal field is eliminated, the dipoles precess about the earth's field and induce an alternating emf in the solenoid.

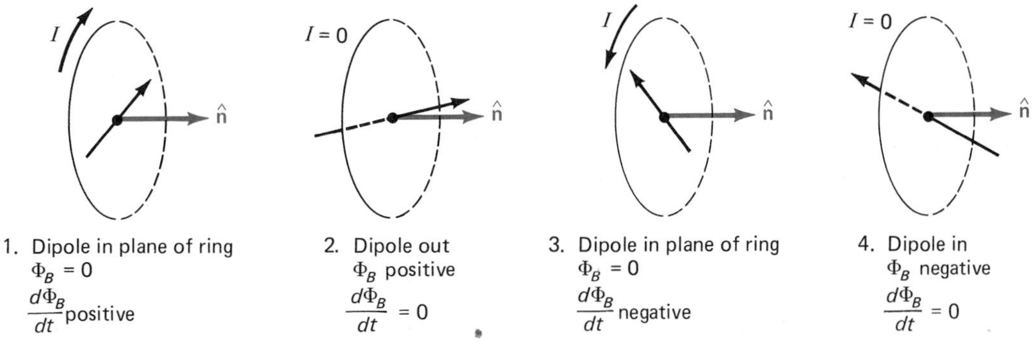

Figure 34.30

A precessing dipole varies the magnetic flux and induces an emf in the ring.

earth, plus the fields created by any magnetized areas of the ocean floor. The precessing proton magnetic dipoles induce an alternating emf in the solenoid, by Faraday's law. The frequency of the induced emf is equal to the precession frequency of the protons, which in turn is proportional to the strength of the remaining magnetic field. (See Eq. 32.13 of Section 32.4.) A measurement of this frequency thus gives oceanographers and geophysicists a measurement of strength of the magnetic field on the ocean floor.

From such magnetic measurements, earth scientists have concluded that portions of the earth's crust are moving relative to one another with speeds of a few centimeters per year. These measurements also document very clearly that the earth's magnetic field has experienced many polarity reversals over time (see Section 36.6).

Eddy Currents

Whenever a conductor or portion of a conductor moves across a magnetic field ("cutting" magnetic field lines), an emf is generated. If this induced emf is not balanced by other emf's, charges will flow. In low-resistance circuits these currents may be quite large. These internal circulating currents are called *eddy currents* because of their resemblance to the eddies in the flow of fluids. As described by Lenz's law, the eddy currents will circulate in such a way as to oppose the motion that induced them. The net effect is a drag, or slowing down. You can see this eddy current drag in action on almost any household watt-hour meter.

The eddy current drag can be dramatically demonstrated by letting a sheet of copper swing into a magnetic field. Eddy currents are generated as shown in Figure 34.31. These eddy currents oppose the external magnetic field. We expect this if we use the right-hand rule to find the direction of the magnetic field produced by the induced current loops. But where does the retarding force, the drag, come in? Remember that the current loop acts like a magnetic dipole, or bar magnet. The sheet of copper *entering* the magnetic field therefore acts like a bar magnet in terms of the repulsion of adjacent like poles. As the sheet of copper leaves the magnetic field, the eddy current loops and the equivalent bar magnet both *reverse*, and we find a net attraction. In each case the motion is opposed—agreeing with Lenz's law and conservation of energy.

Mathematically the retardation follows from the same two vector products that we used in Section 34.3 to show the drag on a conductor moving through a magnetic field. The relevant vectors are shown in Figure 34.32. The charges in the conducting disk moving with velocity **v** experience a force $\mathbf{F}_1 = q\mathbf{v} \times \mathbf{B}$. If only part of the disk is immersed in the field **B**, the charges will flow, giving rise to a current I in the direction of $\mathbf{F}_1$. Then the interaction of field and currents yields a force $d\mathbf{F}_2 = I\,d\mathbf{s} \times \mathbf{B}$ on a current element $I\,d\mathbf{s}$. For **B** perpendicular to the disk and therefore to **v**, $\mathbf{F}_2$ is opposite in direction to **v**. Force $\mathbf{F}_2$ is a retarding force.

The eddy currents in the sheet of copper can be minimized by interrupting their path. A series of slots, or rectangular holes, in the copper sheet does the job very effectively (Figure 34.33). The slots interrupt possible current paths, and thus increase the resistance. Because $P = \mathscr{E}^2/R$ (Section 30.5), an increase in the resistance means that less power is lost.

Eddy currents in the iron cores of transformers (Section 37.6) are minimized by building the cores of thin sheets of iron called *laminations*. The laminated iron core retains the desired magnetic properties but charge does not flow across the laminations.

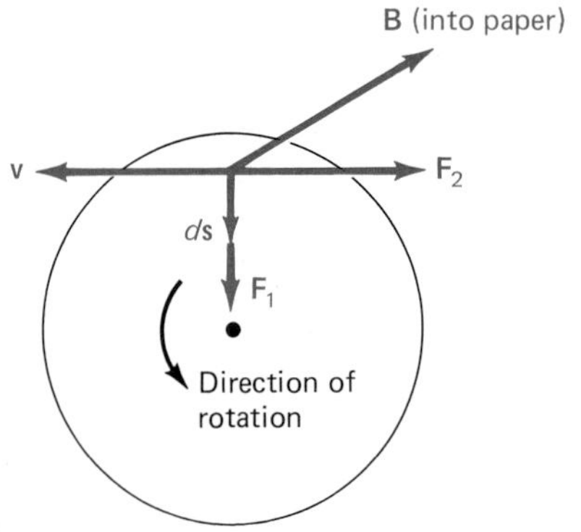

Figure 34.32

Eddy currents and magnetic drag.

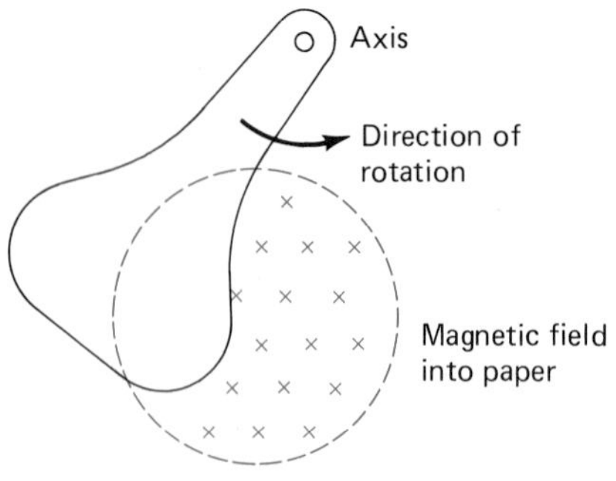

Figure 34.31

A copper pendulum in a magnetic field.

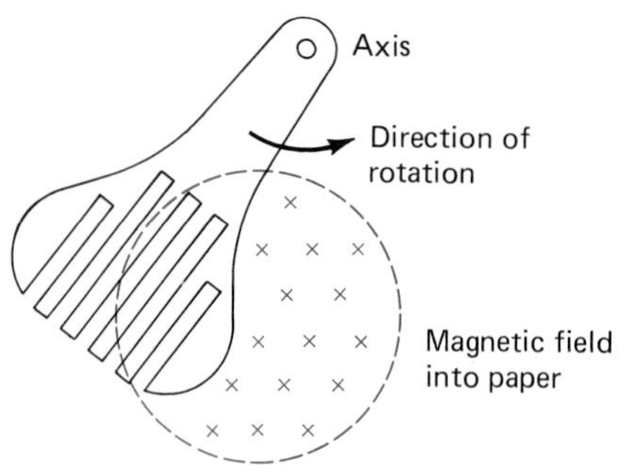

Figure 34.33

Slotted copper pendulum.

Questions

9. A changing magnetic field induces an emf in a metal ring. How would the induced emf change if the resistance of the ring were doubled? Assume that the induced current is so small that its magnetic effects can be neglected.
10. A short metal rod moves through a uniform magnetic field, cutting the field lines. Is an emf induced in the rod? Is there a magnetic force opposing the motion of the rod? Does work have to be done to maintain the motion of the rod?
11. A conducting loop surrounds a magnetic field that is entirely confined to the circular region shown in Figure 34.34. The loop is moved to one side. What eddy currents are induced? In what direction? What electromagnetic force acts on the loop?
12. A strip of sheet copper is partly in and partly out of a magnetic field, as shown in Figure 34.35. The plane of the copper sheet is perpendicular to the field. When you try to pull the copper sheet to the right, out of the field, there is a resisting force. Explain.
13. A bar magnet falls under the influence of gravity along the axis of a long cylindrical copper tube. If air resistance and any friction against the tube walls are neglected, will there be any forces opposing the fall of the magnet? Explain.
14. The falling bar magnet in Question 13 reaches a constant terminal velocity. Explain.

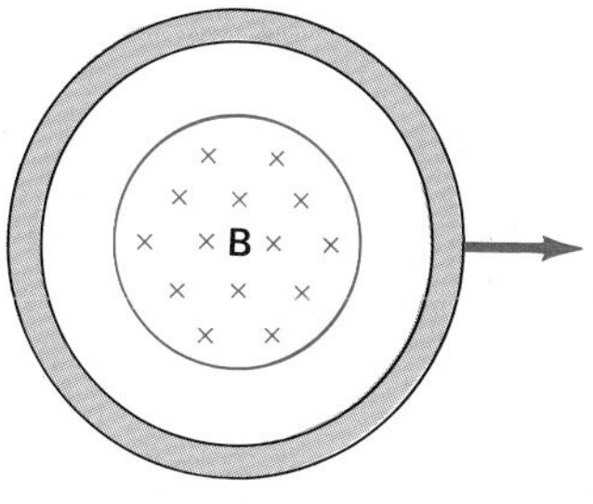

Figure 34.34

A conducting ring and a localized magnetic field.

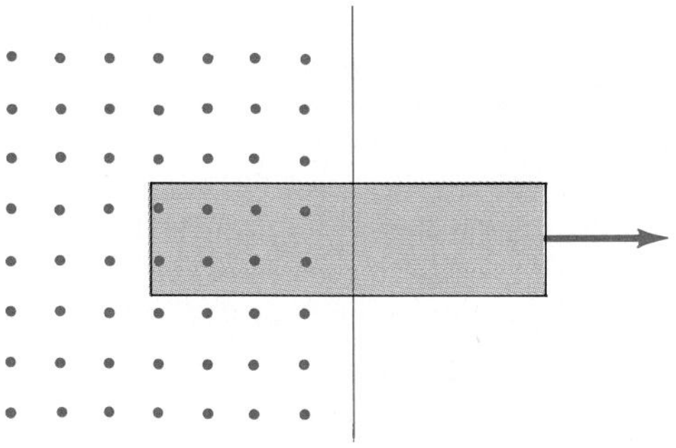

Figure 34.35

A copper plate being pulled out of a magnetic field.

Summary

The charges in a conductor moving through a magnetic field experience a force $\mathbf{F} = q\mathbf{v} \times \mathbf{B}$. To an observer moving with the charges this magnetic field appears to be due to an electrical field. The effect of the motion of the conductor is the creation of an electric field and a potential difference in the moving conductor. The potential difference $\mathscr{E}$ is a form of an electromotive force, or emf.

Faraday's law of electromagnetic induction is

$$\oint \mathbf{E} \cdot d\mathbf{s} = -\frac{d\Phi_B}{dt} \tag{34.11}$$

The total time derivative in Faraday's law includes the possibility of a time-dependent magnetic field and stationary conductors. This possibility can be stated as follows:

A changing magnetic field creates an electric field.

Consequently, our focus shifts from electrostatics to electrodynamics. This induced electric field is not conservative.

The negative sign in Faraday's law is accounted for by Lenz's law, which states that "when a current is induced in a conductor, the direction of the current will be such that the current's magnetic effects oppose the change that induced it." Lenz's law is a form of the law of conservation of energy.

Faraday's magnetic field lines give us a picture of Faraday's electromagnetic induction and of Lenz's law. The induced emf may be associated with the cutting of lines of force by a moving conductor. Lenz's law is illustrated by a distortion of the field lines resulting in a force that opposes the motion of a conductor cutting the field lines.

A moving conductor (or a time-dependent field) may produce internal circulating currents called eddy currents. The production of eddy currents is an illustration of Faraday's law. Eddy currents can be suppressed by interrupting the circuit or reducing the conductivity.

Suggested Reading

H. H. Kolm and R. D. Thornton, Electromagnetic Flight, *Sci. Am.*, 229, 17–25 (October 1973).

Problems

Section 34.1 Motional Electromotive Force

1. Show that the quantities

$$\int_0^l \mathbf{E} \cdot d\mathbf{s} \qquad vBl \quad \text{and} \quad \frac{d\Phi_B}{dt}$$

all have the same dimensions: [mass] · $[\text{length}]^2$ · $[\text{current}]^{-1}$ · $[\text{time}]^{-3}$.

2. The conducting rod of Figure 34.2 is 0.82 m long. It is moving through a field of 0.58 T at 4.31 m/s. Calculate (a) the emf developed and (b) the potential difference.

3. Referring to Figure 34.4 once again, consider the 0.1-m rod AB to be moving to the *left* with a velocity of 8 m/s through a magnetic field **B** of 1.3 T (**B** is into the paper, as shown in Figure 34.4). Rod AB and the horizontal conductors BC and DA are excellent conductors with negligible resistance. In contrast, CD is a section of Nichrome wire with a resistance of 108 Ω. Calculate (a) the emf across the moving rod; (b) the potential difference $V_A - V_B$; (c) the magnitude and direction of the electric current in the closed loop.

4. A spacecraft 10 m wide moves through an interstellar magnetic field $B = 3 \times 10^{-10}$ T. Treat the craft as a metal rod 10 m long moving perpendicular to the field, and calculate the emf developed across the width of the craft at a speed of 3×10^7 m/s.

5. A metal rod of length $l = 0.80$ m moves through the earth's magnetic field with a speed $v = 3.1$ m/s. Take **l**, **v**, and **B** to be mutually perpendicular. Take the magnitude of the earth's field to be 5.0×10^{-5} T. Calculate the emf induced in the moving rod.

6. A car having a radio antenna 1 m long travels 80 km/hr in an area where the earth's magnetic field is 5×10^{-5} T. What is the maximum possible induced emf in the antenna as a result of moving through the earth's magnetic field?

7. A circular loop of wire has a radius of 0.50 m. The loop is in a magnetic field of $B = B_0(1 - 0.031t)$ where $B_0 = 1.12$ T. The positive normal to the loop and the magnetic field are parallel. Calculate (a) the magnetic flux Φ_B through the loop at $t = 0$; (b) the time rate of change of magnetic flux, $d\Phi_B/dt$.

8. Referring to Figure 34.1, let length $AB = 0.16$ m, length $BC = 2.16$ m, $v = 10.2$ m/s, and $B = 0.511$ T. Calculate Φ_B and $d\Phi_B/dt$.

9. A long wire carrying a current I and a rectangular loop of wire whose length is l and width is w lie on the surface of a nonconducting table (Figure 1). The loop is pulled away from the wire with velocity **v**. (a) Why is there a net induced emf around the loop? (b) Indicate the polarities of the induced emf in each side of the wire. (c) Calling r the distance from the wire to the trailing end of the loop, derive a relation for the net induced emf around the loop.

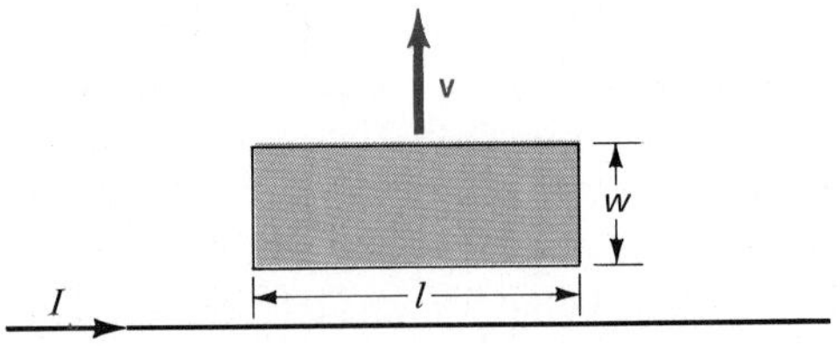

Figure 1

Section 34.2 Faraday's Law of Induction

10. Calculate the electric field that would be seen (measured) by an observer riding on the moving rod of Problem 2.

11. A 1000-turn coil is 10 cm in diameter. What emf is developed across the coil if the magnetic field through the coil is reduced from 6 T to 0 in (a) 1 s; (b) 0.1 s; or (c) 0.0001 s?

12. Figure 2 shows the flux threading a conducting loop as a function of time in milliseconds. Assuming the time dependence to be quadratic, deduce a relation for the induced emf in the loop and evaluate the induced emf for $t = 0$, $t = 2$, and $t = 4$ ms.

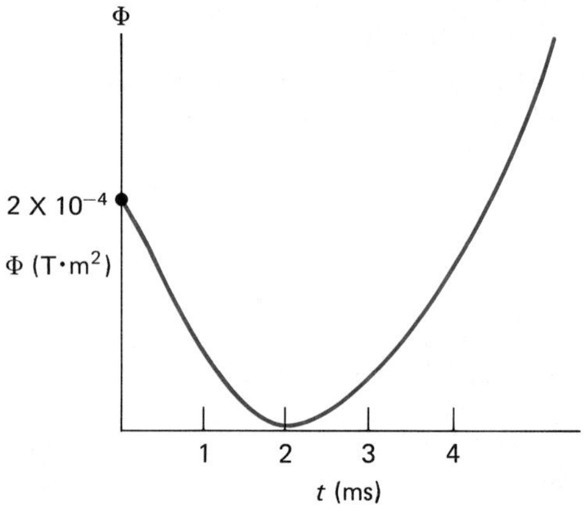

Figure 2

13. A circular loop of wire 5 cm in radius is in a spatially uniform magnetic field, the plane of the circular loop perpendicular to the direction of the field (Figure 3). The magnetic field varies with time: $B(t) = 0.20 + 0.32t$ (t in seconds, B in teslas). (a) Calculate the magnetic flux through the loop and the time rate of change of this flux at $t = 0$. (b) Calculate the emf induced in the loop at time t. (c) If the resistance of the loop is 1.2 Ω, what current exists? (d) At what rate is electrical energy being dissipated in the ring?

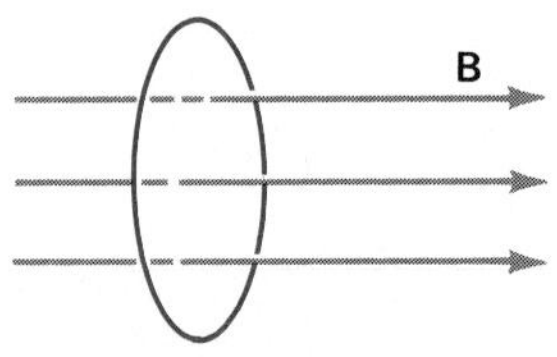

Figure 3

14. Repeat Problem 13 for a loop of radius r_o cm, a field of $B(t) = B_o + B_1 t$, and a loop of resistance of $R\ \Omega$.

15. Continuing Problem 14, if a current of 1.2 A flows in the ring ($r_o = 50$ cm, $R = 0.3\ \Omega$), what is B_1?

16. A circular loop of wire 10 cm in radius is in a uniform magnetic field, the plane of the circular loop perpendicular to the direction of the field. The magnetic field varies with time: $B(t) = B_o \cos \omega t$. (a) Calculate the magnetic flux through the loop and the time rate of change of this flux. (b) Calculate the emf $\mathcal{E}$ induced in the loop. (c) If the resistance of the loop is $R\ \Omega$, what current exists in the loop? (d) At what rate is electrical energy being dissipated in the ring?

17. The axial magnetic flux Φ_B in a given solenoid varies according to the equation

$$\Phi_B = \Phi_o \sin 2\pi\nu t$$

where

$$\Phi_o = 1.30 \times 10^{-4}\ \mathrm{T \cdot m^2}$$

$$\nu = 60\ \mathrm{s^{-1}}$$

and t is in seconds. Wrapped tightly around the outside of the solenoid is a small coil of 12 turns (Figure 4). Calculate the total emf induced in the small coil at $t = 1/60$ s.

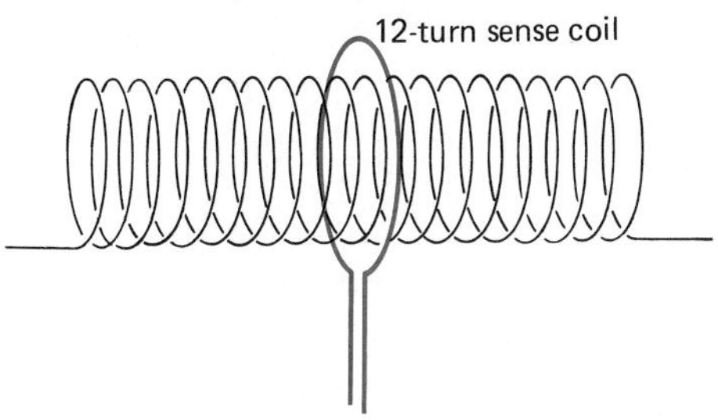

Figure 4

18. A changing magnetic field is confined to a cylindrical region of radius R. The field **B** varies with time but is spatially uniform. The direction of the field everywhere is parallel to the axis of the cylindrical region. Assume that at a radial distance r from the cylindrical axis the electric field is tangent to a circle of radius r. Show that the magnitude of the electric field is given by $E = \frac{1}{2}(R^2/r)(dB/dt)$ for $r \geq R$ and

$$E = \frac{r}{2}\frac{dB}{dt} \text{ for } r \leq R.$$

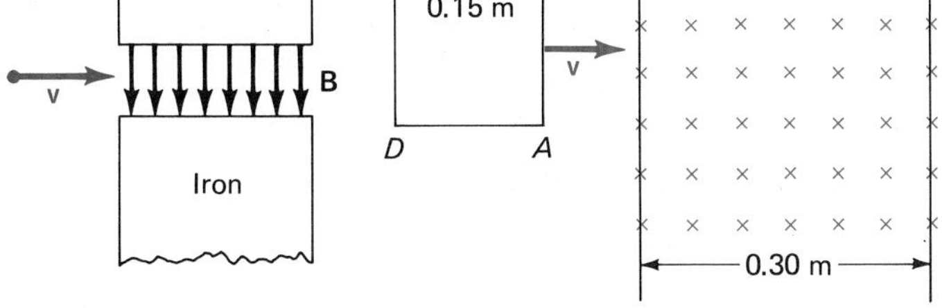

Figure 5

19. In analogy to Example 1 a square loop of copper wire 0.15 m on a side is pulled into, through, and out of a uniform but sharply bounded magnetic field, $B = 1.4$ T, as indicated in Figure 5. The loop, moving with a fixed velocity of 6.0 m/s, enters the magnetic field at $t = 0$. Calculate the induced emf for (a) $t < 0$, (b) $0 < t < 0.025$, (c) $0.025 < t < 0.050$, (d) $0.050 < t < 0.075$, (e) $0.075 < t$. Plot $\mathcal{E}$ versus t. If the boundary of the magnetic field is not mathematically sharp, that is, if there is a fringing field, what will this do to your plot?

20. Referring to Problem 19, what is the direction of the induced emf's in the wire segment AB and in the wire segment CD for (a) $t < 0$; (b) $0 < t < 0.025$; (c) $0.025 < t < 0.050$; (d) $0.050 < t < 0.075$; (e) $0.075 < t$.

21. Referring to Figure 5 of Problem 19, (a) sketch the fields of the magnet and of the current in segment AB separately at $t = 0.012$ s. (b) Sketch the resultant field, superposing the two fields of part (a).

22. Referring to Figure 34.14a, sketch the magnetic field in the vicinity of the circular loop in the y–z plane (a) with the bar magnet approaching but $I = 0$, and (b) with the field of the current I superposed on the bar magnet field.

Section 34.3 Lenz's Law

23. A pair of identical rings, coaxially mounted, are free to move along the axis AB. Ring 1 carries current I defined to be positive, as drawn in Figure 6. Give the direction (+ or −) of the current induced in ring 2 if (a) ring 1 moves toward A and ring 2 is at rest; (b) ring 2 moves toward A and ring 1 is at rest; (c) ring 1 moves toward B and ring 2 is at rest; (d) ring 2 moves toward B and ring 1 is at rest; (e) both rings move toward A with the same speed; (f) ring 1 moves toward B and ring 2 moves toward A with the same speed.

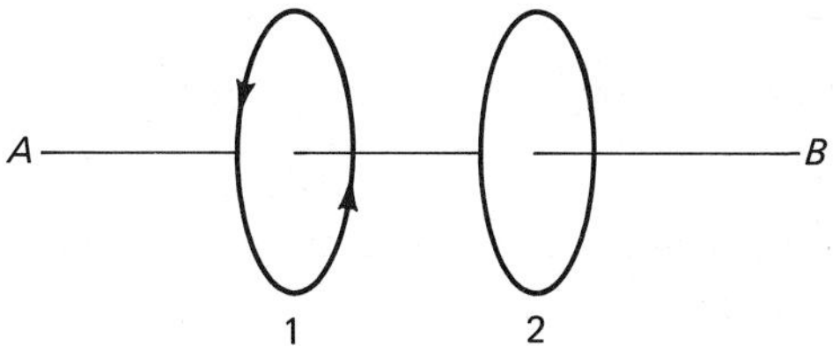

Figure 6

24. For the current loop of Figure 34.17–34.19, let the radial distance of the two conductors shown in cross section be 0.18 m. The angular position is described by sin ωt with $\omega = 2\pi \cdot 60$ rad/s. Calculate the maximum force per coulomb, F/q, for $B = 0.923$ T.

Section 34.4 Applications of Faraday's Law

25. A flip coil 1 cm in radius consists of 80 turns of fine copper wire. The total resistance of the coil, circuit, and galvanometer is 0.20 Ω. The coil is flipped through 180°. (a) If $Q = 0.06$ C, what is the magnetic field, B? (b) What is Q for a magnetic field of 0.5 T, 5000 G?

26. A flip coil whose radius is 10 cm consists of 20 turns of copper wire. The coil is flipped so that its normal, $\hat{n}$, moves from parallel to the magnetic field to perpendicular to the field. With a total resistance of 0.14 Ω in the circuit the measured charge Q is 0.079 C. Calculate B.

27. A metal rod is free to slide along a pair of conducting rails electrically connected by a 200-Ω resistor. If 50 W of mechanical power are expended in pushing the rod through a magnetic field, and all of this power is dissipated in the resistor, what is the induced current in the rails?

28. A rectangular loop generator has dimensions of 0.12 m × 0.37 m. It rotates in a uniform field of 0.682 T with a radian velocity of $2\pi \cdot 60$. (a) Calculate the maximum magnetic flux through the loop. (b) Calculate the maximum value of $d\Phi_B/dt$.

29. Let the rectangular loop generator of Section 34.4 be a square 10 cm on a side. It is rotated at 60 Hz in a uniform field of 0.8 T. Calculate (a) the flux through the loop at time t; (b) the time rate of change of flux through the loop and the voltage induced in the loop; (c) the current induced in the loop for loop resistance 1.0 Ω; (d) the power dissipated in the loop; and (e) the torque that must be exerted to rotate the loop.

30. Continuing Example 1 of Section 34.1, the resistance of the complete circuit is 0.2 Ω. (a) Calculate the electric current I in the circuit. (b) At what rate is electric energy being dissipated as heat? (c) Calculate the retarding force on the moving conductor. (d) Calculate the rate at which mechanical work is being done to move the conductor. Compare the results with part (b).

31. A 20-turn circular coil with a radius of 9 cm rotates in a uniform magnetic field of 0.20 T with a constant frequency of 10 Hz. The axis of rotation (in the plane of the coil) is perpendicular to the magnetic field (Figure 7). The coil ends are joined to form a closed circuit with a total resistance of 1.1 Ω. Calculate (a) the total induced emf in the coil, (b) the current in the coil, and (c) the rate of dissipation of electric energy.

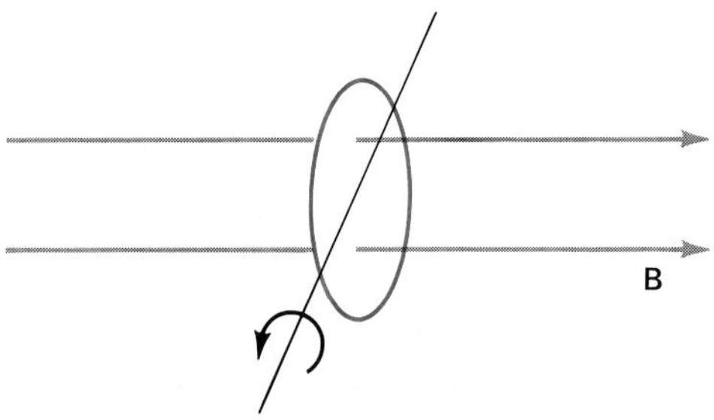

Figure 7

32. Continuing Problem 31, assume that energy is conserved. From the mechanical work being done, calculate the torque required to rotate the coil.

33. A sheet of copper enters a strong magnetic field. Sketch the eddy currents induced. Show the equivalent induced bar magnet and explain why this results in opposition to the motion of the copper sheet.

chapter 35

INDUCTANCE AND INDUCTORS

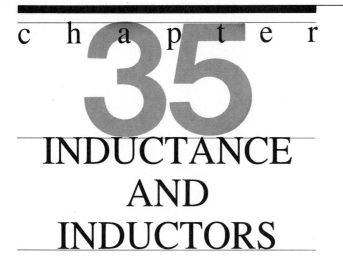

Preview

Resistors and capacitors are two-terminal (two electrical connections) elements that have great practical use in electronic circuits. A resistor converts electric energy to thermal energy, whereas a capacitor stores and releases electric energy. The concepts of resistance (Section 30.2) and capacitance (Section 29.1) help us to evaluate energy storage and energy conversion in electrical circuits. In this chapter we introduce another two-terminal circuit element, the inductor.

An inductor can store magnetic energy. A solenoid is a typical inductor. The concept of inductance helps us to evaluate the amount of magnetic energy stored by a solenoid, and to understand how a solenoid responds to a changing current in an electrical circuit. We use Faraday's law of induction to show that the induced emf in a loop of a conductor is proportional to the time rate of change of current. Inductance thus emerges as a constant of proportionality. Next we relate the stored magnetic energy to the work required to establish the magnetic field around the inductor. Finally, we examine the behavior of inductors in circuits involving resistors and capacitors. In a circuit containing an inductor and a capacitor, magnetic energy and electric energy are transferred back and forth between the two elements in a manner reminiscent of the interchange of kinetic energy and potential energy in a harmonic oscillator. These oscillations in an inductor–capacitor circuit have widespread applications in electronics.

35.1 Self-Inductance and Inductors

Inductors

In Chapter 34 we learned that a change of magnetic flux threading a loop of a conducting material induces an emf that gives rise to a current in the loop (Figure 35.1). The loop might be an intricate connection of components in an integrated electronic circuit or it might be a carefully engineered coil of wire. The current in the loop and the magnetic flux threading the loop are related by a property that we call the **self-inductance*** of the loop. Self-inductance can exist in wires, electrical transmission lines, and arbitrarily shaped loops. For example, self-inductance is a very important characteristic of a coaxial transmission line (Figure 29.7). However, we are usually interested in circuit elements that are designed to *exploit* inductive effects. Such circuit elements are called *inductors* and generally are coils of wire of varied shapes and sizes. The symbol for an inductor is  If the coil is wrapped around an iron core so as to enhance its magnetic effect, it is symbolized by 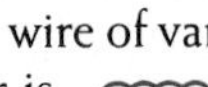. Just as we have described the electrical behavior of resistors and capacitors in electrical circuits, let's now describe the electrical behavior of inductors in electrical circuits.

*Self-inductance pertains to a circuit element for which the *magnetic flux depends only on current in the circuit*. The concept of mutual inductance includes magnetic flux from sources in adjacent circuits. We do not treat mutual inductance formally in this chapter, but we show its implications when we study the transformer in Chapter 37.

Figure 35.1

Relative motion between a conducting loop and a magnetic dipole causes an emf to be induced in the loop. Such an emf is induced whenever there is a change in the magnetic flux threading the loop.

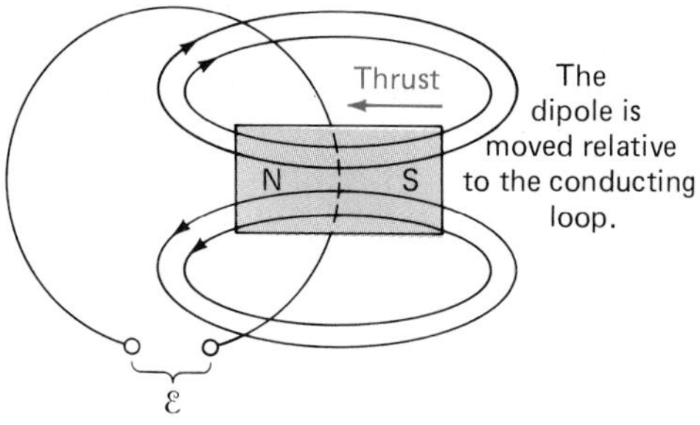

Self-Inductance

Let's consider a loop of a conducting material. The loop contains an electric current. The current produces a magnetic field, **B**. The magnetic field gives rise to a magnetic flux. The magnetic flux* $d\Phi$ threading an area $d\mathbf{A}$ is

$$d\Phi = \mathbf{B} \cdot d\mathbf{A} \tag{35.1}$$

where **B** is the magnetic field at the location of $d\mathbf{A}$. The total magnetic flux threading a loop is determined by integrating $d\Phi$ over the surface defined by the perimeter of the loop.

$$\Phi = \int d\Phi = \int \mathbf{B} \cdot d\mathbf{A} \tag{35.2}$$

In the absence of any external source of magnetic flux (for example, an adjacent coil carrying a current) the Biot–Savart law (Section 33.1) requires that the magnetic field be proportional to the current in the loop. Therefore the flux threading the loop is also proportional to the current (I) in the loop.†

$$\Phi \propto I \tag{35.3}$$

The proportionality constant between magnetic flux and current is called the *self-inductance of the loop*. Self-inductance is given the symbol L, and is defined by

$$\Phi = LI \tag{35.4}$$

Faraday's Law in Terms of Self-Inductance

If the current in the loop in Figure 35.2 is changed by closing the switch, or by varying R or the emf after the switch is closed, the magnetic flux threading the loop changes. This change in magnetic flux is accompanied by an induced emf in the loop, called a *self-induced* emf. At any moment the current in the loop is determined by both the self-induced emf and the potential difference created by the imposed emf (a battery, for example) between the ends of the loop. In accordance with Lenz's law, the self-induced emf in this case (closing the switch) opposes the change that produces it.

For self-induced emf's in a circuit, let's now obtain a form of Faraday's law relating the induced emf to the rate of change of current. We begin by differentiating Eq. 35.4 with respect to time to obtain

$$\frac{d\Phi}{dt} = L\frac{dI}{dt} \tag{35.5}$$

Using Faraday's law of induction (Section 34.2), we can replace $d\Phi/dt$ with the negative of the induced emf ($-\mathscr{E}$) so that

$$\mathscr{E} = -L\frac{dI}{dt} \tag{35.6}$$

According to this equation, the induced emf is proportional to the time rate of change of current in the loop. This is the form of Faraday's law that is applicable to electrical circuits. Note that a constant current, no matter how large, produces zero induced emf, whereas a current changing rapidly produces a large induced emf (Figure 35.3). The units of self-inductance are

$$\text{units of } L = \frac{\text{units of } \mathscr{E}}{\text{units of } dI/dt} = \frac{\text{volts}}{\text{amperes/second}} = \text{ohm second}$$

An ohm second is called a *henry*,* abbreviated H. For most applications, the henry is a rather large unit, and we often use the more convenient units millihenry, mH (10^{-3} H), and microhenry μH (10^{-6} H), instead. Self-inductance, like capacitance, depends on geometric factors. We will

Figure 35.2

If a battery is connected to a conducting loop by way of a switch, then closing the switch causes a current in the loop and magnetic field lines within the loop. This flux change within the loop produces an induced emf in the loop. This induced emf is in addition to any potential difference produced by the battery.

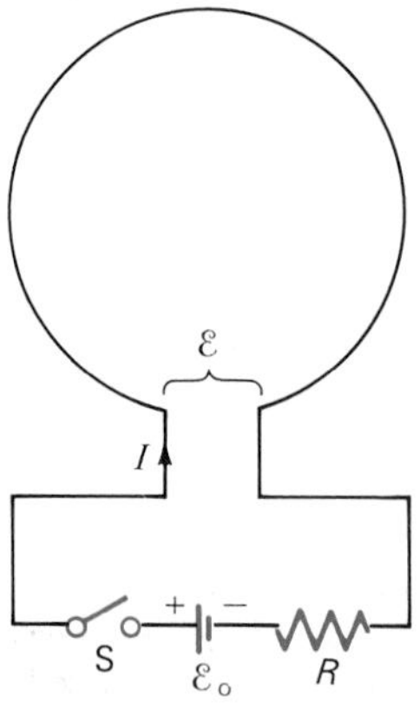

*Because all references to flux in this chapter are to magnetic flux, we can drop the subscript B that we used in Φ_B.

†In Chapter 36, we will see that if magnetic materials are present, the magnetic flux can be a rather complicated function of the current in the inductor, and so the self-inductance also depends on the current. We do not treat these systems in this text.

*Named for Joseph Henry, an American physicist (1797–1878) and a pioneer in the study of electromagnetic phenomena.

Figure 35.3a

Regardless of the size of the current in an inductor, the induced emf is zero if the current is constant.

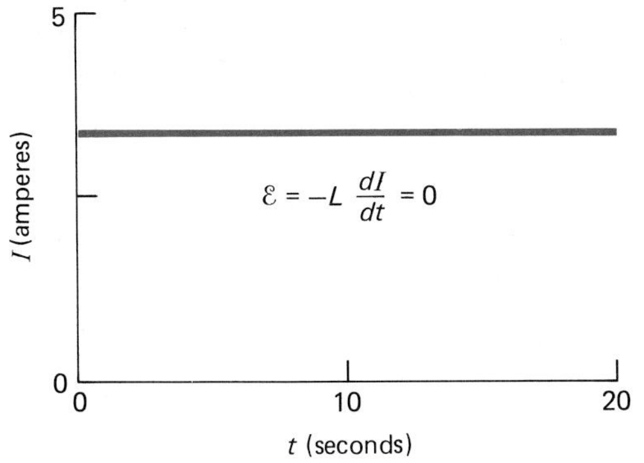

Figure 35.3b

A rapidly changing current in an inductor can produce a significant induced emf even though the currents involved are not particularly large. In this example the inductance is taken to be 0.01 H.

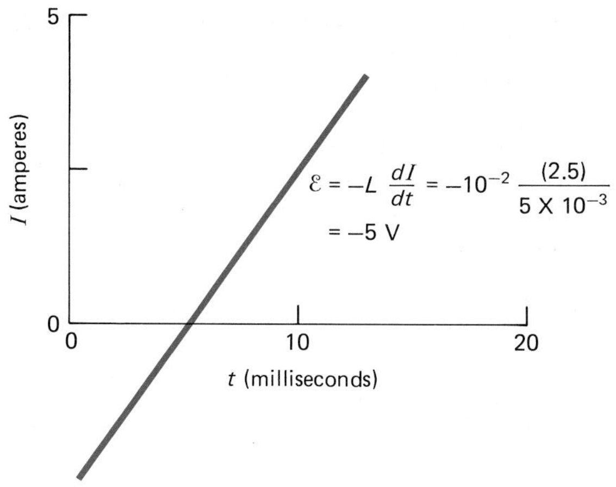

now calculate the self-inductance of a few symmetric systems as illustrations, although in general the value of the self-inductance must be obtained experimentally.

According to Lenz's law (Section 34.3), the negative sign in the equation

$$\mathcal{E} = -L\frac{dI}{dt}$$

states that the direction of the induced emf ($\mathcal{E}$) opposes the direction of the rate of current change in the circuit. For example, if the current is decreasing, the induced emf will tends to keep the current from decreasing. A decreasing current means that $dI/dt < 0$ (Figure 35.4); hence the induced emf will be a positive quantity. If the current is increasing, the induced emf will tend to oppose this increase. An increasing current means that $dI/dt > 0$ (Figure 35.4); hence in this case the induced emf will be a *negative* quantity. If $dI/dt = 0$ at some instant of time (Figure 35.4), then the induced emf is instantaneously zero.

Equation 35.6, $\mathcal{E} = -L\,dI/dt$, has the same mathematical form as Newton's second law, $\mathbf{F} = m\,d\mathbf{v}/dt$, with the self-inductance L appearing in the same position as the inertial mass m. Inertial mass measures the resistance to a change in velocity of a mass; self-inductance measures the resistance to a change in the current in an inductor.

Figure 35.4

This figure shows how current in an inductor might vary with time. At time t_A the slope of the curve is negative ($dI/dt < 0$). Hence the induced emf is positive ($\mathcal{E} > 0$). The induced emf will tend to keep the current from decreasing. At time t_B the slope of the curve is positive ($dI/dt > 0$). Hence the induced emf is negative ($\mathcal{E} < 0$). The induced emf will tend to keep the current from increasing. At time t_C the slope of the curve is zero ($dI/dt = 0$). The induced emf is instantaneously zero.

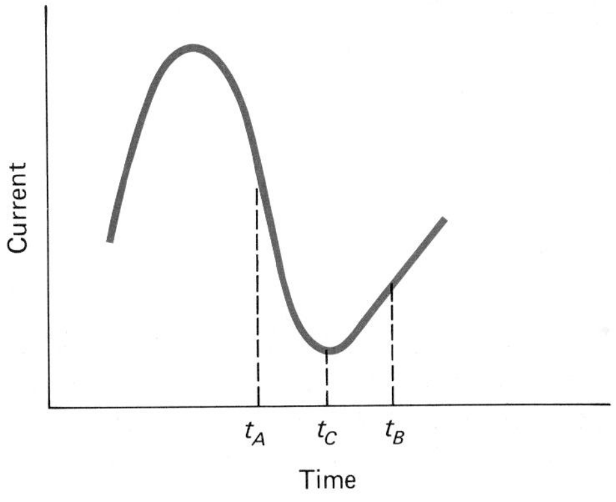

Example 1 Experimental Determination of Inductance

The current in a coil of wire is changed uniformly according to the relation $I = 0.10t$ A, where t is the time in seconds. Experimental measurements show that an induced emf of 0.13 mV is produced across the leads to the coil. The induced emf of 0.13 mV is in addition to any potential difference across the leads that is due to electrical resistance of the coil. What is the self-inductance of the coil?

From the relation $I = 0.10t$ we have

$$\frac{dI}{dt} = 0.10 \text{ A/s}$$

Given that $\mathcal{E} = -0.13 \times 10^{-3}$ V, the self-inductance is

$$L = \frac{\mathcal{E}}{-dI/dt} = \frac{-0.13 \times 10^{-3}}{-0.10}$$

$$= 1.3 \times 10^{-3} \text{ henrys} = 1.3 \text{ mH}$$

Self-Inductance of a Solenoid

The equation $\mathcal{E} = -L\,dI/dt$ is the basis for the measurement of self-inductance regardless of the geometric shape of the conducting loop. But the theoretical calculation of self-inductance is difficult unless the geometry possesses some symmetry. The usual theoretical technique is to calculate the total flux threading a loop (or connected loops in a coil) for some given current and then to divide the flux by the current ($L = \Phi/I$). For example, consider

Figure 35.5

Wire wrapped tightly into a cylindrical form constitutes a solenoid. Shown here is a cross-sectional view of such a solenoid. In the exposed upper wires, current is directed toward the viewer. The magnetic field **B** produced by the current is nearly constant in the interior of the solenoid.

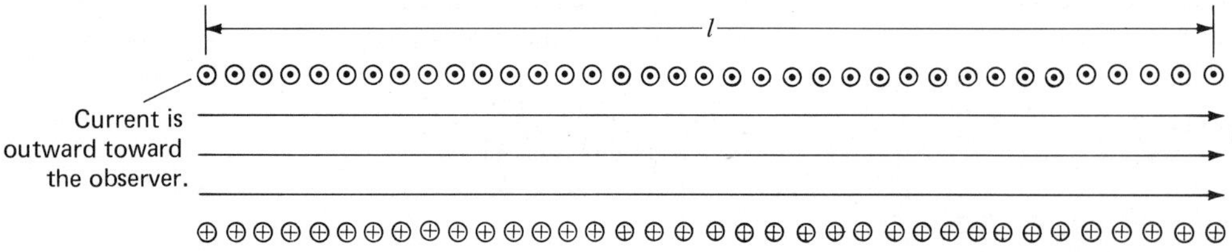

a long solenoid, of length l, containing N closely spaced identical turns of wire (Figure 35.5). We assume the magnetic field to be constant within the solenoid and the magnetic field lines to be parallel to the axis of the solenoid (see Section 33.4). Hence, each loop of the solenoid encircles magnetic flux $\phi = BA$ where A is the cross-sectional area of a single loop. For a solenoid (Section 33.4), we know that $B = \mu_o(NI/l)$. Therefore the flux threading each loop is $\phi = \mu_o(NI/l)A$. Because there are N loops in the solenoid, the total flux encircled is

$$\Phi = N\phi$$

$$= \frac{\mu_o N^2 IA}{l} \quad (35.7)$$

Dividing the total flux by the current in the solenoid gives for the self-inductance

$$L = \frac{\Phi}{I}$$

$$L = \frac{\mu_o N^2 A}{l} \quad (35.8)$$

Note that the self-inductance, L, depends only on μ_o and the geometric properties (N, A, and l) of the solenoid. With A and l fixed, L is proportional to the square of the number of turns.

Self-Inductance of a Coaxial Cable

A coaxial cable (described in Section 29.1) is a straight conductor surrounded by a concentric, cylindrically shaped conductor. Coaxial cables are widely used as conductors for current. For example, currents produced in the pickup of a stereo turntable generally travel to the amplifier through a coaxial cable. We calculated the capacitance of such a cable in Section 29.1. The self-inductance of a coaxial cable is another important characteristic, because it influences the propagation of electrical signals in the cable. Let's now calculate the self-inductance of the coaxial cable shown in Figure 35.6. An emf is connected to one end of the cable and a load resistor is connected to the other end.

Charge flows down the central conductor from the emf and returns to the emf through the outer conductor (Figure 35.6). Thus the inner and outer conductors of the coaxial cable form a conducting loop. A changing current in this loop will induce an emf that acts to oppose the change. We know that the magnetic field a distance r from a long straight wire is (Section 33.2)

$$B = \frac{\mu_o I}{2\pi r}$$

and that the magnetic field lines are circles centered on the wire. Because the magnetic field depends on the radial

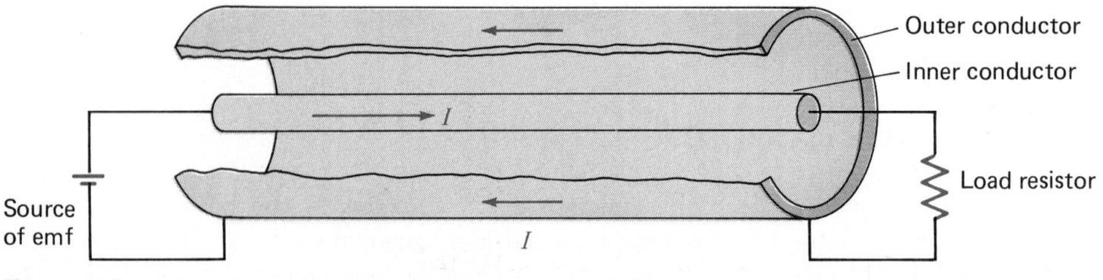

Figure 35.6

A coaxial cable is shown with a section of the outer conductor removed to enable us to view the interior. At the ends of the inner and outer conductors, there is connected a source of emf at one end and a resistor at the other end. Charge flows out of the source, through the inner conductor, and returns in the outer conductor to the source.

Figure 35.7

The magnetic field lines inside a coaxial cable form circles concentric with the inner conductor. To determine the flux we define a strip of area $dA = l\,dr$. The number of lines threading this loop is $B\,dA = Bl\,dr$. We determine the total flux by integrating $Bl\,dr$ from the radius of the inner conductor to the radius of the outer conductor.

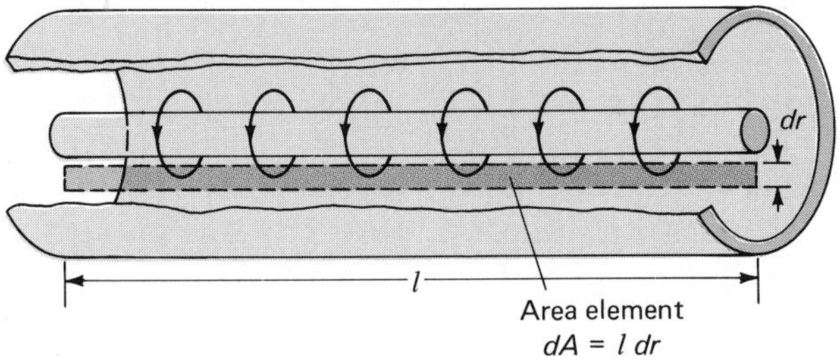

distance from the central conductor, we must perform an integration in order to get the total magnetic flux. Consider an area element (dA) whose length equals the length (l) of the inner conductor and whose width (dr) is measured radially outward from the conductor (Figure 35.7). The circular magnetic field lines intercept this area element perpendicularly. Therefore the number of magnetic lines threading the area element is

$$d\Phi = B\,dA = Bl\,dr$$

Substituting for the value of B at the position of the area element, we have

$$d\Phi = \frac{\mu_o Il\,dr}{2\pi r}$$

Hence the total flux* threading the conducting loop is

$$\Phi = \int d\Phi = \frac{\mu_o Il}{2\pi}\int_{r_1}^{r_2}\frac{dr}{r} = \frac{\mu_o Il}{2\pi}\ln\left(\frac{r_2}{r_1}\right) \qquad (35.9)$$

By definition, the self-inductance is

$$L = \frac{\Phi}{I} = \frac{\mu_o l}{2\pi}\ln\left(\frac{r_2}{r_1}\right) \qquad (35.10)$$

The self-inductance per unit length is

$$\frac{L}{l} = \frac{\mu_o}{2\pi}\ln\left(\frac{r_2}{r_1}\right) \qquad (35.11)$$

The self-inductance of the coaxial cable depends only on μ_o and the cable geometry.

Questions

1. The net self-inductance of two inductors in series is the sum of the two self-inductances if the inductors are far apart. Why?

*This calculation neglects the flux inside the wires themselves. This approximation is justified if $r_2 \gg r_1$.

Figure 35.8

Folded wire wound onto a cylindrical insulator.

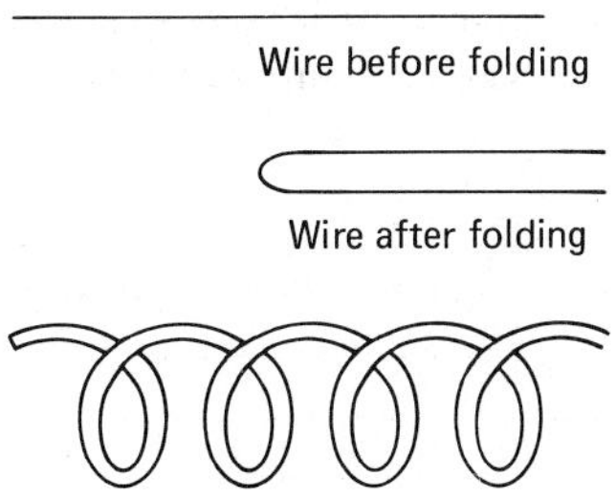

2. A certain length (l) of wire folded into two parallel, adjacent strips of length $l/2$ is wound onto a cylindrical insulator to form a type of wire-wound noninductive resistor (Figure 35.8). Why can this configuration be called noninductive?
3. Suppose that a steady current exists in an inductor that has negligible resistance. If this current is reduced to zero at a constant rate, what quantities determine the induced emf in the inductor?
4. The magnetic flux produced by a certain inductor containing magnetic material can be represented by $\Phi = 10I^2\ \text{T}\cdot\text{m}^2$, where I is the current in the inductor. What problems do you envision in trying to determine the inductance of this inductor?
5. Why should you expect the magnetic field to be zero outside a coaxial cable?
6. If you keep the radius of the inner conductor of a coaxial cable fixed and increase the radius of the outside conductor, why should you expect the self-inductance to increase?
7. The self-inductance of a solenoid is

   $$L = \frac{\mu_o N^2 A}{l}$$

   If the number of turns in the solenoid shown in Figure 35.5 is doubled, by how much will the self-inductance change?
8. If a second layer of turns of wire are wound on a solenoid like that shown in Figure 35.5 by how much will the self-inductance change?

35.2 Circuit Aspects of Inductors

LR Series Circuit

To illustrate how an inductor in an electrical circuit influences the current, let us consider an LR series circuit—a series connection of a battery (strength $\mathscr{E}_o$), a switch, an inductor, and a resistor (Figure 35.9). Because the inductor

Figure 35.9

A source of emf, such as a battery, is connected in series with a switch, an inductor, and a resistor. We call this an *LR* circuit. Initially, the switch is open, so that there is no current in the circuit. When the switch is closed, charge flows in the circuit. An emf is induced in the inductor whenever the current is changing in the circuit.

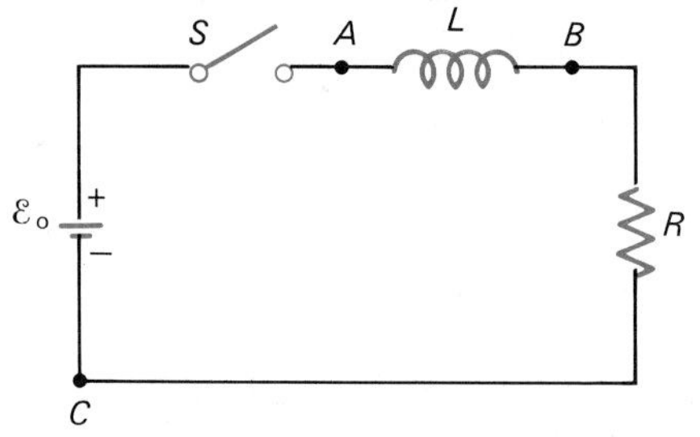

Figure 35.10

Physical arguments suggest that the current in an *LR* circuit (Figure 35.9) changes with time as shown in this graph. Initially, the current is zero. It rises to an equilibrium value equal to $\mathscr{E}_0/R$. The rate of change of the current depends on the inductance of the inductor.

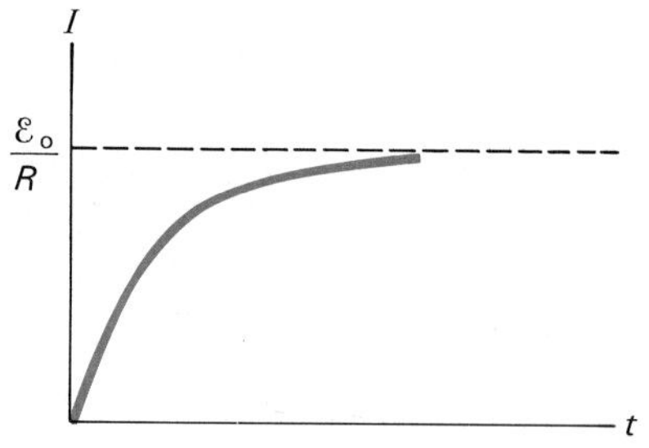

is made of coils of wire it possesses electrical resistance.* Unless stated otherwise, the resistance of the inductor and the internal resistance of the battery are neglected. When the switch is closed, charge begins flowing and the current begins increasing ($dI/dt > 0$). The induced emf across the inductor opposes the change in current, and hence the polarity of the induced emf is such that point *A* is positive relative to point *B*. In other words, *the inductor behaves as if it were a battery (of variable strength) with its terminals connected so as to oppose the action of the emf of the real battery.* This "induced battery" operates as long as the current in the circuit is changing ($dI/dt \neq 0$). The induced emf drops to zero once the current has achieved a constant value. The ultimate current is independent of the value of the inductance. Through its inertial property, the inductance can delay the time required for the current to reach its final value. We can gain further insight into the circuit current by applying Kirchhoff's loop law (Section 31.2). An induced emf is subject to this law, just like any other emf. Starting at point *C* and proceeding clockwise, we get

$$\mathscr{E}_0 - L\frac{dI}{dt} - IR = 0$$

or

$$\mathscr{E}_0 = L\frac{dI}{dt} + IR \qquad (35.12)$$

Carefully note that the polarity for the induced emf ($-L\,dI/dt$) is opposite to the applied emf ($\mathscr{E}_0$). Initially (that is, when $t = 0$) $I = 0$ and the induced emf is equal in magnitude to the applied emf ($\mathscr{E}_0$). For a steady current $dI/dt = 0$ and $I = \mathscr{E}_0/R$. The inductance provides electrical "inertia" that prevents an instantaneous change in current, allowing the current to build up gradually to its maximum value. Figure 35.10 suggests how the current changes from the initial value of zero to the final value of $\mathscr{E}_0/R$.

*Generally this resistance is negligible compared to other resistances in series with the inductor and we neglect it. If it is not negligible, we account for the coil resistance explicitly or include it in the resistance of series-connected resistors. Similarly, we neglect any internal resistance associated with the battery.

Example 2 How to Turn Off a Magnet

If an emf connected to an inductor is switched off suddenly, the induced emf may be sufficiently great to produce an electric discharge across the switch. To illustrate, assume that the components in the circuit in Figure 35.9 have the following values.

$$\mathscr{E}_0 = 12\text{ V} \qquad R = 6\ \Omega \qquad L = 10\text{ H}$$

After the switch has been kept closed for several minutes, the current is

$$I = \frac{\mathscr{E}_0}{R} = \frac{12}{6} = 2\text{ A}$$

If the switch is opened in a time of 10^{-3} s, then the induced emf is

$$\begin{aligned}\mathscr{E} &= -L\frac{dI}{dt}\\ &\cong -L\frac{\Delta I}{\Delta t}\\ &= -10\left(\frac{2}{10^{-3}}\right)\\ &= -20{,}000\text{ V}\end{aligned}$$

Such a potential difference is often sufficient to produce a dangerous discharge across the terminals of the switch. To avoid this, the current in the inductor should be reduced slowly to zero before the switch is opened.

Solution of the *LR* Circuit Equation

To determine the equation for the current in the *LR* series circuit of Figure 35.9 we must solve Eq. 35.12 for *I*. The mathematical form of Eq. 35.12 is identical with the equation for the *RC* series circuit (Eq. 31.26). We can therefore solve Eq. 35.12 by analogy with the *RC* series circuit.

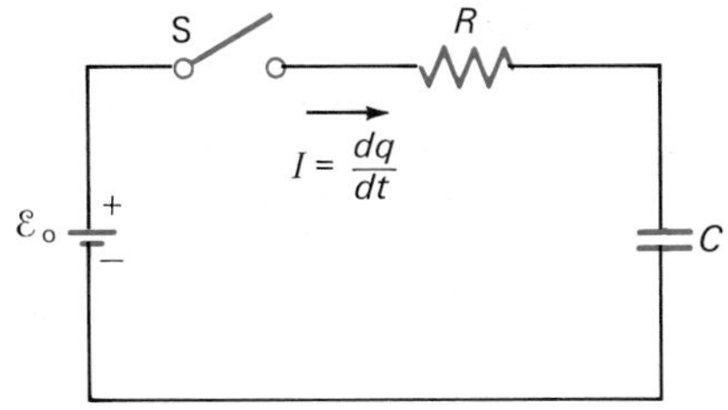

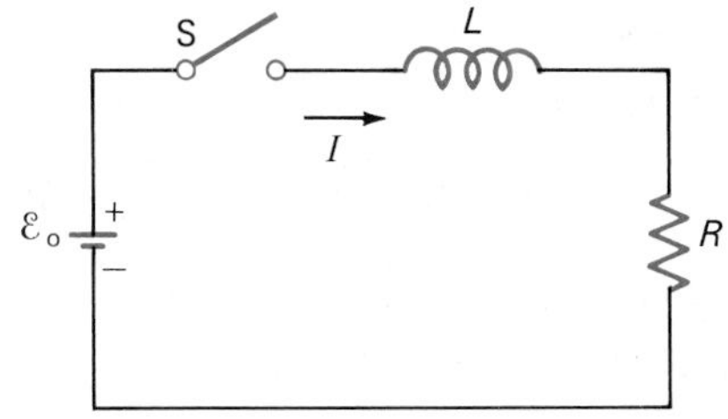

	RC circuit	*LR* circuit
Kirchhoff's loop law	$\mathscr{E}_0 = R\frac{dq}{dt} + \frac{q}{C}$	$\mathscr{E}_0 = L\frac{dI}{dt} + IR$
When the switch is first closed	$t = 0,\ q = 0$	$t = 0,\ I = 0$
	Solution by direct integration (see Section 31.6): $q = C\mathscr{E}_0(1 - e^{-t/RC})$	Solution by comparing terms in the Kirchhoff loop rule equations: I is analogous to q, L to R, R to $\frac{1}{C}$; therefore $I = \frac{\mathscr{E}_0}{R}(1 - e^{-Rt/L})$ (35.13)

Let us test Eq. 35.13 by asking, "Does this solution make sense?" (See Problem Solving Guide in Section 3.2). In particular, we note the following points.

1 The solution is dimensionally consistent because each term in Eq. 35.13 has units of current.

2 At ($t = 0$), the current (I) is zero, as required by the initial conditions.

3 After the switch is closed for a long time ($t \rightarrow \infty$), $I = \mathscr{E}_0/R$, as required from physical arguments discussed earlier.

The ratio L/R, having units of time, is called the *time constant*. To interpret the time constant, consider the current in the circuit after a time equal to L/R has elapsed. Substituting $t = L/R$ into Eq. 35.13, we have for the current in the inductor $I = (\mathscr{E}_0/R)(1 - e^{-1}) = 0.632\mathscr{E}_0/R$. Thus 63.21% of the ultimate current ($\mathscr{E}_0/R$) in the inductor is achieved in a time equal to the time constant. The current approaches its limiting value of $\mathscr{E}_0/R$ as $t \rightarrow \infty$. The time dependence is shown graphically in Figure 35.11. The potential difference across the inductor is $V_L = -L\,dI/dt$. It follows from Eq. 35.13 that

$$V_L = -L\frac{d}{dt}\left[\frac{\mathscr{E}_0}{R}(1 - e^{-Rt/L})\right]$$
$$= -\mathscr{E}_0 e^{-Rt/L} \qquad (35.14)$$

This equation describes the potential difference as a function of time. Initially ($t = 0$), $V_L = -\mathscr{E}_0$, as deduced earlier from physical arguments. As t increases, the poten-

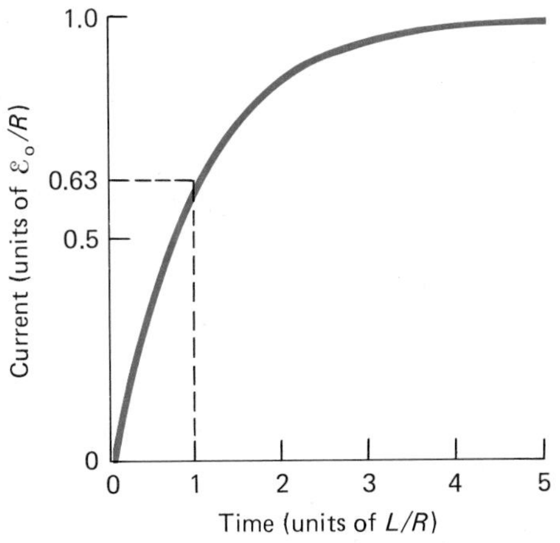

Figure 35.11

When a source of emf is connected to a series *LR* circuit the current changes with time as illustrated in this generalized plot. Time is expressed in units of L/R and current is expressed in units of $\mathscr{E}_0/R$. About 63% of the ultimate current $\mathscr{E}_0/R$ is achieved in a time equal to L/R.

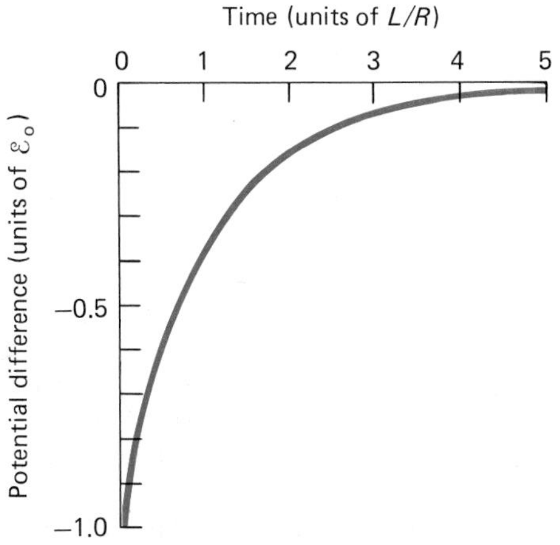

Figure 35.12

When a source of emf is connected to a series *LR* circuit the induced potential difference in the inductor changes with time as illustrated in this generalized plot. Time is expressed in units of L/R and potential difference in units of $\mathscr{E}_0$. Initially the potential difference is $-\mathscr{E}_0$. The potential difference tends to zero as time progresses.

tial difference tends to zero. After a time equal to one time constant (L/R) has elapsed, $V_L = -\mathscr{E}_0 e^{-1} = -0.368\mathscr{E}_0$. Then the potential difference tends to zero as $t \to \infty$. The time behavior of V_L is shown in Figure 35.12. The gradual decrease in the magnitude of the potential difference is a reflection of the inertial effect of the inductor. The larger the inductance, the more slowly the potential difference approaches zero.

Decline of Current in an Inductor

Let us replace the switch in Figure 35.9 with the one shown in Figure 35.13. This new switch is designed to connect terminals 2 and 3 as it breaks the connection between terminals 1 and 3 (see Example 2). In position 1, the resistor and inductor are connected to the emf as in Figure 35.13. In position 2, a closed circuit is formed consisting of only the resistor and the inductor. With the switch in position 2 the emf is removed and Eq. 35.12 reduces to

$$L\frac{dI}{dt} + IR = 0 \qquad (35.15)$$

The solution of Eq. 35.15 is

$$I = I_0 e^{-Rt/L} \qquad (35.16)$$

where I_0 is the current in the circuit at the instant the switch connects terminals 2 and 3 and disconnects terminals 1 and 3. If the switch had connected terminals 1 and 3 long enough for the current to stabilize, then I_0 would equal $\mathscr{E}_0/R$ and Eq. 35.16 would become

$$I = \frac{\mathscr{E}_0}{R} e^{-Rt/L} \qquad (35.17)$$

The current falls exponentially from $\mathscr{E}_0/R$ to zero. In a time equal to L/R, the current has declined to $1/e = 0.368$ of its initial value. Note that the direction of the current in the inductor and the resistor is the same for both positions of the switch. This situation is different from that of the RC circuit.

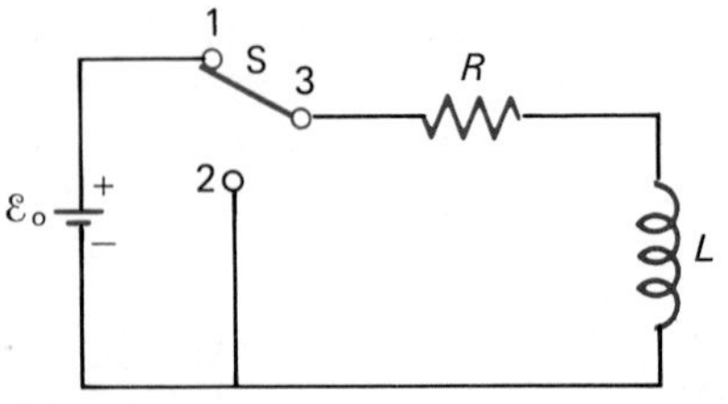

Figure 35.13

With the switch in the position shown here, charge flows through the LR series combination. At equilibrium, the current is $I_0 = \mathscr{E}_0/R$. When the switch is moved to position 2, the current decays exponentially according to $I = \mathscr{E}_0/R\ e^{(-Rt/L)}$.

Figure 35.14

The LR circuit shown in Figure 35.9 is modified to include a resistor (R_2) in parallel with the inductor.

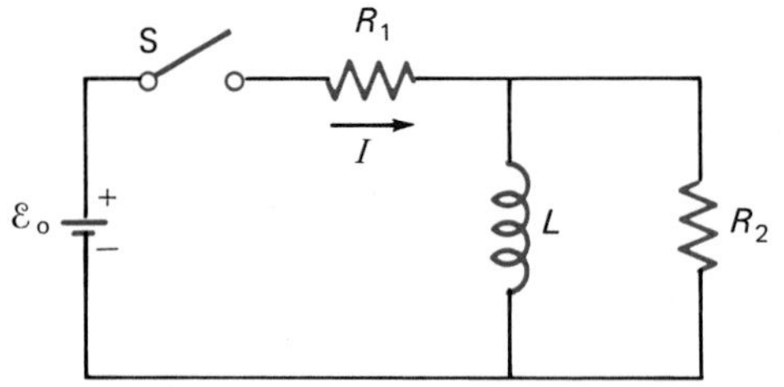

Example 3 Initial and Ultimate Currents in a Resistor–Inductor Circuit

A circuit consisting of a battery, a switch, an inductor, and two resistors is shown in Figure 35.14. Let's determine the current for the limiting cases $t = 0$ (switch is closed) and $t \to \infty$ (switch has been closed for a long time).

Before the switch is closed the current is zero throughout. When the switch is closed, the current in the inductor cannot change instantaneously. At $t = 0$, the charge flowing out of the battery must flow through R_1 and R_2, which effectively are in series when the switch is closed. Hence

$$I\ (\text{at } t = 0) = \frac{\mathscr{E}_0}{R_1 + R_2}$$

When the switch has been closed for a long time, and the magnetic field is established, the zero resistance (assumed) of the inductor is in parallel with R_2. Thus no current is in R_2, and the same current is in R_1 and L. Hence

$$I\ (\text{as } t \to \infty) = \frac{\mathscr{E}_0}{R_1}$$

Questions

9. A light bulb connected to a battery and a switch comes to full brightness nearly instantaneously when the switch is closed. However, if a large inductance is in series with the bulb, several seconds may pass before the bulb achieves full brightness. Why?

10. Figure 35.15 shows a pair of separated electrodes attached to an inductor in a circuit. When the switch is opened, the induced emf across the electrodes can exceed $\mathscr{E}_0$. What makes this possible? (This principle is used to provide a short-lived large potential difference to start a fluorescent lamp.)

11. How would you use a battery of known emf, a resistor of known value, and a means of measuring time and potential difference to help ascertain the inductance of an inductor?

12. An inductor having inductance L and resistance R is connected to a battery having emf $\mathscr{E}_0$. Why doesn't the time required for the current to achieve 90% of its maximum value depend on the value of $\mathscr{E}_0$?

Figure 35.15

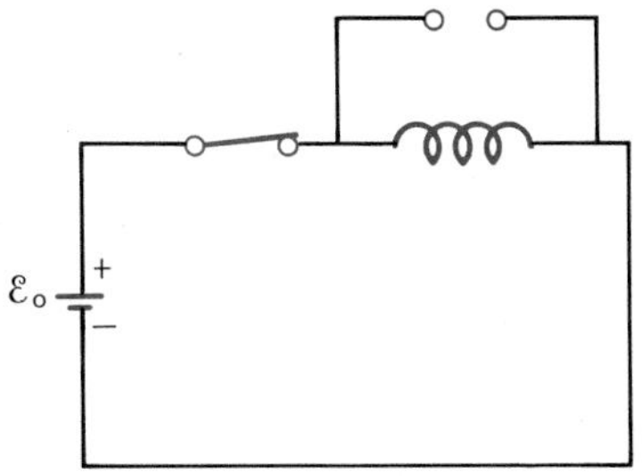

35.3 Energy Stored in the Magnetic Field

Energy Stored by an Inductor

At the beginning of this chapter we said that an inductor can store magnetic energy. By examining the *LR* circuit shown in Figure 35.16, we can now show that energy is stored in the magnetic field of the inductor. We will generalize this observation to show that energy is stored in all magnetic fields, whatever their origin. We begin by multiplying Kirchhoff's loop equation (Eq. 35.12) for this circuit by the current I to obtain

$$\mathcal{E}_o I = I^2 R + LI\frac{dI}{dt} \tag{35.18}$$

Each term in this equation has dimensions of power. For example, $\mathcal{E}_o I$ represents the rate at which energy is delivered by the emf. In a time interval dt the emf will have delivered energy equal to $\mathcal{E}_o I\, dt$. According to Eq. 35.18 this energy may be expressed as

$$\mathcal{E}_o I\, dt = I^2 R\, dt + LI\, dI \tag{35.19}$$

As discussed in Section 30.5, $I^2R\, dt$ is thermal energy produced in time dt in the resistor. To extend the principle of conservation of energy to include magnetic energy, we identify $LI\, dI$ as magnetic energy stored in time dt. The energy converted by the battery ($\mathcal{E}_o I\, dt$) in a time interval dt goes partly into thermal energy ($I^2R\, dt$) and partly into magnetic energy ($LI\, dI$). The total magnetic energy stored

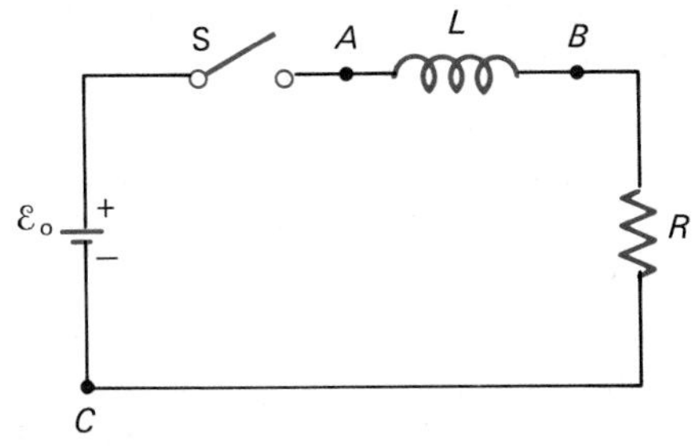

Figure 35.16

from the time the switch is closed ($t = 0$) to the time the current has reached its limiting value ($I = \mathcal{E}_o/R$) is

$$U_M = L\int_0^I I\, dI = \frac{1}{2}LI^2 \tag{35.20}$$

In Section 29.3 we saw that in a parallel-plate capacitor, electric energy is stored in the electric field between the capacitor plates. By analogy we reason that magnetic energy is stored in the magnetic field of the inductor. We also saw that when the electric field is reduced to zero, the capacitor energy is recovered. Similarly, magnetic energy is recovered when the magnetic field is reduced to zero. This suggests another way of looking at the effect of an inductance in opposing a change in the current. If we try to increase the current in an *LR* circuit, the inductance demands that more energy be stored in the increased magnetic field. It takes time to feed in this energy. The demand for stored energy prevents any instantaneous increase in the current. If we try to reduce the current in an *LR* circuit, the energy stored in the magnetic field becomes available, tending to maintain the current.

Magnetic Energy Density

If magnetic energy is stored in the magnetic field of an inductor, we should be able to express the magnetic field energy in terms of the magnetic field B. Using Eq. 33.26, which relates the magnetic field in the interior of a long solenoid to the current in the solenoid windings, we can express the current as

$$I = \frac{Bl}{\mu_o N} \tag{35.21}$$

In Section 35.1 the inductance of the solenoid was determined to be

$$L = \frac{\mu_o N^2 A}{l} \tag{35.22}$$

Hence the magnetic energy associated with the solenoid is

$$\begin{aligned} U_M &= \frac{1}{2}LI^2 \\ &= \frac{1}{2}\left(\frac{\mu_o N^2 A}{l}\right)\left(\frac{Bl}{\mu_o N}\right)^2 \\ &= \frac{1}{2}\left(\frac{1}{\mu_o}\right)(Al)B^2 \end{aligned}$$

Because the magnetic field is essentially confined to the core of the long solenoid, the magnetic energy is confined to the same volume. The volume of the core is Al, and so the magnetic energy density (magnetic energy per unit volume) is

$$u_M = \frac{U_M}{Al} = \left(\frac{1}{2\mu_o}\right)B^2 \tag{35.23}$$

This relation, although derived for a solenoid, is valid for all magnetic fields. To set up a magnetic field requires energy, and we may consider the energy to be stored in the magnetic field. The magnetic field energy density $u_M = (1/2\mu_o)B^2$ is very similar in form to the electric field energy density $u_E = \frac{1}{2}\epsilon_o E^2$ (Eq. 29.19).

Example 4 Magnetic Energy Density of a Superconducting Solenoid

Example 5 of Section 33.4 is the calculation of the magnetic field of a superconducting solenoid. The field is found to be 3.0 T. Let us determine the magnetic energy density of this field.

$$u_M = \left(\frac{1}{2\mu_o}\right)B^2$$

$$= \frac{10^7}{8\pi} \times (3.0)^2$$

$$= 3.58 \times 10^6 \text{ J/m}^3$$

$$= 3.58 \text{ J/cm}^3$$

For comparison, the energy density in gasoline is 3.5×10^4 J/cm^3. Strong forces are required to contain large magnetic energy densities even though magnetic energy densities are well below the energy density of gasoline. Magnets have been known to self-destruct as a result of magnetic forces.

35.4 Oscillations in a Circuit Containing a Capacitor and an Inductor

LC Circuit

We discussed the behavior of an *RC* circuit in Section 31.6 and that of an *LR* circuit in Section 35.2. Now let's consider the *LC* circuit. A charged capacitor connected to an inductor is an interesting and useful configuration. The inductor provides a conducting path, which allows the capacitor to discharge. By discharging, the capacitor converts the electric energy stored in its electric field to magnetic energy in the inductor. Once this transfer of energy is complete, the magnetic field begins to diminish, and the magnetic energy is transformed back to electric energy in the capacitor. In the absence of a resistor, this back-and-forth transfer of energy continues indefinitely. In principle, an *LC* circuit behaves precisely like the spring–mass version of the simple harmonic oscillator (Section 14.1). The initial potential energy of a compressed spring is transferred into kinetic energy of an attached mass. In the absence of energy-dissipating mechanisms (friction), the back-and-forth transfer of kinetic and potential energy continues indefinitely.

We can show the equivalence between a simple harmonic oscillator and an *LC* circuit formally by analyzing Kirchhoff's equation for the *LC* circuit. Let us connect a charged capacitor through a switch to an inductor having zero electrical resistance (Figure 35.17). Charge flows from the capacitor when the switch is closed. The resulting current produces in the inductor a magnetic field and an induced emf that opposes the action of the capacitor. Applying Kirchhoff's loop law, we have

Figure 35.17

A charged capacitor is connected to an inductor and a switch. There is no current as long as the switch is open. Closing the switch allows charges to flow through the inductor.

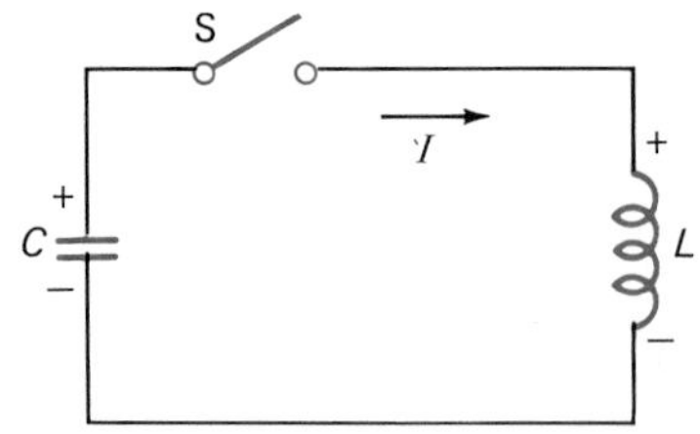

$$\frac{q}{C} - L\frac{dI}{dt} = 0 \qquad (35.24)$$

The charge on the capacitor decreases so that

$$I = -\frac{dq}{dt}$$

Therefore the equation governing the charge q is

$$\frac{q}{C} + L\frac{d^2q}{dt^2} = 0 \qquad (35.25)$$

We arrive at a differential equation identical in form to that deduced for a mass connected to a spring (Section 14.1).

$$kx + m\frac{d^2x}{dt^2} = 0 \qquad \text{spring–mass}$$

$$\frac{1}{C}q + L\frac{d^2q}{dt^2} = 0 \qquad \text{capacitor–inductor}$$

Charge q corresponds to displacement x, $1/C$ corresponds to the spring constant k, and L corresponds to mass m. Calling q_o the initial charge on the capacitor and measuring time from the instant the switch is closed, we have as the solution of Eq. 35.25

$$q = q_o \cos \omega t \qquad (35.26)$$

where $\omega = 1/\sqrt{LC}$. (You can check Eq. 35.26 by differentiating $q_o \cos \omega t$ twice with respect to time and substituting back into Eq. 35.25.) The frequency $\nu = \omega/2\pi$ is called the *natural frequency* of the oscillations. This is the oscillation frequency when there is no electrical resistance in the circuit. It is analogous to the oscillation frequency of a mechanical simple harmonic oscillator when no frictional forces are present. Differentiating Eq. 35.26 to obtain an expression for the current, we have

$$I = -\frac{dq}{dt} = \omega q_o \sin \omega t \qquad (35.27)$$

Equations 35.26 and 35.27 indicate that charge and current will vary with time as shown in Figure 35.18. In a time $\pi/2\omega$ all the initial charge has been removed from the capacitor. The capacitor then begins recharging, but with

Figure 35.18

The time variation of charge stored by the capacitor, and the associated current, in the circuit shown in Figure 35.17. The charge stored is a maximum when the current is zero. We say that the charge stored and current are $\pi/2$ radians out of phase.

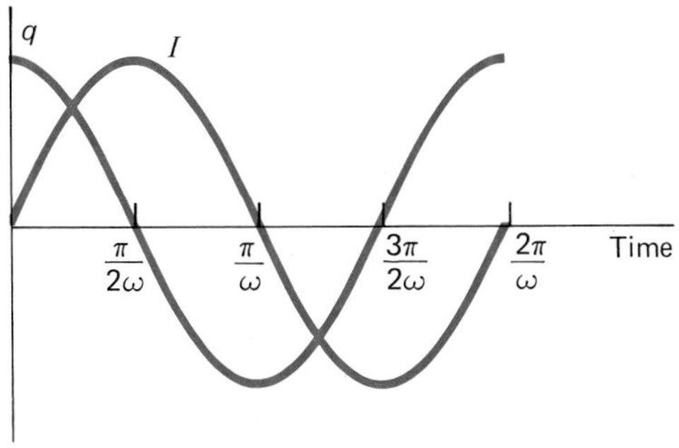

Figure 35.19

Oscilloscope recording of the potential difference across a capacitor for a circuit like that shown in Figure 35.17.

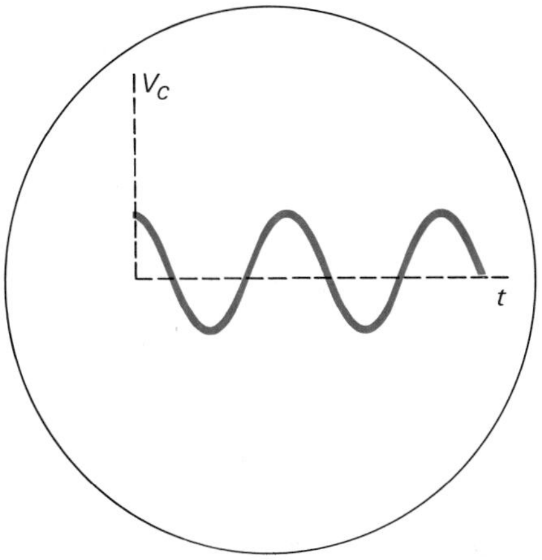

the polarity reversed. In a time π/ω the plate that was initially charged positive has achieved a negative charge $-q_o$. Continuing, the capacitor discharges, recharges, and regains its initial charge and polarity after a time $2\pi/\omega$ has elapsed. An actual oscilloscope recording of the potential difference across the capacitor in an *LC* circuit is shown in Figure 35.19.

Initially the capacitor has energy $U_o = \frac{1}{2}(q_o^2/C)$. At any instant thereafter its electric energy (U_E) is

$$U_E = \frac{1}{2}\frac{q^2}{C} = \left(\frac{1}{2}\frac{q_o^2}{C}\right)\cos^2 \omega t$$

$$= U_o \cos^2 \omega t$$

If we start with zero magnetic energy, the magnetic energy (U_M) stored by the inductor at any instant is

$$U_M = \frac{1}{2}LI^2$$

$$= \frac{1}{2}L\omega^2 q_o^2 \sin^2 \omega t$$

Since

$$\omega^2 = \frac{1}{LC}$$

then

$$U_M = \left(\frac{1}{2}\frac{q_o^2}{C}\right)\sin^2 \omega t$$

$$= U_o \sin^2 \omega t$$

The total energy at any instant is

$$U_{total} = U_E + U_M$$

$$= U_o(\cos^2 \omega t + \sin^2 \omega t)$$

$$= U_o$$

Thus, analogous to a mechanical simple harmonic oscillator, the total energy is conserved.

LCR Circuit

Now that we have examined *RC*, *LR*, and *LC* circuits, let's take a look at the fourth and final combination of our three two-terminal circuit elements: the *LCR* circuit. An *LCR* circuit consists of a charged capacitor connected to a resistor and an inductor (Figure 35.20). Unless the inductor is made of a superconducting material, it has some electrical resistance. Consequently, thermal energy will be produced, thereby diminishing the electric and magnetic energy in the circuit. Eventually the back-and-forth transfer of energy between the inductor and the capacitor vanishes as all the available energy is converted to thermal energy. The behavior is analogous to that of an oscillator in which mechanical energy is converted to thermal energy through frictional forces (Section 14.5). We account for the electrical resistance by including a potential difference term for the resistance in Eq. 35.24.

$$\frac{q}{C} - L\frac{dI}{dt} - IR = 0 \qquad (35.28)$$

Using $I = -dq/dt$, we obtain

$$\frac{q}{C} + L\frac{d^2q}{dt^2} + R\frac{dq}{dt} = 0 \qquad (35.29)$$

Equation 35.29 is identical in form to Eq. 14.37, which describes the motion of a damped harmonic oscillator. For example, the motion of the mass in Figure 35.21, which

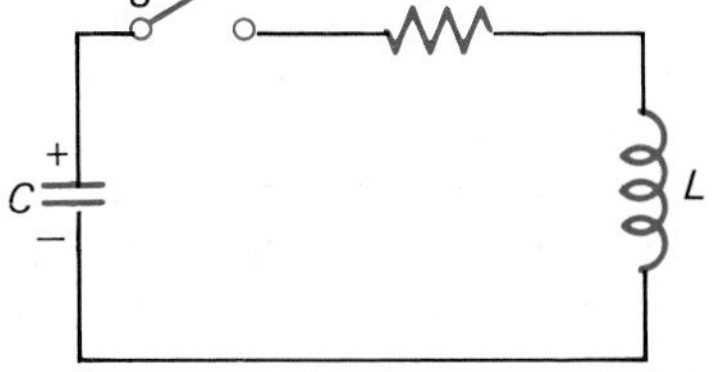

Figure 35.20

The basic *LCR* circuit, which consists of a series connection of an inductor, a capacitor, and a resistor.

Figure 35.21

(a) A mass connected to a spring and moving in a viscous medium constitutes a damped mechanical harmonic oscillator. (b) A series connection of an inductor, a capacitor, and a resistor is the electrical analog of the damped mechanical harmonic oscillator.

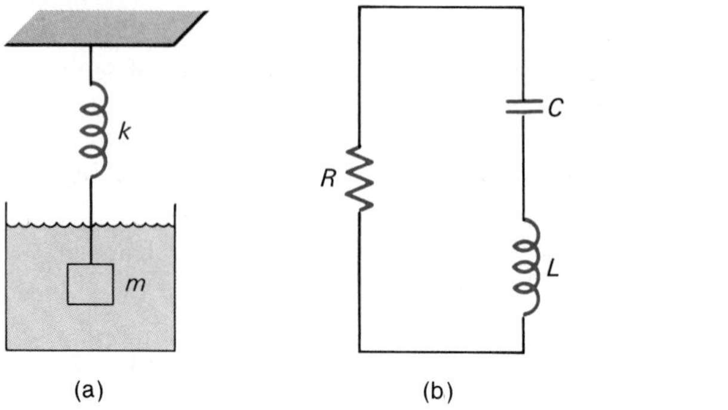

Figure 35.22

The charge on a capacitor in an LCR circuit varies with time as shown in this figure. The decline in the size of the peaks is caused by resistance in the circuit.

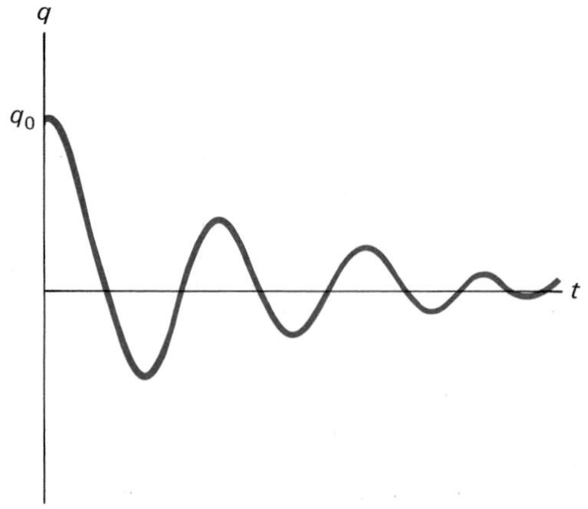

Figure 35.23

Oscillations do not occur in an overdamped LCR circuit; the capacitor simply discharges as shown in this figure. The discharge is most rapid for the critically damped condition.

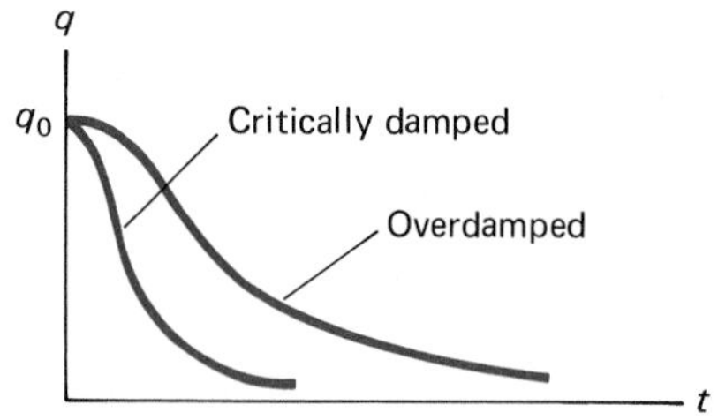

experiences the forces of a connected spring $(-kx)$ and a viscous medium $(-\gamma\, dx/dt)$, is

$$m\frac{d^2x}{dt^2} + \gamma\frac{dx}{dt} + kx = 0$$

For the behavior of the LCR circuit (Figure 35.21b) we again have Eq. 35.29:

$$L\frac{d^2q}{dt^2} + R\frac{dq}{dt} + \frac{q}{C} = 0$$

This equation representing charge on the capacitor as a function of time is analogous to Eq. 14.37, which represents the position of the mass connected to a spring as a function of time. To make the analogy we identify the following:

charge $q \Rightarrow$ displacement x

inductance $L \Rightarrow$ mass m

resistance $R \Rightarrow$ viscous force constant γ

reciprocal of capacitance $\frac{1}{C} \Rightarrow$ spring constant k

Having studied the mechanical oscillator in detail in Chapter 14, we can draw on that presentation to analyze the behavior of the electrical LCR circuit. Assuming that $q = q_o$ at $t = 0$, we have as the solution of Eq. 35.29 for the condition $1/LC > (R/2L)^2$

$$q = q_o e^{-Rt/2L} \cos \omega' t \qquad (35.30)$$

where

$$\omega' = \sqrt{\frac{1}{LC} - \left(\frac{R}{2L}\right)^2}$$

You can check this solution by differentiating Eq. 35.30 and substituting into Eq. 35.29 (see Problem 44). As a plausibility check, note that when $R = 0$, Eq. 35.30 reduces to the result deduced, assuming that the circuit has no electrical resistance. The solution indicates that oscillations still occur but at a lower frequency. The exponential term continually reduces the amplitude of the oscillations until they eventually die out (Figure 35.22). Increasing the resistance makes the damping of the oscillations more severe and reduces the frequency of the oscillations. At a critical value of R such that $(R/2L)^2 = 1/LC$, the circuit no longer oscillates, and the capacitor simply discharges, converting its electric energy directly to thermal energy. As long as $(R/2L)^2 > 1/LC$, the circuit will not oscillate. The circuit is said to be *critically damped* if $(R/2L)^2 = 1/LC$ and *overdamped* if $(R/2L)^2 > 1/LC$ (Figure 35.23). This behavior is completely analogous to the behavior of the damped mechanical oscillator discussed in Section 14.5.

Use of *LCR* Circuits

The analogy between oscillating LCR electrical circuits and oscillating mechanical systems makes it possible to devise electrical analogs of mechanical systems; and because it generally is much easier to take electrical measurements than it is to take mechanical measurements, it is often easier to determine the behavior of a mechanical system by studying its electrical analog. For example, shock absorbers on automobiles are designed to damp out oscillations that may occur when the automobile hits a bump in the road. An engineer can explore the characteristics of a variety of different shock absorbers by studying analog electrical circuits. Electronic analog computers have been developed to facilitate studies like these.

The use of combinations of capacitors and inductors is widespread. They are used to generate controlled oscillations in a variety of electronic circuits. Tuning circuits such as those in radios and television sets are often *LCR* circuits.

Questions

13. Inertial mass m is analogous to self inductance L. Velocity v is analogous to electric current I. What is the mechanical analogy of the energy stored in the magnetic field of an inductor?

14. In the absence of friction, the equation of motion for a mass connected to a spring is $m\, d^2x/dt^2 + kx = 0$. In the absence of resistance, the equation of "motion" for charge in an inductor–capacitor circuit is $L\, d^2q/dt^2 + (1/C)q = 0$. To what extent are the inductance (L) and inertial mass (m) physically similar?

15. The natural frequency of oscillation of a mass connected to a spring is

$$\nu = \frac{1}{2\pi}\sqrt{\frac{k}{m}}$$

By comparing the expressions for the natural frequency of oscillation of charge in an inductor–capacitor circuit, explain why inductance exhibits inertia in an electrical circuit.

16. Figure 35.18 shows that for an inductor–capacitor combination the charge on the capacitor and current in the circuit are out of phase by $\pi/2$ rad. Does this ($\pi/2$)-rad phase difference still prevail if the resistance of the inductor is not zero? (Phase was discussed in Section 17.3).

17. If the inductance of a certain inductor depended on the current in the inductor, could the magnetic energy still be expressible as $U = \frac{1}{2}LI^2$?

Summary

A conducting loop, generally in the form of a coil or wire, designed to exploit inductive effects is called an inductor and symbolized by .

The magnetic flux (Φ) threading a conducting loop as a result of current in the loop is directly proportional to the current (I).

$$\Phi = LI \qquad (35.4)$$

The constant of proportionality L is called the self-inductance of the loop.

A changing current in a conducting loop creates a changing magnetic flux within the loop. The changing magnetic flux induces an emf in the loop. The induced emf is related to self-inductance and the time rate of change of current by

$$\mathscr{E} = -L\frac{dI}{dt} \qquad (35.6)$$

Self-inductance has units of ohm seconds. One ohm second is called a henry, abbreviated H.

Functioning like an electrical inertia, self-inductance measures the resistance to a change in the current in an inductor.

The self-inductance of a long, closely wound solenoid of cross-sectional area A, length l, and N turns is

$$L = \frac{\mu_o N^2 A}{l} \qquad (35.8)$$

The self-inductance of a coaxial cable of length l, inner radius r_1, and outer radius r_2 is

$$L = \frac{\mu_o l}{2\pi}\ln\left(\frac{r_2}{r_1}\right) \qquad (35.10)$$

The current in a series combination of an emf, a resistor, and an inductor is given by

$$I = \frac{\mathscr{E}_o}{R}(1 - e^{-Rt/L}) \qquad (35.13)$$

Having units of seconds, L/R is called the inductive time constant. In a time equal to L/R, the current equals 63.2% of the maximum value ($\mathscr{E}_o/R$).

The magnetic energy stored by an inductor carrying a current I is identified as

$$U_M = \frac{1}{2}LI^2 \qquad (35.20)$$

The magnetic energy density (joules per cubic meter) in the magnetic field is

$$u_M = \left(\frac{1}{2\mu_o}\right)B^2 \qquad (35.23)$$

Given a circuit containing a capacitor and an ideal inductor (zero resistance), there is a regular interchange of electric and magnetic energy. The frequency of these oscillations is given by

$$\nu = \frac{1}{2\pi\sqrt{LC}}$$

Resistance in the circuit produces thermal energy losses and monotonically decreases the amplitude of the oscillations. For $(R/2L)^2 \geq 1/LC$, no oscillations occur and the energy is gradually transformed to thermal energy.

Problems

Section 35.1 Self-Inductance and Inductors

1. When the current in an inductor is 1.2 mA, the magnetic flux threading the turns of wire in the inductor is 2.76×10^{-6} Wb. Determine the inductor's self-inductance.

2. If the self-inductance of a certain circular loop of wire is 2×10^{-4} H, how much magnetic flux threads the loop when the current in the loop is 2.5×10^{-2} A?

3. The total magnetic flux produced by a certain inductor can be represented as $\Phi_B = 0.001I$ Wb where I is the current in the inductor. Determine the inductance of this inductor.

4. A particular loop of wire is found to have a self-inductance of 3.4×10^{-4} H. If the magnetic flux through the loop is 9.1×10^{-6} Wb, what is the current in the loop?

5. A changing current in a 0.1-H inductor produces a constant potential difference of +2 V across the inductor. How does this current depend on time?

6. The current in a 10-mH inductor is observed to vary exponentially according to $I_L = 3.8e^{-0.21t}$ mA if t is expressed in seconds. Deduce the functional form of the induced emf in the inductor.

7. The current in an inductor with $L = 3$ mH is given by

   $$I = 2.2\cos(60t)\text{ A}$$

   where t is in seconds. Determine the maximum emf induced across the inductor.

8. Because of a misprint, a certain textbook reports the inductance of a long solenoid to be $L = \mu_o n^2 A$ where n is the number of turns per unit length of the solenoid and A is the cross-sectional area of the solenoid. Without actually calculating the inductance of a solenoid or looking up the expression for the inductance, show why this expression for inductance is incorrect.

9. Show that the self-inductance per unit volume of a solenoid is $\mu_o n^2$ where n is the number of turns per unit length.

10. Potential difference and current measurements for an inductor in a circuit produce the graphs shown in Figure 1. Estimate the inductance of the inductor from the data presented in the graphs.

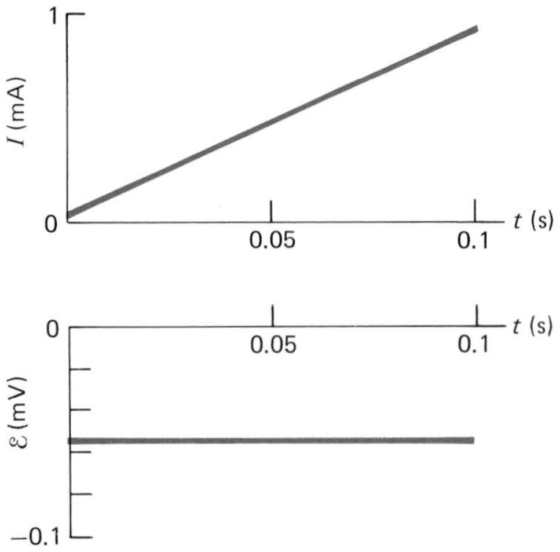

Figure 1

11. The current in a 0.01-H inductor is changing at a constant rate of 10 A/min. Determine the induced emf in the inductor.

12. A manufacturer of superconducting solenoids rates the magnetic field as a function of current of a particular model as 0.1338 T/A. How many turns per meter does this solenoid have?

13. A superconducting solenoid has an inductance of 19 H. How much magnetic flux threads the turns of the solenoid when it carries a current of 120 A?

14. A type of commercial coaxial cable uses the prefix *RG*- to designate a particular construction. For example, type RG-58/U has an inner wire diameter of 0.812 mm and an outer conductor diameter of 3.24 mm. Determine the inductance per unit length of type RG-58/U coaxial cable.

15. A solenoid with tightly formed turns of wire forms a toroid (doughnut) when bent into a circular shape. If the solenoid's length is very large compared to its diameter, then the wires will still be tightly wound after the bending. Following the analysis in Section 35.1 for the determination of the self-inductance of a solenoid, calculate the self-inductance of a toroidal coil.

16. The lead-in wires from a TV antenna are often constructed in the form of two parallel wires (Figure 2).

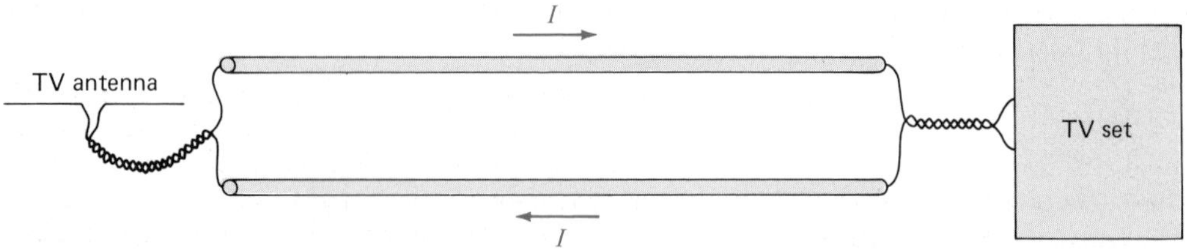

Figure 2

(a) Why does this configuration of conductors have a self-inductance? (b) What constitutes the flux loop for this configuration? (c) Neglecting any magnetic flux inside the wires, show that the self-inductance of a length l of this type of lead-in is

$$L = \left(\frac{\mu_0}{\pi}\right) l \ln\left(\frac{w - a}{a}\right)$$

where a is the radius of the wires and w is the center-to-center separation of the wires.

17. Using the relations $V = IR$ and $\mathcal{E} = -L\, dI/dt$, show that L/R has dimensions of time.

Section 35.2 Circuit Aspects of Inductors

18. An inductor having an intrinsic resistance of 100 Ω and an inductance of 0.25 H is connected to a 10-V battery by a switch that initially is open. After the switch is closed, calculate the time required for the current to increase to 75 mA.

19. A 6-V battery, a switch that initially is open, and an inductance of 100 mH and resistance of 1.5 Ω are all connected in series. (a) While the switch is open, determine the current in the inductor. (b) Determine the current in the inductor 0.12 s after the switch is closed.

20. A series circuit contains only an inductor (L = 0.1 H) and a resistor (R = 6 Ω). At a certain moment the current is 5 A. (a) What is the potential difference across the inductor? (b) Calculate the rate of change of current in amperes per second. (c) The circuit is suddenly broken, dropping from 5 A to zero in a time of 1 ms. Estimate the current while the circuit is being broken.

21. The inductor in the circuit in Figure 3 has negligible resistance. When the switch is opened after having been closed for a long time, the current in the inductor drops to 0.25 A in 0.15 s. What is the inductance of the inductor?

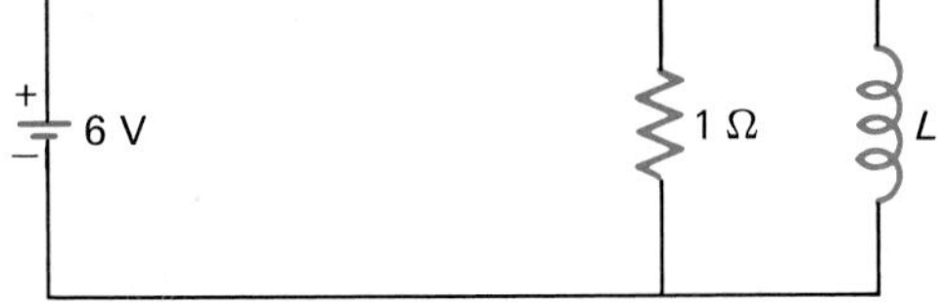

Figure 3

22. Consider the circuit shown in Figure 35.9. With $\mathcal{E}_0$ = 6 V, R = 200 Ω, and L = 3 mH, determine the current when the rate of change of current is 800 A/s.

23. A circuit consisting of a 6-V battery, 0.1-H inductor, and two 1000-Ω resistors is shown in Figure 4. Determine the currents I_1 and I_2 for the limiting cases t = 0 (switch is just closed) and $t \to \infty$ (switch is closed for a long time).

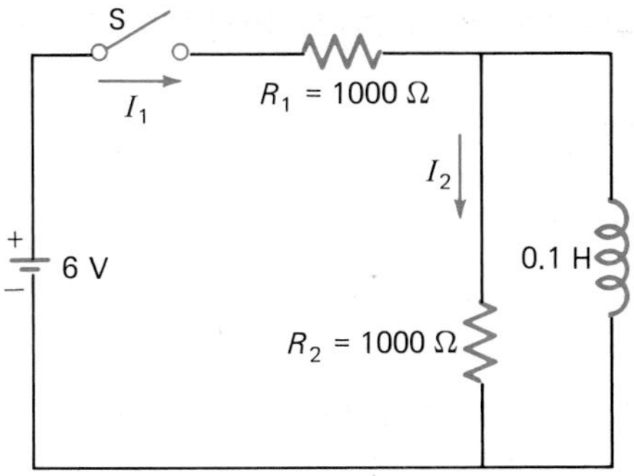

Figure 4

24. Using Kirchhoff's loop rule, show that the potential difference across the switch in Figure 35.9 is $\mathcal{E}_0$ when the switch has been opened for a long time ($t \gg L/R$). If the switch is opened after having been closed for a long time, how is it possible for the potential difference across the switch to exceed $\mathcal{E}_0$?

25. Measurements of the battery current in a circuit similar to that in Figure 4 reveal that the current increases from a minimum of 0.25 A when the switch is first closed to a maximum of 1 A when the switch is closed for a long time. Assuming that the emf of the battery is 6 V, what are the values of R_1 and R_2?

26. (a) If the switch in the circuit in Figure 5 has been open for a long period of time, what is the current in the inductor? (b) Qualitatively, how do you expect the current in the inductor to change when the switch is closed? (c) Using Lenz's law as a guide, determine the polarity of the points A and B after the switch is closed. (d) Argue why the current in the inductor after the switch is closed is

$$I = \left(\frac{\mathcal{E}_0}{r + R}\right) e^{-Rt/L}$$

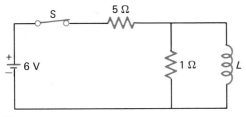

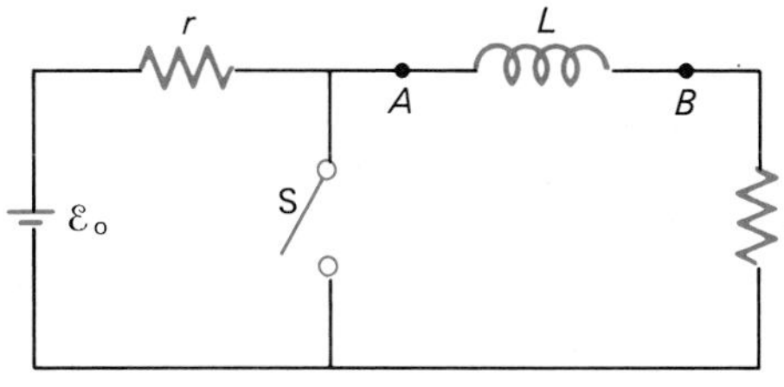

Figure 5

27. Determine the LR time constant for a research magnet having an inductance of 5.3 H and a resistance of 2.1 Ω.

28. The current in an LR circuit (Figure 35.13) drops to one half its initial current in 0.24 s. How long will it take for the current to drop to one tenth of its initial value?

29. An inductor (L), a resistor (R), and an open switch are connected in series with an emf ($\mathcal{E}$). At t = 0 the switch is closed. (a) How many L/R periods are required for the current to rise to 90% of its asymptotic value ($\mathcal{E}/R$)? (b) How many L/R periods are required for the current to rise to 99% of its asymptotic value?

Section 35.3 Energy Stored in the Magnetic Field

30. The magnetic field inside a superconducting solenoid is 4.5 T. The solenoid has an inner diameter of 6.2 cm and a length of 26 cm. (a) Determine the magnetic energy density in the field. (b) Determine the magnetic energy stored in the magnetic field.

31. A superconducting magnet used for research purposes is 20 cm long and 3 cm in diameter. When it produces a magnetic field of 7.0 T, what is (a) the magnetic energy density produced and (b) the total magnetic energy stored? Assume that the magnetic field is contained entirely within the solenoid.

32. The total magnetic energy of a research solenoid is 700,000 J when the current is 240 A. Compute the inductance of the solenoid.

33. Show that the magnetic energy per unit length of a coaxial cable is given by

$$\frac{U_M}{l} = \frac{\mu_o}{4\pi} \ln\left(\frac{r_2}{r_1}\right) I^2$$

where r_1 and r_2 are the radii of the inner and outer conductors and I is the current in the inner (or outer) conductor.

34. Using the result of Problem 35.33, calculate the magnetic energy per unit length of a coaxial cable having an inner radius of 0.406 mm and an outer radius of 3.12 mm, and carrying a current of 0.25 A.

35. An inductor and a resistor are connected in a single loop. The current reduces to zero from I_o with a time constant L/R. How much time elapses before the energy converted to heat equals the energy stored at that moment?

36. The strength of the earth's magnetic field near its surface is about 5×10^{-5} T. Compare the magnetic energy density of the earth's magnetic field near the surface with the magnetic energy density in a superconducting magnet producing a magnetic field of 7 T.

37. A circular wire loop of radius a carries a current I. Calculate the magnetic energy density for points along the axis of rotational symmetry. Show that for axial distance $x \gg a$, the magnetic energy density falls off as the sixth power of the distance.

38. The magnetic field inside a long wire of radius a carrying a current I is (Eq. 33.23)

$$B = \frac{\mu_o I r}{2\pi a^2}$$

where r is the radial distance from the center of the wire. (a) Derive a relation for the total internal magnetic energy per unit length of the wire. (b) Compute the magnetic energy per meter inside a wire 2 mm in diameter that carries a current of 1.2 A.

39. The magnetic field outside a long wire of radius a and carrying a current I is (Eq. 33.11)

$$B = \frac{\mu_o I}{2\pi r}$$

where r is the radial distance from the center of the wire. (a) Derive a relation for the magnetic energy per unit length contained between the surface of the wire and a radial distance R from the center of the wire. (b) Compute the magnetic energy per meter between the surface of a wire 2 mm in diameter and a distance 10 mm from the center of the wire if the wire carries a current of 1.2 A. Compare the magnetic energy per meter with that contained inside the wire (see Problem 38).

40. (a) Compute the electric energy density in a parallel-plate capacitor having a plate separation of 0.01 mm and a potential difference of 2.7 V between the plates. (b) Compare this electric energy density with the magnetic energy density in a solenoid having 125 turns per centimeter and carrying a current of 0.13 A.

Section 35.4 Oscillations in a Circuit Containing a Capacitor and an Inductor

41. A 0.1-μF capacitor initially with a charge of 1 μC is connected in parallel with an inductor having a resistance of 2 Ω and an inductance of 0.1 μH. Make a plot of the charge on the capacitor as a function of time.

42. A 100-μF capacitor initially with a charge of 10 μC is connected in parallel with an inductor having a measurable resistance. After several oscillations in the circuit the charge on the capacitor has decreased to 1 μC at an instant when the current is zero. Account for the energy transformation in the circuit.

43. Consider the circuit shown in Figure 35.17 with L = 2 mH and C = 4 μF. When the switch is closed, determine the frequency of the oscillating current.

44. Show that Eq. 35.30 is a solution of Eq. 35.29.

45. In an LC circuit the maximum energy stored in the capacitor equals 12 J. (a) What is the maximum energy stored in the inductor? (b) What is the total energy stored in the circuit?

46. Consider an ideal LC circuit (Figure 35.17) in which the capacitor has a charge Q_o. At $t = 0$ the switch S is closed. (a) Show that Kirchhoff's loop rule leads to

$$\frac{q}{C} - L\frac{dI}{dt} = 0$$

(b) From part (a) derive the equation

$$\frac{d^2I(t)}{dt^2} + \frac{I(t)}{LC} = 0$$

(c) Show that the equation of part (b) is satisfied by

$$I(t) = I_{max} \sin \omega t$$

with $\omega^2 = 1/LC$. (d) Determine I_{max} in terms of the initial charge Q_o.

chapter 36

MAGNETIC PROPERTIES OF MATTER

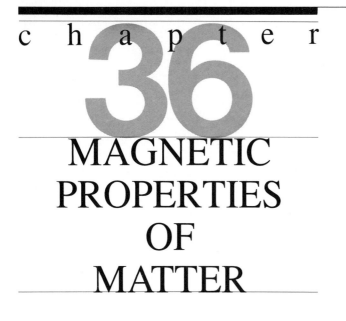

Preview

We begin this study of magnetic properties of matter by making qualitative observations of the behavior of various materials placed in a nonuniform magnetic field. These observations allow us to classify magnetic materials as diamagnetic, paramagnetic, and ferromagnetic. Measurements of the magnetic dipole moment of a wide variety of materials then lead to a quantitative classification of magnetic materials in terms of the magnetic susceptibility. A completely satisfactory explanation of atomic magnetism requires an understanding of modern quantum physics, but some insight is provided in this chapter from a knowledge of magnetic effects produced by circulating currents. We conclude with a discussion of the properties and possible origins of the earth's magnetic field, and a few examples of the use of magnetic materials.

36.1
Behavior of Materials in a Nonuniform Magnetic Field

Two pieces of ordinary iron do not exert magnetic forces on each other. Yet iron and many other materials experience a force when placed in a *nonuniform* magnetic field. We see this, for example, when a small paper clip is attracted to the pole of a permanent magnet, or when an electromagnet in a scrapyard is used to lift an old car or truck. If the magnetic field of this scrapyard magnet were sufficiently nonuniform, the magnet would be able to pick up aluminum beverage cans as well as iron. In contrast, copper in the vicinity of this same magnet would experience a repulsive force (Figure 36.1).

Two striking features emerge from studies of the force exerted on materials by a nonuniform magnetic field. (1) The force exerted on a sample of material by a given nonuniform magnetic field is either very strong (if the sample is iron, for example) or very weak (if the sample is aluminum or copper). (2) Some weakly affected materials will be urged in the same direction as a piece of iron and others will be urged in the opposite direction (as when copper is repulsed). The magnitude and direction of the magnetic force experienced by different materials allow us to classify matter as diamagnetic, paramagnetic, or ferromagnetic. Table 36.1 lists these three categories of matter and includes examples of each.

Figure 36.1

Comparison of the force on an iron, an aluminum, and a copper nail in the magnetic field of a toy bar magnet.

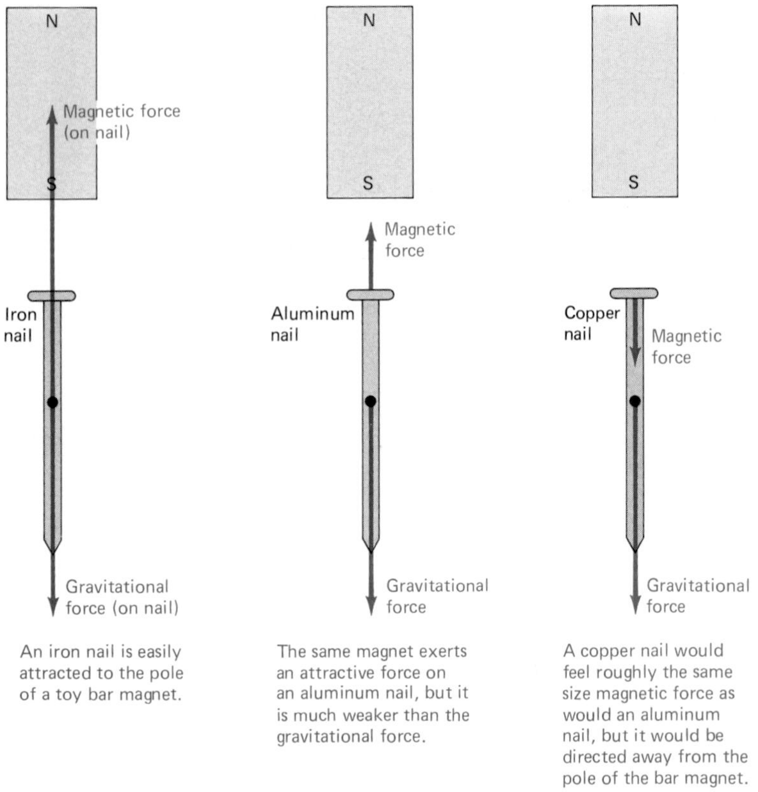

N
Magnetic force (on nail)
S
Iron nail
Gravitational force (on nail)
An iron nail is easily attracted to the pole of a toy bar magnet.
N
S
Magnetic force
Aluminum nail
Gravitational force
The same magnet exerts an attractive force on an aluminum nail, but it is much weaker than the gravitational force.
N
S
Copper nail
Magnetic force
Gravitational force
A copper nail would feel roughly the same size magnetic force as would an aluminum nail, but it would be directed away from the pole of the bar magnet.

To experience a force in a nonuniform magnetic field, a material must possess a magnetic dipole moment (Section 32.4). This moment is either a permanent feature of the material or is induced by the magnetic field. In some cases, an induced magnetic moment can become permanent. An examination of induced magnetic dipole moments provides quantitative information for the classification scheme presented in Table 36.1. By first examining the effect of a magnetic dipole such as a compass needle on the magnetic field that it is placed in, we can clarify our interpretation of the magnetic dipole moment measurements.

Let a material having no initial magnetic moment be placed in a nonuniform magnetic field such as that shown in Figure 36.2. Suppose it is urged toward a region of greater concentration ($dB/dx > 0$) of magnetic field lines (toward the S pole in Figure 36.2). The induced magnetic moment μ must be aligned in the direction of the magnetic field lines (see Section 32.4). Conversely, if a material having no permanent magnetic moment is urged toward a region of lower concentration ($dB/dx < 0$) of magnetic field lines (toward the N pole in Figure 36.2), then the induced magnetic moment must be aligned opposite to the direc-

Table 36.1

Magnetic classification of materials according to the force experienced in a nonuniform field

Magnetic Category of Matter	Interaction	Direction of Force	Examples
diamagnetic	weak	toward a region of weaker magnetic field	copper, bismuth, carbon, gold, lead, silver, zinc
paramagnetic	weak	toward a region of stronger magnetic field	aluminum, chromium, palladium, platinum, potassium, sodium, tungsten (wolfram), manganese, magnesium
ferromagnetic	strong	toward a region of stronger magnetic field	iron, nickel, cobalt, gadolinium, dysprosium

Figure 36.2

A magnetic dipole such as a compass needle will be attracted to a region of greater concentration of magnetic field lines.

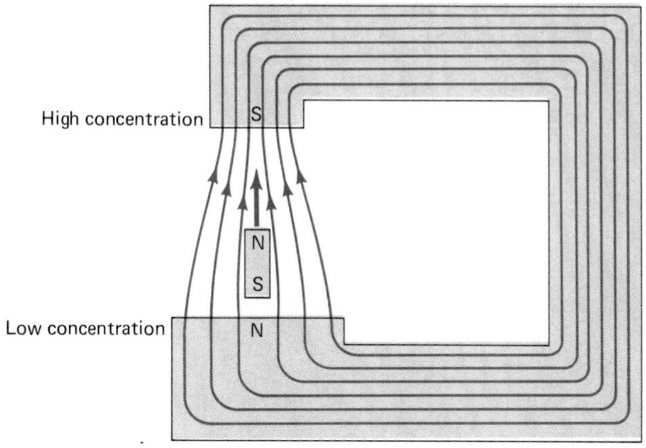

tion of the magnetic field lines. Measurements of the induced magnetic moment confirm these suppositions. In the next section we look at how these measurements are made.

Questions

1. Can a magnetic dipole be induced in a piece of aluminum placed in a uniform magnetic field?
2. Why will a piece of soft iron be pulled into the core of a solenoid when placed near the end of the solenoid? What would happen if the piece of metal were bismuth rather than iron?

36.2 Classification of Magnetic Materials

Magnetic Dipole Moments

We evaluate the magnetic properties of materials in much the same way that we assessed the electrical properties of insulators in Section 29.4. There we studied the alteration of an electric field between the plates of a capacitor when an insulator was inserted between the capacitor plates. Here we determine the modification of a magnetic field when a material of interest is placed in the magnetic field. To avoid complications we want to ensure that the magnetic field is *uniform* in the region occupied by the material being studied. Therefore we use the magnetic field in the interior of a long solenoid, because the magnetic field inside such a solenoid is nearly uniform (Figure 36.3).

We saw in Section 32.4 that a current I in a loop of wire of cross-sectional area A produces a magnetic dipole moment.

$$\boldsymbol{\mu} = I\mathbf{A} \qquad (36.1)$$

The direction of $\boldsymbol{\mu}$ is perpendicular to the plane defined by $\mathbf{A}$ (Figure 36.4). A long solenoid is equivalent to N closely spaced identical loops of wire, each having the same current I. Each loop produces an identical magnetic dipole moment $I\mathbf{A}$ aligned along the axis of the solenoid. Thus the solenoid possesses a net magnetic dipole moment.

$$\boldsymbol{\mu} = NI\mathbf{A} \qquad (36.2)$$

If the solenoid (Figure 36.5) is placed in a magnetic field, then like the needle of a compass, the solenoid experiences a torque

$$\boldsymbol{\tau} = \boldsymbol{\mu} \times \mathbf{B} \qquad (36.3)$$

Figure 36.3

A solenoid consists of closely spaced turns of wire on a cylindrical form. For a long solenoid, the magnetic field is nearly uniform in the interior of the windings.

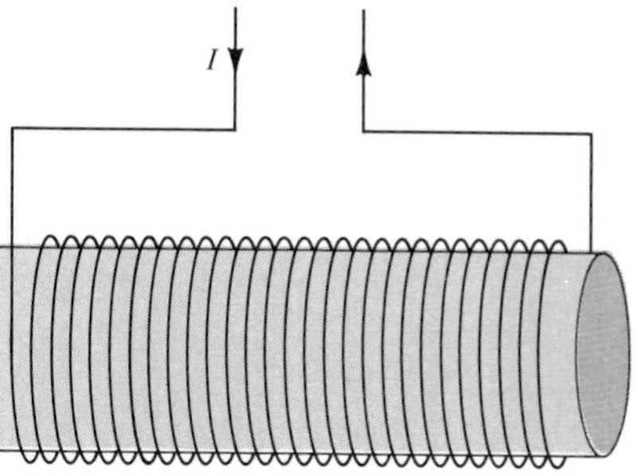

Figure 36.4

A current I in a circular loop of wire produces a magnetic dipole moment $\boldsymbol{\mu} = I\mathbf{A}$ that is oriented perpendicular to the plane of the loop. The direction of $\boldsymbol{\mu}$ is determined by using the right-hand rule.

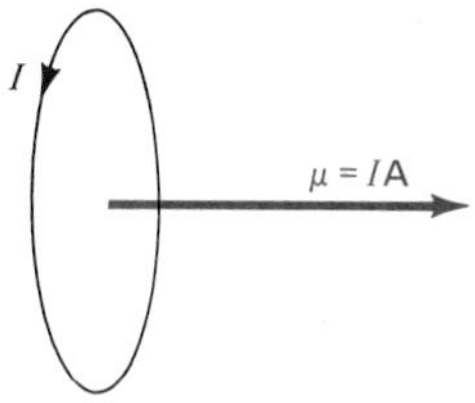

Figure 36.5

A long solenoid is equivalent to N closely spaced loops of wire, each having magnetic dipole $I\mathbf{A}$. Thus the solenoid possesses a magnetic dipole moment $\boldsymbol{\mu} = NI\mathbf{A}$. If placed in a magnetic field, the solenoid experiences a torque $\boldsymbol{\tau} = \boldsymbol{\mu} \times \mathbf{B}$ and rotates on the pivot point.

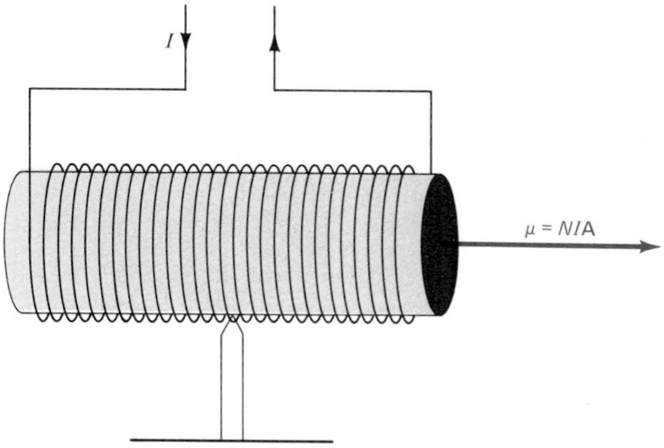

In a vacuum, the magnetic field inside a long solenoid is (Section 33.4)

$$B_E = \mu_o\left(\frac{N}{l}\right)I \qquad (36.4)$$

where l is the length of the solenoid, μ_o is the permeability of free space,* and the subscript E reminds us that the interior of the solenoid is empty. Multiplying and dividing the right-hand side of Eq 36.4 by the cross-sectional area A, we can rewrite Eq. 36.4 in terms of the magnetic dipole moment NIA (Eq. 36.2) as

$$B_E = \mu_o\left(\frac{NIA}{Al}\right)$$

$$= \frac{\mu_o\mu}{V} \qquad (36.5)$$

where $V = Al$ is the volume contained within the windings of the solenoid. If N and A are known, a measurement of the current I determines the magnetic moment μ. The magnetic moment is changed by changing the current. When the inside of the solenoid is filled with the material that we are studying, and a magnetic dipole moment is then induced, the magnetic dipole moment of the solenoid changes and the magnetic field in the sample material also changes. A measurement of the net magnetic field within the solenoid determines how its magnetic dipole moment has been altered by the presence of the material.

Magnetic Field Measurement

A direct measurement of the magnetic field inside a material of interest is difficult because a probe cannot be inserted inside the sample.† Therefore, we generally measure some quantity that is related to magnetic field. One such quantity is magnetic flux. To measure the magnetic flux in a material of interest, we place a coil of wire (called a *sense coil*) around a solenoid that is filled with the material (Figure 36.6). By closing the switch we connect an emf to the solenoid and produce a change of flux in the interior of the sense coil (Figure 36.7). This change in flux is related to the change in magnetic field by

$$\Delta\phi = NA\,\Delta B \qquad (36.6)$$

where N is the number of turns of wire in the sense coil and A is the area defined by the perimeter of a single turn. According to Lenz's law the change in flux is related to the induced emf by

$$\Delta\phi = -\int \mathscr{E}\,dt \qquad (36.7)$$

Equating Eqs. 36.6 and 36.7 we find that

$$\Delta B = -\frac{1}{NA}\int \mathscr{E}\,dt \qquad (36.8)$$

*We are using μ to denote magnetic moment and μ_o to denote the permeability of free space.

†Some internal measurements can be made by using a beam of neutrons. The internal magnetic fields act on the magnetic moment of the neutron.

Figure 36.6

Experimental arrangement for investigating the magnetic properties of matter. When the switch is closed, the ensuing magnetic flux change induces a current in the sense coil. A measurement of the total charge moved through the sense coil leads to a determination of the magnetic field in the sample.

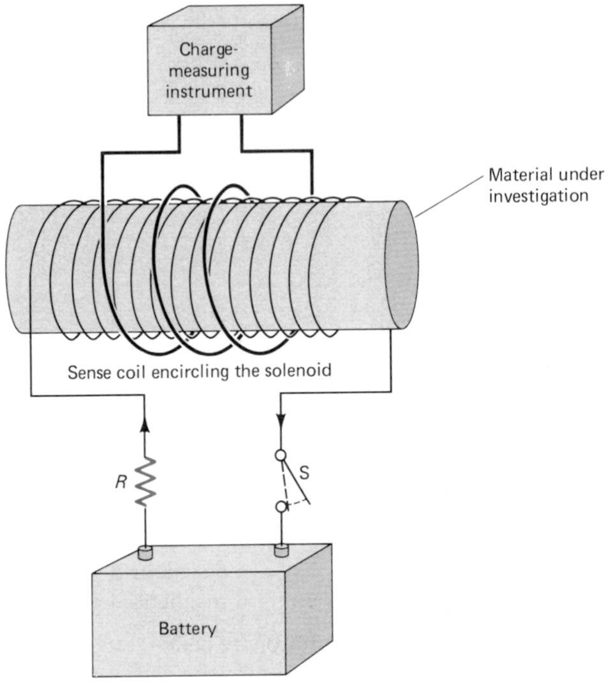

Figure 36.7

(a) With no current in the solenoid, no magnetic field lines thread the turns of the sense coil. (b) With current in the solenoid, magnetic field lines produced by the solenoid thread the turns of the sense coil. Resistor r represents a meter connected to the coil to measure a current or charge. There is no current in the coil so long as the magnetic flux is not changing.

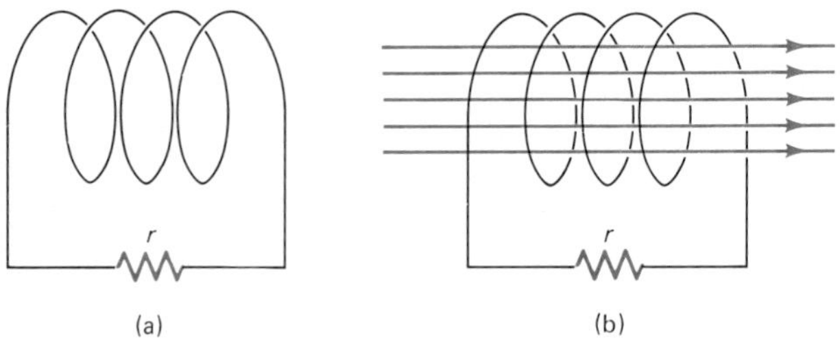

The current produced in the sense coil is related to the induced emf and the resistance of the sense coil by

$$I = \frac{\mathscr{E}}{r} \qquad (36.9)$$

Then Eq. 36.8 can be written

$$\Delta B = -\frac{r}{NA}\int I\,dt \qquad (36.10)$$

The integral $\int I\,dt$ measures the total charge q moved through the sense coil. Thus when r, N, and A are known, a measure of q allows us to deduce the *change* in magnetic field: $\Delta B = B_{after} - B_{before}$, where B_{after} and B_{before} are the magnetic fields after and before the flux change. A variety of ways exist for measuring the charge q, one being to allow the current to charge a capacitor and then to measure the net charge stored by the capacitor.

Classification of Magnetic Materials

If a solenoid is empty, the only contribution to the magnetic moment comes from current in the solenoid, and we have

$$B_E = \frac{\mu_o \mu}{V} \tag{36.11}$$

When the interior of the solenoid is filled with some material, the induced magnetic moments produce an additional contribution to the magnetic moment. We denote this additional contribution by μ_I and generalize Eq. 36.11 to account for the presence of material in the solenoid by writing*

$$\underbrace{B = \frac{\mu_o}{V}}_{\text{net magnetic field arising from current in the coil and induced magnetic dipole moments}}\underbrace{(\mu + \mu_I)}_{\text{contributions to the magnetic dipole moment arising from current in the coil and induced magnetic dipole moments}} \tag{36.12}$$

We find experimentally that the induced magnetic moment (μ_I) depends on current in the solenoid. We write this dependence as

$$\mu_I = \chi_m \mu \tag{36.13}$$

where χ_m is called the **magnetic susceptibility.** Depending on the type of material that is inside the solenoid, χ_m may be constant or may depend on the current I. In terms of magnetic susceptibility, Eq. 36.12 can be written as

$$B = \frac{\mu_o}{V}(1 + \chi_m)\mu \tag{36.14}$$

where the magnetic moment μ is determined from the characteristics of the solenoid and the current in the solenoid (Eq. 36.2). The magnetic field B is determined from measurements of magnetic flux. Thus the magnetic susceptibility is computed from the equation

$$\chi_m = \frac{BV}{\mu_o \mu} - 1 \tag{36.15}$$

The three classes of magnetic materials given in Table 36.1 can now be described in terms of measurements of the magnetic susceptibility as follows.

Diamagnetic materials interact weakly with an imposed magnetic field, weaken the existing magnetic field, and have negative values of χ_m. The magnetic susceptibility χ_m

*The induced magnetic moment for some special materials does not have the same orientation as μ. For these materials, Eq. 36.12 is a vector equation $\mathbf{B} = (\mu_o/V)(\boldsymbol{\mu} + \boldsymbol{\mu}_I)$. We consider only homogeneous materials for which $\boldsymbol{\mu}_I$ has orientation either parallel or antiparallel to $\boldsymbol{\mu}$.

Table 36.2

Magnetic susceptibilities for some paramagnetic and diamagnetic elements*

Element	Magnetic Susceptibility
Paramagnetic	
Aluminum	$+20.7 \times 10^{-6}$
Magnesium	$+11.8 \times 10^{-6}$
Potassium	$+5.82 \times 10^{-6}$
Diamagnetic	
Bismuth	-280.1×10^{-6}
Carbon	-14.1×10^{-6}
Copper	-10.8×10^{-6}
Silver	-23.8×10^{-6}

*The measurements reported here are for a temperature of 300 K, and were taken from a list found in *Handbook of Chemistry and Physics*, 61st Ed., CRC Press, 2000 N.W. 24th St., Boca Raton, FL 33431.

is essentially independent of temperature and solenoid current (I), and is usually on the order of -10^{-5}. Values of χ_m for some representative diamagnetic materials are given in Table 36.2.

Paramagnetic materials interact weakly with an imposed magnetic field, strengthen the existing magnetic field, and have positive values of χ_m. The quantity χ_m depends on temperature, is essentially independent of solenoid current (I), and is usually on the order of $+10^{-5}$. Values of χ_m for some representative paramagnetic materials are also given in Table 36.2.

Ferromagnetic materials interact strongly with an imposed magnetic field, strengthen the existing magnetic field, and have magnetic susceptibilities that depend sensitively on the solenoid current (I).

We commonly call $1 + \chi_m$ the **relative permeability** and denote it by κ_m.

$$\kappa_m = 1 + \chi_m \tag{36.16}$$

so that Eq. 36.14 becomes

$$B = \frac{\mu_o \kappa_m \mu}{V}$$

The combination $\mu_o \kappa_m$ is called the *permeability*. The relative permeability (κ_m) plays the same role for magnetic materials that the dielectric constant (κ_e) plays for dielectric materials. It is customary to describe ferromagnetic materials in terms of the relative permeability rather than the magnetic susceptibility.

Example 1 **Magnetic Force on a Piece of Aluminum in a Nonuniform Magnetic Field**

Let's determine the magnetic force on a sample of aluminum in a nonuniform magnetic field. The magnetic field near the pole face of a permanent magnet in the speaker of a stereo system is about 0.3 T. The strength of the magnetic field diminishes rapidly away from the pole face, and so at about 10 cm away from the face the magnetic field is essentially zero. Thus the space variation of the field (dB/dx) is about 3 T/m. Near the pole face a piece of aluminum having a magnetic susceptibility of $\chi_m = 2.07 \times 10^{-5}$ and dimensions 1 mm × 5 mm × 1 mm has an induced magnetic moment of

$$\mu_I = \chi_m \mu$$

$$= \frac{\chi_m B_E V}{\mu_o}$$

$$= \frac{2.07 \times 10^{-5} \times 0.3\ \text{T} \cdot 5 \times 10^{-9}\ \text{m}^3}{4\pi \times 10^{-7} \dfrac{\text{T}\cdot\text{m}}{\text{A}}}$$

$$= 2.47 \times 10^{-8}\ \text{A}\cdot\text{m}^2$$

The force on the piece of aluminum is (Section 32.5)

$$F = \mu \frac{dB}{dx} = 2.47 \times 10^{-8}\ \text{A}\cdot\text{m}^2 \cdot 3.00\ \text{T/m}$$

$$= 7.41 \times 10^{-8}\ \text{N}$$

The weight of the aluminum is

$$W = mg = \rho V g = 2700 \frac{\text{kg}}{\text{m}^3} (5 \times 10^{-9})\ \text{m}^3\ 9.8 \frac{\text{m}}{\text{s}^2}$$

$$= 1.32 \times 10^{-4}\ \text{N}$$

The gravitational force is over 1000 times greater than the magnetic force and the aluminum is not pulled to the magnet.

The classification of materials into three magnetic categories serves a dual purpose. It provides us with important practical information about the behavior of materials in a magnetic field, and it gives us a framework for understanding the structural laws that give rise to this behavior. The idea that magnetic effects are based on free charges in motion provides insight into the origin of magnetism at the level of atoms. However, only modern quantum physics yields a completely satisfactory explanation of magnetism. But without calling on quantum physics (Chapter 42), let's try to visualize the atomic origin of magnetism in the following sections, bearing in mind that our conclusions are qualitative.

Questions

3. When investigating the magnetic properties of a diamagnetic or paramagnetic material, why is it extremely important to minimize iron impurities in the samples?

4. Two materials, labeled I and II, are placed in the interior of a solenoid. A qualitative plot of measurements of induced magnetic moment (μ_I) and magnetic moment (μ) due only to current in the solenoid is shown in Figure 36.8. Classify these materials.

5. Electrical resistance is defined as the ratio of potential difference across the element to the current in the element ($R = V/I$). For some materials resistance is independent of potential difference and current, but for many materials it is not. In a similar vein, magnetic susceptibility is defined as the ratio of induced magnetic moment to the magnetic moment due to current ($\chi_m = \mu_I/\mu$). Draw an analogy between the general behavior of electrical resistance and magnetic susceptibility.

Figure 36.8

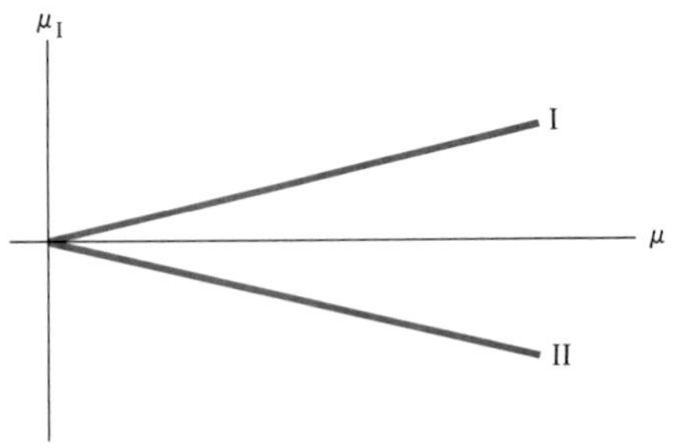

6. An iron nail is placed inside a solenoid, its length parallel to the solenoid axis. A current is then established in the solenoid windings. What happens to the nail? What happens to the nail if the current direction is reversed?

7. Why will an iron nail be attracted to either an N or an S pole of a bar magnet like that shown in Figure 36.1?

36.3 Diamagnetism

The negative magnetic susceptibility of diamagnetic materials can be uderstood if we apply Faraday's law (Section 34.2) and Lenz's law (Section 34.3) to the induction of electron currents in diamagnetic materials.

A change in the magnetic field lines threading a current loop causes a current to be induced in the loop. This is Faraday's law. The magnetic flux produced by the induced current always acts to *oppose the change.* This is Lenz's law. For example, if a bar magnet is thrust into a loop of wire, the magnetic flux threading the loop changes, and a current is induced in the loop (Figure 36.9). The orientation of the magnetic dipole moment created by the circulating current is such as to *oppose* the change in magnetic field lines and the resulting motion of the bar magnet.

Whenever a material is subjected to a magnetic field, magnetic field lines thread the paths of electrons, and the currents and magnetic dipole moments created by the circulating electrons change. According to Lenz's law, these changes oppose the action of the applied magnetic field, and the induced magnetic dipole moments orient oppositely to the applied magnetic field.* In terms of Eq. 36.12, $\mathbf{B} = \mu_o(\boldsymbol{\mu} + \boldsymbol{\mu}_I)V$, the induced magnetic moment ($\boldsymbol{\mu}_I$) and the magnetic moment due to current ($\boldsymbol{\mu}$) have

*This behavior is discussed in more detail in Section 34.3.

Figure 36.9

Thrusting a bar magnet into a loop of wire induces a current in the wire. The direction of the induced magnetic dipole moment is such as to oppose the motion of the bar magnet.

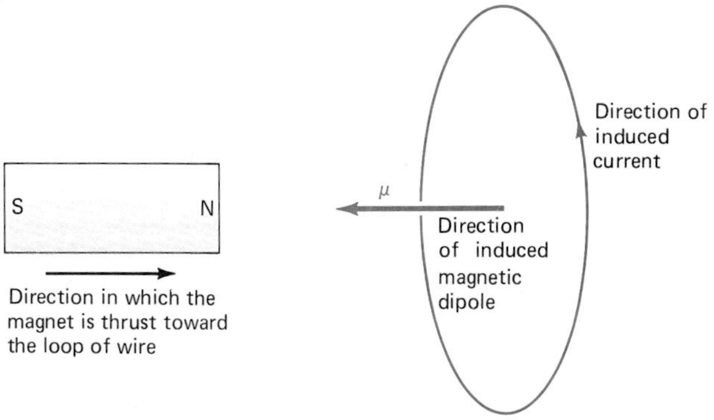

Figure 36.10

A current I on the surface of the superconductor produces a magnetic field $\mathbf{B}_{\text{induced}}$ that just cancels the applied magnetic field **B**.

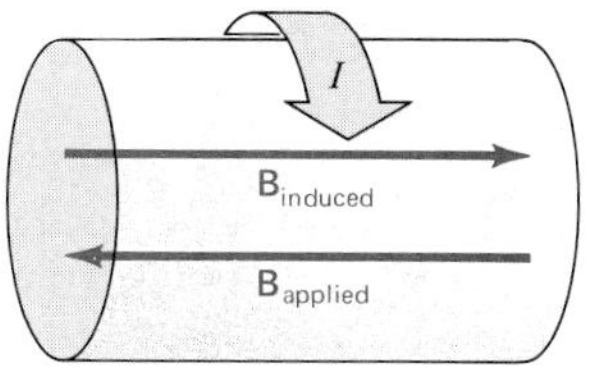

opposite directions. From Eq. 36.13, $\mu_I = \chi_m \mu$, it follows that χ_m is negative, in accordance with our criteria for categorizing magnetic materials. Diamagnetism is a property of all materials, but it is a very weak property and is obscured in materials made of atoms that have permanent magnetic dipole moments.

Diamagnetism in Superconductors

A superconductor is a conductor that loses all electrical resistance below a certain temperature (usually less than 10 K).* Superconductors have particularly interesting diamagnetic properties. For example, the magnetic flux is zero in the interior of a superconductor. In terms of Eq. 36.12, this means that the induced magnetic moment μ_I must exactly cancel the magnetic moment μ.

$$B = \frac{\mu_0}{V}(\mu + \mu_I)$$

$$= 0$$

$$\mu_I = -\mu$$

Because

$$\mu_I = \chi_m \mu$$

it follows that the magnetic susceptibility is

$$\chi_m = -1$$

Superconductors are said to be perfectly diamagnetic.

Example 2 **Surface Currents in Superconductors**

We can envision the induced magnetic dipole moment of a superconductor as arising from an induced current on the surface of the superconductor (Figure 36.10). Let us estimate the size of this surface current in a superconducting cylinder 10 cm long in a magnetic field of 1 T.

The surface current is roughly equivalent to the current in a one-turn solenoid. Hence the magnetic field produced by the surface current (I) is

$$B = \mu_0\left(\frac{I}{l}\right)$$

We want this magnetic field to cancel the applied magnetic field of 1 T. Hence the surface current is

$$I = \frac{Bl}{\mu_0}$$

$$= \frac{1\ \text{T} \times 0.1\ \text{m}}{4\pi \times 10^{-7}\ \dfrac{\text{T}\cdot\text{m}}{\text{A}}}$$

$$= 79{,}600\ \text{A}$$

In the superconductor this enormous current can continue indefinitely because the electrical resistance is zero.

*Superconductors are discussed further in Section 42.7.

Questions

8. The prefix *dia-* means "in opposite directions." Why is dia- an appropriate prefix in the word *diamagnetic*?
9. Why might you expect all materials to exhibit diamagnetism to some extent?

36.4 Paramagnetism

At the atomic level, we view a paramagnetic material as composed of a uniform distribution of atomic magnetic dipoles sufficiently separated so that the magnetic field of any given dipole does not influence any of its neighbors. In the absence of an applied magnetic field, the net magnetic moment of a paramagnetic material is zero because the dipoles are randomly oriented as a result of thermal motions. But when an external magnetic field is applied, the atomic dipoles align themselves with the field, thereby producing a net magnetic moment in the material. This magnetic alignment is opposed by the thermal motions of the atoms, however. At 300 K (room temprature), for example, the random thermal energy of the atomic dipoles is generally more than 100 times larger than the magnetic ordering energy for a fairly large magnetic field of 1 T. Consequently, the thermal motions tend to disrupt the magnetic alignment. *If* the magnetic ordering energy increases and gradually exceeds the random thermal energy, then the alignment will increase. Thus net magnetic moment gradually builds up in the paramagnetic material. If the magnetic ordering energy is sufficiently large, then the magnetic alignment becomes complete. Magnetic alignment can be achieved by lowering the temperature of the sample or by increasing the applied magnetic field.

The magnetic ordering energy of an atom is proportional to the applied magnetic field ($U = -\mu \cdot \mathbf{B}$), and the random thermal energy of an atom is proportional to the Kelvin temperature ($E = \frac{3}{2} kT$). The ratio B/T reflects the relative sizes of the magnetic and thermal energies. For this reason, the magnetization of a sample is generally recorded as a function of the ratio B/T. The experimental results in Figure 36.11 show the general behavior of the magnetization of a paramagnetic sample. As B/T increases, μ_I increases and then levels off. We can understand this behavior if we consider what happens to μ_I if only B is increased while T is held constant. As the magnetic field is increased, dipole alignment becomes more and more nearly complete, and thus μ_I increases. Once the alignment is almost complete, increasing B has little effect on μ_I. The leveling off of μ_I is called *saturation*.

The initial straight-line part of the magnetization curve can be represented by an equation of the form

$$\mu_I = C\left(\frac{B}{T}\right) \tag{36.17}$$

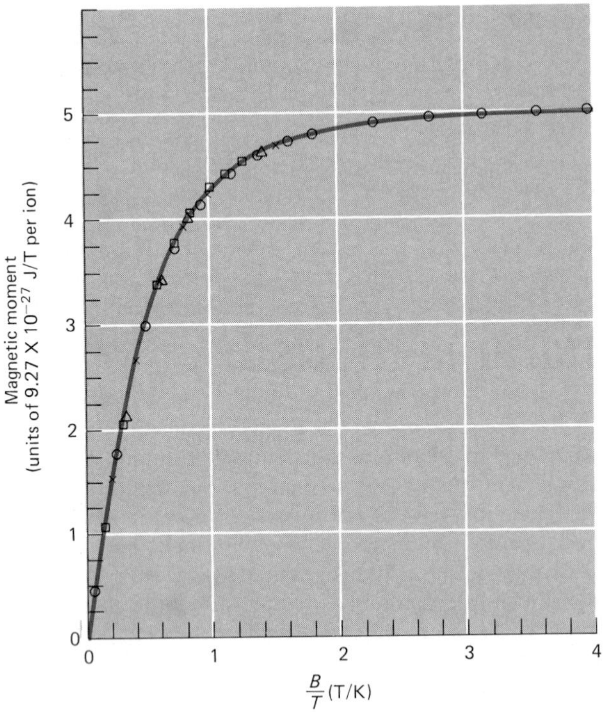

Figure 36.11

Plot of magnetic moment versus B/T for a sample of ferric ammonium alum. The points denote experimental measurements. Different point symbols represent different temperatures for the sample. The solid line is a quantum physics theoretical calculation. Data are from W. E. Henry, Physical Review, 88, 559 (1952).

where C is a constant for a given paramagnetic material. This paramagnetic relation is called **Curie's law.*** Figure 36.11 shows the results of a theoretical calculation of magnetic moment based on quantum physics.

The alignment of atomic magnetic dipoles in a paramagnetic sample enhances the magnetic dipole moment, and the magnetic field (B) is increased from the empty-space magnetic field ($B_E = \mu_o\mu/V$). From Eq. 36.14, $B = (\mu_o/V)(1 + \chi_m)\mu$, it follows that the magnetic susceptibility (χ_m) is positive. Unlike diamagnetism, paramagnetism is not a property of all materials. This is because at least some of the atoms in a material must have a permanent magnetic dipole moment in order for the material to be paramagnetic. If a material *is* paramagnetic, the paramagnetism will usually mask the diamagnetism.

Questions

10. *Para-*, as in *parallel*, is a prefix meaning "alongside." Why is para- an appropriate prefix in the word *paramagnetic*?
11. A small compass is placed at the center of each square of a checkerboard. How could you use this array of compasses to demonstrate the principle of paramagnetism?
12. An aluminum can is not noticeably attracted to a common household magnet. Yet aluminum objects can be magnetically separated from trash. What property of aluminum is exploited in, and what are the magnetic requirements for, this separation?

36.5 Ferromagnetism

Ferromagnetism is exhibited by five elements—iron, nickel, cobalt, dysprosium, and gadolinium—and by some alloys, which usually contain one or more of these five elements. Ferromagnetism relies strongly on the mutual interactions between the magnetic moments of electrons—cooperative interactions that favor a high degree of parallel alignment of the magnetic dipole moments. There is no quantitative classical explanation for either the intrinsic electron magnetic moment or the alignment mechanism; they are quantum mechanical effects (Chapter 42).

There are regions in every ferromagnetic sample that have near perfect alignment of magnetic dipole moments even when there is no applied magnetic field. These experimentally observable regions (Figure 36.12) are called *magnetic domains*. Depending on the structure and type of ferromagnetic material, the volumes of magnetic domains vary from about 10^{-18} to 10^{-12} m^3. Because each cubic centimeter of any solid contains roughly Avogadro's number ($\sim 10^{24}$) of molecules, domain volumes involve between 10^{12} and 10^{18} molecules. But even though the magnetic dipole alignment in a given domain is nearly complete, a ferromagnetic sample will not display a net

*Named for the French chemist Pierre Curie, who first discovered the relation in 1895.

Figure 36.12

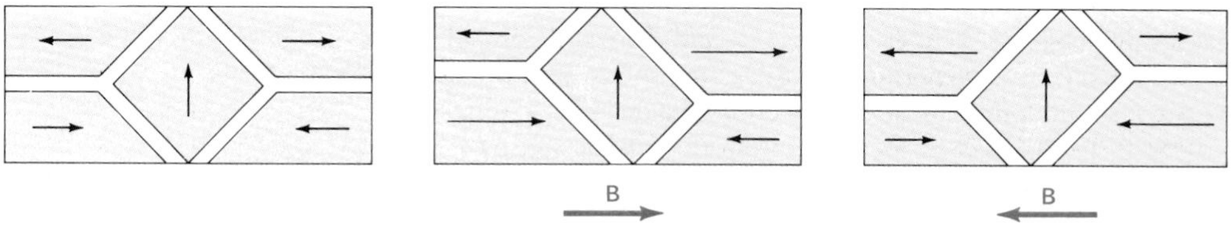

Magnetic domains in a crystalline whisker of iron. The crystal is 0.1 mm thick and there is no applied magnetic field.

Application of a magnetic field causes some domains to grow at the expense of neighboring domains.

Domains having their intrinsic magnetic moments parallel to the applied magnetic field grow when the direction of the applied magnetic field is reversed. Note how the central domain has rotated.

magnetic dipole moment if the domains themselves are randomly oriented. A net magnetic dipole moment in a sample develops only when the domains are aligned by the application of a magnetic field.

There are two types of domain interaction that give rise to a net magnetic dipole moment:

1 Those domains with magnetic dipole moments parallel to the magnetizing field grow at the expense of neighboring domains (Figure 36.12). This effect is responsible for producing a net magnetic dipole moment in a weak applied magnetic field.

2 The magnetic dipole moments of the domains rotate toward alignment with the applied magnetic field. This is the mechanism of magnetic dipole alignment when the applied magnetic field is strong.

For a given ferromagnetic material, ferromagnetism vanishes at a sharply defined temperature called the **Curie temperature.** Above the Curie temperature, the material is only paramagnetic. For example, the Curie temperature of iron is 1043 K (Table 36.3). Because the interior of the earth is at a temperature of about 2000 K, we know that there can be no ferromagnetic contributions to the earth's magnetic field from its molten iron interior.

We saw that the magnetism of a paramagnetic or diamagnetic material vanishes once the material is removed form a magnetic field. However, this may not be the case if a substance is ferromagnetic. To illustrate, let us consider measurements of the magnetic field (B) in an iron sample as a function of the current in a solenoid arrangement like that discussed in Section 36.2. We start with the iron sample unmagnetized. Then according to Eq. 36.10, the first increase in the solenoid current allows us to determine the magnetic field in the iron. The next increase in current allows us to determine the increase in magnetic field. In this way we can record a series of measurements of B and the corresponding values of I. Because the magnetic field with the solenoid interior empty is directly proportional to the current ($B_E = \mu_o NI/l$), we choose to display the measurements as a plot of B versus B_E. Figure 36.13 shows such a set of measurements plotted on a graph. Let's examine each part of this graph.

In Figure 36.13a we see that as the current increases, the magnetic field (B) also increases. The field B tends to saturate, somewhat as does a paramagnetic material (Figure 36.11). The nonlinear relationship between B and B_E means that the magnetic susceptibility is not constant—it depends on the value of B_E. As we will see, it is a rather complicated function.

In Figure 36.13b we see that if B_E is decreased from the value labeled 2, the magnetic field measurements are consistently higher than when the current was increased from the zero value. A remanent magnetic field (or equivalently, a magnetic dipole moment) designated B_r persists even when the current in the solenoid vanishes. The remanent magnetic field results from alignment of magnetic domains. This is one way to produce a permanent magnet.

Figure 36.13c shows that if we now reverse the direction of the current in the solenoid, the magnetic field (B) within the sample is reduced steadily from the remanent field value B_r. At a critical value of B_E, called the *coercive force* (B_c), the magnetic field is zero. The larger the coercive force B_c, the more difficult it is to demagnetize a ferromagnetic sample. *Ferromagnetic materials having a large coercive force are said to be magnetically "hard"; those having a small coercive force are said to be magnetically "soft."* Ordinary iron is magnetically soft and has a coercive force of about 10^{-4} T. A hard magnetic material used in the speaker of a high-fidelity system may have a coercive force 20–50 times that of ordinary iron.

In Figure 36.13d we find that if we keep the same orientation for the magnetic moment μ by maintaining the same current direction in the solenoid and increasing the magnitude of the current, the magnetic field increases. However, the direction is now reversed from the starting direction. Again the magnetic field tends to saturate,

Table 36.3

Curie temperatures of the ferromagnetic elements*

Element	T_c (K)
Iron	1043
Cobalt	1404
Nickel	631
Gadolinium	289
Dysprosium	85

*All values except that for dysprosium were taken from *Handbook of Chemistry and Physics*, 61st ed. CRC Press, 2000 N.W. 24th St., Boca Raton, FL 33431.

Figure 36.13

(a) As B_E is increased, the magnetic field increases. μ and B are oriented in the same direction. (b) Decreasing the current causes the magnetic field to decrease. μ and B are still parallel but for a given B_E, B is different than when B_E was increasing. (c) Reversing the current direction changes the direction of μ. μ and B are no longer parallel. At a critical value of B_E, designated B_c, the magnetic field B becomes zero. (d) As B_E is increased from the critical value B_c, the magnetic field also increases. μ and B are again parallel. (e) Decreasing B_E and subsequently changing the direction of μ, we find the magnetic field returns to the initial saturation value.

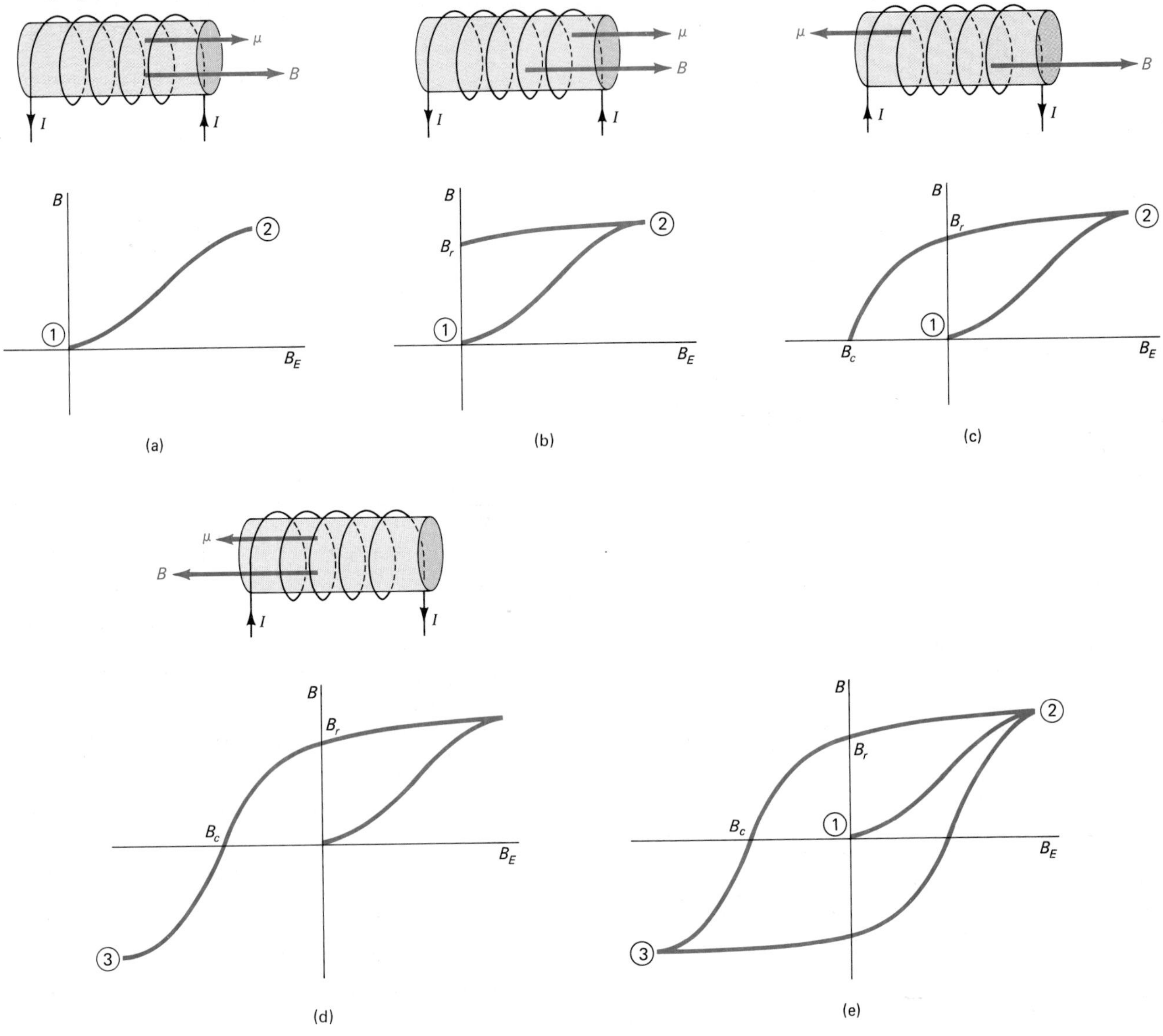

achieving the value labeled 3. This indicates that the alignment of the magnetic domains is approaching completion.

Finally, Figure 36.13e shows that decreasing the magnetic intensity with an ultimate change in the initial direction brings the magnetic field back to its initial saturation value, labeled 2. Subsequent changes in the magnetic intensity retrace this closed loop. Measurements for the steel commonly used in transformers are shown in Table 36.4. Note that the magnetic susceptibility is not constant for a ferromagnetic material. It is generally 10^8 times larger than for a paramagnetic material and depends sensitively on the value of B_E.

The failure of the B versus B_E curve to retrace the initial magnetization curve, labeled 1 to 2 in Figure 36.13, is termed **hysteresis.** A closed curve representing measurements of B and B_E is called a *hysteresis loop*. As a ferromagnetic sample is cycled around a hysteresis loop, irreversible (frictionlike) changes occur in its domain structure.

Table 36.4

Magnetic measurements for ordinary transformer steel*

B (T)	B_E (T)	χ_m
0.2	5.99×10^{-5}	3300
0.4	8.70×10^{-5}	4600
0.6	11.0×10^{-5}	5500
0.8	14.8×10^{-5}	5400
1.0	22.7×10^{-5}	4400
1.2	38.5×10^{-5}	3100
1.4	109×10^{-5}	1300
1.6	430×10^{-5}	370
1.8	$15,000 \times 10^{-5}$	120

*The data are taken from *Handbook of Chemistry and Physics*, 61st Ed., CRC Press, 2000 N.W. 24th Street, Boca Raton, FL 33431.

Figure 36.14

Hysteresis curves for two commercial ferromagnetic materials. The commercial iron is a soft magnetic material. The tungsten steel is a hard magnetic material.

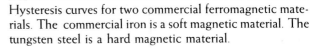

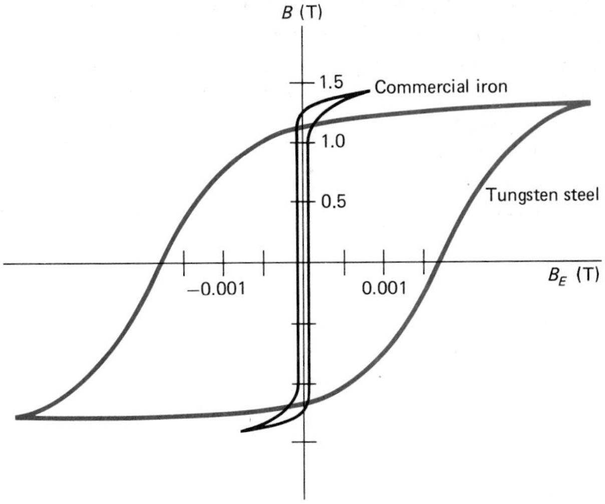

Work is done by the magnetizing field in order to alter the domains, and the temperature of the sample increases. The greater the area of the hysteresis loop, the greater the amount of work required and the larger the temperature rise of the sample.

Use of Magnetically Soft Materials

Many practical electromagnetic devices utilize one or more coils of wire wound around a ferromagnetic medium. The current in the coil often varies cyclically in direction and magnitude. This means that the ferromagnetic medium is continually cycled through a hysteresis loop, producing an energy loss in each cycle. To minimize this energy loss it is desirable to use materials having a hysteresis loop with as small a B–B_E area as possible. This criterion is generally satisfied by magnetically soft materials. Because iron is magnetically soft (Figure 36.14) and has good structural characteristics, it is widely used in electromagnetic devices such as transformers (Section 37.7). Modern materials science has resulted in the development of new alloys with superior magnetic properties. One family of alloys, the metallic glasses, offers the prospect of drastically reducing hysteresis losses.

Use of Magnetically Hard Materials

Magnetically hard materials are characterized by broad hysteresis loops (Figure 36.14) and immobile magnetic domains. They are hard to magnetize and hard to demagnetize. Magnetically hard materials are often used as permanent magnets. For example, magnets in stereo speakers, magnets in mechanical ammeters and voltmeters, and magnets in some types of electric motors are made from magnetically hard materials. Some alloys that consist mostly of aluminum, nickel, cobalt, and iron also make very good permanent magnets. Still better permanent magnets are made of alloys of cobalt and the rare earths (elements with atomic numbers between 58 and 71).

Questions

13. Which would be better to use as a magnetic door-latch, a magnetically hard or magnetically soft ferromagnet?

14. Iron is one of the most versatile materials used by man. Give a use that requires that iron be ferromagnetic and a use that does not require that iron be ferromagnetic.

15. Why is it better to use a magnetically soft material in the tape used in tape recorders and floppy disks?

16. Two ferromagnetic materials have the following properties:

	material 1	material 2
remanent field B_r	0.60 T	0.65 T
coercive force B_c	7.5×10^{-7} T	547 T

Which would make the better permanent magnet?

36.6 The Earth's Magnetic Field

The earth's magnetic field has received more scientific scrutiny than any other magnetic field, and for good reasons. For one , the earth's magnetic field provides an invaluable map for global navigators. But because the field changes continually in both magnitude and direction, constant monitoring is necessary. Questions concerning the origin of the earth's magnetic field and the *reasons* for its continual change have also prompted scientists to investigate. Yet despite intense interest, detailed answers have not been found.

General Properties

Mapping the earth's magnetic field requires local measurements of both the magnitude and direction of the field. The direction is normally established by specifying two angular measurements. First a compass is oriented with its axis of rotation perpendicular to a plane containing the north–south and east–west geographic directions. The angle between the magnetically aligned compass needle and geographic north, called the *angle of declination*, is then recorded. As shown in Figure 36.15, the angle of declination is a measure of the direction of the horizontal component of the earth's magnetic field vector relative to geographic north. If the compass were freely suspended, it would orient itself tangent to the magnetic field lines. The second angle recorded is between the horizontal component and the tangent to the magnetic field lines. This angle is called the *angle of inclination*, or "dip angle." A variety of methods exist for measuring the magnitude of the field. The flip coil discussed in Section 34.4 has long been a favorite and reliable method. When ultimate precision is demanded, the modern SQUID (Superconducting Quantum Interference Device) magnetometer discussed in Chapter 42 is often used instead.

Systematic worldwide measurements of the earth's magnetic field show that to a good first approximation it is representative of the dipole field. The dipole moment axis is tilted at about 11° with respect to the earth's axis of rotation. The dipole moment has a magnitude of 7.94×10^{22} J/T. Although the strength of the field varies on the earth's surface, it is generally around 10^{-4} T.

Solar Wind

Satellite mapping has added to our understanding of the earth's magnetic field. At distances close to the earth, measurements confirm the basic dipole nature of the magnetic field. At distances of about five earth radii, however, space probe measurements record significant departures from a simple dipole field. We find that at these distances the earth's magnetic field appears to have been "blown away" from the sun, as illustrated in Figure 36.16. From satellite measurements we now know that this variation in the earth's magnetic field is a consequence of a "wind" of energetic electrons and protons emanating from the sun. As these particles penetrate the earth's magnetic field they experience a magnetic force ($\mathbf{F} = q\mathbf{v} \times \mathbf{B}$) and most are trapped in spiraling paths around the earth's magnetic field lines. This gives rise to the two Van Allen radiation belts that were discovered in 1958. The first belt, at a mean altitude of about 3000 km, is composed of protons. The second belt, at a mean distance of about 14,000 km, consists of electrons.

Scientific interest in magnetic fields does not end with that of the earth. Space probe measurements have been made of the magnetic fields of Mercury, Venus, Mars, Jupiter, Saturn, Earth's moon, and the sun. This exciting field of study awaits new theories and measurements.

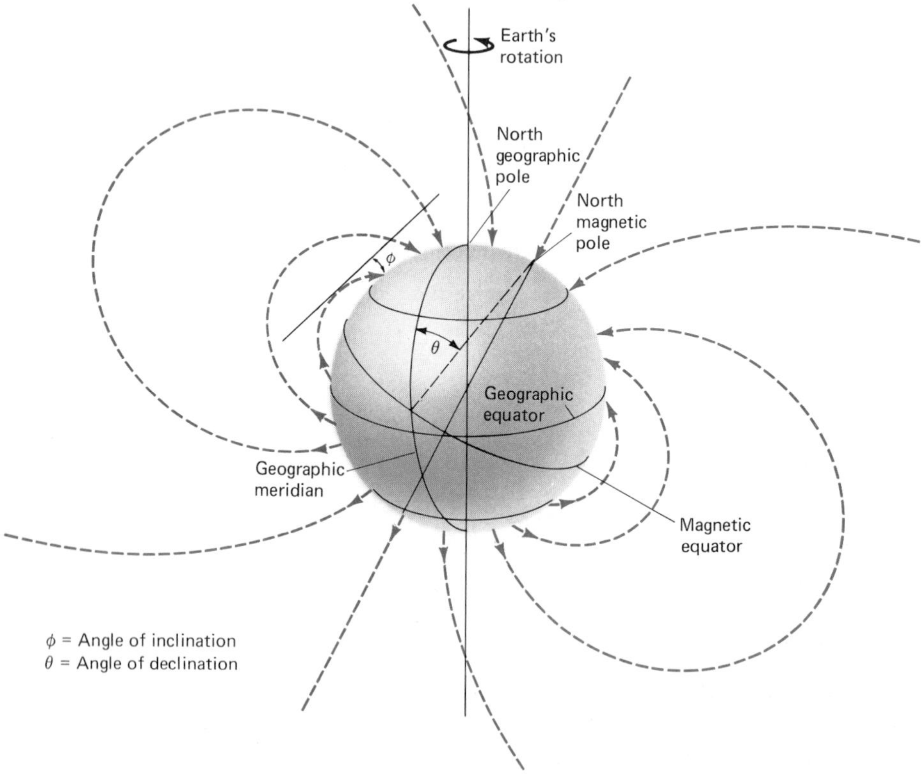

Figure 36.15

Earth's magnetic field line pattern.

Figure 36.16

The effect of the solar wind on Earth's magnetic field line pattern. Downwind from the stream of charged particles the magnetic field lines are blown away from Earth.

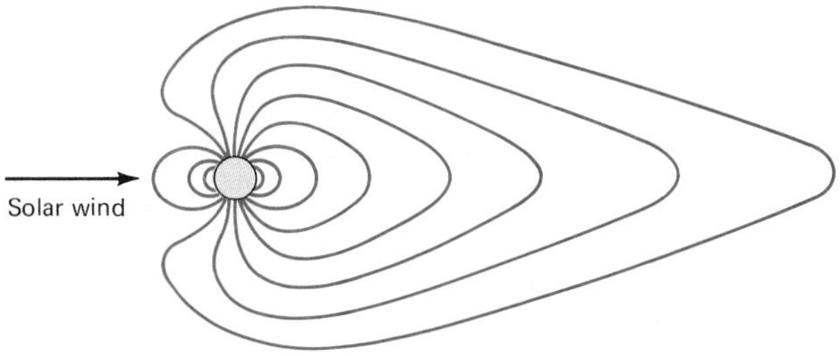

Summary

The magnetic field in the interior of a solenoid produced only by current is related to the magnetic moment μ and volume V of the interior by

$$B_E = \frac{\mu_0 \mu}{V} \quad (36.5)$$

When the interior of a solenoid is filled with a material, magnetic dipole moments may be induced in the medium and we write

$$B = \frac{\mu_o}{V}(\mu + \mu_I) \quad (36.12)$$

where μ_I is the induced magnetic moment.

Magnetic susceptibility χ_m is defined by the relation

$$\mu_I = \chi_m \mu \quad (36.13)$$

Magnetic materials may be classified quantitatively in terms of their magnetic susceptibility:

Magnetic category	Magnetic susceptibility
Diamagnetic	$\chi_m < 0$ and $\chi_m \approx -10^{-5}$
Paramagnetic	$\chi_m > 0$ and $\chi_m \approx +10^{-5}$
Ferromagnetic	$\chi_m > 0$ and χ_m can be large, around 1000, for example, and strongly dependent on B_E.

Diamagnetism is the result of magnetically induced atomic dipoles.

Paramagnetism is due to orientation of weakly interacting magnetic moments associated with electron magnetic moments and circulating electron currents.

Ferromagnetism results from a very strong mutual interaction between magnetic moments. This interaction produces nearly perfect magnetic alignment in regions of the sample called magnetic domains. Permanent ferromagnets are produced by the alignment of magnetic domains.

The extent of magnetic dipole alignment in paramagnetic materials depends on the temperature (T) and strength of the applied magnetic field (B). For a restricted range of the ratio B/T, the magnetic moment (μ_I) of paramagnetic materials obeys Curie's law,

$$\mu_I = C\left(\frac{B}{T}\right) \quad (36.17)$$

where C is a constant.

For a given ferromagnetic material, ferromagnetism vanishes above a well-defined temperature called the Curie temperature. For iron, $T_c = 1043$ K.

The internal magnetic field (B) in a ferromagnetic sample depends sensitively on the magnetizing field (B_E) and on the way B_E is achieved.

Measurements reveal that the earth's magnetic field is similar to that of an elementary magnetic dipole. Although not understood quantitatively, the earth's magnetic field is believed to be caused by circulating interior electric currents.

Suggested Reading

J. J. Becker, Permanent Magnets, *Sci. Am.*, *222*, 92–100 (December 1970).

A. H. Bobeck and H. E. D. Scovil, Magnetic Bubbles, *Sci. Am.*, *224*, 78–90 (June 1971).

P. Dyal and C. W. Parkin, The Magnetism of the Moon, *Sci. Am.*, *225*, 67–73 (August 1971).

U. Essemann and H. Träuble, The Magnetic Structure of Superconductors, *Sci. Am.*, *224*, 74–84 (March 1971).

J. T. Gosling and A. J. Hundhausen, Waves in the Solar Wind, *Sci. Am.*, *237*, 36–43 (March 1977).

H. Kolm, J. Obertueffer, and D. Kelland, High-Gradient Magnetic Separation, *Sci. Am.*, *233*, 46–54 (November 1975).

W. C. Livingston, Magnetic Fields on the Quiet Sun, *Sci. Am.*, *215*, 54–62 (November 1966).

J. A. Van Allen, Radiation Belts Around the Earth, *Sci. Am.*, *200*, 39–47 (March 1959).

Problems

Section 36.2 Classification of Magnetic Materials

1. (a) A current of 1 A exists in a solenoid having 4000 turns per meter. Calculate the magnetic field within the solenoid when it is empty. (b) How much does the magnetic field change when the interior is filled with copper?

2. A long, tightly wound solenoid produces a magnetic field B_E for a current I. When the solenoid core is filled with aluminum, the field in the interior is B. Find the relative change $(B - B_E)/B_E$.

3. The core of a solenoid having 250 turns per meter is filled with a material of unknown composition. When the current in the solenoid is 2 A, measurements reveal that the magnetic field within the core is 0.13 T. Determine the magnetic susceptibility of the material and classify it magnetically.

4. When a vacuum exists in the interior of a solenoid the magnetic field measured in the interior is 0.16 T. Insertion of a certain material in the interior causes the net magnetic field to increase to 0.27 T. (a) Determine the induced magnetic moment per unit volume in the interior of the solenoid. (b) Determine the magnetic susceptibility of the medium and categorize it magnetically.

5. The sense coil surrounding a long solenoid has 120 turns of wire, a resistance of 0.13 Ω, and a mean diameter of 1.21 cm. When the flux in the solenoid is reduced to zero, a net charge of 1.3 mC flows through the sense coil. What was the magnetic field in the interior of the solenoid?

6. One cubic centimeter of aluminum has about 7.84×10^{23} electrons. Suppose each electron has a magnetic moment $\mu = 9.27 \times 10^{-24}\ \mathrm{A \cdot m^2}$, and that these magnetic moments are all parallel. Calculate the resulting dipole moment per unit volume for the aluminum cube.

7. Determine the magnetic field and magnetic moment in the interior of a long solenoid 10.0 cm long and 1.1 cm in diameter that has 68 turns per centimeter and carries a current of 0.56 A. Assume that the interior is a vacuum.

8. The magnetic field in the interior of a solenoid is determined to be 2.6 T. Determine the magnetic moment of this solenoid if the volume of the interior of the solenoid is 63.1 $\mathrm{cm^3}$

9. Figure 1 displays the current induced in the sense coil surrounding a solenoid similar to that shown in Figure 36.6. Determine the total charge moved through the sense coil.

10. With no material in the core of a long solenoid, the magnetic field is precisely 1.27 T. What is the change in the magnetic field in the core when it is filled with (a) aluminum and (b) copper?

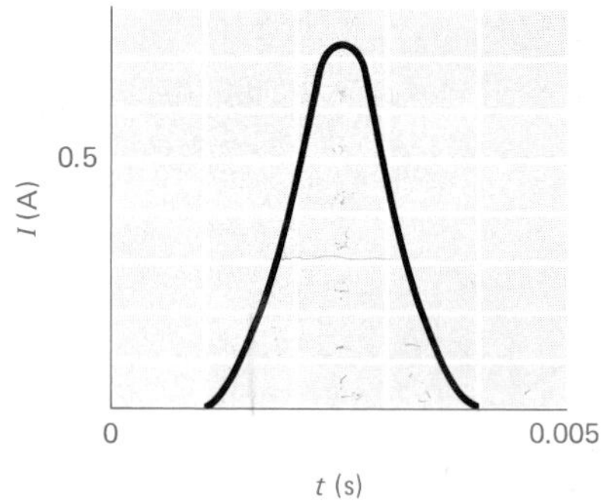

Figure 1

11. A long solenoid has a mean diameter of 13.2 cm, contains 750 closely spaced turns, and carries a current of 0.23 A. Determine the magnetic dipole moment of this solenoid when there is a vacuum in the core.

12. The interior of a long solenoid having 100 turns per centimeter and carrying a current of 1.2 A is filled with aluminum. Calculate the induced dipole moment per unit volume in the aluminum.

13. Determine a relation between B and μ in a material if the magnetic susceptibility (χ_m) is directly proportional to μ.

14. A cube of aluminum 1 cm on a side is placed in a uniform magnetic field of 1 T. Calculate the induced magnetic moment in the cube.

15. Suppose that you want to use an electromagnet to lift a 1-$\mathrm{cm^3}$ cube of aluminum from the floor. Near the end of the magnet the vertical variation of the nonuniform magnetic field is constant, that is dB/dy = constant. If the average magnetic field in the vicinity of the cube is 10 T, determine the value of dB/dy that is required to pick the cube up from the floor.

16. For an electromagnet to pick up a solid piece of material, the upward magnetic force must exceed the downward gravitational force. Show that the variation of magnetic field must be at least $dB/dy = mg/\mu$, where m is the mass and μ is the magnetic dipole moment.

Section 36.3 Diamagnetism

17. A 2.5-cm-long rod of a superconducting material is placed in a uniform magnetic field of 0.54 T with its cylindrical axis along the magnetic field lines. Recalling that a superconductor is perfectly diamagnetic, (a) sketch the orientation of $\mathbf{B}_E$ and μ in the material and the direction of the surface current, and (b) deduce the size of the surface current.

18. A superconducting material is a perfect diamagnet: The field inside the material equals zero. Consider a circular superconducting disk (diameter = 1 cm) placed in an external magnetic field ($B = 2 \times 10^{-3}$ T) parallel to the axis of the disk. What surface current around the disk perimeter would be needed to make $B = 0$ at the center of the disk?

Section 36.4 Paramagnetism

19. Given N magnetic dipoles, each having a dipole moment μ. (a) Organize these N dipoles into an arrangement having a net magnetic dipole moment of 500 μ. (b) Is there any restriction on the size of N? (c) Is your arrangement the only possibility?

20. An aluminum atom has a magnetic dipole moment on the order of 10^{-23} J/T. If all the atomic moments of a sample of aluminum are aligned, estimate the magnetic dipole moment in 1 m^3 of the aluminum.

21. Taking the magnetic dipole moment of an atom in a paramagnetic gas to be 10^{-23} J/T, determine the temperature at which the average thermal energy of an atom is equal to its magnetic energy when the atom is in a magnetic field of 0.6 T.

22. Show that the magnetic susceptibility of a paramagnetic material obeying Curie's law is

$$\chi_m = \frac{C\mu_0}{T}$$

where C is a constant and T is the Kelvin temperature.

23. Evaluate C in Eq. 36.17 for the material depicted in Figure 36.11.

24. A paramagnetic material obeying Curie's law is held at constant temperature. Sketch the induced magnetic moment as a function of an applied magnetic field (B).

25. A paramagnetic material obeying Curie's law is placed in a uniform magnetic field. Make a plot of the induced magnetic moment as a function of the temperature of the sample.

Section 36.5 Ferromagnetism

26. Measurements of the magnetic field (B) and magnetic field due to solenoid current (B_E) are presented in Table 1. Determine the induced magnetic moment per unit volume and magnetic susceptibility χ_m for the values of B_E recorded in the table.

Table 1

B (T)	B_E (T)
0	0
0.094	3.14×10^{-4}
0.38	6.28×10^{-4}
0.70	9.42×10^{-4}
0.90	12.6×10^{-4}
0.99	15.7×10^{-4}
1.05	18.9×10^{-4}

27. The initial magnetization of a ferromagnetic material is approximated by $B = 7.0 \times 10^{-4}\,(B_E/\mu_0)$, where B and B_E are expressed in teslas. Determine the magnetic susceptibility of this material.

28. Magnetic field **B** is defined from the relation $\mathbf{F} = q\mathbf{v} \times \mathbf{B}$. Using this relation and Ampère's law, $\int \mathbf{B} \cdot d\mathbf{s} = \mu_0 I$, show that the product of B and B_E/μ_0 has units of joules per cubic meter.

29. Determine the remanent magnetic field B_r and coercive force B_c for the material whose B versus B_E/μ_0 properties are shown in Figure 2.

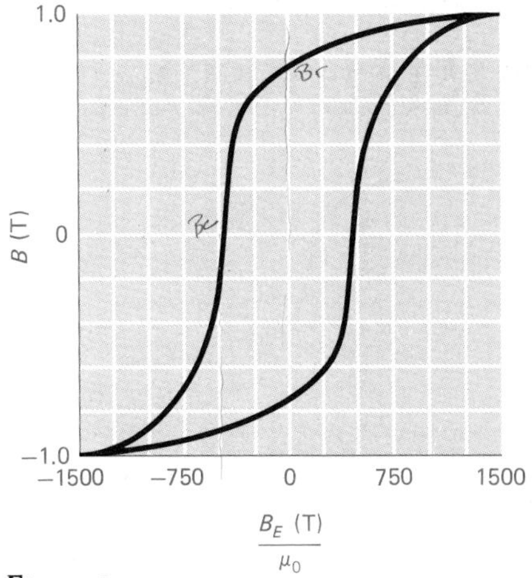

Figure 2

30. The area bounded by a hysteresis loop represents the work required to take the material through a hysteresis cycle. Figure 3 shows an approximation to the real hysteresis loop shown in Figure 2. Determine the work required in joules per cubic meter to take this material through one hysteresis cycle.

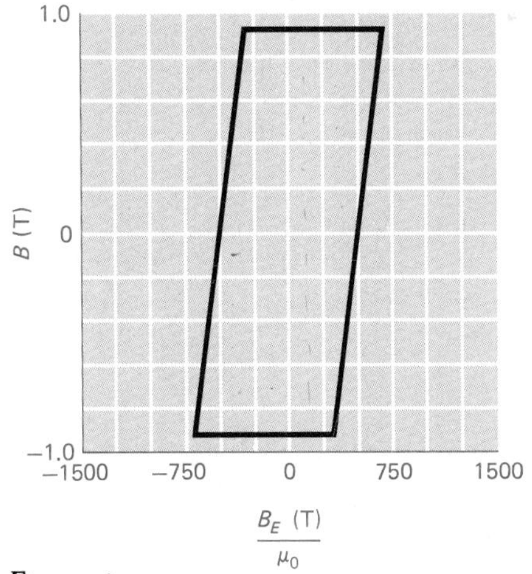

Figure 3

31. A commercial iron used for the cores of electromagnets has the following properties:

remanent field: $B_r = 1.3$ T

coercive field: $B_c = 1.0 \times 10^{-4}$ T

saturation field: $B_s = 2.16$ T

hysteresis loss per cycle: 500 J/m^3 for a maximum field of 1.0 T

Sketch a hysteresis loop for this material that is consistent with its stated properties.

chapter 37

ALTERNATING CURRENTS

Preview

Charge flows in only one direction through an automobile light bulb because the polarity of the emf (battery) that provides the energy source never changes. But in the lights in a home the direction of the current continually reverses because the polarity of the emf (generator) continually reverses. Energy sources having this oscillatory feature, and the currents they produce, are especially useful. Electric energy can be transmitted with minimal loss from an oscillating source to a system where it can then be used. The most useful oscillating electric energy source produces a potential difference that varies sinusoidally with time. We term such an emf, and the currents that it produces, "ac" (pronounced "ay see"). Resistors, capacitors, and inductors are used widely in electrical circuits involving both alternating and direct currents. We now study the behavior of these elements in ac circuits.

We begin by determining the current in, and the potential difference across, a resistor, a capacitor, and an inductor when each is connected to an ac generator. We then introduce reactance as a measure of the extent to which these elements individually limit the ac current. We then investigate series combinations of these elements connected to an ac source in the same way, and develop the idea of impedance as a measure of the extent to which a combination of elements limits an ac current. The series combination of a resistor, capacitor, inductor, and ac source is particularly instructive because it exhibits an electrical resonant behavior that is entirely analogous to the resonant behavior of a driven mechanical system. We end this chapter with a discussion of the transformer, the key to the efficient transmission of electric power.

37.1 Alternating Currents

When a battery is connected to a resistor, charge flows through the resistor in only one direction. If we want to reverse the direction of the current, we must interchange the battery connections. If we could systematically interchange the battery connections, we would produce a current in the resistor that periodically changes direction. We call such a current an *alternating current.*

Sources of potential difference whose polarity changes with time are called *alternating* sources. The types of alternating potential differences are limited only by our imagination and ingenuity, and modern electronics exploits a number of different ones. For example, the sawtooth potential difference shown in Figure 37.1 is used across the horizontal deflector plates of a TV tube to sweep the electron beam back and forth across the screen. Among the many mathematical functions that describe alternating potential differences, $\mathscr{E} \cos \omega t$ (or equivalently, $\mathscr{E} \sin \omega t$), which we discussed in Section 34.4, is particularly significant. $\mathscr{E}$ is the amplitude, and ω is the radian frequency. Both $\mathscr{E}$ and ω are constant for a given physical situation. The radian frequency has the same interpretation as in simple harmonic motion (Section 14.1). It is related to the frequency, ν, expressed in hertz, for example, by

$$\omega = 2\pi\nu \tag{37.1}$$

Figure 37.1

Potential differences whose polarity changes with time are termed *alternating*. The alternating potential differences shown here are used often in electronics.

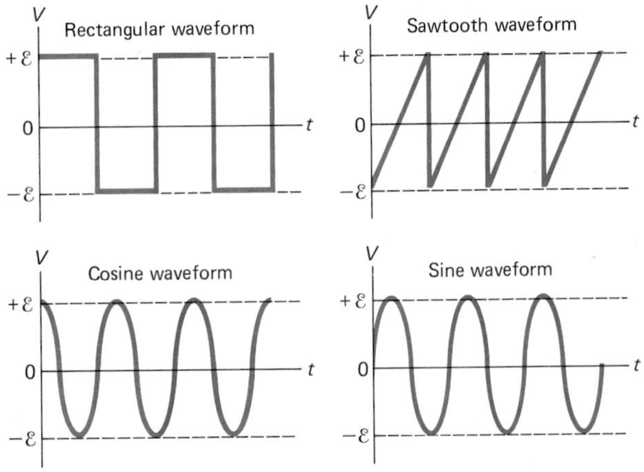

The usefulness of the sinusoidal form $\mathscr{E} \cos \omega t$ is twofold:

1 First, many practical alternating potential differences are represented by $V = \mathscr{E} \cos \omega t$ because it is easily produced by mechanically rotating generators. The potential difference across the electric outlet in an American home has an amplitude ($\mathscr{E}$) of 170 V,* a frequency of 60 Hz, and a radian frequency $\omega = 2\pi 60 = 377$ rad/s. Outlets in European homes generally are furnished with $\mathscr{E} = 340$ V and a frequency of 50 Hz.

2 Second, alternating potential differences that are periodic can be represented as a superposition of sinusoidal functions with appropriate amplitudes and radian frequencies. Figure 37.2 shows such a representation for the rectangular waveform presented in Figure 37.1. In this text we will refer to sinusoidally varying potential differences as *ac potential differences*.†

Potential differences of the form $\mathscr{E} \cos \omega t$ can be produced in several ways. For example, if a coil of wire is rotating in a magnetic field with angular frequency ω, a potential difference of the form $\mathscr{E} \cos \omega t$ develops between the leads to the coil (this principle was discussed in detail in Section 34.4). For a coil of N identical loops, $\mathscr{E}$ is related to N, the radian frequency ω, magnetic field B, and area A of a loop of the coil by

$$\mathscr{E} = N\omega BA \tag{37.2}$$

Electric generators at commercial electric power plants operate by this principle to produce potential differences for use in households and industries. Electronic oscillators that exploit the oscillatory action of charges in circuits containing an inductor and a capacitor also produce potential differences of the form $\mathscr{E} \cos \omega t$. These electronic oscillators play key roles in radio and TV transmitters and receivers. An ac generator, regardless of how the potential difference is produced, is symbolized by ⊖.

*The potential difference quoted usually is 120 V. This is called the *root-mean-square* (rms) value, meaning the square root of the average of the square of the potential difference. We discuss rms values in Section 37.2.

†Technically "ac" is an abbreviation for "alternating current." However, this definition differs from contemporary usage in science and engineering. To be in agreement with this usage we restrict the term ac to mean sinusoidally varying. Accordingly we will use such terms as "ac current" and "ac generator."

Figure 37.2

Periodic waveforms like those shown in Figure 37.1 can be described by a sum of sine and/or cosine terms having appropriate radian frequencies and amplitudes. The plot shown here illustrates how the function

$$V(t) = \frac{4\mathscr{E}}{\pi}\left(\sin \omega t + \frac{1}{3}\sin 3\omega t + \frac{1}{5}\sin 5\omega t + \frac{1}{7}\sin 7\omega t + \frac{1}{9}\sin 9\omega t\right)$$

approximates the rectangular waveform shown in Figure 37.1. The reproduction improves with each additional term added to the sum. Hand-held electronic calculators and personal computers make it easy to calculate these sums of sines and cosines. Try it!

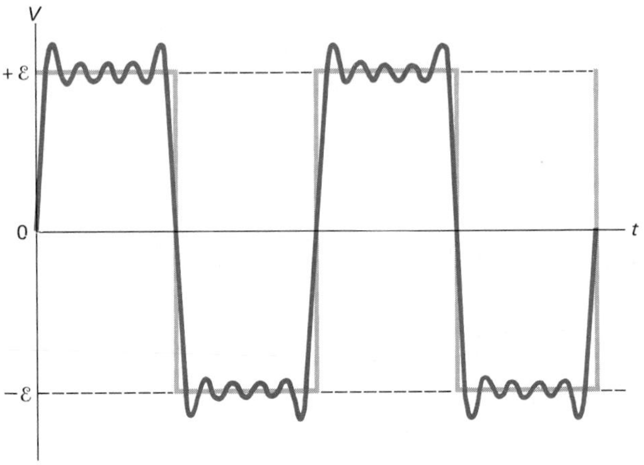

Example 1 An ac Generator

A simple ac generator can be constructed by rotating a single loop of wire inside a horseshoe-shaped magnet that produces a reasonably large magnetic field. Suppose that the magnet produces a field of about 0.1 T, and that the coil is 6 cm long and 3 cm wide. Let us determine the maximum potential difference developed if the coil rotates at 5 rev/s.

Using Eq. 37.2 we have

$$\begin{aligned}\mathscr{E} &= N\omega BA\\ &= (1 \text{ turn})\left(2\pi\,\frac{\text{rad}}{\text{rev}} \times \frac{5 \text{ rev}}{\text{s}}\right)\\ &\quad\times (0.1 \text{ T})(0.06 \times 0.03 \text{ m}^2)\\ &= 5.65 \times 10^{-3} \text{ V}\end{aligned}$$

Although this potential difference is useful for demonstrating a physical principle, it is too small for most other applications. Commercial generators achieve larger values of $\mathscr{E}$ because they have many more loops of wire, a stronger magnetic field, and (usually) a larger rotational frequency.

Questions

1. If a loop of wire is bent around the axis of rotation so that the closed end of the loop forms a right angle, sketch the form of the potential difference developed across the open ends of the loop. Would the potential difference be termed *alternating*? Would the potential difference be termed *ac*?

2. A lighted lamp connected to the ends of a coil rotating in a magnetic field requires energy for its opertion. What is the origin of this energy?

3. An ac generator produces a potential difference that varies sinusoidally with time. Why is it equivalent to represent this variation as either $\mathscr{E} \cos \omega t$ or $\mathscr{E} \sin \omega t$?

37.2
Behavior of a Resistor Connected to an ac Generator

If we connect a resistor to an ac generator as shown in Figure 37.3, we establish in the circuit a current that changes direction with time. The instantaneous value of the current is given by the instantaneous value of the potential difference across the resistor divided by the resistance.

$$I = \frac{V}{R}$$

$$= \frac{\mathscr{E} \cos \omega t}{R} \qquad (37.3)$$

The quantity $\mathscr{E}/R$ has units of volts per ohm, or amperes. It represents the maximum value of the current in the circuit. The current changes direction with time, and so we use positive and negative values of the current to represent the two possible current directions. Substituting I_m, the maximum current in the circuit, for $\mathscr{E}/R$ in Eq. 37.3, we have

$$I = I_m \cos \omega t \qquad (37.4)$$

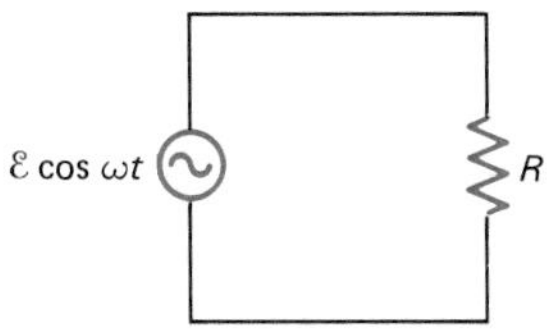

Figure 37.3

Schematic representation of an ac generator connected to a resistor.

Figure 37.4

The time variation of the potential difference between the ends of a resistor and the current in the resistor. Note that the potential difference and current are in phase—the peaks and valleys occur at the same time.

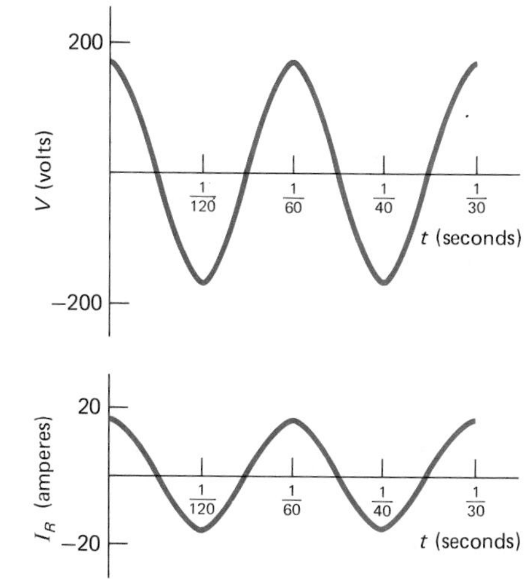

Figure 37.5

The time variation of the thermal power developed in a resistor.

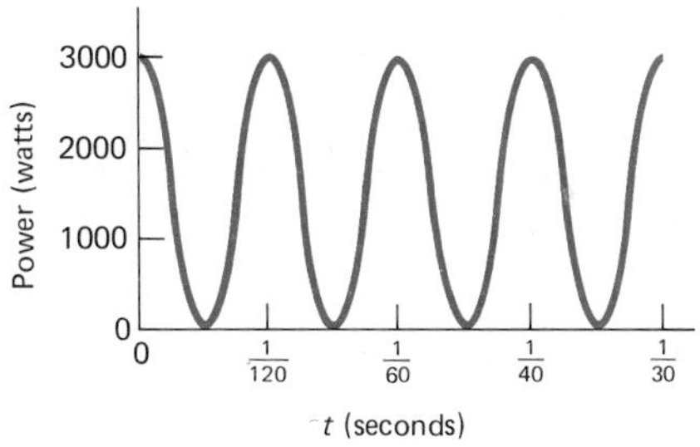

If $\mathscr{E} = 170$ V and $\nu = 60$ Hz (as they do in most wall outlets in the United States), then for $R = 10$ ohms, we get

$$V = 170 \cos(2\pi 60t) \text{ V}$$

and

$$I = \frac{170}{10} \cos(2\pi 60t)$$

$$= 17 \cos(2\pi 60t) \text{ A}$$

Two cycles of V and I are shown in Figure 37.4. Both V and I are proportional to $\cos(2\pi 60t)$. Therefore, their peaks and valleys occur at the same time, and we say that they are *in phase*. The average current is zero over one, two, or any integral number of cycles. The average thermal power developed in the resistor is not zero because the instantaneous power $P = I^2R$ is always positive (Figure 37.5). Because I^2 is periodic, we can determine the average power, P_{av}, by considering a single cycle:

$$P_{av} = (I^2R)_{av} = R(I^2)_{av} = RI_m^2(\cos^2 \omega t)_{av} \qquad (37.5)$$

Figure 37.6

An ac potential difference displayed on an oscilloscope.

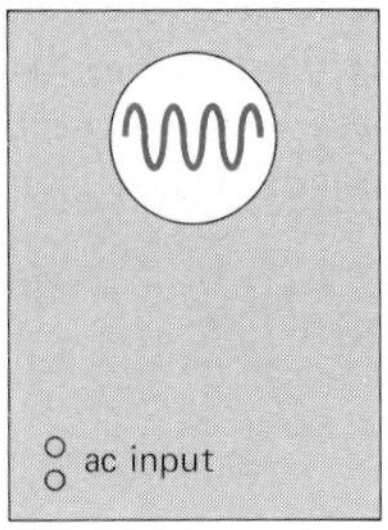

Recalling from Section 9.1 the procedure for determining average values, we have

$$P_{av} = RI_m^2 \left(\frac{\int_0^{2\pi} \cos^2 \omega t \, d(\omega t)}{\int_0^{2\pi} d(\omega t)} \right)$$

$$= R\left(\frac{I_m^2}{2}\right) \qquad (37.6)$$

Note that the same power would be produced by a constant dc current of $I_m/\sqrt{2}$ A in the resistor. It would also result if we were to connect the resistor to a potential difference having a constant value of $\mathcal{E}/\sqrt{2}$ V. The quantities $I_m/\sqrt{2}$ and $\mathcal{E}/\sqrt{2}$ are called the *rms values* of the current and potential difference. The term *rms* is short for root-mean-square, which means "the square root of the mean value of the square of the quantity of interest." For an electric outlet in an American home where $\mathcal{E} = 170$ V, the rms value of the potential difference is $V_{rms} = \mathcal{E}/\sqrt{2} = 120$ V. As mentioned in Section 37.1, this is the value generally quoted for the potential difference.

The fluctuations of an ac potential difference can be measured and observed with an oscilloscope (Figure 37.6). However, the inertia of the moving coil in a galvanometer-type ammeter or voltmeter prevents a measurement of instantaneous current or instantaneous potential difference, and consequently, the use of an oscilloscope is sometimes preferable. But often these values are not required and it is sufficient to measure the rms value. To this end, a galvanometer-type voltmeter can be modified to record an rms potential difference or current. It requires inserting an element called a *rectifier* in series with the meter. A rectifier allows charge to pass in only one direction. The meter therefore records the average value of the current for one direction of the ac current. The average value and the rms value are related by a constant numerical factor, making it possible to compute the rms value from the average value. Most ac meters are calibrated to read the rms value directly.

Example 2 How to Make a Conventional Ammeter Record an rms Value

By connecting a rectifier in series with an ordinary galvanometer-type ammeter, we can convert the latter to an ac ammeter, which reads the rms value of the original (unrectified) current $I = I_m \sin \omega t$. The rectifier eliminates current during the time the current would otherwise be negative. As a result, the current in the meter corresponds to the graph shown in Figure 37.7. The charge flows in only one direction, but because the current is not constant it is called a *pulsating dc*. The meter will record the average value of this rectified current, which is

Figure 37.7

The time variation of the current in a series connection of an ac generator, a resistor, and a rectifier.

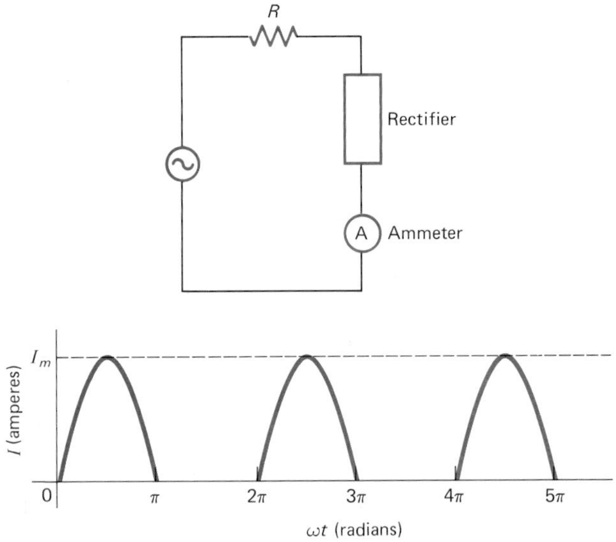

$$I_{av} = \frac{\int_0^{2\pi} I \, d(\omega t)}{\int_0^{2\pi} d(\omega t)}$$

Because

$$I = I_m \sin \omega t \quad \text{for} \quad 0 \le \omega t \le \pi$$

and

$$I = 0 \quad \text{for} \quad \pi < \omega t < 2\pi$$

then

$$I_{av} = \frac{I_m \int_0^{\pi} \sin \omega t \, d(\omega t)}{2\pi} = \frac{I_m[-\cos \omega t]_0^{\pi}}{2\pi} = \frac{I_m}{\pi}$$

The rms value of $I_m \sin \omega t$ was shown to be $I_{rms} = I_m/\sqrt{2}$. The meter determines I_{av}. Hence $I_{rms} = (\pi/\sqrt{2})I_{av}$. Multiplying the average current recorded by the ammeter by $\pi/\sqrt{2}$ yields the rms value of the current.

Questions

4. Why is the rms value of an ac current always less than the peak value?
5. Why would you expect the current in a resistor and the potential difference across the resistor to be in phase?
6. In a light bulb connected to a household outlet, the instantaneous current is zero two times in each cycle of the current. Why isn't the light extinguished during these times of zero current?
7. Consider a series connection of an ac generator, a resistor, and a simple on–off switch. If you could

open the switch during that part of the cycle where the polarity of the ac is negative, what would a plot of current versus time look like? Although our reflexes do not ordinarily permit this, this is exactly the way an elementary rectifier behaves in a circuit.

37.3 Behavior of a Capacitor Connected to an ac Generator

Figure 37.8 shows a capacitor connected to an ac generator. We assume that the capacitor has been connected for a sufficiently long time to avoid any transient effect that might take place when the capacitor is first connected to the generator. The instantaneous charge on the capacitor equals the instantaneous potential difference across the capacitor multiplied by the capacitance ($q = CV$). This follows from the definition of capacitance (Eq. 29.1). Thus

$$q = C\mathscr{E} \cos \omega t \qquad (37.7)$$

Relating current and charge by $I = dq/dt$, we have

$$I = -\omega C\mathscr{E} \sin \omega t \qquad (37.8)$$

In order to compare V and I, let us again take the standard household values $\mathscr{E} = 170$ V (corresponding to 120 V, rms) and

$$\omega = 2\pi(60) \text{ rad/s}$$
$$= 377 \text{ rad/s}$$

and assume that $C = 10^{-5}$ F. Then

$$\omega = 377 \text{ rad/s}$$
$$V = 170 \cos(2\pi 60t) \text{ V}$$
$$I = -170(2\pi 60)10^{-5} \sin(2\pi 60t)$$
$$= -0.641 \sin(2\pi 60t) \text{ A}$$

Two cycles of V and I are shown in Figure 37.9.

Unlike the previous situation for a resistor (Section 37.2), the current I and potential difference V for a capacitor are *not* in phase. The first peak of the current plot occurs one quarter of a cycle *before* the first peak in the potential difference plot. Hence we say that the capacitor current *leads* capacitor potential difference by one quarter of a period. One quarter of a period corresponds to a phase difference of $\pi/2$ rad, or 90°. Accordingly, we also say the potential difference lags the current by 90°.

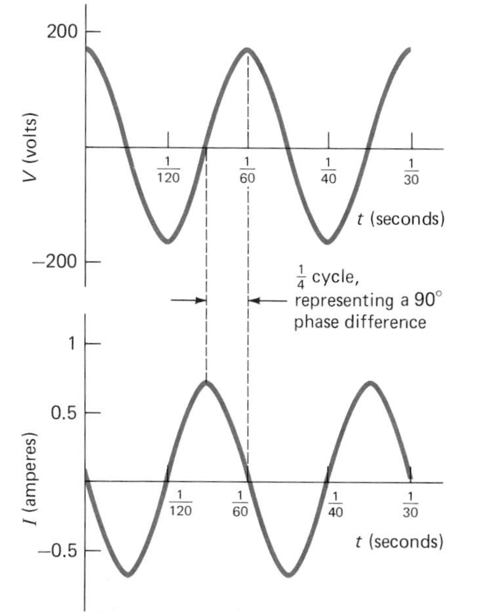

Figure 37.9

The time variation of the potential difference between the plates of a capacitor and the current in the capacitor. Note that the potential difference and the current are not in phase—the peak in the current occurs one-quarter cycle before the corresponding potential difference peak.

Rewriting Eq. 37.8 as

$$I = -\frac{\mathscr{E}}{1/(\omega C)} \sin \omega t \qquad (37.9)$$

and comparing Eqs. 37.3 and 37.9, we see that $1/\omega C$ must have units of resistance. The quantity $1/\omega C$ is called the *capacitive reactance*, symbolized by X_C.

$$X_C = \frac{1}{\omega C} \qquad (37.10)$$

Capacitive reactance is a measure of how the capacitor limits the ac current in the circuit. It depends on both the size of the capacitance and the frequency of the generator (Figure 37.10). The capacitive reactance *decreases* if either frequency or capacitance increases. In the limit as the frequency goes to zero, the capacitive reactance becomes infinite. Zero frequency for the generator implies that there is a dc source, such as a battery. Because no charge actually flows between the plates of a capacitor, the infinite capacitive reactance for zero frequency is consistent with the behavior of a capacitor connected to a dc source. Resistance and capacitive reactance are similar in the sense that both measure limitations to ac currents; but unlike resistance, capacitive reactance depends on the frequency of the ac generator.* The concept of capacitive reactance

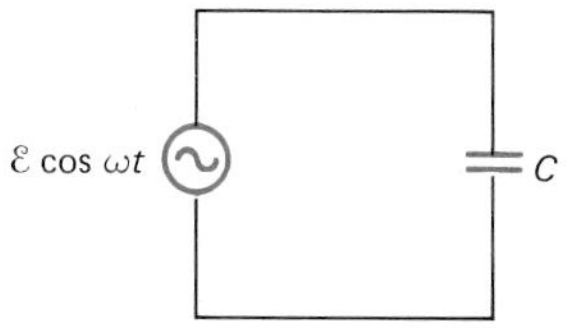

Figure 37.8

Schematic representation of an ac generator connected to a capacitor.

*For sufficiently high frequencies (generally in the hundreds of megahertz region), ac currents are confined to the surface of a conductor. This is called the *skin effect*. At these frequencies the electrical resistance is dependent on frequency. For the range of frequencies considered in this chapter, the skin effect is inconsequential, and the electrical resistance of a conductor is essentially independent of frequency.

Figure 37.10

The reactance of a capacitor ($C = 1/2\pi\ \mu F$) as a function of the frequency measured in hertz. Note that the scales are logarithmic. The capacitive reactance *decreases* as the frequency *increases*.

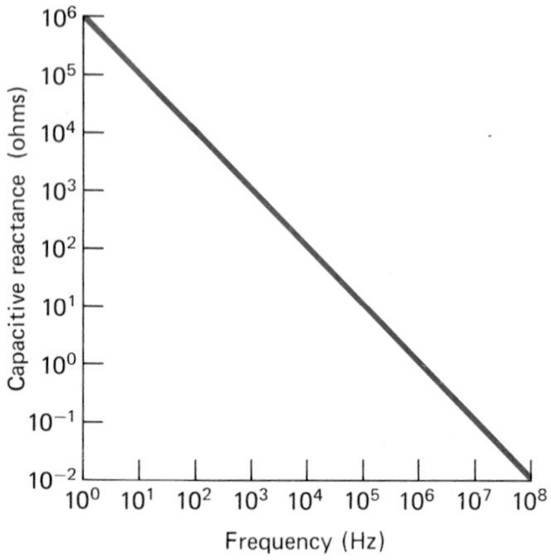

Figure 37.11

The time variation of the potential difference, current, and power for a capacitor connected to an ac generator.

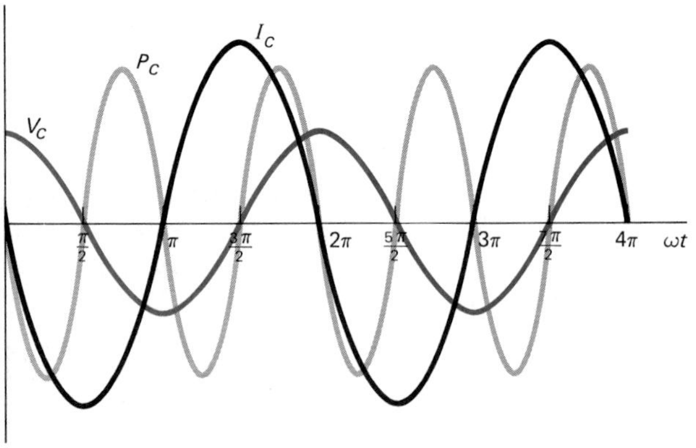

allows us to introduce an equation analogous to the equation $I = V/R$ involving resistance R.

$$I_{rms} = \frac{V_{rms}}{X_C} \qquad (37.11)$$

The instantaneous power delivered to the capacitor is the product of the instantaneous capacitor current and the potential difference.

$$\begin{aligned} P &= VI \\ &= -\omega C\mathcal{E}^2 \sin \omega t \cos \omega t \qquad (37.12) \\ &= -\frac{1}{2}\omega C\mathcal{E}^2 \sin 2\omega t \end{aligned}$$

The sign of P determines the direction of energy flow with time. When P is positive, energy is being stored in the electric field of the capacitor. When P is negative, energy is being released by the capacitor. Graphical representations of V, I, and P are shown in Figure 37.11. While both the current and the potential difference vary with radian frequency ω, the power varies with radian frequency 2ω. The average power is zero. The electric energy stored in the capacitor during a charging cycle is completely recovered when the capacitor discharges. On the average, then, there is no energy stored or lost in the capacitor.

Example 3 **Current in a Circuit Containing a Capacitor Connected to an ac Generator**

A 100-μF capacitor is connected to a 60-Hz ac generator having a peak amplitude of 170 V. Let us determine the current recorded by an rms ac ammeter connected in series with the capacitor.

At this frequency, the capacitive reactance of the capacitor is

$$X_C = \frac{1}{\omega C} = \frac{1}{2\pi(60)\ \frac{\text{rad}}{\text{s}}\ 100 \times 10^{-6}\ \text{F}} = 26.5\ \Omega$$

Assuming that the ammeter has no effect on the current measurement, we have for the instantaneous current in the capacitor

$$\begin{aligned} I &= \frac{-\mathcal{E}}{X_C} \cos \omega t = -\frac{170}{26.5} \cos \omega t \\ &= -6.41 \cos \omega t \text{ A} \end{aligned}$$

The rms current (Section 37.2) is

$$\begin{aligned} I_{rms} &= \frac{I_m}{\sqrt{2}} \\ &= \frac{6.41}{\sqrt{2}} \\ &= 4.53 \text{ A} \end{aligned}$$

Questions

8. Use your knowledge of how a capacitor functions to explain why the current in a capacitor–ac-generator circuit increases as the capacitance increases.
9. A capacitor is connected to an ac generator having a fixed peak value ($\mathcal{E}$) but variable frequency. In terms of the manner in which a capacitor charges and discharges, why would you expect the current to increase as the frequency increases?
10. Why would you expect the average power delivered to a capacitor by an ac generator to be zero?
11. Mathematically, capacitive reactance ($X_C = 1/\omega C$) becomes infinite as the frequency goes to zero. Explain how this makes sense physically.
12. Why do capacitative reactances become small in high-frequency circuits, such as those in a TV set?

37.4
Behavior of an Inductor Connected to an ac Generator

We consider next an ideal (zero-resistance) inductor connected to an ac generator (Figure 37.12). Calling V the potential difference across the inductor, we can write (Eq. 35.6

$$V = L\frac{dI}{dt} = \mathscr{E}\cos\omega t \qquad (37.13)$$

Integration of Eq. 37.13 results in

$$I = \frac{\mathscr{E}}{\omega L}\sin\omega t + \text{constant} \qquad (37.14)$$

The constant (of integration) represents a circuit current not depending on time. Because there is only an ac source, there is no time-independent current. Hence, the constant of integration is zero and we have

$$I = \frac{\mathscr{E}}{\omega L}\sin\omega t \qquad (37.15)$$

To compare V and I let us take $\mathscr{E} = 170$ V and $\omega = 2\pi(60)$ rad/s, and assume that $L = 1$ H. Then

$$V = 170\cos(2\pi 60t)\ \text{V}$$

$$I = \frac{170}{2\pi\cdot 60}\sin(2\pi 60t) = 0.451\sin(2\pi 60t)\ \text{A}$$

Two cycles of V and I are shown in Fgiure 37.13. The inductor current and potential difference are *not in phase*. Instead, the peaks and valleys of current and potential difference occur at different times. The first peak in the potential difference plot occurs one-quarter cycle *before* the first peak in the current plot. We say that the inductor current *lags* the inductor potential difference by $\pi/2$ rad (or 90°). This is what we would expect from Lenz's law. Another way of seeing this is to rewrite Eq. 37.15 as

$$I = \frac{\mathscr{E}}{\omega L}\cos\left(\omega t - \frac{\pi}{2}\right)$$

Because $V = \mathscr{E}\cos\omega t$, the $-\pi/2$ phase difference for I means that I lags V by $\pi/2$ rad. This is in contrast to the current in a capacitor, which *leads* the potential difference. For an inductor, the current *lags* the potential difference.

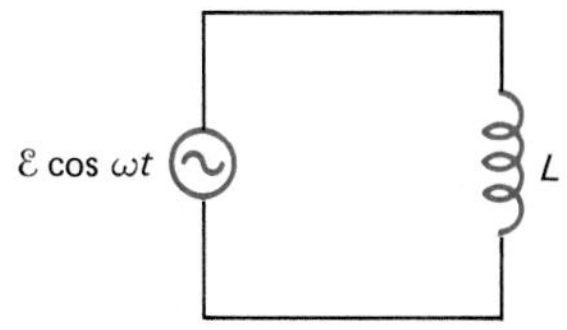

Figure 37.12

Schematic representation of an ideal inductor (zero resistance) connected to an ac generator.

Figure 37.13

The time variation of the potential difference between the ends of an inductor and the current in the inductor. Note that the potential difference and current are not in phase—the peak in the current occurs one-quarter cycle after the corresponding potential difference peak.

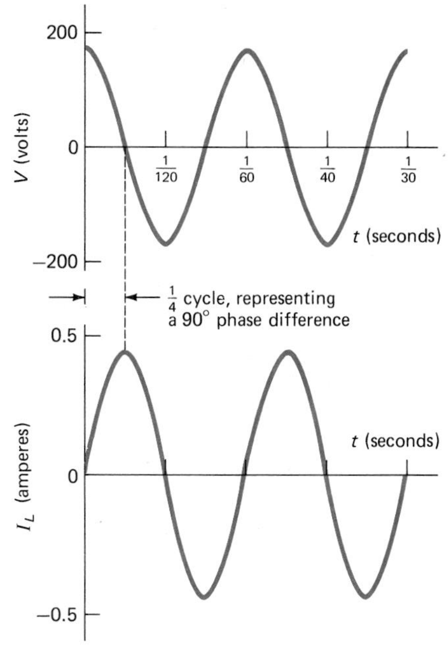

Figure 37.14

The reactance of an inductor ($L = 1/2\pi$ μH) as a function of the frequency measured in hertz. Note that the scales are logarithmic. The inductive reactance *increases* as the frequency *increases*.

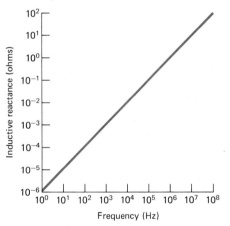

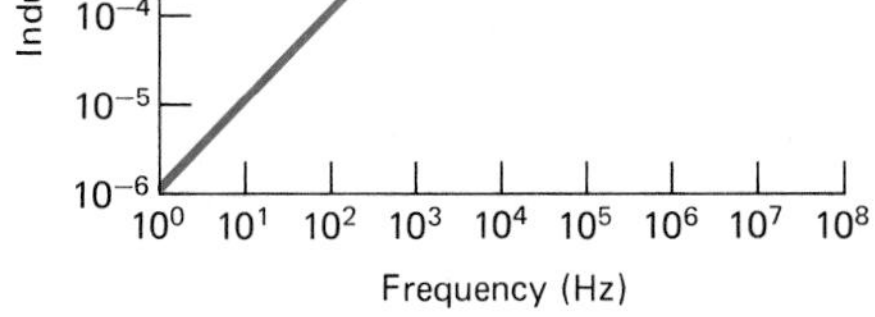

The quantity ωL in Eq. 37.15 has units of resistance and is called the *inductive reactance*, symbolized by X_L.

$$X_L = \omega L \qquad (37.16)$$

As with capacitive reactance, X_C, inductive reactance, X_L, has units of ohms. Inductive reactance is a measure of how the inductor limits the ac current in the circuit. It depends on both the size of the inductance and the frequency of the generator (Figure 37.14). Inductive reactance increases if either frequency or inductance increases. This is just the opposite of capacitive reactance. *In the limit as the frequency*

Table 37.1

	Reactance	Low-Frequency limit	High-Frequency limit
Resistor	$X_R = R$	R	R
Capacitor	$X_C = \dfrac{1}{\omega C}$	∞	0
Inductor	$X_L = \omega L$	0	∞

goes to zero, the inductive reactance goes to zero; in the same limit, capacitive reactance becomes infinite (see Table 37.1). Because inductive effects vanish for a dc source such as a battery, zero inductive reactance for zero frequency is consistent with the behavior of an inductor connected to a dc source.

The concept of inductive reactance allows us to introduce an inductor analog to the equation $I = V/R$ involving resistance R.

$$I_{\rm rms} = \frac{V_{\rm rms}}{X_L} \tag{37.17}$$

The instantaneous power delivered to the inductor is

$$P = VI$$
$$= \frac{\mathscr{E}^2}{\omega L} \sin \omega t \cos \omega t = \frac{\mathscr{E}^2}{2\omega L} \sin 2\omega t \tag{37.18}$$

Graphical representations of V, I, and P are shown in Figure 37.15. Although both the current and the potential difference vary with radian frequency ω, the power varies with radian frequency 2ω. The average power delivered is zero. Energy is alternately stored and released as the magnetic field alternately grows and dwindles.

because P=VI (add V & I graphicly)

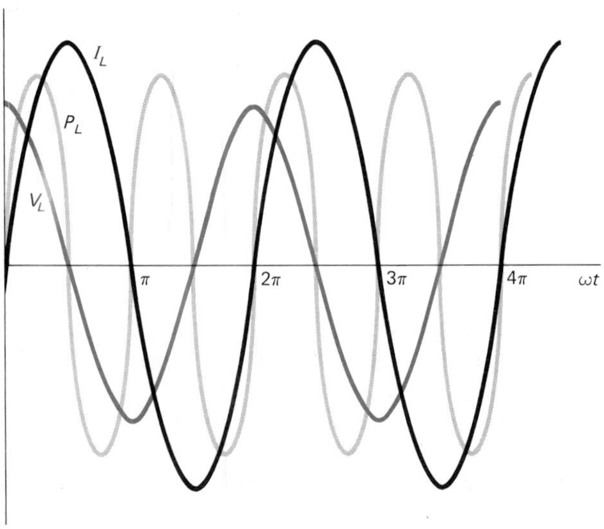

Figure 37.15

The time variation of the potential difference, current, and power for an ideal inductor connected to an ac generator.

Example 4 Inductive Reactance of a Solenoid

A certain solenoid with nothing in its interior has a length of 25 cm and diameter of 2.5 cm, and contains 1000 closely spaced turns. The resistance of the coil was measured to be 1.00 Ω. Let us compare the inductive reactance at 100 Hz with the resistance of the coil.

The inductance of a solenoid whose length is large compared to its diameter is (Section 35.1)

$$L = \frac{\mu_0 N^2 \pi a^2}{l}$$

where N is the number of turns, a is the radius, and l is the length. Hence

$$L = \frac{(4\pi \times 10^{-7})\,\dfrac{\rm H}{\rm m}\,(1000)^2\pi(0.0125)^2\ {\rm m}^2}{0.25\ {\rm m}}$$
$$= 2.47 \times 10^{-3}\ {\rm H}$$

The inductive reactance at a frequency of 100 Hz is

$$X_L = \omega L = 2\pi 100\ \frac{\rm rad}{\rm s}\,(2.47 \times 10^{-3})\ {\rm H}$$
$$= 1.55\ \Omega$$

The inductive reactance of this solenoid at 100 Hz is comparable to the intrinsic (ohmic) resistance. In a circuit diagram it would be shown as

2.47 mH 1.00 Ω

We consider the effect of the R–X_L combination in Section 37.6.

Questions

13. Describe the role of Lenz's law when an ideal inductor is connected to an ac generator.

14. In Section 35.1, self-inductance was characterized as an electrical inertia. Using this as a guide, why would you expect the current in an inductor connected to an ac generator to decrease as the self-inductance increases?

15. A solenoid consisting of several meters of wire becomes warm when it is connected to an ac generator. What characteristic of the solenoid is responsible for the production of this heat energy?

16. Why do inductive reactances become large in high-frequency circuits?

37.5 The *RC* Circuit

In Chapters 29, 30, and 35 we saw how a resistor, a capacitor, and an inductor each functions in a circuit energized by a source, such as a battery, that has a constant potential difference between its output terminals. We based our analyses of these circuits on Kirchhoff's rules, which incorporate two important physical principles, conservation of charge and conservation of energy. We found that, following the connection of a dc source, charge accumulates on capacitor plates and currents build up in inductors. This transient situation leads to a steady-state condition in which capacitors are fully charged and currents in inductors reach steady values. Now let's analyze circuits containing combinations of resistors, capacitors, and inductors connected to an ac generator. Here the currents and potential differences change periodically with time, but the conservation principles (energy and charge) remain valid at any instant. Thus, although our analysis may involve new techniques, it involves no new principles.

Resistor–Capacitor (*RC*) Circuit

Let us consider first a resistor and a capacitor connected to an ac generator as shown in Figure 37.16. This configuration is called a *filter circuit,* which, as we will soon see, is an appropriate name. Filter circuits are important in electronics, and are used in, for example, stereo systems. We are going to analyze the filter circuit in some detail in order to illustrate the physics principles involved, and to develop some analysis methods that we will use later on. We will not need to analyze other circuits in such detail because we will be able to draw on the concepts and methods developed in this analysis.

Consider the instantaneous values of the potential differences across the resistor and capacitor and the current in the circuit. Kirchhoff's loop law for potential differences allows us to write

$$\mathcal{E} \cos \omega t = V_R + V_C \tag{37.19}$$

where V_R and V_C are the potential differences across the resistor and capacitor at some instant of time t. Relating these potential differences to resistance and capacitance by using Eqs. 30.7 and 29.1, we can write

$$\mathcal{E} \cos \omega t = IR + \frac{q}{C} \tag{37.20}$$

Relating current and charge by $I = dq/dt$, we obtain from Eq. 37.20 a differential equation for the charge.

$$\mathcal{E} \cos \omega t = R\frac{dq}{dt} + \frac{q}{C} \tag{37.21}$$

We arrive at an equation that must be solved in order to obtain the charge stored instantaneously by the capacitor.

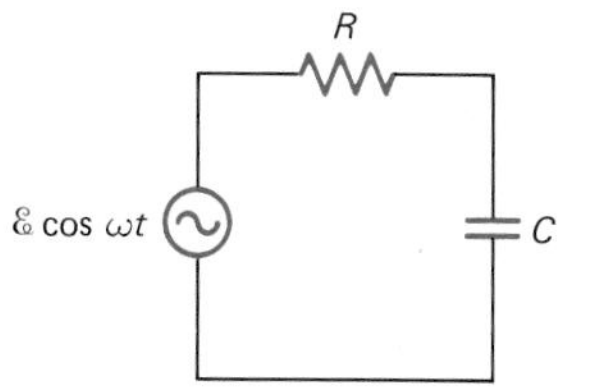

Figure 37.16

Schematic representation of a series connection of a resistor, a capacitor, and an ac generator.

Mathematical Solution of the *RC*, or Filter, Circuit

Apart from the term representing the ac generator, Eq. 37.21 is identical to the equation resulting from connecting a resistor and a capacitor to a battery (Section 31.6). But integrating Eq. 37.21 requires a different technique because of the $\mathcal{E} \cos \omega t$ term. Whereas a dc source, such as a battery, can only cause the capacitor to accumulate charge, the ac generator, whose polarity changes with time, causes the polarity of the charge on the capacitor plates to alternate. Because it is the ac generator that produces these oscillating currents and potential differences, it seems reasonable that their frequency should match the frequency of the ac generator. But our experience with inductors and capacitors connected to an ac generator suggests that the current oscillations may not be in phase with the ac generator. Let us incorporate these two ideas (generator frequency ω and phase shift ϕ) into a possible solution having the form

$$I = I_m \cos(\omega t + \phi) \tag{37.22}$$

where I_m and ϕ are independent of time. When the parameter ϕ is positive, the current leads the potential difference provided by the generator. When ϕ is negative, the current lags the potential difference provided by the generator. This is not the only form for the solution. We could use the sine function and we could use $-\phi$ just as well (see Problem 35). The essential point is that the solution must satisfy the equation involving current, charge, and time.

Because current and charge are related by $I = dq/dt$, integration of Eq. 37.22 yields

$$q = \frac{I_m}{\omega} \sin(\omega t + \phi)$$

for the form of the solution for q. We determine the parameters I_m and ϕ by substituting the assumed solution into Eq. 37.21. This procedure is somewhat like solving a numerical algebraic equation by assuming a number for the solution. To facilitate the substitution we use a trigonometric identity to rewrite the expression for q as

$$q = \frac{I_m}{\omega} \sin \omega t \cos \phi + \frac{I_m}{\omega} \cos \omega t \sin \phi \tag{37.23}$$

Then

$$\frac{dq}{dt} = I_m \cos \omega t \cos \phi - I_m \sin \omega t \sin \phi$$

Substituting for q and dq/dt into Eq. 37.21, we have

$$\mathscr{E}\cos\omega t = RI_m\cos\omega t\cos\phi - RI_m\sin\omega t\sin\phi + \frac{I_m}{\omega C}\sin\omega t\cos\phi + \frac{I_m}{\omega C}\cos\omega t\sin\phi$$

Collecting terms involving $\cos\omega t$ and $\sin\omega t$, we have

$$\left(\mathscr{E} - RI_m\cos\phi - \frac{I_m}{\omega C}\sin\phi\right)\cos\omega t + \left(RI_m\sin\phi - \frac{I_m}{\omega C}\cos\phi\right)\sin\omega t = 0$$

In order that the sum of these two terms be zero for all possible values of the time, the coefficients of $\cos\omega t$ and $\sin\omega t$ must each be zero. For instance, when $\omega t = 0$, then $\sin\omega t = 0$ but $\cos\omega t = 1$. Therefore the coefficient of $\cos\omega t$ must be zero. This gives us two equations.

$$\mathscr{E} = RI_m\cos\phi + \frac{I_m}{\omega C}\sin\phi \qquad (37.24)$$

$$RI_m\sin\phi = \frac{I_m}{\omega C}\cos\phi \qquad (37.25)$$

We start with the second of these two equations because I_m cancels out and ϕ is the only undetermined quantity.

$$\frac{\sin\phi}{\cos\phi} = \tan\phi = \frac{1}{\omega RC} = \frac{X_C}{R} \qquad (37.26)$$

The constant ϕ is determined in terms of known quantities in the circuit. It will help to visualize the relationships if we construct a "useful" triangle (Figure 37.17) having sides $1/\omega C$ and R and hypotenuse $\sqrt{R^2 + (1/\omega C)^2}$. For $\sin\phi$ and $\cos\phi$ we write

$$\sin\phi = \frac{X_C}{\sqrt{R^2 + X_C^2}} \qquad (37.27)$$

$$\cos\phi = \frac{R}{\sqrt{R^2 + X_C^2}} \qquad (37.28)$$

We have identified $X_C = 1/\omega C$ as the capacitive reactance. The quantity $\sqrt{R^2 + X_C^2}$ is called the *impedance* of this circuit. Impedance is given the symbol Z, and so we can write

$$\sin\phi = \frac{X_C}{Z} \qquad \text{and} \qquad \cos\phi = \frac{R}{Z}$$

Notice that $\sin\phi$ is positive. The phase parameter ϕ is positive, indicating that the current leads the generator potential difference in the RC circuit.

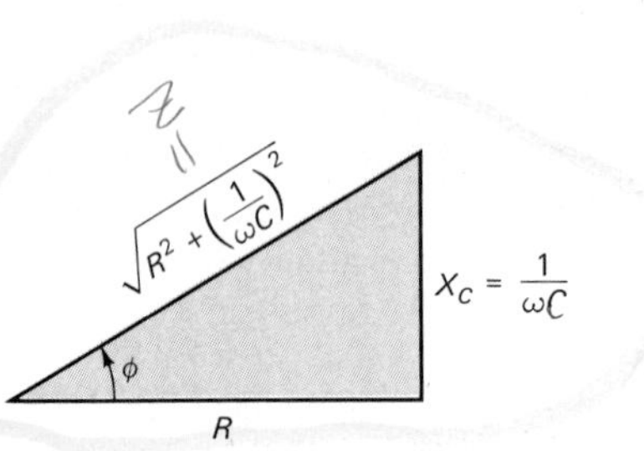

Figure 37.17

A "useful" triangle for an RC circuit. The base is of "length" R and the altitude is of "length" $1/\omega C$. The phase angle ϕ is then given by $\phi = \tan^{-1}[(1/\omega C)/R]$.

Returning to Eq. 37.24, we have

$$\mathscr{E} = RI_m\cos\phi + \frac{I_m}{\omega C}\sin\phi$$

with $\cos\phi$ and $\sin\phi$ now known. Solving for I_m, we have

$$I_m = \frac{\mathscr{E}}{R\cos\phi + \dfrac{\sin\phi}{\omega C}}$$

Substituting for $\sin\phi$ and $\cos\phi$, we obtain

$$I_m = \frac{\mathscr{E}}{\dfrac{R^2}{Z} + \dfrac{1}{\omega^2C^2Z}} = \frac{\mathscr{E}Z}{R^2 + \dfrac{1}{\omega^2C^2}} = \frac{\mathscr{E}}{Z}$$

The solutions for charge (q), Eq. 37.23, and current (I), Eq. 37.22, become

$$q = \frac{\mathscr{E}}{\omega Z}\sin(\omega t + \phi) \qquad (37.29)$$

$$I = \frac{\mathscr{E}}{Z}\cos(\omega t + \phi)$$

You are invited to check these solutions by substituting these expressions for q and I into Eq. 37.20.

Now we consider the consequences for V_C. In particular, we consider the effect of the frequency dependence of the amplitude of V_C.

The potential difference across the capacitor is

$$V_C = \frac{q}{C} = IX_C = \frac{\mathscr{E}}{\omega CZ}\sin(\omega t + \phi)$$

or, with $1/\omega C = X_C$,

$$V_C = \frac{\mathscr{E}X_C}{Z}\sin(\omega t + \phi) \qquad (37.30)$$

The amplitude of this potential difference ($\mathscr{E}X_C/Z$) depends on the frequency of the generator because both X_C and Z depend on ω (Figure 37.18). Substituting $X_C = 1/\omega C$ and $Z = \sqrt{R^2 + 1/\omega^2C^2}$, we have

$$\frac{\mathscr{E}X_C}{Z} = \frac{\mathscr{E}}{\sqrt{(\omega RC)^2 + 1}} \qquad (37.31)$$

We see that as the generator frequency (ω) decreases and becomes zero, the amplitude goes to $\mathscr{E}$; and as the generator frequency increases and becomes infinite, the amplitude goes to zero. This behavior is exploited in electronics.

In an electronics circuit such as a high-fidelity amplifier, an ac source (in effect, a generator) from one part of a circuit provides the input to another part of the circuit. The source may involve many frequencies and it may be desirable to suppress the high-frequency components from the input. If the source is connected to an RC circuit (Figure 37.19), and the input to another part of the circuit is provided by the potential difference across the capacitor, then the potential difference on the input terminals is

Figure 37.18

The amplitude of the potential difference across the capacitor as a function of the dimensionless quantity ωRC. For a given combination of resistance and capacitance, this amplitude *decreases* as the frequency of the generator *increases*.

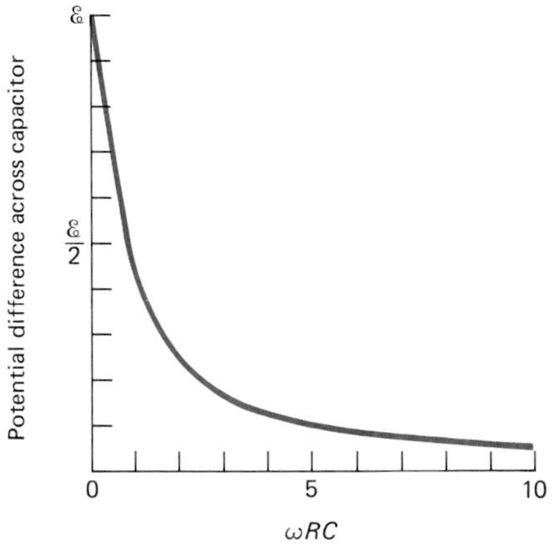

Figure 37.19

Schematic representation of an RC circuit used to suppress high-frequency ac signals.

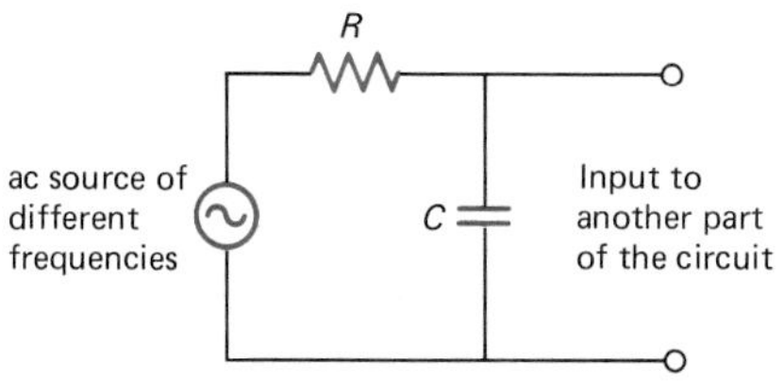

given by Eq. 37.30. For equal amplitudes in the source, the amplitude of the high-frequency component across the capacitor will be smaller. We say that the high-frequency component has been "filtered out" and the RC circuit constitutes a filter circuit for high-frequency ac signals. In qualitative terms, the capacitor is a short circuit for the high-frequency ac signals.

Current in the *RC*, or Filter, Circuit

Using Eq. 37.29 to determine the current in the circuit, we have

$$I = \frac{dq}{dt} = \frac{\mathcal{E}}{Z}\cos(\omega t + \phi) \tag{37.32}$$

and the potential difference across the resistor is

$$V_R = IR = \mathcal{E}\frac{R}{Z}\cos(\omega t + \phi) \tag{37.33}$$

Plausibility Checks

We have labored through considerable algebra to arrive at Eqs. 37.32 and 37.33. Let's do a plausibility check to confirm the results. We can check the two equations by setting $\omega = 0$. Physically this means that the ac generator is replaced by a constant (dc) potential difference $\mathcal{E}$. Therefore our results for the current and potential differences should reduce to those obtained in Section 31.6 for the RC circuit when the steady state (equilibrium) is established. In the steady state, we recognize three features: (1) The current is zero; (2) the potential difference across the resistor is zero; and (3) the potential difference across the capacitor is that of the dc source, namely, $\mathcal{E}$. To show these three features we need only recognize that capacitive reactance becomes infinite as the frequency goes to zero (see Figure 37.10 or Eq. 37.10). In effect, the capacitor behaves as an open circuit. Setting $\omega = 0$ in the equation representing the current (Eq. 37.32) yields

$$I = \frac{\mathcal{E}}{Z}\cos\phi \tag{37.34}$$

Using the "useful" triangle (Figure 37.17) to replace $\cos\phi$ with $R/Z = R/\sqrt{R^2 + X_C^2}$, we have

$$I = \frac{\mathcal{E}}{Z}\left(\frac{R}{Z}\right) = \frac{\mathcal{E}R}{R^2 + X_C^2} \tag{37.35}$$

Now letting X_C become infinite, we see that the current I goes to zero, thus demonstrating the first feature of the steady state. Because the potential difference across the resistor is directly related to the current, the potential difference across the resistor vanishes, demonstrating the second feature. Using Eq. 37.30 and the relation $\sin\phi = X_C/Z$, we obtain the potential difference across the capacitor for zero frequency:

$$V_C = \frac{\mathcal{E}}{Z}X_C\frac{X_C}{Z} = \frac{\mathcal{E}X_C^2}{R^2 + X_C^2} \tag{37.36}$$

Let us write this equation as

$$V_C = \frac{\mathcal{E}}{1 + R^2/X_C^2} \tag{37.37}$$

Then as X_C becomes infinite, the term R^2/X_C^2 vanishes and the potential difference V_C equals $\mathcal{E}$, the strength of the dc source. With this we demonstrate the third feature. Thus our results for the general case ($\omega \neq 0$) are made plausible by the fact that they reduce to the known dc results as $\omega \to 0$.

Maximum and rms Values of Current and Potential Difference

The expressions for current and potential difference bear a strong resemblance to the equation $V = IR$ for resistors. For example, the current is written in the form of the ratio of a potential difference and a quantity having units of ohms. The maximum current in the circuit is just the ratio of the peak generator output ($\mathcal{E}$) divided by the impedance.

$$I_m = \frac{\mathcal{E}}{Z} \tag{37.38}$$

The maximum potential difference across the resistor is

$$V_m = I_m R = \frac{\mathcal{E}}{Z}R \tag{37.39}$$

The maximum potential difference across the capacitor is

$$V_m = \frac{X_C}{Z}\mathcal{E} = I_m X_C \tag{37.40}$$

These expressions for the resistor and the capacitor reveal that the maximum potential difference across an element of interest equals the maximum current multiplied by the reactance (or resistance). Thus the concepts of reactance and impedance allow us to determine maximum values without having to solve a differential equation.

On many occasions, it is only the maximum values that are of interest. On other occasions, it may be the rms values that we want to find. We compute the rms values by using the technique given in Section 37.2. We determine the average of the square of the quantity of interest. When the quantities of interest vary sinusoidally, the results averaged over time correspond identically to those determined in Section 37.2 for a sinusoidal function. Thus

$$I_{\text{rms}} = \frac{1}{\sqrt{2}}\frac{\mathcal{E}}{Z} = \frac{I_m}{\sqrt{2}}, \tag{37.41}$$

$$V_{\text{rms}} = \frac{1}{\sqrt{2}}\frac{R}{Z}\mathcal{E}$$

$$= I_{\text{rms}}R \text{ for the resistor, and} \tag{37.42}$$

$$V_{\text{rms}} = \frac{1}{\sqrt{2}}\frac{X_C}{Z}\mathcal{E}$$

$$= I_{\text{rms}}X_C \text{ for the capacitor,} \tag{37.43}$$

37.6 The *RL* Circuit

Replacing the capacitor in Figure 37.16 with an inductor produces an *RL* filter circuit (Figure 37.20). We can find the current and potential differences of an *RL* circuit just as we found these quantities for the *RC* circuit. Kirchhoff's loop law allows us to write

$$\mathcal{E}\cos\omega t = V_R + V_L \tag{37.44}$$

Knowing that $V_R = IR$ and $V_L = L\,dI/dt$, we can write Eq. 37.44 as

$$\mathcal{E}\cos\omega t = IR + L\frac{dI}{dt} \tag{37.45}$$

Proceeding as we did when solving for the *RC* circuit, we assume a solution of the form $I = I_m\cos(\omega t + \phi)$ and we find

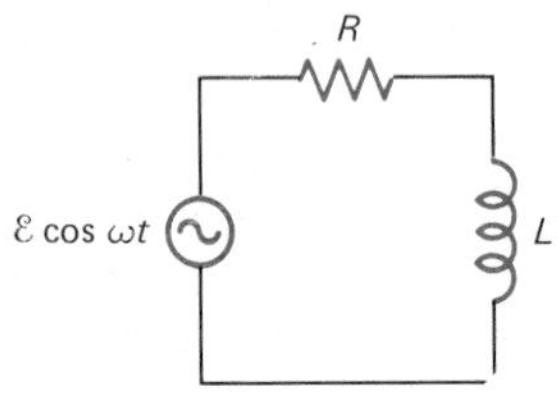

Figure 37.20

Schematic representation of a series connection of a resistor, an inductor, and an ac generator.

Figure 37.21

"Useful" triangle for the *RL* circuit. The base is of "length" R and the altitude is of "length" $-\omega L$. The phase angle ϕ is then given by $\phi = \tan^{-1}(-\omega L/R)$.

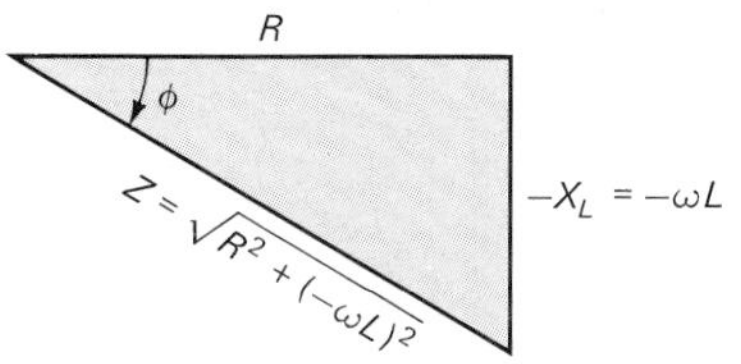

$$I = \frac{\mathcal{E}}{Z}\cos(\omega t + \phi) \tag{37.46}$$

where $Z = \sqrt{R^2 + X_L^2}$ and $\tan\phi = -X_L/R$. Figure 37.21 shows the "useful" triangle for the *RL* circuit. Note that $\sin\phi = -X_L/\sqrt{R^2 + X_L^2}$. The phase parameter ϕ is negative, indicating that the current lags the generator potential difference in the *RL* circuit. The current and potential differences become

$$I = \frac{\mathcal{E}}{Z}\cos(\omega t + \phi) \tag{37.47}$$

$$V_R = IR = \frac{\mathcal{E}}{Z}R\cos(\omega t + \phi) \tag{37.48}$$

$$V_L = L\frac{dI}{dt} = -\mathcal{E}\left(\frac{X_L}{Z}\right)\sin(\omega t + \phi) \tag{37.49}$$

The phase angle ϕ is negative, varying from 0 to $-\pi/2$, corresponding to a frequency variation from 0 to ∞. At low frequencies $X_L \ll R$ and $\phi \approx 0$ so that with Eq. 37.47 the current is

$$I \approx \frac{\mathcal{E}}{Z}\cos\omega t \approx \frac{\mathcal{E}}{R}\cos\omega t \tag{37.50}$$

in agreement with Eq. 37.3 for the purely resistive circuit. The ac generator output and current are in phase. At high frequencies $X_L \gg R$ and $\phi \approx -\pi/2$ so that the current is

$$I \approx \frac{\mathcal{E}}{Z}\cos\left(\omega t - \frac{\pi}{2}\right) \approx \frac{\mathcal{E}}{\omega L}\sin\omega t \tag{37.51}$$

in agreement with Eq. 37.15 for the purely inductive circuit. The current lags the ac generator output ($\mathcal{E}\cos\omega t$) by $\pi/2$ rad (90°). The rms values of current and potential differences are related to the maximum values by

$$I_{\text{rms}} = \frac{1}{\sqrt{2}}\frac{\mathcal{E}}{Z} = \frac{I_m}{\sqrt{2}}, \tag{37.52}$$

$$V_{\text{rms}} = \frac{1}{\sqrt{2}}\frac{\mathcal{E}}{Z}R$$

$$= I_{\text{rms}}R \text{ for the resistor, and} \tag{37.53}$$

$$V_{\text{rms}} = \frac{1}{\sqrt{2}}\frac{\mathcal{E}}{Z}X_L$$

$$= I_{\text{rms}}X_L \text{ for the inductor.} \tag{37.54}$$

With the concept of impedance, we begin to see a certain consistency in the equations and the results.

37.7 The *RLC* Circuit

Let us now conclude our discussion of impedance by considering all three elements—a resistor, an inductor, and a capacitor—connected to an ac generator (Figure 37.22). Such circuits are very useful in electronics. Analysis of this circuit is analogous to that for the *RC* and *RL* circuits. We write

$$\mathscr{E}\cos\omega t = V_R + V_C + V_L$$

$$= R\frac{dq}{dt} + \frac{q}{C} + L\frac{d^2q}{dt^2} \qquad (37.55)$$

This equation is identical in form to that deduced in Section 14.6 for the forced harmonic oscillator. Assuming, as for the *RC* and *RL* circuits, that

$$q = Q\sin(\omega t + \phi)$$

with Q and ϕ undetermined parameters, we find that

$$q = \frac{\mathscr{E}}{\omega Z}\sin(\omega t + \phi)$$

Here we define the impedance Z as

$$Z = \sqrt{R^2 + \left(\frac{1}{\omega C} - \omega L\right)^2}$$

$$= \sqrt{R^2 + (X_C - X_L)^2}$$

and

$$\tan\phi = \frac{X_C - X_L}{R}$$

Figure 37.23 shows the "useful" triangle for an *RLC* circuit. That X_L and X_C tend to cancel in the impedance equation is a consequence of the phase relations discussed earlier.

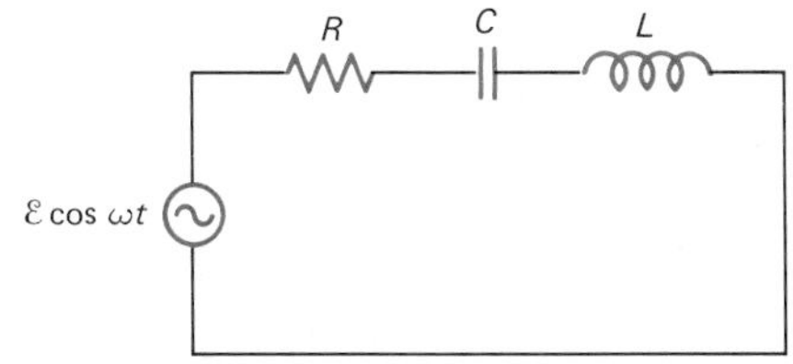

Figure 37.22

Schematic representation of a series connection of a resistor, a capacitor, an inductor, and an ac generator.

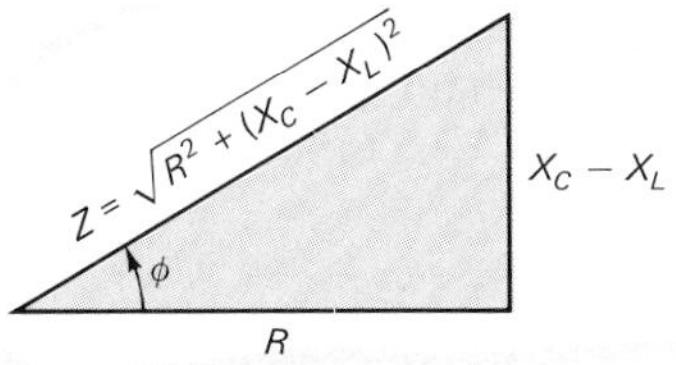

Figure 37.23

A "useful" triangle for an *RLC* circuit. The base is of "length" R and the altitude is of "length" $X_C - X_L$. The phase angle ϕ is then given by $\phi = \tan^{-1}[(X_C - X_L)/R]$.

Note that if $X_L = 0$, meaning that there is *no inductor* in the circuit, then we arrive at the solution we obtained for the *RC* circuit. And if $X_C = 0$, meaning that there is *no capacitor* in the circuit, then we arrive at the solution we obtained for the *RL* circuit. The expressions for current and potential differences for this *RLC* circuit are identical in form to those obtained for the *RC* and *RL* circuit.

$$I = \frac{\mathscr{E}}{Z}\cos(\omega t + \phi) \qquad (37.56)$$

$$V_R = \frac{\mathscr{E}}{Z}R\cos(\omega t + \phi) \qquad (37.57)$$

$$V_C = \frac{\mathscr{E}}{Z}X_C\sin(\omega t + \phi) \qquad (37.58)$$

$$V_L = -\frac{\mathscr{E}}{Z}X_L\sin(\omega t + \phi) \qquad (37.59)$$

Consequently, the relations for the rms values for current and potential differences are unaltered.

$$I_{rms} = \frac{1}{\sqrt{2}}\frac{\mathscr{E}}{Z} = \frac{I_m}{\sqrt{2}}, \qquad (37.60)$$

$$V_{rms} = \frac{1}{\sqrt{2}}\frac{\mathscr{E}}{Z}R$$

$$= I_{rms}R \text{ for the resistor, and} \qquad (37.61)$$

$$V_{rms} = \frac{1}{\sqrt{2}}\frac{\mathscr{E}}{Z}X_C$$

$$= I_{rms}X_C \text{ for the capacitor,} \qquad (37.62)$$

$$V_{rms} = \frac{1}{\sqrt{2}}\frac{\mathscr{E}}{Z}X_L$$

$$= I_{rms}X_L \text{ for the inductor.} \qquad (37.63)$$

These relations involving all three components also include the *RC* and *RL* circuits and the singly connected elements as special cases. For example, removing the inductor and capacitor is equivalent to setting $X_C = 0$ and $X_L = 0$. Then $Z = R$ and $\tan\phi = 0$, implying that $\phi = 0$, and we find that

$$V_R = \mathscr{E}\cos\omega t$$

exactly as in Section 37.2 when only a resistor was connected to the ac generator.

Power in the *RLC* Circuit

A capacitor connected to an ac generator reversibly stores and releases electric energy. There is no net energy delivered by the generator (Section 37.3). Similarly, an inductor connected to an ac generator reversibly stores and releases magnetic energy (Section 37.4). There is no net energy delivered by the generator. However, an ac generator delivers a net amount of energy when connected to a resistor (Section 37.2). The energy is transformed into thermal energy in the resistor. When a resistor, an inductor, and a capacitor are connected in series with an ac generator, it is still only the resistor that causes a net energy transfer. We can confirm this by calculating the power delivered by the generator.

The instantaneous power is the product of the generator output and the current that results.

$$P = VI$$

$$= \mathscr{E} \cos \omega t \left[\frac{\mathscr{E}}{Z} \cos(\omega t + \phi)\right] \qquad (37.64)$$

The phase angle ϕ and radian frequency ω play important roles in the power delivered. If the impedance Z is large at a particular radian frequency, then the power will be small for all values of the time. This is consistent with the idea that impedance measures how the combination of elements impedes (or limits) ac current. Example 5 illustrates the effect of phase angle on the power delivered.

Figure 37.24

Time variation of the power delivered by an ac generator in an RLC circuit having $R = 50\ \Omega$, $L = 1$ H, $C = 200$ μF, $\omega = 100$ rad/s, and $\mathscr{E} = 100$ V.

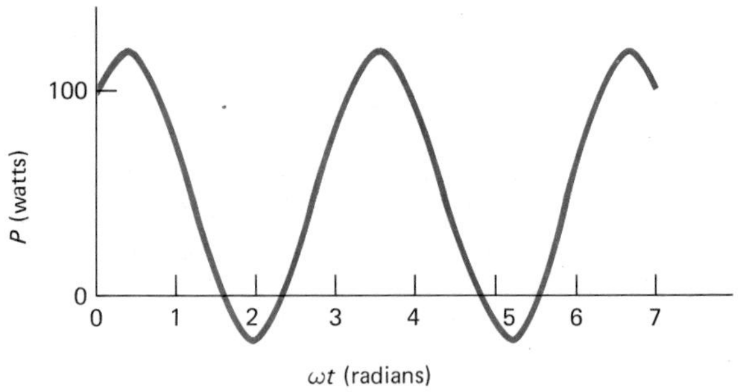

Example 5 Effect of Phase Angle on the Power Delivered by the Generator in an *RLC* Circuit

Let us choose $\omega = 100$ rad/s, $L = 1$ H, $C = 200$ μF, $R = 50\ \Omega$, and $\mathscr{E} = 100$ V, and calculate from Eq. 37.64 the instantaneous power delivered by the generator. The inductive reactance, capacitive reactance, and impedance are

$$X_L = \omega L = 100 \frac{\text{rad}}{\text{s}} \times 1\ \text{H} = 100\ \Omega$$

$$X_C = \frac{1}{\omega C} = \frac{1}{100 \frac{\text{rad}}{\text{s}} \times 200 \times 10^{-6}\ \text{F}} = 50\ \Omega$$

$$Z = \sqrt{R^2 + (X_C - X_L)^2}$$

$$= \sqrt{50^2 + 50^2} = 70.7\ \Omega$$

From the "useful" triangle in Figure 37.23 we determine the phase angle.

$$\tan \phi = \frac{X_C - X_L}{R} = \frac{50\ \Omega - 100\ \Omega}{50\ \Omega} = -1$$

$$\phi = -0.785\ \text{rad}$$

At this frequency the inductive reactance is larger than the capacitive reactance and the current lags the potential difference provided by the generator. The instantaneous power for the given choice of component values is

$$P = \mathscr{E} \cos \omega t \times \frac{\mathscr{E}}{Z} \cos(\omega t + \phi)$$

$$= 100 \cos 100t \times \frac{100}{70.7} \cos(100t - 0.785)$$

In Figure 37.24 the power P is plotted versus ωt for constant ω. Positive values of the power means that electric energy delivered by the generator is being converted into thermal energy by the resistor. Negative values of the power means that energy is being delivered to the generator. It is energy derived from the electric field of the capacitor and magnetic field of the inductor. The average power is positive—the generator must deliver energy to the resistor.

The average power delivered by the generator is determined by calculating the average value of the instantaneous power.

$$P_{av} = \frac{\int_0^{2\pi} VI\, d(\omega t)}{\int_0^{2\pi} d(\omega t)}$$

$$= \frac{\int_0^{2\pi} \mathscr{E} \cos \omega t \frac{\mathscr{E}}{Z} \cos(\omega t + \phi)\, d(\omega t)}{\int_0^{2\pi} d(\omega t)}$$

To facilitate the calculation we make use of the trigonometric identity

$$\cos(\omega t + \phi) = \cos \omega t \cos \phi - \sin \omega t \sin \phi$$

Then

$$P_{av} = \frac{\mathscr{E}^2}{Z} \frac{\int_0^{2\pi} \cos \omega t \cos(\omega t + \phi)\, d(\omega t)}{\int_0^{2\pi} d(\omega t)}$$

$$= \frac{\mathscr{E}^2}{Z} \frac{[\cos \phi \int_0^{2\pi} \cos^2 \omega t\, d(\omega t)}{2\pi}$$

$$- \frac{\sin \phi \int_0^{2\pi} \cos \omega t \sin \omega t\, d(\omega t)]}{2\pi}$$

The first integral equals π. The second integral equals zero. The result is

$$P_{av} = \frac{\mathscr{E}^2}{2Z} \cos \phi \qquad (37.65)$$

The $\cos \phi$ in Eq. 37.65 is called the *power factor*. For given Z and $\mathscr{E}$ the power factor determines how much average power must be delivered by a generator to an RLC circuit. From the "useful" triangle (Figure 37.23) for the RLC circuit it follows that

$$\cos \phi = \frac{R}{Z}$$

Hence the average power may be written

$$P_{av} = \frac{\mathscr{E}^2}{2Z^2} R$$

In terms of the rms current (Eq. 37.60) and rms potential difference the average power is

$$P_{av} = \frac{V_{rms}^2}{Z^2} R \tag{37.66}$$

$$= I_{rms}^2 R \tag{37.67}$$

As anticipated, net power delivered by the generator is a result of resistance in the circuit. If there is no resistance, there is no net power delivered by the generator, regardless of the sizes of the capacitive and inductive reactances. The *RLC* circuit is now an *LC* circuit. Even though the average power delivered is zero, the rms current is not zero. This current produces thermal energy in the transmission lines provided by the power company. An energy penalty accrues even though no net power is delivered to the user. An exactly-zero-resistance situation never occurs, but if the power factor is small, a large rms current may develop, producing thermal energy losses in the transmission lines.

Example 6 Power Factors and Power Companies

An electromagnet having an inductance of 1.5 H and a resistance of 100 Ω is used to operate a valve in a household washing machine. If the power source has an rms output of 115 V at a frequency of 60 Hz, let us determine the power delivered when the electromagnet is operating.

We will use Eq. 37.65 to determine the power delivered. Thus we need the inductive reactance X_L, the impedance Z, and the power factor $\cos \phi$.

$$X_L = \omega L$$

$$= 2\pi(60)\ \frac{\text{rad}}{\text{s}}\ 1.5\ \text{H} = 565\ \Omega$$

$$Z = \sqrt{R^2 + X_L^2}$$

$$= \sqrt{100^2 + 565^2}$$

$$= 574\ \Omega$$

$$\tan \phi = -\frac{X_L}{R}$$

$$\phi = \tan^{-1}\left(-\frac{X_L}{R}\right)$$

$$= \tan^{-1}\left(-\frac{565}{100}\right) = -80.0°$$

$$\cos \phi = \cos(-80.0°) = 0.174$$

With these numbers the average power delivered is

$$P_{av} = \frac{V_{rms}^2}{Z} \cos \phi$$

$$= \frac{(115)^2}{574}\ 0.174$$

$$= 4.01\ \text{W}$$

This is the power that enters into calculation of the user's energy bill.

In order to deliver this power, the power company had to provide an rms current.

$$I_{rms} = \frac{P_{av}}{V_{rms} \cos \phi} = \frac{4.01\ \text{W}}{115\ \text{V}(0.174)}$$

$$= 0.200\ \text{A}$$

It is this current that figures into thermal power losses ($I_{rms}^2 R$) in wires carrying the current. Had the load been purely resistive, the 4.01 W could have been delivered by a current

$$I_{rms} = \frac{P_{av}}{V_{rms}} = \frac{4.01\ \text{W}}{115\ \text{V}}$$

$$= 0.035\ \text{A}$$

A current of 0.200 A is required to deliver 4.01 W of power to this resistor–inductor combination. To deliver the 4.01 W of power to the resistor alone requires a current of 0.035 A. This significant difference is a consequence of the current and the generator output being in phase for the resistor alone but 80.0° out of phase for the resistor–inductor combination.

Questions

17. A commercial capacitor is observed to get warm when connected to an ac generator. What characteristic of the capacitor is responsible for the production of this energy?

18. Why can a manufacturer specify the resistance of a resistor but not the reactance of a capacitor nor of an inductor?

19. If frequency increases, determine whether the impedance of an *RLC* series combination increases or decreases.

20. What is "filtered" by *RC* and *RL* circuits?

21. How is it possible to have zero average power deliverd by a generator in a circuit containing a series combination of a resistor, a capacitor, an inductor, and an ac generator?

37.8 Resonance in a Series *RLC* Circuit

As we saw in the preceding section, reactances depend on the frequency of the power source. This dependence can be used in a variety of electronics circuits. For example, suppose that we want to maximize the ac current in a circuit at a particular generator frequency, as is the case in a radio receiver. The antenna receiving energy from electromagnetic waves emitted by numerous broadcasting stations functions as an ac source. We require that the current be significant in the circuitry connected to the antenna only for a frequency corresponding to the broadcast frequency of a station of interest. One way of meeting this requirement is with a series *RLC* circuit. Let us see how.

Resonance

Resistance is introduced in Section 30.2 as a measure of how a conductor "resists" the flow of charge in a dc circuit. Similarly, impedance measures how a combination of inductors, capacitors, and resistors "impedes" the flow of charge in an ac circuit. Unlike resistance, reactances depend on the frequency of the ac generator. This means that the rms and instantaneous current in a circuit also depend on the frequency of the generator. For the series *RLC* circuit the current increases as the impedance decreases. Because the impedance is given by

$$Z = \sqrt{R^2 + \left(\frac{1}{\omega C} - \omega L\right)^2} \qquad (37.68)$$

the smallest value occurs when

$$\omega L = \frac{1}{\omega C} \qquad (37.69)$$

Solving for the radian frequency, we obtain

$$\omega = \omega_r = \frac{1}{\sqrt{LC}} \qquad (37.70)$$

This condition produces *resonance* in the *RLC* circuit and ω_r is called the *resonant radian frequency*. At resonance, the impedance is equal to R and the current is $I = (\mathscr{E}/R)\cos \omega t$. The current and generator output are in phase at resonance. The current at the resonant frequency is the same as if only the resistor R were connected to the generator.

Power Delivered at Resonance

The resonant frequency equals the natural frequency of oscillation of an *LC* circuit (Section 35.4). Driving the circuit at the natural frequency of oscillation produces a maximum in the amplitude of the current. This situation is completely analogous to driving a mechanical system at its natural frequency of oscillation in order to produce a maximum velocity (Section 14.6). As observed in Eq. 37.65, the average power is always positive (or zero). The average power is zero when the generator frequency (ω) is zero, and approaches zero as the generator frequency becomes very large ($\omega \to \infty$). Thus between $\omega = 0$ and $\omega \to \infty$, a maximum occurs. The maximum power occurs for $\omega = \omega_r$, which is the same condition for maximum current. Evaluating Eq. 37.65 for the maximum power condition, we have

$$P_{max} = \frac{V_{rms}^2}{R} \qquad (37.71)$$

We show shortly that it is advantageous to have $R^2C/L \ll 1$. Taking $R^2C/L = 10^{-3}$, we obtain the average power delivered by the generator for different values of the radian frequency ω shown in Figure 37.25. The power falls off sharply at frequencies above or below the resonant frequency ω_r. The mountain-like appearance of the power versus frequency plot inspires its name, the *resonance peak*. A narrow resonance peak means that power is efficiently delivered to the circuit for only a very narrow range of frequencies.

Figure 37.25

Power delivered to an *RLC* circuit as a function of the radian frequency of the generator for the situation $R^2C/L = 10^{-3}$. The power delivered is a maximum when the generator frequency equals the resonant frequency of the circuit, $\omega = \omega_r = 1/\sqrt{LC}$. For a generator frequency 10% less or 10% greater than the resonant frequency the power delivered is about 97% less than the maximum.

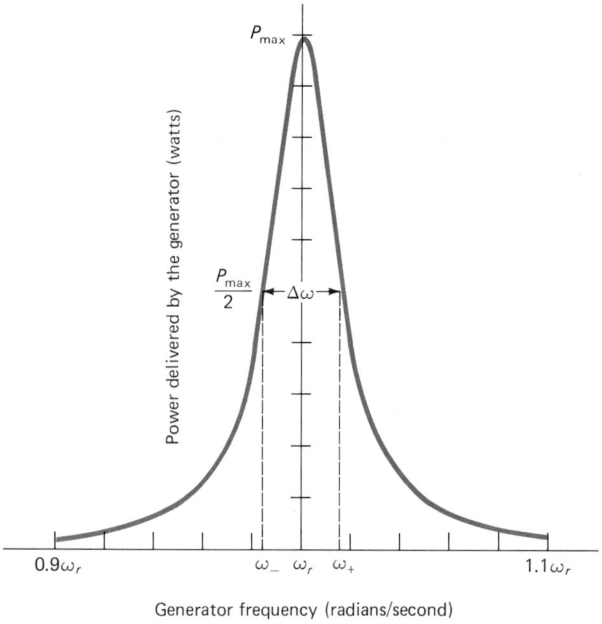

Example 7 Tuning a Radio

A radio is tuned in to a particular station by the adjustment of a capacitive reactance or an inductive reactance to achieve a resonance condition in a resonant circuit. For a particular *RLC* tuning circuit let us determine the current produced at resonance when an rms potential difference of 0.05 V is provided by the antenna. Let us take $R = 150\ \Omega$, $L = 15$ mH, and $C = 3.45$ pF. The resonant frequency for this *RLC* circuit is

$$\omega_r = \frac{1}{\sqrt{LC}}$$

$$= \frac{1}{(15 \times 10^{-3}\ \text{H} \times 3.45 \times 10^{-12}\ \text{F})^{1/2}}$$

$$= 4.40 \times 10^6\ \text{rad/s}$$

$$\nu_r = \frac{\omega_r}{2\pi}$$

$$= 7.00 \times 10^5\ \text{Hz}$$

This frequency is in the AM broadcast band. At resonance the impedance equals the circuit resistance: $Z = R$

$= 150\ \Omega$. Hence the rms current is

$$I_{rms} = \frac{V_{rms}}{R}$$

$$= \frac{0.05\ \text{V}}{150\ \Omega} = 333\ \mu\text{A}$$

To see how the current changes for a slightly different frequency, let us compute the current in the same circuit for an rms input of 0.05 V at a frequency of 7.2×10^5 Hz. First we determine the impedance at this frequency.

$$X_L = \omega L$$

$$= 2\pi \times 7.2 \times 10^5 \frac{\text{rad}}{\text{s}} \times 15 \times 10^{-3}\ \text{H}$$

$$= 6.786 \times 10^4\ \Omega$$

$$X_C = \frac{1}{\omega C}$$

$$= \frac{1}{2\pi \times 7.2 \times 10^5 \frac{\text{rad}}{\text{s}} 3.45 \times 10^{-12}\ \text{F}}$$

$$= 6.407 \times 10^4\ \Omega$$

$$Z = \sqrt{R^2 + (X_C - X_L)^2}$$

$$= \sqrt{150^2 + (6.407 \times 10^4 - 6.785 \times 10^4)^2}$$

$$= 3793\ \Omega$$

Then the rms current for this nonresonant frequency is

$$I_{rms} = \frac{V_{rms}}{Z} = \frac{0.05\ \text{V}}{3793\ \Omega}$$

$$= 13.2\ \mu\text{A}$$

A 0.05-V signal at the resonant frequency of 7.0×10^5 Hz produces a current of 333 μA. A 0.05-V signal at a frequency of 7.2×10^5 Hz produces only 13.2 μA when connected to the same *RLC* circuit. There is a sharp response of the circuit at the resonant frequency.

Width of the Resonant Peak

Resonance is desirable in many situations, and consequently the relation between the circuit parameters and the width of the resonance peak is important. It is customary to determine the frequency width of the resonant peak at the "half-maximum" positions. In Figure 37.25, ω_+ and ω_- define frequencies for which $P = \frac{1}{2}P_{max}$. The difference $\omega_+ - \omega_-$ is called the *half-width* of the resonance.

We evaluate $\omega_+ - \omega_-$ by using Eq. 37.65 for the average power.

$$P_{av} = \left(\frac{V_{rms}^2}{Z^2}\right) R \qquad (37.72)$$

where

$$Z^2 = R^2 + \left(\frac{1}{\omega C} - \omega L\right)^2$$

At resonance $Z^2 = R^2$; at the half-power values the denominator must double; that is, Z^2 increases to $2R^2$. Let us designate Z_+^2 and Z_-^2 as the two values of Z^2 for which the impedance is $2R^2$. Then we require

$$Z_\pm^2 = \left(\frac{1}{\omega C} - \omega L\right)^2 + R^2 = 2R^2 \qquad (37.73)$$

The values of ω corresponding to Z_+^2 and Z_-^2 are solutions to Eq. 37.73. Writing Eq. 37.73 as

$$\left(\frac{1}{\omega C} - \omega L\right)^2 = R^2$$

and taking the square root of both sides we obtain

$$\frac{1}{\omega C} - \omega L = \pm R \qquad (37.74)$$

This gives the quadratic equation

$$\omega^2 \mp \omega\left(\frac{R}{L}\right) - \omega_r^2 = 0$$

where

$$\omega_r^2 = \frac{1}{LC}$$

The roots of this equation are

$$\omega_+ = \frac{1}{2}\frac{R}{L} + \sqrt{\left(\frac{1}{2}\frac{R}{L}\right)^2 + \omega_r^2}$$

$$\omega_- = -\frac{1}{2}\frac{R}{L} + \sqrt{\left(\frac{1}{2}\frac{R}{L}\right)^2 + \omega_r^2}$$

Hence the width of the resonance peak is

$$\Delta\omega = \omega_+ - \omega_-$$

$$= \frac{R}{L} \qquad (37.75)$$

independent of *C*! The width of the resonance peak is small compared to the resonant frequency ω_r if $\Delta\omega/\omega_r \ll 1$. Using Eq. 37.75, and $\omega_r = 1/\sqrt{LC}$, we can write $R\sqrt{LC}/L \ll 1$ or $R^2C/L \ll 1$.

In the limit as the resistance vanishes, the width of the resonance peak goes to zero. Figure 37.26 illustrates how the width of the resonance peak decreases as the ratio R^2C/L decreases.

Example 8 **Half-Width of a Resonance Peak**

An *RLC* circuit has $R = 1\ \Omega$, $L = 1$ mH, and $C = 100$ pF. Let us determine the resonant frequency and the half-width of the resonance peak.

The resonant frequency is determined from Eq. 37.70.

$$\nu_r = \frac{1}{2\pi\sqrt{LC}} = \frac{1}{2\pi\sqrt{10^{-3}\ \text{H} \times 100 \times 10^{-12}\ \text{F}}}$$

$$= 5.03 \times 10^5\ \text{Hz}$$

Figure 37.26

The width of the resonance peak for an *RLC* circuit depends on the parameter $R/\omega_r L$. Each of the three curves shown here corresponds to the same value of inductance and capacitance, and therefore has the same resonant frequency. Only the resistance is changed. As observed, the width of the resonance peak decreases as the parameter $R/\omega_r L$ decreases.

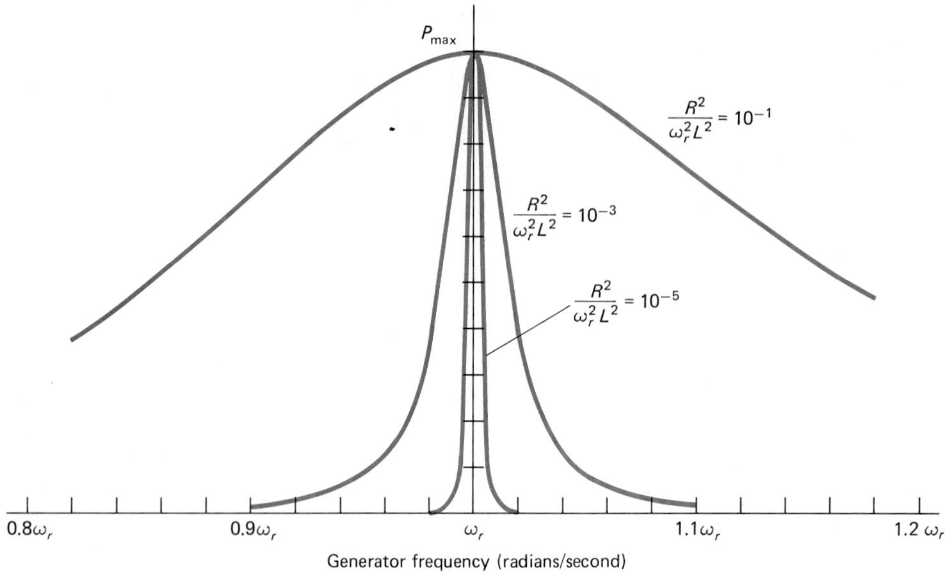

The half-width of the resonance peak is determined from Eq. 37.75.

$$\Delta\omega = 2\pi\Delta\nu = \frac{R}{L}$$

$$\Delta\nu = \frac{1}{2\pi}\left(\frac{R}{L}\right) = \frac{1}{2\pi}\left(\frac{1\ \Omega}{10^{-3}\ \text{H}}\right)$$

$$= 159\ \text{Hz}$$

The sharpness of the peak is measured by the ratio

$$\frac{\Delta\nu}{\nu_r} = \frac{159\ \text{Hz}}{5.03 \times 10^5\ \text{Hz}} = 3.16 \times 10^{-4}$$

As a percentage

$$\frac{\Delta\nu}{\nu_r} = 0.316\%$$

The resonance peak occupies a small fraction of "frequency space."

Questions

22. At resonance, how does the impedance of an *RLC* series combination depend on the resonant frequency?

23. What energy transformations occur in an *RLC* series combination connected to an ac generator?

24. Pulling a pendulum away from its equilibrium position and releasing it leads to several energy transformations. Devise an analogous situation using a resistor, a capacitor, and an inductor.

25. All other parameters being equal, how will the half-width of a resonance change if the inductance is doubled?

37.9 Transformers

Transmission of Electric Power

A modern electric power plant produces about 10^6 kilowatts (kW) of electric power. At this power level, the energy produced in 1 hour is 10^6 kWh. Depending on where the energy is used, it is generally worth from 2¢ to 10¢ per kWh. Thus, 1 hour's output of electrical energy, 10^6 kWh at 5¢ per kWh is worth $50,000. Clearly, selling electric energy is big business, and a power company likes to deliver to its customers all of the electrical energy that is produced at its plant. But unfortunately there are energy losses between the generator and the consumers. Foremost, is the heat lost in the wires of the transmission lines because of the electrical resistance of the wires. As shown in Section 30.5, the thermal power generated in a resistance R due to a current I is

$$P = I^2R \tag{37.76}$$

To minimize the thermal power produced, the current and resistance in the transmission lines are kept as low as practically possible by constructing the lines of copper or aluminum. The resistance can be lowered even more by using cables of larger and larger cross sections, but the cost of the cables increases as the cross section increases, and eventually sets a limit.

Example 9 The Cost of Transmitting Electric Power

Thermal power losses in the transmission wires bringing electricity from an electric power plant to a community are significantly costly. Let us see why.

Suppose that the total length of two wires connecting the generating facility to the community is 50 km and that the wires are made of solid copper having a 1-cm diameter. Using data from Table 30.1 for the resistivity of copper, we obtain as the resistance of the lines

$$R = \rho\left(\frac{l}{A}\right) = \frac{1.72 \times 10^{-8}\ \Omega \times \text{m}\ (50 \times 10^3\ \text{m})}{\dfrac{(0.01)^2\ \text{m}^2}{4}}$$

$$= 11\ \Omega$$

We will show later in this section that in the transmission of 100 MW (about one tenth the output of a large facility) of electric power, the current in a transmission line is often around 500 A. The thermal power produced by a current of 500 A in a line of resistance 11 Ω is

$$P = I^2R$$
$$= (500)^2\ \text{A}^2\ 11\ \Omega$$
$$= 2750\ \text{kW}$$

At 5¢ per kWh this is an hourly cost of $137.50.

In order to minimize the current in the transmission lines and still transmit the desired amount of electrical energy, power companies use devices called *transformers.* With a transformer, ac potential differences can be raised or lowered as needed. At a power plant, potential differences are made large (typically 330 kV) and currents are made small for power transmission. Then at the consumer end of the transmission lines, the potential differences are lowered for safe, practical use. In the home, potential differences can be lowered still further—for the operation of doorbells, for example. It is the ease with which ac potential differences and currents can be changed by transformers that makes ac power particularly useful. In contrast, dc potential differences and current cannot readily be changed.

A diagram of the conventional arrangement for a transformer is shown in Figure 37.27. The purpose of a transformer is to produce a changing flux in one coil of wire, called the *primary winding,* and to have this changing flux intercept a second coil of wire, called the *secondary winding,* causing a potential difference to be induced in the secondary. This same principle is used in the sense coil arrangement described in Section 36.2. The changing flux in the primary winding is produced by an ac source connected to its leads. The transformer core enhances the flux produced by the current in the primary winding and guides this flux into the secondary winding. Ideally all the flux produced in the primary winding threads through the secondary winding. Such a condition is not perfectly realizable but can be approximated if the core is made of a magnetic material such as soft iron to steer the magnetic flux and to minimize hysteresis losses (Section 36.5). The size of the emf induced in the secondary winding depends on the number of turns of wire and the time rate of change of magnetic field lines threading the turns. Let us now formally show how the induced emf in the secondary winding is related to the potential difference impressed on the primary winding by the ac source.

Figure 37.27

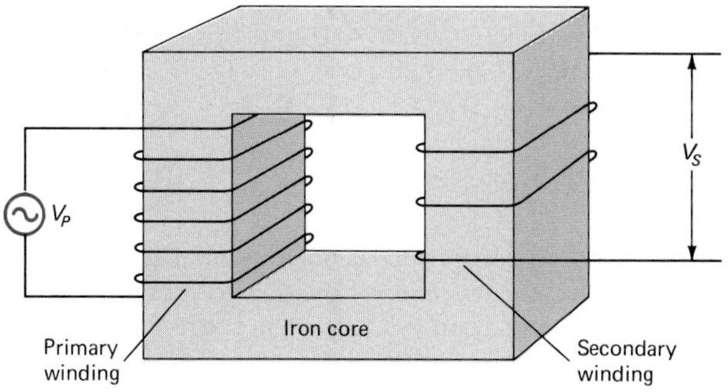

We assume initially that no load, such as a resistor, is connected to the secondary winding. Thus the current (and power) in this winding is zero. However, according to Faraday's law (Section 34.2), a changing magnetic flux in the secondary winding will induce in it an emf. The primary winding of the transformer constitutes an *RL* circuit. Here *R* represents the resistance of the wire in the primary winding. When an ac generator, represented by $\mathscr{E}\cos\omega t$, is connected to the primary winding, our analysis of the *RL* circuit showed (Section 37.5) that

$$\mathscr{E}\cos\omega t = V_R + V_L \tag{37.77}$$

$$= IR + L\frac{dI}{dt} \tag{37.78}$$

Ideally $R = 0$ so that IR is zero regardless of the size of the current. Although the ideal never pertains, in practice IR is small compared to $L\,dI/dt$ so that to a good approximation

$$\mathscr{E}\cos\omega t = L\frac{dI}{dt} \tag{37.79}$$

The definition of self-inductance, $N_p\Phi_p = LI$ (Section 35.1), allows us to write $L\,dI/dt = N_p\,d\Phi_p/dt$ where Φ_p is the magnetic flux and N_p is the number of loops of wire in the primary winding. Thus Eq. 37.79 becomes

$$\mathscr{E}\cos\omega t = N_p\frac{d\Phi_p}{dt} \tag{37.80}$$

To the extent that there is no resistance in the primary winding, and the secondary winding is open circuited, the average power delivered by the ac generator is zero. Assuming that the same magnetic field lines thread the primary and secondary windings we have for the magnitude of the emf induced in the secondary winding having N_s turns of wire

$$\mathscr{E}_s = N_s\frac{d\Phi_p}{dt} \tag{37.81}$$

Dividing Eqs. 37.80 and 37.81, we obtain

$$\frac{\mathscr{E}_s}{\mathscr{E}\cos\omega t} = \frac{N_s}{N_p} \tag{37.82}$$

which yields

$$\mathscr{E}_s = \frac{N_s}{N_p}\mathscr{E}\cos\omega t \tag{37.83}$$

for the induced emf in the secondary winding. The ratio N_s/N_p is called the *turns ratio* of the transformer. If $N_s/N_p > 1$, then the induced emf is greater than the potential difference imposed on the primary winding by the ac generator. Such a transformer is called a *step-up transformer*. If $N_s/N_p < 1$, then the induced emf is less than the potential difference imposed on the primary winding by the ac generator. Such a transformer is called a *step-down transformer*. Thus the induced emf in the secondary winding may be raised or lowered simply by increasing the turns ratio of the transformer.

If we connect a resistor, for example, to the secondary winding, then the secondary circuit is no longer open and the induced emf produces current in the secondary winding. This induced current produces a magnetic flux that, by Lenz's law, reduces the flux produces by the current in the primary circuit. This reduction in flux tends to reduce the potential difference across the primary winding. However, the primary potential difference is established by the applied ac source. Hence the primary current will increase and produce flux that compensates for the flux decrease produced by the secondary current. The power at the secondary winding must be provided by the generator connected to the primary winding. Ideally, the primary and secondary power are equal. Although thermal power losses prevent the ideal situation, the secondary power is seldom less than 90% of the primary power and can be 99% of the primary power in a large, carefully engineered transformer. The predominant energy losses accrue from the resistance of the wire making up the primary and secondary coils, hysteresis losses in the transformer core (Section 36.5), and thermal energy from induced currents in the core. Thermal energy loss in the wires is minimized by using a low-resistivity material such as copper. Hysteresis losses are minimized by using magnetically soft materials such as iron. Thermal energy losses from induced currents in the core are minimized by constructing the core from thin sheets rather than as a solid piece (Section 34.4.). Assuming 100% power transfer from the primary winding to the secondary winding, we can write

$$V_pI_p = V_sI_s$$

$$I_p = \left(\frac{V_s}{V_p}\right)I_s \tag{37.84}$$

If the transformer is ideal, the electric power fed to the primary winding equals the electric power emerging from the secondary winding. If a step-up transformer is used to increase the potential difference across the output of the secondary, then the current in the secondary is necessarily smaller than the primary current in order that the electric power in the primary and secondary be the same. This is the scheme used to minimize the current in the wires of electric power transmission lines, as we see in Example 10.

Example 10 Role of Transformers in Transmitting Electric Power

Let us see how transformers help to minimize thermal energy losses in transmission lines.

In essence, an electric power system consists of a transformer to "step up" the emf at the generating site, a transformer to "step down" the emf at the consumer end, and a transmission line connecting the two facilities (Figure 37.28). At the generating site, the potential difference across the secondary winding of a power transformer is typically 400,000 V, and in some instances it is close to 1,000,000 V. If, for example, the power from the generator is 100 MW, and the potential difference across the secondary winding is 400,000 V, then assuming no energy loss, we have for the current in the secondary winding

$$I = \frac{P}{V} = \frac{10^8\ \text{W}}{4\times10^8\ \text{V}} = 250\ \text{A}$$

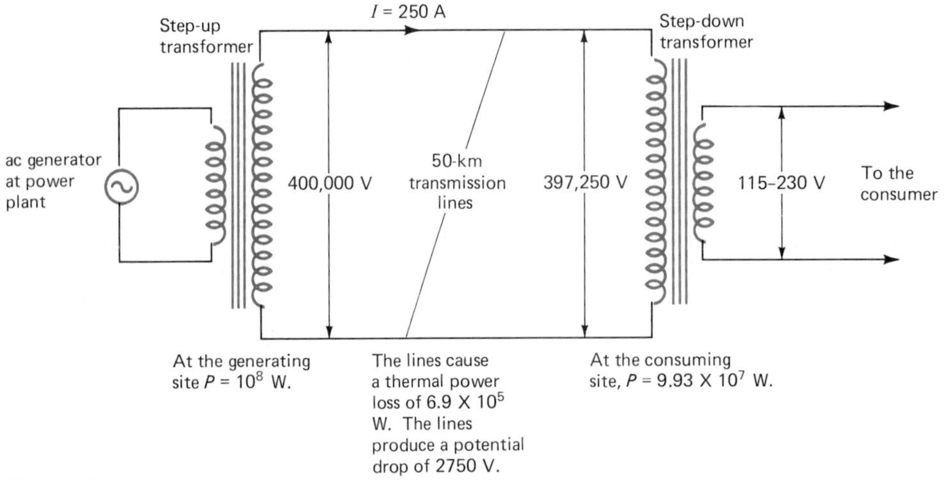

Figure 37.28

Principle of transmission of electric power.

If the resistance of the transmission lines is 11 Ω, as in Example 9, then the thermal power produced in the lines, I^2R, is 0.69 MW. This means that 0.69% of the power delivered by the generator is lost as heat. Because of the resistance of the lines, a potential difference is produced across the ends of each line. This amounts to

$$V = IR$$
$$= 250 \text{ A} \times 11\ \Omega = 2750 \text{ V}$$

Thus the potential difference across the primary winding of the step-down transformer is 400,000 − 2750 = 397,250 V. In practice, at the consumer end, the potential difference is reduced to 115 or 230 V (rms). If the 100 MW had been produced at 230 V at the power plant, then the current would be

$$I = \frac{P}{V} = \frac{10^8 \text{ W}}{230 \text{ V}} = 4.35 \times 10^5 \text{ A}$$

A wire of sufficient size to carry this current would be prohibitively heavy and expensive.

Impedance Matching

Just as a dc source has an internal resistance (Section 31.1), an ac generator has an internal impedance; and just as the maximum transfer of power to a resistor connected to a dc source occurs when the internal resistance of the source equals the resistance of the connected resistor (see Problems for Section 31.2), the maximum transfer of the power delivered by an ac generator to a load also requires a matching of the source and load impedances. Often, for unavoidable reasons, the impedances of the generator and load are very different, but with a properly designed transformer included as part of the load, the load impedance can be matched to the source impedance, with most of the power transfer being to the device connected to the secondary winding.

To illustrate the power transfer principle, let's suppose that the output from an ac source is connected to the primary winding of a transformer, and that a load such as a radio speaker is connected to the secondary winding. Let's also assume that the impedances of the source and the load are purely resistive, and that the transformer is ideal. The current in a resistive load is then

$$I_s = \frac{V_s}{R} \tag{37.85}$$

where V_s is the potential difference between the leads to the secondary. Using Eq. 37.84 we can relate the primary current (I_p) and the secondary current (I_s) by

$$I_p = \left(\frac{V_s}{V_p}\right) I_s \tag{37.84}$$

Multiplying Eq. 37.85 by Eq. 37.86 to eliminate I_s, we obtain

$$I_p = \frac{V_s^2}{V_p R}$$

which can be written

$$I_p = \frac{V_s^2}{V_p^2}\left(\frac{V_p}{R}\right) = \frac{V_p}{(V_p/V_s)^2 R} \tag{37.86}$$

Relating V_p and V_s to the turns ratio N_p/N_s of the transformer (Eq. 37.82), we have

$$I_p = = \frac{V_p}{(N_p/N_s)^2 R} \tag{37.87}$$

In effect, the current produced by the potential difference V_p is due to a resistor of value $(N_p/N_s)^2 R$. By choosing N_p/N_s properly, this effective resistance can be made to match the internal resistance of the ac generator, thereby optimizing the transfer of power to the load. Many stereo amplifiers have an impedance-matching transformer between the amplifier output and the speakers. Stereo enthusiasts know that it is important to connect the speakers to the proper impedance terminals of the audio amplifier. If the mismatch is large, power is dissipated within the amplifier rather than in the speaker and the sound level is diminished.

Example 11 Getting the Most Power from a Stereo Amplifier

A speaker for a stereo system typically has an impedance of 16 Ω. In operation, the speaker is connected to the secondary winding of an impedance-matching transformer. Let us see why this transformer is very important as far as the power delivered to the speaker is concerned.

The output of the amplifier constitutes an ac source for the primary winding of the transformer. Let us assume that the source has an internal resistance of 1600 Ω and produces a 12-V (rms) output. If the speaker were connected directly to the source (Figure 37.29) rather than to the transformer, then the current would be

$$I = \frac{\mathscr{E}_{\text{source}}}{R_{\text{source}} + R_{\text{speaker}}} + \frac{12 \text{ V}}{1600\ \Omega + 16\ \Omega}$$
$$= 0.619 \text{ mA}$$

and the power delivered to the speaker would be

$$P = I^2 R_{\text{speaker}}$$
$$= (0.619 \times 10^{-3})^2 \text{ A}^2 \cdot 16\ \Omega$$
$$= 6.13 \times 10^{-6} \text{ W}$$
$$= 6.13\ \mu\text{W}$$

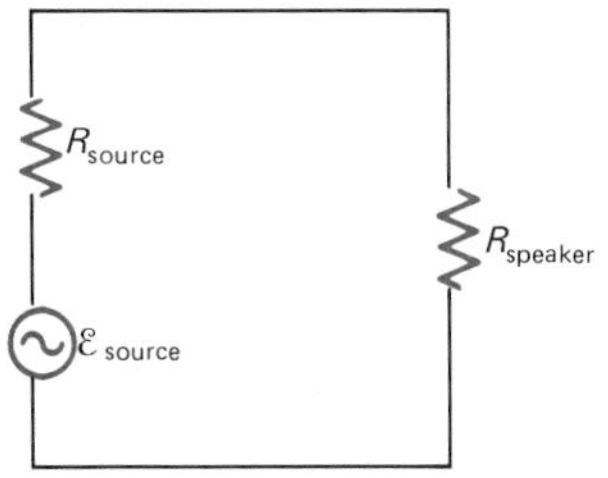

Figure 37.29

Schematic representation of an ac source connected to a speaker of a stereo system. The source has an emf, designated $\mathscr{E}_{\text{source}}$, and an internal resistance, designated R_{source}.

Figure 37.30

The circuit in Figure 37.29 is modified to include a transformer between the ac source and the speaker. Everything to the right of the connections labeled A and B constitutes the load for the source.

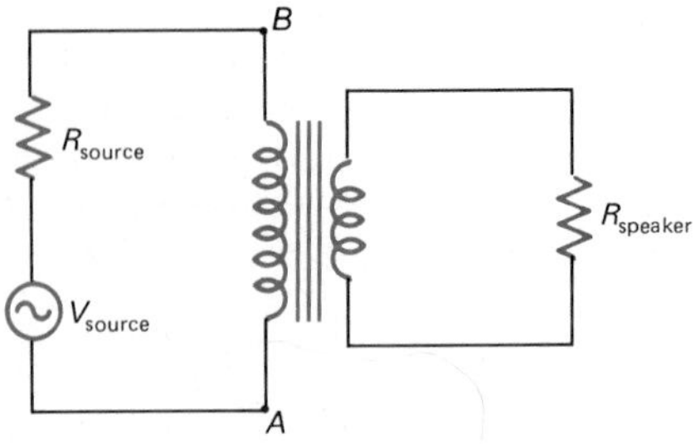

Suppose now that according to Eq. 37.87 we choose an impedance-matching transformer (Figure 37.30) such that

$$\left(\frac{N_p}{N_s}\right)^2 R_{\text{speaker}} = R_{\text{source}}$$

This transformer would have

$$\left(\frac{N_p}{N_s}\right)^2 = \frac{R_{\text{source}}}{R_{\text{speaker}}} = \frac{1600}{16}$$

$$\frac{N_p}{N_s} = 10$$

The transformer is a step-down type having a turns ratio of 10. With this choice the source delivers power to an effective load of 16 Ω. Thus the current produced is

$$I = \frac{\mathscr{E}_{\text{source}}}{R_{\text{source}} + R_{\substack{\text{effective}\\\text{load}}}} = \frac{12\text{ V}}{16\ \Omega + 16\ \Omega}$$

$$= 0.375\text{ A}$$

and the power delivered is

$$P = I^2 R_{\substack{\text{effective}\\\text{load}}} = (0.375)^2\text{ A}^2 \times 16\ \Omega$$

$$= 2.25\text{ W}$$

If we assume that the transformer is 100% efficient, then the power delivered to the speaker is also 2.25 W. Thus the impedance-matching transformer has permitted a significant increase in the power delivered.

Questions

26. What is being transformed by an electrical transformer?

27. In principle, is there any reason why the roles of the primary and secondary windings of a transformer could not be interchanged?

28. Toy electric trains often use a transformer to supply power for the trains and controls. Would this transformer be a step-up or a step-down transformer. Explain.

29. Why does constructing a transformer core from thin sheets tend to minimize energy losses from induced currents in the core?

30. A light bulb in series with an ac generator and the primary winding of a transformer glows dimly when the secondary leads are unconnected. But if the secondary leads are connected to a load such as a resistor, the bulb in the primary winding will brighten. Why?

31. The power supply for a picture tube in a color TV set typically requires 15,000 V ac. How can this potential difference be provided if only 120 V are available at a household electric outlet?

32. If the terminals of a battery are connected to the primary winding of a transformer, why will a steady potential difference not appear across the secondary winding?

Summary

An alternating current is one in which the direction of the current changes with time.

A potential difference that varies sinusoidally with time is termed an ac potential difference. A current that varies sinusoidally with time is called an ac current.

The average ac current in any element in a circuit is zero, and the average ac potential difference across any element in a circuit is zero. The rms values of current and potential difference, meaning the square root of the average of the square, are not zero. For ac currents and ac potential differences, the rms value is always $1/\sqrt{2}$ times the peak value.

The ac current in a resistor is in phase with the ac emf producing the current.

The ac current in a capacitor leads the ac emf producing the current by $\pi/2$ rad.

The ac current in an inductor lags the ac emf producing the current by $\pi/2$ rad.

For resistors, capacitors, and inductors in an ac circuit, current and potential difference are related by

$$I = \frac{V}{X}$$

where I is the rms current, V is the rms potential difference, and X is reactance. For a resistor, X is the ordinary electrical resistance. For a capacitor, the capacitive reactance is

$$X_C = \frac{1}{\omega C} \qquad (37.10)$$

For an inductor, the inductive reactance is

$$X_L = \omega L \qquad (37.16)$$

In terms of the impedance

$$Z = \sqrt{R_2 + \left(\frac{1}{\omega_C} - \omega L\right)^2} \qquad (37.68)$$

the potential difference across each element in a series RLC combination connected to an ac generator $\mathcal{E} \cos \omega t$ is

$$V_R = \frac{\mathcal{E}R}{Z} \cos(\omega t + \phi) \quad (37.56)$$

$$V_L = \frac{-\mathcal{E}X_L}{Z} \sin(\omega t + \phi) \quad (37.59)$$

$$V_C = \frac{\mathcal{E}X_C}{Z} \sin(\omega t + \phi) \quad (37.58)$$

where ϕ is the phase angle, defined as

$$\phi = \tan^{-1}\left(\frac{X_C - X_L}{R}\right) \quad (37.55)$$

The average power delivered by the generator in an RLC circuit is

$$P_{av} = \frac{\mathcal{E}^2}{2Z} \cos \phi \quad (37.65)$$

The quantity $\cos \phi$ is called the power factor.

A series RLC combination is said to be in resonance if the radian frequency of the generator is

$$\omega_r = \frac{1}{\sqrt{LC}} \quad (37.70)$$

The average power in an RLC series circuit as a function of the radian frequency (ω) of the ac generator is

$$P_{av} = \frac{V_{rms}^2 R}{R^2 + \left(\dfrac{1}{\omega C} - \omega L\right)^2} \quad (37.72)$$

and P_{av} is a maximum for $\omega = 1/\sqrt{LC}$. At resonance, $\omega = \omega_r$, and the impedance is a minimum and the average power is a maximum.

The frequency width of the P_{av} versus ω curve at half the maximum value is $\Delta\omega = R/L$. Proper choice of $R/\omega_r L$ allows significant power to be extracted for only a very narrow range of frequencies.

A transformer consists of two magnetically coupled coils, or windings, in close proximity. The transformer permits the amplitude of ac potential differences to be changed with little loss of energy. To a good approximation, the ratio of the potential differences across the two windings of a transformer is

$$\frac{V_p}{V_s} = \frac{N_p}{N_s} \quad (37.83)$$

where N_p and N_s are the numbers of turns in the primary and secondary windings and V_p and V_s are the potential differences across the primary and secondary windings.

Suggested Reading

L. O. Barthold and H. G. Pfeiffer, High-Voltage Power Transmission, *Sci. Am.* 210, 39 (May 1964).

Problems

Section 37.1 Alternating Currents

1. A single loop of wire is rotated through a uniform magnetic field. If the angle θ is measured between a perpendicular to the plane of the loop and the direction of the magnetic field, what is the expression for the induced potential difference in the loop?

2. A 10-turn coil having a mean cross-sectional area of 100 cm^2 is rotated 20 rev/s in a uniform magnetic field of 0.5 T. Calculate the rms potential difference produced in the leads of the coil.

Section 37.2 Current in a Resistor Connected to an ac Generator

3. A toaster having a resistance of 12 Ω is connected to a 115-V (rms), 60-Hz household outlet. Determine the average current, the instantaneous current, and the rms current.

4. A periodic potential difference is described by the relation

$$V = V_m(\sin \omega t + \cos \omega t)$$

where $V_m = 120$ V. (a) Sketch V versus t in the interval $0 \leq \omega t \leq 2\pi$ rad by plotting $V_m \sin \omega t$ and $V_m \cos \omega t$, and graphically adding the two functions. (b) Evaluate V at $\omega t = \pi/4$ rad and compare your result with the values of V at $\omega = 0$ rad and $\omega t = \pi/2$ rad.

5. An electric iron having a resistance of 23 Ω is connected to a 115-V, 60-Hz household outlet. Determine the minimum power, the maximum power, and the average power.

6. Determine the average and rms current in a 6000-W clothes dryer connected to a 230-V (rms), 60-Hz source.

7. An ammeter designed to measure rms current reads 1.83 A when connected to an ac circuit. What are the smallest and largest positive values of the instantaneous current in the circuit?

8. Two 1000-W hair dryers are connected to two 115-V (rms), 60-Hz outlets in a house. What is the total rms current in the two dryers?

9. An electric light bulb dissipates an average power of 100 W when connected to a line described by $V = 170 \sin \omega t$ V. (a) What is the light bulb's resistance? (b) What is the maximum current in the light bulb?

10. A type of rectifier called a bridge rectifier can transform a sinusoidal potential difference to that shown graphically in Figure 1. (a) Show that the average value of the rectified potential difference is

$$V_{av} = \frac{2\mathcal{E}}{\pi} = 0.637\mathcal{E}$$

(b) Show that the rms value of the rectified potential difference is

$$V_{rms} = \frac{\mathscr{E}}{\sqrt{2}} = 0.707\mathscr{E}$$

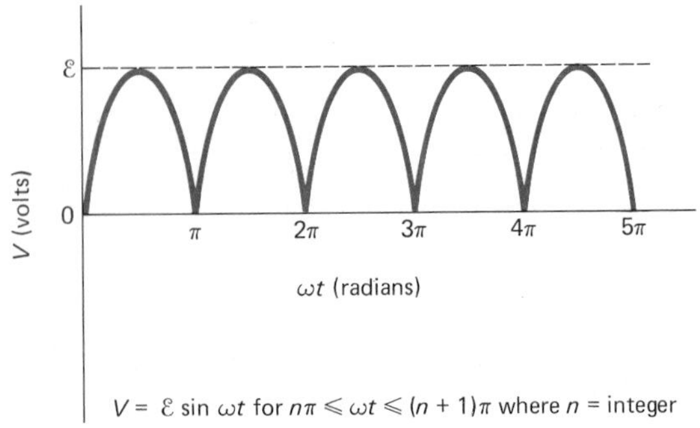

Figure 1

Section 37.3 Behavior of a Capacitor Connected to an ac Generator

11. The current in a 2.5-μF capacitor connected to an ac generator is represented by

$I = -4.71 \sin 377t$ μA

Determine the maximum potential difference across the capacitor.

12. Determine the capacitive reactance of a 3.1-pF capacitor at frequencies of 10^3, 10^6, and 10^9 Hz.

13. Sketch the capacitive reactance of a 12-μF capacitor as a function of (a) frequency (ν) and (b) reciprocal of frequency ($1/\nu$).

14. Calculate the reactance of a capacitor ($C = 2 \times 10^{-6}$ F) at (a) $\nu = 25$ Hz; (b) $\nu = 60$ Hz.

15. At what frequency is the capacitive reactance of a 3.6-pF capacitor equal to 1 Ω?

16. Compute the maximum and rms currents in a 15-pF capacitor subjected to a 10-V (rms), 50-MHz ac generator.

17. The frequency of an ac generator connected to a capacitor is increased by a factor of 10^5. Show that the current in the capacitor increases by a factor of 10^5.

18. Sketch the current in and potential difference across a 5-μF capacitor connected to an ac generator producing a potential difference of the form $V = 10 \cos 300t$ V.

19. An ac source having an output $V = \mathscr{E} \sin \omega t$ is connected to the terminals of a capacitor. Derive an expression for the current in the capacitor and show that the current leads the potential difference by $\pi/2$ rad.

20. Figure 2 shows a plot of the current in a 1.2-nF capacitor connected to an ac generator. Determine the frequency (ν) and peak output ($\mathscr{E}$) of the generator.

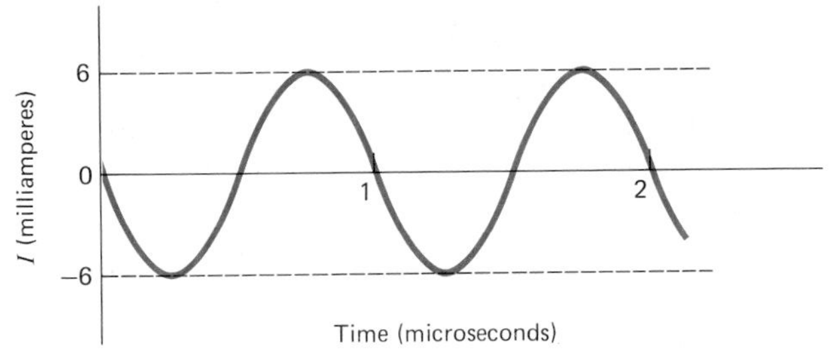

Figure 2

Section 37.4 Behavior of an Inductor Connected to an ac Generator

21. Sketch the inductive reactance of a 12-μH inductor as a function of frequency (ν).

22. At what frequency is the inductive reactance of a 4.8-μH inductor equal to 2.5 Ω?

23. Calculate the rms current in a 1.2-H ideal inductor connected to a 115-V (rms), 60-Hz generator.

24. Calculate the reactance of an inductor (L = 2mH) at (a) 25 Hz and (b) 60 Hz.

25. Calculate the maximum and rms currents in a 22-μH inductor connected to a 5-V (rms), 75-MHz generator.

26. The frequency of an ac generator connected to an inductor is decreased by a factor of 10^3. Show that the current in the inductor increases by a factor of 10^3.

27. Sketch a graph of the current in and potential difference across a 3.2-H inductor connected to an ac generator producing a potential difference of the form $V = 1.6 \cos 120t$ V.

28. Derive an expression for the instantaneous current in an inductor connected to an ac generator producing a potential difference of the form $V = \mathscr{E} \sin \omega t$.

29. If the current in an inductor is represented by $I = (\mathscr{E}/\omega L) \sin \omega t$, determine a representation for the potential difference across the inductor. Compare with Eqs. 37.13 and 37.14.

30. The current in an ideal inductor connected to a 5-V (rms) ac generator can be represented by

$I = 8.0 \sin 200\pi t$ A

Determine the self-inductance of the inductor.

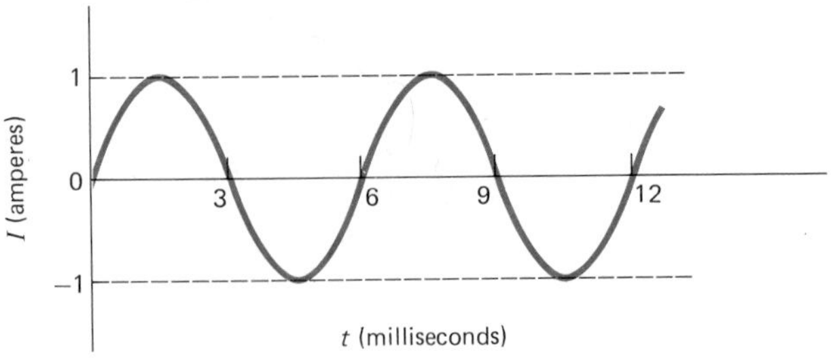

Figure 3

31. Figure 3 shows a plot of the current in an ideal inductor connected to an ac generator producing a peak potential difference ($\mathscr{E}$) of 1 V. Determine (a) the frequency of the generator and (b) the inductance of the inductor.

Section 37.5 The *RC* Circuit

32. A 2.5-V (rms), 100-Hz generator is connected in series with a 0.1-μF capacitor and a 2500-Ω resistor. Determine (a) the impedance of the circuit and (b) the rms current in the circuit.

33. A 1000-Ω resistor and a (2.65×10^{-6})-F capacitor are connected in series to a generator whose maximum output is 170 V. Calculate the maximum potential difference across the capacitor. (Use $\nu = 60$ Hz.)

34. By direct substitution, show that $q = (\mathscr{E}/\omega Z)\sin(\omega t + \phi)$ is a solution to Eq. 37.21.

35. When solving the equation

$$\mathscr{E}\cos\omega t = R\frac{dq}{dt} + \frac{q}{C}$$

we assumed that

$$q = \frac{I_m}{\omega}\sin(\omega t + \phi)$$

and showed that

$$I_m = \frac{\mathscr{E}}{Z} \qquad \tan\phi = \frac{1}{\omega RC} \qquad \text{and}$$

$$Z = \sqrt{R^2 + \left(\frac{1}{\omega C}\right)^2}$$

(a) Show that had we assumed that

$$q = \frac{I_m}{\omega}\sin(\omega t - \phi)$$

we would have achieved an identical result except that

$$\tan\phi = -\frac{1}{\omega RC}$$

(b) Show that we still conclude that the current in the capacitor leads the potential difference provided by the generator.

36. In an *RC* circuit the phase is defined as $\phi = \tan^{-1}(1/\omega RC)$. Show that at sufficiently large frequencies $\phi \approx 0$ and the current can be represented by $I = (\mathscr{E}/Z)\cos\omega t$.

37. In an *RC* circuit having a 10.5-Ω resistor and a 1.5-μF capacitor, at what frequency will the impedance equal 50 Ω?

38. The amplitude of the potential difference across the capacitor in an *RC* circuit can be written (Eq. 37.30)

$$V = \frac{\mathscr{E}X_C}{Z}$$

where $\mathscr{E}$ is the amplitude of the ac generator, X_C is the capacitive reactance, and Z is the impedance of the *RC* combination. Show that this relation for V can also be written as

$$V = \frac{\mathscr{E}}{\sqrt{(\omega RC)^2 + 1}}$$

39. Compute the rms potential difference across a 15-Ω resistor and a 15-μF capactitor in an *RC* combination connected in series with a 10-V (rms), 1-kHz generator.

40. Determine the average power in a series combination of a 100-Ω resistor and a capacitor having a capacitive reactance of 300 Ω at 60 Hz when connected to a 120-V (rms), 60-Hz outlet in a laboratory.

41. A 1000-Ω resistor and a 100-μF capacitor are connected in parallel to a 120-V (rms), 60-Hz ac source. Compare the rms currents in the two elements.

Section 37.6 The *RL* Circuit

42. An *RL* series combination consisting of a 1.5-Ω resistor and a 2.5-mH inductor is connected to a 12.5-V (rms), 400-Hz generator. Determine the impedance of the circuit, the rms current, and the rms potential difference across the resistor and the inductor.

43. An *RL* combination is connected to an ac generator, producing an output of the form $V = \mathscr{E}\cos\omega t$. Why would you expect the current to be of the form

$$I = \frac{\mathscr{E}}{R}\cos\omega t$$

at sufficiently low frequencies?

44. An *RL* combination having a 2.2-Ω resistor has an impedance of 120 Ω at a frequency of 10^3 Hz. What is the self-inductance of the inductor?

45. As a way of determining the self-inductance of a coil used in a research project, a student first connects the coil to a 12-V (dc) potential difference and measures a current of 0.63 A. Then the student connects the coil to a 24-V (rms), 60-Hz potential difference and measures a current (rms) of 0.57 A. What values does the student calculate for the self-inductance?

Section 37.7 The *RLC* Circuit

46. Compute the rms current in an *RLC* circuit having a resistance of 10 Ω, a capacitance of 10 μF, and an inductance of 1.5 mH when connected to a 25-V (rms), 100-Hz generator.

47. Determine the average power in an *RLC* circuit having a resistance of 3.5 Ω, capacitive reactance of 25 Ω, and inductive reactance of 53 Ω when connected to an ac generator having a peak potential difference ($\mathscr{E}$) of 50 V.

48. A 100-mH inductor and a 100-μF capacitor are connected in series with a 100-Hz ac source $V = 100\cos\omega t$ volts. (a) Derive an expression for the instan-

taneous potential difference across each element. (b) Show that the sum of the instantaneous potential differences is $100 \cos \omega t$.

49. If a series combination of a resistor (R), a capacitor (C), and an inductor (L) is connected to an ac generator $V = \mathcal{E} \cos \omega t$, Kirchhoff's loop law tells us that at any instant $V_R + V_C + V_L = \mathcal{E} \cos \omega t$. Show by direct substitution that this equation is solved by

$$V_R = \frac{\mathcal{E}R}{Z} \cos(\omega t + \phi) \qquad V_C = \frac{\mathcal{E}X_C}{Z} \sin(\omega t + \phi)$$

$$V_L = -\frac{\mathcal{E}X_L}{Z} \sin(\omega t + \phi)$$

where $\phi = \tan^{-1}[(1/\omega C - \omega L)/R]$. (*Hint:* Make use of the relations $\cos(\omega t + \phi) = \cos \omega t \cos \phi - \sin \omega t \sin \phi$; $\sin(\omega t + \phi) = \sin \omega t \cos \phi + \cos \omega t \sin \phi$.)

50. An inductor and a capacitor are connected in parallel to an ac potential difference $\mathcal{E} \cos \omega t$. (a) Derive an expression for the instantaneous current in each element. (b) Evaluate the total current for the special case in which $\omega L = 1/\omega C$.

Section 37.8 Resonance in a Series *RLC* Circuit

51. A series *RLC* circuit having $C = 100$ pF resonates at 50 kHz. What is the inductance of the circuit?

52. A resistor ($R = 1000\ \Omega$), a capacitor ($C = 4 \times 10^{-6}$ F), and an inductor ($L = 2 \times 10^{-3}$ H) are connected in series. (a) For a frequency of 60-Hz determine the impedance. (b) Determine the resonant frequency. (c) Determine the impedance at resonance. Calculate the impedance at a frequency (d) 50% above and (e) 50% below resonance.

53. A series *RLC* circuit having $R = 200\ \Omega$, $L = 0.1$ mH, and $C = 0.01\ \mu$F is connected to a 100-V (rms) ac source whose frequency is variable. Determine the thermal power in the resistor at the resonant frequency and at nine tenths of the resonant frequency.

54. If the inductance of a series *RLC* circuit is doubled, what is the percentage change in the resonance frequency?

55. The half-maximum frequency width of a certain *RLC* resonance is 50-Hz. If $R = 20\ \Omega$ and $C = 10$ pF, what is the resonance frequency?

56. Calculate the half-maximum frequency width of the resonance peak of an *RLC* series circuit if $R = 100\ \Omega$, $L = 20\ \mu$H, and $C = 0.1$ pF.

57. For the previous exercise, determine the frequencies at which the average power is one half the maximum average power.

58. An *RLC* series circuit with $R = 1500\ \Omega$ and $C = 0.015\ \mu$F is connected to an ac generator whose frequency is variable. When the frequency is adjusted to 50.5 kHz, the rms current in the circuit is a maximum and is 0.14 A. Determine (a) the inductance L and (b) the rms value of the generator output.

59. An *RLC* series circuit has $L = 1.2$ mH and $C = 1.6$ nF. It is desired to change the resonant frequency by changing the capacitance. What percentage change in the capacitance is required to increase the resonant frequency by 5%?

60. An FM radio frequency of 92.1 MHz produces resonance in an *RLC* series circuit having $R = 2.5\ \Omega$, $L = 1.50\ \mu$H, and $C = 1.99$ pF. (a) If an rms potential difference of 2.5 mV is provided to the circuit by the antenna, determine the rms current at resonance. (b) If 2.5 mV (rms) is provided at a frequency of 90.0 MHz, determine the percentage change in the rms current in the circuit.

61. The power resonance peak in an *RLC* circuit is required to have a width of 4.1 kHz at the half-power values. Determine the ratio of resistance to inductance in the circuit.

62. A plot of average power versus radian frequency for an *RLC* series circuit is shown in Figure 4. The parameters used were $R = 10\ \Omega$, $L = 0.1$ H, and $V_{rms} = 100$ V. (a) Using the graph as a source of information, determine the value of the capacitance. (b) Determine the width of the peak at the points where the power is half the maximum value and show that your result is consistent with $\Delta\omega = R/L$.

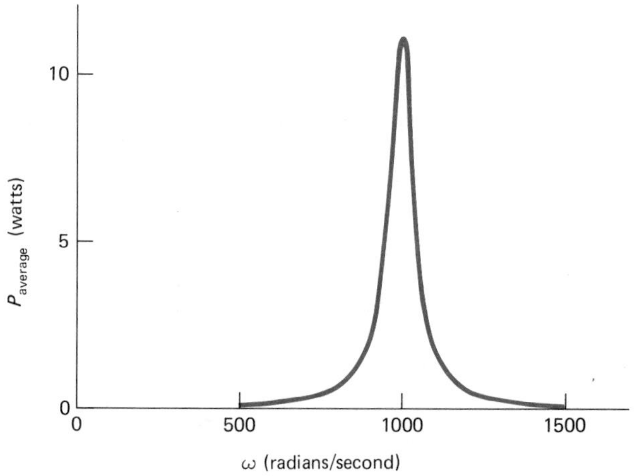

Figure 4

63. The average power in a series *RLC* circuit was shown to be

$$P_{av} = \frac{V_{rms}^2 \omega^2 R}{\omega^2 R^2 + L^2(\omega^2 - \omega_r^2)^2}$$

(a) Show that this can be written in terms of the maximum power as

$$P_{av} = \frac{P_{max}\omega^2}{\omega^2 + (L^2/R^2)(\omega^2 - \omega_r^2)^2}$$

(b) The frequencies for which the power is half the maximum value can be determined by solving the foregoing equation for $P_{av} = P_{max}/2$. Show that this leads to the condition

$$\omega^2 = \frac{L^2}{R^2}(\omega^2 - \omega_r^2)^2$$

(c) This equation, involving the fourth power of ω, means that mathematically there are four possible solutions for ω. Only two of these are physically acceptable. By taking the square root of both sides of the foregoing equation we see that $\pm\omega = (L/R) \times (\omega^2 - \omega_r^2)$. This result yields two quadratic equations for ω. Show that the two physically acceptable solutions to these two quadratic equations are

$$\omega_1 = \frac{L}{2R} + \sqrt{\left(\frac{L}{2R}\right)^2 + \omega_r^2} \quad \text{and}$$

$$\omega_2 = -\frac{L}{2R} + \sqrt{\left(\frac{L}{2R}\right)^2 + \omega_r^2}$$

Hence $\omega_1 - \omega_2 = L/R$, which yields the radian frequency width of the power curve at half-maximum.

Section 37.9 Transformers

64. In Europe, households and businesses are provided with 240-V, 50-Hz current rather than 120-V, 60-Hz current as in the United States. What are the characteristics of a transfomer that would allow you to use an American appliance in Europe? (The frequency is not crucial for this application.)

65. The principle of transmission of electric power is illustrated in Figure 37.28. Recalculate the currents, potential differences, and electic powers shown in this figure assuming that the potential difference across the secondary winding of the step-up transformer is 500,000 V (rms).

66. An ac source producing 10.5 V (rms) has an internal resistance of 12,000 Ω. In order to transfer the maximum amount of power to a resistance of 8 Ω, the source is connected to the primary winding of a transformer and the resistor to the secondary winding. Determine (a) the turns ratio of the transformer and (b) the power delivered to the resistor.

chapter 38

ELECTROMAGNETIC WAVES AND MAXWELL'S EQUATIONS

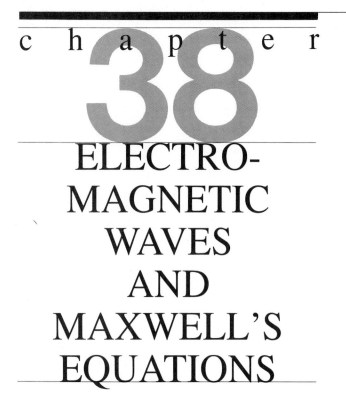

Preview

In 1864, James Clerk Maxwell* brought together the four basic laws governing electric and magnetic fields. He modified one of them (Ampère's law) and thereby achieved a unified theory of electric and magnetic phenomena. The four basic laws are known as Maxwell's equations.

Maxwell found that these four equations have wavelike solutions and showed that the speed of electromagnetic waves equals the speed of light. This led him to conclude that light is an electromagnetic wave. Thus, Maxwell's research not only united electricity and magnetism, it also laid a theoretical foundation for many optical phenomena.

In this chapter we recount the history of the particle and wave theories of light. We use qualitative arguments to suggest that electromagnetic waves should exist and that they should be transverse waves.

We present the modern forms of Maxwell's equations and verify that they have wavelike solutions. We then use Maxwell's equations to show that the speed of electromagnetic waves equals the speed of light.

We develop the idea of intensity as a measure of the rate at which electromagnetic waves transfer energy. The transverse character of electromagnetic waves is revealed by the fact that they can exhibit polarization. Elementary aspects of polarized electromagnetic waves are presented in the final section.

*James Clerk Maxwell (1831–1879), a brilliant English physicist, ranks with Newton and Einstein. Although best known for his research in electromagnetism, Maxwell also made important contributions to the kinetic theory of gases and to our understanding of color vision.

38.1 Introduction

In the 17th century there developed two conflicting views as to the nature of light. Newton favored the "corpuscular" theory—the view that light was a stream of particles. The Dutch physicist Christian Huygens (1629–1695) was the most famous proponent of the wave theory of light. Both theories offered simple explanations of certain optical phenomena, but neither could explain all optical effects. For example, the particle theory could explain the straight-line propagation of light, and the fact that an object may cast a "sharp" shadow. However, the particle theory was unable to explain why the continued loss of particles did not also cause a source of light to lose weight. The wave theory, on the other hand, easily accounted for the transfer of radiant energy without any transfer of matter. Huygens was able to derive the optical laws of reflection and refraction by using the wave theory. But straight-line propagation was not so easily explained, and sharp shadows were still more troublesome for the wave theory. With the technology then available, there was no way to decide whether the wave theory or the particle theory was more satisfactory.

More than 100 years after Newton and Huygens argued the merits of wave versus particle, Thomas Young* performed a famous experiment that demonstrated the *interference* of light. Young made an innovative assumption

*Thomas Young (1773–1829), an English physician, is best known for his experiment demonstrating the interference of light. He also investigated the elastic properties of matter. Young's modulus of elasticity is named in his honor.

Figure 38.1

(a) In longitudinal waves the oscillations of the medium are parallel to the direction of wave travel. (b) In transverse waves the oscillations of the medium are perpendicular to the direction of wave travel.

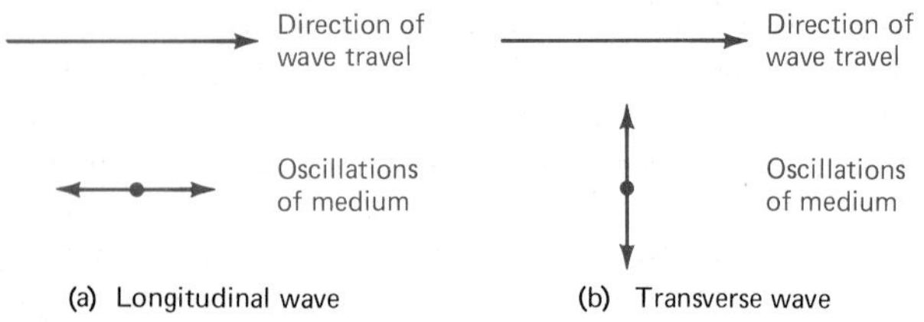

that allowed him to explain diffraction as an interference phenomenon. He assumed that a *principle of superposition* is valid for the light waves. This was a controversial idea at the time, and was accepted only after Young and Fresnel* used it to explain a number of phenomena, such as the coloring of thin films (for example, oil slicks) and the straight-line propagation of waves.

Initially, both Young and Fresnel assumed that light waves are longitudinal. In longitudinal waves the oscillations of the medium are parallel to the direction of wave travel (Figure 38.1a). For example, sound waves in air are longitudinal. Young and Fresnel conceived of an elastic medium—the *luminiferous ether*—as filling all space. The vibrations of the ether propagated as light, just as longitudinal vibrations in air propagate as sound. But the longitudinal wave theory of light contained a fatal flaw: it could not explain *polarization* (Section 38.6), a property exhibited by transverse waves but not by longitudinal waves. In transverse waves the oscillations of the medium are perpendicular to the direction of wave travel (Figure 38.1b). It was Young who eventually realized that the wave theory could be enlarged to accommodate polarization merely by assuming that the vibrations of the ether were tranverse to the direction of wave travel.

Young and Fresnel left behind a legacy—a picture of light as a transverse wave that propagates at great speed through the luminiferous ether. A hint as to the electromagnetic nature of light was later provided by Michael Faraday, who showed that the polarization of light was affected by a strong magnetic field. The stage was then set for Maxwell, who synthesized the empirical laws of electricity and magnetism into a coherent theory of electromagnetism. This theory stands as one of the great intellectual achievements of man. Perhaps the most dramatic prediction of Maxwell's theory was that light propagates as an electromagnetic wave.

Maxwell's theory seemed esoteric at the time of its publication in 1864. (A standard physics textbook published in 1870 does not even mention Maxwell's theory.†) But times change. Today we can look back more than 100 years at a growing parade of scientific and technological developments that are outgrowths of Maxwell's theory. Radio, television, and fiber optics communications are all based on Maxwell's theory.

*Augustin Jean Fresnel (1788–1827), a French scientist, performed many experiments to verify his wave theory predictions.

†Benjamin Silliman, *Principles of Physics*, Ivison, Blakeman, Taylor & Co., Chicago, 1870.

38.2 Electromagnetic Waves

In this section we draw together ideas that we developed earlier in this book in order to build a picture of electromagnetic waves. The picture that emerges is one of wavelike variations in the electric (E) and magnetic (B) fields* generated by an oscillating charge. These field variations travel at the speed of light. The E-field and B-field of the wave are perpendicular to each other and both are perpendicular to the direction of wave travel (Figure 38.2). In Section 38.4 we will make use of Maxwell's equations to show that electromagnetic waves travel at the speed of light.

To begin, we recall the E-field set up by an isolated positive charge at rest (Section 27.2). The E-field lines are directed radially away from a positive charge (Figure 38.3). So long as the charge is at rest it does not set up a B-field. If the charge moves, however, it constitutes an electric current and produces a B-field as well as an E-field. You may recall (Section 33.2) that the direction of the B-field is related to the direction of the current by the right-hand rule. In a plane perpendicular to a steady current the B-field lines are circles, concentric with the current (Figure 38.4).

*Henceforth we refer to these as the E-field and the B-field.

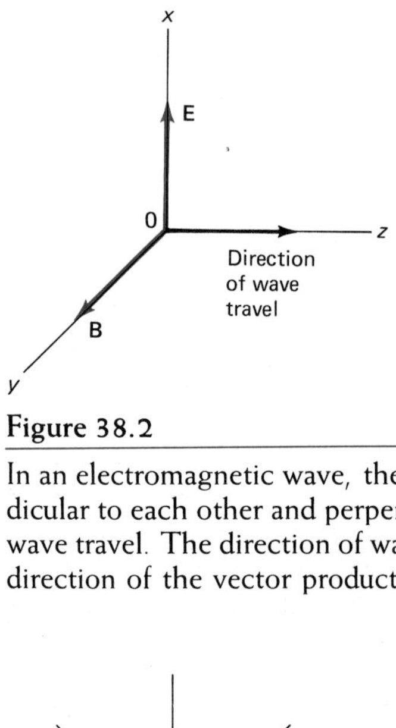

Figure 38.2

In an electromagnetic wave, the fields **E** and **B** are perpendicular to each other and perpendicular to the direction of wave travel. The direction of wave travel is the same as the direction of the vector product **E × B**.

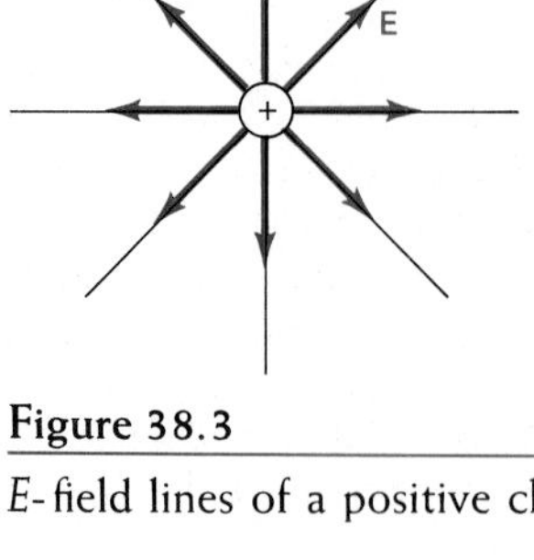

Figure 38.3

E-field lines of a positive charge at rest.

Figure 38.4

B-field lines surrounding a steady current I directed out of the page.

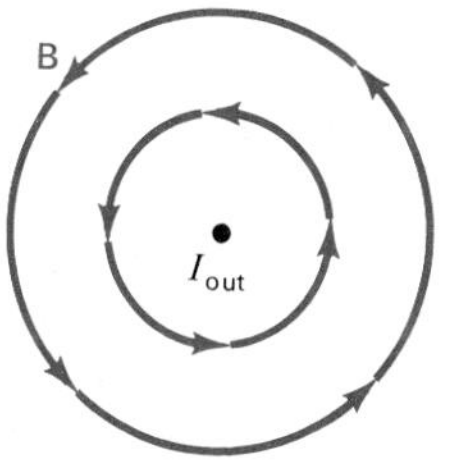

Figure 38.5

As the charge oscillates up and down, changes in the E-field travel outward as a wave. Only one field line is shown.

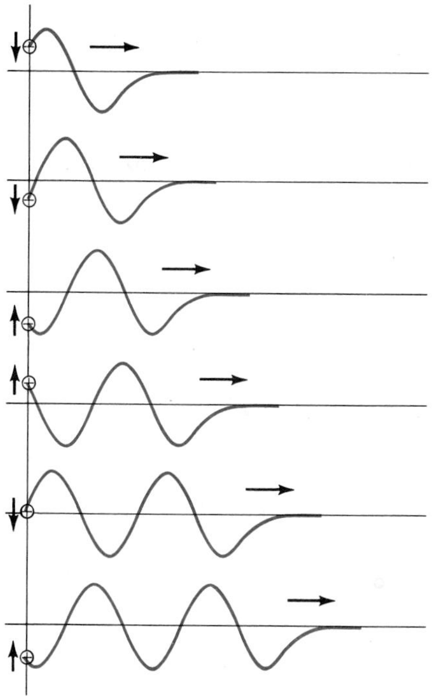

To establish a rudimentary picture of an electromagnetic wave, we envision a single electric charge oscillating up and down. As a matter of convenience we consider a positive charge, although in practice the charge could be negative. Figure 38.5 suggests what happens to *one* of the E-field lines originating on the charge. As the charge oscillates, the E-field line is carried up and down like the end of a rope attached to a moving hand. Ripples in the E-field move outward as a wavelike disturbance. A test charge located some distance from the oscillating charge therefore experiences a time-varying electric force. Note that the E-field of the wave lies in the plane of motion of the charge. Figure 38.5 suggests that an oscillating charge generates an electromagnetic wave. However, the actual E-field set up by an oscillating charge is more complicated than that shown in Figure 38.5 and cannot be represented faithfully by a few lines.

The B-field produced by the oscillating charge is also time varying. As the charge oscillates up and down it constitutes a current that changes direction repeatedly. The B-field set up by this oscillating current travels through space as a wave. The B-field wave accompanies the E-field wave. In fact, the two waves fields are inseparably linked. Accordingly, we often speak of an *electromagnetic wave* rather than of two separate waves. The B-field wave lies in a plane perpendicular to the oscillating current, just as the B-field lies in a plane perpendicular to a steady current. On the other hand, the E-field wave lies in the same plane as the current. Thus, the two planes containing the E-field and B-field of an electromagnetic wave are mutually perpendicular. Furthermore, both the E-field and the B-field of the electromagnetic wave are perpendicular to the direction of wave travel. In other words, electromagnetic waves are *transverse.* The E-field and the B-field travel through space together at a finite speed. As we will see in Section 38.4, electromagnetic waves travel at the speed of light, c. The implication is that light is an electromagnetic wave.

If the charge executes simple harmonic motion with a period T, the electromagnetic wave period is also T. At each point in space the E-field and B-field oscillate with a period T. The wave profile (Figure 38.6) advances at speed c. The distance between wave crests is λ, the wavelength. The time it takes for successive crests to pass a stationary point is one period, T. The basic kinematic equation for waves relates λ, c, and T (Section 17.2).

$$\lambda = cT \tag{38.1}$$

The reciprocal of the period is the frequency, ν (Greek nu).

$$\nu = \frac{1}{T} \tag{38.2}$$

Replacing T by $1/\nu$ in Eq. 38.1 gives an alternate form of the kinematic relation,

$$\nu\lambda = c \tag{38.3}$$

The character of the electromagnetic wave fronts* depends on the nature of the source. Spherical, cylindrical, and plane wave fronts are among the many possible types. The mathematical analysis of electromagnetic waves is simplest for plane waves. Our analysis is based on plane waves. For a plane electromagnetic wave, **E** and **B** lie in a common plane, are mutually perpendicular, and are independent of position within that plane. The wave advances in a direction perpendicular to the plane containing **E** and **B** (see Figure 38.2). For a wave advancing in the positive z-direction we can take **E** to lie along the x-axis and **B** to lie along the y-axis (Figure 38.7). The wave fronts are then parallel to the x–y plane.

The E-field and B-field of a sinusoidal plane wave can be represented by the forms

$$E_x = E_o \sin\left[\frac{2\pi}{\lambda}(z - ct)\right] \tag{38.4}$$

$$B_y = B_o \sin\left[\frac{2\pi}{\lambda}(z - ct)\right] \tag{38.5}$$

*A wave front is a surface of constant phase, such as a wave crest (Section 18.4).

Figure 38.6

The wave profile advances at speed c. The distance between adjacent wave crests is λ, the wavelength. The time for successive crests to pass a stationary point is called the period and is denoted by T. The equation $\lambda = cT$ expresses the kinematic relationship of λ, c, and T.

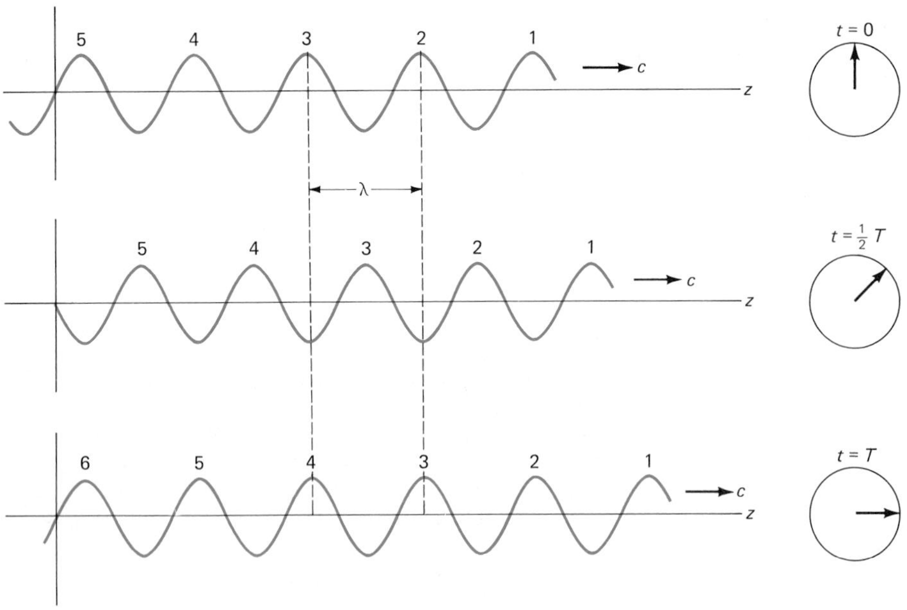

Equations 38.4 and 38.5 represent plane wave fronts in the x–y plane because E_x and B_y are independent of x and y; they depend only on z. E_0 and B_0 are the amplitudes of the waves; λ is the wavelength; and c is the speed (phase velocity). If we choose to introduce the wave number, k,

$$k = \frac{2\pi}{\lambda} \tag{38.6}$$

and radian frequency, ω,

$$\omega = 2\pi\nu \tag{38.7}$$

the form of the fields becomes

$$E_x = E_0 \sin(kz - \omega t) \tag{38.8}$$

$$B_y = B_0 \sin(kz - \omega t) \tag{38.9}$$

Figure 38.8 portrays E_x and B_y at different positions at the same instant of time.

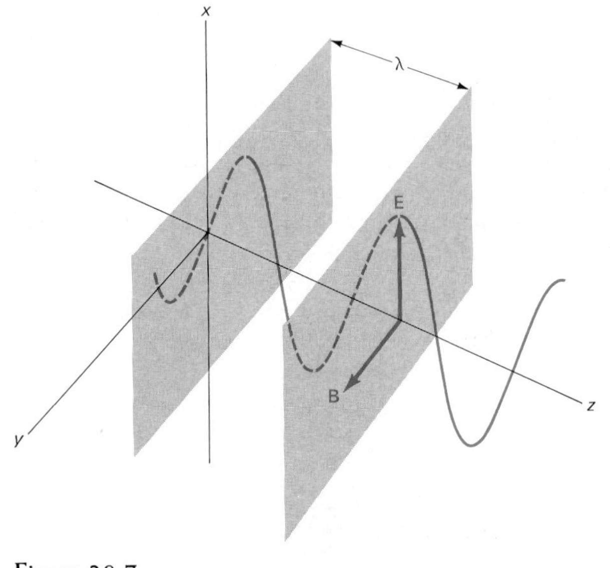

Figure 38.7

Plane electromagnetic waves. The sinusoidal curve indicates the value of the E-field. With **E** along the x-direction and **B** along the y-direction, the wave fronts are parallel to the x–y-plane. Segments of two wave fronts are shown. The sinusoidal variation of the B-field is not shown.

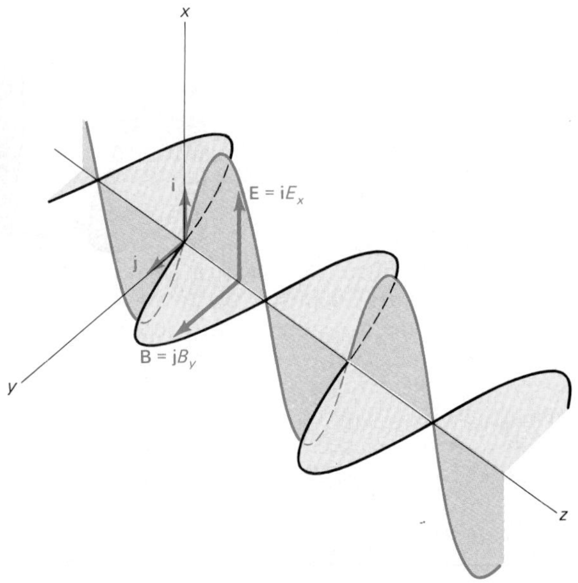

Figure 38.8

The fields E_x and B_y of Eqs. 38.8 and 38.9. The waves advance in the positive z-direction.

We have just constructed a picture of an electromagnetic wave by starting with the notion of a wiggling E-field line attached to an oscillating charge. A rigorous justification of this picture must be based on Maxwell's electromagnetic theory. In the next section we review the four basic equations of electricity and magnetism and show how Maxwell modified one of them (Ampère's law) by adding the so-called displacement current. This modification enabled Maxwell to combine the description of electric and magnetic phenomena. Electrostatics and magnetostatics were no longer isolated and unrelated phenomena. They became special cases of a unified theory of electromagnetism.

38.3 Maxwell's Equations

Let us first review the four basic laws governing the E- and B-fields as we now know them. We can then show how Maxwell modified one of them, thereby creating a unified theory of electric and magnetic phenomena.

Magnetic Flux

The magnetic flux Φ_B is defined as the surface integral of the B-field. In equation form [Section 32.2, Eq. 32.5*]

$$\Phi_B = \int \mathbf{B} \cdot d\mathbf{A} \quad (38.10)$$

Φ_B can be interpreted as the net number of B-field lines "threading" the surface (Figure 38.9).

Two of the basic laws of electromagnetism involve magnetic flux. The first of these states that the net magnetic flux through any *closed* surface is zero [Section 32.2, Eq. 32.6]:

$$\Phi_B = \int_{\substack{\text{closed}\\\text{surface}}} \mathbf{B} \cdot d\mathbf{A} = 0 \quad (38.11)$$

*The section and equation numbers in brackets refer to the original treatment of the topic. New equation numbers are assigned here for convenience.

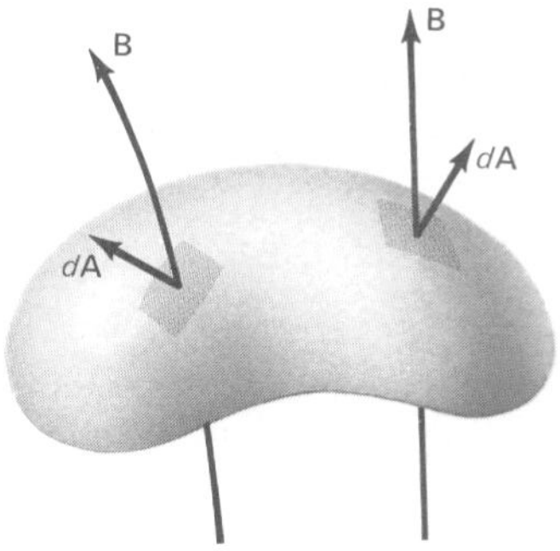

Figure 38.9

Figure 38.10

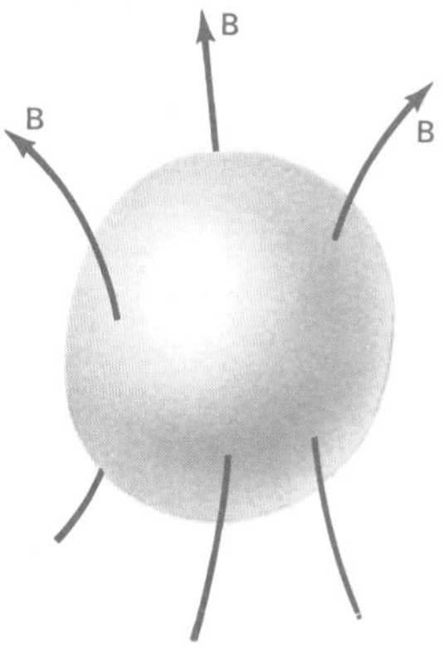

Geometrically, $\Phi_B = 0$ means that equal numbers of B-field lines enter and leave the volume enclosed by the surface (Figure 38.10). Physically, $\Phi_B = 0$ is a consequence of the fact that there are no magnetic monopoles —no isolated magnetic poles on which B-field lines originate or terminate.

Faraday's Law of Induction

The second basic law involving magnetic flux is Faraday's law of induction [Section 34.2, Eq. 34.11]:

$$\oint \mathbf{E} \cdot d\mathbf{s} = -\frac{d}{dt}\int \mathbf{B} \cdot d\mathbf{A} = -\frac{d\Phi_B}{dt} \quad (38.12)$$

The line integral of the E-field extends around the periphery of the surface through which the flux of the B-field is evaluated (Figure 38.11). Faraday's law describes how a changing magnetic flux induces an electric field. Faraday's law is of special significance because it describes a coupling of the E-field and B-field and recognizes that this coupling requires a time variation of the flux. Only when Φ_B changes is there an induced E-field.

Electric Flux, Gauss' Law

The third basic law describing the fields is Gauss' law [Section 27.4, Eq. 27.15], which relates the electric flux (Φ_E) through a closed surface to the net charge enclosed by that surface:

$$\Phi_E = \int_{\substack{\text{closed}\\\text{surface}}} \mathbf{E} \cdot d\mathbf{A} = \frac{q}{\epsilon_0} \quad (38.13)$$

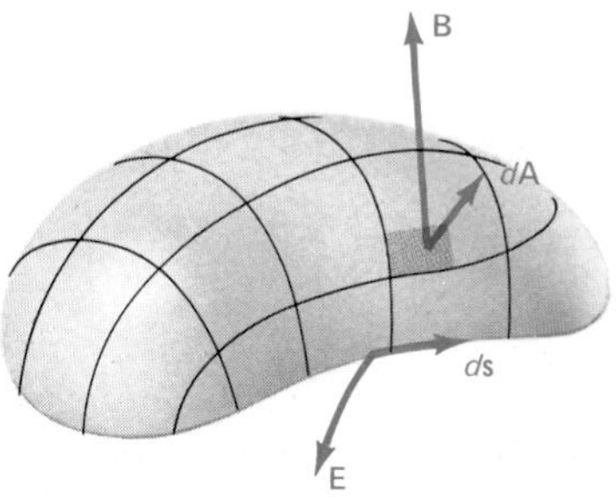

Figure 38.11

Figure 38.12

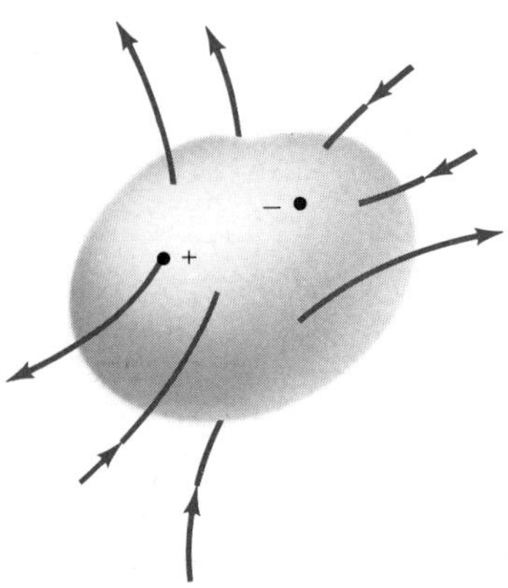

Gauss' law is a quantitative expression of the idea that E-field lines originate on positive charges and terminate on negative charges. If a surface encloses a net positive charge, more E-field lines will leave the surface than enter it ($\Phi_E > 0$). If the surface encloses no net charge, equal numbers of E-field lines enter and leave ($\Phi_E = 0$; Figure 38.12).

Ampère's Law

The fourth basic law describing electromagnetic phenomena is Ampère's law [Section 33.3, Eq. 33.19], which recognizes that a current can set up a B-field. In equation form Ampère's law is

$$\oint \mathbf{B} \cdot d\mathbf{s} = \mu_o I \qquad (38.14)$$

The line integral of $\mathbf{B}$ is around a closed path, and I is the net current encircled by that path (Figure 38.13).

As formulated in Eq. 38.14, Ampère's law is adequate only for phenomena involving *steady* currents. Ampère's law is incomplete when the current is not steady. We can show that this is so by considering the region near a parallel-plate capacitor while the capacitor is charging (Figure 38.14). Charge flows onto one plate and charge flows off the other plate. A current I reaches one plate and a current I leaves the other plate. There is no current between the plates. That is, no charge flows across the space separating the plates. Experiment shows that there is a B-field in the region between the plates as well as on either side of the plates. To a good approximation, the

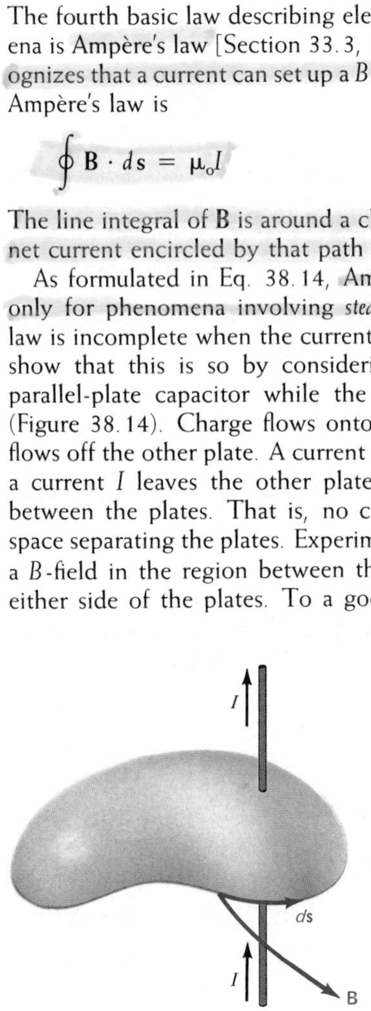

Figure 38.13

Figure 38.14

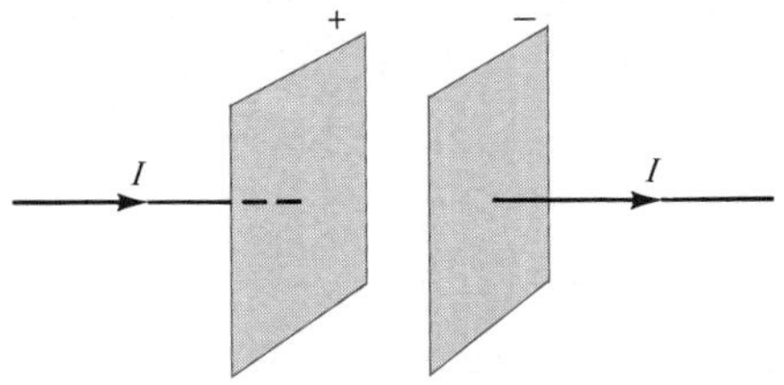

Figure 38.15

Experiment shows that $\oint \mathbf{B} \cdot d\mathbf{s}$ is the same for the loops labeled ①, ②, and ③. Loops ① and ③ encircle a current I, and Eq. 38.14 is satisfied. Loop ② does not encircle any conventional current. That $\oint \mathbf{B} \cdot d\mathbf{s}$ is not zero for loop ② shows that Eq. 38.14 is incomplete.

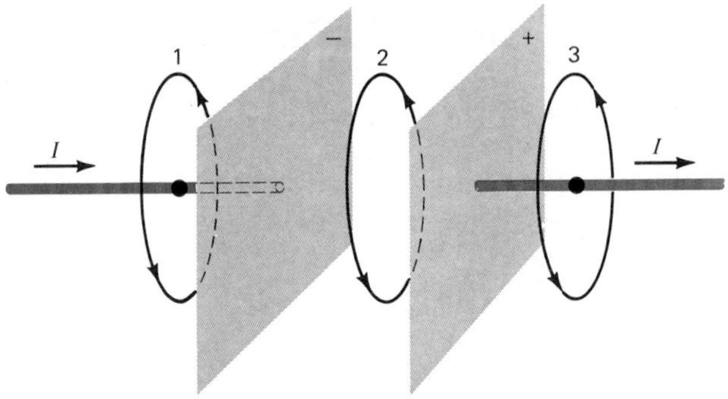

B-field lines are circles, concentric with the current as shown in Figure 38.15. Experiment shows that the value of the line integral $\oint \mathbf{B} \cdot d\mathbf{s}$ is the same around the loops labeled ①, ②, and ③ in the figure. The loops labeled ① and ③ encircle a current I and Eq. 38.14 is satisfied. According to Eq. 38.14, $\oint \mathbf{B} \cdot d\mathbf{s}$ should vanish for loop ②, because no conventional current is encircled. Evidently, Ampère's law, as expressed by Eq. 38.14, is incomplete.

Maxwell's Displacement Current

Maxwell remedied the shortcoming of Eq. 38.14 by adding what is called the *displacement current*. The displacement current is related to the *changing electric field* between the plates of the capacitor. To establish this relation for ourselves we apply Gauss' law to the closed surface shown in Figure 38.16. The surface encloses a net charge q—the charge on one plate of the capacitor. This charge is the source of the E-field between the plates. The charge q is related to the electric flux through the surface by Gauss' law (Eq. 38.13):

$$\frac{q}{\epsilon_o} = \int \mathbf{E} \cdot d\mathbf{A} = \Phi_E$$

The charge q changes with time because the capacitor is charging. As q changes the E-field also changes. The rate at which q changes is

$$\frac{dq}{dt} = \epsilon_o \frac{d}{dt} \int \mathbf{E} \cdot d\mathbf{A} \qquad (38.15)$$

Figure 38.16

The shaded Gaussian surface encloses a charge q, which gives rise to the E-field between the plates. Both q and E change with time as the capacitor charges and discharges.

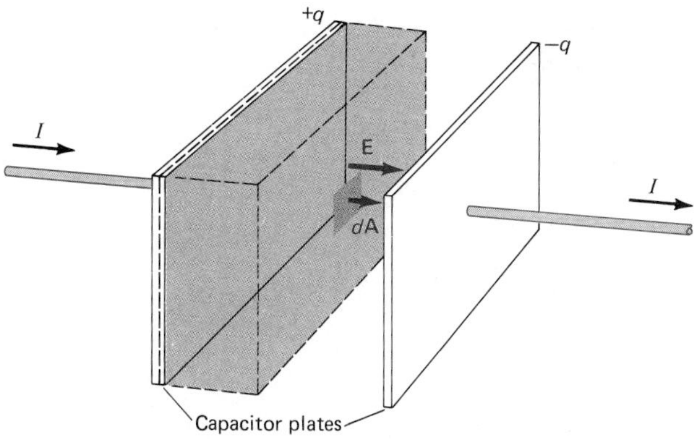

Here dq/dt is the rate at which charge flows onto the capacitor plate; that is, dq/dt equals the current in the wire leading to the capacitor plate.

$$I = \frac{dq}{dt}$$

Maxwell realized that the quantity

$$\epsilon_o \frac{d}{dt}\int \mathbf{E}\cdot d\mathbf{A} \equiv I_{\text{displacement}} \qquad (38.16)$$

which he called a *displacement current*, was equivalent to the current in the wires on either side of the capacitor plates. The displacement current is equivalent to a real current because it produces the same B-field.* Maxwell modified Ampère's law, Eq. 38.14, by adding a displacement current to the conventional current. That is, in Eq. 38.14, he replaced I by $I + I_{\text{displacement}}$. The modified version of Ampère's law is

$$\oint \mathbf{B}\cdot d\mathbf{s} = \mu_o I + \mu_o\epsilon_o \frac{d}{dt}\int \mathbf{E}\cdot d\mathbf{A} \qquad (38.17)$$

The line integral $\oint \mathbf{B}\cdot d\mathbf{s}$ is evaluated around the periphery of the surface over which the electric flux integral $\int \mathbf{E}\cdot d\mathbf{A}$ is evaluated (Figure 38.17). In the region between the capacitor plates there is no conventional current, only a displacement current, and Eq. 38.17 reduces to

$$\oint \mathbf{B}\cdot d\mathbf{s} = \mu_o\epsilon_o \frac{d}{dt}\int \mathbf{E}\cdot d\mathbf{A} \qquad (38.18)$$

This special form of Ampère's law describes the fact that a changing electric flux induces a B-field. Recall that Faraday's law (Eq. 38.12)

$$\oint \mathbf{E}\cdot d\mathbf{s} = -\frac{d}{dt}\int \mathbf{B}\cdot d\mathbf{A} = -\frac{d\Phi_B}{dt}$$

describes how a changing magnetic flux induces an E-field. Collectively, Eqs. 38.17 and 38.12 describe the coupled or interlocking nature of time-varying E- and B-fields. It is

*The phrase *displacement current* is related to the fact that $\epsilon_o\mathbf{E}$, which appears in Eq. 38.14, is sometimes referred to as the *electric displacement*.

Figure 38.17

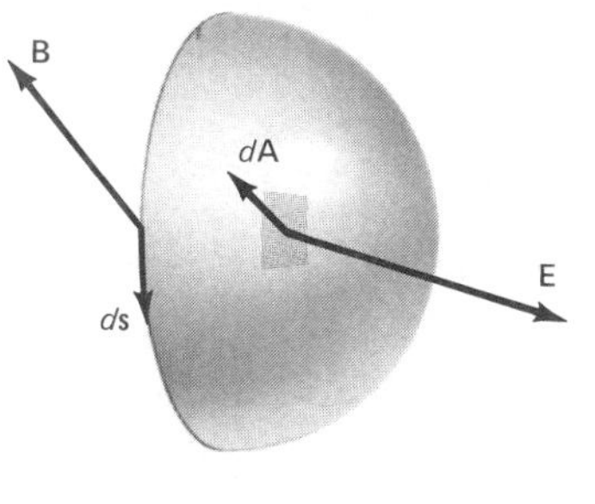

because of this coupling that we speak of the "electromagnetic field."

Maxwell's genius enabled him to discover the underlying unity of electric and magnetic phenomena. To honor his contribution, the set of four basic electromagnetic equations are now called *Maxwell's equations:*

$$\int_{\substack{\text{closed}\\\text{surface}}} \mathbf{B}\cdot d\mathbf{A} = 0 \qquad (38.11)$$

$$\int_{\substack{\text{closed}\\\text{surface}}} \mathbf{E}\cdot d\mathbf{A} = \frac{q}{\epsilon_o} \quad \text{(Gauss)} \qquad (38.13)$$

$$\oint \mathbf{E}\cdot d\mathbf{s} = -\frac{d}{dt}\int \mathbf{B}\cdot d\mathbf{A} \quad \text{(Faraday)} \qquad (38.12)$$

$$\oint \mathbf{B}\cdot d\mathbf{s} = \mu_o I + \mu_o\epsilon_o \frac{d}{dt}\int \mathbf{E}\cdot d\mathbf{A} \quad \text{(Ampère)} \qquad (38.17)$$

Question

1. According to Eq. 38.11, the surface integral of **B** is zero. In Faraday's law, Eq. 38.12, there appears the time derivative of the surface integral of **B**. Won't this derivative automatically equal zero?

38.4 The Speed of Electromagnetic Waves

In this section we verify that Maxwell's equations have wavelike solutions and we determine the speed of the electromagnetic waves. Specifically, we show that electromagnetic waves travel through a vacuum at a speed (c) given by

$$c = \frac{1}{\sqrt{\mu_o\epsilon_o}} \qquad (38.19)$$

where μ_o is the magnetic permeability of empty space (Section 33.1) and ϵ_o is the electric permittivity of empty space (Section 27.1). Inserting the values

$$\epsilon_o = 8.85419 \times 10^{-12}\ \text{C}^2/\text{N}\cdot\text{m}^2 \qquad (38.20)$$

$$\mu_o = 1.25664 \times 10^{-6}\ \text{N/A}^2 \qquad (38.21)$$

into Eq. 38.19 gives

$$c = 2.99793 \times 10^8\ \text{m/s} \qquad (38.22)$$

Figure 38.18

E_x and B_y vary from point to point along the z-axis. But keep in mind that the wave fields **E** and **B** "fill" space.

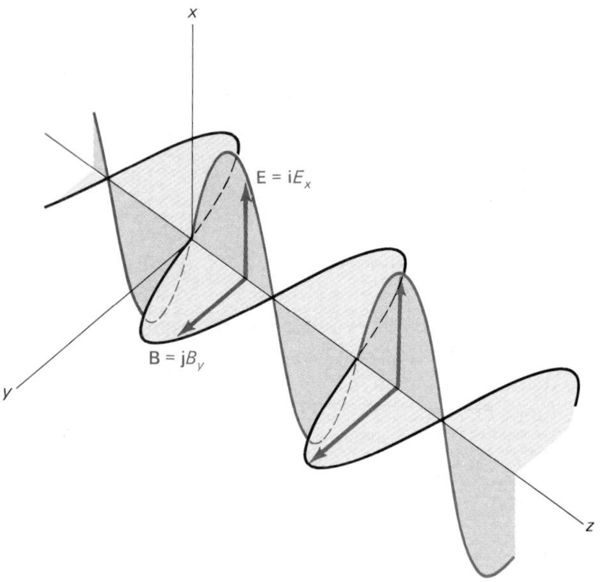

Because this equals the measured speed of light, one immediate implication of Maxwell's equations is that light is an electromagnetic wave.

Let's first verify that two of Maxwell's equations—Faraday's law, Eq. 38.12, and Ampère's law, Eq. 38.17—are satisfied by wavelike E- and B-fields. The E-field and B-field are taken to have the form of the transverse *plane waves* discussed in Section 38.1,

$$E_x = E_o \sin(kz - \omega t) \quad (38.8)$$

$$B_y = B_o \sin(kz - \omega t) \quad (38.9)$$

Equations 38.8 and 38.9 describe a transverse plane electromagnetic wave that advances in the positive z-direction. Figure 38.18 suggests how E_x and B_y vary along the z-axis. Keep in mind that E_x and B_y "fill" space. That is, they are not restricted to points on the z-axis. The speed c of the wave is related to the wave number k and radian frequency ω by*

$$c = \frac{\omega}{k} \quad (38.23)$$

The region under consideration is a vacuum. There are no currents or charges to contend with. The source of the waves lies outside the region we consider.

We have claimed that Maxwell's equations are satisfied by plane waves, with E_x and B_y having the forms specified by Eqs. 38.8 and 38.9. To verify this claim we substitute the plane waveforms of E_x and B_y into Faraday's law, Eq. 38.12, and Ampère's law, Eq. 38.17. The two resulting equations not only uphold the original claim—that Maxwell's equations are satisfied by plane electromagnetic waves—but also predict the speed of the waves.

Faraday's Law: Plane Waves

Figure 38.19 shows the rectangular path used to evaluate the line integral $\oint \mathbf{E} \cdot d\mathbf{s}$. Figure 38.20 shows how the *same* rectangle is used to evaluate the magnetic flux integral $\int \mathbf{B} \cdot d\mathbf{A}$. Figure 38.21 indicates the contributions to $\oint \mathbf{E} \cdot d\mathbf{s}$ made by each of the four legs of the rectangular loop. These add to give

$$\oint \mathbf{E} \cdot d\mathbf{s} = \int_1^2 \mathbf{E} \cdot d\mathbf{s} + \int_2^3 \mathbf{E} \cdot d\mathbf{s} + \int_3^4 \mathbf{E} \cdot d\mathbf{s} + \int_4^1 \mathbf{E} \cdot d\mathbf{s} = E_o h[\sin(kz_2 - \omega t) - \sin(kz_1 - \omega t)] \quad (38.24)$$

$^*c = \nu\lambda = 2\pi\nu\left(\frac{1}{2\pi/\lambda}\right) = \frac{\omega}{k}.$

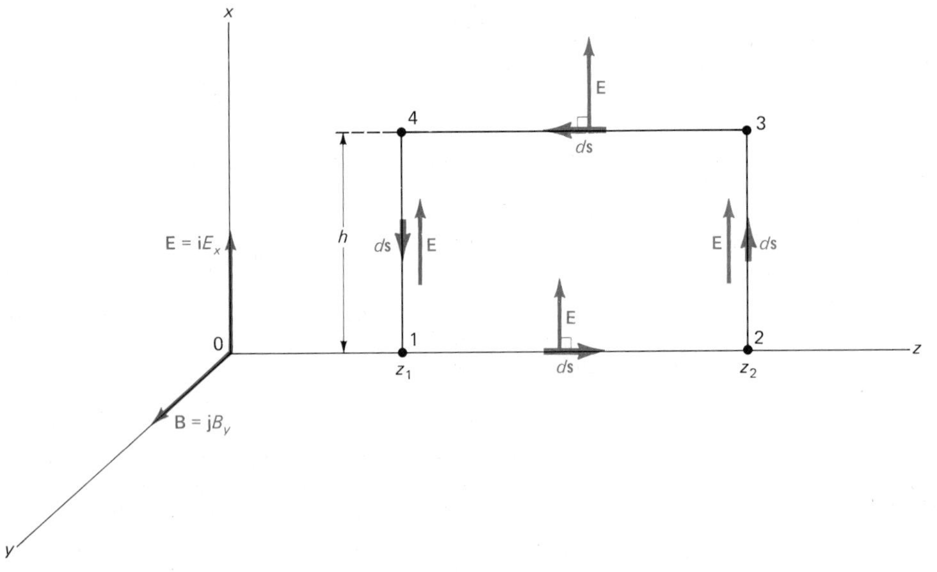

Figure 38.19

A rectangular loop is used to evaluate the line integral $\oint \mathbf{E} \cdot d\mathbf{s}$.

Figure 38.20

The rectangular loop of Figure 38.19 is also used to evaluate the magnetic flux integral $\int \mathbf{B} \cdot d\mathbf{A}$. The vector $d\mathbf{A}$ is parallel to $\mathbf{B}$.

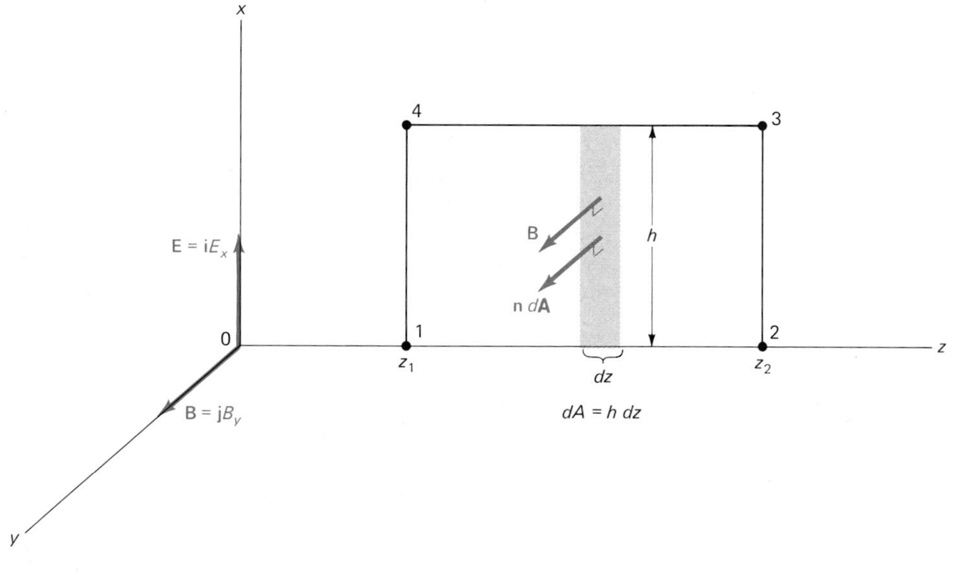

Figure 38.21

Contributions to $\oint \mathbf{E} \cdot d\mathbf{s}$ and $\int \mathbf{B} \cdot d\mathbf{A}$.

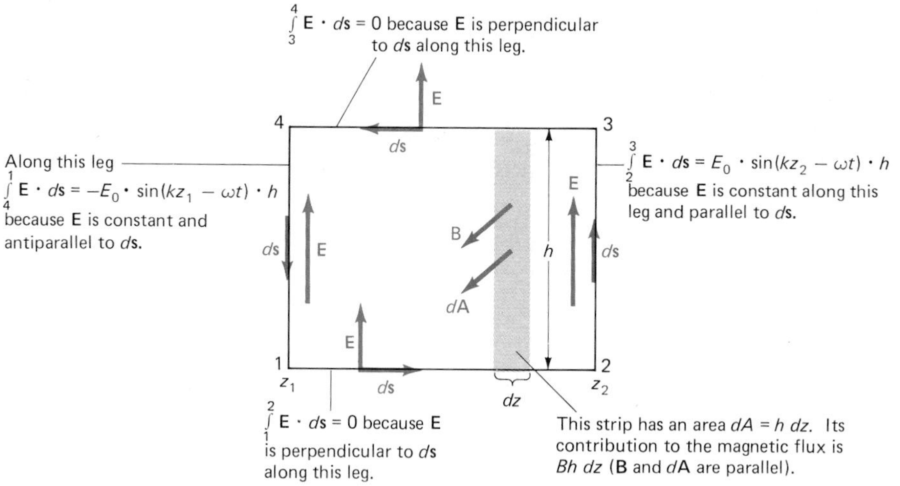

The area of the loop can be divided into a set of strips like the one shown in Figure 38.21. The total magnetic flux threading the loop is the sum (integral) of the contributions made by each strip. The total magnetic flux is

$$\Phi_B = \int \mathbf{B} \cdot d\mathbf{A} = \int_{z_1}^{z_2} B_o h \sin(kz - \omega t)\, dz$$

$$= -\frac{B_o h}{k}[\cos(kz_2 - \omega t) - \cos(kz_1 - \omega t)]$$

Forming the time derivative $-d\Phi_B/dt$, we get

$$-\frac{d\Phi_B}{dt} = +\frac{B_o h \omega}{k}[\sin(kz_2 - \omega t) - \sin(kz_1 - \omega t)] \tag{38.25}$$

Note that the difference of sine terms is the same in Eqs. 38.24 and 38.25. Substituting from Eqs. 38.24 and 38.25 into Faraday's law, Eq. 38.12, and canceling common factors leaves

$$E_o = B_o\left(\frac{\omega}{k}\right) \tag{38.26}$$

Equation 38.23 reminds us that ω/k is the speed c of the waves, so that Eq. 38.26 gives us one relation between the field amplitudes (E_o, B_o) and the speed c:

$$E_o = B_o c \tag{38.27}$$

Equation 38.27 tells us that Faraday's law, Eq. 38.12, is satisfied by plane electromagnetic waves (Eqs. 38.8 and 38.9) provided that the ratio E_o/B_o equals c. The cancellation of the factors $[\sin(kz_2 - \omega t) - \sin(kz_1 - \omega t)]$ is not a trivial result; rather, it shows that the transverse sinusoidal waveforms assumed for E_x and B_y are indeed solutions of Maxwell's equations.

Ampère's Law: Plane Waves

Because the region we are considering is a vacuum, there is no current due to the flow of charge ($I = 0$), and Ampère's law, Eq. 38.17, contains only the displacement current term.

$$\oint \mathbf{B} \cdot d\mathbf{s} = \mu_0 \epsilon_0 \frac{d}{dt} \int \mathbf{E} \cdot d\mathbf{A} \qquad (38.28)$$

Figure 38.22 and 38.23 show the rectangular loop used to evaluate the integrals that appear in Ampère's law. The evaluation of the integrals follows the same lines used for Faraday's law and yields the result

$$B_0 = \mu_0 \epsilon_0 E_0 c \qquad (38.29)$$

Equation 38.29 tells us that Ampère's law, Eq. 38.17, is satisfied by plane electromagnetic waves provided the ratio E_0/B_0 equals $1/\mu_0\epsilon_0 c$.

Equations 38.27 and 38.29 constitute a pair of simultaneous equations for E_0 and B_0. When we solve them simultaneously we obtain an equation for the speed c. First we substitute from Eq. 38.27 into Eq. 38.29 to get

$$B_0 = \mu_0 \epsilon_0 B_0 c^2$$

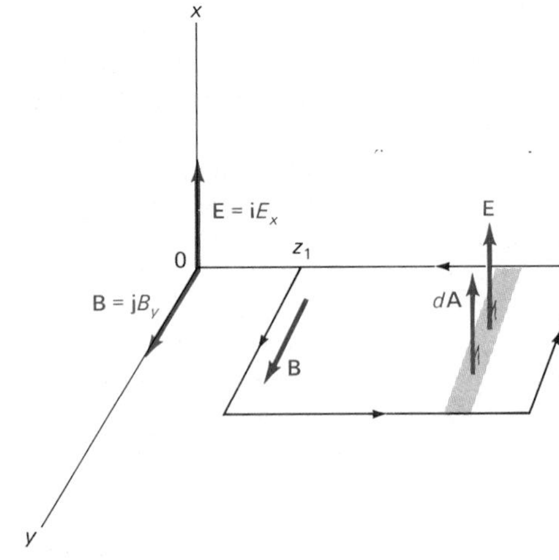

Figure 38.22

Rectangular loop in the y–z plane for evaluating $\mathbf{B} \cdot d\mathbf{s}$ and $\mathbf{E} \cdot d\mathbf{A}$ in Eq. 38.28.

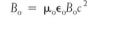

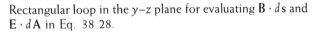

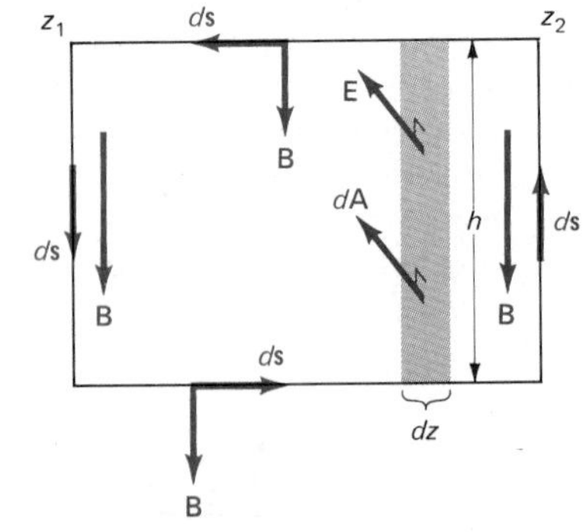

Figure 38.23

The two sides of the rectangular loop along which $\mathbf{B}$ and $d\mathbf{s}$ are perpendicular make no contribution to $\oint \mathbf{B} \cdot d\mathbf{s}$. $\mathbf{E}$ and $d\mathbf{A}$ are parallel.

From this equation we obtain the speed of electromagnetic waves as

$$c = \frac{1}{\sqrt{\mu_0 \epsilon_0}} \qquad (38.19)$$

Equation 38.19 was first obtained by Maxwell. The value of $c = 1/\sqrt{\mu_0\epsilon_0}$ agrees with the measured value of the speed of light. This is remarkable because Eq. 38.19 expresses c in terms of μ_0 and ϵ_0, which are magnetic and electric constants previously unrelated to light. This was one of the triumphs of Maxwell's theory—the implication that light is an electromagnetic wave. The E-field and B-field of the wave are inseparably linked. Their coupling is described quantitatively by Maxwell's equations.

The implications of Maxwell's theory stretch far beyond the suggestion that light is an electromagnetic wave. Maxwell's theory does not restrict the wavelength to the visible spectrum. In fact, Eq. 38.19 predicts that the speed of the waves is independent of wavelength. In principle, then, there could be electromagnetic waves of any wavelength. In 1887, Heinrich Hertz* generated and detected electromagnetic waves with wavelengths of approximately 5 m. He was able to verify Maxwell's predicted speed by measuring both frequency (ν) and wavelength (λ). Their product gave the speed, $\nu\lambda = c$. Hertz's results were in substantial agreement with the speed of visible light. Thus, they confirmed Maxwell's theory.

Maxwell died in 1879, at age 47, and Hertz died in 1894, at age 36. Fate deprived both men of the satisfaction of watching their labors bear fruit. Maxwell's theory still finds new applications, most recently in lasers. Hertz' 5-m waves were soon stretched to 50 m by Marconi and transmitted across the Atlantic Ocean.

Example 1 **B-Field Strength for a Plane Wave**

A plane electromagnetic wave has an E-field strength of $E_0 = 800$ V/m. This value of E_0 equals approximately the amplitude of the E-field of a plane wave that has the same intensity as the sunlight reaching the earth. We wish to determine the strength of the B-field. From Eq. 38.27 we find

$$B_0 = \frac{E_0}{c} = \frac{800 \text{ V/m}}{3 \times 10^8 \text{ m/s}} = 2.67 \times 10^{-6} \text{ T}$$

For comparison, the magnetic field strength of the earth is approximately 10^{-4} T.

The Electromagnetic Spectrum and Its Sources

In 1665, when he was 23, Isaac Newton used a glass prism to disperse a beam of sunlight into a rainbow of colors (Figure 38.24). He concluded that white light is a mixture of colors. The range of color, or *spectrum* as Newton called

*Heinrich Hertz (1857–1894), a brilliant German physicist, carried out his famous experiments between 1887 and 1890. In one series of experiments he used a large parabolic mirror to focus the electromagnetic waves.

Figure 38.24

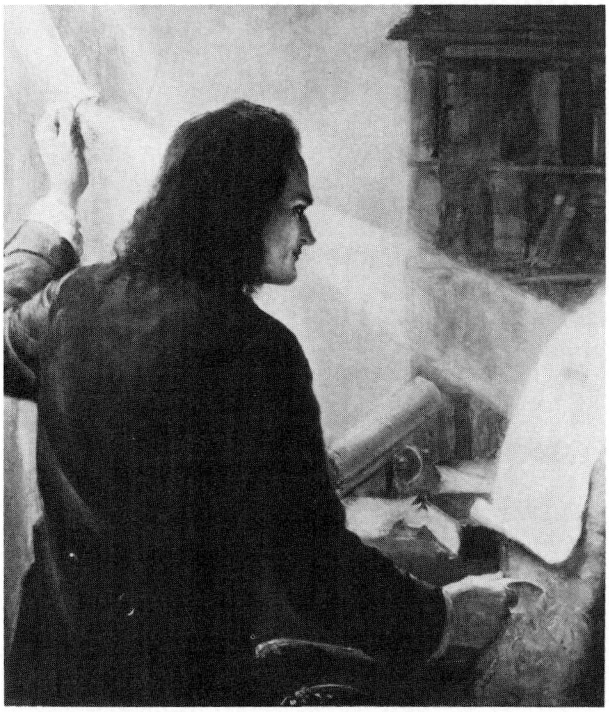

it, ran from deep violet to blue, green, orange, and finally dark red. The full electromagnetic spectrum extends beyond both ends of the visible spectrum and includes radio and television waves, infrared radiation, ultraviolet radiation, x rays, and gamma rays. Figure 38.25 shows the electromagnetic spectrum. The borders separating the various regions of the spectrum are not sharply defined. The range of frequencies and wavelengths explored to date is enormous, yet the spectrum has not been completely spanned.

Frequency (Hertz)	Name of Spectral Region	Wavelength	Sources
10^{20}	Gamma rays	30 fm	Atomic nuclei
			Accelerators
			X-ray tubes
10^{17}	X-rays	3 nm	Atomic electrons
10^{15}	Ultraviolet	300 nm	Lasers
	Visible		Electric arcs
10^{13}	Infrared	30 μm	Hot solids
			Molecules
10^{10} (10 GHz)	Microwaves	3 cm	Magnetron klystron
10^8	Television FM radio	3 m	Electronic currents
10^6 (1 MHz)		300 m	
	AM radio		
10^3 (1 kHz)	Longwave radio	30 km	

Figure 38.25

The electromagnetic spectrum and typical sources. The narrow range of visible wavelength runs from red at about 670 nm to violet at about 430 nm.

Sources of Electromagnetic Waves

If you charge a glass rod by rubbing it, and then wave it up and down periodically, you will generate electromagnetic waves. The frequency of the waves equals the frequency at which you wave the rod. Radiation is produced because the charges are *accelerated*. Charges at rest or in uniform motion do not radiate, even though they may produce electric and magnetic fields. The current in a radio antenna produces electromagnetic waves because the moving charges experience accelerations. When the electron beam in a color TV set hits the screen, some x rays are emitted. The electrons are decelerated via collisions and convert their kinetic energy into radiant energy—x rays. An electron moving along a curved path in a magnetic field is accelerated, and it too emits electromagnetic radiation. The energy carried away by the waves is supplied by the kinetic energy of the electron.

Nature provides us with many examples of the conversion of kinetic energy into electromagnetic energy via the acceleration of electric charges. None is more spectacular than the Crab nebula, the remnant of a supernova (an exploding star). Electrons in the Crab nebula convert kinetic energy into electromagnetic energy as they move through the magnetic field associated with the nebula. The rate at which the Crab nebula radiates electromagnetic energy is enormous—nearly 100,000 times the rate at which our sun radiates.

There are many technological problems related to generating, transmitting, and receiving electromagnetic waves. Quite distinct techniques are used to generate x rays, visible light, and AM radio waves. For example, small crystals are used to control the frequency of the oscillatory electric currents in radio antennas. The crystal is stimulated to vibrate at one of its natural (or resonant) frequencies, and the mechanical vibrations are then used to control the antenna current. The size of such a crystal can be estimated by recalling that the fundamental frequency of a vibrating structure is given by (Section 18.5)

$$\nu_{\text{fund}} \approx \frac{\text{speed of wave}}{\begin{array}{c}\text{largest dimension}\\ \text{of vibrating}\\ \text{structure}\end{array}} = \frac{v}{L} \qquad (38.30)$$

Example 2 Size of an Oscillator Crystal

We want to estimate the characteristic dimension of a crystal that vibrates at 1 MHz (the middle of the AM radio band). We take the speed of sound in the crystal to be 5×10^3 m/s, a typical value for many solids. With $\nu_{fund} = 10^6$ Hz and $v = 5 \times 10^3$ m/s, Eq. 38.30 gives

$$L \simeq \frac{v}{\nu_{fund}} = \frac{5 \times 10^3 \text{ m/s}}{10^6\text{/s}} = 5 \times 10^{-3} \text{ m}$$

Thus, a thin wafer about 5 mm thick would oscillate at the desired frequency.

Crystal oscillators can be used to generate wavelengths as small as several centimeters. Still shorter electromagnetic waves can be generated by klystrons and magnetrons, devices that rely on the coherent accelerated motions of electrons.

For wavelengths of less than about 1 mm, the "classical" sources—accelerated charges—give way to various "quantum" sources: atomic nuclei, atoms, and molecules. Molecular rotations are one source of infrared radiation for example. Electronic transitions in atoms and incandescent solids provide much of the visible light that we use. Changes in nuclear structure often result in the emission of gamma rays (Section 43.2). Figure 38.25 indicates the more prominent sources of electromagnetic radiation across the spectrum.

Questions

2. Which have a longer wavelength, AM radio waves or FM radio waves?
3. The Lorentz force (Section 32.3, Eq. 32.15) on a charge q subject to an E-field and B-field is

   $$\mathbf{F} = q\mathbf{E} + q\mathbf{v} \times \mathbf{B}$$

   If **E** and **B** refer to the fields of a plane electromagnetic wave (Eqs. 38.8 and 38.9), which is larger, the electric force (q**E**) or the magnetic force (q**v** × **B**)? (*Hint:* Refer to Eq. 38.29.)

38.5 Energy Transfer via Electromagnetic Waves

That electromagnetic waves can transfer energy becomes painfully evident to anyone who stays in the sun too long. The rate at which sunlight and other electromagnetic waves transport energy is described by the *wave intensity*. The intensity (I) of a wave is defined as the rate at which it transfers energy divided by the area over which the energy is spread. In other words, intensity is the rate of energy flow per unit area. If an electromagnetic wave carries energy U through an area A* in a time t, the wave intensity is (Figure 38.26)

$$I = U/tA \qquad (38.31)$$

*To be precise, A is the area perpendicular to the direction of the energy flow.

Figure 38.26

The energy U crossing the area A in a time t is contained in a volume Act.

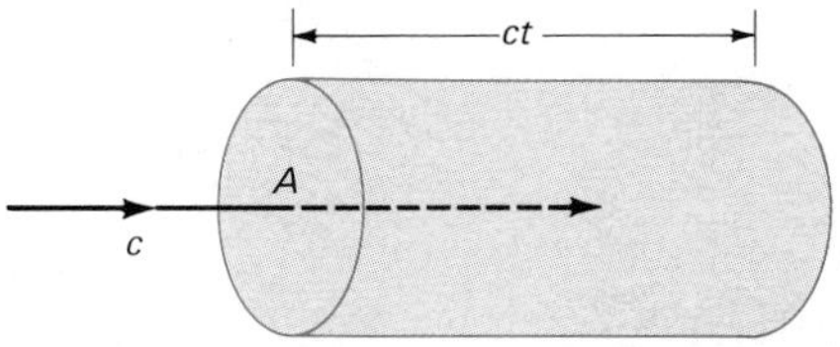

The units of intensity are joules per second per square meter (J/s·m^2) or watts per square meter (W/m^2).

Example 3 Solar Energy and the Intensity of Sunlight

The sun radiates electromagnetic energy at the rate of 3.90×10^{26} W. Its radius is $R_\odot = 6.96 \times 10^8$ m, and so its surface area is

$$A_\odot = 4\pi R_\odot^2 = 6.09 \times 10^{18} \text{ m}^2$$

The intensity of sunlight at the solar surface is

$$I_{surf} = \frac{3.90 \times 10^{26} \text{ W}}{6.09 \times 10^{18} \text{ m}^2} = 5.60 \times 10^7 \text{ W/m}^2$$

As light waves travel outward from the sun their energy is spread over larger and larger areas, causing the intensity to decrease. At a distance of 150 million km from the sun (the distance from the sun to the earth) the energy has spread over an area

$$A = 4\pi(1.50 \times 10^{11} \text{ m})^2 = 2.83 \times 10^{23} \text{ m}^2$$

and the intensity has decreased to

$$I = \frac{3.90 \times 10^{26} \text{ W}}{2.83 \times 10^{23} \text{ m}^2} = 1.38 \times 10^3 \text{ W/m}^2$$

This intensity is referred to as the *solar constant*. The intensity at the surface of the earth is lower because about 40% of the incident solar energy is reflected back into space.

Solar energy reaches each square kilometer of the earth's surface at a rate of roughly 1000 MW. Contemporary power plants deliver electric energy at approximately this same rate—1000 MW. This favorable comparison explains why solar energy is so attractive. But although solar energy is plentiful and arrives free of charge, there are enormous costs in collecting it and converting it to thermal and electric energy.

As we will see in Chapter 42, lasers are usually rated in terms of the power they radiate. When used in this sense, "power" refers to the total electromagnetic energy per second carried away by the laser beam. The small helium–neon lasers used in classroom demonstrations are rated at powers in the range from 0.1 mW to 1.0 mW (1 mW = 10^{-3} W). This power seems inconsequential when compared with that of the sun, 3.90×10^{26} W. However, the intensity of a laser beam may exceed the intensity of sunlight.

Example 4 **Laser Intensity**

A laser beam can be focused on an area approximately equal to the square of its wavelength. For a He–Ne laser, the wavelength is 632.8 nm. The area through which the energy of the beam passes is

$$A = (632.8 \times 10^{-9}\ \mathrm{m})^2 = 4.00 \times 10^{-13}\ \mathrm{m}^2$$

If the laser radiates energy at the rate of 1 mW, the intensity of the focused beam is

$$I = \frac{10^{-3}\ \mathrm{W}}{4.00 \times 10^{-13}\ \mathrm{m}^2} = 2.5 \times 10^{15}\ \mathrm{W/m^2}$$

Note that the laser intensity is many times greater than the intensity of sunlight, even though the laser power is quite small.

Intensity–Field Relationship

To see how intensity is related to the E-field and B-field, we multiply the numerator and denominator of Eq. 38.31 ($I = U/tA$) by c, the speed of light. This gives

$$I = \left(\frac{U}{ctA}\right)c = \left(\frac{U}{V}\right)c \qquad (38.32)$$

The quantity

$$ctA = V \qquad (38.33)$$

is the volume containing the energy U that has crossed the area A in time t (Figure 38.26). the ratio U/V, the electromagnetic energy per unit volume, is called the *energy density*.

$$\frac{U}{V} = \text{energy density} \equiv u \qquad (38.34)$$

The intensity can be expressed as the product of the energy density and the speed of light.

$$I = uc \qquad (38.35)$$

In an electromagnetic wave, both the E-field and the B-field contribute to the energy density, u. The E-field contribution is (Section 29.3)

$$u_E = \frac{\epsilon_o E^2}{2} \qquad (38.36)$$

The B-field contribution to the energy density is (Section 35.3)

$$u_B = \frac{B^2}{2\mu_o} \qquad (38.37)$$

The total energy density is $u_E + u_B$:

$$u = u_E + u_B = \frac{\epsilon_o E^2}{2} + \frac{B^2}{2\mu_o} \qquad (38.38)$$

Because of the wave nature of E and B, the values of E^2 and B^2 vary from point to point and from moment to moment. The effective values of E^2 and B^2 are their time averages,

Figure 38.27

Graphs of $\sin^2\theta$ (dotted) and $\cos^2\theta$ (solid) versus θ. The two graphs are identical in shape, but are shifted by $\pi/2$. It follows that the average values of $\sin^2\theta$ and $\cos^2\theta$ are equal (over any integral multiple of π).

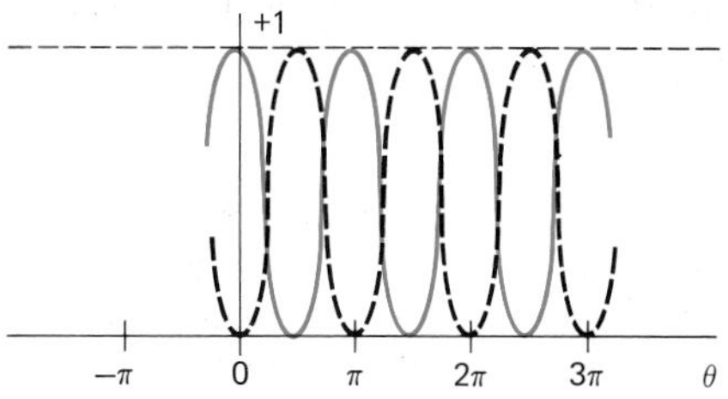

$(E^2)_{av}$ and $(B^2)_{av}$. For sinusoidal plane waves having the forms given by Eqs. 38.8 and 38.9,

$$(E^2)_{av} = E_o^2[\sin^2(kz - \omega t)]_{av} \qquad (38.39)$$

$$(B^2)_{av} = B_o^2[\sin^2(kz - \omega t)]_{av} \qquad (38.40)$$

We can argue by symmetry that $[\sin^2(kz - \omega t)]_{av}$ and $[\cos^2(kz - \omega t)]_{av}$ are equal (see Figure 38.27). Because of the identity

$$\sin^2\theta + \cos^2\theta = 1$$

the average values must both equal $\frac{1}{2}$. Thus

$$(E^2)_{av} = \frac{1}{2}E_o^2 \qquad (B^2)_{av} = \frac{1}{2}B_o^2 \qquad (38.41)$$

The average energy density for a plane electromagnetic wave is therefore

$$u_{av} = \frac{\epsilon_o E_o^2}{4} + \frac{B_o^2}{4\mu_o} \qquad (38.42)$$

We can show that the two terms on the right-hand side of Eq. 38.42 are equal. To do so, we substitute from Eq. 38.27 ($E_o = B_o c$) and Eq. 38.19 ($c = 1/\sqrt{\mu_o\epsilon_o}$) into the second term on the right-hand side of Eq. 38.42. Thus,

$$\frac{B_o^2}{4\mu_o} = \frac{(E_o/c)^2}{4\mu_o} = \left(\frac{E_o^2}{4\mu_o}\right)\mu_o\epsilon_o = \frac{\epsilon_o E_o^2}{4}$$

This result means that the E-field and B-field contribute equally to the energy density and intensity of light waves. Because of this equality the time average intensity can be expressed in terms of ϵ_o and E_o.

$$u_{av} = \frac{\epsilon_o E_o^2}{2} \qquad (38.43)$$

From Eq. 38.35 ($I = uc$) it follows that the time average intensity is

$$I_{av} = \frac{c\epsilon_o E_o^2}{2} \qquad (38.44)$$

Equation 38.44 is the desired relation between average intensity and E-field amplitude.

Example 5 **Floodlight Intensity**

A floodlight is covered with a filter that transmits red light. The E-field of the emerging beam is represented by a sinusoidal plane wave

$$E_x = 36 \sin(1.20 \times 10^7 z - 3.60 \times 10^{15} t) \text{ V/m}$$

Let's calculate the time average intensity of the beam. The amplitude of the E-field is

$$E_o = 36 \text{ V/m}$$

Using Eq. 38.44 we have

$$I_{av} = \frac{c\epsilon_o E_o^2}{2} = \frac{3.00 \times 10^8 \, 8.85 \times 10^{-12}}{2}(36)^2$$

$$= 1.72 \frac{\text{W}}{\text{m}^2}$$

Questions

4. How does a microwave oven cook a piece of food without igniting its plastic packaging? Can you think of an analogous situation involving visible light? (Imagine yourself seated indoors, next to a window, on a very cold but sunny winter day.)
5. Devise an argument that would convince an 8-year-old child that light waves transport energy.
6. Qualitatively, why does the intensity of sunlight decrease with distance form the sun? If distance from the sun is doubled, what happens to the intensity?
7. As a laser beam travels across the width of a classroom, its intensity remains essentially constant. What geometric characteristic of the beam is responsible for the constant intensity?

38.6 Polarization

In a transverse mechanical wave, the vibrations of the medium are perpendicular to the direction of wave travel. In a light wave the E-field and B-field are perpendicular to the direction of propagation; in other words, light waves are transverse waves.

All types of transverse waves can exhibit a preferred direction or alignment, which we call **polarization.** In a polarized mechanical wave the vibration of the medium can be oriented in a chosen direction. For example, waves on a string can be polarized with the vibrations vertical, horizontal, or along some other selected direction. In a mechanical wave, the plane of polarization is defined as the plane containing the direction of vibration and the direction of propagation. In Figure 38.28a the plane of polarization is vertical. In Figure 38.28b it is horizontal. Electromagnetic waves also can exhibit polarization. The plane of polarization of an electromagnetic wave is defined as the plane containing the E-field and the direction of propagation. In Figure 38.29a the plane of polarization is the x–z plane. In Figure 38.29b the plane of polarization is the y–z plane.

Figure 38.28a

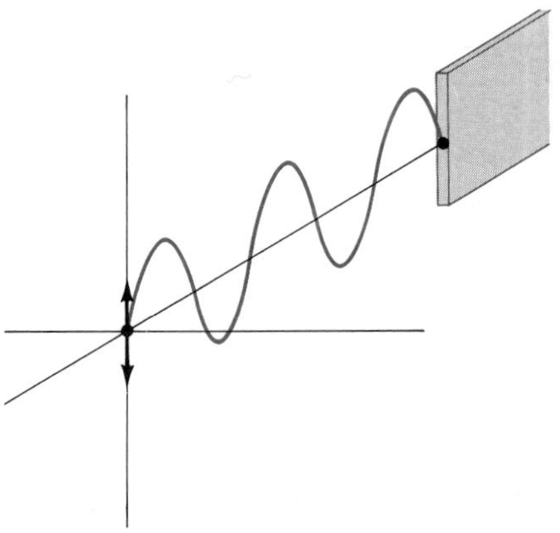

Figure 38.28b

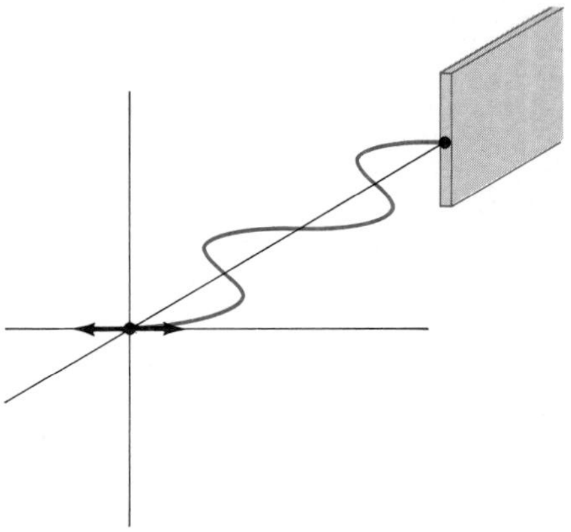

Figure 38.29a

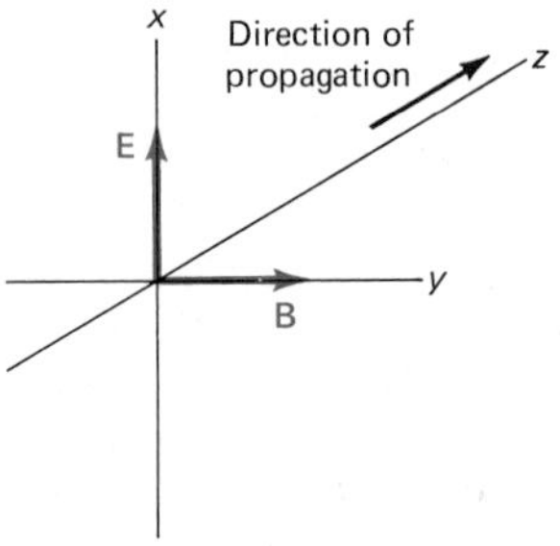

The light streaming from an ordinary light source, such as an incandescent bulb, is a superposition of the waves emitted by an enormous number of atoms and molecules. The E-fields of these many waves are randomly oriented, and as a result, the light is unpolarized (Figure 38.30).

In contrast, a conventional microwave generator produces an electromagnetic wave whose plane of polarization coincides with that of the plane of the dipole antenna (Figure 38.31). In general, microscopic sources composed of many independently radiating atoms, such as an incandescent light, produce unpolarized waves. Macroscopic sources such as radio, TV, or microwave antennas generally produce waves with a definite polarization.

Figure 38.29b

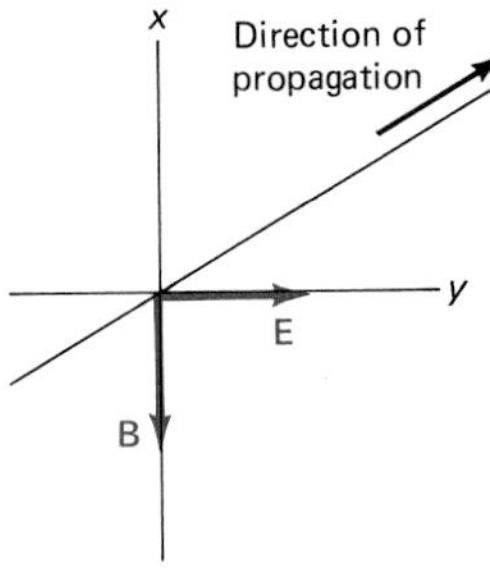

Figure 38.30

The superposition of many light waves with randomly polarized E-fields gives unpolarized light.

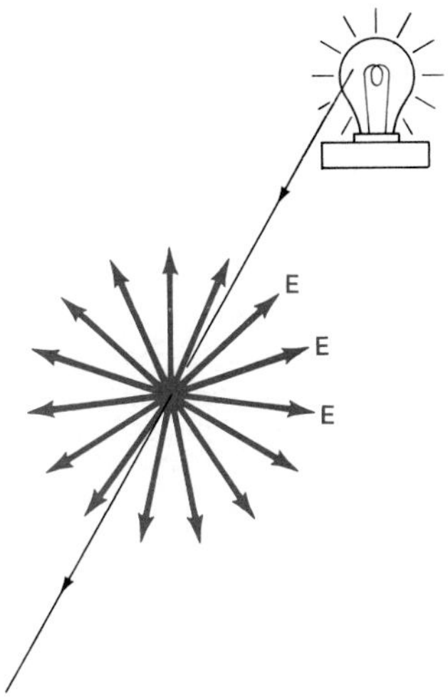

A beam of unpolarized light may be polarized, or partially polarized, in several different ways. Reflection, scattering, refraction, and absorption are all capable of affecting the polarization of light. For example, Polaroid H-sheet* converts unpolarized light by *selective absorption.* The H-sheet is formed from polyvinyl alcohol chemically treated with iodine. The long-chain molecules of the plastic are aligned by stretching the sheet during its fabrication. The aligned molecules absorb almost completely the component of the E-field that is parallel to the axis of molecular alignment. The component of the E-field perpendicular to the axis of molecular alignment is virtually unaffected by absorption. The transmission axis of the sheet is therefore perpendicular to the direction in which the sheet was stretched. A Polaroid sheet thereby acts almost like an ideal polarizer, a filter that transmits only the component of the E-field that is parallel to the characteristic transmission axis. Figure 38.32a and 38.32b illustrate the effect of a Polaroid sheet on the E-field of a beam of unpolarized light, and the way in which the E-field can be analyzed by using a second sheet of Polaroid. The first sheet (the polarizer) converts the unpolarized beam into a plane-polarized beam. If the second sheet (the *analyzer*) has its polarizing axis at right angles to the polarizing axis of the first sheet, then essentially no light is transmitted. But when the polarizer and analyzer axes are aligned (Figure 38.32b) there is an essentially* complete transmission by the analyzer.

A longitudinal wave, such as a sound wave in the air, cannot be polarized. (There is rotational symmetry about

*Edwin H. Land invented a synthetic polarizing material, which he called Polaroid J-sheet, in 1928. Ten years later he invented an improved polarizing material, Polaroid H-sheet. We generally refer to these simply as Polaroids or sheets of Polaroid. Land is perhaps better known for his invention of the Land camera.

Figure 38.31

A microwave antenna. Currents oscillating in the vertical antenna produce an electromagnetic wave with **E** polarized in the same vertical plane. The parabolic reflector gives a directional beam of microwaves.

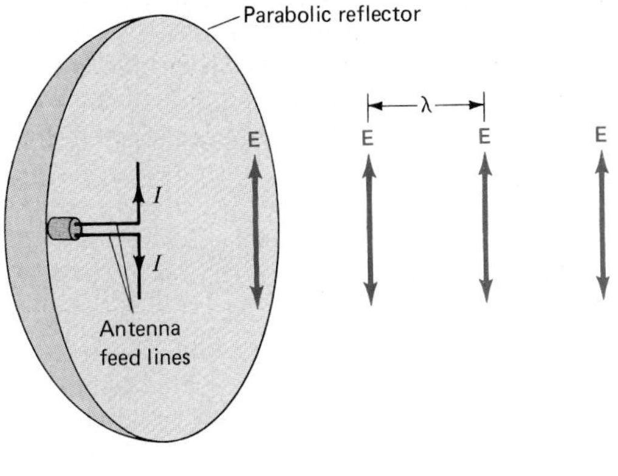

Figure 38.32

(a) With transmission axes crossed, no light passes through the analyzer. (b) With transmission axes parallel, the analyzer transmits the incident linearly polarized light.

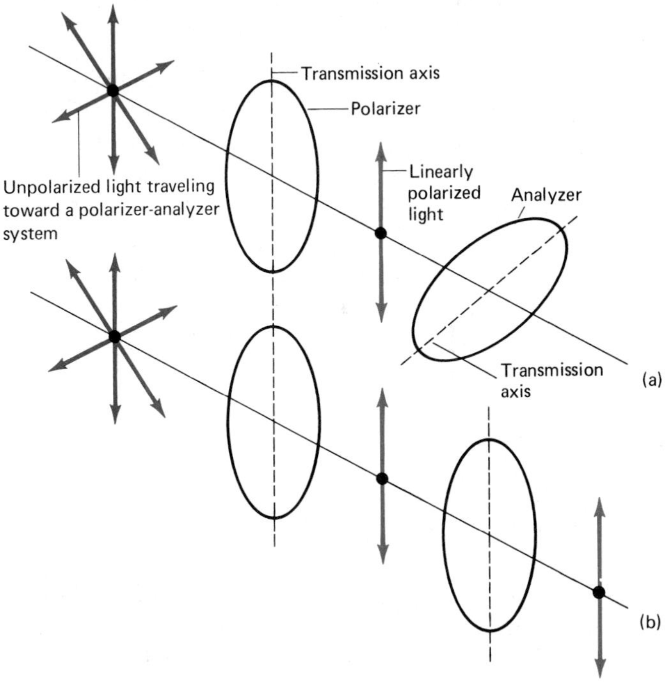

*Reflection slightly alters the intensity of the transmitted light.

the axis of propagation and no transverse direction is singled out.) Only transverse waves can be polarized. Hence, the demonstration of the polarization of light waves, which we will now describe, is a convincing proof of their transverse character.*

When unpolarized light strikes a smooth surface, the reflected light is partially polarized. Specifically, the amplitude of the E-field component that is parallel to the reflecting surface is larger than the amplitude of the component that is perpendicular to the surface. Polaroid sunglasses take advantage of this fact to reduce reflected light, or glare (Figure 38.33a and 38.33b). Most surfaces that cause glare are horizontal (e.g., automobile hoods and water surfaces). The glare is therefore composed predominantly of light polarized in a horizontal plane. The transmission axes of Polaroid sunglasses are vertical, which enables them to absorb over half of the reflected light. You can demonstrate this yourself by viewing reflected glare through one lens of Polaroid glasses. Rotating the lens through 90° will cause the intensity of this transmitted light to increase noticeably.

The vector character of the E-field is important to any quantitative discussion of polarized light. The electric field **E** of the wave can be resolved into two mutually perpendicular components (Figure 38.34). That we can resolve **E** into components means that any transverse electric field can be built up by superimposing two mutually perpendicular E-fields. Thus, when we write the equation for the field **E** of Figure 38.35,

$$\mathbf{E} = \mathbf{i}E_x + \mathbf{j}E_y$$

we are saying, "The field **E** is a superposition of a field $\mathbf{i}E_x$ polarized in the x–z plane, and a field $\mathbf{j}E_y$ polarized in the y–z plane."

*Strictly speaking, a longitudinal wave is polarized along its direction of propagation. However, the polarization of longitudinal waves cannot be controlled or altered, as it can be for transverse waves.

Figure 38.33

Two pictures showing how Polaroid™ sunglasses reduce glare. (a) A scene filmed without a Polariod filter. (b) The same scene filmed through a Polaroid filter, showing a great reduction in glare.

Figure 38.34

The E-field of an electromagnetic wave can be built up by superposing the E-fields of two waves polarized in mutually perpendicular planes.

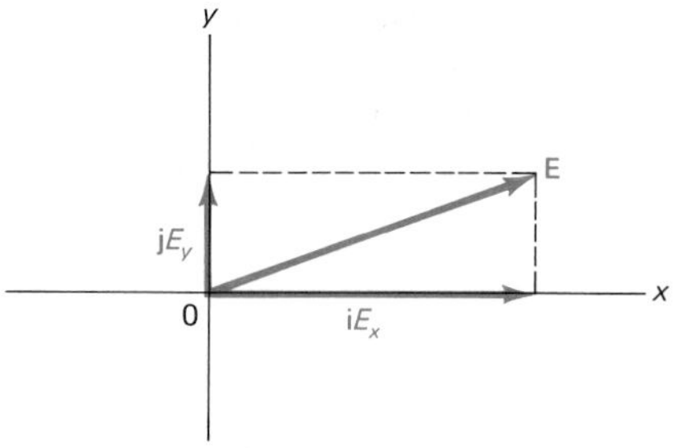

Intensity

Suppose a beam of plane-polarized light strikes a Polaroid sheet such that the angle between the transmission axis and the plane of polarization is θ. Let's determine the relationship between the angle θ and the intensities of the incident and transmitted beams. As Figure 38.35 suggests, we can resolve the incident E-field into two components. If E_0 is the amplitude of the incident E-field, $E_0 \cos\theta$ is the amplitude of the transmitted field. The intensity of the incident beam (I_0) is related to E_0 by Eq. 38.44.

$$I_0 = \frac{c\epsilon_0 E_0^2}{2}$$

The intensity of the transmitted wave (I_t) is given by Eq. 38.44, but with $E_0 \cos\theta$ in place of E_0:

$$I_t = \left(\frac{c\epsilon_0 E_0^2}{2}\right)\cos^2\theta = I_0 \cos^2\theta$$

Thus, the intensity transmitted by the Polaroid is given by

$$I_t = I_0 \cos^2\theta \qquad (38.45)$$

When $\theta = 0°, 180°$ (axes aligned), $I_t = I_0$ and we have complete transmission. When $\theta = 90°, 270°$ (axes perpendicular), $I_t = 0$ and we have complete extinction.

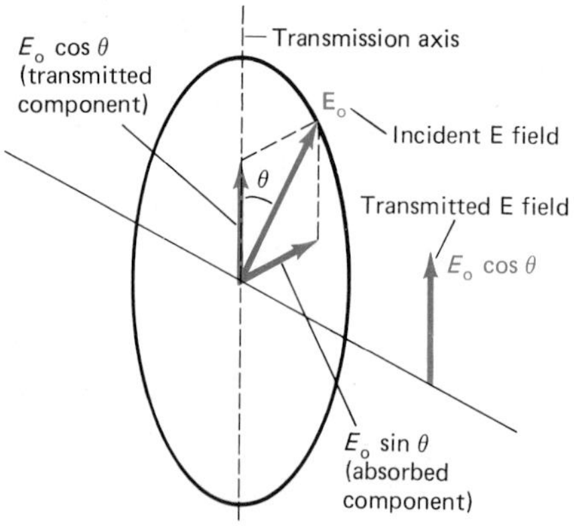

Figure 38.35

The incident E-field may be resolved into components parallel and perpendicular to the transmission axis. The parallel component ($E_0 \cos\theta$) is transmitted. The perpendicular component ($E_0 \sin\theta$) is absorbed by the Polaroid sheet.

Example 6 Polaroids Crossed at 45°

A beam of light passes through a Polaroid sheet (a polarizer). The emergent beam of plane-polarized light strikes a second Polaroid (analyzer). The transmission axis of the analyzer is oriented at 45° with respect to the plane of polarization of the incident beam (Figure 38.36a). Using Eq. 38.45, we can show that the transmitted intensity is one half that of the incident intensity. Thus, with $\theta = 45°$, Eq. 38.45 gives

$$I_t = I_o \cos^2 45° = \frac{1}{2} I_o$$

Half the incident intensity is absorbed and half is transmitted. It should be note that the transmitted beam remains plane polarized; the plane of polarization now coincides with the transmission axis of the analyzer.

If the emergent light is now sent through a second analyzer, with its transmission axis oriented at 45° with respect to the first analyzer (and 90° with respect to the transmission axis of the original polarizer), a transmitted beam will emerge (Figure 38.36b). Surprisingly, if the first analyzer is now removed, the transmitted intensity drops to zero, because we are left with two Polaroids with their transmission axes at 90° (Figure 38.36c).

Equation 38.45 can also be used to show that if a beam of unpolarized light strikes a Polaroid sheet, the intensity of the transmitted light is one half that of the incident light. To prove this statement we first recall that the incident wave is the superposition of a multitude of randomly oriented E-fields. Equation 38.45 applies to each of the component E-fields, but the angle θ ranges from 0° to 360°. Because the orientation is random, all values of θ will occur equally often. As a result, Eq. 38.45 will give the transmitted intensity if we use the average value of $\cos^2 \theta$. As noted in Section 38.5, the average value of $\cos^2 \theta$ is $\frac{1}{2}$. Replacing $\cos^2 \theta$ by $\frac{1}{2}$ in Eq. 38.45 gives us the desired result. One half of the incident intensity is transmitted.

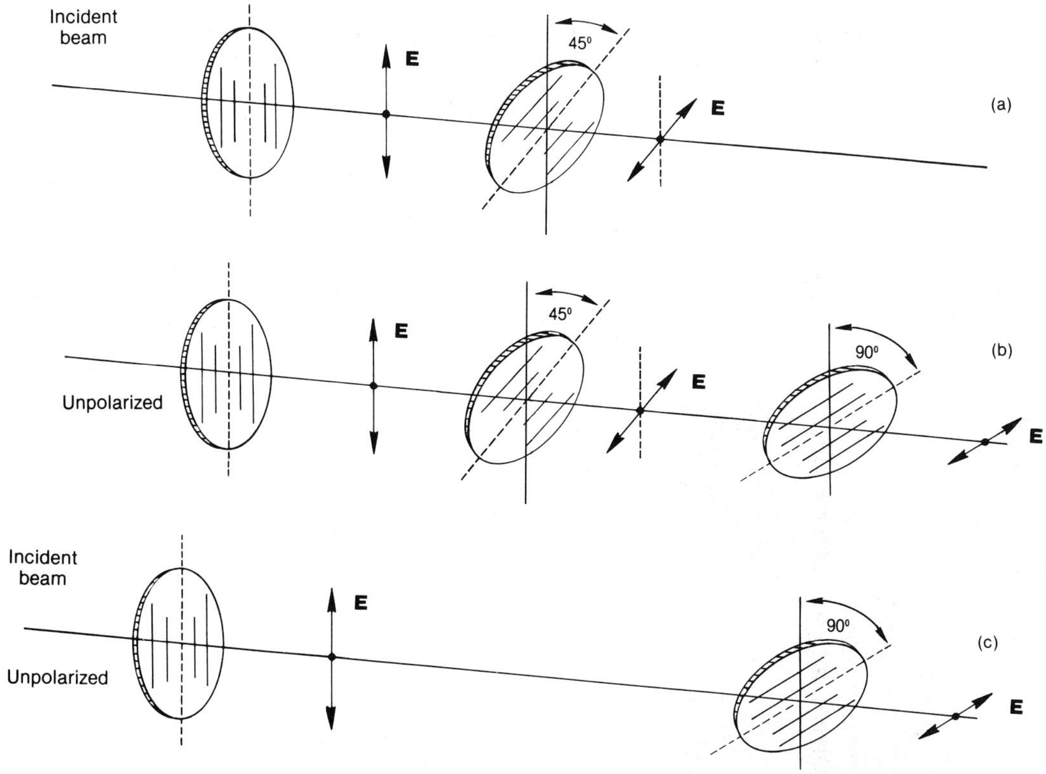

Figure 38.36

(a) Two polaroids with transmission axes inclined at 45°. The transmitted beam is plane polarized. (b) When a third polaroid is put in place with its transmission axis perpendicular to the first polaroid there is still a transmitted beam. (c) When the middle polaroid is removed we are left with two polaroids with their transmission axes perpendicular. The transmitted intensity is zero.

Example 7 A Polarization Puzzle

A beam of light is composed of a mixture of unpolarized and plane-polarized waves. The beam is analyzed with a Polaroid sheet. As the transmssion axis is rotated, the transmitted intensity varies from a maximum I_{max} to a minimum of $\frac{1}{3} I_{max}$. We want to show that the intensities of the polarized and unpolarized beams are equal.

Let I_o and I_p denote the intensities of the unpolarized (I_o) and polarized (I_p) beams that reach the analyzer. The transmitted intensity is

$$I_t = \frac{1}{2} I_o + I_p \cos^2 \theta \qquad (38.46)$$

where θ is the angle between the plane of polarization and the transmission axis. The transmitted intensity is a maximum at $\theta = 0°$ and $180°$, where $\cos^2 \theta = 1$. Thus

$$I_{max} = \frac{1}{2} I_o + I_p \qquad (38.37)$$

The intensity is a minimum at $\theta = 90°$ and $270°$, where $\cos^2 \theta = 0$. Thus

$$I_{min} = \frac{1}{2} I_o$$

Setting $I_{min} = \frac{1}{3} I_{max}$ gives

$$I_o = \frac{2}{3} I_{max}$$

Inserting this result into Eq. 38.47 shows that

$$I_p = \frac{2}{3} I_{max}$$

which establishes the desired result, $I_p = I_o$. The two intensities are equal.

Questions

8. Can sound waves in solids exhibit polarization? Explain.
9. A roadside store offers "nonglare" sunglasses at a bargain price. How might you determine whether or not the lenses are made of a polarizing material?
10. Explain qualitatively how light can partially penetrate three Polaroids placed one after the other with axes at 0°, 45°, 90°, but cannot penetrate two polaroids at 0° and 90°.

Summary

In 1864 James Clerk Maxwell developed a unified theory of electric and magnetic phenomena. A quantitative formulation of this theory is contained in Maxwell's equations:

$$\int_{\substack{\text{closed}\\\text{surface}}} \mathbf{B} \cdot d\mathbf{A} = 0 \qquad (38.11)$$

$$\int_{\substack{\text{closed}\\\text{surface}}} \mathbf{E} \cdot d\mathbf{A} = \frac{q}{\epsilon_o} \qquad (38.13)$$

$$\oint \mathbf{E} \cdot d\mathbf{s} = -\frac{d}{dt} \int \mathbf{B} \cdot d\mathbf{A} \qquad (38.12)$$

$$\oint \mathbf{B} \cdot d\mathbf{s} = \mu_o I + \mu_o \epsilon_o \frac{d}{dt} \int \mathbf{E} \cdot d\mathbf{A} \qquad (38.17)$$

The quantity $\epsilon_o(d/dt) \int \mathbf{E} \cdot d\mathbf{A}$ is Maxwell's displacement current.

Maxwell's equations predict that transverse electromagnetic waves travel through free space at a speed given by

$$c = \frac{1}{\sqrt{\mu_o \epsilon_o}} = 2.99793 \times 10^8 \text{ m/s}$$

Visible light is just one of the many types of electromagnetic waves that travel at this speed.

The intensity (I) of an electromagnetic wave is related to the energy density of the electromagnetic field (u) by

$$I = uc \qquad (38.35)$$

For a plane wave traveling through free space the average intensity is related to the amplitude of the electric field (E_o) by

$$I = \frac{c\epsilon_o E_o^2}{2} \qquad (38.44)$$

Transverse waves can be polarized; the fact that light exhibits polarization indicates that it is a transverse wave.

If plane-polarized light of intensity I_o strikes an analyzer, the intensity of the transmitted light is

$$I_t = I_o \cos^2 \theta \qquad (38.45)$$

where θ is the angle between the plane of polarization and the transmission axis of the analyzer.

Suggested Reading

E. M. Rowe and J. H. Weaver, The Uses of Synchrotron Radiation, *Sci. Am.*, 226, 32 (June 1972).

A. C. S. van Heel and C. H. F. Velzel, *What Is Light?* McGraw-Hill, New York, 1968.

V. F. Weisskopf, How Light Interacts with Matter, *Sci. Am.*, 219, 60 (September 1968).

Problems

Section 38.2 Electromagnetic Waves

1. Radio Station WLW in Cincinnati, Ohio, broadcasts at a frequency of 700 kilohertz (kHz). Radio station WMUB in Oxford, Ohio, broadcasts at a frequency of 88.5 megahertz (MHz). Calculate the wavelengths for these two stations.

2. Radio astronomers learn much about the structure of our galaxy by observing radiation emitted by hydrogen atoms at a frequency of 1420 MHz. What is the wavelength of this radiation?

3. The length characteristic of the size of atoms is 10^{-10} m. What is the frequency of an electromagnetic wave whose wavelength is 10^{-10} m?

4. (a) Red light has a wavelength of 650 nm. What is its frequency? (b) Blue light has a wavelength of 450 nm. What is its frequency?

5. The E-field of a sinusoidal plane wave traveling in a vacuum is represented by

$$E_x = 250 \sin(1.21 \times 10^7 z - 3.63 \times 10^{15} t) \text{ V/m}$$

(a) Determine the frequency of the wave. (b) Determine the wavelength. Express the wavelength in nanometers. If it corresponds to visible light, indicate its color.

6. Police radar operates in the microwave range, at a wavelength of about 6 cm. What is the frequency of a microwave whose wavelength is 6 cm?

7. What is the radian frequency of a microwave for which $\nu = 5.0 \times 10^9$ Hz?

8. From Figure 38.2 it can be seen that the direction of propagation of an electromagnetic wave is in the direction of the vector **E** × **B**. The vectors **E** and **B** reverse their directions every half cycle of the wave. Does **E** × **B** change direction when both **E** and **B** reverse direction? If **E** is directed along the positive y-axis (Figure 38.2), along what (perpendicular) axis must **B** lie for a wave traveling along the positive z-axis?

Section 38.3 Maxwell's Equations

9. Show that the ordinary current (I) and Maxwell's displacement current $[\epsilon_0 (d/dt) \int \mathbf{E} \cdot d\mathbf{A}]$ have the same dimensions. (*Possible hint:* Start with Gauss' law, Eq. 38.13.)

10. Maxwell's equations, as given by Eqs. 38.11, 38.12, 38.13, and 38.17, are in integral form: they relate integrals of **E** and **B**. An alternate formulation, in terms of *derivatives* of the E-field and B-field, is also possible. *Differential* forms of Faraday's law and Ampère's law for empty space are as follows:

$$\frac{dE_x}{dz} = -\frac{dB_y}{dt} \qquad \text{(Faraday)}$$

$$\mu_0 \epsilon_0 \frac{dE_x}{dt} = -\frac{dB_y}{dz} \qquad \text{(Ampère)}$$

Show that the plane waveforms of E_x and B_y given by Eqs. 38.8 and 38.9 satisfy these differential equations provided that

$$\left(\frac{\omega}{k}\right)^2 = \frac{1}{\mu_0 \epsilon_0}$$

Does this lead to the same expression (Eq. 38.19) for the speed, c?

Section 38.4 The Speed of Electromagnetic Waves

11. Show that the *units* of ϵ_0 and μ_0 are such that $1/\sqrt{\mu_0 \epsilon_0}$ has the units of speed.

12. Use the rectangular loop shown in Figures 38.22 and 38.23 to evaluate the integrals in Ampère's law, Eq. 38.28. Show that Eq. 38.29 results.

13. Show that $E_x = E_0 \cos(kz - \omega t)$ and $B_y = B_0 \cos(kz - \omega t)$ satisfy the integral form of Faraday's law, Eq. 38.12, provided that $E_0 = B_0 c$. (*Hint:* Use Figures 38.19, 38.20, and 38.21.)

14. Show that $E_x = E_0 \cos(kz - \omega t)$ and $B_y = B_0 \cos(kz - \omega t)$ satisfy the integral form of Ampère law, Eq. 38.17, with $I = 0$, provided that $B_0 = \mu_0 \epsilon_0 E_0 c$. (*Hint:* Use Figures 38.22 and 38.23.)

15. Show that $E_x = E_0 \cos(kz - \omega t)$ and $B_y = B_0 \cos(kz - \omega t)$ satisfy the differential forms of Faraday's law and Ampère's law (Problem 10) subject to the same condition on ω/k.

16. In a certain plane electromagnetic wave the maximum electric field is 2.1 microvolts/meter. (a) Calculate the maximum electric force (eE) on an electron. (b) Assume the electron of part (a) is moving with velocity **v** perpendicular to **B**. Taking c as an upper limit to the electron's speed, calculate the magnitude of the maximum magnetic force $[q(\mathbf{v} \times \mathbf{B})]$ on the electron.

Section 38.5 Energy Transfer via Electromagnetic Waves

17. A 10-mW laser has a beam diameter of 1.6 mm. (a) What is the intensity of the light, assuming it is uniform across the beam? (b) What is the average energy density of the laser beam?

18. Suppose that the intensity of sunlight reaching the outermost layers of our atmosphere (Example 3) were due to a monochromatic plane wave. Calculate the amplitude of the electric field of the wave.

19. The solar constant is 1.38×10^3 W/m^2. Assume that 60% of the arriving solar energy reaches the earth's surface. *Estimate* the amount of solar energy you absorb in a 60-minute sunbath, assuming you absorb 50% of the incident energy.

20. The solar constant is 1.38×10^3 W/m^2. Assume that 60% of the arriving solar energy reaches the earth's surface. Calculate the rate at which energy reaches a surface area of one square kilometer (10^6 m^2). Would all of the incident energy be absorbed? Could all of the absorbed energy be converted into electric energy?

21. Assume that the laser light in Example 4 is in the form of a plane electromagnetic wave. Determine the field strengths (E_o, B_o) for the intensity of 2.5×10^{15} W/m^2.

22. The B-field of a sinusoidal plane electromagnetic wave is represented by

$$B_y = 1.42 \times 10^{-4} \sin(1.43 \times 10^7 z - 4.29 \times 10^{15} t)$$

Determine the time average intensity of the electromagnetic wave (E-field and B-field combined).

23. A high-power laser has a power output of 1.0 MW. The laser beam is focused onto a spot of 1.0 square millimeter. Calculate (a) the intensity, I; (b) the average energy density, u_A; (c) the maximum electric field, E_o; and (d) the maximum magnetic field, B_o.

24. Pluto is approximately 39 times as far from the sun as is the earth. Use the results of Example 3 to estimate the intensity of sunlight on Pluto.

25. In the text, a symmetry argument was used to show that the average value of $\sin^2 \theta$ equals $\frac{1}{2}$. The average value of $\sin^2 \theta$ is defined by

$$(\sin^2 \theta)_{av} = \frac{\int_0^{2\pi} \sin^2 \theta \, d\theta}{\int_0^{2\pi} d\theta}$$

Evaluate both integrals and show that $(\sin^2 \theta)_{av} = \frac{1}{2}$.

Section 38.6 Polarization

26. A plane-polarized electromagnetic wave with an E-field amplitude of $E_o = 100$ V/m strikes an ideal polarizer. The incident **E** vector makes an angle of 62° with the transmission axis of the polarizer. Determine the amplitude of the transmitted E-field.

27. In an undergraduate laboratory experiment, a beam of plane-polarized light strikes a Polaroid sheet such that the angle between the transmission axis and the plane of polarization is θ. A student measures the ratio of intensities of the transmitted and incident beams and records the data shown in Table 1. By plotting the ratio of intensities versus $\cos^2 \theta$, show that the experimental results are consistent with the relation $I_t = I_o \cos^2 \theta$.

28. A horizontal beam of light is 100% polarized in the vertical direction. The beam passes through three polarizers in succession. The first polarizer's transmission axis is 30° from the vertical, the second is 60° from the vertical, and the third is 90° from the vertical. What is the intensity and what is the polarization of the beam emerging from the third polarizer?

Table 1

I_t/I_o	θ (degrees)
0.97	10
0.87	20
0.76	30
0.59	40
0.40	50
0.26	60
0.12	70
0.03	80

29. A beam of light is known to be a mixture of unpolarized light (intensity I_o) and plane-polarized light (intensity I_p). When the beam is analyzed with a Polaroid sheet, the transmitted intensity varies from a maximum I_{max} to a minimum of $0.60 I_{max}$. Show that $I_p = \frac{1}{3} I_o$.

30. A beam of unpolarized light of intensity 200 W/m^2 strikes a Polaroid (a polarizer). (a) What is the intensity of the transmitted beam? (b) The transmitted beam passes through a second polaroid. The transmission axes of the two Polaroids are oriented at 45°. What is the intensity of the beam transmitted by the second Polaroid? (c) The beam continues, passing through a third Polaroid, whose transmission axis is inclined at an angle of 45° with respect to the second Polaroid and 90° with respect to the first Polaroid. Determine the intensity transmitted by the third Polaroid. (d) The third Polaroid is rotated so that its transmission axis is at right angles (90°) with respect to the second polaroid. Determine the transmitted intensity.

chapter 39

REFLECTION, REFRACTION, AND GEOMETRIC OPTICS

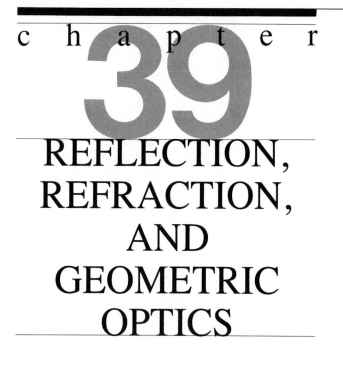

Preview

Optics is now in the midst of its third childhood. By the 17th century, optics had grown from its ancient beginnings into a well-defined field of study. René Descartes, Isaac Newton, and Christian Huygens showed that the behavior of light could be described in quantitative terms, and in their hands optics became a science. But the applications of optics were rather limited: A man could spend his lifetime grinding lenses for spectacles without knowing about the wave or particle view of light, and without contributing anything to the advancement of optical science.

James Clerk Maxwell's electromagnetic theory of light gave new life to optics. The vistas opened by Maxwell led to a flood of discoveries and applications. Electrical engineering, radio, and optical communication systems all have their origins in Maxwell's theory.

By 1950, however, optics had again stagnated; although still useful and respected, it was no longer exciting. But developments that began in the early 1960s have again rejuvenated optics. Today, optics is for the third time at one of the frontiers of science—swept there by laser beams and by the integration of solid-state physics and optical physics. New technologies have sprung up, and both theoretical and applied developments continue at an exhilarating pace. We see many daily reminders of the revolution in optics, such as liquid crystal and LED* displays in watches and calculators, and laser scanners at supermarket checkout counters. Many optical innovations also affect our daily lives even though we are not aware of them. These include remote sensing devices such as weather satellites. The versatile laser is powerful enough to weld auto bodies, yet delicate enough to perform facial surgery.

In this chapter and the two that follow, we develop the basic aspects of optics. In Chapter 42 we present a brief introduction to lasers.

We begin this chapter with a statement of Huygens' principle.† This principle is then used to develop geometric optics, which by and large is concerned with the image-forming properties of mirrors and lenses. The basic concept in geometric optics is that of a light ray—a narrow stream of radiation that moves along a well-defined path. The name *geometric optics* stems from the fact that the optics of light rays involves only geometric considerations. There are no wave equations or electromagnetic field vectors to contend with. We use geometric optics to explain the formation of rainbows, as well as to examine various optical aspects of the human eye, the magnifying glass, the microscope, and the telescope.

*The Light-Emitting Diode is a solid-state light source.

†Christian Huygens (1629–1695), a Dutch scientist, made numerous important contributions to both mechanics and optics. He invented a pendulum clock whose period is independent of the amplitude of the oscillations. He is most famous for his book *Traité de la Lumière*, published in 1690, in which he presented what is now called Huygens' principle and used it to derive the laws of reflection and refraction.

39.1 Huygens' Principle, Refraction, and Dispersion

We can understand many aspects of the propagation of light without using Maxwell's electromagnetic wave theory. We need only know that light behaves like a wave: That the waves are transverse and electromagnetic in character is of secondary importance. In these instances, we can use Huygens' principle, which describes how wave fronts* move:

Each point on a wave front acts as a source of secondary wavelets. The wave front at any subsequent moment is the envelope of the secondary wavelets.

Huygens established this principle long before Maxwell's era. However, it is possible to start with Maxwell's equations and derive Huygens' principle. Thus, Huygens' principle is a special case of Maxwell's theory and is not an independent physical law.

In Figure 39.1 we see how Huygens' principle can be used to follow a plane wave front through a *uniform* medium—a medium in which the speed of light is the same at all points. The wave front remains a plane because

*A wave front is a surface of constant phase, such as a wave crest (Section 18.4).

Figure 39.1

Huygens' principle. Each point on the wave front at time t acts as a source of secondary wavelets. The wave front at time $t + \Delta t$ is the envelope of the secondary wavelets.

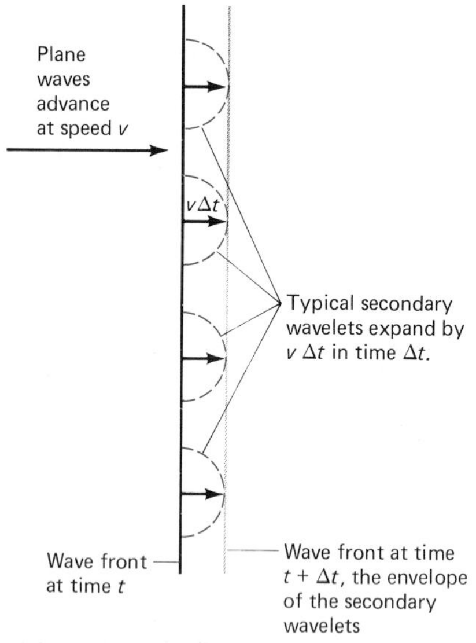

Figure 39.2

The speed of the waves at the top (v_2) is greater than at the bottom (v_1). This causes a "bending" (refraction) of the direction of wave travel. The wave fronts bend toward the region of lower speed.

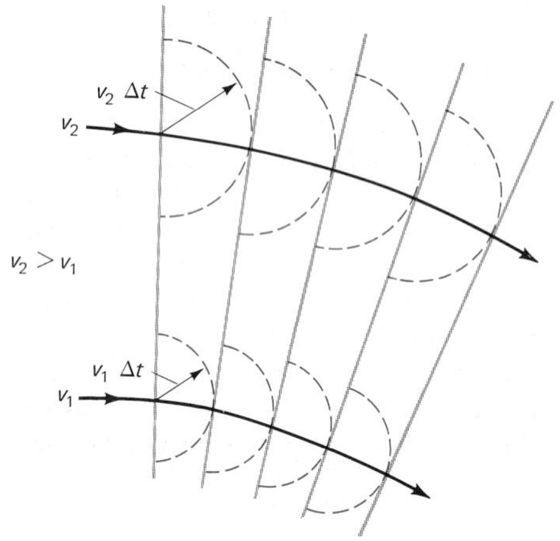

the secondary wavelets expand at the same speed at all points. In Figure 39.2 we see how Huygens' principle can be used to follow a wave through a *nonuniform* medium—one in which the wave speed is not the same at all points. The direction of the wave front changes because the secondary wavelets expand at different speeds. This causes a *bending* of the waves, called *refraction*. The refraction of light by a lens is a familiar example of such bending. Sunlight is refracted by the earth's atmosphere. This refraction is a consequence of the fact that the speed of light decreases slightly as the light travels from the near-vacuum of space toward the denser atomospheric layers near the earth's surface. Refraction in the atmosphere gives rise to many optical phenomena, including rainbows and mirages.

Index of Refraction

The speed of light in a material medium differs from the speed of light in a vacuum. The *index of refraction, n*, is the ratio of the speed of light in a vacuum (c) to the speed of light in the material medium (v),

$$n \equiv \frac{\text{speed in vacuum}}{\text{speed in medium}} = \frac{c}{v} \qquad (39.1)$$

In Table 39.1 we have listed the indices of refraction for several substances. Notice that $n > 1$ in every case. This is because the speed of light in a material medium is always less than its speed in a vacuum. The reduction in speed when light enters a material medium is a consequence of the interaction between the electromagnetic field and the atomic electrons in the material.

Qualitatively, the larger the change in the index of refraction across the boundary between two substances, the greater is the refractive bending of the light as it crosses the boundary. The quantitative relation between the indices of refraction and the refractive bending at a boundary is given by Snell's law, which we will derive in Section 39.2. Snell's law also provides a basis for measuring refractive indices.

Notice, also, that the values in Table 39.1 refer to a particular wavelength. (This wavelength (λ) is 589.3 nm,

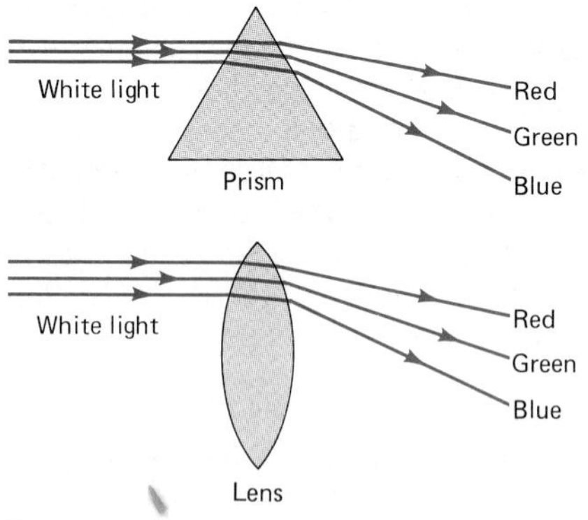

Figure 39.3

White light is dispersed when it passes through a prism or the edge of a lens. The dispersion occurs because the index of refraction varies with wavelength.

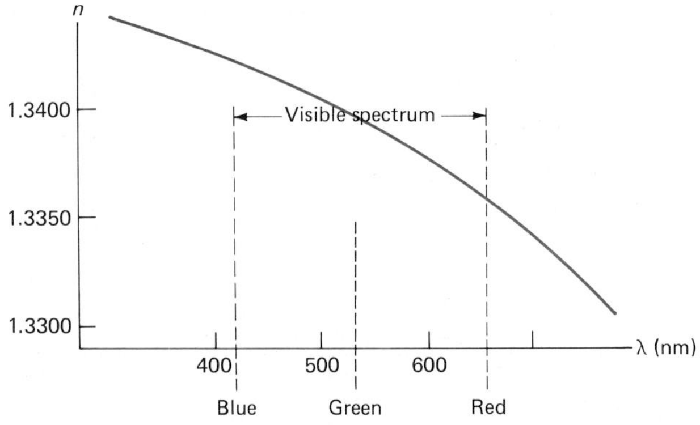

Figure 39.4

Dispersion curve for water. The index of refraction (n) decreases as the wavelength (λ) increases.

Table 39.1

Index of refraction (at λ = 589.3 nm)*

Substance	n	Substance	n
Water (ice)	1.31	Amber	1.56
Water (20°C)	1.33	Emerald	1.58
Acetone	1.36	Crown glass	1.59
Fluorite (CaF_2)	1.43	Flint glass	1.65
Glycerol	1.47	Calcite	1.66
Benzene	1.50	Ruby	1.77
Sodium chloride	1.54	Diamond	2.42

*The index of refraction can be measured more precisely than indicated here. Over the visible spectrum it is generally possible to achieve five-figure precision. Values of n vary with the temperature of the substance as well as with the wavelength of light.

the yellow light characteristic of sodium atoms.) The speed of light in a material medium varies with wavelength, and thus the index of refraction also varies with wavelength. This results in the *dispersion* of white light by a prism (Figure 39.3). The different wavelengths in white light are refracted through different angles, causing a fanning out, or a dispersal, of the beam. Dispersion in a lens produces a defect called *chromatic aberration*. Different wavelengths are focused at slightly different points. This smears the image and alters its color. A graph of the index of refraction (n) versus wavelength (λ) is called a *dispersion curve*. Figure 39.4 shows the dispersion curve for water. Nature gives us a spectacular example of the dispersion of light—the rainbow, which we will study in Section 39.3.

39.2 The Laws of Reflection and Refraction

Huygens' principle can be used to derive two basic optical laws, the law of *reflection* and the law of *refraction*. We will make use of these laws later in this chapter when we study the rainbow and the optics of mirrors and lenses.

Reflection

Consider a plane wave as it strikes a smooth surface.* In general, both a reflected wave and a transmitted wave are formed, as shown in Figure 39.5a. The angle between the surface and the incident wave fronts is called the *angle of incidence*. The angle between the surface and the reflected wave fronts is called the *angle of reflection*. In Figure 39.5b we see that the angles of incidence and reflection can also be measured relative to the normal (perpendicular) to the surface and the direction of travel of the incident and reflected waves. The **law of reflection** states:

The angles of incidence and reflection are equal.

This means that $\theta_i = \theta_r$ in Figure 39.5. We can use Huygens' principle to establish this equality. Secondary wavelets spread out from points A, B, and C on the incident wave front, and travel a distance $v\ \Delta t$ in time Δt. The envelope of these secondary wavelets forms the reflected wave front $A'B'$ and the incident wave front $B'C'$. The two right triangles $AA'B'$ and $B'BA$ are congruent (Figure 39.5a). It follows that the corresponding angles, θ_i and θ_r, are equal, proving the law of reflection. Notice also in Figure 39.5b that the normals to the incident and reflected

*A surface is considered smooth if it gives a clear reflected image.

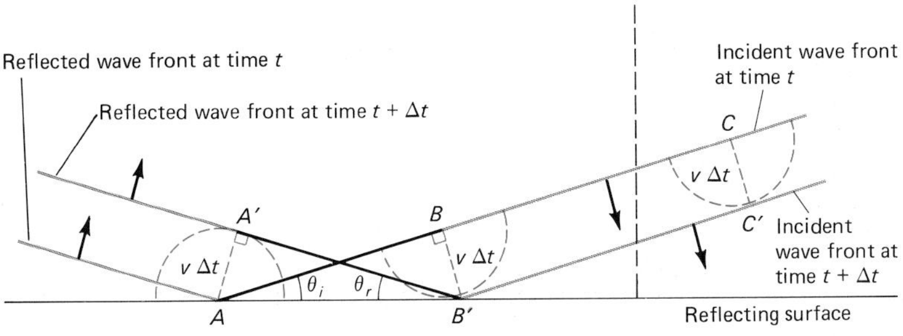

Figure 39.5a

The law of reflection states that the angle of incidence, θ_i, equals the angle of reflection, θ_r. The proof of the equality $\theta_i = \theta_r$ follows from the fact that $AA'B'$ and $B'BA$ are congruent triangles. Their corresponding angles, θ_i and θ_r, are therefore equal.

Figure 39.5b

The angles of incidence and reflection also may be referred to the angles between the normal to the surface and the normals to the incident and reflected wave fronts.

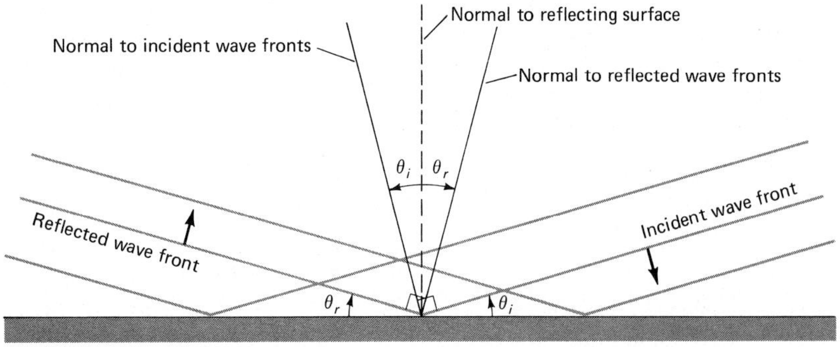

wave fronts make equal angles (θ_i, θ_r) with the normal to the surface. Experiment concurs with the prediction that $\theta_i = \theta_r$. Figure 39.6 shows the equality of θ_i and θ_r for laser light reflected at an air–water boundary.

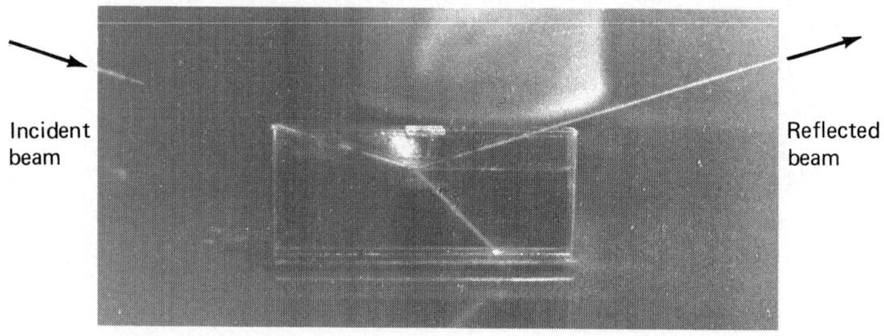

Figure 39.6

The beam of laser light strikes the surface of the water. There is both a reflected and a refracted beam.

Refraction

The refraction of wave fronts is depicted in Figure 39.7. The incident waves travel at speed v_1, cross the boundary, and continue onward at speed v_2. The angle of the incident wave, θ_1, and the angle of the refracted wave, θ_2, are related to the speeds v_1 and v_2 by the **law of refraction:**

$$\frac{v_1}{v_2} = \frac{\sin\theta_1}{\sin\theta_2} \qquad (39.2)$$

To prove the law of refraction, we can follow the secondary wavelets that spread out from points A and B of the incident wave front. The wavelets from B spread out at speed v_1, traveling a distance $v_1\,\Delta t$ in time Δt. During this same time interval, wavelets from A move at speed v_2 and travel a distance $v_2\,\Delta t$. The envelope of the wavelets in medium 2 is the refracted wave front, $A'B'$. Inspection of

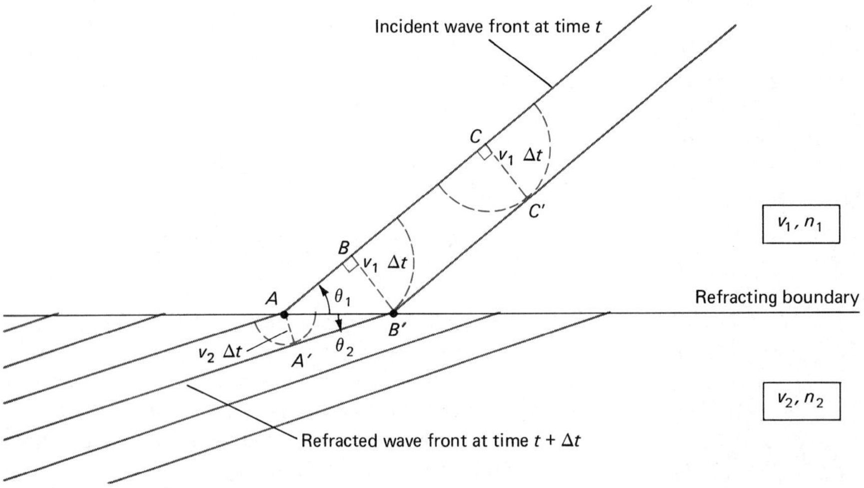

Figure 39.7

The two right triangles ABB' and $B'A'A$ have common hypotenuse, AB'. It follows that $\sin\theta_1/\sin\theta_2 = v_1/v_2$, which is one form of the law of refraction. The relative sizes of the angles θ_1 and θ_2 are appropriate for the situation $v_1 > v_2$ ($n_2 > n_1$).

the right triangle ABB' shows that

$$\frac{v_1 \Delta t}{AB'} = \sin \theta_1 \qquad (39.3)$$

From the right triangle $B'A'A$ we see that

$$\frac{v_2 \Delta t}{AB'} = \sin \theta_2 \qquad (39.4)$$

Dividing equals by equals (Eq. 39.3 and 39.4), we get Eq. 39.2, the law of refraction:

$$\frac{v_1}{v_2} = \frac{\sin \theta_1}{\sin \theta_2}$$

In practice it is more convenient to deal with the indices of refraction than with the speeds. We therefore introduce the indices of refraction, n_1 and n_2 ($n = c/v$),

$$\frac{v_1}{v_2} = \frac{v_1/c}{v_2/c} = \frac{n_2}{n_1} \qquad (39.5)$$

Inserting n_2/n_1 in place of v_1/v_2 in Eq. 39.2 and then cross multiplying, we get

$$n_1 \sin \theta_1 = n_2 \sin \theta_2 \qquad (39.6)$$

This form of the law of refraction is known as **Snell's law.*** Figure 39.6 shows the refraction of laser light across an air–water boundary. You might want to try measuring θ_1 and θ_2 to check whether Snell's law is satisfied ($n_{air} = 1.00$; $n_{water} = 1.33$). Snell's law offers a quick explanation of many optical illusions, such as the apparent bend in partially submerged objects (Figure 39.8).

*Snell is actually credited only with the small-angle form of Eq. 39.6, $n_1\theta_1 \simeq n_2\theta_2$. The small-angle form is appropriate when θ_1 and θ_2 are small enough that the approximations $\sin \theta_1 \simeq \theta_1$, $\sin \theta_2 \simeq \theta_2$ are applicable.

Figure 39.8

Refraction at the air–water boundary causes the partially submerged straw to appear bent.

Figure 39.9

Refraction at the air–water boundary changes the apparent position of the fish.

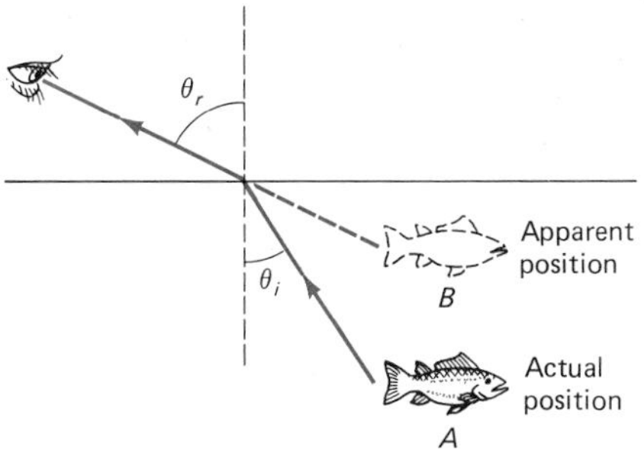

Example 1 **Refraction at an Air–Water Boundary**

Light from a fish reaches the surface of a pond at an angle $\theta_2 = 36.3°$. What is the angle of refraction in the air above? With $n_{air} = 1.00$ and $n_{water} = 1.33$, Snell's law gives

$$1.00 \sin \theta_1 = 1.33 \sin 36.3°$$

from which we find

$$\sin \theta_1 = 0.787$$

and

$$\theta_1 = 51.9°$$

As Figure 39.9 shows, light reaching the observer's eye appears to come from a position B above the actual position A of the fish. In other words, refraction at the air–water boundary changes the apparent position of the fish.

In deriving the laws of reflection and refraction, we have pictured light as a wave. But we can understand many applications of these two laws most easily if we picture light as a *ray*. In the next section, we will discuss light rays and explain how they fit into the overall picture.

39.3 Light Rays and Geometric Optics

The observation that an object can cast a sharp shadow led scientists to the notion of a *light ray*—a narrow stream of radiation that travels in a line, never diverging or converging. We can think of a broad beam of light as a bundle of parallel rays. A shadow is formed when an object reflects or absorbs some of the incident rays.

Geometric optics is the study of light in terms of light rays. Geometric optics is adequate provided the wavelength of light is much smaller than the characteristic dimensions of the particular optical system being considered. For example, geometric optics is satisfactory for

Figure 39.10

Refractions at the two parallel faces result in a "sideways" displacement (d), but do not change the ray direction.

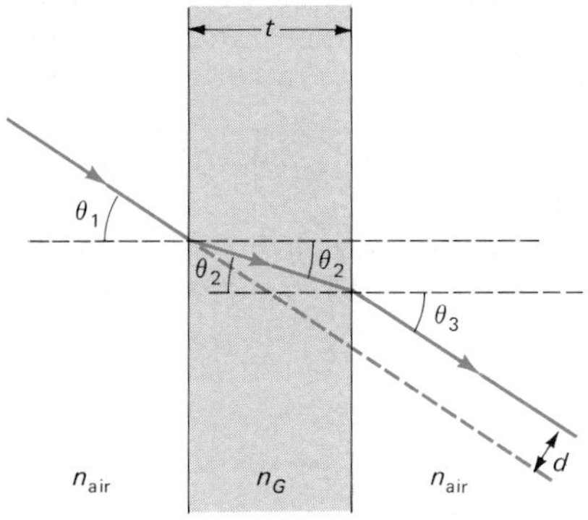

studying the refraction of light by a lens as long as the wavelength is much smaller than the diameter of the lens.

A light ray that travels through a plate glass window is refracted at both faces. The two refractions at the parallel surfaces result in a "sideways" displacement, but do not change the direction of the ray. To show that there is no change in direction, we can apply Snell's law at both faces. At the first face (Figure 39.10) Snells' law takes the form

$$n_{air} \sin\theta_1 = n_G \sin\theta_2 \quad (39.7)$$

with θ_2 being the angle of refraction. Because the faces are parallel, the angle of incidence for the second face equals θ_2, and Snells' law prescribes

$$n_{air} \sin\theta_3 = n_G \sin\theta_2 \quad (39.8)$$

By comparing Eqs. 39.7 and 39.8 we see that $\theta_1 = \theta_3$; the incident and emergent rays are parallel.

The net effect of the two refractions is a parallel displacement of the ray. The distance d between the incident and emergent rays is given by

$$d = \frac{t \cdot \sin(\theta_1 - \theta_2)}{\cos\theta_2} \quad (39.9)$$

where t is the thickness of the glass. For an ordinary windowpane the displacement d is generally not more than a few millimeters.

Example 2 Displacement of a Light Ray by a Window

A ray of light is incident at an angle $\theta_1 = 45°$ on a pane of glass that is 8 mm thick and has an index of refraction of $n_G = 1.50$. Determine the parallel displacement, d, by using Eq. 39.9. First we use Eq. 39.7 to determine θ_2. Taking $n_{air} = 1.00$, we obtain from Eq. 39.7

$$1.00 \sin 45° = 1.50 \sin\theta_2$$

which yields

$$\theta_2 = \arcsin(0.4714) = 28.0°$$

Equation 39.9 then gives

$$d = \frac{8\text{ mm} \cdot \sin(45° - 28.1°)}{\cos 28.1°} = 2.6\text{ mm}$$

Total Internal Reflection

In Figure 39.11 we see a series of light rays that travel through water and strike a water–air boundary. For some light rays the light reaching the boundary divides into a refracted ray, which passes into the air, and a reflected ray. But other light rays exhibit *total internal reflection*. All of the light of these rays is reflected, and no light escapes from the water to the air.

The condition under which light undergoes total internal reflection follows from Snell's law, Eq. 39.6.

$$n_1 \sin\theta_1 = n_2 \sin\theta_2$$

If $n_1 > n_2$, then $\theta_2 > \theta_1$. For example, when light travels from water ($n_1 = 1.33$) to air ($n_2 = 1.00$), the refracted rays are "bent" away from the normal ($\theta_2 > \theta_1$). As Figure 39.12 shows, a limit for refraction is reached when the value of $\sin\theta_2$ is $+1$, corresponding to an angle $\theta_2 = 90°$. The angle of incidence that results in $\theta_2 = 90°$ is called the *critical angle*. With $\theta_1 = \theta_{crit}$, $\theta_2 = 90°$, Snell's law prescribes

$$\sin\theta_{crit} = \frac{n_2}{n_1} \quad (39.10)$$

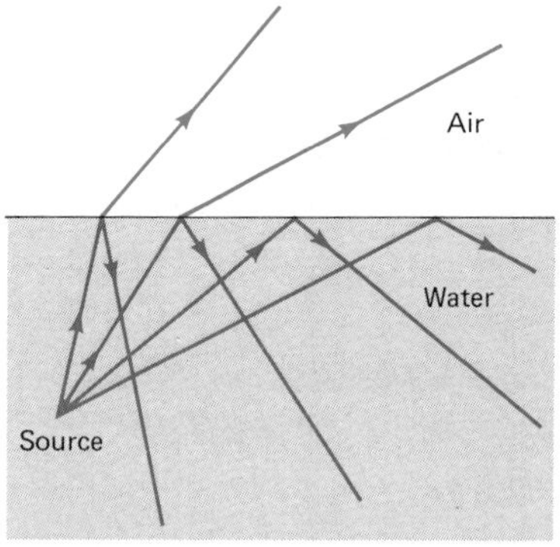

Figure 39.11

Rays from a submerged source reach the air–water boundary. For some rays there is no refracted light. Such rays exhibit total internal reflection.

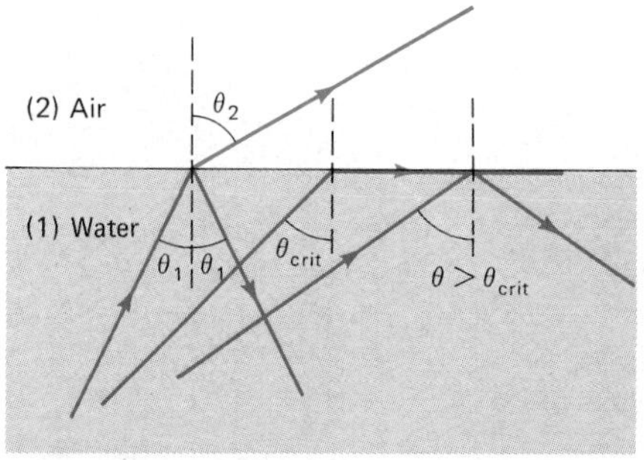

Figure 39.12

Rays originating beneath the surface that strike the water–air boundary at angles greater than the critical angle undergo total internal reflection. For the water–air boundary the critical angle is 48.6°. Rays striking the interface at angles less than the critical angle divide into refracted and reflected rays.

Figure 39.13

A light pipe. Light travels around curved sections via internal reflections. Some rays strike the surface at angles less than the critical angle and are refracted out of the pipe.

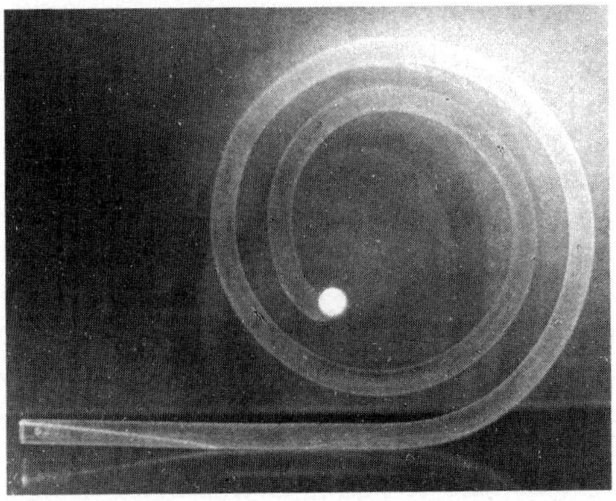

Light incident at angles equal to or exceeding θ_{crit} undergoes total internal reflection; there is no refracted ray. For a water–air boundary the critical angle follows from Eq. 39.10 as

$$\theta_{crit} = \arcsin\left(\frac{1.00}{1.33}\right) = \arcsin(0.75)$$

which gives

$$\theta_{crit} = 48.6° \qquad \textit{(water–air critical angle)}$$

Total internal reflection results in many beautiful optical effects, including the sparkle of a diamond. Most light rays entering a diamond from above undergo total internal reflection one or more times and then come back out. There are also important practical applications of total internal reflection. For example, Figure 39.13 shows a clear plastic "light pipe" through which a laser beam travels via internal reflections. Some rays strike at angles less than the critical angle and light is refracted as well as reflected. Compared to commercially available optical fibers, the pipe of Figure 39.13 is so "porous" it leaks light like a sieve. Commercial fibers have diameters ranging from a few micrometers to several hundred micrometers ($1\mu m = 10^{-6}$ m). They are very flexible and can be tied into knots with no noticeable diminution in their ability to transmit light. These fibers are therefore able to carry beams of light around corners, and are also economically small and lightweight. Fiber optics communication systems using infrared radiation are now in operation.

Rainbows

The index of refraction for a particular material varies with the wavelength, or color, of the light. This phenomenon often results in the dispersion of white light, as in a prism. The rainbow is nature's most beautiful illustration of dispersion. We can understand the formation of a rainbow by applying the laws of reflection and refraction to a spherical raindrop. Figure 39.14 shows the two refractions and one reflection in the raindrop that divert a ray of light from the sun into an observer's eye. The angle of deviation, D, is the total angle through which a ray is turned by refraction and reflection. Notice that in Figure 39.14 *ABC* and *BCE* are isosceles triangles. The deviation of the ray via refraction at A is $\theta_1 - \theta_2$. At B the reflected ray is deviated by $180° - 2\theta_2$. The ray refracted at E is deviated through the angle $\theta_1 - \theta_2$. The total deviation is the sum of these three angles

$$D = 2(\theta_1 - \theta_2) + (180° - 2\theta_2) \qquad (39.11)$$

For a specified angle of incidence (θ_1), we can use Snell's law to determine θ_2. The dispersive action of the raindrop enters through the index of refraction, n. For violet light, $n = 1.3435$. For red light, $n = 1.3318$. For a given value of θ_1, these slightly different values of n result in slightly

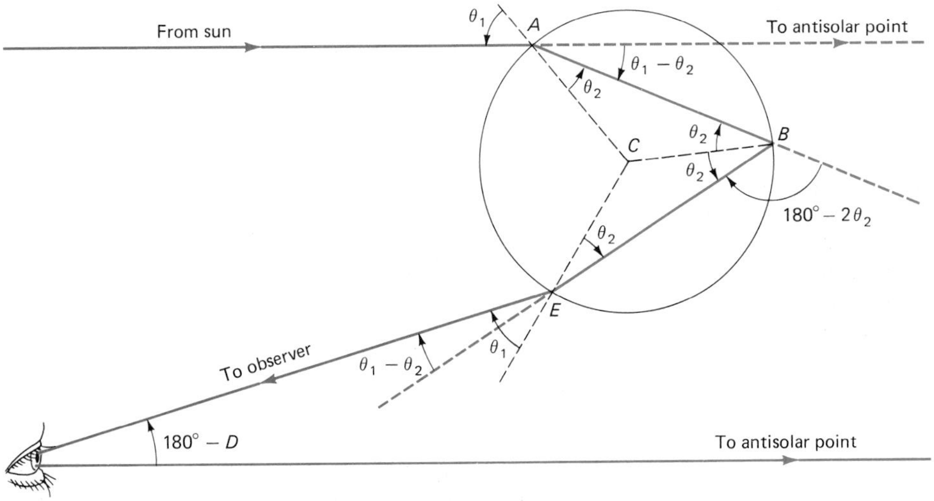

Figure 39.14

Two refractions and one internal reflection deviate the ray through the angle $D = 2(\theta_1 - \theta_2) + (180° - 2\theta_2)$. Light reaches the observer's eye at the angle $180° - D$, measured with respect to the direction of the antisolar point.

different values of D. Different colors emerge at slightly different angles. An observer viewing the refracted light (Figure 39.15) receives different colors from slightly different directions, and thus sees a rainbow.

Light reaches the observer's eye at the angle $180° - D$, measured with respect to the direction of the antisolar point the direction of the incoming sunlight; see Figure 39.16. The locus of points for which $180° - D$ has a particular value is a cone with its apex at the observer's eye. Raindrops that refract a particular color to the observer lie on the portion of the cone that is above the horizon. Light reaching the eye appears to come from points on the arc of a circle—a bow. The lower edge of the bow is violet. It appears at an angle of 40.6° above the antisolar direction (Figure 39.16). The upper edge of the bow is red and it appears at an angle of 42.2°. Thus the angular width of the rainbow is approximately 2°.

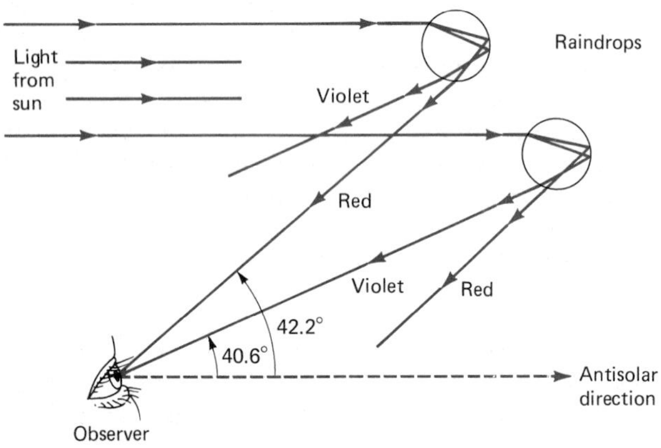

Figure 39.15

Light refracted by the raindrops reaches the observer's eye, producing a rainbow. Red light forms the upper edge of the bow, 42.2° above the antisolar direction. Violet forms the lower edge, 40.6° above the antisolar direction.

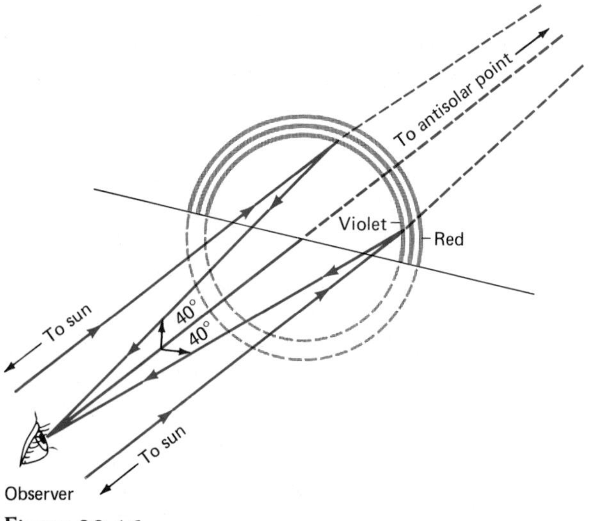

Figure 39.16

A rainbow is formed by rays that reach the observer at angles ranging from approximately 40° (violet) to 42° (red), measured relative to the antisolar direction.

Figure 39.17

A secondary bow is formed by rays that undergo two internal reflections.

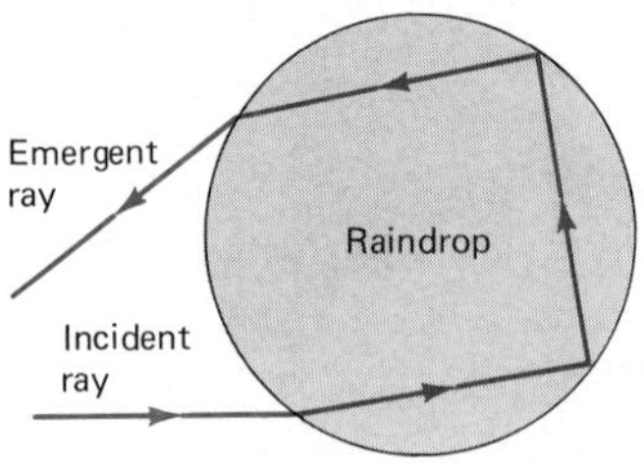

The rainbow we ordinarily view is but one of a series of bows. A secondary bow is formed by rays that undergo two internal reflections, as suggested by Figure 39.17. The secondary bow appears higher in the sky than the primary bow (over angles ranging from 50.7° to 53.6°). A tertiary bow is produced by rays that undergo three internal reflections. Tertiary bows are seldom seen because the observer must face the sun to view them. The normal atmospheric scattering of sunlight ordinarily outshines the triply reflected light of the bow. Internal reflections that occur at angles less than the critical angle are accompanied by refraction. The refractions reduce the intensity of the light that eventually emerges to produce a particular bow.

Questions

1. Suppose that you are under water near the edge of a swimming pool. Does someone standing above water at poolside look shorter or taller to you now than when viewed from the same position when the pool is empty?
2. Fountains are often lit with underwater lights that illuminate the full length of many vertical streams of water. How is this achieved?

39.4 Mirrors

Mirrors and lenses gather and redirect light rays, thereby forming an *image* of some *object*. Image formation by mirrors involves only the law of reflection; the angles of incidence and reflection are equal.

Plane Mirror

The way in which a *plane* mirror forms an image is depicted in Figure 39.18. The full image can be inferred by locating the images of few key points on the object. The object in Figure 39.18 is an upright arrow. Consider two rays leaving the point P at the tip of the arrow. Diverging rays from P strike the mirror and are reflected to the eye of an

Figure 39.18

The upright arrow a distance s in front of the mirror is the object whose image appears behind the mirror. Two rays leave the point P at the tip of the object. These rays are reflected and enter the observer's eye. To the eye the rays appear to come from the point P'. The point P' is the image of the point P. For a plane mirror the image appears to be as far behind the mirror as the object is in front of the mirror.

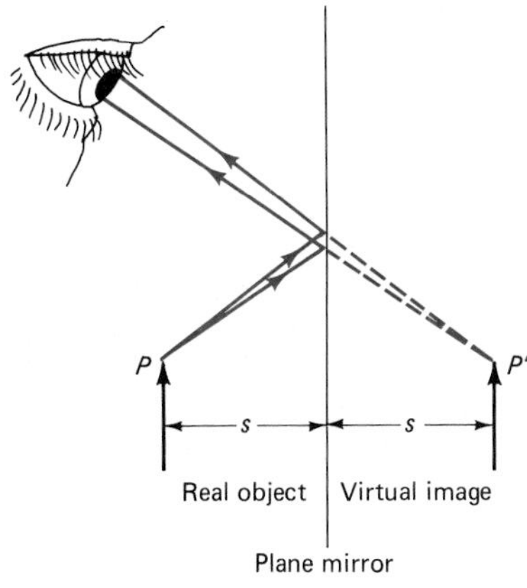

Figure 39.19

A triple mirror in a clothing store lets you see the back of your head by using three mirrors.

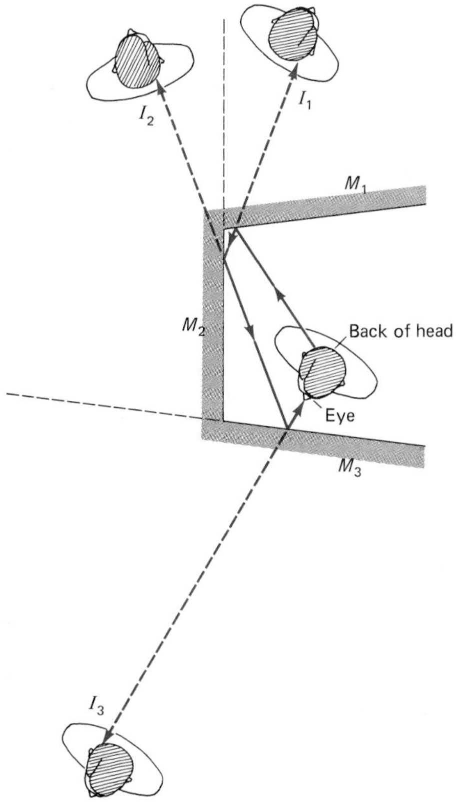

observer. The rays appear to diverge from P'. The *image* of P is the point from which the reflected rays appear to come. Thus, the point P' in Figure 39.18 is the image of the point P. The image is opposite the object and as far behind the mirror as the object is in front. The image P' is called a *virtual* image because no light energy actually passes through that point.

The image formed by a plane mirror is *upright* and *reversed*. The reversal of right and left by a mirror is the reason ambulances often have "ƎƆИA⅃UᗺMA" painted across their front. When seen in a rearview mirror, the word "AMBULANCE" is immediately evident.

Clothing stores often provide triple mirrors that allow you to view your side and back while looking straight ahead. Figure 39.19 shows the path of a light ray as it travels from the back of the head to the eye. Virtual images (I_1, I_2, I_3) are formed at each of the three reflections.

Spherical Mirrors

Let's consider the image-forming action of spherical mirrors, which are formed from segments of a spherical surface and usually have circular edges. Consequently, spherical mirrors have an axis of symmetry, and we can conveniently analyze the image formed by such a mirror by tracing rays in any plane containing the symmetry axis. In many situations, most light rays travel nearly parallel to the spherical mirror axis. A ray is said to be *paraxial* if the angle ϕ between the ray and the symmetry axis is small enough to allow us to use the small-angle approximation (Appendix 4). Provided the angle ϕ is expressed in radians, the small-angle approximation for $\sin\phi$ is

$$\sin\phi \simeq \phi$$

As we will now show, most paraxial rays originating from a given point on an object are focused at a common image point. We will restrict ourselves to paraxial rays in our study of mirrors and lenses.

Point Object

Let's consider first the image of a point object. In Figure 39.20 we illustrate how the image of a point object is located for a concave spherical mirror. A ray from P is reflected at Q, the incident and reflected rays making equal angles (θ) with the normal CQ (C denotes the center of curvature for the spherical mirrror surface). A second ray travels from P along the mirror axis to A and is reflected. The two reflected rays intersect at P', forming a *real* image

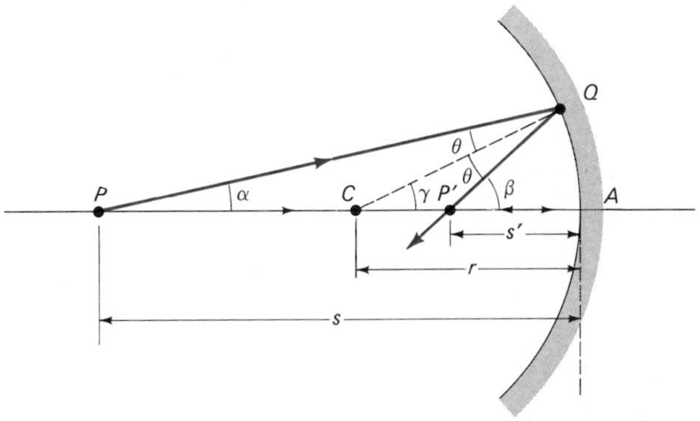

Figure 39.20

Paraxial rays from a point object at P are focused by a concave mirror to give a real image at P'.

of P. An image is termed *real* if light energy passes through the image position. (In contrast, no light energy passes through the position of a virtual image.)

Not all paraxial rays are reflected through the same point, and therefore the image of a point is not a point. In other words, the image is not sharp. This lack of image sharpness caused by the spherical shape of the mirror surface is called spherical aberration. In reflecting telescopes, spherical aberration is avoided by using a mirror with a paraboloidal surface. The reflecting surfaces for automobile headlamps are paraboloids. With the lamp at the focus of the paraboloid, the rays travel outward as a nearly parallel beam.

The basic equation governing spherical mirrors relates three lengths, each measured from the mirror surface along the axis. These are the object distance s, the image distance s', and the radius of curvature of the mirror, r. In Figure 39.20 we see how two paraxial rays, PQP' and PAP' intersect at P' to form an image of P. We can use the geometric proposition that states that the sum of the opposite interior angles of a triangle equals the exterior angle to derive the *spherical mirror equation*. Figure 39.21 shows the triangle PQC of Figure 39.20. The three angles γ, α, and θ are related by

$$\gamma = \alpha + \theta \qquad (39.12)$$

Figure 39.22 shows the triangle PQP' of Figure 39.20. The three angles β, α, and 2θ are related by

$$\beta = \alpha + 2\theta \qquad (39.13)$$

Eliminating the angle θ from this pair of equations gives

$$\alpha + \beta = 2\gamma \qquad (39.14)$$

The angle γ is measured from C and is related to the radius of curvature r and the arc length $\overline{QA}$ by

$$\gamma = \frac{\overline{QA}}{r} \qquad (39.15)$$

When the angles α and β are small (as they must be for paraxial rays) the length $\overline{QB}$ is very nearly equal to the length $\overline{QA}$. Thus

$$\tan\alpha = \frac{\overline{QB}}{s} \simeq \frac{\overline{QA}}{s}$$

and

$$\tan\beta = \frac{\overline{QB}}{s'} \simeq \frac{\overline{QA}}{s'}$$

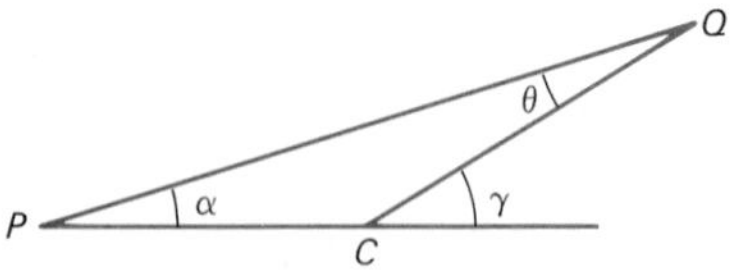

Figure 39.21

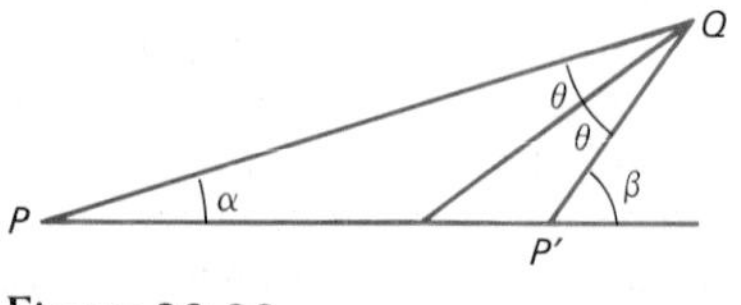

Figure 39.22

Further, by using the small-angle approximation for the tangent (Appendix 4)

$$\tan\alpha \simeq \alpha \qquad \tan\beta \simeq \beta$$

we can write

$$\alpha \simeq \frac{\overline{QA}}{s} \qquad \beta \simeq \frac{\overline{QA}}{s'} \qquad (39.16)$$

Inserting these expressions for α, β, and γ into Eq. 39.14 and then canceling the common factor $\overline{QA}$ gives the *spherical mirror equation:*

$$\frac{1}{s} + \frac{1}{s'} = \frac{2}{r} \qquad (39.17)$$

That the angles α, β, γ, and θ do not appear in the mirror equation shows that all (paraxial) rays from a given object point form an image at the same point. (Remember that Eq. 39.17 is an approximation that is adequate for paraxial rays.)

Incident rays traveling parallel to the mirror axis converge at a common point, called the *focal point*. The *focal length*, f, is defined as the image distance when the object is at infinity ($s \to \infty$ corresponds to incident rays traveling parallel to the mirror axis). For a *concave mirror,*

$$\frac{1}{\infty} + \frac{1}{f} = \frac{2}{r} \qquad (39.18)$$

which shows that the focal length of a concave spherical mirror is half the radius of curvature,

$$f = \frac{1}{2}r \qquad (39.19)$$

We can rewrite the mirror equation in terms of f as

$$\frac{1}{s} + \frac{1}{s'} = \frac{1}{f} \qquad (39.20)$$

As a reminder of the significance of each symbol we re-express Eq. 39.20 as

$$\frac{1}{\text{object distance}} + \frac{1}{\text{image distance}} = \frac{1}{\text{focal length}} \qquad (39.21)$$

Extended Object

So far we have considered only a point object. An extended object can be treated as a set of point objects. Each point on an object that faces a mirror gives rise to an image, and collectively the image points form an extended image. Equation 39.21 applies for each point on the object. If the object distance is essentially the same for all object points, then Eq. 39.21 shows that the image distance is the same for all image points. Stated somewhat differently, if the object points lie in a plane perpendicular to the symmetry axis, then the image points also lie in a plane perpendicular to the symmetry axis. In this case we say that the object plane is being "mapped" onto the image plane.

Figure 39.23

The candle is at P. The image of the candle at P' is real and inverted.

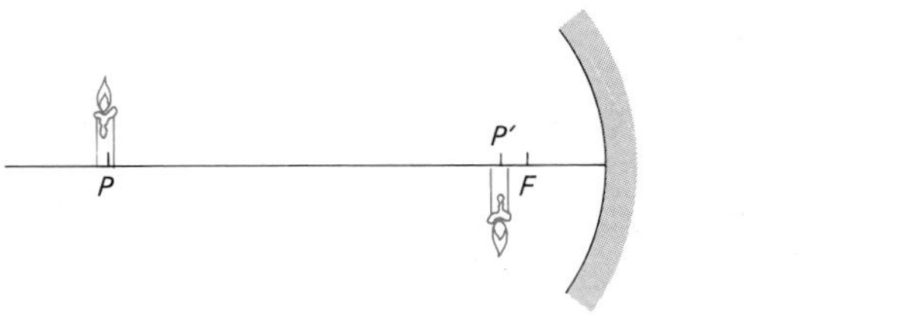

Example 3 **Locating the Image of a Candle**

A concave mirror with a radius of curvature of 1 m is illuminated by a candle located on the symmetry axis 3 m from the mirror (Figure 39.23). Locate the image of the candle.

From Eq. 39.19, the focal length is

$$f = \frac{1}{2}r = 0.50 \text{ m}$$

Equation 39.20 gives

$$\frac{1}{s'} = \frac{1}{f} - \frac{1}{s} = \frac{1}{0.50} - \frac{1}{3} = \frac{5}{3} \text{ m}^{-1}$$

showing that $s' = \frac{3}{5}$ m $= 60$ cm is the mirror-to-image distance. A real image is formed 60 cm in front of the mirror. The image is also inverted.

Optical Reversibility

The *symmetric* way in which s and s' enter the mirror equation is noteworthy:

$$\frac{1}{s} + \frac{1}{s'} = \frac{1}{f}$$

This symmetry illustrates the *principle of optical reversibility,* which states that if any ray is reversed, it will retrace its path through an optical system. Mathematically, we can interchange s and s' and still satisfy the mirror equation. Physically, this interchange means that if the candle in Figure 39.23 is placed at P', its image will be formed at P.

Image Construction

The image of an extended object can be located by following two of the three so-called "easy" rays shown in Figure 39.24. You should try to learn the routes followed by these rays.

① *A ray traveling parallel to the mirror axis.* This ray appears to come from infinity and is therefore reflected through the focal point.

② *A ray passing through the focal point.* This ray is reflected and travels away parallel to the mirror axis.

③ *A ray passing through the center of curvature.* This ray strikes the spherical mirror along a normal and is reflected back along the same line.

Figure 39.24

The three "easy" rays. Any two suffice to locate the image. C is the center of curvature; F is the focal point. ① is the incident ray parallel to the axis that is reflected through F. ② is the incident ray through F that is reflected parallel to the axis. ③ represents incident and reflected rays through C. With the object distance (s) greater than the focal length (f), the image distance (s') is positive, which corresponds to a real image at P'. The image is real and inverted.

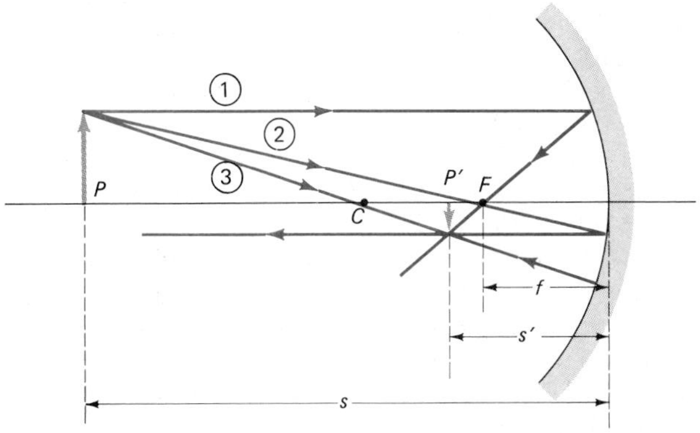

Figure 39.25

When the object distance (s) is less than the focal length (f), a concave mirror forms an upright virtual image. The virtual image corresponds to a negative value of the image distance (s') in the mirror equation, Eq. 39.20.

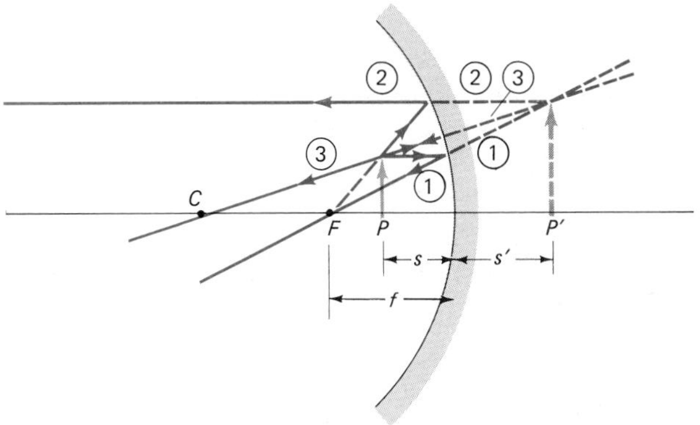

The three easy rays in Figure 39.24 intersect at a common point—the image point of the tip of the arrow. All three rays are shown, even though the intersection of any two is all that is needed to locate the image.

Notice in Figure 39.24 that $s > f$. In other words, the object is *outside* the focal point of the mirror. It follows from the mirror equation, Eq. 39.20,

$$\frac{1}{s} + \frac{1}{s'} = \frac{1}{f}$$

that s' is positive when $s > f$. A positive value of s' means that the image is *real.* As we can see in Figure 39.24, the image is also *inverted.* Figure 39.25 shows a different situation. Now the object is between the focal point and the mirror surface. Mathematically this means that $s < f$, and the mirror equation is satisfied by a negative value of the image distance s'. The negative image distance corresponds to a *virtual image.* As indicated in Figure 39.25, the easy rays show that the virtual image is *upright* and *enlarged*. Easy ray (1) travels toward the mirror parallel to the mirror axis. It is reflected and passes through the focal point. It appears to come from a point behind the mirror surface. Easy ray (2) travels toward the mirror along a line passing through the focal point, F. It is reflected, and then travels away parallel to the mirror axis, apeparing to come from a point behind the mirror. Easy ray (3) strikes the mirror at right angles and is reflected back along its incident direction, passing through the center of curvature, C. Like rays (1) and (2), ray (3) appears to come from a point behind the mirror. The intersection of the three easy rays defines the image of the tip of the arrow. All three easy rays are shown in Figure 39.25, although the intersection of any two is enough to locate an image point. Concave mirrors with long focal lengths magnify the image and are sold as vanity mirrors—the type you might use when shaving whiskers or plucking eyebrows.

Convex mirrors are also properly described by the mirror equation (Eq. 39.20 or 39.21) provided that both s' and r (and $f = \frac{1}{2}r$) are assigned negative values. Figure 39.26 shows the easy rays for a convex mirror. Easy ray (1) travels parallel to the mirror axis and is reflected along a line whose extension passes through the focal point, F. Easy ray (2) travels toward the focal point. After reflection it travels away parallel to the mirror axis. Easy ray (3) travels toward the center of curvature, C, and is reflected back along itself. All three easy rays are shown in Figure 39.26, although again the intersection of any two is sufficient to locate the image and show that it is *virtual* ($s' < 0$) and *upright*. Convex mirrors are often used as rearview mirrors, particularly by truck drivers and other drivers towing wide loads, because they allow the driver to view a wide area.

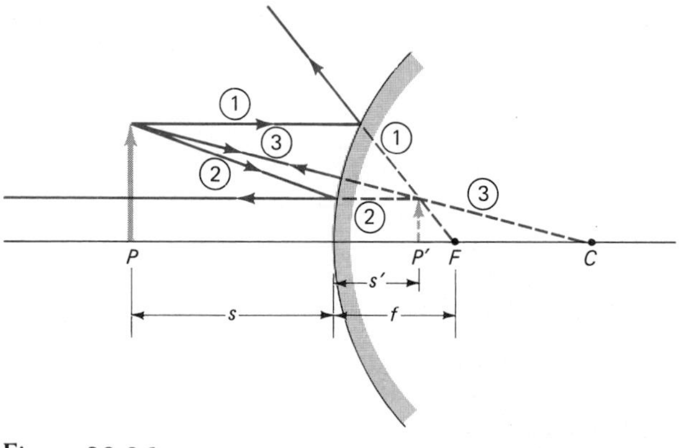

Figure 39.26

The easy rays show that a convex mirror always produces an upright virtual image. The convex mirror is described by the mirror equation,

$$\frac{1}{s} + \frac{1}{s'} = \frac{1}{f},$$

provided s' and f are assigned negative values. The limiting case of a plane mirror corresponds to $f = \frac{1}{2}r \to -\infty$, and gives the result $s = -s'$.

We note one final feature of the mirror equation in the form

$$\frac{1}{s} + \frac{1}{s'} = \frac{2}{r} \qquad (39.17)$$

As the radius of curvature r tends toward infinity, the mirror surface tends toward a *plane*. Thus, the mirror equation for a plane mirror is

$$\frac{1}{s} + \frac{1}{s'} = \frac{1}{\infty} = 0$$

which gives

$$s' = -s$$

This result verifies a conclusion about plane mirrors that we arrived at early in this section: The image of a plane mirror is virtual and is located as far "behind" the mirror surface as the object is in front of it.

Questions

3. Can a virtual image be projected onto a viewing screen? Can a real image?
4. No light energy passes through the position of a virtual image. Could you photograph a virtual image?
5. How many virtual images of the object O will be visible in the right-angle mirror of Figure 39.27? Could the observer position his eye in any way so as to see more virtual images?
6. Many trucks are equipped with dual rearview mirrors: one a plane, and the other convex. What special function does each mirror serve that the other cannot?
7. An object is placed at the focal point of a concave mirror. Where is the image formed?
8. A woman 5 ft 6 in. tall wants to view herself head to toe in a plane mirror. What is the minimum required vertical length of the mirror? Must the mirror be vertically positioned in any special way?
9. Look at your reflected images from both sides of a spoon. Are the images real or virtual?

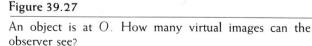

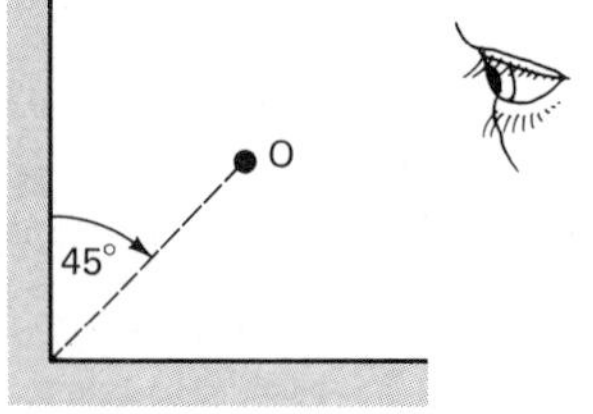

Figure 39.27

An object is at O. How many virtual images can the observer see?

39.5 Lenses

A lens forms an image by refraction. A real or virtual image is formed, depending on the type of lens and on the object-to-lens distance. In this section, we will first derive the equation that governs paraxial rays refracted at a convex spherical surface. We will then illustrate this equation by showing how real and virtual images can be formed. We will do the same for the refraction at a concave spherical surface. Finally, we will combine the results to derive and illustrate the thin lens equation. In this section, the words *convex* and *concave* describe the *surface of a lens,* as viewed from a position *outside* that lens surface. (Figure 39.28 shows three different combinations of lens surfaces.)

Figure 39.29 shows the formation of a real image of a point object by refraction at a convex spherical surface. We want to obtain an equation relating to the object and image distances (s, s'') and the radius of curvature of the refracting surface (r_A). We can obtain this equation by using the same method used in Section 39.4 to derive the mirror equation, but in place of the law of reflection, we use Snell's law to relate the angles θ_1 and θ_2.

$$n_1 \sin\theta_1 = n_2 \sin\theta_2 \qquad (39.6)$$

For paraxial rays $\sin\theta_1 \simeq \theta_1$, $\sin\theta_2 \simeq \theta_2$, and Snell's law becomes

$$n_1\theta_1 = n_2\theta_2 \qquad (39.22)$$

From triangle PQC in Figure 39.29,

$$\alpha + \gamma = \theta_1 \qquad (39.23)$$

From triangle $QP''C$ in Figure 39.29,

$$\theta_2 + \beta = \gamma \qquad (39.24)$$

We can multiply Eq. 39.23 by n_1 and Eq. 39.24 by n_2, and then use $n_1\theta_1 = n_2\theta_2$ to eliminate θ_1 and θ_2. The result is

$$n_1\alpha + n_2\beta = (n_2 - n_1)\gamma \qquad (39.25)$$

Figure 39.28

Three types of lenses, classified according to the shapes of their surfaces.

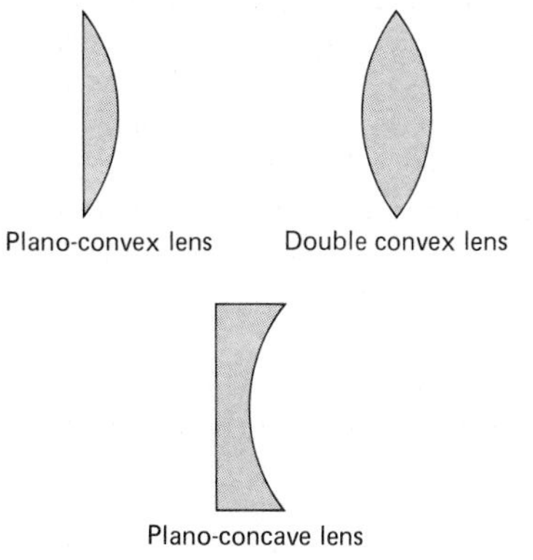

For paraxial rays,

$$\alpha \simeq \frac{\overline{QA}}{s} \qquad \beta \simeq \frac{\overline{QA}}{s''} \qquad \gamma = \frac{\overline{QA}}{r_A} \qquad (39.26)$$

Inserting these expressions for α, β, and γ into Eq. 39.25 and canceling the common factor $\overline{QA}$, we get

$$\frac{n_1}{s} + \frac{n_2}{s''} = (n_2 - n_1)\frac{1}{r_A} \qquad (39.27)$$

For $n_2 > n_1$, the right-hand side of Eq. 39.27 is positive. That the angles α, β, and γ do not appear in Eq. 39.27 means that all paraxial rays for a given s meet to form an image at the same s''. We can also determine the nature of the image from Eq. 39.27. When the object distance s is large enough, the term n_1/s is less than the right-hand side, $(n_2 - n_1)/r_A$. In this case the term n_2/s'' is positive, which means that the image distance s'' is positive. A *positive* value of s'' indicates that a *real* image is formed in the region to the right of the interface (Figure 39.29).

A different type of image is formed when the object distance s is small enough. If s is small enough, n_1/s is greater than $(n_2 - n_1)/r_A$ and Eq. 39.27 is satisfied by a *negative* value of the image distance s''. A *negative* value of s''

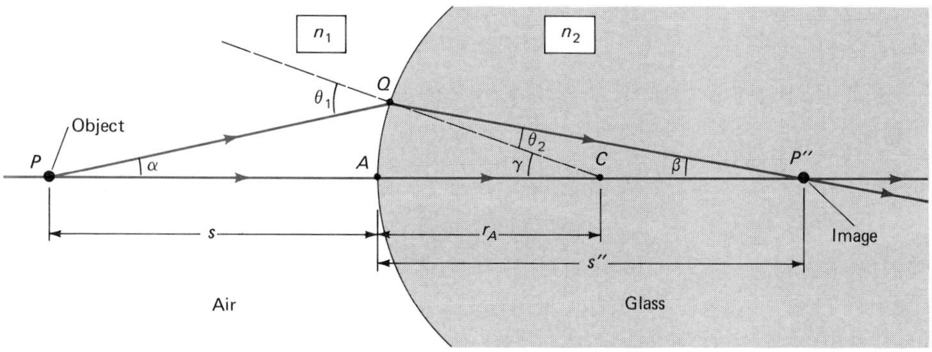

Figure 39.29

The rays PAP'' and PQP'' travel from the object point P and converge to form a real image at P''. The object distance is s. The image distance is s''. The refractive ray bending shown is appropriate for the case where $n_2 > n_1$.

Figure 39.30

The refracted rays PQR and PAC from the object at P, close to a convex surface, appear to diverge from P''. Because no light rays actually diverge from P'' it is a virtual image.

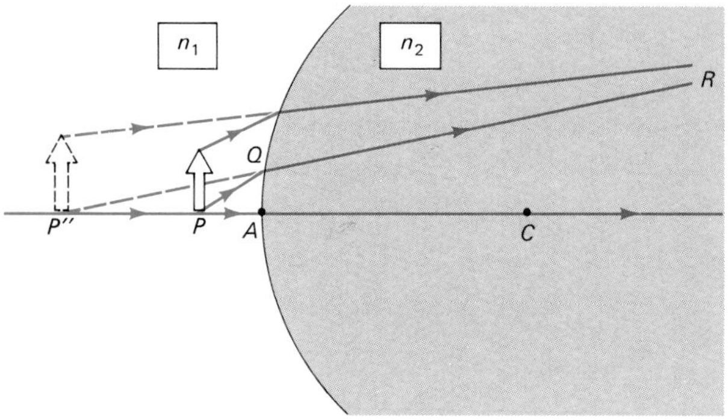

indicates that a *virtual* image is formed to the left of the interface (Figure 39.30). The light in medium 2 appears to have come in a straight line from P''.

Example 4 Locating the Image

A point source of light is placed 6 cm from a convex glass surface ($r_A = 2$ cm, $n_2 = 1.60$). What is the image location? Taking $s = 6$ cm and $n_1 = 1.00$ (air) in Eq. 39.27 gives

$$\frac{1.00}{6} + \frac{1.60}{s''} = \frac{1.60 - 1.00}{2} = 0.30$$

which yields $s'' = +12$ cm. Thus, a *real* image is formed inside the glass, 12 cm beyond the interface (Figure 39.31).

If the source is placed 2 cm from the surface, Eq. 39.27 gives

$$\frac{1.00}{2} + \frac{1.60}{s''} = 0.30$$

The image distance is $s'' = -8$ cm, showing that the light appears to come from a point P'' outside the glass, 8 cm from the interface. In this instance the image is virtual (Figure 39.32). Note that in both cases ($s'' > 0$ and $s'' < 0$) the index of refraction for glass (n_2) is used in the term n_2/s''. This is so even though for $s'' < 0$, the virtual image is located on the *air* side of the interface.

Figure 39.31

A real image is formed by the convex glass surface.

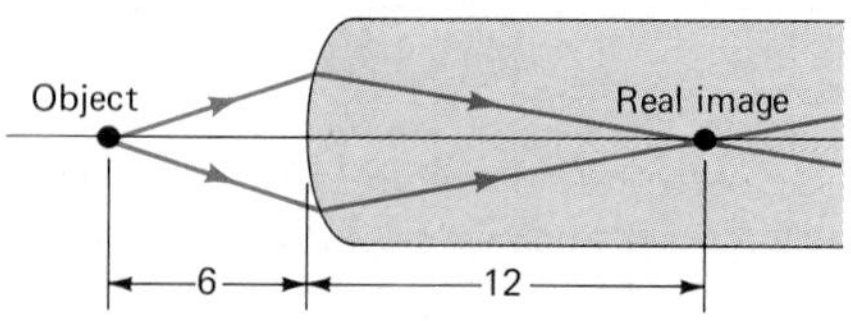

Figure 39.32

With the object sufficiently close to the surface, a virtual image results.

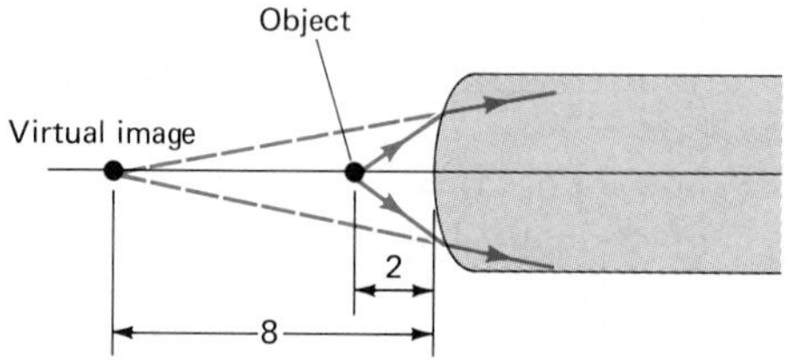

Figure 39.33 shows the refraction of paraxial rays emerging from a convex surface. An object at P''' is imaged at P'. The equation relating s''', s', and r_B can be derived by applying the *principle of optical reversibility*. Reversing the rays creates a situation in which an object at P' is imaged at P''' by a convex surface. This reversed-ray situation is described by Figure 39.29 and Eq. 39.27. If in Figure 39.29 we make the replacements $r_A \rightarrow r_B$, $s \rightarrow s'$, and $s'' \rightarrow s'''$, we get Figure 39.33 (with rays reversed). Making the same replacements in Eq. 39.27, we obtain the equation relating s', s''', and r_B,

$$\frac{n_1}{s'} + \frac{n_2}{s'''} = \frac{n_2 - n_1}{r_B} \qquad (39.28)$$

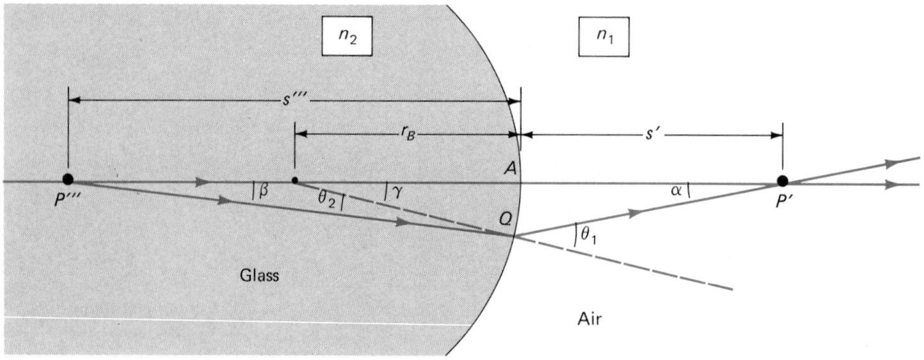

Figure 39.33

Refraction at a convex interface. Rays from the point object at P''' are focused, forming a real image at P'.

Figure 39.34

Rays from the point object at P are refracted at the first surface, forming an image inside the lens at P''. This image acts as a real object. Rays from P'' undergo a second refraction and form a real point image at P', outside the lens.

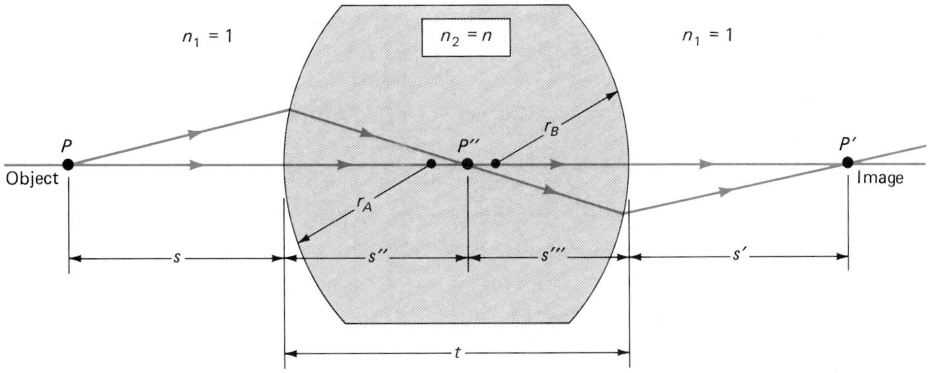

Thin Lens Equation

By combining the preceding analyses of refraction at spherical surfaces we can derive an equation that describes image formation by a lens. The basic idea is that the image formed by the first refracting surface acts as an object for the second refracting surface.

Figure 39.34 shows how a convex surface forms a real image inside a thick lens. This real image serves as an object from which rays diverge, undergo a second refraction, and finally form a real image outside the lens.

A particularly simple result is obtained for a *thin* lens—one for which the thickness t is negligibly small compared with all other lengths (s, s'', r_A, r_B). We will first derive an expression describing the thick lens and then obtain the thin lens equation as the limiting case $t \to 0$. By combining Eqs. 39.27 and 39.28 we get

$$\frac{n_1}{s} + \frac{n_1}{s'} + \frac{n_2}{s''} + \frac{n_2}{s'''} = (n_2 - n_1)\left(\frac{1}{r_A} + \frac{1}{r_B}\right) \quad (39.29)$$

From Figure 39.34 we see that

$$s'' + s''' = t$$

If we set $s''' = t - s''$, the third and fourth terms on the left-hand side of Eq. 39.29 can be combined to give

$$\frac{n_2}{s''} + \frac{n_2}{s'''} = \frac{n_2}{s''} + \frac{n_2}{t - s''} = \frac{n_2 \cdot t}{s''(t - s'')}$$

whereupon Eq. 39.29 becomes

$$\frac{n_1}{s} + \frac{n_1}{s'} + \frac{n_2 t}{s''(t - s'')} = (n_2 - n_1)\left(\frac{1}{r_A} + \frac{1}{r_B}\right) \quad (39.30)$$

If the lens thickness t is much less than the other lengths (s, s', s'', r_A, r_B), then we can take the limit $t \to 0$ in Eq. 39.30. In the limit $t \to 0$, Eq. 39.30 reduces to

$$\frac{n_1}{s} + \frac{n_1}{s'} = (n_2 - n_1)\left(\frac{1}{r_A} + \frac{1}{r_B}\right) \quad (39.31)$$

Dividing through by n_1 gives us one form of the *thin lens equation:*

$$\frac{1}{s} + \frac{1}{s'} = \left(\frac{n_2}{n_1} - 1\right)\left(\frac{1}{r_A} + \frac{1}{r_B}\right) \quad (39.32)$$

In Eq. 39.32, n_2 is the lens index of refraction and n_1 is the index of refraction of the medium surrounding the lens.

$$\frac{n_2}{n_1} = \frac{\text{index of refraction of lens}}{\text{index of refraction of medium surrounding lens}} \quad (39.33)$$

In most situations the medium surrounding the lens is air and n_1 can be set equal to 1. Equation 39.32 describes a thin *double convex* lens. The equation for a thin plano-convex lens follows from Eq. 39.32 with $r_A = \infty$.

A lens has *two* focal points, located at equal distances on either side of the lens. The focal length, f, is the distance from the lens to a focal point. The focal point is defined as the image distance for an object at infinity ($s' \to f$ for $s \to \infty$), or as the object distance for an image at infinity ($s \to f$ for $s' \to \infty$). An image or object "at infinity" corresponds to rays traveling parallel to the lens axis. Figure 39.35 shows one set of parallel rays being focused at one focal point (F_2) and another set of divergent rays leaving the other focal point (F_1) and traveling parallel to the lens axis after being refracted.

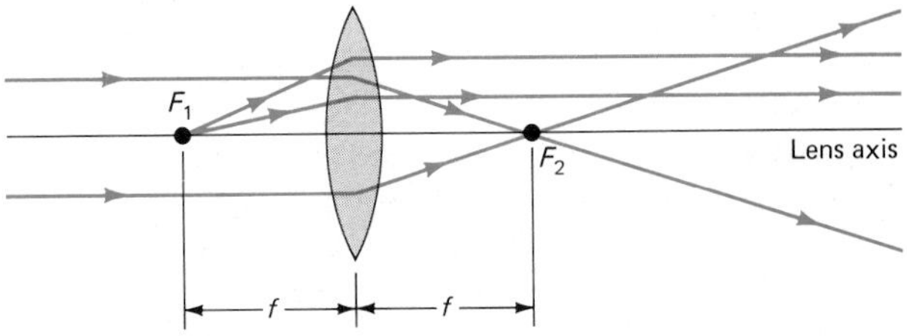

Figure 39.35

A double convex lens. Rays from infinity are focused at a focal point ($s = \infty$; $s' = f > 0$). Rays originating at a focal point are refracted and emerge traveling parallel to the lens axis ($s = f > 0$; $s' = \infty$).

It follows from Eq. 39.32 that the focal length is determined by

$$\frac{1}{f} = \left(\frac{n_2}{n_1} - 1\right)\left(\frac{1}{r_A} + \frac{1}{r_B}\right) \qquad (39.34)$$

Equation 39.34 is often referred to as the *lensmaker's formula.* The focal length for a plano-convex lens follows from Eq. 39.34 by setting $r_A = \infty$. Expressed in terms of f, the thin lens equation, Eq. 39.32, reads

$$\frac{1}{s} + \frac{1}{s'} = \frac{1}{f} \qquad (39.35)$$

Mathematically, this form of the thin lens equation is identical to the mirror equation, Eq. 39.20.

Example 5 **How to Cut the Focal Length in Half**

A plano-convex lens has a radius of curvature of 6 cm. Its index of refraction is $n = 1.55$. The focal length (in air) follows from Eq. 39.32.

$$\frac{1}{f} = (1.55 - 1.00)\left(\frac{1}{\infty} + \frac{1}{6}\right)$$

giving $f = 10.90$ cm. If two identical plano-convex lenses are cemented together, back to back, the resulting double convex lens has a focal length that is one half that of the plano-convex lenses. Thus, with $r_A = r_B = 6$ cm,

$$\frac{1}{f} = (1.55 - 1.00)\left(\frac{1}{6} + \frac{1}{6}\right)$$

This gives $f = 5.45$ cm, which is just half the value for a single plano-convex lens. We can see from Eq. 39.34 that this is a general result. With $r_A = r_B$ (double convex), $1/f$ is twice as great as with $r_A = \infty$ (plano-convex), and thus f is cut in half. Physically, this means that equal ray "bending" occurs at the two identical refracting surfaces.

Figure 39.36

The "easy" rays. ① A ray parallel to the lens axis is refracted and passes through the focal point F_2. ② A ray through the center of the lens undergoes no net refraction. ③ A ray through the focal point F_1 is refracted and continues on parallel to the lens axis.

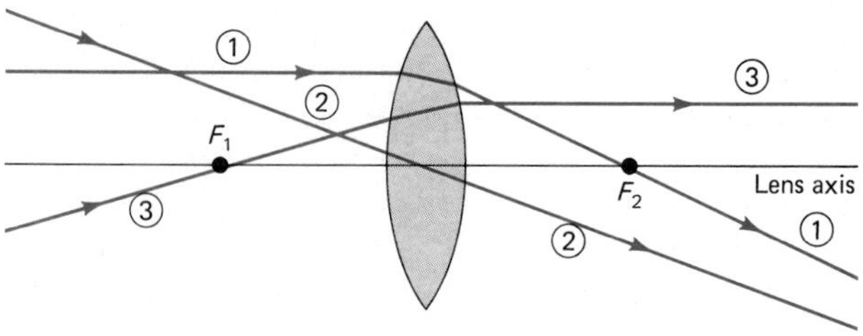

Concave Lens

For a thin lens formed by two concave refracting surfaces, the same analysis that we used for the double convex lens gives us

$$\frac{1}{s} + \frac{1}{s'} = -\left(\frac{n_2}{n_1} - 1\right)\left(\frac{1}{r_A} + \frac{1}{r_B}\right) \qquad (39.36)$$

Equation 39.36 is identical to Eq. 39.32 except for a factor of -1 on the right-hand side. We can rewrite Eq. 39.36 as

$$\frac{1}{s} + \frac{1}{s'} = \frac{1}{f} \qquad (39.37)$$

where f is the focal length, given by another version of the lensmaker's formula,

$$\frac{1}{f} = -\left(\frac{n_2}{n_1} - 1\right)\left(\frac{1}{r_A} + \frac{1}{r_B}\right) \qquad (39.38)$$

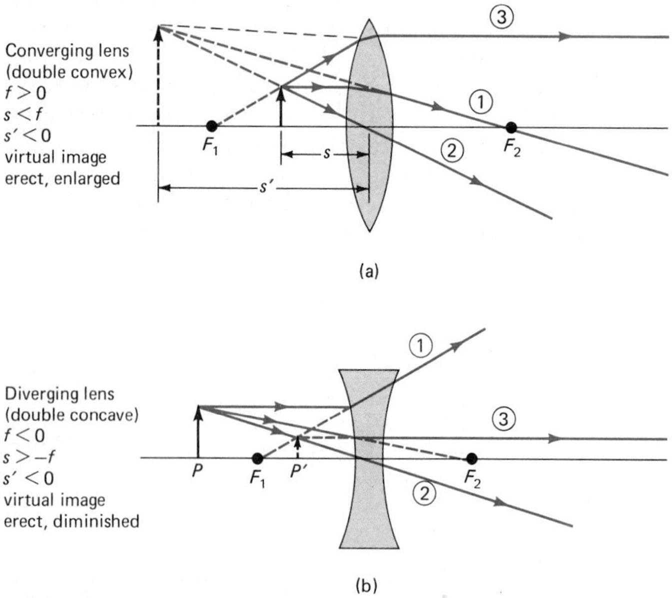

Figure 39.37

The easy rays can be used to locate image points. (a) Converging lens. (b) Diverging lens.

The focal length of the double concave lens is negative. Such a lens causes rays to diverge. The focal length of a plano-concave lens follows from Eq. 39.38 with $r_A = \infty$.

By comparing Eqs. 39.37, 39.35, and 39.20 we see that the two thin lens equations have the same mathematical form as the mirror equation $(1/s + 1/s' = 1/f)$, with f being positive for converging lenses and negative for diverging lenses. The types of images formed by thin lenses can be determined by using a ray diagram. Just as we did with mirrors, we will use "easy" rays.

(1) A ray parallel to the lens axis is refracted and passes through the focal point F_2.

(2) A ray through the center of the lens undergoes no net refraction.

(3) A ray through the focal point F_1 is refracted and continues on parallel to the lens axis.

Figure 39.36 shows how the three easy rays are affected by a lens. Figures 39.37a and 39.37b illustrate the use of the easy rays to determine the images formed by positive ($f > 0$) and negative ($f < 0$) lenses.

Fresnel Lens

In certain applications, a large-diameter, short-focal-length lens is required. For example, large lens arrays are needed in certain types of solar energy collectors. Simple plano-convex lenses are not well suited for such arrays because they are heavy and costly. The Fresnel lens is a cheap, lightweight substitute. The surface of a Fresnel lens is a series of concentric circular ridges, as shown in the cross-sectional view in Figure 39.38. The focusing action of a plano-convex lens occurs at the curved surface. Figure 39.38 illustrates a process in which segments of the curved surface of a plano-convex lens are removed and placed on a thin flat surface. Incident light normally meets the same curved refracting surface as it would encounter on the much bulkier plano-convex lens. Fresnel lenses are also used in lighthouses and in overhead projectors. They collect diverging rays and combine them into a well-defined beam. *Fresnel* lenses with a negative focal length are often used in the rear windows of recreational vehicles or motor homes. They offer the driver a wide-angle view, much of which would otherwise be blocked by the vehicle (Figure 39.39).

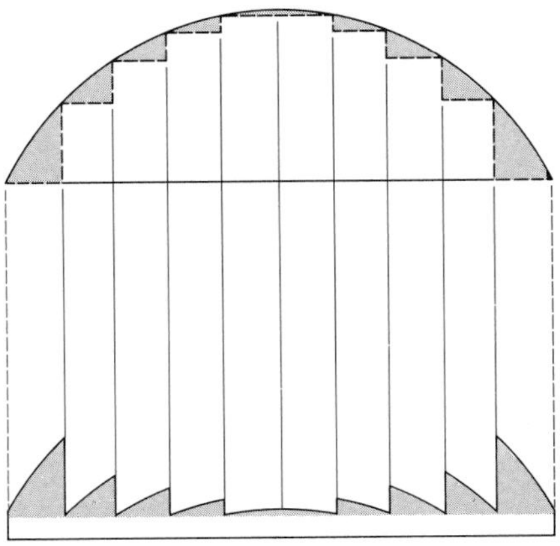

Figure 39.38

Cross-sectional view showing the principle of the Fresnel lens. Segments of the plano-convex lens are combined to produce the Fresnel lens. The important properties of the plano-convex lens (large light-gathering area and short focal length) are preserved by the Fresnel lens. In actual Fresnel lenses there are many more circular ridges. As a consequence, the ridge height varies more gradually than is suggested by the figure. The image quality of Fresnel lenses is generally mediocre because the ridges are not cut or molded precisely.

Figure 39.39

Recreational vehicle window with Fresnel lens.

Chromatic Aberration

The different colors composing white light are refracted through very slightly different angles when they pass through a prism or lens. This dispersion of colors causes *chromatic aberration* in a lens. Blue light undergoes the greatest refractive bending and red light undergoes the least. Consequently, the red light is focused farther from the lens than is the blue light (Figure 39.40), and the result is a smeared image with red edges. Various combinations of lenses are used to eliminate chromatic aberration and other image distortions. Chromatic aberration can be corrected by using a pair of thin lenses (Figure 39.41). One lens is positive, or converging, and the other is negative, or diverging. The combination of these two lenses, which is called an *achromatic doublet,* is converging. The two lenses

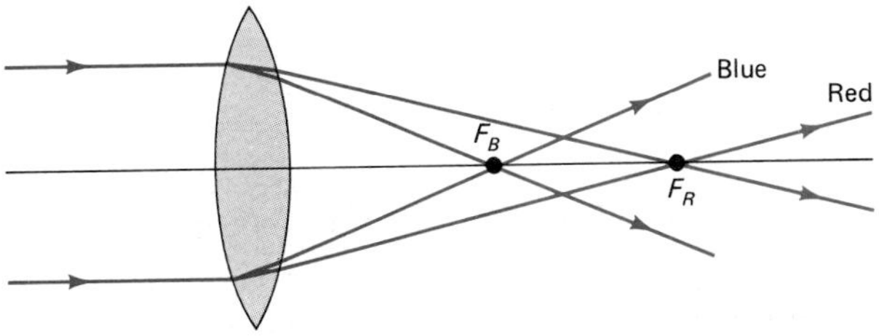

Figure 39.40

Chromatic aberration. Red light experiences the least refractive bending. The result is a smeared image with red edges.

Figure 39.41

The achromatic doublet. The two lenses are made from different types of glass, chosen so that their dispersions cancel.

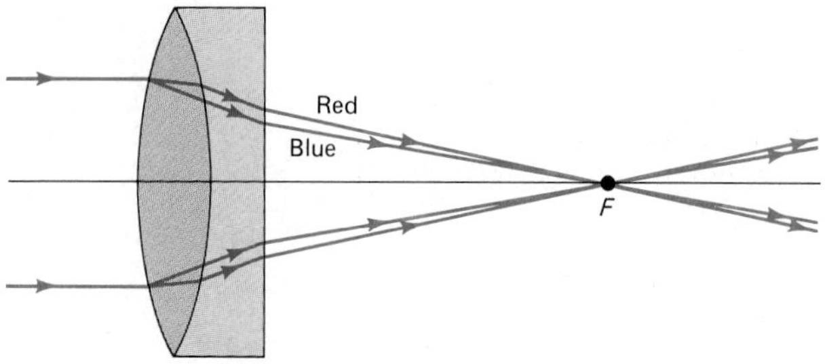

Figure 39.42

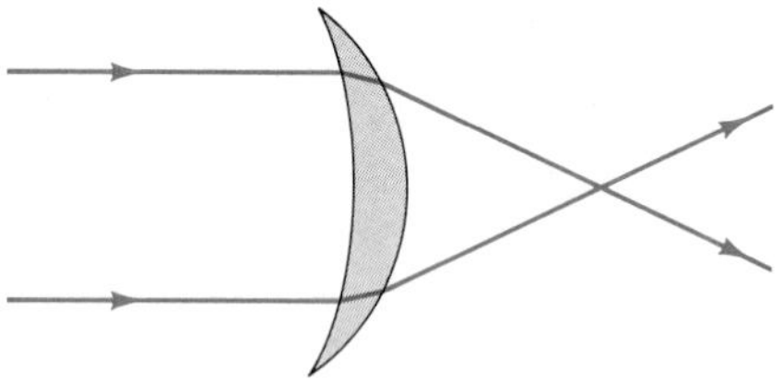

are fashioned from different types of glass, which are chosen so that their dispersions cancel.

Chromatic aberration and other image distortions present a challenge to optical engineers. A remedy for correcting one distortion should not magnify others. Approaches to lens design were changed dramatically by the advent of high-speed computers. The computer allows the optical engineer to simulate the performance of different lens combinations and materials. The effects of minor changes in each component can be evaluated without grinding a single lens.

Questions

10. In Figure 39.42 we see parallel rays entering the concave side of a concave–convex thin lens. The rays come to a focus 6 cm from the lens. Describe what happens to parallel rays that approach from the convex side of the lens.
11. A small air bubble is formed inside a piece of transparent plastic ($n = 1.80$). Does the bubble act like a converging or a diverging lens?
12. Why are magnifying glasses sometimes called "burning glasses"?
13. Suppose that you were handed a lens and a ruler and told to determine the focal length of the lens. How would you proceed?

39.6 Optical Systems

To illustrate and reinforce the ideas developed in the preceding sections, let's now consider four specific optical systems: the human eye, a simple magnifier, the microscope, and the telescope.

The Human Eye

Light enters the eye through the *cornea*, a transparent bulge on the surface of the otherwise opaque shell of the eye, the sclera (see Figure 39.43). Most of the converging action of the eye occurs at the air–cornea boundary, where the index of refraction changes from 1.00 to 1.38. The rays travel onward through a liquid called the *aqueous humor*. They pass through the pupil, which is a variable-diameter opening in the opaque *iris*. (The iris is the colored part of the eye.) The rays undergo further refraction in the *crystalline lens*. The curvature, and thus the focal length, of this lens can be varied by eye muscles. This capacity to vary the curvature is called *accommodation*, and allows the eye to focus on objects at different distances. The light rays then travel onward through the vitreous humor, and are at last focused on a thin layer of photosensitive cells that make up the *retina*.

The retinal cells are of two kinds: The *rods*, numbering over 100 million, are very sensitive to low-intensity light, but do not distinguish colors nor provide sharp images. The *cones*, numbering between 5 and 10 million, sharpen images and are sensitive to colors, but are ineffective at low light intensities. Thus, when you view a magazine in a semidark room, the cones of your retinas are not activated. You may be able to see that the pages carry words and pictures, but you will not be able to distinguish the words. Nor will you be able to tell whether the pictures are in color or in black and white. Increasing the intensity of light brings the cones into play. The retinal signals are relayed to the brain through the *optic nerve*. There are no

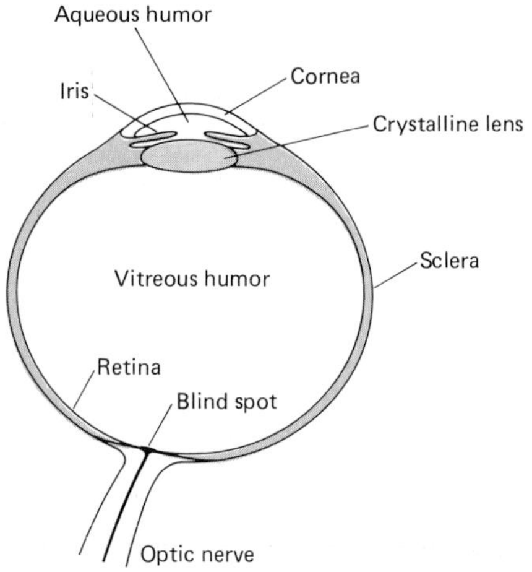

Figure 39.43

The human eye.

Figure 39.44

How to locate your blind spot. Close your left eye and look at the "X." With the page about 15 in. from your eye, move the page toward your eye. The "R" will disappear at some point. Moving the page closer to your eye causes the "R" to reappear and the "L" to disappear.

X **L** **R**

rods or cones in the region where the optic nerve is attached. Accordingly, light focused on this spot is not perceived. This region is called the *blind spot* (see Figure 39.44).

In a normal human eye, light from a distant object is focused on the retina with the lens muscles relaxed. As an object is brought closer, the lens contracts, becoming more convex and acquiring a shorter focal length. The necessity of decreasing f follows from the lens equation, $1/s, + 1/s' = 1/f$. If the image distance is to remain constant, a decrease in the object distance s must be accompanied by a decrease in the focal length. The ability of the eye to vary its focal length is limited. Objects closer than a certain distance appear blurred. The *near point* is the closest point on which the eye can focus. The distance from the eye to the near point ranges from less than 10 cm for youngsters to over 100 cm for older people. (You can assess your own approach to old age by measuring the distance to your near point.)

Two very common eye disorders are nearsightedness (myopia) and farsightedness (hyperopia). In the relaxed myopic eye, parallel rays are focused before they reach the retina. This can be corrected by a spectacle or contact lens with a negative focal length (Figure 39.45). Such a corrective lens diverges the rays slightly, thereby compensating the excessive refraction by the eye. In the farsighted eye, rays are converged toward a focal point behind the retina. This condition is alleviated by a lens with a positive focal length. Such a lens converges rays, compensating for the lack of refraction in the eye.

Contact lenses are thin glass or plastic lenses that adhere to the cornea. We can get a rough idea of how contact lenses alter rays by assuming that the eye is described by the thin lens equation.

$$\frac{1}{s} + \frac{1}{s'} = \frac{1}{f_e}$$

This allows us to treat the eye plus contact lens as a pair of thin lenses in contact. Two thin lenses in contact, with focal lengths f_1 and f_2, behave like a single thin lens with a focal length f given by

$$\frac{1}{f} = \frac{1}{f_1} + \frac{1}{f_2} \qquad (39.39)$$

Figure 39.46 indicates how this statement can be proved.

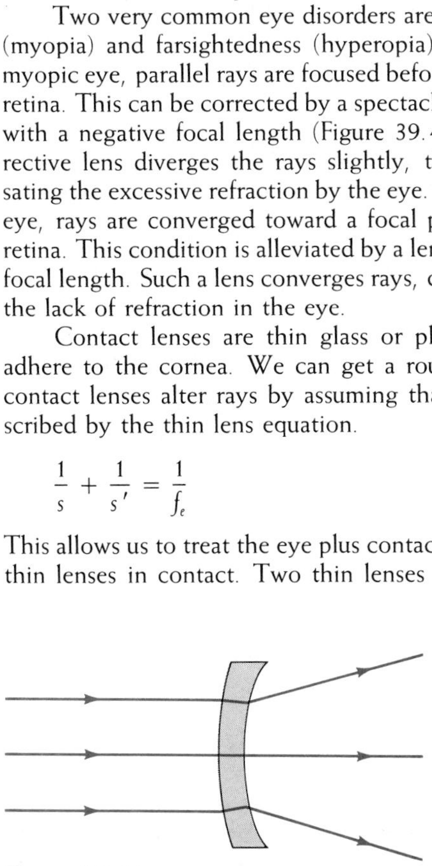

Figure 39.45

Section of a spectacle lens for a myopic (nearsighted) eye. The lens has a negative focal length—it causes the rays to diverge. This compensates the excessive refraction by the eye.

Figure 39.46

For thin lenses the image distance (s'') for the left lens acts as a virtual object distance for the right lens. Adding the two lens equations gives

$$\frac{1}{s} + \frac{1}{s'} = \frac{1}{f_1} + \frac{1}{f_2},$$

showing that the two lenses act like a single lens with a focal length f given by Eq. 39.39,

$$\frac{1}{f} = \frac{1}{f_1} + \frac{1}{f_2}.$$

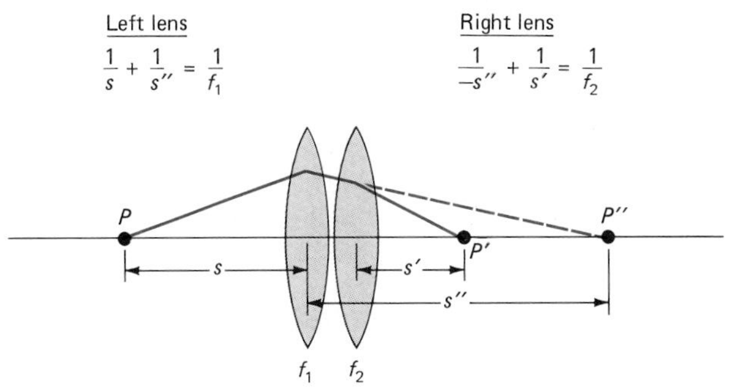

To apply Eq. 39.39 we consider a myopic eye in which parallel rays are focused at a distance $f_2 = 1.69$ cm. The correct focal length, which will focus rays on the retina, is $f = 1.71$ cm. The contact lens must have a focal length f_1 given by Eq. 39.39. Thus,

$$\frac{1}{f_1} = \frac{1}{f} - \frac{1}{f_2} = \frac{1}{1.71} - \frac{1}{1.69}$$

giving $f_1 = -144$ cm. In this case, a weak diverging contact lens will remedy the myopic condition.

Our study of optics so far has been restricted to lenses that are symmetric. If a symmetric lens is rotated about its symmetry axis, the image will not change. If the lens is asymmetric, however, rotation will distort the image. Very often the cornea of the human eye becomes asymmetric, causing what is called *astigmatism*. If you wear glasses, there is a very simple way to determine whether or not you have astigmatism. Remove your glasses and rotate the lenses about a horizontal axis while looking through them. If objects appear to change shape and become distorted as you rotate the lens, you are astigmatic—your cornea, and the correcting lens, are not symmetric about the rotation axis.

The Magnifying Glass

The size of the retinal image is determined by the angle subtended by the object. As an object is moved closer to the eye its apparent size increases (Figure 39.47). Psychologists have developed a test that shows that the eye detects only the angular size of objects. Using one eye, a subject

Figure 39.47

The size of the retinal image increases as the object is moved closer to the eye. The largest clear image is formed with the object at the near point, N. A blurred image results for smaller eye-to-object distances.

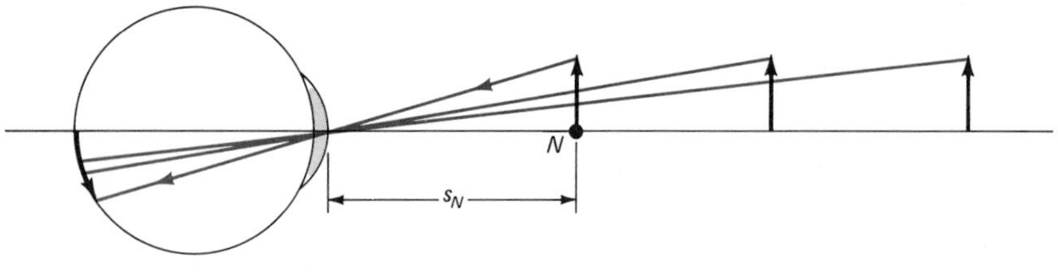

views two automobiles. One is a full-sized vehicle, placed rather far from the subject, and the other is a miniature model. The model automobile is placed much closer to the subject, with the result that both the model and the full-sized car subtend the same angle. With proper lighting, most people perceive both cars to be full sized and far away.

When an object is brought closer than the near point, its apparent size continues to increase, but the image blurs. A simple magnifier consists of a single positive lens ($f > 0$) that permits an object to be brought closer than the near point and still remain in focus. A drop of water magnifies the surface on which it rests. In addition to magnification, it is desirable that a magnifier give an upright image. These conditions require that the object distance be less than the focal length. The greatest usable magnification is achieved when the image is located at the near point. A closer image will appear larger, but blurred. The *angular magnification*, M (also called *magnifying power*), is the ratio of the angles subtended by the magnified image and by the unmagnified image when the object is at the near point. Figure 39.48a shows that the angular size of the unmagnified image is

$$\alpha_o = \frac{y}{S_N} \tag{39.40}$$

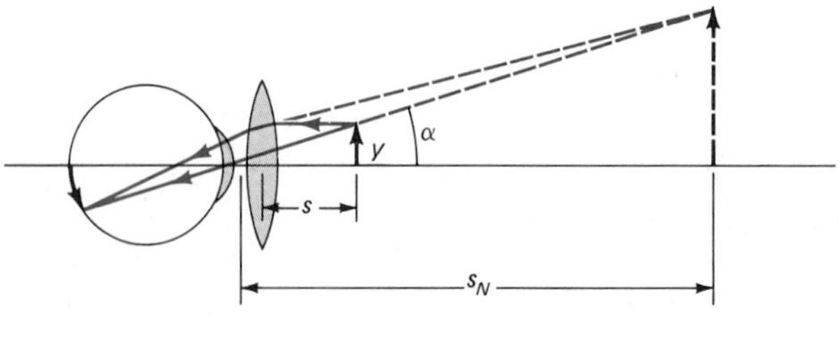

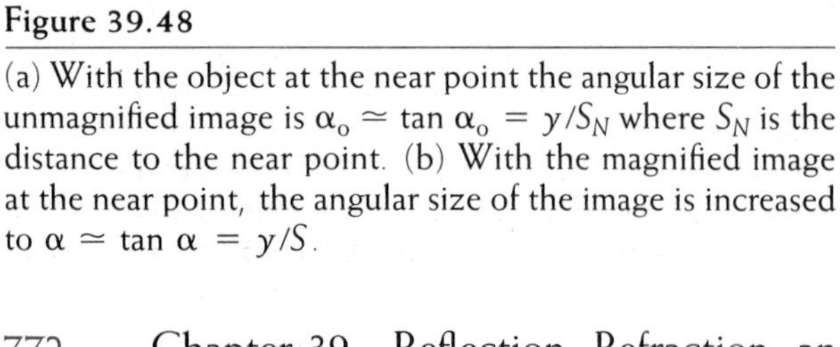

Figure 39.48

(a) With the object at the near point the angular size of the unmagnified image is $\alpha_o \simeq \tan \alpha_o = y/S_N$ where S_N is the distance to the near point. (b) With the magnified image at the near point, the angular size of the image is increased to $\alpha \simeq \tan \alpha = y/S$.

where y is the height of the object and S_N is the distance from the eye to the near point. Figure 39.48b shows that when the magnified virtual image appears at a distance S_N from the magnifier, its angular size is

$$\alpha = \frac{y}{s} \tag{39.41}$$

where s is the object-to-magnifier distance. The maximum angular magnification follows from its definition and Eqs. 39.40 and 39.41.

$$M_{max} = \frac{\alpha}{\alpha_o} = \frac{S_N}{s} \tag{39.42}$$

Using the thin lens equation $1/s + 1/s' = 1/f$, with $s' = -S_N$ for the virtual image distance, gives

$$M_{max} = S_N\left(-\frac{1}{s'} + \frac{1}{f}\right) = 1 + \frac{S_N}{f} \tag{39.43}$$

Equation 39.43 shows that a large magnification requires a lens with a short focal length. Customarily, S_N is taken to be 25 cm in the design of magnifiers and other commercial optical instruments.

Example 6 Focal Length of an 8× Magnifier

We can use Eq. 39.43 to determine the focal length of an "8 power" (8×) magnifier. We take S_N equal to 25 cm. With $M_{max} = 8$ and $S_N = 25$ cm, Eq. 39.43 gives

$$f = \frac{S_N}{M_{max} - 1} = \frac{25 \text{ cm}}{8 - 1} = 3.57 \text{ cm}$$

The Microscope

The microscope was invented by a Dutch optician, Zacharias Jensen, late in the 16th century. In its most elementary form the microscope consists of two lenses that produce a magnified image. Figure 39.49 shows how light rays traverse the instrument. The first lens (the *objective*) focuses rays from a nearby object, giving a real magnified image. The second lens (the eyepiece, or *ocular*) focuses on the image formed by the objective and magnifies it further. The final image is inverted. The thin lens equation is used for both the objective,

$$\frac{1}{s_o} + \frac{1}{s'_o} = \frac{1}{f_o} \tag{39.44}$$

Figure 39.49

Basics of the compound microscope. Rays from the object are focused to form an intermediate image. This image serves as an object for the eyepiece. With the intermediate image positioned at a focal point of the eyepiece rays will enter the eye parallel, allowing relaxed viewing.

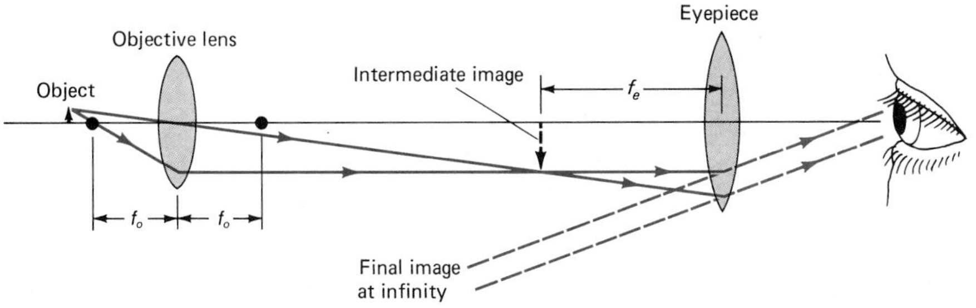

and the eyepiece,

$$\frac{1}{s_e} + \frac{1}{s_e'} = \frac{1}{f_e} \qquad (39.45)$$

(Primes denote image distances.) Both focal lengths, f_o and f_e, are positive.

The image formed by the objective lens serves as the object for the eyepiece. Because of this, the total magnification for the elementary microscope is the product of the magnifications for the objective (M_o) and the eyepiece (M_e):

$$M_{mic} = M_o M_e \qquad (39.46)$$

Let's see how the magnification of the microscope is related to the focal length of the objective and that of the eyepiece. The *linear magnification* of the objective lens is defined as the ratio of the image height and the object height (Figure 39.50).

$$M_o = \frac{y'}{y} \qquad (39.47)$$

From Figure 39.50

$$\tan\theta = \frac{y'}{L} = \frac{y}{f_o}$$

from which we obtain

$$M_o = \frac{L}{f_o} \qquad (39.48)$$

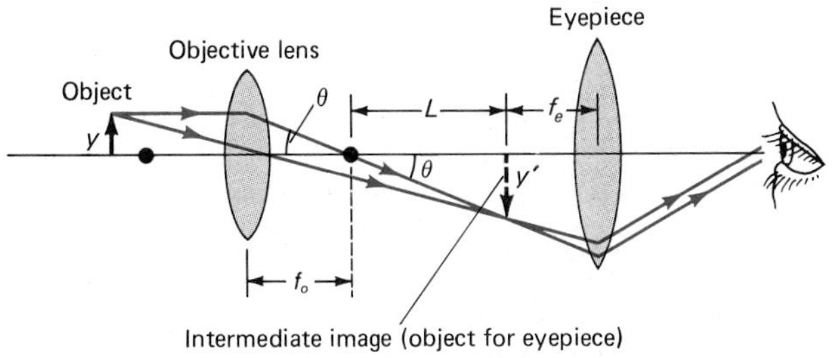

Figure 39.50

The total magnification of the microscope is the product of the magnifications for the objective lens and the eyepiece. The image formed by the objective serves as the object for the eyepiece.

In these equations L is the distance between the second focal point of the objective lens and the first focal point of the eyepiece. The eyepiece magnification is given by Eq. 39.42, in which we set $s = f_e$, because the object of the eyepiece (the intermediate image) is located at the focal point of the eyepiece (Figure 39.50). Thus,

$$M_e = \frac{S_N}{s} = \frac{S_N}{f_e} \qquad (39.49)$$

The microscope magnification is therefore

$$M_{mic} = M_o M_e = \frac{L S_N}{f_o f_e} \qquad (39.50)$$

In most microscopes the distance L is fixed (16 cm), and S_N also has a fixed value (25 cm). Focusing is achieved by varying the object distance. The significant feature of Eq. 39.50 is that the magnification is inversely proportional to the product of the focal lengths of the objective and the eyepiece.

$$M_{mic} \propto \frac{1}{f_o f_e}$$

Both the objective and the eyepiece function as magnifiers. Short focal lengths for both the objective and eyepiece lenses are required in order to achieve a large total magnification.

The Telescope

Credit for inventing the telescope is given to a Dutchman, Hans Lippershey. When Lippershey applied for a patent for his invention in 1608, the Dutch government refused to grant it. Instead, the rights to the telescope were purchased by the government and Lippershey was hired to perform research aimed at improving the instrument. The Dutch government recognized the great military significance of the telescope and wanted the device to remain its military secret. But then as now, *scientific* secrets were difficult to keep. When the knowledge that Lippershey had fashioned a telescope reached Italy, Galileo was able to figure out independently how a combination of

Figure 39.51

Ray path for the refracting telescope. Parallel rays from a distant object are focused in the plane passing through P'', which is also a focal plane of the eyepiece. The image at P'' acts as an object for the eyepiece.

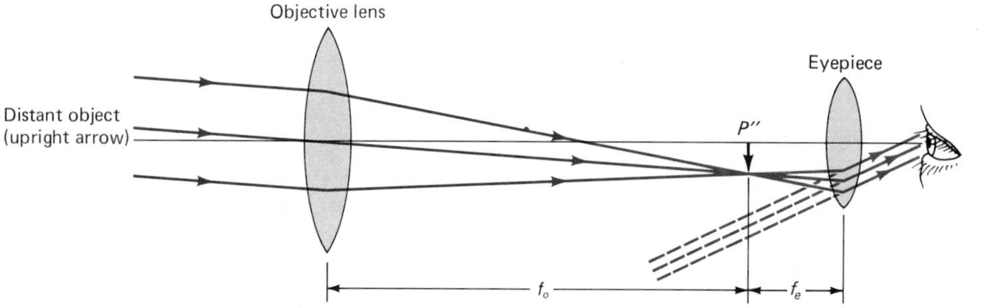

two lenses could be arranged to magnify distant objects.* Galileo then constructed his own telescope and used it to discover four of the moons of Jupiter as well as Saturn's rings.

The simplest refracting telescope (Figure 39.51) consists of two lenses that form a magnified image of a distant object. The object distance (s_o) is nearly infinite by comparison to the focal length (f_o). It follows from the thin lens formula that the image distance (s_o') is very slightly greater than f_o.

$$\frac{1}{s_o} + \frac{1}{s_o'} = \frac{1}{f_o} \qquad s_o' \simeq f_o \text{ when } s_o \gg f_o$$

The eyepiece is positioned so that the image formed by the objective coincides with the focal point of the eyepiece ($s_e = f_e$). This configuration gives an inverted virtual image at infinity.

The magnification of the telescope is given by

$$M_{tel} = \frac{f_o}{f_e} \qquad (39.51)$$

A large magnification can be achieved by using an objective lens with a large focal length. In order to gather light efficiently from faint and distant sources the objective lens must also have a large diameter. Happily, these two requirements are compatible: It is possible to make a large-diameter lens that has a slight curvature and thus a large focal length.

The basic designs of the microscope (Figure 39.50) and the telescope (Figure 39.51) are similar but not identical. The eyepiece performs the same function in both: It magnifies the image formed by the objective lens. The difference between the microscope and the telescope is in their objective lenses. The objective lens of a telescope is *large* so that it can gather many rays of light. The objective lens is not designed to provide magnification. In sharp contrast, the objective lens of the microscope *is* designed to provide magnification. The objective lens of a microscope is *small*. It has a *short* focal length and provides a large magnification (see Eq. 39.50).

The basic optical features of the magnifying glass, the human eye, the microscope, and the telescope can be explained by using geometric optics. In general, the ray picture of geometric optics is satisfactory for phenomena in which relevant dimensions of the optical system (the diameter of a lens, for example) are large by comparison to the wavelength of light. Many important optical instruments (eyeglasses and microscopes, for example) are simply described in terms of light rays. The ray picture is inadequate for describing *interference* phenomena, such as the diffraction of light. Such phenomena must be described in terms of waves. In Chapters 40 and 41 we will examine *physical optics*, which is based on the wave picture of light.

*Galileo also "reinvented" the compound microscope, working without knowledge of Jensen's work.

Questions

14. How might you determine whether a pair of eyeglasses is designed to correct nearsightedness or farsightedness?

15. When you open your eyes under water it is difficult to focus on objects, whether near or far. But if you wear goggles, focusing is easy. Explain.

16. Suppose that you see a newspaper ad that reads "wholesale lot of magnifying glasses, focal length 82 cm." The price quoted is 100 times less than the retail price of a good magnifier. Should you stock up? Why or why not?

Summary

In many situations the progress of light waves can be mapped by using Huygens' principle, which states:

Each point on a wave front acts as a source of secondary wavelets. The wave front at any subsequent moment is the envelope of the secondary wavelets.

The index of refraction (n) is the ratio of the speeds of light in a vacuum and in a particular medium.

$$n = \frac{\text{speed in vacuum}}{\text{speed in medium}} = \frac{c}{v} \qquad (39.1)$$

Huygens' principle can be used to derive the law of reflection:

The angles of incidence and reflection are equal.

and the law of refraction:

$$n_1 \sin \theta_1 = n_2 \sin \theta_2 \qquad \text{(Snell's law)} \qquad (39.6)$$

For phenomena in which the dimensions of the optical system are large by comparison with the wavelength of light, we can picture light rays, and use geometric optics. In particular, ordinary mirrors and lenses can be studied within the framework of geometric optics.

The fundamental equation for spherical mirrors relates the object distance (s), the image distance (s'), and the focal length (f).

$$\frac{1}{s} + \frac{1}{s'} = \frac{1}{f} \qquad (39.20)$$

The basic equation governing lenses has precisely the same form as Eq. 39.20 when the thickness of the lens is negligible in comparison to other characteristic lengths (s, s', f). This is the thin lens equation,

$$\frac{1}{s} + \frac{1}{s'} = \frac{1}{f} \qquad (39.35)$$

with $f > 0$ signifying a converging lens and $f < 0$ signifying a diverging lens.

Suggested Reading

R. C. Jones, How Images Are Detected, *Sci. Am.*, 219, 110 (September 1968).

M. F. Land, Animal Eyes with Mirror Optics, *Sci. Am.*, 239, 126 (December 1978).

D. K. Lynch, Atmospheric Halos, *Sci. Am.*, 238, 144 (April 1978).

F. D. Smith, How Images Are Formed, *Sci. Am.*, 219, 96 (September 1968).

J. Walker, How to Create and Observe a Dozen Rainbows in a Single Drop of Water, *Sci. Am.*, 237, 138 (July 1977).

R. J. Whitaker, Physics of the Rainbow, *Physics Teacher*, 12, 283 (May 1974).

Problems

Section 39.2 The Laws of Reflection and Refraction

1. A ray of light in air making an angle of 40° with the vertical strikes a horizontal water surface, $n = 1.33$. (a) Part of the light is reflected. What is the angle of the reflected ray relative to the vertical? (b) Calculate the angle of the refracted ray relative to the vertical.

2. A ray of light in air making an angle of 40° with the vertical strikes a horizontal liquid surface. The refracted ray makes an angle of 25.4° with the vertical. Calculate the index of refraction of the unknown liquid and, from Table 39.1, suggest what the liquid might be.

3. Figure 1 shows a top view of a square enclosure. The inner surfaces are plane mirrors. A ray of light enters a small hole in the center of one mirror. At what angle θ must the ray enter in order to exit through the hole after being reflected once by each of the other three mirrors?

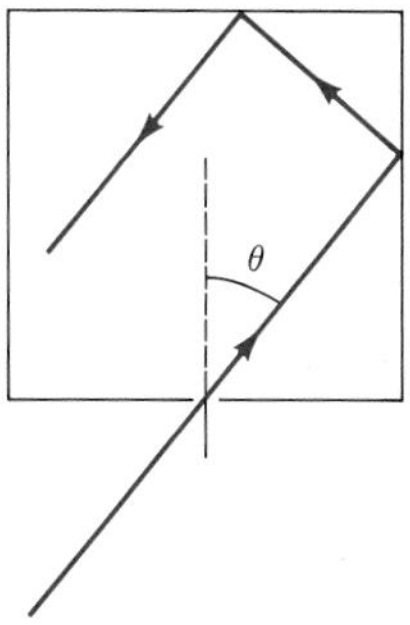

Figure 1

4. For the mirrored enclosure of Problem 3 (Figure 1), are there other values of θ for which the ray can exit after multiple reflections? If so, make a sketch of one of them.

Section 39.3 Light Rays and Geometric Optics

5. Derive Eq. 39.9.

6. A light ray is incident at an angle of 30° on a pane of glass 0.8 cm thick. The index of refraction of the glass is 1.52. Determine the parallel displacement, d, of the emergent ray. Take the index of refraction of air to be 1.

7. A ray of light is incident upon a pane of glass 9.0 mm thick (Figure 2). The light makes an angle of 35° with the vertical. The glass pane has an index of refraction of 1.60. The top and bottom surfaces are parallel. Calculate the separation, δ, of the two reflected rays.

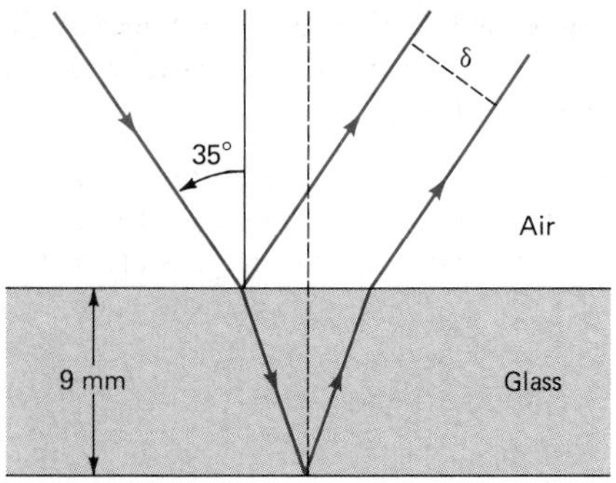

Figure 2

8. Light rays travel upward through the glass bottom of a fish bowl. The index of refraction of the galss is 1.52. Determine the critical angle for the glass–water boundary. Is it larger or smaller than the critical angle for a water–air boundary? The index of refraction of water is 1.33.

9. It is desired to have light enter one face of a prism and emerge from an adjacent face as shown in Figure 3. What is the minimum value for the index of refraction of the glass?

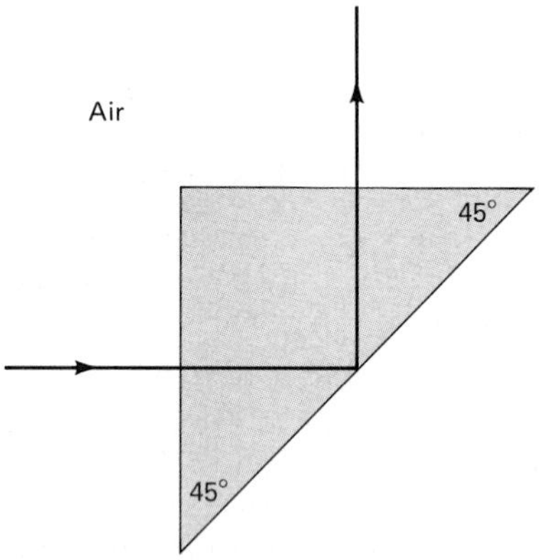

Figure 3

10. The prism shown in Figure 4 has an index of refraction of 1.55. Light is incident at an angle of 20°. Determine the angle θ at which the light emerges.

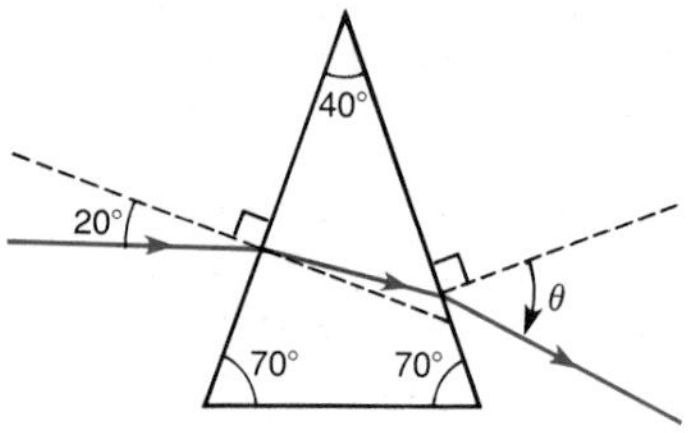

Figure 4

11. (a) A coin rests on the bottom of a pool of water 80 cm deep (Figure 5). When viewed from positions nearly above the coin, the angles of the incident and refracted light are small (θ_1, $\theta_2 \ll 1$ rad). Use the small-angle approximations for sin θ_1 and sin θ_2 in Snell's law to show that the apparent depth of the water is 60 cm. Take the index of refraction for water to be 4/3. (b) For larger angles, does the apparent depth increase or decrease? (c) Can the apparent depth exceed the actual depth?

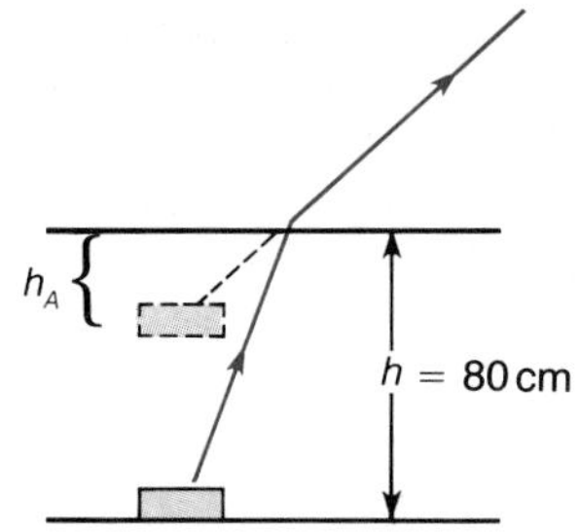

Figure 5

12. The high index of refraction of diamond enables a gem to "sparkle" via internally reflected rays. Such rays would be refracted out of materials with lower values of n, thereby reducing the retroreflected intensity. Compare the critical angle of incidence for diamond and flint glass. For the rays indicated in Figure 6, show that the second reflection is totally internal for diamond, but that refraction occurs for glass.

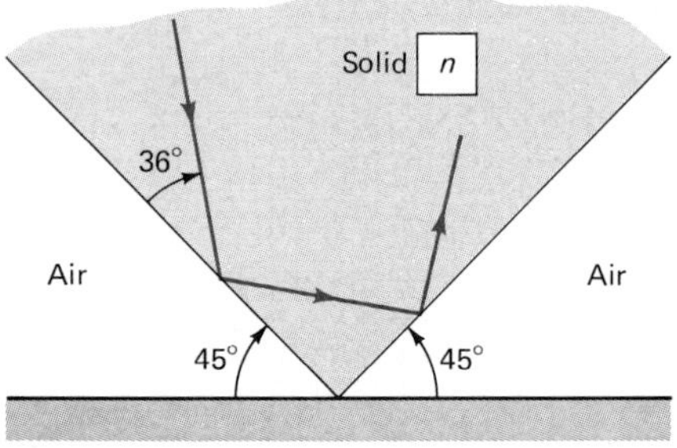

Figure 6

13. The critical angle for a material is 45°. What is its index of refraction?

Section 39.4 Mirrors

14. The relative error made in using the paraxial approximation is given by

$$\frac{\phi - \sin\phi}{\sin\phi}$$

Calculate the relative error for angles of 1°, 3°, 10°, 30°, and 45°. Note that ϕ must first be expressed in radians.

15. (a) Use a ray diagram to show that a plane mirror of length $\frac{1}{2}L$ is long enough for a person of height L to view his entire reflected image. (b) Is there any restriction on the position of the top edge of the mirror? Would the minimum length be increased or decreased if the mirror were (c) slightly concave; (d) slightly convex?

16. The radius of curvature of a convex mirror is 10 cm. What is its focal length? (Remember, the focal length of a convex mirror is *negative.*) The image of an object lies halfway between the mirror surface and the focal point. Use the mirror equation to determine the position of the object. Make a ray sketch.

17. A point object lies at one end of the diameter of a hemispherical mirror (Figure 7. Paraxial rays are focused at a point F located $\frac{1}{2}R$ from the mirror surface. Do rays reflected through large angles cross the mirror axis at points closer to or farther from the mirror than F? Draw one ray that confirms your answer.

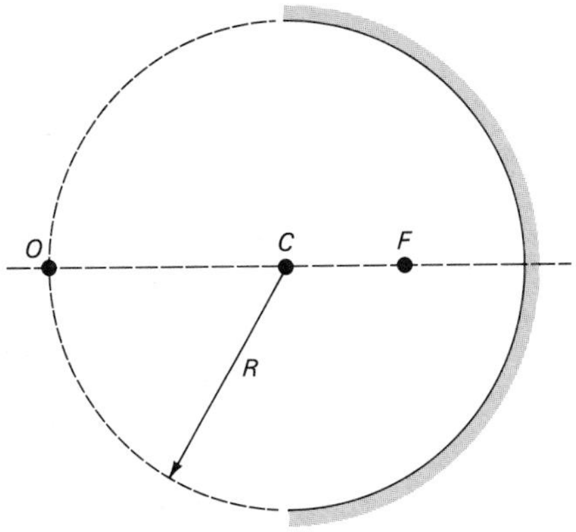

Figure 7

18. Draw a figure similar to Figure 39.20, but for a *convex* spherical mirror with radius of curvature r. Show that paraxial rays are described by the mirror equation

$$\frac{1}{s} + \frac{1}{s'} = \frac{1}{f} \qquad f = \frac{1}{2}r$$

provided that s' and r are assigned negative values.

19. The radius of curvature of a concave spherical mirror is 2.12 m. What is its focal length?

20. A concave mirror has a radius of curvature of 1.2 m. If you stand 5 m in front of the mirror, where is your image formed? Can you look into the mirror and see your image? Explain.

21. A concave mirrror has a focal length of 2.0 m. If you stand 1.0 m in front of it, where is your image formed? Is the image real or virtual? Is it upright or inverted?

22. An object is placed 7.5 cm in front of a concave mirror with a 10-cm radius of curvature. (a) Using the *three* easy rays, construct the image. (b) Is the image real or virtual; erect or inverted? (c) Use the mirror equation to calculate the image distance, s'.

23. An object is placed 0.84 m in front of a concave mirror. A real, inverted image is formed at the same position (0.84 m in front of the mirror). Determine the focal length and radius of curvature of the mirror.

24. An object is 20 m from a convex mirror with a focal length of -0.20 m. Determine the position of the image. Is it upright or inverted? Real or virtual?

25. A convex mirror has a focal length of -0.88 m. One object is 2.1 m in front of the mirror. A second object is 0.62 m in front of the mirror. The first object distance is greater than the magnitude of the focal length ($s > |f|$). The second object distance is less than the magnitude of the focal length ($s < |f|$). Does this difference affect the character of the images? Specifically, are both images real or virtual, upright or inverted? Is either image inverted? Determine the location of both images.

26. An object is located a distance s from a concave mirror of 0.50-m focal length. Calculate the image distance s' and state whether the image is (1) real or virtual and (2) erect or inverted when the distance s is (a) 1.20 m; (b) 0.78 m; (c) 0.36 m.

Section 39.5 Lenses

27. Use a ray diagram to show how a raindrop on the windshield of an automobile gives an inverted real image.

28. (a) Paraxial rays emerge from a point source and are refracted from a thin lens ($f = +5$ cm) away. Where is the image located? Is it real or virtual? (b) The rays travel onward through a second thin lens ($f = -5$ cm) located 20 cm beyond the first. Where is the image located? Is it real or virtual?

29. The surfaces of a symmetric double convex lens are ground to a curvature of 20.3 cm. If the index of refraction is $n = 1.64$, what is the focal length?

30. The image and object for a thin convex lens are on opposite sides of the lens, each 28 cm away from the lens. What is the focal length of the lens?

31. A symmetric double convex lens is designed to have a focal length of 15.0 cm in air. The index of refraction of the glass is 1.48. Calculate the radius of curvature of the convex surfaces.

32. A symmetric double convex lens is designed to have a focal length of 6.0 cm under water. The index of refraction of the glass is 1.54. Calculate the radius of curvature of the convex surfaces. The index of refraction of water is 1.33.

33. A thin double convex lens of glass with index of refraction $n_2 = 1.56$ has a radius of curvature of 6.3 cm on one side and 7.8 cm on the other side. (a) Calculate the focal length of this lens in air. (b) Calculate the focal length in water ($n_1 = 1.33$). (c) Calculate the focal length of this lens when immersed in tetrabromomethane ($n_1 = 2.96$).

34. A thin converging lens has a focal length of 20 cm. Locate and describe the images (real or virtual, upright or inverted) for objects placed at distances of (a) 10 cm, (b) 20 cm, (c) 40 cm, and (d) 80 cm.

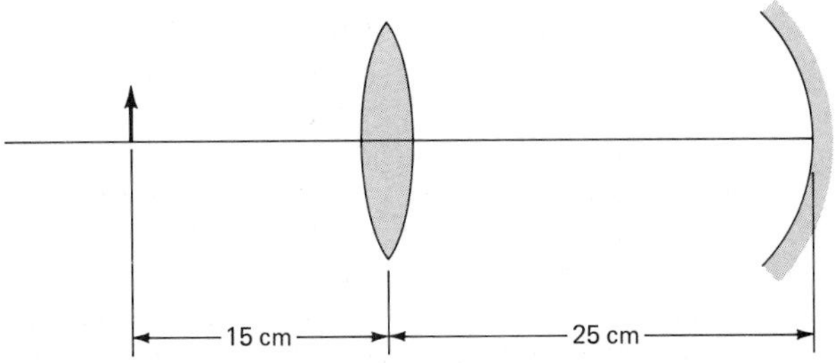

Figure 8

35. A thin convex lens ($f = +10$ cm) is placed 25 cm in front of a concave mirror ($f = +5$ cm; Figure 8). An object is positioned 15 cm in front of the lens. (a) Draw a ray diagram to locate the image. Is the image real or virtual, upright or inverted? (b) Use the mirror and lens equations to locate the position of the image.

36. An object is placed 12 cm to the left of a thin convex lens with a focal length of 7.5 cm. (a) Use easy rays to construct the image. (b) Use the thin lens equation to determine the image distance, s'.

37. A point source of light is placed 8 cm from a convex glass surface ($r = 2$ cm, $n = 1.60$). Determine the image location. Is it a real image or a virtual image? Repeat for an object distance of 1 cm. Sketch rays for both situations.

38. Using glass with an index of refraction $n = 1.52$, what radius of curvature is needed to produce a plano-convex lens with a focal length of 30 cm? What radius of curvature (the same for both faces) would be required to produce a double convex lens with $f = 30$ cm?

Section 39.6 Optical Systems

39. If the distance from the cornea to the retina is 2.12 cm, by how much must the focal length of the eye *change* to follow an object from the near point (25 cm) to infinity? Treat the eye as a thin lens in which all of the refraction occurs at the cornea.

40. A microscope of the type shown in Figure 39.49 has $L = 16$ cm, $f_o = 0.60$ cm, and $f_e = 1.80$ cm. If $S_N = 25$ cm, determine the magnification.

41. The objective lens of a small portable telescope has a focal length of 1.20 m. What focal length eyepiece should be used to achieve 200 power magnification?

42. Determine the magnification of a simple magnifier with a focal length of 2.5 cm.

chapter 40

PHYSICAL OPTICS: INTERFERENCE

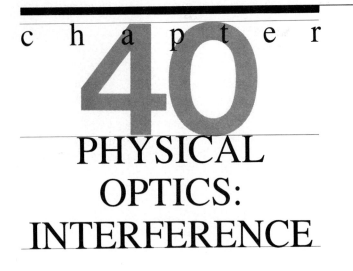

Preview

In Chapter 39 we studied geometric optics, which is based on the ray picture of light. In this chapter we will study physical optics, which pictures light as a wave.

We begin by reviewing the superposition of waves, which was introduced in Chapter 17. A principle of superposition for light was used by Thomas Young to explain his historic demonstration of the interference of light. We describe Young's experiment in Section 40.2.

We then introduce the concept of coherence and use it to explain why light waves exhibit interference in certain circumstances but not in others.

We next study thin-film interference to illustrate the effects of optical interference. The coloring of oil films and the design of antireflecting coatings for camera lenses are but two of the many examples of thin-film interference effects.

Interference can be used to measure extremely small displacements. Distances as small as a fraction of a wavelength of light are routinely measured with optical interfereometers. We complete our investigation of interference with a brief study of optical interferometers. We give particular attention to Michelson's interferometer, an instrument that helped to establish an experimental basis for Einstein's relativity theory.

40.1 Interference

In Section 17.4 we introduced a *principle of superposition* for waves:

A meeting of two or more waves produces a waveform that is the algebraic sum of the waveforms produced by each wave acting separately.

The superposition of waves can result in *interference*, as shown for water waves in Figure 40.1. Circular water waves are generated at two points. At certain positions the superposition of the two circular wave trains results in *destructive interference*. The two waves cancel each other and the surface of the water is undisturbed. At other positions the superposition of the two wave trains results in *constructive interference*. The two waves reinforce one another and produce positions of maximum amplitude.

In general, interference is said to be *constructive* at points where the intensity of the superposed waves is greater than the sum of the intensities of the two separate waves. At points where the intensity is less than the sum of the intensities of the two separate waves, the interference is called *destructive*. In short, *constructive interference increases the intensity; destructive interference reduces the intensity*. The increased intensity over regions of constructive interference is balanced by the decreased intensity over regions of destructive interference.

Figure 40.1

Water waves are generated by two sources. When the two sets of waves meet, they interfere. The interference is constructive at certain points and destructive at other points.

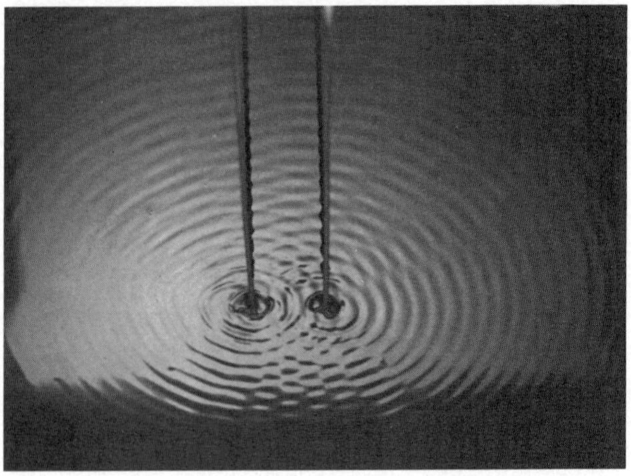

40.2 The Interference of Light: Thomas Young's Experiment

In ordinary circumstances, light does *not* display interference. For example, if two flashlights illuminate the same surface, their combined intensity is simply the sum of the two separate intensities. Likewise, the many floodlamps used to illuminate a stadium produce a total intensity that is the sum of the individual lamp intensities. As we will see later, the absence of interference is a consequence of the *incoherence* of ordinary light sources.

Under the proper conditions, however, light *does* display interference. The interference of light was first demonstrated in 1801 by Thomas Young, a brilliant English physician. Young's experiment, the basic features of which are shown in Figure 40.2, is important because it established the wave picture of light. In the experiment, monochromatic light from a sodium flame reaches a small hole in a screen. The hole acts like a point source, from which secondary wavelets expand in accord with Huygens' principle. Wave fronts from the point source reach two closely spaced slits in a second screen.* Secondary wavelets from these two slits spread out and superpose, interfering constructively at some places and destructively at others. When displayed on a third screen, the regions of constructive and destructive interference appear as an alternating series of bright and dark bands called *interference* fringes. Figure 40.3 shows the interference fringes produced by two closely spaced slits.

*In Young's original experiment there were two pinholes in the second screen rather than slits. The use of narrow slits increases the intensity of the interference pattern.

Figure 40.2

Young's interference experiment. Monochromatic light from a sodium flame reaches a pinhole in a card, from which secondary wavelets expand. The wavelets reach a second card in which there are two narrow slits. Wavelets from the two slits expand in accord with Huygens' principle. The interference of the secondary waves is revealed as a series of bright and dark fringes on the screen.

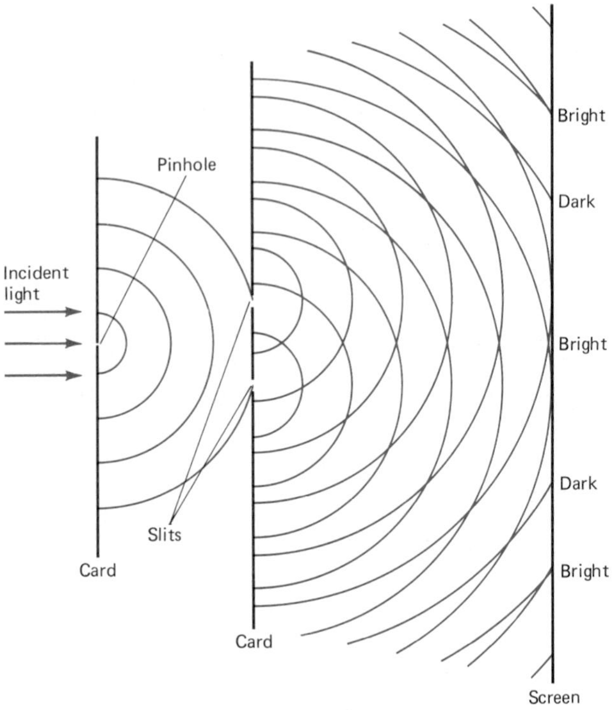

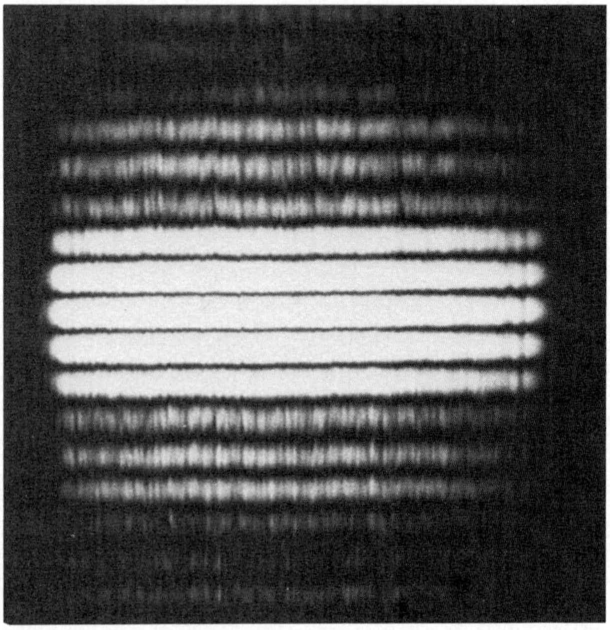

Figure 40.3

Interference fringes produced by light passing through two narrow slits.

We can explain the results of Young's experiment if we describe each wavelet in terms of its electric field, $E(x,t)$. We let E_1 and E_2 denote the E-fields of the wavelets from the two slits. For monochromatic light we can represent E_1 and E_2 in terms of sinusoidal waveforms like the one shown in Figure 40.4a. A mathematical description of the sinusoidal waveform* of Figure 40.4a is given by

$$E_1(x_1,t) = E_o \sin\left[2\pi\left(\frac{x_1}{\lambda} - \nu t\right)\right] \tag{40.1}$$

The quantity λ is the wavelength (the crest-to-crest distance in Figure 40.4a), the quantity ν is the light wave frequency. The quantity in brackets is called the *phase* of the wave and is represented by ϕ_1.

$$\phi_1 = 2\pi\left(\frac{x_1}{\lambda} - \nu t\right) \tag{40.2}$$

The field E_1 can be expressed in terms of the phase as

$$E_1 = E_o \sin \phi_1 \tag{40.3}$$

Likewise,

$$E_2(x_2,t) = E_o \sin \phi_2 \tag{40.4}$$

with

$$\phi_2 = 2\pi\left(\frac{x_2}{\lambda} - \nu t\right) \tag{40.5}$$

*Consult Sections 17.2 and 17.3 for an expanded treatment of sinusoidal waves.

Figure 40.4

(a) A sinusoidal waveform, representing the electric field of a monochromatic light wave. λ is the wavelength. The figure is a graph of the E-field.

$$E_1 = E_o \sin\left[2\pi\left(\frac{x_1}{\lambda} - \nu t\right)\right]$$

versus position along the direction of propagation, x_1, at the moment $t = 0$. (b) The E-field of the wave plotted versus the phase ϕ_1. The angle ϕ_1 is specified in radians. Comparing (a) and (b) shows that a phase difference of π rad corresponds to a shift of one-half wavelength ($\frac{1}{2}\lambda$). In (a), for example, the distance from a wave crest to an adjacent wave trough is $\frac{1}{2}\lambda$. In (b) there is a phase difference of π rad between a crest and an adjacent trough. Likewise, observe that a phase difference of 2π rad corresponds to a linear shift of one full wavelength.

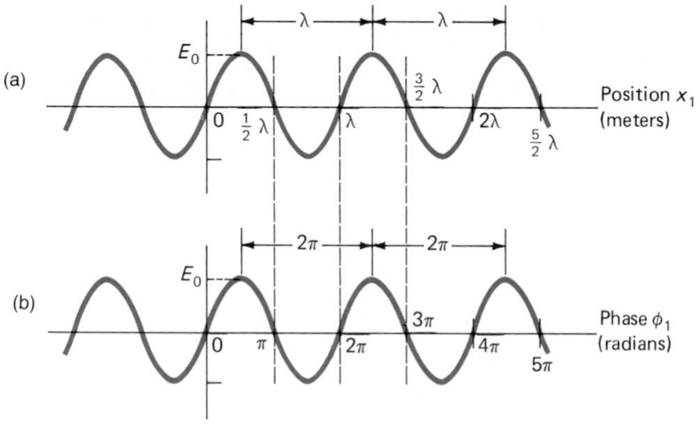

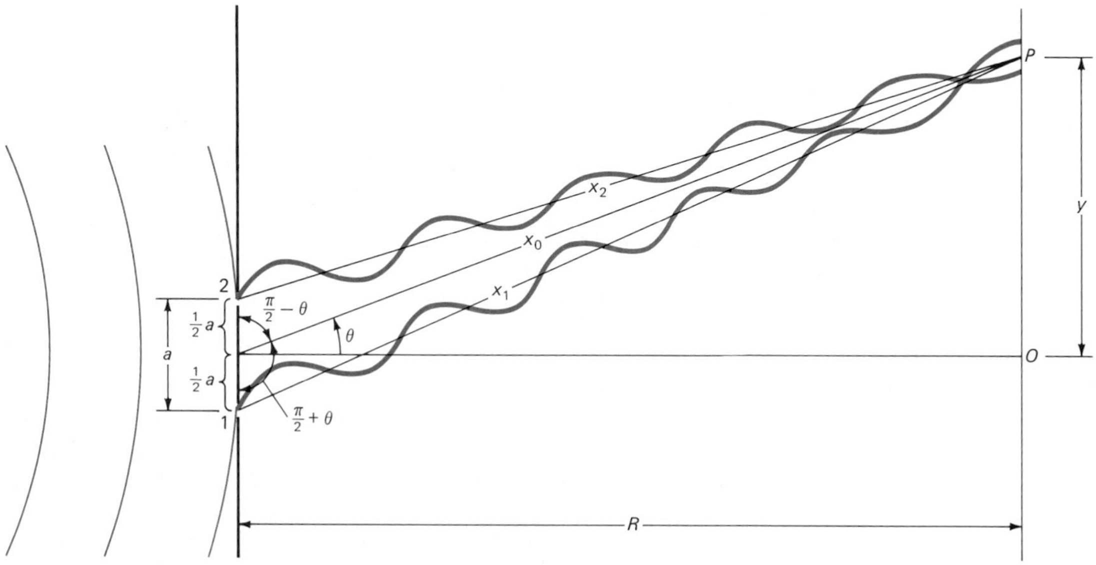

Figure 40.5

Top view of the geometry for the double slit. Waves spread out from the slits 1 and 2 and subsequently interfere. The nature of the interference at point P depends on the path difference, $x_1 - x_2$. If the path difference is an integral number of wavelengths, the waves interfere constructively and P is an intensity maximum. If the path difference is an odd number of half wavelengths, the waves interfere destructively and P marks an intensity minimum.

The distances x_1 and x_2 are measured from origins located at slits 1 and 2. E_1 and E_2 have equal amplitudes (E_o) because we assume that the two slits are identical.

The principle of superposition states that the resultant E-field at points where the two waves superpose is the sum $E_1 + E_2$. At the point P (Figure 40.5) where the two waves superpose the resultant E-field is

$$E = E_1 + E_2 = E_o \sin \phi_1 + E_o \sin \phi_2 \quad (40.6)$$

The phases of the two wavelets differ at P because the slit-to-screen distances (x_1 and x_2) differ.

$$\phi_1 = 2\pi\left(\frac{x_1}{\lambda} - \nu t\right) \quad (40.2)$$

$$\phi_2 = 2\pi\left(\frac{x_2}{\lambda} - \nu t\right) \quad (40.5)$$

The phase difference $\phi_1 - \phi_2$ is independent of time and depends on the difference in the path lengths, $x_1 - x_2$.

$$\phi_1 - \phi_2 = \frac{2\pi}{\lambda}(x_1 - x_2) \quad (40.7)$$

If we analyze the phase relationship between E_1 and E_2, we can determine the positions of maximum and minimum intensity. Referring to Figure 40.5, let's establish the conditions under which the point P is an intensity maximum (a point of constructive interference). If the *path difference* $x_1 - x_2$ is an integral number of wavelengths, the waves superposed at P are in phase. That is, crests of waves from slit 1 meet crests of waves from slit 2, giving a maximum wave amplitude and intensity at P. Thus, P will be a position of maximum intensity provided that

$$x_1 - x_2 = m\lambda$$

$$m = 0, 1, 2, \ldots, \text{ maxima} \quad (40.8)$$

We can transform Eq. 40.8 into an equation that locates the angular position of P (the angle θ in Figure 40.5) or the linear position of P (the distance y in Figure 40.5). We let a denote the slit separation and let x_o denote the distance from a point midway between the two slits to the point P. We can use the law of cosines to relate x_1, x_2, and θ:

$$x_1^2 = x_o^2 + \left(\frac{1}{2}a\right)^2 - x_o a \cos\left(\frac{\pi}{2} + \theta\right) \quad (40.9)$$

$$x_2^2 = x_o^2 + \left(\frac{1}{2}a\right)^2 - x_o a \cos\left(\frac{\pi}{2} - \theta\right) \quad (40.10)$$

Using the identities $\cos(\pi/2 \pm \theta) = \mp \sin \theta$ and subtracting Eqs. 40.9 and 40.10 leads to

$$x_1^2 - x_2^2 = 2x_o a \sin \theta$$

or

$$x_1 - x_2 = \frac{2x_o a \sin \theta}{x_1 + x_2} \quad (40.1)$$

In practice x_1 and x_2 are nearly equal to x_o. Thus, to a very good approximation, the denominator on the right-hand side of Eq. 40.11 is $2x_o$,

$$x_1 + x_2 \simeq 2x_o \quad (40.12)$$

which gives for the path difference

$$x_1 - x_2 = a \sin \theta \quad (40.13)$$

Inserting this result into Eq. 40.8, we get the angular positions of the *intensity maxima*,

$$a \sin \theta = m\lambda \qquad m = 0, 1, 2, \ldots \quad (40.14)$$

The integer m defines the *order* of the intensity maxima. Each value of m corresponds to a definite value of θ for which the interference is constructive. The positions of constructive interference are intensity maxima. The central maximum ($m = 0$) occurs at $\theta = 0$. Note that $\theta = 0$ corresponds to a position directly opposite the point midway between the slits (Figure 40.5). According to Young's wave theory, this point is an intensity maximum—a bright spot. In contrast, the particle picture of light predicts that the $\theta = 0$ position should be a shadow region—a dark spot. Young's experiment thus confirmed the wave theory by demonstrating that the $\theta = 0$ position is a bright spot. In Young's experiment light is behaving like a wave.

A first-order maximum occurs at an angle given by

$$a \sin \theta = \lambda \qquad m = 1 \quad (40.15)$$

Equation 40.15 shows that the first-order maximum occurs at different angles for different wavelengths, unlike the central maximum, which occurs at $\theta = 0$ for all wavelengths. For example, if white light illuminates the slits, then the central maximum is also white. But because the first-order maximum occurs at different positions for different wavelengths, the first-order maximum is a colored fringe pattern. The higher-order maxima also occur at different positions for different wavelengths.

From Figure 40.5 we see that the linear position of an intensity maximum, y, is related to the angular position θ by

$$\sin \theta = \frac{y}{\sqrt{R^2 + y^2}} \quad (40.16)$$

Thus, the linear positions of *intensity maxima* are given by

$$\frac{ay}{\sqrt{R^2 + y^2}} = m\lambda \qquad m = 0, 1, 2, \ldots \quad (40.17)$$

In many instances, the slit-to-screen distance R is much greater than the distance y. In these cases we can replace $\sqrt{R^2 + y^2}$ by R in Eq. 40.17 to obtain the following approximation.

$$\frac{ay}{R} \approx m\lambda \qquad m = 0, 1, 2, \ldots \quad (40.18)$$

If we rewrite this approximation as

$$y \approx \frac{Rm\lambda}{a} \quad (40.19)$$

we see that the position of an intensity maximum is directly proportional to the wavelength. Equations 40.17 and 40.19 show that the wavelength of light can be determined from measurements of a, y, and R.

Example 1 **Wavelength Measurement Using a Double Slit**

Red light illuminates a double-slit system with $a = 0.20$ mm. The first-order maximum occurs at $y = 3.3$ mm on a screen 1.0 m from the slits. We can determine the wavelength by using Eq. 40.19 with $m = 1$.

$$\lambda = \frac{a \cdot y}{R} = 2 \times 10^{-4}\ \text{m} \cdot \frac{3.3 \times 10^{-3}\ \text{m}}{1\ \text{m}}$$

$$= 6.6 \times 10^{-7}\ \text{m}$$

or

$$\lambda = 660\ \text{nm}$$

The remarkable aspect of the double-slit system is that it enables us to measure the wavelength of light—a length far too small to be gauged by any conventional method. To appreciate just how small 660 nm is, consider this: A string of 1500 wavelengths, each 660 nm long, would not span *one millimeter*. Greater precision is obtained by using a multiple-slit system (a diffraction grating, Section 41.4). The interference of light has become a powerful technique in modern science and technology for making many types of delicate measurements.

At points on the screen where waves from slits 1 and 2 arrive one-half wavelength "out of step" there is an *intensity minimum*. When we say that waves are a half wavelength out of step we mean that wave crests from slit 1 meet wave troughs from slit 2 and result in destructive interference. In general, a path difference that is any odd number of half wavelengths results in destructive interference. The angular positions of the intensity minima can be found by using the same procedure used to locate the intensity maxima, Eq. 40.14. The minima are located at angles given by

$$a \sin \theta = \left(m - \frac{1}{2}\right)\lambda \qquad m = 1, 2, \ldots \tag{40.20}$$

The linear positions of the minima are given by

$$y = \frac{R}{a}\left(m - \frac{1}{2}\right)\lambda \qquad m = 1, 2, \ldots \tag{40.21}$$

Example 2 **Casting Shadows with a Double-Slit System**

Light from a helium–cadmium laser ($\lambda = 442$ nm) passes through a double-slit system with $a = 0.40$ mm. Let's determine how far away a screen must be placed in order that a dark fringe appear directly opposite both slits (Figure 40.6).

Taking $m = 1$ and $y = 0.2$ mm in Eq. 40.21 gives

$$R = \frac{2ay}{\lambda} = \frac{2(0.4 \times 10^{-3})(0.2 \times 10^{-3})}{0.442 \times 10^{-6}} = 0.36\ \text{m}$$

or

$$R = 36\ \text{cm}$$

Geometric optics incorrectly predicts bright regions opposite the slits and darkness in between. But as this example shows, interference can produce just the opposite.

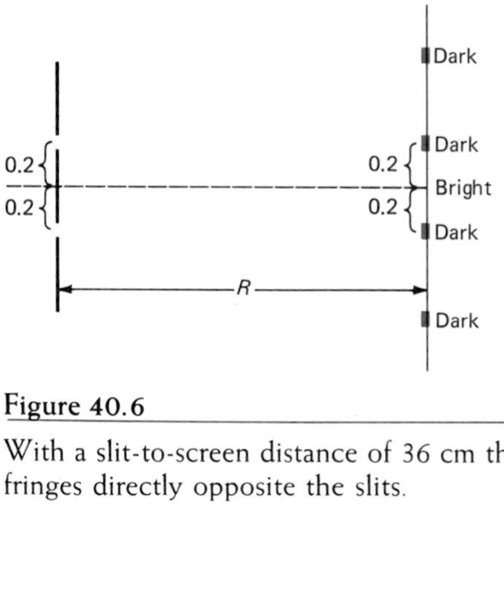

Figure 40.6

With a slit-to-screen distance of 36 cm there will be dark fringes directly opposite the slits.

Questions

1. If the wavelength is held constant, does decreasing the slit separation, a, in Figure 40.5 narrow or spread the interference pattern? That is, do the intensity maxima move closer together or farther apart? If the slit width is held constant, does decreasing the wavelength narrow or spread the interference pattern?
2. Give examples of the interference of (1) sound waves; (2) water waves. Try to think of examples where both constructive and destructive interference are evident.

40.3 Coherence

When we use a light source such as the sun or an ordinary light bulb to demonstrate the double-slit experiment shown in Figure 40.2, we find that the intensity of the interference pattern is quite low. We can increase the intensity greatly by removing the first card in Figure 40.2. This allows light from a broad source (the sun or a light bulb) to fall directly on the two slits. However, such an arrangement destroys the interference pattern—the intensity on the screen no longer shows the maxima and minima characteristic of interference. Evidently, the waves traveling away from the slits have a property that waves coming directly from a broad source do not. This critical property is called **coherence.** The superposition of waves from coherent sources results in interference.

Sources, and the waves they emit, are said to be coherent if they

1 have *equal frequencies*; and

2 maintain a *phase difference* that is *constant in time*.

If either property is lacking, the sources are incoherent, and the waves do not exhibit interference. Some examples of coherent waves and their sources are the following:

1 sound waves from two loudspeakers driven by the same audio oscillator;

2 electromagnetic waves from two microwave "horns" driven by the same oscillator;

3 sound waves from audio headsets, such as those used on commercial airliners;

4 light waves generated by Young's double-slit system;

5 light waves from a laser.

By contrast, the sound waves from two speakers are incoherent if the loudspeakers are driven by two audio oscillators tuned to different frequencies. Or, if two microwave horns are driven by two independent oscillators then the phase difference of the waves will vary randomly with time, and the waves will not interfere. A neon sign is an incoherent source even though some of its light waves have the same frequency as the helium–neon laser light. In the laser (Section 42.6), the light output is the product of neon atoms acting *in concert*—coherently. The incoherent light streaming from a neon sign is the result of the superposition of waves that do not maintain a constant phase difference.

To study coherence in detail we must examine the intensity pattern produced by the superposition of two waves. In Section 38.5 we showed that the average intensity of a monochromtic plane wave is given by

$$I = c\epsilon_o(E^2)_{av} \tag{40.22}$$

where $(\ )_{av}$ denotes a time average over an integral number of cycles of the wave. We can use this expression for the intensity I to illustrate the key aspects of interference.

Consider the superposition of light waves form two sources. The combined E-field is the sum

$$E = E_1 + E_2 \tag{40.23}$$

where E_1 and E_2 are the E-fields produced by the two sources. The time-averaged value* of E^2 is

$$(E^2)_{av} = (E_1^2)_{av} + (E_2^2)_{av} + 2(E_1E_2)_{av} \tag{40.24}$$

The intensity is

$$I = c\epsilon_o(E_1^2)_{av} + c\epsilon_o(E_2^2)_{av} + 2c\epsilon_o(E_1E_2)_{av} \tag{40.25}$$

If we compare this equation to Eq. 40.24, we see that

$$c\epsilon_o(E_1^2)_{av} = I_1 \tag{40.26}$$

and

$$c\epsilon_o(E_2^2)_{av} = I_2 \tag{40.27}$$

are the intensities of the two separate sources. Therefore we can write

$$I = I_1 + I_2 + 2c\epsilon_o(E_1E_2)_{av} \tag{40.28}$$

The product term $2c\epsilon_o(E_1E_2)_{av}$ is called the *interference term*. It represents the interference between the two sources. If

$$(E_1E_2)_{av} = 0 \tag{40.29}$$

the sources are *incoherent* and there is no interference. With $(E_1E_2)_{av} = 0$, Eq. 40.28 becomes

$$I = I_1 + I_2 \tag{40.30}$$

The intensities simply add when the sources are incoherent. For example, when the headlights of an automobile illuminate the same area, their combined intensity is simply the sum of the two separate intensities. The headlights are incoherent sources and there is no interference.

When light sources are coherent, the intensities do not simply add. Coherence alters the intensity. Let's consider two cases where the interference term is not zero, $(E_1E_2)_{av} \neq 0$. In both cases two coherent light waves of equal amplitude are superposed. The interference is constructive in one instance, and destructive in the other.

1 Let $E_1 = E_2$. This means that the two E-fields have the same amplitude, frequency, and phase. The intensities of the individual waves are equal.

$$I_1 = I_2 = c\epsilon_o(E_a^2)_{av} \tag{40.31}$$

and the interference term is

$$2c\epsilon_o(E_1E_2)_{av} = 2I_1 \tag{40.32}$$

The total intensity is

$$I = I_1 + I_2 + 2c\epsilon_o(E_1E_2)_{av} = 4I_1 \tag{40.33}$$

which is twice the intensity that the two sources would produce if they were incoherent. In this instance the coherent waves interfere *constructively*.

2 Let $E_1 = -E_2$. This describes the superposition of two waves with the same amplitude and frequency, but with a constant phase difference of 180° (π). Their superposition results in **complete destructive interference.** The resultant wave amplitude and intensity are zero.

$$E = E_1 + E_2 = 0$$

$$I = 0$$

The superposition of incoherent light waves can never result in zero intensity. The situation in which $I = 0$ requires coherent waves and is the limit of complete destructive interference.

Superposition of Sinusoidal Waves

The two special cases just described illustrate the extremes of interference effects. We can obtain more general results by considering the superposition of two sinusoidal plane waves.

$$E_1 = E_o \sin \phi_1 = E_o \sin\left[2\pi\left(\frac{x_1}{\lambda_1} - \nu_1 t\right)\right] \tag{40.34}$$

$$E_2 = E_o \sin \phi_2 = E_o \sin\left[2\pi\left(\frac{x_2}{\lambda_2} - \nu_2 t\right)\right] \tag{40.35}$$

With $E = E_1 + E_2$ as before, the intensity $I = c\epsilon_o(E^2)_{av}$ is

$$I = c\epsilon_o(E_1^2 + E_2^2 + 2E_1E_2)_{av} \tag{40.36}$$

As before we have the results

$$(E_1^2)_{av} = E_o^2(\sin^2 \phi_1)_{av} = \frac{1}{2}E_o^2$$

$$(E_2^2)_{av} = E_o^2(\sin^2 \phi_2)_{av} = \frac{1}{2}E_o^2$$

We can write

$$(E_1E_2)_{av} = E_o^2(\sin \phi_1 \sin \phi_2)_{av} \tag{40.37}$$

*The time-averaging operation is an integration over time and is therefore linear—the average of a sum equals the sum of the averages.

Inserting these results into Eq. 40.36 and noting that

$$\frac{1}{2} c\epsilon_o E_o^2 = I_1 \quad (40.38)$$

is the intensity of one wave, we can express I as

$$I = 2I_1[1 + 2(\sin \phi_1 \sin \phi_2)_{av}] \quad (40.39)$$

The term $2(\sin \phi_1 \sin \phi_2)_{av}$ incorporates interference effects. Using the trigonometric identity, we write

$$2 \sin \phi_1 \sin \phi_2 = \cos(\phi_1 - \phi_2) - \cos(\phi_1 + \phi_2) \quad (40.40)$$

For our situation,

$$\phi_1 = 2\pi\left(\frac{x_1}{\lambda_1} - \nu_1 t\right) \quad (40.41a)$$

$$\phi_2 = 2\pi\left(\frac{x_2}{\lambda_2} - \nu_2 t\right) \quad (40.41b)$$

and the time average of $\cos(\phi_1 + \phi_2)$ is zero. The time average of $\cos(\phi_1 - \phi_2)$ is zero *unless* $\nu_1 = \nu_2$. Thus, the interference term is zero whenever the waves have different frequencies,

$$2(\sin \phi_1 \sin \phi_2)_{av} = [\cos(\phi_1 - \phi_2)]_{av} = 0$$

$$\nu_1 \neq \nu_2 \quad (40.42)$$

If $\nu_1 \neq \nu_2$, then the sources are incoherent and Eq. 40.39 shows that the intensity is simply the sum of the (equal) intensities of the two waves.

$$I = 2I_1 \quad (40.43)$$

For coherent sources the waves have the same frequency. The frequency ν and the wavelength λ are related to the speed of light c by the kinematic relation (Section 17.2)

$$\nu\lambda = c \quad (40.44)$$

It follows that when $\nu_1 = \nu_2$, the two waves also have equal wavelengths.

$$\lambda_1 = \lambda_2 \equiv \lambda \quad (40.45)$$

With $\nu_1 = \nu_2$, the phase difference $\phi_1 - \phi_2$ is independent of time. Thus

$$\phi_1 - \phi_2 = \frac{2\pi}{\lambda}(x_1 - x_2) \qquad \nu_1 = \nu_2 \quad (40.46)$$

and the interference term reduces to

$$2(\sin \phi_1 \sin \phi_2)_{av} = \cos(\phi_1 - \phi_2)$$

$$= \cos\left[\frac{2\pi}{\lambda}(x_1 - x_2)\right] \quad (40.47)$$

Inserting this result into Eq. 40.39, we get the intensity for coherent sources

$$I = 2I_1\left\{1 + \cos\left[\frac{2\pi}{\lambda}(x_1 - x_2)\right]\right\} \quad (40.48)$$

Equation 40.48 shows how the interference of coherent waves alters the intensity from the value it would have for incoherent waves ($2I_1$). The quantity $\cos[(2\pi/\lambda)(x_1 - x_2)]$ can take on values ranging from -1 to $+1$. When the phase difference $(2\pi/\lambda)(x_1 - x_2)$ is π rad (180°), $\cos[(2\pi/\lambda)(x_1 - x_2)] = -1$ and Eq. 40.48 shows that $I = 0$. Thus we get complete destructive interference when we superpose coherent waves of equal amplitude that differ in phase by π rad. Complete constructive interference occurs when the phase difference is 2π rad (360°). In this case, $\cos[(2\pi/\lambda)(x_1 - x_2)] = +1$ and $I = 4I_1$. These extremes of complete destructive interference and complete constructive interference are the two special cases that we considered earlier, $E_1 = \pm E_2$.

Coherence Length

A wave front is defined as a surface of constant phase. Thus, the notion of a wave front implies one type of coherence—a lateral phase coherence *across* the wave front. A beam of light also displays *longitudinal* coherence—a phase coherence *along* the direction of wave travel. The distance along the direction of propagation over which such phase coherence exists is called the *coherence length.*

If an atom radiates for a time t, the emitted wave train has a length ct, where c is the speed of light (Figure 40.7). The coherence length L_c is given by

$$L_c = ct \quad (40.49)$$

For a typical incoherent light source the coherence length is roughly 1 mm. For example, a sodium atom typically radiates for 4×10^{-12} s. Using Eq. 40.49 we find for the coherence length

$$L_c = ct = \left(3 \times 10^8 \frac{\text{m}}{\text{s}}\right)(4 \times 10^{-12}\ \text{s}) = 1.2\ \text{mm}$$

The light emitted by individual atoms forms short wave trains because the radiation process lasts only a short time. The coherence length is the average length of these wave trains. In a typical interference experiment the wave trains are divided, either physically (by the use of slits) or optically (by reflection). The two sets of wave trains are then sent along different paths that ultimately reunite them. If the paths differ by more than the coherence length, there is no interference because the reunited waves are not parts of the same group of wave trains. They are incoherent.

Light reflected from an oil film can exhibit interference (Figure 40.8). Waves are reflected from both the top and the bottom surfaces of the film, and superimposed when they enter the eye. Interference is evident provided that the thickness of the film does not exceed the coherence length. If the film thickness exceeds the coherence length, then the reflected waves are not parts of the same group of wave trains. There is no phase coherence between the reflected waves and thus no interference.

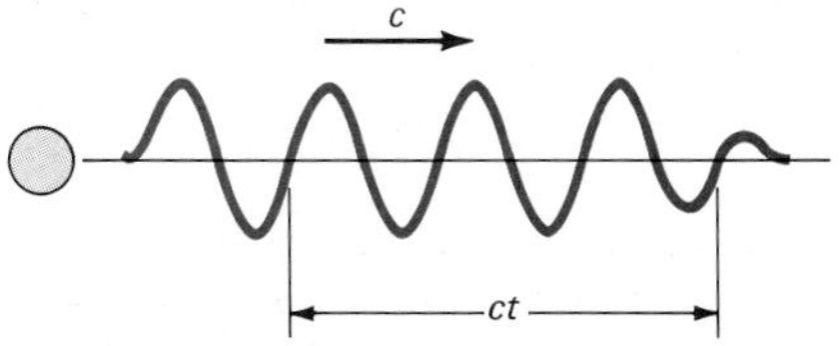

Figure 40.7

An atom that radiates for a time t generates a wave train of length ct. The length of these wave trains is the coherence length.

Figure 40.8

(a) Wave trains (1, 2, 3) of finite length strike an oil film. Coherence extends over the length of each wave train. There is no coherence between different wave trains. Portions of each wave train are reflected by the front and back faces of the film. Thus, each wave train gives rise to two reflected wave trains (F1, B1; F2, B2; F3, B3). The light reflected by the film is a superposition of wave trains reflected from the front and rear faces. In (a) the difference in the optical path lengths of F1 and B1 is small compared to the coherence length (length of wave train). Consequently, the reflected wave exhibits interference because it is a superposition of coherent waves (F1 and B1, F2 and B2, F3 and B3). For the much thicker film in (b) the difference in optical path lengths of F1 and B1 is larger than the coherence length. The reflected wave does not exhibit interference because it is a superposition of incoherent waves (F2 and B1, F3 and B2).

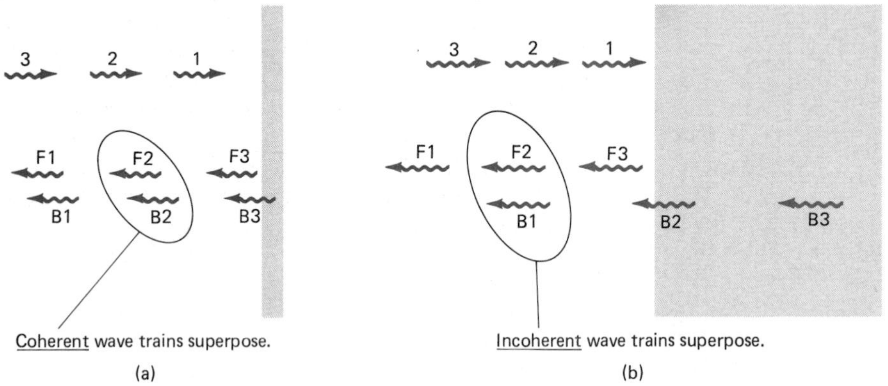

Questions

3. Light waves are reflected from the interior and exterior faces of a department store window. Why don't the two sets of reflected waves show interference effects?
4. Define coherence in your own words. In what way is your definition similar to the nonscientific meaning of the word *coherence*?
5. Which of the following have coherence: (a) raindrops striking a pavement; (b) sound from a corps of drummers; (c) electrons striking a TV screen; (d) your footsteps; (e) the footsteps of marchers in Macy's Thanksgiving Day parade?
6. What coherence length would you expect for a source whose individual atoms radiate for times of roughly 10^{-9} s?

40.4 Thin-Film Interference

Light reflected from oil films, soap bubbles, and other thin films of material often exhibits beautiful interference effects. In practice such interference is most pronounced when the films are very thin—no more than a few times the wavelength of light. Accordingly, we call this interference *thin-film interference.*

The concept of *optical path length* is useful in our study of thin-film interference. If light travels a distance x in a medium of refractive index n, its optical path length is defined as nx.

Optical path length

$$= \text{(index of refraction)(physical path length)} \qquad (40.50)$$

To show why optical path length is a useful concept, we consider the phase difference between two coherent waves that are superposed after traveling different distances (x_1 and x_2) through a medium with an index of refraction n. Their phases differ by

$$\delta = \frac{2\pi}{\lambda_n}(x_1 - x_2) \qquad (40.51)$$

where λ_n is the wavelength of light in the medium. It is convenient to refer all wavelengths to their vacuum value. We can replace λ_n by λ, the corresponding vacuum wavelength, if we also replace $x_1 - x_2$ by $n(x_1 - x_2)$, the difference in optical path lengths. To prove this we first note that when a wave crosses a boundary from one medium to another there is a change in wave speed and wavelength, but not in wave frequency. The frequency equals the number of wave crests per second reaching the boundary in one medium, and it also equals the number per second traveling away from the boundary in the second medium. No wave crests can "get lost" at the boundary—equal numbers must arrive and leave each second. Thus, the wave frequency remains unchanged when a wave crosses a boundary.

Next we compare the basic kinematic relation for a vacuum and for the medium. In a vacuum the frequency ν and wavelength λ are related to the speed of light c by

$$\nu\lambda = c$$

In a material medium the speed of light v differs from c, and the wavelength λ_n differs from its vacuum value, but the frequency remains unchanged.

$$\nu\lambda_n = v \tag{40.52}$$

The ratio

$$n \equiv \frac{c}{v} = \frac{\nu\lambda}{\nu\lambda_n} = \frac{\lambda}{\lambda_n}$$

introduces the index of refraction, n, and shows that

$$\frac{1}{\lambda_n} = \frac{n}{\lambda} \tag{40.53}$$

Inserting this result into Eq. 40.51 shows that we can replace λ_n by λ provided that we also replace the path length difference by the optical path length difference.

$$\delta = \frac{2\pi}{\underbrace{\lambda}_{\text{vacuum wavelength}}}\overbrace{n(x_1 - x_2)}^{\text{optical path length difference}} \tag{40.54}$$

Note that when the optical path lengths differ by one vacuum wavelength, the phases differ by 2π.

Example 3 **Optical Path Length in a Soap Film**

A soap bubble ($n = 1.333$) is 300 nm thick. Green light ($\lambda = 530$ nm) strikes the surface and is partially reflected and partially transmitted. The transmitted wave travels through the film, where it is again partially reflected and partially transmitted. The wave that crosses and returns travels a path of length 600 nm. We determine its optical path length as follows:

$$\text{Optical path length} = n \times 600 \text{ nm} = 800 \text{ nm}$$

The change in phase for the round trip is $(2\pi/\lambda) \times$ optical path length.

$$\delta = \frac{2\pi}{530} \times 800 = 9.48 \text{ rad}$$

Phase Change Accompanying Reflection

Equation 40.51 gives the phase difference resulting from differences in optical path lengths. Additional phase changes may occur as a result of *reflection*. Specifically, we assert that reflection has the following effects:

1 A reflected wave undergoes a phase change of π rad if it travels *faster* than the transmitted wave.

2 A reflected wave undergoes no phase change it if travels slower than the transmitted wave.

The criterion for reflected-wave phase change can also be expressed in terms of indices of refraction. If the incident wave strikes a surface of higher index of refraction, the reflected wave undergoes a phase change of π rad. This follows from the original statement of the criterion and the definition of n, the index of refraction (Section 39.1),

$$n = \frac{c}{v}$$

Thus, a higher n corresponds to a slower wave speed. Likewise, if the incident wave strikes a surface of lower index of refraction, there is no phase change in the reflected wave. For example, consider a light wave that strikes a window pane (Figure 40.9). The wave reflected from the outer surface experiences a phase change of π rad relative to the incident wave because the index of refraction for glass is higher than that of air. If we follow the transmitted wave through the glass, there will be a second reflection at the inner surface, where the light emerges. This reflection produces no phase change because the wave encounters a surface of lower index of refraction (glass –air).

A rigorous proof of the phase change effects of reflection involves Maxwell's equations. However, there are several easier ways to simply *demonstrate* the phase change that accompanies reflection. For example, we can consider a thin soap film suspended vertically in a wire loop (Figure 40.10). Gravity establishes a variable thickness in the film; the thinnest portion is at the top of the loop. In fact, the thickness of the film at the top of the loop is small compared to the wavelength of visible light. Light reflected from such a thin film exhibits destructive interference, as shown by Figure 40.10. Figure 40.11 shows the origin of the destructive interference. Reflected light entering the eye is a superposition of waves reflected from the two air–film boundaries. Light reflected at the front surface undergoes a phase change of π rad. Light reflected from the rear surface experiences no phase change. The round trip across the film and back introduces an insignificant phase change because the film is so thin. Thus, a phase difference of π rad is established between the two reflected waves, and they interfere destructively.

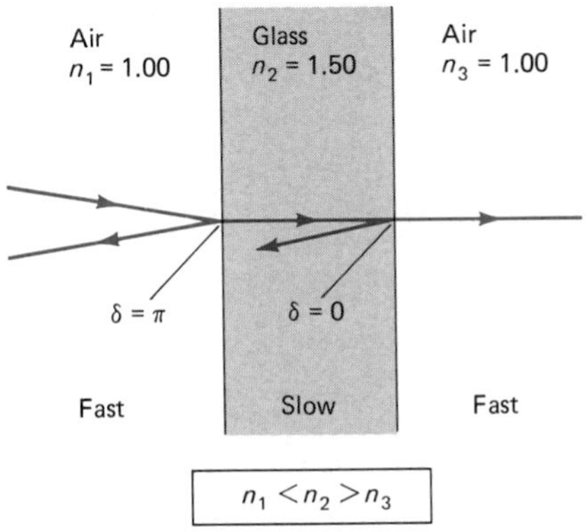

Figure 40.9

At the left (air to glass) interface the reflected wave undergoes a phase change of π rad relative to the incident wave. At the right (glass to air) interface the reflected wave undergoes no phase change.

Figure 40.10

A soap film suspended by a wire loop viewed by reflected light. The upper portion of the film appears black because of destructive interference between waves reflected from the front and rear surfaces of the film.

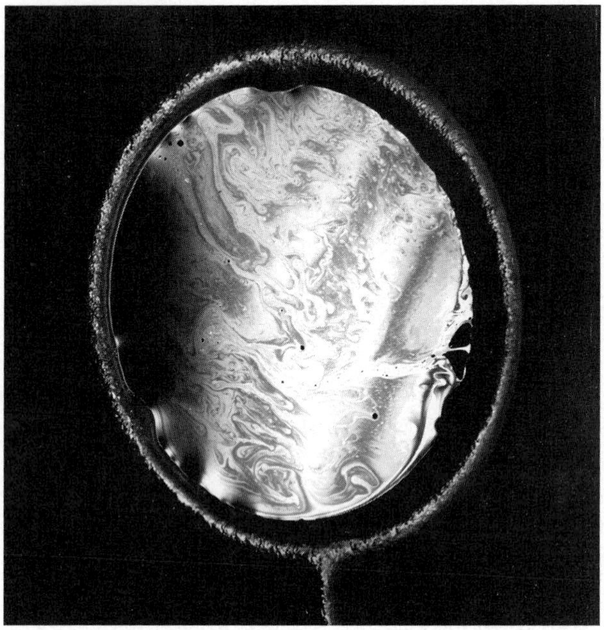

Figure 40.11

The reflection at the front surface introduces a phase shift of π rad (fast to slow). The reflection at the rear surface does not introduce any phase change (slow to fast). The phase difference of π rad between the two reflected waves results in destructive interference.

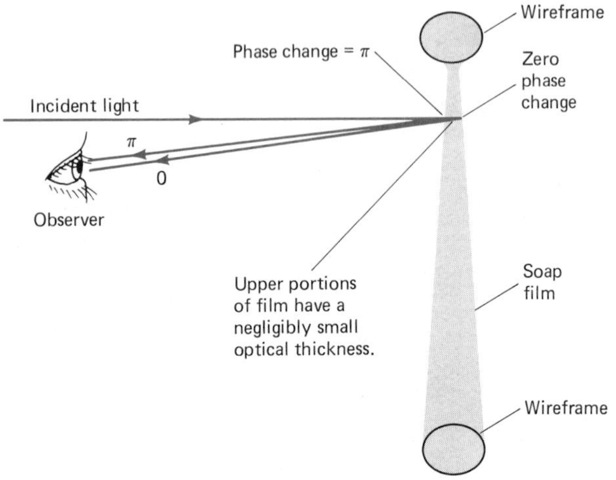

Thin-film interference results from the combined effects of optical path length differences and reflection phase shifts. Consider the situation shown in Figure 40.12. White light is normally incident on a soap bubble. The light that enters the eye is a superposition of waves reflected from the two air–film boundaries. If white light is incident, the reflected light is not white. Certain reflected wavelengths interfere destructively, and these wavelengths are removed from the reflected light. Like-

Figure 40.12

White light is normally incident on a soap film. The eye responds to light formed by the superposition of waves reflected from the two air–film boundaries. The difference in optical path lengths of the two waves is $2nd$. The wave reflected from the upper surface undergoes a phase change of π. The wave reflected from the lower surface experiences no phase change.

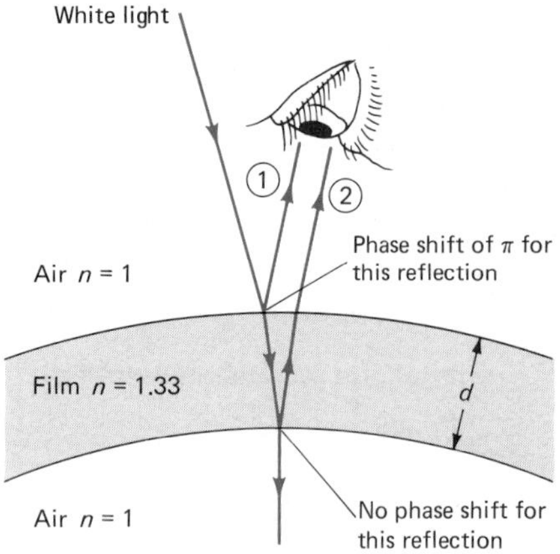

wise, certain reflected wavelengths interfere constructively, and the reflected light is richer in colors corresponding to these wavelengths.

For a film of thickness d the total phase difference between the two waves is

$$\delta = \underbrace{\frac{2\pi}{\lambda} n(2d)}_{\text{due to optical path difference}} + \underbrace{\pi}_{\text{due to air–film reflection from top surface}} \qquad (40.55)$$

The reflected waves interfere constructively when δ is an integral multiple of 2π.

$$\delta = 2\pi,\ 4\pi,\ 6\pi,\ \ldots \qquad \textit{constructive interference} \qquad (40.56)$$

By combining Eqs. 40.55 and 40.56, we get the conditions for constructive interference of the reflected light,

$$2nd = \left(m + \frac{1}{2}\right)\lambda \qquad m = 0, 1, 2, \ldots \qquad (40.57)$$

Wavelengths satisfying Eq. 40.57 are reflected strongly. These same wavelengths are missing from the transmitted light.

The reflected waves interfere destructively when the phase difference δ is an odd multiple of π.

$$\delta = \pi,\ 3\pi,\ 5\pi,\ \ldots \qquad \textit{destructive interference} \qquad (40.58)$$

By combining Eqs. 40.55 and 40.58 we get the conditions for destructive interference of the reflected light,

$$2nd = m\lambda \qquad m = 0, 1, 2, \ldots \qquad (40.59)$$

Wavelengths satisfying Eq. 40.59 are missing from the reflected light. This means they are completely transmitted.

Example 4 Reflected and Transmitted Colors

A soap film has a thickness of 300 nm and an index of refraction of 1.33. What colors are strongly reflected and transmitted by this film? Wavelengths for which interference is constructive are most prominent in the reflected light. Using Eq. 40.57 we can show that these wavelengths are

$$\lambda = \frac{2nd}{m + \frac{1}{2}} = \frac{2(1.33)(300)}{m + \frac{1}{2}} = \frac{800}{m + \frac{1}{2}}\ \text{nm}$$

Taking $m = 0$, 1, and 2, we get $\lambda = 1600$ nm, 533.3 nm, and 320 nm, respectively. Larger values of m give shorter wavelengths. A wavelength of 1600 nm lies in the infrared (IR) region of the spectrum, above the visible range (410–700 nm). The 320-nm wavelength is in the ultraviolet (UV) region, below the visible spectrum. The only wavelength within the visible spectrum that interferes constructively is 533.3 nm, which corresponds to the color green.

Equation 40.59 allows us to determine wavelengths that interfere destructively. These wavelengths are "missing" in the reflected light and hence are strongly transmitted. From Eq. 40.59 we find that for $m = 1$, 2, and 3, the wavelengths are $\lambda = 800/m = 800$ nm, 400 nm, and 266.7 nm, respectively. None of these wavelengths lies in the visible range, although the first two are close to the red and violet ends of the visible spectrum.

The film will have a greenish color when viewed by reflected light because green is strongly reflected and red and blue are transmitted. When viewed in transmitted light the film will appear purple—a mixture of red and blue with very little green.

Antireflection Coatings

A thin film of material can be used to reduce the intensity of reflected light. Camera lenses are often coated with thin films to reduce reflection. Such antireflection coatings result in destructive interference of the reflected waves. A single-layer coating can achieve zero reflectance for only a very narrow range of wavelengths. We will assume that the index of refraction of the coating is intermediate between those of the air and lens. The two most commonly used coatings have refractive indices of 1.38 (magnesium fluoride) and 1.35 (cryolite). In these circumstances, reflections at both surfaces produce phase changes of π rad (Figure 40.13). Therefore, the overall phase difference between the two reflected waves depends on the difference of their optical path lengths. For a coating of thickness d and refractive index n, the optical paths differ by $2nd$ and the phase difference between the two reflected waves is

$$\delta = 2\pi\left(\frac{2nd}{\lambda}\right) \tag{40.60}$$

The reflected waves interfere destructively when $\delta = \pi$; that is, when the two waves are one-half wavelength out of step. Setting $\delta = \pi$ in Eq. 40.60 gives

$$d = \frac{\lambda}{4n} \tag{40.61}$$

which makes the film one-quarter wavelength thick.*

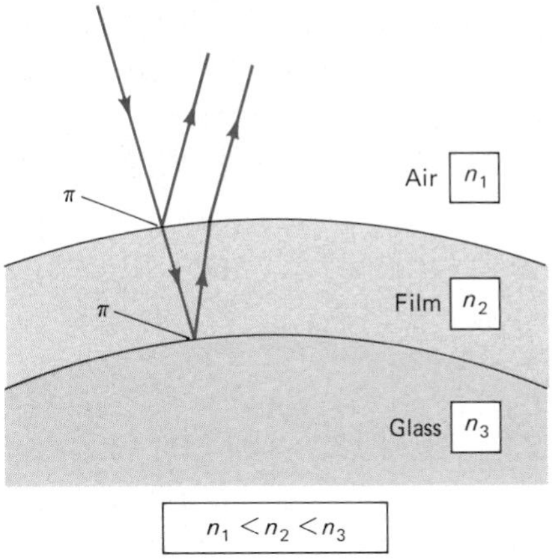

Figure 40.13

Phase changes of π radians occur at both reflecting surfaces. The net phase difference between the reflected waves depends on the difference in optical path lengths ($2nd$) and the wavelength (λ).

Example 5 Thickness of an Antireflection Coating

Let us determine the thickness of an antireflection coating of magnesium fluoride (MgF_2), for which $n = 1.38$. We want the film to give maximum transmission at 550 nm, a wavelength in the middle of the visible spectrum. Equation 40.61 gives for the thickness

$$d = \frac{550}{4(1.38)}\ \text{nm} \approx 99.6\ \text{nm}$$

Thin-film interference can also be used to produce highly reflecting coatings. A film thickness of one-half wavelength produces a path difference of one wavelength and results in constructive interference for the reflected light. Such coatings are particularly useful in fabricating laser mirrors, where high reflectivity for only a single wavelength is desired.

Questions

7. A very thin film (thickness ≪ wavelenth), illuminated by white light and viewed by reflection, may appear black. What condition on the indices of refraction of the film and the medium "behind" it leads to this situation?

8. At night you can see into a well-lighted room from outside, but a person inside the room cannot see you. On a bright sunny day the reverse may be true. Explain.

*Recall that λ is the wavelength in a vacuum; λ/n is the wavelength inside the film.

9. The rear and side windows on passenger vans are sometimes coated so as to act as one-way mirrors. How do such mirrors work? Consider the levels of illumination for observers on both sides of the mirror. Is it possible for any kind of coating to be highly transmitting from one side and highly reflecting from the other (at the same wavelength)? (Do you know of any optical principle that applies?)

10. The light reflected from a coated camera lens has a purple cast. What colors are most completely transmitted by the lens?

40.5 Optical Interferometers

An optical interferometer is an instrument that allows us to measure small changes in optical path length by observing an interference pattern. Basically, an interferometer operates by splitting a beam of light, sending the two beams along different paths, and then recombining the beams to produce an interference pattern. There are many varieties of optical interferometers. We will now describe two, the Rayleigh interferometer and the Michelson interferometer.

The Rayleigh Interferometer

A diagram of the Rayleigh interferometer is shown in Figure 40.14. A slit source of light (S) is placed at the focal point of a lens (L_1). Light emerges from the lens in the form of plane waves that strike a double-slit system (S_1, S_2). The secondary waves emerging from S_1 and S_2 are *coherent* and travel along parallel paths through two identical sealed tubes (T_1, T_2). A second lens (L_2) superposes the waves, which *interfere* and produce a fringe pattern (P).

The Rayleigh interferometer can be used to measure the index of refraction of gases. Suppose, for example, that one tube is filled with a gas having a refractive index n, and the other tube is evacuated. The optical path length through the tube of gas is nx, where x is the length of the tube. The optical path length through the evacuated tube is simply x, because the index of refraction for a vacuum is unity. The difference in optical paths is $(n - 1)x$. If the gas-filled tube is slowly evacuated, the difference in optical path lengths drops to zero and the interference pattern changes. Each time $(n - 1)x$ changes by one wavelength, the interference pattern returns to its original form. Let N denote the number of times the interference pattern returns to its original form as the tube is evacuated (N is called the *number of fringe shifts**). Then $N\lambda$ equals the change in the optical path length difference:

$$N\lambda = (n - 1)Z \qquad (40.62)$$

In the following example we use Eq. 40.62 to determine the index of refraction of air.

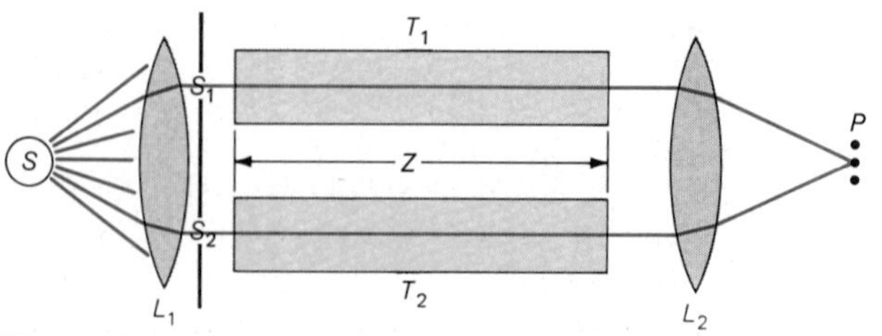

Figure 40.14

The Rayleigh interferometer. A light source (S), a lens (L_1), and a double-slit system (S_1, S_2) generate coherent light waves that travel through the sealed tubes (T_1, T_2). The lens L_2 superposes the waves to produce an interference pattern (P).

Example 6 **Index of Refraction of Air**

The tube length of a Rayleigh refractometer is 10.0 cm. Initially, one tube is filed with air and the other is evacuated. Using light with a wavelength of 656.3 nm we observe 44.5 fringe shifts as the air is removed. Using Eq. 40.62, we can calculate $n - 1$.

$$n - 1 = \frac{N\lambda}{Z} = \frac{44.5 \times 656.3 \times 10^{-9}}{0.100}$$

$$= 2.92 \times 10^{-4}$$

Thus, the index of refraction of air for a wavelength of 656.3 nm is

$$n = 1.000292$$

The Michelson Interferometer

The Michelson interferometer (Figure 40.15) was developed by the American physicist A. A. Michelson. Figure 40.16 indicates how the Michelson interferometer produces an interference pattern. Light from a source strikes a half-silvered mirror (HSM), a glass plate with a thin metallic coating on its back. The thickness of the coating is such that half of the incident intensity is reflected and half is transmitted. The half-silvered mirror thereby serves as a beam splitter, producing two coherent light waves of equal amplitude. The reflected wave travels to a movable mirror (MW) and then retraces its path back to the half-silvered mirror. The transmitted wave passes through a compensator plate (C), strikes a fixed mirror (FM), and then also retraces its path to the half-silvered mirror. The compensator is an unsilvered "twin" of the half-silvered mirror. It is inserted so that both waves make three passes through equal thicknesses of glass. Because of the compensator plate, any difference in optical path lengths comes about as a result of the difference in physical path lengths between the half-silvered mirror and the other two mirrors. The two reflected waves return to the half-silvered mirror, where both are again split into two waves of equal amplitudes. The transmitted portion of one wave and the reflected portion of the other wave are superposed and produce an interference pattern.

*In general N is not an integer. The visual acuity of the human eye is such that changes as small as 1/40 of one wavelength can be measured reliably.

Figure 40.15

A Michelson interferometer.

Figure 40.16

Light paths for the Michelson interferometer. Light from the source (S) reaches the half-silvered mirror (HSM), where it is split into two coherent waves of equal amplitude. The reflected portion ① travels to a movable mirror (MM); the transmitted portion ④ travels to a fixed mirror. The two waves are reflected and retrace their paths, ② and ⑤, to the HSM. The transmitted portion ③ of one wave and the reflected portion ⑥ of the other wave are superposed and produce an interference pattern at the position of the observer.

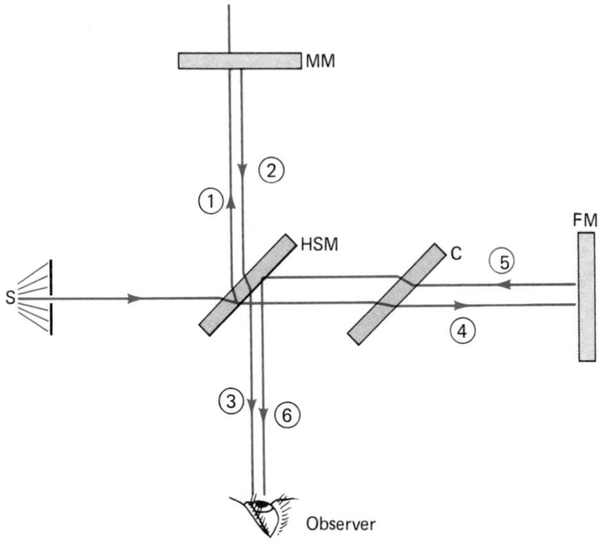

The interference pattern appears as a series of concentric interference fringes. The targetlike pattern of bright and dark fringes shows that positions of constructive and destructive interference are located symmetrically about the axis of the interferometer.

If the movable mirror is displaced a distance d, the optical path length of the waves traveling to and from it is changed by $2d$. If we start with some particular interference pattern, say a dark central region, and increase d until $2d$ equals one wavelength, then the interference pattern returns to its original form. In other words, a relative shift of one wavelength leaves the interference pattern unchanged. Similarly, an increase of $2d$ by two, three, or any integral number of wavelengths will reproduce the original pattern. In general, then, a displacement of the mirror by a distance d will cause the pattern to reproduce itself m times, when

$$2d = m\lambda \qquad (40.63)$$

In order to obtain interference it is necessary that the difference in optical paths not exceed the coherence length. For laser light this is seldom a problem, since coherence lengths of 30 cm or more are typical.

Example 7 **The Wavelength of Helium–Neon Laser Light**

Light from a helium–neon laser is beamed into a Michelson interferometer. The movable mirror is displaced 0.567 mm, causing the interference pattern to reproduce itself 1800 times. The wavelength follows from Eq. 40.63 as

$$\lambda = \frac{2d}{m}$$

with

$$d = 0.567 \times 10^{-3}\ \text{m}$$

$$m = 1800$$

We get

$$\lambda = 630\ \text{nm}$$

Because the displacement d in this example is measured to only three-significant-figure accuracy, the precision of the resulting wavelength is limited. A more precise value for the wavelength of the helium–neon laser light is 632.8 nm.

You should not conclude from Example 7 that Michelson's interferometer is an instrument of low precision. Michelson obtained very high precision when he used his interferometer to compare the standard meter in Sèvres, France, with the wavelength of red light emitted by cadmium atoms. His value for the wavelength of red light has seven significant figures, or in other words, is precise to within one part in 10 million. More important, perhaps, is that Michelson's work led to a definition of the standard meter in terms of the wavelength of light (Section 1.3). Using interferometers, scientists now measure distances in terms of the wavelength of light, accurate to a fraction of one wavelength of light.

Experiments performed with this interferometer by Michelson and E. W. Morley helped establish an experimental basis for Einstein's special relativity theory (Section 19.3). In 1887 Michelson and Morley attempted to measure the speed of the earth relative to the hypothetical medium that propagated light waves. The expected fringe shifts in the interference pattern were never observed. Einstein's special theory of relativity explains the failure of the Michelson–Morley experiment by recognizing that there is no preferred frame of reference for light.

Summary

The superposition of waves can result in interference. The interference is said to be constructive at points where the intensity of the superposed waves is greater than the sum of the intensities of the two separate waves. At points where the intensity is less than the sum of the intensities of the two separate wave the interference is called destructive. The interference of light was first demonstrated by Thomas Young's double-slit experiment. For light of wavelength λ, intensity maxima and minima for slits separated by a distance a are located at angular positions (θ) given by

$$a \sin\theta = \begin{cases} m\lambda & m = 0, 1, 2, \ldots, \text{ maxima} \quad (40.8) \\ \left(m - \frac{1}{2}\right)\lambda & m = 1, 2, \ldots, \text{ minima} \quad (40.20) \end{cases}$$

Sources, and the waves they emit, are said to be coherent if they (1) have equal frequencies and (2) maintain a phase difference that is constant in time. Sources and waves lacking either of these properties are incoherent. When incoherent waves are superposed, the total intensity is simply the sum of the individual wave intensities.

The superposition of light waves reflected from both sides of a thin film can produce optical interference. The constructive or destructive character of optical interference is determined by the difference in the phases of the superposed waves. The optical path length for light moving through a medium with an index of refraction n is given by

optical path length
= (index of refraction) · (physical path length) (40.50)

Two coherent waves that start together and travel different distances (x_1 and x_2) through a medium of refractive index n, and are then superposed, differ in phase by

$$\delta = \frac{2\pi}{\lambda} n(x_1 - x_2) \quad (40.54)$$

In Eq. 40.54, λ is the vacuum wavelength. Additional phase differences may arise at reflecting surfaces:

1 A reflected wave undergoes a phase change of π rad if it travels *faster* than the transmitted wave.

2 A reflected wave undergoes no phase change if it travels slower than the transmitted wave.

Suggested Reading

P. Baumeister and G. Pincus, Optical Interference Coatings, *Sci. Am.*, 223, 59 (December 1970).

R. S. Shankland, Michelson and His Interferometer, *Physics Today*, 27, 36 (April 1974).

J. Walker, The Bright Colors in a Soap Film Are a Lesson in Wave Interference. *Sci. Am.*, 239, 232 (September 1978).

Problems

Section 40.2 The Interference of Light: Thomas Young's Experiment

1. In a double-slit experiment the wavelength is 600 nm and the two slit-to-screen distances (x_1, x_2) differ by 621 nm. Determine the phase difference.

2. Monochromatic light illuminates a double-slit system with a = 0.3 mm. The second-order maximum occurs at y = 4.0 mm on a screen 1 m from the slits. (a) Determine the wavelength. (b) Determine the positions (y) of the first-order and third-order *maxima*. (c) Determine the angular position (θ) of the m = 1 *minimum*.

3. Monochromatic light passes through two slits 0.26 mm apart. The m = 1 maximum is observed at an angle of 0.14°. Calculate the wavelength of the light.

4. Plane light waves with a wavelength of 500 nm are incident on two narrow slits separated by 0.25 mm. The diffraction pattern is viewed on a screen 5 m from the slits. Determine the separation of the central maximum and the first-order (m = 1) intensity minimum.

5. The helium–cadmium laser of Example 2 is replaced by a helium–neon laser (λ = 632.8 nm). What slit-to-screen distance will place the m = 1 dark fringe directly opposite the slits?

6. In a double-slit experiment the second-order maximum occurs at $\theta = 4.4 \times 10^{-3}$ rad. The wavelength is 660 nm. Determine the slit separation.

7. Light of wavelength 486 nm forms a double-slit interference pattern. The slit separation is 0.44 mm and the slit-to-screen distance (R) is 2.40 m. Determine the angular (θ) and linear (y) positions of the first-order intensity maximum.

Section 40.3 Coherence

8. Give two reasons why the following waves are not coherent:

$$E_1 = E_0 \sin(26x - 4t)$$
$$E_2 = E_0 \sin(13x - 2t + \sin 0.01t)$$

9. Two coherent waves are described by

$$E_1 = E_0 \sin\left(\frac{2\pi x_1}{\lambda} - 2\pi\nu t + \frac{\pi}{6}\right)$$
$$E_2 = E_0 \sin\left(\frac{2\pi x_2}{\lambda} - 2\pi\nu t + \frac{\pi}{8}\right)$$

Determine the relationship between x_1 and x_2 that produces constructive interference when the two waves are superposed.

10. The coherence time for a low-power laser used in a classroom demonstration is approximately 10^{-7} s. Estimate the coherence length of the laser light.

Section 40.4 Thin-Film Interference

11. (a) Calculate the wavelength of light in the film of Example 3. (b) In Example 3 we used Eq. 40.54 to calculate the phase change. Show that the same result is obtained when Eq. 40.51 is used.

12. If the antireflection coating of Example 5 is made 90 nm thick, what wavelength of light will receive maximum transmission?

13. A soap film appears green when exposed to white light. The incident and reflected light are normally incident. Determine the *minimum* thickness of the film that can produce the observed effect. Take the wavelength of green light to be 500 nm. The index of refraction of the film is 1.41.

14. Light with a wavelength of 620 nm travels through a film of water (n = 1.33) 0.72×10^{-6} m thick. Determine (a) the optical thickness of the film; (b) the phase change of the light across the film.

15. A thin film of soap and water is held in a wire frame so that its surface is vertical (Figure 1). Gravity causes the thickness of the film to be greatest at the bottom. At the very top, the film appears black (in reflected light; incident light is white). Lower portions of the film appear as colored bands. (a) Explain why the upper portion of the film appears black. (b) The three bands marked 1, 2, 3, in Figure 1 are red, blue, and green—not necessarily in that order. Identify the color of each of the three bands and indicate the reasoning for your ordering. (c) Calculate the thickness of the green band, indicating the reasoning behind any formula you use. The index of refraction of the film is 1.35. Use your own estimate for the wavelength of green light.

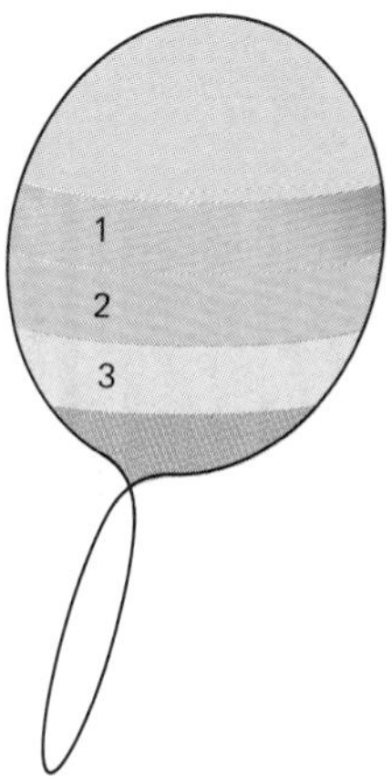

Figure 1

16. A thin film of oil (n_o = 1.38) floats on water (n_w = 1.33). (a) Show that the equation

$$\lambda = \frac{2n_o d}{m + 1/2} \qquad m = 0, 1, 2, \ldots$$

where d is the film thickness, gives the wavelengths that exhibit constructive interference for reflected light. (b) What color is the film at points where d = 300 nm? (The film is viewed from above, and white light is normally incident.)

17. A thin film of cryolite (n = 1.35) is applied to a camera lens (n = 1.5). The coating is designed to reflect wavelengths at the blue end of the visible spectrum and transmit wavelengths in the near infrared. What minimum thickness will give high reflectivity at 450 nm and high transmission at 900 nm?

18. A triangular film of air is formed between two glass plates separated at one side by an optical fiber 1 μm in diameter (1 μm = 10^{-6} m) (Figure 2). The film is illuminated with white light and is viewed from above. (a) Determine the thicknesses of the air film that give constructive interference for yellow light (590 nm). (b) Determine the distances from the end (x) at which constructive interference occurs for yellow light. (c) The air is replaced by water (n = 1.33). Does the spacing of the yellow bands increase or decrease? How many yellow bands are there?

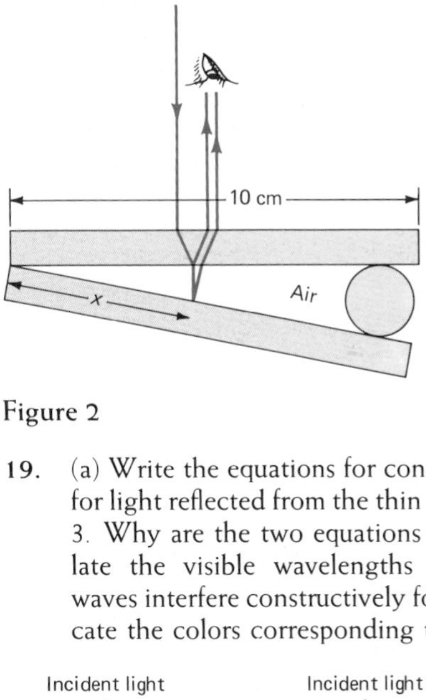

Figure 2

19. (a) Write the equations for constructive interference for light reflected from the thin films shown in Figure 3. Why are the two equations different? (b) Calculate the visible wavelengths for which reflected waves interfere constructively for the two films. Indicate the colors corresponding to each.

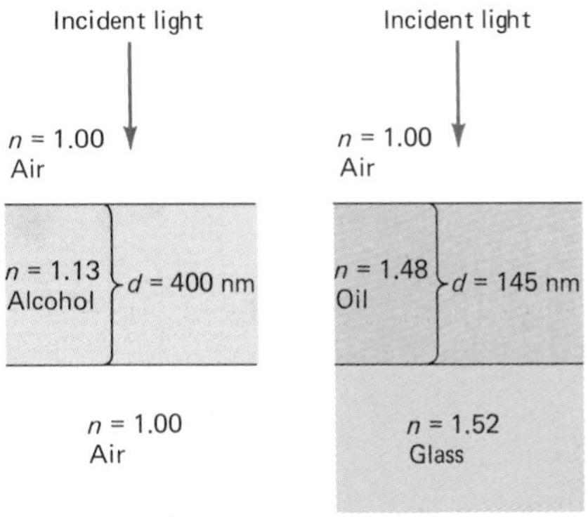

Figure 3

Section 40.5 Optical Interferometers

20. The sealed tubes of a Rayleigh interferometer are each 5.00 cm in length. When light with a wave-

length of 430.8 nm is used, 34.5 fringe shifts are counted as the air is removed from one tube. Determine the index of refraction of air.

21. Using the following data for a Rayleigh interferometer, determine the index of refraction of the gas (hydrogen).

Tube length	11.2 cm
Wavelength	589.2 nm
Number of fringe shifts	25.1

22. Light from a helium–cadmium laser is beamed into a Michelson interferometer. The movable mirror is displaced 0.382 mm, causing the interferometer pattern to reproduce itself 1700 times. Determine the wavelength of the laser light. What color is it?

23. A microwave version of the Michelson interferometer uses waves whose wavelength is 4 cm. The detector receiving the two interfering microwave beams is sensitive to phase differences corresponding to path differences of one tenth of a wavelength. In other words, it can detect a phase difference as small as $\frac{1}{5}\pi$. What is the smallest displacement that can be detected with the interferometer?

24. Reddish light ($\lambda = 632.2$ nm) from a laser is beamed into a Michelson interferometer. How many times will the interference pattern reproduce itself when the movable mirror is moved back exactly 1 mm?

chapter 41

PHYSICAL OPTICS: DIFFRACTION

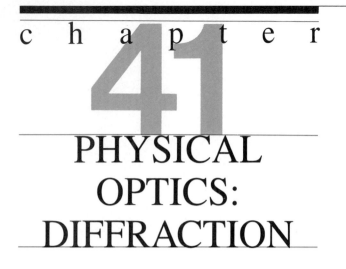

Preview

We saw in Chapter 40 that the superposition of coherent waves results in interference. In this chapter we study another aspect of interference, called diffraction, that can result when a wave strikes an obstacle. When a wave encounters an obstacle or a series of obstacles, portions of the wave are removed, and the remaining portions are scattered. These scattered portions of the wave are called secondary waves. Secondary waves can interfere with each other, and this interference is called diffraction. The diffraction pattern for light waves is a pattern of bright and dark areas corresponding to regions of constructive and destructive interference. The form of the diffraction pattern depends primarily on the size and shape of the obstruction and on the wavelength of the light.

We will consider, in turn, the diffraction of light by a single narrow slit and the diffraction of light by a system of many slits. We will show how a system of many slits, called a diffraction grating, can be used to measure wavelengths with great precision. Finally, we will discuss the use of a very special type of diffraction pattern, called a hologram, to form three-dimensional images.

41.1 Diffraction

When a wave encounters an obstacle, portions of the wave are removed. The remaining portions travel away in all directions, and we say that the wave has been *scattered* by the obstacle. Water waves on a pond are scattered when they strike a partially submerged rock. Sound waves are scattered by interstellar dust grains. Whenever a portion of a wave front is absorbed or reflected, we can expect scattering.

The scattered portions of the wave are called *secondary waves*. From the viewpoint of Huygens' principle (Section 39.1), the effects of scattering can be described in terms of secondary waves. In regions where secondary waves intersect, interference results. The interference of secondary waves is called *diffraction,** and the interference patterns produced are called *diffraction patterns*.

We will discuss diffraction in terms of electromagnetic waves: light. However, all types of waves exhibit diffraction, and thus the ideas and techniques we develop here can be applied to other types of waves. We will confine ourselves to monochromatic (one-wavelength) light waves.†

*All diffraction phenomena are consequences of interference, but all interference phenomena do not produce diffraction. For example, the interference pattern produced by a Michelson interferometer is not a diffraction pattern.

†When many different wavelengths are present, as in white light, the diffraction pattern is a blurred superposition of many patterns.

Figures 41.1–41.4 show the diffraction patterns produced when light encounters various obstacles. The diffraction pattern of a circular disk (Figure 41.1) presents a surprising example of constructive interference and the wave nature of light. Waves scattered by the edges of the disk intersect on the screen, exhibiting constructive interference at some points and destructive interference at other points. The shadow cast by the disk is a circular interference pattern of alternating bright and dark bands. At the center of the shadow is a bright circular spot, indicating constructive interference of the scattered waves reaching that region. The explanation of this bright spot is quite simple: The secondary waves start from the edge of the disk with equal phases. Those reaching the *center* of the shadow travel equal distances—by symmetry. Consequently, they arrive in phase and therefore interfere constructively.

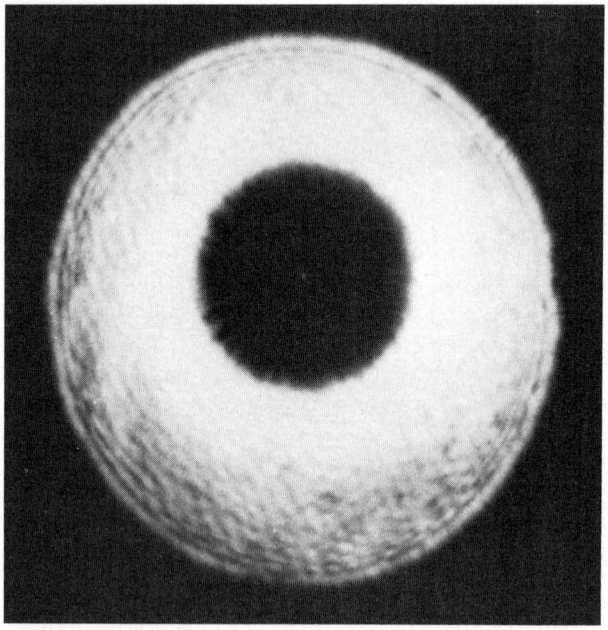

Figure 41.1

Diffraction pattern of a circular disk. Observe the bright spot at the center of the shadow region.

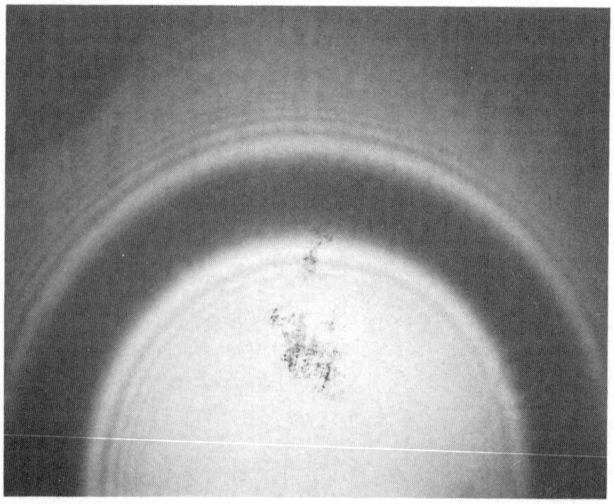

Figure 41.2

Diffraction pattern of a paper clip.

Figure 41.3

Diffraction pattern of the hole in a safety razor blade.

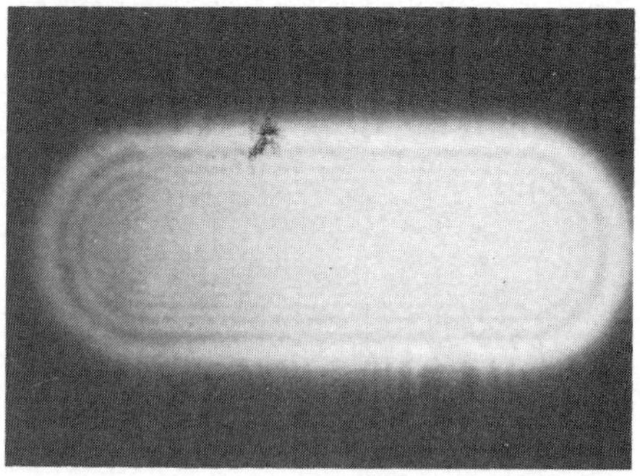

Figure 41.4

Diffraction pattern of a fingertip.

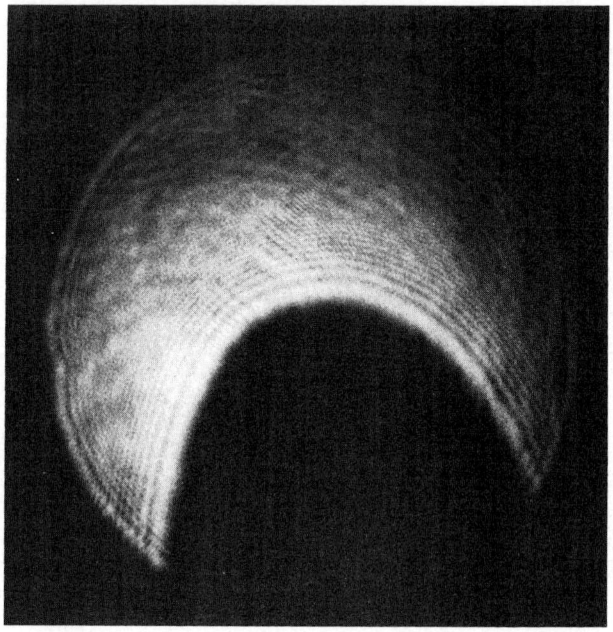

The general approach to determining the diffraction pattern is indicated in Figure 41.5. The wave incident on an aperture Ⓐ is divided into sections. Each section is regarded as a source of secondary waves, in accordance with Huygens' principle. The secondary waves reaching the viewing screen Ⓢ are added, taking into account their in-transit changes of phase.

The most general treatment of diffraction, called *Fresnel* diffraction*, takes into account the curvature of the incident and secondary wave fronts. But under certain conditions, the curvatures of the wave fronts are ignorably small and the waves can be treated as plane waves. Such special conditions produce what we call *Fraunhofer diffrac-*

*Augustin Jean Fresnel (1788–1827), a French engineer (his surname is pronounced fray-*nel*), developed a wave theory of light and performed elaborate optical experiments. In the published reports of his diffraction experiments, Fresnel points out that the results cannot be explained in terms of the particle picture of light. He then shows how the wave picture successfully accounts for his observations.

Figure 41.5

General approach to diffraction. The wave front entering the aperture Ⓐ is divided into sections. Each section is treated as a source of secondary waves. The secondary waves reaching the screen Ⓢ are added, taking account of their in-transit changes of phase.

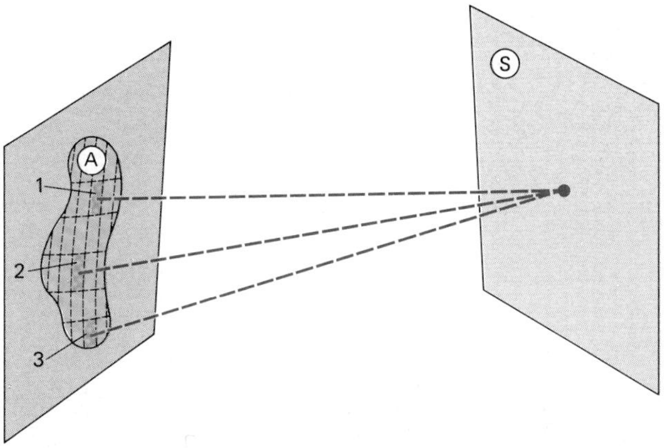

Figure 41.6

Fraunhofer conditions (plane waves incident on the aperture and screen) can be achieved by placing the source and screen at great distances from the aperture.

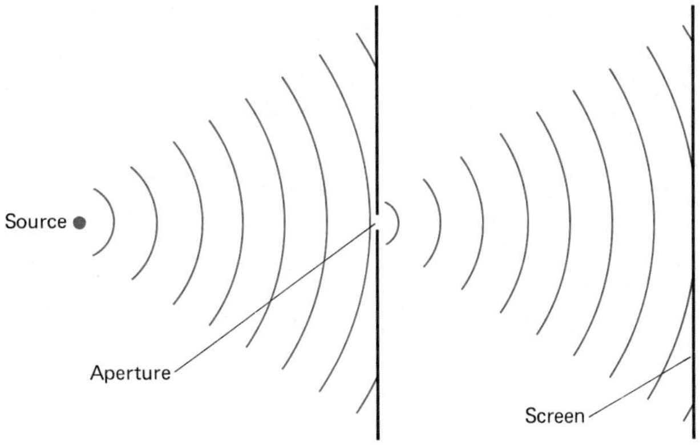

Figure 41.7

An alternate way of achieving Fraunhofer conditions. Lenses on either side of the aperture can be positioned to place the source and screen at infinite (optically) distances from the aperture.

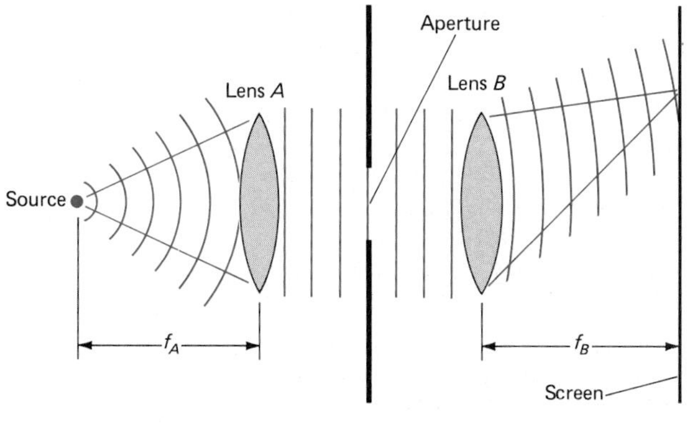

tion. Figures 41.6 and 41.7 show two ways of achieving the plane-wave conditions for Fraunhofer diffraction. The curvature of the wave fronts can be made ignorably small by placing the source and screen great distances from the aperture (Figure 41.6). Plane-wave conditions can also be achieved by inserting lenses on either side of the aperture (Figure 41.7). Waves reaching the aperture will then be plane waves, provided that the source is at the focal point of lens A. (From the standpoint of geometric optics, plane waves correspond to parallel rays.) The screen is placed at the focal point of lens B. This is the optical equivalent of placing the screen a great distance from the aperture. We will investigate Fraunhofer diffraction because it reveals the full range of diffraction phenomena with fewer mathematical complexities than Fresnel diffraction.

41.2 Single-Slit Diffraction Pattern

Let's begin our investigation of Fraunhofer diffraction by analyzing the diffraction of light by a single narrow slit. In Figure 41.8 we see the geometry as viewed from above the slit. Each point on the wave front within the slit acts like a source of secondary waves, as prescribed by Huygens' principle. Waves reaching the screen from different sections of the slit differ in phase because they travel different distances. These phase differences result in interference of the secondary waves and produce a characteristic diffraction pattern. Our goal is to derive an expression for the light intensity on the screen as a function of the angular position on the screen (the angle θ of Figure 41.8).

We take the incident and diffracted waves to be sinusoidal with a wavelength λ and frequency ν. The "wave" in this instance is taken to be the E-field of the electromagnetic wave. The phase of a secondary wave at the screen has the form (recall Eq. 40.2)

$$\phi = 2\pi\left(\frac{x}{\lambda} - \nu t\right) \tag{41.1}$$

where x is the distance from a point in the slit to the screen. In order to make the notation more compact, we introduce the *wave number*, k:

$$k = \frac{2\pi}{\lambda} \tag{41.2}$$

and the *radian frequency*, ω:

$$\omega = 2\pi\nu \tag{41.3}$$

The wave number has units of radians per meter. The radian frequency has units of radians per second. In terms of k and ω the phase, ϕ, of Eq. 41.1 has the form

$$\phi = kx - \omega t \tag{41.4}$$

For example, the distance from the center of the slit to the point P on the screen is x_o. The phase of a wave reaching P from the center of the slit is $kx_o - \omega t$. The E-field of this wave is proportional to the sinusoidal factor $\sin(kx_o - \omega t)$.

Figure 41.8

Geometry for single-slit diffraction, viewed from above the slit. The length y measures distance from the center of the slit and is positive for both halves of the slit.

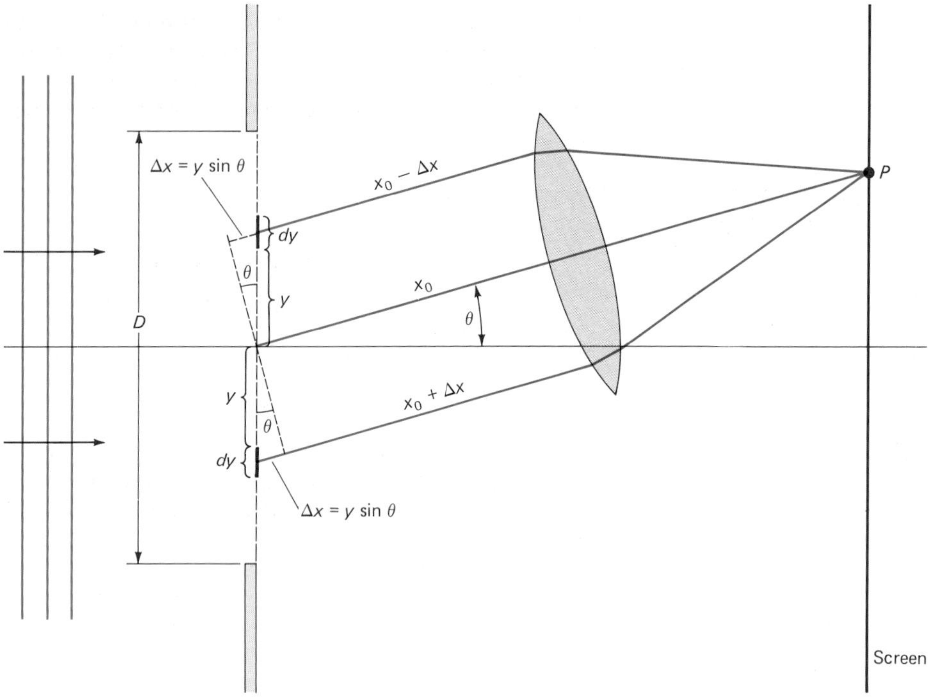

The width of the slit is denoted by D. Figure 41.9 shows a portion of the slit as seen when looking in a direction parallel to the direction of the incident light wave. We can think of the slit as being divided into narrow strips parallel to the length of the slit. Each strip is the source of a secondary wave. We can take advantage of symmetry by first adding the E-fields reaching P from two strips, at positions $\pm y$, equidistant from the center of the slit. These strips are shaded in Figure 41.9. We assume that the E-fields reaching P from the two strips are proportional to the product of three factors:

1 E_o, the amplitude of the wave entering the slit;

2 dy/D, the fraction of the slit covered by a strip of width dy;

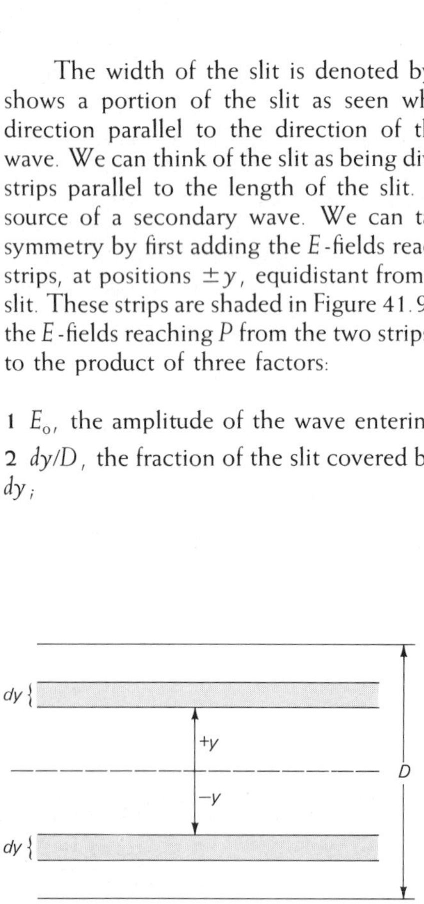

Figure 41.9

The slit has a width D. It is divided—conceptually—into narrow strips of width dy parallel to the length of the slit.

3 $\sin[k(x_o \pm \Delta x) - \omega t]$, the sinusoidal phase factor for waves starting from $\pm y$. As Figure 41.8 indicates, $x_o \pm \Delta x$ are the distances from the strips at $\pm y$ to point P on the screen.

The sum of the two E-fields reaching P from the strips at $\pm y$ is denoted by dE. By introducing a dimensionless proportionality factor, b, we can relate the foregoing three factors and write

$$dE = bE_o \frac{dy}{D} \sin[k(x_o + \Delta x) - \omega t] + bE_o \frac{dy}{D} \sin[k(x_o - \Delta x) - \omega t] \quad (41.5)$$

(The value of b is not important at this time.) Taking $kx_o - \omega t = u$ and $k\,\Delta x = v$, we can use the trigonometric addition theorem $\sin(u + v) + \sin(u - v) = 2 \sin u \cos v$ to convert Eq. 41.5 to the form

$$dE = 2bE_o \frac{dy}{D} \sin(kx_o - \omega t) \cos(k\,\Delta x) \quad (41.6)$$

As Figure 41.8 shows, Δx is related to the positions of the strips by

$$\Delta x = y \sin\theta \quad (41.7)$$

The angle θ is the angular position of the point P on the screen (Figure 41.8). Inserting $\Delta x = y \sin\theta$ into Eq. 41.6 gives the E-field at P produced by waves originating from the strips at $\pm y$.

$$dE = \frac{2bE_o}{D} \sin(kx_o - \omega t) \cos[(k \sin\theta)y]\, dy \quad (41.8)$$

The contributions from all the strips are summed by integrating from $y = 0$ to $y = D/2$. The resultant E-field at P is

$$E_\theta = \int_{\substack{\text{all}\\ \text{strips}}} dE = \frac{2bE_o}{D}\sin(kx_o - \omega t) \times \int_0^{D/2} \cos[(k\sin\theta)y]\,dy \qquad (41.9)$$

The subscript θ is a reminder that the angle θ identifies the angular position of the point P. By following the steps below* we obtain

$$E_\theta = bE_o \sin(kx_o - \omega t)\,\frac{\sin\left(\frac{1}{2}kD\sin\theta\right)}{\frac{1}{2}kD\sin\theta} \qquad (41.10)$$

The dimensionless factor $\frac{1}{2}kD\sin\theta$ appears so frequently that it is convenient to represent it by a single symbol, Φ.

$$\Phi = \frac{1}{2}kD\sin\theta = \frac{\pi D\sin\theta}{\lambda} \qquad (41.11)$$

In terms of Φ we can rewrite Eq. 41.10 for E_θ as

$$E_\theta = bE_o\sin(kx_o - \omega t)\frac{\sin\Phi}{\Phi} \qquad (41.12)$$

The intensity at P is (Section 40.2)

$$I = c\epsilon_o(E_\theta^2)_{av} = c\epsilon_o b^2E_o^2\left(\frac{\sin\Phi}{\Phi}\right)^2[\sin^2(kx_o - \omega t)]_{av}$$

The time average, $[\sin^2(kx_o - \omega t)]_{av}$, is $\frac{1}{2}$. Thus,

$$I_\theta = \frac{1}{2}c\epsilon_o b^2E_o^2\left(\frac{\sin\Phi}{\Phi}\right)^2 \qquad (41.13)$$

The intensity I_θ gives a quantitative description of the interference and is referred to as the *diffraction pattern*. The factor $\frac{1}{2}c\epsilon_o b^2E_o^2$ is the intensity at $\theta = 0$, marked by the point opposite the center of the slit. To prove this to yourself, notice that as the angle θ approaches zero the quantity $\Phi = \pi D\sin\theta/\lambda$ also approaches zero. Noting that

$$\lim_{\Phi\to 0}\frac{\sin\Phi}{\Phi} = 1$$

we see that the $\theta = 0$ limit of Eq. 41.14 is

$$I_{\theta=0} = \frac{1}{2}c\epsilon_o b^2E_o^2 \qquad (41.14)$$

The forward intensity for the single slit is a useful standard and we designate it by a special symbol, I_o:

$$I_o = \frac{1}{2}c\epsilon_o b^2E_o^2 \qquad (41.15)$$

*Equation 41.10 is obtained by the integration

$$\int_0^{D/2}\cos[(k\sin\theta)y]\,dy = \frac{\sin[(k\sin\theta)y]_0^{D/2}}{k\sin\theta} = \frac{\sin\left(\frac{1}{2}kD\sin\theta\right)}{k\sin\theta}$$

Inserting this result into Eq. 41.13 simplifies its form to

$$I_\theta = I_o\left(\frac{\sin\Phi}{\Phi}\right)^2 \qquad (41.16)$$

This expression is the result we have been seeking—the intensity on the screen as a function of angular position. We refer to I_θ as the **single slit diffraction pattern.** The dependence of I_θ on the angle θ enters through the auxiliary quantity, Φ:

$$\Phi = \frac{\pi D\sin\theta}{\lambda} \qquad (41.11)$$

Equation 41.16 can be rearranged to yield

$$\frac{I_\theta}{I_o} = \left(\frac{\sin\Phi}{\Phi}\right)^2 = \left[\frac{\sin\left(\frac{\pi D\sin\theta}{\lambda}\right)}{\frac{\pi D\sin\theta}{\lambda}}\right]^2 \qquad (41.17)$$

The quantity I_θ/I_o is the relative intensity. We can display the form of the diffraction pattern by plotting the relative intensity versus the angle θ. Figure 41.10 is a plot of I_θ/I_o versus θ for the case in which $D/\lambda = 10$ (the slit width is ten times the wavelength). The quantity Φ is also indicated in Figure 41.10. Three significant features of the diffraction pattern can be easily understood as interference effects:

1 The intensity is sharply peaked in the forward ($\theta = 0$) direction. The optical path length from all segments of the slit to the screen is the same for $\theta = 0$. Thus, scattered waves from all segments arrive in phase, giving complete constructive interference and an intensity maximum.

2 The intensity drops to zero at certain angles. As θ increases away from $\theta = 0$, the optical path length from the slit to the screen becomes different for each segment of the slit. The waves from each segment arrive at the screen with different phases. At certain angles their relative phases result in complete destructive interference and zero intensity.

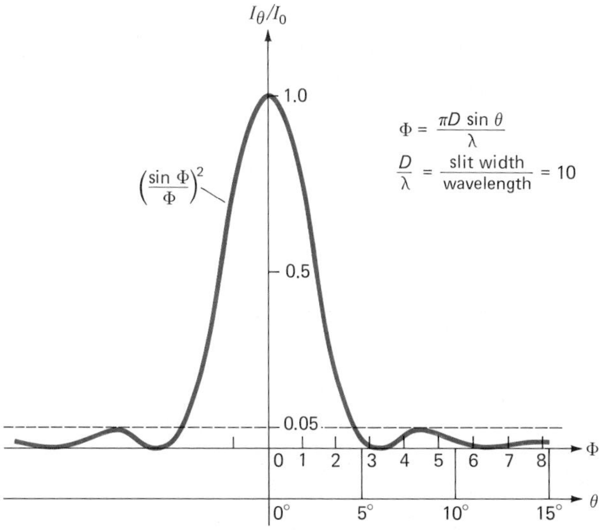

Figure 41.10

Relative intensity I_θ/I_o for the single-slit diffraction pattern. The slit width is ten times the wavelength.

3 Much weaker intensity peaks occur at certain angles. These are the result of incomplete destructive interference. Secondary waves from most segments interfere destructively at the screen. Constructive interference of waves from a small fraction of the segments produces a finite intensity that exhibits peaks.

Now let's determine the features by analyzing the relative intensity,

$$\frac{I_\theta}{I_0} = \left(\frac{\sin\Phi}{\Phi}\right)^2 \qquad (41.17)$$

First we can determine the angular positions of zero intensity by setting $\sin\Phi = 0$. The zeros of $\sin\Phi$ occur at

$$\Phi = m\pi \qquad m = \pm 1, \pm 2, \pm 3, \ldots$$

Setting

$$\Phi = \frac{\pi D \sin\theta}{\lambda} = m\pi$$

shows that the intensity falls to zero at angles θ_m given by

$$\sin\theta_m = m\left(\frac{\lambda}{D}\right) \qquad m = \pm 1, \pm 2, \ldots \qquad (41.18)$$

The plus-or-minus signs show that the zeros of intensity are located symmetrically about the central peak at $\theta = 0$. In particular, the intensity falls to zero on either side of the central maximum, at an angle given by setting $m = 1$ in Eq. 41.18.

$$\sin\theta_1 = \frac{\lambda}{D} \qquad (41.19)$$

The Central Peak

The equation $\sin\theta_1 = \lambda/D$ defines an angular "radius" of the central peak of intensity. That the intensity of the central peak is so much greater than the secondary peaks shows that the bulk of the wave intensity is scattered into a narrow range of angle about the forward direction ($-\lambda/D \lesssim \sin\theta_1 \lesssim +\lambda/D$). In many instances $\lambda \ll D$ and $\sin\theta_1 = \lambda/D \ll 1$, in which case it is legitimate to use the small-angle approximation and replace $\sin\theta_1$ by θ_1.* Thus, when $\lambda \ll D$ the intensity falls to zero on either side of the central peak at an angle

$$\theta_1 = \frac{\lambda}{D} \qquad (41.20)$$

There are two immediate implications of the relation $\theta_1 = \lambda/D$: (1) For a fixed slit width, longer wavelengths are diffracted over a wider range of angle than are shorter wavelengths. For example, red light has a longer wavelength than other colors in the visible spectrum; as a result, when white light is diffracted, the edges of the central peak have a reddish color. (2) For a given wavelength, the angular width of the central peak is increased when the slit width is decreased. In other words, if the slit is narrowed, the energy diffracted into the central peak spreads over a wider range of angle.

*For example, with $\lambda/D = \frac{1}{3}$, Eq. 41.19 is satisfied by $\theta_1 = 0.3398$ rad, whereas Eq. 41.20 yields $\theta_1 = 0.3333$ rad.

Figure 41.11

Single-slit diffraction. W denotes the distance between positions of zero intensity.

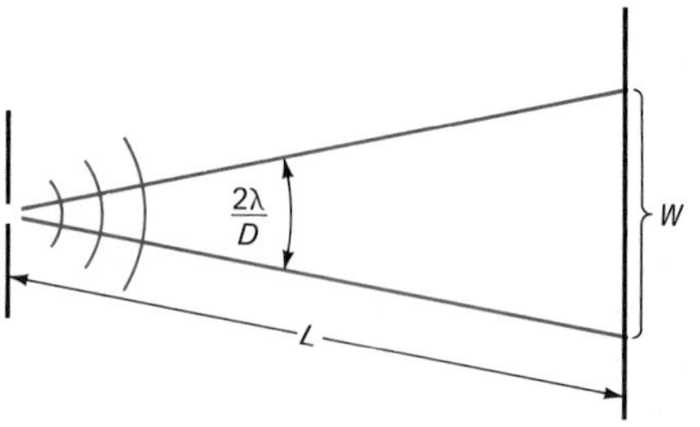

Example 1 **An Estimate of the Wavelength of Laser Light**

A beam of red light from a helium–neon laser is diffracted by a slit 0.50 mm wide. The interference pattern forms on a wall 2.12 m beyond the slit (Figure 41.11). The distance between the positions of zero intensity on either side of the forward peak is 5.1 mm. Let us estimate the wavelength of the laser light. The angular positions of the zero intensity points are given by Eq. 41.18, with $m = \pm 1$.

$$\theta_{\pm 1} = \pm\frac{\lambda}{D}$$

Their angular separation is therefore $2\lambda/D$. At a distance L from the slit their linear separation is approximately

$$W = L\left(\frac{2\lambda}{D}\right)$$

The wavelength is

$$\lambda = \frac{DW}{2L} = \frac{0.50 \times 10^{-3} \cdot 5.1 \times 10^{-3}}{2 \cdot (2.12)}\ \text{m}$$
$$= 6.0 \times 10^{-7}\ \text{m}$$

or

$$\lambda = 600\ \text{nm}$$

A more precise value is 632.8 nm. The single slit diffraction pattern provides an estimate of the wavelength. A more precise measurement, obtained by using a diffraction grating, is described in Section 41.4.

Questions

1. Hold two fingers close together so as to form a narrow slit (Figure 41.12). View a source of light through this slit and explain what you see. Try varying the slit width. (You can also use two pencils or two pieces of chalk to form a narrow slit and then observe the diffraction fringes.)
2. If the wavelength of light diffracted by a single slit is increased by 50%, what change in the slit width will leave the diffraction pattern unchanged?

Figure 41.18

Geometry for diffraction by a grating. The optical path difference for waves from adjacent slits is $a \sin \theta$.

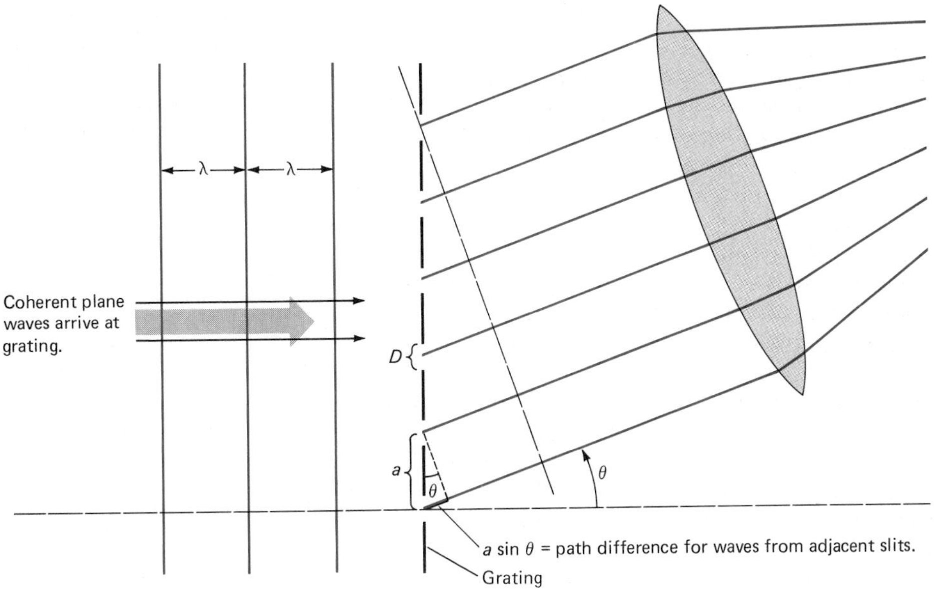
λ
λ
Coherent plane waves arrive at grating.
D
a
θ
θ
a sin θ = path difference for waves from adjacent slits.
Grating

The total E-field at the screen is the sum of the fields arriving from the N slits. We designate this total E field as E_θ and write it as

$$E_\theta = E_{1\theta} + E_{2\theta} + \cdots + E_{N\theta}$$

$$= bE_o \frac{\sin \Phi}{\Phi} \{\sin \phi_o + \sin(\phi_o + \delta) + \cdots + \sin[\phi_o + (N - 1)\delta]\} \qquad (41.27)$$

where

$$\phi_o = kx_o - \omega t \qquad (41.28)$$

$$\delta = ka \sin \theta = \frac{2\pi a \sin \theta}{\lambda} \qquad (41.29)$$

In order to use E_θ to compute the intensity, we first convert the sum of sines in Eq. 41.27 into a product. The mathematical relation needed is

$$\sin \theta_o + \sin(\phi_o + \delta) + \cdots + \sin[\phi_o + (N - 1)\delta]$$

$$= \left(\frac{\sin \frac{1}{2} N\delta}{\sin \frac{1}{2}\delta}\right) \sin\left[\phi_o + \frac{1}{2}(N - 1)\delta\right] \qquad (41.30)$$

When Eq. 41.30 is used in Eq. 41.27, the E-field at point P on the screen becomes

$$E_\theta = bE_o\left(\frac{\sin \Phi}{\Phi}\right)\left(\frac{\sin \frac{1}{2} N\delta}{\sin \frac{1}{2}\delta}\right) \times \sin\left[\phi_o + \frac{1}{2}(N - 1)\delta\right] \qquad (41.31)$$

The intensity is (Section 40.3)

$$I_\theta = c\epsilon_o(E_\theta^2)_{av}$$

$$= c\epsilon_o b^2 E_o^2 \left(\frac{\sin \Phi}{\Phi}\right)^2 \left(\frac{\sin \frac{1}{2} N\delta}{\sin \frac{1}{2}\delta}\right)^2 \times \left\{\sin^2\left[\phi_o + \frac{1}{2}(N - 1)\delta\right]\right\}_{av} \qquad (41.32)$$

Because $\phi_o = kx_o - \omega t$, the time average $\{\sin^2[\phi_o + \frac{1}{2}(N - 1)\delta]\}_{av}$ is $\frac{1}{2}$. Thus, the intensity is

$$I_\theta = \frac{1}{2} c\epsilon_o b^2 E_o^2 \left(\frac{\sin \Phi}{\Phi}\right)^2 \left(\frac{\sin \frac{1}{2} N\delta}{\sin \frac{1}{2}\delta}\right)^2 \qquad (41.33)$$

As a reference for intensity we use I_o, the intensity in the forward ($\theta = 0$) direction for one slit. As we showed for the single slit (Section, 41.2, Eq. 41.15),

$$I_o = \frac{1}{2} c\epsilon_o b^2 E_o^2$$

In terms of I_o, the intensity for the N-slit grating is

$$I_\theta = I_o\left(\frac{\sin \Phi}{\Phi}\right)^2 \left(\frac{\sin \frac{1}{2} N\delta}{\sin \frac{1}{2}\delta}\right)^2 \qquad (41.34)$$

This is the ***N*-slit diffraction pattern.** The angular position of a point on a viewing screen is specified by the angle θ. Both ϕ and δ, which control the value of I_θ, are functions of θ. Thus,

$$\Phi = \frac{\pi D \sin \theta}{\lambda} \qquad (41.11)$$

$$\delta = \frac{2\pi a \sin \theta}{\lambda} \qquad (41.29)$$

It is possible to use Eq. 41.34 to plot I_θ versus θ for any value of N by using Eqs. 41.11 and 41.29 for Φ and δ. However, we will use a somewhat different approach, which is designed to reveal the important features of the diffraction pattern.

First, we observe that when $N = 1$, Eq. 41.34 reduces to the single-slit result, Eq. 41.16. Thus, with $N = 1$ the factor $(\sin \frac{1}{2}N\delta/\sin \frac{1}{2}\delta)^2$ reduces to unity and Eq. 41.34 becomes

$$I_\theta = I_o\left(\frac{\sin \Phi}{\Phi}\right)^2 \qquad (41.16)$$

which is the single-slit intensity derived in Section 41.2.

Next, we observe that I_θ as given by Eq. 41.34 is the product of a single slit intensity factor, $I_o(\sin \Phi/\Phi)^2$, and an *interference factor*, $(\sin \frac{1}{2}N\delta/\sin \frac{1}{2}\delta)^2$. The interference factor describes interference arising from the finite *separation* (a) of the slits. The factor $(\sin \Phi/\Phi)^2$ describes interference arising from the finite *width* (D) of the slits.

Gratings with Large N

The dependence of I_θ on the number of slits, N, is contained in the interference factor $(\sin \frac{1}{2}N\delta/\sin \frac{1}{2}\delta)^2$. We can see the effect of increasing the number of slits by comparing plots of the interference factor for different values of N. Figure 41.19 is a plot of $(\sin \frac{1}{2}N\delta/\sin \frac{1}{2}\delta)^2$ for $N =$ 2, 4, and 6. The large peaks are called *principal maxima*. The effect of increasing N is to increase the intensity of the principal maxima at the expense of the regions in between.

The principal maxima correspond to angles (θ) for which waves from adjacent slits are out of step by an integral number of wavelengths and therefore interfere constructively. The path difference for waves from adjacent slits is $a \sin \theta$. The condition for a principal maximum is therefore

$$a \sin \theta = m\lambda$$

$$m = 0, 1, 2, 3, \ldots \qquad (41.35)$$

Substituting $a \sin \theta = m\lambda$ into $\delta = 2\pi a \sin \theta/\lambda$ shows that the principal maxima correspond to values of δ that are integral multiples of 2π:

$$\delta = 2\pi m \qquad m = 0, 1, 2, 3, \ldots \qquad (41.36)$$

The integer m is called the *order* of the maximum.

The interference factor $(\sin \frac{1}{2}N\delta/\sin \frac{1}{2}\delta)^2$ has the value N^2 at a principal maximum. This is a consquence of complete constructive interference of the waves from all N slits. We can prove this assertion most simply by considering the central maximum at $\theta = 0$. Because $\delta = 2\pi a \sin \theta/\lambda$, both $\frac{1}{2}\delta$ and $\frac{1}{2}N\delta$ approach zero as θ approaches zero. For small δ we can use the small-angle approximation, $\sin \frac{1}{2}\delta \simeq \frac{1}{2}\delta$ and $\sin \frac{1}{2}N\delta \simeq \frac{1}{2}N\delta$. Thus, at $\theta = 0$ the interference factor has the value

$$\lim_{\theta \to 0} \left(\frac{\sin \frac{1}{2}N\delta}{\sin \frac{1}{2}\delta}\right)^2 = \left(\frac{\frac{1}{2}N\delta}{\frac{1}{2}\delta}\right)^2 = N^2 \qquad (41.37)$$

Figure 41.19 confirms numerically that the interference factor has the value N^2 at principal maxima.

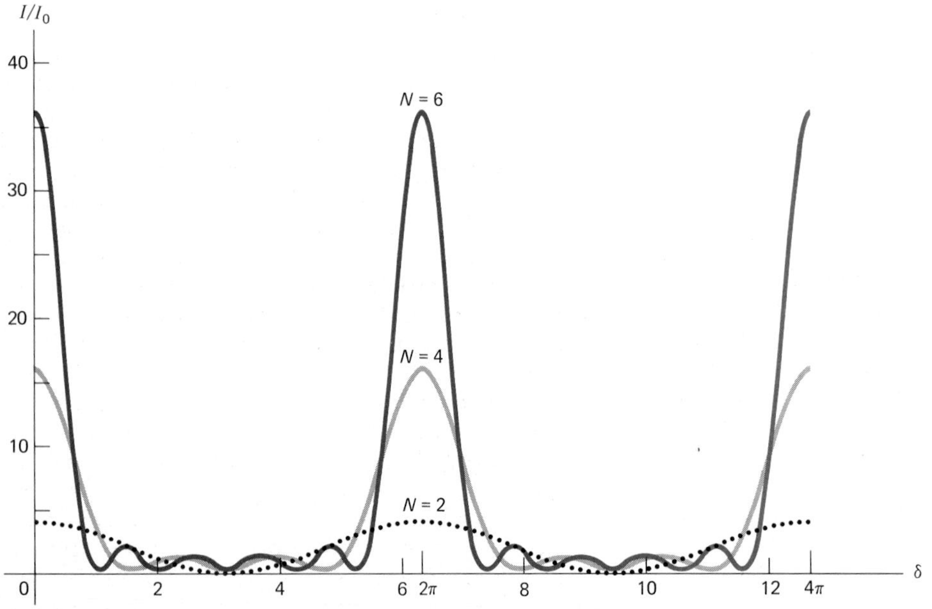

Figure 41.19

The interference factor $(\sin \frac{1}{2}N\delta/\sin \frac{1}{2}\delta)^2$ for $N = 2$, 4, and 6. The principal maxima occur at angles where δ is an integral multiple of 2π. The interference factor has the value N^2 at a principal maximum.

Figure 41.20

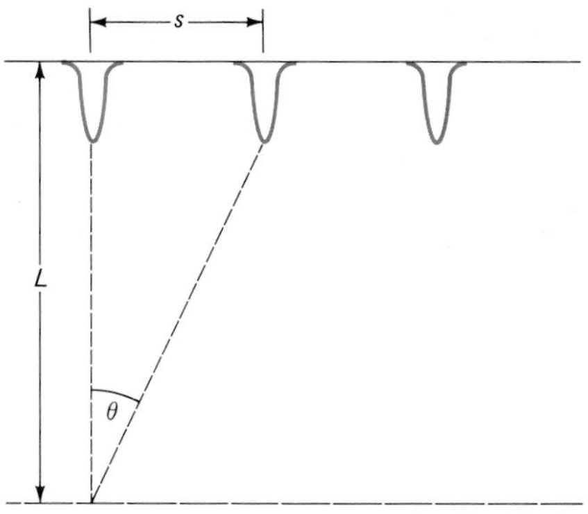

Example 4 **Determining the Wavelength of Laser Light**

In Example 1 we estimated the wavelength of light from a He–Ne laser by using a single slit. A multiple-slit grating provides a much more precise method for measuring the wavelength.

Light from the laser strikes a grating with 5220 lines per centimeter. The zeroth- and first-order principal maxima ($m = 0, 1$) are separated by a distance of $S = 0.639$ m on a wall that is a distance $L = 1.83$ m from the grating. From Figure 41.20

$$\sin\theta = \frac{S}{\sqrt{L^2 + S^2}} = 0.330$$

The slit spacing is the reciprocal of the number of lines per centimeter

$$a = \frac{1}{5220/\text{cm}} = 1.92 \times 10^{-6}\ \text{m} = 1920\ \text{nm}$$

From Eq. 41.35 (with $m = 1$)

$$\lambda = a \sin\theta = (1920\ \text{nm})(0.330) = 633\ \text{nm}$$

If you have access to a laser and replica grating, you should try to perform this measurement.

Intensity Zeros

The positions of zero intensity correspond to angles at which complete destructive interference occurs. We will now locate the angular positions of the intensity zeros, and then demonstrate that the change in the angle θ between a principal maximum and an adjacent zero is inversely proportional to the number of slits (N). For large N, then, the change in angle θ is small and the principal maxima are sharply defined.

The interference factor (and the intensity) is zero provided that

$$\sin\frac{1}{2}N\delta = 0 \quad \text{and} \quad \sin\frac{1}{2}\delta \neq 0 \qquad (41.38)$$

If both $\sin\frac{1}{2}N\delta$ and $\sin\frac{1}{2}\delta$ are zero, we get a principal maximum. At a principal maximum Eq. 41.36 shows that $\delta = 2\pi m$ where m is an integer. The value of $\frac{1}{2}N\delta$ at a principal maximum is πmN. The intensity zero adjacent to a principal maximum occurs when $\frac{1}{2}N\delta$ increases by π, from πmN to $\pi(mN + 1)$. The change in δ between a principal maximum and an adjacent zero is

$$\Delta\delta = \frac{2\pi}{N} \qquad (41.39)$$

Recall (Eq. 41.29) that

$$\delta = \frac{2\pi a \sin\theta}{\lambda}$$

The range of angle ($\Delta\theta$) corresponding to $\Delta\delta = 2\pi/N$ can be found by forming the differential of Eq. 41.29 and replacing $d\delta$ and $d\theta$ by $\Delta\delta$ and $\Delta\theta$. Thus,

$$d\delta = \frac{2\pi a \cos\theta}{\lambda}\, d\theta$$

Setting $d\delta = \Delta\delta = 2\pi/N$ and $d\theta = \Delta\theta$ gives

$$\Delta\theta = \frac{\lambda}{Na\cos\theta} \qquad (41.40)$$

This is our first key result–it specifies $\Delta\theta$ as a functin of N. The quantity $\Delta\theta$ is the difference betwen the values of θ at a principal maximum and at an adjacent zero. As such, $\Delta\theta$ measures the angular radius of the intensity peaks. Equation 41.40 shows that the angular radius of the intensity peaks decreases as N is increased (see also Figure 41.19). Therefore, as the number of slits is increased, light of a given wavelength is concentrated into a smaller range of angle. This makes it possible for a diffraction grating to separate light with different wavelengths. The separation becomes more clear-cut as the value of N is increased. Gratings with thousands of lines per centimeter enable us to measure visible wavelengths with a precision of 0.01 nm–a distance that is about one tenth the characteristic size of an atom!

Resolving Power

The precision with which wavelengths can be measured by using a grating spectroscope can be related to $\Delta\theta = \lambda/Na\cos\theta$. Suppose that light consisting of a mixture of two wavelengths, λ and $\lambda + \Delta\lambda$, strikes the grating. If λ and $\lambda + \Delta\lambda$ are very nearly equal, the corresponding values of θ for which δ takes on any particular value (say $2\pi m$) are very nearly equal. The intensity maxima overlap, and it is not possible to determine that different wavelengths are present. In this instance we say that the wavelengths are *not resolved*. In our discussion of Rayleigh's criterion in the preceding section resolution was a question of detecting two sources that emit the same wavelength, but are located at slightly different *angular positions*. Here, on the other hand, the sources coincide with respect to direction, but have slightly different *wavelengths*. When the difference $\Delta\lambda$ is large enough, the principal maxima do not overlap, and it is clear that two different wavelengths are present. In this instance we say the wavelengths are *clearly resolved*. The arbitrary criterion used to decide the issue of wavelength is this: The difference $\Delta\lambda$ must be large enough to cause the principal maximum for one wavelength to fall on the first position of zero intensity for the other wavelength. (This is Rayleigh's criterion, Section

41.3.) This criterion results in a slight "dimple" in the overall intensity, and shows that two wavelengths are present. The criterion for resolvability can be formulated as follows: If the mth-order intensity peak for the wavelength λ falls at an angle θ, the mth-order maximum for the wavelength $\lambda + \Delta\lambda$ must fall at an angle $\theta + \Delta\theta$, where $\Delta\theta = \lambda/Na \cos\theta$ locates the position of zero intensity for the wavelength λ. We form the differential of the condition for an intensity maximum, $a \sin\theta = m\lambda$, to obtain

$$a \cos\theta \, d\theta = m d\lambda \tag{41.41}$$

We replace $d\theta$ by $\Delta\theta = \lambda/Na \cos\theta$ (Eq. 41.40) and $d\lambda$ by $\Delta\lambda$. This converts Eq. 41.41 to

$$a \cos\theta \frac{\lambda}{Na \cos\theta} = m\,\Delta\lambda$$

Rearranging leaves us with the fundamental equation governing resolvability,

$$\frac{\Delta\lambda}{\lambda} = \frac{1}{mN} \tag{41.42}$$

According to Eq. 41.42 the larger N, the smaller $\Delta\lambda$ and the greater the precision. This, then, is one incentive for having a large number of lines on the grating. A high-quality grating with $N \simeq 100{,}000$ can resolve wavelength differences as small as $\lambda/100{,}00$. For visible light (430–700 nm), this amounts to a precision of better than 0.01 nm. Great precision in measurements of wavelengths is desirable for several reasons. For example, the set of wavelengths (spectrum) of light emitted by each element is unique. No two elements have the same optical spectrum. The spectrum acts like a set of optical "fingerprints" that can be used to identify the emitter, provided that precision measurements of wavelength can be made.

A measure of the precision with which a grating can measure wavelengths is the quantity

$$R \equiv \frac{\lambda}{\Delta\lambda} \tag{41.43}$$

called the *resolving power* of the grating. A large value of R means that the grating can resolve small differences of wavelength. For the mth order of a grating with N slits, Eq. 41.42 shows that

$$R = mN \tag{41.44}$$

Notice that the resolving power is directly proportional to the number of slits—a large grating is desirable. The diffraction pattern contains many images for each wavelength. These multiple images correspond to the different orders $m = 1, 2, 3, \ldots$. Equation 41.44 shows that the resolving power increases with the order m. Two wavelengths that overlap in the first-order spectrum may be resolved in the second- or third-order spectra. In practice, spectra from different orders begin to overlap for values of m larger than 2 or 3. Thus, the resolving power $R = mN$ cannot be improved by using a very high-order spectrum.

Example 5 **Resolving the Sodium Doublet**

The yellow–orange light from a sodium vapor lamp consists primarily of a mixture of two wavelengths, $\lambda = 589.0$ nm and $\lambda + \Delta\lambda = 589.6$ nm. These wavelengths are called the *sodium doublet*. Let us determine the minimum number of slits required to resolve these wavelengths in the second-order spectrum. We use Eq. 41.42

$$\frac{\Delta\lambda}{\lambda} = \frac{1}{mN}$$

with $\Delta\lambda = 0.6$ nm, $\lambda = 589.0$ nm, and $m = 2$, the minimum number of slits is

$$N = \frac{589 \text{ nm}}{0.6 \text{ nm}} \frac{1}{2} \simeq 491$$

The resolving power required for such a grating is

$$R = mN = 982$$

Questions

5. Why doesn't a long picket fence make a good diffraction grating for visible light?
6. When reflection gratings are used with x rays, the incident waves are nearly parallel to the grating surface ("glancing" incidence). The spacing of the slits on such a grating is comparable to, or even larger than, the spacing on an optical transmission grating. In order to have significant diffraction effects the slit spacing must decrease as the wavelength decreases. Because x rays have much shorter wavelengths than does visible light, it would seem that the slit spacing would have to be correspondingly smaller. How does the glancing angle of incidence assist in the production of a quality diffraction pattern?

41.5 Holography

A photographic negative and the positive prints made from it are two-dimensional records of light intensity. The negative contains no information about the phase of the light waves that produced it. Because the phase information is lost, an ordinary photograph gives us a two-dimensional image of a three-dimensional object. If, on the other hand, we record both phase and intensity, we can obtain a three-dimensional image. The process of *holography* allows us to make such images, which are called *holograms*.

We have seen that the superposition of coherent light waves produces an interference pattern. Constructive interference occurs at points where the waves have phases differing by integral multiples of 2π rad, and destructive interference results at points where the phases differ by odd integral multiples of π rad. If the interfering light waves fall on a photographic film, we can record the interference pattern. Such a photographic record contains information about the relative phases of the two waves. It also contains information about the amplitudes of the individual waves.

A hologram is a photographic record of the interference pattern formed by the superposition of two coherent light waves. It can be used to reconstruct the wave fronts that produced it, and this reconstruction gives us a three-dimensional image.

Figure 41.21 shows how the simplest type of hologram is produced. Coherent light from a laser is scattered by a point object, P. The scattered waves then interfere with other waves from the laser, while the film at S records the interference pattern. The pattern is a series of concentric rings that mark alternating regions of constructive and destructive interference. This ring pattern is called a *Gabor zone plate,* in honor of Dennis Gabor, who conceived the principles of holography in 1947. In Figure 41.22 we see that when the Gabor zone plate is illuminated with the

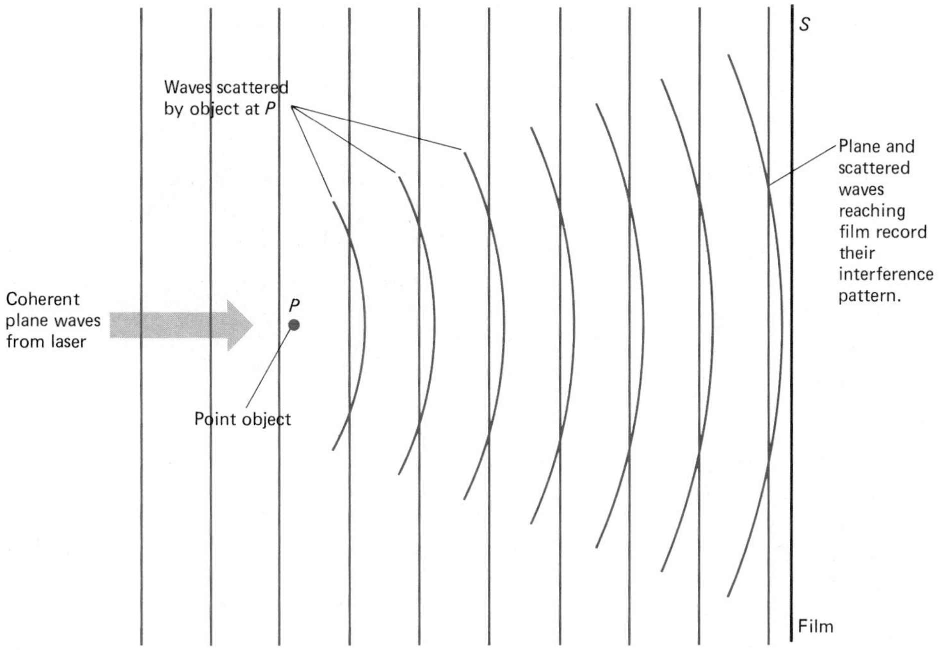

Figure 41.21

A hologram of a point object (P) is formed when waves scattered by the object interfere with unscattered waves. The interference pattern is recorded in the film. For a point object the hologram consists of a set of circular interference rings called a *Gabor zone plate.*

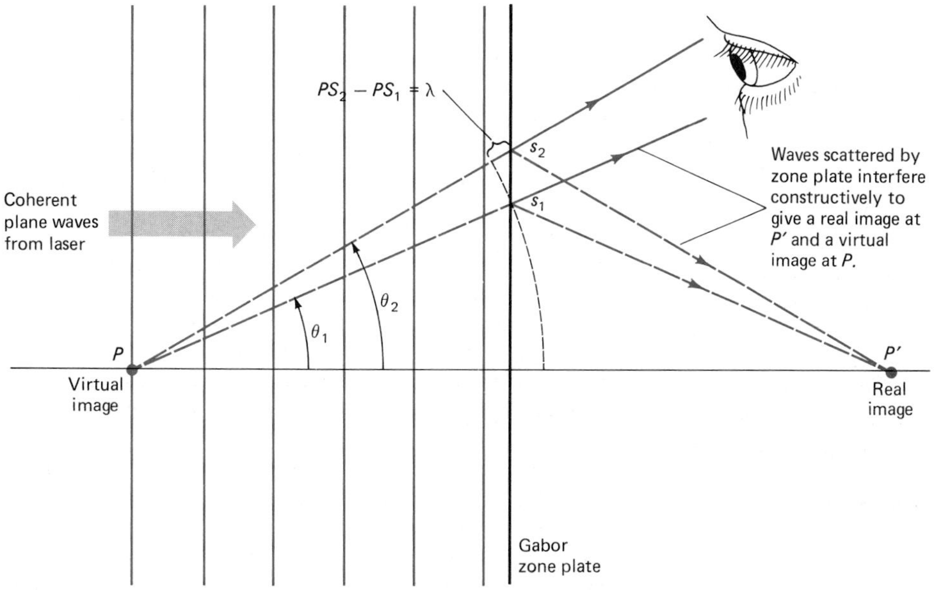

Figure 41.22

When a Gabor zone plate is illuminated with coherent light the scattered waves interfere constructively in certain directions. These waves "reconstruct" the waves that formed the zone plate and thereby produce an image of the object.

same coherent laser light used in its production, it acts like a diffraction grating with circular slits. Secondary waves from the slits interfere constructively in certain directions. In Figure 41.22, S_1 and S_2 mark adjacent positions at which waves from P interfere constructively with the waves arriving directly from the laser. The locations of S_1 and S_2 are such that $\overline{PS_2}$ and $\overline{PS_1}$ differ by one wavelength, λ.

$$\overline{PS_2} - \overline{PS_1} = \lambda$$

Rings of destructive interference occur at positions where the path difference equals an odd number of half wavelengths. The geometry of Figures 41.21 and 41.22 shows that waves diffracted by adjacent slits through the angles θ_1 and θ_2 interfere constructively and form a real image at P'. A second image—this one virtual—is formed behind the hologram at P, the original site of the point object. An observer sees light waves diverging from P. The relative phase of these waves is the same as for the scattered waves that produced the hologram. In effect, then, the diffracted waves are *reconstructions* of the waves that produced the hologram.

An extended object produces a complicated-looking hologram (Figure 41.23) that bears no resemblance to the object. Nevertheless, the diffracted laser light reconstructs the wave fronts that produced the hologram and creates two images—one real and one virtual. Figures 41.24 and 41.25 show how a hologram can be produced and used to display the real and virtual images of an extended object. Recent advances in holography make it possible to view a hologram with ordinary white light. (See the references in the Suggested Reading section.)

In Chapter 40 we introduced the concept of *coherence length*, the length of the wave trains emitted by a source. There is a definite phase relationship between different portions of a single wave train, and the waves that make up the train are coherent. Therefore, the superpositon of different portions of the same wave train results in interference. On the other hand, there is no correlation of phases between different wave trains, and the superposition of portions of different wave trains consquently does not produce interference. In the production of a hologram, wave trains are divided and later superposed. One portion becomes a reference wave and the other becomes a wave scattered by the object. In order for the two waves to interfere, they must be portions of the same wave train. The two waves will interfere provided that their optical

Figure 41.23

A piece of film used to record the holographic interference pattern produced by an extended object. The scale of the holographic interference pattern is too small to be seen. The visible interference fringes are caused by minor imperfections in lenses used to "expand" the laser beam.

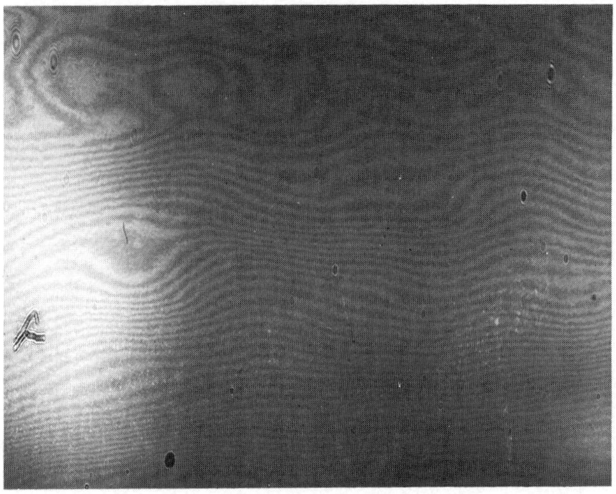

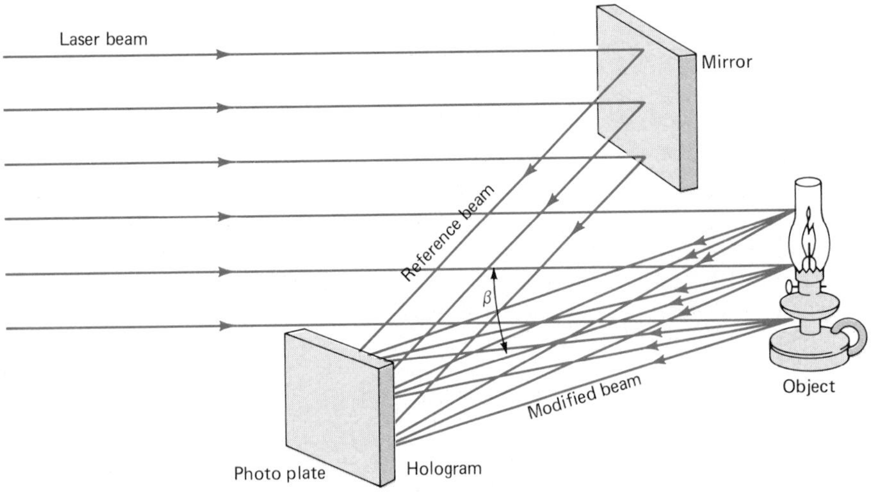

Figure 41.24

Hologram production. Waves reflected from the mirror and from the object interfere at the photo plate. This developed plate is a hologram. (From *Fundamentals of Optics*, 4th ed., by F. A. Jenkins and H. E. White; McGraw–Hill, New York, 1976. Used with permission of McGraw–Hill Book Company.)

Figure 41.25

Reconstruction of wave fronts by a hologram. The hologram diffracts light, producing a real image and a virtual image. (From *Fundamentals of Optics*, 4th ed., by F. A. Jenkins and H. E. White; McGraw–Hill, New York, 1976. Used with permission of McGraw–Hill Book Company.)

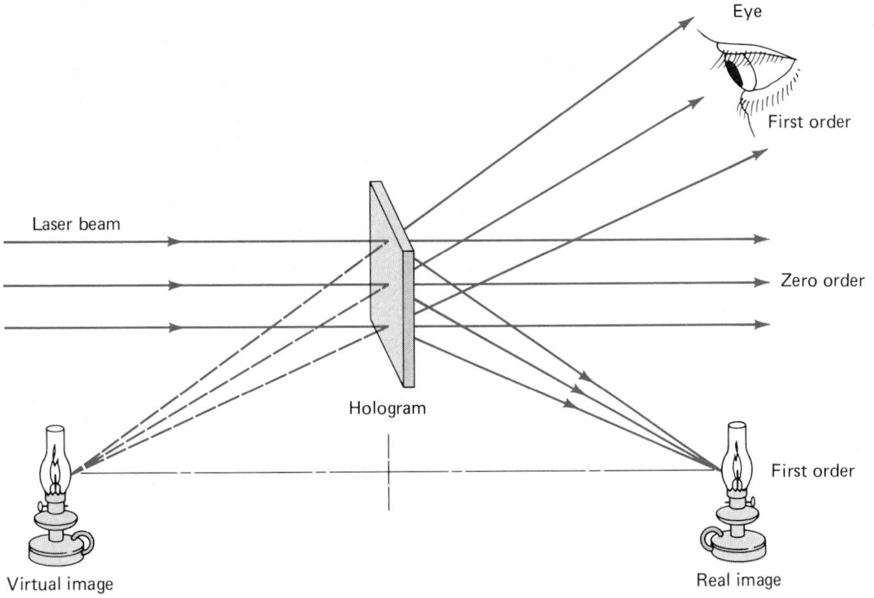

path lengths differ by less than the coherence length. If the optical path difference exceeds the coherence length, the superposed waves do not interfere—they are incoherent.

Before the development of the laser in 1960 the maximum coherence length that could be achieved with a visible light source was approximately 1 mm, and consequently holograms could be produced only for very small objects. The coherence length for lasers, however, can be many meters and allows us to produce holograms of life-sized subjects.

Summary

When a wave encounters an obstacle or a series of obstacles, secondary or scattered waves are generated. The interference of the secondary waves produces a diffraction pattern. When waves of wavelength λ encounter an object or aperture whose characteristic dimension D is large in comparison to λ, most of the diffracted intensity is channeled into a narrow range of angle (2θ) given by

$$2\theta \simeq \frac{2\lambda}{D} \qquad (41.21)$$

Diffraction sets a limit to the smallest angular separation that can be resolved by an image-forming system. For a lens of diameter D used to view light with a wavelength λ, this limit, called the angular resolution of the lens, is given by

$$\text{angular resolution} = 1.22\left(\frac{\lambda}{D}\right) \qquad (41.24)$$

The diffraction pattern of a system of N identical slits is described by the relation

$$I_\theta = I_0\left(\frac{\sin\Phi}{\Phi}\right)^2\left(\frac{\sin\frac{1}{2}N\delta}{\sin\frac{1}{2}\delta}\right)^2 \qquad (41.34)$$

where

$$\delta = \frac{2\pi a\sin\theta}{\lambda} \qquad (41.29)$$

$$\Phi = \frac{\pi D\sin\theta}{\lambda} \qquad (41.11)$$

and a is the slit separation, D the slit width, λ the wavelength, I_θ the intensity at an angle θ, measured relative to the forward direction (the direction of wave travel prior to scattering), and I_0 the intensity in the forward direction for a single slit. A special case of Eq. 41.34 gives the single-slit $(N = 1)$ pattern.

$$I_\theta = I_0\left(\frac{\sin\Phi}{\Phi}\right)^2 \qquad (41.16)$$

A diffraction grating is optically equivalent to a multiple-slit system in which N is very large—typically from 10,000 to 100,000. The angular positions of constructive interference are sharply defined, enabling the grating to measure wavelengths with great precision. The resolving power (R) of a grating is defined as $\lambda/\Delta\lambda$, where $\Delta\lambda$ is the smallest resolvable difference in wavelength for waves with an average wavelength λ. For a grating with N "slits" the resolving power in the mth order is

$$R = mN \qquad (41.44)$$

It is possible to make records of the phase and intensity of a wave in a process called holography. These records, called holograms, are interference patterns. A hologram can be illuminated in such a way that it reconstructs the wave fronts that produced it. This wave-front reconstruction produces a three dimensional image.

Suggested Reading

P. Connes, How Light Is Analyzed, *Sci. Am.*, 219, 72 (September 1968).

E. N. Lieth, White-Light Holograms, *Sci. Am.*, 235, 80 (October 1976).

White-Light Holographic Displays, *Laser Focus*, 13, 12 (July 1977).

Problems

Section 41.2 Single-Slit Diffraction Pattern

1. Monochromatic light falls on a single slit having a width of 6.0×10^{-5} m. The smallest angle at which the intensity falls to zero is 0.011 rad. Determine the wavelength of the light.

2. Monochromatic light is diffracted by a single slit. The slit width is ten times the wavelength of the light. What is the smallest angle for which the intensity is zero?

3. A beam of green light from a helium–cadmium (He–Cd) laser is diffracted by a single slit 0.55 mm wide. The interference pattern forms on a wall 2.06 m beyond the slit. The distance between the positions of zero intensity ($m = \pm 1$) is 4.1 mm. Estimate the wavelength of the laser light. (The He–Cd laser operates at wavelengths of 441.16 nm and 537.8 nm.)

4. The light of a He–Cd laser ($\lambda = 441.16$ nm) passes through a slit whose width s is adjustable. What width slit will result in the first diffraction maximum's appearing at 5°? What is the ratio of D/λ? Compare your result with Figure 41.10.

5. Light with a wavelength of 486.3 nm passes through a single slit. The slit width is 2.2×10^{-5} m. Determine the smallest angle at which the intensity falls to zero.

6. Light of wavelength 589 nm passes through a narrow slit. The intensity falls to zero at an angle of 0.22° on either side of the central peak. Determine the width of the slit.

7. The relative intensity I_θ/I_o for a single slit exhibits peaks between the zeros located at $\Phi = m\pi$. Assuming that the peaks occur midway between zeros, that is, at positions where $\Phi = (m + \frac{1}{2})\pi$ ($m = 1, 2, 3, \ldots$), evaluate I_θ/I_o at the first three secondary peaks ($m = 1, 2, 3$).

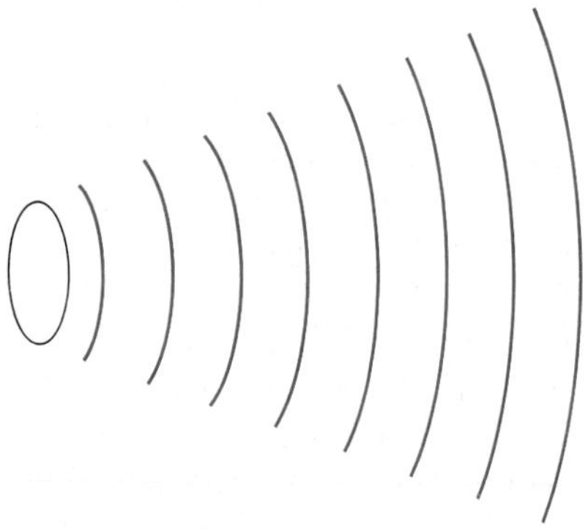

Figure 1

Section 41.3 Diffraction and Angular Resolution

8. Radar waves ($\lambda = 2.6$ cm) pass through a circular opening (Figure 1). The diameter of the opening is 22 cm. Use Eq. 41.21 to determine the approximate angular diameter of the emerging radar beam.

9. A student constructs a reflecting telescope with a 15-cm-diameter mirror. The focal length is 160 cm. Using this telescope, she photographs a star (point source) in blue light, 400 nm. Calculate (a) the angular radius and (b) the linear diameter of the Airy disk.

10. A helium–neon laser emits red light with a wavelength of 632.8 nm. The circular aperture through which the beam emerges has a diameter of 0.50 cm. Estimate the diameter of the beam at a distance 10 km from the laser.

11. The moon is approximately 400,000 km away. Can two lunar craters 50 km apart be resolved by using a telescope with a mirror 6 in. in diameter? Can craters 1 km apart be resolved? Take the wavelength to be 700 nm. Justify your answers with approximate calculations.

12. Grote Reber was a pioneer in radio astronomy. He constructed a radio telescope with a 10-m-diameter receiving dish. What was the telescope's angular resolution for radio waves with a wavelength of 2 m?

13. Estimate the angular resolution of an ordinary camera lens, the diameter of which is 1.6 cm. Take the wavelength to be 600 nm. Could the lens resolve the millimeter markings on a meter stick placed 40 m away?

14. An earth-orbiting satellite designed to inventory crops by photographing them orbits at an altitude of 705.3 km. If the satellite's camera has an angular resolution of 42.5 μrad, what is the minimum separation of two objects which are resolved?

15. When Mars is nearest the earth, the distance separating the two planets is 88.6×10^6 km. Mars is viewed with a telescope whose mirror has a diameter of 30 cm. If the wavelength of the light is 590 nm, what is the angular resolution of the telescope? What is the smallest distance that can be resolved on Mars (the distance between two points that can "just" be resolved)?

16. Compare the angular resolution of a reflecting telescope that has a mirror diameter of 0.60 m and a viewing light with a wavelength of 600 nm to that of the Arecibo radio telescope, which has a diameter of 305 m. The Arecibo telescope can be tuned to a wide range of wavelengths. Take the wavelength to be 21 cm for the purpose of your comparison.

17. A telescope lens 30 cm in diameter has a focal length of 80 cm. The telescope is equipped with a filter that transmits over a narrow range of wavelengths centered on 440 nm. Determine the size of the stellar image.

Section 41.4 Diffraction Gratings

18. Light from an argon laser strikes a grating with 5310 lines per centimeter. The central and first-order principal maxima are separated by a distance of 0.488 m on a wall that is 1.72 m from the grating. Determine the wavelength of the laser light.

19. A diffraction grating with 38,380 lines has a width of 8.211 cm. Determine the resolving power for the second-order spectrum of the grating when it is used to study light with a wavelength of 583 nm.

20. A grating has a width of 0.1062 m. The spacing of lines is 2.313×10^{-6} m. Determine the resolving power for the first-order spectrum.

21. Imagine that you are designing a large diffraction grating that must be able to resolve a $\Delta\lambda = 0.01$ nm for $\lambda = 500$ nm in the second-order spectrum ($m = 2$). How many lines must you have in your grating?

22. A source emits light with wavelengths of 531.62 nm and 531.81 nm. What is the minimum number of lines for a grating that resolves the two wavelengths in the first-order spectrum? Determine the slit spacing for a grating 1.32 cm wide that has the required minimum.

23. The sum

$$\sin\phi_o + \sin(\phi_o + \delta) + \sin(\phi_o + 2\delta) + \cdots + \sin[\phi_o + (N-1)\delta] = \sum_{n=0}^{N-1} \sin(\phi_o + n\delta)$$

occurs in the theory of the N-slit grating. Use the trigonometric identity

$$\sin(\phi_o + n\delta) = \sin\phi_o \cos n\delta + \cos\phi_o \sin n\delta$$

and the summation formulas

$$\sum_{n=0}^{N-1} \cos n\delta = \left(\frac{\sin\frac{1}{2}N\delta}{\sin\frac{1}{2}\delta}\right)\cos(N-1)\frac{\delta}{2}$$

$$\sum_{n=0}^{N-1} \sin n\delta = \left(\frac{\sin\frac{1}{2}N\delta}{\sin\frac{1}{2}\delta}\right)\sin(N-1)\frac{\delta}{2}$$

to derive the result quoted as Eq. 41.30:

$$\sum_{n=0}^{N-1} \sin(\phi_o + n\delta) = \left(\frac{\sin\frac{1}{2}N\delta}{\sin\frac{1}{2}\delta}\right)\sin\left[\phi_o + \frac{1}{2}(N-1)\delta\right]$$

chapter 42

QUANTUM PHYSICS, LASERS, AND SQUIDS

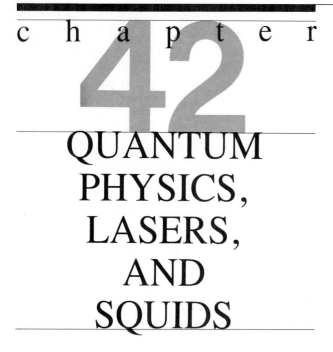

Preview

A systematic development of quantum and nuclear physics would require many volumes, each the size of this text. Accordingly, we will use a different approach. In this chapter and the next, we present condensed accounts of the revolutionary discoveries and ideas that have become benchmarks of 20th-century physics. Additionally, we show how these ideas have been applied. In some instances these applications have created new fields of engineering and technology.

Toward the end of the 19th century it seemed that Newton's mechanics and Maxwell's electromagnetic theory provided a firm basis for a complete understanding of the various branches of physics. Newton's laws adequately described the motions of falling apples and orbiting planets. The theories of heat and sound had been successfully extended into the microscopic domain by using a kinetic theory of matter based on Newtonian mechanics. Maxwell's researches not only unified electrical and magnetic phenomena, but also provided a fundamental starting point for the study of light. To be sure, there were many unanswered questions, but most physicists were confident that the theories of Newton and Maxwell would lead to satisfactory answers.

They were mistaken. The assumptions of Newton and Maxwell were called into question in 1900 by Max Planck. Planck's theory of blackbody radiation ushered in a revolution in scientific and philosophic thought. Albert Einstein and Niels Bohr were the champions of this revolution. They revealed that at the atomic level, familiar physical quantities such as energy and angular momentum are quantized—that is, they can take on only certain discrete values.

The revolution did not end with the discovery of quantization. Newton's mechanics, even when supplemented by quantization, was found to be inadequate for all but the simplest atomic systems. A new quantum theory based on the wave nature of matter—wave mechanics—quickly developed (1924–1926) in the hands of de Broglie, Schrödinger, Heisenberg, and Born. The postulated wave nature of matter was confirmed by the electron interference experiments of Davisson and Germer and of G. P. Thomson. The development of modern physics continued in 1932 with two exciting and fundamental experiments. In America, Anderson discovered the positron—the antimatter mate of the electron. In England, Chadwick identified the neutron, a nuclear particle, and thereby launched a period of fundamental research and development in nuclear physics that continues to affect our lives.

Today's scientists and engineers are still exploiting the discoveries made between the years 1900 and 1932. The ideas of quantum theory have worked their way from the research laboratory into our television sets and our local supermarkets. Our everyday lives are increasingly affected by the products of quantum engineering. In this chapter we study two areas of quantum engineering: quantum optics and squids.

42.1 The Origins of Quantum Physics

Quantum physics originated in 1900 when Max Planck* (Figure 42.1) presented his theory of blackbody radiation (Section 23.3). Planck envisioned the surface of a blackbody as containing innumerable Hertzian oscillators.† The exact physical nature of these microscopic oscillators was not specified by Planck. Each oscillator had a characteristic frequency at which it radiated. An oscillator could absorb electromagnetic radiation of the same frequency it emitted.

Planck found he could explain the observed features of blackbody radiation by assigning discrete energies to the oscillators. For the first time, energy was quantized. In Planck's theory the energy E of an oscillator of frequency ν must be an integral multiple of a quantum of energy, $h\nu$ (Figure 42.2).

$$E = 0,\ h\nu,\ 2h\nu,\ 3h\nu,\ \ldots,\ nh\nu, \qquad (42.1)$$

where n is an integer. The constant h is called *Planck's constant* and has the (modern) value

$$h = 6.6262 \times 10^{-34}\ \mathrm{J \cdot s} \qquad (42.2)$$

Planck explained the blackbody spectrum by quantizing energy—by restricting the oscillator energies to multiples of the value $h\nu$. Despite his success, Planck initially was convinced that energy quantization was nothing but a mathematical artifice—a neat trick that led to the correct blackbody spectrum.

*Max K. E. L. Planck (1858–1947), a German physicist, presented his theory to the German Physical Society on October 19, 1900.

†In 1887 Hertz generated and detected electromagnetic waves, using short segments of wire as radiating and receiving antennas. These became known as *Hertzian oscillators*.

Figure 42.1

Max Planck (right). Quantum theory originated in 1900 with Planck's theory of blackbody radiation. Niels Bohr is on the left.

Figure 42.2

Planck imposed energy quantization on the Hertzian oscillators. The quantized energies open to an oscillator are integral multiples of the energy quantum $h\nu$.

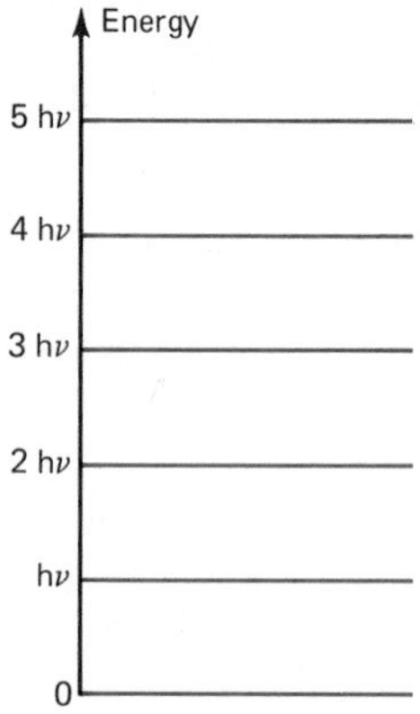

Einstein and Photons

Albert Einstein readily accepted the idea of energy quantization. He went beyond Planck and argued that light quanta were a necessary consequence of Planck's theory. For Einstein, the fact that an oscillator absorbs and emits electromagnetic energy in packets of size $h\nu$ meant that the electromagnetic field could be interpreted as a collection of energy parcels, or *photons* as they are now called.* Thus, Einstein extended Planck's quantization of energy to the radiation field itself. The energy emitted by a light source of frequency ν is carried by photons of energy:

$$E = h\nu \tag{42.3}$$

In 1905, Einstein used the photon theory of light to explain the photoelectric effect, an achievement that earned him the Nobel prize.

*The name *photon* was suggested by G. N. Lewis in 1926.

When ultraviolet radiation strikes a metal surface, electrons are ejected from the metal. This is the photoelectric effect, discovered in 1887 by Hertz during his experiments that verified Maxwell's theory of electromagnetic waves. Today, there are many devices whose operation is based on the photoelectric effect, and that use materials that are photosensitive to visible light. One of the oldest of these is the electric eye, which triggers door openers in elevators.

Attempts to explain the photoelectric effect in terms of light waves were unsuccessful. Einstein successfully explained the photoelectric effect in 1905 by treating the incident light as a flood of *photons*. If ν is the frequency of the light source, $h\nu$ is the energy of the photons. Einstein assumed that when a photon is absorbed by a metal surface its energy is transformed into the work needed to free an electron from the metal and into the kinetic energy of the ejected electron.

$$h\nu = \text{work to free electron} + \text{kinetic energy of ejected electron}$$

The amount of energy required to free an electron depends primarily on the type of surface. The energy needed will also vary slightly, depending on the depth from which the electron originates. If W is the *minimum* energy required to free an electron, then the maximum kinetic energy of the ejected electrons, K_{max}, is related to W by

$$h\nu = W + K_{max} \tag{42.4}$$

The quantity W is called the *work function* of the surface. Typically, W is a few electron volts. To measure K_{max} we can subject the electrons to a repelling electrostatic force (Figure 42.3). The ejected electrons are decelerated as they travel toward the negatively charged plate. If the repelling potential difference is large enough, then not even the most energetic electrons will cross the gap, and the current will drop to zero. The *stopping potential*, V_s, is the potential difference that reduces the current to zero. An electron that is ejected with a kinetic energy K_{max} and momentarily brought to a halt just short of the negatively

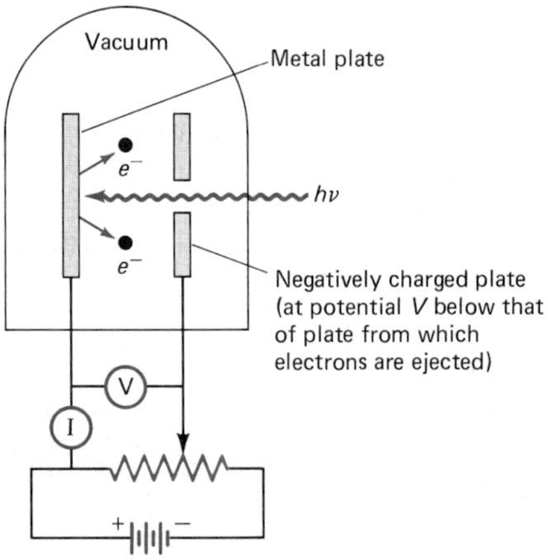

Figure 42.3

The photoelectric effect. Photons eject electrons from a metal plate, causing charge to flow. The charge flow can be reduced to zero (ejected electrons are prevented from reaching the plate at the right) by applying a sufficiently large repelling potential to the plate at the right.

charged plate has converted its kinetic energy into electric potential energy (eV_s). Thus,

$$K_{max} = eV_s \tag{42.5}$$

which can be used in Eq. 42.4 to obtain

$$V_s = \left(\frac{h}{e}\right)\nu - \frac{W}{e} \tag{42.6}$$

Equation 42.6 reveals the key prediction of Einstein's theory: A graph of V_s versus ν would be a straight line whose slope is the same (h/e) for all materials. Experiments carried out by Millikan in 1916 verified Einstein's theory. Millikan demonstrated that the value of h determined from experiments on the photoelectric effect agrees substantially with the value obtained by Planck from blackbody experiments. The photoelectric effect thus provided a second confirmation of the quantum hypothesis. Planck's assumption of energy quantization no longer stood alone, but was joined by Einstein's quanta of radiant energy—photons. Today, quantization and the photon continue to play central roles in modern physics.

The lowest frequency that ejects electrons is called the *threshhold frequency*. The threshhold frequency, ν_o, is related to the work function, W, by

$$h\nu_o = W \tag{42.7}$$

which can be regarded as a limiting case of Eq. 42.4, in which $K_{max} = 0$.

Example 1 **Threshhold Frequency for Silver**

The work function for silver is 4.73 eV. Let's calculate the threshhold frequency for silver. We have

$$\nu_o = \frac{W}{h} = \frac{4.73 \text{ eV} \cdot 1.60 \times 10^{-19} \text{ J/eV}}{6.62 \times 10^{-34} \text{ J} \cdot \text{s}}$$

$$= 1.14 \times 10^{15} \text{ Hz}$$

The corresponding wavelength for this threshhold frequency is

$$\lambda = \frac{c}{\nu_o} = \frac{3 \times 10^8 \text{ m/s}}{1.14 \times 10^{15}\text{/s}} = 2.63 \times 10^{-7} \text{ m}$$

$$= 263 \text{ nm}$$

The wavelength of 263 nm lies in the ultraviolet (UV) portion of the electromagnetic spectrum. Photons of visible light have longer wavelengths and lower frequencies and therefore less energy than UV photons. Photons of visible light cannot eject electrons from silver.

In addition to energy, $E = h\nu$, a photon possesses other particle-like properties. We will now examine three of these: the photon mass, its linear momentum, and its angular momentum.

Photon Mass

As we saw in Section 20.3 any particle that travels at the speed of light and has energy must be a zero-mass particle. The photon possesses energy and travels at the speed of light, and therefore is a zero-mass particle.

$$m_{\text{photon}} = 0 \tag{42.8}$$

When we say that photons are zero-mass particles we mean that they have no existence as objects at rest. For example, when a photon is emitted by an atom, energy stored in the atom is transformed into radiant energy. The photon has no existence *prior* to its emission, but is created at the moment of emission. Likewise, when an atom absorbs a photon, the photon is *destroyed*—it does not survive, "bottled up" inside the atom.

Photon Linear Momentum

A photon possesses linear momentum, which is parallel to the direction of motion of the photon. The magnitude of the linear momentum follows from the general relation between energy, linear momentum, and mass derived in Section 20.3:

$$E = \sqrt{(pc)^2 + (mc^2)^2} \tag{42.9}$$

Taking the mass (m) to be zero, we find that

$$p = \frac{E}{c} \tag{42.10}$$

for the magnitude of the linear momentum. We can obtain a useful expression for the linear momentum of a photon by using $E = h\nu$ (Eq. 42.3) and the kinematic relation for waves $\nu\lambda = c$ (Section 17.2). The expression $p = E/c$ then becomes

$$p = \frac{h}{\lambda} \tag{42.11}$$

Example 2 **The Energy of a Blue Photon**

A light source emits radiation with a wavelength of 440 nm, which appears as blue light. To determine the photon energy by using $E = h\nu$ we use the kinematic relation $\nu\lambda = c$ to obtain

$$E = h\nu = \frac{hc}{\lambda} = \frac{6.62 \times 10^{-34} \text{ J} \cdot \text{s} \cdot 3 \times 10^8 \text{ m/s}}{440 \times 10^{-9} \text{ m}}$$

$$= 4.51 \times 10^{-19} \text{ J}$$

In the more convenient units of electrons volts (1 eV = 1.60×10^{-19} J)

$$E = 2.82 \text{ eV}$$

Photon Angular Momentum

A photon possesses a discrete (quantized) amount of angular momentum. The angular momentum vector must be parallel or antiparallel to the direction of motion of the photon, with a magnitude related to Planck's constant (h) by

$$L_{\text{photon}} = \frac{h}{2\pi} \tag{42.12}$$

This is referred to as the "spin" angular momentum of the photon.

Questions

1. How might you explain the concept of quantization to an 8-year-old child, using money as a basic element?
2. How might the photoelectric effect be used in a burglar alarm system?
3. A source of light that ejects photoelectrons from a metal is replaced by another source of equal intensity (energy per second per unit area), the frequency of which is twice as great. Does the energy of the ejected electrons increase or decrease? Would you expect the *rate* at which electrons are ejected to increase of decrease? (How does the rate of incident photons change when intensity remains constant and frequency is doubled?)
4. Our atmospheric ozone layer protects us from ultraviolet (UV) photons emitted by our sun. Why might UV photons be harmful to human beings, whereas visible light photons are not?

42.2 Rutherford and the Nuclear Atom

The researches of Planck and Einstein introduced quantization into physics. The next thread in the fabric of quantum physics was Rutherford's development of the nuclear atom, which led to Bohr's quantum theory of the atom.

In 1909, Ernest Rutherford (Figure 42.4) and his associates H. Geiger and E. Marsden began analyzing the deflection of α-particles passing through thin metal foils. The α-particles were emitted by radioactive substances and had kinetic energies of several MeV (1 MeV = 10^6 eV). The α-particle was known to have a positive charge and a mass several thousand times that of the electron.* Geiger and Marsden directed a beam of α-particles at a thin gold foil. Gold was chosen as a target because it can be rolled into a very thin film, and this minimizes the number of multiple collisions between an α-particle and the gold atoms. Geiger and Marsden wanted to avoid multiple collisions because successive deflections tend to

*The α-particle was subsequently identified as a doubly ionized helium atom.

Figure 42.4

Ernest Rutherford, originator of the nuclear atom.

Figure 42.5

Top view of the α-scatter apparatus of Geiger and Marsden. The radioactive source (R) gives a collimated beam of α-particles. The microscope (M) is focused on the screen (S). The foil and source are fixed. The chamber and microscope can rotate about the foil.

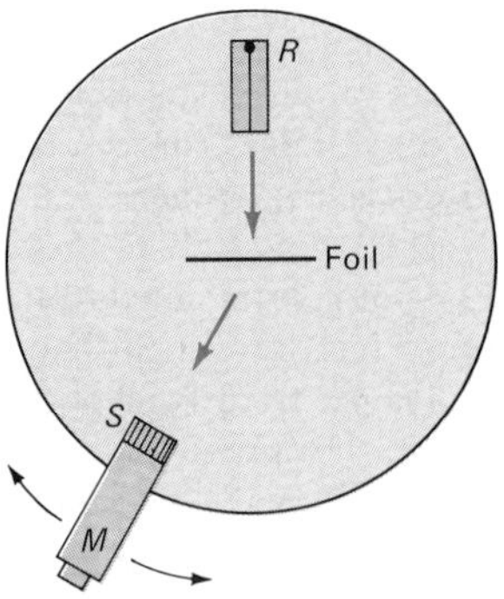

cancel, thereby obscuring the details of atomic structure. If an α-particle collides with just one gold atom, its deflection can be related precisely to the electric force it experiences. The electric force, in turn, depends on the distribution of the negatively charged electrons and the positive charge of the atom. The gold film used by Geiger and Mars den had a thickness of approximately 10^{-6} m. If we view the gold atoms as being stacked in layers, the thickness of the film was about 2000 such layers.

Figure 42.5 indicates the experimental arrangement used by Geiger and Marsden. A movable eyepiece was focused on a small screen that gave off brief flashes of light when struck by an α-particle. Most α-particles suffered very small deflections. But surprisingly, a tiny fraction of the α-particles were scattered backward, through angles as great as 150° (the largest angle observable with their apparatus). Such large deflections were completely unexpected.

In 1911, Rutherford described an atomic model that accounts quantitatively for the alpha scattering observations. Rutherford argued that the alpha could "bounce back" only if it struck an object more massive than itself (Section 9.4). This would be the case if the positive charge and most of the mass of the atom were concentrated in a small portion of the atom rather than spread diffusely throughout the atom. Rutherford called this concentrated region the *nucleus*. That backward scattering is rare suggested that the nucleus was a small target—much smaller than the 1-angstrom (Å) size that characterizes the distribution of atomic electrons (1 Å = 10^{-10} m = 0.1 nm).

Rutherford analyzed the scattering of α-particles by a nucleus, assuming that the only force between the two was the Coulomb force of repulsion between their positive charges. The classic experiments of Geiger and Marsden verified the pattern of scattering predicted by Rutherford (Figure 42.6).

Knowing that the force between a nucleus and an α-particle is electrical, Rutherford was able to estimate the size of the nucleus. His method makes use of energy conservation. Figures 42.7a,b show a head-on collision between an α-particle and a gold nucleus. When the α-particle is outside the gold atom, the Coulomb force of the nucleus is shielded by the atomic electrons. For this

Figure 42.6

Rutherford's theory (solid line) predicted the relative number of alpha particles that will scatter into different ranges of angle. The measurements of Geiger and Marsden (suggested by open circles and solid circles, respectively) confirmed Rutherford's "nuclear atom."

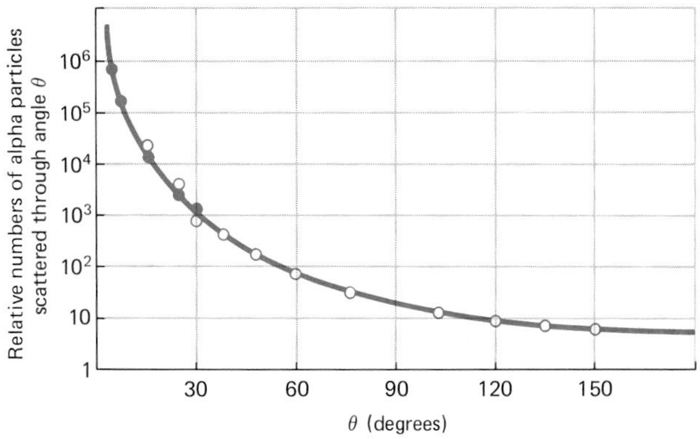

Figure 42.7

(a) An α-particle approaches an Au atom. (b) During the head-on collision the α-particle comes to rest momentarily.

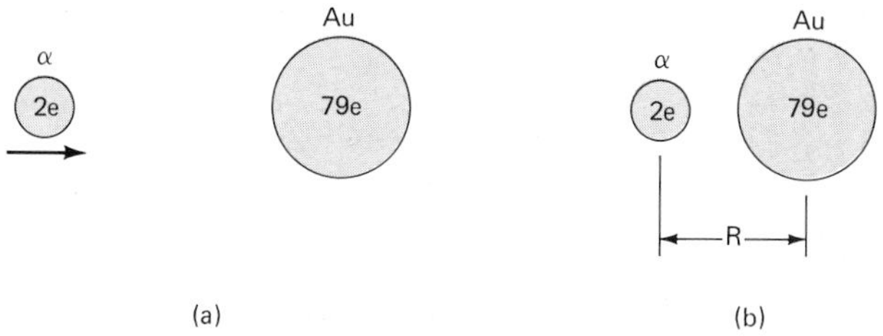

configuration we take the electric potential energy of the system (α + nucleus) to be zero. The total energy of the system is K_α, the kinetic energy of the α-particle. As the α-particle approaches the nucleus, it experiences the repulsive Coulomb force and slows down. In a head-on collision the α-particle is brought to rest momentarily, then repelled backward (180° scatter). At the moment when the α-particle is at rest, the total energy of the system is stored as electric potential energy.* If R is the center-to-center distance between the α-particle (charge $+2e$) and the gold nucleus (charge $+79e$), the potential energy is (see Section 8.2)

$$U = \frac{1}{4\pi\epsilon_0}\left[\frac{(2e)(79e)}{R}\right] \qquad (42.13)$$

The conservation of energy lets us write

$$K_\alpha = U \qquad (42.14)$$

Solving for R gives

$$R = \frac{1}{4\pi\epsilon_0}\left(\frac{158e^2}{K_\alpha}\right) \qquad (42.15).$$

*This assumption ignores the kinetic energy of the recoiling nucleus.

The alphas used by Geiger and Marsden had kinetic energies of 7.7 MeV. The resulting value for R is 3.0×10^{-14} m. As Figure 42.7b suggests, R is an upper limit for the sum of the radii of the α-particle and the gold nucleus. We have seen that the angstrom characterizes the size of the atom. The atomic electrons spread themselves out over a roughly spherical volume 1 Å (10^{-10} m) in radius. Rutherford's figure of 3×10^{-14} m for R shows that the positively charged nucleus is much smaller—about 10,000 times smaller. This concentration of mass and charge explains why large-angle deflections of α-particles are infrequent. As the beam of α-particles races through the gold foil, most will pass a gold nucleus at relatively large distances ($\approx 10^{-10}$ m) and experience only a small deflection. A tiny fraction will experience head-on or nearly head-on collisions and be deflected through large angles.

The experiments of Geiger and Marsden and Rutherford's analysis of the Coulomb scattering helped establish the model of the nuclear atom. The nucleus carries the full positive charge of the atom and most of the mass. The atom's electrons contribute an equal negative charge and an insignificant fraction of the atomic mass. But the positions of the atomic electrons were not revealed by the α-particle scattering. Because of their small mass, the electrons are unable to deflect the α-particles significantly. Thus, observation of the α-particle scattering could not reveal how the electrons are distributed. However, it was clear that the electrons were not distributed in a static configuration. The strong attractive force of the positively charged nucleus would lead to the collapse of any static configuration—the electrons would "fall" into the the nucleus. If, on the other hand, electrons moved about the nucleus in orbits, they could achieve a dynamic stability, just as the motion of the planets about the sun prevents the gravitational collapse of the solar system. And so the planetary model of the atom developed. In the next section we will see how Niels Bohr combined the nuclear atom with the photon concept and quantization to develop a successful model of the hydrogen atom.

42.3 Bohr and the Hydrogen Atom

The idea of an atom in which the electrons whirl about a tiny nucleus intrigued Niels Bohr, a young Danish physicist* (Figure 42.8). Bohr was familiar with Planck's relation ($E = h\nu$) and Einstein's notion of light quanta, and he knew that atoms exhibit a discrete spectrum of wavelengths.

Early in 1913 Bohr was alerted to an empirical relation discovered in 1885 by J. J. Balmer, a Swiss schoolteacher. Balmer's equation described the measured wavelengths of

*For his research on atomic structure, Niels Bohr (1885–1962) received the 1922 Nobel Prize for physics. Subsequently, Bohr made major contributions to the theory of nuclear structure, particularly the theory of nuclear fission.

Figure 42.8

Niels Bohr. Bohr developed the first quantum theory of the atom in 1913.

visible light emitted by hydrogen atoms in terms of a single integer, n.

$$\lambda_n = 364.56 \text{ nm} \frac{n^2}{n^2 - 2^2} \qquad n = 3, 4, 5, \ldots \tag{42.16}$$

Balmer arrived at his formula by a method of trial and error. He systematically tried different combinations of integers, seeking a "simple" numerical relation that predicted the observed wavelengths.

Bohr realized that quantization and the photon concept could be combined to derive the Balmer formula. Bohr's theory of the hydrogen atom ties together the light quanta of Einstein and the nuclear atom of Rutherford. Bohr envisioned the hydrogen atom as an electron moving in a circular orbit about a much more massive nucleus. (In ordinary hydrogen the nucleus consists of one proton.) In Bohr's model, the electron orbits the nucleus the way the earth orbits the sun. The electron is held in orbit by the attractive Coulomb force exerted by the nucleus, just as the earth is held in orbit by the sun's gravity.

Bohr's model raised one troublesome point. According to Maxwell's theory of electromagnetism, an accelerating charge radiates. The electron in orbit about the nucleus experiences a radial acceleration, and therefore should radiate. The continuous loss of energy via radiation would cause the electron to spiral inward toward the nucleus. In short, Maxwell's theory predicted that the planetary atom should collapse. Furthermore, the collapse time could be calculated, and was very brief—about 10^{-10} s. Bohr realized that Maxwell's electromagnetic theory was not adequate to describe fully the dynamics of atomic electrons, but rather than scrap the planetary model, Bohr reached beyond the limited domain of classical physics and postulated the following:

1. *Stationary-state Postulate.* There exist stationary states of the atom in which the orbiting electron does not radiate.
2. *Radiation Postulate.* When the orbiting electron changes from one stationary state to another stationary state of lower energy, a photon is emitted. The energy carried away by the photon ($h\nu$) equals the decrease in the energy of the atom.

If the atom makes a transition from a stationary state of energy E_a to one of energy E_b, the energy decrease is $E_a - E_b$. The radiation postulate can be expressed in equation form as

$$h\nu = E_a - E_b \tag{42.17}$$

3. *Quantization Postulate.* The angular momentum of the atom is quantized in units of $h/2\pi$.

If L denotes the angular momentum of the electron, the quantization postulate can be expressed by the equation

$$L = n\left(\frac{h}{2\pi}\right) \qquad n = 1, 2, 3, \ldots \tag{42.18}$$

Note that the quantization postulate *imposes* quantization on the atom. Quantization did not follow from Bohr's theory as a natural consequence of the classical physics of Newton and Maxwell; instead, it was inserted into the theory.

The quantization of angular momentum, when combined with Newton's second law, leads to the quantization of two properties of the hydrogen atom, the radii of electron orbits and the energies of the stationary states.

Quantization of Orbit Radii

The only orbits that satisfy both Newtonian mechanics and the quantization of angular momentum are those for which the radius is given by

$$r \equiv r_n = n^2\left(\frac{4\pi\epsilon_0}{me^2}\right)\left(\frac{h}{2\pi}\right)^2 \tag{42.19}$$

The quantity $(4\pi\epsilon_0/me^2)(h/2\pi)^2$ is a length, and is called the *Bohr radius*. We denote it by a_0.

$$a_0 = \left(\frac{4\pi\epsilon_0}{me^2}\right)\left(\frac{h}{2\pi}\right)^2 = 5.29177 \times 10^{-11} \text{ m} \tag{42.20}$$

which is approximately 0.5 Å. In terms of a_0 the quantized orbit radii are given by

$$r_n = n^2 a_0 \qquad n = 1, 2, 3, 4 \ldots \tag{42.21}$$

The smallest orbit radius is $r_1 = a_0$. The corresponding diameter of the hydrogen atom is $2a_0$, or approximately 1 Å, the size that characterizes all atoms (Section 25.1). Bohr's theory thus explains the size of atoms and shows that the size is controlled by quantization.

Energy Quantization

The total energy of the hydrogen atom is the sum of the kinetic energy of the orbiting electron and the electric potential energy associated with the Coulomb force that binds the electron and proton. The quantized energies of the atom are given by

$$E = E_n = -\frac{e^2}{8\pi\epsilon_0 a_0 n^2} \qquad n = 1, 2, 3, \ldots \tag{42.22}$$

The quantity E_n is the total (kinetic + potential) energy of the atom when the electron is in the nth orbit. We write this as

$$E_n = -\frac{E_H}{n^2} \qquad n = 1, 2, 3, \ldots \tag{42.23}$$

where E_H is a characteristic energy for the hydrogen atom,

$$E_H = \frac{e^2}{8\pi\epsilon_0 a_0} = 13.6 \text{ eV} \tag{42.24}$$

The quantized energies available to the hydrogen atom are displayed in Figure 42.9. The quantum number n labels the

Figure 42.9

Quantized energy levels of the hydrogen atom. Each energy level is represented by a horizontal line.

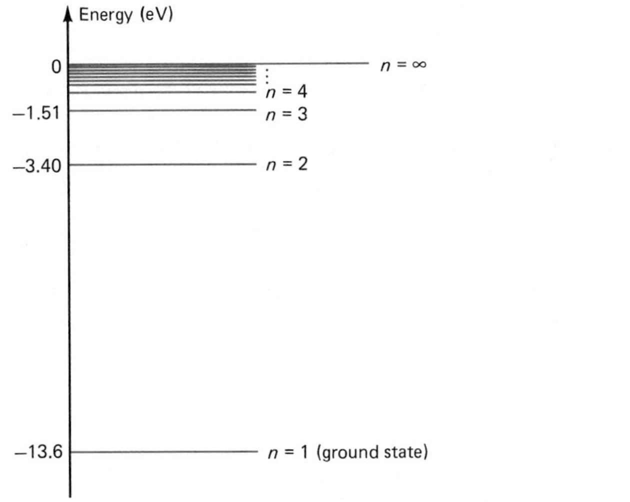

stationary energy states of the atom and the corresponding energy "levels." In Figure 42.10 each energy level is represented by a horizontal line. The state of lowest energy ($n = 1$) is called the *ground state,* and the other states are referred to as *excited states.*

Derivation of Balmer's Formula

When the hydrogen atom makes a transition from an excited state to a state of lower energy, the electron changes orbits and a photon is emitted. Energy is conserved in the overall process. That is, the photon energy ($h\nu$) equals the decrease in the energy of the atom—as prescribed by Bohr's radiation postulate, $h\nu = E_a - E_b$. When the transition terminates on the $n = 2$ level the energy of the emitted photon is

$$h\nu_n = -\frac{E_H}{n^2} + \frac{E_H}{2^2} \qquad n = 3, 4, 5, \ldots \tag{42.25}$$

Replacing ν_n by c/λ_n and solving for the wavelength (λ_n) gives

$$\lambda_n = \frac{4hc}{E_H}\frac{n^2}{n^2 - 2^2} \tag{42.26}$$

Evaluating $4hc/E_H$, we find that

$$\lambda_n = 364.51\left(\frac{n^2}{n^2 - 2^2}\right)$$

$$n = 3, 4, 5, 6, \ldots \tag{42.27}$$

By comparing this equation with Balmer's empirical formula (Eq. 42.16), we see that the two are in substantial agreement. Thus, Bohr's theory provided a theoretical basis for Balmer's empirical formula.

The explanation of Balmer's formula was a great triumph. *It revealed that the atomic world is quantized.* Bohr's original theory of the hydrogen atom has since been replaced, but the quantization of physical properties such as energy and angular momentum survives as an inherent feature of modern quantum theory.

Bohr's theory inspired many experimental investigations. The series of transitions that produced the Balmer wavelengths terminated on the $n = 2$ level (Figure 42.10).

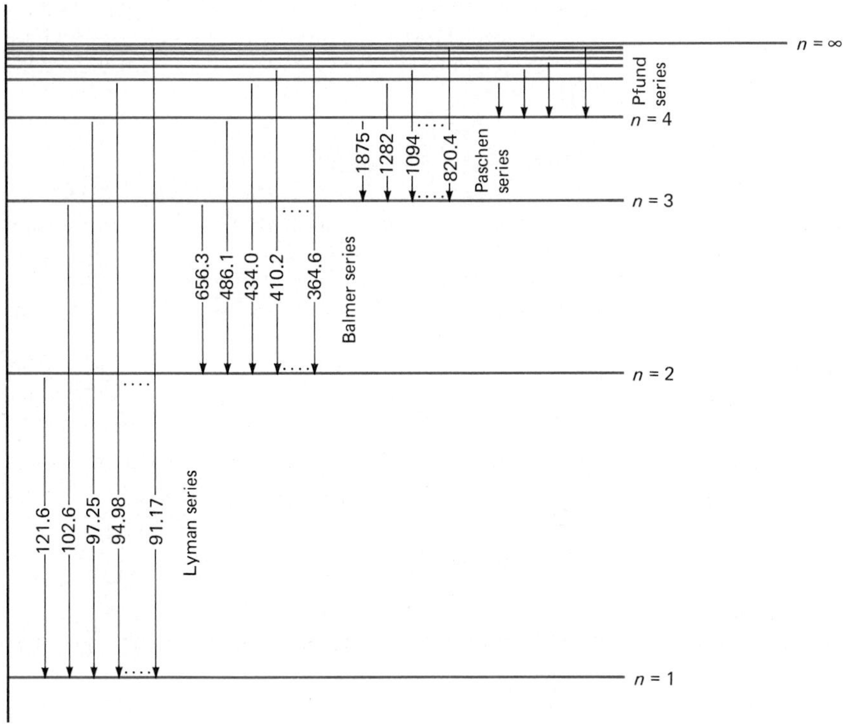

Figure 42.10

Spectral series for the hydrogen atom. Wavelengths are indicated in nanometers.

It seemed likely that other spectral series should be possible, corresponding to transitions that end on the $n = 1, 3, 4, \ldots$ levels. Experiment soon confirmed these expectations. In 1913 Theodore Lyman observed the series of wavelengths that now bear his name. The Lyman series arises from transitions that terminate on the $n = 1$ level (ground state). The wavelengths in the Lyman series lie in the ultraviolet region of the spectrum.

Example 3 **Wavelength of the Lyman α Radiation**

The wavelengths of the Lyman series are labeled L_α, L_β, L_γ, . . . , with L_α being the longest. The L_α radiation is emitted in the $n = 2$ to $n = 1$ transition (Figure 42.10). The wavelength follows from the radiation condition, Eq. 42.17.

$$h\nu = E_a - E_b$$

With $h\nu = hc/\lambda$, $E_a = -E_H/2^2$, and $E_b = -E_H/1^2$, we get

$$\lambda_{L\alpha} = \frac{4}{3}\left(\frac{hc}{E_H}\right)$$

We noted earlier that $4hc/E_H = 365$ nm, and so

$$\lambda_{L\alpha} = 122 \text{ nm}$$

The shortest wavelength in the Lyman series, which corresponds to the $n = \infty$ to $n = 1$ transition, is 91.2 nm. The entire Lyman series lies in the ultraviolet portion of the spectrum. The atmosphere of our sun, composed primarily of hydrogen, emits Lyman radiation. The earth is surrounded by a tenuous cloud of gradually escaping hydrogen and also has a halo of Lyman radiation, largely L_α.

Questions

5. How might an inelastic collision between two hydrogen atoms make it possible for both to subsequently emit light?
6. Suppose you were to construct a Bohr-like model of the helium atom. Would you expect the innermost Bohr orbit to have a radius smaller than, larger than, or equal to the radius of the first Bohr orbit of hydrogen?
7. The three energy levels for a hypothetical atom are shown in Figure 42.11. Identify all possible transitions that result in the emission of light. Which transition produces the longest wavelength? If the atom is in the $n = 2$ level, how many different wavelengths can it absorb?

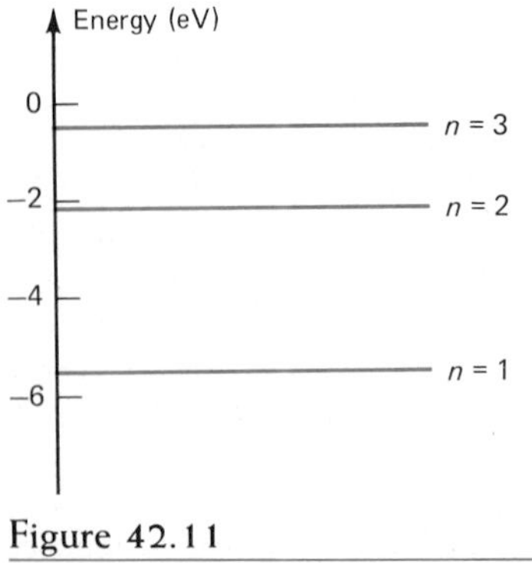

Figure 42.11

42.4 De Broglie and the Wave–Particle Duality

In our study of optics (Chapters 38–41) we treated light as a wave. In this chapter we have seen how light also exhibits *particle-like* behavior. Einstein and Bohr considered light to be composed of particle-like packets of energy called *photons*. The wave and particle interpretations of light are not contradictory, but instead are complementary. In fact, taken together, they enable us to explain the bulk of optical phenomena. The wave picture provides a simple explanation of diffraction and other optical interference phenomena, whereas the particle picture provides a simple explanation of the photoelectric effect and the scattering of x rays by the electrons in atoms. We now recognize that light is neither a wave nor a particle, but has aspects of both. The dual nature of light is one example of the overall wave–particle duality exhibited by nature.

In 1924, a young Frenchman, Louis Victor de Broglie (Figure 42.12), argued that wave–particle duality is a uni-

Figure 42.12

Louis Victor de Broglie. In his 1924 Ph.D. thesis, de Broglie recognized the wave–particle duality as a universal principle.

versal principle and is not restricted to light. Not only do "waves" exhibit particle-like behavior, he argued, but objects that we traditionally regard as particles must also exhibit wavelike properties. For example, we are accustomed to thinking of an electron as a particle; but as we shall see, there are interference-type experiments involving electrons that call for a wave interpretation.

Two quantities used to describe a wave are frequency (ν) and wavelength (λ). Two quantities used to describe a particle are energy (E) and linear momentum (p). The relationships between the wave and particle attributes of light are

$$E = h\nu \tag{42.3}$$

$$p = \frac{h}{\lambda} \tag{42.11}$$

De Broglie postulated Eqs. 42.3 and 42.11 as *universal relationships* between wave (ν, λ) and particle (E, p) attributes.

De Broglie tested his hypothesis by applying it to the electron in a hydrogen atom and found that it leads to the same quantization of angular momentum imposed by Bohr, and in turn, to the same quantized energy levels and spectrum.

The Davisson–Germer Experiment

De Broglie's theory had two points in its favor: (1) It predicted the same quantum structure of the hydrogen atom as did the Bohr theory; and (2) it offered a different viewpoint, which was potentially broader and more general than Bohr's approach. But there were also two shortcomings: (1) There was nothing to indicate the *nature* of the de Broglie waves (specifically, there was no wave equation describing how the waves propagated); and (2) there was no direct experimental evidence to support de Broglie's hypothesis of the wave nature of matter.

Both of these shortcomings were soon remedied. First, in 1925 Erwin Schrödinger set forth a wave equation that described the behavior of de Broglie's waves. (We will examine Schrödinger's equation in the following section.) Then in 1927 Clinton Davisson and Lester Germer performed an experiment that confirmed the wave character of electrons. Their experiment showed that electrons could exhibit *interference*—a clear indication of wave behavior. A beam of electrons accelerated through a potential difference of 54 V was directed toward a nickel crystal. Electrons were scattered in all directions, but their intensity showed a strong maximum at a certain angle. The regularly spaced atoms in the crystal acted like a diffraction grating. The direction of maximum intensity of the scattered electrons corresponded to the direction of *constructive interference* of the electron waves. Davisson and Germer were able to relate the atomic spacing to the electron wavelength, and found that $\lambda = 1.65 \times 10^{-10}$ m. The de Broglie wavelength can also be calculated by using Eq. 42.11 and our knowledge that the measured kinetic energy of the electrons was 54 eV. Thus,

$$\lambda = \frac{h}{p} = \frac{h}{\sqrt{2mK}} \qquad K = \frac{p^2}{2m} \tag{42.28}$$

where K is the kinetic energy and m is the electron mass. With $K = 54$ eV, the calculated de Broglie wavelength is

$$\lambda = \frac{6.62 \times 10^{-34}}{\sqrt{2(9.11 \times 10^{-31})(54 \cdot 1.60 \times 10^{-19})}} = 1.67 \times 10^{-10}\ \text{m}$$

The small difference between the two values of λ can be traced to the fact that the crystal contributes to the overall potential difference experienced by the electrons. Thus, the potential difference differs slightly from the applied 54 V.

Questions

8. If an electron and a photon have the same kinetic energy, which has the shorter de Broglie wavelength?
9. If an electron and a photon have the same de Broglie wavelength, which has the greater kinetic energy?

42.5 Schrödinger's Equation, Probability Waves, and Quantum Mechanical Tunneling

In 1925, Erwin Schrödinger presented an equation that describes de Broglie's waves. The Schrödinger equation is similar in some respects to the equation describing waves on a string (Section 18.1):

$$\frac{\partial^2\psi}{\partial t^2} = v^2\frac{\partial^2\psi}{\partial x^2} \tag{42.29}$$

For waves on a string, the wave function $\psi(x, t)$* describes the displacement of the string from equilibrium. The variable x labels positions along the direction of wave travel and t denotes time. The **Schrödinger wave equation** for stationary states of the hydrogen atom (states of constant energy, E) has the form

$$\left(E + \frac{e^2}{4\pi\epsilon_0 r}\right)\psi = -\frac{h^2}{8\pi^2 m}\left(\frac{\partial^2\psi}{\partial x^2} + \frac{\partial^2\psi}{\partial y^2} + \frac{\partial^2\psi}{\partial z^2}\right) \tag{42.30}$$

The variables x, y, z mark the position of the electron relative to the nucleus, and $r = \sqrt{x^2 + y^2 + z^2}$. The physical significance of the Schrödinger wave function $\psi(x, y, z)$ was not immediately recognized. Yet despite the lack of a physical interpretation of ψ, Schrödinger found that only certain *discrete* (quantized) values of E yield "proper" solutions of the wave equation. A proper solution

*The symbol ψ (pronounced sī) is the 23rd letter of the Greek alphabet.

is one for which the wave function satisfies prescribed boundary conditions.* Remarkably, the quantized energies for hydrogen were the same as those predicted by the Bohr theory.

Schrödinger's wave equation caused great excitement, especially since it was not restricted to the hydrogen atom. Techniques were quickly developed for describing the absorption and emission of radiation when an atom undergoes a transition from one energy state to another. The frequency of the emitted radiation is given by Bohr's radiation condition, $h\nu = E_a - E_b$. The great beauty of Schrödinger's wave equation is that quantization emerges in a natural way—as a consequence of boundary conditions (Section 18.5). It is not necessary to impose quantization, as Bohr did—or as we did with the quantization postulate in Section 42.3.

The Schrödinger wave equation was subsequently solved for numerous atomic and molecular systems, and the optical spectra of many atoms and molecules were successfully interpreted. Schrödinger's wave equation gave birth to a "new" quantum theory—a theory that became known as *wave mechanics*.† But despite its immediate successes and recognition as the key equation governing quantum structure, the Schrödinger equation at first presented one very troublesome point. Neither the equation nor its mathematical solutions reveals the physical meaning of the wave function, ψ. The question physicists asked themselves was "What does ψ measure?"

The answer was given in 1926 by Max Born. Schrödinger's wave is a *probability wave*. Specifically, the square of the magnitude of $\psi(x, y, z)$ for the electron in a hydrogen atom is proportional to the probability that the electron will be located in the vicinity of the point (x, y, z).‡

$$|\psi(x, y, z)|^2\, dV = \text{probability that the particle is in the volume } dV \text{ centered at } (x, y, z) \quad (42.31)$$

The upshot of Born's probability interpretation of ψ is this: The *deterministic* statements of physical laws that characterize classical physics (for example, Newton's laws of motion and Maxwell's electromagnetic theory) are replaced at the atomic level by *probability* statements. For example, the classical description of an electron at rest is quite deterministic. An electron at rest, free of any external forces, will remain at rest. But the quantum description of the electron is different. It says that an electron, free of any external forces, that is within a specified region of space at one moment need not remain in that region. There is a probability that it will leave the region, and quantum theory prescribes how to calculate this probability.

*The role of boundary conditions on the classical wave equation is discussed in Section 18.5. For a wave that represents an electron localized in the vicinity of a nucleus, one boundary condition requires that $\psi \rightarrow 0$ as $r \rightarrow \infty$.

†In today's parlance, the theory that includes the Schrödinger wave equation is referred to as *quantum mechanics*.

‡This interpretation has profound philosophical implications. See *Atomic Physics*, 6th ed., by Max Born, Hafner, New York, 1956 (pp. 94–97); and *Physics for Poets*, 2nd ed., by Robert H. March, McGraw–Hill, New York, 1978 (Chapter 17).

You may be wondering, "Has physics lost touch with reality?" But the concept of physical reality is as elusive as Schrödinger's wave function. Our conception of physical reality has changed as our perception of nature has widened and deepened.* Quantum theory adequately describes reality in the sense that its predictions agree with experiment.

Quantum Mechanical Tunneling

One of the most striking of all quantum phenomena is potential barrier penetration, or *quantum tunneling*. To illustrate the idea of tunneling, consider an electron that moves in response to the potential energy function shown in Figure 42.13† (Section 9.6). The total energy of the electron (E) is the sum of its potential energy (U) and its kinetic energy (K). The total energy E is less than the maximum potential energy (U_B). The peak in the potential energy presents a barrier in the sense that the kinetic energy of an electron decreases as it approaches the point $x = x_A$. According to classical physics, an electron traveling to the right must come to rest momentarily at the turning point at $x = x_A$. At the turning point $E = U(x_A)$, and the kinetic energy is zero. The electron will then move away to the left. In short, the electron is reflected by the potential energy barrier. In the words of classical physics, the region between the turning points (x_A and x_B) is *forbidden* because it corresponds to a negative kinetic energy, $K = (E - U) < 0$. Classically, this is impossible: The kinetic energy, $K = \frac{1}{2}mv^2$, cannot be negative. Quantum mechanics, on the other hand, describes the electron in terms of its wave function, ψ, and the solution of the Schrödinger equation shows that ψ is not zero in the barrier region between x_A and x_B. In other words, there is a definite probability—related to ψ—that an electron can travel from x_A to x_B and continue on. We say that the electron "tunnels" through the region that is "forbidden" by classical mechanics.‡

Quantum tunneling has many practical applications. For example, many solid-state electronic circuits incorporate so-called tunnel diodes, whose operation relies on the quantum tunneling of electrons. Ultrasensitive magnetometers called *squids* (Section 42.7) rely on the quantum tunneling of electrons through a thin insulating barrier that separates two superconductors. One of the earliest applications of quantum mechanics to the nucleus invoked quantum tunneling to explain the type of radioactivity known as α-decay (Section 43.2).

The very existence of life on earth depends on quantum tunneling. The energy that nourishes life on our planet is released via nuclear fusion reactions inside our sun (Section 43.5). One important nuclear fusion reaction

*See *The Nature of Physical Reality*, by Henry Margenau; McGraw–Hill, New York, 1950.

†The potential energy function of Figure 42.13 is *not* typical of the potential energy of an electron in an atom. It has some of the features of the potential energy of electrons in certain types of "solid-state" circuit elements.

‡There are classical analogs of quantum tunneling. For example, light waves may exhibit tunneling. See the discussion of total internal reflection in Section 38.2.

Figure 42.13

The total energy of the particle (E) is less than the maximum potential energy (U_B). The region between $x = x_A$ and $x = x_B$ is "forbidden" by classical physics because it corresponds to a negative kinetic energy.

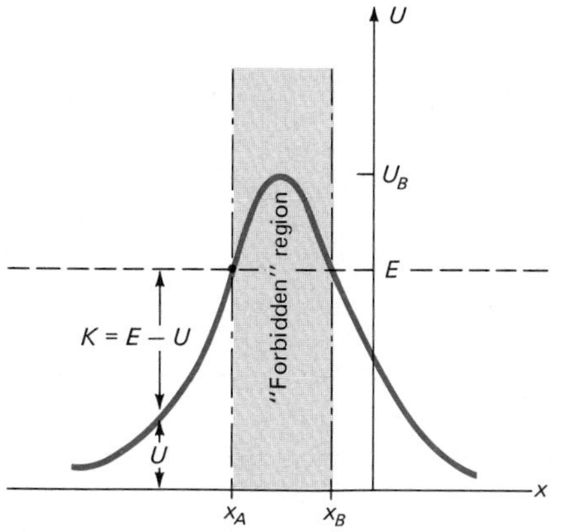

Figure 42.14

The potential energy barrier facing colliding protons.

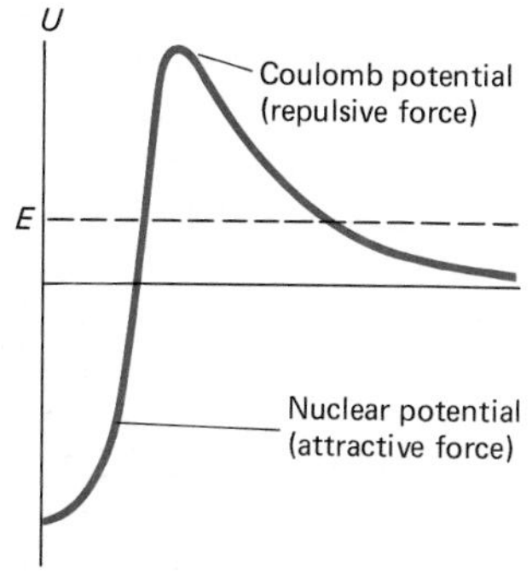

unites two protons. Figure 42.14 shows that the protons face a Coulomb potential energy barrier as they approach one another. The barrier is a consequence of the repulsive Coulomb force between the positive charges of the two protons. To get close enough to experience the attractive nuclear force (Section 43.1), a proton must tunnel through the Coulomb barrier. There is only a small probability that a proton will tunnel through the barrier, and therefore only a tiny fraction of proton–proton collisions result in a fusion reaction. However, this tiny fraction is enough to supply the energy output of the sun, which sustains life on earth. If quantum tunneling were not a physical reality, our sun would never have been able to ignite fusion reactions, nor kindle life.

Questions

10. Figure 42.15 shows the Schrödinger wave function, $\psi(x)$, for a particle confined between the points $x = 0$ and $x = a$. About what point is the particle most probably located?

11. A particle starts inside the potential energy "well" shown in Figure 42.16. Its total energy is E. Describe its motion on the basis of classical physics. From a quantum viewpoint, which side of the barrier is it most likely to tunnel through? (Make an instinctive guess!) On which side would it end, moving with the greater speed?

Figure 42.15

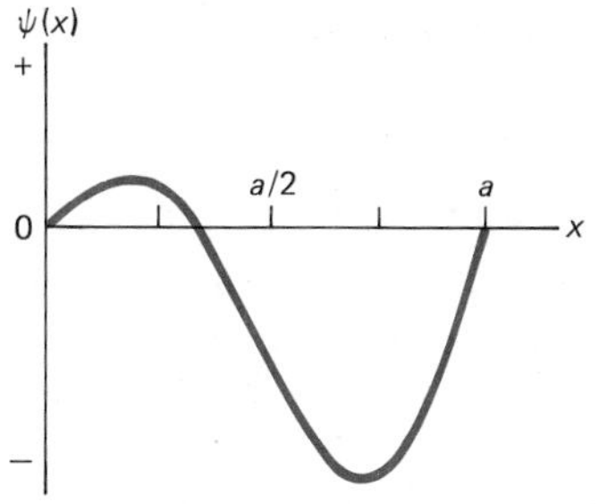

Figure 42.16

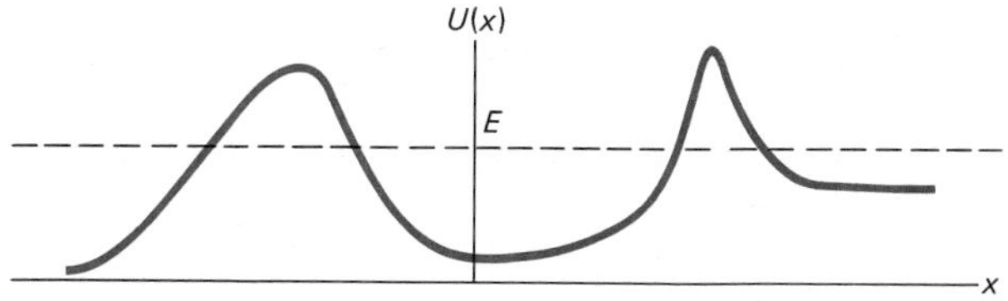

42.6 Quantum Optics and Lasers

The quantum theory of light is called *quantum optics*. As we will now see, the quantum theory of matter and quantum optics are tied together.

Absorption and Emission of Electromagnetic Radiation

One of the basic equations linking quantum optics to the quantum theory of matter is the **Bohr radiation condition.** The radiation condition relates the photon energy to the energy change of an atom or molecule or nucleus.

$$h\nu = E_a - E_b \qquad (42.17)$$

In 1917, Albert Einstein showed that three basic processes are described by Eq. 42.17. These are *absorption, spontaneous emission*, and *stimulated emission*. Figure 42.17 depicts these three processes. In Figure 42.17a an atom in a state of energy E_b absorbs a photon and is thereby raised to a state of energy E_a. In Figure 42.17b an isolated atom in an excited state of energy E_a spontaneously emits a photon. This leaves the atom in a state of lower energy, E_b. The process of stimulated emission is shown in Figure 42.17c. A photon of energy $h\nu = E_a - E_b$ interacts with an atom in a state of energy E_a, and triggers the birth of another photon of energy $h\nu$. The two photons travel off together,

Figure 42.17

Three basic mechanisms by which atoms interact with photons. (a) Absorption: A photon of energy $E_a - E_b$ is absorbed, raising the energy of the atom from E_b to E_a. (b) Spontaneous emission: An atom spontaneously emits a photon of energy $E_a - E_b$. The energy of the atom decreases from E_a to E_b. (c) Stimulated emission: A photon of energy $E_a - E_b$ triggers the emission of its twin. The energy of the emitting atom decreases from E_a to E_b.

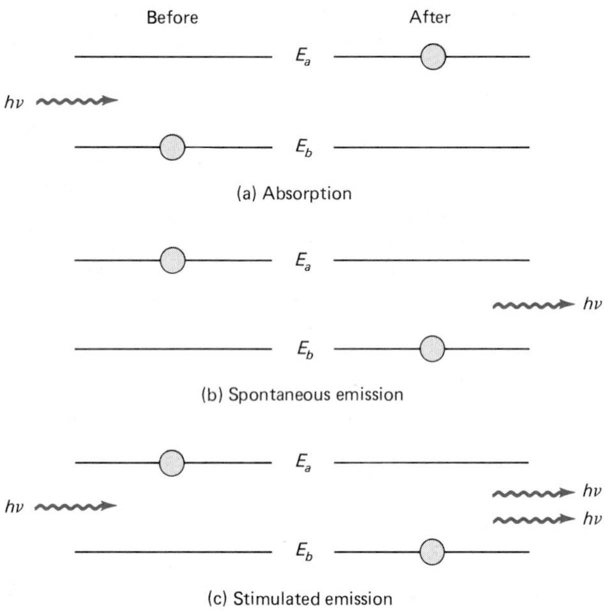

leaving the atom in a level of lower energy. From the wave viewpoint, the excited atom emits a wave that is *coherent* with the stimulating wave. The emitted wave has the same frequency, direction of propagation, and polarization as the stimulating wave. Their superposition results in an amplification of the stimulating wave (the amplitude is increased.)

Lasers

A laser* is a device that produces a highly directional and intense beam of coherent light. Ordinarily, when a beam of light travels through matter, the effects of absorption and spontaneous emission overpower the effects of stimulated emission, and there is a net *reduction* in the intensity of the beam. In a laser, however, conditions are arranged so that stimulated emissions are dominant over absorptions and spontaneous emissions. As a result, the intensity of the beam grows as the beam moves through the matter. Let's first examine the conditions that reduce the intensity of a light beam traveling through matter, and then describe the conditions that increase the intensity in a laser.

Consider a narrow beam of photons entering a slab of material that contains atoms capable of absorbing the photons (Figure 42.18). When a photon is absorbed, the intensity of the beam is reduced. Each atom that absorbs a photon is left in an *excited state*. Its energy is increased by $h\nu$, the photon energy. Ordinarily, the atom remains in an excited state only very briefly—typically about 10^{-8} s. It can return to its former state by emitting a photon of the same frequency as the photon it absorbed. If every photon emitted this way was emitted in the direction of propagation of the beam, no overall change in the beam intensity would result. That is, absorptions followed by spontaneous emissions along the beam would not change the intensity. But the direction of a spontaneously emitted photon is rarely along the direction of the beam. Instead, photons are emitted in all directions, and only a tiny fraction end up traveling with the beam. Thus, absorptions followed by spontaneous emissions result in a *net loss of photons* from the beam and thereby *reduce* the beam intensity (Figure 42.18).

Next, consider a photon in the beam that encounters an atom in an appropriate excited state. If the incident photon should cause the excited atom to emit a photon—

*The word *laser* is an acronym for *l*ight *a*mplification by *s*timulated *e*mission of *r*adiation.

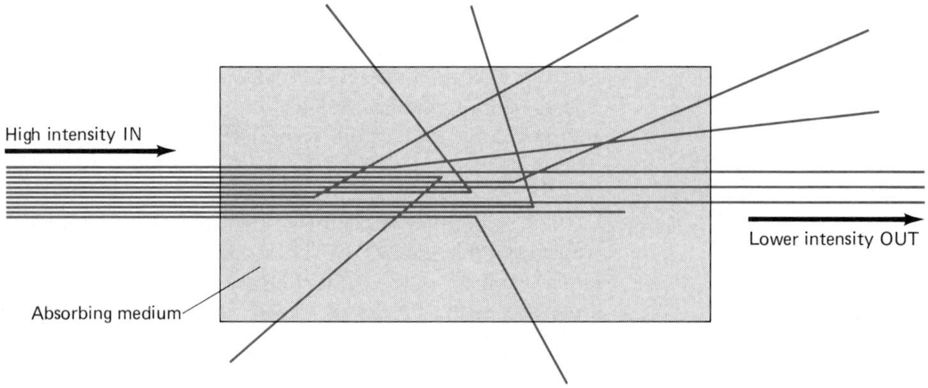

Figure 42.18

A light beam enters a medium capable of absorbing photons. Ordinarily, photon absorption is quickly followed by spontaneous emission. The emitted photons emerge traveling in all directions. Only a tiny fraction end up traveling with the beam. Thus, absorption followed by spontaneous emission reduces the intensity of the emerging beam.

stimulated emission—then the two photons would travel off "hand in hand" in the direction of the beam (Figure 42.17c). Stimulated emissions thereby *increase* the beam intensity.

As you can see, we have two competing effects: Absorptions *reduce* the beam intensity; stimulated emissions *increase* the beam intensity. Normally, the net result of these two processes is a reduced intensity, because there are many more absorptions than there are stimulated emissions. The reason that absorptions normally outweigh stimulated emissions is that most atoms are ordinarily in their lowest energy state, where they can absorb but not emit. In order to have stimulated emissions there must be atoms in excited states, and ordinarily only a tiny fraction of atoms are in excited states.

Population Inversion

There are several methods that make it possible to "invert" the usual population of energy states. A *population inversion* creates a situation where there are relatively many atoms in excited states, capable of emitting light of a particular wavelength, and relatively few in states capable of absorbing light of the same wavelength. In particular, it is possible to add energy selectively to atoms, thereby "pumping" them into a particular excited state. A beam of photons passing through an inverted population is amplified because there are more stimulated emissions than absorptions. Each photon of the incident beam can trigger the emission of others, and each newborn photon can travel on to stimulate further emissions. If a sufficient number of atoms are in excited states, an avalanche of photons can result. Figure 42.19 illustrates how the intensity of a photon beam can grow as it moves through a medium with an inverted population.

The possibility of amplifying light by stimulated emission became evident after Einstein's theoretical study (in 1917) of the basic processes of absorption and emission. The possibility became a reality in 1954, when Gordon, Zeiger, and Townes built a device that amplified microwave radiation via stimulated emission. This type of microwave amplifier is called a *maser* (*m*icrowave *a*mplification by *s*timulated *e*mission of *r*adiation). Then, in 1960 Theodore Maiman first demonstrated the amplification of visible light via stimulated emission. Maiman's laser produced red light with a wavelength of 694.3 nm.

Laser Operation

The two essential components of a laser are an *active medium* and an *optical resonator*. In the helium–neon (He–Ne) laser the active medium is a gaseous mixture of helium and neon. An electric discharge causes some of the atoms to ionize. Thus, the active medium also contains positive ions and electrons. The optical resonator usually consists of a pair of mirrors made highly reflective at the laser wavelength via thin-film interference (Section 40.4). One of the mirrors transmits about 1% of the incident laser intensity. This transmitted light constitutes the laser output beam.

The He–Ne laser is referred to as a *four-level system*. Figure 42.20 shows how four different energy levels are used to achieve a population inversion in a four-level system. Atoms* are pumped (Ⓐ) from the ground state (1) to a band of excited states (2). Transitions (Ⓑ) from the band of excited states populate the upper laser level (3). The transitions from the band of excited states must be *selective:* Transitions to the upper laser level must be more probable than transitions to the lower laser level. In the absence of pumping, both laser levels are essentially empty. Therefore, any selective population of the upper laser level results in a population inversion. Stimulated emissions (Ⓒ) from level 3 to level 4 result in laser action.

Figure 42.21 shows the energy levels for the He–Ne system. The helium plays a role in the pumping scheme. An electric discharge provides energetic electrons that bombard the helium atoms. Inelastic collisions between electrons and helium atoms in the ground state transfer energy to the helium atoms, boosting them into excited states. The energy of the excited state shown for the he-

*Lasers may employ molecules or ions as well as atoms.

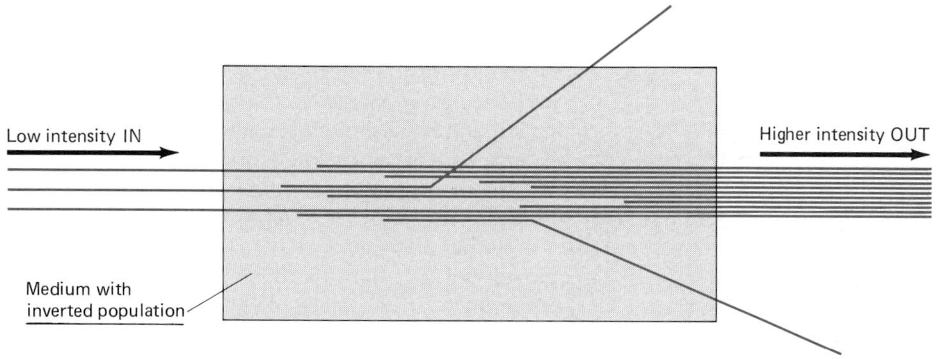

Figure 42.19

A beam of photons enters a medium in which there is an inverted population of atoms. Stimulated emissions outweigh the effects of absorptions. The number of photons emerging exceeds the number of photons entering. This results in an increase in the beam intensity as it travels through the medium.

Figure 42.20

A four-level scheme for achieving an inverted population, Ⓐ, pump transition; Ⓑ, selective transitions to upper laser level; Ⓒ, laser transition.

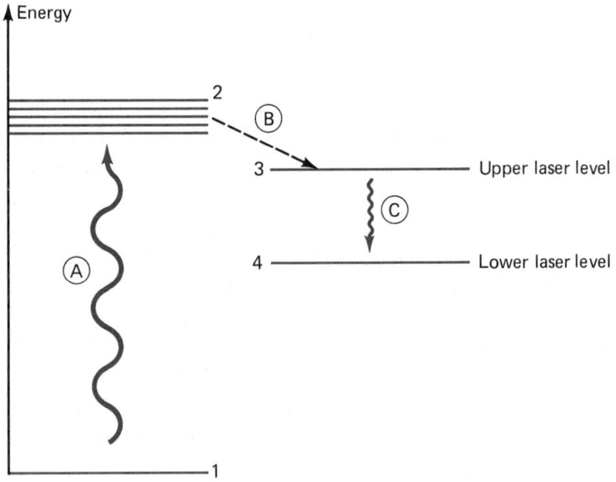

Figure 42.21

Steps involved in producing an inverted population in the helium–neon laser. Ⓐ, electric discharge excitation (inelastic collision transfers energy from electron to helium atom, thereby "pumping" helium atom to excited state); Ⓑ, inelastic collision between excited helium atom and neon atom in ground state "pumps" neon atom to excited states; Ⓒ laser transition. (Neon atom in excited state is stimulated to emit photon.)

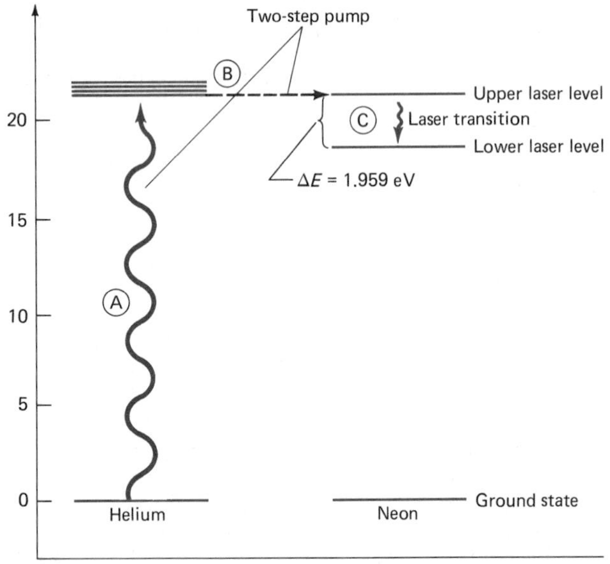

lium atom is a few hundredths of an electron volt greater than the energy needed to excite the upper laser level of the neon atom. Because of this, an inelastic collision between an excited helium atom and a ground-state neon atom can transfer energy to the neon atom. The neon atom is raised to the upper laser level and the helium atom returns to its ground state. The transfer of energy through inelastic collisions is more probable than energy loss through spontaneous emission, making the pumping action efficient. The energy transfer between helium and neon atoms is favored by the very small difference in their excitation energies. The small excess of energy is parceled out as translational kinetic energy between the colliding atoms. This collisional energy transfer process is selective. The excitation energy of the helium atoms is nearly 2 eV higher than the excitation energy of the lower level of the laser transition. This large difference makes it very unlikely that an excited helium atom will transfer its energy to a neon atom and thereby populate the terminal state of the laser transition.

Of course, collisions between the swift electrons and neon atoms in the ground state will populate both laser levels. However, such collisions are not selective. They do not tend to produce a population inversion. The key to the pumping scheme is the tiny energy difference between the two excited states of helium and neon, which favors the upper laser level.

Example 4 **Helium–Neon Laser Wavelength**

Using the energies shown in Figure 42.21 for the upper and lower laser levels, we can calculate the frequency and wavelength of the radiation. The frequency follows from the Bohr radiation condition, Eq. 42.17:

$$h\nu = E_a - E_b$$

For the He–Ne laser, $E_a - E_b = 1.9593$ eV. Thus,

$$\nu = \frac{E_a - E_b}{h} = \frac{1.9593 \text{ eV}}{4.1357 \times 10^{-5} \text{ eV} \cdot \text{s}}$$

$$= 4.7375 \times 10^{14} \text{ Hz}$$

The laser wavelength (λ) follows from the basic kinematic relation $\nu\lambda = c$. Thus,

$$\lambda = \frac{c}{\nu} = \frac{2.9979 \times 10^8}{4.7375 \times 10^{14}} \text{ m} = 6.328 \times 10^{-7} \text{ m}$$

or

$$\lambda = 632.8 \text{ nm}$$

This wavelength gives the laser light a fiery red color.

The laser output is highly directional. This stems from the fact that the amplitude of the light wave builds up via repeated trips through the active medium. Only waves that start out nearly parallel to the axis of the laser cavity are trapped and have the opportunity grow in amplitude. Waves starting in other directions are refracted or reflected out of the active medium. The light that is transmitted is also diffracted as it passes through the exit mirror, and this causes a slight divergence of the laser beam. The high directionality of laser light was demonstrated spectacularly in an experiment in which laser light was beamed to the moon. The light was reflected back to earth by an array of reflectors left by the Apollo 11 astronauts. Optical components expanded the laser beam to give it a diameter of 3 m when it left earth. After traveling nearly 400,000 km to the moon, its diameter was approximately 2 km. In traveling to the moon the beam diameter grew by only 1 m every 200 km.

The lunar reflectors have made it possible for scientists to measure the earth-to-moon distance accurately. These measurements, which have a precision of a few centimeters, confirm the prediction that the moon is slowly spiraling away from the earth because of the effects of tidal friction (Section 12.4). Measurements of continental drifts on earth have also been achieved by using laser signals between the moon and selected points on earth.

Laser Applications

The laser touches our lives in more ways than we realize. The denim in your blue jeans may have been cut to shape by a laser beam. Powerful lasers are used to weld auto frames and anneal large metal pipes. Delicate lasers are used to remove tattoos and weld detached retinas. Low-power lasers are now used routinely by surveyors to establish straight lines.

Laser research pushes frontiers in many directions—toward higher powers and shorter wavelengths. Sophisticated information storage and retrieval systems using lasers are under development. Multiple-laser systems operated in tandem are being tested as part of a broad effort aimed at the development of thermonuclear power reactors (Section 43.6). New methods of pumping lasers are also being developed. The nuclear pumped laser converts the kinetic energy of fission fragments into the energy of excitation of laser atoms. A nuclear pumped laser would make it possible to beam power from one space vehicle to another. The laser seems destined to become an important aspect of future engineering systems, both on earth and in space.

42.7 Squids

Quantum engineering has given us a class of ultrasensitive instruments called *squids*. The word *squid* is an acronym for *s*uperconducting *q*uantum *i*nterference *d*evice. Basically, a squid measures changes in magnetic flux, but it can be converted to an ammeter or voltmeter by using auxiliary circuitry.

The physics underlying squids is almost as old as quantum theory itself. We will begin by sketching the early physics that eventually culminated in the invention of the squid.

Superconductivity

In 1908 Heike Kammerlingh Onnes succeeded in liquefying helium by lowering its temperature to 4.2 K. The liquid helium "bath" furnished Onnes with a new frontier. He could immerse materials in the frigid liquid and study their properties at temperatures never before achieved. Onnes conducted systematic studies of the properties of different electrical conductors. In 1911 he discovered that mercury loses *all* electrical resistance below a temperature of 4.153 K (modern value). This phenomenon is called *superconductivity* (Section 30.3). The temperature below which a material is superconducting is called the *transition temperature*. Since the initial discovery by Onnes, several thousand compounds and more than one third of the elements have been found to exhibit superconductivity.

Meissner Effect

In 1933 Meissner and Ochsenfeld performed an experiment that revealed that a superconductor is not only a perfect conductor, but a perfect diamagnet as well (Section 36.3). They observed that when a tin cylinder has (a) been cooled below its superconducting transition temperature and then (b) been subjected to a magnetic field, the lines of magnetic flux do not penetrate the sample. This is the behavior expected for a perfect conductor. As the magnetic field is turned on, currents are induced that tend to prevent any change in the magnetic flux through the sample (Lenz's law, Section 33.3). In the absence of any resistance, this current assumes whatever value is necessary to prevent flux penetration.

Meissner and Ochsenfeld then reversed the order of (a) and (b). The magnetic field was first applied to the sample in its normal (nonsuperconducting) state (Figure 42.22). Because the magnetic properties of the normal state are very weak, the magnetic flux lines fully penetrated the sample. The material was then cooled below its transition temperature. If zero resistivity were the only attribute of superconductivity, the flux threading the specimen would have remained unchanged as the transition temperature was passed. What Meissner and Ochsenfeld observed, however, was something quite different: They found that the magnetic flux was completely expelled from the sample as it became superconducting (Figure 42.22b). This perfect diamagnetism means that the magnetic field inside a superconductor is zero.*

*The magnetic field does penetrate the surface layers to a depth of about 10^{-7} m.

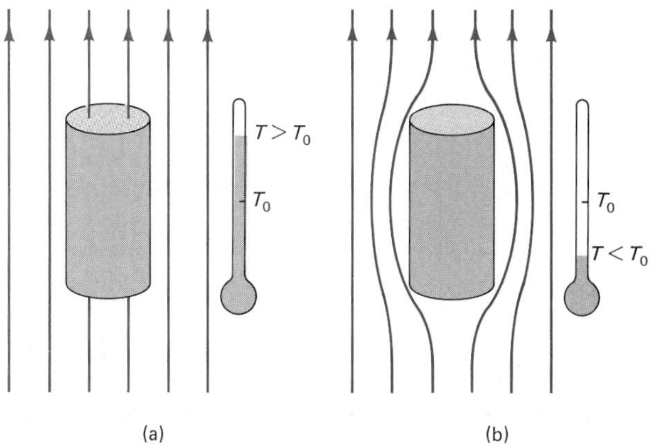

Figure 42.22

The Meissner effect. (a) Lines of flux penetrate the material when it is above its transition temperature (T_0). (b) Flux is expelled when the temperature falls below the transition temperature.

Figure 42.23

Flux trapping. (a) Flux produced by an external field penetrates a ring when the ring is above its transition temperature. (b) If the external field is removed when the ring is below its transition temperature, the flux threading the ring is trapped. Supercurrents induced in the ring maintain the trapped flux.

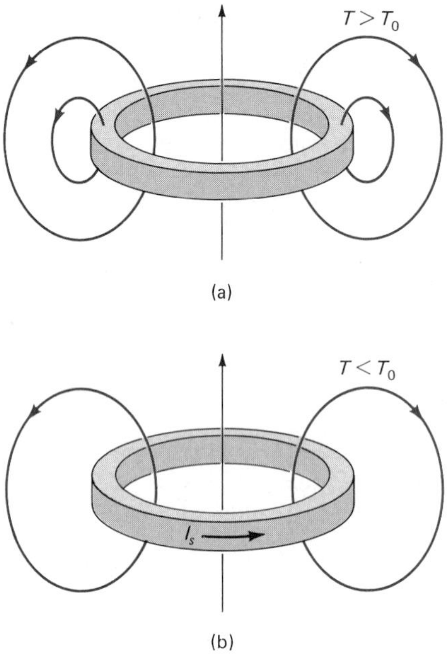

If the solid cylinder is replaced by a ring-shaped superconductor, a phenomenon called *flux trapping* can be observed (Figure 42.23). When the ring is above its transition temperature, an external magnetic field causes flux to penetrate the ring. If the ring is cooled below its transition temperature and the external field is then removed, the flux threading the ring remains unchanged—the flux is trapped.

Cooper Pairs

A quantum theory of superconductivity was developed in 1957 by Bardeen, Cooper, and Schrieffer (Figure 42.24). The BCS theory shows that in a superconductor the motions of a small fraction of the electrons are strongly correlated.* A pair of correlated electrons is treated as a single entity and is called a *Cooper pair.*† When a material becomes superconducting a small fraction of its electrons "condense," forming a group of Cooper pairs. In a superconductor it is the ordered "drift" motions of Cooper pairs that results in the supercurrent. The correlated motion of Cooper pairs proceeds without resistance, thereby "short-circuiting" any normal electric current.

*Quantum mechanically, it is the electron *wave functions* that are *correlated*.

†In 1956, L. N. Cooper showed how such pairs might form in a superconductor.

Figure 42.24

John Bardeen, Leon N. Cooper, and J. Robert Schrieffer received the 1972 Nobel Prize in physics for their superconductivity research.

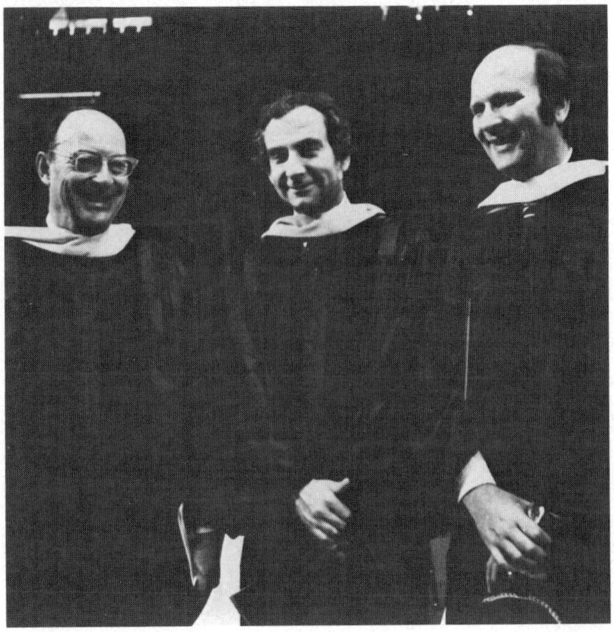

Magnetic Flux Quanta

In 1957 A. A. Abrikosov predicted the existence of magnetic flux quanta. The quantum of magnetic flux is called the *fluxon*. The fluxon (Φ_o) is the ratio of Planck's constant (h) and the magnitude of the electric charge of a Cooper pair ($2e$):

$$\Phi_o = \frac{h}{2e} = 2.0678538 \times 10^{-15}\ \mathrm{T \cdot m^2} \qquad (42.32)$$

The existence of such flux quanta was verified experimentally in 1961 by Fairbank and Deaver and independently by Doll and Nabauer. More recently, the arrangement of flux tubes has been studied directly by using a novel photographic technique.* The quantization of magnetic flux is a special property of superconductors. The magnetic flux threading an ordinary solenoid or transformer coil is not quantized.

Josephson Junctions

We introduced the concept of quantum mechanical tunneling in Section 42.5, where we saw that the wave properties of electrons enable them to penetrate potential energy barriers. Several types of electron tunneling can occur in solids. For example, electrons can tunnel through a very thin layer of insulating material separating two metals. A significant tunnel current can exist only when the de Broglie wavelength of the electron is comparable to or greater than the barrier thickness. In practice, this requires very thin barriers, generally with thicknesses less than 50 Å (500 nm).

*See the "Suggested Reading" section at end of this chapter.

Figure 42.25

Brian Josephson developed a theory describing the tunneling of supercurrents. Josephson shared the 1973 Nobel Prize for physics.

In 1962, an English graduate student, Brian Josephson (Figure 42.25), developed a theory describing the tunneling of supercurrents through a junction separating *two superconductors*. These tunneling junctions between two superconductors are called *Josephson junctions*. At a Josephson junction, Cooper pairs tunnel through the oxide layer separating the two superconductors.

A novel feature of the tunneling supercurrent is that a steady supercurrent can exist *without any potential difference across the junction*. For a given junction area there is a maximum tunnel supercurrent (I_c) that can exist without setting up a potential difference across the junction. If the current through the junction exceeds I_c a potential difference appears across the junction. The tunneling is then due to individual electrons. In other words, if $I > I_c$, the Cooper pairs are broken.

The maximum supercurrent that can tunnel is strongly influenced by an applied magnetic field. More specifically, the supercurrent exhibits interference effects related to the quantization of magnetic flux.

Supercurrent Interference Effects

Josephson showed that a constant magnetic field, acting at right angles to the supercurrent through the junction, results in interference that limits the critical current. Specifically, the critical current has the form

Figure 42.26

Critical current versus magnetic flux (in units of Φ_o) for a Josephson junction. The quantum interference of the de Broglie waves of Cooper pairs produces a pattern similar to the E-field amplitude produced by the optical interference in a single slit.

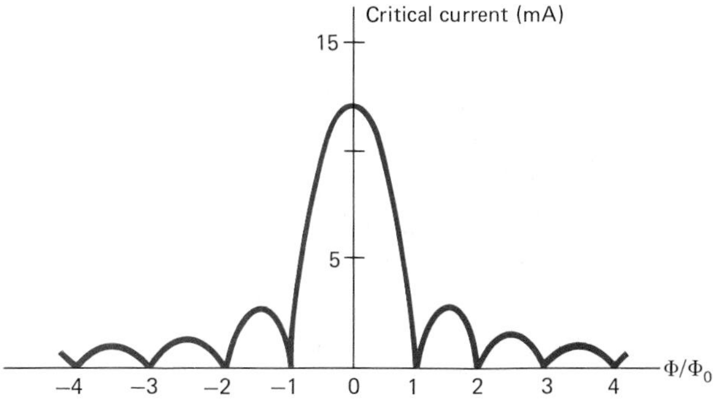

$$I_c = I_o \frac{|\sin(\pi\Phi/\Phi_o)|}{(\pi\Phi/\Phi_o)} \qquad (42.33)$$

where Φ is the magnetic flux through the junction and Φ_o is the fluxon, the quantum of magnetic flux,

$$\Phi_o = \frac{h}{2e} = 2.0678538 \times 10^{-15}\ \mathrm{T\cdot m^2} \qquad (42.32)$$

Figure 42.26 is a graph of I_c versus Φ/Φ_o. The graph shows that I_c is zero whenever the flux threading the junction is an integral number of fluxons. The mathematical form of I_c is identical to that of the magnitude of the E-field for single-slit diffraction (Section 41.2).* There are certain analogies between the optical and supercurrent interference effects. In both cases, interference results from the superposition of coherent waves having different phases. In optical interference, the waves are electromagnetic and the phase difference results because of differences in optical path lengths. In supercurrent interference, the waves are the de Broglie waves of Cooper pairs and the phase difference is a consequence of the applied magnetic field.

In a typical junction, the area penetrated by the applied magnetic field is roughly $10^{-11}\ \mathrm{m^2}$. A change in the field of $\Delta B = 2 \times 10^{-4}$ T gives a flux change of

$$\Delta\Phi = \Delta B \cdot A = 2 \times 10^{-15}\ \mathrm{T\cdot m^2} \approx \Phi_o$$

The $|\sin(\pi\Phi/\Phi_o)|$ factor in Eq. 42.33 shows that the critical current goes from one zero to the next each time the flux changes by Φ_o. If the magnetic field through the junction is varied by several gauss, the critical current should exhibit variations having the form described by Eq. 42.33. In 1963 John Rowell studied the critical current through a lead–lead-oxide–lead junction operated at a temperature of 1.3 K. The critical current varied with the applied magnetic field strength in the manner displayed in Figure 42.27. Rowell's experiment was a striking confirmation of the quantum interference predicted by Josephson.

*The symbols Φ and Φ_o in Eq. 42.33 represent magnetic fluxes. The same symbols used in Section 41.2 to describe the E-field for single-slit diffraction have completely different meanings.

Figure 42.27

The single-slit interference pattern for the critical current in a Josephson junction (lead–lead oxide–lead) operated at $T = 1.3$ K. [After J. M. Rowell, *Phys. Rev. Lett.* 11, 200 (1963).]

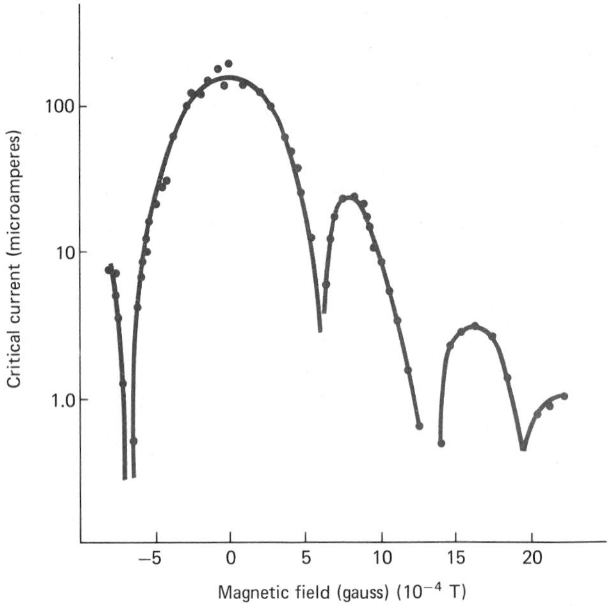

Figure 42.28

The double-slit interference of supercurrents. A supercurrent I_s is divided and sent along two different paths. The coherent supercurrents tunnel through Josephson junctions and exhibit interference when they reunite. An external magnetic field **B** is applied parallel to the surface of the glass substrate. The phase difference between the supercurrents is proportional to the magnetic flux through the area A ($\Phi = BA$). A schematic representation of the double-junction device is shown at the upper right.

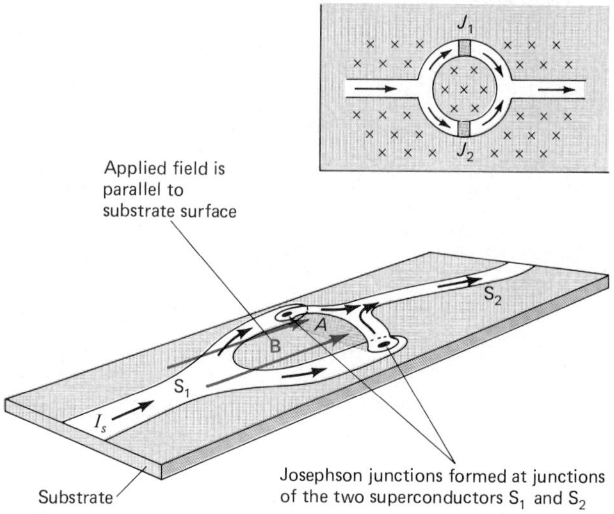

In 1964, a research group headed by James Mercereau performed the supercurrent analog of the double slit interference experiment (Young's experiment, Section 40.2). A supercurrent is divided (see Figure 42.28) and sent along two different paths. The supercurrents tunnel through separate junctions and are reunited. Josephson's theory predicts that the phase difference between the reunited currents is directly proportional to the magnetic flux through the loop formed by the two supercurrent paths. This loop area, the area A in Figure 42.28, can be made much larger than the junction area. The change in magnetic field necessary to cause a flux change of one fluxon is therefore much smaller. Thus, a double-junction device has a greater sensitivity than one using a single junction. Areas on the order of 10^{-6} m^2 were typical of some of the early devices. An area of 10^{-6} m^2 requires a field change of approximately 2×10^{-9} T to cause the flux to change by one fluxon.* Sensitivities approaching (10^{-15} T) are possible with more sophisticated devices.

Observing the interference of supercurrents allows us to count flux quanta in a direct way and has led to the development of the ultrasensitive measuring instruments called squids.

Squids

The heart of a squid is a *superconducting ring*, which is simply a closed loop of superconducting material. The magnetic flux threading a superconducting ring is *quantized*. The total flux Φ must be an integral number of fluxons ($\Phi_o = 2.07 \times 10^{-15}$ T · m^2).

$$\Phi = n\Phi_o \qquad n = \text{integer} \qquad (42.34)$$

In order for a superconducting ring to be a useful part of the squid, it must be able to exchange energy with other parts of the instrument. This is accomplished by introducing a *weak link* in the ring. A weak link is a region that has a much lower critical current than the rest of the superconducting ring. The weak link is also a region into which an applied magnetic field can penetrate. Figure 42.29 shows one way of producing a weak link.

Because of the constrictive nature of the weak link, the current density is greater in the link than in any other portion of the ring. If the current in the link exceeds the critical value, the link becomes normal—it is no longer superconducting. This break in the superconducting path around the ring allows fluxons to move through the link. The weak link thus acts as a "gate" through which fluxons enter and leave the ring. The weak link can be "tailored" to let a single fluxon pass through before reverting to the superconducting state. When a fluxon passes through the link, the link is resistive, and the potential drop across it can be measured with a voltmeter. This makes it possible to count individual fluxons with a squid.

In practice, the weak link of the squid is operated in the "resistive mode," in which the link is in the normal state while the rest of the ring is superconducting. Flux can therefore pass *continuously* through the link. The counting

$$^*\Delta B = \frac{\Phi_o}{A} \simeq \frac{2 \times 10^{-15}\ \text{T} \cdot \text{m}^2}{10^{-6}\ \text{m}^2} = 2 \times 10^{-9}\ \text{T}$$

Figure 42.29

A thin circular film of superconducting material is deposited on a substrate. A narrow constriction in the film (the weak link) is created by scraping away portions of the film.

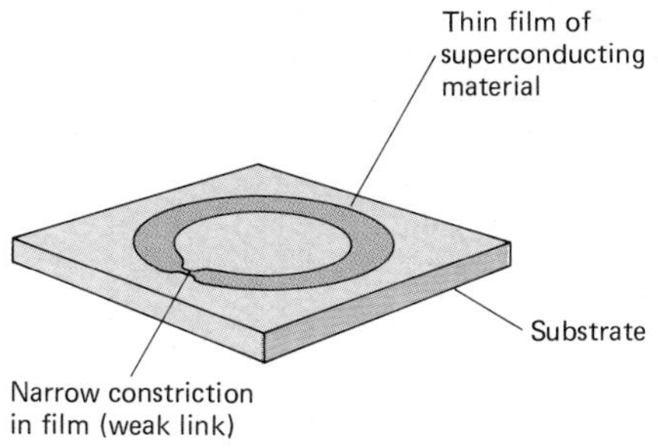

Figure 42.30

The critical current in the weak link varies periodically with the total flux through the squid area (area enclosed by the superconducting ring). The current executes one cycle each time the flux changes by one fluxon (Φ_0). The maximum variation in the critical current is given by $\Delta I_c = \Phi_0/L$. The potential difference across the link shows the same periodic variations as does I_c.

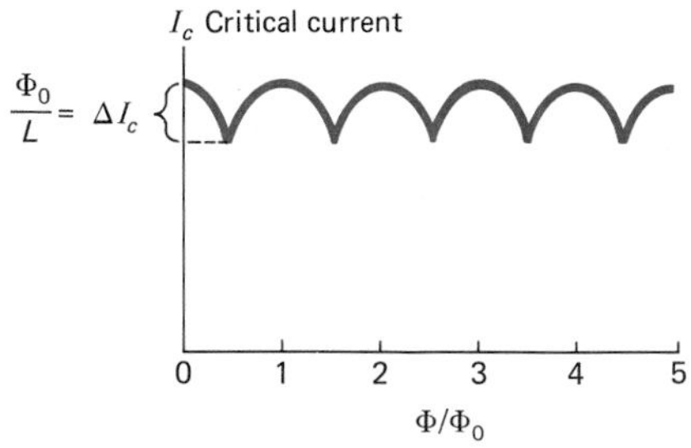

Figure 42.31

As the external field ($\mathbf{B}_{ex}$) is turned on, the flux through the squid area (A) changes, causing periodic changes in the squid current (I_s). Variations in the squid current induce an emf in the adjacent coil (inductance L). The induced emf $V(\Phi)$ has the same flux-dependent period as the squid current.

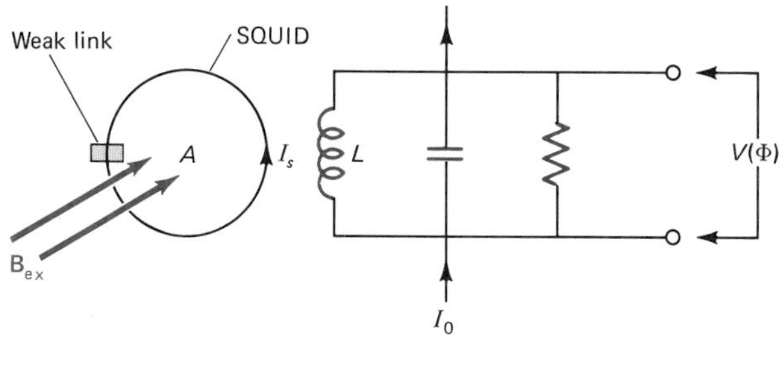

of fluxons then depends on the periodic relationship between the flux and the critical current. The critical current repeats as the applied flux passes through the values $\Phi = 0, \Phi_0, 2\Phi_0, 3\Phi_0, \ldots$ (Figure 42.30). The potential difference across the link shows the same periodic behavior of the critical squid current and potential difference, and converts these into "counts" of fluxons. The potential difference across the link can be measured directly and used to drive an electronic counting circuit. Such a circuit is used to count the number of periodic variations in the squid current, each of which corresponds to a flux change of one fluxon (Φ_0). Counting rates of 2000 fluxons/s have been achieved with such digital squids. The squid current can be sensed by a nearby coil (Figure 42.31). Periodic variations in the squid current set up a changing magnetic field that produces similar variations in the flux threading the coil. The induced current and emf in the coil have the same period as the squid current. The emf induced in the coil acts to drive electronic counting circuitry.

As we have seen, the squid measures flux changes. The relationship between flux and magnetic field strength

$$\Phi = BA \tag{42.35}$$

makes it possible to convert flux measurements into B-field measurements. The sensitivity of squids is determined by the largest area (A) that can be used and the minimum flux change that can be detected. If $\Delta\Phi$ is the smallest detectable change in Φ, then the smallest change in the magnetic field (ΔB) follows from Eq. 42.35 as

$$\Delta B = \frac{\Delta\Phi}{A} \tag{42.36}$$

The smallest flux change that can be detected has steadily decreased with advances in the design of squids. Changes as small as $\Delta\Phi \simeq 0.01\Phi_0$ have been detected. The largest area A that can be used is limited by the self-inductance (L) of the squid. The larger A, the larger the inductance L.* The largest inductance that can be used is approximately 10^{-8} H. This sets a limit for A of about 1 cm^2. The corresponding magnetic field sensitivity follows from Eq. 42.36. With

$$\Delta\Phi = 0.01\Phi_0 = 2 \times 10^{-17}\ \text{T} \cdot \text{m}^2$$

$$A = 10^{-4}\ \text{m}^2$$

we find

$$\Delta B = \frac{2 \times 10^{-17}\ \text{T} \cdot \text{m}^2}{10^{-4}\ \text{m}^2} = 2 \times 10^{-13}\ \text{T}$$

The earth's B-field is on the order of 10^{-4} T. Fluctuations in the earth's field of about 10^{-9} T are caused by irregularities in the ionosphere. In order to achieve a sensitivity of $\Delta B \simeq 10^{-13}$ T, it is essential that the squid be shielded from fluctuations in the earth's field.†

The Squid as an Ammeter

We can convert a squid into a sensitive ammeter by coupling it to a superconducting coil placed near the squid (Figure 42.32). A current I in the coil produces a field that tends to change the flux through the squid. The expression

*Remember, inductance measures "electrical inertia." The larger the squid, the larger its electrical inertia.

†Very elaborate shielding from stray electric fields is also required. The first generation of squids was very fragile; even a slight electrostatic spark between an experimenter's fingertip and the apparatus could "burn out" the weak link.

Figure 42.32

The current I in the superconducting coil produces a flux $\Phi = LI$ which threads both the coil and the squid. Changes in the current cause changes in the flux. The relation $\Delta\Phi = L\Delta I$ relates the smallest detectable change in current (ΔI) to the smallest detectable change in squid flux ($\Delta\Phi$).

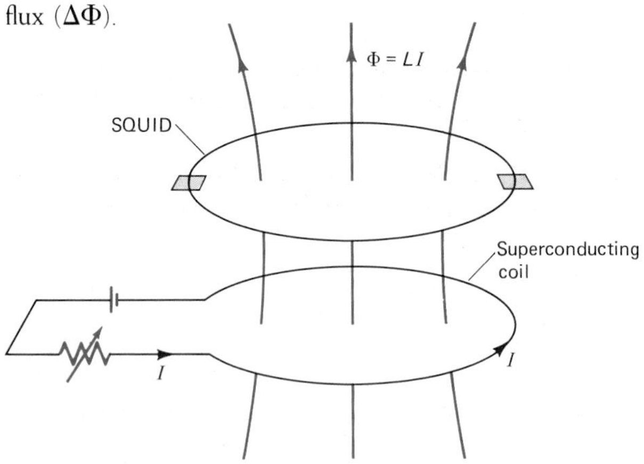

$LI = \Phi$ lets us relate the smallest detectable change in current (ΔI) and the minimum flux change that the squid can sense ($\Delta\Phi$):

$$\Delta I = \frac{\Delta\Phi}{L} \tag{42.37}$$

With

$$\Delta\Phi \simeq 0.01\Phi_0 = 2 \times 10^{-17}\ \text{T} \cdot \text{m}^2$$

$$L = 10^{-8}\ \text{H}$$

We find

$$\Delta I \simeq \frac{2 \times 10^{-17}}{10^{-8}} = 2 \times 10^{-9}\ \text{A}$$

Squids have revolutionized the technology of electrical and magnetic measurements. Geologists are adapting quantum magnetometers to measurements of rock magnetism and continental drift. Applications of squids to the fields of biology, medicine, and psychology are being pursued actively. The heart and the brain generate tiny electric currents, which in turn set up weak magnetic fields. Magnetic fields of about 10^{-14} T are generated by the human heart, and the human brain generates magnetic fields of about 10^{-15} T. Squids are able to detect even these feeble fields. Ultimately, squids may replace the equipment now used to make electrocardiograms (EKG's) and electroencephalograms (EEG's).

The squid owes its existence to the quantization of magnetic flux in a superconducting ring. It is a product of quantum engineering, the branch of engineering that also produced the transistor and the laser. It is safe to predict that the squid will find many applications that promote our well-being.

Summary

Planck adopted a quantum viewpoint to explain the spectrum of blackbody radiation. Einstein treated light as a particle—the photon—in order to develop a theory of the photoelectric effect. The energy of a photon emitted by a source of frequency ν is

$$E = h\nu \tag{42.3}$$

where h is a constant, called Planck's constant. The rest mass of the photon is zero. It exists only as an entity in motion. The linear momentum of the photon is parallel to its direction of motion. The magnitude of the linear momentum is related to the wavelength λ of the source by

$$p = \frac{h}{\lambda} \tag{42.11}$$

The intrinsic spin angular momentum of the photon is

$$L_{\text{photon}} = \frac{h}{2\pi} \tag{42.12}$$

Rutherford's analysis of α-particle scatter led him to conclude that the positive charge and most of the mass of an atom were concentrated in a small central region, which he called the nucleus. The linear dimension of the nucleus is approximately 10,000 times smaller than that of the neutral atom.

Bohr conceived a quantum theory for the hydrogen atom. Energy and angular momentum are quantized in the Bohr theory, which successfully explains Balmer's empirical formula for the spectrum of hydrogen. Historically, the Bohr theory is important because it marks the recognition of the atomic domain as the realm of quantum phenomena. The shortcomings and successes of the Bohr theory spurred the development of a "new" quantum theory.

De Broglie recognized the wave–particle duality of light as an example of a general principle. He postulated relationships between the wave (ν, λ) and particle (E, p) attributes:

$$E = h\nu \tag{42.3}$$

$$p = \frac{h}{\lambda} \tag{42.11}$$

Davisson and Germer confirmed de Broglie's wave theory by demonstrating experimentally that electrons can show interference.

Erwin Schrödinger developed an equation that describes the de Broglie waves. Quantization emerged naturally from the solutions of Schrödinger's equation. Max Born showed that the Schrödinger wave function, ψ, was a probability wave.

$$|\psi(x, y, z)|^2\, dV = \text{probability that particle is in the volume } dV \text{ centered at } (x, y, z) \tag{42.31}$$

Quantum mechanical tunneling is an important consequence of the wave nature of matter.

Quantum physics has given birth to the quantum engineering fields of quantum optics and squids. Most noteworthy in the area of quantum optics is the laser, a device that produces a highly directional and intense beam of coherent light. Key elements in laser operations are the *stimulated emission* of radiation from an *inverted population* of atoms or molecules. The inverted population is maintained by a *pumping* mechanism that selectively raises atoms to excited states.

The phenomenon of superconductivity has made possible the development of *squids*, a class of measuring instruments based on the quantum interference of supercurrents. The squid current and voltage vary periodically with the magnetic flux threading the squid. This flux is quantized, the quantum of magnetic flux being the fluxon, Φ_o.

$$\Phi_o = \frac{h}{2e} = 2.0678538 \times 10^{-55}\ \mathrm{T \cdot m^2} \qquad (42.32)$$

The periodic nature of the current and voltage makes it possible to accurately count the flux change through the squid. A squid can be used as a fluxmeter or adapted as a magnetometer, ammeter, or voltmeter.

Suggested Reading

P. W. Anderson, How Josephson Discovered His Effect, *Physics Today*, 23, 23 (November 1970).

E. M. Breinan, B. H. Kear, and C. M. Banas, Processing Materials with Lasers, *Physics Today*, 29, 44 (November 1976).

J. Clarke, Electronics with Superconducting Junctions, *Physics Today*, 24, 30 (August 1971).

U. Essman and J. E. Gordon, The Magnetic Structure of Superconductors, *Sci. Am.*, 224, 75 (March 1971).

J. J. Ewing, Rare-Gas Halide Lasers, *Physics Today*, 31, 32 (May 1978).

J. E. Faller and E. J. Wampler, The Lunar Laser Reflector, *Sci. Am.* 222, 38 (March 1970).

G. Feinberg, Light, *Sci. Am.*, 219, 50 (September 1968).

T. H. Geballe, New Superconductors, *Sci. Am.* 225, 22 (November 1971).

R. K. Gehrenbeck, Electron Diffraction: 50 Years Ago, *Physics Today*, 31, 34 (January 1978).

S. Goldhaber and M. M. Nieto, The Mass of the Photon, *Sci. Am.*, 234, 86 (May 1976).

S. A. Goudsmit and G. E. Uhlenbeck, Fifty Years of Spin, *Physics Today*, 29, 40 (June 1976).

D. N. Langenberg, D. J. Scalapino, and B. N. Taylor, The Josephson Effects, *Sci. Am.* 214, 30 (May 1966).

C. Long, Superconducting Ring Fluxmeters, *Physics Teacher*, 13, 532 (December 1975).

B. T. Matthias, The Search for High-Temperature Superconductors, *Physics Today*, 24, 21 (August 1971).

C. K. N. Patel, High-Power Carbon Dioxide Lasers, *Sci. Am.*, 219, 22 (August 1968).

A. L. Schawlow, Laser Light, *Sci. Am.*, 219, 120 (September 1968).

Superconductivity, Selected Reprints, American Institute of Physics, New York, 1964. (Contains reprint of Onnes' report describing the discovery of superconductivity.)

V. Vali, Measuring Earth Strains by Laser, *Sci. Am.*, 221, 88 (December 1969).

Problems

Section 42.1 The Origins of Quantum Physics

1. The work function of platinum is 6.30 eV. Determine the threshold frequency and the corresponding wavelength.

2. The threshold frequency for gold is 1.16×10^{15} Hz. Determine the work function for gold. Express your results in electron volts.

3. The following data were obtained in a low-precision photoelectric experiment.

Wavelength	Stopping Potential
$\lambda_1 = 552$ nm	$V_1 = 1.8$ V
$\lambda_2 = 410$ nm	$V_2 = 2.8$ V

(a) Show that Planck's constant is given in terms of these data by

$$h = \frac{e}{c}(V_2 - V_1)\frac{\lambda_1 \lambda_2}{\lambda_1 - \lambda_2}$$

(b) Evaluate h and compare it with the more precise value 6.62×10^{-34} J · s.

4. What is the wavelength associated with a photon with an energy of 1 eV?

Section 42.2 Rutherford and the Nuclear Atom

5. A bowling ball weighs 16 pounds. A BB weighs 0.012 oz. Compare the mass ratio of a bowling ball to a BB with the mass ratio of an alpha particle to an electron. Is the mass ratio for alpha scatter by an electron comparable to the scatter of a bowling ball by a BB?

6. An alpha particle with a kinetic energy of 8.2 MeV collides head on with a uranium nucleus (which carries a charge 92 times that of the proton). Assume that the uranium nucleus does not recoil, and determine the distance of closest approach (R) between the alpha and the nucleus.

7. A gold atom has a radius of 2.6×10^{-10} m. The gold nucleus has a radius of 8.6×10^{-15} m. (a) What fraction of the total cross-sectional area of the atom is "blocked" by the nuclear cross-sectional area? (b) Approximately how many atoms thick was the gold film used by Geiger and Marsden? (c) Estimate the fraction of the incident α-particles that suffered a head-on collision with a gold nucleus.

Section 42.3 Bohr and the Hydrogen Atom

8. Consult Appendix 2 and verify the claim, following Eq. 42.26, that

$$\frac{4hc}{E_H} = 364.51 \text{ nm}$$

9. Two hydrogen atoms collide head on and end up with zero kinetic energy. Each then emits a photon with a wavelength of 121.6 nm ($n = 2$ to $n = 1$ transition). How fast were they traveling before the collision? (The mass of the hydrogen atom is 1.67×10^{-27} kg.)

10. Calculate the energy of a Lyman alpha photon ($n = 2$ to $n = 1$). Express your result in electron volts.

11. A hydrogen atom undergoes a transition from the $n = 105$ level to the $n = 104$ level. Show that the wavelength of the emitted radiation is a few centimeters. (This places it in the radio-frequency portion of the electromagnetic spectrum.)

Section 42.4 De Broglie and the Wave–Particle Duality

12. Calculate your de Broglie wavelength, assuming that you move at a speed of 1 m/s. How does your result compare with the size of an atom?

13. The neutral nuclear particle, called the *neutron*, has a mass of 1.67×10^{-27} kg. Neutrons emitted in nuclear reactions can be slowed down via collisions with matter. They are referred to as thermal neutrons once they come into thermal equilibrium with their surroundings. The average kinetic energy ($3kT/2$) of a thermal neutron is approximately 0.04 eV. Calculate the de Broglie wavelength of a neutron with a kinetic energy of 0.04 eV. How does it compare with the characteristic atomic spacing in a crystal? Would you expect thermal neutrons to exhibit diffraction effects when scattered by a crystal?

14. Assume that the de Broglie wavelength of 1.65×10^{-10} m measured by Davisson and Germer is correct. Calculate the kinetic energy of the electrons (in eV). If the applied potential difference is 54.0 V, what potential difference is contributed by the crystal?

Section 42.6 Quantum Optics and Lasers

15. Water is pumped from a reservoir G to a tank A at a rate of 600 liters/min (Figure 1). The water leaks out of A through holes. One hole returns water to G at a rate of $6N_A$ liters/min, where N_A is the number of liters in A. The other hole leaks water into tank B at a rate of $54N_A$ liters/min. Water leaks from B to C at a rate of $2N_B$ liters/min and from C to G at a rate of $30N_C$ liters/min. Tanks A, B, and C are all empty when pumping begins. (a) Determine the *steady-state* content of tank A [the number of liters in A when it reaches a constant (steady) value]. (b) Determine the steady-state contents of B and C. (c) Draw a diagram showing the steady-state rate of transfer into and out of A, B, C, and G.

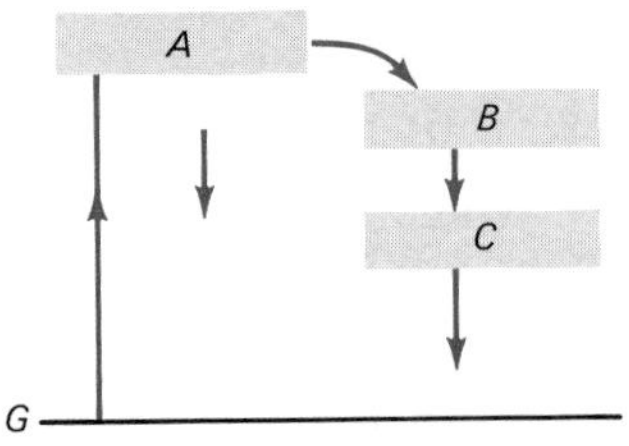

Figure 1

16. The helium–neon system is capable of lasing at several different infrared wavelengths. The most prominent infrared wavelength is 3.3913 μm. Determine the energy difference (in electron volts) between the upper and lower laser levels for this wavelength.

17. The carbon dioxide (CO_2) laser is one of the most powerful lasers developed. The energy difference between the two laser levels is 0.117 eV. Determine the frequency and wavelength of the radiation. In what portion of the electromagnetic spectrum is this radiation?

Section 42.7 Squids

18. The quantum of magnetic flux, the fluxon, is defined by

$$\Phi_o = \frac{h}{2e}$$

Show that h/e has the units of magnetic flux ($T \cdot m^2$).

19. Estimate the area of a ring that would fit one of your fingers, and calculate the magnetic flux through the ring due to the earth's magnetic field (take $B_{earth} = 5.8 \times 10^{-5}$ T). If this flux were quantized, how many fluxons would the ring enclose?

20. A squid operating as an ammeter can detect a flux change of $0.03\Phi_o$. The inductance of the superconducting coil is 6.0×10^{-8} H. What is the minimum change in current that the ammeter can detect?

chapter 43

NUCLEAR STRUCTURE AND NUCLEAR TECHNOLOGY

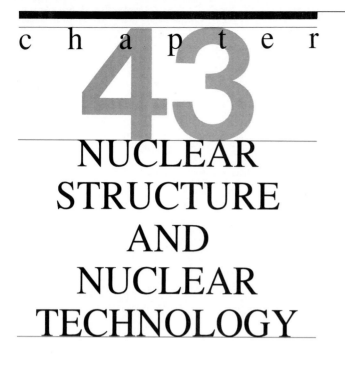

Preview

Nuclear terms evoke a variety of personal reactions. To the nuclear particle physicist, nuclear research is the most exciting and most challenging of all human investigative endeavors. Some people see electricity produced by nuclear power plants as an answer to universal energy problems. A patient whose malady is diagnosed by a radioactive tracer technique may owe his life to this application of nuclear physics principles. But these and other benefits of nuclear technology are accompanied by risks. Regardless of one's "nuclear" feelings, nuclear physics and nuclear technology have an impact on all of our lives. The first step in controlling nuclear technology is understanding the underlying physical ideas.

Nuclear physics emerged as a distinct branch of physics during the first three decades of this century, and many brilliant minds contributed to its development. In this chapter, we introduce the basic neutron–proton model of the nucleus upon which modern nuclear models are based. Our goal is to gain an understanding of this model and the important role it plays in 20th-century physics and technology, which are so much a part of our modern world.

43.1 The Neutron–Proton Model of the Nucleus

Ingredients

In 1932, while investigating products produced by the bombardment of beryllium with alpha particles, James Chadwick discovered the neutron. As the name suggests, the neutron is electrically neutral. It has a mass slightly greater than that of the proton. In atomic mass units* (Section 32.3) the neutron and proton masses are

$$m_n = 1.008665 \text{ u}$$

$$m_p = 1.007277 \text{ u}$$

Also in 1932, John Cockcroft and Ernest Walton used their newly constructed proton accelerator to bombard the element lithium and verified the Einstein relation, $E = mc^2$, for the equivalence of mass and energy. Applying the principle of conservation of energy, they accounted for the kinetic energy of alpha particles produced in the proton–lithium interactions by assuming that some of the mass of the reactants was converted to energy according to $E = mc^2$. Using these experimental results, Werner Heisenberg introduced the **neutron–proton model of the nucleus.** This model has been refined in many ways, but its original key ingredients remain unaltered.

*$1 \text{ u} = 1.66 \times 10^{-27}$ kg.

In Heisenberg's model of the nucleus, neutrons and protons (referred to collectively as *nucleons*) are bound together by the *strong nuclear force*. This force is the strongest of the four fundamental forces of nature (Section 3.1). It has the unusual characteristic of being essentially zero when neutrons and protons are separated by distances greater than about 3×10^{-15} m. This feature is dramatically different from either the gravitational or electrical force, which (in principle at least) extend to infinity.

Protons in the nucleus account for the nuclear charge. We call the number of protons the *atomic number* and give it the symbol Z. The number of neutrons is denoted by N. The number of neutrons and protons, $A = N + Z$, accounts for the nuclear mass, and is called the *mass number*. An electrically neutral atom possesses Z extranuclear electrons and Z protons in the nucleus. When we are interested in the specific nuclear properties of an atom it is customary to attach the mass number A to the chemical symbol that identifies the atom. Atoms having the same atomic number but different neutron number are called *isotopes*. Isotopes form a family of atoms sharing the same chemical symbol.* The notation ^{235}U for a particular isotope of uranium means that the nucleus contains

1. a sum total of $A = 235$ neutrons and protons;
2. $Z = 92$ protons, as indicated by the symbol U; and
3. $N = A - Z = 235 - 92 = 143$ neutrons.

For emphasis, the neutron and proton numbers are sometimes designated explicitly as $^{235}_{92}U_{143}$.

Unless stated otherwise, the notation ^{A}X denotes the neutral atom of the element having chemical symbol X. On occasion we are interested in the properties of bare nuclei, that is, atoms with the electrons removed. We use the same notation in this case, but indicate explicitly that we are dealing with bare nuclei. For example, ^{239}Pu(bare nucleus) means the nucleus of the plutonium atom having 94 protons in 145 neutrons. When designating masses, we use a lowercase letter to denote the mass of a bare nucleus and an uppercase letter to denote the mass of a neutral atom. For example, m_N symbolizes the mass of a bare nitrogen nucleus; M_N denotes the mass of a neutral nitrogen atom. The difference in mass of a bare nucleus and a neutral atom is small. For nitrogen, $(M_N - m_N)/M_N = 0.0003$.

Nuclear Binding Energy

The equivalence of mass and energy and the conservation of energy play central roles in understanding how neutrons and protons are bound together in a nucleus. For example deuterium, ^2H, is a hydrogen isotope with one proton and one neutron in its nucleus. When a neutron and proton meet and interact, they can produce a deuteron (a deuterium nucleus), but in so doing they release a photon having 2.225 MeV of energy. This process can be represented as

$$n + p \rightarrow d + \gamma \qquad (43.1)$$

*The chemical elements, their symbols, and their atomic numbers are presented in Appendix 5.

The energy of the γ-ray photon (γ) is derived from the conversion of some of the mass of the neutron and proton. Conservation of energy then requires the deuteron mass to be less than the combined masses of a free neutron and proton. Separating a deuteron into a free neutron and proton requires that energy be added to the deuteron in order to make up the mass difference between the neutron and proton and a deuteron. One way of adding the required energy is to bombard deuterons with energetic photons. Symbolically,

$$\gamma + d \rightarrow n + p \qquad (43.2)$$

Both the formation of a deuteron according to Eq. 43.1 and the separation of a deuteron according to Eq. 43.2 are routine events in many physics laboratories.

Example 1 **Conservation of Energy in a Nuclear Reaction**

A photon is emitted in the capture of a neutron by a proton to form a deuteron, the bare nucleus of the ^2H atom. Using conservation of energy at the nuclear level, we can calculate the photon energy. The reaction is written

$$n + p \rightarrow d + \gamma$$

The kinetic energies of the neutron, proton, and deuteron are ignorably small. Hence only the rest mass energy of these particles need to be considered. Using the neutron and proton masses presented in Section 43.1, we have

$$\begin{aligned} m_p &= 1.007277 \text{ u} \\ m_n &= 1.008665 \text{ u} \\ \hline m_p + m_n &= 2.015942 \text{ u} \end{aligned}$$

From magnetic spectrometer and nuclear reaction measurements we find*

$$m_d = 2.013553 \text{ u}$$

Hence the combined masses of the neutron and proton exceed the deuteron mass by 0.002389 u. The rest mass energy of one atomic mass unit is 931.48 MeV (Appendix 2). Hence the energy associated with the 0.002389-u mass difference is

$$0.002389 \text{ u} \cdot 931.48 \frac{\text{MeV}}{\text{u}} = 2.225 \text{ MeV}$$

It is this 2.225-MeV rest mass energy difference that provides the energy for the photon and the negligibly small energy of the recoiling deuteron.

Measurement of the photon energy confirms the conservation of energy principle. Once the conservation principle is established, it can be used to deduce nuclear masses from a measurement of photon energies. For example, if the neutron mass were an unknown, then a knowledge of the masses of the proton and deuteron and a measurement of the photon energy would permit a determination of the mass of the neutron.

---

*The atomic mass scale based on the mass of the ^{12}C isotope is presented in Section 32.3. Some selected atomic masses measured in atomic mass units are presented in Table 43.1.

Table 43.1

Selected atomic masses (neutral atoms) in atomic mass unit*

Isotope	Z	Mass	Isotope	Z	Mass	Isotope	Z	Mass
n	0	1.008665	^{16}O	8	15.994915	^{210}Po	84	209.982883
^{1}H	1	1.007825	^{17}O	8	16.999133	^{234}Th	90	234.043635
^{2}H	1	2.014103	^{27}Al	13	26.981540	^{233}U	92	233.039654
^{3}H	1	3.016050	^{40}Ar	18	39.962383	^{234}U	92	234.040976
^{3}He	2	3.016030	^{40}K	19	39.963999	^{235}U	92	235.043945
^{4}He	2	4.002603	^{40}Ca	20	39.962591	^{238}U	92	238.050817
^{6}Li	3	6.015124	^{60}Co	27	59.933809	^{239}U	92	239.054326
^{7}Li	3	7.016005	^{60}Ni	28	59.930778	^{239}Np	93	239.052954
^{7}Be	4	7.016930	^{75}As	33	74.921598	^{238}Pu	94	238.049583
^{8}Be	4	8.005305	^{90}Sr	38	89.907751	^{239}Pu	94	239.052177
^{9}Be	4	9.012183	^{91}Kr	36	90.923241	^{240}Pu	94	240.053828
^{11}B	5	11.009306	^{95}Sr	38	94.918903	^{252}Cf	98	252.081654
^{12}B	5	12.014354	^{137}Cs	55	136.907072			
^{11}C	6	11.011434	^{137}Ba	56	136.905812			
^{12}C	6	12.000000	^{140}Ba	56	139.910636			
^{13}C	6	13.003355	^{151}Nd	60	150.923886			
^{14}C	6	14.003242	^{197}Au	79	196.966547			
^{14}N	7	14.003074	^{210}Bi	83	209.984130			

*A complete list may be found in *Nuclear Data Tables*, *9*, 267 (1971), by A. H. Wapstra and N. B. Gove.

The **binding energy** (B.E.) of a nucleus of mass m containing Z protons and N neutrons is

$$\begin{aligned} \text{B.E.} &= \text{energy to take apart the nucleus} \\ &= (mc^2)_{\text{pieces}} - (mc^2)_{\text{assembled nucleus}} \\ &= Zm_pc^2 + Nm_nc^2 - mc^2 \qquad (43.3) \end{aligned}$$

In order to calculate the binding energy from Eq. 43.3, we must know the mass of the nucleus. However, it is generally easier to measure the mass of the neutral atom than it is to measure the mass of the nucleus. Equation 43.3 is therefore re-expressed in terms of atomic masses by adding to and subtracting from the right-hand side of the equation the mass of Z electrons:

$$\text{B.E.} = Zm_pc^2 + Zm_ec^2 + Nm_nc^2 - (mc^2 + Zm_ec^2)$$

This reduces to

$$\text{B.E.} = ZM_{\text{H}}c^2 + Nm_nc^2 - Mc^2 \qquad (43.4)$$

where $M_{\text{H}} \cong m_p + m_e$ is the mass of the hydrogen atom and $M = m + Zm_e$ is the mass of the atom. This final expression is only approximate because it neglects the binding energies of the atomic electrons, but for most purposes the atomic binding energies involved are small compared to the nuclear energies, and Eq. 43.4 is sufficiently accurate. For example, the electron binding energy in deuterium is 13.6 eV (Section 42.3), as compared to a binding energy of 2.225 MeV for the deuteron.

The binding energy increases systematically as the number of nucleons in a nucleus increases. But if we divide the binding energy by the number of nucleons in order to obtain the binding energy per nucleon, the result is approximately the same for most elements—around 8 MeV/nucleon (Figure 43.1). The binding energy per nucleon is generally 10^4–10^6 times greater than the binding energy per electron in an atom. This huge difference is a reflection of the enormous difference between the strength of the nuclear force binding neutrons and protons in the nucleus and that of the electric force binding electrons in the atom.

For large and small mass numbers the binding energy per nucleon is less than the nominal value of 8 MeV/nucleon. This is an extremely important consideration in nuclear technology. A conventional nuclear reactor derives energy by literally splitting (fissioning) nuclei having mass numbers around 235. Since the binding energy per nucleon of the fragments is larger than that of the original nucleus, the energy difference is liberated. From Figure 43.1 we see that the difference in binding energy per nucleon for $A = 240$ and $A = 120$ is about 1 MeV/nucleon. Hence the splitting of a single $A = 240$ nucleus liberates about

$$\frac{1 \text{ MeV}}{\text{nucleon}} \cdot 240 \text{ nucleons} = 240 \text{ MeV}$$

This energy is very large compared to a typical chemical reaction that liberates a few electron volts of energy.

Figure 43.1

The binding energy per nucleon for representative nuclei as determined from Eq. 43.4 plotted against mass number A. Note the orderly variation for mass numbers greater than about 10. That the binding energy per nucleon is nearly constant (~ 8 MeV per nucleon) is a consequence of the short-range nature of the nuclear force that binds nucleons together.

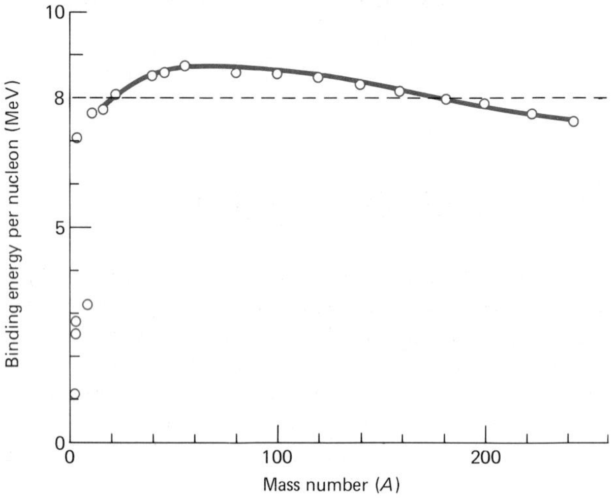

At the low-mass-number end of the binding energy curve it is possible to combine two nuclei into a more massive nucleus having a higher binding energy per nucleon, thereby causing an energy release. This process is called **nuclear fusion.** Energy is released in the core of a star by nuclear fusion reactions. On earth nuclear fusion has yet to be exploited as a practical energy source, but as we will see in Section 43.5, there is intense interest in controlled fusion reactions.

Example 2 **Binding Energy of ^{235}U**

The isotopic form of uranium used in contemporary nuclear reactors is ^{235}U. Its binding energy is an important consideration in energy production in nuclear reactors. To see how binding energies are computed, let's determine the binding energy of ^{235}U.

The atomic number of uranium is $Z = 92$. Because ^{235}U has $A = 235$ nucleons, the number of neutrons is

$$N = A - Z$$

$$= 235 - 92 = 143$$

The binding energy of ^{235}U as determined from Eq. 43.4 is

$$\text{B.E.} = ZM_{H}c^2 + Nm_nc^2 - M_{U}c^2$$

From Table 43.1 we find that

$$M_H = 1.007825 \text{ u}$$

$$m_n = 1.008665 \text{ u}$$

$$M_U = 235.043945 \text{ u}$$

The rest mass energies of the hydrogen atom and the neutron are

$$M_Hc^2 = 1.007825 \text{ u} \cdot 931.48 \frac{\text{MeV}}{\text{u}} = 938.77 \text{ MeV}$$

$$m_nc^2 = 1.008665 \text{ u} \cdot 931.48 \frac{\text{MeV}}{\text{u}} = 939.55 \text{ MeV}$$

Hence the binding energy is

$$\text{B.E.} = 92(938.77) + 143(939.55)$$

$$- 235.043945(931.48)$$

$$= 1780 \text{ MeV}$$

This is the minimum energy that must be given to the ^{235}U nucleus in order to separate it into 92 protons and 143 neutrons. The binding energy per nucleon is

$$\frac{\text{B.E.}}{A} = \frac{1780 \text{ MeV}}{235 \text{ nucleons}}$$

$$= 7.59 \text{ MeV/nucleon}$$

As expected, the binding energy per nucleon is slightly less than 8 MeV. It is this "slightly less" that accounts for the energy release in nuclear fission.

Nuclear Size

To a first approximation we can view a nucleus as a sphere of constant density. In this spherical model of the nucleus, volume is proportional to the number of nucleons.

$$V \sim A \tag{43.5}$$

Because the volume of a sphere is proportional to the cube of its radius (R), it follows that

$$R^3 \sim A$$

$$R \sim A^{1/3} \tag{43.6}$$

Introducing a proportionality constant with dimensions of length, we write

$$R = r_oA^{1/3} \tag{43.7}$$

Experiments reveal that r_o has the value

$$r_o = 1.2 \times 10^{-15} \text{ m} \tag{43.8}$$

In Chapter 1, 10^{-15} was denoted by the prefix *femto-*. Thus, 10^{-15} m is 1 femtometer. Nuclear physicists call 10^{-15} m a *fermi,** and symbolize it by fm. With this notation

$$r_o = 1.2 \text{ fm}$$

and

$$R = 1.2A^{1/3} \text{ fm} \tag{43.9}$$

Although the details of nuclear structure are needed for some physics calculations, the spherical model of the nucleus suffices for our considerations here.

*Named for Enrico Fermi (1901–1954), an Italian physicist and a pioneer in nuclear physics and nuclear technology.

Example 3 Nuclear Radii

Because the nuclear radius as given by Eq. 43.7 varies as the cube root of the mass number A, the difference between the radii of light and heavy nuclei is fairly small. To illustrate, let us determine the radii of ^{7}Li and ^{239}Pu.

For ^{7}Li,

$$r = r_0 A^{1/3}$$
$$= 1.2(7^{1/3})$$
$$= 2.3 \text{ fm}$$

For ^{239}Pu, $r = 7.4$ fm. Although the mass number of ^{239}Pu is about 34 times larger than the mass of ^{7}Li, their radii differ by only about a factor of 3.

Most of the mass of an atom is concentrated in its nucleus. Because the volume of the nucleus is very small compared to the volume of the atom, the mass density of nuclear matter is very large. In Example 4 we estimate the mass density of nuclear matter.

Example 4 Mass Density of Nuclear Matter

To demonstrate the use of our nuclear model to estimate the mass density of nuclear matter, we consider the nucleus of the aluminum atom. The mass of an aluminum atom in atomic mass units is 27.0. The mass in kilograms of 1 atomic mass unit is (Section 32.3) 1.66×10^{-27} kg. Hence the mass of an aluminum atom is

$$27.0 \text{ u} \cdot 1.66 \times 10^{-27} \frac{\text{kg}}{\text{u}} = 4.48 \times 10^{-26} \text{ kg}$$

According to our model, the volume of an aluminum nucleus is

$$V = \tfrac{4}{3}\pi R^3$$
$$= \tfrac{4}{3}\pi(1.2A^{1/3} \times 10^{-15})^3$$
$$= \tfrac{4}{3}\pi[1.2(3) \times 10^{-15}]^3$$
$$= 1.95 \times 10^{-43} \text{ m}^3$$

Because the great bulk of the mass of an atom resides in its nucleus, the density of an aluminum nucleus is approximately

$$\rho = \frac{4.48 \times 10^{-26} \text{ kg}}{1.95 \times 10^{-43} \text{ m}^3} = 2.3 \times 10^{17} \text{ kg/m}^3$$

Such a density is incomprehensible in the everyday world, where densities are on the order of 10^3 kg/m^3, but it is a reality in the astrophysical world of neutron stars.

Questions

1. What is meant by the notation ^{239}Pu?
2. How many neutrons are contained in the sodium isotope designated ^{24}Na?
3. A hydrogen atom is a bound system of an electron and a proton. Why would you expect a difference between the mass of a hydrogen atom and the combined masses of a proton and an electron?
4. If you had a table of nuclear masses, how could you *compute* the "ionization" energy for removing a proton from a nucleus?
5. If a nucleus had a diameter about the thickness of a dime (about 1 mm), which of the following would be the atomic diameter: 1 m, 100 m, or 100 km?

43.2 Nuclear Stability

The light that we see emanating from neon signs and gas lasers is a result of photon emission by energetically unstable atoms in the gas. Much like unstable atoms, nuclei of atoms can also be energetically unstable, and a wide variety of unstable nuclei occur naturally. Through emission of photons or particles that carry away energy, unstable nuclei proceed to more stable configurations. Both stable and unstable nuclei play important roles in nuclear physics (Figure 43.2). We begin this section by examining some features of stable nuclei.

Proceeding from the smallest nuclei, we find empirically that the stable nuclei tend to have nearly equal numbers of neutrons and protons. Helium has two stable isotopes: ^{4}He and ^{3}He. Oxygen has three stable isotopes: ^{16}O, ^{17}O, and ^{18}O. The lighter nuclei are all characterized by this feature of $Z \approx N$. But as the mass number increases, we find that stable nuclei have an increasing excess of neutrons (Figure 43.3). When ^{209}Bi is reached, the neutron excess is $N - Z = 43$. A nucleon (proton or neutron) is attracted to another nucleon (proton or neutron) by the strong nuclear force. We refer to these interactions as $n \leftrightarrow n$, $p \leftrightarrow p$, and $n \leftrightarrow p$. Experiment shows that these three interactions have nearly the same strength. The neutron excess in nuclei gives rise to specifically attractive nuclear forces that counteract the repulsive electrical forces between protons.

The importance of $n \leftrightarrow p$ and $n \leftrightarrow n$ interactions in nuclear stability is clearly seen by evaluating the electric energy of the protons in the nucleus. We saw in Section 28.3 that the electric potential energy of two protons separated by a distance r is $U = (1/4\pi\epsilon_0)(e^2/r)$. This energy is positive. It represents work done against the repulsive electric force between protons, a force that tends to disrupt the nucleus. Protons in a nucleus are not fixed in position, and the electric energy of any two protons changes with time. There is some average electric energy that is determined by the average of the reciprocal of the separation of the two protons.

$$U_{av} = \frac{1}{4\pi\epsilon_0} e^2 \left(\frac{1}{r}\right)_{av} \qquad (43.10)$$

A precise determination of $(1/r)_{av}$ is difficult, but it is reasonable to expect that it is roughly the reciprocal of the radius of the nucleus. Each proton interacts with every other proton in the nucleus, and the total electric energy is given by multiplying U_{av} by the total number of pairs of

Figure 43.2

Shown here is a portion of a chart of nuclides. This chart summarizes the properties of nuclei and is very useful for following pictorially the disintegration of unstable nuclei. The entire chart includes over 1000 nuclei having atomic numbers (Z) ranging from 0 to 105. The meaning of the notation for the block denoting ^{14}C is

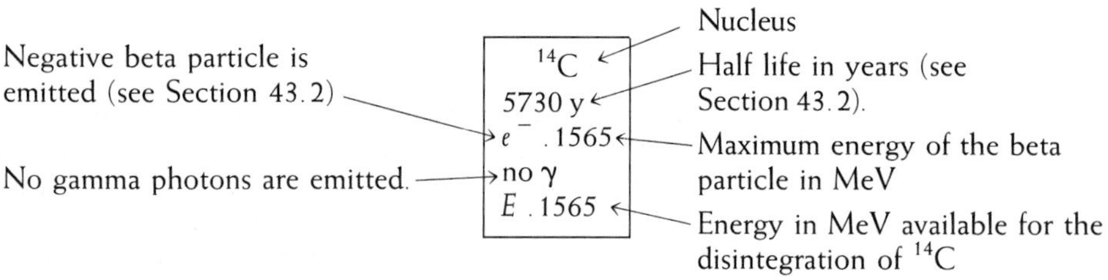
Nucleus
Half life in years (see Section 43.2).
Maximum energy of the beta particle in MeV
Energy in MeV available for the disintegration of 14C
Negative beta particle is emitted (see Section 43.2)
No gamma photons are emitted.
14C
5730 y
e− .1565
no γ
E .1565

The data shown in this portion of the chart were taken from *Table of Isotopes*, 7th ed., Wiley, New York. Copies of the chart can be purchased from the Superintendent of Documents, U. S. Government Printing Office, Washington, D.C.

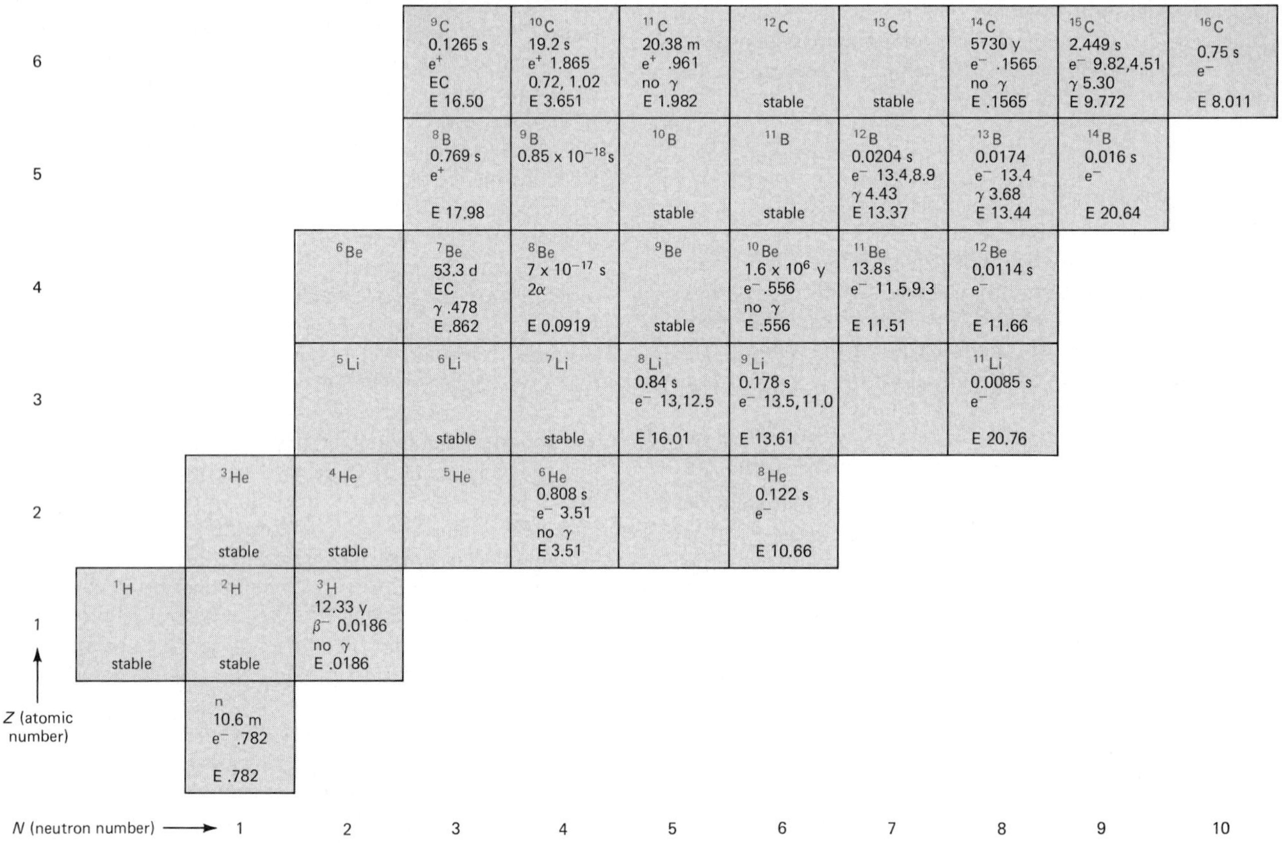
9C 0.1265 s e+ EC E 16.50
10C 19.2 s e+ 1.865 0.72, 1.02 E 3.651
11C 20.38 m e+ .961 no γ E 1.982
12C stable
13C stable
14C 5730 y e− .1565 no γ E .1565
15C 2.449 s e− 9.82,4.51 γ 5.30 E 9.772
16C 0.75 s e− E 8.011
8B 0.769 s e+ E 17.98
9B 0.85 x 10−18 s
10B stable
11B stable
12B 0.0204 s e− 13.4,8.9 γ 4.43 E 13.37
13B 0.0174 e− 13.4 γ 3.68 E 13.44
14B 0.016 s e− E 20.64
6Be
7Be 53.3 d EC γ .478 E .862
8Be 7 x 10−17 s 2α E 0.0919
9Be stable
10Be 1.6 x 106 y e− .556 no γ E .556
11Be 13.8s e− 11.5,9.3 E 11.51
12Be 0.0114 s e− E 11.66
5Li
6Li stable
7Li stable
8Li 0.84 s e− 13,12.5 E 16.01
9Li 0.178 s e− 13.5,11.0 E 13.61
11Li 0.0085 s e− E 20.76
3He stable
4He stable
5He
6He 0.808 s e− 3.51 no γ E 3.51
8He 0.122 s e− E 10.66
1H stable
2H stable
3H 12.33 y β− 0.0186 no γ E .0186
n 10.6 m e− .782 E .782
6
5
4
3
2
1
Z (atomic number)
N (neutron number)
1
2
3
4
5
6
7
8
9
10

protons. Given Z protons, there are $Z(Z-1)/2$ pairs. (Convince yourself of this by identifying the number of pairs in, say, three or four objects.) The total electric potential energy is then

$$U_{tot} = \frac{1}{4\pi\epsilon_0}\frac{Z(Z-1)}{2}e^2\left(\frac{1}{R}\right) \qquad (43.11)$$

where we have taken $(1/r)_{av} = 1/R$, with R being the nuclear radius (Eq. 43.9). Note the the total energy per proton (U_{total}/Z) increases as Z increases. For helium, $Z = 2$, $R \approx 2 \times 10^{-15}$ m, and

$$U_{tot} = (8.988 \times 10^9)\frac{2(2-1)}{2}\frac{(1.602 \times 10^{-19})^2}{2 \times 10^{-15}}$$
$$= 1.153 \times 10^{-13}\text{ J}$$
$$= 0.72\text{ MeV}$$

On a nuclear scale, this energy is not particularly large. Because the electrical energy varies as $Z(Z-1) \approx Z^2$, it becomes important for heavier nuclei (large Z). For ura-

Figure 43.3

Plot of the proton and neutron numbers of selected stable nuclei. For low mass numbers, there is a tendency for stable nuclei to have equal numbers of protons and neutrons (see Figure 43.2). But as the mass number increases there is a clear tendency for the neutron number to exceed the proton number.

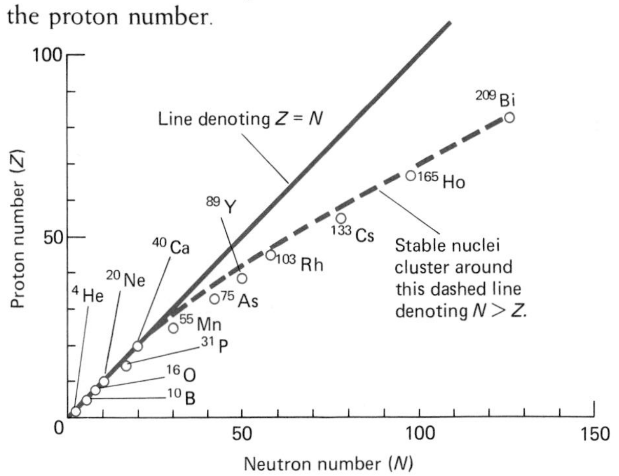

nium, $Z = 92$, $R \approx 8 \times 10^{-15}$ m, and

$$U_{tot} \approx 750 \text{ MeV}$$

The electric potential energy of a nucleus is positive, but the nuclear potential energy is negative. Like a mass confined in a hole where the gravitational potential is zero at the surface, the total nuclear energy, kinetic plus potential, must be negative for stability. An excess of neutrons with no electric interactions, but providing binding through the strong nuclear force, helps achieve stability in a nucleus. That is why heavy stable nuclei have more neutrons than protons (Figure 43.3). For example, ^{235}U has 143 neutrons as compared to 92 protons. As the proton number Z increases, the electric potential energy increases accordingly. However, the nuclear energy tends to level off, since all nucleons (protons and neutrons) do not interact with all other nucleons because of the short-range nature of the strong nuclear force. This places an upper limit on the proton number for stable nuclei. There is no completely stable nucleus in nature having Z greater than that of bismuth, $Z = 83$.

Three types of radiation from unstable nuclei play especially important roles in nuclear physics and nuclear technology. Termed alpha, beta, and gamma, these three kinds of radiations provide clues to the detailed structure of nuclei. Their properties have been exploited in a wide variety of technological applications that we will briefly describe at the end of this chapter. Let us look a little more closely at the origin and properties of these three types of radiation.

Alpha-Particle Emission

Nuclear instability may lead to a spontaneous fracture of the nucleus with the emission of an energetic particle (or particles). When this occurs, the nucleus is said to *decay*, or disintegrate. Alpha-particle emission is a spontaneous nuclear decay of this type. An alpha particle is identical to the nucleus of a helium atom having mass number 4 ($^4_2He_2$). According to the neutron–proton model of the nucleus, alpha-particle emission is the escape of a bound configuration of two protons and two neutrons. For example, $^{238}_{92}U_{146}$ is a natural alpha-particle emitter and decays as follows:

$$^{238}_{92}U_{146} \rightarrow {}^4_2He_2 + {}^{234}_{90}Th_{144} \tag{43.12}$$

Note that charge is conserved; initially 92 protons are in the ^{238}U nucleus, and finally 2 protons end up in the alpha particle and 90 protons end up in the thorium nucleus (Figure 43.4). From the energy conservation standpoint, rest mass energy plays a central role. When a uranium nucleus at rest decays, kinetic energy is imparted to the alpha particle and residual thorium nucleus. Conservation of energy requires that the uranium rest mass energy be greater than the combined rest mass energy of the alpha particle and thorium nucleus. The difference in rest mass energies is converted into kinetic energy, which is shared by the alpha particle and the thorium nucleus. The thorium nucleus has kinetic energy and may be in an elevated ("excited") energy state (Figure 43.5).

Applying conservation of energy formally, we develop a condition that alpha-particle emission be possible.

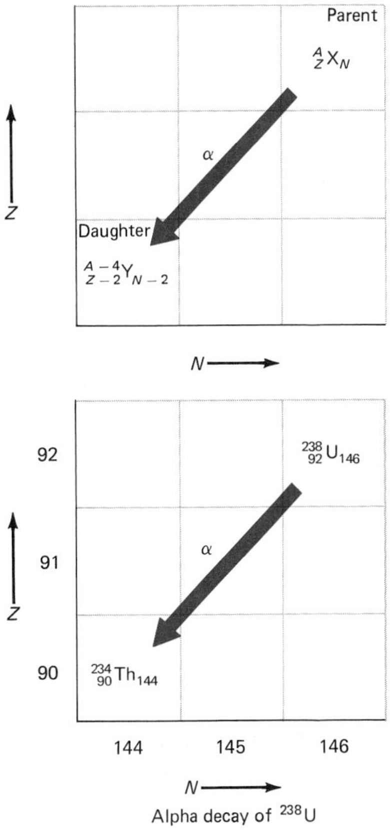

Figure 43.4

On a chart of nuclides, alpha emission is represented by a diagonal arrow pointing from a parent nucleus having Z protons and N neutrons to a daughter nucleus having $Z - 2$ protons and $N - 2$ neutrons.

Figure 43.5

Energy diagram for the alpha-particle decay of ^{238}U. If, for example, ^{238}U emits an alpha particle labeled α_1, then ^{234}Th is left in an excited state. The ^{234}Th nucleus normally loses this excess energy by emitting gamma photons.

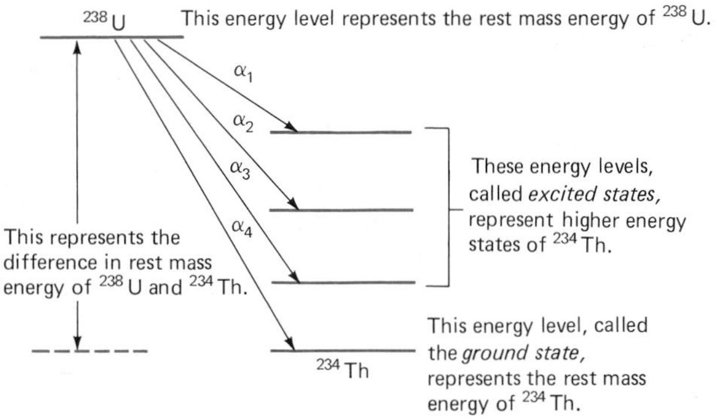

Including rest mass energy, kinetic energy, and excitation energy, we have

$$\text{energy before} = \text{energy after}$$

$$M_U c^2 = M_{He} c^2 + M_{Th} c^2 + K_\alpha + K_{Th} + E_{\text{excitation}}$$

where $E_{\text{excitation}}$ represents energy in excess of that of a thorium nucleus in its lowest energy state. Separating the rest mass energy terms, we obtain

$$K_\alpha + K_{Th} + E_{\text{excitation}} = M_U c^2 - M_{He} c^2 - M_{Th} c^2 \quad (43.13)$$

Because $K_\alpha + K_{Th} + E_{\text{excitation}}$ is a positive quantity, then

$$M_U c^2 > (M_{He} c^2 + M_{Th} c^2)$$

or

$$M_U > M_{He} + M_{Th} \quad (43.14)$$

The mass of the parent nucleus (uranium in this example) must exceed the combined masses of the neutral helium atom and the daughter nucleus (thorium). To show this explicitly, let us examine the masses of the atoms involved in the alpha decay of ^{238}U. From Table 43.1 we find

$$M_U = 238.050817 \text{ u} \qquad M_{He} = 4.002603 \text{ u}$$

$$M_{Th} = 234.043635 \text{ u}$$

$$M_{He} + M_{Th} = 238.046238 \text{ u}$$

$$M_U - (M_{He} + M_{Th}) = 0.004579 \text{ u}$$

The rest mass of ^{238}U differs from the combined rest masses of ^{234}Th and ^{4}He by 0.004579 u. In energy terms, this difference represents an energy difference of

$$0.004579 \text{ u} \cdot 931.48 \frac{\text{MeV}}{\text{u}} = 4.265 \text{ MeV}$$

This energy is shared by the alpha particle and ^{234}Th. The conservation of linear momentum dictates the linear momentum (Section 9.3), and therefore the kinetic energy, garnered by each particle. As shown in Example 5, the alpha-particle energy is 4.19 MeV.

Example 5 Two-Particle Breakup

We want to prove that if (1) energy is conserved and (2) linear momentum is conserved, then each fragment in the breakup of a stationary particle into two pieces has a definite (unique) kinetic energy. For purposes of illustration, let us consider the alpha emitter ^{238}U.

$$^{238}U \rightarrow \alpha + {}^{234}Th$$

Before the emission the parent nucleus ^{238}U at rest has zero linear momentum.

$$\mathbf{P}_{\text{before}} = 0$$

The emission of the alpha particle is a consequence of forces internal to the ^{238}U nucleus. According to the conservation of linear momentum, the linear momentum is still zero after the emission of the alpha particle.

$$\mathbf{P}_{\text{after}} = 0 = \mathbf{P}_\alpha + \mathbf{P}_r = m_\alpha \mathbf{v}_\alpha + m_r \mathbf{v}_r$$

The subscripts α and r designate the alpha particle and the ^{234}Th daughter nucleus from the decay. The energy for the alpha particle and the daughter comes from internal energy of the parent ^{238}U nucleus. Calling this internal energy U and invoking the conservation of energy (nonrelativistic form), we have

$$\tfrac{1}{2} m_\alpha v_\alpha^2 + \tfrac{1}{2} m_r v_r^2 = U$$

Using the conservation of linear momentum equation to replace v_r, we have

$$\tfrac{1}{2} m_\alpha v_\alpha^2 + \tfrac{1}{2} m_r \left(\frac{m_\alpha v_\alpha}{m_r}\right)^2 = U$$

$$\tfrac{1}{2} m_\alpha v_\alpha^2 = \left(\frac{m_r}{m_\alpha + m_r}\right) U$$

Thus for a given internal energy U, the kinetic energy of the alpha particle is determined uniquely. For the alpha emitter in question, $U = 4.265$ MeV. Hence the kinetic energy of the alpha particle is

$$\tfrac{1}{2} m_\alpha v_\alpha^2 = \left(\frac{234.04}{234.04 + 4.00}\right) 4.265 = 4.19 \text{ MeV}$$

In the two-body breakup of any system at rest the energies of the fragments are unique. If more than two fragments are produced, the sharing of linear momentum and energy can occur in a variety of ways, and the kinetic energy of a given fragment is not unique. We see this, for example, in the emission of beta particles by unstable nuclei.

Mass considerations severely limit the number of alpha-particle emitters. There are a few nuclei having mass numbers (A) between 140 and 190 that are natural alpha-particle emitters. The majority of alpha emitters have $A > 200$. For example, 14 isotopes of uranium are unstable to alpha decay.

Beta-Particle Emission

There exist two types of beta particles. They are identical in mass but differ in the sign of electric charge. The negative beta particle is identical to an electron. The positive counterpart is termed a *positron*. Explanation of the emission of beta particles is quite different from that of alpha-particle emission. In alpha decay, alpha particles are emitted with discrete energies. The radium isotope ^{224}Ra emits alpha particles having energies of 5.447 and 5.684 MeV. Beta particles are emitted with a distribution of energies. For example, ^{210}Bi emits beta particles having a continuous distribution of energy varying from 0 to 1.16 MeV (Figure 43.6). This distribution of beta-particle energies is a consequence of the simultaneous emission of a companion called the *neutrino* (or *antineutrino*, as the case may be).

Antineutrinos (and neutrinos) have energy and momentum but they lack charge and mass.* Neutrino interaction with matter is extremely weak, and we do not consider neutrino radiation in this discussion. The energy available for beta decay is apportioned to the beta particle, the neutrino, and the nucleus remaining after the decay. Linear momentum is conserved in beta decay but there is no unique direction of emission for the beta particle, the neutrino, and the nucleus. Consequently, the energy and linear momentum of the products depend on the directions the products take following beta decay.

It is not the strong nuclear force that manifests beta decay. Rather it is the fourth and final fundamental force we shall encounter—the weak nuclear force (Section 3.1). In relative strength, the weak nuclear force falls between the gravitational and electromagnetic forces. At the time of emission of an electron from a nucleus, a neutron is transformed into a proton, an electron (the negative beta particle), and an antineutrino (Figure 43.7). The proton remains in the parent nucleus, but the electron and antineutrino escape. The energy and linear momentum of each of the three particles depend on the directions in which they are emitted. However, conservation of energy dictates that the total energy be constant in any disintegration. Symbolically, the process is

$$n \rightarrow p + e^- + \bar{\nu} \qquad (43.15)$$

This does not mean that a neutron consists of a proton, an electron, and an antineutrino that separate at the moment of decay. The product particles are created at the moment of decay, and at the same time the neutron disappears. Although this seems strange at first, it is very much akin to photon emission by an atom. The photon is actually created at the time of emission (Section 42.3).

The *triton* (a tritium nucleus, $^3_1H_2$) is unstable and undergoes a beta decay typical of electron emission. One of its neutrons is transformed, leaving a proton behind (Figure 43.8). Symbolically, the disintegration of the bare nucleus is

$$^3_1H_2 \rightarrow {}^3_2He_1 + e^- + \bar{\nu} \qquad \text{(bare nuclei)} \qquad (43.16)$$

You should note carefully that because the nuclear remnant ($^3_2He_1$) has two protons it is no longer a triton, but is now a helium isotope. By capturing two electrons, this nuclear remnant is transformed into a neutral helium atom. This species of helium is actually produced in the laboratory through the beta decay of tritons. This transmutation of elements was the elusive goal of alchemists for centuries. Today it is done routinely.

Total charge is conserved in beta decay. In the decay of tritium (Eq. 43.16) we start with +1 unit in the triton and end up with +2 units in the helium nucleus and −1 unit in the electron. Unlike what occurs in alpha decay, individual proton and neutron numbers are not conserved. Contemporary physics views neutrons and protons as belonging to a family called *baryons*. In beta decay, the number of baryons is conserved.

The mechanisms for alpha-particle emission and beta-particle emission are very different. However, the energy available for both types of reaction comes from the difference in rest mass energies of the parent nucleus and products of the reaction. The calculation of the energy available is done in basically the same way for both processes (Example 6).

*Some neutrinos may have a very small mass, no more than one ten-thousandth the mass of an electron.

Figure 43.7

Schematic representation of the beta decay process. A neutron disintegrates, producing a beta particle and antineutrino that escape and a proton that remains in the nucleus.

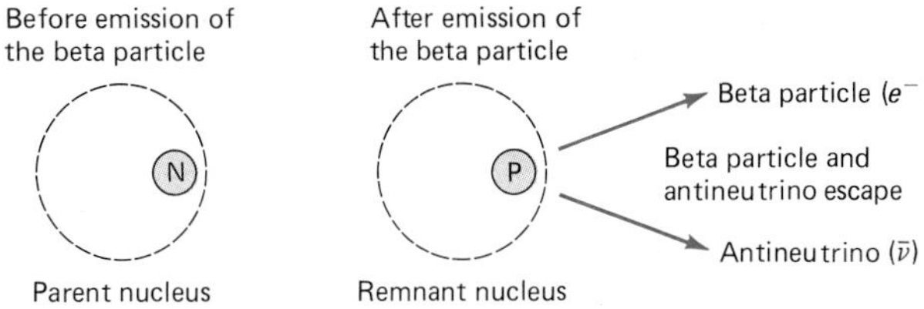

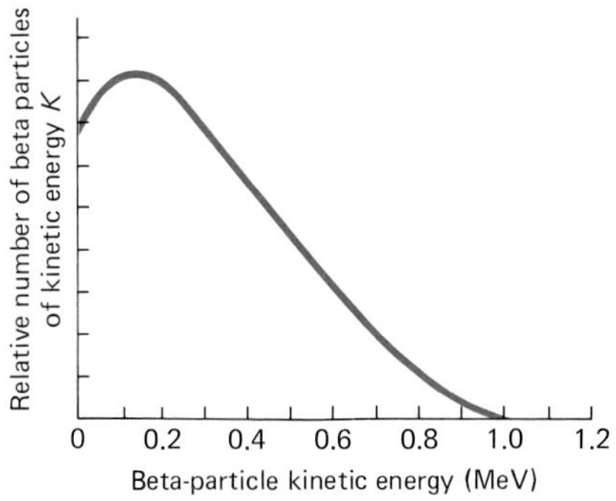

Figure 43.6

In a beta decay experiment, the number of beta particles having a certain energy is determined. This number is plotted against the value of the energy as shown in this graph. The energies are found to vary from zero up to some maximum that is determined by the type of beta-particle emitter. The shape of the curve will change for different types of beta-particle emitters.

Figure 43.8

On a chart of nuclides, negative beta-particle (e^-) emission is represented by a diagonal arrow pointing from a parent nucleus having Z protons and N neutrons to a daughter nucleus having $Z + 1$ protons and $N - 1$ neutrons. The mass number A does not change.

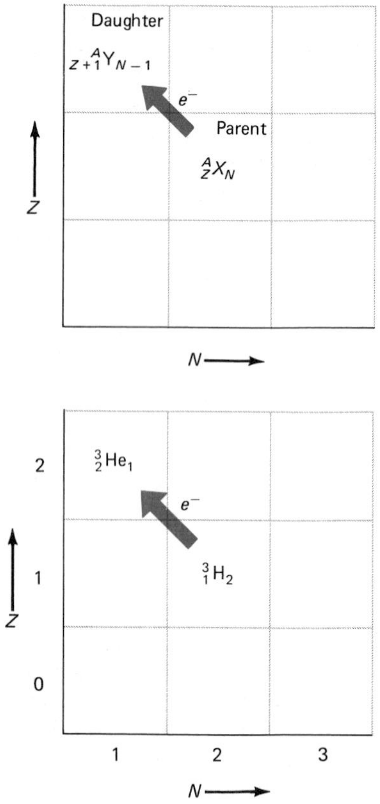

Beta decay of ^{3}H (tritium)

Example 6 **Kinetic Energy Available in Beta Decay**

One of the most useful techniques for determining the age of very old carbon-based materials (such as wood) is based on the radioactive decay of ^{14}C. Using energy conservation, we want to determine the amount of kinetic energy available to the beta particle, antineutrino, and remnant nucleus in the beta decay of ^{14}C. In terms of the bare nuclei we depict the process as

$$^{14}_{6}C_8 \rightarrow {}^{14}_{7}N_7 + e^- + \bar{\nu} \qquad \text{(bare nuclei)}$$

The energy available (E) for the reaction products comes from the difference in total mass of the initial and final systems.

$$E = (m_C - m_N - m_e)c^2$$

Algebraically this equation is unaltered if we add and subtract the rest mass energy of six electrons.

$$E = (m_C + 6m_e - m_N - 6m_e - m_e)c^2$$
$$= (m_C + 6m_e - m_N - 7m_e)c^2$$

To within the few electron volts difference between the binding energies of six electrons in ^{14}C and seven electrons in ^{14}N, $(m_C + 6m_e)c^2$ is the rest mass energy of the ^{14}C atom and $(m_N + 7m_e)c^2$ is the rest mass energy of the ^{14}N atom. In terms of the masses of the neutral atoms, the energy available is

$$E = (M_C - M_N)c^2$$

From Table 43.1, we find

$$M_C = 14.003242 \text{ u}$$
$$M_N = 14.003074 \text{ u}$$
$$\Delta M = 0.000168 \text{ u}$$
$$E = \Delta M c^2$$
$$= 0.000168 \text{ u} \cdot 931.48 \frac{\text{MeV}}{\text{u}}$$
$$= 0.156 \text{ MeV}$$

This decay energy represents the maximum kinetic energy of the beta particle. If a beta particle is emitted with this kinetic energy, it has all the available kinetic energy. The antineutrino gets none. On the other hand, if the antineutrino is emitted with energy $E_{\bar{\nu}}$, then that much less kinetic energy is available for the beta particle. By observing many beta decays, we may determine the distribution of beta-particle kinetic energies and present the results in a manner similar to that shown in Figure 43.6.

Carl Anderson discovered the positron in 1932 while investigating cosmic rays (particles emanating from space). The mechanics of positron emission is much like that of electron emission. Drawing on the internal (nuclear) energy of a nucleus, a proton may be transformed into a neutron, positron, and neutrino. The neutron remains in the nucleus but the positron and neutrino escape from the nucleus (Figure 43.9). In terms of bare nuclei, the positron decay of $^{11}_{6}C_5$ would be depicted as

$$^{11}_{6}C_5 \rightarrow {}^{11}_{5}B_6 + e^+ + \nu \qquad \text{(bare nuclei)} \qquad (43.17)$$

Note that charge and number of baryons are conserved. Momentum and energy are also conserved. With the addition of five orbital electrons, ^{11}B becomes a neutral atom chemically different from the ^{11}C atom.

If a proton within a nucleus transforms as described, we might wonder why a proton, which is a hydrogen nucleus, doesn't disintegrate according to $p \rightarrow n + e^+ + \nu$. The reason is simply that it is energetically impossible. The mass of the proton is less than the combined masses of the disintegration products. Such is not the case with a neutron. An isolated neutron can, and does, decay into a proton, an electron, and an antineutrino. The slight mass difference between the neutron and proton permits an isolated neutron, but not an isolated proton, to decay. A proton in a nucleus may be able to transform, because every nucleon in a nucleus is able to contribute energetically in beta decay. A proton in a nucleus may be able to "borrow" enough energy to permit it to decay.

Figure 43.9

On a chart of nuclides, positive beta-particle (^{+}e) emission is represented by a diagonal arrow pointing from a parent nucleus having Z protons and N neutrons to a daughter nucleus having $Z - 1$ protons and $N + 1$ neutrons. The mass number A does not change.

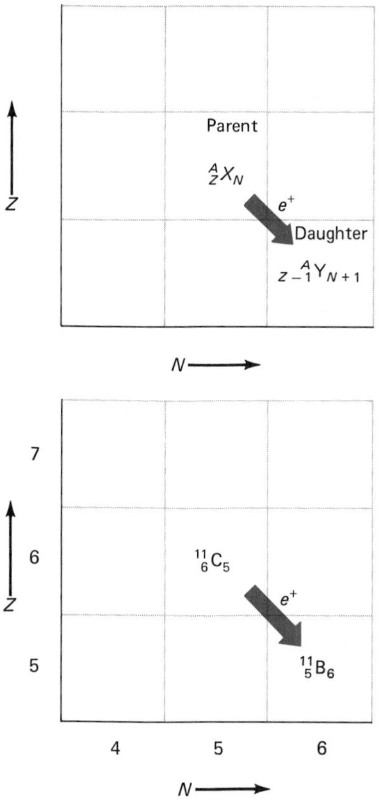

Beta decay of ^{11}C

Figure 43.10

The energy states of a nucleus are discrete. A possible energy value for a state is indicated by a horizontal line. The higher the position of the line, the larger the energy value. Absorption of energy is required for a transition to a higher energy state. This is indicated by an arrow pointing up. A release of energy is required for a transition to a lower energy state. This is indicated by an arrow pointing down. Shown here are some of the measured energy states for ^{45}Sc. Each state is characterized by a definite energy measured in keV. Transition from a higher energy state to a lower energy state involves the emission of a gamma photon having energy equal to the energy difference of the two states minus the recoil energy of the nucleus.

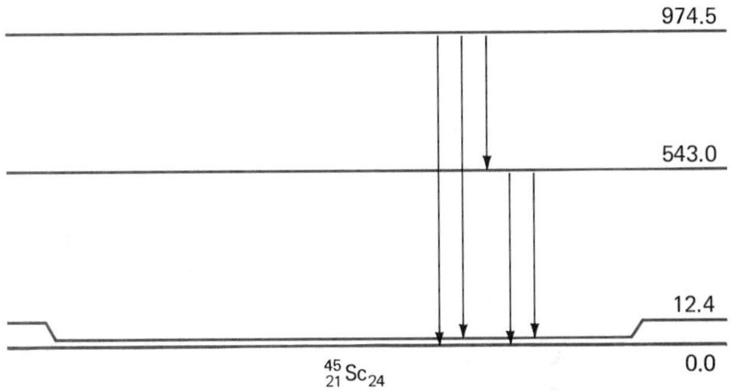

Gamma Emission

Few aspects of physics are more intriguing than quantum phenomena. And, as we saw in Chapter 42, scientists and engineers have developed a variety of quantum technologies. It is satisfying that a relatively simple theory can be formulated to account for some quantum phenomena involving atoms. Bohr's theory of hydrogen-like atoms is a prime example. The Bohr model is mathematically tractable because of the mathematical simplicity of the electrical force between electrons and the nucleus. The strong nuclear force between protons and neutrons in the nucleus is much more complicated. This is one reason why no counterpart of the Bohr model has been discovered for the nucleus. Nevertheless, many of the same concepts apply to the nucleus. (1) Angular momentum is quantized in terms of the same fundamental unit ($h/2\pi$). (2) Energies are discrete, just as for the atom as a whole. But whereas atomic energies are on the order of electron volts, nuclear energies are on the order of millions of electron volts. The larger nuclear energies reflect the greater strength of the strong nuclear force in comparison with the electrical force. Nuclear energy states are measured routinely, and theories based on the nuclear force are offered for their calculation. But no theory can yet account for all the details of the energy states of even a single nucleus. The four lowest measured energy levels of the ^{45}Sc nucleus are shown in Figure 43.10. (3) Like an atom, a nucleus can make a transition to a higher energy state by absorbing energy. Once in a higher state it can change to a lower energy state by giving up energy. As in the atomic case, this energy release is often in the form of a photon of electromagnetic energy, the energy being given by

$$h\nu = E_{\text{higher}} - E_{\text{lower}} \tag{43.18}$$

Equation 43.18 is Bohr's radiation postulate applied to nuclei. It is a direct consequence of conservation of energy.

Energy differences between atomic states are of the order of electron volts (eV). The energy differences between states in the nucleus are generally of the order of millions of electron volts (MeV). Therefore, nuclear photons (gamma rays) are much more energetic than atomic photons. For example, Figure 43.10 shows two energy levels in ^{45}Sc at 974.5 and 543.0 keV. The energy difference between these two states is

$$\begin{aligned} E_{\text{higher}} - E_{\text{lower}} &= 974.5 - 543.0 \\ &= 431.5 \text{ keV} \end{aligned}$$

This energy is in the gamma region of the electromagnetic spectrum. For comparison, a photon of green light has an energy of 2.4 eV = 0.0024 keV. The wavelength of green light is about 1.8×10^5 times larger than the wavelength of gamma radiation composed of 431.5-keV photons.

Radioactive Decay Law

For a given nucleus, any disintegration process consistent with the conservation principles of physics is possible. But even if a given disintegration is possible, the conservation principles say nothing about how frequently the event occurs. Given a sample of tritium atoms, for example, we do not find that all the nuclei simultaneously undergo beta decay. Some decay in fractions of a second. Others survive extraordinarily long times, perhaps many years, before disintegrating. When we detect beta particles by using a Geiger counter that produces an audible click when it registers an impinging electron, we hear random clicks, indicating that the time interval between the reception of any two beta particles is unpredictable.

When studying nuclear disintegrations we generally measure the number of disintegrations (Δn) occurring in some well-defined time interval (Δt). The ratio $\Delta n/\Delta t$ measures the average disintegration rate over the time interval Δt, and approximates the instantaneous decay rate dn/dt. If the measurement is repeated successively and the time for each measurement is noted, we arrive at a set of data representing decay rate as a function of time. Generally, the data are displayed as a plot of the logarithm of the decay rate versus time, as shown in Figure 43.11.

The decay rate versus time feature of radioactivity can be derived quite simply if we assume that nuclei decay independently of each other and independently of how long a given nucleus has existed. Then the decay rate is directly proportional to the number of nuclei present. Let n denote the number of nuclei yet to decay and let dn/dt denote the rate at which n changes. Because n decreases with time, dn/dt is negative.

Invoking the assumption that the decay rate is directly proportional to n, we have

$$\frac{dn}{dt} = -\lambda n \tag{43.19}$$

The proportionality factor λ represents the probability (per unit time) that any particular nucleus will decay. It is independent of the history of the sample of unstable nuclei. Equation 43.19 can be integrated to yield

$$n = n_o e^{-\lambda t} \tag{43.20}$$

where n_o is the number of nuclei present at the initial time ($t = 0$).* You may check Eq. 43.20 by differentiating it to see if you recover Eq. 43.19. The instantaneous decay rate is

$$\frac{dn}{dt} = -\lambda n_o e^{-\lambda t}$$

which yields for the initial ($t = 0$) decay rate

$$\left(\frac{dn}{dt}\right)_{t=0} = -\lambda n_o$$

Hence we can write the decay rate at any time as

$$\frac{dn}{dt} = \left(\frac{dn}{dt}\right)_{t=0} e^{-\lambda t}$$

To the extent that the instantaneous decay rate, called the *activity*, can be approximated by the experimentally measured activity $a = \Delta n/\Delta t$, we have

$$a = a_o e^{-\lambda t} \tag{43.21}$$

$$\ln a = \ln a_o - \lambda t \tag{43.22}$$

which is precisely the form of the experimentally derived relation. This result holds for any type of nuclear disintegration—alpha, beta, gamma, and so on.

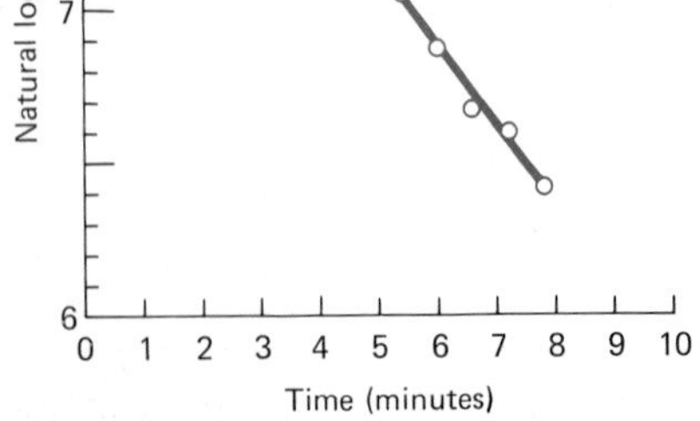

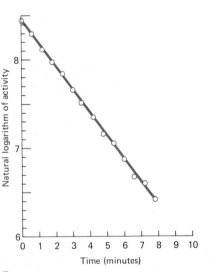

Figure 43.11

Here the data are plotted as the natural logarithm of the decay rate (activity) versus the time of the measurement. The straight line shown has the mathematical form $\ln a = 8.44 - 0.262t$, where a is the activity and t is the time.

Half-Life

The half-life ($\tau_{1/2}$) is the time required for the decay rate of a sample of unstable nuclei to decrease by a factor of 2. If a_o is the initial decay rate, then the decay rate after the elapse of one half-life is $a_o/2$. Substituting into Eq. 43.21 we have

$$\frac{a_o}{2} = a_o e^{-\lambda\tau_{1/2}}$$

$$\frac{1}{2} = e^{-\lambda\tau_{1/2}}$$

$$\ln\frac{1}{2} = -\lambda\tau_{1/2} \qquad \text{or}$$

$$\tau_{1/2} = \frac{0.693}{\lambda} \tag{43.23}$$

The half-life can be determined by an inspection of the experimental data. Or if λ is determined by fitting Eq. 43.22 to the data, the half-life can be calculated from Eq. 43.23. There are very few physical parameters in nature spanning a greater range of values than the half-life for nuclear disintegration. The reported half-life of only 2×10^{-21} s for ^{5}He is not much greater than the time required for a light beam to travel a distance equal to the diameter of a nucleus. An extremely important nucleus in nuclear technology, ^{239}Pu, has a half-life of 24,390 yr. A sampling of alpha decay half-lives is presented in Table 43.2.

*The two assumptions listed for nuclei have counterparts in many other physical systems. As a consequence, exponential relations analogous to Eq. 43.20 occur frequently in physics.

Table 43.2

Some natural alpha-particle emitters and their half-lives*

Uranium Isotopes		A = 140–190		Technically Significant Emitters	
Mass Number	Half-life	Isotope	Half-life	Isotope	Half-life
227	1.1 min	^{146}Sm	1.03×10^8 yr	^{238}Pu	87.74 yr
228	9.1 min	^{142}Ce	$>5 \times 10^{16}$ yr	^{239}Pu	2.44×10^4 yr
229	58 min	^{144}Nd	2.1×10^{15} yr	^{239}Np	2.35 days
230	20.8 days	^{150}Gd	1.8×10^6 yr	^{232}Th	1.41×10^{10} yr
232	72 yr	^{152}Gd	1.1×10^{14} yr	^{226}Ra	1.60×10^3 yr
233	1.592×10^5 yr	^{156}Yb	24 s	^{210}Po	138.38 days
234	2.45×10^5 yr	^{174}Hf	2.0×10^{15} yr		
235	7.038×10^8 yr	^{185}Au	4.3 min		
236	2.342×10^7 yr				
238	4.468×10^9 yr				

*Column 1 indicates the enormous range of half-lives encountered in the uranium isotopes; column 2 lists some of the alpha-particle emitters in the mass region from $A = 140$ to $A = 190$; column 3 shows some alpha-particle emitters of particular interest in nuclear technology. Chemical names for the symbols are given in Appendix 5. The data are from *Table of Isotopes*, 7th ed., C. Michael Lederer and Virginia S. Shirley, eds., Wiley, New York. 1978.

Example 7 **Experimental Determination of the Half-Life for Nuclear Decay**

The radioactive element ^{137}Ba has a relatively short half-life and can be easily extracted from a solution containing radioactive cesium (^{137}Cs). We want to determine the half-life for the decay of ^{137}Ba by using the experimental data presented in Figure 43.11.

The logarithmic plot shown in Figure 43.11 is fitted by

$$\ln a = 8.44 - 0.262t$$

Thus the decay constant λ is 0.262 per minute and the half-life is

$$\tau_{1/2} = \frac{0.693}{\lambda} = \frac{0.693}{0.262}$$
$$= 2.64 \text{ min}$$

The reported half-life for ^{137}Ba is 2.55 min. The difference reflects experimental uncertainties.

The Curie

Disintegration rates are usually expressed in a unit called a *curie*, symbolized by Ci. Based on the radioactivity of a gram of radium, one curie is defined as 3.7×10^{10} disintegrations per second. Gamma radiation sources like those used in cancer treatment, or the spent fuel elements from nuclear reactors, are generally on the order of hundreds or thousands of curies. Sources encountered in nuclear physics research are usually in the millicurie (1 mCi $= 10^{-3}$ Ci) range. Demonstration sources used in classrooms are typically microcuries (1μCi $= 10^{-6}$ Ci). The radioactive content of samples of material from our environment is typically on the order of picocuries (1pCi $= 10^{-12}$ Ci). Sensitive instruments have made possible the measurement of extremely small amounts of radioactivity.

Questions

6. If the electric energy of a stable nucleus is +700 MeV, what is the least amount of specifically nuclear energy possessed by the nucleus?
7. What prevents a ^{12}C nucleus from decaying spontaneously into three alpha particles according to the following reaction?

$$^{12}\text{C} \rightarrow {}^4\text{He} + {}^4\text{He} + {}^4\text{He}$$

8. When a nucleus at rest decays spontaneously into two fragments, why does the fragment of smaller mass acquire the greater amount of kinetic energy?
9. A positron and an electron are deflected in a uniform magnetic field as shown below. If the magnetic field is directed into the plane of the paper, which track is due to the electron?

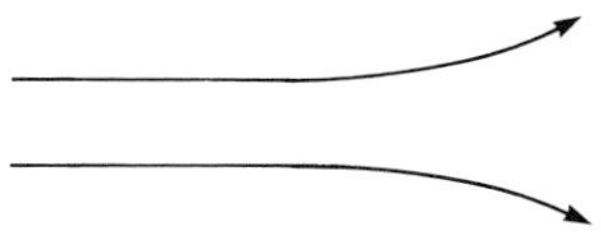

10. Why can a proton in a nucleus, but not a free proton, transform into a neutron, a positron, and a neutrino?
11. A radioactive nucleus above the line of stability in Figure 43.3 may undergo beta decay, producing a more stable nucleus closer to the line of stability. Would the decay be by positron emission or by electron emission?

43.3 Radioactive Dating

Archaeologists are interested in determining the age of artifacts, and geologists are interested in determining the age of geologic samples. Both use dating techniques that exploit the half-life of selected radioactive atoms. Let us consider first the carbon-14 method used by archaeologists.

The Carbon-14 Method

The earth's atmosphere is 78% nitrogen, and 99.63% of this nitrogen is of the isotopic form ^{14}N. There exists in the atmosphere a small flux of energetic neutrons produced from nuclear reactions initiated by protons from outer space (cosmic rays). These neutrons interact with ^{14}N nuclei in the atmosphere, producing ^{14}C according to the reaction

$$^{14}N + n \rightarrow {}^{14}C + p \tag{43.24}$$

Since ^{14}C is unstable, it disintegrates, emitting an electron and an antineutrino according to the beta decay reaction

$$^{14}C \rightarrow {}^{14}N + e^- + \bar{\nu} \tag{43.25}$$

The half-life for this decay is 5730 yr. Because the production and the disintegration processes have been operating for much longer than 5730 yr, the rate of production of ^{14}C equals its rate of disintegration, and an equilibrium concentration of ^{14}C is achieved in the atmosphere, approximately one ^{14}C atom for every 10^{12} carbon atoms. ^{14}C enters into chemical reactions in the same way as ^{12}C and ^{13}C. In particular, ^{14}C forms gaseous carbon dioxide (CO_2) just as readily as do the stable isotopes ^{12}C and ^{13}C. The ratio of ^{14}C to total carbon in a living plant or animal consuming CO_2 remains constant. But when either dies, the consumption of CO_2 ceases. Accumulation of ^{14}C in the solid parts of the organism (bones, for example) ceases, while decay of ^{14}C proceeds as usual. Consequently the ^{14}C content in the dead organism decreases through radioactive decay, whereas the stable ^{12}C and ^{13}C content remains fixed. The equilibrium ratio of ^{14}C atoms to the total carbon atoms in the dead plant or animal decreases in a manner determined by the time since death and by the ^{14}C half-life. By measuring this ratio we can determine the time that has elapsed since the organism's death.

We can use the time scale for the decay of ^{14}C and the radioactive decay law (Section 43.2) to determine the age of dead plant or animal material. If we call N_o the number of ^{14}C atoms per gram of total carbon present when the plant dies, then from Eq. 43.19 the initial rate of decay per gram (I_o) of total carbon is*

$$I_o = -\lambda N_o \tag{43.26}$$

where λ is the disintegration constant which is related to the half-life by $\lambda = \ln 2/\tau_{1/2}$. In a contemporary sample of an organic material, measurements reveal that the ^{14}C activity amounts to 13.6 disintegrations per minute per gram of total carbon. As time progresses, the ^{14}C content diminishes through radioactive decay. At some time t later the rate of decay per gram will be

$$I = -\lambda N \tag{43.27}$$

Using the radioactive decay law (Eq. 43.20), we can write Eq. 43.26 as

$$I = -\lambda N_o e^{-\lambda t} \tag{43.28}$$

Dividing Eq. 43.28 by Eq. 43.26, we obtain

$$\frac{I}{I_o} = e^{-\lambda t} \tag{43.29}$$

If we assume that the rate of decay per gram (I_o) in a present-day organic sample is the same as when the sample was living, then measurements of I and I_o and a knowledge of the decay constant λ allow us to calculate the age of the material from

$$t = -\frac{1}{\lambda}\ln\frac{I}{I_o} = -\frac{\tau_{1/2}\ln(I/I_o)}{\ln 2} \tag{43.30}$$

Example 8 **The ^{14}C Dating Game**

The specific disintegration rate in a wooden archaeological artifact is measured to be 10.5 disintegrations/hr · gm. From this measurement, let us estimate the age of the artifact by means of the ^{14}C dating method.

The determination follows from Eq. 43.30. Making the assumption that the specific disintegration rate at the time the tree or other plant from which the artifact was made died is the same as at the present, we have

$$\frac{I}{I_o} = \frac{10.5\ \dfrac{\text{disintegrations}}{\text{hr}\cdot\text{gm}}}{13.6\ \dfrac{\text{disintegrations}}{\text{min}\cdot\text{gm}}}\ \frac{1\ \text{hr}}{60\ \text{min}}$$

$$= 0.0129$$

Relating the disintegration constant λ to the half-life $\tau_{1/2}$, we have

$$\lambda = \frac{\ln 2}{5730\ \text{yr}}$$

Thus from Eq. 43.30

$$t = \frac{-5730}{\ln 2}\ln\frac{I}{I_o} = \frac{-5730}{\ln 2}\ln 0.0129$$

$$= 36{,}000\ \text{yr}$$

The plant died 36,000 yr ago.

Although simple in principle, the ^{14}C dating scheme is tedious in practice. Generally the carbon content is oxidized to produce gaseous carbon dioxide (CO_2). The radioactivity of the carbon dioxide gas is measured with an instrument that detects the beta particles emitted by ^{14}C atoms. This instrument is usually a metal cylindrical container in which an electric field is established between a central wire and the outer cylinder. When a beta particle

*In Section 43.2, we used the symbol a (for activity) to denote the rate of decay (dn/dt) of a radioactive sample. Here, the symbol I denotes the rate of decay *per gram*. Activity (a) and specific disintegration rate (I) are related but have differing measuring units.

collides with and ionizes carbon dioxide molecules in the container, the positive carbon dioxide ions and free electrons are accelerated by the electric field. The motions of these charges give rise to a tiny electric current that can be recorded. Thus the beta particle triggers a sequence of events that signal its presence. Because there are extraneous nuclear radiations from cosmic sources and building materials that interfere with the ^{14}C radiation of interest, the experimental apparatus must be carefully shielded in order to minimize these background interferences. Even in a comparatively young sample the counting rates are low, requiring on the order of 10 hr of counting time. These complications limit this form of ^{14}C dating to ages of less than about 50,000 yr.

An alternate ^{14}C dating scheme has been developed that employs a cyclotron (Section 32.3) which counts atoms rather than beta particles emitted during the decay of nuclei. A cyclotron accelerates positive ions having the same charge-to-mass ratio (q/m). An ion is produced by stripping an electron from an atom. If the cyclotron is adjusted to accelerate singly charged ions of ^{14}C, then it will exclude ions of ^{12}C and ^{13}C from the beam. Thus when the atoms of a material of interest are ionized, the cyclotron separates them according to their mass and they can be counted as they emerge. By adjusting the cyclotron conditions sequentially to accelerate ^{14}C ions, ^{12}C ions, and ^{13}C ions, the researcher can determine the ^{14}C-to-total-carbon ratio. Because the procedure counts atoms rather than disintegrations, it takes advantage of the still rather large reservoir of ^{14}C atoms in a sample. Milligram samples are generally required, and the range of age determinations can be extended to around 100,000 yr. The precision of the method is limited by the fact that other ions having the same charge-to-mass ratio as ^{14}C, particularly ^{14}N, are accelerated along with ^{14}C. Separating ^{14}N ions from ^{14}C is an experimental obstacle that has to be overcome. The cyclotron technique also lends itself to other long-lived radioactive nuclei, such as ^{10}Be (beryllium), which has a half-life of 1.5 million yr.

Geologic time scales are typically much larger than archaeological ages. A radioactive dating scheme for geologic samples therefore requires unstable nuclei with correspondingly long half-lives. Several naturally occurring radioactive nuclei are used. We will discuss a method involving ^{87}Rb (rubidium).

The Rubidium–Strontium Method

Geologists who generally deal with inorganic material (material not containing carbon) cannot use the carbon-14 method. One method that they use instead to date samples is the rubidium–strontium method. ^{87}Rb is unstable to beta decay and disintegrates by emitting an electron according to the reaction

$$^{87}\text{Rb} \rightarrow {}^{87}\text{Sr} + e^- + \bar{\nu}$$

The half-life for the decay is 5.0×10^{10} yr, about ten times the age of the earth. If at the time a rock solidified it contained N_o rubidium atoms per gram, then at a time t later the number per gram (N_{Rb}) remaining is

$$N_{Rb} = N_o e^{-\lambda t} \qquad (43.31)$$

Figure 43.12

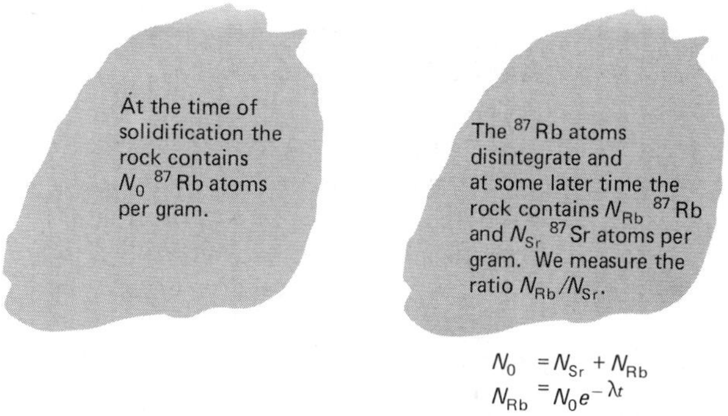

At the time of solidification the rock contains N_0 ^{87}Rb atoms per gram.
The ^{87}Rb atoms disintegrate and at some later time the rock contains N_{Rb} ^{87}Rb and N_{Sr} ^{87}Sr atoms per gram. We measure the ratio N_{Rb}/N_{Sr}.
$N_0 = N_{Sr} + N_{Rb}$
$N_{Rb} = N_0 e^{-\lambda t}$

where

$$\lambda = 1.39 \times 10^{-11}\ \text{yr}^{-1}$$

Because each ^{87}Rb decay forms a stable ^{87}Sr nucleus, the number of ^{87}Sr atoms per gram accumulating in this period is

$$N_{Sr} = N_o - N_{Rb}$$
$$= N_o(1 - e^{-\lambda t}) \qquad (43.32)$$

Using Eq. 43.31 to write N_o in terms of the number of ^{87}Rb atoms per gram present at the time of the measurement, we have

$$N_{Sr} = N_{Rb} e^{\lambda t}(1 - e^{-\lambda t})$$
$$= N_{Rb}(e^{\lambda t} - 1) \qquad (43.33)$$

By measuring the ratio N_{Sr}/N_{Rb} and knowing the decay constant λ, we can calculate the time since solidification from the melt of the rock (Figure 43.12). Generally the ratio N_{Sr}/N_{Rb} is determined with a mass spectrometer (Section 32.3). This radioactive dating method is one of several similar methods widely used for dating geologic materials.

Example 9 Dating a Meteorite

One technique using the ^{87}Rb dating method measures the ratio of the concentrations of ^{87}Rb and ^{86}Sr and the ratio of the concentrations of ^{87}Sr and ^{86}Sr. Since ^{86}Sr is common to both of these measurements, the ratio

$$\frac{N(^{87}\text{Rb})}{N(^{86}\text{Sr})} \div \frac{N(^{87}\text{Sr})}{N(^{86}\text{Sr})} = \frac{N(^{87}\text{Rb})}{N(^{87}\text{Sr})}$$

which is the quantity needed to determine a sample age. In a certain meteorite sample it is found that

$$\frac{N(^{87}\text{Sr})}{N(^{86}\text{Sr})} \div \frac{N(^{87}\text{Rb})}{N(^{86}\text{Sr})} = 0.0664$$

Let us determine the age of the meteorite from this measurement.

Using Eq. 43.33 we have

$$\frac{N_{Sr}}{N_{Rb}} = 0.0664 = e^{\lambda t} - 1$$

Therefore

$$e^{\lambda t} = 1 + 0.0664$$
$$= 1.0664$$
$$t = \frac{\ln 1.0664}{\lambda}$$
$$= \frac{\ln 1.0664}{1.39 \times 10^{-11}}$$
$$= 4.63 \times 10^{9} \text{ yr}$$

Questions

12. Suppose that the natural ^{14}C disintegration rate per gram of total carbon 5000 yr ago was in fact less than what it is now. If we want to date an artifact that is known to be 5000 yr old, would the assumption that the natural disintegration rate per gram of total carbon was the same as now yield an age greater than or less than 5000 yr?

13. Why does a measurement of the ^{14}C disintegration rate per gram of total carbon indirectly determine the concentration of ^{14}C atoms?

14. The rubidium–strontium dating method requires a measurement of the ratio of ^{87}Sr and ^{87}Rb atoms. If a given rock sample was separated into two unequal parts and the ^{87}Sr content was measured in one and the ^{87}Rb content was measured in the other, how could the desired rubidium–strontium ratio be determined?

43.4 Neutron Activation Analysis

Recall from Section 42.2 that every atom has its unique set of energy states. When an atom in an excited state makes a transition to a lower energy state by emitting a photon, the energy of the photon is determined by the two energy states of the emitting atom. By measuring the wavelengths of such photons it is possible to infer the chemical species emitting the radiation. Atomic spectroscopic analysis techniques exploit this principle. An analogous situation exists with nuclei. Like an atom, the energy states of a nucleus are unique. When a nucleus in an excited state makes a transition to another energy state by emitting a photon, the energy of the photon is determined by the two energy states of the emitting nucleus. By measuring the photon energies it is possible to identify the nuclei emitting the radiation. **Neutron activation analysis** takes advantage of this principle. "Neutron activation" means that the nuclei have been made radioactive by bombardment with neutrons. "Analysis" means the measurement of the energies of photons released in the radioactive decay processes.

The capture of a neutron produces an isotope of the capturing nucleus. For example, the capture of a neutron by ^{75}As produces ^{76}As according to the neutron capture reaction

$$^{75}As + n \rightarrow {}^{76}As \qquad (43.34)$$

The nucleus formed by the capture may be unstable as a result of an excess of neutrons, in which case it will disintegrate by emitting an electron and an antineutrino according to the beta decay reaction*

$$^{76}As \rightarrow {}^{76}Se + e^{-} + \bar{\nu}$$

The product nucleus of the beta decay is generally in an excited energy state, and decays by emitting gamma photons. Figure 43.13 shows this sequence of events on an energy diagram. The experiment involves the following steps: (1) By measuring the gamma photon energies we first identify the emitting nucleus (Figure 43.14). (2) Knowing that the nucleus was formed by beta decay, we then identify the isotope that led to the beta decay. (3) Knowing that this isotope was formed by neutron capture, we finally work backward to identify the nucleus that

*Typical neutron activation reactions are (a) neutron capture

$$^{75}As + n \rightarrow {}^{76}As$$

(b) beta emission

$$^{76}As \rightarrow {}^{76}Se^{*} + e^{-} + \bar{\nu}$$

and (c) gamma emission

$$^{76}Se^{*} \rightarrow {}^{76}Se + \gamma$$

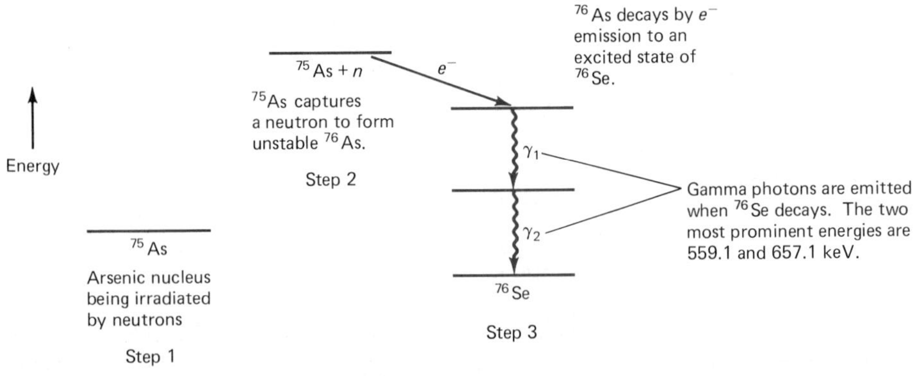

Figure 43.13

The activation of ^{75}As by neutron capture proceeds in the sequence of Steps 1, 2, and 3. The analysis procedure reverses these three steps.

Figure 43.14

Recordings of gamma photons emitted form neutron-activated sediment samples from a lake. The samples are being analyzed for their arsenic content. The horizontal scale records the photon energy; the vertical (logarithmic) scale represents the number of photons having a given energy. Each peak in the spectrum identifies a gamma photon source in the sample. The size of the photon peak is proportional to the concentration of the isotope emitting the photons. One photon source is identified as being initiated by ^{76}As. The relative sizes of the ^{76}As peaks in the standard and unknown samples enable us to deduce the arsenic content in the unknown.

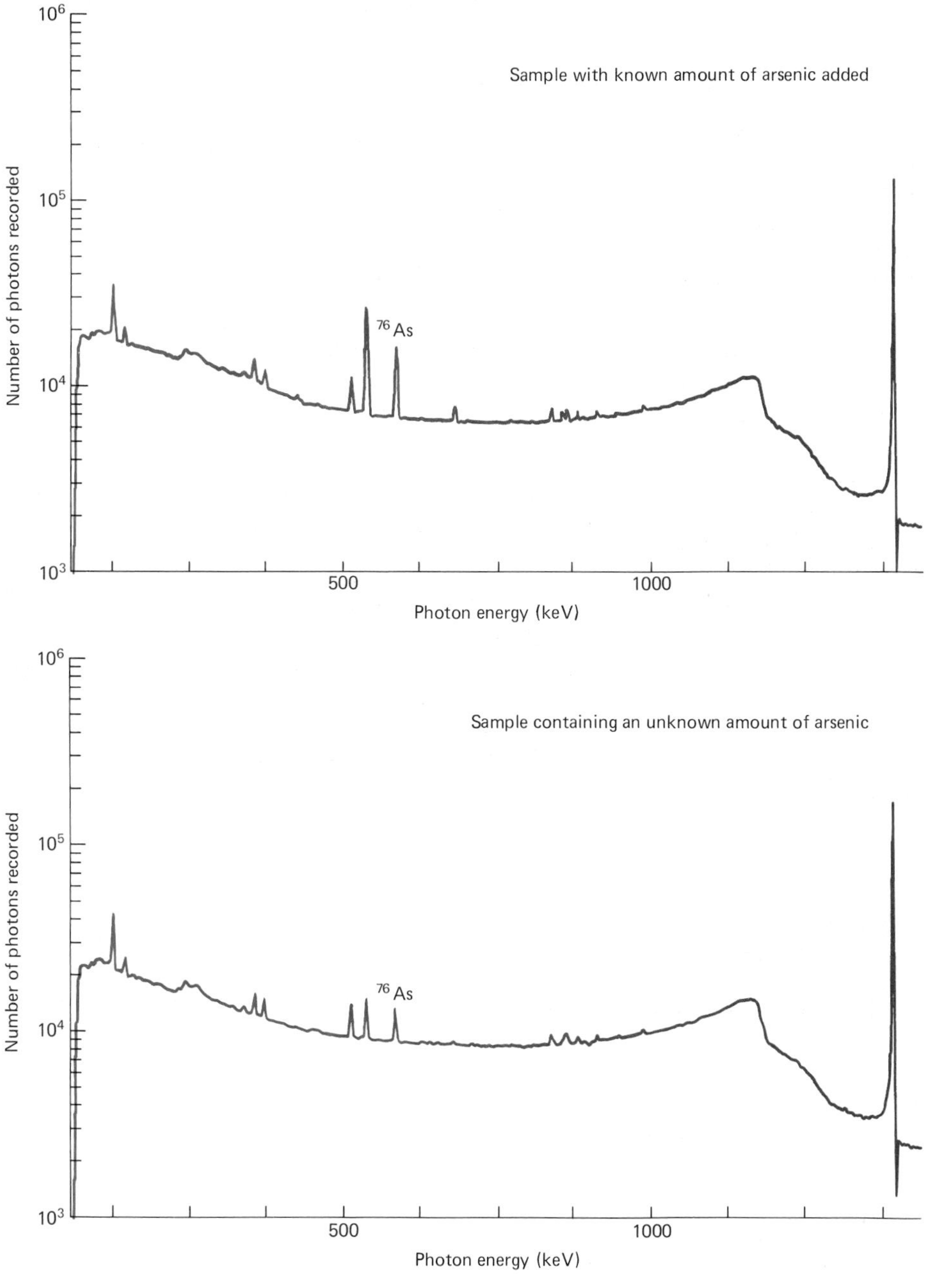

originated the sequence. Thus we not only can identify the initial isotope, but can also determine its concentration.

If we want to determine the concentration of an element in some material of interest, we prepare two identical samples of the material. Each of these samples contains C_1 atoms of interest per gram of the sample. To one of these samples we add a known concentration of C_o atoms of interest per gram of the sample. Both samples are irradiated simultaneously with a "rain" of neutrons—from a nuclear reactor, for example. We then measure the gamma photon intensities for the two irradiated materials. The ratio of the activities of the gamma photons of interest is equal to the ratio of the concentrations of the isotopes of interest.

$$\frac{I_1}{I_0} = \frac{C_1}{C_0 + C_1} \qquad (43.35)$$

Here I_1 is the activity in the unaltered sample and I_0 is the activity in the sample with the known number of atoms added. The only unknown in Eq. 43.35 is C_1, the concentration of the element of interest in the sample. By exposing the two samples to the same neutron source and taking the ratio of the induced activities, we need not know

1 how many neutrons actually impinged on the samples, nor

2 the probability that the neutrons are captured by the atoms of interest.

This is a very important feature of neutron activation analysis.

It is the great sensitivity of neutron activation analysis that makes the technique so useful. In many instances it is possible, while using only a fragment of a sample, to determine concentrations equivalent to 1 atom of interest in the presence of 1 billion other atoms. Environmentalists, criminologists, and analytical chemists, all of whom must often analyze samples for trace amounts (very small quantities) of impurities, use the neutron activation analysis technique to great advantage.

Example 10 Accountability

A metal-working company claims the arsenic content in its wastewater is less than 75 μg/liter (1 liter = 10^{-3} m^3) of water. Let us see how the neutron activation analysis technique can check the company's accountability.

Two 5-liter samples of the effluent are prepared. Into one is dissolved 0.00136 gm of arsenic pentoxide (As_2O_5), the molecular weight of which is 229.1. The atomic weight of arsenic is 74.96. The added amount of arsenic is thus

$$\frac{2(74.96)}{229.1} \cdot 0.00136 \text{ gm} = 8.90 \times 10^{-4} \text{ gm}$$

The added arsenic concentration is

$$\frac{8.90 \times 10^{-4}}{5} = 1.78 \times 10^{-4} \text{ gm/liter}$$

After neutron irradiation of 150 milliliters of both samples, the selenium gamma activities in the two samples are

standard $I_0 = 531$ counts/min

specimen $I_1 = 93$ counts/min

Using Eq. 43.35, we have

$$\frac{I_1}{I_0} = \frac{C_1}{C_0 + C_1}$$

Solving for C_1, we have

$$C_1 = \frac{I_1 C_0}{I_0 - I_1}$$

$$= \frac{93}{531 - 93} \cdot 1.78 \times 10^{-4} \frac{\text{gm}}{\text{liter}}$$

$$= 38 \frac{\mu\text{g}}{\text{liter}}$$

In this case, the measured concentration verifies the company's statement.

Question

15. In neutron activation analysis, the activities in the samples with known and unknown concentrations of the atoms of interest are measured at different times. What role does the half-life of the atoms of interest play in these measurements?

43.5 Nuclear Energy

All societies depend on thermal energy—for warmth, cooking, and drying, for example. Industrial societies need thermal energy to drive steam turbines as well. A nuclear reactor is one source of thermal energy. At present, the heat from nuclear reactors is used almost exclusively to provide steam for turbines that drive electric generators at central electric power stations.

Nuclear Fission

The separation of a nucleus into two fairly equally massive fragments is termed **nuclear fission.** In a contemporary nuclear reactor, energy is liberated by inducing nuclear fission of ^{235}U with neutrons. A typical neutron-induced nuclear fission reaction liberating 174 MeV of energy is depicted as

$$n + {}^{235}\text{U} \rightarrow {}^{143}\text{Ba} + {}^{90}\text{Kr} + 3n \qquad (43.36)$$

Products other than ^{143}Ba and ^{90}Kr may be produced, but the energy released is always about 200 MeV. For comparison, the chemical oxidation of a carbon atom in coal according to the reaction $C + O_2 \rightarrow CO_2$ liberates 4.1 eV, about 50 million times less energy than that obtained from the fission reaction. This large factor is the major attraction of nuclear energy. The fission energy is distributed among several particles, but the kinetic energy of the fission fragments (^{143}Ba and ^{90}Kr in Eq. 43.36) accounts for more than 80% of it (Table 43.3). The fission fragments are relatively massive and charged, and they readily transfer their kinetic energy to the medium through which they move. In other words, the medium (*fuel elements*) is *heated.* A nuclear reactor converts nuclear energy into thermal energy.

Table 43.3

	Mev
Kinetic energy of fission fragments	165
Instantaneous gamma-ray energy	7
Kinetic energy of fission neutrons	5
Beta particles from fission products	7
Gamma rays from fission products	6
Neutrinos from fission products (not available for power production)	10
Total energy per fission	200

In a nuclear power plant producing electricity, the thermal energy of the fission fragments is transferred to water that circulates in contact with ^{235}U fuel elements containing the ^{235}U atoms. Either directly or indirectly through a heat exchanger, water is converted to steam for a turbine that provides rotational energy for an electric generator that produces the electricity (Figure 43.15). Apart from the way the heat is provided for the turbine, a nuclear electric power plant is similar to the more familiar coal-burning electric power plant. Although simple in principle, few technological accomplishments rival the nuclear reactor. It could not have been developed without an understanding of nuclear physics principles.

Prior to the formation of the reaction products and the liberation of energy in a neutron-induced nuclear fission reaction, a short-lived nucleus is formed from the capture of the neutron by the target nucleus. This nucleus is an isotope of the target nucleus because the number of protons is unaltered by the capture of a neutron. For example, if ^{235}U captures a neutron, ^{236}U is formed. Symbolically

$$^{235}U + n \rightarrow (^{236}U) \tag{43.37}$$

Likewise, neutron capture by ^{238}U produces ^{239}U.

$$^{238}U + n \rightarrow (^{239}U) \tag{43.38}$$

There appears to be little difference in these two neutron capture reactions. However, it turns out that ^{235}U is a useful nuclear reactor fuel and ^{238}U is not. Let us see why.

Short-range nuclear forces behaving like surface tension tend to bind a nucleus into a nearly spherical configuration. The repulsive electric force between protons opposes this binding. Thus, competition between the repulsive electric forces and attractive surface tension forces determines whether or not nuclear fission occurs. The surface tension energy is proportional to the surface area of the nucleus. For a spherical nucleus the area is $4\pi R^2$, hence the surface tension energy varies as R^2. The electric energy of a heavy nucleus is proportional to Z^2/R (Section 43.2). The relative size of these two forces is reflected in the ratio of the nuclear parameters describing the competing forces.

$$\frac{U_{\text{electric}}}{U_{\text{surface}}} \sim \frac{Z^2/R}{R^2} \sim \frac{Z^2}{R^3} \tag{43.39}$$

Relating R to mass number A through the relation $R = r_0 A^{1/3}$, we have

$$\frac{U_{\text{electric}}}{U_{\text{surface}}} \sim \frac{Z^2}{A} \tag{43.40}$$

As the ratio Z^2/A increases, a point is reached where the electric force dominates the surface tension force. Niels Bohr and John Wheeler first recognized this fact and formulated a successful theory of nuclear fission with Z^2/A as a key parameter. For ^{236}U, $Z^2/A = 35.9$ and for ^{239}U, $Z^2/A = 35.4$. The slightly higher value for ^{236}U and its larger excitation energy (Figure 43.16) makes the fission of ^{236}U more likely than the fission of ^{239}U.

Nature has provided only a single naturally occurring atomic species, ^{235}U, with a significant induced fission probability. But through nuclear transmutations, we can make other fissionable nuclei that do not occur naturally. Most noteworthy from the nuclear fission standpoint are ^{239}Pu (plutonium) and ^{233}U. ^{240}Pu ($Z^2/A = 36.8$) and ^{234}U ($Z^2/A = 36.2$) formed from the capture of a neutron by ^{239}Pu and ^{233}U, respectively, readily fission.

Large values of Z^2/A for the transuranic (beyond uranium) elements help limit the stability of these elements by producing spontaneous nuclear fission.

On the average, the fission of ^{235}U produces 2.5 neutrons. Therefore, on the average, a given fission reaction is capable of providing the neutrons for succeeding neutron-induced fission reactions. If this occurs, there results a self-sustaining chain reaction of fissions, which is what makes possible the nuclear reactor.

Achieving a self-sustaining chain reaction involves several considerations. When a neutron impinges on a fissionable nucleus, there is no guarantee that it will induce

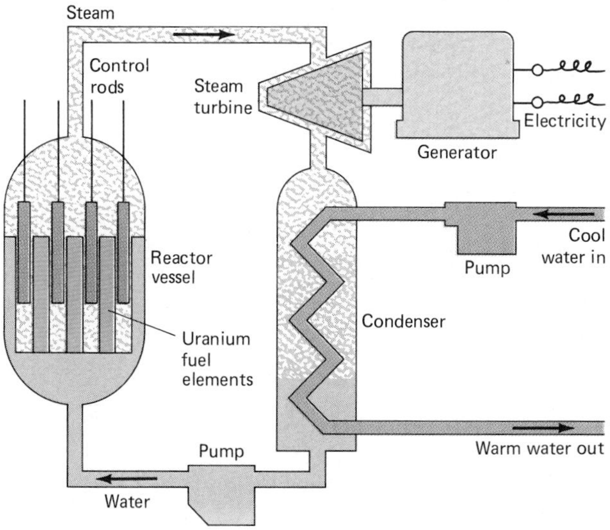

Figure 43.15

Schematic illustration of energy conversion and energy transfer in a nuclear fission reactor. Energy liberated from nuclear fission reactions causes the temperature of the uranium fuel elements to rise. Water circulating around the hot fuel elements is vaporized and channeled to a steam turbine. After passing through the turbine, the steam is condensed and returned to the reactor, completing the cycle. The heat of condensation liberated at the condenser is removed by circulating water through the condenser.

Figure 43.16

Schematic illustration of the energy available for the fission of ^{236}U and ^{239}U. The excitation energy of ^{236}U amounts of 6.546 MeV; that of ^{239}U amounts to 4.803 MeV. The larger excitation of ^{236}U coupled with its larger value of Z^2/A is sufficient to cause ^{236}U but not ^{239}U to fission.

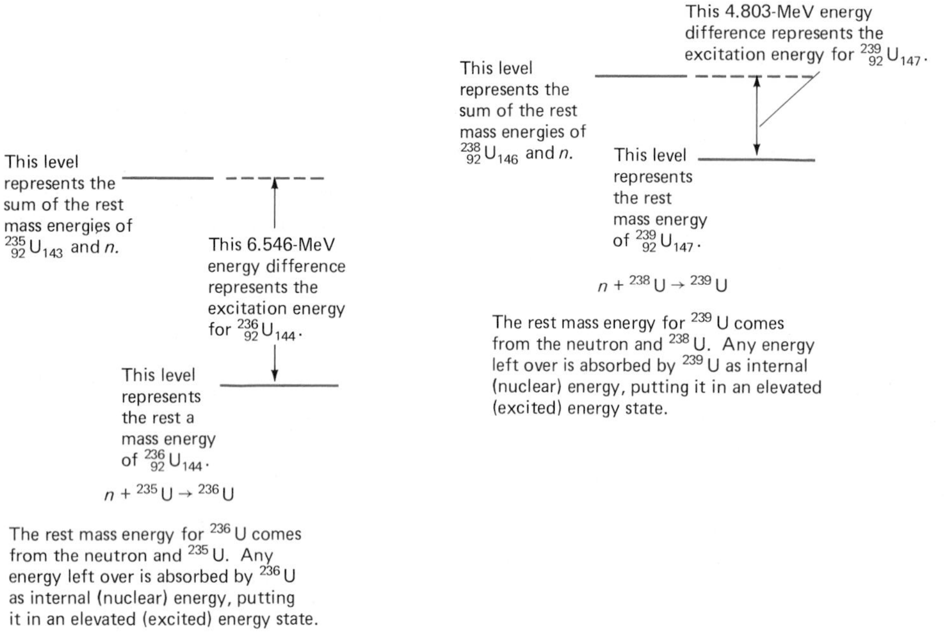
This level represents the sum of the rest mass energies of $^{235}_{92}U_{143}$ and n.
This 6.546-MeV energy difference represents the excitation energy for $^{236}_{92}U_{144}$.
This level represents the rest a mass energy of $^{236}_{92}U_{144}$.
$n + {}^{235}U \rightarrow {}^{236}U$
The rest mass energy for ^{236}U comes from the neutron and ^{235}U. Any energy left over is absorbed by ^{236}U as internal (nuclear) energy, putting it in an elevated (excited) energy state.
This level represents the sum of the rest mass energies of $^{238}_{92}U_{146}$ and n.
This 4.803-MeV energy difference represents the excitation energy for $^{239}_{92}U_{147}$.
This level represents the rest mass energy of $^{239}_{92}U_{147}$.
$n + {}^{238}U \rightarrow {}^{239}U$
The rest mass energy for ^{239}U comes from the neutron and ^{238}U. Any energy left over is absorbed by ^{239}U as internal (nuclear) energy, putting it in an elevated (excited) energy state.

fission. Other processes compete with fission. For instance, the neutron may simply scatter elastically, like a golf ball bouncing off a bowling ball. It may scatter and transfer energy to the target nucleus, leaving it in one of its higher energy states. Or it may induce a nuclear reaction that transmutes the target nucleus into a different nuclear species. Even if the target nucleus captures the neutron, there is no guarantee that the intermediate nucleus will fission; it may simply lose its energy by emission of one or more gamma photons. Each of these possible results occurs with a particular probability, determined by details of the structure of the target nucleus and the kinetic energy of the neutron.

Figure 43.17 shows the measured neutron-induced fission probabilities for ^{235}U and ^{238}U. In both cases the probabilities depend sensitively on neutron energy. For ^{238}U the probabilities are negligibly small for neutron energies less than about 1 MeV. As discussed earlier, there is insufficient energy in the nucleus (^{239}U) to stimulate nuclear fission until the kinetic energy of the neutron is greater than about 1 MeV.

Note the large, sharp fluctuations in the probability for energies between 1 eV and 10^3 eV. These sharp increases are referred to as *resonances*. At a resonance energy, the nucleus readily absorbs energy. The resonance is much like a mechanical or electrical system that readily absorbs energy at a sharply defined resonant frequency. ^{238}U also exhibits a resonant behavior for neutron capture for neutron energies between 1 eV and 10^3 eV. However, the capture does not induce fission. Rather, the excited nucleus loses the absorbed energy by gamma photon emission.

The capture probability for ^{235}U (to be followed by fission) increases steadily as the neutron energy decreases. The lower-energy limit for the neutrons is determined by the ambient temperature. In a practical situation this temperature is about 300 K and the lower-energy limit is about

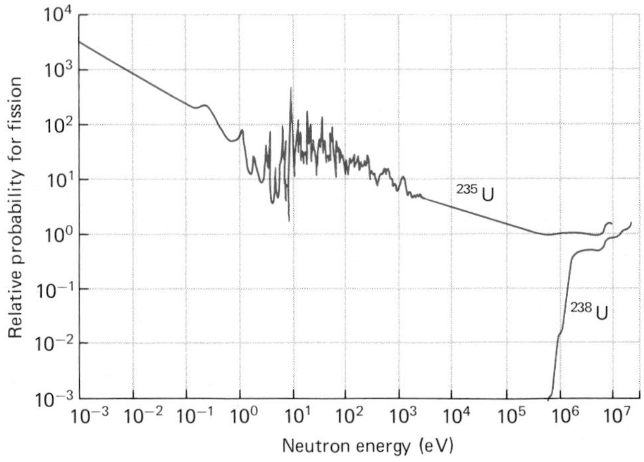
Relative probability for fission
10^4
10^3
10^2
10^1
10^0
10^{-1}
10^{-2}
10^{-3}
^{235}U
^{238}U
10^{-3} 10^{-2} 10^{-1} 10^0 10^1 10^2 10^3 10^4 10^5 10^6 10^7
Neutron energy (eV)

Figure 43.17

The relative probability for the neutron-induced fission of ^{235}U and ^{238}U. Both the probability and energy scales are logarithmic. The fluctuations near the center of the plot are very large; typically they represent orders-of-magnitude changes in the probability. Although the fission probability appears to be significant for both ^{235}U and ^{238}U for neutron energies greater than 1 MeV, the probabilities are very small compared to the fission probability of ^{235}U for low-energy neutrons.

1/40 eV. Neutrons having these energies are referred to as *slow* (or *thermal*) neutrons. Neutrons with energy greater than 10 keV are termed *fast* neutrons. Figure 43.17 shows that

1 with ^{235}U, fission is most easily induced with slow neutrons;

2 with ^{238}U, fission is induced *only* with (very) fast neutrons; and

3 the ^{235}U fission probability induced by slow neutrons is about three orders of magnitude larger than the ^{238}U fission probability induced by fast neutrons.

It happens that the neutrons produced in the fissioning of ^{235}U are very fast. The probability of these fast neutrons' producing fissions is relatively low. Therefore, most contemporary nuclear reactors provide a means of slowing down (reducing the energy of) the fission neutrons in order to produce a self-sustaining chain reaction. When the neutron speed is moderated, the reactor fuel need only contain a few percent of ^{235}U, the remainder being ^{238}U, which is not an appropriate fuel. This is in contrast to a nuclear fission bomb using ^{235}U, in which the fuel is nearly 100% ^{235}U, and there is no neutron moderation. The combination of neutron moderation and a very dilute concentration of ^{235}U in a nuclear reactor prevents a nuclear-bomblike explosion.

The rate of nuclear fission reactions is controlled in nuclear reactors by inserting rods between the fuel elements. These rods generally contain boron or cadmium, both of which have a very high affinity for capturing neutrons. The position of the rods determines the number of neutrons available for inducing nuclear fission reactions. In turn, the number of neutron-induced fission reactions determines the energy output of the reactor.

Nuclear Fusion

The binding energy per nucleon curve (Figure 43.1) peaks at a nucleon number of about 56. This means that nuclei having $A = 56$ are more tightly bound than either less massive or more massive nuclei. Nuclear fission of nuclei with $A \approx 240$ releases energy because the fragments have mass numbers closer to the $A = 56$ peak and are therefore more tightly bound. Nuclear reactions between light nuclei that produce more massive and more tightly bound nuclei also release energy. The difference in binding energy is released as kinetic energy of the reaction products. This process is called *nuclear fusion* because nuclei are fused together to form more tightly bound nuclei. In our sun, the initial reaction involves the fusion of two protons into a deuteron (d), a positron (e^+), and a neutrino (ν) according to the reaction

$$p + p \rightarrow d + e^+ + \nu \qquad (43.41)$$

Example 11 Energy Released in the Proton–Proton Fusion Reaction

Our very existence depends on energy from the sun. Although a variety of nuclear reactions are involved in energy production in the sun, proton–proton fusion reactions trigger the major reactions. Let us determine the energy released in a proton–proton fusion reaction.

The reaction involving bare nuclei is represented as

$$p + p \rightarrow d + e^+ + \nu \qquad \text{(bare nuclei)}$$

To make neutral atoms out of the two protons we need to add two electrons to both sides of the equation.

$$(p + e^-) + (p + e^-) \rightarrow (d + e^-) + e^- + e^+ + \nu$$

In terms of neutral atoms this equation becomes

$$^1H + {}^1H \rightarrow {}^2H + e^- + e^+ + \nu$$

The energy released is the mass energy difference between the left- and right-hand sides of the equation. From Table 43.1 we have

$$2M_H = 2(1.007825)$$
$$= 2.01565 \text{ u}$$
$$M_{2H} + m_{e^-} + m_{e^+} = 2.014103 + 0.000549 + 0.000549$$
$$= 2.015201 \text{ u}$$

The mass difference (initial minus final) is given by

$$\Delta m = 2M_H - M_{2H} - m_{e^-} - m_{e^+}$$
$$= 0.000449 \text{ u}$$

Converting from mass to energy units we have

$$\Delta m \cdot c^2 = 0.000449 \text{ u} \cdot 931.48 \frac{\text{MeV}}{\text{u}}$$
$$= 0.42 \text{ MeV}$$

The fusion of two protons yields 0.42 MeV of energy. In addition, in a very short time following the fusion of two protons the positron and an electron will interact, converting their mass energy into photons having a total energy of 1.02 MeV. The total energy release is then 1.44 MeV.

The total energy released in a proton–proton fusion reaction is small compared to that released in a nuclear fusion reaction (about 200 MeV). But the energy released per nucleon in a fusion reaction is comparable to the 1 MeV/nucleon released in a nuclear fission reaction. In this sense, nuclear fusion is a very significant energy source.

Each $p + p$ reaction releases 0.42 MeV of energy. When the positron interacts with an electron, an additional 1.02 MeV of energy is released in the form of gamma photons. The production of a positron and a neutrino in the $p + p$ reaction (Eq. 43.41) signals that the reaction involves the weak nuclear force. Such reactions proceed extremely slowly, and consequently it is difficult to liberate significant amounts of energy by means of $p + p$ reactions unless a large number of protons participate. The sun is a prolific energy source because the density of solar hydrogen is large.

Numerous other nuclear fusion reactions are possible, but three are of special importance. They are

$$^2H + {}^2H \rightarrow {}^3He + n + 3.3 \text{ MeV} \qquad (43.42)$$

$$^2H + {}^2H \rightarrow {}^3H + p + 4.0 \text{ Mev} \qquad (43.43)$$

$$^2H + {}^3H \rightarrow {}^4He + n + 17.6 \text{ MeV} \qquad (43.44)$$

It now appears that the first generation of nuclear fusion reactors will be based on these reactions. The first two of these reactions are termed DD (deuteron–deuteron) reactions. The third is termed a DT (deuteron–triton) reaction.

Producing these reactions on an individual basis is not especially difficult. Deuterons from an accelerator can be made to impinge on deuterium atoms in a suitable target, and fusion reactions will proceed at a definite rate. But such a scheme is not practical as a nuclear reactor because the investment in energy to accelerate the deuterons far exceeds the energy derived from the fusion reactions. If nuclear fusion reactions are to be a useful energy source, a technology must be devised that derives more energy than is used to initiate the reactions. Although hopes are high for doing this ultimately, it has not yet been done.

In order for a nuclear fusion reaction to occur, the reacting nuclei must come within the range of the nuclear force. Because of the electrical repulsion between the positively charged nuclei, the latter can approach one another closely enough to fuse only if they have sufficient kinetic energy. The energy barrier (see Figure 42.14, Section 42.5) that the nuclei face is just the electric potential energy of the two interacting nuclei evaluated at a separation where the nuclear force comes into play. This distance is roughly 3 fm. The electric potential energy of two $Z = 1$ nuclei separated by 3 fm is

$$U = \frac{1}{4\pi\epsilon_0}\frac{e^2}{r} = \frac{(9.0 \times 10^9)(1.602 \times 10^{-19})^2}{3 \times 10^{-15}\text{ m}}$$
$$= 7.7 \times 10^{-14}\text{ J}$$
$$= 481\text{ keV}$$

This energy is easily achieved by modest particle accelerators, which is why fusion reactions can be achieved on a laboratory scale. In principle this energy could be provided thermally. The required temperature may be estimated from the expression relating the average kinetic energy of a molecule to the Kelvin temperature—$E_{av} = \frac{3}{2}kT$ (see Section 25.3). If we assume that each nucleus furnishes one half of the energy required to surmount the electric potential energy barrier, then

$$\tfrac{3}{2}kT \approx 240\text{ keV}$$
$$T \approx \frac{2(240 \times 10^3\text{ eV})(1.602 \times 10^{-19}\text{ J/ev})}{3(1.38 \times 10^{-23}\text{ J/K})}$$
$$\approx 2 \times 10^9\text{ K}$$

The actual temperatures required to kindle fusion are significantly lower because some nuclei have energies substantially larger than the average and some reactions proceed by quantum tunneling of the electric potential energy barrier. This reduces the required temperature by more than a factor of 10. Nevertheless, it is no trivial task to create a temperature on the order of 10^8 K. This temperature, called the *ignition temperature,* depends on the particular nuclear fusion reaction being considered. For the DD and DT reactions the ignition temperature is around 4×10^8 K and 4×10^7 K, respectively. Because the ignition temperature is lower for the DT reaction and the energy release per reaction is larger, the DT reaction is receiving the most attention in nuclear fusion energy research.

At the ignition temperature, the thermal energy of a particle greatly exceeds the atomic binding energy of an electron. Hence, a gaseous mixture of fusible atoms becomes an electrically neutral sea of unattached negative electrons and positive nuclei. Such a system is called a *plasma.* If the plasma is allowed to come into contact with any material structure, it cools and loses energy and is not able to participate in the induction of a nuclear fusion reaction. Therefore, a containment scheme must be devised to prevent the plasma from contacting any material framework. As we shall see, most schemes attempt to do this magnetically. In the following paragraphs, we derive a criterion for achieving an energy profit in a nuclear fusion energy system.

To ignite the fusion reactions, energy must be added to heat the plasma to the ignition temperature. The energy invested is proportional to the total number of particles present in the plasma. This is no different than trying to raise the temperature of an ordinary gas by transferring heat to it. Considering a unit volume, the energy required is proportional to the number density (n) of the nuclei.

$$\text{Energy invested} \propto n$$
$$= Bn \tag{43.45}$$

where B is a proportionality constant. A given nucleus may possibly interact with any of the remaining nuclei in the plasma. Hence the collision rate of any given nucleus will be proportional to the density of the particles. Including the collisions of all ions, the total collision rate is proportional to the square of the ion density. The rate at which energy is produced (power) is proportional to the square of the density of particles. Since energy is related to the product of power and time, the energy liberated by the colliding atoms is proportional to the square of the ion density and the time τ during which the ions interact.

$$\text{Energy produced} = \text{power} \times \text{time}$$
$$= Cn^2\tau \tag{43.46}$$

where C is a second proportionality constant. Net energy production requires that the energy produced be greater than the energy invested:

$$Cn^2\tau > Bn$$

or

$$n\tau > \frac{B}{C}$$

The factors B and C include the details of the fusion mechanism. The quantity τ is called the *confinement time,* and $n\tau$ is called the *confinement parameter.* For the DT reactions a theoretical analysis shows that

$$\frac{B}{C} = 10^{14}\,\frac{\text{ions}\cdot\text{s}}{\text{cm}^3}$$

and so

$$n\tau > 10^{14}\,\frac{\text{ions}\cdot\text{s}}{\text{cm}^3} \tag{43.47}$$

This condition for the confinement parameter is termed the *Lawson criterion.* *

*It is named for the British physicist John D. Lawson.

A practical energy-producing nuclear fusion system must satisfy two basic criteria:

1 the creation of a sufficiently high ignition temperature —this ensures sufficient thermal energy for the interacting nuclei to overcome the electric potential energy of the two charges of the nuclei;

2 achievement of the Lawson criterion—this ensures a net gain in energy from the nucleon fusion reactions.

Although the Lawson criterion is simple, the physics involved in achieving the criterion on a large scale is far from simple. We need such measurements as the probabilities for the interacting nuclei and we need models for the energy transfer mechanisms associated with the nuclear fusion mechanism. Besides the conduction and convection heat transfer processes, we must consider radiative energy losses from rapidly moving charged particles within the plasma. Contemporary nuclear fusion research focuses on achieving the Lawson criterion. In roughly 30 years of research, progress has been slow and success elusive. But there are encouraging results, and the work continues.

Questions

16. The nuclear reaction $n + {}^9\text{Be} \rightarrow {}^8\text{Be} + 2n$ produces enough neutrons to make possible a self-sustaining chain reaction. For energy reasons, however, it will not work. Explain.

17. Why isn't ${}^{239}\text{Np}$ (neptunium) a useful nuclear reactor fuel?

Summary

The nucleus of the atom can be thought of as a bound system of neutrons and protons that, collectively, are called nucleons. A proton has one positive unit (1.602×10^{-19} C) of charge and a mass of 1.007277 u. A neutron has no net charge and a mass of 1.008665 u.

We characterize a given nucleus by the symbolism

$${}^A_Z\text{X}_N$$

where Z is the atomic (proton) number, N is the neutron number, $A = Z + N$ is the mass (protons and neutrons) number, and X denotes the chemical symbol for the nucleus.

For many purposes we can consider the nucleus to be a sphere of constant density having a radius

$$r = 1.2A^{1/3} \text{ fm} \tag{43.9}$$

$$1 \text{ fm} = 10^{-15} \text{ m}$$

The binding energy of a nucleus is the minimum energy that must be added to a nucleus to separate it into its constituent neutrons and protons. In terms of mass energy of the neutral atoms involved

$$\text{binding energy} = ZM_{\text{H}}c^2 + Nm_n c^2 - Mc^2 \tag{43.4}$$

where M_{H} is the mass of the hydrogen atom, m_n is the mass of the neutron, and M is the mass of the neutral atom. For nucleon numbers greater than about 40, the binding energy per nucleon is about 8 MeV/nucleon. As the number of protons in stable nuclei increases, the neutron number gets progressively larger than the proton number.

In all nuclear reactions, the conservation laws involving energy, linear momentum, angular momentum, and charge apply.

In a spontaneous disintegration of a nucleus, rest mass energy considerations require that the combined masses of the disintegration products be less than the mass of the disintegrating nucleus. The rest mass energy is transformed into kinetic energy, which is shared by the reaction products.

An alpha particle is a bound system of two protons and two neutrons. It is identical to the ${}^4\text{He}$ nucleus. Alpha particles are emitted spontaneously by a number of nuclei usually having $A > 140$.

Beta particles are charged particles emitted spontaneously by some unstable nuclei. Some beta particles are negatively charged, others are positively charged. The negative beta particle (e^-) has properties identical to an electron. It is a result of the transformation by the weak nuclear force of a neutron into a proton, an electron, and an antineutrino. Symbolically, $n \rightarrow p + e^- + \bar{\nu}$. The positive beta particle (e^+), called a positron, is identical to an electron except that it possesses one positive unit of charge. It is a result of the transformation by the weak nuclear force of a proton into a neutron, a positron, and a neutrino. Symbolically, $p \rightarrow n + e^+ + \nu$.

Like the energies associated with an atom, the energies of a nucleus are discrete. A nucleus may absorb energy and make a transition to a state of higher energy. It may release energy and move to a lower energy state. Often this energy release is in the form of a photon having frequency

$$\nu = \frac{E_2 - E_1}{h} \tag{43.18}$$

This process is called gamma emission.

The half-life ($\tau_{1/2}$) of a radioactive sample is defined as the time required for one half of the radioactive nuclei to disintegrate. In terms of the decay constant λ,

$$\tau_{1/2} = \frac{0.693}{\lambda} \tag{43.23}$$

Nuclear disintegration rates are measured in units called curies, symbolized by Ci. One Ci $= 3.7 \times 10^{10}$ disintegrations per second.

The half-life ($\tau_{1/2}$) for the decay of selected unstable nuclei is used to determine the age of materials of archaeological and geologic interest. The decay of ${}^{14}\text{C}$, which has a half-life of 5730 yr, is used to determine the age of carbon-based archaeological artifacts. The pertinent equation for the age evaluation is

$$t = -\frac{1}{\lambda} \ln \frac{I}{I_o} \tag{43.30}$$

where $\lambda = 0.693/\tau_{1/2}$ is the decay constant and I and I_o are, respectively, the activities per gram at the time of measurement and at the time of termination of photosynthesis in the object.

The age of geologic materials can be deduced from the decay of ${}^{87}\text{Rb}$ to ${}^{87}\text{Sr}$. We measure the ratio of ${}^{87}\text{Sr}$ and

^{87}Rb atoms in the sample of interest and determine the age from the relation

$$\frac{N_{Sr}}{N_{Rb}} = e^{\lambda t} - 1 \qquad (43.33)$$

Neutron activation is the process of making materials radioactive by neutron irradiation. By analyzing the energies and numbers of the gamma photons emitted by the radioactive sample we can identify and assess trace amounts of atomic species of interest. The algebraic relation required is

$$\frac{I_1}{I_o} = \frac{C_1}{C_1 + C_o} \qquad (43.35)$$

where I_1 and I_o are the measured activities in the sample and standard and C_1 and C_o are the atomic concentrations in the sample and standard.

A nuclear reactor is basically a source of heat. It derives energy from ^{235}U or ^{239}Pu according to the neutron-induced nuclear fission reaction

$$n + {}^{235}\text{U} \rightarrow \text{X} + \text{Y} + \text{neutrons} + \text{energy}$$

where X and Y denote the fission fragments.

Combining two nuclei to form nuclei having a combined mass less than that of the reactants is termed nuclear fusion. The mass difference is converted to energy according to the equation $E = mc^2$. Deuterium (^{2}H) and tritium (^{3}H) combine according to the nuclear fusion reaction

$${}^2\text{H} + {}^3\text{H} \rightarrow {}^4\text{He} + n + 17.6 \text{ MeV}$$

Because of the significant energy release and relatively high probability for fusing, the DT reaction is being explored for use in practical nuclear fusion reactors.

Suggested Reading

Nuclear Physics

G. F. Bertsch, Vibrations of the Atomic Nucleus, *Sci. Am.*, 248, 62–73 (May 1983).

W. Brandt, Positrons as a Probe of the Solid State, *Sci. Am.*, 233, 34–42 (July 1975).

D. Allan Bromley, Nuclear Molecules, *Sci. Am.*, 239, 58–69 (December 1978).

J. Cerny and A. M. Poskanzer, Exotic Light Nuclei, *Sci. Am.*, 238, 60–72 (June 1978).

M. Jacob and P. Landsoff, The Inner Structure of the Proton, *Sci. Am.*, 243, 66–75 (March 1980).

Positron–Emission Tomography, Michel M. Ter-Pogossian, Marcus E. Raichle, and Burton E. Sobel, *Sci. Am.*, 243, 170–181 (October 1980).

The Decay of the Proton, S. Weinberg, *Sci. Am.*, 244, 64–75 (June 1981).

Nuclear Technology

Radioactive Dating

J. D. Macdougall, Fission-Track Dating, *Sci. Am.*, 235, 114–122 (December 1976).

S. Moorbath, The Oldest Rocks and the Growth of Continents, *Sci. Am.*, 236, 92–104 (March 1977).

C. Renfrew, Carbon-14 and the Prehistory of Europe, *Sci. Am.*, 225, 63–72 (October 1971).

D. N. Schramm, The Age of the Elements, *Sci. Am.*, 230, 69–77 (January 1974).

Neutron Activation Analysis

W. H. Wahl and H. H. Kramer, Neutron Activation Analysis, *Sci. Am.*, 216, 68–82 (April 1967).

Nuclear Reactors

H. W. Agnew, Gas-Cooled Nuclear Power Reactors, *Sci. Am.*, 244, 55–63 (June 1981).

W. P. Beggington, The Reprocessing of Nuclear Fuels, *Sci. Am.*, 235, 30–41 (December 1976).

H. A. Bethe, The Necessity of Fission Power, *Sci. Am.*, 234, 21–31 (January 1976).

B. L. Cohen, The Disposal of Radioactive Wastes from Fission Reactors, *Sci. Am.*, 236, 21–31 (June 1977).

B. Coppi and J. Rem, The Tokamak Approach in Fusion Research, *Sci. Am.*, 227, 65–75 (July 1972).

G. A. Cowan, A Natural Fission Reactor, *Sci. Am.*, 235, 36–47 (July 1976).

J. L. Emmett, J. Nuckolls, and L. Wood, Fusion Power by Laser Implosion, *Sci. Am.*, 230, 24–37 (June 1974).

S. A. Fetter and K. Tsipis, Catastrophic Releases of Radioactivity, *Sci. Am.*, 244, 41–47 (April 1981).

H. P. Furth, Progress toward a Tokamak Fusion Reactor, *Sci. Am.*, 241, 50–61 (August 1979).

O. Hahn, The Discovery of Fission, *Sci. Am.*, 198, 76–84 (February 1958).

H. W. Lewis, The Safety of Fission Reactors, *Sci. Am.*, 243, 53–65 (March 1980).

R. Golub, W. Mampe, J. M. Pendlebury, and P. Ageron, Ultracold Neutrons, *Sci. Am.*, 240, 134–154 (June 1979).

H. C. McIntyre, Natural-Uranium Heavy-Water Reactors, *Sci. Am.*, 233, 17–27 (October 1975).

D. R. Olander, The Gas Centrifuge, *Sci. Am.*, 239, 37–53 (August 1978).

G. Yonas, Fusion Power with Particle Beams, *Sci. Am.*, 239, 50–61 (November 1978).

R. N. Zare, Laser Separation of Isotopes, *Sci. Am.*, 236, 86–97 (February 1977).

Problems

Atomic mass data for these problems may be found in Table 43.1 in Section 43.1.

Section 43.1 The Neutron–Proton Model of the Nucleus

1. James Chadwick observed that radiation from the interaction of alpha particles with beryllium caused protons to be ejected from paraffin with an energy of about 6 MeV. Conceivably this ejection could have been the result of a collision of a photon with a proton. In such a collision the photon will transfer the maximum amount of energy to the proton when the collision is head on and the photon scatters backward. (a) Assuming a head-on collision of a photon with a proton intially at rest, use nonrelativistic linear momentum and energy conservation principles to show that

$$E_\gamma = \sqrt{\frac{E_p m_p c^2}{2}} + \frac{E_p}{2}$$

where E_p and E_γ are the proton and photon energies and m_p is the proton mass. (b) Using the equation derived in part (a), determine the photon energy for a proton energy of 6 MeV. Does it seem reasonable that such an energetic proton could be ejected from paraffin in Chadwick's experiment?

2. Determine and compare the binding energy per nucleon of ^{40}Ar, ^{90}Sr, ^{137}Ba, and ^{238}U.

3. Determine the nuclear radii of ^{60}Co, ^{131}I, and ^{235}U.

4. (a) Assuming that a nucleus is a sphere having a radius given by Eq. 43.7, show that the number of nucleons per unit volume is $3/4\pi r_o^3$. (b) Determine the number of nucleons per cubic fermi by using $r_o = 1.2$ fm.

5. If a nucleus is spherical, its volume can be calculated by using $r = r_o A^{1/3}$. Since nearly all the mass of an atom resides in the nucleus, the mass of a nucleus can be estimated from a knowledge of the mass of the neutral atom (Table 43.1). Estimate in kilograms per cubic meter the density of nuclear matter in gold. Compare your result with the density of nuclear matter in aluminum (Example 4).

6. Using the data in Example 4 for the density of nuclear matter, estimate the length in centimeters of a side of a cube of nuclear matter having a mass of 1 metric ton. Compare this length with the thickness of a dime, which is about 1 mm.

7. A beryllium nucleus has four protons ($Z = 4$). Label these protons p_1, p_2, p_3, and p_4; determine the number (n) of possible pairs of protons; and show that your result is consistent with the formula $n = [Z(Z - 1)]/2$.

Section 43.2 Nuclear Stability

8. $^5_2He_3$ decays by neutron emission. Write the nuclear equation for this disintegration.

9. Beryllium 8 ($^8_4Be_4$) is a very special case among light nuclei because it disintegrates by alpha-particle emission. (a) Write the nuclear equation for this disintegration. (b) If the atomic masses of ^8Be and an alpha particle are 8.005305 u and 4.002603 u, respectively, show that this disintegration is energetically possible.

10. Determine the maximum beta-particle energy in the beta decay of the neutron.

11. A ^{238}U nucleus at rest emits an alpha particle, leaving the daughter ^{234}Th nucleus in its ground state. The total energy liberated is 4.265 MeV. Using the conservation of linear momentum, show that the alpha-particle energy is 4.19 MeV.

12. In positron decay, a nuclear proton is transformed into a neutron, a positron, and a neutrino. (a) Complete the following nuclear equation for the positron decay of the $^{11}_6C_5$ nucleus.

$$^{11}_6C_5 \rightarrow$$

(b) The nuclear equation in part (a) involves only nuclear masses. Maintaining a balance of charge on both sides of the equation, change the nuclear equation into one involving masses of the neutral atoms by adding the same number of electrons to both sides of the equation. (c) Using energy conservation principles, determine the maximum kinetic energy of the positron.

13. Plutonium 239 ($^{239}_{94}Pu_{145}$) decays by alpha-particle emission according to

$$^{239}_{94}Pu_{145} \rightarrow {}^{235}_{92}U_{143} + {}^4_2He_2$$

Determine the energy available for this decay.

14. The activity of a certain radioactive sample declines by a factor of 10 in 4 min. What is the half-life of this radioactive species? Of the nuclei listed in Table 43.2, which is mostly likely to be involved in this sample?

15. The mean lifetime of an unstable nucleus is related to the disintegration constant λ by $\bar{t} = 1/\lambda$ (see Problem 18). Determine the mean lifetime of a free neutron and a ^{239}Pu nucleus. The half-life of a neutron is 11.0 min. The half-life of ^{239}Pu is 24,390 yr.

16. A disintegrating nucleus is often referred to as a *parent*, and the end nuclear product is called the *daughter*. For example, in the decay of tritium, the parent is $^3_1H_2$ and the daughter is $^3_2He_1$. If the decay of the parent is described by $n = n_o e^{-\lambda t}$, what is the corresponding equation for the growth of the daughter? Make a simple sketch of the decay and growth curves.

17. Determine the disintegration rate (activity) of 1 gm of ^{60}Co. The half-life of ^{60}Co is 5.24 yr.

18. In the disintegration of a radioactive sample having n_0 nuclei at the beginning, the number (n) remaining after time t is $n = n_0 e^{-\lambda t}$. Thus a plot of t versus n is as shown in Figure 1. Some nuclei disintegrate immediately; others live (in principle) forever. From the scheme for calculating average values (Section 10.1) the average lifetime of a nucleus is then

$$\bar{t} = \frac{\int_{n_0}^{0} t\, dn}{\int_{n_0}^{0} dn}$$

(a) Using the relation $n = n_0 e^{-\lambda t}$, show that

$$\bar{t} = \frac{\int_0^{\infty} t e^{-\lambda t}\, dt}{\int_0^{\infty} e^{-\lambda t}\, dt}$$

(b) Evaluate the integrals in part (a) to show that $\bar{t} = 1/\lambda$.

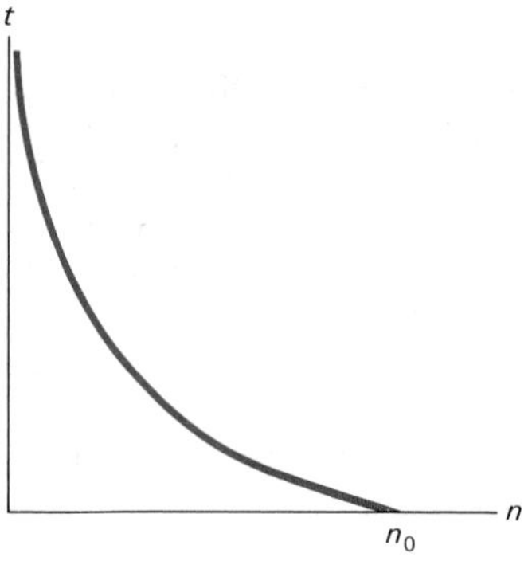

Figure 1

19. If the daughter (remnant) nucleus in a radioactive decay process is stable, then the rate of decay of the parent nuclei (dn_p/dt) is exactly equal to the rate of growth of daughter nuclei (dn_d/dt). But if the daughter is radioactive, then the rate of change of the daughter is a combination of growth and decay rates. Show that the rate of change of the daughter is zero when $\lambda_d n_d = \lambda_p n_p$, where λ_d and λ_p are the decay constants of the daughter and parent, respectively.

Section 43.3 Radioactive Dating

20. The ^{14}C specific disintegration rate in an artifact is measured to be 23.1 disintegrations/hr·gm. Determine the age of the artifact.

21. The specific disintegration rate in an artifact is measured to be 35.1 ± 0.5 disintegrations/hr·gm. Estimate the age and the uncertainty in the age by using the ^{14}C dating method.

22. By the ^{14}C dating method the age of an artifact was determined to be 22,500 yr. Determine the activity per gram of total carbon in the sample.

23. In a contemporary carbon-based material the ^{14}C specific activity is 13.6 disintegrations/min·gm of total carbon. (a) Show that there are 5.9×10^{10} ^{14}C atoms for each gram of total carbon. (b) How many ^{14}C atoms will have disintegrated for each gram of total carbon in a sample that is 5730 yr old?

24. (a) Calling R the ratio of daughter to parent atoms, show that the age in the rubidium–strontium method is given by

$$t = \frac{1}{\lambda} \ln(1 + R)$$

(b) Under what conditions would the approximate relation

$$t \approx \frac{R}{\lambda}$$

be valid? (c) The age determination in Example 9 utilized the expression in part (a). Show that the approximate equation in part (b) gives essentially the same result.

25. In a piece of rock from the moon the ^{87}Rb content is assessed to be 1.82×10^{10} atoms per gram of material. In a piece of the same rock the ^{87}Sr content is found to be 1.07×10^9 atoms per gram. (a) What assumption would you have to make in order to determine the age of the rock from these measurements? (b) Making the required assumption about the data, determine the age of the rock.

Section 43.4 Neutron Activation Analysis

26. If the sodium activity in a neutron-irradiated material is 59,300 disintegrations per second, how many sodium atoms are in the sample? (The half-life of the sodium isotope is 15.0 hr.)

27. In a neutron activation analysis experiment involving the determination of aluminum, a measurement of the activity in a sample in which 1.25 mg of aluminum had been inserted was 386 disintegrations per second. Ten minutes later the activity in an identical sample having no aluminum purposely inserted was found to be 5.90 disintegrations per second. Determine the unknown aluminum concentration. (^{28}Al has a half-life of 2.30 min.)

28. When a material of interest is irradiated by neutrons, radioactive atoms are produced continually and some decay according to their given half-life. (a) If radioactive atoms are produced at a constant rate R and their decay is governed by the conventional radioactive decay law, show that the number of radioactive atoms accumulated after an irradiation time t is

$$n = \frac{R}{\lambda}(1 - e^{-\lambda t})$$

(b) What is the maximum number of radioactive atoms that can be produced?

29. Figure 2 shows in an idealized way data recorded for the activities of two neutron-irradiated samples. One sample has been doped with 5.23×10^{16} lanthanum atoms. The data for the doped sample required 1 min of counting time. The data for the other sample required 10 min of counting time. Determine the number of atoms in the unknown sample.

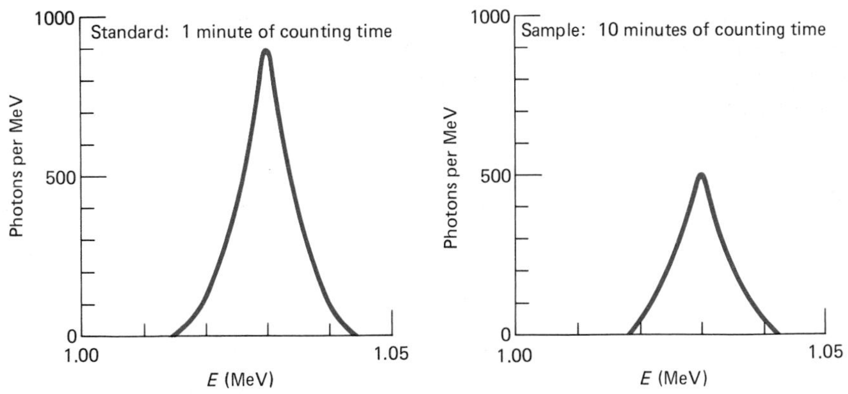

Figure 2

Section 43.5 Nuclear Energy

30. When ^{239}Pu captures a neutron to form ^{240}Pu, how much energy is available for the excitation of ^{240}Pu?

31. How much energy is released in the following neutron-induced fission reaction?

$$n + {}^{233}U \rightarrow {}^{91}Kr + {}^{140}Ba + 3n$$

32. Determine the difference in excitation energy for the capture of a neutron by ^{233}U and ^{238}U. Which of these two reactions would most likely lead to nuclear fission?

33. Suppose that a neutron-induced fission reaction produces two neutrons, and that the two neutrons in each reaction proceed to induce fission reactions. (a) If the first reaction is called the first generation, what is the total number of neutrons produced in eight generations? (b) Derive a relation for the neutrons produced in n generations.

34. Both ^{235}U and ^{238}U are radioactive. The half-life of ^{235}U is 7.13×10^8 yr and the half-life of ^{238}U is 4.5×10^9 yr. In natural uranium there is one ^{235}U atom for every 140 ^{235}U atoms. (a) What would this ratio have been 10^9 yr ago? (b) What will it be 10^9 yr from now?

35. A modern nuclear power plant produces about 1000 MW of electric power. The overall energy conversion efficiency of the plant is generally close to 33%. If the plant were operated at full capacity for a year and all the energy were derived from the fissioning of ^{235}U (each reaction producing 190 MeV of energy), how many kilograms of ^{235}U would be required?

36. If all the energy from the fissioning of 1 gm of ^{235}U could be converted to electrical energy, and each reaction produces 190 MeV of. energy, how long could this energy keep a 200-W bulb operating?

37. About 200 MeV of energy is liberated in the fission of a ^{235}U nucleus. (a) Show that the complete fissioning of 1 kg of ^{235}U liberates about 8.2×10^{13} J. (b) On the average, a city of 1 million people requires 2000 MW of electric power. If nuclear energy is converted to electrical energy at an efficiency of 30%, determine the mass of ^{235}U that would have to be fissioned to furnish the daily energy of a city of 1 million people.

38. (a) Neutrons are reduced in energy through elastic collisions with appropriate nuclei, such as protons. The energy lost depends on how the neutron strikes the scatterer. If a neutron collides with a more massive nucleus at rest and scatters at an angle of 90° measured relative to its incident direction, use nonrelativistic conservation of linear momentum and energy to show that the ratio of its kinetic energy after the collision (K) to its kinetic energy before the collision (K_o) is

$$\frac{K}{K_o} = \frac{M - m}{M + m}$$

where m is the mass of the neutron and M is the mass of the scatterer. (b) How many 90° scatterings must a neutron make with ^{12}C nuclei in order to reduce its energy from 2 MeV to 0.2 eV?

39. In Section 43.5 we reasoned that

$$\frac{U_{electric}}{U_{surface}} \approx \frac{Z^2}{A}$$

where Z is the atomic number of a nucleus and A is its mass number. If a nucleus is split into two nuclei, each having atomic number $Z/2$ and essentially equal mass, how will the ratio of the Coulomb energy to the surface energy for the two nuclei compare with that of the parent?

40. About 1 of every 7000 hydrogen isotopes in water is deuterium. (a) If all the deuterium atoms in 1 liter of water are fused in pairs according to the reaction depicted in Eq. 43.42. how many joules of energy would be liberated? (b) Burning gasoline produces about 3.4×10^7 J/liter. Compare the energy obtainable from the fusion of the deuterium in a liter of water with the energy liberated from the burning of a liter of gasoline.

41. Assuming that a deuteron and a triton are at rest when they fuse according to Eq. 43.44, use nonrelativistic conservation principles to determine the energy acquired by the neutron.

42. Using conservation principles and atomic mass data from Table 43.1, verify the fusion energies stated in Eqs. 43.42, 43.43, and 43.44.

appendix 1

TABLE OF SYMBOLS

Mechanics

Quantity	Symbol	Unit	SI Units	Section Reference
Acceleration	**a**	m/s^2	m/s^2	4.4, 5.2
Angle	α, β, γ, θ, ϕ	degree, radian	----	12.1
Angular acceleration	α	$radians/s^2$	s^{-2}	12.1
Angular frequency	ω	radians/s	s^{-1}	12.1
Angular impulse	**A**	N · m · s	N · m · s	13.3
Angular momentum	**L**	$kg \cdot m^2/s$	$kg \cdot m^2/s$	11.1, 12.2
Coefficient of viscosity	η	$N \cdot s/m^2$	kg/m · s	6.8, 16.6
Displacement, length	**s**, *x*	METER	m	1.3, 4.1
Total energy	*E*	joule	$kg \cdot m^2/s^2$	8.3, 20.3
Rest mass energy	E_o	joule	$kg \cdot m^2/s^2$	20.3
Kinetic energy	*K*	joule	$kg \cdot m^2/s^2$	7.4, 12.2
Potential energy	*U*	joule	$kg \cdot m^2/s^2$	8.2
Force	**F**, **f**	newton	$kg \cdot m/s^2$	3.1, 6.3, 6.4
Frequency	ν	hertz	s^{-1}	14.1
Kinetic coefficient of friction	μ_K	----	----	6.8
Linear impulse	**J**	N · s	kg · m/s	9.1
Linear momentum	**p**	kg · m/s	kg · m/s	9.2, 20.2
Mass	*m*, *M*	KILOGRAM	kg	1.5, 6.4, 6.6
Mass density	ρ	kg/m^3	kg/m^3	6.6
Moment of inertia	*I*	$kg \cdot m^2$	$kg \cdot m^2$	12.2, 12.3
Period	*T*	SECOND	s	5.2
Position vector	**r**	METER	m	4.1
Power	*P*	watt	$kg \cdot m^2/s^3$	7.5
Spring constant	*k*	N/m	kg/s^2	6.8, 14.3
Time	*t*	SECOND	s	1.4
Torque	τ	N · m	$kg \cdot m^2/s^2$	3.3, 11.1
Velocity	**v**, **u**	m/s	m/s	4.2, 4.3
Weight	*W*	newton	$kg \cdot m/s^2$	3.1, 6.7
Work	*W*	joule	$kg \cdot m^2/s^2$	7.1–7.3

Mechanics of Solids and Fluids

Quantity	Symbol	Unit	SI Units	Section Reference
Bulk modulus	B	pascal	$kg/m \cdot s^2$	15.5
Buoyant force	B	newton	$kg \cdot m/s^2$	16.1
Compressibility	K	1/pascal	$m \cdot s^2/kg$	15.5
Discharge rate	Q	m^3/s	m^3/s	16.4
Fluid velocity	$\mathbf{u}$	m/s	m/s	16.3
Mass density	ρ	kg/m^3	kg/m^3	15.1, 16.3
Pressure	P	pascal	$kg/m \cdot s^2$	15.2
Reynolds number	R	----	----	16.7
Shear modulus	μ	pascal	$kg/m \cdot s^2$	15.6
Shear strain	θ	----	----	15.2
Shear stress	S	pascal	$kg/m \cdot s^2$	15.2
Spring constant	k	newton/meter	kg/s^2	15.4
Tensile strain	$\Delta l/l$	----	----	15.3
Tensile stress	F_t/A	pascal	$kg/m \cdot s^2$	15.3
Viscosity (dynamic)	η	$N \cdot s/m^2$	$kg/m \cdot s$	16.6
Viscosity (kinematic)	ν	m^2/s	m^2/s	16.6
Volume strain	$\Delta V/V$	----	----	15.5, 16.2
Young's modulus	Y	pascal	$kg/m \cdot s^2$	15.3

Waves

Quantity	Symbol	Unit	SI Units	Section Reference
Amplitude	A	METER	m	17.2
Energy density	u	J/m^3	$kg/m \cdot s^2$	18.4
Frequency	ν	Hz	s^{-1}	17.2
Frequency (radian)	ω	radian/s	s^{-1}	17.2
Intensity	I	W/m^2	kg/s^3	18.4
Intensity level	β	dB	----	18.4
Period	T	SECOND	s	17.2
Phase	ϕ	radian	----	17.3
Wavelength	λ	METER	m	17.2
Wave number	k	radian/m	m^{-1}	17.2

Optics

Quantity	Symbol	Unit	SI Units	Section Reference
Angle of deviation	D	degree, radian	----	39.3
Angular size of image	α	degree, radian	----	39.6
Energy density	u	J/m^3	$kg/m \cdot s^2$	38.5
Focal length	f	METER	m	39.4
Index of refraction	n	----	----	39.1
Intensity	I	W/m^2	kg/s^3	38.5
Magnification	M	----	----	39.6
Phase difference	δ	radian	----	40.3
Resolving power	R	----	----	41.4
Wavelength	λ	METER	m	42.1

Quantum and Nuclear Physics

Quantity	Symbol	Unit	SI Units	Section Reference
Activity	a	----	----	43.2
Activity per unit mass	I	kg^{-1}	kg^{-1}	43.3
Atomic mass unit	u	KILOGRAM	kg	32.3, 43.1
Atomic number	Z	----	----	43.1
Curie	Ci	s^{-1}	s^{-1}	43.2
Flux quantum	Φ_o	Wb, $T \cdot m^2$	$kg \cdot m^2/A \cdot s^2$	42.7
Half-life	$\tau_{1/2}$	SECOND	s	43.2
Mass number	A	----	----	43.1
Neutron number	N	----	----	43.1
Nuclear decay constant	λ	s^{-1}	s^{-1}	43.2
Planck's constant	h	$J \cdot s$	$kg \cdot m^2/s$	42.1
Reproduction constant	k	----	----	43.5
Schrödinger wave function	ψ	$m^{-3/2}$	$m^{-3/2}$	42.5
Work function	W	joule	$kg \cdot m^2/s^2$	42.1

appendix 2

PHYSICAL CONSTANTS*

For computations in most problems, values should be rounded to *three* significant figures.

Quantity	Value	Symbol
Gravitation constant	6.6720×10^{-11} N · m²/kg²	G
Acceleration of gravity at surface of earth (values differ slightly with location)	9.80 m/s²	g
Speed of light in vacuum	2.99792×10^{8} m/s	c
Boltzmann constant	1.38066×10^{-23} J/K 8.61733×10^{-5} eV/K	k
Gas constant	8.31441 J/K	R
Avogadro's number	6.02205×10^{23}	N_A
Joule constant (mechanical equivalent of heat)	4.186 joule = 1 calorie	J
Volume of one mole of ideal gas under standard conditions (Pressure = 1 atm Temperature = 0°C)	2.24138×10^{-2} m³	V_o
Blackbody energy density constant	7.56464×10^{-16} J/m³ · K⁴	a
Coulomb force constant	8.98755×10^{9} N · m²/C²	$k_e = \frac{1}{4\pi\epsilon_o}$
Permittivity of free space	8.85419×10^{-12} C²/N · m²	ϵ_o
Electron charge to mass ratio	1.75880×10^{11} C/kg	e/m_e
Planck's constant	6.62618×10^{-34} J · s 4.13570×10^{-15} eV · s	h
$\frac{\text{Planck's constant}}{2\pi}$ ("h cross")	1.05459×10^{-34} J · s 6.58218×10^{-16} eV · s	$\hbar \equiv \frac{h}{2\pi}$
Ratio of electron charge and Planck's constant	2.41797×10^{12} Hz/volt	e/h

*Based largely on The International System of Units, E. A. Mechtly. NASA report SP-7012 (1973), National Aeronautics and Space Administration, Washington, D.C., and, The Least-Squares Adjustment of the Fundamental Constants, E. R. Cohen and B. N. Taylor. *J. Physical and Chemical Reference Data*, Vol. 2, #4, 663–734 (1973).

Quantity	Value	Symbol
Bohr radius	5.29177×10^{-11} m	a_o
Rydberg constant	$1.097373 \times 10^{7}\ m^{-1}$	R_∞
Classical electron radius	2.81794×10^{-15} m	r_e
Compton wavelength of electron	2.42631×10^{-12} m	λ_c
Fine structure constant	7.2973506×10^{-3}	α
Inverse of fine structure constant	137.03604	α^{-1}
Stefan–Boltzmann constant	$5.67032 \times 10^{-8}\ W/m^2 \cdot K^4$	σ
Electron magnetic moment	9.28483×10^{-24} J/T	μ_e
Electron charge	1.60219×10^{-19} C	e
Faraday constant	9.64846×10^{4} C/mol	F
Permeability of free space	$1.25664 \times 10^{-6}\ T \cdot m/A$	μ_o
Permeability constant	$10^{-7}\ T \cdot m/A$	$\frac{\mu_o}{4\pi}$
Electron rest mass	9.10953×10^{-31} kg	m_e
Unified atomic mass unit	1.660531×10^{-27} kg	u
Proton rest mass	1.67265×10^{-27} kg 1.00728 u	m_p
Hydrogen atom mass	1.67352×10^{-27} kg 1.00782 u	m_H
Neutron rest mass	1.67495×10^{-27} kg 1.00866 u	m_n
Rest mass energy of electron	511,003 eV	m_ec^2
Rest mass energy of one unified atomic mass unit	931.48×10^{6} eV	uc^2
Rest mass energy of proton	938.26×10^{6} eV	m_pc^2
Rest mass energy of neutron	939.55×10^{6} eV	m_nc^2
Rest mass energy of hydrogen atom	938.77×10^{6} eV	M_Hc^2
Ratio of proton mass and electron mass	1836.151	m_p/m_e
Ground state energy of hydrogen (infinite nuclear mass)	13.6058 eV 2.17991×10^{-18} J	E_H
Bohr magneton	9.27408×10^{-24} J/T	μ_B
Proton magnetic moment	1.41062×10^{-26} J/T	μ_p
Nuclear magneton	5.05082×10^{-27} J/T	μ_n
Magnetic flux quantum	$2.06785 \times 10^{-15}\ T \cdot m^2$	Φ_o
Hubble constant	$1.8 \times 10^{-18}\ s^{-1}$ 17 km/s per million light years	H

Astrophysical Data

Quantity	Value
Earth	
Mass	5.98×10^{24} kg
Radius (equatorial)	6.38×10^{6} m
Radius (polar)	6.37×10^{6} m
Sidereal day (time for one rotation with respect to fixed stars)	8.6164×10^{4} s
Solar day (time for one rotation with respect to sun)	8.6400×10^{4} s
Sidereal year (time for earth to complete one orbit about sun, with respect to fixed stars)	3.1558150×10^{7} s
Average distance from sun	1.50×10^{11} m
Moon	
Mass	7.35×10^{22} kg
Radius	1.74×10^{6} m
Sidereal period	2.3606×10^{6} s
Average distance from earth	3.84×10^{8} m
Sun	
Mass	1.99×10^{30} kg
Radius	6.96×10^{8} m
Sidereal period (at equator)	2.1868×10^{6} s
Distance from center of galaxy	3×10^{20} m
Time required for sun to complete one orbit of galaxy	7.5×10^{15} s (250 millions years)
Milky Way Galaxy	
Mass	2.8×10^{41} kg (1.4×10^{11} solar masses)
Diameter (disk)	8×10^{20} m (25,000 parsecs)
Thickness (disk)	6×10^{19} m (2,000 parsecs)
Universe	
Mass	3×10^{51} kg
Radius	10^{26} m
Average mass density	2×10^{-28} kg/m^3
Age (upper limit)	18×10^{9} years
Hubble constant	$1.8 \times 10^{-18}\ \mathrm{s}^{-1} = 17 \dfrac{\mathrm{km/s}}{\mathrm{M\,l.y.}}$

appendix 3

CONVERSION FACTORS

Nearly all calculations in this text employ SI units. This table of conversion factors is designed primarily to convert non-SI units to SI units. Non-SI units are presented in the left column; mostly SI units appear in the right column. To convert non-SI units to SI units, multiply by the conversion factor in the right column.

Y in SI units = Y in non-SI units × conversion factor

Example: Convert 2.12 parsecs to meters

$$Y\ (\text{m}) = 2.12\ \text{parsec} \cdot 3.086 \times 10^{6}\ \text{m/parsec}$$
$$= 6.54 \times 10^{16}\ \text{m}$$

Numbers for the conversion factors were taken from The International System of Units, Physical Constants and Conversion Factors, Second Revision, National Aeronautics and Space Administration Publication SP-7012 (1973).

Length

1 fermi (femtometer, fm)	= 10^{-15} meters (m)
1 angstrom (Å)	= 10^{-10} meters (m)
1 nanometer (nm)	= 10^{-9} meters (m)
1 micrometer (μm)	= 10^{-6} meters (m)
1 centimeter (cm)	= 10^{-2} meters (m)
1 inch (in.)	= 0.0254 meters (m) – Exactly
1 kilometer (km)	= 10^{3} meters (m)
1 mile (mi)	= 1.609×10^{3} meters (m)
1 mile (mi)	= 5280 feet (ft)
1 astronomical unit (AU)	= 1.496×10^{11} meters (m)
1 light year (l.y.)	= 9.461×10^{15} meters (m)
1 parsec	= 3.086×10^{16} meters (m)

Volume

1 gallon (U.S. liquid)	= 3.785×10^{-3} cubic meter (m^3)

Time

1 minute (min)	= 60 seconds (s)
1 hour (hr)	= 60 minutes (min)
1 hour (hr)	= 3600 seconds (s)
1 sidereal day	= 86,164 seconds (s)
1 sidereal year (yr)	= 3.156×10^{7} seconds (s)

Speed

1 foot/second	= 0.3048 meter/second (m/s)
1 mile/hour	= 0.4470 meter/second (m/s)
1 kilometer/hour	= 0.2778 meter/second (m/s)

Mass

1 atomic mass unit (u)	= 1.66053×10^{-27} kilograms (kg)
1 gram (gm)	= 10^{-3} kilograms (kg)

Force

1 pound (lb)	= 4.448 newtons (N)
1 dyne (dyn)	= 10^{-5} newtons (N)

Pressure

1 pascal (Pa)	= 1 newton/meter^2 (N/m^2)
1 pound/inch^2 (lb · in^{-2}, psi)	= 6.895×10^{3} newtons/meter^2 (N/m^2, Pa)
1 millimeter of mercury at 0°C	= 1.333×10^{2} newtons/meter^2 (N/m^2, Pa)
1 torricelli (torr)	= 1.333×10^{2} newtons/meter^2 (N/m^2, Pa)
1 bar	= 10^{5} newtons/meter^2 (N/m^2, Pa)
1 atmosphere (atm)	= 1.013×10^{5} newtons/meters (N/m^2, Pa)
1 atmosphere (atm)	= 14.70 pounds/inch^2 (lb/in^2, psi)

Energy

1 electron volt (eV)	$= 1.602 \times 10^{-19}$ joules (J)
1 erg	$= 10^{-7}$ joules (J)
1 foot · pound (ft · lb)	= 1.356 joules (J)
1 calorie (cal)	= 4.186 joules (J)
1 British thermal unit (Btu)	$= 1.055 \times 10^{3}$ joules (J)
1 kilocalorie (kcal)	$= 4.186 \times 10^{3}$ joules (J)
1 kilowatt · hour (kWh)	$= 3.600 \times 10^{6}$ joules (J)

Power

1 foot · pound/second ($ft \cdot lb \cdot s^{-1}$)	= 1.356 watts (W)
1 horsepower (hp)	= 746 watts (W)

Plane Angle

1 degree (deg)	$= 1.745 \times 10^{-2}$ radians
1 minute (min)	= 60 seconds (s)
1 degree (deg)	= 60 minutes (min)
1 degree (deg)	= 3600 seconds (s)

Solid Angle

1 sphere = 4π steradians = 12.57 steradians

Magnetic Field

1 gauss (G) = 10^{-4} tesla (T)

appendix

4 MATHEMATICS

Powers of 10

$10^m \cdot 10^n = 10^{m+n}$
$10^m/10^n = 10^{m-n}$
$(10^m)^n = 10^{m \cdot n}$
$10^3 = 1000$ $\quad 10^6 = 1,000,000$
$10^{-3} = 0.001$ $\quad 10^{-6} = 0.000,001$

Geometry

Area of a circle πr^2
Area of a sphere $4\pi r^2$
Volume of a sphere $\frac{4}{3}\pi r^3$
$\pi = 3.141592\ldots$

Algebra

The roots of a quadratic equation:

$ax^2 + bx + c = 0$ are $x = \dfrac{-b \pm \sqrt{b^2 - 4ac}}{2a}$

$x^2 + 2\beta x + \gamma = 0$ are $x = -\beta \pm \sqrt{\beta^2 - \gamma}$

Trigonometry

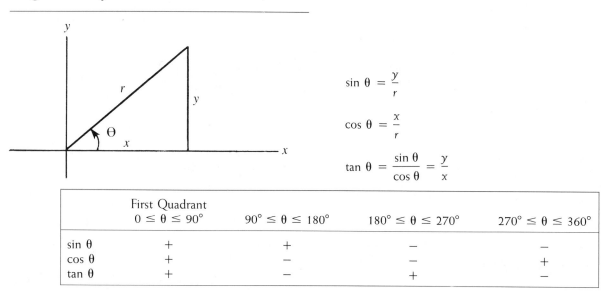

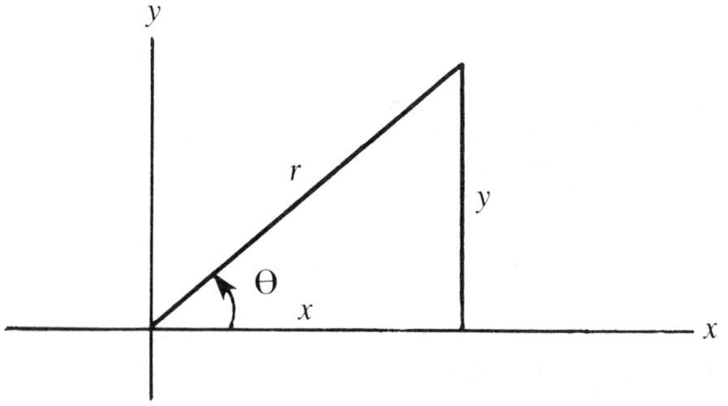

$$\sin\theta = \frac{y}{r}$$

$$\cos\theta = \frac{x}{r}$$

$$\tan\theta = \frac{\sin\theta}{\cos\theta} = \frac{y}{x}$$

	First Quadrant $0 \le \theta \le 90°$	$90° \le \theta \le 180°$	$180° \le \theta \le 270°$	$270° \le \theta \le 360°$
$\sin\theta$	+	+	−	−
$\cos\theta$	+	−	−	+
$\tan\theta$	+	−	+	−

Binomial Expansion

$$(1 \pm x)^n = 1 \pm nx + \frac{n(n-1)}{1 \cdot 2}x^2 \pm \frac{n(n-1)(n-2)}{1 \cdot 2 \cdot 3}x^3 + \ldots$$

The series terminates and becomes a polynomial if n is a positive integer. If n is a fraction or a negative integer, the expansion is an infinite series and we must require $|x| < 1$ for convergence.

Special Cases

$n = \frac{1}{2}$ $\quad (1 \pm x)^{1/2} = 1 + \frac{1}{2}x - \frac{1}{8}x^2 \pm \frac{1}{16}x^3 - \ldots$
$n = -\frac{1}{2}$ $\quad (1 \pm x)^{-1/2} = 1 \mp \frac{1}{2}x + \frac{3}{8}x^2 \mp \frac{5}{16}x^3 + \ldots$
$n = -1$ $\quad (1 \pm x)^{-1} = 1 \mp x + x^2 \mp x^3 + \ldots$
$n = -\frac{3}{2}$ $\quad (1 \pm x)^{-3/2} = 1 \mp \frac{3}{2}x + \frac{15}{8}x^2 \mp \frac{35}{16}x^3 + \ldots$

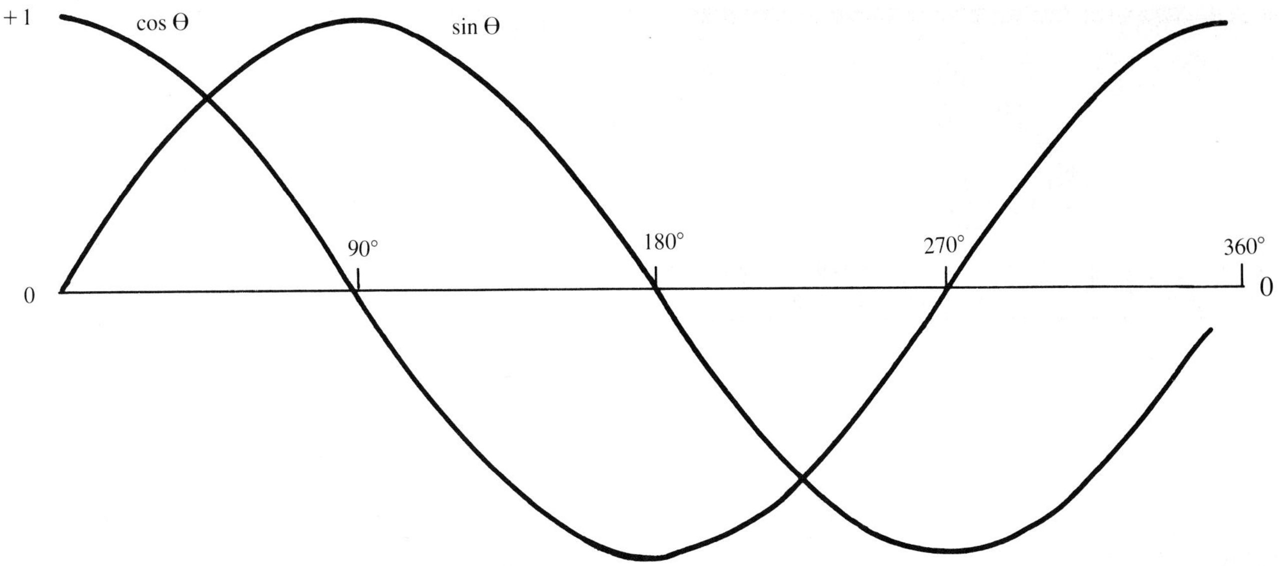

Special Values

	0°	30°	45°	60°	90°
sin θ	0	$\frac{1}{2}$	$\frac{1}{\sqrt{2}} = 0.707$	$\frac{\sqrt{3}}{2} = 0.866$	1
cos θ	1	$\frac{\sqrt{3}}{2} = 0.866$	$\frac{1}{\sqrt{2}} = 0.707$	$\frac{1}{2}$	0
tan θ	0	$\frac{1}{\sqrt{3}} = 0.577$	1	$\sqrt{3} = 1.732$	∞

Special Properties

$\sin(-\theta) = -\sin\theta \qquad \cos(-\theta) = +\cos\theta$

$\sin\left(\theta \pm \frac{\pi}{2}\right) = \pm\cos\theta \qquad \cos\left(\theta \pm \frac{\pi}{2}\right) = \mp\sin\theta$

$\sin(\theta \pm \pi) = -\sin\theta \qquad \cos(\theta \pm \pi) = -\cos\theta$

Trigonometric Formulas

$\sin^2\theta + \cos^2\theta = 1$ — This corresponds to the Pythagorean theorem $x^2 + y^2 = r^2$ for a right triangle.

$\sin 2\theta = 2\sin\theta\cos\theta$

$\cos 2\theta = \cos^2\theta - \sin^2\theta$

$\sin(\alpha \pm \beta) = \sin\alpha\cos\beta \pm \cos\alpha\sin\beta$

$\cos(\alpha \pm \beta) = \cos\alpha\cos\beta \mp \sin\alpha\sin\beta$

$\sin\alpha \pm \sin\beta = 2\sin\frac{1}{2}(\alpha \pm \beta)\cos\frac{1}{2}(\alpha \mp \beta)$

$\cos\alpha + \cos\beta = 2\cos\frac{1}{2}(\alpha + \beta)\cos\frac{1}{2}(\alpha - \beta)$

$\cos\alpha - \cos\beta = 2\sin\frac{1}{2}(\alpha + \beta)\sin\frac{1}{2}(\beta - \alpha)$

$\sin\alpha\sin\beta = \frac{1}{2}[\cos(\alpha - \beta) - \cos(\alpha + \beta)]$

$\cos\alpha\cos\beta = \frac{1}{2}[\cos(\alpha - \beta) + \cos(\alpha + \beta)]$

$\sin\alpha\cos\beta = \frac{1}{2}[\sin(\alpha - \beta) + \sin(\alpha + \beta)]$

$\sin\theta + \sin 2\theta + \sin 3\theta + \ldots + \sin N\theta$

$$= \frac{\sin[\frac{1}{2}(N + 1)\theta]\sin(\frac{1}{2}N\theta)}{\sin(\frac{1}{2}\theta)}$$

Law of Cosines

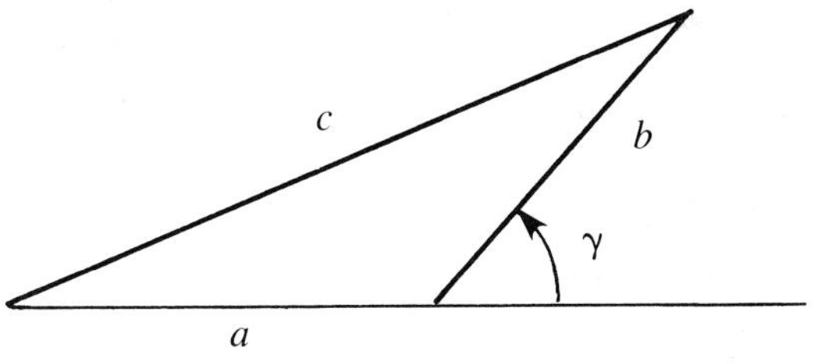

$c^2 = a^2 + b^2 + 2ab\cos\gamma$
For $\gamma = 90°$ (right triangle)
$c^2 = a^2 + b^2$, Pythagorean theorem

Series Expansions

$$\sin x = x - \frac{x^3}{3!} + \frac{x^5}{5!} - \ldots$$

$$\cos x = 1 - \frac{x^2}{2!} + \frac{x^4}{4!} - \ldots$$

where $n! \equiv 1 \cdot 2 \cdot 3 \cdot \ldots (n - 1) \cdot n$

Exponentials, Logarithms

$e = 2.718281\ldots$

$e^{\ln x} = x$

$e^{-\ln x} = 1/x$

$e^{0} = 1$

$e^{\infty} = \infty$

$e^{-\infty} = 0$

$$e^{\pm x} = 1 \pm x + \frac{x^2}{2!} \pm \frac{x^3}{3!} + \ldots$$

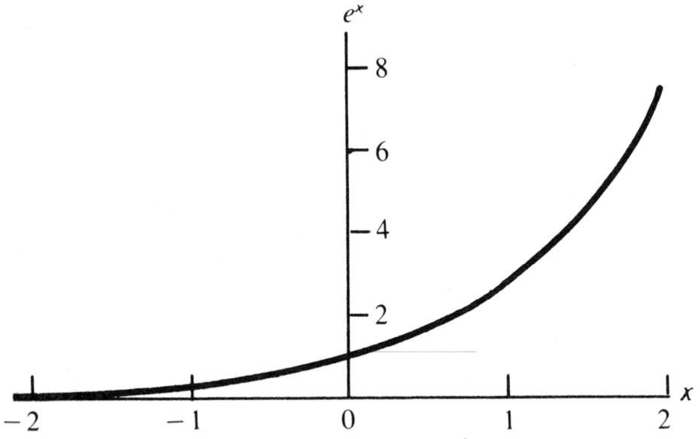

$\ln e = 1$

$\ln e^x = x$

$\ln(xy) = \ln x + \ln y$

$\ln(x/y) = \ln x - \ln y$

$\ln a = 2.3026 \log_{10} a$

$\log_{10} a = 0.43429 \ln a$

$\ln 1 = 0$

$\ln 0 = -\infty$

$\ln \infty = \infty$

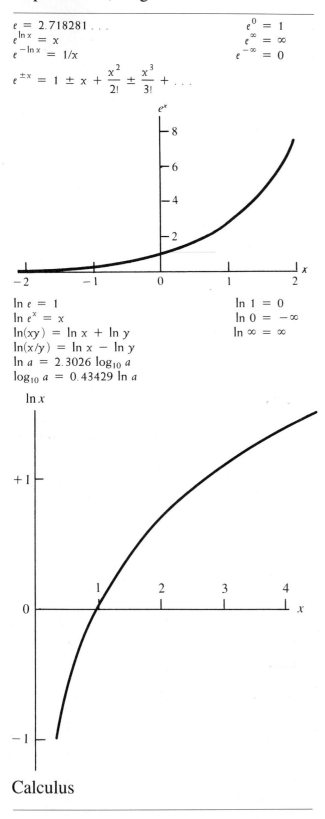

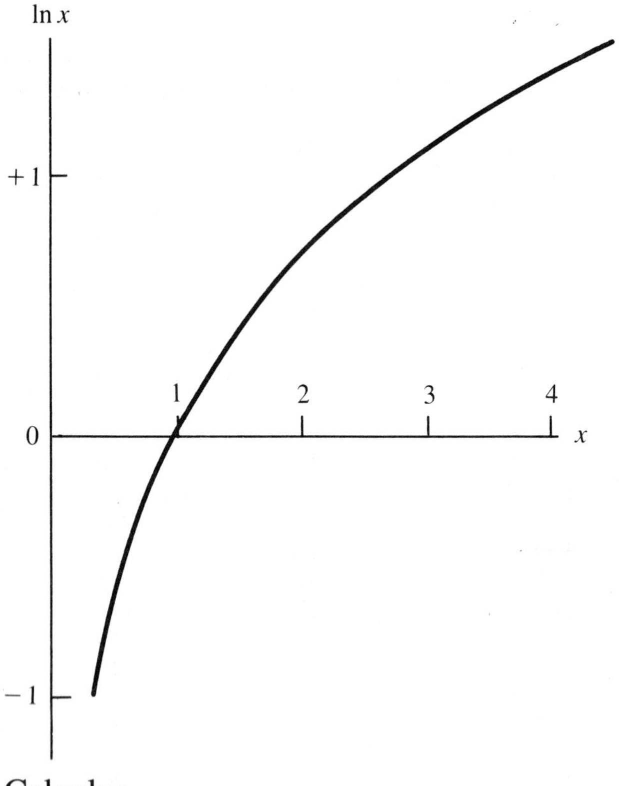

Calculus

$$\frac{d}{dx}[f(x)\,g(x)] = \frac{df}{dx}g + f\frac{dg}{dx}$$

If $f(u)$ is a function of u and $u = u(x)$ is a function of x,

$$\frac{d}{dx}f(u) = \frac{df(u)}{du} \cdot \frac{du(x)}{dx} \quad (\textit{chain rule}).$$

Taylor Series

$$f(x_o + h) = f(x_o) + h\frac{df}{dx}\bigg|_{x_o} + \frac{h^2}{2!}\frac{d^2f}{dx^2}\bigg|_{x_o} + \ldots$$

This Taylor series may be used to derive the sin x, cos x, e^x series and the binomial expansion given previously.

$$\int_a^b f(x)\frac{dg(x)}{dx}\,dx = f(x)\,g(x)\Big|_a^b - \int_a^b \frac{df(x)}{dx}\,g(x)\,dx$$

$$\frac{d}{dx}(x^n) = nx^{n-1}$$

$$\frac{d}{dx}(x^2 \pm a^2)^n = 2\,nx(x^2 \pm a^2)^{n-1}$$

$$\frac{d}{dx}(\sin \omega x) = \omega \cos \omega x$$

$$\frac{d}{dx}(\cos \omega x) = -\omega \sin \omega x$$

$$\frac{d}{dx}(e^{ax}) = a\,e^{ax}$$

$$\frac{d}{dx}(\ln x) = \frac{1}{x}$$

Indefinite Integrals

$$\int x^n\,dx = \frac{x^{n+1}}{n+1} + C,\ n \neq -1$$

$$\int \frac{dx}{x} = \ln x + C$$

$$\int \frac{dx}{(x^2 \pm a^2)^{3/2}} = \frac{\pm x}{a^2(x^2 \pm a^2)^{1/2}} + C$$

$$\int \frac{dx}{\sqrt{a^2 - x^2}} = \arcsin\left(\frac{x}{a}\right)$$

$$\int \frac{dx}{x^2\sqrt{x^2 - a^2}} = \frac{\sqrt{x^2 - a^2}}{a^2 x}$$

$$\int \frac{dx}{x^2 + a^2} = \frac{1}{a}\tan^{-1}\left(\frac{x}{a}\right)$$

$$\int \sin ax\,dx = -\frac{1}{a}\cos ax + C$$

$$\int \cos ax\,dx = \frac{1}{a}\sin ax + C$$

$$\int e^{ax}\,dx = \frac{1}{a}e^{ax} + C$$

$$\int \sin^2 ax\,dx = \frac{x}{2} - \frac{\sin 2ax}{4a}$$

$$\int \cos^2 ax\,dx = \frac{x}{2} + \frac{\sin 2ax}{4a}$$

Handbook of Chemistry and Physics, 61st Ed., CRC Press, 2000 N.W. 24th Street, Boca Raton, FL 33431.
H. B. Dwight, *Tables of Integrals and Other Mathematical Data*, 4th Ed., New York: The Macmillan Co., 1961.

appendix 5

THE ELEMENTS

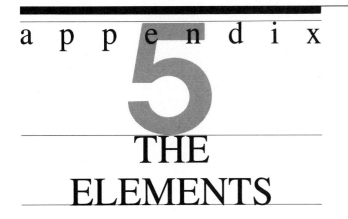

Name	Symbol	Atomic Number	Name	Symbol	Atomic Number
Actinium	Ac	89	Mercury	Hg	80
Aluminum	Al	13	Molybdenum	Mo	42
Americium	Am	95	Neodymium	Nd	60
Antimony	Sb	51	Neon	Ne	10
Argon	Ar	18	Neptunium	Np	93
Arsenic	As	33	Nickel	Ni	28
Astatine	At	85	Niobium	Nb	41
Barium	Ba	56	Nitrogen	N	7
Berkelium	Bk	97	Nobelium	No	102
Beryllium	Be	4	Osmium	Os	76
Bismuth	Bi	83	Oxygen	O	8
Boron	B	5	Palladium	Pd	46
Bromine	Br	35	Phosphorus	P	15
Cadmium	Cd	48	Platinum	Pt	78
Calcium	Ca	20	Plutonium	Pu	94
Californium	Cf	98	Polonium	Po	84
Carbon	C	6	Potassium	K	19
Cerium	Ce	58	Praseodymium	Pr	59
Cesium	Cs	55	Promethium	Pm	61
Chlorine	Cl	17	Protactinium	Pa	91
Chromium	Cr	24	Radium	Ra	88
Cobalt	Co	27	Radon	Rn	86
Copper	Cu	29	Rhenium	Re	75
Curium	Cm	96	Rhodium	Rh	45
Dysprosium	Dy	66	Rubidium	Rb	37
Einsteinium	Es	99	Ruthenium	Ru	44
Erbium	Er	68	Rutherfordium*	Rf	104
Europium	Eu	63	Kurchatovium**	Ku	104
Fermium	Fm	100	Samarium	Sm	62
Fluorine	F	9	Scandium	Sc	21
Francium	Fr	87	Selenium	Se	34

Name	Symbol	Atomic Number	Name	Symbol	Atomic Number
Gadolinium	Gd	64	Silicon	Si	14
Gallium	Ga	31	Silver	Ag	47
Germanium	Ge	32	Sodium	Na	11
Gold	Au	79	Strontium	Sr	38
Hafnium	Hf	72	Sulfur	S	16
Hahnium*	Ha	105	Tantalum	Ta	73
Helium	He	2	Technetium	Tc	43
Holmium	Ho	67	Tellurium	Te	52
Hydrogen	H	1	Terbium	Tb	65
Indium	In	49	Thallium	Tl	81
Iodine	I	53	Thorium	Th	90
Iridium	Ir	77	Thulium	Tm	69
Iron	Fe	26	Tin	Sn	50
Krypton	Kr	36	Titanium	Ti	22
Lanthanum	La	57	Tungsten	W	74
Lawrencium	Lw	103	Unnilhexium***	Unh	106
Lead	Pb	82	Uranium	U	92
Lithium	Li	3	Vanadium	V	23
Lutetium	Lu	71	Xenon	Xe	54
Magnesium	Mg	12	Ytterbium	Yb	70
Manganese	Mn	25	Yttrium	Y	39
Mendelevium	Md	101	Zinc	Zn	30
			Zirconium	Zr	40

*American-proposed name
**Soviet-proposed name
***No name proposed as yet

ANSWERS TO ODD-NUMBERED PROBLEMS

Chapter 1

1. (a) 4×10^7 m; (b) 6.37×10^6 m
3. (a) 3.80×10^8 m; (b) 2.53 s
5. 1650.76373
7. (a) Nineteen years is approximately 6.00×10^8 s.
(b) Atomic clock precision is about 1 part in 10^{12}, or about 6×10^{-4} s for a 19-yr time interval.
9. (a) 5.95×10^{24} kg; (b) 5.52×10^3 kg/m^3
11. m^3/kg · s^2
13. $v \sim a^{1/2} x^{1/2}$
15. (a) $v \sim r^{-1/2}$; (b) $T \sim r^{3/2}$
17. 49.7 mi/hr
19. 1 acre = 0.405 hectare
21. 9.81 m/s^2
23. 224 s = 3 min 44 s
25. 53.6 m/s^2

Chapter 2

3. Analytic results: $|\mathbf{F}_1 + \mathbf{F}_2| = 9.54$ N, 57° above $+x$-axis
5. Minimum = 0 N, maximum = 200 N.
7. (b) Plane with respect to air (188, 68.4) km/hr; air with respect to ground (10.4, −59.1) km/hr; plane with respect to ground (198, 9.3) km/hr
9. B = 7.81; $\alpha = 59.2°$; $\beta = 39.8°$; $\gamma = 67.4°$
11. $\hat{\mathbf{r}} = (0.866, 0.500, 0)$
13. (b) $v_x = 2.73$ km/hr; $v_y = 0$
15. (−3, 3, 0)
17. (7.66, −4.23, 4.85)
19. $v_x = 12.3$ m/s; $v_y = 17.0$ m/s
21. 52.3 mi/hr
23. 1
25. 7
27. 516 N·m
31. $\mathbf{M}\cdot\mathbf{N} = 13$; $\mathbf{M} \times \mathbf{N} = -7\mathbf{i} + 16\mathbf{j} - 10\mathbf{k}$
33. 5.20**k**
35. $1.98 \times 10^{-15}\mathbf{i} - 2.08 \times 10^{-14}\mathbf{j}$ N
37. $-5.86\mathbf{i} - 1.25\mathbf{j} + 99.1\mathbf{k}$

Chapter 3

1. (b) 10.1 cm; (c) 11.2 cm
3. 1.04 cm
5. 0.01 N
7. 100 N
9. T = 283 N; P = 200 N
11. (b) $T_2 = T_1 + f$; (c) $T_1 = T_2 + f$; (d) $T_3 = W_1 + W_2 + F$
13. (b) $T_1 = T_2 = 1440$ N; (c) $T_3 = 1420$ N
15. (a) $T_1 = T_2 = 7.64$ N; (b) $T_1 = 8.76$ N, $T_2 = 6.46$ N
17. 160 N·m, counterclockwise
19. (a) 20 N·m, counterclockwise; (b) 10 N; (c) + for (a), − for (b); (d) 2 m for both
21. (a) 900 N; (b) 90 N·m
23. (a) 8.91 N; (b) 5.35 N·m
25. (a) All but the vertical components of the wall force have zero lever arms and hence exert zero torque. Because the net torque must be zero the vertical component of the wall force must be zero. (b) 866 N; (c) 1000 N
27. (a) Max and Min for both T_1 and T_2 are 1050 N and 250 N; (b) 4 m
31. T denotes tension in horizontal beams, C the compression in the upper (long) beams, and S the cable tension. In terms of the weight, W = 200,000 N,
T = W/(4 tan θ), C = W/(4 sin θ), S = W/2.
(a) T = 86,600 N, C = 100,000 N, S = 100,000 N
(b) T = 28,900 N, C = 57,700 N, S = 100,000 N
35. H = 50 N, V = 467 N, $T_1 = 583$ N
37. (a) center of middle ball; (b) $4R/5$ to left of center of middle ball; (c) $8R/5$ to left of center of middle ball
39. $x = y = z = 0.75$ cm
41. $x = 1.75$ cm, $y = 0.75$ cm
43. Remains stationary. Maximum restoring torque exceeds maximum tipping torque.
45. (a) $\tan\theta = a/b$

Chapter 4

1. (a) For (0, 0, 0), $\mathbf{r}_i = (10, -2, 7)$ m, $\mathbf{r}_f = (8, -3, 8)$ m; for (2, 1, 3), $\mathbf{r}_i = (8, -3, 4)$ m, $\mathbf{r}_f = (6, -4, 5)$ m. (b) $\Delta\mathbf{r} = \mathbf{r}_f - \mathbf{r}_i = (-2, -1, 1)$ m for both observers.
3. First base: $\mathbf{r} = 90\mathbf{i}$ ft; second base: $\mathbf{r} = 90(\mathbf{i} + \mathbf{j})$ ft; third base: $\mathbf{r} = 90\mathbf{j}$ ft
5. (a) (13, 4) m; (b) (1, −12) m
7. (a) 65.4 km/hr at 23.4° north of east; (b) 65.4 km/hr; (c) 86 km/hr
9. (Approximate answers deduced from graph.) (a) 0.58 m/s; (b) 1.2 m/s
11. (a) −0.81 m/s; (b) −1.22 m/s; (c) −1.54 m/s
13. (Approximate answers deduced from graph.) (a) 0.8 m/s; (b) ±0.1 m/s
15. **NOTE ERRATUM**: Reference should be to Fig 4.8, not Fig. 4.7. (Approximate answer deduced from graph.) −55 m/s
17. 1.18 m/s
19. (Approximate answers deduced from graph.) (a) 0.14 km/min; (b) 1.1 km/min; (c) 1.2 km/min; (d) 0.14 km/min
21. (a) 4 m/s^2; (b) 12 m/s^2
23. 29.5 m/s^2; (b) 24.5 m/s^2
25. first gear, 13 mi/hr·s
27. 0.8 m/s^2
29. (a) $t = v_o/g$; (b) height $= v_o^2/2g$; (c) $t = v_o/g$, same as rise time; (d) $-v_o$, same speed but opposite direction as initial velocity.
31. (a) 120 m/s; (b) 240 m/s; (c) 3 m/s^2
33. 405 m
35. Only choice (d) has the correct dimension of time.
37. 180 frames/s
39. (a) $4.9[2/(1 - 0.81) - 1]$ m = 46.7 m; (b) 19 s; (c) 2.46 m/s; (d) 4.90 m/s
43. (a) 20 s; (b) 278 m; (c) 100 km/hr
45. (a) 9.80 m/s^2; (b) 46.5 m/s; (c) 110 m
47. (a) $2(t^2 + 1)^{1/2}$ m/s directed at $\theta = \sin^{-1}(t^2 + 1)^{-1/2}$ measured from across-stream direction toward downstream direction. (b) 2 m/s^2, across-stream direction
49. (a) 50 mi/hr, 53.1° north of east; (b) 50 mi/hr, 53.1° south of west
51. 45 mi/hr

Chapter 5

3. (a) 8.94 m/s; (b) 63.4° clockwise from $+x$-direction
9. 24.2 m/s
11. 1.60 km, 28.9 s
13. 2.99° and 87.01°
15. 63.9°, 27.3 m/s
17. (a) 28.0 m/s; (b) 1.01 s
21. 1.41 hours
23. 1.48 m/s^2
25. 0.224 m/s^2

Chapter 6

1. Not force-free motion because average velocity changes.
3. 9.8 N
5. 8.66 N, east
7. (a) $a_{max} = 25$ m/s^2, east; $a_{min} = 5$ m/s^2, west; (b) 41.4° north of west, 41.4° south of west; (c) 7.2° west of north, 7.2° west of south
9. $190\mathbf{i} - 310\mathbf{j}$ m/s^2
11. (a) 6 N, east; (b) 6 kg
13. $F_1 = 9$ N, $F_2 = 6$ N
15. (a) −2.5 m/s^2; (b) 2500 N
17. 9.83×10^{-23} m/s^2
19. 9.35 m/s^2
21. (a) $F_{moon}/F_{earth} = 3.38 \times 10^{-6}$, $F_{sun}/F_{earth} = 6.01 \times 10^{-4}$; (b) decrease, 0.483 N
23. (a) 346,000 km; (b) 261,000 km
25. $r_{IV} = 3.30 \times 10^6$ m
27. 1210 kg
29. (a) 3.9×10^{17} kg/m^3; (b) 1.3×10^{12} m/s^2; (c) 0.60 ms; (d) 1.2×10^8 m/s
31. 30 lbs
33. 3.58×10^7 m = 22,200 mi
35. (a) 475 km; (b) 24.7 lbs
37. (a) 100 N/m; (b) 1 m/s^2; (c) −6 m/s^2
39. 210 N = 47.1 lbs
41. 0.598 m
43. (a) 28.4 m; (b) 256 m
47. (a) −9 m/s^2; (b) −9000 N; (c) 0.918; (d) −1080 N; (e) 417 m
49. 37 m
51. 43 years
53. water, 1.0×10^{-3} kg/m·s; glycerin, 0.74 kg/m·s; sucrose, 2700 kg/m·s
55. (a) $T_2 = 18$ N, $T_1 = 6$ N; (b) $T_2 = 18$ N, $T_1 = 12$ N
57. (a) 14.3°; (b) 4.42 s
59. 3 N
63. (a) 491 N; (b) 50 kg; (c) 2.00 m/s^2
65. 6.01×10^{24} kg
67. 35,800 km = 22,200 mi
69. (b) 1.05×10^{32} kg

Chapter 7

1. (a) 0.0328 J; (b) −0.0328 J
3. (a) zero; (b) -5.13×10^4 J; (c) 5.13×10^4 J
5. (a) zero; (b) 78.4 J; (c) 78.4 J; (d) −78.4 J; (e) −78.4 J
7. (a) 25 J; (b) −25 J; (c) zero
9. (a) −74 kJ; (b) −74.1 kJ
11. (a) 0.53 J; (b) 0.20 J; (c) 0.60 J; (d) zero
13. (a) 6 J; (b) 9 J; (c) −9 J; (d) zero; (e) zero
15. (a) -1.4×10^{-21} J; (b) -1.37×10^{-21} J
17. (a) positive; (b) negative; (c) zero
19. (a) $Gm_1m_2[(1/r_2) - (1/r_1)]$; (b) $Gm_1m_2[(1/r_1) - (1/r_2)]$; (c) zero
21. (a) −5 J; (b) −5 J; (c) −5 J

23. 38.9 km/hr
25. 5.93×10^7 m/s
27. 2.66×10^{33} J
29. (a) $v(0) = 32.4$ m/s, $v(10) = 16.2$ m/s, $v(20) = 0$, $v(30) = -16.2$ m/s, $v(40) = -32.4$ m/s; (b) $K(0) = 262$ J, $K(10) = 65.6$ J, $K(20) = 0$, $K(30) = 65.6$ J, $K(40) = 262$ J
31. (a) 30 cm; (b) 0.049 J; (c) -0.0245 J; (d) 0.0245 J; (e) 0.0245 J
33. (a) 1.34×10^7 m/s; (b) 1.23×10^7 m/s; (c) 1.03×10^7 m/s; (d) 3.79×10^{-14} m
35. 221 m
37. (a) 4.43 m/s; (b) 7.67 m/s; (c) 7.67 m/s; (d) 3.43 N
39. (c) 0.864 m/s; (d) 14 m/s; (e) 14 m/s, same as (d)
41. 2.34×10^9 J
43. 48.1 hp
45. (a) $v = 2.5t$ m/s (t in s); (b) 25 kW; (c) $3125t^2$ J; (d) 50 kW; (e) 25 kW; (f) 50 kW
47. 746 N = 167 lbs
49. (a) 2.30 kW; (b) 62.2 kW
53. $mg/4$
55. $r_2 = 0.235$ m

Chapter 8

3. (a) zero; (b) -10 J; (c) 10 J; (d) -10 J
5. (a) 825 J; (b) -825 J; (c) zero
7. 10^3 N/m
9. 8.22 J
11. -1.45 MJ
13. (b) 33.9 J; (c) 2.61 kJ; (d) 8.92 kJ
15. -1.92×10^{-57} J
17. (b) 6.24×10^{10} J; (c) 1.27×10^7 m
19. Analytic results: (a) $r_B = 1.22 \times 10^{-14}$ m, $(r_A - r_B) \to \infty$; (b) $r_B = 1.08 \times 10^{-14}$ m, $r_A - r_B = 8.92 \times 10^{-14}$ m; (c) $r_B = 7.56 \times 10^{-15}$ m, $r_A - r_B = 1.24 \times 10^{-14}$ m
21. Analytic results: 3.66×10^{-10} m and 4.46×10^{-10} m
23. 0.447 m/s
25. (a) 9.07 m/s; (b) $\theta_1 = 180°$, $\theta_2 = \theta_3 = 90°$
27. 3.61 m/s
29. (a) 20.4 m; (b) 147 m
31. (a) 8.28 m/s; (b) 10.4 m/s
33. (a) 22.3 m/s; (b) 28.4 m/s
37. 618 km/s
39. (a) 1.7×10^{-21} J; (b) 2.4×10^{-21} J; (c) 1.0×10^{-21} J
41. (b) $F(-3) = 30$ N, $F(1) = -10$ N, $F(2) = -20$ N
43. 9.6×10^{-11} N ($r = 3.5 \times 10^{-10}$ m), -9.5×10^{-12} N ($r = 4.4 \times 10^{-10}$ m), -2.0×10^{-12} N ($r = 6 \times 10^{-10}$ m)
45. (a) 5.18 mJ; (b) 0.0864 N
47. 63.0 N, directed down the plane
49. $2a/r^3$
51. (a) 4×10^{-10} m; (b) 3.93×10^{-10} m
53. (Approximate values deduced from graph.) $r = 2.4$ m, stable; $r = 5$ m, unstable
55. $x = 0$ is unstable. Displacement to right or left of $x = 0$ gives rise to a force that leads to unstable motion to the left ($x < 0$).
57. (Approximate results deduced from graph.) Turning points: r = 1.2 m and r = 3.3 m; maximum kinetic energy = 17 J
59. (Approximate values deduced from graph.) (a) 0.6 m, 4.3 m, 6.2 m; (b) allowed: $0.6\text{m} \le r \le 4.3$ m, $r \ge 6.2$ m. Otherwise, forbidden.
61. (a) particle experiences net force; (b) bounded, oscillatory; (c) x_C; (d) x_B, x_D

Chapter 9

1. 12 N·s
3. 2.5 N·s
5. 60 N·s
7. (Approximate result deduced from graph.) 160 N·s
9. $-kA/\omega$
15. (a) $-\omega mA$; (b) zero
19. 48.8 m/s
23. (a) 12 N·s; (b) 2.40 m/s
25. $\frac{1}{2}J(\mathbf{i} + \mathbf{j}\sqrt{3})$
27. (a) away from spacecraft; (b) 0.503 m/s
29. $-v(\mathbf{i} + 2\mathbf{j})$
31. 8.64 m/s, 40.6° north of $+x$-axis
33. 4.5 km/hr
39. 11.2 units
41. (a) -1; (b) $-8/9$; (c) $-16/25$; (d) $-48/169$; (e) $-92/576$
43. 301 m/s
45. (a) $2v$; (b) zero

Chapter 10

1. position 1
3. (a) (0,0) m; (b) (0,0) m; (c) (0.50, 0) m
5. 2.88×10^{-14} m from proton, toward electron.
7. $x_{cm} = 0$, $y_{cm} = -0.543d$
9. Center of mass is on longer rod, 18 cm from weld.
11. $(0, 4.70 \times 10^{-11}, 3.15 \times 10^{-11})$ m
13. $a/4$ directly above center of base.
15. (a) 9 m; (b) 1 m, opposite to displacement of man.
17. 79,900 kg
19. 2.45 cm
21. 2.32×10^5 m/s, opposite to alpha velocity
23. (a) 310 km/s; (b) -1.70 mm
25. $v_{center} = 3.6$ m/s, $v_{goalie} = -4.4$ m/s
27. (a) before = after = 160 J; (b) before = after = 80 J
29. (a) zero; (b) 2.32×10^5 m/s
31. 0.030
37. 63.3°
39. $P_{boy} = (3/4)mu_o$; $P_{girl} = (5/12)mu_o$; $P_{car} = -\frac{1}{2}mu_o$ (when boy jumps; does not include momentum of girl); $P_{car} = -(7/6)mu_o$ (when girl jumps)
41. 0.632
43. 6.91 km/s

Chapter 11

1. 5.31×10^7 kg·m²/s
3. $L = 3.33 \times 10^6$ kg·m²/s for all angles. (a) $r \rightarrow \infty$ as $\theta \rightarrow 90°$; (b) $r = 141$ m; (c) $r = 100$ m
5. $3.53\ \hbar$
7. 2.68×10^{40} kg·m²/s
15. $-22.0\mathbf{k}$ kg·m²/s
17. (a) 0.0522 kg·m²/s; (b) 0.608 m/s
19. 2.4×10^3 kg·m²/s, out of page. The angular momentum is the same for origins at A, B, and C.
23. (a) 16**k** kg·m²/s; (b) 16**k** kg·m²/s; (c) 16**k** kg·m²/s
25. 7.78**i** m/s
27. Length of day = 27.2 hrs (13.3% increase)
29. 0.0123
31. **NOTE ERRATA**: Asume that the cord between the crates remains along the radial direction, so that the crates orbit the earth with equal periods. The equation should read $T \cong (3/2)GM_E mh/r_1^3$ (*Hint:* The tension is less than the difference between the gravitational forces because the crates have different accelerations.) (b) 4.22×10^{-5} N; (c) 1.67×10^{-13} N; (d) 4.16×10^9 m

Chapter 12

1. 8.11°; (b) 0.142 rad
3. 377 rad/s
5. hour hand: 1.45×10^{-4} rad/s; minute hand: 1.75×10^{-3} rad/s; second hand: 0.105 rad/s
7. (a) 0.447 rad/s; (b) -0.0318 rad/s²; (c) 14.1 s
9. $\pi - 2\theta$
11. 0.443 m/s
13. 31.4 rad/s²
15. 105 revolutions
19. 6.25×10^{-3}
21. 1.05×10^{-34} kg·m²/s
23. 4.43×10^{-47} kg·m²
25. (a) 9.72 kJ; (b) 9.72 kJ
27. earth: 2.58×10^{29} J; moon: 3.15×10^{23} J
29. (a) $\frac{1}{2}ma^2$; (b) $\frac{1}{2}ma^2$; (c) ma^2
31. 310 rad/s
33. $I_{square} = 2ML^2/3$; $I_{cross} = 4ML^2/3$
35. $2MR^2$
37. (a) 1.07×10^{40} kg·m²; (b) 1.08×10^{40} kg·m²
39. $(1/12)M(a^2 + b^2)$
41. $I_x = \frac{1}{2}MR^2(1 - 8/\pi^2)$; $I_y = \frac{1}{2}MR^2$; $I_z = MR^2(1 - 4/\pi^2)$
49. (a) 0.0105 kg·m²; (b) 0.0105 kg·m²
53. (a) $\omega/3$; (b) $K_f = K_i/3 = (1/12)MR^2\omega^2$
55. (a) 31.4 rad/s; (b) 126 kg·m²/s; (c) 1.5 s

Chapter 13

1. (a) 60 N·m; (b) 1.57 rad; (c) 94.2 J; (d) 12.1 rad/s
3. 12.8 hours
5. 1630 rad/s
7. (a) 245 J; (b) 234 N·m; (c) 23.4 kW
9. (a) 7.29×10^{-5} rad/s; (b) -8.46×10^{-15} rad/s
11. (a) 2.15×10^{16} N·m; (b) 1.57 TW
13. 2/5
15. 2/7 ($K_f = 5K_i/7$)
17. 7.43 R
19. 0.729
23. 390 rad/s = 62.1 rev/s
25. 0.209 rad/s

Chapter 14

1. (a) 2.45 N/m; (b) 35 rad/s; (c) 4.08 cm
3. 62 gm
5. (a) 0.00508%; (b) 0.510%; (c) 57.1%
7. (a) 38.6 N/m; (b) 1.32 kg
9. 0.707 A
13. 0.384 J
17. (a) 4.2; (b) 0.5 rad/s; (c) 12.6 s; (d) 3.14 rad
19. (a) 1.53 s; (b) 0.581 m
21. (a) $T = 2\pi(2R/g)^{1/2}$; (b) 0.898 s ($R = 0.1$ m), 8.98 s ($R = 10$ m)
25. 3.36 N/m
27. 6,370 km (1 earth radius above surface)
29. frequency = 1 Hz; length = 0.248 m
33. (a) 5.84×10^8 mi/yr; (b) 3.67×10^9 mi/yr²
37. $A = 4.22 \times 10^8$ m; $\omega = 3.55$ rad/day
39. **NOTE ERRATA**: In line 3 the damping constant units are kg/s, not s^{-1} In (e) and (f), ignore the words "average" and "per cycle." (a) 9.80 m; (b) 0.513 m/s; (c) 9.21 J; (d) 3.29 J; (e) -2.63×10^{-3} W; (f) -9.40×10^{-4} W
41. (a) underdamped; (b) 0.790 kg/s
45. (a) 2 s; (b) one oscillation for one particle and two for the other
47. 2.79 s

Chapter 15

1. 1 $kg/m^3 = 10^{-3}$ gm/cm^3
3. (a) 924 $cm^3 = 9.24 \times 10^{-4}$ m^3; (b) 6.04 cm
5. For a mass of 80 kg and a density equal to that of water (10^3 kg/m^3) the volume is 0.080 m^3.
7. (a) 43.8 psi; (b) 2.98 atm; (c) 3.02×10^5 Pa
9. 2.22×10^{-5}
11. -0.03
13. (a) 4.99×10^{14} m^2; (b) 4.99×10^{19} N; (c) 5.09×10^{18} kg; (d) 1.06×10^{44}
15. 7.6×10^{11} Pa
17. 6.6 cm
19. (a) 0.254 N; (b) 1.53×10^{-3} J
21. (a) 2.26×10^3 N; (b) 1.70 J
25. $F/mg = 2.6 \times 10^3$
27. $B = 3.4 \times 10^8$ Pa; $K = 3.0 \times 10^{-9}$/Pa
29. Approximately 6%
31. 392 atm
33. 2.1×10^{-3}

Chapter 16

1. (a) 2.99×10^{-29} m^3; (b) 3.34×10^7; (c) No. Ten molecules represent less than 1 part per million of the number in the volume.
5. floats between Cognac and Curacao
7. volume = 90 m^3; mass of helium = 16.1 kg. A typical 1-car garage has a volume of approximately 150 m^3, about the same as the 90 m^3 blimp.
9. 1.50×10^7 Pa (including 0.01×10^7 Pa from atmospheric pressure)
11. 220 km
13. 1.96×10^5 Pa = 1.94 atm; gauge pressure. (Absolute pressure on diver is approximately one atmosphere greater.)
15. 7.9 km (Mt. Everest rises 8.89 km above sea level.)
17. (a) 1.25 m/s; (b) 9.8×10^{-5} m^3/s; (c) 0.098 kg/s
19. 0.4 m/s
21. 1.6 m^3/s
23. 1.42×10^5 Pa
25. 9.8×10^4 Pa
27. 4.50×10^5 Pa = 4.46 atm
29. (a) On axis of pipe, $y = 0$; (b) on pipe surface, $y = R$; (c) on pipe surface, $y = R$.
33. (a) 2.5×10^{-3} cm/s; (b) 4×10^7 s = 463 days
35. 3.75 m/s
37. 970, laminar
41. 25,000, turbulent

Chapter 17

1. a, b, c
3. amplitude = 1.80 cm; wavelength = 1.02 m
5. 30 cm in the $+x$ direction
7. 429 nm
9. wavelength = 5.03 m; speed = 2.40 m/s
13. (a) -0.20 radian; (b) 1.79 radian
15. 0.868 radian
17. 325 nm
19. The x' origin is 0.70 m to the left of the x origin.
21. (a) $3\sin(\pi x/L) \cdot \cos(\pi t/T) = (3/2)\sin(\pi x/L + \pi t/T) + (3/2)\sin(\pi x/L - \pi t/T)$
(b) wavelength = 6 m; (c) period = 4 s; (d) frequency = 0.25 Hz; (e) velocities = +1.5 m/s and -1.5 m/s
25. 254 Hz and 253 Hz
27. 432 Hz to 448 Hz
29. (a) 535 Hz; (b) 445 Hz
31. $\nu = \nu_0(1 + u/v)$; (b) Eq. 17.32 gives 21.9 Hz. whereas the result of (a) gives 20.9 Hz.

Chapter 18

3. (a) 77.5 m/s; (b) 0.103 s
5. 115 m/s
11. 422 m/s
13. 3.83 m/s = 8.56 mph
15. For Al, $(v_l - v_t)/v_t = 1.03$; for Cu, $(v_l - v_t)/v_t = 1.00$.
17. 147 nm to 7.35 cm
19. 0.052 mm
21. 10^7 J/m^3
23. (a) 10^{-2} W/m^2; (b) 3.33×10^{-11} J/m^3
25. (a) 100 W/m^2; (b) 200 Pa = 1.98×10^{-3} atm
27. (b) 3.8×10^{18} kWh; 3.9×10^{-7} times U.S. consumption
29. (a) 3.98×10^{-7} W/m^2; (b) 7.96×10^{-7} W/m^2
31. 1.38×10^{11}, approximately 28 times the population of the earth
35. 63.4 m/s
37. Approximately 2×10^{-3} Hz

Chapter 19

1. 8×10^{-14} s
3. (b) $v/c < 0.128$; (c) approximately 85×10^6 mph
7. 1.5c (Galilean); 0.96c (Einstein, SRT)
9. 0.357c
13. $L = (30 - 3.35 \times 10^{-11})$ m; contraction = 3.35×10^{-8} mm
19. **NOTE ERRATUM**: In the last line the exponent should be +1/2 not −1/2. 1.54 ns slow
21. 3.61×10^{-10} s; 11.4 ms/yr
23. No, it can travel only about 4.8 km in 16 μs.

Chapter 20

1. 1000
3. (a) 1.52×10^8 m/s; (b) 1.36×10^8 m/s
5. (a) 1.76×10^{17} m/s^2; (b) 2.96×10^8 m/s
7. 6.19×10^{11} m/s^2
11. $0.99999726c$
13. (a) 0.9938; (b) 4
15. $v/c < 0.082$
19. 1.42 MeV/c $= 7.58 \times 10^{-22}$ kg·m/s
21. (a) 1770 MeV; (b) 0.848
25. 2×10^{-10}
29. 600 MeV
33. Comparing the solutions for (c) and (d) with the exact solutions given in Problem 20.34 shows the effects of assuming $(1 - v_p^2/c^2)^{1/2} \simeq 1$. (b) $v_\pi/c = 0.617$; (c) $v_p/c = 0.1159$; (d) $K_\pi = 0.271m_\pi c^2 = 37.8$ MeV, $K_p = 0.0456m_\pi c^2 = 6.4$ MeV

Chapter 21

1. (a) 375.3 K, 374.2 K, 373.7 K; (b) 373.1 K (linear extrapolation)
3. -459.67°F
5. 574.59 K = 574.59°F
7. The numerator and denominator both approach zero. Their ratio approaches a finite limit.
9. 100 quads = 263 megatons
11. 0.234 kJ/kg·C°
13. 1.68×10^{15} kJ, 52.3 yr
15. 225 kJ
17. 3.22×10^3 m^3
19. 1.8×10^{-3} C°; "splashed" water acquires "ordered" kinetic energy
21. 21 C°
23. 1.92×10^6 J
25. 1.2×10^7 J; -1.2×10^7 J during compression
27. 1.86×10^5 J
29. 2.2×10^7 J
31. 7.00 kJ
33. (a) $W < 0$; counterclockwise area greater than clockwise area. (b) $W < 0$; same reason (as a). (c) $W = 0$; equal areas, one corresponding to + work, other to − work.

Chapter 22

1. 2.171 cm
3. 116°C; (b) −100°C
5. Upward. Zinc expands more than aluminum.
7. 86,390 (in one *solar* day of 86,400 s)
9. $V_i = 1.11 \times 10^{-2}$ m^3; $V_f = 1.97 \times 10^{-2}$ m^3
11. (a) 1.20×10^{-2} m^3; (b) 2.4×10^{-2} m^3; (c) 1.82×10^{-2} m^3; (d) 222 K
13. 80.5 K. This is below the boiling point of oxygen, implying that some oxygen in the air might liquefy. Because air at a pressure of 100 atm does not behave like an ideal gas, there may or may not be any liquefaction of oxygen.
15. $T_f = 193$ K, $P_f = 0.859$ atm
21. (a) 33.4 s; (b) 105 s; (c) 564 s
23. 17.9 W
25. 878 kJ
27. 561 kJ/kg
29. (a) 2.25×10^8 J; (b) yes; (c) yes. The available kinetic energy is approximately 80 times the energy needed to vaporize the comet. Thus, the comet would be vaporized and super-heated steam would burn the surroundings.

Chapter 23

1. 20.4 kW
3. 16 cm
5. 45°C
7. Heat flow through the thinner window is greater by a factor of 3.25.
9. (a) 63.1 W/m; (b) 50 W/m; (c) 43.1 W/m
11. (q/L) for 1 cm, 3 cm, 10 cm are in the ratios 1 : 0.50 : 0.29
13. 90 W
15. Approximately 1400 Calories
17. Greater T for silver paint; 423 K
19. 544 kcal, about 18% of the 3000 kcal daily intake.
21. 24.0 W/m^2
23. Approximately 2200 K
25. $2.8\ R_{\text{earth}}$

Chapter 24

1. Don't invest. Claim violates law of conservation of energy.
3. (a) 60 kJ; (b) 0.40
5. 1/3
7. (a) 15 kJ; (b) 1/3
9. 18.9 kWh
11. 0.21
13. (a) 120 kJ; (b) 4
15. (a) $1 \rightarrow 2$, 693 kJ absorbed; $2 \rightarrow 3$, no heat exchange; $3 \rightarrow 4$, 478 kJ ejected; $4 \rightarrow 1$ no heat exchange; (b) 0.319
17. Reducing T_L by 25 K produces greater increase in efficiency.
19. 0.120
21. No. The Carnot efficiency is 0.014, and is an upper limit to actual efficiency for engines operating between 0°C and 4°C.
23. 967 K
25. 4.88 kJ/kg·K
27. 3.77 J/K
29. 202 gm
31. 1.73 kJ/K
33. (a) 20°C = 293 K; (c) $\Delta S = 4.88$ J/K > 0; (d) mixing is irreversible, as implied by the $\Delta S > 0$ result of (c).

Chapter 25

1. (a) 1.99×10^{-26} kg; (b) 1.67×10^{-27} kg
5. 0.05 Pa
7. Approximately 7 m
9. 4.2 mm
11. $v_{mean} = 2$ km/s; $v_{rms} = 2.16$ km/s
13. $v_{rms} = 2.22$ km/s $\simeq (1/5)v_{escape}$. A small but significant fraction are moving outward at speeds greater tha v_{escape} and thus escape. Over millions of years any particular hydrogen atom has a good chance of escape.
15. (a) 1.96×10^7 m/s; (b) 2.82×10^3 m/s
17. (a) $v_{proton} = 1.57 \times 10^6$ m/s; (b) $v_{deuteron} = 1.11 \times 10^6$ m/s; (c) SRT not required; kinetic energy less than 0.1 % of rest mass energy
21. $v_{rms} > v_{mp} > v_{sound}$
23. (a) 300 K; (b) 100 K
25. The Maxwellian distributions are the same for all four because all have the same value of m/T.
27. (a) 82; (b) 610; (c) 760; (d) zero

Chapter 26

3. Approximately 6×10^8
5. (a) 2.62×10^{24}; (b) 2.39 per billion
7. $(+5.82 \times 10^{-6}$ C and -2.31×10^{-6} C) or $(-5.82 \times 10^{-6}$ C and $+2.31 \times 10^{-6}$ C)
9. 1.05×10^{-5} C
11. 0.999 m
13. (a) 8.22×10^{-8} N; (b) 2.19×10^6 m/s
15. 8.34×10^9 N $= 9.36 \times 10^5$ tons
17. 1.16×10^3 N, repulsive
19. $q_{earth} = -5.15 \times 10^{14}$ C; $q_{sun} = 1.71 \times 10^{20}$ C; (b) 8.61×10^{-11} C/kg
21. (a) 82.9 N; (b) 4.16×10^{27} m/s^2
25. $F_e/F_g = 4.83 \times 10^{-11}$
27. 0.018 N, directed at 45° above $+x$-axis
29. (a) -4 μC; (b) unstable
33. (a) $\mathbf{F} = -0.018[(1-y)^{-2} + (1+y)^{-2}]\,\mathbf{j}$ N, $|y| < 1$ m
    $\mathbf{F} = +0.018[(y-1)^{-2} - (y+1)^{-2}]\,\mathbf{j}$ N, $y > 1$ m
    (b) $\mathbf{F} = -0.036(1 + x^2)^{-3/2}\,\mathbf{j}$ N
35. $\theta = 81.7°$, $x = 0.939a$
37. (b) period $= 0.0788(m/q_o)^{1/2}$
39. $[k_e q_o \lambda L/a(a - L)]\,\mathbf{i}$
41. $(2k_e q_o \lambda/y_o)\,\mathbf{j}$
43. (a) $\mathbf{F} = (k_e q_o Q x/a^3)\,\mathbf{i}$; (b) period $= 2\pi(ma^3/k_e Q|q_o|)^{1/2}$

Chapter 27

1. (a) 5.58×10^{-11} N/C, down; (b) 1.02×10^{-7} N/C, up; (c) impossible; neutron has no electric charge
3. (a) 2.0×10^{-6} C/kg; (b) 2.49×10^{-13} electrons per atom
5. (a) 5×10^5 N/C; (b) negative x-direction
7. 5.90×10^{-3} N/kg
9. (b) $x = +a_o/\sqrt{2}$
13. (b) E at the midpoint is four times E at point B.
19. (a) $E_x = 6qa^2/4\pi\epsilon_o x^4$; (b) $E_y = -3qa^2/4\pi\epsilon_o y^4$
21. 1660
23. 4.34×10^3 N·m^2/C
25. 0.149k N·m^2/C
29. 28.2 N·m^2/C
31. 1.35×10^{-7} C
33. (a) 1.33×10^{-9} C/m^2; (b) 8.31×10^5/cm^2
37. (a) 2.93×10^{21} N/C; (b) 1.18×10^{17} N/C
41. $E = kr^2/4\epsilon_o$, $r \leq R$; $E = kR^2/4\epsilon_o r^4$, $r \geq R$. Field is radial.
43. 4.77×10^4 N·m^2/C
47. (b) $E = \lambda/2\pi\epsilon_o r_b$ radial
49. 1.96×10^{10} m/s^2, downward
51. (a) $v = 5.93 \times 10^5\sqrt{E}$ m/s; (b) 2.56×10^3 N/C
53. $s = mv^2/2eE$
55. 5.37×10^{-11} m
57. 1.2×10^{-20} N·m
59. (a) $10^{-2} \sin\theta$ N·m; (b) 2×10^{-2} J
61. (a) 4.94×10^{-26} N; (b) 1.65 m/s^2

Chapter 28

1. (a) 8×10^{-3} J; (b) 8×10^{-3} J; (c) 8×10^{-3} J; (d) -4×10^4 V
5. (a) $-2aqE$; (b) $2aqE$
7. (a) 1.28×10^3 V/m; (b) 25.6 V
9. 1.36×10^5 V/m; (b) 3.25 mm
11. (a) 10^4 V/m; (b) $-2000x$ V/m
13. 0.08 J/s
15. (a) 1.60×10^{-18} J, 1.88×10^6 m/s; (b) 1.60×10^{-13} J, 2.82×10^8 m/s
17. (a) 0.99999999967; (b) 6.21×10^5 V/m
19. Zero work
21. (a) 0; (b) 0; (c) 5.27×10^3 V
23. 1.15×10^{-13} J $= 0.719$ MeV
27. $(2eQ/4\pi\epsilon_o m_p a)^{1/2}$; m_p = proton mass
29. 0.0270 J
33. $(\sigma/2\epsilon_o)[(x^2 + a^2)^{1/2} - |x|]$; $\sigma = Q/\pi a_o^2$ = charge density
35. 7.86×10^6 V
37. (a) $E_r = aq\cos\theta/\pi\epsilon_o r^3$; (b) dipole aligned along z-axis. Rotation of coordinate system (or observer's position) about z-axis changes ϕ but leaves V unchanged, so V cannot depend on ϕ.
41. (a) $V(r_2) - V(r_1) = (\lambda/2\pi\epsilon_o)\ln(r_2/r_1)$; (b) no. There is an infinite charge on the line. Infinite work is required to move a test charge from finite r to infinity. Hence, choosing $V = 0$ at $r \to \infty$ would make $V(r)$ infinite for all finite values of r.
43. 2.70 m
44. (a) 6.70 mm; (b) -25 V
47. (a) $r = 8.99 \times 10^9$ m $= 1410\ r_{earth}$; (b) $\sigma = 9.85 \times 10^{-22}$ C/m^2, 6.15×10^{-3} protons/m^2
49. (a) 1.77×10^{-9} C/m^2; (b) 200 V/m

Chapter 29

1. 0.1 pF
3. (a) 1 μC; (b) 5 μC; (c) 25 μC
5. 709 μF
7. (a) 0.09 C; (b) 300 V; (c) capacitor allows short duration flash
9. 1.13×10^8 m^2
11. 45.8×10^{-3} F
17. $C = C_1C_2C_3C_4/(C_1C_2C_3 + C_1C_2C_4 + C_1C_3C_4 + C_2C_3C_4)$
19. 4.43 μF
21. 45 μF
23. 480,000
25. 10
27. $C_{min} = C_{series} = 2$ pF; $C_{max} = C_{parallel} = 72$ pF
29. (a) 8 V; (b) 128 μC; (c) 5.33 μF
31. (a) 21.4 μJ; (b) 72 μJ; (c) 288 μJ
33. Both are doubled
35. 55.9 μC, 4.65 V
39. 0.0737 J (half of which is stored in the capacitor)
43. 1.83×10^{20} V/m; 1.49×10^{29} J/m^3
45. Total charge conserved. Total linear momentum conserved. Math analogs.
47. 0.0376 μF
49. 1.06 kJ/m^3
51. Five 6 pF in series gives joint C of 1.2 pF. Each of the five can sustain 200 V, so the series combination can withstand 1000 V.
53. 10.1 V
55. 0.133 μC; 0.620 μJ
57. (a) 500 J; (b) 5 J
61. 19.7 pF/ft, within 10% of quoted value
63. (a) Same potential drop over slab and air space, indicative of parallel combination.

Chapter 30

1. 6.25×10^{18}/s
3. 152 C
5. 1.74×10^3 A
7. 8.44×10^{11} protons/s
9. 0.282 mm
11. 1.57×10^{-5} A
13. 2.08 mC
15. Rate = Mnv
19. 0.06 ohm
21. 1.57 A
23. $d_{Al} = 1.28$ mm
25. $d_1/d_2 = 1/2$
27. 4.36×10^{-3} ohm, ignorably small
29. 4.52 Ω
31. 10^{-14} $(\Omega \cdot m)^{-1}$
33. 1.13 V/m
37. (a) (Celsius degrees)$^{-1}$ or K^{-1}
39. (a) added term gives one more adjustable parameter (β) to fit data. (b) (Celsius degrees)$^{-2}$ or K^{-2}; (c) 6×10^{-8}/(Celsius degrees)^2
41. (c) $R_{Fe} = 9.1$ ohms, $R_C = 90.9$ ohms
43. (Approximate answers deduced from graph.) (a) $R = 1100$ ohms, $dV/dI = 290$ ohms; (b) $R = 470$ ohms, $dV/dI = 67$ ohms
45. (a) 10^4 ohms; (b) -10^3 ohms
47. (a) 5×10^7 W; (b) 55 J
49. 0.773 m
51. (a) 133 ohms; (b) 9.40 m
53. (a) 3.6×10^5 J; (b) 100 W
55. $I_{Al} = 55.6$ A, $I_{Cu} = 273$ A, $P_{Al} = 55.6$ W, $P_{Cu} = 273$ W
57. (a) 6.25×10^{12}; (b) 6 J; (c) 6 W
59. 15.8 mA
61. 22.4 mA and 22.4 V
63. 0.162 mm

Chapter 31

1. $R_{ext} = 10$ ohms; $R_{int} = 4$ ohms
3. (a) 2 A; (b) 0.5 ohm
5. (a) 4.32 MJ; (b) could melt and vaporize
7. (b) ohm/ampere; (c) $I = -(R/2k) + [(R/2k)^2 + \mathcal{E}/k]^{1/2}$
9. (Approximate answers deduced from graph.) (b) $\mathcal{E} = 10$ V, $r = 1.1$ ohms
13. 3.33 kΩ and 6.67 kΩ
15. 25 kΩ
17. Single bulb brightens. Bulb in parallel dims.
19. (b) $N = 1$, 50%, $N = 10$, 9.1%, $N = 100$, 0.99%
21. starter, 171 A; battery, 0.28 A
23. 0.327 A
25. $I_2 = 0.8$ A, $I_4 = 0.2$ A, $I_5 = 0.4$ A, $I_7 = 1$ A
27. (a) 11.1 V; (b) 17.0 W; (c) 0.8 V increase and 2.3 W decrease
31. (b) one; (c) $I_1 = I_3 = 5$ A, $I_2 = 0$
33. (a) $I_1 = 1.42$ A, $I_2 = -0.16$ A, $I_3 = 1.26$ A
(b) $V_1 = 5.68$ V, $V_2 = -0.32$ V, $V_3 = 6.32$ V
(c) $P_1 = 8.08$ W, $P_2 = 0.05$ W, $P_3 = 7.98$ W
(d) $P_{in} = 12I_1 + 6I_2 = 16.1$ W $= P_1 + P_2 + P_3$
35. (a) 4 A; (b) 8 A; (c) −2 A
37. (7/5)R, (2/3)R, (3/4)R
41. 0.23 s
43. 40%
45. 6.4 μF
49. 3 J
51. (Approximate answer deduced from graph.) 7.2 megohms

Chapter 32

1. 0.10 T in negative y-direction; $\mathbf{B} = -0.10\mathbf{j}$ T
3. 1.59×10^{-13} N
7. vertically downward (the direction of $\mathbf{F} \times \mathbf{v}$)
9. (a) $-(8.0\mathbf{k} + 12.8\mathbf{i}) \times 10^{-19}$ N; (b) $-(4.79\mathbf{k} + 7.66\mathbf{i}) \times 10^{8}$ m/s
11. 2.84×10^{-12} N (nonrelativistic treatment of motion)
13. 71 km; no (not directly, possibly if trapped).
15. 5×10^{5} m/s
17. 4.27 cm
19. 1.24×10^{-2} T
21. (a) 0.1%; (b) 0.2%
23. (a) $p_{alpha} = 2p_p$; (b) $v_{alpha} = v_p/2$; $K_{alpha} = K_p$
25. 2.80×10^{11} Hz
27. 21.2 J
29. 0.101 T
31. 0.461 T
33. 2.5 T
35. (b) 2.48 mm/s; (c) $5.86 \times 10^{28}/\text{m}^3$
37. 0.112**i** N
39. 7.1 J
41. (b) 0.243 radian = 13.9°
45. (a) 1,3; (b) 2,6 (stable, near zero), 1,3 (unstable, near zero)

Chapter 33

5. $\sqrt{2}\,\mu_o I/\pi a$
7. (a) $2\mu_o I a^2/\pi z^3$; (b) field has dipole character ($\sim z^{-3}$) with magnetic moment $\mu = I \cdot A = 4Ia^2$
11. 1.51×10^{-6} T·m²
13. (a) 1.88×10^{-4} T; (b) ±7.66 cm
15. $8I/3$, counterclockwise
17. (a) $\mu_o I/4a$; (b) $\mu_o I/4a$
19. $B = (\mu_o I s_o/2\pi r_o r_1)$
21. 2.0×10^{19} A·m²
23. (a) $8\pi \times 10^{-7}$ T·m; (b) $-8\pi \times 10^{-7}$ T·m; (c) zero
25. (b) $-\mathbf{j}(\mu_o I/\pi)[a/(a^2 + y^2)]$
27. $-\mathbf{i}(\mu_o I/\pi)[y/(a^2 + y^2)]$
29. (a) zero; (b) $(\mu_o I/2\pi r)[(r^2 - r_1^2)/(r_2^2 - r_1^2)]$; (c) $\mu_o I/2\pi r$
31. 260 A
35. The integral $\int_{-\infty}^{+\infty} B_x(x,0,0)\,dx$ can be interpreted as $\oint \mathbf{B} \cdot d\mathbf{s}$ around a path that consists of the x-axis (from $-\infty$ to $+\infty$) and a semicircular path of infinite radius (along which $B = 0$). This path encloses the current I, and in accord with Ampere's, law one verifies $\oint \mathbf{B} \cdot d\mathbf{s} = \mu_o I$.
37. 1.80 T
39. 12 layers; total length approximately 121 m.
45. (a) 9.6×10^{-6} N, directed toward a long wire; (b) zero ($\boldsymbol{\mu}_{rectangle}$ and $\mathbf{B}_{wire}$ are parallel)
47. 0.78 A
49. 13,000 N (≃ 1.5 tons)

Chapter 34

3. (a) 1.04 V; (b) 1.04 V; (c) 9.63 mA, clockwise
5. 1.24×10^{-4} V
7. (a) 0.880 T·m²; (b) −0.0273 V
9. (a) B decreases with distance from the wire and so the flux through the loop decreases as the coil moves away. (b) Left ends are positive, right ends are negative. Induced current is counterclockwise. (c) $[\mu_o I\, l\, w\, v/2\pi r(r + w)]$
11. (a) 47.1 V; (b) 471 V; (c) 471,000 V
13. (a) 1.57×10^{-3} T·m², 2.51×10^{-3} V; (b) -2.51×10^{-3} V; (c) 2.09 mA; (d) 5.26×10^{-6} W
15. 0.458 T/s
17. −0.588 V
19. (a) zero; (b) 1.26 V; (c) zero; (d) −1.26 V; (e) zero. A fringing field would "smooth" the emf discontinuities.
23. (a) +; (b) −; (c) −; (d) +; (e) no current; (f) −
25. (a) 0.239 T; (b) 0.126 C
27. 0.5 A
29. (a) $0.008 \cos 377t$ T·m²; (b) $-d\Phi_B/dt = \mathscr{E} = 3.02 \sin 377t$ T·m²/s or volts; (c) $3.02 \sin 377t$ A; (d) $9.10 \sin^2 377t$ W; (e) $0.0241 \sin^2 377t$ N·m
31. (a) $6.40 \sin 62.8t$ V; (b) $5.81 \sin 62.8t$ A; (c) $37.2 \sin^2 62.8t$ W

Chapter 35

1. 2.3 mH
3. 1 mH
5. $I = -20t +$ constant A (t in seconds)
7. 0.396 V
11. −1.67 mV
13. 2280 Wb (or T·m²)
15. $L_{toroid} = L_{solenoid} = \mu_o N^2 A/l = \mu_o n^2 Al$
19. (a) 0; (b) 3.34 A
21. 96 mH
23. $t = 0+$, $I_1 = I_2 = 3$ mA; $t \to \infty$, $I_1 = 6$ mA, $I_2 = 0$
25. $R_1 = 6$ ohms, $R_2 = 18$ ohms
27. 2.52 s
29. (a) 2.30; (b) 4.61
31. (a) 1.95×10^{7} J/m³; (b) 2.76×10^{3} J
35. $(L/2R)\ln 2 = 0.347(L/R)$
37. $u_M = \mu_o I^2 a^4/8(a^2 + x^2)^3 \underset{x \gg a}{\longrightarrow} \mu_o I^2 a^4/8x^6$
39. (a) $U_M/l = (\mu_o I^2/4\pi)\ln(R/a)$; (b) 3.32×10^{-7} J/m, ratio of energy per meter outside/inside = 9.21
41. Critically damped
43. 1.78 kHz
45. (a) 12 J; (b) 12 J

Chapter 36

1. (a) 5.03×10^{-3} T; (b) decreases by 5.43×10^{-8} T
3. 206, ferromagnetic
5. 0.0122 T
7. $B = 4.79 \times 10^{-3}$ T, $\mu = 0.0362$ J/T
9. (Approximate answer deduced from graph.) 1.05 mC
11. 2.36 J/T (equivalent units are $A \cdot m^2$)
13. Based on Eq. 36.13 [$\mu_I = \chi_m \mu$] and $\chi_m = a\mu$, $B = \mu_o(1 + a\mu)\mu/V$. Based on the more general relation, $d\mu_I = \chi_m d\mu$, $B = \mu_o(1 + \frac{1}{2}a\mu)\mu/V$; $\chi_m = a\mu$, a = constant.
15. 161 T/m
17. 10,700 A
19. (a) $\frac{1}{2}N + 250$ in one direction, $\frac{1}{2}N - 250$ in opposite direction; (b) $N \geq 500$; (c) no, other orientations of oppositely directed dipoles can yield same total magnetic moment.
21. 0.30 K
23. (Approximate answer deduced from graph.) 7×10^{-26} $J \cdot K/T^2$
27. 560
29. (Approximate answers deduced from graph.) $B_r = 0.72$ T; $B_c = 9.42 \times 10^{-4}$ T

Chapter 37

1. $\omega BA \sin \omega t = \omega BA \sin \theta$
3. $I_{av} = 0$; $I(t) = 13.6 \cos 377t$ A; $I_{rms} = 9.58$ A
5. $P_{min} = 0$; $P_{max} = 1.15$ kW; $P_{av} = 575$ W
7. $I_{min} = 0$; $I_{max} = 2.59$ A
9. (a) 145 ohms; (b) 1.17 A
11. 5.00 mV
15. 44.2 GHz = 4.42×10^{10} Hz
23. 0.254 A
25. $I_{max} = 0.682$ mA; $I_{rms} = 0.482$ mA
29. $V = \mathcal{E} \cos \omega t$
31. (a) 167 Hz; (b) 0.955 mH.
33. 120 V
37. 2.17 kHz
39. $V_R = 8.16$ V; $V_C = 5.77$ V
41. $I_C = 4.52$ A; $I_R = 0.12$ A
    At sufficiently low frequencies the inductor does not affect the current. With $L = 0$ the current is $I = (\mathcal{E}/R) \cos \omega t$.
45. 99.7 mH
47. 5.50 W
51. 0.101 H
53. $P_{res} = 50$ W; $P(\nu = 0.9\nu_r) = 49.5$ W
55. 199 kHz
57. 112.1 MHz, 112.9 MHz
59. 9.3% decrease
61. 2.58×10^4
65. $I = 200$ A; $V_{primary} = 498{,}000$ V; $P_{lost} = 0.442$ MW; $P_{consumer} = 99.6$ MW

Chapter 38

1. 429 m (WLW), 3.39 m (WMUB)
3. 3×10^{18} Hz
5. (a) 5.78×10^{14} Hz; (b) 519 nm, green
7. 3.14×10^{10} radians/s
17. (a) 4.97 kW/m^2; (b) 1.66×10^{-5} J/m^3
19. Between 10^5 J and 10^6 J. For an absorbing area of 0.4 m^2 the calculated result is approximately 6×10^5 J.
21. $E_o = 1.37 \times 10^9$ V/m; $B_o = 4.57$ T
23. (a) 10^{12} W/m^2; (b) 3.33 kJ/m^3; (c) 27.4 MV/m; (d) 0.0915 T

Chapter 39

1. (a) 40°; (b) 28.9°
3. 45°
7. 5.66 mm
9. 1.414
11. (b) decrease; (c) no
13. 1.414
15. (b) Yes. It must be at a level half between the eyes and the top of the head. (c) Decreased; (d) increased
17. Closer to the mirror surface than F
19. 1.06 m
21. 2.0 m behind mirror; virtual, upright
23. $f = 0.42$ m; $r = 0.84$ m
25. Both images are virtual and upright. The images are located 0.62 m behind the mirror (for $s = 2.1$ m) and 0.36 m behind the mirror (for $s = 0.62$ m).
29. 15.9 cm
31. 14.4 cm
35. (a) 6.22 cm; (b) 20.2 cm; (c) −7.37 cm
37. For $s = 8$ cm, real image is formed 9.14 cm inside glass. For $s = 1$ cm virtual image is formed 2.29 cm in front of glass–air interface.
39. 0.17 cm
41. 6 mm

Chapter 40

1. 6.50 radians
3. 635 nm
5. 25.3 cm
7. 1.1×10^{-3} radian, 2.7 mm
9. $x_2 - x_1 = \lambda(m + 1/48)$
11. (a) 398 nm
13. 88.6 nm
15. (a) Waves reflected from the front and rear surfaces of the film differ in phase by π and therefore interfere destructively. (b) 1—blue, 2—green, 3—red; (c) 98 nm for a wavelength of 530 nm.
17. 167 nm
19. (b) 603 nm (orange); 429 nm (violet)
21. 1.000132
23. 2 mm

Chapter 41

1. 660 nm
3. 547 nm
5. 0.022 radian = 1.3°
7. 0.045 (m = 1); 0.0162 (m = 2); 0.00827 (m = 3)
9. (a) 3.25×10^{-6} radian; (b) 1.04×10^{-5} m
11. 50 km is resolved; 1 km is not resolved
13. 4.6×10^{-5} radian; markings not resolved
15. 2.4×10^{-6}; approximately 210 km
17. 2.86 μm
19. 76,760
21. 25,000

Chapter 42

1. 1.52×10^{15} Hz; 197 nm
3. (b) 8.5×10^{-34} J·s
5. $M_{bowling}/M_{BB}$ = 21,000; $M_{alpha}/M_{electron}$ = 7,500; fair analogy
7. (a) 1.1×10^{-9}; (b) 1900; (c) 1/500,000
9. 4.43×10^4 m/s
11. 5.22 cm
13. 1.44×10^{-10} m
15. (a) 10 liters; (b) 270 liters; (c) 18 liters
17. 2.83×10^{13} Hz; 10.6 μm; infrared
19. 8.8×10^6

Chapter 43

1. (b) 56 MeV
3. 4.7 fm (^{60}Co); 6.1 fm (^{131}I); 7.4 fm (^{235}U)
5. 2.29×10^{17} kg/m^3
7. 6 [= 4(4 − 1)/2]
9. (a) $^8\text{Be} \rightarrow {^4\text{He}} + {^4\text{He}}$; (b) [8.005305 − 2(4.002603)] > 0
11. 4.19 MeV
13. 5.24 MeV
15. 15.9 min (neutron); 35,200 yr (^{239}Pu)
17. 1140 Ci
21. 26,000 ± 100 yr
23. (b) 2.95×10^{10}
25. (a) No escape of Sr or Rb and no decay of Sr; (b) 4.11×10^9
27. 0.564 mg
29. (Approximate result based on graph.); 3×10^{15} atoms
31. 176 MeV
33. (a) 510; (b) $2^{n+1} - 2$
35. 1220 kg
37. (b) 7.0 kg
39. $(Z^2/A)_{\text{after fission}} = \frac{1}{2}(Z^2/A)_{\text{before fission}}$
41. 14.1 MeV

Photo Credits

15.6 Courtesy of NASA.
16.3 Photo by George Holten/Photo Researchers, Inc.
16.4 Courtesy of Educational Materials Division, Fisher Scientific Company.
17.2 Photo by Bernice Abbott/Photo Researchers, Inc.
19.10 Einstein Archives, Courtesy AIP, Neils Bohr Library.
21.19 Photo by Ed Finley/Photo Researchers, Inc.
23.12 Courtesy of Educational Materials Division, Fisher Scientific Company.
26.1 The Granger Collection, New York.
26.2 Courtesy of Deutsches Museum, Munich.
28.11 Fundamental Photographs, New York.
29.18 Courtesy of Central Scientific Company, Franklin Park, IL.
30.10 Courtesy of Central Scientific Company, Franklin Park, IL.
31.1 Photo by Keith Gunnar/Photo Researchers, Inc.
31.32 Courtesy of Simpson Electric Company, Elgin, IL.
32.3(a) From J. B. Marion, *Physics in the Modern World*, 2/e, 1981 Academic Press, New York. Used with permission.
32.3(b) Fundamental Photographs, New York.
32.5 Fundamental Photographs, New York.
32.6 Courtesy of Simpson Electric Company, Elgin, IL.
32.14 Courtesy of Brookhaven National Laboratory.
32.24 From J. B. Marion, *Physics in the Modern World*, 2/e, 1981 Academic Press, New York. Used with permission.
33.5 Fundamental Photographs, New York.
33.11 From J. B. Marion, *Physics in the Modern World*, 2/e, 1981 Academic Press, New York. Used with permission.
38.24 Courtesy of Baush & Lomb Optical Company.
38.34 Courtesy of Polaroid Corporation.
38.26 Courtesy of Mount Wilson & Palomar Observatories.
39.8 Fundamental Photographs, New York.
40.1 Fundamental Photographs, New York.
40.10 Fundamental Photographs, New York.
41.14 From M. Cagent, M. Francon, J. Thrien, *Atlas of Optical Phenomena*, 1962, Springer-Verlag, New York. Used with permission.
42.1 Courtesy AIP, Niels Bohr Library.
42.12 Courtesy AIP, Niels Bohr Library.
42.24 Courtesy AIP, Niels Bohr Library.
42.25 Courtesy AIP, Meggars Gallery of Nobel Laureates.

INDEX

Entries followed by 'p' refer to a problem.
Entries followed by 't' refer to a table.

A 4
B 5
C 6
D 7
E 8
F 9
G 0
H 1